LIST OF ELEMENTS WITH THEIR SYMBOLS AND ATOMIC WEIGHTS

Element	Symbol	Atomic number	Atomic weight	Element	Symbol	Atomic number	Atomic weight	Element	Symbol	Atomic number	Atomic weight
Actinium	Ac	89	227.03[a]	Helium	He	2	4.002602	Rhenium	Re	75	186.207
Aluminum	Al	13	26.981538	Holmium	Ho	67	164.93032	Rhodium	Rh	45	102.90550
Americium	Am	95	243.06[a]	Hydrogen	H	1	1.00794	Rubidium	Rb	37	85.4678
Antimony	Sb	51	121.760	Indium	In	49	114.818	Ruthenium	Ru	44	101.07
Argon	Ar	18	39.948	Iodine	I	53	126.90447	Rutherfordium	Rf	104	261.11[a]
Arsenic	As	33	74.92160	Iridium	Ir	77	192.217	Samarium	Sm	62	150.36
Astatine	At	85	209.99[a]	Iron	Fe	26	55.845	Scandium	Sc	21	44.955910
Barium	Ba	56	137.327	Krypton	Kr	36	83.80	Seaborgium	Sg	106	266[a]
Berkelium	Bk	97	247.07[a]	Lanthanum	La	57	138.9055	Selenium	Se	34	78.96
Beryllium	Be	4	9.012182	Lawrencium	Lr	103	262.11[a]	Silicon	Si	14	28.0855
Bismuth	Bi	83	208.98038	Lead	Pb	82	207.2	Silver	Ag	47	107.8682
Bohrium	Bh	107	264.12[a]	Lithium	Li	3	6.941	Sodium	Na	11	22.989770
Boron	B	5	10.811	Lutetium	Lu	71	174.967	Strontium	Sr	38	87.62
Bromine	Br	35	79.904	Magnesium	Mg	12	24.3050	Sulfur	S	16	32.065
Cadmium	Cd	48	112.411	Manganese	Mn	25	54.938049	Tantalum	Ta	73	180.9479
Calcium	Ca	20	40.078	Meitnerium	Mt	109	268.14[a]	Technetium	Tc	43	98[a]
Californium	Cf	98	251.08[a]	Mendelevium	Md	101	258.10[a]	Tellurium	Te	52	127.60
Carbon	C	6	12.0107	Mercury	Hg	80	200.59	Terbium	Tb	65	158.92534
Cerium	Ce	58	140.116	Molybdenum	Mo	42	95.94	Thallium	Tl	81	204.3833
Cesium	Cs	55	132.90545	Neodymium	Nd	60	144.24	Thorium	Th	90	232.0381
Chlorine	Cl	17	35.453	Neon	Ne	10	20.1797	Thulium	Tm	69	168.93421
Chromium	Cr	24	51.9961	Neptunium	Np	93	237.05[a]	Tin	Sn	50	118.710
Cobalt	Co	27	58.933200	Nickel	Ni	28	58.6934	Titanium	Ti	22	47.867
Copper	Cu	29	63.546	Niobium	Nb	41	92.90638	Tungsten	W	74	183.84
Curium	Cm	96	247.07[a]	Nitrogen	N	7	14.0067	Uranium	U	92	238.02891
Dubnium	Db	105	262.11[a]	Nobelium	No	102	259.10[a]	Vanadium	V	23	50.9415
Dysprosium	Dy	66	162.50	Osmium	Os	76	190.23	Xenon	Xe	54	131.293
Einsteinium	Es	99	252.08[a]	Oxygen	O	8	15.9994	Ytterbium	Yb	70	173.04
Erbium	Er	68	167.259	Palladium	Pd	46	106.42	Yttrium	Y	39	88.90585
Europium	Eu	63	151.964	Phosphorus	P	15	30.973761	Zinc	Zn	30	65.39
Fermium	Fm	100	257.10[a]	Platinum	Pt	78	195.078	Zirconium	Zr	40	91.224
Fluorine	F	9	18.9984032	Plutonium	Pu	94	244.06[a]	*[b]		110	271.15[a]
Francium	Fr	87	223.02[a]	Polonium	Po	84	208.98[a]	*[b]		111	272.15[a]
Gadolinium	Gd	64	157.25	Potassium	K	19	39.0983	*[b]		112	277[a]
Gallium	Ga	31	69.723	Praseodymium	Pr	59	140.90765	*[b]		113	284[a]
Germanium	Ge	32	72.64	Promethium	Pm	61	145[a]	*[b]		114	289[a]
Gold	Au	79	196.96655	Protactinium	Pa	91	231.03588	*[b]		115	288[a]
Hafnium	Hf	72	178.49	Radium	Ra	88	226.03[a]	*[b]		116	292[a]
Hassium	Hs	108	269.13[a]	Radon	Rn	86	222.02[a]				

[a]Massof longest-lived or most important isotope.

[b]The names of elements 110 and above have not yet been decided.

In an effort to save students money, the publisher has worked with the Chemistry faculty at LSU to reduce the cost of this textbook by omitting Chapters 21-25. These missing chapters are still reflected in the Table of Contents.

Should you need to reference Chapters 21-25, you may do so by accessing them with your One Key Access Code.

CHEMISTRY
THE CENTRAL SCIENCE

Theodore L. Brown ♦ H. Eugene LeMay, Jr. ♦ Bruce E. Bursten ♦ Catherine J. Murphy

Taken from:
Chemistry: The Central Science, Tenth Edition
by Theodore L. Brown, H. Eugene LeMay, Jr., Bruce E. Bursten, and Catherine J. Murphy

PEARSON Custom Publishing

PEARSON Prentice Hall

Taken from:

Chemistry: The Central Science
by Theodore L. Brown, H. Eugene LeMay, Jr., Bruce E. Bursten, and Catherine J. Murphy

Published by Prentice Hall, Inc.
Upper Saddle River, New Jersey 07458

This special edition published in cooperation with Pearson Custom Publishing.

Printed in the United States of America

10 9 8 7 6 5 4 3 2 1

ISBN 0-536-94345-1

2005180039

KC

Please visit our web site at *www.pearsoncustom.com*

PEARSON Custom Publishing

PEARSON CUSTOM PUBLISHING
75 Arlington Street, Suite 300, Boston, MA 02116
A Pearson Education Company

To our students, whose enthusiasm
and curiosity have often inspired us,
and whose questions and suggestions
have sometimes taught us.

Contents

1 | Introduction: Matter and Measurement 1

2 | Atoms, Molecules, and Ions 36

9 | Molecular Geometry and Bonding Theories 344

10 | Gases 398

24 | Chemistry of Coordination Compounds 1024

25 | The Chemistry of Life: Organic and Biological Chemistry 1062

Appendices

Preface

To the Instructor

Philosophy

This is the tenth edition of a text that has enjoyed unprecedented success over its many editions. It is fair to ask why there needs to be yet another edition. The answer in part lies in the nature of chemistry itself, a dynamic science in a process of continual discovery. New research leads to new applications of chemistry in other fields of science and in technology. A textbook that purports to introduce chemistry to students who have only a limited prior acquaintance with it should reflect that dynamic, changing character. We want the text to convey the excitement that scientists experience in making new discoveries and contributing to our understanding of the physical world.

In addition, the teaching of chemistry is also continuously changing. New ideas about how to present chemistry are being offered by teachers of chemistry, and many of these new ideas are reflected in how the textbook is organized and the ways in which individual topics are presented. In addition, new technologies and new devices to assist students in learning lead to new ways of presenting learning materials: the Internet, computer-based classroom projection tools, and more effective means of testing, to name just a few. All of these factors impact on how the text and the accompanying supplementary materials are modified from one edition to the next.

Our aim in revising the text has been to ensure that the text remains a central, indispensable learning tool for the student. It is the one device that can be carried everywhere and used at any time, and as such, it is a one-stop source of all the information that the student is likely to need for learning, skill development, reference, and test preparation. We believe that students are more enthusiastic about learning chemistry when they see its importance to their own goals and interests. With this in mind, we have highlighted many important applications of chemistry in everyday life. At the same time, the text provides the background in modern chemistry that students need to serve their professional interests and, as appropriate, to prepare for more advanced chemistry courses.

If the text is to support your role as teacher effectively, it must be addressed to the students. We have done our best to keep our writing clear and interesting and the book attractive and well-illustrated. Furthermore, we have provided numerous in-text study aids for students, including carefully placed descriptions of problem-solving strategies. Together, we have logged many years of teaching experience. We hope this is evident in our pacing and choice of examples.

A textbook is only as useful to students as the instructor permits it to be. This book is loaded with many features that can help students learn and that can guide them as they acquire both conceptual understanding and problem-solving skills. But the text and all the supplementary materials provided to support its use must work in concert with the instructor. There is a great deal for the students to use here, too much for all of it to be absorbed by any one student. You, the instructor, are the guide to a proper use of the book. Only with your active help will the students be able to fully utilize all that the text and its supplements offer. Students care about grades, of course, but with encouragement, they can also care about learning just because the subject matter is interesting. Please consider emphasizing features of the book that can materially enhance student appreciation of chemistry, such as the *Chemistry at Work* and *Chemistry and Life* boxes that show how chemistry impacts modern life and its relationship to health and life processes. Learn to use, and urge students to use, the rich Internet resources available. Emphasize conceptual understanding, and place less emphasis on simple manipulative, algorithmic problem-solving. Spending less time on solving a variety of gas law problems, for example, can open up opportunities to talk about chemistry and the environment.

Organization and Contents

In the present edition the first five chapters give a largely macroscopic, phenomenological view of chemistry. The basic concepts introduced—such as nomenclature, stoichiometry, and thermochemistry—provide necessary background for many of the laboratory experiments usually performed in general chemistry. We believe that an early introduction to thermochemistry is desirable because so much of our understanding of chemical processes is based on considerations of energy change. Thermochemistry is also important when we come to a discussion of bond enthalpies. We have continued to refine our approach to teaching thermodynamics in general chemistry. It is no easy matter to walk the narrow pathway between—on the one hand—trying to teach too much at too high a level and—on the other—resorting to oversimplifications. As with the book as a whole, the emphasis has been on imparting conceptual understanding, as opposed to presenting equations into which students are supposed to plug numbers.

The next four chapters (Chapters 6–9) deal with electronic structure and bonding. Here we have made several significant changes, particularly in the presentation of atomic orbitals. New *Closer Look* boxes deal with radial probability functions and the nature of antibonding orbitals. The focus of the text then changes to the next level of the organization of matter: the states of matter (Chapters 10 and 11) and solutions (Chapter 13). Also included in this section is an applications chapter on the chemistry of modern materials (Chapter 12), which builds on the student's understanding of chemical bonding and intermolecular interactions. This chapter has received a major revision, in keeping with the rapid pace of change in technology. It has been reorganized to emphasize the classification of materials according to their uses (materials for structure, materials for electronics, etc.). In addition, new topics such as light-emitting polymers and materials for nanotechnology have been added.

The next several chapters examine the factors that determine the speed and extent of chemical reactions: kinetics (Chapter 14), equilibria (Chapters 15–17), thermodynamics (Chapter 19), and electrochemistry (Chapter 20). Also in this section is a chapter on environmental chemistry (Chapter 18), in which the concepts developed in preceding chapters are applied to a discussion of the atmosphere and hydrosphere. We have further revised and refined our introduction to equilibrium constants in Chapter 15. The chapter on thermodynamics has been carefully reworked to give students a better sense of how the macroscopic and microscopic views of entropy are connected.

After a discussion of nuclear chemistry (Chapter 21), the final chapters survey the chemistry of nonmetals, metals, organic chemistry, and biochemistry (Chapters 22–25). These chapters are developed in a parallel fashion and can be treated in any order.

Our chapter sequence provides a fairly standard organization, but we recognize that not everyone teaches all the topics in exactly the order we have chosen. We have therefore made sure that instructors can make common changes in teaching sequence with no loss in student comprehension. In particular, many instructors prefer to introduce gases (Chapter 10) after stoichiometry or after thermochemistry rather than with states of matter. The chapter on gases has been written to permit this change with *no* disruption in the flow of material. It is also possible to treat the balancing of redox equations (Sections 20.1 and 20.2) earlier, after the introduction of redox reactions in Section 4.4. Finally, some instructors like to cover organic chemistry (Chapter 25) right after bonding (Chapter 9). With the exception of the discussion of stereochemistry (which is introduced in Section 24.3), this, too, is a seamless move.

We have introduced students to descriptive organic and inorganic chemistry by integrating examples throughout the text. You will find pertinent and relevant examples of "real" chemistry woven into all the chapters as a means to illustrate principles and applications. Some chapters, of course, more directly address the properties of elements and their compounds, especially Chapters 4, 7, 12, 18, and 22–25. We also incorporate descriptive organic and inorganic chemistry in the end-of-chapter exercises.

Changes in this edition

Some of the changes in the tenth edition made in individual chapters have already been mentioned. More broadly, we have introduced a number of new features that are general throughout the text. *Chemistry: The Central Science* has traditionally been valued for its clarity of writing, its scientific accuracy and currency, its strong end-of-chapter exercises, and its consistency in level of coverage. In making changes, we have made sure not to compromise those characteristics. At the same time, we have responded to feedback received from the faculty and students who used the ninth edition. Sections that have seemed most difficult to students have in many cases been rewritten and augmented with improved artwork. In order to make the text easier for students to use, we have continued to employ an open, clean design in the layout of the book. Illustrations that lend themselves to a more schematic, bolder presentation of the underlying principles have been introduced or revised from earlier versions. The art program in general has been strengthened, to better convey the beauty, excitement, and concepts of chemistry to students. The chapter-opening photos have been integrated into the introduction to each chapter, and thus made more relevant to the chapter's contents.

We have continued to use the What's Ahead overview at the opening of each chapter, introduced in the ninth edition. *Concept links* ( ) continue to provide easy-to-see cross-references to pertinent material covered earlier in the text. The essays titled *Strategies in Chemistry*, which provide advice to students on problem solving and "thinking like a chemist," continue to be an important feature. A new in-chapter feature is the *Give It Some Thought* exercises. These are informal, rather sharply focused questions that give students opportunities to test whether they are actually "getting it" as they read along. We have added more conceptual exercises to the end-of-chapter exercises. A new category of end-of-chapter exercises, *Visualizing Concepts*, has been added to every chapter. These exercises are designed to facilitate concept understanding through use of models, graphs, and other visual materials. They precede the regular end-of-chapter exercises and are identified in each case with the relevant chapter section number. New Multi-Focus Graphics have been added. These graphics depict topics in macroscopic, microscopic, symbolic and conceptual representation so students learn to see chemistry the way scientists do, from a variety of perspectives. The *Integrative Exercises*, which give students the opportunity to solve more challenging problems that integrate concepts from the present chapter with those of previous chapters, have also been increased in number.

New essays in our well-received *Chemistry at Work* and *Chemistry and Life* series emphasize world events, scientific discoveries, and medical breakthroughs that have occurred since publication of the ninth edition. We maintain our focus on the positive aspects of chemistry, without neglecting the problems that can arise in an increasingly technological world. Our goal is to help students appreciate the real-world perspective of chemistry and the ways in which chemistry affects their lives.

You'll also find that we've

- Revised the end-of-chapter Exercises, with particular focus on the black-numbered exercises (those not answered in the Appendix).
- Integrated more conceptual questions into the end-of-chapter material. For the convenience of instructors, these are identified by the **CQ** annotation in the Annotated Instructor's Edition, but not in the student edition of the text.
- Added new eLaboratory Exercises in the end-of-chapter material. These exercises allow students to develop and practice the skills they learn from the chapter in a simulated laboratory, available in software that accompanies the text. The realistic setting helps students see how problem-solving in chemistry applies in practice.
- Updated the eMedia Exercises in the end-of-chapter material. These exercises take advantage of the integrated media components and extend students' understanding, using the advantages that interactive, media-rich presentations offer.
- Continued the practice of using an eMedia Activity icon in the margins to indicate where students can extend understanding of a concept or topic by looking at an activity located on the Web site or the Accelerator CD-ROM.
- Carried the stepwise Analyze, Plan, Solve, Check problem-solving strategy into nearly all of the Sample Exercises of the book to provide additional guidance in problem solving.
- Expanded the use of dual-column problem-solving strategies in many Sample Exercises to more clearly outline the process underlying mathematical calculations, thereby helping students to better perform mathematical calculations.

To the Student

Chemistry: The Central Science, Tenth Edition, has been written to introduce you to modern chemistry. As authors, we have, in effect, been engaged by your instructor to help you learn chemistry. Based on the comments of students and instructors who have used this book in its previous editions, we believe that we have done that job well. Of course, we expect the text to continue to evolve through future editions. We invite you to write to us to tell us what you like about the book so that we will know where we have helped you most. Also, we would like to learn of any shortcomings, so that we might further improve the book in subsequent editions. Our addresses are given at the end of the Preface.

Advice for learning and studying chemistry

Learning chemistry requires both the assimilation of many new concepts and the development of analytical skills. In this text we have provided you with numerous tools to help you succeed in both. We have provided details of the features of this text in the "walk-through" on pages xxxii–xxxv. You will find it helpful to examine those features.

If you are going to succeed in your course in chemistry, you will have to develop good study habits. Science courses, and chemistry in particular, make different demands on your learning skills than other types of courses. We offer the following tips for success in your study of chemistry:

Don't fall behind! As your chemistry course moves along, new topics will build on material already presented. If you don't keep up in your reading and problem solving, you will find it much harder to follow the lectures and discussions on current topics. "Cramming" just before an exam has been shown to be an ineffective way to study any subject, chemistry included.

Focus your study. The amount of information you will be expected to learn can sometimes seem overwhelming. It is essential to recognize those concepts and skills that are particularly important. Pay attention to what your instructor is emphasizing. As you work through the Sample Exercises and homework assignments, try to see what general principles and skills they deal with. Use the What's Ahead feature at the beginning of each chapter to help orient you to what is important in each chapter. A single reading of a chapter will simply not be enough for successful learning of chapter concepts and problem-solving skills. You will need to go over assigned materials more than once. Don't skip the Give It Some Thought features, Sample Exercises, and Practice Exercises. They are your guide to whether you are actually learning the material.

Keep good lecture notes. Your lecture notes will provide you with a clear and concise record of what your instructor regards as the most important material to learn. Use your lecture notes in conjunction with this text; that's your best way to determine which material to study.

Skim topics in the text before they are covered in lecture. Reviewing a topic before lecture will make it easier for you to take good notes. First read the introduction and Summary, then quickly read through the chapter, skipping Sample Exercises and supplemental sections. Pay attention to the titles of sections and subsections, which give you a feeling for the scope of topics. Try to avoid thinking that you must learn and understand everything right away.

After lecture, carefully read the topics covered in class. As you read, pay attention to the concepts presented and to the application of these concepts in the Sample Exercises. Once you think you understand a Sample Exercise, test your understanding by working the accompanying Practice Exercise.

Learn the language of chemistry. As you study chemistry, you will encounter many new words. It is important to pay attention to these words and to know their meanings or the entities to which they refer. Knowing how to identify chemical substances from their names is an important skill; it can help you avoid painful mistakes on examinations. For example, "chlorine" and "chloride" refer to very different things.

Attempt the assigned end-of-chapter exercises. Working the exercises that have been selected by your instructor provides necessary practice in recalling and using the essential ideas of the chapter. You cannot learn merely by observing; you must be a participant. In particular, try to resist checking the *Student Solutions Manual* (if you have one) until you have made a sincere effort to solve the exercise yourself. If you really get stuck on an exercise, however, get help from your instructor, your teaching assistant, or another student. Spending more than 20 minutes on a single exercise is rarely effective unless you know that it is particularly challenging.

Make use of the online resources. Some things are more easily learned by discovery, and others are best shown in three dimensions. Use the Companion Website with GradeTracker or OneKey course management materials to this text to get the most out of your time in chemistry.

The bottom line is to work hard, study effectively, and use the tools that are available to you, including this textbook. We want to help you learn more about the world of chemistry and why chemistry is the central science. If you learn chemistry well, you can be the life of the party, impress your friends and parents, and . . . well, also pass the course with a good grade.

Acknowledgments

Many people contributed hard work and talent to bring this edition to life. Colleagues helped us immensely by sharing their insights, providing suggestions for improvements and reviewing draft materials. We would especially like to thank the following individuals who have contributed to this and prior editions:

Tenth Edition Reviewers

S.K. Airee	University of Tennessee, Martin
Patricia Amateis	Virginia Polytechnic Institute and State University
Sandra Anderson	University of Wisconsin, Parkside
Socorro Arteaga	El Paso Community College
Rosemary Bartoszek-Loza	The Ohio State University
Amy Beilstein	Centre College
Victor Berner	New Mexico Junior College
Narayan Bhat	University of Texas, Pan American
Salah M. Blaih	Kent State University, Trumbull
Leon Borowski	Diablo Valley College
Simon Bott	University of Houston
Karen Brewer	Virginia Polytechnic Institute and State University
Carmela Byrnes	Texas A&M University (grad student)
Kim Calvo	University of Akron
Elaine Carter	Los Angeles City College
Robert Carter	University of Massachusetts, Boston
Ann Cartwright	San Jacinto Central College
David Cedeno	Illinois State University
Paul Chirik	Cornell University
Beverly Clement	Blinn College
John Collins	Broward Community College
Enriqueta Cortez	Southern Texas Community College
Nancy De Luca	University of Massachusetts, Lowell North Campus
Daniel Domin	Tennessee State University
James Donaldson	University of Toronto
Bill Donovan	University of Akron
Ronald Duchovic	Indiana University-Purdue University at Fort Wayne
Joseph Ellison	United States Military Academy
James Farrar	University of Rochester
Gregory Ferrence	Illinois State University
Jennifer Firestine	Lindenwood University
Michelle Fossum	Laney College
David Frank	California State University, Fresno
Cheryl Frech	University of Central Oklahoma
Ewa Fredette	Moraine Valley College
Kenneth French	Blinn College
Karen Frindell	Santa Rosa Junior College
Paul Gilletti	Mesa Community College
Eric Goll	Brookdale Community College
James Gordon	Central Methodist College
John Hagadorn	University of Colorado, Boulder
Daniel Haworth	Marquette University
Inna Hefley	Blinn College
David Henderson	Trinity College
Paul Higgs	Barry University
Carl Hoeger	University of California, San Diego
Deborah Hokien	Marywood University
Janet Johannessen	County College of Morris
Milton Johnston	University of South Florida
Andrew Jones	Southern Alberta Institute of Technology
Booker Juma	Fayetteville State University
Ismail Kady	East Tennessee State University
Steven Keller	University of Missouri, Columbia
David Kort	George Mason University
William R. Lammela	Nazareth College
Robley Light	Florida State University
Patrick Lloyd	Kingsborough Community College
Encarnacion Lopez	Miami Dade College, Wolfson
Marcus T. McEllistrem	University of Wisconsin, Eau Claire
Craig McLauchlan	Illinois State University
Larry Manno	Triton College
Pam Marks	Arizona State University
Albert H. Martin	Moravian College
Przemyslaw Maslak	Pennsylvania State University, University Park
Armin Mayr	El Paso Community College
Joseph Merola	Virginia Polytechnic Institute and State University
Stephen Mezyk	California State University, Long Beach
Shelley Minteer	St. Louis University
Tracy Morkin	Emory University
Kathy Nabona	Austin Community College
Al Nichols	Jacksonville State University
Mark Ott	Jackson Community College
Sandra Patrick	Malaspina University College
Tammi Pavelec	Lindenwood University
Albert Payton	Broward Community College
Gita Perkins	Estrella Mountain Community College
John Pfeffer	Highline Community College
Lou Pignolet	University of Minnesota
Bernard Powell	University of Texas, San Antonio
Jeffrey Rahn	Eastern Washington University
Steve Rathbone	Blinn College
Scott Reeve	Arkansas State University
John Reissner	University of North Carolina, Pembroke
Thomas Ridgway	University of Cincinnati
Lenore Rodicio	Miami Dade College, Wolfson
Amy Rogers	College of Charleston
Jimmy Rogers	University of Texas at Arlington
Steven Rowley	Middlesex Community College
Theodore Sakano	Rockland Community College
Mark Schraf	West Virginia University
Paula Secondo	Western Connecticut State University
Vince Sollimo	Burlington Community College
David Soriano	University of Pittsburgh, Bradford
Michael Tubergen	Kent State University
James Tyrell	Southern Illinois University
Michael Van Stipdonk	Wichita State University
Philip Verhalen	Panola College
Ann Verner	University of Toronto at Scarborough
Edward Vickner	Gloucester County Community College
John Vincent	University of Alabama, Tuscaloosa
Tony Wallner	Barry University
Lichang Wang	Southern Illinois University
Thomas R. Webb	Auburn University
Paul Wenthold	Purdue University
Thao Yang	University of Wisconsin, Eau Claire

Tenth Edition Accuracy Checkers

Richard Langley	Stephen F. Austin State University
Arthur Low	Tarleton State University
Stephen Mezyk	California State University, Long Beach
Kathy Thrush	Villanova University

Special Thanks to Those Who Provided Valuable Feedback to the Authors and/or Publisher

James Birk	Arizona State University
Rik Blumenthal	Auburn University
Paul Chirik	Cornell University
Roger DeKock	Calvin College
David Harris	University of Illinois, Urbana-Champaign
Daniel Haworth	Marquette University
Alan D. Isaacson	Miami University
Friedrich Koknat	Youngstown State University
Frank L. Lambert	Professor Emeritus, Occidental College
Tom H. Ridgway	University of Cincinnati
Thomas R. Webb	Auburn University
John Wheeler	University of California, San Diego

Ninth Edition Reviewers

John Arnold	University of California, Berkeley
Merrill Blackman (Col.)	United States Military Academy
Daeg Scott Brenner	Clark University
Gregory Alan Brewer	Catholic University of America
Gary Buckley	Cameron University
Gene O. Carlisle	Texas A&M University
Dana Chatellier	University of Delaware
William Cleaver	University of Vermont
Elzbieta Cool	University of Calgary
Dwaine Davis	Forsyth Tech Community College
Angel deDios	Georgetown University
John Farrar	University of St. Francis
Clark L. Fields	University of Northern Colorado
Jan M. Fleischner	The College of New Jersey
Peter Gold	Pennsylvania State University
Michael Greenlief	University of Missouri
John M. Halpin	New York University
Robin Horner	Fayetteville Tech Community College
Roger K. House	Moraine Valley College
William Jensen	South Dakota State University
Siam Kahmis	University of Pittsburgh
John W. Kenney	Eastern New Mexico University
George P. Kreishman	University of Cincinnati
Paul Kreiss	Anne Arundel Community College
David Lehmpuhl	University of Southern Colorado
Gary L. Lyon	Louisiana State University
Albert H. Martin	Moravian College
William A. Meena	Rock Valley College
Massoud Miri	Rochester Institute of Technology
Eric Miller	San Juan College
Mohammad Moharerrzadeh	Bowie State University
Kathleen E. Murphy	Daemen College
Robert T. Paine	University of New Mexico
Albert Payton	Broward Community College
Kim Percell	Cape Fear Community College
Nancy Peterson	North Central College
James P. Schneider	Portland Community College
Eugene Stevens	Binghamton University
James Symes	Cosumnes River College
Edmund Tisko	University of Nebraska at Omaha
Richard S. Treptow	Chicago State University
Claudia Turro	The Ohio State University
Maria Vogt	Bloomfield College
Sarah West	University of Notre Dame
Linda M. Wilkes	University at Southern Colorado
Darren L. Williams	West Texas A&M University
Troy D. Wood	SUNY Buffalo
David Zax	Cornell University

Ninth Edition Accuracy Checkers

Boyd Beck	Snow College
B. Edward Cain	Rochester Institute of Technology
Thomas Edgar Crumm	Indiana University of Pennsylvania
Angel deDios	Georgetown University
David Easter	Southwest Texas State University
Jeffrey Madura	Duquesne University
Hilary L. Maybaum	ThinkQuest, Inc.
Robert T. Paine	Rochester Institute of Technology
Christopher J. Peeples	University of Tulsa
Jimmy R. Rogers	University of Texas at Arlington
Iwao Teraoka	Polytechnic University
Richard Treptow	Chicago State University
Maria Vogt	Bloomfield College

Eighth Edition Reviewers

John J. Alexander	University of Cincinnati
Robert Allendoerfer	SUNY Buffalo
Boyd Beck	Snow College
James A. Boiani	College at Geneseo-SUNY
Kevin L. Bray	Washington State University
Edward Brown	Lee University
Donald L. Campbell	University of Wisconsin, Eau Claire
Stanton Ching	Connecticut College
Robert D. Cloney	Fordham University
Edward Werner Cook	Tunxis Community Technical College
John M. DeKorte	Glendale Community College
Roger Frampton	Tidewater Community College
Joe Franek	University of Minnesota
John I. Gelder	Oklahoma State University
Thomas J. Greenbowe	Iowa State University
Eric P. Grimsrud	Montana State University
Marie Hankins	University of Southern Indiana
Robert M. Hanson	St. Olaf College
Gary G. Hoffman	Florida International University
Robin Horner	Fayetteville Tech Community College
Donald Kleinfelter	University of Tennessee, Knoxville
Manickham Krishnamurthy	Howard University
Brian D. Kybett	University of Regina
William R. Lammela	Nazareth College
John T. Landrum	Florida International University
N. Dale Ledford	University of South Alabama
Ernestine Lee	Utah State University
Donald E. Linn, Jr.	Indiana University-Purdue University Indianapolis
David Lippmann	Southwest Texas State

Ramón López de la Vega	Florida International University	Jeffrey A. Rahn	Eastern Washington University
Preston J. MacDougall	Middle Tennessee State University	Mark G. Rockley	Oklahoma State University
Asoka Marasinghe	Moorhead State University	Jimmy Rogers	University of Texas at Arlington
Earl L. Mark	ITT Technical Institute	James E. Russo	Whitman College
William A. Meena	Rock Valley College	Michael J. Sanger	University of Northern Iowa
Gordon Miller	Iowa State University	Jerry L. Sarquis	Miami University
Massoud (Matt) Miri	Rochester Institute of Technology	Gray Scrimgeour	University of Toronto
Kathleen E. Murphy	Daemon College	Richard Treptow	Chicago State University
Ross Nord	Eastern Michigan University	Laurence Werbelow	New Mexico Institute of Mining and Technology
Robert H. Paine	Rochester Institute of Technology	Troy D. Wood	SUNY Buffalo
Mary Jane Patterson	Brazosport College		
Robert C. Pfaff	Saint Joseph's College		

Eighth Edition Accuracy Reviewers

Leslie Kinsland	University of Louisiana, Lafayette	Robert H. Paine	Rochester Institute of Technology
Albert Martin	Moravian College	Richard Perkins	University of Louisiana, Lafayette
Robert Nelson	Georgia Southern University		

Special Thanks to Others Involved in the Review of the Previous Edition Text and Its Various Components

Patricia Amateis	Virginia Polytechnic Institute and State University	Barbara Mowery	Yorktown, VA
Randy Hall	Louisiana State University	Helen Richter	University of Akron
Daniel Haworth	Marquette University	Mark Rockley	Oklahoma State University
Neil Kestner	Louisiana State University	David Shinn	University of Hawaii at Hilo
Albert Martin	Moravian College	John Vincent	University of Alabama
		Karen Weichelman	University of Louisiana-Lafayette

We would also like to express our gratitude to our colleagues at Prentice Hall whose imagination and commitment have done a great deal to shape this edition: Nicole Folchetti, our Executive Chemistry Editor, who brought energy and many creative ideas to our collaboration; Ray Mullaney, our developmental editorial manager, who has been with us for a long time and who continues to infuse a sense of style, accuracy, and quality into the project; John Challice, Editor in Chief, who continues to support us in all the right ways; Kathleen Schiaparelli, Executive Managing Editor, for her support and encouragement; Irene Nunes, our developmental editor, whose sharp, good-humored scrutiny of our draft materials has kept us consistent and on-message; Fay Ahuja, our copy editor, for her keen eye and excellent memory; Rosaria Cassinese, who worked successfully with a difficult and demanding production schedule; Paul Draper, our media editor, and Margaret Trombley, media development editor, who have taken on increasingly complex assignments as the electronic components of our project have gained in importance; Yvonne Gerin, our photo researcher, who has always been there with the right stuff; Kristen Kaiser, project manager, and Angela Sica, editorial assistant, for their valuable contributions to the new edition. We offer a special thanks to all the students and faculty who have sent us comments and suggestions about Chemistry: The Central Science, Ninth Edition. You will see that many of your suggestions have been taken into account in preparing this new edition. Finally, we are happy to thank out families and friends for their love, support and patience as we brought this tenth edition to completion.

Theodore L. Brown
School of Chemical Sciences
University of Illinois, Urbana-Champaign
Urbana, IL 61801
tlbrown@uiuc.edu

H. Eugene LeMay, Jr.
Department of Chemistry
University of Nevada
Reno, NV 89557
lemay@unr.edu

Bruce E. Bursten
Department of Chemistry
The Ohio State University
Columbus, OH 43210
bursten.1@osu.edu

Catherine J. Murphy
Department of Chemistry and Biochemistry
University of South Carolina
Columbia, SC 29208
murphy@mail.chem.sc.edu

A Student's Guide to Using this Text

THE FOLLOWING PAGES WALK YOU THROUGH some of the main features of this text and its integrated media components. This learning system was designed with you, the student, in mind. We hope you enjoy your study of chemistry—the central science.

Building a Conceptual Framework

"What's Ahead" Sections At the beginning of each chapter, reading the "What's Ahead" sections will give you a sense of direction for studying the chapter and help you to recognize key ideas and relationships of the topics within the chapter.

"Give It Some Thought" Questions These new, informal, rather sharply focused questions will give you opportunities to test whether you are actually "getting it" as you study the material. Give these questions some real thought before you check your responses against the answers at the back of the book.

"Visualizing Concepts" Exercises Just before the end-of-chapter exercises, there are new Visualizing Concepts exercises that ask you to consider concepts through the use of models, graphs, and other visual materials. These exercises will help you develop a conceptual understanding of the key ideas in the chapter. There are additional conceptual exercises among the end-of-chapter exercises.

GIVE IT SOME THOUGHT

Why is the choice of indicator more crucial for a weak acid–strong base titration than for a strong acid–strong base titration?

VISUALIZING CONCEPTS

16.1 CQ (a) Identify the Brønsted–Lowry acid and the Brønsted–Lowry base in the following reaction:

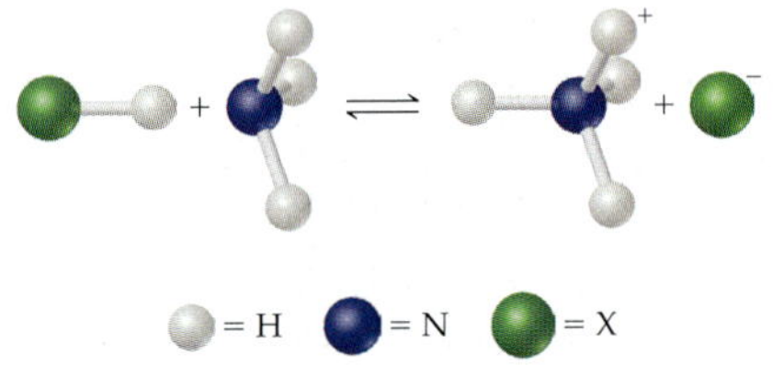

(b) Identify the Lewis acid and the Lewis base in the reaction. [Sections 16.2 and 16.11]

16.2 CQ The following diagrams represent aqueous solutions of two monoprotic acids, HA (A = X or Y). The water

Problem Solving

Learning effective problem-solving skills is one of your most important goals in this course. To help you solve problems with confidence, the text integrates problem-solving pedagogy.

Worked Solutions demonstrate the strategy and thought process involved in solving each exercise.

Analyze/Plan/Solve/Check Theme provides a consistent framework for helping you to understand what you are being asked to solve, to plan how you will solve each problem, to work your way through the solution, and to check to make sure that your answer is correct.

Dual-Column Problem-Solving Strategies found in selected Sample Exercises provide an explanation of the thought process involved in each step of a mathematical calculation to give you a conceptual understanding of those calculations.

SAMPLE EXERCISE 3.13 | Calculating an Empirical Formula

Ascorbic acid (vitamin C) contains 40.92% C, 4.58% H, and 54.50% O by mass. What is the empirical formula of ascorbic acid?

Solution

Analyze: We are to determine an empirical formula of a compound from the mass percentages of its elements.

Plan: The strategy for determining the empirical formula involves the three steps given in Figure 3.11.

Solve: We *first* assume, for simplicity, that we have exactly 100 g of material (although any mass can be used). In 100 g of ascorbic acid, we have

40.92 g C, 4.58 g H, and 54.50 g O.

Second, we calculate the number of moles of each element:

$$\text{Moles C} = (40.92\ \cancel{\text{g C}})\left(\frac{1\text{ mol C}}{12.01\ \cancel{\text{g C}}}\right) = 3.407\text{ mol C}$$

$$\text{Moles H} = (4.58\ \cancel{\text{g H}})\left(\frac{1\text{ mol H}}{1.008\ \cancel{\text{g H}}}\right) = 4.54\text{ mol H}$$

$$\text{Moles O} = (54.50\ \cancel{\text{g O}})\left(\frac{1\text{ mol O}}{16.00\ \cancel{\text{g O}}}\right) = 3.406\text{ mol O}$$

Third, we determine the simplest whole-number ratio of moles by dividing each number of moles by the smallest number of moles, 3.406:

$$\text{C}: \frac{3.407}{3.406} = 1.000 \qquad \text{H}: \frac{4.54}{3.406} = 1.33 \qquad \text{O}: \frac{3.406}{3.406} = 1.000$$

The ratio for H is too far from 1 to attribute the difference to experimental error; in fact, it is quite close to $1\frac{1}{3}$. This suggests that if we multiply the ratio by 3, we will obtain whole numbers:

$$\text{C:H:O} = 3(1:1.33:1) = 3:4:3$$

The whole-number mole ratio gives us the subscripts for the empirical formula:

$C_3H_4O_3$

Check: It is reassuring that the subscripts are moderately sized whole numbers. Otherwise, we have little by which to judge the reasonableness of our answer.

PRACTICE EXERCISE

A 5.325-g sample of methyl benzoate, a compound used in the manufacture of perfumes, is found to contain 3.758 g of carbon, 0.316 g of hydrogen, and 1.251 g of oxygen. What is the empirical formula of this substance?

Answer: C_4H_4O

Practice Exercises include answers but not solutions, thereby giving you the opportunity to test your knowledge and get instant feedback.

Strategies in Chemistry Strategies in Chemistry boxes teach you ways to analyze information and organize thoughts, helping to improve your problem-solving and critical-thinking abilities.

STRATEGIES IN CHEMISTRY | Using Enthalpy as a Guide

If you hold a brick in the air and let it go, it will fall as the force of gravity pulls it toward Earth. A process that is thermodynamically favored to happen, such as a falling brick, is called a *spontaneous* process. A spontaneous process can be either fast or slow. Speed is not the issue in thermodynamics.

Many chemical processes are thermodynamically favored, or spontaneous, too. By "spontaneous," we don't mean that the reaction will form products without any intervention. That can be the case, but often some energy must be imparted to get the process started. The enthalpy change in a reaction gives one indication as to whether it is likely to be spontaneous. The combustion of $H_2(g)$ and $O_2(g)$, for example, is a highly exothermic process:

$$H_2(g) + \frac{1}{2} O_2(g) \longrightarrow H_2O(g) \qquad \Delta H = -242 \text{ kJ}$$

Hydrogen gas and oxygen gas can exist together in a volume indefinitely without noticeable reaction occurring, as in Figure 5.14(a). Once initiated, however, energy is rapidly transferred from the system (the reactants) to the surroundings. As the reaction proceeds, large amounts of heat are released, which greatly increases the temperature of the reactants and the products. The system then loses enthalpy by transferring the heat to the surroundings. (Recall that the first law of thermodynamic ... surrou

Enthalpy change is not the only consideration in the spontaneity of reactions, however, nor is it a foolproof guide. For example, the melting of ice is an endothermic process:

$$H_2O(s) \longrightarrow H_2O(l) \qquad \Delta H = +6.01 \text{ kJ}$$

Even though this process is endothermic, it is spontaneous at temperatures above the freezing point of water (0°C). The reverse process, the freezing of water to ice, is spontaneous at temperatures below 0°C. Thus, we know that ice at room temperature will melt and that water put into a freezer at −20°C will turn into ice; both of these processes are spontaneous under different conditions even though they are the reverse of one another. In Chapter 19 we will address the spontaneity of processes more fully. We will see why a process can be spontaneous at one temperature, but not at another, as is the case for the conversion of water to ice.

Despite these complicating factors, however, you should pay attention to the enthalpy changes in reactions. As a general observation, when the enthalpy change is large, it is the dominant factor in determining spontaneity. Thus, reactions for which ΔH is *large* and *negative* tend to be spontaneous. Reactions for which ΔH is *large* and *positive* tend to be spontaneous only in the reverse direction. There are a number of ways in which the enthalpy of a reaction can be estimated; from those estimates the likelihood of the reaction being ther

End-of-Chapter Exercises

- *Visualizing Concepts* exercises, identified by the relevant chapter section number, help to build a conceptual understanding of the chapter as a whole.
- The next section of exercises is grouped by topic. They are presented in matched pairs, giving you multiple opportunities to test each concept. *Additional Exercises* follow the paired exercises and are not categorized, because many of these exercises draw on multiple concepts from within the chapter.
- *Integrative Exercises*, which are included among the exercises at the end of Chapters 3-25, connect concepts from the current chapter with those from previous chapters. These exercises will help you gain a deeper understanding of how chemistry fits together. In addition, they serve as an overall review of key concepts.

SAMPLE INTEGRATIVE EXERCISE | Putting Concepts Together

Boron, atomic number 5, occurs naturally as two isotopes, ^{10}B and ^{11}B, with natural abundances of 19.9% and 80.1%, respectively. **(a)** In what ways do the two isotopes differ from each other? Does the electronic configuration of ^{10}B differ from that of ^{11}B? **(b)** Draw the orbital diagram for an atom of ^{11}B. Which electrons are the valence electrons? **(c)** Indicate three major ways in which the 1*s* electrons in boron differ from the 2*s* electrons in the atom. **(d)** Elemental boron reacts with fluorine to form BF_3, a gas. Write a balanced chemical equation for the reaction of solid boron with fluorine gas. **(e)** ΔH_f° for $BF_3(g)$ is −1135.6 kJ mol^{-1}. Calculate the standard enthalpy change in the reaction of boron with fluorine. **(f)** When BCl_3, also a gas at room temperature, comes into contact with water, the two react to form hydrochloric acid and boric acid, H_3BO_3, a very weak acid in water. Write a balanced net ionic equation for this reaction.

Solution (a) The two nuclides of boron differ in the number of neutrons in the nucleus. (Sections 2.3 and 2.4) Each of the nuclides contains five protons, but ^{10}B contains five neutrons, whereas ^{11}B contains six neutrons. The two isotopes of boron have identical electron configurations, $1s^22s^22p^1$, because each has five electrons.

(b) The complete orbital diagram is

- The *Sample Integrative Exercise* at the end of the chapter shows you how to analyze and solve problems that encompass more than one concept.

eMEDIA EXERCISES

These exercises make use of the interactive objects available online in OneKey or the Companion Website, and on your Accelerator CD. Access to these resources comes in your MediaPak.

6.105 The **Electromagnetic Spectrum** activity (*6.1*) allows you to choose a color in the visible spectrum and see its wavelength, frequency, and energy per photon. **(a)** What is the wavelength range of blue light? **(b)** What are the ranges of its frequency and energy per photon? **(c)** Exercise 6.25 indicates that a type of sunburn is caused by light with wavelength ~325 nm. Would you expect any of the visible wavelengths to cause sunburn? Explain.

6.106 In the **Flame Tests for Metals** movie (*6.3*) the characteristic color of the flame is produced by emissions at several visible wavelengths, with the most intense spectral lines dominating the color. For instance, the most intense visible lines in the spectrum of lithium occur at ~671 nm. **(a)** What color is light of ... length? **(b)** At what approximate wavelen... you expect to find the most intense lines in ... spectrum of potassium? **(c)** Based on the ... would you expect the intensity of visible ... spectrum of potassium to compare to those ... trum of lithium? **(d)** Would it be possible to

- Answers are provided in the back of the book for red-numbered exercises. More challenging exercises are indicated by brackets around the exercise number.
- *eMedia Exercises* are answered by using the movies and simulations indicated in the margins by the eMedia Activity icon and available in the student MediaPak. These questions foster your understanding through visual and interactive learning styles.
- *eLaboratory Exercises* are answered using the Virtual ChemLab software in your student MediaPak. These are chemistry questions posed in a simulated laboratory setting.

Visualization

One of the challenges facing you in general chemistry is the often abstract nature of the subject. First, chemistry relies on a symbolic language based on chemical formulas and chemical equations. Second, chemistry is based on the behavior of molecules and atoms—particles far too small to see.

This text has been designed expressly to help you better visualize the chemistry you need to learn and thus help you to succeed in your course. Spend time with the illustrations in the text— they'll help you understand the chemistry concepts being discussed.

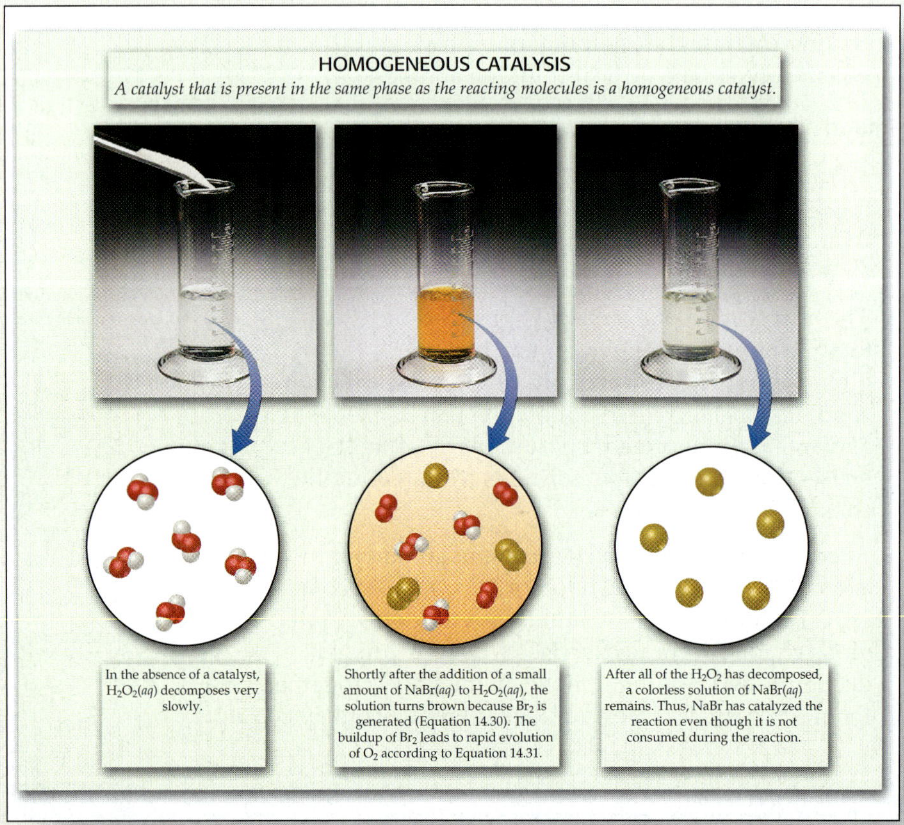

Multi-Focus Graphics Scientists see chemical concepts from a multitude of perspectives. Concepts in the text that require a mental picture of different representations are presented through new Multi-Focus graphics. In addition to a textual statement of key information, these new graphics include a representation from a variety of perspectives, including macroscopic, microscopic, symbolic, and conceptual.

Molecular Illustrations Computer-generated renditions of molecules and materials provide visual representations of matter at the atomic level. These drawings help you visualize molecules in three dimensions and enhance your understanding of molecular architecture.

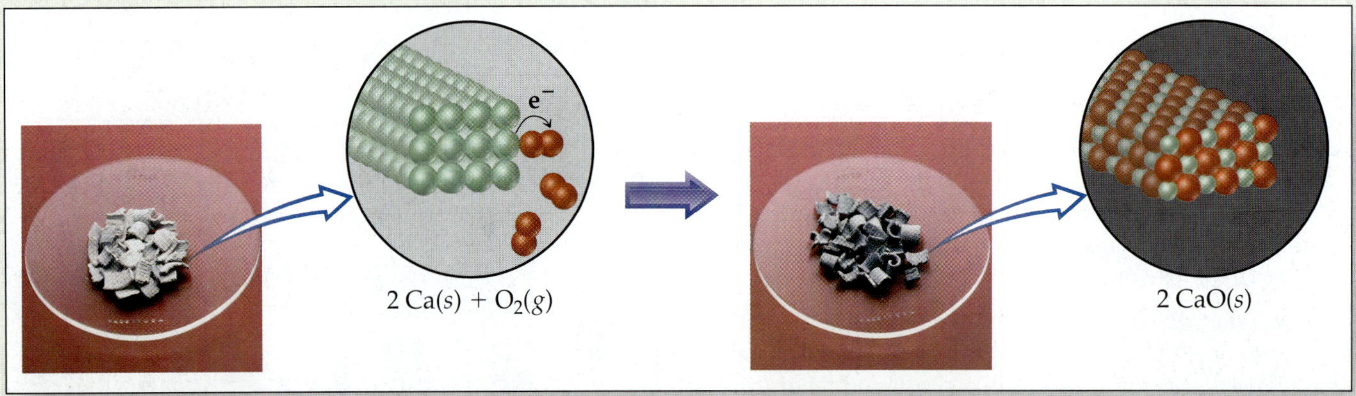

Applications

Why is chemistry "the central science"? In this text and in your course, you will recognize the importance of chemistry for your major areas of study and in the world around you. You will see why chemistry is fundamental to understanding our environment, for new developments in technology, and in our society.

CHEMISTRY AND LIFE | How Much Perchlorate Is Too Much?

Since the 1950s both NASA and the Pentagon have used ammonium perchlorate, NH_4ClO_4, as a rocket fuel or in munitions. The result is that traces of perchlorate ion are found in groundwater in many regions of the United States, with levels ranging from about 4 to 100 ppb. Perchlorate is known to suppress hormone levels in humans by acting in the thyroid gland, and there is concern that perchlorate may especially impair the development of fetuses and young children. However, it is disputed whether the amounts found in drinking water are sufficiently high to cause health problems. The Environmental Protection Agency has proposed a recommended limit of 1 ppb in drinking water. This figure is vigorously disputed by the Pentagon and by industries that have done government work that involves perchlorates. They argue for a standard limit of 200 ppb, which would likely exempt most contamination from cleanup. California has set an interim safety level of 2 to 6 ppb and has ordered water systems not to exceed 40 ppb.

Removal of perchlorate ion from water supplies is not an easy matter. Although perchlorate is an oxidizing agent, the ClO_4^- ion is quite stable in aqueous solution. One promising avenue is to employ biological reduction, using microorganisms. While research continues on the best means of reducing perchlorate levels in drinking water, agencies of the federal government continue to disagree about appropriate target levels and about who must shoulder the costs.

Related Exercises: 22.39, 22.40, 22.102

CHEMISTRY AT WORK | Cell Phone Tower Range

Cell phone towers increasingly dot both rural and urban landscapes (Figure 12.8 ▼), yet it can be difficult to maintain contact with a tower in the midst of a telephone conversation. A cell phone communicates with the system by receiving a signal from the tower's transmitter and transmitting signals back to it. Although the tower's transmitter may be quite powerful, the cell phone has very limited power. As the distance from the tower increases or intervening structures interfere, the cell phone signal becomes too weak to be distinguished from general electronic noise.

The amplifiers in the tower's receiver have electronic filters that discriminate between the desired incoming signal and other electronic signals. The sharper the filter, the higher the discrimination between one channel and another, and therefore, the higher the capability of detecting the desired signal with clarity. Filters can be made from a high-temperature superconducting oxide that when cooled below its T_c provides much sharper filtering capabilities than conventional filters. By incorporating such filters in the tower receivers, communications companies can extend the range of the tower by up to a factor of 2, which saves construction costs and improves the reliability of communication.

Superconducting technology is now being deployed in PC-sized boxes located in cell phone base stations (the little building at the foot of the tower). The filters are formed from a ceramic oxide, typically $YBa_2Cu_3O_7$ or $Tl_2Ba_2CaCu_2O_8$. The requisite cooling is provided by a mechanical cooling device, which is a small refrigeration unit capable of cooling the filter below its T_c (Figure 12.9 ▼).

▲ **Figure 12.8 Wireless communications tower.** Towers like this (here, disguised as a tree) enable cell phone communicaton.

▲ **Figure 12.9 superconducting f** object to the left is t at a temperature be

Chemistry and Life and Chemistry at Work

Chemistry and Life and Chemistry at Work essays emphasize chemistry's connection to world events, scientific discoveries, and medical breakthroughs. Your instructor may ask you to read these boxes or you may just want to read them on your own. They'll show you a new way of looking at the world around you.

A Closer Look A Closer Look essays supplement the chapter material by covering high-interest topics in more detail.

A CLOSER LOOK | Probability Density and Radial Probability Functions

The quantum mechanical description of the hydrogen atom requires that we talk about the position of the electron in the atom in a somewhat unfamiliar way. In classical physics, we can exactly pinpoint the position and velocity of an orbiting object, such as a planet orbiting a star. Under quantum mechanics, however, we must describe the position of the electron in the hydrogen atom in terms of probabilities rather than an exact location—an exact answer would violate the uncertainty principle, which we have seen becomes important when considering subatomic particles. The information we need about the probability of finding the electron is contained in the wave functions, ψ, that are obtained when Schrödinger's equation is solved. Remember that there are an infinite number of wave functions (orbitals) for the hydrogen atom, but the electron can occupy only one of them at any given time. Here we will discuss briefly how we can use the orbitals to obtain radial probability functions, such as those in Figure 6.18.

In Section 6.5 we stated that the square of the wave function, ψ^2, gives the probability that the electron is at any one given point in space—recall that this quantity is called the *probability density* for the point. For a spherically symmetric *s* orbital, the value of ψ depends only on the distance from the nucleus, *r*. Let's consider a straight line outward from the nucleus, as shown in Figure 6.20 ▼. The probability of finding the electron at distance *r* from the nucleus along that line is

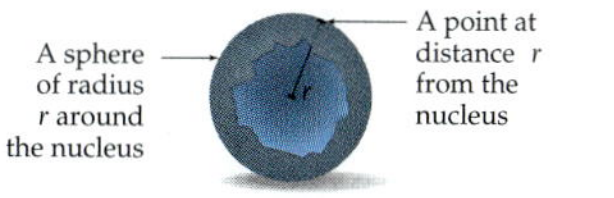

▲ **Figure 6.20 Probability at a point.** The probability density, $\psi(r)^2$, gives the probability that the electron will be found at *a specific point* at distance *r* from the nucleus. The radial probability function, $4\pi r^2\psi(r)^2$, gives the probability that the electron will be found at *any* point distance *r* from the nucleus—in other words, at any point on the sphere of radius *r*.

$[\psi(r)]^2$ where $\psi(r)$ is the value of ψ at distance *r*. Figure 6.21 ▼ shows plots of $[\psi(r)]^2$ as a function of *r* for the 1*s*, 2*s*, and 3*s* orbitals of the hydrogen atom.

You will notice that the plots in Figure 6.21 look distinctly different from the radial probability functions plotted in Figure 6.18. These two types of plots for the *s* orbitals are very closely related, but they provide somewhat different information. The probability density, $[\psi(r)]^2$, tells us the probability of finding the electron at *a specific* point in space that is at distance *r* from the nucleus. The radial probability function, which we will denote $P(r)$, tells us the probability of finding the electron at *any* point that is distance *r* from the nucleus—in other words, to get $P(r)$ we need to "add up" the probabilities of finding the electron over all the points at distance *r* from the nucleus. The difference between these descriptions may seem rather subtle, but mathematics provides us with a precise way to connect them.

As shown in Figure 6.20, the collection of points at distance *r* from the nucleus is simply a sphere of radius *r*. The probability density at every point on that sphere is $[\psi(r)]^2$. To add up all of the individual probability densities requires the use of calculus and is beyond the scope of this text (in the language of calculus "we integrate the probability density over the surface of the sphere"). The result we obtain is easy to describe, however. The radial probability function at distance *r*, $P(r)$, is simply the probability density at distance *r*, $[\psi(r)]^2$, multiplied by the surface area of the sphere, which is given by the formula $4\pi r^2$:

$$P(r) = 4\pi r^2[\psi(r)]^2$$

Thus, the plots of $P(r)$ in Figure 6.18 are equal to the plots of $[\psi(r)]^2$ in Figure 6.21 multiplied by $4\pi r^2$. The fact that $4\pi r^2$ increases rapidly as we move away from the nucleus makes the two sets of plots look very different. For example, the plot of $[\psi(r)]^2$ for the 3*s* orbital (Figure 6.21) shows that the function generally gets smaller the farther we go from the nucleus. But when we multiply by $4\pi r^2$, we see peaks that get larger

Resources and How They Can Help You

Instructors and students engage in a multitude of activities in the course of teaching and of learning general chemistry. Prentice Hall understands the challenges involved in each of these steps and has provided a wide spectrum of resources that are specifically designed to help solve those problems. The flowchart below outlines the resources that are available, based on events that occur throughout the course. Following the flowchart is a list of these resources with descriptions. As always, you may contact your Prentice Hall representative for help determining the best solution for your needs.

INSTRUCTOR

Prepare

Prepare your lecture – ***The Instructor's Resource Center (IRC)*** on CD/DVD contains lecture outlines from the ***Instructor's Resource Manual***, pre-built PowerPoint lecture presentations, and a searchable library of all images from the textbook and all interactive animations, manipulable molecules, and movies for you to incorporate in your lecture.

Prepare your students – Help students prepare for the lecture by assigning a pre-lecture reading comprehension assignment using ***PH GradeAssist*** or our ***OneKey*** course management tools.

Lecture

Illustrate concepts in class – Using the art-only version, or adapting our pre-built lecture presentations from the ***IRC***, you can take advantage of the high-quality art and lecture guidelines from the book. Use our dynamic animations of microscopic processes, our video clips of difficult demonstrations, our 3-D molecular models, and the "what-if" power of our simulations to explain concepts engagingly.

Interact in class – Our support for ***Classroom Response Systems (CRS)***, including hardware discounts from our two CRS partners and "clicker" questions, lets you poll students in class and collect their responses quickly and anonymously. You can quickly determine what misconceptions are tripping up your students, and you will engage them as they discuss the answers to questions carefully designed for response systems.

STUDENT

Prepare

Read – Prepare for lecture by reading the chapter in the ***Textbook***, whether you purchase the print version or the online ***WebBook*** version.

Focus – If you elected to purchase the ***WebBook,*** use the diagnostic Custom Study Plan to determine where you are weak and what readings in the textbook address those areas, or use the practice questions on the ***Companion Website with GradeTracker*** to check whether you understand the basic concepts before the lecture.

Catch up – The ***Math Toolkit*** will help you brush up on basic math skills, so that you can focus on understanding the chemistry.

Lecture

Follow the lecture – The ***Lecture Notebook*** contains the same text images that your instructor will use in lecture, allowing you to take notes quickly and efficiently. All of the interactive eMedia assets your instructor uses in the classroom are also available to you on the ***Student Accelerator CD***, on the ***Companion Website with GradeTracker***, and in ***OneKey***.

Participate in lecture – Classroom Response Systems let you "click" answers to questions posed in class, letting your instructor know on the fly whether you have misunderstood a concept or are confused. Though you might discuss the questions with fellow students, your response is anonymous and not shared with your peers.

Guide student learning outside of class – Assemble practice problem sets in ***PH GradeAssist*** or ***OneKey*** that your students can work repeatedly, getting instantaneous feedback on each attempt until they establish mastery. You may elect to track student progress or allow them work on their own.

Communicate outside of class – Use ***OneKey*** course management tools to hold virtual office hours, to set up online study groups for your students, or to post your lecture notes or sample quizzes for them.

Prepare for lab – The ***Virtual ChemLab Workbook*** is a realistic, simulated laboratory environment where students can get a feel for what to expect in a real wet lab, or conduct experiments that are not included in your department's lab program. Choose a lab sequence – Any of our general chemistry ***Lab Manuals*** has a wide selection of carefully designed laboratory experiments to follow your syllabus, or you can use our ***Catalyst*** system to select exactly the experiments you will use from any of our lab manuals and create a customized lab manual for your course.

Assign homework – Use ***PH GradeAssist*** or ***OneKey*** course management to assign homework online that is automatically graded, is easily configured, and stores student scores in a versatile grade book. PH GradeAssist question banks include textbook questions, interactive eMedia questions, and other book-specific questions. You can randomize the questions so that each student receives a slightly different assignment.

Create exams – Our ***TestGen*** test-creation software lets you choose questions from deep question banks and format them flexibly for printed exams or for online delivery in ***OneKey*** course management.

Practice & Support

Experiment

Assess

Practice – Test your understanding by working through practice problems that provide instant feedback on your correct and incorrect answers. Repeat these assignments until you are comfortable with the material. The ***Companion Website with GradeTracker*** provides practice questions, and your instructor may provide additional practice assignments with ***PH GradeAssist*** or ***OneKey*** that preview homework assignments and quizzes. If you are using the ***WebBook***, the Custom Study Plan will tell you where you need additional practice.

Get help – The ***Student Study Guide*** provides a problem-oriented aid and summary of key points for each chapter. The ***Student Solutions Manual*** contains detailed, worked solutions to textbook problems. The textbook eLaboratory Exercises and the ***Virtual ChemLab Workbook*** contain problems in a simulated laboratory setting that help you see how textbook examples work in real life. If your instructor is using ***OneKey***, you can work in virtual study groups with other students in your class or ask questions of your instructor, anytime and anywhere.

Prepare for lab – Optimize your valuable time in the laboratory by knowing what to expect when you arrive. Use the ***Virtual ChemLab Workbook*** to try out an experiment identical or similar to the experiment in the ***Lab Manual***.

Homework, quizzes, and exams – Using ***PH GradeAssist*** or ***OneKey***, your instructor can assign homework that gives you instantaneous feedback on how you are doing, gives you another try to improve your performance, and gives you repeated practice with randomized problems to help you prepare for exams. Keep track of your own progress in the grade book, so that you know how you are doing day by day.

List of Resources

For Students

Student MediaPak. (0-13-134451-X) Every copy of the text comes with a Student MediaPak, which includes access to the Companion Website with GradeTracker available at http://chem.prenhall.com/brown, the Accelerator CD, and the Virtual ChemLab Workbook and CD.

Student MediaPak for OneKey. Assigned by an instructor who prefers more robust, customizable web resources, includes access to a OneKey course and access to the PH GradeAssist online homework system. It also contains the Accelerator CD and the Virtual ChemLab Workbook and CD.

Accelerator CD serves as a resource of media animations and movies and works in conjunction with PH online resources to speed up download and interaction with the media. The Accelerator CD contains all the same movies, animations, simulations, and 3-D molecular models found on the Student MediaPak websites.

Virtual ChemLab Workbook and CD contains a full virtual lab environment as well as the pre-arranged labs that are referenced in the workbook and at the end of the chapter in the textbook. Virtual ChemLab can be run directly from the CD or installed on the student's computer.

WebBook, a complete, electronic version of the textbook, with a diagnostic study-plan quiz and embedded interactive objects, that also supports student highlighting, bookmarking, and jotted notes.

Solutions to Red Exercises (0-13-146486-8) Prepared by Roxy Wilson of University of Illinois–Urbana-Champaign. Full solutions to all of the red-numbered exercises in the text are provided. (Short answers to red exercises are found in the appendix of the text).

Solutions to Black Exercises (0-13-146485-X) Prepared by Roxy Wilson of University of Illinois–Urbana-Champaign. Full solutions to all of the black-numbered exercises in the text are provided.

Student's Guide (0-13-146484-1) Prepared by James C. Hill of California State University. This book assists students through the text material with chapter overviews, learning objectives, review of key terms, cumulative chapter review quizzes, and self-tests. Included are answers to all Student's Guide exercises. Chapter summaries are correlated to those in the Instructor's Resource Manual.

Math Review Toolkit (0-13-186052-6) Prepared by Gary Long of Virginia Polytechnic Institute. This book reinforces the skills necessary to succeed in chemistry and is included at no additional charge if ordered in a package with the text. It is keyed specifically to chapters in *Chemistry: The Central Science, Tenth Edition*, and includes additional mathematics review, problem-solving tools and examples, and a section on writing for the laboratory.

Student Lecture Notebook (0-13-146481-7) This lecture notebook contains the art from the text with note-taking sections to obviate the need for students to spend time re-drawing figures in lecture and instead lets them concentrate on taking notes.

Companion Website with GradeTracker contains problems, movies, animations, and hundreds of molecules in the molecular gallery. The Companion website can be accessed at http://chem.prenhall.com/brown.

For Instructors On Qualified Adoptions

Annotated Instructor's Edition (0-13-146489-2) Prepared by Linda Brunauer of Santa Clara University. This special instructor's edition provides marginal notes and information for instructors and TAs, including Instructor Resource Center icons, suggested lecture demonstrations, teaching tips, and background references from the chemical education literature for key topics.

Full Solutions Manual (0-13-146491-4) Prepared by Roxy Wilson of University of Illinois–Urbana-Champaign. This full solutions manual contains all end-of-chapter exercises in the text. With an instructor's permission, this manual may be made available to students.

OneKey's online course management content is all you need to plan and administer your course, and includes the best teaching and learning resources all in one place. Conveniently organized by textbook chapter, these compiled resources help you save time and help your students reinforce and apply what they have learned in class. Available resources include Summary with Key Terms, Tools, Research Navigator web research center, Math Toolkit, and media-enriched WebBook sections to reference while solving problems. Assessment content includes Practice Problems, Current Topics, MCAT Practice Questions, and the Test Item File, plus access to the PH GradeAssist online homework system. Resources from the Instructor's Resource Center on CD/DVD are also included. OneKey is available for institutions using WebCT or Blackboard, and is also available in our nationally hosted CourseCompass course management system. If desired, WebCT and Blackboard cartridges containing only the Test Item File are also available for download. Visit http://cms.prenhall.com for details.

PH GradeAssist online homework system that allows students unlimited practice with problems that are algorithmically generated and media-enhanced. Instructors can administer quizzes and assignments, control the content and assignment parameters, and view performance statistics with the built-in gradebook. Many questions are linked to individual sections of the WebBook.

Instructor's Resource Center on CD-DVD (0-13-146487-6) Prepared by Linda Brunauer of Santa Clara University and Karen Wiechelman of The University of Louisiana-Lafayette. This lecture resource provides a fully searchable and integrated collection of resources to help you make efficient and effective use of your lecture preparation time, as well as to enhance your classroom presentations and assessment efforts. This resource features almost all the art from the text, including tables; two pre-built PowerPoint™ presentations; PDF files of the art for high-resolution printing; all interactive and dynamic media objects from the student MediaPak; the Instructor's Resource Manual Word™ files, and a set of "clicker" questions for use with Classroom Response Systems. This CD also features a search-engine tool that enables you find relevant resources via key terms, learning objectives, figure numbers, and resource type (e.g., simulations). This CD/DVD set also contains TestGen test-generation software and a TestGen version of the Test Item File that enables professors to create and tailor exams to their needs, or to create online quizzes for delivery in WebCT, Blackboard, or CourseCompass.

Classroom Response Systems and Questions

Prepared by Dr. Donna Narsavage-Heald of Siena College and Troy Wood of the University of Buffalo. We offer a variety of options for ***Classroom Response Systems (CRS)***, including hardware discounts and textbook-specific "clicker" questions. Please contact your PH representative for more information.

Test Item File (0-13-146496-5) Prepared by Joseph P. Laurino and Donald Cannon, both of the University of Tampa. The Test Item File now provides a selection of more than 4000 test questions.

Instructor's Resource Manual (0-13-146495-7) Prepared by Linda Brunauer of Santa Clara University and Karen Wiechelman of The University of Louisiana-Lafayette. This useful guide describes all the different resources available to instructors and shows how to integrate them into your course. Organized by chapter, this manual offers detailed lecture outlines and complete descriptions of all available lecture demonstrations, the interactive media assets, common student misconceptions, and much more.

Transparencies (0-13-146490-6) Over 300 full-color transparencies chosen from the text put principles into visual perspective and save you time when preparing lectures.

Laboratory Experiments (0-13-146479-5) Prepared by John H. Nelson and Kenneth C. Kemp, both of The University of Nevada. This manual contains 43 finely tuned experiments chosen to introduce students to basic lab techniques and to illustrate core chemical principles. Pre-lab questions and post-lab questions have been revised and include detachable report sheets. Safety and disposal information has been expanded and includes waste management information. This new edition has been revised to correlate more tightly with the text. **Annotated Instructor's Edition to *Laboratory Experiments*** (0-13-146488-4) Prepared by John H. Nelson and Kenneth C. Kemp, both of The University of Nevada. This AIE combines the full student lab manual with front and back appendices covering the proper disposal of chemical waste, safety instructions for the lab, descriptions of standard lab equipment and materials, answers to questions, and more. This laboratory book, as well as a wealth of other Prentice Hall lab experiments, are available in a fully customizable format. Catalyst allows you to view, select, and organize lab experiments from all of our lab offerings, and add your own content as well. Visit http://www.prenhall.com/catalyst or contact your Prentice Hall representative for more information.

WebAssign is a Prentice Hall partner online homework system. For information on this system, contact your PH representative.

Options for Ordering

Text or WebBook with Student MediaPak. The MediaPak comes with every new copy of the text, and contains all the digital resources that students will need, including access to the Companion Website with GradeTracker, Student Accelerator CD, and the Virtual ChemLab Workbook and CD.

Text or WebBook with MediaPak for OneKey. This MediaPak requires selection by instructors and includes the Student Accelerator CD, the Virtual ChemLab Workbook and CD, as well as a OneKey and PHGradeAssist Access Code. OneKey with PHGradeAssist is available in WebCT (0-13-168632-1), Blackboard (0-13-168633-X), and CourseCompass (0-13-168635-6).

About the Authors

THEODORE L. BROWN received his Ph.D. from Michigan State University in 1956. Since then, he has been a member of the faculty of the University of Illinois, Urbana-Champaign, where he is now Professor of Chemistry, Emeritus. He served as Vice Chancellor for Research, and Dean, The Graduate College, from 1980 to 1986, and as Founding Director of the Arnold and Mabel Beckman Institute for Advanced Science and Technology from 1987 to 1993. Professor Brown has been an Alfred P. Sloan Foundation Research Fellow and has been awarded a Guggenheim Fellowship. In 1972 he was awarded the American Chemical Society Award for Research in Inorganic Chemistry, and received the American Chemical Society Award for Distinguished Service in the Advancement of Inorganic Chemistry in 1993. He has been elected a Fellow of both the American Association for the Advancement of Science and the American Academy of Arts and Sciences.

H. EUGENE LEMAY, JR., received his B.S. degree in Chemistry from Pacific Lutheran University (Washington) and his Ph.D. in Chemistry in 1966 from the University of Illinois (Urbana). He then joined the faculty of the University of Nevada, Reno, where he is currently Professor of Chemistry, Emeritus. He has enjoyed Visiting Professorships at the University of North Carolina at Chapel Hill, at the University College of Wales in Great Britain, and at the University of California, Los Angeles. Professor LeMay is a popular and effective teacher, who has taught thousands of students during more than 35 years of university teaching. Known for the clarity of his lectures and his sense of humor, he has received several teaching awards, including the University Distinguished Teacher of the Year Award (1991) and the first Regents' Teaching Award given by the State of Nevada Board of Regents (1997).

BRUCE E. BURSTEN received his Ph.D. in Chemistry from the University of Wisconsin in 1978. After two years as a National Science Foundation Postdoctoral Fellow at Texas A&M University, he joined the faculty of The Ohio State University, where he is currently Distinguished University Professor. Professor Bursten has been a Camille and Henry Dreyfus Foundation Teacher-Scholar and an Alfred P. Sloan Foundation Research Fellow, and he has been elected a Fellow of the American Association for the Advancement of Science. At Ohio State he has received the University Distinguished Teaching Award in 1982 and 1996, the Arts and Sciences Student Council Outstanding Teaching Award in 1984, and the University Distinguished Scholar Award in 1990. In addition to his teaching activities, Professor Bursten's research program focuses on compounds of the transition-metal and actinide elements.

Contributing Author

CATHERINE J. MURPHY received two B.S. degrees, one in Chemistry and one in Biochemistry, from the University of Illinois, Urbana-Champaign, in 1986. She received her Ph.D. in Chemistry from the University of Wisconsin in 1990. She was a National Science Foundation and National Institutes of Health Postdoctoral Fellow at the California Institute of Technology from 1990 to 1993. In 1993, she joined the faculty of the University of South Carolina, Columbia, where she is currently the Guy F. Lipscomb Professor of Chemistry. Professor Murphy has been honored for both research and teaching as a Camille Dreyfus Teacher-Scholar, an Alfred P. Sloan Foundation Research Fellow, a Cottrell Scholar of the Research Corporation, a National Science Foundation CAREER Award winner and a subsequent NSF Award for Special Creativity. She has also received a USC Mortar Board Excellence in Teaching Award, the USC Golden Key Faculty Award for Creative Integration of Research and Undergraduate Teaching, the USC Michael J. Mungo Undergraduate Teaching Award, and the USC Outstanding Undergraduate Research Mentor Award. Professor Murphy's research program focuses on the synthesis and optical properties of inorganic nanomaterials, and on the local structure and dynamics of the DNA double helix.

CHAPTER

1

WHAT'S AHEAD

- We begin by providing a very brief perspective of what chemistry is about and why it is useful to learn chemistry. *(Section 1.1)*
- Next, we examine some fundamental ways to classify materials, distinguishing between *pure substances* and *mixtures* and noting that there are two fundamentally different kinds of pure substances: *elements* and *compounds*. *(Section 1.2)*
- We then consider some of the different kinds of characteristics, or *properties*, that we use to characterize, identify, and separate substances. *(Section 1.3)*
- Many properties rely on quantitative measurements, involving both numbers and units. The units of measurement used throughout science are those of the *metric system*, a decimal system of measurement. *(Section 1.4)*
- The uncertainties inherent in all measured quantities and those obtained from calculations involving measured quantities are expressed by the number of *significant figures* used to report the number. *(Section 1.5)*
- Units as well as numbers are carried through calculations, and obtaining correct units for the result of a calculation is an important way to check whether the calculation is correct. *(Section 1.6)*

Introduction: Matter and Measurement

HAVE YOU EVER WONDERED why ice melts and water evaporates? Why leaves turn colors in the fall and how a battery generates electricity? Why keeping foods cold slows their spoilage and how our bodies use food to maintain life? Chemistry supplies answers to these questions and countless others like them. **Chemistry** is the study of the properties of materials and the changes that materials undergo. One of the joys of learning chemistry is seeing how chemical principles operate in all aspects of our lives, from everyday activities like lighting a match to more far-reaching matters like the development of drugs to cure cancer. Chemical principles also operate in the far reaches of our galaxy (chapter-opening photo) as well as both within and around us.

You are just beginning the journey of learning chemistry. In a sense, this text is your guide on this journey. Throughout your studies, we hope that you will find this text enjoyable as well as educational. As you study, keep in mind that the chemical facts and concepts you are asked to learn are not ends in themselves, but tools to help you better understand the world around you. This first chapter lays a foundation for our studies by providing an overview of what chemistry is about and dealing with some fundamental concepts of matter and scientific measurements. The list to the left, entitled "What's Ahead," gives a brief overview of some of the ideas that we will be considering in this chapter.

HUBBLE SPACE TELESCOPE IMAGE of the Omega Nebula, which resides 5,500 light-years away in the constellation Sagittarius. Ultraviolet radiation from young, massive stars in the region cause atoms to emit light. Hydrogen, the simplest and most plentiful element, occurs as molecules in cool regions, as atoms in hotter regions, and as ions in the hottest regions. The processes that occur within stars are responsible for creating other chemical elements from hydrogen.

1.1 | The Study of Chemistry

Before traveling to an unfamiliar city, you might look at a map to get some sense of where you are heading. Because chemistry may be unfamiliar to you, it's useful to get a general idea of what lies ahead before you embark on your journey. In fact, you might even ask why you are taking the trip.

The Atomic and Molecular Perspective of Chemistry

Chemistry involves studying the properties and behavior of matter. **Matter** is the physical material of the universe; it is anything that has mass and occupies space. A **property** is any characteristic that allows us to recognize a particular type of matter and to distinguish it from other types. This book, your body, the clothes you are wearing, and the air you are breathing are all samples of matter. Not all forms of matter are so common or so familiar, but countless experiments have shown that the tremendous variety of matter in our world is due to combinations of only about 100 very basic, or elementary, substances called **elements**. As we proceed through this text, we will seek to relate the properties of matter to its composition, that is, to the particular elements it contains.

Chemistry also provides a background to understanding the properties of matter in terms of **atoms**, the almost infinitesimally small building blocks of matter. Each element is composed of a unique kind of atom. We will see that the properties of matter relate not only to the kinds of atoms it contains (*composition*), but also to the arrangements of these atoms (*structure*).

3-D MODELS Oxygen, Water, Carbon Dioxide, Ethanol, Ethylene Glycol, Aspirin

Atoms can combine to form **molecules** in which two or more atoms are joined together in specific shapes. Throughout this text you will see molecules represented using colored spheres to show how their component atoms connect to each other (Figure 1.1 ▼). The color merely provides a convenient way to distinguish between the atoms of different elements. As examples, compare the molecules of ethanol and ethylene glycol, which are depicted in Figure 1.1. Notice that these molecules differ somewhat in composition. Ethanol contains one red sphere, which represents one oxygen atom, whereas ethylene glycol contains two.

Even apparently minor differences in the composition or structure of molecules can cause profound differences in their properties. Ethanol, also called grain alcohol, is the alcohol in beverages such as beer and wine. Ethylene glycol, on the other hand, is a viscous liquid used as automobile antifreeze. The properties of these two substances differ in a great number of ways, including the temperatures at which they freeze and boil. One of the challenges that chemists undertake is to alter the composition or structure of molecules in a controlled way, creating new substances with different properties.

Every change in the observable world—from boiling water to the changes that occur as our bodies combat invading viruses—has its basis in the world of atoms and molecules. Thus, as we proceed with our study of chemistry, we will find

▼ **Figure 1.1 Molecular models.** The white, dark gray, and red spheres represent atoms of hydrogen, carbon, and oxygen, respectively.

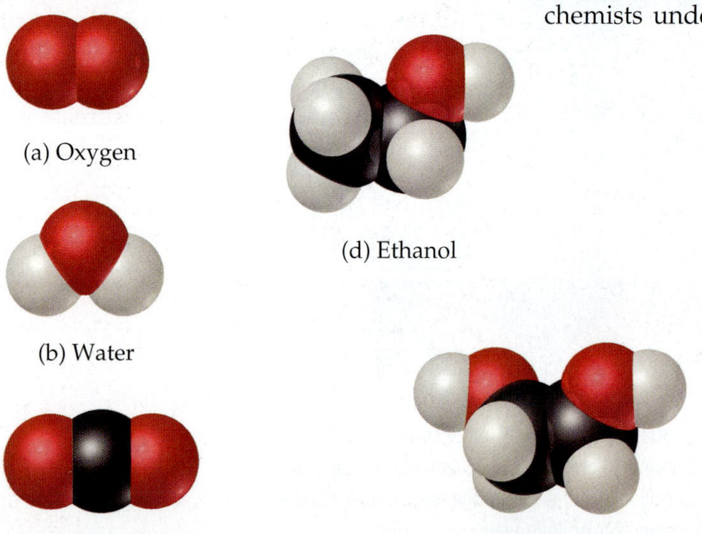

(a) Oxygen

(b) Water

(c) Carbon dioxide

(d) Ethanol

(e) Ethylene glycol

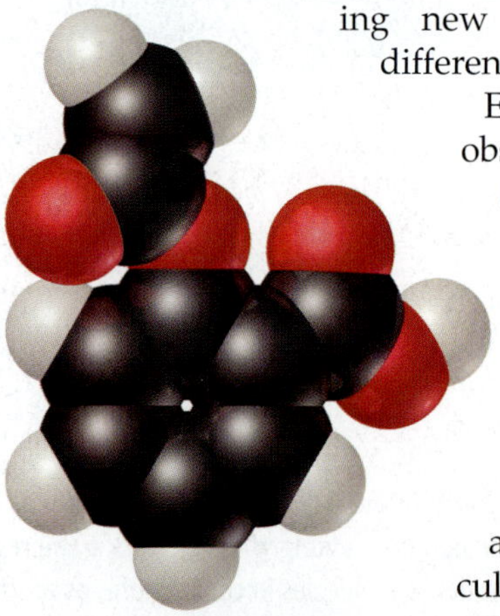

(f) Aspirin

ourselves thinking in two realms, the *macroscopic* realm of ordinary-sized objects (*macro* = large) and the *submicroscopic* realm of atoms and molecules. We make our observations in the macroscopic world—in the laboratory and in our everyday surroundings. In order to understand that world, however, we must visualize how atoms and molecules behave at the submicroscopic level. Chemistry is the science that seeks to understand the properties and behavior of matter by studying the properties and behavior of atoms and molecules.

GIVE IT SOME THOUGHT

(a) In round numbers, about how many elements are there? (b) What submicroscopic particles are the building blocks of matter?

Why Study Chemistry?

Chemistry provides important understanding of our world and how it works. It is an extremely practical science that greatly impacts our daily living. Indeed, chemistry lies near the heart of many matters of public concern: improvement of health care, conservation of natural resources, protection of the environment, and provision of our everyday needs for food, clothing, and shelter. Using chemistry, we have discovered pharmaceutical chemicals that enhance our health and prolong our lives. We have increased food production through the development of fertilizers and pesticides. We have developed plastics and other materials that are used in almost every facet of our lives. Unfortunately, some chemicals also have the potential to harm our health or the environment. It is in our best interest as educated citizens and consumers to understand the profound effects, both positive and negative, that chemicals have on our lives and to strike an informed balance about their uses.

Most of you are studying chemistry, however, not merely to satisfy your curiosity or to become more informed consumers or citizens, but because it is an essential part of your curriculum. Your major might be biology, engineering, agriculture, geology, or some other field. Why do so many diverse subjects share an essential tie to chemistry? The answer is that chemistry, by its very nature, is the *central science*, central to a fundamental understanding of other sciences and technologies. For example, our interactions with the material world raise basic questions about the materials around us. What are their compositions and properties? How do they interact with us and our environment? How, why, and when do they undergo change? These questions are important whether the material is part of high-tech computer chips, an aged pigment used by a Renaissance painter, or the DNA that transmits genetic information in our bodies (Figure 1.2 ▼).

By studying chemistry, you will learn to use the powerful language and ideas that have evolved to describe and enhance our understanding of matter.

(a) (b) (c)

◀ **Figure 1.2 Chemistry helps us better understand materials.** (a) A microscopic view of an EPROM (Erasable Programmable Read-Only Memory) silicon microchip. (b) A Renaissance painting, *Young Girl Reading*, by Vittore Carpaccio (1472–1526). (c) A long strand of DNA that has spilled out of the damaged cell wall of a bacterium.

CHEMISTRY AT WORK | Chemistry and the Chemical Industry

Many people are familiar with common household chemicals such as those shown in Figure 1.3 ▶, but few realize the size and importance of the chemical industry. Worldwide sales of chemicals and related products manufactured in the United States total over $450 billion annually. The chemical industry employs over 10% of all scientists and engineers and is a major contributor to the U.S. economy.

Vast amounts of chemicals are produced each year and serve as raw materials for a variety of uses, including the manufacture of metals, plastics, fertilizers, pharmaceuticals, fuels, paints, adhesives, pesticides, synthetic fibers, microprocessor chips, and numerous other products. Table 1.1 ▼ lists the top ten chemicals produced in the United States. We will discuss many of these substances and their uses as the course progresses.

People who have degrees in chemistry hold a variety of positions in industry, government, and academia. Those who work in the chemical industry find positions as laboratory chemists, carrying out experiments to develop new products (research and development), analyzing materials (quality control), or assisting customers in using products (sales and service). Those with more experience or training may work as managers or company directors. There are also alternate careers that a chemistry degree prepares you for such as teaching, medicine, biomedical research, information science, environmental work, technical sales, work with government regulatory agencies, and patent law.

▲ **Figure 1.3 Household chemicals.** Many common supermarket products have very simple chemical compositions.

TABLE 1.1 The Top Ten Chemicals Produced by the Chemical Industry in 2003[a]

Rank	Chemical	Formula	2003 Production (billions of pounds)	Principal End Uses
1	Sulfuric acid	H_2SO_4	82	Fertilizers, chemical manufacturing
2	Nitrogen	N_2	75	Fertilizers
3	Oxygen	O_2	61	Steel, welding
4	Ethylene	C_2H_4	51	Plastics, antifreeze
5	Lime	CaO	38	Paper, cement, steel
6	Propylene	C_3H_6	31	Plastics
7	Ammonia	NH_3	24	Fertilizers
8	Chlorine	Cl_2	24	Bleaches, plastics, water purification
9	Phosphoric acid	H_3PO_4	24	Fertilizers
10	Sodium hydroxide	NaOH	19	Aluminum production, soap

[a]Most data from *Chemical and Engineering News*, July 5, 2004, pp. 51, 54.

The language of chemistry is a universal scientific language that is widely used in other disciplines. Furthermore, an understanding of the behavior of atoms and molecules provides powerful insights in other areas of modern science, technology, and engineering. For this reason, chemistry will probably play a significant role in your future. You will be better prepared for the future if you increase your understanding of chemical principles, and it is our goal to help you achieve this end.

1.2 | Classifications of Matter

Let's begin our study of chemistry by examining some fundamental ways in which matter is classified and described. Two principal ways of classifying matter are according to its physical state (as a gas, liquid, or solid) and according to its composition (as an element, compound, or mixture).

States of Matter

A sample of matter can be a gas, a liquid, or a solid. These three forms of matter are called the **states of matter**. The states of matter differ in some of their simple observable properties. A **gas** (also known as *vapor*) has no fixed volume or shape; rather, it conforms to the volume and shape of its container. A gas can be compressed to occupy a smaller volume, or it can expand to occupy a larger one. A **liquid** has a distinct volume independent of its container but has no specific shape: It assumes the shape of the portion of the container that it occupies. A **solid** has both a definite shape and a definite volume. Neither liquids nor solids can be compressed to any appreciable extent.

The properties of the states can be understood on the molecular level (Figure 1.4 ▶). In a gas the molecules are far apart and are moving at high speeds, colliding repeatedly with each other and with the walls of the container. In a liquid the molecules are packed more closely together, but still move rapidly, allowing them to slide over each other; thus, liquids pour easily. In a solid the molecules are held tightly together, usually in definite arrangements, in which the molecules can wiggle only slightly in their otherwise fixed positions.

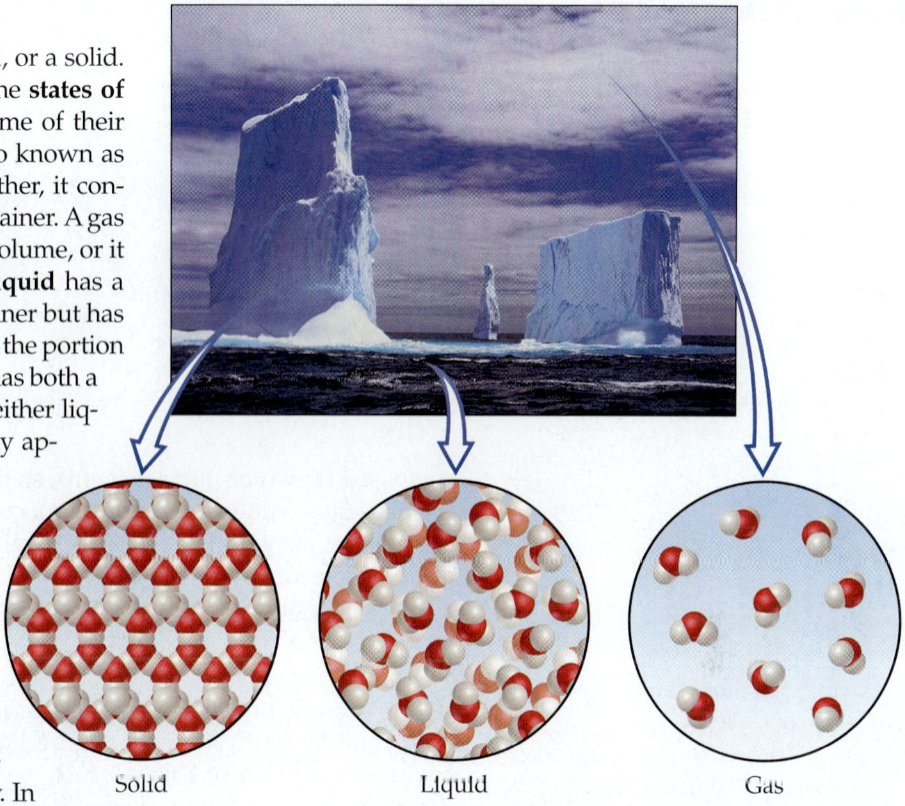

▲ **Figure 1.4 The three physical states of water—water vapor, liquid water, and ice.** In this photo we see both the liquid and solid states of water. We cannot see water vapor. What we see when we look at steam or clouds is tiny droplets of liquid water dispersed in the atmosphere. The molecular views show that the molecules in the solid are arranged in a more orderly way than in the liquid. The molecules in the gas are much farther apart than those in the liquid or the solid.

ACTIVITY
Phases of Water

Pure Substances

Most forms of matter that we encounter—for example, the air we breathe (a gas), gasoline for cars (a liquid), and the sidewalk on which we walk (a solid)—are not chemically pure. We can, however, resolve, or separate, these forms of matter into different pure substances. A **pure substance** (usually referred to simply as a *substance*) is matter that has distinct properties and a composition that doesn't vary from sample to sample. Water and ordinary table salt (sodium chloride), the primary components of seawater, are examples of pure substances.

All substances are either elements or compounds. **Elements** cannot be decomposed into simpler substances. On the molecular level, each element is composed of only one kind of atom [Figure 1.5(a and b) ▼]. **Compounds** are substances

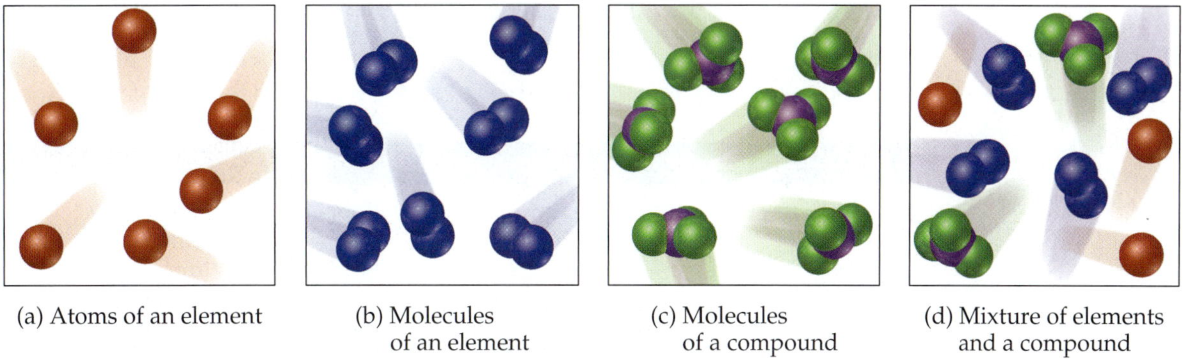

▲ **Figure 1.5 Molecular comparison of element, compounds, and mixtures.** Each element contains a unique kind of atom. Elements might consist of individual atoms, as in (a), or molecules, as in (b). Compounds contain two or more different atoms chemically joined together, as in (c). A mixture contains the individual units of its components, shown in (d) as both atoms and molecules.

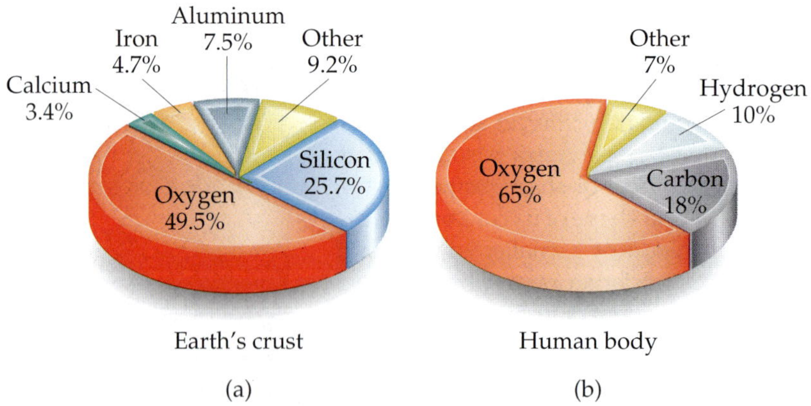

▶ **Figure 1.6 Relative abundances of elements.** Elements in percent by mass in (a) Earth's crust (including oceans and atmosphere) and (b) the human body.

composed of two or more elements, so they contain two or more kinds of atoms [Figure 1.5(c)]. Water, for example, is a compound composed of two elements, hydrogen and oxygen. Figure 1.5(d) shows a mixture of substances. **Mixtures** are combinations of two or more substances in which each substance retains its own chemical identity.

Elements

At the present time 116 elements are known. These elements vary widely in their abundance, as shown in Figure 1.6 ▲. For example, only five elements account for over 90% of Earth's crust: oxygen, silicon, aluminum, iron, and calcium. In contrast, just three elements (oxygen, carbon, and hydrogen) account for over 90% of the mass of the human body.

Some of the more common elements are listed in Table 1.2 ▼, along with the chemical abbreviations—or chemical *symbols*—used to denote them. All the known elements and their symbols are listed on the front inside cover of this text. The table in which the symbol for each element is enclosed in a box is called the *periodic table*. In the periodic table the elements are arranged in vertical columns so that closely related elements are grouped together. We describe this important tool in more detail in Section 2.5.

ACTIVITY
Phases of the Elements

The symbol for each element consists of one or two letters, with the first letter capitalized. These symbols are often derived from the English name for the element, but sometimes they are derived from a foreign name instead (last column in Table 1.2). You will need to know these symbols and to learn others as we encounter them in the text.

GIVE IT SOME THOUGHT

Which element is most abundant in both Earth's crust and in the human body? What is the symbol for this element?

TABLE 1.2 Some Common Elements and Their Symbols

Carbon	C	Aluminum	Al	Copper	Cu (from *cuprum*)
Fluorine	F	Bromine	Br	Iron	Fe (from *ferrum*)
Hydrogen	H	Calcium	Ca	Lead	Pb (from *plumbum*)
Iodine	I	Chlorine	Cl	Mercury	Hg (from *hydrargyrum*)
Nitrogen	N	Helium	He	Potassium	K (from *kalium*)
Oxygen	O	Lithium	Li	Silver	Ag (from *argentum*)
Phosphorus	P	Magnesium	Mg	Sodium	Na (from *natrium*)
Sulfur	S	Silicon	Si	Tin	Sn (from *stannum*)

Compounds

Most elements can interact with other elements to form compounds. For example, consider the fact that when hydrogen gas burns in oxygen gas, the elements hydrogen and oxygen combine to form the compound water. Conversely, water can be decomposed into its component elements by passing an electrical current through it, as shown in Figure 1.7 ▼. Pure water, regardless of its source, consists of 11% hydrogen and 89% oxygen by mass. This macroscopic composition corresponds to the molecular composition, which consists of two hydrogen atoms combined with one oxygen atom:

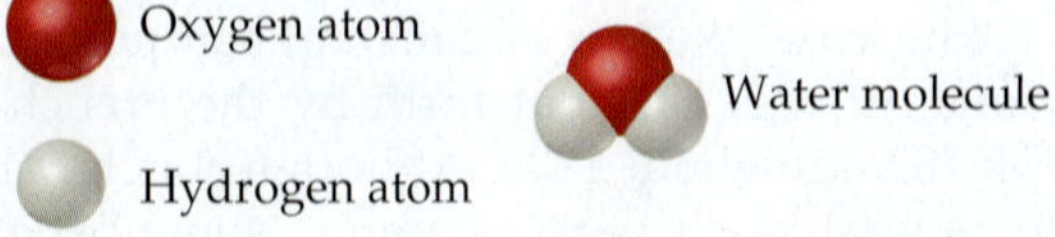

The elements hydrogen and oxygen themselves exist naturally as diatomic (two-atom) molecules:

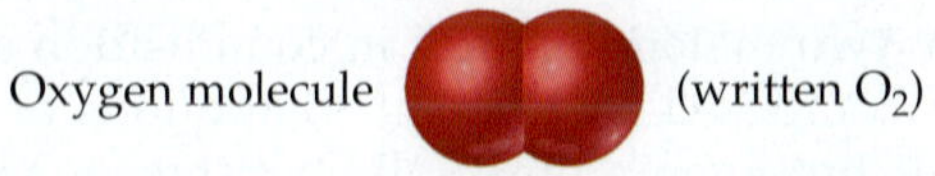

As seen in Table 1.3 ▶, the properties of water bear no resemblance to the properties of its component elements. Hydrogen, oxygen, and water are each unique substances, a consequence of the uniqueness of their respective molecules.

ANIMATION
Electrolysis of Water

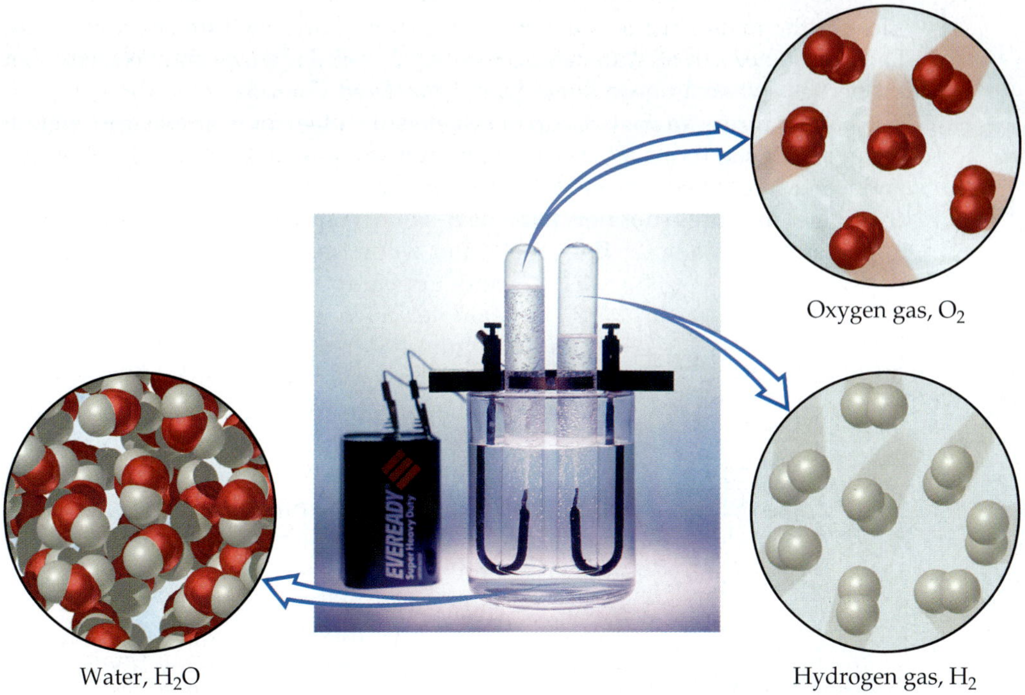

▲ **Figure 1.7 Electrolysis of water.** Water decomposes into its component elements, hydrogen and oxygen, when a direct electrical current is passed through it. The volume of hydrogen, which is collected in the right tube of the apparatus, is twice the volume of oxygen, which is collected in the left tube.

TABLE 1.3 Comparison of Water, Hydrogen, and Oxygen

	Water	Hydrogen	Oxygen
State[a]	Liquid	Gas	Gas
Normal boiling point	100°C	−253°C	−183°C
Density[a]	1.00 g/mL	0.084 g/L	1.33 g/L
Flammable	No	Yes	No

[a] At room temperature and atmospheric pressure. (See Section 10.2.)

The observation that the elemental composition of a pure compound is always the same is known as the **law of constant composition** (or the **law of definite proportions**). It was first put forth by the French chemist Joseph Louis Proust (1754–1826) in about 1800. Although this law has been known for 200 years, the general belief persists among some people that a fundamental difference exists between compounds prepared in the laboratory and the corresponding compounds found in nature. However, a pure compound has the same composition and properties regardless of its source. Both chemists and nature must use the same elements and operate under the same natural laws. When two materials differ in composition and properties, we know that they are composed of different compounds or that they differ in purity. Occasionally, however, a pure solid substance may have more than one crystalline form, and these may have slightly different properties.

GIVE IT SOME THOUGHT

Hydrogen, oxygen, and water are all composed of molecules. What is it about the molecules of water that makes water a compound?

Mixtures

Most of the matter we encounter consists of mixtures of different substances. Each substance in a mixture retains its own chemical identity and hence its own properties. Whereas pure substances have fixed compositions, the compositions of mixtures can vary. A cup of sweetened coffee, for example, can contain either a little sugar or a lot. The substances making up a mixture (such as sugar and water) are called *components* of the mixture.

Some mixtures do not have the same composition, properties, and appearance throughout. Both rocks and wood, for example, vary in texture and appearance throughout any typical sample. Such mixtures are *heterogeneous* [Figure 1.8(a) ◀]. Mixtures that are uniform throughout are *homogeneous*. Air is a homogeneous mixture of the gaseous substances nitrogen, oxygen, and smaller amounts of other substances. The nitrogen in air has all the properties that pure nitrogen does because both the pure substance and the mixture contain the same nitrogen molecules. Salt, sugar, and many other substances dissolve in water to form homogeneous mixtures [Figure 1.8(b)]. Homogeneous mixtures are also called **solutions**. Figure 1.9 ▶ summarizes the classification of matter into elements, compounds, and mixtures.

▼ **Figure 1.8 Mixtures.** (a) Many common materials, including rocks, are heterogeneous. This close-up photo is of *malachite*, a copper mineral. (b) Homogeneous mixtures are called solutions. Many substances, including the blue solid shown in this photo (copper sulfate), dissolve in water to form solutions.

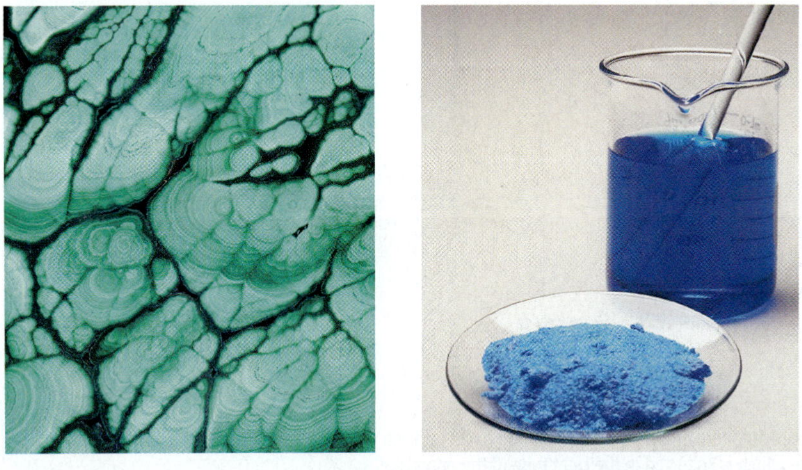

(a) (b)

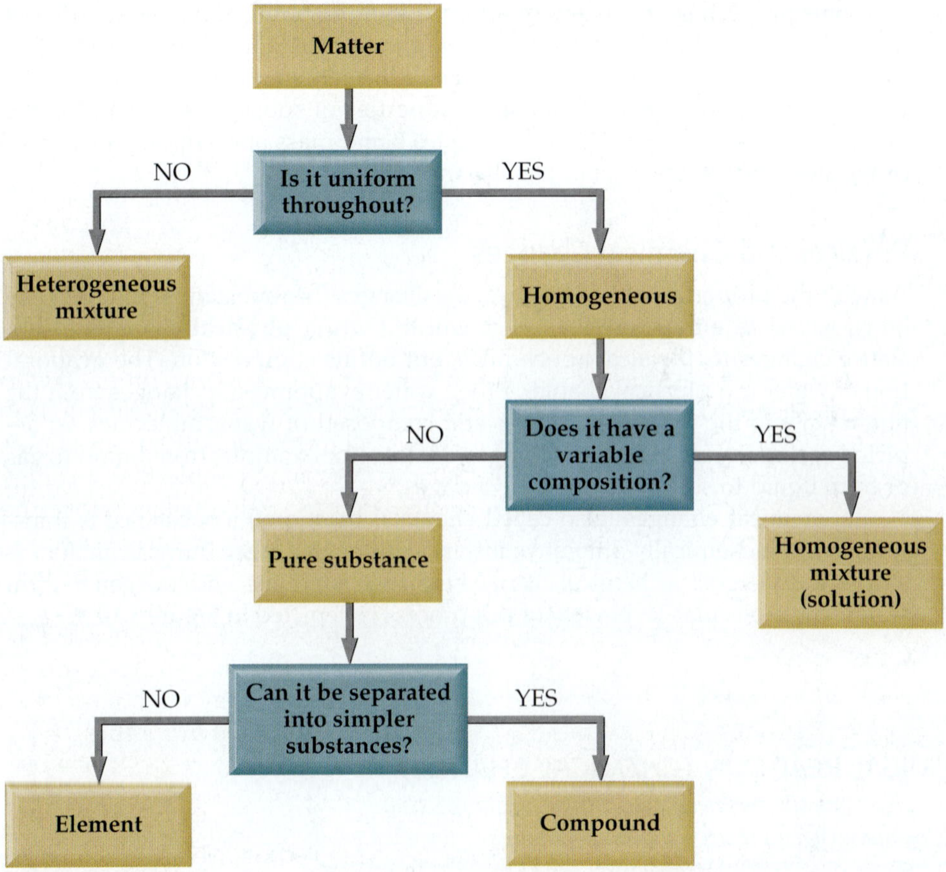

◀ **Figure 1.9 Classification of matter.** At the chemical level all matter is classified ultimately as either elements or compounds.

ACTIVITY
Classification of Matter

SAMPLE EXERCISE 1.1 | Distinguishing among elements, compounds, and mixtures

"White gold," used in jewelry, contains two elements, gold and palladium. Two different samples of white gold differ in the relative amounts of gold and palladium that they contain. Both are uniform in composition throughout. Without knowing any more about the materials, use Figure 1.9 to characterize and classify white gold.

Solution Because the material is uniform throughout, it is homogeneous. Because its composition differs for the two samples, it cannot be a compound. Instead, it must be a homogeneous mixture. Gold and palladium can be said to form a solid solution with one another.

PRACTICE EXERCISE

Aspirin is composed of 60.0% carbon, 4.5% hydrogen, and 35.5% oxygen by mass, regardless of its source. Use Figure 1.9 to characterize and classify aspirin.
Answer: It is a compound because it has constant composition and can be separated into several elements.

1.3 | Properties of Matter

Every substance has a unique set of properties. For example, the properties listed in Table 1.3 allow us to distinguish hydrogen, oxygen, and water from one another. The properties of matter can be categorized as physical or chemical. **Physical properties** can be measured without changing the identity and composition of the substance. These properties include color, odor, density, melting point, boiling point, and hardness. **Chemical properties** describe the way a substance may change, or *react*, to form other substances. A common chemical property is flammability, the ability of a substance to burn in the presence of oxygen.

Some properties—such as temperature, melting point, and density—do not depend on the amount of the sample being examined. These properties, called **intensive properties**, are particularly useful in chemistry because many can be used to *identify* substances. **Extensive properties** of substances depend on the quantity of the sample, with two examples being mass and volume. Extensive properties relate to the *amount* of substance present.

Physical and Chemical Changes

ANIMATION
Changes of State

As with the properties of a substance, the changes that substances undergo can be classified as either physical or chemical. During **physical changes** a substance changes its physical appearance, but not its composition. The evaporation of water is a physical change. When water evaporates, it changes from the liquid state to the gas state, but it is still composed of water molecules, as depicted earlier in Figure 1.4. All **changes of state** (for example, from liquid to gas or from liquid to solid) are physical changes.

In **chemical changes** (also called **chemical reactions**) a substance is transformed into a chemically different substance. When hydrogen burns in air, for example, it undergoes a chemical change because it combines with oxygen to form water. The molecular-level view of this process is depicted in Figure 1.10 ▼.

▼ **Figure 1.10 A chemical reaction.**

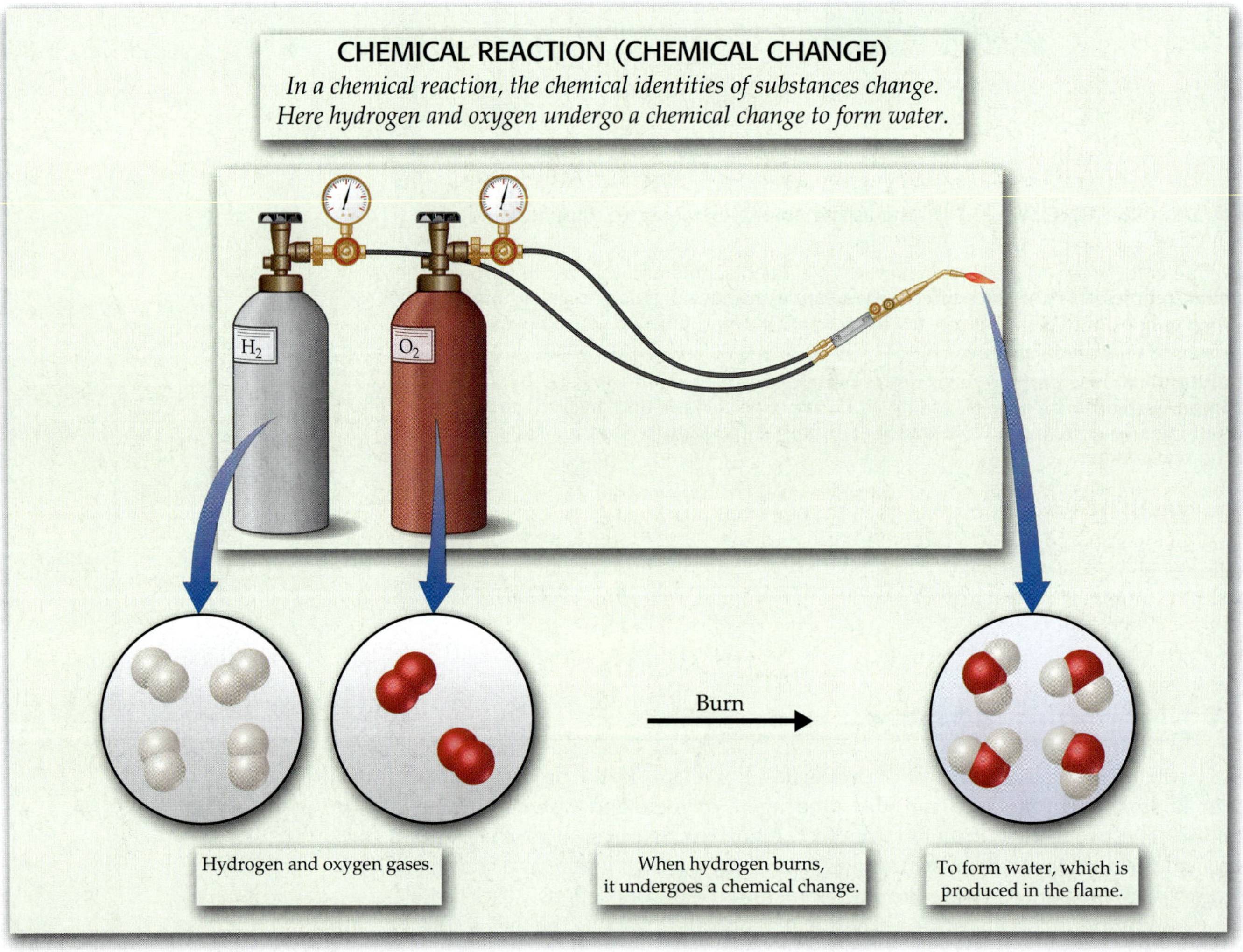

(a) (b) (c)

▲ **Figure 1.11 The chemical reaction between a copper penny and nitric acid.** The dissolved copper produces the blue-green solution; the reddish brown gas produced is nitrogen dioxide.

Chemical changes can be dramatic. In the account that follows, Ira Remsen, author of a popular chemistry text published in 1901, describes his first experiences with chemical reactions. The chemical reaction that he observed is shown in Figure 1.11 ▲.

> While reading a textbook of chemistry, I came upon the statement "nitric acid acts upon copper," and I determined to see what this meant. Having located some nitric acid, I had only to learn what the words "act upon" meant. In the interest of knowledge I was even willing to sacrifice one of the few copper cents then in my possession. I put one of them on the table, opened a bottle labeled "nitric acid," poured some of the liquid on the copper, and prepared to make an observation. But what was this wonderful thing which I beheld? The cent was already changed, and it was no small change either. A greenish-blue liquid foamed and fumed over the cent and over the table. The air became colored dark red. How could I stop this? I tried by picking the cent up and throwing it out the window. I learned another fact: nitric acid acts upon fingers. The pain led to another unpremeditated experiment. I drew my fingers across my trousers and discovered nitric acid acts upon trousers. That was the most impressive experiment I have ever performed. I tell of it even now with interest. It was a revelation to me. Plainly the only way to learn about such remarkable kinds of action is to see the results, to experiment, to work in the laboratory.

GIVE IT SOME THOUGHT

Which of the following is a physical change, and which is a chemical change? Explain. (a) Plants use carbon dioxide and water to make sugar. (b) Water vapor in the air on a cold day forms frost.

MOVIE
Sodium and Potassium in Water

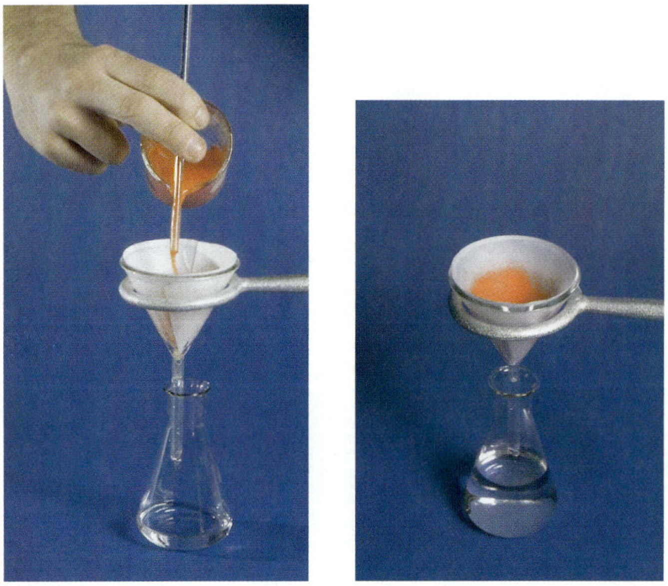

(a) (b)

▲ **Figure 1.12 Separation by filtration.** A mixture of a solid and a liquid is poured through a porous medium, in this case filter paper. The liquid passes through the paper while the solid remains on the paper.

Separation of Mixtures

Because each component of a mixture retains its own properties, we can separate a mixture into its components by taking advantage of the differences in their properties. For example, a heterogeneous mixture of iron filings and gold filings could be sorted individually by color into iron and gold. A less tedious approach would be to use a magnet to attract the iron filings, leaving the gold ones behind. We can also take advantage of an important chemical difference between these two metals: Many acids dissolve iron, but not gold. Thus, if we put our mixture into an appropriate acid, the iron would dissolve and the gold would be left behind. The two could then be separated by *filtration*, a procedure illustrated in Figure 1.12 ◀. We would have to use other chemical reactions, which we will learn about later, to transform the dissolved iron back into metal.

An important method of separating the components of a homogeneous mixture is *distillation*, a process that depends on the different abilities of substances to form gases. For example, if we boil a solution of salt and water, the water evaporates, forming a gas, and the salt is left behind. The gaseous water can be converted back to a liquid on the walls of a condenser, as shown in the apparatus depicted in Figure 1.13 ▼.

The differing abilities of substances to adhere to the surfaces of various solids such as paper and starch can also be used to separate mixtures. This is the basis of *chromatography* (literally "the writing of colors"), a technique that can give beautiful and dramatic results. An example of the chromatographic separation of ink is shown in Figure 1.14 ▶.

MOVIES
Mixtures and Compounds, Paper Chromatography of Ink

▶ **Figure 1.13 Distillation.** A simple apparatus for the separation of a sodium chloride solution (salt water) into its components. Boiling the solution evaporates the water, which is condensed, then collected in the receiving flask. After all the water has boiled away, pure sodium chloride remains in the boiling flask.

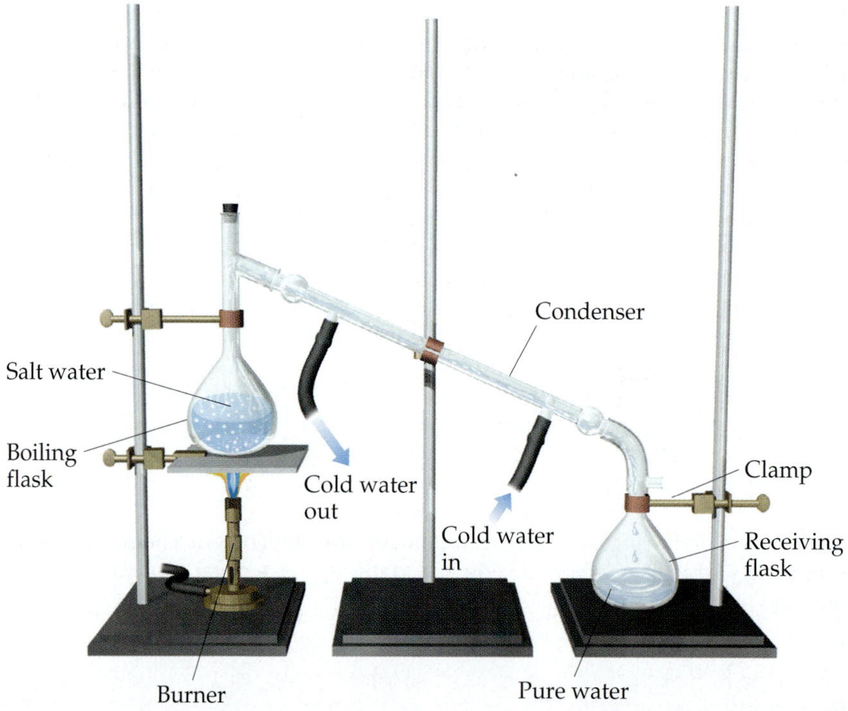

(a) (b) (c)

▲ **Figure 1.14 Separation of ink into components by paper chromatography.** (a) Water begins to move up the paper. (b) Water moves past the ink spot, dissolving different components of the ink at different rates. (c) Water has separated the ink into its several different components.

A CLOSER LOOK | The Scientific Method

Although two scientists rarely approach the same problem in exactly the same way, there are guidelines for the practice of science that have come to be known as the **scientific method**. These guidelines are outlined in Figure 1.15 ▼. We begin by collecting information, or *data*, by observation and experiment. The collection of information, however, is not the ultimate goal. The goal is to find a pattern or sense of order in our observations and to understand the origin of this order.

As we perform our experiments, we may begin to see patterns that lead us to a *tentative explanation*, or **hypothesis**, that guides us in planning further experiments. Eventually, we may be able to tie together a great number of observations in a single statement or equation called a scientific law. A **scientific law** *is a concise verbal statement or a mathematical equation that summarizes a broad variety of observations and experiences*. We tend to think of the laws of nature as the basic rules under which nature operates. However, it is not so much that matter obeys the laws of nature, but rather that the laws of nature describe the behavior of matter.

At many stages of our studies we may propose explanations of why nature behaves in a particular way. If a hypothesis is sufficiently general and is continually effective in predicting facts yet to be observed, it is called a theory. A **theory** *is an explanation of the general causes of certain phenomena, with considerable evidence or facts to support it*. For example, Einstein's theory of relativity was a revolutionary new way of thinking about space and time. It was more than just a simple hypothesis, however, because it could be used to make predictions that could be tested experimentally. When these experiments were conducted, the results were generally in agreement with the predictions and were not explainable by earlier theories. Thus, the theory of relativity was supported, but not proven. Indeed, theories can never be proven to be absolutely correct.

As we proceed through this text, we will rarely have the opportunity to discuss the doubts, conflicts, clashes of personalities, and revolutions of perception that have led to our present ideas. We need to be aware that just because we can spell out the results of science so concisely and neatly in textbooks does not mean that scientific progress is smooth, certain, and predictable. Some of the ideas we present in this text took centuries to develop and involved large numbers of scientists. We gain our view of the natural world by standing on the shoulders of the scientists who came before us. Take advantage of this view. As you study, exercise your imagination. Don't be afraid to ask daring questions when they occur to you. You may be fascinated by what you discover!

Related Exercise: 1.59

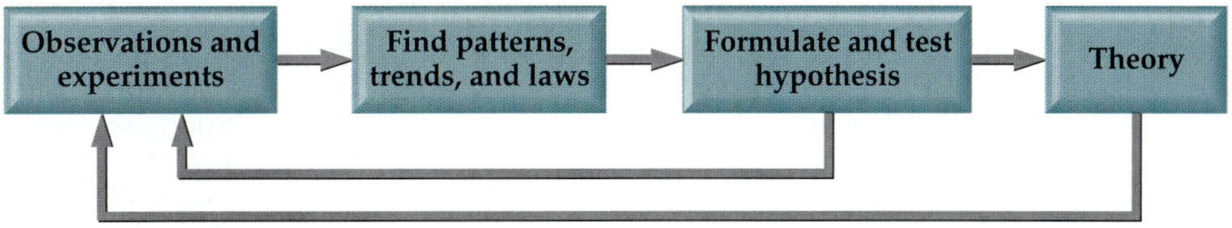

▲ **Figure 1.15 The scientific method.** The scientific method is a general approach to problems that involves making observations, seeking patterns in the observations, formulating hypotheses to explain the observations, and testing these hypotheses by further experiments. Those hypotheses that withstand such tests and prove themselves useful in explaining and predicting behavior become known as theories.

1.4 | Units of Measurement

Many properties of matter are *quantitative*; that is, they are associated with numbers. When a number represents a measured quantity, the units of that quantity must always be specified. To say that the length of a pencil is 17.5 is meaningless. Expressing the number with its units, 17.5 centimeters (cm), properly specifies the length. The units used for scientific measurements are those of the **metric system**.

The metric system, which was first developed in France during the late eighteenth century, is used as the system of measurement in most countries throughout the world. The United States has traditionally used the English system, although use of the metric system has become more common. For example, the contents of most canned goods and soft drinks in grocery stores are now given in metric as well as in English units as shown in Figure 1.16 ◀.

▲ **Figure 1.16 Metric units.** Metric measurements are increasingly common in the United States, as exemplified by the volume printed on this one-liter soda bottle.

SI Units

In 1960 an international agreement was reached specifying a particular choice of metric units for use in scientific measurements. These preferred units are called **SI units**, after the French *Système International d'Unités*. This system has seven *base units* from which all other units are derived. Table 1.4 ▼ lists these base units and their symbols. In this chapter we will consider the base units for length, mass, and temperature.

TABLE 1.4 SI Base Units

Physical Quantity	Name of Unit	Abbreviation
Mass	Kilogram	kg
Length	Meter	m
Time	Second	s[a]
Temperature	Kelvin	K
Amount of substance	Mole	mol
Electric current	Ampere	A
Luminous intensity	Candela	cd

[a]The abbreviation sec is frequently used.

In the metric system, prefixes are used to indicate decimal fractions or multiples of various units. For example, the prefix *milli-* represents a 10^{-3} fraction of a unit: A milligram (mg) is 10^{-3} gram (g), a millimeter (mm) is 10^{-3} meter (m), and so forth. Table 1.5 ▼ presents the prefixes commonly encountered in chemistry. In using SI units and in working problems throughout this text, you must be comfortable using exponential notation. If you are unfamiliar with exponential notation or want to review it, refer to Appendix A.1.

TABLE 1.5 Selected Prefixes Used in the Metric System

Prefix	Abbreviation	Meaning	Example
Giga	G	10^9	1 gigameter (Gm) $= 1 \times 10^9$ m
Mega	M	10^6	1 megameter (Mm) $= 1 \times 10^6$ m
Kilo	k	10^3	1 kilometer (km) $= 1 \times 10^3$ m
Deci	d	10^{-1}	1 decimeter (dm) = 0.1 m
Centi	c	10^{-2}	1 centimeter (cm) = 0.01 m
Milli	m	10^{-3}	1 millimeter (mm) = 0.001 m
Micro	μ[a]	10^{-6}	1 micrometer (μm) $= 1 \times 10^{-6}$ m
Nano	n	10^{-9}	1 nanometer (nm) $= 1 \times 10^{-9}$ m
Pico	p	10^{-12}	1 picometer (pm) $= 1 \times 10^{-12}$ m
Femto	f	10^{-15}	1 femtometer (fm) $= 1 \times 10^{-15}$ m

[a]This is the Greek letter mu (pronounced "mew").

Although non-SI units are being phased out, there are still some that are commonly used by scientists. Whenever we first encounter a non-SI unit in the text, the proper SI unit will also be given.

GIVE IT SOME THOUGHT

Which of the following quantities is the smallest: 1 mg, 1 μg, or 1 pg?

Length and Mass

The SI base unit of *length* is the meter (m), a distance only slightly longer than a yard. The relations between the English and metric system units that we will use most frequently in this text appear on the back inside cover. We will discuss how to convert English units into metric units, and vice versa, in Section 1.6.

Mass* is a measure of the amount of material in an object. The SI base unit of mass is the kilogram (kg), which is equal to about 2.2 pounds (lb). This base unit is unusual because it uses a prefix, *kilo-*, instead of the word *gram* alone. We obtain other units for mass by adding prefixes to the word *gram*.

SAMPLE EXERCISE 1.2 | Using metric prefixes

What is the name given to the unit that equals **(a)** 10^{-9} gram, **(b)** 10^{-6} second, **(c)** 10^{-3} meter?

Solution In each case we can refer to Table 1.5, finding the prefix related to each of the decimal fractions: **(a)** nanogram, ng, **(b)** microsecond, μs, **(c)** millimeter, mm.

PRACTICE EXERCISE

(a) What decimal fraction of a second is a picosecond, ps? **(b)** Express the measurement 6.0×10^3 m using a prefix to replace the power of ten. **(c)** Use exponential notation to express 3.76 mg in grams.
Answers: **(a)** 10^{-12} second, **(b)** 6.0 km, **(c)** 3.76×10^{-3} g

Temperature

Temperature is a measure of the hotness or coldness of an object. Indeed, temperature is a physical property that determines the direction of heat flow. Heat always flows spontaneously from a substance at higher temperature to one at lower temperature. Thus, we feel the influx of heat when we touch a hot object, and we know that the object is at a higher temperature than our hand.

The temperature scales commonly employed in scientific studies are the Celsius and Kelvin scales. The **Celsius scale** is also the everyday scale of temperature in most countries (Figure 1.17 ▶). It was originally based on the assignment of 0°C to the freezing point of water and 100°C to its boiling point at sea level (Figure 1.18 ▶).

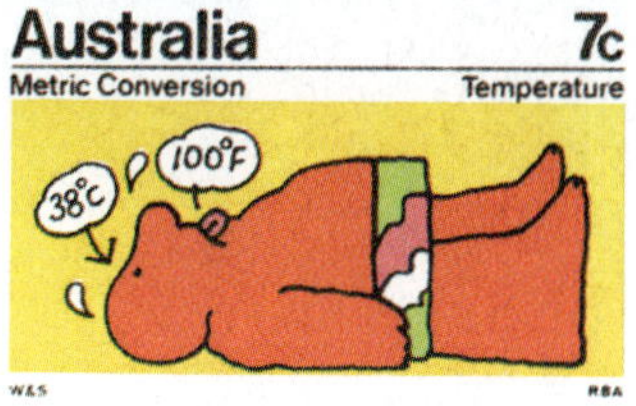

▲ **Figure 1.17 Australian stamp.** Many countries employ the Celsius temperature scale in everyday use, as illustrated by this stamp.

The **Kelvin scale** is the SI temperature scale, and the SI unit of temperature is the kelvin (K). Historically, the Kelvin scale was based on the properties of gases; its origins will be considered in Chapter 10. Zero on this scale is the lowest attainable temperature, −273.15°C, a temperature referred to as *absolute zero*. Both the Celsius and Kelvin scales have equal-sized units—that is, a kelvin is the same size as a degree Celsius. Thus, the Kelvin and Celsius scales are related as follows:

$$K = °C + 273.15 \qquad [1.1]$$

* Mass and weight are not interchangeable terms and are often incorrectly thought to be the same. The weight of an object is the force that its mass exerts due to gravity. In space, where gravitational forces are very weak, an astronaut can be weightless, but he or she cannot be massless. In fact, the astronaut's mass in space is the same as it is on Earth.

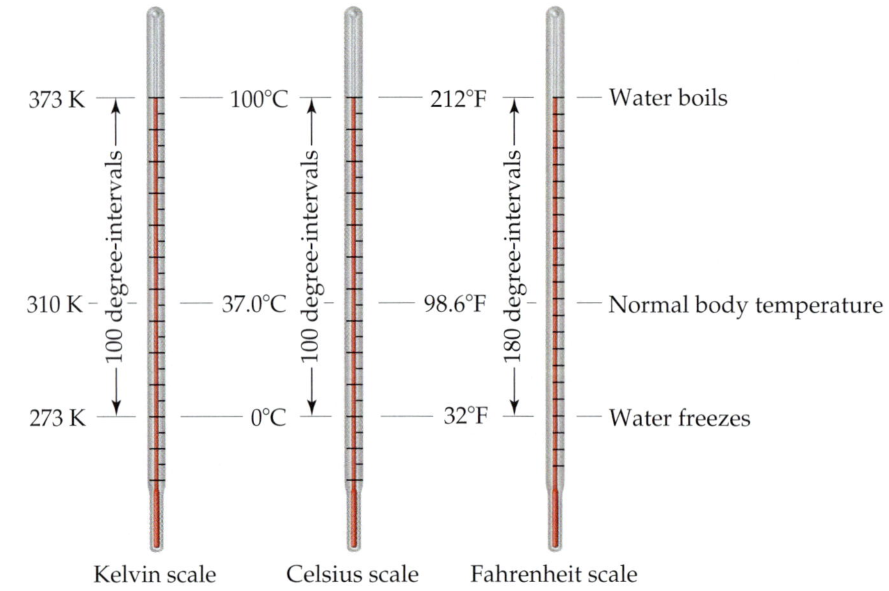

▶ **Figure 1.18 Comparison of the Kelvin, Celsius, and Fahrenheit temperature scales.** The freezing point and boiling point of water as well as normal human body temperature is indicated on each of the scales.

ACTIVITY
Temperature

The freezing point of water, 0°C, is 273.15 K (Figure 1.18). Notice that we do not use a degree sign (°) with temperatures on the Kelvin scale.

The common temperature scale in the United States is the *Fahrenheit scale*, which is not generally used in scientific studies. On the Fahrenheit scale, water freezes at 32°F and boils at 212°F. The Fahrenheit and Celsius scales are related as follows:

$$°C = \frac{5}{9}(°F - 32) \text{ or } °F = \frac{9}{5}(°C) + 32 \qquad [1.2]$$

SAMPLE EXERCISE 1.3 | Converting units of temperature

If a weather forecaster predicts that the temperature for the day will reach 31°C, what is the predicted temperature **(a)** in K, **(b)** in °F?

Solution **(a)** Using Equation 1.1, we have $K = 31 + 273 = 304$ K

(b) Using Equation 1.2, we have $°F = \frac{9}{5}(31) + 32 = 56 + 32 = 88°F$

PRACTICE EXERCISE

Ethylene glycol, the major ingredient in antifreeze, freezes at −11.5°C. What is the freezing point in **(a)** K, **(b)** °F?
Answers: **(a)** 261.7 K, **(b)** 11.3°F

Derived SI Units

The SI base units in Table 1.4 are used to derive the units of other quantities. To do so, we use the defining equation for the quantity, substituting the appropriate base units. For example, speed is defined as the ratio of distance traveled to elapsed time. Thus, the SI unit for speed is the SI unit for distance (length) divided by the SI unit for time, m/s, which we read as "meters per second." We will encounter many derived units, such as those for force, pressure, and energy, later in this text. In this chapter we examine the derived units for volume and density.

Volume

The *volume* of a cube is given by its length cubed, $(\text{length})^3$. Thus, the SI unit of volume is the SI unit of length raised to the third power. The cubic meter, or m^3, is the volume of a cube that is 1 m on each edge. Smaller units, such as cubic centimeters, cm^3 (sometimes written as cc), are frequently used in chemistry. Another unit of volume commonly used in chemistry is the *liter* (L), which equals a cubic decimeter, dm^3, and is slightly larger than a quart. The liter is the first metric unit we have encountered that is *not* an SI unit. There are 1000 milliliters (mL) in a liter (Figure 1.19 ▶), and each milliliter is the same volume as a cubic centimeter: 1 mL = 1 cm^3. The terms *milliliter* and *cubic centimeter* are used interchangeably in expressing volume.

The devices used most frequently in chemistry to measure volume are illustrated in Figure 1.20 ▼. Syringes, burets, and pipets deliver liquids with more precision than graduated cylinders. Volumetric flasks are used to contain specific volumes of liquid.

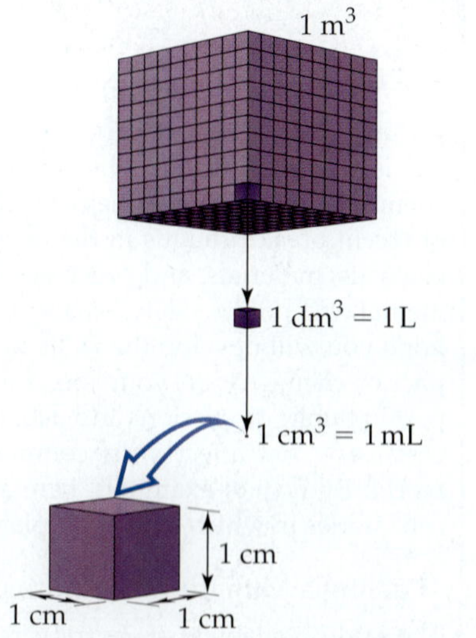

▲ **Figure 1.19 Volume relationships.** The volume occupied by a cube that is 1 m on each edge is a cubic meter, 1 m^3 (top). Each cubic meter contains 1000 dm^3 (middle). A liter is the same volume as a cubic decimeter, 1 L = 1 dm^3. Each cubic decimeter contains 1000 cubic centimeters, 1 dm^3 = 1000 cm^3. Each cubic centimeter equals a milliliter, 1 cm^3 = 1 mL (bottom).

Density

Density is a property of matter that is widely used to characterize substances. It is defined as the amount of mass in a unit volume of the substance:

$$\text{Density} = \frac{\text{mass}}{\text{volume}} \quad [1.3]$$

The densities of solids and liquids are commonly expressed in units of grams per cubic centimeter (g/cm^3) or grams per milliliter (g/mL). The densities of some common substances are listed in Table 1.6 ▼. It is no coincidence that the density of water is 1.00 g/mL; the gram was originally defined as the mass of 1 mL of water at a specific temperature. Because most substances change volume when heated or cooled, densities are temperature dependent.

TABLE 1.6 Densities of Some Selected Substances at 25°C

Substance	Density (g/cm^3)
Air	0.001
Balsa wood	0.16
Ethanol	0.79
Water	1.00
Ethylene glycol	1.09
Table sugar	1.59
Table salt	2.16
Iron	7.9
Gold	19.32

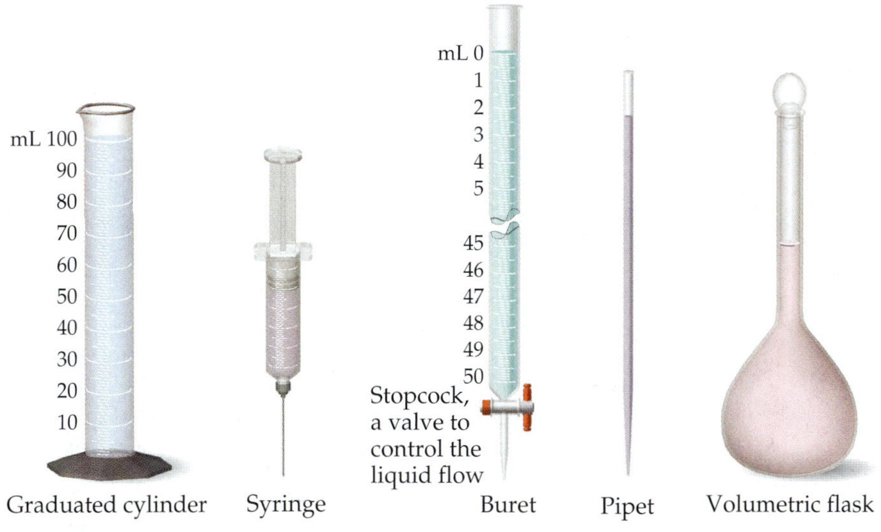

◀ **Figure 1.20 Common volumetric glassware.** The graduated cylinder, syringe, and buret are used in laboratories to deliver variable volumes of liquid; the pipet is used to deliver a specific volume of liquid. The volumetric flask contains a specific volume of liquid when filled to the mark.

CHEMISTRY AT WORK | Chemistry in the News

Chemistry is a very lively, active field of science. Because it is so central to our lives, there are reports on matters of chemical significance in the news nearly every day. Some tell of recent breakthroughs in the development of new pharmaceuticals, materials, and processes. Others deal with environmental and public safety issues. As you study chemistry, we hope you will develop the skills to better understand the impact of chemistry on your life. You need these skills to take part in public discussions and debates about matters related to chemistry that affect your community, the nation, and the world. By way of examples, here are summaries of a few recent stories in which chemistry plays a role.

"Fueling a Miniature Revolution"

The ready availability of electric cars has been delayed for years by problems in finding a suitable energy source. The batteries that are available at reasonable cost are too heavy, and they can propel a car only a limited distance before needing to be recharged. The fuel cell, an electrical device in which fuels are used to directly generate electricity, is an alternative to a battery. Experimental vehicles powered by fuel cells, such as the car shown in Figure 1.21 ▼, are now commonly displayed on the pages of magazines and newspapers, offering hope that someday such vehicles will be a practical means of transportation.

The development of fuel cells for transportation has caused some to propose that fuel cells might also be practical for small electronic devices such as cell phones and notebook computers. As these devices become smaller, faster, and more capable, their power demands increase. Even the best batteries are hard-pressed to meet the need for lightweight, portable power sources that deliver power for an extended period of time. Many people have had the unfortunate experience of having a battery go dead during a cell-phone call or the inconvenience of having to carry extra batteries or charging devices while traveling.

▲ **Figure 1.21 Cutaway view of car powered by fuel cells.**

In a fuel cell, the fuel flows through the cell on demand. Once the fuel is consumed, more can be added to make the cell produce more energy. In contrast, batteries are sealed systems, which means that once their chemical reactions are complete, the user has to either recharge the batteries or buy new ones. Small-scale fuel cells for portable electronic devices would be the size of conventional battery packs but weigh considerably less. Until now, successful fuel cells have required the use of hydrogen as a fuel. One of the keys to the commercial development of small cells is the development of practical systems, which convert a hydrogen-containing fuel such as methanol into the hydrogen that the cell can use.

"Biting into Research for New Dental Materials"

Rapid changes are occurring in the materials that dentists use. Research has yielded new and improved ways to bond restorative materials to teeth, improvements in the durability and appearance of materials used to restore or protect teeth, and improved treatments to reduce tooth decay.

For example, new materials are being developed that do not merely plug the cavities caused by decay and the dentist's drill, but are biologically active and stimulate the tooth to repair itself. The new materials contain amorphous calcium phosphate, which releases calcium and phosphate to tooth structures in proportions that form the mineral found naturally in teeth. The composition is not useful for large cavities because it is not as strong as conventional filling materials, but it is useful for fillings in small cavities and to seal pits and fissures in teeth where bits of food can accumulate and cause decay.

"Nanotechnology: Hype and Hope"

The past 15 years have witnessed an explosion of relatively inexpensive equipment and techniques for probing and manipulating materials on the nanometer-length scale. These capabilities have led to optimistic forecasts of futuristic nanotechnologies including molecular-scale machines and robots that can manipulate matter with atomic precision. Many believe that such futuristic visions are mere hype, while others express the conviction that they can be realized.

Nanoscale materials exhibit chemical and physical properties that are different from the properties of bulk materials. For example, carbon can be made to form tubular structures as shown in Figure 1.22 ▶. These tubes, called nanotubes, resemble a cylindrical roll of chicken wire. When nanotubes are perfectly formed, they conduct electricity like a metal.

Scientists have learned that the electric and optical properties of certain nanometer-size particles can be tuned by adjusting the particle size or shape. These particles are therefore of interest for applications in various kinds of electronic devices. For example, scientists at IBM's T. J. Watson Research Center recently generated an electric current in a single carbon nanotube by shining light on it. By developing the means to control light and electrical current, scientists aim to produce tiny optoelectronic devices. Although such applications are still years from commercial fruition, they nevertheless offer the promise of dramatically changing not only the size of

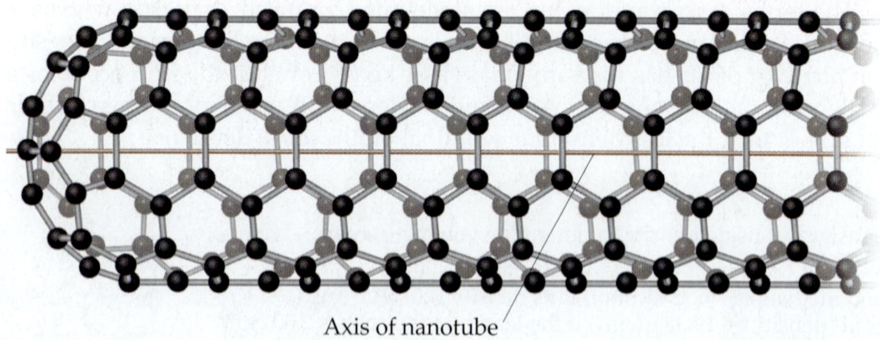

◀ **Figure 1.22 A section of carbon nanotube.** Each intersection in the network represents a carbon atom chemically joined to three others.

electronic devices and many other items but also the way they are manufactured. The results obtained at the IBM labs also suggest that such devices might be assembled from simpler, smaller components such as individual molecules and other nanostructures. This approach is similar to the one nature uses to construct complex biological architectures.

"Got Mercury?"

The Food and Drug Administration (FDA) and the Environmental Protection Agency (EPA) recently issued a joint advisory about fish and shellfish consumption for people at risk from exposure to high mercury levels, namely pregnant women, nursing mothers, and young children. According to the advisory, people in these populations should not eat shark, swordfish, king mackerel, or tilefish, and should limit their intake of other fish as well. The concern is based on the fact that nearly all fish contain at least traces of methylmercury (Figure 1.23 ▼), a biologically active form of mercury.

Tiny amounts of mercury vapor occur in the atmosphere, resulting from both natural sources and pollution. In the United States, coal-fired power plants are the largest unregulated source of mercury that is due to human activities. Mercury in the atmosphere is converted to simple mercury-containing compounds and returned to Earth's surface in rainwater. The mercury compounds are converted to methylmercury by aquatic bacteria. Plankton take up some of the methylmercury, fish eat the plankton, and bigger fish eat them. The resultant accumulation of mercury through the food chain can produce mercury levels a million times greater in fish than in the surrounding waters.

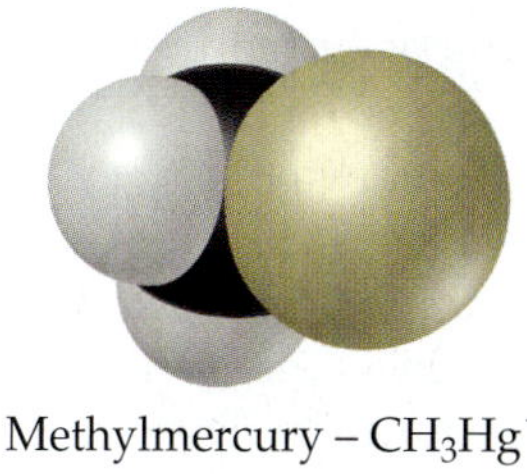

Methylmercury – CH_3Hg^+

▲ **Figure 1.23 Molecular model of methylmercury.**

When we eat fish, the methylmercury is absorbed from our gastrointestinal tract and distributed widely throughout our body. Methylmercury produces the greatest damage in the brain and central nervous system where it causes loss of nerve cells, which in turn results in neurological problems. The symptoms of acute mercury poisoning include numbness, difficulty in speaking, and loss of coordination. Fortunately, adults are able to release about 1% of their methylmercury burden daily by a number of different pathways. For infants, however, these mechanisms are not fully developed, leading to greater potential damage.

"New Elements Created"

Two new elements were reported in February 2004 as a result of studies performed at the Joint Institute for Nuclear Research (JINR) in Dubna, Russia. JINR scientists and their collaborators from Lawrence Livermore National Lab in California announced that they had produced 4 atoms of a new superheavy element (element 115) by striking a target of americium atoms (element 95) with a highly energetic beam of calcium atoms in a device called a particle accelerator.

The atoms of element 115 last only 90 milliseconds (90 ms) before decomposing to yet another new element (element 113). Element 113 is also unstable, lasting a little over a second. Although these times seem very short, they are actually long compared with other unstable nuclei of comparable mass. How long a nucleus lasts before falling apart is a measure of its stability. Theory suggests that elements in the vicinity of element 114 in the periodic table should be more stable than other heavy nuclei. Scientists, however, caution that these results will need to be confirmed by others before they are fully accepted.

When reporting densities, the temperature should be specified. If no temperature is reported, we usually assume that the temperature is 25°C, close to normal room temperature.

The terms *density* and *weight* are sometimes confused. A person who says that iron weighs more than air generally means that iron has a higher density than air; 1 kg of air has the same mass as 1 kg of iron, but the iron occupies a smaller volume, thereby giving it a higher density. If we combine two liquids that do not mix, the less dense one will float on the more dense one.

SAMPLE EXERCISE 1.4 | Determining density and using density to determine volume or mass

(a) Calculate the density of mercury if 1.00×10^2 g occupies a volume of 7.36 cm^3.
(b) Calculate the volume of 65.0 g of the liquid methanol (wood alcohol) if its density is 0.791 g/mL.
(c) What is the mass in grams of a cube of gold (density = 19.32 g/cm^3) if the length of the cube is 2.00 cm?

Solution

(a) We are given mass and volume, so Equation 1.3 yields

$$\text{Density} = \frac{\text{mass}}{\text{volume}} = \frac{1.00 \times 10^2\ \text{g}}{7.36\ \text{cm}^3} = 13.6\ \text{g/cm}^3$$

(b) Solving Equation 1.3 for volume and then using the given mass and density gives

$$\text{Volume} = \frac{\text{mass}}{\text{density}} = \frac{65.0\ \text{g}}{0.791\ \text{g/mL}} = 82.2\ \text{mL}$$

(c) We can calculate the mass from the volume of the cube and its density. The volume of a cube is given by its length cubed:

$$\text{Volume} = (2.00\ \text{cm})^3 = (2.00)^3\ \text{cm}^3 = 8.00\ \text{cm}^3$$

Solving Equation 1.3 for mass and substituting the volume and density of the cube, we have

$$\text{Mass} = \text{volume} \times \text{density} = (8.00\ \text{cm}^3)(19.32\ \text{g/cm}^3) = 155\ \text{g}$$

PRACTICE EXERCISE

(a) Calculate the density of a 374.5-g sample of copper if it has a volume of 41.8 cm^3. **(b)** A student needs 15.0 g of ethanol for an experiment. If the density of ethanol is 0.789 g/mL, how many milliliters of ethanol are needed? **(c)** What is the mass, in grams, of 25.0 mL of mercury (density = 13.6 g/mL)?
Answers: **(a)** 8.96 g/cm^3, **(b)** 19.0 mL, **(c)** 340 g

1.5 | Uncertainty in Measurement

ACTIVITY
Uncertainty in Measurement

There are two kinds of numbers in scientific work: *exact numbers* (those whose values are known exactly) and *inexact numbers* (those whose values have some uncertainty). Most of the exact numbers that we will encounter in this course have defined values. For example, there are exactly 12 eggs in a dozen, exactly 1000 g in a kilogram, and exactly 2.54 cm in an inch. The number 1 in any conversion factor between units, as in 1 m = 100 cm or 1 kg = 2.2046 lb, is also an exact number. Exact numbers can also result from counting numbers of objects. For example, we can count the exact number of marbles in a jar or the exact number of people in a classroom.

Numbers obtained by measurement are always *inexact*. There are always inherent limitations in the equipment used to measure quantities (equipment errors), and there are differences in how different people make the same measurement (human errors). Suppose that ten students with ten balances are given the same dime and told to determine its mass. The ten measurements will probably vary slightly from one another for various reasons. The balances might be calibrated slightly differently, and there might be differences in how each student reads the mass from the balance. Remember: *Uncertainties always exist in measured quantities*. Counting very large numbers of objects usually has some associated error as well. Consider, for example, how difficult it is to obtain accurate census information for a city or vote counts for an election.

GIVE IT SOME THOUGHT

Which of the following is an inexact quantity: (a) the number of people in your chemistry class, (b) the mass of a penny, (c) the number of grams in a kilogram?

Precision and Accuracy

The terms precision and accuracy are often used in discussing the uncertainties of measured values. **Precision** is a measure of how closely individual measurements agree with one another. **Accuracy** refers to how closely individual measurements agree with the correct, or "true," value. The analogy of darts stuck in a dartboard pictured in Figure 1.24 ▶ illustrates the difference between these two concepts.

In the laboratory we often perform several different "trials" of the same experiment. We gain confidence in the accuracy of our measurements if we obtain nearly the same value each time. Figure 1.24 should remind us, however, that precise measurements could be inaccurate. For example, if a very sensitive balance is poorly calibrated, the masses we measure will be consistently either high or low. They will be inaccurate even if they are precise.

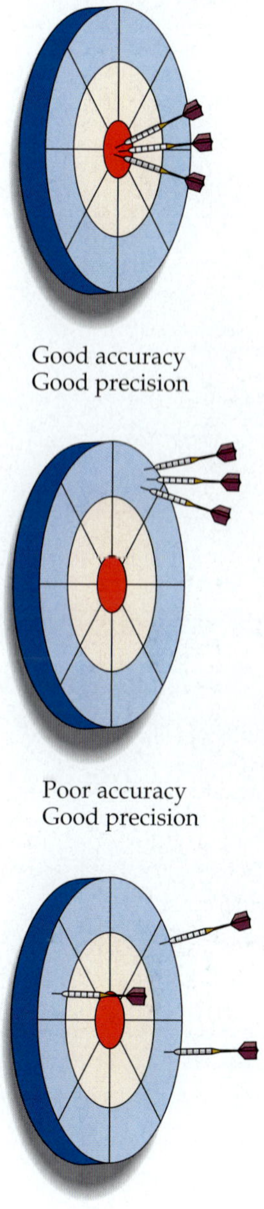

▲ **Figure 1.24 Precision and accuracy.** The distribution of darts on a target illustrates the difference between accuracy and precision.

Significant Figures

Suppose you determine the mass of a dime on a balance capable of measuring to the nearest 0.0001 g. You could report the mass as 2.2405 ± 0.0001 g. The ± notation (read "plus or minus") expresses the magnitude of the uncertainty of your measurement. In much scientific work we drop the ± notation with the understanding that there is always some uncertainty in the last digit of the measured quantity. That is, *measured quantities are generally reported in such a way that only the last digit is uncertain.*

Figure 1.25 ▼ shows a thermometer with its liquid column between the scale marks. We can read the certain digits from the scale and estimate the uncertain one. From the scale marks on the thermometer, we see that the liquid is between the 25°C and 30°C marks. We might estimate the temperature to be 27°C, being somewhat uncertain of the second digit of our measurement.

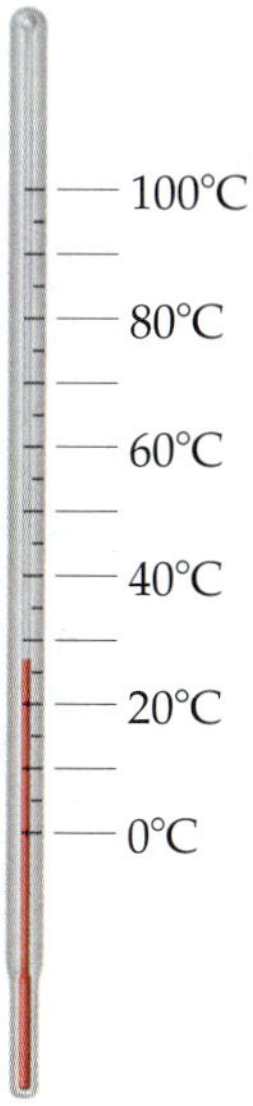

◀ **Figure 1.25 Significant figures in measurements.** The thermometer has markings every 5°C. The temperature is between 25°C and 30°C and is approximately 27°C. The two significant figures in the measurement include the second digit, which is estimated by reading between the scale marks.

All digits of a measured quantity, including the uncertain one, are called **significant figures**. A measured mass reported as 2.2 g has two significant figures, whereas one reported as 2.2405 g has five significant figures. The greater the number of significant figures, the greater is the certainty implied for the measurement. When multiple measurements are made of a quantity, the results can be averaged, and the number of significant figures estimated by using statistical methods.

SAMPLE EXERCISE 1.5 | Relating significant figures to the uncertainty of a measurement

What difference exists between the measured values 4.0 g and 4.00 g?

Solution Many people would say there is no difference, but a scientist would note the difference in the number of significant figures in the two measurements. The value 4.0 has two significant figures, while 4.00 has three. This difference implies that the first measurement has more uncertainty. A mass of 4.0 g indicates that the uncertainty is in the first decimal place of the measurement. Thus, the mass might be anything between 3.9 and 4.1 g, which we can represent as 4.0 ± 0.1 g. A measurement of 4.00 g implies that the uncertainty is in the second decimal place. Thus, the mass might be anything between 3.99 and 4.01 g, which we can represent as 4.00 ± 0.01 g. Without further information, we cannot be sure whether the difference in uncertainties of the two measurements reflects the precision or accuracy of the measurement.

PRACTICE EXERCISE

A balance has a precision of ± 0.001 g. A sample that has a mass of about 25 g is placed on this balance. How many significant figures should be reported for this measurement?

Answer: five, as in the measurement 24.995 g

To determine the number of significant figures in a reported measurement, read the number from left to right, counting the digits starting with the first digit that is not zero. *In any measurement that is properly reported, all nonzero digits are significant*. Zeros, however, can be used either as part of the measured value or merely to locate the decimal point. Thus, zeros may or may not be significant, depending on how they appear in the number. The following guidelines describe the different situations involving zeros:

1. Zeros *between* nonzero digits are always significant—1005 kg (four significant figures); 1.03 cm (three significant figures).

©2006 by Sidney Harris

2. Zeros *at the beginning* of a number are never significant; they merely indicate the position of the decimal point—0.02 g (one significant figure); 0.0026 cm (two significant figures).
3. Zeros *at the end* of a number are significant if the number contains a decimal point—0.0200 g (three significant figures); 3.0 cm (two significant figures).

A problem arises when a number ends with zeros but contains no decimal point. In such cases, it is normally assumed that the zeros are not significant. Exponential notation (Appendix A) can be used to clearly indicate whether zeros at the end of a number are significant. For example, a mass of 10,300 g can be written in exponential notation showing three, four, or five significant figures depending on how the measurement is obtained:

1.03×10^4 g (three significant figures)
1.030×10^4 g (four significant figures)
1.0300×10^4 g (five significant figures)

In these numbers all the zeros to the right of the decimal point are significant (rules 1 and 3). (The exponential term does not add to the number of significant figures.)

SAMPLE EXERCISE 1.6 | Determining the number of significant figures in a measurement

How many significant figures are in each of the following numbers (assume that each number is a measured quantity): **(a)** 4.003, **(b)** 6.023×10^{23}, **(c)** 5000?

Solution **(a)** Four; the zeros are significant figures. **(b)** Four; the exponential term does not add to the number of significant figures. **(c)** One. We assume that the zeros are not significant when there is no decimal point shown. If the number has more significant figures, a decimal point should be employed or the number written in exponential notation. Thus, 5000. has four significant figures, whereas 5.00×10^3 has three.

PRACTICE EXERCISE

How many significant figures are in each of the following measurements: **(a)** 3.549 g, **(b)** 2.3×10^4 cm, **(c)** 0.00134 m^3?
Answers: **(a)** four, **(b)** two, **(c)** three

ACTIVITY
Significant Figures

Significant Figures in Calculations

When carrying measured quantities through calculations, *the least certain measurement limits the certainty of the calculated quantity and thereby determines the number of significant figures in the final answer.* The final answer should be reported with only one uncertain digit. To keep track of significant figures in calculations, we will make frequent use of two rules, one for multiplication and division and another for addition and subtraction.

1. *For multiplication and division,* the result contains the same number of significant figures as the measurement with the fewest significant figures. When the result contains more than the correct number of significant figures, it must be rounded off. For example, the area of a rectangle whose measured edge lengths are 6.221 cm and 5.2 cm should be reported as 32 cm^2 even though a calculator shows the product of 6.221 and 5.2 to have more digits:

$$\text{Area} = (6.221\ \text{cm})(5.2\ \text{cm}) = 32.3492\ \text{cm}^2 \Rightarrow \text{round off to } 32\ \text{cm}^2$$

We round off to two significant figures because the least precise number—5.2 cm—has only two significant figures.

2. *For addition and subtraction*, the result has the same number of decimal places as the measurement with the fewest decimal places. Consider the following example in which the uncertain digits appear in color:

This number limits	20.42	← two decimal places
the number of significant	1.322	← three decimal places
figures in the result ⟶	83.1	← one decimal place
	104.842	← round off to 104. 8 (one decimal place)

We report the result as 104.8 because 83.1 has only one decimal place.

Notice that for multiplication and division, significant figures are counted. For addition and subtraction, decimal places are counted. In determining the final answer for a calculated quantity, exact numbers can be treated as if they have an infinite number of significant figures. This rule applies to many definitions between units. Thus, when we say, "There are 12 inches in 1 foot," the number 12 is exact, and we need not worry about the number of significant figures in it.

In *rounding off numbers*, look at the leftmost digit to be removed:

- If the leftmost digit removed is less than 5, the preceding number is left unchanged. Thus, rounding 7.248 to two significant figures gives 7.2.
- If the leftmost digit removed is 5 or greater, the preceding number is increased by 1. Rounding 4.735 to three significant figures gives 4.74, and rounding 2.376 to two significant figures gives 2.4.*

SAMPLE EXERCISE 1.7 | Determining the number of significant figures in a calculated quantity

The width, length, and height of a small box are 15.5 cm, 27.3 cm, and 5.4 cm, respectively. Calculate the volume of the box, using the correct number of significant figures in your answer.

Solution The volume of a box is determined by the product of its width, length, and height. In reporting the product, we can show only as many significant figures as given in the dimension with the fewest significant figures, that for the height (two significant figures):

$$\text{Volume} = \text{width} \times \text{length} \times \text{height}$$

$$= (15.5\text{ cm})(27.3\text{ cm})(5.4\text{ cm}) = 2285.01\text{ cm}^3 \Rightarrow 2.3 \times 10^3\text{ cm}^3$$

When we use a calculator to do this calculation, the display shows 2285.01, which we must round off to two significant figures. Because the resulting number is 2300, it is best reported in exponential notation, 2.3×10^3, to clearly indicate two significant figures.

PRACTICE EXERCISE

It takes 10.5 s for a sprinter to run 100.00 m. Calculate the average speed of the sprinter in meters per second, and express the result to the correct number of significant figures.

Answer: 9.52 m/s (3 significant figures)

* Your instructor may want you to use a slight variation on the rule when the leftmost digit to be removed is exactly 5, with no following digits or only zeros. One common practice is to round up to the next higher number if that number will be even, and down to the next lower number otherwise. Thus, 4.7350 would be rounded to 4.74, and 4.7450 would also be rounded to 4.74.

SAMPLE EXERCISE 1.8 | **Determining the number of significant figures in a calculated quantity**

A gas at 25°C fills a container whose volume is $1.05 \times 10^3\ cm^3$. The container plus gas have a mass of 837.6 g. The container, when emptied of all gas, has a mass of 836.2 g. What is the density of the gas at 25°C?

Solution

To calculate the density, we must know both the mass and the volume of the gas. The mass of the gas is just the difference in the masses of the full and empty container:

$$(837.6 - 836.2)\ g = 1.4\ g$$

In subtracting numbers, we determine the number of significant figures in our result by counting decimal places in each quantity. In this case each quantity has one decimal place. Thus, the mass of the gas, 1.4 g, has one decimal place.

Using the volume given in the question, $1.05 \times 10^3\ cm^3$, and the definition of density, we have

$$\text{Density} = \frac{\text{mass}}{\text{volume}} = \frac{1.4\ g}{1.05 \times 10^3\ cm^3}$$
$$= 1.3 \times 10^{-3}\ g/cm^3 = 0.0013\ g/cm^3$$

In dividing numbers, we determine the number of significant figures in our result by counting the number of significant figures in each quantity. There are two significant figures in our answer, corresponding to the smaller number of significant figures in the two numbers that form the ratio.

PRACTICE EXERCISE

To how many significant figures should the mass of the container be measured (with and without the gas) in Sample Exercise 1.8 in order for the density to be calculated to three significant figures?

Answer: five (In order for the difference in the two masses to have three significant figures, there must be two decimal places in the masses of the filled and empty containers.)

When a calculation involves two or more steps and you write down answers for intermediate steps, retain at least one additional digit—past the number of significant figures—for the intermediate answers. This procedure ensures that small errors from rounding at each step do not combine to affect the final result. When using a calculator, you may enter the numbers one after another, rounding only the final answer. Accumulated rounding-off errors may account for small differences among results you obtain and answers given in the text for numerical problems.

1.6 | Dimensional Analysis

Throughout the text we use an approach called **dimensional analysis** as an aid in problem solving. In dimensional analysis we carry units through all calculations. Units are multiplied together, divided into each other, or "canceled." Using dimensional analysis helps ensure that the solutions to problems yield the proper units. Moreover, it provides a systematic way of solving many numerical problems and of checking our solutions for possible errors.

The key to using dimensional analysis is the correct use of conversion factors to change one unit into another. A **conversion factor** is a fraction whose numerator and denominator are the same quantity expressed in different units. For example, 2.54 cm and 1 in. are the same length, 2.54 cm = 1 in. This relationship allows us to write two conversion factors:

$$\frac{2.54\ \text{cm}}{1\ \text{in.}} \quad \text{and} \quad \frac{1\ \text{in.}}{2.54\ \text{cm}}$$

We use the first of these factors to convert inches to centimeters. For example, the length in centimeters of an object that is 8.50 in. long is given by

Desired unit

$$\text{Number of centimeters} = (8.50\ \cancel{\text{in.}})\ \frac{2.54\ \text{cm}}{1\ \cancel{\text{in.}}} = 21.6\ \text{cm}$$

Given unit

The unit inches in the denominator of the conversion factor cancels the unit inches in the given data (8.50 *inches*). The unit centimeters in the numerator of the conversion factor becomes the unit of the final answer. Because the numerator and denominator of a conversion factor are equal, multiplying any quantity by a conversion factor is equivalent to multiplying by the number 1 and so does not change the intrinsic value of the quantity. The length 8.50 in. is the same as the length 21.6 cm.

In general, we begin any conversion by examining the units of the given data and the units we desire. We then ask ourselves what conversion factors we have available to take us from the units of the given quantity to those of the desired one. When we multiply a quantity by a conversion factor, the units multiply and divide as follows:

$$\cancel{\text{Given unit}} \times \frac{\text{desired unit}}{\cancel{\text{given unit}}} = \text{desired unit}$$

If the desired units are not obtained in a calculation, then an error must have been made somewhere. Careful inspection of units often reveals the source of the error.

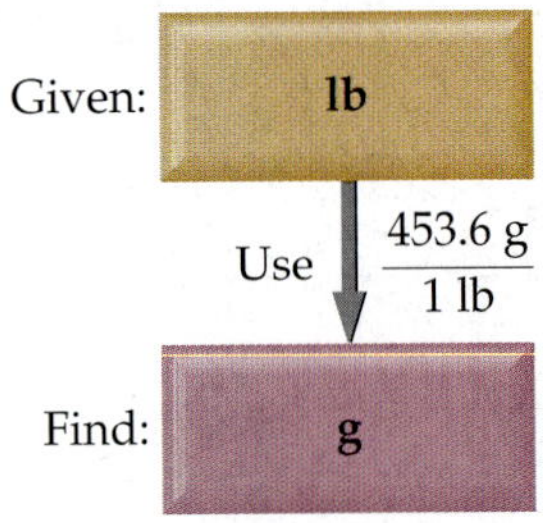

SAMPLE EXERCISE 1.9 | Converting units

If a woman has a mass of 115 lb, what is her mass in grams? (Use the relationships between units given on the back inside cover of the text.)

Solution Because we want to change from lb to g, we look for a relationship between these units of mass. From the back inside cover we have 1 lb = 453.6 g. In order to cancel pounds and leave grams, we write the conversion factor with grams in the numerator and pounds in the denominator:

$$\text{Mass in grams} = (115\ \cancel{\text{lb}})\left(\frac{453.6\ \text{g}}{1\ \cancel{\text{lb}}}\right) = 5.22 \times 10^4\ \text{g}$$

The answer can be given to only three significant figures, the number of significant figures in 115 lb. The process we have used is diagrammed in the margin.

PRACTICE EXERCISE

By using a conversion factor from the back inside cover, determine the length in kilometers of a 500.0-mi automobile race.
Answer: 804.7 km

GIVE IT SOME THOUGHT

How do we determine how many digits to use in conversion factors, such as the one between pounds and grams in Sample Exercise 1.9?

STRATEGIES IN CHEMISTRY | Estimating Answers

A friend once remarked cynically that calculators let you get the wrong answer more quickly. What he was implying by that remark was that unless you have the correct strategy for solving a problem and have punched in the correct numbers, the answer will be incorrect. If you learn to *estimate* answers, however, you will be able to check whether the answers to your calculations are reasonable.

The idea is to make a rough calculation using numbers that are rounded off in such a way that the arithmetic can be easily performed without a calculator. This approach is often referred to as making a "ballpark" estimate, meaning that while it doesn't give an exact answer, it gives one that is roughly the right size. By working with units using dimensional analysis and by estimating answers, we can readily check the reasonableness of our answers to calculations.

Using Two or More Conversion Factors

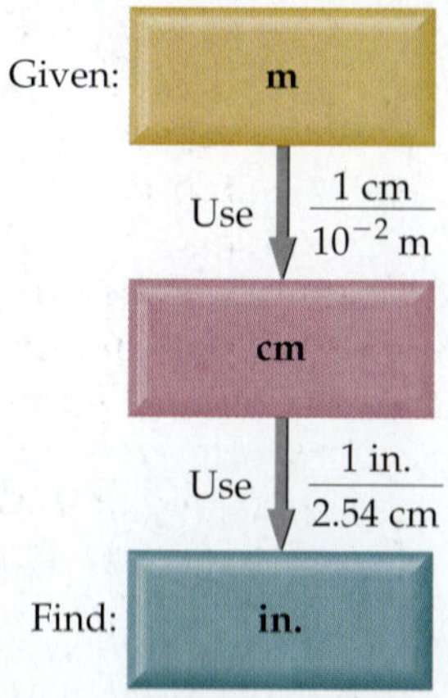

It is often necessary to use several conversion factors in solving a problem. As an example, let's convert the length of an 8.00-m rod to inches. The table on the back inside cover doesn't give the relationship between meters and inches. It *does* give the relationship between centimeters and inches (1 in. = 2.54 cm), though, and from our knowledge of metric prefixes we know that 1 cm $= 10^{-2}$ m. Thus, we can convert step by step, first from meters to centimeters, and then from centimeters to inches as diagrammed in the margin.

Combining the given quantity (8.00 m) and the two conversion factors, we have

$$\text{Number of inches} = (8.00\ \text{m})\left(\frac{100\ \text{cm}}{1\ \text{m}}\right)\left(\frac{1\ \text{in.}}{2.54\ \text{cm}}\right) = 315\ \text{in.}$$

The first conversion factor is applied to cancel meters and convert the length to centimeters. Thus, meters are written in the denominator and centimeters in the numerator. The second conversion factor is written to cancel centimeters, so it has centimeters in the denominator and inches, the desired unit, in the numerator.

SAMPLE EXERCISE 1.10 | Converting units using two or more conversion factors

The average speed of a nitrogen molecule in air at 25°C is 515 m/s. Convert this speed to miles per hour.

Solution To go from the given units, m/s, to the desired units, mi/hr, we must convert meters to miles and seconds to hours. From the relationships given on the back inside cover of the book, we find that 1 mi = 1.6093 km. From our knowledge of metric prefixes we know that 1 km $= 10^3$ m. Thus, we can convert m to km and then convert km to mi. From our knowledge of time we know that 60 s = 1 min and 60 min = 1 hr. Thus, we can convert s to min and then convert min to hr.

Applying first the conversions for distance and then those for time, we can set up one long equation in which unwanted units are canceled:

$$\text{Speed in mi/hr} = \left(515\frac{\text{m}}{\text{s}}\right)\left(\frac{1\ \text{km}}{10^3\ \text{m}}\right)\left(\frac{1\ \text{mi}}{1.6093\ \text{km}}\right)\left(\frac{60\ \text{s}}{1\ \text{min}}\right)\left(\frac{60\ \text{min}}{1\ \text{hr}}\right)$$
$$= 1.15 \times 10^3\ \text{mi/hr}$$

Our answer has the desired units. We can check our calculation, using the estimating procedure described in the previous "Strategies" box. The given speed is about 500 m/s. Dividing by 1000 converts m to km, giving 0.5 km/s. Because 1 mi is about 1.6 km, this speed corresponds to $0.5/1.6 = 0.3$ mi/s. Multiplying by 60 gives about $0.3 \times 60 = 20$ mi/min. Multiplying again by 60 gives $20 \times 60 = 1200$ mi/hr. The approximate solution (about 1200 mi/hr) and the detailed solution (1150 mi/hr) are reasonably close. The answer to the detailed solution has three significant figures, corresponding to the number of significant figures in the given speed in m/s.

PRACTICE EXERCISE

A car travels 28 mi per gallon of gasoline. How many kilometers per liter will it go?
Answer: 12 km/L

Conversions Involving Volume

The conversion factors previously noted convert from one unit of a given measure to another unit of the same measure, such as from length to length. We also have conversion factors that convert from one measure to a different one. The density of a substance, for example, can be treated as a conversion factor between mass and volume. Suppose that we want to know the mass in grams of two cubic inches (2.00 in.^{3}) of gold, which has a density of 19.3 g/cm^3. The density gives us the following factors:

$$\frac{19.3\ \text{g}}{1\ \text{cm}^3} \quad \text{and} \quad \frac{1\ \text{cm}^3}{19.3\ \text{g}}$$

Because the answer we want is a mass in grams, we can see that we will use the first of these factors, which has mass in grams in the numerator. To use this factor, however, we must first convert cubic inches to cubic centimeters. The relationship between in.^{3} and cm^3 is not given on the back inside cover, but the relationship between inches and centimeters is given: 1 in. = 2.54 cm (exactly). Cubing both sides of this equation gives (1 in.)^3 = (2.54 cm)^3 from which we write the desired conversion factor:

$$\frac{(2.54\ \text{cm})^3}{(1\ \text{in.})^3} = \frac{(2.54)^3\ \text{cm}^3}{(1)^3\ \text{in.}^3} = \frac{16.39\ \text{cm}^3}{1\ \text{in.}^3}$$

Notice that both the numbers and the units are cubed. Also, because 2.54 is an exact number, we can retain as many digits of (2.54)^3 as we need. We have used four, one more than the number of digits in the density (19.3 g/cm^3). Applying our conversion factors, we can now solve the problem:

$$\text{Mass in grams} = (2.00\ \cancel{\text{in.}^3})\left(\frac{16.39\ \cancel{\text{cm}^3}}{1\ \cancel{\text{in.}^3}}\right)\left(\frac{19.3\ \text{g}}{1\ \cancel{\text{cm}^3}}\right) = 633\ \text{g}$$

The procedure is diagrammed below. The final answer is reported to three significant figures, the same number of significant figures as in 2.00 and 19.3.

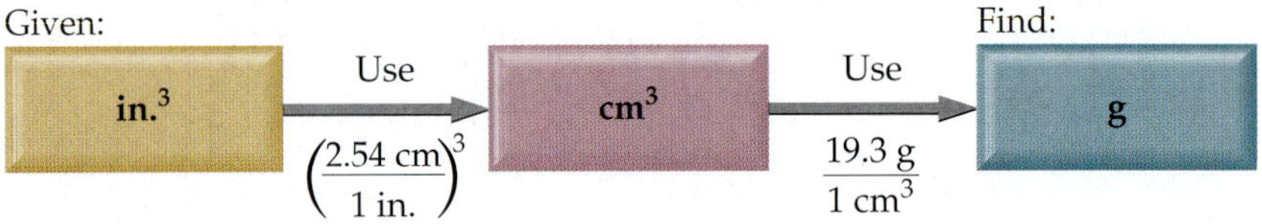

SAMPLE EXERCISE 1.11 | Converting volume units

Earth's oceans contain approximately 1.36×10^9 km^3 of water. Calculate the volume in liters.

Solution This problem involves conversion of km^3 to L. From the back inside cover of the text we find 1 L = 10^{-3} m^3, but there is no relationship listed involving km^3. From our knowledge of metric prefixes, however, we have 1 km = 10^3 m, and we can use this relationship between lengths to write the desired conversion factor between volumes:

$$\left(\frac{10^3\ \text{m}}{1\ \text{km}}\right)^3 = \frac{10^9\ \text{m}^3}{1\ \text{km}^3}$$

Thus, converting from km^3 to m^3 to L, we have

$$\text{Volume in liters} = (1.36 \times 10^9\ \cancel{\text{km}^3})\left(\frac{10^9\ \cancel{\text{m}^3}}{1\ \cancel{\text{km}^3}}\right)\left(\frac{1\ \text{L}}{10^{-3}\ \cancel{\text{m}^3}}\right) = 1.36 \times 10^{21}\ \text{L}$$

PRACTICE EXERCISE

If the volume of an object is reported as 5.0 ft^3, what is the volume in cubic meters?
Answer: 0.14 m^3

SAMPLE EXERCISE 1.12 | Conversions involving density

What is the mass in grams of 1.00 gal of water? The density of water is 1.00 g/mL.

Solution Before we begin solving this exercise, we note the following:

1. We are given 1.00 gal of water (the known, or given, quantity) and asked to calculate its mass in grams (the unknown).
2. We have the following conversion factors either given, commonly known, or available on the back inside cover of the text:

$$\frac{1.00\text{ g water}}{1\text{ mL water}} \quad \frac{1\text{ L}}{1000\text{ mL}} \quad \frac{1\text{ L}}{1.057\text{ qt}} \quad \frac{1\text{ gal}}{4\text{ qt}}$$

The first of these conversion factors must be used as written (with grams in the numerator) to give the desired result, whereas the last conversion factor must be inverted in order to cancel gallons:

$$\text{Mass in grams} = (1.00\ \cancel{\text{gal}})\left(\frac{4\ \cancel{\text{qt}}}{1\ \cancel{\text{gal}}}\right)\left(\frac{1\ \cancel{\text{L}}}{1.057\ \cancel{\text{qt}}}\right)\left(\frac{1000\ \cancel{\text{mL}}}{1\ \cancel{\text{L}}}\right)\left(\frac{1.00\text{ g}}{1\ \cancel{\text{mL}}}\right)$$

$$= 3.78 \times 10^3\text{ g water}$$

The units of our final answer are appropriate, and we've also taken care of our significant figures. We can further check our calculation by the estimation procedure. We can round 1.057 off to 1. Focusing on the numbers that don't equal 1 then gives merely $4 \times 1000 = 4000$ g, in agreement with the detailed calculation.

PRACTICE EXERCISE

The density of benzene is 0.879 g/mL. Calculate the mass in grams of 1.00 qt of benzene.

Answer: 832 g

STRATEGIES IN CHEMISTRY | The Importance of Practice

If you've ever played a musical instrument or participated in athletics, you know that the keys to success are practice and discipline. You can't learn to play a piano merely by listening to music, and you can't learn how to play basketball merely by watching games on television. Likewise, you can't learn chemistry by merely watching your instructor do it. Simply reading this book, listening to lectures, or reviewing notes will not usually be sufficient when exam time comes around. Your task is not merely to understand how someone else uses chemistry, but to be able to do it yourself. That takes practice on a regular basis, and anything that you have to do on a regular basis requires self-discipline until it becomes a habit.

Throughout the book, we have provided sample exercises in which the solutions are shown in detail. A practice exercise, for which only the answer is given, accompanies each sample exercise. It is important that you use these exercises as learning aids. End-of-chapter exercises provide additional questions to help you understand the material in the chapter. Red numbers indicate exercises for which answers are given at the back of the book. A review of basic mathematics is given in Appendix A.

The practice exercises in this text and the homework assignments given by your instructor provide the minimal practice that you will need to succeed in your chemistry course. Only by working all the assigned problems will you face the full range of difficulty and coverage that your instructor expects you to master for exams. There is no substitute for a determined and perhaps lengthy effort to work problems on your own. If you do get stuck on a problem, however, get help from your instructor, a teaching assistant, a tutor, or a fellow student. Spending an inordinate amount of time on a single exercise is rarely effective unless you know that it is particularly challenging and requires extensive thought and effort.

SUMMARY AND KEY TERMS

Introduction and Section 1.1 **Chemistry** is the study of the composition, structure, properties, and changes of **matter**. The composition of matter relates to the kinds of **elements** it contains. The structure of matter relates to the ways the **atoms** of these elements are arranged. A **property** is any characteristic that gives a sample of matter its unique identity. A **molecule** is an entity composed of two or more atoms with the atoms attached to one another in a specific **way**.

Section 1.2 Matter exists in three physical states, **gas**, **liquid**, and **solid**, which are known as the **states of matter**. There are two kinds of **pure substances: elements** and **compounds**. Each element has a single kind of atom and is represented by a chemical symbol consisting of one or two letters, with the first letter capitalized. Compounds are composed of two or more elements joined chemically. The **law of constant composition**, also called the **law of definite proportions**, states that the elemental composition of a pure compound is always the same. Most matter consists of a mixture of substances. **Mixtures** have variable compositions and can be either homogeneous or heterogeneous; homogeneous mixtures are called **solutions**.

Section 1.3 Each substance has a unique set of **physical properties** and **chemical properties** that can be used to identify it. During a **physical change**, matter does not change its composition. **Changes of state** are physical changes. In a **chemical change (chemical reaction)** a substance is transformed into a chemically different substance. **Intensive properties** are independent of the amount of matter examined and are used to identify substances. **Extensive properties** relate to the amount of substance present. Differences in physical and chemical properties are used to separate substances.

The **scientific method** is a dynamic process used to answer questions about our physical world. Observations and experiments lead to **scientific laws**, general rules that summarize how nature behaves. Observations also lead to tentative explanations or **hypotheses**. As a hypothesis is tested and refined, a **theory** may be developed.

Section 1.4 Measurements in chemistry are made using the **metric system**. Special emphasis is placed on a particular set of metric units called **SI units**, which are based on the meter, the kilogram, and the second as the basic units of length, **mass**, and time, respectively. The metric system employs a set of prefixes to indicate decimal fractions or multiples of the base units. The SI temperature scale is the **Kelvin scale**, although the **Celsius scale** is frequently used as well. **Density** is an important property that equals mass divided by volume.

Section 1.5 All measured quantities are inexact to some extent. The **precision** of a measurement indicates how closely different measurements of a quantity agree with one another. The **accuracy** of a measurement indicates how well a measurement agrees with the accepted or "true" value. The **significant figures** in a measured quantity include one estimated digit, the last digit of the measurement. The significant figures indicate the extent of the uncertainty of the measurement. Certain rules must be followed so that a calculation involving measured quantities is reported with the appropriate number of significant figures.

Section 1.6 In the **dimensional analysis** approach to problem solving, we keep track of units as we carry measurements through calculations. The units are multiplied together, divided into each other, or canceled like algebraic quantities. Obtaining the proper units for the final result is an important means of checking the method of calculation. When converting units and when carrying out several other types of problems, **conversion factors** can be used. These factors are ratios constructed from valid relations between equivalent quantities.

VISUALIZING CONCEPTS

Visualizing Concepts

The exercises in this section are intended to probe your understanding of key concepts rather than your ability to utilize formulas and perform calculations. Those exercises with red exercise numbers have answers in the back of the book.

1.1 Which of the following figures represents **(a)** a pure element, **(b)** a mixture of two elements, **(c)** a pure compound, **(d)** a mixture of an element and a compound? (More than one picture might fit each description.)

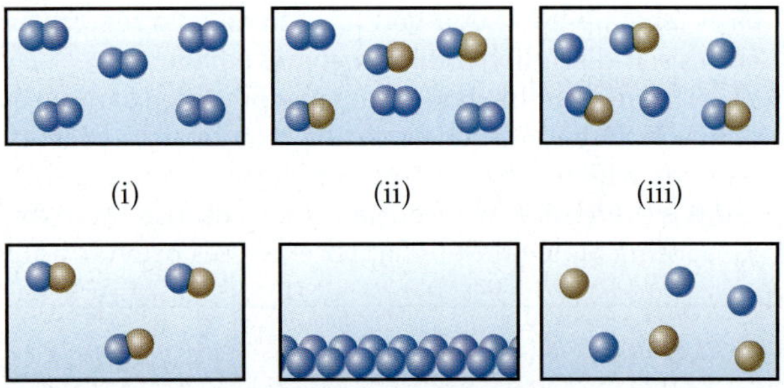

(i) (ii) (iii)

(iv) (v) (vi)

1.2 Does the following diagram represent a chemical or physical change? How do you know?

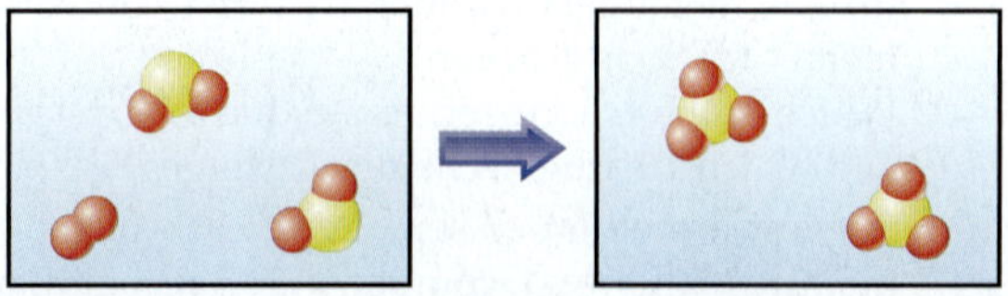

1.3 Identify each of the following as measurements of length, area, volume, mass, density, time, or temperature: **(a)** 5 ns, **(b)** 5.5 kg/m^3, **(c)** 0.88 pm, **(d)** 540 km^2, **(e)** 173 K, **(f)** 2 mm^3, **(g)** 23°C.

1.4 The following dartboards illustrate the types of errors often seen when one measurement is repeated several times. The bull's-eye represents the "true value," and the darts represent the experimental measurements. Which board best represents each of the following scenarios: **(a)** measurements both accurate and precise, **(b)** measurements precise but inaccurate, **(c)** measurements imprecise but yield an accurate average?

(i) (ii) (iii)

1.5 **(a)** What is the length of the pencil in the following figure if the scale reads in centimeters? How many significant figures are there in this measurement? **(b)** An oven thermometer with a circular scale reading degrees Fahrenheit is shown. What temperature does the scale indicate? How many significant figures are in the measurement?

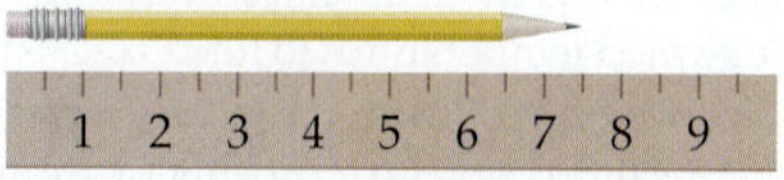

1.6 What is wrong with the following statement? Twenty years ago an ancient artifact was determined to be 1900 years old. It must now be 1920 years old.

1.7 When you convert units, how do you decide which part of the conversion factor is in the numerator and which is in the denominator?

1.8 Draw a logic map indicating the steps you would take to convert miles per hour to kilometers per second. Write down the conversion factor for each step, as done on page 28.

EXERCISES

Classification and Properties of Matter

The following exercises are divided into sections that deal with specific topics in the chapter. These exercises are grouped in pairs, with the answer given in the back of the book to the odd-numbered one, as indicated by the red exercise number. Those exercises whose number appears in brackets are more challenging than the nonbracketed exercises.

1.9 Classify each of the following as a pure substance or a mixture; if a mixture, indicate whether it is homogeneous or heterogeneous: **(a)** rice pudding, **(b)** seawater, **(c)** magnesium, **(d)** gasoline.

1.10 Classify each of the following as a pure substance or a mixture; if a mixture, indicate whether it is homogeneous or heterogeneous: **(a)** air, **(b)** tomato juice, **(c)** iodine crystals, **(d)** sand.

1.11 Give the chemical symbols for the following elements: **(a)** sulfur, **(b)** potassium, **(c)** chlorine, **(d)** copper, **(e)** silicon, **(f)** nitrogen, **(g)** calcium, **(h)** helium.

1.12 Give the chemical symbol for each of the following elements: **(a)** carbon, **(b)** sodium, **(c)** fluorine, **(d)** iron, **(e)** phosphorus, **(f)** argon, **(g)** nickel, **(h)** silver.

1.13 Name the chemical elements represented by the following symbols: **(a)** Li, **(b)** Al, **(c)** Pb, **(d)** S, **(e)** Br, **(f)** Sn, **(g)** Cr, **(h)** Zn.

1.14 Name each of the following elements: **(a)** Co, **(b)** I, **(c)** Kr, **(d)** Hg, **(e)** As, **(f)** Ti, **(g)** K, **(h)** Ge.

1.15 A solid white substance A is heated strongly in the absence of air. It decomposes to form a new white substance B and a gas C. The gas has exactly the same properties as the product obtained when carbon is burned in an excess of oxygen. Based on these observations, can we determine whether solids A and B and the gas C are elements or compounds? Explain your conclusions for each substance.

1.16 In 1807 the English chemist Humphry Davy passed an electric current through molten potassium hydroxide and isolated a bright, shiny reactive substance. He claimed the discovery of a new element, which he named potassium. In those days, before the advent of modern instruments, what was the basis on which one could claim that a substance was an element?

1.17 In the process of attempting to characterize a substance, a chemist makes the following observations: The substance is a silvery white, lustrous metal. It melts at 649°C and boils at 1105°C. Its density at 20°C is 1.738 g/cm^3. The substance burns in air, producing an intense white light. It reacts with chlorine to give a brittle white solid. The substance can be pounded into thin sheets or drawn into wires. It is a good conductor of electricity. Which of these characteristics are physical properties, and which are chemical properties?

1.18 Read the following description of the element zinc, and indicate which are physical properties and which are chemical properties. Zinc is a silver–gray-colored metal that melts at 420°C. When zinc granules are added to dilute sulfuric acid, hydrogen is given off and the metal dissolves. Zinc has a hardness on the Mohs scale of 2.5 and a density of 7.13 g/cm^3 at 25°C. It reacts slowly with oxygen gas at elevated temperatures to form zinc oxide, ZnO.

1.19 Label each of the following as either a physical process or a chemical process: **(a)** corrosion of aluminum metal, **(b)** melting of ice, **(c)** pulverizing an aspirin, **(d)** digesting a candy bar, **(e)** explosion of nitroglycerin.

1.20 A match is lit and held under a cold piece of metal. The following observations are made: **(a)** The match burns. **(b)** The metal gets warmer. **(c)** Water condenses on the metal. **(d)** Soot (carbon) is deposited on the metal. Which of these occurrences are due to physical changes, and which are due to chemical changes?

1.21 Suggest a method of separating each of the following mixtures into two components: **(a)** sugar and sand, **(b)** iron and sulfur.

1.22 A beaker contains a clear, colorless liquid. If it is water, how could you determine whether it contained dissolved table salt? Do *not* taste it!

Units and Measurement

1.23 What decimal power do the following abbreviations represent: **(a)** d, **(b)** c, **(c)** f, **(d)** μ, **(e)** M, **(f)** k, **(g)** n, **(h)** m, **(i)** p?

1.24 Use appropriate metric prefixes to write the following measurements without use of exponents: **(a)** 6.35×10^{-2} L, **(b)** 6.5×10^{-6} s, **(c)** 9.5×10^{-4} m, **(d)** 4.23×10^{-9} m^3, **(e)** 12.5×10^{-8} kg, **(f)** 3.5×10^{-10} g, **(g)** 6.54×10^{9} fs.

1.25 Perform the following conversions: **(a)** 25.5 mg to g, **(b)** 4.0×10^{-10} m to nm, **(c)** 0.575 mm to μm.

1.26 Convert **(a)** 9.5×10^{-2} kg to g, **(b)** 0.0023 μm to nm; **(c)** 7.25×10^{-4} s to ms.

1.27 **(a)** A sample of carbon tetrachloride, a liquid once used in dry cleaning, has a mass of 39.73 g and a volume of 25.0 mL at 25°C. What is its density at this temperature? Will carbon tetrachloride float on water? (Materials that are less dense than water will float.) **(b)** The density of platinum is 21.45 g/cm^3 at 20°C. Calculate the mass of 75.00 cm^3 of platinum at this temperature. **(c)** The density of magnesium is 1.738 g/cm^3 at 20°C. What is the volume of 87.50 g of this metal at this temperature?

1.28 **(a)** A cube of osmium metal 1.500 cm on a side has a mass of 76.31 g at 25°C. What is its density in g/cm^3 at this temperature? **(b)** The density of titanium metal is 4.51 g/cm^3 at 25°C. What mass of titanium displaces 65.8 mL of water at 25°C? **(c)** The density of benzene at 15°C is 0.8787 g/mL. Calculate the mass of 0.1500 L of benzene at this temperature.

1.29 **(a)** To identify a liquid substance, a student determined its density. Using a graduated cylinder, she measured out a 45-mL sample of the substance. She then measured the mass of the sample, finding that it weighed 38.5 g. She knew that the substance had to be either isopropyl alcohol (density 0.785 g/mL) or toluene (density 0.866 g/mL). What are the calculated density and the probable identity of the substance? **(b)** An experiment requires 45.0 g of ethylene glycol, a liquid whose density is 1.114 g/mL. Rather than weigh the sample on a balance, a chemist chooses to dispense the liquid using a graduated cylinder. What volume of the liquid should he use? **(c)** A cubic piece of metal measures 5.00 cm on each edge. If the metal is nickel, whose density is 8.90 g/cm^3, what is the mass of the cube?

1.30 **(a)** After the label fell off a bottle containing a clear liquid believed to be benzene, a chemist measured the density of the liquid to verify its identity. A 25.0-mL portion of the liquid had a mass of 21.95 g. A chemistry handbook lists the density of benzene at 15°C as 0.8787 g/mL. Is the calculated density in agreement with the tabulated value? **(b)** An experiment requires 15.0 g of cyclohexane, whose density at 25°C is 0.7781 g/mL. What volume of cyclohexane should be used? **(c)** A spherical ball of lead has a diameter of 5.0 cm. What is the mass of the sphere if lead has a density of 11.34 g/cm^3? (The volume of a sphere is $\left(\frac{4}{3}\right)\pi r^3$ where r is the radius.)

[1.31] Gold can be hammered into extremely thin sheets called gold leaf. If a 200-mg piece of gold (density = 19.32 g/cm^3) is hammered into a sheet measuring 2.4×1.0 ft, what is the average thickness of the sheet in meters? How might the thickness be expressed without exponential notation, using an appropriate metric prefix?

[1.32] A cylindrical rod formed from silicon is 16.8 cm long and has a mass of 2.17 kg. The density of silicon is 2.33 g/cm^3. What is the diameter of the cylinder? (The volume of a cylinder is given by $\pi r^2 h$, where r is the radius, and h is its length.)

1.33 Make the following conversions: **(a)** 62°F to °C, **(b)** 216.7°C to °F, **(c)** 233°C to K, **(d)** 315 K to °F, **(e)** 2500°F to K.

1.34 **(a)** The temperature on a warm summer day is 87°F. What is the temperature in °C? **(b)** Many scientific data are reported at 25°C. What is this temperature in kelvins and in degrees Fahrenheit? **(c)** Suppose that a recipe calls for an oven temperature of 175°F. Convert this temperature to degrees Celsius and to kelvins. **(d)** The melting point of sodium bromide (a salt) is 755°C. Calculate this temperature in °F and in kelvins. **(e)** Neon, the gaseous element used to make electronic signs, has a melting point of −248.6°C and a boiling point of −246.1°C. Convert these temperatures to kelvins.

Uncertainty in Measurement

1.35 Indicate which of the following are exact numbers: **(a)** the mass of a paper clip, **(b)** the surface area of a dime, **(c)** the number of inches in a mile, **(d)** the number of ounces in a pound, **(e)** the number of microseconds in a week, **(f)** the number of pages in this book.

1.36 Indicate which of the following are exact numbers: **(a)** the mass of a 32-oz can of coffee, **(b)** the number of students in your chemistry class, **(c)** the temperature of the surface of the sun, **(d)** the mass of a postage stamp, **(e)** the number of milliliters in a cubic meter of water, **(f)** the average height of students in your school.

1.37 What is the number of significant figures in each of the following measured quantities? **(a)** 358 kg, **(b)** 0.054 s, **(c)** 6.3050 cm, **(d)** 0.0105 L, **(e)** 7.0500×10^{-3} m^3.

1.38 Indicate the number of significant figures in each of the following measured quantities: **(a)** 3.7745 km, **(b)** 205 m^2, **(c)** 1.700 cm, **(d)** 350.0 K, **(e)** 307.080 g.

1.39 Round each of the following numbers to four significant figures, and express the result in standard exponential notation: **(a)** 102.53070, **(b)** 656,980, **(c)** 0.008543210, **(d)** 0.000257870, **(e)** −0.0357202.

1.40 **(a)** The diameter of Earth at the equator is 7926.381 mi. Round this number to three significant figures, and express it in standard exponential notation. **(b)** The circumference of Earth through the poles is 40,008 km. Round this number to four significant figures, and express it in standard exponential notation.

1.41 Carry out the following operations, and express the answers with the appropriate number of significant figures: **(a)** 12.0550 + 9.05, **(b)** 257.2 − 19.789, **(c)** $(6.21 \times 10^3)(0.1050)$, **(d)** 0.0577/0.753.

1.42 Carry out the following operations, and express the answer with the appropriate number of significant figures:
(a) 320.55 − (6104.5/2.3),
(b) $[(285.3 \times 10^5) - (1.200 \times 10^3)] \times 2.8954$,
(c) $(0.0045 \times 20{,}000.0) + (2813 \times 12)$,
(d) $863 \times [1255 - (3.45 \times 108)]$.

Dimensional Analysis

1.43 Using the information on the back inside cover and your knowledge of English units, write down the conversion factors needed to convert **(a)** cm to ft, **(b)** in.^{3} to cm^3.

1.44 Using the information on the back inside cover, write down the conversion factors needed to convert **(a)** mi to km, **(b)** qt to L.

1.45 Perform the following conversions: **(a)** 0.076 L to mL, **(b)** 5.0×10^{-8} m to nm, **(c)** 6.88×10^5 ns to s, **(d)** 0.50 lb to g, **(e)** 1.55 kg/m^3 to g/L, **(f)** 5.850 gal/hr to L/s.

1.46 **(a)** The speed of light in a vacuum is 2.998×10^8 m/s. Calculate its speed in km/hr. **(b)** The Sears Tower in Chicago is 1454 ft tall. Calculate its height in meters. **(c)** The Vehicle Assembly Building at the Kennedy Space Center in Florida has a volume of 3,666,500 m^3. Convert this volume to liters, and express the result in standard exponential notation. **(d)** An individual suffering from a high cholesterol level in her blood has 232 mg of cholesterol per 100 mL of blood. If the total blood volume of the individual is 5.2 L, how many grams of total blood cholesterol does the individual's body contain?

1.47 Perform the following conversions: **(a)** 5.00 days to s, **(b)** 0.0550 mi to m, **(c)** $1.89/gal to dollars per liter, **(d)** 0.510 in./ms to km/hr, **(e)** 22.50 gal/min to L/s, **(f)** 0.02500 ft^3 to cm^3.

1.48 Carry out the following conversions: **(a)** 0.105 in. to mm, **(b)** 0.870 qt to mL, **(c)** 8.75 μm/s to km/hr, **(d)** 4.733 yd^3 to m^3, **(e)** $3.99/lb to dollars per kg, **(f)** 8.75 lb/ft^3 to g/mL.

1.49 **(a)** How many liters of wine can be held in a wine barrel whose capacity is 31 gal? **(b)** The recommended adult dose of Elixophyllin®, a drug used to treat asthma, is 6 mg/kg of body mass. Calculate the dose in milligrams for a 150-lb person. **(c)** If an automobile is able to travel 254 mi on 11.2 gal of gasoline, what is the gas mileage in km/L? **(d)** A pound of coffee beans yields 50 cups of coffee (4 cups = 1 qt). How many milliliters of coffee can be obtained from 1 g of coffee beans?

1.50 **(a)** If an electric car is capable of going 225 km on a single charge, how many charges will it need to travel from Boston, Massachusetts, to Miami, Florida, a distance of 1486 mi, assuming that the trip begins with a full charge? **(b)** If a migrating loon flies at an average speed of 14 m/s, what is its average speed in mi/hr? **(c)** What is the engine piston displacement in liters of an engine whose displacement is listed as 450 in.^{3}? **(d)** In March 1989 the *Exxon Valdez* ran aground and spilled 240,000 barrels of crude petroleum off the coast of Alaska. One barrel of petroleum is equal to 42 gal. How many liters of petroleum were spilled?

1.51 The density of air at ordinary atmospheric pressure and 25°C is 1.19 g/L. What is the mass, in kilograms, of the air in a room that measures $12.5 \times 15.5 \times 8.0$ ft?

1.52 The concentration of carbon monoxide in an urban apartment is 48 μg/m^3. What mass of carbon monoxide in grams is present in a room measuring $9.0 \times 14.5 \times 18.8$ ft?

1.53 By using estimation techniques, arrange these items in order from shortest to longest: a 57-cm length of string, a 14-in. long shoe, and a 1.1-m length of pipe.

1.54 By using estimation techniques, determine which of the following is the heaviest and which is the lightest: a 5-lb bag of potatoes, a 5-kg bag of sugar, or 1 gal of water (density = 1.0 g/mL).

[1.55] The Morgan silver dollar has a mass of 26.73 g. By law, it was required to contain 90% silver, with the remainder being copper. **(a)** When the coin was minted in the late 1800s, silver was worth $1.18 per troy ounce (31.1 g). At this price, what is the value of the silver in the silver dollar? **(b)** Today, silver sells for about $5.30 per troy ounce. How many Morgan silver dollars are required to obtain $25.00 worth of pure silver?

[1.56] A copper refinery produces a copper ingot weighing 150 lb. If the copper is drawn into wire whose diameter is 8.25 mm, how many feet of copper can be obtained from the ingot? The density of copper is 8.94 g/cm^3. (Assume that the wire is a cylinder whose volume is $V = \pi r^2 h$, where r is its radius and h is its height or length.)

Additional Exercises

The exercises in this section are not divided by category, although they are roughly in the order of the topics in the chapter. They are not paired.

1.57 What is meant by the terms composition and structure when referring to matter?

1.58 **(a)** Classify each of the following as a pure substance, a solution, or a heterogeneous mixture: a gold coin, a cup of coffee, a wood plank. **(b)** What ambiguities are there in answering part (a) from the descriptions given?

1.59 **(a)** What is the difference between a hypothesis and a theory? **(b)** Explain the difference between a theory and a scientific law. Which addresses how matter behaves, and which addresses why it behaves that way?

1.60 A sample of ascorbic acid (vitamin C) is synthesized in the laboratory. It contains 1.50 g of carbon and 2.00 g of oxygen. Another sample of ascorbic acid isolated from citrus fruits contains 6.35 g of carbon. How many grams of oxygen does it contain? Which law are you assuming in answering this question?

1.61 Two students determine the percentage of lead in a sample as a laboratory exercise. The true percentage is 22.52%. The students' results for three determinations are as follows:

1. 22.52, 22.48, 22.54

2. 22.64, 22.58, 22.62

(a) Calculate the average percentage for each set of data, and tell which set is the more accurate based on the average. **(b)** Precision can be judged by examining the average of the deviations from the average value for that data set. (Calculate the average value for each data set, then calculate the average value of the absolute deviations of each measurement from the average.) Which set is more precise?

1.62 Is the use of significant figures in each of the following statements appropriate? Why or why not? **(a)** The 2000 circulation of *National Geographic* was 8,783,752. **(b)** In July 2003 the population of Cook County, Illinois, was 5,351,552. **(c)** The average annual rainfall in San Diego, California, is 20.54 cm. **(d)** In the United States, 0.621% of the population has the surname Brown.

1.63 What type of quantity (for example, length, volume, density) do the following units indicate: **(a)** mL, **(b)** cm^2, **(c)** mm^3, **(d)** mg/L, **(e)** ps, **(f)** nm, **(g)** K?

1.64 Give the derived SI units for each of the following quantities in base SI units: **(a)** acceleration = distance/time^2; **(b)** force = mass × acceleration; **(c)** work = force × distance; **(d)** pressure = force/area; **(e)** power = work/time.

1.65 The liquid substances mercury (density = 13.5 g/mL), water (1.00 g/mL), and cyclohexane (0.778 g/mL) do not form a solution when mixed, but separate in distinct layers. Sketch how the liquids would position themselves in a test tube.

1.66 A 40-lb container of peat moss measures 14 × 20 × 30 in. A 40-lb container of topsoil has a volume of 1.9 gal. **(a)** Calculate the average densities of peat moss and topsoil in units of g/cm^3. Would it be correct to say that peat moss is "lighter" than topsoil? Explain. **(b)** How many bags of the peat moss are needed to cover an area measuring 10. ft by 20. ft to a depth of 2.0 in.?

1.67 Small spheres of equal mass are made of lead (density = 11.3 g/cm^3), silver (10.5 g/cm^3), and aluminum (2.70 g/cm^3). Which sphere has the largest diameter, and which has the smallest?

1.68 The annual production of sodium hydroxide in the United States in 1999 was 23.2 billion pounds. **(a)** How many grams of sodium hydroxide were produced in that year? **(b)** The density of sodium hydroxide is 2.130 g/cm^3. How many cubic kilometers were produced?

1.69 **(a)** You are given a bottle that contains 4.59 cm^3 of a metallic solid. The total mass of the bottle and solid is 35.66 g. The empty bottle weighs 14.23 g. What is the density of the solid? **(b)** Mercury is traded by the "flask," a unit that has a mass of 34.5 kg. What is the volume of a flask of mercury if the density of mercury is 13.5 g/mL? **(c)** A thief plans to steal a gold sphere with a radius of 28.9 cm from a museum. If the gold has a density of 19.3 g/cm^3, what is the mass of the sphere? [The volume of a sphere is $V = (4/3)\pi r^3$.] Is he likely to be able to walk off with it unassisted?

1.70 Automobile batteries contain sulfuric acid, which is commonly referred to as "battery acid." Calculate the number of grams of sulfuric acid in 0.500 L of battery acid if the solution has a density of 1.28 g/mL and is 38.1% sulfuric acid by mass.

[1.71] A 32.65-g sample of a solid is placed in a flask. Toluene, in which the solid is insoluble, is added to the flask so that the total volume of solid and liquid together is 50.00 mL. The solid and toluene together weigh 58.58 g. The density of toluene at the temperature of the experiment is 0.864 g/mL. What is the density of the solid?

[1.72] Suppose you decide to define your own temperature scale using the freezing point (−11.5°C) and boiling point (197.6°C) of ethylene glycol. If you set the freezing point as 0°G and the boiling point as 100°G, what is the freezing point of water on this new scale?

1.73 Recently, one of the text authors completed an exactly 10-mi road race, in a time of 1 hr, 25 min, and 14 s. **(a)** What was the runner's average speed in miles per hour? **(b)** What was the runner's pace in minutes and seconds per mile?

1.74 The distance from Earth to the Moon is approximately 240,000 mi. **(a)** What is this distance in meters? **(b)** The *Concorde SST* has an air speed of about 2400 km/hr. If the *Concorde* could fly to the Moon, how many seconds would it take?

1.75 The U.S. quarter has a mass of 5.67 g and is approximately 1.55 mm thick. **(a)** How many quarters would have to be stacked to reach 575 ft, the height of the Washington Monument? **(b)** How much would this stack weigh? **(c)** How much money would this stack contain? **(d)** In 2004 the national debt was $7.2 trillion. How many stacks like the one described would be necessary to pay off this debt?

1.76 In the United States, water used for irrigation is measured in acre-feet. An acre-foot of water covers an acre to a depth of exactly 1 ft. An acre is 4840 yd^2. An acre-foot is enough water to supply two typical households for 1.00 yr. **(a)** If desalinated water costs $2480 per acre-foot, how much does desalinated water cost per liter? **(b)** How much would it cost one household per day if it were the only source of water?

[1.77] A package of aluminum foil contains 50 ft^2 of foil, which weighs approximately 8.0 oz. Aluminum has a density of 2.70 g/cm^3. What is the approximate thickness of the foil in millimeters?

[1.78] A 15.0-cm long cylindrical glass tube, sealed at one end, is filled with ethanol. The mass of ethanol needed to fill the tube is found to be 11.86 g. The density of ethanol is 0.789 g/mL. Calculate the inner diameter of the tube in centimeters.

[1.79] Gold is alloyed (mixed) with other metals to increase its hardness in making jewelry. **(a)** Consider a piece of gold jewelry that weighs 9.85 g and has a volume of 0.675 cm^3. The jewelry contains only gold and silver, which have densities of 19.3 g/cm^3 and 10.5 g/cm^3, respectively. Assuming that the total volume of the jewelry is the sum of the volumes of the gold and silver that it contains, calculate the percentage of gold (by mass) in the jewelry. **(b)** The relative amount of gold in an alloy is commonly expressed in units of carats. Pure gold is 24-carat, and the percentage of gold in an alloy is given as a percentage of this value. For example, an alloy that is 50% gold is 12-carat. State the purity of the gold jewelry in carats.

[1.80] Suppose you are given a sample of a homogeneous liquid. What would you do to determine whether it is a solution or a pure substance?

[1.81] Chromatography (Figure 1.14) is a simple, but reliable, method for separating a mixture into its constituent substances. Suppose you are using chromatography to separate a mixture of two substances. How would you know whether the separation is successful? Can you propose a means of quantifying how good or how poor the separation is?

[1.82] You are assigned the task of separating a desired granular material, with a density of 3.62 g/cm^3, from an undesired granular material that has a density of 2.04 g/cm^3. You want to do this by shaking the mixture in a liquid in which the heavier material will fall to the bottom and the lighter material will float. A solid will float on any liquid that is more dense. Using a handbook of chemistry, find the densities of the following substances: carbon tetrachloride, hexane, benzene, and methylene iodide. Which of these liquids will serve your purpose, assuming no chemical interaction between the liquid and the solids?

[1.83] The concepts of accuracy and precision are not always easy to grasp. Here are two sets of studies: **(a)** The mass of a secondary weight standard is determined by weighing it on a very precise balance under carefully controlled laboratory conditions. The average of 18 different weight measurements is taken as the weight of the standard. **(b)** A group of 10,000 males between the ages of 50 and 55 is surveyed to ascertain a relationship between calorie intake and blood cholesterol level. The survey questionnaire is quite detailed, asking the respondents about what they eat, smoking and drinking habits, and so on. The results are reported as showing that for men of comparable lifestyles, there is a 40% chance of the blood cholesterol level being above 230 for those who consume more than 40 calories per gram of body weight per day, as compared with those who consume fewer than 30 calories per gram of body weight per day.

Discuss and compare these two studies in terms of the precision and accuracy of the result in each case. How do the two studies differ in nature in ways that affect the accuracy and precision of the results? What makes for high precision and accuracy in any given study? In each of these studies, what factors might not be controlled that could affect the accuracy and precision? What steps can be taken generally to attain higher precision and accuracy?

eMEDIA EXERCISES

These exercises make use of the interactive objects available online in OneKey or the Companion Website, and on your Accelerator CD. Access to these resources comes in your MediaPak.

1.84 Experiment with the **Phases of the Elements** activity (*1.2*). **(a)** How many elements are liquids at room temperature, and what are they? **(b)** Choose two temperatures—one higher and one lower than room temperature—and determine how many elements are liquids at those temperatures.

1.85 Watch the **Electrolysis of Water** animation (*1.2*). **(a)** How can you tell from this experiment that water is a compound and not an element? **(b)** If you were to perform a similar experiment using liquid bromine instead of liquid water in the apparatus, what would you expect to happen?

1.86 The **Changes of State** animation (*1.3*) shows what happens to a solid when it is heated. **(a)** Describe the changes that occur. **(b)** Is the change from solid to liquid a chemical change or a physical change? **(c)** Is the change from liquid to gas a chemical change or a physical change? **(d)** Is enough information given to determine whether the original solid is an element, a compound, or a mixture? Explain.

1.87 **(a)** Use the **Significant Figures** activity (*1.5*) to verify your answers to Exercises 1.41 and 1.42. **(b)** Is it possible for the sum of a column of numbers, each containing two significant figures, to have more than two significant figures? Explain. **(c)** How many significant figures should there be in the answer to the following calculation? $(35.2 - 30.1) \times 1.23 =$ _____.

CHAPTER

2

WHAT'S AHEAD

- We begin by giving a brief history of the notion of *atoms*—the smallest pieces of matter. *(Section 2.1)*
- Next, we look at some of the key experiments that led to the discovery of *electrons* and to the *nuclear model* of the atom. *(Section 2.2)*
- We then discuss the modern theory of atomic structure, including the ideas of *atomic numbers, mass numbers*, and *isotopes*. *(Section 2.3)*
- We introduce the concept of *atomic weights* and how they relate to the masses of individual atoms. *(Section 2.4)*
- Our discussion of atoms leads to the organization of the elements into the *periodic table*, in which elements are put in order of increasing atomic number and grouped by chemical similarity. *(Section 2.5)*
- Our understanding of atoms allows us to discuss the assemblies of atoms called *molecules* and how their compositions are represented by *empirical and molecular formulas*. *(Section 2.6)*
- We learn that atoms can gain or lose electrons to form *ions*, and we look at how to use the periodic table to predict the charges on ions and the empirical formulas of *ionic compounds*. *(Section 2.7)*
- We will see the systematic way in which substances are named, called *nomenclature*, and how this nomenclature is applied to inorganic compounds. *(Section 2.8)*
- Finally, we introduce some basic ideas of *organic chemistry*, which is the chemistry of the element carbon. *(Section 2.9)*

Atoms, Molecules, and Ions

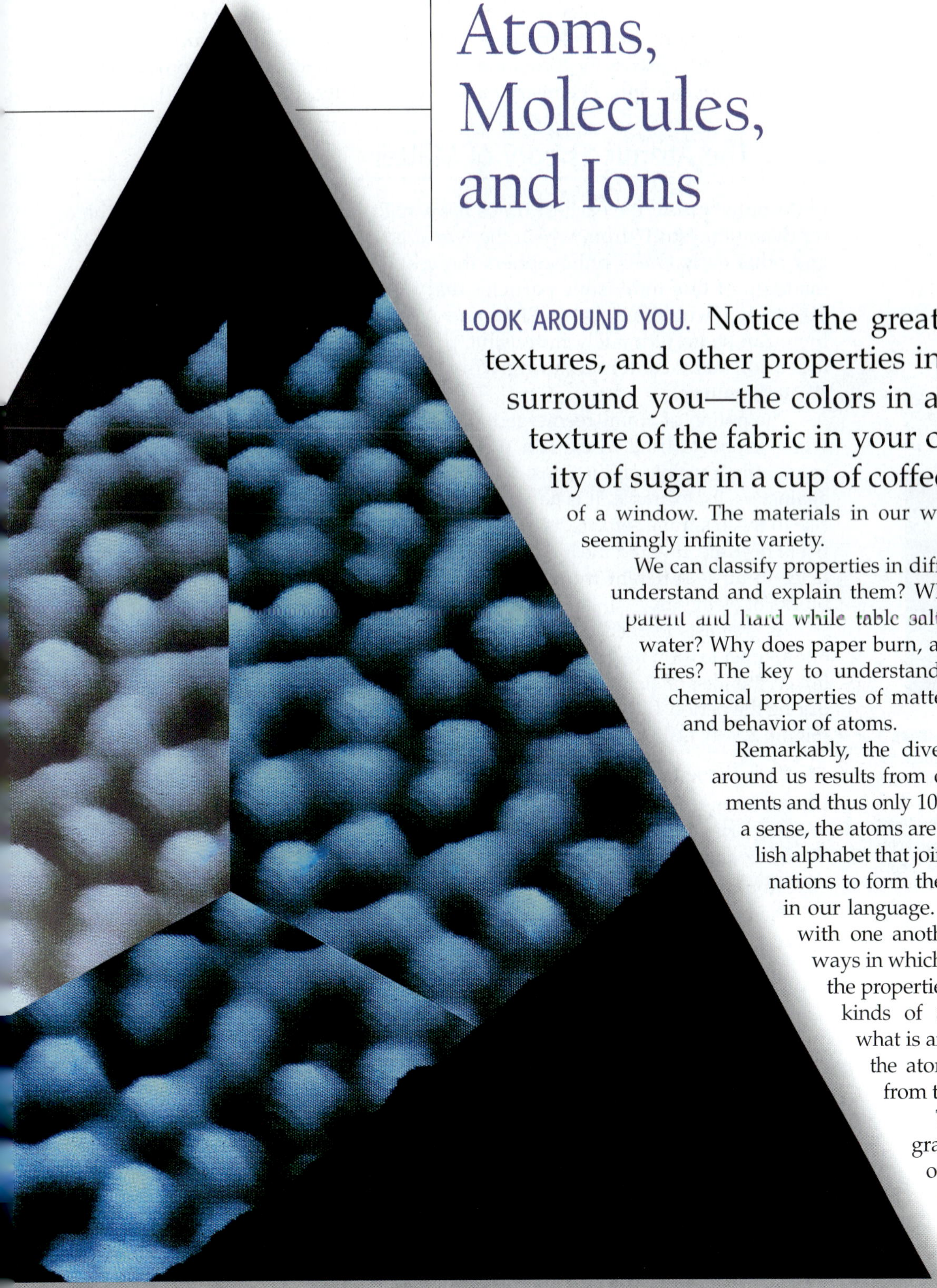

LOOK AROUND YOU. Notice the great variety of colors, textures, and other properties in the materials that surround you—the colors in a garden scene, the texture of the fabric in your clothes, the solubility of sugar in a cup of coffee, the transparency of a window. The materials in our world exhibit a striking and seemingly infinite variety.

We can classify properties in different ways, but how do we understand and explain them? What makes diamonds transparent and hard while table salt is brittle and dissolves in water? Why does paper burn, and why does water quench fires? The key to understanding both the physical and chemical properties of matter is found in the structure and behavior of atoms.

Remarkably, the diversity of properties we see around us results from only about 100 different elements and thus only 100 different kinds of atoms. In a sense, the atoms are like the 26 letters of the English alphabet that join together in different combinations to form the immense number of words in our language. But how do atoms combine with one another? What rules govern the ways in which they can combine? How do the properties of a substance relate to the kinds of atoms it contains? Indeed, what is an atom like, and what makes the atoms of one element different from those of another?

The chapter-opening photograph is an image of a collection of atoms, but as striking as this picture is, it doesn't reveal enough to help us understand the properties of matter. Fortunately, we can use a variety

INDIVIDUAL CARBON ATOMS IN A SAMPLE OF GRAPHITE as viewed by a technique known as scanning tunneling microscopy (STM). Notice the hexagonal arrangements of the atoms in both the STM image and in the adjacent molecular model.

of experimental techniques to probe the atom to gain a clearer understanding of what it is like. In this chapter we begin to explore the fascinating world of atoms that we discover by such experiments. We will examine the basic structure of the atom and briefly discuss the formation of molecules and ions, thereby providing a foundation for exploring chemistry more deeply in later chapters.

2.1 | The Atomic Theory of Matter

Philosophers from the earliest times have speculated about the nature of the fundamental "stuff" from which the world is made. Democritus (460–370 BC) and other early Greek philosophers thought that the material world must be made up of tiny indivisible particles that they called *atomos*, meaning "indivisible or uncuttable." Later, Plato and Aristotle formulated the notion that there can be no ultimately indivisible particles. The "atomic" view of matter faded for many centuries during which Aristotelean philosophy dominated Western culture.

The notion of atoms reemerged in Europe during the seventeenth century, when scientists tried to explain the properties of gases. Air is composed of something invisible and in constant motion; we can feel the motion of the wind against us, for example. It is natural to think of tiny invisible particles as giving rise to these familiar effects. Isaac Newton (1642–1727), the most famous scientist of his time, favored the idea of atoms, but thinking of atoms as invisible particles in air is different from thinking of atoms as the fundamental building blocks of elements.

▲ **Figure 2.1 John Dalton (1766–1844).** Dalton was the son of a poor English weaver. He began teaching at the age of 12. He spent most of his years in Manchester, where he taught both grammar school and college. His lifelong interest in meteorology led him to study gases and hence to chemistry and eventually to the atomic theory.

As chemists learned to measure the amounts of elements that reacted with one another to form new substances, the ground was laid for an atomic theory that linked the idea of elements with the idea of atoms. That theory came into being during the period 1803–1807 in the work of an English schoolteacher, John Dalton (Figure 2.1 ◀). Dalton's atomic theory involved the following postulates:

1. Each element is composed of extremely small particles called atoms.
2. All atoms of a given element are identical to one another in mass and other properties, but the atoms of one element are different from the atoms of all other elements.
3. Atoms of an element are not changed into atoms of a different element by chemical reactions; atoms are neither created nor destroyed in chemical reactions.
4. Compounds are formed when atoms of more than one element combine; a given compound always has the same relative number and kind of atoms.

According to Dalton's atomic theory, **atoms** are the smallest particles of an element that retain the chemical identity of the element. (Section 1.1) As noted in the postulates of Dalton's theory, an element is composed of only one kind of atom, whereas a compound contains atoms of two or more elements.

Dalton's theory explains several simple laws of chemical combination that were known in his time. One of these was the *law of constant composition*: In a given compound, the relative numbers and kinds of atoms are constant. (Section 1.2) This law is the basis of Dalton's Postulate 4. Another fundamental chemical law was the *law of conservation of mass* (also known as the *law of conservation of matter*): The total mass of materials present after a chemical reaction is the same as the total mass present before the reaction. This law is the basis for Postulate 3. Dalton proposed that atoms always retain their identities and that atoms taking part in a chemical reaction rearrange to give new chemical combinations.

ACTIVITY
Law of Multiple Proportions

ANIMATION
Multiple Proportions

A good theory should not only explain the known facts but should also predict new ones. Dalton used his theory to deduce the *law of multiple proportions*: If two elements A and B combine to form more than one compound, the masses of B that can combine with a given mass of A are in the ratio of small whole numbers. We can illustrate this law by considering the substances water and hydrogen peroxide, both of which consist of the elements hydrogen and oxygen. In forming water, 8.0 g of oxygen combines with 1.0 g of hydrogen. In forming hydrogen peroxide, 16.0 g of oxygen combines with 1.0 g of hydrogen. In other words, the ratio of the mass of oxygen per gram of hydrogen in the two compounds is 2:1. Using the atomic theory, we can conclude that hydrogen peroxide contains twice as many atoms of oxygen per hydrogen atom as does water.

GIVE IT SOME THOUGHT

One compound of carbon and oxygen contains 1.333 g of oxygen per gram of carbon, whereas a second compound contains 2.666 g of oxygen per gram of carbon. (a) What chemical law do these data illustrate? (b) If the first compound has an equal number of oxygen and carbon atoms, what can we conclude about the composition of the second compound?

2.2 | The Discovery of Atomic Structure

Dalton reached his conclusion about atoms on the basis of chemical observations in the macroscopic world of the laboratory. Neither he nor those who followed him during the century after his work was published had direct evidence for the existence of atoms. Today, however, we can use powerful instruments to measure the properties of individual atoms and even provide images of them (Figure 2.2 ▶).

As scientists began to develop methods for more detailed probing of the nature of matter, the atom, which was supposed to be indivisible, began to show signs of a more complex structure: We now know that the atom is composed of still smaller **subatomic particles**. Before we summarize the current model of atomic structure, we will briefly consider a few of the landmark discoveries that led to that model. We'll see that the atom is composed in part of electrically charged particles, some with a positive (+) charge and some with a negative (−) charge. As we discuss the development of our current model of the atom, keep in mind a simple statement of the behavior of charged particles: *Particles with the same charge repel one another, whereas particles with unlike charges are attracted to one another.*

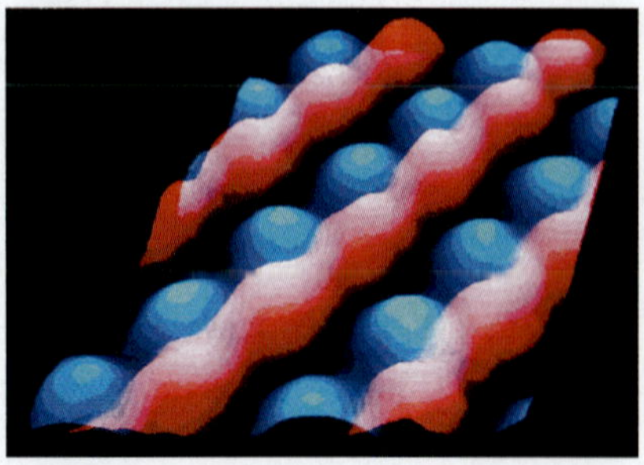

▲ **Figure 2.2 An image of the surface of the semiconductor GaAs (gallium arsenide).** This image was obtained by a technique called scanning tunneling microscopy. The color was added to the image by computer to distinguish the gallium atoms (blue spheres) from the arsenic atoms (red spheres).

Cathode Rays and Electrons

In the mid-1800s, scientists began to study electrical discharge through partially evacuated tubes (tubes that had been pumped almost empty of air), such as those shown in Figure 2.3 ▼. A high voltage produces radiation within the tube.

▼ **Figure 2.3 Cathode-ray tube.** (a) In a cathode-ray tube, electrons move from the negative electrode (cathode) to the positive electrode (anode). (b) A photo of a cathode-ray tube containing a fluorescent screen to show the path of the cathode rays. (c) The path of the cathode rays is deflected by the presence of a magnet.

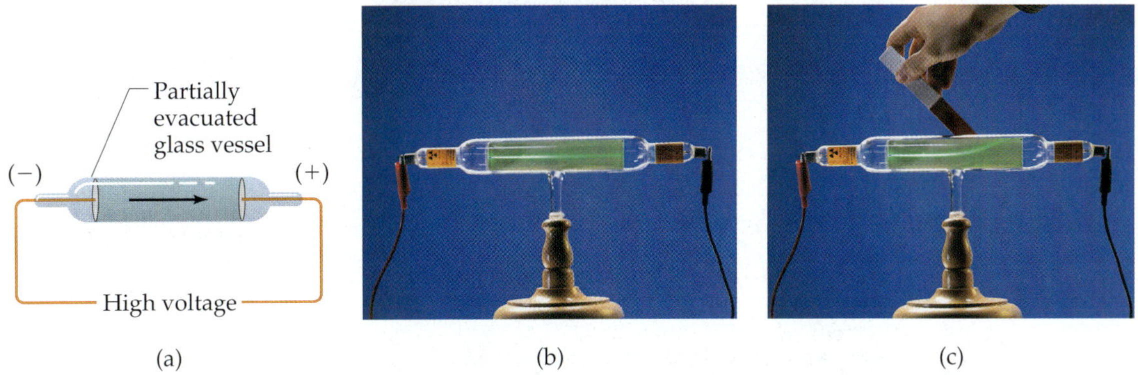

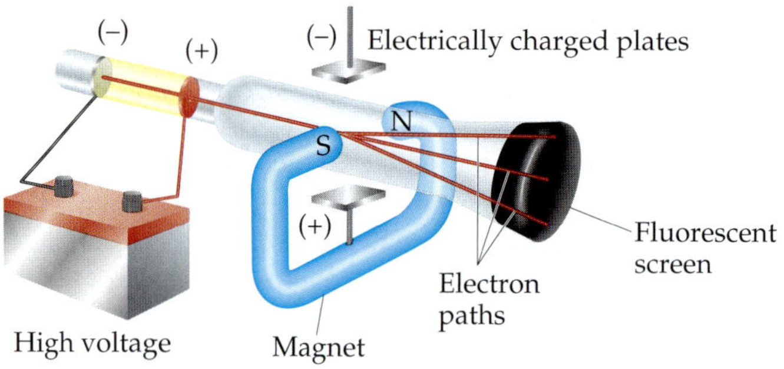

▶ **Figure 2.4 Cathode-ray tube with perpendicular magnetic and electric fields.** The cathode rays (electrons) originate from the negative plate on the left and are accelerated toward the positive plate, which has a hole in its center. A beam of electrons passes through the hole and is then deflected by the magnetic and electric fields. The three paths result from different strengths of the magnetic and electric fields. The charge-to-mass ratio of the electron can be determined by measuring the effects that the magnetic and electric fields have on the direction of the beam.

VCL Thompson Cathode-Ray Experiment

This radiation became known as **cathode rays** because it originated from the negative electrode, or cathode. Although the rays themselves could not be seen, their movement could be detected because the rays cause certain materials, including glass, to *fluoresce*, or give off light. (Television picture tubes are cathode-ray tubes; a television picture is the result of fluorescence from the television screen.)

Scientists held conflicting views about the nature of the cathode rays. It was not initially clear whether the rays were an invisible stream of particles or a new form of radiation. Experiments showed that cathode rays are deflected by electric or magnetic fields in a way consistent with their being a stream of negative electrical charge [Figure 2.3(c)]. The British scientist J. J. Thomson observed many properties of the rays, including the fact that the rays are the same regardless of the identity of the cathode material. In a paper published in 1897, he summarized his observations and concluded that cathode rays are streams of negatively charged particles. Thomson's paper is generally accepted as the "discovery" of what later became known as the *electron*.

ANIMATION Millikan Oil-Drop Experiment

Thomson constructed a cathode-ray tube having a fluorescent screen at one end, such as that shown in Figure 2.4 ▲, so that he could quantitatively measure the effects of electric and magnetic fields on the thin stream of electrons passing through a hole in the positively charged electrode. These measurements made it possible to calculate a value of 1.76×10^8 coulombs per gram for the ratio of the electron's electrical charge to its mass.*

VCL Millikan Oil-Drop Experiment

Once the charge-to-mass ratio of the electron was known, measuring either the charge or the mass of an electron would also yield the value of the other quantity. In 1909, Robert Millikan (1868–1953) of the University of Chicago succeeded in measuring the charge of an electron by performing a series of experiments described in Figure 2.5 ▼. He then calculated the mass of the electron by

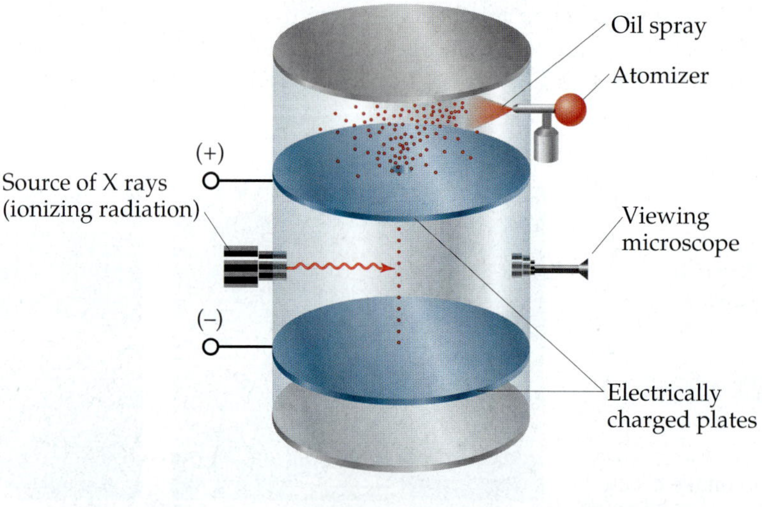

▶ **Figure 2.5 Millikan's oil-drop experiment.** A representation of the apparatus Millikan used to measure the charge of the electron. Small drops of oil, which had picked up extra electrons, were allowed to fall between two electrically charged plates. Millikan monitored the drops, measuring how the voltage on the plates affected their rate of fall. From these data he calculated the charges on the drops. His experiment showed that the charges were always integral multiples of 1.602×10^{-19} C, which he deduced was the charge of a single electron.

* The coulomb (C) is the SI unit for electrical charge.

using his experimental value for the charge, 1.60×10^{-19} C, and Thomson's charge-to-mass ratio, 1.76×10^{8} C/g:

$$\text{Electron mass} = \frac{1.60 \times 10^{-19}\ \text{C}}{1.76 \times 10^{8}\ \text{C/g}} = 9.10 \times 10^{-28}\ \text{g}$$

This result agrees well with the presently accepted value for the mass of the electron, 9.10938×10^{-28} g. This mass is about 2000 times smaller than that of hydrogen, the lightest atom.

Radioactivity

In 1896 the French scientist Henri Becquerel (1852–1908) was studying a uranium compound when he discovered that it spontaneously emits high-energy radiation. This spontaneous emission of radiation is called **radioactivity**. At Becquerel's suggestion Marie Curie (Figure 2.6 ▶) and her husband, Pierre, began experiments to isolate the radioactive components of the compound.

Further study of the nature of radioactivity, principally by the British scientist Ernest Rutherford (Figure 2.7 ▶), revealed three types of radiation: alpha (α), beta (β), and gamma (γ) radiation. Each type differs in its response to an electric field, as shown in Figure 2.8 ▼. The paths of both α and β radiation are bent by the electric field, although in opposite directions, whereas γ radiation is unaffected.

Rutherford showed that both α and β rays consist of fast-moving particles, which were called α and β particles. In fact, β particles are high-speed electrons and can be considered the radioactive equivalent of cathode rays. They are therefore attracted to a positively charged plate. The α particles have a positive charge and are therefore attracted toward a negative plate. In units of the charge of the electron, β particles have a charge of $1-$ and α particles a charge of $2+$. Each α particle is considerably heavier than an electron, having a mass about 7300 times that of an electron. Gamma radiation is high-energy radiation similar to X rays; it does not consist of particles and carries no charge. We will discuss radioactivity in greater detail in Chapter 21.

The Nuclear Atom

With the growing evidence that the atom is composed of smaller particles, attention was given to how the particles fit together. In the early 1900s Thomson reasoned that because electrons contribute only a very small fraction of the mass of an atom, they probably were responsible for an equally small fraction of the atom's size. He proposed that the atom consisted of a uniform positive sphere of matter in which the electrons were embedded, as shown

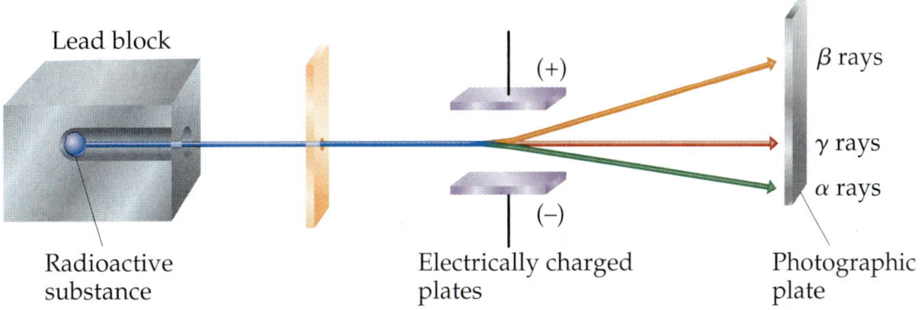

▲ **Figure 2.8 Behavior of alpha (α), beta (β), and gamma (γ) rays in an electric field.** The α rays consist of positively charged particles and are therefore attracted to the negatively charged plate. The β rays consist of negatively charged particles and are attracted to the positively charged plate. The γ rays, which carry no charge, are unaffected by the electric field.

▲ **Figure 2.6 Marie Sklodowska Curie (1867–1934).** When M. Curie presented her doctoral thesis, it was described as the greatest single contribution of any doctoral thesis in the history of science. Among other things, two new elements, polonium and radium, had been discovered. In 1903 Henri Becquerel, M. Curie, and her husband, Pierre, were jointly awarded the Nobel Prize in physics. In 1911 M. Curie won a second Nobel Prize, this time in chemistry.

▲ **Figure 2.7 Ernest Rutherford (1871–1937).** Rutherford, whom Einstein called "the second Newton," was born and educated in New Zealand. In 1895 he was the first overseas student ever to be awarded a position at the Cavendish Laboratory at Cambridge University in England, where he worked with J. J. Thomson. In 1898 he joined the faculty of McGill University in Montreal. While at McGill, Rutherford did his research on radioactivity that led to his being awarded the 1908 Nobel Prize in chemistry. In 1907 Rutherford moved back to England to be a faculty member at Manchester University, where in 1910 he performed his famous α-particle scattering experiments that led to the nuclear model of the atom. In 1992 his native New Zealand honored Rutherford by putting his likeness, along with his Nobel Prize medal, on their $100 currency note.

ACTIVITY
Separation of Alpha, Beta, and Gamma Rays

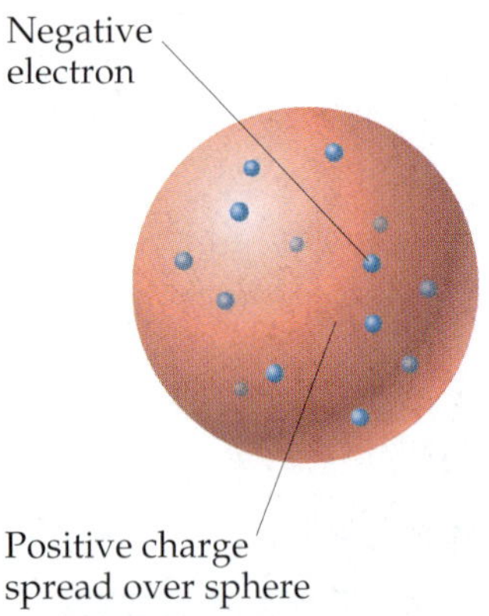

▲ **Figure 2.9 J. J. Thomson's "plum-pudding" model of the atom.** Thomson pictured the small electrons to be embedded in the atom much like raisins in a pudding or like seeds in a watermelon. Ernest Rutherford proved this model wrong.

in Figure 2.9 ◀. This so-called "plum-pudding" model, named after a traditional English dessert, was very short-lived.

In 1910, Rutherford and his coworkers performed an experiment that disproved Thomson's model. Rutherford was studying the angles at which α particles were deflected, or *scattered*, as they passed through a thin gold foil a few thousand atoms thick (Figure 2.10 ▼). He and his coworkers discovered that almost all the α particles passed directly through the foil without deflection. A small percentage were found to be slightly deflected, on the order of 1 degree, consistent with Thomson's plum-pudding model. Just for the sake of completeness, Rutherford suggested that Ernest Marsden, an undergraduate student working in the laboratory, look for evidence of scattering at large angles. To everyone's surprise, a small amount of scattering was observed at large angles. Some particles were even scattered back in the direction from which they had come. The explanation for these results was not immediately obvious, but they were clearly inconsistent with Thomson's plum-pudding model.

By 1911, Rutherford was able to explain these observations. He postulated that most of the mass of each gold atom in his foil and all of its positive charge reside in a very small, extremely dense region, which he called the **nucleus**. He postulated further that most of the total volume of an atom is empty space in which electrons move around the nucleus. In the α-scattering experiment, most α particles passed directly through the foil because they did not encounter the minute nucleus of any gold atom; they merely passed through the empty space making up the greatest part of all the atoms in the foil. Occasionally, however, an α particle came close to a gold nucleus. The repulsion between the highly charged gold nucleus and the α particle was strong enough to deflect the less massive α particle, as shown in Figure 2.11 ▶.

ANIMATION
Rutherford Experiment: Nuclear Atom

Subsequent experimental studies led to the discovery of both positive particles (*protons*) and neutral particles (*neutrons*) in the nucleus. Protons were discovered in 1919 by Rutherford. Neutrons were discovered in 1932 by the British scientist James Chadwick (1891–1972). We examine these particles more closely in Section 2.3.

▶ **Figure 2.10 Rutherford's experiment on the scattering of α particles.** The red lines represent the paths of the α particles. When the incoming beam strikes the gold foil, most particles pass straight through the foil but some are scattered.

VCL Rutherford's Backscattering Experiment

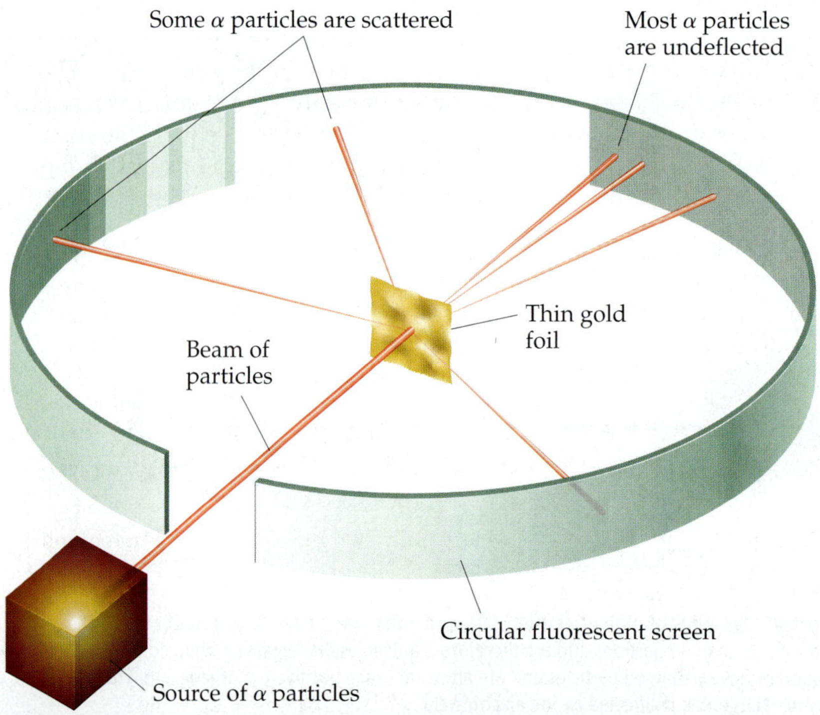

GIVE IT SOME THOUGHT

What happens to most of the α particles that strike the gold foil in Rutherford's experiment, and why do they behave that way?

2.3 | The Modern View of Atomic Structure

Since the time of Rutherford, physicists have learned much about the detailed composition of atomic nuclei. In the course of these discoveries, the list of particles that make up nuclei has grown long and continues to increase. As chemists, however, we can take a very simple view of the atom because only three subatomic particles—the **proton**, **neutron**, and **electron**—have a bearing on chemical behavior.

The charge of an electron is -1.602×10^{-19} C, and that of a proton is $+1.602 \times 10^{-19}$ C. The quantity 1.602×10^{-19} C is called the **electronic charge**. For convenience, the charges of atomic and subatomic particles are usually expressed as multiples of this charge rather than in coulombs. Thus, the charge of the electron is 1−, and that of the proton is 1+. Neutrons are uncharged and are therefore electrically neutral (which is how they received their name). *Every atom has an equal number of electrons and protons, and so atoms have no net electrical charge.*

Protons and neutrons reside together in the nucleus of the atom, which, as Rutherford proposed, is extremely small. The vast majority of an atom's volume is the space in which the electrons reside. The electrons are attracted to the protons in the nucleus by the force that exists between particles of opposite electrical charge. In later chapters we will see that the strength of the attractive forces between electrons and nuclei can be used to explain many of the differences between different elements.

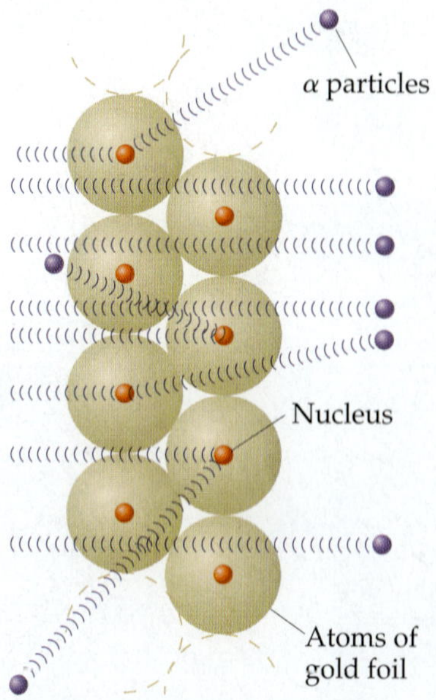

▲ **Figure 2.11 Rutherford's model explaining the scattering of α particles.** The gold foil is several thousand atoms thick. Because most of the volume of each atom is empty space, most α particles pass through the foil without deflection. When an α particle passes very close to a gold nucleus, however, it is repelled, causing its path to be altered.

GIVE IT SOME THOUGHT

(a) If an atom has 15 protons, how many electrons does it have? (b) Where do the protons reside in an atom?

Atoms have extremely small masses. The mass of the heaviest known atom, for example, is on the order of 4×10^{-22} g. Because it would be cumbersome to express such small masses in grams, we use instead the **atomic mass unit**, or amu.* One amu equals 1.66054×10^{-24} g. The masses of the proton and neutron are very nearly equal, and both are much greater than that of the electron: A proton has a mass of 1.0073 amu, a neutron 1.0087 amu, and an electron 5.486×10^{-4} amu. Because it would take 1836 electrons to equal the mass of 1 proton, the nucleus contains most of the mass of an atom. Table 2.1 ▼ summarizes the charges and masses of the subatomic particles. We will have more to say about atomic masses in Section 2.4.

TABLE 2.1 Comparison of the Proton, Neutron, and Electron

Particle	Charge	Mass (amu)
Proton	Positive (1+)	1.0073
Neutron	None (neutral)	1.0087
Electron	Negative (1−)	5.486×10^{-4}

* The SI abbreviation for the atomic mass unit is u. We will use the more common abbreviation amu.

Atoms are also extremely small. Most atoms have diameters between 1×10^{-10} m and 5×10^{-10} m, or 100–500 pm. A convenient, although non-SI, unit of length used to express atomic dimensions is the **angstrom** (Å). One angstrom equals 10^{-10} m. Thus, atoms have diameters on the order of 1–5 Å. The diameter of a chlorine atom, for example, is 200 pm, or 2.0 Å. Both picometers and angstroms are commonly used to express the dimensions of atoms and molecules.

SAMPLE EXERCISE 2.1 | Illustrating the Size of an Atom

The diameter of a U.S. penny is 19 mm. The diameter of a silver atom, by comparison, is only 2.88 Å. How many silver atoms could be arranged side by side in a straight line across the diameter of a penny?

Solution The unknown is the number of silver (Ag) atoms. We use the relationship 1 Ag atom = 2.88 Å as a conversion factor relating the number of atoms and distance. Thus, we can start with the diameter of the penny, first converting this distance into angstroms and then using the diameter of the Ag atom to convert distance to the number of Ag atoms:

$$\text{Ag atoms} = (19\ \cancel{\text{mm}})\left(\frac{10^{-3}\ \cancel{\text{m}}}{1\ \cancel{\text{mm}}}\right)\left(\frac{1\ \cancel{\text{Å}}}{10^{-10}\ \cancel{\text{m}}}\right)\left(\frac{1\ \text{Ag atom}}{2.88\ \cancel{\text{Å}}}\right) = 6.6 \times 10^{7}\ \text{Ag atoms}$$

That is, 66 million silver atoms could sit side by side across a penny!

PRACTICE EXERCISE

The diameter of a carbon atom is 1.54 Å. **(a)** Express this diameter in picometers. **(b)** How many carbon atoms could be aligned side by side in a straight line across the width of a pencil line that is 0.20 mm wide?
Answers: **(a)** 154 pm, **(b)** 1.3×10^{6} C atoms

The diameters of atomic nuclei are on the order of 10^{-4} Å, only a small fraction of the diameter of the atom as a whole. You can appreciate the relative sizes of the atom and its nucleus by imagining that if the atom were as large as a football stadium, the nucleus would be the size of a small marble. Because the tiny nucleus carries most of the mass of the atom in such a small volume, it has an incredible density—on the order of 10^{13}–10^{14} g/cm^3. A matchbox full of material of such density would weigh over 2.5 billion tons! Astrophysicists have suggested that the interior of a collapsed star may approach this density.

An illustration of the atom that incorporates the features we have just discussed is shown in Figure 2.12 ▼. The electrons, which take up most of the volume of the atom, play the major role in chemical reactions. The significance of representing the region containing the electrons as an indistinct cloud will become clear in later chapters when we consider the energies and spatial arrangements of the electrons.

▼ **Figure 2.12 The structure of the atom.** The nucleus, which contains protons and neutrons, is the location of virtually all the mass of the atom. The rest of the atom is the space in which the light, negatively charged electrons reside.

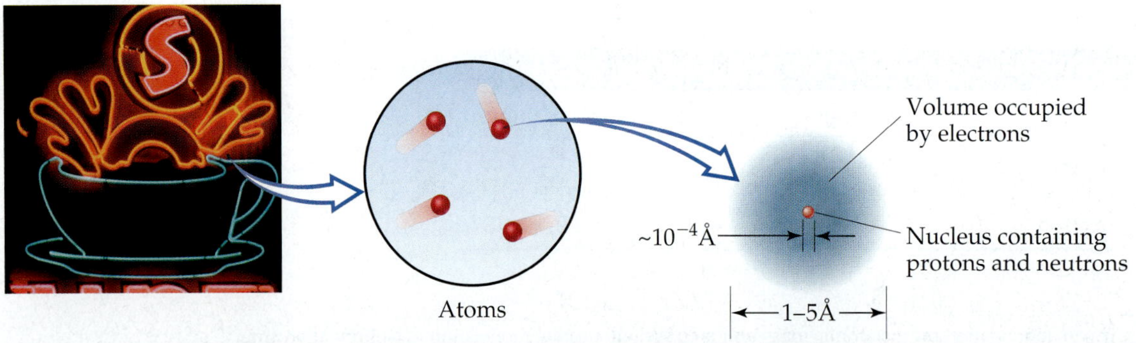

A CLOSER LOOK | Basic Forces

There are four basic forces, or interactions, known in nature: gravitational, electromagnetic, strong nuclear, and weak nuclear. *Gravitational forces* are attractive forces that act between all objects in proportion to their masses. Gravitational forces between atoms or between subatomic particles are so small that they are of no chemical significance.

Electromagnetic forces are attractive or repulsive forces that act between either electrically charged or magnetic objects. Electric and magnetic forces are intimately related. Electric forces are of fundamental importance in understanding the chemical behavior of atoms. The magnitude of the electric force between two charged particles is given by *Coulomb's law*: $F = kQ_1Q_2/d^2$, where Q_1 and Q_2 are the magnitudes of the charges on the two particles, d is the distance between their centers, and k is a constant determined by the units for Q and d. A negative value for the force indicates attraction, whereas a positive value indicates repulsion.

All nuclei except those of hydrogen atoms contain two or more protons. Because like charges repel, electrical repulsion would cause the protons to fly apart if a stronger attractive force did not keep them together. This force is called the *strong nuclear force*. It acts between subatomic particles, as in the nucleus. At this distance, the strong nuclear force is stronger than the electric force, and as a result the nucleus holds together. The *weak nuclear force* is weaker than the electric force but stronger than the gravitational force. We are aware of its existence only because it shows itself in certain types of radioactivity.

Related Exercises: 2.78(b) and 2.83

Atomic Numbers, Mass Numbers, and Isotopes

What makes an atom of one element different from an atom of another element? For example, how does an atom of carbon differ from an atom of oxygen? The significant difference is in their subatomic compositions. The atoms of each element have a characteristic number of protons. Indeed, the number of protons in the nucleus of an atom of any particular element is called that element's **atomic number**. Because an atom has no net electrical charge, the number of electrons it contains must equal the number of protons. All atoms of carbon, for example, have six protons and six electrons, whereas all atoms of oxygen have eight protons and eight electrons. Thus, carbon has atomic number 6, whereas oxygen has atomic number 8. The atomic number of each element is listed with the name and symbol of the element on the front inside cover of the text.

Atoms of a given element can differ in the number of neutrons they contain and consequently in mass. For example, most atoms of carbon have six neutrons, although some have more and some have less. The symbol $^{12}_{6}C$ (read "carbon twelve," carbon-12) represents the carbon atom containing six protons and six neutrons. The atomic number is shown by the subscript, and the superscript, called the **mass number**, is the total number of protons plus neutrons in the atom:

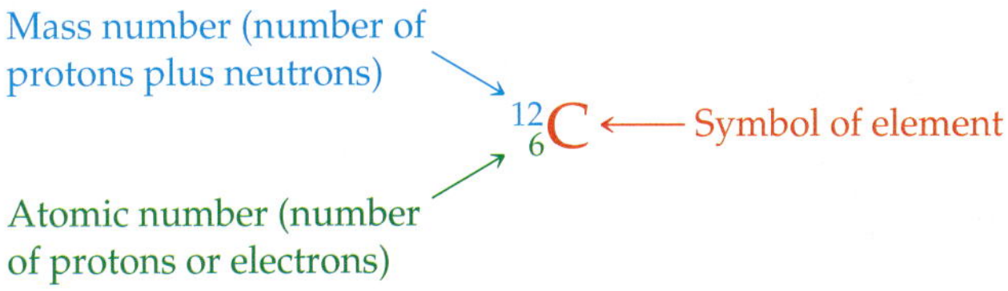

Because all atoms of a given element have the same atomic number, the subscript is redundant and is often omitted. Thus, the symbol for carbon-12 can be represented simply as ^{12}C. As one more example of this notation, atoms that contain six protons and eight neutrons have a mass number of 14 and are represented as $^{14}_{6}C$ or ^{14}C and referred to as carbon-14.

Atoms with identical atomic numbers but different mass numbers (that is, same number of protons but different numbers of neutrons) are called **isotopes**

ACTIVITY
Isotopes of Hydrogen

TABLE 2.2 Some Isotopes of Carbon[a]

Symbol	Number of Protons	Number of Electrons	Number of Neutrons
^{11}C	6	6	5
^{12}C	6	6	6
^{13}C	6	6	7
^{14}C	6	6	8

Almost 99% of the carbon found in nature is ^{12}C.

of one another. Several isotopes of carbon are listed in Table 2.2 ▲. We will generally use the notation with superscripts only when referring to a particular isotope of an element.

SAMPLE EXERCISE 2.2 | Determining the Number of Subatomic Particles in Atoms

How many protons, neutrons, and electrons are in **(a)** an atom of ^{197}Au; **(b)** an atom of strontium-90?

Solution **(a)** The superscript 197 is the mass number, the sum of the number of protons plus the number of neutrons. According to the list of elements given on the front inside cover, gold has an atomic number of 79. Consequently, an atom of ^{197}Au has 79 protons, 79 electrons, and 197 − 79 = 118 neutrons. **(b)** The atomic number of strontium (listed on the front inside cover) is 38. Thus, all atoms of this element have 38 protons and 38 electrons. The strontium-90 isotope has 90 − 38 = 52 neutrons.

PRACTICE EXERCISE

How many protons, neutrons, and electrons are in **(a)** a ^{138}Ba atom, **(b)** an atom of phosphorus-31?
Answer: **(a)** 56 protons, 56 electrons, and 82 neutrons; **(b)** 15 protons, 15 electrons, and 16 neutrons.

SAMPLE EXERCISE 2.3 | Writing Symbols for Atoms

Magnesium has three isotopes, with mass numbers 24, 25, and 26. **(a)** Write the complete chemical symbol (superscript and subscript) for each of them. **(b)** How many neutrons are in an atom of each isotope?

Solution **(a)** Magnesium has atomic number 12, and so all atoms of magnesium contain 12 protons and 12 electrons. The three isotopes are therefore represented by $^{24}_{12}Mg$, $^{25}_{12}Mg$, and $^{26}_{12}Mg$. **(b)** The number of neutrons in each isotope is the mass number minus the number of protons. The numbers of neutrons in an atom of each isotope are therefore 12, 13, and 14, respectively.

PRACTICE EXERCISE

Give the complete chemical symbol for the atom that contains 82 protons, 82 electrons, and 126 neutrons.
Answer: $^{208}_{82}Pb$

2.4 | Atomic Weights

Atoms are small pieces of matter, and so they have mass. In this section we will discuss the mass scale used for atoms and introduce the concept of *atomic weights*. In Section 3.3 we will extend these concepts to show how atomic masses are used to determine the masses of compounds and *molecular weights*.

The Atomic Mass Scale

Although scientists of the nineteenth century knew nothing about subatomic particles, they were aware that atoms of different elements have different masses. They found, for example, that each 100.0 g of water contains 11.1 g of hydrogen and 88.9 g of oxygen. Thus, water contains 88.9/11.1 = 8 times as much oxygen, by mass, as hydrogen. Once scientists understood that water contains two hydrogen atoms for each oxygen, they concluded that an oxygen atom must have $2 \times 8 = 16$ times as much mass as a hydrogen atom. Hydrogen, the lightest atom, was arbitrarily assigned a relative mass of 1 (no units), and atomic masses of other elements were at first determined relative to this value. Thus, oxygen was assigned an atomic mass of 16.

Today we can determine the masses of individual atoms with a high degree of accuracy. For example, we know that the ^{1}H atom has a mass of 1.6735×10^{-24} g and the ^{16}O atom has a mass of 2.6560×10^{-23} g. As we noted in Section 2.3, it is convenient to use the *atomic mass unit* (amu) when dealing with these extremely small masses:

$$1 \text{ amu} = 1.66054 \times 10^{-24} \text{ g and } 1 \text{ g} = 6.02214 \times 10^{23} \text{ amu}$$

The atomic mass unit is presently defined by assigning a mass of exactly 12 amu to an atom of the ^{12}C isotope of carbon. In these units, an ^{1}H atom has a mass of 1.0078 amu and an ^{16}O atom has a mass of 15.9949 amu.

Average Atomic Masses

Most elements occur in nature as mixtures of isotopes. We can determine the *average atomic mass* of an element by using the masses of its various isotopes and their relative abundances. Naturally occurring carbon, for example, is composed of 98.93% ^{12}C and 1.07% ^{13}C. The masses of these isotopes are 12 amu (exactly) and 13.00335 amu, respectively. We calculate the average atomic mass of carbon from the fractional abundance of each isotope and the mass of that isotope:

$$(0.9893)(12 \text{ amu}) + (0.0107)(13.00335 \text{ amu}) = 12.01 \text{ amu}$$

The average atomic mass of each element (expressed in atomic mass units) is also known as its **atomic weight**. Although the term *average atomic mass* is more proper, the term *atomic weight* is more common. The atomic weights of the elements are listed in both the periodic table and the table of elements on the inside front cover of this text.

GIVE IT SOME THOUGHT

A particular atom of chromium has a mass of 52.94 amu, whereas the atomic weight of chromium is 51.99 amu. Explain the difference in the two masses.

SAMPLE EXERCISE 2.4 | Calculating the Atomic Weight of an Element from Isotopic Abundances

Naturally occurring chlorine is 75.78% ^{35}Cl, which has an atomic mass of 34.969 amu, and 24.22% ^{37}Cl, which has an atomic mass of 36.966 amu. Calculate the average atomic mass (that is, the atomic weight) of chlorine.

Solution The average atomic mass is found by multiplying the abundance of each isotope by its atomic mass and summing these products. Because 75.78% = 0.7578 and 24.22% = 0.2422, we have

$$\begin{aligned}\text{Average atomic mass} &= (0.7578)(34.969 \text{ amu}) + (0.2422)(36.966 \text{ amu}) \\ &= 26.50 \text{ amu} + 8.953 \text{ amu} \\ &= 35.45 \text{ amu}\end{aligned}$$

This answer makes sense: The average atomic mass of Cl is between the masses of the two isotopes and is closer to the value of ^{35}Cl, which is the more abundant isotope.

PRACTICE EXERCISE

Three isotopes of silicon occur in nature: ^{28}Si (92.23%), which has an atomic mass of 27.97693 amu; ^{29}Si (4.68%), which has an atomic mass of 28.97649 amu; and ^{30}Si (3.09%), which has an atomic mass of 29.97377 amu. Calculate the atomic weight of silicon.
Answer: 28.09 amu

ACTIVITY
Mass Spectrometer

A CLOSER LOOK | The Mass Spectrometer

The most direct and accurate means for determining atomic and molecular weights is provided by the **mass spectrometer** (Figure 2.13 ▼). A gaseous sample is introduced at *A* and bombarded by a stream of high-energy electrons at *B*. Collisions between the electrons and the atoms or molecules of the gas produce positively charged particles, mostly with a 1+ charge. These charged particles are accelerated toward a negatively charged wire grid (*C*). After they pass through the grid, the particles encounter two slits that allow only a narrow beam of particles to pass. This beam then passes between the poles of a magnet, which deflects the particles into a curved path, much as electrons are deflected by a magnetic field (Figure 2.4). For charged particles with the same charge, the extent of deflection depends on mass—the more massive the particle, the less the deflection. The particles are thereby separated according to their masses. By changing the strength of the magnetic field or the accelerating voltage on the negatively charged grid, charged particles of various masses can be selected to enter the detector at the end of the instrument.

A graph of the intensity of the detector signal versus particle atomic mass is called a *mass spectrum*. The mass spectrum of chlorine atoms, shown in Figure 2.14 ▼, reveals the presence of two isotopes. Analysis of a mass spectrum gives both the masses of the charged particles reaching the detector and their relative abundances. The abundances are obtained from the signal intensities. Knowing the atomic mass and the abundance of each isotope allows us to calculate the atomic weight of an element, as shown in Sample Exercise 2.4.

Mass spectrometers are used extensively today to identify chemical compounds and analyze mixtures of substances. Any molecule that loses electrons falls apart, forming an array of positively charged fragments. The mass spectrometer measures the masses of these fragments, producing a chemical "fingerprint" of the molecule and providing clues about how the atoms were connected in the original molecule. Thus, a chemist might use this technique to determine the molecular structure of a newly synthesized compound or to identify a pollutant in the environment.

Related Exercises: 2.31, 2.32, 2.33(b), 2.34, 2.86, and 2.87

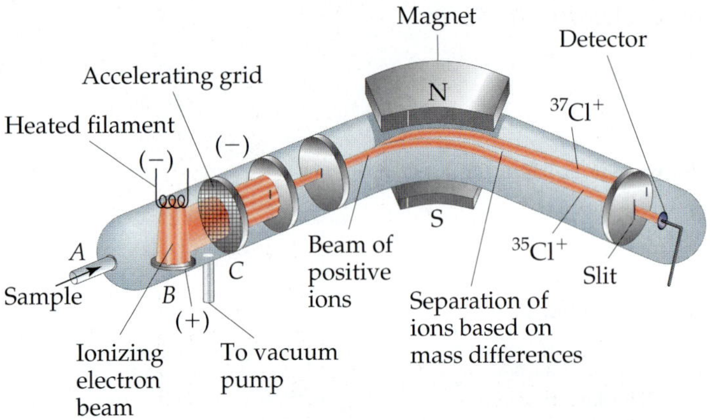

▲ **Figure 2.13 A mass spectrometer.** Cl atoms are introduced on the left side of the spectrometer and are ionized to form Cl^+ ions, which are then directed through a magnetic field. The paths of the ions of the two isotopes of Cl diverge as they pass through the magnetic field. As drawn, the spectrometer is tuned to detect $^{35}Cl^+$ ions. The heavier $^{37}Cl^+$ ions are not deflected enough for them to reach the detector.

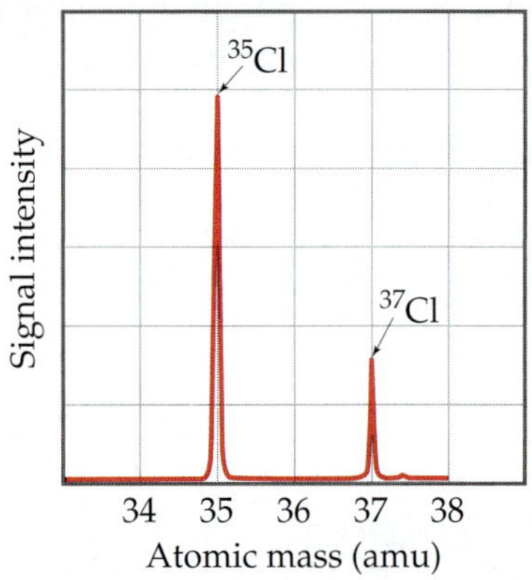

▲ **Figure 2.14 Mass spectrum of atomic chlorine.** The fractional abundances of the ^{35}Cl and ^{37}Cl isotopes of chlorine are indicated by the relative signal intensities of the beams reaching the detector of the mass spectrometer.

2.5 | The Periodic Table

Dalton's atomic theory set the stage for a vigorous growth in chemical experimentation during the early 1800s. As the body of chemical observations grew and the list of known elements expanded, attempts were made to find regular patterns in chemical behavior. These efforts culminated in the development of the periodic table in 1869. We will have much to say about the periodic table in later chapters, but it is so important and useful that you should become acquainted with it now. You will quickly learn that *the periodic table is the most significant tool that chemists use for organizing and remembering chemical facts.*

Many elements show very strong similarities to one another. The elements lithium (Li), sodium (Na), and potassium (K) are all soft, very reactive metals, for example; and the elements helium (He), neon (Ne), and argon (Ar) are all very nonreactive gases. If the elements are arranged in order of increasing atomic number, their chemical and physical properties are found to show a repeating, or periodic, pattern. For example, each of the soft, reactive metals—lithium, sodium, and potassium—comes immediately after one of the nonreactive gases—helium, neon, and argon—as shown in Figure 2.15 ▼.

The arrangement of elements in order of increasing atomic number, with elements having similar properties placed in vertical columns, is known as the **periodic table**. The periodic table is shown in Figure 2.16 ▶ and is also given on the front inside cover of the text. For each element in the table, the atomic number and atomic symbol are given, and the atomic weight is often given as well, as in the following typical entry for potassium:

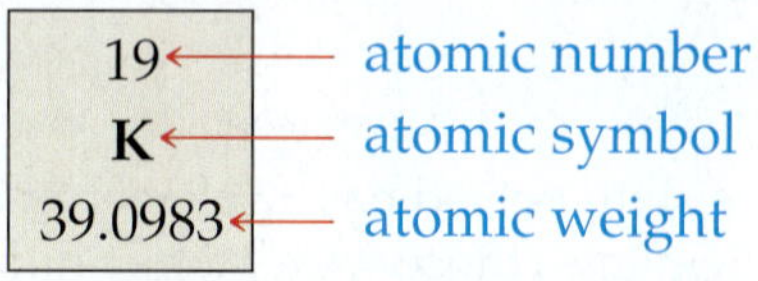

You may notice slight variations in periodic tables from one book to another or between those in the lecture hall and in the text. These are simply matters of style, or they might concern the particular information included; there are no fundamental differences.

The horizontal rows of the periodic table are called **periods**. The first period consists of only two elements, hydrogen (H) and helium (He). The second and third periods, which begin with lithium (Li) and sodium (Na), respectively, consist of eight elements each. The fourth and fifth periods

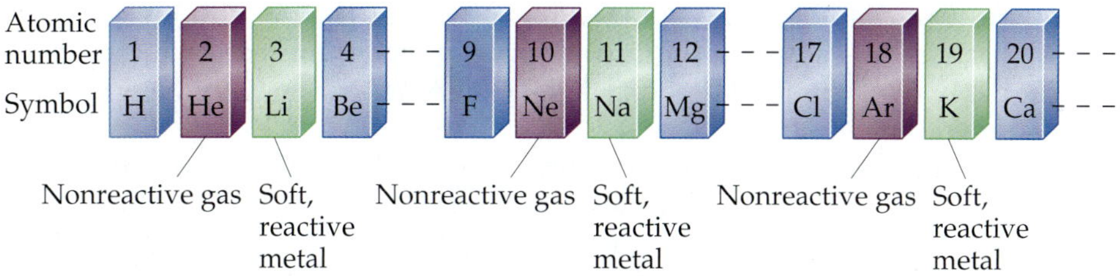

▲ **Figure 2.15 Arranging the elements by atomic number reveals a periodic pattern of properties.** This periodic pattern is the basis of the periodic table.

Period	1A 1	2A 2	3B 3	4B 4	5B 5	6B 6	7B 7	8B 8	8B 9	8B 10	1B 11	2B 12	3A 13	4A 14	5A 15	6A 16	7A 17	8A 18
1	1 H																	2 He
2	3 Li	4 Be											5 B	6 C	7 N	8 O	9 F	10 Ne
3	11 Na	12 Mg											13 Al	14 Si	15 P	16 S	17 Cl	18 Ar
4	19 K	20 Ca	21 Sc	22 Ti	23 V	24 Cr	25 Mn	26 Fe	27 Co	28 Ni	29 Cu	30 Zn	31 Ga	32 Ge	33 As	34 Se	35 Br	36 Kr
5	37 Rb	38 Sr	39 Y	40 Zr	41 Nb	42 Mo	43 Tc	44 Ru	45 Rh	46 Pd	47 Ag	48 Cd	49 In	50 Sn	51 Sb	52 Te	53 I	54 Xe
6	55 Cs	56 Ba	71 Lu	72 Hf	73 Ta	74 W	75 Re	76 Os	77 Ir	78 Pt	79 Au	80 Hg	81 Tl	82 Pb	83 Bi	84 Po	85 At	86 Rn
7	87 Fr	88 Ra	103 Lr	104 Rf	105 Db	106 Sg	107 Bh	108 Hs	109 Mt	110	111	112	113	114	115	116		

57 La	58 Ce	59 Pr	60 Nd	61 Pm	62 Sm	63 Eu	64 Gd	65 Tb	66 Dy	67 Ho	68 Er	69 Tm	70 Yb
89 Ac	90 Th	91 Pa	92 U	93 Np	94 Pu	95 Am	96 Cm	97 Bk	98 Cf	99 Es	100 Fm	101 Md	102 No

Metals

Metalloids

Nonmetals

▲ **Figure 2.16 Periodic table of the elements.** Different colors are used to show the division of the elements into metals, metalloids, and nonmetals.

ACTIVITY
Periodic Table

contain 18 elements. The sixth period has 32 elements, but in order for it to fit on a page, 14 of these elements (those with atomic numbers 57–70) appear at the bottom of the table. The seventh and last period is incomplete, but it also has 14 of its members placed in a row at the bottom of the table.

The vertical columns of the periodic table are called **groups**. The way in which the groups are labeled is somewhat arbitrary, and three labeling schemes are in common use, two of which are shown in Figure 2.16. The top set of labels, which have A and B designations, is widely used in North America. Roman numerals, rather than Arabic ones, are often employed in this scheme. Group 7A, for example, is often labeled VIIA. Europeans use a similar convention that numbers the columns from 1A through 8A and then from 1B through 8B, thereby giving the label 7B (or VIIB) instead of 7A to the group headed by fluorine (F). In an effort to eliminate this confusion, the International Union of Pure and Applied Chemistry (IUPAC) has proposed a convention that numbers the groups from 1 through 18 with no A or B designations, as shown in the lower set of labels at the top of the table in Figure 2.16. We will use the traditional North American convention with Arabic numerals.

Elements that belong to the same group often exhibit similarities in physical and chemical properties. For example, the "coinage metals"—copper (Cu), silver (Ag), and gold (Au)—all belong to group 1B. As their name suggests, the coinage metals are used throughout the world to make coins. Many other groups in the periodic table also have names, as listed in Table 2.3 ▶.

TABLE 2.3 Names of Some Groups in the Periodic Table

Group	Name	Elements
1A	Alkali metals	Li, Na, K, Rb, Cs, Fr
2A	Alkaline earth metals	Be, Mg, Ca, Sr, Ba, Ra
6A	Chalcogens	O, S, Se, Te, Po
7A	Halogens	F, Cl, Br, I, At
8A	Noble gases (or rare gases)	He, Ne, Ar, Kr, Xe, Rn

We will learn in Chapters 6 and 7 that the elements in a group of the periodic table have similar properties because they have the same arrangement of electrons at the periphery of their atoms. However, we need not wait until then to make good use of the periodic table; after all, the table was invented by chemists who knew nothing about electrons! We can use the table, as they intended, to correlate the behaviors of elements and to aid in remembering many facts. You will find it helpful to refer to the periodic table frequently when studying the remainder of this chapter.

Except for hydrogen, all the elements on the left side and in the middle of the periodic table are **metallic elements**, or **metals**. The majority of elements are metallic, and they all share many characteristic properties, such as luster and high electrical and heat conductivity. All metals, with the exception of mercury (Hg), are solids at room temperature. The metals are separated from the **nonmetallic elements**, or **nonmetals**, by a diagonal steplike line that runs from boron (B) to astatine (At), as shown in Figure 2.16. Hydrogen, although on the left side of the periodic table, is a nonmetal. At room temperature some of the nonmetals are gaseous, some are solid, and one is liquid. Nonmetals generally differ from the metals in appearance (Figure 2.17 ▶) and in other physical properties. Many of the elements that lie along the line that separates metals from nonmetals, such as antimony (Sb), have properties that fall between those of metals and those of nonmetals. These elements are often referred to as **metalloids**.

▲ **Figure 2.17 Some familiar examples of metals and nonmetals.** The nonmetals (from bottom left) are sulfur (yellow powder), iodine (dark, shiny crystals), bromine (reddish brown liquid and vapor in glass vial), and three samples of carbon (black charcoal powder, diamonds, and graphite in the pencil lead). The metals are in the form of an aluminum wrench, copper pipe, lead shot, silver coins, and gold nuggets.

GIVE IT SOME THOUGHT

Chlorine is a halogen. Locate this element in the periodic table. (a) What is its symbol? (b) In what period and in what group is the element located? (c) What is its atomic number? (d) Is chlorine a metal or nonmetal?

SAMPLE EXERCISE 2.5 | Using the Periodic Table

Which two of the following elements would you expect to show the greatest similarity in chemical and physical properties: B, Ca, F, He, Mg, P?

Solution Elements that are in the same group of the periodic table are most likely to exhibit similar chemical and physical properties. We therefore expect that Ca and Mg should be most alike because they are in the same group (2A, the alkaline earth metals).

PRACTICE EXERCISE

Locate Na (sodium) and Br (bromine) on the periodic table. Give the atomic number of each, and label each a metal, metalloid, or nonmetal.
Answer: Na, atomic number 11, is a metal; Br, atomic number 35, is a nonmetal.

A CLOSER LOOK | Glenn Seaborg and Seaborgium

Prior to 1940 the periodic table ended at uranium, element number 92. Since that time, no scientist has had a greater effect on the periodic table than Glenn Seaborg. Seaborg (Figure 2.18 ▶) became a faculty member in the chemistry department at the University of California, Berkeley, in 1937. In 1940 he and his colleagues Edwin McMillan, Arthur Wahl, and Joseph Kennedy succeeded in isolating plutonium (Pu) as a product of the reaction between uranium and neutrons. We will talk about reactions of this type, called *nuclear reactions*, in Chapter 21. We will also discuss the key role that plutonium plays in nuclear fission reactions, such as those that occur in nuclear power plants and atomic bombs.

During the period 1944 through 1958, Seaborg and his coworkers also identified various products of nuclear reactions as being the elements having atomic numbers 95 through 102. All these elements are radioactive and are not found in nature; they can be synthesized only via nuclear reactions. For their efforts in identifying the elements beyond uranium (the *transuranium* elements), McMillan and Seaborg shared the 1951 Nobel Prize in chemistry.

From 1961 to 1971, Seaborg served as the chairman of the U.S. Atomic Energy Commission (now the Department of Energy). In this position he had an important role in establishing international treaties to limit the testing of nuclear weapons. Upon his return to Berkeley he was part of the team that in 1974 first identified element number 106; that discovery was corroborated by another team at Berkeley in 1993. In 1994, to honor Seaborg's many contributions to the discovery of new elements, the American Chemical Society proposed that element number 106 be named "seaborgium," with a proposed symbol of Sg. After several years of controversy about whether an element should be named after a living person, the name seaborgium was officially adopted by the IUPAC in 1997, and Seaborg became the first person to have an element named after him while he was still alive.

Related Exercise: 2.90

▲ **Figure 2.18 Glenn Seaborg (1912–1999).** The photograph shows Seaborg at Berkeley in 1941 using a Geiger counter to try to detect radiation produced by plutonium. Geiger counters will be discussed in Section 21.5.

2.6 | Molecules and Molecular Compounds

The atom is the smallest representative sample of an element, but only the noble-gas elements are normally found in nature as isolated atoms. Most matter is composed of molecules or ions, both of which are formed from atoms. We examine molecules here and ions in Section 2.7.

A **molecule** is an assembly of two or more atoms tightly bound together. The resultant "package" of atoms behaves in many ways as a single, distinct object, just as a television set composed of many parts can be recognized as a single object. We will discuss the forces that hold the atoms together (the chemical bonds) in Chapters 8 and 9.

Molecules and Chemical Formulas

Many elements are found in nature in molecular form; that is, two or more of the same type of atom are bound together. For example, the oxygen normally found in air consists of molecules that contain two oxygen atoms. We represent this molecular form of oxygen by the **chemical formula** O_2 (read "oh two"). The subscript in the formula tells us that two oxygen atoms are present in each molecule. A molecule that is made up of two atoms is called a **diatomic molecule**. Oxygen also exists in another molecular form known as *ozone*. Molecules of ozone consist of three oxygen atoms, making the chemical formula for this substance O_3. Even though "normal" oxygen (O_2) and ozone

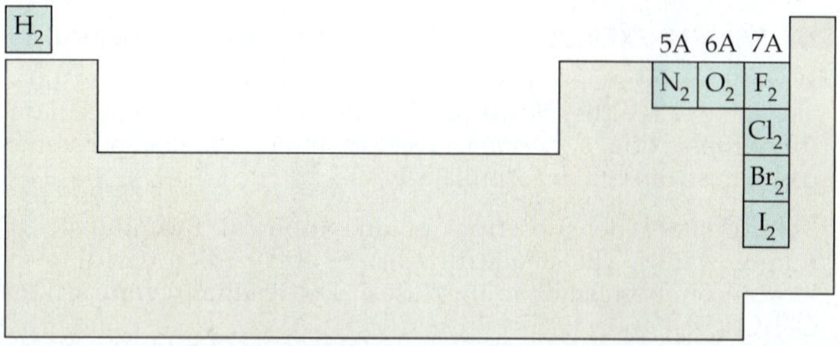

◀ Figure 2.19 Diatomic molecules. Seven common elements exist as diatomic molecules at room temperature.

are both composed only of oxygen atoms, they exhibit very different chemical and physical properties. For example, O_2 is essential for life, but O_3 is toxic; O_2 is odorless, whereas O_3 has a sharp, pungent smell.

The elements that normally occur as diatomic molecules are hydrogen, oxygen, nitrogen, and the halogens. Their locations in the periodic table are shown in Figure 2.19 ▲. When we speak of the substance hydrogen, we mean H_2 unless we explicitly indicate otherwise. Likewise, when we speak of oxygen, nitrogen, or any of the halogens, we are referring to O_2, N_2, F_2, Cl_2, Br_2, or I_2. Thus, the properties of oxygen and hydrogen listed in Table 1.3 are those of O_2 and H_2. Other, less common forms of these elements behave much differently.

Compounds that are composed of molecules are called **molecular compounds** and contain more than one type of atom. A molecule of water, for example, consists of two hydrogen atoms and one oxygen atom. It is therefore represented by the chemical formula H_2O. Lack of a subscript on the O indicates one atom of O per water molecule. Another compound composed of these same elements (in different relative proportions) is hydrogen peroxide, H_2O_2. The properties of hydrogen peroxide are very different from the properties of water.

Several common molecules are shown in Figure 2.20 ▶. Notice how the composition of each compound is given by its chemical formula. Notice also that these substances are composed only of nonmetallic elements. *Most molecular substances that we will encounter contain only nonmetals.*

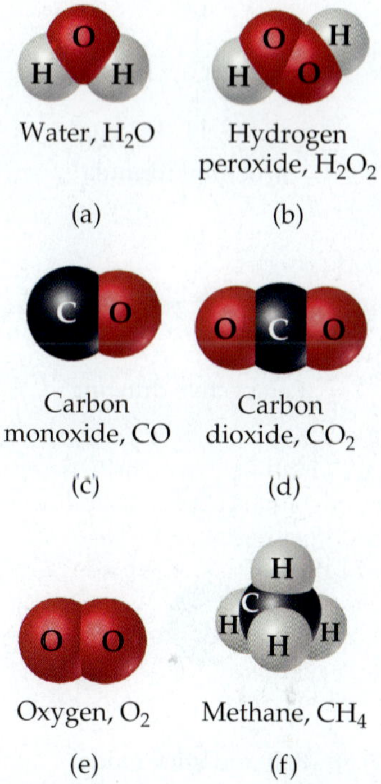

▲ Figure 2.20 Molecular models of some simple molecules. Notice how the chemical formulas of these substances correspond to their compositions.

Molecular and Empirical Formulas

Chemical formulas that indicate the actual numbers and types of atoms in a molecule are called **molecular formulas**. (The formulas in Figure 2.20 are molecular formulas.) Chemical formulas that give only the relative number of atoms of each type in a molecule are called **empirical formulas**. The subscripts in an empirical formula are always the smallest possible whole-number ratios. The molecular formula for hydrogen peroxide is H_2O_2, for example, whereas its empirical formula is HO. The molecular formula for ethylene is C_2H_4, and its empirical formula is CH_2. For many substances, the molecular formula and the empirical formula are identical, as in the case of water, H_2O.

Molecular formulas provide more information about molecules than do empirical formulas. Whenever we know the molecular formula of a compound, we can determine its empirical formula. The converse is not true, however; if we know the empirical formula of a substance, we can't determine its molecular formula unless we have more information. So why do chemists bother with empirical formulas? As we will see in Chapter 3, certain common methods of analyzing substances lead to the empirical formula only. Once the empirical formula is known, however, additional experiments can give the information needed to convert the empirical formula to the molecular one. In addition, there are substances, such as the most common forms of elemental carbon, that don't exist as isolated molecules. For these substances, we must rely on empirical formulas. Thus, all the common forms of elemental carbon are represented by the element's chemical symbol, C, which is the empirical formula for all the forms.

SAMPLE EXERCISE 2.6 | Relating Empirical and Molecular Formulas

Write the empirical formulas for the following molecules: **(a)** glucose, a substance also known as either blood sugar or dextrose, whose molecular formula is $C_6H_{12}O_6$; **(b)** nitrous oxide, a substance used as an anesthetic and commonly called laughing gas, whose molecular formula is N_2O.

Solution (a) The subscripts of an empirical formula are the smallest whole-number ratios. The smallest ratios are obtained by dividing each subscript by the largest common factor, in this case 6. The resultant empirical formula for glucose is CH_2O.

(b) Because the subscripts in N_2O are already the lowest integral numbers, the empirical formula for nitrous oxide is the same as its molecular formula, N_2O.

PRACTICE EXERCISE

Give the empirical formula for the substance called *diborane*, whose molecular formula is B_2H_6.

Answer: BH_3

ACTIVITY
Representations of Methane

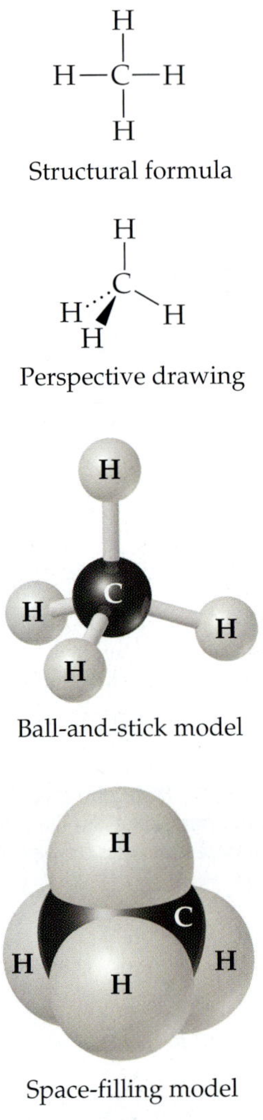

▲ **Figure 2.21 Different representations of the methane (CH_4) molecule.** Structural formulas, perspective drawings, ball-and-stick models, and space-filling models each help us visualize the ways atoms are attached to each other in molecules. In the perspective drawing, solid lines represent bonds in the plane of the paper, the solid wedge represents a bond that extends out from the plane of the paper, and dashed lines represent bonds behind the paper.

Picturing Molecules

The molecular formula of a substance summarizes the composition of the substance but does not show how the atoms come together to form the molecule. The **structural formula** of a substance shows which atoms are attached to which within the molecule. For example, the structural formulas for water, hydrogen peroxide, and methane (CH_4) can be written as follows:

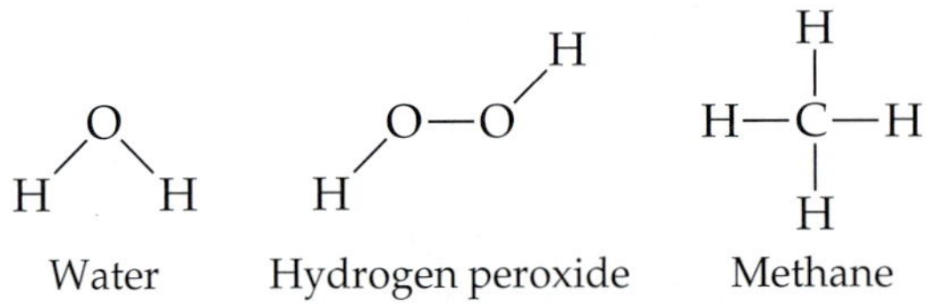

The atoms are represented by their chemical symbols, and lines are used to represent the bonds that hold the atoms together.

A structural formula usually does not depict the actual geometry of the molecule, that is, the actual angles at which atoms are joined together. A structural formula can be written as a *perspective drawing*, however, to give some sense of three-dimensional shape, as shown in Figure 2.21 ◀.

Scientists also rely on various models to help visualize molecules. *Ball-and-stick models* show atoms as spheres and bonds as sticks. This type of model has the advantage of accurately representing the angles at which the atoms are attached to one another within the molecule (Figure 2.21). All atoms may be represented by balls of the same size, or the relative sizes of the balls may reflect the relative sizes of the atoms. Sometimes the chemical symbols of the elements are superimposed on the balls, but often the atoms are identified simply by color.

A *space-filling model* depicts what the molecule would look like if the atoms were scaled up in size (Figure 2.21). These models show the relative sizes of the atoms, but the angles between atoms, which help define their molecular geometry, are often more difficult to see than in ball-and-stick models. As in ball-and-stick models, the identities of the atoms are indicated by their colors, but they may also be labeled with the element's symbol.

GIVE IT SOME THOUGHT

The structural formula for the substance ethane is shown here:

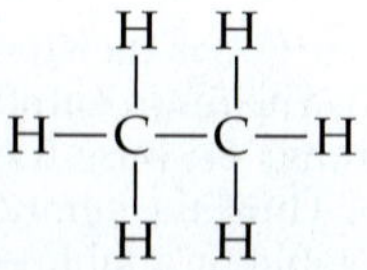

(a) What is the molecular formula for ethane? (b) What is its empirical formula? (c) Which kind of molecular model would most clearly show the angles between atoms?

2.7 | Ions and Ionic Compounds

The nucleus of an atom is unchanged by chemical processes, but atoms can readily gain or lose electrons. If electrons are removed from or added to a neutral atom, a charged particle called an **ion** is formed. An ion with a positive charge is called a **cation** (pronounced CAT-ion); a negatively charged ion is called an **anion** (AN-ion).

To see how ions form, consider the sodium atom, which has 11 protons and 11 electrons. This atom easily loses one electron. The resulting cation has 11 protons and 10 electrons, which means it has a net charge of 1+.

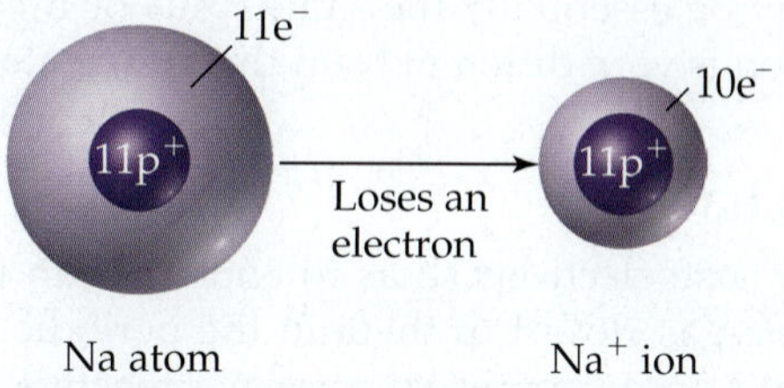

The net charge on an ion is represented by a superscript; the superscripts +, 2+, and 3+, for instance, mean a net charge resulting from the loss of one, two, and three electrons, respectively. The superscripts −, 2−, and 3− represent net charges resulting from the gain of one, two, and three electrons, respectively. Chlorine, with 17 protons and 17 electrons, for example, can gain an electron in chemical reactions, producing the Cl^- ion:

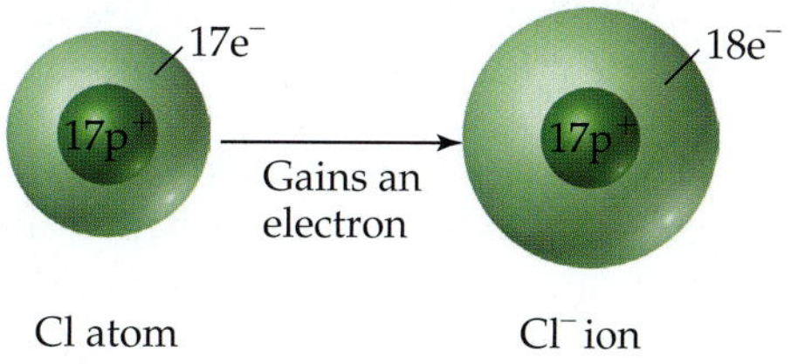

In general, metal atoms tend to lose electrons to form cations, whereas nonmetal atoms tend to gain electrons to form anions.

SAMPLE EXERCISE 2.7 | Writing Chemical Symbols for Ions

Give the chemical symbol, including mass number, for each of the following ions: **(a)** The ion with 22 protons, 26 neutrons, and 19 electrons; **(b)** the ion of sulfur that has 16 neutrons and 18 electrons.

Solution (a) The number of protons (22) is the atomic number of the element, which means this element is titanium (Ti). The mass number of this isotope is 22 + 26 = 48 (the sum of the protons and neutrons). Because the ion has three more protons than electrons, it has a net charge of 3+. Thus, the symbol for the ion is $^{48}Ti^{3+}$.

(b) By referring to a periodic table or a table of elements, we see that sulfur (S) has an atomic number of 16. Thus, each atom or ion of sulfur must contain 16 protons. We are told that the ion also has 16 neutrons, meaning the mass number of the ion is 16 + 16 = 32. Because the ion has 16 protons and 18 electrons, its net charge is 2−. Thus, the symbol for the ion is $^{32}S^{2-}$.

In general, we will focus on the net charges of ions and ignore their mass numbers unless the circumstances dictate that we specify a certain isotope.

PRACTICE EXERCISE

How many protons and electrons does the Se^{2-} ion possess?
Answer: 34 protons and 36 electrons

In addition to simple ions, such as Na^+ and Cl^-, there are **polyatomic ions**, such as NO_3^- (nitrate ion) and SO_4^{2-} (sulfate ion). These latter ions consist of atoms joined as in a molecule, but they have a net positive or negative charge. We will consider further examples of polyatomic ions in Section 2.8.

It is important to realize that the chemical properties of ions are very different from the chemical properties of the atoms from which the ions are derived. The difference is like the change from Dr. Jekyll to Mr. Hyde: Although a given atom and its ion may be essentially the same (plus or minus a few electrons), the behavior of the ion is very different from that of the atom.

Predicting Ionic Charges

Many atoms gain or lose electrons so as to end up with the same number of electrons as the noble gas closest to them in the periodic table. The members of the noble-gas family are chemically very nonreactive and form very few compounds. We might deduce that this is because their electron arrangements are very stable. Nearby elements can obtain these same stable arrangements by losing or gaining electrons. For example, loss of one electron from an atom of sodium leaves it with the same number of electrons as the neutral neon atom (atomic number 10). Similarly, when chlorine gains an electron, it ends up with 18, the same number of electrons as in argon (atomic number 18). We will use this simple observation to explain the formation of ions until Chapter 8, where we discuss chemical bonding.

SAMPLE EXERCISE 2.8 | Predicting the Charges of Ions

Predict the charge expected for the most stable ion of barium and for the most stable ion of oxygen.

Solution We will assume that these elements form ions that have the same number of electrons as the nearest noble-gas atom. From the periodic table, we see that barium has atomic number 56. The nearest noble gas is xenon, atomic number 54. Barium can attain a stable arrangement of 54 electrons by losing two of its electrons, forming the Ba^{2+} cation.

Oxygen has atomic number 8. The nearest noble gas is neon, atomic number 10. Oxygen can attain this stable electron arrangement by gaining two electrons, thereby forming the O^{2-} anion.

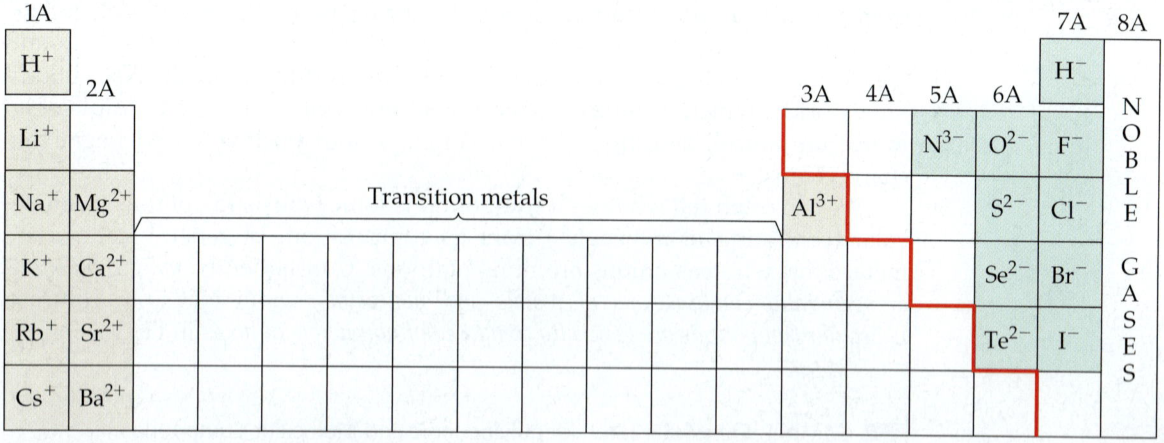

▲ **Figure 2.22 Charges of some common ions.** Notice that the steplike line that divides metals from nonmetals also separates cations from anions.

PRACTICE EXERCISE

Predict the charge expected for the most stable ion of aluminum and for the most stable ion of fluorine.
Answer: 3+ and 1−

The periodic table is very useful for remembering the charges of ions, especially those of the elements on the left and right sides of the table. As Figure 2.22 ▲ shows, the charges of these ions relate in a simple way to their positions in the table. On the left side of the table, for example, the group 1A elements (the alkali metals) form 1+ ions, and the group 2A elements (the alkaline earths) form 2+ ions. On the other side of the table the group 7A elements (the halogens) form 1− ions, and the group 6A elements form 2− ions. As we will see later in the text, many of the other groups do not lend themselves to such simple rules.

Ionic Compounds

A great deal of chemical activity involves the transfer of electrons from one substance to another, and, as we just saw, ions form when one or more electrons transfer from one neutral atom to another. Figure 2.23 ▼ shows that when elemental sodium is allowed to react with elemental chlorine, an electron transfers

▼ **Figure 2.23 The formation of an ionic compound.** (a) The transfer of an electron from a neutral Na atom to a neutral Cl atom leads to the formation of an Na^+ ion and a Cl^- ion. (b) Arrangement of these ions in solid sodium chloride (NaCl). (c) A sample of sodium chloride crystals.

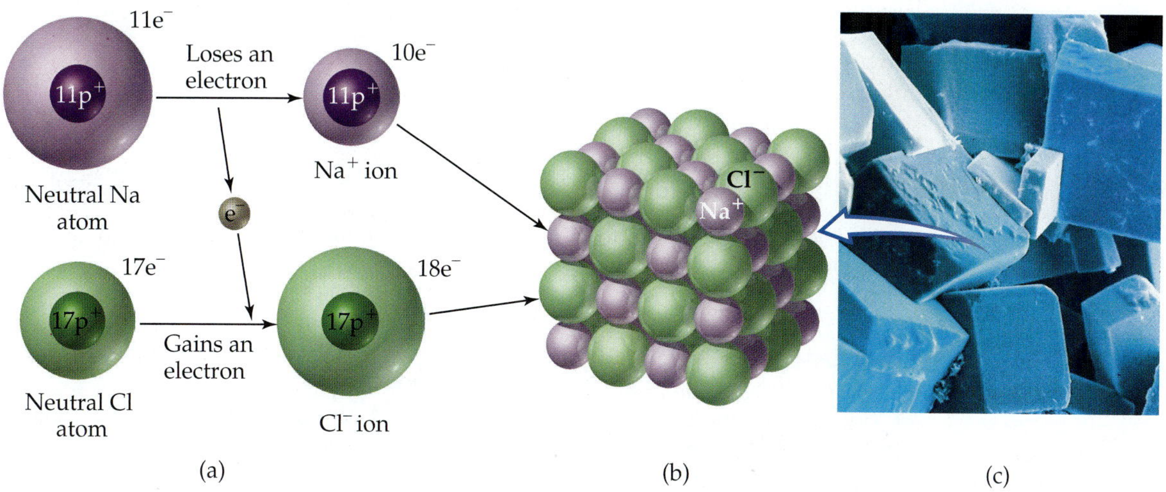

from a neutral sodium atom to a neutral chlorine atom. We are left with an Na^+ ion and a Cl^- ion. Because objects of opposite charge attract, the Na^+ and the Cl^- ions bind together to form the compound sodium chloride (NaCl). Sodium chloride, which we know better as common table salt, is an example of an **ionic compound**, a compound that contains both positively and negatively charged ions.

We can often tell whether a compound is ionic (consisting of ions) or molecular (consisting of molecules) from its composition. In general, cations are metal ions, whereas anions are nonmetal ions. Consequently, *ionic compounds are generally combinations of metals and nonmetals*, as in NaCl. In contrast, *molecular compounds are generally composed of nonmetals only*, as in H_2O.

SAMPLE EXERCISE 2.9 | Identifying Ionic and Molecular Compounds

Which of the following compounds would you expect to be ionic: N_2O, Na_2O, $CaCl_2$, SF_4?

Solution We would predict that Na_2O and $CaCl_2$ are ionic compounds because they are composed of a metal combined with a nonmetal. The other two compounds, composed entirely of nonmetals, are predicted (correctly) to be molecular compounds.

PRACTICE EXERCISE

Which of the following compounds are molecular: CBr_4, FeS, P_4O_6, PbF_2?
Answer: CBr_4 and P_4O_6

The ions in ionic compounds are arranged in three-dimensional structures. The arrangement of Na^+ and Cl^- ions in NaCl is shown in Figure 2.23. Because there is no discrete molecule of NaCl, we are able to write only an empirical formula for this substance. In fact, only empirical formulas can be written for most ionic compounds.

We can readily write the empirical formula for an ionic compound if we know the charges of the ions of which the compound is composed. Chemical compounds are always electrically neutral. Consequently, the ions in an ionic compound always occur in such a ratio that the total positive charge equals the total negative charge. Thus, there is one Na^+ to one Cl^- (giving NaCl), one Ba^{2+} to two Cl^- (giving $BaCl_2$), and so forth.

As you consider these and other examples, you will see that if the charges on the cation and anion are equal, the subscript on each ion will be 1. If the charges are not equal, the charge on one ion (without its sign) will become the subscript on the other ion. For example, the ionic compound formed from Mg (which forms Mg^{2+} ions) and N (which forms N^{3-} ions) is Mg_3N_2:

$$Mg^{2+} \quad N^{3-} \longrightarrow Mg_3N_2$$

GIVE IT SOME THOUGHT

Why don't we write the formula for the compound formed by Ca^{2+} and O^{2-} as Ca_2O_2?

SAMPLE EXERCISE 2.10 | Using Ionic Charge to Write Empirical Formulas for Ionic Compounds

What are the empirical formulas of the compounds formed by **(a)** Al^{3+} and Cl^- ions, **(b)** Al^{3+} and O^{2-} ions, **(c)** Mg^{2+} and NO_3^- ions?

Solution **(a)** Three Cl^- ions are required to balance the charge of one Al^{3+} ion. Thus, the formula is $AlCl_3$.

(b) Two Al^{3+} ions are required to balance the charge of three O^{2-} ions (that is, the total positive charge is 6+, and the total negative charge is 6−). Thus, the formula is Al_2O_3.

(c) Two NO_3^- ions are needed to balance the charge of one Mg^{2+}. Thus, the formula is $Mg(NO_3)_2$. In this case the formula for the entire polyatomic ion NO_3^- must be enclosed in parentheses so that it is clear that the subscript 2 applies to all the atoms of that ion.

PRACTICE EXERCISE

Write the empirical formulas for the compounds formed by the following ions: **(a)** Na^+ and PO_4^{3-}, **(b)** Zn^{2+} and SO_4^{2-}, **(c)** Fe^{3+} and CO_3^{2-}.
Answers: **(a)** Na_3PO_4, **(b)** $ZnSO_4$, **(c)** $Fe_2(CO_3)_3$

CHEMISTRY AND LIFE | Elements Required by Living Organisms

Figure 2.24 ▼ shows the elements that are essential for life. More than 97% of the mass of most organisms comprises just six elements—oxygen, carbon, hydrogen, nitrogen, phosphorus, and sulfur. Water (H_2O) is the most common compound in living organisms, accounting for at least 70% of the mass of most cells. Carbon is the most prevalent element (by mass) in the solid components of cells. Carbon atoms are found in a vast variety of organic molecules, in which the carbon atoms are bonded to other carbon atoms or to atoms of other elements, principally H, O, N, P, and S. All proteins, for example, contain the following group of atoms that occurs repeatedly within the molecules:

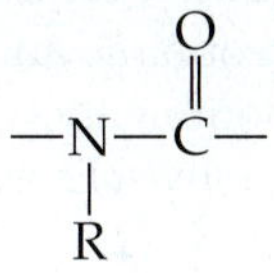

(R is either an H atom or a combination of atoms such as CH_3.)

In addition, 23 more elements have been found in various living organisms. Five are ions that are required by all organisms: Ca^{2+}, Cl^-, Mg^{2+}, K^+, and Na^+. Calcium ions, for example, are necessary for the formation of bone and for the transmission of signals in the nervous system, such as those that trigger the contraction of cardiac muscles, causing the heart to beat. Many other elements are needed in only very small quantities and consequently are called *trace* elements. For example, trace quantities of copper are required in the diet of humans to aid in the synthesis of hemoglobin.

▼ **Figure 2.24 Biologically essential elements.** The elements that are essential for life are indicated by colors. Red denotes the six most abundant elements in living systems (hydrogen, carbon, nitrogen, oxygen, phosphorus, and sulfur). Blue indicates the five next most abundant elements. Green indicates the elements needed in only trace amounts.

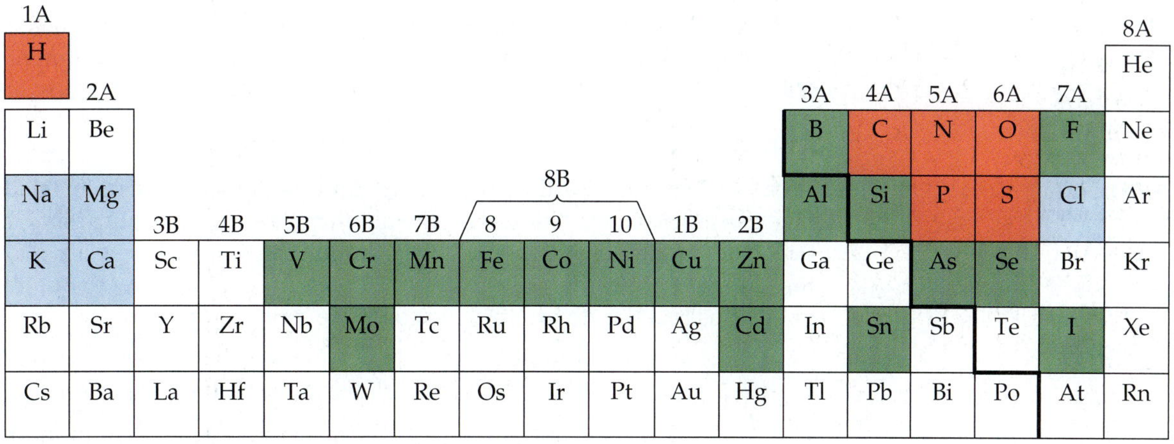

STRATEGIES IN CHEMISTRY | Pattern Recognition

Someone once said that drinking at the fountain of knowledge in a chemistry course is like drinking from a fire hydrant. Indeed, the pace can sometimes seem brisk. More to the point, however, we can drown in the facts if we don't see the general patterns. The value of recognizing patterns and learning rules and generalizations is that the patterns, rules, and generalizations free us from learning (or trying to memorize) many individual facts. The patterns, rules, and generalizations tie ideas together so that we don't get lost in the details.

Many students struggle with chemistry because they don't see how the topics relate to one another, how ideas connect together. They therefore treat every idea and problem as being unique instead of as an example or application of a general rule, procedure, or relationship. You can avoid this pitfall by remembering the following: Begin to notice the structure of the topic you are studying. Pay attention to the trends and rules given to summarize a large body of information. Notice, for example, how atomic structure helps us understand the existence of isotopes (as seen in Table 2.2) and how the periodic table aids us in remembering the charges of ions (as seen in Figure 2.22). You may surprise yourself by observing patterns that are not even explicitly spelled out yet. Perhaps you've even noticed certain trends in chemical formulas. Moving across the periodic table from element 11 (Na), we find that the elements form compounds with F having the following compositions: NaF, MgF_2, and AlF_3. Does this trend continue? Do SiF_4, PF_5, and SF_6 exist? Indeed they do. If you have picked up on trends like this from the scraps of information you've seen so far, then you're ahead of the game and you've already prepared yourself for some topics we will address in later chapters.

2.8 | Naming Inorganic Compounds

To obtain information about a particular substance, you must know its chemical formula and name. The names and formulas of compounds are essential vocabulary in chemistry. The system used in naming substances is called **chemical nomenclature** from the Latin words *nomen* (name) and *calare* (to call).

There are now more than 19 million known chemical substances. Naming them all would be a hopelessly complicated task if each had a special name independent of all others. Many important substances that have been known for a long time, such as water (H_2O) and ammonia (NH_3), do have individual, traditional names (so-called "common" names). For most substances, however, we rely on a systematic set of rules that leads to an informative and unique name for each substance, a name based on the composition of the substance.

The rules for chemical nomenclature are based on the division of substances into categories. The major division is between organic and inorganic compounds. *Organic compounds* contain carbon, usually in combination with hydrogen, oxygen, nitrogen, or sulfur. All others are *inorganic compounds*. Early chemists associated organic compounds with plants and animals, and they associated inorganic compounds with the nonliving portion of our world. Although this distinction between living and nonliving matter is no longer pertinent, the classification between organic and inorganic compounds continues to be useful. In this section we consider the basic rules for naming inorganic compounds, and in Section 2.9 we will introduce the names of some simple organic compounds. Among inorganic compounds, we will consider three categories: ionic compounds, molecular compounds, and acids.

Names and Formulas of Ionic Compounds

ACTIVITIES
Naming Cations, Naming Anions

Recall from Section 2.7 that ionic compounds usually consist of metal ions combined with nonmetal ions. The metals form the positive ions, and the nonmetals form the negative ions. Let's examine the naming of positive ions, then the naming of negative ones. After that, we will consider how to put the names of the ions together to identify the complete ionic compound.

1. ***Positive Ions (Cations)***
   (a) *Cations formed from metal atoms have the same name as the metal:*

   Na^+ sodium ion Zn^{2+} zinc ion Al^{3+} aluminum ion

   Ions formed from a single atom are called *monatomic ions*.

(b) *If a metal can form different cations, the positive charge is indicated by a Roman numeral in parentheses following the name of the metal:*

Fe^{2+}	iron(II) ion	Cu^{+}	copper(I) ion
Fe^{3+}	iron(III) ion	Cu^{2+}	copper(II) ion

Ions of the same element that have different charges exhibit different properties, such as different colors (Figure 2.25 ▶).

Most of the metals that can form more than one cation are *transition metals*, elements that occur in the middle block of elements, from group 3B to group 2B in the periodic table. The charges of these ions are indicated by Roman numerals. The metals that form only one cation are those of group 1A (Na^{+}, K^{+}, and Rb^{+}) and group 2A (Mg^{2+}, Ca^{2+}, Sr^{2+}, and Ba^{2+}), as well as Al^{3+} (group 3A) and two transition-metal ions: Ag^{+} (group 1B) and Zn^{2+} (group 2B). Charges are not expressed explicitly when naming these ions. However, if there is any doubt in your mind whether a metal forms more than one cation, use a Roman numeral to indicate the charge. It is never wrong to do so, even though it may be unnecessary.

An older method still widely used for distinguishing between two differently charged ions of a metal is to apply the ending *-ous* or *-ic*. These endings represent the lower and higher charged ions, respectively. They are added to the root of the element's Latin name:

Fe^{2+}	ferrous ion	Cu^{+}	cuprous ion
Fe^{3+}	ferric ion	Cu^{2+}	cupric ion

Although we will only rarely use these older names in this text, you might encounter them elsewhere.

(c) *Cations formed from nonmetal atoms have names that end in* -ium:

NH_4^{+}	ammonium ion	H_3O^{+}	hydronium ion

These two ions are the only ions of this kind that we will encounter frequently in the text. They are both polyatomic. The vast majority of cations are monatomic metal ions.

The names and formulas of some common cations are shown in Table 2.4 ▶ and are also included in a table of common ions that is placed in the back inside cover of the text. The ions listed on the left in Table 2.4 are the monatomic ions that do not have variable charges. Those listed on the right are either polyatomic cations or cations with variable charges. The Hg_2^{2+} ion is unusual because this metal ion is not monatomic. It is called the mercury(I) ion because it can be thought of as two Hg^{+} ions fused together. The cations that you will encounter most frequently are shown in boldface. These are the ones you should learn first.

▲ **Figure 2.25 Ions of the same element with different charges exhibit different properties.** Compounds containing ions of the same element but with different charge can be very different in appearance and properties. Both substances shown are complex compounds of iron that also contain K^{+} and CN^{-} ions. The one on the left is potassium ferrocyanide, which contains Fe(II) bound to CN^{-} ions. The one on the right is potassium ferricyanide, which contains Fe(III) bound to CN^{-} ions. Both substances are used extensively in blueprinting and other dyeing processes.

GIVE IT SOME THOUGHT

Why is CrO named using a Roman numeral, chromium(II) oxide, whereas CaO is named without a Roman numeral in the name, calcium oxide?

2. *Negative Ions (Anions)*

(a) *The names of monatomic anions are formed by replacing the ending of the name of the element with* -ide:

H^{-}	hydride ion	O^{2-}	oxide ion	N^{3-}	nitride ion

A few simple polyatomic anions also have names ending in *-ide*:

OH^{-}	hydroxide ion	CN^{-}	cyanide ion	O_2^{2-}	peroxide ion

TABLE 2.4 Common Cations*

Charge	Formula	Name	Formula	Name
1+	**H^+**	**Hydrogen ion**	**NH_4^+**	**Ammonium ion**
	Li^+	Lithium ion	Cu^+	Copper(I) or cuprous ion
	Na^+	**Sodium ion**		
	K^+	**Potassium ion**		
	Cs^+	Cesium ion		
	Ag^+	**Silver ion**		
2+	**Mg^{2+}**	**Magnesium ion**	Co^{2+}	Cobalt(II) or cobaltous ion
	Ca^{2+}	**Calcium ion**	**Cu^{2+}**	**Copper(II)** or cupric ion
	Sr^{2+}	Strontium ion	**Fe^{2+}**	**Iron(II)** or ferrous ion
	Ba^{2+}	Barium ion	Mn^{2+}	Manganese(II) or manganous ion
	Zn^{2+}	**Zinc ion**	Hg_2^{2+}	Mercury(I) or mercurous ion
	Cd^{2+}	Cadmium ion	**Hg^{2+}**	**Mercury(II)** or mercuric ion
			Ni^{2+}	Nickel(II) or nickelous ion
			Pb^{2+}	**Lead(II)** or plumbous ion
			Sn^{2+}	Tin(II) or stannous ion
3+	**Al^{3+}**	**Aluminum ion**	Cr^{3+}	Chromium(III) or chromic ion
			Fe^{3+}	**Iron(III)** or ferric ion

*The most common ions are in boldface.

ACTIVITY
Polyatomic Ions

(b) *Polyatomic anions containing oxygen have names ending in* -ate *or* -ite. These anions are called **oxyanions**. The ending *-ate* is used for the most common oxyanion of an element. The ending *-ite* is used for an oxyanion that has the same charge but one O atom fewer:

NO_3^-	nitrate ion	SO_4^{2-}	sulfate ion
NO_2^-	nitrite ion	SO_3^{2-}	sulfite ion

Prefixes are used when the series of oxyanions of an element extends to four members, as with the halogens. The prefix *per-* indicates one more O atom than the oxyanion ending in *-ate*; the prefix *hypo-* indicates one O atom fewer than the oxyanion ending in *-ite*:

ClO_4^-	perchlorate ion (one more O atom than chlorate)
ClO_3^-	**chlorate ion**
ClO_2^-	chlorite ion (one O atom fewer than chlorate)
ClO^-	hypochlorite ion (one O atom fewer than chlorite)

These rules are summarized in Figure 2.26 ▼.

▼ **Figure 2.26 Summary of the procedure for naming anions.** The root of the name (such as "chlor" for chlorine) goes in the blank.

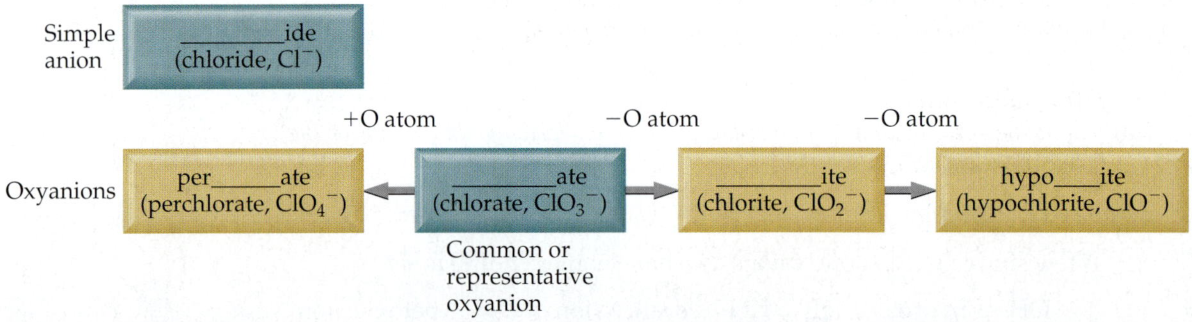

	4A	5A	6A	7A
2	CO_3^{2-} Carbonate ion	NO_3^- Nitrate ion		
3		PO_4^{3-} Phosphate ion	SO_4^{2-} Sulfate ion	ClO_4^- Perchlorate ion

◀ **Figure 2.27 Common oxyanions.** The composition and charges of common oxyanions are related to their location in the periodic table.

GIVE IT SOME THOUGHT

What information is conveyed by the endings *-ide*, *-ate*, and *-ite* in the name of an anion?

Students often have a hard time remembering the number of oxygen atoms in the various oxyanions and the charges of these ions. Figure 2.27 ▲ lists the oxyanions of C, N, P, S, and Cl that contain the maximum number of O atoms. There is a periodic pattern to these formulas that can help you remember them. Notice that C and N, which are in the second period of the periodic table, have only three O atoms each, whereas P, S, and Cl, which are in the third period, have four O atoms each. If we begin at the lower right side of the figure, with Cl, we see that the charges increase from right to left, from 1− for Cl (ClO_4^-) to 3− for P (PO_4^{3-}). In the second period the charges also increase from right to left, from 1− for N (NO_3^-) to 2− for C (CO_3^{2-}). Each anion shown in Figure 2.27 has a name ending in *-ate*. The ClO_4^- ion also has a *per-* prefix. If you know the rules summarized in Figure 2.26 and the names and formulas of the five oxyanions in Figure 2.27, you can deduce the names for the other oxyanions of these elements.

GIVE IT SOME THOUGHT

Predict the formulas for the borate ion and silicate ion, assuming that they contain a single B and Si atom, respectively, and follow the trends shown in Figure 2.27.

SAMPLE EXERCISE 2.11 | Determining the Formula of an Oxyanion from Its Name

Based on the formula for the sulfate ion, predict the formula for **(a)** the selenate ion and **(b)** the selenite ion. (Sulfur and selenium are both members of group 6A and form analogous oxyanions.)

Solution (a) The sulfate ion is SO_4^{2-}. The analogous selenate ion is therefore SeO_4^{2-}.

(b) The ending *-ite* indicates an oxyanion with the same charge but one O atom fewer than the corresponding oxyanion that ends in *-ate*. Thus, the formula for the selenite ion is SeO_3^{2-}.

PRACTICE EXERCISE

The formula for the bromate ion is analogous to that for the chlorate ion. Write the formula for the hypobromite and perbromate ions.
Answer: BrO^- and BrO_4^-

(c) *Anions derived by adding* H^+ *to an oxyanion are named by adding as a prefix the word* hydrogen *or* dihydrogen, *as appropriate:*

CO_3^{2-}	carbonate ion	PO_4^{3-}	phosphate ion
HCO_3^-	hydrogen carbonate ion	$H_2PO_4^-$	dihydrogen phosphate ion

Notice that each H^+ reduces the negative charge of the parent anion by one. An older method for naming some of these ions is to use the prefix *bi-*. Thus, the HCO_3^- ion is commonly called the bicarbonate ion, and HSO_4^- is sometimes called the bisulfate ion.

TABLE 2.5 Common Anions*

Charge	Formula	Name	Formula	Name
1−	H^-	Hydride ion	$\mathbf{C_2H_3O_2^-}$	Acetate ion
	$\mathbf{F^-}$	**Fluoride ion**	ClO_3^-	Chlorate ion
	$\mathbf{Cl^-}$	**Chloride ion**	$\mathbf{ClO_4^-}$	**Perchlorate ion**
	$\mathbf{Br^-}$	**Bromide ion**	$\mathbf{NO_3^-}$	**Nitrate ion**
	$\mathbf{I^-}$	**Iodide ion**	MnO_4^-	Permanganate ion
	CN^-	Cyanide ion		
	$\mathbf{OH^-}$	**Hydroxide ion**		
2−	$\mathbf{O^{2-}}$	**Oxide ion**	$\mathbf{CO_3^{2-}}$	**Carbonate ion**
	O_2^{2-}	Peroxide ion	CrO_4^{2-}	Chromate ion
	$\mathbf{S^{2-}}$	**Sulfide ion**	$Cr_2O_7^{2-}$	Dichromate ion
			$\mathbf{SO_4^{2-}}$	**Sulfate ion**
3−	N^{3-}	Nitride ion	$\mathbf{PO_4^{3-}}$	**Phosphate ion**

*The most common ions are in boldface.

The names and formulas of the common anions are listed in Table 2.5 ▲ and on the back inside cover of the text. Those anions whose names end in *-ide* are listed on the left portion of Table 2.5, and those whose names end in *-ate* are listed on the right. The most common of these ions are shown in boldface. These are the ones you should learn first. The formulas of the ions whose names end with *-ite* can be derived from those ending in *-ate* by removing an O atom. Notice the location of the monatomic ions in the periodic table. Those of group 7A always have a 1− charge (F^-, Cl^-, Br^-, and I^-), and those of group 6A have a 2− charge (O^{2-} and S^{2-}).

VCL Names and Formulas of Ionic Compounds

3. *Ionic Compounds*
Names of ionic compounds consist of the cation name followed by the anion name:

$CaCl_2$	calcium chloride
$Al(NO_3)_3$	aluminum nitrate
$Cu(ClO_4)_2$	copper(II) perchlorate (or cupric perchlorate)

In the chemical formulas for aluminum nitrate and copper(II) perchlorate, parentheses followed by the appropriate subscript are used because the compounds contain two or more polyatomic ions.

ACTIVITY
Ionic Compounds

SAMPLE EXERCISE 2.12 | Determining the Names of Ionic Compounds from Their Formulas

Name the following compounds: **(a)** K_2SO_4, **(b)** $Ba(OH)_2$, **(c)** $FeCl_3$.

Solution Each compound is ionic and is named using the guidelines we have already discussed. In naming ionic compounds, it is important to recognize polyatomic ions and to determine the charge of cations with variable charge. **(a)** The cation in this compound is K^+, and the anion is SO_4^{2-}. (If you thought the compound contained S^{2-} and O^{2-} ions, you failed to recognize the polyatomic sulfate ion.) Putting together the names of the ions, we have the name of the compound, potassium sulfate. **(b)** In this case the compound is composed of Ba^{2+} and OH^- ions. Ba^{2+} is the barium ion and OH^- is the hydroxide ion. Thus, the compound is called barium hydroxide. **(c)** You must determine the charge of Fe in this compound because an iron atom can form more than one cation. Because the compound contains three Cl^- ions, the cation must be Fe^{3+}, which is the iron(III), or ferric, ion. The Cl^- ion is the chloride ion. Thus, the compound is iron(III) chloride or ferric chloride.

PRACTICE EXERCISE

Name the following compounds: **(a)** NH_4Br, **(b)** Cr_2O_3, **(c)** $Co(NO_3)_2$.
Answers: **(a)** ammonium bromide, **(b)** chromium(III) oxide, **(c)** cobalt(II) nitrate

SAMPLE EXERCISE 2.13 | Determining the Formulas of Ionic Compounds from Their Names

Write the chemical formulas for the following compounds: **(a)** potassium sulfide, **(b)** calcium hydrogen carbonate, **(c)** nickel(II) perchlorate.

Solution In going from the name of an ionic compound to its chemical formula, you must know the charges of the ions to determine the subscripts. **(a)** The potassium ion is K^+, and the sulfide ion is S^{2-}. Because ionic compounds are electrically neutral, two K^+ ions are required to balance the charge of one S^{2-} ion, giving the empirical formula of the compound, K_2S. **(b)** The calcium ion is Ca^{2+}. The carbonate ion is CO_3^{2-}, so the hydrogen carbonate ion is HCO_3^-. Two HCO_3^- ions are needed to balance the positive charge of Ca^{2+}, giving $Ca(HCO_3)_2$. **(c)** The nickel(II) ion is Ni^{2+}. The perchlorate ion is ClO_4^-. Two ClO_4^- ions are required to balance the charge on one Ni^{2+} ion, giving $Ni(ClO_4)_2$.

PRACTICE EXERCISE

Give the chemical formula for **(a)** magnesium sulfate, **(b)** silver sulfide, **(c)** lead(II) nitrate.
Answers: **(a)** $MgSO_4$, **(b)** Ag_2S, **(c)** $Pb(NO_3)_2$

Names and Formulas of Acids

Acids are an important class of hydrogen-containing compounds and are named in a special way. For our present purposes, an *acid* is a substance whose molecules yield hydrogen ions (H^+) when dissolved in water. When we encounter the chemical formula for an acid at this stage of the course, it will be written with H as the first element, as in HCl and H_2SO_4.

We can consider an acid to be composed of an anion connected to enough H^+ ions to neutralize, or balance, the anion's charge. Thus, the SO_4^{2-} ion requires two H^+ ions, forming H_2SO_4. The name of an acid is related to the name of its anion, as summarized in Figure 2.28 ▼.

1. *Acids containing anions whose names end in* -ide *are named by changing the* -ide *ending to* -ic, *adding the prefix* hydro- *to this anion name, and then following with the word* acid, *as in the following examples:*

Anion	Corresponding Acid
Cl^- (chloride)	HCl (hydrochloric acid)
S^{2-} (sulfide)	H_2S (hydrosulfuric acid)

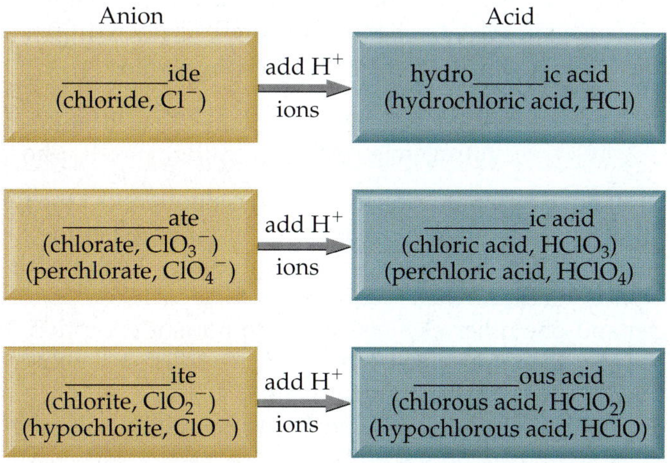

◀ **Figure 2.28 Relating names of anions and acids.** Summary of the way in which anion names and acid names are related. The prefixes *per-* and *hypo-* are retained in going from the anion to the acid.

2. *Acids containing anions whose names end in* -ate *or* -ite *are named by changing* -ate *to* -ic *and* -ite *to* -ous, *and then adding the word* acid. Prefixes in the anion name are retained in the name of the acid. These rules are illustrated by the oxyacids of chlorine:

Anion	Corresponding Acid
ClO_4^- (perchlorate)	$HClO_4$ (perchloric acid)
ClO_3^- (chlorate)	$HClO_3$ (chloric acid)
ClO_2^- (chlorite)	$HClO_2$ (chlorous acid)
ClO^- (hypochlorite)	$HClO$ (hypochlorous acid)

SAMPLE EXERCISE 2.14 | Relating the Names and Formulas of Acids

Name the following acids: **(a)** HCN, **(b)** HNO_3, **(c)** H_2SO_4, **(d)** H_2SO_3.

Solution **(a)** The anion from which this acid is derived is CN^-, the cyanide ion. Because this ion has an *-ide* ending, the acid is given a *hydro-* prefix and an *-ic* ending: hydrocyanic acid. Only water solutions of HCN are referred to as hydrocyanic acid: The pure compound, which is a gas under normal conditions, is called hydrogen cyanide. Both hydrocyanic acid and hydrogen cyanide are *extremely* toxic. **(b)** Because NO_3^- is the nitrate ion, HNO_3 is called nitric acid (the *-ate* ending of the anion is replaced with an *-ic* ending in naming the acid). **(c)** Because SO_4^{2-} is the sulfate ion, H_2SO_4 is called sulfuric acid. **(d)** Because SO_3^{2-} is the sulfite ion, H_2SO_3 is sulfurous acid (the *-ite* ending of the anion is replaced with an *-ous* ending).

PRACTICE EXERCISE

Give the chemical formulas for **(a)** hydrobromic acid, **(b)** carbonic acid.
Answers: **(a)** HBr, **(b)** H_2CO_3

Names and Formulas of Binary Molecular Compounds

The procedures used for naming *binary* (two-element) molecular compounds are similar to those used for naming ionic compounds:

1. *The name of the element farther to the left in the periodic table is usually written first.* An exception to this rule occurs in the case of compounds that contain oxygen. Oxygen is always written last except when combined with fluorine.
2. *If both elements are in the same group in the periodic table, the one having the higher atomic number is named first.*
3. *The name of the second element is given an* -ide *ending.*
4. *Greek prefixes (Table 2.6 ◀) are used to indicate the number of atoms of each element.* The prefix *mono-* is never used with the first element. When the prefix ends in *a* or *o* and the name of the second element begins with a vowel (such as *oxide*), the *a* or *o* of the prefix is often dropped.

The following examples illustrate these rules:

Cl_2O	dichlorine monoxide	NF_3	nitrogen trifluoride
N_2O_4	dinitrogen tetroxide	P_4S_{10}	tetraphosphorus decasulfide

TABLE 2.6 Prefixes Used in Naming Binary Compounds Formed Between Nonmetals

Prefix	*Meaning*
Mono-	1
Di-	2
Tri-	3
Tetra-	4
Penta-	5
Hexa-	6
Hepta-	7
Octa-	8
Nona-	9
Deca-	10

It is important to realize that you cannot predict the formulas of most molecular substances in the same way that you predict the formulas of ionic compounds. That is why we name molecular compounds using prefixes that explicitly indicate their composition. Molecular compounds that contain hydrogen and one other element are an important exception, however. These compounds can be treated as if they were neutral substances containing H^+ ions and anions. Thus, you can predict that the substance whose name is hydrogen chloride has the formula HCl, containing one H^+ to balance the charge of one Cl^-.

(The name hydrogen chloride is used only for the pure compound; water solutions of HCl are called hydrochloric acid.) Similarly, the formula for hydrogen sulfide is H_2S because two H^+ are needed to balance the charge on S^{2-}.

SAMPLE EXERCISE 2.15 | Relating the Names and Formulas of Binary Molecular Compounds

Name the following compounds: **(a)** SO_2, **(b)** PCl_5, **(c)** N_2O_3.

Solution The compounds consist entirely of nonmetals, so they are probably molecular rather than ionic. Using the prefixes in Table 2.6, we have **(a)** sulfur dioxide, **(b)** phosphorus pentachloride, and **(c)** dinitrogen trioxide.

PRACTICE EXERCISE

Give the chemical formula for **(a)** silicon tetrabromide, **(b)** disulfur dichloride.
Answers: **(a)** $SiBr_4$, **(b)** S_2Cl_2

2.9 | Some Simple Organic Compounds

The study of compounds of carbon is called **organic chemistry**, and, as noted earlier in the chapter, compounds that contain carbon and hydrogen, often in combination with oxygen, nitrogen, or other elements, are called *organic compounds*. We will examine organic compounds and organic chemistry in some detail in Chapter 25. You will see a number of organic compounds throughout this text; many of them have practical applications or are relevant to the chemistry of biological systems. Here we present a very brief introduction to some of the simplest organic compounds so as to provide you with a sense of what these molecules look like and how they are named.

Alkanes

Compounds that contain only carbon and hydrogen are called **hydrocarbons**. In the most basic class of hydrocarbons, each carbon atom is bonded to four other atoms. These compounds are called **alkanes**. The three simplest alkanes, which contain one, two, and three carbon atoms, respectively, are methane (CH_4), ethane (C_2H_6), and propane (C_3H_8). The structural formulas of these three alkanes are as follows:

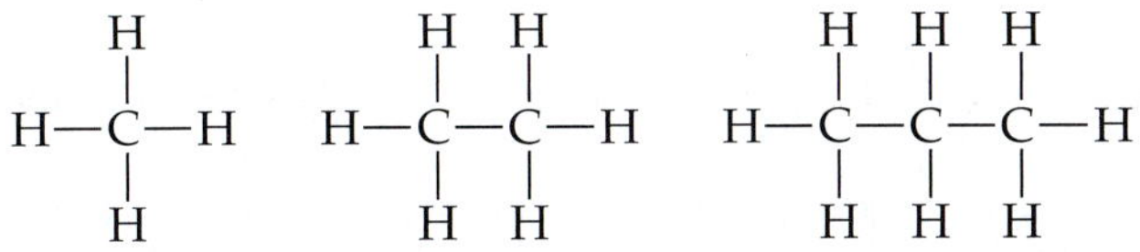

Methane Ethane Propane

3-D MODELS
Methane, Ethane, Propane

Longer alkanes can be made by adding additional carbon atoms to the "skeleton" of the molecule.

Although the hydrocarbons are binary molecular compounds, they are not named like the inorganic binary compounds discussed in Section 2.8. Instead, each alkane has a name that ends in *-ane*. For alkanes with five or more carbon atoms, the names are derived from prefixes like those in Table 2.6. An alkane with eight carbon atoms, for example, is called *octane* (C_8H_{18}), where the *octa-* prefix for eight is combined with the *-ane* ending for an alkane. Gasoline consists primarily of octanes, as will be discussed in Chapter 25.

3-D MODELS
Methanol, Ethanol, 1-Propanol, 2-Propanol

Some Derivatives of Alkanes

Other classes of organic compounds are obtained when hydrogen atoms of alkanes are replaced with *functional groups*, which are specific groups of atoms. An **alcohol**, for example, is obtained by replacing an H atom of an alkane with an —OH group. The name of the alcohol is derived from that of the alkane by adding an *-ol* ending:

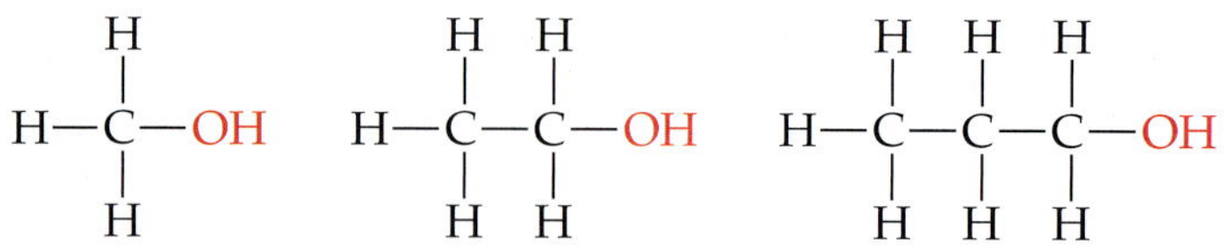

Methanol Ethanol 1-Propanol

Alcohols have properties that are very different from the properties of the alkanes from which the alcohols are obtained. For example, methane, ethane, and propane are all colorless gases under normal conditions, whereas methanol, ethanol, and propanol are colorless liquids. We will discuss the reasons for these differences in properties in Chapter 11.

The prefix "1" in the name 1-propanol indicates that the replacement of H with OH has occurred at one of the "outer" carbon atoms rather than the "middle" carbon atom. A different compound called 2-propanol (also known as isopropyl alcohol) is obtained if the OH functional group is attached to the middle carbon atom:

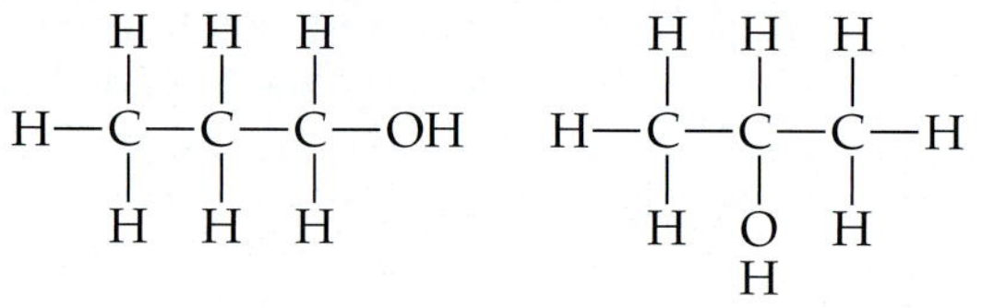

1-Propanol 2-Propanol

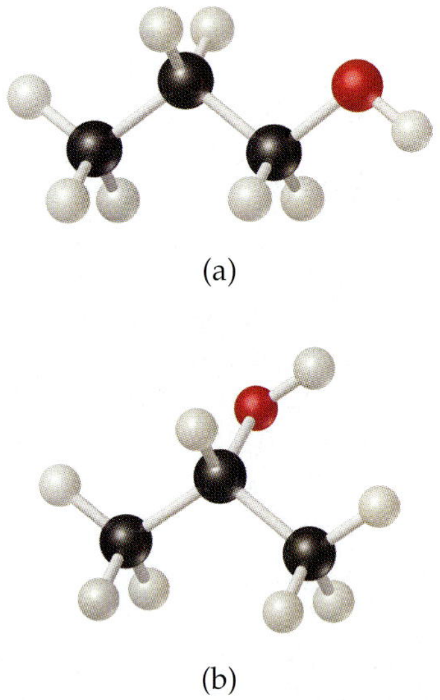

▲ **Figure 2.29 The two forms of propanol (C_3H_7OH).** (a) 1-Propanol, in which the OH group is attached to one of the end carbon atoms, and (b) 2-propanol, in which the OH group is attached to the middle carbon atom.

Ball-and-stick models of these two molecules are presented in Figure 2.29 ◀.

Much of the richness of organic chemistry is possible because organic compounds can form long chains of carbon–carbon bonds. The series of alkanes that begins with methane, ethane, and propane and the series of alcohols that begins with methanol, ethanol, and propanol can both be extended for as long as we desire, in principle. The properties of alkanes and alcohols change as the chains get longer. Octanes, which are alkanes with eight carbon atoms, are liquids under normal conditions. If the alkane series is extended to tens of thousands of carbon atoms, we obtain *polyethylene*, a solid substance that is used to make thousands of plastic products, such as plastic bags, food containers, and laboratory equipment.

SAMPLE EXERCISE 2.16 | Writing Structural and Molecular Formulas for Hydrocarbons

Consider the alkane called *pentane*. **(a)** Assuming that the carbon atoms are in a straight line, write a structural formula for pentane. **(b)** What is the molecular formula for pentane?

Solution (a) Alkanes contain only carbon and hydrogen, and each carbon atom is attached to four other atoms. Because the name pentane contains the prefix *penta-* for five (Table 2.6), we can assume that pentane contains five carbon atoms bonded in a chain. If we then add enough hydrogen atoms to make four bonds to each carbon atom, we obtain the following structural formula:

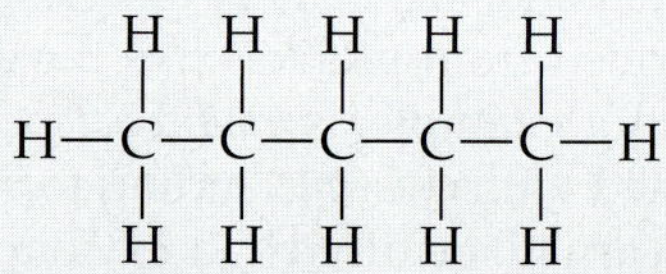

This form of pentane is often called *n*-pentane, where the *n*- stands for "normal" because all five carbon atoms are in one line in the structural formula.

(b) Once the structural formula is written, we can determine the molecular formula by counting the atoms present. Thus, *n*-pentane has the formula C_5H_{12}.

PRACTICE EXERCISE

Butane is the alkane with four carbon atoms. **(a)** What is the molecular formula of butane? **(b)** What are the name and molecular formula of an alcohol derived from butane?
Answers: **(a)** C_4H_{10}, **(b)** butanol, $C_4H_{10}O$ or C_4H_9OH

SUMMARY AND KEY TERMS

Sections 2.1 and 2.2 **Atoms** are the basic building blocks of matter; they are the smallest units of an element that can combine with other elements. Atoms are composed of even smaller particles, called **subatomic particles**. Some of these subatomic particles are charged and follow the usual behavior of charged particles: Particles with the same charge repel one another, whereas particles with unlike charges are attracted to one another. We considered some of the important experiments that led to the discovery and characterization of subatomic particles. Thomson's experiments on the behavior of **cathode rays** in magnetic and electric fields led to the discovery of the electron and allowed its charge-to-mass ratio to be measured. Millikan's oil-drop experiment determined the charge of the electron. Becquerel's discovery of **radioactivity**, the spontaneous emission of radiation by atoms, gave further evidence that the atom has a substructure. Rutherford's studies of how thin metal foils scatter α particles showed that the atom has a dense, positively charged **nucleus**.

Section 2.3 Atoms have a nucleus that contains **protons** and **neutrons**; **electrons** move in the space around the nucleus. The magnitude of the charge of the electron, 1.602×10^{-19} C, is called the **electronic charge**. The charges of particles are usually represented as multiples of this charge; thus, an electron has a 1− charge, and a proton has a 1+ charge. The masses of atoms are usually expressed in terms of **atomic mass units** (1 amu = 1.66054×10^{-24} g). The dimensions of atoms are often expressed in units of **angstroms** (1 Å = 10^{-10} m).

Elements can be classified by **atomic number**, the number of protons in the nucleus of an atom. All atoms of a given element have the same atomic number. The **mass number** of an atom is the sum of the numbers of protons and neutrons. Atoms of the same element that differ in mass number are known as **isotopes**.

Section 2.4 The atomic mass scale is defined by assigning a mass of exactly 12 amu to a ^{12}C atom. The **atomic weight** (average atomic mass) of an element can be calculated from the relative abundances and masses of that element's isotopes. The **mass spectrometer** provides the most direct and accurate means of experimentally measuring atomic (and molecular) weights.

Section 2.5 The **periodic table** is an arrangement of the elements in order of increasing atomic number. Elements with similar properties are placed in vertical columns. The elements in a column are known as a periodic **group**. The elements in a horizontal row are known as a **period**. The **metallic elements (metals)**, which comprise the majority of the elements, dominate the left side and the middle of the table; the **nonmetallic elements (nonmetals)** are located on the upper right side. Many of the elements that lie along the line that separates metals from nonmetals are **metalloids**.

Section 2.6 Atoms can combine to form **molecules**. Compounds composed of molecules (**molecular compounds**) usually contain only nonmetallic elements. A molecule that contains two atoms is called a **diatomic molecule**. The composition of a substance is given by its **chemical formula**. A molecular substance can be represented by its **empirical formula**, which gives the relative numbers of atoms of each kind. It is usually represented by its **molecular formula**, however, which gives the actual numbers of each type of atom in a molecule. **Structural formulas** show the order in which the atoms in a molecule are connected. Ball-and-stick models and space-filling models are often used to represent molecules.

Section 2.7 Atoms can either gain or lose electrons, forming charged particles called **ions**. Metals tend to lose electrons, becoming positively charged ions (**cations**). Nonmetals tend to gain electrons, forming negatively charged ions (**anions**). Because **ionic compounds** are electrically neutral, containing both cations and anions, they usually contain both metallic and nonmetallic elements. Atoms that are joined together, as in a molecule, but carry a net charge are called **polyatomic ions**. The chemical formulas used for ionic compounds are empirical formulas, which can be written readily if the charges of the ions are known. The total positive charge of the cations in an ionic compound equals the total negative charge of the anions.

Section 2.8 The set of rules for naming chemical compounds is called **chemical nomenclature**. We studied the systematic rules used for naming three classes of inorganic substances: ionic compounds, acids, and binary molecular compounds. In naming an ionic compound, the cation is named first and then the anion. Cations formed from metal atoms have the same name as the metal. If the metal can form cations of differing charges, the charge is given using Roman numerals. Monatomic anions have names ending in *-ide*. Polyatomic anions containing oxygen and another element **(oxyanions)** have names ending in *-ate* or *-ite*.

Section 2.9 **Organic chemistry** is the study of compounds that contain carbon. The simplest class of organic molecules are the **hydrocarbons**, which contain only carbon and hydrogen. Hydrocarbons in which each carbon atom is attached to four other atoms are called **alkanes**. Alkanes have names that end in *-ane*, such as methane and ethane. Other organic compounds are formed when an H atom of a hydrocarbon is replaced with a functional group. An **alcohol**, for example, is a compound in which an H atom of a hydrocarbon is replaced by an OH functional group. Alcohols have names that end in *-ol*, such as methanol and ethanol.

VISUALIZING CONCEPTS

2.1 A charged particle is caused to move between two electrically charged plates, as shown below.

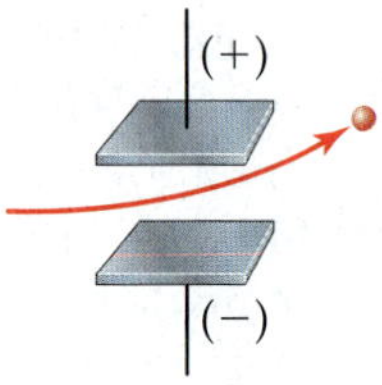

(a) Why does the path of the charged particle bend? **(b)** What is the sign of the electrical charge on the particle? **(c)** As the charge on the plates is increased, would you expect the bending to increase, decrease, or stay the same? **(d)** As the mass of the particle is increased while the speed of the particles remains the same, would you expect the bending to increase, decrease, or stay the same? [Section 2.2]

2.2 Four of the boxes in the following periodic table are colored. Which of these are metals and which are nonmetals? Which one is an alkaline earth metal? Which one is a noble gas? [Section 2.5]

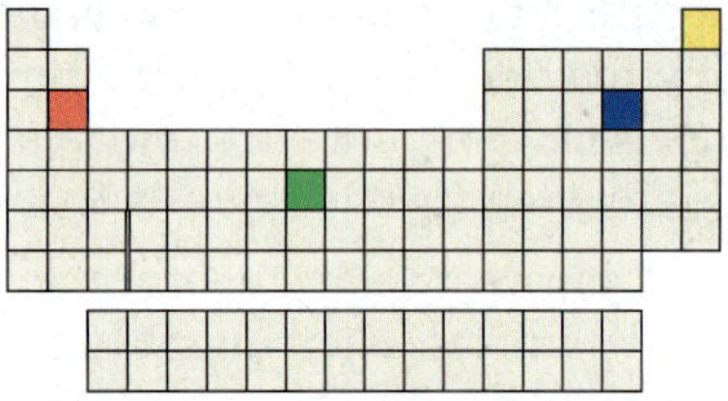

2.3 Does the following drawing represent a neutral atom or an ion? Write its complete chemical symbol including mass number, atomic number, and net charge (if any). [Sections 2.3 and 2.7]

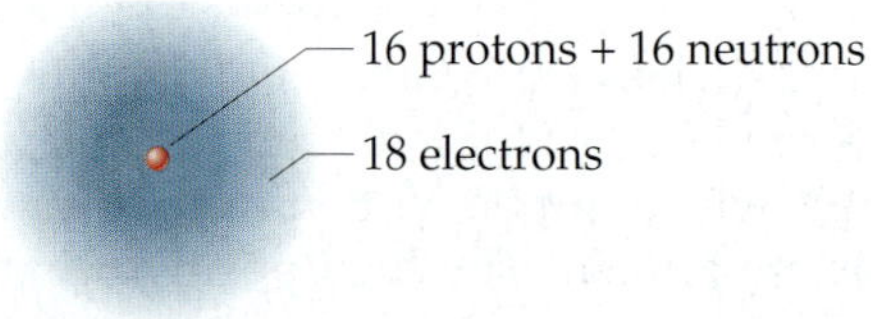

2.4 Which of the following diagrams is most likely to represent an ionic compound, and which a molecular one? Explain your choice. [Sections 2.6 and 2.7]

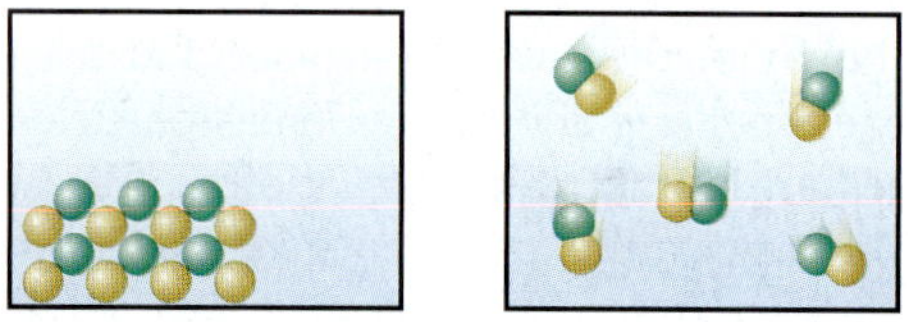

(i) (ii)

2.5 Write the chemical formula for the following compound. Is the compound ionic or molecular? Name the compound. [Sections 2.6 and 2.8]

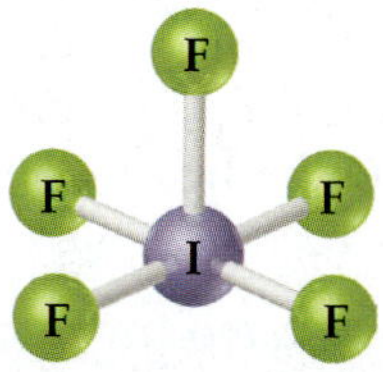

2.6 The following diagram represents an ionic compound in which the cations are indicated by the red spheres and the anions are indicated by the blue spheres. Which of the following formulas is consistent with the drawing: KBr, K_2SO_4, $Ca(NO_3)_2$, $Fe_2(SO_4)_3$? Name the compound. [Sections 2.7 and 2.8]

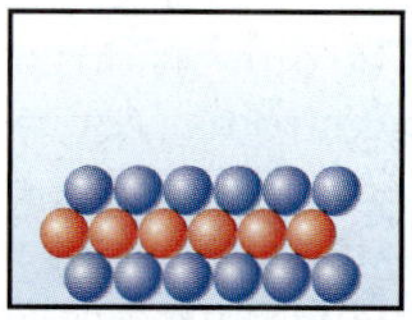

EXERCISES

Atomic Theory and the Discovery of Atomic Structure

2.7 How does Dalton's atomic theory account for the fact that when 1.000 g of water is decomposed into its elements, 0.111 g of hydrogen and 0.889 g of oxygen are obtained regardless of the source of the water?

2.8 Hydrogen sulfide is composed of two elements: hydrogen and sulfur. In an experiment, 6.500 g of hydrogen sulfide is fully decomposed into its elements. **(a)** If 0.384 g of hydrogen is obtained in this experiment, how many grams of sulfur must be obtained? **(b)** What fundamental law does this experiment demonstrate? **(c)** How is this law explained by Dalton's atomic theory?

2.9 A chemist finds that 30.82 g of nitrogen will react with 17.60 g, 35.20 g, 70.40 g, or 88.00 g of oxygen to form four different compounds. **(a)** Calculate the mass of oxygen per gram of nitrogen in each compound. **(b)** How do the numbers in part (a) support Dalton's atomic theory?

2.10 In a series of experiments, a chemist prepared three different compounds that contain only iodine and fluorine and determined the mass of each element in each compound:

Compound	Mass of Iodine (g)	Mass of Fluorine (g)
1	4.75	3.56
2	7.64	3.43
3	9.41	9.86

(a) Calculate the mass of fluorine per gram of iodine in each compound. **(b)** How do the numbers in part (a) support the atomic theory?

2.11 Summarize the evidence used by J. J. Thomson to argue that cathode rays consist of negatively charged particles.

2.12 An unknown particle is caused to move between two electrically charged plates, as illustrated in Figure 2.8. Its path is deflected in the opposite direction from that of a beta particle, and it is deflected by a smaller magnitude. What can you conclude about the charge and mass of this unknown particle?

2.13 **(a)** Figure 2.5 shows the apparatus used in the Millikan oil-drop experiment with the positively charged plate above the negatively charged plate. What do you think would be the effect on the rate of oil drops descending if the charges on the plates were reversed (negative above positive)? **(b)** In his original series of experiments, Millikan measured the charge on 58 separate oil drops. Why do you suppose he chose so many drops before reaching his final conclusions?

2.14 Millikan determined the charge on the electron by studying the static charges on oil drops falling in an electric field. A student carried out this experiment using several oil drops for her measurements and calculated the charges on the drops. She obtained the following data:

Droplet	Calculated Charge (C)
A	1.60×10^{-19}
B	3.15×10^{-19}
C	4.81×10^{-19}
D	6.31×10^{-19}

(a) What is the significance of the fact that the droplets carried different charges? **(b)** What conclusion can the student draw from these data regarding the charge of the electron? **(c)** What value (and to how many significant figures) should she report for the electronic charge?

Modern View of Atomic Structure; Atomic Weights

2.15 The radius of an atom of krypton (Kr) is about 1.9 Å. **(a)** Express this distance in nanometers (nm) and in picometers (pm). **(b)** How many krypton atoms would have to be lined up to span 1.0 mm? **(c)** If the atom is assumed to be a sphere, what is the volume in cm^3 of a single Kr atom?

2.16 An atom of tin (Sn) has a diameter of about 2.8×10^{-8} cm. **(a)** What is the radius of a tin atom in angstroms (Å) and in meters (m)? **(b)** How many Sn atoms would have to be placed side by side to span a distance of 6.0 μm? **(c)** If the atom is assumed to be a sphere, what is the volume in m^3 of a single Sn atom?

2.17 Answer the following questions without referring to Table 2.1: **(a)** What are the main subatomic particles that make up the atom? **(b)** What is the charge, in units of the electronic charge, of each of the particles? **(c)** Which of the particles is the most massive? Which is the least massive?

2.18 Determine whether each of the following statements is true or false; if false, correct the statement to make it true: **(a)** The nucleus has most of the mass and comprises most of the volume of an atom; **(b)** every atom of a given element has the same number of protons; **(c)** the number of electrons in an atom equals the number of neutrons in the atom; **(d)** the protons in the nucleus of the helium atom are held together by a force called the strong nuclear force.

2.19 **(a)** Define atomic number and mass number. **(b)** Which of these can vary without changing the identity of the element?

2.20 **(a)** Which two of the following are isotopes of the same element: $^{31}_{16}X$, $^{31}_{15}X$, $^{32}_{16}X$? **(b)** What is the identity of the element whose isotopes you have selected?

2.21 How many protons, neutrons, and electrons are in the following atoms: **(a)** ^{40}Ar, **(b)** ^{65}Zn, **(c)** ^{70}Ga, **(d)** ^{80}Br, **(e)** ^{184}W, **(f)** ^{243}Am.

2.22 Each of the following isotopes is used in medicine. Indicate the number of protons and neutrons in each isotope: **(a)** phosphorus-32, **(b)** chromium-51, **(c)** cobalt-60, **(d)** technetium-99, **(e)** iodine-131; **(f)** thallium-201.

2.23 Fill in the gaps in the following table, assuming each column represents a neutral atom:

Symbol	^{52}Cr			
Protons		25			82
Neutrons		30	64		
Electrons			48	86	
Mass no.				222	207

2.24 Fill in the gaps in the following table, assuming each column represents a neutral atom:

Symbol	^{121}Sb				
Protons		45			94
Neutrons		58	50		
Electrons			38	52	
Mass no.				127	239

2.25 Write the correct symbol, with both superscript and subscript, for each of the following. Use the list of elements on the front inside cover as needed: **(a)** the isotope of platinum that contains 118 neutrons, **(b)** the isotope of krypton with mass number 84, **(c)** the isotope of arsenic with mass number 75, **(d)** the isotope of magnesium that has an equal number of protons and neutrons.

2.26 One way in which Earth's evolution as a planet can be understood is by measuring the amounts of certain isotopes in rocks. One quantity recently measured is the ratio of ^{129}Xe to ^{130}Xe in some minerals. In what way do these two isotopes differ from one another, and in what respects are they the same?

2.27 **(a)** What isotope is used as the standard in establishing the atomic mass scale? **(b)** The atomic weight of boron is reported as 10.81, yet no atom of boron has the mass of 10.81 amu. Explain.

2.28 **(a)** What is the mass in amu of a carbon-12 atom? **(b)** Why is the atomic weight of carbon reported as 12.011 in the table of elements and the periodic table in the front inside cover of this text?

2.29 Only two isotopes of copper occur naturally, ^{63}Cu (atomic mass = 62.9296 amu; abundance 69.17%) and ^{65}Cu (atomic mass = 64.9278 amu; abundance 30.83%). Calculate the atomic weight (average atomic mass) of copper.

2.30 The element lead (Pb) consists of four naturally occurring isotopes with atomic masses 203.97302, 205.97444, 206.97587, and 207.97663 amu. The relative abundances of these four isotopes are 1.4, 24.1, 22.1, and 52.4%, respectively. From these data, calculate the atomic weight of lead.

2.31 **(a)** In what fundamental way is mass spectrometry related to Thomson's cathode-ray experiments (Figure 2.4)? **(b)** What are the labels on the axes of a mass spectrum? **(c)** In order to measure the mass spectrum of an atom, the atom must first lose or gain one or more electrons. Why is this so?

2.32 **(a)** The mass spectrometer in Figure 2.13 has a magnet as one of its components. What is the purpose of the magnet? **(b)** The atomic weight of Cl is 35.5 amu. However, the mass spectrum of Cl (Figure 2.14) does not show a peak at this mass. Explain. **(c)** A mass spectrum of phosphorus (P) atoms shows only a single peak at a mass of 31. What can you conclude from this observation?

2.33 Naturally occurring magnesium has the following isotopic abundances:

Isotope	Abundance	Atomic mass (amu)
^{24}Mg	78.99%	23.98504
^{25}Mg	10.00%	24.98584
^{26}Mg	11.01%	25.98259

(a) What is the average atomic mass of Mg? **(b)** Sketch the mass spectrum of Mg.

2.34 Mass spectrometry is more often applied to molecules than to atoms. We will see in Chapter 3 that the *molecular weight* of a molecule is the sum of the atomic weights of the atoms in the molecule. The mass spectrum of H_2 is taken under conditions that prevent decomposition into H atoms. The two naturally occurring isotopes of hydrogen are ^{1}H (atomic mass = 1.00783 amu; abundance 99.9885%) and ^{2}H (atomic mass = 2.01410 amu; abundance 0.0115%). **(a)** How many peaks will the mass spectrum have? **(b)** Give the relative atomic masses of each of these peaks. **(c)** Which peak will be the largest, and which the smallest?

The Periodic Table; Molecules and Ions

2.35 For each of the following elements, write its chemical symbol, locate it in the periodic table, and indicate whether it is a metal, metalloid, or nonmetal: **(a)** chromium, **(b)** helium, **(c)** phosphorus, **(d)** zinc, **(e)** magnesium, **(f)** bromine, **(g)** arsenic.

2.36 Locate each of the following elements in the periodic table; indicate whether it is a metal, metalloid, or nonmetal; and give the name of the element: **(a)** Na, **(b)** Ti, **(c)** Ga, **(d)** U, **(e)** Pd, **(f)** Se, **(g)** Kr.

2.37 For each of the following elements, write its chemical symbol, determine the name of the group to which it belongs (Table 2.3), and indicate whether it is a metal, metalloid, or nonmetal: **(a)** potassium, **(b)** iodine, **(c)** magnesium, **(d)** argon, **(e)** sulfur.

2.38 The elements of group 4A show an interesting change in properties moving down the group. Give the name and chemical symbol of each element in the group, and label it as a nonmetal, metalloid, or metal.

2.39 What can we tell about a compound when we know the empirical formula? What additional information is conveyed by the molecular formula? By the structural formula? Explain in each case.

2.40 Two compounds have the same empirical formula. One substance is a gas, the other is a viscous liquid. How is it possible for two substances with the same empirical formula to have markedly different properties?

2.41 Write the empirical formula corresponding to each of the following molecular formulas: **(a)** Al_2Br_6, **(b)** C_8H_{10}, **(c)** $C_4H_8O_2$, **(d)** P_4O_{10}, **(e)** $C_6H_4Cl_2$, **(f)** $B_3N_3H_6$.

2.42 Determine the molecular and empirical formulas of the following: **(a)** The organic solvent *benzene*, which has six carbon atoms and six hydrogen atoms; **(b)** the compound *silicon tetrachloride*, which has a silicon atom and four chlorine atoms and is used in the manufacture of computer chips; **(c)** the reactive substance *diborane*, which has two boron atoms and six hydrogen atoms; **(d)** the sugar called *glucose*, which has six carbon atoms, twelve hydrogen atoms, and six oxygen atoms.

2.43 How many hydrogen atoms are in each of the following: **(a)** C_2H_5OH, **(b)** $Ca(CH_3COO)_2$, **(c)** $(NH_4)_3PO_4$?

2.44 How many of the indicated atoms are represented by each chemical formula: **(a)** carbon atoms in $C_2H_5COOCH_3$, **(b)** oxygen atoms in $Ca(ClO_3)_2$, **(c)** hydrogen atoms in $(NH_4)_2HPO_4$?

2.45 Write the molecular and structural formulas for the compounds represented by the following molecular models:

(a)

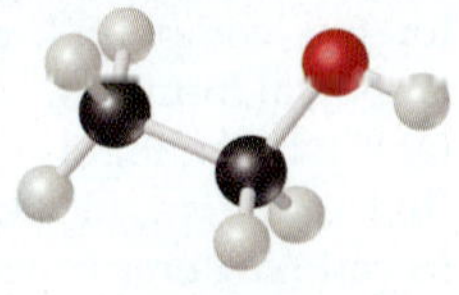
(b)

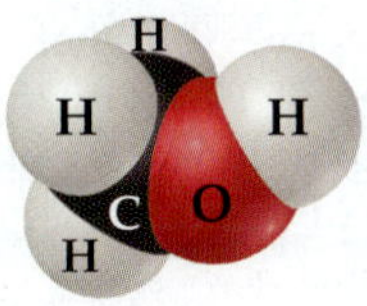

(c)

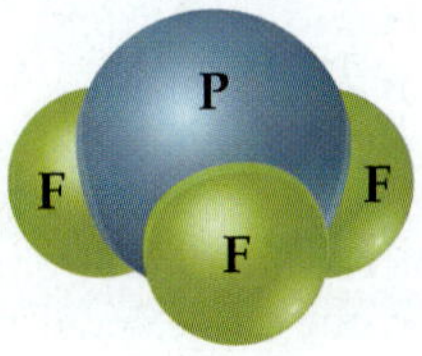

(d)

2.46 Write the molecular and structural formulas for the compounds represented by the following models:

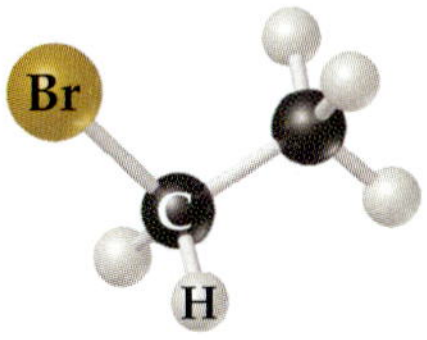

(a)

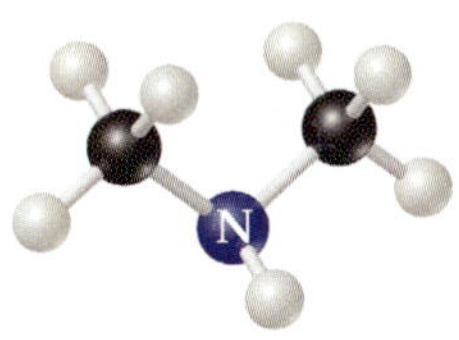

(b)

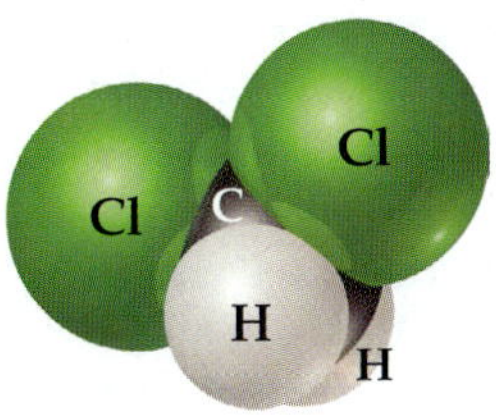

(c)

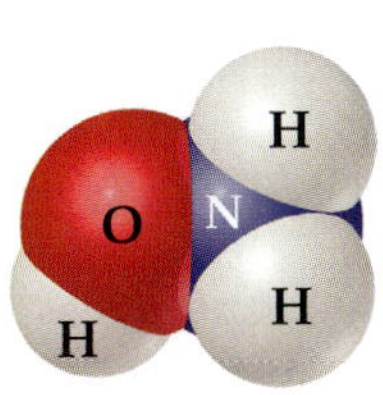

(d)

2.47 Fill in the gaps in the following table:

Symbol	$^{59}Co^{3+}$			
Protons		34	76	80
Neutrons		46	116	120
Electrons		36		78
Net charge			2+	

2.48 Fill in the gaps in the following table:

Symbol	$^{75}As^{3-}$			
Protons		28	53	
Neutrons		31	74	118
Electrons		26		76
Net charge			1−	3+

2.49 Each of the following elements is capable of forming an ion in chemical reactions. By referring to the periodic table, predict the charge of the most stable ion of each: **(a)** Mg, **(b)** Al, **(c)** K, **(d)** S, **(e)** F.

2.50 Using the periodic table, predict the charges of the ions of the following elements: **(a)** Sr, **(b)** Sc, **(c)** P, **(d)** I, **(e)** Se.

2.51 Using the periodic table to guide you, predict the chemical formula and name of the compound formed by the following elements: **(a)** Ga and F, **(b)** Li and H, **(c)** Al and I, **(d)** K and S.

2.52 The most common charge associated with silver in its compounds is 1+. Indicate the chemical formulas you would expect for compounds formed between Ag and **(a)** iodine, **(b)** sulfur, **(c)** fluorine.

2.53 Predict the chemical formula for the ionic compound formed by **(a)** Ca^{2+} and Br^-, **(b)** K^+ and $CO_3{}^{2-}$, **(c)** Al^{3+} and $C_2H_3O_2{}^-$, **(d)** $NH_4{}^+$ and $SO_4{}^{2-}$, **(e)** Mg^{2+} and $PO_4{}^{3-}$.

2.54 Predict the chemical formulas of the compounds formed by the following pairs of ions: **(a)** Cu^+ and S^{2-}, **(b)** Fe^{3+} and O^{2-}, **(c)** $Hg_2{}^{2+}$ and $CO_3{}^{2-}$, **(d)** Ca^{2+} and $AsO_4{}^{3-}$, **(e)** $NH_4{}^+$ and $CO_3{}^{2-}$.

2.55 Predict whether each of the following compounds is molecular or ionic: **(a)** B_2H_6, **(b)** CH_3OH, **(c)** $LiNO_3$, **(d)** Sc_2O_3, **(e)** CsBr, **(f)** NOCl, **(g)** NF_3, **(h)** Ag_2SO_4.

2.56 Which of the following are ionic, and which are molecular? **(a)** PF_5, **(b)** NaI, **(c)** SCl_2, **(d)** $Ca(NO_3)_2$, **(e)** $FeCl_3$, **(f)** LaP, **(g)** $CoCO_3$, **(h)** N_2O_4.

Naming Inorganic Compounds; Organic Molecules

2.57 Give the chemical formula for **(a)** chlorite ion, **(b)** chloride ion, **(c)** chlorate ion, **(d)** perchlorate ion, **(e)** hypochlorite ion.

2.58 Selenium, an element required nutritionally in trace quantities, forms compounds analogous to sulfur. Name the following ions: **(a)** SeO_4^{2-}, **(b)** Se^{2-}, **(c)** HSe^-, **(d)** $HSeO_3^-$.

2.59 Name the following ionic compounds: **(a)** MgO, **(b)** $AlCl_3$, **(c)** Li_3PO_4, **(d)** $Ba(ClO_4)_2$, **(e)** $Cu(NO_3)_2$, **(f)** $Fe(OH)_2$, **(g)** $Ca(C_2H_3O_2)_2$, **(h)** $Cr_2(CO_3)_3$, **(i)** K_2CrO_4, **(j)** $(NH_4)_2SO_4$.

2.60 Name the following ionic compounds: **(a)** Li_2O, **(b)** NaClO, **(c)** $Sr(CN)_2$, **(d)** $Cr(OH)_3$, **(e)** $Fe_2(CO_3)_3$, **(f)** $Co(NO_3)_2$, **(g)** $(NH_4)_2SO_3$, **(h)** NaH_2PO_4, **(i)** $KMnO_4$, **(j)** $Ag_2Cr_2O_7$.

2.61 Write the chemical formulas for the following compounds: **(a)** aluminum hydroxide, **(b)** potassium sulfate, **(c)** copper(I) oxide, **(d)** zinc nitrate, **(e)** mercury(II) bromide, **(f)** iron(III) carbonate, **(g)** sodium hypobromite.

2.62 Give the chemical formula for each of the following ionic compounds: **(a)** sodium phosphate, **(b)** cobalt(II) nitrate, **(c)** barium bromate, **(d)** copper(II) perchlorate, **(e)** magnesium hydrogen carbonate, **(f)** chromium(III) acetate, **(g)** potassium dichromate.

2.63 Give the name or chemical formula, as appropriate, for each of the following acids: **(a)** $HBrO_3$, **(b)** HBr, **(c)** H_3PO_4, **(d)** hypochlorous acid, **(e)** iodic acid, **(f)** sulfurous acid.

2.64 Provide the name or chemical formula, as appropriate, for each of the following acids: **(a)** hydrobromic acid, **(b)** hydrosulfuric acid, **(c)** nitrous acid, **(d)** H_2CO_3, **(e)** $HClO_3$, **(f)** $HC_2H_3O_2$.

2.65 Give the name or chemical formula, as appropriate, for each of the following binary molecular substances: **(a)** SF_6, **(b)** IF_5, **(c)** XeO_3, **(d)** dinitrogen tetroxide, **(e)** hydrogen cyanide, **(f)** tetraphosphorus hexasulfide.

2.66 The oxides of nitrogen are very important ingredients in determining urban air pollution. Name each of the following compounds: **(a)** N_2O, **(b)** NO, **(c)** NO_2, **(d)** N_2O_5, **(e)** N_2O_4.

2.67 Write the chemical formula for each substance mentioned in the following word descriptions (use the front inside cover to find the symbols for the elements you don't know). **(a)** Zinc carbonate can be heated to form zinc oxide and carbon dioxide. **(b)** On treatment with hydrofluoric acid, silicon dioxide forms silicon tetrafluoride and water. **(c)** Sulfur dioxide reacts with water to form sulfurous acid. **(d)** The substance phosphorus trihydride, commonly called phosphine, is a toxic gas. **(e)** Perchloric acid reacts with cadmium to form cadmium(II) perchlorate. **(f)** Vanadium(III) bromide is a colored solid.

2.68 Assume that you encounter the following phrases in your reading. What is the chemical formula for each substance mentioned? **(a)** Sodium hydrogen carbonate is used as a deodorant. **(b)** Calcium hypochlorite is used in some bleaching solutions. **(c)** Hydrogen cyanide is a very poisonous gas. **(d)** Magnesium hydroxide is used as a cathartic. **(e)** Tin(II) fluoride has been used as a fluoride additive in toothpastes. **(f)** When cadmium sulfide is treated with sulfuric acid, fumes of hydrogen sulfide are given off.

2.69 **(a)** What is a hydrocarbon? **(b)** Butane is the alkane with a chain of four carbon atoms. Write a structural formula for this compound, and determine its molecular and empirical formulas.

2.70 **(a)** What ending is used for the names of alkanes? **(b)** Hexane is an alkane whose structural formula has all its carbon atoms in a straight chain. Draw the structural formula for this compound, and determine its molecular and empirical formulas. [*Hint:* You might need to refer to Table 2.6.]

2.71 **(a)** What is a functional group? **(b)** What functional group characterizes an alcohol? **(c)** With reference to Exercise 2.69, write a structural formula for 1-butanol, the alcohol derived from butane, by making a substitution on one of the end carbon atoms.

2.72 **(a)** What do ethane and ethanol have in common? **(b)** How does 1-propanol differ from propane?

Additional Exercises

2.73 Describe a major contribution to science made by each of the following scientists: **(a)** Dalton, **(b)** Thomson, **(c)** Millikan, **(d)** Rutherford.

2.74 What is radioactivity? Indicate whether you agree or disagree with the following statement, and indicate your reasons: Henri Becquerel's discovery of radioactivity shows that the atom is not indivisible, as had been believed for so long.

2.75 How did Rutherford interpret the following observations made during his α-particle scattering experiments? **(a)** Most α particles were not appreciably deflected as they passed through the gold foil. **(b)** A few α particles were deflected at very large angles. **(c)** What differences would you expect if beryllium foil were used instead of gold foil in the α-particle scattering experiment?

[2.76] Suppose a scientist repeats the Millikan oil-drop experiment, but reports the charges on the drops using an unusual (and imaginary) unit called the *warmomb* (wa). He obtains the following data for four of the drops:

Droplet	Calculated Charge (wa)
A	3.84×10^{-8}
B	4.80×10^{-8}
C	2.88×10^{-8}
D	8.64×10^{-8}

(a) If all the droplets were the same size, which would fall most slowly through the apparatus? **(b)** From these data, what is the best choice for the charge of the electron in warmombs? **(c)** Based on your answer to part (b), how many electrons are there on each of the droplets? **(d)** What is the conversion factor between warmombs and coulombs?

2.77 The natural abundance of ^3He is 0.000137%. **(a)** How many protons, neutrons, and electrons are in an atom of ^3He? **(b)** Based on the sum of the masses of their subatomic particles, which is expected to be more massive, an atom of ^3He or an atom of ^3H (which is also called *tritium*)? **(c)** Based on your answer for part (b), what would need to be the precision of a mass spectrometer that is able to differentiate between peaks that are due to $^3He^+$ and $^3H^+$?

[2.78] An α particle is the nucleus of an ^4He atom. **(a)** How many protons and neutrons are in an α particle? **(b)** What force holds the protons and neutrons together in the α particle? **(c)** What is the charge on an α particle in units of electronic charge? **(d)** The charge-to-mass ratio of an α particle is 4.8224×10^4 C/g. Based on the charge on the particle, calculate its mass in grams and in amu. **(e)** By using the data in Table 2.1, compare your answer for part (d) with the sum of the masses of the individual subatomic particles. Can you explain the difference in mass? (If not, we will discuss such mass differences further in Chapter 21.)

[2.79] A cube of gold that is 1.00 cm on a side has a mass of 19.3 g. A single gold atom has a mass of 197.0 amu. **(a)** How many gold atoms are in the cube? **(b)** From the information given, estimate the diameter in Å of a single gold atom. **(c)** What assumptions did you make in arriving at your answer for part (b)?

[2.80] The diameter of a rubidium atom is 4.95 Å. We will consider two different ways of placing the atoms on a surface. In arrangement A, all the atoms are lined up with one another. Arrangement B is called a *close-packed* arrangement because the atoms sit in the "depressions" formed by the previous row of atoms:

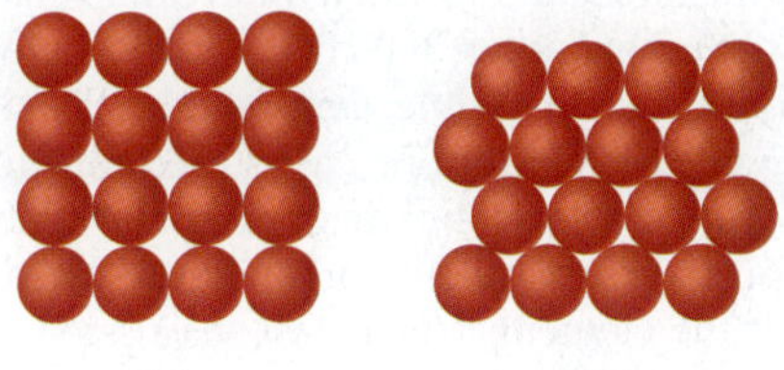

(a) (b)

(a) Using arrangement A, how many Rb atoms could be placed on a square surface that is 1.0 cm on a side? **(b)** How many Rb atoms could be placed on a square surface that is 1.0 cm on a side, using arrangement B? **(c)** By what factor has the number of atoms on the surface increased in going to arrangement B from arrangement A? If extended to three dimensions, which arrangement would lead to a greater density for Rb metal?

[2.81] **(a)** Assuming the dimensions of the nucleus and atom shown in Figure 2.12, what fraction of the *volume* of the atom is taken up by the nucleus? **(b)** Using the mass of the proton from Table 2.1 and assuming its diameter is 1.0×10^{-15} m, calculate the density of a proton in g/cm^3.

2.82 The element oxygen has three naturally occurring isotopes, with 8, 9, and 10 neutrons in the nucleus, respectively. **(a)** Write the full chemical symbols for these three isotopes. **(b)** Describe the similarities and differences between the three kinds of atoms of oxygen.

2.83 Use Coulomb's law, $F = kQ_1Q_2/d^2$, to calculate the electric force on an electron ($Q = -1.6 \times 10^{-19}$ C) exerted by a single proton if the particles are 0.53×10^{-10} m apart. The constant k in Coulomb's law is 9.0×10^9 N·m^2/C^2. (The unit abbreviated N is the Newton, the SI unit of force.)

2.84 Gallium (Ga) consists of two naturally occurring isotopes with masses of 68.926 and 70.925 amu. **(a)** How many protons and neutrons are in the nucleus of each isotope? Write the complete atomic symbol for each, showing the atomic number and mass number. **(b)** The average atomic mass of Ga is 69.72 amu. Calculate the abundance of each isotope.

[2.85] Using a suitable reference such as the *CRC Handbook of Chemistry and Physics* or http://www.webelements.com, look up the following information for nickel: **(a)** the number of known isotopes, **(b)** the atomic masses (in amu) and the natural abundance of the five most abundant isotopes.

[2.86] There are two different isotopes of bromine atoms. Under normal conditions, elemental bromine consists of Br_2 molecules (Figure 2.19), and the mass of a Br_2 molecule is the sum of the masses of the two atoms in the molecule. The mass spectrum of Br_2 consists of three peaks:

Mass (amu)	Relative Size
157.836	0.2569
159.834	0.4999
161.832	0.2431

(a) What is the origin of each peak (of what isotopes does each consist)? **(b)** What is the mass of each isotope? **(c)** Determine the average molecular mass of a Br_2 molecule. **(d)** Determine the average atomic mass of a bromine atom. **(e)** Calculate the abundances of the two isotopes.

2.87 It is common in mass spectrometry to assume that the mass of a cation is the same as that of its parent atom. **(a)** Using data in Table 2.1, determine the number of significant figures that must be reported before the difference in mass of ^1H and $^1H^+$ is significant. **(b)** What percentage of the mass of an ^1H atom does the electron represent?

2.88 *Bronze* is a metallic alloy often used in decorative applications and in sculpture. A typical bronze consists of copper, tin, and zinc, with lesser amounts of phosphorus and lead. Locate each of these five elements in the periodic table, write their symbols, and identify the group of the periodic table to which they belong.

2.89 From the following list of elements—Ar, H, Ga, Al, Ca, Br, Ge, K, O—pick the one that best fits each description; use each element only once: **(a)** an alkali metal, **(b)** an alkaline earth metal, **(c)** a noble gas, **(d)** a halogen, **(e)** a metalloid, **(f)** a nonmetal listed in group 1A, **(g)** a metal that forms a 3+ ion, **(h)** a nonmetal that forms a 2− ion, **(i)** an element that resembles aluminum.

2.90 The first atoms of seaborgium (Sg) were identified in 1974. The longest-lived isotope of Sg has a mass number of 266. **(a)** How many protons, electrons, and neutrons are in an ^{266}Sg atom? **(b)** Atoms of Sg are very unstable, and it is therefore difficult to study this element's properties. Based on the position of Sg in the periodic table, what element should it most closely resemble in its chemical properties?

2.91 From the molecular structures shown here, identify the one that corresponds to each of the following species: **(a)** chlorine gas; **(b)** propane, **(c)** nitrate ion; **(d)** sulfur trioxide; **(e)** methyl chloride, CH_3Cl.

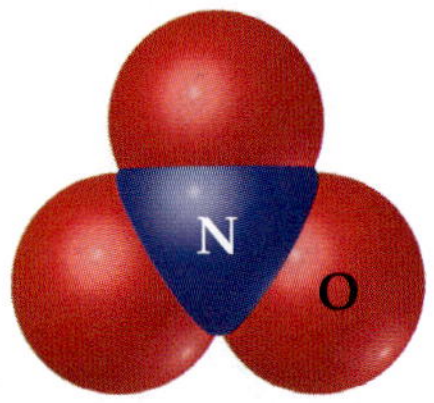

(i)

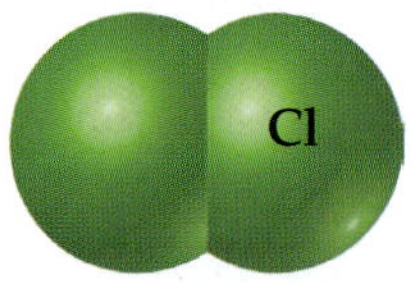

(ii)

(iii)

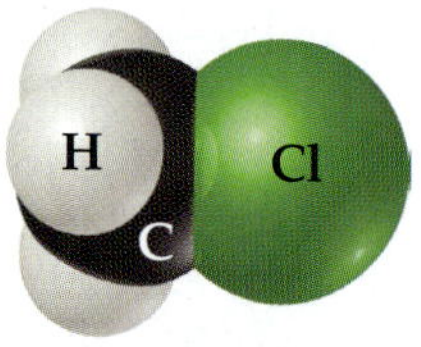

(iv)

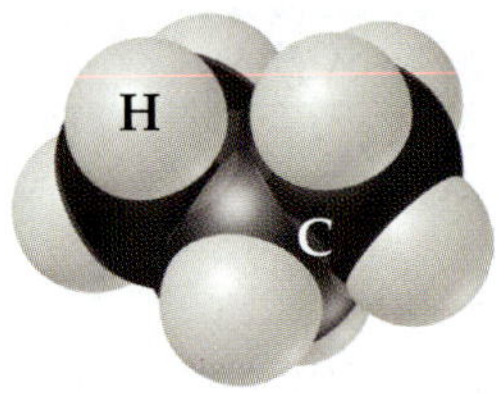

(v)

2.92 Name each of the following oxides. Assuming that the compounds are ionic, what charge is associated with the metallic element in each case? **(a)** NiO, **(b)** MnO_2, **(c)** Cr_2O_3, **(d)** MoO_3.

2.93 Iodic acid has the molecular formula HIO_3. Write the formulas for the following: **(a)** the iodate anion, **(b)** the periodate anion, **(c)** the hypoiodite anion, **(d)** hypoiodous acid, **(e)** periodic acid.

2.94 Elements in the same group of the periodic table often form oxyanions with the same general formula. The anions are also named in a similar fashion. Based on these observations, suggest a chemical formula or name, as appropriate, for each of the following ions: **(a)** BrO_4^-, **(b)** SeO_3^{2-}, **(c)** arsenate ion, **(d)** hydrogen tellurate ion.

2.95 Give the chemical names of each of the following familiar compounds: **(a)** NaCl (table salt), **(b)** $NaHCO_3$ (baking soda), **(c)** NaOCl (in many bleaches), **(d)** NaOH (caustic soda), **(e)** $(NH_4)_2CO_3$ (smelling salts), **(f)** $CaSO_4$ (plaster of Paris).

2.96 Many familiar substances have common, unsystematic names. For each of the following, give the correct systematic name: **(a)** saltpeter, KNO_3; **(b)** soda ash, Na_2CO_3; **(c)** lime, CaO; **(d)** muriatic acid, HCl; **(e)** Epsom salts, $MgSO_4$; **(f)** milk of magnesia, $Mg(OH)_2$.

2.97 Many ions and compounds have very similar names, and there is great potential for confusing them. Write the correct chemical formulas to distinguish between **(a)** calcium sulfide and calcium hydrogen sulfide, **(b)** hydrobromic acid and bromic acid, **(c)** aluminum nitride and aluminum nitrite, **(d)** iron(II) oxide and iron(III) oxide, **(e)** ammonia and ammonium ion, **(f)** potassium sulfite and potassium bisulfite, **(g)** mercurous chloride and mercuric chloride, **(h)** chloric acid and perchloric acid.

[2.98] Using the *CRC Handbook of Chemistry and Physics*, find the density, melting point, and boiling point for **(a)** PF_3, **(b)** $SiCl_4$, **(c)** ethanol, C_2H_6O.

[2.99] The compound *cyclohexane* is an alkane in which six carbon atoms form a ring. The partial structural formula of the compound is as follows:

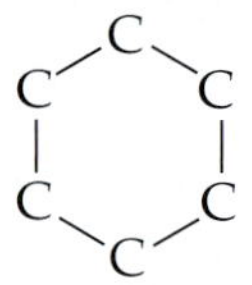

(a) Complete the structural formula for cyclohexane. **(b)** Is the molecular formula for cyclohexane the same as that for *n*-hexane, in which the carbon atoms are in a straight line? If possible, comment on the source of any differences. **(c)** Propose a structural formula for *cyclohexanol*, the alcohol derived from cyclohexane.

2.100 The periodic table helps organize the chemical behaviors of the elements. As a class discussion or as a short essay, describe how the table is organized, and mention as many ways as you can think of in which the position of an element in the table relates to the chemical and physical properties of the element.

eMEDIA EXERCISES

These exercises make use of the interactive objects available online in OneKey or the Companion Website, and on your Accelerator CD. Access to these resources comes in your MediaPak.

2.101 **(a)** After watching the **Multiple Proportions** animation (*2.1*), show how the oxygen-to-hydrogen mass ratios of H_2O and H_2O_2 illustrate the law of multiple proportions. **(b)** Refer to Exercise 2.9, and sketch molecular models (similar to those in the movie) of the three N- and O-containing compounds in the exercise. **(c)** Do the same thing for Exercise 2.10, sketching models of the I- and F-containing molecules.

2.102 Prior to Rutherford's gold-foil experiment, the mass and positively charged particles of an atom were thought to be evenly distributed throughout the volume of the atom. **(a)** Watch the animation of the **Rutherford Experiment: Nuclear Atom** (*2.2*), and describe how the experimental results would have been different if the earlier model had been correct. **(b)** What specific feature of the modern view of atomic structure was illuminated by Rutherford's experiment?

2.103 The **Separation of Alpha, Beta, and Gamma Rays** animation (*2.2*) shows how three different types of radioactive emissions behave in the presence of an electric field. **(a)** Which of the three types of radiation does not consist of a stream of charged particles? **(b)** In Exercise 2.9 you explained why α and β rays are deflected in opposite directions. In the movie, the difference in the *magnitude* of deflection of α versus β particles is attributed primarily to a difference in mass. How much more massive are α particles than β particles? What factors other than mass influence the magnitude of deflection?

2.104 Give the correct formula and name for the ionic compound formed by each of the indicated combinations: NH_4^+ and Al^{3+} each with Br^-, OH^-, S^{2-}, CO_3^{2-}, NO_3^-, and ClO_4^-. Use the **Ionic Compounds** activity (*2.7*) to check your answers.

eLABORATORY EXERCISES

These exercises allow students to apply the concepts and skills for this chapter in the simulated laboratory of Virtual ChemLab. Worksheets for each of these exercises are found in the Virtual ChemLab Workbook in your MediaPak.

2.106 ***(VCL 2-1) Thomson Cathode-Ray Tube Experiment***
In 1897, Thomson showed that if you could measure how much a beam of electrons was bent in an electric field and measure the magnetic field required to move the beam of electrons back to center, you could determine the charge-to-mass ratio (q/m_e) for the electrons. In this problem, you will duplicate the Thomson cathode-ray tube experiment and calculate the charge-to-mass ratio of an electron.

2.107 ***(VCL 2-2) Millikan Oil-Drop Experiment***
In 1909, Robert Millikan and his graduate student Harvey Fletcher showed that they could measure the charge of an electron by making very small oil drops, depositing electrons on these drops, and then applying an electric field. You will get a chance to repeat their experiments and, using the results from the Thomson assignment, be able to experimentally calculate the mass of an electron.

2.108 ***(VCL 2-3) Rutherford's Backscattering Experiment***
A key experiment in understanding the nature of atomic structure was completed by Ernest Rutherford in 1911. He set up an experiment that directed a beam of alpha particles (helium nuclei) through a gold foil and then onto a detector screen. According to the plum-pudding atomic model, electrons float around inside a cloud of positive charge. Based on this model, Rutherford expected that almost all the alpha particles should pass through the gold foil and not be deflected. A few of the alpha particles would experience a slight deflection because of the attraction to the negative electrons. Imagine his surprise when a few alpha particles deflected at all angles, even nearly straight backward. You will have a chance to repeat this experiment and observe the number of alpha particles that are deflected at different angles. You will also be able to experiment with other metals and understand why Rutherford selected gold.

2.109 ***(VCL 2-4) Investigating the Properties of Alpha and Beta Particles***
As scientists began investigating the properties of atoms, their first discovery was that they could extract negatively charged particles. They called these particles electrons, but they are also known as beta particles in the context of nuclear decay. Robert Millikan used beta particles in his famous oil-drop experiment. Another particle ejected during nuclear decay is the alpha particle. An alpha particle is a helium nucleus, or a helium atom without its two electrons. Consequently, an alpha particle is positively charged. Ernest Rutherford used alpha particles in his gold-foil experiment. In this assignment you will investigate the propeties of alpha and beta particles using the virtual laboratory.

2.110 ***(VCL 2-5) Names and Formulas of Ionic Compounds***
In this problem you will go into the virtual laboratory and make a series of ionic compounds containing the cations Ag^+, Pb^+, Ca^{2+}, Fe^{3+}, and Cu^{2+}; observe the reactions and identify the color of the compounds formed; write the chemical formulas; and write the chemical names.

CHAPTER 3

WHAT'S AHEAD

- We begin by considering how we can use chemical formulas to write equations that represent chemical reactions. *(Section 3.1)*
- We then examine some simple kinds of chemical reactions: *combination reactions, decomposition reactions*, and *combustion reactions. (Section 3.2)*
- We then use chemical formulas to relate the masses of substances to the numbers of atoms, molecules, or ions contained in the substances, a relationship that leads to the crucially important concept of a mole. A *mole* is 6.022×10^{23} objects (atoms, molecules, ions, or whatever). *(Sections 3.3 and 3.4)*
- We will apply the mole concept to determine chemical formulas from the masses of each element in a given quantity of a compound. *(Section 3.5)*
- We will use the quantitative information inherent in chemical formulas and equations together with the mole concept to predict the amounts of substances consumed and/or produced in chemical reactions. *(Section 3.6)*
- A special situation arises when one of the reactants is used up before the others and the reaction therefore stops, leaving some of the excess starting material unreacted. *(Section 3.7)*

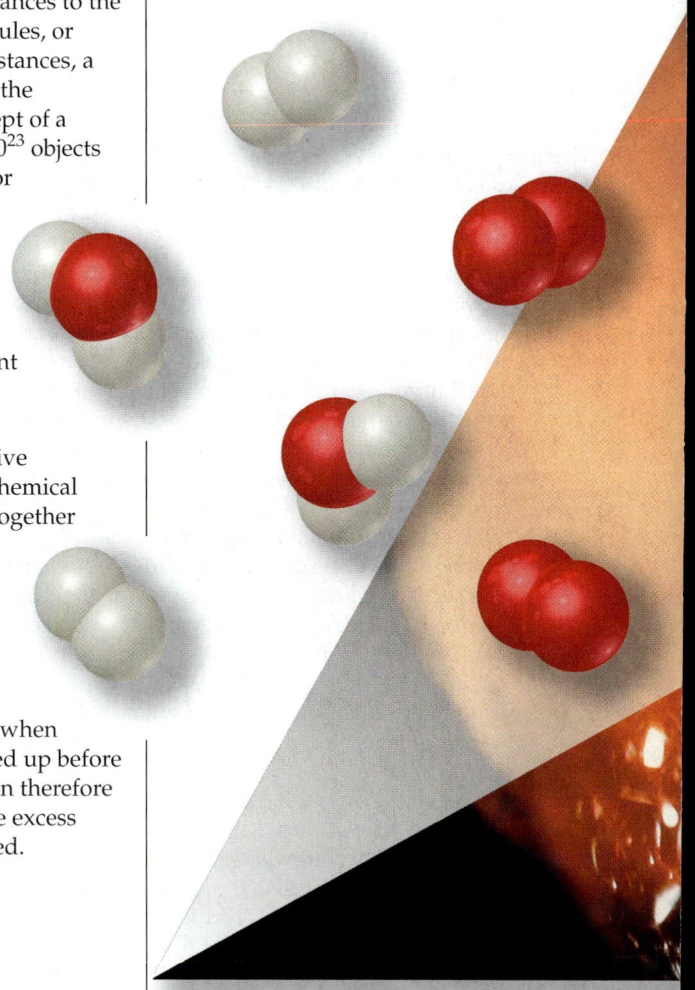

Stoichiometry: Calculations with Chemical Formulas and Equations

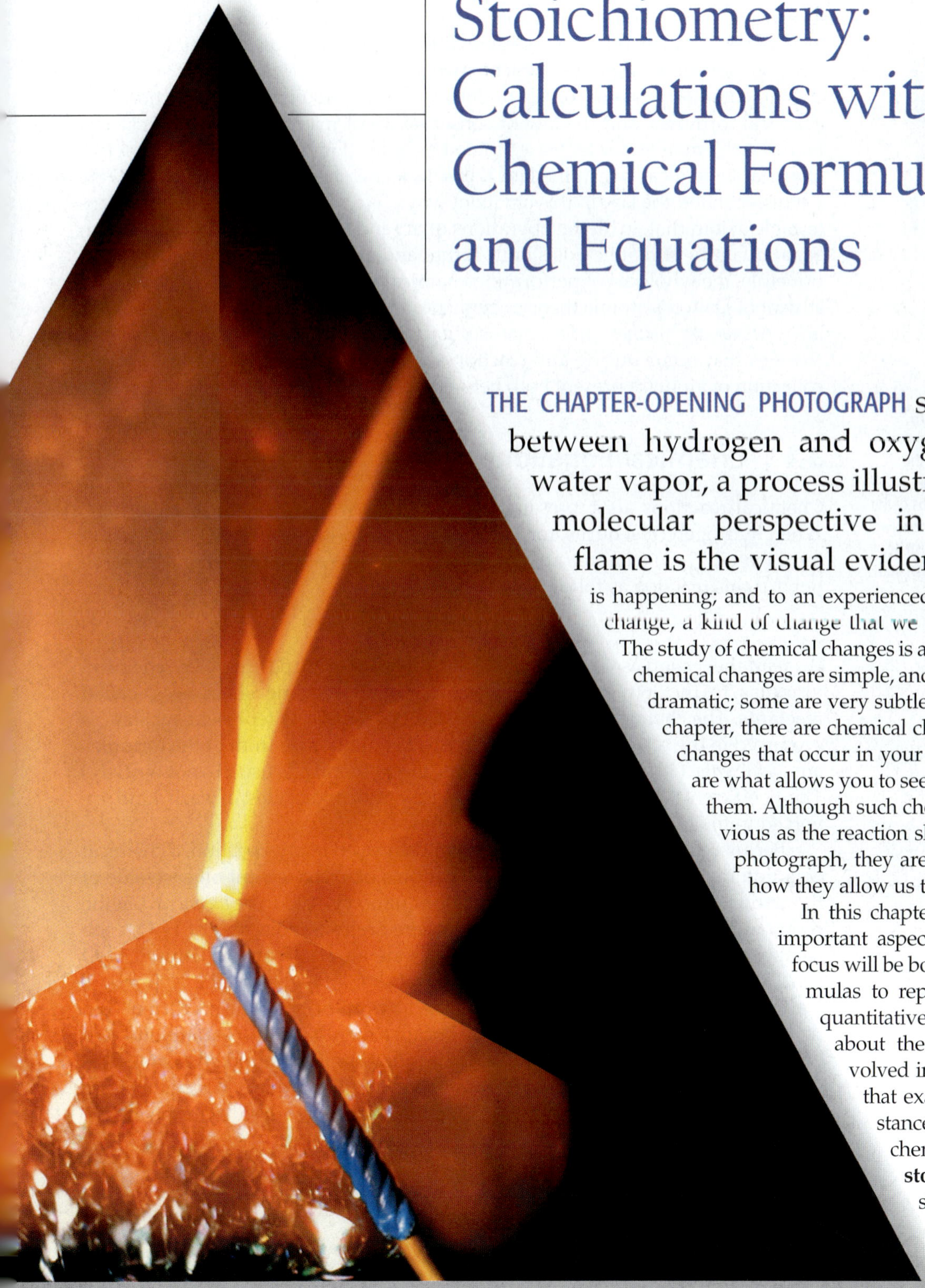

THE CHAPTER-OPENING PHOTOGRAPH shows the reaction between hydrogen and oxygen gases to form water vapor, a process illustrated earlier from a molecular perspective in Figure 1.10. The flame is the visual evidence that something is happening; and to an experienced eye, it indicates a chemical change, a kind of change that we also call a chemical reaction. The study of chemical changes is at the heart of chemistry. Some chemical changes are simple, and some are complex. Some are dramatic; some are very subtle. Even as you sit reading this chapter, there are chemical changes occurring in you. The changes that occur in your eyes and brain, for example, are what allows you to see these words and think about them. Although such chemical changes are not as obvious as the reaction shown in the chapter-opening photograph, they are nevertheless remarkable for how they allow us to function.

In this chapter we begin to explore some important aspects of chemical changes. Our focus will be both on the use of chemical formulas to represent reactions and on the quantitative information we can obtain about the amounts of substances involved in reactions. The area of study that examines the quantities of substances consumed and produced in chemical reactions is known as **stoichiometry** (pronounced stoy-key-OM-uh-tree), a name derived from the Greek *stoicheion* ("element") and *metron* ("measure"). This study provides an

COMBUSTION OF HYDROGEN GAS. The gas is bubbled through a soap solution forming hydrogen-filled bubbles. As the bubbles float upwards, they are ignited by a candle on a long pole. The red-orange flame is due to the combustion of the soap bubbles as the hydrogen reacts with oxygen in the air.

▲ **Figure 3.1 Antoine Lavoisier (1734–1794).** Lavoisier conducted many important studies on combustion reactions. Unfortunately, his career was cut short by the French Revolution. He was a member of the French nobility and a tax collector. He was guillotined in 1794 during the final months of the Reign of Terror. He is now generally considered to be the father of modern chemistry because he conducted carefully controlled experiments and used quantitative measurements.

MOVIE
Formation of water

essential set of tools that are widely used in chemistry. Such diverse problems as measuring the concentration of ozone in the atmosphere, determining the potential yield of gold from an ore, and assessing different processes for converting coal into gaseous fuels all use aspects of stoichiometry.

Stoichiometry is built on an understanding of atomic masses (Section 2.4), chemical formulas, and the law of conservation of mass. ∞ (Section 2.1) The French nobleman and scientist Antoine Lavoisier (Figure 3.1 ◀) discovered this important chemical law in the late 1700s. In a chemistry text published in 1789, Lavoisier stated the law in this eloquent way: "We may lay it down as an incontestable axiom that, in all the operations of art and nature, nothing is created; an equal quantity of matter exists both before and after the experiment. Upon this principle, the whole art of performing chemical experiments depends." With the advent of Dalton's atomic theory, chemists came to understand the basis for this law: *Atoms are neither created nor destroyed during any chemical reaction.* The changes that occur during any reaction merely rearrange the atoms. The same collection of atoms is present both before and after the reaction.

3.1 | Chemical Equations

Chemical reactions are represented in a concise way by **chemical equations**. When hydrogen (H_2) burns, for example, it reacts with oxygen (O_2) in the air to form water (H_2O) (chapter-opening photograph). We write the chemical equation for this reaction as follows:

$$2\,H_2 + O_2 \longrightarrow 2\,H_2O \qquad [3.1]$$

We read the + sign as "reacts with" and the arrow as "produces." The chemical formulas to the left of the arrow represent the starting substances, called **reactants**. The chemical formulas to the right of the arrow represent substances produced in the reaction, called **products**. The numbers in front of the formulas are *coefficients*. (As in algebraic equations, the numeral 1 is usually not written.) The coefficients indicate the relative numbers of molecules of each kind involved in the reaction.

Because atoms are neither created nor destroyed in any reaction, a chemical equation must have an equal number of atoms of each element on each side of the arrow. When this condition is met, the equation is said to be *balanced*. On the right side of Equation 3.1, for example, there are two molecules of H_2O, each composed of two atoms of hydrogen and one atom of oxygen. Thus, $2\,H_2O$ (read "two molecules of water") contains $2 \times 2 = 4$ H atoms and $2 \times 1 = 2$ O. Notice that the number of atoms is obtained by multiplying the coefficient and the subscripts in the chemical formula. Because there are four H atoms and two O atoms on each side of the equation, the equation is balanced. We can represent the balanced equation by the following molecular models, which illustrate that the number of atoms of each kind is the same on both sides of the arrow:

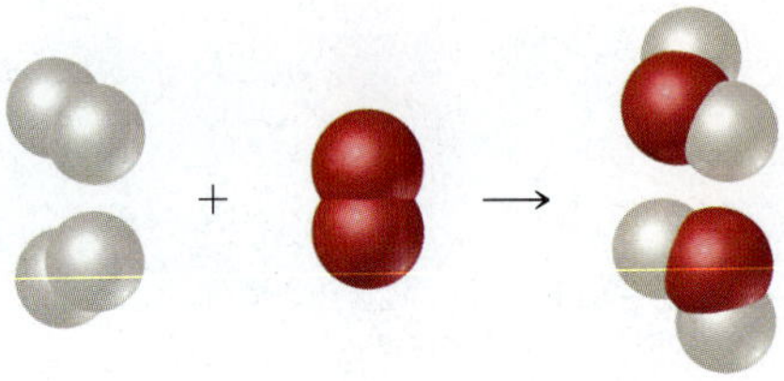

GIVE IT SOME THOUGHT

How many atoms of Mg, O, and H are represented by $3\,Mg(OH)_2$?

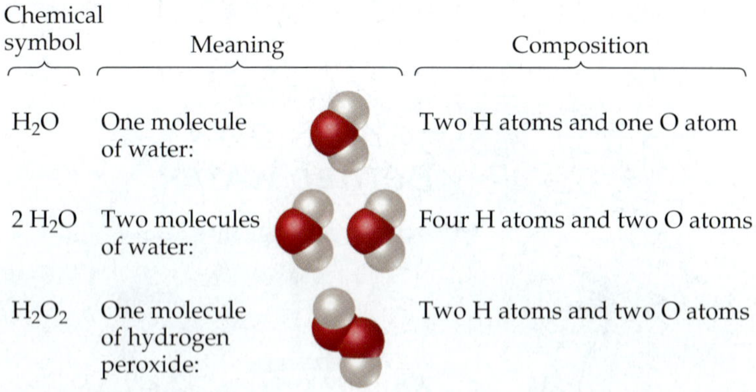

Figure 3.2 The difference between a subscript in a chemical formula and a coefficient in front of the formula. Notice how adding the coefficient 2 in front of the formula (line 2) has a different effect on the implied composition than adding the subscript 2 to the formula (in line 3). The number of atoms of each type (listed under composition) is obtained by multiplying the coefficient and the subscript associated with each element in the formula.

Balancing Equations

Once we know the formulas of the reactants and products in a reaction, we can write the unbalanced equation. We then balance the equation by determining the coefficients that provide equal numbers of each type of atom on each side of the equation. For most purposes, a balanced equation should contain the smallest possible whole-number coefficients.

In balancing equations, it is important to understand the difference between a coefficient in front of a formula and a subscript in a formula. Refer to Figure 3.2 ▲. Notice that changing a subscript in a formula—from H_2O to H_2O_2, for example—changes the identity of the chemical. The substance H_2O_2, hydrogen peroxide, is quite different from the substance H_2O, water. *Subscripts should never be changed in balancing an equation.* In contrast, placing a coefficient in front of a formula changes only the *amount* of the substance and not its *identity*. Thus, $2\ H_2O$ means two molecules of water, $3\ H_2O$ means three molecules of water, and so forth.

ACTIVITY
Reading a Chemical Equation

To illustrate the process of balancing equations, consider the reaction that occurs when methane (CH_4), the principal component of natural gas, burns in air to produce carbon dioxide gas (CO_2) and water vapor (H_2O) (Figure 3.3 ▼). Both of these products contain oxygen atoms that come from O_2 in the air. Thus, O_2 is a reactant, and the unbalanced equation is

$$CH_4 + O_2 \longrightarrow CO_2 + H_2O \quad \text{(unbalanced)} \qquad [3.2]$$

It is usually best to balance first those elements that occur in the fewest chemical formulas on each side of the equation. In our example both C and H

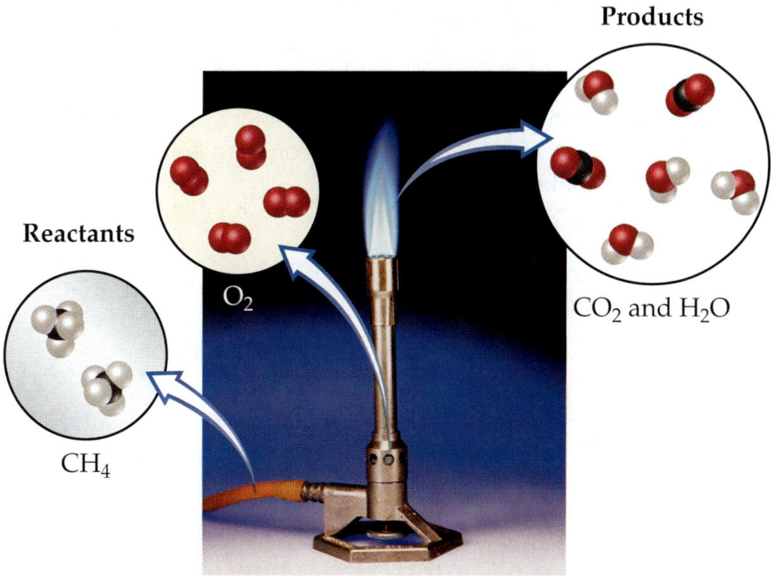

Figure 3.3 Methane reacts with oxygen to produce the flame in a Bunsen burner. The methane (CH_4) in natural gas and oxygen (O_2) from the air are the reactants in the reaction, while carbon dioxide (CO_2) and water (H_2O) are the products.

appear in only one reactant and, separately, in one product each, so we begin by focusing on CH_4. Let's consider first carbon and then hydrogen.

One molecule of the reactant CH_4 contains the same number of C atoms (one) as one molecule of the product CO_2. The coefficients for these substances *must* be the same, therefore, and we choose them both to be 1 as we start the balancing process. However, one molecule of CH_4 contains more H atoms (four) than one molecule of the product H_2O (two). If we place a coefficient 2 in front of H_2O, there will be four H atoms on each side of the equation:

ACTIVITY
Reading a Balanced Chemical Equation

$$CH_4 + O_2 \longrightarrow CO_2 + 2\,H_2O \qquad \text{(unbalanced)} \qquad [3.3]$$

At this stage the products have more O atoms (four—two from CO_2 and two from 2 H_2O) than the reactants (two). If we place a coefficient 2 in front of the reactant O_2, we complete the balancing by making the number of O atoms equal on both sides of the equation:

$$CH_4 + 2\,O_2 \longrightarrow CO_2 + 2\,H_2O \qquad \text{(balanced)} \qquad [3.4]$$

The molecular view of the balanced equation is shown in Figure 3.4 ▼. We see that there are one C, four H, and four O atoms on both sides of the arrow, indicating the equation is balanced.

▶ **Figure 3.4 Balanced chemical equation for the combustion of CH_4.** The drawings of the molecules involved call attention to the conservation of atoms through the reaction.

ACTIVITY
Counting Atoms

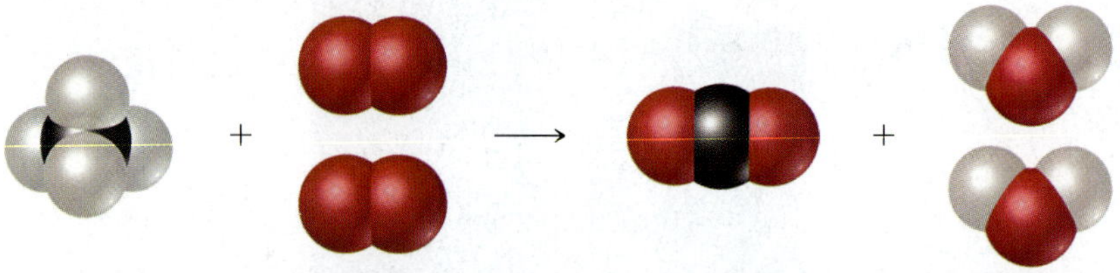

$$CH_4 \quad + \quad 2\,O_2 \quad \longrightarrow \quad CO_2 \quad + \quad 2\,H_2O$$

$$\binom{1\,C}{4\,H} \qquad (4\,O) \qquad \binom{1\,C}{2\,O} \qquad \binom{2\,O}{4\,H}$$

The approach we have taken in arriving at balanced Equation 3.4 is largely trial and error. We balance each kind of atom in succession, adjusting coefficients as necessary. This approach works for most chemical equations.

SAMPLE EXERCISE 3.1 | Interpreting and Balancing Chemical Equations

The following diagram represents a chemical reaction in which the red spheres are oxygen atoms and the blue spheres are nitrogen atoms. **(a)** Write the chemical formulas for the reactants and products. **(b)** Write a balanced equation for the reaction. **(c)** Is the diagram consistent with the law of conservation of mass?

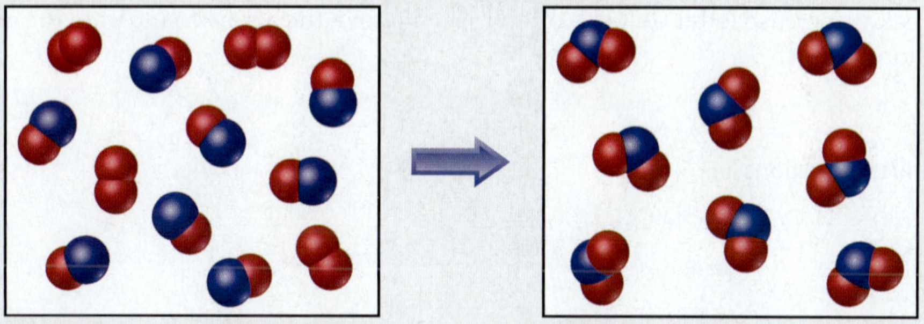

Solution (a) The left box, which represents the reactants, contains two kinds of molecules, those composed of two oxygen atoms (O_2) and those composed of one nitrogen atom and one oxygen atom (NO). The right box, which represents the products, contains only molecules composed of one nitrogen atom and two oxygen atoms (NO_2).

(b) The unbalanced chemical equation is

$$O_2 + NO \longrightarrow NO_2 \qquad \text{(unbalanced)}$$

In this equation there are three O atoms on the left side of the arrow and two O atoms on the right side. We can increase the number of O atoms by placing a coefficient 2 on the product side:

$$O_2 + NO \longrightarrow 2\,NO_2 \qquad \text{(unbalanced)}$$

Now there are two N atoms and four O atoms on the right. Placing a coefficient 2 in front of NO brings both the N atoms and O atoms into balance:

$$O_2 + 2\,NO \longrightarrow 2\,NO_2 \qquad \text{(balanced)}$$

(c) The left box (reactants) contains four O_2 molecules and eight NO molecules. Thus, the molecular ratio is one O_2 for each two NO as required by the balanced equation. The right box (products) contains eight NO_2 molecules. The number of NO_2 molecules on the right equals the number of NO molecules on the left as the balanced equation requires. Counting the atoms, we find eight N atoms in the eight NO molecules in the box on the left. There are also $4 \times 2 = 8$ O atoms in the O_2 molecules and eight O atoms in the NO molecules, giving a total of 16 O atoms. In the box on the right, we find eight N atoms and $8 \times 2 = 16$ O atoms in the eight NO_2 molecules. Because there are equal numbers of both N and O atoms in the two boxes, the drawing is consistent with the law of conservation of mass.

ACTIVITY
Balancing Equations

PRACTICE EXERCISE

In order to be consistent with the law of conservation of mass, how many NH_3 molecules should be shown in the right box of the following diagram?

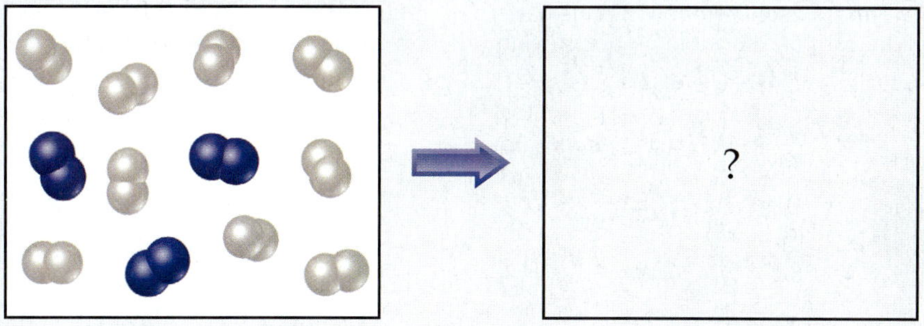

Answer: Six NH_3 molecules

Indicating the States of Reactants and Products

Additional information is often added to the formulas in balanced equations to indicate the physical state of each reactant and product. We use the symbols (*g*), (*l*), (*s*), and (*aq*) for gas, liquid, solid, and aqueous (water) solution, respectively. Thus, Equation 3.4 can be written

$$CH_4(g) + 2\,O_2(g) \longrightarrow CO_2(g) + 2\,H_2O(g) \qquad [3.5]$$

Sometimes the conditions (such as temperature or pressure) under which the reaction proceeds appear above or below the reaction arrow. The symbol Δ (the Greek uppercase letter delta) is often placed above the arrow to indicate the addition of heat.

MOVIE
Sodium and Potassium in Water

SAMPLE EXERCISE 3.2 | Balancing Chemical Equations

Balance this equation:

$$Na(s) + H_2O(l) \longrightarrow NaOH(aq) + H_2(g)$$

Solution We begin by counting the atoms of each kind on both sides of the arrow. The Na and O atoms are balanced (one Na and one O on each side), but there are two H atoms on the left and three H atoms on the right. Thus, we need to increase the number of H atoms on the left. As a trial beginning in our effort to balance H, let's place a coefficient 2 in front of H_2O:

$$Na(s) + 2\,H_2O(l) \longrightarrow NaOH(aq) + H_2(g)$$

Beginning this way doesn't balance H, but introducing the coefficient 2 does increase the number of H atoms among the reactants, which we need to do. The fact that it causes O to be unbalanced is something we will take care of after we balance H. Now that we have 2 H_2O on the left, we can balance H by putting a coefficient 2 in front of NaOH on the right:

$$Na(s) + 2\,H_2O(l) \longrightarrow 2\,NaOH(aq) + H_2(g)$$

Balancing H in this way fortuitously brings O into balance, but notice that Na is now unbalanced, with one on the left but two on the right. To rebalance Na, we put a coefficient 2 in front of the reactant:

$$2\,Na(s) + 2\,H_2O(l) \longrightarrow 2\,NaOH(aq) + H_2(g)$$

Finally, we check the number of atoms of each element and find that we have two Na atoms, four H atoms, and two O atoms on each side of the equation. The equation is balanced.

Comment: Notice that in balancing this equation, we moved back and forth placing a coefficient in front of H_2O, then NaOH, and finally Na. In balancing equations, we often find ourselves following this pattern of moving back and forth from one side of the arrow to the other, placing coefficients first in front of a formula on one side and then in front of a formula on the other side until the equation is balanced.

PRACTICE EXERCISE

Balance the following equations by providing the missing coefficients:

(a) $\_\,Fe(s) + \_\,O_2(g) \longrightarrow \_\,Fe_2O_3(s)$

(b) $\_\,C_2H_4(g) + \_\,O_2(g) \longrightarrow \_\,CO_2(g) + \_\,H_2O(g)$

(c) $\_\,Al(s) + \_\,HCl(aq) \longrightarrow \_\,AlCl_3(aq) + \_\,H_2(g)$

Answers: **(a)** 4, 3, 2; **(b)** 1, 3, 2, 2; **(c)** 2, 6, 2, 3

3.2 | Some Simple Patterns of Chemical Reactivity

In this section we examine three simple kinds of reactions that we will see frequently throughout this chapter. Our first reason for examining these reactions is merely to become better acquainted with chemical reactions and their balanced equations. Our second reason is to consider how we might predict the products of some of these reactions knowing only their reactants. The key to predicting the products formed by a given combination of reactants is recognizing general patterns of chemical reactivity. Recognizing a pattern of reactivity for a class of substances gives you a broader understanding than merely memorizing a large number of unrelated reactions.

Combination and Decomposition Reactions

Table 3.1 ▼ summarizes two simple types of reactions, combination and decomposition reactions. In **combination reactions** two or more substances react to form one product. There are many examples of such reactions, especially those in which elements combine to form compounds. For example, magnesium metal burns in air with a dazzling brilliance to produce magnesium oxide, as shown in Figure 3.5 ▶:

$$2\,Mg(s) + O_2(g) \longrightarrow 2\,MgO(s) \quad [3.6]$$

This reaction is used to produce the bright flame generated by flares.

When a combination reaction occurs between a metal and a nonmetal, as in Equation 3.6, the product is an ionic solid. Recall that the formula of an ionic compound can be determined from the charges of the ions involved ∞ (Section 2.7). When magnesium reacts with oxygen, for example, the magnesium loses electrons and forms the magnesium ion, Mg^{2+}. The oxygen gains electrons and forms the oxide ion, O^{2-}. Thus, the reaction product is MgO. You should be able to recognize when a reaction is a combination reaction and to predict the products of a combination reaction in which the reactants are a metal and a nonmetal.

GIVE IT SOME THOUGHT

When Na and S combine in a combination reaction, what is the chemical formula of the product?

TABLE 3.1 Combination and Decomposition Reactions

Combination Reactions	
$A + B \longrightarrow C$ $C(s) + O_2(g) \longrightarrow CO_2(g)$ $N_2(g) + 3\,H_2(g) \longrightarrow 2\,NH_3(g)$ $CaO(s) + H_2O(l) \longrightarrow Ca(OH)_2(s)$	Two reactants combine to form a single product. Many elements react with one another in this fashion to form compounds.
Decomposition Reactions	
$C \longrightarrow A + B$ $2\,KClO_3(s) \longrightarrow 2\,KCl(s) + 3\,O_2(g)$ $PbCO_3(s) \longrightarrow PbO(s) + CO_2(g)$ $Cu(OH)_2(s) \longrightarrow CuO(s) + H_2O(l)$	A single reactant breaks apart to form two or more substances. Many compounds react this way when heated.

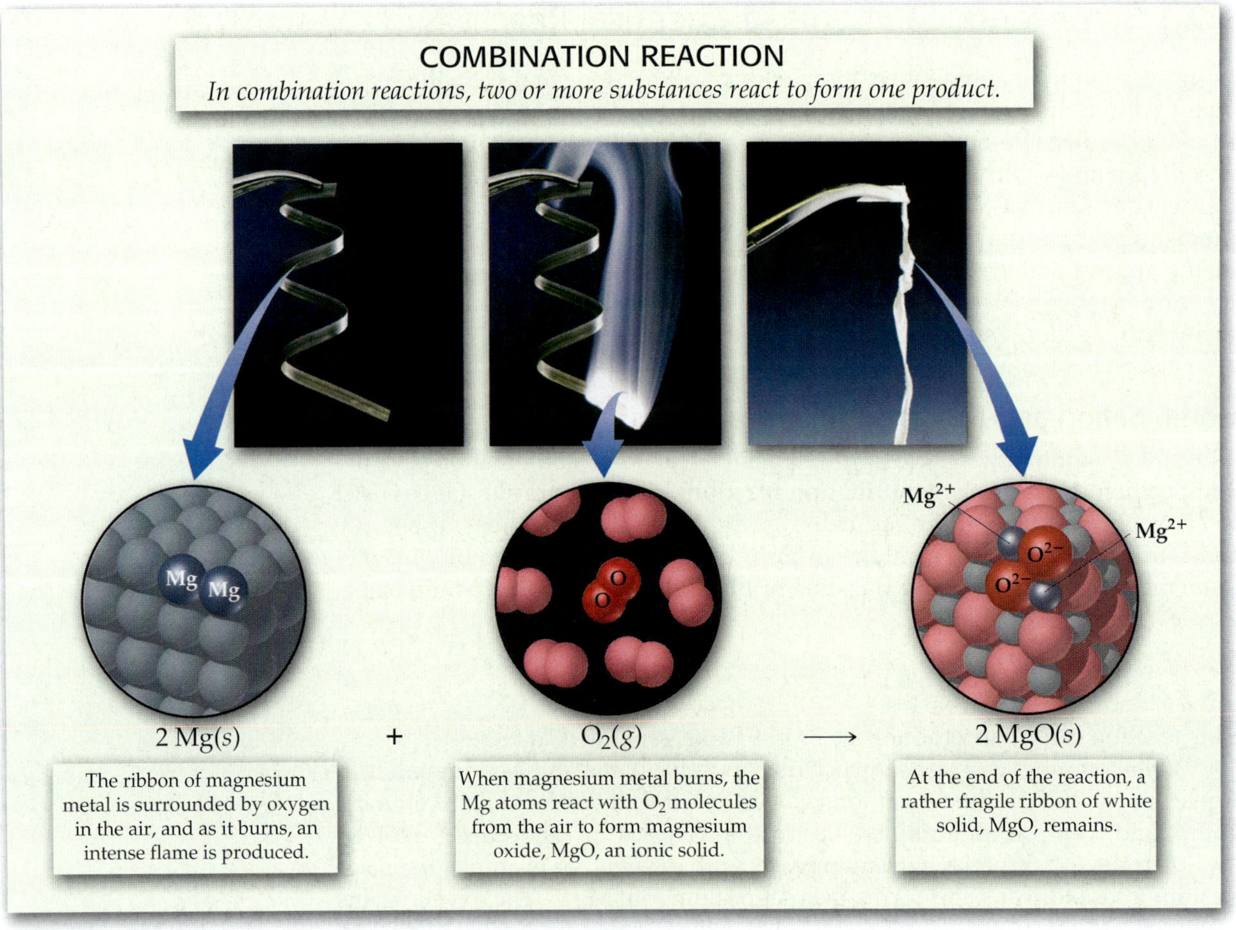

▲ **Figure 3.5 Combustion of magnesium metal in air.**

MOVIE
Reactions with Oxygen

ANIMATION
Air Bags

▲ **Figure 3.6 An automobile air bag.** The decomposition of sodium azide, $NaN_3(s)$, is used to inflate automobile air bags. When properly ignited, the NaN_3 decomposes rapidly, forming nitrogen gas, $N_2(g)$, which expands the air bag.

In a **decomposition reaction** one substance undergoes a reaction to produce two or more other substances. Many compounds undergo decomposition reactions when heated. For example, many metal carbonates decompose to form metal oxides and carbon dioxide when heated:

$$CaCO_3(s) \longrightarrow CaO(s) + CO_2(g) \qquad [3.7]$$

The decomposition of $CaCO_3$ is an important commercial process. Limestone or seashells, which are both primarily $CaCO_3$, are heated to prepare CaO, which is known as lime or quicklime. About 2×10^{10} kg (20 million tons) of CaO is used in the United States each year, principally in making glass, in obtaining iron from its ores, and in making mortar to bind bricks.

The decomposition of sodium azide (NaN_3) rapidly releases $N_2(g)$, so this reaction is used to inflate safety air bags in automobiles (Figure 3.6 ◀):

$$2\,NaN_3(s) \longrightarrow 2\,Na(s) + 3\,N_2(g) \qquad [3.8]$$

The system is designed so that an impact ignites a detonator cap, which in turn causes NaN_3 to decompose explosively. A small quantity of NaN_3 (about 100 g) forms a large quantity of gas (about 50 L). We will consider the volumes of gases produced in chemical reactions in Section 10.5.

SAMPLE EXERCISE 3.3 | Writing Balanced Equations for Combination and Decomposition Reactions

Write balanced equations for the following reactions: **(a)** The combination reaction that occurs when lithium metal and fluorine gas react. **(b)** The decomposition reaction that occurs when solid barium carbonate is heated. (Two products form: a solid and a gas.)

Solution (a) The symbol for lithium is Li. With the exception of mercury, all metals are solids at room temperature. Fluorine occurs as a diatomic molecule (see Figure 2.19). Thus, the reactants are Li(s) and $F_2(g)$. The product will consist of a metal and a nonmetal, so we expect it to be an ionic solid. Lithium ions have a 1+ charge, Li^+, whereas fluoride ions have a 1− charge, F^-. Thus, the chemical formula for the product is LiF. The balanced chemical equation is

$$2\,Li(s) + F_2(g) \longrightarrow 2\,LiF(s)$$

(b) The chemical formula for barium carbonate is $BaCO_3$. As noted in the text, many metal carbonates decompose to form metal oxides and carbon dioxide when heated. In Equation 3.7, for example, $CaCO_3$ decomposes to form CaO and CO_2. Thus, we would expect that $BaCO_3$ decomposes to form BaO and CO_2. Barium and calcium are both in group 2A in the periodic table, moreover, which further suggests they would react in the same way:

$$BaCO_3(s) \longrightarrow BaO(s) + CO_2(g)$$

PRACTICE EXERCISE

Write balanced chemical equations for the following reactions: **(a)** Solid mercury(II) sulfide decomposes into its component elements when heated. **(b)** The surface of aluminum metal undergoes a combination reaction with oxygen in the air.
Answers: **(a)** $HgS(s) \longrightarrow Hg(l) + S(s)$; **(b)** $4\,Al(s) + 3\,O_2(g) \longrightarrow 2\,Al_2O_3(s)$

Combustion in Air

Combustion reactions are rapid reactions that produce a flame. Most of the combustion reactions we observe involve O_2 from air as a reactant. Equation 3.5 and Practice Exercise 3.1(b) illustrate a general class of reactions involving the burning, or combustion, of hydrocarbon compounds (compounds that contain only carbon and hydrogen, such as CH_4 and C_2H_4). (Section 2.9)

When hydrocarbons are combusted in air, they react with O_2 to form CO_2 and H_2O.* The number of molecules of O_2 required in the reaction and the number of molecules of CO_2 and H_2O formed depend on the composition of the hydrocarbon, which acts as the fuel in the reaction. For example, the combustion of propane (C_3H_8), a gas used for cooking and home heating, is described by the following equation:

$$C_3H_8(g) + 5\,O_2(g) \longrightarrow 3\,CO_2(g) + 4\,H_2O(g) \qquad [3.9]$$

The state of the water, $H_2O(g)$ or $H_2O(l)$, depends on the conditions of the reaction. Water vapor, $H_2O(g)$, is formed at high temperature in an open container. The blue flame produced when propane burns is shown in Figure 3.7 ▶.

Combustion of oxygen-containing derivatives of hydrocarbons, such as CH_3OH, also produces CO_2 and H_2O. The simple rule that hydrocarbons and related oxygen-containing derivatives of hydrocarbons form CO_2 and H_2O when they burn in air summarizes the behavior of about 3 million compounds. Many substances that our bodies use as energy sources, such as the sugar glucose ($C_6H_{12}O_6$), similarly react in our bodies with O_2 to form CO_2 and H_2O.

▲ **Figure 3.7 Propane burning in air.** The liquid propane, C_3H_8, vaporizes and mixes with air as it escapes through the nozzle. The combustion reaction of C_3H_8 and O_2 produces a blue flame.

* When there is an insufficient quantity of O_2 present, carbon monoxide (CO) will be produced along with the CO_2; this is called *incomplete* combustion. If the amount of O_2 is severely restricted, fine particles of carbon that we call soot will be produced. *Complete* combustion produces only CO_2 and H_2O. Unless specifically stated to the contrary, we will always take *combustion* to mean *complete combustion*.

In our bodies, however, the reactions take place in a series of steps that occur at body temperature. The reactions are then described as *oxidation reactions* rather than combustion reactions.

SAMPLE EXERCISE 3.4 | Writing Balanced Equations for Combustion Reactions

Write the balanced equation for the reaction that occurs when methanol, $CH_3OH(l)$, is burned in air.

Solution When any compound containing C, H, and O is combusted, it reacts with the $O_2(g)$ in air to produce $CO_2(g)$ and $H_2O(g)$. Thus, the unbalanced equation is

$$CH_3OH(l) + O_2(g) \longrightarrow CO_2(g) + H_2O(g)$$

The C atoms are balanced with one on each side of the arrow. Because CH_3OH has four H atoms, we place a coefficient 2 in front of H_2O to balance the H atoms:

$$CH_3OH(l) + O_2(g) \longrightarrow CO_2(g) + 2\,H_2O(g)$$

This balances H but gives four O atoms in the products. Because there are only three O atoms in the reactants (one in CH_3OH and two in O_2), we are not finished yet. We can place the fractional coefficient $\frac{3}{2}$ in front of O_2 to give a total of four O atoms in the reactants (there are $\frac{3}{2} \times 2 = 3$ O atoms in $\frac{3}{2}O_2$):

$$CH_3OH(l) + \tfrac{3}{2}O_2(g) \longrightarrow CO_2(g) + 2\,H_2O(g)$$

Although the equation is now balanced, it is not in its most conventional form because it contains a fractional coefficient. If we multiply each side of the equation by 2, we will remove the fraction and achieve the following balanced equation:

$$2\,CH_3OH(l) + 3\,O_2(g) \longrightarrow 2\,CO_2(g) + 4\,H_2O(g)$$

PRACTICE EXERCISE

Write the balanced equation for the reaction that occurs when ethanol, $C_2H_5OH(l)$, is burned in air.
Answer: $C_2H_5OH(l) + 3\,O_2(g) \longrightarrow 2\,CO_2(g) + 3\,H_2O(g)$

3.3 | Formula Weights

Chemical formulas and chemical equations both have a *quantitative* significance; the subscripts in formulas and the coefficients in equations represent precise quantities. The formula H_2O indicates that a molecule of this substance contains exactly two atoms of hydrogen and one atom of oxygen. Similarly, the coefficients in a balanced chemical equation indicate the relative quantities of reactants and products. But how do we relate the numbers of atoms or molecules to the amounts we measure out in the laboratory? Although we cannot directly count atoms or molecules, we can indirectly determine their numbers if we know their masses. Therefore, before we can pursue the quantitative aspects of chemical formulas or equations, we must examine the masses of atoms and molecules, which we do in this section and the next.

Formula and Molecular Weights

The **formula weight** of a substance is the sum of the atomic weights of each atom in its chemical formula. Using atomic masses from a periodic table, we find, for example, that the formula weight of sulfuric acid (H_2SO_4) is 98.1 amu:*

* The abbreviation AW is used for atomic weight, FW for formula weight, and MW for molecular weight.

$$\begin{aligned}\text{FW of } H_2SO_4 &= 2(\text{AW of H}) + (\text{AW of S}) + 4(\text{AW of O})\\ &= 2(1.0 \text{ amu}) + 32.1 \text{ amu} + 4(16.0 \text{ amu})\\ &= 98.1 \text{ amu}\end{aligned}$$

For convenience, we have rounded off all the atomic weights to one place beyond the decimal point. We will round off the atomic weights in this way for most problems.

If the chemical formula is merely the chemical symbol of an element, such as Na, then the formula weight equals the atomic weight of the element. If the chemical formula is that of a molecule, then the formula weight is also called the **molecular weight**. The molecular weight of glucose ($C_6H_{12}O_6$), for example, is

$$\text{MW of } C_6H_{12}O_6 = 6(12.0 \text{ amu}) + 12(1.0 \text{ amu}) + 6(16.0 \text{ amu}) = 180.0 \text{ amu}$$

Because ionic substances, such as NaCl, exist as three-dimensional arrays of ions (Figure 2.23), it is inappropriate to speak of molecules of NaCl. Instead, we speak of *formula units*, represented by the chemical formula of the substance. The formula unit of NaCl consists of one Na^+ ion and one Cl^- ion. Thus, the formula weight of NaCl is the mass of one formula unit:

$$\text{FW of NaCl} = 23.0 \text{ amu} + 35.5 \text{ amu} = 58.5 \text{ amu}$$

SAMPLE EXERCISE 3.5 | Calculating Formula Weights

Calculate the formula weight of **(a)** sucrose, $C_{12}H_{22}O_{11}$ (table sugar), and **(b)** calcium nitrate, $Ca(NO_3)_2$.

Solution **(a)** By adding the atomic weights of the atoms in sucrose, we find it to have a formula weight of 342.0 amu:

$$\begin{aligned}12 \text{ C atoms} &= 12(12.0 \text{ amu}) = 44.0 \text{ amu}\\ 22 \text{ H atoms} &= 22(1.0 \text{ amu}) = 22.0 \text{ amu}\\ 11 \text{ O atoms} &= 11(16.0 \text{ amu}) = \underline{176.0 \text{ amu}}\\ & \quad\quad\quad\quad\quad\quad\; 342.0 \text{ amu}\end{aligned}$$

(b) If a chemical formula has parentheses, the subscript outside the parentheses is a multiplier for all atoms inside. Thus, for $Ca(NO_3)_2$, we have

$$\begin{aligned}1 \text{ Ca atom} &= 1(40.1 \text{ amu}) = 40.1 \text{ amu}\\ 2 \text{ N atoms} &= 2(14.0 \text{ amu}) = 28.0 \text{ amu}\\ 6 \text{ O atoms} &= 6(16.0 \text{ amu}) = \underline{96.0 \text{ amu}}\\ & \quad\quad\quad\quad\quad\quad\; 164.1 \text{ amu}\end{aligned}$$

PRACTICE EXERCISE

Calculate the formula weight of **(a)** $Al(OH)_3$ and **(b)** CH_3OH.
Answers: **(a)** 78.0 amu, **(b)** 32.0 amu

Percentage Composition from Formulas

Occasionally we must calculate the *percentage composition* of a compound (that is, the percentage by mass contributed by each element in the substance). For example, in order to verify the purity of a compound, we may wish to compare the calculated percentage composition of the substance with that found experimentally. Calculating percentage composition is a straightforward matter if the chemical formula is known. The calculation depends on the formula weight of the substance, the atomic weight of the element of interest, and the number of atoms of that element in the chemical formula:

ACTIVITY
Molecular Weight and Weight Percent

$$\%\ \text{element} = \frac{(\text{number of atoms of that element})(\text{atomic weight of element})}{\text{formula weight of compound}} \times 100\% \qquad [3.10]$$

SAMPLE EXERCISE 3.6 | Calculating Percentage Composition

Calculate the percentage of carbon, hydrogen, and oxygen (by mass) in $C_{12}H_{22}O_{11}$.

Solution Let's examine this question using the problem-solving steps in the "Strategies in Chemistry: Problem Solving" essay.

Analyze: We are given a chemical formula, $C_{12}H_{22}O_{11}$, and asked to calculate the percentage by mass of its component elements (C, H, and O).

Plan: We can use Equation 3.10, relying on a periodic table to obtain the atomic weight of each component element. The atomic weights are first used to determine the formula weight of the compound. (The formula weight of $C_{12}H_{22}O_{11}$, 342.0 amu, was calculated in Sample Exercise 3.5.) We must then do three calculations, one for each element.

Solve: Using Equation 3.10, we have

$$\%C = \frac{(12)(12.0\ \text{amu})}{342.0\ \text{amu}} \times 100\% = 42.1\%$$

$$\%H = \frac{(22)(1.0\ \text{amu})}{342.0\ \text{amu}} \times 100\% = 6.4\%$$

$$\%O = \frac{(11)(16.0\ \text{amu})}{342.0\ \text{amu}} \times 100\% = 51.5\%$$

Check: The percentages of the individual elements must add up to 100%, which they do in this case. We could have used more significant figures for our atomic weights, giving more significant figures for our percentage composition, but we have adhered to our suggested guideline of rounding atomic weights to one digit beyond the decimal point.

PRACTICE EXERCISE

Calculate the percentage of nitrogen, by mass, in $Ca(NO_3)_2$.
Answer: 17.1%

3.4 | Avogadro's Number and the Mole

Even the smallest samples that we deal with in the laboratory contain enormous numbers of atoms, ions, or molecules. For example, a teaspoon of water (about 5 mL) contains 2×10^{23} water molecules, a number so large that it almost defies comprehension. Chemists, therefore, have devised a special counting unit for describing such large numbers of atoms or molecules.

STRATEGIES IN CHEMISTRY | Problem Solving

The key to success in problem solving is practice. As you practice, you will find that you can improve your skills by following these steps:

Step 1: Analyze the problem. Read the problem carefully for understanding. What does it say? Draw any picture or diagram that will help you visualize the problem. Write down both the data you are given and the quantity that you need to obtain (the unknown).

Step 2: Develop a plan for solving the problem. Consider the possible paths between the given information and the unknown. What principles or equations relate the known data to the unknown? Recognize that some data may not be given explicitly in the problem; you may be expected to know certain quantities (such as Avogadro's number, which we will soon discuss) or look them up in tables (such as atomic weights). Recognize also that your plan may involve either a single step or a series of steps with intermediate answers.

Step 3: Solve the problem. Use the known information and suitable equations or relationships to solve for the unknown. Dimensional analysis (Section 1.6) is a very useful tool for solving a great number of problems. Be careful with significant figures, signs, and units.

Step 4: Check the solution. Read the problem again to make sure you have found all the solutions asked for in the problem. Does your answer make sense? That is, is the answer outrageously large or small, or is it in the ballpark? Finally, are the units and significant figures correct?

In everyday life we use counting units like dozen (12 objects) and gross (144 objects) to deal with modestly large quantities. In chemistry the unit for dealing with the number of atoms, ions, or molecules in a common-sized sample is the **mole**, abbreviated mol.* A mole is the amount of matter that contains as many objects (atoms, molecules, or whatever objects we are considering) as the number of atoms in exactly 12 g of isotopically pure ^{12}C. From experiments, scientists have determined this number to be 6.0221421×10^{23}. Scientists call this number **Avogadro's number**, in honor of Amedeo Avogadro (1776–1856), an Italian scientist. For most purposes we will use 6.02×10^{23} or 6.022×10^{23} for Avogadro's number throughout the text.

A mole of atoms, a mole of molecules, or a mole of anything else all contain Avogadro's number of these objects:

$$1 \text{ mol } ^{12}\text{C atoms} = 6.02 \times 10^{23} \, ^{12}\text{C atoms}$$

$$1 \text{ mol } H_2O \text{ molecules} = 6.02 \times 10^{23} \, H_2O \text{ molecules}$$

$$1 \text{ mol } NO_3^- \text{ ions} = 6.02 \times 10^{23} \, NO_3^- \text{ ions}$$

Avogadro's number is so large that it is difficult to imagine. Spreading 6.02×10^{23} marbles over the entire surface of Earth would produce a layer about 3 mi thick. If Avogadro's number of pennies were placed side by side in a straight line, they would encircle Earth 300 trillion (3×10^{14}) times.

SAMPLE EXERCISE 3.7 | Estimating Numbers of Atoms

Without using a calculator, arrange the following samples in order of increasing numbers of carbon atoms: 12 g ^{12}C, 1 mol C_2H_2, 9×10^{23} molecules of CO_2.

Solution

Analyze: We are given amounts of different substances expressed in grams, moles, and number of molecules and asked to arrange the samples in order of increasing numbers of C atoms.

Plan: To determine the number of C atoms in each sample, we must convert g ^{12}C, mol C_2H_2, and molecules CO_2 all to numbers of C atoms. To do this converting, we use the definition of mole and Avogadro's number.

Solve: A mole is defined as the amount of matter that contains as many units of the matter as there are C atoms in exactly 12 g of ^{12}C. Thus, 12 g of ^{12}C contains 1 mol of C atoms (that is, 6.02×10^{23} C atoms). In 1 mol C_2H_2, there are 6×10^{23} C_2H_2 molecules. Because there are two C atoms in each C_2H_2 molecule, this sample contains 12×10^{23} C atoms. Because each CO_2 molecule contains one C atom, the sample of CO_2 contains 9×10^{23} C atoms. Hence, the order is 12 g ^{12}C (6×10^{23} C atoms) $<$ 9×10^{23} CO_2 molecules (9×10^{23} C atoms) $<$ 1 mol C_2H_2 (12×10^{23} C atoms).

Check: We can check our results by comparing the number of moles of C atoms in each sample because the number of moles is proportional to the number of atoms. Thus, 12 g of ^{12}C is 1 mol C; 1 mol of C_2H_2 contains 2 mol C, and 9×10^{23} molecules of CO_2 contain 1.5 mol C, giving the same order as above: 12 g ^{12}C (1 mol C) $<$ 9×10^{23} CO_2 molecules (1.5 mol C) $<$ 1 mol C_2H_2 (2 mol C).

PRACTICE EXERCISE

Without using a calculator, arrange the following samples in order of increasing number of O atoms: 1 mol H_2O, 1 mol CO_2, 3×10^{23} molecules O_3.

Answer: 1 mol H_2O (6×10^{23} O atoms) $<$ 3×10^{23} molecules O_3 (9×10^{23} O atoms) $<$ 1 mol CO_2 (12×10^{23} O atoms)

* The term *mole* comes from the Latin word *moles*, meaning "a mass." The term *molecule* is the diminutive form of this word and means "a small mass."

SAMPLE EXERCISE 3.8 | Converting Moles to Number of Atoms

Calculate the number of H atoms in 0.350 mol of $C_6H_{12}O_6$.

Solution

Analyze: We are given both the amount of a substance (0.350 mol) and its chemical formula ($C_6H_{12}O_6$). The unknown is the number of H atoms in the sample.

Plan: Avogadro's number provides the conversion factor between the number of moles of $C_6H_{12}O_6$ and the number of molecules of $C_6H_{12}O_6$. Once we know the number of molecules of $C_6H_{12}O_6$, we can use the chemical formula, which tells us that each molecule of $C_6H_{12}O_6$ contains 12 H atoms. Thus, we convert moles of $C_6H_{12}O_6$ to molecules of $C_6H_{12}O_6$ and then determine the number of atoms of H from the number of molecules of $C_6H_{12}O_6$:

$$\text{Moles } C_6H_{12}O_6 \longrightarrow \text{molecules } C_6H_{12}O_6 \longrightarrow \text{atoms H}$$

Solve:

$$\text{H atoms} = (0.350 \text{ mol } C_6H_{12}O_6)\left(\frac{6.02 \times 10^{23} \text{ molecules } C_6H_{12}O_6}{1 \text{ mol } C_6H_{12}O_6}\right)\left(\frac{12 \text{ H atoms}}{1 \text{ molecule } C_6H_{12}O_6}\right)$$

$$= 2.53 \times 10^{24} \text{ H atoms}$$

Check: The magnitude of our answer is reasonable: It is a large number about the magnitude of Avogadro's number. We can also make the following ballpark calculation: Multiplying $0.35 \times 6 \times 10^{23}$ gives about 2×10^{23} molecules. Multiplying this result by 12 gives $24 \times 10^{23} = 2.4 \times 10^{24}$ H atoms, which agrees with the previous, more detailed calculation. Because we were asked for the number of H atoms, the units of our answer are correct. The given data had three significant figures, so our answer has three significant figures.

PRACTICE EXERCISE

How many oxygen atoms are in **(a)** 0.25 mol $Ca(NO_3)_2$ and **(b)** 1.50 mol of sodium carbonate?

Answers: **(a)** 9.0×10^{23}, **(b)** 2.71×10^{24}

Molar Mass

A dozen is the same number (12) whether we have a dozen eggs or a dozen elephants. Clearly, however, a dozen eggs does not have the same mass as a dozen elephants. Similarly, a mole is always the *same number* (6.02×10^{23}), but 1-mole samples of different substances will have *different masses.* Compare, for example, 1 mol of ^{12}C and 1 mol of ^{24}Mg. A single ^{12}C atom has a mass of 12 amu, whereas a single ^{24}Mg atom is twice as massive, 24 amu (to two significant figures). Because a mole always has the same number of particles, a mole of ^{24}Mg must be twice as massive as a mole of ^{12}C. Because a mole of ^{12}C has a mass of 12 g (by definition), then a mole of ^{24}Mg must have a mass of 24 g. This example illustrates a general rule relating the mass of an atom to the mass of Avogadro's number (1 mol) of these atoms: *The mass of a single atom of an element (in amu) is numerically equal to the mass (in grams) of 1 mol of that element.* This statement is true regardless of the element:

$$1 \text{ atom of } ^{12}C \text{ has a mass of 12 amu} \Rightarrow 1 \text{ mol } ^{12}C \text{ has a mass of 12 g}$$

$$1 \text{ atom of Cl has an atomic weight of 35.5 amu} \Rightarrow 1 \text{ mol Cl has a mass of 35.5 g}$$

$$1 \text{ atom of Au has an atomic weight of 197 amu} \Rightarrow 1 \text{ mol Au has a mass of 197 g}$$

Notice that when we are dealing with a particular isotope of an element, we use the mass of that isotope; otherwise we use the atomic weight (the average atomic mass) of the element.

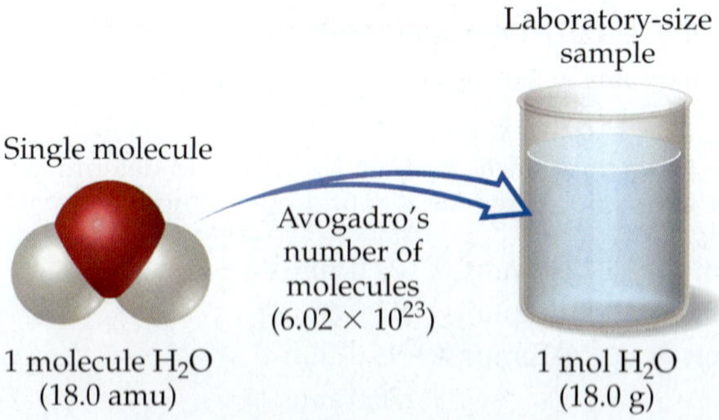

▲ **Figure 3.8 Comparing the mass of 1 molecule H_2O and 1 mol H_2O.** Notice that the masses are numerically equal but have different units (18.0 amu compared to 18.0 g) representing the huge difference in mass.

For other kinds of substances, the same numerical relationship exists between the formula weight (in amu) and the mass (in grams) of one mole of that substance:

1 H_2O molecule has a mass of 18.0 amu ⇒ 1 mol H_2O has a mass of 18.0 g

1 NO_3^- ion has a mass of 62.0 amu ⇒ 1 mol NO_3^- has a mass of 62.0 g

1 NaCl unit has a mass of 58.5 amu ⇒ 1 mol NaCl has a mass of 58.5 g

Figure 3.8 ▲ illustrates the relationship between the mass of a single molecule of H_2O and that of a mole of H_2O.

The mass in grams of one mole of a substance (that is, the mass in grams per mol) is called the **molar mass** of the substance. *The molar mass (in g/mol) of any substance is always numerically equal to its formula weight (in amu).* The substance NaCl, for example, has formula weight of 58.5 amu and a molar mass of 58.5 g/mol. Further examples of mole relationships are shown in Table 3.2 ▼. Figure 3.9 ▶ shows 1-mole quantities of several common substances.

The entries in Table 3.2 for N and N_2 point out the importance of stating the chemical form of a substance exactly when we use the mole concept. Suppose you read that 1 mol of nitrogen is produced in a particular reaction. You might interpret this statement to mean 1 mol of nitrogen atoms (14.0 g). Unless otherwise stated, however, what is probably meant is 1 mol of nitrogen molecules, N_2 (28.0 g), because N_2 is the usual chemical form of the element. To avoid ambiguity, it is important to state explicitly the chemical form being discussed. Using the chemical formula N_2 avoids ambiguity.

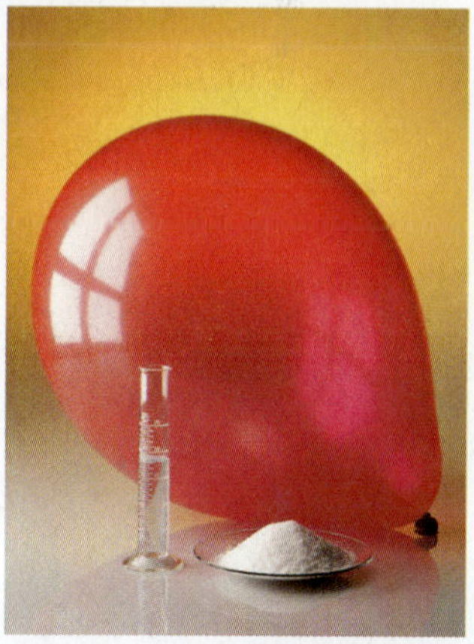

▲ **Figure 3.9 One mole each of a solid, a liquid, and a gas.** One mole of NaCl, the solid, has a mass of 58.45 g. One mole of H_2O, the liquid, has a mass of 18.0 g and occupies a volume of 18.0 mL. One mole of O_2, the gas, has a mass of 32.0 g and occupies a balloon whose diameter is 35 cm.

TABLE 3.2 Mole Relationships

Name of substance	Formula	Formula Weight (amu)	Molar Mass (g/mol)	Number and Kind of Particles in One Mole
Atomic nitrogen	N	14.0	14.0	6.022×10^{23} N atoms
Molecular nitrogen	N_2	28.0	28.0	6.022×10^{23} N_2 molecules; $2(6.022 \times 10^{23})$ N atoms
Silver	Ag	107.9	107.9	6.022×10^{23} Ag atoms
Silver ions	Ag^+	107.9[a]	107.9	6.022×10^{23} Ag^+ ions
Barium chloride	$BaCl_2$	208.2	208.2	6.022×10^{23} $BaCl_2$ units; 6.022×10^{23} Ba^{2+} ions; $2(6.022 \times 10^{23})$ Cl^- ions

[a]Recall that the electron has negligible mass; thus, ions and atoms have essentially the same mass.

SAMPLE EXERCISE 3.9 | Calculating Molar Mass

What is the mass in grams of 1.000 mol of glucose, $C_6H_{12}O_6$?

Solution

Analyze: We are given a chemical formula and asked to determine its molar mass.

Plan: The molar mass of a substance is found by adding the atomic weights of its component atoms.

Solve:

$$\begin{aligned} 6\text{ C atoms} &= 6(12.0\text{ amu}) = 72.0\text{ amu} \\ 12\text{ H atoms} &= 12(1.0\text{ amu}) = 12.0\text{ amu} \\ 6\text{ O atoms} &= 6(16.0\text{ amu}) = \underline{96.0\text{ amu}} \\ & \qquad\qquad\qquad\quad 180.0\text{ amu} \end{aligned}$$

Because glucose has a formula weight of 180.0 amu, one mole of this substance has a mass of 180.0 g. In other words, $C_6H_{12}O_6$ has a molar mass of 180.0 g/mol.

Check: The magnitude of our answer seems reasonable, and g/mol is the appropriate unit for the molar mass.

Comment: Glucose is sometimes called dextrose. Also known as blood sugar, glucose is found widely in nature, occurring, for example, in honey and fruits. Other types of sugars used as food are converted into glucose in the stomach or liver before they are used by the body as energy sources. Because glucose requires no conversion, it is often given intravenously to patients who need immediate nourishment.

PRACTICE EXERCISE

Calculate the molar mass of $Ca(NO_3)_2$.

Answer: 164.1 g/mol

Interconverting Masses and Moles

Conversions of mass to moles and of moles to mass are frequently encountered in calculations using the mole concept. These calculations are made easy through dimensional analysis, as shown in Sample Exercises 3.10 and 3.11.

SAMPLE EXERCISE 3.10 | Converting Grams to Moles

Calculate the number of moles of glucose ($C_6H_{12}O_6$) in 5.380 g of $C_6H_{12}O_6$.

Solution

Analyze: We are given the number of grams of a substance and its chemical formula and asked to calculate the number of moles.

Plan: The molar mass of a substance provides the factor for converting grams to moles. The molar mass of $C_6H_{12}O_6$ is 180.0 g/mol (Sample Exercise 3.9).

Solve: Using 1 mol $C_6H_{12}O_6$ = 180.0 g $C_6H_{12}O_6$ to write the appropriate conversion factor, we have

$$\text{Moles } C_6H_{12}O_6 = (5.380\ \cancel{\text{g } C_6H_{12}O_6})\left(\frac{1\text{ mol } C_6H_{12}O_6}{180.0\ \cancel{\text{g } C_6H_{12}O_6}}\right) = 0.02989\text{ mol } C_6H_{12}O_6$$

Check: Because 5.380 g is less than the molar mass, it is reasonable that our answer is less than one mole. The units of our answer (mol) are appropriate. The original data had four significant figures, so our answer has four significant figures.

PRACTICE EXERCISE

How many moles of sodium bicarbonate ($NaHCO_3$) are there in 508 g of $NaHCO_3$?

Answer: 6.05 mol $NaHCO_3$

SAMPLE EXERCISE 3.11 | Converting Moles to Grams

Calculate the mass, in grams, of 0.433 mol of calcium nitrate.

Solution

Analyze: We are given the number of moles and name of a substance and asked to calculate the number of grams in the sample.

Plan: In order to convert moles to grams, we need the molar mass, which we can calculate using the chemical formula and atomic weights.

Solve: Because the calcium ion is Ca^{2+} and the nitrate ion is NO_3^-, calcium nitrate is $Ca(NO_3)_2$. Adding the atomic weights of the elements in the compound gives a formula weight of 164.1 amu. Using 1 mol $Ca(NO_3)_2$ = 164.1 g $Ca(NO_3)_2$ to write the appropriate conversion factor, we have

$$\text{Grams } Ca(NO_3)_2 = (0.433 \text{ mol } Ca(NO_3)_2)\left(\frac{164.1 \text{ g } Ca(NO_3)_2}{1 \text{ mol } Ca(NO_3)_2}\right) = 71.1 \text{ g } Ca(NO_3)_2$$

Check: The number of moles is less than 1, so the number of grams must be less than the molar mass, 164.1 g. Using rounded numbers to estimate, we have $0.5 \times 150 = 75$ g. Thus, the magnitude of our answer is reasonable. Both the units (g) and the number of significant figures (3) are correct.

PRACTICE EXERCISE

What is the mass, in grams, of **(a)** 6.33 mol of $NaHCO_3$ and **(b)** 3.0×10^{-5} mol of sulfuric acid?

Answers: **(a)** 532 g, **(b)** 2.9×10^{-3} g

Interconverting Masses and Numbers of Particles

The mole concept provides the bridge between masses and numbers of particles. To illustrate how we can interconvert masses and numbers of particles, let's calculate the number of copper atoms in an old copper penny. Such a penny weighs about 3 g, and we'll assume that it is 100% copper:

$$\text{Cu atoms} = (3 \text{ g Cu})\left(\frac{1 \text{ mol Cu}}{63.5 \text{ g Cu}}\right)\left(\frac{6.02 \times 10^{23} \text{ Cu atoms}}{1 \text{ mol Cu}}\right)$$
$$= 3 \times 10^{22} \text{ Cu atoms}$$

Notice how dimensional analysis (Section 1.6) provides a straightforward route from grams to numbers of atoms. The molar mass and Avogadro's number are used as conversion factors to convert grams $\longrightarrow$ moles $\longrightarrow$ atoms. Notice also that our answer is a very large number. Any time you calculate the number of atoms, molecules, or ions in an ordinary sample of matter, you can expect the answer to be very large. In contrast, the number of moles in a sample will usually be much smaller, often less than 1. The general procedure for interconverting mass and number of formula units (atoms, molecules, ions, or whatever is represented by the chemical formula) of a substance is summarized in Figure 3.10 ▼.

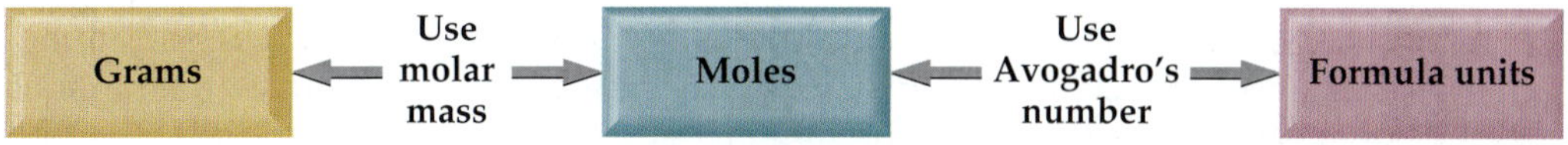

▲ **Figure 3.10 Procedure for interconverting the mass and the number of formula units of a substance.** The number of moles of the substance is central to the calculation; thus, the mole concept can be thought of as the bridge between the mass of a substance in grams and the number of formula units.

SAMPLE EXERCISE 3.12 | Calculating the Number of Molecules and Number of Atoms from Mass

(a) How many glucose molecules are in 5.23 g of $C_6H_{12}O_6$? **(b)** How many oxygen atoms are in this sample?

Solution

Analyze: We are given the number of grams and chemical formula and asked to calculate (a) the number of molecules and (b) the number of O atoms in the sample.

(a) Plan: The strategy for determining the number of molecules in a given quantity of a substance is summarized in Figure 3.10. We must convert 5.23 g $C_6H_{12}O_6$ to moles $C_6H_{12}O_6$, which can then be converted to molecules $C_6H_{12}O_6$. The first conversion uses the molar mass of $C_6H_{12}O_6$: 1 mol $C_6H_{12}O_6$ = 180.0 g $C_6H_{12}O_6$. The second conversion uses Avogadro's number.

Solve: Molecules $C_6H_{12}O_6$

$$= (5.23\ \text{g}\ C_6H_{12}O_6)\left(\frac{1\ \text{mol}\ C_6H_{12}O_6}{180.0\ \text{g}\ C_6H_{12}O_6}\right)\left(\frac{6.022 \times 10^{23}\ \text{molecules}\ C_6H_{12}O_6}{1\ \text{mol}\ C_6H_{12}O_6}\right)$$

$$= 1.75 \times 10^{22}\ \text{molecules}\ C_6H_{12}O_6$$

Check: The magnitude of the answer is reasonable. Because the mass we began with is less than a mole, there should be less than 6.02×10^{23} molecules. We can make a ballpark estimate of the answer: $5/200 = 2.5 \times 10^{-2}$ mol; $2.5 \times 10^{-2} \times 6 \times 10^{23} = 15 \times 10^{21} = 1.5 \times 10^{22}$ molecules. The units (molecules) and significant figures (three) are appropriate.

(b) Plan: To determine the number of O atoms, we use the fact that there are six O atoms in each molecule of $C_6H_{12}O_6$. Thus, multiplying the number of molecules $C_6H_{12}O_6$ by the factor (6 atoms O/1 molecule $C_6H_{12}O_6$) gives the number of O atoms.

Solve:

$$\text{Atoms O} = (1.75 \times 10^{22}\ \text{molecules}\ C_6H_{12}O_6)\left(\frac{6\ \text{atoms O}}{1\ \text{molecule}\ C_6H_{12}O_6}\right)$$

$$= 1.05 \times 10^{23}\ \text{atoms O}$$

Check: The answer is simply 6 times as large as the answer to part (a). The number of significant figures (three) and the units (atoms O) are correct.

PRACTICE EXERCISE

(a) How many nitric acid molecules are in 4.20 g of HNO_3? **(b)** How many O atoms are in this sample?

Answers: **(a)** 4.01×10^{22} molecules HNO_3, **(b)** 1.20×10^{23} atoms O

3.5 | Empirical Formulas from Analyses

3-D MODELS
Water, Hydrogen Peroxide

As we learned in Section 2.6, the empirical formula for a substance tells us the relative number of atoms of each element it contains. Thus, the empirical formula H_2O indicates that water contains two H atoms for each O atom. This ratio also applies on the molar level: 1 mol of H_2O contains 2 mol of H atoms and 1 mol of O atoms. Conversely, the ratio of the number of moles of each element in a compound gives the subscripts in a compound's empirical formula. Thus, the mole concept provides a way of calculating the empirical formulas of chemical substances, as shown in the following examples.

Mercury and chlorine combine to form a compound that is 73.9% mercury and 26.1% chlorine by mass. This means that if we had a 100.0-g sample of the solid, it would contain 73.9 g of mercury (Hg) and 26.1 g of chlorine (Cl). (Any size sample can be used in problems of this type, but we will generally use 100.0 g to simplify the calculation of mass from percentage.) Using the atomic weights of the elements to give us molar masses, we can calculate the number of moles of each element in the sample:

$$(73.9\ \text{g Hg})\left(\frac{1\ \text{mol Hg}}{200.6\ \text{g Hg}}\right) = 0.368\ \text{mol Hg}$$

$$(26.1\ \text{g Cl})\left(\frac{1\ \text{mol Cl}}{35.5\ \text{g Cl}}\right) = 0.735\ \text{mol Cl}$$

We then divide the larger number of moles (0.735) by the smaller (0.368) to obtain a Cl:Hg mole ratio of 1.99:1:

$$\frac{\text{moles of Cl}}{\text{moles of Hg}} = \frac{0.735\ \text{mol Cl}}{0.368\ \text{mol Hg}} = \frac{1.99\ \text{mol Cl}}{1\ \text{mol Hg}}$$

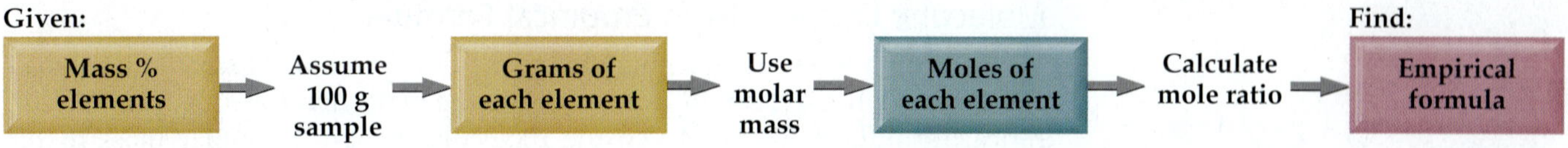

▲ **Figure 3.11 Procedure for calculating an empirical formula from percentage composition.** The central part of the calculation is determining the number of moles of each element in the compound. The procedure is also summarized as "percent to mass, mass to mole, divide by small, multiply 'til whole."

Because of experimental errors, the results may not lead to exact integers for the ratios of moles. The number 1.99 is very close to 2, so we can confidently conclude that the empirical formula for the compound is $HgCl_2$. This is the empirical formula because its subscripts are the smallest integers that express the *ratios* of atoms present in the compound. (Section 2.6) The general procedure for determining empirical formulas is outlined in Figure 3.11 ▲.

ACTIVITY
Empirical Formula Determination: C_8H_6O

SAMPLE EXERCISE 3.13 | Calculating an Empirical Formula

Ascorbic acid (vitamin C) contains 40.92% C, 4.58% H, and 54.50% O by mass. What is the empirical formula of ascorbic acid?

Solution

Analyze: We are to determine an empirical formula of a compound from the mass percentages of its elements.

Plan: The strategy for determining the empirical formula involves the three steps given in Figure 3.11.

Solve: We *first* assume, for simplicity, that we have exactly 100 g of material (although any mass can be used). In 100 g of ascorbic acid, we have

40.92 g C, 4.58 g H, and 54.50 g O.

Second, we calculate the number of moles of each element:

$$\text{Moles C} = (40.92\ \cancel{\text{g C}})\left(\frac{1\ \text{mol C}}{12.01\ \cancel{\text{g C}}}\right) = 3.407\ \text{mol C}$$

$$\text{Moles H} = (4.58\ \cancel{\text{g H}})\left(\frac{1\ \text{mol H}}{1.008\ \cancel{\text{g H}}}\right) = 4.54\ \text{mol H}$$

$$\text{Moles O} = (54.50\ \cancel{\text{g O}})\left(\frac{1\ \text{mol O}}{16.00\ \cancel{\text{g O}}}\right) = 3.406\ \text{mol O}$$

Third, we determine the simplest whole-number ratio of moles by dividing each number of moles by the smallest number of moles, 3.406:

$$\text{C:}\frac{3.407}{3.406} = 1.000 \qquad \text{H:}\frac{4.54}{3.406} = 1.33 \qquad \text{O:}\frac{3.406}{3.406} = 1.000$$

The ratio for H is too far from 1 to attribute the difference to experimental error; in fact, it is quite close to $1\frac{1}{3}$. This suggests that if we multiply the ratio by 3, we will obtain whole numbers:

$$\text{C:H:O} = 3(1:1.33:1) = 3:4:3$$

The whole-number mole ratio gives us the subscripts for the empirical formula:

$$C_3H_4O_3$$

Check: It is reassuring that the subscripts are moderately sized whole numbers. Otherwise, we have little by which to judge the reasonableness of our answer.

PRACTICE EXERCISE

A 5.325-g sample of methyl benzoate, a compound used in the manufacture of perfumes, is found to contain 3.758 g of carbon, 0.316 g of hydrogen, and 1.251 g of oxygen. What is the empirical formula of this substance?

Answer: C_4H_4O

Molecular Formula from Empirical Formula

For any compound, the formula obtained from percentage compositions is always the empirical formula. We can obtain the molecular formula from the empirical formula if we are given the molecular weight or molar mass of the compound. *The subscripts in the molecular formula of a substance are always a whole-number multiple of the corresponding subscripts in its empirical formula.* (Section 2.6) This multiple can be found by comparing the empirical formula weight with the molecular weight:

$$\text{Whole-number multiple} = \frac{\text{molecular weight}}{\text{empirical formula weight}} \quad [3.11]$$

In Sample Exercise 3.13, for example, the empirical formula of ascorbic acid was determined to be $C_3H_4O_3$, giving an empirical formula weight of 3(12.0 amu) + 4(1.0 amu) + 3(16.0 amu) = 88.0 amu. The experimentally determined molecular weight is 176 amu. Thus, the molecular weight is 2 times the empirical formula weight (176/88.0 = 2.00), and the molecular formula must therefore have twice as many of each kind of atom as the empirical formula. Consequently, we multiply the subscripts in the empirical formula by 2 to obtain the molecular formula: $C_6H_8O_6$.

SAMPLE EXERCISE 3.14 | Determining a Molecular Formula

Mesitylene, a hydrocarbon that occurs in small amounts in crude oil, has an empirical formula of C_3H_4. The experimentally determined molecular weight of this substance is 121 amu. What is the molecular formula of mesitylene?

Solution

Analyze: We are given an empirical formula and molecular weight and asked to determine a molecular formula.

Plan: The subscripts in the molecular formula of a compound are whole-number multiples of the subscripts in its empirical formula. To find the appropriate multiple, we must compare the molecular weight with the formula weight of the empirical formula.

Solve: First, we calculate the formula weight of the empirical formula, C_3H_4:

$$3(12.0\ \text{amu}) + 4(1.0\ \text{amu}) = 40.0\ \text{amu}$$

Next, we divide the molecular weight by the empirical formula weight to obtain the multiple used to multiply the subscripts in C_3H_4:

$$\frac{\text{Molecular weight}}{\text{Empirical formula weight}} = \frac{121}{40.0} = 3.02$$

Only whole-number ratios make physical sense because we must be dealing with whole atoms. The 3.02 in this case results from a small experimental error in the molecular weight. We therefore multiply each subscript in the empirical formula by 3 to give the molecular formula: C_9H_{12}.

Check: We can have confidence in the result because dividing the molecular weight by the formula weight yields nearly a whole number.

PRACTICE EXERCISE

Ethylene glycol, the substance used in automobile antifreeze, is composed of 38.7% C, 9.7% H, and 51.6% O by mass. Its molar mass is 62.1 g/mol. **(a)** What is the empirical formula of ethylene glycol? **(b)** What is its molecular formula?

Answers: **(a)** CH_3O, **(b)** $C_2H_6O_2$

Combustion Analysis

The empirical formula of a compound is based on experiments that give the number of moles of each element in a sample of the compound. That is why we use the word "empirical," which means "based on observation and experiment." Chemists have devised a number of experimental techniques to determine

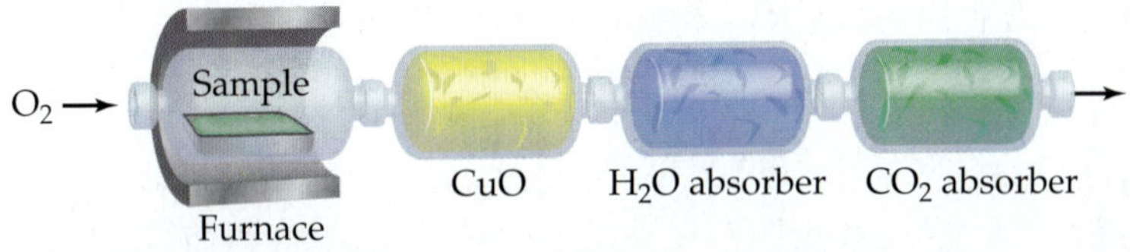

▲ **Figure 3.12 Apparatus to determine percentages of carbon and hydrogen in a compound.** The compound is combusted to form CO_2 and H_2O. Copper oxide helps to oxidize traces of carbon and carbon monoxide to carbon dioxide and to oxidize hydrogen to water.

empirical formulas. One of these is combustion analysis, which is commonly used for compounds containing principally carbon and hydrogen as their component elements.

When a compound containing carbon and hydrogen is completely combusted in an apparatus such as that shown in Figure 3.12 ▲ the carbon in the compound is converted to CO_2, and the hydrogen is converted to H_2O. (Section 3.2) The amounts of CO_2 and H_2O produced are determined by measuring the mass increase in the CO_2 and H_2O absorbers. From the masses of CO_2 and H_2O we can calculate the number of moles of C and H in the original compound and thereby the empirical formula. If a third element is present in the compound, its mass can be determined by subtracting the masses of C and H from the compound's original mass. Sample Exercise 3.15 shows how to determine the empirical formula of a compound containing C, H, and O.

SAMPLE EXERCISE 3.15 | Determining Empirical Formula by Combustion Analysis

Isopropyl alcohol, a substance sold as rubbing alcohol, is composed of C, H, and O. Combustion of 0.255 g of isopropyl alcohol produces 0.561 g of CO_2 and 0.306 g of H_2O. Determine the empirical formula of isopropyl alcohol.

Solution

Analyze: We are told that isopropyl alcohol contains C, H, and O atoms and given the quantities of CO_2 and H_2O produced when a given quantity of the alcohol is combusted. We must use this information to determine the empirical formula for isopropyl alcohol, a task that requires us to calculate the number of moles of C, H, and O in the sample.

Plan: We can use the mole concept to calculate the number of grams of C present in the CO_2 and the number of grams of H present in the H_2O. These are the quantities of C and H present in the isopropyl alcohol before combustion. The number of grams of O in the compound equals the mass of the isopropyl alcohol minus the sum of the C and H masses. Once we have the number of grams of C, H, and O in the sample, we can then proceed as in Sample Exercise 3.13: Calculate the number of moles of each element, and determine the mole ratio, which gives the subscripts in the empirical formula.

Solve: To calculate the number of grams of C, we first use the molar mass of CO_2, 1 mol CO_2 = 44.0 g CO_2, to convert grams of CO_2 to moles of CO_2. Because there is only 1 C atom in each CO_2 molecule, there is 1 mol of C atoms per mole of CO_2 molecules. This fact allows us to convert the moles of CO_2 to moles of C. Finally, we use the molar mass of C, 1 mol C = 12.0 g C, to convert moles of C to grams of C. Combining the three conversion factors, we have:

$$\text{Grams C} = (0.561\ \cancel{\text{g CO}_2})\left(\frac{1\ \cancel{\text{mol CO}_2}}{44.0\ \cancel{\text{g CO}_2}}\right)\left(\frac{1\ \cancel{\text{mol C}}}{1\ \cancel{\text{mol CO}_2}}\right)\left(\frac{12.0\text{ g C}}{1\ \cancel{\text{mol C}}}\right) = 0.153\text{ g C}$$

The calculation of the number of grams of H from the grams of H_2O is similar, although we must remember that there are 2 mol of H atoms per 1 mol of H_2O molecules:

$$\text{Grams H} = (0.306\ \cancel{\text{g H}_2\text{O}})\left(\frac{1\ \cancel{\text{mol H}_2\text{O}}}{18.0\ \cancel{\text{g H}_2\text{O}}}\right)\left(\frac{2\ \cancel{\text{mol H}}}{1\ \cancel{\text{mol H}_2\text{O}}}\right)\left(\frac{1.01\text{ g H}}{1\ \cancel{\text{mol H}}}\right) = 0.0343\text{ g H}$$

The total mass of the sample, 0.255 g, is the sum of the masses of the C, H, and O. Thus, we can calculate the mass of O as follows:

$$\text{Mass of O} = \text{mass of sample} - (\text{mass of C} + \text{mass of H})$$
$$= 0.255\text{ g} - (0.153\text{ g} + 0.0343\text{ g}) = 0.068\text{ g O}$$

We then calculate the number of moles of C, H, and O in the sample:

$$\text{Moles C} = (0.153\ \cancel{\text{g C}})\left(\frac{1\ \text{mol C}}{12.0\ \cancel{\text{g C}}}\right) = 0.0128\ \text{mol C}$$

$$\text{Moles H} = (0.0343\ \cancel{\text{g H}})\left(\frac{1\ \text{mol H}}{1.01\ \cancel{\text{g H}}}\right) = 0.0340\ \text{mol H}$$

$$\text{Moles O} = (0.068\ \cancel{\text{g O}})\left(\frac{1\ \text{mol O}}{16.0\ \cancel{\text{g O}}}\right) = 0.0043\ \text{mol O}$$

To find the empirical formula, we must compare the relative number of moles of each element in the sample. The relative number of moles of each element is found by dividing each number by the smallest number, 0.0043. The mole ratio of C : H : O so obtained is 2.98 : 7.91 : 1.00. The first two numbers are very close to the whole numbers 3 and 8, giving the empirical formula C_3H_8O.

Check: The subscripts work out to be moderately sized whole numbers, as expected.

PRACTICE EXERCISE

(a) Caproic acid, which is responsible for the foul odor of dirty socks, is composed of C, H, and O atoms. Combustion of a 0.225-g sample of this compound produces 0.512 g CO_2 and 0.209 g H_2O. What is the empirical formula of caproic acid? **(b)** Caproic acid has a molar mass of 116 g/mol. What is its molecular formula?

Answers: **(a)** C_3H_6O, **(b)** $C_6H_{12}O_2$

GIVE IT SOME THOUGHT

In Sample Exercise 3.15, how do you explain the fact that the ratios C : H : O are 2.98 : 7.91 : 1.00, rather than exact integers 3 : 8 : 1?

3.6 | Quantitative Information from Balanced Equations

The coefficients in a chemical equation represent the relative numbers of molecules involved in a reaction. The mole concept allows us to convert this information to the masses of the substances. Consider the following balanced equation:

$$2\,H_2(g) + O_2(g) \longrightarrow 2\,H_2O(l) \qquad [3.12]$$

The coefficients indicate that two molecules of H_2 react with each molecule of O_2 to form two molecules of H_2O. It follows that the relative numbers of moles are identical to the relative numbers of molecules:

$2\,H_2(g)$	+	$O_2(g)$	$\longrightarrow$	$2\,H_2O(l)$
2 molecules		1 molecule		2 molecules
$2(6.02 \times 10^{23}$ molecules)		$1(6.02 \times 10^{23}$ molecules)		$2(6.02 \times 10^{23}$ molecules)
2 mol		1 mol		2 mol

We can generalize this observation for all balanced chemical equations: *The coefficients in a balanced chemical equation indicate both the relative numbers of molecules (or formula units) involved in the reaction and the relative numbers of moles.* Table 3.3 ▼ further summarizes this result and also shows how it corresponds

TABLE 3.3 Information from a Balanced Equation

Equation:	$\mathbf{2\,H_2(g)}$	**+**	$\mathbf{O_2(g)}$	$\longrightarrow$	$\mathbf{2\,H_2O(l)}$
Molecules:	2 molecules H_2	+	1 molecule O_2	$\longrightarrow$	2 molecules H_2O

Mass (amu):	4.0 amu H_2	+	32.0 amu O_2	$\longrightarrow$	36.0 amu H_2O
Amount (mol):	2 mol H_2	+	1 mol O_2	$\longrightarrow$	2 mol H_2O
Mass (g):	4.0 g H_2	+	32.0 g O_2	$\longrightarrow$	36.0 g H_2O

to the law of conservation of mass. Notice that the total mass of the reactants (4.0 g + 32.0 g) equals the total mass of the products (36.0 g).

The quantities 2 mol H_2, 1 mol O_2, and 2 mol H_2O, which are given by the coefficients in Equation 3.12, are called *stoichiometrically equivalent quantities.* The relationship between these quantities can be represented as

$$2 \text{ mol } H_2 \hat{=} 1 \text{ mol } O_2 \hat{=} 2 \text{ mol } H_2O$$

where the $\hat{=}$ symbol means "is stoichiometrically equivalent to." In other words, Equation 3.12 shows 2 mol of H_2 and 1 mol of O_2 forming 2 mol of H_2O. These stoichiometric relations can be used to convert between quantities of reactants and products in a chemical reaction. For example, the number of moles of H_2O produced from 1.57 mol of O_2 can be calculated as follows:

$$\text{Moles } H_2O = (1.57 \cancel{\text{ mol } O_2})\left(\frac{2 \text{ mol } H_2O}{1 \cancel{\text{ mol } O_2}}\right) = 3.14 \text{ mol } H_2O$$

GIVE IT SOME THOUGHT

When 1.57 mol O_2 reacts with H_2 to form H_2O, how many moles of H_2 are consumed in the process?

As an additional example, consider the combustion of butane (C_4H_{10}), the fuel in disposable cigarette lighters:

$$2\, C_4H_{10}(l) + 13\, O_2(g) \longrightarrow 8\, CO_2(g) + 10\, H_2O(g) \qquad [3.13]$$

Let's calculate the mass of CO_2 produced when 1.00 g of C_4H_{10} is burned. The coefficients in Equation 3.13 tell how the amount of C_4H_{10} consumed is related to the amount of CO_2 produced: 2 mol C_4H_{10} $\hat{=}$ 8 mol CO_2. In order to use this relationship, however, we must use the molar mass of C_4H_{10} to convert grams of C_4H_{10} to moles of C_4H_{10}. Because 1 mol C_4H_{10} = 58.0 g C_4H_{10}, we have

$$\text{Moles } C_4H_{10} = (1.00 \cancel{\text{ g } C_4H_{10}})\left(\frac{1 \text{ mol } C_4H_{10}}{58.0 \cancel{\text{ g } C_4H_{10}}}\right)$$

$$= 1.72 \times 10^{-2} \text{ mol } C_4H_{10}$$

We can then use the stoichiometric factor from the balanced equation, 2 mol C_4H_{10} $\hat{=}$ 8 mol CO_2, to calculate moles of CO_2:

$$\text{Moles } CO_2 = (1.72 \times 10^{-2} \cancel{\text{ mol } C_4H_{10}})\left(\frac{8 \text{ mol } CO_2}{2 \cancel{\text{ mol } C_4H_{10}}}\right)$$

$$= 6.88 \times 10^{-2} \text{ mol } CO_2$$

Finally, we can calculate the mass of the CO_2, in grams, using the molar mass of CO_2 (1 mol CO_2 = 44.0 g CO_2):

$$\text{Grams } CO_2 = (6.88 \times 10^{-2} \cancel{\text{ mol } CO_2})\left(\frac{44.0 \text{ g } CO_2}{1 \cancel{\text{ mol } CO_2}}\right)$$

$$= 3.03 \text{ g } CO_2$$

Thus, the conversion sequence is

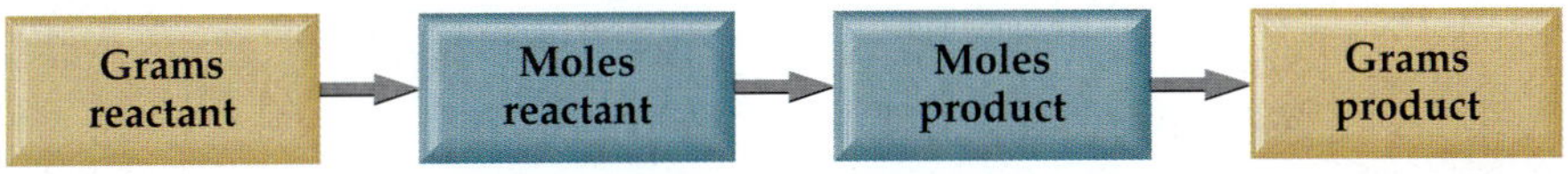

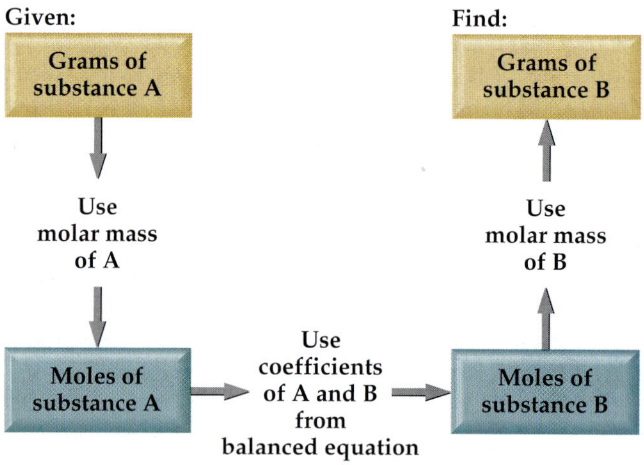

▶ **Figure 3.13 The procedure for calculating amounts of reactants or products in a reaction.** The number of grams of a reactant consumed or of a product formed in a reaction can be calculated, starting with the number of grams of one of the other reactants or products. Notice how molar masses and the coefficients in the balanced equation are used.

ACTIVITY
Stoichiometry Calculation

These steps can be combined in a single sequence of factors:

$$\text{Grams } CO_2 = (1.00 \text{ g } C_4H_{10})\left(\frac{1 \text{ mol } C_4H_{10}}{58.0 \text{ g } C_4H_{10}}\right)\left(\frac{8 \text{ mol } CO_2}{2 \text{ mol } C_4H_{10}}\right)\left(\frac{44.0 \text{ g } CO_2}{1 \text{ mol } CO_2}\right)$$

$$= 3.03 \text{ g } CO_2$$

Similarly, we can calculate the amount of O_2 consumed or H_2O produced in this reaction. For example, to calculate the amount of O_2 consumed, we again rely on the coefficients in the balanced equation to give us the appropriate stoichiometric factor: 2 mol $C_4H_{10} \simeq$ 13 mol O_2:

$$\text{Grams } O_2 = (1.00 \text{ g } C_4H_{10})\left(\frac{1 \text{ mol } C_4H_{10}}{58.0 \text{ g } C_4H_{10}}\right)\left(\frac{13 \text{ mol } O_2}{2 \text{ mol } C_4H_{10}}\right)\left(\frac{32.0 \text{ g } O_2}{1 \text{ mol } O_2}\right)$$

$$= 3.59 \text{ g } O_2$$

Figure 3.13 ▲ summarizes the general approach used to calculate the quantities of substances consumed or produced in chemical reactions. The balanced chemical equation provides the relative numbers of moles of reactants and products involved in the reaction.

SAMPLE EXERCISE 3.16 | Calculating Amounts of Reactants and Products

How many grams of water are produced in the oxidation of 1.00 g of glucose, $C_6H_{12}O_6$?

$$C_6H_{12}O_6(s) + 6\,O_2(g) \longrightarrow 6\,CO_2(g) + 6\,H_2O(l)$$

Solution

Analyze: We are given the mass of a reactant and are asked to determine the mass of a product in the given equation.

Plan: The general strategy, as outlined in Figure 3.13, requires three steps. First, the amount of $C_6H_{12}O_6$ must be converted from grams to moles. We can then use the balanced equation, which relates the moles of $C_6H_{12}O_6$ to the moles of H_2O: 1 mol $C_6H_{12}O_6 \simeq$ 6 mol H_2O. Finally, the moles of H_2O must be converted to grams.

Solve: First, use the molar mass of $C_6H_{12}O_6$ to convert from grams $C_6H_{12}O_6$ to moles $C_6H_{12}O_6$:

$$\text{Moles } C_6H_{12}O_6 = (1.00 \text{ g } C_6H_{12}O_6)\left(\frac{1 \text{ mol } C_6H_{12}O_6}{180.0 \text{ g } C_6H_{12}O_6}\right)$$

Second, use the balanced equation to convert moles of $C_6H_{12}O_6$ to moles of H_2O:

$$\text{Moles } H_2O = (1.00 \text{ g } C_6H_{12}O_6)\left(\frac{1 \text{ mol } C_6H_{12}O_6}{180.0 \text{ g } C_6H_{12}O_6}\right)\left(\frac{6 \text{ mol } H_2O}{1 \text{ mol } C_6H_{12}O_6}\right)$$

Third, use the molar mass of H_2O to convert from moles of H_2O to grams of H_2O:

$$\text{Grams } H_2O = (1.00 \text{ g } C_6H_{12}O_6)\left(\frac{1 \text{ mol } C_6H_{12}O_6}{180.0 \text{ g } C_6H_{12}O_6}\right)\left(\frac{6 \text{ mol } H_2O}{1 \text{ mol } C_6H_{12}O_6}\right)\left(\frac{18.0 \text{ g } H_2O}{1 \text{ mol } H_2O}\right)$$

$$= 0.600 \text{ g } H_2O$$

The steps can be summarized in a diagram like that in Figure 3.13:

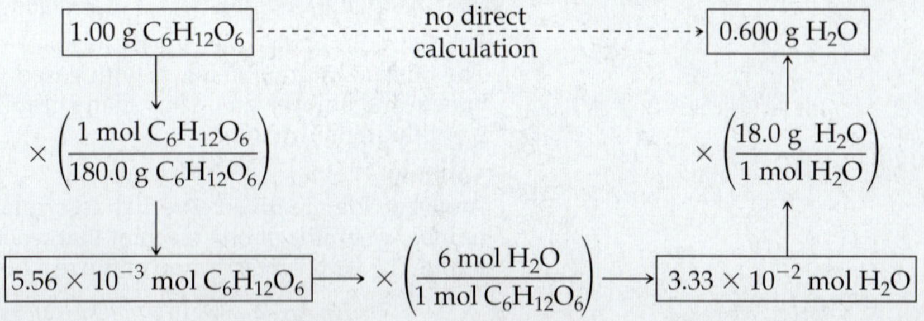

Check: An estimate of the magnitude of our answer, $18/180 = 0.1$ and $0.1 \times 6 = 0.6$, agrees with the exact calculation. The units, grams H_2O, are correct. The initial data had three significant figures, so three significant figures for the answer is correct.

Comment: An average person ingests 2 L of water daily and eliminates 2.4 L. The difference between 2 L and 2.4 L is produced in the metabolism of foodstuffs, such as in the oxidation of glucose. (*Metabolism* is a general term used to describe all the chemical processes of a living animal or plant.) The desert rat (kangaroo rat), on the other hand, apparently never drinks water. It survives on its metabolic water.

PRACTICE EXERCISE

The decomposition of $KClO_3$ is commonly used to prepare small amounts of O_2 in the laboratory: $2\,KClO_3(s) \longrightarrow 2\,KCl(s) + 3\,O_2(g)$. How many grams of O_2 can be prepared from 4.50 g of $KClO_3$?

Answer: 1.77 g

CHEMISTRY AT WORK | CO_2 and the Greenhouse Effect

Coal and petroleum provide the fuels that we use to generate electricity and power our industrial machinery. These fuels are composed primarily of hydrocarbons and other carbon-containing substances. As we have seen, the combustion of 1.00 g of C_4H_{10} produces 3.03 g of CO_2. Similarly, a gallon (3.78 L) of gasoline (density = 0.70 g/mL and approximate composition C_8H_{18}) produces about 8 kg (18 lb) of CO_2. Combustion of such fuels releases about 20 billion tons of CO_2 into the atmosphere annually.

Much CO_2 is absorbed into oceans or used by plants in photosynthesis. Nevertheless, we are now generating CO_2 much faster than it is being absorbed. Chemists have monitored atmospheric CO_2 concentrations since 1958. Analysis of air trapped in ice cores taken from Antarctica and Greenland makes it possible to determine the atmospheric levels of CO_2 during the past 160,000 years. These measurements reveal that the level of CO_2 remained fairly constant from the last Ice Age, some 10,000 years ago, until roughly the beginning of the Industrial Revolution, about 300 years ago. Since that time the concentration of CO_2 has increased by about 25% (Figure 3.14 ▼).

Although CO_2 is a minor component of the atmosphere, it plays a significant role by absorbing radiant heat, acting much like the glass of a greenhouse. For this reason, we often refer to CO_2 and other heat-trapping gases as greenhouse gases, and we call the warming caused by these gases the *greenhouse effect*. Most atmospheric scientists believe that the accumulation of CO_2 and other heat-trapping gases has begun to change the climate of our planet. Still, it is recognized that the factors affecting climate are complex and incompletely understood.

We will examine the greenhouse effect more closely in Chapter 18.

Related Exercises: 3.64, 3.96, 3.99, and 3.103

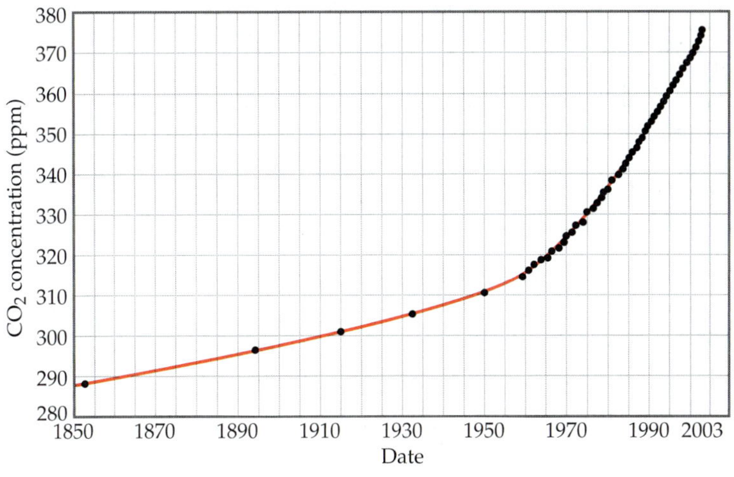

◀ **Figure 3.14 Increasing concentration of atmospheric CO_2.** The world-wide concentration of CO_2 has increased from about 290 ppm to over 370 ppm over the past 150 years. The concentration in ppm is the number of molecules of CO_2 per million (10^6) molecules of air. Data before 1958 came from analyses of air trapped in bubbles of glacial ice.

SAMPLE EXERCISE 3.17 | Calculating Amounts of Reactants and Products

Solid lithium hydroxide is used in space vehicles to remove exhaled carbon dioxide. The lithium hydroxide reacts with gaseous carbon dioxide to form solid lithium carbonate and liquid water. How many grams of carbon dioxide can be absorbed by 1.00 g of lithium hydroxide?

Solution

Analyze: We are given a verbal description of a reaction and asked to calculate the number of grams of one reactant that reacts with 1.00 g of another.

Plan: The verbal description of the reaction can be used to write a balanced equation:

$$2\,LiOH(s) + CO_2(g) \longrightarrow Li_2CO_3(s) + H_2O(l)$$

We are given the grams of LiOH and asked to calculate grams of CO_2. We can accomplish this task by using the following sequence of conversions:

$$\text{Grams LiOH} \longrightarrow \text{moles LiOH} \longrightarrow \text{moles } CO_2 \longrightarrow \text{grams } CO_2.$$

The conversion from grams of LiOH to moles of LiOH requires the molar mass of LiOH (6.94 + 16.00 + 1.01 = 23.95 g/mol). The conversion of moles of LiOH to moles of CO_2 is based on the balanced chemical equation: 2 mol LiOH $\simeq$ 1 mol CO_2. To convert the number of moles of CO_2 to grams, we must use the molar mass of CO_2: 12.01 + 2(16.00) = 44.01 g/mol.

Solve:

$$(1.00\ \cancel{\text{g LiOH}})\left(\frac{1\ \cancel{\text{mol LiOH}}}{23.95\ \cancel{\text{g LiOH}}}\right)\left(\frac{1\ \cancel{\text{mol } CO_2}}{2\ \cancel{\text{mol LiOH}}}\right)\left(\frac{44.01\text{ g } CO_2}{1\ \cancel{\text{mol } CO_2}}\right) = 0.919\text{ g } CO_2$$

Check: Notice that $23.95 \approx 24$, $24 \times 2 = 48$, and 44/48 is slightly less than 1. Thus, the magnitude of the answer is reasonable based on the amount of starting LiOH; the significant figures and units are appropriate, too.

PRACTICE EXERCISE

Propane, C_3H_8, is a common fuel used for cooking and home heating. What mass of O_2 is consumed in the combustion of 1.00 g of propane?

Answer: 3.64 g

3.7 | Limiting Reactants

Suppose you wish to make several sandwiches using one slice of cheese and two slices of bread for each sandwich. Using Bd = bread, Ch = cheese, and Bd_2Ch = sandwich, the recipe for making a sandwich can be represented like a chemical equation:

$$2\,Bd + Ch \longrightarrow Bd_2Ch$$

If you have 10 slices of bread and 7 slices of cheese, you will be able to make only five sandwiches before you run out of bread. You will have 2 slices of cheese left over. The amount of available bread limits the number of sandwiches.

An analogous situation occurs in chemical reactions when one of the reactants is used up before the others. The reaction stops as soon as any one of the reactants is totally consumed, leaving the excess reactants as leftovers. Suppose, for example, that we have a mixture of 10 mol H_2 and 7 mol O_2, which react to form water:

$$2\,H_2(g) + O_2(g) \longrightarrow 2\,H_2O(g)$$

Because 2 mol $H_2 \simeq$ 1 mol O_2, the number of moles of O_2 needed to react with all the H_2 is

$$\text{Moles } O_2 = (10\ \cancel{\text{mol } H_2})\left(\frac{1\text{ mol } O_2}{2\ \cancel{\text{mol } H_2}}\right) = 5\text{ mol } O_2$$

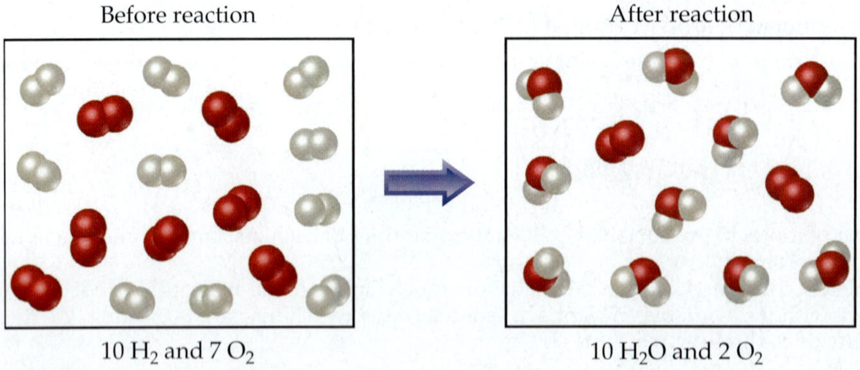

▲ **Figure 3.15 Example illustrating a limiting reactant.** Because the H_2 is completely consumed, it is the limiting reagent in this case. Because there is a stoichiometric excess of O_2, some is left over at the end of the reaction. The amount of H_2O formed is related directly to the amount of H_2 consumed.

Because 7 mol O_2 was available at the start of the reaction, 7 mol O_2 − 5 mol O_2 = 2 mol O_2 will still be present when all the H_2 is consumed. The example we have considered is depicted on a molecular level in Figure 3.15 ▲.

The reactant that is completely consumed in a reaction is called either the **limiting reactant** or *limiting reagent* because it determines, or limits, the amount of product formed. The other reactants are sometimes called either *excess reactants* or *excess reagents*. In our example, H_2 is the limiting reactant, which means that once all the H_2 has been consumed, the reaction stops. O_2 is the excess reactant, and some is left over when the reaction stops.

There are no restrictions on the starting amounts of the reactants in any reaction. Indeed, many reactions are carried out using an excess of one reagent. The quantities of reactants consumed and the quantities of products formed, however, are restricted by the quantity of the limiting reactant. When a combustion reaction takes place in the open air, oxygen is plentiful and is therefore the excess reactant. You may have had the unfortunate experience of running out of gasoline while driving. The car stops because you've run out of the limiting reactant in the combustion reaction, the fuel.

ANIMATION Limiting Reagent

Before we leave our present example, let's summarize the data in a tabular form:

	$2\,H_2(g)$	+	$O_2(g)$	$\longrightarrow$	$2\,H_2O(g)$
Initial quantities:	10 mol		7 mol		0 mol
Change (reaction):	−10 mol		−5 mol		+10 mol
Final quantities:	0 mol		2 mol		10 mol

The initial amounts of the reactants are what we started with (10 mol H_2 and 7 mol O_2). The second line in the table (change) summarizes the amounts of the reactants consumed and the amount of the product formed in the reaction. These quantities are restricted by the quantity of the limiting reactant and depend on the coefficients in the balanced equation. The mole ratio $H_2{:}O_2{:}H_2O = 10{:}5{:}10$ conforms to the ratio of the coefficients in the balanced equation, 2:1:2. The changes are negative for the reactants because they are consumed during the reaction and positive for the product because it is formed during the reaction. Finally, the quantities in the third line of the table (final quantities) depend on the initial quantities and their changes, and these entries are found by adding the entries for the initial quantity and change for each column. There is none of the limiting reactant (H_2) left at the end of the reaction. All that remains is 2 mol O_2 and 10 mol H_2O.

ACTIVITY Limiting Reagents

SAMPLE EXERCISE 3.18 | Calculating the Amount of Product Formed from a Limiting Reactant

The most important commercial process for converting N_2 from the air into nitrogen-containing compounds is based on the reaction of N_2 and H_2 to form ammonia (NH_3):

$$N_2(g) + 3\,H_2(g) \longrightarrow 2\,NH_3(g)$$

How many moles of NH_3 can be formed from 3.0 mol of N_2 and 6.0 mol of H_2?

Solution

Analyze: We are asked to calculate the number of moles of product, NH_3, given the quantities of each reactant, N_2 and H_2, available in a reaction. Thus, this is a limiting reactant problem.

Plan: If we assume that one reactant is completely consumed, we can calculate how much of the second reactant is needed in the reaction. By comparing the calculated quantity with the available amount, we can determine which reactant is limiting. We then proceed with the calculation, using the quantity of the limiting reactant.

Solve: The number of moles of H_2 needed for complete consumption of 3.0 mol of N_2 is

$$\text{Moles } H_2 = (3.0\ \cancel{\text{mol } N_2})\left(\frac{3\ \text{mol } H_2}{1\ \cancel{\text{mol } N_2}}\right) = 9.0\ \text{mol } H_2$$

Because only 6.0 mol H_2 is available, we will run out of H_2 before the N_2 is gone, and H_2 will be the limiting reactant. We use the quantity of the limiting reactant, H_2, to calculate the quantity of NH_3 produced:

$$\text{Moles } NH_3 = (6.0\ \cancel{\text{mol } H_2})\left(\frac{2\ \text{mol } NH_3}{3\ \cancel{\text{mol } H_2}}\right) = 4.0\ \text{mol } NH_3$$

Comment: The table on the right summarizes this example:

	$N_2(g)$	+ $3\,H_2(g)$	$\longrightarrow 2\,NH_3(g)$
Initial quantities:	3.0 mol	6.0 mol	0 mol
Change (reaction):	−2.0 mol	−6.0 mol	+4.0 mol
Final quantities:	1.0 mol	0 mol	4.0 mol

Notice that we can calculate not only the number of moles of NH_3 formed but also the number of moles of each of the reactants remaining after the reaction. Notice also that although the number of moles of H_2 present at the beginning of the reaction is greater than the number of moles of N_2 present, the H_2 is nevertheless the limiting reactant because of its larger coefficient in the balanced equation.

Check: The summarizing table shows that the mole ratio of reactants used and product formed conforms to the coefficients in the balanced equation, 1:3:2. Also, because H_2 is the limiting reactant, it is completely consumed in the reaction, leaving 0 mol at the end. Because 2.0 mol H_2 has two significant figures, our answer has two significant figures.

PRACTICE EXERCISE

Consider the reaction $2\,Al(s) + 3\,Cl_2(g) \longrightarrow 2\,AlCl_3(s)$. A mixture of 1.50 mol of Al and 3.00 mol of Cl_2 is allowed to react. **(a)** Which is the limiting reactant? **(b)** How many moles of $AlCl_3$ are formed? **(c)** How many moles of the excess reactant remain at the end of the reaction?

Answers: **(a)** Al, **(b)** 1.50 mol, **(c)** 0.75 mol Cl_2

SAMPLE EXERCISE 3.19 | Calculating the Amount of Product Formed from a Limiting Reactant

Consider the following reaction:

$$2\,Na_3PO_4(aq) + 3\,Ba(NO_3)_2(aq) \longrightarrow Ba_3(PO_4)_2(s) + 6\,NaNO_3(aq)$$

Suppose a solution containing 3.50 g of Na_3PO_4 is mixed with a solution containing 6.40 g of $Ba(NO_3)_2$. How many grams of $Ba_3(PO_4)_2$ can be formed?

Solution

Analyze: We are asked to calculate the amount of a product, given the amounts of two reactants, so this is a limiting reactant problem.

Plan: We must first identify the limiting reagent. To do so, we can calculate the number of moles of each reactant and compare their ratio with that required by the balanced equation. We then use the quantity of the limiting reagent to calculate the mass of $Ba_3(PO_4)_2$ that forms.

Solve: From the balanced equation, we have the following stoichiometric relations:

$$2\ \text{mol } Na_3PO_4 \simeq 3\ \text{mol } Ba(NO_3)_2 \simeq 1\ \text{mol } Ba_3(PO_4)_2$$

Using the molar mass of each substance, we can calculate the number of moles of each reactant:

$$\text{Moles } Na_3PO_4 = (3.50\ \text{g } Na_3PO_4)\left(\frac{1\ \text{mol } Na_3PO_4}{164\ \text{g } Na_3PO_4}\right) = 0.0213\ \text{mol } Na_3PO_4$$

$$\text{Moles } Ba(NO_3)_2 = (6.40\ \text{g } Ba(NO_3)_2)\left(\frac{1\ \text{mol } Ba(NO_3)_2}{261\ \text{g } Ba(NO_3)_2}\right) = 0.0245\ \text{mol } Ba(NO_3)_2$$

Thus, there are slightly more moles of $Ba(NO_3)_2$ than moles of Na_3PO_4. The coefficients in the balanced equation indicate, however, that the reaction requires 3 mol $Ba(NO_3)_2$ for each 2 mol Na_3PO_4. [That is, 1.5 times more moles of $Ba(NO_3)_2$ are needed than moles of Na_3PO_4.] Thus, there is insufficient $Ba(NO_3)_2$ to completely consume the Na_3PO_4. That means that $Ba(NO_3)_2$ is the limiting reagent. We therefore use the quantity of $Ba(NO_3)_2$ to calculate the quantity of product formed. We can begin this calculation with the grams of $Ba(NO_3)_2$, but we can save a step by starting with the moles of $Ba(NO_3)_2$ that were calculated previously in the exercise:

$$\text{Grams } Ba_3(PO_4)_2 = (0.0245\ \text{mol } Ba(NO_3)_2)\left(\frac{1\ \text{mol } Ba_3(PO_4)_2}{3\ \text{mol } Ba(NO_3)_2}\right)\left(\frac{602\ \text{g } Ba_3(PO_4)_2}{1\ \text{mol } Ba_3(PO_4)_2}\right)$$

$$= 4.92\ \text{g } Ba_3(PO_4)_2$$

Check: The magnitude of the answer seems reasonable: Starting with the numbers in the two conversion factors on the right, we have $600/3 = 200$; $200 \times 0.025 = 5$. The units are correct, and the number of significant figures (three) corresponds to the number in the quantity of $Ba(NO_3)_2$.

Comment: The quantity of the limiting reagent, $Ba(NO_3)_2$, can also be used to determine the quantity of $NaNO_3$ formed (4.16 g) and the quantity of Na_3PO_4 used (2.67 g). The number of grams of the excess reagent, Na_3PO_4, remaining at the end of the reaction equals the starting amount minus the amount consumed in the reaction, $3.50\ \text{g} - 2.67\ \text{g} = 0.82\ \text{g}$.

PRACTICE EXERCISE

A strip of zinc metal having a mass of 2.00 g is placed in an aqueous solution containing 2.50 g of silver nitrate, causing the following reaction to occur:

$$Zn(s) + 2\ AgNO_3(aq) \longrightarrow 2\ Ag(s) + Zn(NO_3)_2(aq)$$

(a) Which reactant is limiting? **(b)** How many grams of Ag will form? **(c)** How many grams of $Zn(NO_3)_2$ will form? **(d)** How many grams of the excess reactant will be left at the end of the reaction?

Answers: **(a)** $AgNO_3$, **(b)** 1.59 g, **(c)** 1.39 g, **(d)** 1.52 g Zn

Theoretical Yields

The quantity of product that is calculated to form when all of the limiting reactant reacts is called the **theoretical yield**. The amount of product actually obtained in a reaction is called the *actual yield*. The actual yield is almost always less than (and can never be greater than) the theoretical yield. There are many reasons for this difference. Part of the reactants may not react, for example, or they may react in a way different from that desired (side reactions). In addition, it is not always possible to recover all of the product from the reaction mixture. The **percent yield** of a reaction relates the actual yield to the theoretical (calculated) yield:

$$\text{Percent yield} = \frac{\text{actual yield}}{\text{theoretical yield}} \times 100\% \qquad [3.14]$$

In the experiment described in Sample Exercise 3.19, for example, we calculated that 4.92 g of $Ba_3(PO_4)_2$ should form when 3.50 g of Na_3PO_4 is mixed with 6.40 g of $Ba(NO_3)_2$. The 4.92 g is the theoretical yield of $Ba_3(PO_4)_2$ in the reaction. If the actual yield turned out to be 4.70 g, the percent yield would be

$$\frac{4.70\ \text{g}}{4.92\ \text{g}} \times 100\% = 95.5\%$$

STRATEGIES IN CHEMISTRY | How to Take a Test

At about this time in your study of chemistry, you are likely to face your first hour-long examination. The best way to prepare for the exam is to study and do homework diligently and to make sure you get help from the instructor on any material that is unclear or confusing. (See the advice for learning and studying chemistry presented in the preface of the book.) We present here some general guidelines for taking tests.

Depending on the nature of your course, the exam could consist of a variety of different types of questions. Let's consider some of the more common types and how they can best be addressed.

1. **Multiple-choice questions**
In large-enrollment courses, the most common kind of testing device is the multiple-choice question. You are given the problem and usually presented with four or five possible answers from which you must select the correct one. The first thing to realize is that the instructor has written the question so that all of the answer choices appear at first glance to be correct. (There would be little point in offering choices you could tell were wrong even without knowing much about the concept being tested.) Thus, you should not jump to the conclusion that because one of the choices looks correct, it must be so.

If a multiple-choice question involves a calculation, perform the calculation, quickly double-check your work, and *only then* compare your answer with the choices. If you find a match, you have probably found the correct answer. Keep in mind, though, that your instructor has anticipated the most common errors one can make in solving a given problem and has probably listed the incorrect answers resulting from those errors. Thus, always double-check your reasoning and make sure to use dimensional analysis to arrive at the correct answer, with the correct units.

In multiple-choice questions that don't involve calculations, one way to proceed if you are not sure of the correct choice is to eliminate all the choices you know for sure to be incorrect. Additionally, the reasoning you used in eliminating incorrect choices will help you in reasoning about which is the one correct choice.

SAMPLE EXERCISE 3.20 | Calculating the Theoretical Yield and Percent Yield for a Reaction

Adipic acid, $H_2C_6H_8O_4$, is used to produce nylon. The acid is made commercially by a controlled reaction between cyclohexane (C_6H_{12}) and O_2:

$$2\,C_6H_{12}(l) + 5\,O_2(g) \longrightarrow 2\,H_2C_6H_8O_4(l) + 2\,H_2O(g)$$

(a) Assume that you carry out this reaction starting with 25.0 g of cyclohexane and that cyclohexane is the limiting reactant. What is the theoretical yield of adipic acid?

(b) If you obtain 33.5 g of adipic acid from your reaction, what is the percent yield of adipic acid?

Solution

Analyze: We are given a chemical equation and the quantity of the limiting reactant (25.0 g of C_6H_{12}). We are asked first to calculate the theoretical yield of a product ($H_2C_6H_8O_4$) and then to calculate its percent yield if only 33.5 g of the substance is actually obtained.

Plan:

(a) The theoretical yield, which is the calculated quantity of adipic acid formed in the reaction, can be calculated using the following sequence of conversions:

$$\text{g } C_6H_{12} \longrightarrow \text{mol } C_6H_{12} \longrightarrow \text{mol } H_2C_6H_8O_4 \longrightarrow \text{g } H_2C_6H_8O_4.$$

(b) The percent yield is calculated by comparing the actual yield (33.5 g) to the theoretical yield using Equation 3.14.

Solve:

$$\textbf{(a)}\ \text{Grams } H_2C_6H_8O_4 = (25.0\ \cancel{\text{g } C_6H_{12}})\left(\frac{1\ \cancel{\text{mol } C_6H_{12}}}{84.0\ \cancel{\text{g } C_6H_{12}}}\right) \times \left(\frac{2\ \cancel{\text{mol } H_2C_6H_8O_4}}{2\ \cancel{\text{mol } C_6H_{12}}}\right)\left(\frac{146.0\text{ g } H_2C_6H_8O_4}{1\ \cancel{\text{mol } H_2C_6H_8O_4}}\right) = 43.5\text{ g } H_2C_6H_8O_4$$

$$\textbf{(b)}\ \text{Percent yield} = \frac{\text{actual yield}}{\text{theoretical yield}} \times 100\% = \frac{33.5\text{ g}}{43.5\text{ g}} \times 100\% = 77.0\%$$

Check: Our answer in (a) has the appropriate magnitude, units, and significant figures. In (b) the answer is less than 100% as necessary.

2. **Calculations in which you must show your work**
Your instructor may present you with a numerical problem in which you are to show your work in arriving at a solution. In questions of this kind, you may receive partial credit even if you don't arrive at the correct answer, depending on whether the instructor can follow your line of reasoning. It is important, therefore, to be as neat and organized as you can be, given the pressures of exam taking. It is helpful in approaching such questions to take a few moments to think about the direction you are going to take in solving the problem, and you may even want to write a few words or a diagram on the test paper to indicate your plan of attack. Then write out your calculations as neatly as you can. Show the units for every number you write down, and use dimensional analysis as much as you can, showing how units cancel.
3. **Questions requiring drawings**
Sometimes a test question will require you to draw a chemical structure, a diagram related to chemical bonding, or a figure showing some kind of chemical process. Questions of this kind will come later in the course, but it is useful to talk about them here. (You should review this box before each exam you take, to remind yourself of good exam-taking practices.) Be sure to label your drawing as completely as possible.
4. **Other types of questions**
Other exam questions you might encounter include true-false questions and ones in which you are given a list and asked to indicate which members of the list match some criterion given in the question. Often students answer such questions incorrectly because, in their haste, they misunderstand the nature of the question. Whatever the question form, ask yourself this: What is the instructor testing here? What material am I supposed to know that this question covers?
Finally, if you find that you simply don't see how to arrive at a reasoned response to a question, don't linger over the question. Put a check next to it and go on to the next one. If time permits, you can come back to the unanswered questions, but lingering over a question when nothing is coming to mind is wasting time you may need to finish the exam.

PRACTICE EXERCISE

Imagine that you are working on ways to improve the process by which iron ore containing Fe_2O_3 is converted into iron. In your tests you carry out the following reaction on a small scale:

$$Fe_2O_3(s) + 3\,CO(g) \longrightarrow 2\,Fe(s) + 3\,CO_2(g)$$

(a) If you start with 150 g of Fe_2O_3 as the limiting reagent, what is the theoretical yield of Fe? **(b)** If the actual yield of Fe in your test was 87.9 g, what was the percent yield?
Answers: **(a)** 105 g Fe, **(b)** 83.7%

SUMMARY AND KEY TERMS

Introduction and Section 3.1 The study of the quantitative relationships between chemical formulas and chemical equations is known as **stoichiometry**. One of the important concepts of stoichiometry is the law of conservation of mass, which states that the total mass of the products of a chemical reaction is the same as the total mass of the reactants. The same numbers of atoms of each type are present before and after a chemical reaction. A balanced **chemical equation** shows equal numbers of atoms of each element on each side of the equation. Equations are balanced by placing coefficients in front of the chemical formulas for the **reactants** and **products** of a reaction, *not* by changing the subscripts in chemical formulas.

Section 3.2 Among the reaction types described in this chapter are (1) **combination reactions**, in which two reactants combine to form one product; (2) **decomposition reactions**, in which a single reactant forms two or more products; and (3) **combustion reactions** in oxygen, in which a hydrocarbon or related compound reacts with O_2 to form CO_2 and H_2O.

Section 3.3 Much quantitative information can be determined from chemical formulas and balanced chemical equations by using atomic weights. The **formula weight** of a compound equals the sum of the atomic weights of the atoms in its formula. If the formula is a molecular formula, the formula weight is also called the **molecular weight**. Atomic weights and formula weights can be used to determine the elemental composition of a compound.

Section 3.4 A **mole** of any substance is **Avogadro's number** (6.02×10^{23}) of formula units of that substance. The mass of a mole of atoms, molecules, or ions (the **molar mass**) equals the formula weight of that material expressed in grams. The mass of one molecule of H_2O, for example, is 18 amu, so the mass of 1 mol of H_2O is 18 g. That is, the molar mass of H_2O is 18 g/mol.

Section 3.5 The empirical formula of any substance can be determined from its percent composition by calculating the relative number of moles of each atom in 100 g of the substance. If the substance is molecular in nature, its molecular formula can be determined from the empirical formula if the molecular weight is also known.

Sections 3.6 and 3.7 The mole concept can be used to calculate the relative quantities of reactants and products involved in chemical reactions. The coefficients in a balanced equation give the relative numbers of moles of the reactants and products. To calculate the number of grams of a product from the number of grams of a reactant, therefore, first convert grams of reactant to moles of reactant. We then use the coefficients in the balanced equation to convert the number of moles of reactant to moles of product. Finally, we convert moles of product to grams of product.

A **limiting reactant** is completely consumed in a reaction. When it is used up, the reaction stops, thus limiting the quantities of products formed. The **theoretical yield** of a reaction is the quantity of product calculated to form when all of the limiting reagent reacts. The actual yield of a reaction is always less than the theoretical yield. The **percent yield** compares the actual and theoretical yields.

VISUALIZING CONCEPTS

3.1 The reaction between reactant A (blue spheres) and reactant B (red spheres) is shown in the following diagram:

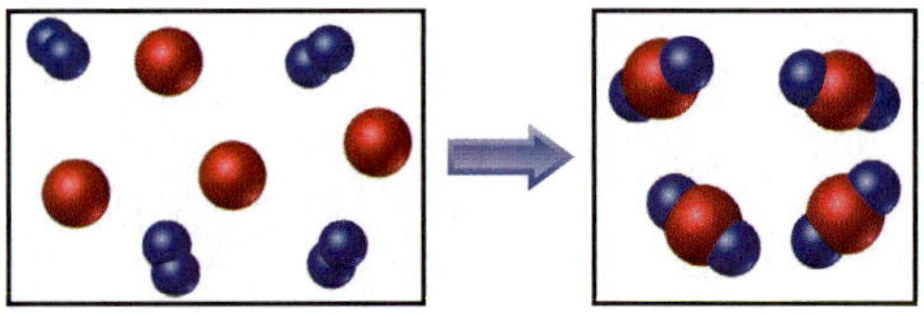

Based on this diagram, which equation best describes the reaction? [Section 3.1]

(a) $A_2 + B \longrightarrow A_2B$ **(b)** $A_2 + 4\,B \longrightarrow 2\,AB_2$
(c) $2\,A + B_4 \longrightarrow 2\,AB_2$ **(d)** $A + B_2 \longrightarrow AB_2$

3.2 Under appropriate experimental conditions, H_2 and CO undergo a combination reaction to form CH_3OH. The drawing below represents a sample of H_2. Make a corresponding drawing of the CO needed to react completely with the H_2. How did you arrive at the number of CO molecules in your drawing? [Section 3.2]

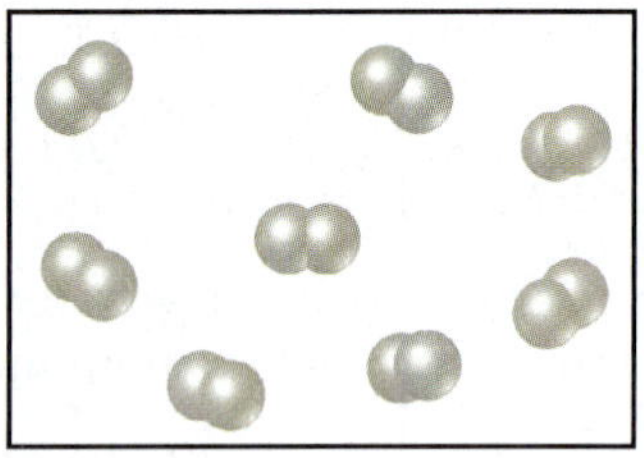

3.3 The following diagram represents the collection of elements formed by decomposition reaction. **(a)** If the blue spheres represent N atoms and the red ones represent O atoms, what was the empirical formula of the original compound? **(b)** Could you draw a diagram representing the molecules of the compound that had been decomposed? Why or why not? [Section 3.2]

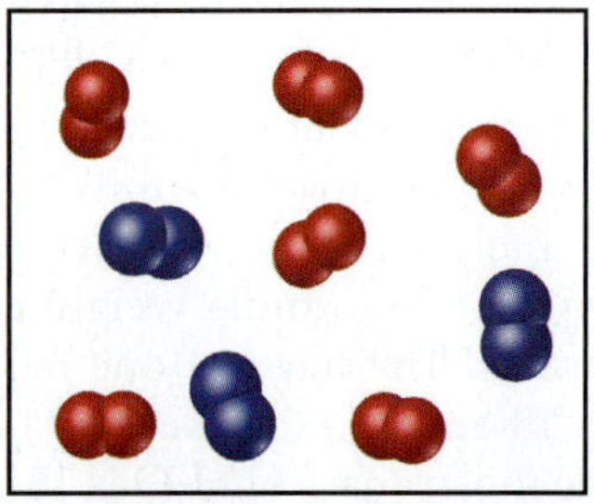

3.4 The following diagram represents the collection of CO_2 and H_2O molecules formed by complete combustion of a hydrocarbon. What is the empirical formula of the hydrocarbon? [Section 3.2]

3.5 Glycine, an amino acid used by organisms to make proteins, is represented by the molecular model below. **(a)** Write its molecular formula. **(b)** Determine its molecular mass. **(c)** Calculate the percent nitrogen by mass in glycine. [Section 3.3 and 3.5]

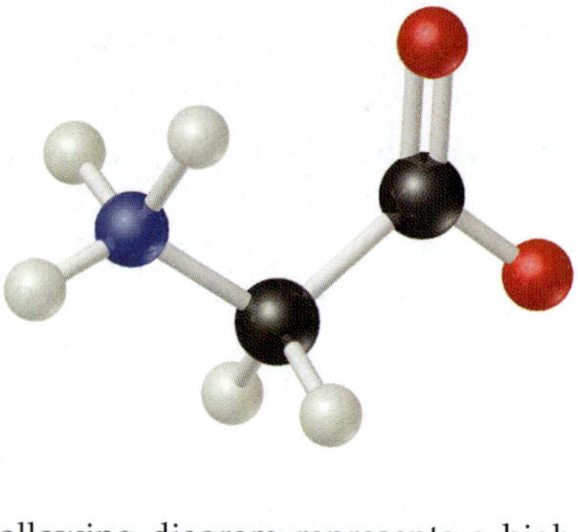

3.6 The following diagram represents a high-temperature reaction between CH_4 and H_2O. Based on this reaction, how many moles of each product can be obtained starting with 4.0 mol CH_4? [Section 3.6]

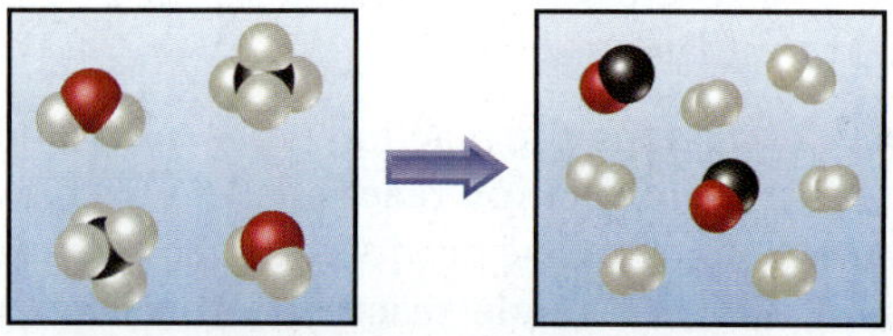

3.7 Nitrogen (N_2) and hydrogen (H_2) react to form ammonia (NH_3). Consider the mixture of N_2 and H_2 shown in the accompanying diagram. The blue spheres represent N, and the white ones represent H. Draw a representation of the product mixture, assuming that the reaction goes to completion. How did you arrive at your representation? What is the limiting reactant in this case? [Section 3.7]

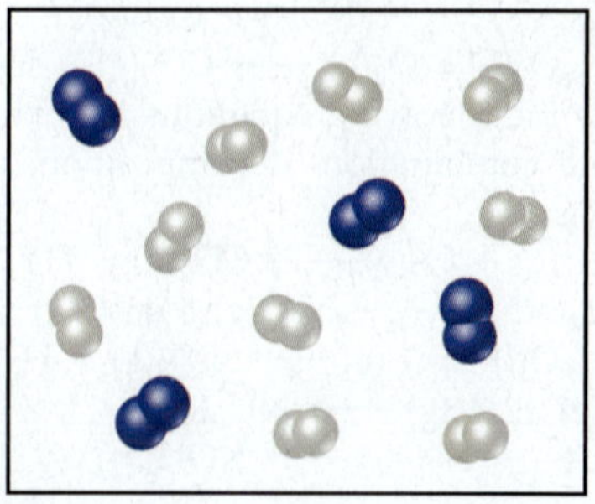

3.8 Nitrogen monoxide and oxygen react to form nitrogen dioxide. Consider the mixture of NO and O_2 shown in the accompanying diagram. The blue spheres represent N, and the red ones represent O. **(a)** Draw a representation of the product mixture, assuming that the reaction goes to completion. What is the limiting reactant in this case? **(b)** How many NO_2 molecules would you draw as products if the reaction had a percent yield of 75%? [Section 3.7]

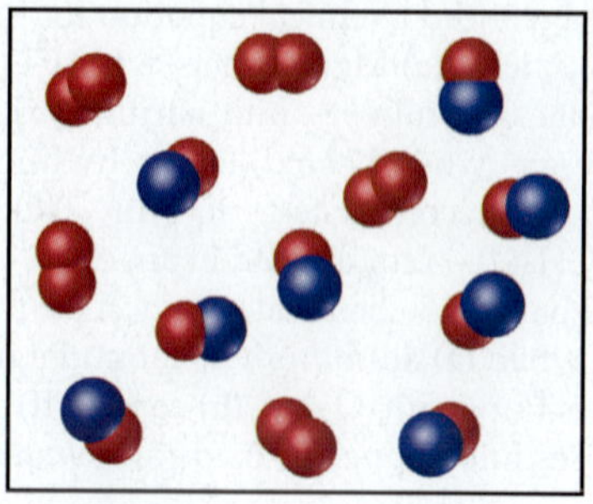

EXERCISES

Balancing Chemical Equations

3.9 **(a)** What scientific principle or law is used in the process of balancing chemical equations? **(b)** In balancing equations, why shouldn't subscripts in chemical formulas be changed? **(c)** What are the symbols used to represent gases, liquids, solids, and aqueous solutions in chemical equations?

3.10 **(a)** What is the difference between adding a subscript 2 to the end of the formula for CO to give CO_2 and adding a coefficient in front of the formula to give 2 CO? **(b)** Is the following chemical equation, as written, consistent with the law of conservation of mass?

$$3\,Mg(OH)_2(s) + 2\,H_3PO_4(aq) \longrightarrow Mg_3(PO_4)_2(s) + 6\,H_2O(l)$$

Why or why not?

3.11 Balance the following equations:
(a) $CO(g) + O_2(g) \longrightarrow CO_2(g)$
(b) $N_2O_5(g) + H_2O(l) \longrightarrow HNO_3(aq)$
(c) $CH_4(g) + Cl_2(g) \longrightarrow CCl_4(l) + HCl(g)$
(d) $Al_4C_3(s) + H_2O(l) \longrightarrow Al(OH)_3(s) + CH_4(g)$
(e) $C_5H_{10}O_2(l) + O_2(g) \longrightarrow CO_2(g) + H_2O(g)$
(f) $Fe(OH)_3(s) + H_2SO_4(aq) \longrightarrow Fe_2(SO_4)_3(aq) + H_2O(l)$
(g) $Mg_3N_2(s) + H_2SO_4(aq) \longrightarrow MgSO_4(aq) + (NH_4)_2SO_4(aq)$

3.12 Balance the following equations:
(a) $Li(s) + N_2(g) \longrightarrow Li_3N(s)$
(b) $La_2O_3(s) + H_2O(l) \longrightarrow La(OH)_3(aq)$
(c) $NH_4NO_3(s) \longrightarrow N_2(g) + O_2(g) + H_2O(g)$
(d) $Ca_3P_2(s) + H_2O(l) \longrightarrow Ca(OH)_2(aq) + PH_3(g)$
(e) $Ca(OH)_2(aq) + H_3PO_4(aq) \longrightarrow Ca_3(PO_4)_2(s) + H_2O(l)$
(f) $AgNO_3(aq) + Na_2SO_4(aq) \longrightarrow Ag_2SO_4(s) + NaNO_3(aq)$
(g) $CH_3NH_2(g) + O_2(g) \longrightarrow CO_2(g) + H_2O(g) + N_2(g)$

3.13 Write balanced chemical equations to correspond to each of the following descriptions: **(a)** Solid calcium carbide, CaC_2, reacts with water to form an aqueous solution of calcium hydroxide and acetylene gas, C_2H_2. **(b)** When solid potassium chlorate is heated, it decomposes to form solid potassium chloride and oxygen gas. **(c)** Solid zinc metal reacts with sulfuric acid to form hydrogen gas and an aqueous solution of zinc sulfate. **(d)** When liquid phosphorus trichloride is added to water, it reacts to form aqueous phosphorous acid, $H_3PO_3(aq)$, and aqueous hydrochloric acid. **(e)** When hydrogen sulfide gas is passed over solid hot iron(III) hydroxide, the resultant reaction produces solid iron(III) sulfide and gaseous water.

3.14 Convert these descriptions into balanced equations: **(a)** When sulfur trioxide gas reacts with water, a solution of sulfuric acid forms. **(b)** Boron sulfide, $B_2S_3(s)$, reacts violently with water to form dissolved boric acid, H_3BO_3, and hydrogen sulfide gas. **(c)** Phosphine, $PH_3(g)$, combusts in oxygen gas to form gaseous water and solid tetraphosphorus decoxide. **(d)** When solid mercury(II) nitrate is heated, it decomposes to form solid mercury(II) oxide, gaseous nitrogen dioxide, and oxygen. **(e)** Copper metal reacts with hot concentrated sulfuric acid solution to form aqueous copper(II) sulfate, sulfur dioxide gas, and water.

Patterns of Chemical Reactivity

3.15 **(a)** When the metallic element sodium combines with the nonmetallic element bromine, $Br_2(l)$, how can you determine the chemical formula of the product? How do you know whether the product is a solid, liquid, or gas at room temperature? Write the balanced chemical equation for the reaction. **(b)** When a hydrocarbon burns in air, what reactant besides the hydrocarbon is involved in the reaction? What products are formed? Write a balanced chemical equation for the combustion of benzene, $C_6H_6(l)$, in air.

3.16 **(a)** Determine the chemical formula of the product formed when the metallic element calcium combines with the nonmetallic element oxygen, O_2. Write the balanced chemical equation for the reaction. **(b)** What products form when a compound containing C, H, and O is completely combusted in air? Write a balanced chemical equation for the combustion of acetone, $C_3H_6O(l)$, in air.

3.17 Write a balanced chemical equation for the reaction that occurs when **(a)** $Mg(s)$ reacts with $Cl_2(g)$; **(b)** barium carbonate decomposes into barium oxide and carbon dioxide gas when heated; **(c)** the hydrocarbon styrene, $C_8H_8(l)$, is combusted in air; **(d)** dimethylether, $CH_3OCH_3(g)$, is combusted in air.

3.18 Write a balanced chemical equation for the reaction that occurs when **(a)** aluminum metal undergoes a combination reaction with $O_2(g)$; **(b)** copper(II) hydroxide decomposes into copper(II) oxide and water when heated; **(c)** heptane, $C_7H_{16}(l)$, burns in air; **(d)** the gasoline additive MTBE (methyl tert-butyl ether), $C_5H_{12}O(l)$, burns in air.

3.19 Balance the following equations, and indicate whether they are combination, decomposition, or combustion reactions:
(a) $Al(s) + Cl_2(g) \longrightarrow AlCl_3(s)$
(b) $C_2H_4(g) + O_2(g) \longrightarrow CO_2(g) + H_2O(g)$
(c) $Li(s) + N_2(g) \longrightarrow Li_3N(s)$
(d) $PbCO_3(s) \longrightarrow PbO(s) + CO_2(g)$
(e) $C_7H_8O_2(l) + O_2(g) \longrightarrow CO_2(g) + H_2O(g)$

3.20 Balance the following equations, and indicate whether they are combination, decomposition, or combustion reactions:
(a) $C_3H_6(g) + O_2(g) \longrightarrow CO_2(g) + H_2O(g)$
(b) $NH_4NO_3(s) \longrightarrow N_2O(g) + H_2O(g)$
(c) $C_5H_6O(l) + O_2(g) \longrightarrow CO_2(g) + H_2O(g)$
(d) $N_2(g) + H_2(g) \longrightarrow NH_3(g)$
(e) $K_2O(s) + H_2O(l) \longrightarrow KOH(aq)$

Formula Weights

3.21 Determine the formula weights of each of the following compounds: **(a)** N_2O_5, **(b)** $CuSO_4$, **(c)** $(NH_4)_3PO_4$, **(d)** $Ca(HCO_3)_2$, **(e)** aluminum sulfide, **(f)** iron(III) sulfate, **(g)** disilicon hexabromide.

3.22 Determine the formula weights of each of the following compounds: **(a)** nitrous oxide, N_2O, known as laughing gas and used as an anesthetic in dentistry; **(b)** benzoic acid, $HC_7H_5O_2$, a substance used as a food preservative; **(c)** $Mg(OH)_2$, the active ingredient in milk of magnesia; **(d)** urea, $(NH_2)_2CO$, a compound used as a nitrogen fertilizer; **(e)** isopentyl acetate, $CH_3CO_2C_5H_{11}$, responsible for the odor of bananas.

3.23 Calculate the percentage by mass of oxygen in the following compounds: **(a)** SO_3; **(b)** CH_3COOCH_3; **(c)** $Cr(NO_3)_3$; **(d)** sodium sulfate; **(e)** ammonium nitrate.

3.24 Calculate the percentage by mass of the indicated element in the following compounds: **(a)** carbon in acetylene, C_2H_2, a gas used in welding; **(b)** hydrogen in ascorbic acid, $HC_6H_7O_6$, also known as vitamin C; **(c)** hydrogen in ammonium sulfate, $(NH_4)_2SO_4$, a substance used as a nitrogen fertilizer; **(d)** platinum in $PtCl_2(NH_3)_2$, a chemotherapy agent called cisplatin; **(e)** oxygen in the female sex hormone estradiol, $C_{18}H_{24}O_2$; **(f)** carbon in capsaicin, $C_{18}H_{27}NO_3$, the compound that gives the hot taste to chili peppers.

3.25 Based on the following structural formulas, calculate the percentage of carbon by mass present in each compound:

(a) 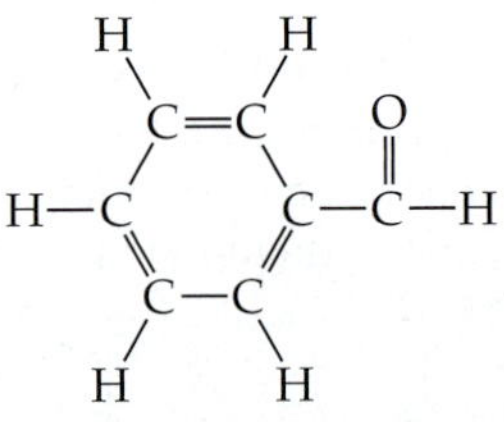
Benzaldehyde (almond fragrance)

(b) 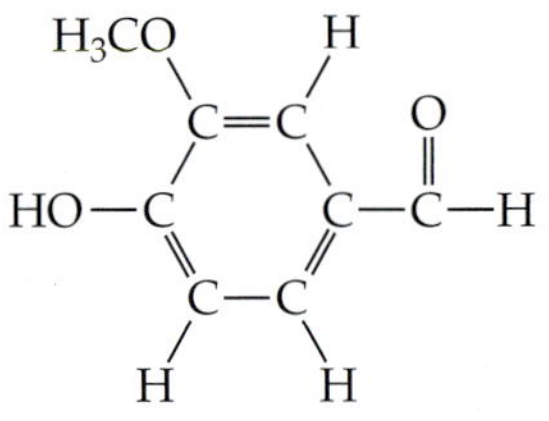
Vanillin (vanilla flavor)

(c) 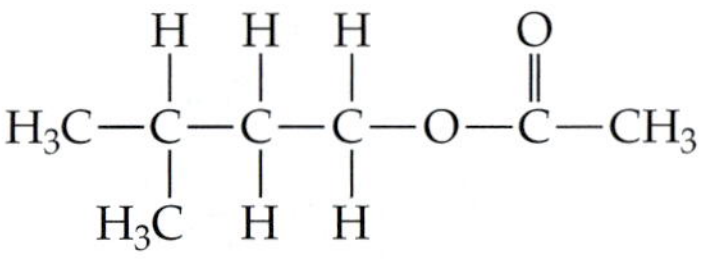
Isopentyl acetate (banana flavor)

3.26 Calculate the percentage of carbon by mass in each of the compounds represented by the following models:

(a)

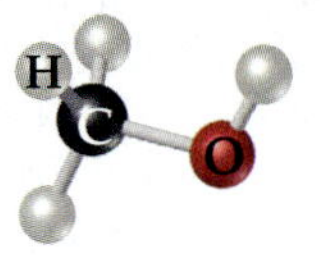
(b)

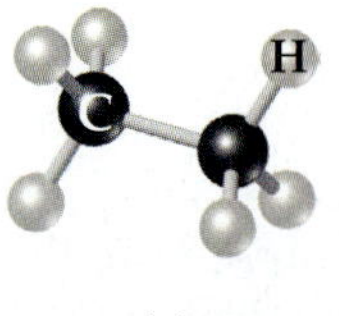
(c)

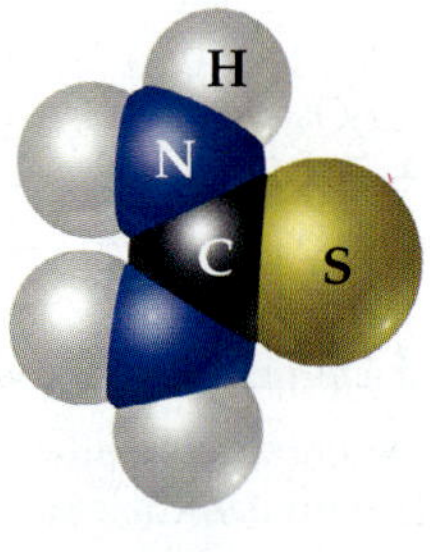
(d)

Avogadro's Number and the Mole

3.27 **(a)** What is Avogadro's number, and how is it related to the mole? **(b)** What is the relationship between the formula weight of a substance and its molar mass?

3.28 **(a)** What is the mass, in grams, of a mole of ^{12}C? **(b)** How many carbon atoms are present in a mole of ^{12}C?

3.29 Without doing any detailed calculations (but using a periodic table to give atomic weights), rank the following samples in order of increasing number of atoms: 0.50 mol H_2O, 23 g Na, 6.0×10^{23} N_2 molecules.

3.30 Without doing any detailed calculations (but using a periodic table to give atomic weights), rank the following samples in order of increasing number of atoms: 3.0×10^{23} molecules of H_2O_2, 2.0 mol CH_4, 32 g O_2.

3.31 What is the mass, in kilograms, of an Avogadro's number of Olympic shot-put balls if each one has a mass of 16 lb? How does this compare with the mass of Earth, 5.98×10^{24} kg?

3.32 If Avogadro's number of pennies is divided equally among the 292 million men, women, and children in the United States, how many dollars would each receive? How does this compare with the national debt of the United States, which was $7.0 trillion at the time of the writing of this text?

3.33 Calculate the following quantities:
(a) mass, in grams, of 0.773 mol CaH_2
(b) moles of $Mg(NO_3)_2$ in 5.35 g of this substance
(c) number of molecules in 0.0305 mol CH_3OH
(d) number of C atoms in 0.585 mol C_4H_{10}

3.34 Calculate the following quantities:
(a) mass, in grams, of 1.906×10^{-2} mol BaI_2
(b) number of moles of NH_4Cl in 48.3 g of this substance
(c) number of molecules in 0.05752 mol $HCHO_2$
(d) number of O atoms in 4.88×10^{-3} mol $Al(NO_3)_3$

3.35 **(a)** What is the mass, in grams, of 2.50×10^{-3} mol of ammonium phosphate?
(b) How many moles of chloride ions are in 0.2550 g of aluminum chloride?
(c) What is the mass, in grams, of 7.70×10^{20} molecules of caffeine, $C_8H_{10}N_4O_2$?
(d) What is the molar mass of cholesterol if 0.00105 mol weighs 0.406 g?

3.36 **(a)** What is the mass, in grams, of 0.0714 mol of iron(III) sulfate?
(b) How many moles of ammonium ions are in 8.776 g of ammonium carbonate?
(c) What is the mass, in grams, of 6.52×10^{21} molecules of aspirin, $C_9H_8O_4$?
(d) What is the molar mass of diazepam (Valium®) if 0.05570 mol weighs 15.86 g?

3.37 The molecular formula of allicin, the compound responsible for the characteristic smell of garlic, is $C_6H_{10}OS_2$. **(a)** What is the molar mass of allicin? **(b)** How many moles of allicin are present in 5.00 mg of this substance? **(c)** How many molecules of allicin are in 5.00 mg of this substance? **(d)** How many S atoms are present in 5.00 mg of allicin?

3.38 The molecular formula of aspartame, the artificial sweetener marketed as NutraSweet®, is $C_{14}H_{18}N_2O_5$. **(a)** What is the molar mass of aspartame? **(b)** How many moles of aspartame are present in 1.00 mg of aspartame? **(c)** How many molecules of aspartame are present in 1.00 mg of aspartame? **(d)** How many hydrogen atoms are present in 1.00 mg of aspartame?

3.39 A sample of glucose, $C_6H_{12}O_6$, contains 1.250×10^{21} atoms of carbon. **(a)** How many atoms of hydrogen does it contain? **(b)** How many molecules of glucose does it contain? **(c)** How many moles of glucose does it contain? **(d)** What is the mass of this sample in grams?

3.40 A sample of the male sex hormone testosterone, $C_{19}H_{28}O_2$, contains 7.08×10^{20} atoms of hydrogen. **(a)** How many atoms of carbon does it contain? **(b)** How many molecules of testosterone does it contain? **(c)** How many moles of testosterone does it contain? **(d)** What is the mass of this sample in grams?

3.41 The allowable concentration level of vinyl chloride, C_2H_3Cl, in the atmosphere in a chemical plant is 2.0×10^{-6} g/L. How many moles of vinyl chloride in each liter does this represent? How many molecules per liter?

3.42 At least 25 μg of tetrahydrocannabinol (THC), the active ingredient in marijuana, is required to produce intoxication. The molecular formula of THC is $C_{21}H_{30}O_2$. How many moles of THC does this 25 μg represent? How many molecules?

Empirical Formulas

3.43 Give the empirical formula of each of the following compounds if a sample contains **(a)** 0.0130 mol C, 0.0390 mol H, and 0.0065 mol O; **(b)** 11.66 g iron and 5.01 g oxygen; **(c)** 40.0% C, 6.7% H, and 53.3% O by mass.

3.44 Determine the empirical formula of each of the following compounds if a sample contains **(a)** 0.104 mol K, 0.052 mol C, and 0.156 mol O; **(b)** 5.28 g Sn and 3.37 g F; **(c)** 87.5% N and 12.5% H by mass.

3.45 Determine the empirical formulas of the compounds with the following compositions by mass:
(a) 10.4% C, 27.8% S, and 61.7% Cl
(b) 21.7% C, 9.6% O, and 68.7% F
(c) 32.79% Na, 13.02% Al, and 54.19% F

3.46 Determine the empirical formulas of the compounds with the following compositions by mass:
(a) 55.3% K, 14.6% P, and 30.1% O
(b) 24.5% Na, 14.9% Si, and 60.6% F
(c) 62.1% C, 5.21% H, 12.1% N, and 20.7% O

3.47 What is the molecular formula of each of the following compounds?
(a) empirical formula CH_2, molar mass = 84 g/mol
(b) empirical formula NH_2Cl, molar mass = 51.5 g/mol

3.48 What is the molecular formula of each of the following compounds?
(a) empirical formula HCO_2, molar mass = 90.0 g/mol
(b) empirical formula C_2H_4O, molar mass = 88 g/mol

3.49 Determine the empirical and molecular formulas of each of the following substances:
(a) Styrene, a compound substance used to make Styrofoam® cups and insulation, contains 92.3% C and 7.7% H by mass and has a molar mass of 104 g/mol.
(b) Caffeine, a stimulant found in coffee, contains 49.5% C, 5.15% H, 28.9% N, and 16.5% O by mass and has a molar mass about 195 g/mol.
(c) Monosodium glutamate (MSG), a flavor enhancer in certain foods, contains 35.51% C, 4.77% H, 37.85% O, 8.29% N, and 13.60% Na, and has a molar mass of 169 g/mol.

3.50 Determine the empirical and molecular formulas of each of the following substances:
(a) Ibuprofen, a headache remedy, contains 75.69% C, 8.80% H, and 15.51% O by mass, and has a molar mass about 206 g/mol.
(b) Cadaverine, a foul smelling substance produced by the action of bacteria on meat, contains 58.55% C, 13.81% H, and 27.40% N by mass; its molar mass is 102.2 g/mol.
(c) Epinephrine (adrenaline), a hormone secreted into the bloodstream in times of danger or stress, contains 59.0% C, 7.1% H, 26.2% O, and 7.7% N by mass; its MW is about 180 amu.

3.51 (a) Combustion analysis of toluene, a common organic solvent, gives 5.86 mg of CO_2 and 1.37 mg of H_2O. If the compound contains only carbon and hydrogen, what is its empirical formula? (b) Menthol, the substance we can smell in mentholated cough drops, is composed of C, H, and O. A 0.1005-g sample of menthol is combusted, producing 0.2829 g of CO_2 and 0.1159 g of H_2O. What is the empirical formula for menthol? If the compound has a molar mass of 156 g/mol, what is its molecular formula?

3.52 (a) The characteristic odor of pineapple is due to ethyl butyrate, a compound containing carbon, hydrogen, and oxygen. Combustion of 2.78 mg of ethyl butyrate produces 6.32 mg of CO_2 and 2.58 mg of H_2O. What is the empirical formula of the compound? (b) Nicotine, a component of tobacco, is composed of C, H, and N. A 5.250-mg sample of nicotine was combusted, producing 14.242 mg of CO_2 and 4.083 mg of H_2O. What is the empirical formula for nicotine? If the substance has a molar mass of 60 ± 5 g/mol, what is its molecular formula?

3.53 Washing soda, a compound used to prepare hard water for washing laundry, is a hydrate, which means that a certain number of water molecules are included in the solid structure. Its formula can be written as $Na_2CO_3 \cdot xH_2O$, where x is the number of moles of H_2O per mole of Na_2CO_3. When a 2.558-g sample of washing soda is heated at 25°C, all the water of hydration is lost, leaving 0.948 g of Na_2CO_3. What is the value of x?

3.54 Epsom salts, a strong laxative used in veterinary medicine, is a hydrate, which means that a certain number of water molecules are included in the solid structure. The formula for Epsom salts can be written as $MgSO_4 \cdot xH_2O$, where x indicates the number of moles of H_2O per mole of $MgSO_4$. When 5.061 g of this hydrate is heated to 250°C, all the water of hydration is lost, leaving 2.472 g of $MgSO_4$. What is the value of x?

Calculations Based on Chemical Equations

3.55 Why is it essential to use balanced chemical equations when determining the quantity of a product formed from a given quantity of a reactant?

3.56 What parts of balanced chemical equations give information about the relative numbers of moles of reactants and products involved in a reaction?

3.57 Hydrofluoric acid, HF(*aq*), cannot be stored in glass bottles because compounds called silicates in the glass are attacked by the HF(*aq*). Sodium silicate (Na_2SiO_3), for example, reacts as follows:

$$Na_2SiO_3(s) + 8\,HF(aq) \longrightarrow H_2SiF_6(aq) + 2\,NaF(aq) + 3\,H_2O(l)$$

(a) How many moles of HF are needed to react with 0.300 mol of Na_2SiO_3?
(b) How many grams of NaF form when 0.500 mol of HF reacts with excess Na_2SiO_3?
(c) How many grams of Na_2SiO_3 can react with 0.800 g of HF?

3.58 The fermentation of glucose ($C_6H_{12}O_6$) produces ethyl alcohol (C_2H_5OH) and CO_2:

$$C_6H_{12}O_6(aq) \longrightarrow 2\,C_2H_5OH(aq) + 2\,CO_2(g)$$

(a) How many moles of CO_2 are produced when 0.400 mol of $C_6H_{12}O_6$ reacts in this fashion?
(b) How many grams of $C_6H_{12}O_6$ are needed to form 7.50 g of C_2H_5OH?
(c) How many grams of CO_2 form when 7.50 g of C_2H_5OH are produced?

3.59 Several brands of antacids use $Al(OH)_3$ to react with stomach acid, which contains primarily HCl:

$$Al(OH)_3(s) + HCl(aq) \longrightarrow AlCl_3(aq) + H_2O(l)$$

(a) Balance this equation.
(b) Calculate the number of grams of HCl that can react with 0.500 g of $Al(OH)_3$.
(c) Calculate the number of grams of $AlCl_3$ and the number of grams of H_2O formed when 0.500 g of $Al(OH)_3$ reaacts.
(d) Show that your calculations in parts (b) and (c) are consistent with the law of conservation of mass.

3.60 An iron ore sample contains Fe_2O_3 together with other substances. Reaction of the ore with CO produces iron metal:

$$Fe_2O_3(s) + CO(g) \longrightarrow Fe(s) + CO_2(g)$$

(a) Balance this equation.
(b) Calculate the number of grams of CO that can react with 0.150 kg of Fe_2O_3.
(c) Calculate the number of grams of Fe and the number of grams of CO_2 formed when 0.150 kg of Fe_2O_3 reacts.
(d) Show that your calculations in parts (b) and (c) are consistent with the law of conservation of mass.

3.61 Aluminum sulfide reacts with water to form aluminum hydroxide and hydrogen sulfide. **(a)** Write the balanced chemical equation for this reaction. **(b)** How many grams of aluminum hydroxide are obtained from 6.75 g of aluminum sulfide?

3.62 Calcium hydride reacts with water to form calcium hydroxide and hydrogen gas. **(a)** Write a balanced chemical equation for the reaction. **(b)** How many grams of calcium hydride are needed to form 8.500 g of hydrogen?

3.63 Automotive air bags inflate when sodium azide, NaN_3, rapidly decomposes to its component elements:

$$2\,NaN_3(s) \longrightarrow 2\,Na(s) + 3\,N_2(g)$$

(a) How many moles of N_2 are produced by the decomposition of 1.50 mol of NaN_3?
(b) How many grams of NaN_3 are required to form 10.0 g of nitrogen gas?
(c) How many grams of NaN_3 are required to produce 10.0 ft^3 of nitrogen gas if the gas has a density of 1.25 g/L?

3.64 The complete combustion of octane, C_8H_{18}, a component of gasoline, proceeds as follows:

$$2\,C_8H_{18}(l) + 25\,O_2(g) \longrightarrow 16\,CO_2(g) + 18\,H_2O(g)$$

(a) How many moles of O_2 are needed to burn 1.25 mol of C_8H_{18}?
(b) How many grams of O_2 are needed to burn 10.0 g of C_8H_{18}?
(c) Octane has a density of 0.692 g/mL at 20°C. How many grams of O_2 are required to burn 1.00 gal of C_8H_{18}?

3.65 A piece of aluminum foil 1.00 cm square and 0.550 mm thick is allowed to react with bromine to form aluminum bromide as shown in the accompanying photo.

(a) How many moles of aluminum were used? (The density of aluminum is 2.699 g/cm^3.) **(b)** How many grams of aluminum bromide form, assuming that the aluminum reacts completely?

3.66 Detonation of nitroglycerin proceeds as follows:

$$4\,C_3H_5N_3O_9(l) \longrightarrow 12\,CO_2(g) + 6\,N_2(g) + O_2(g) + 10\,H_2O(g)$$

(a) If a sample containing 2.00 mL of nitroglycerin (density = 1.592 g/mL) is detonated, how many total moles of gas are produced? **(b)** If each mole of gas occupies 55 L under the conditions of the explosion, how many liters of gas are produced? **(c)** How many grams of N_2 are produced in the detonation?

Limiting Reactants; Theoretical Yields

3.67 **(a)** Define the terms *limiting reactant* and *excess reactant*. **(b)** Why are the amounts of products formed in a reaction determined only by the amount of the limiting reactant?

3.68 **(a)** Define the terms *theoretical yield*, *actual yield*, and *percent yield*. **(b)** Why is the actual yield in a reaction almost always less than the theoretical yield?

3.69 A manufacturer of bicycles has 4815 wheels, 2305 frames, and 2255 handlebars. **(a)** How many bicycles can be manufactured using these parts? **(b)** How many parts of each kind are left over? **(c)** Which part limits the production of bicycles?

3.70 A bottling plant has 121,515 bottles with a capacity of 355 mL, 122,500 caps, and 40,875 L of beverage. **(a)** How many bottles can be filled and capped? **(b)** How much of each item is left over? **(c)** Which component limits the production?

3.71 Sodium hydroxide reacts with carbon dioxide as follows:

$$2\,NaOH(s) + CO_2(g) \longrightarrow Na_2CO_3(s) + H_2O(l)$$

Which reagent is the limiting reactant when 1.85 mol NaOH and 1.00 mol CO_2 are allowed to react? How many moles of Na_2CO_3 can be produced? How many moles of the excess reactant remain after the completion of the reaction?

3.72 Aluminum hydroxide reacts with sulfuric acid as follows:

$$2\,Al(OH)_3(s) + 3\,H_2SO_4(aq) \longrightarrow Al_2(SO_4)_3(aq) + 6\,H_2O(l)$$

Which reagent is the limiting reactant when 0.500 mol $Al(OH)_3$ and 0.500 mol H_2SO_4 are allowed to react? How many moles of $Al_2(SO_4)_3$ can form under these conditions? How many moles of the excess reactant remain after the completion of the reaction?

3.73 The fizz produced when an Alka-Seltzer® tablet is dissolved in water is due to the reaction between sodium bicarbonate ($NaHCO_3$) and citric acid ($H_3C_6H_5O_7$):

$$3\,NaHCO_3(aq) + H_3C_6H_5O_7(aq) \longrightarrow 3\,CO_2(g) + 3\,H_2O(l) + Na_3C_6H_5O_7(aq)$$

In a certain experiment 1.00 g of sodium bicarbonate and 1.00 g of citric acid are allowed to react. **(a)** Which is the limiting reactant? **(b)** How many grams of carbon dioxide form? **(c)** How many grams of the excess reactant remain after the limiting reactant is completely consumed?

3.74 One of the steps in the commercial process for converting ammonia to nitric acid is the conversion of NH_3 to NO:

$$4\,NH_3(g) + 5\,O_2(g) \longrightarrow 4\,NO(g) + 6\,H_2O(g)$$

In a certain experiment, 1.50 g of NH_3 reacts with 2.75 g of O_2. **(a)** Which is the limiting reactant? **(b)** How many grams of NO and of H_2O form? **(c)** How many grams of the excess reactant remain after the limiting reactant is completely consumed? **(d)** Show that your calculations in parts (b) and (c) are consistent with the law of conservation of mass.

3.75 Solutions of sodium carbonate and silver nitrate react to form solid silver carbonate and a solution of sodium nitrate. A solution containing 3.50 g of sodium carbonate is mixed with one containing 5.00 g of silver nitrate. How many grams of sodium carbonate, silver nitrate, silver carbonate, and sodium nitrate are present after the reaction is complete?

3.76 Solutions of sulfuric acid and lead(II) acetate react to form solid lead(II) sulfate and a solution of acetic acid. If 7.50 g of sulfuric acid and 7.50 g of lead(II) acetate are mixed, calculate the number of grams of sulfuric acid, lead(II) acetate, lead(II) sulfate, and acetic acid present in the mixture after the reaction is complete.

3.77 When benzene (C_6H_6) reacts with bromine (Br_2), bromobenzene (C_6H_5Br) is obtained:

$$C_6H_6 + Br_2 \longrightarrow C_6H_5Br + HBr$$

(a) What is the theoretical yield of bromobenzene in this reaction when 30.0 g of benzene reacts with 65.0 g of bromine? **(b)** If the actual yield of bromobenzene was 56.7 g, what was the percentage yield?

3.78 When ethane (C_2H_6) reacts with chlorine (Cl_2), the main product is C_2H_5Cl; but other products containing Cl, such as $C_2H_4Cl_2$, are also obtained in small quantities. The formation of these other products reduces the yield of C_2H_5Cl. **(a)** Calculate the theoretical yield of C_2H_5Cl when 125 g of C_2H_6 reacts with 255 g of Cl_2, assuming that C_2H_6 and Cl_2 react only to form C_2H_5Cl and HCl. **(b)** Calculate the percent yield of C_2H_5Cl if the reaction produces 206 g of C_2H_5Cl.

3.79 Lithium and nitrogen react to produce lithium nitride:

$$6\,Li(s) + N_2(g) \longrightarrow 2\,Li_3N(s)$$

If 5.00 g of each reactant undergoes a reaction with a 88.5% yield, how many grams of Li_3N are obtained from the reaction?

3.80 When hydrogen sulfide gas is bubbled into a solution of sodium hydroxide, the reaction forms sodium sulfide and water. How many grams of sodium sulfide are formed if 1.50 g of hydrogen sulfide is bubbled into a solution containing 2.00 g of sodium hydroxide, assuming that the sodium sulfide is made in 92.0% yield?

Additional Exercises

3.81 Write the balanced chemical equation for **(a)** the complete combustion of butyric acid, $HC_4H_7O_2(l)$, a compound produced when butter becomes rancid; **(b)** the decomposition of solid nickel(II) hydroxide into solid nickel(II) oxide and water vapor; **(c)** the combination reaction between zinc metal and chlorine gas.

3.82 The effectiveness of nitrogen fertilizers depends on both their ability to deliver nitrogen to plants and the amount of nitrogen they can deliver. Four common nitrogen-containing fertilizers are ammonia, ammonium nitrate, ammonium sulfate, and urea [$(NH_2)_2CO$]. Rank these fertilizers in terms of the mass percentage nitrogen they contain.

3.83 **(a)** Diamond is a natural form of pure carbon. How many moles of carbon are in a 1.25-carat diamond (1 carat = 0.200 g)? How many atoms are in this diamond? **(b)** The molecular formula of acetylsalicylic acid (aspirin), one of the most common pain relievers, is $HC_9H_7O_4$. How many moles of $HC_9H_7O_4$ are in a 0.500-g tablet of aspirin? How many molecules of $HC_9H_7O_4$ are in this tablet?

3.84 **(a)** One molecule of the antibiotic known as penicillin G has a mass of 5.342×10^{-21} g. What is the molar mass of penicillin G? **(b)** Hemoglobin, the oxygen-carrying protein in red blood cells, has four iron atoms per molecule and contains 0.340% iron by mass. Calculate the molar mass of hemoglobin.

3.85 Very small crystals composed of 1000 to 100,000 atoms, called quantum dots, are being investigated for use in electronic devices.

(a) Calculate the mass in grams of a quantum dot consisting of 10,000 atoms of silicon.

(b) Assuming that the silicon in the dot has a density of 2.3 g/cm^3, calculate its volume.

(c) Assuming that the dot has the shape of a cube, calculate the length of each edge of the cube.

3.86 Serotonin is a compound that conducts nerve impulses in the brain. It contains 68.2 mass percent C, 6.86 mass percent H, 15.9 mass percent N, and 9.08 mass percent O. Its molar mass is 176 g/mol. Determine its molecular formula.

3.87 The koala dines exclusively on eucalyptus leaves. Its digestive system detoxifies the eucalyptus oil, a poison to other animals. The chief constituent in eucalyptus oil is a substance called eucalyptol, which contains 77.87% C, 11.76% H, and the remainder O. **(a)** What is the empirical formula for this substance? **(b)** A mass spectrum of eucalyptol shows a peak at about 154 amu. What is the molecular formula of the substance?

3.88 Vanillin, the dominant flavoring in vanilla, contains C, H, and O. When 1.05 g of this substance is completely combusted, 2.43 g of CO_2 and 0.50 g of H_2O are produced. What is the empirical formula of vanillin?

[3.89] An organic compound was found to contain only C, H, and Cl. When a 1.50-g sample of the compound was completely combusted in air, 3.52 g of CO_2 was formed. In a separate experiment the chlorine in a 1.00-g sample of the compound was converted to 1.27 g of AgCl. Determine the empirical formula of the compound.

[3.90] An oxybromate compound, $KBrO_x$, where x is unknown, is analyzed and found to contain 52.92% Br. What is the value of x?

[3.91] An element X forms an iodide (XI_3) and a chloride (XCl_3). The iodide is quantitatively converted to the chloride when it is heated in a stream of chlorine:

$$2\ XI_3 + 3\ Cl_2 \longrightarrow 2\ XCl_3 + 3\ I_2$$

If 0.5000 g of XI_3 is treated, 0.2360 g of XCl_3 is obtained. **(a)** Calculate the atomic weight of the element X. **(b)** Identify the element X.

3.92 If 1.5 mol of each of the following compounds is completely combusted in oxygen, which one will produce the largest number of moles of H_2O? Which will produce the least? Explain. C_2H_5OH, C_3H_8, $CH_3CH_2COCH_3$

3.93 A method used by the Environmental Protection Agency (EPA) for determining the concentration of ozone in air is to pass the air sample through a "bubbler" containing sodium iodide, which removes the ozone according to the following equation:

$$O_3(g) + 2\ NaI(aq) + H_2O(l) \longrightarrow O_2(g) + I_2(s) + 2\ NaOH(aq)$$

(a) How many moles of sodium iodide are needed to remove 3.8×10^{-5} mol O_3? **(b)** How many grams of sodium iodide are needed to remove 0.550 mg of O_3?

3.94 A chemical plant uses electrical energy to decompose aqueous solutions of NaCl to give Cl_2, H_2, and NaOH:

$$2\ NaCl(aq) + 2\ H_2O(l) \longrightarrow 2\ NaOH(aq) + H_2(g) + Cl_2(g)$$

If the plant produces 1.5×10^6 kg (1500 metric tons) of Cl_2 daily, estimate the quantities of H_2 and NaOH produced.

3.95 The fat stored in the hump of a camel is a source of both energy and water. Calculate the mass of H_2O produced by metabolism of 1.0 kg of fat, assuming the fat consists entirely of tristearin ($C_{57}H_{110}O_6$), a typical animal fat, and assuming that during metabolism, tristearin reacts with O_2 to form only CO_2 and H_2O.

[3.96] When hydrocarbons are burned in a limited amount of air, both CO and CO_2 form. When 0.450 g of a particular hydrocarbon was burned in air, 0.467 g of CO, 0.733 g of CO_2, and 0.450 g of H_2O were formed. **(a)** What is the empirical formula of the compound? **(b)** How many grams of O_2 were used in the reaction? **(c)** How many grams would have been required for complete combustion?

3.97 A mixture of $N_2(g)$ and $H_2(g)$ reacts in a closed container to form ammonia, $NH_3(g)$. The reaction ceases before either reactant has been totally consumed. At this stage 2.0 mol N_2, 2.0 mol H_2, and 2.0 mol NH_3 are present. How many moles of N_2 and H_2 were present originally?

[3.98] A mixture containing $KClO_3$, K_2CO_3, $KHCO_3$, and KCl was heated, producing CO_2, O_2, and H_2O gases according to the following equations:

$$2\ KClO_3(s) \longrightarrow 2\ KCl(s) + 3\ O_2(g)$$

$$2\ KHCO_3(s) \longrightarrow K_2O(s) + H_2O(g) + 2\ CO_2(g)$$

$$K_2CO_3(s) \longrightarrow K_2O(s) + CO_2(g)$$

The KCl does not react under the conditions of the reaction. If 100.0 g of the mixture produces 1.80 g of H_2O, 13.20 g of CO_2, and 4.00 g of O_2, what was the composition of the original mixture? (Assume complete decomposition of the mixture.)

3.99 When a mixture of 10.0 g of acetylene (C_2H_2) and 10.0 g of oxygen (O_2) is ignited, the resultant combustion reaction produces CO_2 and H_2O. **(a)** Write the balanced chemical equation for this reaction. **(b)** Which is the limiting reactant? **(c)** How many grams of C_2H_2, O_2, CO_2, and H_2O are present after the reaction is complete?

3.100 Aspirin ($C_9H_8O_4$) is produced from salicylic acid ($C_7H_6O_3$) and acetic anhydride ($C_4H_6O_3$):

$$C_7H_6O_3 + C_4H_6O_3 \longrightarrow C_9H_8O_4 + HC_2H_3O_2$$

(a) How much salicylic acid is required to produce 1.5×10^2 kg of aspirin, assuming that all of the salicylic acid is converted to aspirin? **(b)** How much salicylic acid would be required if only 80% of the salicylic acid is converted to aspirin? **(c)** What is the theoretical yield of aspirin if 185 kg of salicylic acid is allowed to react with 125 kg of acetic anhydride? **(d)** If the situation described in part (c) produces 182 kg of aspirin, what is the percentage yield?

Integrative Exercises

(These exercises require skills from earlier chapters as well as skills from the present chapter.)

3.101 Consider a sample of calcium carbonate in the form of a cube measuring 2.005 in. on each edge. If the sample has a density of 2.71 g/cm^3, how many oxygen atoms does it contain?

3.102 **(a)** You are given a cube of silver metal that measures 1.000 cm on each edge. The density of silver is 0.49 g/cm^3. How many atoms are in this cube? **(b)** Because atoms are spherical, they cannot occupy all of the space of the cube. The silver atoms pack in the solid in such a way that 74% of the volume of the solid is actually filled with the silver atoms. Calculate the volume of a single silver atom. **(c)** Using the volume of a silver atom and the formula for the volume of a sphere, calculate the radius in angstroms of a silver atom.

3.103 If an automobile travels 225 mi with a gas mileage of 20.5 mi/gal, how many kilograms of CO_2 are produced? Assume that the gasoline is composed of octane, $C_8H_{18}(l)$, whose density is 0.69 g/mL.

3.104 In 1865 a chemist reported that he had reacted a weighed amount of pure silver with nitric acid and had recovered all the silver as pure silver nitrate. The mass ratio of silver to silver nitrate was found to be 0.634985. Using only this ratio and the presently accepted values for the atomic weights of silver and oxygen, calculate the atomic weight of nitrogen. Compare this calculated atomic weight with the currently accepted value.

3.105 A particular coal contains 2.5% sulfur by mass. When this coal is burned, the sulfur is converted into sulfur dioxide gas. The sulfur dioxide reacts with calcium oxide to form solid calcium sulfite. **(a)** Write the balanced chemical equation for the reaction. **(b)** If the coal is burned in a power plant that uses 2000 tons of coal per day, what is the daily production of calcium sulfite?

3.106 Copper is an excellent electrical conductor widely used in making electric circuits. In producing a printed circuit board for the electronics industry, a layer of copper is laminated on a plastic board. A circuit pattern is then printed on the board using a chemically resistant polymer. The board is then exposed to a chemical bath that reacts with the exposed copper, leaving the desired copper circuit, which has been protected by the overlaying polymer. Finally, a solvent removes the polymer. One reaction used to remove the exposed copper from the circuit board is

$$Cu(s) + Cu(NH_3)_4Cl_2(aq) + 4\,NH_3(aq) \longrightarrow 2\,Cu(NH_3)_4Cl(aq)$$

A plant needs to produce 5000 circuit boards, each with a surface area measuring 2.0 in. × 3.0 in. The boards are covered with a 0.65-mm layer of copper. In subsequent processing, 85% of the copper is removed. Copper has a density of 8.96 g/cm^3. Calculate the masses of $Cu(NH_3)_4Cl_2$ and NH_3 needed to produce the circuit boards, assuming that the reaction used gives a 97% yield.

3.107 Hydrogen cyanide, HCN, is a poisonous gas. The lethal dose is approximately 300 mg HCN per kilogram of air when inhaled. **(a)** Calculate the amount of HCN that gives the lethal dose in a small laboratory room measuring 12 × 15 × 8.0 ft. The density of air at 26°C is 0.00118 g/cm^3. **(b)** If the HCN is formed by reaction of NaCN with an acid such as H_2SO_4, what mass of NaCN gives the lethal dose in the room?

$$2\,NaCN(s) + H_2SO_4(aq) \longrightarrow Na_2SO_4(aq) + 2\,HCN(g)$$

(c) HCN forms when synthetic fibers containing Orlon® or Acrilan® burn. Acrilan® has an empirical formula of CH_2CHCN, so HCN is 50.9% of the formula by mass. A rug measures 12 × 15 ft and contains 30 oz of Acrilan® fibers per square yard of carpet. If the rug burns, will a lethal dose of HCN be generated in the room? Assume that the yield of HCN from the fibers is 20% and that the carpet is 50% consumed.

eMEDIA EXERCISES

These exercises make use of the interactive objects available online in OneKey or the Companion Website, and on your Accelerator CD. Access to these resources comes in your MediaPak.

3.108 **(a)** Balance the three reactions available in the **Balancing Equations** activity (*3.1*). **(b)** If, in the case of the reduction of Fe_2O_3 you were to multiply each coefficient in the balanced equation by the same number, would the reaction still be balanced? **(c)** Would it still be balanced if you were to square each coefficient? **(d)** What would the coefficients be in the balanced equation if the reduction reaction produced carbon *mon*oxide instead of carbon *di*oxide?

3.109 Calculate the percentage composition of each compound in Exercise 3.22. Use the **Molecular Weight and Weight Percent** activity (*3.3*) to check your answers.

3.110 Consider the reaction of zinc metal with hydrochloric acid shown in the animation **Limiting Reagent** (*3.7*). **(a)** What would the limiting reagent have been if 100 mg of Zn had been combined with 2.0×10^{-3} mol of HCl? **(b)** What volume of hydrogen gas would the reaction have produced? **(c)** For the purposes of the limiting reagent experiment shown in the movie, why is it important that the hydrogen gas have a low solubility in water? How would the apparent yield of a reaction be affected if the evolved gas were very soluble in (or reactive with) water?

3.111 Write the balanced equation and predict the masses of products and remaining reactant for each of the following combinations. Use the **Limiting Reagents** (*3.7*) activity to check your answers. **(a)** 50 g $Pb(NO_3)_2$ and 55 g K_2CrO_4 to form $PbCrO_4$ and KNO_3; **(b)** 150 g $FeCl_2$ and 125 g Na_2S to form FeS and NaCl; **(c)** 96 g $Ca(NO_3)_2$ and 62 g Na_2CO_3 to form $CaCO_3$ and $NaNO_3$.

3.112 **(a)** In the reaction between $FeCl_3$ and NaOH to produce $Fe(OH)_3$ and NaCl, what mass of sodium hydroxide would be required to completely consume 50 g of iron(III) chloride? **(b)** Use the **Limiting Reagents** activity (*3.7*) to combine 50 g $FeCl_3$ with as close as you can get to the stoichiometric amount of sodium hydroxide, making sure that you add enough to completely consume the $FeCl_3$. How much sodium hydroxide is left over?

eLABORATORY EXERCISES

These exercises allow students to apply the concepts and skills for this chapter in the simulated laboratory of Virtual ChemLab. Worksheets for each of these exercises are found in the Virtual ChemLab Workbook in your MediaPak.

3.113 *(VCL 3-1) Counting Atoms*
It is a common problem for students to be asked to calculate the number of atoms in a pure elemental material or compound, given the mass of the sample. In this problem you will go into the virtual laboratory, retrieve a sample of gold (Au), use an analytical balance to measure its mass, and then calculate the number of atoms in the sample.

3.114 *(VCL 3-2) Counting Atoms*
In this problem you will go into the virtual laboratory, retrieve a sample of lead (Pb) and of uranium (U), use an analytical balance to measure the mass of each sample, and then calculate the number of atoms in each sample.

3.115 *(VCL 3-3) Counting Atoms*
In this problem you will go into the virtual laboratory; retrieve samples of erbium (Er), sodium (Na), tungsten (W), and a fourth sample of your choice; use an analytical balance to measure the mass of each sample; and then calculate the number of atoms in each sample.

3.116 *(VCL 3-4) Counting Molecules*
In this problem you will go into the virtual laboratory and retrieve a bottle of NaCl, weigh out approximately 1 g of the sample on an analytical balance, and then calculate the number of atoms of Na and Cl in the sample.

3.117 *(VCL 3-5) Counting Molecules*
In this problem you will go into the virtual laboratory and retrieve a bottle of sugar, weigh out approximately 1 g of the sample on an analytical balance, and then calculate the number of atoms of C, H, and O in the sample. You will also repeat this with a bottle of NH_4Cl.

3.118 *(VCL 3-6) Counting Protons, Neutrons, and Electrons*
In this problem you will enter the virtual laboratory and weigh out an isotopically pure sample of scandium (^{45}Sc), calculate the number of moles in the sample, and then calculate the number of protons, neutrons, and electrons.

3.119 *(VCL 3-7) Counting Protons, Neutrons, and Electrons*
In this problem you will enter the virtual laboratory and weigh out an isotopically pure sample of bismuth (^{209}Bi), calculate the number of moles in the sample, and then calculate the number of protons, neutrons, and electrons in $^{209}Bi^{5+}$.

3.120 *(VCL 3-8) Writing Balanced Precipitation Reactions*
It is an important skill that students be able to write balanced net ionic reactions for reactions that are observed in the laboratory. In this problem you will go into the virtual laboratory and perform a series of precipitation reactions using Ag^+, Pb^{2+}, and Sb^{3+}. After observing the reactions, you will write the net ionic equations representing these reactions and then balance them.

CHAPTER 4

WHAT'S AHEAD

- We begin by examining the nature of the substances dissolved in water, whether they exist in water as ions, molecules, or some mixture of the two. This information is necessary to understand the nature of reactants in aqueous solutions. *(Section 4.1)*

  Three major types of chemical processes occur in aqueous solution:
- *Precipitation reactions* are those in which soluble reactants yield an insoluble product. *(Section 4.2)*
- *Acid-base reactions* are those in which H^+ ions are transferred between reactants. *(Section 4.3)*
- *Oxidation-reduction reactions* are those in which electrons are transferred between reactants. *(Section 4.4)*
- Reactions between ions can be represented by *ionic equations* that show, for example, how ions can combine to form precipitates, or how they are removed from the solution or changed in some other way. *(Section 4.2)*
- After examining the common types of chemical reactions and how they are recognized and described, we consider how the *concentrations* of solutions can be expressed. *(Section 4.5)*
- We conclude the chapter by examining how the concepts of stoichiometry and concentration can be used to determine the amounts or concentrations of various substances. *(Section 4.6)*

Aqueous Reactions and Solution Stoichiometry

A VIEW OF THE PACIFIC OCEAN from along the California coastline.

THE WATERS OF THE PACIFIC OCEAN, seen in this chapter-opening photograph of the California coast, are part of the World Ocean that covers almost two-thirds of our planet. Water has been the key to much of Earth's evolutionary history. Life itself almost certainly originated in water, and the need for water by all forms of life has helped determine diverse biological structures. Your own body is about 60% water by mass. We will see repeatedly throughout this text that water possesses many unusual properties essential to supporting life on Earth.

The waters of the World Ocean may not appear to be any different from those of Lake Tahoe or the water that flows from your kitchen faucet, but a taste of seawater is all it takes to demonstrate that there is an important difference. Water has an exceptional ability to dissolve a wide variety of substances. Water in nature—whether it is drinking water from the tap, water from a clear mountain stream, or seawater—invariably contains a variety of dissolved substances. Solutions in which water is the dissolving medium are called **aqueous solutions**. Seawater differs from what we call "freshwater" in having a much higher total concentration of dissolved ionic substances.

Water is the medium for most of the chemical reactions that take place within us and around us. Nutrients dissolved in blood are carried to our cells, where they enter into reactions that help keep us alive. Automobile parts rust when they come into frequent contact with aqueous solutions that contain various

▲ **Figure 4.1 Limestone cave.** When CO_2 dissolves in water, the resulting solution is slightly acidic. Limestone caves are formed by the dissolving action of this acidic solution acting on $CaCO_3$ in the limestone.

dissolved substances. Spectacular limestone caves (Figure 4.1 ◀) are formed by the dissolving action of underground water containing carbon dioxide, $CO_2(aq)$:

$$CaCO_3(s) + H_2O(l) + CO_2(aq) \longrightarrow Ca(HCO_3)_2(aq) \quad [4.1]$$

We saw in Chapter 3 a few simple types of chemical reactions and how they are described. In this chapter we continue to examine chemical reactions by focusing on aqueous solutions. A great deal of important chemistry occurs in aqueous solutions, and we need to learn the vocabulary and concepts used to describe and understand this chemistry. In addition, we will extend the concepts of stoichiometry that we learned in Chapter 3 by considering how solution concentrations can be expressed and used.

4.1 | General Properties of Aqueous Solutions

A *solution* is a homogeneous mixture of two or more substances. (Section 1.2) The substance present in greatest quantity is usually called the **solvent**. The other substances in the solution are known as the **solutes**; they are said to be dissolved in the solvent. When a small amount of sodium chloride (NaCl) is dissolved in a large quantity of water, for example, the water is the solvent and the sodium chloride is the solute.

Electrolytic Properties

Imagine preparing two aqueous solutions—one by dissolving a teaspoon of table salt (sodium chloride) in a cup of water and the other by dissolving a teaspoon of table sugar (sucrose) in a cup of water. Both solutions are clear and colorless. How do they differ? One way, which might not be immediately obvious, is in their electrical conductivity: The salt solution is a good conductor of electricity, whereas the sugar solution is not.

Whether or not a solution conducts electricity can be determined by using a device such as that shown in Figure 4.2 ▶. To light the bulb, an electric current must flow between the two electrodes that are immersed in the solution. Although water itself is a poor conductor of electricity, the presence of ions causes aqueous solutions to become good conductors. Ions carry electrical charge from one electrode to the other, completing the electrical circuit. Thus, the conductivity of NaCl solutions indicates the presence of ions in the solution, and the lack of conductivity of sucrose solutions indicates the absence of ions. When NaCl dissolves in water, the solution contains Na^+ and Cl^- ions, each surrounded by water molecules. When sucrose ($C_{12}H_{22}O_{11}$) dissolves in water, the solution contains only neutral sucrose molecules surrounded by water molecules.

ANIMATION
Electrolytes and Nonelectrolytes

A substance (such as NaCl) whose aqueous solutions contain ions is called an **electrolyte**. A substance (such as $C_{12}H_{22}O_{11}$) that does not form ions in solution is called a **nonelectrolyte**. The difference between NaCl and $C_{12}H_{22}O_{11}$ arises largely because NaCl is ionic, whereas $C_{12}H_{22}O_{11}$ is molecular.

Ionic Compounds in Water

Recall from Section 2.7 and especially Figure 2.23 that solid NaCl consists of an orderly arrangement of Na^+ and Cl^- ions. When NaCl dissolves in water, each ion separates from the solid structure and disperses throughout the solution as shown in Figure 4.3(a) ▶. The ionic solid *dissociates* into its component ions as it dissolves.

Water is a very effective solvent for ionic compounds. Although water is an electrically neutral molecule, one end of the molecule (the O atom) is rich in electrons and thus possesses a partial negative charge, denoted by $\delta-$.

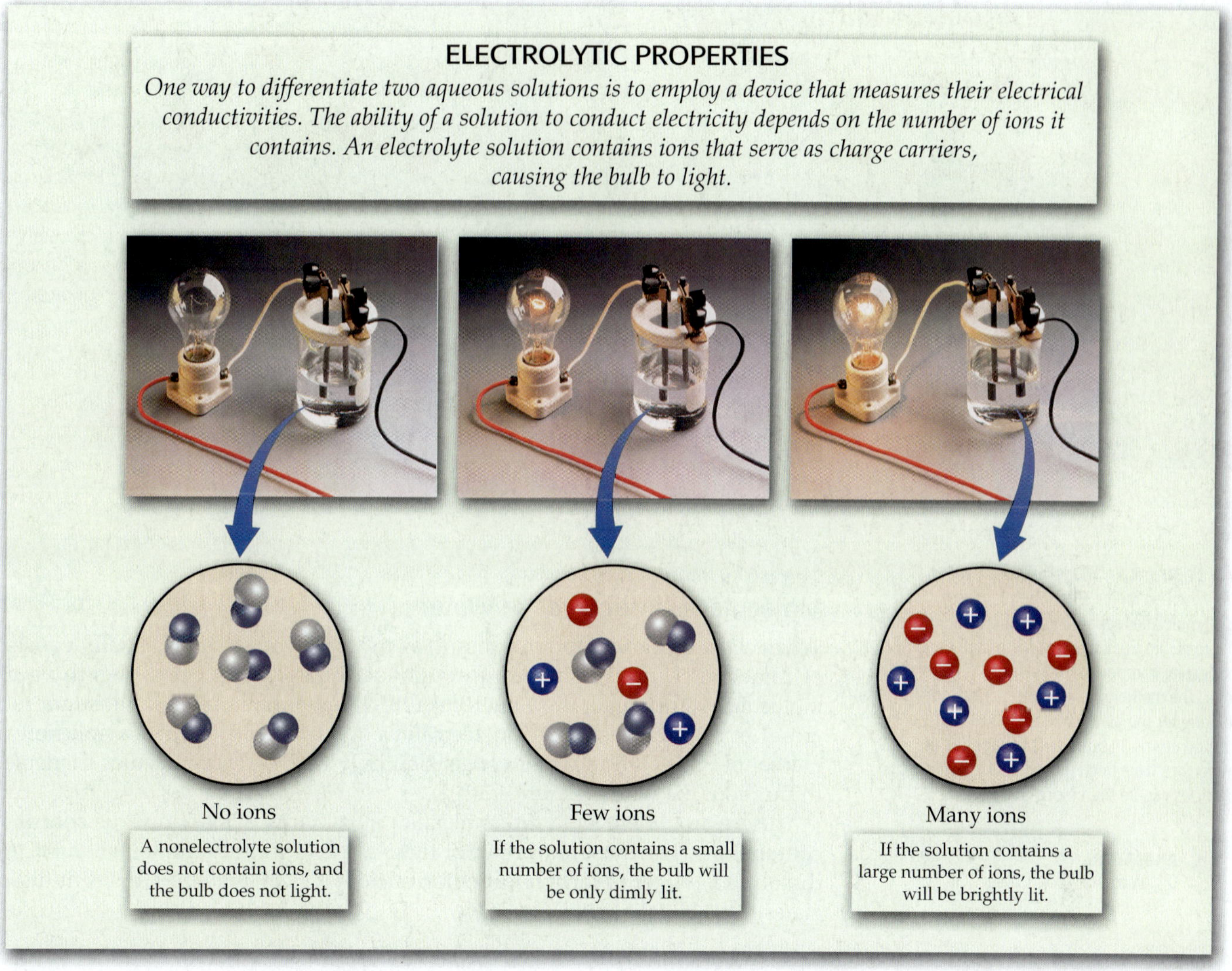

▲ **Figure 4.2 Measuring ion concentrations using conductivity.**

The other end (the H atoms) has a partial positive charge, denoted by $\delta+$. Positive ions (cations) are attracted by the negative end of H_2O, and negative ions (anions) are attracted by the positive end.

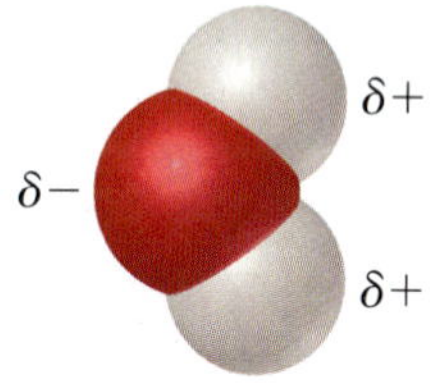

As an ionic compound dissolves, the ions become surrounded by H_2O molecules as shown in Figure 4.3(a). The ions are said to be solvated. The **solvation** process helps stabilize the ions in solution and prevents cations and anions from recombining. Furthermore, because the ions and their shells of surrounding water molecules are free to move about, the ions become dispersed uniformly throughout the solution.

We can usually predict the nature of the ions present in a solution of an ionic compound from the chemical name of the substance. Sodium sulfate (Na_2SO_4), for example, dissociates into sodium ions (Na^+) and sulfate ions (SO_4^{2-}). You must remember the formulas and charges of common ions (Tables 2.4 and 2.5) to understand the forms in which ionic compounds exist in aqueous solution.

GIVE IT SOME THOUGHT

What dissolved species are present in a solution of (a) KCN, (b) $NaClO_4$?

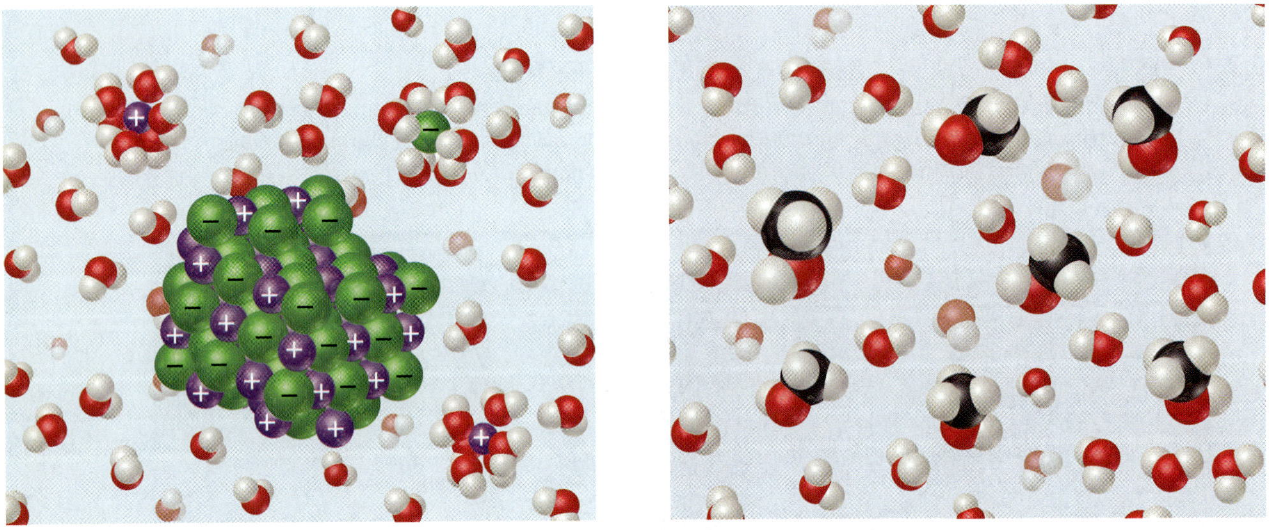

(a) (b)

▲ **Figure 4.3 Dissolution in water.** (a) When an ionic compound dissolves in water, H_2O molecules separate, surround, and disperse the ions into the liquid. (b) Methanol, CH_3OH, a molecular compound, dissolves without forming ions. The methanol molecules can be found by looking for the black spheres, which represent carbon atoms. In both parts (a) and (b) the water molecules have been moved apart so the solute particles can be seen more clearly.

ANIMATION
Dissolution of NaCl in Water

Molecular Compounds in Water

When a molecular compound dissolves in water, the solution usually consists of intact molecules dispersed throughout the solution. Consequently, most molecular compounds are nonelectrolytes. As we have seen, table sugar (sucrose) is an example of a nonelectrolyte. As another example, a solution of methanol (CH_3OH) in water consists entirely of CH_3OH molecules dispersed throughout the water [Figure 4.3(b)].

There are, however, a few molecular substances whose aqueous solutions contain ions. The most important of these are acids. For example, when $HCl(g)$ dissolves in water to form hydrochloric acid, $HCl(aq)$, it *ionizes;* that is, it dissociates into $H^+(aq)$ and $Cl^-(aq)$ ions.

MOVIE
Strong and Weak Electrolytes

VCL Strong and Weak Electrolytes

Strong and Weak Electrolytes

There are two categories of electrolytes, strong and weak, which differ in the extent to which they conduct electricity. **Strong electrolytes** are those solutes that exist in solution completely or nearly completely as ions. Essentially all soluble ionic compounds (such as NaCl) and a few molecular compounds (such as HCl) are strong electrolytes. **Weak electrolytes** are those solutes that exist in solution mostly in the form of molecules with only a small fraction in the form of ions. For example, in a solution of acetic acid ($HC_2H_3O_2$) most of the solute is present as $HC_2H_3O_2$ molecules. Only a small fraction (about 1%) of the $HC_2H_3O_2$ is present as $H^+(aq)$ and $C_2H_3O_2^-(aq)$ ions.

We must be careful not to confuse the extent to which an electrolyte dissolves with whether it is strong or weak. For example, $HC_2H_3O_2$ is extremely soluble in water but is a weak electrolyte. $Ba(OH)_2$, on the other hand, is not very soluble, but the amount of the substance that does dissolve dissociates almost completely, so $Ba(OH)_2$ is a strong electrolyte.

When a weak electrolyte such as acetic acid ionizes in solution, we write the reaction in the following manner:

$$HC_2H_3O_2(aq) \rightleftharpoons H^+(aq) + C_2H_3O_2^-(aq) \quad [4.2]$$

The double arrow means that the reaction is significant in both directions. At any given moment some $HC_2H_3O_2$ molecules are ionizing to form H^+ and $C_2H_3O_2^-$. At the same time, H^+ and $C_2H_3O_2^-$ ions are recombining to form $HC_2H_3O_2$. The balance between these opposing processes determines the relative numbers of ions and neutral molecules. This balance produces a state of **chemical equilibrium** that varies from one weak electrolyte to another. Chemical equilibria are extremely important, and we will devote Chapters 15–17 to examining them in detail.

Chemists use a double arrow to represent the ionization of weak electrolytes and a single arrow to represent the ionization of strong electrolytes. Because HCl is a strong electrolyte, we write the equation for the ionization of HCl as follows:

$$HCl(aq) \longrightarrow H^+(aq) + Cl^-(aq) \quad [4.3]$$

The absence of a reverse arrow indicates that the H^+ and Cl^- ions have no tendency to recombine in water to form HCl molecules.

In the sections ahead we will begin to look more closely at how we can use the composition of a compound to predict whether it is a strong electrolyte, weak electrolyte, or nonelectrolyte. For the moment, it is important only to remember that *soluble ionic compounds are strong electrolytes.* We identify ionic compounds as being ones composed of metals and nonmetals [such as NaCl, $FeSO_4$, and $Al(NO_3)_3$], or compounds containing the ammonium ion, NH_4^+ [such as NH_4Br and $(NH_4)_2CO_3$].

GIVE IT SOME THOUGHT

Which solute will cause the lightbulb in the experiment shown in Figure 4.2 to glow more brightly, CH_3OH or $MgBr_2$?

SAMPLE EXERCISE 4.1 | Relating Relative Numbers of Anions and Cations to Chemical Formulas

The diagram on the right represents an aqueous solution of one of the following compounds: $MgCl_2$, KCl, or K_2SO_4. Which solution does it best represent?

Solution

Analyze: We are asked to associate the charged spheres in the diagram with ions present in a solution of an ionic substance.

Plan: We examine the ionic substances given in the problem to determine the relative numbers and charges of the ions that each contains. We then correlate these charged ionic species with the ones shown in the diagram.

Solve: The diagram shows twice as many cations as anions, consistent with the formulation K_2SO_4.

Check: Notice that the total net charge in the diagram is zero, as it must be if it is to represent an ionic substance.

PRACTICE EXERCISE

If you were to draw diagrams (such as that shown on the right) representing aqueous solutions of each of the following ionic compounds, how many anions would you show if the diagram contained six cations? **(a)** $NiSO_4$, **(b)** $Ca(NO_3)_2$, **(c)** Na_3PO_4, **(d)** $Al_2(SO_4)_3$

Answers: **(a)** 6, **(b)** 12, **(c)** 2, **(d)** 9

4.2 | Precipitation Reactions

VCL Precipitation Reactions

MOVIE Precipitation Reactions

Figure 4.4 ▼ shows two clear solutions being mixed, one containing lead nitrate [$Pb(NO_3)_2$] and the other containing potassium iodide (KI). The reaction between these two solutes produces an insoluble yellow product. Reactions that result in the formation of an insoluble product are known as **precipitation reactions**. A **precipitate** is an insoluble solid formed by a reaction in solution. In Figure 4.4 the precipitate is lead iodide (PbI_2), a compound that has a very low solubility in water:

$$Pb(NO_3)_2(aq) + 2\,KI(aq) \longrightarrow PbI_2(s) + 2\,KNO_3(aq) \quad [4.4]$$

The other product of this reaction, potassium nitrate, remains in solution.

Precipitation reactions occur when certain pairs of oppositely charged ions attract each other so strongly that they form an insoluble ionic solid. To predict whether certain combinations of ions form insoluble compounds, we must consider some guidelines concerning the solubilities of common ionic compounds.

▼ **Figure 4.4 A Precipitation Reaction.**

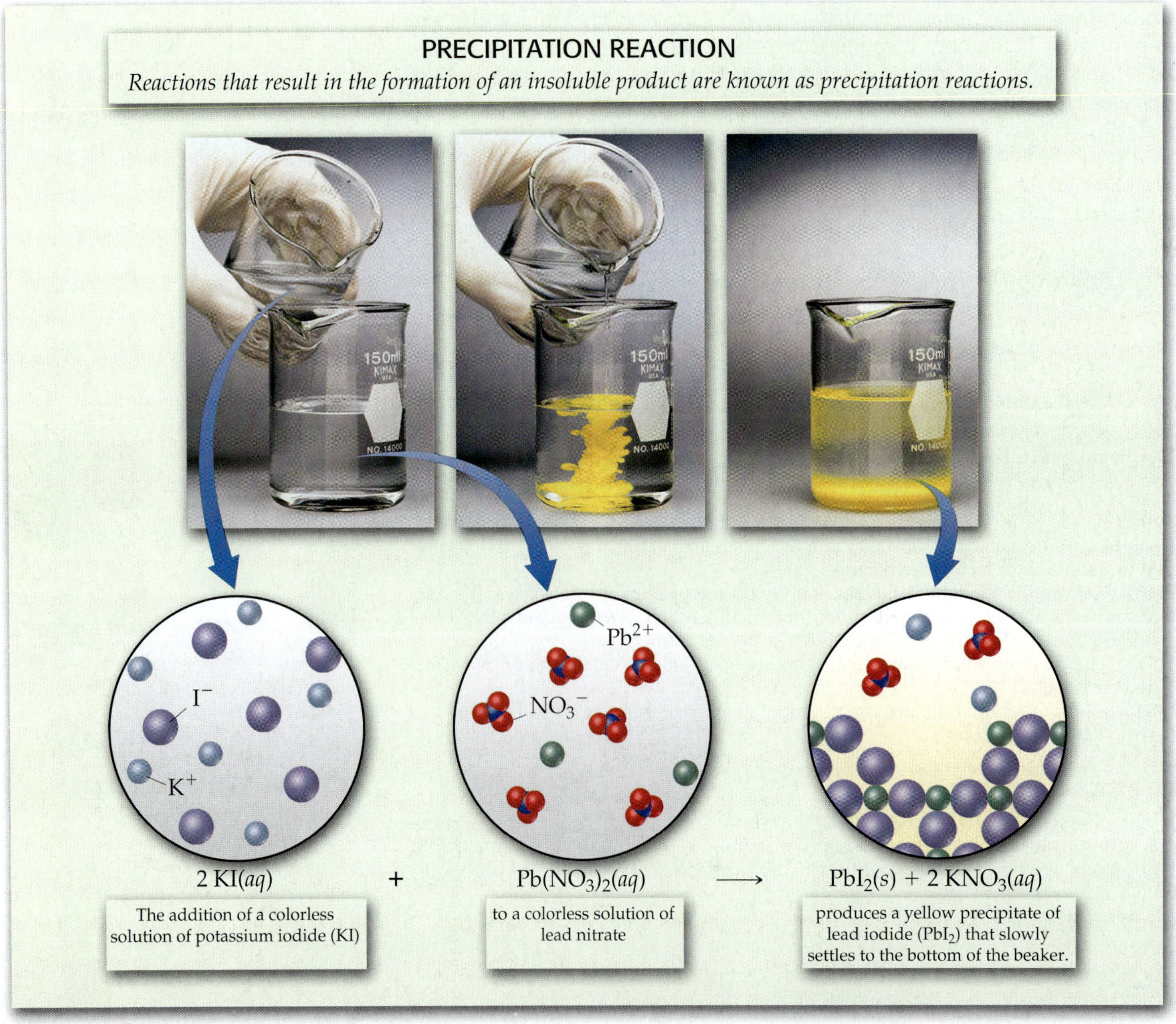

TABLE 4.1 Solubility Guidelines for Common Ionic Compounds in Water

Soluble Ionic Compounds		Important Exceptions
Compounds containing	NO_3^-	None
	$C_2H_3O_2^-$	None
	Cl^-	Compounds of Ag^+, Hg_2^{2+}, and Pb^{2+}
	Br^-	Compounds of Ag^+, Hg_2^{2+}, and Pb^{2+}
	I^-	Compounds of Ag^+, Hg_2^{2+}, and Pb^{2+}
	SO_4^{2-}	Compounds of Sr^{2+}, Ba^{2+}, Hg_2^{2+}, and Pb^{2+}

Insoluble Ionic Compounds		Important Exceptions
Compounds containing	S^{2-}	Compounds of NH_4^+, the alkali metal cations, and Ca^{2+}, Sr^{2+}, and Ba^{2+}
	CO_3^{2-}	Compounds of NH_4^+ and the alkali metal cations
	PO_4^{3-}	Compounds of NH_4^+ and the alkali metal cations
	OH^-	Compounds of the alkali metal cations, and NH_4^+, Ca^{2+}, Sr^{2+}, and Ba^{2+}

Solubility Guidelines for Ionic Compounds

The **solubility** of a substance at a given temperature is the amount of that substance that can be dissolved in a given quantity of solvent at that temperature. For instance, only 1.2×10^{-3} mol of PbI_2 dissolves in a liter of water at 25°C. In our discussions any substance with a solubility less than 0.01 mol/L will be referred to as *insoluble*. In those cases the attraction between the oppositely charged ions in the solid is too great for the water molecules to separate them to any significant extent, and the substance remains largely undissolved.

ACTIVITY Ionic Compounds

Unfortunately, there are no rules based on simple physical properties such as ionic charge to guide us in predicting whether a particular ionic compound will be soluble. Experimental observations, however, have led to guidelines for predicting solubility for ionic compounds. For example, experiments show that all common ionic compounds that contain the nitrate anion, NO_3^-, are soluble in water. Table 4.1 ▲ summarizes the solubility guidelines for common ionic compounds. The table is organized according to the anion in the compound, but it reveals many important facts about cations. Note that *all common ionic compounds of the alkali metal ions (group 1A of the periodic table) and of the ammonium ion* (NH_4^+) *are soluble in water.*

SAMPLE EXERCISE 4.2 | Using Solubility Rules

Classify the following ionic compounds as soluble or insoluble in water: **(a)** sodium carbonate (Na_2CO_3), **(b)** lead sulfate ($PbSO_4$).

Solution

Analyze: We are given the names and formulas of two ionic compounds and asked to predict whether they are soluble or insoluble in water.

Plan: We can use Table 4.1 to answer the question. Thus, we need to focus on the anion in each compound because the table is organized by anions.

Solve: **(a)** According to Table 4.1, most carbonates are insoluble, but carbonates of the alkali metal cations (such as sodium ion) are an exception to this rule and are soluble. Thus, Na_2CO_3 is soluble in water.

(b) Table 4.1 indicates that although most sulfates are water soluble, the sulfate of Pb^{2+} is an exception. Thus, $PbSO_4$ is insoluble in water.

PRACTICE EXERCISE

Classify the following compounds as soluble or insoluble in water: **(a)** cobalt(II) hydroxide, **(b)** barium nitrate, **(c)** ammonium phosphate.
Answers: **(a)** insoluble, **(b)** soluble, **(c)** soluble

To predict whether a precipitate forms when we mix aqueous solutions of two strong electrolytes, we must (1) note the ions present in the reactants, (2) consider the possible combinations of the cations and anions, and (3) use Table 4.1 to determine if any of these combinations is insoluble. For example, will a precipitate form when solutions of $Mg(NO_3)_2$ and NaOH are mixed? Both $Mg(NO_3)_2$ and NaOH are soluble ionic compounds, and they are both strong electrolytes. Mixing $Mg(NO_3)_2(aq)$ and $NaOH(aq)$ first produces a solution containing Mg^{2+}, NO_3^-, Na^+, and OH^- ions. Will either of the cations interact with either of the anions to form an insoluble compound? In addition to the reactants, the other possible interactions are Mg^{2+} with OH^- and Na^+ with NO_3^-. From Table 4.1 we see that hydroxides are generally insoluble. Because Mg^{2+} is not an exception, $Mg(OH)_2$ is insoluble and will thus form a precipitate. $NaNO_3$, however, is soluble, so Na^+ and NO_3^- will remain in solution. The balanced equation for the precipitation reaction is

$$Mg(NO_3)_2(aq) + 2\,NaOH(aq) \longrightarrow Mg(OH)_2(s) + 2\,NaNO_3(aq) \quad [4.5]$$

Exchange (Metathesis) Reactions

Notice in Equation 4.5 that the cations in the two reactants exchange anions—Mg^{2+} ends up with OH^-, and Na^+ ends up with NO_3^-. The chemical formulas of the products are based on the charges of the ions—two OH^- ions are needed to give a neutral compound with Mg^{2+}, and one NO_3^- ion is needed to give a neutral compound with Na^+. (Section 2.7) It is only after the chemical formulas of the products have been determined that the equation can be balanced.

Reactions in which positive ions and negative ions appear to exchange partners conform to the following general equation:

$$AX + BY \longrightarrow AY + BX \quad [4.6]$$

Example: $AgNO_3(aq) + KCl(aq) \longrightarrow AgCl(s) + KNO_3(aq)$

Such reactions are known as **exchange reactions**, or **metathesis reactions** (meh-TATH-eh-sis, which is the Greek word for "to transpose"). Precipitation reactions conform to this pattern, as do many acid-base reactions, as we will see in Section 4.3.

To complete and balance a metathesis equation, we follow these steps:

1. Use the chemical formulas of the reactants to determine the ions that are present.
2. Write the chemical formulas of the products by combining the cation from one reactant with the anion of the other. (Use the charges of the ions to determine the subscripts in the chemical formulas.)
3. Finally, balance the equation.

SAMPLE EXERCISE 4.3 | Predicting a Metathesis Reaction

(a) Predict the identity of the precipitate that forms when solutions of $BaCl_2$ and K_2SO_4 are mixed.
(b) Write the balanced chemical equation for the reaction.

Solution

Analyze: We are given two ionic reactants and asked to predict the insoluble product that they form.

Plan: We need to write down the ions present in the reactants and to exchange the anions between the two cations. Once we have written the chemical formulas for these products, we can use Table 4.1 to determine which is insoluble in water. Knowing the products also allows us to write the equation for the reaction.

Solve: **(a)** The reactants contain Ba^{2+}, Cl^-, K^+, and SO_4^{2-} ions. If we exchange the anions, we will have $BaSO_4$ and KCl. According to Table 4.1, most compounds of SO_4^{2-} are soluble but those of Ba^{2+} are not. Thus, $BaSO_4$ is insoluble and will precipitate from solution. KCl, on the other hand, is soluble.

(b) From part (a) we know the chemical formulas of the products, $BaSO_4$ and KCl. The balanced equation with phase labels shown is

$$BaCl_2(aq) + K_2SO_4(aq) \longrightarrow BaSO_4(s) + 2\,KCl(aq)$$

PRACTICE EXERCISE

(a) What compound precipitates when solutions of $Fe_2(SO_4)_3$ and LiOH are mixed? **(b)** Write a balanced equation for the reaction. **(c)** Will a precipitate form when solutions of $Ba(NO_3)_2$ and KOH are mixed?

Answers: **(a)** $Fe(OH)_3$; **(b)** $Fe_2(SO_4)_3(aq) + 6\,LiOH(aq) \longrightarrow 2\,Fe(OH)_3(s) + 3\,Li_2SO_4(aq)$; **(c)** no (both possible products are water soluble)

Ionic Equations

In writing chemical equations for reactions in aqueous solution, it is often useful to indicate explicitly whether the dissolved substances are present predominantly as ions or as molecules. Let's reconsider the precipitation reaction between $Pb(NO_3)_2$ and 2 KI, shown previously in Figure 4.4:

$$Pb(NO_3)_2(aq) + 2\,KI(aq) \longrightarrow PbI_2(s) + 2\,KNO_3(aq)$$

An equation written in this fashion, showing the complete chemical formulas of the reactants and products, is called a **molecular equation** because it shows the chemical formulas of the reactants and products without indicating their ionic character. Because $Pb(NO_3)_2$, KI, and KNO_3 are all soluble ionic compounds and therefore strong electrolytes, we can write the chemical equation to indicate explicitly the ions that are in the solution:

$$Pb^{2+}(aq) + 2\,NO_3^-(aq) + 2\,K^+(aq) + 2\,I^-(aq) \longrightarrow PbI_2(s) + 2\,K^+(aq) + 2\,NO_3^-(aq) \quad [4.7]$$

An equation written in this form, with all soluble strong electrolytes shown as ions, is known as a **complete ionic equation**.

Notice that $K^+(aq)$ and $NO_3^-(aq)$ appear on both sides of Equation 4.7. Ions that appear in identical forms among both the reactants and products of a complete ionic equation are called **spectator ions**. They are present but play no direct role in the reaction. When spectator ions are omitted from the equation (they cancel out like algebraic quantities), we are left with the **net ionic equation**:

$$Pb^{2+}(aq) + 2\,I^-(aq) \longrightarrow PbI_2(s) \quad [4.8]$$

A net ionic equation includes only the ions and molecules directly involved in the reaction. Charge is conserved in reactions, so the sum of the charges of the ions must be the same on both sides of a balanced net ionic equation. In this case the 2+ charge of the cation and the two 1− charges of the anions add to give zero, the charge of the electrically neutral product. *If every ion in a complete ionic equation is a spectator, then no reaction occurs.*

Net ionic equations are widely used to illustrate the similarities between large numbers of reactions involving electrolytes. For example, Equation 4.8 expresses the essential feature of the precipitation reaction between any strong electrolyte containing Pb^{2+} and any strong electrolyte containing I^-: The $Pb^{2+}(aq)$ and $I^-(aq)$ ions combine to form a precipitate of PbI_2. Thus, a net ionic equation demonstrates that more than one set of reactants can lead to the same net reaction. Aqueous solutions of KI and MgI_2 share many chemical similarities because both contain I^- ions. The complete equation, on the other hand, identifies the actual reactants that participate in a reaction.

The following steps summarize the procedure for writing net ionic equations:

1. Write a balanced molecular equation for the reaction.
2. Rewrite the equation to show the ions that form in solution when each soluble strong electrolyte dissociates into its component ions. *Only strong electrolytes dissolved in aqueous solution are written in ionic form.*
3. Identify and cancel spectator ions.

ACTIVITY
Writing a Net Ionic Equation

SAMPLE EXERCISE 4.4 | Writing a Net Ionic Equation

Write the net ionic equation for the precipitation reaction that occurs when solutions of calcium chloride and sodium carbonate are mixed.

Solution

Analyze: Our task is to write a net ionic equation for a precipitation reaction, given the names of the reactants present in solution.

Plan: We first need to write the chemical formulas of the reactants and products and to determine which product is insoluble. Then we write and balance the molecular equation. Next, we write each soluble strong electrolyte as separated ions to obtain the complete ionic equation. Finally, we eliminate the spectator ions to obtain the net ionic equation.

Solve: Calcium chloride is composed of calcium ions, Ca^{2+}, and chloride ions, Cl^-; hence an aqueous solution of the substance is $CaCl_2(aq)$. Sodium carbonate is composed of Na^+ ions and CO_3^{2-} ions; hence an aqueous solution of the compound is $Na_2CO_3(aq)$. In the molecular equations for precipitation reactions, the anions and cations appear to exchange partners. Thus, we put Ca^{2+} and CO_3^{2-} together to give $CaCO_3$ and Na^+ and Cl^- together to give NaCl. According to the solubility guidelines in Table 4.1, $CaCO_3$ is insoluble and NaCl is soluble. The balanced molecular equation is

$$CaCl_2(aq) + Na_2CO_3(aq) \longrightarrow CaCO_3(s) + 2\,NaCl(aq)$$

In a complete ionic equation, *only* dissolved strong electrolytes (such as soluble ionic compounds) are written as separate ions. As the (aq) designations remind us, $CaCl_2$, Na_2CO_3, and NaCl are all dissolved in the solution. Furthermore, they are all strong electrolytes. $CaCO_3$ is an ionic compound, but it is not soluble. We do not write the formula of any insoluble compound as its component ions. Thus, the complete ionic equation is

$$Ca^{2+}(aq) + 2\,Cl^-(aq) + 2\,Na^+(aq) + CO_3^{2-}(aq) \longrightarrow CaCO_3(s) + 2\,Na^+(aq) + 2\,Cl^-(aq)$$

Cl^- and Na^+ are spectator ions. Canceling them gives the following net ionic equation:

$$Ca^{2+}(aq) + CO_3^{2-}(aq) \longrightarrow CaCO_3(s)$$

Check: We can check our result by confirming that both the elements and the electric charge are balanced. Each side has one Ca, one C, and three O, and the net charge on each side equals 0.

Comment: If none of the ions in an ionic equation is removed from solution or changed in some way, then they all are spectator ions and a reaction does not occur.

PRACTICE EXERCISE

Write the net ionic equation for the precipitation reaction that occurs when aqueous solutions of silver nitrate and potassium phosphate are mixed.

Answers: $3\,Ag^+(aq) + PO_4^{3-}(aq) \longrightarrow Ag_3PO_4(s)$

4.3 | Acid-Base Reactions

Many acids and bases are industrial and household substances (Figure 4.5 ▶), and some are important components of biological fluids. Hydrochloric acid, for example, is not only an important industrial chemical but also the main constituent of gastric juice in your stomach. Acids and bases also happen to be common electrolytes.

▲ **Figure 4.5 Some common household acids (left) and bases (right).**

Acids

Acids are substances that ionize in aqueous solutions to form hydrogen ions, thereby increasing the concentration of $H^+(aq)$ ions. Because a hydrogen atom consists of a proton and an electron, H^+ is simply a proton. Thus, acids are often called proton donors. Molecular models of three common acids, HCl, HNO_3, and $HC_2H_3O_2$, are shown in the margin.

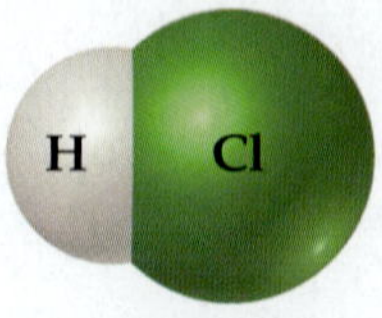

HCl

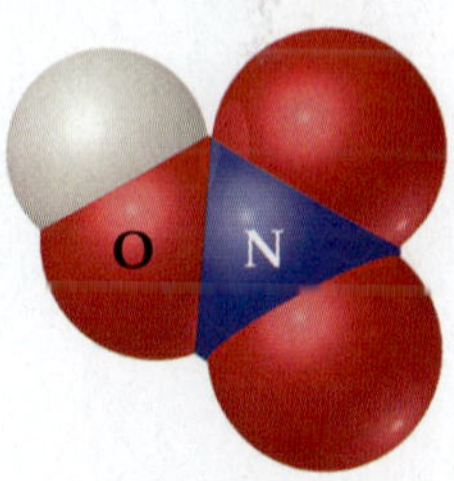

HNO_3

$HC_2H_3O_2$

Just as cations are surrounded and bound by water molecules (see Figure 4.3[a]), the proton is also solvated by water molecules. The nature of the proton in water is discussed in detail in Section 16.2. In writing chemical equations involving the proton in water, we represent it simply as $H^+(aq)$.

Molecules of different acids can ionize to form different numbers of H^+ ions. Both HCl and HNO_3 are *monoprotic* acids, which yield one H^+ per molecule of acid. Sulfuric acid, H_2SO_4, is a *diprotic* acid, one that yields two H^+ per molecule of acid. The ionization of H_2SO_4 and other diprotic acids occurs in two steps:

$$H_2SO_4(aq) \longrightarrow H^+(aq) + HSO_4^-(aq) \quad [4.9]$$

$$HSO_4^-(aq) \rightleftharpoons H^+(aq) + SO_4^{2-}(aq) \quad [4.10]$$

Although H_2SO_4 is a strong electrolyte, only the first ionization is complete. Thus, aqueous solutions of sulfuric acid contain a mixture of $H^+(aq)$, $HSO_4^-(aq)$, and $SO_4^{2-}(aq)$.

Bases

Bases are substances that accept (react with) H^+ ions. Bases produce hydroxide ions (OH^-) when they dissolve in water. Ionic hydroxide compounds such as NaOH, KOH, and $Ca(OH)_2$ are among the most common bases. When dissolved in water, they dissociate into their component ions, introducing OH^- ions into the solution.

Compounds that do not contain OH^- ions can also be bases. For example, ammonia (NH_3) is a common base. When added to water, it accepts an H^+ ion from the water molecule and thereby produces an OH^- ion (Figure 4.6 ▶):

$$NH_3(aq) + H_2O(l) \rightleftharpoons NH_4^+(aq) + OH^-(aq) \quad [4.11]$$

Because only a small fraction of the NH_3 (about 1%) forms NH_4^+ and OH^- ions, ammonia is a weak electrolyte.

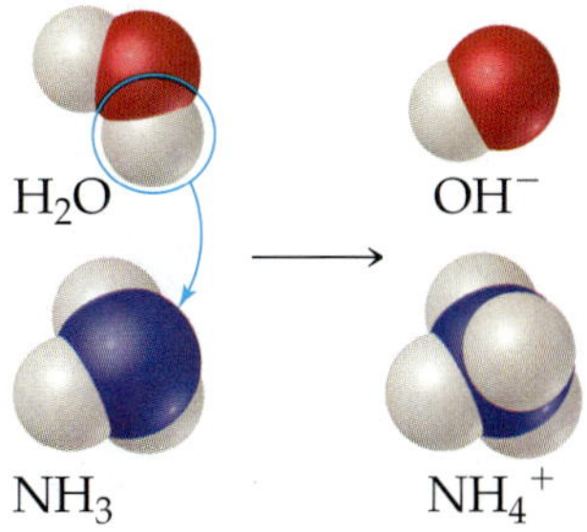

▲ **Figure 4.6 Hydrogen ion transfer.** An H_2O molecule acts as a proton donor (acid), and NH_3 as a proton acceptor (base). Only a fraction of the NH_3 reacts with H_2O; NH_3 is a weak electrolyte.

Strong and Weak Acids and Bases

Acids and bases that are strong electrolytes (completely ionized in solution) are called **strong acids** and **strong bases**. Those that are weak electrolytes (partly ionized) are called **weak acids** and **weak bases**. Strong acids are more reactive than weak acids when the reactivity depends only on the concentration of $H^+(aq)$. The reactivity of an acid, however, can depend on the anion as well as on $H^+(aq)$. For example, hydrofluoric acid (HF) is a weak acid (only partly ionized in aqueous solution), but it is very reactive and vigorously attacks many substances, including glass. This reactivity is due to the combined action of $H^+(aq)$ and $F^-(aq)$.

ANIMATION
Introduction to Aqueous Acids

ANIMATION
Introduction to Aqueous Bases

TABLE 4.2 Common Strong Acids and Bases

Strong Acids	Strong Bases
Hydrochloric, HCl	Group 1A metal hydroxides (LiOH, NaOH, KOH, RbOH, CsOH)
Hydrobromic, HBr	Heavy group 2A metal hydroxides [$Ca(OH)_2$, $Sr(OH)_2$, $Ba(OH)_2$]
Hydroiodic, HI	
Chloric, $HClO_3$	
Perchloric, $HClO_4$	
Nitric, HNO_3	
Sulfuric, H_2SO_4	

Table 4.2 ▲ lists the common strong acids and bases. You should commit these to memory. As you examine this table, notice that some of the most common acids, such as HCl, HNO_3, and H_2SO_4, are strong. Three of the strong acids are the hydrogen compounds of the halogen family. (HF, however, is a weak acid.) The list of strong acids is very short. Most acids are weak. The only common strong bases are the hydroxides of Li^+, Na^+, K^+, Rb^+, and Cs^+ (the alkali metals, group 1A) and the hydroxides of Ca^{2+}, Sr^{2+}, and Ba^{2+} (the heavy alkaline earths, group 2A). These are the common soluble metal hydroxides. Most other metal hydroxides are insoluble in water. The most common weak base is NH_3, which reacts with water to form OH^- ions (Equation 4.11).

GIVE IT SOME THOUGHT

Which of the following is a strong acid: H_2SO_3, HBr, $HC_2H_3O_2$?

SAMPLE EXERCISE 4.5 | Comparing Acid Strengths

The following diagrams represent aqueous solutions of three acids (HX, HY, and HZ) with water molecules omitted for clarity. Rank them from strongest to weakest.

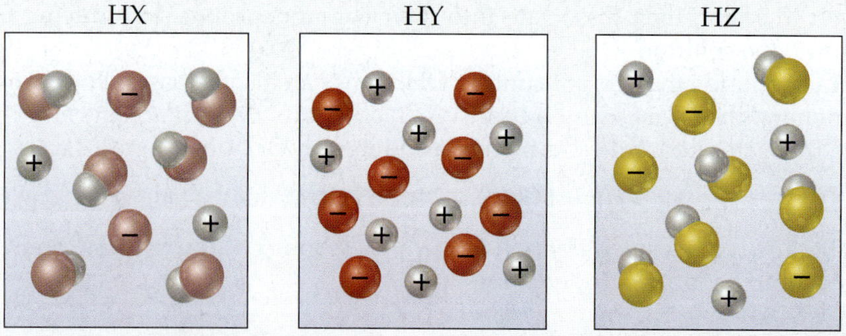

Solution

Analyze: We are asked to rank three acids from strongest to weakest, based on schematic drawings of their solutions.

Plan: We can examine the drawings to determine the relative numbers of uncharged molecular species present. The strongest acid is the one with the most H^+ ions and fewest undissociated acid molecules in solution. The weakest is the one with the largest number of undissociated molecules.

Solve: The order is HY > HZ > HX. HY is a strong acid because it is totally ionized (no HY molecules in solution), whereas both HX and HZ are weak acids, whose solutions consist of a mixture of molecules and ions. Because HZ contains more H^+ ions and fewer molecules than HX, it is a stronger acid.

PRACTICE EXERCISE

Imagine a diagram showing ten Na^+ ions and ten OH^- ions. If this solution were mixed with the one pictured on the previous page for HY, what would the diagram look like that represents the solution after any possible reaction? (H^+ ions will react with OH^- ions to form H_2O.)
Answer: The final diagram would show ten Na^+ ions, two OH^- ions, eight Y^- ions, and eight H_2O molecules.

TABLE 4.3 Summary of the Electrolytic Behavior of Common Soluble Ionic and Molecular Compounds

	Strong Electrolyte	Weak Electrolyte	Nonelectrolyte
Ionic	All	None	None
Molecular	Strong acids (see Table 4.2)	Weak acids (H . . .)	
		Weak bases (NH_3)	All other compounds

Identifying Strong and Weak Electrolytes

If we remember the common strong acids and bases (Table 4.2) and also remember that NH_3 is a weak base, we can make reasonable predictions about the electrolytic strength of a great number of water-soluble substances. Table 4.3 ▲ summarizes our observations about electrolytes. To classify a soluble substance as a strong electrolyte, weak electrolyte, or nonelectrolyte, we simply work our way down and across this table. We first ask ourselves whether the substance is ionic or molecular. If it is ionic, it is a strong electrolyte because the second column of Table 4.3 tells us that all ionic compounds are strong electrolytes. If the substance we want to classify is molecular, we ask whether it is an acid or a base. (Does it have H first in the chemical formula?) If it is an acid, we rely on the memorized list from Table 4.2 to determine whether it is a strong or weak electrolyte: All strong acids are strong electrolytes, and all weak acids are weak electrolytes. If an acid is not listed in Table 4.2, it is probably a weak acid and therefore a weak electrolyte. For example, H_3PO_4, H_2SO_3, and $HC_7H_5O_2$ are not listed in Table 4.2 and are weak acids. If the substance we want to classify is a base, we again turn to Table 4.2 to determine whether it is one of the listed strong bases. NH_3 is the only molecular base that we consider in this chapter, and Table 4.3 tells us it is a weak electrolyte. (There are compounds called amines that are related to NH_3 and are also molecular bases, but we will not consider them until Chapter 16.) Finally, any molecular substance that we encounter in this chapter that is not an acid or NH_3 is probably a nonelectrolyte.

SAMPLE EXERCISE 4.6 | Identifying Strong, Weak, and Nonelectrolytes

Classify each of the following dissolved substances as a strong electrolyte, weak electrolyte, or nonelectrolyte: $CaCl_2$, HNO_3, C_2H_5OH (ethanol), $HCHO_2$ (formic acid), KOH.

Solution
Analyze: We are given several chemical formulas and asked to classify each substance as a strong electrolyte, weak electrolyte, or nonelectrolyte.
Plan: The approach we take is outlined in Table 4.3. We can predict whether a substance is ionic or molecular, based on its composition. As we saw in Section 2.7, most ionic compounds we encounter in this text are composed of a metal and a nonmetal, whereas most molecular compounds are composed only of nonmetals.
Solve: Two compounds fit the criteria for ionic compounds: $CaCl_2$ and KOH. As Table 2.3 tells us that all ionic compounds are strong electrolytes, that is how we classify these two substances. The three remaining compounds are molecular.

Two, HNO_3 and $HCHO_2$, are acids. Nitric acid, HNO_3, is a common strong acid, as shown in Table 4.2, and therefore is a strong electrolyte. Because most acids are weak acids, our best guess would be that $HCHO_2$ is a weak acid (weak electrolyte). This is correct. The remaining molecular compound, C_2H_5OH, is neither an acid nor a base, so it is a nonelectrolyte.

Comment: Although C_2H_5OH has an OH group, it is not a metal hydroxide; thus, it is not a base. Rather, it is a member of a class of organic compounds that have C—OH bonds, which are known as alcohols. (Section 2.9)

PRACTICE EXERCISE

Consider solutions in which 0.1 mol of each of the following compounds is dissolved in 1 L of water: $Ca(NO_3)_2$ (calcium nitrate), $C_6H_{12}O_6$ (glucose), $NaC_2H_3O_2$ (sodium acetate), and $HC_2H_3O_2$ (acetic acid). Rank the solutions in order of increasing electrical conductivity, based on the fact that the greater the number of ions in solution, the greater the conductivity.

Answers: $C_6H_{12}O_6$ (nonelectrolyte) $<$ $HC_2H_3O_2$ (weak electrolyte, existing mainly in the form of molecules with few ions) $<$ $NaC_2H_3O_2$ (strong electrolyte that provides two ions, Na^+ and $C_2H_3O_2^-$) $<$ $Ca(NO_3)_2$ (strong electrolyte that provides three ions, Ca^{2+} and 2 NO_3^-)

Neutralization Reactions and Salts

▲ **Figure 4.7 The acid-base indicator bromthymol blue.** The indicator is blue in basic solution and yellow in acidic solution. The left flask shows the indicator in the presence of a base, aqueous ammonia (labeled as ammonium hydroxide). The right flask shows the indicator in the presence of hydrochloric acid, HCl.

The properties of acidic solutions are quite different from those of basic solutions. Acids have a sour taste, whereas bases have a bitter taste.* Acids can change the colors of certain dyes in a specific way that differs from the effect of a base (Figure 4.7 ◀). The dye known as litmus, for example, is changed from blue to red by an acid, and from red to blue by a base. In addition, acidic and basic solutions differ in chemical properties in several important ways that we will explore in this chapter and in later chapters.

When a solution of an acid and a solution of a base are mixed, a **neutralization reaction** occurs. The products of the reaction have none of the characteristic properties of either the acidic solution or the basic solution. For example, when hydrochloric acid is mixed with a solution of sodium hydroxide, the following reaction occurs:

$$\underset{\text{(acid)}}{HCl(aq)} + \underset{\text{(base)}}{NaOH(aq)} \longrightarrow \underset{\text{(water)}}{H_2O(l)} + \underset{\text{(salt)0}}{NaCl(aq)} \quad [4.12]$$

Water and table salt, NaCl, are the products of the reaction. By analogy to this reaction, the term **salt** has come to mean any ionic compound whose cation comes from a base (for example, Na^+ from NaOH) and whose anion comes from an acid (for example, Cl^- from HCl). In general, *a neutralization reaction between an acid and a metal hydroxide produces water and a salt.*

Because HCl, NaOH, and NaCl are all soluble strong electrolytes, the complete ionic equation associated with Equation 4.12 is

$$H^+(aq) + Cl^-(aq) + Na^+(aq) + OH^-(aq) \longrightarrow H_2O(l) + Na^+(aq) + Cl^-(aq) \quad [4.13]$$

Therefore, the net ionic equation is

$$H^+(aq) + OH^-(aq) \longrightarrow H_2O(l) \quad [4.14]$$

Equation 4.14 summarizes the essential feature of the neutralization reaction between any strong acid and any strong base: $H^+(aq)$ and $OH^-(aq)$ ions combine to form H_2O.

* Tasting chemical solutions is not a good practice. However, we have all had acids such as ascorbic acid (vitamin C), acetylsalicylic acid (aspirin), and citric acid (in citrus fruits) in our mouths, and we are familiar with their characteristic sour taste. Soaps, which are basic, have the characteristic bitter taste of bases.

(a) (b) (c)

▲ **Figure 4.8 Reaction of $Mg(OH)_2(s)$ with acid.** (a) Milk of magnesia is a suspension of magnesium hydroxide, $Mg(OH)_2(s)$, in water. (b) The magnesium hydroxide dissolves upon the addition of hydrochloric acid, $HCl(aq)$. (c) The final clear solution contains soluble $MgCl_2(aq)$, shown in Equation 4.15.

Figure 4.8 ▲ shows the reaction between hydrochloric acid and the base $Mg(OH)_2$, which is insoluble in water. A milky white suspension of $Mg(OH)_2$ called milk of magnesia is seen dissolving as the neutralization reaction occurs:

Molecular equation:

$$Mg(OH)_2(s) + 2\,HCl(aq) \longrightarrow MgCl_2(aq) + 2\,H_2O(l) \qquad [4.15]$$

Net ionic equation:

$$Mg(OH)_2(s) + 2\,H^+(aq) \longrightarrow Mg^{2+}(aq) + 2\,H_2O(l) \qquad [4.16]$$

Notice that the OH^- ions (this time in a solid reactant) and H^+ ions combine to form H_2O. Because the ions exchange partners, neutralization reactions between acids and metal hydroxides are also metathesis reactions.

ANIMATION
Dissolution of $Mg(OH)_2$ by Acid

SAMPLE EXERCISE 4.7 | Writing Chemical Equations for a Neutralization Reaction

(a) Write a balanced molecular equation for the reaction between aqueous solutions of acetic acid ($HC_2H_3O_2$) and barium hydroxide [$Ba(OH)_2$]. **(b)** Write the net ionic equation for this reaction.

Solution

Analyze: We are given the chemical formulas for an acid and a base and asked to write a balanced molecular equation and then a net ionic equation for their neutralization reaction.

Plan: As Equation 4.12 and the italicized statement that follows it indicate, neutralization reactions form two products, H_2O and a salt. We examine the cation of the base and the anion of the acid to determine the composition of the salt.

Solve: **(a)** The salt will contain the cation of the base (Ba^{2+}) and the anion of the acid ($C_2H_3O_2^-$). Thus, the formula of the salt is $Ba(C_2H_3O_2)_2$. According to the solubility guidelines in Table 4.1, this compound is soluble. The unbalanced molecular equation for the neutralization reaction is

$$HC_2H_3O_2(aq) + Ba(OH)_2(aq) \longrightarrow H_2O(l) + Ba(C_2H_3O_2)_2(aq)$$

To balance this molecular equation, we must provide two molecules of $HC_2H_3O_2$ to furnish the two $C_2H_3O_2^-$ ions and to supply the two H^+ ions needed to combine with the two OH^- ions of the base. The balanced molecular equation is

$$2\,HC_2H_3O_2(aq) + Ba(OH)_2(aq) \longrightarrow 2\,H_2O(l) + Ba(C_2H_3O_2)_2(aq)$$

(b) To write the net ionic equation, we must determine whether each compound in aqueous solution is a strong electrolyte. $HC_2H_3O_2$ is a weak electrolyte (weak acid), $Ba(OH)_2$ is a strong electrolyte, and $Ba(C_2H_3O_2)_2$ is also a strong electrolyte (ionic compound). Thus, the complete ionic equation is

$$2\,HC_2H_3O_2(aq) + Ba^{2+}(aq) + 2\,OH^-(aq) \longrightarrow 2\,H_2O(l) + Ba^{2+}(aq) + 2\,C_2H_3O_2^-(aq)$$

Eliminating the spectator ions gives

$$2\,HC_2H_3O_2(aq) + 2\,OH^-(aq) \longrightarrow 2\,H_2O(l) + 2\,C_2H_3O_2^-(aq)$$

Simplifying the coefficients gives the net ionic equation:

$$HC_2H_3O_2(aq) + OH^-(aq) \longrightarrow H_2O(l) + C_2H_3O_2^-(aq)$$

Check: We can determine whether the molecular equation is correctly balanced by counting the number of atoms of each kind on both sides of the arrow. (There are ten H, six O, four C, and one Ba on each side.) However, it is often easier to check equations by counting groups: There are two $C_2H_3O_2$ groups, as well as one Ba, and four additional H atoms and two additional O atoms on each side of the equation. The net ionic equation checks out because the numbers of each kind of element and the net charge are the same on both sides of the equation.

PRACTICE EXERCISE

(a) Write a balanced molecular equation for the reaction of carbonic acid (H_2CO_3) and potassium hydroxide (KOH). **(b)** Write the net ionic equation for this reaction.
Answers: **(a)** $H_2CO_3(aq) + 2\,KOH(aq) \longrightarrow 2\,H_2O(l) + K_2CO_3(aq)$; **(b)** $H_2CO_3(aq) + 2\,OH^-(aq) \longrightarrow 2\,H_2O(l) + CO_3^{2-}(aq)$. ($H_2CO_3$ is a weak acid and therefore a weak electrolyte, whereas KOH, a strong base, and K_2CO_3, an ionic compound, are strong electrolytes.)

Acid-Base Reactions with Gas Formation

There are many bases besides OH^- that react with H^+ to form molecular compounds. Two of these that you might encounter in the laboratory are the sulfide ion and the carbonate ion. Both of these anions react with acids to form gases that have low solubilities in water. Hydrogen sulfide (H_2S), the substance that gives rotten eggs their foul odor, forms when an acid such as HCl(*aq*) reacts with a metal sulfide such as Na_2S:

Molecular equation: $$2\,HCl(aq) + Na_2S(aq) \longrightarrow H_2S(g) + 2\,NaCl(aq) \quad [4.17]$$

Net ionic equation: $$2\,H^+(aq) + S^{2-}(aq) \longrightarrow H_2S(g) \quad [4.18]$$

Carbonates and bicarbonates react with acids to form CO_2 gas. Reaction of CO_3^{2-} or HCO_3^- with an acid first gives carbonic acid (H_2CO_3). For example, when hydrochloric acid is added to sodium bicarbonate, the following reaction occurs:

$$HCl(aq) + NaHCO_3(aq) \longrightarrow NaCl(aq) + H_2CO_3(aq) \quad [4.19]$$

Carbonic acid is unstable; if present in solution in sufficient concentrations, it decomposes to form CO_2, which escapes from the solution as a gas.

$$H_2CO_3(aq) \longrightarrow H_2O(l) + CO_2(g) \quad [4.20]$$

▲ **Figure 4.9 Carbonates react with acids to form carbon dioxide gas.** Here $NaHCO_3$ (white solid) reacts with hydrochloric acid; the bubbles contain CO_2.

The decomposition of H_2CO_3 produces bubbles of CO_2 gas, as shown in Figure 4.9 ◄. The overall reaction is summarized by the following equations:

Molecular equation:
$$HCl(aq) + NaHCO_3(aq) \longrightarrow NaCl(aq) + H_2O(l) + CO_2(g) \quad [4.21]$$

Net ionic equation: $$H^+(aq) + HCO_3^-(aq) \longrightarrow H_2O(l) + CO_2(g) \quad [4.22]$$

Both $NaHCO_3$ and Na_2CO_3 are used as acid neutralizers in acid spills. The bicarbonate or carbonate salt is added until the fizzing due to the formation of

CHEMISTRY AT WORK | Antacids

The stomach secretes acids to help digest foods. These acids, which include hydrochloric acid, contain about 0.1 mol of H^+ per liter of solution. The stomach and digestive tract are normally protected from the corrosive effects of stomach acid by a mucosal lining. Holes can develop in this lining, however, allowing the acid to attack the underlying tissue, causing painful damage. These holes, known as ulcers, can be caused by the secretion of excess acids or by a weakness in the digestive lining. Recent studies indicate, however, that many ulcers are caused by bacterial infection. Between 10 and 20% of Americans suffer from ulcers at some point in their lives, and many others experience occasional indigestion or heartburn that is due to digestive acids entering the esophagus.

We can address the problem of excess stomach acid in two simple ways: (1) removing the excess acid, or (2) decreasing the production of acid. Those substances that remove excess acid are called *antacids*, whereas those that decrease the production of acid are called *acid inhibitors*. Figure 4.10 ▶ shows several common, over-the-counter drugs of both types.

TABLE 4.4 Some Common Antacids

Commercial Name	Acid-Neutralizing Agents
Alka-Seltzer®	$NaHCO_3$
Amphojel®	$Al(OH)_3$
Di-Gel®	$Mg(OH)_2$ and $CaCO_3$
Milk of Magnesia	$Mg(OH)_2$
Maalox®	$Mg(OH)_2$ and $Al(OH)_3$
Mylanta®	$Mg(OH)_2$ and $Al(OH)_3$
Rolaids®	$NaAl(OH)_2CO_3$
Tums®	$CaCO_3$

Antacids are simple bases that neutralize digestive acids. Their ability to neutralize acids is due to the hydroxide, carbonate, or bicarbonate ions they contain. Table 4.4 ◀ lists the active ingredients in some antacids.

The newer generation of antiulcer drugs, such as Tagamet® and Zantac®, are acid inhibitors. They act on acid-producing cells in the lining of the stomach. Formulations that control acid in this way are now available as over-the-counter drugs.

Related Exercise: 4.93

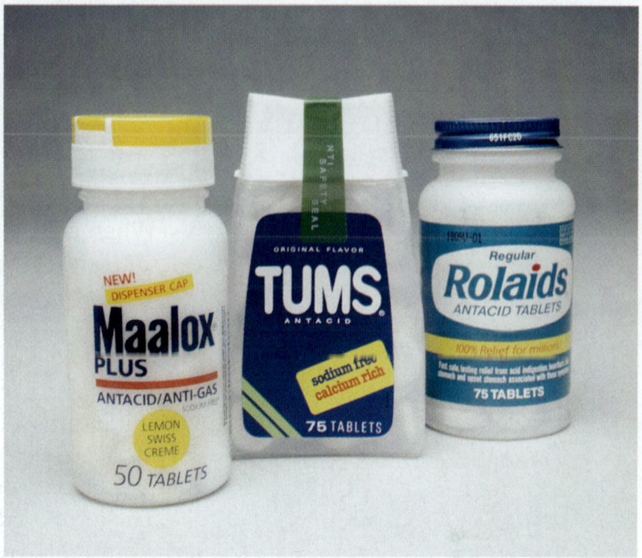

▲ **Figure 4.10 Antacids.** These products all serve as acid-neutralizing agents in the stomach.

$CO_2(g)$ stops. Sometimes sodium bicarbonate is used as an antacid to soothe an upset stomach. In that case the HCO_3^- reacts with stomach acid to form $CO_2(g)$. The fizz when Alka-Seltzer® tablets are added to water arises from the reaction of sodium bicarbonate and citric acid.

GIVE IT SOME THOUGHT

By analogy to examples already given in the text, predict what gas forms when $Na_2SO_3(s)$ is treated with $HCl(aq)$.

4.4 | Oxidation-Reduction Reactions

In precipitation reactions cations and anions come together to form an insoluble ionic compound. In neutralization reactions H^+ ions and OH^- ions come together to form H_2O molecules. Now let's consider a third important kind of reaction, one in which electrons are transferred between reactants. Such reactions are called **oxidation-reduction**, or *redox*, **reactions**.

▶ **Figure 4.11 Corrosion of iron.** Corrosion of iron is caused by chemical attack of oxygen and water on exposed metal surfaces. Corrosion is even more rapid in salt water.

Oxidation and Reduction

The corrosion of iron (rusting) and of other metals, such as the corrosion of the terminals of an automobile battery, are familiar processes. What we call *corrosion* is the conversion of a metal into a metal compound by a reaction between the metal and some substance in its environment. Rusting, as shown in Figure 4.11 ▲, involves the reaction of oxygen with iron in the presence of water.

When a metal corrodes, it loses electrons and forms cations. For example, calcium is vigorously attacked by acids to form calcium ions:

MOVIE
Reduction of CuO

$$Ca(s) + 2\,H^+(aq) \longrightarrow Ca^{2+}(aq) + H_2(g) \qquad [4.23]$$

ANIMATIONS
Oxidation-Reduction Reactions - Part I
Oxidation-Reduction Reactions - Part II

When an atom, ion, or molecule has become more positively charged (that is, when it has lost electrons), we say that it has been oxidized. *Loss of electrons by a substance is called* **oxidation**. Thus, Ca, which has no net charge, is *oxidized* (undergoes oxidation) in Equation 4.23, forming Ca^{2+}.

The term oxidation is used because the first reactions of this sort to be studied thoroughly were reactions with oxygen. Many metals react directly with O_2 in air to form metal oxides. In these reactions the metal loses electrons to oxygen, forming an ionic compound of the metal ion and oxide ion. For example, when calcium metal is exposed to air, the bright metallic surface of the metal tarnishes as CaO forms:

$$2\,Ca(s) + O_2(g) \longrightarrow 2\,CaO(s) \qquad [4.24]$$

As Ca is oxidized in Equation 4.24, oxygen is transformed from neutral O_2 to two O^{2-} ions (Figure 4.12 ▼). When an atom, ion, or molecule has become more negatively charged (gained electrons), we say that it is *reduced*. *Gain of electrons by a substance is called* **reduction**. When one reactant loses

▼ **Figure 4.12 Oxidation of calcium metal by molecular oxygen.** The oxidation involves transfer of electrons from the metal to O_2, eventually leading to formation of CaO.

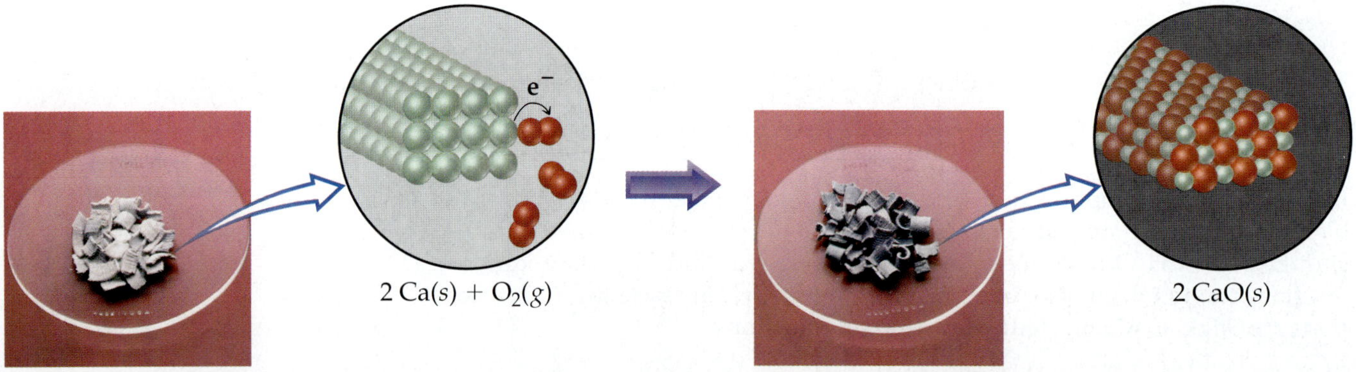

electrons, another reactant must gain them; the oxidation of one substance is always accompanied by the reduction of another as electrons are transferred between them.

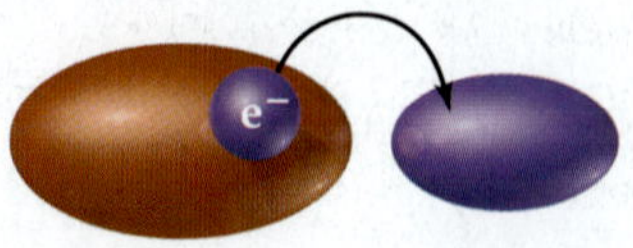

Substance **oxidized** (loses electron) Substance **reduced** (gains electron)

Oxidation Numbers

Before we can properly identify an oxidation-reduction reaction, we must have a kind of bookkeeping system, a way of keeping track of the electrons gained by the substance reduced and those lost by the substance oxidized. The concept of oxidation numbers (also called *oxidation states*) was devised as a way of doing this. Each atom in a neutral molecule or charged species is assigned an **oxidation number**, which is the actual charge for a monatomic ion; otherwise, the oxidation number is the hypothetical charge assigned to the atom, assuming that the electrons are *completely* held by one atom or the other. The oxidation numbers of certain atoms change in an oxidation-reduction reaction. Oxidation occurs when there is an increase in oxidation number, whereas reduction occurs when there is a decrease in oxidation number.

ACTIVITY Oxidation Numbers I

We use the following rules for assigning oxidation numbers:

1. *For an atom in its* ***elemental form****, the oxidation number is always zero.* Thus, each H atom in the H_2 molecule has an oxidation number of 0, and each P atom in the P_4 molecule has an oxidation number of 0.
2. *For any* ***monatomic ion*** *the oxidation number equals the charge on the ion.* Thus, K^+ has an oxidation number of +1, S^{2-} has an oxidation state of −2, and so forth. The alkali metal ions (group 1A) always have a 1+ charge, and therefore the alkali metals always have an oxidation number of +1 in their compounds. Similarly, the alkaline earth metals (group 2A) are always +2, and aluminum (group 3A) is always +3 in its compounds. (In writing oxidation numbers, we will write the sign before the number to distinguish them from the actual electronic charges, which we write with the number first.)

ACTIVITY Oxidation Numbers II

3. ***Nonmetals*** usually have negative oxidation numbers, although they can sometimes be positive:
   - **(a)** *The oxidation number of* ***oxygen*** *is usually* −2 in both ionic and molecular compounds. The major exception is in compounds called peroxides, which contain the O_2^{2-} ion, giving each oxygen an oxidation number of −1.
   - **(b)** *The oxidation number of* ***hydrogen*** *is +1 when bonded to nonmetals and −1 when bonded to metals.*
   - **(c)** *The oxidation number of* ***fluorine*** *is −1 in all compounds.* The other ***halogens*** have an oxidation number of −1 in most binary compounds. When combined with oxygen, as in oxyanions, however, they have positive oxidation states.
4. ***The sum of the oxidation numbers*** *of all atoms in a neutral compound is zero. The sum of the oxidation numbers in a polyatomic ion equals the charge of the ion.* For example, in the hydronium ion, H_3O^+, the oxidation number of each hydrogen is +1 and that of oxygen is −2. Thus, the sum of the oxidation numbers is $3(+1) + (-2) = +1$, which equals the net charge of the ion. This rule is very useful in obtaining the oxidation number of one atom in a compound or ion if you know the oxidation numbers of the other atoms, as illustrated in Sample Exercise 4.8.

GIVE IT SOME THOUGHT

(a) What noble gas element has the same number of electrons as the fluoride ion?
(b) What is the oxidation number of that species?

SAMPLE EXERCISE 4.8 | Determining Oxidation Numbers

Determine the oxidation number of sulfur in each of the following: **(a)** H_2S, **(b)** S_8, **(c)** SCl_2, **(d)** Na_2SO_3, **(e)** SO_4^{2-}.

Solution

Analyze: We are asked to determine the oxidation number of sulfur in two molecular species, in the elemental form, and in two ionic substances.

Plan: In each species the sum of oxidation numbers of all the atoms must equal the charge on the species. We will use the rules outlined above to assign oxidation numbers.

Solve: **(a)** When bonded to a nonmetal, hydrogen has an oxidation number of +1 (rule 3b). Because the H_2S molecule is neutral, the sum of the oxidation numbers must equal zero (rule 4). Letting x equal the oxidation number of S, we have $2(+1) + x = 0$. Thus, S has an oxidation number of −2.

(b) Because this is an elemental form of sulfur, the oxidation number of S is 0 (rule 1).

(c) Because this is a binary compound, we expect chlorine to have an oxidation number of −1 (rule 3c). The sum of the oxidation numbers must equal zero (rule 4). Letting x equal the oxidation number of S, we have $x + 2(-1) = 0$. Consequently, the oxidation number of S must be +2.

(d) Sodium, an alkali metal, always has an oxidation number of +1 in its compounds (rule 2). Oxygen has a common oxidation state of −2 (rule 3a). Letting x equal the oxidation number of S, we have $2(+1) + x + 3(-2) = 0$. Therefore, the oxidation number of S in this compound is +4.

(e) The oxidation state of O is −2 (rule 3a). The sum of the oxidation numbers equals −2, the net charge of the SO_4^{2-} ion (rule 4). Thus, we have $x + 4(-2) = -2$. From this relation we conclude that the oxidation number of S in this ion is +6.

Comment: These examples illustrate that the oxidation number of a given element depends on the compound in which it occurs. The oxidation numbers of sulfur, as seen in these examples, range from −2 to +6.

PRACTICE EXERCISE

What is the oxidation state of the boldfaced element in each of the following: **(a)** $\mathbf{P}_2O_5$, **(b)** Na**H**, **(c)** $\mathbf{Cr}_2O_7^{2-}$, **(d)** **Sn**Br_4, **(e)** Ba$\mathbf{O}_2$?

Answers: **(a)** +5, **(b)** −1, **(c)** +6, **(d)** +4, **(e)** −1

Oxidation of Metals by Acids and Salts

MOVIE
Oxidation-Reduction Chemistry of Tin and Zinc

There are many kinds of redox reactions. For example, combustion reactions are redox reactions because elemental oxygen is converted to compounds of oxygen. (Section 3.2) In this chapter we consider the redox reactions between metals and either acids or salts. In Chapter 20 we will examine more complex kinds of redox reactions.

The reaction of a metal with either an acid or a metal salt conforms to the following general pattern:

$$A + BX \longrightarrow AX + B \qquad [4.25]$$

Examples:

$$Zn(s) + 2\,HBr(aq) \longrightarrow ZnBr_2(aq) + H_2(g)$$

$$Mn(s) + Pb(NO_3)_2(aq) \longrightarrow Mn(NO_3)_2(aq) + Pb(s)$$

These reactions are called **displacement reactions** because the ion in solution is displaced or replaced through oxidation of an element.

Many metals undergo displacement reactions with acids, producing salts and hydrogen gas. For example, magnesium metal reacts with hydrochloric acid to form magnesium chloride and hydrogen gas (Figure 4.13 ▶). To show that oxidation and reduction have occurred, the oxidation number for each atom is shown below the chemical equation for this reaction:

$$\underset{0}{Mg(s)} + 2\,\underset{+1\;-1}{HCl(aq)} \longrightarrow \underset{+2\;-1}{MgCl_2(aq)} + \underset{0}{H_2(g)}, \qquad [4.26]$$

▲ **Figure 4.13 Reaction of magnesium with acid.** The bubbles are due to hydrogen gas.

Notice that the oxidation number of Mg changes from 0 to +2. The increase in the oxidation number indicates that the atom has lost electrons and has therefore been oxidized. The H^+ ion of the acid decreases in oxidation number from +1 to 0, indicating that this ion has gained electrons and has therefore been reduced. The oxidation number of the Cl^- ion remains −1, and it is a spectator ion in the reaction. The net ionic equation is as follows:

$$Mg(s) + 2\,H^+(aq) \longrightarrow Mg^{2+}(aq) + H_2(g) \qquad [4.27]$$

ACTIVITY
Precipitation, Redox, and Neutralization Reactions

Metals can also be oxidized by aqueous solutions of various salts. Iron metal, for example, is oxidized to Fe^{2+} by aqueous solutions of Ni^{2+}, such as $Ni(NO_3)_2(aq)$:

Molecular equation: $$Fe(s) + Ni(NO_3)_2(aq) \longrightarrow Fe(NO_3)_2(aq) + Ni(s) \qquad [4.28]$$

Net ionic equation: $$Fe(s) + Ni^{2+}(aq) \longrightarrow Fe^{2+}(aq) + Ni(s) \qquad [4.29]$$

The oxidation of Fe to form Fe^{2+} in this reaction is accompanied by the reduction of Ni^{2+} to Ni. Remember: *Whenever one substance is oxidized, some other substance must be reduced.*

SAMPLE EXERCISE 4.9 | Writing Molecular and Net Ionic Equations for Oxidation-Reduction Reactions

Write the balanced molecular and net ionic equations for the reaction of aluminum with hydrobromic acid.

Solution

Analyze: We must write two equations—molecular and net ionic—for the redox reaction between a metal and an acid.

Plan: Metals react with acids to form salts and H_2 gas. To write the balanced equations, we must write the chemical formulas for the two reactants and then determine the formula of the salt. The salt is composed of the cation formed by the metal and the anion of the acid.

Solve: The formulas of the given reactants are Al and HBr. The cation formed by Al is Al^{3+}, and the anion from hydrobromic acid is Br^-. Thus, the salt formed in the reaction is $AlBr_3$. Writing the reactants and products and then balancing the equation gives this molecular equation:

$$2\,Al(s) + 6\,HBr(aq) \longrightarrow 2\,AlBr_3(aq) + 3\,H_2(g)$$

Both HBr and $AlBr_3$ are soluble strong electrolytes. Thus, the complete ionic equation is

$$2\,Al(s) + 6\,H^+(aq) + 6\,Br^-(aq) \longrightarrow 2\,Al^{3+}(aq) + 6\,Br^-(aq) + 3\,H_2(g)$$

Because Br^- is a spectator ion, the net ionic equation is

$$2\,Al(s) + 6\,H^+(aq) \longrightarrow 2\,Al^{3+}(aq) + 3\,H_2(g)$$

Comment: The substance oxidized is the aluminum metal because its oxidation state changes from 0 in the metal to +3 in the cation, thereby increasing in oxidation number. The H^+ is reduced because its oxidation state changes from +1 in the acid to 0 in H_2.

PRACTICE EXERCISE

(a) Write the balanced molecular and net ionic equations for the reaction between magnesium and cobalt(II) sulfate. **(b)** What is oxidized and what is reduced in the reaction?

Answers: **(a)** $Mg(s) + CoSO_4(aq) \longrightarrow MgSO_4(aq) + Co(s)$; $Mg(s) + Co^{2+}(aq) \longrightarrow Mg^{2+}(aq) + Co(s)$; **(b)** Mg is oxidized and Co^{2+} is reduced.

The Activity Series

Can we predict whether a certain metal will be oxidized either by an acid or by a particular salt? This question is of practical importance as well as chemical interest. According to Equation 4.28, for example, it would be unwise to store a solution of nickel nitrate in an iron container because the solution would dissolve the container. When a metal is oxidized, it appears to be eaten away as it reacts to form various compounds. Extensive oxidation can lead to the failure of metal machinery parts or the deterioration of metal structures.

Different metals vary in the ease with which they are oxidized. Zn is oxidized by aqueous solutions of Cu^{2+}, for example, but Ag is not. Zn, therefore, loses electrons more readily than Ag; that is, Zn is easier to oxidize than Ag.

A list of metals arranged in order of decreasing ease of oxidation is called an **activity series**. Table 4.5 ▶ gives the activity series in aqueous solution for many of the most common metals. Hydrogen is also included in the table. The metals at the top of the table, such as the alkali metals and the alkaline earth metals, are most easily oxidized; that is, they react most readily to form compounds. They are called the *active metals*. The metals at the bottom of the activity series, such as the transition elements from groups 8B and 1B, are very stable and form compounds less readily. These metals, which are used to make coins and jewelry, are called *noble metals* because of their low reactivity.

The activity series can be used to predict the outcome of reactions between metals and either metal salts or acids. *Any metal on the list can be oxidized by the ions of elements below it.* For example, copper is above silver in the series. Thus, copper metal will be oxidized by silver ions, as pictured in Figure 4.14 ▶:

$$Cu(s) + 2\,Ag^+(aq) \longrightarrow Cu^{2+}(aq) + 2\,Ag(s) \qquad [4.30]$$

The oxidation of copper to copper ions is accompanied by the reduction of silver ions to silver metal. The silver metal is evident on the surface of the copper wires in Figure 4.14(b) and (c). The copper(II) nitrate produces a blue color in the solution, which is most evident in part (c).

TABLE 4.5 Activity Series of Metals in Aqueous Solution

Metal	Oxidation Reaction
Lithium	$Li(s) \longrightarrow Li^{+}(aq) + e^{-}$
Potassium	$K(s) \longrightarrow K^{+}(aq) + e^{-}$
Barium	$Ba(s) \longrightarrow Ba^{2+}(aq) + 2e^{-}$
Calcium	$Ca(s) \longrightarrow Ca^{2+}(aq) + 2e^{-}$
Sodium	$Na(s) \longrightarrow Na^{+}(aq) + e^{-}$
Magnesium	$Mg(s) \longrightarrow Mg^{2+}(aq) + 2e^{-}$
Aluminum	$Al(s) \longrightarrow Al^{3+}(aq) + 3e^{-}$
Manganese	$Mn(s) \longrightarrow Mn^{2+}(aq) + 2e^{-}$
Zinc	$Zn(s) \longrightarrow Zn^{2+}(aq) + 2e^{-}$
Chromium	$Cr(s) \longrightarrow Cr^{3+}(aq) + 3e^{-}$
Iron	$Fe(s) \longrightarrow Fe^{2+}(aq) + 2e^{-}$
Cobalt	$Co(s) \longrightarrow Co^{2+}(aq) + 2e^{-}$
Nickel	$Ni(s) \longrightarrow Ni^{2+}(aq) + 2e^{-}$
Tin	$Sn(s) \longrightarrow Sn^{2+}(aq) + 2e^{-}$
Lead	$Pb(s) \longrightarrow Pb^{2+}(aq) + 2e^{-}$
Hydrogen	$H_2(g) \longrightarrow 2\,H^{+}(aq) + 2e^{-}$
Copper	$Cu(s) \longrightarrow Cu^{2+}(aq) + 2e^{-}$
Silver	$Ag(s) \longrightarrow Ag^{+}(aq) + e^{-}$
Mercury	$Hg(l) \longrightarrow Hg^{2+}(aq) + 2e^{-}$
Platinum	$Pt(s) \longrightarrow Pt^{2+}(aq) + 2e^{-}$
Gold	$Au(s) \longrightarrow Au^{3+}(aq) + 3e^{-}$

↑ Ease of oxidation increases

MOVIE
Formation of Silver Crystals

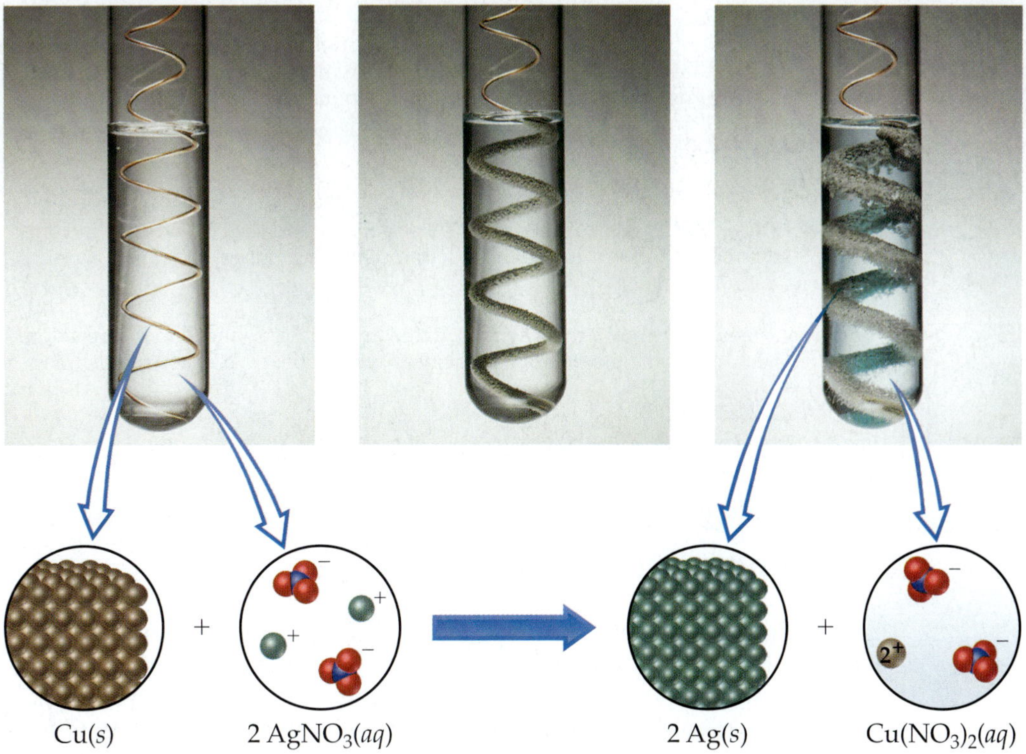

▲ **Figure 4.14 Reaction of copper with silver ion.** When copper metal is placed in a solution of silver nitrate, a redox reaction occurs, forming silver metal and a blue solution of copper(II) nitrate.

GIVE IT SOME THOUGHT

Which is the more easily reduced, $Mg^{2+}(aq)$ or $Ni^{2+}(aq)$?

Only those metals above hydrogen in the activity series are able to react with acids to form H_2. For example, Ni reacts with HCl(*aq*) to form H_2:

$$Ni(s) + 2\,HCl(aq) \longrightarrow NiCl_2(aq) + H_2(g) \quad [4.31]$$

Because elements below hydrogen in the activity series are not oxidized by H^+, Cu does not react with HCl(*aq*). Interestingly, copper does react with nitric acid, as shown previously in Figure 1.11. This reaction, however, is not a simple oxidation of Cu by the H^+ ions of the acid. Instead, the metal is oxidized to Cu^{2+} by the nitrate ion of the acid, accompanied by the formation of brown nitrogen dioxide, $NO_2(g)$:

$$Cu(s) + 4\,HNO_3(aq) \longrightarrow Cu(NO_3)_2(aq) + 2\,H_2O(l) + 2\,NO_2(g) \quad [4.32]$$

What substance is reduced as copper is oxidized in Equation 4.32? In this case the NO_2 results from the reduction of NO_3^-. We will examine reactions of this type in more detail in Chapter 20.

SAMPLE EXERCISE 4.10 | Determining When an Oxidation-Reduction Reaction Can Occur

Will an aqueous solution of iron(II) chloride oxidize magnesium metal? If so, write the balanced molecular and net ionic equations for the reaction.

Solution

Analyze: We are given two substances—an aqueous salt, $FeCl_2$, and a metal, Mg—and asked if they react with each other.

Plan: A reaction will occur if Mg is above Fe in the activity series, Table 4.5. If the reaction occurs, the Fe^{2+} ion in $FeCl_2$ will be reduced to Fe, and the elemental Mg will be oxidized to Mg^{2+}.

Solve: Because Mg is above Fe in the table, the reaction will occur. To write the formula for the salt that is produced in the reaction, we must remember the charges on common ions. Magnesium is always present in compounds as Mg^{2+}; the chloride ion is Cl^-. The magnesium salt formed in the reaction is $MgCl_2$, meaning the balanced molecular equation is

$$Mg(s) + FeCl_2(aq) \longrightarrow MgCl_2(aq) + Fe(s)$$

Both $FeCl_2$ and $MgCl_2$ are soluble strong electrolytes and can be written in ionic form. Cl^-, then, is a spectator ion in the reaction. The net ionic equation is

$$Mg(s) + Fe^{2+}(aq) \longrightarrow Mg^{2+}(aq) + Fe(s)$$

The net ionic equation shows that Mg is oxidized and Fe^{2+} is reduced in this reaction.

Check: Note that the net ionic equation is balanced with respect to both charge and mass.

PRACTICE EXERCISE

Which of the following metals will be oxidized by $Pb(NO_3)_2$: Zn, Cu, Fe?
Answer: Zn and Fe

4.5 | Concentrations of Solutions

The behavior of solutions often depends not only on the nature of the solutes but also on their concentrations. Scientists use the term **concentration** to designate the amount of solute dissolved in a given quantity of solvent or quantity of solution. The concept of concentration is intuitive: The greater the amount of solute dissolved in a certain amount of solvent, the more concentrated the resulting solution. In chemistry we often need to express the concentrations of solutions quantitatively.

A CLOSER LOOK | The Aura of Gold

Gold has been known since the earliest records of human existence. Throughout history people have cherished gold, have fought for it, and have died for it.

The physical and chemical properties of gold serve to make it a special metal. First, its intrinsic beauty and rarity make it precious. Second, gold is soft and can be easily formed into artistic objects, jewelry, and coins (Figure 4.15 ▶). Third, gold is one of the least active metals (Table 4.5). It is not oxidized in air and does not react with water. It is unreactive toward basic solutions and nearly all acidic solutions. As a result, gold can be found in nature as a pure element rather than combined with oxygen or other elements, which accounts for its early discovery.

Many of the early studies of the reactions of gold arose from the practice of alchemy, in which people attempted to turn cheap metals, such as lead, into gold. Alchemists discovered that gold can be dissolved in a 3:1 mixture of concentrated hydrochloric and nitric acids, known as aqua regia ("royal water"). The action of nitric acid on gold is similar to that on copper (Equation 4.32) in that the nitrate ion, rather than H^+, oxidizes the metal to Au^{3+}. The Cl^- ions interact with Au^{3+} to form highly stable $AuCl_4^-$ ions. The net ionic equation for the reaction of gold with aqua regia is

$$Au(s) + NO_3^-(aq) + 4\,H^+(aq) + 4\,Cl^-(aq) \longrightarrow AuCl_4^-(aq) + 2\,H_2O(l) + NO(g)$$

All the gold ever mined would easily fit in a cube 19 m on a side and weighing about 1.1×10^8 kg (125,000 tons). More than 90% of this amount has been produced since the beginning of the California gold rush of 1848. Each year, worldwide production of gold amounts to about 1.8×10^6 kg (2000 tons). By contrast, over 1.5×10^{10} kg (16 million tons) of aluminum are produced annually. Gold is used mainly in jewelry (73%), coins (10%), and electronics (9%). Its use in electronics relies on its excellent conductivity and its corrosion resistance. Gold is used, for example, to plate contacts in electrical switches, relays, and connections. A typical touch-tone telephone contains 33 gold-plated contacts. Gold is also used in computers and other microelectronic devices where fine gold wire is used to link components.

Because of its resistance to corrosion by acids and other substances found in saliva, gold is an ideal metal for dental crowns and caps, which accounts for about 3% of the annual use of the element. The pure metal is too soft to use in dentistry, so it is combined with other metals to form alloys.

Related Exercise: 4.95

◀ **Figure 4.15 Portrait of Pharaoh Tutankhamun (1346–1337 BC) made of gold and precious stones.** This highly prized article is from the inner coffin of the tomb of Tutankhamun.

STRATEGIES IN CHEMISTRY | Analyzing Chemical Reactions

In this chapter you have been introduced to a great number of chemical reactions. A major difficulty that students face in trying to master material of this sort is gaining a "feel" for what happens when chemicals are allowed to react. In fact, you might marvel at the ease with which your professor or teaching assistant can figure out the results of a chemical reaction. One of our goals in this textbook is to help you become more adept at predicting the outcome of reactions. The key to gaining this "chemical intuition" is understanding how to categorize reactions.

There are so many individual reactions in chemistry that memorizing them all is a futile task. It is far more fruitful to try to use pattern recognition to determine the general category of a reaction, such as metathesis or oxidation-reduction. Thus, when you are faced with the challenge of predicting the outcome of a chemical reaction, ask yourself the following pertinent questions:

- What are the reactants in the reaction?
- Are they electrolytes or nonelectrolytes?
- Are they acids and bases?
- If the reactants are electrolytes, will metathesis produce a precipitate? Water? A gas?
- If metathesis cannot occur, can the reactants possibly engage in an oxidation-reduction reaction? This requires that there be both a reactant that can be oxidized and one that can be reduced.

By asking questions such as these, you should be able to predict what might happen during the reaction. You might not always be entirely correct, but if you keep your wits about you, you will not be far off. As you gain experience with chemical reactions, you will begin to look for reactants that might not be immediately obvious, such as water from the solution or oxygen from the atmosphere.

One of the greatest tools available to us in chemistry is experimentation. If you perform an experiment in which two solutions are mixed, you can make observations that help you understand what is happening. For example, using the information in Table 4.1 to predict whether a precipitate will form is not nearly as exciting as actually seeing the precipitate form, as in Figure 4.4. Careful observations in the laboratory portion of the course will make your lecture material easier to master.

(a) (b) (c) (d)

▲ **Figure 4.16 Procedure for preparation of 0.250 L of 1.00 *M* solution of $CuSO_4$.** (a) Weigh out 0.250 mol (39.9 g) of $CuSO_4$ (formula weight = 159.6 amu). (b) Put the $CuSO_4$ (solute) into a 250-mL volumetric flask, and add a small quantity of water. (c) Dissolve the solute by swirling the flask. (d) Add more water until the solution just reaches the calibration mark etched on the neck of the flask. Shake the stoppered flask to ensure complete mixing.

ANIMATION
Solution Formation from a Solid

Molarity

Molarity (symbol *M*) expresses the concentration of a solution as the number of moles of solute in a liter of solution (soln):

ANIMATION
Dissolution of $KMnO_4$

$$\text{Molarity} = \frac{\text{moles solute}}{\text{volume of solution in liters}} \quad [4.33]$$

A 1.00 molar solution (written 1.00 *M*) contains 1.00 mol of solute in every liter of solution. Figure 4.16 ▲ shows the preparation of 250.0 mL of a 1.00 *M* solution of $CuSO_4$ by using a volumetric flask that is calibrated to hold exactly 250.0 mL. First, 0.250 mol of $CuSO_4$ (39.9 g) is weighed out and placed in the volumetric flask. Water is added to dissolve the salt, and the resultant solution is diluted to a total volume of 250.0 mL. The molarity of the solution is $(0.250 \text{ mol } CuSO_4)/(0.250 \text{ L soln}) = 1.00\ M$.

SAMPLE EXERCISE 4.11 | Calculating Molarity

Calculate the molarity of a solution made by dissolving 23.4 g of sodium sulfate (Na_2SO_4) in enough water to form 125 mL of solution.

Solution

Analyze: We are given the number of grams of solute (23.4 g), its chemical formula (Na_2SO_4), and the volume of the solution (125 mL), and we are asked to calculate the molarity of the solution.

Plan: We can calculate molarity using Equation 4.33. To do so, we must convert the number of grams of solute to moles and the volume of the solution from milliliters to liters.

Solve: The number of moles of Na_2SO_4 is obtained by using its molar mass:

$$\text{Moles } Na_2SO_4 = (23.4 \text{ g } Na_2SO_4)\left(\frac{1 \text{ mol } Na_2SO_4}{142 \text{ g } Na_2SO_4}\right) = 0.165 \text{ mol } Na_2SO_4$$

Converting the volume of the solution to liters:

$$\text{Liters soln} = (125 \text{ mL})\left(\frac{1 \text{ L}}{1000 \text{ mL}}\right) = 0.125 \text{ L}$$

Thus, the molarity is

$$\text{Molarity} = \frac{0.165 \text{ mol } Na_2SO_4}{0.125 \text{ L soln}} = 1.32 \frac{\text{mol } Na_2SO_4}{\text{L soln}} = 1.32\ M$$

Check: Because the numerator is only slightly larger than the denominator, it's reasonable for the answer to be a little over 1 *M*. The units (mol/L) are appropriate for molarity, and three significant figures are appropriate for the answer because each of the initial pieces of data had three significant figures.

PRACTICE EXERCISE

Calculate the molarity of a solution made by dissolving 5.00 g of glucose ($C_6H_{12}O_6$) in sufficient water to form exactly 100 mL of solution.
Answer: 0.278 *M*

Expressing the Concentration of an Electrolyte

When an ionic compound dissolves, the relative concentrations of the ions introduced into the solution depend on the chemical formula of the compound. For example, a 1.0 *M* solution of NaCl is 1.0 *M* in Na^+ ions and 1.0 *M* in Cl^- ions. Similarly, a 1.0 *M* solution of Na_2SO_4 is 2.0 *M* in Na^+ ions and 1.0 *M* in SO_4^{2-} ions. Thus, the concentration of an electrolyte solution can be specified either in terms of the compound used to make the solution (1.0 *M* Na_2SO_4) or in terms of the ions that the solution contains (2.0 *M* Na^+ and 1.0 *M* SO_4^{2-}).

SAMPLE EXERCISE 4.12 | Calculating Molar Concentrations of Ions

What are the molar concentrations of each of the ions present in a 0.025 *M* aqueous solution of calcium nitrate?

Solution

Analyze: We are given the concentration of the ionic compound used to make the solution and asked to determine the concentrations of the ions in the solution.
Plan: We can use the subscripts in the chemical formula of the compound to determine the relative concentrations of the ions.
Solve: Calcium nitrate is composed of calcium ions (Ca^{2+}) and nitrate ions (NO_3^-), so its chemical formula is $Ca(NO_3)_2$. Because there are two NO_3^- ions for each Ca^{2+} ion in the compound, each mole of $Ca(NO_3)_2$ that dissolves dissociates into 1 mol of Ca^{2+} and 2 mol of NO_3^-. Thus, a solution that is 0.025 *M* in $Ca(NO_3)_2$ is 0.025 *M* in Ca^{2+} and $2 \times 0.025\ M = 0.050\ M$ in NO_3^-.
Check: The concentration of NO_3^- ions is twice that of Ca^{2+} ions, as the subscript 2 after the NO_3^- in the chemical formula $Ca(NO_3)_2$ suggests it should be.

PRACTICE EXERCISE

What is the molar concentration of K^+ ions in a 0.015 *M* solution of potassium carbonate?
Answer: 0.030 *M* K^+

Interconverting Molarity, Moles, and Volume

The definition of molarity (Equation 4.33) contains three quantities—molarity, moles solute, and liters of solution. If we know any two of these, we can calculate the third. For example, if we know the molarity of a solution, we can calculate the number of moles of solute in a given volume. Molarity, therefore, is a conversion factor between volume of solution and moles of solute. Calculation of the number of moles of HNO_3 in 2.0 L of 0.200 *M* HNO_3 solution illustrates the conversion of volume to moles:

$$\text{Moles HNO}_3 = (2.0\ \cancel{\text{L soln}})\left(\frac{0.200\ \text{mol HNO}_3}{1\ \cancel{\text{L soln}}}\right)$$
$$= 0.40\ \text{mol HNO}_3$$

Dimensional analysis can be used in this conversion if we express molarity as moles/liter soln. To obtain moles, therefore, we multiply liters and molarity: moles = liters × molarity = $\cancel{\text{liters}} \times \text{moles}/\cancel{\text{liter}}$.

To illustrate the conversion of moles to volume, let's calculate the volume of 0.30 M HNO_3 solution required to supply 2.0 mol of HNO_3:

$$\text{Liters soln} = (2.0 \cancel{\text{mol HNO}_3})\left(\frac{1 \text{ L soln}}{0.30 \cancel{\text{mol HNO}_3}}\right) = 6.7 \text{ L soln}$$

In this case we must use the reciprocal of molarity in the conversion: liters = moles $\times$ $1/M$ = $\cancel{\text{moles}}$ $\times$ liters/$\cancel{\text{mole}}$.

SAMPLE EXERCISE 4.13 | Using Molarity to Calculate Grams of Solute

How many grams of Na_2SO_4 are required to make 0.350 L of 0.500 M Na_2SO_4?

Solution

Analyze: We are given the volume of the solution (0.350 L), its concentration (0.500 M), and the identity of the solute (Na_2SO_4) and asked to calculate the number of grams of the solute in the solution.

Plan: We can use the definition of molarity (Equation 4.33) to determine the number of moles of solute, and then convert moles to grams using the molar mass of the solute.

$$M_{Na_2SO_4} = \frac{\text{moles Na}_2\text{SO}_4}{\text{liters soln}}$$

Solve: Calculating the moles of Na_2SO_4 using the molarity and volume of solution gives

$$M_{Na_2SO_4} = \frac{\text{moles Na}_2\text{SO}_4}{\text{liters soln}}$$

$$\begin{aligned}\text{Moles Na}_2\text{SO}_4 &= \text{liters soln} \times M_{Na_2SO_4} \\ &= (0.350 \cancel{\text{L soln}})\left(\frac{0.500 \text{ mol Na}_2\text{SO}_4}{1 \cancel{\text{L soln}}}\right) \\ &= 0.175 \text{ mol Na}_2\text{SO}_4\end{aligned}$$

Because each mole of Na_2SO_4 weighs 142 g, the required number of grams of Na_2SO_4 is

$$\text{Grams Na}_2\text{SO}_4 = (0.175 \cancel{\text{mol Na}_2\text{SO}_4})\left(\frac{142 \text{ g Na}_2\text{SO}_4}{1 \cancel{\text{mol Na}_2\text{SO}_4}}\right) = 24.9 \text{ g Na}_2\text{SO}_4$$

Check: The magnitude of the answer, the units, and the number of significant figures are all appropriate.

PRACTICE EXERCISE

(a) How many grams of Na_2SO_4 are there in 15 mL of 0.50 M Na_2SO_4? **(b)** How many milliliters of 0.50 M Na_2SO_4 solution are needed to provide 0.038 mol of this salt?
Answers: **(a)** 1.1 g, **(b)** 76 mL

Dilution

ANIMATION
Solution Formation by Dilution

Solutions that are used routinely in the laboratory are often purchased or prepared in concentrated form (called *stock solutions*). Hydrochloric acid, for example, is purchased as a 12 M solution (concentrated HCl). Solutions of lower concentrations can then be obtained by adding water, a process called **dilution**.*

To illustrate the preparation of a dilute solution from a concentrated one, suppose we wanted to prepare 250.0 mL (that is, 0.250 L) of 0.100 M $CuSO_4$ solution by diluting a stock solution containing 1.00 M $CuSO_4$. When solvent is added to dilute a solution, the number of moles of solute remains unchanged.

$$\text{Moles solute before dilution} = \text{moles solute after dilution} \qquad [4.34]$$

* In diluting a concentrated acid or base, the acid or base should be added to water and then further diluted by adding more water. Adding water directly to concentrated acid or base can cause spattering because of the intense heat generated.

CHEMISTRY AND LIFE | Drinking Too Much Water Can Kill You

For a long time it was thought that dehydration was a potential danger for people engaged in extended vigorous activity. Thus, athletes were encouraged to drinks lots of water while engaged in active sport. The trend toward extensive hydration has spread throughout society; many people carry water bottles everywhere and dutifully keep well hydrated.

It turns out, though, that in some circumstances, drinking too much water is a greater danger than not drinking enough. Excess water consumption can lead to *hyponatremia*, a condition in which the concentration of sodium ion in the blood is too low. In the past decade at least four marathon runners have died from hyponatremia-related trauma, and dozens more have become seriously ill. For example, a first-time marathoner named Hillary Bellamy, running in the Marine Corps marathon in 2003, collapsed near mile 22 and died the next day. One physician who treated her said that she died from hyponatremia-induced brain swelling, the result of drinking too much water before and during the race.

The normal blood sodium level is 135 to 145 m*M*. When that level drops to as low as 125 m*M*, dizziness and confusion set in. A concentration below 120 m*M* can be critical. Low sodium level in the blood causes brain tissue to swell. Dangerously low levels can occur in a marathon runner or other active athlete who is sweating out salt at the same time that excessive salt-free water is being drunk to compensate for water loss. The condition affects women more than men because of their generally different body composition and patterns of metabolism. Drinking a sport drink, such as Gatorade, which contains some electrolytes, helps to prevent hyponatremia. (Figure 4.17 ▶)

Contrary to popular belief, dehydration is not as likely as overhydration to present a life-threatening situation, though it can contribute to heat stroke when the temperature is high. Athletes frequently lose several pounds in the course of extreme workouts, all in the form of water loss, with no lasting adverse effects. When, for instance, Amby Burfoot ran in the Boston Marathon in 1968, his body weight went from 138 to 129 pounds during the race. He lost 6.5% of his body weight while winning the men's competition that year. Weight losses of this magnitude are typical of elite marathon runners, who produce tremendous amounts of heat and sweat and can't afford to slow down for much drinking.

Related Exercises: 4.63, 4.64

▲ **Figure 4.17 Water stations.** To help prevent overhydration, the number of water stations such as this one has been reduced in many marathon events.

Because we know both the volume and concentration of the dilute solution, we can calculate the number of moles of $CuSO_4$ it contains.

$$\text{Moles } CuSO_4 \text{ in dil soln} = (0.250\ \cancel{\text{L soln}})\left(0.100\frac{\text{mol } CuSO_4}{\cancel{\text{L soln}}}\right)$$
$$= 0.0250 \text{ mol } CuSO_4$$

Now we can calculate the volume of the concentrated solution needed to provide 0.0250 mol $CuSO_4$:

$$\text{Liters of conc soln} = (0.0250\ \cancel{\text{mol } CuSO_4})\left(\frac{1 \text{ L soln}}{1.00\ \cancel{\text{mol } CuSO_4}}\right) = 0.0250 \text{ L}$$

Thus, this dilution is achieved by withdrawing 0.0250 L (that is, 25.0 mL) of the 1.00 *M* solution using a pipet, adding it to a 250-mL volumetric flask, and then diluting it to a final volume of 250.0 mL, as shown in Figure 4.18 ▶. Notice that the diluted solution is less intensely colored than the concentrated one.

In laboratory situations, calculations of this sort are often made very quickly with a simple equation that can be derived by remembering that the number of moles of solute is the same in both the concentrated and dilute solutions and that moles = molarity × liters:

$$\text{Moles solute in conc soln} = \text{moles solute in dil soln}$$
$$M_{conc} \times V_{conc} = M_{dil} \times V_{dil} \qquad [4.35]$$

(a) (b) (c)

▶ **Figure 4.18 Procedure for preparing 250 mL of 0.100 *M* $CuSO_4$ by dilution of 1.00 *M* $CuSO_4$.** (a) Draw 25.0 mL of the 1.00 *M* solution into a pipet. (b) Add this to a 250-mL volumetric flask. (c) Add water to dilute the solution to a total volume of 250 mL.

GIVE IT SOME THOUGHT

How is the molarity of a 0.50 *M* KBr solution changed when water is added to double its volume?

The molarity of the more concentrated stock solution (M_{conc}) is always larger than the molarity of the dilute solution (M_{dil}). Because the volume of the solution increases upon dilution, V_{dil} is always larger than V_{conc}. Although Equation 4.35 is derived in terms of liters, any volume unit can be used as long as that same unit is used on both sides of the equation. For example, in the calculation we did for the $CuSO_4$ solution, we have

$$(1.00\ M)(V_{conc}) = (0.100\ M)(250\ \text{mL})$$

Solving for V_{conc} gives $V_{conc} = 25.0$ mL as before.

SAMPLE EXERCISE 4.14 | Preparing a Solution by Dilution

How many milliliters of 3.0 *M* H_2SO_4 are needed to make 450 mL of 0.10 *M* H_2SO_4?

Solution

Analyze: We need to dilute a concentrated solution. We are given the molarity of a more concentrated solution (3.0 *M*) and the volume and molarity of a more dilute one containing the same solute (450 mL of 0.10 *M* solution). We must calculate the volume of the concentrated solution needed to prepare the dilute solution.

Plan: We can calculate the number of moles of solute, H_2SO_4, in the dilute solution and then calculate the volume of the concentrated solution needed to supply this amount of solute. Alternatively, we can directly apply Equation 4.35. Let's compare the two methods.

Solve: Calculating the moles of H_2SO_4 in the dilute solution:

$$\text{Moles } H_2SO_4 \text{ in dilute solution} = (0.450\ \cancel{\text{L soln}})\left(\frac{0.10\ \text{mol } H_2SO_4}{1\ \cancel{\text{L soln}}}\right)$$
$$= 0.045\ \text{mol } H_2SO_4$$

Calculating the volume of the concentrated solution that contains 0.045 mol H_2SO_4:

$$\text{L conc soln} = (0.045\ \cancel{\text{mol } H_2SO_4})\left(\frac{1\ \text{L soln}}{3.0\ \cancel{\text{mol } H_2SO_4}}\right) = 0.015\ \text{L soln}$$

Converting liters to milliliters gives 15 mL.

If we apply Equation 4.35, we get the same result:

$$(3.0\ M)(V_{conc}) = (0.10\ M)(450\ mL)$$

$$V_{conc} = \frac{(0.10\ \cancel{M})(450\ mL)}{3.0\ \cancel{M}} = 15\ mL$$

Either way, we see that if we start with 15 mL of 3.0 *M* H_2SO_4 and dilute it to a total volume of 450 mL, the desired 0.10 *M* solution will be obtained.

Check: The calculated volume seems reasonable because a small volume of concentrated solution is used to prepare a large volume of dilute solution.

PRACTICE EXERCISE

(a) What volume of 2.50 *M* lead(II) nitrate solution contains 0.0500 mol of Pb^{2+}? **(b)** How many milliliters of 5.0 *M* $K_2Cr_2O_7$ solution must be diluted to prepare 250 mL of 0.10 *M* solution? **(c)** If 10.0 mL of a 10.0 *M* stock solution of NaOH is diluted to 250 mL, what is the concentration of the resulting stock solution?
Answers: **(a)** 0.0200 L = 20.0 mL, **(b)** 5.0 mL, **(c)** 0.40 *M*

4.6 | Solution Stoichiometry and Chemical Analysis

Imagine that you have to determine the concentrations of several ions in a sample of lake water. Although many instrumental methods have been developed for such analyses, chemical reactions such as those discussed in this chapter continue to be used. In Chapter 3 we learned that if you know the chemical equation and the amount of one reactant consumed in the reaction, you can calculate the quantities of other reactants and products. In this section we briefly explore such analyses of solutions.

Recall that the coefficients in a balanced equation give the relative number of moles of reactants and products. (Section 3.6) To use this information, we must convert the quantities of substances involved in a reaction into moles. When we are dealing with grams of substances, as we were in Chapter 3, we use the molar mass to achieve this conversion. When we are working with solutions of known molarity, however, we use molarity and volume to determine the number of moles (moles solute = $M \times L$). Figure 4.19 ▼ summarizes this approach to using stoichiometry.

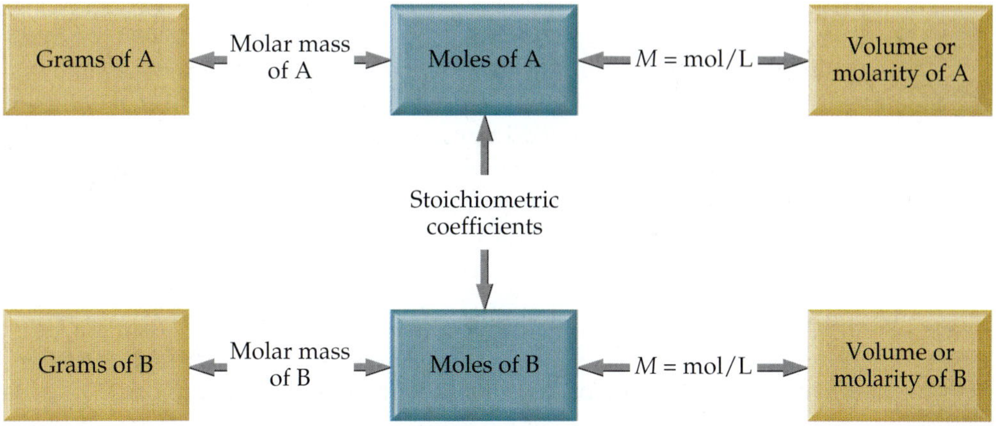

▲ **Figure 4.19 Problem-solving procedure.** Outline of the procedure used to solve stoichiometry problems that involve measured (laboratory) units of mass, solution concentration (molarity), or volume.

SAMPLE EXERCISE 4.15 | Using Mass Relations in a Neutralization Reaction

How many grams of $Ca(OH)_2$ are needed to neutralize 25.0 mL of 0.100 *M* HNO_3?

Solution

Analyze: The reactants are an acid, HNO_3, and a base, $Ca(OH)_2$. The volume and molarity of HNO_3 are given, and we are asked how many grams of $Ca(OH)_2$ are needed to neutralize this quantity of HNO_3.

Plan: We can use the molarity and volume of the HNO_3 solution to calculate the number of moles of HNO_3. We then use the balanced equation to relate the moles of HNO_3 to moles of $Ca(OH)_2$. Finally, we can convert moles of $Ca(OH)_2$ to grams. These steps can be summarized as follows:

$$L_{HNO_3} \times M_{HNO_3} \Rightarrow \text{mol } HNO_3 \Rightarrow \text{mol } Ca(OH)_2 \Rightarrow \text{g } Ca(OH)_2$$

Solve: The product of the molar concentration of a solution and its volume in liters gives the number of moles of solute:

$$\text{Moles } HNO_3 = L_{HNO_3} \times M_{HNO_3} = (0.0250\ \cancel{L})\left(0.100 \frac{\text{mol } HNO_3}{\cancel{L}}\right)$$

$$= 2.50 \times 10^{-3} \text{ mol } HNO_3$$

Because this is an acid-base neutralization reaction, HNO_3 and $Ca(OH)_2$ react to form H_2O and the salt containing Ca^{2+} and NO_3^-:

$$2\ HNO_3(aq) + Ca(OH)_2(s) \longrightarrow 2\ H_2O(l) + Ca(NO_3)_2(aq)$$

Thus, 2 mol $HNO_3 \simeq$ 1 mol $Ca(OH)_2$. Therefore,

$$\text{Grams } Ca(OH)_2 = (2.50 \times 10^{-3}\ \cancel{\text{mol } HNO_3})\left(\frac{1\ \cancel{\text{mol } Ca(OH)_2}}{2\ \cancel{\text{mol } HNO_3}}\right)\left(\frac{74.1 \text{ g } Ca(OH)_2}{1\ \cancel{\text{mol } Ca(OH)_2}}\right)$$

$$= 0.0926 \text{ g } Ca(OH)_2$$

Check: The size of the answer is reasonable. A small volume of dilute acid will require only a small amount of base to neutralize it.

PRACTICE EXERCISE

(a) How many grams of NaOH are needed to neutralize 20.0 mL of 0.150 *M* H_2SO_4 solution? **(b)** How many liters of 0.500 *M* $HCl(aq)$ are needed to react completely with 0.100 mol of $Pb(NO_3)_2(aq)$, forming a precipitate of $PbCl_2(s)$?

Answers: **(a)** 0.240 g, **(b)** 0.400 L

Titrations

To determine the concentration of a particular solute in a solution, chemists often carry out a **titration**, which involves combining a sample of the solution with a reagent solution of known concentration, called a **standard solution**. Titrations can be conducted using acid-base, precipitation, or oxidation-reduction reactions. Suppose we have an HCl solution of unknown concentration and an NaOH solution we know to be 0.100 *M*. To determine the concentration of the HCl solution, we take a specific volume of that solution, say 20.00 mL. We then slowly add the standard NaOH solution to it until the neutralization reaction between the HCl and NaOH is complete. The point at which stoichiometrically equivalent quantities are brought together is known as the **equivalence point** of the titration.

ANIMATION
Acid-Base Titration

VCL Concepts in Acid-Base Titrations

GIVE IT SOME THOUGHT

25.00 mL of a 0.100 *M* HBr solution is titrated with a 0.200 *M* NaOH solution. How many mL of the NaOH solution are required to reach the equivalence point?

In order to titrate an unknown with a standard solution, there must be some way to determine when the equivalence point of the titration has been

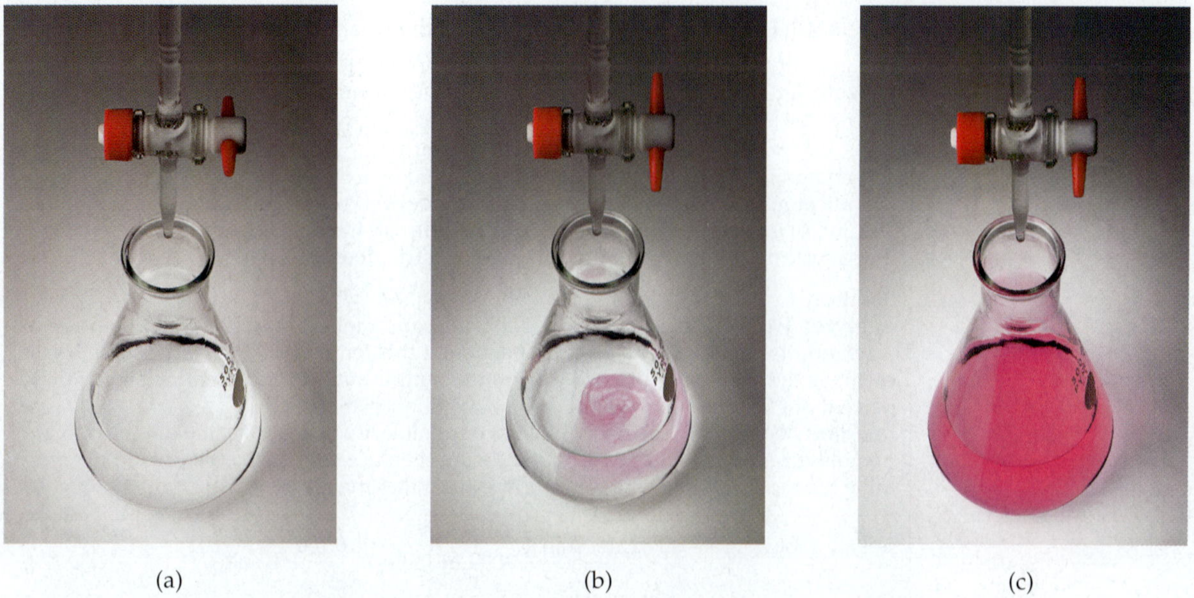

▲ **Figure 4.20 Change in appearance of a solution containing phenolphthalein indicator as base is added.** Before the end point, the solution is colorless (a). As the end point is approached, a pale pink color forms where the base is added (b). At the end point, this pale pink color extends throughout the solution after mixing. As even more base is added, the intensity of the pink color increases (c).

reached. In acid-base titrations, dyes known as acid-base **indicators** are used for this purpose. For example, the dye known as phenolphthalein is colorless in acidic solution but is pink in basic solution. If we add phenolphthalein to an unknown solution of acid, the solution will be colorless, as seen in Figure 4.20(a) ▲. We can then add standard base from a buret until the solution barely turns from colorless to pink, as seen in Figure 4.20(b). This color change indicates that the acid has been neutralized and the drop of base that caused the solution to become colored has no acid to react with. The solution therefore becomes basic, and the dye turns pink. The color change signals the *end point* of the titration, which usually coincides very nearly with the equivalence point. Care must be taken to choose indicators whose end points correspond to the equivalence point of the titration. We will consider this matter in Chapter 17. The titration procedure is summarized in Figure 4.21 ▼.

ACTIVITY
Acid-Base Titration

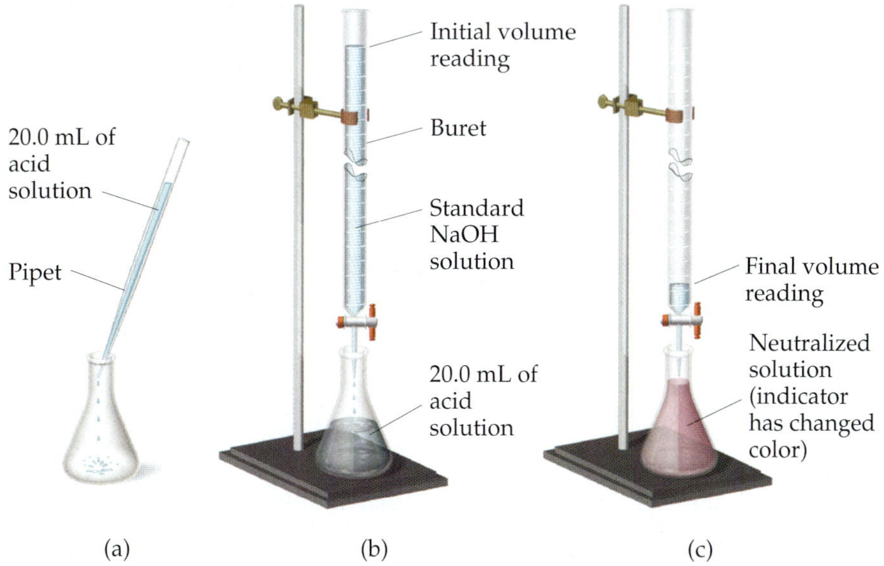

◀ **Figure 4.21 Procedure for titrating an acid against a standardized solution of NaOH.** (a) A known quantity of acid is added to a flask. (b) An acid-base indicator is added, and standardized NaOH is added from a buret. (c) The equivalence point is signaled by a color change in the indicator.

SAMPLE EXERCISE 4.16 | Determining the Quantity of Solute by Titration

The quantity of Cl^- in a municipal water supply is determined by titrating the sample with Ag^+. The reaction taking place during the titration is

$$Ag^+(aq) + Cl^-(aq) \longrightarrow AgCl(s)$$

The end point in this type of titration is marked by a change in color of a special type of indicator. **(a)** How many grams of chloride ion are in a sample of the water if 20.2 mL of 0.100 M Ag^+ is needed to react with all the chloride in the sample? **(b)** If the sample has a mass of 10.0 g, what percent Cl^- does it contain?

Solution

Analyze: We are given the volume (20.2 mL) and molarity (0.100 M) of a solution of Ag^+ and the chemical equation for reaction of this ion with Cl^-. We are asked first to calculate the number of grams of Cl^- in the sample and, second, to calculate the mass percent of Cl^- in the sample.

(a) Plan: We begin by using the volume and molarity of Ag^+ to calculate the number of moles of Ag^+ used in the titration. We can then use the balanced equation to determine the moles of Cl^- in the sample and from that the grams of Cl^-.

Solve: $$\text{Moles } Ag^+ = (20.2\ \cancel{\text{mL soln}})\left(\frac{1\ \cancel{\text{L soln}}}{1000\ \cancel{\text{mL soln}}}\right)\left(0.100\frac{\text{mol } Ag^+}{\cancel{\text{L soln}}}\right)$$
$$= 2.02 \times 10^{-3}\ \text{mol } Ag^+$$

From the balanced equation we see that 1 mol $Ag^+ \simeq$ 1 mol Cl^-. Using this information and the molar mass of Cl, we have

$$\text{Grams } Cl^- = (2.02 \times 10^{-3}\ \cancel{\text{mol } Ag^+})\left(\frac{1\ \cancel{\text{mol } Cl^-}}{1\ \cancel{\text{mol } Ag^+}}\right)\left(\frac{35.5\text{ g } Cl^-}{1\ \cancel{\text{mol } Cl^-}}\right)$$
$$= 7.17 \times 10^{-2}\text{ g } Cl^-$$

(b) Plan: To calculate the percentage of Cl^- in the sample, we compare the number of grams of Cl^- in the sample, 7.17×10^{-2} g, with the original mass of the sample, 10.0 g.

Solve: $$\text{Percent } Cl^- = \frac{7.17 \times 10^{-3}\text{ g}}{10.0\text{ g}} \times 100\% = 0.717\%\ Cl^-$$

Comment: Chloride ion is one of the most common ions in water and sewage. Ocean water contains 1.92% Cl^-. Whether water containing Cl^- tastes salty depends on the other ions present. If the only accompanying ions are Na^+, a salty taste may be detected with as little as 0.03% Cl^-.

PRACTICE EXERCISE

A sample of an iron ore is dissolved in acid, and the iron is converted to Fe^{2+}. The sample is then titrated with 47.20 mL of 0.02240 M MnO_4^- solution. The oxidation-reduction reaction that occurs during titration is as follows:

$$MnO_4^-(aq) + 5\,Fe^{2+}(aq) + 8\,H^+(aq) \longrightarrow Mn^{2+}(aq) + 5\,Fe^{3+}(aq) + 4\,H_2O(l)$$

(a) How many moles of MnO_4^- were added to the solution? **(b)** How many moles of Fe^{2+} were in the sample? **(c)** How many grams of iron were in the sample? **(d)** If the sample had a mass of 0.8890 g, what is the percentage of iron in the sample?
Answers: **(a)** 1.057×10^{-3} mol MnO_4^-, **(b)** 5.286×10^{-3} mol Fe^{2+}, **(c)** 0.2952 g, **(d)** 33.21%

SAMPLE EXERCISE 4.17 | Determining Solution Concentration Via an Acid-Base Titration

One commercial method used to peel potatoes is to soak them in a solution of NaOH for a short time, remove them from the NaOH, and spray off the peel. The concentration of NaOH is normally in the range of 3 to 6 M. The NaOH is analyzed periodically. In one such analysis, 45.7 mL of 0.500 M H_2SO_4 is required to neutralize a 20.0-mL sample of NaOH solution. What is the concentration of the NaOH solution?

Solution

Analyze: We are given the volume (45.7 mL) and molarity (0.500 M) of an H_2SO_4 solution that reacts completely with a 20.0-mL sample of NaOH. We are asked to calculate the molarity of the NaOH solution.

Plan: We can use the volume and molarity of the H_2SO_4 to calculate the number of moles of this substance. Then, we can use this quantity and the balanced equation for the reaction to calculate the number of moles of NaOH. Finally, we can use the moles of NaOH and the volume of this solution to calculate molarity.

Solve: The number of moles of H_2SO_4 is given by the product of the volume and molarity of this solution:

$$\text{Moles } H_2SO_4 = (45.7\ \text{mL soln})\left(\frac{1\ \text{L soln}}{1000\ \text{mL soln}}\right)\left(0.500\frac{\text{mol } H_2SO_4}{\text{L soln}}\right)$$

$$= 2.28 \times 10^{-2}\ \text{mol } H_2SO_4$$

Acids react with metal hydroxides to form water and a salt. Thus, the balanced equation for the neutralization reaction is

$$H_2SO_4(aq) + 2\ NaOH(aq) \longrightarrow 2\ H_2O(l) + Na_2SO_4(aq)$$

According to the balanced equation, 1 mol $H_2SO_4 \simeq$ 2 mol NaOH. Therefore,

$$\text{Moles NaOH} = (2.28 \times 10^{-2}\ \text{mol } H_2SO_4)\left(\frac{2\ \text{mol NaOH}}{1\ \text{mol } H_2SO_4}\right)$$

$$= 4.56 \times 10^{-2}\ \text{mol NaOH}$$

Knowing the number of moles of NaOH present in 20.0 mL of solution allows us to calculate the molarity of this solution:

$$\text{Molarity NaOH} = \frac{\text{mol NaOH}}{\text{L soln}} = \left(\frac{4.56 \times 10^{-2}\ \text{mol NaOH}}{20.0\ \text{mL soln}}\right)\left(\frac{1000\ \text{mL soln}}{1\ \text{L soln}}\right)$$

$$= 2.28\frac{\text{mol NaOH}}{\text{L soln}} = 2.28\ M$$

PRACTICE EXERCISE

What is the molarity of an NaOH solution if 48.0 mL is needed to neutralize 35.0 mL of 0.144 M H_2SO_4?

Answers: 0.210 M

SAMPLE INTEGRATIVE EXERCISE | Putting Concepts Together

Note: Integrative exercises require skills from earlier chapters as well as ones from the present chapter.

A sample of 70.5 mg of potassium phosphate is added to 15.0 mL of 0.050 M silver nitrate, resulting in the formation of a precipitate. **(a)** Write the molecular equation for the reaction. **(b)** What is the limiting reactant in the reaction? **(c)** Calculate the theoretical yield, in grams, of the precipitate that forms.

Solution **(a)** Potassium phosphate and silver nitrate are both ionic compounds. Potassium phosphate contains K^+ and PO_4^{3-} ions, so its chemical formula is K_3PO_4. Silver nitrate contains Ag^+ and NO_3^- ions, so its chemical formula is $AgNO_3$. Because both reactants are strong electrolytes, the solution contains K^+, PO_4^{3-}, Ag^+, and NO_3^- ions before the reaction occurs. According to the solubility guidelines in Table 4.1, Ag^+ and PO_4^{3-} form an insoluble compound, so Ag_3PO_4 will precipitate from the solution. In contrast, K^+ and NO_3^- will remain in solution because KNO_3 is water soluble. Thus, the balanced molecular equation for the reaction is

$$K_3PO_4(aq) + 3\ AgNO_3(aq) \longrightarrow Ag_3PO_4(s) + 3\ KNO_3(aq)$$

(b) To determine the limiting reactant, we must examine the number of moles of each reactant. (Section 3.7) The number of moles of K_3PO_4 is calculated from the mass of the sample using the molar mass as a conversion factor. (Section 3.4) The molar mass of K_3PO_4 is 3(39.1) + 31.0 + 4(16.0) = 212.3 g/mol. Converting milligrams to grams and then to moles, we have

$$(70.5\ \text{mg } K_3PO_4)\left(\frac{10^{-3}\ \text{g } K_3PO_4}{1\ \text{mg } K_3PO_4}\right)\left(\frac{1\ \text{mol } K_3PO_4}{212.3\ \text{g } K_3PO_4}\right) = 3.32 \times 10^{-4}\ \text{mol } K_3PO_4$$

We determine the number of moles of $AgNO_3$ from the volume and molarity of the solution. (Section 4.5) Converting milliliters to liters and then to moles, we have

$$(15.0\ \cancel{\text{mL}})\left(\frac{10^{-3}\ \cancel{\text{L}}}{1\ \cancel{\text{mL}}}\right)\left(\frac{0.050\ \text{mol AgNO}_3}{\cancel{\text{L}}}\right) = 7.5 \times 10^{-4}\ \text{mol AgNO}_3$$

Comparing the amounts of the two reactants, we find that there are $(7.5 \times 10^{-4})/(3.32 \times 10^{-4}) = 2.3$ times as many moles of $AgNO_3$ as there are moles of K_3PO_4. According to the balanced equation, however, 1 mol K_3PO_4 requires 3 mol of $AgNO_3$. Thus, there is insufficient $AgNO_3$ to consume the K_3PO_4, and $AgNO_3$ is the limiting reactant.

(c) The precipitate is Ag_3PO_4, whose molar mass is $3(107.9) + 31.0 + 4(16.0) = 418.7$ g/mol. To calculate the number of grams of Ag_3PO_4 that could be produced in this reaction (the theoretical yield), we use the number of moles of the limiting reactant, converting mol $AgNO_3 \Rightarrow$ mol $Ag_3PO_4 \Rightarrow$ g Ag_3PO_4. We use the coefficients in the balanced equation to convert moles of $AgNO_3$ to moles Ag_3PO_4, and we use the molar mass of Ag_3PO_4 to convert the number of moles of this substance to grams.

$$(7.5 \times 10^{-4}\ \cancel{\text{mol AgNO}_3})\left(\frac{1\ \cancel{\text{mol Ag}_3\text{PO}_4}}{3\ \cancel{\text{mol AgNO}_3}}\right)\left(\frac{418.7\ \text{g Ag}_3\text{PO}_4}{1\ \cancel{\text{mol Ag}_3\text{NO}_4}}\right) = 0.10\ \text{g Ag}_3\text{PO}_4$$

The answer has only two significant figures because the quantity of $AgNO_3$ is given to only two significant figures.

SUMMARY AND KEY TERMS

Introduction and Section 4.1 Solutions in which water is the dissolving medium are called **aqueous solutions**. The component of the solution that is in the greater quantity is the **solvent**. The other components are **solutes**.

Any substance whose aqueous solution contains ions is called an **electrolyte**. Any substance that forms a solution containing no ions is a **nonelectrolyte**. Those electrolytes that are present in solution entirely as ions are **strong electrolytes**, whereas those that are present partly as ions and partly as molecules are **weak electrolytes**. Ionic compounds dissociate into ions when they dissolve, and they are strong electrolytes. The solubility of ionic substances is made possible by **solvation**, the interaction of ions with polar solvent molecules. Most molecular compounds are nonelectrolytes, although some are weak electrolytes and a few are strong electrolytes. When representing the ionization of a weak electrolyte in solution, a double arrow is used, indicating that the forward and reverse reactions can achieve a chemical balance called a **chemical equilibrium**.

Section 4.2 **Precipitation reactions** are those in which an insoluble product, called a **precipitate**, forms. Solubility guidelines help determine whether or not an ionic compound will be soluble in water. (The **solubility** of a substance is the amount that dissolves in a given quantity of solvent.) Reactions such as precipitation reactions, in which cations and anions appear to exchange partners, are called **exchange reactions**, or **metathesis reactions**.

Chemical equations can be written to show whether dissolved substances are present in solution predominantly as ions or molecules. When the complete chemical formulas of all reactants and products are used, the equation is called a **molecular equation**. A **complete ionic equation** shows all dissolved strong electrolytes as their component ions. In a **net ionic equation**, those ions that go through the reaction unchanged (**spectator ions**) are omitted.

Section 4.3 Acids and bases are important electrolytes. **Acids** are proton donors; they increase the concentration of $H^+(aq)$ in aqueous solutions to which they are added. **Bases** are proton acceptors; they increase the concentration of $OH^-(aq)$ in aqueous solutions. Those acids and bases that are strong electrolytes are called **strong acids** and **strong bases**, respectively. Those that are weak electrolytes are **weak acids** and **weak bases**. When solutions of acids and bases are mixed, a **neutralization reaction** results. The neutralization reaction between an acid and a metal hydroxide produces water and a **salt**. Gases can also be formed as a result of acid-base reactions. The reaction of a sulfide with an acid forms $H_2S(g)$; the reaction between a carbonate and an acid forms $CO_2(g)$.

Section 4.4 **Oxidation** is the loss of electrons by a substance, whereas **reduction** is the gain of electrons by a substance. **Oxidation numbers** help us keep track of electrons during chemical reactions and are assigned to atoms by using specific rules. The oxidation of an element results in an increase in its oxidation number, whereas reduction is accompanied by a decrease in oxidation number. Oxidation is always accompanied by reduction, giving **oxidation-reduction**, or redox, **reactions**.

Many metals are oxidized by O_2, acids, and salts. The redox reactions between metals and acids and between metals and salts are called **displacement reactions**.

The products of these displacement reactions are always an element (H_2 or a metal) and a salt. Comparing such reactions allows us to rank metals according to their ease of oxidation. A list of metals arranged in order of decreasing ease of oxidation is called an **activity series**. Any metal on the list can be oxidized by ions of metals (or H^+) below it in the series.

Section 4.5 The composition of a solution expresses the relative quantities of solvent and solutes that it contains. One of the common ways to express the **concentration** of a solute in a solution is in terms of molarity. The **molarity** of a solution is the number of moles of solute per liter of solution. Molarity makes it possible to interconvert solution volume and number of moles of solute. Solutions of known molarity can be formed either by weighing out the solute and diluting it to a known volume or by the **dilution** of a more concentrated solution of known concentration (a stock solution). Adding solvent to the solution (the process of dilution) decreases the concentration of the solute without changing the number of moles of solute in the solution ($M_{conc} \times V_{conc} = M_{dil} \times V_{dil}$).

Section 4.6 In the process called **titration**, we combine a solution of known concentration (a **standard solution**) with a solution of unknown concentration in order to determine the unknown concentration or the quantity of solute in the unknown. The point in the titration at which stoichiometrically equivalent quantities of reactants are brought together is called the **equivalence point**. An **indicator** can be used to show the end point of the titration, which coincides closely with the equivalence point.

VISUALIZING CONCEPTS

4.1 Which of the following schematic drawings best describes a solution of Li_2SO_4 in water (water molecules not shown for simplicity)? [Section 4.1]

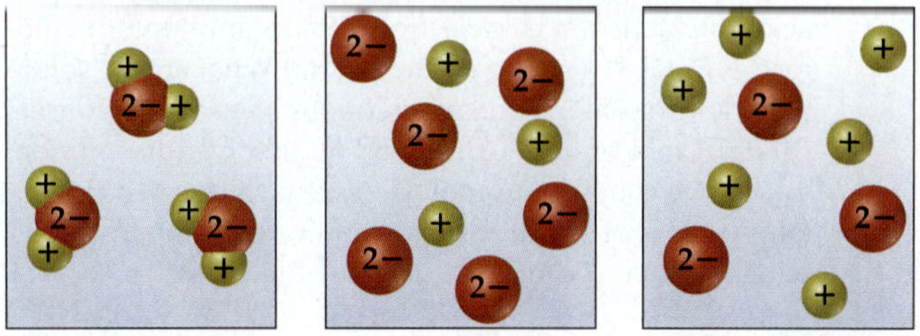

(a) (b) (c)

4.2 Methanol, CH_3OH, and hydrogen chloride, HCl, are both molecular substances, yet an aqueous solution of methanol does not conduct an electrical current, whereas a solution of HCl does conduct. Account for this difference. [Section 4.1]

4.3 Aqueous solutions of three different substances, AX, AY, and AZ, are represented by the three diagrams below. Identify each substance as a strong electrolyte, weak electrolyte, or nonelectrolyte. [Section 4.1]

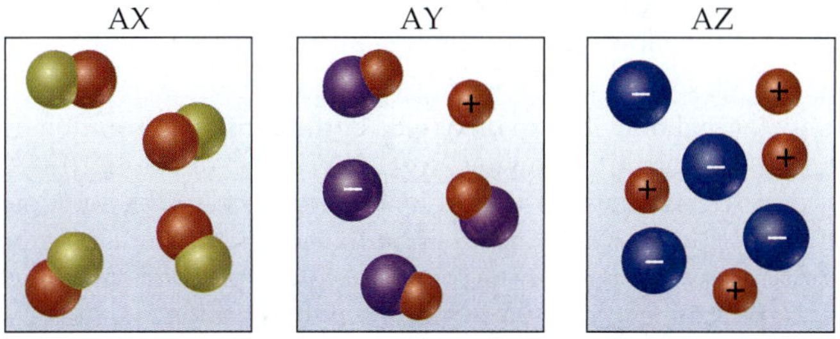

(a) (b) (c)

4.4 A 0.1 *M* solution of acetic acid, $HC_2H_3O_2$, causes the lightbulb in the apparatus of Figure 4.2 to glow about as brightly as a 0.001 *M* solution of HBr. How do you account for this fact? [Section 4.1]

4.5 You are presented with three white solids, A, B and C, which are glucose (a sugar substance), NaOH, and AgBr. Solid A dissolves in water to form a conducting solution. B is not soluble in water. C dissolves in water to form a nonconducting solution. Identify A, B and C. [Section 4.2]

4.6 We have seen that ions in aqueous solution are stabilized by the attractions between the ions and the water molecules. Why then do some pairs of ions in solution form precipitates? [Section 4.2]

4.7 Which of the following ions will *always* be a spectator ion in a precipitation reaction? **(a)** Cl^-, **(b)** NO_3^-, **(c)** NH_4^+, **(d)** S^{2-}, **(e)** SO_4^{2-}. Explain briefly. [Section 4.2]

4.8 The labels have fallen off two bottles, one containing $Mg(NO_3)_2$ and the other containing $Pb(NO_3)_2$. You have a bottle of dilute H_2SO_4. How could you use it to test a portion of each solution to identify which solution is which? [Section 4.2]

4.9 Which of the following chemical equations is represented schematically in the drawing below? [Section 4.2]

(a) $BaCl_2(aq) + Na_2(SO)_4(aq) \longrightarrow BaSO_4(s) + 2\,NaCl(aq)$

(b) $SrCO_3(s) + 2\,HBr(aq) \longrightarrow SrBr_2(aq) + H_2O(l) + CO_2(g)$

(c) $2\,NaOH(aq) + CdCl_2(aq) \longrightarrow Cd(OH)_2(s) + 2\,NaCl(aq)$

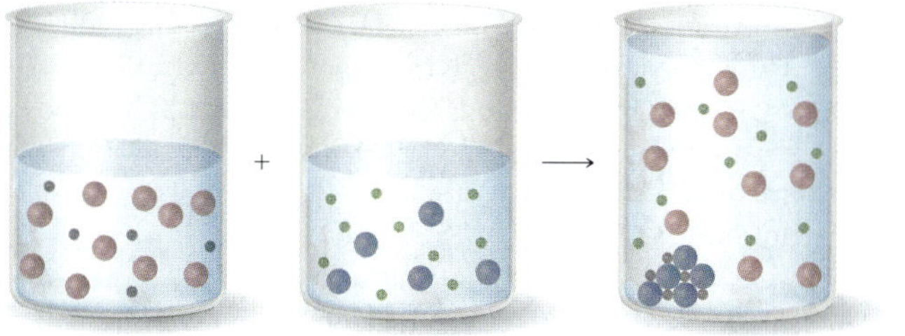

4.10 Which of the three solutions shown schematically to the right corresponds to the result of each of the following reactions: (Water molecules are excluded for clarity) [Section 4.3]

(a) $Ag_2O(s) + 2\,HCl(aq) \longrightarrow 2\,AgCl(s) + H_2O(l)$
(b) $NaOH(aq) + HCl(aq) \longrightarrow NaCl(aq) + H_2O(l)$
(c) $AgNO_3(aq) + KCl(aq) \longrightarrow AgCl(s) + KCl(aq)$

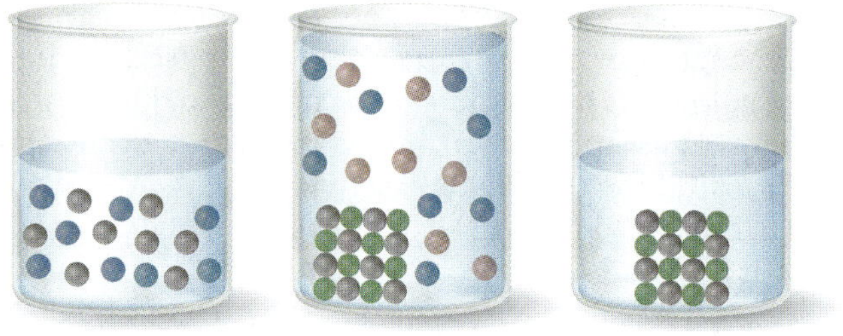

(i) (ii) (iii)

EXERCISES

Electrolytes

4.11 When asked what causes electrolyte solutions to conduct electricity, a student responds that it is due to the movement of electrons through the solution. Is the student correct? If not, what is the correct response?

4.12 When methanol, CH_3OH, is dissolved in water, a nonconducting solution results. When acetic acid, $HC_2H_3O_2$, dissolves in water, the solution is weakly conducting and acidic in nature. Describe what happens upon dissolution in the two cases, and account for the different results.

4.13 We have learned in this chapter that many ionic solids dissolve in water as strong electrolytes, that is, as separated ions in solution. What properties of water facilitate this process?

4.14 What does it mean to say that ions are hydrated when an ionic substance dissolves in water?

4.15 Specify what ions are present in solution upon dissolving each of the following substances in water: **(a)** $ZnCl_2$, **(b)** HNO_3, **(c)** $(NH_4)_2SO_4$, **(d)** $Ca(OH)_2$.

4.16 Specify what ions are present upon dissolving each of the following substances in water: **(a)** MgI_2, **(b)** $Al(NO_3)_3$, **(c)** $HClO_4$, **(d)** $KC_2H_3O_2$.

4.17 Formic acid, $HCHO_2$, is a weak electrolyte. What solute particles are present in an aqueous solution of this compound? Write the chemical equation for the ionization of $HCHO_2$.

4.18 Acetone, CH_3COCH_3, is a nonelectrolyte; hypochlorous acid, HClO, is a weak electrolyte; and ammonium chloride, NH_4Cl, is a strong electrolyte. **(a)** What are the solute particles present in aqueous solutions of each compound? **(b)** If 0.1 mol of each compound is dissolved in solution, which one contains 0.2 mol of solute particles, which contains 0.1 mol of solute particles, and which contains somewhere between 0.1 and 0.2 mol of solute particles?

Precipitation Reactions and Net Ionic Equations

4.19 Using solubility guidelines, predict whether each of the following compounds is soluble or insoluble in water: **(a)** $NiCl_2$, **(b)** Ag_2S, **(c)** Cs_3PO_4, **(d)** $SrCO_3$, **(e)** $PbSO_4$.

4.20 Predict whether each of the following compounds is soluble in water: **(a)** $Ni(OH)_2$, **(b)** $PbBr_2$, **(c)** $Ba(NO_3)_2$, **(d)** $AlPO_4$, **(e)** $AgC_2H_3O_2$.

4.21 Will precipitation occur when the following solutions are mixed? If so, write a balanced chemical equation for the reaction. **(a)** Na_2CO_3 and $AgNO_3$, **(b)** $NaNO_3$ and $NiSO_4$, **(c)** $FeSO_4$ and $Pb(NO_3)_2$.

4.22 Identify the precipitate (if any) that forms when the following solutions are mixed, and write a balanced equation for each reaction. **(a)** $Ni(NO_3)_2$ and NaOH, **(b)** NaOH and K_2SO_4, **(c)** Na_2S and $Cu(C_2H_3O_2)_2$.

4.23 Name the spectator ions in any reactions that may be involved when each of the following pairs of solutions are mixed.

(a) $Na_2CO_3(aq)$ and $MgSO_4(aq)$
(b) $Pb(NO_3)_2(aq)$ and $Na_2S(aq)$
(c) $(NH_4)_3PO_4(aq)$ and $CaCl_2(aq)$

4.24 Write balanced net ionic equations for the reactions that occur in each of the following cases. Identify the spectator ion or ions in each reaction.

(a) $Cr_2(SO_4)_3(aq) + (NH_4)_2CO_3(aq) \longrightarrow$
(b) $Ba(NO_3)_2(aq) + K_2SO_4(aq) \longrightarrow$
(c) $Fe(NO_3)_2(aq) + KOH(aq) \longrightarrow$

4.25 Separate samples of a solution of an unknown salt are treated with dilute solutions of HBr, H_2SO_4, and NaOH. A precipitate forms in all three cases. Which of the following cations could the solution contain: K^+; Pb^{2+}; Ba^{2+}?

4.26 Separate samples of a solution of an unknown ionic compound are treated with dilute $AgNO_3$, $Pb(NO_3)_2$, and $BaCl_2$. Precipitates form in all three cases. Which of the following could be the anion of the unknown salt: Br^-; CO_3^{2-}; NO_3^-?

4.27 You know that an unlabeled bottle contains a solution of one of the following: $AgNO_3$, $CaCl_2$, or $Al_2(SO_4)_3$. A friend suggests that you test a portion of the solution with $Ba(NO_3)_2$ and then with NaCl solutions. Explain how these two tests together would be sufficient to determine which salt is present in the solution.

4.28 Three solutions are mixed together to form a single solution. One contains 0.2 mol $Pb(C_2H_3O_2)_2$, the second contains 0.1 mol Na_2S, and the third contains 0.1 mol $CaCl_2$. **(a)** Write the net ionic equations for the precipitation reaction or reactions that occur. **(b)** What are the spectator ions in the solution?

Acid-Base Reactions

4.29 Which of the following solutions has the largest concentration of solvated protons: **(a)** 0.1 M LiOH, **(b)** 0.1 M HI, **(c)** 0.5 M methyl alcohol (CH_3OH)? Explain.

4.30 Which of the following solutions is the most basic? **(a)** 0.5 M NH_3, **(b)** 0.1 M KOH, **(c)** 0.1 M $Ca(OH)_2$. Explain.

4.31 What is the difference between **(a)** a monoprotic acid and a diprotic acid, **(b)** a weak acid and a strong acid, **(c)** an acid and a base?

4.32 Explain the following observations: **(a)** NH_3 contains no OH^- ions, and yet its aqueous solutions are basic; **(b)** HF is called a weak acid, and yet it is very reactive; **(c)** although sulfuric acid is a strong electrolyte, an aqueous solution of H_2SO_4 contains more HSO_4^- ions than SO_4^{2-} ions.

4.33 It is said that $HClO_4$ is a strong acid, whereas $HClO_2$ is a weak acid. What does this mean in terms of the extent to which the two substances are ionized in solution?

4.34 What is the relationship between the solubility rules in Table 4.1 and the list of strong bases in Table 4.2? Another way of asking this question is, why is $Cd(OH)_2$, for example, not listed as a strong base in Table 4.2?

4.35 Label each of the following substances as an acid, base, salt, or none of the above. Indicate whether the substance exists in aqueous solution entirely in molecular form, entirely as ions, or as a mixture of molecules and ions. **(a)** HF; **(b)** acetonitrile, CH_3CN; **(c)** $NaClO_4$; **(d)** $Ba(OH)_2$.

4.36 An aqueous solution of an unknown solute is tested with litmus paper and found to be acidic. The solution is weakly conducting compared with a solution of NaCl of the same concentration. Which of the following substances could the unknown be: KOH, NH_3, HNO_3, $KClO_2$, H_3PO_3, CH_3COCH_3 (acetone)?

4.37 Classify each of the following substances as a nonelectrolyte, weak electrolyte, or strong electrolyte in water: **(a)** H_2SO_3, **(b)** C_2H_5OH (ethanol), **(c)** NH_3, **(d)** $KClO_3$, **(e)** $Cu(NO_3)_2$.

4.38 Classify each of the following aqueous solutions as a nonelectrolyte, weak electrolyte, or strong electrolyte: **(a)** $HClO_4$, **(b)** HNO_3, **(c)** NH_4Cl, **(d)** CH_3COCH_3 (acetone), **(e)** $CoSO_4$, **(f)** $C_{12}H_{22}O_{11}$ (sucrose).

4.39 Complete and balance the following molecular equations, and then write the net ionic equation for each:
(a) $HBr(aq) + Ca(OH)_2(aq) \longrightarrow$
(b) $Cu(OH)_2(s) + HClO_4(aq) \longrightarrow$
(c) $Al(OH)_3(s) + HNO_3(aq) \longrightarrow$

4.40 Write the balanced molecular and net ionic equations for each of the following neutralization reactions:
(a) Aqueous acetic acid is neutralized by aqueous potassium hydroxide.
(b) Solid chromium(III) hydroxide reacts with nitric acid.
(c) Aqueous hypochlorous acid and aqueous calcium hydroxide react.

4.41 Write balanced molecular and net ionic equations for the following reactions, and identify the gas formed in each: **(a)** solid cadmium sulfide reacts with an aqueous solution of sulfuric acid; **(b)** solid magnesium carbonate reacts with an aqueous solution of perchloric acid.

4.42 Because the oxide ion is basic, metal oxides react readily with acids. **(a)** Write the net ionic equation for the following reaction:

$$FeO(s) + 2\,HClO_4(aq) \longrightarrow Fe(ClO_4)_2(aq) + H_2O(l).$$

(b) Based on the equation in part (a), write the net ionic equation for the reaction that occurs between NiO(s) and an aqueous solution of nitric acid.

4.43 Write a balanced molecular equation and a net ionic equation for the reaction that occurs when **(a)** solid $CaCO_3$ reacts with an aqueous solution of nitric acid; **(b)** solid iron(II) sulfide reacts with an aqueous solution of hydrobromic acid.

4.44 As K_2O dissolves in water, the oxide ion reacts with water molecules to form hydroxide ions. Write the molecular and net ionic equations for this reaction. Based on the definitions of acid and base, what ion is the base in this reaction? What is the acid? What is the spectator ion in the reaction?

Oxidation-Reduction Reactions

4.45 Define oxidation and reduction in terms of **(a)** electron transfer and **(b)** oxidation numbers.

4.46 Can oxidation occur without accompanying reduction? Explain.

4.47 Which circled region of the periodic table shown here contains the most readily oxidized elements? Which contains the least readily oxidized?

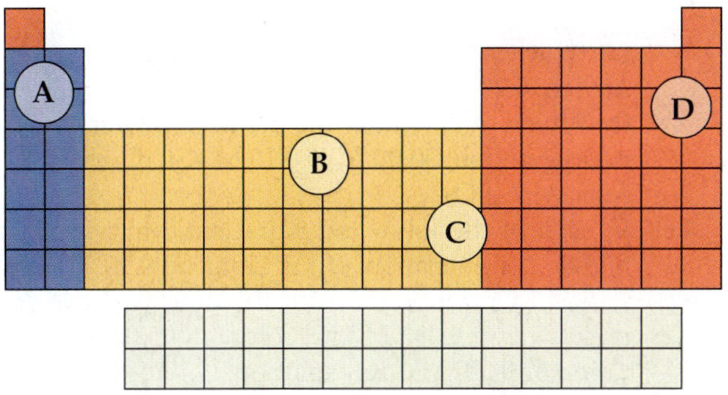

4.48 From the elements listed in Table 4.5, select an element that lies in region A of the periodic table shown to the left and an element that lies in region C. Write a balanced oxidation-reduction equation that shows the oxidation of one metal and reduction of an ion of the other. You will need to decide which element is oxidized and which is reduced.

4.49 Determine the oxidation number for the indicated element in each of the following substances: **(a)** S in SO_2, **(b)** C in $COCl_2$, **(c)** Mn in MnO_4^-, **(d)** Br in HBrO, **(e)** As in As_4, **(f)** O in K_2O_2.

4.50 Determine the oxidation number for the indicated element in each of the following compounds: **(a)** Ti in TiO_2, **(b)** Sn in $SnCl_3^-$, **(c)** C in $C_2O_4^{2-}$, **(d)** N in N_2H_4, **(e)** N in HNO_2, **(f)** Cr in $Cr_2O_7^{2-}$.

4.51 Which element is oxidized and which is reduced in the following reactions?
(a) $Ni(s) + Cl_2(g) \longrightarrow NiCl_2(s)$
(b) $3\,Fe(NO_3)_2(aq) + 2\,Al(s) \longrightarrow 3\,Fe(s) + 2\,Al(NO_3)_3(aq)$

(c) $Cl_2(aq) + 2\,NaI(aq) \longrightarrow I_2(aq) + 2\,NaCl(aq)$
(d) $PbS(s) + 4\,H_2O_2(aq) \longrightarrow PbSO_4(s) + 4\,H_2O(l)$

4.52 Which of the following are redox reactions? For those that are, indicate which element is oxidized and which is reduced. For those that are not, indicate whether they are precipitation or acid-base reactions.
(a) $Cu(OH)_2(s) + 2\,HNO_3(aq) \longrightarrow Cu(NO_3)_2(aq) + 2\,H_2O(l)$
(b) $Fe_2O_3(s) + 3\,CO(g) \longrightarrow 2\,Fe(s) + 3\,CO_2(g)$
(c) $Sr(NO_3)_2(aq) + H_2SO_4(aq) \longrightarrow SrSO_4(s) + 2\,HNO_3(aq)$
(d) $4\,Zn(s) + 10\,H^+(aq) + 2\,NO_3^-(aq) \longrightarrow 4\,Zn^{2+}(aq) + N_2O(g) + 5\,H_2O(l)$

4.53 Write balanced molecular and net ionic equations for the reactions of **(a)** manganese with dilute sulfuric acid; **(b)** chromium with hydrobromic acid; **(c)** tin with hydrochloric acid; **(d)** aluminum with formic acid, $HCHO_2$.

4.54 Write balanced molecular and net ionic equations for the reactions of **(a)** hydrochloric acid with nickel; **(b)** dilute sulfuric acid with iron; **(c)** hydrobromic acid with magnesium; **(d)** acetic acid, $HC_2H_3O_2$, with zinc.

4.55 Using the activity series (Table 4.5), write balanced chemical equations for the following reactions. If no reaction occurs, simply write NR. **(a)** Iron metal is added to a solution of copper(II) nitrate; **(b)** zinc metal is added to a solution of magnesium sulfate; **(c)** hydrobromic acid is added to tin metal; **(d)** hydrogen gas is bubbled through an aqueous solution of nickel(II) chloride; **(e)** aluminum metal is added to a solution of cobalt(II) sulfate.

4.56 Based on the activity series (Table 4.5), what is the outcome (if any) of each of the following reactions?
(a) $Mn(s) + NiCl_2(aq) \longrightarrow$
(b) $Cu(s) + Cr(C_2H_3O_2)_3(aq) \longrightarrow$
(c) $Cr(s) + NiSO_4(aq) \longrightarrow$
(d) $Pt(s) + HBr(aq) \longrightarrow$
(e) $H_2(g) + CuCl_2(aq) \longrightarrow$

4.57 The metal cadmium tends to form Cd^{2+} ions. The following observations are made: (i) When a strip of zinc metal is placed in $CdCl_2(aq)$, cadmium metal is deposited on the strip. (ii) When a strip of cadmium metal is placed in $Ni(NO_3)_2(aq)$, nickel metal is deposited on the strip. **(a)** Write net ionic equations to explain each of the observations made above. **(b)** What can you conclude about the position of cadmium in the activity series? **(c)** What experiments would you need to perform to locate more precisely the position of cadmium in the activity series?

4.58 **(a)** Use the following reactions to prepare an activity series for the halogens:

$$Br_2(aq) + 2\,NaI(aq) \longrightarrow 2\,NaBr(aq) + I_2(aq);$$

$$Cl_2(aq) + 2\,NaBr(aq) \longrightarrow 2\,NaCl(aq) + Br_2(aq).$$

(b) Relate the positions of the halogens in the periodic table with their locations in this activity series. **(c)** Predict whether a reaction occurs when the following reagents are mixed: $Cl_2(aq)$ and $KI(aq)$; $Br_2(aq)$ and $LiCl(aq)$.

Solution Composition; Molarity

4.59 **(a)** Is the concentration of a solution an intensive or an extensive property? **(b)** What is the difference between 0.50 mol HCl and 0.50 *M* HCl?

4.60 **(a)** Suppose you prepare 500 mL of a 0.10 *M* solution of some salt and then spill some of it. What happens to the concentration of the solution left in the container? **(b)** A certain volume of a 0.50 *M* solution contains 4.5 g of a salt. What mass of the salt is present in the same volume of a 2.50 *M* solution?

4.61 **(a)** Calculate the molarity of a solution that contains 0.0345 mol NH_4Cl in exactly 400 mL of solution. **(b)** How many moles of HNO_3 are present in 35.0 mL of a 2.20 *M* solution of nitric acid? **(c)** How many milliliters of 1.50 *M* KOH solution are needed to provide 0.125 mol of KOH?

4.62 **(a)** Calculate the molarity of a solution made by dissolving 0.145 mol Na_2SO_4 in enough water to form exactly 750 mL of solution. **(b)** How many moles of $KMnO_4$ are present in 125 mL of a 0.0850 *M* solution? **(c)** How many milliliters of 11.6 *M* HCl solution are needed to obtain 0.255 mol of HCl?

4.63 The average adult human male has a total blood volume of 5.0 L. If the concentration of sodium ion in this average individual is 0.135 *M*, what is the mass of sodium ion circulating in the blood?

4.64 A person suffering from hyponatremia has a sodium ion concentration in the blood of 0.118 *M* and a total blood volume of 4.6 L. What mass of sodium chloride would need to be added to the blood to bring the sodium ion concentration up to 0.138 *M*, assuming no change in blood volume?

4.65 Calculate **(a)** the number of grams of solute in 0.250 L of 0.150 *M* KBr, **(b)** the molar concentration of a solution containing 4.75 g of $Ca(NO_3)_2$ in 0.200 L, **(c)** the volume of 1.50 *M* Na_3PO_4 in milliliters that contains 5.00 g of solute.

4.66 **(a)** How many grams of solute are present in 50.0 mL of 0.360 *M* $K_2Cr_2O_7$? **(b)** If 4.28 g of $(NH_4)_2SO_4$ is dissolved in enough water to form 300 mL of solution, what is the molarity of the solution? **(c)** How many milliliters of 0.240 *M* $CuSO_4$ contain 2.25 g of solute?

4.67 **(a)** Which will have the highest concentration of potassium ion: 0.20 *M* KCl, 0.15 *M* K_2CrO_4, or 0.080 *M* K_3PO_4? **(b)** Which will contain the greater number of moles of potassium ion: 30.0 mL of 0.15 M K_2CrO_4 or 25.0 mL of 0.080 *M* K_3PO_4?

4.68 In each of the following pairs, indicate which has the higher concentration of Cl^- ion: **(a)** 0.10 *M* $CaCl_2$ or 0.15 *M* KCl solution, **(b)** 100 mL of 0.10 *M* KCl solution or 400 mL of 0.080 *M* LiCl solution, **(c)** 0.050 *M* HCl solution or 0.020 *M* $CdCl_2$ solution.

4.69 Indicate the concentration of each ion or molecule present in the following solutions: **(a)** 0.22 *M* NaOH, **(b)** 0.16 *M* $CaBr_2$, **(c)** 0.15 *M* CH_3OH, **(d)** a mixture of 40.0 mL of 0.15 *M* $KClO_3$ and 35.0 mL of 0.22 *M* Na_2SO_4. Assume the volumes are additive.

4.70 Indicate the concentration of each ion present in the solution formed by mixing **(a)** 16.0 mL of 0.130 *M* HCl and 12.0 mL of 0.600 *M* HCl, **(b)** 18.0 mL of 0.200 *M* Na_2SO_4 and 15.0 mL of 0.150 *M* KCl, **(c)** 2.38 g of NaCl in 50.0 mL of 0.400 *M* $CaCl_2$ solution. (Assume that the volumes are additive.)

4.71 **(a)** You have a stock solution of 14.8 *M* NH_3. How many milliliters of this solution should you dilute to make 100.0 mL of 0.250 *M* NH_3? **(b)** If you take a 10.0-mL portion of the stock solution and dilute it to a total volume of 0.250 L, what will be the concentration of the final solution?

4.72 **(a)** How many milliliters of a stock solution of 10.0 *M* HNO_3 would you have to use to prepare 0.350 L of 0.400 *M* HNO_3? **(b)** If you dilute 25.0 mL of the stock solution to a final volume of 0.500 L, what will be the concentration of the diluted solution?

4.73 **(a)** Starting with solid sucrose, $C_{12}H_{22}O_{11}$, describe how you would prepare 125 mL of 0.150 *M* sucrose solution. **(b)** Describe how you would prepare 400.0 mL of 0.100 *M* $C_{12}H_{22}O_{11}$ starting with 2.00 L of 1.50 *M* $C_{12}H_{22}O_{11}$.

4.74 **(a)** How would you prepare 250.0 mL of 0.150 *M* $AgNO_3$ solution starting with pure solute? **(b)** An experiment calls for you to use 100 mL of 0.50 *M* HNO_3 solution. All you have available is a bottle of 6.0 *M* HNO_3. How would you prepare the desired solution?

[4.75] Pure acetic acid, known as glacial acetic acid, is a liquid with a density of 1.049 g/mL at 25°C. Calculate the molarity of a solution of acetic acid made by dissolving 20.00 mL of glacial acetic acid at 25°C in enough water to make 250.0 mL of solution.

[4.76] Glycerol, $C_3H_8O_3$, is a substance used extensively in the manufacture of cosmetics, foodstuffs, antifreeze, and plastics. Glycerol is a water-soluble liquid with a density of 1.2656 g/L at 15°C. Calculate the molarity of a solution of glycerol made by dissolving 50.000 mL glycerol at 15°C in enough water to make 250.00 mL of solution.

Solution Stoichiometry; Titrations

4.77 What mass of NaCl is needed to precipitate the silver ions from 20.0 mL of 0.100 *M* $AgNO_3$ solution?

4.78 What mass of NaOH is needed to precipitate the Cd^{2+} ions from 25.0 mL of 0.500 *M* $Cd(NO_3)_2$ solution?

4.79 **(a)** What volume of 0.115 *M* $HClO_4$ solution is needed to neutralize 50.00 mL of 0.0875 *M* NaOH? **(b)** What volume of 0.128 *M* HCl is needed to neutralize 2.87 g of $Mg(OH)_2$? **(c)** If 25.8 mL of $AgNO_3$ is needed to precipitate all the Cl^- ions in a 785-mg sample of KCl (forming AgCl), what is the molarity of the $AgNO_3$ solution? **(d)** If 45.3 mL of 0.108 *M* HCl solution is needed to neutralize a solution of KOH, how many grams of KOH must be present in the solution?

4.80 **(a)** How many milliliters of 0.120 *M* HCl are needed to completely neutralize 50.0 mL of 0.101 *M* $Ba(OH)_2$ solution? **(b)** How many milliliters of 0.125 *M* H_2SO_4 are needed to neutralize 0.200 g of NaOH? **(c)** If 55.8 mL of $BaCl_2$ solution is needed to precipitate all the sulfate ion in a 752-mg sample of Na_2SO_4, what is the molarity of the solution? **(d)** If 42.7 mL of 0.208 *M* HCl solution is needed to neutralize a solution of $Ca(OH)_2$, how many grams of $Ca(OH)_2$ must be in the solution?

4.81 Some sulfuric acid is spilled on a lab bench. It can be neutralized by sprinkling sodium bicarbonate on it and then mopping up the resultant solution. The sodium bicarbonate reacts with sulfuric acid as follows:

$$2\,NaHCO_3(s) + H_2SO_4(aq) \longrightarrow Na_2SO_4(aq) + 2\,H_2O(l) + 2\,CO_2(g)$$

Sodium bicarbonate is added until the fizzing due to the formation of $CO_2(g)$ stops. If 27 mL of 6.0*M* H_2SO_4 was spilled, what is the minimum mass of $NaHCO_3$ that must be added to the spill to neutralize the acid?

4.82 The distinctive odor of vinegar is due to acetic acid, $HC_2H_3O_2$, which reacts with sodium hydroxide in the following fashion:

$$HC_2H_3O_2(aq) + NaOH(aq) \longrightarrow H_2O(l) + NaC_2H_3O_2(aq)$$

If 3.45 mL of vinegar needs 42.5 mL of 0.115 *M* NaOH to reach the equivalence point in a titration, how many grams of acetic acid are in a 1.00-qt sample of this vinegar?

4.83 A sample of solid $Ca(OH)_2$ is stirred in water at 30°C until the solution contains as much dissolved $Ca(OH)_2$ as it can hold. A 100-mL sample of this solution is withdrawn and titrated with 5.00×10^{-2} *M* HBr. It requires 48.8 mL of the acid solution for neutralization. What is the molarity of the $Ca(OH)_2$ solution? What is the solubility of $Ca(OH)_2$ in water, at 30°C in grams of $Ca(OH)_2$ per 100 mL of solution?

4.84 In the laboratory 6.82 g of $Sr(NO_3)_2$ is dissolved in enough water to form 0.500 L. A 0.100-L sample is withdrawn from this stock solution and titrated with a 0.0335 *M* solution of Na_2CrO_4. What volume of Na_2CrO_4 solution is needed to precipitate all the $Sr^{2+}(aq)$ as $SrCrO_4$?

4.85 A solution of 100.0 mL of 0.200 *M* KOH is mixed with a solution of 200.0 mL of 0.150 *M* $NiSO_4$. **(a)** Write the balanced chemical equation for the reaction that occurs. **(b)** What precipitate forms? **(c)** What is the limiting reactant? **(d)** How many grams of this precipitate form? **(e)** What is the concentration of each ion that remains in solution?

4.86 A solution is made by mixing 12.0 g of NaOH and 75.0 mL of 0.200 *M* HNO_3. **(a)** Write a balanced equation for the reaction that occurs between the solutes. **(b)** Calculate the concentration of each ion remaining in solution. **(c)** Is the resultant solution acidic or basic?

[4.87] A 0.5895-g sample of impure magnesium hydroxide is dissolved in 100.0 mL of 0.2050 M HCl solution. The excess acid then needs 19.85 mL of 0.1020 M NaOH for neutralization. Calculate the percent by mass of magnesium hydroxide in the sample, assuming that it is the only substance reacting with the HCl solution.

[4.88] A 1.248-g sample of limestone rock is pulverized and then treated with 30.00 mL of 1.035 M HCl solution. The excess acid then requires 11.56 mL of 1.010 M NaOH for neutralization. Calculate the percent by mass of calcium carbonate in the rock, assuming that it is the only substance reacting with the HCl solution.

Additional Exercises

[4.89] Suppose it were possible to instantaneously form a solution of uniformly dispersed K^+ and Br^- ions in a solvent such as mineral oil. What would you expect to occur? How does this compare with what would happen if the same uniform dispersion of ions were formed in water? How do you account for the difference?

4.90 The accompanying photo shows the reaction between a solution of $Cd(NO_3)_2$ and one of Na_2S. What is the identity of the precipitate? What ions remain in solution? Write the net ionic equation for the reaction.

4.91 Suppose you have a solution that might contain any or all of the following cations: Ni^{2+}, Ag^+, Sr^{2+}, and Mn^{2+}. Addition of HCl solution causes a precipitate to form. After filtering off the precipitate, H_2SO_4 solution is added to the resultant solution and another precipitate forms. This is filtered off, and a solution of NaOH is added to the resulting solution. No precipitate is observed. Which ions are present in each of the precipitates? Which of the four ions listed above must be absent from the original solution?

4.92 You choose to investigate some of the solubility guidelines for two ions not listed in Table 4.1, the chromate ion (CrO_4^{2-}) and the oxalate ion ($C_2O_4^{2-}$). You are given 0.01 M solutions (A, B, C, D) of four water-soluble salts:

Solution	Solute	Color of Solution
A	Na_2CrO_4	Yellow
B	$(NH_4)_2C_2O_4$	Colorless
C	$AgNO_3$	Colorless
D	$CaCl_2$	Colorless

When these solutions are mixed, the following observations are made:

Expt Number	Solutions Mixed	Result
1	A + B	No precipitate, yellow solution
2	A + C	Red precipitate forms
3	A + D	No precipitate, yellow solution
4	B + C	White precipitate forms
5	B + D	White precipitate forms
6	C + D	White precipitate forms

(a) Write a net ionic equation for the reaction that occurs in each of the experiments. **(b)** Identify the precipitate formed, if any, in each of the experiments. **(c)** Based on these limited observations, which ion tends to form the more soluble salts, chromate or oxalate?

4.93 Antacids are often used to relieve pain and promote healing in the treatment of mild ulcers. Write balanced net ionic equations for the reactions between the $HCl(aq)$ in the stomach and each of the following substances used in various antacids: **(a)** $Al(OH)_3(s)$, **(b)** $Mg(OH)_2(s)$, **(c)** $MgCO_3(s)$, **(d)** $NaAl(CO_3)(OH)_2(s)$, **(e)** $CaCO_3(s)$.

[4.94] Salts of the sulfite ion, SO_3^{2-}, react with acids in a way similar to that of carbonates. **(a)** Predict the chemical formula, and name the weak acid that forms when the sulfite ion reacts with acids. **(b)** The acid formed in part (a) decomposes to form water and a gas. Predict the molecular formula, and name the gas formed. **(c)** Use a source book such as the *CRC Handbook of Chemistry and Physics* to confirm that the substance in part (b) is a gas under normal room-temperature conditions. **(d)** Write balanced net ionic equations of the reaction of $HCl(aq)$ with (i) $Na_2SO_3(aq)$, (ii) $Ag_2SO_3(s)$, (iii) $KHSO_3(s)$, and (iv) $ZnSO_3(aq)$.

4.95 The commercial production of nitric acid involves the following chemical reactions:

$$4\,NH_3(g) + 5\,O_2(g) \longrightarrow 4\,NO(g) + 6\,H_2O(g)$$

$$2\,NO(g) + O_2(g) \longrightarrow 2\,NO_2(g)$$

$$3\,NO_2(g) + H_2O(l) \longrightarrow 2\,HNO_3(aq) + NO(g)$$

(a) Which of these reactions are redox reactions? **(b)** In each redox reaction identify the element undergoing oxidation and the element undergoing reduction.

4.96 Use Table 4.5 to predict which of the following ions can be reduced to their metal forms by reacting with zinc: **(a)** $Na^+(aq)$, **(b)** $Pb^{2+}(aq)$, **(c)** $Mg^{2+}(aq)$, **(d)** $Fe^{2+}(aq)$, **(e)** $Cu^{2+}(aq)$, **(f)** $Al^{3+}(aq)$. Write the balanced net ionic equation for each reaction that occurs.

[4.97] Lanthanum metal forms cations with a charge of 3+. Consider the following observations about the chemistry of lanthanum: When lanthanum metal is exposed to air, a white solid (compound A) is formed that contains lanthanum and one other element. When lanthanum metal is added to water, gas bubbles are observed and a different white solid (compound B) is formed. Both A and B dissolve in hydrochloric acid to give a clear solution. When either of these solution is evaporated, a soluble white solid (compound C) remains. If compound C is dissolved in water and sulfuric acid is added, a white precipitate (compound D) forms. **(a)** Propose identities for the substances A, B, C, and D. **(b)** Write net ionic equations for all the reactions described. **(c)** Based on the preceding observations, what can be said about the position of lanthanum in the activity series (Table 4.5)?

4.98 A 35.0-mL sample of 1.00 *M* KBr and a 60.0-mL sample of 0.600 *M* KBr are mixed. The solution is then heated to evaporate water until the total volume is 50.0 mL. What is the molarity of the KBr in the final solution?

4.99 Calculate the molarity of the solution produced by mixing **(a)** 40.0 mL of 0.160 *M* NaCl and 65.0 mL of 0.150 *M* NaCl, **(b)** 32.5 mL of 0.750 *M* NaOH and 26.8 mL of 0.750 *M* NaOH. (Assume that the volumes are additive.)

4.100 Using modern analytical techniques, it is possible to detect sodium ions in concentrations as low as 50 pg/mL. What is this detection limit expressed in **(a)** molarity of Na^+, **(b)** Na^+ ions per cubic centimeter?

4.101 Hard water contains Ca^{2+}, Mg^{2+}, and Fe^{2+}, which interfere with the action of soap and leave an insoluble coating on the insides of containers and pipes when heated. Water softeners replace these ions with Na^+. If 1.0×10^3 L of hard water contains 0.010 *M* Ca^{2+} and 0.0050 *M* Mg^{2+}, how many moles of Na^+ are needed to replace these ions?

4.102 Tartaric acid, $H_2C_4H_4O_6$, has two acidic hydrogens. The acid is often present in wines and precipitates from solution as the wine ages. A solution containing an unknown concentration of the acid is titrated with NaOH. It requires 22.62 mL of 0.2000 *M* NaOH solution to titrate both acidic protons in 40.00 mL of the tartaric acid solution. Write a balanced net ionic equation for the neutralization reaction, and calculate the molarity of the tartaric acid solution.

4.103 The concentration of hydrogen peroxide in a solution is determined by titrating a 10.0-mL sample of the solution with permanganate ion.

$$2\,MnO_4^-(aq) + 5\,H_2O_2(aq) + 6\,H^+(aq) \longrightarrow 2\,Mn^{2+}(aq) + 5\,O_2(g) + 8\,H_2O(l)$$

If it takes 16.8 mL of 0.124 *M* MnO_4^- solution to reach the equivalence point, what is the molarity of the hydrogen peroxide solution?

[4.104] A solid sample of $Zn(OH)_2$ is added to 0.400 L of 0.500 *M* aqueous HBr. The solution that remains is still acidic. It is then titrated with 0.500 *M* NaOH solution, and it takes 98.5 mL of the NaOH solution to reach the equivalence point. What mass of $Zn(OH)_2$ was added to the HBr solution?

Integrative Exercises

4.105 A solution of sodium chlorate is prepared by dissolving 1.28 g of the salt in water to form 1.00 L of solution. To form a solution of sodium chlorite of the same molarity and volume, what mass of sodium chlorite would be needed?

4.106 **(a)** By titration, 15.0 mL of 0.1008 *M* sodium hydroxide is needed to neutralize a 0.2053-g sample of an organic acid. What is the molar mass of the acid if it is monoprotic? **(b)** An elemental analysis of the acid indicates that it is composed of 5.89% H, 70.6% C, and 23.5% O by mass. What is its molecular formula?

4.107 A 3.455-g sample of a mixture was analyzed for barium ion by adding a small excess of sulfuric acid to an aqueous solution of the sample. The resultant reaction produced a precipitate of barium sulfate, which was collected by filtration, washed, dried, and weighed. If 0.2815 g of barium sulfate was obtained, what was the mass percentage of barium in the sample?

[4.108] A tanker truck carrying 5.0×10^3 kg of concentrated sulfuric acid solution tips over and spills its load. If the sulfuric acid is 95.0% H_2SO_4 by mass and has a density of 1.84 gm/L, how many kilograms of sodium carbonate must be added to neutralize the acid?

4.109 A sample of 5.53 g of $Mg(OH)_2$ is added to 25.0 mL of 0.200 *M* HNO_3. **(a)** Write the chemical equation for the reaction that occurs. **(b)** Which is the limiting reactant in the reaction? **(c)** How many moles of $Mg(OH)_2$, HNO_3, and $Mg(NO_3)_2$ are present after the reaction is complete?

4.110 A sample of 1.50 g of lead(II) nitrate is mixed with 125 mL of 0.100 *M* sodium sulfate solution. **(a)** Write the chemical equation for the reaction that occurs. **(b)** Which is the limiting reactant in the reaction? **(c)** What are the concentrations of all ions that remain in solution after the reaction is complete?

4.111 A mixture contains 76.5% NaCl, 6.5% $MgCl_2$, and 17.0% Na_2SO_4 by mass. What is the molarity of Cl^- ions in a solution formed by dissolving 7.50 g of the mixture in enough water to form 500.0 mL of solution?

[4.112] The average concentration of bromide ion in seawater is 65 mg of bromide ion per kg of seawater. What is the molarity of the bromide ion if the density of the seawater is 1.025 g/mL?

[4.113] The mass percentage of chloride ion in a 25.00-mL sample of seawater was determined by titrating the sample with silver nitrate, precipitating silver chloride. It took 42.58 mL of 0.2997 *M* silver nitrate solution to reach the equivalence point in the titration. What is the mass percentage of chloride ion in the seawater if its density is 1.025 g/mL?

4.114 The arsenic in a 1.22-g sample of a pesticide was converted to AsO_4^{3-} by suitable chemical treatment. It was then titrated using Ag^+ to form Ag_3AsO_4 as a precipitate. **(a)** What is the oxidation state of As in AsO_4^{3-}? **(b)** Name Ag_3AsO_4 by analogy to the corresponding compound containing phosphorus in place of arsenic. **(c)** If it took 25.0 mL of 0.102 *M* Ag^+ to reach the equivalence point in this titration, what is the mass percentage of arsenic in the pesticide?

[4.115] The new standard for arsenate in drinking water, mandated by the Safe Drinking Water Act, requires that by January, 2006, public water supplies must contain no greater than 10 parts per billion (ppb) arsenic. Assuming that this arsenic is present as arsenate, AsO_4^{3-}, what mass of sodium arsenate would be present in a 1.00-L sample of drinking water that just meets the standard?

[4.116] Federal regulations set an upper limit of 50 parts per million (ppm) of NH_3 in the air in a work environment (that is, 50 molecules of $NH_3(g)$ for every million molecules in the air). Air from a manufacturing operation was drawn through a solution containing 1.00×10^2 mL of 0.0105 *M* HCl. The NH_3 reacts with HCl as follows:

$$NH_3(aq) + HCl(aq) \longrightarrow NH_4Cl(aq)$$

After drawing air through the acid solution for 10.0 min at a rate of 10.0 L/min, the acid was titrated. The remaining acid needed 13.1 mL of 0.0588 *M* NaOH to reach the equivalence point. **(a)** How many grams of NH_3 were drawn into the acid solution? **(b)** How many ppm of NH_3 were in the air? (Air has a density of 1.20 g/L and an average molar mass of 29.0 g/mol under the conditions of the experiment.) **(c)** Is this manufacturer in compliance with regulations?

eMEDIA EXERCISES

These exercises make use of the interactive objects available online in OneKey or the Companion Website, and on your Accelerator CD. Access to these resources comes in your MediaPak.

4.117 The **Electrolytes and Non-Electrolytes** animation (*4.1*) and the **Introduction to Aqueous Acids** and **Introduction to Aqueous Bases** animations (*4.3*) illustrate the behavior of various substances in aqueous solution. For each of the seven substances mentioned in the animations, write the chemical equation that corresponds to dissolution in water. (The chemical formula of sugar is $C_{12}H_{22}O_{11}$.) Where appropriate, use the double arrow notation.

4.118 In the **Strong and Weak Electrolytes** movie (*4.1*), the lightbulb glows brightly when the beaker contains aqueous hydrochloric acid, but relatively dimly when the beaker contains aqueous acetic acid. **(a)** For each of the compounds in Exercise 4.3, would you expect an aqueous solution to cause the bulb to light? If so, how brightly? **(b)** Consider the use of aqueous solutions of each of the following compounds in the apparatus shown in the demonstration. For each compound, tell whether you would expect the lightbulb to glow brightly, dimly, or not at all: H_2CO_3, C_2H_5OH, NH_4Cl, CaF_2, and HF.

4.119 **(a)** Use the solubility rules to predict what precipitate, if any, will form as the result of each combination. (i) $Na_2CO_3(aq)$ and $Fe(NO_3)_2(aq)$, (ii) $NH_4NO_3(aq)$ and $K_2SO_4(aq)$, (iii) $AlBr_3(aq)$ and $Fe_2(SO_4)_3(aq)$, (iv) $H_2SO_4(aq)$ and $Pb(NO_3)_2(aq)$, (v) $Na_2S(aq)$ and $(NH_4)_2SO_4(aq)$. Use the **Ionic Compounds** activity (*4.2*) to check your answers. **(b)** For each combination that produces a precipitate, write a balanced net ionic equation. **(c)** When $NH_4Cl(aq)$ and $Pb(NO_3)_2(aq)$ are combined, a precipitate forms. What ions are still present in the solution in significant concentration after the precipitation? Explain.

4.120 In the **Oxidation-Reduction Chemistry of Tin and Zinc** movie (*4.4*), zinc is oxidized by a solution containing tin ions. **(a)** Write the equation corresponding to this redox reaction. **(b)** In addition to the reaction between zinc metal and tin ions, there is another process occurring. Write the net ionic equation corresponding to this process. (Refer to Exercise 4.54.)

4.121 After watching the **Solution Formation from a Solid** animation (*4.5*), answer the following questions: **(a)** If we neglect to account for the mass of the weighing paper, how would our calculated concentration differ from the actual concentration of the solution? **(b)** Describe the process of preparing an aqueous solution of known concentration, starting with a solid. **(c)** Why is it necessary to make the solution as described in the animation, rather than simply filling the flask up to the mark with water and then adding the solute? **(d)** Describe how you would prepare the solution in part (a) starting with the concentrated stock solution in the **Solution Formation by Dilution** animation (*4.5*).

4.122 Use the **Acid-Base Titration** activity (*4.6*) to determine the concentration of an unknown acid by adding 0.40 *M* NaOH in increments of 1.0 mL. Repeat the titration, adding increments of 0.10 mL of base near the end point. Once more, repeat the titration, adding increments of 0.05 mL of base near the end point. If your acid is dilute enough, repeat the titration three more times, using 0.10 *M* NaOH in 1.0-mL, 0.50-mL, and 0.05-mL increments. **(a)** Tabulate the acid concentrations that you calculate from your titration data. Are the values all the same? If not, why not? **(b)** Which value do you consider to be most accurate and why?

4.123 **(a)** What is the maximum concentration of monoprotic acid that could be titrated in the **Acid-Base Titration** activity *(4.6)* using 0.05 *M* NaOH? **(b)** What is the maximum concentration of a diprotic acid that could be titrated in this simulation using 0.10 *M* base? **(c)** All of the acid-base indicators available in the simulation change color within a pH range of ~4 to 10.5. What effect would the use of an indicator such as metacresol purple have on the experimentally determined value of an unknown acid's concentration? (Metacresol purple changes colors in the pH range of ~1.2 to 2.8.)

eLABORATORY EXERCISES

These exercises allow students to apply the concepts and skills for this chapter in the simulated laboratory of Virtual ChemLab. Worksheets for each of these exercises are found in the Virtual ChemLab Workbook in your MediaPak.

4.124 ***(VCL 4-1) Strong and Weak Electrolytes***
As discussed in the chapter, strong electrolytes are compounds that fully dissociate in water and weak electrolytes are ones that do so only to a small extent. A method for measuring the degree of dissociation is to measure the electrical conductivity of the solution. In this assignment you will measure the conductivity of several solutions and make some important conclusions.

4.125 ***(VCL 4-2) Precipitation Reactions***
A precipitation reaction is defined as one in which a soluble cation and a soluble anion combine to form an insoluble combination product. In this assignment you will mix several different combinations of cations and anions and observe those that form precipitates and those that do not. You will also write full balanced equations and net ionic equations for these reactions.

4.126 ***(VCL 4-3) Concepts in Acid-Base Titrations***
Titrations provide a method of quantitatively measuring the concentration of an unknown solution. In an acid-base titration this is done by delivering a titrant of known concentration into an analyte of known volume. The equivalence point of the titration, or the point where the analyte has been completely consumed by the titrant, is identified by the point where the pH changes rapidly over a small volume of titrant delivered. In this assignment you will titrate a strong acid, HCl, with a strong base, NaOH, and observe the pH of the solution as a function of the volume of titrant delivered.

4.127 ***(VCL 4-4) Predicting the Equivalence Point***
To make a titration more efficient and more accurate, it is often important to be able to predict the equivalence point for the titration. In this assignment you will be given 0.3000 *M* HCl and 0.3000 *M* NaOH. You will predict the equivalence point and then perform the titration to check your prediction.

4.128 ***(VCL 4-5) Predicting the Equivalence Point***
To make a titration more efficient and more accurate, it is often important to be able to predict the equivalence point for the titration. In this assignment you will be given 0.1276 *M* HCl and 0.1475 *M* NaOH. You will predict the equivalence point and then perform the titration to check your prediction.

4.129 ***(VCL 4-6) Predicting the Equivalence Point***
To make a titration more efficient and more accurate, it is often important to be able to predict the equivalence point for the titration. In this assignment you will be given 0.1033 *M* CH_3COOH (acetic acid, or HAc) and 0.1104 *M* NaOH. You will predict the equivalence point and then perform the titration to check your prediction.

4.130 ***(VCL 4-7) Acid-Base Titration: Practice***
Titrations provide a method of quantitatively measuring the concentration of an unknown solution. In an acid-base titration this is done by delivering a titrant of known concentration into an analyte of known volume. In this assignment you will titrate a 0.3000 *M* solution of NaOH into 25 mL of 0.3000 *M* HCl. Although in this situation you know the concentration of both NaOH and HCl, this will give you practice in performing a titration and calculating the concentration of the analyte, which in this case is HCl.

4.131 ***(VCL 4-8) Acid-Base Titration: Unknown HCl***
Titrations provide a method of quantitatively measuring the concentration of an unknown solution. In an acid-base titration this is done by delivering a titrant of known concentration into an analyte of known volume. In this assignment you will titrate a 0.2564 *M* solution of NaOH into 25 mL of an unknown concentration of HCl and calculate the concentration of the HCl solution.

CHAPTER

5

WHAT'S AHEAD

- We will consider the nature of *energy* and the forms it takes, notably *kinetic energy* and *potential energy*. We will also discuss the units used in measuring energy and the fact that energy can be used to accomplish *work* or transfer *heat*. To study energy changes, we focus on a particular part of the universe, which we call the *system*. Everything else is called the *surroundings*. *(Section 5.1)*
- We'll then explore the *first law of thermodynamics*: Energy cannot be created or destroyed, but it can be transformed from one form to another or transferred between systems and surroundings. *(Section 5.2)*
- The energy possessed by a system is called its *internal energy*. Internal energy, E, is a *state function* because its value depends only on the state, or condition, of a system now, not on how the system came to be in that state. *(Section 5.2)*
- A related state function—*enthalpy*, H—is useful because the change in enthalpy, ΔH, measures the quantity of heat energy gained or lost by a system in a process occurring under a constant pressure. *(Sections 5.3 and 5.4)*
- We will examine *calorimetry*, which is the technique we use to measure heat changes in chemical processes. *(Section 5.5)*
- Then we will discuss how to establish standard values for enthalpy changes in chemical reactions and how to use them to calculate enthalpy changes for reactions (ΔH_{rxn}). *(Sections 5.6 and 5.7)*
- Finally, we'll examine foods and fuels as sources of energy and discuss some related health and social issues. *(Section 5.8)*

Thermochemistry

MODERN SOCIETY DEPENDS ON ENERGY for its existence. Energy is used to drive our machinery and appliances, to power our transportation vehicles, and to keep us warm in the winter and cool in the summer. It's not just modern society, however, that depends on energy. Energy is necessary for all of life. Plants, such as those in the chapter-opening photograph, use solar energy to carry out photosynthesis, allowing the plants to grow. The plants, in turn, provide food, from which we humans derive the energy we need to move, to maintain body temperature, and to carry out bodily functions. What exactly is energy, though, and what principles are involved in its many transactions and transformations, such as those from the sun to plants to animals?

In this chapter we begin to explore energy and its changes. In doing so, we are motivated partly by the fact that energy changes invariably accompany chemical reactions. Indeed, sometimes we use a chemical reaction specifically to obtain energy, as when we burn fossil fuels (coal, oil, gasoline, or natural gas). Thus, energy is very much a chemical topic. Nearly all of the energy on which we depend is derived from chemical reactions, whether those reactions are associated with the combustion of fossil fuels, the discharge of a battery, or the metabolism of our foods. Thus, if we are to properly understand chemistry, we must also understand the energy changes that accompany chemical reactions.

The study of energy and its transformations

WILD FLOWERS. Plants utilize solar energy to carry out photosynthesis of carbohydrates, which provide the energy needed for plant growth. Among the molecules produced by plants are O_2 and ethylene, C_2H_4.

is known as **thermodynamics** (Greek: *thérme-*, "heat"; *dy'namis*, "power"). This area of study began during the Industrial Revolution as scientists and engineers studied the relationships among heat, work, and the energy content of fuels in an effort to maximize the performance of steam engines. Today thermodynamics continues to be enormously important in all areas of science and engineering. In this chapter we will examine an aspect of thermodynamics that involves the relationships between chemical reactions and energy changes involving heat. This portion of thermodynamics is called **thermochemistry**. We will discuss additional aspects of thermodynamics in Chapter 19.

(a)

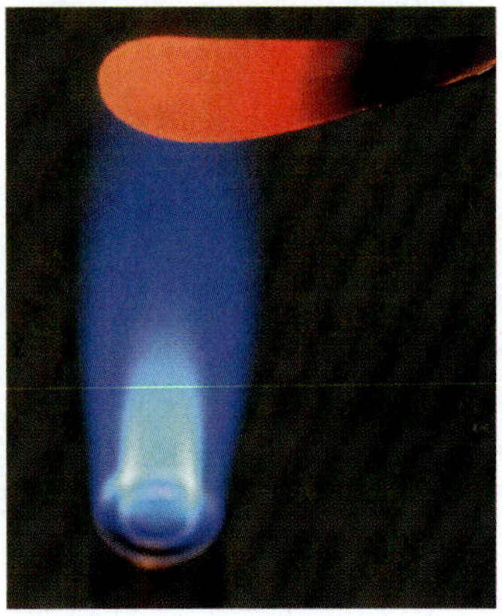

(b)

▲ **Figure 5.1 Work and heat.** Energy can be used to achieve two basic types of tasks: (a) Work is energy used to cause an object with mass to move. (b) Heat is energy used to cause the temperature of an object to increase.

5.1 | The Nature of Energy

Although the idea of energy is a familiar one, it is a bit challenging to deal with the concept in a precise way. **Energy** is commonly defined as *the capacity to do work or to transfer heat.* This definition requires us to understand the concepts of work and heat. We can think of **work** as being *energy used to cause an object with mass to move*, and **heat** as being *the energy used to cause the temperature of an object to increase* (Figure 5.1 ◀). Now let's consider each of these concepts more closely to give them fuller meaning. We begin by examining the ways in which matter can possess energy and how that energy can be transferred from one piece of matter to another.

Kinetic Energy and Potential Energy

Objects, whether they are tennis balls or molecules, can possess **kinetic energy**, the energy of *motion*. The magnitude of the kinetic energy, E_k, of an object depends on its mass, m, and speed, v:

$$E_k = \frac{1}{2}mv^2 \qquad [5.1]$$

Equation 5.1 shows that the kinetic energy increases as the speed of an object increases. For example, a car moving at 55 miles per hour (mph) has greater kinetic energy than it does at 40 mph. For a given speed, moreover, the kinetic energy increases with increasing mass. Thus, a large sport-utility vehicle traveling at 55 mph has greater kinetic energy than a small sedan traveling at the same speed, because the SUV has greater mass than the sedan. Atoms and molecules have mass and are in motion. They therefore possess kinetic energy.

An object can also possess another form of energy, called **potential energy**, by virtue of its *position* relative to other objects. Potential energy arises when there is a force operating on an object. A **force** is any kind of push or pull exerted on an object. The most familiar force is the pull of gravity. Think of a cyclist poised at the top of a hill, as illustrated in Figure 5.2 ▼. Gravity acts upon her and her bicycle, exerting a force directed toward the center of Earth. At the top of

▶ **Figure 5.2 Potential energy and kinetic energy.** (a) A bicycle at the top of a hill has a high potential energy relative to the bottom of the hill. (b) As the bicycle proceeds down the hill, the potential energy is converted into kinetic energy.

(a) (b)

the hill the cyclist and her bicycle possess a certain potential energy by virtue of their elevation. The potential energy, E_p, is given by the equation $E_p = mgh$, where m is the mass of the object in question (in this case the cyclist and bicycle), h is the height of the object relative to some reference height, and g is the gravitational constant, 9.8 m/s^2. Once in motion, without any further effort on her part, the cyclist gains speed as the bicycle rolls down the hill. Her potential energy decreases as she moves downward, but the energy does not simply disappear. It is converted to other forms of energy, principally kinetic energy, the energy of motion. This example illustrates that forms of energy are interconvertible.

Gravity is an important kind of force for large objects, such as the cyclist and Earth. Chemistry, however, deals mostly with extremely small objects—atoms and molecules—and gravitational forces play a negligible role in the ways these submicroscopic objects interact with one another. More important are forces that arise from electrical charges. One of the most important forms of potential energy in chemistry is *electrostatic potential energy*, which arises from the interactions between charged particles. The electrostatic potential energy, E_{el}, is proportional to the electrical charges on the two interacting objects, Q_1 and Q_2, and inversely proportional to the distance separating them:

$$E_{el} = \frac{\kappa Q_1 Q_2}{d} \qquad [5.2]$$

Here κ is simply a constant of proportionality, 8.99×10^9 J-m/C^2. [C is the coulomb, a unit of electrical charge (Section 2.2), and J is the joule, a unit of energy that we will soon discuss.] When dealing with molecular-level objects, the electrical charges Q_1 and Q_2 are typically on the order of magnitude of the charge of the electron (1.60×10^{-19} C). When Q_1 and Q_2 have the same sign (for example, both are positive), the two charges repel one another, pushing them apart, and E_{el} is positive. When they have opposite signs, they attract one another, pulling them toward each other, and E_{el} is negative. The lower the energy of a system, the more stable it is. Thus, the more strongly opposite charges interact, the more stable the system.

One of our goals in chemistry is to relate the energy changes that we see in our macroscopic world to the kinetic or potential energy of substances at the atomic or molecular level. Many substances—fuels, for example—release energy when they react. The *chemical energy* of these substances is due to the potential energy stored in the arrangements of the atoms of the substance. Likewise, we will see that the energy a substance possesses because of its temperature (its *thermal energy*) is associated with the kinetic energy of the molecules in the substance.

GIVE IT SOME THOUGHT

What are the terms for the energy an object possesses (a) because of its motion, (b) because of its position? What terms are used to describe changes of energy associated with (c) temperature changes, (d) moving an object?

Units of Energy

The SI unit for energy is the **joule** (pronounced "jool"), J, in honor of James Joule (1818–1889), a British scientist who investigated work and heat: J = 1 kg-m^2/s^2. A mass of 2 kg moving at a speed of 1 m/s possesses a kinetic energy of 1 J:

$$E_k = \tfrac{1}{2}mv^2 = \tfrac{1}{2}(2\text{ kg})(1\text{ m/s})^2 = 1\text{ kg-m}^2/\text{s}^2 = 1\text{ J}$$

A joule is not a large amount of energy, and we will often use *kilojoules* (kJ) in discussing the energies associated with chemical reactions.

Traditionally, energy changes accompanying chemical reactions have been expressed in calories, a non-SI unit still widely used in chemistry, biology, and biochemistry. A **calorie** (cal) was originally defined as the amount of energy required to raise the temperature of 1 g of water from 14.5°C to 15.5°C. It is now defined in terms of the joule:

$$1 \text{ cal} = 4.184 \text{ J (exactly)}$$

A related energy unit used in nutrition is the nutritional *Calorie* (note that this unit is capitalized): 1 Cal = 1000 cal = 1 kcal.

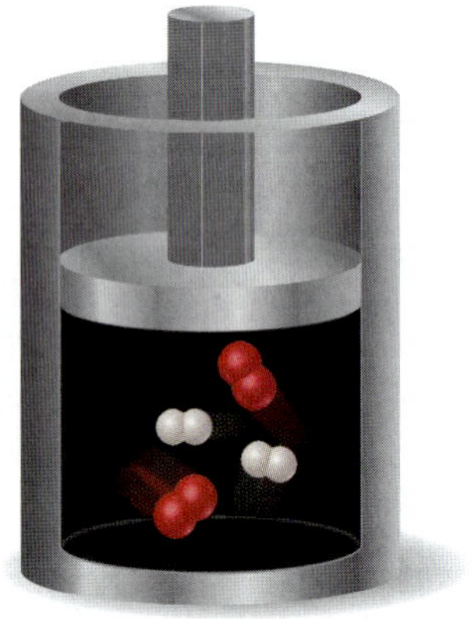

▲ **Figure 5.3 A closed system and its surroundings.** Hydrogen and oxygen gases are confined in a cylinder with a movable piston. If we are interested only in the properties of these gases, the gases are the system and the cylinder and piston are part of the surroundings. Because the system can exchange energy (in the form of heat and work) but not matter with its surroundings, it is a closed system.

System and Surroundings

When we analyze energy changes, we need to focus our attention on a limited and well-defined part of the universe to keep track of the energy changes that occur. The portion we single out for study is called the **system**; everything else is called the **surroundings**. When we study the energy change that accompanies a chemical reaction in the laboratory, the reactants and products usually constitute the system. The container and everything beyond it are then considered the surroundings. The systems we can most readily study are called *closed systems*. A closed system can exchange energy but not matter with its surroundings. For example, consider a mixture of hydrogen gas, H_2, and oxygen gas, O_2, in a cylinder, as illustrated in Figure 5.3 ◄. The system in this case is just the hydrogen and oxygen; the cylinder, piston, and everything beyond them (including us) are the surroundings. If the hydrogen and oxygen react to form water, energy is liberated:

$$2\,H_2(g) + O_2(g) \longrightarrow 2\,H_2O(g) + \text{energy}$$

Although the chemical form of the hydrogen and oxygen atoms in the system is changed by this reaction, the system has not lost or gained mass; it undergoes no exchange of matter with its surroundings. However, it does exchange energy with its surroundings in the form of *work* and *heat*. These are quantities that we can measure, as we will now discuss.

Transferring Energy: Work and Heat

Figure 5.1 illustrates the two ways that we experience energy changes in our everyday lives—in the form of work or heat. In Figure 5.1(a) energy is transferred from the tennis racquet to the ball, changing the direction and speed of the ball's movement. In Figure 5.1(b) energy is transferred in the form of heat. Indeed, energy is transferred between systems and surroundings in two general ways, as work or heat.

Energy used to cause an object to move against a force is called *work*. Thus, we can define work, w, as the energy transferred when an object is moved by a force. The magnitude of this work equals the product of the force, F, and the distance, d, that the object is moved:

$$w = F \times d \qquad [5.3]$$

We perform work, for example, when we lift an object against the force of gravity or when we bring two like charges closer together. If we define the object as the system, then we—as part of the surroundings—are performing work on that system, transferring energy to it.

The other way in which energy is transferred is as heat. *Heat* is the energy transferred from a hotter object to a colder one. Or stating this idea in a slightly more abstract but nevertheless useful way, we can define heat as the energy transferred between a system and its surroundings as a result of their difference in temperature. A combustion reaction, such as the burning of natural gas illustrated in Figure 5.1(b), releases the chemical energy stored in the molecules of the fuel. (Section 3.2) If we define the substances involved in the reaction as the system and everything else as the surroundings, we find that the released energy causes the temperature of the system to increase. Energy in the form of heat is then transferred from the hotter system to the cooler surroundings.

We will end this section with one more example of energy changes that illustrates some of the concepts that we have discussed. Consider a ball of modeling clay, which we will define as the system. If we lift the ball to the top of a wall, as shown in Figure 5.4(a) ▶, we are doing work against the force of gravity. The energy we transfer to the ball by doing work on it increases its potential energy because the ball is now at a greater height. If the ball now rolls off the wall, as in Figure 5.4(b), its downward speed increases as its potential energy is converted to kinetic energy. When the clay ball strikes the ground [Figure 5.4(c)], it stops moving and its kinetic energy goes to zero. Some of the kinetic energy is used to do work in squashing the ball, and the rest is dissipated to the surroundings as heat during the collision with the ground. The "bookkeeping" of the various transfers of energy between the system and the surroundings as work and heat is the focus of Section 5.2.

▲ **Figure 5.4 A ball of clay can be used to show energy interconversions.** (a) At the top of the wall, the ball has potential energy that is due to gravity. (b) As the ball falls, its potential energy is converted into kinetic energy. (c) When the ball strikes the ground, some of the kinetic energy is used to do work in squashing the ball; the rest is released as heat.

SAMPLE EXERCISE 5.1 | Describing and Calculating Energy Changes

A bowler lifts a 5.4-kg (12-lb) bowling ball from ground level to a height of 1.6 m (5.2 feet) and then drops the ball back to the ground. **(a)** What happens to the potential energy of the bowling ball as it is raised from the ground? **(b)** What quantity of work, in J, is used to raise the ball? **(c)** After the ball is dropped, it gains kinetic energy. If we assume that all of the work done in part (b) has been converted to kinetic energy by the time the ball strikes the ground, what is the speed of the ball at the instant just before it hits the ground? (Note: The force due to gravity is $F = m \times g$, where m is the mass of the object and g is the gravitational constant; $g = 9.8\ \text{m/s}^2$.)

Solution

Analyze: We need to relate the potential energy of the bowling ball to its position relative to the ground. We then need to establish the relationship between work and the change in potential energy of the ball. Finally, we need to connect the change in potential energy when the ball is dropped with the kinetic energy attained by the ball.

Plan: We can calculate the work done in lifting the ball by using Equation 5.3: $w = F \times d$. We can calculate the kinetic energy of the ball at the moment of impact using Equation 5.1 and from that calculate the speed v.

Solve: **(a)** Because the bowling ball is raised to a greater height above the ground, its potential energy increases.

(b) The ball has a mass of 5.4 kg, and it is lifted a distance of 1.6 m. To calculate the work performed to raise the ball, we use both Equation 5.3 and $F = m \times g$ for the force that is due to gravity:

$$w = F \times d = m \times g \times d = (5.4\ \text{kg})(9.8\ \text{m/s}^2)(1.6\ \text{m}) = 85\ \text{kg-m}^2/\text{s}^2 = 85\ \text{J}$$

Thus, the bowler has done 85 J of work to lift the ball to a height of 1.6 m.

(c) When the ball is dropped, its potential energy is converted to kinetic energy. At the instant just before the ball hits the ground, we assume that the kinetic energy is equal to the work done in part (b), 85 J:

$$E_k = \tfrac{1}{2}mv^2 = 85\ \text{J} = 85\ \text{kg-m}^2/\text{s}^2$$

We can now solve this equation for v:

$$v^2 = \left(\frac{2E_k}{m}\right) = \left(\frac{2(85\ \text{kg-m}^2/\text{s}^2)}{5.4\ \text{kg}}\right) = 31.5\ \text{m}^2/\text{s}^2$$

$$v = \sqrt{31.5\ \text{m}^2/\text{s}^2} = 5.6\ \text{m/s}$$

Check: Work must be done in part (b) to increase the potential energy of the ball, which is in accord with our experiences. The units are appropriate in both parts (b) and (c). The work is in units of J and the speed in units of m/s. In part (c) we have carried an additional digit in the intermediate calculation involving the square root, but we report the final value to only two significant figures, as appropriate.

Comment: A speed of 1 m/s is roughly 2 mph, so the bowling ball has a speed greater than 10 mph upon impact.

PRACTICE EXERCISE

What is the kinetic energy, in J, of **(a)** an Ar atom moving with a speed of 650 m/s, **(b)** a mole of Ar atoms moving with a speed of 650 m/s? (Hint: 1 amu $= 1.66 \times 10^{-27}$ kg)
Answers: **(a)** 1.4×10^{-20} J, **(b)** 8.4×10^{3} J

5.2 | The First Law of Thermodynamics

We have seen that the potential energy of a system can be converted into kinetic energy, and vice versa. We have also seen that energy can be transferred back and forth between a system and its surroundings in the forms of work and heat. In general, energy can be converted from one form to another, and it can be transferred from one place to another. All of these transactions proceed in accord with one of the most important observations in science—that energy can be neither created nor destroyed. This universal truth, known as the **first law of thermodynamics**, can be summarized by a simple statement: *Energy is conserved.* Any energy that is lost by the system must be gained by the surroundings, and vice versa. To apply the first law quantitatively, let's first define the energy of a system more precisely.

Internal Energy

The **internal energy** of a system is the sum of *all* the kinetic and potential energies of all its components. For the system in Figure 5.3, for example, the internal energy includes the motions of the H_2 and O_2 molecules through space, their rotations and internal vibrations. It also includes the energies of the nuclei of each atom and of the component electrons. We represent the internal energy with the symbol E. (Some texts, particularly more advanced ones, use the symbol U.) We generally don't know the actual numerical value of E. What we can hope to know, however, is ΔE (read "delta E"),* the change in E that accompanies a change in the system.

Imagine that we start with a system with an initial internal energy, E_{initial}. The system then undergoes a change, which might involve work being done or heat being transferred. After the change, the final internal energy of the system is E_{final}. We define the *change* in internal energy, ΔE, as the difference between E_{final} and E_{initial}:

$$\Delta E = E_{\text{final}} - E_{\text{initial}} \qquad [5.4]$$

We don't really need to know the actual values of E_{final} or E_{initial} for the system. To apply the first law of thermodynamics, we need only the value of ΔE.

Thermodynamic quantities such as ΔE have three parts: a number and a unit that together give the magnitude of the change, and a sign that gives the direction. A *positive* value of ΔE results when $E_{\text{final}} > E_{\text{initial}}$, indicating the system has gained energy from its surroundings. A *negative* value of ΔE is obtained when $E_{\text{final}} < E_{\text{initial}}$, indicating the system has lost energy to its surroundings. Notice that we are taking the system's point of view rather than that of the surroundings in discussing the energy changes. We need to remember, however, that any change in the energy of the system is accompanied by an opposite change in the energy of the surroundings. These features of energy changes are summarized in Figure 5.5 ▶.

* The symbol Δ is commonly used to denote *change*. For example, a change in height, h, can be represented by Δh.

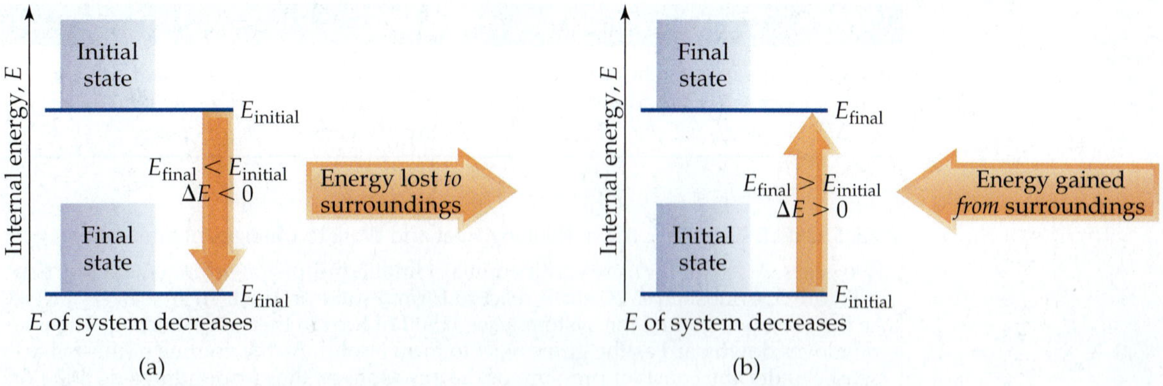

▲ **Figure 5.5 Changes in internal energy.** (a) When a system loses energy, that energy is released to the surroundings. The loss of energy is represented by an arrow that points downward between the initial and final states of the system. In this case, the energy change of the system, $\Delta E = E_{final} - E_{initial}$, is negative. (b) When a system gains energy, that energy is gained from the surroundings. In this case, the gain of energy is represented by an arrow that points upward between the initial and final states of the system, and the energy change of the system is positive. Notice in both (a) and (b) that the vertical arrow originates at the initial state and has its head at the final state.

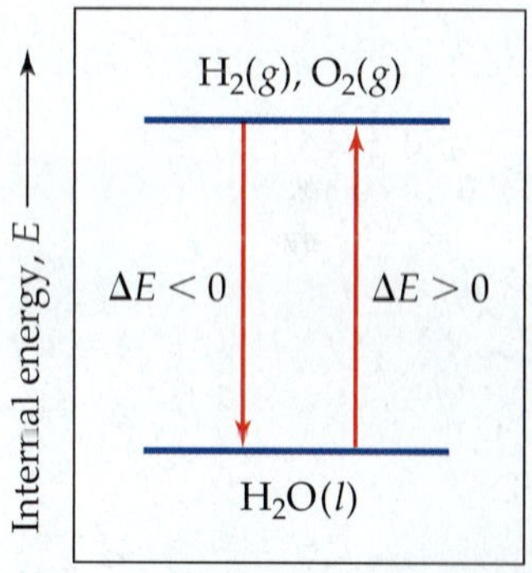

▲ **Figure 5.6 Energy diagram for the interconversion of $H_2(g)$, $O_2(g)$ and $H_2O(l)$.** A system composed of $H_2(g)$ and $O_2(g)$ has a greater internal energy than one composed of $H_2O(l)$. The system loses energy ($\Delta E < 0$) when H_2 and O_2 are converted to H_2O. It gains energy ($\Delta E > 0$) when H_2O is decomposed into H_2 and O_2.

In a chemical reaction the initial state of the system refers to the reactants, and the final state refers to the products. When hydrogen and oxygen form water at a given temperature, the system loses energy to the surroundings as heat. Because heat is lost from the system, the internal energy of the products (final state) is less than that of the reactants (initial state), and ΔE for the process is negative. Thus, the *energy diagram* in Figure 5.6 ▶ shows that the internal energy of the mixture of H_2 and O_2 is greater than that of H_2O.

Relating ΔE to Heat and Work

As we noted in Section 5.1, a system may exchange energy with its surroundings as heat or as work. The internal energy of a system changes in magnitude as heat is added to or removed from the system or as work is done on it or by it. If we think of internal energy as being the system's bank account of energy, we see that deposits or withdrawals can be made either in terms of heat or in terms of work. Deposits increase the energy of the system (positive ΔE), whereas withdrawals decrease the energy of the system (negative ΔE).

We can use these ideas to write a very useful algebraic expression of the first law of thermodynamics. When a system undergoes any chemical or physical change, the magnitude and sign of the accompanying change in internal energy, ΔE, is given by the heat added to or liberated from the system, q, plus the work done on or by the system, w:

$$\Delta E = q + w \qquad [5.5]$$

When heat is added to a system or work is done on a system, its internal energy increases. Therefore, when heat is transferred to the system from the surroundings, q has a positive value. Adding heat to the system is like making a deposit to the energy account—the total amount of energy goes up. Likewise, when work is done on the system by the surroundings, w has a positive value (Figure 5.7 ▶). That too is a deposit, increasing the internal energy of the system. Conversely, both the heat lost by the system to the surroundings and the work done by the system on the surroundings have negative values; that is, they lower the internal energy of the system. They are energy withdrawals and as a result, lower the total amount of energy in the energy account. The sign conventions for q, w, and ΔE are summarized in Table 5.1 ▶. Notice that any energy entering the system as either heat or work carries a positive sign.

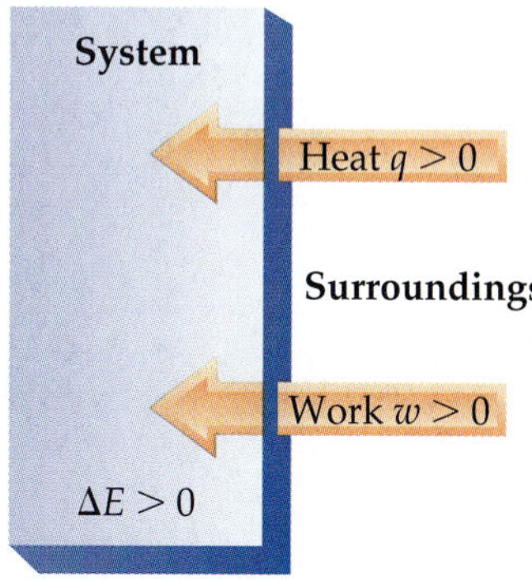

▲ **Figure 5.7 Sign conventions for heat and work.** Heat, q, absorbed by a system and work, w, done on a system are both positive quantities. Both increase the internal energy, E, of the system, causing ΔE, which equals $q + w$, to also be a positive quantity.

ACTIVITY
Sign Conventions for q and w

TABLE 5.1 Sign Conventions for q, w, and ΔE

For q	+ means system *gains* heat	− means system *loses* heat
For w	+ means work done *on* system	− means work done *by* system
For ΔE	+ means *net gain* of energy by system	− means *net loss* of energy by system

SAMPLE EXERCISE 5.2 | Relating Heat and Work to Changes of Internal Energy

Two gases, A(g) and B(g), are confined in a cylinder-and-piston arrangement like that in Figure 5.3. Substances A and B react to form a solid product: $A(g) + B(g) \rightarrow C(s)$. As the reactions occurs, the system loses 1150 J of heat to the sourrounding. The piston moves downward as the gases react to form a solid. As the volume of the gas decreases under the constant pressure of the atmosphere, the surroundings do 480 J of work on the system. What is the change in the internal energy of the system?

Solution

Analyze: The question asks us to determine ΔE, given information about q and w.
Plan: We first determine the signs of q and w (Table 5.1) and then use Equation 5.5, $\Delta E = q + w$, to calculate ΔE.
Solve: Heat is transferred from the system to the surroundings, and work is done on the system by the surroundings, so q is negative and w is positive: $q = -1150$ J and $w = 480$ kJ. Thus, ΔE is

$$\Delta E = q + w = (-1150 \text{ J}) + (480 \text{ J}) = -670 \text{ J}$$

The negative value of ΔE tells us that a net quantity of 670 J of energy has been transferred from the system to the surroundings.
Comment: You can think of this change as a decrease of 670 J in the net value of the system's energy bank account (hence the negative sign); 1150 J is withdrawn in the form of heat, while 480 J is deposited in the form of work. Notice that as the volume of the gases decreases, work is being done on the system by the surroundings resulting in a deposit of energy.

PRACTICE EXERCISE

Calculate the change in the internal energy of the system for a process in which the system absorbs 140 J of heat from the surroundings and does 85 J of work on the surroundings.
Answer: +55 J

Endothermic and Exothermic Processes

When a process occurs in which the system absorbs heat, the process is called **endothermic**. (*Endo-* is a prefix meaning "into.") During an endothermic process, such as the melting of ice, heat flows *into* the system from its surroundings. If we, as part of the surroundings, touch a container in which ice is melting, it feels cold to us because heat has passed from our hands to the container.

A process in which the system loses heat is called **exothermic**. (*Exo-* is a prefix meaning "out of.") During an exothermic process, such as the combustion of gasoline, heat *exits* or flows *out* of the system and into the surroundings. Figure 5.8 ▶ shows two examples of chemical reactions, one endothermic and the other highly exothermic. In the endothermic process shown in Figure 5.8(a), the temperature in the beaker decreases. In this case the system consists of the chemical reactants and products. The solvent in which they are dissolved is part of the surroundings. Heat flows from the solvent, as part of the surroundings, into the system as reactants are converted to products. Thus, the temperature of the solution drops.

State Functions

Although we usually have no way of knowing the precise value of the internal energy of a system, E, it does have a fixed value for a given set of conditions.

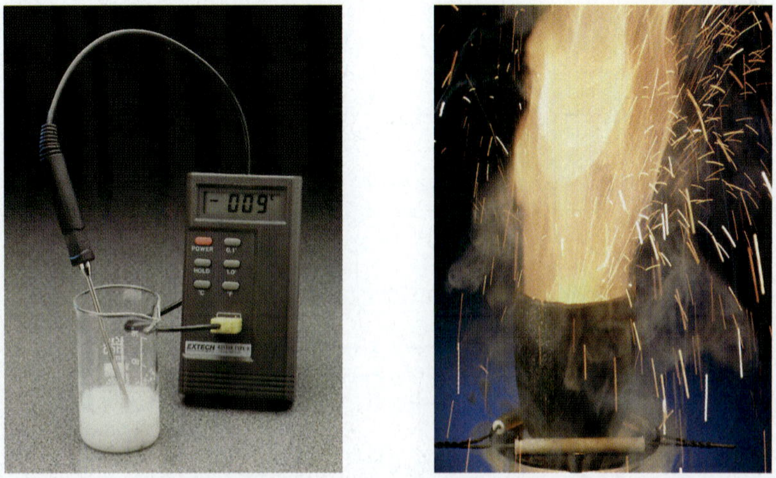

(a) (b)

◀ **Figure 5.8 Examples of endothermic and exothermic reactions.** (a) When ammonium thiocyanate and barium hydroxide octahydrate are mixed at room temperature, an endothermic reaction occurs:
$2\ NH_4SCN(s) + Ba(OH)_2 \cdot 8\ H_2O(s) \longrightarrow Ba(SCN)_2(aq) + 2\ NH_3(aq) + 10\ H_2O(l)$.
As a result, the temperature of the system drops from about 20°C to −9°C. (b) The reaction of powdered aluminum with Fe_2O_3 (the thermite reaction) is highly exothermic. The reaction proceeds vigorously to form Al_2O_3 and molten iron:
$2\ Al(s) + Fe_2O_3(s) \longrightarrow Al_2O_3(s) + 2\ Fe(l)$.

The conditions that influence internal energy include the temperature and pressure. Furthermore, the total internal energy of a system is proportional to the total quantity of matter in the system because energy is an extensive property. (Section 1.3)

MOVIE
Thermite

ACTIVITY
Dissolution of Ammonium Nitrate

Suppose we define our system as 50 g of water at 25°C, as in Figure 5.9 ▼. The system could have arrived at this state by cooling 50 g of water from 100°C or by melting 50 g of ice and subsequently warming the water to 25°C. The internal energy of the water at 25°C is the same in either case. Internal energy is an example of a **state function**, a property of a system that is determined by specifying the system's condition, or state, (in terms of temperature, pressure, location, and so forth). *The value of a state function depends only on the present state of the system, not on the path the system took to reach that state.* Because E is a state function, ΔE depends only on the initial and final states of the system, not on how the change occurs.

An analogy may help to explain the difference between quantities that are state functions and those that are not. Suppose you are traveling between Chicago and Denver. Chicago is 596 ft above sea level; Denver is 5280 ft above sea level. No matter what route you take, the altitude change will be 4684 ft. The distance you travel, however, will depend on your route. Altitude is analogous to a state function because the change in altitude is independent of the path taken. Distance traveled is not a state function.

Some thermodynamic quantities, such as E, are state functions. Others, such as q and w, are not. Although $\Delta E = q + w$ does not depend on how the change occurs, the specific amounts of heat and work produced depend on

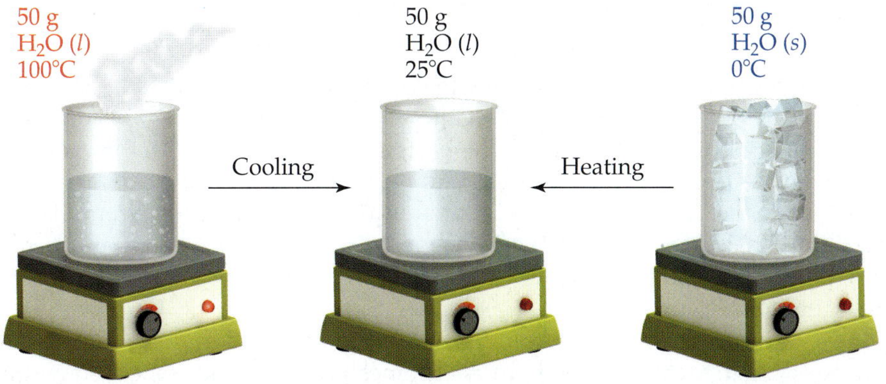

◀ **Figure 5.9 Internal energy, *E*, a state function.** *E* depends only on the present state of the system and not on the path by which it arrived at that state. The internal energy of 50 g of water at 25°C is the same whether the water is cooled from a higher temperature to 25°C or warmed from a lower temperature to 25°C.

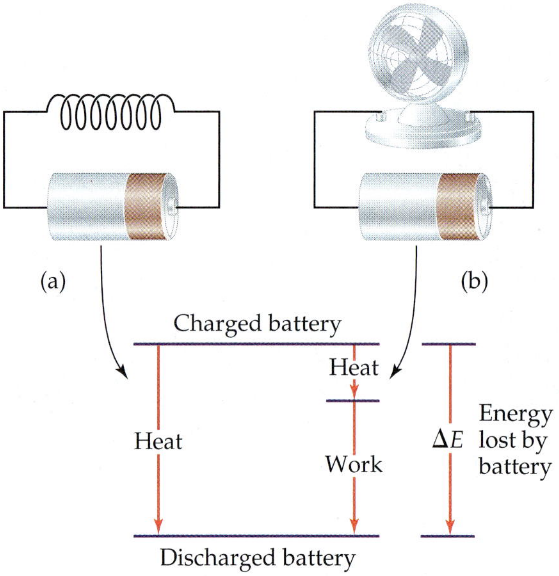

▶ **Figure 5.10 Internal energy is a state function, but heat and work are not.** The amounts of heat and work transferred between the system and the surroundings depend on the way in which the system goes from one state to another. (a) A battery shorted out by a wire loses energy to the surroundings only as heat; no work is performed by the system. (b) A battery discharged through a motor loses energy as work (to make the fan turn) and also loses energy as heat. Now, however, the amount of heat lost is much less than in (a). The value of ΔE is the same for both processes even though the values of q and w in (a) are different from the values of q and w in (b).

the way in which the change is carried out, analogous to the choice of travel route between Chicago and Denver. Nevertheless, if changing the path by which a system goes from an initial state to a final state increases the value of q, that path change will also decrease the value of w by exactly the same amount. The result is that the value for ΔE for the two paths will be the same.

We can illustrate this principle with the example of a flashlight battery as our system. In Figure 5.10 ▲, we consider two possible ways of discharging the battery at constant temperature. If the battery is shorted out by a coil of wire, no work is accomplished because nothing is moved against a force. All the energy is lost from the battery in the form of heat. (The wire coil will get warmer and release heat to the surrounding air.) On the other hand, if the battery is used to make a small motor turn, the discharge of the battery produces work. Some heat will be released as well, although not as much as when the battery is shorted out. The magnitudes of q and w are different for these two cases. If the initial and final states of the battery are identical in both cases, however, then $\Delta E = q + w$ must be the same in both cases because E is a state function. Thus, ΔE depends only on the initial and final states of the system, regardless of how the transfers of energy occur in terms of heat and work.

GIVE IT SOME THOUGHT

In what ways is the balance in your checkbook a state function?

5.3 | Enthalpy

The chemical and physical changes that occur around us, such as photosynthesis in the leaves of a plant, the evaporation of water from a lake, or a reaction in an open beaker in a laboratory, occur at essentially constant atmospheric pressure. The changes can result in the release or absorption of heat or can be accompanied by work that is done by or on the system. It is the heat flow that is easiest to measure, so we will begin to focus on that aspect of reactions. Nevertheless, we still need to account for any work that accompanies the process.

Most commonly, the only kind of work produced by chemical or physical changes open to the atmosphere is the mechanical work associated with a

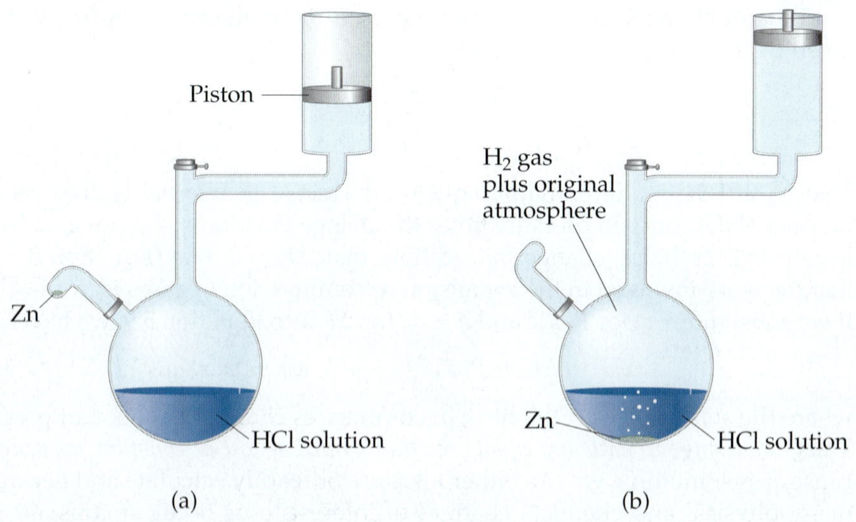

◀ **Figure 5.11 A system that does work on its surroundings.** (a) An apparatus for studying the reaction of zinc metal with hydrochloric acid at constant pressure. The piston is free to move up and down in its cylinder to maintain a constant pressure equal to atmospheric pressure inside the apparatus. Notice the pellets of zinc in the L-shaped arm on the left. When this arm is rotated, the pellets will fall into the main container and the reaction will begin. (b) When zinc is added to the acid solution, hydrogen gas is evolved. The hydrogen gas does work on the surroundings, raising the piston against atmospheric pressure to maintain constant pressure inside the reaction vessel.

change in the volume of the system. Consider, for example, the reaction of zinc metal with hydrochloric acid solution:

$$Zn(s) + 2\,H^+(aq) \longrightarrow Zn^{2+}(aq) + H_2(g) \qquad [5.6]$$

If we carry out this reaction in the laboratory hood in an open beaker, we can see the evolution of hydrogen gas, but it may not be so obvious that work is being done. Still, the hydrogen gas that is being produced must expand against the existing atmosphere, which requires the system to do work. We can see this better by conducting the reaction in a closed vessel at constant pressure, as illustrated in Figure 5.11 ▲. In this apparatus the piston moves up or down to maintain a constant pressure in the reaction vessel. If we assume for simplicity that the piston has no mass, the pressure in the apparatus is the same as the atmospheric pressure outside the apparatus. As the reaction proceeds, H_2 gas forms, and the piston rises. The gas within the flask is thus doing work on the surroundings by lifting the piston against the force of atmospheric pressure that presses down on it.

The work involved in the expansion or compression of gases is called **pressure-volume work** (or *P-V* work). When the pressure is constant, as in our example, the sign and magnitude of the pressure-volume work is given by

$$w = -P\,\Delta V \qquad [5.7]$$

where P is pressure and ΔV is the change in volume of the system ($\Delta V = V_{final} - V_{initial}$). The negative sign in Equation 5.7 is necessary to comform to the sign conventions given in Table 5.1. Thus, when the volume expands, ΔV is a positive quantity and w is a negative quantity. That is, energy leaves the system as work, indicating that work is done *by* the system *on* the surroundings. When a gas is compressed, ΔV is a negative quantity (the volume decreases), which makes w a positive quantity. That is, energy enters the system as work, indicating that work is done on the system by the surroundings. The "Closer Look" box discusses pressure-volume work in more detail, but all you really need to keep in mind for now is Equation 5.7, which applies to processes occurring at constant pressure. We will take up the properties of gases in more detail in Chapter 10.

ANIMATION
Work of Gas Expansion

A thermodynamic function called **enthalpy** (from the Greek word *enthalpein*, meaning "to warm") accounts for heat flow in processes occurring at constant pressure when no forms of work are performed other than *P-V* work. Enthalpy, which we denote by the symbol H, equals the internal energy plus the product of the pressure and volume of the system:

$$H = E + PV \qquad [5.8]$$

Enthalpy is a state function because internal energy, pressure, and volume are all state functions.

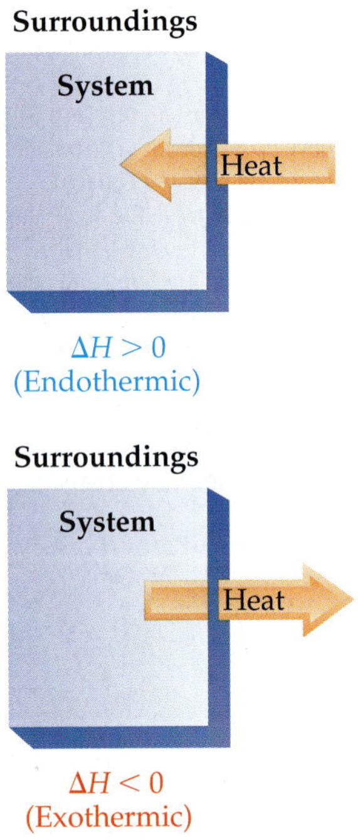

▲ **Figure 5.12 Endothermic and exothermic processes.** (a) If the system absorbs heat (endothermic process), ΔH will be positive ($\Delta H > 0$). (b) If the system loses heat (exothermic process), ΔH will be negative ($\Delta H < 0$).

When a change occurs at constant pressure, the change in enthalpy, ΔH, is given by the following relationship:

$$\Delta H = \Delta(E + PV) = \Delta E + P\,\Delta V \qquad [5.9]$$

That is, the change in enthalpy equals the change in internal energy plus the product of the constant pressure times the change in volume. We can gain further insight into enthalpy change by recalling that $\Delta E = q + w$ (Equation 5.5) and that the work involved in the expansion or compression of gases is $w = -P\,\Delta V$. If we substitute $-w$ for $P\,\Delta V$ and $q + w$ for ΔE into Equation 5.9, we have

$$\Delta H = \Delta E + P\,\Delta V = (q_P + w) - w = q_P \qquad [5.10]$$

where the subscript P on the heat, q, emphasizes changes at constant pressure. Thus, *the change in enthalpy equals the heat gained or lost at constant pressure.* Because q_P is something we can either measure or readily calculate and because so many physical and chemical changes of interest to us occur at constant pressure, enthalpy is a more useful function than internal energy. For most reactions the difference in ΔH and ΔE is small because $P\,\Delta V$ is small.

When ΔH is positive (that is, when q_P is positive), the system has gained heat from the surroundings (Table 5.1), which is an endothermic process. When ΔH is negative, the system has released heat to the surroundings, which is an exothermic process. These cases are diagrammed in Figure 5.12 ◀. Because H is a state function, ΔH (which equals q_P) depends only on the initial and final states of the system, not on how the change occurs. At first glance this statement might seem to contradict our earlier discussion in Section 5.2, in which we said that q is *not* a state function. There is no contradiction, however, because the relationship between ΔH and heat (q_P) has the special limitations that only P-V work is involved and the pressure is constant.

GIVE IT SOME THOUGHT

What is the advantage of using enthalpy rather than internal energy to describe energy changes in reactions?

SAMPLE EXERCISE 5.3 | Determining the Sign of ΔH

Indicate the sign of the enthalpy change, ΔH, in each of the following processes carried out under atmospheric pressure, and indicate whether the process is endothermic or exothermic: **(a)** An ice cube melts; **(b)** 1 g of butane (C_4H_{10}) is combusted in sufficient oxygen to give complete combustion to CO_2 and H_2O.

Solution

Analyze: Our goal is to determine whether ΔH is positive or negative for each process. Because each process appears to occur at constant pressure, the enthalpy change of each one equals the amount of heat absorbed or evolved, $\Delta H = q_P$.

Plan: We must predict whether heat is absorbed or released by the system in each process. Processes in which heat is absorbed are endothermic and have a positive sign for ΔH; those in which heat is evolved are exothermic and have a negative sign for ΔH.

Solve: In (a) the water that makes up the ice cube is the system. The ice cube absorbs heat from the surroundings as it melts, so ΔH is positive and the process is endothermic. In (b) the system is the 1 g of butane and the oxygen required to combust it. The combustion of butane in oxygen gives off heat, so ΔH is negative and the process is exothermic.

PRACTICE EXERCISE

Suppose we confine 1 g of butane and sufficient oxygen to completely combust it in a cylinder like that in Figure 5.13. The cylinder is perfectly insulating, so no heat can escape to the surroundings. A spark initiates combustion of the butane, which forms carbon dioxide and water vapor. If we used this apparatus to measure the enthalpy change in the reaction, would the piston rise, fall, or stay the same?

Answer: The piston must move so as to maintain a constant pressure in the cylinder. The products contain more molecules of gas than the reactants, as shown by the balanced equation

$$2\,C_4H_{10}(g) + 13\,O_2(g) \longrightarrow 8\,CO_2(g) + 10\,H_2O(g)$$

As a result, the piston would rise to make room for the additional molecules of gas. Heat is given off, moreover, so the piston would also rise an additional amount to accommodate the expansion of the gases because of the temperature increase.

A CLOSER LOOK | Energy, Enthalpy, and *P*-*V* Work

In chemistry we are interested mainly in two types of work: electrical work and mechanical work done by expanding gases. We will focus here only on the latter, called pressure-volume, or *P*-*V*, work. Expanding gases in the cylinder of an automobile engine do *P*-*V* work on the piston, and this work eventually turns the wheels. Expanding gases from an open reaction vessel do *P*-*V* work on the atmosphere. This work accomplishes nothing in a practical sense, but we must keep track of all work, useful or not, when monitoring the energy changes of a system.

Consider a gas confined to a cylinder with a movable piston of cross-sectional area A (Figure 5.13 ▶). A downward force, F, acts on the piston. The *pressure*, P, on the gas is the force per area: $P = F/A$. We will assume that the piston is weightless and that the only pressure acting on it is the *atmospheric pressure* that is due to the weight of Earth's atmosphere, which we will assume to be constant.

Suppose the gas in the cylinder expands and the piston moves a distance, Δh. From Equation 5.3, the magnitude of the work done by the system equals the distance moved times the force acting on the piston:

$$\text{Magnitude of work} = \text{force} \times \text{distance} = F \times \Delta h \qquad [5.11]$$

We can rearrange the definition of pressure, $P = F/A$, to $F = P \times A$. In addition, the volume change, ΔV, resulting from the movement of the piston, is the product of the cross-sectional area of the piston and the distance it moves: $\Delta V = A \times \Delta h$. Substituting into Equation 5.11,

$$\text{Magnitude of work} = F \times \Delta h = P \times A \times \Delta h$$
$$= P \times \Delta V$$

Because the system (the confined gas) is doing work on the surroundings, the work is a negative quantity:

$$w = -P\,\Delta V \qquad [5.12]$$

Now, if *P*-*V* work is the only work that can be done, we can substitute Equation 5.12 into Equation 5.5 to give

$$\Delta E = q + w = q - P\,\Delta V \qquad [5.13]$$

When a reaction is carried out in a constant-volume container ($\Delta V = 0$), the heat transferred equals the change in internal energy:

$$\Delta E = q_V \qquad \text{(constant volume)} \qquad [5.14]$$

The subscript V indicates that the volume is constant.

Most reactions are run under constant-pressure conditions. In this case Equation 5.13 becomes

$$\Delta E = q_P - P\,\Delta V \text{ or}$$

$$q_P = \Delta E + P\,\Delta V \qquad \text{(constant pressure)} \qquad [5.15]$$

But we see from Equation 5.9 that the right-hand side of Equation 5.15 is just the enthalpy change under constant-pressure conditions. Thus, $\Delta H = q_P$, as we saw earlier in Equation 5.10.

In summary, the change in internal energy measures the heat gained or lost at constant volume; the change in enthalpy measures the heat gained or lost at constant pressure. The difference between ΔE and ΔH is the amount of *P*-*V* work done by the system when the process occurs at constant pressure, $-P\,\Delta V$. The volume change accompanying many reactions is close to zero, which makes $P\,\Delta V$, and therefore the difference between ΔE and ΔH, small. It is generally satisfactory to use ΔH as the measure of energy changes during most chemical processes.

Related Exercises: 5.41, 5.42, 5.43, 5.44

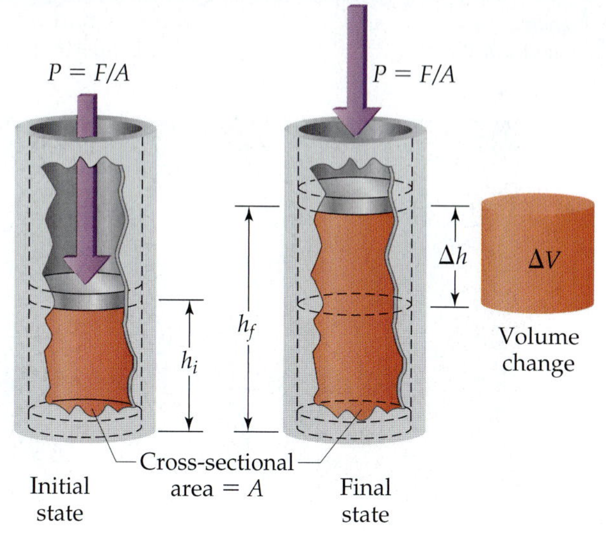

▲ **Figure 5.13 Pressure-volume work.** A piston moving upward, expanding the volume of the system against an external pressure, P, does work on the surroundings. The amount of work done by the system on the surroundings is $w = -P\,\Delta V$.

VCL Endothermic versus Exothermic

5.4 | Enthalpies of Reaction

Because $\Delta H = H_{\text{final}} - H_{\text{initial}}$, the enthalpy change for a chemical reaction is given by the enthalpy of the products minus the enthalpy of the reactants:

$$\Delta H = H_{\text{products}} - H_{\text{reactants}} \quad [5.16]$$

The enthalpy change that accompanies a reaction is called the **enthalpy of reaction**, or merely the *heat of reaction*, and is sometimes written ΔH_{rxn}, where "rxn" is a commonly used abbreviation for "reaction."

The combustion of hydrogen is shown in Figure 5.14 ▼. When the reaction is controlled so that 2 mol $H_2(g)$ burn to form 2 mol $H_2O(g)$ at a constant pressure, the system releases 483.6 kJ of heat. We can summarize this information as

ACTIVITY
Enthalpy of Solution

$$2\,H_2(g) + O_2(g) \longrightarrow 2\,H_2O(g) \qquad \Delta H = -483.6\text{ kJ} \quad [5.17]$$

ΔH is negative, so this reaction is exothermic. Notice that ΔH is reported at the end of the balanced equation, without explicitly mentioning the amounts of chemicals involved. In such cases the coefficients in the balanced equation represent the number of moles of reactants and products producing the associated enthalpy change. Balanced chemical equations that show the associated enthalpy change in this way are called *thermochemical equations*.

GIVE IT SOME THOUGHT

What information is summarized by the coefficients in a *thermochemical equation*?

The enthalpy change accompanying a reaction may also be represented in an *enthalpy diagram* such as that shown in Figure 5.14(c). Because the combustion of $H_2(g)$ is exothermic, the enthalpy of the products in the reaction is lower than the enthalpy of the reactants. The enthalpy of the system is lower after the reaction because energy has been lost in the form of heat released to the surroundings.

The reaction of hydrogen with oxygen is highly exothermic (ΔH is negative and has a large magnitude), and it occurs rapidly once it starts. It can occur with explosive violence, too, as demonstrated by the disastrous explosions of the German airship *Hindenburg* in 1937 (Figure 5.15 ▶) and the space shuttle *Challenger* in 1986.

MOVIE
Formation of Water

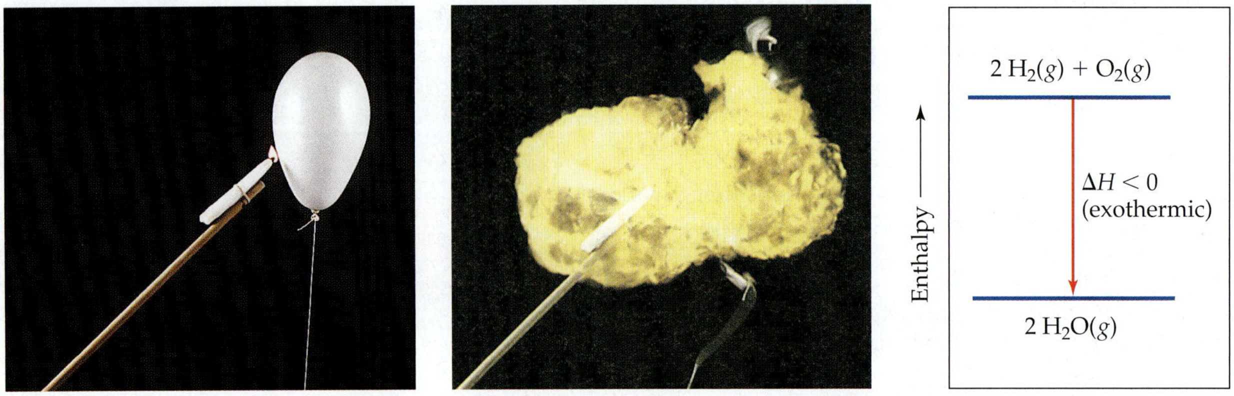

(a) (b) (c)

▲ **Figure 5.14 Exothermic reaction of hydrogen with oxygen.** (a) A candle is held near a balloon filled with hydrogen gas and oxygen gas. (b) The $H_2(g)$ ignites, reacting with $O_2(g)$ to form $H_2O(g)$. The resultant explosion produces a ball of flame. The system gives off heat to its surroundings. (c) The enthalpy diagram for this reaction, showing its exothermic character.

◀ **Figure 5.15 The burning of the hydrogen-filled airship *Hindenburg*.** This photograph was taken only 22 seconds after the first explosion occurred. This tragedy, which occurred in Lakehurst, New Jersey, on May 6, 1937, led to the discontinuation of hydrogen as a buoyant gas in such craft. Modern-day blimps are filled with helium, which is not as buoyant as hydrogen but is not flammable.

The following guidelines are helpful when using thermochemical equations and enthalpy diagrams:

1. *Enthalpy is an extensive property.* The magnitude of ΔH, therefore, is directly proportional to the amount of reactant consumed in the process. For the combustion of methane to form carbon dioxide and liquid water, for example, 890 kJ of heat is produced when 1 mol of CH_4 is burned in a constant-pressure system:

$$CH_4(g) + 2\,O_2(g) \longrightarrow CO_2(g) + 2\,H_2O(l) \qquad \Delta H = -890\text{ kJ} \qquad [5.18]$$

Because the combustion of 1 mol of CH_4 with 2 mol of O_2 releases 890 kJ of heat, the combustion of 2 mol of CH_4 with 4 mol of O_2 releases twice as much heat, 1780 kJ.

2. *The enthalpy change for a reaction is equal in magnitude, but opposite in sign, to ΔH for the reverse reaction.* For example, if we could reverse Equation 5.18 so that $CH_4(g)$ and $O_2(g)$ formed from $CO_2(g)$ and $H_2O(l)$, ΔH for the process would be +890 kJ:

$$CO_2(g) + 2\,H_2O(l) \longrightarrow CH_4(g) + 2\,O_2(g) \qquad \Delta H = 890\text{ kJ} \qquad [5.19]$$

When we reverse a reaction, we reverse the roles of the products and the reactants; thus, the reactants in a reaction become the products of the reverse reaction, and so forth. From Equation 5.16, we can see that reversing the products and reactants leads to the same magnitude, but a change in sign for ΔH_{rxn}. This relationship is diagrammed for Equations 5.18 and 5.19 in Figure 5.16 ▶.

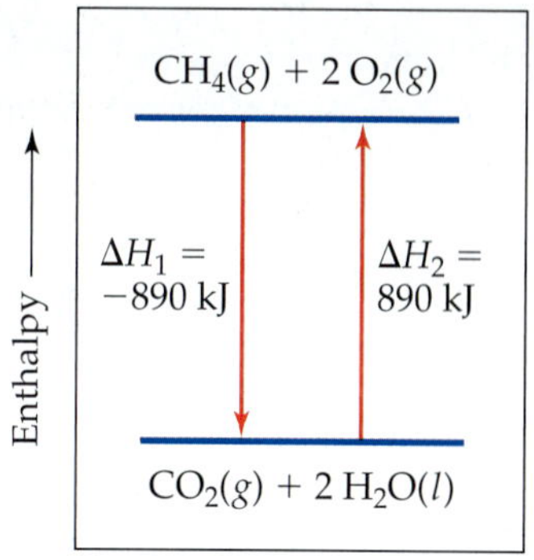

▲ **Figure 5.16 ΔH for a reverse reaction.** Reversing a reaction changes the sign but not the magnitude of the enthalpy change: $\Delta H_2 = -\Delta H_1$.

3. *The enthalpy change for a reaction depends on the state of the reactants and products.* If the product in the combustion of methane (Equation 5.18) were gaseous H_2O instead of liquid H_2O, ΔH_{rxn} would be -802 kJ instead of -890 kJ. Less heat would be available for transfer to the surroundings because the enthalpy of $H_2O(g)$ is greater than that of $H_2O(l)$. One way to see this is to imagine that the product is initially liquid water. The liquid water must be converted to water vapor, and the conversion of 2 mol $H_2O(l)$ to 2 mol $H_2O(g)$ is an endothermic process that absorbs 88 kJ:

$$2\,H_2O(l) \longrightarrow 2\,H_2O(g) \qquad \Delta H = +88\text{ kJ} \qquad [5.20]$$

Thus, it is important to specify the states of the reactants and products in thermochemical equations. In addition, we will generally assume that the reactants and products are both at the same temperature, 25°C, unless otherwise indicated.

SAMPLE EXERCISE 5.4 | Relating ΔH to Quantities of Reactants and Products

How much heat is released when 4.50 g of methane gas is burned in a constant-pressure system? (Use the information given in Equation 5.18.)

Solution

Analyze: Our goal is to use a thermochemical equation to calculate the heat produced when a specific amount of methane gas is combusted. According to Equation 5.18, 890 kJ is released by the system when 1 mol CH_4 is burned at constant pressure ($\Delta H = -890$ kJ).

Plan: Equation 5.18 provides us with a stoichiometric conversion factor: 1 mol $CH_4 \simeq -890$ kJ. Thus, we can convert moles of CH_4 to kJ of energy. First, however, we must convert grams of CH_4 to moles of CH_4. Thus, the conversion sequence is grams CH_4 (given) $\rightarrow$ moles $CH_4 \rightarrow$ kJ (unknown to be found).

Solve: By adding the atomic weights of C and 4 H, we have 1 mol CH_4 = 16.0 g CH_4. Thus, we can use the appropriate conversion factors to convert grams of CH_4 to moles of CH_4 to kilojoules:

$$\text{Heat} = (4.50 \text{ g } CH_4)\left(\frac{1 \text{ mol } CH_4}{16.0 \text{ g } CH_4}\right)\left(\frac{-890 \text{ kJ}}{1 \text{ mol } CH_4}\right) = -250 \text{ kJ}$$

The negative sign indicates that 250 kJ is released by the system into the surroundings.

PRACTICE EXERCISE

Hydrogen peroxide can decompose to water and oxygen by the following reaction:

$$2\,H_2O_2(l) \longrightarrow 2\,H_2O(l) + O_2(g) \qquad \Delta H = -196 \text{ kJ}$$

Calculate the value of q when 5.00 g of $H_2O_2(l)$ decomposes at constant pressure.
Answer: -14.4 kJ

There are many situations in which it is valuable to know the enthalpy change associated with a given chemical process. As we will see in the following sections, ΔH can be determined directly by experiment or calculated from the known enthalpy changes of other reactions by invoking the first law of thermodynamics.

5.5 | Calorimetry

The value of ΔH can be determined experimentally by measuring the heat flow accompanying a reaction at constant pressure. Typically, we can determine the magnitude of the heat flow by measuring the magnitude of the temperature change the heat flow produces. The measurement of heat flow is **calorimetry**; a device used to measure heat flow is a **calorimeter**.

Heat Capacity and Specific Heat

The more heat an object gains, the hotter it gets. All substances change temperature when they are heated, but the magnitude of the temperature change produced by a given quantity of heat varies from substance to substance. The temperature change experienced by an object when it absorbs a certain amount of heat is determined by its **heat capacity**, C. The heat capacity of an object is the amount of heat required to raise its temperature by 1 K (or 1°C). The greater the heat capacity, the greater the heat required to produce a given increase in temperature.

STRATEGIES IN CHEMISTRY | Using Enthalpy as a Guide

If you hold a brick in the air and let it go, it will fall as the force of gravity pulls it toward Earth. A process that is thermodynamically favored to happen, such as a falling brick, is called a *spontaneous* process. A spontaneous process can be either fast or slow. Speed is not the issue in thermodynamics.

Many chemical processes are thermodynamically favored, or spontaneous, too. By "spontaneous," we don't mean that the reaction will form products without any intervention. That can be the case, but often some energy must be imparted to get the process started. The enthalpy change in a reaction gives one indication as to whether it is likely to be spontaneous. The combustion of $H_2(g)$ and $O_2(g)$, for example, is a highly exothermic process:

$$H_2(g) + \tfrac{1}{2}O_2(g) \longrightarrow H_2O(g) \qquad \Delta H = -242 \text{ kJ}$$

Hydrogen gas and oxygen gas can exist together in a volume indefinitely without noticeable reaction occurring, as in Figure 5.14(a). Once initiated, however, energy is rapidly transferred from the system (the reactants) to the surroundings. As the reaction proceeds, large amounts of heat are released, which greatly increases the temperature of the reactants and the products. The system then loses enthalpy by transferring the heat to the surroundings. (Recall that the first law of thermodynamics indicates that the total energy of the system plus the surroundings will not change; energy is conserved.)

Enthalpy change is not the only consideration in the spontaneity of reactions, however, nor is it a foolproof guide. For example, the melting of ice is an endothermic process:

$$H_2O(s) \longrightarrow H_2O(l) \qquad \Delta H = +6.01 \text{ kJ}$$

Even though this process is endothermic, it is spontaneous at temperatures above the freezing point of water (0°C). The reverse process, the freezing of water to ice, is spontaneous at temperatures below 0°C. Thus, we know that ice at room temperature will melt and that water put into a freezer at −20°C will turn into ice; both of these processes are spontaneous under different conditions even though they are the reverse of one another. In Chapter 19 we will address the spontaneity of processes more fully. We will see why a process can be spontaneous at one temperature, but not at another, as is the case for the conversion of water to ice.

Despite these complicating factors, however, you should pay attention to the enthalpy changes in reactions. As a general observation, when the enthalpy change is large, it is the dominant factor in determining spontaneity. Thus, reactions for which ΔH is *large* and *negative* tend to be spontaneous. Reactions for which ΔH is *large* and *positive* tend to be spontaneous only in the reverse direction. There are a number of ways in which the enthalpy of a reaction can be estimated; from these estimates, the likelihood of the reaction being thermodynamically favorable can be predicted.

Related Exercises: 5.45, 5.46

For pure substances the heat capacity is usually given for a specified amount of the substance. The heat capacity of one mole of a substance is called its **molar heat capacity**, C_{molar}. The heat capacity of one gram of a substance is called its *specific heat capacity*, or merely its **specific heat** (Figure 5.17 ▶). The specific heat, s, of a substance can be determined experimentally by measuring the temperature change, ΔT, that a known mass, m, of the substance undergoes when it gains or loses a specific quantity of heat, q:

$$\text{Specific heat} = \frac{\text{(quantity of heat transferred)}}{\text{(grams of substance)} \times \text{(temperature change)}}$$

$$s = \frac{q}{m \times \Delta T} \qquad [5.21]$$

For example, 209 J is required to increase the temperature of 50.0 g of water by 1.00 K. Thus, the specific heat of water is

$$s = \frac{209 \text{ J}}{(50.0 \text{ g})(1.00 \text{ K})} = 4.18 \frac{\text{J}}{\text{g-K}}$$

A temperature change in kelvins is equal in magnitude to the temperature change in degrees Celsius: ΔT in K = ΔT in °C. (Section 1.4) When the sample gains heat (positive q), the temperature of the sample increases (positive ΔT). Rearranging Equation 5.21, we get

$$q = s \times m \times \Delta T \qquad [5.22]$$

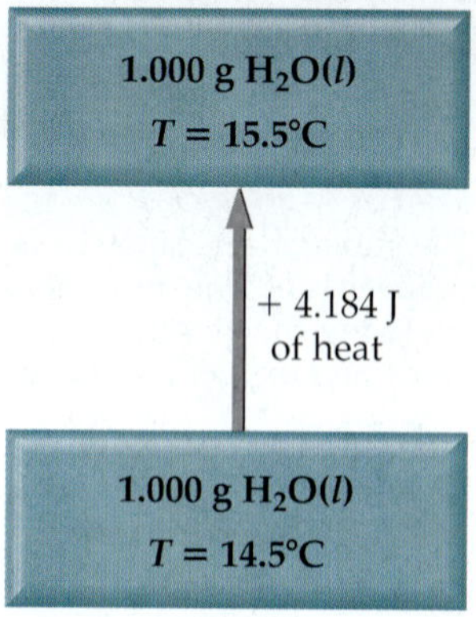

▲ **Figure 5.17 Specific heat of water.** Specific heat indicates the amount of heat that must be added to one gram of a substance to raise its temperature by 1 K (or 1°C). Specific heats can vary slightly with temperature, so for precise measurements the temperature is specified. The specific heat of $H_2O(l)$ at 14.5°C is 4.184 J/g-K; the addition of 4.184 J of heat to 1 g of liquid water at this temperature raises the temperature to 15.5°C. This amount of energy defines the calorie: 1 cal = 4.184 J.

TABLE 5.2 Specific Heats of Some Substances at 298 K

Elements		Compounds	
Substance	Specific Heat (J/g-K)	Substance	Specific Heat (J/g-K)
$N_2(g)$	1.04	$H_2O(l)$	4.18
$Al(s)$	0.90	$CH_4(g)$	2.20
$Fe(s)$	0.45	$CO_2(g)$	0.84
$Hg(l)$	0.14	$CaCO_3(s)$	0.82

Thus, we can calculate the quantity of heat that a substance has gained or lost by using its specific heat together with its measured mass and temperature change.

The specific heats of several substances are listed in Table 5.2 ▲. Notice that the specific heat of liquid water is higher than those of the other substances listed. For example, it is about 5 times as great as that of aluminum metal. The high specific heat of water affects Earth's climate because it keeps the temperatures of the oceans relatively resistant to change. It also is very important in maintaining a constant temperature in our bodies, as will be discussed in the "Chemistry and Life" box later in this chapter.

VCL Specific Heat of a Pb

SAMPLE EXERCISE 5.5 | Relating Heat, Temperature Change, and Heat Capacity

(a) How much heat is needed to warm 250 g of water (about 1 cup) from 22°C (about room temperature) to near its boiling point, 98°C? The specific heat of water is 4.18 J/g-K. **(b)** What is the molar heat capacity of water?

Solution

Analyze: In part **(a)** we must find the total quantity of heat needed to warm the sample of water, given the mass of water (m), its temperature change (ΔT), and its specific heat (s). In part **(b)** we must calculate the molar heat capacity (heat capacity per mole) of water from its specific heat (heat capacity per gram).

Plan: **(a)** Given s, m, and ΔT, we can calculate the quantity of heat, q, using Equation 5.22. **(b)** We can use the molar mass of water and dimensional analysis to convert from heat capacity per gram to heat capacity per mole.

Solve: **(a)** The water undergoes a temperature change of

$$\Delta T = 98°C - 22°C = 76°C = 76\ K$$

Using Equation 5.22, we have

$$q = s \times m \times \Delta T$$
$$= (4.18\ J/g\text{-}K)(250\ g)(76\ K) = 7.9 \times 10^4\ J$$

(b) The molar heat capacity is the heat capacity of one mole of substance. Using the atomic weights of hydrogen and oxygen, we have

$$1\ mol\ H_2O = 18.0\ g\ H_2O$$

From the specific heat given in part (a), we have

$$C_{molar} = (4.18\ J/g\text{-}K)\left(\frac{18.0\ g}{1\ mol}\right) = 75.2\ J/mol\text{-}K$$

PRACTICE EXERCISE

(a) Large beds of rocks are used in some solar-heated homes to store heat. Assume that the specific heat of the rocks is 0.082 J/g-K. Calculate the quantity of heat absorbed by 50.0 kg of rocks if their temperature increases by 12.0°C. **(b)** What temperature change would these rocks undergo if they emitted 450 kJ of heat?

Answers: **(a)** 4.9×10^5 J, **(b)** 11 K = 11°C decrease

Constant-Pressure Calorimetry

The techniques and equipment employed in calorimetry depend on the nature of the process being studied. For many reactions, such as those occurring in solution, it is easy to control pressure so that ΔH is measured directly. (Recall that $\Delta H = q_P$.) Although the calorimeters used for highly accurate work are precision instruments, a very simple "coffee-cup" calorimeter, as shown in Figure 5.18 ▶, is often used in general chemistry labs to illustrate the principles of calorimetry. Because the calorimeter is not sealed, the reaction occurs under the essentially constant pressure of the atmosphere.

There is no physical boundary between the system and the surroundings in this case. The reactants and products of the reaction are the system, and the water in which they dissolve as well as the calorimeter are part of the surroundings. If we assume that the calorimeter perfectly prevents the gain or loss of heat from the solution, the heat gained by the solution must be produced from the chemical reaction under study. In other words, the heat produced by the reaction, q_{rxn}, is entirely absorbed by the solution; it does not escape the calorimeter. (We also assume that the calorimeter itself does not absorb heat. In the case of the coffee-cup calorimeter, this is a reasonable approximation because the calorimeter has a very low thermal conductivity and heat capacity.) For an exothermic reaction, heat is "lost" by the reaction and "gained" by the solution, so the temperature of the solution rises. The opposite occurs for an endothermic reaction. The heat gained by the solution, q_{soln}, is therefore equal in magnitude to q_{rxn} but and opposite in sign: $q_{soln} = -q_{rxn}$. The value of q_{soln} is readily calculated from the mass of the solution, its specific heat, and the temperature change:

$$q_{soln} = (\text{specific heat of solution}) \times (\text{grams of solution}) \times \Delta T = -q_{rxn} \qquad [5.23]$$

For dilute aqueous solutions the specific heat of the solution will be approximately the same as that of water, 4.18 J/g-K.

Equation 5.23 makes it possible to calculate q_{rxn} from the temperature change of the solution in which the reaction occurs. A temperature increase ($\Delta T > 0$) means the reaction is exothermic ($q_{rxn} < 0$).

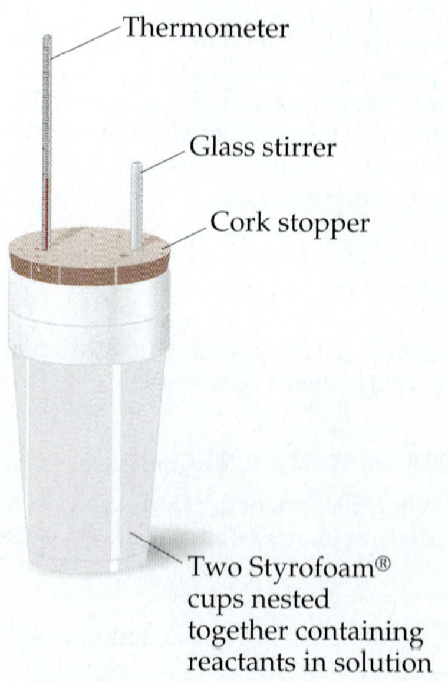

▲ **Figure 5.18 Coffee-cup calorimeter.** This simple apparatus is used to measure heat accompanying reactions at constant pressure.

ACTIVITY Calorimetry

GIVE IT SOME THOUGHT

(a) How are the energy changes of a system and its surroundings related? (b) How is the heat gained or lost by a system related to the heat gained or lost by its surroundings?

VCL Heat of Reaction: NaOH(*aq*) + HCl(*aq*)

SAMPLE EXERCISE 5.6 | Measuring ΔH Using a Coffee-Cup Calorimeter

When a student mixes 50 mL of 1.0 *M* HCl and 50 mL of 1.0 *M* NaOH in a coffee-cup calorimeter, the temperature of the resultant solution increases from 21.0°C to 27.5°C. Calculate the enthalpy change for the reaction in kJ/mol HCl, assuming that the calorimeter loses only a negligible quantity of heat, that the total volume of the solution is 100 mL, that its density is 1.0 g/mL, and that its specific heat is 4.18 J/g-K.

Solution

Analyze: Mixing solutions of HCl and NaOH results in an acid-base reaction:

$$HCl(aq) + NaOH(aq) \longrightarrow H_2O(l) + NaCl(aq)$$

We need to calculate the heat produced by this reaction per mole of HCl, given the temperature increase of the solution, the number of moles of HCl and NaOH involved, and the density and specific heat of the solution.

Plan: The total heat evolved can be calculated using Equation 5.23. The number of moles of HCl consumed in the reaction must be calculated from the volume and molarity of this substance, and this amount then used to determine the heat produced per mol HCl.

Solve: Because the total volume of the solution is 100 mL, its mass is

$$(100\ mL)(1.0\ g/mL) = 100\ g$$

The temperature change is

$$27.5°C - 21.0°C = 6.5°C = 6.5\ K$$

Because the temperature increases, the reaction must be exothermic (that is, q must have a negative sign):

$$q_{rxn} = -s \times m \times \Delta T$$
$$= -(4.18\ J/g\text{-}K)(100\ g)(6.5\ K) = -2.7 \times 10^3\ J = -2.7\ kJ$$

Because the process occurs at constant pressure,

$$\Delta H = q_P = -2.7\ kJ$$

To express the enthalpy change on a molar basis, we use the fact that the number of moles of HCl and NaOH is given by the product of the respective solution volumes (50 mL = 0.050 L) and concentrations:

$$(0.050\ \text{L})(1.0\ \text{mol/L}) = 0.050\ \text{mol}$$

Thus, the enthalpy change per mole of HCl (or NaOH) is

$$\Delta H = -2.7\ \text{kJ}/0.050\ \text{mol} = -54\ \text{kJ/mol}$$

Check: ΔH is negative (exothermic), which is expected for the reaction of an acid with a base. The molar magnitude of the heat evolved seems reasonable.

PRACTICE EXERCISE

When 50.0 mL of 0.100 *M* $AgNO_3$ and 50.0 mL of 0.100 *M* HCl are mixed in a constant-pressure calorimeter, the temperature of the mixture increases from 22.20°C to 23.11°C. The temperature increase is caused by the following reaction:

$$AgNO_3(aq) + HCl(aq) \longrightarrow AgCl(s) + HNO_3(aq)$$

Calculate ΔH for this reaction in kJ/mol $AgNO_3$, assuming that the combined solution has a mass of 100.0 g and a specific heat of 4.18 J/g-°C.

Answer: −68,000 J/mol = −68 kJ/mol

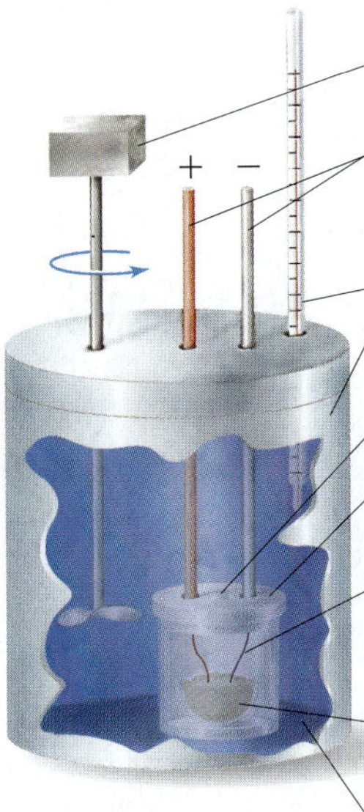

▲ **Figure 5.19 Bomb calorimeter.** This device is used to measure heat accompanying combustion reactions at constant volume.

VCL Heat of Combustion: Chicken Fat

Bomb Calorimetry (Constant-Volume Calorimetry)

Calorimetry can be used to study the chemical potential energy stored in substances. One of the most important types of reactions studied using calorimetry is combustion, in which a compound (usually an organic compound) reacts completely with excess oxygen. (Section 3.2) Combustion reactions are most conveniently studied using a **bomb calorimeter**, a device shown schematically in Figure 5.19 ◀. The substance to be studied is placed in a small cup within a sealed vessel called a *bomb*. The bomb, which is designed to withstand high pressures, has an inlet valve for adding oxygen and electrical contacts to initiate the combustion. After the sample has been placed in the bomb, the bomb is sealed and pressurized with oxygen. It is then placed in the calorimeter, which is essentially an insulated container, and covered with an accurately measured quantity of water. When all the components within the calorimeter have come to the same temperature, the combustion reaction is initiated by passing an electrical current through a fine wire that is in contact with the sample. When the wire gets sufficiently hot, the sample ignites.

Heat is released when combustion occurs. This heat is absorbed by the calorimeter contents, causing a rise in the temperature of the water. The temperature of the water is very carefully measured before reaction and then after reaction when the contents of the calorimeter have again arrived at a common temperature.

To calculate the heat of combustion from the measured temperature increase in the bomb calorimeter, we must know the total heat capacity of the calorimeter, C_{cal}. This quantity is determined by combusting a sample that releases a known quantity of heat and measuring the resulting temperature change. For example, the combustion of exactly 1 g of benzoic acid, $C_7H_6O_2$, in a bomb calorimeter produces 26.38 kJ of heat. Suppose 1.000 g of benzoic acid is combusted in a calorimeter, and it increases the temperature by 4.857°C. The heat capacity of the calorimeter is then given by $C_{cal} = 26.38\ \text{kJ}/4.857°\text{C} = 5.431\ \text{kJ}/°\text{C}$. Once we know the value of C_{cal}, we can measure temperature changes produced by other reactions, and from these we can calculate the heat evolved in the reaction, q_{rxn}:

$$q_{rxn} = -C_{cal} \times \Delta T \qquad [5.24]$$

SAMPLE EXERCISE 5.7 | Measuring q_{rxn} Using a Bomb Calorimeter

Methylhydrazine (CH_6N_2) is commonly used as a liquid rocket fuel. The combustion of methylhydrazine with oxygen produces $N_2(g)$, $CO_2(g)$, and $H_2O(l)$:

$$2\,CH_6N_2(l) + 5\,O_2(g) \longrightarrow 2\,N_2(g) + 2\,CO_2(g) + 6\,H_2O(l)$$

When 4.00 g of methylhydrazine is combusted in a bomb calorimeter, the temperature of the calorimeter increases from 25.00°C to 39.50°C. In a separate experiment the heat capacity of the calorimeter is measured to be 7.794 kJ/°C. What is the heat of reaction for the combustion of a mole of CH_6N_2 in this calorimeter?

Solution

Analyze: We are given a temperature change and the total heat capacity of the calorimeter. We are also given the amount of reactant combusted. Our goal is to calculate the enthalpy change per mole for combustion of the reactant.

Plan: We will first calculate the heat evolved for the combustion of the 4.00-g sample. We will then convert this heat to a molar quantity.

Solve: For combustion of the 4.00-g sample of methylhydrazine, the temperature change of the calorimeter is

$$\Delta T = (39.50°C - 25.00°C) = 14.50°C$$

We can use this value and the value for C_{cal} to calculate the heat of reaction (Equation 5.24):

$$q_{rxn} = -C_{cal} \times \Delta T = -(7.794\ kJ/°C)(14.50°C) = -113.0\ kJ$$

We can readily convert this value to the heat of reaction for a mole of CH_6N_2:

$$\left(\frac{-113.0\ kJ}{4.00\ g\ CH_6N_2}\right) \times \left(\frac{46.1\ g\ CH_6N_2}{1\ mol\ CH_6N_2}\right) = -1.30 \times 10^3\ kJ/mol\ CH_6N_2$$

Check: The units cancel properly, and the sign of the answer is negative as it should be for an exothermic reaction.

PRACTICE EXERCISE

A 0.5865-g sample of lactic acid ($HC_3H_5O_3$) is burned in a calorimeter whose heat capacity is 4.812 kJ/°C. The temperature increases from 23.10°C to 24.95°C. Calculate the heat of combustion of lactic acid **(a)** per gram and **(b)** per mole.

Answers: **(a)** −15.2 kJ/g, **(b)** −1370 kJ/mol

Because the reactions in a bomb calorimeter are carried out under constant-volume conditions, the heat transferred corresponds to the change in internal energy, ΔE, rather than the change in enthalpy, ΔH (Equation 5.14). For most reactions, however, the difference between ΔE and ΔH is very small. For the reaction discussed in Sample Exercise 5.7, for example, the difference between ΔE and ΔH is only about 1 kJ/mol—a difference of less than 0.1%. It is possible to correct the measured heat changes to obtain ΔH values, and these form the basis of the tables of enthalpy change that we will see in the following sections. However, we need not concern ourselves with how these small corrections are made.

5.6 | Hess's Law

VCL Hess's Law

Many enthalpies of reaction have been measured and tabulated. In this section and the next we will see that it is often possible to calculate the ΔH for a reaction from the tabulated ΔH values of other reactions. Thus, it is not necessary to make calorimetric measurements for all reactions.

Because enthalpy is a state function, the enthalpy change, ΔH, associated with any chemical process depends only on the amount of matter that undergoes change and on the nature of the initial state of the reactants and the final state of the products. This means that if a particular reaction can be carried out in one step or in a series of steps, the sum of the enthalpy changes associated with the individual steps must be the same as the enthalpy change associated with the one-step process. As an example, the combustion of methane gas, $CH_4(g)$, to form $CO_2(g)$ and liquid water can be thought of as occurring in two steps: (1) the combustion of $CH_4(g)$ to form $CO_2(g)$ and gaseous water, $H_2O(g)$,

CHEMISTRY AND LIFE | The Regulation of Human Body Temperature

For most of us, the question "Are you running a fever?" was one of our first introductions to medical diagnosis. Indeed, a deviation in body temperature of only a few degrees indicates that something is amiss. In the laboratory you may have tried to maintain a solution at a constant temperature, only to find how difficult it can be. Yet our bodies manage to maintain a near-constant temperature in spite of widely varying weather, levels of physical activity, and periods of high metabolic activity (such as after a meal). How does the human body manage this task, and how does it relate to some of the topics we have discussed in this chapter?

Maintaining a near-constant temperature is one of the primary physiological functions of the human body. Normal body temperature generally ranges from 35.8°C to 37.2°C (96.5°F–99°F). This very narrow temperature range is essential to proper muscle function and to the control of the rates of the biochemical reactions in the body. You will learn more about the effects of temperature on reaction rates in Chapter 14. The temperature is regulated by a portion of the human brain stem called the *hypothalamus*. The hypothalamus acts as a thermostat for body temperature. When body temperature rises above the high end of the normal range, the hypothalamus triggers mechanisms to lower the temperature. It likewise triggers mechanisms to increase the temperature if body temperature drops too low.

To understand qualitatively how the body's heating and cooling mechanisms operate, we can view the body as a thermodynamic system. The body increases its internal energy content by ingesting foods from the surroundings. (Section 5.8) The foods, such as glucose ($C_6H_{12}O_6$), are metabolized—a process that is essentially controlled oxidation to CO_2 and H_2O:

$$C_6H_{12}O_6(s) + 6\,O_2(g) \longrightarrow 6\,CO_2(g) + 6\,H_2O(l) \qquad \Delta H = -2803\text{ kJ}$$

Roughly 40% of the energy produced is ultimately used to do work in the form of muscle contractions and nerve cell activities. The remainder of the energy is released as heat, part of which is used to maintain body temperature. When the body produces too much heat, as in times of heavy physical exertion, it dissipates the excess to the surroundings.

Heat is transferred from the body to its surroundings primarily by *radiation*, *convection*, and *evaporation*. Radiation is the direct loss of heat from the body to cooler surroundings, much as a hot stovetop radiates heat to its surroundings. Convection is heat loss by virtue of heating air that is in contact with the body. The heated air rises and is replaced with cooler air, and the process continues. Warm clothing, which usually consists of insulating layers of material with "dead air" in between, decreases convective heat loss in cold weather. Evaporative cooling occurs when perspiration is generated at the skin surface by the sweat glands. Heat is removed from the body as the perspiration evaporates into the surroundings. Perspiration is predominantly water, so the process involved is the endothermic conversion of liquid water into water vapor:

$$H_2O(l) \longrightarrow H_2O(g) \qquad \Delta H = +44.0\text{ kJ}$$

The speed with which evaporative cooling occurs decreases as the atmospheric humidity increases, which is why people seem to be more sweaty and uncomfortable on hot and humid days.

When the hypothalamus senses that the body temperature has risen too high, it increases heat loss from the body in two principal ways. First, it increases the flow of blood near the surface of the skin, which allows for increased radiational and convective cooling. The reddish "flushed" appearance of a hot individual is the result of this increased subsurface blood flow. Second, the hypothalamus stimulates the secretion of perspiration from the sweat glands, which increases evaporative cooling. During periods of extreme activity, the amount of liquid secreted as perspiration can be as high as 2 to 4 liters per hour. As a result, water must be replenished to the body during these periods (Figure 5.20 ▼). If the body loses too much fluid through perspiration, it will no longer be able to cool itself and blood volume decreases, which can lead to *heat exhaustion* or the more serious and potentially fatal *heat stroke*, during which the body temperature can rise to as high as 41°C to 45°C (106°F–113°F). On the other hand, replenishing water without replenishing the electrolytes that are lost during perspiration can also lead to serious problems as pointed out in the "Chemistry in Life" box in Section 4.5. In fact, in some circumstances, drinking too much water is a greater danger than not drinking enough.

When body temperature drops too low, the hypothalamus decreases the blood flow to the surface of the skin, thereby decreasing heat loss. It also triggers small involuntary contractions of the muscles; the biochemical reactions that generate the energy to do this work also generate more heat for the body. When these contractions get large enough—as when the body feels a chill—a *shiver* results. If the body is unable to maintain a temperature above 35°C (95°F), the very dangerous condition called *hypothermia* can result.

The ability of the human body to maintain its temperature by "tuning" the amount of heat it generates and transfers to its surroundings is truly remarkable. If you take courses in human anatomy and physiology, you will see many other applications of thermochemistry and thermodynamics to the ways in which the human body works.

◀ **Figure 5.20 Marathon runner drinking water.** Runners must constantly replenish the water in their bodies that is lost through perspiration.

and (2) the condensation of gaseous water to form liquid water, $H_2O(l)$. The enthalpy change for the overall process is simply the sum of the enthalpy changes for these two steps:

$$\begin{array}{llll} & CH_4(g) + 2\,O_2(g) & \longrightarrow CO_2(g) + 2\,H_2O(g) & \Delta H = -802\text{ kJ} \\ \text{(Add)} & 2\,H_2O(g) & \longrightarrow 2\,H_2O(l) & \Delta H = -\ \ 88\text{ kJ} \\ \hline & CH_4(g) + 2\,O_2(g) + 2\,H_2O(g) & \longrightarrow CO_2(g) + 2\,H_2O(l) + 2\,H_2O(g) & \Delta H = -890\text{ kJ} \end{array}$$

The net equation is

$$CH_4(g) + 2\,O_2(g) \longrightarrow CO_2(g) + 2\,H_2O(l) \qquad \Delta H = -890\text{ kJ}$$

To obtain the net equation, the sum of the reactants of the two equations is placed on one side of the arrow and the sum of the products on the other side. Because $2\,H_2O(g)$ occurs on both sides, it can be canceled like an algebraic quantity that appears on both sides of an equal sign.

Hess's law states that *if a reaction is carried out in a series of steps, ΔH for the overall reaction will equal the sum of the enthalpy changes for the individual steps.* The overall enthalpy change for the process is independent of the number of steps or the particular nature of the path by which the reaction is carried out. This principle is a consequence of the fact that enthalpy is a state function, and ΔH is therefore independent of the path between the initial and final states. We can therefore calculate ΔH for any process, as long as we find a route for which ΔH is known for each step. This means that a relatively small number of experimental measurements can be used to calculate ΔH for a vast number of different reactions.

Hess's law provides a useful means of calculating energy changes that are difficult to measure directly. For instance, it is impossible to measure directly the enthalpy for the combustion of carbon to form carbon monoxide. Combustion of 1 mol of carbon with 0.5 mol of O_2 produces not only CO but also CO_2, leaving some carbon unreacted. However, solid carbon and carbon monoxide can both be completely burned in O_2 to produce CO_2. We can use the enthalpy changes of these reactions to calculate the heat of combustion of C to CO, as shown in Sample Exercise 5.8.

GIVE IT SOME THOUGHT

What effect do the following changes have on ΔH for a reaction: (a) reversing a reaction, (b) multiplying coefficients by 2?

SAMPLE EXERCISE 5.8 | Using Hess's Law to Calculate ΔH

The enthalpy of reaction for the combustion of C to CO_2 is -393.5 kJ/mol C, and the enthalpy for the combustion of CO to CO_2 is -283.0 kJ/mol CO:

$$\text{(1)} \qquad C(s) + O_2(g) \longrightarrow CO_2(g) \qquad \Delta H_1 = -393.5\text{ kJ}$$

$$\text{(2)} \qquad CO(g) + \tfrac{1}{2}\,O_2(g) \longrightarrow CO_2(g) \qquad \Delta H_2 = -283.0\text{ kJ}$$

Using these data, calculate the enthalpy for the combustion of C to CO:

$$\text{(3)} \qquad C(s) + \tfrac{1}{2}\,O_2(g) \longrightarrow CO(g) \qquad \Delta H_3 = ?$$

Solution

Analyze: We are given two thermochemical equations, and our goal is to combine them in such a way as to obtain the third equation and its enthalpy.

Plan: We will use Hess's law. In doing so, we first note the numbers of moles of substances among the reactants and products in the target equation, (3). We then manipulate equations (1) and (2) to give the same number of moles of these substances, so that when the resulting equations are added, we obtain the target equation. At the same time, we keep track of the enthalpy changes, which we add.

Solve: In order to use equations (1) and (2), we arrange them so that C(*s*) is on the reactant side and CO(*g*) is on the product side of the arrow, as in the target reaction, equation (3). Because equation (1) has C(*s*) as a reactant, we can use that equation just as it is. We need to turn equation (2) around, however, so that CO(*g*) is a product. Remember that when reactions are turned around, the sign of ΔH is reversed. We arrange the two equations so that they can be added to give the desired equation:

$$\begin{array}{lr} C(s) + O_2(g) \longrightarrow CO_2(g) & \Delta H_1 = -393.5\ \text{kJ} \\ CO_2(g) \longrightarrow CO(g) + \tfrac{1}{2}O_2(g) & -\Delta H_2 = 283.0\ \text{kJ} \\ \hline C(s) + \tfrac{1}{2}O_2(g) \longrightarrow CO(g) & \Delta H_3 = -110.5\ \text{kJ} \end{array}$$

When we add the two equations, $CO_2(g)$ appears on both sides of the arrow and therefore cancels out. Likewise, $\frac{1}{2}O_2(g)$ is eliminated from each side.

Comment: It is sometimes useful to add subscripts to the enthalpy changes, as we have done here, to keep track of the associations between the chemical reactions and their ΔH values.

PRACTICE EXERCISE

Carbon occurs in two forms, graphite and diamond. The enthalpy of the combustion of graphite is −393.5 kJ/mol and that of diamond is −395.4 kJ/mol:

$$C(\textit{graphite}) + O_2(g) \longrightarrow CO_2(g) \qquad \Delta H_1 = -393.5\ \text{kJ}$$

$$C(\textit{diamond}) + O_2(g) \longrightarrow CO_2(g) \qquad \Delta H_2 = -395.4\ \text{kJ}$$

Calculate ΔH for the conversion of graphite to diamond:

$$C(\textit{graphite}) \longrightarrow C(\textit{diamond}) \qquad \Delta H_3 = ?$$

Answer: $\Delta H_3 = +1.9$ kJ

SAMPLE EXERCISE 5.9 | Using Three Equations with Hess's Law to Calculate ΔH

Calculate ΔH for the reaction

$$2\,C(s) + H_2(g) \longrightarrow C_2H_2(g)$$

given the following chemical equations and their respective enthalpy changes:

$$\begin{array}{lr} C_2H_2(g) + \tfrac{5}{2}O_2(g) \longrightarrow 2\,CO_2(g) + H_2O(l) & \Delta H = -1299.6\ \text{kJ} \\ C(s) + O_2(g) \longrightarrow CO_2(g) & \Delta H = -393.5\ \text{kJ} \\ H_2(g) + \tfrac{1}{2}O_2(g) \longrightarrow H_2O(l) & \Delta H = -285.8\ \text{kJ} \end{array}$$

Solution

Analyze: We are given a chemical equation and asked to calculate its ΔH using three chemical equations and their associated enthalpy changes.

Plan: We will use Hess's law, summing the three equations or their reverses and multiplying each by an appropriate coefficient so that they add to give the net equation for the reaction of interest. At the same time, we keep track of the ΔH values, reversing their signs if the reactions are reversed and multiplying them by whatever coefficient is employed in the equation.

Solve: Because the target equation has C_2H_2 as a product, we turn the first equation around; the sign of ΔH is therefore changed. The desired equation has 2 C(*s*) as a reactant, so we multiply the second equation and its ΔH by 2. Because the target equation has H_2 as a reactant, we keep the third equation as it is. We then add the three equations and their enthalpy changes in accordance with Hess's law:

$$\begin{array}{lr} 2\,CO_2(g) + H_2O(l) \longrightarrow C_2H_2(g) + \tfrac{5}{2}O_2(g) & \Delta H = 1299.6\ \text{kJ} \\ 2\,C(s) + 2\,O_2(g) \longrightarrow 2\,CO_2(g) & \Delta H = -787.0\ \text{kJ} \\ H_2(g) + \tfrac{1}{2}O_2(g) \longrightarrow H_2O(l) & \Delta H = -285.8\ \text{kJ} \\ \hline 2\,C(s) + H_2(g) \longrightarrow C_2H_2(g) & \Delta H = 226.8\ \text{kJ} \end{array}$$

When the equations are added, there are 2 CO_2, $\frac{5}{2}O_2$, and H_2O on both sides of the arrow. These are canceled in writing the net equation.

Check: The procedure must be correct because we obtained the correct net equation. In cases like this you should go back over the numerical manipulations of the ΔH values to ensure that you did not make an inadvertent error with signs.

PRACTICE EXERCISE

Calculate ΔH for the reaction

$$NO(g) + O(g) \longrightarrow NO_2(g)$$

given the following information:

$$NO(g) + O_3(g) \longrightarrow NO_2(g) + O_2(g) \qquad \Delta H = -198.9 \text{ kJ}$$
$$O_3(g) \longrightarrow \tfrac{3}{2} O_2(g) \qquad \Delta H = -142.3 \text{ kJ}$$
$$O_2(g) \longrightarrow 2\,O(g) \qquad \Delta H = 495.0 \text{ kJ}$$

Answer: -304.1 kJ

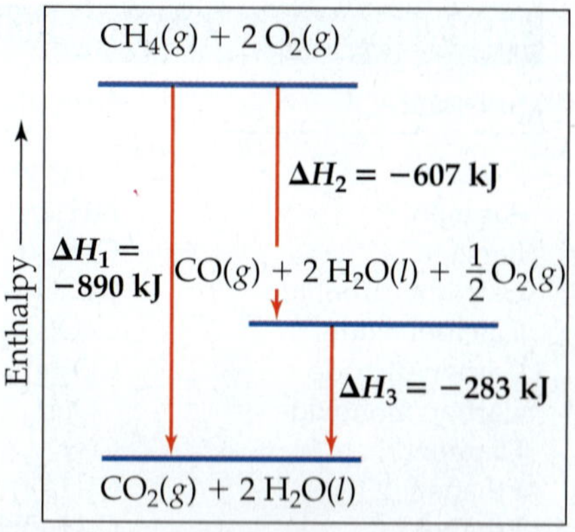

▲ **Figure 5.21 An enthalpy diagram illustrating Hess's law.** The quantity of heat generated by combustion of 1 mol CH_4 is independent of whether the reaction takes place in one or more steps: $\Delta H_1 = \Delta H_2 + \Delta H_3$.

The key point of these examples is that H is a state function, so *for a particular set of reactants and products, ΔH is the same whether the reaction takes place in one step or in a series of steps.* For example, consider the reaction of methane (CH_4) and oxygen (O_2) to form CO_2 and H_2O. We can envision the reaction forming CO_2 directly or with the initial formation of CO, which is then combusted to CO_2. These two paths are compared in Figure 5.21 ▶. Because H is a state function, both paths *must* produce the same value of ΔH. In the enthalpy diagram, that means $\Delta H_1 = \Delta H_2 + \Delta H_3$.

5.7 | Enthalpies of Formation

By using the methods we have just discussed, we can calculate the enthalpy changes for a great many reactions from tabulated ΔH values. Many experimental data are tabulated according to the type of process. For example, extensive tables exist of *enthalpies of vaporization* (ΔH for converting liquids to gases), *enthalpies of fusion* (ΔH for melting solids), *enthalpies of combustion* (ΔH for combusting a substance in oxygen), and so forth. A particularly important process used for tabulating thermochemical data is the formation of a compound from its constituent elements. The enthalpy change associated with this process is called the **enthalpy of formation** (or *heat of formation*) and is labeled ΔH_f, where the subscript f indicates that the substance has been *formed* from its component elements.

The magnitude of any enthalpy change depends on the conditions of temperature, pressure, and state (gas, liquid, or solid crystalline form) of the reactants and products. In order to compare the enthalpies of different reactions, we must define a set of conditions, called a *standard state*, at which most enthalpies are tabulated. The standard state of a substance is its pure form at atmospheric pressure (1 atm; ∞ Section 10.2) and the temperature of interest, which we usually choose to be 298 K (25°C).* The **standard enthalpy change** of a reaction is defined as the enthalpy change when all reactants and products are in their standard states. We denote a standard enthalpy change as $\Delta H°$, where the superscript ° indicates standard-state conditions.

MOVIE
Formation of Water

The **standard enthalpy of formation** of a compound, ΔH_f°, is the change in enthalpy for the reaction that forms one mole of the compound from its elements, with all substances in their standard states. We usually report ΔH_f° values at 298 K. If an element exists in more than one form under standard

MOVIE
Formation of Aluminium Bromide

* The definition of the standard state for gases has been changed to 1 bar (1 atm = 1.013 bar), a slightly lower pressure than the value of 1 atm that is used for the data in this text. For most purposes, this change makes very little difference in the standard changes.

TABLE 5.3 Standard Enthalpies of Formation, ΔH_f°, at 298 K

Substance	Formula	ΔH_f° (kJ/mol)	Substance	Formula	ΔH_f° (kJ/mol)
Acetylene	$C_2H_2(g)$	226.7	Hydrogen chloride	$HCl(g)$	−92.30
Ammonia	$NH_3(g)$	−46.19	Hydrogen fluoride	$HF(g)$	−268.60
Benzene	$C_6H_6(l)$	49.0	Hydrogen iodide	$HI(g)$	25.9
Calcium carbonate	$CaCO_3(s)$	−1207.1	Methane	$CH_4(g)$	−74.80
Calcium oxide	$CaO(s)$	−635.5	Methanol	$CH_3OH(l)$	−238.6
Carbon dioxide	$CO_2(g)$	−393.5	Propane	$C_3H_8(g)$	−103.85
Carbon monoxide	$CO(g)$	−110.5	Silver chloride	$AgCl(s)$	−127.0
Diamond	$C(s)$	1.88	Sodium bicarbonate	$NaHCO_3(s)$	−947.7
Ethane	$C_2H_6(g)$	−84.68	Sodium carbonate	$Na_2CO_3(s)$	−1130.9
Ethanol	$C_2H_5OH(l)$	−277.7	Sodium chloride	$NaCl(s)$	−410.9
Ethylene	$C_2H_4(g)$	52.30	Sucrose	$C_{12}H_{22}O_{11}(s)$	−2221
Glucose	$C_6H_{12}O_6(s)$	−1273	Water	$H_2O(l)$	−285.8
Hydrogen bromide	$HBr(g)$	−36.23	Water vapor	$H_2O(g)$	−241.8

conditions, the most stable form of the element is usually used for the formation reaction. For example, the standard enthalpy of formation for ethanol, C_2H_5OH, is the enthalpy change for the following reaction:

$$2\,C(graphite) + 3\,H_2(g) + \tfrac{1}{2}\,O_2(g) \longrightarrow C_2H_5OH(l) \qquad \Delta H_f^\circ = -277.7\text{ kJ} \qquad [5.25]$$

The elemental source of oxygen is O_2, not O or O_3, because O_2 is the stable form of oxygen at 298 K and standard atmospheric pressure. Similarly, the elemental source of carbon is graphite and not diamond, because graphite is more stable (lower energy) at 298 K and standard atmospheric pressure (see Practice Exercise 5.8). Likewise, the most stable form of hydrogen under standard conditions is $H_2(g)$, so this is used as the source of hydrogen in Equation 5.25.

The stoichiometry of formation reactions always indicates that one mole of the desired substance is produced, as in Equation 5.25. As a result, enthalpies of formation are reported in kJ/mol of the substance. Several standard enthalpies of formation are given in Table 5.3 ▲. A more complete table is provided in Appendix C. By definition, *the standard enthalpy of formation of the most stable form of any element is zero* because there is no formation reaction needed when the element is already in its standard state. Thus, the values of ΔH_f° for C(graphite), $H_2(g)$, $O_2(g)$, and the standard states of other elements are zero by definition.

GIVE IT SOME THOUGHT

In Table 5.3, the standard enthalpy of formation of $C_2H_2(g)$ is listed as 226.7 kJ/mol. Write the thermochemical equation associated with ΔH_f° for this substance.

SAMPLE EXERCISE 5.10 | Identifying Equations Associated with Enthalpies of Formation

For which of the following reactions at 25°C would the enthalpy change represent a standard enthalpy of formation? For those where it does not, what changes would need to be made in the reaction conditions?

(a) $2\,Na(s) + \tfrac{1}{2}\,O_2(g) \longrightarrow Na_2O(s)$

(b) $2\,K(l) + Cl_2(g) \longrightarrow 2\,KCl(s)$

(c) $C_6H_{12}O_6(s) \longrightarrow 6\,C(diamond) + 6\,H_2(g) + 3\,O_2(g)$

Solution

Analyze: The standard enthalpy of formation is represented by a reaction in which each reactant is an element in its standard state and the product is one mole of the compound.

Plan: To solve these problems, we need to examine each equation to determine, first of all, whether the reaction is one in which a substance is formed from the elements. Next, we need to determine whether the reactant elements are in their standard states.

Solve: In (a) Na_2O is formed from the elements sodium and oxygen in their proper states, a solid and O_2 gas, respectively. Therefore, the enthalpy change for reaction (a) corresponds to a standard enthalpy of formation.

In (b) potassium is given as a liquid. It must be changed to the solid form, its standard state at room temperature. Furthermore, two moles of product are formed, so the enthalpy change for the reaction as written is twice the standard enthalpy of formation of KCl(*s*). The proper equation for the formation reaction is

$$K(s) + \frac{1}{2}Cl(g) \longrightarrow KCl(s)$$

Reaction (c) does not form a substance from its elements. Instead, a substance decomposes to its elements, so this reaction must be reversed. Next, the element carbon is given as diamond, whereas graphite is the lowest-energy solid form of carbon at room temperature and 1 atm pressure. The equation that correctly represents the enthalpy of formation of glucose from its elements is

$$6\,C(graphite) + 6\,H_2(g) + 3\,O_2(g) \longrightarrow C_6H_{12}O_6(s)$$

PRACTICE EXERCISE

Write the equation corresponding to the standard enthalpy of formation of liquid carbon tetrachloride (CCl_4).

Answer: $C(s) + 2\,Cl_2(g) \longrightarrow CCl_4(l)$

Using Enthalpies of Formation to Calculate Enthalpies of Reaction

Tabulations of ΔH_f°, such as those in Table 5.3 and Appendix C, have many important uses. As we will see in this section, we can use Hess's law to calculate the standard enthalpy change for any reaction for which we know the ΔH_f° values for all reactants and products. For example, consider the combustion of propane gas, $C_3H_8(g)$, with oxygen to form $CO_2(g)$ and $H_2O(l)$ under standard conditions:

$$C_3H_8(g) + 5\,O_2(g) \longrightarrow 3\,CO_2(g) + 4\,H_2O(l)$$

We can write this equation as the sum of three formation reactions:

$$C_3H_8(g) \longrightarrow 3\,C(s) + 4\,H_2(g) \qquad \Delta H_1 = -\Delta H_f^\circ[C_3H_8(g)] \qquad [5.26]$$

$$3\,C(s) + 3\,O_2(g) \longrightarrow 3\,CO_2(g) \qquad \Delta H_2 = 3\Delta H_f^\circ[CO_2(g)] \qquad [5.27]$$

$$4\,H_2(g) + 2\,O_2(g) \longrightarrow 4\,H_2O(l) \qquad \Delta H_3 = 4\Delta H_f^\circ[H_2O(l)] \qquad [5.28]$$

$$C_3H_8(g) + 5\,O_2(g) \longrightarrow 3\,CO_2(g) + 4\,H_2O(l) \qquad \Delta H_{rxn}^\circ = \Delta H_1 + \Delta H_2 + \Delta H_3 \qquad [5.29]$$

From Hess's law we can write the standard enthalpy change for the overall reaction, Equation 5.29, as the sum of the enthalpy changes for the processes in Equations 5.26 through 5.28. We can then use values from Table 5.3 to compute a numerical value for ΔH° for the overall reaction:

$$\begin{aligned}\Delta H^\circ{}_{rxn} &= \Delta H_1 + \Delta H_2 + \Delta H_3 \\ &= -\Delta H_f^\circ[C_3H_8(g)] + 3\Delta H_f^\circ[CO_2(g)] + 4\Delta H_f^\circ[H_2O(l)] \\ &= -(-103.85\text{ kJ}) + 3(-393.5\text{ kJ}) + 4(-285.8\text{ kJ}) = -2220\text{ kJ} \qquad [5.30]\end{aligned}$$

MOVIE Nitrogen Triiodide

MOVIE Thermite

Several aspects of this calculation depend on the guidelines we discussed in Section 5.4.

1. Equation 5.26 is the reverse of the formation reaction for $C_3H_8(g)$, so the enthalpy change for this reaction is $-\Delta H_f^\circ[C_3H_8(g)]$.
2. Equation 5.27 is the formation reaction for 3 mol of $CO_2(g)$. Because enthalpy is an extensive property, the enthalpy change for this step is $3\Delta H_f^\circ[CO_2(g)]$. Similarly, the enthalpy change for Equation 5.28 is $4\Delta H_f^\circ[H_2O(l)]$. The reaction specifies that $H_2O(l)$ was produced, so be careful to use the value of ΔH_f° for $H_2O(l)$, not $H_2O(g)$.
3. We assume that the stoichiometric coefficients in the balanced equation represent moles. For Equation 5.29, therefore, the value $\Delta H_{rxn}^\circ = -2220$ kJ represents the enthalpy change for the reaction of 1 mol C_3H_8 and 5 mol O_2 to form 3 mol CO_2 and 4 mol H_2O. The product of the number of moles and the enthalpy change in kJ/mol has the units kJ: (number of moles) × (ΔH_f° in kJ/mol) = kJ. We therefore report ΔH_{rxn}° in kJ.

Figure 5.22 ▼ presents an enthalpy diagram for Equation 5.29, showing how it can be broken into steps involving formation reactions.

We can break down any reaction into formation reactions as we have done here. When we do, we obtain the general result that the standard enthalpy change of a reaction is the sum of the standard enthalpies of formation of the products minus the standard enthalpies of formation of the reactants:

$$\Delta H^\circ_{rxn} = \sum n\Delta H_f^\circ(\text{products}) - \sum m\Delta H_f^\circ(\text{reactants}) \qquad [5.31]$$

The symbol Σ (sigma) means "the sum of," and n and m are the stoichiometric coefficients of the chemical equation. The first term in Equation 5.31 represents the formation reactions of the products, which are written in the "forward" direction, that is, elements reacting to form products. This term is analogous to Equations 5.27 and 5.28 in the previous example. The second term represents the reverse of the formation reactions of the reactants, as in Equation 5.26, which is why the ΔH_f° values have a minus sign in front of them.

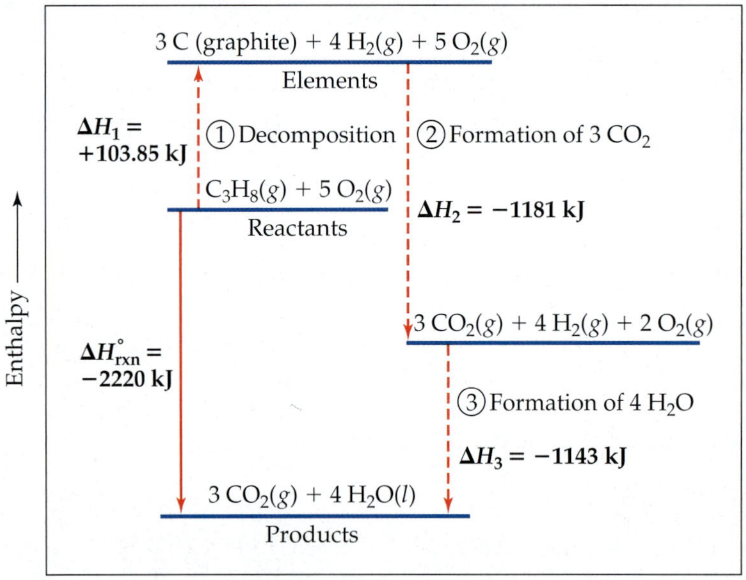

▶ **Figure 5.22 An enthalpy diagram relating the enthalpy change for a reaction to enthalpies of formation.** Enthalpy diagram for the combustion of 1 mol of propane gas, $C_3H_8(g)$. The overall reaction is $C_3H_8(g) + 5\,O_2(g) \longrightarrow 3\,CO_2(g) + 4\,H_2O(l)$. We can imagine this reaction as occurring in three steps. First, $C_3H_8(g)$ is decomposed to its elements, so $\Delta H_1 = -\Delta H_f^\circ[C_3H_8(g)]$. Second, 3 mol $CO_2(g)$ are formed, so $\Delta H_2 = 3\Delta H_f^\circ[CO_2(g)]$. Finally, 4 mol $H_2O(l)$ are formed, so $\Delta H_3 = 4\Delta H_f^\circ[H_2O(l)]$. Hess's law tells us that $\Delta H_{rxn}^\circ = \Delta H_1 + \Delta H_2 + \Delta H_3$. This same result is given by Equation 5.30 because $\Delta H_f^\circ[O_2(g)] = 0$.

SAMPLE EXERCISE 5.11 | Calculating an Enthalpy of Reaction from Enthalpies of Formation

(a) Calculate the standard enthalpy change for the combustion of 1 mol of benzene, $C_6H_6(l)$, to $CO_2(g)$ and $H_2O(l)$. **(b)** Compare the quantity of heat produced by combustion of 1.00 g propane to that produced by 1.00 g benzene.

Solution

Analyze: **(a)** We are given a reaction [combustion of $C_6H_6(l)$ to form $CO_2(g)$ and $H_2O(l)$] and asked to calculate its standard enthalpy change, $\Delta H°$. **(b)** We then need to compare the quantity of heat produced by combustion of 1.00 g C_6H_6 with that produced by 1.00 g of C_3H_8, whose combustion was treated above in the text.

Plan: **(a)** We need to write the balanced equation for the combustion of C_6H_6. We then look up ΔH_f° values in Appendix C or in Table 5.3 and apply Equation 5.31 to calculate the enthalpy change for the reaction. **(b)** We use the molar mass of C_6H_6 to change the enthalpy change per mole to that per gram. We similarly use the molar mass of C_3H_8 and the enthalpy change per mole calculated in the text above to calculate the enthalpy change per gram of that substance.

Solve: **(a)** We know that a combustion reaction involves $O_2(g)$ as a reactant. Thus, the balanced equation for the combustion reaction of 1 mol $C_6H_6(l)$ is

$$C_6H_6(l) + \tfrac{15}{2}\,O_2(g) \longrightarrow 6\,CO_2(g) + 3\,H_2O(l)$$

We can calculate $\Delta H°$ for this reaction by using Equation 5.31 and data in Table 5.3. Remember to multiply the ΔH_f° value for each substance in the reaction by that substance's stoichiometric coefficient. Recall also that $\Delta H_f^\circ = 0$ for any element in its most stable form under standard conditions, so $\Delta H_f^\circ[O_2(g)] = 0$:

$$\begin{aligned}\Delta H^\circ_{rxn} &= [6\Delta H_f^\circ(CO_2) + 3\Delta H_f^\circ(H_2O)] - [\Delta H_f^\circ(C_6H_6) + \tfrac{15}{2}\,\Delta H_f^\circ(O_2)]\\ &= [6(-393.5\text{ kJ}) + 3(-285.8\text{ kJ})] - \left[(49.0\text{ kJ}) + \tfrac{15}{2}(0\text{ kJ})\right]\\ &= (-2361 - 857.4 - 49.0)\text{ kJ}\\ &= -3267\text{ kJ}\end{aligned}$$

(b) From the example worked in the text, $\Delta H° = -2220$ kJ for the combustion of 1 mol of propane. In part (a) of this exercise we determined that $\Delta H° = -3267$ kJ for the combustion of 1 mol benzene. To determine the heat of combustion per gram of each substance, we use the molar masses to convert moles to grams:

$$C_3H_8(g): \quad (-2220\text{ kJ/mol})(1\text{ mol}/44.1\text{ g}) = -50.3\text{ kJ/g}$$

$$C_6H_6(l): \quad (-3267\text{ kJ/mol})(1\text{ mol}/78.1\text{ g}) = -41.8\text{ kJ/g}$$

Comment: Both propane and benzene are hydrocarbons. As a rule, the energy obtained from the combustion of a gram of hydrocarbon is between 40 and 50 kJ.

PRACTICE EXERCISE

Using the standard enthalpies of formation listed in Table 5.3, calculate the enthalpy change for the combustion of 1 mol of ethanol:

$$C_2H_5OH(l) + 3\,O_2(g) \longrightarrow 2\,CO_2(g) + 3\,H_2O(l)$$

Answer: −1367 kJ

SAMPLE EXERCISE 5.12 | Calculating an Enthalpy of Formation Using an Enthalpy of Reaction

The standard enthalpy change for the reaction

$$CaCO_3(s) \longrightarrow CaO(s) + CO_2(g)$$

is 178.1 kJ. From the values for the standard enthalpies of formation of $CaO(s)$ and $CO_2(g)$ given in Table 5.3, calculate the standard enthalpy of formation of $CaCO_3(s)$.

Solution

Analyze: We need to obtain $\Delta H_f^\circ(CaCO_3)$.

Plan: We begin by writing the expression for the standard enthalpy change for the reaction:

$$\Delta H^\circ_{rxn} = [\Delta H_f^\circ(CaO) + \Delta H_f^\circ(CO_2)] - \Delta H_f^\circ(CaCO_3)$$

Solve: Inserting the known values from Table 5.3 or Appendix C, we have

$$78.1 \text{ kJ} = -635.5 \text{ kJ} - 393.5 \text{ kJ} - \Delta H_f^\circ(\text{CaCO}_3)$$

Solving for $\Delta H_f^\circ(\text{CaCO}_3)$ gives

$$\Delta H_f^\circ(\text{CaCO}_3) = -1207.1 \text{ kJ/mol}$$

Check: We expect the enthalpy of formation of a stable solid such as calcium carbonate to be negative, as obtained.

PRACTICE EXERCISE

Given the following standard enthalpy change, use the standard enthalpies of formation in Table 5.3 to calculate the standard enthalpy of formation of CuO(*s*):

$$\text{CuO}(s) + \text{H}_2(g) \longrightarrow \text{Cu}(s) + \text{H}_2\text{O}(l) \qquad \Delta H^\circ = -129.7 \text{ kJ}$$

Answer: −156.1 kJ/mol

5.8 | Foods and Fuels

Most chemical reactions used for the production of heat are combustion reactions. The energy released when one gram of a material is combusted is often called its **fuel value**. Because fuel values represent the heat *released* in a combustion, fuel values are reported as positive numbers. The fuel value of any food or fuel can be measured by calorimetry.

Foods

Most of the energy our bodies need comes from carbohydrates and fats. The forms of carbohydrate known as starch are decomposed in the intestines into glucose, $C_6H_{12}O_6$. Glucose is soluble in blood, and in the human body it is known as blood sugar. It is transported by the blood to cells, where it reacts with O_2 in a series of steps, eventually producing $CO_2(g)$, $H_2O(l)$, and energy:

$$\text{C}_6\text{H}_{12}\text{O}_6(s) + 6\,\text{O}_2(g) \longrightarrow 6\,\text{CO}_2(g) + 6\,\text{H}_2\text{O}(l) \qquad \Delta H^\circ = -2803 \text{ kJ}$$

The breakdown of carbohydrates is rapid, so their energy is quickly supplied to the body. However, the body stores only a very small amount of carbohydrates. The average fuel value of carbohydrates is 17 kJ/g (4 kcal/g).

Like carbohydrates, fats produce CO_2 and H_2O when metabolized and when subjected to combustion in a bomb calorimeter. The reaction of tristearin, $C_{57}H_{110}O_6$, a typical fat, is as follows:

$$2\,\text{C}_{57}\text{H}_{110}\text{O}_6(s) + 163\,\text{O}_2(g) \longrightarrow 114\,\text{CO}_2(g) + 110\,\text{H}_2\text{O}(l) \quad \Delta H^\circ = -75{,}520 \text{ kJ}$$

The body uses the chemical energy from foods to maintain body temperature (see the "Chemistry and Life" box in Section 5.5), to contract muscles, and to construct and repair tissues. Any excess energy is stored as fats. Fats are well suited to serve as the body's energy reserve for at least two reasons: (1) They are insoluble in water, which facilitates storage in the body, and (2) they produce more energy per gram than either proteins or carbohydrates, which makes them efficient energy sources on a mass basis. The average fuel value of fats is 38 kJ/g (9 kcal/g).

The metabolism of proteins in the body produces less energy than combustion in a calorimeter because the products are different. Proteins contain nitrogen, which is released in the bomb calorimeter as N_2. In the body this nitrogen ends up mainly as urea, $(NH_2)_2CO$. Proteins are used by the body mainly as building materials for organ walls, skin, hair, muscle, and so forth. On average, the metabolism of proteins produces 17 kJ/g (4 kcal/g), the same as for carbohydrates.

TABLE 5.4 Compositions and Fuel Values of Some Common Foods

	Approximate Composition (% by mass)			Fuel Value	
	Carbohydrate	Fat	Protein	kJ/g	kcal/g (Cal/g)
Carbohydrate	100	–	–	17	4
Fat	–	100	–	38	9
Protein	–	–	100	17	4
Apples	13	0.5	0.4	2.5	0.59
Beer*	1.2	–	0.3	1.8	0.42
Bread	52	3	9	12	2.8
Cheese	4	37	28	20	4.7
Eggs	0.7	10	13	6.0	1.4
Fudge	81	11	2	18	4.4
Green beans	7.0	–	1.9	1.5	0.38
Hamburger	–	30	22	15	3.6
Milk (whole)	5.0	4.0	3.3	3.0	0.74
Peanuts	22	39	26	23	5.5

*Beers typically contain 3.5% ethanol, which has fuel value.

The fuel values for a variety of common foods are shown in Table 5.4 ▲. Labels on packaged foods show the amounts of carbohydrate, fat, and protein contained in an average serving, as well as the amount of energy supplied by a serving (Figure 5.23 ▶). The amount of energy our bodies require varies considerably depending on such factors as weight, age, and muscular activity. About 100 kJ per kilogram of body weight per day is required to keep the body functioning at a minimal level. An average 70-kg (154-lb) person expends about 800 kJ/hr when doing light work, such as slow walking or light gardening. Strenuous activity, such as running, often requires 2000 kJ/hr or more. When the energy value, or caloric content, of our food exceeds the energy we expend, our body stores the surplus as fat.

▲ **Figure 5.23 Labels of processed foods showing nutritional information.** Such labels give information about the quantities of different nutrients and the energy value (caloric value) in an average serving.

GIVE IT SOME THOUGHT

Which releases the greatest amount energy per gram upon metabolism, carbohydrates, proteins, or fats?

SAMPLE EXERCISE 5.13 | Comparing Fuel Values

A plant such as celery contains carbohydrates in the form of starch and cellulose. These two kinds of carbohydrates have essentially the same fuel values when combusted in a bomb calorimeter. When we consume celery, however, our bodies receive fuel value from the starch only. What can we conclude about the difference between starch and cellulose as foods?

Solution If cellulose does not provide fuel value, we must conclude that it is not converted in the body into CO_2 and H_2O, as starch is. A slight, but critical, difference in the structures of starch and cellulose explains why only starch is broken down into glucose in the body. Cellulose passes through without undergoing significant chemical change. It serves as fiber, or roughage, in the diet, but provides no caloric value.

PRACTICE EXERCISE

The nutritional label on a bottle of canola oil indicates that 10 g of the oil has an energy value of 86 kcal. A similar label on a bottle of pancake syrup indicates that 60 mL (about 60 g) has an energy value of 200 kcal. Account for the difference.

Answer: The oil has a fuel value of 8.6 kcal/g, whereas the syrup has a fuel value of about 3.3 kcal/g. The higher fuel value for the canola oil arises because the oil is essentially pure fat, whereas the syrup is a solution of sugars (carbohydrates) in water. The oil has a higher fuel value per gram; in addition, the syrup is diluted by water.

SAMPLE EXERCISE 5.14 | Estimating the Fuel Value of a Food from Its Composition

(a) A 28-g (1-oz) serving of a popular breakfast cereal served with 120 mL of skim milk provides 8 g protein, 26 g carbohydrates, and 2 g fat. Using the average fuel values of these kinds of substances, estimate the energy value (caloric content) of this serving. **(b)** A person of average weight uses about 100 Cal/mi when running or jogging. How many servings of this cereal provide the energy value requirements for running 3 mi?

Solution

(a) Analyze: The energy value of the serving will be the sum of the energy values of the protein, carbohydrates, and fat.

Plan: We are given the masses of the protein, carbohydrates, and fat contained in the combined cereal/milk serving. We can use the data in Table 5.4 to convert these masses to their energy values, which we can sum to get the total energy value.

Solve:
$$(8\text{ g protein})\left(\frac{17\text{ kJ}}{1\text{ g protein}}\right) + (26\text{ g carbohydrate})\left(\frac{17\text{ kJ}}{1\text{ g carbohydrate}}\right) +$$
$$(2\text{ g fat})\left(\frac{38\text{ kJ}}{1\text{ g fat}}\right) = 650\text{ kJ (to two significant figures)}$$

This corresponds to 160 kcal:

$$(650\text{ kJ})\left(\frac{1\text{ kcal}}{4.18\text{ kJ}}\right) = 160\text{ kcal}$$

Recall that the dietary Calorie is equivalent to 1 kcal. Thus, the serving provides 160 Cal.

(b) Analyze: Here we are faced with the reverse problem, calculating the quantity of food that provides a specific energy value.

Plan: The problem statement provides a conversion factor between Calories and miles. The answer to part (a) provides us with a conversion factor between servings and Calories.

Solve: We can use these factors in a straightforward dimensional analysis to determine the number of servings needed, rounded to the nearest whole number:

$$\text{Servings} = (3\text{ mi})\left(\frac{100\text{ Cal}}{1\text{ mi}}\right)\left(\frac{1\text{ serving}}{160\text{ Cal}}\right) = 2\text{ servings}$$

PRACTICE EXERCISE

(a) Dry red beans contain 62% carbohydrate, 22% protein, and 1.5% fat. Estimate the fuel value of these beans. **(b)** Very light activity like reading or watching television uses about 7 kJ/min. How many minutes of such activity can be sustained by the energy provided by a serving of chicken noodle soup containing 13 g protein, 15 g carbohydrate, and 5 g fat?

Answers: **(a)** 15 kJ/g, **(b)** 95 min

Fuels

The elemental compositions and fuel values of several common fuels are compared in Table 5.5 ▶. During the complete combustion of fuels, carbon is converted to CO_2 and hydrogen is converted to H_2O, both of which have large negative enthalpies of formation. Consequently, the greater the percentage of carbon and hydrogen in a fuel, the higher its fuel value. Compare, for example, the compositions and fuel values of bituminous coal and wood. The coal has a higher fuel value because of its greater carbon content.

In 2002 the United States consumed 1.03×10^{17} kJ of energy. This value corresponds to an average daily energy consumption per person of 1.0×10^{6} kJ, which is roughly 100 times greater than the per capita food-energy needs. We are a very energy-intensive society. Although our population is only about 4.5% of the world's population, the United States accounts for nearly one-fourth of the total world energy consumption. Figure 5.24 ▶ illustrates the sources of this energy.

TABLE 5.5 Fuel Values and Compositions of Some Common Fuels

	Approximate Elemental Composition (mass %)			
	C	H	O	Fuel Value (kJ/g)
Wood (pine)	50	6	44	18
Anthracite coal (Pennsylvania)	82	1	2	31
Bituminous coal (Pennsylvania)	77	5	7	32
Charcoal	100	0	0	34
Crude oil (Texas)	85	12	0	45
Gasoline	85	15	0	48
Natural gas	70	23	0	49
Hydrogen	0	100	0	142

Coal, petroleum, and natural gas, which are our major sources of energy, are known as **fossil fuels**. All have formed over millions of years from the decomposition of plants and animals and are being depleted far more rapidly than they are being formed. **Natural gas** consists of gaseous hydrocarbons, compounds of hydrogen and carbon. It contains primarily methane (CH_4), with small amounts of ethane (C_2H_6), propane (C_3H_8), and butane (C_4H_{10}). We determined the fuel value of propane in Sample Exercise 5.11. **Petroleum** is a liquid composed of hundreds of compounds. Most of these compounds are hydrocarbons, with the remainder being chiefly organic compounds containing sulfur, nitrogen, or oxygen. **Coal**, which is solid, contains hydrocarbons of high molecular weight as well as compounds containing sulfur, oxygen, or nitrogen. Coal is the most abundant fossil fuel; it constitutes 80% of the fossil fuel reserves of the United States and 90% of those of the world. However, the use of coal presents a number of problems. Coal is a complex mixture of substances, and it contains components that cause air pollution. When coal is combusted, the sulfur it contains is converted mainly to sulfur dioxide, SO_2, a very troublesome air pollutant. Because coal is a solid, recovery from its underground deposits is expensive and often dangerous. Furthermore, coal deposits are not always close to locations of high-energy use, so there are often substantial shipping costs.

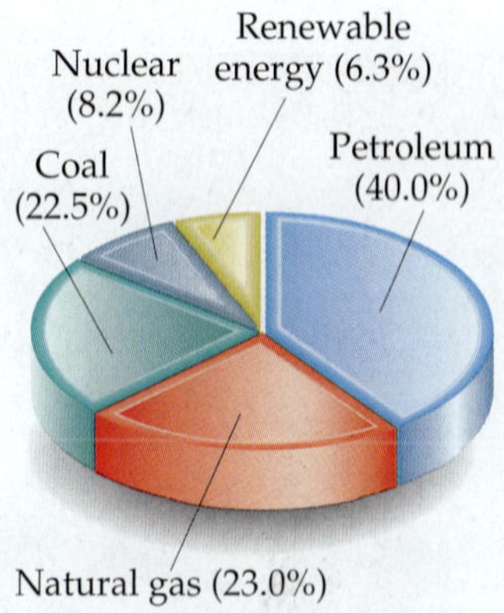

▲ **Figure 5.24 Sources of energy consumed in the United States.** In 2002 the United States consumed a total of 1.0×10^{17} kJ of energy.

One promising way to utilize coal reserves is to use them to produce a mixture of gaseous hydrocarbons called *syngas* (for "*syn*thesis *gas*"). In this process, called *coal gasification*, the coal typically is pulverized and treated with superheated steam. Sulfur-containing compounds, water, and carbon dioxide can be removed from the products, leading to a gaseous mixture of CH_4, H_2, and CO, all of which have high fuel values:

$$\text{Coal} + \text{steam} \xrightarrow{\text{conversion}} \text{complex mixture} \xrightarrow{\text{purification}} \text{mixture of } CH_4, H_2, CO \text{ (syngas)}$$

Because it is gaseous, syngas can be easily transported in pipelines. Additionally, because much of the sulfur in coal is removed during the gasification process, combustion of syngas causes less air pollution than burning coal. For these reasons, the economical conversion of coal and petroleum into "cleaner" fuels such as syngas and hydrogen is a very active area of current research in chemistry and engineering.

Other Energy Sources

Nuclear energy is energy that is released in either the splitting or the fusion (combining) of the nuclei of atoms. Nuclear power is currently used to produce about 22% of the electric power in the United States and comprises about 8% of the total U.S. energy production (Figure 5.24). Nuclear energy is, in principle, free of the polluting emissions that are a major problem in the

CHEMISTRY AT WORK | The Hybrid Car

The hybrid cars now entering the automobile marketplace nicely illustrate the convertibility of energy from one form to another. Hybrid cars run on either gasoline or electricity. The so-called "full hybrids" are cars capable of running on battery power alone at low speeds (Figure 5.25 ▼). In the full hybrid cars, an electrical engine is capable of driving the car at lower speeds. The "mild hybrid" cars are best described as electrically assisted gasoline engines.

▲ **Figure 5.25 A hybrid car.** The Ford Escape Hybrid is an example of a hybrid gasoline-electric vehicle.

Full hybrid cars are more efficient than the mild hybrid designs, but the full hybrids are more costly to produce and require more technological advances than the mild hybrid versions. The mild hybrids are likely to be more widely produced and sold within the next several years. Let's consider how they operate and some of the interesting thermodynamic considerations they incorporate.

Figure 5.26 ▶ shows a schematic diagram of the power system for a mild hybrid car. In addition to the 12-volt battery that is standard on conventional autos, the hybrid car carries a 42-volt battery pack. The electrical energy from this battery pack is not employed to move the car directly; an electrical engine capable of doing that, as in the full hybrids, requires from 150 to 300 volts. In the mild hybrid cars the added electrical source is employed to run various auxiliary devices that would otherwise be run off the gasoline engine, such as water pump, power steering, and air systems. To save on energy, when the hybrid car comes to a stop, the engine shuts off. It restarts automatically when the driver

generation of energy from fossil fuels. However, nuclear power plants produce radioactive waste products, and their use has therefore been fraught with controversy. We will discuss issues related to the production of nuclear energy in Chapter 21.

Fossil fuel and nuclear energy are *nonrenewable* sources of energy; the fuels used are limited resources that we are consuming at a much greater rate than they are regenerated. Eventually these fuels will be expended, although estimates vary greatly as to when this will occur. Because nonrenewable sources of energy will eventually be used up, there is a great deal of research into sources of **renewable energy**, energy sources that are essentially inexhaustible. Renewable energy sources include *solar energy* from the Sun, *wind energy* harnessed by windmills, *geothermal energy* from the heat stored in the mass of Earth, *hydroelectric energy* from flowing rivers, and *biomass energy* from crops, such as trees and corn, and from biological waste matter. Currently, renewable sources provide about 6.3% of the U.S. annual energy consumption, with hydroelectric (2.7%) and biomass (3.5%) sources as the major contributors.

Providing our future energy needs will most certainly depend on developing the technology to harness solar energy with greater efficiency. Solar energy is the world's largest energy source. On a clear day about 1 kJ of solar energy reaches each square meter of Earth's surface every second. The solar energy that falls on only 0.1% of U.S. land area is equivalent to all the energy that this nation currently uses. Harnessing this energy is difficult because it is dilute (it is distributed over a wide area) and it varies with time of day and weather conditions. The effective use of solar energy will depend on the presses the accelerator. This feature saves fuel that would otherwise be used to keep the engine idling at traffic lights and other stopping situations.

The idea is that the added electrical system will improve overall fuel efficiency of the car. The added battery, moreover, is not supposed to need recharging from an external power source. Where, then, can the improved fuel efficiency come from? Clearly, if the battery pack is to continue to operate auxiliary devices such as the water pump, it must be recharged. We can think of it this way: The source of the voltage that the battery develops is a chemical reaction. Recharging the battery thus represents a conversion of mechanical energy into chemical potential energy. The recharging occurs in part through the agency of an alternator, which runs off the engine and provides a recharging voltage. In the mild hybrid car, the braking system serves as an additional source of mechanical energy for recharging. When the brakes are applied in a conventional car, the car's kinetic energy is converted through the brake pads in the wheels into heat, so no useful work is done. In the hybrid car, some of the car's kinetic energy is used to recharge the battery when the brakes are applied. Thus, kinetic energy that would otherwise be dissipated as heat is partially converted into useful work. Overall, the mild hybrid cars are expected to yield 10–20% improvements in fuel economy as compared with similar conventional cars.

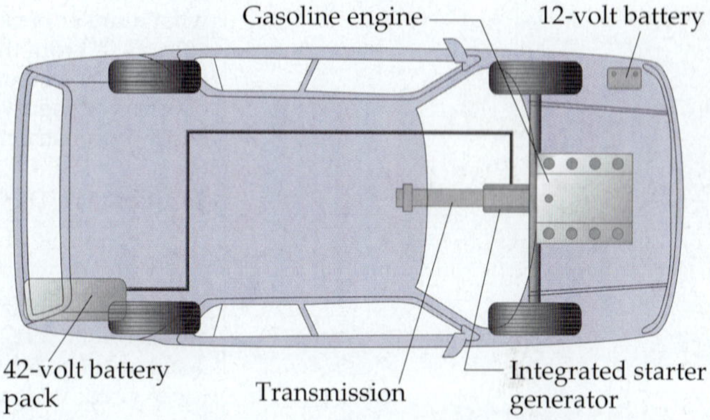

▲ **Figure 5.26 Schematic diagram of a mild hybrid car.** The 42-volt battery pack provides energy for operating several auxiliary functions. It is recharged from the engine and through the braking system.

development of some means of storing the collected energy for use at a later time. Any practical means for doing this will almost certainly involve use of an endothermic chemical process that can be later reversed to release heat. One such reaction is the following:

$$CH_4(g) + H_2O(g) + \text{heat} \rightleftharpoons CO(g) + 3\,H_2(g)$$

This reaction proceeds in the forward direction at high temperatures, which can be obtained in a solar furnace. The CO and H_2 formed in the reaction could then be stored and allowed to react later, with the heat released being put to useful work.

A survey taken about 20 years ago at Walt Disney's EPCOT Center revealed that nearly 30% of the visitors expected that solar energy would be the principal source of energy in the United States in the year 2000. The future of solar energy has proven to be a lot like the Sun itself: big and bright but farther away than it seems. Nevertheless, important progress has been made in recent years. Perhaps the most direct way to make use of the Sun's energy is to convert it directly into electricity by use of photovoltaic devices, sometimes called *solar cells*. The efficiencies of solar energy conversion by use of such devices have increased dramatically during the past few years as a result of intensive research efforts. Photovoltaics are vital to the generation of power for the space station. More significant for our Earth-bound concerns, the unit costs of solar panels have been steadily declining, even as their efficiencies have improved dramatically. Now that the year 2000 has come and gone, when do you think solar energy will become a principal source of energy in the United States?

SAMPLE INTEGRATIVE EXERCISE | Putting Concepts Together

Trinitroglycerin, $C_3H_5N_3O_9$ (usually referred to simply as nitroglycerin), has been widely used as an explosive. Alfred Nobel used it to make dynamite in 1866. Rather surprisingly, it also is used as a medication, to relieve angina (chest pains resulting from partially blocked arteries to the heart) by dilating the blood vessels. The enthalpy of decomposition at 1 atm pressure of trinitroglycerin to form nitrogen gas, carbon dioxide gas, liquid water, and oxygen gas at 25°C is −1541.4 kJ/mol. **(a)** Write a balanced chemical equation for the decomposition of trinitroglycerin. **(b)** Calculate the standard heat of formation of trinitroglycerin. **(c)** A standard dose of trinitroglycerin for relief of angina is 0.60 mg. Assuming that the sample is eventually oxidized in the body (not explosively, though!) to nitrogen gas, carbon dioxide gas, and liquid water, what number of calories is released? **(d)** One common form of trinitroglycerin melts at about 3°C. From this information and the formula for the substance, would you expect it to be a molecular or ionic compound? Explain. **(e)** Describe the various conversions of forms of energy when trinitroglycerin is used as an explosive to break rockfaces in highway construction.

Solution **(a)** The general form of the equation we must balance is

$$C_3H_5N_3O_9(l) \longrightarrow N_2(g) + CO_2(g) + H_2O(l) + O_2(g)$$

We go about balancing in the usual way. To obtain an even number of nitrogen atoms on the left, we multiply the formula for $C_3H_5N_3O_9(s)$ by 2. This then gives us 3 mol of $N_2(g)$, 6 mol of $CO_2(g)$ and 5 mol of $H_2O(l)$. Everything is balanced except for oxygen. We have an odd number of oxygen atoms on the right. We can balance the oxygen by adding $\frac{1}{2}$ mol of $O_2(g)$ on the right:

$$2\,C_3H_5N_3O_9(l) \longrightarrow 3\,N_2(g) + 6\,CO_2(g) + 5\,H_2O(l) + \tfrac{1}{2}\,O_2(g)$$

We multiply through by 2 to convert all coefficients to whole numbers:

$$4\,C_3H_5N_3O_9(l) \longrightarrow 6\,N_2(g) + 12\,CO_2(g) + 10\,H_2O(l) + O_2(g)$$

(At the temperature of the explosion, water is a gas. It is the rapid expansion of the gaseous products that creates the force of an explosion.)

(b) The heat of formation is the enthalpy change in the balanced chemical equation:

$$3\,C(s) + \tfrac{3}{2}\,N_2(g) + \tfrac{5}{2}\,H_2(g) + \tfrac{9}{2}\,O_2(g) \longrightarrow C_3H_5N_3O_9(l) \qquad \Delta H_f^\circ = ?$$

We can obtain the value of ΔH_f° by using the equation for the heat of decomposition of trinitroglycerin:

$$4\,C_3H_5N_3O_9(l) \longrightarrow 6\,N_2(g) + 12\,CO_2(g) + 10\,H_2O(l) + O_2(g)$$

The enthalpy change in this reaction is 4(−1541.4 kJ) = −6155.6 kJ. [We need to multiply by 4 because there are 4 mol of $C_3H_5N_3O_9(l)$ in the balanced equation.] This enthalpy change is given by the sum of the heats of formation of the products minus the heats of formation of the reactants, each multiplied by its coefficient in the balanced equation:

$$-6155.6\text{ kJ} = \{6\Delta H_f^\circ[N_2(g)] + 12\Delta H_f^\circ[CO_2(g)] + 10\Delta H_f^\circ[H_2O(l)] + \Delta H_f^\circ[O_2(g)]\} - 4\Delta H_f^\circ[C_3H_5N_3O_9(l)].$$

The ΔH_f° values for $N_2(g)$ and $O_2(g)$ are zero, by definition. We look up the values for $H_2O(l)$ and $CO_2(g)$ from Table 5.3 and find that

$$-6155.6\text{ kJ} = 12(-393.5\text{ kJ}) + 10(-285.8\text{ kJ}) - 4\Delta H_f^\circ(C_3H_5N_3O_9(l))$$

$$\Delta H_f^\circ(C_3H_5N_3O_9(l)) = -353.6\text{ kJ/mol}$$

(c) We know that on oxidation 1 mol of $C_3H_5N_3O_9(l)$ yields 1541.4 kJ. We need to calculate the number of moles of $C_3H_5N_3O_9(l)$ in 0.60 mg:

$$0.60 \times 10^{-3}\text{ g } C_3H_5N_3O_9\left(\frac{1\text{ mol } C_3H_5N_3O_9}{227\text{ g } C_3H_5N_3O_9}\right)\left(\frac{1541.4\text{ kJ}}{1\text{ mol } C_3H_5N_3O_9}\right) = 4.1 \times 10^{-3}\text{ kJ}$$

$$= 4.1\text{ J}$$

(d) Because trinitroglycerin melts below room temperature, we expect that it is a molecular compound. With few exceptions, ionic substances are generally hard, crystalline materials that melt at high temperatures. (Sections 2.5 and 2.6) Also, the molecular formula suggests that it is likely to be a molecular substance. All the elements of which it is composed are nonmetals.

(e) The energy stored in trinitroglycerin is chemical potential energy. When the substance reacts explosively, it forms substances such as carbon dioxide, water, and nitrogen gas, which are of lower potential energy. In the course of the chemical transformation, energy is released in the form of heat; the gaseous reaction products are very hot. This very high heat energy is transferred to the surroundings; the gases expand against the surroundings, which may be solid materials. Work is done in moving the solid materials and imparting kinetic energy to them. For example, a chunk of rock might be impelled upward. It has been given kinetic energy by transfer of energy from the hot, expanding gases. As the rock rises, its kinetic energy is transformed into potential energy. Eventually, it again acquires kinetic energy as it falls to Earth. When it strikes Earth, its kinetic energy is converted largely to thermal energy, though some work may be done on the surroundings as well.

SUMMARY AND KEY TERMS

Introduction and Section 5.1 **Thermodynamics** is the study of energy and its transformations. In this chapter we have focused on **thermochemistry**, the transformations of energy—especially heat—during chemical reactions.

An object can possess energy in two forms: **Kinetic energy** is the energy due to motion of the object, and **potential energy** is the energy that an object possesses by virtue of its position relative to other objects. An electron in motion near a proton, for example, has kinetic energy because of its motion and potential energy because of its electrostatic attraction to the proton. The SI unit of energy is the **joule** (J): $1\ J = 1\ kg\text{-}m^2/s^2$. Another common energy unit is the **calorie** (cal), which was originally defined as the quantity of energy necessary to increase the temperature of 1 g of water by 1°C: 1 cal = 4.184 J.

When we study thermodynamic properties, we define a specific amount of matter as the **system**. Everything outside the system is the **surroundings**. When we study a chemical reaction, the system is generally the reactants and products. A closed system can exchange energy, but not matter, with the surroundings. Energy can be transferred between the system and the surroundings as work or heat. **Work** is the energy expended to move an object against a **force**. **Heat** is the energy that is transferred from a hotter object to a colder one. **Energy** is the capacity to do work or to transfer heat.

Section 5.2 The **internal energy** of a system is the sum of all the kinetic and potential energies of its component parts. The internal energy of a system can change because of energy transferred between the system and the surroundings. The **first law of thermodynamics** states that the change in the internal energy of a system, ΔE, is the sum of the heat, q, transferred into or out of the system and the work, w, done on or by the system: $\Delta E = q + w$. Both q and w have a sign that indicates the direction of energy transfer. When heat is transferred from the surroundings to the system, $q > 0$. Likewise, when the surroundings do work on the system, $w > 0$. In an **endothermic** process the system absorbs heat from the surroundings; in an **exothermic** process the system releases heat to the surroundings.

The internal energy, E, is a **state function**. The value of any state function depends only on the state or condition of the system and not on the details of how it came to be in that state. The heat, q, and the work, w, are not state functions; their values depend on the particular way in which a system changes its state.

Sections 5.3 and 5.4 When a gas is produced or consumed in a chemical reaction occurring at constant pressure, the system may perform **pressure-volume (*P*-*V*) work** against the prevailing pressure. For this reason, we define a new state function called **enthalpy**, H, which is related to energy: $H = E + PV$. In systems where only pressure-volume work that is due to gases is involved, the change in the enthalpy of a system, ΔH, equals the heat gained or lost by the system at constant pressure, $\Delta H = q_P$. For an endothermic process, $\Delta H > 0$; for an exothermic process, $\Delta H < 0$.

Every substance has a characteristic enthalpy. In a chemical process, the **enthalpy of reaction** is the enthalpy of the products minus the enthalpy of the reactants: $\Delta H_{rxn} = H\,(\text{products}) - H\,(\text{reactants})$. Enthalpies of reaction follow some simple rules: (1) The enthalpy of reaction is proportional to the amount of reactant that reacts. (2) Reversing a reaction changes the sign of ΔH. (3) The enthalpy of reaction depends on the physical states of the reactants and products.

Section 5.5 The amount of heat transferred between the system and the surroundings is measured experimentally by **calorimetry**. A **calorimeter** measures the temperature change accompanying a process. The temperature change of a calorimeter depends on its **heat**

capacity, the amount of heat required to raise its temperature by 1 K. The heat capacity for one mole of a pure substance is called its **molar heat capacity**; for one gram of the substance, we use the term **specific heat**. Water has a very high specific heat, 4.18 J/g-K. The amount of heat, q, absorbed by a substance is the product of its specific heat (s), its mass, and its temperature change: $q = s \times m \times \Delta T$.

If a calorimetry experiment is carried out under a constant pressure, the heat transferred provides a direct measure of the enthalpy change of the reaction. Constant-volume calorimetry is carried out in a vessel of fixed volume called a **bomb calorimeter**. Bomb calorimeters are used to measure the heat evolved in combustion reactions. The heat transferred under constant-volume conditions is equal to ΔE. However, corrections can be applied to ΔE values to yield enthalpies of combustion.

Section 5.6 Because enthalpy is a state function, ΔH depends only on the initial and final states of the system. Thus, the enthalpy change of a process is the same whether the process is carried out in one step or in a series of steps. **Hess's law** states that if a reaction is carried out in a series of steps, ΔH for the reaction will be equal to the sum of the enthalpy changes for the steps. We can therefore calculate ΔH for any process, as long as we can write the process as a series of steps for which ΔH is known.

Section 5.7 The **enthalpy of formation**, ΔH_f, of a substance is the enthalpy change for the reaction in which the substance is formed from its constituent elements. The **standard enthalpy change** of a reaction, $\Delta H°$, is the enthalpy change when all reactants and products are at 1 atm pressure and a specific temperature, usually 298 K (25° C). Combining these ideas, the **standard enthalpy of formation**, ΔH_f°, of a substance is the change in enthalpy for the reaction that forms one mole of the substance from its elements in their most stable form with all reactants and products at 1 atm pressure and usually 298 K. For any element in its most stable state at 298 K and 1 atm pressure, $\Delta H_f^\circ = 0$. The standard enthalpy change for any reaction can be readily calculated from the standard enthalpies of formation of the reactants and products in the reaction:

$$\Delta H^\circ_{rxn} = \sum n\Delta H_f^\circ(\text{products}) - \sum m\Delta H_f^\circ(\text{reactants})$$

Section 5.8 The **fuel value** of a substance is the heat released when one gram of the substance is combusted. Different types of foods have different fuel values and differing abilities to be stored in the body. The most common fuels are hydrocarbons that are found as **fossil fuels**, such as **natural gas**, **petroleum**, and **coal**. Coal is the most abundant fossil fuel, but the sulfur present in most coals causes air pollution. Coal gasification is one possible way to use existing resources as sources of cleaner energy in the future. Sources of **renewable energy** include solar energy, wind energy, biomass, and hydroelectric energy. Nuclear power does not utilize fossil fuels but does create serious waste-disposal problems.

VISUALIZING CONCEPTS

5.1 Imagine a book that is falling from a shelf. At a particular moment during its fall, it has a kinetic energy of 13 J and a potential energy with respect to the floor of 72 J. How do its kinetic energy and its potential energy change as it continues to fall? What is its total kinetic energy at the instant just before it strikes the floor? [Section 5.1]

5.2 Consider the energy diagram shown below. **(a)** Does this diagram represent an increase or decrease in the internal energy of the system? **(b)** What sign is given to ΔE for this process? **(c)** If there is no work associated with the process, is it exothermic or endothermic? [Section 5.2]

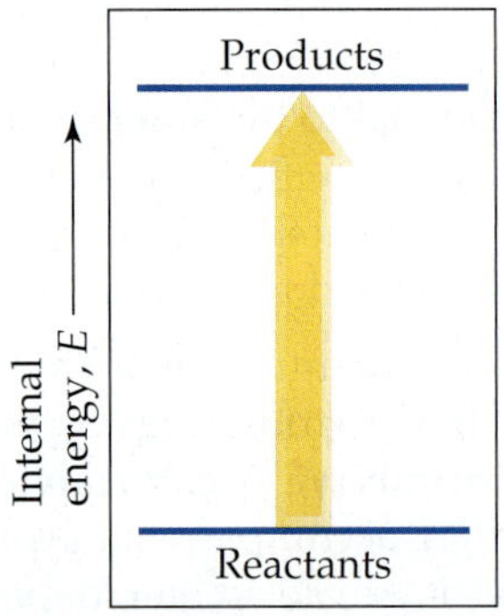

5.3 The contents of the closed box in each of the following illustrations represent a system, and the arrows show the changes to the system during some process. The lengths of the arrows represent the relative magnitudes of q and w. **(a)** Which of these processes is endothermic? **(b)** For which of these processes, if any, is $\Delta E < 0$? **(c)** For which process, if any, is there a net gain in internal energy? [Section 5.2]

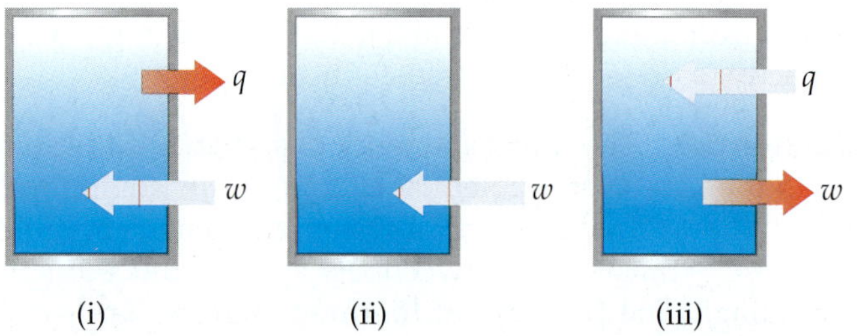

5.4 Imagine that you are climbing a mountain. **(a)** Is the distance you travel to the top a state function? Why or why not? **(b)** Is the change in elevation between your base camp and the peak a state function? Why or why not? [Section 5.2]

5.5 In the cylinder diagrammed below, a chemical process occurs at constant temperature and pressure. Is the sign of w indicated by this change positive or negative? If the process is endothermic, does the internal energy of the system within the cylinder increase or decrease during the change? Is ΔE positive or negative? [Sections 5.2 and 5.3]

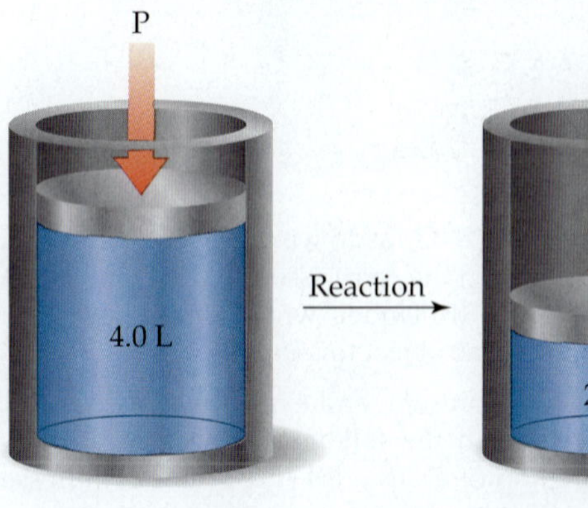

5.6 Imagine a container placed in a tub of water, as depicted in the following diagram. **(a)** If the contents of the container are the system and heat is able to flow through the container walls, what qualitative changes will occur in the temperatures of the system and in its surroundings? What is the sign of q associated with each change? From the system's perspective, is the process exothermic or endothermic? **(b)** If neither the volume nor the pressure of the system changes during the process, how is the change in internal energy related to the change in enthalpy? [Sections 5.2 and 5.3]

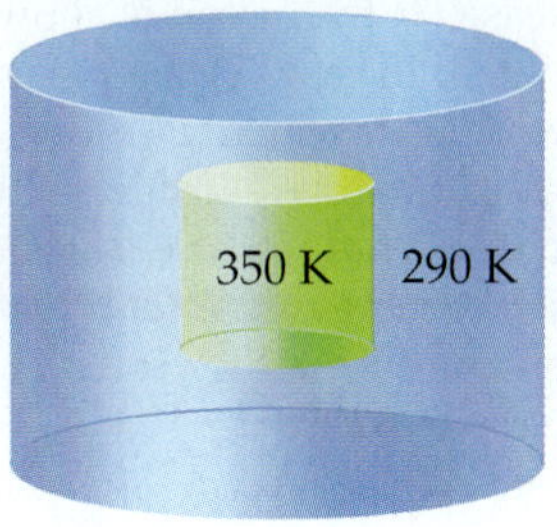

5.7 A gas-phase reaction was run in an apparatus designed to maintain a constant pressure. **(a)** Write a balanced chemical equation for the reaction depicted, and predict whether w is positive, negative, or zero. **(b)** Using data from Appendix C, determine ΔH for the formation of one mole of the product. Why is this enthalpy change called the enthalpy of formation of the involved product? [Sections 5.3 and 5.7]

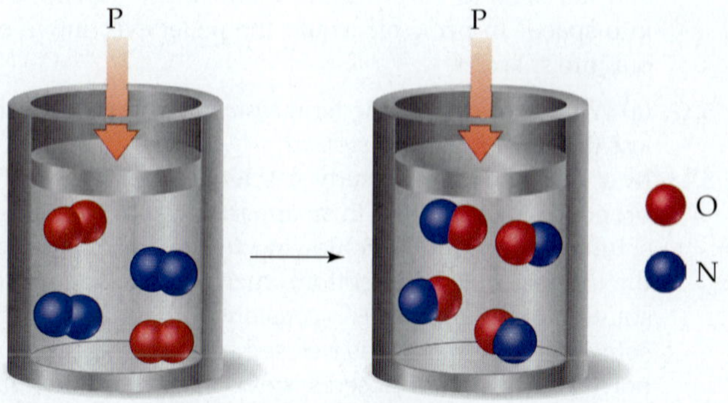

5.8 Consider the two diagrams below. **(a)** Based on (*i*), write an equation showing how ΔH_A is related to ΔH_B and ΔH_C. How do both diagram (*i*) and your equation relate to the fact that enthalpy is a state function? **(b)** Based on (*ii*), write an equation relating ΔH_Z to the other enthalpy changes in the diagram. How does diagram (*ii*) relate to Hess's law? [Section 5.6]

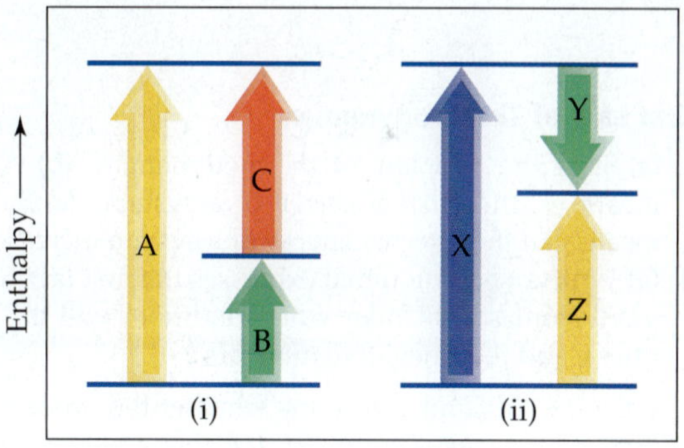

EXERCISES

The Nature of Energy

5.9 In what two ways can an object possess energy? How do these two ways differ from one another?

5.10 Suppose you toss a tennis ball upward. **(a)** Does the kinetic energy of the ball increase or decrease as it moves higher? **(b)** What happens to the potential energy of the ball as it moves higher? **(c)** If the same amount of energy were imparted to a ball the same size as a tennis ball, but of twice the mass, how high would it go in comparison to the tennis ball? Explain your answers.

5.11 **(a)** Calculate the kinetic energy in joules of a 45-g golf ball moving at 61 m/s. **(b)** Convert this energy to calories. **(c)** What happens to this energy when the ball lands in a sand trap?

5.12 **(a)** What is the kinetic energy in joules of a 950-lb motorcycle moving at 68 mph? **(b)** By what factor will the kinetic energy change if the speed of the motorcycle is decreased to 34 mph? **(c)** Where does the kinetic energy of the motorcycle go when the rider brakes to a stop?

5.13 In much engineering work it is common to use the British thermal unit (Btu). A Btu is the amount of heat required to raise the temperature of 1 lb of water by 1°F. Calculate the number of joules in a Btu.

5.14 A watt is a measure of power (the rate of energy change) equal to 1 J/s. **(a)** Calculate the number of joules in a kilowatt-hour. **(b)** An adult person radiates heat to the surroundings at about the same rate as a 100-watt electric incandescent lightbulb. What is the total amount of energy in kcal radiated to the surroundings by an adult in 24 hours?

5.15 Describe the source of the energy and the nature of the energy conversions involved when a 100-watt electric lightbulb radiates energy to its surroundings. Compare this with the energy source and energy conversions involved when an adult person radiates energy to the surroundings.

5.16 Suppose that a pellet is shot from an air gun straight up into the air. Why does the pellet eventually stop rising and fall back to Earth, rather than simply moving out into space? In principle, could the pellet ever move on out into space?

5.17 **(a)** What is meant by the term *system* in thermodynamics? **(b)** What is a *closed system*?

5.18 In a thermodynamic study a scientist focuses on the properties of a solution in an apparatus as illustrated. A solution is continuously flowing into the apparatus at the top and out at the bottom, such that the amount of solution in the apparatus is constant with time. **(a)** Is the solution in the apparatus a closed system? Why or why not? **(b)** If it is not a closed system, what could be done to make it a closed system?

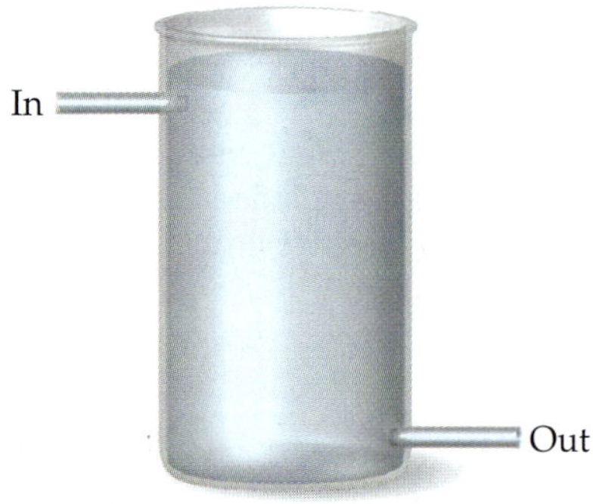

5.19 **(a)** What is work? **(b)** How do we determine the amount of work done, given the force associated with the work?

5.20 **(a)** What is heat? **(b)** Under what conditions is heat transferred from one object to another?

5.21 Identify the force present, and explain whether work is being performed in the following cases: **(a)** You lift a pencil off the top of a desk. **(b)** A spring is compressed to half its normal length.

5.22 Identify the force present, and explain whether work is done when **(a)** a positively charged particle moves in a circle at a fixed distance from a negatively charged particle; **(b)** an iron nail is pulled off a magnet.

The First Law of Thermodynamics

5.23 **(a)** State the first law of thermodynamics. **(b)** What is meant by the *internal energy* of a system? **(c)** By what means can the internal energy of a system increase?

5.24 **(a)** Write an equation that expresses the first law of thermodynamics. **(b)** Under what conditions will the quantities q and w be negative numbers?

5.25 Calculate ΔE, and determine whether the process is endothermic or exothermic for the following cases: **(a)** A system absorbs 85 kJ of heat from its surroundings while doing 29 kJ of work on the surroundings; **(b)** $q = 1.50$ kJ and $w = -657$ J; **(c)** the system releases 57.5 kJ of heat while doing 13.5 kJ of work on the surroundings.

5.26 For the following processes, calculate the change in internal energy of the system and determine whether the process is endothermic or exothermic: **(a)** A balloon is heated by adding 900 J of heat. It expands, doing 422 J of work on the atmosphere. **(b)** A 50-g sample of water is cooled from 30°C to 15°C, thereby losing approximately 3140 J of heat. **(c)** A chemical reaction releases 8.65 kJ of heat and does no work on the surroundings.

5.27 A gas is confined to a cylinder fitted with a piston and an electrical heater, as shown in the accompanying illustration. Suppose that current is supplied to the heater so that 100 J of energy is added. Consider two different situations. In case (1) the piston is allowed to move as the energy is added. In case (2) the piston is fixed so that it cannot move. **(a)** In which case does the gas have the higher temperature after addition of the electrical energy? Explain. **(b)** What can you say about the values of q and w in each of these cases? **(c)** What can you say about the relative values of ΔE for the system (the gas in the cylinder) in the two cases?

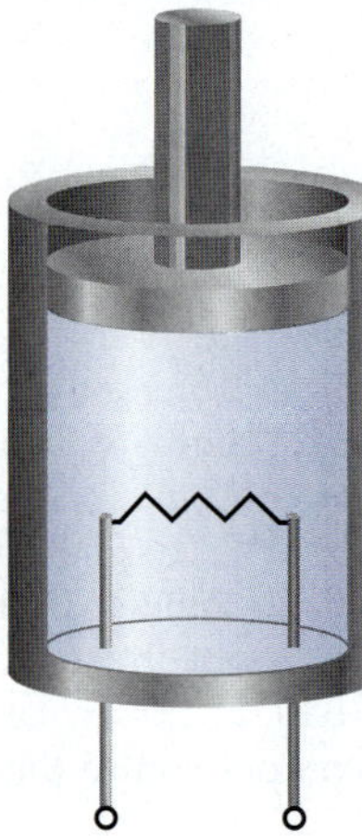

5.28 Consider a system consisting of two oppositely charged spheres hanging by strings and separated by a distance r_1, as shown in the accompanying illustration. Suppose they are separated to a larger distance r_2, by moving them apart along a track. **(a)** What change, if any, has occurred in the potential energy of the system? **(b)** What effect, if any, does this process have on the value of ΔE? **(c)** What can you say about q and w for this process?

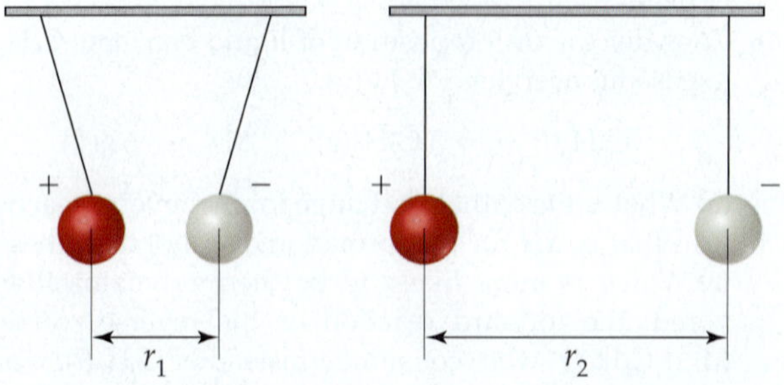

5.29 **(a)** What is meant by the term *state function?* **(b)** Give an example of a quantity that is a state function and one that is not. **(c)** Is work a state function? Why or why not?

5.30 Indicate which of the following is independent of the path by which a change occurs: **(a)** the change in potential energy when a book is transferred from table to shelf, **(b)** the heat evolved when a cube of sugar is oxidized to $CO_2(g)$ and $H_2O(g)$, **(c)** the work accomplished in burning a gallon of gasoline.

Enthalpy

5.31 **(a)** Why is the change in enthalpy usually easier to measure than the change in internal energy? **(b)** For a given process at constant pressure, ΔH is negative. Is the process endothermic or exothermic?

5.32 **(a)** Under what condition will the enthalpy change of a process equal the amount of heat transferred into or out of the system? **(a)** During a constant-pressure process the system absorbs heat from the surroundings. Does the enthalpy of the system increase or decrease during the process?

5.33 The complete combustion of acetic acid, $HC_2H_3O_2(l)$, to form $H_2O(l)$ and $CO_2(g)$ at constant pressure releases 871.7 kJ of heat per mole of $HC_2H_3O_2$. **(a)** Write a balanced thermochemical equation for this reaction. **(b)** Draw an enthalpy diagram for the reaction.

5.34 The decomposition of zinc carbonate, $ZnCO_3(s)$, into zinc oxide, $ZnO(s)$, and $CO_2(g)$ at constant pressure requires the addition of 71.5 kJ of heat per mole of $ZnCO_3$. **(a)** Write a balanced thermochemical equation for the reaction. **(b)** Draw an enthalpy diagram for the reaction.

5.35 Consider the following reaction, which occurs at room temperature and pressure:

$$2\,Cl(g) \longrightarrow Cl_2(g) \qquad \Delta H = -243.4\ \text{kJ}$$

Which has the higher enthalpy under these conditions, $2\,Cl(g)$ or $Cl_2(g)$?

5.36 Without referring to tables, predict which of the following has the higher enthalpy in each case: **(a)** 1 mol $CO_2(s)$ or 1 mol $CO_2(g)$ at the same temperature, **(b)** 2 mol of hydrogen atoms or 1 mol of H_2, **(c)** 1 mol $H_2(g)$ and 0.5 mol $O_2(g)$ at 25°C or 1 mol $H_2O(g)$ at 25°C, **(d)** 1 mol $N_2(g)$ at 100°C or 1 mol $N_2(g)$ at 300°C.

5.37 Consider the following reaction:

$$2\,Mg(s) + O_2(g) \longrightarrow 2\,MgO(s) \qquad \Delta H = -1204\ \text{kJ}$$

(a) Is this reaction exothermic or endothermic? **(b)** Calculate the amount of heat transferred when 2.4 g of $Mg(s)$ reacts at constant pressure. **(c)** How many grams of MgO are produced during an enthalpy change of −96.0 kJ? **(d)** How many kilojoules of heat are absorbed when 7.50 g of $MgO(s)$ is decomposed into $Mg(s)$ and $O_2(g)$ at constant pressure?

5.38 Consider the following reaction:

$$CH_3OH(g) \longrightarrow CO(g) + 2\,H_2(g) \qquad \Delta H = +90.7\ \text{kJ}$$

(a) Is heat absorbed or evolved in the course of this reaction? **(b)** Calculate the amount of heat transferred when 45.0 g of $CH_3OH(g)$ is decomposed by this reaction at constant pressure. **(c)** For a given sample of CH_3OH, the enthalpy change on reaction is 18.5 kJ. How many grams of hydrogen gas are produced? **(d)** What is the value of ΔH for the reverse of the previous reaction? How many kilojoules of heat are released when 27.0 g of $CO(g)$ reacts completely with $H_2(g)$ to form $CH_3OH(g)$ at constant pressure?

5.39 When solutions containing silver ions and chloride ions are mixed, silver chloride precipitates:

$$Ag^+(aq) + Cl^-(aq) \longrightarrow AgCl(s) \qquad \Delta H = -65.5\ \text{kJ}$$

(a) Calculate ΔH for formation of 0.200 mol of AgCl by this reaction. **(b)** Calculate ΔH for the formation of 2.50 g of AgCl. **(c)** Calculate ΔH when 0.150 mmol of AgCl dissolves in water.

5.40 At one time, a common means of forming small quantities of oxygen gas in the laboratory was to heat $KClO_3$:

$$2\,KClO_3(s) \longrightarrow 2\,KCl(s) + 3\,O_2(g) \qquad \Delta H = -89.4\ \text{kJ}$$

For this reaction, calculate ΔH for the formation of **(a)** 0.855 mol of O_2 and **(b)** 10.75 g of KCl. **(c)** The decomposition of $KClO_3$ proceeds spontaneously when it is heated. Do you think that the reverse reaction, the formation of $KClO_3$ from KCl and O_2, is likely to be feasible under ordinary conditions? Explain your answer.

5.41 You are given ΔH for a process that occurs at constant pressure. What additional information is needed to determine ΔE for the process?

5.42 Suppose that the gas-phase reaction $2\ NO(g) + O_2(g) \longrightarrow 2\ NO_2(g)$, were carried out in a constant-volume container at constant temperature. Would the measured heat change represent ΔH or ΔE? If there is a difference, which quantity is larger for this reaction? Explain.

5.43 A gas is confined to a cylinder under constant atmospheric pressure, as illustrated in Figure 5.3. When the gas undergoes a particular chemical reaction, it releases 79 kJ of heat to its surroundings and does 18 kJ of *P-V* work on its surroundings. What are the values of ΔH and ΔE for this process?

5.44 A gas is confined to a cylinder under constant atmospheric pressure, as illustrated in Figure 5.3. When 418 J of heat is added to the gas, it expands and does 107 J of work on the surroundings. What are the values of ΔH and ΔE for this process?

5.45 Consider the combustion of liquid methanol, $CH_3OH(l)$:

$$CH_3OH(l) + \tfrac{3}{2}\ O_2(g) \longrightarrow CO_2(g) + 2\ H_2O(l) \qquad \Delta H = -726.5\ \text{kJ}$$

(a) What is the enthalpy change for the reverse reaction? **(b)** Balance the forward reaction with whole-number coefficients. What is ΔH for the reaction represented by this equation? **(c)** Which is more likely to be thermodynamically favored, the forward reaction or the reverse reaction? **(d)** If the reaction were written to produce $H_2O(g)$ instead of $H_2O(l)$, would you expect the magnitude of ΔH to increase, decrease, or stay the same? Explain.

5.46 Consider the decomposition of liquid benzene, $C_6H_6(l)$, to gaseous acetylene, $C_2H_2(g)$:

$$C_6H_6(l) \longrightarrow 3\ C_2H_2(g) \qquad \Delta H = +630\ \text{kJ}$$

(a) What is the enthalpy change for the reverse reaction? **(b)** What is ΔH for the formation of 1 mol of acetylene? **(c)** Which is more likely to be thermodynamically favored, the forward reaction or the reverse reaction? **(d)** If $C_6H_6(g)$ were consumed instead of $C_6H_6(l)$, would you expect the magnitude of ΔH to increase, decrease, or stay the same? Explain.

Calorimetry

5.47 **(a)** What are the units of heat capacity? **(b)** What are the units of specific heat? **(c)** If you know the specific heat of copper, what additional information do you need to calculate the heat capacity of a particular piece of copper pipe?

5.48 Two solid objects, A and B, are placed in boiling water and allowed to come to temperature there. Each is then lifted out and placed in separate beakers containing 1000 g water at 10.0°C. Object A increases the water temperature by 3.50°C; B increases the water temperature by 2.60°C. **(a)** Which object has the larger heat capacity? **(b)** What can you say about the specific heats of A and B?

5.49 **(a)** What is the specific heat of liquid water? **(b)** What is the molar heat capacity of liquid water? **(c)** What is the heat capacity of 185 g of liquid water? **(d)** How many kJ of heat are needed to raise the temperature of 10.00 kg of liquid water from 24.6°C to 46.2°C?

5.50 **(a)** Which substance in Table 5.2 requires the smallest amount of energy to increase the temperature of 50.0 g of that substance by 10 K? **(b)** Calculate the energy needed for this temperature change.

5.51 The specific heat of iron metal is 0.450 J/g-K. How many J of heat are necessary to raise the temperature of a 1.05-kg block of iron from 25.0°C to 88.5°C?

5.52 The specific heat of ethylene glycol is 2.42 J/g-K. How many J of heat are needed to raise the temperature of 62.0 g of ethylene glycol from 15.2°C to 40.8°C?

5.53 When a 9.55-g sample of solid sodium hydroxide dissolves in 100.0 g of water in a coffee-cup calorimeter (Figure 5.18), the temperature rises from 23.6°C to 47.4°C. Calculate ΔH (in kJ/mol NaOH) for the solution process

$$NaOH(s) \longrightarrow Na^+(aq) + OH^-(aq)$$

Assume that the specific heat of the solution is the same as that of pure water.

5.54 **(a)** When a 3.88-g sample of solid ammonium nitrate dissolves in 60.0 g of water in a coffee-cup calorimeter (Figure 5.18), the temperature drops from 23.0°C to 18.4°C. Calculate ΔH (in kJ/mol NH_4NO_3) for the solution process

$$NH_4NO_3(s) \longrightarrow NH_4^+(aq) + NO_3^-(aq)$$

Assume that the specific heat of the solution is the same as that of pure water. **(b)** Is this process endothermic or exothermic?

5.55 A 2.200-g sample of quinone ($C_6H_4O_2$) is burned in a bomb calorimeter whose total heat capacity is 7.854 kJ/°C. The temperature of the calorimeter increases from 23.44°C to 30.57°C. What is the heat of combustion per gram of quinone? Per mole of quinone?

5.56 A 1.800-g sample of phenol (C_6H_5OH) was burned in a bomb calorimeter whose total heat capacity is 11.66 kJ/°C. The temperature of the calorimeter plus contents increased from 21.36°C to 26.37°C. **(a)** Write a balanced chemical equation for the bomb calorimeter reaction. **(b)** What is the heat of combustion per gram of phenol? Per mole of phenol?

5.57 Under constant-volume conditions the heat of combustion of glucose ($C_6H_{12}O_6$) is 15.57 kJ/g. A 2.500-g sample of glucose is burned in a bomb calorimeter. The temperature of the calorimeter increased from 20.55°C to 23.25°C. **(a)** What is the total heat capacity of the calorimeter? **(b)** If the size of the glucose sample had been exactly twice as large, what would the temperature change of the calorimeter have been?

5.58 Under constant-volume conditions the heat of combustion of benzoic acid ($HC_7H_5O_2$) is 26.38 kJ/g. A 1.640-g sample of benzoic acid is burned in a bomb calorimeter. The temperature of the calorimeter increases from 22.25°C to 27.20°C. **(a)** What is the total heat capacity of the calorimeter? **(b)** A 1.320-g sample of a new organic substance is combusted in the same calorimeter. The temperature of the calorimeter increases from 22.14°C to 26.82°C. What is the heat of combustion per gram of the new substance? **(c)** Suppose that in changing samples, a portion of the water in the calorimeter were lost. In what way, if any, would this change the heat capacity of the calorimeter?

Hess's Law

5.59 What is the connection between Hess's law and the fact that H is a state function?

5.60 Consider the following hypothetical reactions:

$$A \longrightarrow B \qquad \Delta H = +30 \text{ kJ}$$
$$B \longrightarrow C \qquad \Delta H = +60 \text{ kJ}$$

(a) Use Hess's law to calculate the enthalpy change for the reaction $A \longrightarrow C$. **(b)** Construct an enthalpy diagram for substances A, B, and C, and show how Hess's law applies.

5.61 Calculate the enthalpy change for the reaction

$$P_4O_6(s) + 2\,O_2(g) \longrightarrow P_4O_{10}(s)$$

given the following enthalpies of reaction:

$$P_4(s) + 3\,O_2(g) \longrightarrow P_4O_6(s) \qquad \Delta H = -1640.1 \text{ kJ}$$
$$P_4(s) + 5\,O_2(g) \longrightarrow P_4O_{10}(s) \qquad \Delta H = -2940.1 \text{ kJ}$$

5.62 From the enthalpies of reaction

$$2\,H_2(g) + O_2(g) \longrightarrow 2\,H_2O(g) \qquad \Delta H = -483.6 \text{ kJ}$$
$$3\,O_2(g) \longrightarrow 2\,O_3(g) \qquad \Delta H = +284.6 \text{ kJ}$$

calculate the heat of the reaction

$$3\,H_2(g) + O_3(g) \longrightarrow 3\,H_2O(g)$$

5.63 From the enthalpies of reaction

$$H_2(g) + F_2(g) \longrightarrow 2\,HF(g) \qquad \Delta H = -537 \text{ kJ}$$
$$C(s) + 2\,F_2(g) \longrightarrow CF_4(g) \qquad \Delta H = -680 \text{ kJ}$$
$$2\,C(s) + 2\,H_2(g) \longrightarrow C_2H_4(g) \qquad \Delta H = +52.3 \text{ kJ}$$

calculate ΔH for the reaction of ethylene with F_2:

$$C_2H_4(g) + 6\,F_2(g) \longrightarrow 2\,CF_4(g) + 4\,HF(g)$$

5.64 Given the data

$$N_2(g) + O_2(g) \longrightarrow 2\,NO(g) \qquad \Delta H = +180.7 \text{ kJ}$$
$$2\,NO(g) + O_2(g) \longrightarrow 2\,NO_2(g) \qquad \Delta H = -113.1 \text{ kJ}$$
$$2\,N_2O(g) \longrightarrow 2\,N_2(g) + O_2(g) \qquad \Delta H = -163.2 \text{ kJ}$$

use Hess's law to calculate ΔH for the reaction

$$N_2O(g) + NO_2(g) \longrightarrow 3\,NO(g)$$

Enthalpies of Formation

5.65 **(a)** What is meant by the term *standard conditions*, with reference to enthalpy changes? **(b)** What is meant by the term *enthalpy of formation*? **(c)** What is meant by the term *standard enthalpy of formation*?

5.66 **(a)** Why are tables of standard enthalpies of formation so useful? **(b)** What is the value of the standard enthalpy of formation of an element in its most stable form? **(c)** Write the chemical equation for the reaction whose enthalpy change is the standard enthalpy of formation of glucose, $C_6H_{12}O_6(s)$, $\Delta H_f^\circ[C_6H_{12}O_6]$.

5.67 For each of the following compounds, write a balanced thermochemical equation depicting the formation of one mole of the compound from its elements in their standard states and use Appendix C to obtain the value of ΔH_f°: **(a)** $NH_3(g)$, **(b)** $SO_2(g)$, **(c)** $RbClO_3(s)$, **(d)** $NH_4NO_3(s)$.

5.68 Write balanced equations that describe the formation of the following compounds from their elements in their standard states, and use Appendix C to obtain the values of their standard enthalpies of formation: **(a)** $HBr(g)$, **(b)** $AgNO_3(s)$, **(c)** $Hg_2Cl_2(s)$, **(d)** $C_2H_5OH(l)$.

5.69 The following is known as the thermite reaction [Figure 5.8(b)]:

$$2\,Al(s) + Fe_2O_3(s) \longrightarrow Al_2O_3(s) + 2\,Fe(s)$$

This highly exothermic reaction is used for welding massive units, such as propellers for large ships. Using enthalpies of formation in Appendix C, calculate ΔH° for this reaction.

5.70 Many cigarette lighters contain liquid butane, $C_4H_{10}(l)$. Using enthalpies of formation, calculate the quantity of heat produced when 1.0 g of butane is completely combusted in air.

5.71 Using values from Appendix C, calculate the standard enthalpy change for each of the following reactions:
(a) $2\,SO_2(g) + O_2(g) \longrightarrow 2\,SO_3(g)$
(b) $Mg(OH)_2(s) \longrightarrow MgO(s) + H_2O(l)$
(c) $CH_4(g) + 4\,Cl_2(g) \longrightarrow CCl_4(l) + 4\,HCl(g)$
(d) $SiCl_4(l) + 2\,H_2O(l) \longrightarrow SiO_2(s) + 4\,HCl(g)$

5.72 Using values from Appendix C, calculate the value of ΔH° for each of the following reactions:
(a) $N_2O_4(g) + 4\,H_2(g) \longrightarrow N_2(g) + 4\,H_2O(g)$
(b) $2\,KOH(s) + CO_2(g) \longrightarrow K_2CO_3(s) + H_2O(g)$
(c) $SO_2(g) + 2\,H_2S(g) \longrightarrow \frac{3}{8}\,S_8(s) + 2\,H_2O(g)$
(d) $Fe_2O_3(s) + 6\,HCl(g) \longrightarrow 2\,FeCl_3(s) + 3\,H_2O(g)$

5.73 Complete combustion of 1 mol of acetone (C_3H_6O) liberates 1790 kJ:

$$C_3H_6O(l) + 4\,O_2(g) \longrightarrow 3\,CO_2(g) + 3\,H_2O(l)$$
$$\Delta H^\circ = -1790 \text{ kJ}$$

Using this information together with data from Appendix C, calculate the enthalpy of formation of acetone.

5.74 Calcium carbide (CaC_2) reacts with water to form acetylene (C_2H_2) and $Ca(OH)_2$. From the following enthalpy of reaction data and data in Appendix C, calculate ΔH_f° for $CaC_2(s)$:

$$CaC_2(s) + 2\,H_2O(l) \longrightarrow Ca(OH)_2(s) + C_2H_2(g) \qquad \Delta H^\circ = -127.2\text{ kJ}$$

5.75 Calculate the standard enthalpy of formation of solid $Mg(OH)_2$, given the following data:

$$2\,Mg(s) + O_2(g) \longrightarrow 2\,MgO(s) \qquad \Delta H^\circ = -1203.6\text{ kJ}$$
$$Mg(OH)_2(s) \longrightarrow MgO(s) + H_2O(l) \qquad \Delta H^\circ = +37.1\text{ kJ}$$
$$2\,H_2(g) + O_2(g) \longrightarrow 2\,H_2O(l) \qquad \Delta H^\circ = -571.7\text{ kJ}$$

5.76 **(a)** Calculate the standard enthalpy of formation of gaseous diborane (B_2H_6) using the following thermochemical information:

$$4\,B(s) + 3\,O_2(g) \longrightarrow 2\,B_2O_3(s) \qquad \Delta H^\circ = -2509.1\text{ kJ}$$
$$2\,H_2(g) + O_2(g) \longrightarrow 2\,H_2O(l) \qquad \Delta H^\circ = -571.7\text{ kJ}$$
$$B_2H_6(g) + 3\,O_2(g) \longrightarrow B_2O_3(s) + 3\,H_2O(l) \qquad \Delta H^\circ = -2147.5\text{ kJ}$$

(b) Pentaborane (B_5H_9) is another boron hydride. What experiment or experiments would you need to perform to yield the data necessary to calculate the heat of formation of $B_5H_9(l)$? Explain by writing out and summing any applicable chemical reactions.

5.77 Gasoline is composed primarily of hydrocarbons, including many with eight carbon atoms, called *octanes*. One of the cleanest-burning octanes is a compound called 2,3,4-trimethylpentane, which has the following structural formula:

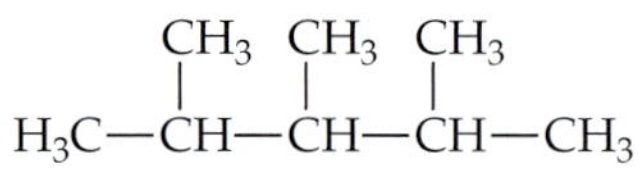

The complete combustion of one mole of this compound to $CO_2(g)$ and $H_2O(g)$ leads to $\Delta H^\circ = -5069$ kJ. **(a)** Write a balanced equation for the combustion of 1 mol of $C_8H_{18}(l)$. **(b)** Write a balanced equation for the formation of $C_8H_{18}(l)$ from its elements. **(c)** By using the information in this problem and data in Table 5.3, calculate ΔH_f° for 2,3,4-trimethylpentane.

5.78 Naphthalene ($C_{10}H_8$) is a solid aromatic compound often sold as mothballs. The complete combustion of this substance to yield $CO_2(g)$ and $H_2O(l)$ at 25°C yields 5154 kJ/mol. **(a)** Write balanced equations for the formation of naphthalene from the elements and for its combustion. **(b)** Calculate the standard enthalpy of formation of naphthalene.

Foods and Fuels

5.79 **(a)** What is meant by the term *fuel value*? **(b)** Which is a greater source of energy as food, 5 g of fat or 9 g of carbohydrate?

5.80 **(a)** Why are fats well suited for energy storage in the human body? **(b)** A particular chip snack food is composed of 12% protein, 14% fat, and the rest carbohydrate. What percentage of the calorie content of this food is fat? **(c)** How many grams of protein provide the same fuel value as 25 g of fat?

5.81 A serving of condensed cream of mushroom soup contains 7 g fat, 9 g carbohydrate, and 1 g protein. Estimate the number of Calories in a serving.

5.82 A pound of plain M&M® candies contains 96 g fat, 320 g carbohydrate, and 21 g protein. What is the fuel value in kJ in a 42-g (about 1.5 oz) serving? How many Calories does it provide?

5.83 The heat of combustion of fructose, $C_6H_{12}O_6$, is −2812 kJ/mol. If a fresh golden delicious apple weighing 4.23 oz (120 g) contains 16.0 g of fructose, what caloric content does the fructose contribute to the apple?

5.84 The heat of combustion of ethanol, $C_2H_5OH(l)$, is −1367 kJ/mol. A batch of Sauvignon Blanc wine contains 10.6% ethanol by mass. Assuming the density of the wine to be 1.0 g/mL, what caloric content does the alcohol (ethanol) in a 6-oz glass of wine (177 mL) have?

5.85 The standard enthalpies of formation of gaseous propyne (C_3H_4), propylene (C_3H_6), and propane (C_3H_8) are +185.4, +20.4, and −103.8 kJ/mol, respectively. **(a)** Calculate the heat evolved per mole on combustion of each substance to yield $CO_2(g)$ and $H_2O(g)$. **(b)** Calculate the heat evolved on combustion of 1 kg of each substance. **(c)** Which is the most efficient fuel in terms of heat evolved per unit mass?

5.86 It is interesting to compare the "fuel value" of a hydrocarbon in a world where fluorine rather than oxygen is the combustion agent. The enthalpy of formation of $CF_4(g)$ is −679.9 kJ/mol. Which of the following two reactions is the more exothermic?

$$CH_4(g) + 2\,O_2(g) \longrightarrow CO_2(g) + 2\,H_2O(g)$$
$$CH_4(g) + 4\,F_2(g) \longrightarrow CF_4(g) + 4\,HF(g)$$

Additional Exercises

5.87 At 20°C (approximately room temperature) the average velocity of N_2 molecules in air is 1050 mph. **(a)** What is the average speed in m/s? **(b)** What is the kinetic energy (in J) of an N_2 molecule moving at this speed? **(c)** What is the total kinetic energy of 1 mol of N_2 molecules moving at this speed?

5.88 Suppose an Olympic diver who weighs 52.0 kg executes a straight dive from a 10-m platform. At the apex of the dive, the diver is 10.8 m above the surface of the water. **(a)** What is the potential energy of the diver at the apex of the dive, relative to the surface of the water? **(b)** Assuming that all the potential energy of the diver is converted into kinetic energy at the surface of the water, at what speed in m/s will the diver enter the water? **(c)** Does the diver do work on entering the water? Explain.

5.89 When a mole of Dry Ice, $CO_2(s)$, is converted to $CO_2(g)$ at atmospheric pressure and $-78°C$, the heat absorbed by the system exceeds the increase in internal energy of the CO_2. Why is this so? What happens to the remaining energy?

5.90 The air bags that provide protection in autos in the event of an accident expand as a result of a rapid chemical reaction. From the viewpoint of the chemical reactants as the system, what do you expect for the signs of q and w in this process?

[5.91] An aluminum can of a soft drink is placed in a freezer. Later, the can is found to be split open and its contents frozen. Work was done on the can in splitting it open. Where did the energy for this work come from?

[5.92] A sample of gas is contained in a cylinder-and-piston arrangement. It undergoes the change in state shown in the drawing. **(a)** Assume first that the cylinder and piston are perfect thermal insulators that do not allow heat to be transferred. What is the value of q for the state change? What is the sign of w for the state change? What can be said about ΔE for the state change? **(b)** Now assume that the cylinder and piston are made up of a thermal conductor such as a metal. During the state change, the cylinder gets warmer to the touch. What is the sign of q for the state change in this case? Describe the difference in the state of the system at the end of the process in the two cases. What can you say about the relative values of ΔE?

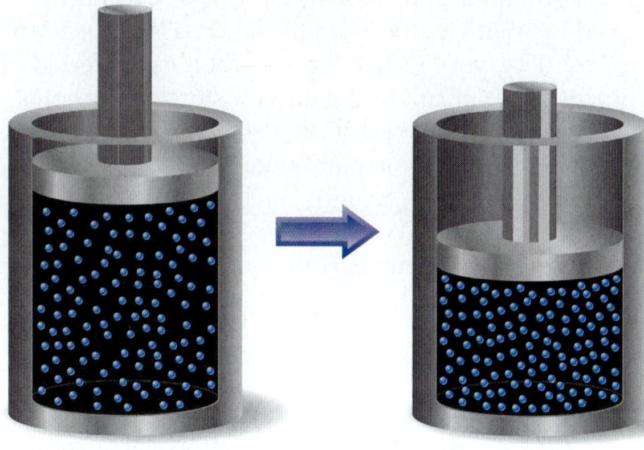

[5.93] Limestone stalactites and stalagmites are formed in caves by the following reaction:

$$Ca^{2+}(aq) + 2\,HCO_3^-(aq) \longrightarrow CaCO_3(s) + CO_2(g) + H_2O(l)$$

If 1 mol of $CaCO_3$ forms at 298 K under 1 atm pressure, the reaction performs 2.47 kJ of P-V work, pushing back the atmosphere as the gaseous CO_2 forms. At the same time, 38.95 kJ of heat is absorbed from the environment. What are the values of ΔH and of ΔE for this reaction?

[5.94] Consider the systems shown in Figure 5.10. In one case the battery becomes completely discharged by running the current through a heater, and in the other by running a fan. Both processes occur at constant pressure. In both cases the change in state of the system is the same: The battery goes from being fully charged to being fully discharged. Yet in one case the heat evolved is large, and in the other it is small. Is the enthalpy change the same in the two cases? If not, how can enthalpy be considered a state function? If it is, what can you say about the relationship between enthalpy change and q in this case, as compared with others that we have considered?

5.95 A house is designed to have passive solar energy features. Brickwork is to be incorporated into the interior of the house to act as a heat absorber. Each brick weighs approximately 1.8 kg. The specific heat of the brick is 0.85 J/g-K. How many bricks must be incorporated into the interior of the house to provide the same total heat capacity as 1.7×10^3 gal of water?

[5.96] A coffee-cup calorimeter of the type shown in Figure 5.18 contains 150.0 g of water at 25.1°C. A 121.0-g block of copper metal is heated to 100.4°C by putting it in a beaker of boiling water. The specific heat of Cu(s) is 0.385 J/g-K. The Cu is added to the calorimeter, and after a time the contents of the cup reach a constant temperature of 30.1°C. **(a)** Determine the amount of heat, in J, lost by the copper block. **(b)** Determine the amount of heat gained by the water. The specific heat of water is 4.18 J/g-K. **(c)** The difference between your answers for (a) and (b) is due to heat loss through the Styrofoam® cups and the heat necessary to raise the temperature of the inner wall of the apparatus. The heat capacity of the calorimeter is the amount of heat necessary to raise the temperature of the apparatus (the cups and the stopper) by 1 K. Calculate the heat capacity of the calorimeter in J/K. **(d)** What would be the final temperature of the system if all the heat lost by the copper block were absorbed by the water in the calorimeter?

[5.97] **(a)** When a 0.235-g sample of benzoic acid is combusted in a bomb calorimeter, the temperature rises 1.642°C. When a 0.265-g sample of caffeine, $C_8H_{10}O_2N_4$, is burned, the temperature rises 1.525°C. Using the value 26.38 kJ/g for the heat of combustion of benzoic acid, calculate the heat of combustion per mole of caffeine at constant volume. **(b)** Assuming that there is an uncertainty of 0.002°C in each temperature reading and that the masses of samples are measured to 0.001 g, what is the estimated uncertainty in the value calculated for the heat of combustion per mole of caffeine?

5.98 How many grams of methane [$CH_4(g)$] must be combusted to heat 1.00 kg of water from 25.0°C to 90.0°C, assuming $H_2O(l)$ as a product and 100% efficiency in heat transfer?

5.99 Burning methane in oxygen can produce three different carbon-containing products: soot (very fine particles of graphite), CO(g), and $CO_2(g)$. **(a)** Write three balanced equations for the reaction of methane gas with oxygen to produce these three products. In each case assume that $H_2O(l)$ is the only other product. **(b)** Determine the standard enthalpies for the reactions in part (a). **(c)** Why, when the oxygen supply is adequate, is $CO_2(g)$ the predominant carbon-containing product of the combustion of methane?

5.100 From the following data for three prospective fuels, calculate which could provide the most energy per unit volume:

Fuel	Density at 20°C (g/cm^3)	Molar Enthalpy of Combustion (kJ/mol)
Nitroethane, $C_2H_5NO_2(l)$	1.052	−1368
Ethanol, $C_2H_5OH(l)$	0.789	−1367
Methylhydrazine, $CH_6N_2(l)$	0.874	−1305

5.101 The hydrocarbons acetylene (C_2H_2) and benzene (C_6H_6) have the same empirical formula. Benzene is an "aromatic" hydrocarbon, one that is unusually stable because of its structure. **(a)** By using the data in Appendix C, determine the standard enthalpy change for the reaction $3\,C_2H_2(g) \longrightarrow C_6H_6(l)$. **(b)** Which has greater enthalpy, 3 mol of acetylene gas or 1 mol of liquid benzene? **(c)** Determine the fuel value in kJ/g for acetylene and benzene.

[5.102] Ammonia (NH_3) boils at −33°C; at this temperature it has a density of 0.81 g/cm^3. The enthalpy of formation of $NH_3(g)$ is −46.2 kJ/mol, and the enthalpy of vaporization of $NH_3(l)$ is 23.2 kJ/mol. Calculate the enthalpy change when 1 L of liquid NH_3 is burned in air to give $N_2(g)$ and $H_2O(g)$. How does this compare with ΔH for the complete combustion of 1 L of liquid methanol, $CH_3OH(l)$? For $CH_3OH(l)$, the density at 25°C is 0.792 g/cm^3, and ΔH_f° equals −239 kJ/mol.

[5.103] Three common hydrocarbons that contain four carbons are listed here, along with their standard enthalpies of formation:

Hydrocarbon	Formula	ΔH_f° (kJ/mol)
1,3-Butadiene	$C_4H_6(g)$	111.9
1-Butene	$C_4H_8(g)$	1.2
n-Butane	$C_4H_{10}(g)$	−124.7

(a) For each of these substances, calculate the molar enthalpy of combustion to $CO_2(g)$ and $H_2O(l)$. **(b)** Calculate the fuel value in kJ/g for each of these compounds. **(c)** For each hydrocarbon, determine the percentage of hydrogen by mass. **(d)** By comparing your answers for parts (b) and (c), propose a relationship between hydrogen content and fuel value in hydrocarbons.

5.104 The two common sugars, glucose ($C_6H_{12}O_6$) and sucrose ($C_{12}H_{22}O_{11}$), are both carbohydrates. Their standard enthalpies of formation are given in Table 5.3. Using these data, **(a)** calculate the molar enthalpy of combustion to $CO_2(g)$ and $H_2O(l)$ for the two sugars; **(b)** calculate the enthalpy of combustion per gram of each sugar; **(c)** determine how your answers to part (b) compare to the average fuel value of carbohydrates discussed in Section 5.8.

5.105 A 200-lb man decides to add to his exercise routine by walking up three flights of stairs (45 ft) 20 times per day. He figures that the work required to increase his potential energy in this way will permit him to eat an extra order of French fries, at 245 Cal, without adding to his weight. Is he correct in this assumption?

5.106 The sun supplies about 1.0 kilowatt of energy for each square meter of surface area (1.0 kW/m^2, where a watt = 1 J/s). Plants produce the equivalent of about 0.20 g of sucrose ($C_{12}H_{22}O_{11}$) per hour per square meter. Assuming that the sucrose is produced as follows, calculate the percentage of sunlight used to produce sucrose.

$$12\,CO_2(g) + 11\,H_2O(l) \longrightarrow C_{12}H_{22}O_{11} + 12\,O_2(g) \qquad \Delta H = 5645\text{ kJ}$$

[5.107] It is estimated that the net amount of carbon dioxide fixed by photosynthesis on the landmass of the Earth is 5.5×10^{16} g/yr of CO_2. All this carbon is converted into glucose. **(a)** Calculate the energy stored by photosynthesis on land per year in kJ. **(b)** Calculate the average rate of conversion of solar energy into plant energy in MW (1 W = 1 J/s). A large nuclear power plant produces about 10^3 MW. The energy of how many such nuclear power plants is equivalent to the solar energy conversion?

Integrative Exercises

5.108 Consider the combustion of a single molecule of $CH_4(g)$ forming $H_2O(l)$ as a product. **(a)** How much energy, in J, is produced during this reaction? **(b)** A typical X-ray photon has an energy of 8 keV. How does the energy of combustion compare to the energy of the X-ray photon?

5.109 Consider the dissolving of NaCl in water, illustrated in Figure 4.3. Let's say that the system consists of 0.1 mol NaCl and 1 L of water. Considering that the NaCl readily dissolves in the water and that the ions are strongly stabilized by the water molecules, as shown in the figure, is it safe to conclude that the dissolution of NaCl in water results in a lower enthalpy for the system? Explain your response. What experimental evidence would you examine to test this question?

5.110 Consider the following unbalanced oxidation-reduction reactions in aqueous solution:

$$Ag^+(aq) + Li(s) \longrightarrow Ag(s) + Li^+(aq)$$
$$Fe(s) + Na^+(aq) \longrightarrow Fe^{2+}(aq) + Na(s)$$
$$K(s) + H_2O(l) \longrightarrow KOH(aq) + H_2(g)$$

(a) Balance each of the reactions. **(b)** By using data in Appendix C, calculate ΔH° for each of the reactions. **(c)** Based on the values you obtain for ΔH°, which of the reactions would you expect to be favorable? Which would you expect to be unfavorable? **(d)** Use the activity series to predict which of these reactions should occur. ∞ (Section 4.4) Are these results in accord with your conclusion in part (c) of this problem?

[5.111] Consider the following acid-neutralization reactions involving the strong base $NaOH(aq)$:

$$HNO_3(aq) + NaOH(aq) \longrightarrow NaNO_3(aq) + H_2O(l)$$

$$HCl(aq) + NaOH(aq) \longrightarrow NaCl(aq) + H_2O(l)$$

$$NH_4^+(aq) + NaOH(aq) \longrightarrow NH_3(aq) + Na^+(aq) + H_2O(l)$$

(a) By using data in Appendix C, calculate $\Delta H°$ for each of the reactions. **(b)** As we saw in Section 4.3, nitric acid and hydrochloric acid are strong acids. Write net ionic equations for the neutralization of these acids. **(c)** Compare the values of $\Delta H°$ for the first two reactions. What can you conclude? **(d)** In the third equation $NH_4^+(aq)$ is acting as an acid. Based on the value of $\Delta H°$ for this reaction, do you think it is a strong or a weak acid? Explain.

5.112 Consider two solutions, the first being 50.0 mL of 1.00 *M* $CuSO_4$ and the other 50.0 mL of 2.00 *M* KOH. When the two solutions are mixed in a constant-pressure calorimeter, a precipitate forms and the temperature of the mixture rises from 21.5°C to 27.7°C. **(a)** Before mixing, how many grams of Cu are present in the solution of $CuSO_4$? **(b)** Predict the identity of the precipitate in the reaction. **(c)** Write complete and net ionic equations for the reaction that occurs when the two solutions are mixed. **(d)** From the calorimetric data, calculate ΔH for the reaction that occurs on mixing. Assume that the calorimeter absorbs only a negligible quantity of heat, that the total volume of the solution is 100.0 mL, and that the specific heat and density of the solution after mixing are the same as that of pure water.

5.113 The metathesis reaction between $AgNO_3(aq)$ and $NaCl(aq)$ proceeds as follows:

$$AgNO_3(aq) + NaCl(aq) \longrightarrow NaNO_3(aq) + AgCl(s)$$

(a) By using Appendix C, calculate $\Delta H°$ for the net ionic equation of this reaction. **(b)** What would you expect for the value of $\Delta H°$ of the overall molecular equation compared to that for the net ionic equation? Explain. **(c)** Use the results from (a) and (b) along with data in Appendix C to determine the value of $\Delta H_f°$ for $AgNO_3(aq)$.

[5.114] A sample of a hydrocarbon is combusted completely in $O_2(g)$ to produce 21.83 g $CO_2(g)$, 4.47 g $H_2O(g)$, and 311 kJ of heat. **(a)** What is the mass of the hydrocarbon sample that was combusted? **(b)** What is the empirical formula of the hydrocarbon? **(c)** Calculate the value of $\Delta H_f°$ per empirical-formula unit of the hydrocarbon. **(d)** Do you think that the hydrocarbon is one of those listed in Appendix C? Explain your answer.

5.115 The methane molecule, CH_4, has the geometry shown in Figure 2.21. Imagine a hypothetical process in which the methane molecule is "expanded," by simultaneously extending all four C—H bonds to infinity. We then have the process

$$CH_4(g) \longrightarrow C(g) + 4\,H(g)$$

(a) Compare this process with the reverse of the reaction that represents the standard enthalpy of formation. **(b)** Calculate the enthalpy change in each case. Which is the more endothermic process? What accounts for the difference in $\Delta H°$ values? **(c)** Suppose that 3.45 g $CH_4(g)$ is reacted with 1.22 g $F_2(g)$, forming $CF_4(g)$ and $HF(g)$ as sole products. What is the limiting reagent in this reaction? Assuming that the reaction occurs at constant pressure, what amount of heat is evolved?

eMEDIA EXERCISES

These exercises make use of the interactive objects available online in OneKey or the Companion Website, and on your Accelerator CD. Access to these resources comes in your MediaPak.

5.116 The **Enthalpy of Solution** activity (*5.4*) allows you to dissolve five different compounds in water. For each compound, tell whether the process is endothermic or exothermic.

5.117 Using the calculated ΔH from Exercise 5.53, determine how much sodium hydroxide must be dissolved in 100.0 g of water in order for the temperature to rise by 25.0°C. Use the **Enthalpy of Solution** activity (*5.4*) to verify your answer.

5.118 **(a)** Which combination (what compound, mass of compound, and mass of water) in the **Enthalpy of Solution** activity (*5.4*) produces the largest temperature change? **(b)** Suppose a weighed quantity of a sparingly soluble salt were used for a calorimetry experiment similar to those in the simulation. After measuring the temperature change, you discover that not all of the salt has dissolved. Discuss the errors that would arise in the calculated value of the molar enthalpy of solution as a result of this complication.

5.119 **(a)** Using the heat of combustion given in Exercise 5.58, calculate the temperature increase for the combustion of a 450-mg sample of benzoic acid in a calorimeter with a heat capacity of 420 J/°C that contains 500 g of water. Use the **Calorimetry** activity (*5.5*) to check your prediction. **(b)** Perform calorimetry experiments using the other compounds available in the simulation. For each compound (excluding nitroglycerin), determine the molar heat of combustion. **(c)** Which compound has the highest heat of combustion per mole? **(d)** Which compound has the highest heat of combustion per *gram*?

5.120 **(a)** Using the heat of combustion for sucrose determined in the previous exercise, calculate the amount of heat that would be evolved by the combustion of 228 mg of sucrose. **(b)** What temperature change would be observed if this combustion were carried out with 735 grams of water in the calorimeter? Use the **Calorimetry** activity (*5.5*) to verify your answer.

5.121 Carry out a series of experiments with the **Calorimetry** activity (*5.5*), using 500 grams of water in the calorimeter. Vary the amount of sucrose, using values of 50 mg,

150 mg, 250 mg, and 450 mg. **(a)** Graph your data with mg sucrose on the *x*-axis and temperature change on the *y*-axis. **(b)** Is the temperature change proportional to the mass of sucrose combusted?

5.122 Repeat the experiment of the previous exercise with a constant 255 mg of sucrose, but vary the amount of water, using values of 600 g, 700 g, 800 g, and 900 g. **(a)** Again, graph your data with grams of water on the *x*-axis and temperature change on the *y*-axis. **(b)** Is the temperature change proportional to the mass of water in the calorimeter? If your answer to part (b) is different from your answer to part (b) of the previous exercise, explain why.

5.123 The enthalpy of combustion of nitroglycerin is −1541 kJ/mol. **(a)** Using this and the data from experiments conducted in the **Calorimetry** activity (*5.5*), determine the molar mass of nitroglycerin. **(b)** Explain how your answer to part (a) would be affected by the use of a poorly insulated calorimeter.

5.124 Compare the **Thermite** and **Formation of Water** movies (*5.7*). **(a)** Use the information provided to calculate the ΔH° for the formation of water, and compare it to the ΔH° calculated for the thermite reaction in Exercise 5.69. **(b)** Comment on the apparent relationship between the nature of a reaction and the magnitude of its ΔH°.

VCL eLABORATORY EXERCISES

These exercises allow students to apply the concepts and skills for this chapter in the simulated laboratory of Virtual ChemLab. Worksheets for each of these exercises are found in the Virtual ChemLab Workbook in your MediaPak.

5.125 ***(VCL 5-1) Endothermic Versus Exothermic***
In various chemical processes, such as reactions and the dissolving of salts, heat is either absorbed or given off. We call these events either an endothermic (heat in) or exothermic (heat out) process. It is usual to detect these heat events by measuring the temperature change associated with the process. In this problem you will dissolve several salts in water and measure the resulting temperature change and then make deductions about the nature of the process.

5.126 ***(VCL 5-2) Heat of Solution: NH_4NO_3***
Have you ever used one of those "instant cold packs" that looks like a plastic bag filled with liquid? If you hit the bag and shake it up, it gets extremely cold, but why does it do that? The liquid inside the cold pack is water, and in the water is another plastic bag or tube containing NH_4NO_3. When you hit the cold pack, it breaks the tube so that the water mixes with the NH_4NO_3. The dissolving of a salt, such as NH_4NO_3, in water is called dissolution, and the heat associated with the dissolving process is called the enthalpy of solution. In this problem you will take a sample of NH_4NO_3, dissolve it in water, and from measuring the change in temperature, calculate the enthalpy of solution for NH_4NO_3.

5.127 ***(VCL 5-3) Specific Heat of Al***
On a sunny day the water in a swimming pool may warm up a degree or two while the concrete around the pool may become too hot to walk on with bare feet. This may seem strange since the water and concrete are being heated by the same source—the Sun. This evidence suggests it takes more heat to raise the temperature of some substances than others, which is true. The amount of heat required to raise the temperature of 1 g of a substance by 1°C is called the *specific heat capacity* or *specific heat* of that substance. Water, for instance, has a specific heat of 4.18 J/K-g. This value is high in comparison with the specific heats for other materials, such as concrete or metals. In this experiment you will use a simple calorimeter and your knowledge of the specific heat of water to measure the specific heat of aluminum (Al).

5.128 ***(VCL 5-4) Specific Heat of Pb***
In this problem you will repeat the previous specific heat experiment except measure the specific heat of Pb.

5.129 ***(VCL 5-5) Heat of Combustion: Chicken Fat***
The heat of combustion (ΔH_{comb}) is the heat of reaction for the complete burning (reacting with O_2) of one mole of a substance to form CO_2 and H_2O. Calorimetry experiments that measure the heat of combustion can be performed at constant volume using a device called a bomb calorimeter. In a bomb calorimeter a sample is burned in a constant-volume chamber in the presence of oxygen at high pressure. The heat that is released warms the water surrounding the chamber. By measuring the temperature increase of the water, it is possible to calculate the quantity of heat released during the combustion reaction. In this assignment you will calculate the heat of combustion of chicken fat.

5.130 ***(VCL 5-6) Heat of Combustion: Sugar***
In this assignment you will measure the heat of combustion of sugar as outlined in the heat of combustion problem on chicken fat.

5.131 ***(VCL 5-7) Heat of Combustion: TNT***
In this assignment you will measure the heat of combustion of 2,4,6-trinitrotoluene (TNT) as outlined in the heat of combustion problem on chicken fat.

5.132 ***(VCL 5-8) Heat of Formation: Ethanol***
The heat of formation is the heat of reaction for the formation of a compound from the elements. The heat of formation can be determined by measuring the heat of combustion for the compound and then using Hess's law to convert the heat of combustion to a heat of formation. Calorimetry experiments that measure the heat of combustion can be performed at constant volume, using a device called a bomb calorimeter. In a bomb calorimeter a sample is burned in a constant-volume chamber in the presence of oxygen at high pressure. The heat that is released warms the water surrounding the chamber. By measuring the temperature increase of the water, it is possible to calculate the quantity of heat

released during the combustion reaction. In this assignment you will first measure the heat of combustion of ethanol (ethyl alcohol, C_2H_5OH) and convert the heat of combustion to a heat of formation.

5.133 ***(VCL 5-9) Heat of Formation: Aspirin***
In this assignment you will measure the heat of combustion of aspirin and then calculate the heat of formation as outlined in the heat of formation problem for ethanol.

5.134 ***(VCL 5-10) Heat of Reaction: NaOH(aq) + HCl(aq)***
Energy that is either absorbed or released in a chemical reaction is known as the enthalpy of reaction (ΔH_{rxn}). An *endothermic* reaction absorbs energy, and the enthalpy of reaction is positive. An *exothermic* reaction gives off energy, and the enthalpy of reaction is negative. Most chemical reactions are exothermic. In this problem you will measure the amount of heat released when aqueous solutions of NaOH and HCl are mixed in a neutralization reaction to form water and a salt, and calculate the heat of reaction.

5.135 ***(VCL 5-11) Heat of Reaction: MgO(s) + HCl(aq)***
In this assignment you will measure the heat of reaction for $MgO(s) + HCl(aq)$ as outlined in the heat of reaction problem for $NaOH(aq)$ and $HCl(aq)$.

5.136 ***(VCL 5-12) Hess's Law***
In this experiment you will measure the amount of heat released in these three related exothermic reactions:

$$NaOH(s) \longrightarrow Na^+(aq) + OH^-(aq) \quad \Delta H_1$$

$$NaOH(s) + H^+(aq) + Cl^-(aq) \longrightarrow H_2O + Na^+(aq) + Cl^-(aq) \quad \Delta H_2$$

$$Na^+(aq) + OH^-(aq) + H^+(aq) + Cl^-(aq) \longrightarrow H_2O + Na^+(aq) + Cl^-(aq) \quad \Delta H_3$$

After determining the heats of reaction (ΔH_1, ΔH_2 and ΔH_3), you will then analyze your data and verify Hess's law, or $\Delta H_1 + \Delta H_3 = \Delta H_2$.

CHAPTER

6

WHAT'S AHEAD

- Light (radiant energy, or *electromagnetic radiation*) has wavelike properties and so is characterized by *wavelength*, *frequency*, and *speed*. *(Section 6.1)*
- Studies of the radiation given off by hot objects and of the way in which light striking a metal surface can free electrons indicate that electromagnetic radiation also has particle-like properties and can be described in terms of *photons*. *(Section 6.2)*
- The fact that atoms give off characteristic colors of light (*line spectra*) provides clues about how electrons are arranged in atoms, leading to two important ideas: Electrons exist only in certain energy levels around a nucleus, and energy is involved in moving an electron from one level to another. *(Section 6.3)*
- Because matter also has wavelike properties, it is impossible to determine simultaneously the exact position and exact motion of an electron in an atom (*Heisenberg's uncertainty principle*). *(Section 6.4)*
- The arrangement of electrons in atoms is described by quantum mechanics in terms of *atomic orbitals*. *(Sections 6.5, 6.6, and 6.7)*
- Knowing the energies of orbitals as well as some fundamental characteristics of electrons described by *Hund's rule* allows us to determine how electrons are distributed among various orbitals in an atom (*electron configurations*). *(Section 6.8)*
- The electron configuration of an atom is related to the location of the element in the periodic table. *(Section 6.9)*

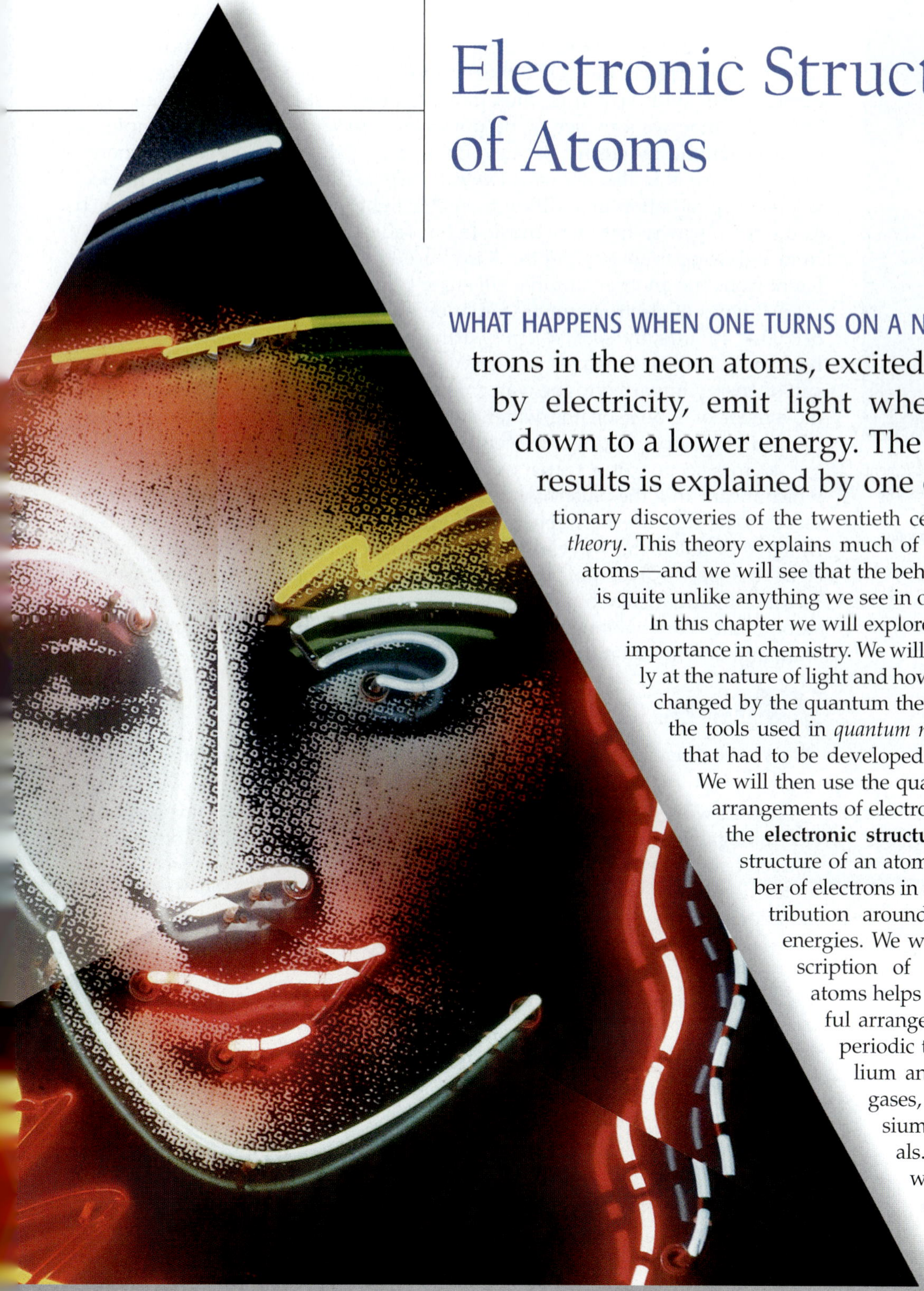

THE GLASS TUBES OF NEON LIGHTS contain various gases that can be excited by electricity. Light is produced when electrically excited atoms return to their lowest-energy states.

Electronic Structure of Atoms

WHAT HAPPENS WHEN ONE TURNS ON A NEON LIGHT? The electrons in the neon atoms, excited to a higher energy by electricity, emit light when they drop back down to a lower energy. The pleasing glow that results is explained by one of the most revolutionary discoveries of the twentieth century, namely, the *quantum theory*. This theory explains much of the behavior of electrons in atoms—and we will see that the behavior of subatomic electrons is quite unlike anything we see in our macroscopic world.

In this chapter we will explore the quantum theory and its importance in chemistry. We will begin by looking more closely at the nature of light and how our description of light was changed by the quantum theory. We will explore some of the tools used in *quantum mechanics*, the "new" physics that had to be developed to describe atoms correctly. We will then use the quantum theory to describe the arrangements of electrons in atoms, or what we call the **electronic structure** of atoms. The electronic structure of an atom refers not only to the number of electrons in an atom but also to their distribution around the nucleus and to their energies. We will see that the quantum description of the electronic structure of atoms helps us to understand the beautiful arrangement of the elements in the periodic table—why, for example, helium and neon are both unreactive gases, whereas sodium and potassium are both soft, reactive metals. In the chapters that follow, we will see how the concepts of quantum theory are used to explain trends in the periodic table and the formation of bonds between atoms.

6.1 | The Wave Nature of Light

Much of our present understanding of the electronic structure of atoms has come from analysis of the light either emitted or absorbed by substances. To understand electronic structure, therefore, we must first learn more about light. The light that we can see with our eyes, *visible light*, is an example of **electromagnetic radiation**. Because electromagnetic radiation carries energy through space, it is also known as *radiant energy*. There are many types of electromagnetic radiation in addition to visible light. These different forms—such as the radio waves that carry music to our radios, the infrared radiation (heat) from a glowing fireplace, and the X rays used by a dentist—may *seem* very different from one another, but they all share certain fundamental characteristics.

All types of electromagnetic radiation move through a vacuum at a speed of 3.00×10^8 m/s, the *speed of light*. Furthermore, all have wavelike characteristics similar to those of waves that move through water. Water waves are the result of energy imparted to the water, perhaps by the dropping of a stone or the movement of a boat on the water surface (Figure 6.1 ◀). This energy is expressed as the up-and-down movements of the water.

▲ **Figure 6.1 Water waves.** The movement of the boat through the water forms waves. The regular variation of the peaks and troughs enables us to sense the motion, or *propagation*, of the waves.

A cross section of a water wave (Figure 6.2 ◀) shows that it is periodic, which means that the pattern of peaks and troughs repeats itself at regular intervals. The distance between two adjacent peaks (or between two adjacent troughs) is called the **wavelength**. The number of complete wavelengths, or *cycles*, that pass a given point each second is the **frequency** of the wave. We can measure the frequency of a water wave by counting the number of times per second that a cork bobbing on the water moves through a complete cycle of upward and downward motion.

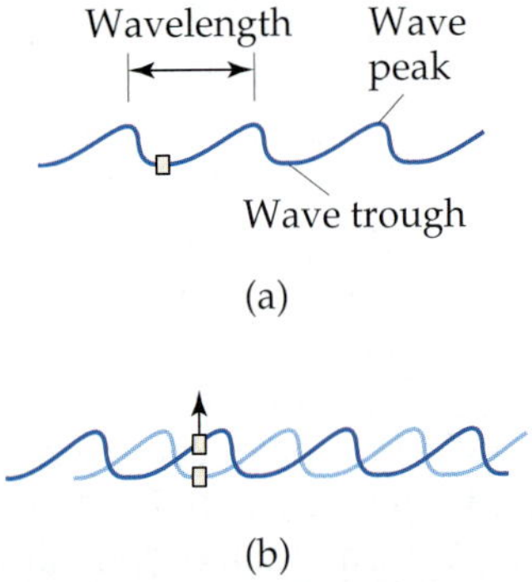

▲ **Figure 6.2 Characteristics of water waves.** (a) The distance between corresponding points on each wave is called the *wavelength*. In this drawing, the two corresponding points are two peaks, but they could be any other two corresponding points, such as two adjacent troughs. (b) The number of times per second that the cork bobs up and down is called the *frequency* of the wave.

GIVE IT SOME THOUGHT

What is the difference between visible light and electromagnetic radiation?

Just as with water waves, we can assign a frequency and wavelength to electromagnetic waves, as illustrated in Figure 6.3 ▼. These and all other wave characteristics of electromagnetic radiation are due to the periodic oscillations in the intensities of the electric and magnetic fields associated with the radiation.

▼ **Figure 6.3 Characteristics of electromagnetic waves.** Radiant energy has wave characteristics; it consists of electromagnetic waves. Notice that the shorter the wavelength, λ, the higher the frequency, ν. The wavelength in (b) is half as long as that in (a), and the frequency of the wave in (b) is therefore twice as great as the frequency in (a). The *amplitude* of the wave relates to the intensity of the radiation. It is the maximum extent of the oscillation of the wave. In these diagrams amplitude is measured as the vertical distance from the midline of the wave to its peak. The waves in (a) and (b) have the same amplitude. The wave in (c) has the same frequency as that in (b), but its amplitude is lower.

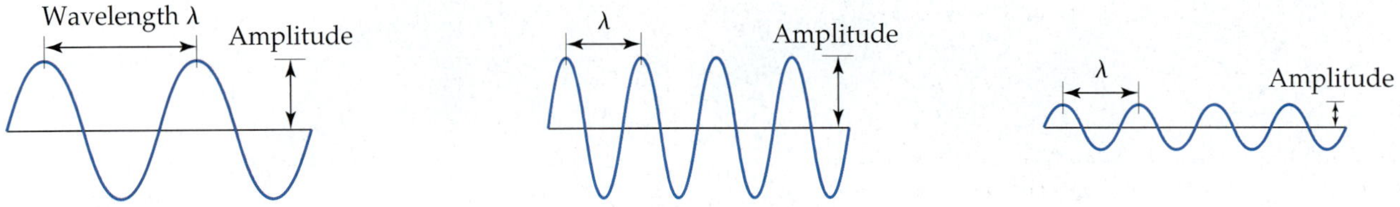

(a) Two complete cycles of wavelength λ

(b) Wavelength half of that in (a); frequency twice as great as in (a)

(c) Same frequency as (b), smaller amplitude

The speed of water waves can vary depending on how they are created—for example, the waves produced by a speedboat travel faster than those produced by a rowboat. In contrast, all electromagnetic radiation moves at the same speed, namely the speed of light. As a result, the wavelength and frequency of electromagnetic radiation are always related in a straightforward way. If the wavelength is long, there will be fewer cycles of the wave passing a point per second; thus, the frequency will be low. Conversely, for a wave to have a high frequency, the distance between the peaks of the wave must be small (short wavelength). This inverse relationship between the frequency and the wavelength of electromagnetic radiation can be expressed by the equation

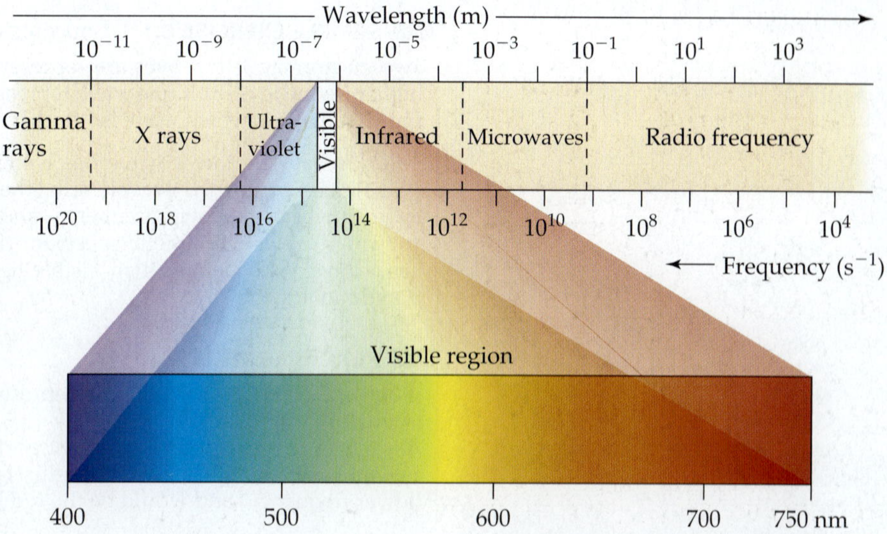

▲ **Figure 6.4 The electromagnetic spectrum.** Wavelengths in the spectrum range from very short gamma rays to very long radio waves. Notice that the color of visible light can be expressed quantitatively by wavelength.

$$\nu\lambda = c \qquad [6.1]$$

where ν (nu) is the frequency, λ (lambda) is the wavelength, and c is the speed of light.

Why do different forms of electromagnetic radiation have different properties? Their differences are due to their different wavelengths, which are expressed in units of length. Figure 6.4 ▲ shows the various types of electromagnetic radiation arranged in order of increasing wavelength, a display called the *electromagnetic spectrum*. Notice that the wavelengths span an enormous range. The wavelengths of gamma rays are similar to the diameters of atomic nuclei, whereas the wavelengths of radio waves can be longer than a football field. Notice also that visible light, which corresponds to wavelengths of about 400 to 700 nm (4×10^{-9} m to 7×10^{-9} m), is an extremely small portion of the electromagnetic spectrum. We can see visible light because of the chemical reactions it triggers in our eyes. The unit of length normally chosen to express wavelength depends on the type of radiation, as shown in Table 6.1 ▼.

ACTIVITY
Electromagnetic Spectrum

Frequency is expressed in cycles per second, a unit also called a *hertz* (Hz). Because it is understood that cycles are involved, the units of frequency are normally given simply as "per second," which is denoted by s^{-1} or /s. For example, a frequency of 820 kilohertz (kHz), a typical frequency for an AM radio station, could be written either as 820,000 s^{-1} or as 820,000/s.

TABLE 6.1 Common Wavelength Units for Electromagnetic Radiation

Unit	Symbol	Length (m)	Type of Radiation
Angstrom	Å	10^{-10}	X ray
Nanometer	nm	10^{-9}	Ultraviolet, visible
Micrometer	μm	10^{-6}	Infrared
Millimeter	mm	10^{-3}	Infrared
Centimeter	cm	10^{-2}	Microwave
Meter	m	1	TV, radio

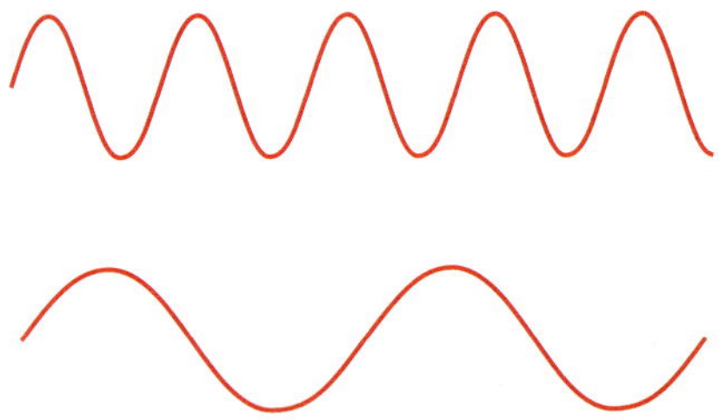

SAMPLE EXERCISE 6.1 | Concepts of Wavelength and Frequency

Two electromagnetic waves are represented in the margin. **(a)** Which wave has the higher frequency? **(b)** If one wave represents visible light and the other represents infrared radiation, which wave is which?

Solution (a) The lower wave has a longer wavelength (greater distance between peaks). The longer the wavelength, the lower the frequency ($\nu = c/\lambda$). Thus, the lower wave has the lower frequency, and the upper one has the higher frequency.

(b) The electromagnetic spectrum (Figure 6.4) indicates that infrared radiation has a longer wavelength than visible light. Thus, the lower wave would be the infrared radiation.

PRACTICE EXERCISE

If one of the waves in the margin represents blue light and the other red light, which is which?

Answer: The expanded visible-light portion of Figure 6.4 tells you that red light has a longer wavelength than blue light. The lower wave has the longer wavelength (lower frequency) and would be the red light.

SAMPLE EXERCISE 6.2 | Calculating Frequency from Wavelength

The yellow light given off by a sodium vapor lamp used for public lighting has a wavelength of 589 nm. What is the frequency of this radiation?

Solution

Analyze: We are given the wavelength, λ, of the radiation and asked to calculate its frequency, ν.

Plan: The relationship between the wavelength (which is given) and the frequency (which is the unknown) is given by Equation 6.1. We can solve this equation for ν and then use the values of λ and c to obtain a numerical answer. (The speed of light, c, is a fundamental constant whose value is given in the text or in the table of fundamental constants on the back inside cover.)

Solve: Solving Equation 6.1 for frequency gives $\nu = c/\lambda$. When we insert the values for c and λ, we note that the units of length in these two quantities are different. We can convert the wavelength from nanometers to meters, so the units cancel:

$$\nu = \frac{c}{\lambda} = \left(\frac{3.00 \times 10^8 \text{ m/s}}{589 \text{ nm}}\right)\left(\frac{1 \text{ nm}}{10^{-9} \text{ m}}\right) = 5.09 \times 10^{14} \text{ s}^{-1}$$

Check: The high frequency is reasonable because of the short wavelength. The units are proper because frequency has units of "per second," or s^{-1}.

PRACTICE EXERCISE

(a) A laser used in eye surgery to fuse detached retinas produces radiation with a wavelength of 640.0 nm. Calculate the frequency of this radiation. **(b)** An FM radio station broadcasts electromagnetic radiation at a frequency of 103.4 MHz (megahertz; $\text{MHz} = 10^6 \text{ s}^{-1}$). Calculate the wavelength of this radiation.

Answers: **(a)** $4.688 \times 10^{14} \text{ s}^{-1}$, **(b)** 2.901 m

GIVE IT SOME THOUGHT

Human skin is not penetrated by visible light but is penetrated by X rays. Which travels faster, visible light or X rays?

6.2 | Quantized Energy and Photons

Although the wave model of light explains many aspects of its behavior, there are several phenomena this model can't explain. Three of these are particularly pertinent to our understanding of how electromagnetic radiation and atoms interact: (1) the emission of light from hot objects (referred to as *blackbody radiation* because the objects studied appear black before heating), (2) the emission of electrons from metal surfaces on which light shines (the *photoelectric effect*), and (3) the emission of light from electronically excited gas atoms (*emission spectra*). We examine the first two here and the third in Section 6.3.

Hot Objects and the Quantization of Energy

When solids are heated, they emit radiation, as seen in the red glow of an electric stove burner and the bright white light of a tungsten lightbulb. The wavelength distribution of the radiation depends on temperature, a red-hot object being cooler than a white-hot one (Figure 6.5 ▶). In the late 1800s a number of physicists were studying this phenomenon, trying to understand the relationship between the temperature and the intensity and wavelengths of the emitted radiation. The prevailing laws of physics could not account for the observations.

▲ **Figure 6.5 Color as a function of temperature.** The color and intensity of the light emitted by a hot object depend on the temperature of the object. The temperature is highest at the center of this pour of molten steel. As a result, the light emitted from the center is most intense and of shortest wavelength.

VCL Blackbody Radiation

In 1900 a German physicist named Max Planck (1858–1947) solved the problem by making a daring assumption: He assumed that energy can be either released or absorbed by atoms only in discrete "chunks" of some minimum size. Planck gave the name **quantum** (meaning "fixed amount") to the smallest quantity of energy that can be emitted or absorbed as electromagnetic radiation. He proposed that the energy, E, of a single quantum equals a constant times the frequency of the radiation:

$$E = h\nu \qquad [6.2]$$

The constant h is called **Planck's constant** and has a value of 6.626×10^{-34} joule-seconds (J-s). According to Planck's theory, matter is allowed to emit and absorb energy only in whole-number multiples of $h\nu$, such as $h\nu$, $2h\nu$, $3h\nu$, and so forth. If the quantity of energy emitted by an atom is $3h\nu$, for example, we say that three quanta of energy have been emitted (quanta being the plural of quantum). Because the energy can be released only in specific amounts, we say that the allowed energies are *quantized*—their values are restricted to certain quantities. Planck's revolutionary proposal that energy is quantized was proved correct, and he was awarded the 1918 Nobel Prize in Physics for his work on the quantum theory.

If the notion of quantized energies seems strange, it might be helpful to draw an analogy by comparing a ramp and a staircase (Figure 6.6 ▼). As you walk up a ramp, your potential energy increases in a uniform, continuous manner. When you climb a staircase, you can step only *on* individual stairs, not *between* them, so that your potential energy is restricted to certain values and is therefore quantized.

If Planck's quantum theory is correct, why aren't its effects more obvious in our daily lives? Why do energy changes seem continuous rather than quantized, or "jagged"? Notice that Planck's constant is an extremely small number. Thus, a quantum of energy, $h\nu$, is an extremely small amount. Planck's rules regarding the gain or loss of energy are always the same, whether we are concerned with objects on the size scale of our ordinary experience or with microscopic objects. With everyday macroscopic objects, however, the gain or loss of a single quantum of energy goes completely unnoticed. In contrast, when dealing with matter at the atomic level, the impact of quantized energies is far more significant.

▲ **Figure 6.6 A model for quantized energy.** The potential energy of a person walking up a ramp (a) increases in a uniform, continuous manner, whereas that of a person walking up steps (b) increases in a stepwise, quantized manner.

GIVE IT SOME THOUGHT

The temperature of stars is gauged by their colors. For example, red stars have a lower temperature than blue-white stars. How is this temperature scale consistent with Planck's assumption?

The Photoelectric Effect and Photons

A few years after Planck presented his theory, scientists began to see its applicability to a great many experimental observations. It soon became apparent that Planck's theory had within it the seeds of a revolution in the way the physical world is viewed. In 1905, Albert Einstein (1879–1955) used Planck's quantum theory to explain the **photoelectric effect**, which is illustrated in Figure 6.7 ▼. Experiments had shown that light shining on a clean metal surface causes the surface to emit electrons. For each metal, there is a minimum frequency of light below which no electrons are emitted. For example, light with a frequency of $4.60 \times 10^{14}\ s^{-1}$ or greater will cause cesium metal to emit electrons, but light of lower frequency has no effect.

VCL Photoelectric Effect

To explain the photoelectric effect, Einstein assumed that the radiant energy striking the metal surface is behaving not like a wave but rather as if it were a stream of tiny energy packets. Each energy packet, called a **photon**, behaves like a tiny particle. Extending Planck's quantum theory, Einstein deduced that each photon must have an energy equal to Planck's constant times the frequency of the light:

ANIMATION Photoelectric Effect

$$\text{Energy of photon} = E = h\nu \qquad [6.3]$$

Thus, radiant energy itself is quantized.

Under the right conditions, a photon can strike a metal surface and be absorbed. When this happens, the photon can transfer its energy to an electron in the metal. A certain amount of energy—called the *work function*—is required for an electron to overcome the attractive forces that hold it in the metal. If the photons of the radiation impinging on the metal have less energy than the work function, electrons do not acquire sufficient energy to escape from the metal surface, even if the light beam is intense. If the photons of radiation have sufficient energy, electrons are emitted from the metal. If the photons have more than the minimum energy required to free electrons, the excess energy appears as the kinetic energy of the emitted electrons.

To better understand what a photon is, imagine that you have a light source that produces radiation with a single wavelength. Further suppose that you could switch the light on and off faster and faster to provide ever-smaller bursts of energy. Einstein's photon theory tells us that you would eventually come to a smallest energy burst, given by $E = h\nu$. This smallest burst of energy consists of a single photon of light.

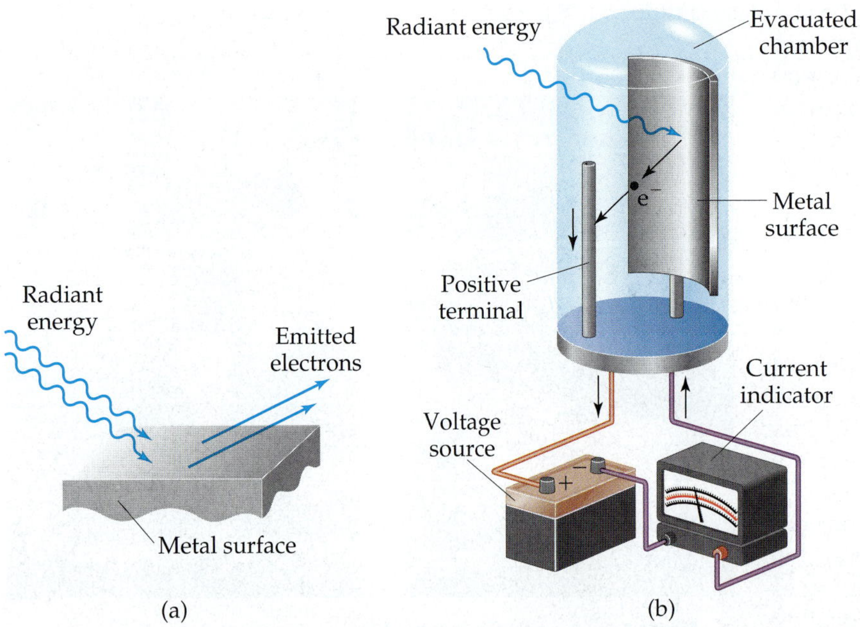

▼ **Figure 6.7 The photoelectric effect.** When photons of sufficiently high energy strike a metal surface, electrons are emitted from the metal, as in (a). The photoelectric effect is the basis of the photocell shown in (b). The emitted electrons are drawn toward the positive terminal. As a result, current flows in the circuit. Photocells are used in photographic light meters as well as in numerous other electronic devices.

SAMPLE EXERCISE 6.3 | Energy of a Photon

Calculate the energy of one photon of yellow light whose wavelength is 589 nm.

Solution

Analyze: Our task is to calculate the energy, E, of a photon, given $\lambda = 589$ nm.

Plan: We can use Equation 6.1 to convert the wavelength to frequency:

$$\nu = c/\lambda$$

We can then use Equation 6.3 to calculate energy:

$$E = h\nu$$

Solve: The frequency, ν, is calculated from the given wavelength, as shown in Sample Exercise 6.2:

$$\nu = c/\lambda = 5.09 \times 10^{14}\ \text{s}^{-1}$$

The value of Planck's constant, h, is given both in the text and in the table of physical constants on the inside front cover of the text, and so we can easily calculate E:

$$E = (6.626 \times 10^{-34}\ \text{J-s})(5.09 \times 10^{14}\ \text{s}^{-1}) = 3.37 \times 10^{-19}\ \text{J}$$

Comment: If one photon of radiant energy supplies 3.37×10^{-19} J, then one mole of these photons will supply

$$(6.02 \times 10^{23}\ \text{photons/mol})(3.37 \times 10^{-19}\ \text{J/photon}) = 2.03 \times 10^{5}\ \text{J/mol}$$

This is the magnitude of enthalpies of reactions (Section 5.4), so radiation can break chemical bonds, producing what are called *photochemical reactions*.

PRACTICE EXERCISE

(a) A laser emits light with a frequency of $4.69 \times 10^{14}\ \text{s}^{-1}$. What is the energy of one photon of the radiation from this laser? **(b)** If the laser emits a pulse of energy containing 5.0×10^{17} photons of this radiation, what is the total energy of that pulse? **(c)** If the laser emits 1.3×10^{-2} J of energy during a pulse, how many photons are emitted during the pulse?

Answers: **(a)** 3.11×10^{-19} J, **(b)** 0.16 J, **(c)** 4.2×10^{16} photons

The idea that the energy of light depends on its frequency helps us understand the diverse effects that different kinds of electromagnetic radiation have on matter. For example, the high frequency (short wavelength) of X rays (Figure 6.4) causes X-ray photons to have high energy, sufficient to cause tissue damage and even cancer. Thus, signs are normally posted around X-ray equipment warning us of high-energy radiation.

Although Einstein's theory of light as a stream of particles rather than a wave explains the photoelectric effect and a great many other observations, it does pose a dilemma. Is light a wave, or does it consist of particles? The only way to resolve this dilemma is to adopt what might seem to be a bizarre position: We must consider that light possesses both wavelike and particle-like characteristics and, depending on the situation, will behave more like a wave or more like particles. We will soon see that this dual nature is also characteristic of matter.

GIVE IT SOME THOUGHT

Suppose that yellow visible light can be used to eject electrons from a certain metal surface. What would happen if ultraviolet light were used instead?

▲ **Figure 6.8 Quantum giants.** Niels Bohr (right) with Albert Einstein. Bohr (1885–1962) made major contributions to the quantum theory. From 1911 to 1913 Bohr studied in England, working first with J. J. Thomson at Cambridge University and then with Ernest Rutherford at the University of Manchester. He published his quantum theory of the atom in 1914 and was awarded the Nobel Prize in Physics in 1922.

▲ **Figure 6.9 Monochromatic radiation.** Lasers produce light of one wavelenght, which we call *monocromatic light*. Different lasers produce light of different wavelenght. The photo shows beams from a variety of lasers that produce visible light of different colors. Other lasers produce light that is not visible, including infrared and ultraviolet light.

6.3 | Line Spectra and the Bohr Model

The work of Planck and Einstein paved the way for understanding how electrons are arranged in atoms. In 1913 the Danish physicist Niels Bohr (Figure 6.8 ◀) offered a theoretical explanation of *line spectra*, another phenomenon that had puzzled scientists in the nineteenth century. Let's first examine this phenomenon and then consider how Bohr used the ideas of Planck and Einstein.

Line Spectra

A particular source of radiant energy may emit a single wavelength, as in the light from a laser (Figure 6.9 ◀). Radiation composed of a single wavelength is said to be *monochromatic*. However, most common radiation sources, including lightbulbs and stars, produce radiation containing many different wavelengths. When radiation from such sources is separated into its different wavelength components, a **spectrum** is produced. Figure 6.10 ▼ shows how a prism spreads light from a lightbulb into its component wavelengths. The spectrum so produced consists of a continuous range of colors: Violet merges into blue, blue into green, and so forth, with no blank spots. This rainbow of colors, containing light of all wavelengths, is called a **continuous spectrum**. The most familiar example of a continuous spectrum is the rainbow produced when raindrops or mist acts as a prism for sunlight.

Not all radiation sources produce a continuous spectrum. When different gases are placed under reduced pressure in a tube and a high voltage is applied, the gases emit different colors of light (Figure 6.11 ▶). The light emitted by neon gas is the familiar red-orange glow of many "neon" lights, whereas sodium vapor emits the yellow light characteristic of some modern streetlights. When light coming from such tubes is passed through a prism, only a few wavelengths are present in the resultant spectra, as shown in Figure 6.12 ▶. Each wavelength is represented by a colored line in one of these spectra. The

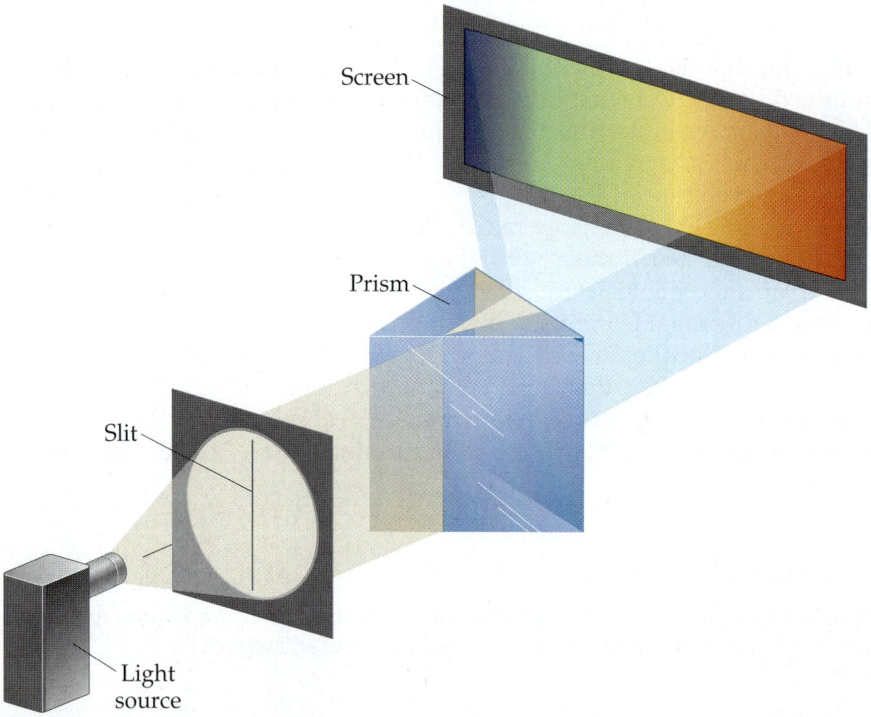

▶ **Figure 6.10 Creating a spectrum.** A continuous visible spectrum is produced when a narrow beam of white light is passed through a prism. The white light could be sunlight or light from an incandescent lamp.

colored lines are separated by black regions, which correspond to wavelengths that are absent from the light. A spectrum containing radiation of only specific wavelengths is called a **line spectrum**.

When scientists first detected the line spectrum of hydrogen in the mid-1800s, they were fascinated by its simplicity. At that time, only the four lines in the visible portion of the spectrum were observed, as shown in Figure 6.12. In 1885 a Swiss schoolteacher named Johann Balmer showed that the wavelengths of these four visible lines of hydrogen fit an intriguingly simple formula. Later, additional lines were found to occur in the ultraviolet and infrared regions of the hydrogen spectrum. Soon Balmer's equation was extended to a more general one, called the *Rydberg equation*, which allowed the calculation of the wavelengths of all the spectral lines of hydrogen:

$$\frac{1}{\lambda} = (R_H)\left(\frac{1}{n_1^2} - \frac{1}{n_2^2}\right) \qquad [6.4]$$

In this formula λ is the wavelength of a spectral line, R_H is the *Rydberg constant* ($1.096776 \times 10^7\ \text{m}^{-1}$), and n_1 and n_2 are positive integers, with n_2 being larger than n_1. How could the remarkable simplicity of this equation be explained? It took nearly 30 more years to answer this question.

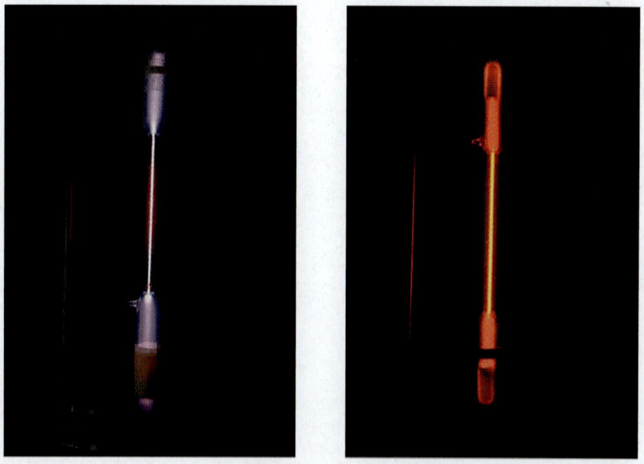

▲ **Figure 6.11 Atomic emission.** Different gases emit light of different characteristic colors upon excitation by an electrical discharge: (a) hydrogen, (b) neon.

VCL Atomic Emission Spectra

VCL The Rydberg Equation

MOVIE
Flame Tests for Metals

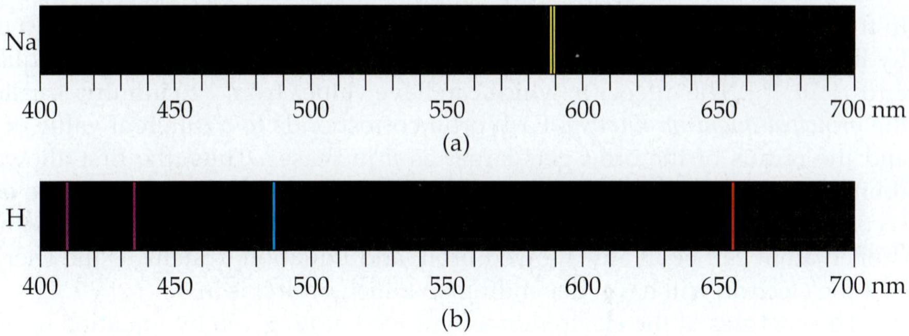

◄ **Figure 6.12 Line spectra.** Spectra obtained from the electrical discharge from (a) Na, (b) H. Light of only a few specific wavelenghts are produced, as shown by colored lines in the spectra.

Bohr's Model

After Rutherford discovered the nuclear nature of the atom (Section 2.2), scientists thought of the atom as a "microscopic solar system" in which electrons orbited the nucleus. To explain the line spectrum of hydrogen, Bohr started by assuming that electrons move in circular orbits around the nucleus. According to classical physics, however, an electrically charged particle (such as an electron) that moves in a circular path should continuously lose energy by emitting electromagnetic radiation. As the electron loses energy, it should spiral into the nucleus. This spiraling obviously does not happen since hydrogen atoms are stable, and so what could explain how these atoms seem to violate the laws of physics? Bohr approached this problem in much the same way that Planck had approached the problem of the nature of the radiation emitted by hot objects: He assumed that the prevailing laws of physics were inadequate to describe all aspects of atoms. Furthermore, he adopted Planck's idea that energies are quantized.

ACTIVITY
Bohr Model

Bohr based his model on three postulates:

1. Only orbits of certain radii, corresponding to certain definite energies, are permitted for the electron in a hydrogen atom.
2. An electron in a permitted orbit has a specific energy and is in an "allowed" energy state. An electron in an allowed energy state will not radiate energy and therefore will not spiral into the nucleus.
3. Energy is emitted or absorbed by the electron only as the electron changes from one allowed energy state to another. This energy is emitted or absorbed as a photon, $E = h\nu$.

GIVE IT SOME THOUGHT

Before reading further about the details of Bohr's model, speculate as to how they explain the fact that hydrogen gas emits a line spectrum (Figure 6.12) rather than a continuous spectrum.

The Energy States of the Hydrogen Atom

Starting with his three postulates and using classical equations for motion and for interacting electrical charges, Bohr calculated the energies corresponding to each allowed orbit for the electron in the hydrogen atom. These energies fit the formula

$$E = (-hcR_H)\left(\frac{1}{n^2}\right) = (-2.18 \times 10^{-18}\ \text{J})\left(\frac{1}{n^2}\right) \quad [6.5]$$

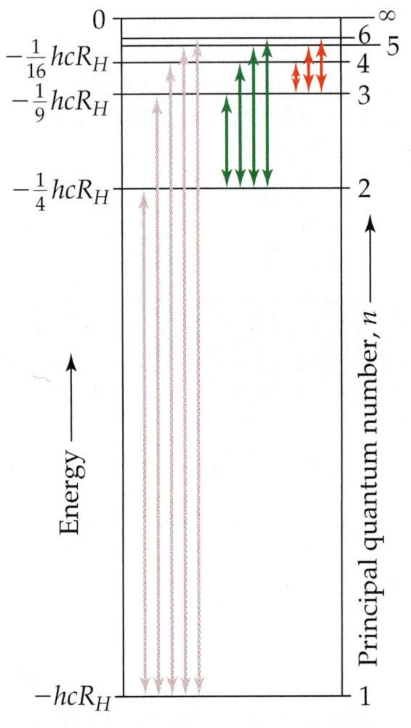

▲ **Figure 6.13 Energy levels in the hydrogen atom from the Bohr model.** The arrows refer to the transitions of the electron from one allowed energy state to another. The states shown are those for which $n = 1$ through $n = 6$ and the state for $n = \infty$, for which the energy, E, equals zero.

In this equation, h, c, and R_H are Planck's constant, the speed of light, and the Rydberg constant, respectively. The product of these three constants equals 2.18×10^{-18} J. The integer n, which can have values from 1 to infinity, is called the *principal quantum number*. Each orbit corresponds to a different value of n, and the radius of the orbit gets larger as n increases. Thus, the first allowed orbit (the one closest to the nucleus) has $n = 1$, the next allowed orbit (the one second closest to the nucleus) has $n = 2$, and so forth. The electron in the hydrogen atom can be in any allowed orbit, and Equation 6.5 tells us the energy that the electron will have, depending on which orbit it is in.

The energies of the electron of a hydrogen atom given by Equation 6.5 are negative for all values of n. The lower (more negative) the energy is, the more stable the atom will be. The energy is lowest (most negative) for $n = 1$. As n gets larger, the energy becomes successively less negative and therefore increases. We can liken the situation to a ladder in which the rungs are numbered from the bottom rung on up. The higher one climbs the ladder (the greater the value of n), the higher the energy. The lowest energy state ($n = 1$, analogous to the bottom rung) is called the **ground state** of the atom. When the electron is in a higher energy (less negative) orbit—$n = 2$ or higher—the atom is said to be in an **excited state**. Figure 6.13 ◀ shows the energy of the electron in a hydrogen atom for several values of n.

What happens to the orbit radius and the energy as n becomes infinitely large? The radius increases as n^2, so we reach a point at which the electron is completely separated from the nucleus. When $n = \infty$, the energy is zero:

$$E = (-2.18 \times 10^{-18}\ \text{J})\left(\frac{1}{\infty^2}\right) = 0$$

Thus, the state in which the electron is removed from the nucleus is the reference, or zero-energy, state of the hydrogen atom. This zero-energy state is *higher* in energy than the states with negative energies.

In his third postulate, Bohr assumed that the electron could "jump" from one allowed energy state to another by either absorbing or emitting photons whose radiant energy corresponds exactly to the energy difference between the two states. Energy must be absorbed for an electron to move to a higher energy state (one with a higher value of n). Conversely, radiant energy is emitted when the electron jumps to a lower energy state (one with a lower value of n). Thus, if the electron jumps from an initial state that has energy E_i to a final state of energy E_f, the change in energy is

$$\Delta E = E_f - E_i = E_{\text{photon}} = h\nu \qquad [6.6]$$

Bohr's model of the hydrogen atom states, therefore, that only the specific frequencies of light that satisfy Equation 6.6 can be absorbed or emitted by the atom.

Substituting the energy expression in Equation 6.5 into Equation 6.6 and recalling that $\nu = c/\lambda$, we have

$$\Delta E = h\nu = \frac{hc}{\lambda} = (-2.18 \times 10^{-18}\ \text{J})\left(\frac{1}{n_f^2} - \frac{1}{n_i^2}\right) \qquad [6.7]$$

In this equation n_i and n_f are the principal quantum numbers of the initial and final states of the atom, respectively. If n_f is smaller than n_i, the electron moves closer to the nucleus and ΔE is a negative number, indicating that the atom releases energy. For example, if the electron moves from $n_i = 3$ to $n_f = 1$, we have

$$\Delta E = (-2.18 \times 10^{-18}\ \text{J})\left(\frac{1}{1^2} - \frac{1}{3^2}\right) = (-2.18 \times 10^{-18}\ \text{J})\left(\frac{8}{9}\right) = -1.94 \times 10^{-18}\ \text{J}$$

Knowing the energy for the emitted photon, we can calculate either its frequency or its wavelength. For the wavelength, we have

$$\lambda = \frac{c}{\nu} = \frac{hc}{\Delta E} = \frac{(6.63 \times 10^{-34}\ \text{J-s})(3.00 \times 10^{8}\ \text{m/s})}{1.94 \times 10^{-18}\ \text{J}} = 1.03 \times 10^{-7}\ \text{m}$$

We have not included the negative sign of the energy in this calculation because wavelength and frequency are always reported as positive quantities. The direction of energy flow is indicated by saying that a photon of wavelength 1.03×10^{-7} has been *emitted*.

If we solve Equation 6.7 for $1/\lambda$, we find that this equation derived from Bohr's theory corresponds to the Rydberg equation, Equation 6.4, which was obtained using experimental data:

$$\frac{1}{\lambda} = \frac{-hcR_H}{hc}\left(\frac{1}{n_f^2} - \frac{1}{n_i^2}\right) = R_H\left(\frac{1}{n_i^2} - \frac{1}{n_f^2}\right)$$

Thus, the existence of discrete spectral lines can be attributed to the quantized jumps of electrons between energy levels.

GIVE IT SOME THOUGHT

As the electron in a hydrogen atom jumps from the $n = 3$ orbit to the $n = 7$ orbit, does it absorb energy or emit energy?

SAMPLE EXERCISE 6.4 | Electronic Transitions in the Hydrogen Atom

Using Figure 6.13, predict which of the following electronic transitions produces the spectral line having the longest wavelength: $n = 2$ to $n = 1$, $n = 3$ to $n = 2$, or $n = 4$ to $n = 3$.

Solution The wavelength increases as frequency decreases ($\lambda = c/\nu$). Hence the longest wavelength will be associated with the lowest frequency. According to Planck's equation, $E = h\nu$, the lowest frequency is associated with the lowest energy. In Figure 6.13 the shortest vertical line represents the smallest energy change. Thus, the $n = 4$ to $n = 3$ transition produces the longest wavelength (lowest frequency) line.

PRACTICE EXERCISE

Indicate whether each of the following electronic transitions emits energy or requires the absorption of energy: **(a)** $n = 3$ to $n = 1$; **(b)** $n = 2$ to $n = 4$.
Answers: **(a)** emits energy, **(b)** requires absorption of energy

Limitations of the Bohr Model

While the Bohr model offers an explanation for the line spectrum of the hydrogen atom, it cannot explain the spectra of other atoms, except in a rather crude way. Furthermore, there is a problem with describing an electron merely as a small particle circling about the nucleus. As we will see in Section 6.4, the electron exhibits wavelike properties, a fact that any acceptable model of electronic structure must accommodate. As it turns out, the Bohr model was only an important step along the way toward the development of a more comprehensive model. What is most significant about Bohr's model is that it introduces two important ideas that are also incorporated into our current model: (1) Electrons exist only in certain discrete energy levels, which are described by quantum numbers, and (2) energy is involved in moving an electron from one level to another. We will now start to develop the successor to the Bohr model, which requires that we take a closer look at the behavior of matter.

6.4 | The Wave Behavior of Matter

In the years following the development of Bohr's model for the hydrogen atom, the dual nature of radiant energy became a familiar concept. Depending on the experimental circumstances, radiation appears to have either a wavelike or a particle-like (photon) character. Louis de Broglie (1892–1987), who was working on his Ph.D. thesis in physics at the Sorbonne in Paris, boldly extended this idea. If radiant energy could, under appropriate conditions, behave as though it were a stream of particles, could matter, under appropriate conditions, possibly show the properties of a wave? Suppose that the electron orbiting the nucleus of a hydrogen atom could be thought of not as a particle but rather as a wave, with a characteristic wavelength. De Broglie suggested that, in its movement about the nucleus, the electron has associated with it a particular wavelength. He went on to propose that the characteristic wavelength of the electron or of any other particle depends on its mass, m, and on its velocity, v:

$$\lambda = \frac{h}{mv} \qquad [6.8]$$

(h is Planck's constant.) The quantity mv for any object is called its **momentum**. De Broglie used the term **matter waves** to describe the wave characteristics of material particles.

Because de Broglie's hypothesis is applicable to all matter, any object of mass m and velocity v would give rise to a characteristic matter wave. However, Equation 6.8 indicates that the wavelength associated with an object of ordinary size, such as a golf ball, is so tiny as to be completely out of the range of any possible observation. This is not so for an electron because its mass is so small, as we see in Sample Exercise 6.5.

SAMPLE EXERCISE 6.5 | Matter Waves

What is the wavelength of an electron moving with a speed of 5.97×10^6 m/s? (The mass of the electron is 9.11×10^{-28} g.)

Solution

Analyze: We are given the mass, m, and velocity, v, of the electron, and we must calculate its de Broglie wavelength, λ.

Plan: The wavelength of a moving particle is given by Equation 6.8, so λ is calculated by inserting the known quantities h, m, and v. In doing so, however, we must pay attention to units.

Solve: Using the value of Planck's constant,

$$h = 6.63 \times 10^{-34} \text{ J-s}$$

and recalling that

$$1 \text{ J} = 1 \text{ kg-m}^2/\text{s}^2$$

we have the following:

$$\lambda = \frac{h}{mv}$$

$$= \frac{(6.63 \times 10^{-34} \text{ J-s})}{(9.11 \times 10^{-28} \text{ g})(5.97 \times 10^6 \text{ m/s})}\left(\frac{1 \text{ kg-m}^2/\text{s}^2}{1 \text{ J}}\right)\left(\frac{10^3 \text{ g}}{1 \text{ kg}}\right)$$

$$= 1.22 \times 10^{-10} \text{ m} = 0.122 \text{ nm}$$

Comment: By comparing this value with the wavelengths of electromagnetic radiation shown in Figure 6.4, we see that the wavelength of this electron is about the same as that of X rays.

PRACTICE EXERCISE

Calculate the velocity of a neutron whose de Broglie wavelength is 500 pm. The mass of a neutron is given in the table on the back inside cover of the text.

Answer: 7.92×10^2 m/s

Within a few years after de Broglie published his theory, the wave properties of the electron were demonstrated experimentally. Electrons passed through a crystal were diffracted by the crystal, just as X rays are diffracted. Thus, a stream of moving electrons exhibits the same kinds of wave behavior as electromagnetic radiation.

The technique of electron diffraction has been highly developed. In the electron microscope, for instance, the wave characteristics of electrons are used to obtain pictures of tiny objects. This microscope is an important tool for studying surface phenomena at very high magnifications. Figure 6.14 ▶ is a photograph of an electron microscope image, which demonstrates that tiny particles of matter can behave as waves.

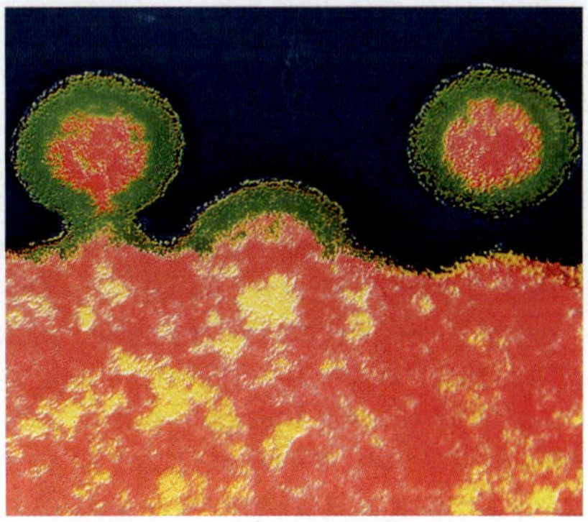

▲ **Figure 6.14 Electrons as waves.** Color-enhanced electron microscope image of human immunodeficiency viruses (HIV) budding from an infected T-lymphocyte human blood cell. In an electron microscope the wave behavior of a stream of electrons is utilized in analogy to the way that a conventional microscope uses the wave behavior of a beam of light.

GIVE IT SOME THOUGHT

A baseball pitcher throws a fastball that moves at 95 miles per hour. Does that moving baseball generate matter waves? If so, can we observe them?

The Uncertainty Principle

The discovery of the wave properties of matter raised some new and interesting questions about classical physics. Consider, for example, a ball rolling down a ramp. Using the equations of classical physics, we can calculate its position, direction of motion, and speed at any time, with great accuracy. Can we do the same for an electron, which exhibits wave properties? A wave extends in space, and therefore its location is not precisely defined. We might therefore anticipate that it is impossible to determine exactly where an electron is located at a specific time.

VCL Heisenberg Uncertainty Principle

▲ **Figure 6.15 Werner Heisenberg (1901–1976).** During his postdoctoral assistantship with Niels Bohr, Heisenberg formulated his famous uncertainty principle. At the age of 25, he became the chair in theoretical physics at the University of Leipzig. At 32 he was one of the youngest scientists to receive the Nobel Prize.

The German physicist Werner Heisenberg (Figure 6.15 ◀) proposed that the dual nature of matter places a fundamental limitation on how precisely we can know both the location and the momentum of any object. The limitation becomes important only when we deal with matter at the subatomic level (that is, with masses as small as that of an electron). Heisenberg's principle is called the **uncertainty principle**. When applied to the electrons in an atom, this principle states that it is inherently impossible for us to know simultaneously both the exact momentum of the electron and its exact location in space.

Heisenberg mathematically related the uncertainty of the position (Δx) and the uncertainty in momentum $\Delta(mv)$ to a quantity involving Planck's constant:

$$\Delta x \cdot \Delta(mv) \geq \frac{h}{4\pi} \qquad [6.9]$$

A brief calculation illustrates the dramatic implications of the uncertainty principle. The electron has a mass of 9.11×10^{-31} kg and moves at an average speed of about 5×10^{6} m/s in a hydrogen atom. Let's assume that we know the speed to an uncertainty of 1% [that is, an uncertainty of $(0.01)(5 \times 10^{6}\ \text{m/s}) = 5 \times 10^{4}\ \text{m/s}$] and that this is the only important source of uncertainty in the momentum, so that $\Delta(mv) = m\Delta v$. We can then use Equation 6.9 to calculate the uncertainty in the position of the electron:

$$\Delta x \geq \frac{h}{4\pi m \Delta v} = \frac{(6.63 \times 10^{-34}\ \text{J-s})}{4\pi(9.11 \times 10^{-31}\ \text{kg})(5 \times 10^{4}\ \text{m/s})} = 1 \times 10^{-9}\ \text{m}$$

Because the diameter of a hydrogen atom is only about 1×10^{-10} m, the uncertainty is an order of magnitude greater than the size of the atom. Thus, we have essentially no idea of where the electron is located within the atom. On the other hand, if we were to repeat the calculation with an object of ordinary mass

A CLOSER LOOK | Measurement and the Uncertainty Principle

Whenever any measurement is made, some uncertainty exists. Our experience with objects of ordinary dimensions, like balls or trains or laboratory equipment, indicates that the uncertainty of a measurement can be decreased by using more precise instruments. In fact, we might expect that the uncertainty in a measurement can be made indefinitely small. However, the uncertainty principle states that there is an actual limit to the accuracy of measurements. This limit is not a restriction on how well instruments can be made; rather, it is inherent in nature. This limit has no practical consequences when dealing with ordinary-sized objects, but its implications are enormous when dealing with subatomic particles, such as electrons.

To measure an object, we must disturb it, at least a little, with our measuring device. Imagine using a flashlight to locate a large rubber ball in a dark room. You see the ball when the light from the flashlight bounces off the ball and strikes your eyes. When a beam of photons strikes an object of this size, it does not alter its position or momentum to any practical extent. Imagine, however, that you wish to locate an electron by similarly bouncing light off it into some detector. Objects can be located to an accuracy no greater than the wavelength of the radiation used. Thus, if we want an accurate position measurement for an electron, we must use a short wavelength. This means that photons of high energy must be employed. The more energy the photons have, the more momentum they impart to the electron when they strike it, which changes the electron's motion in an unpredictable way. The attempt to measure accurately the electron's position introduces considerable uncertainty in its momentum; the act of measuring the electron's position at one moment makes our knowledge of its future position inaccurate.

Suppose, then, that we use photons of longer wavelength. Because these photons have lower energy, the momentum of the electron is not so appreciably changed during measurement, but its position will be correspondingly less accurately known. This is the essence of the uncertainty principle: *There is an uncertainty in simultaneously knowing either the position or the momentum of the electron that cannot be reduced beyond a certain minimum level.* The more accurately one is known, the less accurately the other is known. Although we can never know the exact position and momentum of the electron, we can talk about the probability of its being at certain locations in space. In Section 6.5 we introduce a model of the atom that provides the probability of finding electrons of specific energies at certain positions in atoms.

Related Exercises: 6.45 and 6.46

such as a tennis ball, the uncertainty would be so small that it would be inconsequential. In that case, m is large and Δx is out of the realm of measurement and therefore of no practical consequence.

De Broglie's hypothesis and Heisenberg's uncertainty principle set the stage for a new and more broadly applicable theory of atomic structure. In this new approach, any attempt to define precisely the instantaneous location and momentum of the electron is abandoned. The wave nature of the electron is recognized, and its behavior is described in terms appropriate to waves. The result is a model that precisely describes the energy of the electron while describing its location not precisely, but in terms of probabilities.

GIVE IT SOME THOUGHT

What is the principal reason that the uncertainty principle seems very important when discussing electrons and other subatomic particles, but seems rather unimportant in our macroscopic world?

6.5 | Quantum Mechanics and Atomic Orbitals

In 1926 the Austrian physicist Erwin Schrödinger (1887–1961) proposed an equation, now known as Schrödinger's wave equation, that incorporates both the wavelike behavior and the particle-like behavior of the electron. His work opened a new way of dealing with subatomic particles, known as either *quantum mechanics* or *wave mechanics*. The application of Schrödinger's equation requires advanced calculus, and we will not be concerned with the details of his approach. We will, however, qualitatively consider the results he obtained, because they give us a powerful new way to view electronic structure. Let's begin by examining the electronic structure of the simplest atom, hydrogen.

Solving Schrödinger's equation leads to a series of mathematical functions called **wave functions** that describe the electron. These wave functions are usually represented by the symbol ψ (the Greek lowercase letter *psi*). Although the wave function itself has no direct physical meaning, the square of the wave function, ψ^2, provides information about an electron's location when the electron is in an allowed energy state.

For the hydrogen atom, the allowed energies are the same as those predicted by the Bohr model. However, the Bohr model assumes that the electron is in a circular orbit of some particular radius about the nucleus. In the quantum mechanical model, the electron's location cannot be described so simply. According to the uncertainty principle, if we know the momentum of the electron with high accuracy, our simultaneous knowledge of its location is very uncertain. Thus, we cannot hope to specify the exact location of an individual electron around the nucleus. Rather, we must be content with a kind of statistical knowledge. In the quantum mechanical model, we therefore speak of the *probability* that the electron will be in a certain region of space at a given instant. As it turns out, the square of the wave function, ψ^2, at a given point in space represents the probability that the electron will be found at that location. For this reason, ψ^2 is called either the **probability density** or the **electron density**.

One way of representing the probability of finding the electron in various regions of an atom is shown in Figure 6.16 ▶. In this figure the density of the dots represents the probability of finding the electron. The regions with a high density of dots correspond to relatively large values for ψ^2 and are therefore regions where there is a high probability of finding the electron. In Section 6.6 we will say more about the ways in which we can represent electron density.

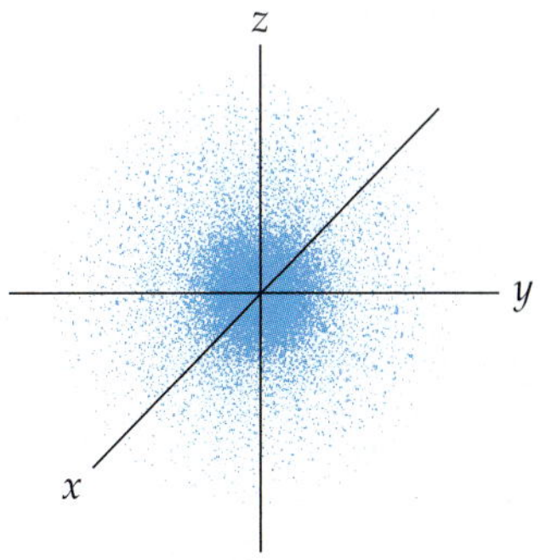

▲ **Figure 6.16 Electron-density distribution.** This rendering represents the probability of where in the space surrounding the nucleus the electron is to be found in a hydrogen atom in its ground state.

GIVE IT SOME THOUGHT

Is there a difference between stating, "The electron is located at a particular point in space" and "There is a high probability that the electron is located at a particular point in space"?

Orbitals and Quantum Numbers

The solution to Schrödinger's equation for the hydrogen atom yields a set of wave functions and corresponding energies. These wave functions are called **orbitals**. Each orbital describes a specific distribution of electron density in space, as given by the orbital's probability density. Each orbital therefore has a characteristic energy and shape. For example, the lowest-energy orbital in the hydrogen atom has an energy of -2.18×10^{-18} J and the shape illustrated in Figure 6.16. Note that an *orbital* (quantum mechanical model) is not the same as an *orbit* (Bohr model). The quantum mechanical model doesn't refer to orbits because the motion of the electron in an atom cannot be precisely measured or tracked (Heisenberg uncertainty principle).

The Bohr model introduced a single quantum number, n, to describe an orbit. The quantum mechanical model uses three quantum numbers, n, l, and m_l, to describe an orbital. Let's consider what information we obtain from each of these and how they are interrelated.

ACTIVITY
Quantum Numbers

1. The *principal quantum number*, n, can have positive integral values of 1, 2, 3, and so forth. As n increases, the orbital becomes larger, and the electron spends more time farther from the nucleus. An increase in n also means that the electron has a higher energy and is therefore less tightly bound to the nucleus. For the hydrogen atom, $E_n = -(2.18 \times 10^{-18}\ \text{J})(1/n^2)$, as in the Bohr model.
2. The second quantum number—the *azimuthal quantum number*, l—can have integral values from 0 to $n - 1$ for each value of n. This quantum number defines the shape of the orbital. (We will consider these shapes in Section 6.6.) The value of l for a particular orbital is generally designated by the letters s, p, d, and f,* corresponding to l values of 0, 1, 2, and 3, respectively, as summarized here:

Value of l	0	1	2	3
Letter used	s	p	d	f

3. The *magnetic quantum number*, m_l, can have integral values between $-l$ and l, including zero. This quantum number describes the orientation of the orbital in space, as we will discuss in Section 6.6.

Notice that because the value of n can be any positive integer, there is an infinite number of orbitals for the hydrogen atom. The electron in a hydrogen atom is described by only one of these orbitals at any given time—we say that the electron *occupies* a certain orbital. The remaining orbitals are *unoccupied* for that particular state of the hydrogen atom. We will see that we are mainly interested in the orbitals of the hydrogen atom with small values of n.

GIVE IT SOME THOUGHT

What is the difference between an *orbit* (Bohr model) and an *orbital* (quantum mechanical model)?

* The letters s, p, d, and f come from the words sharp, principal, diffuse, and fundamental, which were used to describe certain features of spectra before quantum mechanics was developed.

TABLE 6.2 Relationship Among Values of n, l, and m_l Through $n = 4$

n	Possible Values of l	Subshell Designation	Possible Values of m_l	Number of Orbitals in Subshell	Total Number of Orbitals in Shell
1	0	1s	0	1	1
2	0	2s	0	1	
	1	2p	1, 0, −1	3	4
3	0	3s	0	1	
	1	3p	1, 0, −1	3	
	2	3d	2, 1, 0, −1, −2	5	9
4	0	4s	0	1	
	1	4p	1, 0, −1	3	
	2	4d	2, 1, 0, −1, −2	5	
	3	4f	3, 2, 1, 0, −1, −2, −3	7	16

The collection of orbitals with the same value of n is called an **electron shell**. For example, all the orbitals that have $n = 3$ are said to be in the third shell. Further, the set of orbitals that have the same n and l values is called a **subshell**. Each subshell is designated by a number (the value of n) and a letter (s, p, d, or f, corresponding to the value of l). For example, the orbitals that have $n = 3$ and $l = 2$ are called 3d orbitals and are in the 3d subshell.

Table 6.2 ▲ summarizes the possible values of the quantum numbers l and m_l for values of n through $n = 4$. The restrictions on the possible values of the quantum numbers give rise to the following very important observations:

1. The shell with principal quantum number n will consist of exactly n subshells. Each subshell corresponds to a different allowed value of l from 0 to $n - 1$. Thus, the first shell ($n = 1$) consists of only one subshell, the 1s ($l = 0$); the second shell ($n = 2$) consists of two subshells, the 2s ($l = 0$) and 2p ($l = 1$); the third shell consists of three subshells, 3s, 3p, and 3d, and so forth.
2. Each subshell consists of a specific number of orbitals. Each orbital corresponds to a different allowed value of m_l. For a given value of l, there are $2l + 1$ allowed values of m_l, ranging from $-l$ to $+l$. Thus, each s ($l = 0$) subshell consists of one orbital; each p ($l = 1$) subshell consists of three orbitals; each d ($l = 2$) subshell consists of five orbitals, and so forth.
3. The total number of orbitals in a shell is n^2, where n is the principal quantum number of the shell. The resulting number of orbitals for the shells—1, 4, 9, 16—is related to a pattern seen in the periodic table: We see that the number of elements in the rows of the periodic table—2, 8, 18, and 32—equal twice these numbers. We will discuss this relationship further in Section 6.9.

Figure 6.17 ▶ shows the relative energies of the hydrogen atom orbitals through $n = 3$. Each box represents an orbital; orbitals of the same subshell, such as the 2p, are grouped together. When the electron occupies the lowest-energy orbital (1s), the hydrogen atom is said to be in its *ground state*. When the electron occupies any other orbital, the atom is in an *excited state*. At ordinary temperatures, essentially all hydrogen atoms are in the ground state. The electron can be excited to a higher-energy orbital by absorption of a photon of appropriate energy.

GIVE IT SOME THOUGHT

In Figure 6.17, why is the energy difference between the $n = 1$ and $n = 2$ levels so much greater than the energy difference between the $n = 2$ and $n = 3$ levels?

▼ Figure 6.17 Orbital energy levels in the hydrogen atom. Each box represents an orbital. Note that all orbitals with the same value for the principal quantum number, n, have the same energy. This is true only in one-electron systems, such as the hydrogen atom.

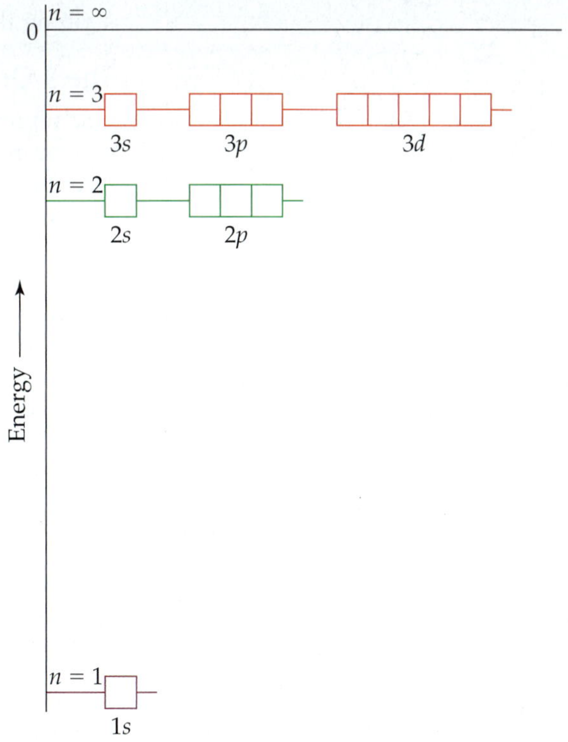

SAMPLE EXERCISE 6.6 | **Subshells of the Hydrogen Atom**

(a) Without referring to Table 6.2, predict the number of subshells in the fourth shell, that is, for $n = 4$. **(b)** Give the label for each of these subshells. **(c)** How many orbitals are in each of these subshells?

Analyze and Plan: We are given the value of the principal quantum number, n. We need to determine the allowed values of l and m_l for this given value of n and then count the number of orbitals in each subshell.

Solution **(a)** There are four subshells in the fourth shell, corresponding to the four possible values of l (0, 1, 2, and 3).

(b) These subshells are labeled 4*s*, 4*p*, 4*d*, and 4*f*. The number given in the designation of a subshell is the principal quantum number, n; the letter designates the value of the azimuthal quantum number, l: for $l = 0$, *s*; for $l = 1$, *p*; for $l = 2$, *d*; for $l = 3$, *f*.

(c) There is one 4*s* orbital (when $l = 0$, there is only one possible value of m_l: 0). There are three 4*p* orbitals (when $l = 1$, there are three possible values of m_l: 1, 0, and −1). There are five 4*d* orbitals (when $l = 2$, there are five allowed values of m_l: 2, 1, 0, −1, −2). There are seven 4*f* orbitals (when $l = 3$, there are seven permitted values of m_l: 3, 2, 1, 0, −1, −2, −3).

PRACTICE EXERCISE

(a) What is the designation for the subshell with $n = 5$ and $l = 1$? **(b)** How many orbitals are in this subshell? **(c)** Indicate the values of m_l for each of these orbitals.
Answers: **(a)** 5*p*; **(b)** 3; **(c)** 1, 0, −1

6.6 | Representations of Orbitals

In our discussion of orbitals so far, we have emphasized their energies. But the wave function also provides information about the electron's location in space when it occupies an orbital. Let's examine the ways that we can picture the orbitals. In doing so, we will examine some important aspects of the electron-density distributions of the orbitals. First, we will look at the three-dimensional shape of the orbital—is it spherical, for example, or does it have directionality? Second, we will examine how the probability density changes as we move on a straight line farther and farther from the nucleus. Finally, we will look at the typical three-dimensional sketches that chemists use in describing the orbitals.

The *s* Orbitals

One representation of the lowest-energy orbital of the hydrogen atom, the 1*s*, is shown in Figure 6.16. This type of drawing, which shows the distribution of electron density around the nucleus, is one of the several ways we use to help us visualize orbitals. The first thing that we notice about the electron density for the 1*s* orbital is that it is *spherically symmetric*—in other words, the electron density at a given distance from the nucleus is the same regardless of the direction in which we proceed from the nucleus. All of the other *s* orbitals (2*s*, 3*s*, 4*s*, and so forth) are spherically symmetric as well.

So what is different about the *s* orbitals having different n quantum numbers? For example, how does the electron-density distribution of the hydrogen atom change when the electron is excited from the 1*s* orbital to the 2*s* orbital? To address questions like this, we must take a look at the *radial probability density*, that is, the probability that we will find the electron at a specific distance from the nucleus. In Figure 6.18 ▶ we have plotted the radial probability density for the 1*s* orbital as a function of r, the distance from the nucleus—the resulting curve is the **radial probability function** for the 1*s* orbital. (Radial probability functions are described more fully in the "A Closer Look" box in this section.) We see that the probability of finding the electron rises rapidly as we move away from the nucleus, maximizing at a distance of 0.529 Å from the nucleus, and then falls off rapidly. Thus, when the electron occupies the 1*s* orbital, it is *most likely* to be found

ANIMATION
Radial Electron Distribution

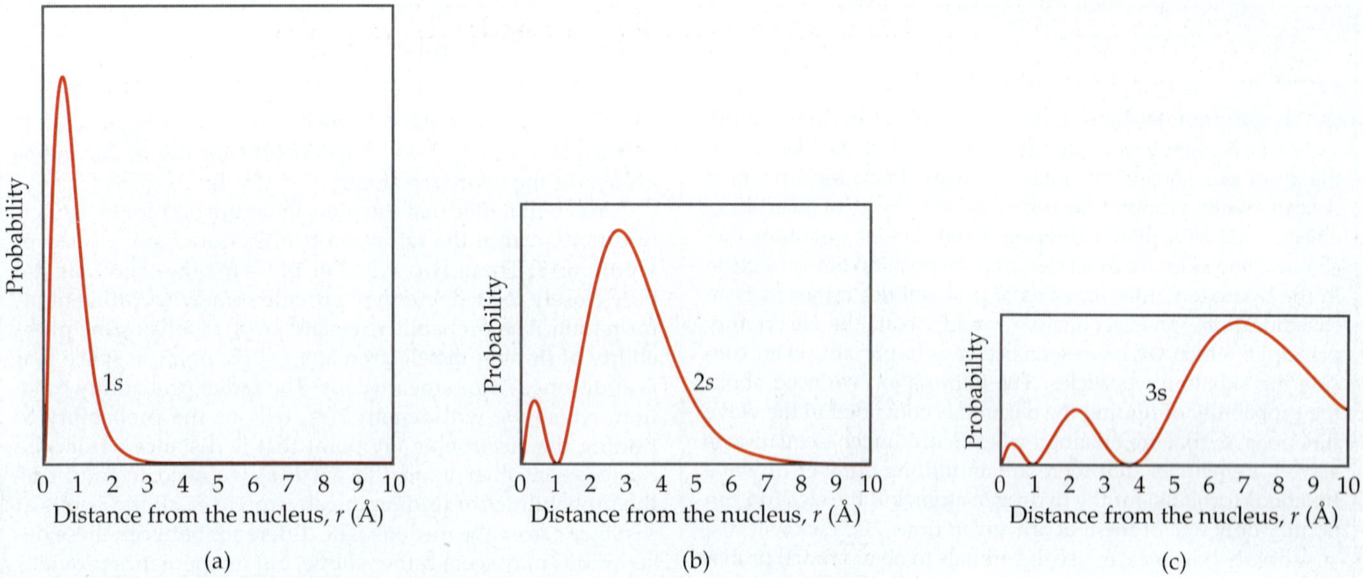

▲ **Figure 6.18 Radial probability functions for the 1*s*, 2*s*, and 3*s* orbitals.** These plots show the probability of finding the electron as a function of distance from the nucleus. As *n* increases, the most likely distance at which to find the electron moves farther from the nucleus, similar to the Bohr model. In the 2*s* and 3*s* orbitals the radial probability function drops to zero at certain distances from the nucleus but then rises again. The points at which the probability is zero are called *nodes*.

0.529 Å from the nucleus*—we still use the probabilistic description, consistent with the uncertainty principle. Notice also that the probability of finding the electron at a distance greater than 3 Å from the nucleus is essentially zero.

Figure 6.18(b) shows the radial probability function for the 2*s* orbital of the hydrogen atom. We can see three significant differences between this plot and that for the 1*s* orbital: (1) There are two separate maxima in the radial probability function for the 2*s* orbital, namely a small peak at about $r = 0.5$ Å and a much larger peak at about $r = 3$ Å; (2) Between these two peaks is a point at which the function goes to zero (at about $r = 1$ Å). An intermediate point at which a probability function goes to zero is called a **node**. There is a zero probability of finding the electron at a distance corresponding to a node, even though the electron might be found at shorter or longer distances; (3) The radial probability function for the 2*s* orbital is significantly broader (more spread out) than that for the 1*s* orbital. Thus, for the 2*s* orbital, there is a larger range of distances from the nucleus, at which there is a high probability of finding the electron than for the 1*s* orbital. This trend continues for the 3*s* orbital, as shown in Figure 6.18(c). The radial probability function for the 3*s* orbital has three peaks of increasing size, with the largest peak maximizing even farther from the nucleus (at about $r = 7$ Å) at which it has two nodes and is even more spread out.

The radial probability functions in Figure 6.18 tell us that as *n* increases, there is also an increase in the most likely distance from the nucleus to find the electron. In other words, the size of the orbital increases with increasing *n*, just as it did in the Bohr model.

One widely used method of representing orbitals is to display a boundary surface that encloses some substantial portion, say 90%, of the total electron density for the orbital. For the *s* orbitals, these contour representations are spheres. The contour representations of the 1*s*, 2*s*, and 3*s* orbitals are shown in Figure 6.19 ▶. They all have the same shape, but they differ in size. Although

▲ **Figure 6.19 Contour representations of the 1*s*, 2*s*, and 3*s* orbitals.** The relative radii of the spheres correspond to a 90% probability of finding the electron within each sphere.

* In the quantum mechanical model, the most probable distance at which to find the electron in the 1*s* orbital—0.529 Å—is identical to the radius of the orbit predicted by Bohr for $n = 1$. The distance 0.529 Å is often called the *Bohr radius*.

A CLOSER LOOK | Probability Density and Radial Probability Functions

The quantum mechanical description of the hydrogen atom requires that we talk about the position of the electron in the atom in a somewhat unfamiliar way. In classical physics, we can exactly pinpoint the position and velocity of an orbiting object, such as a planet orbiting a star. Under quantum mechanics, however, we must describe the position of the electron in the hydrogen atom in terms of probabilities rather than an exact location—an exact answer would violate the uncertainty principle, which we have seen becomes important when considering subatomic particles. The information we need about the probability of finding the electron is contained in the wave functions, ψ, that are obtained when Schrödinger's equation is solved. Remember that there are an infinite number of wave functions (orbitals) for the hydrogen atom, but the electron can occupy only one of them at any given time. Here we will discuss briefly how we can use the orbitals to obtain radial probability functions, such as those in Figure 6.18.

In Section 6.5 we stated that the square of the wave function, ψ^2, gives the probability that the electron is at any one given point in space—recall that this quantity is called the *probability density* for the point. For a spherically symmetric *s* orbital, the value of ψ depends only on the distance from the nucleus, *r*. Let's consider a straight line outward from the nucleus, as shown in Figure 6.20 ▼. The probability of finding the electron at distance *r* from the nucleus along that line is $[\psi(r)]^2$ where $\psi(r)$ is the value of ψ at distance *r*. Figure 6.21 ▼ shows plots of $[\psi(r)]^2$ as a function of *r* for the 1*s*, 2*s*, and 3*s* orbitals of the hydrogen atom.

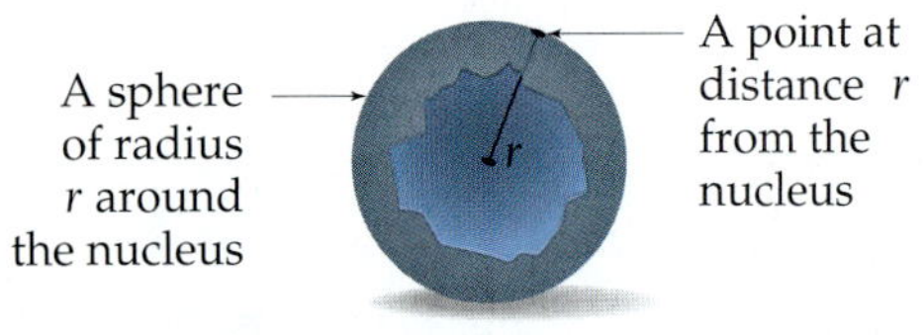

▲ **Figure 6.20 Probability at a point.** The probability density, $\psi(r)^2$, gives the probability that the electron will be found at *a specific point* at distance *r* from the nucleus. The radial probability function, $4\pi r^2\psi(r)^2$, gives the probability that the electron will be found at *any* point distance *r* from the nucleus—in other words, at any point on the sphere of radius *r*.

You will notice that the plots in Figure 6.21 look distinctly different from the radial probability functions plotted in Figure 6.18. These two types of plots for the *s* orbitals are very closely related, but they provide somewhat different information. The probability density, $[\psi(r)]^2$, tells us the probability of finding the electron at *a specific* point in space that is at distance *r* from the nucleus. The radial probability function, which we will denote $P(r)$, tells us the probability of finding the electron at *any* point that is distance *r* from the nucleus—in other words, to get $P(r)$ we need to "add up" the probabilities of finding the electron over all the points at distance *r* from the nucleus. The difference between these descriptions may seem rather subtle, but mathematics provides us with a precise way to connect them.

As shown in Figure 6.20, the collection of points at distance *r* from the nucleus is simply a sphere of radius *r*. The probability density at every point on that sphere is $[\psi(r)]^2$. To add up all of the individual probability densities requires the use of calculus and is beyond the scope of this text (in the language of calculus "we integrate the probability density over the surface of the sphere"). The result we obtain is easy to describe, however. The radial probability function at distance *r*, $P(r)$, is simply the probability density at distance *r*, $[\psi(r)]^2$, multiplied by the surface area of the sphere, which is given by the formula $4\pi r^2$:

$$P(r) = 4\pi r^2[\psi(r)]^2$$

Thus, the plots of $P(r)$ in Figure 6.18 are equal to the plots of $[\psi(r)]^2$ in Figure 6.21 multiplied by $4\pi r^2$. The fact that $4\pi r^2$ increases rapidly as we move away from the nucleus makes the two sets of plots look very different. For example, the plot of $[\psi(r)]^2$ for the 3*s* orbital (Figure 6.21) shows that the function generally gets smaller the farther we go from the nucleus. But when we multiply by $4\pi r^2$, we see peaks that get larger and larger as we move away from the nucleus (Figure 6.18). We will see that the radial probability functions in Figure 6.18 provide us with the more useful information because they tell us the probability for finding the electron at *all* points distance *r* from the nucleus, not just one particular point.

Related Exercises: 6.48, 6.57, 6.58, and 6.91

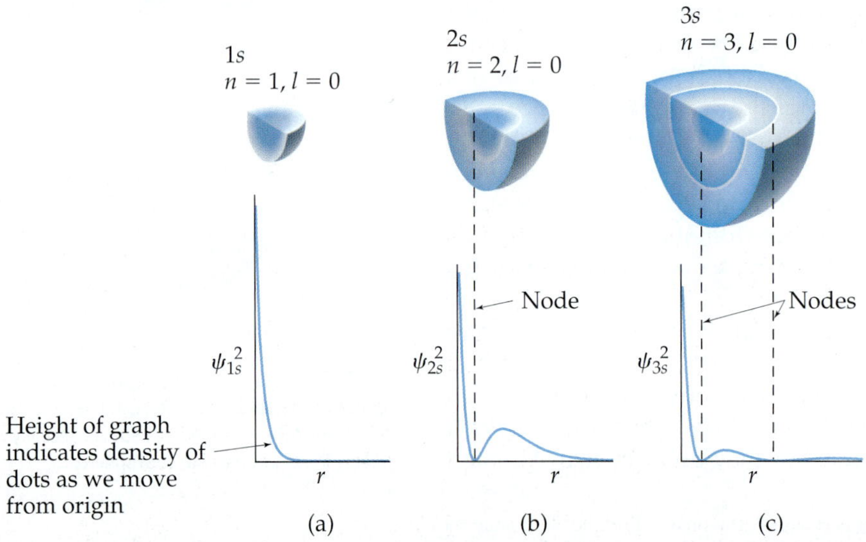

◀ **Figure 6.21 Probability density distribution in 1*s*, 2*s*, and 3*s* orbitals.** The lower part of the figure shows how the probability density, $\psi(r)^2$, varies as a function of distance *r* from the nucleus. The upper part of the figure shows a cutaway of the spherical electron density in each of the *s* orbitals.

the details of how the electron density varies within the contour representation are lost in these representations, this is not a serious disadvantage. For more qualitative discussions, the most important features of orbitals are their relative sizes and their shapes, which are adequately displayed by contour representations.

GIVE IT SOME THOUGHT

How many maxima would you expected to find in the radial probability function for the 4*s* orbital of the hydrogen atom? How many nodes would you expect in the 4*s* radial probability function?

The *p* Orbitals

The distribution of electron density for a 2*p* orbital is shown in Figure 6.22(a) ▶. As we can see from this figure, the electron density is not distributed in a spherically symmetric fashion as in an *s* orbital. Instead, the electron density is concentrated in two regions on either side of the nucleus, separated by a node at the nucleus. We say that this dumbbell-shaped orbital has two *lobes*. It is useful to recall that we are making no statement of how the electron is moving within the orbital; the only thing Figure 6.22(a) portrays is the *averaged* distribution of the electron density in a 2*p* orbital.

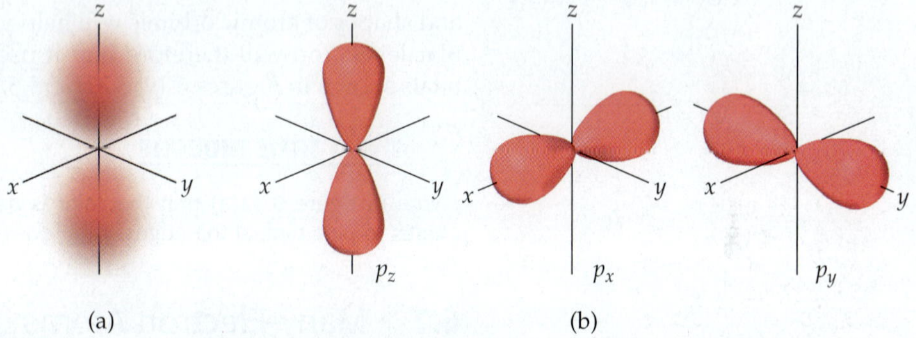

▼ **Figure 6.22 The *p* orbitals.** (a) Electron-density distribution of a 2*p* orbital. (b) Contour representations of the three *p* orbitals. Note that the subscript on the orbital label indicates the axis along which the orbital lies.

Beginning with the $n = 2$ shell, each shell has three *p* orbitals. Thus, there are three 2*p* orbitals, three 3*p* orbitals, and so forth. Each set of *p* orbitals has the dumbbell shapes shown in Figure 6.22(a) for the 2*p* orbitals. For each value of *n*, the three *p* orbitals have the same size and shape but differ from one another in spatial orientation. We usually represent *p* orbitals by drawing the shape and orientation of their wave functions, as shown in Figure 6.22(b). It is convenient to label these as the p_x, p_y, and p_z orbitals. The letter subscript indicates the Cartesian axis along which the orbital is oriented.* Like *s* orbitals, *p* orbitals increase in size as we move from 2*p* to 3*p* to 4*p*, and so forth.

The *d* and *f* Orbitals

When *n* is 3 or greater, we encounter the *d* orbitals (for which $l = 2$). There are five 3*d* orbitals, five 4*d* orbitals, and so forth. The different *d* orbitals in a given shell have different shapes and orientations in space, as shown in Figure 6.23 ▶. Four of the *d*-orbital contour representations have a "four-leaf clover" shape, and each lies primarily in a plane. The d_{xy}, d_{xz}, and d_{yz} lie in the *xy*, *xz*, and *yz* planes,

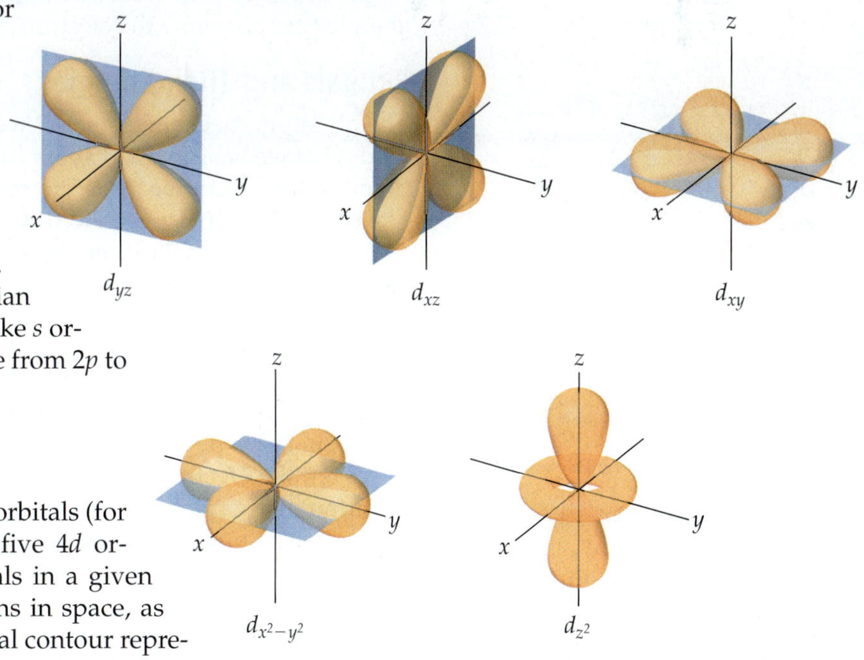

▲ **Figure 6.23 Contour representations of the five *d* orbitals.**

* We cannot make a simple correspondence between the subscripts (*x*, *y*, and *z*) and the allowed m_l values (1, 0, and −1). To explain why this is so is beyond the scope of an introductory text.

respectively, with the lobes oriented *between* the axes. The lobes of the $d_{x^2-y^2}$ orbital also lie in the *xy* plane, but the lobes lie *along* the *x* and *y* axes. The d_{z^2} orbital looks very different from the other four: It has two lobes along the *z* axis and a "doughnut" in the *xy* plane. Even though the d_{z^2} orbital looks different from the other *d* orbitals, it has the same energy as the other four *d* orbitals. The representations in Figure 6.23 are commonly used for all *d* orbitals, regardless of principal quantum number.

When *n* is 4 or greater, there are seven equivalent *f* orbitals (for which $l = 3$). The shapes of the *f* orbitals are even more complicated than those of the *d* orbitals and are not presented here. As you will see in the next section, however, you must be aware of *f* orbitals as we consider the electronic structure of atoms in the lower part of the periodic table.

In many instances later in the text you will find that knowing the number and shapes of atomic orbitals will help you understand chemistry at the molecular level. You will therefore find it useful to memorize the shapes of the orbitals shown in Figures 6.19, 6.22, and 6.23.

GIVE IT SOME THOUGHT

Note in Figure 6.22(a) that the color is deep pink in the interior of each lobe but fades to pale pink at the edges. What does this change in color represent?

6.7 | Many-Electron Atoms

One of our goals in this chapter has been to determine the electronic structures of atoms. So far, we have seen that quantum mechanics leads to a very elegant description of the hydrogen atom. This atom, however, has only one electron. How must our description of the electronic structure of atoms change when we consider atoms with two or more electrons (a *many-electron* atom)? To describe these atoms, we must consider not only the nature of orbitals and their relative energies but also how the electrons populate the available orbitals.

Orbitals and Their Energies

The quantum mechanical model would not be very useful if we could not extend what we have learned about hydrogen to other atoms. Fortunately, we can describe the electronic structure of a many-electron atom in terms of orbitals like those of the hydrogen atom. Thus, we can continue to designate orbitals as $1s$, $2p_x$, and so forth. Further, these orbitals have the same general shapes as the corresponding hydrogen orbitals.

Although the shapes of the orbitals for many-electron atoms are the same as those for hydrogen, the presence of more than one electron greatly changes the energies of the orbitals. In hydrogen the energy of an orbital depends only on its principal quantum number, *n* (Figure 6.17); the $3s$, $3p$, and $3d$ subshells all have the same energy, for instance. In a many-electron atom, however, the electron-electron repulsions cause the different subshells to be at different energies, as shown in Figure 6.24 ◀. To understand why this is so, we must consider the forces between the electrons and how these forces are affected by the shapes of the orbitals. We will, however, forgo this analysis until Chapter 7.

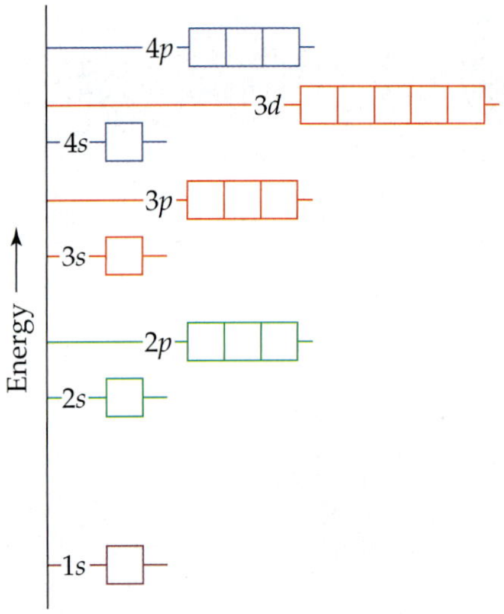

▲ **Figure 6.24 Orbital energy levels in many-electron atoms.** In a many-electron atom, the energies of the subshells in each shell follow the order $ns < np < nd < nf$. As in Figure 6.17, each box represents an orbital.

The important idea is this: *In a many-electron atom, for a given value of n, the energy of an orbital increases with increasing value of l.* You can see this illustrated in Figure 6.24. Notice, for example, that the $n = 3$ orbitals (red) increase in energy in the order $3s < 3p < 3d$. Figure 6.24 is a *qualitative* energy-level diagram; the exact energies of the orbitals and their spacings differ from one atom to another. Notice that all orbitals of a given subshell (such as the five $3d$ orbitals) still have the same energy as one another, just as they do in the hydrogen atom. Orbitals with the same energy are said to be **degenerate**.

GIVE IT SOME THOUGHT

For a many-electron atom, can we predict unambiguously whether the $4s$ orbital is lower in energy or higher in energy than the $3d$ orbitals?

Electron Spin and the Pauli Exclusion Principle

We have now seen that we can use hydrogen-like orbitals to describe many-electron atoms. What, however, determines which orbitals the electrons reside in? That is, how do the electrons of a many-electron atom populate the available orbitals? To answer this question, we must consider an additional property of the electron.

When scientists studied the line spectra of many-electron atoms in great detail, they noticed a very puzzling feature: Lines that were originally thought to be single were actually closely spaced pairs. This meant, in essence, that there were twice as many energy levels as there were "supposed" to be. In 1925 the Dutch physicists George Uhlenbeck and Samuel Goudsmit proposed a solution to this dilemma. They postulated that electrons have an intrinsic property, called **electron spin**, that causes each electron to behave as if it were a tiny sphere spinning on its own axis.

By now it probably does not surprise you to learn that electron spin is quantized. This observation led to the assignment of a new quantum number for the electron, in addition to n, l, and m_l that we have already discussed. This new quantum number, the **spin magnetic quantum number**, is denoted m_s (the subscript s stands for *spin*). Two possible values are allowed for m_s, $+\frac{1}{2}$ or $-\frac{1}{2}$, which was first interpreted as indicating the two opposite directions in which the electron can spin. A spinning charge produces a magnetic field. The two opposite directions of spin therefore produce oppositely directed magnetic fields, as shown in Figure 6.25 ▶.* These two opposite magnetic fields lead to the splitting of spectral lines into closely spaced pairs.

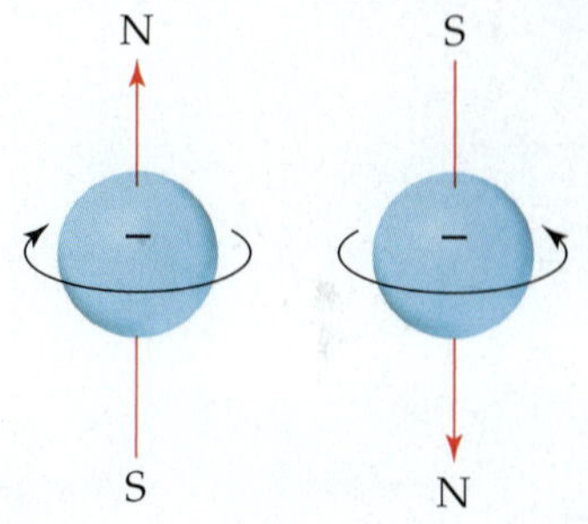

▲ **Figure 6.25 Electron spin.** The electron behaves as if it were spinning about an axis, thereby generating a magnetic field whose direction depends on the direction of spin. The two directions for the magnetic field correspond to the two possible values for the spin quantum number, m_s.

Electron spin is crucial for understanding the electronic structures of atoms. In 1925 the Austrian-born physicist Wolfgang Pauli (1900–1958) discovered the principle that governs the arrangements of electrons in many-electron atoms. The **Pauli exclusion principle** states that *no two electrons in an atom can have the same set of four quantum numbers* n, l, m_l, and m_s. For a given orbital ($1s$, $2p_z$, and so forth), the values of n, l, and m_l are fixed. Thus, if we want to put more than one electron in an orbital *and* satisfy the Pauli exclusion principle, our only choice is to assign different m_s values to the electrons. Because there are only two such values, we conclude that *an orbital can hold a maximum of two electrons and they must have opposite spins*. This restriction allows us to index the electrons in an atom, giving their quantum numbers and thereby defining the region in space where each electron is most likely to be found. It also provides the key to one of the great problems in chemistry—understanding the structure of the periodic table of the elements. We will discuss these issues in the next two sections.

* As we discussed earlier, the electron has both particle-like and wavelike properties. Thus, the picture of an electron as a spinning charged sphere is, strictly speaking, just a useful pictorial representation that helps us understand the two directions of magnetic field that an electron can possess.

A CLOSER LOOK | Experimental Evidence for Electron Spin

Even before electron spin had been proposed, there was experimental evidence that electrons had an additional property that needed explanation. In 1921, Otto Stern and Walter Gerlach succeeded in separating a beam of neutral atoms into two groups by passing them through a nonhomogeneous magnetic field. Their experiment is diagrammed in Figure 6.26 ▶. Let's assume that they used a beam of hydrogen atoms (in actuality, they used silver atoms, which contain just one unpaired electron). We would normally expect that neutral atoms would not be affected by a magnetic field. However, the magnetic field arising from the electron's spin interacts with the magnet's field, deflecting the atom from its straight-line path. As shown in Figure 6.26, the magnetic field splits the beam in two, suggesting that there are two (and only two) equivalent values for the electron's own magnetic field. The Stern–Gerlach experiment could be readily interpreted once it was realized that there are exactly two values for the spin of the electron. These values will produce equal magnetic fields that are opposite in direction.

Related Exercise: 6.94

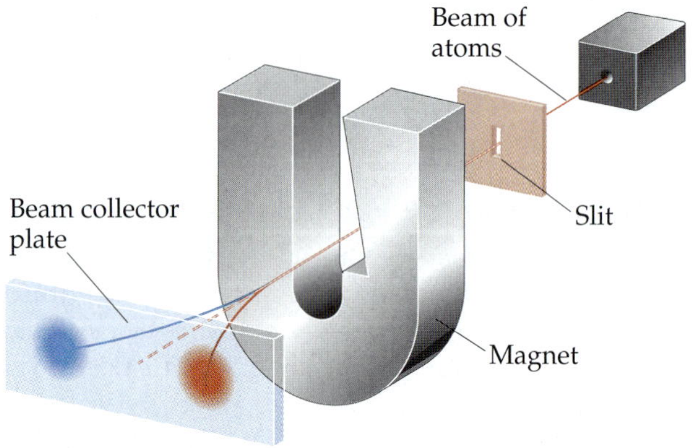

▲ **Figure 6.26 The Stern–Gerlach experiment.** Atoms in which the electron spin quantum number (m_s) of the unpaired electron is $+\frac{1}{2}$ are deflected in one direction, and those in which m_s is $-\frac{1}{2}$ are deflected in the other.

6.8 | Electron Configurations

Armed with a knowledge of the relative energies of orbitals and the Pauli exclusion principle, we are now in a position to consider the arrangements of electrons in atoms. The way in which the electrons are distributed among the various orbitals of an atom is called the **electron configuration** of the atom. The most stable electron configuration of an atom—the ground state—is that in which the electrons are in the lowest possible energy states. If there were no restrictions on the possible values for the quantum numbers of the electrons, all the electrons would crowd into the 1*s* orbital because it is the lowest in energy (Figure 6.24). The Pauli exclusion principle tells us, however, that there can be at most two electrons in any single orbital. Thus, *the orbitals are filled in order of increasing energy, with no more than two electrons per orbital.* For example, consider the lithium atom, which has three electrons. (Recall that the number of electrons in a neutral atom equals its atomic number.) The 1*s* orbital can accommodate two of the electrons. The third one goes into the next lowest energy orbital, the 2*s*.

We can represent any electron configuration by writing the symbol for the occupied subshell and adding a superscript to indicate the number of electrons in that subshell. For example, for lithium we write $1s^22s^1$ (read "1*s* two, 2*s* one"). We can also show the arrangement of the electrons as

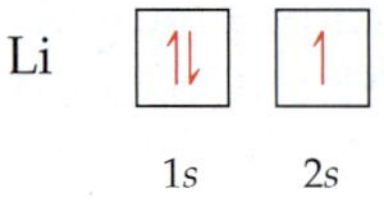

In this kind of representation, which we call an *orbital diagram*, each orbital is denoted by a box and each electron by a half arrow. A half arrow pointing up (↿) represents an electron with a positive spin magnetic quantum number $\left(m_s = +\frac{1}{2}\right)$, and a half arrow pointing down (⇂) represents an electron with a negative spin magnetic quantum number $\left(m_s = -\frac{1}{2}\right)$. This pictorial representation of electron spin is quite convenient. In fact, chemists and physicists often

CHEMISTRY AND LIFE | Nuclear Spin and Magnetic Resonance Imaging

A major challenge facing medical diagnosis is seeing inside the human body from the outside. Until recently, this was accomplished primarily by using X rays to image human bones, muscles, and organs. However, there are several drawbacks to using X rays for medical imaging. First, X rays do not give well-resolved images of overlapping physiological structures. Moreover, because damaged or diseased tissue often yields the same image as healthy tissue, X rays frequently fail to detect illness or injuries. Finally, X rays are high-energy radiation that can cause physiological harm, even in low doses.

In the 1980s a new technique called *magnetic resonance imaging* (MRI) moved to the forefront of medical imaging technology. The foundation of MRI is a phenomenon called nuclear magnetic resonance (NMR), which was discovered in the mid-1940s. Today NMR has become one of the most important spectroscopic methods used in chemistry. It is based on the observation that, like electrons, the nuclei of many elements possess an intrinsic spin. Like electron spin, nuclear spin is quantized. For example, the nucleus of ^1H (a proton) has two possible magnetic nuclear spin quantum numbers, $+\frac{1}{2}$ and $-\frac{1}{2}$. The hydrogen nucleus is the most common one studied by NMR.

A spinning hydrogen nucleus acts like a tiny magnet. In the absence of external effects, the two spin states have the same energy. However, when the nuclei are placed in an external magnetic field, they can align either parallel or opposed (antiparallel) to the field, depending on their spin. The parallel alignment is lower in energy than the antiparallel one by a certain amount, ΔE (Figure 6.27 ▼). If the nuclei are irradiated with photons with energy equal to ΔE, the spin of the nuclei can be "flipped," that is, excited from the parallel to the antiparallel alignment. Detection of the flipping of nuclei between the two spin states leads to an NMR spectrum. The radiation used in an NMR experiment is in the radiofrequency range, typically 100 to 900 MHz.

Because hydrogen is a major constituent of aqueous body fluids and fatty tissue, the hydrogen nucleus is the most convenient one for study by MRI. In MRI a person's body is placed in a strong magnetic field. By irradiating the body with pulses of radiofrequency radiation and using sophisticated detection techniques, tissue can be imaged at specific depths within the body, giving pictures with spectacular detail (Figure 6.28 ▼). The ability to sample at different depths allows medical technicians to construct a three-dimensional picture of the body.

MRI has none of the disadvantages of X rays. Diseased tissue appears very different from healthy tissue, resolving overlapping structures at different depths in the body is much easier, and the radiofrequency radiation is not harmful to humans in the doses used. The technique has had such a profound influence on the modern practice of medicine that Paul Lauterbur, a chemist, and Peter Mansfield, a physicist, were awarded the 2003 Nobel Prize in Physiology or Medicine for their discoveries concerning MRI. The major drawback of this technique is expense: The current cost of a new MRI instrument for clinical applications is over $1.5 million.

Related Exercises: 6.94 and 6.95

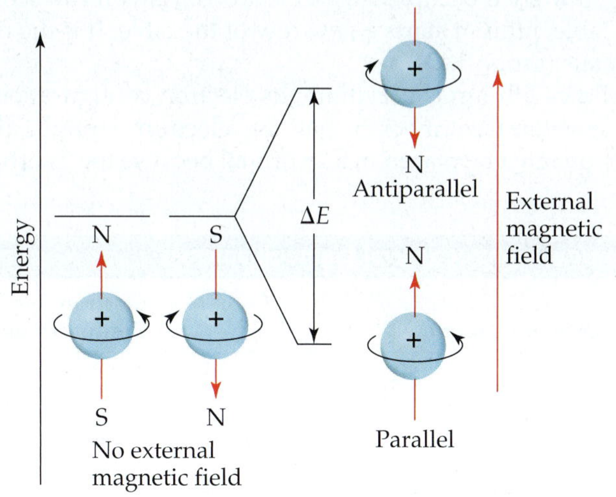

▲ **Figure 6.27 Nuclear spin.** Like electron spin, nuclear spin generates a small magnetic field and has two allowed values. In the absence of an external magnetic field (left), the two spin states have the same energy. If an external magnetic field is applied (right), the parallel alignment of the nuclear magnetic field is lower in energy than the antiparallel alignment. The energy difference, ΔE, is in the radio frequency portion of the electromagnetic spectrum.

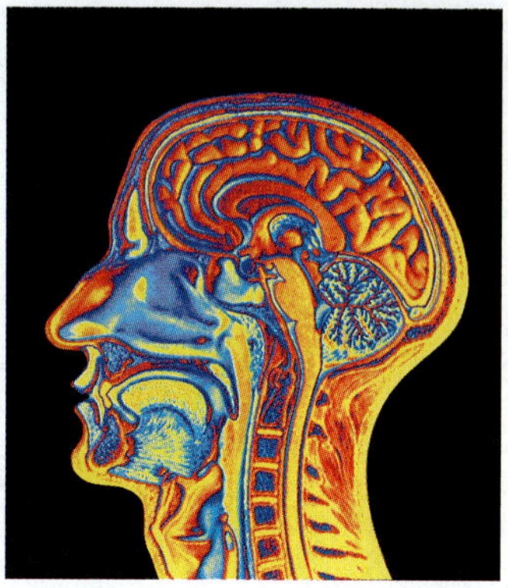

◀ **Figure 6.28 MRI image.** This image of a human head, obtained using MRI, shows the structures of a normal brain, airways, and facial tissues.

refer to electrons as "spin-up" and "spin-down" rather than specifying the value for m_s.

Electrons having opposite spins are said to be *paired* when they are in the same orbital (↑↓). An *unpaired electron* is one not accompanied by a partner of opposite spin. In the lithium atom the two electrons in the 1*s* orbital are paired and the electron in the 2*s* orbital is unpaired.

Hund's Rule

Consider now how the electron configurations of the elements change as we move from element to element across the periodic table. Hydrogen has one electron, which occupies the 1*s* orbital in its ground state.

ANIMATION Electron Configurations

H [↑] : $1s^1$
1*s*

The choice of a spin-up electron here is arbitrary; we could equally well show the ground state with one spin-down electron in the 1*s* orbital. It is customary, however, to show unpaired electrons with their spins up.

The next element, helium, has two electrons. Because two electrons with opposite spins can occupy an orbital, both of helium's electrons are in the 1*s* orbital.

He [↑↓] : $1s^2$
1*s*

The two electrons present in helium complete the filling of the first shell. This arrangement represents a very stable configuration, as is evidenced by the chemical inertness of helium.

The electron configurations of lithium and several elements that follow it in the periodic table are shown in Table 6.3 ▼. For the third electron of lithium, the change in principal quantum number represents a large jump in energy and a corresponding jump in the average distance of the electron from the nucleus. It represents the start of a new shell occupied with electrons. As you can see by examining the periodic table, lithium starts a new row of the table. It is the first member of the alkali metals (group 1A).

The element that follows lithium is beryllium; its electron configuration is $1s^22s^2$ (Table 6.3). Boron, atomic number 5, has the electron configuration $1s^22s^22p^1$. The fifth electron must be placed in a 2*p* orbital because the 2*s* orbital

ACTIVITY Electron Configuration

TABLE 6.3 Electron Configurations of Several Lighter Elements

Element	Total Electrons	Orbital Diagram 1*s*	2*s*	2*p*	3*s*	Electron Configuration
Li	3	↑↓	↑	[ ][ ][ ]	[ ]	$1s^22s^1$
Be	4	↑↓	↑↓	[ ][ ][ ]	[ ]	$1s^22s^2$
B	5	↑↓	↑↓	[↑][ ][ ]	[ ]	$1s^22s^22p^1$
C	6	↑↓	↑↓	[↑][↑][ ]	[ ]	$1s^22s^22p^2$
N	7	↑↓	↑↓	[↑][↑][↑]	[ ]	$1s^22s^22p^3$
Ne	10	↑↓	↑↓	[↑↓][↑↓][↑↓]	[ ]	$1s^22s^22p^6$
Na	11	↑↓	↑↓	[↑↓][↑↓][↑↓]	↑	$1s^22s^22p^63s^1$

is filled. Because all the three $2p$ orbitals are of equal energy, it doesn't matter which $2p$ orbital is occupied.

With the next element, carbon, we encounter a new situation. We know that the sixth electron must go into a $2p$ orbital. However, does this new electron go into the $2p$ orbital that already has one electron, or into one of the other two $2p$ orbitals? This question is answered by **Hund's rule**, which states that *for degenerate orbitals, the lowest energy is attained when the number of electrons with the same spin is maximized*. This means that electrons will occupy orbitals singly to the maximum extent possible and that these single electrons in a given subshell will all have the same spin magnetic quantum number. Electrons arranged in this way are said to have *parallel spins*. For a carbon atom to achieve its lowest energy, therefore, the two $2p$ electrons will have the same spin. In order for this to happen, the electrons must be in different $2p$ orbitals, as shown in Table 6.3. Thus, a carbon atom in its ground state has two unpaired electrons. Similarly, for nitrogen in its ground state, Hund's rule requires that the three $2p$ electrons singly occupy each of the three $2p$ orbitals. This is the only way that all three electrons can have the same spin. For oxygen and fluorine, we place four and five electrons, respectively, in the $2p$ orbitals. To achieve this, we pair up electrons in the $2p$ orbitals, as we will see in Sample Exercise 6.7.

Hund's rule is based in part on the fact that electrons repel one another. By occupying different orbitals, the electrons remain as far as possible from one another, thus minimizing electron-electron repulsions.

SAMPLE EXERCISE 6.7 | Orbital Diagrams and Electron Configurations

Draw the orbital diagram for the electron configuration of oxygen, atomic number 8. How many unpaired electrons does an oxygen atom possess?

Solution

Analyze and Plan: Because oxygen has an atomic number of 8, each oxygen atom has 8 electrons. Figure 6.24 shows the ordering of orbitals. The electrons (represented as arrows) are placed in the orbitals (represented as boxes) beginning with the lowest-energy orbital, the $1s$. Each orbital can hold a maximum of two electrons (the Pauli exclusion principle). Because the $2p$ orbitals are degenerate, we place one electron in each of these orbitals (spin-up) before pairing any electrons (Hund's rule).

Solve: Two electrons each go into the $1s$ and $2s$ orbitals with their spins paired. This leaves four electrons for the three degenerate $2p$ orbitals. Following Hund's rule, we put one electron into each $2p$ orbital until all three orbitals have one electron each. The fourth electron is then paired up with one of the three electrons already in a $2p$ orbital, so that the representation is

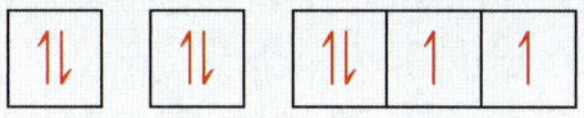

1s 2s 2p

The corresponding electron configuration is written $1s^2 2s^2 2p^4$. The atom has two unpaired electrons.

PRACTICE EXERCISE

(a) Write the electron configuration for phosphorus, element 15. **(b)** How many unpaired electrons does a phosphorus atom possess?

Answers: **(a)** $1s^2 2s^2 2p^6 3s^2 3p^3$, **(b)** three

Condensed Electron Configurations

The filling of the $2p$ subshell is complete at neon (Table 6.3), which has a stable configuration with eight electrons (an *octet*) in the outermost occupied shell. The next element, sodium, atomic number 11, marks the beginning of a new row of the periodic table. Sodium has a single $3s$ electron beyond the stable configuration of neon. We can therefore abbreviate the electron configuration of sodium as

$$\text{Na:} \quad [\text{Ne}]3s^1$$

The symbol [Ne] represents the electron configuration of the ten electrons of neon, $1s^22s^22p^6$. Writing the electron configuration as $[\text{Ne}]3s^1$ helps focus attention on the outermost electrons of the atom, which are the ones largely responsible for the chemical behavior of an element.

We can generalize what we have just done for the electron configuration of sodium: In writing the *condensed electron configuration* of an element, the electron configuration of the nearest noble-gas element of lower atomic number is represented by its chemical symbol in brackets. For example, we can write the electron configuration of lithium as

$$\text{Li:} \quad [\text{He}]2s^1$$

We refer to the electrons represented by the symbol for a noble gas as the *noble-gas core* of the atom. More usually, these inner-shell electrons are referred to merely as the **core electrons**. The electrons given after the noble-gas core are called the *outer-shell electrons*. The outer-shell electrons include the electrons involved in chemical bonding, which are called the **valence electrons.** For ligher elements (those with atomic number of 30 or less), all of the outer-shell electrons are valence electrons. As we will discuss later, many of the heavier elements have completely filled subshells in their outer-shell electrons that are not involved in bonding and are therefore not considered valence electrons.

By comparing the condensed electron configuration of lithium with that of sodium, we can appreciate why these two elements are so similar chemically: They have the same type of electron configuration in the outermost occupied shell. Indeed, all the members of the alkali metal group (1A) have a single *s* valence electron beyond a noble-gas configuration.

Transition Metals

The noble-gas element argon marks the end of the row started by sodium. The configuration for argon is $1s^22s^22p^63s^23p^6$. The element following argon in the periodic table is potassium (K), atomic number 19. In all its chemical properties, potassium is clearly a member of the alkali metal group. The experimental facts about the properties of potassium leave no doubt that the outermost electron of this element occupies an *s* orbital. But this means that the electron that has the highest energy has *not* gone into a 3*d* orbital, which we might have expected it to do. Here the ordering of energy levels is such that the 4*s* orbital is lower in energy than the 3*d* (Figure 6.24). Hence, the condensed electron configuration of potassium is

$$\text{K:} \quad [\text{Ar}]4s^1$$

Following complete filling of the 4*s* orbital (this occurs in the calcium atom), the next set of orbitals to be filled is the 3*d*. (You will find it helpful as we go along to refer often to the periodic table on the front inside cover.) Beginning with scandium and extending through zinc, electrons are added to the five 3*d* orbitals until they are completely filled. Thus, the fourth row of the periodic table is ten elements wider than the two previous rows. These ten elements are known as either **transition elements** or **transition metals**. Note the position of these elements in the periodic table.

In deriving the electron configurations of the transition elements, the orbitals are filled in accordance with Hund's rule—electrons are added to the 3*d* orbitals singly until all five orbitals have one electron each. Additional electrons are then placed in the 3*d* orbitals with spin pairing until the shell is completely filled. The condensed electron configurations and the corresponding orbital diagram representations of two transition elements are as follows:

		4s	3d
Mn: $[Ar]4s^23d^5$	or [Ar]	⇅	↑ ↑ ↑ ↑ ↑
Zn: $[Ar]4s^23d^{10}$	or [Ar]	⇅	⇅ ⇅ ⇅ ⇅ ⇅

Once all the 3*d* orbitals have been filled with two electrons each, the 4*p* orbitals begin to be occupied until the completed octet of outer electrons ($4s^24p^6$) is reached with krypton (Kr), atomic number 36, another of the noble gases. Rubidium (Rb) marks the beginning of the fifth row. Refer again to the periodic table on the front inside cover. Notice that this row is in every respect like the preceding one, except that the value for n is greater by 1.

GIVE IT SOME THOUGHT

Based on the structure of the periodic table, which becomes occupied first, the 6*s* orbital or the 5*d* orbitals?

The Lanthanides and Actinides

The sixth row of the periodic table begins similarly to the preceding one: one electron in the 6*s* orbital of cesium (Cs) and two electrons in the 6*s* orbital of barium (Ba). Notice, however, that the periodic table then has a break, and the subsequent set of elements (elements 57–70) is placed below the main portion of the table. It is at this place that we begin to encounter a new set of orbitals, the 4*f*.

There are seven degenerate 4*f* orbitals, corresponding to the seven allowed values of m_l, ranging from 3 to -3. Thus, it takes 14 electrons to fill the 4*f* orbitals completely. The 14 elements corresponding to the filling of the 4*f* orbitals are known as either the **lanthanide elements** or the **rare earth elements**. They are set below the other elements to avoid making the periodic table unduly wide. The properties of the lanthanide elements are all quite similar, and these elements occur together in nature. For many years it was virtually impossible to separate them from one another.

Because the energies of the 4*f* and 5*d* orbitals are very close to each other, the electron configurations of some of the lanthanides involve 5*d* electrons. For example, the elements lanthanum (La), cerium (Ce), and praseodymium (Pr) have the following electron configurations:

$[Xe]6s^25d^1$	$[Xe]6s^25d^14f^1$	$[Xe]6s^24f^3$
Lanthanum	Cerium	Praseodymium

Because La has a single 5*d* electron, it is sometimes placed below yttrium (Y) as the first member of the third series of transition elements, and Ce is then placed as the first member of the lanthanides. Based on their chemistry, however, La can be considered the first element in the lanthanide series. Arranged this way, there are fewer apparent exceptions to the regular filling of the 4*f* orbitals among the subsequent members of the series.

After the lanthanide series, the third transition element series is completed by the filling of the 5*d* orbitals, followed by the filling of the 6*p* orbitals. This brings us to radon (Rn), heaviest of the known noble-gas elements.

The final row of the periodic table begins by filling the 7*s* orbitals. The **actinide elements**, of which uranium (U, element 92) and plutonium (Pu, element 94) are the best known, are then built up by completing the 5*f* orbitals. The actinide elements are radioactive, and most of them are not found in nature.

6.9 | Electron Configurations and the Periodic Table

TABLE 6.4 Electron Configurations of the Group 2A and 3A Elements

Group 2A	
Be	$[He]2s^2$
Mg	$[Ne]3s^2$
Ca	$[Ar]4s^2$
Sr	$[Kr]5s^2$
Ba	$[Xe]6s^2$
Ra	$[Rn]7s^2$
Group 3A	
B	$[He]2s^22p^1$
Al	$[Ne]3s^23p^1$
Ga	$[Ar]3d^{10}4s^24p^1$
In	$[Kr]4d^{10}5s^25p^1$
Tl	$[Xe]4f^{14}5d^{10}6s^26p^1$

Our rather brief survey of electron configurations of the elements has taken us through the periodic table. We have seen that the electron configurations of the elements are related to their locations in the periodic table. The periodic table is structured so that elements with the same pattern of outer-shell (valence) electron configuration are arranged in columns. For example, the electron configurations for the elements in groups 2A and 3A are given in Table 6.4 ◀. We see that the 2A elements all have ns^2 outer configurations, while the 3A elements all have ns^2np^1 configurations.

Earlier, in Table 6.2, we saw that the total number of orbitals in each shell is equal to n^2: 1, 4, 9, or 16. Because each orbital can hold two electrons, each shell can accommodate up to $2n^2$ electrons: 2, 8, 18, or 32. The structure of the periodic table reflects this orbital structure. The first row has two elements, the second and third rows have eight elements, the fourth and fifth rows have 18 elements, and the sixth row has 32 elements (including the lanthanide metals). Some of the numbers repeat because we reach the end of a row of the periodic table before a shell completely fills. For example, the third row has eight elements, which corresponds to filling the 3*s* and 3*p* orbitals. The remaining orbitals of the third shell, the 3*d* orbitals, do not begin to fill until the fourth row of the periodic table (and after the 4*s* orbital is filled). Likewise, the 4*d* orbitals don't begin to fill until the fifth row of the table, and the 4*f* orbitals don't begin filling until the sixth row.

All these observations are evident in the structure of the periodic table. For this reason, we will emphasize that *the periodic table is your best guide to the order in which orbitals are filled*. You can easily write the electron configuration of an element based on its location in the periodic table. The pattern is summarized in Figure 6.29 ◀. Notice that the elements can be grouped by the *type* of orbital into which the electrons are placed. On the left are *two* columns of elements, depicted in blue. These elements, known as the alkali metals (group 1A) and alkaline earth metals (group 2A), are those in which the valence *s* orbitals are being filled. On the right is a pink block of *six* columns. These are the elements in which the valence *p* orbitals are being filled. The *s* block and the *p* block of the periodic table together are the **representative elements**, which are sometimes called the **main-group elements**.

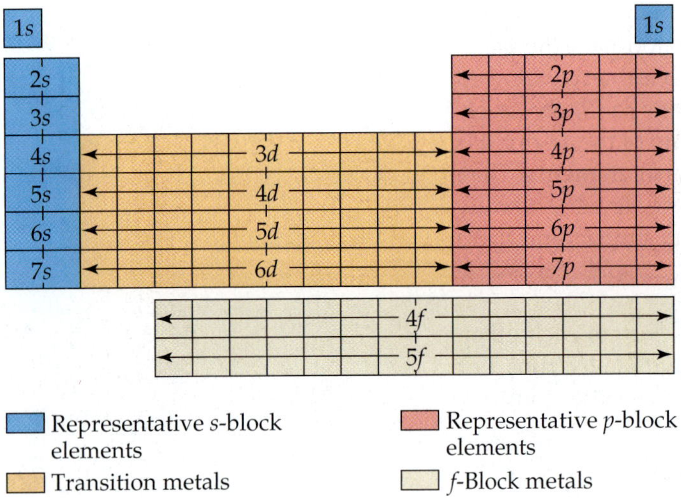

▲ **Figure 6.29 Regions of the periodic table.** This block diagram of the periodic table shows the order in which electrons are added to orbitals as we move through the table from beginning to end.

In the middle of Figure 6.29 is a gold block of *ten* columns containing the transition metals. These are the elements in which the valence *d* orbitals are being filled. Below the main portion of the table are two tan rows containing *14* columns. These elements are often referred to as the ***f*-block metals** because they are the ones in which the valence *f* orbitals are being filled. Recall that the numbers 2, 6, 10, and 14 are precisely the number of electrons that can fill the *s*, *p*, *d*, and *f* subshells, respectively. Recall also that the 1*s* subshell is the first *s* subshell, the 2*p* is the first *p* subshell, the 3*d* is the first *d* subshell, and the 4*f* is the first *f* subshell.

SAMPLE EXERCISE 6.8 | Electron Configurations for a Group

What is the characteristic valence electron configuration of the group 7A elements, the halogens?

Solution

Analyze and Plan: We first locate the halogens in the periodic table, write the electron configurations for the first two elements, and then determine the general similarity between them.

Solve: The first member of the halogen group is fluorine, atomic number 9. The condensed electron configuration for fluorine is

$$\text{F:}\quad [\text{He}]2s^2 2p^5$$

Similarly, that for chlorine, the second halogen, is

$$\text{Cl:}\quad [\text{Ne}]3s^2 3p^5$$

From these two examples, we see that the characteristic valence electron configuration of a halogen is ns^2np^5, where n ranges from 2 in the case of fluorine to 6 in the case of astatine.

PRACTICE EXERCISE

Which family of elements is characterized by an ns^2np^2 electron configuration in the outermost occupied shell?
Answer: group 4A

SAMPLE EXERCISE 6.9 | Electron Configurations from the Periodic Table

(a) Write the electron configuration for bismuth, element number 83. **(b)** Write the condensed electron configuration for this element. **(c)** How many unpaired electrons does each atom of bismuth possess?

Solution (a) We write the electron configuration by moving across the periodic table one row at a time and writing the occupancies of the orbital corresponding to each row (refer to Figure 6.29).

First row	$1s^2$
Second row	$2s^2 2p^6$
Third row	$3s^2 3p^6$
Fourth row	$4s^2 3d^{10} 4p^6$
Fifth row	$5s^2 4d^{10} 5p^6$
Sixth row	$6s^2 4f^{14} 5d^{10} 6p^3$
Total:	$1s^2 2s^2 2p^6 3s^2 3p^6 3d^{10} 4s^2 4p^6 4d^{10} 4f^{14} 5s^2 5p^6 5d^{10} 6s^2 6p^3$

Note that 3 is the lowest possible value that n may have for a d orbital and that 4 is the lowest possible value of n for an f orbital.

The total of the superscripted numbers should equal the atomic number of bismuth, 83. The electrons may be listed, as shown above in the "Total" row, in the order of increasing principal quantum number. However, it is equally correct to list the orbitals in the order in which they are read from Figure 6.29: $1s^2 2s^2 2p^6 3s^2 3p^6 4s^2 3d^{10} 4p^6 5s^2 4d^{10} 5p^6 6s^2 4f^{14} 5d^{10} 6p^3$.

(b) We write the condensed electron configuration by locating bismuth on the periodic table and then moving *backward* to the nearest noble gas, which is Xe, element 54. Thus, the noble-gas core is [Xe]. The outer electrons are then read from the periodic table as before. Moving from Xe to Cs, element 55, we find ourselves in the sixth row. Moving across this row to Bi gives us the outer electrons. Thus, the abbreviated electron configuration is $[\text{Xe}]6s^2 4f^{14} 5d^{10} 6p^3$ or $[\text{Xe}]4f^{14} 5d^{10} 6s^2 6p^3$.

(c) We can see from the abbreviated electron configuration that the only partially occupied subshell is the $6p$. The orbital diagram representation for this subshell is

In accordance with Hund's rule, the three $6p$ electrons occupy the three $6p$ orbitals singly, with their spins parallel. Thus, there are three unpaired electrons in each atom of bismuth.

PRACTICE EXERCISE

Use the periodic table to write the condensed electron configurations for **(a)** Co (atomic number 27), **(b)** Te (atomic number 52).
Answers: **(a)** $[\text{Ar}]4s^2 3d^7$ or $[\text{Ar}]3d^7 4s^2$, **(b)** $[\text{Kr}]5s^2 4d^{10} 5p^4$ or $[\text{Kr}]4d^{10} 5s^2 5p^4$

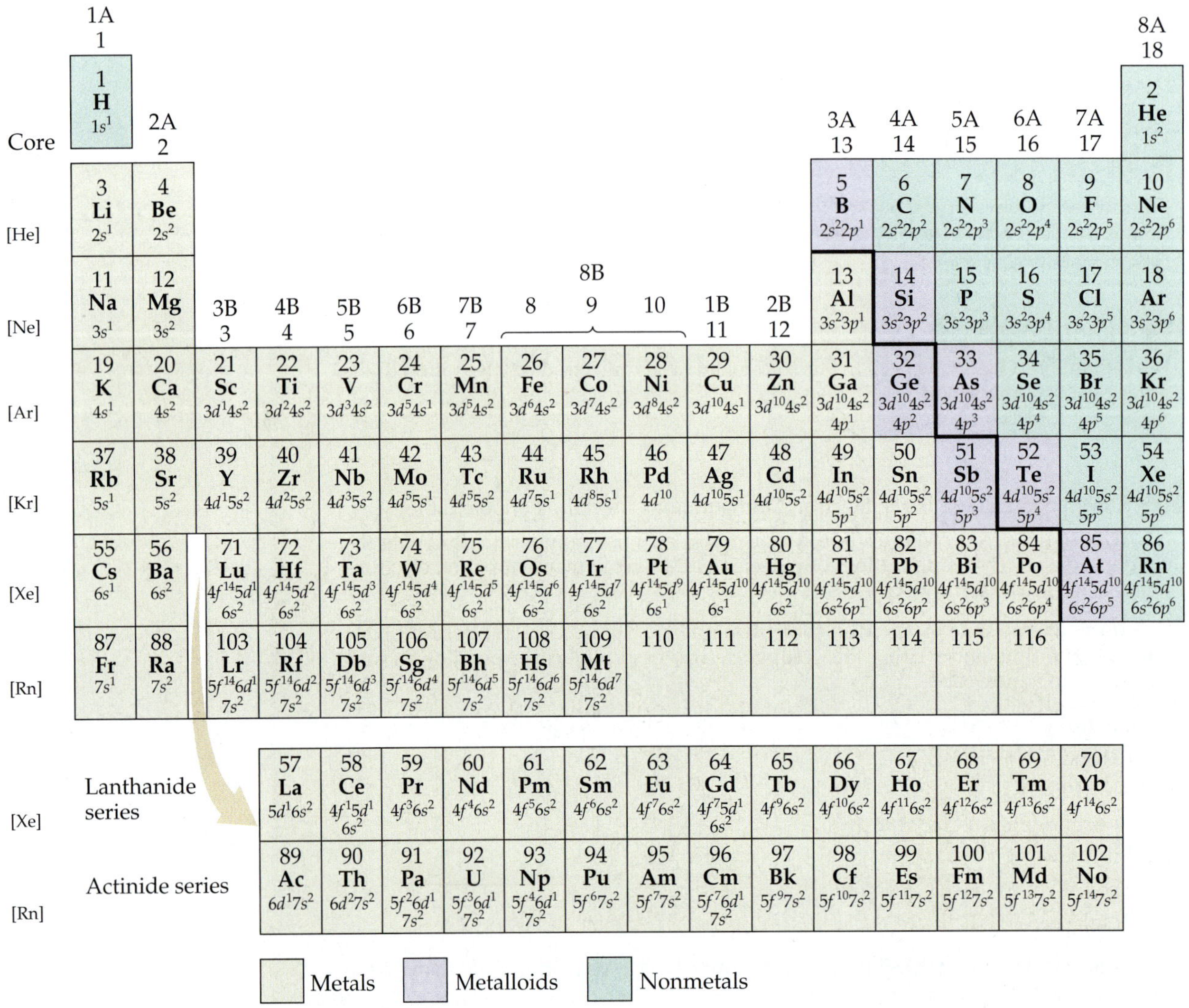

▲ **Figure 6.30 Valence electron configurations of the elements.**

ACTIVITY Periodic Table

Figure 6.30 ▲ gives the valence ground-state electron configurations for all the elements. You can use this figure to check your answers as you practice writing electron configurations. We have written these configurations with orbitals listed in order of increasing principal quantum number. As we saw in Sample Exercise 6.9, the orbitals can also be listed in order of filling, as they would be read off the periodic table.

The electron configurations in Figure 6.30 allow us to reexamine the concept of *valence electrons*. Notice, for example, that as we proceed from Cl ([Ne]$3s^23p^5$) to Br ([Ar]$3d^{10}4s^24p^5$) we have added a complete subshell of 3*d* electrons to the outer-shell electrons beyond the noble-gas core of Ar. Although the 3*d* electrons are outer-shell electrons, they are not involved in chemical bonding and are therefore not considered valence electrons. Thus, we consider only the 4*s* and 4*p* electrons of Br to be valence electrons. Similarly, if we compare the electron configuration of Ag and Au, Au has a completely full $4f^{14}$ subshell beyond its noble-gas core, but those 4*f* electrons are not involved in bonding. In general, *for representative elements we do not consider completely full* d *or* f *subshells to be among the valence electrons*, and *for transition elements we likewise do not consider a completely full* f *subshell to be among the valence electrons*.

Anomalous Electron Configurations

If you inspect Figure 6.30 closely, you will see that the electron configurations of certain elements appear to violate the rules we have just discussed. For example, the electron configuration of chromium is $[Ar]3d^54s^1$ rather than the $[Ar]3d^44s^2$ configuration we might have expected. Similarly, the configuration of copper is $[Ar]3d^{10}4s^1$ instead of $[Ar]3d^94s^2$. This anomalous behavior is largely a consequence of the closeness of the 3*d* and 4*s* orbital energies. It frequently occurs when there are enough electrons to lead to precisely half-filled sets of degenerate orbitals (as in chromium) or to completely fill a *d* subshell (as in copper). There are a few similar cases among the heavier transition metals (those with partially filled 4*d* or 5*d* orbitals) and among the *f*-block metals. Although these minor departures from the expected are interesting, they are not of great chemical significance.

GIVE IT SOME THOUGHT

The elements Ni, Pd, and Pt are all in the same group. By examining the electron configurations for these elements in Figure 6.30, what can you conclude about the relative energies of the *nd* and $(n + 1)s$ orbitals for this group?

SAMPLE INTEGRATIVE EXERCISE | Putting Concepts Together

Boron, atomic number 5, occurs naturally as two isotopes, ^{10}B and ^{11}B, with natural abundances of 19.9% and 80.1%, respectively. **(a)** In what ways do the two isotopes differ from each other? Does the electronic configuration of ^{10}B differ from that of ^{11}B? **(b)** Draw the orbital diagram for an atom of ^{11}B. Which electrons are the valence electrons? **(c)** Indicate three major ways in which the 1*s* electrons in boron differ from the 2*s* electrons in the atom. **(d)** Elemental boron reacts with fluorine to form BF_3, a gas. Write a balanced chemical equation for the reaction of solid boron with fluorine gas. **(e)** $\Delta H_f°$ for $BF_3(g)$ is -1135.6 kJ mol^{-1}. Calculate the standard enthalpy change in the reaction of boron with fluorine. **(f)** When BCl_3, also a gas at room temperature, comes into contact with water, the two react to form hydrochloric acid and boric acid, H_3BO_3, a very weak acid in water. Write a balanced net ionic equation for this reaction.

Solution (a) The two nuclides of boron differ in the number of neutrons in the nucleus. (Sections 2.3 and 2.4) Each of the nuclides contains five protons, but ^{10}B contains five neutrons, whereas ^{11}B contains six neutrons. The two isotopes of boron have identical electron configurations, $1s^22s^22p^1$, because each has five electrons.

(b) The complete orbital diagram is

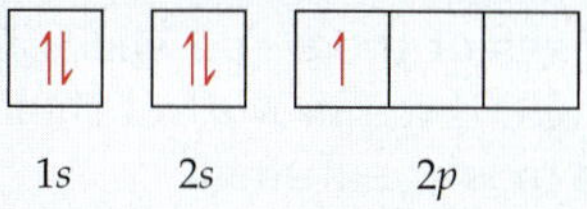

The valence electrons are the ones in the outermost occupied shell, the $2s^2$ and $2p^1$ electrons. The $1s^2$ electrons constitute the core electrons, which we represent as [He] when we write the condensed electron configuration, $[He]2s^22p^1$.

(c) The 1*s* and 2*s* orbitals are both spherical, but they differ in three important respects: First, the 1*s* orbital is lower in energy than the 2*s* orbital. Second, the average distance of the 2*s* electrons from the nucleus is greater than that of the 1*s* electrons, so the 1*s* orbital is smaller than the 2*s*. Third, the 2*s* orbital has one node, whereas the 1*s* orbital has no nodes (Figure 6.18).

(d) The balanced chemical equation is

$$2\,B(s) + 3\,F_2(g) \longrightarrow 2\,BF_3(g)$$

(e) $\Delta H° = 2(-1135.6) - [0 + 0] = -2271.2$ kJ. The reaction is strongly exothermic.

(f) $BCl_3(g) + 3\,H_2O(l) \longrightarrow H_3BO_3(aq) + 3\,H^+(aq) + 3\,Cl^-(aq)$. Note that because H_3BO_3 is a very weak acid, its chemical formula is written in molecular form, as discussed in Section 4.3.

SUMMARY AND KEY TERMS

Introduction and Section 6.1 The **electronic structure** of an atom describes the energies and arrangement of electrons around the atom. Much of what is known about the electronic structure of atoms was obtained by observing the interaction of light with matter. Visible light and other forms of **electromagnetic radiation** (also known as radiant energy) move through a vacuum at the speed of light, $c = 3.00 \times 10^8$ m/s. Electromagnetic radiation has both electric and magnetic components that vary periodically in wavelike fashion. The wave characteristics of radiant energy allow it to be described in terms of **wavelength**, λ, and **frequency**, ν, which are interrelated: $\lambda\nu = c$.

Section 6.2 Planck proposed that the minimum amount of radiant energy that an object can gain or lose is related to the frequency of the radiation: $E = h\nu$. This smallest quantity is called a **quantum** of energy. The constant h is called **Planck's constant:** $h = 6.626 \times 10^{-34}$ J-s. In the quantum theory, energy is quantized, meaning that it can have only certain allowed values. Einstein used the quantum theory to explain the **photoelectric effect**, the emission of electrons from metal surfaces by light. He proposed that light behaves as if it consists of quantized energy packets called **photons**. Each photon carries energy $E = h\nu$.

Section 6.3 Dispersion of radiation into its component wavelengths produces a **spectrum**. If the spectrum contains all wavelengths, it is called a **continuous spectrum**; if it contains only certain specific wavelengths, the spectrum is called a **line spectrum**. The radiation emitted by excited hydrogen atoms forms a line spectrum; the frequencies observed in the spectrum follow a simple mathematical relationship that involves small integers.

Bohr proposed a model of the hydrogen atom that explains its line spectrum. In this model the energy of the hydrogen atom depends on the value of a number n, called the quantum number. The value of n must be a positive integer $(1, 2, 3, \ldots)$, and each value of n corresponds to a different specific energy, E_n. The energy of the atom increases as n increases. The lowest energy is achieved for $n = 1$; this is called the **ground state** of the hydrogen atom. Other values of n correspond to **excited states** of the atom. Light is emitted when the electron drops from a higher energy state to a lower energy state; light must be absorbed to excite the electron from a lower energy state to a higher one. The frequency of light emitted or absorbed must be such that $h\nu$ equals the difference in energy between two allowed states of the atom.

Section 6.4 De Broglie proposed that matter, such as electrons, should exhibit wavelike properties; this hypothesis of **matter waves** was proved experimentally by observing the diffraction of electrons. An object has a characteristic wavelength that depends on its **momentum**, mv: $\lambda = h/mv$. Discovery of the wave properties of the electron led to Heisenberg's **uncertainty principle**, which states that there is an inherent limit to the accuracy with which the position and momentum of a particle can be measured simultaneously.

Section 6.5 In the quantum mechanical model of the hydrogen atom, the behavior of the electron is described by mathematical functions called **wave functions**, denoted with the Greek letter ψ. Each allowed wave function has a precisely known energy, but the location of the electron cannot be determined exactly; rather, the probability of it being at a particular point in space is given by the **probability density**, ψ^2. The **electron density** distribution is a map of the probability of finding the electron at all points in space.

The allowed wave functions of the hydrogen atom are called **orbitals**. An orbital is described by a combination of an integer and a letter, corresponding to values of three quantum numbers for the orbital. The principal quantum number, n, is indicated by the integers 1, 2, 3, ... This quantum number relates most directly to the size and energy of the orbital. The azimuthal quantum number, l, is indicated by the letters s, p, d, f, and so on, corresponding to the values of 0, 1, 2, 3, ... The l quantum number defines the shape of the orbital. For a given value of n, l can have integer values ranging from 0 to $n - 1$. The magnetic quantum number, m_l, relates to the orientation of the orbital in space. For a given value of l, m_l can have integral values ranging from $-l$ to l. Cartesian labels can be used to label the orientations of the orbitals. For example, the three $3p$ orbitals are designated $3p_x$, $3p_y$, and $3p_z$, with the subscripts indicating the axis along which the orbital is oriented.

An **electron shell** is the set of all orbitals with the same value of n, such as $3s$, $3p$, and $3d$. In the hydrogen atom all the orbitals in an electron shell have the same energy. A **subshell** is the set of one or more orbitals with the same n and l values; for example, $3s$, $3p$, and $3d$ are each subshells of the $n = 3$ shell. There is one orbital in an s subshell, three in a p subshell, five in a d subshell, and seven in an f subshell.

Section 6.6 Contour representations are useful for visualizing the spatial characteristics (shapes) of the orbitals. Represented this way, s orbitals appear as spheres that increase in size as n increases. The **radial probability function** tells us the probability that the electron will be found at a certain distance from the nucleus. The wave function for each p orbital has two lobes on opposite sides of the nucleus. They are oriented along the x-, y-, and z-axes. Four of the d orbitals appear as shapes with four lobes around the nucleus; the fifth one, the d_{z^2} orbital, is represented as two lobes along the z-axis and a "doughnut" in the xy plane. Regions in which the wave function is zero are called **nodes**. There is zero probability that the electron will be found at a node.

Section 6.7 In many-electron atoms, different subshells of the same electron shell have different energies. For a given value of n, the energy of the subshells increases as the value of l increases: $ns < np < nd < nf$. Orbitals within the same subshell are **degenerate**, meaning they have the same energy.

Electrons have an intrinsic property called **electron spin**, which is quantized. The **spin magnetic quantum number**, m_s, can have two possible values, $+\frac{1}{2}$ and $-\frac{1}{2}$, which can be envisioned as the two directions of an electron spinning about an axis. The **Pauli exclusion principle** states that no two electrons in an atom can have the same values for n, l, m_l, and m_s. This principle places a limit of two on the number of electrons that can occupy any one atomic orbital. These two electrons differ in their value of m_s.

Sections 6.8 and 6.9 The **electron configuration** of an atom describes how the electrons are distributed among the orbitals of the atom. The ground-state electron configurations are generally obtained by placing the electrons in the atomic orbitals of lowest possible energy with the restriction that each orbital can hold no more than two electrons. When electrons occupy a subshell with more than one degenerate orbital, such as the $2p$ subshell, **Hund's rule** states that the lowest energy is attained by maximizing the number of electrons with the same electron spin. For example, in the ground-state electron configuration of carbon, the two $2p$ electrons have the same spin and must occupy two different $2p$ orbitals.

Elements in any given group in the periodic table have the same type of electron arrangements in their outermost shells. For example, the electron configurations of the halogens fluorine and chlorine are $[He]2s^22p^5$ and $[Ne]3s^23p^5$, respectively. The outer-shell electrons are those that lie outside the orbitals occupied in the next lowest noble-gas element. The outer-shell electrons that are involved in chemical bonding are the **valence electrons** of an atom; for the elements with atomic number 30 or less, all the outer-shell electrons are valence electrons. The electrons that are not valence electrons are called **core electrons**.

The periodic table is partitioned into different types of elements, based on their electron configurations. Those elements in which the outermost subshell is an s or p subshell are called the **representative** (or **main-group**) **elements**. The alkali metals (group 1A), halogens (group 7A), and noble gases (group 8A) are representative elements. Those elements in which a d subshell is being filled are called the **transition elements** (or **transition metals**). The elements in which the $4f$ subshell is being filled are called the **lanthanide** (or **rare earth**) **elements**. The **actinide elements** are those in which the $5f$ subshell is being filled. The lanthanide and actinide elements are collectively referred to as the ***f*-block metals**. These elements are shown as two rows of 14 elements below the main part of the periodic table. The structure of the periodic table, summarized in Figure 6.29, allows us to write the electron configuration of an element from its position in the periodic table.

VISUALIZING CONCEPTS

6.1 Consider the water wave shown here. **(a)** How could you measure the speed of this wave? **(b)** How would you determine the wavelength of the wave? **(c)** Given the speed and wavelength of the wave, how could you determine the frequency of the wave? **(d)** Suggest an independent experiment to determine the frequency of the wave. [Section 6.1]

6.2 A popular kitchen appliance produces electromagnetic radiation with a wavelength of 1 cm. With reference to Figure 6.4, answer the following: **(a)** Would the radiation produced by the appliance be visible by the human eye? **(b)** If the radiation is not visible, do photons of this radiation have more or less energy than photons of visible light? **(c)** Propose an identity to the kitchen appliance. [Section 6.1]

6.3 As shown in the photograph below, an electric stove burner on its highest setting exhibits an orange glow. **(a)** When the burner setting is changed to low, the burner continues to produce heat but the orange glow disappears. How can this observation be explained with reference to one of the fundamental observations that led to the notion of quanta? **(b)** Suppose that the energy provided to the burner could be increased beyond the highest setting of the stove. What would we expect to observe with regard to visible light emitted by the burner? [Section 6.2]

6.4 The familiar phenomenon of a rainbow results from the diffraction of sunlight through raindrops. **(a)** Does the wavelength of light increase or decrease as we proceed outward from the innermost band of the rainbow? **(b)** Does the frequency of light increase or decrease as we proceed outward? **(c)** Suppose that instead of sunlight, the visible light from a hydrogen discharge tube (Figure 6.11) was used as the light source. What do you think the resulting "hydrogen discharge rainbow" would look like? [Section 6.3]

6.5 A certain quantum mechanical system has the energy levels shown in the diagram below. The energy levels are indexed by a single quantum number n that is an integer. **(a)** As drawn, which quantum numbers are involved in the transition that requires the most energy? **(b)** Which quantum numbers are involved in the transition that requires the least energy? **(c)** Based on the drawing, put the following in order of increasing wavelength of the light absorbed or emitted during the transition: (*i*) $n = 1$ to $n = 2$; (*ii*) $n = 3$ to $n = 2$; (*iii*) $n = 2$ to $n = 4$; (*iv*) $n = 3$ to $n = 1$. [Section 6.3]

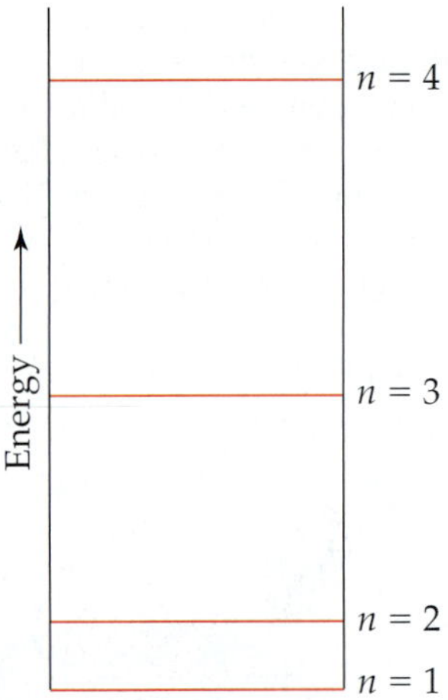

6.6 Consider a fictitious one-dimensional system with one electron. The wave function for the electron, drawn below, is $\psi(x) = \sin x$ from $x = 0$ to $x = 2\pi$. **(a)** Sketch the probability density, $\psi^2(x)$, from $x = 0$ to $x = 2\pi$. **(b)** At what value or values of x will there be the greatest probability of finding the electron? **(c)** What is the probability that the electron will be found at $x = \pi$? What is such a point in a wave function called? [Section 6.5]

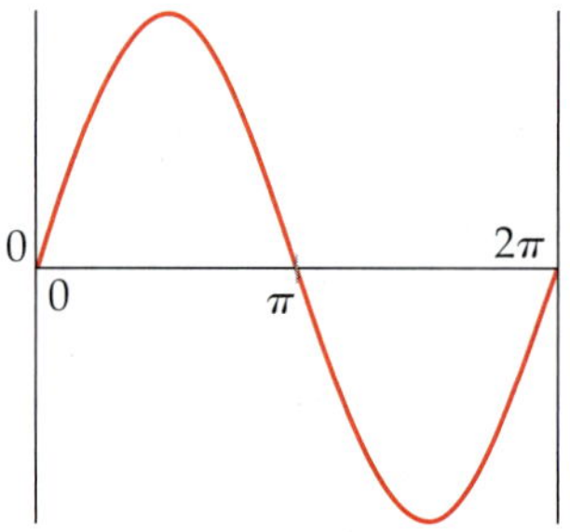

6.7 The contour representation of one of the orbitals for the $n = 3$ shell of a hydrogen atom is shown below. **(a)** What is the quantum number l for this orbital? **(b)** How do we label this orbital? **(c)** How would you modify this sketch to show the analogous orbital for the $n = 4$ shell? [Section 6.6]

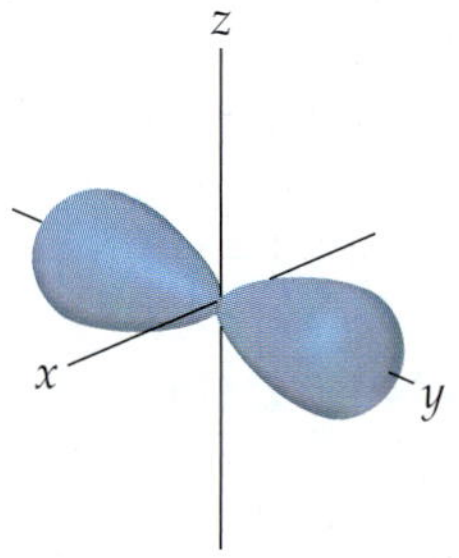

6.8 The drawing below shows part of the orbital diagram for an element. **(a)** As drawn, the drawing is *incorrect*. Why? **(b)** How would you correct the drawing without changing the number of electrons? **(c)** To which group in the periodic table does the element belong? [Section 6.8]

EXERCISES

Radiant Energy

6.9 What are the basic SI units for **(a)** the wavelength of light, **(b)** the frequency of light, **(c)** the speed of light?

6.10 **(a)** What is the relationship between the wavelength and the frequency of radiant energy? **(b)** Ozone in the upper atmosphere absorbs energy in the 210–230-nm range of the spectrum. In what region of the electromagnetic spectrum does this radiation occur?

6.11 Label each of the following statements as true or false. For those that are false, correct the statement. **(a)** Visible light is a form of electromagnetic radiation. **(b)** The frequency of radiation increases as the wavelength increases. **(c)** Ultraviolet light has longer wavelengths than visible light. **(d)** Electromagnetic radiation and sound waves travel at the same speed.

6.12 Determine which of the following statements are false, and correct them. **(a)** Electromagnetic radiation is incapable of passing through water. **(b)** Electromagnetic radiation travels through a vacuum at a constant speed, regardless of wavelength. **(c)** Infrared light has higher frequencies than visible light. **(d)** The glow from a fireplace, the energy within a microwave oven, and a foghorn blast are all forms of electromagnetic radiation.

6.13 Arrange the following kinds of electromagnetic radiation in order of increasing wavelength: infrared, green light, red light, radio waves, X rays, ultraviolet light.

6.14 List the following types of electromagnetic radiation in order of increasing wavelength: **(a)** the gamma rays produced by a radioactive nuclide used in medical imaging; **(b)** radiation from an FM radio station at 93.1 MHz on the dial; **(c)** a radio signal from an AM radio station at 680 kHz on the dial; **(d)** the yellow light from sodium vapor streetlights; **(e)** the red light of a light-emitting diode, such as in a calculator display.

6.15 **(a)** What is the frequency of radiation that has a wavelength of 955 μm? **(b)** What is the wavelength of radiation that has a frequency of $5.50 \times 10^{14}\ s^{-1}$? **(c)** Would the radiations in part (a) or part (b) be visible to the human eye? **(d)** What distance does electromagnetic radiation travel in 50.0 μs?

6.16 **(a)** What is the frequency of radiation whose wavelength is 10.0 Å? **(b)** What is the wavelength of radiation that has a frequency of $7.6 \times 10^{10}\ s^{-1}$? **(c)** Would the radiations in part (a) or part (b) be detected by an X-ray detector? **(d)** What distance does electromagnetic radiation travel in 25.5 fs?

6.17 Excited mercury atoms emit light strongly at a wavelength of 436 nm. What is the frequency of this radiation? Using Figure 6.4, predict the color associated with this wavelength.

6.18 An argon ion laser emits light at 489 nm. What is the frequency of this radiation? Is this emission in the visible spectrum? If yes, what color is it?

Quantized Energy and Photons

6.19 **(a)** What does it mean when we say energy is quantized? **(b)** Why don't we notice the quantization of energy in everyday activities?

6.20 Einstein's 1905 paper on the photoelectric effect was the first important application of Planck's quantum hypothesis. Describe Planck's original hypothesis, and explain how Einstein made use of it in his theory of the photoelectric effect.

6.21 **(a)** Calculate the smallest increment of energy (a quantum) that can be emitted or absorbed at a wavelength of 438 nm. **(b)** Calculate the energy of a photon of frequency $6.75 \times 10^{12}\ s^{-1}$. **(c)** What wavelength of radiation has photons of energy 2.87×10^{-18} J? In what portion of the electromagnetic spectrum would this radiation be found?

6.22 **(a)** Calculate the smallest increment of energy that can be emitted or absorbed at a wavelength of 10.8 mm. **(b)** Calculate the energy of a photon from an FM radio station at a frequency of 101.1 MHz. **(c)** For what frequency of radiation will a mole of photons have energy 24.7 kJ? In what region of the electromagnetic spectrum would this radiation be found?

6.23 **(a)** Calculate and compare the energy of a photon of wavelength 3.3 μm with that of wavelength 0.154 nm. **(b)** Use Figure 6.4 to identify the region of the electromagnetic spectrum to which each belongs.

6.24 An AM radio station broadcasts at 1010 kHz, and its FM partner broadcasts at 98.3 MHz. Calculate and compare the energy of the photons emitted by these two radio stations.

6.25 One type of sunburn occurs on exposure to UV light of wavelength in the vicinity of 325 nm. **(a)** What is the energy of a photon of this wavelength? **(b)** What is the energy of a mole of these photons? **(c)** How many photons are in a 1.00 mJ burst of this radiation?

6.26 The energy from radiation can be used to cause the rupture of chemical bonds. A minimum energy of 941 kJ/mol is required to break the nitrogen-nitrogen bond in N_2. What is the longest wavelength of radiation that possesses the necessary energy to break the bond? What type of electromagnetic radiation is this?

6.27 A diode laser emits at a wavelength of 987 nm. **(a)** In what portion of the electromagnetic spectrum is this radiation found? **(b)** All of its output energy is absorbed in a detector that measures a total energy of 0.52 J over a period of 32 s. How many photons per second are being emitted by the laser?

6.28 A stellar object is emitting radiation at 3.55 mm. **(a)** What type of electromagnetic spectrum is this radiation? **(b)** If the detector is capturing 3.2×10^8 photons per second at this wavelength, what is the total energy of the photons detected in one hour?

6.29 Molybdenum metal must absorb radiation with a minimum frequency of $1.09 \times 10^{15}\ s^{-1}$ before it can emit an electron from its surface via the photoelectric effect. **(a)** What is the minimum energy needed to produce this effect? **(b)** What wavelength radiation will provide a photon of this energy? **(c)** If molybdenum is irradiated with light of wavelength of 120 nm, what is the maximum possible kinetic energy of the emitted electrons?

6.30 It requires a photon with a minimum energy of 4.41×10^{-19} J to emit electrons from sodium metal. **(a)** What is the minimum frequency of light necessary to emit electrons from sodium via the photoelectric effect? **(b)** What is the wavelength of this light? **(c)** If sodium is irradiated with light of 439 nm, what is the maximum possible kinetic energy of the emitted electrons? **(d)** What is the maximum number of electrons that can be freed by a burst of light whose total energy is 1.00 μJ?

Bohr's Model; Matter Waves

6.31 Explain how the existence of line spectra is consistent with Bohr's theory of quantized energies for the electron in the hydrogen atom.

6.32 (a) In terms of the Bohr theory of the hydrogen atom, what process is occurring when excited hydrogen atoms emit radiant energy of certain wavelengths and only those wavelengths? (b) Does a hydrogen atom "expand" or "contract" as it moves from its ground state to an excited state?

6.33 Is energy emitted or absorbed when the following electronic transitions occur in hydrogen? (a) from $n = 4$ to $n = 2$, (b) from an orbit of radius 2.12 Å to one of radius 8.46 Å, (c) an electron adds to the H^+ ion and ends up in the $n = 3$ shell.

6.34 Indicate whether energy is emitted or absorbed when the following electronic transitions occur in hydrogen: (a) from $n = 2$ to $n = 6$, (b) from an orbit of radius 4.76 Å to one of radius 0.529 Å, (c) from the $n = 6$ to the $n = 9$ state.

6.35 Using Equation 6.5, calculate the energy of an electron in the hydrogen atom when $n = 2$ and when $n = 6$. Calculate the wavelength of the radiation released when an electron moves from $n = 6$ to $n = 2$. Is this line in the visible region of the electromagnetic spectrum? If so, what color is it?

6.36 For each of the following electronic transitions in the hydrogen atom, calculate the energy, frequency, and wavelength of the associated radiation, and determine whether the radiation is emitted or absorbed during the transition: (a) from $n = 4$ to $n = 1$, (b) from $n = 5$ to $n = 2$, (c) from $n = 3$ to $n = 6$. Does any of these transitions emit or absorb visible light?

6.37 The visible emission lines observed by Balmer all involved $n_f = 2$. (a) Explain why only the lines with $n_f = 2$ were observed in the visible region of the electromagnetic spectrum. (b) Calculate the wavelengths of the first three lines in the Balmer series—those for which $n_i = 3$, 4, and 5—and identify these lines in the emission spectrum shown in Figure 6.12.

6.38 The Lyman series of emission lines of the hydrogen atom are those for which $n_f = 1$. (a) Determine the region of the electromagnetic spectrum in which the lines of the Lyman series are observed. (b) Calculate the wavelengths of the first three lines in the Lyman series—those for which $n_i = 2$, 3, and 4.

[6.39] One of the emission lines of the hydrogen atom has a wavelength of 93.8 nm. (a) In what region of the electromagnetic spectrum is this emission found? (b) Determine the initial and final values of n associated with this emission.

[6.40] The hydrogen atom can absorb light of wavelength 2626 nm. (a) In what region of the electromagnetic spectrum is this absorption found? (b) Determine the initial and final values of n associated with this absorption.

6.41 Use the de Broglie relationship to determine the wavelengths of the following objects: (a) an 85-kg person skiing at 50 km/hr, (b) a 10.0-g bullet fired at 250 m/s, (c) a lithium atom moving at 2.5×10^5 m/s.

6.42 Among the elementary subatomic particles of physics is the muon, which decays within a few nanoseconds after formation. The muon has a rest mass 206.8 times that of an electron. Calculate the de Broglie wavelength associated with a muon traveling at a velocity of 8.85×10^5 cm/s.

6.43 Neutron diffraction is an important technique for determining the structures of molecules. Calculate the velocity of a neutron that has a characteristic wavelength of 0.955 Å. (Refer to the inside cover for the mass of the neutron.)

6.44 The electron microscope has been widely used to obtain highly magnified images of biological and other types of materials. When an electron is accelerated through a particular potential field, it attains a speed of 9.38×10^6 m/s. What is the characteristic wavelength of this electron? Is the wavelength comparable to the size of atoms?

6.45 Using Heisenberg's uncertainty principle, calculate the uncertainty in the position of (a) a 1.50-mg mosquito moving at a speed of 1.40 m/s if the speed is known to within ± 0.01 m/s; (b) a proton moving at a speed of $(5.00 \pm 0.01) \times 10^4$ m/s. (The mass of a proton is given in the table of fundamental constants in the inside cover of the text.)

6.46 Calculate the uncertainty in the position of (a) an electron moving at a speed of $(3.00 \pm 0.01) \times 10^5$ m/s, (b) a neutron moving at this same speed. (The masses of an electron and a neutron are given in the table of fundamental constants in the inside cover of the text.) (c) What are the implications of these calculations to our model of the atom?

Quantum Mechanics and Atomic Orbitals

6.47 (a) Why does the Bohr model of the hydrogen atom violate the uncertainty principle? (b) In what way is the description of the electron using a wave function consistent with de Broglie's hypothesis? (c) What is meant by the term *probability density*? Given the wave function, how do we find the probability density at a certain point in space?

6.48 (a) According to the Bohr model, an electron in the ground state of a hydrogen atom orbits the nucleus at a specific radius of 0.53 Å. In the quantum mechanical description of the hydrogen atom, the most probable distance of the electron from the nucleus is 0.53 Å. Why are these two statements different? (b) Why is the use of Schrödinger's wave equation to describe the location of a particle very different from the description obtained from classical physics? (c) In the quantum mechanical description of an electron, what is the physical significance of the square of the wave function, ψ^2?

6.49 (a) For $n = 4$, what are the possible values of l? (b) For $l = 2$, what are the possible values of m_l?

6.50 How many possible values for l and m_l are there when (a) $n = 3$; (b) $n = 5$?

6.51 Give the numerical values of n and l corresponding to each of the following designations: **(a)** $3p$, **(b)** $2s$, **(c)** $4f$, **(d)** $5d$.

6.52 Give the values for n, l, and m_l for **(a)** each orbital in the $2p$ subshell, **(b)** each orbital in the $5d$ subshell.

6.53 Which of the following represent impossible combinations of n and l: **(a)** $1p$, **(b)** $4s$, **(c)** $5f$, **(d)** $2d$?

6.54 Which of the following are permissible sets of quantum numbers for an electron in a hydrogen atom: **(a)** $n = 2$, $l = 1$, $m_l = 1$; **(b)** $n = 1$, $l = 0$, $m_l = -1$; **(c)** $n = 4$, $l = 2$, $m_l = -2$; **(d)** $n = 3$, $l = 3$, $m_l = 0$? For those combinations that are permissible, write the appropriate designation for the subshell to which the orbital belongs (that is, $1s$, and so on).

6.55 Sketch the shape and orientation of the following types of orbitals: **(a)** s, **(b)** p_z, **(c)** d_{xy}.

6.56 Sketch the shape and orientation of the following types of orbitals: **(a)** p_x, **(b)** d_{z^2}, **(c)** $d_{x^2-y^2}$.

6.57 **(a)** What are the similarities and differences between the $1s$ and $2s$ orbitals of the hydrogen atom? **(b)** In what sense does a $2p$ orbital have directional character? Compare the "directional" characteristics of the p_x and $d_{x^2-y^2}$ orbitals (that is, in what direction or region of space is the electron density concentrated?). **(c)** What can you say about the average distance from the nucleus of an electron in a $2s$ orbital as compared with a $3s$ orbital? **(d)** For the hydrogen atom, list the following orbitals in order of increasing energy (that is, most stable ones first): $4f$, $6s$, $3d$, $1s$, $2p$.

6.58 **(a)** With reference to Figure 6.18, what is the relationship between the number of nodes in an s orbital and the value of the principal quantum number? **(b)** Identify the number of nodes; that is, identify places where the electron density is zero, in the $2p_x$ orbital; in the $3s$ orbital. **(c)** What information is obtained from the radial probability functions in Figure 6.18? **(d)** For the hydrogen atom, list the following orbitals in order of increasing energy: $3s$, $2s$, $2p$, $5s$, $4d$.

Many-Electron Atoms and Electron Configurations

6.59 For a given value of the principal quantum number, n, how do the energies of the s, p, d, and f subshells vary for **(a)** hydrogen, **(b)** a many-electron atom?

6.60 **(a)** The average distance from the nucleus of a $3s$ electron in a chlorine atom is smaller than that for a $3p$ electron. In light of this fact, which orbital is higher in energy? **(b)** Would you expect it to require more or less energy to remove a $3s$ electron from the chlorine atom, as compared with a $2p$ electron? Explain.

6.61 **(a)** What are the possible values of the electron spin quantum number? **(b)** What piece of experimental equipment can be used to distinguish electrons that have different values of the electron spin quantum number? **(c)** Two electrons in an atom both occupy the $1s$ orbital. What quantity must be different for the two electrons? What principle governs the answer to this question?

6.62 **(a)** State the Pauli exclusion principle in your own words. **(b)** The Pauli exclusion principle is, in an important sense, the key to understanding the periodic table. Explain why.

6.63 What is the maximum number of electrons that can occupy each of the following subshells: **(a)** $3p$, **(b)** $5d$, **(c)** $2s$, **(d)** $4f$?

6.64 What is the maximum number of electrons in an atom that can have the following quantum numbers: **(a)** $n = 2$, $m_s = -\frac{1}{2}$; **(b)** $n = 5$, $l = 3$; **(c)** $n = 4$, $l = 3$, $m_l = -3$; **(d)** $n = 4$, $l = 1$, $m_l = 1$.

6.65 **(a)** What does each box in an orbital diagram represent? **(b)** What quantity is represented by the direction (either up or down) of the half arrows in an orbital diagram? **(c)** Is Hund's rule needed to write the electron configuration of beryllium? Explain.

6.66 **(a)** What are "valence electrons"? **(b)** What are "unpaired electrons"? **(c)** How many valence electrons does a P atom possess? How many of these are unpaired?

6.67 Write the condensed electron configurations for the following atoms, using the appropriate noble-gas core abbreviations: **(a)** Cs, **(b)** Ni, **(c)** Se, **(d)** Cd, **(e)** Ac, **(f)** Pb.

6.68 Write the condensed electron configurations for the following atoms, and indicate how many unpaired electrons each has: **(a)** Ga, **(b)** Ca, **(c)** V, **(d)** I, **(e)** Y, **(f)** Pt, **(g)** Lu.

6.69 Meitnerium, Mt, element 109, named after Lisa Meitner, is a transition metal expected to have the same outer-electron configuration as iridium. By using this observation (and without looking at Figure 6.30), write the electron configuration of meitnerium. Use [Rn] to represent the first 86 electrons of the electron configuration.

6.70 In 1999 it was reported that element 118 had been made artificially. This report was retracted in 2001, and since then there have been no other claims that element 118 can be made. **(a)** Write the electron configuration for element 118. **(b)** How many unpaired electrons would you expect an atom of element 118 to have? **(c)** To which group of elements in the periodic table would you expect element 118 to belong?

6.71 Identify the specific element that corresponds to each of the following electron configurations: **(a)** $1s^2 2s^2 2p^6 3s^2$, **(b)** $[Ne]3s^2 3p^1$, **(c)** $[Ar]4s^1 3d^5$, **(d)** $[Kr]5s^2 4d^{10} 5p^4$.

6.72 Identify the group of elements that corresponds to each of the following generalized electron configurations:
(a) [noble gas] $ns^2 np^5$
(b) [noble gas] $ns^2(n-1)d^2$
(c) [noble gas] $ns^2(n-1)d^{10}np^1$
(d) [noble gas] $ns^2(n-2)f^6$

6.73 What is wrong with the following electron configurations for atoms in their ground states? **(a)** $1s^2 2s^2 3s^1$, **(b)** $[Ne]2s^2 2p^3$, **(c)** $[Ne]3s^2 3d^5$.

6.74 The following electron configurations represent excited states. Identify the element, and write its ground-state condensed electron configuration. **(a)** $1s^2 2s^2 3p^2 4p^1$, **(b)** $[Ar]3d^{10} 4s^1 4p^4 5s^1$, **(c)** $[Kr]4d^6 5s^2 5p^1$.

Additional Exercises

6.75 Consider the two waves shown here, which we will consider to represent two electromagnetic radiations:
(a) What is the wavelength of wave A? Of wave B?
(b) What is the frequency of wave A? Of wave B?
(c) Identify the regions of the electromagnetic spectrum to which waves A and B belong.

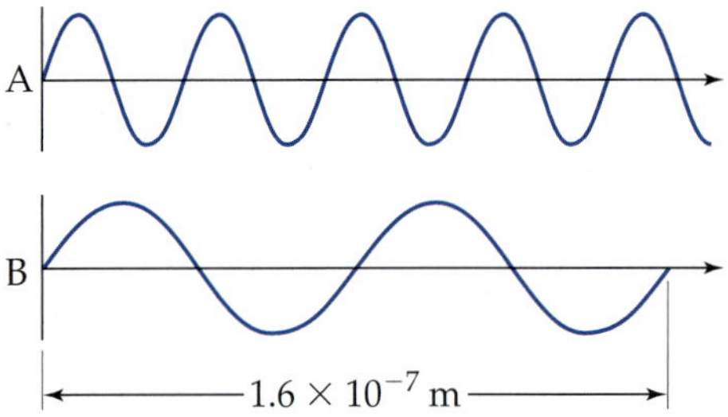

6.76 Certain elements emit light of a specific wavelength when they are burned. Historically, chemists used such emission wavelengths to determine whether specific elements were present in a sample. Some characteristic wavelengths for some of the elements are

Ag	328.1 nm	Fe	372.0 nm
Au	267.6 nm	K	404.7 nm
Ba	455.4 nm	Mg	285.2 nm
Ca	422.7 nm	Na	589.6 nm
Cu	324.8 nm	Ni	341.5 nm

(a) Determine which elements emit radiation in the visible part of the spectrum. **(b)** Which element emits photons of highest energy? Of lowest energy? **(c)** When burned, a sample of an unknown substance is found to emit light of frequency $6.59 \times 10^{14}\ s^{-1}$. Which of these elements is probably in the sample?

6.77 In June 2004, the Cassini–Huygens spacecraft began orbiting Saturn and transmitting images to Earth. The closest distance between Saturn and Earth is 746 million miles. What is the minimum amount of time it takes for the transmitted signals to travel from the spacecraft to Earth?

6.78 The rays of the Sun that cause tanning and burning are in the ultraviolet portion of the electromagnetic spectrum. These rays are categorized by wavelength: So-called UV-A radiation has wavelengths in the range of 320–380 nm, whereas UV-B radiation has wavelengths in the range of 290–320 nm. **(a)** Calculate the frequency of light that has a wavelength of 320 nm. **(b)** Calculate the energy of a mole of 320-nm photons. **(c)** Which are more energetic, photons of UV-A radiation or photons of UV-B radiation? **(d)** The UV-B radiation from the Sun is considered a greater cause of sunburn in humans than is UV-A radiation. Is this observation consistent with your answer to part (c)?

6.79 The watt is the derived SI unit of power, the measure of energy per unit time: 1 W = 1 J/s. A semiconductor laser in a CD player has an output wavelength of 780 nm and a power level of 0.10 mW. How many photons strike the CD surface during the playing of a CD 69 minutes in length?

6.80 Carotenoids, present in all organisms capable of photosynthesis, extend the range of light absorbed by the organism. They exhibit maximal capacity for absorption of light in the range of 440–470 nm. Calculate the energy in kilojoules represented by absorption of an Avogadro's number of photons of wavelength 455 nm.

[6.81] A photocell, such as the one illustrated in Figure 6.7(b), is a device used to measure the intensity of light. In a certain experiment, when light of wavelength 630 nm is directed onto the photocell, electrons are emitted at the rate of 2.6×10^{-12} C/s. Assume that each photon that impinges on the photocell emits one electron. How many photons per second are striking the photocell? How much energy per second is the photocell absorbing?

6.82 The light-sensitive substance in black-and-white photographic film is AgBr. Photons provide the energy necessary to transfer an electron from Br^- to Ag^+ to produce Ag and Br and thereby darken the film. **(a)** If a minimum energy of 2.00×10^5 J/mol is needed for this process, what is the minimum energy needed by each photon? **(b)** Calculate the wavelength of the light necessary to provide photons of this energy. **(c)** Explain why this film can be handled in a darkroom under red light.

[6.83] In an experiment to study the photoelectric effect, a scientist measures the kinetic energy of ejected electrons as a function of the frequency of radiation hitting a metal surface. She obtains the following plot:

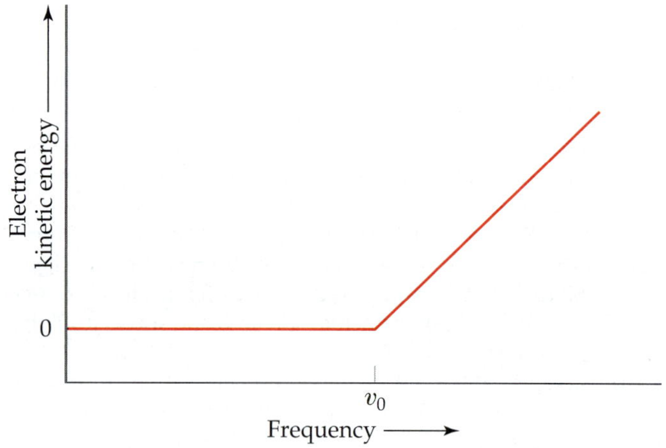

The point labeled "ν_0" corresponds to light with a wavelength of 680 nm. **(a)** What is the value of ν_0 in s^{-1}? **(b)** What is the value of the work function of the metal in units of kJ/mol of ejected electrons? **(c)** What happens when the metal is irradiated with light of frequency less than ν_0? **(d)** Note that when the frequency of the light is greater than ν_0, the plot shows a straight line with a nonzero slope. Why is this the case? **(e)** Can you determine the slope of the line segment discussed in part (d)? Explain.

6.84 The series of emission lines of the hydrogen atom for which $n_f = 3$ is called the *Paschen series*. **(a)** Determine the region of the electromagnetic spectrum in which the lines of the Paschen series are observed. **(b)** Calculate the wavelengths of the first three lines in the Paschen series—those for which $n_i = 4, 5,$ and 6.

6.85 When the spectrum of light from the Sun is examined in high resolution in an experiment similar to that illustrated in Figure 6.10, dark lines are evident. These are called Fraunhofer lines, after the scientist who studied them extensively in the early nineteenth century. Altogether, about 25,000 lines have been identified in the solar spectrum between 2950 Å and 10,000 Å. The Fraunhofer lines are attributed to absorption of certain wavelengths of the Sun's "white" light by gaseous elements in the Sun's atmosphere. **(a)** Describe the process that causes absorption of specific wavelengths of light from the solar spectrum. **(b)** If a scientist wanted to know which Fraunhofer lines belonged to a given element, say neon, what experiments could she conduct here on Earth to provide data?

[6.86] Bohr's model can be used for hydrogen-like ions—ions that have only one electron, such as He^+ and Li^{2+}. **(a)** Why is the Bohr model applicable to He^+ ions but not to neutral He atoms? **(b)** The ground-state energies of H, He^+, and Li^{2+} are tabulated as follows:

Atom or ion	H	He^+	Li^{2+}
Ground-state energy	2.18×10^{-18} J	8.72×10^{-18} J	1.96×10^{-17} J

By examining these numbers, propose a relationship between the ground-state energy of hydrogen-like systems and the nuclear charge, *Z*. **(c)** Use the relationship you derive in part (b) to predict the ground-state energy of the C^{5+} ion.

6.87 Under appropriate conditions, molybdenum emits X rays that have a characteristic wavelength of 0.711 Å. These X rays are used in diffraction experiments to determine the structures of molecules. How fast would an electron have to be moving in order to have the same wavelength as these X rays?

[6.88] An electron is accelerated through an electric potential to a kinetic energy of 18.6 keV. What is its characteristic wavelength? [*Hint:* Recall that the kinetic energy of a moving object is $E = \frac{1}{2}mv^2$, where m is the mass of the object and v is the speed of the object.]

6.89 What is the difference between an *orbit* (Bohr model of the hydrogen atom) and an *orbital* (quantum mechanical model of the hydrogen atom)?

6.90 Which of the quantum numbers governs **(a)** the shape of an orbital, **(b)** the energy of an orbital, **(c)** the spin properties of the electron, **(d)** the spatial orientation of the orbital?

[6.91] Consider the discussion of radial probability functions in the "A Closer Look" box in Section 6.6. **(a)** What is the difference between the probability density as a function of r and the radial probability function as a function of r? **(b)** What is the significance of the term $4\pi r^2$ in the radial probability functions for the s orbitals? **(c)** Based on Figures 6.18 and 6.21, make sketches of what you think the probability density as a function of r and the radial probability function would look like for the 4s orbital of the hydrogen atom.

6.92 The "magic numbers" in the periodic table are the atomic numbers of elements with high stability (the noble gases): 2, 10, 18, 36, 54, and 86. In terms of allowed values of orbitals and spin quantum numbers, explain why these electron arrangements correspond to special stability.

[6.93] For non-spherically symmetric orbitals, the contour representations (as in Figures 6.22 and 6.23) suggest where nodal planes exist (that is, where the electron density is zero). For example, the p_x orbital has a node wherever $x = 0$; this equation is satisfied by all points on the yz plane, so this plane is called a nodal plane of the p_x orbital. **(a)** Determine the nodal plane of the p_z orbital. **(b)** What are the two nodal planes of the d_{xy} orbital? **(c)** What are the two nodal planes of the $d_{x^2-y^2}$ orbital?

[6.94] As noted in Figure 6.25, the spin of an electron generates a magnetic field, with spin-up and spin-down electrons having opposite fields. In the absence of a magnetic field, a spin-up and a spin-down electron have the same energy. **(a)** Why do you think that the use of a magnet was important in the discovery of electron spin (see the "A Closer Look" box in Section 6.7)? **(b)** Imagine that the two spinning electrons in Figure 6.25 were placed between the poles of a horseshoe magnet, with the north pole of the magnet at the top of the figure. Based on what you know about magnets, would you expect the left or right electron in the figure to have the lower energy? **(c)** A phenomenon called *electron spin resonance* (ESR) is closely related to nuclear magnetic resonance. In ESR a compound with an unpaired electron is placed in a magnetic field, which causes the unpaired electron to have two different energy states analogous to Figure 6.27. ESR uses microwave radiation to excite the unpaired electron from one state to the other. Based on your reading of the "Chemistry and Life" box in Section 6.7, does an ESR experiment require photons of greater or lesser energy than an NMR experiment?

[6.95] The "Chemistry and Life" box in Section 6.7 described the techniques called NMR and MRI. **(a)** Instruments for obtaining MRI data are typically labeled with a frequency, such as 600 MHz. Why do you suppose this label is relevant to the experiment? **(b)** What is the value of ΔE in Figure 6.27 that would correspond to the absorption of a photon of radiation with frequency 450 MHz? **(c)** In general, the stronger the magnetic field, the greater the information obtained from an NMR or MRI experiment. Why do you suppose this is the case?

[6.96] Suppose that the spin quantum number, m_s, could have *three* allowed values instead of two. How would this affect the number of elements in the first four rows of the periodic table?

6.97 Using only a periodic table as a guide, write the condensed electron configurations for the following atoms: **(a)** Se, **(b)** Rh, **(c)** Si, **(d)** Hg, **(e)** Hf.

6.98 Scientists have speculated that element 126 might have a moderate stability, allowing it to be synthesized and characterized. Predict what the condensed electron configuration of this element might be.

Integrative Exercises

[6.99] Microwave ovens use microwave radiation to heat food. The microwaves are absorbed by moisture in the food, which is transferred to other components of the food. As the water becomes hotter, so does the food. Suppose that the microwave radiation has a wavelength of 11.2 cm. How many photons are required to heat 200 mL of coffee from 23°C to 60°C?

6.100 The stratospheric ozone (O_3) layer helps to protect us from harmful ultraviolet radiation. It does so by absorbing ultraviolet light and falling apart into an O_2 molecule and an oxygen atom, a process known as photodissociation.

$$O_3(g) \longrightarrow O_2(g) + O(g)$$

Use the data in Appendix C to calculate the enthalpy change for this reaction. What is the maximum wavelength a photon can have if it is to possess sufficient energy to cause this dissociation? In what portion of the spectrum does this wavelength occur?

6.101 The discovery of hafnium, element number 72, provided a controversial episode in chemistry. G. Urbain, a French chemist, claimed in 1911 to have isolated an element number 72 from a sample of rare earth (elements 58–71) compounds. However, Niels Bohr believed that hafnium was more likely to be found along with zirconium than with the rare earths. D. Coster and G. von Hevesy, working in Bohr's laboratory in Copenhagen, showed in 1922 that element 72 was present in a sample of Norwegian zircon, an ore of zirconium. (The name hafnium comes from the Latin name for Copenhagen, *Hafnia*). **(a)** How would you use electron configuration arguments to justify Bohr's prediction? **(b)** Zirconium, hafnium's neighbor in group 4B, can be produced as a metal by reduction of solid $ZrCl_4$ with molten sodium metal. Write a balanced chemical equation for the reaction. Is this an oxidation-reduction reaction? If yes, what is reduced and what is oxidized? **(c)** Solid zirconium dioxide, ZrO_2, is reacted with chlorine gas in the presence of carbon. The products of the reaction are $ZrCl_4$ and two gases, CO_2 and CO in the ratio 1 : 2. Write a balanced chemical equation for the reaction. Starting with a 55.4-g sample of ZrO_2, calculate the mass of $ZrCl_4$ formed, assuming that ZrO_2, is the limiting reagent and assuming 100% yield. **(d)** Using their electron configurations, account for the fact that Zr and Hf form chlorides MCl_4 and oxides MO_2.

6.102 **(a)** Account for formation of the following series of oxides in terms of the electron configurations of the elements and the discussion of ionic compounds in Section 2.7: K_2O, CaO, Sc_2O_3, TiO_2, V_2O_5, CrO_3. **(b)** Name these oxides. **(c)** Consider the metal oxides whose enthalpies of formation (in kJ mol^{-1}) are listed here.

Oxide	$K_2O(s)$	$CaO(s)$	$TiO_2(s)$	$V_2O_5(s)$
ΔH_f°	−363.2	−635.1	−938.7	−1550.6

Calculate the enthalpy changes in the following general reaction for each case:

$$M_nO_m(s) + H_2(g) \longrightarrow nM(s) + mH_2O(g)$$

(You will need to write the balanced equation for each case, then compute ΔH°.) **(d)** Based on the data given, estimate a value of ΔH_f° for $Sc_2O_3(s)$.

6.103 The first 25 years of the twentieth century were momentous for the rapid pace of change in scientists' understanding of the nature of matter. **(a)** How did Rutherford's experiments on the scattering of α particles by a gold foil set the stage for Bohr's theory of the hydrogen atom? **(b)** In what ways is de Broglie's hypothesis, as it applies to electrons, consistent with J. J. Thomson's conclusion that the electron has mass? In what sense is it consistent with proposals that preceded Thomson's work, that the cathode rays are a wave phenomenon?

[6.104] The two most common isotopes of uranium are ^{235}U and ^{238}U. **(a)** Compare the number of protons, the number of electrons, and the number of neutrons in atoms of these two isotopes. **(b)** Using the periodic table in the front inside cover, write the electron configuration for a U atom. **(c)** Compare your answer to part (b) to the electron configuration given in Figure 6.30. How can you explain any differences between these two electron configurations? **(d)** ^{238}U undergoes radioactive decay to ^{234}Th. How many protons, electrons, and neutrons are gained or lost by the ^{238}U atom during this process? **(e)** Examine the electron configuration for Th in Figure 6.30. Are you surprised by what you find? Explain.

eMEDIA EXERCISES

These exercises make use of the interactive objects available online in OneKey or the Companion Website, and on your Accelerator CD. Access to these resources comes in your MediaPak.

6.105 The **Electromagnetic Spectrum** activity (*6.1*) allows you to choose a color in the visible spectrum and see its wavelength, frequency, and energy per photon. **(a)** What is the wavelength range of blue light? **(b)** What are the ranges of its frequency and energy per photon? **(c)** Exercise 6.25 indicates that a type of sunburn is caused by light with wavelength ~325 nm. Would you expect any of the visible wavelengths to cause sunburn? Explain.

6.106 In the **Flame Tests for Metals** movie (*6.3*) the characteristic color of the flame is produced by emissions at several visible wavelengths, with the most intense spectral lines dominating the color. For instance, the most intense visible lines in the spectrum of lithium occur at ~671 nm. **(a)** What color is light of this wavelength? **(b)** At what approximate wavelength would you expect to find the most intense lines in the visible spectrum of potassium? **(c)** Based on the movie, how would you expect the intensity of visible lines in the spectrum of potassium to compare to those in the spectrum of lithium? **(d)** Would it be possible to verify the presence of individual metals using flame color if several metal salts were mixed together? If not, explain why not.

6.107 In the **Radial Electron Distribution** animation (*6.6*) the radial electron density plots of helium, neon, and argon are all placed on the same graph. **(a)** Explain why the maximum for the helium plot and the first maximum for each of the other two occur at significantly different distances from the nucleus. **(b)** Based on how far the *outermost* maximum for each plot is from the nucleus, predict which pair would have the greater difference between their first ionization energies (the energy required to remove completely an electron from the outermost shell): helium and neon, or neon and argon. Explain your reasoning.

6.108 The electron configuration given in Exercise 6.71(c) is one of a handful of examples of an *s* electron being "stolen" in order to fill or half-fill a *d* subshell. (As seen in the **Electron Configurations** animation (*6.8*), there is special stability associated with filled and half-filled subshells.) **(a)** Using the **Electron Configuration** activity (*6.8*), identify at least three more examples like this. **(b)** Are there any instances of *both s* electrons being stolen in order to fill a *d* subshell? If so, name the element(s). **(c)** Why is this phenomenon not observed in *p*-block elements? That is, why is chlorine's electron configuration $[Ne]3s^2 3p^5$ rather than $[Ne]3s^1 3p^6$?

VCL eLABORATORY EXERCISES

These exercises allow students to apply the concepts and skills for this chapter in the simulated laboratory of Virtual ChemLab. Worksheets for each of these exercises are found in the Virtual ChemLab Workbook in your MediaPak.

6.109 ***(VCL 6-1) Blackbody Radiation***
In the early 1900s several experimental results appeared to be in conflict with classical physics. One of these experiments was the study of blackbody radiation. A blackbody is a solid (such as a piece of iron) that does not emit light at low temperatures, but when heated, the "blackbody" begins to emit first red and then orange light and at higher temperatures eventually becomes white-hot. The intensity of the emitted light is also a function of temperature. In this problem you will make observations similar to those of Max Planck (1858–1947), who, through his study of blackbody radiation, found an explanation that revolutionized how scientists think about radiated energy.

6.110 ***(VCL 6-2) Photoelectric Effect***
Though Albert Einstein is most famous for $E = mc^2$ and his work in describing relativity in mechanics, his Nobel Prize was for understanding a very simple experiment. It was long understood that if you directed light of a certain wavelength at a piece of metal, it would emit electrons. In classical theory the energy of the light was thought to be based on its intensity and not its frequency. However, the results of the photoelectric effect contradicted classical theory. These inconsistencies led Einstein to suggest that we need to think of light as being composed of particles (photons) and not just as waves. In this experiment you will reproduce a photoelectric experiment and show that the energy (E) of a photon of light is related to its frequency and not its intensity.

6.111 ***(VCL 6-3) The Rydberg Equation***
When a sample of gas is excited by applying a large alternating electric field, the gas emits light at certain discrete wavelengths, but the classical picture of atoms would allow electrons to be at any energy level and thus the energy would be continuous over all wavelengths. In the late 1800s two scientists, Johann Balmer and Johannes Rydberg, developed a mathematical equation that correlated the wavelength of the emitted light for certain gases such as H_2. Later, Niels Bohr's concept of quantized "jumps" by electrons between orbits was shown to be consistent with the Rydberg equation. In this assignment you will measure the wavelengths of the lines in the hydrogen emission spectra and graphically determine the value of the Rydberg constant, R_H.

6.112 ***(VCL 6-4) Atomic Emission Spectra***
When a sample of gas is excited by applying a large alternating electric field, the gas emits light at certain discrete wavelengths. The intensity and wavelength of the light that is emitted is called the atomic emission spectrum and is unique for each gas. In this assignment you will measure the emission spectra for several gases and make some observations about the differences in their spectra.

6.113 ***(VCL 6-5) Heisenberg Uncertainty Principle***
It has long been known that if you shine light through narrow slits that are spaced at small intervals, the light will form a diffraction pattern. A diffraction pattern is a series of light and dark patterns caused by wave interference. The wave interference can be either constructive (light) or destructive (dark). In this experiment you will shine a laser through a device with two slits where the spacing can be adjusted, and you will investigate the patterns that will be made at a distance from the slits. From this you will be able to study the effect of the Heisenberg uncertainty principle on the diffraction of light.

CHAPTER 7

WHAT'S AHEAD

- Our discussion begins with a brief history of the periodic table. *(Section 7.1)*
- We will see that many properties of atoms depend both on the net attraction of the outer electrons to the nucleus (an attraction that is due to the *effective nuclear charge*) and on the average distance of those electrons from the nucleus. *(Sections 7.2 and 7.3)*
- We will examine periodic trends in the *atomic size*, *ionization energy*, and *electron affinity* of atoms. *(Sections 7.3, 7.4, and 7.5)*
- We will also examine the sizes of ions and their electron configurations. *(Sections 7.3 and 7.4)*
- We will explore some of the differences in the physical and chemical properties of metals, nonmetals, and metalloids. *(Section 7.6)*
- Finally, we will discuss some periodic trends in the chemistry of the active metals (groups 1A and 2A), of hydrogen, and of several other nonmetals (groups 6A to 8A). *(Sections 7.7 and 7.8)*

Periodic Properties of the Elements

THE BEAUTY OF AN IMPRESSIONIST OIL PAINTING, such as the Monet masterpiece shown here, begins with chemistry. Colorful inorganic salts are suspended in various organic media that contain hydrocarbons and other molecular substances. Indeed, the great painters had a gift for using compounds of elements that span nearly the entire periodic table.

Today the periodic table is still the most significant tool chemists have for organizing and remembering chemical facts. As we saw in Chapter 6, the periodic nature of the table arises from the repeating patterns in the electron configurations of the elements. Elements in the same column of the table contain the same number of electrons in their **valence orbitals**, the occupied orbitals that hold the electrons involved in bonding. For example, O ($[He]2s^2 2p^4$) and S ($[Ne]3s^2 3p^4$) are both members of group 6A; the similarity of the electron distribution in their valence *s* and *p* orbitals leads to similarities in the properties of these two elements. When we compare O and S, however, it is apparent that they exhibit differences as well, not the least of which is that oxygen is a colorless gas at room temperature, whereas sulfur is a yellow solid (Figure 7.1 ▶)! One of the major differences between atoms of these two elements is their electron configurations: The outermost electrons of O are in the second shell, whereas those of S are in the third shell. We will see that electron configurations can be used to explain differences as well as similarities in the properties of elements.

OIL PAINTS CONTAIN PIGMENTS, which are usually highly colored salts, suspended in an organic carrier composed of a variety of heavy hydrocarbon molecules. This painting, by the famous French Impressionist Claude Monet (1840–1926), is entitled *La rue Montorgueil, fête du 30 juin 1878.*

Claude Monet (1840–1926) "Rue Montorgueil in Paris, Festival of 30 June 1878" 1878. Herve Lewandowski/Reunion des Musees Nationaux/Art Resource, NY.

▲ **Figure 7.1 Oxygen and sulfur.** Because they are both group 6A elements, oxygen and sulfur have many chemical similarities. They also have many differences, however, including the forms they take at room temperature. Oxygen consists of O_2 molecules that appear as a colorless gas (shown here enclosed in a glass container on the left). In contrast, sulfur consists of S_8 molecules that form a yellow solid.

In this chapter we explore how some of the important properties of elements change as we move across a row or down a column of the periodic table. In many cases the trends within a row or column allow us to make predictions about the physical and chemical properties of the elements.

7.1 | Development of the Periodic Table

The discovery of the chemical elements has been an ongoing process since ancient times (Figure 7.2 ▼). Certain elements, such as gold, appear in nature in elemental form and were thus discovered thousands of years ago. In contrast, some elements are radioactive and intrinsically unstable. We know about them only because of technology developed in the twentieth century.

The majority of the elements, although stable, are dispersed widely in nature and are incorporated into numerous compounds. For centuries, therefore, scientists were unaware of their existence. In the early nineteenth century, advances in chemistry made it easier to isolate elements from their compounds. As a result, the number of known elements more than doubled from 31 in 1800 to 63 by 1865.

As the number of known elements increased, scientists began to investigate the possibilities of classifying them in useful ways. In 1869, Dmitri Mendeleev in Russia and Lothar Meyer in Germany published nearly identical classification schemes. Both scientists noted that similar chemical and physical properties recur periodically when the elements are arranged in order of increasing atomic weight. Scientists at that time had no knowledge of atomic numbers. Atomic weights, however, generally increase with increasing atomic number, so both Mendeleev and Meyer fortuitously arranged the elements in proper sequence. The tables of elements advanced by Mendeleev and Meyer were the forerunners of the modern periodic table.

Although Mendeleev and Meyer came to essentially the same conclusion about the periodicity of elemental properties, Mendeleev is given credit for advancing his ideas more vigorously and stimulating much new work in chemistry. His insistence that elements with similar characteristics be listed in the same family forced him to leave several blank spaces in his table. For example,

▶ **Figure 7.2 Discovering the elements.** Periodic table showing the dates of discovery of the elements.

ANIMATION
Periodic Properties

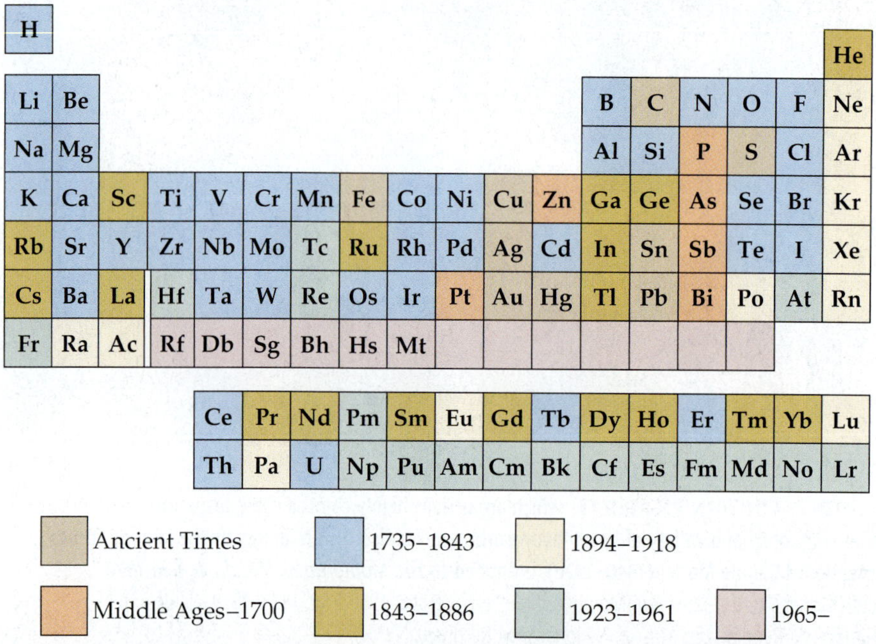

ACTIVITY
Periodic Table

TABLE 7.1 Comparison of the Properties of Eka-Silicon Predicted by Mendeleev with the Observed Properties of Germanium

Property	Mendeleev's Predictions for Eka-Silicon (made in 1871)	Observed Properties of Germanium (discovered in 1886)
Atomic weight	72	72.59
Density (g/cm^3)	5.5	5.35
Specific heat (J/g-k)	0.305	0.309
Melting point (°C)	High	947
Color	Dark gray	Grayish white
Formula of oxide	XO_2	GeO_2
Density of oxide (g/cm^3)	4.7	4.70
Formula of chloride	XCl_4	$GeCl_4$
Boiling point of chloride (°C)	A little under 100	84

both gallium (Ga) and germanium (Ge) were unknown at that time. Mendeleev boldly predicted their existence and properties, referring to them as *eka-aluminum* ("under" aluminum) and *eka-silicon* ("under" silicon), respectively, after the elements under which they appear in the periodic table. When these elements were discovered, their properties closely matched those predicted by Mendeleev, as shown in Table 7.1 ▲.

In 1913, two years after Rutherford proposed the nuclear model of the atom (Section 2.2), an English physicist named Henry Moseley (1887–1915) developed the concept of atomic numbers. Moseley determined that the frequencies of X rays emitted as different elements were bombarded with high-energy electrons. He found that each element produces X rays of a unique frequency; furthermore, he found that the frequency generally increased as the atomic mass increased. He arranged the X-ray frequencies in order by assigning a unique whole number, called an *atomic number*, to each element. Moseley correctly identified the atomic number as the number of protons in the nucleus of the atom and the number of electrons in the atom. (Section 2.3)

The concept of atomic number clarified some problems in the early version of the periodic table, which was based on atomic weights. For example, the atomic weight of Ar (atomic number 18) is greater than that of K (atomic number 19). However, when the elements are arranged in order of increasing atomic number, rather than increasing atomic weight, Ar and K appear in their correct places in the table. Moseley's studies also made it possible to identify "holes" in the periodic table, which led to the discovery of other previously unknown elements.

GIVE IT SOME THOUGHT

Arranging the elements by atomic weight leads to a slightly different order than arranging them by atomic number. How can this happen?

7.2 | Effective Nuclear Charge

Because electrons are negatively charged, they are attracted to nuclei, which are positively charged. Many of the properties of atoms depend not only on their electron configurations but also on how strongly their outer electrons are attracted to the nucleus. Coulomb's law tells us that the strength of the interaction between two electrical charges depends on the signs and magnitudes of the charges and on the distance between them. (Section 2.3) Thus, the force of attraction between an electron and the nucleus depends on the magnitude of the net nuclear charge acting on the electron and on the average distance between the nucleus and the electron. The force of attraction increases as the nuclear charge increases and decreases as the electron moves farther from the nucleus.

In a many-electron atom, each electron is simultaneously attracted to the nucleus and repelled by the other electrons. In general, there are so many electron-electron repulsions that we cannot analyze the situation exactly. We can, however, estimate the net attraction of each electron to the nucleus by considering how it interacts with the *average* environment created by the nucleus and the other electrons in the atom. This approach allows us to treat each electron individually as though it were moving in the net electric field created by the nucleus and the electron density of the other electrons. We can view this net electric field as if it results from a single positive charge located at the nucleus, called the **effective nuclear charge,** Z_{eff}. It is important to realize that the effective nuclear charge acting on an electron in an atom is smaller than the *actual* nuclear charge because the effective nuclear charge also accounts for the repulsion of the electron by the other electrons in the atom—in other words, $Z_{eff} < Z$. Let's consider how we can get a sense of the magnitude of Z_{eff} for an electron in an atom.

ANIMATION
Effective Nuclear Charge

A valence electron in an atom is attracted to the nucleus of the atom and is repelled by the other electrons in the atom. In particular, the electron density that is due to the inner (core) electrons is particularly effective at partially canceling the attraction of the valence electron to the nucleus—we say that the inner electrons partially *shield* or *screen* the outer electrons from the attraction of the nucleus. We can therefore write a simple relationship between the effective nuclear charge, Z_{eff}, and the number of protons in the nucleus, Z:

$$Z_{eff} = Z - S \qquad [7.1]$$

The factor S is a positive number called the *screening constant*, and it represents the portion of the nuclear charge that is screened from the valence electron by the other electrons in the atom. Because the core electrons are most effective at screening a valence electron from the nucleus, *the value of S is usually close to the number of core electrons in an atom.* Electrons in the same valence shell hardly screen one another but do affect the value of S slightly.

▼ **Figure 7.3 Effective nuclear charge.** (a) The effective nuclear charge experienced by the valence electron in sodium depends mostly on the 11+ charge of the nucleus and the 10− charge of the neon core. If the neon core were totally effective in shielding the valence electron from the nucleus, then the valence electron would experience an effective nuclear charge of 1+. (b) The 3*s* electron has some probability of being inside the Ne core. As a consequence of this "penetration," the core is not totally effective in screening the 3*s* electron from the nucleus. Thus, the effective nuclear charge experienced by the 3*s* electron is somewhat greater than 1+.

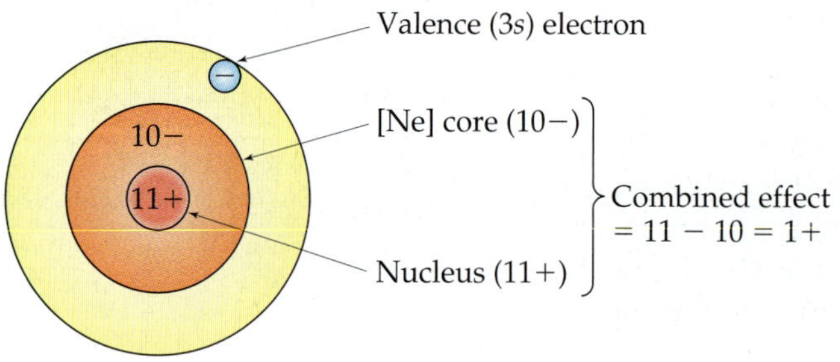

(a)

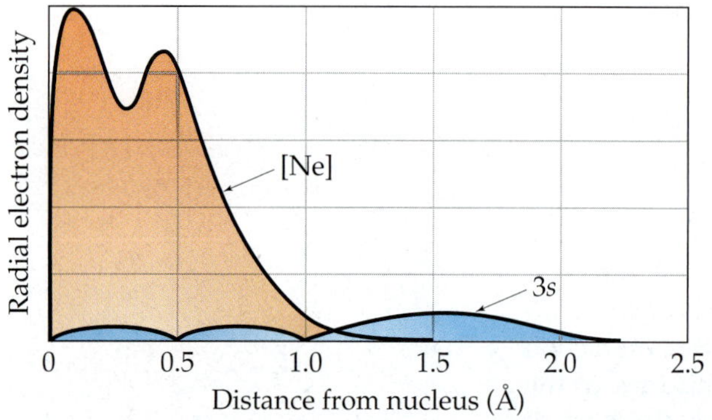

(b)

Let's take a look at an Na atom to see what we would expect for the magnitude of Z_{eff}. Sodium (atomic number 11) has a condensed electron configuration of $[Ne]3s^1$. The nuclear charge of the atom is 11+, and the Ne inner core consists of 10 electrons ($1s^2 2s^2 2p^6$). Very roughly then, we would expect the 3*s* valence electron of the Na atom to experience an effective nuclear charge of about 11 − 10 = 1+, as pictured in a simplified way in Figure 7.3(a) ◄. The situation is a bit more complicated because of the electron distributions of atomic orbitals. ∞ (Section 6.6) Recall that a 3*s* electron has a small probability of being found close to the nucleus and inside the core electrons, as shown in Figure 7.3(b). Thus, there is a probability that the 3*s* electron will experience a greater attraction than our simple model suggests, which will increase the value of Z_{eff} somewhat. Indeed, more detailed calculations (which are beyond the scope of our discussion) indicate that the effective nuclear charge acting on the 3*s* electron in Na is 2.5+.

The notion of effective nuclear charge also explains an important effect we noted in Section 6.7, namely, that for a many-electron atom the energies of orbitals with the same n value increase with increasing l value. For example, consider a carbon atom, for which the electron configuration is $1s^22s^22p^2$. The energy of the $2p$ orbital ($l = 1$) is slightly higher than that of the $2s$ orbital ($l = 0$) even though both of these orbitals are in the $n = 2$ shell (Figure 6.24). The reason that these orbitals have different energy in a many-electron atom is due to the radial probability functions for the orbitals, shown in Figure 7.4 ▶. Notice that the $2s$ probability function has a small peak fairly close to the nucleus, whereas the $2p$ probability function does not. As a result, an electron in the $2s$ orbital is less effectively screened by the core orbitals than is an electron in the $2p$ orbital. In other words, the electron in the $2s$ orbital has a higher effective nuclear charge than one in the $2p$ orbital. The greater attraction between the $2s$ electron and the nucleus leads to a lower energy for the $2s$ orbital than for the $2p$ orbital. The same reasoning explains the general trend in orbital energies ($ns < np < nd$) in many-electron atoms.

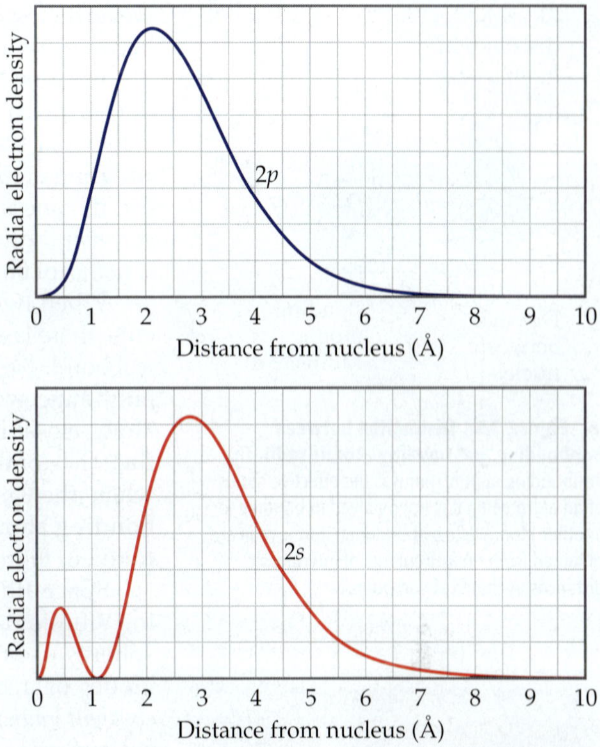

▲ **Figure 7.4 2s and 2p radial functions.** The radial probability function for the $2s$ orbital of the hydrogen atom (red curve) shows a "bump" of probability close to the nucleus, whereas that for the $2p$ orbital (blue curve) does not. As a result, an electron in the $2s$ orbital for a many-electron atom "sees" more of the nuclear charge than does an electron in the $2p$ orbital—the effective nuclear charge experienced by the $2s$ electron is greater than that for the $2p$ electron. This difference leads to our observation that in a many-electron atom the orbitals for a given n value increase in energy with increasing l value, that is, ns is lower in energy than np, which is lower in energy than nd (Figure 6.24).

Finally, let's examine the trends in Z_{eff} for valence electrons as we move from one element to another in the periodic table. The effective nuclear charge increases as we move across any row (period) of the table. Although the number of core electrons stays the same as we move across the row, the actual nuclear charge increases. The valence electrons added to counterbalance the increasing nuclear charge shield one another very ineffectively. Thus, the effective nuclear charge increases steadily. For example, the $1s^2$ core electrons of lithium ($1s^22s^1$) shield the $2s$ valence electron from the 3+ nucleus fairly efficiently. Consequently, the outer electron experiences an effective nuclear charge of roughly $3 - 2 = 1+$. For beryllium ($1s^22s^2$) the effective nuclear charge experienced by each $2s$ valence electron is larger; in this case, the inner $1s^2$ electrons are shielding a 4+ nucleus, and each $2s$ electron only partially shields the other from the nucleus. Consequently, the effective nuclear charge experienced by each $2s$ electron is about $4 - 2 = 2+$.

Going down a column, the effective nuclear charge experienced by valence electrons changes far less than it does across a row. For example, we would expect the effective nuclear charge for the outer electrons in lithium and sodium to be about the same, roughly $3 - 2 = 1+$ for lithium and $11 - 10 = 1+$ for sodium. In fact, however, the effective nuclear charge increases slightly as we go down a family because larger electron cores are less able to screen the outer electrons from the nuclear charge. We saw earlier, for example, that the value for sodium is 2.5+. Nevertheless, the small change in effective nuclear charge that occurs moving down a column is generally of less importance than the increase that occurs when moving across a row.

GIVE IT SOME THOUGHT

Which would you expect to experience a greater effective nuclear charge, a $2p$ electron of an Ne atom or a $3s$ electron of an Na atom?

7.3 | Sizes of Atoms and Ions

One of the important properties of an atom or an ion is its size. We often think of atoms and ions as hard, spherical objects. According to the quantum mechanical model, however, atoms and ions do not have sharply defined boundaries at which the electron distribution becomes zero. ∞ (Section 6.5)

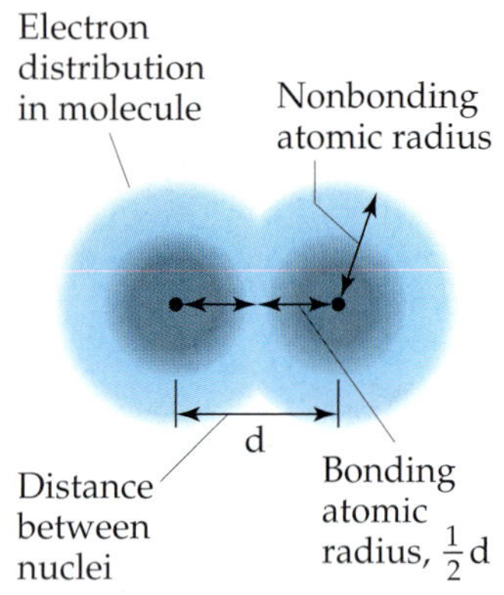

▲ **Figure 7.5 Distinction between nonbonding and bonding atomic radii.** The nonbonding atomic radius is the effective radius of an atom when it is not involved in bonding to another atom. Values of bonding atomic radii are obtained from measurements of interatomic distances in chemical compounds.

Nevertheless, we can define atomic size in several ways, based on the distances between atoms in various situations.

Imagine a collection of argon atoms in the gas phase. When two atoms collide with each other in the course of their motions, they ricochet apart—somewhat like billiard balls. This happens because the electron clouds of the colliding atoms cannot penetrate each other to any significant extent. The closest distances separating the nuclei during such collisions determine the *apparent* radii of the argon atoms. We might call this radius the *nonbonding atomic radius* of an atom.

When two atoms are chemically bonded to each other, as in the Cl_2 molecule, there is an attractive interaction between the two atoms leading to a chemical bond. We will discuss the nature of such bonding in Chapter 8. For now, the only thing we need to realize is that this attractive interaction brings the two atoms closer together than they would be in a nonbonding collision. We can define an atomic radius based on the distances separating the nuclei of atoms when they are chemically bonded to each other. This distance, called the **bonding atomic radius**, is shorter than the nonbonding atomic radius, as illustrated in Figure 7.5 ◀.

Space-filling models, such as those in Figures 1.1 and 2.20, use nonbonding atomic radii (also called *van der Waals radii*) to determine the size of the spheres used to represent atoms of different elements. When a space-filling model of a molecule is being built, the bonding atomic radii (also called *covalent radii*) are used to determine how much the spheres interpenetrate each other to represent the correct distance between the centers of two adjacent atoms in the molecule.

▼ **Figure 7.6 Trends in atomic radii.** Bonding atomic radii for the first 54 elements of the periodic table. The height of the bar for each element is proportional to its radius, giving a "relief map" view of the radii.

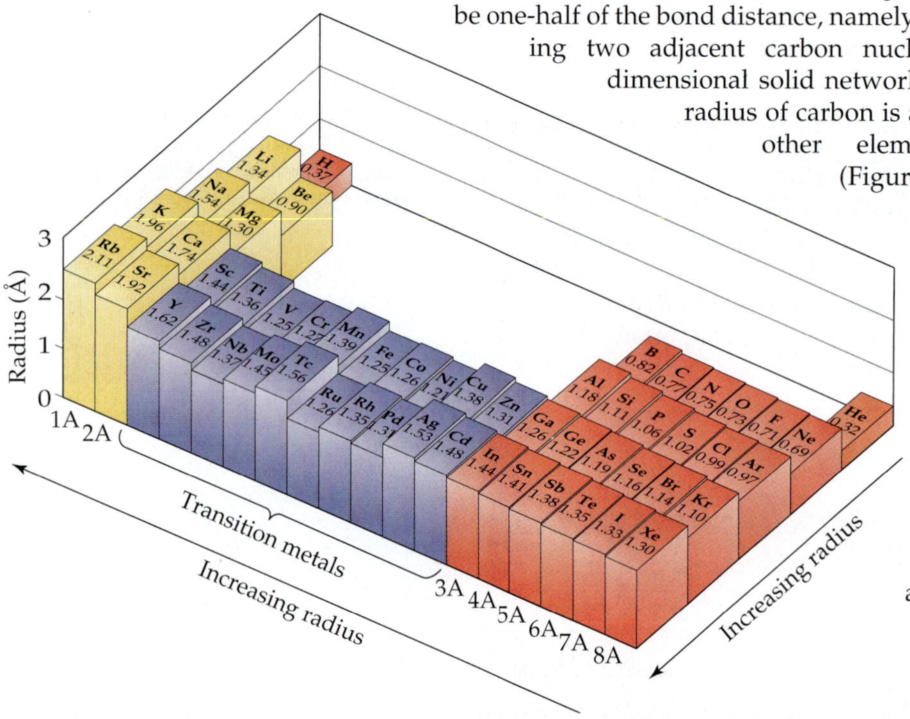

Scientists have developed a variety of methods for measuring the distances separating nuclei in molecules. From observations of these distances in many molecules, each element can be assigned a bonding atomic radius. For example, in the I_2 molecule, the distance separating the iodine nuclei is observed to be 2.66 Å.* We can define the bonding atomic radius of iodine on this basis to be one-half of the bond distance, namely 1.33 Å. Similarly, the distance separating two adjacent carbon nuclei in diamond, which is a three-dimensional solid network, is 1.54 Å; thus, the bonding atomic radius of carbon is assigned the value 0.77 Å. The radii of other elements can be similarly defined (Figure 7.6 ◀). (For helium and neon, the bonding radii must be estimated because there are no known compounds of these elements.)

Knowing atomic radii allows us to estimate the bond lengths between different elements in molecules. For example, the Cl—Cl bond length in Cl_2 is 1.99 Å, so a radius of 0.99 Å is assigned to Cl. In the compound CCl_4 the measured length of the C—Cl bond is 1.77 Å, very close to the sum (0.77 + 0.99 Å) of the atomic radii of C and Cl.

* *Remember:* The angstrom (1 Å $= 10^{-10}$ m) is a convenient metric unit for atomic measurements of length. The angstrom is *not* an SI unit. The most commonly used SI unit for such measurements is the picometer (1 pm $= 10^{-12}$ m; 1 Å = 100 pm).

SAMPLE EXERCISE 7.1 | Bond Lengths in a Molecule

Natural gas used in home heating and cooking is odorless. Because natural gas leaks pose the danger of explosion or suffocation, various smelly substances are added to the gas to allow detection of a leak. One such substance is methyl mercaptan, CH_3SH, whose structure is shown in the margin. Use Figure 7.6 to predict the lengths of the C—S, C—H, and S—H bonds in this molecule.

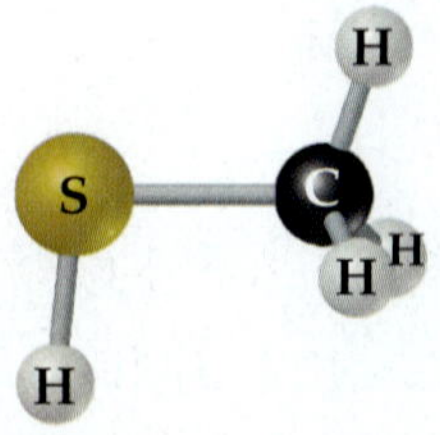

Solution

Analyze and Plan: We are given three bonds and the list of bonding atomic radii. We will assume that each bond length is the sum of the radii of the two atoms involved.

Solve: Using radii for C, S, and H from Figure 7.6, we predict

$$\text{C—S bond length} = \text{radius of C} + \text{radius of S}$$
$$= 0.77 \text{ Å} + 1.02 \text{ Å} = 1.79 \text{ Å}$$
$$\text{C—H bond length} = 0.77 \text{ Å} + 0.37 \text{ Å} = 1.14 \text{ Å}$$
$$\text{S—H bond length} = 1.02 \text{ Å} + 0.37 \text{ Å} = 1.39 \text{ Å}$$

Check: The experimentally determined bond lengths in methyl mercaptan are C—S = 1.82 Å, C—H = 1.10 Å, and S—H = 1.33 Å. (In general, the lengths of bonds involving hydrogen show larger deviations from the values predicted by the sum of the atomic radii than do those bonds involving larger atoms.)

Comment: Notice that the estimated bond lengths using bonding atomic radii are close to, but not exact matches of, the experimental bond lengths. Atomic radii must be used with some caution in estimating bond lengths. In Chapter 8 we will examine some of the average lengths of common types of bonds.

PRACTICE EXERCISE

Using Figure 7.6, predict which will be greater, the P—Br bond length in PBr_3 or the As—Cl bond length in $AsCl_3$.

Answer: P—Br

ANIMATION
Periodic Trends: Atomic Radii

Periodic Trends in Atomic Radii

If we examine the "relief map" of atomic radii shown in Figure 7.6, we observe two interesting trends in the data:

1. Within each column (group), atomic radius tends to increase from top to bottom. This trend results primarily from the increase in the principal quantum number (n) of the outer electrons. As we go down a column, the outer electrons have a greater probability of being farther from the nucleus, causing the atom to increase in size.
2. Within each row (period), atomic radius tends to decrease from left to right. The major factor influencing this trend is the increase in the effective nuclear charge (Z_{eff}) as we move across a row. The increasing effective nuclear charge steadily draws the valence electrons closer to the nucleus, causing the atomic radius to decrease.

GIVE IT SOME THOUGHT

As we proceed across a row of the periodic table, atomic weight increases but atomic radius decreases. Are these trends a contradiction?

SAMPLE EXERCISE 7.2 | Atomic Radii

Referring to a periodic table, arrange (as much as possible) the following atoms in order of increasing size: ${}_{15}P$, ${}_{16}S$, ${}_{33}As$, ${}_{34}Se$. (Atomic numbers are given for the elements to help you locate them quickly in the periodic table.)

Solution

Analyze and Plan: We are given the chemical symbols for four elements. We can use their relative positions in the periodic table and the two periodic trends just described to predict the relative order of their atomic radii.

Solve: Notice that P and S are in the same row of the periodic table, with S to the right of P. Therefore, we expect the radius of S to be smaller than that of P. (Radii decrease as we move from left to right.) Likewise, the radius of Se is expected to be smaller than that of As. We also notice that As is directly below P and that Se is directly below S. We expect, therefore, that the radius of As is greater than that of P and the radius of Se is greater than that of S. From these observations, we predict S < P, P < As, S < Se, and Se < As. We can therefore conclude that S has the smallest radius of the four elements and that As has the largest radius.

Using just the two trends described above, we cannot determine whether P or Se has the larger radius; to go from P to Se in the periodic table, we must move down (radius tends to increase) and to the right (radius tends to decrease). In Figure 7.6 we see that the radius of Se (1.16 Å) is greater than that of P (1.06 Å). If you examine the figure carefully, you will discover that for the representative elements the increase in radius moving down a column tends to be the greater effect. There are exceptions, however.

Check: From Figure 7.6, we have S (1.02 Å) < P (1.06 Å) < Se (1.16 Å) < As (1.19 Å).

Comment: Note that the trends we have just discussed are for the representative elements. You will see in Figure 7.6 that the transition elements do not show a regular decrease from left to right in a row.

PRACTICE EXERCISE

Arrange the following atoms in order of increasing atomic radius: Na, Be, Mg.
Answer: Be < Mg < Na

ANIMATION
Gain and Loss of Electrons

Periodic Trends in Ionic Radii

The radii of ions are based on the distances between ions in ionic compounds. Like the size of an atom, the size of an ion depends on its nuclear charge, on the number of electrons it possesses, and on the orbitals in which the valence electrons reside. The formation of a cation vacates the most spatially extended occupied orbitals in an atom and also decreases the number of electron-electron repulsions. As a consequence, *cations are smaller than their parent atoms*, as illustrated in Figure 7.7 ►. The opposite is true of anions. When electrons are added to a neutral atom to form an anion, the increased electron-electron repulsions cause the electrons to spread out more in space. Thus, *anions are larger than their parent atoms*.

For ions carrying the same charge, size increases as we go down a column in the periodic table. This trend is also seen in Figure 7.7. As the principal quantum number of the outermost occupied orbital of an ion increases, the radius of the ion increases.

SAMPLE EXERCISE 7.3 | Atomic and Ionic Radii

Arrange these atoms and ions in order of decreasing size: Mg^{2+}, Ca^{2+}, and Ca.

Solution Cations are smaller than their parent atoms, and so the Ca^{2+} ion is smaller than the Ca atom. Because Ca is below Mg in group 2A of the periodic table, Ca^{2+} is larger than Mg^{2+}. Consequently, Ca > Ca^{2+} > Mg^{2+}.

PRACTICE EXERCISE

Which of the following atoms and ions is largest: S^{2-}, S, O^{2-}?
Answer: S^{2-}

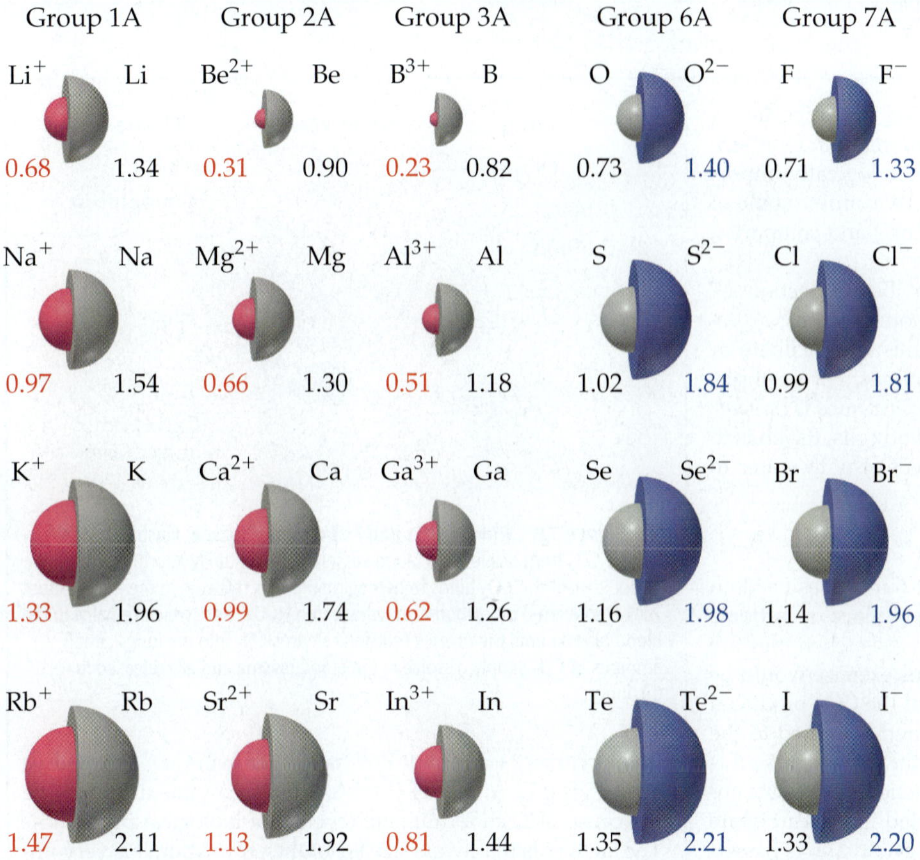

Figure 7.7 Cation and anion size. Comparisons of the radii, in Å, of neutral atoms and ions for several of the groups of representative elements. Neutral atoms are shown in gray, cations in red, and anions in blue.

An **isoelectronic series** is a group of ions all containing the same number of electrons. For example, each ion in the isoelectronic series O^{2-}, F^-, Na^+, Mg^{2+}, Al^{3+} has 10 electrons. In any isoelectronic series we can list the members in order of increasing atomic number, and therefore nuclear charge increases as we move through the series. (Recall that the charge on the nucleus of an atom or monatomic ion is given by the atomic number of the element.) Because the number of electrons remains constant, the radius of the ion decreases with increasing nuclear charge, as the electrons are more strongly attracted to the nucleus:

—— Increasing nuclear charge ⟶

O^{2-}	F^-	Na^+	Mg^{2+}	Al^{3+}
1.40 Å	1.33 Å	0.97 Å	0.66 Å	0.51 Å

—— Decreasing ionic radius ⟶

Notice the positions of these elements in the periodic table and also their atomic numbers. The nonmetal anions precede the noble gas Ne in the table. The metal cations follow Ne. Oxygen, the largest ion in this isoelectronic series, has the lowest atomic number, 8. Aluminum, the smallest of these ions, has the highest atomic number, 13.

CHEMISTRY AND LIFE | Ionic Size Makes a BIG Difference

Ionic size plays a major role in determining the properties of ions in solution. For example, a small difference in ionic size is often sufficient for one metal ion to be biologically important and another not to be. To illustrate, let's examine some of the biological chemistry of the zinc ion (Zn^{2+}) and compare it with the cadmium ion (Cd^{2+}).

Recall from the "Chemistry and Life" box in Section 2.7 that zinc is needed in our diets in trace amounts. Zinc is an essential part of several enzymes, the proteins that facilitate or regulate the speeds of key biological reactions. For example, one of the most important zinc-containing enzymes is *carbonic anhydrase*. This enzyme is found in red blood cells. Its job is to facilitate the reaction of carbon dioxide (CO_2) with water to form the bicarbonate ion (HCO_3^-):

$$CO_2(aq) + H_2O(l) \longrightarrow HCO_3^-(aq) + H^+(aq) \quad [7.2]$$

You might be surprised to know that our bodies need an enzyme for such a simple reaction. In the absence of carbonic anhydrase, however, the CO_2 produced in cells when they are oxidizing glucose or other fuels in vigorous exercise would be cleared out much too slowly. About 20% of the CO_2 produced by cell metabolism binds to hemoglobin and is carried to the lungs, where it is expelled. About 70% of the CO_2 produced is converted to bicarbonate ion through the action of carbonic anhydrase. When the CO_2 has been converted into bicarbonate ion, it diffuses into the blood plasma and eventually is passed into the lungs in the reverse of Equation 7.2. These processes are illustrated in Figure 7.8 ▶. In the absence of zinc, the carbonic anhydrase would be inactive and serious imbalances would result in the amount of CO_2 present in blood.

Zinc is also found in several other enzymes, including some found in the liver and kidneys. It is obviously essential to life. By contrast, cadmium, zinc's neighbor in group 2B, is extremely toxic to humans. But why are two elements so different?

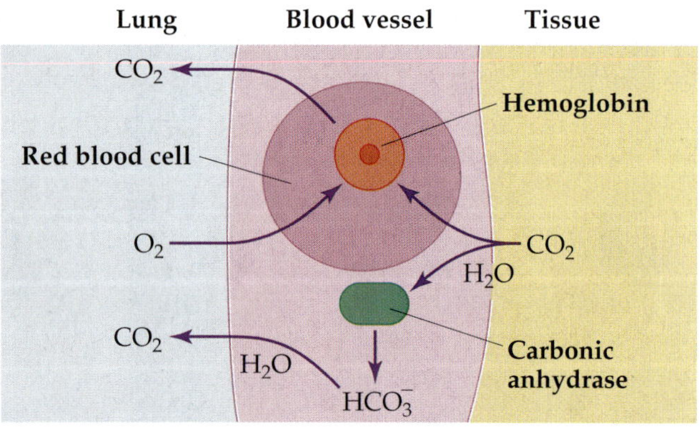

▲ **Figure 7.8 Ridding the body of carbon dioxide.** Illustration of the flow of CO_2 from tissues into blood vessels and eventually into the lungs. About 20% of the CO_2 binds to hemoglobin and is released in the lungs. About 70% is converted by carbonic anhydrase into HCO_3^- ion, which remains in the blood plasma until the reverse reaction releases CO_2 into the lungs. Small amounts of CO_2 simply dissolve in the blood plasma and are released in the lungs.

Both occur as 2+ ions, but Zn^{2+} is smaller than Cd^{2+}. The radius of Zn^{2+} is 0.74 Å; that of Cd^{2+} is 0.95 Å. Can this difference be the cause of such a dramatic reversal of biological properties? The answer is that while size is not the only factor, it is very important. In the carbonic anhydrase enzyme the Zn^{2+} ion is found electrostatically bonded to atoms on the protein, as shown in Figure 7.9 ▼. It turns out that Cd^{2+} binds in this same place preferentially over Zn^{2+}, thus displacing it. When Cd^{2+} is present instead of Zn^{2+}, however, the reaction of CO_2 with water is not facilitated. More seriously, Cd^{2+} inhibits reactions that are essential to the kidneys' functioning.

Related Exercises: 7.28, 7.88, 7.89, and 7.92

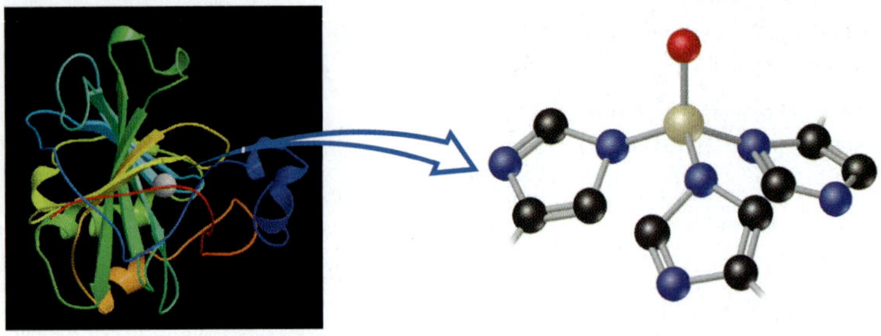

◀ **Figure 7.9 A zinc-containing enzyme.** The enzyme called carbonic anhydrase (left) catalyzes the reaction between CO_2 and water to form HCO_3^-. The ribbon represents the folding of the protein chain. The "active site" of the enzyme (represented by the ball-and-stick model) is where the reaction occurs. (H atoms have been excluded from this model for clarity.) The red sphere represents the oxygen of a water molecule that is bound to the zinc ion (gold sphere) at the center of the active site. The water molecule is replaced by CO_2 in the reaction. The bonds coming off the five-member rings attach the active site to the protein.

SAMPLE EXERCISE 7.4 | Ionic Radii in an Isoelectronic Series

Arrange the ions K^+, Cl^-, Ca^{2+}, and S^{2-} in order of decreasing size.

Solution First, we note that this is an isoelectronic series of ions, with all ions having 18 electrons. In such a series, size decreases as the nuclear charge (atomic number) of the ion increases. The atomic numbers of the ions are S (16), Cl (17), K (19), and Ca (20). Thus, the ions decrease in size in the order $S^{2-} > Cl^- > K^+ > Ca^{2+}$.

PRACTICE EXERCISE

Which of the following ions is largest, Rb^+, Sr^{2+}, or Y^{3+}?

Answer: Rb^+

7.4 | Ionization Energy

The ease with which electrons can be removed from an atom or ion has a major impact on chemical behavior. The **ionization energy** of an atom or ion is the minimum energy required to remove an electron from the ground state of the isolated gaseous atom or ion. The *first ionization energy*, I_1, is the energy needed to remove the first electron from a neutral atom. For example, the first ionization energy for the sodium atom is the energy required for the process

ANIMATION
Ionization Energy

$$Na(g) \longrightarrow Na^+(g) + e^- \quad [7.3]$$

The *second ionization energy*, I_2, is the energy needed to remove the second electron, and so forth, for successive removals of additional electrons. Thus, I_2 for the sodium atom is the energy associated with the process

$$Na^+(g) \longrightarrow Na^{2+}(g) + e^- \quad [7.4]$$

The greater the ionization energy, the more difficult it is to remove an electron.

GIVE IT SOME THOUGHT

Light can be used to ionize atoms and ions, as in Equations 7.3 and 7.4. What concept discussed in Chapter 6 can be related to the ionization of atoms and molecules?

Variations in Successive Ionization Energies

Ionization energies for the elements sodium through argon are listed in Table 7.2 ▼. Notice that the values for a given element increase as successive electrons are removed: $I_1 < I_2 < I_3$, and so forth. This trend exists because with each successive removal, an electron is being pulled away from an increasingly more positive ion, requiring increasingly more energy.

ACTIVITY
Ionization Energies

A second important feature shown in Table 7.2 is the sharp increase in ionization energy that occurs when an inner-shell electron is removed. For example, consider silicon, whose electron configuration is $1s^22s^22p^63s^23p^2$ or $[Ne]3s^23p^2$. The ionization energies increase steadily from 786 kJ/mol to 4360 kJ/mol for the loss of the four electrons in the outer 3s and 3p subshells. Removal of the fifth electron, which comes from the 2p subshell, requires a great deal more energy: 16,100 kJ/mol. The large increase occurs because the 2p electron is much more likely to be found close to the nucleus than are the four $n = 3$ electrons, and therefore the 2p electron experiences a much greater effective nuclear charge than do the 3s and 3p electrons.

TABLE 7.2 Successive Values of Ionization Energies, *I*, for the Elements Sodium Through Argon (kJ/mol)

Element	I_1	I_2	I_3	I_4	I_5	I_6	I_7
Na	495	4562			(inner-shell electrons)		
Mg	738	1451	7733				
Al	578	1817	2745	11,577			
Si	786	1577	3232	4356	16,091		
P	1012	1907	2914	4964	6274	21,267	
S	1000	2252	3357	4556	7004	8496	27,107
Cl	1251	2298	3822	5159	6542	9362	11,018
Ar	1521	2666	3931	5771	7238	8781	11,995

GIVE IT SOME THOUGHT

Which would you expect to be greater, I_1 for a boron atom or I_2 for a carbon atom?

Every element exhibits a large increase in ionization energy when electrons are removed from its noble-gas core. This observation supports the idea that only the outermost electrons, those beyond the noble-gas core, are involved in the sharing and transfer of electrons that give rise to chemical bonding and reactions. The inner electrons are too tightly bound to the nucleus to be lost from the atom or even shared with another atom.

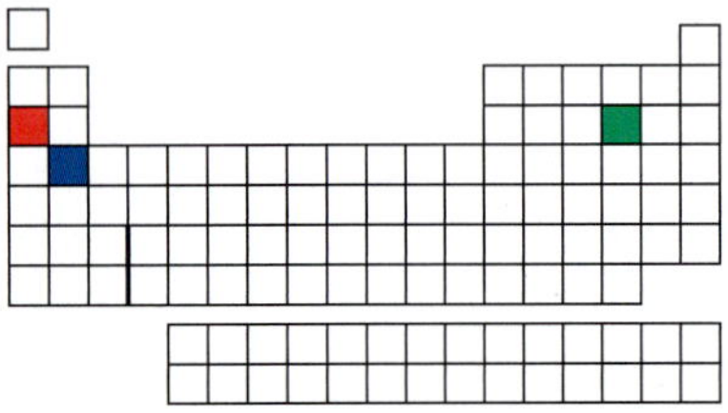

SAMPLE EXERCISE 7.5 | Trends in Ionization Energy

Three elements are indicated in the periodic table in the margin. Based on their locations, predict the one with the largest second ionization energy.

Solution

Analyze and Plan: The locations of the elements in the periodic table allow us to predict the electron configurations. The greatest ionization energies involve removal of core electrons. Thus, we should look first for an element with only one electron in the outermost occupied shell.

Solve: The element in group 1A (Na), indicated by the red box, has only one valence electron. The second ionization energy of this element is associated, therefore, with the removal of a core electron. The other elements indicated, S (green box) and Ca (blue box), have two or more valence electrons. Thus, Na should have the largest second ionization energy.

Check: If we consult a chemistry handbook, we find the following values for the second ionization energies (I_2) of the respective elements: Ca (1,145 kJ/mol) < S (2,252 kJ/mol) < Na (4,562 kJ/mol).

PRACTICE EXERCISE

Which will have the greater third ionization energy, Ca or S?

Answer: Ca

Periodic Trends in First Ionization Energies

We have seen that the ionization energy for a given element increases as we remove successive electrons. What trends do we observe in ionization energy as we move from one element to another in the periodic table? Figure 7.10 ◀ shows a graph of I_1 versus atomic number for the first 54 elements. The important trends are as follows:

1. Within each row (period) of the table, I_1 generally increases with increasing atomic number. The alkali metals show the lowest ionization energy in each row, and the noble gases the highest. There are slight irregularities in this trend that we will discuss shortly.
2. Within each column (group) of the table, the ionization energy generally decreases with increasing atomic number. For example, the ionization energies of the noble gases follow the order He > Ne > Ar > Kr > Xe.
3. The representative elements show a larger range of values of I_1 than do the transition-metal elements. Generally,

▼ **Figure 7.10 First ionization energy versus atomic number.** The red dots mark the beginning of a period (alkali metals), the blue dots mark the end of a period (noble gases), and the black dots indicate other representative elements. Green dots are used for the transition metals.

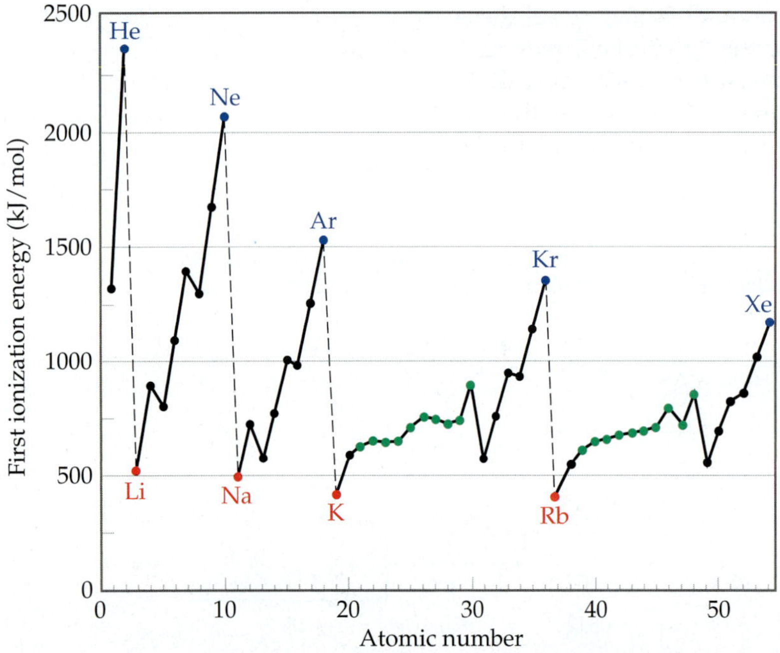

the ionization energies of the transition metals increase slowly as we proceed from left to right in a period. The *f*-block metals, which are not shown in Figure 7.10, also show only a small variation in the values of I_1.

The periodic trends in the first ionization energies of the representative elements are further illustrated in Figure 7.11 ▶.

In general, smaller atoms have higher ionization energies. The same factors that influence atomic size also influence ionization energies. The energy needed to remove an electron from the outermost occupied shell depends on both the effective nuclear charge and the average distance of the electron from the nucleus. Either increasing the effective nuclear charge or decreasing the distance from the nucleus increases the attraction between the electron and the nucleus. As this attraction increases, it becomes harder to remove the electron and, thus, the ionization energy increases. As we move across a period, there is both an increase in effective nuclear charge and a decrease in atomic radius, causing the ionization energy to increase. As we move down a column, however, the atomic radius increases, while the effective nuclear charge changes little. Thus, the attraction between the nucleus and the electron decreases, causing the ionization energy to decrease.

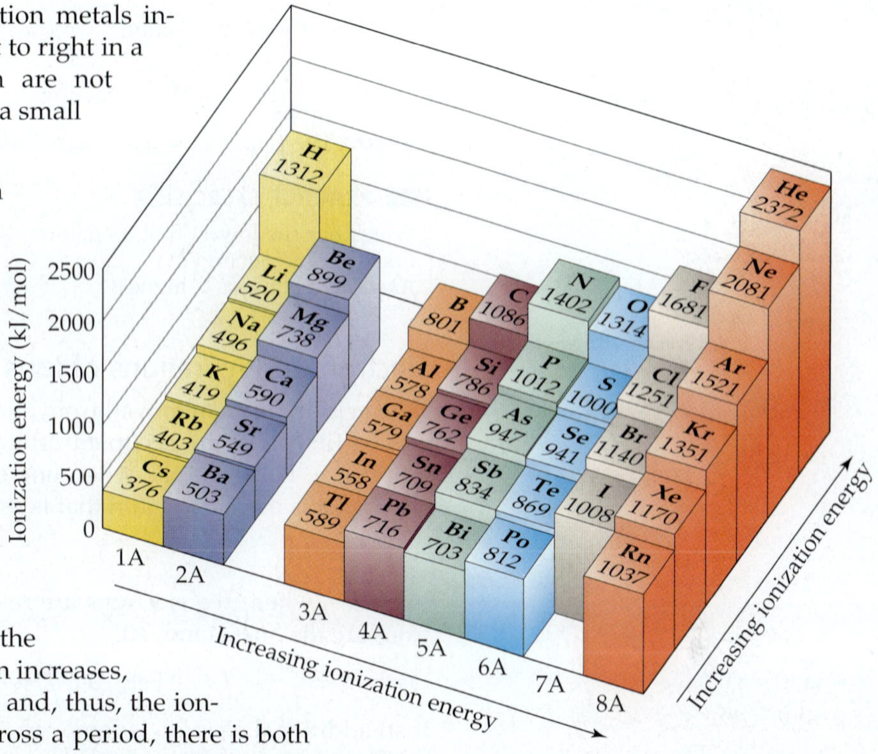

▲ **Figure 7.11 Trends in first ionization energy.** First ionization energies for the representative elements in the first six periods. The ionization energy generally increases from left to right and decreases from top to bottom. The ionization energy of astatine has not been determined.

The irregularities within a given row are somewhat more subtle but still readily explained. For example, the decrease in ionization energy from beryllium ([He]$2s^2$) to boron ([He]$2s^2 2p^1$), shown in Figures 7.10 and 7.11, occurs because the third valence electron of B must occupy the $2p$ subshell, which is empty for Be. Recall that, as we discussed earlier, the $2p$ subshell is at a higher energy than the $2s$ (Figure 6.22). The decrease in ionization energy on going from nitrogen ([He]$2s^2 2p^3$) to oxygen ([He]$2s^2 2p^4$) is because of repulsion of paired electrons in the p^4 configuration. Remember that according to Hund's rule, each electron in the p^3 configuration resides in a different p orbital, which minimizes the electron-electron repulsion among the three $2p$ electrons. (Section 6.8)

ANIMATION
Periodic Trends: Ionization Energies

SAMPLE EXERCISE 7.6 | Periodic Trends in Ionization Energy

Referring to a periodic table, arrange the following atoms in order of increasing first ionization energy: Ne, Na, P, Ar, K.

Solution

Analyze and Plan: We are given the chemical symbols for five elements. In order to rank them according to increasing first ionization energy, we need to locate each element in the periodic table. We can then use their relative positions and the trends in first ionization energies to predict their order.

Solve: Ionization energy increases as we move left to right across a row. It decreases as we move from the top of a group to the bottom. Because Na, P, and Ar are in the same row of the periodic table, we expect I_1 to vary in the order Na $<$ P $<$ Ar.

Because Ne is above Ar in group 8A, we expect Ne to have the greater first ionization energy: Ar $<$ Ne. Similarly, K is the alkali metal directly below Na in group 1A, and so we expect I_1 for K to be less than that of Na: K $<$ Na.

From these observations, we conclude that the ionization energies follow the order

$$K < Na < P < Ar < Ne$$

Check: The values shown in Figure 7.11 confirm this prediction.

PRACTICE EXERCISE

Which has the lowest first ionization energy, B, Al, C, or Si? Which has the highest first ionization energy?
Answer: Al lowest, C highest

Electron Configurations of Ions

When electrons are removed from an atom to form a cation, they are always removed first from the occupied orbitals having the largest principal quantum number, n. For example, when one electron is removed from a lithium atom ($1s^22s^1$), it is the $2s^1$ electron that is removed:

$$\text{Li } (1s^22s^1) \Rightarrow \text{Li}^+ \ (1s^2)$$

Likewise, when two electrons are removed from Fe ($[\text{Ar}]3d^64s^2$), the $4s^2$ electrons are the ones removed:

$$\text{Fe } ([\text{Ar}]3d^64s^2) \Rightarrow \text{Fe}^{2+} \ ([\text{Ar}]3d^6)$$

If an additional electron is removed, forming Fe^{3+}, it now comes from a $3d$ orbital because all the orbitals with $n = 4$ are empty:

$$\text{Fe}^{2+} \ ([\text{Ar}]3d^6) \Rightarrow \text{Fe}^{3+} \ ([\text{Ar}]3d^5)$$

It may seem odd that the $4s$ electrons are removed before the $3d$ electrons in forming transition-metal cations. After all, in writing electron configurations, we added the $4s$ electrons before the $3d$ ones. In writing electron configurations for atoms, however, we are going through an imaginary process in which we move through the periodic table from one element to another. In doing so, we are adding not only an electron to an orbital but also a proton to the nucleus, in order to change the identity of the element. In ionization, we do not reverse this process because no protons are being removed.

When electrons are added to an atom to form an anion, they are added to the empty or partially filled orbital having the lowest value of n. For example, when an electron is added to a fluorine atom to form the F^- ion, the electron goes into the one remaining vacancy in the $2p$ subshell:

$$\text{F } (1s^22s^22p^5) \Rightarrow \text{F}^- \ (1s^22s^22p^6)$$

GIVE IT SOME THOUGHT

Would Cr^{3+} and V^{2+} have the same or different electron configurations?

SAMPLE EXERCISE 7.7 | Electron Configurations of Ions

Write the electron configuration for **(a)** Ca^{2+}, **(b)** Co^{3+}, and **(c)** S^{2-}.

Solution

Analyze and Plan: We are asked to write electron configurations for three ions. To do so, we first write the electron configuration of the parent atom. We then remove electrons to form cations or add electrons to form anions. Electrons are first removed from the orbitals having the highest value of n. They are added to the empty or partially filled orbitals having the lowest value of n.

Solve: (a) Calcium (atomic number 20) has the electron configuration

$$\text{Ca: } [\text{Ar}]4s^2$$

To form a 2+ ion, the two outer electrons must be removed, giving an ion that is isoelectronic with Ar:

$$Ca^{2+}\text{: [Ar]}$$

(b) Cobalt (atomic number 27) has the electron configuration

$$\text{Co: [Ar]}3d^7 4s^2$$

To form a 3+ ion, three electrons must be removed. As discussed in the text preceding this Sample Exercise, the 4*s* electrons are removed before the 3*d* electrons. Consequently, the electron configuration for Co^{3+} is

$$Co^{3+}\text{: [Ar]}3d^6$$

(c) Sulfur (atomic number 16) has the electron configuration

$$\text{S: [Ne]}3s^2 3p^4$$

To form a 2− ion, two electrons must be added. There is room for two additional electrons in the 3*p* orbitals. Thus, the S^{2-} electron configuration is

$$S^{2-}\text{: [Ne]}3s^2 3p^6 = \text{[Ar]}$$

Comment: Remember that many of the common ions of the representative elements, such as Ca^{2+} and S^{2-}, have the same number of electrons as the closest noble gas. (Section 2.7)

PRACTICE EXERCISE

Write the electron configuration for **(a)** Ga^{3+}, **(b)** Cr^{3+}, and **(c)** Br^-.
Answers: **(a)** $[Ar]3d^{10}$, **(b)** $[Ar]3d^3$, **(c)** $[Ar]3d^{10}4s^2 4p^6 = [Kr]$

7.5 | Electron Affinities

The first ionization energy of an atom is a measure of the energy change associated with removing an electron from the atom to form a positively charged ion. For example, the first ionization energy of Cl(*g*), 1251 kJ/mol, is the energy change associated with the process

$$\textit{Ionization energy:}\quad \underset{[Ne]3s^2 3p^5}{Cl(g)} \longrightarrow \underset{[Ne]3s^2 3p^4}{Cl^+(g)} + e^- \qquad \Delta E = 1251 \text{ kJ/mol} \qquad [7.5]$$

The positive value of the ionization energy means that energy must be put into the atom in order to remove the electron.

In addition, most atoms can gain electrons to form negatively charged ions. The energy change that occurs when an electron is added to a gaseous atom is called the **electron affinity** because it measures the attraction, or *affinity*, of the atom for the added electron. For most atoms, energy is released when an electron is added. For example, the addition of an electron to a chlorine atom is accompanied by an energy change of −349 kJ/mol, the negative sign indicating that energy is released during the process. We therefore say that the electron affinity of Cl is −349 kJ/mol:*

ANIMATION
Electron Affinity

$$\textit{Electron affinity:}\quad \underset{[Ne]3s^2 3p^5}{Cl(g)} + e^- \longrightarrow \underset{[Ne]3s^2 3p^6}{Cl^-(g)} \qquad \Delta E = -349 \text{ kJ/mol} \qquad [7.6]$$

It is important to understand the difference between ionization energy and electron affinity: Ionization energy measures the ease with which an atom *loses*

* Two sign conventions are used for electron affinity. In most introductory texts, including this one, the thermodynamic sign convention is used: A negative sign indicates that the addition of an electron is an exothermic process, as in the electron affinity given for chlorine, −349 kJ/mol. Historically, however, electron affinity has been defined as the energy released when an electron is added to a gaseous atom or ion. Because 349 kJ/mol is *released* when an electron is added to Cl(*g*), the electron affinity by this convention would be +349 kJ/mol.

an electron, whereas electron affinity measures the ease with which an atom *gains* an electron.

The greater the attraction between a given atom and an added electron, the more negative the atom's electron affinity. For some elements, such as the noble gases, the electron affinity has a positive value, meaning that the anion is higher in energy than are the separated atom and electron:

$$\underset{[\text{Ne}]3s^2 3p^6}{\text{Ar}(g)} + e^- \longrightarrow \underset{[\text{Ne}]3s^2 3p^6 4s^1}{\text{Ar}^-(g)} \qquad \Delta E > 0 \qquad [7.7]$$

ANIMATION
Periodic Trends: Electron Affinity

The fact that the electron affinity is a positive number means that an electron will not attach itself to an Ar atom; the Ar^- ion is unstable and does not form.

Figure 7.12 ◀ shows the electron affinities for the representative elements in the first five rows of the periodic table. Notice that the trends in electron affinity as we proceed through the periodic table are not as evident as they were for ionization energy. The halogens, which are one electron shy of a filled *p* subshell, have the most-negative electron affinities. By gaining an electron, a halogen atom forms a stable negative ion that has a noble-gas configuration (Equation 7.6). The addition of an electron to a noble gas, however, would require that the electron reside in a higher-energy subshell that is empty in the neutral atom (Equation 7.7). Because occupying a higher-energy subshell is energetically very unfavorable, the electron affinity is highly positive. The electron affinities of Be and Mg are positive for the same reason; the added electron would reside in a previously empty *p* subshell that is higher in energy.

1A	2A	3A	4A	5A	6A	7A	8A
H −73							He > 0
Li −60	Be > 0	B −27	C −122	N > 0	O −141	F −328	Ne > 0
Na −53	Mg > 0	Al −43	Si −134	P −72	S −200	Cl −349	Ar > 0
K −48	Ca −2	Ga −30	Ge −119	As −78	Se −195	Br −325	Kr > 0
Rb −47	Sr −5	In −30	Sn −107	Sb −103	Te −190	I −295	Xe > 0

▲ **Figure 7.12 Electron affinity.** Electron affinities in kJ/mol for the representative elements in the first five rows of the periodic table. The more negative the electron affinity, the greater the attraction of the atom for an electron. An electron affinity > 0 indicates that the negative ion is higher in energy than the separated atom and electron.

The electron affinities of the group 5A elements (N, P, As, Sb) are also interesting. Because these elements have half-filled *p* subshells, the added electron must be put in an orbital that is already occupied, resulting in larger electron-electron repulsions. Consequently, these elements have electron affinities that are either positive (N) or less negative than their neighbors to the left (P, As, Sb).

Electron affinities do not change greatly as we move down a group. For example, consider the electron affinities of the halogens (Figure 7.12). For F, the added electron goes into a 2*p* orbital, for Cl a 3*p* orbital, for Br a 4*p* orbital, and so forth. As we proceed from F to I, therefore, the average distance between the added electron and the nucleus steadily increases, causing the electron-nucleus attraction to decrease. The orbital that holds the outermost electron is increasingly spread out, however, as we proceed from F to I, thereby reducing the electron-electron repulsions. A lower electron-nucleus attraction is thus counterbalanced by lower electron-electron repulsions.

GIVE IT SOME THOUGHT

Suppose you were asked for a value for the first ionization energy of a $Cl^-(g)$ ion. What is the relationship between this quantity and the electron affinity of Cl(*g*) ?

7.6 | Metals, Nonmetals, and Metalloids

Atomic radii, ionization energies, and electron affinities are properties of individual atoms. With the exception of the noble gases, however, none of the elements exists in nature as individual atoms. To get a broader understanding of the properties of elements, we must also examine periodic trends in properties that involve large collections of atoms.

The elements can be broadly grouped into the categories of metals, nonmetals, and metalloids. (Section 2.5) This classification is shown in

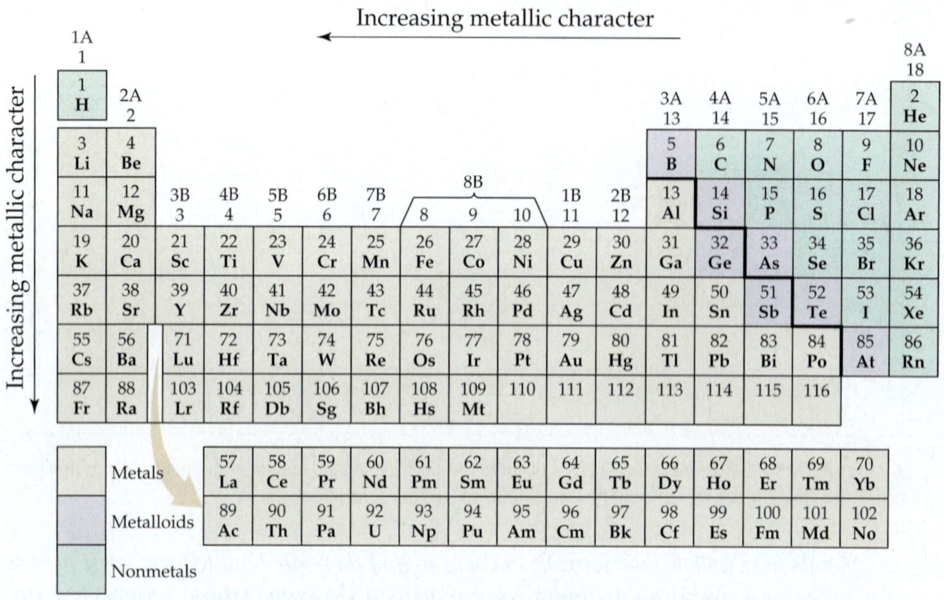

▲ **Figure 7.13 Metals, metalloids, and nonmetals.** The majority of elements are metals. Metallic character increases from right to left across a period and also increases from top to bottom in a group.

Figure 7.13 ▲. Roughly three-quarters of the elements are metals, situated in the left and middle portions of the table. The nonmetals are located at the top right corner, and the metalloids lie between the metals and nonmetals. Hydrogen, which is located at the top left corner, is a nonmetal. This is why we set off hydrogen from the remaining group 1A elements in Figure 7.13 by inserting a space between the H box and the Li box. Some of the distinguishing properties of metals and nonmetals are summarized in Table 7.3 ▼.

The more an element exhibits the physical and chemical properties of metals, the greater its **metallic character**. As indicated in Figure 7.13, metallic character generally increases as we proceed down a column of the periodic table and increases as we proceed from right to left in a row. Let's now examine the close relationships that exist between electron configurations and the properties of metals, nonmetals, and metalloids.

Metals

Most metallic elements exhibit the shiny luster that we associate with metals (Figure 7.14 ▶). Metals conduct heat and electricity. They are malleable (can be pounded into thin sheets) and ductile (can be drawn into wires). All are solids at room temperature except mercury (melting point = −39°C), which is a liquid. Two melt at slightly above room temperature, cesium at 28.4°C and gallium at 29.8°C. At the other extreme, many metals melt at very high temperatures. For example, chromium melts at 1900°C.

▲ **Figure 7.14 The luster of metals.** Metallic objects are readily recognized by their characteristic shiny luster.

TABLE 7.3 Characteristic Properties of Metals and Nonmetals

Metals	Nonmetals
Have a shiny luster; various colors, although most are silvery	Do not have a luster; various colors
Solids are malleable and ductile	Solids are usually brittle; some are hard, some are soft
Good conductors of heat and electricity	Poor conductors of heat and electricity
Most metal oxides are ionic solids that are basic	Most nonmetal oxides are molecular substances that form acidic solutions
Tend to form cations in aqueous solution	Tend to form anions or oxyanions in aqueous solution

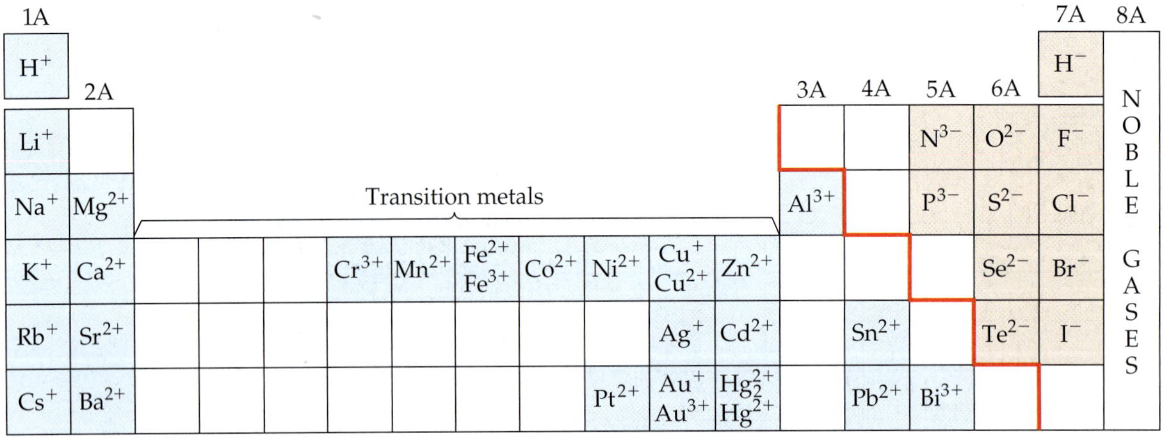

▲ **Figure 7.15 Common ions.** Charges of some common ions found in ionic compounds. Notice that the steplike line that divides metals from nonmetals also separates cations from anions.

Metals tend to have low ionization energies and therefore tend to form positive ions relatively easily. As a result, metals are oxidized (lose electrons) when they undergo chemical reactions. The relative ease of oxidation of common metals was discussed earlier, in Section 4.6. As we noted there, many metals are oxidized by a variety of common substances, including O_2 and acids.

Figure 7.15 ▲ shows the charges of some common ions of both metals and nonmetals. As we noted in Section 2.7, the charge on any alkali metal ion is always 1+, and that on any alkaline earth metal is always 2+ in their compounds. In atoms in both these groups, the outer *s* electrons are easily lost, yielding a noble-gas electron configuration. The charge on transition-metal ions does not follow an obvious pattern. Many transition-metal ions carry a charge of 2+, but 1+ and 3+ are also encountered. One of the characteristic features of the transition metals is their ability to form more than one positive ion. For example, iron may be 2+ in some compounds and 3+ in others.

GIVE IT SOME THOUGHT

Based on periodic trends discussed in this chapter, can you see a general relationship between the trends in metallic character and those for ionization energy?

Compounds of metals with nonmetals tend to be ionic substances. For example, most metal oxides and halides are ionic solids. To illustrate, the reaction between nickel metal and oxygen produces nickel oxide, an ionic solid containing Ni^{2+} and O^{2-} ions:

$$2\,Ni(s) + O_2(g) \longrightarrow 2\,NiO(s) \qquad [7.8]$$

The oxides are particularly important because of the great abundance of oxygen in our environment.

Most metal oxides are basic. Those that dissolve in water react to form metal hydroxides, as in the following examples:

$$\text{Metal oxide} + \text{water} \longrightarrow \text{metal hydroxide}$$

$$Na_2O(s) + H_2O(l) \longrightarrow 2\,NaOH(aq) \qquad [7.9]$$

$$CaO(s) + H_2O(l) \longrightarrow Ca(OH)_2(aq) \qquad [7.10]$$

The basicity of metal oxides is due to the oxide ion, which reacts with water according to the net ionic equation

$$O^{2-}(aq) + H_2O(l) \longrightarrow 2\,OH^-(aq) \qquad [7.11]$$

NiO

(a) (b)

◀ **Figure 7.16 Metal oxides react with acids.** (a) Nickel oxide (NiO), nitric acid (HNO_3), and water. (b) NiO is insoluble in water, but reacts with HNO_3 to give a green solution of the salt $Ni(NO_3)_2$.

ANIMATION
Periodic Trends: Acid-Base Behavior of Oxides

Metal oxides also demonstrate their basicity by reacting with acids to form a salt plus water, as illustrated in Figure 7.16 ▲:

$$\text{Metal oxide} + \text{acid} \longrightarrow \text{salt} + \text{water}$$

$$NiO(s) + 2\,HCl(aq) \longrightarrow NiCl_2(aq) + H_2O(l) \qquad [7.12]$$

In contrast, we will soon see that nonmetal oxides are acidic, dissolving in water to form acidic solutions and reacting with bases to form salts.

SAMPLE EXERCISE 7.8 | Metal Oxides

(a) Would you expect aluminum oxide to be a solid, liquid, or gas at room temperature? **(b)** Write the balanced chemical equation for the reaction of aluminum oxide with nitric acid.

Solution

Analyze and Plan: We are asked about one physical property of aluminum oxide—its state at room temperature—and one chemical property—how it reacts with nitric acid.

Solve: **(a)** Because aluminum oxide is the oxide of a metal, we would expect it to be an ionic solid. Indeed it is, with the very high melting point of 2072°C.

(b) In its compounds, aluminum has a 3+ charge, Al^{3+}; the oxide ion is O^{2-}. Consequently, the formula of aluminum oxide is Al_2O_3. Metal oxides tend to be basic and therefore to react with acids to form a salt plus water. In this case the salt is aluminum nitrate, $Al(NO_3)_3$. The balanced chemical equation is

$$Al_2O_3(s) + 6\,HNO_3(aq) \longrightarrow 2\,Al(NO_3)_3(aq) + 3\,H_2O(l)$$

PRACTICE EXERCISE

Write the balanced chemical equation for the reaction between copper(II) oxide and sulfuric acid.
Answer: $CuO(s) + H_2SO_4(aq) \longrightarrow CuSO_4(aq) + H_2O(l)$

Nonmetals

Nonmetals vary greatly in appearance (Figure 7.17 ▶). They are not lustrous and generally are poor conductors of heat and electricity. Their melting points are generally lower than those of metals (although diamond, a form of carbon, melts at 3570°C). Under ordinary conditions, seven nonmetals exist as diatomic molecules. Five of these are gases (H_2, N_2, O_2, F_2, and Cl_2), one is a liquid (Br_2), and one is a volatile solid (I_2). The remaining nonmetals are solids that can be either hard, such as diamond, or soft, such as sulfur.

Because of their electron affinities, nonmetals tend to gain electrons when they react with metals. For example, the reaction of aluminum

▼ **Figure 7.17 The diversity of nonmetals.** Nonmetallic elements are diverse in their appearances. Shown here are (clockwise from left) carbon in the form of graphite, sulfur, white phosphorus (stored under water), and iodine.

with bromine produces aluminum bromide, an ionic compound containing the aluminum ion, Al^{3+}, and the bromide ion, Br^-:

$$2\,Al(s) + 3\,Br_2(l) \longrightarrow 2\,AlBr_3(s) \qquad [7.13]$$

A nonmetal typically will gain enough electrons to fill its outermost occupied *p* subshell, giving a noble-gas electron configuration. For example, the bromine atom gains one electron to fill its 4*p* subshell:

$$Br\ ([Ar]4s^23d^{10}4p^5) \Rightarrow Br^-\ ([Ar]4s^23d^{10}4p^6)$$

Compounds composed entirely of nonmetals are molecular substances. For example, the oxides, halides, and hydrides of the nonmetals are molecular substances that tend to be gases, liquids, or low-melting solids at room temperature.

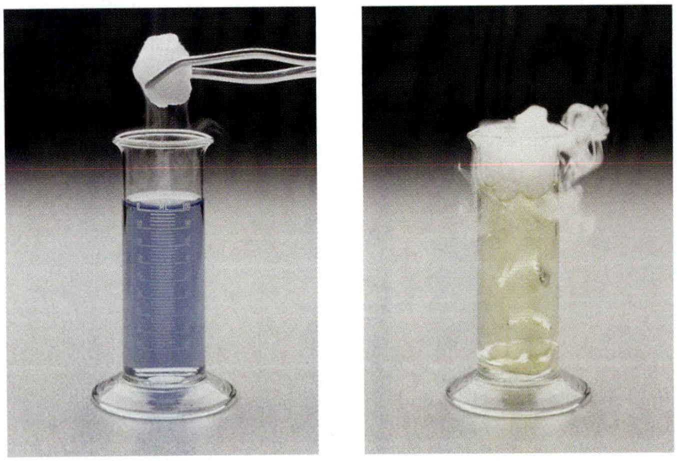

(a) (b)

▲ **Figure 7.18 The reaction of CO_2 with water.** (a) The water has been made slightly basic and contains a few drops of bromthymol blue, an acid-base indicator that is blue in basic solution. (b) Upon the addition of a piece of solid carbon dioxide, $CO_2(s)$, the color changes to yellow, indicating an acidic solution. The mist is due to water droplets condensed from the air by the cold CO_2 gas.

Most nonmetal oxides are acidic; those that dissolve in water react to form acids, as in the following examples:

$$\text{Nonmetal oxide} + \text{water} \longrightarrow \text{acid}$$

$$CO_2(g) + H_2O(l) \longrightarrow H_2CO_3(aq) \qquad [7.14]$$

$$P_4O_{10}(s) + 6\,H_2O(l) \longrightarrow 4\,H_3PO_4(aq) \qquad [7.15]$$

The reaction of carbon dioxide with water (Figure 7.18 ◀) accounts for the acidity of carbonated water and, to some extent, rainwater. Because sulfur is present in oil and coal, combustion of these common fuels produces sulfur dioxide and sulfur trioxide. These substances dissolve in water to produce *acid rain*, a major pollution problem in many parts of the world. Like acids, most nonmetal oxides dissolve in basic solutions to form a salt plus water:

$$\text{Nonmetal oxide} + \text{base} \longrightarrow \text{salt} + \text{water}$$

$$CO_2(g) + 2\,NaOH(aq) \longrightarrow Na_2CO_3(aq) + H_2O(l) \qquad [7.16]$$

GIVE IT SOME THOUGHT

A compound ACl_3 (A is an element) has a melting point of $-112°C$. Would you expect the compound to be a molecular or ionic substance? Is element A more likely to be scandium (Sc) or phosphorus (P)?

SAMPLE EXERCISE 7.9 | Nonmetal Oxides

Write the balanced chemical equations for the reactions of solid selenium dioxide with **(a)** water, **(b)** aqueous sodium hydroxide.

Solution

Analyze and Plan: We first note that selenium (Se) is a nonmetal. We therefore need to write chemical equations for the reaction of a nonmetal oxide, first with water and then with a base, NaOH. Nonmetal oxides are acidic, reacting with water to form an acid and with bases to form a salt and water.

Solve: (a) Selenium dioxide is SeO_2. Its reaction with water is like that of carbon dioxide (Equation 7.14):

$$SeO_2(s) + H_2O(l) \longrightarrow H_2SeO_3(aq)$$

(It doesn't matter that SeO_2 is a solid and CO_2 is a gas; the point is that both are water-soluble nonmetal oxides.)

(b) The reaction with sodium hydroxide is like the reaction summarized by Equation 7.16:

$$SeO_2(s) + 2\,NaOH(aq) \longrightarrow Na_2SeO_3(aq) + H_2O(l)$$

PRACTICE EXERCISE

Write the balanced chemical equation for the reaction of solid tetraphosphorus hexoxide with water.

Answer: $P_4O_6(s) + 6\,H_2O(l) \longrightarrow 4\,H_3PO_3(aq)$

▲ **Figure 7.19 Elemental silicon.** Silicon is an example of a metalloid. Although it looks metallic, silicon is brittle and is a poor thermal and electrical conductor as compared to metals. Large crystals of silicon are sliced into thin wafers for use in integrated circuits.

Metalloids

Metalloids have properties intermediate between those of metals and those of nonmetals. They may have *some* characteristic metallic properties but lack others. For example, silicon *looks* like a metal (Figure 7.19 ▶), but it is brittle rather than malleable and is a much poorer conductor of heat and electricity than are metals. Compounds of metalloids can have characteristics of the compounds of metals or nonmetals, depending on the specific compound.

Several of the metalloids, most notably silicon, are electrical semiconductors and are the principal elements used in the manufacture of integrated circuits and computer chips.

7.7 | Group Trends for the Active Metals

Our discussion of atomic radius, ionization energy, electron affinity, and metallic character gives some idea of the way the periodic table can be used to organize and remember facts. Not only do elements in a group possess general similarities, but there are also trends as we move through a group or from one group to another. In this section we will use the periodic table and our knowledge of electron configurations to examine the chemistry of the **alkali metals** (group 1A) and the **alkaline earth metals** (group 2A).

▲ **Figure 7.20 Alkali metals.** Sodium and the other alkali metals are soft enough to be cut with a knife. The shiny metallic surface quickly tarnishes as the metal reacts with oxygen in the air.

Group 1A: The Alkali Metals

The alkali metals are soft metallic solids (Figure 7.20 ▶). All have characteristic metallic properties such as a silvery, metallic luster and high thermal and electrical conductivities. The name *alkali* comes from an Arabic word meaning "ashes." Many compounds of sodium and potassium, two alkali metals, were isolated from wood ashes by early chemists.

Sodium and potassium are among the most abundant elements in Earth's crust, in seawater, and in biological systems. We all have sodium ions in our bodies. If we ingest too much, however, it can raise our blood pressure. Potassium is also prevalent in our bodies; a 140-pound person contains about 130 g of potassium, as K^+ ions in intracellular fluids. Plants require potassium for growth and development (Figure 7.21 ▶).

Some of the physical and chemical properties of the alkali metals are given in Table 7.4 ▼. The elements have low densities and melting points, and these properties vary in a fairly regular way with increasing atomic number. We can

▲ **Figure 7.21 Elements in fertilizers.** Fertilizers contain large quantities of potassium, phosphorus, and nitrogen to meet the needs of growing plants.

TABLE 7.4 Some Properties of the Alkali Metals

Element	Electron Configuration	Melting Point (°C)	Density (g/cm^3)	Atomic Radius (Å)	I_1 (kJ/mol)
Lithium	$[He]2s^1$	181	0.53	1.34	520
Sodium	$[Ne]3s^1$	98	0.97	1.54	496
Potassium	$[Ar]4s^1$	63	0.86	1.96	419
Rubidium	$[Kr]5s^1$	39	1.53	2.11	403
Cesium	$[Xe]6s^1$	28	1.88	2.25	376

(a) (b) (c)

▲ **Figure 7.22 The alkali metals react vigorously with water.** (a) The reaction of lithium is evidenced by the bubbling of escaping hydrogen gas. (b) The reaction of sodium is more rapid and is so exothermic that the hydrogen gas produced burns in air. (c) Potassium reacts almost explosively.

MOVIE
Sodium and Potassium in Water

also see some of the usual trends as we move down the group, such as increasing atomic radius and decreasing first ionization energy. For each row of the periodic table, the alkali metal has the lowest I_1 value (Figure 7.10), which reflects the relative ease with which its outer *s* electron can be removed. As a result, the alkali metals are all very reactive, readily losing one electron to form ions carrying a 1+ charge. (Section 4.4)

The alkali metals exist in nature only as compounds. The metals combine directly with most nonmetals. For example, they react with hydrogen to form hydrides and with sulfur to form sulfides:

$$2\,M(s) + H_2(g) \longrightarrow 2\,MH(s) \qquad [7.17]$$

$$2\,M(s) + S(s) \longrightarrow M_2S(s) \qquad [7.18]$$

(The symbol M in Equations 7.17 and 7.18 represents any one of the alkali metals.) In hydrides of the alkali metals (LiH, NaH, and so forth), hydrogen is present as H^-, called the **hydride ion**. The hydride ion, which is a hydrogen atom that has *gained* an electron, is distinct from the hydrogen ion, H^+, formed when a hydrogen atom *loses* its electron.

The alkali metals react vigorously with water, producing hydrogen gas and a solution of an alkali metal hydroxide:

$$2\,M(s) + 2\,H_2O(l) \longrightarrow 2\,MOH(aq) + H_2(g) \qquad [7.19]$$

These reactions are very exothermic. In many cases enough heat is generated to ignite the H_2, producing a fire or sometimes even an explosion (Figure 7.22 ▲). This reaction is most violent for the heavier members of the group, in keeping with their weaker hold on the single valence electron.

The reactions between the alkali metals and oxygen are more complex. When oxygen reacts with metals, metal oxides, which contain the O^{2-} ion, are usually formed. Indeed, lithium shows this reactivity:

$$4\,Li(s) + O_2(g) \longrightarrow \underset{\text{lithium oxide}}{2\,Li_2O(s)} \qquad [7.20]$$

When dissolved in water, Li_2O and other soluble metal oxides react with water to form hydroxide ions from the reaction of O^{2-} ions with H_2O (Equation 7.11).

In contrast, the other alkali metals all react with oxygen to form metal *peroxides*, which contain the $O_2{}^{2-}$ ion. For example, sodium forms sodium peroxide, Na_2O_2:

$$2\,Na(s) + O_2(g) \longrightarrow \underset{\text{sodium peroxide}}{Na_2O_2(s)} \qquad [7.21]$$

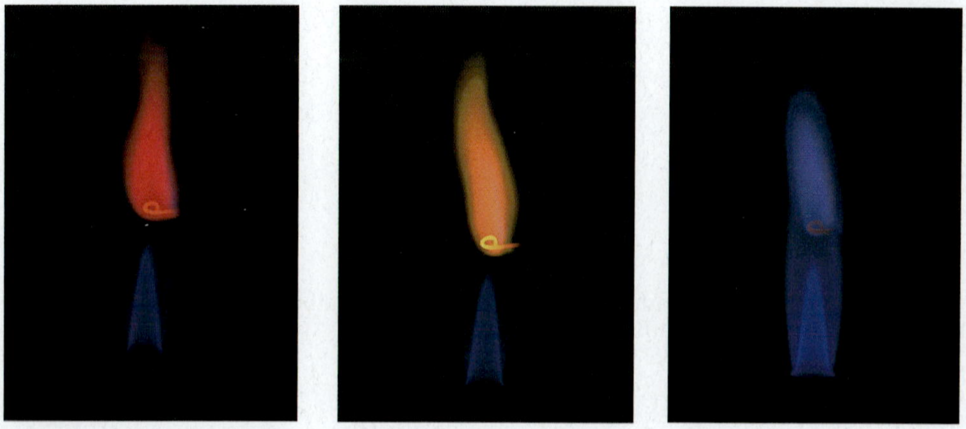

(a) (b) (c)

▲ **Figure 7.23 Flame tests.** (a) Li (crimson red), (b) Na (yellow), and (c) K (lilac).

MOVIE
Flame Tests for Metals

VCL Flame Test for Metals

Potassium, rubidium, and cesium also form compounds that contain the O_2^- ion, which we call the *superoxide ion*. For example, potassium forms potassium superoxide, KO_2:

$$K(s) + O_2(g) \longrightarrow KO_2(s) \quad [7.22]$$

potassium superoxide

You should realize that the reactions shown in Equations 7.21 and 7.22 are somewhat surprising; in most cases, the reaction of oxygen with a metal forms the metal oxide.

As is evident from Equations 7.19 through 7.22, the alkali metals are extremely reactive toward water and oxygen. Because of this, the metals are usually stored submerged in a liquid hydrocarbon, such as mineral oil or kerosene.

Although alkali metal ions are colorless, each emits a characteristic color when placed in a flame (Figure 7.23 ▲). The ions are reduced to gaseous metal atoms in the central region of the flame. The high temperature of the flame then excites the valence electron to a higher-energy orbital, causing the atom to be in an excited state. The atom then emits energy in the form of visible light as the electron falls back into the lower-energy orbital and the atom returns to its ground state. Sodium, for instance, gives a yellow flame because of emission at 589 nm. This wavelength is produced when the excited valence electron drops from the 3*p* subshell to the lower-energy 3*s* subshell. The characteristic yellow emission of sodium is the basis for sodium vapor lamps (Figure 7.24 ▶).

▲ **Figure 7.24 Light from sodium.** Sodium vapor lamps, which are used for commercial and highway lighting, have a yellow glow because of the emission from excited sodium atoms.

VCL Emission Spectra of Sodium and Mercury

GIVE IT SOME THOUGHT

Cesium metal tends to be the most reactive of the stable alkali metals (francium, element number 87, is radioactive and has not been extensively studied). What *atomic* property of Cs is most responsible for its high reactivity?

SAMPLE EXERCISE 7.10 | Reactions of an Alkali Metal

Write a balanced equation that predicts the reaction of cesium metal with **(a)** $Cl_2(g)$, **(b)** $H_2O(l)$, **(c)** $H_2(g)$.

Solution

Analyze and Plan: Cesium is an alkali metal (atomic number 55). We therefore expect that its chemistry will be dominated by oxidation of the metal to Cs^+ ions. Further, we recognize that Cs is far down the periodic table, which means it will be among the most active of all metals and will probably react with all three of the substances listed.

Solve: The reaction between Cs and Cl_2 is a simple combination reaction between two elements, one a metal and the other a nonmetal, forming the ionic compound CsCl:

$$2\,Cs(s) + Cl_2(g) \longrightarrow 2\,CsCl(s)$$

By analogy to Equations 7.19 and 7.17, respectively, we predict the reactions of cesium with water and hydrogen to proceed as follows:

$$2\,Cs(s) + 2\,H_2O(l) \longrightarrow 2\,CsOH(aq) + H_2(g)$$
$$2\,Cs(s) + H_2(g) \longrightarrow 2\,CsH(s)$$

In each case cesium forms a Cs^+ ion in its compounds. The chloride (Cl^-), hydroxide (OH^-), and hydride (H^-) ions are all 1− ions, which means the final products have 1:1 stoichiometry with Cs^+.

PRACTICE EXERCISE

Write a balanced equation for the reaction between potassium metal and elemental sulfur.
Answer: $2\,K(s) + S(s) \longrightarrow K_2S(s)$

CHEMISTRY AND LIFE | The Improbable Development of Lithium Drugs

The alkali metal ions tend to play a rather unexciting role in most chemical reactions in general chemistry. As noted in Section 4.2, all salts of the alkali metal ions are soluble, and the ions are spectators in most aqueous reactions (except for those involving the alkali metals in their elemental form, such as in Equations 7.17 through 7.22).

The alkali metal ions play an important role in human physiology, however. Sodium and potassium ions are major components of blood plasma and intracellular fluid, respectively, with average concentrations on the order of 0.1 *M*. These electrolytes serve as vital charge carriers in normal cellular function, and, as we will discuss further in Chapter 20, they are two of the principal ions involved in regulation of the heart.

In contrast, the lithium ion (Li^+) has no known function in normal human physiology. Since the discovery of lithium in 1817, however, salts of the element were thought to possess almost mystical healing powers; there were claims that it was an ingredient in ancient "fountain of youth" formulas. In 1927, Mr. C. L. Grigg began marketing a lithium-containing soft drink with the unwieldy name "Bib-Label Lithiated Lemon-Lime Soda." Grigg soon gave his lithiated beverage a much simpler name: Seven-Up® (Figure 7.25 ▶).

Because of concerns of the Food and Drug Administration, lithium was removed from Seven-Up® in the early 1950s. At nearly the same time, it was found that the lithium ion has a remarkable therapeutic effect on the mental disorder called *bipolar affective disorder*, or *manic-depressive illness*. Over 1 million Americans suffer from this psychosis, undergoing severe mood swings from deep depression to a manic euphoria. The lithium ion smooths out these mood swings, allowing the bipolar patient to function more effectively in daily life.

The antipsychotic action of Li^+ was discovered by accident in the late 1940s by Australian psychiatrist John Cade as he was researching the use of uric acid—a component of urine—to treat manic-depressive illness. He administered the acid to manic laboratory animals in the form of its most soluble salt, lithium urate, and found that many of the manic symptoms seemed to disappear. Later studies showed that uric acid has no role in the therapeutic effects observed; rather, the seemingly innocuous Li^+ ions were responsible. Because lithium overdose can cause severe side effects in humans, including kidney failure and death, lithium salts were not approved as antipsychotic drugs for humans until 1970. Today Li^+ is usually administered orally in the form of Li_2CO_3, which is the active ingredient in prescription drugs such as Eskalith®. Lithium drugs are effective for about 70% of the bipolar patients who take it.

In this age of sophisticated drug design and biotechnology, the simple lithium ion is still the most effective treatment of a destructive psychological disorder. Remarkably, in spite of intensive research, scientists still don't fully understand the biochemical action of lithium that leads to its therapeutic effects. Because of its similarity to the Na^+ ion, the Li+ ion is incorporated into blood plasma, where it can affect the behavior of both nerve cells and muscle cells. The Li^+ ion has a smaller radius than the Na^+ ion (Figure 7.7), so its interaction with molecules in cells is somewhat different. Other studies indicate that Li^+ alters the function of certain neurotransmitters, which might lead to its effectiveness as an antipsychotic drug.

▲ **Figure 7.25 Lithium no more.** The soft drink Seven-Up® originally contained lithium citrate, the lithium salt of citric acid. The lithium was claimed to give the beverage healthful benefits, including "an abundance of energy, enthusiasm, a clear complexion, lustrous hair, and shining eyes!" The lithium was removed from the beverage in the early 1950s, about the same time that the antipsychotic action of Li^+ was discovered.

TABLE 7.5 Some Properties of the Alkaline Earth Metals

Element	Electron Configuration	Melting Point (°C)	Density (g/cm^3)	Atomic Radius (Å)	I_1 (kJ/mol)
Beryllium	$[He]2s^2$	1287	1.85	0.90	899
Magnesium	$[Ne]3s^2$	650	1.74	1.30	738
Calcium	$[Ar]4s^2$	842	1.55	1.74	590
Strontium	$[Kr]5s^2$	777	2.63	1.92	549
Barium	$[Xe]6s^2$	727	3.51	1.98	503

Group 2A: The Alkaline Earth Metals

Like the alkali metals, the group 2A elements are all solids at room temperature and have typical metallic properties, some of which are listed in Table 7.5 ▲. Compared with the alkali metals, the alkaline earth metals are harder and more dense, and they melt at higher temperatures.

The first ionization energies of the alkaline earth elements are low, but not as low as those of the alkali metals. Consequently, the alkaline earths are less reactive than their alkali metal neighbors. As we noted in Section 7.4, the ease with which the elements lose electrons decreases as we move across the periodic table from left to right and increases as we move down a group. Thus, beryllium and magnesium, the lightest members of the alkaline earth metals, are the least reactive.

The trend of increasing reactivity within the group is shown by the way the alkaline earths behave in the presence of water. Beryllium does not react with water or steam, even when heated red-hot. Magnesium does not react with liquid water but does react with steam to form magnesium oxide and hydrogen:

$$Mg(s) + H_2O(g) \longrightarrow MgO(s) + H_2(g) \qquad [7.23]$$

Calcium and the elements below it react readily with water at room temperature (although more slowly than the alkali metals adjacent to them in the periodic table), as shown in Figure 7.26 ▶. The reaction between calcium and water, for example, is

$$Ca(s) + 2\,H_2O(l) \longrightarrow Ca(OH)_2(aq) + H_2(g) \qquad [7.24]$$

The reactions represented in Equations 7.23 and 7.24 illustrate the dominant pattern in the reactivity of the alkaline earth elements—the tendency to lose their two outer *s* electrons and form 2+ ions. For example, magnesium reacts with chlorine at room temperature to form $MgCl_2$ and burns with dazzling brilliance in air to give MgO (Figure 3.5):

$$Mg(s) + Cl_2(g) \longrightarrow MgCl_2(s) \qquad [7.25]$$

$$2\,Mg(s) + O_2(g) \longrightarrow 2\,MgO(s) \qquad [7.26]$$

▲ **Figure 7.26 Elemental calcium solution.** Calcium metal reacts with water to form hydrogen gas and aqueous calcium hydroxide, $Ca(OH)_2(aq)$.

In the presence of O_2, magnesium metal is protected from many chemicals by a thin surface coating of water-insoluble MgO. Thus, even though it is high in the activity series (Section 4.4), Mg can be incorporated into lightweight structural alloys used in, for example, automobile wheels. The heavier alkaline earth metals (Ca, Sr, and Ba) are even more reactive toward nonmetals than is magnesium.

The heavier alkaline earth ions give off characteristic colors when strongly heated in a flame. The colored flame produced by calcium is brick red, that of strontium is crimson red, and that of barium is green. Strontium salts produce the brilliant red color in fireworks, and barium salts produce the green color.

Both magnesium and calcium are essential for living organisms (Figure 2.24). Calcium is particularly important for growth and maintenance of

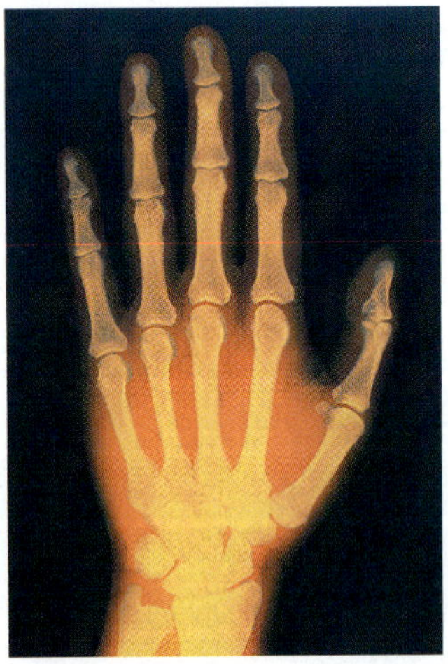

▲ **Figure 7.27 Calcium in the body.** This X-ray photograph shows the bone structure of the human hand. The primary mineral in bone and teeth is hydroxyapatite, $Ca_5(PO_4)_3OH$, in which calcium is present as Ca^{2+}.

▲ **Figure 7.28 Exploring Saturn.** The Cassini–Huygens spacecraft entered orbit around Saturn in June 2004. The probe will use advanced scientific instruments to explore several aspects of Saturn and its largest moon, Titan, including the metallic hydrogen composition of the planet. This photograph of Saturn was taken by Cassini–Huygens in May 2004 at a distance of 28.2 million kilometers (17.6 million miles) from the planet.

bones and teeth (Figure 7.27 ◀). In humans 99% of the calcium is found in the skeletal system.

GIVE IT SOME THOUGHT

Calcium carbonate, $CaCO_3$, is often used as a dietary calcium supplement for bone health. Although $CaCO_3(s)$ is insoluble in water (Table 4.1), it can be taken orally to allow for the delivery of $Ca^{2+}(aq)$ ions to the musculoskeletal system. Why is this the case? [*Hint:* Recall the reactions of metal carbonates that were discussed in Section 4.3]

7.8 | Group Trends for Selected Nonmetals

Hydrogen

Hydrogen, the first element in the periodic table, has a $1s^1$ electron configuration, and for this reason its usual position in the table is above the alkali metals. However, hydrogen does not truly belong to any particular group. Unlike the alkali metals, hydrogen is a nonmetal that occurs as a colorless diatomic gas, $H_2(g)$, under most conditions. Nevertheless, hydrogen can be metallic under tremendous pressures. The interiors of the planets Jupiter and Saturn, for example, are believed to consist of a rocky core surrounded by a thick shell of metallic hydrogen. The metallic hydrogen is in turn surrounded by a layer of liquid molecular hydrogen with gaseous hydrogen occurring above that near the surface. The Cassini–Huygens satellite mission, launched in 1997, is designed to explore Saturn and its largest moon, Titan, including their chemical compositions. The spacecraft successfully entered orbit around Saturn in June 2004 and is already using telescopy and spectroscopy to provide spectacular images and chemical analysis (Figure 7.28 ◀).

Owing to the complete absence of nuclear shielding of its sole electron, the ionization energy of hydrogen, 1312 kJ/mol, is markedly higher than that of any of the alkali metals (Figure 7.10). In fact, it is comparable to the I_1 values of other nonmetals, such as oxygen and chlorine. As a result, hydrogen has less tendency to lose an electron than do the alkali metals. Whereas the alkali metals readily lose their valence electron to nonmetals to form ionic compounds, hydrogen shares its electron with nonmetals and thereby forms molecular compounds. The reactions between hydrogen and nonmetals can be quite exothermic, as evidenced by the combustion reaction between hydrogen and oxygen to form water (Figure 5.14):

$$2\,H_2(g) + O_2(g) \longrightarrow 2\,H_2O(l) \qquad \Delta H^\circ = -571.7\text{ kJ} \qquad [7.27]$$

We have also seen (Equation 7.17) that hydrogen reacts with active metals to form solid metal hydrides, which contain the hydride ion, H^-. The fact that hydrogen can gain an electron further illustrates that it is not truly a member of the alkali metal family. In fact, this property of hydrogen suggests a slight resemblance between hydrogen and the halogens.

In addition to its ability to form covalent bonds and metal hydrides, probably the most important characteristic of hydrogen is its ability to lose its electron to form a cation. Indeed, the aqueous chemistry of hydrogen is dominated by the $H^+(aq)$ ion, which we first encountered in Chapter 4. ∞ (Section 4.1) We will study this important ion in greater detail in Chapter 16.

Group 6A: The Oxygen Group

As we proceed down group 6A, there is a change from nonmetallic to metallic character (Figure 7.13). Oxygen, sulfur, and selenium are typical nonmetals. Tellurium has some metallic properties and is classified as a metalloid. Polonium, which is radioactive and quite rare, is a metal.

TABLE 7.6 Some Properties of the Group 6A Elements

Element	Electron Configuration	Melting Point (°C)	Density	Atomic Radius (Å)	I_1 (kJ/mol)
Oxygen	$[He]2s^2 2p^4$	−218	1.43 g/L	0.73	1314
Sulfur	$[Ne]3s^2 3p^4$	115	1.96 g/cm^3	1.02	1000
Selenium	$[Ar]3d^{10}4s^2 4p^4$	221	4.82 g/cm^3	1.16	941
Tellurium	$[Kr]4d^{10}5s^2 5p^4$	450	6.24 g/cm^3	1.35	869
Polonium	$[Xe]4f^{14}5d^{10}6s^2 6p^4$	254	9.20 g/cm^3	—	812

Oxygen is a colorless gas at room temperature; all of the other members of group 6A are solids. Some of the physical properties of the group 6A elements are given in Table 7.6 ▲.

As we saw in Section 2.6, oxygen is encountered in two molecular forms, O_2 and O_3. The O_2 form is the common one. People generally mean O_2 when they say "oxygen," although the name *dioxygen* is more descriptive. The O_3 form is called **ozone**. The two forms of oxygen are examples of *allotropes*. Allotropes are different forms of the same element in the same state. (In this case both forms are gases.) About 21% of dry air consists of O_2 molecules. Ozone, which is toxic and has a pungent odor, is present in very small amounts in the upper atmosphere and in polluted air. It is also formed from O_2 in electrical discharges, such as in lightning storms:

$$3\,O_2(g) \longrightarrow 2\,O_3(g) \qquad \Delta H^\circ = 284.6\text{ kJ} \qquad [7.28]$$

This reaction is endothermic, telling us that O_3 is less stable than O_2.

Oxygen has a great tendency to attract electrons from other elements (to *oxidize* them). Oxygen in combination with a metal is almost always present as the oxide ion, O^{2-}. This ion has a noble-gas configuration and is particularly stable. As shown in Equation 7.27, the formation of nonmetal oxides is also often very exothermic and thus energetically favorable.

In our discussion of the alkali metals, we noted two less common oxygen anions—the peroxide (O_2^{2-}) ion and the superoxide (O_2^{-}) ion. Compounds of these ions often react with themselves to produce an oxide and O_2. For example, aqueous hydrogen peroxide, H_2O_2, slowly decomposes into water and O_2 at room temperature:

$$2\,H_2O_2(aq) \longrightarrow 2\,H_2O(l) + O_2(g) \qquad \Delta H^\circ = -196.1\text{ kJ} \qquad [7.29]$$

For this reason, bottles of aqueous hydrogen peroxide are topped with caps that are able to release the $O_2(g)$ produced before the pressure inside becomes too great (Figure 7.29 ▶).

▲ **Figure 7.29 Hydrogen peroxide solution.** Bottles of this common antiseptic are topped with a cap that allows any excess pressure created by $O_2(g)$ to be released from the bottle. Hydrogen peroxide is often stored in dark-colored or opaque bottles to minimize exposure to light, which accelerates its decomposition.

After oxygen, the most important member of group 6A is sulfur. This element also exists in several allotropic forms, the most common and stable of which is the yellow solid having the molecular formula S_8. This molecule consists of an eight-membered ring of sulfur atoms, as shown in Figure 7.30 ▶. Even though solid sulfur consists of S_8 rings, we usually write it simply as S(*s*) in chemical equations to simplify the stoichiometric coefficients.

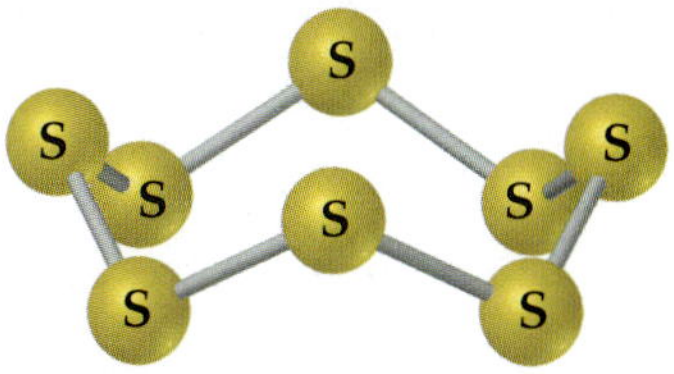

▲ **Figure 7.30 Elemental sulfur.** At room temperature, the most common allotropic form of sulfur is an eight-member ring, S_8.

Like oxygen, sulfur has a tendency to gain electrons from other elements to form sulfides, which contain the S^{2-} ion. In fact, most sulfur in nature is present as metal sulfides. Sulfur is below oxygen in the periodic table, and the tendency of sulfur to form sulfide anions is not as great as that of oxygen to form oxide ions. As a result, the chemistry of sulfur is more complex than that of oxygen. In fact, sulfur and its compounds (including those in coal and petroleum) can be burned in oxygen. The main product is sulfur dioxide, a major pollutant:

$$S(s) + O_2(g) \longrightarrow SO_2(g) \qquad [7.30]$$

We will discuss the environmental aspects of sulfur oxide chemistry in greater depth in Chapter 18.

TABLE 7.7 Some Properties of the Halogens

Element	Electron Configuration	Melting Point (°C)	Density	Atomic Radius (Å)	I_1 (kJ/mol)
Fluorine	$[He]2s^2 2p^5$	−220	1.69 g/L	0.71	1681
Chlorine	$[Ne]3s^2 3p^5$	−102	3.21 g/L	0.99	1251
Bromine	$[Ar]3d^{10} 4s^2 4p^5$	−7.3	3.12 g/cm^3	1.14	1140
Iodine	$[Kr]4d^{10} 5s^2 5p^5$	114	4.94 g/cm^3	1.33	1008

Group 7A: The Halogens

MOVIE
Physical Properties of the Halogens

The group 7A elements are known as the **halogens**, after the Greek words *halos* and *gennao*, meaning "salt formers." Some of the properties of these elements are given in Table 7.7 ▲. Astatine, which is both extremely rare and radioactive, is omitted because many of its properties are not yet known.

Unlike the group 6A elements, all the halogens are typical nonmetals. Their melting and boiling points increase with increasing atomic number. Fluorine and chlorine are gases at room temperature, bromine is a liquid, and iodine is a solid. Each element consists of diatomic molecules: F_2, Cl_2, Br_2, and I_2. Fluorine gas is pale yellow; chlorine gas is yellow-green; bromine liquid is reddish brown and readily forms a reddish brown vapor; and solid iodine is grayish black and readily forms a violet vapor (Figure 7.31 ◀).

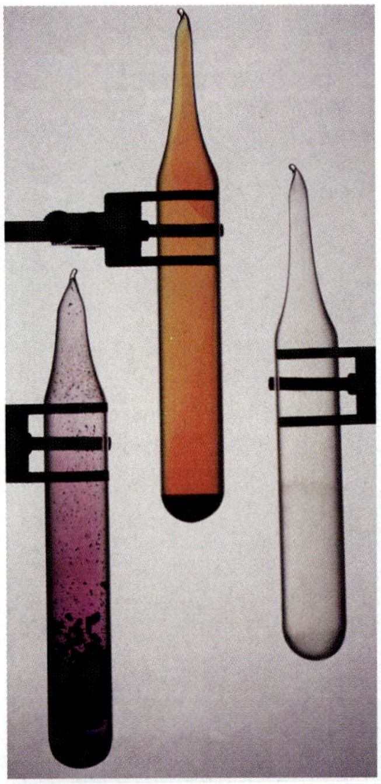

▲ **Figure 7.31 Elemental halogens.** All three of these elements—from left to right, iodine (I_2), bromine (Br_2), and chlorine (Cl_2)—exist as diatomic molecules.

The halogens have highly negative electron affinities (Figure 7.12). Thus, it is not surprising that the chemistry of the halogens is dominated by their tendency to gain electrons from other elements to form halide ions, X^-. (In many equations X is used to indicate any one of the halogen elements.) Fluorine and chlorine are more reactive than bromine and iodine. In fact, fluorine removes electrons from almost any substance with which it comes into contact, including water, and usually does so very exothermically, as in the following examples:

$$2\,H_2O(l) + 2\,F_2(g) \longrightarrow 4\,HF(aq) + O_2(g) \qquad \Delta H = -758.9\text{ kJ} \qquad [7.31]$$

$$SiO_2(s) + 2\,F_2(g) \longrightarrow SiF_4(g) + O_2(g) \qquad \Delta H = -704.0\text{ kJ} \qquad [7.32]$$

As a result, fluorine gas is difficult and dangerous to use in the laboratory, requiring specialized equipment.

Chlorine is the most industrially useful of the halogens. In 2003 total production was 26 billion pounds, making it the eighth most produced chemical in the United States. Unlike fluorine, chlorine reacts slowly with water to form relatively stable aqueous solutions of HCl and HOCl (hypochlorous acid):

$$Cl_2(g) + H_2O(l) \longrightarrow HCl(aq) + HOCl(aq) \qquad [7.33]$$

Chlorine is often added to drinking water and swimming pools, where the HOCl(*aq*) that is generated serves as a disinfectant.

The .halogens react directly with most metals to form ionic halides. The halogens also react with hydrogen to form gaseous hydrogen halide compounds:

$$H_2(g) + X_2 \longrightarrow 2\,HX(g) \qquad [7.34]$$

These compounds are all very soluble in water and dissolve to form the hydrohalic acids. As we discussed in Section 4.3, HCl(*aq*), HBr(*aq*), and HI(*aq*) are strong acids, whereas HF(*aq*) is a weak acid.

GIVE IT SOME THOUGHT

Can you use data in Table 7.7 to provide estimates for the atomic radius and first ionization energy of an astatine atom?

TABLE 7.8 Some Properties of the Noble Gases

Element	Electron Configuration	Boiling Point (K)	Density (g/L)	Atomic Radius* (Å)	I_1 (kJ/mol)
Helium	$1s^2$	4.2	0.18	0.32	2372
Neon	$[He]2s^22p^6$	27.1	0.90	0.69	2081
Argon	$[Ne]3s^23p^6$	87.3	1.78	0.97	1521
Krypton	$[Ar]3d^{10}4s^24p^6$	120	3.75	1.10	1351
Xenon	$[Kr]4d^{10}5s^25p^6$	165	5.90	1.30	1170
Radon	$[Xe]4f^{14}5d^{10}6s^26p^6$	211	9.73	1.45	1037

*Only the heaviest of the noble-gas elements form chemical compounds. Thus, the atomic radii for the lighter noble-gas elements are estimated values.

Group 8A: The Noble Gases

The group 8A elements, known as the **noble gases**, are all nonmetals that are gases at room temperature. They are all *monatomic* (that is, they consist of single atoms rather than molecules). Some physical properties of the noble-gas elements are listed in Table 7.8 ▲. The high radioactivity of radon (Rn, atomic number 86) has limited the study of its reaction chemistry and some of its properties.

The noble gases have completely filled *s* and *p* subshells. All elements of group 8A have large first ionization energies, and we see the expected decrease as we move down the column. Because the noble gases possess such stable electron configurations, they are exceptionally unreactive. In fact, until the early 1960s the elements were called the *inert gases* because they were thought to be incapable of forming chemical compounds. In 1962, Neil Bartlett at the University of British Columbia reasoned that the ionization energy of Xe might be low enough to allow it to form compounds. In order for this to happen, Xe would have to react with a substance with an extremely high ability to remove electrons from other substances, such as fluorine. Bartlett synthesized the first noble-gas compound by combining Xe with the fluorine-containing compound PtF_6. Xenon also reacts directly with $F_2(g)$ to form the molecular compounds XeF_2, XeF_4, and XeF_6 (Figure 7.32 ▶). Krypton has a higher I_1 value than xenon and is therefore less reactive. In fact, only a single stable compound of krypton is known, KrF_2. In 2000, Finnish scientists reported the first neutral molecule that contains argon, the HArF molecule, which is stable only at low temperatures.

▲ **Figure 7.32 A compound of xenon.** Crystals of XeF_4, which is one of the very few compounds that contain a group 8A element.

SAMPLE INTEGRATIVE EXERCISE | Putting Concepts Together

The element bismuth (Bi, atomic number 83) is the heaviest member of group 5A. A salt of the element, bismuth subsalicylate, is the active ingredient in Pepto-Bismol®, an over-the-counter medication for gastric distress.

(a) The covalent atomic radii of thallium (Tl) and lead (Pb) are 1.48 Å and 1.47 Å, respectively. Using these values and those in Figure 7.6, predict the covalent atomic radius of the element bismuth (Bi). Explain your answer.

(b) What accounts for the general increase in atomic radius going down the group 5A elements?

(c) Another major use of bismuth has been as an ingredient in low-melting metal alloys, such as those used in fire sprinkler systems and in typesetting. The element itself is a brittle white crystalline solid. How do these characteristics fit with the fact that bismuth is in the same periodic group with such nonmetallic elements as nitrogen and phosphorus?

(d) Bi_2O_3 is a basic oxide. Write a balanced chemical equation for its reaction with dilute nitric acid. If 6.77 g of Bi_2O_3 is dissolved in dilute acidic solution to make up 500 mL of solution, what is the molarity of the solution of Bi^{3+} ion?

(e) ^{209}Bi is the heaviest stable isotope of any element. How many protons and neutrons are present in this nucleus?

(f) The density of Bi at 25°C is 9.808 g/cm^3. How many Bi atoms are present in a cube of the element that is 5.00 cm on each edge? How many moles of the element are present?

Solution **(a)** Note that there is a rather steady decrease in radius for the elements in the row preceding the one we are considering, that is, in the series In–Sn–Sb. It is reasonable to expect a decrease of about 0.02 Å in moving from Pb to Bi, leading to an estimate of 1.45 Å. The tabulated value is 1.46 Å.

(b) The general increase in radius with increasing atomic number in the group 5A elements occurs because additional shells of electrons are being added, with corresponding increases in nuclear charge. The core electrons in each case largely shield the outermost electrons from the nucleus, so the effective nuclear charge does not vary greatly as we go to higher atomic numbers. However, the principal quantum number, n, of the outermost electrons steadily increases, with a corresponding increase in orbital radius.

(c) The contrast between the properties of bismuth and those of nitrogen and phosphorus illustrates the general rule that there is a trend toward increased metallic character as we move down in a given group. Bismuth, in fact, is a metal. The increased metallic character occurs because the outermost electrons are more readily lost in bonding, a trend that is consistent with lower ionization energy.

(d) Following the procedures described in Section 4.2 for writing molecular and net ionic equations, we have the following:

Molecular equation: $Bi_2O_3(s) + 6\,HNO_3(aq) \longrightarrow 2\,Bi(NO_3)_3(aq) + 3\,H_2O(l)$

Net ionic equation: $Bi_2O_3(s) + 6\,H^+(aq) \longrightarrow 2\,Bi^{3+}(aq) + 3\,H_2O(l)$

In the net ionic equation, nitric acid is a strong acid and $Bi(NO_3)_3$ is a soluble salt, so we need show only the reaction of the solid with the hydrogen ion forming the $Bi^{3+}(aq)$ ion and water.

To calculate the concentration of the solution, we proceed as follows (Section 4.5):

$$\frac{6.77\text{ g }Bi_2O_3}{0.500\text{ L soln}} \times \frac{1\text{ mol }Bi_2O_3}{466.0\text{ g }Bi_2O_3} \times \frac{2\text{ mol }Bi^{3+}}{1\text{ mol }Bi_2O_3} = \frac{0.0581\text{ mol }Bi^{3+}}{\text{L soln}} = 0.0581\,M$$

(e) We can proceed as in Section 2.3. Bismuth is element 83; there are therefore 83 protons in the nucleus. Because the atomic mass number is 209, there are $209 - 83 = 126$ neutrons in the nucleus.

(f) We proceed as in Sections 1.4 and 3.4: The volume of the cube is $(5.00)^3$ $cm^3 = 125\ cm^3$. Then we have

$$125\text{ cm}^3\text{ Bi} \times \frac{9.780\text{ g Bi}}{1\text{ cm}^3} \times \frac{1\text{ mol Bi}}{209.0\text{ g Bi}} = 5.87\text{ mol Bi}$$

$$5.87\text{ mol Bi} \times \frac{6.022 \times 10^{23}\text{ atom Bi}}{1\text{ mol Bi}} = 3.54 \times 10^{24}\text{ atoms Bi}$$

SUMMARY AND KEY TERMS

Introduction and Section 7.1 The periodic table was first developed by Mendeleev and Meyer on the basis of the similarity in chemical and physical properties exhibited by certain elements. Moseley established that each element has a unique atomic number, which added more order to the periodic table. We now recognize that elements in the same column of the periodic table have the same number of electrons in their **valence orbitals**. This similarity in valence electronic structure leads to the similarities among elements in the same group. The differences among elements in the same group arise because their valence orbitals are in different shells.

Section 7.2 Many properties of atoms are due to the average distance of the outer electrons from the nucleus and to the **effective nuclear charge** experienced by these electrons. The core electrons are very effective in screening the outer electrons from the full charge of the nucleus, whereas electrons in the same shell do not screen each other very effectively at all. As a result, the effective nuclear charge experienced by valence electrons increases as we move left to right across a period.

Section 7.3 The size of an atom can be gauged by its **bonding atomic radius**, based on measurements of the distances separating atoms in their chemical compounds. In general, atomic radii increase as we go down a column in the periodic table and decrease as we proceed left to right across a row.

Cations are smaller than their parent atoms; anions are larger than their parent atoms. For ions of the same charge, size increases going down a column of the periodic table. An **isoelectronic series** is a series of ions that has the same number of electrons. For such a series, size decreases with increasing nuclear charge as the electrons are attracted more strongly to the nucleus.

Section 7.4 The first **ionization energy** of an atom is the minimum energy needed to remove an electron from the atom in the gas phase, forming a cation. The second ionization energy is the energy needed to remove a second electron, and so forth. Ionization energies show a sharp increase after all the valence electrons have been removed, because of the much higher effective nuclear charge experienced by the core electrons. The first ionization energies of the elements show periodic trends that are opposite those seen for atomic radii, with smaller atoms having higher first ionization energies. Thus, first ionization energies decrease as we go down a column and increase as we proceed left to right across a row.

We can write electron configurations for ions by first writing the electron configuration of the neutral atom and then removing or adding the appropriate number of electrons. Electrons are removed first from the orbitals with the largest value of n. Electrons are added to orbitals with the lowest value of n.

Section 7.5 The **electron affinity** of an element is the energy change upon adding an electron to an atom in the gas phase, forming an anion. A negative electron affinity means that the anion is stable; a positive electron affinity means that the anion is not stable relative to the separated atom and electron. In general, electron affinities become more negative as we proceed from left to right across the periodic table. The halogens have the most-negative electron affinities. The electron affinities of the noble gases are all positive because the added electron would have to occupy a new, higher-energy subshell.

Section 7.6 The elements can be categorized as metals, nonmetals, and metalloids. Most elements are metals; they occupy the left side and the middle of the periodic table. Nonmetals appear in the upper-right section of the table. Metalloids occupy a narrow band between the metals and nonmetals. The tendency of an element to exhibit the properties of metals, called the **metallic character**, increases as we proceed down a column and decreases as we proceed from left to right across a row.

Metals have a characteristic luster, and they are good conductors of heat and electricity. When metals react with nonmetals, the metal atoms are oxidized to cations and ionic substances are generally formed. Most metal oxides are basic; they react with acids to form salts and water.

Nonmetals lack metallic luster and are poor conductors of heat and electricity. Several are gases at room temperature. Compounds composed entirely of nonmetals are generally molecular. Nonmetals usually form anions in their reactions with metals. Nonmetal oxides are acidic; they react with bases to form salts and water. Metalloids have properties that are intermediate between those of metals and nonmetals.

Section 7.7 The periodic properties of the elements can help us understand the properties of groups of the representative elements. The **alkali metals** (group 1A) are soft metals with low densities and low melting points. They have the lowest ionization energies of the elements. As a result, they are very reactive toward nonmetals, easily losing their outer s electron to form 1+ ions. The **alkaline earth metals** (group 2A) are harder and more dense and have higher melting points than the alkali metals. They are also very reactive toward nonmetals, although not as reactive as the alkali metals. The alkaline earth metals readily lose their two outer s electrons to form 2+ ions. Both alkali and alkaline earth metals react with hydrogen to form ionic substances that contain the **hydride ion**, H^-.

Section 7.8 Hydrogen is a nonmetal with properties that are distinct from any of the groups of the periodic table. It forms molecular compounds with other nonmetals, such as oxygen and the halogens.

Oxygen and sulfur are the most important elements in group 6A. Oxygen is usually found as a diatomic molecule, O_2. **Ozone**, O_3, is an important allotrope of oxygen. Oxygen has a strong tendency to gain electrons from other elements, thus oxidizing them. In combination with metals, oxygen is usually found as the oxide ion, O^{2-}, although salts of the peroxide ion, $O_2{}^{2-}$, and superoxide ion, $O_2{}^-$, are sometimes formed. Elemental sulfur is most commonly found as S_8 molecules. In combination with metals, it is most often found as the sulfide ion, S^{2-}.

The **halogens** (group 7A) are nonmetals that exist as diatomic molecules. The halogens have the most negative electron affinities of the elements. Thus their chemistry is dominated by a tendency to form 1− ions, especially in reactions with metals.

The **noble gases** (group 8A) are nonmetals that exist as monoatomic gases. They are very unreactive because they have completely filled s and p subshells. Only the heaviest noble gases are known to form compounds, and they do so only with very active nonmetals, such as fluorine.

VISUALIZING CONCEPTS

7.1 We can draw an analogy between the attraction of an electron to a nucleus and seeing a lightbulb—in essence, the more nuclear charge the electron "sees," the greater the attraction. **(a)** Within this analogy, discuss how the shielding by core electrons is analogous to putting a frosted-glass lampshade between the lightbulb and your eyes, as shown in the illustration. **(b)** Explain how we could mimic moving to the right in a row of the periodic table by changing the wattage of the lightbulb. **(c)** How would you change the wattage of the bulb and/or the frosted glass to mimic the effect of moving down a column of the periodic table? [Section 7.2]

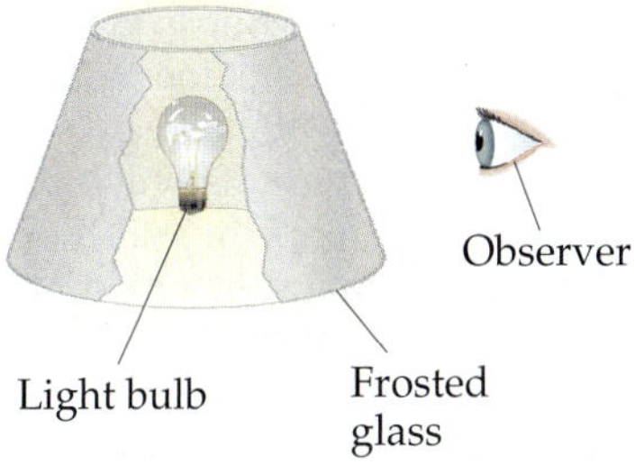

7.2 Neon has atomic number 10. What are the similarities and differences in describing the radius of a neon atom and the radius of the billiard ball illustrated here? Could you use billiard balls to illustrate the concept of bonding atomic radius? Explain. [Section 7.3]

7.3 Consider the A_2X_4 molecule depicted below, where A and X are elements. The A—A bond length in this molecule is d_1, and the four A—X bond lengths are each d_2. **(a)** In terms of d_1 and d_2, how could you define the bonding atomic radii of atoms A and X? **(b)** In terms of d_1 and d_2, what would you predict for the X—X bond length of an X_2 molecule? [Section 7.3]

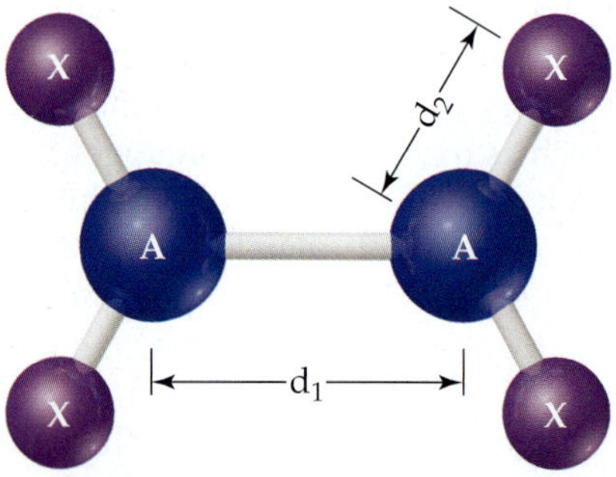

7.4 Make a simple sketch of the shape of the main part of the periodic table, as shown. **(a)** Ignoring H and He, write a single straight arrow from the element with the smallest bonding atomic radius to the element with the largest. **(b)** Ignoring H and He, write a single straight arrow from the element with the smallest first ionization energy to the element with the largest. **(c)** What significant observation can you make from the arrows you drew in parts (a) and (b)? [Sections 7.3 and 7.4]

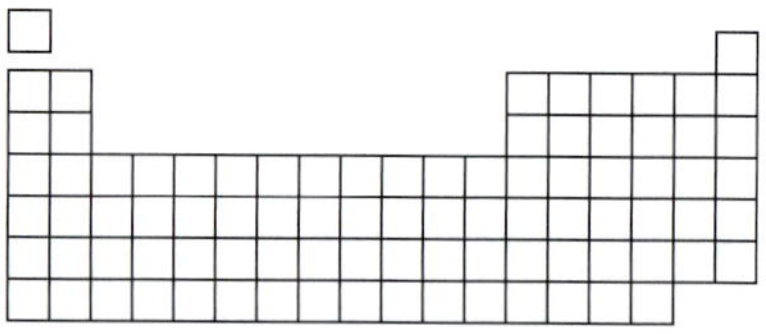

7.5 In the chemical process called *electron transfer*, an electron is transferred from one atom or molecule to another (we will talk about electron transfer extensively in Chapter 20). A simple electron transfer reaction is

$$A(g) + A(g) \longrightarrow A^{+}(g) + A^{-}(g)$$

In terms of the ionization energy and electron affinity of atom A, what is the energy change for this reaction? [Sections 7.4 and 7.5]

7.6 An element X reacts with $F_2(g)$ to form the molecular product shown below. **(a)** Write a balanced equation for this reaction (do not worry about the phases for X and the product). **(b)** Do you think that X is a metal or nonmetal? Explain. [Section 7.6]

EXERCISES

Periodic Table; Effective Nuclear Charge

7.7 Why did Mendeleev leave blanks in his early version of the periodic table? How did he predict the properties of the elements that belonged in those blanks?

7.8 **(a)** In the period from about 1800 to about 1865, the atomic weights of many elements were accurately measured. Why was this important to Mendeleev's formulation of the periodic table? **(b)** What property of the atom did Moseley associate with the wavelength of X rays emitted from an element in his experiments? In what ways did this affect the meaning of the periodic table?

7.9 **(a)** What is meant by the term *effective nuclear charge*? **(b)** How does the effective nuclear charge experienced by the valence electrons of an atom vary going from left to right across a period of the periodic table?

7.10 **(a)** How is the concept of effective nuclear charge used to simplify the numerous electron-electron repulsions in a many-electron atom? **(b)** Which experiences a greater effective nuclear charge in a Be atom, the 1*s* electrons or the 2*s* electrons? Explain.

7.11 **(a)** If each core electron were totally effective in screening the valence electrons from the full charge of the nucleus and the valence electrons provided no screening for each other, what would be the values of the screening constant, *S*, and the effective nuclear charge, Z_{eff}, for the 4*s* electron in a potassium atom? **(b)** Detailed calculations show that the value of Z_{eff} for a K atom is 3.49+. Explain the difference between this number and the one you obtained in part (a).

7.12 **(a)** If the core electrons were totally effective at shielding the valence electrons from the full charge of the nucleus and the valence electrons provided no shielding for each other, what would be the values of *S* and Z_{eff} for a 3*p* electron in a sulfur atom? **(b)** Detailed calculations indicate that the value of *S* for a 3*p* electron in a sulfur atom is 10.52. Can you explain any difference between this number and the one you obtained in part (a)?

7.13 Which will experience the greater effective nuclear charge, the electrons in the $n = 3$ shell in Ar or the $n = 3$ shell in Kr? Which will be closer to the nucleus? Explain.

7.14 Arrange the following atoms in order of increasing effective nuclear charge experienced by the electrons in the $n = 3$ electron shell: K, Mg, P, Rh, and Ti. Explain the basis for your order.

Atomic and Ionic Radii

7.15 Because an exact outer boundary cannot be measured or even calculated for an atom, how are atomic radii determined? What is the difference between a bonding radius and a nonbonding radius?

7.16 **(a)** Why does the quantum mechanical description of many-electron atoms make it difficult to define a precise atomic radius? **(b)** When nonbonded atoms come up against one another, what determines how closely the nuclear centers can approach?

7.17 The distance between W atoms in tungsten metal is 2.74 Å. What is the atomic radius of a tungsten atom in this environment? (This radius is called the *metallic radius*.)

7.18 Based on the radii presented in Figure 7.6, predict the distance between Ge atoms in solid germanium.

7.19 Estimate the As—I bond length from the data in Figure 7.6, and compare your value to the experimental As—I bond length in arsenic triiodide, AsI_3, 2.55 Å.

7.20 The experimental Bi—I bond length in bismuth triiodide, BiI_3, is 2.81 Å. Based on this value and data in Figure 7.6, predict the atomic radius of Bi.

7.21 How do the sizes of atoms change as we move **(a)** from left to right across a row in the periodic table, **(b)** from top to bottom in a group in the periodic table? **(c)** Arrange the following atoms in order of increasing atomic radius: F, P, S, As.

7.22 **(a)** Among the nonmetallic elements, the change in atomic radius in moving one place left or right in a row is smaller than the change in moving one row up or down. Explain these observations. **(b)** Arrange the following atoms in order of increasing atomic radius: Si, S, Ge, Se.

7.23 Using only the periodic table, arrange each set of atoms in order of increasing radius: **(a)** Ca, Mg, Be; **(b)** Ga, Br, Ge; **(c)** Al, Tl, Si.

7.24 Using only the periodic table, arrange each set of atoms in order of increasing radius: **(a)** Cs, K, Rb; **(b)** In, Te, Sn; **(c)** P, Cl, Sr.

7.25 **(a)** Why are monatomic cations smaller than their corresponding neutral atoms? **(b)** Why are monatomic anions larger than their corresponding neutral atoms? **(c)** Why does the size of ions increase as one proceeds down a column in the periodic table?

7.26 Explain the following variations in atomic or ionic radii: **(a)** $I^- > I > I^+$, **(b)** $Ca^{2+} > Mg^{2+} > Be^{2+}$, **(c)** $Fe > Fe^{2+} > Fe^{3+}$.

7.27 Consider a reaction represented by the following spheres:

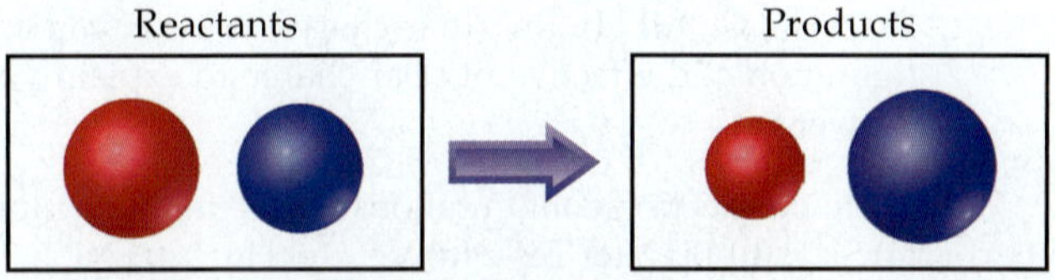

Which sphere represents a metal and which a nonmetal? Explain.

7.28 Consider the following spheres:

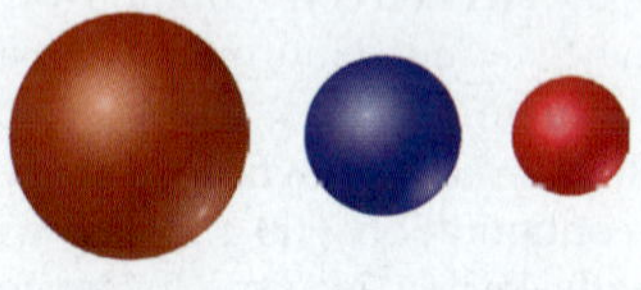

Which one represents Ca, which Ca^{2+}, and which Mg^{2+}?

7.29 **(a)** What is an isoelectronic series? **(b)** Which neutral atom is isoelectronic with each of the following ions: (*i*) N^{3-}, (*ii*) Ba^{2+}, (*iii*) Se^{2-}, (*iv*) Bi^{3+}?

7.30 Select the ions or atoms from the following sets that are isoelectronic with each other: **(a)** Na^+, Sr^{2+}, Br^-; **(b)** Y^{3+}, Br^-, Kr; **(c)** N^{3-}, P^{3-}, Ti^{4+}; **(d)** Fe^{3+}, Co^{3+}, Mn^{2+}.

7.31 **(a)** Why do the radii of isoelectronic ions decrease with increasing nuclear charge? **(b)** Which experiences the greatest effective nuclear charge, a 2*p* electron in F^-, a 2*p* electron in Ne, or a 2*p* electron in Na^+?

7.32 Consider S, Cl, and K and their most common ions. **(a)** List the atoms in order of increasing size. **(b)** List the ions in order of increasing size. **(c)** Explain any differences in the orders of the atomic and ionic sizes.

7.33 For each of the following sets of atoms and ions, arrange the members in order of increasing size: **(a)** Se^{2-}, Te^{2-}, Se; **(b)** Co^{3+}, Fe^{2+}, Fe^{3+}; **(c)** Ca, Ti^{4+}, Sc^{3+}; **(d)** Be^{2+}, Na^+, Ne.

7.34 For each of the following statements, provide an explanation: **(a)** Cl^- is larger than Cl; **(b)** S^{2-} is larger than O^{2-}; **(c)** K^+ is larger than Ca^{2+}.

Ionization Energies; Electron Affinities

7.35 Write equations that show the processes that describe the first, second, and third ionization energies of a boron atom.

7.36 Write equations that show the process for **(a)** the first two ionization energies of tin and **(b)** the fourth ionization energy of titanium.

7.37 **(a)** Why are ionization energies always positive quantities? **(b)** Why does F have a larger first ionization energy than O? **(c)** Why is the second ionization energy of an atom always greater than its first ionization energy?

7.38 **(a)** Why does Li have a larger first ionization energy than Na? **(b)** The difference between the third and fourth ionization energies of scandium is much larger than the difference between the third and fourth ionization energies of titanium. Why? **(c)** Why does Li have a much larger second ionization energy than Be?

7.39 **(a)** What is the general relationship between the size of an atom and its first ionization energy? **(b)** Which element in the periodic table has the largest ionization energy? Which has the smallest?

7.40 **(a)** What is the trend in first ionization energies as one proceeds down the group 7A elements? Explain how this trend relates to the variation in atomic radii. **(b)** What is the trend in first ionization energies as one moves across the fourth period from K to Kr? How does this trend compare with the trend in atomic sizes?

7.41 Based on their positions in the periodic table, predict which atom of the following pairs will have the larger first ionization energy: **(a)** Cl, Ar; **(b)** Be, Ca; **(c)** K, Co; **(d)** S, Ge; **(e)** Sn, Te.

7.42 For each of the following pairs, indicate which element has the larger first ionization energy: **(a)** Rb, Mo; **(b)** N, P; **(c)** Ga, Cl; **(d)** Pb, Rn. (In each case use electron configuration and effective nuclear charge to explain your answer.)

7.43 Write the electron configurations for the following ions: **(a)** Si^{2+}, **(b)** Bi^{3+}, **(c)** Te^{2-}, **(d)** V^{3+}, **(e)** Hg^{2+}, **(f)** Ni^{2+}.

7.44 Write electron configurations for the following ions, and determine which have noble-gas configurations: **(a)** Mn^{3+}, **(b)** Se^{2-}, **(c)** Sc^{3+}, **(d)** Ru^{2+}, **(e)** Tl^{+}, **(f)** Au^{+}.

7.45 Write the electron configuration for **(a)** the Mn^{2+} ion and **(b)** the Si^{2-} ion. How many unpaired electrons does each contain?

7.46 Identify the element whose ions have the following electron configurations: **(a)** a 2+ ion with $[Ar]3d^9$, **(b)** a 1+ ion with $[Xe]4f^{14}5d^{10}6s^2$. How many unpaired electrons does each ion contain?

7.47 Write equations, including electron configurations beneath the species involved, that explain the difference between the first ionization energy of Se(*g*) and the electron affinity of Se(*g*).

7.48 The first ionization energy of Ar and the electron affinity of Ar are both positive values. What is the significance of the positive value in each case?

7.49 The electron affinity of lithium is a negative value, whereas the electron affinity of beryllium is a positive value. Use electron configurations to account for this observation.

7.50 While the electron affinity of bromine is a negative quantity, it is positive for Kr. Use the electron configurations of the two elements to explain the difference.

7.51 What is the relationship between the ionization energy of an anion with a 1− charge such as F^- and the electron affinity of the neutral atom, F?

7.52 Write an equation for the process that corresponds to the electron affinity of the Mg^+ ion. Also write the electron configurations of the species involved. What process does this electron affinity equation correspond to? What is the magnitude of the energy change in the process? [*Hint:* The answer is in Table 7.2.]

Properties of Metals and Nonmetals

7.53 How are metallic character and first ionization energy related?

7.54 Arrange the following pure solid elements in order of increasing electrical conductivity: Ge, Ca, S, and Si. Explain the reasoning you used.

7.55 For each of the following pairs, which element will have the greater metallic character: **(a)** Li or Be, **(b)** Li or Na, **(c)** Sn or P, **(d)** Al or B?

7.56 **(a)** What data can you cite from this chapter to support a prediction that the metallic character of the group 5A elements will increase with increasing atomic number? **(b)** *Nonmetallic character* is the opposite of metallic character—nonmetallic character decreases as metallic character increases. Arrange the following elements in order of increasing nonmetallic character: Se, Ag, Sn, F, and C.

7.57 Predict whether each of the following oxides is ionic or molecular: SO_2, MgO, Li_2O, P_2O_5, Y_2O_3, N_2O, and XeO_3. Explain the reasons for your choices.

7.58 When metal oxides react with water, the oxygen generally ends up as the hydroxide ion, separate from the metal. In contrast, when nonmetallic oxides react with water, the oxygen ends up as part of the nonmetal species. (For example, upon reaction of CO_2 with water, the oxygen remains on the carbon in H_2CO_3.) **(a)** Give two examples each from metals and nonmetals to support these generalizations. **(b)** What connection is there between this contrasting behavior of metal and nonmetal oxides and ionization energies?

7.59 **(a)** What is meant by the terms acidic oxide and basic oxide? **(b)** How can we predict whether an oxide will be acidic or basic, based on its composition?

7.60 Arrange the following oxides in order of increasing acidity: CO_2, CaO, Al_2O_3, SO_3, SiO_2, and P_2O_5.

7.61 Chlorine reacts with oxygen to form Cl_2O_7. **(a)** What is the name of this product (see Table 2.6)? **(b)** Write a balanced equation for the formation $Cl_2O_7(l)$ from the elements. **(c)** Under usual conditions, Cl_2O_7 is a colorless liquid with a boiling point of 81°C. Is this boiling point expected or surprising? **(d)** Would you expect Cl_2O_7 to be more reactive toward $H^+(aq)$ or $OH^-(aq)$? Explain.

[7.62] An element X reacts with oxygen to form XO_2 and with chlorine to form XCl_4. XO_2 is a white solid that melts at high temperatures (above 1000°C). Under usual conditions, XCl_4 is a colorless liquid with a boiling point of 58°C. **(a)** XCl_4 reacts with water to form XO_2 and another product. What is the likely identity of the other product? **(b)** Do you think that element X is a metal, nonmetal, or metalloid? Explain. **(c)** By using a sourcebook such as the *CRC Handbook of Chemistry and Physics*, try to determine the identity of element X.

[7.63] Write balanced equations for the following reactions: **(a)** barium oxide with water, **(b)** iron(II) oxide with perchloric acid, **(c)** sulfur trioxide with water, **(d)** carbon dioxide with aqueous sodium hydroxide.

7.64 Write balanced equations for the following reactions: **(a)** potassium oxide with water, **(b)** diphosphorus trioxide with water, **(c)** chromium(III) oxide with dilute hydrochloric acid, **(d)** selenium dioxide with aqueous potassium hydroxide.

Group Trends in Metals and Nonmetals

7.65 Compare the elements sodium and magnesium with respect to the following properties: **(a)** electron configuration, **(b)** most common ionic charge, **(c)** first ionization energy, **(d)** reactivity toward water, **(e)** atomic radius. Account for the differences between the two elements.

7.66 **(a)** Compare the electron configurations and atomic radii (see Figure 7.6) of rubidium and silver. In what respects are their electronic configurations similar? Account for the difference in radii of the two elements. **(b)** As with rubidium, silver is most commonly found as the 1+ ion, Ag^+. However, silver is far less reactive. Explain these observations.

7.67 **(a)** Why is calcium generally more reactive than magnesium? **(b)** Why is calcium generally less reactive than potassium?

7.68 **(a)** Why is cesium more reactive toward water than is lithium? **(b)** One of the alkali metals reacts with oxygen to form a solid white substance. When this substance is dissolved in water, the solution gives a positive test for hydrogen peroxide, H_2O_2. When the solution is tested in a burner flame, a lilac-purple flame is produced. What is the likely identity of the metal? **(c)** Write a balanced chemical equation for reaction of the white substance with water.

7.69 Write a balanced equation for the reaction that occurs in each of the following cases: **(a)** Potassium metal burns in an atmosphere of chlorine gas. **(b)** Strontium oxide is added to water. **(c)** A fresh surface of lithium metal is exposed to oxygen gas. **(d)** Sodium metal is reacted with molten sulfur.

7.70 Write a balanced equation for the reaction that occurs in each of the following cases: **(a)** Potassium is added to water. **(b)** Barium is added to water. **(c)** Lithium is heated in nitrogen, forming lithium nitride. **(d)** Magnesium burns in oxygen.

7.71 Use electron configurations to explain why hydrogen exhibits properties similar to those of both Li and F.

7.72 **(a)** As described in Section 7.7, the alkali metals react with hydrogen to form hydrides and react with halogens—for example, fluorine—to form halides. Compare the roles of hydrogen and the halogen in these reactions. In what sense are the forms of hydrogen and halogen in the products alike? **(b)** Write balanced equations for the reaction of fluorine with calcium and for the reaction of hydrogen with calcium. What are the similarities among the products of these reactions?

7.73 Compare the elements fluorine and chlorine with respect to the following properties: **(a)** electron configuration, **(b)** most common ionic charge, **(c)** first ionization energy, **(d)** reactivity toward water, **(e)** electron affinity, **(f)** atomic radius. Account for the differences between the two elements.

7.74 Little is known about the properties of astatine, At, because of its rarity and high radioactivity. Nevertheless, it is possible for us to make many predictions about its properties. **(a)** Do you expect the element to be a gas, liquid, or solid at room temperature? Explain. **(b)** What is the chemical formula of the compound it forms with Na?

7.75 Until the early 1960s the group 8A elements were called the inert gases. Why was this name given? Why is it inappropriate?

7.76 Why does xenon react with fluorine whereas neon does not?

7.77 Write a balanced equation for the reaction that occurs in each of the following cases: **(a)** Ozone decomposes to dioxygen. **(b)** Xenon reacts with fluorine. (Write three different equations.) **(c)** Sulfur reacts with hydrogen gas. **(d)** Fluorine reacts with water.

7.78 Write a balanced equation for the reaction that occurs in each of the following cases: **(a)** Chlorine reacts with water. **(b)** Barium metal is heated in an atmosphere of hydrogen gas. **(c)** Lithium reacts with sulfur. **(d)** Fluorine reacts with magnesium metal.

Additional Exercises

7.79 Consider the stable elements through lead ($Z = 82$). In how many instances are the atomic weights of the elements in the reverse order relative to the atomic numbers of the elements? What is the explanation for these cases?

7.80 In 1871, Mendeleev predicted the existence of an element that he called eka-aluminum, which would have the following properties: atomic weight of about 68 amu, density of about 5.9 g/cm^3, low melting point, high boiling point, and an oxide with stoichiometry M_2O_3. **(a)** In 1875 the element predicted by Mendeleev was discovered. By what name is the element known? **(b)** Use a reference such as the *CRC Handbook of Chemistry and Physics* or WebElements.com to check the accuracy of Mendeleev's predictions.

7.81 **(a)** Which will have the lower energy, a 4*s* or a 4*p* electron in an As atom? **(b)** How can we use the concept of effective nuclear charge to explain your answer to part (a)?

7.82 **(a)** If the core electrons were totally effective at shielding the valence electrons and the valence electrons provided no shielding for each other, what would be the effective nuclear charge acting on the valence electron in P? **(b)** Detailed calculations indicate that the effective nuclear charge is 5.6+ for the 3*s* electrons and 4.9+ for the 3*p* electrons. Why are the values for the 3*s* and 3*p* electrons different? **(c)** If you remove a single electron from a P atom, which orbital will it come from? Explain.

7.83 Nearly all the mass of an atom is in the nucleus, which has a very small radius. When atoms bond together (for example, two fluorine atoms in F_2), why is the distance separating the nuclei so much larger than the radii of the nuclei?

[7.84] Consider the change in effective nuclear charge experienced by a 2*p* electron as we proceed from C to N. **(a)** Based on a simple model in which core electrons screen the valence electrons completely and valence electrons do not screen other valence electrons, what do you predict for the change in Z_{eff} from C to N? **(b)** The actual calculated change in Z_{eff} from C to N is 0.70+. How can we explain the difference between this number and the one obtained in part (a)? **(c)** The calculated change in Z_{eff} from N to O is smaller than that from C to N. Can you provide an explanation for this observation?

7.85 As we move across a period of the periodic table, why do the sizes of the transition elements change more gradually than those of the representative elements?

7.86 In the series of group 5A hydrides, of general formula MH_3, the measured bond distances are P—H, 1.419 Å; As—H, 1.519 Å; Sb—H, 1.707 Å. **(a)** Compare these values with those estimated by use of the atomic radii in Figure 7.6. **(b)** Explain the steady increase in M—H bond distance in this series in terms of the electronic configurations of the M atoms.

7.87 It is possible to produce compounds of the form $GeClH_3$, $GeCl_2H_2$, and $GeCl_3H$. What values do you predict for the Ge—H and Ge—Cl bond lengths in these compounds?

7.88 Note from the following table that the increase in atomic radius in moving from Zr to Hf is smaller than in moving from Y to La. Suggest an explanation for this effect.

Atomic Radii (Å)

Sc 1.44	Ti 1.36
Y 1.62	Zr 1.48
La 1.69	Hf 1.50

7.89 The "Chemistry and Life" box on ionic size in Section 7.3 compares the ionic radii of Zn^{2+} and Cd^{2+}. **(a)** The 2+ ion of which other element seems the most obvious one to compare to Zn^{2+} and Cd^{2+}? **(b)** With reference to Figure 2.24, is the element in part (a) essential for life? **(c)** Estimate the ionic radius of the 2+ ion of the element in part (a). Explain any assumptions you have made. **(d)** Would you expect the 2+ ion of the element in part (a) to be physiologically more similar to Zn^{2+} or to Cd^{2+}? **(e)** Use a sourcebook or a Web search to determine whether the element in part (a) is toxic to humans.

[7.90] The ionic substance strontium oxide, SrO, forms from the direct reaction of strontium metal with molecular oxygen. The arrangement of the ions in solid SrO is analogous to that in solid NaCl (see Figure 2.23) and is shown below. **(a)** Write a balanced equation for the formation of SrO(*s*) from the elements. **(b)** In the figure, do the large spheres represent Sr^{2+} ions or O^{2-} ions? Explain. **(c)** Based on the ionic radii in Figure 7.7, predict the length of the side of the cube in the figure. **(d)** The experimental density of SrO is 5.10 g/cm^3. Given your answer to part (c), what is the number of formula units of SrO that are contained in the cube in the figure? (We will examine structures like those in the figure more closely in Chapter 11.)

7.91 Explain the variation in ionization energies of carbon, as displayed in the following graph:

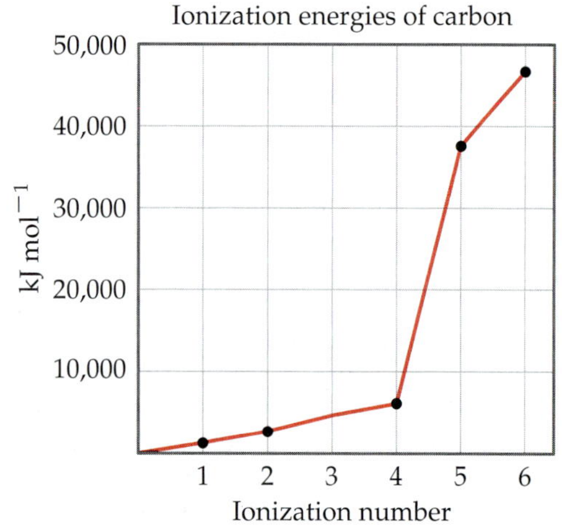

[7.92] Listed here are the atomic and ionic (2+) radii for calcium and zinc:

Radii (Å)

Ca 1.74	Ca^{2+} 0.99
Zn 1.31	Zn^{2+} 0.74

(a) Explain why the ionic radius in each case is smaller than the atomic radius. **(b)** Why is the atomic radius of calcium larger than that of zinc? **(c)** Suggest a reason why the difference in the ionic radii is much less than the difference in the atomic radii.

7.93 Do you agree with the following statement? "A negative value for the electron affinity of an atom occurs when the outermost electrons only incompletely shield one another from the nucleus." If not, change it to make it more nearly correct in your view. Apply either the statement as given or your revised statement to explain why the electron affinity of bromine is −325 kJ/mol and that for its neighbor Kr is > 0.

7.94 Use orbital diagrams to illustrate what happens when an oxygen atom gains two electrons. Why is it extremely difficult to add a third electron to the atom?

[7.95] Use electron configurations to explain the following observations: **(a)** The first ionization energy of phosphorus is greater than that of sulfur. **(b)** The electron affinity of nitrogen is lower (less negative) than those of both carbon and oxygen. **(c)** The second ionization energy of oxygen is greater than that of fluorine. **(d)** The third ionization energy of manganese is greater than those of both chromium and iron.

7.96 The following table gives the electron affinities, in kJ/mol, for the group 1B and group 2B metals:

Cu −119	Zn > 0
Ag −126	Cd > 0
Au −223	Hg > 0

(a) Why are the electron affinities of the group 2B elements greater than zero? (b) Why do the electron affinities of the group 1B elements become more negative as we move down the group? [*Hint:* Examine the trends in the electron affinity of other groups as we proceed down the periodic table.]

7.97 Hydrogen is an unusual element because it behaves in some ways like the alkali metal elements and in other ways like a nonmetal. Its properties can be explained in part by its electron configuration and by the values for its ionization energy and electron affinity. **(a)** Explain why the electron affinity of hydrogen is much closer to the values for the alkali elements than for the halogens. **(b)** Is the following statement true? "Hydrogen has the smallest bonding atomic radius of any element that forms chemical compounds." If not, correct it. If it is, explain in terms of electron configurations. **(c)** Explain why the ionization energy of hydrogen is closer to the values for the halogens than for the alkali metals.

[7.98] The first ionization energy of the oxygen molecule is the energy required for the following process:

$$O_2(g) \longrightarrow O_2^{+}(g) + e^-$$

The energy needed for this process is 1175 kJ/mol, very similar to the first ionization energy of Xe. Would you expect O_2 to react with F_2? If so, suggest a product or products of this reaction.

7.99 Based on your reading of this chapter, arrange the following in order of increasing melting point: K, Br_2, Mg, and O_2. Explain the factors that determine the order.

[7.100] There are certain similarities in properties that exist between the first member of any periodic family and the element located below it and to the right in the periodic table. For example, in some ways Li resembles Mg, Be resembles Al, and so forth. This observation is called the diagonal relationship. Using what we have learned in this chapter, offer a possible explanation for this relationship.

[7.101] The elements at the bottom of groups 1A, 2A, 6A, 7A, and 8A—Fr, Ra, Po, At, and Rn—are all radioactive. As a result, much less is known about their physical and chemical properties than those of the elements above them. Based on what we have learned in this chapter, which of these five elements would you expect **(a)** to have the most metallic character, **(b)** to have the least metallic (that is, the most nonmetallic) character, **(c)** to have the largest first ionization energy, **(d)** to have the smallest first ionization energy, **(e)** to have the greatest (most negative) electron affinity, **(f)** to have the largest atomic radius, **(g)** to resemble least in appearance the element immediately above it, **(h)** to have the highest melting point, **(i)** to react most exothermically with water?

[7.102] A historian discovers a nineteenth-century notebook in which some observations, dated 1822, on a substance thought to be a new element, were recorded. Here are some of the data recorded in the notebook: Ductile, silver-white, metallic looking. Softer than lead. Unaffected by water. Stable in air. Melting point: 153°C. Density: 7.3 g/cm^3. Electrical conductivity: 20% that of copper. Hardness: About 1% as hard as iron. When 4.20 g of the unknown is heated in an excess of oxygen, 5.08 g of a white solid is formed. The solid could be sublimed by heating to over 800°C. **(a)** Using information in the text and a handbook of chemistry, and making allowances for possible variations in numbers from current values, identify the element reported. **(b)** Write a balanced chemical equation for the reaction with oxygen. **(c)** Judging from Figure 7.2, might this nineteenth-century investigator have been the first to discover a new element?

[7.103] It has been discovered in recent years that many organic compounds that contain chlorine, including dioxins, which had been thought to be entirely of man-made origin, are formed in natural processes. More than 3000 natural organohalogen compounds, most involving chlorine and bromine, are known. These compounds, in which the halogen is attached to carbon, are nearly all nonionic materials. Why are these materials typically not ionic, as are the more abundant inorganic halogen compounds found in nature?

Integrative Exercises

[7.104] Moseley established the concept of atomic number by studying X rays emitted by the elements. The X rays emitted by some of the elements have the following wavelengths:

Element	Wavelength (Å)
Ne	14.610
Ca	3.358
Zn	1.435
Zr	0.786
Sn	0.491

(a) Calculate the frequency, ν, of the X rays emitted by each of the elements, in Hz. **(b)** Using graph paper (or suitable computer software), plot the square root of ν versus the atomic number of the element. What do you observe about the plot? **(c)** Explain how the plot in part (b) allowed Moseley to predict the existence of undiscovered elements. **(d)** Use the result from part (b) to predict the X-ray wavelength emitted by iron. **(e)** A particular element emits X rays with a wavelength of 0.980 Å. What element do you think it is?

[7.105] **(a)** Write the electron configuration for Li, and estimate the effective nuclear charge experienced by the valence electron. **(b)** The energy of an electron in a one-electron atom or ion equals $(-2.18 \times 10^{-18}\,\text{J})\left(\frac{Z^2}{n^2}\right)$, where Z is the nuclear charge and n is the principal quantum number of the electron. Estimate the first ionization energy of Li. **(c)** Compare the result of your calculation with the value reported in Table 7.4, and explain the difference. **(d)** What value of the effective nuclear charge gives the proper value for the ionization energy? Does this agree with your explanation in (c)?

[7.106] One way to measure ionization energies is photoelectron spectroscopy (PES), a technique based on the photoelectric effect. (Section 6.2) In PES, monochromatic light is directed onto a sample, causing electrons to be emitted. The kinetic energy of the emitted electrons is measured. The difference between the energy of the photons and the kinetic energy of the electrons corresponds to the energy needed to remove the electrons (that is, the ionization energy). Suppose that a PES experiment is performed in which mercury vapor is irradiated with ultraviolet light of wavelength 58.4 nm. **(a)** What is the energy of a photon of this light, in eV? **(b)** Write an equation that shows the process corresponding to the first ionization energy of Hg. **(c)** The kinetic energy of the emitted electrons is measured to be 10.75 eV. What is the first ionization energy of Hg, in kJ/mol? **(d)** With reference to Figure 7.11, determine which of the halogen elements has a first ionization energy closest to that of mercury.

7.107 Consider the gas-phase transfer of an electron from a sodium atom to a chlorine atom:

$$Na(g) + Cl(g) \longrightarrow Na^+(g) + Cl^-(g)$$

(a) Write this reaction as the sum of two reactions, one that relates to an ionization energy and one that relates to an electron affinity. **(b)** Use the result from part (a), data in this chapter, and Hess's law to calculate the enthalpy of the above reaction. Is the reaction exothermic or endothermic? **(c)** The reaction between sodium metal and chlorine gas is highly exothermic and produces $NaCl(s)$, whose structure was discussed in Section 2.6. Comment on this observation relative to the calculated enthalpy for the aforementioned gas-phase reaction.

[7.108] When magnesium metal is burned in air (Figure 3.5), two products are produced. One is magnesium oxide, MgO. The other is the product of the reaction of Mg with molecular nitrogen, magnesium nitride. When water is added to magnesium nitride, it reacts to form magnesium oxide and ammonia gas. **(a)** Based on the charge of the nitride ion (Table 2.5), predict the formula of magnesium nitride. **(b)** Write a balanced equation for the reaction of magnesium nitride with water. What is the driving force for this reaction? **(c)** In an experiment a piece of magnesium ribbon is burned in air in a crucible. The mass of the mixture of MgO and magnesium nitride after burning is 0.470 g. Water is added to the crucible, further reaction occurs, and the crucible is heated to dryness until the final product is 0.486 g of MgO. What was the mass percentage of magnesium nitride in the mixture obtained after the initial burning? **(d)** Magnesium nitride can also be formed by reaction of the metal with ammonia at high temperature. Write a balanced equation for this reaction. If a 6.3-g Mg ribbon reacts with 2.57 g $NH_3(g)$ and the reaction goes to completion, which component is the limiting reactant? What mass of $H_2(g)$ is formed in the reaction? **(e)** The standard enthalpy of formation of solid magnesium nitride is $-461.08 \text{ kJ mol}^{-1}$. Calculate the standard enthalpy change for the reaction between magnesium metal and ammonia gas.

7.109 **(a)** The experimental Bi—Br bond length in bismuth tribromide, $BiBr_3$, is 2.63 Å. Based on this value and the data in Figure 7.6, predict the atomic radius of Bi. **(b)** Bismuth tribromide is soluble in acidic solution. It is formed by treating solid bismuth(III) oxide with aqueous hydrobromic acid. Write a balanced chemical equation for this reaction. **(c)** While bismuth(III) oxide is soluble in acidic solutions, it is insoluble in basic solutions such as $NaOH(aq)$. On the basis of these properties, is bismuth characterized as a metallic, metalloid, or nonmetallic element? **(d)** Treating bismuth with fluorine gas forms BiF_5. Use the electron configuration of Bi to explain the formation of a compound with this formulation. **(e)** While it is possible to form BiF_5 in the manner just described, pentahalides of bismuth are not known for the other halogens. Explain why the pentahalide might form with fluorine, but not with the other halogens. How does the behavior of bismuth relate to the fact that xenon reacts with fluorine to form compounds, but not with the other halogens?

7.110 Potassium superoxide, KO_2, is often used in oxygen masks (such as those used by firefighters) because KO_2 reacts with CO_2 to release molecular oxygen. Experiments indicate that 2 mol of $KO_2(s)$ react with each mole of $CO_2(g)$. **(a)** The products of the reaction are $K_2CO_3(s)$ and $O_2(g)$. Write a balanced equation for the reaction between $KO_2(s)$ and $CO_2(g)$. **(b)** Indicate the oxidation number for each atom involved in the reaction in part (a). What elements are being oxidized and reduced? **(c)** What mass of $KO_2(s)$ is needed to consume 18.0 g $CO_2(g)$? What mass of $O_2(g)$ is produced during this reaction?

eMEDIA EXERCISES

These exercises make use of the interactive objects available online in OneKey or the Companion Website, and on your Accelerator CD. Access to these resources comes in your MediaPak.

7.111 The **Periodic Trends: Atomic Radii** animation (*7.3*) describes the trends in the sizes of atoms on the periodic table—from left to right and from top to bottom. **(a)** What factors influence atomic radius? **(b)** Based on the factors that influence the atomic radius, explain why the radius of gallium is smaller than that of aluminum.

7.112 The **Gain and Loss of Electrons** animation (*7.3*) illustrates how addition or subtraction of an electron affects the size of an atom. The first ionization of aluminum produces the Al^+ ion, which is smaller than the neutral Al atom. The second ionization of aluminum produces the Al^{2+} ion, which is smaller still. The third ionization of aluminum produces the Al^{3+} ion, and the 3+ cation is even smaller than the 2+ cation. Of the first, second, and third ionizations, which would you expect to cause the biggest change in size? Explain your reasoning.

7.113 According to the information given in the **Periodic Trends: Ionization Energy** animation (*7.4*), you might expect fluorine and chlorine to have two of the highest ionization energies among the representative elements. Explain why those two elements are almost always found in nature as ions.

7.114 Although the **Periodic Trends: Ionization Energy** animation (*7.4*) depicts clear trends in the magnitudes of first ionization energies, the movie shows that the trend from left to right across the periodic table is not smooth. **(a)** Based on electron configurations, explain why this is so. **(b)** For each pair, predict which one will have the higher first ionization energy: N, O; Be, B; Ca, Ga; P, S.

7.115 **(a)** What happens to the size of the chlorine atom in the **Ionization Energy** animation (*7.4*) when the first electron is removed? **(b)** How do you account for this change in size? **(c)** Based on your answer to part (b), how would you expect the size of the chlorine atom to change if an electron were *added* rather than removed?

7.116 **(a)** Based on the information in the **Periodic Trends: Acid-Base Behavior of Oxides** animation (*7.6*), which of the following compounds would you expect to form a basic solution with water: NaO, NO, N_2O, K_2O, CO_2? **(b)** Which oxide in each pair would produce the most acidic solution, and why? CO_2, CO; NO, N_2O; N_2O_5, N_2O_4.

7.117 Watch the **Sodium and Potassium in Water** movie (*7.7*), and note the differences between the two reactions shown. **(a)** What property of the elements involved is responsible for the observed differences? **(b)** How would you expect the reaction of Rb and water to differ from the reactions in the demonstration? Describe what you would expect to happen.

eLABORATORY EXERCISES

These exercises allow students to apply the concepts and skills for this chapter in the simulated laboratory of Virtual ChemLab. Worksheets for each of these exercises are found in the Virtual ChemLab Workbook in your MediaPak.

7.119 ***(VCL 7-1) Flame Tests for Metals***

Have you ever wondered why a candle flame is yellow? The characteristic yellow of the flame comes from the glow of burning carbon fragments. The carbon fragments are produced by the incomplete combustion reaction of the wick and candle wax. When elements, such as carbon, are heated to high temperatures, some of their electrons are excited to higher energy levels. When these excited electrons fall back to lower energy levels, they release excess energy in packages of light called photons. The color of the emitted light depends on the individual energy level spacing in the atom. When heated, each element emits a characteristic pattern of photons, which is useful for identifying the element. The characteristic colors of light produced when substances are heated in the flame of a gas burner are the basis for flame tests of several elements. In this assignment you will perform flame tests that are used to identify several metallic elements.

7.120 ***(VCL 7-2) Emission Spectra of Sodium and Mercury***

In the 1800s scientists found that when a sample of gas is excited by an alternating electric field, light with only certain discrete wavelengths is emitted. This property allowed for the development of spectroscopic techniques that can be used in the identification and analysis of elements and compounds. Even though scientists found spectroscopy very useful, they could not explain why the spectrum was not continuous. The explanation of this was left to Niels Bohr, a Danish physicist, who first proposed that energy levels of electrons are quantized and excited electrons can fall only to discrete energy levels. This assignment illustrates the measurements that helped Bohr develop his original quantum model and also illustrates some practical uses for this science by measuring the emission spectra for mercury and sodium. Mercury vapor is used in fluorescent lights, and sodium vapor in street lighting.

CHAPTER 8

WHAT'S AHEAD

- We begin with a brief discussion of chemical bond types and introduce *Lewis symbols*, a way of showing the valence electrons in atoms and ions. *(Section 8.1)*
- *Ionic bonding* results from the essentially complete transfer of one or more electrons from one atom to another. We will study the energetics of formation of ionic substances and describe the *lattice energy* of these substances. *(Section 8.2)*
- *Covalent bonding* involves the sharing of one or more electron pairs between atoms, as needed to attain an *octet* of electrons about each atom. *(Section 8.3)*
- We then see that covalent bonds can be represented by diagrams called *Lewis structures* and that covalent bonds can be *single, double,* or *triple.* *(Section 8.3)*
- *Electronegativity* is defined as the ability of an atom to attract electrons to itself in a bonding situation. Electron pairs will in general be shared unequally between atoms of differing electronegativity, leading to *polar covalent bonds.* *(Section 8.4)*
- We will examine more closely some of the rules we use for drawing Lewis structures, including the assignment of *formal charges* to atoms in molecules. *(Section 8.5)*
- It may occur that more than one equivalent Lewis structure can be drawn for a molecule or polyatomic ion. The actual structure in such cases is a blend of two or more contributing Lewis structures, called *resonance structures*. *(Section 8.6)*
- *Exceptions to the octet rule* include a relatively few cases in which there are fewer than 8 electrons in the valence-shell orbitals. By contrast, the structures of many compounds of third-period and heavier elements can best be described by assuming that the valence-shell orbitals hold more than an octet of electrons. *(Section 8.7)*
- The strengths of covalent bonds vary with the number of shared electron pairs as well as other factors. We can use *average bond enthalpy* values, which are average values of the energy needed to break a specific type of bond, to estimate the enthalpies of reaction in cases where thermodynamic data such as heats of formation are unavailable. *(Section 8.8)*

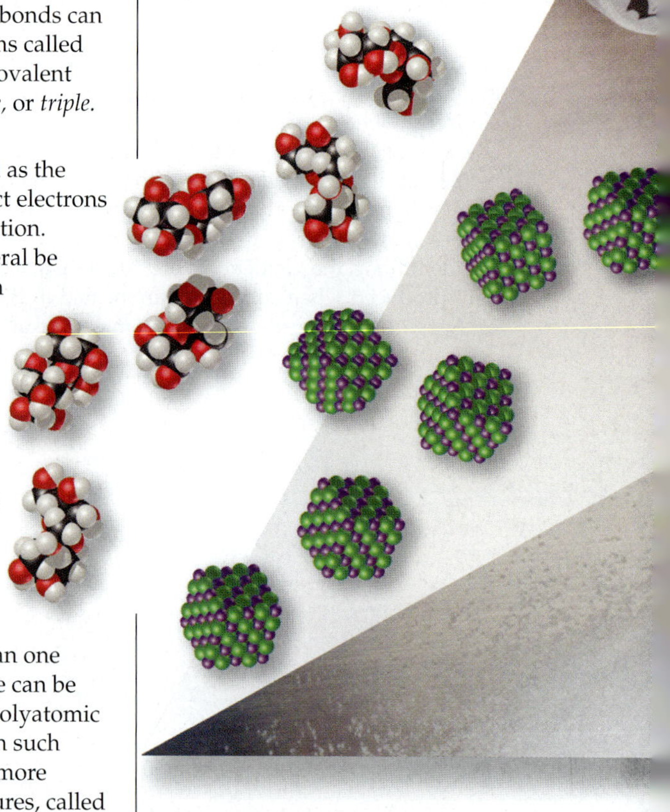

Basic Concepts of Chemical Bonding

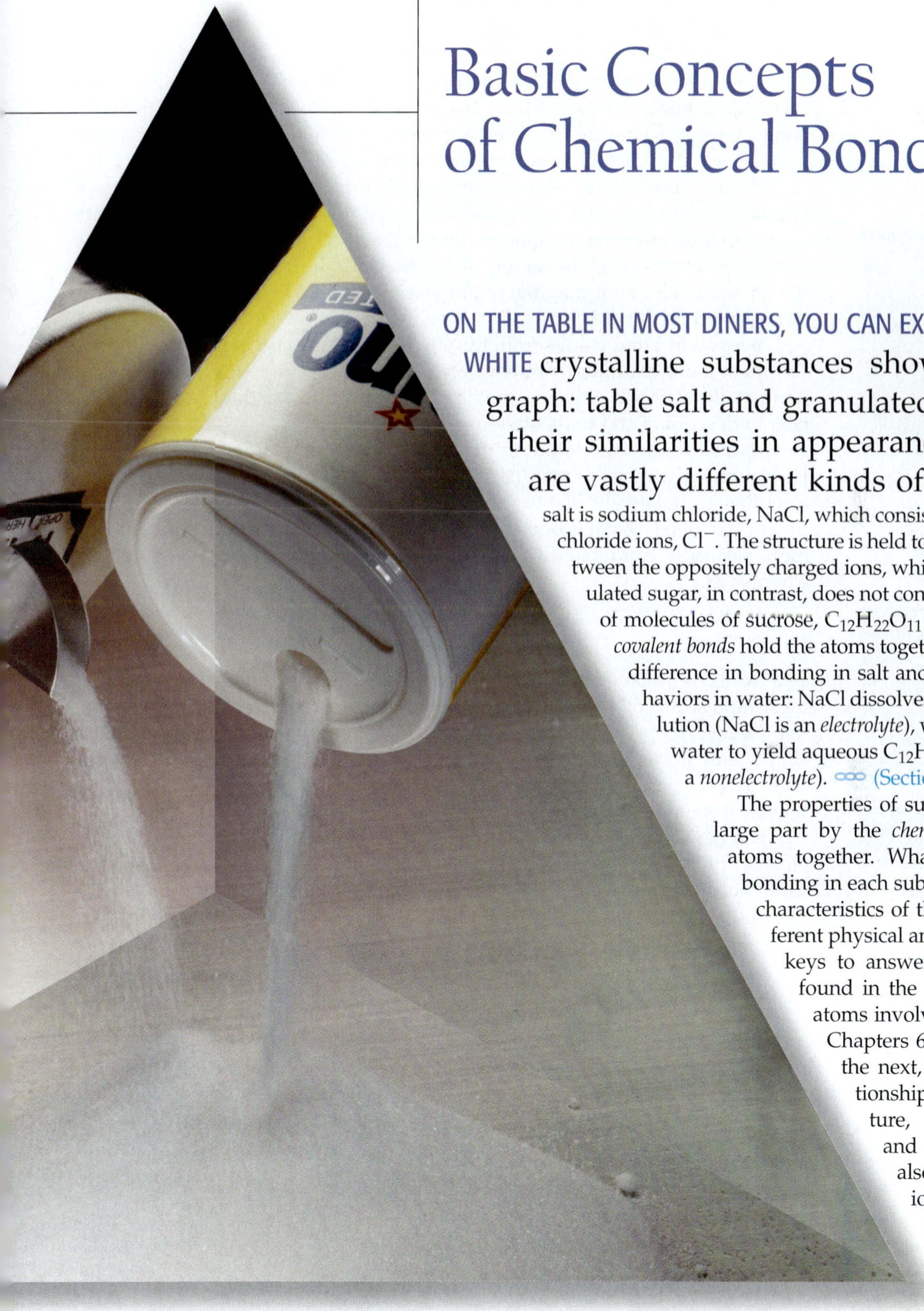

ALTHOUGH TABLE SALT AND SUGAR are both white crystalline substances, they involve distinctly different chemical bonds. Table salt has ionic bonds between Na^+ and Cl^- ions, whereas table sugar (sucrose, $C_{12}H_{22}O_{11}$) is a molecular substance with covalent bonds.

ON THE TABLE IN MOST DINERS, YOU CAN EXPECT TO FIND THE TWO WHITE crystalline substances shown in the photograph: table salt and granulated sugar. In spite of their similarities in appearance, salt and sugar are vastly different kinds of substances. Table salt is sodium chloride, NaCl, which consists of sodium ions, Na^+, and chloride ions, Cl^-. The structure is held together by the attractions between the oppositely charged ions, which we call *ionic bonds*. Granulated sugar, in contrast, does not contain ions. Instead, it consists of molecules of sucrose, $C_{12}H_{22}O_{11}$, in which attractions called *covalent bonds* hold the atoms together. One consequence of the difference in bonding in salt and sugar is their different behaviors in water: NaCl dissolves in water to yield ions in solution (NaCl is an *electrolyte*), whereas sucrose dissolves in water to yield aqueous $C_{12}H_{22}O_{11}$ molecules (sucrose is a *nonelectrolyte*). (Section 4.1)

The properties of substances are determined in large part by the *chemical bonds* that hold their atoms together. What determines the type of bonding in each substance, and just how do the characteristics of these bonds give rise to different physical and chemical properties? The keys to answering the first question are found in the electronic structures of the atoms involved, which we discussed in Chapters 6 and 7. In this chapter and the next, we will examine the relationships among electronic structure, chemical bonding forces, and chemical bond type. We'll also see how the properties of ionic and covalent substances arise from the distributions of electronic charge within atoms, ions, and molecules.

8.1 | Chemical Bonds, Lewis Symbols, and the Octet Rule

Whenever two atoms or ions are strongly attached to each other, we say there is a **chemical bond** between them. There are three general types of chemical bonds: ionic, covalent, and metallic. Figure 8.1 ◀ shows examples of substances in which we find each of these types of attractive forces.

The term **ionic bond** refers to electrostatic forces that exist between ions of opposite charge. Ions may be formed from atoms by the transfer of one or more electrons from one atom to another. Ionic substances generally result from the interaction of metals on the far left side of the periodic table with nonmetals on the far right side (excluding the noble gases, group 8A). Ionic bonding will be discussed in Section 8.2.

A **covalent bond** results from the sharing of electrons between two atoms. The most familiar examples of covalent bonding are seen in the interactions of nonmetallic elements with one another. We devote much of this chapter and the next to describing and understanding covalent bonds.

Metallic bonds are found in metals, such as copper, iron, and aluminum. Each atom in a metal is bonded to several neighboring atoms. The bonding electrons are relatively free to move throughout the three-dimensional structure of the metal. Metallic bonds give rise to such typical metallic properties as high electrical conductivity and luster. We will examine these bonds in Chapter 23.

ACTIVITY
Periodic Trends: Lewis Symbols

▲ **Figure 8.1 Chemical bonds.** Examples of substances in which (a) ionic, (b) covalent, and (c) metallic bonds are found.

Lewis Symbols

The electrons involved in chemical bonding are the *valence electrons*, which, for most atoms, are those residing in the outermost occupied shell of an atom. ∞ (Section 6.8) The American chemist G. N. Lewis (1875–1946) suggested a simple way of showing the valence electrons in an atom and tracking them in the course of bond formation, using what are now known as *Lewis electron-dot symbols* or merely Lewis symbols.

The **Lewis symbol** for an element consists of the chemical symbol for the element plus a dot for each valence electron. Sulfur, for example, has the electron configuration [Ne]$3s^2 3p^4$; its Lewis symbol therefore shows six valence electrons:

$$\cdot\ddot{\underset{\cdot\cdot}{\mathrm{S}}}\cdot$$

The dots are placed on the four sides of the atomic symbol: the top, the bottom, and the left and right sides. Each side can accommodate up to two electrons. All four sides of the symbol are equivalent, which means that the choice of which side to place two electrons versus one electron is arbitrary.

The electron configurations and Lewis symbols for the representative elements of the second and third rows of the periodic table are shown in Table 8.1 ▶. Notice that the number of valence electrons in any representative element is the same as the group number of the element. For example, the Lewis symbols for oxygen and sulfur, members of group 6A, both show six dots.

GIVE IT SOME THOUGHT

Which of the following three possible Lewis symbols for Cl is correct?

$$:\ddot{\underset{\cdot\cdot}{\mathrm{Cl}}}\cdot \quad :\ddot{\underset{\cdot}{\mathrm{Cl}}}: \quad :\dot{\underset{\cdot}{\mathrm{Cl}}}\cdot$$

TABLE 8.1 Lewis Symbols

Element	Electron Configuration	Lewis Symbol	Element	Electron Configuration	Lewis Symbol
Li	$[He]2s^1$	Li·	Na	$[Ne]3s^1$	Na·
Be	$[He]2s^2$	·Be·	Mg	$[Ne]3s^2$	·Mg·
B	$[He]2s^22p^1$	·Ḃ·	Al	$[Ne]3s^23p^1$	·Ȧl·
C	$[He]2s^22p^2$	·Ċ̣·	Si	$[Ne]3s^23p^2$	·Ṩi·
N	$[He]2s^22p^3$	·Ṇ̇:	P	$[Ne]3s^23p^3$	·Ṗ̣:
O	$[He]2s^22p^4$	:Ọ̇:	S	$[Ne]3s^23p^4$	:Ṩ:
F	$[He]2s^22p^5$	·F̤̈:	Cl	$[Ne]3s^23p^5$	·C̤̈l:
Ne	$[He]2s^22p^6$	:N̤̈e:	Ar	$[Ne]3s^23p^6$	:Ä̤r:

The Octet Rule

ACTIVITY
Octet Rule

Atoms often gain, lose, or share electrons to achieve the same number of electrons as the noble gas closest to them in the periodic table. The noble gases have very stable electron arrangements, as evidenced by their high ionization energies, low affinity for additional electrons, and general lack of chemical reactivity. (Section 7.8) Because all noble gases (except He) have eight valence electrons, many atoms undergoing reactions also end up with eight valence electrons. This observation has led to a guideline known as the **octet rule:** *Atoms tend to gain, lose, or share electrons until they are surrounded by eight valence electrons.*

An octet of electrons consists of full *s* and *p* subshells in an atom. In terms of Lewis symbols, an octet can be thought of as four pairs of valence electrons arranged around the atom, as in the Lewis symbol for Ne in Table 8.1. There are many exceptions to the octet rule, but it provides a useful framework for introducing many important concepts of bonding.

8.2 | Ionic Bonding

When sodium metal, Na(*s*), is brought into contact with chlorine gas, $Cl_2(g)$, a violent reaction ensues (Figure 8.2 ▶). The product of this very exothermic reaction is sodium chloride, NaCl(*s*):

$$Na(s) + \tfrac{1}{2}Cl_2(g) \longrightarrow NaCl(s) \qquad \Delta H_f^\circ = -410.9 \text{ kJ} \qquad [8.1]$$

Sodium chloride is composed of Na^+ and Cl^- ions, which are arranged in a regular three-dimensional array, as shown in Figure 8.3 ▶.

The formation of Na^+ from Na and Cl^- from Cl_2 indicates that an electron has been lost by a sodium atom and gained by a chlorine atom—we can envision an *electron transfer* from the Na atom to the Cl atom. Two of the atomic properties we discussed in Chapter 7 give us an indication of how readily electron transfer occurs: the ionization energy, which indicates how easily an electron can be removed from an atom, and the electron affinity, which measures how much an atom wants to gain an electron. (Sections 7.4 and 7.5) Electron transfer to form oppositely charged ions occurs when one of the atoms readily gives up an electron (low ionization energy) and the other atom readily gains an electron (high electron affinity). Thus, NaCl is a typical ionic compound because it consists of a metal of low ionization energy and a nonmetal of high electron affinity. Using Lewis electron-dot symbols (and showing a chlorine atom rather than the Cl_2 molecule), we can represent this reaction as follows:

$$\text{Na}\cdot + \cdot\ddot{\underset{\cdot\cdot}{\text{Cl}}}: \longrightarrow \text{Na}^+ + [:\ddot{\underset{\cdot\cdot}{\text{Cl}}}:]^- \qquad [8.2]$$

MOVIE
Formation of Sodium Chloride

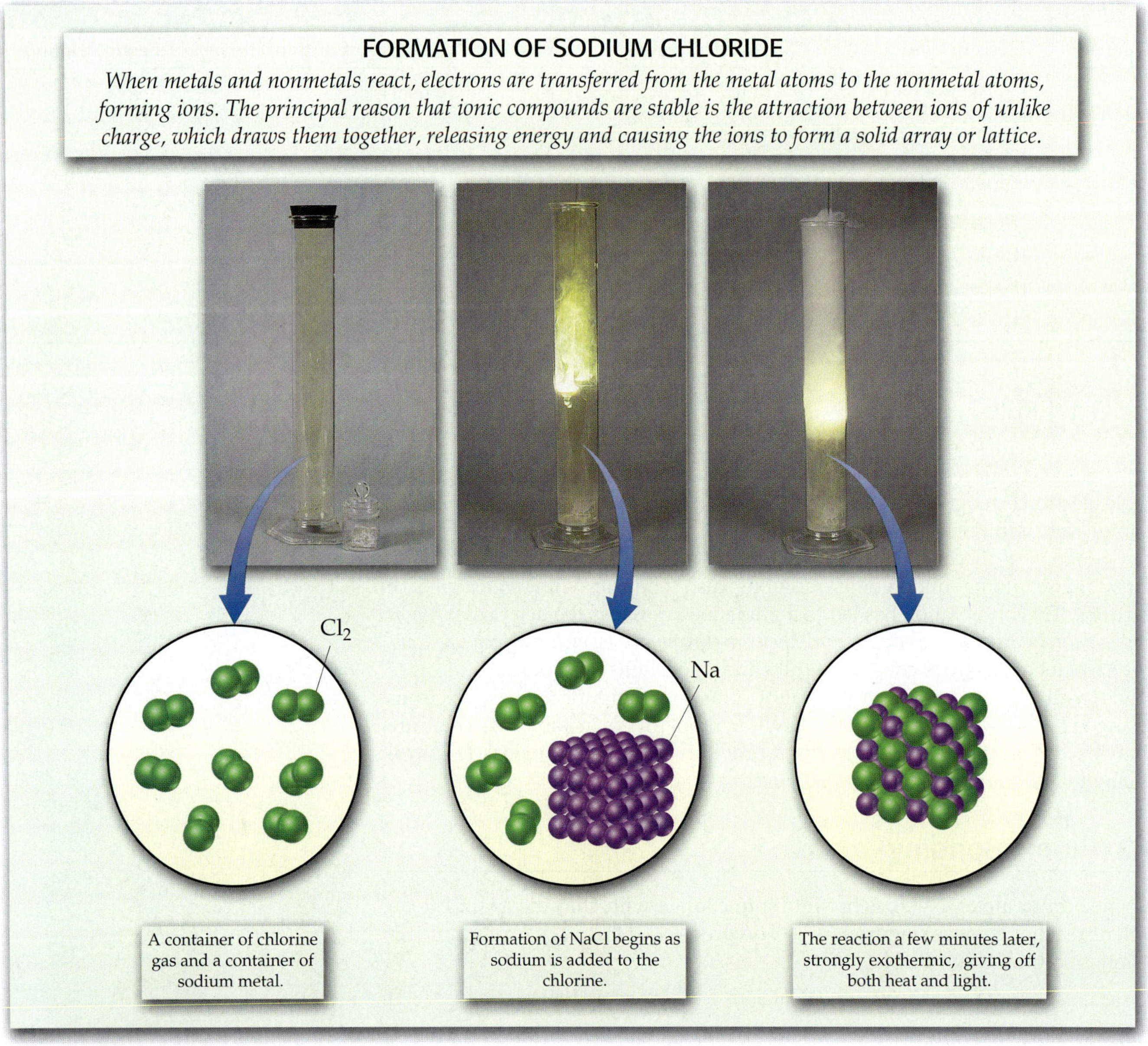

▲ **Figure 8.2 Formation of sodium chloride.**

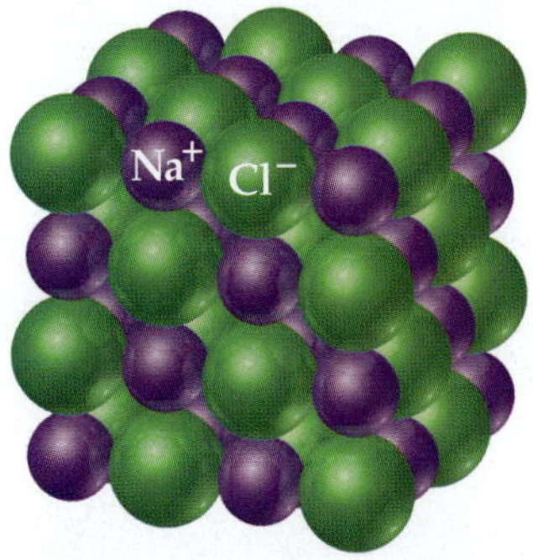

▲ **Figure 8.3 The crystal structure of sodium chloride.** In this three-dimensional array of ions, each Na^+ ion is surrounded by six Cl^- ions, and each Cl^- ion is surrounded by six Na^+ ions.

The arrow indicates the transfer of an electron from the Na atom to the Cl atom. Each ion has an octet of electrons, the octet on Na^+ being the $2s^22p^6$ electrons that lie below the single 3*s* valence electron of the Na atom. We've put a bracket around the chloride ion to emphasize that all eight electrons are located exclusively on the Cl^- ion.

GIVE IT SOME THOUGHT

Describe the electron transfers that occur in the formation of magnesium fluoride from elemental magnesium and fluorine.

Energetics of Ionic Bond Formation

As seen in Figure 8.2, the reaction of sodium with chlorine is *very* exothermic. In fact, Equation 8.1 is the reaction for the formation of NaCl(*s*) from its elements, so that the enthalpy change for the reaction is ΔH_f° for NaCl(*s*). In Appendix C we see that the heat of formation of other ionic substances is also quite negative. What factors make the formation of ionic compounds so exothermic?

In Equation 8.2 we represented the formation of NaCl as the transfer of an electron from Na to Cl. Recall from our discussion of ionization energies, however, that the loss of electrons from an atom is always an endothermic process. (Section 7.4) Removing an electron from Na(*g*) to form $Na^+(g)$, for instance, requires 496 kJ/mol. Conversely, when a nonmetal gains an electron, the process is generally exothermic, as seen from the negative electron affinities of the elements. (Section 7.5) Adding an electron to Cl(*g*), for example, releases 349 kJ/mol. If the transfer of an electron from one atom to another were the only factor in forming an ionic bond, the overall process would rarely be exothermic. For example, removing an electron from Na(*g*) and adding it to Cl(*g*) is an endothermic process that requires 496 − 349 = 147 kJ/mol. This endothermic process corresponds to the formation of sodium and chloride ions that are infinitely far apart—in other words, the positive energy change assumes that the ions are not interacting with one another, which is quite different from the situation in ionic solids.

ACTIVITY
Bond Enthalpy

The principal reason that ionic compounds are stable is the attraction between ions of unlike charge. This attraction draws the ions together, releasing energy and causing the ions to form a solid array, or lattice, such as that shown for NaCl in Figure 8.3. A measure of just how much stabilization results from arranging oppositely charged ions in an ionic solid is given by the **lattice energy**, which is *the energy required to completely separate a mole of a solid ionic compound into its gaseous ions.*

To get a picture of this process for NaCl, imagine that the structure shown in Figure 8.3 expands from within, so that the distances between the ions increase until the ions are very far apart. This process requires 788 kJ/mol, which is the value of the lattice energy:

$$NaCl(s) \longrightarrow Na^+(g) + Cl^-(g) \qquad \Delta H_{lattice} = +788 \text{ kJ/mol} \qquad [8.3]$$

Notice that this process is highly endothermic. The reverse process—the coming together of $Na(g)^+$ and $Cl(g)^-$ to form NaCl(*s*)—is therefore highly exothermic ($\Delta H = -788$ kJ/mol).

ACTIVITY
Coulomb's Law

Table 8.2 ▶ lists the lattice energies of NaCl and other ionic compounds. All are large positive values, indicating that the ions are strongly attracted to one another in these solids. The energy released by the attraction between ions of unlike charge more than makes up for the endothermic nature of ionization energies, making the formation of ionic compounds an exothermic process. The strong attractions also cause most ionic materials to be hard and brittle, with high melting points—for example, NaCl melts at 801°C.

The magnitude of the lattice energy of a solid depends on the charges of the ions, their sizes, and their arrangement in the solid. We saw in Chapter 5 that the potential energy of two interacting charged particles is given by

$$E_{el} = \frac{\kappa Q_1 Q_2}{d} \qquad [8.4]$$

TABLE 8.2 Lattice Energies for Some Ionic Compounds

Compound	Lattice Energy (kJ/mol)	Compound	Lattice Energy (kJ/mol)
LiF	1030	$MgCl_2$	2326
LiCl	834	$SrCl_2$	2127
LiI	730		
NaF	910	MgO	3795
NaCl	788	CaO	3414
NaBr	732	SrO	3217
NaI	682		
KF	808	ScN	7547
KCl	701		
KBr	671		
CsCl	657		
CsI	600		

In this equation Q_1 and Q_2 are the charges on the particles, d is the distance between their centers, and κ is a constant, 8.99×10^9 J-m/C^2. (Section 5.1) Equation 8.4 indicates that the attractive interaction between two oppositely charged ions increases as the magnitudes of their charges increase and as the distance between their centers decreases. Thus, *for a given arrangement of ions, the lattice energy increases as the charges on the ions increase and as their radii decrease*. The magnitude of lattice energies depends primarily on the ionic charges because ionic radii do not vary over a very wide range.

SAMPLE EXERCISE 8.1 | Magnitudes of Lattice Energies

Without consulting Table 8.2, arrange the following ionic compounds in order of increasing lattice energy: NaF, CsI, and CaO.

Solution

Analyze: From the formulas for three ionic compounds, we must determine their relative lattice energies.

Plan: We need to determine the charges and relative sizes of the ions in the compounds. We can then use Equation 8.4 qualitatively to determine the relative energies, knowing that the larger the ionic charges, the greater the energy and the farther apart the ions are, the lower the energy.

Solve: NaF consists of Na^+ and F^- ions, CsI of Cs^+ and I^- ions, and CaO of Ca^{2+} and O^{2-} ions. Because the product of the charges, Q_1Q_2, appears in the numerator of Equation 8.4, the lattice energy will increase dramatically when the charges of the ions increase. Thus, we expect the lattice energy of CaO, which has 2+ and 2− ions, to be the greatest of the three.

The ionic charges in NaF and CsI are the same. As a result, the difference in their lattice energies will depend on the difference in the distance between the centers of the ions in their lattice. Because ionic size increases as we go down a group in the periodic table (Section 7.3), we know that Cs^+ is larger than Na^+ and I^- is larger than F^-. Therefore the distance between the Na^+ and F^- ions in NaF will be less than the distance between the Cs^+ and I^- ions in CsI. As a result, the lattice energy of NaF should be greater than that of CsI. In order of increasing energy, therefore, we have CsI < NaF < CaO.

Check: Table 8.2 confirms our predicted order is correct.

PRACTICE EXERCISE

Which substance would you expect to have the greatest lattice energy, AgCl, CuO, or CrN?

Answer: CrN

Electron Configurations of Ions of the Representative Elements

We began considering the electron configurations of ions in Section 7.4. In light of our examination of ionic bonding, we will continue with that discussion here. The energetics of ionic bond formation helps explain why many ions tend to have noble-gas electron configurations. For example, sodium readily loses one electron to form Na^+, which has the same electron configuration as Ne:

$$\text{Na} \quad 1s^22s^22p^63s^1 = [\text{Ne}]3s^1$$

$$\text{Na}^+ \quad 1s^22s^22p^6 = [\text{Ne}]$$

Even though lattice energy increases with increasing ionic charge, we never find ionic compounds that contain Na^{2+} ions. The second electron removed would have to come from an inner shell of the sodium atom, and removing electrons from an inner shell requires a very large amount of energy. (Section 7.4) The increase in lattice energy is not enough to compensate for the energy needed to remove an inner-shell electron. Thus, sodium and the other group 1A metals are found in ionic substances only as 1+ ions.

Similarly, the addition of electrons to nonmetals is either exothermic or only slightly endothermic as long as the electrons are being added to the valence shell. Thus, a Cl atom easily adds an electron to form Cl^-, which has the same electron configuration as Ar:

$$\text{Cl} \quad 1s^22s^22p^63s^23p^5 = [\text{Ne}]3s^23p^5$$

$$\text{Cl}^- \quad 1s^22s^22p^63s^23p^6 = [\text{Ne}]3s^23p^6 = [\text{Ar}]$$

ACTIVITY
Ion Electron Configurations

In order to form a Cl^{2-} ion, the second electron would have to be added to the next higher shell of the Cl atom, which is energetically very unfavorable. Therefore, we never observe Cl^{2-} ions in ionic compounds.

Based on these concepts, we expect that ionic compounds of the representative metals from groups 1A, 2A, and 3A will contain cations with charges of 1+, 2+, and 3+, respectively. Likewise, ionic compounds of the representative nonmetals of groups 5A, 6A, and 7A usually contain anions of charge 3−, 2−, and 1−, respectively. Although we rarely find ionic compounds of the nonmetals from group 4A (C, Si, and Ge), the heaviest elements in group 4A (Sn and Pb) are metals and are usually found as 2+ cations in ionic compounds: Sn^{2+} and Pb^{2+}. This behavior is consistent with the increasing metallic character found as one proceeds down a column in the periodic table. (Section 7.6)

SAMPLE EXERCISE 8.2 | Charges on Ions

Predict the ion generally formed by **(a)** Sr, **(b)** S, **(c)** Al.

Solution

Analyze: We must decide how many electrons are most likely to be gained or lost by atoms of Sr, S, and Al.

Plan: In each case we can use the element's position in the periodic table to predict whether it will form a cation or an anion. We can then use its electron configuration to determine the ion that is likely to be formed.

Solve: **(a)** Strontium is a metal in group 2A and will therefore form a cation. Its electron configuration is $[Kr]5s^2$, and so we expect that the two valence electrons can be lost easily to give an Sr^{2+} ion. **(b)** Sulfur is a nonmetal in group 6A and will thus tend to be found as an anion. Its electron configuration ($[Ne]3s^23p^4$) is two electrons short of a noble-gas configuration. Thus, we expect that sulfur tends to form S^{2-} ions. **(c)** Aluminum is a metal in group 3A. We therefore expect it to form Al^{3+} ions.

Check: The ionic charges we predict here are confirmed in Tables 2.4 and 2.5.

PRACTICE EXERCISE

Predict the charges on the ions formed when magnesium reacts with nitrogen.
Answer: Mg^{2+} and N^{3-}

Transition-Metal Ions

Because ionization energies increase rapidly for each successive electron removed, the lattice energies of ionic compounds are generally large enough to compensate for the loss of up to only three electrons from atoms. Thus, we find cations with charges of 1+, 2+, or 3+ in ionic compounds. Most transition metals, however, have more than three electrons beyond a noble-gas core. Silver, for example, has a $[Kr]4d^{10}5s^1$ electron configuration. Metals of group 1B (Cu, Ag, Au) often occur as 1+ ions (as in CuBr and AgCl). In forming Ag^+, the 5*s* electron is lost, leaving a completely filled 4*d* subshell. As in this example, transition metals generally do not form ions that have a noble-gas configuration. The octet rule, although useful, is clearly limited in scope.

Recall from our discussion in Section 7.4 that when a positive ion is formed from an atom, electrons are always lost first from the subshell having the largest value of *n*. Thus, *in forming ions, transition metals lose the valence-shell* s *electrons first, then as many* d *electrons as are required to reach the charge of the ion*. Let's consider Fe, which has the electron configuration $[Ar]3d^64s^2$. In forming the Fe^{2+} ion, the two 4*s* electrons are lost, leading to an $[Ar]3d^6$ configuration. Removal of an additional electron gives the Fe^{3+} ion, whose electron configuration is $[Ar]3d^5$.

A CLOSER LOOK | Calculation of Lattice Energies: The Born–Haber Cycle

Lattice energy is a useful concept because it relates directly to the stability of an ionic solid. Unfortunately, the lattice energy cannot be determined directly by experiment. It can, however, be calculated by envisioning the formation of an ionic compound as occurring in a series of well-defined steps. We can then use Hess's law (Section 5.6) to put these steps together in a way that gives us the lattice energy for the compound. By so doing, we construct a **Born–Haber cycle**, a thermochemical cycle named after the German scientists Max Born (1882–1970) and Fritz Haber (1868–1934), who introduced it to analyze the factors contributing to the stability of ionic compounds.

In the Born–Haber cycle for NaCl, we consider the formation of NaCl(*s*) from the elements Na(*s*) and $Cl_2(g)$ by two different routes, as shown in Figure 8.4 ▶. The enthalpy change for the direct route (red arrow) is the heat of formation of NaCl(*s*):

$$\text{Na}(s) + \tfrac{1}{2}\text{Cl}_2(g) \longrightarrow \text{NaCl}(s) \qquad \Delta H_f^\circ[\text{NaCl}(s)] = -411 \text{ kJ} \quad [8.5]$$

The indirect route consists of five steps, shown by the green arrows in Figure 8.4. First, we generate gaseous atoms of sodium by vaporizing sodium metal. Then we form gaseous atoms of chlorine by breaking the bonds in the Cl_2 molecules. The enthalpy changes for these processes are available to us as enthalpies of formation (Appendix C):

$$\text{Na}(s) \longrightarrow \text{Na}(g) \qquad \Delta H_f^\circ[\text{Na}(g)] = 108 \text{ kJ} \quad [8.6]$$

$$\tfrac{1}{2}\text{Cl}_2(g) \longrightarrow \text{Cl}(g) \qquad \Delta H_f^\circ[\text{Cl}(g)] = 122 \text{ kJ} \quad [8.7]$$

Both of these processes are endothermic; energy is required to generate gaseous sodium and chlorine atoms.

In the next two steps we remove the electron from Na(*g*) to form $Na^+(g)$ and then add the electron to Cl(*g*) to form $Cl^-(g)$. The enthalpy changes for these processes equal the first ionization energy of Na, I_1(Na), and the electron affinity of Cl, denoted *E*(Cl), respectively: (Sections 7.4, 7.5)

$$\text{Na}(g) \longrightarrow \text{Na}^+(g) + e^- \qquad \Delta H = I_1(\text{Na}) = 496 \text{ kJ} \quad [8.8]$$

$$\text{Cl}(g) + e^- \longrightarrow \text{Cl}^-(g) \qquad \Delta H = E(\text{Cl}) = -349 \text{ kJ} \quad [8.9]$$

Finally, we combine the gaseous sodium and chloride ions to form solid sodium chloride. Because this process is just the reverse of the lattice energy (breaking a solid into gaseous ions), the enthalpy change is the negative of the lattice energy, the quantity that we want to determine:

$$\text{Na}^+(g) + \text{Cl}^-(g) \longrightarrow \text{NaCl}(s) \qquad \Delta H = -\Delta H_{\text{lattice}} = ? \quad [8.10]$$

The sum of the five steps in the indirect path gives us NaCl(*s*) from Na(*s*) and $\frac{1}{2}Cl_2(g)$. Thus, from Hess's law we know that the sum of the enthalpy changes for these five steps equals that for the direct path, indicated by the red arrow, Equation 8.5:

$$\Delta H_f^\circ[\text{NaCl(s)}] = \Delta H_f^\circ[\text{Na}(g)] + \Delta H_f^\circ[\text{Cl}(g)] + I_1(\text{Na}) + E(\text{Cl}) - \Delta H_{\text{lattice}}$$

$$-411 \text{ kJ} = 108 \text{ kJ} + 122 \text{ kJ} + 496 \text{ kJ} - 349 \text{ kJ} - \Delta H_{\text{lattice}}$$

Solving for $\Delta H_{\text{lattice}}$:

$$\Delta H_{\text{lattice}} = 108 \text{ kJ} + 122 \text{ kJ} + 496 \text{ kJ} - 349 \text{ kJ} + 411 \text{ kJ} = 788 \text{ kJ}$$

Thus, the lattice energy of NaCl is 788 kJ/mol.
Related Exercises: 8.26, 8.27, and 8.28

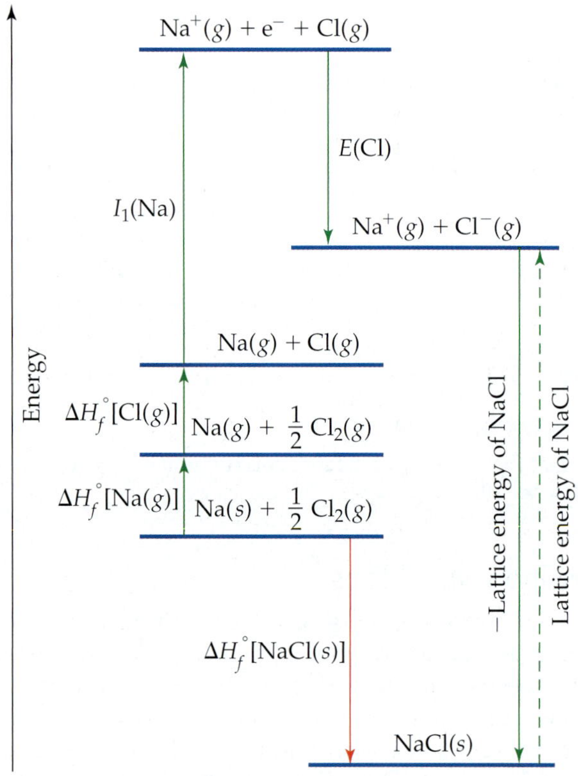

▲ **Figure 8.4 The Born-Haber cycle.** This representation shows the energetic relationships in the formation of ionic solids from the elements. By Hess's law, the enthalpy of formation of NaCl(*s*) from elemental sodium and chlorine (Equation 8.5) is equal to the sum of the energies of several individual steps (Equations 8.6 through 8.10).

GIVE IT SOME THOUGHT

Which element forms a 1+ ion that has the electron configuration [Kr]$4d^8$?

Polyatomic Ions

Let's now briefly reconsider Tables 2.4 and 2.5, which list the common ions. (Section 2.8) Several cations and many common anions are polyatomic. Examples include the ammonium ion, NH_4^+, and the carbonate ion, CO_3^{2-}. In polyatomic ions, two or more atoms are bound together by predominantly covalent bonds. They form a stable grouping that carries a charge, either positive or negative. We will examine the covalent bonding forces in these ions in Chapter 9. For now, the only thing you need to understand about any polyatomic ion is that the group of atoms as a whole acts as a charged species when the ion forms an ionic compound with an ion of opposite charge.

8.3 | Covalent Bonding

Ionic substances possess several characteristic properties. They are usually brittle substances with high melting points. They are usually crystalline, meaning that the solids have flat surfaces that make characteristic angles with one another. Ionic crystals often can be cleaved; that is, they break apart along smooth, flat surfaces. These characteristics result from electrostatic forces that maintain the ions in a rigid, well-defined, three-dimensional arrangement such as that shown in Figure 8.3.

The vast majority of chemical substances do not have the characteristics of ionic materials. Most of the substances with which we come into daily contact—such as water—tend to be gases, liquids, or solids with low melting points. Many, such as gasoline, vaporize readily. Many are pliable in their solid forms—for example, plastic bags and paraffin.

For the very large class of substances that do not behave like ionic substances, we need a different model for the bonding between atoms. G. N. Lewis reasoned that atoms might acquire a noble-gas electron configuration by sharing electrons with other atoms. As we noted in Section 8.1, a chemical bond formed by sharing a pair of electrons is called a *covalent bond*.

The hydrogen molecule, H_2, provides the simplest example of a covalent bond. When two hydrogen atoms are close to each other, electrostatic interactions occur between them. The two positively charged nuclei repel each other, the two negatively charged electrons repel each other, and the nuclei and electrons attract each other, as shown in Figure 8.5 ▶. Because the H_2 molecule exists as a stable entity, the attractive forces must exceed the repulsive ones. Why is this so?

By using quantum mechanical methods analogous to those employed for atoms (Section 6.5), it is possible to calculate the distribution of electron density in molecules. Such a calculation for H_2 shows that the attractions between the nuclei and the electrons cause electron density to concentrate between the nuclei, as shown in Figure 8.5(b). As a result, the overall electrostatic interactions are attractive. Thus, the atoms in H_2 are held together principally because the two nuclei are electrostatically attracted to the concentration of negative charge between them. In essence, the shared pair of electrons in any covalent bond acts as a kind of "glue" to bind atoms together.

ANIMATION
H_2 Bond Formation

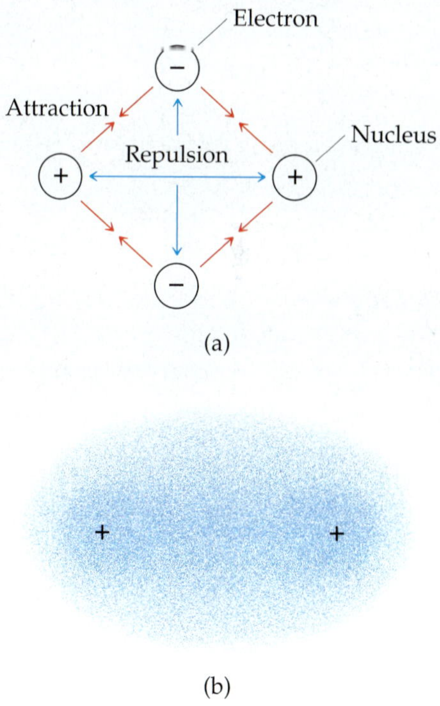

▲ **Figure 8.5 The covalent bond in H_2.** (a) The attractions and repulsions among electrons and nuclei in the hydrogen molecule. (b) Electron distribution in the H_2 molecule. The concentration of electron density between the nuclei leads to a net attractive force that constitutes the covalent bond holding the molecule together.

GIVE IT SOME THOUGHT

Because it is less stable than two separated He atoms, the He_2 molecule does not exist. What are the attractive forces in He_2? What are the repulsive forces in He_2? Which are greater, the attractive or the repulsive forces?

Lewis Structures

The formation of covalent bonds can be represented using Lewis symbols for the constituent atoms. The formation of the H_2 molecule from two H atoms, for example, can be represented as

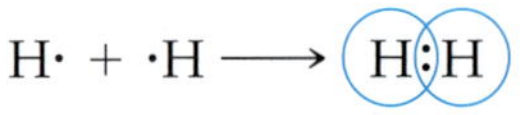

In this way, each hydrogen atom acquires a second electron, achieving the stable, two-electron, noble-gas electron configuration of helium.

The formation of a bond between two Cl atoms to give a Cl_2 molecule can be represented in a similar way:

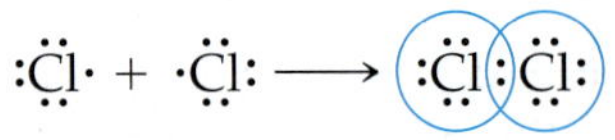

By sharing the bonding electron pair, each chlorine atom has eight electrons (an octet) in its valence shell. It thus achieves the noble-gas electron configuration of argon.

The structures shown here for H_2 and Cl_2 are called **Lewis structures** (or Lewis electron-dot structures). In writing Lewis structures, we usually show each electron pair shared between atoms as a line and the unshared electron pairs as dots. Written this way, the Lewis structures for H_2 and Cl_2 are

For the nonmetals, the number of valence electrons in a neutral atom is the same as the group number. Therefore, one might predict that 7A elements, such as F, would form one covalent bond to achieve an octet; 6A elements, such as O, would form two covalent bonds; 5A elements, such as N, would form three covalent bonds; and 4A elements, such as C, would form four covalent bonds. These predictions are borne out in many compounds. For example, consider the simple hydrogen compounds of the nonmetals of the second row of the periodic table:

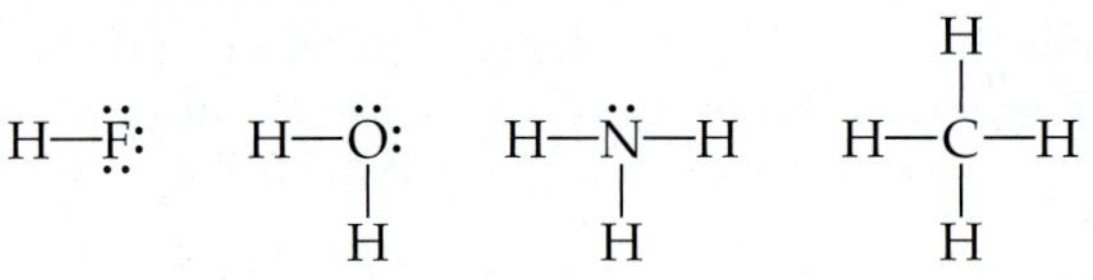

Thus, the Lewis model succeeds in accounting for the compositions of many compounds of nonmetals, in which covalent bonding predominates.

SAMPLE EXERCISE 8.3 | Lewis Structure of a Compound

Given the Lewis symbols for the elements nitrogen and fluorine shown in Table 8.1, predict the formula of the stable binary compound (a compound composed of two elements) formed when nitrogen reacts with fluorine, and draw its Lewis structure.

Solution

Analyze: The Lewis symbols for nitrogen and fluorine reveal that nitrogen has five valence electrons and fluorine has seven.

Plan: We need to find a combination of the two elements that results in an octet of electrons around each atom in the compound. Nitrogen requires three additional electrons to complete its octet, whereas fluorine requires but one. Sharing a pair of electrons between one N atom and one F atom will result in an octet of electrons for fluorine but not for nitrogen. We therefore need to figure out a way to get two more electrons for the N atom.

Solve: Nitrogen must share a pair of electrons with three fluorine atoms to complete its octet. Thus, the Lewis structure for the resulting compound, NF_3, is

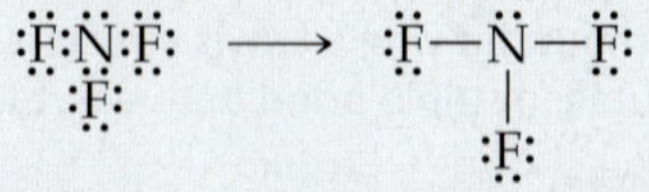

Check: The Lewis structure on the left shows that each atom is surrounded by an octet of electrons. Once you are accustomed to thinking of each line in a Lewis structure as representing *two* electrons, you can just as easily use the structure on the right to check for octets.

PRACTICE EXERCISE

Compare the Lewis symbol for neon with the Lewis structure for methane, CH_4. In what important way are the electron arrangements about neon and carbon alike? In what important respect are they different?

Answer: Both atoms have an octet of electrons about them. However, the electrons about neon are unshared electron pairs, whereas those about carbon are shared with four hydrogen atoms.

Multiple Bonds

The sharing of a pair of electrons constitutes a single covalent bond, generally referred to simply as a **single bond**. In many molecules, atoms attain complete octets by sharing more than one pair of electrons. When two electron pairs are shared, two lines are drawn, representing a **double bond**. In carbon dioxide, for example, bonding occurs between carbon, with four valence electrons, and oxygen, with six:

$$:\ddot{O}: + \cdot\dot{C}\cdot + :\ddot{O}: \longrightarrow \ddot{O}::C::\ddot{O} \quad (\text{or } \ddot{O}{=}C{=}\ddot{O})$$

As the diagram shows, each oxygen acquires an octet of electrons by sharing two electron pairs with carbon. Carbon, on the other hand, acquires an octet of electrons by sharing two pairs with two oxygen atoms.

A **triple bond** corresponds to the sharing of three pairs of electrons, such as in the N_2 molecule:

$$:\dot{N}\cdot + \cdot\dot{N}: \longrightarrow :N:::N: \quad (\text{or } :N{\equiv}N:)$$

Because each nitrogen atom possesses five electrons in its valence shell, three electron pairs must be shared to achieve the octet configuration.

The properties of N_2 are in complete accord with its Lewis structure. Nitrogen is a diatomic gas with exceptionally low reactivity that results from the very stable nitrogen–nitrogen bond. Study of the structure of N_2 reveals that the nitrogen atoms are separated by only 1.10 Å. The short N—N bond distance is a result of the triple bond between the atoms. From structure studies of many different substances in which nitrogen atoms share one or two electron pairs, we have learned that the average distance between bonded nitrogen atoms varies with the number of shared electron pairs:

N—N	N=N	N≡N
1.47 Å	1.24 Å	1.10 Å

As a general rule, the distance between bonded atoms decreases as the number of shared electron pairs increases. The distance between the nuclei of the atoms involved in a bond is called the **bond length** for the bond. We first encountered bond lengths in Section 7.3 in our discussion of atomic radii, and we will discuss them further in Section 8.8.

GIVE IT SOME THOUGHT

The C—O bond length in carbon monoxide, CO, is 1.13 Å, whereas the C—O bond length in CO_2 is 1.24 Å. Without drawing a Lewis structure, do you think that carbon monoxide has a single, double, or triple bond between the C and O atoms?

8.4 | Bond Polarity and Electronegativity

When two identical atoms bond, as in Cl_2 or N_2, the electron pairs must be shared equally. In ionic compounds, on the other hand, such as NaCl, there is essentially no sharing of electrons, which means that NaCl is best described as composed of Na^+ and Cl^- ions. The 3*s* electron of the Na atom is, in effect, transferred completely to chlorine. The bonds occurring in most covalent substances fall somewhere between these extremes.

ANIMATION
Periodic Trends: Electronegativity

The concept of **bond polarity** helps describe the sharing of electrons between atoms. A **nonpolar covalent bond** is one in which the electrons are shared equally between two atoms, as in the Cl_2 and N_2 examples we just cited. In a **polar covalent bond**, one of the atoms exerts a greater attraction for the bonding electrons than the other. If the difference in relative ability to attract electrons is large enough, an ionic bond is formed.

Electronegativity

We use a quantity called electronegativity to estimate whether a given bond will be nonpolar covalent, polar covalent, or ionic. **Electronegativity** is defined as the ability of an atom *in a molecule* to attract electrons to itself. The greater an atom's electronegativity, the greater is its ability to attract electrons to itself. The electronegativity of an atom in a molecule is related to its ionization energy and electron affinity, which are properties of isolated atoms. The *ionization energy* measures how strongly an atom holds on to its electrons. (Section 7.4) Likewise, the *electron affinity* is a measure of how strongly an atom attracts additional electrons. (Section 7.5) An atom with a very negative electron affinity and high ionization energy will both attract electrons from other atoms and resist having its electrons attracted away; it will be highly electronegative. Numerical estimates of electronegativity can be based on a variety of properties, not just ionization energy and electron affinity.

The first and most widely used electronegativity scale was developed by the American chemist Linus Pauling (1901–1994), who based his scale on thermochemical data. Figure 8.6 ◀ shows Pauling's electronegativity values for many of the elements. The values are unitless. Fluorine, the most electronegative element, has an electronegativity of 4.0. The least electronegative element, cesium, has an electronegativity of 0.7. The values for all other elements lie between these two extremes.

▼ **Figure 8.6 Electronegativities of the elements.** Electronegativity generally increases from left to right across a period and decreases from top to bottom down a group.

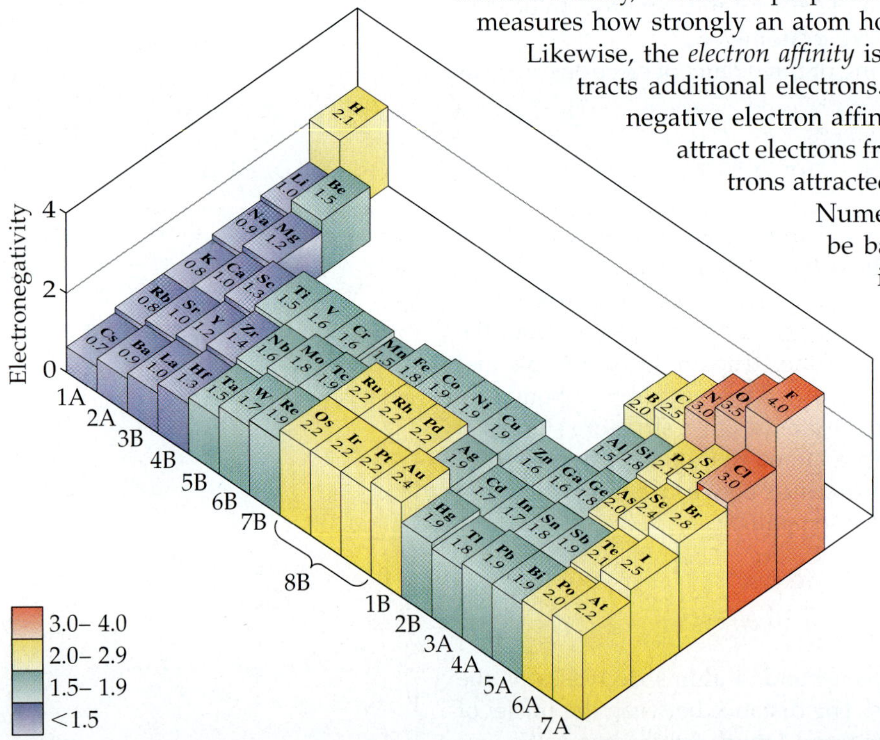

Within each period there is generally a steady increase in electronegativity from left to right; that is, from the most metallic to the most nonmetallic elements. With some exceptions (especially within the transition metals), electronegativity decreases with increasing atomic number in any one group. This is what we might expect because we know that ionization energies tend to decrease with increasing atomic number in a group and electron affinities don't change very much. You do not need to memorize numerical values for electronegativity. Instead, you should know the periodic trends so that you can predict which of two elements is more electronegative.

GIVE IT SOME THOUGHT

How does the *electronegativity* of an element differ from its *electron affinity*?

Electronegativity and Bond Polarity

We can use the difference in electronegativity between two atoms to gauge the polarity of the bonding between them. Consider these three fluorine-containing compounds:

Compound	F_2	HF	LiF
Electronegativity difference	$4.0 - 4.0 = 0$	$4.0 - 2.1 = 1.9$	$4.0 - 1.0 = 3.0$
Type of bond	Nonpolar covalent	Polar covalent	Ionic

In F_2 the electrons are shared equally between the fluorine atoms, and thus the covalent bond is *nonpolar*. In general, a nonpolar covalent bond results when the electronegativities of the bonded atoms are equal.

In HF the fluorine atom has a greater electronegativity than the hydrogen atom, with the result that the sharing of electrons is unequal—the bond is polar. In general, a polar covalent bond results when the atoms differ in electronegativity. In HF the more electronegative fluorine atom attracts electron density away from the less electronegative hydrogen atom, leaving a partial positive charge on the hydrogen atom and a partial negative charge on the fluorine atom. We can represent this charge distribution as

ACTIVITY
Electronegativity Differences and Bond Types

$$\overset{\delta+}{\text{H}}-\overset{\delta-}{\text{F}}$$

The $\delta+$ and $\delta-$ (read "delta plus" and "delta minus") symbolize the partial positive and negative charges, respectively.

In LiF the electronegativity difference is very large, meaning that the electron density is shifted far toward F. In the three-dimensional structure of LiF, analogous to that shown for NaCl in Figure 8.3, the transfer of electronic charge from Li to F is essentially complete. The resultant bond is therefore *ionic*. This shift of electron density toward the more electronegative atom can be seen in the results of calculations of electron density distributions. For the three species in our example, the calculated electron density distributions are shown in Figure 8.7 ▶. The regions of space that have relatively higher electron density are shown in red, and those with a relatively lower electron density are shown in blue. You can see that in F_2 the distribution is symmetrical, in HF it is clearly shifted

F_2 HF LiF

▼ **Figure 8.7 Electron density distribution.** This computer-generated rendering shows the calculated electron-density distribution on the surface of the F_2, HF, and LiF molecules. The regions of relatively low electron density (net positive charge) appear blue, those of relatively high electron density (net negative charge) appear red, and regions that are close to electrically neutral appear green.

toward fluorine, and in LiF the shift is even greater.* These examples illustrate, therefore, that *the greater the difference in electronegativity between two atoms, the more polar their bond*. The nonpolar covalent bond lies at one end of a continuum of bond types, and the ionic bond lies at the other end. In between is a broad range of polar covalent bonds, differing in the extent to which there is unequal sharing of electrons.

GIVE IT SOME THOUGHT

The difference in the electronegativity of two elements is 0.7. Would you expect a bond between these elements to be nonpolar, polar covalent, or ionic?

SAMPLE EXERCISE 8.4 | Bond Polarity

In each case, which bond is more polar: **(a)** B—Cl or C—Cl, **(b)** P—F or P—Cl? Indicate in each case which atom has the partial negative charge.

Solution

Analyze: We are asked to determine relative bond polarities, given nothing but the atoms involved in the bonds.

Plan: We need to know electronegativity values for all the atoms involved, which we can get from Figure 8.6. Alternatively, because we are not asked for quantitative answers, we can use the periodic table and our knowledge of electron-affinity trends to answer the question.

Solve: **(a)** Using Figure 8.6: The difference in the electronegativities of chlorine and boron is $3.0 - 2.0 = 1.0$; the difference between chlorine and carbon is $3.0 - 2.5 = 0.5$. Consequently, the B—Cl bond is more polar; the chlorine atom carries the partial negative charge because it has a higher electronegativity.

Using the periodic table: Because boron is to the left of carbon in the periodic table, we predict that boron has the lower electronegativity. Chlorine, being on the right side of the table, has a higher electronegativity. The more polar bond will be the one between the atoms having the lowest electronegativity (boron) and the highest electronegativity (chlorine).

(b) The electronegativities are P = 2.1, F = 4.0, Cl = 3.0. Consequently, the P—F bond will be more polar than the P—Cl bond. You should compare the electronegativity differences for the two bonds to verify this prediction. The fluorine atom carries the partial negative charge. We reach the same conclusion by noting that fluorine is above chlorine in the periodic table, and so fluorine should be more electronegative and will form the more polar bond with P.

PRACTICE EXERCISE

Which of the following bonds is most polar: S—Cl, S—Br, Se—Cl, Se—Br?
Answer: Se—Cl

Dipole Moments

The difference in electronegativity between H and F leads to a polar covalent bond in the HF molecule. As a consequence, there is a concentration of negative charge on the more electronegative F atom, leaving the less electronegative H atom at the positive end of the molecule. A molecule such as HF, in which the centers of positive and negative charge do not coincide, is said to be a **polar molecule**. Thus, we describe not only bonds as being polar and nonpolar but also entire molecules.

ACTIVITY Molecular Polarity

We can indicate the polarity of the HF molecule in two ways:

$$\overset{\delta+}{\mathrm{H}}-\overset{\delta-}{\mathrm{F}} \quad \text{or} \quad \overset{\longleftrightarrow}{\mathrm{H}-\mathrm{F}}$$

* The calculated electron density distribution for LiF is for an isolated LiF "molecule," not the ionic solid. While the bond in this isolated diatomic system is very polar, it is not 100% ionic, as is the bonding in solid lithium fluoride. The solid state promotes a more complete electron transfer from Li to F because of the stabilizing effects of the ionic lattice.

Recall from the preceding subsection that "δ+" and "δ−" indicate the partial positive and negative charges on the H and F atoms. In the notation on the right, the arrow denotes the shift in electron density toward the fluorine atom. The crossed end of the arrow can be thought of as a plus sign that designates the positive end of the molecule.

Polarity helps determine many of the properties of substances that we observe at the macroscopic level, in the laboratory and in everyday life. Polar molecules align themselves with respect to one another, with the negative end of one molecule and the positive end of another attracting each other. Polar molecules are likewise attracted to ions. The negative end of a polar molecule is attracted to a positive ion, and the positive end is attracted to a negative ion. These interactions account for many properties of liquids, solids, and solutions, as you will see in Chapters 11, 12, and 13.

How can we quantify the polarity of a molecule? Whenever two electrical charges of equal magnitude but opposite sign are separated by a distance, a **dipole** is established. The quantitative measure of the magnitude of a dipole is called its **dipole moment**, denoted μ. If two equal and opposite charges $Q+$ and $Q-$ are separated by a distance r, the magnitude of the dipole moment is the product of Q and r (Figure 8.8 ▶):

$$\mu = Qr \qquad [8.11]$$

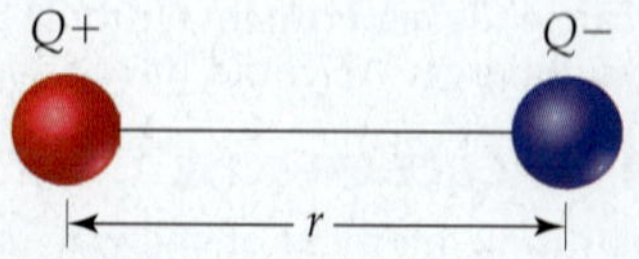

▲ **Figure 8.8 Dipole and dipole moment.** When charges of equal magnitude and opposite sign $Q+$ and $Q-$ are separated by a distance r, a dipole is produced. The size of the dipole is given by the dipole moment, μ, which is the product of the charge separated and the distance of separation between the charge centers: $\mu = Qr$.

The dipole moment increases as the magnitude of charge that is separated increases and as the distance between the charges increases. For a nonpolar molecule, such as F_2, the dipole moment is zero because there is no charge separation.

GIVE IT SOME THOUGHT

The molecules chlorine monofluoride, ClF, and iodine monofluoride, IF, are examples of *interhalogen* compounds—compounds that contain bonds between different halogen elements. Which of these molecules will have the larger dipole moment?

Dipole moments are usually reported in *debyes* (D), a unit that equals 3.34×10^{-30} coulomb-meters (C-m). For molecules, we usually measure charge in units of the electronic charge e, 1.60×10^{-19} C, and distance in units of angstroms. Suppose that two charges 1+ and 1− (in units of e) are separated by a distance of 1.00 Å. The dipole moment produced is

$$\mu = Qr = (1.60 \times 10^{-19}\ \mathrm{C})(1.00\ \text{Å})\left(\frac{10^{-10}\ \mathrm{m}}{1\ \text{Å}}\right)\left(\frac{1\ \mathrm{D}}{3.34 \times 10^{-30}\ \mathrm{C\text{-}m}}\right) = 4.79\ \mathrm{D}$$

Measurement of the dipole moments can provide us with valuable information about the charge distributions in molecules, as illustrated in Sample Exercise 8.5.

SAMPLE EXERCISE 8.5 | Dipole Moments of Diatomic Molecules

The bond length in the HCl molecule is 1.27 Å. **(a)** Calculate the dipole moment, in debyes, that would result if the charges on the H and Cl atoms were 1+ and 1−, respectively. **(b)** The experimentally measured dipole moment of HCl(*g*) is 1.08 D. What magnitude of charge, in units of *e*, on the H and Cl atoms would lead to this dipole moment?

Solution

Analyze and Plan: We are asked in (a) to calculate the dipole moment of HCl that would result if there was a full charge transferred from H to Cl. We can use Equation 8.11 to obtain this result. In (b), we are given the actual dipole moment for the molecule and will use that value to calculate the actual partial charges on the H and Cl atoms.

Solve: **(a)** The charge on each atom is the electronic charge, $e = 1.60 \times 10^{-19}$ C. The separation is 1.27 Å. The dipole moment is therefore

$$\mu = Qr = (1.60 \times 10^{-19}\ \mathrm{C})(1.27\ \text{Å})\left(\frac{10^{-10}\ \mathrm{m}}{1\ \text{Å}}\right)\left(\frac{1\ \mathrm{D}}{3.34 \times 10^{-30}\ \mathrm{C\text{-}m}}\right) = 6.08\ \mathrm{D}$$

(b) We know the value of μ, 1.08 D, and the value of r, 1.27 Å, and we want to calculate the value of Q:

$$Q = \frac{\mu}{r} = \frac{(1.08\ \text{D})\left(\dfrac{3.34 \times 10^{-30}\ \text{C-m}}{1\ \text{D}}\right)}{(1.27\ \text{Å})\left(\dfrac{10^{-10}\ \text{m}}{1\ \text{Å}}\right)} = 2.84 \times 10^{-20}\ \text{C}$$

We can readily convert this charge to units of e:

$$\text{Charge in } e = (2.84 \times 10^{-20}\ \text{C})\left(\frac{1\ e}{1.60 \times 10^{-19}\ \text{C}}\right) = 0.178e$$

Thus, the experimental dipole moment indicates that the charge separation in the HCl molecule is

$$\overset{0.178+}{\text{H}} - \overset{0.178-}{\text{Cl}}$$

Because the experimental dipole moment is less than that calculated in part (a), the charges on the atoms are less than a full electronic charge. We could have anticipated this because the H—Cl bond is polar covalent rather than ionic.

PRACTICE EXERCISE

The dipole moment of chlorine monofluoride, ClF(g), is 0.88 D. The bond length of the molecule is 1.63 Å. **(a)** Which atom is expected to have the partial negative charge? **(b)** What is the charge on that atom, in units of e?
Answers: **(a)** F, **(b)** 0.11−

TABLE 8.3 Bond Lengths, Electronegativity Differences, and Dipole Moments of the Hydrogen Halides

Compound	Bond Length (Å)	Electronegativity Difference	Dipole Moment (D)
HF	0.92	1.9	1.82
HCl	1.27	0.9	1.08
HBr	1.41	0.7	0.82
HI	1.61	0.4	0.44

Table 8.3 ◀ presents the bond lengths and dipole moments of the hydrogen halides. Notice that as we proceed from HF to HI, the electronegativity difference decreases and the bond length increases. The first effect decreases the amount of charge separated and causes the dipole moment to decrease from HF to HI, even though the bond length is increasing. We can "observe" the varying degree of electronic charge shift in these substances from computer-generated renderings based on calculations of electron distribution, as shown in Figure 8.9 ▼. For these molecules, the change in the electronegativity difference has a greater effect on the dipole moment than does the change in bond length.

GIVE IT SOME THOUGHT

How do you interpret the fact that there is no red in the HBr and HI representations in Figure 8.9?

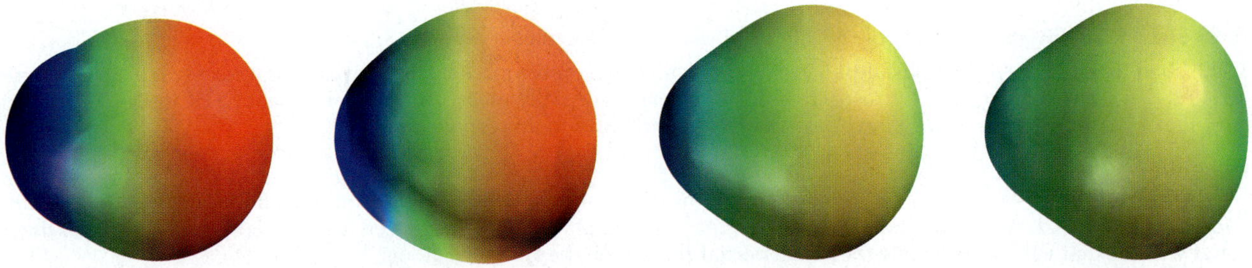

HF HCl HBr HI

▲ **Figure 8.9 Charge separation in the hydrogen halides.** Blue represents regions of lowest electron density, red regions of highest electron density. In HF the strongly electronegative F pulls much of the electron density away from H. In HI the I, being much less electronegative than F, does not attract the shared electrons as strongly, and consequently there is far less polarization of the bond.

Bond Types and Nomenclature

This is a good point for a brief interlude about nomenclature. We saw in Section 2.8 that there are two general approaches to naming binary compounds: one used for ionic compounds and the other for molecular ones. In both approaches, the name of the less electronegative element is given first. The name of the more electronegative element then follows, modified to have an *-ide* ending. Compounds that are ionic are given names based on their component ions, including the charge of the cation if that charge is variable. Those that are molecular are named using the prefixes listed in Table 2.6 to indicate the number of atoms of each kind in the substance (except when there is only one atom, in which case the prefix *mono-* is frequently not used):

Ionic		*Molecular*	
MgH_2	Magnesium hydride	H_2S	Dihydrogen sulfide
FeF_2	Iron(II) fluoride	OF_2	Oxygen difluoride
Mn_2O_3	Manganese(III) oxide	Cl_2O_3	Dichlorine trioxide

The dividing line between the two approaches, however, is not always clear, and both approaches are often applied to the same substances. The compound TiO_2, for example, which is a commercially important white paint pigment, is sometimes referred to as titanium(IV) oxide but is more commonly called titanium dioxide. The roman numeral in the first name is the oxidation number of the titanium. ∞ (Section 4.4)

One reason for the overlap in the two approaches to nomenclature is that many compounds of metals with higher oxidation numbers have properties more similar to molecular compounds (which contain covalent bonds) than to ionic compounds (which contain ionic bonds). Historically, these metal compounds have been named using both conventions even though naming them as molecular compounds is more correct based on their properties. For example, ionic compounds are solids with very high melting points, but $SnCl_4$, which is called either tin tetrachloride or tin(IV) chloride, is a colorless liquid that freezes at −33°C and boils at 114°C. As another example, the compound Mn_2O_7, which is called either dimanganese heptoxide or manganese(VII) oxide, is a green liquid that freezes at 5.9°C. In short, whenever you see the formula of a compound containing a metal in a high oxidation state (usually above +3), you should not be surprised if it exhibits the general properties of molecular, rather than ionic, compounds.

GIVE IT SOME THOUGHT

The compounds MoO_3 and OsO_4 are more properly named molybdenum(VI) oxide and osmium tetroxide, respectively. Which of these compounds do you think has the higher melting point?

8.5 | Drawing Lewis Structures

Lewis structures can help us understand the bonding in many compounds and are frequently used when discussing the properties of molecules. For this reason, drawing Lewis structures is an important skill that you should practice. To do so, you should follow a regular procedure. First we'll outline the procedure, and then we'll go through several examples.

1. *Sum the valence electrons from all atoms.* (Use the periodic table as necessary to help you determine the number of valence electrons in each atom.) For an anion, add one electron to the total for each negative charge. For a cation, subtract one electron from the total for each positive charge. Don't worry about keeping track of which electrons come from which atoms. Only the total number is important.

ACTIVITIES
Writing Lewis Structures I, Writing Lewis Structures II

2. *Write the symbols for the atoms to show which atoms are attached to which, and connect them with a single bond* (a dash, representing *two* electrons). Chemical formulas are often written in the order in which the atoms are connected in the molecule or ion; the formula HCN, for example, tells you that the carbon atom is bonded to the H and to the N. When a central atom has a group of other atoms bonded to it, the central atom is usually written first, as in CO_3^{2-} and SF_4. It also helps to remember that the central atom is generally less electronegative than the atoms surrounding it. In other cases you may need more information before you can draw the Lewis structure.
3. *Complete the octets around all the atoms bonded to the central atom.* (Remember, however, that you use only a single pair of electrons around hydrogen.)
4. *Place any leftover electrons on the central atom,* even if doing so results in more than an octet of electrons around the atom. In Section 8.7 we will discuss molecules that don't adhere to the octet rule.
5. *If there are not enough electrons to give the central atom an octet, try multiple bonds.* Use one or more of the unshared pairs of electrons on the atoms bonded to the central atom to form double or triple bonds.

SAMPLE EXERCISE 8.6 | Drawing Lewis Structures

Draw the Lewis structure for phosphorus trichloride, PCl_3.

Solution

Analyze and Plan We are asked to draw a Lewis structure from a molecular formula. Our plan is to follow the five-step procedure just described.

Solve: First, we sum the valence electrons. Phosphorus (group 5A) has five valence electrons, and each chlorine (group 7A) has seven. The total number of valence electrons is therefore

$$5 + (3 \times 7) = 26$$

Second, we arrange the atoms to show which atom is connected to which, and we draw a single bond between them. There are various ways the atoms might be arranged. In binary (two-element) compounds, however, the first element listed in the chemical formula is generally surrounded by the remaining atoms. Thus, we begin with a skeleton structure that shows a single bond between the phosphorus atom and each chlorine atom:

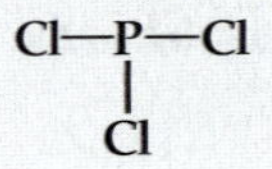

(It is not crucial to place the atoms in exactly this arrangement.)

Third, we complete the octets on the atoms bonded to the central atom. Placing octets around each Cl atom accounts for 24 electrons (remember, each line in our structure represents *two* electrons):

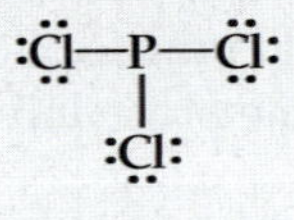

Fourth, we place the remaining two electrons on the central atom, completing the octet around it:

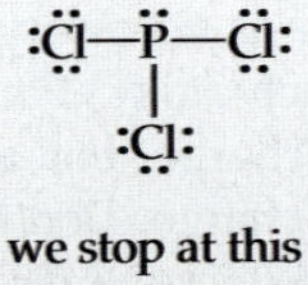

This structure gives each atom an octet, so we stop at this point. (Remember that in achieving an octet, the bonding electrons are counted for both atoms.)

PRACTICE EXERCISE

(a) How many valence electrons should appear in the Lewis structure for CH_2Cl_2?
(b) Draw the Lewis structure.

Answers: **(a)** 20, **(b)** 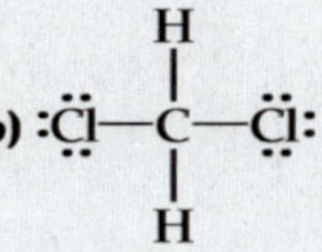

SAMPLE EXERCISE 8.7 | Lewis Structures with Multiple Bonds

Draw the Lewis structure for HCN.

Solution Hydrogen has one valence electron, carbon (group 4A) has four, and nitrogen (group 5A) has five. The total number of valence electrons is therefore $1 + 4 + 5 = 10$. In principle, there are different ways in which we might choose to arrange the atoms. Because hydrogen can accommodate only one electron pair, it always has only one single bond associated with it in any compound. Therefore, C—H—N is an impossible arrangement. The remaining two possibilities are H—C—N and H—N—C. The first is the arrangement found experimentally. You might have guessed this to be the atomic arrangement because the formula is written with the atoms in this order. Thus, we begin with a skeleton structure that shows single bonds between hydrogen, carbon, and nitrogen:

H—C—N

These two bonds account for four electrons. If we then place the remaining six electrons around N to give it an octet, we do not achieve an octet on C:

H—C—N̈:

We therefore try a double bond between C and N, using one of the unshared pairs of electrons we placed on N. Again, there are fewer than eight electrons on C, and so we next try a triple bond. This structure gives an octet around both C and N:

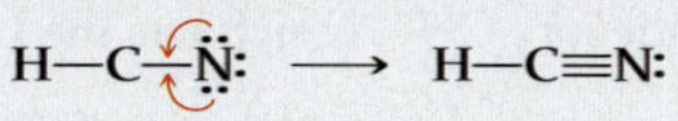

We see that the octet rule is satisfied for the C and N atoms, and the H atom has two electrons around it, so this appears to be a correct Lewis structure.

PRACTICE EXERCISE

Draw the Lewis structure for **(a)** NO^+ ion, **(b)** C_2H_4.

Answers: **(a)** $[:N{\equiv}O:]^+$, **(b)** 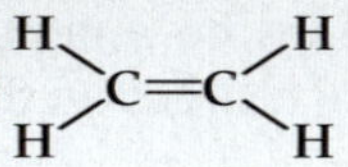

SAMPLE EXERCISE 8.8 | Lewis Structure for a Polyatomic Ion

Draw the Lewis structure for the BrO_3^- ion.

Solution Bromine (group 7A) has seven valence electrons, and oxygen (group 6A) has six. We must now add one more electron to our sum to account for the 1− charge of the ion. The total number of valence electrons is therefore $7 + (3 \times 6) + 1 = 26$. For oxyanions—$BrO_3^-$, SO_4^{2-}, NO_3^-, CO_3^{2-}, and so forth—the oxygen atoms surround the central nonmetal atoms. After following this format and then putting in the single bonds and distributing the unshared electron pairs, we have

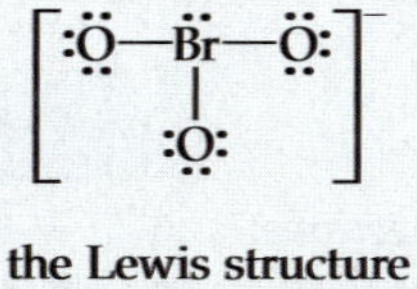

Notice here and elsewhere that the Lewis structure for an ion is written in brackets with the charge shown outside the brackets at the upper right.

PRACTICE EXERCISE

Draw the Lewis structure for **(a)** ClO_2^- ion, **(b)** PO_4^{3-} ion.

Answers: **(a)** 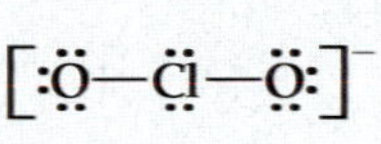 **(b)** 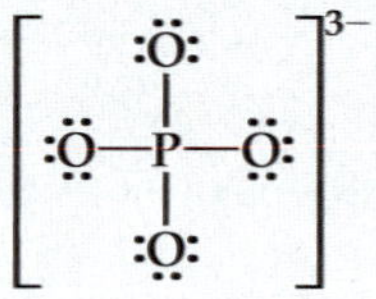

ANIMATION
Formal Charges

Formal Charge

When we draw a Lewis structure, we are describing how the electrons are distributed in a molecule (or polyatomic ion). In some instances we can draw several different Lewis structures that all obey the octet rule. How do we decide which one is the most reasonable? One approach is to do some "bookkeeping" of the valence electrons to determine the formal charge of each atom in each Lewis structure. The **formal charge** of any atom in a molecule is the charge the atom would have if all the atoms in the molecule had the same electronegativity (that is, if each bonding electron pair in the molecule were shared equally between its two atoms).

To calculate the formal charge on any atom in a Lewis structure, we assign the electrons to the atom as follows:

1. *All* unshared (nonbonding) electrons are assigned to the atom on which they are found.
2. For any bond—single, double, or triple—*half* of the bonding electrons are assigned to each atom in the bond.

The formal charge of each atom is then calculated *by subtracting the number of electrons assigned to the atom from the number of valence electrons in the isolated atom.*

Let's illustrate this procedure by calculating the formal charges on the C and N atoms in the cyanide ion, CN^-, which has the Lewis structure

$$[:C\equiv N:]^-$$

For the C atom, there are 2 nonbonding electrons and 3 electrons from the 6 in the triple bond $\left(\frac{1}{2} \times 6 = 3\right)$, for a total of 5. The number of valence electrons on a neutral C atom is 4. Thus, the formal charge on C is $4 - 5 = -1$. For N, there are 2 nonbonding electrons and 3 electrons from the triple bond. Because the number of valence electrons on a neutral N atom is 5, its formal charge is $5 - 5 = 0$. Thus, the formal charges on the atoms in the Lewis structure of CN^- are

$$[:\overset{-1}{C}\equiv \overset{0}{N}:]^-$$

Notice that the sum of the formal charges equals the overall charge on the ion, 1−. In general, the formal charges on a neutral molecule add to zero, whereas those on an ion add to give the overall charge on the ion.

The concept of formal charge can help us choose between alternative Lewis structures. We will see how this is done by considering the CO_2 molecule. As shown in Section 8.3, CO_2 is represented as having two double bonds. The octet rule is also obeyed, however, in a Lewis structure having one single bond and one triple bond. Calculating the formal charge for each atom in these structures, we have

	Ö	=C=	Ö	:Ö	–C≡	O:
Valence electrons:	6	4	6	6	4	6
−(Electrons assigned to atom):	6	4	6	7	4	5
Formal charge:	0	0	0	−1	0	+1

Note that in both cases the formal charges add up to zero, as they must because CO_2 is a neutral molecule. So, which is the correct structure? With both choices following all our rules, how do we decide? As a general rule, when several Lewis structures are possible, we will use the following guidelines to choose the most correct one:

1. *We generally choose the Lewis structure in which the atoms bear formal charges closest to zero.*
2. *We generally choose the Lewis structure in which any negative charges reside on the more electronegative atoms.*

Thus, the first Lewis structure of CO_2 is preferred because the atoms carry no formal charges and so satisfy the first guideline.

Although the concept of formal charge helps us choose between alternative Lewis structures, it is very important that you remember that *formal charges do not represent real charges on atoms.* These charges are just a bookkeeping convention. The actual charge distributions in molecules and ions are determined not by formal charges but by a number of factors, including the electronegativity differences between atoms.

GIVE IT SOME THOUGHT

Suppose that a Lewis structure for a neutral fluorine-containing molecule results in a formal charge on the fluorine atom of +1. What conclusion would you draw?

SAMPLE EXERCISE 8.9 | Lewis Structures and Formal Charges

The following are three possible Lewis structures for the thiocyanate ion, NCS^-:

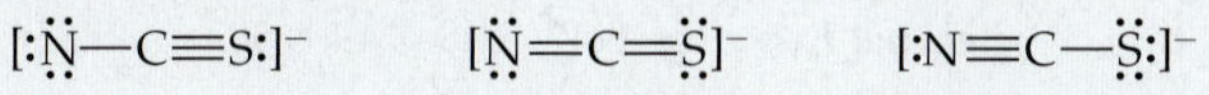

(a) Determine the formal charges of the atoms in each structure. **(b)** Which Lewis structure is the preferred one?

Solution (a) Neutral N, C, and S atoms have five, four, and six valence electrons, respectively. We can determine the following formal charges in the three structures by using the rules we just discussed:

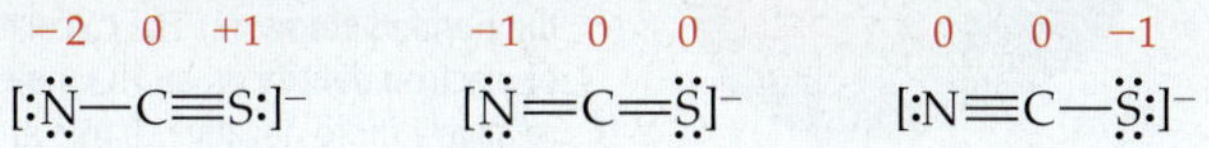

As they must, the formal charges in all three structures sum to 1−, the overall charge of the ion.

(b) We will use the guidelines for the best Lewis structure to determine which of the three structures is likely the most correct. As discussed in Section 8.4, N is more electronegative than C or S. Therefore, we expect that any negative formal charge will reside on the N atom (guideline 2). Further, we usually choose the Lewis structure that produces the formal charges of smallest magnitude (guideline 1). For these two reasons, the middle structure is the preferred Lewis structure of the NCS^- ion.

PRACTICE EXERCISE

The cyanate ion (NCO^-), like the thiocyanate ion, has three possible Lewis structures. **(a)** Draw these three Lewis structures, and assign formal charges to the atoms in each structure. **(b)** Which Lewis structure is the preferred one?

Answers: **(a)**

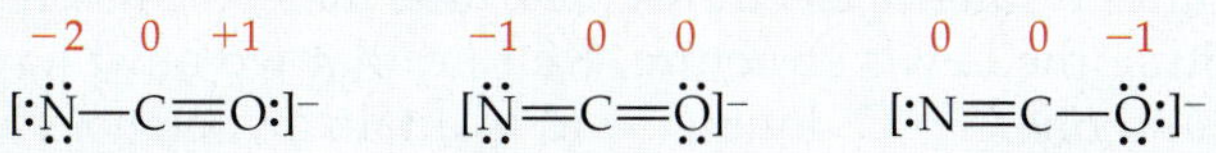

(i) (ii) (iii)

(b) Structure (iii), which places a negative charge on oxygen, the most electronegative of the three elements, is the preferred Lewis structure.

A CLOSER LOOK | Oxidation Numbers, Formal Charges, and Actual Partial Charges

In Chapter 4 we introduced the rules for assigning *oxidation numbers* to atoms. The concept of electronegativity is the basis of these numbers. The oxidation number of an atom is the charge it would have if its bonds were completely ionic. That is, in determining the oxidation number, all shared electrons are counted with the more electronegative atom. For example, consider the Lewis structure of HCl shown in Figure 8.10(a) ▼. To assign oxidation numbers, the pair of electrons in the covalent bond between the atoms is assigned to the more electronegative Cl atom. This procedure gives Cl eight valence-shell electrons, one more than the neutral atom. Thus, it is assigned an oxidation number of -1. Hydrogen has no valence electrons when they are counted this way, giving it an oxidation number of $+1$.

In this section we have just considered another way of counting electrons that gives rise to *formal charges*. The formal charge is assigned by completely ignoring electronegativity and assigning the electrons in bonds equally between the bonded atoms. Consider again the HCl molecule, but this time divide the bonding pair of electrons equally between H and Cl as shown in Figure 8.10(b). In this case Cl has seven assigned electrons, the same as that of the neutral Cl atom. Thus, the formal charge of Cl in this compound is 0. Likewise, the formal charge of H is also 0.

Neither the oxidation number nor the formal charge gives an accurate depiction of the actual charges on atoms. Oxidation numbers overstate the role of electronegativity, and formal charges ignore it completely. It seems reasonable that electrons in covalent bonds should be apportioned according to the relative electronegativities of the bonded atoms. From Figure 8.6 we see that Cl has an electronegativity of 3.0 while that of H is 2.1. The more electronegative Cl atom might therefore be expected to have roughly $3.0/(3.0 + 2.1) = 0.59$ of the electrical charge in the bonding pair, whereas the H atom has $2.1/(3.0 + 2.1) = 0.41$ of the charge. Because the bond consists of two electrons, the Cl atom's share is $0.59 \times 2e = 1.18e$, or $0.18e$ more than the neutral Cl atom. This gives rise to a partial charge of 0.18− on Cl and 0.18+ on H (notice again that we place the + and − signs before the magnitude when speaking about oxidation numbers and formal charges but after the magnitude when talking about actual charges).

The dipole moment of HCl gives an experimental measure of the partial charges on each atom. In Sample Exercise 8.5 we saw that the dipole moment of HCl indicates a charge separation with a partial charge of 0.178+ on H and 0.178− on Cl, in remarkably good agreement with our simple approximation based on electronegativities. Although that type of calculation provides "ballpark" numbers for the magnitude of charge on atoms, the relationship between electronegativities and charge separation is generally more complicated. As we have already seen, computer programs employing quantum mechanical principles have been developed to calculate the partial charges on atoms, even in complex molecules. Figure 8.10(c) shows a graphical representation of the charge distribution in HCl.

Related Exercises: 8.47, 8.48, 8.49, 8.50, and 8.85

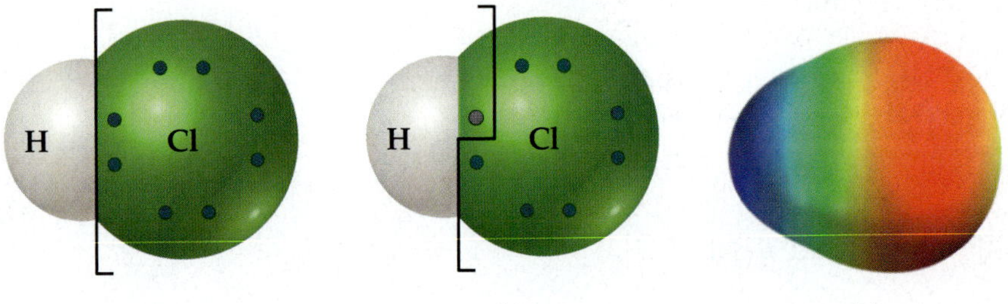

(a) (b) (c)

◀ **Figure 8.10 Oxidation number and formal charge.** (a) The oxidation number for any atom in a molecule is determined by assigning all shared electrons to the more electronegative atom (in this case Cl). (b) Formal charges are derived by dividing all shared electron pairs equally between the bonded atoms. (c) The calculated distribution of electron density on an HCl molecule. Regions of relatively more negative charge are red; those of more positive charge are blue. Negative charge is clearly localized on the chlorine atom.

8.6 | Resonance Structures

We sometimes encounter molecules and ions in which the experimentally determined arrangement of atoms is not adequately described by a single Lewis structure. Consider a molecule of ozone, O_3, which is a bent molecule with two equal O—O bond lengths (Figure 8.11 ▶). Because each oxygen atom contributes 6 valence electrons, the ozone molecule has 18 valence electrons. In writing the Lewis structure, we find that we must have one O—O single bond and one O—O double bond to attain an octet of electrons about each atom:

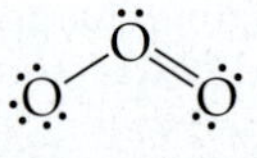

However, this structure cannot by itself be correct because it requires that one O—O bond be different from the other, contrary to the observed structure—we would expect the O=O double bond to be shorter than the O—O single bond. (Section 8.3) In drawing the Lewis structure, however, we could just as easily have put the O=O bond on the left:

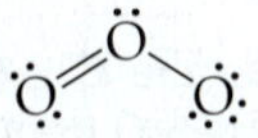

The placement of the atoms in these two alternative but completely equivalent Lewis structures for ozone is the same, but the placement of the electrons is different. Lewis structures of this sort are called **resonance structures**. To describe the structure of ozone properly, we write both Lewis structures and use a double-headed arrow to indicate that the real molecule is described by an average of the two resonance structures:

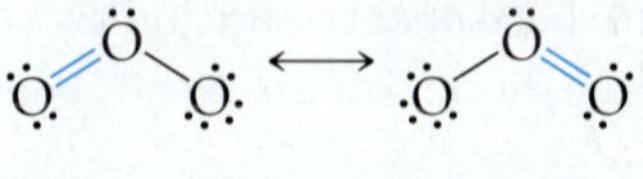

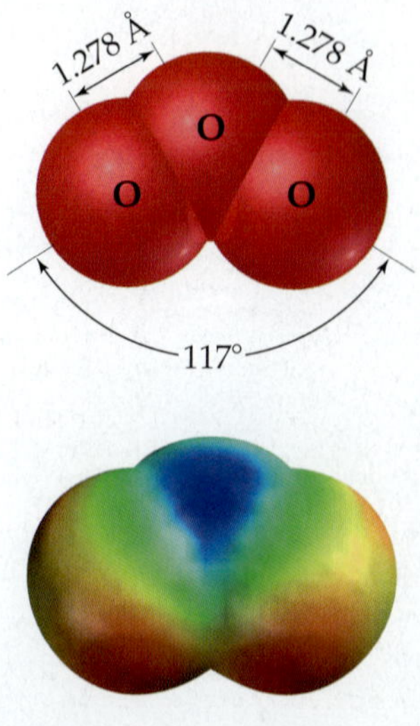

▲ **Figure 8.11 Ozone.** Molecular structure (top) and electron-distribution diagram (bottom) for the ozone molecule, O_3.

To understand why certain molecules require more than one resonance structure, we can draw an analogy to the mixing of paint (Figure 8.12 ▶). Blue and yellow are both primary colors of paint pigment. An equal blend of blue and yellow pigments produces green pigment. We can't describe green paint in terms of a single primary color, yet it still has its own identity. Green paint does not oscillate between its two primary colors: It is not blue part of the time and yellow the rest of the time. Similarly, molecules such as ozone cannot be described by a single Lewis structure in which the electrons are "locked into" a particular arrangement.

The true arrangement of the electrons in molecules such as O_3 must be considered as a blend of two (or more) Lewis structures. By analogy to the green paint, the molecule has its own identity separate from the individual resonance structures. For example, the ozone molecule always has two equivalent O—O bonds whose lengths are intermediate between the lengths of an oxygen–oxygen single bond and an oxygen–oxygen double bond. Another way of looking at it is to say that the rules for drawing Lewis structures don't allow us to have a single structure that adequately represents the ozone molecule. For example, there are no rules for drawing half-bonds. We can get around this limitation by drawing two equivalent Lewis structures that, when averaged, amount to something very much like what is observed experimentally.

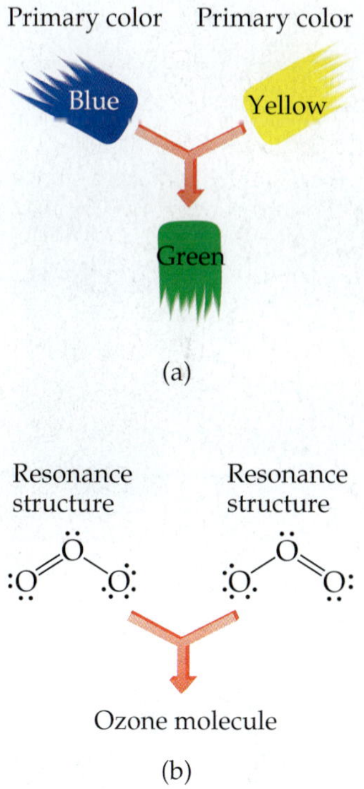

▲ **Figure 8.12 Resonance.** Describing a molecule as a blend of different resonance structures is similar to describing a paint color as a blend of primary colors. (a) Green paint is a blend of blue and yellow. We cannot describe green as a single primary color. (b) The ozone molecule is a blend of two resonance structures. We cannot describe the ozone molecule in terms of a single Lewis structure.

GIVE IT SOME THOUGHT

The O—O bonds in ozone are often described as "one-and-a-half" bonds. Is this description consistent with the idea of resonance?

As an additional example of resonance structures, consider the nitrate ion, NO_3^-, for which three equivalent Lewis structures can be drawn:

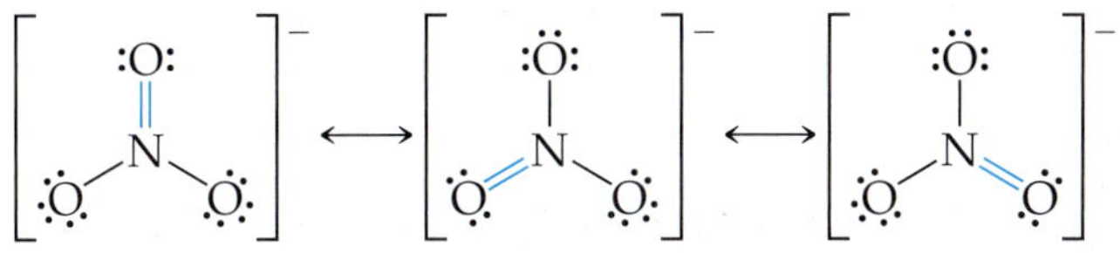

Notice that the arrangement of atoms is the same in each structure; only the placement of electrons differs. In writing resonance structures, the same atoms must be bonded to each other in all structures, so that the only differences are in the arrangements of electrons. All three Lewis structures taken together adequately describe the nitrate ion, in which all three N—O bond lengths are the same.

GIVE IT SOME THOUGHT

In the same sense that we describe the O—O bonds in O_3 as "one-and-a-half" bonds, how would you describe the N—O bonds in NO_3^-?

In some instances all the possible Lewis structures for a species may not be equivalent to one another; instead, one or more may represent a more stable arrangement than other possibilities. We will encounter examples of this as we proceed.

SAMPLE EXERCISE 8.10 | Resonance Structures

Which is predicted to have the shorter sulfur–oxygen bonds, SO_3 or SO_3^{2-}?

Solution The sulfur atom has six valence electrons, as does oxygen. Thus, SO_3 contains 24 valence electrons. In writing the Lewis structure, we see that three equivalent resonance structures can be drawn:

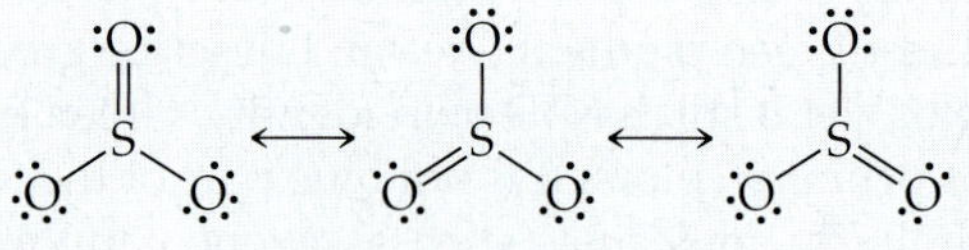

As was the case for NO_3^-, the actual structure of SO_3 is an equal blend of all three. Thus, each S—O bond distance should be about one-third of the way between that of a single and that of a double bond (see the immediately preceding Give It Some Thought exercise). That is, they should be shorter than single bonds but not as short as double bonds.

The SO_3^{2-} ion has 26 electrons, which leads to a Lewis structure in which all the S—O bonds are single bonds:

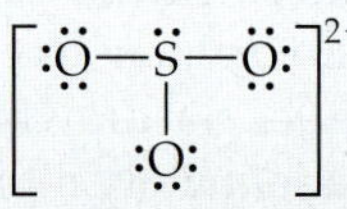

There are no other reasonable Lewis structures for this ion—it can be described quite well by a single Lewis structure rather than by multiple resonance structures.

Our analysis of the Lewis structures leads us to conclude that SO_3 should have the shorter S—O bonds and SO_3^{2-} the longer ones. This conclusion is correct: The experimentally measured S—O bond lengths are 1.42 Å in SO_3 and 1.51 Å in SO_3^{2-}.

PRACTICE EXERCISE

Draw two equivalent resonance structures for the formate ion, HCO_2^-.

Answer:

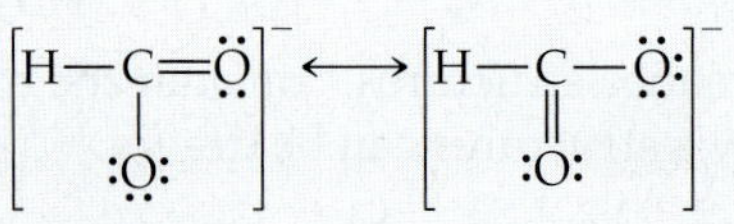

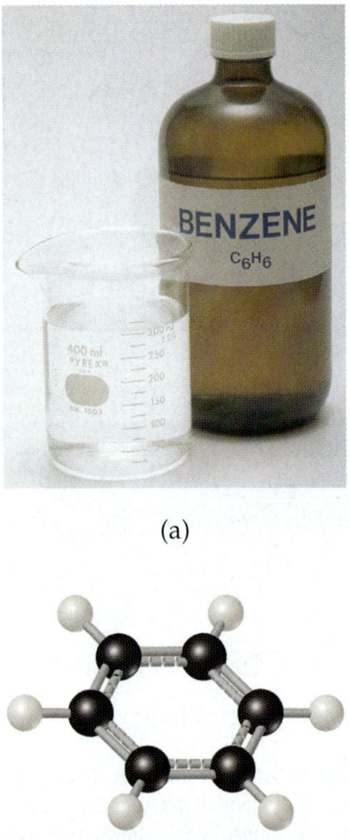

(a)

(b)

▲ **Figure 8.13 Benzene, an "aromatic" organic compound.** (a) Benzene is obtained from the distillation of fossil fuels. More than 4 billion pounds of benzene is produced annually in the United States. Because benzene is a carcinogen, its use is closely regulated. (b) The benzene molecule is a regular hexagon of carbon atoms with a hydrogen atom bonded to each one.

Resonance in Benzene

Resonance is an extremely important concept in describing the bonding in organic molecules, particularly in the ones called *aromatic* molecules. Aromatic organic molecules include the hydrocarbon called *benzene*, which has the molecular formula C_6H_6 (Figure 8.13 ◀). The six C atoms of benzene are bonded in a hexagonal ring, and one H atom is bonded to each C atom.

We can write two equivalent Lewis structures for benzene, each of which satisfies the octet rule. These two structures are in resonance:

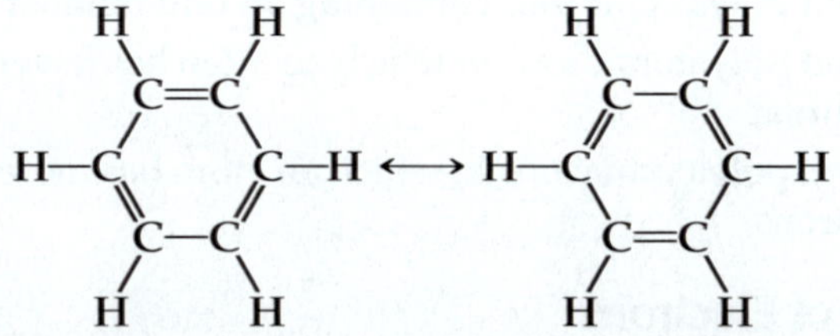

Each resonance structure shows three C—C single bonds and three C=C double bonds, but the double bonds are in different places in the two structures. The experimental structure of benzene shows that all six C—C bonds are of equal length, 1.40 Å, intermediate between the typical bond lengths for a C—C single bond (1.54 Å) and a C=C double bond (1.34 Å).

Benzene is commonly represented by omitting the hydrogen atoms attached to carbon and showing only the carbon–carbon framework with the vertices unlabeled. In this convention, the resonance in the benzene molecule is represented either by two structures separated by a double-headed arrow, as with our other examples, or by a shorthand notation in which we draw a hexagon with a circle in it:

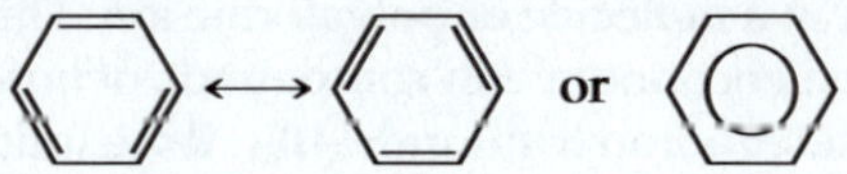

The shorthand notation on the right reminds us that benzene is a blend of two resonance structures—it emphasizes that the C=C double bonds cannot be assigned to specific edges of the hexagon. Chemists use both representations of benzene interchangeably.

The bonding arrangement in benzene confers special stability to the molecule. As a result, literally millions of organic compounds contain the six-membered rings characteristic of benzene. Many of these compounds are important in biochemistry, in pharmaceuticals, and in the production of modern materials. We will say more about the bonding in benzene in Chapter 9 and about its unusual stability in Chapter 25.

GIVE IT SOME THOUGHT

Each Lewis structure of benzene has three C=C double bonds. Another hydrocarbon containing three C=C double bonds is *hexatriene*, C_6H_8. A Lewis structure of hexatriene is

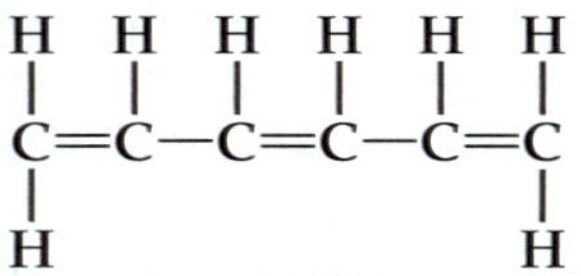

Would you expect hexatriene to have multiple resonance structures like benzene? If not, why is this molecule different from benzene with respect to resonance?

8.7 | Exceptions to the Octet Rule

The octet rule is so simple and useful in introducing the basic concepts of bonding that you might assume that it is always obeyed. In Section 8.2, however, we noted its limitation in dealing with ionic compounds of the transition metals.

The octet rule also fails in many situations involving covalent bonding. These exceptions to the octet rule are of three main types:

1. Molecules and polyatomic ions containing an odd number of electrons
2. Molecules and polyatomic ions in which an atom has fewer than an octet of valence electrons
3. Molecules and polyatomic ions in which an atom has more than an octet of valence electrons

Odd Number of Electrons

In the vast majority of molecules and polyatomic ions, the total number of valence electrons is even, and complete pairing of electrons occurs. In a few molecules and polyatomic ions, however, such as ClO_2, NO, NO_2, and O_2^-, the number of valence electrons is odd. Complete pairing of these electrons is impossible, and an octet around each atom cannot be achieved. For example, NO contains $5 + 6 = 11$ valence electrons. The two most important Lewis structures for this molecule are

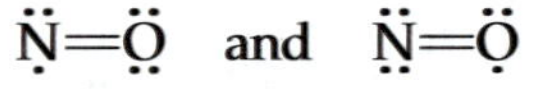

Less than an Octet of Valence Electrons

A second type of exception occurs when there are fewer than eight valence electrons around an atom in a molecule or polyatomic ion. This situation is also relatively rare, most often encountered in compounds of boron and beryllium. As an example, let's consider boron trifluoride, BF_3. If we follow the first four steps of the procedure at the beginning of Section 8.5 for drawing Lewis structures, we obtain the structure

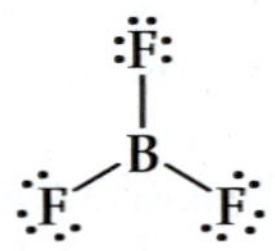

3-D MODELS
BF_3, BF_3NH_3

There are only six electrons around the boron atom. In this Lewis structure the formal charges on both the B and the F atoms are zero. We could complete the octet around boron by forming a double bond (step 5). In so doing, we see that there are three equivalent resonance structures (the formal charges on each atom are shown in red):

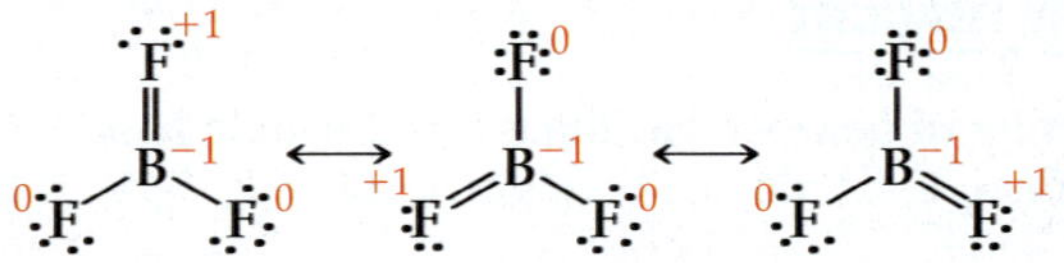

These Lewis structures force a fluorine atom to share additional electrons with the boron atom, which is inconsistent with the high electronegativity of fluorine. In fact, the formal charges tell us that this is an unfavorable situation: In each of the Lewis structures, the F atom involved in the B=F double bond has a formal charge of +1, while the less electronegative B atom has a formal charge of −1. Thus, the Lewis structures in which there is a B=F double bond are less important than the one in which there are fewer than an octet of valence electrons around boron:

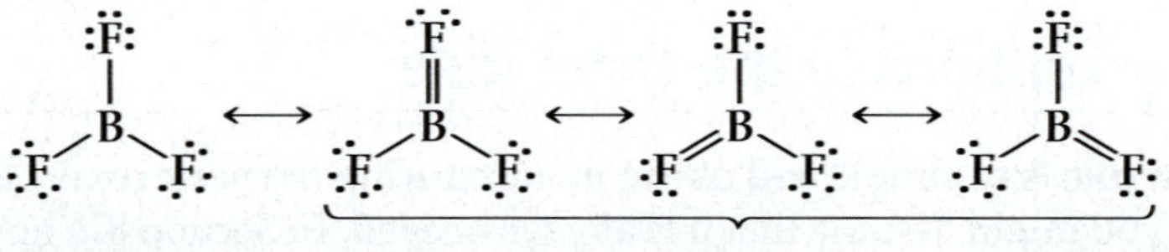

Most important Less important

We usually represent BF_3 solely by the leftmost resonance structure, in which there are only six valence electrons around boron. The chemical behavior of BF_3 is consistent with this representation. In particular, BF_3 reacts very energetically with molecules having an unshared pair of electrons that can be used to form a bond with boron. For example, it reacts with ammonia, NH_3, to form the compound NH_3BF_3:

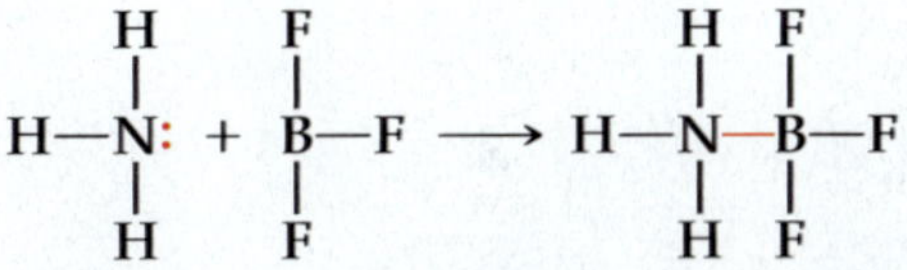

In this stable compound, boron has an octet of valence electrons.

More than an Octet of Valence Electrons

The third and largest class of exceptions consists of molecules or polyatomic ions in which there are more than eight electrons in the valence shell of an atom. When we draw the Lewis structure for PCl_5, for example, we are forced to "expand" the valence shell and place ten electrons around the central phosphorus atom:

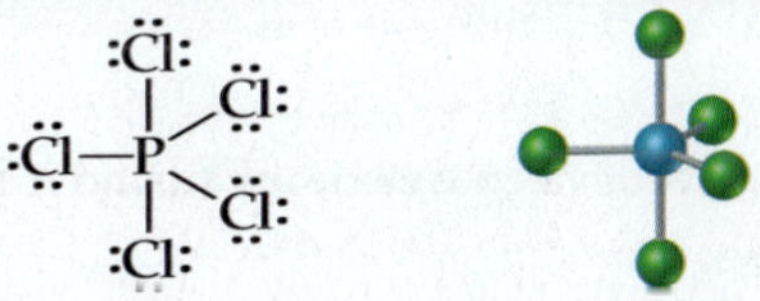

Other examples of molecules and ions with "expanded" valence shells are SF_4, AsF_6^-, and ICl_4^-. The corresponding molecules with a second-period atom bonded to the halogen atom, such as NCl_5 and OF_4, do *not* exist. Let's take a look at why expanded valence shells are observed only for elements in period 3 and beyond in the periodic table.

Elements of the second period have only the 2*s* and 2*p* valence orbitals available for bonding. Because these orbitals can hold a maximum of eight electrons, we never find more than an octet of electrons around elements from the second period. Elements from the third period and beyond, however, have *ns*, *np*, and unfilled *nd* orbitals that can be used in bonding. For example, the orbital diagram for the valence shell of a phosphorus atom is

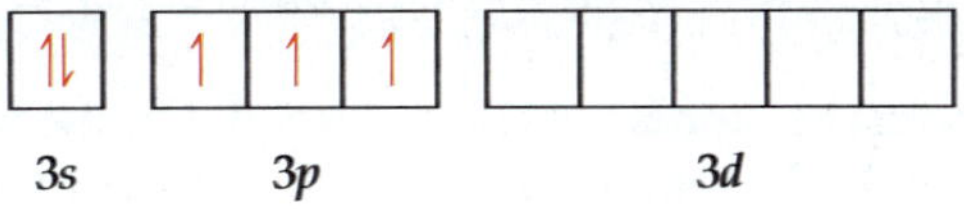

Although third-period elements often satisfy the octet rule, as in PCl_3, they also often exceed an octet by seeming to use their empty *d* orbitals to accommodate additional electrons.*

Size also plays an important role in determining whether an atom in a molecule or polyatomic ion can accommodate more than eight electrons in its valence shell. The larger the central atom, the larger the number of atoms that can surround it. The number of molecules and ions with expanded valence shells therefore increases with increasing size of the central atom. The size of the surrounding atoms is also important. Expanded valence shells occur most often when the central atom is bonded to the smallest and most electronegative atoms, such as F, Cl, and O.

* On the basis of theoretical calculations, some chemists have questioned whether valence *d* orbitals are actually used in the bonding of molecules and ions with expanded valence shells. Nevertheless, the presence of valence *d* orbitals in period 3 and beyond provides the simplest explanation of this phenomenon, especially within the scope of a general chemistry textbook.

SAMPLE EXERCISE 8.11 | **Lewis Structure for an Ion with an Expanded Valence Shell**

Draw the Lewis structure for ICl_4^-.

Solution Iodine (group 7A) has seven valence electrons; each chlorine (group 7A) also has seven; an extra electron is added to account for the 1− charge of the ion. Therefore, the total number of valence electrons is

$$7 + 4(7) + 1 = 36$$

The I atom is the central atom in the ion. Putting eight electrons around each Cl atom (including a pair of electrons between I and each Cl to represent the single bond between these atoms) requires $8 \times 4 = 32$ electrons.

We are thus left with $36 - 32 = 4$ electrons to be placed on the larger iodine:

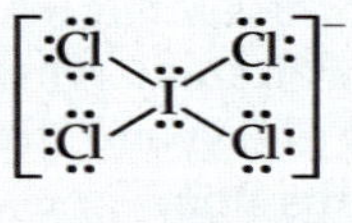

Iodine has 12 valence electrons around it, four more than needed for an octet.

PRACTICE EXERCISE

(a) Which of the following atoms is never found with more than an octet of valence electrons around it: S, C, P, Br? **(b)** Draw the Lewis structure for XeF_2.

Answers: **(a)** C, **(b)** :F̤—X̤e—F̤:

At times you may see Lewis structures written with an expanded valence shell even though structures can be written with an octet. For example, consider these Lewis structures for the phosphate ion, PO_4^{3-}:

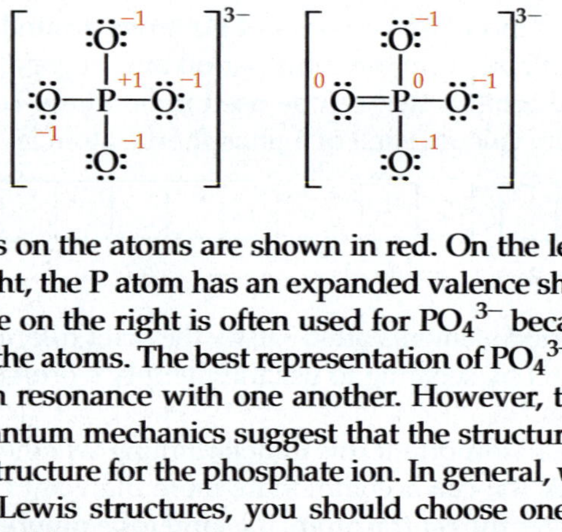

The formal charges on the atoms are shown in red. On the left, the P atom has an octet; on the right, the P atom has an expanded valence shell of five electron pairs. The structure on the right is often used for PO_4^{3-} because it has smaller formal charges on the atoms. The best representation of PO_4^{3-} is a series of such Lewis structures in resonance with one another. However, theoretical calculations based on quantum mechanics suggest that the structure on the left is the best single Lewis structure for the phosphate ion. In general, when choosing between alternative Lewis structures, you should choose one that satisfies the octet rule if it is possible to do so.

8.8 | Strengths of Covalent Bonds

The stability of a molecule is related to the strengths of the covalent bonds it contains. The strength of a covalent bond between two atoms is determined by the energy required to break that bond. It is easiest to relate bond strength to the enthalpy change in reactions in which bonds are broken. ∞ (Section 5.4)

The **bond enthalpy** is the enthalpy change, ΔH, for the breaking of a particular bond in one mole of a gaseous substance. For example, the bond enthalpy for the bond between chlorine atoms in the Cl_2 molecule is the enthalpy change when 1 mol of Cl_2 is dissociated into chlorine atoms:

$$:\ddot{\underset{..}{Cl}}-\ddot{\underset{..}{Cl}}:(g) \longrightarrow 2\,:\ddot{\underset{..}{Cl}}\cdot(g)$$

We use the designation *D*(bond type) to represent bond enthalpies.

It is relatively simple to assign bond enthalpies to bonds that are found in diatomic molecules, such as the Cl—Cl bond in Cl_2 or the H—Br bond in HBr: The bond enthalpy is just the energy required to break the diatomic molecule into its component atoms. Many important bonds, such as the C—H bond, exist only in polyatomic molecules. For these types of bonds, we usually utilize *average* bond enthalpies. For example, the enthalpy change for the following process in which a methane molecule is decomposed to its five atoms (a process called *atomization*) can be used to define an average bond enthalpy for the C—H bond:

$$\mathrm{CH_4}(g) \longrightarrow \cdot\dot{\underset{\cdot}{C}}\cdot(g) + 4\,\mathrm{H}\cdot(g) \qquad \Delta H = 1660\ \mathrm{kJ}$$

Because there are four equivalent C—H bonds in methane, the heat of atomization is equal to the sum of the bond enthalpies of the four C—H bonds. Therefore, the average C—H bond enthalpy for CH_4 is $D(\mathrm{C{-}H}) = (1660/4)$ kJ/mol = 415 kJ/mol.

The bond enthalpy for a given set of atoms, say C—H, depends on the rest of the molecule of which the atom pair is a part. However, the variation from one molecule to another is generally small, which supports the idea that bonding electron pairs are localized between atoms. If we consider C—H bond enthalpies in many different compounds, we find that the average bond enthalpy is 413 kJ/mol, which compares closely with the 415 kJ/mol value calculated from CH_4.

GIVE IT SOME THOUGHT

The hydrocarbon *ethane*, C_2H_6, was first introduced in Section 2.9. How could you use the enthalpy of atomization of $C_2H_6(g)$ along with the value of *D*(C—H) to provide an estimate for *D*(C—C)?

Table 8.4 ▶ lists several average bond enthalpies. *The bond enthalpy is always a positive quantity*; energy is always required to break chemical bonds. Conversely, energy is always released when a bond forms between two gaseous atoms or molecular fragments. The greater the bond enthalpy, the stronger the bond.

A molecule with strong chemical bonds generally has less tendency to undergo chemical change than does one with weak bonds. This relationship between strong bonding and chemical stability helps explain the chemical form in which many elements are found in nature. For example, Si—O bonds are among the strongest ones that silicon forms. It should not be surprising, therefore, that SiO_2 and other substances containing Si—O bonds (silicates) are so common; it is estimated that over 90% of Earth's crust is composed of SiO_2 and silicates.

TABLE 8.4 Average Bond Enthalpies (kJ/mol)

Single Bonds

C—H	413	N—H	391	O—H	463	F—F	155
C—C	348	N—N	163	O—O	146		
C—N	293	N—O	201	O—F	190	Cl—F	253
C—O	358	N—F	272	O—Cl	203	Cl—Cl	242
C—F	485	N—Cl	200	O—I	234		
C—Cl	328	N—Br	243			Br—F	237
C—Br	276			S—H	339	Br—Cl	218
C—I	240	H—H	436	S—F	327	Br—Br	193
C—S	259	H—F	567	S—Cl	253		
		H—Cl	431	S—Br	218	I—Cl	208
Si—H	323	H—Br	366	S—S	266	I—Br	175
Si—Si	226	H—I	299			I—I	151
Si—C	301						
Si—O	368						
Si—Cl	464						

Multiple Bonds

C=C	614	N=N	418	O_2	495		
C≡C	839	N≡N	941				
C=N	615	N=O	607	S=O	523		
C≡N	891			S=S	418		
C=O	799						
C≡O	1072						

Bond Enthalpies and the Enthalpies of Reactions

We can use average bond enthalpies to estimate the enthalpies of reactions in which bonds are broken and new bonds are formed. This procedure allows us to estimate quickly whether a given reaction will be endothermic ($\Delta H > 0$) or exothermic ($\Delta H < 0$) even if we do not know ΔH_f° for all the chemical species involved.

Our strategy for estimating reaction enthalpies is a straightforward application of Hess's law. (Section 5.6) We use the fact that breaking bonds is always an endothermic process and bond formation is always exothermic. We therefore imagine that the reaction occurs in two steps: (1) We supply enough energy to break those bonds in the reactants that are not present in the products. In this step the enthalpy of the system is increased by the sum of the bond enthalpies of the bonds that are broken. (2) We form the bonds in the products that were not present in the reactants. This step releases energy and therefore lowers the enthalpy of the system by the sum of the bond enthalpies of the bonds that are formed. The enthalpy of the reaction, ΔH_{rxn}, is estimated as the sum of the bond enthalpies of the bonds broken minus the sum of the bond enthalpies of the bonds formed:

$$\Delta H_{rxn} = \Sigma(\text{bond enthalpies of bonds broken}) - \Sigma(\text{bond enthalpies of bonds formed}) \quad [8.12]$$

Consider, for example, the gas-phase reaction between methane, CH_4, and chlorine to produce methyl chloride, CH_3Cl, and hydrogen chloride, HCl:

$$\text{H—CH}_3(g) + \text{Cl—Cl}(g) \longrightarrow \text{Cl—CH}_3(g) + \text{H—Cl}(g) \quad \Delta H_{rxn} = ? \quad [8.13]$$

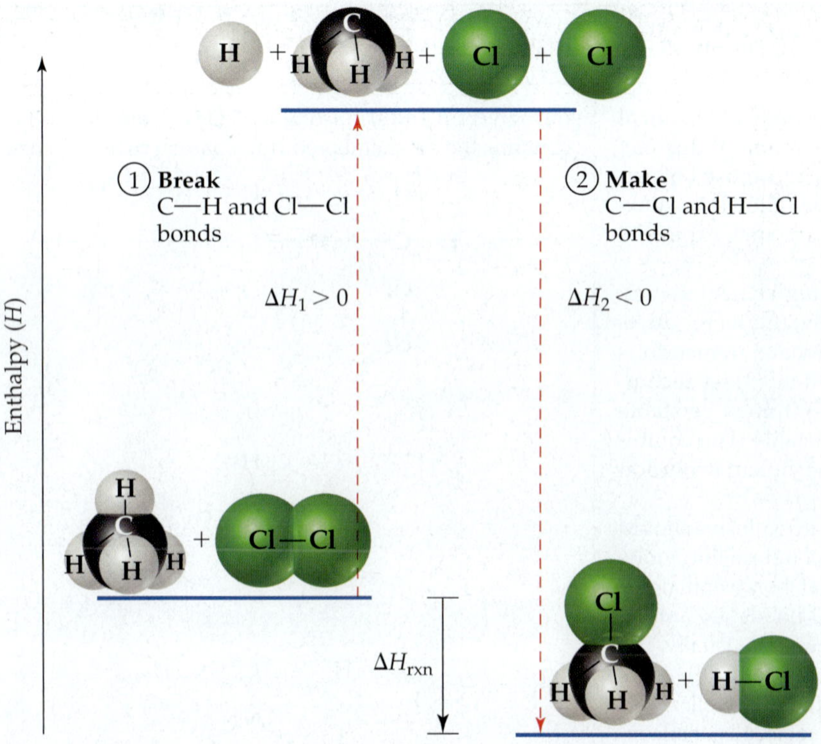

◀ **Figure 8.14 Using bond enthalpies to calculate ΔH_{rxn}.** Average bond enthalpies are used to estimate ΔH_{rxn} for the reaction in Equation 8.13. Breaking the C—H and Cl—Cl bonds produces a positive enthalpy change (ΔH_1), whereas making the C—Cl and H—Cl bonds causes a negative enthalpy change (ΔH_2). The values of ΔH_1 and ΔH_2 are estimated from the values in Table 8.4. From Hess's law, $\Delta H_{rxn} = \Delta H_1 + \Delta H_2$.

ACTIVITY
Bond Enthalpy

Our two-step procedure is outlined in Figure 8.14 ▲. We note that in the course of this reaction the following bonds are broken and made:

Bonds broken: 1 mol C—H, 1 mol Cl—Cl

Bonds made: 1 mol C—Cl, 1 mol H—Cl

We first supply enough energy to break the C—H and Cl—Cl bonds, which will raise the enthalpy of the system. We then form the C—Cl and H—Cl bonds, which will release energy and lower the enthalpy of the system. By using Equation 8.12 and the data in Table 8.4, we estimate the enthalpy of the reaction as

$$\begin{aligned}\Delta H_{rxn} &= [D(\text{C—H}) + D(\text{Cl—Cl})] - [D(\text{C—Cl}) + D(\text{H—Cl})] \\ &= (413\text{ kJ} + 242\text{ kJ}) - (328\text{ kJ} + 431\text{ kJ}) = -104\text{ kJ}\end{aligned}$$

The reaction is exothermic because the bonds in the products (especially the H—Cl bond) are stronger than the bonds in the reactants (especially the Cl—Cl bond).

We usually use bond enthalpies to estimate ΔH_{rxn} only if we do not have the needed ΔH_f° values readily at hand. For the above reaction, we cannot calculate ΔH_{rxn} from ΔH_f° values and Hess's law because the value of ΔH_f° for $CH_3Cl(g)$ is not given in Appendix C. If we obtain the value of ΔH_f° for $CH_3Cl(g)$ from another source (such as the *CRC Handbook of Chemistry and Physics*) and use Equation 5.31, we find that $\Delta H_{rxn} = -99.8$ kJ for the reaction in Equation 8.13. Thus, the use of average bond enthalpies provides a reasonably accurate estimate of the actual reaction enthalpy change.

It is important to remember that bond enthalpies are derived for *gaseous* molecules and that they are often *averaged* values. Nonetheless, average bond enthalpies are useful for estimating reaction enthalpies quickly, especially for gas-phase reactions.

CHEMISTRY AT WORK | Explosives and Alfred Nobel

Enormous amounts of energy can be stored in chemical bonds. Perhaps the most graphic illustration of this fact is seen in certain molecular substances that are used as explosives. Our discussion of bond enthalpies allows us to examine more closely some of the properties of such explosive substances.

An explosive must have the following characteristics: (1) It must decompose very exothermically; (2) the products of its decomposition must be gaseous, so that a tremendous gas pressure accompanies the decomposition; (3) its decomposition must occur very rapidly; and (4) it must be stable enough so that it can be detonated predictably. The combination of the first three effects leads to the violent evolution of heat and gases.

To give the most exothermic reaction, an explosive should have weak chemical bonds and should decompose into molecules with very strong bonds. Looking at bond enthalpies (Table 8.4), the N≡N, C≡O, and C=O bonds are among the strongest. Not surprisingly, explosives are usually designed to produce the gaseous products $N_2(g)$, $CO(g)$, and $CO_2(g)$. Water vapor is nearly always produced as well.

Many common explosives are organic molecules that contain nitro (NO_2) or nitrate (NO_3) groups attached to a carbon skeleton. The structures of two of the most familiar explosives, nitroglycerin and trinitrotoluene (TNT), are shown here. TNT contains the six-membered ring characteristic of benzene.

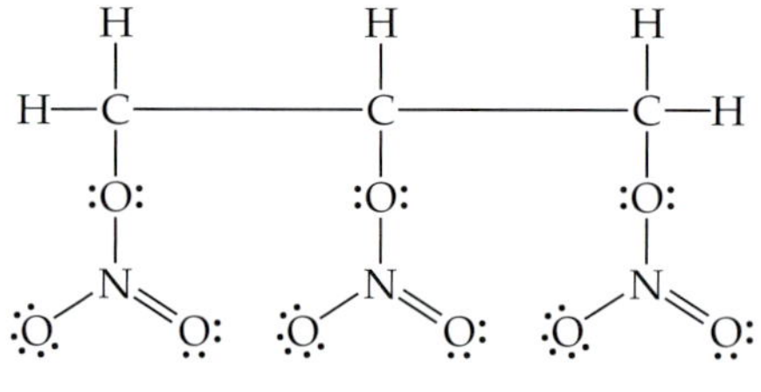

Nitroglycerin

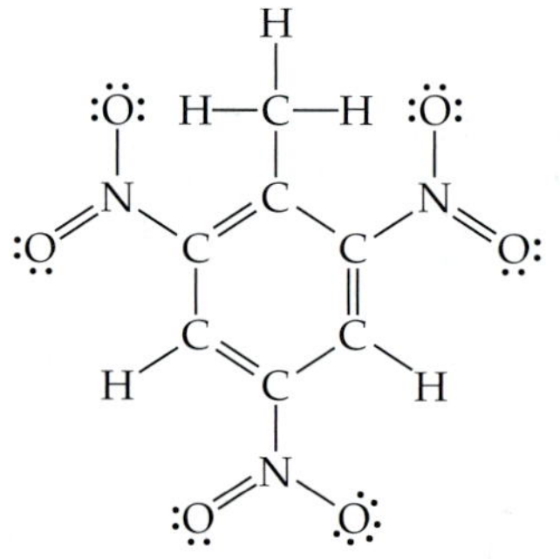

TNT

SAMPLE EXERCISE 8.12 | Using Average Bond Enthalpies

Using Table 8.4, estimate ΔH for the following reaction (where we explicitly show the bonds involved in the reactants and products):

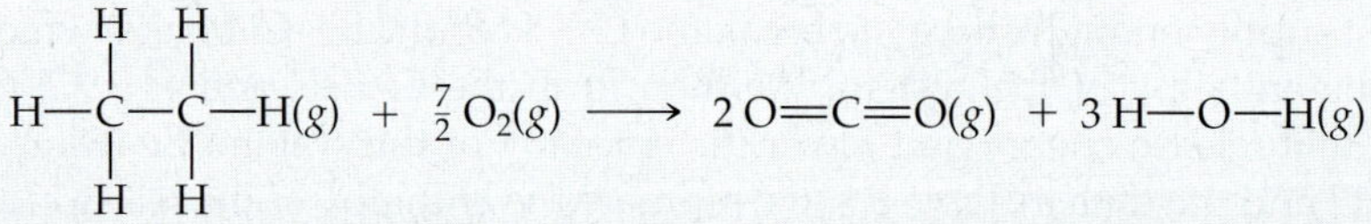

Solution

Analyze: We are asked to estimate the enthalpy change for a chemical process by using average bond enthalpies for the bonds that are broken in the reactants and formed in the products.

Plan: Among the reactants, we must break six C—H bonds and a C—C bond in C_2H_6; we also break $\frac{7}{2}$ O_2 bonds. Among the products, we form four C=O bonds (two in each CO_2) and six O—H bonds (two in each H_2O).

Solve: Using Equation 8.12 and data from Table 8.4, we have

$$\begin{aligned}\Delta H &= 6D(\text{C—H}) + D(\text{C—C}) + \tfrac{7}{2}D(\text{O}_2) - 4D(\text{C=O}) - 6D(\text{O—H}) \\ &= 6(413\text{ kJ}) + 348\text{ kJ} + \tfrac{7}{2}(495\text{ kJ}) - 4(799\text{ kJ}) - 6(463\text{ kJ}) \\ &= 4558\text{ kJ} - 5974\text{ kJ} \\ &= -1416\text{ kJ}\end{aligned}$$

Check: This estimate can be compared with the value of −1428 kJ calculated from more accurate thermochemical data; the agreement is good.

PRACTICE EXERCISE

Using Table 8.4, estimate ΔH for the reaction

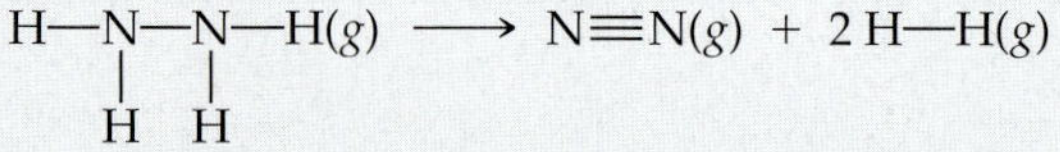

Answer: −86 kJ

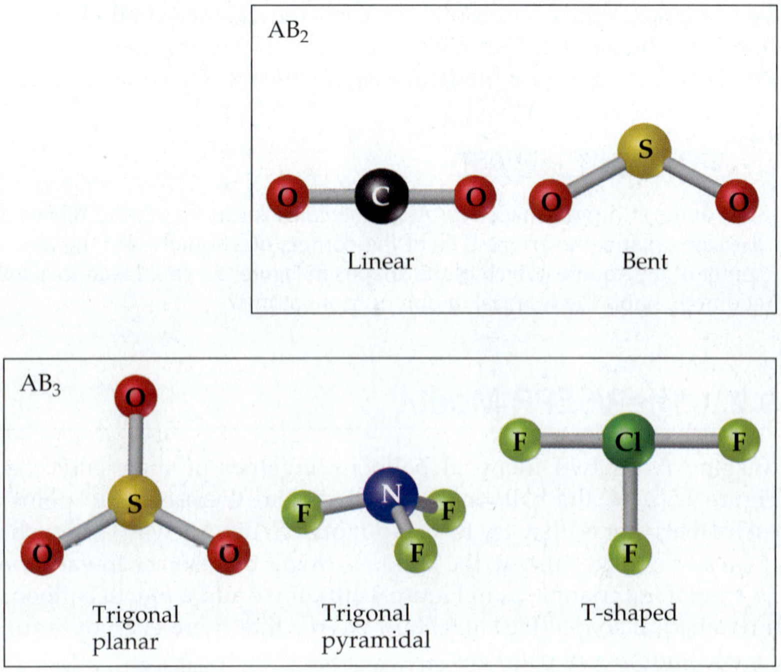

Figure 9.2 Shapes of AB_2 and AB_3 molecules. Top: AB_2 molecules can be either linear or bent. Bottom: Three possible shapes for AB_3 molecules.

3-D MODELS CO_2, SO_2, SO_3, NF_3, ClF_3

CO_2 is linear and SO_2 is bent. For AB_3 molecules, the two most common shapes place the B atoms at the corners of an equilateral triangle. If the A atom lies in the same plane as the B atoms, the shape is called *trigonal planar*. If the A atom lies above the plane of the B atoms, the shape is called *trigonal pyramidal* (a pyramid with an equilateral triangle as its base). For example, SO_3 is trigonal planar and NF_3 is trigonal pyramidal. Some AB_3 molecules, such as ClF_3, exhibit the more unusual *T shape* shown in Figure 9.2.

The shape of any particular AB_n molecule can usually be derived from one of the five basic geometric structures shown in Figure 9.3 ▶. Starting with a tetrahedron, for example, we can remove atoms successively from the corners as shown in Figure 9.4 ▼. When an atom is removed from one corner of the tetrahedron, the remaining fragment has a trigonal-pyramidal geometry such as that found for NF_3. When two atoms are removed, a bent geometry results.

Why do so many AB_n molecules have shapes related to the basic structures in Figure 9.3, and can we predict these shapes? When A is a representative element (one of the elements from the *s* block or *p* block of the periodic table),

Figure 9.3 Shapes of AB_n molecules. For molecules whose formula is of the general form AB_n, there are five fundamental shapes.

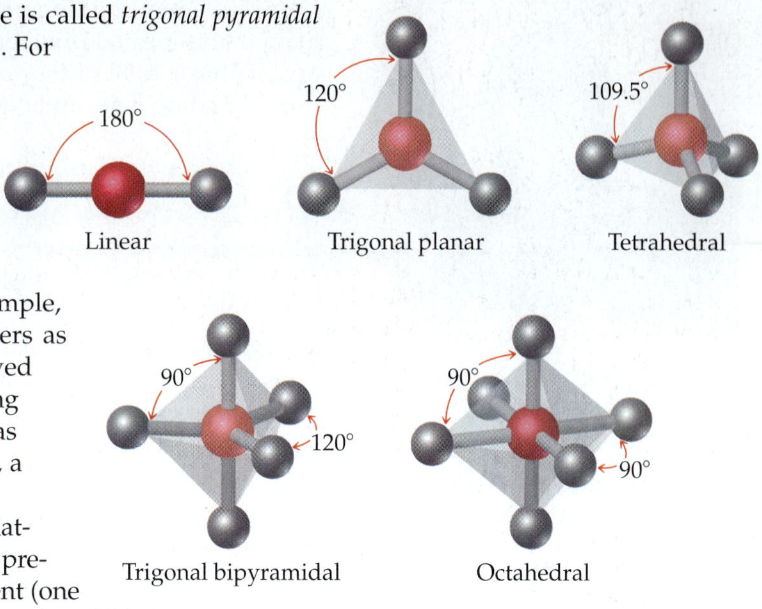

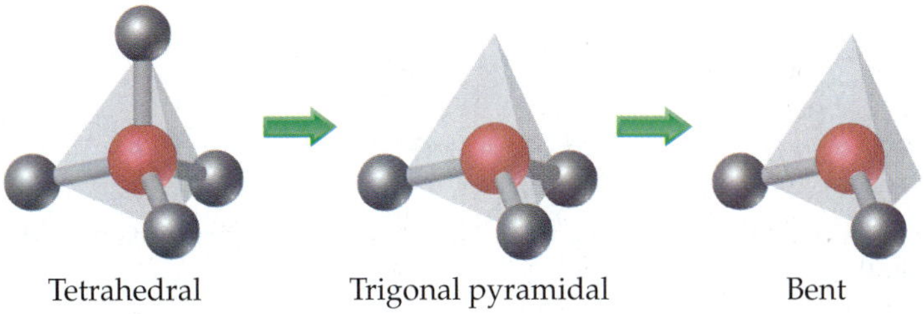

Figure 9.4 Derivatives from the AB_n geometries. Additional molecular shapes can be obtained by removing corner atoms from the basic shapes shown in Figure 9.3. Here we begin with a tetrahedron and successively remove corners, producing first a trigonal-pyramidal geometry and then a bent geometry, each having ideal bond angles of 109.5°. Molecular shape is meaningful only when there are at least three atoms. If there are only two, they must be arranged next to each other, and no special name is given to describe the molecule.

we can answer these questions by using the **valence-shell electron-pair repulsion (VSEPR) model**. Although the name is rather imposing, the model is quite simple, and it has useful predictive capabilities, as we will see in Section 9.2.

GIVE IT SOME THOUGHT

One of the common shapes for AB_4 molecules is *square planar*: All five atoms lie in the same plane, the atoms B lie at the corners of a square, and the atom A is at the center of the square. Which of the shapes in Figure 9.3 could lead to a square-planar geometry upon the removal of one or more atoms?

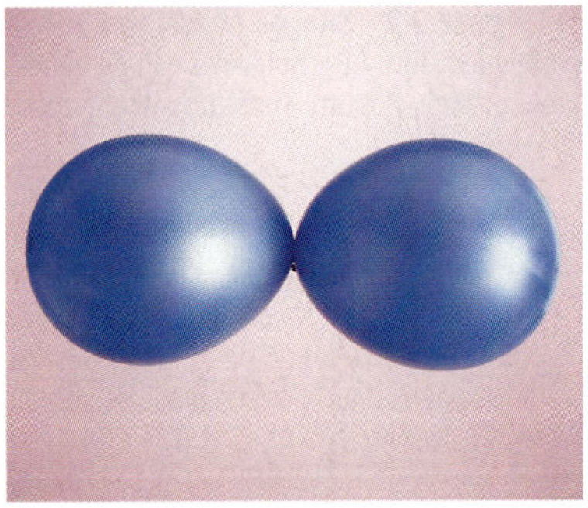

(a) Two balloons adopt a linear arrangement.

(b) Three balloons adopt a trigonal-planar arrangement.

(c) Four balloons adopt a tetrahedral arrangement.

▲ **Figure 9.5 A balloon analogy for electron domains.** Balloons tied together at their ends naturally adopt their lowest-energy arrangement.

ANIMATION
VSEPR

9.2 | The VSEPR Model

Imagine tying two identical balloons together at their ends. As shown in Figure 9.5(a) ◀, the balloons naturally orient themselves to point away from each other; that is, they try to "get out of each other's way" as much as possible. If we add a third balloon, the balloons orient themselves toward the vertices of an equilateral triangle as in Figure 9.5(b). If we add a fourth balloon, they adopt a tetrahedral shape [Figure 9.5(c)]. We see that there is an optimum geometry for each number of balloons.

In some ways the electrons in molecules behave like the balloons in Figure 9.5. We have seen that a single covalent bond is formed between two atoms when a pair of electrons occupies the space between the atoms. (Section 8.3) A **bonding pair** of electrons thus defines a region in which the electrons will most likely be found. We will refer to such a region as an **electron domain**. Likewise, a **nonbonding pair** (or *lone pair*) of electrons defines an electron domain that is located principally on one atom. For example, the Lewis structure of NH_3 has a total of four electron domains around the central nitrogen atom (three bonding pairs and one nonbonding pair):

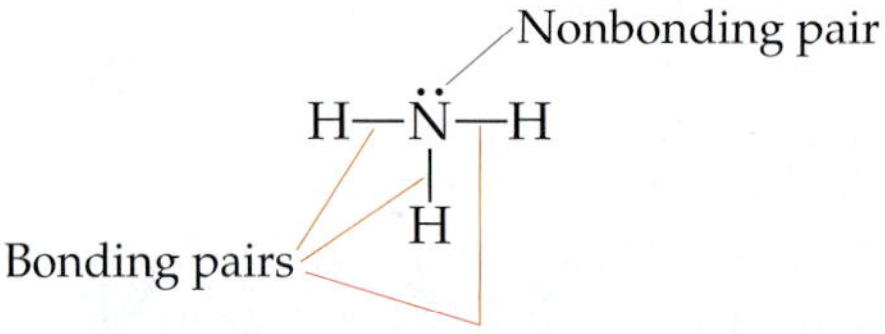

Each multiple bond in a molecule also constitutes a single electron domain. Thus, the following resonance structure for O_3 has three electron domains around the central oxygen atom (a single bond, a double bond, and a nonbonding pair of electrons):

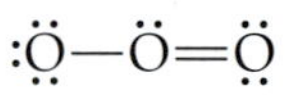

In general, then, *each nonbonding pair, single bond, or multiple bond produces an electron domain around the central atom.*

GIVE IT SOME THOUGHT

An AB_3 molecule has the resonance structure

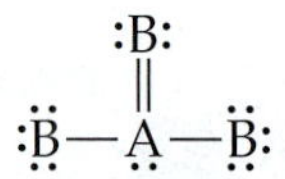

Does this Lewis structure follow the octet rule? How many electron domains are there around the atom A?

Because electron domains are negatively charged, they repel one another. Therefore, like the balloons in Figure 9.5, electron domains try to stay out of one another's way. *The best arrangement of a given number of electron domains is the one that minimizes the repulsions among them.* This simple idea is the basis of the VSEPR model. In fact, the analogy between electron domains and balloons is so close that the same preferred geometries are found in both cases. Thus, like the balloons in Figure 9.5, two electron domains are arranged *linearly*, three domains are arranged in a *trigonal-planar* fashion, and four are arranged *tetrahedrally*. These arrangements, together with those for five electron domains (*trigonal bipyramidal*) and six electron domains (*octahedral*), are summarized in Table 9.1 ▼. If you compare the geometries in Table 9.1 with those in Figure 9.3, you will see that they are the same. *The shapes of different AB_n molecules or ions depend on the number of electron domains surrounding the central A atom.*

TABLE 9.1 Electron-Domain Geometries as a Function of the Number of Electron Domains

Number of Electron Domains	Arrangement of Electron Domains	Electron-Domain Geometry	Predicted Bond Angles
2	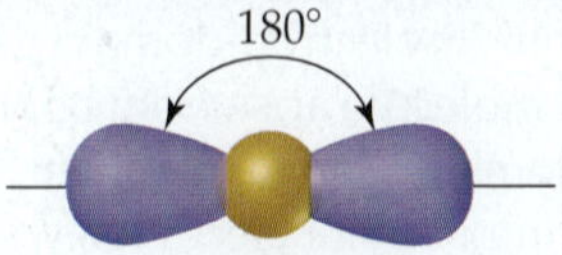	Linear	180°
3	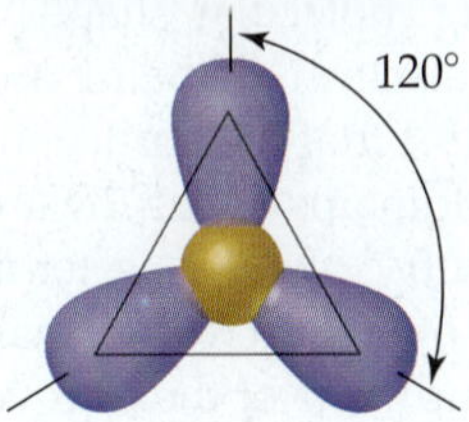	Trigonal planar	120°
4	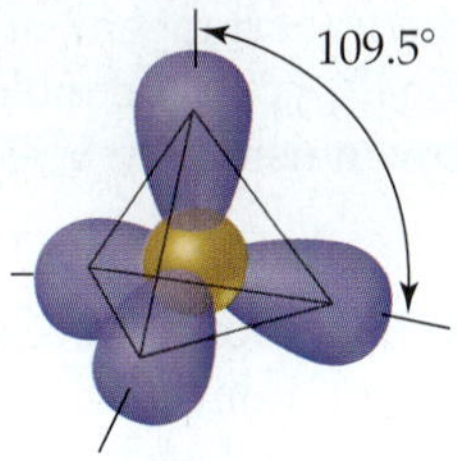	Tetrahedral	109.5°
5	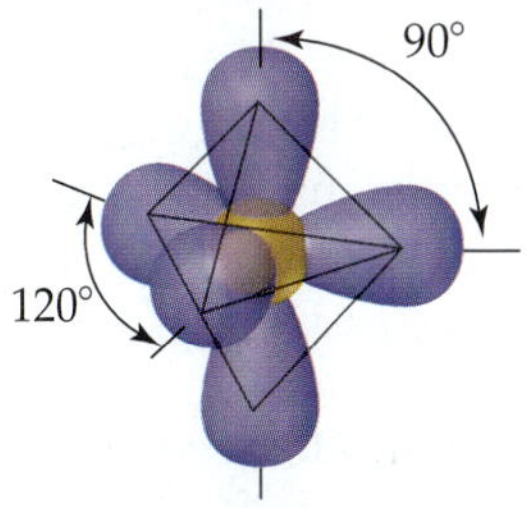	Trigonal bipyramidal	120° 90°
6	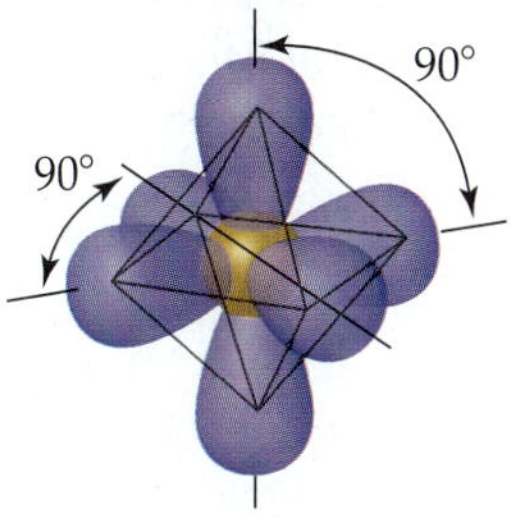	Octahedral	90°

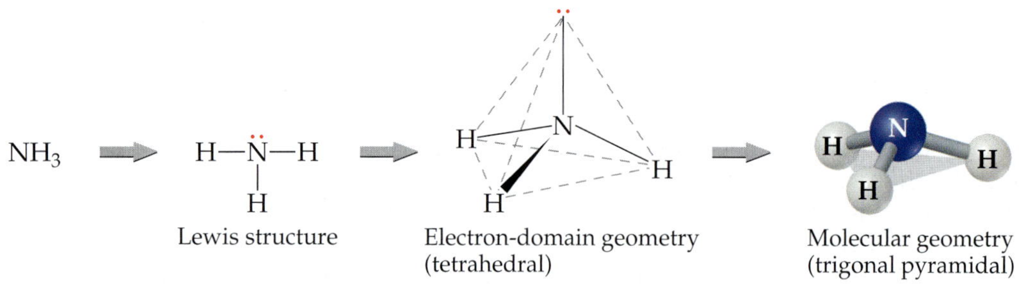

▲ **Figure 9.6 The molecular geometry of NH_3.** The geometry is predicted by first drawing the Lewis structure, then using the VSEPR model to determine the electron-domain geometry, and finally focusing on the atoms themselves to describe the molecular geometry.

The arrangement of electron domains about the central atom of an AB_n molecule or ion is called its **electron-domain geometry**. In contrast, the **molecular geometry** is the arrangement of *only the atoms* in a molecule or ion—any nonbonding pairs are not part of the description of the molecular geometry. In the VSEPR model, we predict the electron-domain geometry, and from knowing how many domains are due to nonbonding pairs, we can then predict the molecular geometry of a molecule or ion from its electron-domain geometry.

When all the electron domains in a molecule arise from bonds, the molecular geometry is identical to the electron-domain geometry. When, however, one or more of the domains involve nonbonding pairs of electrons, we must remember to ignore those domains when predicting molecular shape. Consider the NH_3 molecule, for instance, which has four electron domains around the nitrogen atom (Figure 9.6 ▲). We know from Table 9.1 that the repulsions among four electron domains are minimized when the domains point toward the vertices of a tetrahedron—the electron-domain geometry of NH_3 is tetrahedral. We know from the Lewis structure of NH_3 that one of the electron domains is due to a nonbonding pair of electrons, which will occupy one of the four vertices of the tetrahedron. Hence the molecular geometry of NH_3 is trigonal pyramidal, as shown in Figure 9.6 . Notice that it is the tetrahedral arrangement of the four electron domains that leads us to predict the trigonal-pyramidal molecular geometry.

We can generalize the steps we follow in using the VSEPR model to predict the shapes of molecules or ions:

1. Draw the *Lewis structure* of the molecule or ion, and count the total number of electron domains around the central atom. Each nonbonding electron pair, each single bond, each double bond, and each triple bond counts as an electron domain.
2. Determine the *electron-domain geometry* by arranging the electron domains about the central atom so that the repulsions among them are minimized, as shown in Table 9.1.
3. Use the arrangement of the bonded atoms to determine the *molecular geometry*.

Figure 9.6 shows how these steps are applied to predict the geometry of the NH_3 molecule. Because the trigonal-pyramidal molecular geometry is based on a tetrahedral electron-domain geometry, the *ideal bond angles* are 109.5°. As we will soon see, bond angles deviate from the ideal angles when the surrounding atoms and electron domains are not identical.

Let's apply these steps to determine the shape of the CO_2 molecule. We first draw its Lewis structure, which reveals two electron domains (two double bonds) around the central carbon:

$$:\ddot{O}=C=\ddot{O}:$$

Two electron domains will arrange themselves to give a linear electron-domain geometry (Table 9.1). Because neither of the domains is a nonbonding pair of electrons, the molecular geometry is also linear and the O—C—O bond angle is 180°.

Table 9.2 ▼ summarizes the possible molecular geometries when an AB_n molecule has four or fewer electron domains about A. These geometries are important because they include all the commonly occurring shapes found for molecules or ions that obey the octet rule.

3-D MODELS
VSEPR: Basic Molecular Configurations

TABLE 9.2 Electron-Domain Geometries and Molecular Shapes for Molecules with Two, Three, and Four Electron Domains Around the Central Atom

Number of Electron Domains	Electron-Domain Geometry	Bonding Domains	Nonbonding Domains	Molecular Geometry	Example
2	 Linear	2	0	 Linear	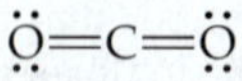
3	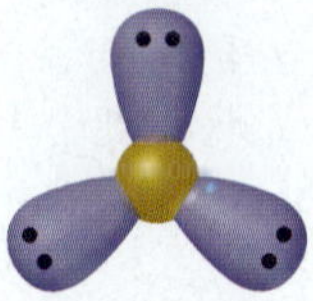 Trigonal planar	3	0	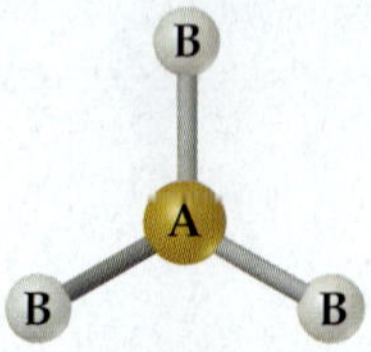 Trigonal planar	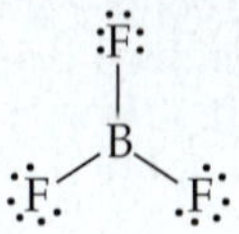
		2	1	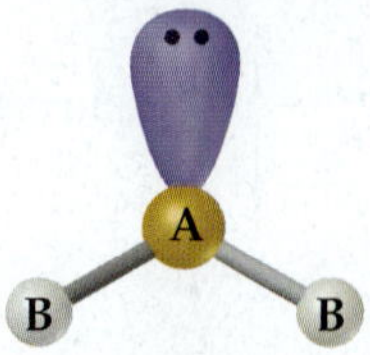 Bent	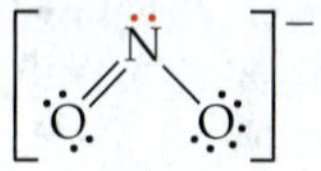
4	 Tetrahedral	4	0	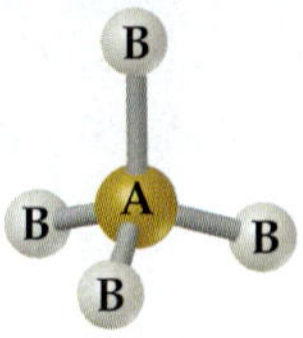 Tetrahedral	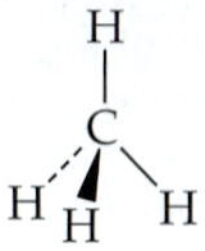
		3	1	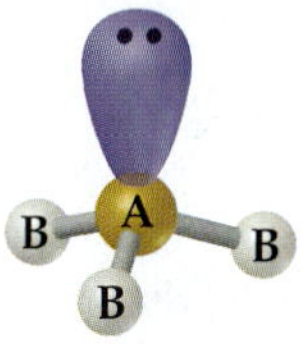 Trigonal pyramidal	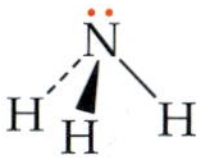
		2	2	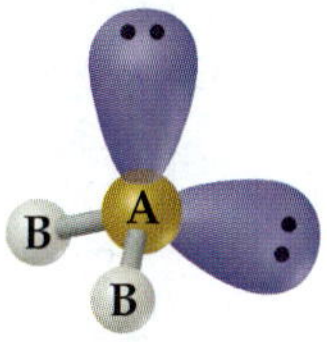 Bent	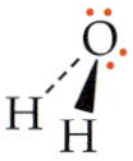

SAMPLE EXERCISE 9.1 | Using the VSEPR Model

Use the VSEPR model to predict the molecular geometry of **(a)** O_3, **(b)** $SnCl_3^-$.

Solution

Analyze: We are given the molecular formulas of a molecule and a polyatomic ion, both conforming to the general formula AB_n and both having a central atom from the *p* block of the periodic table.

Plan: To predict the molecular geometries of these species, we first draw their Lewis structures and then count the number of electron domains around the central atom. The number of electron domains gives the electron-domain geometry. We then obtain the molecular geometry from the arrangement of the domains that are due to bonds.

Solve: (a) We can draw two resonance structures for O_3:

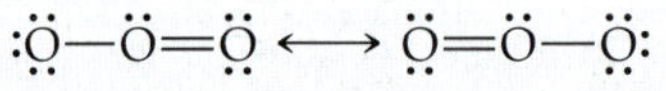

Because of resonance, the bonds between the central O atom and the outer O atoms are of equal length. In both resonance structures the central O atom is bonded to the two outer O atoms and has one nonbonding pair. Thus, there are three electron domains about the central O atoms. (Remember that a double bond counts as a single electron domain.) The best arrangement of three electron domains is trigonal planar (Table 9.1). Two of the domains are from bonds, and one is due to a nonbonding pair, so the molecule has a bent shape with an ideal bond angle of 120° (Table 9.2).

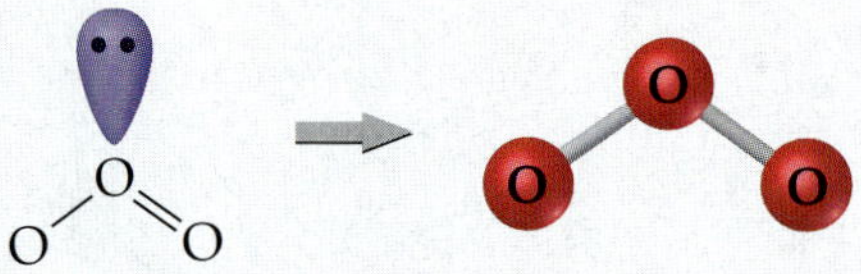

As this example illustrates, when a molecule exhibits resonance, any one of the resonance structures can be used to predict the molecular geometry.

(b) The Lewis structure for the $SnCl_3^-$ ion is

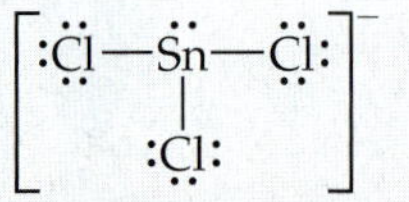

The central Sn atom is bonded to the three Cl atoms and has one nonbonding pair. Therefore, the Sn atom has four electron domains around it. The resulting electron-domain geometry is tetrahedral (Table 9.1) with one of the corners occupied by a nonbonding pair of electrons. The molecular geometry is thus trigonal pyramidal (Table 9.2), like that of NH_3.

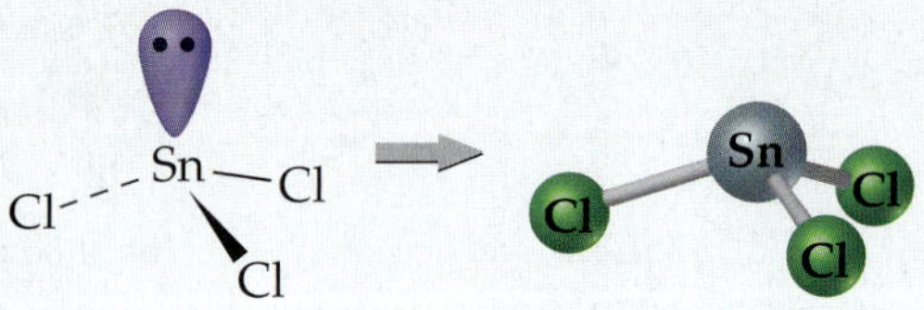

PRACTICE EXERCISE

Predict the electron-domain geometry and the molecular geometry for **(a)** $SeCl_2$, **(b)** CO_3^{2-}.

Answers: **(a)** tetrahedral, bent; **(b)** trigonal planar, trigonal planar

The Effect of Nonbonding Electrons and Multiple Bonds on Bond Angles

We can refine the VSEPR model to predict and explain slight distortions of molecules from the ideal geometries summarized in Table 9.2. For example, consider methane (CH_4), ammonia (NH_3), and water (H_2O). All three have tetrahedral electron-domain geometries, but their bond angles differ slightly:

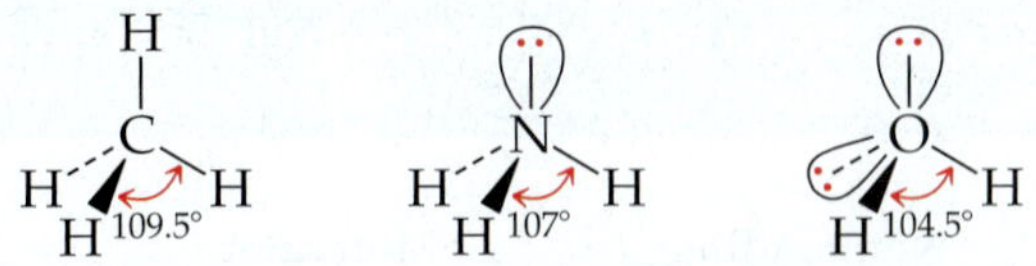

Notice that the bond angles decrease as the number of nonbonding electron pairs increases. A bonding pair of electrons is attracted by both nuclei of the bonded atoms. By contrast, a nonbonding pair is attracted primarily by only one nucleus. Because a nonbonding pair experiences less nuclear attraction, its electron domain is spread out more in space than is the electron domain for a bonding pair, as shown in Figure 9.7 ▶. As a result, *electron domains for nonbonding electron pairs exert greater repulsive forces on adjacent electron domains and thus tend to compress the bond angles.* Using the analogy in Figure 9.5, we can envision the domains for nonbonding electron pairs as represented by balloons that are slightly larger and slightly fatter than those for bonding pairs.

Because multiple bonds contain a higher electronic-charge density than single bonds, multiple bonds also represent larger electron domains ("fatter balloons"). Consider the Lewis structure of *phosgene*, Cl_2CO:

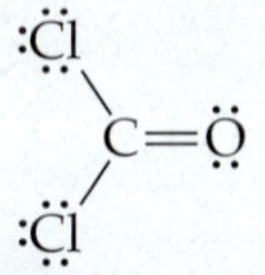

Because the central carbon atom is surrounded by three electron domains, we might expect a trigonal-planar geometry with 120° bond angles. The double bond seems to act much like a nonbonding pair of electrons, however, reducing the Cl—C—Cl bond angle from the ideal angle of 120° to an actual angle of 111.4°:

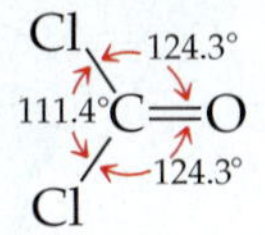

In general, *electron domains for multiple bonds exert a greater repulsive force on adjacent electron domains than do electron domains for single bonds.*

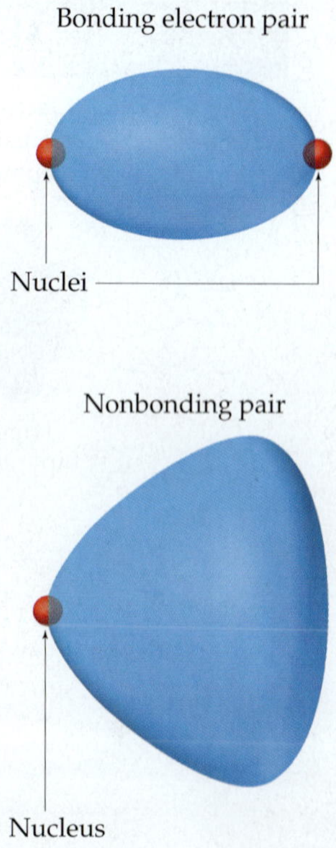

▲ **Figure 9.7 Relative "sizes" of bonding and nonbonding electron domains.**

GIVE IT SOME THOUGHT

One of the resonance structures of the nitrate ion, NO_3^-, is

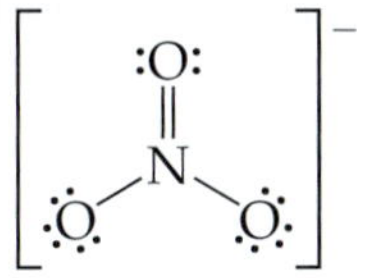

The bond angles in this ion are exactly 120°. Is this observation consistent with the above discussion of the effect of multiple bonds on bond angles?

Molecules with Expanded Valence Shells

Our discussion of the VSEPR model thus far has involved molecules with no more than an octet of electrons around the central atom. Recall, however, that when the central atom of a molecule is from the third period of the periodic table and beyond, that atom may have more than four electron pairs around it. (Section 8.7) Molecules with five or six electron domains around the central atom display a variety of molecular geometries based on the *trigonal-bipyramidal* (five electron domains) or the *octahedral* (six electron domains) electron-domain geometries, as shown in Table 9.3 ▶.

TABLE 9.3 Electron-Domain Geometries and Molecular Shapes for Molecules with Five and Six Electron Domains Around the Central Atom

Total Electron Domains	Electron-Domain Geometry	Bonding Domains	Nonbonding Domains	Molecular Geometry	Example
5	 Trigonal bipyramidal	5	0	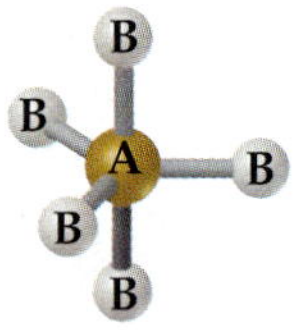 Trigonal bipyramidal	PCl_5
		4	1	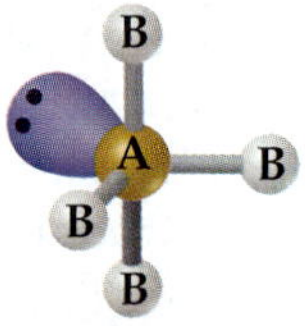 Seesaw	SF_4
		3	2	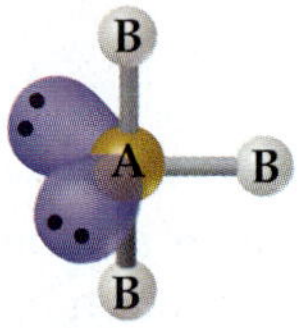 T-shaped	ClF_3
		2	3	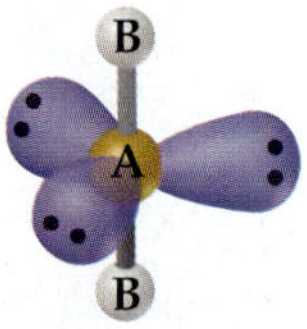 Linear	XeF_2
6	 Octahedral	6	0	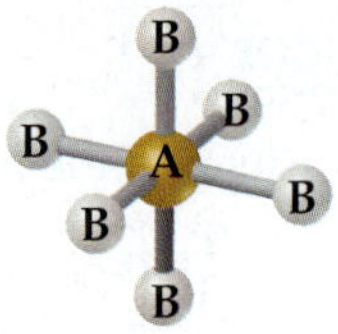 Octahedral	SF_6
		5	1	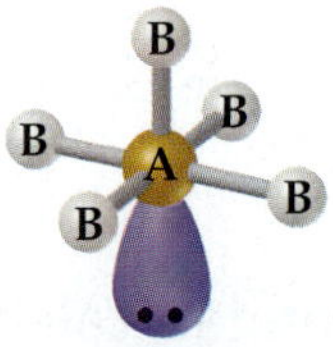 Square pyramidal	BrF_5
		4	2	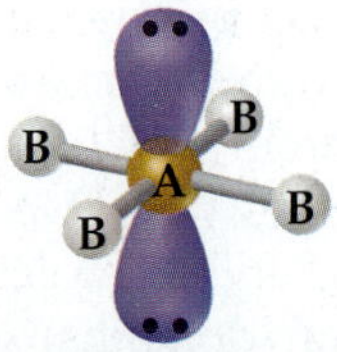 Square planar	XeF_4

The most stable electron-domain geometry for five electron domains is the trigonal bipyramid (two trigonal pyramids sharing a base). Unlike the arrangements we have seen to this point, the electron domains in a trigonal bipyramid can point toward two geometrically distinct types of positions. Two of the five domains point toward what are called *axial positions*, and the remaining three domains point toward *equatorial positions* (Figure 9.8 ▶). Each axial domain makes a 90° angle with any equatorial domain. Each equatorial domain makes a 120° angle with either of the other two equatorial domains and a 90° angle with either axial domain.

Suppose a molecule has five electron domains, one or more of which originates from a nonbonding pair. Will the electron domains from the nonbonding pairs occupy axial or equatorial positions? In order to answer this question, we must determine which location minimizes the repulsions between the electron domains. Repulsions between domains are much greater when they are situated 90° from each other than when they are at 120°. An equatorial domain is 90° from only two other domains (the two axial domains). By contrast, an axial domain is situated 90° from *three* other domains (the three equatorial domains). Hence, an equatorial domain experiences less repulsion than an axial domain. Because the domains from nonbonding pairs exert larger repulsions than those from bonding pairs, they always occupy the equatorial positions in a trigonal bipyramid.

The most stable electron-domain geometry for six electron domains is the *octahedron*. As shown in Figure 9.9 ▶, an octahedron is a polyhedron with six vertices and eight faces, each of which is an equilateral triangle. If an atom has six electron domains around it, that atom can be visualized as being at the center of the octahedron with the electron domains pointing toward the six vertices. All the bond angles in an octahedron are 90°, and all six vertices are equivalent. Therefore, if an atom has five bonding electron domains and one nonbonding domain, we can point the nonbonding domain at any of the six vertices of the octahedron. The result is always a *square-pyramidal* molecular geometry. When there are two nonbonding electron domains, however, their repulsions are minimized by pointing them toward opposite sides of the octahedron, producing a *square-planar* molecular geometry, as shown in Table 9.3.

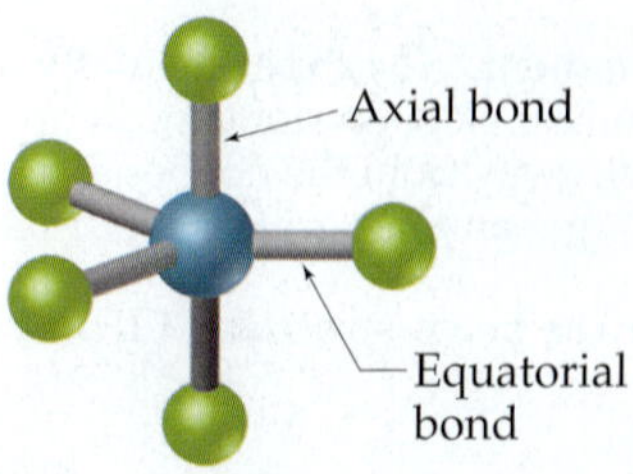

▲ **Figure 9.8 Trigonal-bipyramidal geometry.** Five electron domains arrange themselves around a central atom as a trigonal bipyramid. The three *equatorial* electron domains define an equilateral triangle. The two *axial* domains lie above and below the plane of the triangle. If a molecule has nonbonding electron domains, they will occupy the equatorial positions.

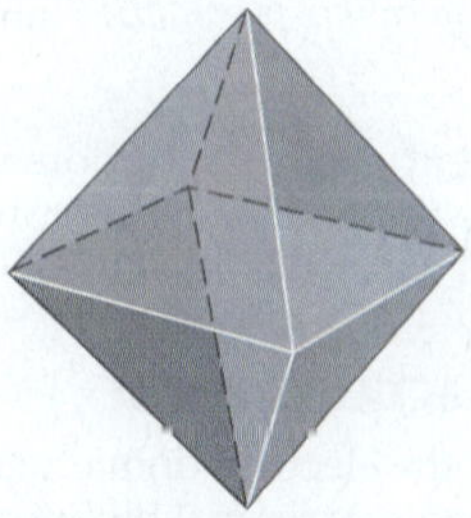

▲ **Figure 9.9 An octahedron.** The octahedron is an object with eight faces and six vertices. Each face is an equilateral triangle.

SAMPLE EXERCISE 9.2 | Molecular Geometries of Molecules with Expanded Valence Shells

Use the VSEPR model to predict the molecular geometry of **(a)** SF_4, **(b)** IF_5.

Solution

Analyze: The molecules are of the AB_n type with a central atom from the *p* block of the periodic table.

Plan: We can predict their structures by first drawing Lewis structures and then using the VSEPR model to determine the electron-domain geometry and molecular geometry.

Solve: (a) The Lewis structure for SF_4 is

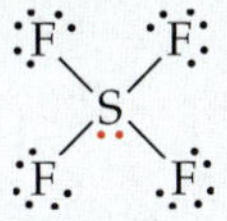

The sulfur has five electron domains around it: four from the S—F bonds and one from the nonbonding pair. Each domain points toward a vertex of a trigonal bipyramid. The domain from the nonbonding pair will point toward an equatorial position. The four bonds point toward the remaining four positions, resulting in a molecular geometry that is described as seesaw-shaped:

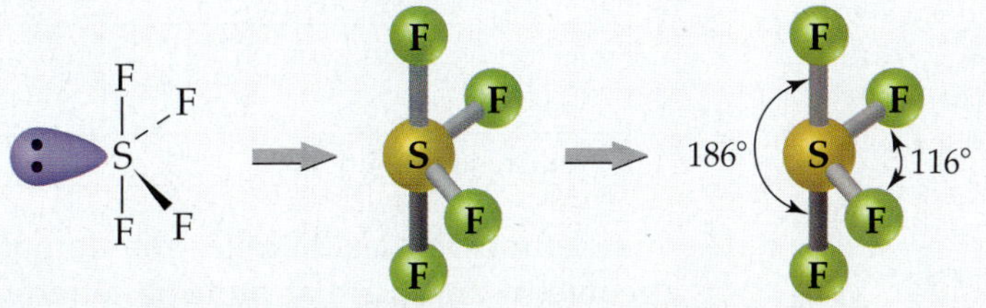

Comment: The experimentally observed structure is shown on the previous page on the right, and we can infer that the nonbonding electron domain occupies an equatorial position, as predicted. The axial and equatorial S—F bonds are slightly bent back away from the nonbonding domain, suggesting that the bonding domains are "pushed" by the nonbonding domain, which is larger and has greater repulsion (Figure 9.7).

(b) The Lewis structure of IF_5 is

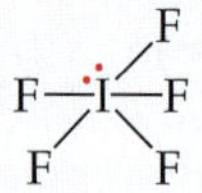

(There are three lone pairs on each of the F atoms, but they are not shown.)

The iodine has six electron domains around it, one of which is from a nonbonding pair. The electron-domain geometry is therefore octahedral, with one position occupied by the nonbonding electron pair. The resulting molecular geometry is therefore *square pyramidal* (Table 9.3):

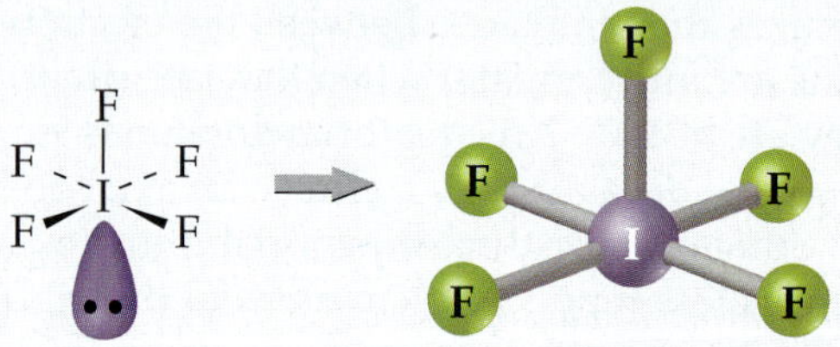

Comment: Because the domain for the nonbonding pair is larger than the other domains, the four F atoms in the base of the pyramid are tipped up slightly toward the F atom on top. Experimentally, it is found that the angle between the base and top F atoms is 82°, smaller than the ideal 90° angle of an octahedron.

PRACTICE EXERCISE

Predict the electron-domain geometry and molecular geometry of **(a)** ClF_3, **(b)** ICl_4^-.
Answers: **(a)** trigonal bipyramidal, T-shaped; **(b)** octahedral, square planar

Shapes of Larger Molecules

Although the molecules and ions whose structures we have thus far considered contain only a single central atom, the VSEPR model can be extended to more complex molecules. Consider the acetic acid molecule, whose Lewis structure is

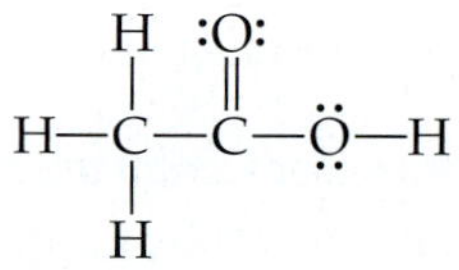

Acetic acid has three interior atoms: the leftmost C atom, the central C atom, and the rightmost O atom. We can use the VSEPR model to predict the geometry about each of these atoms individually:

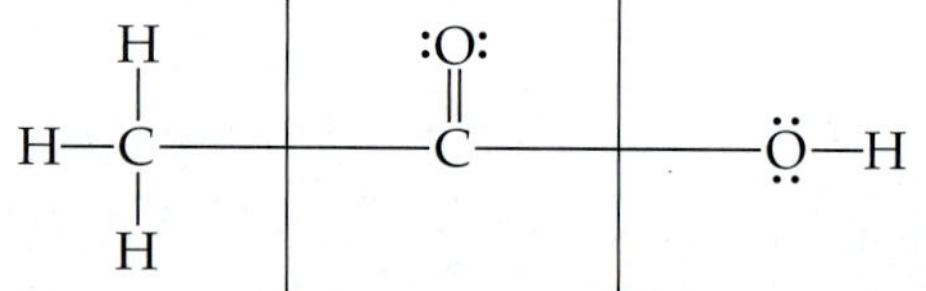

Number of electron domains	4	3	4
Electron-domain geometry	Tetrahedral	Trigonal planar	Tetrahedral
Predicted bond angles	109.5°	120°	109.5°

The leftmost C has four electron domains (all from bonding pairs), and so the geometry around that atom is tetrahedral. The central C has three electron domains (counting the double bond as one domain). Thus, the geometry around that atom is trigonal planar. The O atom has four electron domains (two from bonding pairs and two from nonbonding pairs), so its electron-domain geometry is tetrahedral and the molecular geometry around O is bent. The bond angles about the central C atom and the O atom are expected to deviate slightly from the ideal values of 120° and 109.5°, because of the spatial demands of multiple bonds and nonbonding electron pairs. The structure of the acetic acid molecule is shown in Figure 9.10 ▶.

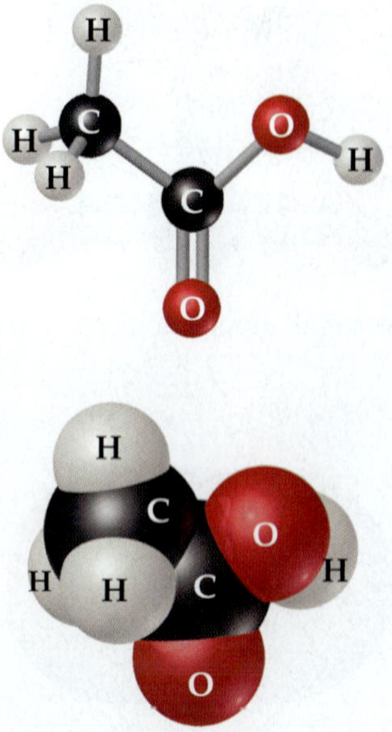

▲ **Figure 9.10 Ball-and-stick (top) and space-filling (bottom) representations of acetic acid, $HC_2H_3O_2$.**

SAMPLE EXERCISE 9.3 | Predicting Bond Angles

Eyedrops for dry eyes usually contain a water-soluble polymer called *poly(vinyl alcohol)*, which is based on the unstable organic molecule called *vinyl alcohol*:

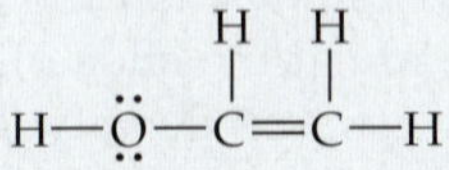

Predict the approximate values for the H—O—C and O—C—C bond angles in vinyl alcohol.

Solution

Analyze We are given a molecular structure and asked to determine two bond angles in the structure.

Plan: To predict a particular bond angle, we consider the middle atom of the angle and determine the number of electron domains surrounding that atom. The ideal angle corresponds to the electron-domain geometry around the atom. The angle will be compressed somewhat by nonbonding electrons or multiple bonds.

Solve: For the H—O—C bond angle, there are four electron domains around the middle O atom (two bonding and two nonbonding). The electron-domain geometry around O is therefore tetrahedral, which gives an ideal angle of 109.5°. The H—O—C angle will be compressed somewhat by the nonbonding pairs, so we expect this angle to be slightly less than 109.5°.

To predict the O—C—C bond angle, we must examine the leftmost C atom, which is the central atom for this angle. There are three atoms bonded to this C atom and no nonbonding pairs, and so it has three electron domains about it. The predicted electron-domain geometry is trigonal planar, resulting in an ideal bond angle of 120°. Because of the larger size of the C=C domain, however, the O—C—C bond angle should be slightly greater than 120°.

PRACTICE EXERCISE

Predict the H—C—H and C—C—C bond angles in the following molecule, called *propyne*:

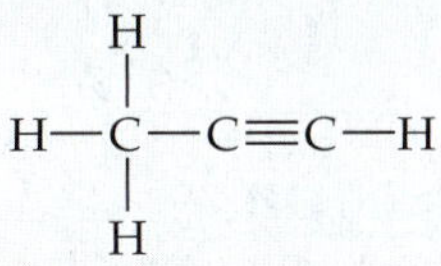

Answers: 109.5°, 180°

9.3 | Molecular Shape and Molecular Polarity

We now have a sense of the shapes that molecules adopt and why they do so. We will spend the rest of this chapter looking more closely at the ways in which electrons are shared to form the bonds between atoms in molecules. We will begin by returning to a topic that we first discussed in Section 8.4, namely *bond polarity* and *dipole moments*.

Recall that bond polarity is a measure of how equally the electrons in a bond are shared between the two atoms of the bond: As the difference in electronegativity between the two atoms increases, so does the bond polarity. (Section 8.4) We saw that the dipole moment of a diatomic molecule is a quantitative measure of the amount of charge separation in the molecule. The charge separation in molecules has a significant effect on physical and chemical properties. We will see in Chapter 11, for example, how molecular polarity affects boiling points, melting points, and other physical properties.

For a molecule that consists of more than two atoms, *the dipole moment depends on both the polarities of the individual bonds and the geometry of the molecule.* For each bond in the molecule, we can consider the **bond dipole**, which is the dipole moment that is due only to the two atoms in that bond. Consider the linear CO_2 molecule, for example. As shown in Figure 9.11(a) ◀, each C=O bond is polar, and because the C=O bonds are identical, the bond dipoles are equal in magnitude. A plot of the electron density of the CO_2 molecule, shown in Figure 9.11(b), clearly shows that the bonds are polar—the regions of high electron density (red) are at the ends of the molecule, on the oxygen atoms; and the regions of low electron density (blue) are in the center, on the carbon atom. But what can we say about the *overall* dipole moment of the CO_2 molecule?

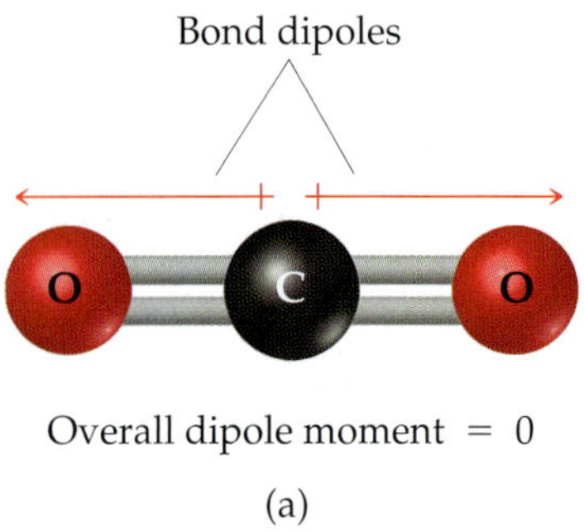

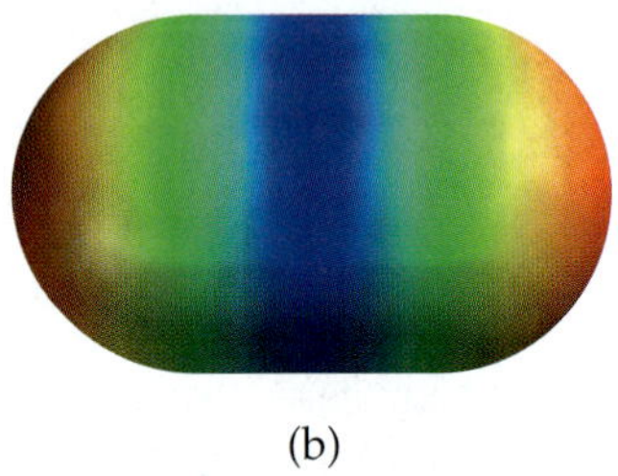

▲ **Figure 9.11 CO_2, a nonpolar molecule.** (a) The overall dipole moment of a molecule is the sum of its bond dipoles. In CO_2 the bond dipoles are equal in magnitude, but exactly oppose each other. The overall dipole moment is zero, therefore, making the molecule nonpolar. (b) The electron-density model shows that the regions of higher electron density (red) are at the ends of the molecule while the region of lower electron density (blue) is at the center.

Bond dipoles and dipole moments are *vector* quantities; that is, they have both a magnitude and a direction. The *overall* dipole moment of a polyatomic molecule is the vector sum of its bond dipoles. Both the magnitudes *and* the directions of the bond dipoles must be considered when summing these vectors. The two bond dipoles in CO_2, although equal in magnitude, are exactly opposite in direction. Adding them together is the same as adding two numbers that are equal in magnitude but opposite in sign, such as $100 + (-100)$: The bond dipoles, like the numbers, "cancel" each other. Therefore, the overall dipole moment of CO_2 is zero, even though the individual bonds are polar. Thus, the geometry of the molecule dictates that the overall dipole moment be zero, making CO_2 a *nonpolar* molecule.

Now let's consider H_2O, which is a bent molecule with two polar bonds (Figure 9.12 ◀). Again, the two bonds in the molecule are identical, and so the bond dipoles are equal in magnitude. Because the molecule is bent, however, the bond dipoles do not directly oppose each other and therefore do not cancel each other. Hence, the H_2O molecule has an overall nonzero dipole moment ($\mu = 1.85$ D). Because H_2O has a nonzero dipole moment, it is a *polar* molecule. The oxygen atom carries a partial negative charge, and the hydrogen atoms each have a partial positive charge, as shown by the electron-density model in Figure 9.12(b).

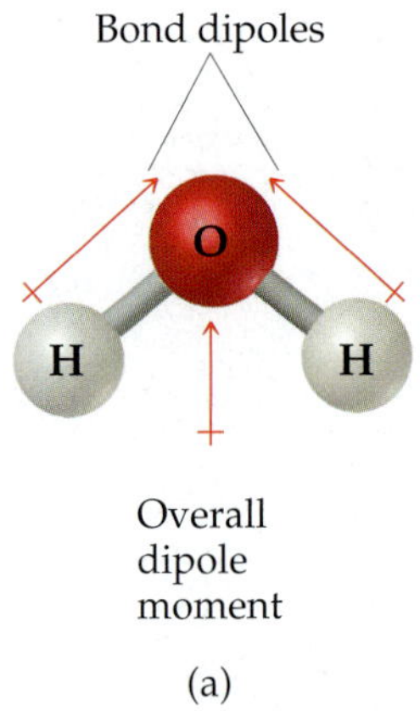

▲ **Figure 9.12 The dipole moment of a bent molecule.** (a) In H_2O the bond dipoles are equal in magnitude, but do not exactly oppose each other. The molecule has a nonzero dipole moment overall, making the molecule polar. (b) The electron-density model shows that one end of the molecule has more electron density (the oxygen end) while the other end has less electron density (the hydrogens).

GIVE IT SOME THOUGHT

The molecule O=C=S has a Lewis structure analogous to that of CO_2 and is a linear molecule. Will it necessarily have a zero dipole moment like CO_2?

Figure 9.13 ▶ shows examples of polar and nonpolar molecules, all of which have polar bonds. The molecules in which the central atom is symmetrically surrounded by identical atoms (BF_3 and CCl_4) are nonpolar. For AB_n molecules in which all the B atoms are the same, certain symmetrical shapes—linear (AB_2), trigonal planar (AB_3), tetrahedral and square planar (AB_4), trigonal bipyramidal (AB_5), and octahedral (AB_6)—must lead to nonpolar molecules even though the individual bonds might be polar.

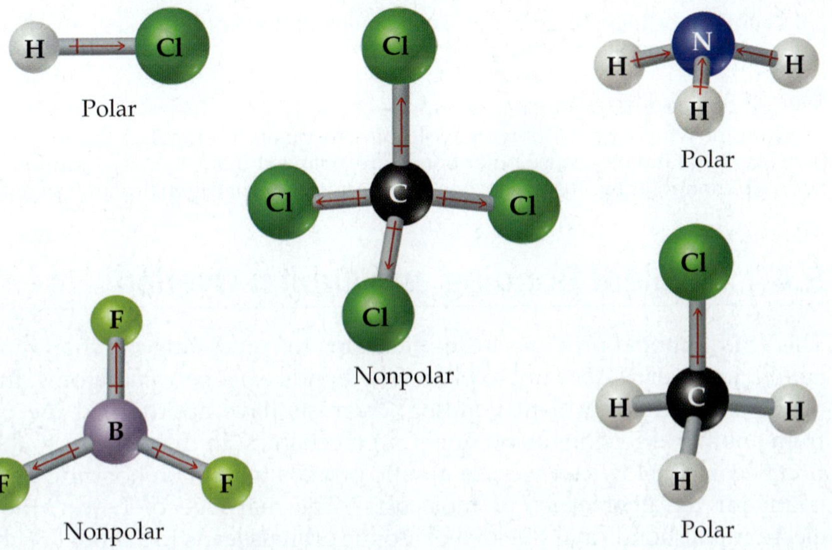

◀ **Figure 9.13 Molecules containing polar bonds.** Two of these molecules have a zero dipole moment because their bond dipoles cancel one another, while the other molecules are polar.

ACTIVITY
Molecular Polarity

SAMPLE EXERCISE 9.4 | Polarity of Molecules

Predict whether the following molecules are polar or nonpolar: **(a)** BrCl, **(b)** SO_2, **(c)** SF_6.

Solution

Analyze: We are given the molecular formulas of several substances and asked to predict whether the molecules are polar.

Plan: If the molecule contains only two atoms, it will be polar if the atoms differ in electronegativity. If it contains three or more atoms, its polarity depends on both its molecular geometry and the polarity of its bonds. Thus, we must draw a Lewis structure for each molecule containing three or more atoms and determine its molecular geometry. We then use the relative electronegativities of the atoms in each bond to determine the direction of the bond dipoles. Finally, we see if the bond dipoles cancel each other to give a nonpolar molecule or reinforce each other to give a polar one.

Solve: **(a)** Chlorine is more electronegative than bromine. All diatomic molecules with polar bonds are polar molecules. Consequently, BrCl will be polar, with chlorine carrying the partial negative charge:

Br—Cl

The actual dipole moment of BrCl, as determined by experimental measurement, is $\mu = 0.57$ D.

(b) Because oxygen is more electronegative than sulfur, SO_2 has polar bonds. Three resonance forms can be written for SO_2:

$$:\ddot{O}-\ddot{S}=\ddot{O}: \longleftrightarrow :\ddot{O}=\ddot{S}-\ddot{O}: \longleftrightarrow :\ddot{O}=\ddot{S}=\ddot{O}:$$

For each of these, the VSEPR model predicts a bent geometry. Because the molecule is bent, the bond dipoles do not cancel and the molecule is polar:

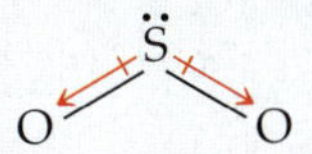

Experimentally, the dipole moment of SO_2 is $\mu = 1.63$ D.

(c) Fluorine is more electronegative than sulfur, so the bond dipoles point toward fluorine. The six S—F bonds are arranged octahedrally around the central sulfur:

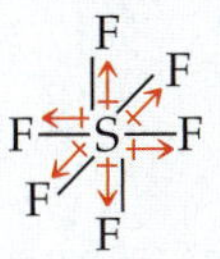

Because the octahedral geometry is symmetrical, the bond dipoles cancel, and the molecule is nonpolar, meaning that $\mu = 0$.

PRACTICE EXERCISE

Determine whether the following molecules are polar or nonpolar: **(a)** NF_3, **(b)** BCl_3.
Answers: **(a)** polar because polar bonds are arranged in a trigonal-pyramidal geometry, **(b)** nonpolar because polar bonds are arranged in a trigonal-planar geometry

9.4 | Covalent Bonding and Orbital Overlap

The VSEPR model provides a simple means for predicting the shapes of molecules. However, it does not explain why bonds exist between atoms. In developing theories of covalent bonding, chemists have approached the problem from another direction, using quantum mechanics. In this approach, the question becomes, How can we use atomic orbitals to explain bonding and to account for the geometries of molecules? The marriage of Lewis's notion of electron-pair bonds and the idea of atomic orbitals leads to a model of chemical bonding called **valence-bond theory**. By extending this approach to include the ways in which atomic orbitals can mix with one another, we obtain a picture that corresponds nicely to the VSEPR model.

In the Lewis theory, covalent bonding occurs when atoms share electrons. Such sharing concentrates electron density between the nuclei. In the valence-bond theory, the buildup of electron density between two nuclei is visualized as occurring when a valence atomic orbital of one atom merges with that of another atom. The orbitals are then said to share a region of space, or to **overlap**. The overlap of orbitals allows two electrons of opposite spin to share the common space between the nuclei, forming a covalent bond.

The coming together of two H atoms to form H_2 is depicted in Figure 9.14(a) ▼. Each atom has a single electron in a 1*s* orbital. As the orbitals overlap, electron density is concentrated between the nuclei. Because the electrons in the overlap region are simultaneously attracted to both nuclei, they hold the atoms together, forming a covalent bond.

The idea of orbital overlap producing a covalent bond applies equally well to other molecules. In HCl, for example, chlorine has the electron configuration $[Ne]3s^2 3p^5$. All the valence orbitals of chlorine are full except one 3*p* orbital, which contains a single electron. This electron pairs up with the single electron of H to form a covalent bond. Figure 9.14(b) shows the overlap of the 3*p* orbital of Cl with the 1*s* orbital of H. Likewise, we can explain the covalent bond in the Cl_2 molecule in terms of the overlap of the 3*p* orbital of one atom with the 3*p* orbital of another, as shown in Figure 9.14(c).

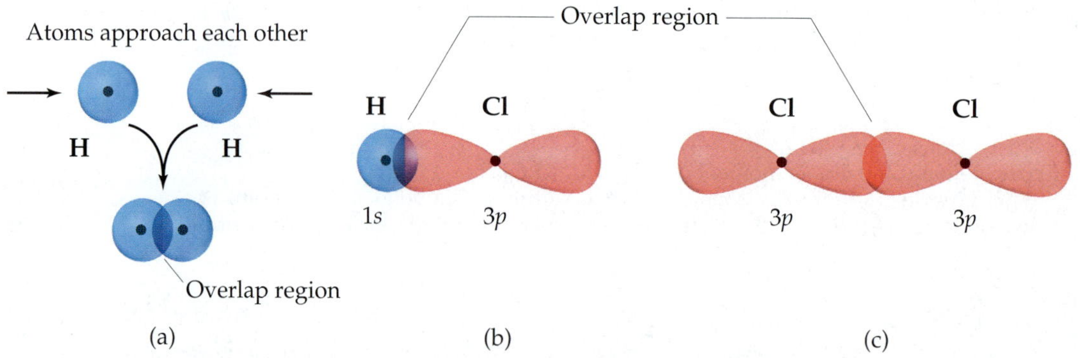

▲ **Figure 9.14 The overlap of orbitals to form covalent bonds.** (a) The bond in H_2 results from the overlap of two 1*s* orbitals from two H atoms. (b) The bond in HCl results from the overlap of a 1*s* orbital of H and one of the lobes of a 3*p* orbital of Cl. (c) The bond in Cl_2 results from the overlap of two 3*p* orbitals from two Cl atoms.

There is always an optimum distance between the two bonded nuclei in any covalent bond. Figure 9.15 ▶ shows how the potential energy of the system changes as two H atoms come together to form an H_2 molecule. As the distance between the atoms decreases, the overlap between their 1*s* orbitals increases. Because of the resultant increase in electron density between the nuclei, the potential energy of the system decreases. That is, the strength of the bond increases, as shown by the decrease in the energy on the curve. However, the curve also shows that as the atoms come very close together, the energy increases rapidly. This rapid increase is due mainly to the electrostatic repulsion between the nuclei, which becomes significant at short internuclear distances. The internuclear distance at a minimum of the potential-energy curve corresponds to the observed bond length. Thus, the observed bond length is the distance at which the attractive forces between unlike charges (electrons and nuclei) are balanced by the repulsive forces between like charges (electron-electron and nucleus-nucleus).

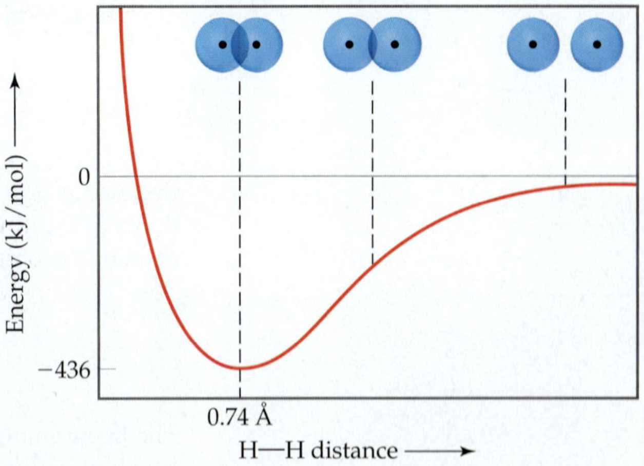

▲ **Figure 9.15 Formation of the H_2 molecule.** Plot of the change in potential energy as two hydrogen atoms come together to form the H_2 molecule. The minimum in the energy, at 0.74 Å, represents the equilibrium bond distance. The energy at that point, −436 kJ/mol, corresponds to the energy change for formation of the H—H bond.

GIVE IT SOME THOUGHT

Can orbital overlap explain why the bond length in Cl_2 is longer than the bond length in F_2?

9.5 | Hybrid Orbitals

ANIMATION
H_2 Bond Formation

Although the idea of orbital overlap allows us to understand the formation of covalent bonds, it is not always easy to extend these ideas to polyatomic molecules. When we apply valence-bond theory to polyatomic molecules, we must explain both the formation of electron-pair bonds *and* the observed geometries of the molecules.

To explain geometries, we often assume that the atomic orbitals on an atom mix to form new orbitals called **hybrid orbitals**. The shape of any hybrid orbital is different from the shapes of the original atomic orbitals. The process of mixing atomic orbitals as atoms approach each other to form bonds is called **hybridization**. The total number of atomic orbitals on an atom remains constant, however, and so the number of hybrid orbitals on an atom equals the number of atomic orbitals mixed.

Let's examine the common types of hybridization. As we do so, notice the connection between the type of hybridization and the five basic electron-domain geometries—linear, trigonal planar, tetrahedral, trigonal bipyramidal, and octahedral—predicted by the VSEPR model.

ANIMATION
Hybridization

sp Hybrid Orbitals

To illustrate the process of hybridization, consider the BeF_2 molecule, which is generated when solid BeF_2 is heated to high temperatures. The Lewis structure of BeF_2 is

:F̤̈—Be—F̤̈:

The VSEPR model correctly predicts that BeF_2 is linear with two identical Be—F bonds, but how can we use valence-bond theory to describe the bonding? The electron configuration of F ($1s^22s^22p^5$) indicates there is an unpaired electron in a 2*p* orbital. This 2*p* electron can be paired with an unpaired electron from the Be atom to form a polar covalent bond. Which orbitals on the Be atom, however, overlap with those on the F atoms to form the Be—F bonds?

The orbital diagram for a ground-state Be atom is

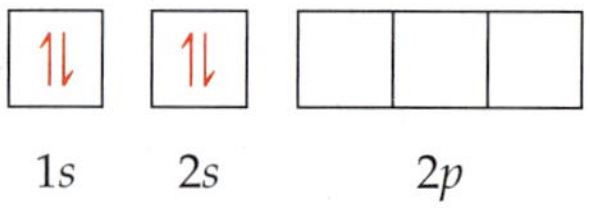

Because it has no unpaired electrons, the Be atom in its ground state is incapable of forming bonds with the fluorine atoms. It could form two bonds, however, by "promoting" one of the 2*s* electrons to a 2*p* orbital:

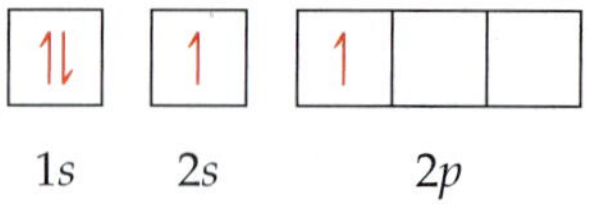

The Be atom now has two unpaired electrons and can therefore form two polar covalent bonds with the F atoms. The two bonds would not be identical, however, because a Be 2*s* orbital would be used to form one of the bonds and a 2*p* orbital would be used for the other. Therefore, although the promotion of an electron allows two Be—F bonds to form, we still haven't explained the structure of BeF_2.

We can solve this dilemma by "mixing" the 2*s* orbital and one of the 2*p* orbitals to generate two new orbitals, as shown in Figure 9.16 ▼. Like *p* orbitals, each of the new orbitals has two lobes. Unlike *p* orbitals, however, one lobe is much larger than the other. The two new orbitals are identical in shape, but their large lobes point in opposite directions. We have created two hybrid orbitals. In this case we have hybridized one *s* and one *p* orbital, so we call each hybrid an *sp* hybrid orbital. *According to the valence-bond model, a linear arrangement of electron domains implies* sp *hybridization.*

For the Be atom of BeF_2, we write the orbital diagram for the formation of two *sp* hybrid orbitals as follows:

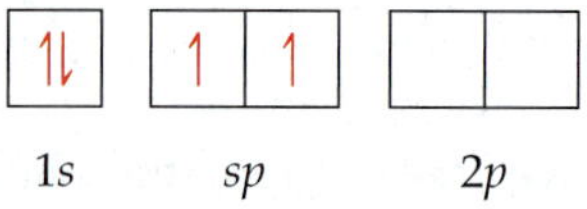

The electrons in the *sp* hybrid orbitals can form two-electron bonds with the two fluorine atoms (Figure 9.17 ▶). Because the *sp* hybrid orbitals are equivalent but point in opposite directions, BeF_2 has two identical bonds and a linear geometry. The remaining two 2*p* orbitals remain unhybridized.

Our first step in constructing the *sp* hybrids, namely the promotion of a 2*s* electron to a 2*p* orbital in Be, requires energy. Why, then, do the orbitals hybridize? Hybrid orbitals have one large lobe and can therefore be directed at other atoms better than can unhybridized atomic orbitals. Hence, they can overlap more strongly with the orbitals of other atoms than atomic orbitals, and

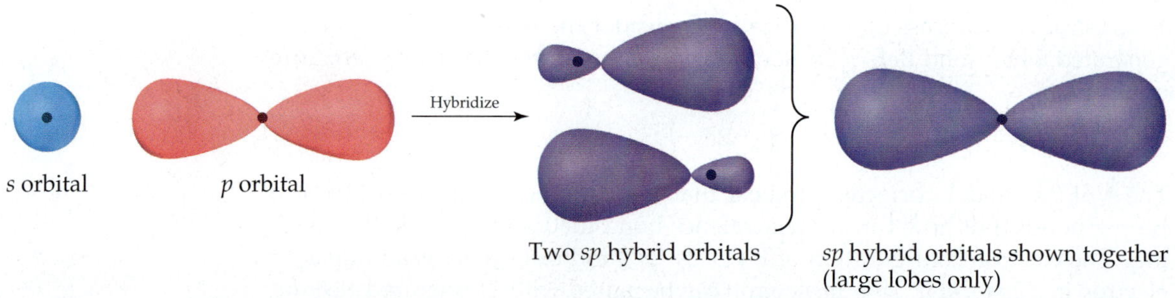

▲ **Figure 9.16 Formation of *sp* hybrid orbitals.** One *s* orbital and one *p* orbital can hybridize to form two equivalent *sp* hybrid orbitals. The two hybrid orbitals have their large lobes pointing in opposite directions, 180° apart.

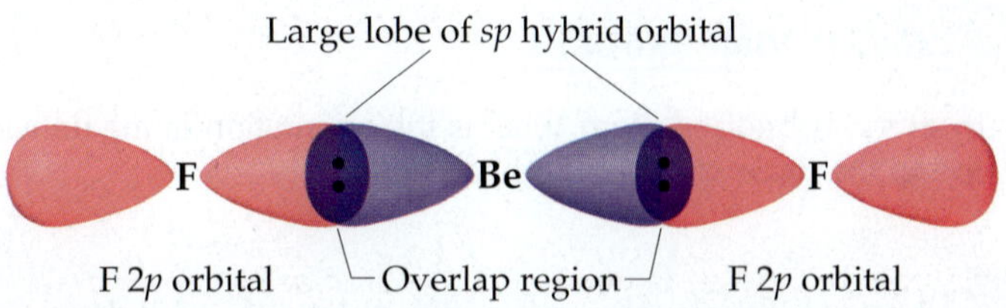

◀ **Figure 9.17 Formation of two equivalent Be—F bonds in BeF_2.** Each *sp* hybrid orbital on Be overlaps with a 2*p* orbital on F to form a bond. The two bonds are equivalent to each other and form an angle of 180°.

stronger bonds result. The energy released by the formation of bonds more than offsets the energy that must be expended to promote electrons.

GIVE IT SOME THOUGHT

Suppose that two unhybridized 2*p* orbitals were used to make the Be—F bonds in BeF_2. Would the two bonds be equivalent to each other? What would be the expected F—Be—F bond angle?

sp^2 and sp^3 Hybrid Orbitals

Whenever we mix a certain number of atomic orbitals, we get the same number of hybrid orbitals. Each of these hybrid orbitals is equivalent to the others but points in a different direction. Thus, mixing one 2*s* and one 2*p* orbital yields two equivalent *sp* hybrid orbitals that point in opposite directions (Figure 9.16). Other combinations of atomic orbitals can be hybridized to obtain different geometries. In BF_3, for example, a 2*s* electron on the B atom can be promoted to a vacant 2*p* orbital. Mixing the 2*s* and two of the 2*p* orbitals yields three equivalent sp^2 (pronounced "s-p-two") hybrid orbitals:

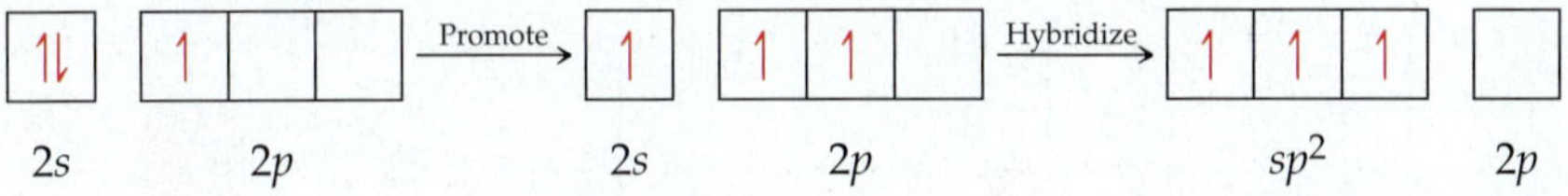

The three sp^2 hybrid orbitals lie in the same plane, 120° apart from one another (Figure 9.18 ▼). They are used to make three equivalent bonds with the three fluorine atoms, leading to the trigonal-planar geometry of BF_3. Notice that an unfilled 2*p* orbital remains unhybridized. This unhybridized orbital will be important when we discuss double bonds in Section 9.6.

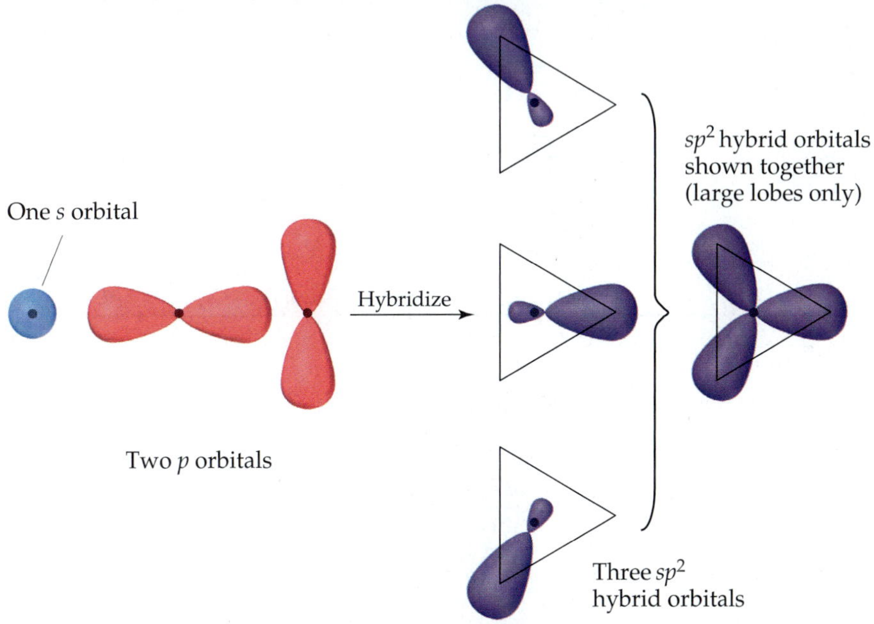

◀ **Figure 9.18 Formation of sp^2 hybrid orbitals.** One *s* orbital and two *p* orbitals can hybridize to form three equivalent sp^2 hybrid orbitals. The large lobes of the hybrid orbitals point toward the corners of an equilateral triangle.

GIVE IT SOME THOUGHT

In an sp^2 hybridized atom, what is the orientation of the unhybridized p orbital relative to the three sp^2 hybrid orbitals?

An s orbital can also mix with all three p orbitals in the same subshell. For example, the carbon atom in CH_4 forms four equivalent bonds with the four hydrogen atoms. We envision this process as resulting from the mixing of the 2s and all three 2p atomic orbitals of carbon to create four equivalent sp^3 (pronounced "s-p-three") hybrid orbitals:

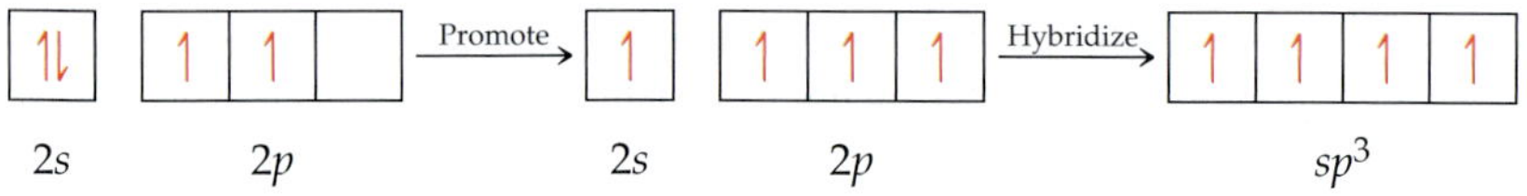

Each sp^3 hybrid orbital has a large lobe that points toward a vertex of a tetrahedron, as shown in Figure 9.19 ▼. These hybrid orbitals can be used to form two-electron bonds by overlap with the atomic orbitals of another atom, such as H. Thus, using valence-bond theory, we can describe the bonding in CH_4 as the overlap of four equivalent sp^3 hybrid orbitals on C with the 1s orbitals of the four H atoms to form four equivalent bonds.

The idea of hybridization is used in a similar way to describe the bonding in molecules containing nonbonding pairs of electrons. In H_2O, for example,

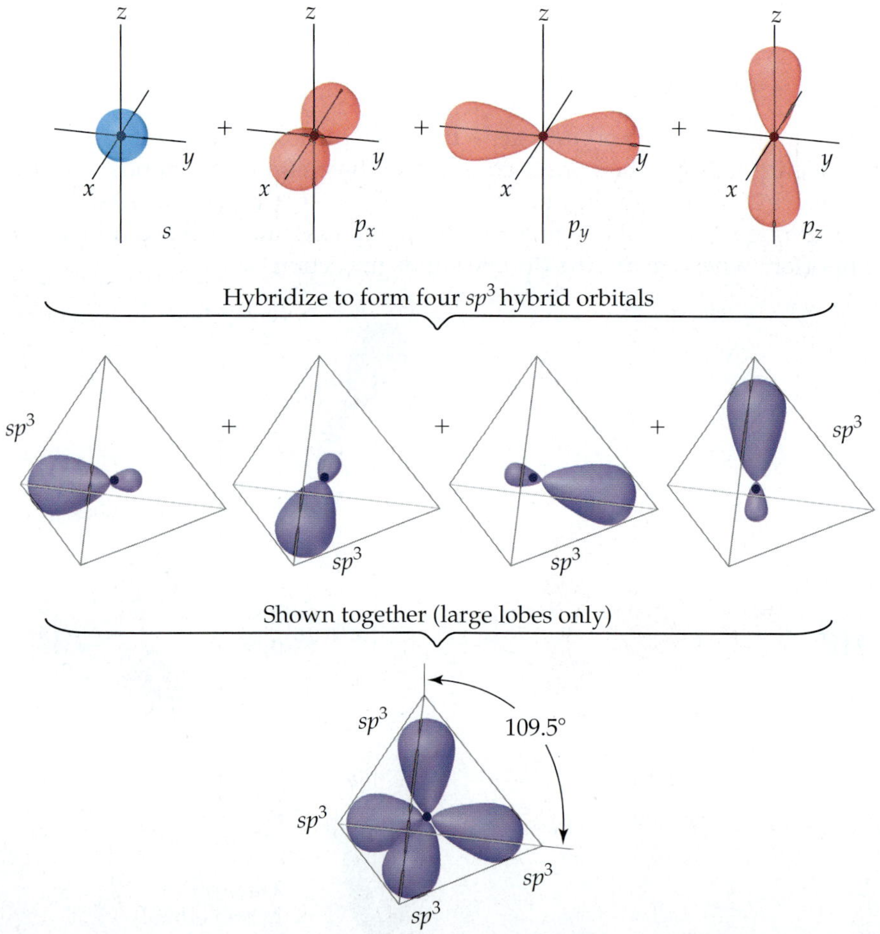

▶ **Figure 9.19 Formation of sp^3 hybrid orbitals.** One s orbital and three p orbitals can hybridize to form four equivalent sp^3 hybrid orbitals. The large lobes of the hybrid orbitals point toward the corners of a tetrahedron.

the electron-domain geometry around the central O atom is approximately tetrahedral. Thus, the four electron pairs can be envisioned as occupying sp^3 hybrid orbitals. Two of the hybrid orbitals contain nonbonding pairs of electrons, while the other two are used to form bonds with hydrogen atoms, as shown in Figure 9.20 ▶.

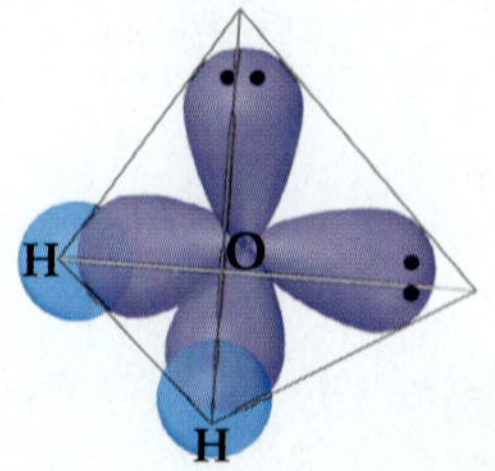

▲ **Figure 9.20 Valence-bond description of H_2O.** The bonding in a water molecule can be envisioned as sp^3 hybridization of the orbitals on O. Two of the four hybrid orbitals overlap with 1*s* orbitals of H to form covalent bonds. The other two hybrid orbitals are occupied by nonbonding pairs of electrons.

Hybridization Involving *d* Orbitals

Atoms in the third period and beyond can also use *d* orbitals to form hybrid orbitals. Mixing one *s* orbital, three *p* orbitals, and one *d* orbital leads to five sp^3d hybrid orbitals. These hybrid orbitals are directed toward the vertices of a trigonal bipyramid. The formation of sp^3d hybrids is exemplified by the phosphorus atom in PF_5:

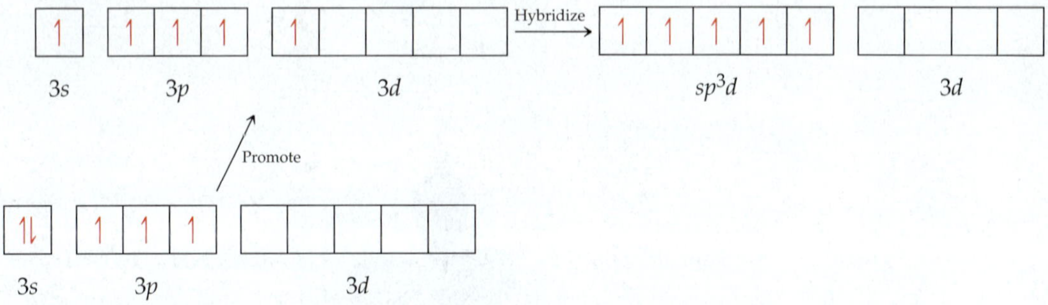

Similarly, mixing one *s* orbital, three *p* orbitals, and two *d* orbitals gives six sp^3d^2 hybrid orbitals, which are directed toward the vertices of an octahedron. The use of *d* orbitals in constructing hybrid orbitals nicely corresponds to the notion of an expanded valence shell. ∞ (Section 8.7) The geometric arrangements characteristic of hybrid orbitals are summarized in Table 9.4 ▶.

ACTIVITY
s-p Hybridization

Summary

Hybrid orbitals provide a convenient model for using valence-bond theory to describe covalent bonds in molecules whose geometries conform to the electron-domain geometries predicted by the VSEPR model. The picture of hybrid orbitals has limited predictive value; that is, we cannot say in advance that the nitrogen atom in NH_3 uses sp^3 hybrid orbitals. When we know the electron-domain geometry, however, we can employ hybridization to describe the atomic orbitals used by the central atom in bonding.

The following steps allow us to predict the hybrid orbitals used by an atom in bonding:

1. Draw the *Lewis structure* for the molecule or ion.
2. Determine the electron-domain geometry using the *VSEPR model.*
3. Specify the *hybrid orbitals* needed to accommodate the electron pairs based on their geometric arrangement (Table 9.4).

These steps are illustrated in Figure 9.21 ▼, which shows how the hybridization employed by N in NH_3 is determined.

▼ **Figure 9.21 Bonding in NH_3.** The hybrid orbitals used by N in the NH_3 molecule are predicted by first drawing the Lewis structure, then using the VSEPR model to determine the electron-domain geometry, and then specifying the hybrid orbitals that correspond to that geometry. This is essentially the same procedure as that used to determine molecular structure (Figure 9.6), except we focus on the orbitals used to make bonds and to hold nonbonding pairs.

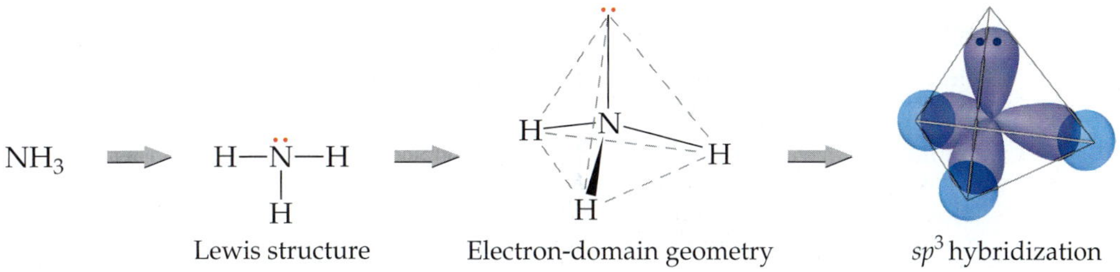

TABLE 9.4 Geometric Arrangements Characteristic of Hybrid Orbital Sets

Atomic Orbital Set	Hybrid Orbital Set	Geometry	Examples
s,p	Two *sp*	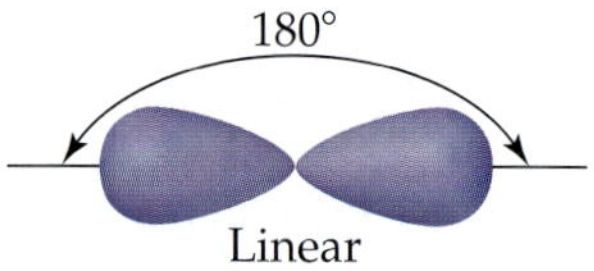 Linear	BeF_2, $HgCl_2$
s,p,p	Three sp^2	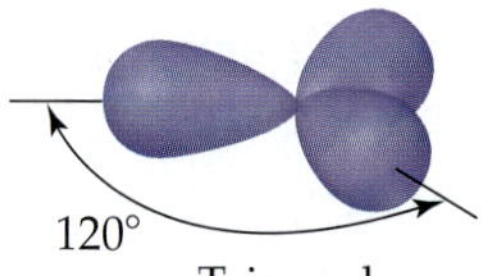 Trigonal planar	BF_3, SO_3
s,p,p,p	Four sp^3	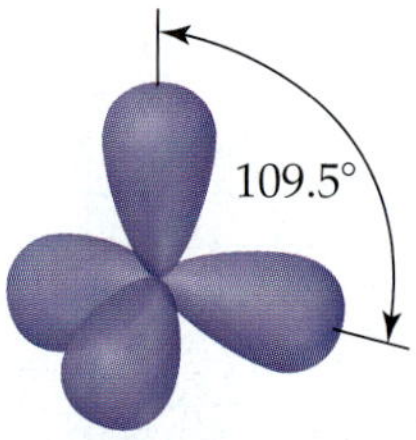 Tetrahedral	CH_4, NH_3, H_2O, NH_4^+
s,p,p,p,d	Five sp^3d	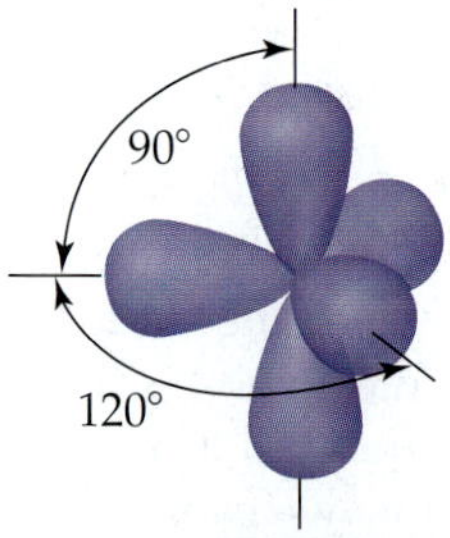 Trigonal bipyramidal	PF_5, SF_4, BrF_3
s,p,p,p,d,d	Six sp^3d^2	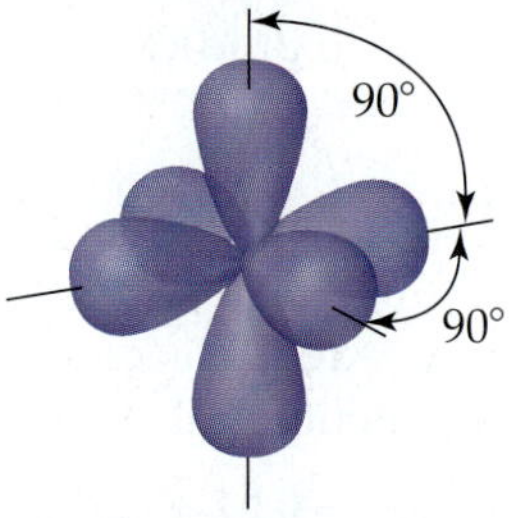 Octahedral	SF_6, ClF_5, XeF_4, PF_6^-

SAMPLE EXERCISE 9.5 | Hybridization

Indicate the hybridization of orbitals employed by the central atom in **(a)** NH_2^-, **(b)** SF_4 (see Sample Exercise 9.2).

Solution

Analyze: We are given two chemical formulas—one for a polyatomic anion and one for a molecular compound—and asked to describe the type of hybrid orbitals surrounding the central atom in each case.

Plan: To determine the hybrid orbitals used by an atom in bonding, we must know the electron-domain geometry around the atom. Thus, we first draw the Lewis structure to determine the number of electron domains around the central atom. The hybridization conforms to the number and geometry of electron domains around the central atom as predicted by the VSEPR model.
Solve: **(a)** The Lewis structure of NH_2^- is

$$\left[H:\ddot{\underset{\cdot\cdot}{N}}:H\right]^-$$

Because there are four electron domains around N, the electron-domain geometry is tetrahedral. The hybridization that gives a tetrahedral electron-domain geometry is sp^3 (Table 9.4). Two of the sp^3 hybrid orbitals contain nonbonding pairs of electrons, and the other two are used to make two-electron bonds with the hydrogen atoms.

(b) The Lewis structure and electron-domain geometry of SF_4 are shown in Sample Exercise 9.2. There are five electron domains around S, giving rise to a trigonal-bipyramidal electron-domain geometry. With an expanded octet of ten electrons, a *d* orbital on the sulfur must be used. The trigonal-bipyramidal electron-domain geometry corresponds to sp^3d hybridization (Table 9.4). One of the hybrid orbitals that points in an equatorial direction contains a nonbonding pair of electrons; the other four are used to form the S—F bonds.

PRACTICE EXERCISE

Predict the electron-domain geometry and the hybridization of the central atom in **(a)** SO_3^{2-}, **(b)** SF_6.
Answers: **(a)** tetrahedral, sp^3; **(b)** octahedral, sp^3d^2

9.6 | Multiple Bonds

In the covalent bonds we have considered thus far, the electron density is concentrated symmetrically about the line connecting the nuclei (the *internuclear axis*). In other words, the line joining the two nuclei passes through the middle of the overlap region. These bonds are all called **sigma (σ) bonds**. The overlap of two *s* orbitals as in H_2 [Figure 9.14(a)], the overlap of an *s* and a *p* orbital as in HCl [Figure 9.14(b)], the overlap between two *p* orbitals as in Cl_2 [Figure 9.14(c)], and the overlap of a *p* orbital with an *sp* hybrid orbital as in BeF_2 (Figure 9.17) are all examples of σ bonds.

To describe multiple bonding, we must consider a second kind of bond, one that results from the overlap between two *p* orbitals oriented perpendicularly to the internuclear axis (Figure 9.22 ▶). This sideways overlap of *p* orbitals produces a **pi (π) bond**. A π bond is a covalent bond in which the overlap regions lie above and below the internuclear axis. Unlike in a σ bond, in a π bond there is no probability of finding the electron on the internuclear axis. Because the *p* orbitals in a π bond overlap sideways rather than directly facing each other, the total overlap in a π bond tends to be less than that in a σ bond. As a consequence, π bonds are generally weaker than σ bonds.

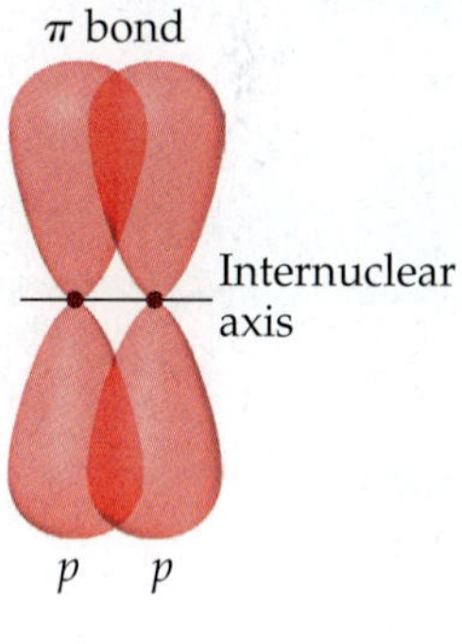

▲ **Figure 9.22 The π bond.** When two *p* orbitals overlap in a sideways fashion, the result is a π bond. Note that the two regions of overlap constitute a *single* π bond.

In almost all cases, single bonds are σ bonds. A double bond consists of one σ bond and one π bond, and a triple bond consists of one σ bond and two π bonds:

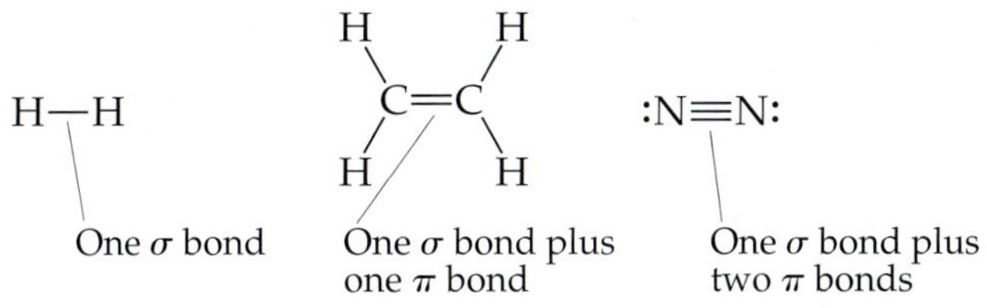

▲ **Figure 9.23 The molecular geometry of ethylene.** Ethylene, C_2H_4, has one C—C σ bond and one C—C π bond.

To see how these ideas are used, consider ethylene (C_2H_4), which possesses a C═C double bond. The bond angles in ethylene are all approximately 120° (Figure 9.23 ◀), suggesting that each carbon atom uses sp^2 hybrid orbitals (Figure 9.18) to form σ bonds with the other carbon and with two hydrogens. Because carbon has four valence electrons, after sp^2 hybridization one electron in each of the carbon atoms remains in the *unhybridized* $2p$ orbital:

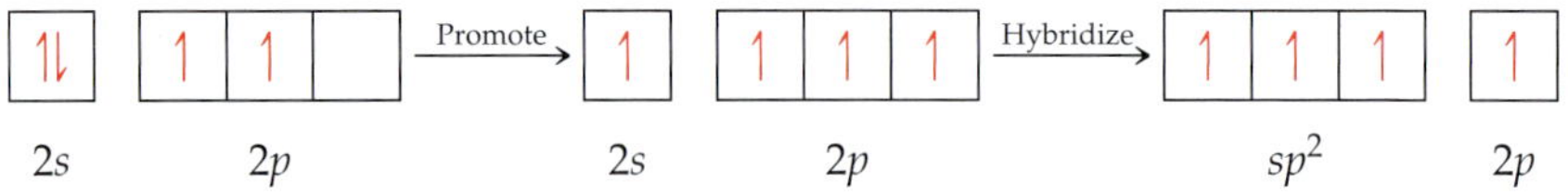

The unhybridized $2p$ orbital is directed perpendicular to the plane that contains the three sp^2 hybrid orbitals.

Each sp^2 hybrid orbital on a carbon atom contains one electron. Figure 9.24 ◀ shows how the four C—H σ bonds are formed by overlap of sp^2 hybrid orbitals on C with the $1s$ orbitals on each H atom. We use eight electrons to form these four electron-pair bonds. The C—C σ bond is formed by the overlap of two sp^2 hybrid orbitals, one on each carbon atom, and requires two more electrons. Thus 10 of the 12 valence electrons in the C_2H_4 molecule are used to form five σ bonds.

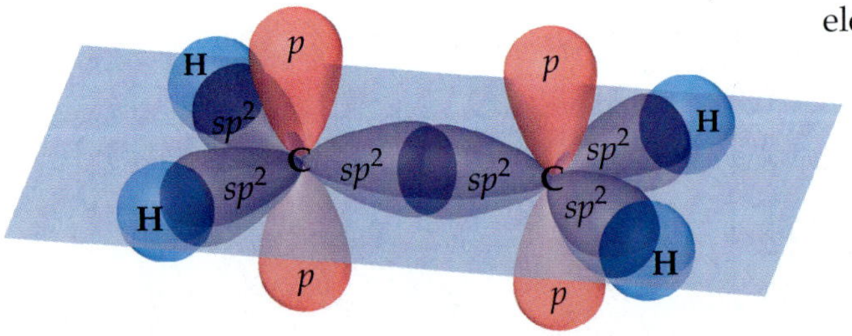

▲ **Figure 9.24 Hybridization in ethylene.** The σ bonding framework is formed from sp^2 hybrid orbitals on the carbon atoms. The unhybridized $2p$ orbitals on the C atoms are used to make a π bond.

The remaining two valence electrons reside in the unhybridized $2p$ orbitals, one electron on each carbon atom. These two $2p$ orbitals can overlap with each other in a sideways fashion, as shown in Figure 9.25 ▼. The resultant electron density is concentrated above and below the C—C bond axis, which means this is a π bond (Figure 9.22). Thus, the C═C double bond in ethylene consists of one σ bond and one π bond.

Although we cannot experimentally observe a π bond directly (all we can observe are the positions of the atoms), the structure of ethylene provides strong support for its presence. First, the C—C bond length in ethylene (1.34 Å) is much shorter than in compounds with C—C single bonds (1.54 Å), consistent with the presence of a stronger C═C double bond. Second, all six atoms in C_2H_4 lie in the same plane. The $2p$ orbitals that make up the π bond can achieve a good overlap only when the two CH_2 fragments lie in the same plane. If the π bond were absent, there would be no reason for the two CH_2 fragments of ethylene to lie in the same plane. Because π bonds require that portions of a molecule be planar, they can introduce rigidity into molecules.

ACTIVITY
Multiple Bonds

GIVE IT SOME THOUGHT

The molecule called *diazine* has the formula N_2H_2 and the Lewis structure

$$\mathrm{H{-}\ddot{N}{=}\ddot{N}{-}H}$$

Would you expect diazine to be a linear molecule (all four atoms on the same line)? If not, would you expect the molecule to be planar (all four atoms in the same plane)?

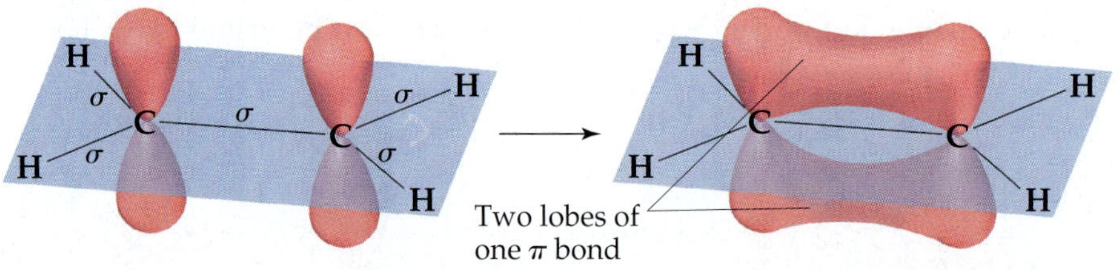

▶ **Figure 9.25 The π bond in ethylene.** The unhybridized $2p$ orbitals on each C atom overlap to form a π bond. The electron density in the π bond is above and below the bond axis, whereas in the σ bonds the electron density lies directly along the bond axes. As noted in Figure 9.22, the two lobes constitute one π bond.

Triple bonds can also be explained by using hybrid orbitals. Acetylene (C_2H_2), for example, is a linear molecule containing a triple bond: H—C≡C—H. The linear geometry suggests that each carbon atom uses *sp* hybrid orbitals to form σ bonds with the other carbon and one hydrogen. Each carbon atom thus has two remaining unhybridized 2*p* orbitals at right angles to each other and to the axis of the *sp* hybrid set (Figure 9.26 ▶). These *p* orbitals overlap to form a pair of π bonds. Thus, the triple bond in acetylene consists of one σ bond and two π bonds.

Although it is possible to make π bonds from *d* orbitals, the only π bonds we will consider are those formed by the overlap of *p* orbitals. These π bonds can form only if unhybridized *p* orbitals are present on the bonded atoms. Therefore, only atoms having *sp* or sp^2 hybridization can be involved in such π bonding. Further, double and triple bonds (and hence π bonds) are more common in molecules made up of small atoms, especially C, N, and O. Larger atoms, such as S, P, and Si, form π bonds less readily.

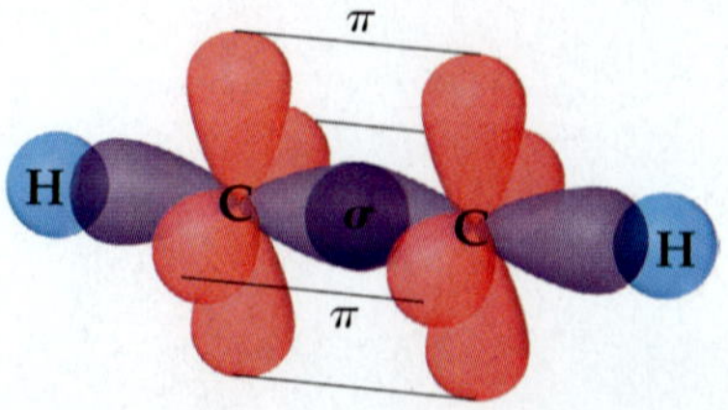

▲ **Figure 9.26 Formation of two π bonds.** In acetylene, C_2H_2, the overlap of two sets of unhybridized carbon 2*p* orbitals leads to the formation of two π bonds.

SAMPLE EXERCISE 9.6 | Describing σ and π Bonds in a Molecule

Formaldehyde has the Lewis structure

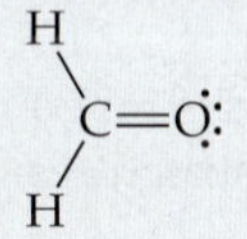

Describe how the bonds in formaldehyde are formed in terms of overlaps of appropriate hybridized and unhybridized orbitals.

Solution

Analyze: We are asked to describe the bonding in formaldehyde in terms of orbital overlaps.

Plan: Single bonds will be of the σ type, whereas double bonds will consist of one σ bond and one π bond. The ways in which these bonds form can be deduced from the geometry of the molecule, which we predict using the VSEPR model.

Solve: The C atom has three electron domains around it, which suggests a trigonal-planar geometry with bond angles of about 120°. This geometry implies sp^2 hybrid orbitals on C (Table 9.4). These hybrids are used to make the two C—H and one C—O σ bonds to C. There remains an unhybridized 2*p* orbital on carbon, perpendicular to the plane of the three sp^2 hybrids.

The O atom also has three electron domains around it, and so we will assume that it has sp^2 hybridization as well. One of these hybrids participates in the C—O σ bond, while the other two hybrids hold the two nonbonding electron pairs of the O atom. Like the C atom, therefore, the O atom has an unhybridized 2*p* orbital that is perpendicular to the plane of the molecule. The unhybridized 2*p* orbitals on the C and O atoms overlap to form a C—O π bond, as illustrated in Figure 9.27 ▼.

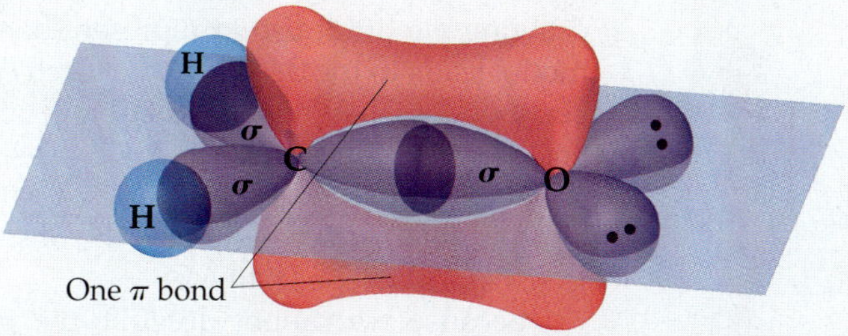

▲ **Figure 9.27 Formation of σ and π bonds in formaldehyde, H_2CO.**

PRACTICE EXERCISE

Consider the acetonitrile molecule:

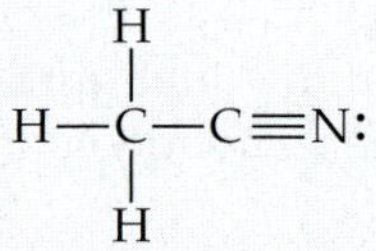

(a) Predict the bond angles around each carbon atom; **(b)** describe the hybridization at each of the carbon atoms; **(c)** determine the total number of σ and π bonds in the molecule.

Answers: **(a)** approximately 109° around the left C and 180° on the right C; **(b)** sp^3, sp; **(c)** five σ bonds and two π bonds

Delocalized π Bonding

In the molecules we have discussed thus far in this section, the bonding electrons are *localized*. By this we mean that the σ and π electrons are associated totally with the two atoms that form the bond. In many molecules, however, we cannot adequately describe the bonding as being entirely localized. This situation arises particularly in molecules that have two or more resonance structures involving π bonds.

One molecule that cannot be described with localized π bonds is benzene (C_6H_6), which has two resonance structures: (Section 8.6)

To describe the bonding in benzene using hybrid orbitals, we first choose a hybridization scheme consistent with the geometry of the molecule. Because each carbon is surrounded by three atoms at 120° angles, the appropriate hybrid set is sp^2. Six localized C—C σ bonds and six localized C—H σ bonds are formed from the sp^2 hybrid orbitals, as shown in Figure 9.28(a) ◀. This leaves a $2p$ orbital on each carbon that is oriented perpendicularly to the plane of the molecule. The situation is very much like that in ethylene, except we now have six carbon $2p$ orbitals arranged in a ring [Figure 9.28(b)]. Each of the unhybridized $2p$ orbitals is occupied by one electron, leaving a total of six electrons to be accounted for by π bonding.

We could envision using the unhybridized $2p$ orbitals of benzene to form three localized π bonds. As shown in Figure 9.29(a) and (b) ▶, there are two equivalent ways to make these localized bonds, and each corresponds to one of the resonance structures of the molecule. A representation that reflects *both* resonance structures has the six π electrons "smeared out" among all six carbon atoms, as shown in Figure 9.29(c). Notice how this figure corresponds to the "circle-in-a-hexagon" drawing we often use to represent benzene. This model leads to the description of each carbon–carbon bond as having identical bond lengths that are between the length of a C—C single bond (1.54 Å) and the length of a C=C double bond (1.34 Å), consistent with the observed bond lengths in benzene (1.40 Å).

Because we cannot describe the π bonds in benzene as individual electron-pair bonds between neighboring atoms, we say that the π bonds are **delocalized** among the six carbon atoms. Delocalization of the electrons in its π bonds gives

▼ **Figure 9.28 The σ and π bond networks in benzene, C_6H_6.** (a) The C—C and C—H σ bonds all lie in the plane of the molecule and are formed by using carbon sp^2 hybrid orbitals. (b) Each carbon atom has an unhybridized $2p$ orbital that lies perpendicular to the molecular plane. These six $2p$ orbitals are the *π orbitals* of benzene.

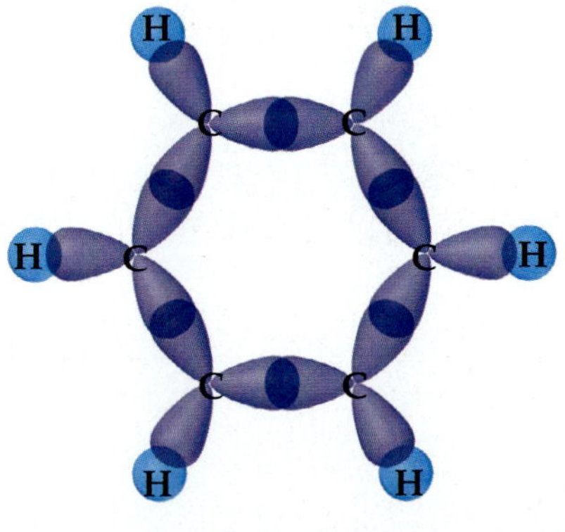

(a) σ bonds

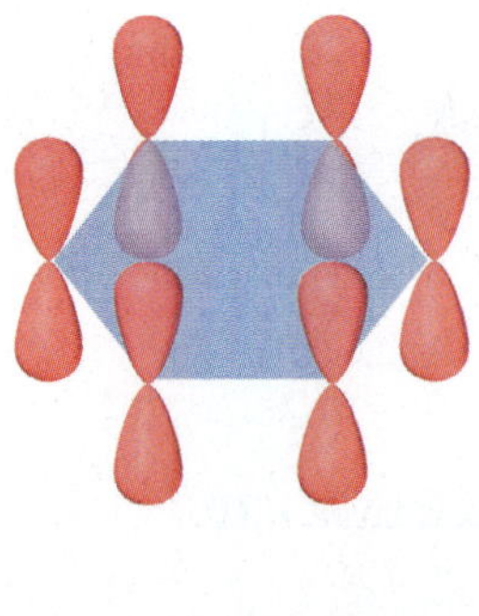

(b) $2p$ atomic orbitals

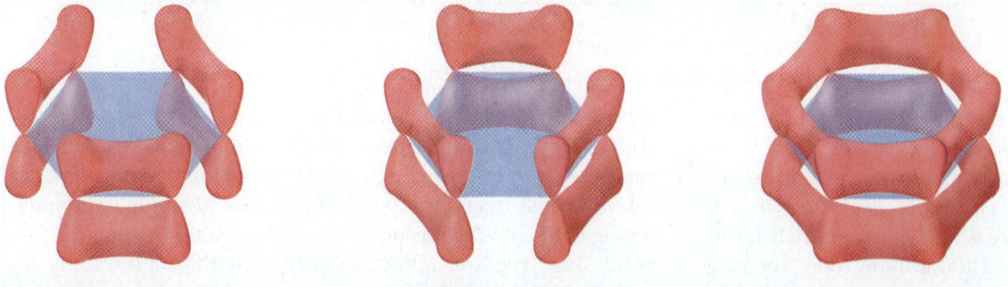

(a) Localized π bonds (b) Localized π bonds (c) Delocalized π bonds

▲ **Figure 9.29 Delocalized π bonds.** The six $2p$ orbitals of benzene, shown in Figure 9.28(b), can be used to make C—C π bonds. (a, b) Two equivalent ways to make localized π bonds. These π bonds correspond to the two resonance structures for benzene. (c) A representation of the smearing out, or delocalization, of the three C—C π bonds among the six C atoms.

benzene a special stability, as will be discussed in Section 25.4. Delocalization of π bonds is also responsible for the color of many organic molecules. (See the "Chemistry at Work" box on organic dyes at the end of this chapter.) If you take a course in organic chemistry, you will see many examples of how electron delocalization influences the properties of organic molecules.

SAMPLE EXERCISE 9.7 | Delocalized Bonding

Describe the bonding in the nitrate ion, NO_3^-. Does this ion have delocalized π bonds?

Solution

Analyze: Given the chemical formula for a polyatomic anion, we are asked to describe the bonding and determine whether the ion has delocalized π bonds.

Plan: Our first step in describing the bonding in NO_3^- is to construct appropriate Lewis structures. If there are multiple resonance structures that involve the placement of the double bonds in different locations, that suggests that the π component of the double bonds is delocalized.

Solve: In Section 8.6 we saw that NO_3^- has three resonance structures:

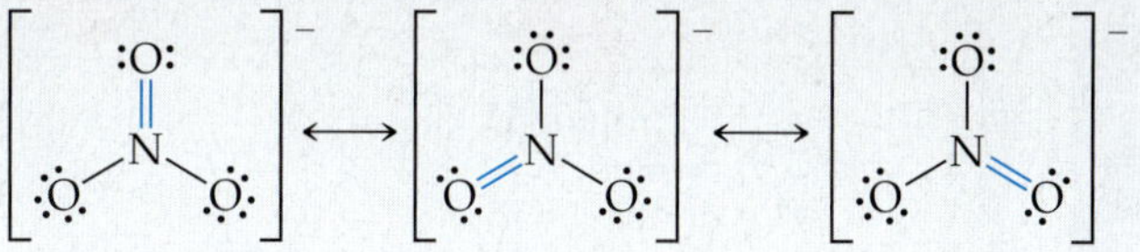

In each of these structures the electron-domain geometry at nitrogen is trigonal planar, which implies sp^2 hybridization of the N atom. The sp^2 hybrid orbitals are used to construct the three N—O σ bonds that are present in each of the resonance structures.

The unhybridized $2p$ orbital on the N atom can be used to make π bonds. For any one of the three resonance structures shown, we might imagine a single localized N—O π bond formed by the overlap of the unhybridized $2p$ orbital on N and a $2p$ orbital on one of the O atoms, as shown in Figure 9.30(a) ▶. Because each resonance structure contributes equally to the observed structure of NO_3^-, however, we represent the π bonding as spread out, or delocalized, over the three N—O bonds, as shown in Figure 9.30(b).

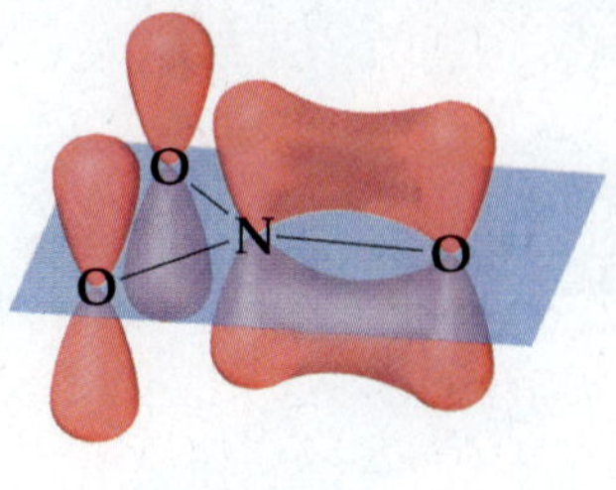

(a) N—O π bond in one of the resonance structures of NO_3^-.

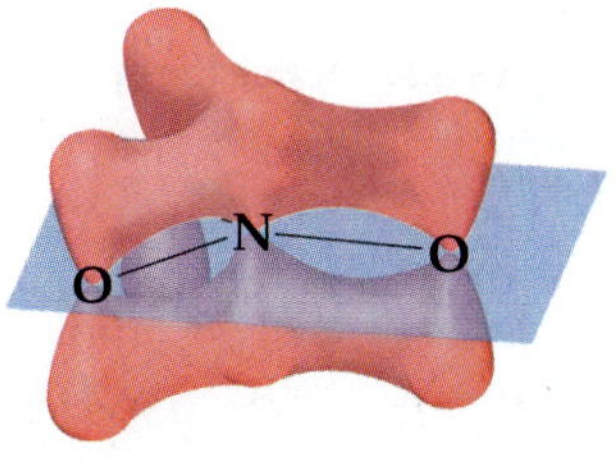

(b) Delocalization of the π bonds in the NO_3^- ion.

▲ **Figure 9.30 Localized and delocalized π bonds in NO_3^-.**

PRACTICE EXERCISE

Which of the following molecules or ions will exhibit delocalized bonding: SO_3, SO_3^{2-}, H_2CO, O_3, NH_4^+?

Answer: SO_3 and O_3, as indicated by the presence of two or more resonance structures involving π bonding for each of these molecules

CHEMISTRY AND LIFE | The Chemistry of Vision

In recent years scientists have begun to understand the complex chemistry of vision. Vision begins when light is focused by the lens onto the retina, the layer of cells lining the interior of the eyeball. The retina contains *photoreceptor* cells known as rods and cones (Figure 9.31 ▼). The human retina contains about 3 million cones and 100 million rods. The rods are sensitive to dim light and are used in night vision. The cones are sensitive to colors. The tops of the rods and cones contain a molecule called *rhodopsin*. Rhodopsin consists of a protein, called *opsin*, bonded to a reddish purple pigment called *retinal*. Structural changes around a double bond in the retinal portion of the molecule trigger a series of chemical reactions that result in vision.

Double bonds between atoms are stronger than single bonds between the same atoms (Table 8.4). For example, a C=C double bond is stronger [D(C=C) = 614 kJ/mol] than a C—C single bond [D(C—C) = 348 kJ/mol], though not twice as strong. Our recent discussions now allow us to appreciate another aspect of double bonds: the stiffness or rigidity that they introduce into molecules.

Imagine taking the $—CH_2$ group of the ethylene molecule and rotating it relative to the other $—CH_2$ group as shown in Figure 9.32 ▶. This rotation destroys the overlap of p orbitals, breaking the π bond, a process that requires considerable energy. Thus, the presence of a double bond restricts the rotation of the bonds in a molecule. In contrast, molecules can rotate almost freely around the bond axis in single (σ) bonds because this motion has no effect on the orbital overlap for a σ bond. This rotation allows molecules with single bonds to twist and fold almost as if their atoms were attached by hinges.

Our vision depends on the rigidity of double bonds in retinal. In its normal form, retinal is held rigid by its double bonds, as shown on the left in Figure 9.33 ▼. Light entering the eye is absorbed by rhodopsin, and the energy is used to break the π-bond portion of the indicated double bond. The molecule then rotates around this bond, changing its geometry. The retinal then separates from the opsin, triggering the reactions that produce a nerve impulse that the brain interprets as the sensation of vision. It takes as few as five closely spaced molecules reacting in this fashion to produce the sensation of vision. Thus, only five photons of light are necessary to stimulate the eye.

The retinal slowly reverts to its original form and reattaches to the opsin. The slowness of this process helps explain why intense bright light causes temporary blindness. The light causes all the retinal to separate from opsin, leaving no further molecules to absorb light.

Related Exercises: 9.99 and 9.100

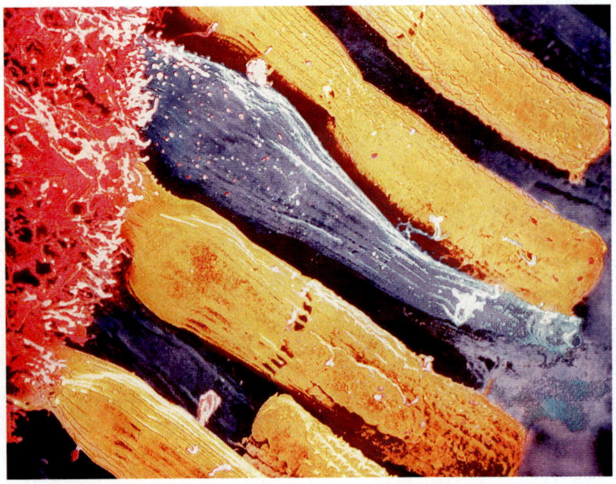

▲ **Figure 9.31 Inside the eye.** A color-enhanced scanning electron micrograph of the rods (yellow) and cones (blue) in the retina of the human eye.

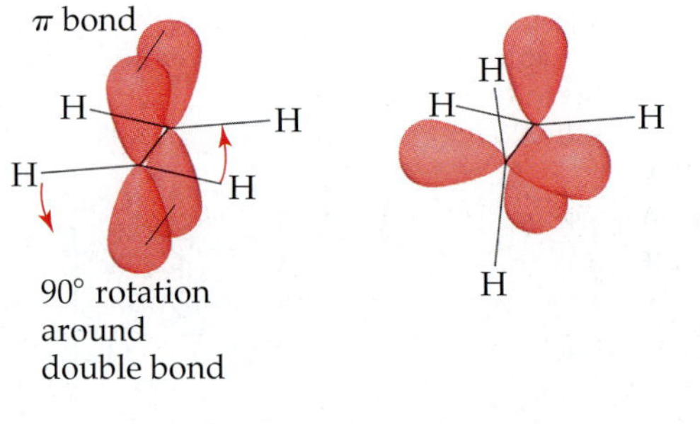

▲ **Figure 9.32 Rotation about the carbon–carbon double bond in ethylene.** The overlap of the p orbitals that form the π bond is lost in the rotation. For this reason, rotation about double bonds requires the absorption of energy.

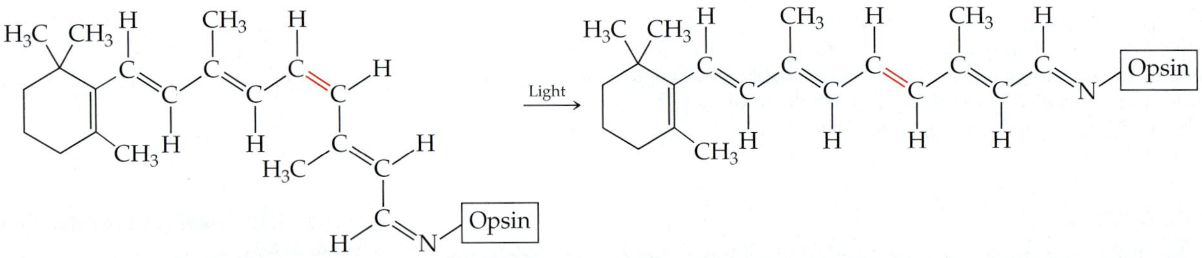

▲ **Figure 9.33 The chemical basis of vision.** When rhodopsin absorbs visible light, the π component of the double bond shown in red breaks, allowing rotation that produces a change in molecular geometry.

General Conclusions

On the basis of the examples we've seen, we can draw a few helpful conclusions for using the concept of hybrid orbitals to describe molecular structures:

1. Every pair of bonded atoms shares one or more pairs of electrons. In every bond at least one pair of electrons is localized in the space between the atoms in a σ bond. The appropriate set of hybrid orbitals used to form the σ bonds between an atom and its neighbors is determined by the observed geometry of the molecule. The correlation between the set of hybrid orbitals and the geometry about an atom is given in Table 9.4.
2. The electrons in σ bonds are localized in the region between two bonded atoms and do not make a significant contribution to the bonding between any other two atoms.
3. When atoms share more than one pair of electrons, one pair is used to form a σ bond and the additional pairs form π bonds. The centers of charge density in a π bond lie above and below the bond axis.
4. Molecules with two or more resonance structures can have π bonds that extend over more than two bonded atoms. Electrons in π bonds that extend over more than two atoms are delocalized.

9.7 | Molecular Orbitals

Valence-bond theory and hybrid orbitals allow us to move in a straightforward way from Lewis structures to rationalizing the observed geometries of molecules in terms of atomic orbitals. For example, we can use this theory to understand why methane has the formula CH_4, how the carbon and hydrogen atomic orbitals are used to form electron-pair bonds, and why the arrangement of the C—H bonds about the central carbon is tetrahedral. This model, however, does not explain all aspects of bonding. It is not successful, for example, in describing the excited states of molecules, which we must understand in order to explain how molecules absorb light, giving them color.

Some aspects of bonding are better explained by an alternative model called **molecular orbital theory**. In Chapter 6 we saw that electrons in atoms can be described by certain wave functions, which we call atomic orbitals. In a similar way molecular orbital theory describes the electrons in molecules by using specific wave functions called **molecular orbitals**. Chemists use the abbreviation **MO** for molecular orbital.

Molecular orbitals have many of the same characteristics as atomic orbitals. For example, an MO can hold a maximum of two electrons (with opposite spins), it has a definite energy, and we can visualize its electron-density distribution by using a contour representation, as we did when we discussed atomic orbitals. Unlike atomic orbitals, however, MOs are associated with the entire molecule, not with a single atom.

The Hydrogen Molecule

To get a sense of the approach taken in MO theory, we will begin with the simplest molecule: the hydrogen molecule, H_2. We will use the two 1*s* atomic orbitals (one on each H atom) to "build" molecular orbitals for the H_2 molecule. *Whenever two atomic orbitals overlap, two molecular orbitals form.* Thus, the overlap of the 1*s* orbitals of two hydrogen atoms to form H_2 produces two MOs (Figure 9.34 ▶).

The lower-energy MO of H_2 concentrates electron density between the two hydrogen nuclei and is called the **bonding molecular orbital**. This sausage-shaped

▼ **Figure 9.34 The molecular orbitals of H_2.** The combination of two H 1*s* atomic orbitals forms two molecular orbitals (MOs) of H_2. In the bonding MO, σ_{1s}, the atomic orbitals combine constructively, leading to a buildup of electron density between the nuclei. In the antibonding MO, σ^*_{1s}, the orbitals combine destructively in the bonding region. Note that the σ^*_{1s} MO has a node between the two nuclei.

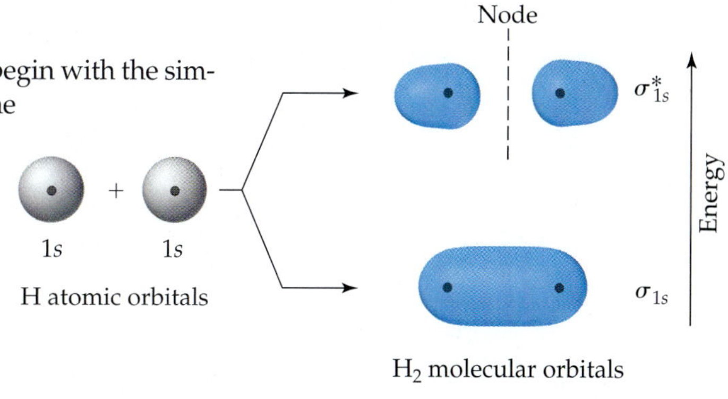

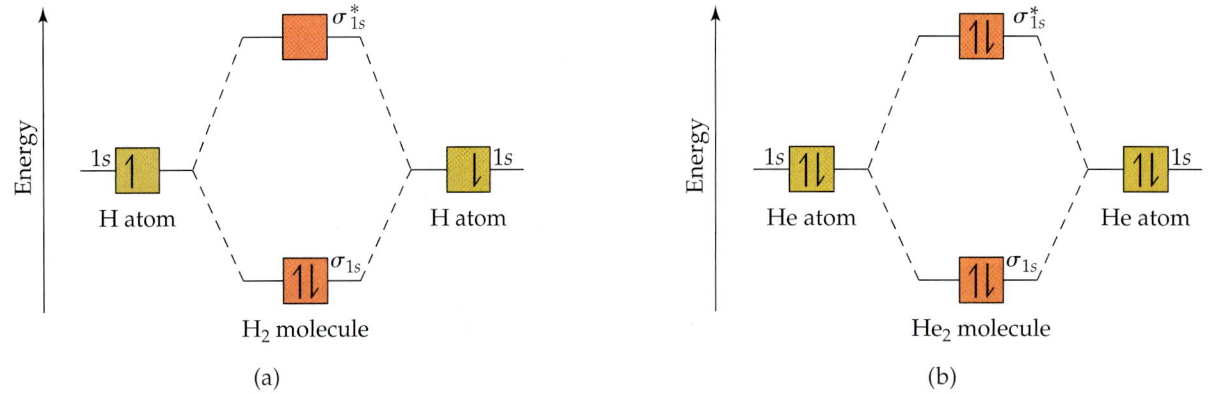

▲ **Figure 9.35 Energy-level diagrams for H_2 and He_2.** (a) The two electrons in the H_2 molecule occupy the σ_{1s} bonding MO. (b) In the (hypothetical) He_2 molecule, both the σ_{1s} bonding MO and the σ^*_{1s} antibonding MO are occupied by two electrons.

ANIMATION
Molecular Orbital Theory

MO results from summing the two atomic orbitals so that the atomic orbital wave functions enhance each other in the bond region. Because an electron in this MO is strongly attracted to both nuclei, the electron is more stable (in other words, it has lower energy) than it is in the 1*s* atomic orbital of an isolated hydrogen atom. Further, because it concentrates electron density between the nuclei, the bonding MO holds the atoms together in a covalent bond.

The higher-energy MO in Figure 9.34 has very little electron density between the nuclei and is called the **antibonding molecular orbital**. Instead of enhancing each other in the region between the nuclei, the atomic orbitals cancel each other in this region, and the greatest electron density is on opposite sides of the nuclei. Thus, this MO excludes electrons from the very region in which a bond must be formed. An electron in this MO is repelled from the bonding region and is therefore less stable (in other words, it has higher energy) than it is in the 1*s* atomic orbital of a hydrogen atom.

The electron density in both the bonding MO and the antibonding MO of H_2 is centered about the internuclear axis, an imaginary line passing through the two nuclei. MOs of this type are called **sigma (σ) molecular orbitals**. The bonding sigma MO of H_2 is labeled σ_{1s}, the subscript indicating that the MO is formed from two 1*s* orbitals. The antibonding sigma MO of H_2 is labeled σ^*_{1s} (read "sigma-star-one-*s*"), the asterisk denoting that the MO is antibonding.

The interaction between two 1*s* atomic orbitals and the molecular orbitals that result can be represented by an **energy-level diagram** (also called a **molecular orbital diagram**), like those in Figure 9.35 ▲. Such diagrams show the interacting atomic orbitals in the left and right columns and the MOs in the middle column. Note that the bonding molecular orbital, σ_{1s}, is lower in energy than the atomic 1*s* orbitals, whereas the antibonding orbital, σ^*_{1s}, is higher in energy than the 1*s* orbitals. Like atomic orbitals, each MO can accommodate two electrons with their spins paired (Pauli exclusion principle). (Section 6.7)

The molecular orbital diagram of the H_2 molecule is shown in Figure 9.35(a). Each H atom has one electron, and so there are two electrons in H_2. These two electrons occupy the lower-energy bonding (σ_{1s}) MO, and their spins are paired. Electrons occupying a bonding molecular orbital are called *bonding electrons*. Because the σ_{1s} MO is lower in energy than the isolated 1*s* atomic orbitals, the H_2 molecule is more stable than the two separate H atoms.

In contrast, the hypothetical He_2 molecule requires four electrons to fill its molecular orbitals, as in Figure 9.35(b). Because only two electrons can be put in the σ_{1s} MO, the other two must be placed in the σ^*_{1s} MO. The energy decrease from the two electrons in the bonding MO is offset by the energy increase from the two electrons in the antibonding MO.* Hence, He_2 is an unstable molecule. Molecular orbital theory correctly predicts that hydrogen forms diatomic molecules but helium does not.

GIVE IT SOME THOUGHT

Suppose light is used to excite one of the electrons in the H_2 molecule from the σ_{1s} MO to the σ^*_{1s} MO. Would you expect the H atoms to remain bonded to each other, or would the molecule fall apart?

Bond Order

In molecular orbital theory the stability of a covalent bond is related to its **bond order**, defined as half the difference between the number of bonding electrons and the number of antibonding electrons:

$$\text{Bond order} = \tfrac{1}{2}(\text{no. of bonding electrons} - \text{no. of antibonding electrons})$$

We take half the difference because we are used to thinking of bonds as pairs of electrons. *A bond order of 1 represents a single bond, a bond order of 2 represents a double bond, and a bond order of 3 represents a triple bond.* Because MO theory also treats molecules containing an odd number of electrons, bond orders of 1/2, 3/2, or 5/2 are possible.

Because H_2 has two bonding electrons and zero antibonding electrons [Figure 9.35(a)], it has a bond order of 1. Because He_2 has two bonding and two antibonding electrons [Figure 9.35(b)], it has a bond order of 0. A bond order of 0 means that no bond exists.

GIVE IT SOME THOUGHT

What is the bond order in the excited H_2 molecule described in the immediately preceding Give It Some Thought exercise?

SAMPLE EXERCISE 9.8 | Bond Order

What is the bond order of the He_2^+ ion? Would you expect this ion to be stable relative to the separated He atom and He^+ ion?

Solution

Analyze: We will determine the bond order for the He_2^+ ion and use it to predict whether the ion is stable.

Plan: To determine the bond order, we must determine the number of electrons in the molecule and how these electrons populate the available MOs. The valence electrons of He are in the 1s orbital, and the 1s orbitals combine to give an MO diagram like that for H_2 or He_2 (Figure 9.35). If the bond order is greater than 0, we expect a bond to exist and the ion is stable.

Solve: The energy-level diagram for the He_2^+ ion is shown in Figure 9.36 ▶. This ion has three electrons. Two are placed in the bonding orbital, the third in the antibonding orbital. Thus, the bond order is

$$\text{Bond order} = \tfrac{1}{2}(2 - 1) = \tfrac{1}{2}$$

Because the bond order is greater than 0, the He_2^+ ion is predicted to be stable relative to the separated He and He^+. Formation of He_2^+ in the gas phase has been demonstrated in laboratory experiments.

PRACTICE EXERCISE

Determine the bond order of the H_2^- ion.
Answer: $\frac{1}{2}$

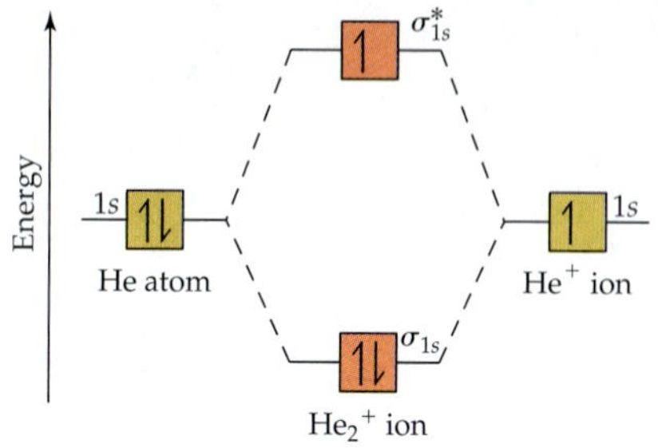

▲ **Figure 9.36 Energy-level diagram for the He_2^+ ion.**

* In fact, antibonding MOs are slightly more unfavorable than bonding MOs are favorable. Thus, whenever there is an equal number of electrons in bonding and antibonding orbitals, the energy of the molecule is slightly higher than that for the separated atoms and no bond is formed.

9.8 | Second-Row Diatomic Molecules

Just as we treated the bonding in H_2 by using molecular orbital theory, we can consider the MO description of other diatomic molecules. Initially we will restrict our discussion to *homonuclear* diatomic molecules (those composed of two identical atoms) of elements in the second row of the periodic table. As we will see, the procedure for determining the distribution of electrons in these molecules closely follows the one we used for H_2.

Second-row atoms have valence 2*s* and 2*p* orbitals, and we need to consider how they interact to form MOs. The following rules summarize some of the guiding principles for the formation of MOs and for how they are populated by electrons:

1. The number of MOs formed equals the number of atomic orbitals combined.
2. Atomic orbitals combine most effectively with other atomic orbitals of similar energy.
3. The effectiveness with which two atomic orbitals combine is proportional to their overlap; that is, as the overlap increases, the energy of the bonding MO is lowered and the energy of the antibonding MO is raised.
4. Each MO can accommodate, at most, two electrons, with their spins paired (Pauli exclusion principle).
5. When MOs of the same energy are populated, one electron enters each orbital (with the same spin) before spin pairing occurs (Hund's rule).

Molecular Orbitals for Li_2 and Be_2

Lithium, the first element of the second period, has a $1s^22s^1$ electron configuration. When lithium metal is heated above its boiling point (1342°C), Li_2 molecules are found in the vapor phase. The Lewis structure for Li_2 indicates an Li—Li single bond. We will now use MOs to describe the bonding in Li_2.

Because the 1*s* and 2*s* orbitals of Li are so different in energy, we may assume that the 1*s* orbital on one Li atom interacts only with the 1*s* orbital on the other atom (rule 2). Likewise, the 2*s* orbitals interact only with each other. The resulting energy-level diagram is shown in Figure 9.37 ◀. Notice that combining four atomic orbitals produces four MOs (rule 1).

The 1*s* orbitals of Li combine to form σ_{1s} and σ^*_{1s} bonding and antibonding MOs, as they did for H_2. The 2*s* orbitals interact with one another in exactly the same way, producing bonding (σ_{2s}) and antibonding (σ^*_{2s}) MOs. Because the 2*s* orbitals of Li extend farther from the nucleus than the 1*s* orbitals do, the 2*s* orbitals overlap more effectively. As a result, the energy separation between the σ_{2s} and σ^*_{2s} orbitals is greater than that for the 1*s*-based MOs. The 1*s* orbitals of Li are so much lower in energy than the 2*s* orbitals, however, that the σ^*_{1s} antibonding MO is still well below the σ_{2s} bonding MO.

▼ Figure 9.37 Energy-level diagram for the Li_2 molecule.

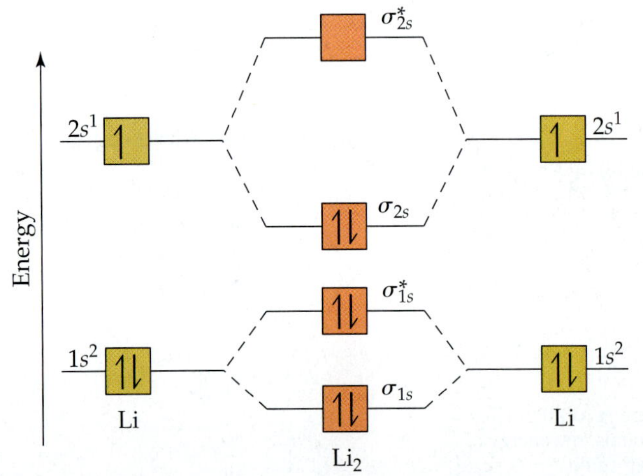

Each Li atom has three electrons, so six electrons must be placed in the MOs of Li_2. As shown in Figure 9.37, these occupy the σ_{1s}, σ^*_{1s}, and σ_{2s} MOs, each with two electrons. There are four electrons in bonding orbitals and two in antibonding orbitals, so the bond order equals $\frac{1}{2}(4 - 2) = 1$. The molecule has a single bond, in accord with its Lewis structure.

Because both the σ_{1s} and σ^*_{1s} MOs of Li_2 are completely filled, the 1*s* orbitals contribute almost nothing to the bonding. The single bond in Li_2 is due essentially to the interaction of the valence 2*s* orbitals on the Li atoms. This example illustrates the general rule that *core electrons usually do not con-*

tribute significantly to bonding in molecule formation. The rule is equivalent to using only the valence electrons when drawing Lewis structures. Thus, we need not consider further the 1*s* orbitals while discussing the other second-row diatomic molecules.

The MO description of Be_2 follows readily from the energy-level diagram for Li_2. Each Be atom has four electrons ($1s^2 2s^2$), so we must place eight electrons in molecular orbitals. Thus, we completely fill the σ_{1s}, σ^*_{1s}, σ_{2s}, and σ^*_{2s} MOs. We have an equal number of bonding and antibonding electrons, so the bond order equals 0. Consistent with this analysis, Be_2 does not exist.

GIVE IT SOME THOUGHT

Would you expect Be_2^+ to be a stable ion?

Molecular Orbitals from 2*p* Atomic Orbitals

Before we can consider the remaining second-row molecules, we must look at the MOs that result from combining 2*p* atomic orbitals. The interactions between *p* orbitals are shown in Figure 9.38 ▼, where we have arbitrarily chosen the internuclear axis to be the *z*-axis. The $2p_z$ orbitals face each other in a "head-to-head" fashion. Just as we did for the *s* orbitals, we can combine the $2p_z$ orbitals in two ways. One combination concentrates electron density between the nuclei and is therefore a bonding molecular orbital. The other combination excludes electron density from the bonding region; it is an antibonding molecular orbital. In each of these MOs the electron density lies along the line through the nuclei, so they are σ molecular orbitals: σ_{2p} and σ^*_{2p}.

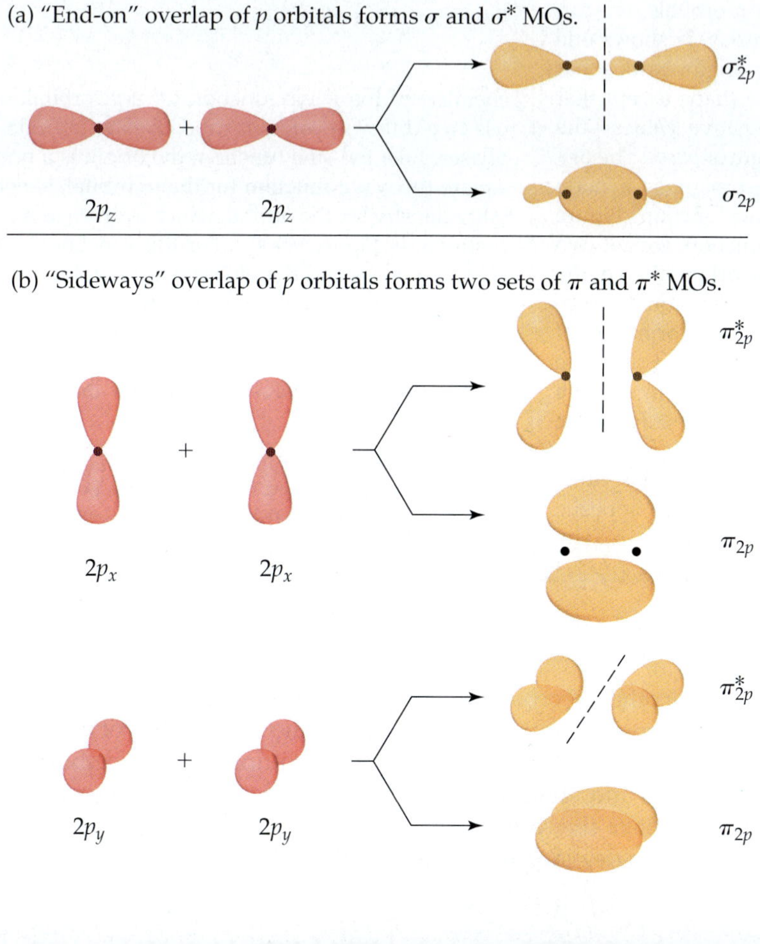

◀ **Figure 9.38 Contour representations of the molecular orbitals formed by 2*p* orbitals.** Each time we combine two atomic orbitals, we obtain two MOs: one bonding and one antibonding.

A CLOSER LOOK | Phases in Atomic and Molecular Orbitals

Our discussion of atomic orbitals in Chapter 6 and molecular orbitals in this chapter highlights some of the most important applications of quantum mechanics in chemistry. In the quantum mechanical treatment of electrons in atoms and molecules, we are mainly interested in obtaining two characteristics of the electrons—namely, their energies and their distribution in space. Recall that solving Schrödinger's wave equation yields the electron's energy, E, and wave function, ψ, but that ψ itself does not have a direct physical meaning. (Section 6.5) The contour representations of atomic and molecular orbitals that we have presented thus far are based on the square of the wave function, ψ^2 (the *probability density*), which gives the probability of finding the electron at a given point in space.

Because probability densities are squares of functions, their values must be nonnegative (zero or positive) at all points in space. Remember that points where the probability density is zero are called *nodes*: At a node there is a zero probability of finding the electron. Consider, for example, the p_x orbital presented in Figure 6.22. For this orbital, the probability density at any point on the plane defined by the y- and z-axes is zero—we say that the yz plane is a *nodal plane* of the orbital. Likewise, the σ^*_{1s} MO of H_2 shown in Figure 9.34 has a nodal plane that is perpendicular to the line connecting the H atoms and midway between them. Why do these nodes arise? To answer that question, we must take a closer look at the wave functions for the orbitals.

In order to understand the nodes in p orbitals, we can draw an analogy to a sine function. Figure 9.39 ▶ shows one cycle of the function $\sin x$ centered on the origin. Notice that the two halves of the wave have the same shape except that one has positive values and the other negative values—the two halves of the function differ in their sign, or *phase*. The origin, which is where the function changes sign, is a node ($\sin 0 = 0$). What happens if we square this function? Figure 9.39 (b) shows that when we square the function, we get two peaks that look the same on each side of the origin. Both of the peaks are positive because squaring a negative number produces a positive number—we lose the phase information of the function upon squaring it. Note also that because $\sin 0 = 0$, it follows that $\sin^2 0 = 0$. Thus, the origin is still a node even though the function is positive on both sides of the origin.

The wave function for a p orbital is much like a sine function inasmuch as it has two equal parts that have opposite phases. Figure 9.40(a) ▶ gives a typical representation used by chemists of the wave function for a p_x orbital.* We typically use two different colors for the lobes to indicate the different phases. Like the sine function, the origin is a node. When we square the wave function for the p_x orbital, we get the probability density for the orbital, which is given as a contour representation in Figure 9.40(b). For this function, both lobes have the same phase and therefore the same color.

The lobes of the wave functions for the d orbitals also have different phases. For example, the wave function for a d_{xy} orbital has four lobes that alternate in phase, as shown in Figure 9.40(c). The wave functions for the d_{xz}, d_{yz}, and $d_{x^2-y^2}$

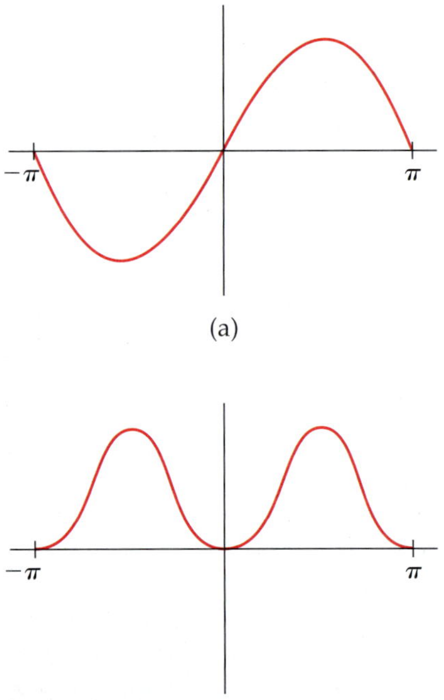

(b)

▲ **Figure 9.39 Phases in the sine function.** (a) A full cycle of a sine wave has two equivalent halves that are opposite in sign (phase). The origin is a node. (b) Squaring the sine function produces two equivalent positive peaks. The origin is still a node.

* The mathematical development of this three-dimensional function (and its square) is beyond the scope of this book, and, as is typically done by chemists, we have used lobes that are the same shape as in Figure 6.22.

The other $2p$ orbitals overlap sideways and thus concentrate electron density on opposite sides of the line through the nuclei. MOs of this type are called **pi (π) molecular orbitals**. We get one π bonding MO by combining the $2p_x$ atomic orbitals and another from the $2p_y$ atomic orbitals. These two π_{2p} molecular orbitals have the same energy; in other words, they are degenerate. Likewise, we get two degenerate π^*_{2p} antibonding MOs.

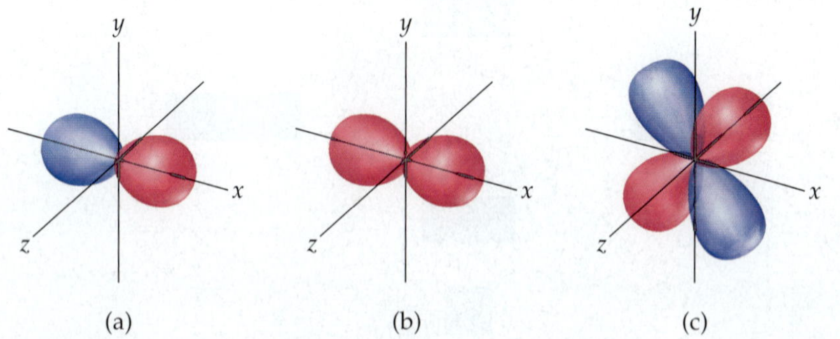

◀ **Figure 9.40 Atomic orbital wave functions.** (a) The wave function for a *p* orbital has two equivalent lobes that have opposite sign. (b) The probability density for a *p* orbital has two lobes that have the same sign. (c) The wave function for the d_{xy} orbital has lobes of alternating sign.

orbitals likewise have lobes of alternating phase. In the wave function for the d_{z^2} orbital, the "doughnut" has a phase opposite that of the two large lobes.

We encounter a very similar situation with molecular orbitals. The wave function of the σ^*_{1s} MO of H_2 is constructed by adding the wave function for a 1*s* orbital on one atom to the wave function for a 1*s* orbital on the other atom, but with the two orbitals having different phases. Chemists generally sketch this wave function by simply drawing the contour representations of the 1*s* orbitals in different colors, as shown in Figure 9.41(a) ▶. The fact that the 1*s* orbitals have different phases causes a node halfway between the atoms. Notice how this wave function has similarities to the sine function and to a *p* orbital—in each of these cases we have two parts of the function of opposite phase separated by a node. When we square the wave function of the σ^*_{1s} MO, we get the probability density representation given in Figure 9.34—notice that we once again lose the phase information when we look at the probability density.

The same phenomenon occurs when *p* orbitals form bonding and antibonding MOs. Figures 9.41(b) and 9.41(c) show the atomic orbital wave functions oriented to form the σ_{2p} and σ^*_{2p} MOs of a homonuclear diatomic molecule. In the bonding σ_{2p} orbital, the lobes of the same phase point at one another, which concentrates the electron density between the nuclei. In contrast, to construct the antibonding σ^*_{2p} MO, the 2*p* orbitals are lined up so that the lobes that point at one another have different phases. Because they have a different phase, the mixing of these orbitals leads to the exclusion of electron density in the region between the nuclei and the formation of a node halfway between the nuclei, both of which are important characteristics of antibonding MOs. We presented the probability densities for these MOs in Figure 9.38.

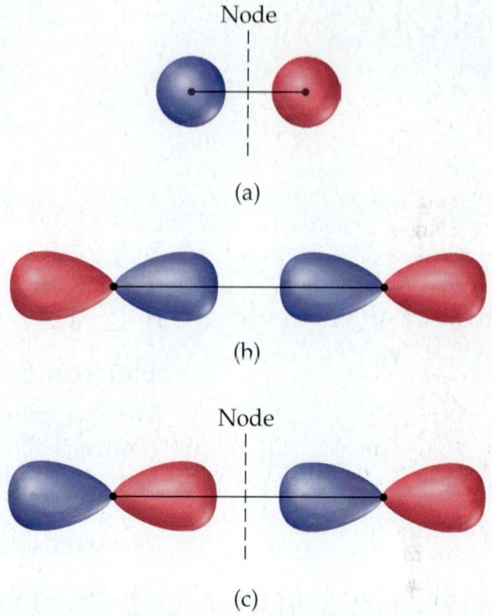

▲ **Figure 9.41 Molecular orbitals from atomic orbital wave functions.** (a) The mixing of two 1*s* orbitals that are opposite in phase produces a σ^*_{1s} MO. (b) The mixing of two 2*p* orbitals pointing at each other with lobes of the same phase produces a σ_{2p} MO. (c) The mixing of two 2*p* orbitals pointing at each other with lobes of opposite phase produces a σ^*_{2p} MO. Notice that a node is generated in (a) and (c).

Our brief discussion has given you only an introduction to the mathematical subtleties of atomic and molecular orbitals. The wave functions of atomic and molecular orbitals are used by chemists to understand many aspects of chemical bonding and spectroscopy, and you will probably see orbitals drawn in color to show phases if you take a course in organic chemistry. *Related Exercises:* 9.89 and 9.103

The $2p_z$ orbitals on two atoms point directly at each other. Hence, the overlap of two $2p_z$ orbitals is greater than that for two $2p_x$ or $2p_y$ orbitals. From rule 3 we therefore expect the σ_{2p} MO to be lower in energy (more stable) than the π_{2p} MOs. Similarly, the σ^*_{2p} MO should be higher in energy (less stable) than the π^*_{2p} MOs.

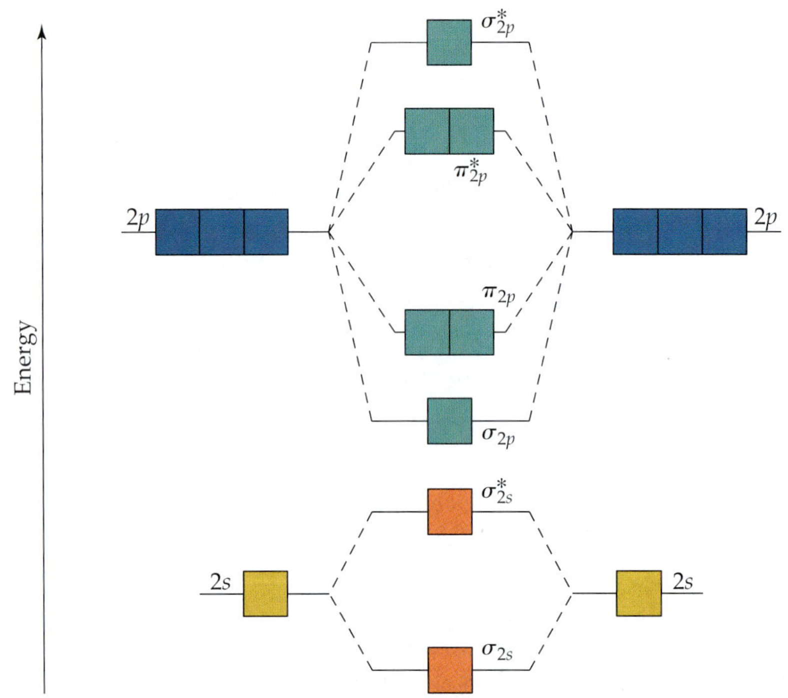

▶ **Figure 9.42 Energy-level diagram for MOs of second-row homonuclear diatomic molecules.** The diagram assumes no interaction between the 2*s* atomic orbital on one atom and the 2*p* atomic orbitals on the other atom, and experiment shows that it fits only for O_2, F_2, and Ne_2.

Electron Configurations for B_2 Through Ne_2

We have thus far independently considered the MOs that result from *s* orbitals (Figure 9.37) and from *p* orbitals (Figure 9.38). We can combine these results to construct an energy-level diagram (Figure 9.42 ▲) for homonuclear diatomic molecules of the elements boron through neon, all of which have valence 2*s* and 2*p* atomic orbitals. The following features of the diagram are notable:

1. The 2*s* atomic orbitals are lower in energy than the 2*p* atomic orbitals. (Section 6.7) Consequently, both of the molecular orbitals that result from the 2*s* orbitals, the bonding σ_{2s} and antibonding σ^*_{2s} are lower in energy than the lowest-energy MO derived from the 2*p* atomic orbitals.
2. The overlap of the two $2p_z$ orbitals is greater than that of the two $2p_x$ or $2p_y$ orbitals. As a result, the bonding σ_{2p} MO is lower in energy than the π_{2p} MOs, and the antibonding σ^*_{2p} MO is higher in energy than the π^*_{2p} MOs.
3. Both the π_{2p} and π^*_{2p} molecular orbitals are *doubly degenerate*; that is, there are two degenerate MOs of each type.

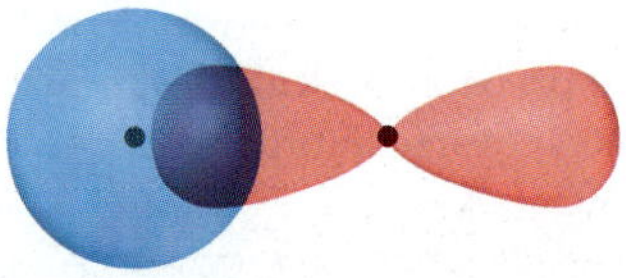

▲ **Figure 9.43 Interaction of 2*s* and 2*p* atomic orbitals.** The 2*s* orbital on one atom of a diatomic molecule can overlap with the $2p_z$ orbital on the other atom. These 2*s*-2*p* interactions can alter the energetic ordering of the MOs of the molecule.

Before we can add electrons to the energy-level diagram in Figure 9.42, there is one more effect we must consider. We have constructed the diagram assuming there is no interaction between the 2*s* orbital on one atom and the 2*p* orbitals on the other. In fact, such interactions can and do take place. Figure 9.43 ◀ shows the overlap of a 2*s* orbital on one of the atoms with a $2p_z$ orbital on the other. These interactions affect the energies of the σ_{2s} and σ_{2p} molecular orbitals in such a way that these MOs move further apart in energy, the σ_{2s} falling and the σ_{2p} rising in energy (Figure 9.44 ▶). These 2*s*-2*p* interactions are strong enough that the energetic ordering of the MOs can be altered: *For B_2, C_2, and N_2, the σ_{2p} MO is above the π_{2p} MOs in energy. For O_2, F_2, and Ne_2, the σ_{2p} MO is below the π_{2p} MOs.*

Given the energy ordering of the molecular orbitals, it is a simple matter to determine the electron configurations for the second-row diatomic molecules B_2 through Ne_2. For example, a boron atom has three valence electrons.

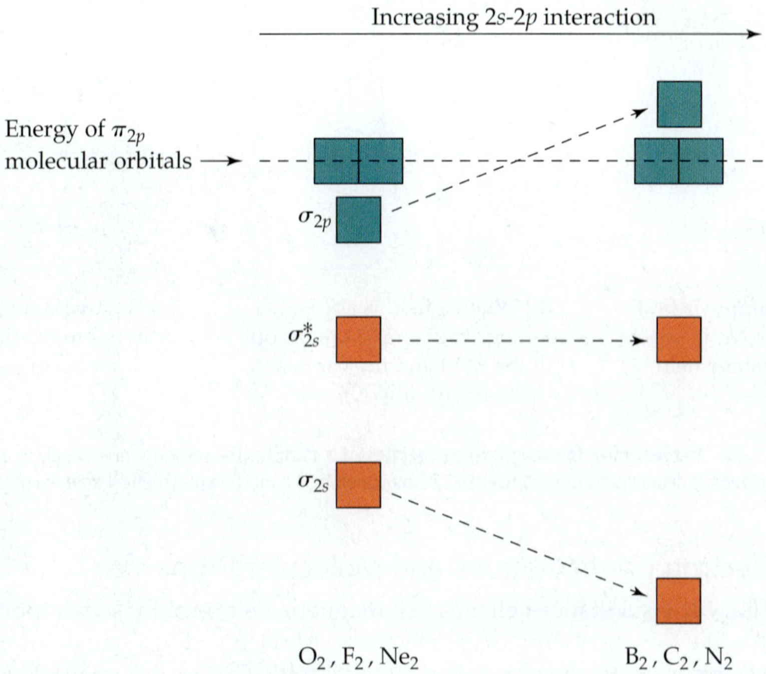

◀ **Figure 9.44 The effect of 2*s*-2*p* interactions.** When the 2*s* and 2*p* orbitals interact, the σ_{2s} MO falls in energy and the σ_{2p} MO rises in energy. For O_2, F_2, and Ne_2, the interaction is small, and the σ_{2p} MO remains below the π_{2p} MOs, as in Figure 9.42. For B_2, C_2, and N_2, the 2*s*-2*p* interaction is great enough that the σ_{2p} MO rises above the π_{2p} MOs, as shown on the right.

(Remember that we are ignoring the inner-shell 1*s* electrons.) Thus, for B_2 we must place six electrons in MOs. Four of these fully occupy the σ_{2s} and σ^*_{2s} MOs, leading to no net bonding. The last two electrons are put in the π_{2p} bonding MOs; one electron is put in one π_{2p} MO and the other electron is put in the other π_{2p} MO, with the two electrons having the same spin. Therefore, B_2 has a bond order of 1. Each time we move one element to the right in the second row, two more electrons must be placed in the diagram. For example, on moving to C_2, we have two more electrons than in B_2, and these electrons are also placed in the π_{2p} MOs, completely filling them. The electron configurations and bond orders for the diatomic molecules B_2 through Ne_2 are given in Figure 9.45 ▼.

Large 2*s*-2*p* interaction				Small 2*s*-2*p* interaction			
	B_2	C_2	N_2		O_2	F_2	Ne_2
σ^*_{2p}				σ^*_{2p}			↑↓
π^*_{2p}				π^*_{2p}	↑ ↑	↑↓ ↑↓	↑↓ ↑↓
σ_{2p}			↑↓	π_{2p}	↑↓ ↑↓	↑↓ ↑↓	↑↓ ↑↓
π_{2p}	↑ ↑	↑↓ ↑↓	↑↓ ↑↓	σ_{2p}	↑↓	↑↓	↑↓
σ^*_{2s}	↑↓	↑↓	↑↓	σ^*_{2s}	↑↓	↑↓	↑↓
σ_{2s}	↑↓	↑↓	↑↓	σ_{2s}	↑↓	↑↓	↑↓
Bond order	1	2	3		2	1	0
Bond enthalpy (kJ/mol)	290	620	941		495	155	—
Bond length (Å)	1.59	1.31	1.10		1.21	1.43	—
Magnetic behavior	Paramagnetic	Diamagnetic	Diamagnetic		Paramagnetic	Diamagnetic	—

▲ **Figure 9.45 The second-row diatomic molecules.** Molecular orbital electron configurations and some experimental data for several second-row diatomic molecules.

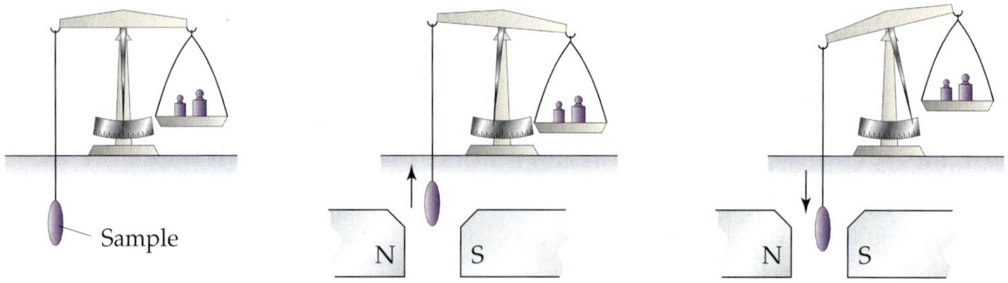

(a) The sample is first weighed in the absence of a magnetic field.

(b) When a field is applied, a diamagnetic sample moves out of the field and thus appears to have a lower mass.

(c) A paramagnetic sample is drawn into the field and thus appears to gain mass.

▲ **Figure 9.46 Determining the magnetic properties of a sample.** The response of a sample to a magnetic field indicates whether it is diamagnetic or paramagnetic. Paramagnetism is a much stronger effect than diamagnetism.

Electron Configurations and Molecular Properties

The way a substance behaves in a magnetic field provides an important insight into the arrangements of its electrons. Molecules with one or more unpaired electrons are attracted into a magnetic field. The more unpaired electrons in a species, the stronger the force of attraction. This type of magnetic behavior is called **paramagnetism**.

Substances with no unpaired electrons are weakly repelled from a magnetic field. This property is called **diamagnetism**. Diamagnetism is a much weaker effect than paramagnetism. A straightforward method for measuring the magnetic properties of a substance, illustrated in Figure 9.46 ▲, involves weighing the substance in the presence and absence of a magnetic field. If the substance is paramagnetic, it will appear to weigh more in the magnetic field; if it is diamagnetic, it will appear to weigh less. The magnetic behaviors observed for the diatomic molecules of the second-row elements agree with the electron configurations shown in Figure 9.45.

GIVE IT SOME THOUGHT

Figure 9.45 indicates that the C_2 molecule is diamagnetic. Would that be expected if the σ_{2p} MO were lower in energy than the π_{2p} MOs?

The electron configurations can also be related to the bond distances and bond enthalpies of the molecules. (Section 8.8) As bond orders increase, bond distances decrease and bond enthalpies increase. N_2, for example, whose bond order is 3, has a short bond distance and a large bond enthalpy. The N_2 molecule does not react readily with other substances to form nitrogen compounds. The high bond order of the molecule helps explain its exceptional stability. We should also note, however, that molecules with the same bond orders do *not* have the same bond distances and bond enthalpies. Bond order is only one factor influencing these properties. Other factors include the nuclear charges and the extent of orbital overlap.

Bonding in the dioxygen molecule, O_2, is especially interesting. Its Lewis structure shows a double bond and complete pairing of electrons:

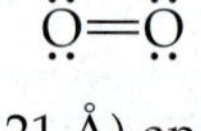

The short O—O bond distance (1.21 Å) and the relatively high bond enthalpy (495 kJ/mol) are in agreement with the presence of a double bond. However, the molecule is found to contain two unpaired electrons. The paramagnetism of

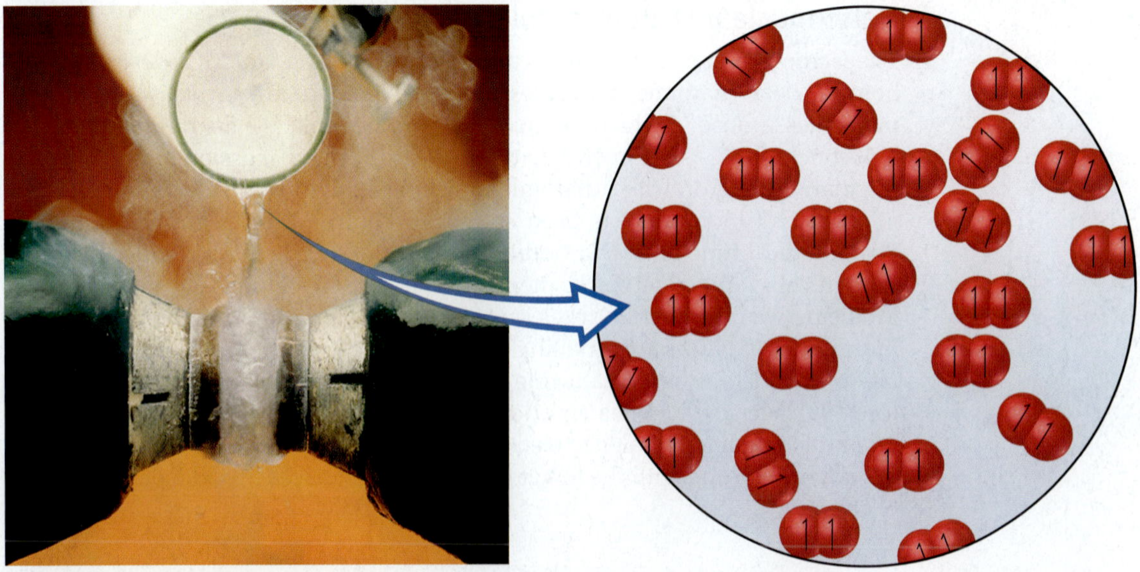

▲ **Figure 9.47 Paramagnetism of O_2.** Liquid O_2 being poured between the poles of a magnet. Because each O_2 molecule contains two unpaired electrons, O_2 is paramagnetic. It is therefore attracted into the magnetic field and "sticks" between the magnetic poles.

O_2 is demonstrated in Figure 9.47 ▲. Although the Lewis structure fails to account for the paramagnetism of O_2, molecular orbital theory correctly predicts that there are two unpaired electrons in the π^*_{2p} orbital of the molecule (Figure 9.45). The MO description also correctly indicates a bond order of 2.

Going from O_2 to F_2, we add two more electrons, completely filling the π^*_{2p} MOs. Thus, F_2 is expected to be diamagnetic and have an F—F single bond, in accord with its Lewis structure. Finally, the addition of two more electrons to make Ne_2 fills all the bonding and antibonding MOs; therefore, the bond order of Ne_2 is zero, and the molecule is not expected to exist.

SAMPLE EXERCISE 9.9 | Molecular Orbitals of a Second-Row Diatomic Ion

Predict the following properties of O_2^+: **(a)** number of unpaired electrons, **(b)** bond order, **(c)** bond enthalpy and bond length.

Solution

Analyze: Our task is to predict several properties of the cation O_2^+.

Plan: We will use the MO description of O_2^+ to determine the desired properties. We must first determine the number of electrons in O_2^+ and then draw its MO energy diagram. The unpaired electrons are those without a partner of opposite spin. The bond order is one-half the difference between the number of bonding and antibonding electrons. After calculating the bond order, we can use the data in Figure 9.45 to estimate the bond enthalpy and bond length.

Solve: **(a)** The O_2^+ ion has 11 valence electrons, one fewer than O_2. The electron removed from O_2 to form O_2^+ is one of the two unpaired π^* electrons (see Figure 9.45). Therefore, O_2^+ has just one unpaired electron.

(b) The molecule has eight bonding electrons (the same as O_2) and three antibonding electrons (one fewer than O_2). Thus, its bond order is

$$\frac{1}{2}(8 - 3) = 2\frac{1}{2}$$

(c) The bond order of O_2^+ is between that for O_2 (bond order 2) and N_2 (bond order 3). Thus, the bond enthalpy and bond length should be about midway between those for O_2 and N_2, approximately 700 kJ/mol and 1.15 Å, respectively. The experimental bond enthalpy and bond length of the ion are 625 kJ/mol and 1.123 Å, respectively.

PRACTICE EXERCISE

Predict the magnetic properties and bond orders of **(a)** the peroxide ion, O_2^{2-}; **(b)** the acetylide ion, C_2^{2-}.

Answers: **(a)** diamagnetic, 1; **(b)** diamagnetic, 3

Heteronuclear Diatomic Molecules

The same principles we have used in developing an MO description of homonuclear diatomic molecules can be extended to *heteronuclear* diatomic molecules—those in which the two atoms in the molecule are not the same. We will conclude this section on MO theory with a brief discussion of the MOs of a fascinating heteronuclear diatomic molecule—the nitric oxide, NO, molecule.

The NO molecule has been shown to control several important human physiological functions. Our bodies use it, for example, to relax muscles, to kill foreign cells, and to reinforce memory. The 1998 Nobel Prize in Physiology or Medicine was awarded to three scientists for their research that uncovered the importance of NO as a "signalling" molecule in the cardiovascular system. That NO plays such an important role in human metabolism was unsuspected before 1987 because NO has an odd number of electrons and is highly reactive. The molecule has 11 valence electrons, and two possible Lewis structures can be drawn. The one with the lower formal charges places the odd electron on the N atom:

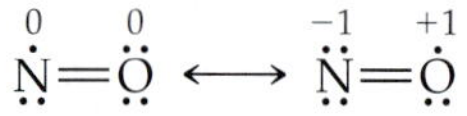

Both structures indicate the presence of a double bond, but when compared with the molecules in Figure 9.45, the experimental bond length of NO (1.15 Å) suggests a bond order greater than two. How do we treat NO using the MO model?

If the atoms in a heteronuclear diatomic molecule do not differ too greatly in their electronegativities, the description of their MOs will resemble those for homonuclear diatomics, with one important modification: The atomic energies of the more electronegative atom will be lower in energy than those of the less electronegative element. The MO diagram for NO is shown in Figure 9.48 ◀, and you can see that the 2*s* and 2*p* atomic orbitals of oxygen (more electronegative) are slightly lower than those of nitrogen (less electronegative). We see that the MO energy-level diagram is much like that of a homonuclear diatomic molecule—because the 2*s* and 2*p* orbitals on the two atoms interact, the same types of MOs are produced.

▼ Figure 9.48 The MO energy-level diagram for NO.

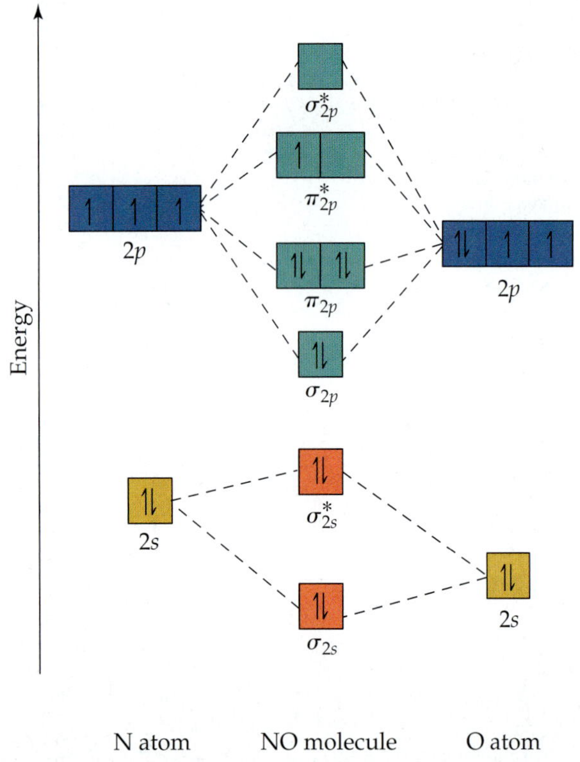

There is one other important change in the MOs when we consider heteronuclear molecules. The MOs that result are still a mix of the atomic orbitals from both atoms, but in general *an MO will have a greater contribution from the atomic orbital to which it is closer in energy*. In the case of NO, for example, the σ_{2s} bonding MO is closer in energy to the O 2*s* atomic orbital than to the N 2*s* atomic orbital. As a result, the σ_{2s} MO has a slightly greater contribution from O than from N—the orbital is no longer an equal mixture of the two atoms, as was the case for the homonuclear diatomic molecules. Similarly, the σ^*_{2s} antibonding MO is weighted more heavily toward the N atom because that MO is closest in energy to the N 2*s* atomic orbital.

We complete the MO diagram for NO by filling the MOs in Figure 9.48 with the 11 valence electrons. We see that there are eight bonding and three antibonding electrons, giving a bond order of $\frac{1}{2}(8 - 3) = 2\frac{1}{2}$, which agrees better with experiment than the Lewis structures do. The unpaired electron resides in one of the π^*_{2p} MOs, which are more heavily weighted toward the N atom. This description is consistent with the left Lewis structure above (the one preferred on the basis of formal charge), which places the unpaired electron on the N atom.

CHEMISTRY AT WORK | Organic Dyes

The chemistry of color has fascinated people since ancient times. The brilliant colors around you—those of your clothes, the photographs in this book, the foods you eat—are due to the selective absorption of light by chemicals. Light excites electrons in molecules. In a molecular orbital picture, we can envision light exciting an electron from a filled molecular orbital to an empty one at higher energy. Because the MOs have definite energies, only light of the proper wavelengths can excite electrons. The situation is analogous to that of atomic line spectra. ooo (Section 6.3) If the appropriate wavelength for exciting electrons is in the visible portion of the electromagnetic spectrum, the substance will appear colored: Certain wavelengths of white light are absorbed, others are not. A red traffic light appears red because only red light is transmitted through the lens. The other wavelengths of visible light are absorbed by it.

In using molecular orbital theory to discuss the absorptions of light by molecules, we can focus on two MOs in particular. The *highest occupied molecular orbital* (HOMO) is the MO of highest energy that has electrons in it. The *lowest unoccupied molecular orbital* (LUMO) is the MO of lowest energy that does not have electrons in it. In N_2, for example, the HOMO is the π_{2p} MO and the LUMO is the π^*_{2p} MO (Figure 9.45). The energy difference between the HOMO and the LUMO—known as the HOMO-LUMO gap—is related to the minimum energy needed to excite an electron in the molecule. Colorless or white substances usually have such a large HOMO-LUMO gap that visible light is not energetic enough to excite an electron to the higher level. The minimum energy needed to excite an electron in N_2 corresponds to light with a wavelength of less than 200 nm, which is far into the ultraviolet part of the spectrum (Figure 6.4). As a result, N_2 cannot absorb any visible light and is therefore colorless.

Many rich colors are due to *organic dyes*, organic molecules that strongly absorb selected wavelengths of visible light. Organic dyes are most familiar as the substances that are used to provide vibrant colors to fabrics. They are also used in color photographic film and in new high-tech applications such as the recordable compact discs called CD-Rs (Figure 9.49 ▼). In a CD-R a thin layer of a transparent colored organic dye is sandwiched between a reflective surface and a clear, rigid polymer backing. Data are "burned" onto the CD-R by use of a laser. When the laser strikes the dye, the dye molecules absorb the light, change structure, and become opaque. The selective production of these opaque "pits" in the CD-R gives it the capacity to store data in binary ("clear" or "opaque") form. Because the structure of the dye is irreversibly altered during the writing of data on the disc, data can be written only once to any given part of the disc.

Organic dyes contain extensively delocalized π electrons. The molecules contain atoms that are predominantly sp^2 hybridized, like the carbon atoms in benzene (Figure 9.28). This leaves one unhybridized p orbital on each atom to form π bonds with neighboring atoms. The p orbitals are arranged so that electrons can be delocalized throughout the entire molecule; we say that the π bonds are *conjugated*. The HOMO-LUMO gap in such molecules decreases as the number of conjugated double bonds increases. Butadiene, (C_4H_6), for example, has alternating carbon–carbon double and single bonds:

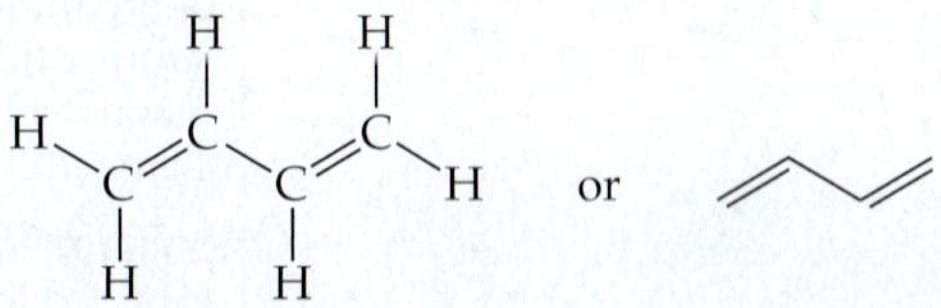

The representation on the right is the shorthand notation that chemists use for organic molecules. There are implicit carbon atoms at the ends of the three straight segments, and there are implicitly sufficient hydrogen atoms to make a total of four bonds at each carbon. Butadiene is planar, so the unhybridized p orbitals on the carbons are pointing in the same direction. The π electrons are delocalized among the four carbon atoms.

Because butadiene has only two conjugated double bonds, it still has a fairly large HOMO-LUMO gap. Butadiene absorbs light at 217 nm, still well into the ultraviolet part of

◄ **Figure 9.49 Organic dyes.** The color in organic dyes is the result of electrons being excited from one molecular orbital to another. Organic dyes have a variety of useful applications, from the production of colorful fabrics (left) to the production of laser-recordable compact discs (CD-Rs) for computer data storage (right).

continues on next page

continued

the spectrum. It is therefore colorless. If we keep adding new conjugated double bonds, however, the HOMO-LUMO gap keeps shrinking until visible light is absorbed. β-Carotene, for example, is the substance chiefly responsible for the bright orange color of carrots.

Because β-carotene contains 11 conjugated double bonds, its π electrons are very extensively delocalized. It absorbs light of wavelength 500 nm, in the middle of the visible part of the spectrum. The human body converts β-carotene into vitamin A, which in turn is converted into retinal, a component of *rhodopsin*, found in the retinas of your eyes. (See the "Chemistry and Life" box in Section 9.6.) The absorption of visible light by rhodopsin is a major reason why "visible" light is indeed visible. Thus, there seems to be a good basis for the maxim that eating carrots is good for your eyesight.

Related Exercises: 9.93 and 9.104

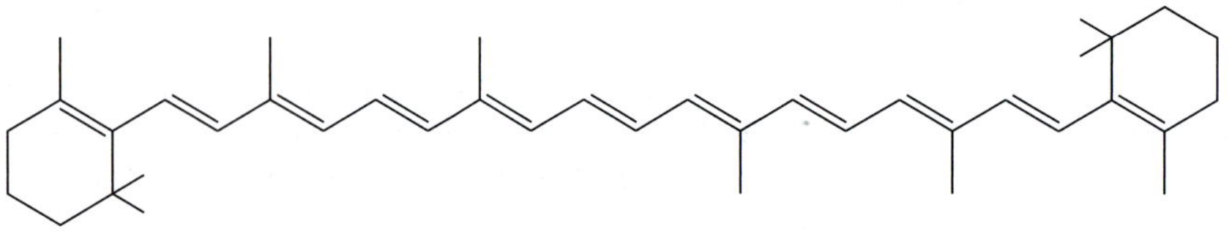

β-carotene

SAMPLE INTEGRATIVE EXERCISE | Putting Concepts Together

Elemental sulfur is a yellow solid that consists of S_8 molecules. The structure of the S_8 molecule is a puckered eight-membered ring (Figure 7.30). Heating elemental sulfur to high temperatures produces gaseous S_2 molecules:

$$S_8(s) \longrightarrow 4\,S_2(g)$$

(a) With respect to electronic structure, which element in the second row of the periodic table is most similar to sulfur? **(b)** Use the VSEPR model to predict the S—S—S bond angles in S_8 and the hybridization at S in S_8. **(c)** Use MO theory to predict the sulfur–sulfur bond order in S_2. Is the molecule expected to be diamagnetic or paramagnetic? **(d)** Use average bond enthalpies (Table 8.4) to estimate the enthalpy change for the reaction just described. Is the reaction exothermic or endothermic?

Solution **(a)** Sulfur is a group 6A element with an [Ne]$3s^2 3p^4$ electron configuration. It is expected to be most similar electronically to oxygen (electron configuration, [He]$2s^2 2p^4$), which is immediately above it in the periodic table. (Chapter 7, Introduction)
(b) The Lewis structure of S_8 is

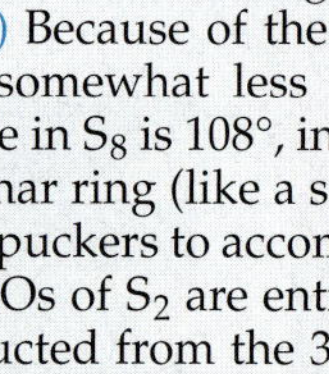

There is a single bond between each pair of S atoms and two nonbonding electron pairs on each S atom. Thus, we see four electron domains around each S atom, and we would expect a tetrahedral electron-domain geometry corresponding to sp^3 hybridization. (Sections 9.2, 9.5) Because of the nonbonding pairs, we would expect the S—S—S angles to be somewhat less than 109°, the tetrahedral angle. Experimentally, the S—S—S angle in S_8 is 108°, in good agreement with this prediction. Interestingly, if S_8 were a planar ring (like a stop sign), it would have S—S—S angles of 135°. Instead, the S_8 ring puckers to accommodate the smaller angles dictated by sp^3 hybridization. **(c)** The MOs of S_2 are entirely analogous to those of O_2, although the MOs for S_2 are constructed from the 3s and 3p atomic orbitals of sulfur. Further, S_2 has the same number of valence electrons as O_2. Thus, by analogy to our discussion of O_2, we would expect S_2 to have a bond order of 2 (a double bond) and to be paramagnetic with two unpaired electrons in the π^*_{3p} molecular orbitals of S_2. (Section 9.8) **(d)** We are considering the reaction in which an S_8 molecule falls apart into four S_2 molecules. From parts (b) and (c), we see that S_8 has S—S single bonds and S_2 has S=S double bonds. During the course of the reaction, therefore, we are breaking eight S—S single bonds and forming four S=S double bonds. We can estimate the enthalpy of the reaction by using Equation 8.12 and the average bond enthalpies in Table 8.4:

$$\Delta H_{rxn} = 8\,D(\text{S—S}) - 4\,D(\text{S=S}) = 8(266\text{ kJ}) - 4(418\text{ kJ}) = +456\text{ kJ}$$

Because $\Delta H_{rxn} > 0$, the reaction is endothermic. (Section 5.4) The very positive value of ΔH_{rxn} suggests that high temperatures are required to cause the reaction to occur.

SUMMARY AND KEY TERMS

Introduction and Section 9.1 The three-dimensional shapes and sizes of molecules are determined by their **bond angles** and bond lengths. Molecules with a central atom A surrounded by n atoms B, denoted AB_n, adopt a number of different geometric shapes, depending on the value of n and on the particular atoms involved. In the overwhelming majority of cases, these geometries are related to five basic shapes (linear, trigonal pyramidal, tetrahedral, trigonal bipyramidal, and octahedral).

Section 9.2 The **valence-shell electron-pair repulsion (VSEPR) model** rationalizes molecular geometries based on the repulsions between **electron domains**, which are regions about a central atom in which electrons are likely to be found. **Bonding pairs** of electrons, which are those involved in making bonds, and **nonbonding pairs** of electrons, also called lone pairs, both create electron domains around an atom. According to the VSEPR model, electron domains orient themselves to minimize electrostatic repulsions; that is, they remain as far apart as possible. Electron domains from nonbonding pairs exert slightly greater repulsions than those from bonding pairs, which leads to certain preferred positions for nonbonding pairs and to the departure of bond angles from idealized values. Electron domains from multiple bonds exert slightly greater repulsions than those from single bonds. The arrangement of electron domains around a central atom is called the **electron-domain geometry**; the arrangement of atoms is called the **molecular geometry**.

Section 9.3 The dipole moment of a polyatomic molecule depends on the vector sum of the dipole moments associated with the individual bonds, called the **bond dipoles**. Certain molecular shapes, such as linear AB_2 and trigonal planar AB_3, assure that the bond dipoles cancel, producing a nonpolar molecule, which is one whose dipole moment is zero. In other shapes, such as bent AB_2 and trigonal pyramidal AB_3, the bond dipoles do not cancel and the molecule will be polar (that is, it will have a nonzero dipole moment).

Section 9.4 **Valence-bond theory** is an extension of Lewis's notion of electron-pair bonds. In valence-bond theory, covalent bonds are formed when atomic orbitals on neighboring atoms **overlap** one another. The overlap region is a favorable one for the two electrons because of their attraction to two nuclei. The greater the overlap between two orbitals, the stronger the bond that is formed.

Section 9.5 To extend the ideas of valence-bond theory to polyatomic molecules, we must envision mixing s, p, and sometimes d orbitals to form **hybrid orbitals**. The process of **hybridization** leads to hybrid atomic orbitals that have a large lobe directed to overlap with orbitals on another atom to make a bond. Hybrid orbitals can also accommodate nonbonding pairs. A particular mode of hybridization can be associated with each of the five common electron-domain geometries (linear = sp; trigonal planar = sp^2; tetrahedral = sp^3; trigonal bipyramidal = sp^3d; and octahedral = sp^3d^2).

Section 9.6 Covalent bonds in which the electron density lies along the line connecting the atoms (the internuclear axis) are called **sigma (σ) bonds**. Bonds can also be formed from the sideways overlap of p orbitals. Such a bond is called a **pi (π) bond**. A double bond, such as that in C_2H_4, consists of one σ bond and one π bond; a triple bond, such as that in C_2H_2, consists of one σ and two π bonds. The formation of a π bond requires that molecules adopt a specific orientation; the two CH_2 groups in C_2H_4, for example, must lie in the same plane. As a result, the presence of π bonds introduces rigidity into molecules. In molecules that have multiple bonds and more than one resonance structure, such as C_6H_6, the π bonds are **delocalized**; that is, the π bonds are spread among several atoms.

Section 9.7 **Molecular orbital theory** is another model used to describe the bonding in molecules. In this model the electrons exist in allowed energy states called **molecular orbitals (MOs)**. These orbitals can be spread among all the atoms of a molecule. Like an atomic orbital, a molecular orbital has a definite energy and can hold two electrons of opposite spin. The combination of two atomic orbitals leads to the formation of two MOs, one at lower energy and one at higher energy relative to the energy of the atomic orbitals. The lower-energy MO concentrates charge density in the region between the nuclei and is called a **bonding molecular orbital**. The higher-energy MO excludes electrons from the region between the nuclei and is called an **antibonding molecular orbital**. Occupation of bonding MOs favors bond formation, whereas occupation of antibonding MOs is unfavorable. The bonding and antibonding MOs formed by the combination of s orbitals are **sigma (σ) molecular orbitals**; like σ bonds, they lie on the internuclear axis.

The combination of atomic orbitals and the relative energies of the molecular orbitals are shown by an **energy-level** (or **molecular orbital**) **diagram**. When the appropriate number of electrons are put into the MOs, we can calculate the **bond order** of a bond, which is half the difference between the number of electrons in bonding MOs and the number of electrons in antibonding MOs. A bond order of 1 corresponds to a single bond, and so forth. Bond orders can be fractional numbers.

Section 9.8 Electrons in core orbitals do not contribute to the bonding between atoms, so a molecular orbital description usually needs to consider only electrons in the outermost electron subshells. In order to describe the MOs of second-row homonuclear diatomic molecules, we need to consider the MOs that can form by the combination of p orbitals. The p orbitals that point directly at one another can form σ bonding and σ^* antibonding MOs. The p orbitals that are oriented perpendicular to the internuclear axis combine to form **pi (π) molecular orbitals**. In diatomic molecules the π molecular orbitals occur as pairs of degenerate (same energy) bonding and antibonding MOs. The σ_{2p} bonding MO is expected to be lower in energy than the π_{2p} bonding MOs because of larger orbital overlap. This ordering is reversed in B_2, C_2, and N_2 because of interaction between the 2s and 2p atomic orbitals.

The molecular orbital description of second-row diatomic molecules leads to bond orders in accord with the Lewis structures of these molecules. Further, the model predicts correctly that O_2 should exhibit **paramagnetism**, an attraction of a molecule by a magnetic field due to unpaired electrons. Those molecules in which all the electrons are paired exhibit **diamagnetism**, a weak repulsion from a magnetic field.

VISUALIZING CONCEPTS

9.1 A certain AB_4 molecule has a "seesaw" shape:

From which of the fundamental geometries shown in Figure 9.3 could you remove one or more atoms to create a molecule having this seesaw shape? [Section 9.1]

9.2 **(a)** If the three balloons shown below are all the same size, what angle is formed between the red one and the green one? **(b)** If additional air is added to the blue balloon so that it gets larger, what happens to the angle between the red and green balloons? **(c)** What aspect of the VSEPR model is illustrated by part (b)? [Section 9.2]

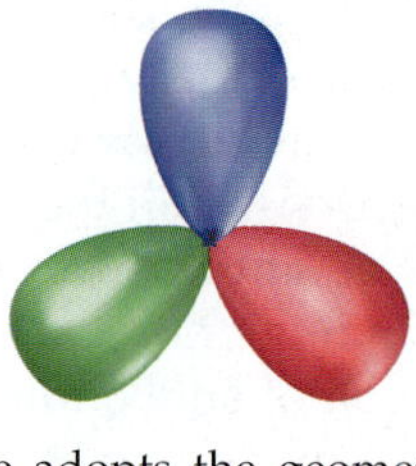

9.3 An AB_5 molecule adopts the geometry shown below. **(a)** What is the name of this geometry? **(b)** Do you think there are any nonbonding electron pairs on atom A? Why or why not? **(c)** Suppose the atoms B are halogen atoms. Can you determine uniquely to which group in the periodic table atom A belongs? [Section 9.2]

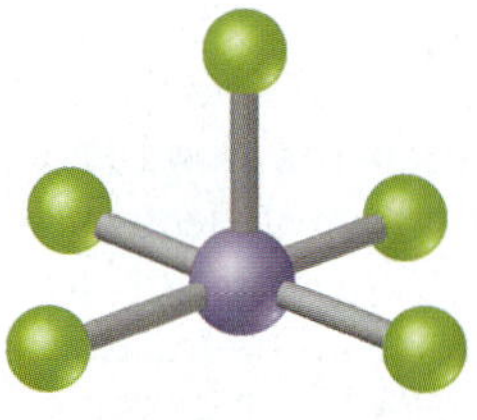

9.4 The molecule shown here is *difluoromethane* (CH_2F_2), which is used as a refrigerant called R-32. **(a)** Based on the structure, how many electron domains surround the C atom in this molecule? **(b)** Would the molecule have a nonzero dipole moment? **(c)** If the molecule is polar, in what direction will the overall dipole moment vector point in the molecule? [Sections 9.2 and 9.3]

9.5 The plot below shows the potential energy of two Cl atoms as a function of the distance between them. **(a)** To what does an energy of zero correspond in this diagram? **(b)** According to the valence-bond model, why does the energy decrease as the Cl atoms move from a large separation to a smaller one? **(c)** What is the significance of the Cl-Cl distance at the minimum point in the plot? **(d)** Why does the energy rise at Cl-Cl distances less than that at the minimum point in the plot? [Section 9.4]

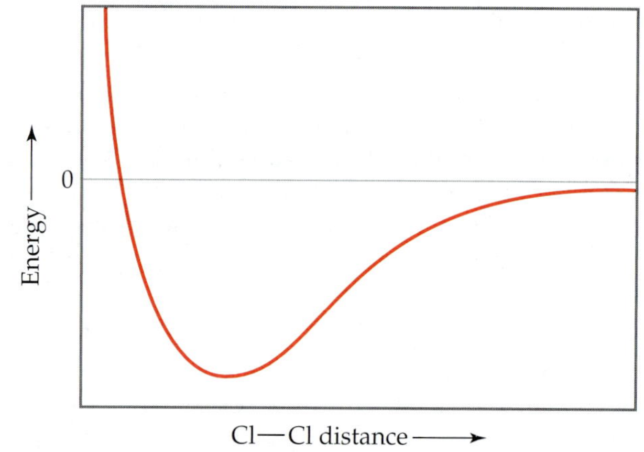

9.6 Shown below are three pairs of hybrid orbitals, with each set at a characteristic angle. For each pair, determine the type or types of hybridization that could lead to hybrid orbitals at the specified angle. [Section 9.5]

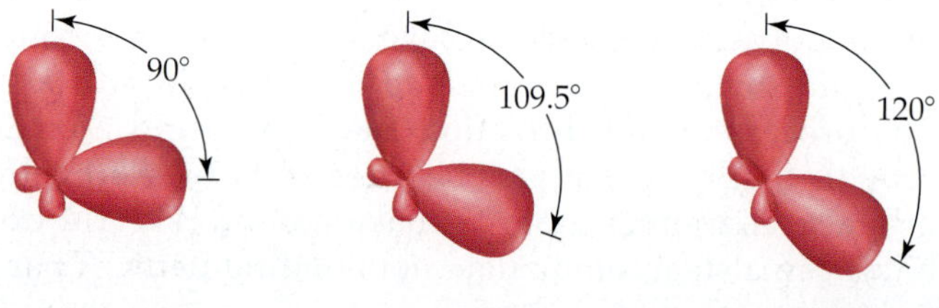

9.7 The orbital diagram below presents the final step in the formation of hybrid orbitals by a silicon atom. **(a)** Do you think one or more electrons have been promoted? Why or why not? **(b)** What type of hybrid orbitals are being produced in this hybridization? [Section 9.5]

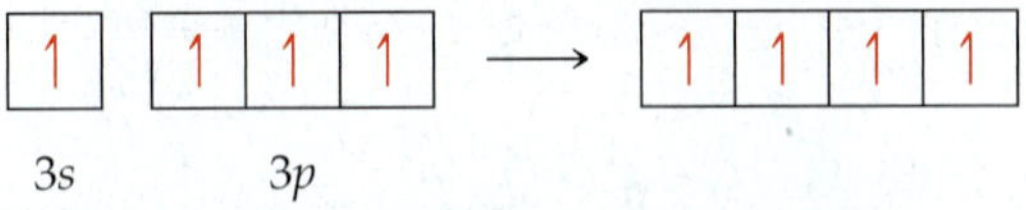

9.8 Consider the hydrocarbon drawn below. **(a)** What is the hybridization at each carbon atom in the molecule? **(b)** How many σ bonds are there in the molecule? **(c)** How many π bonds? [Section 9.6]

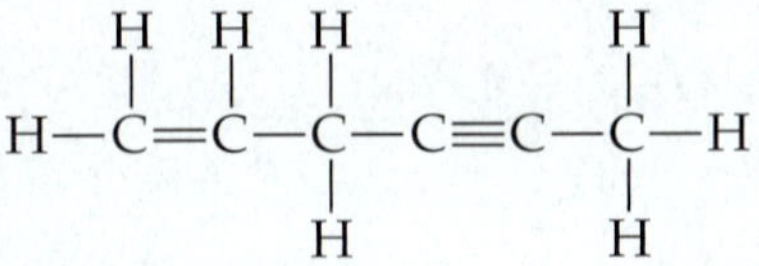

9.9 For each of the following contour representations of molecular orbitals, identify (i) the atomic orbitals (*s* or *p*) used to construct the MO, (ii) the type of MO (σ or π), and (iii) whether the MO is bonding or antibonding. [Sections 9.7 and 9.8]

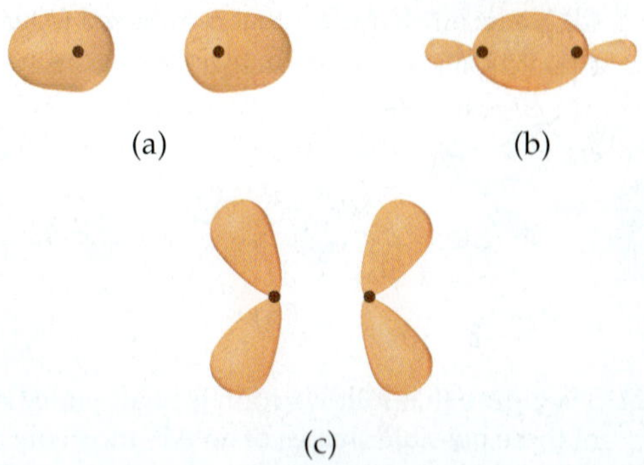

9.10 The diagram below shows the highest occupied MOs of a neutral molecule CX, where element X is in the same row of the periodic table as C. **(a)** Based on the number of electrons, can you determine the identity of X? **(b)** Would the molecule be diamagnetic or paramagnetic? **(c)** Consider the π_{2p} MOs of the molecule. Would you expect them to have a greater atomic orbital contribution from C, have a greater atomic orbital contribution from X, or be an equal mixture of atomic orbitals from the two atoms? [Section 9.8]

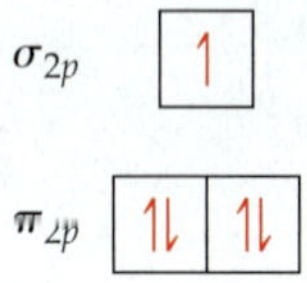

EXERCISES

Molecular Shapes; the VSEPR Model

9.11 **(a)** An AB_2 molecule is described as linear, and the A—B bond length is specified. Does this information completely describe the geometry of the molecule? **(b)** The molecules BF_3 and SO_3 are both described as trigonal planar. Does this information completely define the bond angles of these molecules?

9.12 **(a)** Methane (CH_4) and the perchlorate ion (ClO_4^-) are both described as tetrahedral. What does this indicate about their bond angles? **(b)** The NH_3 molecule is trigonal pyramidal. How many parameters need to be specified to define its geometry completely?

9.13 **(a)** What is meant by the term *electron domain*? **(b)** Explain in what way electron domains behave like the balloons in Figure 9.5. Why do they do so?

9.14 **(a)** How does one determine the number of electron domains in a molecule or ion? **(b)** What is the difference between a *bonding electron domain* and a *nonbonding electron domain*?

9.15 Describe the characteristic electron-domain geometry of each of the following numbers of electron domains about a central atom: **(a)** 3, **(b)** 4, **(c)** 5, **(d)** 6.

9.16 Indicate the number of electron domains about a central atom, given the following angles between them: **(a)** 120°, **(b)** 180°, **(c)** 109.5°, **(d)** 90°.

9.17 What is the difference between the electron-domain geometry and the molecular geometry of a molecule? Use the water molecule as an example in your discussion.

9.18 An AB_3 molecule is described as having a trigonal-bipyramidal electron-domain geometry. How many nonbonding domains are on atom A? Explain.

9.19 Give the electron-domain and molecular geometries of a molecule that has the following electron domains on its central atom: **(a)** four bonding domains and no nonbonding domains, **(b)** three bonding domains and two nonbonding domains, **(c)** five bonding domains and one nonbonding domain.

9.20 What are the electron-domain and molecular geometries of a molecule that has the following electron domains on its central atom? **(a)** Three bonding domains and no nonbonding domains, **(b)** three bonding domains and one nonbonding domain, **(c)** two bonding domains and three nonbonding domains.

9.21 Draw the Lewis structure for each of the following molecules or ions, and predict their electron-domain and molecular geometries: **(a)** PF_3, **(b)** CH_3^+, **(c)** BrF_3, **(d)** ClO_4^-, **(e)** XeF_2, **(f)** BrO_2^-.

9.22 Give the electron-domain and molecular geometries for the following molecules and ions: **(a)** HCN, **(b)** SO_3^{2-}, **(c)** SF_4, **(d)** PF_6^-, **(e)** NH_3Cl^+, **(f)** N_3^-.

9.23 The figure that follows shows ball-and-stick drawings of three possible shapes of an AF_3 molecule. **(a)** For each shape, give the electron-domain geometry on which the molecular geometry is based. **(b)** For each shape, how many nonbonding electron domains are there on atom A? **(c)** Which of the following elements will lead to an AF_3 molecule with the shape in (ii): Li, B, N, Al, P, Cl?

(d) Name an element A that is expected to lead to the AF_3 structure shown in (iii). Explain your reasoning.

(i) (ii) (iii)

9.24 The figure that follows contains ball-and-stick drawings of three possible shapes of an AF_4 molecule. **(a)** For each shape, give the electron-domain geometry on which the molecular geometry is based. **(b)** For each shape, how many nonbonding electron domains are there on atom A? **(c)** Which of the following elements will lead to an AF_4 molecule with the shape in (iii): Be, C, S, Se, Si, Xe? **(d)** Name an element A that is expected to lead to the AF_4 structure shown in (i).

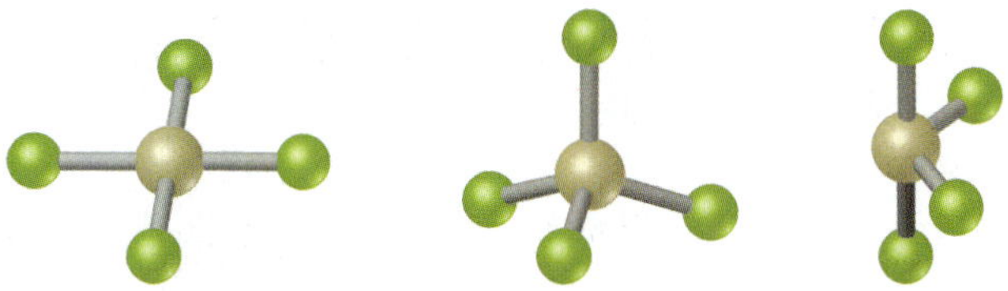

(i) (ii) (iii)

9.25 Give the approximate values for the indicated bond angles in the following molecules:

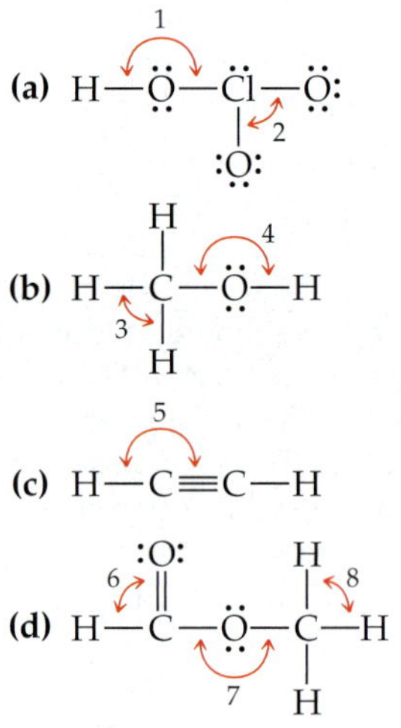

9.26 Give approximate values for the indicated bond angles in the following molecules:

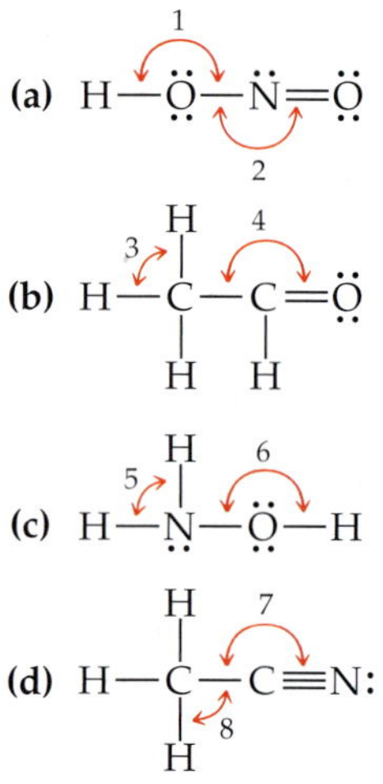

9.27 The three species NH_2^-, NH_3, and NH_4^+ have H—N—H bond angles of 105°, 107°, and 109°, respectively. Explain this variation in bond angles.

9.28 Predict the trend in the F(axial)—A—F(equatorial) bond angle in the following AF_n molecules: PF_5, SF_4, and ClF_3.

9.29 **(a)** Explain why BrF_4^- is square planar, whereas BF_4^- is tetrahedral. **(b)** In which of these molecules, CF_4 or SF_4, do you think the actual bond angle is closest to the ideal angle predicted by the VSEPR model? Explain briefly.

9.30 **(a)** Explain why the following ions have different bond angles: ClO_2^- and NO_2^-. Predict the bond angle in each case. **(b)** Given that the spatial requirement of a nonbonding pair of electrons is greater than that of a bonding pair, explain why the XeF_2 molecule is linear and not bent.

Polarity of Polyatomic Molecules

9.31 Does SO_2 have a dipole moment? If so, in which direction does the net dipole point?

9.32 The PH_3 molecule is polar. How does this offer experimental proof that the molecule cannot be planar?

9.33 **(a)** Consider the AF_3 molecules in Exercise 9.23. Which of these will have a nonzero dipole moment? Explain. **(b)** Which of the AF_4 molecules in Exercise 9.24 will have a zero dipole moment?

9.34 **(a)** What conditions must be met if a molecule with polar bonds is nonpolar? **(b)** What geometries will give nonpolar molecules for AB_2, AB_3, and AB_4 geometries?

9.35 Predict whether each of the following molecules is polar or nonpolar: **(a)** CCl_4, **(b)** NH_3, **(c)** SF_4, **(d)** XeF_4, **(e)** CH_3Br, **(f)** GaH_3.

9.36 Predict whether each of the following molecules is polar or nonpolar: **(a)** IF, **(b)** CS_2, **(c)** SO_3, **(d)** PCl_3, **(e)** SF_6, **(f)** IF_5.

9.37 Dichloroethylene ($C_2H_2Cl_2$) has three forms (isomers), each of which is a different substance. A pure sample of one of these substances is found experimentally to have a dipole moment of zero. Can we determine which of the three substances was measured?

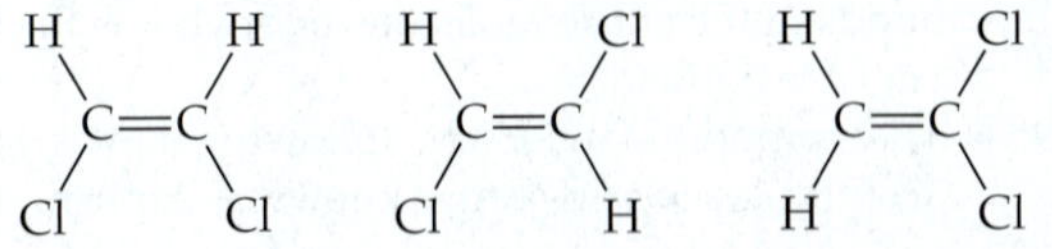

9.38 Dichlorobenzene, $C_6H_4Cl_2$, exists in three forms (isomers), called *ortho, meta,* and *para*:

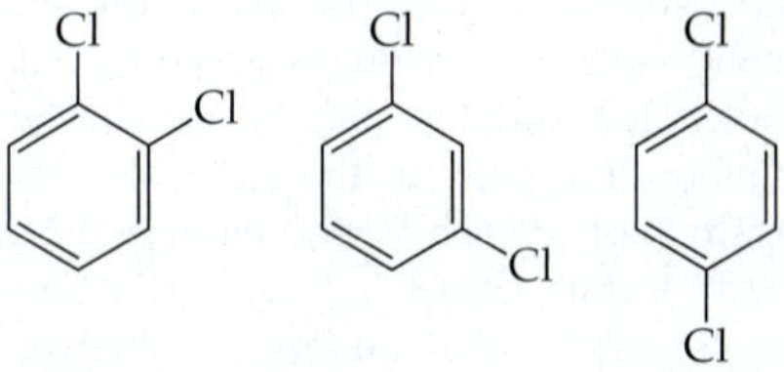

ortho *meta* *para*

Which of these would have a nonzero dipole moment? Explain.

Orbital Overlap; Hybrid Orbitals

9.39 **(a)** What is meant by the term *orbital overlap*? **(b)** What is the significance of overlapping orbitals in valence-bond theory? **(c)** What two fundamental concepts are incorporated in valence-bond theory?

9.40 Draw sketches illustrating the overlap between the following orbitals on two atoms: **(a)** the 2*s* orbital on each, **(b)** the $2p_z$ orbital on each (assume that the atoms are on the *z*-axis), **(c)** the 2*s* orbital on one and the $2p_z$ orbital on the other.

9.41 Consider the bonding in an MgH_2 molecule. **(a)** Draw a Lewis structure for the molecule, and predict its molecular geometry. **(b)** Why is it necessary to promote an electron before forming hybrid orbitals for the Mg atom? **(c)** What hybridization scheme is used in MgH_2? **(d)** Sketch one of the two-electron bonds between an Mg hybrid orbital and an H 1*s* atomic orbital.

9.42 **(a)** Draw a Lewis structure for silane (SiH_4), and predict its molecular geometry. **(b)** Is it necessary to promote an electron before forming hybrid orbitals for the Si atom? **(c)** What type of hybridization exists in SiH_4? **(d)** In one diagram, sketch two of the two-electron bonds formed between a hybrid orbital on Si and an H 1*s* orbital. How would the other Si—H bonds be oriented relative to your sketch?

9.43 Indicate the hybridization and bond angles associated with each of the following electron-domain geometries: **(a)** linear, **(b)** tetrahedral, **(c)** trigonal planar, **(d)** octahedral, **(e)** trigonal bipyramidal.

9.44 What is the designation for the hybrid orbitals formed from each of the following combinations of atomic orbitals: **(a)** one *s* and two *p*; **(b)** one *s*, three *p*, and one *d*; **(c)** one *s*, three *p*, and two *d*? What characteristic bond angles are associated with each?

9.45 **(a)** Starting with the orbital diagram of a boron atom, describe the steps needed to construct hybrid orbitals appropriate to describe the bonding in BF_3. **(b)** What is the name given to the hybrid orbitals constructed in (a)? **(c)** On one origin, sketch the large lobes of the hybrid orbitals constructed in part (a). **(d)** Are there any valence atomic orbitals of B that are left unhybridized? If so, how are they oriented relative to the hybrid orbitals?

9.46 **(a)** Starting with the orbital diagram of a sulfur atom, describe the steps needed to construct hybrid orbitals appropriate to describe the bonding in SF_2. **(b)** What is the name given to the hybrid orbitals constructed in (a)? **(c)** On one origin, sketch the large lobes of the hybrid orbitals constructed in part (a). **(d)** Would the hybridization scheme in part (a) be appropriate for SF_4? Explain.

9.47 Indicate the hybridization of the central atom in **(a)** BCl_3, **(b)** $AlCl_4^-$, **(c)** CS_2, **(d)** KrF_2, **(e)** PF_6^-.

9.48 What set of hybrid orbitals is used by the central atom in **(a)** $SiCl_4$, **(b)** HCN, **(c)** SO_3, **(d)** ICl_2^-, **(e)** BrF_4^-?

Multiple Bonds

9.49 **(a)** Sketch a σ bond that is constructed from *p* orbitals. **(b)** Sketch a π bond that is constructed from *p* orbitals. **(c)** Which is generally the stronger, a σ bond or a π bond? Explain.

9.50 **(a)** If the valence atomic orbitals of an atom are *sp* hybridized, how many unhybridized *p* orbitals remain in the valence shell? How many π bonds can the atom form? **(b)** How many σ and π bonds are generally part of a triple bond? **(c)** How do multiple bonds introduce rigidity into molecules?

9.51 **(a)** Draw Lewis structures for methane (CH_4) and formaldehyde (H_2CO). **(b)** What is the hybridization at the carbon atom in CH_4 and H_2CO? **(c)** The carbon atom in CH_4 cannot participate in multiple bonding, whereas that in H_2CO can. Explain this observation by using the hybridization at the carbon atom.

9.52 The nitrogen atoms in N_2 participate in multiple bonding, whereas those in hydrazine, N_2H_4, do not. How can you explain this observation in light of the hybridization at the nitrogen atoms in the two molecules?

9.53 Propylene, C_3H_6, is a gas that is used to form the important polymer called polypropylene. Its Lewis structure is

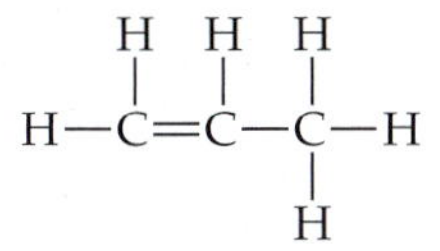

(a) What is the total number of valence electrons in the propylene molecule? (b) How many valence electrons are used to make σ bonds in the molecule? (c) How many valence electrons are used to make π bonds in the molecule? (d) How many valence electrons remain in nonbonding pairs in the molecule? (e) What is the hybridization at each carbon atom in the molecule?

9.54 Ethyl acetate, $C_4H_8O_2$, is a fragrant substance used both as a solvent and as an aroma enhancer. Its Lewis structure is

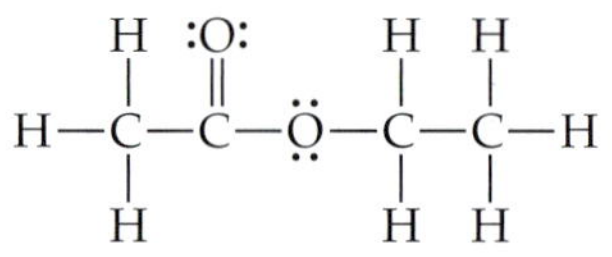

(a) What is the hybridization at each of the carbon atoms of the molecule? (b) What is the total number of valence electrons in ethyl acetate? (c) How many of the valence electrons are used to make σ bonds in the molecule? (d) How many valence electrons are used to make π bonds? (e) How many valence electrons remain in nonbonding pairs in the molecule?

9.55 Consider the Lewis structure for glycine, the simplest amino acid:

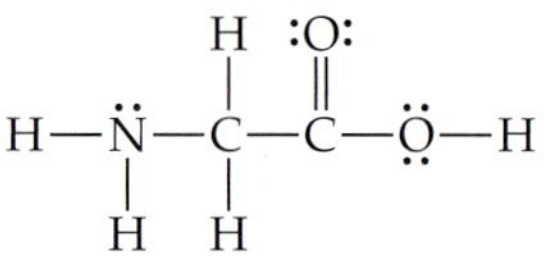

(a) What are the approximate bond angles about each of the two carbon atoms, and what are the hybridizations of the orbitals on each of them? (b) What are the hybridizations of the orbitals on the two oxygens and the nitrogen atom, and what are the approximate bond angles at the nitrogen? (c) What is the total number of σ bonds in the entire molecule, and what is the total number of π bonds?

9.56 The compound with the following Lewis structure is acetylsalicylic acid, better known as aspirin:

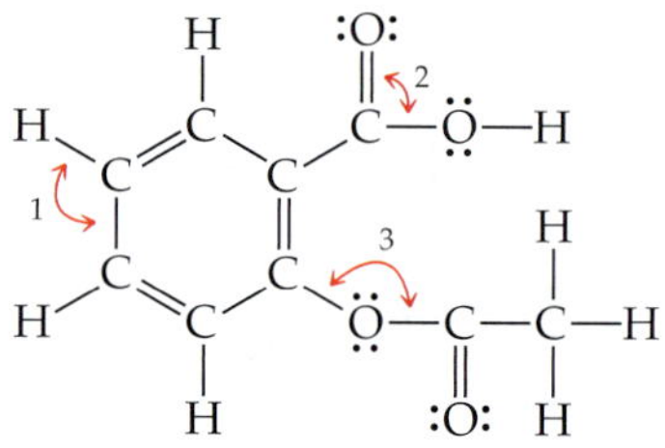

(a) What are the approximate values of the bond angles labeled 1, 2, and 3? (b) What hybrid orbitals are used about the central atom of each of these angles? (c) How many σ bonds are in the molecule?

9.57 (a) What is the difference between a localized π bond and a delocalized one? (b) How can you determine whether a molecule or ion will exhibit delocalized π bonding? (c) Is the π bond in NO_2^- localized or delocalized?

9.58 (a) Write a single Lewis structure for SO_3, and determine the hybridization at the S atom. (b) Are there other equivalent Lewis structures for the molecule? (c) Would you expect SO_3 to exhibit delocalized π bonding? Explain.

Molecular Orbitals

9.59 (a) What are the similarities and differences between atomic orbitals and molecular orbitals? (b) Why is the bonding molecular orbital of H_2 at lower energy than the electron in a hydrogen atom? (c) How many electrons can be placed into each MO of a molecule?

9.60 (a) Why is the antibonding molecular orbital of H_2 at higher energy than the electron in a hydrogen atom? (b) Does the Pauli exclusion principle (Section 6.7) apply to MOs? Explain. (c) If two *p* orbitals of one atom combine with two *p* orbitals of another atom, how many MOs result? Explain.

9.61 Consider the H_2^+ ion. (a) Sketch the molecular orbitals of the ion, and draw its energy-level diagram. (b) How many electrons are there in the H_2^+ ion? (c) Write the electron configuration of the ion in terms of its MOs. (d) What is the bond order in H_2^+? (e) Suppose that the ion is excited by light so that an electron moves from a lower-energy to a higher-energy MO. Would you expect the excited-state H_2^+ ion to be stable or to fall apart? Explain.

9.62 (a) Sketch the molecular orbitals of the H_2^- ion, and draw its energy-level diagram. (b) Write the electron configuration of the ion in terms of its MOs. (c) Calculate the bond order in H_2^-. (d) Suppose that the ion is excited by light, so that an electron moves from a lower-energy to a higher-energy molecular orbital. Would you expect the excited-state H_2^- ion to be stable? Explain.

9.63 (a) Sketch the σ and σ^* molecular orbitals that can result from the combination of two $2p_z$ atomic orbitals. (b) Sketch the π and π^* MOs that result from the combination of two $2p_x$ atomic orbitals. (c) Place the MOs from parts (a) and (b) in order of increasing energy, assuming no mixing of 2*s* and 2*p* orbitals.

9.64 (a) What is the probability of finding an electron on the internuclear axis if the electron occupies a π molecular orbital? (b) For a homonuclear diatomic molecule, what similarities and differences are there between the π_{2p} MO made from the $2p_x$ atomic orbitals and the π_{2p} MO made from the $2p_y$ atomic orbitals? (c) Why are the π_{2p} MOs lower in energy than the π^*_{2p} MOs?

9.65 (a) What are the relationships among bond order, bond length, and bond energy? (b) According to molecular orbital theory, would either Be_2 or Be_2^+ be expected to exist? Explain.

9.66 Explain the following: (a) The *peroxide* ion, O_2^{2-}, has a longer bond than the *superoxide* ion, O_2^-. (b) The magnetic properties of B_2 are consistent with the π_{2p} MOs being lower in energy than the σ_{2p} MO.

9.67 **(a)** What is meant by the term *diamagnetism*? **(b)** How does a diamagnetic substance respond to a magnetic field? **(c)** Which of the following ions would you expect to be diamagnetic: N_2^{2-}, O_2^{2-}, Be_2^{2+}, C_2^{-}?

9.68 **(a)** What is meant by the term *paramagnetism*? **(b)** How can one determine experimentally whether a substance is paramagnetic? **(c)** Which of the following ions would you expect to be paramagnetic: O_2^{+}, N_2^{2-}, Li_2^{+}, O_2^{2-}? If the ion is paramagnetic, how many unpaired electrons does it possess?

9.69 Using Figures 9.37 and 9.45 as guides, give the molecular orbital electron configuration for **(a)** B_2^{+}, **(b)** Li_2^{+}, **(c)** N_2^{+}, **(d)** Ne_2^{2+}. In each case indicate whether the addition of an electron to the ion would increase or decrease the bond order of the species.

9.70 If we assume that the energy-level diagrams for homonuclear diatomic molecules shown in Figure 9.42 can be applied to heteronuclear diatomic molecules and ions, predict the bond order and magnetic behavior of **(a)** CO^{+}, **(b)** NO^{-}, **(c)** OF^{+}, **(d)** NeF^{+}.

9.71 Determine the electron configurations for CN^{+}, CN, and CN^{-}. Calculate the bond order for each, and indicate which ones are paramagnetic.

9.72 **(a)** The nitric oxide molecule, NO, readily loses one electron to form the NO^{+} ion. Why is this consistent with the electronic structure of NO? **(b)** Predict the order of the N—O bond strengths in NO, NO^{+}, and NO^{-}, and describe the magnetic properties of each. **(c)** With what neutral homonuclear diatomic molecules are the NO^{+} and NO^{-} ions isoelectronic (same number of electrons)?

[9.73] Consider the molecular orbitals of the P_2 molecule. (Assume that the MOs of diatomics from the third row of the periodic table are analogous to those from the second row.) **(a)** Which valence atomic orbitals of P are used to construct the MOs of P_2? **(b)** The figure below shows a sketch of one of the MOs for P_2. What is the label for this MO? **(c)** For the P_2 molecule, how many electrons occupy the MO in the figure? **(d)** Is P_2 expected to be diamagnetic or paramagnetic? Explain.

[9.74] The iodine bromide molecule, IBr, is an *interhalogen compound*. Assume that the molecular orbitals of IBr are analogous to the homonuclear diatomic molecule F_2. **(a)** Which valence atomic orbitals of I and of Br are used to construct the MOs of IBr? **(b)** What is the bond order of the IBr molecule? **(c)** One of the valence MOs of IBr is sketched below. Why are the atomic orbital contributions to this MO different in size? **(d)** What is the label for the MO? **(e)** For the IBr molecule, how many electrons occupy the MO?

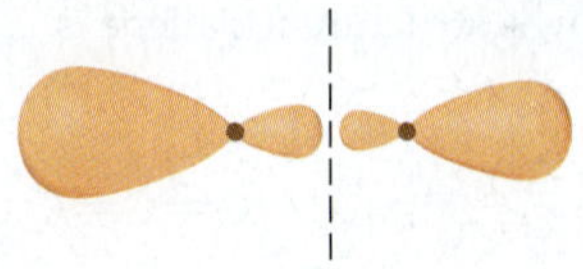

Additional Exercises

9.75 **(a)** What is the physical basis for the VSEPR model? **(b)** When applying the VSEPR model, we count a double or triple bond as a single electron domain. Why is this justified?

9.76 The molecules SiF_4, SF_4, and XeF_4 all have molecular formulas of the type AF_4, but the molecules have different molecular geometries. Predict the shape of each molecule, and explain why the shapes differ.

[9.77] The vertices of a tetrahedron correspond to four alternating corners of a cube. By using analytical geometry, demonstrate that the angle made by connecting two of the vertices to a point at the center of the cube is 109.5°, the characteristic angle for tetrahedral molecules.

9.78 Consider the molecule PF_4Cl. **(a)** Draw a Lewis structure for the molecule, and predict its electron-domain geometry. **(b)** Which would you expect to produce a larger electron domain, a P—F bond or a P—Cl bond? Explain. **(c)** Predict the molecular geometry of PF_4Cl. How did your answer for part (b) influence your answer here in part (c)? **(d)** Would you expect the molecule to distort from its ideal electron domain geometry? If so, how would it distort?

9.79 From their Lewis structures, determine the number of σ and π bonds in each of the following molecules or ions: **(a)** CO_2; **(b)** thiocyanate ion, NCS^{-}; **(c)** formaldehyde, H_2CO; **(d)** formic acid, HCO(OH), which has one H and two O atoms attached to C.

9.80 The lactic acid molecule, $CH_3CH(OH)CO(OH)$, gives sour milk its unpleasant, sour taste. **(a)** Draw the Lewis structure for the molecule, assuming that carbon always forms four bonds in its stable compounds. **(b)** How many π and how many σ bonds are in the molecule? **(c)** Which CO bond is shortest in the molecule? **(d)** What is the hybridization of atomic orbitals around each carbon atom associated with that short bond? **(e)** What are the approximate bond angles around each carbon atom in the molecule?

9.81 The PF_3 molecule has a dipole moment of 1.03 D, but BF_3 has a dipole moment of zero. How can you explain the difference?

9.82 There are two compounds of the formula $Pt(NH_3)_2Cl_2$:

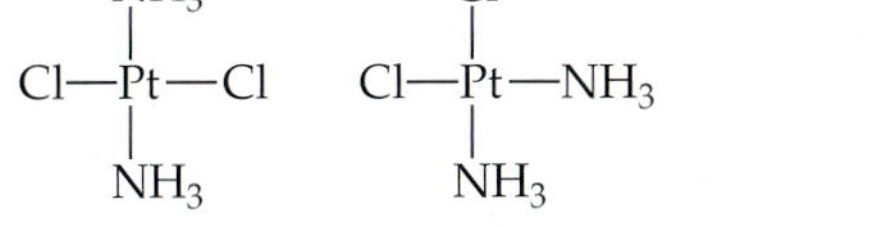

The compound on the right, *cisplatin*, is used in cancer therapy. Both compounds have a square-planar geometry. Which compound has a nonzero dipole moment?

[9.83] The O—H bond lengths in the water molecule (H_2O) are 0.96 Å, and the H—O—H angle is 104.5°. The dipole moment of the water molecule is 1.85 D. **(a)** In what directions do the bond dipoles of the O—H bonds point? In what direction does the dipole moment vector of the water molecule point? **(b)** Calculate the magnitude of the bond dipole of the O—H bonds. (*Note:* You will need to use vector addition to do this.) **(c)** Compare your answer from part (b) to the dipole moments of the hydrogen halides (Table 8.3). Is your answer in accord with the relative electronegativity of oxygen?

[9.84] The reaction of three molecules of fluorine gas with an Xe atom produces the substance xenon hexafluoride, XeF_6:

$$Xe(g) + 3\,F_2(g) \longrightarrow XeF_6(s)$$

(a) Draw a Lewis structure for XeF_6. **(b)** If you try to use the VSEPR model to predict the molecular geometry of XeF_6, you run into a problem. What is it? **(c)** What could you do to resolve the difficulty in part (b)? **(d)** Suggest a hybridization scheme for the Xe atom in XeF_6. **(e)** The molecule IF_7 has a pentagonal-bipyramidal structure (five equatorial fluorine atoms at the vertices of a regular pentagon and two axial fluorine atoms). Based on the structure of IF_7, suggest a structure for XeF_6.

[9.85] The Lewis structure for allene is

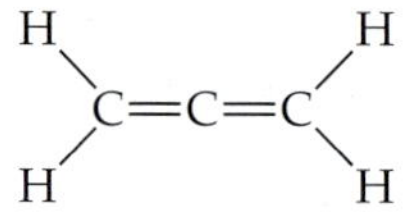

Make a sketch of the structure of this molecule that is analogous to Figure 9.27. In addition, answer the following three questions: **(a)** Is the molecule planar? **(b)** Does it have a nonzero dipole moment? **(c)** Would the bonding in allene be described as delocalized? Explain.

[9.86] The azide ion, N_3^-, is linear with two N—N bonds of equal length, 1.16 Å. **(a)** Draw a Lewis structure for the azide ion. **(b)** With reference to Table 8.5, is the observed N—N bond length consistent with your Lewis structure? **(c)** What hybridization scheme would you expect at each of the nitrogen atoms in N_3^-? **(d)** Show which hybridized and unhybridized orbitals are involved in the formation of σ and π bonds in N_3^-. **(e)** It is often observed that σ bonds that involve an *sp* hybrid orbital are shorter than those that involve only sp^2 or sp^3 hybrid orbitals. Can you propose a reason for this? Is this observation applicable to the observed bond lengths in N_3^-?

[9.87] In ozone, O_3, the two oxygen atoms on the ends of the molecule are equivalent to one another. **(a)** What is the best choice of hybridization scheme for the atoms of ozone? **(b)** For one of the resonance forms of ozone, which of the orbitals are used to make bonds and which are used to hold nonbonding pairs of electrons? **(c)** Which of the orbitals can be used to delocalize the π electrons? **(d)** How many electrons are delocalized in the π system of ozone?

9.88 Butadiene, C_4H_6, is a planar molecule that has the following carbon–carbon bond lengths:

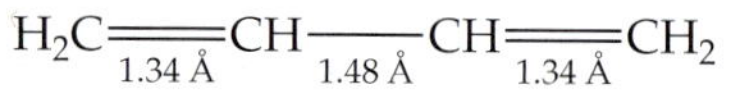

(a) Predict the bond angles around each of the carbon atoms, and sketch the molecule. **(b)** Compare the bond lengths to the average bond lengths listed in Table 8.5. Can you explain any differences?

[9.89] The sketches below show the atomic orbital wave functions (with phases) used to construct some of the MOs of a homonuclear diatomic molecule. For each sketch, determine the MO that will result from mixing the atomic orbital wave functions as drawn. Use the same labels for the MOs as in Figure 9.42.

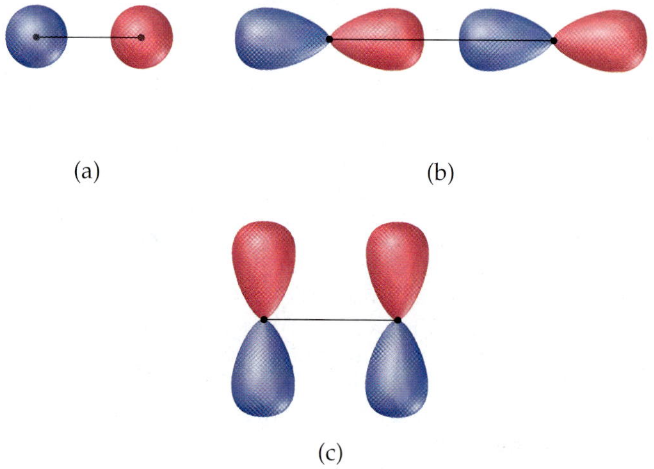

[9.90] The *cyclopentadienide ion* has the formula $C_5H_5^-$. The ion consists of a regular pentagon of C atoms, each bonded to two C neighbors, with a hydrogen atom bonded to each C atom. All the atoms lie in the same plane. **(a)** Draw a Lewis structure for the ion. According to your structure, do all five C atoms have the same hybridization? Explain. **(b)** Chemists generally view this ion as having sp^2 hybridization at each C atom. Is that view consistent with your answer to part (a)? **(c)** Your Lewis structure should show one nonbonding pair of electrons. Under the assumption of part (b), in what type of orbital must this nonbonding pair reside? **(d)** Are there resonance structures equivalent to the Lewis structure you drew in part (a)? If so, how many? **(e)** The ion is often drawn as a pentagon enclosing a circle. Is this representation consistent with your answer to part (d)? Explain. **(f)** Both benzene and the cyclopentadienide ion are often described as systems containing six π electrons. What do you think is meant by this description?

9.91 Write the electron configuration for the first excited state for N_2 (that is, the state with the highest-energy electron moved to the next available energy level). What differences do you expect in the properties of N_2 in its ground state and its first excited state?

9.92 Figure 9.46 shows how the magnetic properties of a compound can be measured experimentally. When such measurements are made, the sample is generally covered by an atmosphere of pure nitrogen gas rather than air. Why do you suppose this is done?

[9.93] *Azo dyes* are organic dyes that are used for many applications, such as the coloring of fabrics. Many azo dyes are derivatives of the organic substance *azobenzene*, $C_{12}H_{10}N_2$. A closely related substance is *hydrazobenzene*, $C_{12}H_{12}N_2$. The Lewis structures of these two substances are

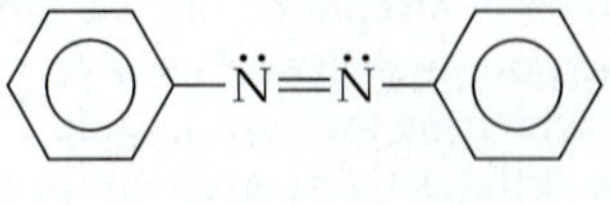

Azobenzene

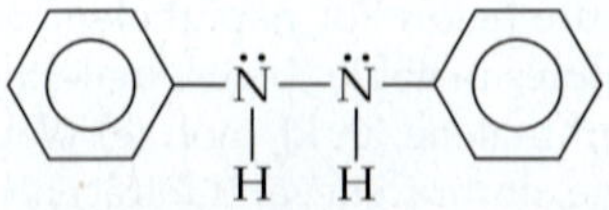

Hydrazobenzene

(Recall the shorthand notation used for organic molecules.) **(a)** What is the hybridization at the N atom in each of the substances? **(b)** How many unhybridized atomic orbitals are there on the N and the C atoms in each of the substances? **(c)** Predict the N—N—C angles in each of the substances. **(d)** Azobenzene is said to have greater delocalization of its π electrons than hydrazobenzene. Discuss this statement in light of your answers to (a) and (b). **(e)** All the atoms of azobenzene lie in one plane, whereas those of hydrazobenzene do not. Is this observation consistent with the statement in part (d)? **(f)** Azobenzene is an intense red-orange color, whereas hydrazobenzene is nearly colorless. Discuss this observation with reference to the "Chemistry at Work" box on organic dyes.

9.94 **(a)** Using only the valence atomic orbitals of a hydrogen atom and a fluorine atom, how many MOs would you expect for the HF molecule? **(b)** How many of the MOs from part (a) would be occupied by electrons? **(c)** Do you think the MO diagram shown in Figure 9.48 could be used to describe the MOs of the HF molecule? Why or why not?

[9.95] Carbon monoxide, CO, is isoelectronic to N_2. **(a)** Draw a Lewis structure for CO that satisfies the octet rule. **(b)** Assume that the diagram in Figure 9.48 can be used to describe the MOs of CO. What is the predicted bond order for CO? Is this answer in accord with the Lewis structure you drew in part (a)? **(c)** Experimentally, it is found that the highest-energy electrons in CO reside in a σ-type MO. Is that observation consistent with Figure 9.48? If not, what modification needs to be made to the diagram? How does this modification relate to Figure 9.44? **(d)** Would you expect the π_{2p} MOs of CO to have equal atomic orbital contributions from the C and O atoms? If not, which atom would have the greater contribution?

Integrative Exercises

9.96 A compound composed of 2.1% H, 29.8% N, and 68.1% O has a molar mass of approximately 50 g/mol. **(a)** What is the molecular formula of the compound? **(b)** What is its Lewis structure if H is bonded to O? **(c)** What is the geometry of the molecule? **(d)** What is the hybridization of the orbitals around the N atom? **(e)** How many σ and how many π bonds are there in the molecule?

9.97 Sulfur tetrafluoride (SF_4) reacts slowly with O_2 to form sulfur tetrafluoride monoxide (OSF_4) according to the following unbalanced reaction:

$$SF_4(g) + O_2(g) \longrightarrow OSF_4(g)$$

The O atom and the four F atoms in OSF_4 are bonded to a central S atom. **(a)** Balance the equation. **(b)** Write a Lewis structure of OSF_4 in which the formal charges of all atoms are zero. **(c)** Use average bond enthalpies (Table 8.4) to estimate the enthalpy of the reaction. Is it endothermic or exothermic? **(d)** Determine the electron-domain geometry of OSF_4, and write two possible molecular geometries for the molecule based on this electron-domain geometry. **(e)** Which of the molecular geometries in part (d) is more likely to be observed for the molecule? Explain.

[9.98] The phosphorus trihalides (PX_3) show the following variation in the bond angle X—P—X: PF_3, 96.3°; PCl_3, 100.3°; PBr_3, 101.0°; PI_3, 102°. The trend is generally attributed to the change in the electronegativity of the halogen. **(a)** Assuming that all electron domains exhibit the same repulsion, what value of the X—P—X angle is predicted by the VSEPR model? **(b)** What is the general trend in the X—P—X angle as the electronegativity increases? **(c)** Using the VSEPR model, explain the observed trend in X—P—X angle as the electronegativity of X changes. **(d)** Based on your answer to part (c), predict the structure of $PBrCl_4$.

[9.99] The molecule 2-butene, C_4H_8, can undergo a geometric change called *cis-trans isomerization*:

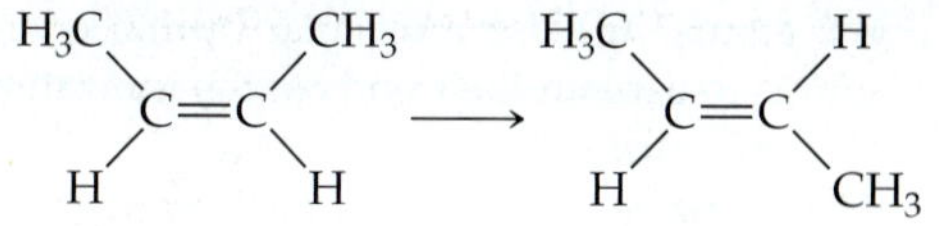

cis-2-butene *trans*-2-butene

As discussed in the "Chemistry and Life" box on the chemistry of vision, such transformations can be induced by light and are the key to human vision. **(a)** What is the hybridization at the two central carbon atoms of 2-butene? **(b)** The isomerization occurs by rotation about the central C—C bond. With reference to Figure 9.32, explain why the π bond between the two central carbon atoms is destroyed halfway through the rotation from *cis*- to *trans*-2-butene. **(c)** Based on average bond enthalpies (Table 8.4), how much energy per molecule must be supplied to break the C—C π bond? **(d)** What is the longest wavelength of light that will provide photons of sufficient energy to break the C—C π bond and cause the isomerization? **(e)** Is the wavelength in your answer to part (d) in the visible portion of the electromagnetic spectrum? Comment on the importance of this result for human vision.

9.100 **(a)** Compare the bond enthalpies (Table 8.4) of the carbon–carbon single, double, and triple bonds to deduce an average π-bond contribution to the enthalpy. What fraction of a single bond does this quantity represent? **(b)** Make a similar comparison of nitrogen–nitrogen bonds. What do you observe? **(c)** Write Lewis structures of N_2H_4, N_2H_2, and N_2, and determine the hybridization around nitrogen in each case. **(d)** Propose a reason for the large difference in your observations of parts (a) and (b).

9.101 Use average bond enthalpies (Table 8.4) to estimate ΔH for the atomization of benzene, C_6H_6:

$$C_6H_6(g) \longrightarrow 6\,C(g) + 6\,H(g)$$

Compare the value to that obtained by using ΔH_f° data given in Appendix C and Hess's law. To what do you attribute the large discrepancy in the two values?

[9.102] For both atoms and molecules, ionization energies (Section 7.4) are related to the energies of orbitals: The lower the energy of the orbital, the greater the ionization energy. The first ionization energy of a molecule is therefore a measure of the energy of the highest occupied molecular orbital (HOMO; see "Chemistry at Work" box on organic dyes). The first ionization energies of several diatomic molecules are given in electron-volts in the following table:

Molecule	I_1 (eV)
H_2	15.4
N_2	15.6
O_2	12.1
F_2	15.7

(a) Convert these ionization energies to kJ/mol. **(b)** On the same piece of graph paper, plot I_1 for the H, N, O, and F atoms (Figure 7.11) and I_1 for the molecules listed. **(c)** Do the ionization energies of the molecules follow the same periodic trends as the ionization energies of the atoms? **(d)** Use molecular orbital energy-level diagrams to explain the trends in the ionization energies of the molecules.

[9.103] Many compounds of the transition-metal elements contain direct bonds between metal atoms. We will assume that the z-axis is defined as the metal–metal bond axis. **(a)** Which of the $3d$ orbitals (Figure 6.23) can be used to make a σ bond between metal atoms? **(b)** Sketch the σ_{3d} bonding and σ_{3d}^* antibonding MOs. **(c)** With reference to the "Closer Look" box on the phases of orbitals, explain why a node is generated in the σ_{3d}^* MO. **(d)** Sketch the energy-level diagram for the Sc_2 molecule, assuming that only the $3d$ orbital from part (a) is important. **(e)** What is the bond order in Sc_2?

[9.104] The organic molecules shown below are derivatives of benzene in which additional six-membered rings are "fused" at the edges of the hexagons. The compounds are shown in the usual abbreviated method for organic molecules. **(a)** Determine the empirical formula of benzene and of these three compounds. **(b)** Suppose you are given a sample of one of the compounds. Could combustion analysis be used to determine unambiguously which of the compounds it is? **(c)** Naphthalene, which is the active ingredient in mothballs, is a white solid. Write a balanced equation for the combustion of naphthalene to $CO_2(g)$ and $H_2O(g)$. **(d)** Using the Lewis structure below for naphthalene and the average bond enthalpies in Table 8.4, estimate the heat of combustion of naphthalene, in kJ/mol. **(e)** Would you expect naphthalene, anthracene, and tetracene to have multiple resonance structures? If so, draw the additional resonance structures for naphthalene. **(f)** Benzene, naphthalene, and anthracene are colorless. Tetracene is orange. With reference to the "Chemistry at Work" box on organic dyes, how can this observation be explained?

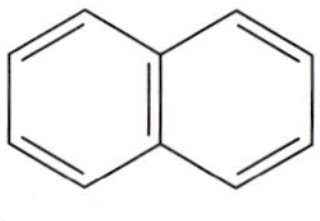

Naphthalene

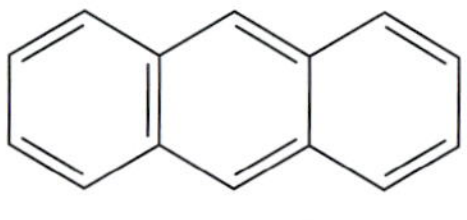

Anthracene

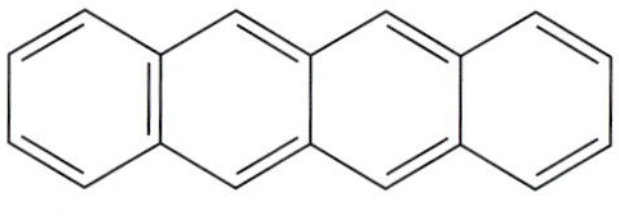

Tetracene

[9.105] As noted in Section 7.8, the chemistry of astatine (At) is far less developed than that of the other halogen elements. **(a)** In the periodic table the atomic weight of At is written as (210). Why are the parentheses used? How is this nomenclature related to the difficulty of studying astatine? **(b)** Write the complete electron configuration for a neutral At atom. **(c)** Although At_2 is not known, the interhalogen compound AtI has been characterized. Would this compound be expected to have a covalent, polar covalent, or ionic bond? Explain. **(d)** The reaction of AtI with I^- forms the AtI_2^- ion. Use the VSEPR method to predict the geometry of this ion. **(e)** Suppose we construct the molecular orbitals of the unknown At_2 molecule. What bond order is predicted for the molecule? What type of MO is the highest-occupied MO of the molecule?

eMEDIA EXERCISES

These exercises make use of the interactive objects available online in OneKey or the Companion Website, and on your Accelerator CD. Access to these resources comes in your MediaPak.

9.106 The **VSEPR** animation (*9.2*) shows the arrangements of electron domains around a central atom. **(a)** In which of the arrangements are the bond angles not all equal? **(b)** In which of the arrangements are all positions unequal with regard to placement of a single lone pair? **(c)** Are there any electron-domain geometries for which all positions are equal with regard to placement of the first lone pair, but not with regard to placement of the second? Explain.

9.107 In the trigonal-bipyramidal arrangement of electron domains shown in the **VSEPR** animation (*9.2*), a single lone pair will occupy an equatorial position rather than an axial position. **(a)** Explain why this is so. **(b)** Is it possible for a molecule with trigonal-bipyramidal electron-domain geometry to have linear molecular geometry? Explain your answer. **(c)** Is it possible for a molecule with trigonal-bipyramidal electron-domain geometry to have a trigonal-planar molecular geometry? If so, how? If not, why not?

9.108 The 3-D model set **VSEPR: Basic Molecular Configurations** tutorial (*9.2*) gives examples of electron-domain and molecular geometries for AB_n molecules with and without lone pairs. It also allows you to manipulate and view the molecules from any angle. For each of the examples in the set, give a different example with the same electron-domain and molecular geometries. Write the formula, draw the Lewis-dot structure, and give the molecular geometry for each of your examples.

9.109 The **Molecular Polarity** activity (*9.3*) allows you to experiment with various central and terminal atoms in AB_2 and AB_3 molecules. For the case of AB_3, where the terminal atoms are all the same element, use geometry to show quantitatively how a molecule with three polar bonds can be nonpolar overall.

9.110 Exercise 9.51 asks about the hybridization at the carbon atom of the molecule H_2CO. The **Hybridization** animation (*9.5*) illustrates the formation of *sp*, sp^2, and sp^3 hybrid orbitals. **(a)** Beginning with the electron configuration of a ground-state carbon atom, diagram and describe the process by which hybridization leads to the formation of the σ bonds in this molecule. **(b)** Using your diagrams from part (a), describe the process by which the π bond forms.

9.111 In Exercise 9.45 you described the steps needed to construct hybrid orbitals to account for the bonding in BF_3. Examine the **Hybridization** animation (*9.5*), and use it to explain why promotion and hybridization are necessary to describe the B—F bonds in BF_3.

CHAPTER 10

WHAT'S AHEAD

- We will compare the distinguishing characteristics of gases with those of liquids and solids. *(Section 10.1)*
- We'll study gas *pressure*, how it is measured, and the units used to express it, as well as consider Earth's atmosphere and the pressure it exerts. *(Section 10.2)*
- The state of a gas can be expressed in terms of its volume, pressure, temperature, and quantity of gas. We will examine several empirical relationships that relate these variables to each other. Put together, these empirical relationships yield the *ideal-gas equation*, $PV = nRT$. *(Sections 10.3 and 10.4)*
- While the ideal-gas equation is not obeyed exactly by any real gas, most gases come quite close to obeying it at temperature and pressure conditions of greatest interest. Consequently, we can use the ideal-gas equation to make many useful calculations. *(Sections 10.4, 10.5, and 10.6)*
- In the *kinetic-molecular theory* of gases, the atoms or molecules that make up the gas are assumed to be point masses that move about with an average kinetic energy that is proportional to the gas temperature. *(Section 10.7)*
- The kinetic-molecular theory leads to the ideal-gas equation and helps us account for such gas properties as *effusion* through tiny openings and also for *diffusion*. *(Section 10.8)*
- Real gases deviate from ideal behavior, primarily because the gas molecules have finite volume and because attractive forces exist between molecules. The *van der Waals equation* gives a more accurate account of real gas behavior at high pressures and low temperatures. *(Section 10.9)*

Gases

IN THE PAST SEVERAL CHAPTERS we have learned about electronic structures of atoms and about how atoms combine to form molecules and ionic substances. In everyday life, however, we don't have direct experiences with atoms. Instead, we encounter matter as collections of enormous numbers of atoms or molecules that make up gases, liquids, and solids. In the atmosphere it's the action of such large collections of atoms and molecules that are responsible for our weather—the gentle breezes and the gales, the humidity and the rain. Tornados, such as the one shown in the chapter-opening photo, form when moist, warm air at lower elevations converges with cooler, dry air above. The resultant air flows produce winds that can approach speeds up to 500 km/hr (300 mph).

It was his interest in the weather that motivated John Dalton (Section 2.1) to study gases and eventually led him to propose the atomic theory of matter. We now know that the properties of gases, liquids, and solids are readily understood in terms of the behavior of their component atoms, ions, and molecules. In this chapter we will examine the physical properties of gases and consider how we can understand these properties in terms of the behavior of gas molecules. In Chapter 11 we will turn our attention to the physical properties of liquids and solids.

In many ways gases are the most easily understood form of matter. Even though different gaseous substances may have very different *chemical* properties, they behave quite similarly as far as their *physical* properties are concerned.

A TORNADO IS A VIOLENT, rotating column of air characterized by a funnel-shaped cloud. A tornado can be up to 100 m in diameter with wind speeds approaching 500 km/hr.

For example, we live in an atmosphere composed of a mixture of gases that we refer to as air. We breathe air to absorb oxygen, O_2, which supports human life. Air also contains nitrogen, N_2, which has very different chemical properties from oxygen, yet this mixture behaves physically as one gaseous material. The relative simplicity of the gas state affords a good starting point as we seek to understand the properties of the states of matter.

10.1 | Characteristics of Gases

In considering the characteristics of gases, there is no more appropriate place to begin than with Earth's atmosphere, vital to all life on the planet. Air is a complex mixture of several substances, either atomic in nature or consisting of small molecules. Air consists primarily, however, of N_2 (78%) and O_2 (21%), with small amounts of several other gases, including Ar (0.9%).

In addition to O_2 and N_2, a few other elements (H_2, F_2, Cl_2) exist as gases under ordinary conditions of temperature and pressure. The noble gases (He, Ne, Ar, Kr, and Xe) are all monatomic gases. Many molecular compounds are also gases. Table 10.1 ▼ lists a few of the more common gaseous compounds. Notice that all of these gases are composed entirely of nonmetallic elements. Furthermore, all have simple molecular formulas and, therefore, low molar masses. Substances that are liquids or solids under ordinary conditions can also exist in the gaseous state, too, where they are often referred to as **vapors**. The substance H_2O, for example, can exist as liquid water, solid ice, or water vapor.

Gases differ significantly from solids and liquids in several respects. For example, a gas expands spontaneously to fill its container. Consequently, the volume of a gas equals the volume of the container in which it is held. Gases also are highly compressible: When pressure is applied to a gas, its volume readily decreases. Solids and liquids, on the other hand, do not expand to fill their containers, and solids and liquids are not readily compressible.

Gases form homogeneous mixtures with each other regardless of the identities or relative proportions of the component gases. The atmosphere serves as an excellent example. As a further example, when water and gasoline are mixed, the two liquids remain as separate layers. In contrast, the water vapor and gasoline vapors above the liquids form a homogeneous gas mixture. The characteristic properties of gases arise because the individual molecules are relatively far apart. In the air we breathe, for example, the molecules take up only about 0.1% of the total volume, with the rest being empty space. Thus, each molecule behaves largely as though the others were not present. As a result, different gases behave similarly, even though they are made up of different

TABLE 10.1 Some Common Compounds That Are Gases at Room Temperature

Formula	Name	Characteristics
HCN	Hydrogen cyanide	Very toxic, slight odor of bitter almonds
H_2S	Hydrogen sulfide	Very toxic, odor of rotten eggs
CO	Carbon monoxide	Toxic, colorless, odorless
CO_2	Carbon dioxide	Colorless, odorless
CH_4	Methane	Colorless, odorless, flammable
C_2H_4	Ethylene	Colorless; ripens fruit
C_3H_8	Propane	Colorless; bottled gas
N_2O	Nitrous oxide	Colorless, sweet odor; laughing gas
NO_2	Nitrogen dioxide	Toxic, red-brown, irritating odor
NH_3	Ammonia	Colorless, pungent odor
SO_2	Sulfur dioxide	Colorless, irritating odor

molecules. In contrast, the individual molecules in a liquid are close together and occupy perhaps 70% of the total space. The attractive forces among the molecules keep the liquid together.

GIVE IT SOME THOUGHT

What is the major reason that physical properties do not differ much from one gaseous substance to another?

10.2 | Pressure

Among the most readily measured properties of a gas are its temperature, volume, and pressure. It is not surprising, therefore, that many early studies of gases focused on relationships among these properties. We have already discussed volume and temperature. (Section 1.4) Let's now consider the concept of pressure.

In general terms, **pressure** conveys the idea of a force, a push that tends to move something in a given direction. Pressure, P, is, in fact, the force, F, that acts on a given area, A.

$$P = \frac{F}{A} \qquad [10.1]$$

Gases exert a pressure on any surface with which they are in contact. The gas in an inflated balloon, for example, exerts a pressure on the inside surface of the balloon.

Atmospheric Pressure and the Barometer

You and I, coconuts and nitrogen molecules, all experience an attractive force that pulls toward the center of Earth. When a coconut comes loose from its place at the top of the tree, for example, the gravitational attractive force causes it to be accelerated toward Earth, increasing in speed as its potential energy is converted into kinetic energy. (Section 5.1) The atoms and molecules of the atmosphere experience a gravitational acceleration, too. Because the gas particles have such tiny masses, however, their thermal energies of motion (their kinetic energies) override the gravitational forces, so all the molecules that make up the atmosphere don't just pile up in a thin layer at Earth's surface. Nevertheless, gravity does operate, and it causes the atmosphere as a whole to press down on Earth's surface, creating an atmospheric pressure.

You can demonstrate the existence of atmospheric pressure to yourself with an empty plastic bottle of the sort used to hold water or soft drinks. If you suck on the mouth of the empty bottle, chances are you can cause it to partially cave in. When you break the partial vacuum you have created, the bottle pops out to its original shape. What causes the bottle to cave in when the pressure inside it is reduced, even by the relatively small amount you can manage with your lungs? The gas molecules in the atmosphere are exerting a force on the outside of the bottle that is greater than the force within the bottle when some of the gas has been sucked out.

We can calculate the magnitude of the atmospheric pressure as follows: The force, F, exerted by any object is the product of its mass, m, times its acceleration, a; that is, $F = ma$. The acceleration given by Earth's gravity to any object located near Earth's surface is 9.8 m/s^2. Now imagine a column of air 1 m^2 in cross section extending through the entire atmosphere. That column has a mass of roughly 10,000 kg (Figure 10.1 ▶). The force exerted by the column on the Earth's surface is

$$F = (10{,}000\ \text{kg})(9.8\ \text{m/s}^2) = 1 \times 10^5\ \text{kg-m/s}^2 = 1 \times 10^5\ \text{N}$$

▼ **Figure 10.1 Atmospheric pressure.** Illustration of the manner in which Earth's atmosphere exerts pressure at the surface of the planet. The mass of a column of atmosphere exactly 1 m^2 in cross-sectional area and extending to the upper atmosphere exerts a force of 1.01×10^5 N.

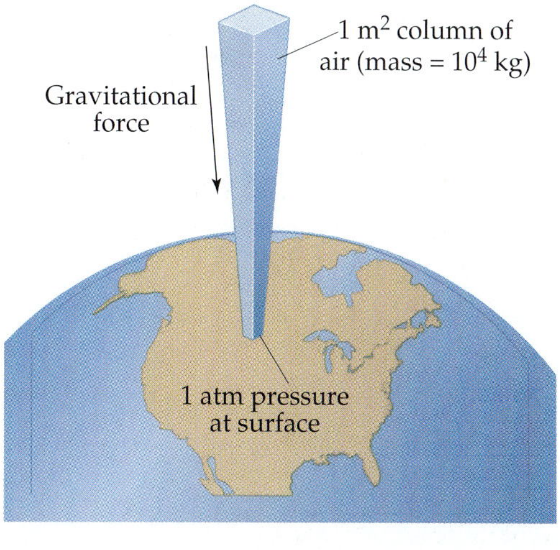

The SI unit for force is kg-m/s^2 and is called the *newton* (N): 1 N = 1 kg-m/s^2. The pressure exerted by the column is the force divided by the cross-sectional area, A, over which the force is applied. Because our air column has a cross-sectional area of 1 m^2, we have

$$P = \frac{F}{A} = \frac{1 \times 10^5\ \text{N}}{1\ \text{m}^2} = 1 \times 10^5\ \text{N/m}^2 = 1 \times 10^5\ \text{Pa} = 1 \times 10^2\ \text{kPa}$$

The SI unit of pressure is N/m^2. It is given the name **pascal** (Pa) after Blaise Pascal (1623–1662), a French mathematician and scientist who studied pressure: 1 Pa = 1 N/m^2. A related unit sometimes used to report pressures is the **bar**, which equals 10^5 Pa. Atmospheric pressure at sea level is about 100 kPa, or 1 bar. The actual atmospheric pressure at any location depends on weather conditions and altitude.

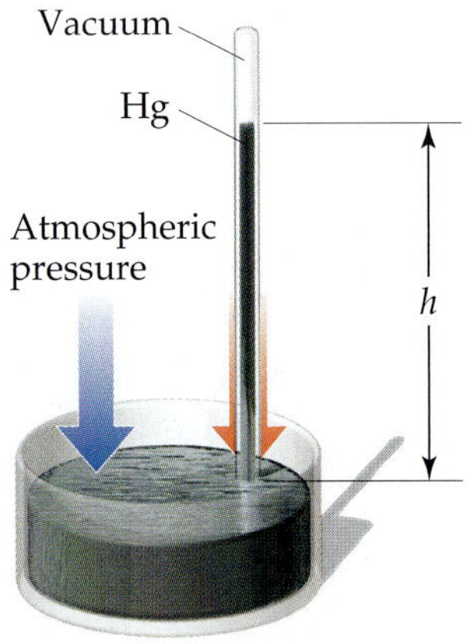

▲ **Figure 10.2 A mercury barometer.** The pressure of the atmosphere on the surface of the mercury (represented by the blue arrow) equals the pressure of the column of mercury (red arrow).

In the early part of the seventeenth century it was widely believed that the atmosphere had no weight. Evangelista Torricelli (1608–1647), who was a student of Galileo's, invented the *barometer* (Figure 10.2 ◀) to show that it did. A glass tube more than 760 mm long that is closed at one end is completely filled with mercury and inverted into a dish that contains additional mercury. Care must be taken so that no air gets into the tube. Some of the mercury flows out when the tube is inverted, but a column of mercury remains in the tube. Torricelli argued that the mercury surface in the dish experiences the full force, or weight, of Earth's atmosphere. Because there is no air (and therefore no atmospheric pressure) above the mercury in the tube, the mercury is pushed up the tube until the pressure at the base of the tube that is due to the mass of the mercury column balances the atmospheric pressure. Thus, the height, h, of the mercury column is a measure of the atmosphere's pressure, so it will change as the atmospheric pressure changes.

Torricelli's proposed explanation met with fierce opposition. Some argued that there could not possibly be a vacuum at the top of the tube. They said, "Nature does not permit a vacuum!" But Torricelli also had his supporters. Blaise Pascal, for example, had one of the barometers carried to the top of Puy de Dome, a volcanic mountain in central France, and compared its readings with a duplicate barometer kept at the foot of the mountain. As the barometer ascended, the height of the mercury column diminished, as expected, because the amount of atmosphere pressing down on the surface decreases as one moves higher. These and other experiments by scientists eventually prevailed, and the idea that the atmosphere has weight became accepted over a period of many years.

GIVE IT SOME THOUGHT

What happens to the height of the mercury column in a mercury barometer as you move to higher altitude, and why?

Standard atmospheric pressure, which corresponds to the typical pressure at sea level, is the pressure sufficient to support a column of mercury 760 mm high. In SI units, this pressure equals 1.01325 × 10^5 Pa. Standard atmospheric pressure defines some common non-SI units used to express gas pressures, such as the **atmosphere** (atm) and the *millimeter of mercury* (mm Hg). The latter unit is also called the **torr**, after Torricelli.

$$1\ \text{atm} = 760\ \text{mm Hg} = 760\ \text{torr} = 1.01325 \times 10^5\ \text{Pa} = 101.325\ \text{kPa}$$

Note that the units mm Hg and torr are equivalent: 1 torr = 1 mm Hg.

We will usually express gas pressure in units of atm, Pa (or kPa), or torr, so you should be comfortable converting gas pressures from one set of units to another.

SAMPLE EXERCISE 10.1 | Converting Units of Pressure

(a) Convert 0.357 atm to torr. **(b)** Convert 6.6×10^{-2} torr to atm. **(c)** Convert 147.2 kPa to torr.

Solution

Analyze: In each case we are given the pressure in one unit and asked to convert it to another unit. Our task, therefore, is to choose the appropriate conversion factors.

Plan: We can use dimensional analysis to perform the desired conversions.

Solve: (a) To convert atmospheres to torr, we use the relationship 760 torr = 1 atm:

$$(0.357 \cancel{\text{atm}})\left(\frac{760 \text{ torr}}{1 \cancel{\text{atm}}}\right) = 271 \text{ torr}$$

Note that the units cancel in the required manner.

(b) We use the same relationship as in part (a). To get the appropriate units to cancel, we must use the conversion factor as follows:

$$(6.6 \times 10^{-2} \cancel{\text{torr}})\left(\frac{1 \text{ atm}}{760 \cancel{\text{torr}}}\right) = 8.7 \times 10^{-5} \text{ atm}$$

(c) The relationship 760 torr = 101.325 kPa allows us to write an appropriate conversion factor for this problem:

$$(147.2 \cancel{\text{kPa}})\left(\frac{760 \text{ torr}}{101.325 \cancel{\text{kPa}}}\right) = 1104 \text{ torr}$$

Check: In each case look at the magnitude of the answer and compare it with the starting value. The torr is a much smaller unit than the atmosphere, so we expect the *numerical* answer to be larger than the starting quantity in (a), smaller in (b). In (c) notice that there are nearly 8 torr per kPa, so the numerical answer in torr should be about 8 times larger than its value in kPa, consistent with our calculation.

PRACTICE EXERCISE

(a) In countries that use the metric system, such as Canada, atmospheric pressure in weather reports is given in units of kPa. Convert a pressure of 745 torr to kPa. **(b)** An English unit of pressure sometimes used in engineering is pounds per square inch (lb/in.^{2}), or psi: 1 atm = 14.7 lb/in.^{2}. If a pressure is reported as 91.5 psi, express the measurement in atmospheres.

Answer: **(a)** 99.3 kPa, **(b)** 6.22 atm

We can use various devices to measure the pressures of enclosed gases. Tire gauges, for example, measure the pressure of air in automobile and bicycle tires. In laboratories we sometimes use a device called a *manometer*. A manometer operates on a principle similar to that of a barometer, as shown in Sample Exercise 10.2.

ACTIVITY
Manometer

SAMPLE EXERCISE 10.2 | Using a Manometer to Measure Gas Pressure

On a certain day the barometer in a laboratory indicates that the atmospheric pressure is 764.7 torr. A sample of gas is placed in a flask attached to an open-end mercury manometer, shown in Figure 10.3 ▶. A meter stick is used to measure the height of the mercury above the bottom of the manometer. The level of mercury in the open-end arm of the manometer has a height of 136.4 mm, and the mercury in the arm that is in contact with the gas has a height of 103.8 mm. What is the pressure of the gas **(a)** in atmospheres, **(b)** in kPa?

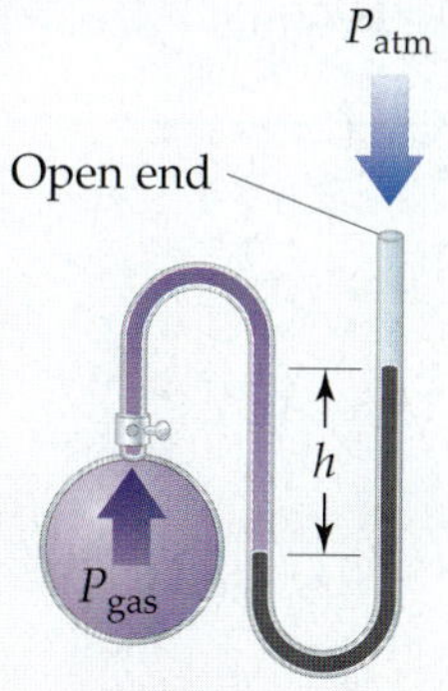

$$P_{gas} = P_{atm} + P_h$$

▲ **Figure 10.3 A mercury manometer.** This device is sometimes employed in the laboratory to measure gas pressures near atmospheric pressure.

Solution

Analyze: We are given the atmospheric pressure (764.7 torr) and the heights of the mercury in the two arms of the manometer and asked to determine the gas pressure in the flask. We know that this pressure must be greater than atmospheric because the manometer level on the flask side (103.8 mm) is lower than that on the side open to the atmosphere (136.4 mm), as indicated in Figure 10.3.

Plan: We'll use the difference in height between the two arms (h in Figure 10.3) to obtain the amount by which the pressure of the gas exceeds atmospheric pressure. Because an open-end mercury manometer is used, the height difference directly measures the pressure difference in mm Hg or torr between the gas and the atmosphere.

Solve: **(a)** The pressure of the gas equals the atmospheric pressure plus h:

$$P_{gas} = P_{atm} + h$$
$$= 764.7\ \text{torr} + (136.4\ \text{torr} - 103.8\ \text{torr})$$
$$= 797.3\ \text{torr}$$

We convert the pressure of the gas to atmospheres:

$$P_{gas} = (797.3\ \text{torr})\left(\frac{1\ \text{atm}}{760\ \text{torr}}\right) = 1.049\ \text{atm}$$

(b) To calculate the pressure in kPa, we employ the conversion factor between atmospheres and kPa:

$$1.049\ \text{atm}\left(\frac{101.3\ \text{kPa}}{1\ \text{atm}}\right) = 106.3\ \text{kPa}$$

Check: The calculated pressure is a bit more than one atmosphere. This makes sense because we anticipated that the pressure in the flask would be greater than the pressure of the atmosphere acting on the manometer, which is a bit greater than one standard atmosphere.

PRACTICE EXERCISE

Convert a pressure of 0.975 atm into Pa and kPa.
Answers: 9.88×10^4 Pa and 98.8 kPa

CHEMISTRY AND LIFE | Blood Pressure

The human heart pumps blood to the parts of the body through arteries, and the blood returns to the heart through veins. When your blood pressure is measured, two values are reported, such as 120/80 (120 over 80), which is a normal reading. The first measurement is the *systolic pressure*, the maximum pressure when the heart is pumping. The second is the *diastolic pressure*, the pressure when the heart is in the resting part of its pumping cycle. The units associated with these pressure measurements are torr.

Blood pressure is measured using a pressure gauge attached to a closed, air-filled jacket or cuff that is applied like a tourniquet to the arm (Figure 10.4 ▶). The pressure gauge may be a mercury manometer or some other device. The air pressure in the cuff is increased using a small pump until it is above the systolic pressure and prevents the flow of blood. The air pressure inside the cuff is then slowly reduced until blood just begins to pulse through the artery, as detected by the use of a stethoscope. At this point the pressure in the cuff equals the pressure that the blood exerts inside the arteries. Reading the gauge gives the systolic pressure. The pressure in the cuff is then reduced further until the blood flows freely. The pressure at this point is the diastolic pressure.

Hypertension is the presence of abnormally high blood pressure. The usual criterion for hypertension is a blood pressure greater than 140/90, although recent studies suggest that health risks increase for systolic readings above 120. Hypertension significantly increases the workload on the heart and also places a stress on the walls of the blood vessels throughout the body. These effects increase the risk of aneurysms, heart attacks, and strokes.

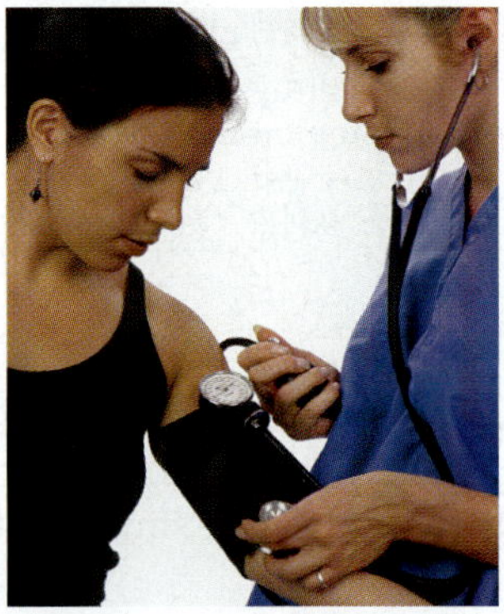

▲ **Figure 10.4 Measuring blood pressure.**

10.3 | The Gas Laws

Experiments with a large number of gases reveal that four variables are needed to define the physical condition, or *state*, of a gas: temperature, T, pressure, P, volume, V, and the amount of gas, which is usually expressed as the number of moles, n. The equations that express the relationships among T, P, V, and n are known as the *gas laws*.

The Pressure-Volume Relationship: Boyle's Law

If the pressure on a balloon is decreased, the balloon expands. That is why weather balloons expand as they rise through the atmosphere (Figure 10.5 ▶). Conversely, when a volume of gas is compressed, the pressure of the gas increases. British chemist Robert Boyle (1627–1691) first investigated the relationship between the pressure of a gas and its volume.

◀ **Figure 10.5 An application of the pressure-volume relationship.** The volume of gas in this weather balloon will increase as it ascends into the high atmosphere, where the atmospheric pressure is lower than on Earth's surface.

VCL Boyle's Law: Pressure and Volume

To perform his gas experiments, Boyle used a J-shaped tube like that shown in Figure 10.6 ▶. In the tube on the left, a quantity of gas is trapped behind a column of mercury. Boyle changed the pressure on the gas by adding mercury to the tube. He found that the volume of the gas decreased as the pressure increased. For example, doubling the pressure caused the gas volume to decrease to half its original value.

Boyle's law, which summarizes these observations, states that *the volume of a fixed quantity of gas maintained at constant temperature is inversely proportional to the pressure.* When two measurements are inversely proportional, one gets smaller as the other gets larger. Boyle's law can be expressed mathematically as

$$V = \text{constant} \times \frac{1}{P} \quad \text{or} \quad PV = \text{constant} \qquad [10.2]$$

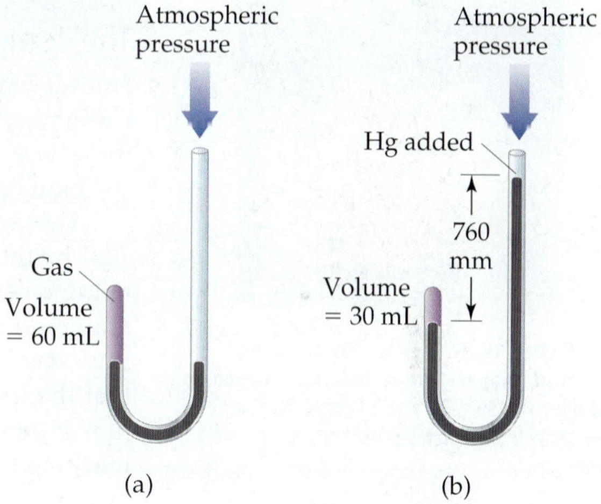

▲ **Figure 10.6 An illustration of Boyle's experiment relating pressure and volume.** In (a) the volume of the gas trapped in the J-tube is 60 mL when the gas pressure is 760 torr. When additional mercury is added, as shown in (b), the trapped gas is compressed. The volume is 30 mL when its total pressure is 1520 torr, corresponding to atmospheric pressure plus the pressure exerted by the 760-mm column of mercury.

The value of the constant depends on the temperature and the amount of gas in the sample. The graph of V versus P in Figure 10.7(a) ▼ shows the type of curve obtained for a given quantity of gas at a fixed temperature. A linear relationship is obtained when V is plotted versus $1/P$ [Figure 10.7(b)].

ANIMATION *P–V* Relationships

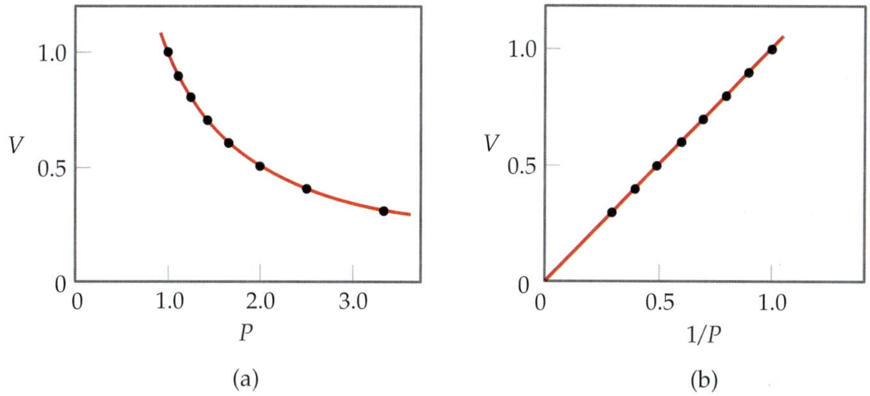

◀ **Figure 10.7 Graphs based on Boyle's law.** (a) Volume versus pressure, (b) volume versus $1/P$.

Simple though it is, Boyle's law occupies a special place in the history of science. Boyle was the first to carry out a series of experiments in which one variable was systematically changed to determine the effect on another variable. The data from the experiments were then employed to establish an empirical relationship, a "law."

We apply Boyle's law every time we breathe. The volume of the lungs is governed by the rib cage, which can expand and contract, and the diaphragm, a muscle beneath the lungs. Inhalation occurs when the rib cage expands and the diaphragm moves downward. Both of these actions increase the volume of the lungs, thus decreasing the gas pressure inside the lungs. The atmospheric pressure then forces air into the lungs until the pressure in the lungs equals atmospheric pressure. Exhalation reverses the process: The rib cage contracts and the diaphragm moves up, both of which decrease the volume of the lungs. Air is forced out of the lungs by the resulting increase in pressure.

▲ **Figure 10.8 An illustration of the effect of temperature on volume.** As liquid nitrogen (−196°C) is poured over a balloon, the gas in the balloon is cooled and its volume decreases.

VCL Charles's Law: Temperature and Volume

GIVE IT SOME THOUGHT

What happens to the volume of a gas if you double its pressure, say from 1 atm to 2 atm, while its temperature is held constant?

The Temperature-Volume Relationship: Charles's Law

Hot-air balloons rise because air expands as it is heated. The warm air in the balloon is less dense than the surrounding cool air at the same pressure. This difference in density causes the balloon to ascend. Conversely, a balloon will shrink when the gas in it is cooled, as seen in Figure 10.8 ◀.

The relationship between gas volume and temperature was discovered in 1787 by the French scientist Jacques Charles (1746–1823). Charles found that the volume of a fixed quantity of gas at constant pressure increases linearly with temperature. Some typical data are shown in Figure 10.9 ▼. Notice that the extrapolated (extended) line (which is dashed) passes through −273°C. Note also that the gas is predicted to have zero volume at this temperature. This condition is never realized, however, because all gases liquefy or solidify before reaching this temperature.

In 1848 William Thomson (1824–1907), a British physicist whose title was Lord Kelvin, proposed an absolute-temperature scale, now known as the Kelvin scale. On this scale 0 K, which is called *absolute zero*, equals −273.15°C. (Section 1.4) In terms of the Kelvin scale, **Charles's law** can be stated as follows: *The volume of a fixed amount of gas maintained at constant pressure is directly*

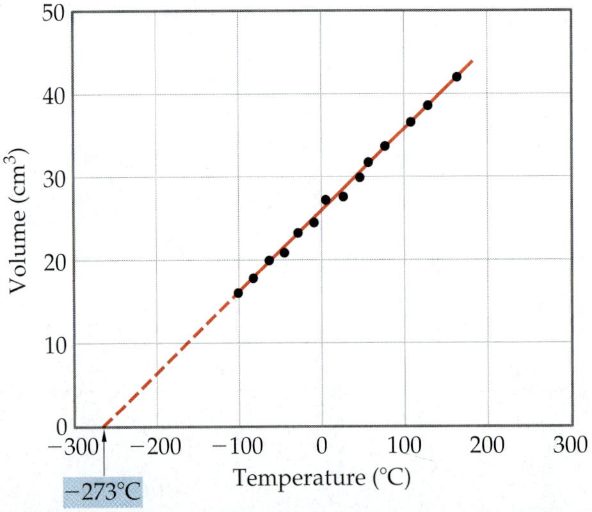

▶ **Figure 10.9 Graph based on Charles's law.** At constant pressure, the volume of an enclosed gas increases as the temperature increases. The dashed line is an extrapolation to temperatures at which the substance is no longer a gas.

proportional to its absolute temperature. Thus, doubling the absolute temperature, say from 200 K to 400 K, causes the gas volume to double. Mathematically, Charles's law takes the following form:

$$V = \text{constant} \times T \quad \text{or} \quad \frac{V}{T} = \text{constant} \qquad [10.3]$$

The value of the constant depends on the pressure and amount of gas.

GIVE IT SOME THOUGHT

Does the volume of a fixed quantity of gas decrease to half its original value when the temperature is lowered from 100°C to 50°C?

The Quantity-Volume Relationship: Avogadro's Law

As we add gas to a balloon, the balloon expands. The volume of a gas is affected not only by pressure and temperature but also by the amount of gas. The relationship between the quantity of a gas and its volume follows from the work of Joseph Louis Gay-Lussac (1778–1823) and Amedeo Avogadro (1776–1856).

Gay-Lussac is one of those extraordinary figures in the history of science who could truly be called an adventurer. He was interested in lighter-than-air balloons, and in 1804 he made an ascent to 23,000 ft—an exploit that held the altitude record for several decades. To better control lighter-than-air balloons, Gay-Lussac carried out several experiments on the properties of gases. In 1808 he observed the *law of combining volumes*: At a given pressure and temperature, the volumes of gases that react with one another are in the ratios of small whole numbers. For example, two volumes of hydrogen gas react with one volume of oxygen gas to form two volumes of water vapor, as shown in Figure 10.10 ▼.

Observation — Two volumes hydrogen — One volume oxygen — Two volumes water vapor

Explanation

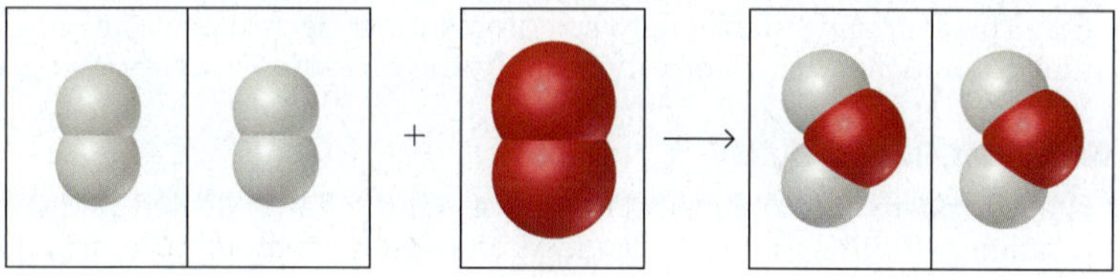

Equation $2\,H_2(g) + O_2(g) \longrightarrow 2\,H_2O(g)$

◄ **Figure 10.10 The law of combining volumes.** Gay-Lussac's experimental observation of combining volumes is shown together with Avogadro's explanation of this phenomenon.

Three years later Amedeo Avogadro interpreted Gay-Lussac's observation by proposing what is now known as **Avogadro's hypothesis**: *Equal volumes of gases at the same temperature and pressure contain equal numbers of molecules*. For example, experiments show that 22.4 L of any gas at 0°C and 1 atm contain 6.02×10^{23} gas molecules (that is, 1 mol), as depicted in Figure 10.11 ▼.

	He	N_2	CH_4
Volume	22.4 L	22.4 L	22.4 L
Pressure	1 atm	1 atm	1 atm
Temperature	0°C	0°C	0°C
Mass of gas	4.00 g	28.0 g	16.0 g
Number of gas molecules	6.02×10^{23}	6.02×10^{23}	6.02×10^{23}

◄ **Figure 10.11 A comparison illustrating Avogadro's hypothesis.** Note that helium gas consists of helium atoms. Each gas has the same volume, temperature, and pressure and thus contains the same number of molecules. Because a molecule of one substance differs in mass from a molecule of another, the masses of gas in the three containers differ.

VCL Avogadro's Law: Moles and Volume

Avogadro's law follows from Avogadro's hypothesis: *The volume of a gas maintained at constant temperature and pressure is directly proportional to the number of moles of the gas*. That is,

$$V = \text{constant} \times n \qquad [10.4]$$

Thus, doubling the number of moles of gas will cause the volume to double if T and P remain constant.

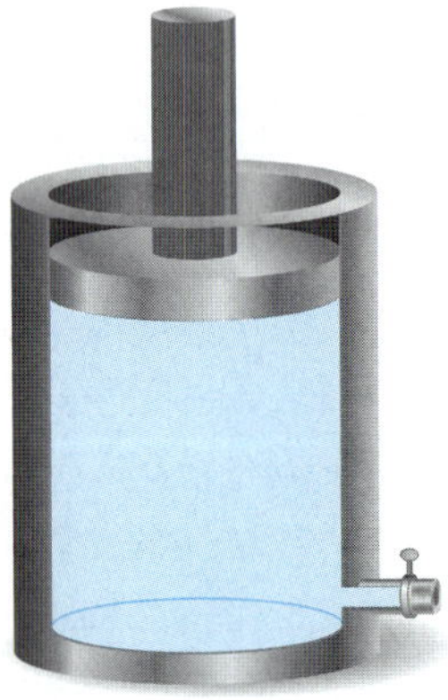

▲ **Figure 10.12 Cylinder with piston and gas inlet valve.**

SAMPLE EXERCISE 10.3 | Evaluating the Effects of Changes in P, V, n, and T on a Gas

Suppose we have a gas confined to a cylinder as shown in Figure 10.12 ◀. Consider the following changes: **(a)** Heat the gas from 298 K to 360 K, while maintaining the piston in the position shown in the drawing. **(b)** Move the piston to reduce the volume of gas from 1 L to 0.5 L. **(c)** Inject additional gas through the gas inlet valve. Indicate how each of these changes will affect the average distance between molecules, the pressure of the gas, and the number of moles of gas present in the cylinder.

Solution

Analyze: We need to think how each of three different changes in the system affects (1) the distance between molecules, (2) the pressure of the gas, and (3) the number of moles of gas in the cylinder.

Plan: We'll use our understanding of the gas laws and the general properties of gases to analyze each situation.

Solve: **(a)** Heating the gas while maintaining the position of the piston will cause no change in the number of molecules per unit volume. Thus, the distance between molecules and the total moles of gas remain the same. The increase in temperature, however, will cause the pressure to increase (Charles's law).

(b) Moving the piston compresses the same quantity of gas into a smaller volume. The total number of molecules of gas, and thus the total number of moles, remains the same. The average distance between molecules, however, must decrease because of the smaller volume in which the gas is confined. The reduction in volume causes the pressure to increase (Boyle's law).

(c) Injecting more gas into the cylinder while keeping the volume and temperature the same will result in more molecules and thus a greater number of moles of gas. The average distance between atoms must decrease because their number per unit volume increases. Correspondingly, the pressure increases (Avogadro's law).

PRACTICE EXERCISE

What happens to the density of a gas as **(a)** the gas is heated in a constant-volume container; **(b)** the gas is compressed at constant temperature; **(c)** additional gas is added to a constant-volume container?

Answer: **(a)** no change, **(b)** increase, **(c)** increase

10.4 | The Ideal-Gas Equation

ACTIVITY Gas Laws

VCL Derivation of the Ideal-Gas Law

In Section 10.3 we examined three historically important gas laws that describe the relationships between the four variables, P, V, T, and n, that define the state of a gas. Each law was obtained by holding two variables constant in order to see how the remaining two variables affect each other. We can express each law as a proportionality relationship. Using the symbol $\propto$, which is read "is proportional to," we have

$$\textit{Boyle's law:} \quad V \propto \frac{1}{P} \quad (\text{constant } n, T)$$

$$\textit{Charles's law:} \quad V \propto T \quad (\text{constant } n, P)$$

$$\textit{Avogadro's law:} \quad V \propto n \quad (\text{constant } P, T)$$

We can combine these relationships to make a more general gas law.

$$V \propto \frac{nT}{P}$$

If we call the proportionality constant R, we obtain

$$V = R\left(\frac{nT}{P}\right)$$

Rearranging, we have this relationship in its more familiar form:

$$PV = nRT \qquad [10.5]$$

This equation is known as the **ideal-gas equation**. An **ideal gas** is a hypothetical gas whose pressure, volume, and temperature behavior is completely described by the ideal-gas equation.

The term R in the ideal-gas equation is called the **gas constant**. The value and units of R depend on the units of P, V, n, and T. Temperature must *always* be expressed as an absolute temperature when used in the ideal-gas equation. The quantity of gas, n, is normally expressed in moles. The units chosen for pressure and volume are most often atm and liters, respectively. However, other units can be used. In most countries other than the United States, the SI unit of Pa (or kPa) is most commonly employed. Table 10.2 ▶ shows the numerical value for R in various units. As we saw in the "Closer Look" box on P-V work in Section 5.3, the product PV has the units of energy. Therefore, the units of R can include joules or calories. In working problems with the ideal-gas equation, the units of P, V, n, and T must agree with the units in the gas constant. In this chapter we will most often use the value $R = 0.08206$ L-atm/mol-K (four significant figures) or 0.0821 L-atm/mol-K (three significant figures) whenever we use the ideal-gas equation, consistent with the units of atm for pressure. Use of the value $R = 8.314$ J/mol-K, consistent with the units of Pa for pressure, is also very common.

TABLE 10.2 Numerical Values of the Gas Constant, R, in Various Units

Units	Numerical Value
L-atm/mol-K	0.08206
J/mol-K*	8.314
cal/mol-K	1.987
m^3-Pa/mol-K*	8.314
L-torr/mol-K	62.36

*SI unit.

Suppose we have 1.000 mol of an ideal gas at 1.000 atm and 0.00°C (273.15 K). According to the ideal-gas equation, the volume of the gas is

$$V = \frac{nRT}{P} = \frac{(1.000\ \text{mol})(0.08206\ \text{L-atm/mol-K})(273.15\ \text{K})}{1.000\ \text{atm}} = 22.41\ \text{L}$$

The conditions 0°C and 1 atm are referred to as the **standard temperature and pressure (STP)**. Many properties of gases are tabulated for these conditions. The volume occupied by one mole of ideal gas at STP, 22.41 L, is known as the *molar volume* of an ideal gas at STP.

GIVE IT SOME THOUGHT

How many molecules are in 22.41 L of an ideal gas at STP?

The ideal-gas equation accounts adequately for the properties of most gases under a wide variety of circumstances. It is not exactly correct, however, for any real gas. Thus, the measured volume, V, for given conditions of P, n, and T, might differ from the volume calculated from $PV = nRT$. To illustrate, the measured molar volumes of real gases at STP are compared with the calculated volume of an ideal gas in Figure 10.13 ▼. While these real gases don't match the ideal gas behavior exactly, the differences are so small that we can ignore them for all but the most accurate work. We'll have more to say about the differences between ideal and real gases in Section 10.9.

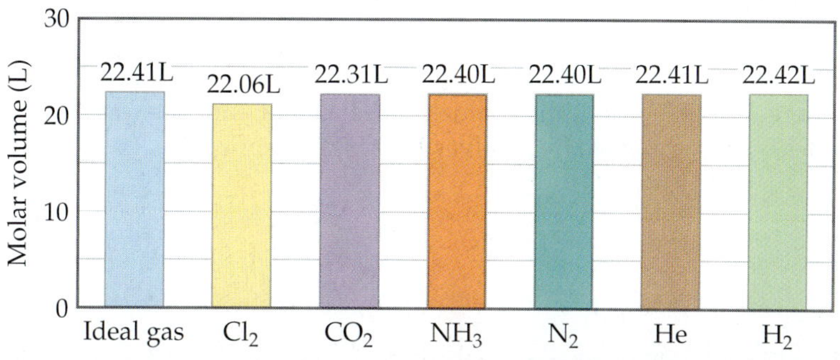

◀ **Figure 10.13 Comparison of molar volumes at STP.** One mole of an ideal gas at STP occupies a volume of 22.41 L. One mole of various real gases at STP occupies close to this ideal volume.

SAMPLE EXERCISE 10.4 | Using the Ideal-Gas Equation

Calcium carbonate, $CaCO_3(s)$, decomposes upon heating to give $CaO(s)$ and $CO_2(g)$. A sample of $CaCO_3$ is decomposed, and the carbon dioxide is collected in a 250-mL flask. After the decomposition is complete, the gas has a pressure of 1.3 atm at a temperature of 31°C. How many moles of CO_2 gas were generated?

Solution

Analyze: We are given the volume (250 mL), pressure (1.3 atm), and temperature (31°C) of a sample of CO_2 gas and asked to calculate the number of moles of CO_2 in the sample.

Plan: Because we are given V, P, and T, we can solve the ideal-gas equation for the unknown quantity, n.

Solve: In analyzing and solving gas-law problems, it is helpful to tabulate the information given in the problems and then to convert the values to units that are consistent with those for R (0.0821 L-atm/mol-K). In this case the given values are

$$V = 250 \text{ mL} = 0.250 \text{ L}$$
$$P = 1.3 \text{ atm}$$
$$T = 31°\text{C} = (31 + 273) \text{ K} = 304 \text{ K}$$

Remember: *Absolute temperature must always be used when the ideal-gas equation is solved.*

We now rearrange the ideal-gas equation (Equation 10.5) to solve for n.

$$n = \frac{PV}{RT}$$

$$n = \frac{(1.3 \text{ atm})(0.250 \text{ L})}{(0.0821 \text{ L-atm/mol-K})(304 \text{ K})} = 0.013 \text{ mol } CO_2$$

Check: Appropriate units cancel, thus ensuring that we have properly rearranged the ideal-gas equation and have converted to the correct units.

PRACTICE EXERCISE

Tennis balls are usually filled with air or N_2 gas to a pressure above atmospheric pressure to increase their "bounce." If a particular tennis ball has a volume of 144 cm^3 and contains 0.33 g of N_2 gas, what is the pressure inside the ball at 24°C?

Answer: 2.0 atm

Relating the Ideal-Gas Equation and the Gas Laws

The simple gas laws that we discussed in Section 10.3, such as Boyle's law, are special cases of the ideal-gas equation. For example, when the quantity of gas and the temperature are held constant, n and T have fixed values. Therefore, the product nRT is the product of three constants and must itself be a constant.

$$PV = nRT = \text{constant} \quad \text{or} \quad PV = \text{constant} \qquad [10.6]$$

Thus, we have Boyle's law. We see that if n and T are constant, the individual values of P and V can change, but the product PV must remain constant.

We can use Boyle's law to determine how the volume of a gas changes when its pressure changes. For example, if a metal cylinder holds 50.0 L of O_2 gas at 18.5 atm and 21°C, what volume will the gas occupy if the temperature is maintained at 21°C while the pressure is reduced to 1.00 atm? Because the product PV is a constant when a gas is held at constant n and T, we know that

$$P_1V_1 = P_2V_2 \qquad [10.7]$$

where P_1 and V_1 are initial values and P_2 and V_2 are final values. Dividing both sides of this equation by P_2 gives the final volume, V_2.

$$V_2 = V_1 \times \frac{P_1}{P_2}$$

STRATEGIES IN CHEMISTRY | Calculations Involving Many Variables

In chemistry and throughout your studies of science and math, you may encounter problems that involve several experimentally measured variables as well as several different physical constants. In this chapter we encounter a variety of problems based on the ideal-gas equation, which consists of four experimental quantities—P, V, n, and T—and one constant, R. Depending on the type of problem, we might need to solve for any of the four quantities.

To avoid any difficulty extracting the necessary information from problems when many variables are involved, we suggest you take the following steps as you analyze, plan, and solve such problems:

1. *Tabulate information.* Read the problems carefully to determine which quantity is the unknown and which quantities are given. Every time you encounter a numerical value, jot it down. In many cases, constructing a table of the given information will be useful.

2. *Convert to consistent units.* As you have already seen, we can use several different units to express the same quantity. Make certain that quantities are converted to the proper units by using the correct conversion factors. In using the ideal-gas equation, for example, we usually use the value of R that has units of L-atm/mol-K. If you are given a pressure in torr, you will need to convert it to atmospheres.

3. *If a single equation relates the variables, rearrange the equation to solve for the unknown.* Make certain that you are comfortable using algebra to solve the equation for the desired variable. In the case of the ideal-gas equation the following algebraic rearrangements will all be used at one time or another:

$$P = \frac{nRT}{V}; \quad V = \frac{nRT}{P}; \quad n = \frac{PV}{RT}; \quad T = \frac{PV}{nR}$$

4. *Use dimensional analysis.* Carry the units through your calculation. Use of dimensional analysis enables you to check that you have solved the equation correctly. If the units of the quantities in the equation cancel properly to give the units of the desired variable, you have probably used the equation correctly.

Sometimes you will not be given values for the necessary variables directly. Rather, you will be given the values of other quantities that can be used to determine the needed variables. For example, suppose you are trying to use the ideal-gas equation to calculate the pressure of a gas. You are given the temperature of the gas, but you are not given explicit values for n and V. However, the problem states that "the sample of gas contains 0.15 mol of gas per liter." We can turn this statement into the expression

$$\frac{n}{V} = 0.15 \text{ mol/L}$$

Solving the ideal-gas equation for pressure yields

$$P = \frac{nRT}{V} = \left(\frac{n}{V}\right)RT$$

Thus, we can solve the equation even though we are not given specific values for n and V. We will examine how to use the density and molar mass of a gas in this fashion in Section 10.5.

As we have continuously stressed, the most important thing you can do to become proficient at solving problems is to practice by using practice exercises and assigned exercises at the end of each chapter. By using systematic procedures, such as those described here, you should be able to minimize difficulties in solving problems involving many variables.

Substituting the given quantities into this equation gives

$$V_2 = (50.0 \text{ L})\left(\frac{18.5 \text{ atm}}{1.00 \text{ atm}}\right) = 925 \text{ L}$$

The answer is reasonable because gases expand as their pressures are decreased.

In a similar way we can start with the ideal-gas equation and derive relationships between any other two variables, V and T (Charles's law), n and V (Avogadro's law), or P and T. Sample Exercise 10.5 illustrates how these relationships can be derived and used.

SAMPLE EXERCISE 10.5 | Calculating the Effect of Temperature Changes on Pressure

The gas pressure in an aerosol can is 1.5 atm at 25°C. Assuming that the gas inside obeys the ideal-gas equation, what would the pressure be if the can were heated to 450°C?

Solution

Analyze: We are given the initial pressure (1.5 atm) and temperature (25°C) of the gas and asked for the pressure at a higher temperature (450°C).

Plan: The volume and number of moles of gas do not change, so we must use a relationship connecting pressure and temperature. Converting temperature to the Kelvin scale and tabulating the given information, we have

	P	T
Initial	1.5 atm	298 K
Final	P_2	723 K

Solve: In order to determine how P and T are related, we start with the ideal-gas equation and isolate the quantities that don't change (n, V, and R) on one side and the variables (P and T) on the other side.

$$\frac{P}{T} = \frac{nR}{V} = \text{constant}$$

Because the quotient P/T is a constant, we can write

$$\frac{P_1}{T_1} = \frac{P_2}{T_2}$$

(where the subscripts 1 and 2 represent the initial and final states, respectively). Rearranging to solve for P_2 and substituting the given data gives

$$P_2 = P_1 \times \frac{T_2}{T_1}$$

$$P_2 = (1.5 \text{ atm})\left(\frac{723 \text{ K}}{298 \text{ K}}\right) = 3.6 \text{ atm}$$

Check: This answer is intuitively reasonable—increasing the temperature of a gas increases its pressure.
Comment: It is evident from this example why aerosol cans carry a warning not to incinerate.

PRACTICE EXERCISE

A large natural-gas storage tank is arranged so that the pressure is maintained at 2.20 atm. On a cold day in December when the temperature is $-15°C$ (4°F), the volume of gas in the tank is 28,500 ft^3. What is the volume of the same quantity of gas on a warm July day when the temperature is 31°C (88°F)?
Answer: 33,600 ft^3

We are often faced with the situation in which P, V, and T all change for a fixed number of moles of gas. Because n is constant under these circumstances, the ideal-gas equation gives

$$\frac{PV}{T} = nR = \text{constant}$$

If we represent the initial and final conditions of pressure, temperature, and volume by subscripts 1 and 2, respectively, we can write

$$\frac{P_1V_1}{T_1} = \frac{P_2V_2}{T_2} \qquad [10.8]$$

SAMPLE EXERCISE 10.6 | Calculating the Effect of Changing *P* and *T* on the Volume of a Gas

An inflated balloon has a volume of 6.0 L at sea level (1.0 atm) and is allowed to ascend in altitude until the pressure is 0.45 atm. During ascent the temperature of the gas falls from 22°C to $-21°C$. Calculate the volume of the balloon at its final altitude.

Solution

Analyze: We need to determine a new volume for a gas sample in a situation where both pressure and temperature change.
Plan: Let's again proceed by converting temperature to the Kelvin scale and tabulating the given information.

	P	V	T
Initial	1.0 atm	6.0 L	295 K
Final	0.45 atm	V_2	252 K

Because n is constant, we can use Equation 10.8.

Solve: Rearranging Equation 10.8 to solve for V_2 gives

$$V_2 = V_1 \times \frac{P_1}{P_2} \times \frac{T_2}{T_1} = (6.0\text{ L})\left(\frac{1.0\text{ atm}}{0.45\text{ atm}}\right)\left(\frac{252\text{ K}}{295\text{ K}}\right) = 11\text{ L}$$

Check: The result appears reasonable. Notice that the calculation involves multiplying the initial volume by a ratio of pressures and a ratio of temperatures. Intuitively, we expect that decreasing pressure will cause the volume to increase. Similarly, decreasing temperature should cause the volume to decrease. Note that the difference in pressures is more dramatic than the difference in temperatures. Thus, we should expect the effect of the pressure change to predominate in determining the final volume, as it does.

PRACTICE EXERCISE

A 0.50-mol sample of oxygen gas is confined at 0°C in a cylinder with a movable piston, such as that shown in Figure 10.12. The gas has an initial pressure of 1.0 atm. The gas is then compressed by the piston so that its final volume is half the initial volume. The final pressure of the gas is 2.2 atm. What is the final temperature of the gas in degrees Celsius?

Answer: 27°C

10.5 | Further Applications of the Ideal-Gas Equation

The ideal-gas equation can be used to determine many relationships involving the physical properties of gases. In this section we use it first to define the relationship between the density of a gas and its molar mass, and then to calculate the volumes of gases formed or consumed in chemical reactions.

Gas Densities and Molar Mass

The ideal-gas equation allows us to calculate gas density from the molar mass, pressure, and temperature of the gas. Recall that density has the units of mass per unit volume ($d = m/V$) (Section 1.4). We can arrange the gas equation to obtain similar units, moles per unit volume, n/V:

ACTIVITY Density of Gases

$$\frac{n}{V} = \frac{P}{RT}$$

If we multiply both sides of this equation by the molar mass, $\mathcal{M}$, which is the number of grams in one mole of a substance, we obtain the following relationship:

$$\frac{n\mathcal{M}}{V} = \frac{P\mathcal{M}}{RT} \qquad [10.9]$$

The product of the quantities n/V and $\mathcal{M}$ equals the density in g/L, as seen from their units:

$$\frac{\text{Moles}}{\text{Liter}} \times \frac{\text{grams}}{\text{mole}} = \frac{\text{grams}}{\text{liter}}$$

Thus, the density, d, of the gas is given by the expression on the right in Equation 10.9:

$$d = \frac{P\mathcal{M}}{RT} \qquad [10.10]$$

From Equation 10.10 we see that the density of a gas depends on its pressure, molar mass, and temperature. The higher the molar mass and pressure, the more dense the gas; the higher the temperature, the less dense the gas. Although gases form homogeneous mixtures regardless of their identities, a less

▲ **Figure 10.14 A CO_2 fire extinguisher.** The CO_2 gas from a fire extinguisher is denser than air. The CO_2 cools significantly as it emerges from the tank. Water vapor in the air is condensed by the cool CO_2 gas and forms a white fog accompanying the colorless CO_2.

dense gas will lie above a more dense one in the absence of mixing. For example, CO_2 has a higher molar mass than N_2 or O_2 and is therefore more dense than air. When CO_2 is released from a CO_2 fire extinguisher, as shown in Figure 10.14 ◀, it blankets a fire, preventing O_2 from reaching the combustible material. The fact that a hotter gas is less dense than a cooler one explains why hot air rises. The difference between the densities of hot and cold air is responsible for the lift of hot-air balloons. It is also responsible for many phenomena in weather, such as the formation of large thunderhead clouds during thunderstorms.

GIVE IT SOME THOUGHT

Is water vapor more or less dense than N_2 under the same conditions of temperature and pressure?

SAMPLE EXERCISE 10.7 | Calculating Gas Density

What is the density of carbon tetrachloride vapor at 714 torr and 125°C?

Solution

Analyze: We are asked to calculate the density of a gas given its name, its pressure, and its temperature. From the name we can write the chemical formula of the substance and determine its molar mass.

Plan: We can use Equation 10.10 to calculate the density. Before we can use that equation, however, we need to convert the given quantities to the appropriate units. We must convert temperature to the Kelvin scale and pressure to atmospheres. The molar mass of CCl_4 is $12.0 + (4)(35.5) = 154.0$ g/mol.

Solve: Using Equation 10.10, we have

$$d = \frac{(714\ \text{torr})(1\ \text{atm}/760\ \text{torr})(154.0\ \text{g/mol})}{(0.0821\ \text{L-atm/mol-K})(398\ K)} = 4.43\ \text{g/L}$$

Check: If we divide the molar mass (g/mol) by the density (g/L), we end up with L/mol. The numerical value is roughly $154/4.4 = 35$. That is in the right ballpark for the molar volume of a gas heated to 125°C at near atmospheric pressure, so our answer is reasonable.

PRACTICE EXERCISE

The mean molar mass of the atmosphere at the surface of Titan, Saturn's largest moon, is 28.6 g/mol. The surface temperature is 95 K, and the pressure is 1.6 atm. Assuming ideal behavior, calculate the density of Titan's atmosphere.

Answer: 5.9 g/L

Equation 10.10 can be rearranged to solve for the molar mass of a gas:

$$\mathcal{M} = \frac{dRT}{P} \qquad [10.11]$$

Thus, we can use the experimentally measured density of a gas to determine the molar mass of the gas molecules, as shown in Sample Exercise 10.8.

SAMPLE EXERCISE 10.8 | Calculating the Molar Mass of a Gas

A series of measurements are made in order to determine the molar mass of an unknown gas. First, a large flask is evacuated and found to weigh 134.567 g. It is then filled with the gas to a pressure of 735 torr at 31°C and reweighed; its mass is now 137.456 g. Finally, the flask is filled with water at 31°C and found to weigh 1067.9 g. (The density of the water at this temperature is 0.997 g/mL.) Assuming that the ideal-gas equation applies, calculate the molar mass of the unknown gas.

Solution

Analyze: We are given the temperature (31°C) and pressure (735 torr) for a gas, together with information to determine its volume and mass, and we are asked to calculate its molar mass.

Plan: We need to use the mass information given to calculate the volume of the container and the mass of the gas within it. From this we calculate the gas density and then apply Equation 10.11 to calculate the molar mass of the gas.

Solve: The mass of the gas is the difference between the mass of the flask filled with gas and that of the empty (evacuated) flask:

$$137.456 \text{ g} - 134.567 \text{ g} = 2.889 \text{ g}$$

The volume of the gas equals the volume of water that the flask can hold. The volume of water is calculated from its mass and density. The mass of the water is the difference between the masses of the full and empty flask:

$$1067.9 \text{ g} - 134.567 \text{ g} = 933.3 \text{ g}$$

By rearranging the equation for density ($d = m/V$), we have

$$V = \frac{m}{d} = \frac{(933.3 \text{ g})}{(0.997 \text{ g/mL})} = 936 \text{ mL}$$

Knowing the mass of the gas (2.889 g) and its volume (936 mL), we can calculate the density of the gas:

$$2.889 \text{ g}/0.936 \text{ L} = 3.09 \text{ g/L}$$

After converting pressure to atmospheres and temperature to kelvins, we can use Equation 10.11 to calculate the molar mass:

$$\mathcal{M} = \frac{dRT}{P}$$

$$= \frac{(3.09 \text{ g/L})(0.0821 \text{ L-atm/mol-K})(304 \text{ K})}{(735/760) \text{ atm}}$$

$$= 79.7 \text{ g/mol}$$

Check: The units work out appropriately, and the value of molar mass obtained is reasonable for a substance that is gaseous near room temperature.

PRACTICE EXERCISE

Calculate the average molar mass of dry air if it has a density of 1.17 g/L at 21°C and 740.0 torr.
Answer: 29.0 g/mol

Volumes of Gases in Chemical Reactions

Understanding the properties of gases is important because gases are often reactants or products in chemical reactions. For this reason we are often faced with calculating the volumes of gases consumed or produced in reactions. We have seen that the coefficients in balanced chemical equations tell us the relative amounts (in moles) of reactants and products in a reaction. The number of moles of a gas, in turn, is related to P, V, and T.

ANIMATION
Air Bags

SAMPLE EXERCISE 10.9 | Relating the Volume of a Gas to the Amount of Another Substance in a Reaction

The safety air bags in automobiles are inflated by nitrogen gas generated by the rapid decomposition of sodium azide, NaN_3:

$$2\,NaN_3(s) \longrightarrow 2\,Na(s) + 3\,N_2(g)$$

If an air bag has a volume of 36 L and is to be filled with nitrogen gas at a pressure of 1.15 atm at a temperature of 26.0°C, how many grams of NaN_3 must be decomposed?

Solution

Analyze: This is a multistep problem. We are given the volume, pressure, and temperature of the N_2 gas and the chemical equation for the reaction by which the N_2 is generated. We must use this information to calculate the number of grams of NaN_3 needed to obtain the necessary N_2.

Plan: We need to use the gas data (P, V, and T) and the ideal-gas equation to calculate the number of moles of N_2 gas that should be formed for the air bag to operate correctly. We can then use the balanced equation to determine the number of moles of NaN_3. Finally, we can convert the moles of NaN_3 to grams.

Gas data → mol N_2 → mol NaN_3 → g NaN_3

Solve: The number of moles of N_2 is determined using the ideal-gas equation:

$$n = \frac{PV}{RT} = \frac{(1.15\ \text{atm})(36\ \text{L})}{(0.0821\ \text{L-atm/mol-K})(299\ \text{K})} = 1.7\ \text{mol N}_2$$

From here we use the coefficients in the balanced equation to calculate the number of moles of NaN_3.

$$(1.7\ \text{mol N}_2)\left(\frac{2\ \text{mol NaN}_3}{3\ \text{mol N}_2}\right) = 1.1\ \text{mol NaN}_3$$

Finally, using the molar mass of NaN_3, we convert moles of NaN_3 to grams:

$$(1.1\ \text{mol NaN}_3)\left(\frac{65.0\ \text{g NaN}_3}{1\ \text{mol NaN}_3}\right) = 72\ \text{g NaN}_3$$

Check: The best way to check our approach is to make sure the units cancel properly at each step in the calculation, leaving us with the correct units in the answer, g NaN_3.

PRACTICE EXERCISE

In the first step in the industrial process for making nitric acid, ammonia reacts with oxygen in the presence of a suitable catalyst to form nitric oxide and water vapor:

$$4\,NH_3(g) + 5\,O_2(g) \longrightarrow 4\,NO(g) + 6\,H_2O(g)$$

How many liters of $NH_3(g)$ at 850°C and 5.00 atm are required to react with 1.00 mol of $O_2(g)$ in this reaction?
Answer: 14.8 L

CHEMISTRY AT WORK | Gas Pipelines

Most people are quite unaware of the vast network of underground pipelines that undergirds the developed world. Pipelines are used to move massive quantities of liquids and gases over considerable distances. For example, pipelines move natural gas (methane) from huge natural-gas fields in Siberia to Western Europe. Natural gas from Algeria is moved to Italy through a pipeline 120 cm in diameter and 2500 km in length that stretches across the Mediterranean Sea at depths up to 600 m. In the United States the pipeline systems consist of trunk lines, large-diameter pipes for long-distance transport, with branch lines of lower diameter and lower pressure for local transport to and from the trunk lines.

Essentially all substances that are gases at STP are transported commercially by pipeline, including ammonia, carbon dioxide, carbon monoxide, chlorine, ethane, helium, hydrogen, and methane. The largest volume transport by far, though, is natural gas. (Figure 10.15 ▶) The methane-rich gas from oil and gas wells is processed to remove particulates, water, and various gaseous impurities such as hydrogen sulfide and carbon dioxide. The gas is then compressed to pressures ranging from 3.5 MPa (35 atm) to 10 MPa (100 atm), depending on the age and diameter of the pipe. The long-distance pipelines are about 40 cm in diameter and made of steel. Pressure is maintained by large compressor stations along the pipeline, spaced at 50- to 100-mile intervals.

Recall from Figure 5.24 that natural gas is a major source of energy for the United States. To meet this demand, methane must be transported from source wells throughout the United States and Canada to all parts of the nation. The total length of pipeline for natural-gas transport in the United States is about 6×10^5 km, and growing. The United States is divided into seven regions exceeds 2.7×10^{12} L (measured at STP), which is almost 100 billion cubic feet per day! The volume of the pipelines themselves would be entirely inadequate for managing the enormous quantities of natural gas that are placed into and taken out of the system on a continuing basis. For this reason, underground storage facilities, such as salt caverns and other natural formations, are employed to hold large quantities of gas.
Related Exercise: 10.108

▲ **Figure 10.15 A natural gas pipeline junction.**

10.6 | Gas Mixtures and Partial Pressures

VCL Dalton's Law of Partial Pressure

Thus far we have considered only the behavior of pure gases—those that consist of only one substance in the gaseous state. How do we deal with gases composed of a mixture of two or more different substances? While studying the properties of air, John Dalton (Section 2.1) observed that the *total pressure of a mixture of gases equals the sum of the pressures that each would exert if it were present alone*. The pressure exerted by a particular component of a mixture of gases is called the **partial pressure** of that gas, and Dalton's observation is known as **Dalton's law of partial pressures**.

GIVE IT SOME THOUGHT

How is the pressure exerted by N_2 gas affected when some O_2 is introduced into a container if the temperature and volume remain constant?

ACTIVITY Partial Pressures

If we let P_t be the total pressure of a mixture of gases and P_1, P_2, P_3, and so forth be the partial pressures of the individual gases, we can write Dalton's law as follows:

$$P_t = P_1 + P_2 + P_3 + \cdots \qquad [10.12]$$

This equation implies that each gas in the mixture behaves independently of the others, as we can see by the following analysis. Let n_1, n_2, n_3, and so forth be the number of moles of each of the gases in the mixture and n_t be the total number of moles of gas ($n_t = n_1 + n_2 + n_3 + \cdots$).

If each of the gases obeys the ideal-gas equation, we can write

$$P_1 = n_1\left(\frac{RT}{V}\right); \quad P_2 = n_2\left(\frac{RT}{V}\right); \quad P_3 = n_3\left(\frac{RT}{V}\right); \quad \text{and so forth.}$$

All the gases in the mixture are at the same temperature and occupy the same volume. Therefore, by substituting into Equation 10.12, we obtain

$$P_t = (n_1 + n_2 + n_3 + \cdots)\frac{RT}{V} = n_t\left(\frac{RT}{V}\right) \qquad [10.13]$$

That is, the total pressure at constant temperature and constant volume is determined by the total number of moles of gas present, whether that total represents just one substance or a mixture.

SAMPLE EXERCISE 10.10 | Applying Dalton's Law of Partial Pressures

A gaseous mixture made from 6.00 g O_2 and 9.00 g CH_4 is placed in a 15.0-L vessel at 0°C. What is the partial pressure of each gas, and what is the total pressure in the vessel?

Solution

Analyze: We need to calculate the pressure for two different gases in the same volume and at the same temperature.

Plan: Because each gas behaves independently, we can use the ideal-gas equation to calculate the pressure that each would exert if the other were not present. The total pressure is the sum of these two partial pressures.

Solve: We must first convert the mass of each gas to moles:

$$n_{O_2} = (6.00 \text{ g } O_2)\left(\frac{1 \text{ mol } O_2}{32.0 \text{ g } O_2}\right) = 0.188 \text{ mol } O_2$$

$$n_{CH_4} = (9.00 \text{ g } CH_4)\left(\frac{1 \text{ mol } CH_4}{16.0 \text{ g } CH_4}\right) = 0.563 \text{ mol } CH_4$$

We can now use the ideal-gas equation to calculate the partial pressure of each gas:

$$P_{O_2} = \frac{n_{O_2}RT}{V} = \frac{(0.188\ \text{mol})(0.0821\ \text{L-atm/mol-K})(273\ \text{K})}{15.0\ \text{L}} = 0.281\ \text{atm}$$

$$P_{CH_4} = \frac{n_{CH_4}RT}{V} = \frac{(0.563\ \text{mol})(0.0821\ \text{L-atm/mol-K})(273\ \text{K})}{15.0\ \text{L}} = 0.841\ \text{atm}$$

According to Dalton's law (Equation 10.12), the total pressure in the vessel is the sum of the partial pressures:

$$P_t = P_{O_2} + P_{CH_4} = 0.281\ \text{atm} + 0.841\ \text{atm} = 1.122\ \text{atm}$$

Check: Performing rough estimates is good practice, even when you may not feel that you need to do it to check an answer. In this case a pressure of roughly 1 atm seems right for a mixture of about 0.2 mol O_2 (6/32) and a bit more than 0.5 mol CH_4 (9/16), together in a 15-L volume, because one mole of an ideal gas at 1 atm pressure and 0°C occupies about 22 L.

PRACTICE EXERCISE

What is the total pressure exerted by a mixture of 2.00 g of H_2 and 8.00 g of N_2 at 273 K in a 10.0-L vessel?
Answer: 2.86 atm

Partial Pressures and Mole Fractions

Because each gas in a mixture behaves independently, we can relate the amount of a given gas in a mixture to its partial pressure. For an ideal gas, $P = nRT/V$, and so we can write

$$\frac{P_1}{P_t} = \frac{n_1RT/V}{n_tRT/V} = \frac{n_1}{n_t} \qquad [10.14]$$

The ratio n_1/n_t is called the mole fraction of gas 1, which we denote X_1. The **mole fraction**, X, is a dimensionless number that expresses the ratio of the number of moles of one component to the total number of moles in the mixture. We can rearrange Equation 10.14 to give

$$P_1 = \left(\frac{n_1}{n_t}\right)P_t = X_1P_t \qquad [10.15]$$

Thus, the partial pressure of a gas in a mixture is its mole fraction times the total pressure.

The mole fraction of N_2 in air is 0.78 (that is, 78% of the molecules in air are N_2). If the total barometric pressure is 760 torr, then the partial pressure of N_2 is

$$P_{N_2} = (0.78)(760\ \text{torr}) = 590\ \text{torr}$$

This result makes intuitive sense: Because N_2 makes up 78% of the mixture, it contributes 78% of the total pressure.

SAMPLE EXERCISE 10.11 | Relating Mole Fractions and Partial Pressures

A study of the effects of certain gases on plant growth requires a synthetic atmosphere composed of 1.5 mol percent CO_2, 18.0 mol percent O_2, and 80.5 mol percent Ar. **(a)** Calculate the partial pressure of O_2 in the mixture if the total pressure of the atmosphere is to be 745 torr. **(b)** If this atmosphere is to be held in a 120-L space at 295 K, how many moles of O_2 are needed?

Solution

Analyze: **(a)** We first need to calculate the partial pressure of O_2 given its mole percentage and the total pressure of the mixture. **(b)** We need to calculate the number of moles of O_2 in the mixture given its volume (120 L), temperature (745 torr), and partial pressure (from part (a)).

Plan: **(a)** We will calculate the partial pressures using Equation 10.15. **(b)** We will then use P_{O_2}, V, and T together with the ideal-gas equation to calculate the number of moles of O_2, n_{O_2}.

Solve: **(a)** The mole percent is just the mole fraction times 100. Therefore, the mole fraction of O_2 is 0.180. Using Equation 10.15, we have

$$P_{O_2} = (0.180)(745 \text{ torr}) = 134 \text{ torr}$$

(b) Tabulating the given variables and changing them to appropriate units, we have

$$P_{O_2} = (134 \text{ torr})\left(\frac{1 \text{ atm}}{760 \text{ torr}}\right) = 0.176 \text{ atm}$$

$$V = 120 \text{ L}$$

$$n_{O_2} = ?$$

$$R = 0.0821 \frac{\text{L-atm}}{\text{mol-K}}$$

$$T = 295 \text{ K}$$

Solving the ideal-gas equation for n_{O_2}, we have

$$n_{O_2} = P_{O_2}\left(\frac{V}{RT}\right)$$

$$= (0.176 \text{ atm})\frac{120 \text{ L}}{(0.0821 \text{ L-atm/mol-K})(295 \text{ K})} = 0.872 \text{ mol}$$

Check: The units check out satisfactorily, and the answer seems to be the right order of magnitude.

PRACTICE EXERCISE

From data gathered by *Voyager 1*, scientists have estimated the composition of the atmosphere of Titan, Saturn's largest moon. The total pressure on the surface of Titan is 1220 torr. The atmosphere consists of 82 mol percent N_2, 12 mol percent Ar, and 6.0 mol percent CH_4. Calculate the partial pressure of each of these gases in Titan's atmosphere.

Answer: 1.0×10^3 torr N_2, 1.5×10^2 torr Ar, and 73 torr CH_4

Collecting Gases over Water

An experiment that is often encountered in general chemistry laboratories involves determining the number of moles of gas collected from a chemical reaction. Sometimes this gas is collected over water. For example, solid potassium chlorate, $KClO_3$, can be decomposed by heating it in a test tube in an arrangement such as that shown in Figure 10.16 ▼. The balanced equation for the reaction is

$$2\ KClO_3(s) \longrightarrow 2\ KCl(s) + 3\ O_2(g) \qquad [10.16]$$

The oxygen gas is collected in a bottle that is initially filled with water and inverted in a water pan.

The volume of gas collected is measured by raising or lowering the bottle as necessary until the water levels inside and outside the bottle are the same. When this condition is met, the pressure inside the bottle is equal to the atmospheric pressure outside. The total pressure inside is the sum of the pressure of gas collected and the pressure of water vapor in equilibrium with liquid water.

$$P_{total} = P_{gas} + P_{H_2O} \qquad [10.17]$$

The pressure exerted by water vapor, P_{H_2O}, at various temperatures is listed in Appendix B.

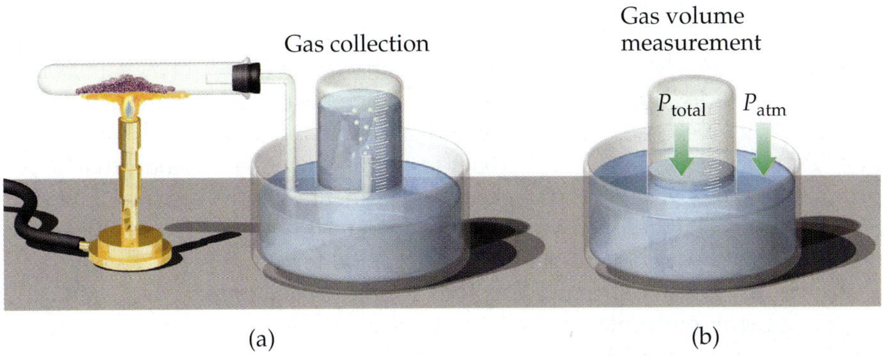

◀ **Figure 10.16 Collecting a water-insoluble gas over water.** (a) A solid is heated, releasing a gas, which is bubbled through water into a collection bottle. (b) When the gas has been collected, the bottle is raised or lowered so that the water levels inside and outside the bottle are equal. The total pressure of the gases inside the bottle is then equal to the atmospheric pressure.

SAMPLE EXERCISE 10.12 | Calculating the Amount of Gas Collected over Water

A sample of $KClO_3$ is partially decomposed (Equation 10.16), producing O_2 gas that is collected over water as in Figure 10.16. The volume of gas collected is 0.250 L at 26°C and 765 torr total pressure. **(a)** How many moles of O_2 are collected? **(b)** How many grams of $KClO_3$ were decomposed?

Solution

(a) Analyze: We need to calculate the number of moles of O_2 gas in a container that also contains water vapor.

Plan: If we tabulate the information presented, we will see that values are given for V and T. In order to use the ideal-gas equation to calculate the unknown, n_{O_2}, we also must know the partial pressure of O_2 in the system. We can calculate the partial pressure of O_2 from the total pressure (765 torr) and the vapor pressure of water.

Solve: The partial pressure of the O_2 gas is the difference between the total pressure, 765 torr, and the pressure of the water vapor at 26°C, 25 torr (Appendix B):

$$P_{O_2} = 765 \text{ torr} - 25 \text{ torr} = 740 \text{ torr}$$

We can use the ideal-gas equation to calculate the number of moles of O_2:

$$n_{O_2} = \frac{P_{O_2}V}{RT} = \frac{(740 \text{ torr})(1 \text{ atm}/760 \text{ torr})(0.250 \text{ L})}{(0.0821 \text{ L-atm/mol-K})(299 \text{ K})} = 9.92 \times 10^{-3} \text{ mol } O_2$$

(b) Analyze: We now need to calculate the number of moles of reactant $KClO_3$ decomposed.

Plan: We can use the number of moles of O_2 formed and the balanced chemical equation to determine the number of moles of $KClO_3$ decomposed, which we can then convert to grams of $KClO_3$.

Solve: From Equation 10.16, we have 2 mol $KClO_3 \simeq 3$ mol O_2. The molar mass of $KClO_3$ is 122.6 g/mol. Thus, we can convert the moles of O_2 that we found in part (a) to moles of $KClO_3$ and then to grams of $KClO_3$:

$$(9.92 \times 10^{-3} \text{ mol } O_2)\left(\frac{2 \text{ mol } KClO_3}{3 \text{ mol } O_2}\right)\left(\frac{122.6 \text{ g } KClO_3}{1 \text{ mol } KClO_3}\right) = 0.811 \text{ g } KClO_3$$

Check: As always, we make sure that the units cancel appropriately in the calculations. In addition, the numbers of moles of O_2 and $KClO_3$ seem reasonable, given the small volume of gas collected.

Comment: Many chemical compounds that react with water and water vapor would be degraded by exposure to wet gas. Thus, in research laboratories gases are often dried by passing wet gas over a substance that absorbs water (a *desiccant*), such as calcium sulfate, $CaSO_4$. Calcium sulfate crystals are sold as a desiccant under the trade name Drierite™.

PRACTICE EXERCISE

Ammonium nitrite, NH_4NO_2, decomposes upon heating to form N_2 gas:

$$NH_4NO_2(s) \longrightarrow N_2(g) + 2\, H_2O(l)$$

When a sample of NH_4NO_2 is decomposed in a test tube, as in Figure 10.16, 511 mL of N_2 gas is collected over water at 26°C and 745 torr total pressure. How many grams of NH_4NO_2 were decomposed?

Answer: 1.26 g

10.7 | Kinetic-Molecular Theory

The ideal-gas equation describes *how* gases behave, but it doesn't explain *why* they behave as they do. Why does a gas expand when heated at constant pressure? Or, why does its pressure increase when the gas is compressed at constant temperature? To understand the physical properties of gases, we need a model that helps us picture what happens to gas particles as experimental conditions such as pressure or temperature change. Such a model, known as the **kinetic-molecular theory**, was developed over a period of about 100 years, culminating in 1857 when Rudolf Clausius (1822–1888) published a complete and satisfactory form of the theory.

The kinetic-molecular theory (the theory of moving molecules) is summarized by the following statements:

1. Gases consist of large numbers of molecules that are in continuous, random motion. (The word *molecule* is used here to designate the smallest particle of any gas; some gases, such as the noble gases, consist of individual atoms.)
2. The combined volume of all the molecules of the gas is negligible relative to the total volume in which the gas is contained.
3. Attractive and repulsive forces between gas molecules are negligible.
4. Energy can be transferred between molecules during collisions, but the *average* kinetic energy of the molecules does not change with time, as long as the temperature of the gas remains constant. In other words, the collisions are perfectly elastic.
5. The average kinetic energy of the molecules is proportional to the absolute temperature. At any given temperature the molecules of all gases have the same average kinetic energy.

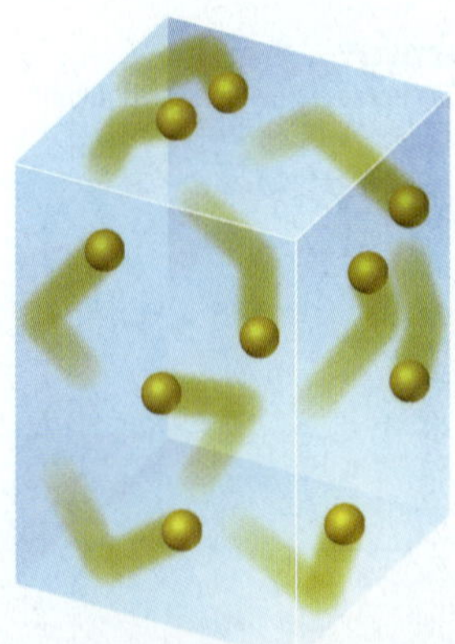

▲ **Figure 10.17 The molecular origin of gas pressure.** The pressure exerted by a gas is caused by collisions of the gas molecules with the walls of their container.

ANIMATION
Kinetic Energy of a Gas

The kinetic-molecular theory explains both pressure and temperature at the molecular level. The pressure of a gas is caused by collisions of the molecules with the walls of the container, as shown in Figure 10.17 ▶. The magnitude of the pressure is determined by both how often and how forcefully the molecules strike the walls.

The absolute temperature of a gas is a measure of the *average* kinetic energy of its molecules. If two different gases are at the same temperature, their molecules have the same average kinetic energy (statement 5 of the kinetic-molecular theory). If the absolute temperature of a gas is doubled, the average kinetic energy of its molecules doubles. Thus, molecular motion increases with increasing temperature.

Although the molecules in a sample of gas have an *average* kinetic energy and hence an average speed, the individual molecules move at varying speeds. The moving molecules collide frequently with other molecules. Momentum is conserved in each collision, but one of the colliding molecules might be deflected off at high speed while the other is nearly stopped. The result is that the molecules at any instant have a wide range of speeds. Figure 10.18 ▶ illustrates the distribution of molecular speeds for nitrogen gas at 0°C (blue line) and at 100°C (red line). The curve shows us the fraction of molecules moving at each speed. At higher temperatures, a larger fraction of molecules moves at greater speeds; the distribution curve has shifted to the right toward higher speeds and hence toward higher average kinetic energy. The peak of each curve represents the most probable speed (the speed of the largest number of molecules). Notice that the blue curve (0°C) has a peak at about 4×10^2 m/s, whereas the red curve (100°C) has a peak at a higher speed, about 5×10^2 m/s.

ACTIVITY
Motions of a Gas

▼ **Figure 10.18 The effect of temperature on molecular speeds.** Distribution of molecular speeds for nitrogen at 0°C (blue line) and 100°C (red line). Increasing temperature increases both the most probable speed (curve maximum) and the rms speed, *u*, which is indicated by the vertical dashed line.

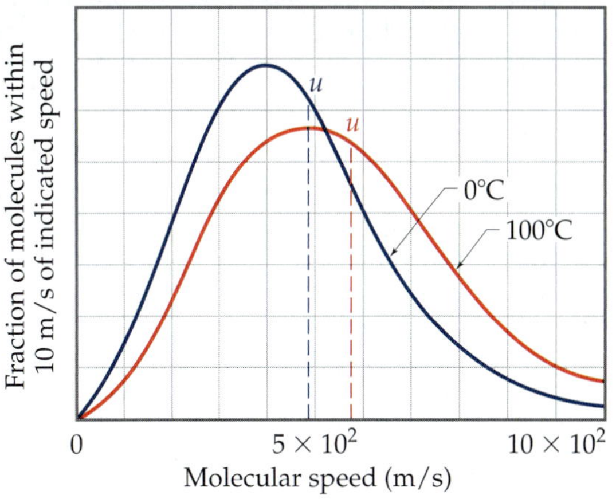

Figure 10.18 also shows the value of the **root-mean-square (rms) speed**, u, of the molecules at each temperature. This quantity is the speed of a molecule possessing average kinetic energy. The rms speed is not quite the same as the average (mean) speed. The difference between the two, however, is small.* Notice that the rms speed is higher at 100°C than at 0°C.

The rms speed is important because the average kinetic energy of the gas molecules in a sample, ϵ, is related directly to u^2:

$$\epsilon = \tfrac{1}{2}mu^2 \qquad [10.18]$$

where m is the mass of the molecule. Mass does not change with temperature. Thus, the increase in the average kinetic energy as the

* To illustrate the difference between rms speed and average speed, suppose that we have four objects with speeds of 4.0, 6.0, 10.0, and 12.0 m/s. Their average speed is $\frac{1}{4}(4.0 + 6.0 + 10.0 + 12.0) = 8.0$ m/s. The rms speed, u, however, is the square root of the average squared speeds of the molecules:

$$\sqrt{\tfrac{1}{4}(4.0^2 + 6.0^2 + 10.0^2 + 12.0^2)} = \sqrt{74.0} = 8.6 \text{ m/s}$$

For an ideal gas the average speed equals $0.921 \times u$. Thus, the average speed is directly proportional to the rms speed, and the two are in fact nearly equal.

ACTIVITY
Gas Phase Boltzmann Distribution

temperature increases implies that the rms speed (and also the average speed) of molecules likewise increases as temperature increases.

GIVE IT SOME THOUGHT

Consider three samples of gas: HCl at 298 K, H_2 at 298 K, and O_2 at 350 K. Compare the average kinetic energies of the molecules in the three samples.

Application to the Gas Laws

The empirical observations of gas properties as expressed in the various gas laws are readily understood in terms of the kinetic-molecular theory. The following examples illustrate this point:

1. *Effect of a volume increase at constant temperature:* A constant temperature means that the average kinetic energy of the gas molecules remains unchanged. This in turn means that the rms speed of the molecules, u, is unchanged. If the volume is increased, however, the molecules must move a longer distance between collisions. Consequently, there are fewer collisions per unit time with the container walls, and pressure decreases. Thus, the model accounts in a simple way for Boyle's law.
2. *Effect of a temperature increase at constant volume:* An increase in temperature means an increase in the average kinetic energy of the molecules and thus an increase in u. If there is no change in volume, there will be more collisions with the walls per unit time. Furthermore, the change in momentum in each collision increases (the molecules strike the walls more forcefully). Hence, the model explains the observed pressure increase.

A CLOSER LOOK | The Ideal-Gas Equation

Beginning with the five statements given in the text for the kinetic-molecular theory, it is possible to derive the ideal-gas equation. Rather than proceed through a derivation, let's consider in somewhat qualitative terms how the ideal-gas equation might follow. As we have seen, pressure is force per unit area. (Section 10.2) The total force of the molecular collisions on the walls and hence the pressure produced by these collisions depend both on how strongly the molecules strike the walls (impulse imparted per collision) and on the rate at which these collisions occur:

$$P \propto \text{impulse imparted per collision} \times \text{rate of collisions}$$

For a molecule traveling at the rms speed, u, the impulse imparted by a collision with a wall depends on the momentum of the molecule; that is, it depends on the product of its mass and speed, mu. The rate of collisions is proportional to both the number of molecules per unit volume, n/V, and their speed, u. If there are more molecules in a container, there will be more frequent collisions with the container walls. As the molecular speed increases or the volume of the container decreases, the time required for molecules to traverse the distance from one wall to another is reduced, and the molecules collide more frequently with the walls. Thus, we have

$$P \propto mu \times \frac{n}{V} \times u \propto \frac{nmu^2}{V} \quad [10.19]$$

Because the average kinetic energy, $\frac{1}{2}mu^2$, is proportional to temperature, we have $mu^2 \propto T$. Making this substitution into Equation 10.19 gives

$$P \propto \frac{n(mu^2)}{V} \propto \frac{nT}{V} \quad [10.20]$$

Let's now convert the proportionality sign to an equal sign by expressing n as the number of moles of gas; we then insert a proportionality constant—R, the gas constant:

$$P = \frac{nRT}{V} \quad [10.21]$$

This expression is the ideal-gas equation.

An eminent Swiss mathematician, Daniel Bernoulli (1700–1782), conceived of a model for gases that was, for all practical purposes, the same as the kinetic theory model. From this model, Bernoulli derived Boyle's law and the ideal-gas equation. His was one of the first examples in science of developing a mathematical model from a set of assumptions, or hypothetical statements. However, Bernoulli's work on this subject was completely ignored, only to be rediscovered a hundred years later by Clausius and others. It was ignored because it conflicted with popular beliefs and was in conflict with Isaac Newton's incorrect model for gases. Those idols of the times had to fall before the way was clear for the kinetic-molecular theory. As this story illustrates, science is not a straight road running from here to the "truth." The road is built by humans, so it zigs and zags.

Related Exercises: 10.69, 10.70, 10.71 and 10.72

SAMPLE EXERCISE 10.13 | Applying the Kinetic-Molecular Theory

A sample of O_2 gas initially at STP is compressed to a smaller volume at constant temperature. What effect does this change have on **(a)** the average kinetic energy of O_2 molecules, **(b)** the average speed of O_2 molecules, **(c)** the total number of collisions of O_2 molecules with the container walls in a unit time, **(d)** the number of collisions of O_2 molecules with a unit area of container wall per unit time?

Solution

Analyze: We need to apply the concepts of the kinetic-molecular theory to a situation in which a gas is compressed at constant temperature.

Plan: We will determine how each of the quantities in (a)–(d) is affected by the change in volume at constant temperature.

Solve: **(a)** The average kinetic energy of the O_2 molecules is determined only by temperature. Thus the average kinetic energy is unchanged by the compression of O_2 at constant temperature. **(b)** If the average kinetic energy of O_2 molecules doesn't change, the average speed remains constant. **(c)** The total number of collisions with the container walls per unit time must increase because the molecules are moving within a smaller volume but with the same average speed as before. Under these conditions they must encounter a wall more frequently. **(d)** The number of collisions with a unit area of wall per unit time increases because the total number of collisions with the walls per unit time increases and the area of the walls decreases.

Check: In a conceptual exercise of this kind, there is no numerical answer to check. All we can check in such cases is our reasoning in the course of solving the problem.

PRACTICE EXERCISE

How is the rms speed of N_2 molecules in a gas sample changed by **(a)** an increase in temperature, **(b)** an increase in volume, **(c)** mixing with a sample of Ar at the same temperature?

Answers: **(a)** increases, **(b)** no effect, **(c)** no effect

10.8 | Molecular Effusion and Diffusion

According to the kinetic-molecular theory, the average kinetic energy of *any* collection of gas molecules, $\frac{1}{2}mu^2$, has a specific value at a given temperature. Thus, a gas composed of light particles, such as He, will have the same average kinetic energy as one composed of much heavier particles, such as Xe, provided the two gases are at the same temperature. The mass, m, of the particles in the lighter gas is smaller than that in the heavier gas. Consequently, the particles of the lighter gas must have a higher rms speed, u, than the particles of the heavier one. The following equation, which expresses this fact quantitatively, can be derived from kinetic-molecular theory:

ACTIVITY
Gas Phase Boltzmann Distribution

$$u = \sqrt{\frac{3RT}{\mathcal{M}}} \qquad [10.22]$$

Because the molar mass, $\mathcal{M}$, appears in the denominator, the less massive the gas molecules, the higher the rms speed, u. Figure 10.19 ▼ shows the distribution

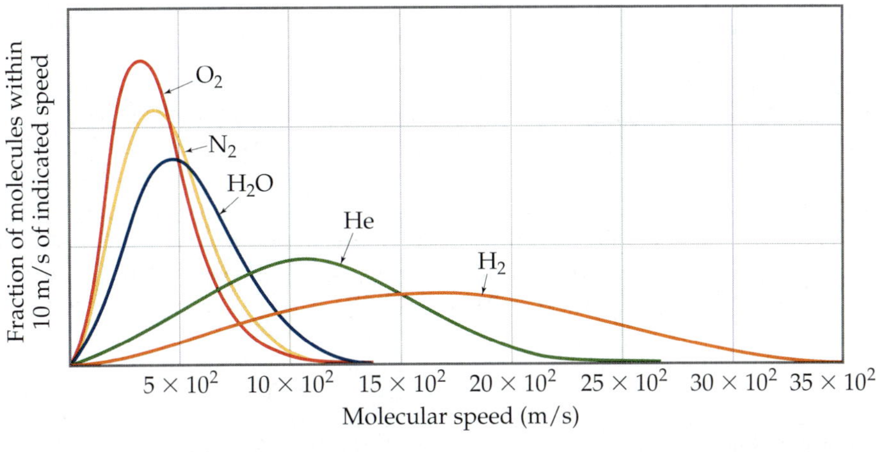

◀ **Figure 10.19 The effect of molecular mass on molecular speeds.** The distributions of molecular speeds for different gases are compared at 25°C. The molecules with lower molecular masses have higher rms speeds.

of molecular speeds for several gases at 25°C. Notice how the distributions are shifted toward higher speeds for gases of lower molar masses.

SAMPLE EXERCISE 10.14 | Calculating a Root-Mean-Square Speed

Calculate the rms speed, u, of an N_2 molecule at 25°C.

Solution

Analyze: We are given the identity of the gas and the temperature, the two quantities we need to calculate the rms speed.

Plan: We will calculate the rms speed using Equation 10.22.

Solve: In using Equation 10.22, we should convert each quantity to SI units so that all the units are compatible. We will also use R in units of J/mol-K (Table 10.2) in order to make the units cancel correctly.

$$T = 25 + 273 = 298 \text{ K}$$

$$\mathcal{M} = 28.0 \text{ g/mol} = 28.0 \times 10^{-3} \text{ kg/mol}$$

$$R = 8.314 \text{ J/mol-K} = 8.314 \text{ kg-m}^2/\text{s}^2\text{-mol-K}$$ (These units follow from the fact that $1 \text{ J} = 1 \text{ kg-m}^2/\text{s}^2$)

$$u = \sqrt{\frac{3(8.314 \text{ kg-m}^2/\text{s}^2\text{-mol-K})(298 \text{ K})}{28.0 \times 10^{-3} \text{ kg/mol}}} = 5.15 \times 10^2 \text{ m/s}$$

Comment: This corresponds to a speed of 1150 mi/hr. Because the average molecular weight of air molecules is slightly greater than that of N_2, the rms speed of air molecules is a little slower than that for N_2. The speed at which sound propagates through air is about 350 m/s, a value about two-thirds the average rms speed for air molecules.

PRACTICE EXERCISE

What is the rms speed of an He atom at 25°C?

Answer: 1.36×10^3 m/s

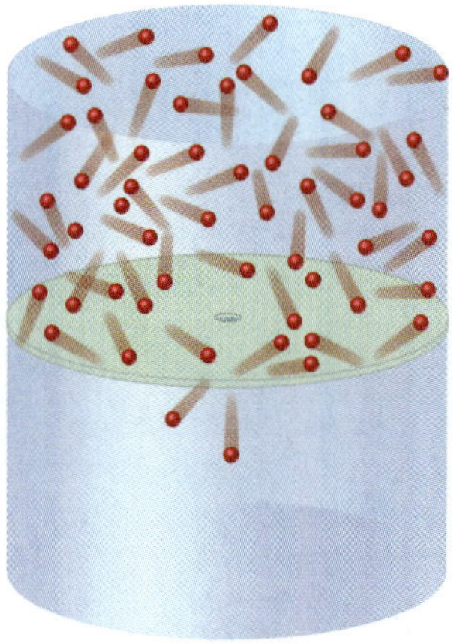

▲ **Figure 10.20 Effusion.** The top half of this cylinder is filled with a gas, and the bottom half is an evacuated space. Gas molecules effuse through a pinhole in the partitioning wall only when they happen to hit the hole.

ACTIVITY
Gas Diffusion and Effusion

The dependence of molecular speeds on mass has several interesting consequences. The first phenomenon is **effusion**, which is the escape of gas molecules through a tiny hole into an evacuated space as shown in Figure 10.20 ◄. The second is **diffusion**, which is the spread of one substance throughout a space or throughout a second substance. For example, the molecules of a perfume diffuse throughout a room.

Graham's Law of Effusion

In 1846 Thomas Graham (1805–1869) discovered that the effusion rate of a gas is inversely proportional to the square root of its molar mass. Assume that we have two gases at the same temperature and pressure in containers with identical pinholes. If the rates of effusion of the two substances are r_1 and r_2 and their respective molar masses are $\mathcal{M}_1$ and $\mathcal{M}_2$, **Graham's law** states

$$\frac{r_1}{r_2} = \sqrt{\frac{\mathcal{M}_2}{\mathcal{M}_1}} \quad [10.23]$$

Equation 10.23 compares the *rates* of effusion of two different gases under identical conditions, and it indicates that the lighter gas effuses more rapidly.

Figure 10.20 illustrates the basis of Graham's law. The only way for a molecule to escape from its container is for it to "hit" the hole in the partitioning wall. The faster the molecules are moving, the greater is the likelihood that a molecule will hit the hole and effuse. This implies that the rate of effusion is directly proportional to the rms speed of the molecules. Because R and T are constant, we have, from Equation 10.22

$$\frac{r_1}{r_2} = \frac{u_1}{u_2} = \sqrt{\frac{3RT/\mathcal{M}_1}{3RT/\mathcal{M}_2}} = \sqrt{\frac{\mathcal{M}_2}{\mathcal{M}_1}} \quad [10.24]$$

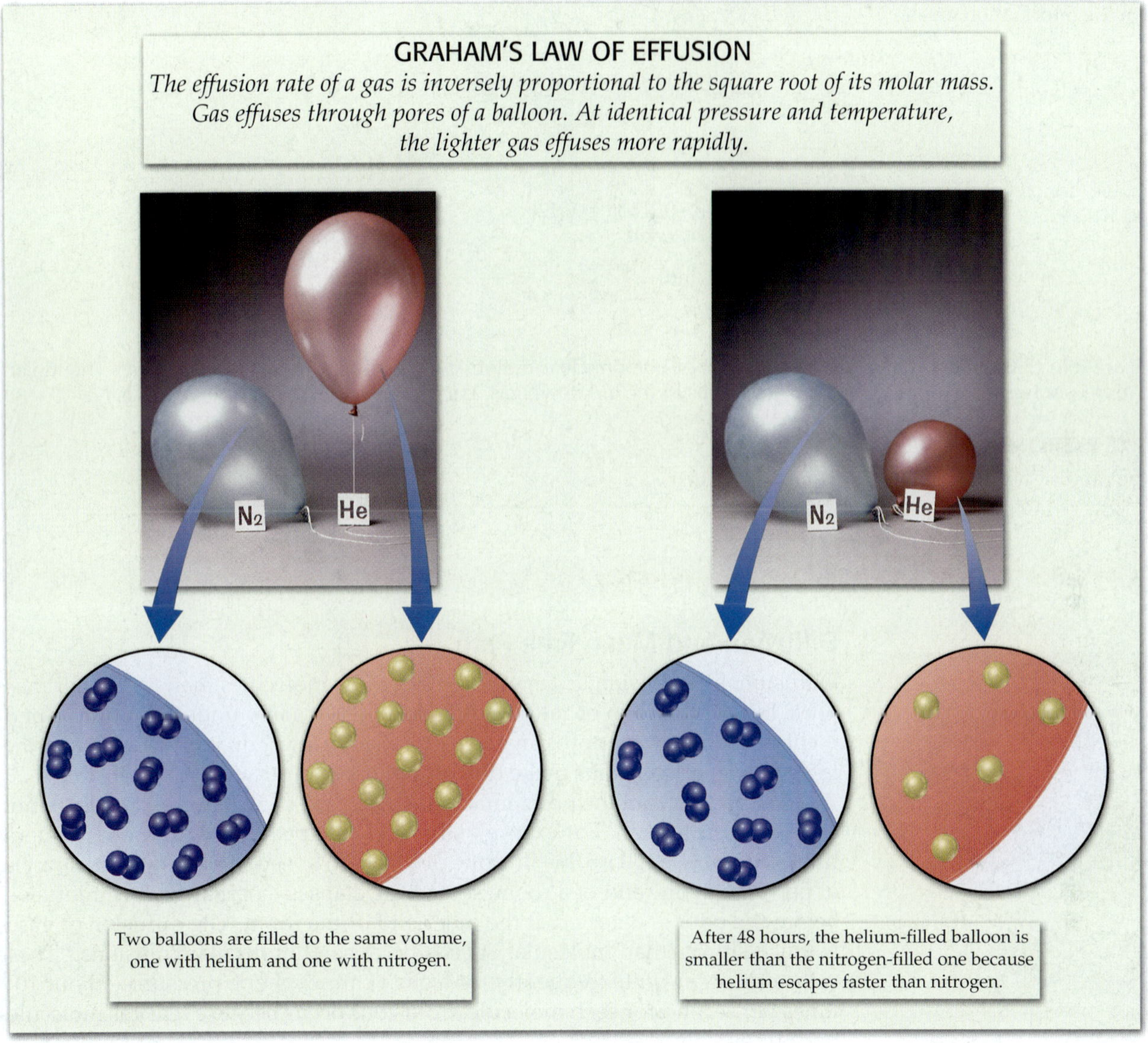

▲ **Figure 10.21 An illustration of Graham's law.**

As expected from Graham's law, helium escapes from containers through tiny pinhole leaks more rapidly than other gases of higher molecular weight (Figure 10.21 ▲).

SAMPLE EXERCISE 10.15 | Applying Graham's Law

An unknown gas composed of homonuclear diatomic molecules effuses at a rate that is only 0.355 times that of O_2 at the same temperature. Calculate the molar mass of the unknown, and identify it.

Solution

Analyze: We are given the rate of effusion of an unknown gas relative to that of O_2, and we are asked to find the molar mass and identity of the unknown. Thus, we need to connect relative rates of effusion to relative molar masses.

Plan: We can use Graham's law of effusion, Equation 10.23, to determine the molar mass of the unknown gas. If we let r_x and $\mathcal{M}_x$ represent the rate of effusion and molar mass of the unknown gas, Equation 10.23 can be written as follows:

$$\frac{r_x}{r_{O_2}} = \sqrt{\frac{\mathcal{M}_{O_2}}{\mathcal{M}_x}}$$

Solve: From the information given,

$$r_x = 0.355 \times r_{O_2}$$

Thus,

$$\frac{r_x}{r_{O_2}} = 0.355 = \sqrt{\frac{32.0\ \text{g/mol}}{\mathcal{M}_x}}$$

We now solve for the unknown molar mass, $\mathcal{M}_x$.

$$\frac{32.0\ \text{g/mol}}{\mathcal{M}_x} = (0.355)^2 = 0.126$$

$$\mathcal{M}_x = \frac{32.0\ \text{g/mol}}{0.126} = 254\ \text{g/mol}$$

Because we are told that the unknown gas is composed of homonuclear diatomic molecules, it must be an element. The molar mass must represent twice the atomic weight of the atoms in the unknown gas. We conclude that the unknown gas is I_2.

PRACTICE EXERCISE

Calculate the ratio of the effusion rates of N_2 and O_2, r_{N_2}/r_{O_2}.

Answer: $r_{N_2}/r_{O_2} = 1.07$

Diffusion and Mean Free Path

Diffusion, like effusion, is faster for lower mass molecules than for higher mass ones. In fact, the ratio of rates of diffusion of two gases under identical experimental conditions is approximated by Graham's law, Equation 10.23. Nevertheless, molecular collisions make diffusion more complicated than effusion.

We can see from the horizontal scale in Figure 10.19 that the speeds of molecules are quite high. For example, the average speed of N_2 at room temperature is 515 m/s (1150 mi/hr). In spite of this high speed, if someone opens a vial of perfume at one end of a room, some time elapses—perhaps a few minutes—before the odor is detected at the other end of the room. The diffusion of gases is much slower than molecular speeds because of molecular collisions.* These collisions occur quite frequently for a gas at atmospheric pressure—about 10^{10} times per second for each molecule. Collisions occur because real gas molecules have finite volumes.

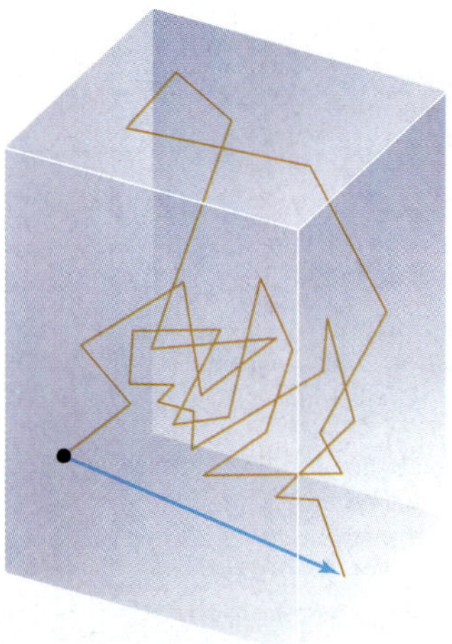

▲ **Figure 10.22 Diffusion of a gas molecule.** For clarity, no other gas molecules in the container are shown. The path of the molecule of interest begins at the dot. Each short segment of line represents travel between collisions. The blue arrow indicates the net distance traveled by the molecule.

MOVIE
Diffusion of Bromine Vapor

Because of molecular collisions, the direction of motion of a gas molecule is constantly changing. Therefore, the diffusion of a molecule from one point to another consists of many short, straight-line segments as collisions buffet it around in random directions, as depicted in Figure 10.22 ◀. First the molecule moves in one direction, then in another; one instant at high speed, the next at low speed.

The average distance traveled by a molecule between collisions is called the **mean free path** of the molecule. The mean free path varies with pressure as the following analogy illustrates. Imagine walking through a shopping mall. When the mall is very crowded (high pressure), the average distance you can walk before bumping into someone is short (short mean free path). When the mall is empty (low pressure), you can walk a long way (long mean free path) before bumping into someone. The mean free path for air molecules at sea level is about 60 nm (6×10^{-8} m). At about 100 km in altitude, where the air density is much lower, the mean free path is about 10 cm, about 1 million times longer than at Earth's surface.

* The rate at which the perfume moves across the room also depends on how well stirred the air is from temperature gradients and the movement of people. Nevertheless, even with the aid of these factors, it still takes much longer for the molecules to traverse the room than one would expect from the rms speed alone.

CHEMISTRY AT WORK | Gas Separations

The fact that lighter molecules move at higher average speeds than more massive ones has many interesting consequences and applications. For example, the effort to develop the atomic bomb during World War II required scientists to separate the relatively low-abundance uranium isotope ^{235}U (0.7%) from the much more abundant ^{238}U (99.3%). This was accomplished by converting the uranium into a volatile compound, UF_6, that was then allowed to pass through porous barriers. Because of the diameters of the pores, this is not a simple effusion. Nevertheless, the dependence on molar mass is essentially the same. The slight difference in molar mass between the two hexafluorides, $^{235}UF_6$ and $^{238}UF_6$, caused the molecules to move at slightly different rates:

$$\frac{r_{235}}{r_{238}} = \sqrt{\frac{352.04}{349.03}} = 1.0043$$

Thus, the gas initially appearing on the opposite side of the barrier was very slightly enriched in the lighter molecule. The diffusion process was repeated thousands of times, leading to a nearly complete separation of the two isotopes of uranium.
Related Exercises: 10.77 and 10.78

GIVE IT SOME THOUGHT

Will the following changes increase, decrease, or have no effect on the mean free path of the gas molecules in a sample of gas? (a) Increasing pressure, (b) increasing temperature?

10.9 | Real Gases: Deviations from Ideal Behavior

Although the ideal-gas equation is a very useful description of gases, all real gases fail to obey the relationship to some degree. The extent to which a real gas departs from ideal behavior can be seen by rearranging the ideal-gas equation to solve for n:

$$\frac{PV}{RT} = n \qquad [10.25]$$

For one mole of ideal gas ($n = 1$) the quantity PV/RT equals 1 at all pressures. In Figure 10.23 ▼ PV/RT is plotted as a function of P for one mole of several different gases. At high pressures the deviation from ideal behavior ($PV/RT = 1$) is large and is different for each gas. *Real gases, therefore, do not behave ideally at high pressure.* At lower pressures (usually below 10 atm), however, the deviation from ideal behavior is small, and we can use the ideal-gas equation without generating serious error.

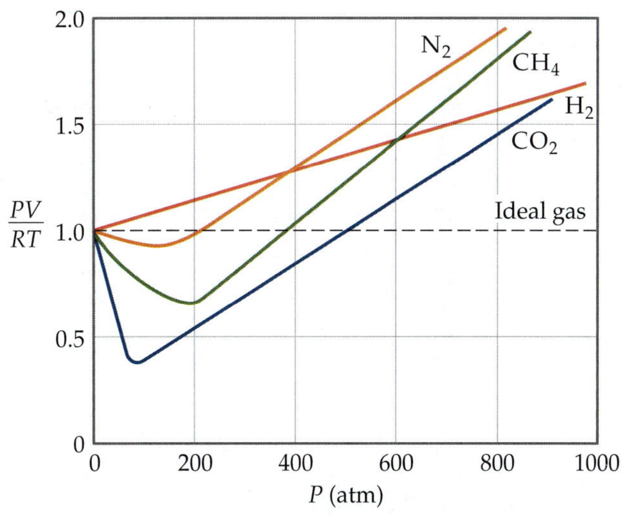

◀ **Figure 10.23 The effect of pressure on the behavior of several gases.** The ratios of PV/RT versus pressure are compared for one mole of several gases at 300 K. The data for CO_2 are at 313 K because under high pressure CO_2 liquefies at 300 K. The dashed horizontal line shows the behavior of an ideal gas.

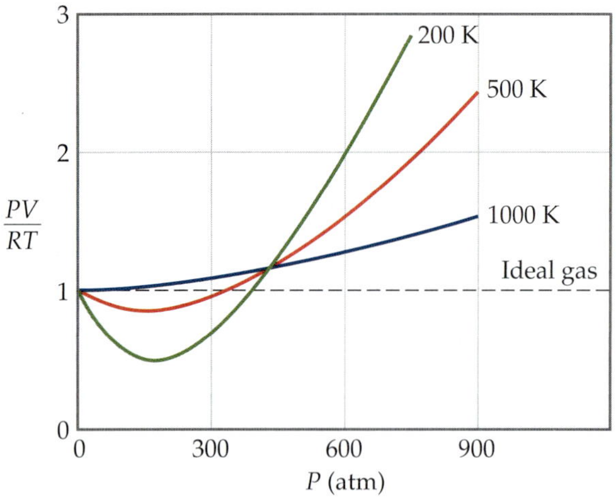

▲ **Figure 10.24 The effect of temperature and pressure on the behavior of nitrogen gas.** The ratios of *PV/RT* versus pressure are shown for 1 mol of nitrogen gas at three temperatures. As temperature increases, the gas more closely approaches ideal behavior, which is represented by the dashed horizontal line.

The deviation from ideal behavior also depends on temperature. Figure 10.24 ◀ shows graphs of PV/RT versus P for 1 mol of N_2 at three temperatures. As temperature increases, the behavior of the gas more nearly approaches that of the ideal gas. In general, *the deviations from ideal behavior increase as temperature decreases*, becoming significant near the temperature at which the gas is converted into a liquid.

GIVE IT SOME THOUGHT

Would you expect helium gas to deviate from ideal behavior more at (a) 100 K and 1 atm, (b) 100 K and 5 atm, or (c) 300 K and 2 atm?

The basic assumptions of the kinetic-molecular theory give us insight into why real gases deviate from ideal behavior. The molecules of an ideal gas are assumed to occupy no space and have no attractions for one another. *Real molecules, however, do have finite volumes, and they do attract one another.* As shown in Figure 10.25 ◀, the free, unoccupied space in which molecules can move is somewhat less than the container volume. At relatively low pressures the volume of the gas molecules is negligible relative to the container volume. Thus, the free volume available to the molecules is essentially the entire volume of the container. As the pressure increases, however, the free space in which the molecules can move becomes a smaller fraction of the container volume. Under these conditions, therefore, gas volumes tend to be slightly greater than those predicted by the ideal-gas equation.

In addition, the attractive forces between molecules come into play at short distances, as when molecules are crowded together at high pressures. Because of these attractive forces, the impact of a given molecule with the wall of the container is lessened. If we could stop the action in a gas, the positions of the molecules might resemble the illustration in Figure 10.26 ▼. The molecule about to make contact with the wall experiences the attractive forces of nearby molecules. These attractions lessen the force with which the molecule hits the wall. As a result, the pressure is less than that of an ideal gas. This effect serves to decrease PV/RT below its ideal value, as seen in Figure 10.23. When the pressure is sufficiently high, however, the volume effects dominate and PV/RT increases to above the ideal value.

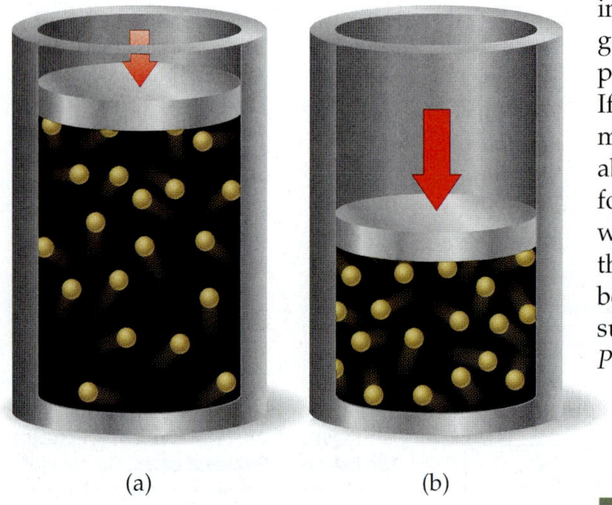

▲ **Figure 10.25 Comparing the volume of gas molecules to the container volume.** In (a), at low pressure, the combined volume of the gas molecules is small relative to the container volume, and we can approximate the empty space between molecules as being equal to the container volume. In (b), at high pressure, the combined volume of the gas molecules is a larger fraction of the total space available. Now we must account for the volume of the molecules themselves in determining the empty space available for the motion of the gas molecules.

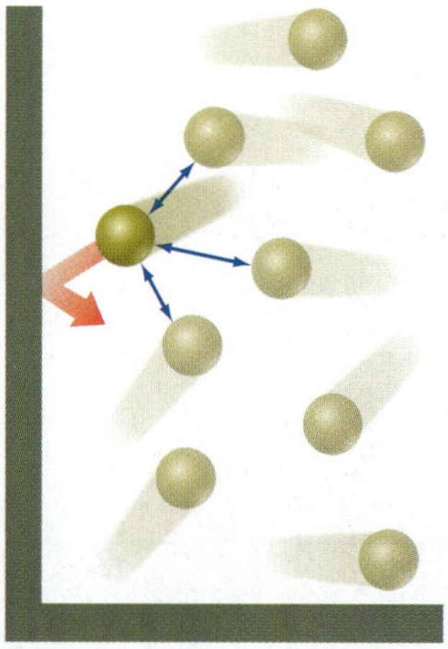

◀ **Figure 10.26 The effect of intermolecular forces on gas pressure.** The molecule that is about to strike the wall experiences attractive forces from nearby gas molecules, and its impact on the wall is thereby lessened. The lessened impact means the molecule exerts a lower-than-expected pressure on the wall. The attractive forces become significant only under high-pressure conditions, when the average distance between molecules is small.

Temperature determines how effective attractive forces between gas molecules are. As a gas is cooled, the average kinetic energy of the molecules decreases but intermolecular attractions remain constant. In a sense, cooling a gas deprives molecules of the energy they need to overcome their mutual attractive influence. The effects of temperature shown in Figure 10.24 illustrate this point very well. As temperature increases, the negative departure of PV/RT from ideal-gas behavior disappears. The difference that remains at high temperature stems mainly from the effect of the finite volumes of the molecules.

GIVE IT SOME THOUGHT

List two reasons why gases deviate from ideal behavior.

The van der Waals Equation

VCL Ideal versus Real Gases

Engineers and scientists who work with gases at high pressures often cannot use the ideal-gas equation to predict the pressure-volume properties of gases because departures from ideal behavior are too large. One useful equation developed to predict the behavior of real gases was proposed by the Dutch scientist Johannes van der Waals (1837–1923).

The ideal-gas equation predicts that the pressure of a gas is

$$P = \frac{nRT}{V} \quad \text{(ideal gas)}$$

ACTIVITY Nonideal Gas Behavior

Van der Waals recognized that for a real gas this expression would have to be corrected for the finite volume occupied by the gas molecules and for the attractive forces between the gas molecules. He introduced two constants, a and b, to make these corrections.

$$P = \frac{nRT}{V - nb} - \frac{n^2a}{V^2} \qquad [10.26]$$

Correction for volume of molecules (the nb term); Correction for molecular attraction (the n^2a/V^2 term)

The volume of the container, V, is decreased by the factor nb, to account for the small but finite volume occupied by the gas molecules themselves (Figure 10.25). Thus, the free volume available to the gas molecules is $V - nb$. The van der Waals constant b is a measure of the actual intrinsic volume occupied by a mole of gas molecules; b has units of L/mol. The pressure is in turn decreased by the factor n^2a/V^2, which accounts for the attractive forces between the gas molecules (Figure 10.26). The unusual form of this correction results because the attractive forces between pairs of molecules increase as the square of the number of molecules per unit volume $(n/V)^2$. Hence, the van der Waals constant a has units of L^2-atm/mol^2. The magnitude of a reflects how strongly the gas molecules attract each other.

Equation 10.26 is generally rearranged to give the following form of the **van der Waals equation**:

$$\left(P + \frac{n^2a}{V^2}\right)(V - nb) = nRT \qquad [10.27]$$

The van der Waals constants a and b are different for each gas. Values of these constants for several gases are listed in Table 10.3 ▶. Note that the values of both a and b generally increase with an increase in mass of the molecule and with an increase in the complexity of its structure. Larger, more massive molecules not only have larger volumes, they also tend to have greater intermolecular attractive forces.

TABLE 10.3 van der Waals Constants for Gas Molecules

Substance	a (L^2-atm/mol^2)	b (L/mol)
He	0.0341	0.02370
Ne	0.211	0.0171
Ar	1.34	0.0322
Kr	2.32	0.0398
Xe	4.19	0.0510
H_2	0.244	0.0266
N_2	1.39	0.0391
O_2	1.36	0.0318
Cl_2	6.49	0.0562
H_2O	5.46	0.0305
CH_4	2.25	0.0428
CO_2	3.59	0.0427
CCl_4	20.4	0.1383

SAMPLE EXERCISE 10.16 | Using the van Der Waals Equation

If 1.000 mol of an ideal gas were confined to 22.41 L at 0.0°C, it would exert a pressure of 1.000 atm. Use the van der Waals equation and the constants in Table 10.3 to estimate the pressure exerted by 1.000 mol of $Cl_2(g)$ in 22.41 L at 0.0°C.

Solution

Analyze: The quantity we need to solve for is pressure. Because we will use the van der Waals equation, we must identify the appropriate values for the constants that appear there.

Plan: Using Equation 10.26, we have

$$P = \frac{nRT}{V - nb} - \frac{n^2a}{V^2}$$

Solve: Substituting $n = 1.000$ mol, $R = 0.08206$ L-atm/mol-K, $T = 273.2$ K, $V = 22.41$ L, $a = 6.49$ L^2-atm/mol^2, and $b = 0.0562$ L/mol:

$$P = \frac{(1.000 \text{ mol})(0.08206 \text{ L-atm/mol-K})(273.2 \text{ K})}{22.41 \text{ L} - (1.000 \text{ mol})(0.0562 \text{ L/mol})} - \frac{(1.000 \text{ mol})^2(6.49 \text{ L}^2\text{-atm/mol}^2)}{(22.41 \text{ L})^2}$$

$$= 1.003 \text{ atm} - 0.013 \text{ atm} = 0.990 \text{ atm}$$

Check: We expect a pressure not far from 1.000 atm, which would be the value for an ideal gas, so our answer seems very reasonable.

Comment: Notice that the first term, 1.003 atm, is the pressure corrected for molecular volume. This value is higher than the ideal value, 1.000 atm, because the volume in which the molecules are free to move is smaller than the container volume, 22.41 L. Thus, the molecules must collide more frequently with the container walls. The second factor, 0.013 atm, corrects for intermolecular forces. The intermolecular attractions between molecules reduce the pressure to 0.990 atm. We can conclude, therefore, that the intermolecular attractions are the main cause of the slight deviation of $Cl_2(g)$ from ideal behavior under the stated experimental conditions.

PRACTICE EXERCISE

Consider a sample of 1.000 mol of $CO_2(g)$ confined to a volume of 3.000 L at 0.0°C. Calculate the pressure of the gas using **(a)** the ideal-gas equation and **(b)** the van der Waals equation.

Answers: **(a)** 7.473 atm, **(b)** 7.182 atm

SAMPLE INTEGRATIVE EXERCISE | Putting Concepts Together

Cyanogen, a highly toxic gas, is composed of 46.2% C and 53.8% N by mass. At 25°C and 751 torr, 1.05 g of cyanogen occupies 0.500 L. **(a)** What is the molecular formula of cyanogen? **(b)** Predict its molecular structure. **(c)** Predict the polarity of the compound.

Solution

Analyze: First we need to determine the molecular formula of a compound from elemental analysis data and data on the properties of the gaseous substance. Thus, we have two separate calculations to do.

(a) Plan: We can use the percentage composition of the compound to calculate its empirical formula. (Section 3.5) Then we can determine the molecular formula by comparing the mass of the empirical formula with the molar mass. (Section 3.5)

Solve: To determine the empirical formula, we assume that we have a 100-g sample of the compound and then calculate the number of moles of each element in the sample:

$$\text{Moles C} = (46.2 \text{ g C})\left(\frac{1 \text{ mol C}}{12.01 \text{ g C}}\right) = 3.85 \text{ mol C}$$

$$\text{Moles N} = (53.8 \text{ g N})\left(\frac{1 \text{ mol N}}{14.01 \text{ g N}}\right) = 3.84 \text{ mol N}$$

Because the ratio of the moles of the two elements is essentially 1 : 1, the empirical formula is CN.

To determine the molar mass of the compound, we use Equation 10.11.

$$\mathcal{M} = \frac{dRT}{P} = \frac{(1.05 \text{ g}/0.500 \text{ L})(0.0821 \text{ L-atm/mol-K})(298 \text{ K})}{(751/760) \text{ atm}} = 52.0 \text{ g/mol}$$

The molar mass associated with the empirical formula, CN, is 12.0 + 14.0 = 26.0 g/mol. Dividing the molar mass of the compound by that of its empirical formula gives (52.0 g/mol)/(26.0 g/mol) = 2.00. Thus, the molecule has twice as many atoms of each element as the empirical formula, giving the molecular formula C_2N_2.

(b) Plan: To determine the molecular structure of the molecule, we must first determine its Lewis structure. (Section 8.5) We can then use the VSEPR model to predict the structure. (Section 9.2)

Solve: The molecule has 2(4) + 2(5) = 18 valence-shell electrons. By trial and error, we seek a Lewis structure with 18 valence electrons in which each atom has an octet and in which the formal charges are as low as possible. The following structure meets these criteria:

$$:N\equiv C-C\equiv N:$$

(This structure has zero formal charges on each atom.)

The Lewis structure shows that each atom has two electron domains. (Each nitrogen has a nonbonding pair of electrons and a triple bond, whereas each carbon has a triple bond and a single bond.) Thus the electron-domain geometry around each atom is linear, causing the overall molecule to be linear.

(c) Plan: To determine the polarity of the molecule, we must examine the polarity of the individual bonds and the overall geometry of the molecule.

Solve: Because the molecule is linear, we expect the two dipoles created by the polarity in the carbon–nitrogen bond to cancel each other, leaving the molecule with no dipole moment.

SUMMARY AND KEY TERMS

Section 10.1 Substances that are gases at room temperatures tend to be molecular substances with low molar masses. Air, a mixture composed mainly of N_2 and O_2, is the most common gas we encounter. Some liquids and solids can also exist in the gaseous state, where they are known as vapors. Gases are compressible; they mix in all proportions because their component molecules are far apart.

Section 10.2 To describe the state, or condition, of a gas, we must specify four variables: pressure (P), volume (V), temperature (T), and quantity (n). Volume is usually measured in liters, temperature in kelvins, and quantity of gas in moles. **Pressure** is the force per unit area. It is expressed in SI units as **pascals**, Pa (1 Pa = 1 N/m^2 = 1 kg/m-s^2). A related unit, the **bar**, equals 10^5 Pa. In chemistry, **standard atmospheric pressure** is used to define the **atmosphere** (atm) and the **torr** (also called the millimeter of mercury). One atmosphere of pressure equals 101.325 kPa, or 760 torr. A barometer is often used to measure the atmospheric pressure. A manometer can be used to measure the pressure of enclosed gases.

Sections 10.3 and 10.4 Studies have revealed several simple gas laws: For a constant quantity of gas at constant temperature, the volume of the gas is inversely proportional to the pressure (**Boyle's law**). For a fixed quantity of gas at constant pressure, the volume is directly proportional to its absolute temperature (**Charles's law**). Equal volumes of gases at the same temperature and pressure contain equal numbers of molecules (**Avogadro's hypothesis**). For a gas at constant temperature and pressure, the volume of the gas is directly proportional to the number of moles of gas (**Avogadro's law**). Each of these gas laws is a special case of the ideal-gas equation.

The **ideal-gas equation**, $PV = nRT$, is the equation of state for an **ideal gas**. The term R in this equation is the **gas constant**. We can use the ideal-gas equation to calculate variations in one variable when one or more of the others are changed. Most gases at pressures of about 1 atm and temperatures near 273 K and above obey the ideal-gas equation reasonably well. The conditions of 273 K (0°C) and 1 atm are known as the **standard temperature and pressure (STP)**.

Sections 10.5 and 10.6 Using the ideal-gas equation, we can relate the density of a gas to its molar mass: $\mathcal{M} = dRT/P$. We can also use the ideal-gas equation to solve problems involving gases as reactants or products in chemical reactions. In all applications of the ideal-gas equation we must remember to convert temperatures to the absolute-temperature scale (the Kelvin scale).

In gas mixtures the total pressure is the sum of the **partial pressures** that each gas would exert if it were present alone under the same conditions (**Dalton's law of partial pressures**). The partial pressure of a component of a mixture is equal to its mole fraction times the total pressure: $P_1 = X_1P_t$. The **mole fraction** is the ratio of the moles of one component of a mixture to the total moles of all components. In calculating the quantity of a gas collected over water, correction must be made for the partial pressure of water vapor in the gas mixture.

Section 10.7 The **kinetic-molecular theory** accounts for the properties of an ideal gas in terms of a set of statements about the nature of gases. Briefly, these statements are as follows: Molecules are in continuous chaotic motion; the volume of gas molecules is negligible compared to the volume of their container; the gas molecules have no attractive forces for one another; their collisions are elastic; and the average kinetic energy of the gas molecules is proportional to the absolute temperature.

The molecules of a gas do not all have the same kinetic energy at a given instant. Their speeds are distributed over a wide range; the distribution varies with the molar mass of the gas and with temperature. The **root-mean-square (rms) speed**, u, varies in proportion to the square root of the absolute temperature and inversely with the square root of the molar mass: $u = \sqrt{3RT/\mathcal{M}}$.

Section 10.8 It follows from kinetic-molecular theory that the rate at which a gas undergoes **effusion** (escapes through a tiny hole into a vacuum) is inversely proportional to the square root of its molar mass (**Graham's law**). The **diffusion** of one gas through the space occupied by a second gas is another phenomenon related to the speeds at which molecules move. Because molecules undergo frequent collisions with one another, the **mean free path**—the mean distance traveled between collisions—is short. Collisions between molecules limit the rate at which a gas molecule can diffuse.

Section 10.9 Departures from ideal behavior increase in magnitude as pressure increases and as temperature decreases. The extent of nonideality of a real gas can be seen by examining the quantity PV/RT for one mole of the gas as a function of pressure; for an ideal gas, this quantity is exactly 1 at all pressures. Real gases depart from ideal behavior because the molecules possess finite volume and because the molecules experience attractive forces for one another. The **van der Waals equation** is an equation of state for gases that modifies the ideal-gas equation to account for intrinsic molecular volume and intermolecular forces.

VISUALIZING CONCEPTS

10.1 Assume that you have a sample of gas in a container with a movable piston such as the one in the drawing. **(a)** Redraw the container to show what it might look like if the temperature of the gas is increased from 300 K to 500 K while the pressure is kept constant. **(b)** Redraw the container to show what it might look like if the pressure on the piston is increased from 1.0 atm to 2.0 atm while the temperature is kept constant. [Section 10.3]

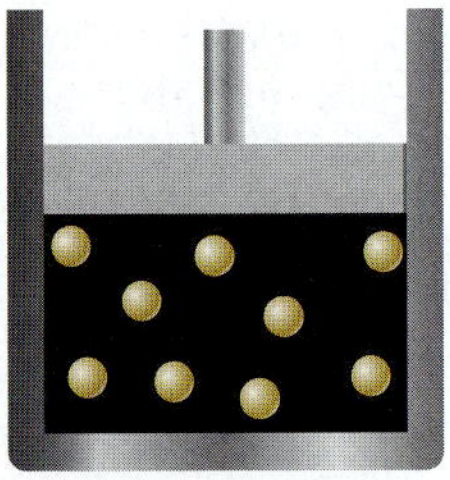

10.2 Consider the sample of gas depicted below. What would the drawing look like if the volume and temperature remained constant while you removed enough of the gas to decrease the pressure by a factor of 2? [Section 10.3]

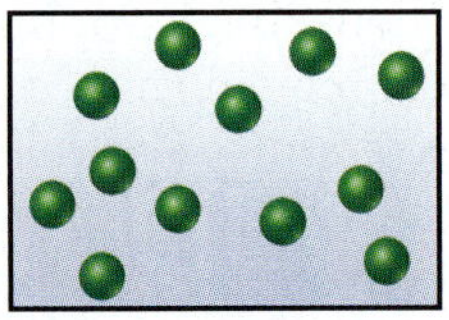

10.3 Consider the following reaction:

$$2\,CO(g) + O_2(g) \longrightarrow 2\,CO_2(g)$$

Imagine that this reaction occurs in a container that has a piston that moves to allow a constant pressure to be maintained when the reaction occurs at constant temperature. **(a)** What happens to the volume of the container as a result of the reaction? Explain. **(b)** If the piston is not allowed to move, what happens to the pressure as a result of the reaction? [Sections 10.3 and 10.5]

10.4 Consider the apparatus below, which shows gases in two containers and one empty container. When the stopcocks are opened and the gases allowed to mix at constant temperature, what is the distribution of atoms in each container? Assume that the containers are of equal volume and ignore the volume of the tubing connecting them. Which gas has the greater partial pressure after the stopcocks are opened? [Section 10.6]

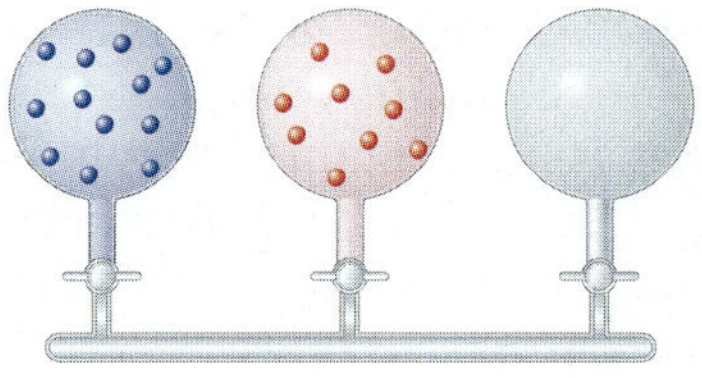

10.5 The drawing below represents a mixture of three different gases. **(a)** Rank the three components in order of increasing partial pressure. **(b)** if the total pressure of the mixture is 0.90 atm, calculate the partial pressure of each gas. [Section 10.6]

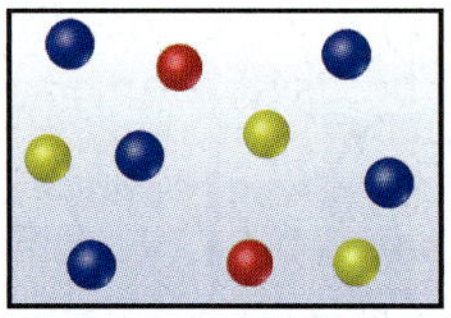

10.6 On a single plot, qualitatively sketch the distribution of molecular speeds for **(a)** Kr(g) at −50°C, **(b)** Kr(g) at 0°C, **(c)** Ar(g) at 0°C. [Section 10.7]

10.7 Consider the drawing below. **(a)** If the curves A and B refer to two different gases, He and O_2, at the same temperature, which is which? Explain. **(b)** If A and B refer to the same gas at two different temperatures, which represents the higher temperature? [Section 10.7]

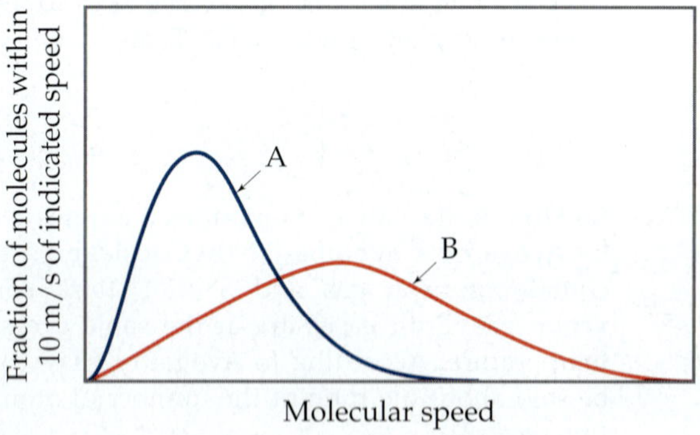

10.8 Consider the following samples of gases:

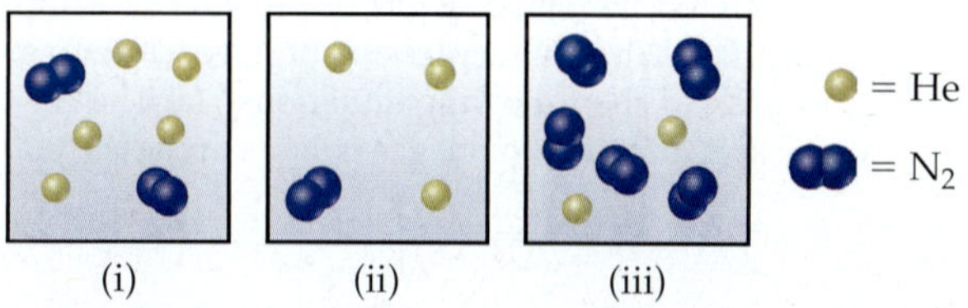

If the three samples are all at the same temperature, rank them with respect to **(a)** total pressure, **(b)** partial pressure of helium, **(c)** density, **(d)** average kinetic energy of particles. [Section 10.7]

EXERCISES

Gas Characteristics; Pressure

10.9 How does a gas differ from a liquid with respect to each of the following properties: **(a)** density, **(b)** compressibility, **(c)** ability to mix with other substances of the same phase to form homogeneous mixtures?

10.10 **(a)** Both a liquid and a gas are moved to larger containers. How does their behavior differ? Explain the difference in molecular terms. **(b)** Although water and carbon tetrachloride, $CCl_4(l)$, do not mix, their vapors form homogeneous mixtures. Explain. **(c)** The densities of gases are generally reported in units of g/L, whereas those for liquids are reported as g/mL. Explain the molecular basis for this difference.

10.11 Suppose that a woman weighing 130 lb and wearing high-heeled shoes momentarily places all her weight on the heel of one foot. If the area of the heel is 0.50 in.^{2}, calculate the pressure exerted on the underlying surface in kilopascals.

10.12 A set of bookshelves rests on a hard floor surface on four legs, each having a cross-sectional dimension of 3.0×4.1 cm in contact with the floor. The total mass of the shelves plus the books stacked on them is 262 kg. Calculate the pressure in pascals exerted by the shelf footings on the surface.

10.13 **(a)** How high in meters must a column of water be to exert a pressure equal to that of a 760-mm column of mercury? The density of water is 1.0 g/mL, whereas that of mercury is 13.6 g/mL. **(b)** What is the pressure in atmospheres on the body of a diver if he is 36 ft below the surface of the water when atmospheric pressure at the surface is 0.95 atm?

10.14 The compound 1-iodododecane is a nonvolatile liquid with a density of 1.20 g/mL. The density of mercury is 13.6 g/mL. What do you predict for the height of a barometer column based on 1-iodododecane, when the atmospheric pressure is 752 torr?

10.15 Each of the following statements concerns a mercury barometer such as that shown in Figure 10.2. Identify any incorrect statements, and correct them. **(a)** The tube must be 1 cm^2 in cross-sectional area. **(b)** At equilibrium the force of gravity per unit area acting on the mercury column at the level of the outside mercury equals the force of gravity per unit area acting on the atmosphere. **(c)** The column of mercury is held up by the vacuum at the top of the column.

10.16 Suppose you make a mercury barometer using a glass tube about 50 cm in length, closed at one end. What would you expect to see if the tube is filled with mercury and inverted in a mercury dish, as in Figure 10.2? Explain.

10.17 The typical atmospheric pressure on top of Mt. Everest (29,028 ft) is about 265 torr. Convert this pressure to **(a)** atm, **(b)** mm Hg, **(c)** pascals, **(d)** bars.

10.18 Perform the following conversions: **(a)** 0.850 atm to torr, **(b)** 785 torr to kilopascals, **(c)** 655 mm Hg to atmospheres, **(d)** 1.323×10^5 Pa to atmospheres, **(e)** 2.50 atm to bars.

10.19 In the United States, barometric pressures are reported in inches of mercury (in. Hg). On a beautiful summer day in Chicago the barometric pressure is 30.45 in. Hg. **(a)** Convert this pressure to torr. **(b)** A meteorologist explains the nice weather by referring to a "high-pressure area." In light of your answer to part (a), explain why this term makes sense.

10.20 **(a)** On Titan, the largest moon of Saturn, the atmospheric pressure is 1.63105 Pa. What is the atmospheric pressure of Titan in atm? **(b)** On Venus the surface atmospheric pressure is about 90 Earth atmospheres. What is the Venusian atmospheric pressure in kilopascals?

10.21 If the atmospheric pressure is 0.985 atm, what is the pressure of the enclosed gas in each of the three cases depicted in the drawing?

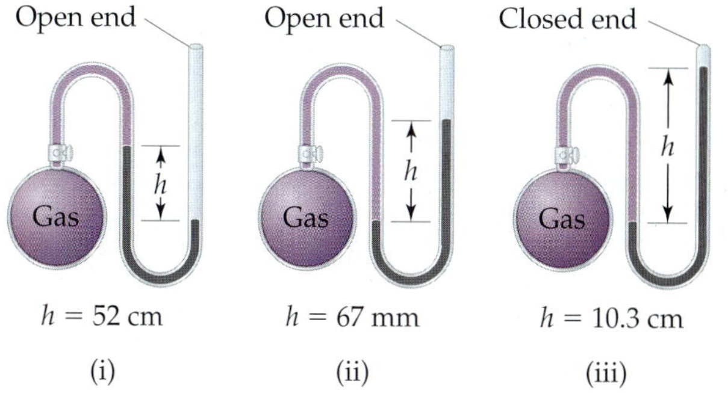

10.22 An open-end manometer containing mercury is connected to a container of gas, as depicted in Sample Exercise 10.2. What is the pressure of the enclosed gas in torr in each of the following situations? **(a)** The mercury in the arm attached to the gas is 10.7 cm higher than in the one open to the atmosphere; atmospheric pressure is 0.977 atm. **(b)** The mercury in the arm attached to the gas is 9.5 mm lower than in the one open to the atmosphere; atmospheric pressure is 1.02 atm.

The Gas Laws

10.23 Assume that you have a cylinder with a movable piston. What would happen to the gas pressure inside the cylinder if you do the following? **(a)** Decrease the volume to one-fourth the original volume while holding the temperature constant. **(b)** Reduce the Kelvin temperature to half its original value while holding the volume constant. **(c)** Reduce the amount of gas to half while keeping the volume and temperature constant.

10.24 A fixed quantity of gas at 23°C exhibits a pressure of 735 torr and occupies a volume of 5.22 L. **(a)** Use Boyle's law to calculate the volume the gas will occupy if the pressure is increased to 1.88 atm while the temperature is held constant. **(b)** Use Charles's law to calculate the volume the gas will occupy if the temperature is increased to 165°C while the pressure is held constant.

10.25 **(a)** How is the law of combining volumes explained by Avogadro's hypothesis? **(b)** Consider a 1.0-L flask containing neon gas and a 1.5-L flask containing xenon gas. Both gases are at the same pressure and temperature. According to Avogadro's law, what can be said about the ratio of the number of atoms in the two flasks?

10.26 Nitrogen and hydrogen gases react to form ammonia gas as follows:

$$N_2(g) + 3\,H_2(g) \longrightarrow 2\,NH_3(g)$$

At a certain temperature and pressure, 1.2 L of N_2 reacts with 3.6 L of H_2. If all the N_2 and H_2 are consumed, what volume of NH_3, at the same temperature and pressure, will be produced?

The Ideal-Gas Equation

10.27 **(a)** Write the ideal-gas equation, and give the units used for each term in the equation when $R = 0.0821$ L-atm/mol-K. **(b)** What is an ideal gas?

10.28 **(a)** What conditions are represented by the abbreviation STP? **(b)** What is the molar volume of an ideal gas at STP? **(c)** Room temperature is often assumed to be 25°C. Calculate the molar volume of an ideal gas at room temperature and 1 atm pressure.

10.29 Suppose you are given two 1-L flasks and told that one contains a gas of molar mass 30, the other a gas of molar mass 60, both at the same temperature. The pressure in flask A is X atm, and the mass of gas in the flask is 1.2 g. The pressure in flask B is 0.5X atm, and the mass of gas in that flask is 1.2 g. Which flask contains gas of molar mass 30, and which contains the gas of molar mass 60?

10.30 Suppose you are given two flasks at the same temperature, one of volume 2 L and the other of volume 3 L. The 2-L flask contains 4.8 g of gas, and the gas pressure is X atm. The 3-L flask contains 0.36 g of gas, and the gas pressure is 0.1X. Do the two gases have the same molar mass? If not, which contains the gas of higher molar mass?

10.31 Complete the following table for an ideal gas:

P	*V*	*n*	*T*
2.00 atm	1.00 L	0.500 mol	? K
0.300 atm	0.250 L	? mol	27°C
650 torr	? L	0.333 mol	350 K
? atm	585 mL	0.250 mol	295 K

10.32 Calculate each of the following quantities for an ideal gas: **(a)** the volume of the gas, in liters, if 1.75 mol has a pressure of 0.985 atm at a temperature of −6°C; **(b)** the absolute temperature of the gas at which 3.33×10^{-3} mol occupies 255 mL at 720 torr; **(c)** the pressure, in atmospheres, if 0.0467 mol occupies 413 mL at 122°C; **(d)** the quantity of gas, in moles, if 67.5 L at 54°C has a pressure of 11.25 kPa.

10.33 The *Hindenburg* was a hydrogen-filled dirigible that exploded in 1937. If the *Hindenburg* held $2.0 \times 10^5\ m^3$ of hydrogen gas at 23°C and 1.0 atm, what mass of hydrogen was present?

10.34 A neon sign is made of glass tubing whose inside diameter is 2.5 cm and whose length is 5.5 m. If the sign contains neon at a pressure of 1.78 torr at 35°C, how many grams of neon are in the sign? (The volume of a cylinder is $\pi r^2 h$.)

10.35 Calculate the number of molecules in a deep breath of air whose volume is 2.50 L at body temperature, 37°C, and a pressure of 735 torr.

10.36 If the pressure exerted by ozone, O_3, in the stratosphere is 3.0×10^{-3} atm and the temperature is 250 K, how many ozone molecules are in a liter?

10.37 A scuba diver's tank contains 0.29 kg of O_2 compressed into a volume of 2.3 L. **(a)** Calculate the gas pressure inside the tank at 9°C. **(b)** What volume would this oxygen occupy at 26°C and 0.95 atm?

10.38 An aerosol spray can with a volume of 250 mL contains 2.30 g of propane gas (C_3H_8) as a propellant. **(a)** If the can is at 23°C, what is the pressure in the can? **(b)** What volume would the propane occupy at STP? **(c)** The can says that exposure to temperatures above 130°F may cause the can to burst. What is the pressure in the can at this temperature?

10.39 Chlorine is widely used to purify municipal water supplies and to treat swimming pool waters. Suppose that the volume of a particular sample of Cl_2 gas is 8.70 L at

895 torr and 24°C. **(a)** How many grams of Cl_2 are in the sample? **(b)** What volume will the Cl_2 occupy at STP? **(c)** At what temperature will the volume be 15.00 L if the pressure is 8.76×10^2 torr? **(d)** At what pressure will the volume equal 6.00 L if the temperature is 58°C?

10.40 Many gases are shipped in high-pressure containers. Consider a steel tank whose volume is 65.0 L and which contains O_2 gas at a pressure of 16,500 kPa at 23°C. **(a)** What mass of O_2 does the tank contain? **(b)** What volume would the gas occupy at STP? **(c)** At what temperature would the pressure in the tank equal 150.0 atm? **(d)** What would be the pressure of the gas, in kPa, if it were transferred to a container at 24°C whose volume is 55.0 L?

10.41 In an experiment reported in the scientific literature, male cockroaches were made to run at different speeds on a miniature treadmill while their oxygen consumption was measured. In one hour the average cockroach running at 0.08 km/hr consumed 0.8 mL of O_2 at 1 atm pressure and 24°C per gram of insect weight. **(a)** How many moles of O_2 would be consumed in 1 hr by a 5.2-g cockroach moving at this speed? **(b)** This same cockroach is caught by a child and placed in a 1-qt fruit jar with a tight lid. Assuming the same level of continuous activity as in the research, will the cockroach consume more than 20% of the available O_2 in a 48-hr period? (Air is 21 mol percent O_2)

10.42 After the large eruption of Mount St. Helens in 1980, gas samples from the volcano were taken by sampling the downwind gas plume. The unfiltered gas samples were passed over a gold-coated wire coil to absorb mercury (Hg) present in the gas. The mercury was recovered from the coil by heating it, and then analyzed. In one particular set of experiments scientists found a mercury vapor level of 1800 ng of Hg per cubic meter in the plume, at a gas temperature of 10°C. Calculate **(a)** the partial pressure of Hg vapor in the plume, **(b)** the number of Hg atoms per cubic meter in the gas, **(c)** the total mass of Hg emitted per day by the volcano if the daily plume volume was 1600 km^3.

Further Applications of the Ideal-Gas Equation

10.43 Which gas is most dense at 1.00 atm and 298 K? **(a)** CO_2, **(b)** N_2O, **(c)** Cl_2. Explain.

10.44 Which gas is least dense at 1.00 atm and 298 K? **(a)** SO_2, **(b)** HBr, **(c)** CO_2. Explain.

10.45 Which of the following statements best explains why a closed balloon filled with helium gas rises in air?
(a) Helium is a monatomic gas, whereas nearly all the molecules that make up air, such as nitrogen and oxygen, are diatomic.
(b) The average speed of helium atoms is higher than the average speed of air molecules, and the higher speed of collisions with the balloon walls propels the balloon upward.
(c) Because the helium atoms are of lower mass than the average air molecule, the helium gas is less dense than air. The balloon thus weighs less than the air displaced by its volume.
(d) Because helium has a lower molar mass than the average air molecule, the helium atoms are in faster motion. This means that the temperature of the helium is higher than the air temperature. Hot gases tend to rise.

10.46 Which of the following statements best explains why nitrogen gas at STP is less dense than Xe gas at STP?
(a) Because Xe is a noble gas, there is less tendency for the Xe atoms to repel one another, so they pack more densely in the gas state.
(b) Xe atoms have a higher mass than N_2 molecules. Because both gases at STP have the same number of molecules per unit volume, the Xe gas must be denser.
(c) The Xe atoms are larger than N_2 molecules and thus take up a larger fraction of the space occupied by the gas.
(d) Because the Xe atoms are much more massive than the N_2 molecules, they move more slowly and thus exert less upward force on the gas container and make the gas appear denser.

10.47 **(a)** Calculate the density of NO_2 gas at 0.970 atm and 35°C. **(b)** Calculate the molar mass of a gas if 2.50 g occupies 0.875 L at 685 torr and 35°C.

10.48 **(a)** Calculate the density of sulfur hexafluoride gas at 678 torr and 28°C. **(b)** Calculate the molar mass of a vapor that has a density of 7.135 g/L at 12°C and 743 torr.

10.49 In the Dumas-bulb technique for determining the molar mass of an unknown liquid, you vaporize the sample of a liquid that boils below 100°C in a boiling-water bath and determine the mass of vapor required to fill the bulb (see drawing). From the following data, calculate the molar mass of the unknown liquid: mass of unknown vapor, 1.012 g; volume of bulb, 354 cm^3; pressure, 742 torr; temperature, 99°C.

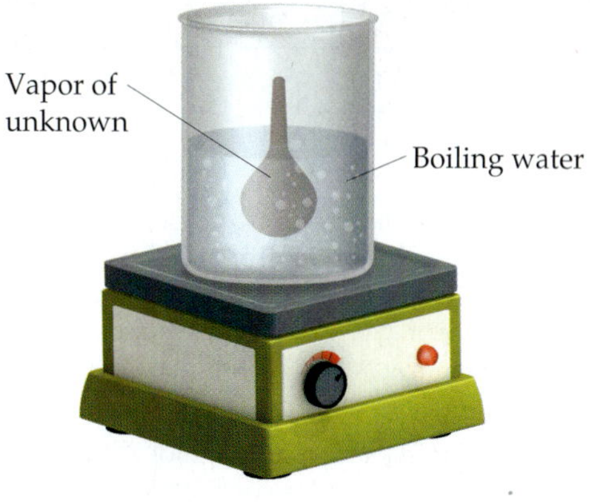

10.50 The molar mass of a volatile substance was determined by the Dumas-bulb method described in Exercise 10.49. The unknown vapor had a mass of 0.846 g; the volume of the bulb was 354 cm^3, pressure 752 torr, and temperature 100°C. Calculate the molar mass of the unknown vapor.

10.51 Magnesium can be used as a "getter" in evacuated enclosures, to react with the last traces of oxygen. (The magnesium is usually heated by passing an electric current through a wire or ribbon of the metal.) If an enclosure of

0.382 L has a partial pressure of O_2 of 3.5×10^{-6} torr at 27°C, what mass of magnesium will react according to the following equation?

$$2\ Mg(s) + O_2(g) \longrightarrow 2\ MgO(s)$$

10.52 Calcium hydride, CaH_2, reacts with water to form hydrogen gas:

$$CaH_2(s) + 2\ H_2O(l) \longrightarrow Ca(OH)_2(aq) + 2\ H_2(g)$$

This reaction is sometimes used to inflate life rafts, weather balloons, and the like, where a simple, compact means of generating H_2 is desired. How many grams of CaH_2 are needed to generate 53.5 L of H_2 gas if the pressure of H_2 is 814 torr at 21°C?

10.53 The metabolic oxidation of glucose, $C_6H_{12}O_6$, in our bodies produces CO_2, which is expelled from our lungs as a gas:

$$C_6H_{12}O_6(aq) + 6\ O_2(g) \longrightarrow 6\ CO_2(g) + 6\ H_2O(l)$$

Calculate the volume of dry CO_2 produced at body temperature (37°C) and 0.970 atm when 24.5 g of glucose is consumed in this reaction.

10.54 Consider the following unbalanced chemical equation:

$$C_5H_{12}(l) + O_2(g) \longrightarrow CO_2(g) + H_2O(g)$$

What volume of oxygen gas, measured at 23°C and 0.980 atm, is needed to react with 2.50 g of C_5H_{12}? What volume of each product is produced under the same conditions?

10.55 Hydrogen gas is produced when zinc reacts with sulfuric acid:

$$Zn(s) + H_2SO_4(aq) \longrightarrow ZnSO_4(aq) + H_2(g)$$

If 159 mL of wet H_2 is collected over water at 24°C and a barometric pressure of 738 torr, how many grams of Zn have been consumed? (The vapor pressure of water is tabulated in Appendix B.)

10.56 Acetylene gas, $C_2H_2(g)$, can be prepared by the reaction of calcium carbide with water:

$$CaC_2(s) + 2\ H_2O(l) \longrightarrow Ca(OH)_2(s) + C_2H_2(g)$$

Calculate the volume of C_2H_2 that is collected over water at 26°C by reaction of 0.887 g of CaC_2 if the total pressure of the gas is 726 torr. (The vapor pressure of water is tabulated in Appendix B.)

Partial Pressures

10.57 Consider the apparatus shown in the drawing. **(a)** When the stopcock between the two containers is opened and the gases allowed to mix, how does the volume occupied by the N_2 gas change? What is the partial pressure of N_2 after mixing? **(b)** How does the volume of the O_2 gas change when the gases mix? What is the partial pressure of O_2 in the mixture? **(c)** What is the total pressure in the container after the gases mix?

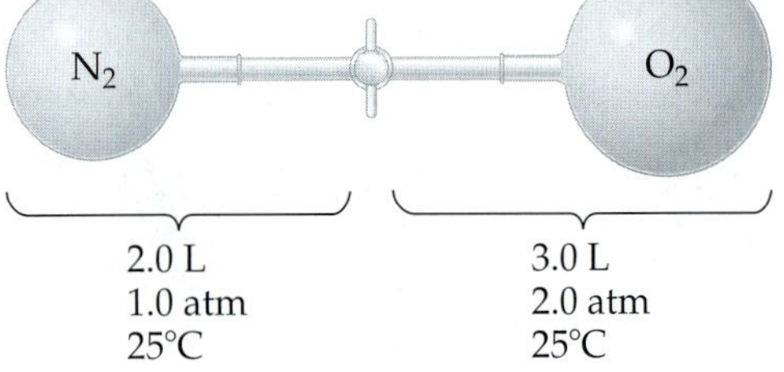

10.58 Consider a mixture of two gases, A and B, confined to a closed vessel. A quantity of a third gas, C, is added to the same vessel at the same temperature. How does the addition of gas C affect the following: **(a)** the partial pressure of gas A, **(b)** the total pressure in the vessel, **(c)** the mole fraction of gas B?

10.59 A mixture containing 0.538 mol He(*g*), 0.315 mol Ne(*g*), and 0.103 mol Ar(*g*) is confined in a 7.00-L vessel at 25°C. **(a)** Calculate the partial pressure of each of the gases in the mixture. **(b)** Calculate the total pressure of the mixture.

10.60 A mixture containing 2.50 g each of $CH_4(g)$, $C_2H_4(g)$, and $C_4H_{10}(g)$ is contained in a 2.00-L flask at a temperature of 15°C. **(a)** Calculate the partial pressure of each of the gases in the mixture. **(b)** Calculate the total pressure of the mixture.

10.61 A piece of solid carbon dioxide with a mass of 5.50 g is placed in a 10.0-L vessel that already contains air at 705 torr and 24°C. After the carbon dioxide has totally vaporized, what is the partial pressure of carbon dioxide and the total pressure in the container at 24°C?

10.62 A sample of 4.00 mL of diethylether ($C_2H_5OC_2H_5$; density = 0.7134 g/mL) is introduced into a 5.00-L vessel that already contains a mixture of N_2 and O_2, whose partial pressures are $P_{N_2} = 0.751$ atm and $P_{O_2} = 0.208$ atm. The temperature is held at 35.0°C, and the diethylether totally evaporates. **(a)** Calculate the partial pressure of the diethylether. **(b)** Calculate the total pressure in the container.

10.63 A mixture of gases contains 0.75 mol N_2, 0.30 mol O_2, and 0.15 mol CO_2. If the total pressure of the mixture is 1.56 atm, what is the partial pressure of each component?

10.64 A mixture of gases contains 10.25 g of N_2, 2.05 g of H_2, and 7.63 g of NH_3. If the total pressure of the mixture is 2.35 atm, what is the partial pressure of each component?

10.65 At an underwater depth of 250 ft, the pressure is 8.38 atm. What should the mole percent of oxygen be in the diving gas for the partial pressure of oxygen in the mixture to be 0.21 atm, the same as in air at 1 atm?

10.66 **(a)** What are the mole fractions of each component in a mixture of 5.08 g of O_2, 7.17 g of N_2, and 1.32 g of H_2? **(b)** What is the partial pressure in atm of each component of this mixture if it is held in a 12.40-L vessel at 15°C?

10.67 A quantity of N_2 gas originally held at 4.75 atm pressure in a 1.00-L container at 26°C is transferred to a 10.0-L container at 20°C. A quantity of O_2 gas originally at 5.25 atm and 26°C in a 5.00-L container is transferred to this same container. What is the total pressure in the new container?

10.68 A sample of 4.00 g of $SO_2(g)$ originally in a 5.00-L vessel at 26°C is transferred to a 10.0-L vessel at 25°C. A sample of 2.35 g $N_2(g)$ originally in a 2.50-L vessel at 20°C is transferred to this same 10.0-L vessel. **(a)** What is the partial pressure of $SO_2(g)$ in the larger container? **(b)** What is the partial pressure of $N_2(g)$ in this vessel? **(c)** What is the total pressure in the vessel?

Kinetic-Molecular Theory; Graham's Law

10.69 What change or changes in the state of a gas bring about each of the following effects? **(a)** The number of impacts per unit time on a given container wall increases. **(b)** The average energy of impact of molecules with the wall of the container decreases. **(c)** The average distance between gas molecules increases. **(d)** The average speed of molecules in the gas mixture is increased.

10.70 Indicate which of the following statements regarding the kinetic-molecular theory of gases are correct. For those that are false, formulate a correct version of the statement. **(a)** The average kinetic energy of a collection of gas molecules at a given temperature is proportional to $m^{1/2}$. **(b)** The gas molecules are assumed to exert no forces on each other. **(c)** All the molecules of a gas at a given temperature have the same kinetic energy. **(d)** The volume of the gas molecules is negligible in comparison to the total volume in which the gas is contained.

10.71 What property or properties of gases can you point to that support the assumption that most of the volume in a gas is empty space?

10.72 Newton had an incorrect theory of gases in which he assumed that all gas molecules repel one another and the walls of their container. Thus, the molecules of a gas are statically and uniformly distributed, trying to get as far apart as possible from one another and the vessel walls. This repulsion gives rise to pressure. Explain why Charles's law argues for the kinetic-molecular theory and against Newton's model.

10.73 Vessel A contains $CO(g)$ at 0°C and 1 atm. Vessel B contains $SO_2(g)$ at 20°C and 0.5 atm. The two vessels have the same volume. **(a)** Which vessel contains more molecules? **(b)** Which contains more mass? **(c)** In which vessel is the average kinetic energy of molecules higher? **(d)** In which vessel is the rms speed of molecules higher?

10.74 Suppose you have two 1-L flasks, one containing N_2 at STP, the other containing CH_4 at STP. How do these systems compare with respect to **(a)** number of molecules, **(b)** density, **(c)** average kinetic energy of the molecules, **(d)** rate of effusion through a pinhole leak?

10.75 **(a)** Place the following gases in order of increasing average molecular speed at 25°C: Ne, HBr, SO_2, NF_3, CO. **(b)** Calculate the rms speed of NF_3 molecules at 25°C.

10.76 **(a)** Place the following gases in order of increasing average molecular speed at 300 K: CO, SF_6, H_2S, Cl_2, HBr. **(b)** Calculate and compare the rms speeds of CO and Cl_2 molecules at 300 K.

10.77 Hydrogen has two naturally occurring isotopes, ^1H and ^2H. Chlorine also has two naturally occurring isotopes, ^{35}Cl and ^{37}Cl. Thus, hydrogen chloride gas consists of four distinct types of molecules: $^1H^{35}Cl$, $^1H^{37}Cl$, $^2H^{35}Cl$, and $^2H^{37}Cl$. Place these four molecules in order of increasing rate of effusion.

10.78 As discussed in the "Chemistry at Work" box in Section 10.8, enriched uranium is produced via gaseous diffusion of UF_6. Suppose a process were developed to allow diffusion of gaseous uranium atoms, $U(g)$. Calculate the ratio of diffusion rates for ^{235}U and ^{238}U, and compare it to the ratio for UF_6 given in the essay.

10.79 Arsenic(III) sulfide sublimes readily, even below its melting point of 320°C. The molecules of the vapor phase are found to effuse through a tiny hole at 0.28 times the rate of effusion of Ar atoms under the same conditions of temperature and pressure. What is the molecular formula of arsenic(III) sulfide in the gas phase?

10.80 A gas of unknown molecular mass was allowed to effuse through a small opening under constant-pressure conditions. It required 105 s for 1.0 L of the gas to effuse. Under identical experimental conditions it required 31 s for 1.0 L of O_2 gas to effuse. Calculate the molar mass of the unknown gas. (Remember that the faster the rate of effusion, the shorter the time required for effusion of 1.0 L; that is, rate and time are inversely proportional.)

Nonideal-Gas Behavior

10.81 **(a)** List two experimental conditions under which gases deviate from ideal behavior. **(b)** List two reasons the gases deviate from ideal behavior. **(c)** Explain how the function PV/RT can be used to show how gases behave nonideally.

10.82 The planet Jupiter has a mass 318 times that of Earth, and its surface temperature is 140 K. Mercury has a mass 0.05 times that of Earth, and its surface temperature is between 600 K and 700 K. On which planet is the atmosphere more likely to obey the ideal-gas law? Explain.

10.83 Based on their respective van der Waals constants (Table 10.3), is Ar or CO_2 expected to behave more nearly like an ideal gas at high pressures? Explain.

10.84 Briefly explain the significance of the constants a and b in the van der Waals equation.

10.85 Calculate the pressure that CCl_4 will exert at 40°C if 1.00 mol occupies 28.0 L, assuming that **(a)** CCl_4 obeys the ideal-gas equation; **(b)** CCl_4 obeys the van der Waals equation. (Values for the van der Waals constants are given in Table 10.3.)

10.86 It turns out that the van der Waals constant b equals four times the total volume actually occupied by the molecules of a mole of gas. Using this figure, calculate the fraction of the volume in a container actually occupied by Ar atoms **(a)** at STP, **(b)** at 100 atm pressure and 0°C. (Assume for simplicity that the ideal-gas equation still holds.)

Additional Exercises

10.87 Suppose the mercury used to make a barometer has a few small droplets of water trapped in it that rise to the top of the mercury in the tube. Will the barometer show the correct atmospheric pressure? Explain.

10.88 Suppose that when Torricelli had his great idea for constructing a mercury manometer, he had rushed into the laboratory and found the following items of glass:

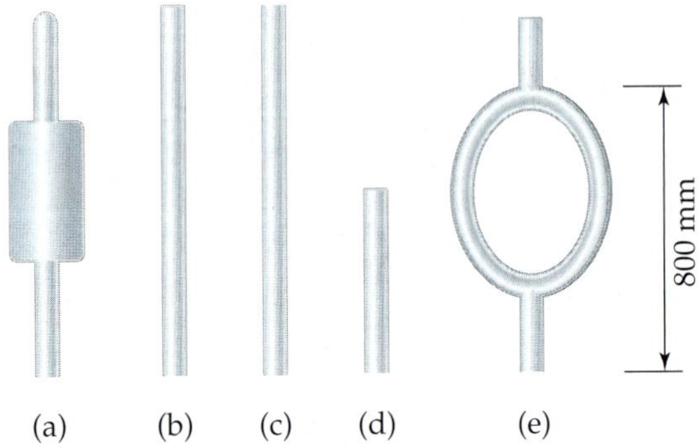

Which of these would have been satisfactory for his use in forming the first manometer? Explain why the unsatisfactory ones would not have worked.

10.89 A gas bubble with a volume of 1.0 mm^3 originates at the bottom of a lake where the pressure is 3.0 atm. Calculate its volume when the bubble reaches the surface of the lake where the pressure is 695 torr, assuming that the temperature doesn't change.

10.90 A 15.0-L tank is filled with helium gas at a pressure of 1.00×10^2 atm. How many balloons (each 2.00 L) can be inflated to a pressure of 1.00 atm, assuming that the temperature remains constant and that the tank cannot be emptied below 1.00 atm?

10.91 To minimize the rate of evaporation of the tungsten filament, 1.4×10^{-5} mol of argon is placed in a 600-cm^3 lightbulb. What is the pressure of argon in the lightbulb at 23°C?

10.92 Propane, C_3H_8, liquefies under modest pressure, allowing a large amount to be stored in a container. **(a)** Calculate the number of moles of propane gas in a 110-L container at 3.00 atm and 27°C. **(b)** Calculate the number of moles of liquid propane that can be stored in the same volume if the density of the liquid is 0.590 g/mL. **(c)** Calculate the ratio of the number of moles of liquid to moles of gas. Discuss this ratio in light of the kinetic-molecular theory of gases.

10.93 What is the total mass (in grams) of O_2 in a room measuring 10.0 ft $\times$ 8.0 ft $\times$ 8.0 ft if the air in the room is at STP and contains 20.95% O_2?

10.94 Nickel carbonyl, $Ni(CO)_4$, is one of the most toxic substances known. The present maximum allowable concentration in laboratory air during an 8-hr workday is 1 part in 10^9 by volume, which means that there is one mole of $Ni(CO)_4$ for every 10^9 moles of gas. Assume 24°C and 1.00 atm pressure. What mass of $Ni(CO)_4$ is allowable in a laboratory that is 54 m^2 in area, with a ceiling height of 3.1 m?

10.95 Consider the arrangement of bulbs shown in the drawing. Each of the bulbs contains a gas at the pressure shown. What is the pressure of the system when all the stopcocks are opened, assuming that the temperature remains constant? (We can neglect the volume of the capillary tubing connecting the bulbs.)

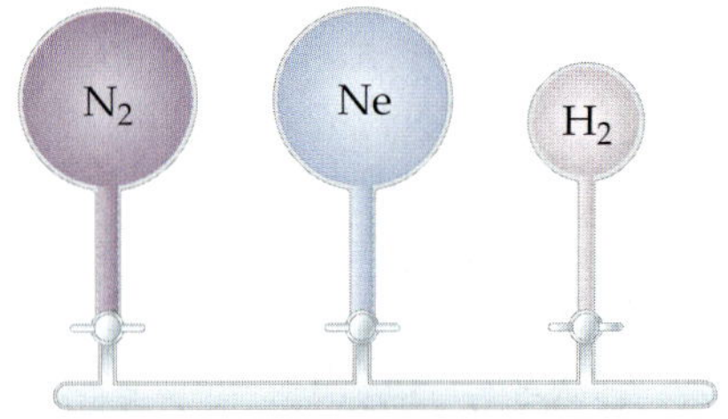

Volume	1.0 L	1.0 L	0.5 L
Pressure	265 torr	800 torr	532 torr

10.96 Assume that a single cylinder of an automobile engine has a volume of 524 cm^3. **(a)** If the cylinder is full of air at 74°C and 0.980 atm, how many moles of O_2 are present? (The mole fraction of O_2 in dry air is 0.2095.) **(b)** How many grams of C_8H_{18} could be combusted by this quantity of O_2, assuming complete combustion with formation of CO_2 and H_2O?

[10.97] Ammonia, $NH_3(g)$, and hydrogen chloride, $HCl(g)$, react to form solid ammonium chloride, $NH_4Cl(s)$:

$$NH_3(g) + HCl(g) \rightarrow NH_4Cl(s)$$

Two 2.00-L flasks at 25°C are connected by a stopcock, as shown in the drawing. One flask contains 5.00 g $NH_3(g)$, and the other contains 5.00 g $HCl(g)$.

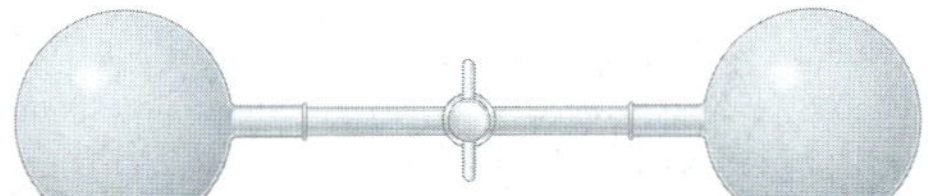

When the stopcock is opened, the gases react until one is completely consumed. **(a)** Which gas will remain in the system after the reaction is complete? **(b)** What will be the final pressure of the system after the reaction is complete? (Neglect the volume of the ammonium chloride formed.)

10.98 A sample of 1.42 g of helium and an unweighed quantity of O_2 are mixed in a flask at room temperature. The partial pressure of helium in the flask is 42.5 torr, and the partial pressure of oxygen is 158 torr. What is the mass of the oxygen in the container?

[10.99] A gaseous mixture of O_2 and Kr has a density of 1.104 g/L at 435 torr and 300 K. What is the mole percent O_2 in the mixture?

[10.100] A glass vessel fitted with a stopcock has a mass of 337.428 g when evacuated. When filled with Ar, it has a mass of 339.854 g. When evacuated and refilled with a mixture of Ne and Ar, under the same conditions of temperature and pressure, it weighs 339.076 g. What is the mole percent of Ne in the gas mixture?

[10.101] The density of a gas of unknown molar mass was measured as a function of pressure at 0°C, as in the table below. **(a)** Determine a precise molar mass for the gas. [*Hint:* Graph d/P versus P.] **(b)** Why is d/P not a constant as a function of pressure?

Pressure (atm)	1.00	0.666	0.500	0.333	0.250
Density (g/L)	2.3074	1.5263	1.1401	0.7571	0.5660

10.102 Does the effect of intermolecular attraction on the properties of a gas become more significant or less significant if **(a)** the gas is compressed to a smaller volume at constant temperature; **(b)** the temperature of the gas is increased at constant volume.

10.103 For nearly all real gases, the quantity PV/RT decreases below the value of 1, which characterizes an ideal gas, as pressure on the gas increases. At much higher pressures, however, PV/RT increases and rises above the value of 1. **(a)** Explain the initial drop in value of PV/RT below 1 and the fact that it rises above 1 for still higher pressures. **(b)** The effects we have just noted are smaller for gases at higher temperature. Why is this so?

[10.104] Large amounts of nitrogen gas are used in the manufacture of ammonia, principally for use in fertilizers. Suppose 120.00 kg of $N_2(g)$ is stored in a 1100.0-L metal cylinder at 280°C. **(a)** Calculate the pressure of the gas, assuming ideal-gas behavior. **(b)** By using data in Table 10.3, calculate the pressure of the gas according to the van der Waals equation. **(c)** Under the conditions of this problem, which correction dominates, the one for finite volume of gas molecules or the one for attractive interactions?

Integrative Exercises

10.105 Cyclopropane, a gas used with oxygen as a general anesthetic, is composed of 85.7% C and 14.3% H by mass. **(a)** If 1.56 g of cyclopropane has a volume of 1.00 L at 0.984 atm and 50.0°C, what is the molecular formula of cyclopropane? **(b)** Judging from its molecular formula, would you expect cyclopropane to deviate more or less than Ar from ideal-gas behavior at moderately high pressures and room temperature? Explain.

10.106 Consider the combustion reaction between 25.0 mL of liquid methanol (density = 0.850 g/mL) and 12.5 L of oxygen gas measured at STP. The products of the reaction are $CO_2(g)$ and $H_2O(g)$. Calculate the number of moles of H_2O formed if the reaction goes to completion.

[10.107] A 4.00-g sample of a mixture of CaO and BaO is placed in a 1.00-L vessel containing CO_2 gas at a pressure of 730 torr and a temperature of 25°C. The CO_2 reacts with the CaO and BaO, forming $CaCO_3$ and $BaCO_3$. When the reaction is complete, the pressure of the remaining CO_2 is 150 torr. **(a)** Calculate the number of moles of CO_2 that have reacted. **(b)** Calculate the mass percentage of CaO in the mixture.

10.108 In the "Chemistry at Work" box on pipelines, Section 10.5, it is mentioned that the total deliverability of natural gas (methane, CH_4) to the various regions of the United States is on the order of 2.7×10^{12} L per day, measured at STP. Calculate the total enthalpy change for combustion of this quantity of methane. (*Note*: Less than this amount of methane is actually combusted daily. Some of the delivered gas is passed through to other regions.)

[10.109] A gas forms when elemental sulfur is heated carefully with AgF. The initial product boils at 15°C. Experiments on several samples yielded a gas density of 0.803 ± 0.010 g/L for the gas at 150 mm pressure and 32°C. When the gas reacts with water, all the fluorine is converted to aqueous HF. Other products are elemental sulfur, S_8, and other sulfur-containing compounds. A 480-mL sample of the dry gas at 126 mm pressure and 28°C, when reacted with 80 mL of water, yielded a 0.081 *M* solution of HF. The initial gaseous product undergoes a transformation over a period of time to a second compound with the same empirical and molecular formula, which boils at −10°C. **(a)** Determine the empirical and molecular formulas of the first compound formed. **(b)** Draw at least two reasonable Lewis structures that represent the initial compound and the one into which it is transformed over time. **(c)** Describe the likely geometries of these compounds, and estimate the single bond distances, given that the S—S bond distance in S_8 is 2.04 Å and the F—F distance in F_2 is 1.43 Å

10.110 Chlorine dioxide gas (ClO_2) is used as a commercial bleaching agent. It bleaches materials by oxidizing them. In the course of these reactions, the ClO_2 is itself reduced. **(a)** What is the Lewis structure for ClO_2? **(b)** Why do you think that ClO_2 is reduced so readily? **(c)** When a ClO_2 molecule gains an electron, the chlorite ion, ClO_2^-, forms. Draw the Lewis structure for ClO_2^-. **(d)** Predict the O—Cl—O bond angle in the ClO_2^- ion. **(e)** One method of preparing ClO_2 is by the reaction of chlorine and sodium chlorite:

$$Cl_2(g) + 2\,NaClO_2(s) \longrightarrow 2\,ClO_2(g) + 2\,NaCl(s).$$

If you allow 10.0 g of $NaClO_2$ to react with 2.00 L of chlorine gas at a pressure of 1.50 atm at 21°C, how many grams of ClO_2 can be prepared?

[10.111] Natural gas is very abundant in many Middle Eastern oil fields. However, the costs of shipping the gas to markets in other parts of the world are high because it is necessary to liquefy the gas, which is mainly methane and thus has a boiling point at atmospheric pressure of −164°C. One possible strategy is to oxidize the methane to methanol, CH_3OH, which has a boiling point of 65° C and can therefore be shipped more readily. Suppose that 10.7×10^9 ft^3 of methane at atmospheric pressure and 25°C are oxidized to methanol. **(a)** What volume of methanol is formed if the density of CH_3OH is 0.791 g/mL? **(b)** Write balanced chemical equations for the oxidations of methane and methanol to $CO_2(g)$ and $H_2O(l)$. Calculate the total enthalpy change for complete combustion of the 10.7×10^9 ft^3 of methane described above and for complete combustion of the equivalent amount of methanol, as calculated in part (a).

(c) Methane, when liquefied, has a density of 0.466 g/mL; the density of methanol at 25°C is 0.791 g/mL. Compare the enthalpy change upon combustion of a unit volume of liquid methane and liquid methanol. From the standpoint of energy production, which substance has the higher enthalpy of combustion per unit volume?

[10.112] Gaseous iodine pentafluoride, IF_5, can be prepared by the reaction of solid iodine and gaseous fluorine:

$$I_2(s) + 5\,F_2(g) \longrightarrow 2\,IF_5(g)$$

A 5.00-L flask containing 10.0 g I_2 is charged with 10.0 g F_2, and the reaction proceeds until one of the reagents is completely consumed. After the reaction is complete, the temperature in the flask is 125°C. **(a)** What is the partial pressure of IF_5 in the flask? **(b)** What is the mole fraction of IF_5 in the flask?

[10.113] A 6.53-g sample of a mixture of magnesium carbonate and calcium carbonate is treated with excess hydrochloric acid. The resulting reaction produces 1.72 L of carbon dioxide gas at 28°C and 743 torr pressure. **(a)** Write balanced chemical equations for the reactions that occur between hydrochloric acid and each component of the mixture. **(b)** Calculate the total number of moles of carbon dioxide that forms from these reactions. **(c)** Assuming that the reactions are complete, calculate the percentage by mass of magnesium carbonate in the mixture.

eMEDIA EXERCISES

These exercises make use of the interactive objects available online in OneKey or the Companion Website, and on your Accelerator CD. Access to these resources comes in your MediaPak.

10.114 Using the **Gas Laws** activity (*10.3*), select a mass and a pressure to be held constant, and compare the volumes of N_2 and Xe at various temperatures. Under identical conditions—the same mass at the same pressure and temperature—are the volumes of N_2 and Xe equal? If not, explain why.

10.115 The ***P–V* Relationships** animation (*10.3*) illustrates Boyle's law and points out that this law holds only when temperature is constant. **(a)** Reproduce the pressure versus volume graph presented in the movie. **(b)** Using the ideal-gas equation, deduce and superimpose on your graph from part (a) the line you would expect on the *P–V* plot at a temperature higher than the original and at a temperature lower than the original. **(c)** Do the same for the *V* versus $1/P$ plot.

10.116 Automobile air bags are inflated by the explosive decomposition of sodium azide, as shown in the **Air Bags** animation (*10.5*). **(a)** If an air bag is to be inflated with 40.0 L of gas, initially at 110°C and 1.05 atm pressure, what mass of sodium azide must be available for decomposition? **(b)** What does the notation O(*s*) stand for in the decomposition reaction? **(c)** Why is it important that the reactants include an oxidant to react with the sodium metal produced by the decomposition?

10.117 Use the **Density of Gases** activity (*10.5*) to compare the densities of two different gases at the same pressure and temperature. Explain in terms of kinetic-molecular theory why the molar mass of a gas is necessary to calculate the density of a gas, but not to determine its pressure.

$$d = \frac{P\mathcal{M}}{RT} \qquad P = \frac{nRT}{V}$$

10.118 The relative speeds of helium and neon atoms are shown in the **Kinetic Energy of a Gas** animation (*10.7*). **(a)** If the average kinetic energies of both gases are the same at a given temperature, determine how much faster helium atoms are moving (on average) than neon atoms. **(b)** How does the fact that average kinetic energy of a gas is directly proportional to absolute temperature explain Boyle's observation that pressure decreases with increasing volume at constant temperature? **(c)** How does it explain Charles's observation that pressure increases with increasing temperature at constant volume?

eLABORATORY EXERCISES

These exercises allow students to apply the concepts and skills for this chapter in the simulated laboratory of Virtual ChemLab. Worksheets for each of these exercises are found in the Virtual ChemLab Workbook in your MediaPak.

10.119 ***(VCL 10-1) Boyle's Law: Pressure and Volume***
Robert Boyle, a philosopher and theologian, studied the properties of gases in the seventeenth century. He noticed that gases behave similarly to springs; when compressed or expanded, they tend to spring back to their original volume. He published his findings in 1662 in a monograph entitled *The Spring of the Air and Its Effects*. You will make observations similar to those of Robert Boyle and learn about the relationship between the pressure and volume of an ideal gas.

10.120 ***(VCL 10-2) Charles's Law: Temperature and Volume***
Charles's law was discovered by Joseph Louis Gay-Lussac in 1802, based on unpublished work done by Jacques Charles in about 1787. Charles had found that a number of gases expand to the same extent over the same 80° temperature interval. You will be observing the relationship between the temperature and volume of a gas similar to that studied by Charles.

10.121 ***(VCL 10-3) Avogadro's Law: Moles and Volume***
In 1808, Joseph Gay-Lussac observed the *law of combining values*, which states that the volumes of gases that react with one another react in the ratio of small whole numbers. Three years later, Amedeo Avogadro built upon this observation by proposing what is now known

as *Avogadro's hypothesis*: Equal volumes of gases at the same temperature and pressure contain equal numbers of molecules. Avogadro's law, which states the relationship between moles and volume, followed from his hypothesis. You will be observing the same principle that Avogadro stated 200 years ago.

10.122 ***(VCL 10-4) Derivation of the Ideal-Gas Law***
An ideal gas is a hypothetical gas whose pressure, volume, and temperature follow the relationship $PV = nRT$. Ideal gases do not actually exist, although all real gases can behave like an ideal gas at certain temperatures and pressures. All gases can be described to some extent using the ideal-gas law, and it is important in our understanding of how all gases behave. In this assignment you will derive the ideal-gas law from experimental observations.

10.123 ***(VCL 10-5) Dalton's Law of Partial Pressure***
Dalton's law of partial pressure, named after its discoverer John Dalton, describes the behavior of gas mixtures. It states that the total pressure of the gas, P_{tot}, is the sum of the partial pressures of each gas, or the sum of the pressures that each gas would exert if it were alone in the container. In this assignment you will become more familiar with Dalton's law.

10.124 ***(VCL 10-6) Ideal versus Real Gases***
At room temperature and normal atmospheric pressures, real gases behave similarly to ideal gases, but real gases can deviate significantly from ideal-gas behavior at extreme conditions such as high temperatures and high pressures. It is not always easy to find an effective means to show the deviation from ideal-gas behavior for real gases. An effective but simple method to see these deviations is to calculate the value of R, the ideal-gas constant, for real gases, *assuming* they follow the ideal-gas law, and then compare the value for R against the actual value of 0.08205 L-atm/mol-K. In this assignment you will measure P, V, T, and n for various real gases under different conditions and then calculate the value for R and compare it to its actual value.

10.125 ***(VCL 10-7) The Effect of Mass on Pressure***
An understanding of pressure is an integral part of our understanding of the behavior of gases. Pressure is defined as the force per unit area exerted by a gas or other medium. The pressure of a gas is affected by many variables, such as temperature, external pressure, volume, moles of a gas, and other factors. This assignment will help you become more familiar with pressure and the effect of adding mass to a frictionless, massless piston.

CHAPTER

11

WHAT'S AHEAD

- We begin with a brief comparison of solids, liquids, and gases from a molecular perspective, which reveals the important roles that temperature and *intermolecular forces* play in determining the physical state of a substance. *(Section 11.1)*
- We then examine the main kinds of intermolecular forces that occur within and between substances: *ion-dipole forces, dipole-dipole forces, London dispersion forces*, and *hydrogen bonds*. *(Section 11.2)*
- We will learn that the nature and strength of the intermolecular forces between molecules are largely responsible for many properties of liquids, including their *viscosity*, which is a measure of a liquid's resistance to flow, and *surface tension*, which is a measure of a liquid's resistance to increasing its surface area. *(Section 11.3)*
- We'll explore *phase changes*, the transitions of matter between the gaseous, liquid, and solid states, and the energy changes that accompany phase changes. *(Section 11.4)*
- We'll examine the *dynamic equilibrium* that exists between a liquid and its gaseous state and introduce the idea of *vapor pressure*. A liquid boils when its vapor pressure equals the pressure acting on the surface of the liquid. *(Section 11.5)*
- In a *phase diagram* the equilibria among the gaseous, liquid, and solid phases are displayed graphically. *(Section 11.6)*
- Finally, we'll examine solids. Orderly arrangements of molecules or ions in three dimensions characterize *crystalline solids*. We'll examine how the structure of a crystalline solid can be conveyed in terms of its *unit cell* and how simple molecules and ions are most efficiently arranged in three dimensions. *(Section 11.7)*
- Solids can be characterized and classified according to the attractive forces between their component atoms, molecules, or ions. We examine four such classes: *molecular solids, covalent-network solids, ionic solids*, and *metallic solids*. *(Section 11.8)*

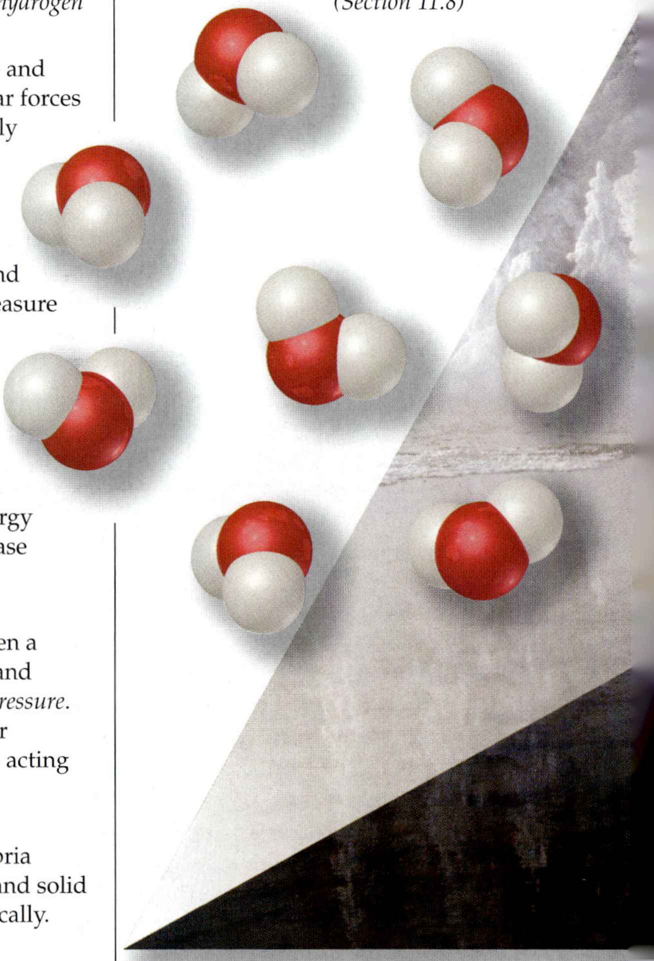

Intermolecular Forces, Liquids, and Solids

SITTING IN A HOT SPRING ON A SNOWY day is not something many of us have experienced. If we were in the hot spring shown in the chapter-opening photo, however, we would be surrounded simultaneously by all three phases of water—gas, liquid, and solid. The water vapor—or humidity—in the air, the water in the hot spring, and the surrounding snow are all forms of the same substance, H_2O. They all have the same chemical properties. Their physical properties differ greatly, however, because the physical properties of a substance depend on its physical state. In Chapter 10 we discussed the gaseous state in some detail. In this chapter we turn our attention to the physical properties of liquids and solids and to the phase changes that occur between the three states of matter.

Many of the substances that we will consider in this chapter are molecular. In fact, virtually all substances that are liquids at room temperature are molecular substances. The intramolecular forces *within* molecules that give rise to covalent bonding influence molecular shape, bond energies, and many aspects of chemical behavior. The physical properties of molecular liquids and solids, however, are due largely to **intermolecular forces**, the forces that exist *between* molecules. We learned in Section 10.9 that attractions between gas molecules lead to deviations from ideal-gas behavior. But how do these intermolecular attractions arise? By understanding the nature and strength of intermolecular forces, we can begin to relate the composition and structure of molecules to their physical properties.

THE WATER IN HOT SPRINGS is at least 5–10°C warmer than the mean annual air temperature where they are located. There are over 1000 hot springs in the United States, the majority in the West. Some hot springs are popular for bathing while others are far too hot for people to safely enter them.

TABLE 11.1 Some Characteristic Properties of the States of Matter

Gas	Assumes both the volume and shape of its container Is compressible Flows readily Diffusion within a gas occurs rapidly
Liquid	Assumes the shape of the portion of the container it occupies Does not expand to fill container Is virtually incompressible Flows readily Diffusion within a liquid occurs slowly
Solid	Retains its own shape and volume Is virtually incompressible Does not flow Diffusion within a solid occurs extremely slowly

MOVIE
Physical Properties of the Halogens

11.1 | A Molecular Comparison of Gases, Liquids, and Solids

Some of the characteristic properties of gases, liquids, and solids are listed in Table 11.1 ▲. These properties can be understood in terms of the energy of motion (kinetic energy) of the particles of each state compared to the intermolecular forces between those particles. As we learned from the kinetic-molecular theory of gases in Chapter 10, the average kinetic energy, which is related to the particle's average speed, is proportional to the absolute temperature.

Gases consist of a collection of widely separated molecules in constant, chaotic motion. The average energy of the attractions between the molecules is much smaller than their average kinetic energy. The lack of strong attractive forces between molecules allows a gas to expand to fill its container.

In liquids the intermolecular attractive forces are strong enough to hold molecules close together. Thus, liquids are much denser and far less compressible than gases. Unlike gases, liquids have a definite volume, independent of the size and shape of their container. The attractive forces in liquids are not strong enough, however, to keep the molecules from moving past one another. Thus, any liquid can be poured, and it assumes the shape of whatever portion of its container it occupies.

In solids the intermolecular attractive forces are strong enough not only to hold molecules close together, but to virtually lock them in place. Solids, like liquids, are not very compressible because the molecules have little free space between them. Because the particles in a solid or liquid are fairly close together compared with those of a gas, we often refer to solids and liquids as *condensed phases*. Often the molecules of a solid take up positions in a highly regular pattern. Solids that possess highly ordered structures are said to be *crystalline*. (The transition from a liquid to a crystalline solid is rather like the change that occurs on a military parade ground when the troops are called to formation.) Because the particles of a solid are not free to undergo long-range movement, solids are rigid. Keep in mind, however, that the units that form the solid, whether ions or molecules, possess thermal energy and vibrate in place. This vibrational motion increases in amplitude as a solid is heated. In fact, the energy may increase to the point that the solid either melts or sublimes.

Figure 11.1 ▶ compares the three states of matter. The particles that compose the substance can be individual atoms, as in Ar; molecules, as in H_2O; or ions, as in NaCl. *The state of a substance depends largely on the balance between the kinetic energies of the particles and the interparticle energies of attraction.* The kinetic energies, which depend on temperature, tend to keep the particles apart and moving. The interparticle attractions tend to draw the particles together. Sub-

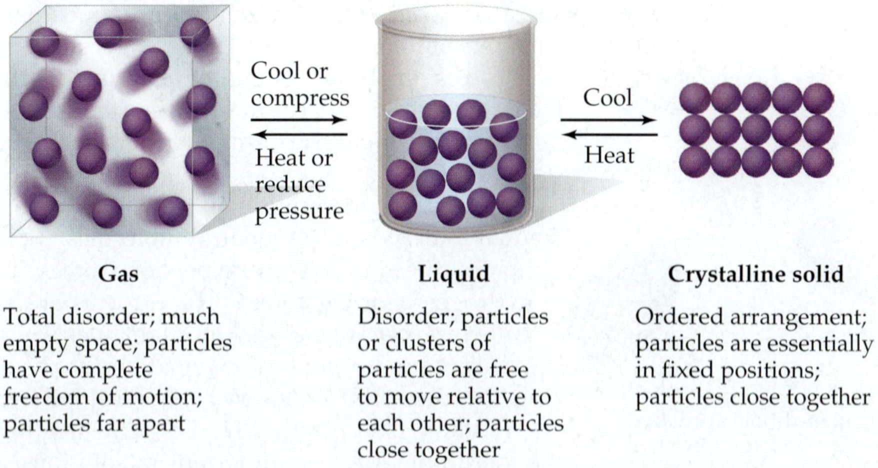

◀ **Figure 11.1 Molecular-level comparison of gases, liquids, and solids.** The particles can be atoms, ions, or molecules. The density of particles in the gas phase is exaggerated compared with most real situations.

ACTIVITY
Phases of Matter

stances that are gases at room temperature have weaker interparticle attractions than those that are liquids; substances that are liquids have weaker interparticle attractions than those that are solids.

We can change a substance from one state to another by heating or cooling, which changes the average kinetic energy of the particles. NaCl, for example, which is a solid at room temperature, melts at 801°C and boils at 1413°C under 1 atm pressure. N_2O, on the other hand, which is a gas at room temperature, liquefies at −88.5°C and solidifies at −90.8°C under 1 atm pressure. As the temperature of a gas decreases, the average kinetic energy of its particles decreases, allowing the attractions between the particles to first draw the particles close together, forming a liquid, and then virtually locking them in place, forming a solid.

Increasing the pressure on a gas forces the molecules closer together, which in turn increases the strength of the intermolecular forces of attraction. Propane (C_3H_8) is a gas at room temperature and 1 atm pressure, whereas liquefied propane (LP) gas is a liquid at room temperature because it is stored under much higher pressure.

GIVE IT SOME THOUGHT

How does the energy of attraction between particles compare with their kinetic energies in (a) a gas, (b) a solid?

11.2 | Intermolecular Forces

The strengths of intermolecular forces of different substances vary over a wide range, but they are generally much weaker than ionic or covalent bonds (Figure 11.2 ▶). Less energy, therefore, is required to vaporize, or evaporate, a liquid or to melt a solid than to break covalent bonds in molecules. For example, only 16 kJ/mol is required to overcome the intermolecular attractions between HCl molecules in liquid HCl in order to vaporize it. In contrast, the energy required to break the covalent bond to dissociate HCl into H and Cl atoms is 431 kJ/mol. Thus, when a molecular substance like HCl changes from solid to liquid to gas, the molecules themselves remain intact.

▼ **Figure 11.2 Intermolecular attraction.** Comparison of a covalent bond (an intramolecular force) and an intermolecular attraction. Because intermolecular attractions are weaker than covalent bonds, they are usually represented by dashes or dots.

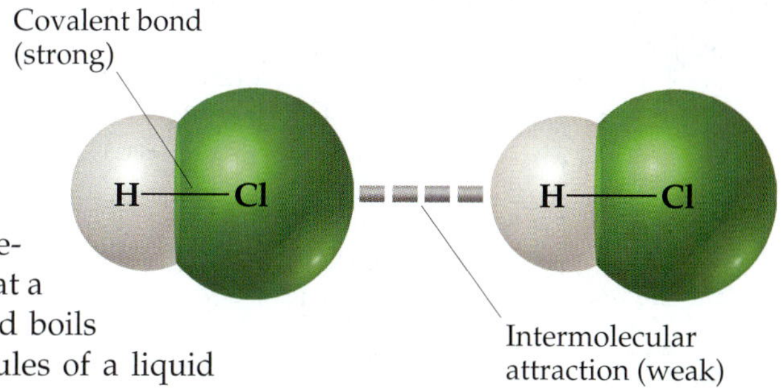

Many properties of liquids, including their *boiling points*, reflect the strengths of the intermolecular forces. For example, because the forces between HCl molecules are so weak, HCl boils at a very low temperature, −85°C at atmospheric pressure. A liquid boils when bubbles of its vapor form within the liquid. The molecules of a liquid must overcome their attractive forces in order to separate and form a vapor. The stronger the attractive forces, the higher is the temperature at which the liquid boils. Similarly, the *melting points* of solids increase as the strengths of the intermolecular forces increase.

Three types of intermolecular attractive forces are known to exist between neutral molecules: dipole-dipole forces, London dispersion forces, and hydrogen-bonding forces. These forces are also called *van der Waals forces* after Johannes van der Waals, who developed the equation for predicting the deviation of gases from ideal behavior. (Section 10.9) Another kind of attractive force, the ion-dipole force, is important in solutions. All four forces are electrostatic in nature, involving attractions between positive and negative species. All tend to be less than 15% as strong as covalent or ionic bonds.

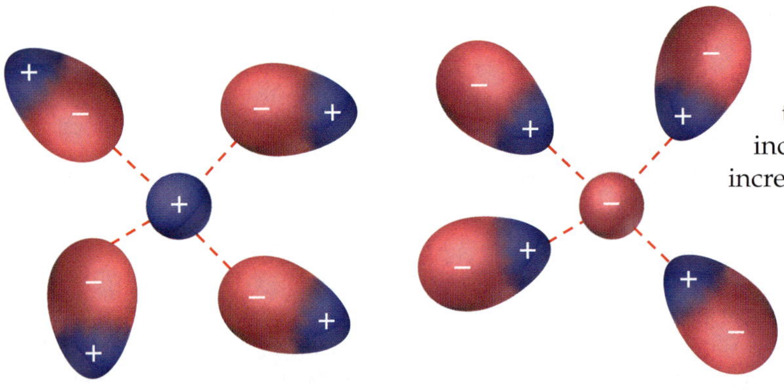

Cation-dipole attractions

(a)

Anion-dipole attractions

(b)

▲ **Figure 11.3 Ion-dipole attractions.** Illustration of the preferred orientations of polar molecules toward ions. The negative end of the dipoles is oriented toward a cation (a), and the positive end of the dipoles is oriented toward an anion (b).

Ion-Dipole Forces

An **ion-dipole force** exists between an ion and the partial charge on the end of a polar molecule. Polar molecules are dipoles; they have a positive end and a negative end. (Section 9.3) HCl is a polar molecule, for example, because the electronegativities of the H and Cl atoms differ.

Positive ions are attracted to the negative end of a dipole, whereas negative ions are attracted to the positive end, as shown in Figure 11.3 ◀. The magnitude of the attraction increases as either the charge of the ion or the magnitude of the dipole moment increases. Ion-dipole forces are especially important for solutions of ionic substances in polar liquids, such as a solution of NaCl in water. (Section 4.1) We will discuss these solutions in more detail in Section 13.1.

GIVE IT SOME THOUGHT

In which of the following mixtures do you encounter ion-dipole forces: CH_3OH in water or $Ca(NO_3)_2$ in water?

Dipole-Dipole Forces

Neutral polar molecules attract each other when the positive end of one molecule is near the negative end of another, as in Figure 11.4(a) ◀. These **dipole-dipole forces** are effective only when polar molecules are very close together, and they are generally weaker than ion-dipole forces.

In liquids polar molecules are free to move with respect to one another. As shown in Figure 11.4(b), they will sometimes be in an orientation that is attractive (red solid lines) and sometimes in an orientation that is repulsive (blue dashed lines). Two molecules that are attracting each other spend more time near each other than do two that are repelling each other. Thus, the overall effect is a net attraction. When we examine various liquids, we find that *for molecules of approximately equal mass and size, the strengths of intermolecular attractions increase with increasing polarity*. We can see this trend in Table 11.2 ▶, which lists several substances with similar molecular weights but different dipole moments. Notice that the boiling point increases as the dipole moment increases. For dipole-dipole forces to operate, the molecules must be able to get close together in the correct

▼ **Figure 11.4 Dipole-dipole attractions.** The interaction of many dipoles in a condensed state. There are both repulsive interactions between like charges and attractive interactions between unlike charges, but the attractive interactions predominate.

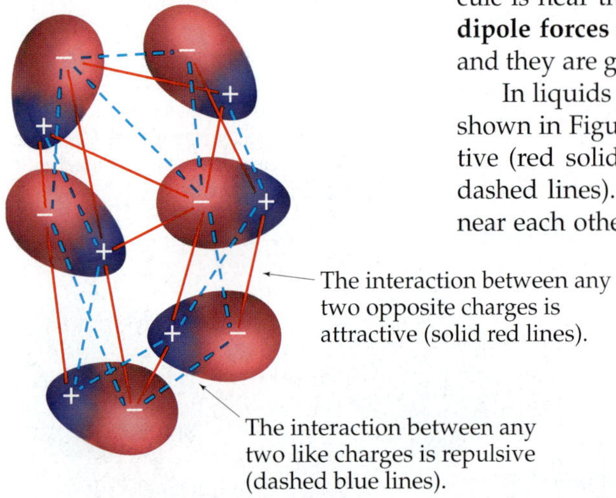

TABLE 11.2 Molecular Weights, Dipole Moments, and Boiling Points of Several Simple Organic Substances

Substance	Molecular Weight (amu)	Dipole Moment μ (D)	Boiling Point (K)
Propane, $CH_3CH_2CH_3$	44	0.1	231
Dimethyl ether, CH_3OCH_3	46	1.3	248
Methyl chloride, CH_3Cl	50	1.9	249
Acetaldehyde, CH_3CHO	44	2.7	294
Acetonitrile, CH_3CN	41	3.9	355

orientation. For molecules of comparable polarity, therefore, those with smaller molecular volumes generally experience higher dipole-dipole attractive forces.

GIVE IT SOME THOUGHT

For which of the substances in Table 11.2 are the dipole-dipole attractive forces greatest?

London Dispersion Forces

There can be no dipole-dipole forces between nonpolar atoms and molecules. There must be some kind of attractive interactions, however, because nonpolar gases can be liquefied. The origin of this attraction was first proposed in 1930 by Fritz London, a German-American physicist. London recognized that the motion of electrons in an atom or molecule can create an *instantaneous*, or momentary, dipole moment.

In a collection of helium atoms, for example, the *average* distribution of the electrons about each nucleus is spherically symmetrical. The atoms are nonpolar and possess no permanent dipole moment. The *instantaneous* distribution of the electrons, however, can be different from the average distribution. If we could freeze the motion of the electrons in a helium atom at any given instant, both electrons could be on one side of the nucleus. At just that instant, then, the atom would have an instantaneous dipole moment.

Because electrons repel one another, the motions of electrons on one atom influence the motions of electrons on its neighbors. Thus, the temporary dipole on one atom can induce a similar temporary dipole on an adjacent atom, causing the atoms to be attracted to each other as shown in Figure 11.5 ▼. This attractive interaction is called the **London dispersion force** (or merely the *dispersion force*). This force, like dipole-dipole forces, is significant only when molecules are very close together.

The strength of the dispersion force depends on the ease with which the charge distribution in a molecule can be distorted to induce a momentary dipole. The ease with which the electron distribution in a molecule is distorted is called its **polarizability**. We can think of the polarizability of a molecule as a measure of the "squashiness" of its electron cloud; the greater the polarizability

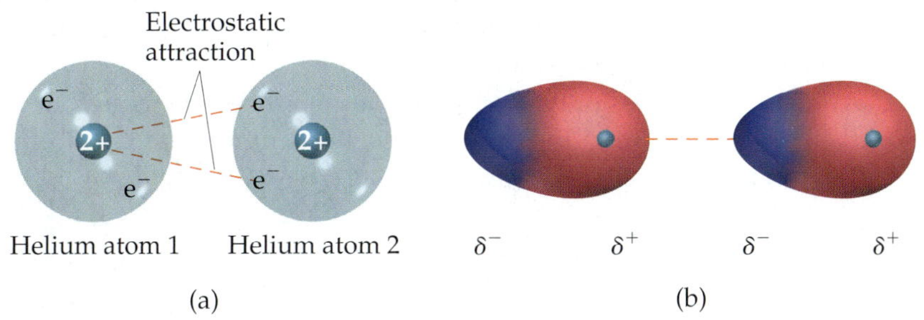

◀ **Figure 11.5 Dispersion forces.** On the average, the charge distribution in the helium atoms is spherical, as represented by the spheres in (a). At a particular instant, however, there can be a nonspherical arrangement of the electrons, as indicated by the location of the electrons (e^-) in (a) and by the nonspherical shape of the electron cloud in (b). The nonspherical electron distributions produce momentary dipoles and allow momentary electrostatic attractions between atoms that are called London dispersion forces or merely dispersion forces.

TABLE 11.3 Boiling Points of the Halogens and the Noble Gases

Halogen	Molecular Weight (amu)	Boiling Point (K)	Noble Gas	Molecular Weight (amu)	Boiling Point (K)
F_2	38.0	85.1	He	4.0	4.6
Cl_2	71.0	238.6	Ne	20.2	27.3
Br_2	159.8	332.0	Ar	39.9	87.5
I_2	253.8	457.6	Kr	83.8	120.9
			Xe	131.3	166.1

of the molecule, the more easily its electron cloud can be distorted to give a momentary dipole. Therefore, more polarizable molecules have stronger dispersion forces. In general, larger molecules tend to have greater polarizabilities because they have a greater number of electrons and their electrons are farther from the nuclei. The strength of the dispersion forces, therefore, tends to increase with increasing molecular size. Because molecular size and mass generally parallel each other, *dispersion forces tend to increase in strength with increasing molecular weight*. Thus, the boiling points of the halogens and the noble gases increase with increasing molecular weight (Table 11.3 ▲).

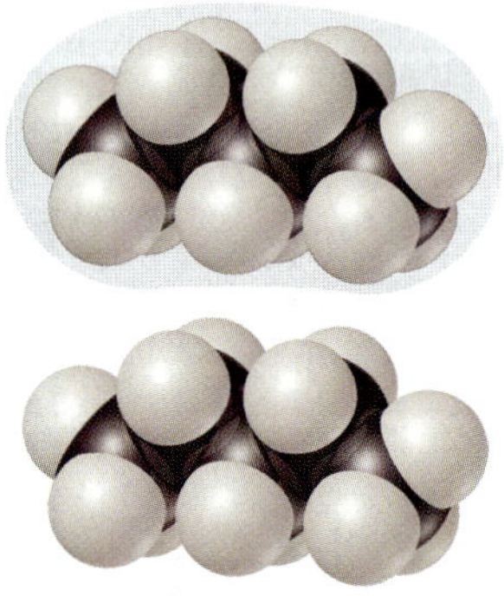

n-Pentane
(bp = 309.4 K)

Neopentane
(bp = 282.7 K)

▲ **Figure 11.6 Molecular shape affects intermolecular attraction.** The *n*-pentane molecules make more contact with each other than do the neopentane molecules. Thus, *n*-pentane has the greater intermolecular attractive forces and therefore has the higher boiling point (bp).

GIVE IT SOME THOUGHT

List the substances CCl_4, CBr_4, and CH_4 in order of increasing (a) polarizability, (b) strength of dispersion forces.

The shapes of molecules also influence the magnitudes of dispersion forces. For example, *n*-pentane* and neopentane, illustrated in Figure 11.6 ◀, have the same molecular formula (C_5H_{12}), yet the boiling point of *n*-pentane is 27 K higher than that of neopentane. The difference can be traced to the different shapes of the two molecules. The overall attraction between molecules is greater for *n*-pentane because the molecules can come in contact over the entire length of the long, somewhat cylindrically shaped molecule. Less contact is possible between the more compact and nearly spherical molecules of neopentane.

Dispersion forces operate between all molecules, whether they are polar or nonpolar. Polar molecules experience dipole-dipole interactions, but they also experience dispersion forces at the same time. In fact, dispersion forces between polar molecules commonly contribute more to intermolecular attractions than do dipole-dipole forces. In liquid HCl, for example, dispersion forces are estimated to account for more than 80% of the total attraction between molecules; dipole-dipole attractions account for the rest.

When comparing the relative strengths of intermolecular attractions in two substances, the following generalizations should be considered:

1. When the molecules of two substances have comparable molecular weights and shapes, dispersion forces are approximately equal in the two substances. In this case differences in the magnitudes of the attractive forces are due to differences in the strengths of dipole-dipole attractions, with the more polar molecules having the stronger attractions.
2. When the molecules of two substances differ widely in molecular weights, dispersion forces tend to be decisive in determining which substance has the stronger intermolecular attractions. In this case differences in the mag-

* The *n* in *n*-pentane is an abbreviation for the word *normal*. A normal hydrocarbon is one in which the carbon atoms are arranged in a straight chain. ∞ (Section 2.9)

nitudes of the attractive forces can usually be associated with differences in molecular weights, with the substance made up of the more massive molecule having the strongest attractions.

The hydrogen bond, which we consider after Sample Exercise 11.1, is a special type of dipole-dipole attraction that is typically stronger than dispersion forces.

SAMPLE EXERCISE 11.1 | Comparing Intermolecular Forces

The dipole moments of acetonitrile, CH_3CN, and methyl iodide, CH_3I, are 3.9 D and 1.62 D, respectively. **(a)** Which of these substances will have the greater dipole-dipole attractions among its molecules? **(b)** Which of these substances will have the greater London dispersion attractions? **(c)** The boiling points of CH_3CN and CH_3I are 354.8 K and 315.6 K, respectively. Which substance has the greater overall attractive forces?

Solution **(a)** Dipole-dipole attractions increase in magnitude as the dipole moment of the molecule increases. Thus, CH_3CN molecules attract each other by stronger dipole-dipole forces than CH_3I molecules do. **(b)** When molecules differ in their molecular weights, the more massive molecule generally has the stronger dispersion attractions. In this case CH_3I (142.0 amu) is much more massive than CH_3CN (41.0 amu), so the dispersion forces will be stronger for CH_3I. **(c)** Because CH_3CN has the higher boiling point, we can conclude that more energy is required to overcome attractive forces between CH_3CN molecules. Thus, the total intermolecular attractions are stronger for CH_3CN, suggesting that dipole-dipole forces are decisive when comparing these two substances. Nevertheless, dispersion forces play an important role in determining the properties of CH_3I.

PRACTICE EXERCISE

Of Br_2, Ne, HCl, HBr, and N_2, which is likely to have **(a)** the largest intermolecular dispersion forces, **(b)** the largest dipole-dipole attractive forces?
Answers: **(a)** Br_2 (largest molecular weight), **(b)** HCl (largest polarity)

ANIMATION
Hydrogen Bonding

Hydrogen Bonding

Figure 11.7 ▶ shows the boiling points of the simple hydrogen compounds of group 4A and 6A elements. In general, the boiling point increases with increasing molecular weight, owing to increased dispersion forces. The notable exception to this trend is H_2O, whose boiling point is much higher than we would expect on the basis of its molecular weight. This observation indicates that there are stronger intermolecular attractions between H_2O molecules than there are between other molecules in the same group. The compounds NH_3 and HF likewise have abnormally high boiling points. In fact, these compounds have many characteristics that distinguish them from other substances of similar molecular weight and polarity. For example, water has a high melting point, a high specific heat, and a high heat of vaporization. Each of these properties indicates that the intermolecular forces between H_2O molecules are abnormally strong.

The strong intermolecular attractions in H_2O result from hydrogen bonding. **Hydrogen bonding** *is a special type of intermolecular attraction between the hydrogen atom in a polar bond (particularly an H—F, H—O, or H—N bond) and an unshared electron pair on a nearby small electronegative ion or atom (usually an F, O, or N atom in another molecule)*. For example, a hydrogen bond exists between the H atom in an HF molecule and the F atom of an adjacent HF molecule, F—H···F—H (where the dots represent the hydrogen

▼ **Figure 11.7 Boiling point as a function of molecular weight.** The boiling points of the group 4A (bottom) and 6A (top) hydrides are shown as a function of molecular weight. In general, the boiling points increase with increasing molecular weight, because of the increasing strength of dispersion forces. The very strong hydrogen-bonding forces between H_2O molecules, however, cause water to have an unusually high boiling point.

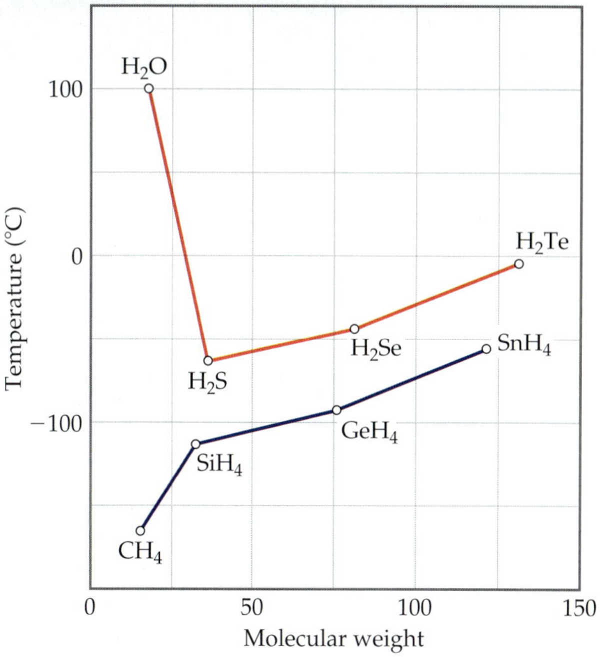

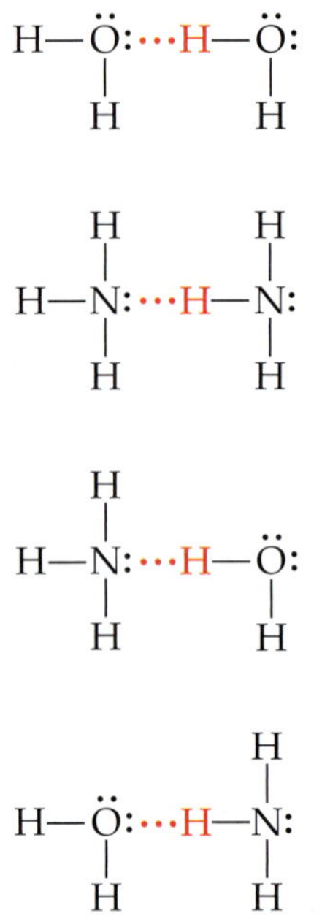

▲ **Figure 11.8 Examples of hydrogen bonding.** The solid lines represent covalent bonds; the red dotted lines represent hydrogen bonds.

ACTIVITY
Hydrogen Bonding in Water

bond between the molecules). Several additional examples are shown in Figure 11.8 ◀.

Hydrogen bonds can be considered unique dipole-dipole attractions. Because F, N, and O are so electronegative, a bond between hydrogen and any of these three elements is quite polar, with hydrogen at the positive end:

$$\overset{\longleftarrow\!\!+}{\text{N—H}} \qquad \overset{\longleftarrow\!\!+}{\text{O—H}} \qquad \overset{\longleftarrow\!\!+}{\text{F—H}}$$

The hydrogen atom has no inner core of electrons. Thus, the positive side of the bond dipole has the concentrated charge of the partially exposed, nearly bare proton of the hydrogen nucleus. This positive charge is attracted to the negative charge of an electronegative atom in a nearby molecule. Because the electron-poor hydrogen is so small, it can approach an electronegative atom very closely and thus interact strongly with it.

SAMPLE EXERCISE 11.2 | Identifying Substances that Can Form Hydrogen Bonds

In which of the following substances is hydrogen bonding likely to play an important role in determining physical properties: methane (CH_4), hydrazine (H_2NNH_2), methyl fluoride (CH_3F), or hydrogen sulfide (H_2S)?

Solution

Analyze: We are given the chemical formulas of four substances and asked to predict whether they can participate in hydrogen bonding. All of these compounds contain H, but hydrogen bonding usually occurs only when the hydrogen is covalently bonded to N, O, or F.

Plan: We can analyze each formula to see if it contains N, O, or F directly bonded to H. There also needs to be an unshared pair of electrons on an electronegative atom (usually N, O, or F) in a nearby molecule, which can be revealed by drawing the Lewis structure for the molecule.

Solve: The criteria listed above eliminate CH_4 and H_2S, which do not contain H bonded to N, O, or F. They also eliminate CH_3F, whose Lewis structure shows a central C atom surrounded by three H atoms and an F atom. (Carbon always forms four bonds, whereas hydrogen and fluorine form one each.) Because the molecule contains a C—F bond and not an H—F bond, it does not form hydrogen bonds. In H_2NNH_2, however, we find N—H bonds. If we draw the Lewis structure for the molecule, we see that there is a nonbonding pair of electrons on each N atom. Therefore, hydrogen bonds can exist between the molecules as depicted below.

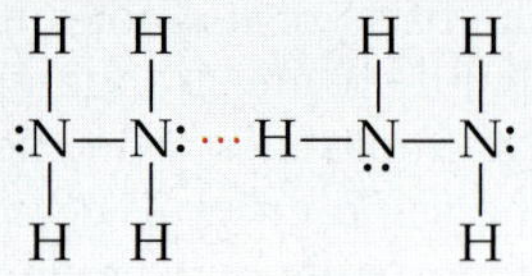

Check: While we can generally identify substances that participate in hydrogen bonding based on their containing N, O, or F covalently bonded to H, drawing the Lewis structure for the interaction, as shown above, provides a way to check the prediction.

PRACTICE EXERCISE

In which of the following substances is significant hydrogen bonding possible: methylene chloride (CH_2Cl_2), phosphine (PH_3), hydrogen peroxide (HOOH), or acetone (CH_3COCH_3)?
Answer: HOOH

The energies of hydrogen bonds vary from about 4 kJ/mol to 25 kJ/mol or so. Thus, they are much weaker than ordinary chemical bonds (see Table 8.4). Nevertheless, because hydrogen bonds are generally stronger than dipole-dipole or dispersion forces, they play important roles in many chemical systems, including those of biological significance. For example, hydrogen bonds help stabilize the structures of proteins, which are key parts of skin, muscles, and other structural components of animal tissues (see Section 25.9). They are also responsible for the way that DNA is able to carry genetic information. (Section 25.11)

One of the remarkable consequences of hydrogen bonding is found when the densities of ice and liquid water are compared. In most substances the molecules in the solid are more densely packed than in the liquid. Thus, the solid phase is denser than the liquid phase (Figure 11.9 ▶). By contrast, the density of ice at 0°C (0.917 g/mL) is less than that of liquid water at 0°C (1.00 g/mL), so ice floats on liquid water (Figure 11.9).

The lower density of ice compared to that of water can be understood in terms of hydrogen-bonding interactions between H_2O molecules. In ice the H_2O molecules assume an ordered, open arrangement as shown in Figure 11.10 ▼. This arrangement optimizes the hydrogen bonding interactions between molecules, with each H_2O molecule forming hydrogen bonds to four other H_2O molecules. These hydrogen bonds, however, create the open cavities shown in the structure. When the ice melts, the motions of the molecules cause

▲ **Figure 11.9 Comparing densities of liquid and solid phases.** As with most other substances, the solid phase of paraffin is denser than the liquid phase, and the solid therefore sinks below the surface of the liquid paraffin in the beaker on the left. In contrast, the solid phase of water, ice, is less dense than its liquid phase (right beaker), causing the ice to float on the water.

(a)

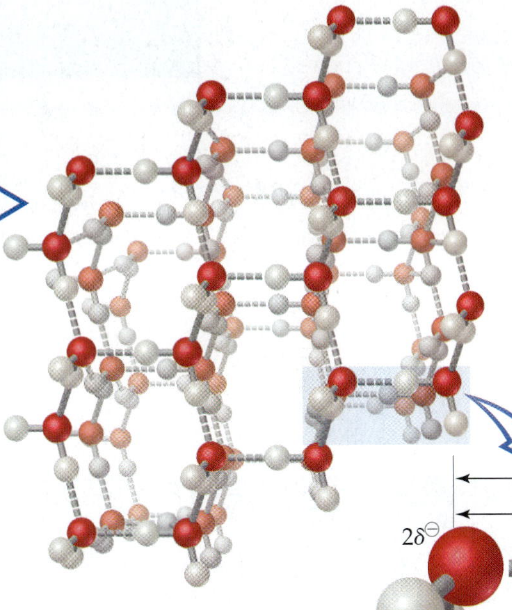

(b)

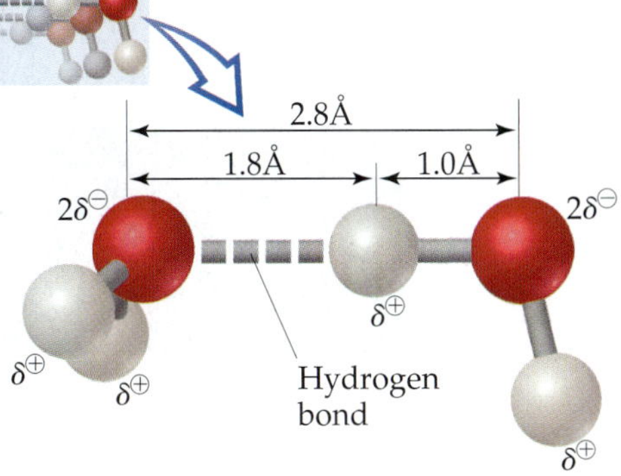

(c)

▲ **Figure 11.10 Hydrogen bonding in ice.** (a) The hexagonal shape is characteristic of snowflakes. (b) The arrangement of H_2O molecules in ice. Each hydrogen atom on one H_2O molecule is oriented toward a nonbonding pair of electrons on an adjacent H_2O molecule. As a result, ice has an open, hexagonal arrangement of H_2O molecules. (c) Hydrogen bonding between two water molecules. The distances shown are those found in ice.

▲ **Figure 11.11 Expansion of water upon freezing.** Water is one of the few substances that expand upon freezing. The expansion is due to the open structure of ice relative to that of liquid water.

3-D MODEL
Ice

the structure to collapse. The hydrogen bonding in the liquid is more random than in ice, but it is strong enough to hold the molecules close together. Consequently, liquid water has a more dense structure than ice, meaning that a given mass of water occupies a smaller volume than the same mass of ice.

The lower density of ice compared to liquid water profoundly affects life on Earth. Because ice floats (Figure 11.9), it covers the top of the water when a lake freezes in cold weather, thereby insulating the water below. If ice were more dense than water, ice forming at the top of a lake would sink to the bottom, and the lake could freeze solid. Most aquatic life could not survive under these conditions. The expansion of water upon freezing (Figure 11.11 ◀) is also what causes water pipes to break in freezing weather.

GIVE IT SOME THOUGHT

What is unusual about the relative densities of liquid water and ice?

Comparing Intermolecular Forces

We can identify the intermolecular forces that are operative in a substance by considering its composition and structure. *Dispersion forces are found in all substances.* The strengths of these forces increase with increased molecular weight and depend on molecular shapes. Dipole-dipole forces add to the effect of dispersion forces and are found in polar molecules. Hydrogen bonds, which require H atoms bonded to F, O, or N, also add to the effect of dispersion forces. Hydrogen bonds tend to be the strongest type of intermolecular force. None of these intermolecular forces, however, is as strong as ordinary ionic or covalent bonds. Figure 11.12 ▼ presents a systematic way of identifying the kinds of intermolecular forces in a particular system, including ion-dipole and ion-ion forces.

ACTIVITY
Intermolecular Forces

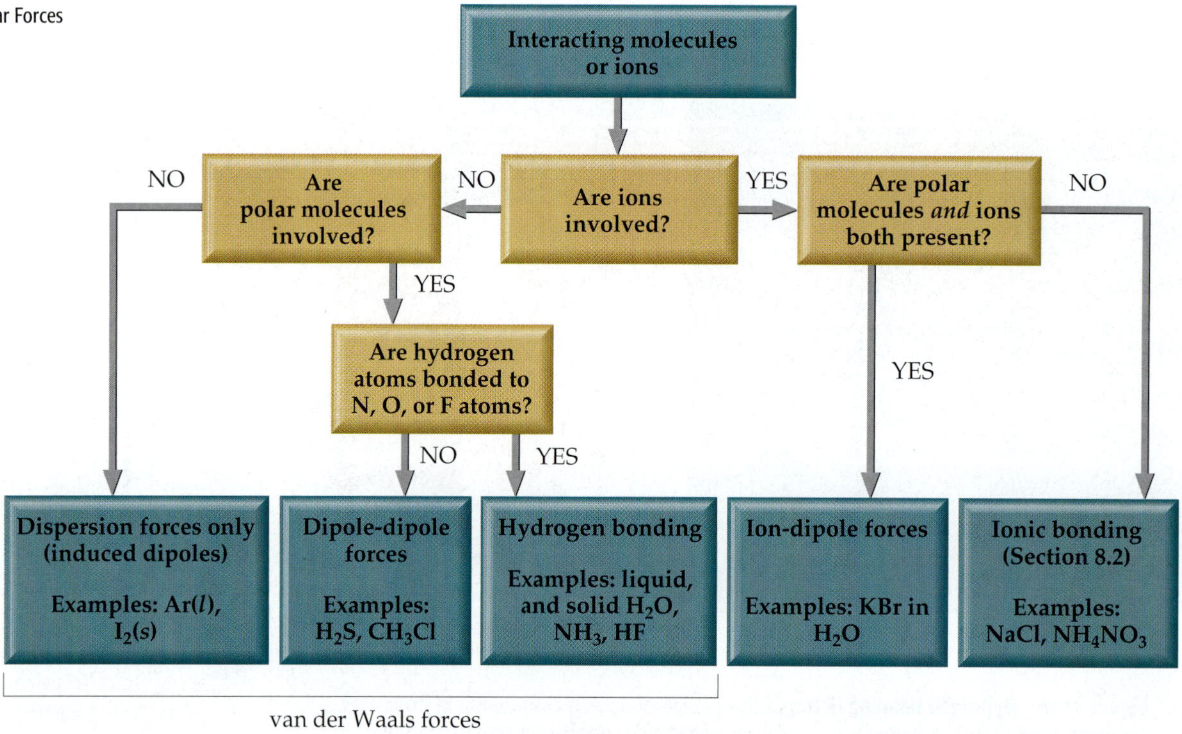

▲ **Figure 11.12 Flowchart for determining intermolecular forces.** London dispersion forces occur in all instances. The strengths of the other forces generally increase proceeding from left to right across the chart.

SAMPLE EXERCISE 11.3 | Predicting the Types and Relative Strengths of Intermolecular Forces

List the substances $BaCl_2$, H_2, CO, HF, and Ne in order of increasing boiling points.

Solution

Analyze: We need to relate the properties of the listed substances to boiling point.
Plan: The boiling point depends in part on the attractive forces in the liquid. We need to order these according to the relative strengths of the different kinds of forces.
Solve: The attractive forces are stronger for ionic substances than for molecular ones, so $BaCl_2$ should have the highest boiling point. The intermolecular forces of the remaining substances depend on molecular weight, polarity, and hydrogen bonding. The molecular weights are H_2 (2), CO (28), HF (20), and Ne (20). The boiling point of H_2 should be the lowest because it is nonpolar and has the lowest molecular weight. The molecular weights of CO, HF, and Ne are roughly the same. Because HF can hydrogen bond, however, it should have the highest boiling point of the three. Next is CO, which is slightly polar and has the highest molecular weight. Finally, Ne, which is nonpolar, should have the lowest boiling point of these three. The predicted order of boiling points is therefore

$$H_2 < Ne < CO < HF < BaCl_2$$

Check: The actual normal boiling points are H_2 (20 K), Ne (27 K), CO (83 K), HF (293 K), and $BaCl_2$ (1813 K), in agreement with our predictions.

PRACTICE EXERCISE

(a) Identify the intermolecular forces present in the following substances, and **(b)** select the substance with the highest boiling point: CH_3CH_3, CH_3OH, and CH_3CH_2OH.
Answers: **(a)** CH_3CH_3 has only dispersion forces, whereas the other two substances have both dispersion forces and hydrogen bonds; **(b)** CH_3CH_2OH

11.3 | Some Properties of Liquids

The intermolecular forces we have just discussed can help us understand many familiar properties of liquids and solids. In this section we examine two important properties of liquids: viscosity and surface tension.

Viscosity

Some liquids, such as molasses and motor oil, flow very slowly; others, such as water and gasoline, flow easily. The resistance of a liquid to flow is called its **viscosity**. The greater a liquid's viscosity, the more slowly it flows. Viscosity can be measured by timing how long it takes a certain amount of the liquid to flow through a thin tube under gravitational force. More viscous liquids take longer (Figure 11.13 ▶). Viscosity can also be determined by measuring the rate at which steel spheres fall through the liquid. The spheres fall more slowly as the viscosity increases.

Viscosity is related to the ease with which individual molecules of the liquid can move with respect to one another. It thus depends on the attractive forces between molecules and on whether structural features exist that cause the molecules to become entangled. For a series of related compounds, therefore, viscosity increases with molecular weight, as illustrated in Table 11.4 ▼.

TABLE 11.4 Viscosities of a Series of Hydrocarbons at 20°C

Substance	Formula	Viscosity (kg/m-s)
Hexane	$CH_3CH_2CH_2CH_2CH_2CH_3$	3.26×10^{-4}
Heptane	$CH_3CH_2CH_2CH_2CH_2CH_2CH_3$	4.09×10^{-4}
Octane	$CH_3CH_2CH_2CH_2CH_2CH_2CH_2CH_3$	5.42×10^{-4}
Nonane	$CH_3CH_2CH_2CH_2CH_2CH_2CH_2CH_2CH_3$	7.11×10^{-4}
Decane	$CH_3CH_2CH_2CH_2CH_2CH_2CH_2CH_2CH_2CH_3$	1.42×10^{-3}

▲ **Figure 11.13 Comparing viscosities.** The Society of Automotive Engineers (SAE) has established numbers to indicate the viscosity of motor oils. The higher the number, the greater the viscosity is at any given temperature. The SAE 40 motor oil on the left is more viscous and flows more slowly than the less viscous SAE 10 oil on the right.

▲ **Figure 11.14 Surface tension.** Surface tension permits an insect such as the water strider to "walk" on water.

▲ **Figure 11.15 Molecular-level view of surface and interior intermolecular forces in a liquid.** Molecules at the surface are attracted only by other surface molecules and by molecules below the surface. The result is a net downward attraction into the interior of the liquid. Molecules in the interior experience attractions in all directions, resulting in no net attraction in any direction.

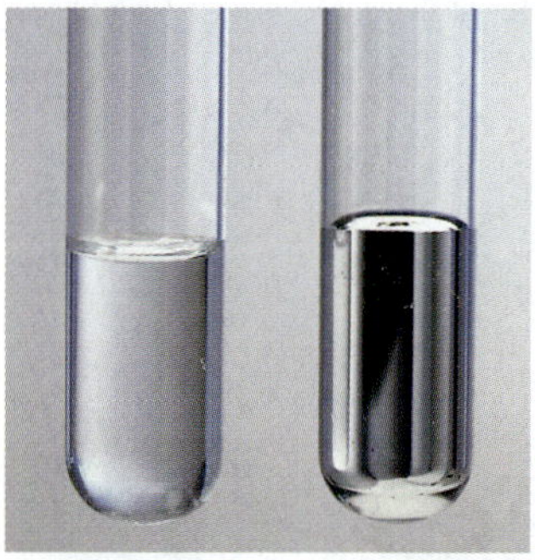

▲ **Figure 11.16 Two meniscus shapes.** The water meniscus in a glass tube compared with the mercury meniscus in a similar tube. Water wets the glass, and the bottom of the meniscus is below the level of the water-glass contact line, giving a U-shape to the water surface. Mercury does not wet glass and the meniscus is above the mercury-glass contact line, giving an inverted U-shape to the mercury surface.

(The SI units for viscosity are kg/m-s.). For any given substance, viscosity decreases with increasing temperature. Octane, for example, has a viscosity of 7.06×10^{-4} kg/m-s at 0°C, and of 4.33×10^{-4} kg/m-s at 40°C. At higher temperatures the greater average kinetic energy of the molecules more easily overcomes the attractive forces between molecules.

Surface Tension

The surface of water behaves almost as if it had an elastic skin, as evidenced by the ability of certain insects to "walk" on water (Figure 11.14 ◀). This behavior is due to an imbalance of intermolecular forces at the surface of the liquid, as shown in Figure 11.15 ◀. Notice that molecules in the interior are attracted equally in all directions, whereas those at the surface experience a net inward force. The resultant inward force pulls molecules from the surface into the interior, thereby reducing the surface area and making the molecules at the surface pack closely together. Because spheres have the smallest surface area for their volume, water droplets assume an almost spherical shape. Similarly, water tends to "bead up" on a newly waxed car because there is little or no attraction between the polar water molecules and the nonpolar wax molecules.

A measure of the inward forces that must be overcome in order to expand the surface area of a liquid is given by its surface tension. **Surface tension** is the energy required to increase the surface area of a liquid by a unit amount. For example, the surface tension of water at 20°C is 7.29×10^{-2} J/m^2, which means that an energy of 7.29×10^{-2} J must be supplied to increase the surface area of a given amount of water by 1 m^2. Water has a high surface tension because of its strong hydrogen bonds. The surface tension of mercury is even higher (4.6×10^{-1} J/m^2) because of even stronger metallic bonds between the atoms of mercury.

GIVE IT SOME THOUGHT

How do viscosity and surface tension change (a) as temperature increases, (b) as intermolecular forces of attraction become stronger?

Intermolecular forces that bind similar molecules to one another, such as the hydrogen bonding in water, are called *cohesive forces*. Intermolecular forces that bind a substance to a surface are called *adhesive forces*. Water placed in a glass tube adheres to the glass because the adhesive forces between the water and glass are even greater than the cohesive forces between water molecules. The curved upper surface, or *meniscus*, of the water is therefore U-shaped (Figure 11.16 ◀). For mercury, however, the meniscus is curved downward where the mercury contacts the glass. In this case the cohesive forces between the mercury atoms are much greater than the adhesive forces between the mercury atoms and the glass.

When a small-diameter glass tube, or capillary, is placed in water, water rises in the tube. The rise of liquids up very narrow tubes is called **capillary action**. The adhesive forces between the liquid and the walls of the tube tend to increase the surface area of the liquid. The surface tension of the liquid tends to reduce the area, thereby pulling the liquid up the tube. The liquid climbs until the adhesive and cohesive forces are balanced by the force of gravity on the liquid. Capillary action helps water and dissolved nutrients move upward through plants.

GIVE IT SOME THOUGHT

Do the viscosity and surface tension of a substance reflect adhesive forces or cohesive forces of attraction?

11.4 | Phase Changes

Water left uncovered in a glass for several days evaporates. An ice cube left in a warm room quickly melts. Solid CO_2 (sold as Dry Ice *sublimes* at room temperature; that is, it changes directly from the solid to the vapor state. In general, each state of matter can change into either of the other two states. Figure 11.17 ▶ shows the name associated with each of these transformations. These transformations are called either **phase changes** or changes of state.

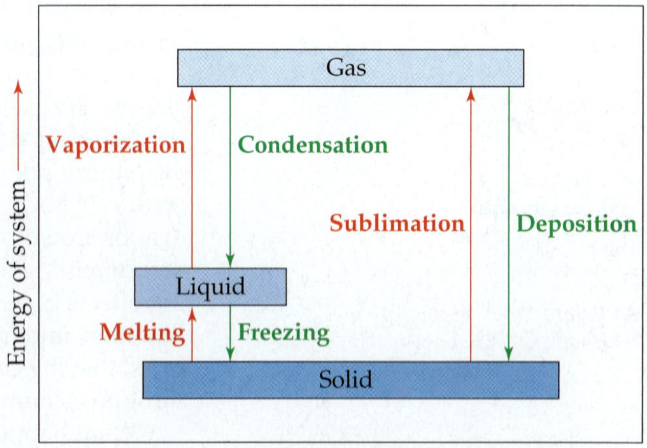

▲ **Figure 11.17 Phase changes and the names associated with them.** Those changes indicated by red arrows and names are endothermic, whereas those in green are exothermic.

VCL Heat of Fusion of Water

Energy Changes Accompanying Phase Changes

Every phase change is accompanied by a change in the energy of the system. In a solid lattice, for example, the molecules or ions are in more or less fixed positions with respect to one another and closely arranged to minimize the energy of the system. As the temperature of the solid increases, the units of the solid vibrate about their equilibrium positions with increasingly energetic motion. When the solid melts, the units that made up the solid are freed to move with respect to one another, which ordinarily means that their average separations increase. This melting process is called *fusion*. The increased freedom of motion of the molecules or ions comes at a price, measured by the **heat of fusion**, or enthalpy of fusion, denoted ΔH_{fus}. The heat of fusion of ice, for example, is 6.01 kJ/mol.

As the temperature of the liquid phase increases, the molecules of the liquid move about with increasing energy. One measure of this increasing energy is that the concentration of gas-phase molecules over the liquid increases with temperature. These molecules exert a pressure called the vapor pressure. We will explore vapor pressure in Section 11.5. For now we just need to understand that the vapor pressure increases with increasing temperature until it equals the external pressure over the liquid, typically atmospheric pressure. At this point the liquid boils; the molecules of the liquid move into the gaseous state, where they are widely separated. The energy required to cause this transition is called the **heat of vaporization**, or enthalpy of vaporization, denoted ΔH_{vap}. For water, the heat of vaporization is 40.7 kJ/mol.

VCL Heat of Vaporization of Water

Figure 11.18 ▼ shows the comparative values of ΔH_{fus} and ΔH_{vap} for four substances. ΔH_{vap} values tend to be larger than ΔH_{fus} because in the transition from the liquid to the vapor state, the molecules must essentially sever all their intermolecular attractive interactions; whereas in melting, many of these attractive interactions remain.

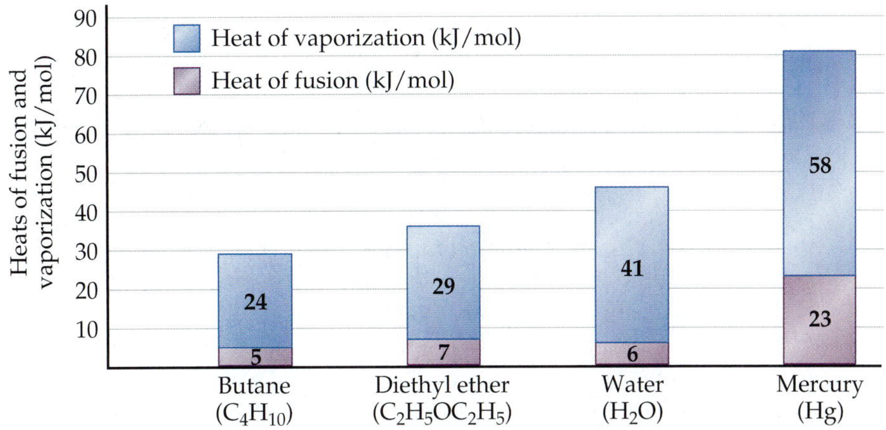

◀ **Figure 11.18 Comparing enthalpy changes for fusion and vaporization.** Heats of fusion (violet bars) and heats of vaporization (blue) for several substances. Notice that the heat of vaporization for a substance is always larger than its heat of fusion. The heat of sublimation is the sum of the heats of vaporization and fusion.

The molecules of a solid can be transformed directly into the gaseous state. The enthalpy change required for this transition is called the **heat of sublimation**, denoted ΔH_{sub}. For the substances shown in Figure 11.18, ΔH_{sub} is the sum of ΔH_{fus} and ΔH_{vap}. Thus, ΔH_{sub} for water is approximately 47 kJ/mol.

Phase changes of matter show up in important ways in our everyday experiences. We use ice cubes to cool our liquid drinks; the heat of fusion of ice cools the liquid in which the ice is immersed. We feel cool when we step out of a swimming pool or a warm shower because the heat of vaporization is drawn from our bodies as the water evaporates from our skin. Our bodies use the evaporation of water from skin to regulate body temperature, especially when we exercise vigorously in warm weather. A refrigerator also relies on the cooling effects of vaporization. Its mechanism contains an enclosed gas that can be liquefied under pressure. The liquid absorbs heat as it subsequently evaporates, thereby cooling the interior of the refrigerator. The vapor is then recycled through a compressor.

ANIMATION Changes of State

What happens to the heat absorbed when the liquid refrigerant vaporizes? According to the first law of thermodynamics (Section 5.2), the heat absorbed by the liquid in vaporizing must be released when the reverse process, condensation of the vapor to form the liquid, occurs. As the refrigerator compresses the vapor and liquid is formed, the heat released is dissipated through cooling coils in the back of the refrigerator. Just as the heat of condensation is equal in magnitude to the heat of vaporization and has the opposite sign, so also the *heat of deposition* is exothermic to the same degree that the heat of sublimation is endothermic; and the *heat of freezing* is exothermic to the same degree that the heat of fusion is endothermic. These relationships, shown in Figure 11.17, are consequences of the first law of thermodynamics.

GIVE IT SOME THOUGHT

What is the name of the phase change that occurs when ice left at room temperature changes to liquid water? Is this change exothermic or endothermic?

Heating Curves

What happens when we heat a sample of ice that is initially at −25°C and 1 atm pressure? The addition of heat causes the temperature of the ice to increase. As long as the temperature is below 0°C, the sample remains frozen. When the temperature reaches 0°C, the ice begins to melt. Because melting is an endothermic process, the heat we add at 0°C is used to convert ice to water and the temperature remains constant until all the ice has melted. Once we reach this point, the further addition of heat causes the temperature of the liquid water to increase.

▼ **Figure 11.19 Heating curve for water.** This graph indicates the changes that occur when 1.00 mol of water is heated from −25°C to 125°C at a constant pressure of 1 atm. Blue lines show the heating of one phase from a lower temperature to a higher one. Red lines show the conversion of one phase to another at constant temperature.

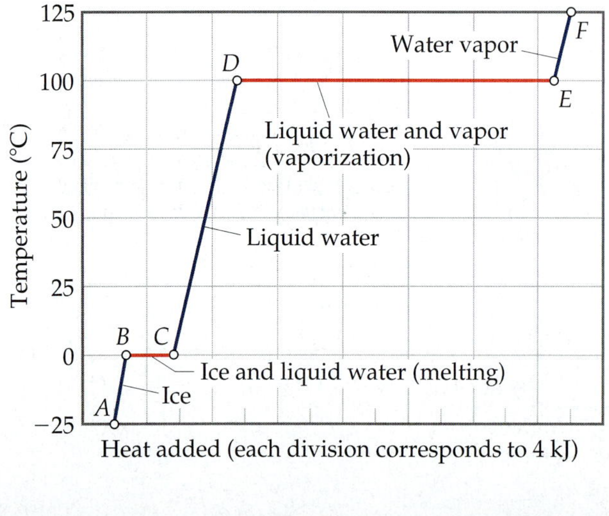

A graph of the temperature of the system versus the amount of heat added is called a *heating curve*. Figure 11.19 ◀ shows a heating curve for transforming ice at −25°C to steam at 125°C under a constant pressure of 1 atm. Heating the ice from −25°C to 0°C is represented by the line segment *AB* in Figure 11.19, while converting the ice at 0°C to water at 0°C is the horizontal segment *BC*. Additional heat increases the temperature of the water until the temperature reaches 100°C (segment *CD*). The heat is then used to convert water to steam at a constant temperature of 100°C (segment *DE*). Once all the water has been converted to steam, the steam is heated to its final temperature of 125°C (segment *EF*).

We can calculate the enthalpy change of the system for each of the segments of the heating curve. In segments *AB*, *CD*, and *EF* we are heating a single phase from one temperature to another. As we saw in Section 5.5, the amount of heat needed to raise the temperature of a substance is given by the product of the specific heat, mass, and temperature change (Equation 5.22). The greater the specific heat of a substance, the more heat we must add to accomplish a certain temperature increase. Because the specific heat of water is greater than that of ice, the slope of segment *CD* is less than that of segment *AB*; we must add more heat to water to achieve a 1°C temperature change than is needed to warm the same quantity of ice by 1°C.

ACTIVITY
Heating Curves

In segments *BC* and *DE* we are converting one phase to another at a constant temperature. The temperature remains constant during these phase changes because the added energy is used to overcome the attractive forces between molecules rather than to increase their average kinetic energy. For segment *BC*, in which ice is converted to water, the enthalpy change can be calculated by using ΔH_{fus}, while for segment *DE* we can use ΔH_{vap}. In Sample Exercise 11.4 we calculate the total enthalpy change for the heating curve in Figure 11.19.

SAMPLE EXERCISE 11.4 | Calculating ΔH for Temperature and Phase Changes

Calculate the enthalpy change upon converting 1.00 mol of ice at −25°C to water vapor (steam) at 125°C under a constant pressure of 1 atm. The specific heats of ice, water, and steam are 2.09 J/g-K, 4.18 J/g-K, and 1.84 J/g-K, respectively. For H_2O, $\Delta H_{fus} = 6.01$ kJ/mol and $\Delta H_{vap} = 40.67$ kJ/mol.

Solution

Analyze: Our goal is to calculate the total heat required to convert 1 mol of ice at −25°C to steam at 125°C.

Plan: We can calculate the enthalpy change for each segment and then sum them to get the total enthalpy change (Hess's law, Section 5.6).

Solve: For segment *AB* in Figure 11.19, we are adding enough heat to ice to increase its temperature by 25°C. A temperature change of 25°C is the same as a temperature change of 25 K, so we can use the specific heat of ice to calculate the enthalpy change during this process:

$$AB: \quad \Delta H = (1.00\ \text{mol})(18.0\ \text{g/mol})(2.09\ \text{J/g-K})(25\ \text{K}) = 940\ \text{J} = 0.94\ \text{kJ}$$

For segment *BC* in Figure 11.19, in which we convert ice to water at 0°C, we can use the molar enthalpy of fusion directly:

$$BC: \quad \Delta H = (1.00\ \text{mol})(6.01\ \text{kJ/mol}) = 6.01\ \text{kJ}$$

The enthalpy changes for segments *CD*, *DE*, and *EF* can be calculated in similar fashion:

$$CD: \quad \Delta H = (1.00\ \text{mol})(18.0\ \text{g/mol})(4.18\ \text{J/g-K})(100\ \text{K}) = 7520\ \text{J} = 7.52\ \text{kJ}$$

$$DE: \quad \Delta H = (1.00\ \text{mol})(40.67\ \text{kJ/mol}) = 40.7\ \text{kJ}$$

$$EF: \quad \Delta H = (1.00\ \text{mol})(18.0\ \text{g/mol})(1.84\ \text{J/g-K})(25\ \text{K}) = 830\ \text{J} = 0.83\ \text{kJ}$$

The total enthalpy change is the sum of the changes of the individual steps:

$$\Delta H = 0.94\ \text{kJ} + 6.01\ \text{kJ} + 7.52\ \text{kJ} + 40.7\ \text{kJ} + 0.83\ \text{kJ} = 56.0\ \text{kJ}$$

Check: The components of the total energy change are reasonable in comparison with the lengths of the horizontal segments of the lines in Figure 11.19. Notice that the largest component is the heat of vaporization.

PRACTICE EXERCISE

What is the enthalpy change during the process in which 100.0 g of water at 50.0°C is cooled to ice at −30.0°C? (Use the specific heats and enthalpies for phase changes given in Sample Exercise 11.4.)
Answer: −20.9 kJ − 33.4 kJ − 6.27 kJ = −60.6 kJ

Cooling a substance has the opposite effect of heating it. Thus, if we start with water vapor and begin to cool it, we would move right to left through the events shown in Figure 11.19. We would first lower the temperature of the vapor ($F \longrightarrow E$), then condense it ($E \longrightarrow D$), and so forth. Sometimes as we remove heat from a liquid, we can temporarily cool it below its freezing point without forming a solid. This phenomenon is called *supercooling*. Supercooling occurs when heat is removed from a liquid so rapidly that the molecules literally have no time to assume the ordered structure of a solid. A supercooled liquid is unstable; particles of dust entering the solution or gentle stirring is often sufficient to cause the substance to solidify quickly.

Critical Temperature and Pressure

A gas normally liquefies at some point when pressure is applied to it. Suppose we have a cylinder with a piston, containing water vapor at 100°C. If we increase the pressure on the water vapor, liquid water will form when the pressure is 760 torr. On the other hand, if the temperature is 110°C, the liquid phase does not form until the pressure is 1075 torr. At 374°C the liquid phase forms only at 1.655×10^5 torr (217.7 atm). Above this temperature no amount of pressure will cause a distinct liquid phase to form. Instead, as pressure increases, the gas merely becomes steadily more compressed. The highest temperature at which a distinct liquid phase can form is referred to as the **critical temperature**. The **critical pressure** is the pressure required to bring about liquefaction at this critical temperature.

The critical temperature is the highest temperature at which a liquid can exist. Above the critical temperature, the motional energies of the molecules are greater than the attractive forces that lead to the liquid state regardless of how much the substance is compressed to bring the molecules closer together. The greater the intermolecular forces, the greater the critical temperature of a substance.

The critical temperatures and pressures are listed for several substances in Table 11.5 ▼. Notice that nonpolar, low molecular weight substances, which have weak intermolecular attractions, have lower critical temperatures and pressures than those that are polar or of higher molecular weight. Notice also that water and ammonia have exceptionally high critical temperatures and pressures as a consequence of strong intermolecular hydrogen-bonding forces.

The critical temperatures and pressures of substances are often of considerable importance to engineers and other people working with gases, because they provide information about the conditions under which gases liquefy. Sometimes we want to liquefy a gas; other times we want to avoid liquefying it. It is useless to try to liquefy a gas by applying pressure if the gas is above its critical temperature. For example, O_2 has a critical temperature of 154.4 K. It must be cooled below this temperature before it can be liquefied by pressure. In contrast, ammonia has a critical temperature of 405.6 K. Thus, it can be liquefied at room temperature (approximately 295 K) by compressing the gas to a sufficient pressure.

TABLE 11.5 Critical Temperatures and Pressures of Selected Substances

Substance	Critical Temperature (K)	Critical Pressure (atm)
Ammonia, NH_3	405.6	111.5
Phosphine, PH_3	324.4	64.5
Argon, Ar	150.9	48
Carbon dioxide, CO_2	304.3	73.0
Nitrogen, N_2	126.1	33.5
Oxygen, O_2	154.4	49.7
Propane, $CH_3CH_2CH_3$	370.0	42.0
Water, H_2O	647.6	217.7
Hydrogen sulfide, H_2S	373.5	88.9

CHEMISTRY AT WORK | Supercritical Fluid Extraction

At ordinary pressures, a substance above its critical temperature behaves as an ordinary gas. However, as pressure increases up to several hundred atmospheres, its character changes. Like a gas, it still expands to fill the confines of its container, but its density approaches that of a liquid. (For example, the critical temperature of water is 647.6 K, and its critical pressure is 217.7 atm. At this temperature and pressure, the density of water is 0.4 g/mL). A substance at temperatures and pressures higher than its critical temperature and pressure is better considered a *supercritical fluid* rather than a gas.

Like liquids, supercritical fluids can behave as solvents, dissolving a wide range of substances. Using *supercritical fluid extraction*, the components of mixtures can be separated. The solvent power of a supercritical fluid increases as its density increases. Conversely, lowering its density (by either decreasing pressure or increasing temperature) causes the supercritical fluid and the dissolved material to separate. Figure 11.20 ▼ shows the solubility of a typical nonpolar organic solid, naphthalene ($C_{10}H_8$), in supercritical carbon dioxide at 45°C. The solubility of naphthalene is essentially zero below the critical pressure of 73 atm. Solubility increases rapidly, however, with increasing pressure (and, thus, increasing density of the supercritical fluid).

By appropriate manipulation of pressure, supercritical fluid extraction has been used successfully to separate complex mixtures in the chemical, food, pharmaceutical, and energy industries. Supercritical carbon dioxide, for example, is environmentally benign because there are no problems disposing of solvent and there are no toxic residues resulting from the process. In addition, supercritical CO_2 is inexpensive compared to solvents other than water. A process for removing caffeine from green coffee beans by extraction with supercritical CO_2, diagrammed in Figure 11.21 ▼, has been in commercial operation for several years. At the proper temperature and pressure the supercritical CO_2 removes caffeine from the beans by dissolution, but leaves the flavor and aroma components, producing decaffeinated coffee. Other applications of supercritical CO_2 extraction include the extraction of the essential flavor elements from hops for use in brewing, and the isolation of the flavor components of herbs and spices. (See also Section 18.7.)

Related Exercises: 11.41, 11.42, 11.87

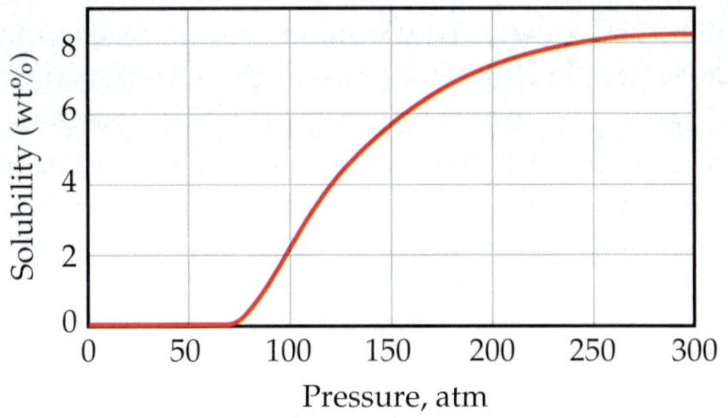

▲ **Figure 11.20 Solubility of naphthalene ($C_{10}H_8$) in supercritical carbon dioxide at 45°C.**

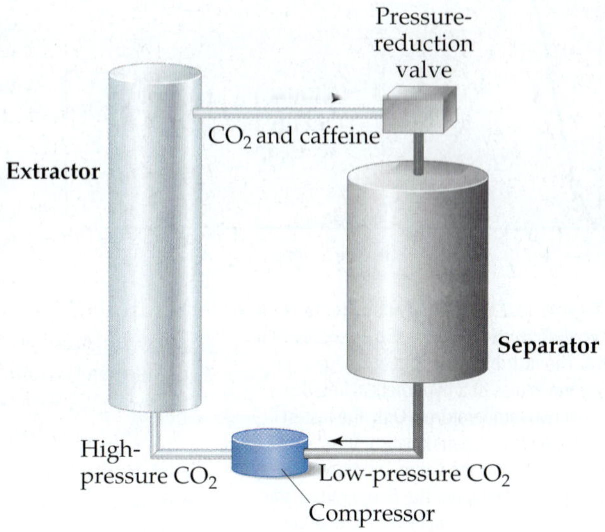

▲ **Figure 11.21 Diagram of a supercritical fluid extraction process.** The material to be processed is placed in the extractor. The desired material dissolves in supercritical CO_2 at high pressure, then is precipitated in the separator when the CO_2 pressure is reduced. The carbon dioxide is then recycled through the compressor with a fresh batch of material in the extractor.

11.5 | Vapor Pressure

Molecules can escape from the surface of a liquid into the gas phase by evaporation. Suppose we conduct an experiment in which we place a quantity of ethanol (C_2H_5OH) in an evacuated, closed container such as that in Figure 11.22 ▶. The ethanol will quickly begin to evaporate. As a result, the pressure exerted by the vapor in the space above the liquid will begin to increase. After a short time the pressure of the vapor will attain a constant value, which we call the **vapor pressure** of the substance.

▶ **Figure 11.22 Illustration of the equilibrium vapor pressure over a liquid.** In (a) we imagine that no molecules exist in the gas phase; there is zero pressure in the cell. In (b) the rate at which molecules leave the surface equals the rate at which gas molecules pass into the liquid phase. These equal rates produce a stable vapor pressure that does not change as long as the temperature remains constant.

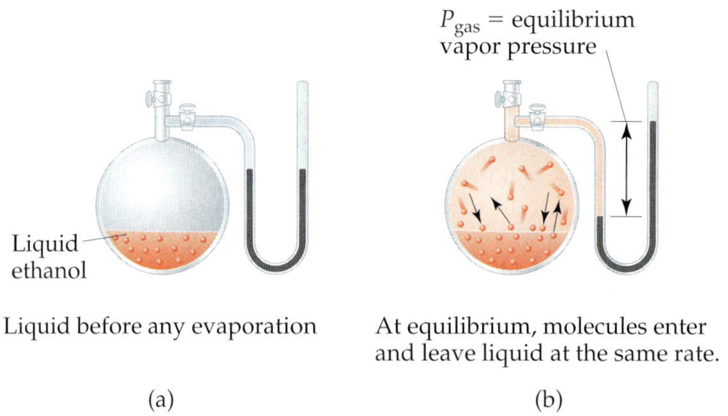

Explaining Vapor Pressure on the Molecular Level

The molecules of a liquid move at various speeds. Figure 11.23 ◀ shows the distribution of kinetic energies of the particles at the surface of a liquid at two temperatures. The distribution curves are like those shown earlier for gases (Figures 10.18 and 10.19). At any instant some of the molecules on the surface of the liquid possess sufficient kinetic energy to overcome the attractive forces of their neighbors and escape into the gas phase. The weaker the attractive forces, the larger is the number of molecules that are able to escape and therefore the higher is the vapor pressure.

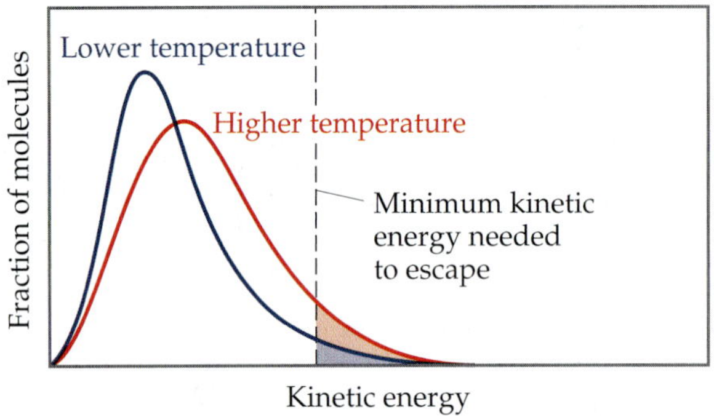

▲ **Figure 11.23 The effect of temperature on the distribution of kinetic energies in a liquid.** The distribution of kinetic energies of surface molecules of a hypothetical liquid are shown at two temperatures. Only the fastest molecules have sufficient kinetic energy to escape the liquid and enter the vapor, as shown by the shaded areas. The higher the temperature, the larger is the fraction of molecules with enough energy to escape from the liquid into the vapor phase.

At any particular temperature the movement of molecules from the liquid to the gas phase goes on continuously. As the number of gas-phase molecules increases, however, the probability increases that a molecule in the gas phase will strike the liquid surface and be recaptured by the liquid, as shown in Figure 11.22(b). Eventually, the rate at which molecules return to the liquid exactly equals the rate at which they escape. The number of molecules in the gas phase then reaches a steady value, and the pressure of the vapor at this stage becomes constant.

The condition in which two opposing processes are occurring simultaneously at equal rates is called a **dynamic equilibrium**, but is usually referred to merely as an *equilibrium*. A liquid and its vapor are in dynamic equilibrium when evaporation and condensation occur at equal rates. It may appear that nothing is occurring at equilibrium, because there is no net change in the system. In fact, a great deal is happening; molecules continuously pass from the liquid state to the gas state and from the gas state to the liquid state. All equilibria between different states of matter possess this dynamic character. *The vapor pressure of a liquid is the pressure exerted by its vapor when the liquid and vapor states are in dynamic equilibrium.*

ANIMATION
Vapor Pressure vs. Temperature

Volatility, Vapor Pressure, and Temperature

When vaporization occurs in an open container, as when water evaporates from a bowl, the vapor spreads away from the liquid. Little, if any, is recaptured at the surface of the liquid. Equilibrium never occurs, and the vapor continues to form until the liquid evaporates to dryness. Substances with high vapor pressure (such as gasoline) evaporate more quickly than substances with low vapor pressure (such as motor oil). Liquids that evaporate readily are said to be **volatile**.

Hot water evaporates more quickly than cold water because vapor pressure increases with increasing temperature. We see this effect in Figure 11.23: As the temperature of a liquid is increased, the molecules move more energetically and a greater fraction can therefore escape more readily from their neighbors. Figure 11.24 ▶ depicts the variation in vapor pressure with temperature for four common substances that differ greatly in volatility. Note that the vapor pressure in all cases increases nonlinearly with increasing temperature.

GIVE IT SOME THOUGHT

Predict (a) the relative vapor pressures of CCl_4 and CBr_4 at 25°C and (b) the relative volatilities of these two substances at this temperature.

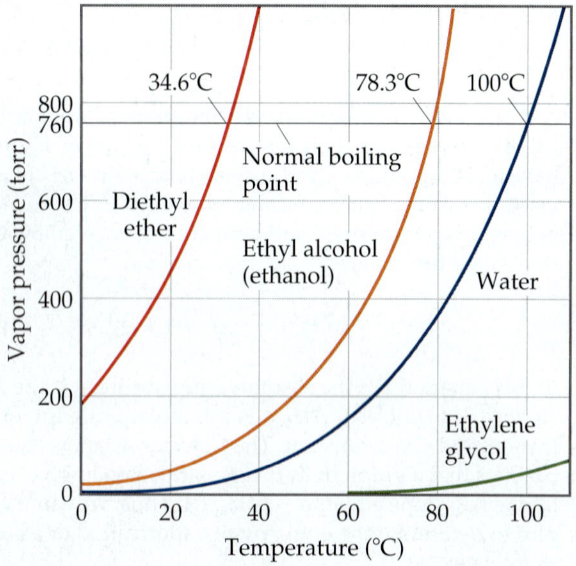

▲ **Figure 11.24 Vapor pressure for four common liquids as a function of temperature.** The temperature at which the vapor pressure is 760 torr is the normal boiling point of each liquid.

Vapor Pressure and Boiling Point

A liquid boils when its vapor pressure equals the external pressure acting on the surface of the liquid. At this point bubbles of vapor are able to form within the liquid. The temperature at which a given liquid boils increases with increasing external pressure. The boiling point of a liquid at 1 atm pressure is called its **normal boiling point**. From Figure 11.24 we see that the normal boiling point of water is 100°C.

The boiling point is important to many processes that involve heating liquids, including cooking. The time required to cook food depends on the temperature. As long as water is present, the maximum temperature of the food being cooked is the boiling point of water. Pressure cookers work by allowing steam to escape only when it exceeds a predetermined pressure; the pressure above the water can therefore increase above atmospheric pressure. The higher pressure causes the water to boil at a higher temperature, thereby allowing the food to get hotter and to cook more rapidly. The effect of pressure on boiling point also explains why it takes longer to cook food at higher elevations than at sea level. The atmospheric pressure is lower at higher altitudes, so water boils at a lower temperature.

ACTIVITY
Equilibrium Vapor Pressure

SAMPLE EXERCISE 11.5 | Relating Boiling Point to Vapor Pressure

Use Figure 11.24 to estimate the boiling point of diethyl ether under an external pressure of 0.80 atm.

Solution

Analyze: We are asked to read a graph of vapor pressure versus temperature to determine the boiling point of a substance at a particular pressure. The boiling point is the temperature at which the vapor pressure is equal to the external pressure.

Plan: We need to convert 0.80 atm to torr because that is the pressure scale on the graph. We estimate the location of that pressure on the graph, move horizontally to the vapor pressure curve, and then drop vertically from the curve to estimate the temperature.

Solve: The pressure equals (0.80 atm)(760 torr/atm) = 610 torr. From Figure 11.24 we see that the boiling point at this pressure is about 27°C, which is close to room temperature.

Comment: We can make a flask of diethyl ether boil at room temperature by using a vacuum pump to lower the pressure above the liquid to about 0.8 atm.

PRACTICE EXERCISE

At what external pressure will ethanol have a boiling point of 60°C?
Answer: about 340 torr (0.45 atm)

A CLOSER LOOK | The Clausius–Clapeyron Equation

You might have noticed that the plots of the variation of vapor pressure with temperature shown in Figure 11.24 have a distinct shape: Each curves sharply upward to a higher vapor pressure with increasing temperature. The relationship between vapor pressure and temperature is given by an equation called the *Clausius–Clapeyron equation*:

$$\ln P = \frac{-\Delta H_{vap}}{RT} + C \qquad [11.1]$$

In this equation T is the absolute temperature, R is the gas constant (8.314 J/mol-K), ΔH_{vap} is the molar enthalpy of vaporization, and C is a constant. The Clausius–Clapeyron equation predicts that a graph of $\ln P$ versus $1/T$ should give a straight line with a slope equal to $-\Delta H_{vap}/R$. Thus, we can use such a plot to determine the enthalpy of vaporization of a substance as follows:

$$\Delta H_{vap} = -\text{slope} \times R$$

As an example of the application of the Clausius–Clapeyron equation, the vapor-pressure data for ethanol shown in Figure 11.24 are graphed as $\ln P$ versus $1/T$ in Figure 11.25 ▶. The data lie on a straight line with a negative slope. We can use the slope of the line to determine ΔH_{vap} for ethanol. We can also extrapolate the line to obtain values for the vapor pressure of ethanol at temperatures above and below the temperature range for which we have data.

Related Exercises: 11.89, 11.90, 11.91

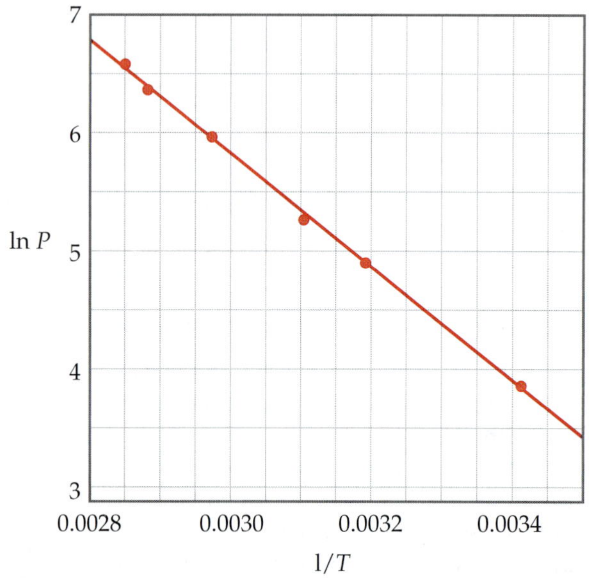

▲ **Figure 11.25 Linear graph of vapor pressure data for ethanol.** The Clausius–Clapeyron equation, Equation 11.1, indicates that a graph of $\ln P$ versus $1/T$ gives a straight line whose slope equals $-\Delta H_{vap}/R$. The slope of this line gives $\Delta H_{vap} = 38.56$ kJ/mol.

11.6 | Phase Diagrams

The equilibrium between a liquid and its vapor is not the only dynamic equilibrium that can exist between states of matter. Under appropriate conditions of temperature and pressure, a solid can be in equilibrium with its liquid state or even with its vapor state. A **phase diagram** is a graphical way to summarize the conditions under which equilibria exist between the different states of matter. Such a diagram also allows us to predict the phase of a substance that is stable at any given temperature and pressure.

The general form of a phase diagram for a substance that can exist in any one of the three phases of matter is shown in Figure 11.26 ▶. The diagram is a two-dimensional graph, with pressure and temperature as the axes. It contains three important curves, each of which represents the conditions of temperature and pressure at which the various phases can coexist at equilibrium. The only substance present in the system is the one whose phase diagram is under consideration. The pressure shown in the diagram is either the pressure applied to the system or the pressure generated by the substance itself. The curves may be described as follows:

1. The line from A to B is the vapor-pressure curve of the liquid. It represents the equilibrium between the liquid and gas phases. The point on this curve where the vapor pressure is 1 atm is the normal boiling point of the substance. The vapor-pressure curve ends at the *critical point* (B), which is at the critical temperature and critical pressure of the substance. Beyond the critical point the liquid and gas phases become indistinguishable from each other.

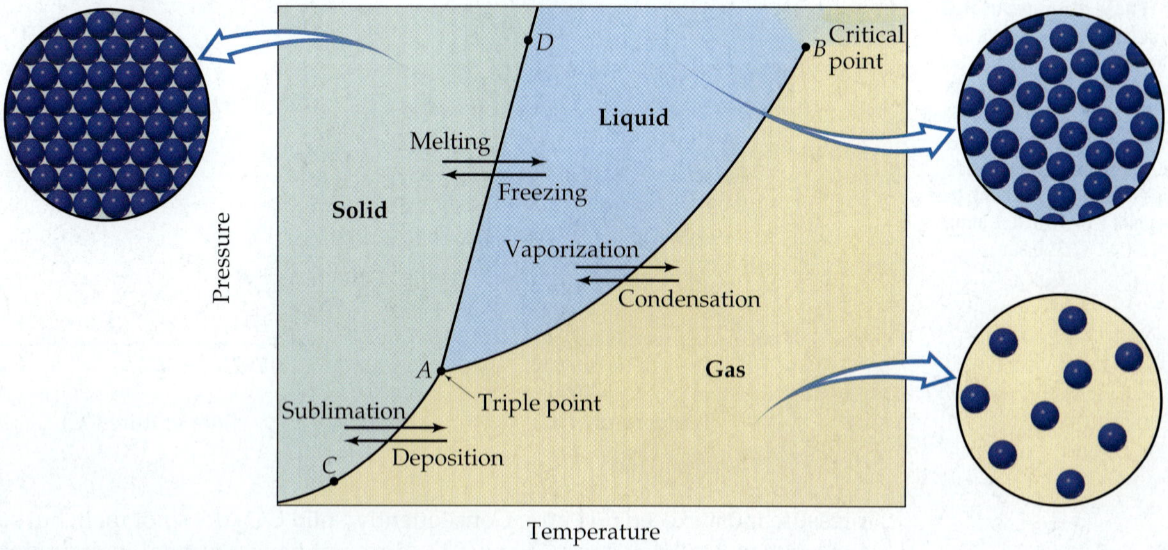

▲ **Figure 11.26 Phase diagram for a three-phase system.** In this generic diagram, the substance being investigated can exist as a solid, a liquid, or a gas, depending on pressure and temperature.

2. The line *AC* represents the variation in the vapor pressure of the solid as it sublimes at different temperatures.
3. The line from *A* through *D* represents the change in melting point of the solid with increasing pressure. This line usually slopes slightly to the right as pressure increases, because for most substances the solid form is denser than the liquid form. An increase in pressure usually favors the more compact solid phase; thus, higher temperatures are required to melt the solid at higher pressures. The *melting point* of a substance is identical to its *freezing point*. The two differ only in the direction from which the phase change is approached. The melting point at 1 atm is the **normal melting point**.

ACTIVITY
Phase Diagram

Point *A*, where the three curves intersect, is known as the **triple point**. All three phases are in equilibrium at this temperature and pressure. Any other point on the three curves represents an equilibrium between two phases. Any point on the diagram that does not fall on a line corresponds to conditions under which only one phase is present. The gas phase, for example, is stable at low pressures and high temperatures, whereas the solid phase is stable at low temperatures and high pressures. Liquids are stable in the region between the other two.

GIVE IT SOME THOUGHT

On a phase diagram, which way does the line representing melting point usually slope, and why?

The Phase Diagrams of H_2O and CO_2

Figure 11.27 ▶ shows the phase diagrams of H_2O and CO_2. The solid-liquid equilibrium (melting point) line of CO_2 follows the typical behavior, slanting to the right with increasing pressure, telling you that its melting point increases with increasing pressure. In contrast, the melting point line of H_2O is atypical, slanting to the left with increasing pressure, indicating that for water the melting point *decreases* with increasing pressure. As seen in Figure 11.11, water is among the very few substances whose liquid form is more compact than its solid form. ∞ (Section 11.2)

The triple point of H_2O (0.0098°C and 4.58 torr) is at much lower pressure than that of CO_2 (−56.4°C and 5.11 atm). For CO_2 to exist as a liquid, the

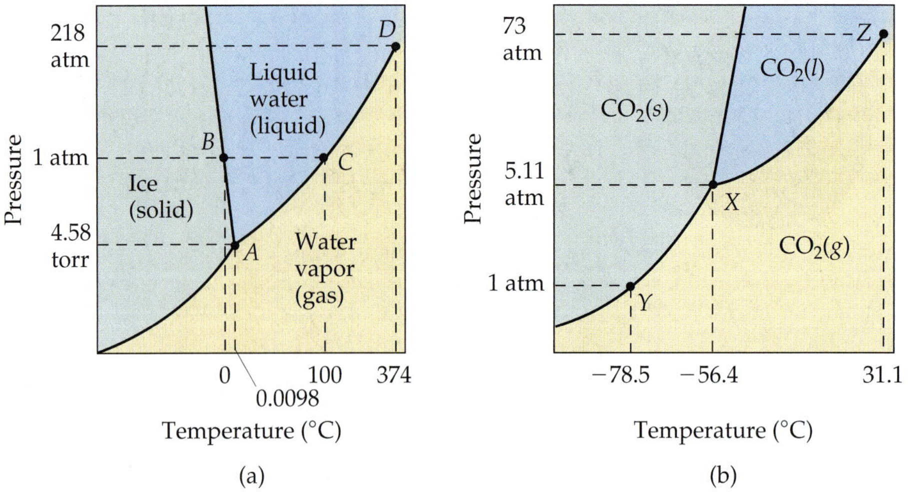

▶ **Figure 11.27 Phase diagrams of H_2O and CO_2.** The axes are not drawn to scale in either case. In (a), for water, note the triple point *A* (0.0098°C, 4.58 torr), the normal melting (or freezing) point *B* (0°C, 1 atm), the normal boiling point *C* (100°C, 1 atm), and the critical point *D* (374.4°C, 217.7 atm). In (b), for carbon dioxide, note the triple point *X* (−56.4°C, 5.11 atm), the normal sublimation point *Y* (−78.5°C, 1 atm), and the critical point *Z* (31.1°C, 73.0 atm).

ANIMATION
Phase Diagram of Water

pressure must exceed 5.11 atm. Consequently, solid CO_2 does not melt, but sublimes when heated at 1 atm. Thus, CO_2 does not have a normal melting point; instead, it has a normal sublimation point, −78.5°C. Because CO_2 sublimes rather than melts as it absorbs energy at ordinary pressures, solid CO_2 (Dry Ice) is a convenient coolant. For water (ice) to sublime, however, its vapor pressure must be below 4.58 torr. Food is freeze-dried by placing frozen food in a low-pressure chamber (below 4.58 torr) so that the ice in it sublimes.

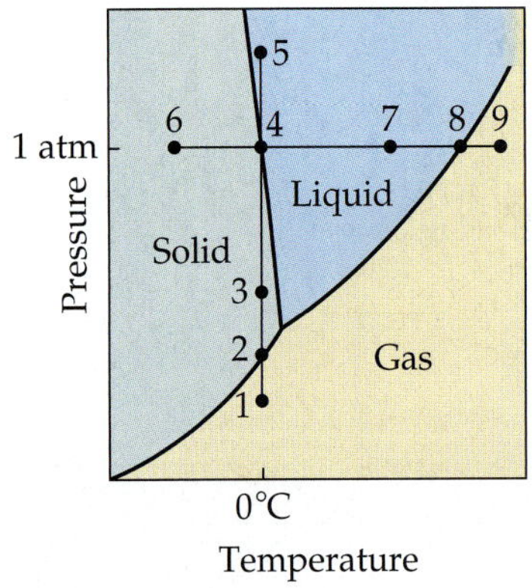

▲ **Figure 11.28 Phase diagram of H_2O.**

SAMPLE EXERCISE 11.6 | Interpreting a Phase Diagram

Referring to Figure 11.28 ◀, describe any changes in the phases present when H_2O is **(a)** kept at 0°C while the pressure is increased from that at point 1 to that at point 5 (vertical line), **(b)** kept at 1.00 atm while the temperature is increased from that at point 6 to that at point 9 (horizontal line).

Solution

Analyze: We are asked to use the phase diagram provided to deduce what phase changes might occur when specific pressure and temperature changes are brought about.

Plan: Trace the path indicated on the phase diagram, and note what phases and phase changes occur.

Solve: **(a)** At point 1, H_2O exists totally as a vapor. At point 2 a solid-vapor equilibrium exists. Above that pressure, at point 3, all the H_2O is converted to a solid. At point 4 some of the solid melts and an equilibrium between solid and liquid is achieved. At still higher pressures all the H_2O melts, so only the liquid phase is present at point 5. **(b)** At point 6 the H_2O exists entirely as a solid. When the temperature reaches point 4, the solid begins to melt and an equilibrium exists between the solid and liquid phases. At an even higher temperature, point 7, the solid has been converted entirely to a liquid. A liquid-vapor equilibrium exists at point 8. Upon further heating to point 9, the H_2O is converted entirely to the vapor phase.

Check: The indicated phases and phase changes are consistent with our knowledge of the properties of water.

PRACTICE EXERCISE

Using Figure 11.27(b), describe what happens when the following changes are made in a CO_2 sample initially at 1 atm and −60°C: **(a)** Pressure increases at constant temperature to 60 atm. **(b)** Temperature increases from −60°C to −20°C at constant 60 atm pressure.

Answers: **(a)** $CO_2(g) \longrightarrow CO_2(s)$; **(b)** $CO_2(s) \longrightarrow CO_2(l)$

11.7 | Structures of Solids

Throughout the remainder of this chapter we will focus on how the properties of solids relate to their structures and bonding. Solids can be either crystalline or amorphous (noncrystalline). In a **crystalline solid** the atoms, ions, or mole-

▲ **Figure 11.29 Crystalline solids.** Crystalline solids come in a variety of forms and colors: (a) pyrite (fool's gold), (b) fluorite, (c) amethyst.

cules are ordered in well-defined arrangements. These solids usually have flat surfaces, or *faces*, that make definite angles with one another. The orderly stacks of particles that produce these faces also cause the solids to have highly regular shapes (Figure 11.29 ▲). Quartz and diamond are crystalline solids.

An **amorphous solid** (from the Greek words for "without form") is a solid whose particles have no orderly structure. These solids lack well-defined faces and shapes. Many amorphous solids are mixtures of molecules that do not stack together well. Most others are composed of large, complicated molecules. Familiar amorphous solids include rubber and glass.

Quartz (SiO_2) is a crystalline solid with a three-dimensional structure like that shown in Figure 11.30 ▼(a). When quartz melts (at about 1600°C), it becomes a viscous, tacky liquid. Although the silicon–oxygen network remains

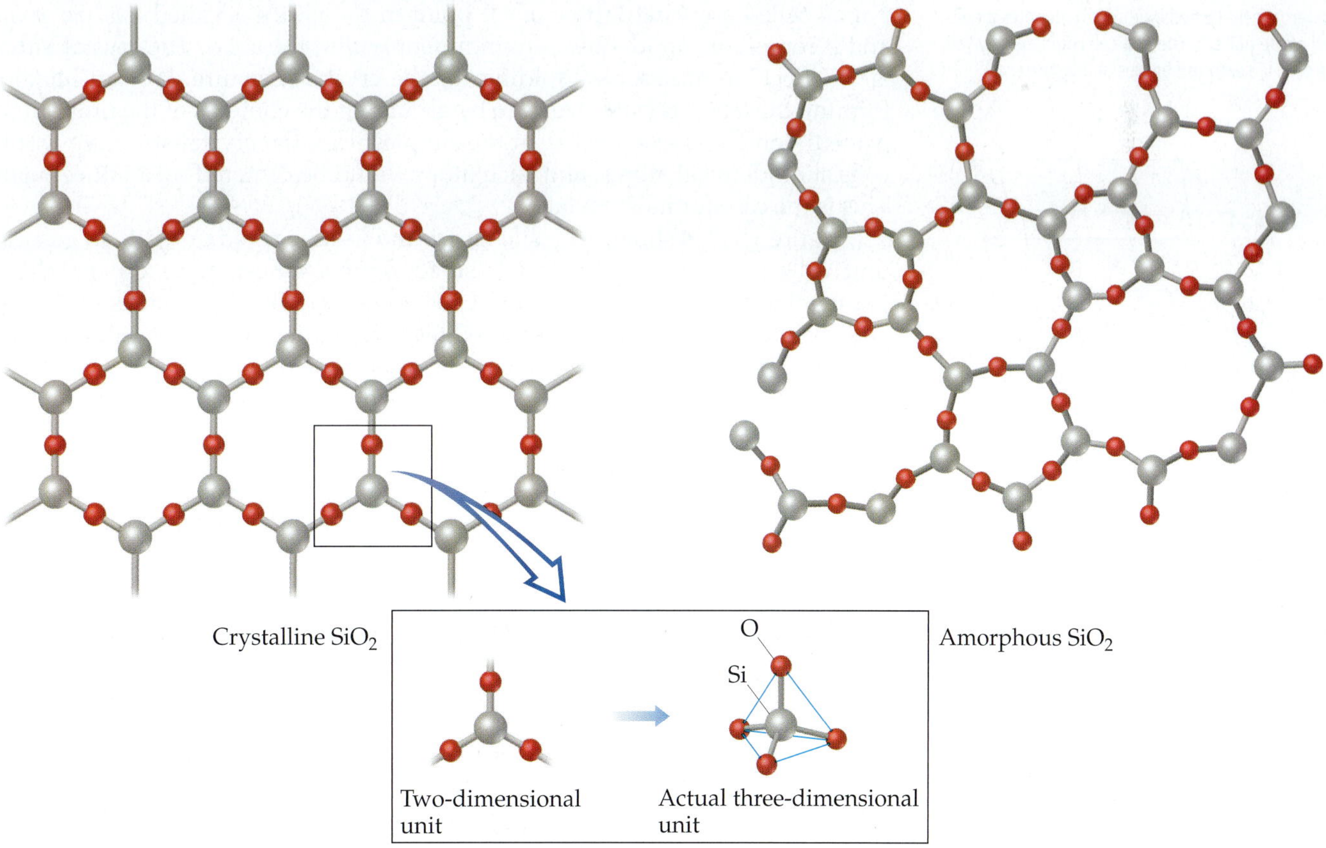

▲ **Figure 11.30 Schematic comparisons of crystalline SiO_2 (quartz) and amorphous SiO_2 (quartz glass).** The structures are actually three-dimensional and not planar as drawn. The two-dimensional unit shown as the basic building block of the structure (silicon and three oxygens) actually has four oxygens, the fourth coming out of the plane of the paper and capable of bonding to other silicon atoms. The actual three-dimensional building block is shown.

largely intact, many Si—O bonds are broken, and the rigid order of the quartz is lost. If the melt is rapidly cooled, the atoms are unable to return to an orderly arrangement. As a result, an amorphous solid known either as quartz glass or as silica glass results [Figure 11.30(b)].

Because the particles of an amorphous solid lack any long-range order, intermolecular forces vary in strength throughout a sample. Thus, amorphous solids do not melt at specific temperatures. Instead, they soften over a temperature range as intermolecular forces of various strengths are overcome. A crystalline solid, in contrast, melts at a specific temperature.

GIVE IT SOME THOUGHT

What is the general difference in the melting behaviors of crystalline and amorphous solids?

Unit Cells

The characteristic order of crystalline solids allows us to convey a picture of an entire crystal by looking at only a small part of it. We can think of the solid as being built up by stacking together identical building blocks, much as a brick wall is formed by stacking rows of individual "identical" bricks. The repeating unit of a solid, the crystalline "brick," is known as the **unit cell**. A simple two-dimensional example appears in the sheet of wallpaper shown in Figure 11.31 ◀. There are several ways of choosing a unit cell, but the choice is usually the smallest unit cell that shows clearly the symmetry characteristic of the entire pattern.

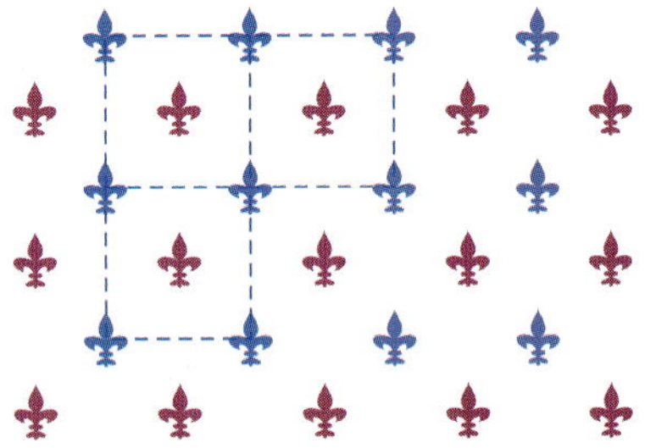

▲ **Figure 11.31 A two-dimensional analog of a lattice and its unit cell.** The wallpaper design shows a characteristic repeat pattern. Each dashed blue square denotes a unit cell of the pattern. The unit cell could equally well be selected with red figures at the corners.

A crystalline solid can be represented by a three-dimensional array of points called a **crystal lattice**. Each point in the lattice is called a *lattice point*, and it represents an identical environment within the solid. The crystal lattice is, in effect, an abstract scaffolding for the crystal structure. We can imagine forming the entire crystal structure by arranging the contents of the unit cell repeatedly on the crystal lattice. In the simplest case the crystal structure would consist of identical atoms, and each atom would be centered on a lattice point. This is the case for most metals.

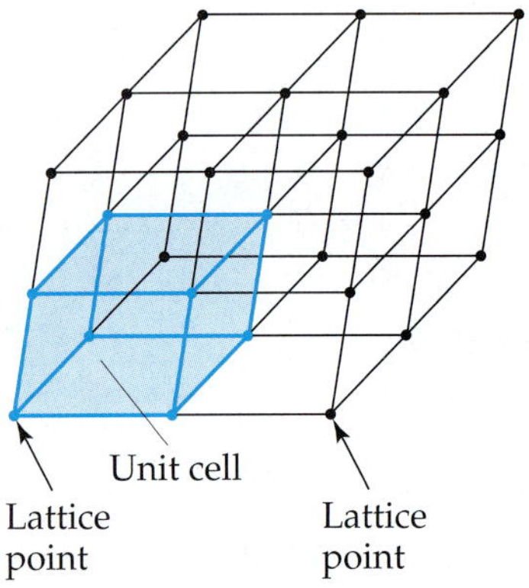

▲ **Figure 11.32 Part of a simple crystal lattice and its associated unit cell.** A lattice is an array of points that define the positions of particles in a crystalline solid. Each lattice point represents an identical environment in the solid. The points here are shown connected by lines to help convey the three-dimensional character of the lattice and to help us see the unit cell.

Figure 11.32 ◀ shows a crystal lattice and its associated unit cell. In general, unit cells are parallelepipeds (six-sided figures whose faces are parallelograms). Each unit cell can be described by the lengths of the edges of the cell and by the angles between these edges. The lattices of all crystalline compounds can be described by seven basic types of unit cells. The simplest of these is the cubic unit cell, in which all the sides are equal in length and all the angles are 90°.

There are three kinds of cubic unit cells, as illustrated in Figure 11.33 ▼. When lattice points are at the corners only, the unit cell is called **primitive cubic** (or *simple cubic*). When a lattice point also occurs at the center of the unit cell,

3-D MODELS
Primitive Cubic, Body-Centered Cubic, Face-Centered Cubic

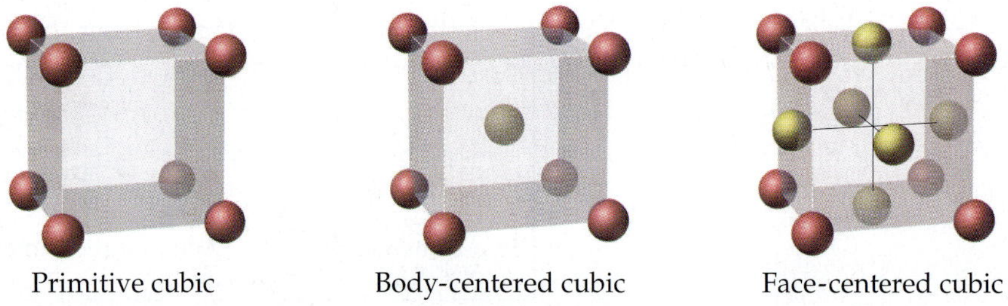

▲ **Figure 11.33 The three types of unit cells found in cubic lattices.** For clarity, the corner spheres are red and the body-centered and face-centered ones are yellow. Each sphere represents a lattice point (an identical environment in the solid).

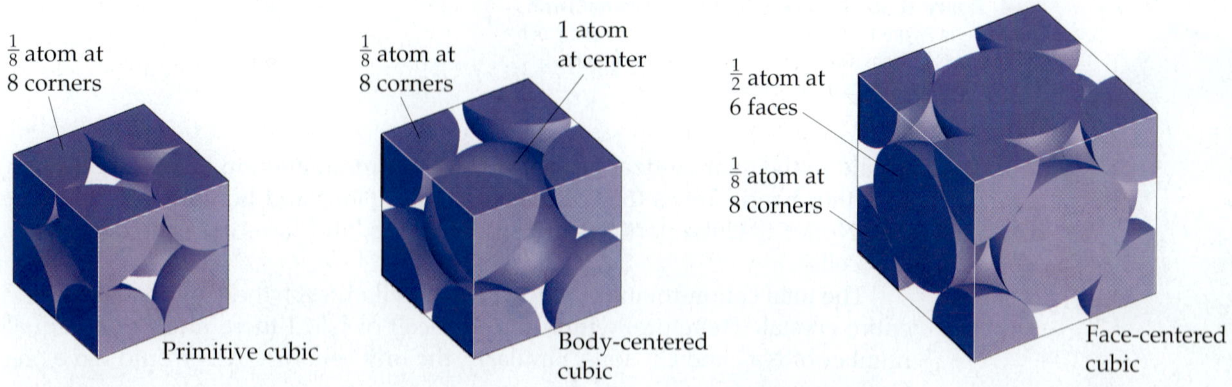

▲ **Figure 11.34 Space-filling view of cubic unit cells.** Only the portion of each atom that belongs to the unit cell is shown.

the cell is **body-centered cubic**. When the cell has lattice points at the center of each face, as well as at each corner, it is **face-centered cubic**.

The simplest crystal structures are cubic unit cells with only one atom centered at each lattice point. Most metals have such structures. Nickel, for example, has a face-centered cubic unit cell, whereas sodium has a body-centered cubic one. Figure 11.34 ▲ shows how atoms fill the cubic unit cells. Notice that the atoms on the corners and faces do not lie wholly within the unit cell. Instead, these atoms are shared between unit cells. Table 11.6 ▶ summarizes the fraction of an atom that occupies a unit cell when atoms are shared between unit cells.

TABLE 11.6 Fraction of an Atom that Occupies a Unit Cell for Various Positions in the Unit Cell

Position in Unit Cell	Fraction in Unit Cell
Center	1
Face	$\frac{1}{2}$
Edge	$\frac{1}{4}$
Corner	$\frac{1}{8}$

3-D MODEL Salt

GIVE IT SOME THOUGHT

For a structure consisting of identical atoms, how many atoms are contained in the body-centered cubic unit cell?

The Crystal Structure of Sodium Chloride

In the crystal structure of NaCl (Figure 11.35 ▼) we can center either the Na^+ ions or the Cl^- ions on the lattice points of a face-centered cubic unit cell. Thus, we can describe the structure as being face-centered cubic.

In Figure 11.35 the Na^+ and Cl^- ions have been moved apart so the symmetry of the structure can be seen more clearly. In this representation no attention

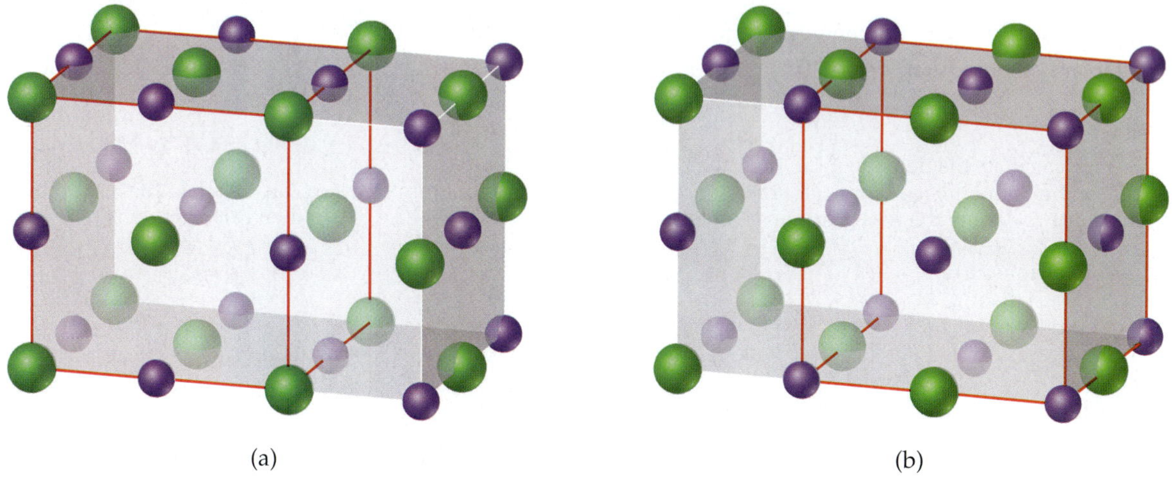

▲ **Figure 11.35 Two ways of defining the unit cell of NaCl.** A representation of an NaCl crystal lattice can show either (a) Cl^- ions (green spheres) or (b) Na^+ ions (purple spheres) at the lattice points of the unit cell. In both cases, the red lines define the unit cell. Both of these choices for the unit cell are acceptable; both have the same volume, and in both cases identical points are arranged in a face-centered cubic fashion.

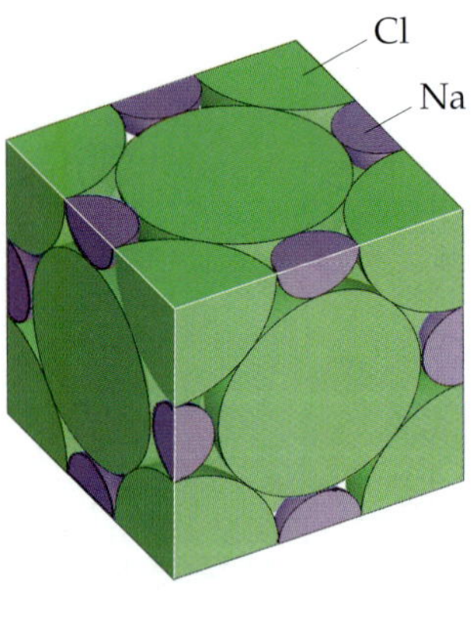

◀ **Figure 11.36 Relative size of ions in an NaCl unit cell.** As in Figure 11.35, purple represents Na^+ ions and green represents Cl^- ions. Only portions of most of the ions lie within the boundaries of the single unit cell.

is paid to the relative sizes of the ions. The representation in Figure 11.36 ◀, on the other hand, shows the relative sizes of the ions and how they fill the unit cell. Notice that the particles at corners, edges, and faces are shared by other unit cells.

The total cation-to-anion ratio of a unit cell must be the same as that for the entire crystal. Therefore, within the unit cell of NaCl there must be an equal number of Na^+ and Cl^- ions. Similarly, the unit cell for $CaCl_2$ would have one Ca^{2+} for every two Cl^-, and so forth.

SAMPLE EXERCISE 11.7 | Determining the Contents of a Unit Cell

Determine the net number of Na^+ and Cl^- ions in the NaCl unit cell (Figure 11.36).

Solution

Analyze: We must sum the various contributing elements to determine the number of Na^+ and Cl^- ions within the unit cell.

Plan: To find the total number of ions of each type, we must identify the different locations within the unit cell and determine the fraction of the ion that lies within the unit cell boundaries.

Solve: There is one-fourth of an Na^+ on each edge, a whole Na^+ in the center of the cube (refer also to Figure 11.35), one-eighth of a Cl^- on each corner, and one-half of a Cl^- on each face. Thus, we have the following:

$$Na^+\text{: } \left(\tfrac{1}{4}\ Na^+ \text{ per edge}\right)(12 \text{ edges}) = 3\ Na^+$$

$$(1\ Na^+ \text{ per center})(1 \text{ center}) = 1\ Na^+$$

$$Cl^-\text{: } \left(\tfrac{1}{8}\ Cl^- \text{ per corner}\right)(8 \text{ corners}) = 1\ Cl^-$$

$$\left(\tfrac{1}{2}\ Cl^- \text{ per face}\right)(6 \text{ faces}) = 3\ Cl^-$$

Thus, the unit cell contains $4\ Na^+$ and $4\ Cl^-$

Check: This result agrees with the compound's stoichiometry: $1\ Na^+$ for each Cl^-

PRACTICE EXERCISE

The element iron crystallizes in a form called α-iron, which has a body-centered cubic unit cell. How many iron atoms are in the unit cell?

Answer: two

SAMPLE EXERCISE 11.8 | Using the Contents and Dimensions of a Unit Cell to Calculate Density

The geometric arrangement of ions in crystals of LiF is the same as that in NaCl. The unit cell of LiF is 4.02 Å on an edge. Calculate the density of LiF.

Solution

Analyze: We are asked to calculate the density of LiF from the size of the unit cell.

Plan: We need to determine the number of formula units of LiF within the unit cell. From that, we can calculate the total mass within the unit cell. Because we know the mass and can calculate the volume of the unit cell, we can then calculate density.

Solve: The arrangement of ions in LiF is the same as that in NaCl (Sample Exercise 11.7), so a unit cell of LiF contains $4\ Li^+$ ions and $4\ F^-$ ions.

Density measures mass per unit volume. Thus, we can calculate the density of LiF from the mass contained in a unit cell and the volume of the unit cell. The mass contained in one unit cell is

$$4(6.94 \text{ amu}) + 4(19.0 \text{ amu}) = 103.8 \text{ amu}$$

The volume of a cube of length a on an edge is a^3, so the volume of the unit cell is $(4.02\ \text{Å})^3$. We can now calculate the density, converting to the common units of g/cm^3:

$$\text{Density} = \frac{(103.8\ \text{amu})}{(4.02\ \text{Å})^3}\left(\frac{1\ \text{g}}{6.02 \times 10^{23}\ \text{amu}}\right)\left(\frac{1\ \text{Å}}{10^{-8}\ \text{cm}}\right)^3 = 2.65\ \text{g/cm}^3$$

Check: This value agrees with that found by simple density measurements, $2.640\ g/cm^3$ at 20°C. The size and contents of the unit cell are therefore consistent with the macroscopic density of the substance.

PRACTICE EXERCISE

The body-centered cubic unit cell of a particular crystalline form of iron is 2.8664 Å on each side. Calculate the density of this form of iron.
Answer $7.8753\ g/cm^3$

Close Packing of Spheres

MOVIE
Close Packing

The structures adopted by crystalline solids are those that bring particles in closest contact to maximize the attractive forces between them. In many cases the particles that make up the solids are spherical or approximately so. Such is the case for atoms in metallic solids. It is therefore instructive to consider how equal-sized spheres can pack most efficiently (that is, with the minimum amount of empty space).

The most efficient arrangement of a layer of equal-sized spheres is shown in Figure 11.37 ▼(a). Each sphere is surrounded by six others in the layer. A second layer of spheres can be placed in the depressions on top of the first layer. A third layer can then be added above the second with the spheres sitting in the depressions of the second layer. However, there are two types of depressions for this third layer, and they result in different structures, as shown in Figure 11.37(b) and (c).

If the spheres of the third layer are placed in line with those of the first layer, as shown in Figure 11.37(b), the structure is known as **hexagonal close packing**. The third layer repeats the first layer, the fourth layer repeats the second layer, and so forth, giving a layer sequence that we denote ABAB.

The spheres of the third layer, however, can be placed so they do not sit above the spheres in the first layer. The resulting structure, shown in Figure 11.37(c), is known as **cubic close packing**. In this case it is the fourth layer that repeats the first layer, and the layer sequence is ABCA. Although it cannot be seen in Figure 11.37(c), the unit cell of the cubic close-packed structure is face-centered cubic.

ANIMATION
Close Packing

In both of the close-packed structures, each sphere has 12 equidistant nearest neighbors: six in one plane, three above that plane, and three below. We say that each sphere has a **coordination number** of 12. The coordination number is the number of particles immediately surrounding a particle in the crystal

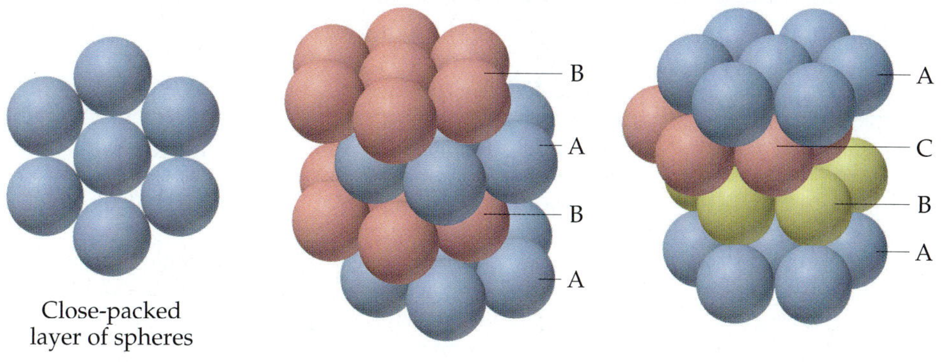

Close-packed layer of spheres

(a) (b) (c)

◀ **Figure 11.37 Close packing of equal-sized spheres.** (a) Close packing of a single layer of equal-sized spheres. (b) In the hexagonal close-packed structure the atoms in the third layer lie directly over those in the first layer. The order of layers is ABAB. (c) In the cubic close-packed structure the atoms in the third layer are not over those in the first layer. Instead, they are offset a bit, and it is the fourth layer that lies directly over the first. Thus, the order of layers is ABCA.

structure. In both types of close packing, 74% of the total volume of the structure is occupied by spheres; 26% is empty space between the spheres. By comparison, each sphere in the body-centered cubic structure has a coordination number of 8, and only 68% of the space is occupied. In the primitive cubic structure the coordination number is 6, and only 52% of the space is occupied.

When unequal-sized spheres are packed in a lattice, the larger particles sometimes assume one of the close-packed arrangements, with smaller particles occupying the holes between the large spheres. In Li_2O, for example, the larger oxide ions assume a cubic close-packed structure, and the smaller Li^+ ions occupy small cavities that exist between oxide ions.

GIVE IT SOME THOUGHT

Based on the information given above for close-packed structures and structures with cubic unit cells, what qualitative relationship exists between coordination numbers and packing efficiencies?

11.8 | Bonding in Solids

The physical properties of crystalline solids, such as melting point and hardness, depend both on the arrangements of particles and on the attractive forces between them. Table 11.7 ▼ classifies solids according to the types of forces between particles in solids.

Molecular Solids

Molecular solids consist of atoms or molecules held together by intermolecular forces (dipole-dipole forces, London dispersion forces, and hydrogen bonds). Because these forces are weak, molecular solids are soft. Furthermore, they normally have relatively low melting points (usually below 200°C). Most substances that are gases or liquids at room temperature form molecular solids at low temperature. Examples include Ar, H_2O, and CO_2.

The properties of molecular solids depend not only on the strengths of the forces that exist between molecules, but also on the abilities of the molecules to

TABLE 11.7 Types of Crystalline Solids

Type of Solid	Form of Unit Particles	Forces Between Particles	Properties	Examples
Molecular	Atoms or molecules	London dispersion forces, dipole-dipole forces, hydrogen bonds	Fairly soft, low to moderately high melting point, poor thermal and electrical conduction	Argon, Ar; methane, CH_4; sucrose, $C_{12}H_{22}O_{11}$; Dry Ice, CO_2
Covalent-network	Atoms connected in a network of covalent bonds	Covalent bonds	Very hard, very high melting point, often poor thermal and electrical conduction	Diamond, C; quartz, SiO_2
Ionic	Positive and negative ions	Electrostatic attractions	Hard and brittle, high melting point, poor thermal and electrical conduction	Typical salts—for example, NaCl, $Ca(NO_3)_2$
Metallic	Atoms	Metallic bonds	Soft to very hard, low to very high melting point, excellent thermal and electrical conduction, malleable and ductile	All metallic elements—for example, Cu, Fe, Al, Pt

A CLOSER LOOK | X-Ray Diffraction by Crystals

When light waves pass through a narrow slit, they are scattered in such a way that the wave seems to spread out. This physical phenomenon is called *diffraction*. When light passes through many evenly spaced narrow slits (a *diffraction grating*), the scattered waves interact to form a series of light and dark bands, known as a diffraction pattern. The most effective diffraction of light occurs when the wavelength of the light and the width of the slits are similar in magnitude.

The spacing of the layers of atoms in solid crystals is usually about 2–20 Å. The wavelengths of X rays are also in this range. Thus, a crystal can serve as an effective diffraction grating for X rays. X-ray diffraction results from the scattering of X rays by a regular arrangement of atoms, molecules, or ions. Much of what we know about crystal structures has been obtained from studies of X-ray diffraction by crystals, a technique known as *X-ray crystallography*. Figure 11.38 ▼ depicts the diffraction of a beam of X rays as it passes through a crystal. The diffracted X rays were formerly detected by a photographic film. Today, crystallographers use an *array detector*, a device analogous to that used in digital cameras, to capture and measure the intensities of the diffracted rays. The diffraction pattern of spots on the detector in Figure 11.38 depends on the particular arrangement of atoms in the crystal. Thus, different types of crystals give rise to different diffraction patterns. In 1913 the English scientists William and Lawrence Bragg (father and son) determined for the first time how the spacing of layers in crystals leads to different X-ray diffraction patterns. By measuring the intensities of the diffracted beams and the angles at which they are diffracted, it is possible to reason backward to the structure that must have given rise to the pattern. One of the most famous X-ray diffraction patterns is the one for crystals of the genetic material DNA (Figure 11.39 ▼), first obtained in the early 1950s. Working from photographs such as this one, Francis Crick, Rosalind Franklin, James Watson, and Maurice Wilkins determined the double-helix structure of DNA, one of the most important discoveries in molecular biology.

Today X-ray crystallography is used extensively to determine the structures of molecules in crystals. The instruments used to measure X-ray diffraction, known as *X-ray diffractometers*, are now computer-controlled, making the collection of diffraction data highly automated. The diffraction pattern of a crystal can be determined very accurately and quickly (sometimes in a matter of hours) even though thousands of diffraction points are measured. Computer programs are then used to analyze the diffraction data and determine the arrangement and structure of the molecules in the crystal.

Related Exercises: 11.95, 11.96

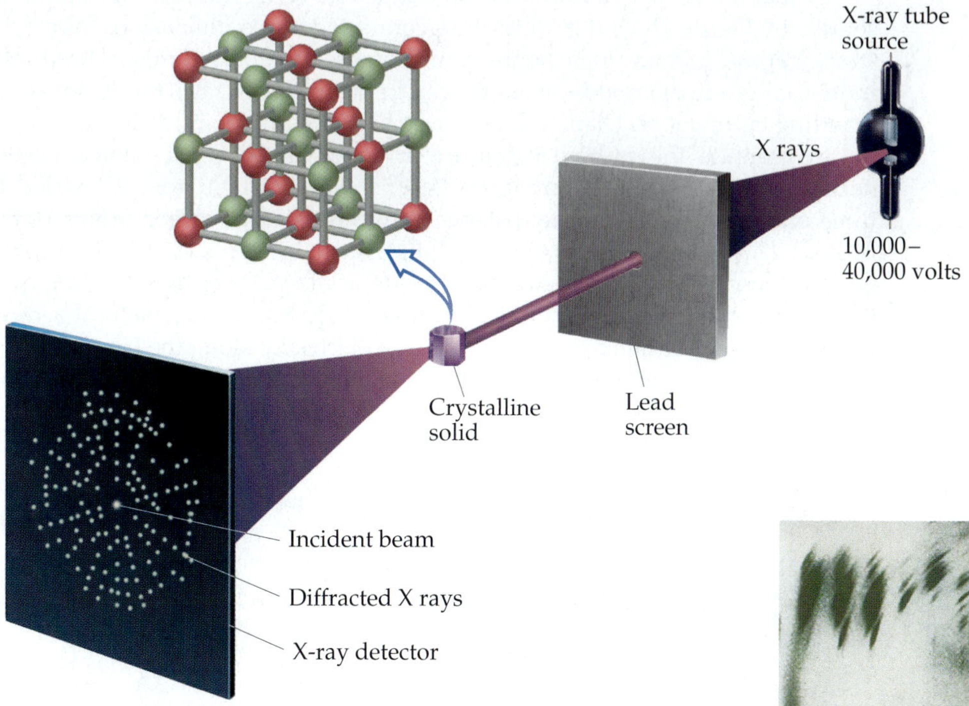

◀ **Figure 11.38 Diffraction of X rays by a crystal.** In X-ray crystallography, an X-ray beam is diffracted by a crystal. The diffraction pattern can be recorded as spots where the diffracted X rays strike a detector, which records the positions and intensities of the spots.

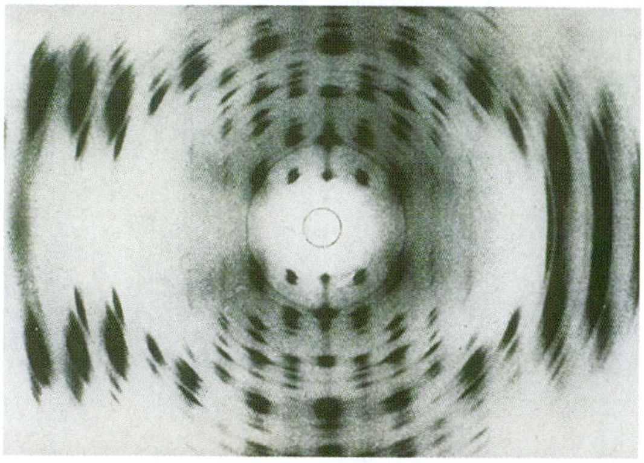

▶ **Figure 11.39 The X-ray diffraction photograph of one form of crystalline DNA.** This photograph was taken in the early 1950s. From the pattern of dark spots, the double-helical shape of the DNA molecule was deduced.

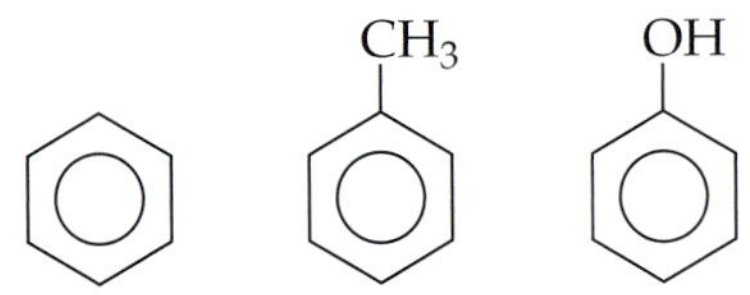

	Benzene	Toluene	Phenol
Melting point (°C)	5	−95	43
Boiling point (°C)	80	111	182

▲ **Figure 11.40 Comparative melting and boiling points for benzene, toluene, and phenol.**

pack efficiently in three dimensions. Benzene (C_6H_6), for example, is a highly symmetrical planar molecule. (Section 8.6) It has a higher melting point than toluene, a compound in which one of the hydrogen atoms of benzene has been replaced by a CH_3 group (Figure 11.40 ◀). The lower symmetry of toluene molecules prevents them from packing as efficiently as benzene molecules. As a result, the intermolecular forces that depend on close contact are not as effective and the melting point is lower. In contrast, the boiling point of toluene is higher than that of benzene, indicating that the intermolecular attractive forces are larger in liquid toluene than in liquid benzene. Both the melting and boiling points of phenol, another substituted benzene shown in Figure 11.40, are higher than those of benzene because the OH group of phenol can form hydrogen bonds.

GIVE IT SOME THOUGHT

Which of the following substances would you expect to form molecular solids: Co, C_6H_6, or K_2O?

Covalent-Network Solids

Covalent-network solids consist of atoms held together in large networks or chains by covalent bonds. Because covalent bonds are much stronger than intermolecular forces, these solids are much harder and have higher melting points than molecular solids. Diamond and graphite, two allotropes of carbon, are covalent-network solids. Other examples include quartz, SiO_2; silicon carbide, SiC; and boron nitride, BN.

3-D MODELS
Graphite, Diamond, Fullerene

In diamond each carbon atom is bonded to four other carbon atoms as shown in Figure 11.41(a) ▼. This interconnected three-dimensional array of strong carbon–carbon single bonds contributes to diamond's unusual hardness. Industrial-grade diamonds are employed in the blades of saws for the most demanding cutting jobs. Diamond also has a high melting point, 3550°C.

In graphite the carbon atoms are arranged in layers of interconnected hexagonal rings as shown in Figure 11.41(b). Each carbon atom is bonded to three others in the layer. The distance between adjacent carbon atoms in the plane, 1.42 Å, is very close to the C—C distance in benzene, 1.395 Å. In fact, the bonding resembles that of benzene, with delocalized π bonds extending over the layers. (Section 9.6) Electrons move freely through the delocalized orbitals, making graphite a good conductor of electricity along the layers. (If you have ever taken apart a flashlight battery, you know that the central electrode in the battery is made of graphite.) The layers, which are separated by 3.41 Å, are

(a) Diamond (b) Graphite

▶ **Figure 11.41 Structures of (a) diamond and (b) graphite.** The blue color in (b) is added to emphasize the planarity of the carbon layers.

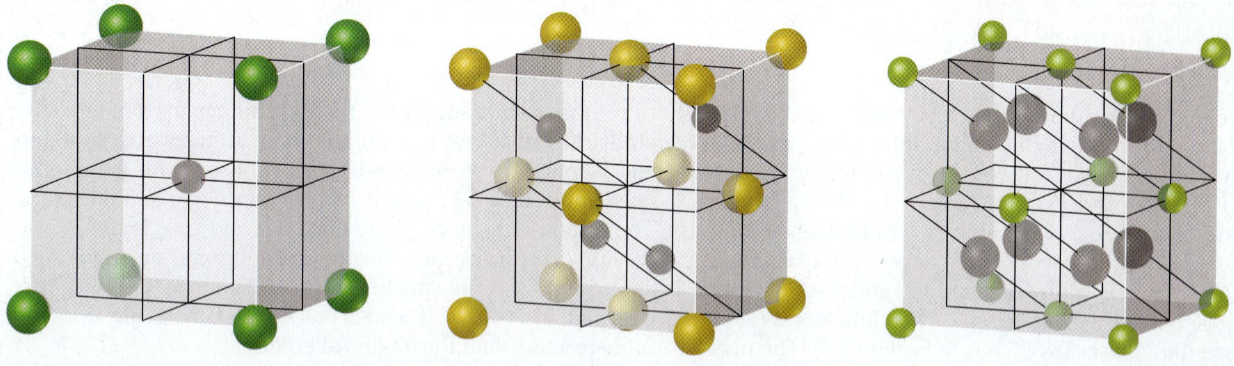

(a) CsCl (b) ZnS (c) CaF_2

▲ **Figure 11.42 Unit cells of some common ionic structures.** The ZnS shown in (b) is called zinc blende, and the CaF_2 of (c) is known as fluorite.

held together by weak dispersion forces. The layers readily slide past one another when rubbed, giving graphite a greasy feel. Graphite is used as a lubricant and in the "lead" in pencils.

Ionic Solids

Ionic solids consist of ions held together by ionic bonds. (Section 8.2) The strength of an ionic bond depends greatly on the charges of the ions. Thus, NaCl, in which the ions have charges of 1+ and 1−, has a melting point of 801°C, whereas MgO, in which the charges are 2+ and 2−, melts at 2852°C.

The structures of simple ionic solids can be classified as a few basic types. The NaCl structure is a representative example of one type. Other compounds that possess this same structure include LiF, KCl, AgCl, and CaO. Three other common types of crystal structures are shown in Figure 11.42 ▲.

The structure adopted by an ionic solid depends largely on the charges and relative sizes of the ions. In the NaCl structure, for example, the Na^+ ions have a coordination number of 6 because each Na^+ ion is surrounded by six nearest neighbor Cl^- ions. In the CsCl structure [Figure 11.42(a)], by comparison, the Cl^- ions adopt a primitive cubic arrangement with each Cs^+ ion surrounded by eight Cl^- ions. The increase in the coordination number as the alkali metal ion changes from Na^+ to Cs^+ is a consequence of the larger size of Cs^+ compared to Na^+.

In the zinc blende (ZnS) structure [Figure 11.42(b)], the S^{2-} ions adopt a face-centered cubic arrangement, with the smaller Zn^{2+} ions arranged so they are each surrounded tetrahedrally by four S^{2-} ions (compare with Figure 11.33). CuCl also adopts this structure.

In the fluorite (CaF_2) structure [Figure 11.42(c)], the Ca^{2+} ions are shown in a face-centered cubic arrangement. As required by the chemical formula of the substance, there are twice as many F^- ions (gray) in the unit cell as there are Ca^{2+} ions. Other compounds that have the fluorite structure include $BaCl_2$ and PbF_2.

Metallic Solids

Metallic solids consist entirely of metal atoms. Metallic solids usually have hexagonal close-packed, cubic close-packed (face-centered cubic), or body-centered cubic structures. Thus, each atom typically has 8 or 12 adjacent atoms.

The bonding in metals is too strong to be due to London dispersion forces, and yet there are not enough valence electrons for ordinary covalent bonds between atoms. The bonding is due to valence electrons that are delocalized throughout the entire solid. In fact, we can visualize the metal as an array of

A CLOSER LOOK | Buckyball

Until the mid-1980s pure solid carbon was thought to exist in only two forms: diamond and graphite, both of which are covalent-network solids. In 1985 a group of researchers led by Richard Smalley and Robert Curl of Rice University in Houston and Harry Kroto of the University of Sussex in England made a startling discovery. They vaporized a sample of graphite with an intense pulse of laser light and used a stream of helium gas to carry the vaporized carbon into a mass spectrometer (see the "Closer Look" box in Section 2.4). The mass spectrum showed peaks corresponding to clusters of carbon atoms, with a particularly strong peak corresponding to molecules composed of 60 carbon atoms, C_{60}.

Because C_{60} clusters were so preferentially formed, the group proposed a radically different form of carbon, namely, C_{60} molecules that were nearly spherical in shape. They proposed that the carbon atoms of C_{60} form a "ball" with 32 faces, of which 12 are pentagons and 20 are hexagons (Figure 11.43 ▼), exactly like a soccer ball. The shape of this molecule is reminiscent of the geodesic dome invented by the U.S. engineer and philosopher R. Buckminster Fuller, so C_{60} was whimsically named "buckminsterfullerene," or "buckyball" for short. Since the discovery of C_{60}, other related molecules of carbon atoms have been discovered. These molecules are now known as fullerenes.

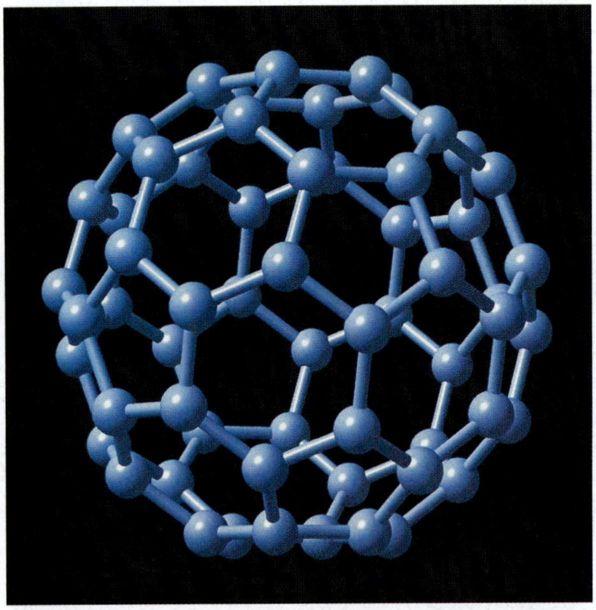

▲ **Figure 11.43 The buckminsterfullerene molecule, C_{60}.** The molecule has a highly symmetric structure in which the 60 carbon atoms sit at the vertices of a truncated icosahedron–the same geometry as a soccer ball.

Appreciable amounts of buckyball can be prepared by electrically evaporating graphite in an atmosphere of helium gas. About 14% of the resulting soot consists of C_{60} and a related molecule, C_{70}, which has a more elongated structure. The carbon-rich gases from which C_{60} and C_{70} condense also contain other fullerenes, mostly with more carbon atoms such as C_{76} and C_{84}. The smallest possible fullerene, C_{20}, was first detected in 2000. This small ball-shaped molecule is much more reactive than the larger fullerenes.

Because the fullerenes are composed of individual molecules, they dissolve in various organic solvents, whereas diamond and graphite do not (Figure 11.44 ▼). This solubility permits the fullerenes to be separated from the other components of soot and even from each other. It also allows the study of their reactions in solution. Study of these substances has led to the discovery of some very interesting chemistry. For example, it is possible to place a metal atom inside a buckyball, generating a molecule in which a metal atom is completely enclosed by the carbon sphere. The C_{60} molecules also react with potassium to give K_3C_{60}, which contains a face-centered cubic array of buckyballs with K^+ ions in the cavities between them. This compound is a superconductor at 18 K (Section 12.5), suggesting the possibility that other fullerenes may also have interesting electrical, magnetic, or optical properties. For their discovery and pioneering work with fullerenes, Professors Smalley, Curl, and Kroto were awarded the 1996 Nobel Prize in chemistry.

Related Exercise: 11.97

▲ **Figure 11.44 Solutions of fullerenes.** Unlike diamond and graphite, the new molecular forms of carbon can be dissolved in organic solvents. The orange solution on the left is a solution of C_{70} in *n*-hexane, which is a colorless liquid. The magenta solution on the right is a solution of buckyball, C_{60}, in *n*-hexane.

positive ions immersed in a sea of delocalized valence electrons, as shown in Figure 11.45 ▶.

Metals vary greatly in the strength of their bonding, as shown by their wide range of physical properties such as hardness and melting point. In general, however, the strength of the bonding increases as the number of electrons available for bonding increases. Thus, sodium, which has only one valence electron per atom, melts at 97.5°C, whereas chromium, with six electrons beyond the noble-gas core, melts at 1890°C. The mobility of the electrons explains why metals are good conductors of heat and electricity. The bonding and properties of metals will be examined more closely in Chapter 23.

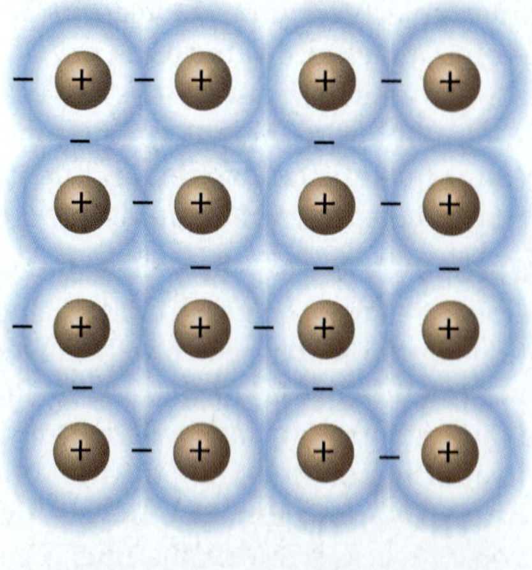

▲ **Figure 11.45 Representation of a cross section of a metal.** Each sphere represents the nucleus and inner-core electrons of a metal atom. The surrounding blue "fog" represents the mobile sea of electrons that binds the atoms together.

SAMPLE INTEGRATIVE EXERCISE | Putting Concepts Together

The substance CS_2 has a melting point of −110.8°C and a boiling point of 46.3°C. Its density at 20°C is 1.26 g/cm^3. It is highly inflammable. **(a)** What is the name of this compound? **(b)** If you were going to look up the properties of this substance in the *CRC Handbook of Chemistry and Physics*, would you look under the physical properties of inorganic or organic compounds? Explain. **(c)** How would you classify $CS_2(s)$ as to type of crystalline solid? **(d)** Write a balanced equation for the combustion of this compound in air. (You will have to decide on the most likely oxidation products.) **(e)** The critical temperature and pressure for CS_2 are 552 K and 78 atm, respectively. Compare these values with those for CO_2 (Table 11.5), and discuss the possible origins of the differences. **(f)** Would you expect the density of CS_2 at 40°C to be greater or less than at 20°C? What accounts for the difference?

Solution **(a)** The compound is named carbon disulfide, in analogy with the naming of other binary molecular compounds. (Section 2.8) **(b)** The substance will be listed as an inorganic compound. It contains no carbon–carbon bonds, or C—H bonds, which are the usual structural features of organic compounds. **(c)** Because $CS_2(s)$ consists of individual CS_2 molecules, it would be a molecular solid in the classification scheme of Table 11.7. **(d)** The most likely products of the combustion will be CO_2 and SO_2. (Sections 3.2 and 7.8) Under some conditions SO_3 might be formed, but this would be the less likely outcome. Thus, we have the following equation for combustion:

$$CS_2(l) + 3\,O_2(g) \longrightarrow CO_2(g) + 2\,SO_2(g)$$

(e) The critical temperature and pressure of CS_2 (552 K and 78 atm) are both higher than those given for CO_2 in Table 11.5 (304 K and 73 atm). The difference in critical temperatures is especially notable. The higher values for CS_2 arise from the greater London dispersion attractions between the CS_2 molecules compared with CO_2. These greater attractions are due to the larger size of the sulfur compared to oxygen and therefore its greater polarizability. **(f)** The density would be lower at the higher temperature. Density decreases with increasing temperature because the molecules possess higher kinetic energies. Their more energetic movements result in larger average spacings per molecule, which translate into lower densities.

SUMMARY AND KEY TERMS

Introduction and Section 11.1 Substances that are gases or liquids at room temperature are usually composed of molecules. In gases the intermolecular attractive forces are negligible compared to the kinetic energies of the molecules; thus, the molecules are widely separated and undergo constant, chaotic motion. In liquids the **intermolecular forces** are strong enough to keep the molecules in close proximity; nevertheless, the molecules are free to move with respect to one another. In solids the interparticle attractive forces are strong enough to restrain molecular motion and to force the particles to occupy specific locations in a three-dimensional arrangement.

Section 11.2 Three types of intermolecular forces exist between neutral molecules: **dipole-dipole forces, London dispersion forces**, and **hydrogen bonding. Ion-dipole forces** are important in solutions in which ionic compounds are dissolved in polar solvents. London dispersion forces operate between all molecules. The relative strengths of the dipole-dipole and dispersion forces depend on the polarity, **polarizability**, size, and shape of the molecule. Dipole-dipole forces increase in strength with increasing polarity. Dispersion forces increase in strength with increasing molecular weight, although molecular shape is also an important factor. Hydrogen bonding occurs in compounds containing O—H, N—H, and F—H bonds. Hydrogen bonds are generally stronger than dipole-dipole or dispersion forces.

Section 11.3 The stronger the intermolecular forces, the greater is the **viscosity**, or resistance to flow, of a liquid. The surface tension of a liquid also increases as intermolecular forces increase in strength. **Surface tension** is a measure of the tendency of a liquid to maintain a minimum surface area. The adhesion of a liquid to the walls of a narrow tube and the cohesion of the liquid account for **capillary action** and the formation of a meniscus at the surface of a liquid.

Section 11.4 A substance may exist in more than one state of matter, or phase. **Phase changes** are transformations from one phase to another. Changes of a solid to liquid (melting), solid to gas (sublimation), and liquid to gas (vaporization) are all endothermic processes. Thus, the **heat of fusion** (melting), the **heat of sublimation**, and the **heat of vaporization** are all positive quantities. The reverse processes are exothermic. A gas cannot be liquefied by application of pressure if the temperature is above its **critical temperature**. The pressure required to liquefy a gas at its critical temperature is called the **critical pressure**.

Section 11.5 The **vapor pressure** of a liquid indicates the tendency of the liquid to evaporate. The vapor pressure is the partial pressure of the vapor when it is in **dynamic equilibrium** with the liquid. At equilibrium the rate of transfer of molecules from the liquid to the vapor equals the rate of transfer from the vapor to the liquid. The higher the vapor pressure of a liquid, the more readily it evaporates and the more **volatile** it is. Vapor pressure increases nonlinearly with temperature. Boiling occurs when the vapor pressure equals the external pressure. The **normal boiling point** is the temperature at which the vapor pressure equals 1 atm.

Section 11.6 The equilibria between the solid, liquid, and gas phases of a substance as a function of temperature and pressure are displayed on a **phase diagram**. Equilibria between any two phases are indicated by a line. The line through the melting point usually slopes slightly to the right as pressure increases, because the solid is usually more dense than the liquid. The melting point at 1 atm is the **normal melting point**. The point on the diagram at which all three phases coexist in equilibrium is called the **triple point**.

Section 11.7 In a **crystalline solid**, particles are arranged in a regularly repeating pattern. An **amorphous solid** is one whose particles show no such order. The essential structural features of a crystalline solid can be represented by its **unit cell**, the smallest part of the crystal that can, by simple displacement, reproduce the three-dimensional structure. The three-dimensional structures of a crystal can also be represented by its **crystal lattice**. The points in a crystal lattice represent positions in the structure where there are identical environments. The simplest unit cells are cubic. There are three kinds of cubic unit cells: **primitive cubic, body-centered cubic**, and **face-centered cubic**.

Many solids have a close-packed structure in which spherical particles are arranged so as to leave the minimal amount of empty space. Two closely related forms of close packing, **cubic close packing** and **hexagonal close packing**, are possible. In both, each sphere has a **coordination number** of 12.

Section 11.8 The properties of solids depend both on the arrangements of particles and on the attractive forces between them. **Molecular solids**, which consist of atoms or molecules held together by intermolecular forces, are soft and low melting. **Covalent-network solids**, which consist of atoms held together by covalent bonds that extend throughout the solid, are hard and high melting. **Ionic solids** are hard and brittle and have high melting points. **Metallic solids**, which consist of metal cations held together by a sea of electrons, exhibit a wide range of properties.

VISUALIZING CONCEPTS

11.1 Does the following diagram best describe a crystalline solid, liquid, or gas? Explain. [Section 11.1]

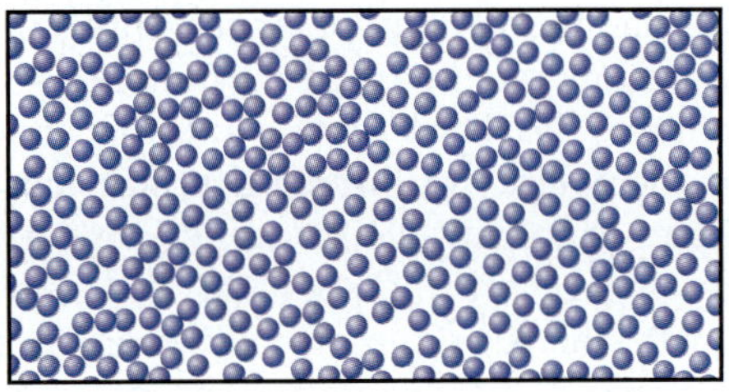

11.2 What kind of intermolecular attractive force is shown in each of the following cases? [Section 11.2]

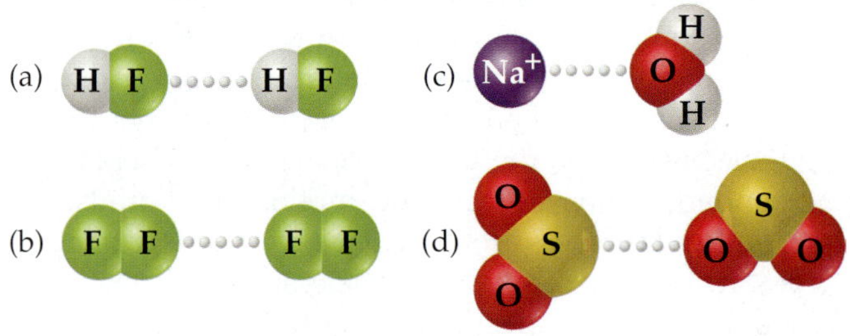

11.3 The molecular models of glycerol and 1-propanol are given here.

(a) Glycerol (b) 1-propanol

Do you expect the viscosity of glycerol to be larger or smaller than that of 1-propanol? Explain. [Section 11.3]

11.4 Using the following graph, determine **(a)** the approximate vapor pressure of CS_2 at 30°C, **(b)** the temperature at which the vapor pressure equals 300 torr, **(c)** the normal boiling point of CS_2. [Section 11.5]

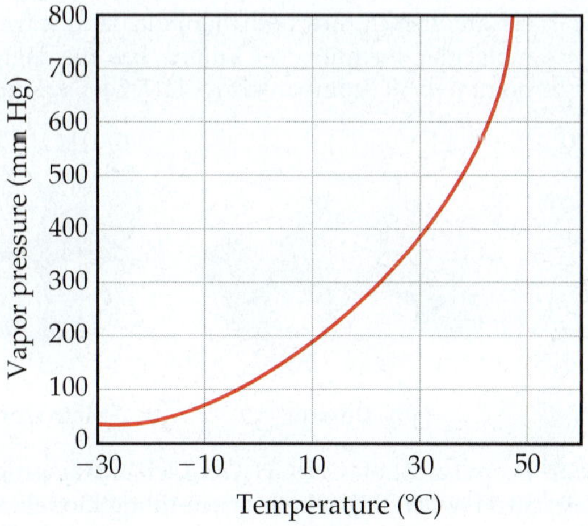

11.5 The following molecules have the same molecular formula (C_3H_8O), yet they have different normal boiling points, as shown. Rationalize the difference in boiling points. [Sections 11.2 and 11.5]

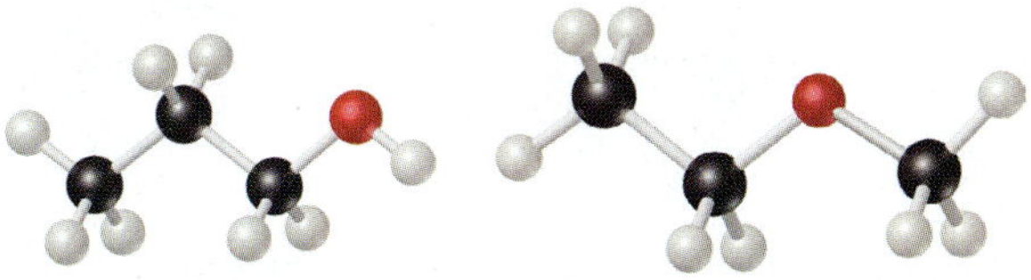

(a) Propanol 97.2°C (b) Ethyl methyl ether 10.8°C

11.6 The phase diagram of a hypothetical substance is shown below.

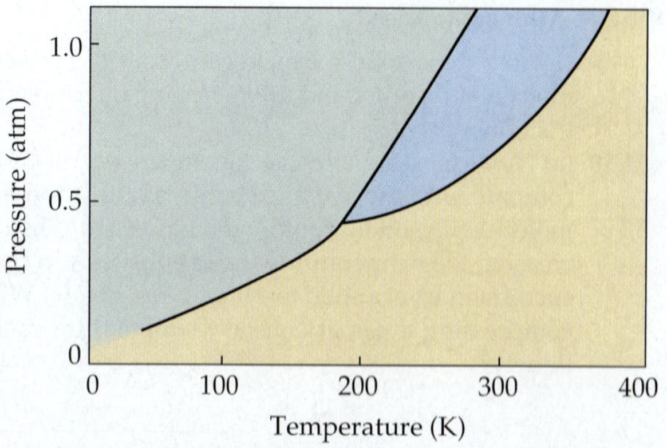

(a) Estimate the normal boiling point and freezing point of the substance.

(b) What is the physical state of the substance under the following conditions?
(i) $T = 150$ K, $P = 0.2$ atm;
(ii) $T = 100$ K, $P = 0.8$ atm;
(iii) $T = 300$ K, $P = 1.0$ atm. [Section 11.6]

11.7 Niobium(II) oxide crystallizes in the following cubic unit cell.

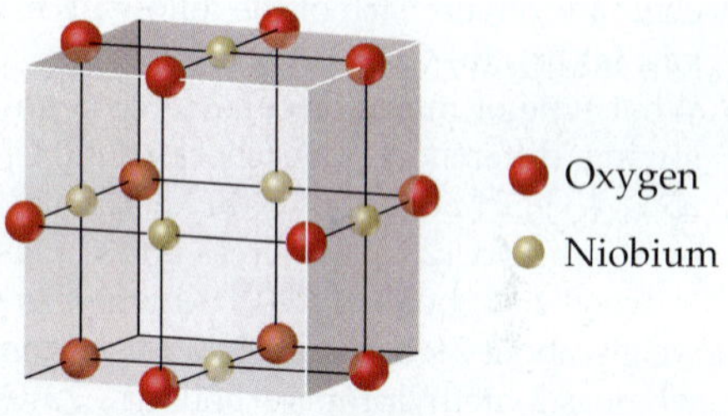

(a) How many niobium atoms and how many oxygen atoms are within the unit cell?

(b) What is the empirical formula of niobium oxide?

(c) Is this a molecular, covalent-network, or ionic solid? [Sections 11.7 and 11.8]

11.8 **(a)** What kind of packing arrangement is seen in the accompanying photo? **(b)** What is the coordination number of each orange in the interior of the stack? **(c)** If each orange represents an argon atom, what category of solid is represented? [Sections 11.7 and 11.8]

EXERCISES

Kinetic-Molecular Theory

11.9 List the three states of matter in order of **(a)** increasing molecular disorder and **(b)** increasing intermolecular attractions.

11.10 **(a)** How does the average kinetic energy of molecules compare with the average energy of attraction between molecules in solids, liquids, and gases? **(b)** Why does increasing the temperature cause a substance to change in succession from a solid to a liquid to a gas? **(c)** Why does compressing a gas at constant temperature cause it to liquefy?

11.11 **(a)** Why are gases more compressible than liquids? **(b)** Why are the liquid and solid forms of a substance referred to as *condensed phases*? **(c)** Why do liquids have a greater ability to flow than solids?

11.12 Benzoic acid, C_6H_5COOH, melts at 122°C. The density in the liquid state at 130°C is 1.08 g/cm^3. The density of solid benzoic acid at 15°C is 1.266 g/cm^3. **(a)** In which of these two states is the average distance between molecules the greater? **(b)** Explain the difference in densities at the two temperatures in terms of the kinetic-molecular theory.

Intermolecular Forces

11.13 Which type of intermolecular attractive force operates between **(a)** all molecules, **(b)** polar molecules, **(c)** the hydrogen atom of a polar bond and a nearby small electronegative atom?

11.14 What type(s) of intermolecular force is (are) common to **(a)** Xe and methanol (CH_3OH), **(b)** CH_3OH and acetonitrile (CH_3CN), **(c)** NH_3 and HF?

11.15 Describe the intermolecular forces that must be overcome to convert each of the following from a liquid to a gas: **(a)** Br_2, **(b)** CH_3OH, **(c)** H_2S.

11.16 What type of intermolecular force accounts for the following differences in each case? **(a)** CH_3OH boils at 65°C, CH_3SH boils at 6°C. **(b)** Xe is liquid at atmospheric pressure and 120 K, whereas Ar is a gas. **(c)** Kr, atomic weight 84, boils at 120.9 K, whereas Cl_2, molecular weight about 71, boils at 238 K. **(d)** Acetone boils at 56°C, whereas 2-methylpropane boils at −12°C.

Acetone 2-Methylpropane

11.17 **(a)** What is meant by the term polarizability? **(b)** Which of the following atoms would you expect to be most polarizable: O, S, Se, or Te? Explain. **(c)** Put the following molecules in order of increasing polarizability: $GeCl_4$, CH_4, $SiCl_4$, SiH_4, and $GeBr_4$. **(d)** Predict the order of boiling points of the substances in part (c).

11.18 **(a)** Why does the strength of dispersion forces increase with increasing polarizability? **(b)** Account for the steady increase in boiling point of the noble-gas elements with increasing atomic weight (Table 11.3). **(c)** What general rule of thumb applies to the relationship between dispersion forces and molecular weight? **(d)** Comment on whether the following statement is correct: "All other factors being the same, dispersion forces between molecules increase with the number of electrons in the molecules."

11.19 Which member of the following pairs has the larger London dispersion forces: **(a)** H_2O or H_2S, **(b)** CO_2 or CO, **(c)** CH_4 or SiH_4?

11.20 Which member of the following pairs has the stronger intermolecular dispersion forces: **(a)** Br_2 or O_2, **(b)** CH_3CH_2SH or $CH_3CH_2CH_2SH$, **(c)** $CH_3CH_2CH_2Cl$ or $(CH_3)_2CHCl$?

11.21 Butane and 2-methylpropane, whose space-filling models are shown, are both nonpolar and have the same molecular formula, yet butane has the higher boiling point (−0.5°C compared to −11.7°C). Explain.

(a) Butane (b) 2-Methylpropane

11.22 Propyl alcohol ($CH_3CH_2CH_2OH$) and isopropyl alcohol [$(CH_3)_2CHOH$], whose space-filling models are shown, have boiling points of 97.2°C and 82.5°C, respectively. Explain why the boiling point of propyl alcohol is higher, even though both have the molecular formula of C_3H_8O.

(a) Propyl alcohol (b) Isopropyl alcohol

11.23 **(a)** What molecular features must a molecule have to participate in hydrogen bonding with other molecules of the same kind? **(b)** Which of the following molecules can form hydrogen bonds with other molecules of the same kind: CH_3F, CH_3NH_2, CH_3OH, CH_3Br?

11.24 Ethylene glycol ($HOCH_2CH_2OH$), the major substance in antifreeze, has a normal boiling point of 198°C. By comparison, ethyl alcohol (CH_3CH_2OH) boils at 78°C at atmospheric pressure. Ethylene glycol dimethyl ether ($CH_3OCH_2CH_2OCH_3$) has a normal boiling point of

83°C, and ethyl methyl ether ($CH_3CH_2OCH_3$) has a normal boiling point of 11°C. **(a)** Explain why replacement of a hydrogen on the oxygen by CH_3 generally results in a lower boiling point. **(b)** What are the major factors responsible for the difference in boiling points of the two ethers?

11.25 Rationalize the difference in boiling points between the members of the following pairs of substances: **(a)** HF (20°C) and HCl (−85°C), **(b)** $CHCl_3$ (61°C) and $CHBr_3$ (150°C), **(c)** Br_2 (59°C) and ICl (97°C).

11.26 Identify the types of intermolecular forces present in each of the following substances, and select the substance in each pair that has the higher boiling point: **(a)** C_6H_{14} or C_8H_{18}, **(b)** C_3H_8 or CH_3OCH_3, **(c)** HOOH or HSSH, **(d)** NH_2NH_2 or CH_3CH_3.

11.27 How are the following observations related to the ability of water to form hydrogen bonds? **(a)** Ice is less dense than liquid water. **(b)** Water has a high specific heat, meaning it requires a large amount of heat to produce a temperature increase of 1 K.

11.28 The following quote about ammonia (NH_3) is from a textbook of inorganic chemistry: "It is estimated that 26% of the hydrogen bonding in NH_3 breaks down on melting, 7% on warming from the melting to the boiling point, and the final 67% on transfer to the gas phase at the boiling point." From the standpoint of the kinetic-molecular theory, explain **(a)** why there is a decrease of hydrogen-bonding energy on melting and **(b)** why most of the loss in hydrogen bonding occurs in the transition from the liquid to the vapor state.

Viscosity and Surface Tension

11.29 **(a)** Why do surface tension and viscosity decrease with increasing temperature? **(b)** Why do substances with high surface tensions also tend to have high viscosities?

11.30 **(a)** Distinguish between adhesive forces and cohesive forces. **(b)** What adhesive and what cohesive forces are involved when a paper towel absorbs water? **(c)** Explain the cause for the U-shaped meniscus formed when water is in a glass tube.

11.31 Explain the following observations: **(a)** The surface tension of $CHBr_3$ is greater than that of $CHCl_3$. **(b)** As temperature increases, oil flows faster through a narrow tube. **(c)** Raindrops that collect on a waxed automobile hood take on a nearly spherical shape.

11.32 Hydrazine (NH_2NH_2), hydrogen peroxide (HOOH), and water (H_2O) all have exceptionally high surface tensions in comparison with other substances of comparable molecular weights. **(a)** Draw the Lewis structures for these three compounds. **(b)** What structural property do these substances have in common, and how might that account for the high surface tensions?

Changes of State

11.33 Name the phase transition in each of the following situations, and indicate whether it is exothermic or endothermic: **(a)** When water is cooled, it turns to ice. **(b)** Wet clothes dry on a warm summer day. **(c)** Frost appears on a window on a cold winter day.

11.34 Name the phase transition in each of the following situations, and indicate whether it is exothermic or endothermic: **(a)** Bromine vapor turns to bromine liquid as it is cooled. **(b)** Crystals of iodine disappear from an evaporating dish as they stand in a fume hood. **(c)** Rubbing alcohol in an open container slowly disappears. **(d)** Molten lava from a volcano turns into solid rock.

11.35 Explain why the heat of fusion of any substance is generally lower than its heat of vaporization.

11.36 Ethyl chloride (C_2H_5Cl) boils at 12°C. When liquid C_2H_5Cl under pressure is sprayed on a room temperature surface in air, the surface is cooled considerably. **(a)** What does this observation tell us about the enthalpy content of $C_2H_5Cl(g)$ as compared with $C_2H_5Cl(l)$? **(b)** In terms of the kinetic-molecular theory, what is the origin of this difference?

11.37 For many years drinking water has been cooled in hot climates by evaporating it from the surfaces of canvas bags or porous clay pots. How many grams of water can be cooled from 35°C to 22°C by the evaporation of 50 g of water? (The heat of vaporization of water in this temperature range is 2.4 kJ/g. The specific heat of water is 4.18 J/g-K.)

11.38 Compounds like CCl_2F_2 are known as chlorofluorocarbons, or CFCs. These compounds were once widely used as refrigerants but are now being replaced by compounds that are believed to be less harmful to the environment. The heat of vaporization of CCl_2F_2 is 289 J/g. What mass of this substance must evaporate in order to freeze 100 g of water initially at 18°C? (The heat of fusion of water is 334 J/g; the specific heat of water is 4.18 J/g-K.)

11.39 Ethanol (C_2H_5OH) melts at −114°C and boils at 78°C. The enthalpy of fusion of ethanol is 5.02 kJ/mol, and its enthalpy of vaporization is 38.56 kJ/mol. The specific heats of solid and liquid ethanol are 0.97 J/g-K and 2.3 J/g-K, respectively. How much heat is required to convert 75.0 g of ethanol at −120°C to the vapor phase at 78°C?

11.40 The fluorocarbon compound $C_2Cl_3F_3$ has a normal boiling point of 47.6°C. The specific heats of $C_2Cl_3F_3(l)$ and $C_2Cl_3F_3(g)$ are 0.91 J/g-K and 0.67 J/g-K, respectively. The heat of vaporization for the compound is 27.49 kJ/mol. Calculate the heat required to convert 25.0 g of $C_2Cl_3F_3$ from a liquid at 5.00°C to a gas at 82.00°C.

11.41 **(a)** What is the significance of the critical pressure of a substance? **(b)** What happens to the critical temperature of a series of compounds as the force of attraction between molecules increases? **(c)** Which of the substances listed in Table 11.5 can be liquefied at the temperature of liquid nitrogen (−196°C)?

11.42 The critical temperatures (K) and pressures (atm) of a series of halogenated methanes are as follows:

Compound	CCl_3F	CCl_2F_2	$CClF_3$	CF_4
Critical Temperature	471	385	302	227
Critical Pressure	43.5	40.6	38.2	37.0

(a) What in general can you say about the variation in intermolecular forces in this series? **(b)** What specific kinds of intermolecular forces are most likely to account for most of the variation in critical parameters in this series?

Vapor Pressure and Boiling Point

11.43 Explain how each of the following affects the vapor pressure of a liquid: **(a)** volume of the liquid, **(b)** surface area, **(c)** intermolecular attractive forces, **(d)** temperature.

11.44 A liquid that has an equilibrium vapor pressure of 130 torr at 25°C is placed into a 1-L vessel like that shown in Figure 11.22. What is the pressure difference shown on the manometer, and what is the composition of the gas in the vessel, under each of the following conditions: **(a)** Two hundred mL of the liquid is introduced into the vessel and frozen at the bottom. The vessel is evacuated and sealed, and the liquid is allowed to warm to 25°C. **(b)** Two hundred milliliters of the liquid is added to the vessel at 25°C under atmospheric pressure, and after a few minutes the vessel is closed off. **(c)** A few mL of the liquid is introduced into the vessel at 25°C while it has a pressure of 1 atm of air in it, without allowing any of the air to escape. After a few minutes a few drops of liquid remain in the vessel.

11.45 **(a)** Place the following substances in order of increasing volatility: CH_4, CBr_4, CH_2Cl_2, CH_3Cl, $CHBr_3$, and CH_2Br_2. Explain. **(b)** How do the boiling points vary through this series?

11.46 PCl_3 and $AsCl_3$ are similar substances, with similar geometries and bonding modes. **(a)** Which of these two substances would you expect to be the more volatile at room temperature? **(b)** Which substance would you expect to have the higher boiling point? **(c)** In which substance would the kinetic energies of the molecules be greater at 40°C, a temperature well below either substance's boiling point? **(d)** In which substance would you expect the intermolecular forces to be larger?

11.47 **(a)** Two pans of water are on different burners of a stove. One pan of water is boiling vigorously, while the other is boiling gently. What can be said about the temperature of the water in the two pans? **(b)** A large container of water and a small one are at the same temperature. What can be said about the relative vapor pressures of the water in the two containers?

11.48 Explain the following observations: **(a)** Water evaporates more quickly on a hot, dry day than on a hot, humid day. **(b)** It takes longer to cook hard-boiled eggs at high altitudes than at lower altitudes.

11.49 **(a)** Use the vapor-pressure curve in Figure 11.24 to estimate the boiling point of diethyl ether at 400 torr.
(b) Use the vapor-pressure curve in Figure 11.24 to estimate the external pressure under which ethyl alcohol will boil at 70°C.
(c) Using the vapor-pressure table in Appendix B, determine the boiling point of water when the external pressure is 25 torr.
(d) Suppose the pressure inside a pressure cooker reaches 1.2 atm. By using the vapor-pressure table in Appendix B, estimate the temperature at which water will boil in this cooker.

11.50 Reno, Nevada, is about 4500 ft above sea level.
(a) If the barometric pressure is 680 mm Hg in Reno, at what temperature will water boil? Refer to Appendix B.
(b) What can you say about the average kinetic energies of the water molecules at the boiling point in Reno as compared with those at the boiling point in Chicago, where the barometric pressure is 752 mm Hg? If you believe the average kinetic energies are different, explain how it is that boiling of water can occur at two different average kinetic energies of the water molecules.

Phase Diagrams

11.51 **(a)** What is the significance of the critical point in a phase diagram? **(b)** Why does the line that separates the gas and liquid phases end at the critical point?

11.52 **(a)** What is the significance of the triple point in a phase diagram? **(b)** Could you measure the triple point of water by measuring the temperature in a vessel in which water vapor, liquid water, and ice are in equilibrium under one atmosphere of air? Explain.

11.53 Refer to Figure 11.27(a), and describe all the phase changes that would occur in each of the following cases: **(a)** Water vapor originally at 1.0×10^{-3} atm and −0.10°C is slowly compressed at constant temperature until the final pressure is 10 atm. **(b)** Water originally at 100.0°C and 0.50 atm is cooled at constant pressure until the temperature is −10°C.

11.54 Refer to Figure 11.27(b), and describe the phase changes (and the temperatures at which they occur) when CO_2 is heated from −80°C to −20°C at **(a)** a constant pressure of 3 atm, **(b)** a constant pressure of 6 atm.

11.55 The normal melting and boiling points of xenon are −112°C and −107°C, respectively. Its triple point is at −121°C and 282 torr, and its critical point is at 16.6°C and 57.6 atm. **(a)** Sketch the phase diagram for Xe, showing the four points given and indicating the area in which each phase is stable. **(b)** Which is denser, Xe(*s*) or Xe(*l*)? Explain. **(c)** If Xe gas is cooled under an external pressure of 100 torr, will it undergo condensation or deposition? Explain.

11.56 The normal melting and boiling points of O_2 are $-218°C$ and $-183°C$, respectively. Its triple point is at $-219°C$ and 1.14 torr, and its critical point is at $-119°C$ and 49.8 atm. **(a)** Sketch the phase diagram for O_2, showing the four points given and indicating the area in which each phase is stable. **(b)** Will $O_2(s)$ float on $O_2(l)$? Explain. **(c)** As it is heated, will solid O_2 sublime or melt under a pressure of 1 atm?

Structures of Solids

11.57 How does an amorphous solid differ from a crystalline one? Give an example of an amorphous solid.

11.58 Amorphous silica has a density of about 2.2 g/cm^3, whereas the density of crystalline quartz is 2.65 g/cm^3. Account for this difference in densities.

11.59 Perovskite, a mineral composed of Ca, O, and Ti, has the cubic unit cell shown in the drawing. What is the chemical formula of this mineral?

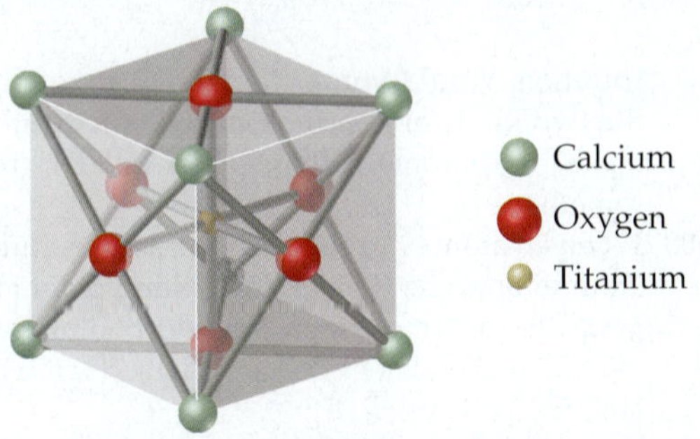

11.60 Rutile is a mineral composed of Ti and O. Its unit cell, shown in the drawing, contains Ti atoms at each corner and a Ti atom at the center of the cell. Four O atoms are on the opposite faces of the cell, and two are entirely within the cell. **(a)** What is the chemical formula of this mineral? **(b)** What is the nature of the bonding that holds the solid together?

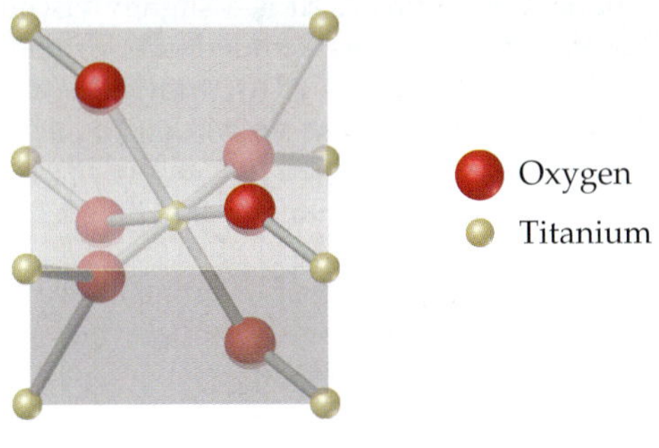

11.61 Iridium crystallizes in a face-centered cubic unit cell that has an edge length of 3.833 Å. The atom in the center of the face is in contact with the corner atoms, as shown in the drawing. **(a)** Calculate the atomic radius of an iridium atom. **(b)** Calculate the density of iridium metal.

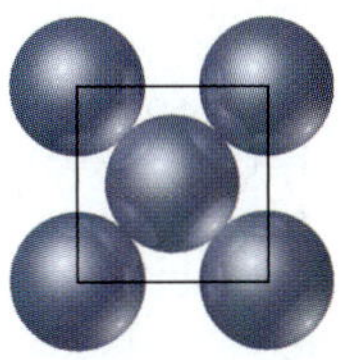

11.62 Aluminum metal crystallizes in a cubic close-packed structure (face-centered cubic cell, Figure 11.34). **(a)** How many aluminum atoms are in a unit cell? **(b)** What is the coordination number of each aluminum atom? **(c)** Assume that the aluminum atoms can be represented as spheres, as shown in the drawing for Exercise 11.61. If each Al atom has a radius of 1.43 Å, what is the length of a side of the unit cell? **(d)** Calculate the density of aluminum metal.

11.63 An element crystallizes in a body-centered cubic lattice. The edge of the unit cell is 2.86 Å, and the density of the crystal is 7.92 g/cm^3. Calculate the atomic weight of the element.

11.64 KCl has the same structure as NaCl. The length of the unit cell is 628 pm. The density of KCl is 1.984 g/cm^3, and its formula mass is 74.55 amu. Using this information, calculate Avogadro's number.

11.65 What is the coordination number of each sphere in **(a)** a three-dimensional, close-packed array of equal-sized spheres; **(b)** a primitive cubic structure; **(c)** a body-centered cubic lattice?

11.66 What is the coordination number of **(a)** Na^+ in the NaCl structure, Figure 11.35; **(b)** Zn^{2+} in the ZnS unit cell, Figure 11.42(b); **(c)** Ca^{2+} in the CaF_2 unit cell, Figure 11.42(c)?

11.67 Clausthalite is a mineral composed of lead selenide (PbSe). The mineral adopts a NaCl-type structure. The density of PbSe at 25°C is 8.27 g/cm^3. Calculate the length of an edge of the PbSe unit cell.

11.68 Nickel oxide (NiO) crystallizes in the NaCl type of crystal structure (Figure 11.35). The length of the unit cell of NiO is 4.18 Å. Calculate the density of NiO.

11.69 The mineral uraninite (UO_2) adopts a fluorite structure [Figure 11.42(c)] in which the length of an edge of the unit cell is 5.468 Å. **(a)** Will the uranium ions be represented by the larger or the smaller spheres in Figure 11.42(c)? Explain. **(b)** Calculate the density of uraninite.

11.70 A particular form of cinnabar (HgS) adopts the zinc blende structure, Figure 11.42(b). The length of the unit cell side is 5.852 Å. **(a)** Calculate the density of HgS in this form. **(b)** The mineral tiemmanite (HgSe) also forms a solid phase with the zinc blende structure. The length of the unit cell side in this mineral is 6.085 Å. What accounts for the larger unit cell length in tiemmanite? **(c)** Which of the two substances has the higher density? How do you account for the difference in densities?

Bonding in Solids

11.71 What kinds of attractive forces exist between particles in **(a)** molecular crystals, **(b)** covalent-network crystals, **(c)** ionic crystals, **(d)** metallic crystals?

11.72 Indicate the type of crystal (molecular, metallic, covalent-network, or ionic) each of the following would form upon solidification: **(a)** $CaCO_3$, **(b)** Pt, **(c)** ZrO_2 (melting point, 2677°C), **(d)** Kr, **(e)** benzene, **(f)** I_2.

11.73 Covalent bonding occurs in both molecular and covalent-network solids. Why do these two kinds of solids differ so greatly in their hardness and melting points?

11.74 Which type (or types) of crystalline solid is characterized by each of the following: **(a)** high mobility of electrons throughout the solid; **(b)** softness, relatively low melting point; **(c)** high melting point and poor electrical conductivity; **(d)** network of covalent bonds; **(e)** charged particles throughout the solid.

11.75 A white substance melts with some decomposition at 730°C. As a solid, it is a nonconductor of electricity, but it dissolves in water to form a conducting solution. Which type of solid (Table 11.7) might the substance be?

11.76 You are given a white substance that sublimes at 3000°C; the solid is a nonconductor of electricity and is insoluble in water. Which type of solid (Table 11.7) might this substance be?

11.77 For each of the following pairs of substances, predict which will have the higher melting point and indicate why: **(a)** Ar, Xe; **(b)** SiO_2, CO_2; **(c)** KBr, Br_2; **(d)** C_6Cl_6, C_6H_6.

11.78 For each of the following pairs of substances, predict which will have the higher melting point, and indicate why: **(a)** HF, HCl; **(b)** C (graphite), CH_4; **(c)** KBr, Br_2; **(d)** LiF, MgF_2.

Additional Exercises

11.79 As the intermolecular attractive forces between molecules increase in magnitude, do you expect each of the following to increase or decrease in magnitude? **(a)** vapor pressure, **(b)** heat of vaporization, **(c)** boiling point, **(d)** freezing point, **(e)** viscosity, **(f)** surface tension, **(g)** critical temperature.

11.80 Suppose you have two colorless molecular liquids, one boiling at −84°C, the other at 34°C, and both at atmospheric pressure. Which of the following statements is correct? For those that are not correct, modify the statement so that it is correct. **(a)** The higher-boiling liquid has greater total intermolecular forces than the other. **(b)** The lower boiling liquid must consist of nonpolar molecules. **(c)** The lower-boiling liquid has a lower molecular weight than the higher-boiling liquid. **(d)** The two liquids have identical vapor pressures at their normal boiling points. **(e)** At 34°C both liquids have vapor pressures of 760 mm Hg.

11.81 Two isomers of the planar compound 1,2-dichloroethylene are shown here, along with their melting and boiling points.

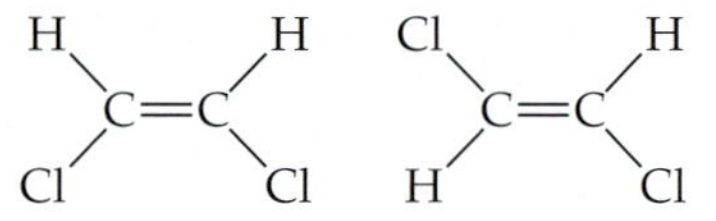

cis isomer   trans isomer

	cis isomer	trans isomer
Melting point (°C)	−80.5	−49.8
Boiling point (°C)	60.3	47.5

(a) Which of the two isomers will have the stronger dipole-dipole forces? Is this prediction borne out by the data presented here? **(b)** Based on the data presented here, which isomer packs more efficiently in the solid phase?

11.82 In dichloromethane, CH_2Cl_2 (μ = 1.60 D), the dispersion force contribution to the intermolecular attractive forces is about 5 times larger than the dipole-dipole contribution. Would you expect the relative importance of the two kinds of intermolecular attractive forces to differ **(a)** in dibromomethane (μ = 1.43 D), **(b)** in difluoromethane (μ = 1.93 D)? Explain.

11.83 When an atom or group of atoms is substituted for an H atom in benzene (C_6H_6), the boiling point changes. Explain the order of the following boiling points: C_6H_6 (80°C), C_6H_5Cl (132°C), C_6H_5Br (156°C), C_6H_5OH (182°C).

11.84 Trimethylamine [$(CH_3)_3N$] boils at 3°C; propylamine ($CH_3CH_2CH_2NH_2$) boils at 49°C. **(a)** What accounts for the difference in boiling points? **(b)** Propylamine is entirely miscible with water; trimethylamine has reasonably high solubility in water. What accounts for these data, considering that isobutane [$(CH_3)_3CH$] is considerably less soluble than trimethylamine?

11.85 Ethylene glycol [$CH_2(OH)CH_2(OH)$] is the major component of antifreeze. It is a slightly viscous liquid, not very volatile at room temperature, with a boiling point of 198°C. Pentane (C_5H_{12}), which has about the same molecular weight, is a nonviscous liquid that is highly volatile at room temperature and whose boiling point is 36.1°C. Explain the differences in the physical properties of the two substances.

[11.86] Using the following list of normal boiling points for a series of hydrocarbons, estimate the normal boiling point for octane, C_8H_{18}: propane (C_3H_8, −42.1°C), butane (C_4H_{10}, −0.5°C), pentane (C_5H_{12}, 36.1°C), hexane (C_6H_{14}, 68.7°C), heptane (C_7H_{16}, 98.4°C). Explain the trend in the boiling points.

[11.87] Notice in Figure 11.21 that there is a pressure-reduction valve in the line just before the supercritical CO_2 and dissolved caffeine enter the separator. Use Figure 11.20 to explain the function of this valve in the overall process.

11.88 **(a)** When you exercise vigorously, you sweat. How does this help your body cool? **(b)** A flask of water is connected to a vacuum pump. A few moments after the pump is turned on, the water begins to boil. After a few minutes, the water begins to freeze. Explain why these processes occur.

[11.89] The following table gives the vapor pressure of hexafluorobenzene (C_6F_6) as a function of temperature:

Temperature (K)	Vapor Pressure (torr)
280.0	32.42
300.0	92.47
320.0	225.1
330.0	334.4
340.0	482.9

(a) By plotting these data in a suitable fashion, determine whether the Clausius–Clapeyron equation is obeyed. If it is obeyed, use your plot to determine ΔH_{vap} for C_6F_6. **(b)** Use these data to determine the boiling point of the compound.

[11.90] Suppose the vapor pressure of a substance is measured at two different temperatures. **(a)** By using the Clausius–Clapeyron equation, Equation 11.1, derive the following relationship between the vapor pressures, P_1 and P_2, and the absolute temperatures at which they were measured, T_1 and T_2:

$$\ln\frac{P_1}{P_2} = -\frac{\Delta H_{vap}}{R}\left(\frac{1}{T_1} - \frac{1}{T_2}\right)$$

(b) The melting point of potassium is 63.2°C. Molten potassium has a vapor pressure of 10.00 torr at 443°C and a vapor pressure of 400.0 torr at 708°C. Use these data and the equation in part (a) to calculate the heat of vaporization of liquid potassium. **(c)** By using the equation in part (a) and the data given in part (b), calculate the boiling point of potassium. **(d)** Calculate the vapor pressure of liquid potassium at 100°C.

[11.91] The following data present the temperatures at which certain vapor pressures are achieved for dichloromethane (CH_2Cl_2) and methyl iodide (CH_3I):

Vapor Pressure (torr):	10.0	40.0	100.0	400.0
T for CH_2Cl_2 (°C):	−43.3	−22.3	−6.3	24.1
T for CH_3I (°C):	−45.8	−24.2	−7.0	25.3

(a) Which of the two substances is expected to have the greater dipole-dipole forces? Which is expected to have the greater London dispersion forces? Based on your answers, explain why it is difficult to predict which compound would be more volatile. **(b)** Which compound would you expect to have the higher boiling point? Check your answer in a reference book such as the *CRC Handbook of Chemistry and Physics*. **(c)** The order of volatility of these two substances changes as the temperature is increased. What quantity must be different for the two substances in order for this phenomenon to occur? **(d)** Substantiate your answer for part (c) by drawing an appropriate graph.

11.92 The elements xenon and gold both have solid-state structures with face-centered cubic unit cells, yet Xe melts at −112°C and gold melts at 1064°C. Account for these greatly different melting points.

11.93 Gold crystallizes in a face-centered cubic unit cell that has an edge length of 4.078 Å. The atom in the center of the face is in contact with the corner atoms, as shown in the drawing for Exercise 11.61. **(a)** Calculate the apparent radius of a gold atom in this structure. **(b)** Calculate the density of gold metal.

11.94 **(a)** Consider the cubic unit cells (Figure 11.33) with an atom located at each lattice point. Calculate the net number of atoms in (i) a primitive cubic unit cell, (ii) a body-centered cubic unit cell, (iii) a face-centered cubic unit cell. **(b)** Why can't $CaCl_2$ have the same crystal structure as NaCl?

11.95 In a typical X-ray crystallography experiment, X rays of wavelength $\lambda = 0.71$ Å are generated by bombarding molybdenum metal with an energetic beam of electrons. Why are these X rays more effectively diffracted by crystals than is visible light?

11.96 In their study of X-ray diffraction, William and Lawrence Bragg determined that the relationship among the wavelength of the radiation (λ), the angle at which the radiation is diffracted (θ), and the distance between the layers of atoms in the crystal that cause the diffraction (d) is given by $n\lambda = 2d \sin\theta$. X rays from a copper X-ray tube ($\lambda = 1.54$ Å) are diffracted at an angle of 14.22 degrees by crystalline silicon. Using the Bragg equation, calculate the interplanar spacing in the crystal, assuming $n = 1$ (first-order diffraction).

[11.97] **(a)** The density of diamond [Figure 11.41(a)] is 3.5 g/cm^3, and that of graphite [Figure 11.41(b)] is 2.3 g/cm^3. Based on the structure of buckminsterfullerene (Figure 11.43), what would you expect its density to be relative to these other forms of carbon? **(b)** X-ray diffraction studies of buckminsterfullerene show that it has a face-centered cubic lattice of C_{60} molecules. The length of a side of the unit cell is 14.2 Å. Calculate the density of buckminsterfullerene.

Integrative Exercises

11.98 Spinel is a mineral that contains 37.9% Al, 17.1% Mg, and 45.0% O, by mass, and has a density of 3.57 g/cm^3. The unit cell is cubic, with an edge length of 809 pm. How many atoms of each type are in the unit cell?

11.99 **(a)** At the molecular level, what factor is responsible for the steady increase in viscosity with increasing molecular weight in the hydrocarbon series shown in Table 11.4? **(b)** Although the viscosity varies over a factor of more than two in the series from hexane to nonane, the surface tension at 25°C increases by only about 20% in the same series. How do you account for this? **(c)** *n*-Octyl alcohol, $CH_3CH_2CH_2CH_2CH_2CH_2CH_2CH_2OH$, has a viscosity of 1.01×10^{-2} kg m^{-1}s^{-1}, much higher than nonane, which has about the same molecular weight. What accounts for this difference? How does your answer relate to the difference in normal boiling points for these two substances?

11.100 Acetone, $(CH_3)_2CO$, is widely used as an industrial solvent. **(a)** Draw the Lewis structure for the acetone molecule, and predict the geometry around each carbon atom. **(b)** Is the acetone molecule polar or nonpolar?

(c) What kinds of intermolecular attractive forces exist between acetone molecules? **(d)** 1-Propanol, $CH_3CH_2CH_2OH$, has a molecular weight that is very similar to that of acetone, yet acetone boils at 56.5°C and 1-propanol boils at 97.2°C. Explain the difference.

11.101 The table shown here lists the molar heats of vaporization for several organic compounds. Use specific examples from this list to illustrate how the heat of vaporization varies with **(a)** molar mass, **(b)** molecular shape, **(c)** molecular polarity, **(d)** hydrogen-bonding interactions. Explain these comparisons in terms of the nature of the intermolecular forces at work. (You may find it helpful to draw out the structural formula for each compound.)

Compound	Heat of Vaporization (kJ/mol)
$CH_3CH_2CH_3$	19.0
$CH_3CH_2CH_2CH_2CH_3$	27.6
$CH_3CHBrCH_3$	31.8
CH_3COCH_3	32.0
$CH_3CH_2CH_2Br$	33.6
$CH_3CH_2CH_2OH$	47.3

11.102 Liquid butane, C_4H_{10}, is stored in cylinders, to be used as a fuel. The normal boiling point of butane is listed as −0.5°C. **(a)** Suppose the tank is standing in the sun and reaches a temperature of 46°C. Would you expect the pressure in the tank to be greater or less than atmospheric pressure? How does the pressure within the tank depend on how much liquid butane is in it? **(b)** Suppose the valve to the tank is opened and a few liters of butane are allowed to escape rapidly. What do you expect would happen to the temperature of the remaining liquid butane in the tank? Explain. **(c)** How much heat must be added to vaporize 155 g of butane if its heat of vaporization is 21.3 kJ/mol? What volume does this much butane occupy at 755 torr and 35°C?

[11.103] Using information in Appendices B and C, calculate the minimum number of grams of $C_3H_8(g)$ that must be combusted to provide the energy necessary to convert 2.50 kg of H_2O from its solid form at −14.0°C to its liquid form at 60.0°C.

11.104 In a certain type of nuclear reactor, liquid sodium metal is employed as a circulating coolant in a closed system, protected from contact with air or water. Much like the coolant that circulates in an automobile engine, the liquid sodium carries heat from the hot reactor core to heat exchangers. **(a)** What properties of the liquid sodium are of special importance in this application? **(b)** The viscosity of liquid sodium varies with temperature as follows:

Temperature (° C)	Viscosity (kg $m^{-1}s^{-1}$)
100	7.05×10^{-4}
200	4.50×10^{-4}
300	3.45×10^{-4}
600	2.10×10^{-4}

What forces within the liquid sodium are likely to be the major contributors to the viscosity? Why does viscosity decrease with increasing temperature?

11.105 The vapor pressure of a volatile liquid can be determined by slowly bubbling a known volume of gas through it at a known temperature and pressure. In an experiment, 5.00 L of N_2 gas is passed through 7.2146 g of liquid benzene, C_6H_6, at 26.0°C. The liquid remaining after the experiment weighs 5.1493 g. Assuming that the gas becomes saturated with benzene vapor and that the total gas volume and temperature remain constant, what is the vapor pressure of the benzene in torr?

11.106 The relative humidity of air equals the ratio of the partial pressure of water in the air to the equilibrium vapor pressure of water at the same temperature. If the relative humidity of the air is 45% and its temperature is 23°C, how many molecules of water are present in a room measuring 14 × 9.0 × 8.6 m?

[11.107] Use a reference source such as the *CRC Handbook of Chemistry and Physics* to compare the melting and boiling points of the following pairs of inorganic substances: **(a)** W and WF_6, **(b)** SO_2 and SF_4, **(c)** SiO_2 and $SiCl_4$. Account for the major differences observed in terms of likely structures and bonding.

eMEDIA EXERCISES

These exercises make use of the interactive objects available online in OneKey or the Companion Website, and on your Accelerator CD. Access to these resources comes in your MediaPak.

11.108 The **Changes of State** animation (*11.4*) shows the heating curve of a substance that melts and then vaporizes. **(a)** Using the information in Figure 11.27(b), sketch the heating curve of carbon dioxide at atmospheric pressure, starting at −100°C and ending at 30°C. **(b)** Sketch the same heating curve as it would appear at a pressure of 10 atm. **(c)** Under what conditions would the heating curve of water resemble the one you sketched for part (a)?

11.109 Using the **Equilibrium Vapor Pressure** activity (*11.5*), compare the vapor pressures of methanol, ethanol, acetic acid, water, and benzene. **(a)** Which compound appears to have the strongest intermolecular forces at 100°C? **(b)** If one compound has a higher equilibrium vapor pressure than another at a particular temperature, will it necessarily have a higher equilibrium vapor pressure at all temperatures? If not, give an example of two compounds whose vapor-pressure curves cross and at what temperature they have roughly the same vapor pressure.

11.110 Ethanol and acetic acid have very similar molar masses, and both exhibit hydrogen bonding. In which of these two compounds is hydrogen bonding a more significant component of the overall intermolecular forces? Support your answer using data from the **Equilibrium Vapor Pressure** activity (*11.5*).

11.111 Water is a substance with some very unusual properties. Based on the structure of the **Ice** 3-D Model (*11.2*), explain **(a)** why ice is less dense than water and **(b)** why increasing pressure, within a certain temperature range, causes ice to melt. **(c)** Why is it not possible to convert ice to liquid water by applying pressure at very low temperatures?

11.112 Compare the structure of the **Ice** 3-D Model (*11.2*) with the structure of the **Diamond** 3-D Model (*11.8*). If you rotate the structures just right, you will see distinct similarities, namely sp^3 hybridization (of oxygen and carbon, respectively) and the hexagonal arrangement of atoms. Given this similarity, explain why ice can be melted by applying pressure whereas diamond cannot.

VCL eLABORATORY EXERCISES

These exercises allow students to apply the concepts and skills for this chapter in the simulated laboratory of Virtual ChemLab. Worksheets for each of these exercises are found in the Virtual ChemLab Workbook in your MediaPak.

11.113 ***(VCL 11-1) Heat of Fusion of Water***
The molar heat of fusion for a substance, ΔH_{fus}, is the heat required to transform one mole of the substance from the solid phase into the liquid phase. In this assignment you will use a simple coffee-cup calorimeter and a thermometer to measure the molar heat of fusion for water.

11.114 ***(VCL 11-2) Heat of Vaporization of Water***
A graph of the vapor pressure of a liquid as a function of temperature has a characteristic shape that can be represented fairly accurately with the *Clausius–Clapeyron* equation. In this assignment you will measure the vapor pressure of water at several temperatures and use this data to estimate the heat of vaporization of water using the *Clausius–Clapeyron* equation.

11.115 ***(VCL 11-3) The Boiling Point of Water at High Altitude***
The relationship between the equilibrium vapor pressure of a liquid or solid and temperature is given by the *Clausius–Clapeyron* equation. In this assignment you will measure the vapor pressure of water at a given temperature and use this data and the *Clausius–Clapeyron* equation to calculate the boiling point of water on the top of Mt. Denali in Alaska.

CHAPTER 12

WHAT'S AHEAD

- We will characterize materials according to their ability to conduct electricity—*metals*, which are very conductive; *semiconductors*, which are not very conductive; *insulators*, which are not at all conductive; and *superconductors*, which are perfectly conductive under certain conditions. *(Section 12.1)*
- The structural properties of materials ultimately depend on the bonding and intermolecular forces within them. Materials that are used for structural purposes can be either soft or hard. We will learn about *polymers* and plastics as examples of soft materials and about *ceramics* and metals as examples of hard materials. *(Section 12.2)*
- Materials used in biomedical applications must meet special requirements to be compatible with living organisms. We will see several examples of biomaterials and how they work in such applications as heart valves, artificial tissues, and sutures. *(Section 12.3)*
- A major use of semiconductors is in electronics, in the form of "silicon chips." We will learn why silicon is the choice for many electronic applications and gain some understanding of how these devices are made. *(Section 12.4)*
- Some materials are valuable for their optical properties—they either interact with light in a unique way or produce light when exposed to external stimulation. We will learn about the materials that underlie LCDs (*liquid crystal displays*) and LEDs (*light-emitting diodes*), two common types of display technologies. *(Section 12.5)*
- *Nanotechnology* is an emerging field that crosses the boundaries between chemistry, physics, biology, engineering, and medicine. We will learn that the properties of materials on the 1–100-nm length scale are different from bulk properties. *(Section 12.6)*

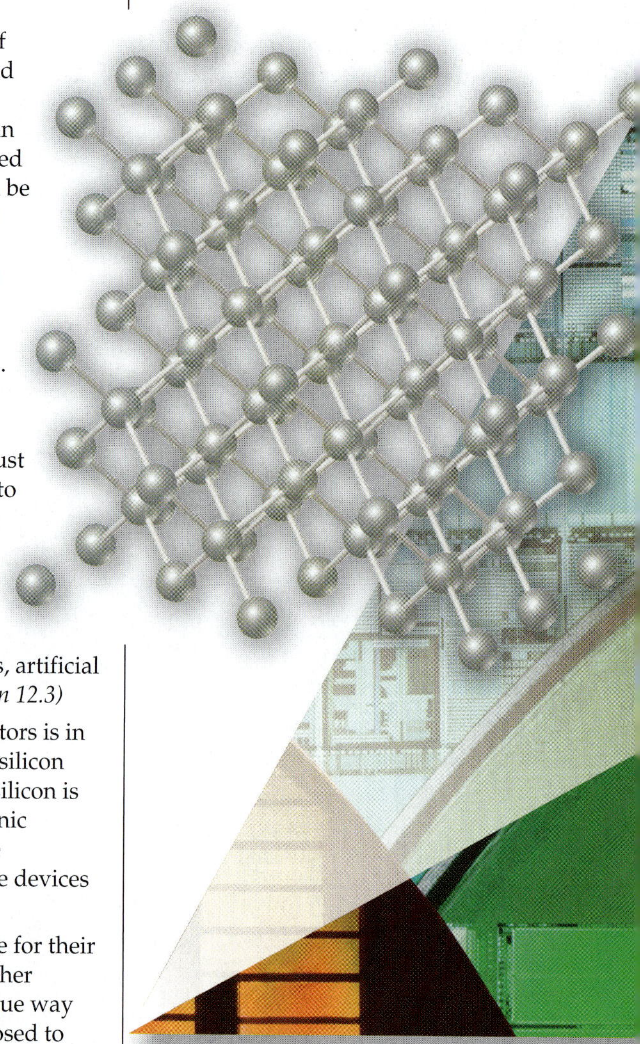

Modern Materials

THE MODERN WORLD IS FULL OF "STUFF"—materials that are used to make computers, compact discs, cell phones, contact lenses, skis, furniture, and a host of other things. Chemists have contributed to the discovery and development of new materials by inventing entirely new substances and by developing the means for processing naturally occurring materials to form fibers, films, coatings, adhesives, and substances with special electrical, magnetic, or optical properties. In this chapter we will discuss classes of materials, their properties, and their applications in modern society. Our aim is to show how we can understand the physical or chemical properties of materials by applying the principles discussed in earlier chapters.

This chapter demonstrates the important point that *the observable, macroscopic properties of materials are the result of atomic- and molecular-level structures and processes.*

HERE, THERE AND EVERYWHERE. The element silicon is the primary component of computer processor chips and commercial solar panels.

12.1 | Classes of Materials

When we say "material," we generally mean a substance or mixture of substances that is held together by strong chemical bonds throughout the whole sample—in other words, covalent-network solids, ionic solids, and metallic solids. (Section 11.8) In addition to these three types of materials, we learned in Chapter 11 about molecular solids, materials held together not by strong chemical bonds but by weaker London dispersion forces, dipole-dipole interactions, and hydrogen bonds. Two common types of soft materials, polymers and liquid crystals, are molecular solids in which the intermolecular forces give rise to unusual properties that we will investigate in this chapter.

Metals and Semiconductors

We can classify materials in a number of ways, but one way is based on the bonding in the material. Recall that atomic orbitals can combine in a molecule to make molecular orbitals that extend over the whole molecule. (Section 9.7) A given molecular orbital can contain zero, one, or two electrons, depending on how many electrons the molecule has. The number of molecular orbitals in a molecule is the same number as the number of atomic orbitals used to make the molecular orbitals.

In the case of many, many atoms the atomic orbitals combine to give many, many molecular orbitals. In the bulk solids we are considering, there are so many molecular orbitals that the energy separation between them becomes vanishingly small and continuous **bands** of energy states are formed (Figure 12.1 ▼). Different atomic orbitals make bands with different energies, so the band structure of a solid consists of a set of bands separated by energy gaps.

Based on their band structure, we can classify materials as **metals, semiconductors**, or **insulators** (Figure 12.2 ▼; Table 12.1 ▶). Metals are very electrically conductive, shiny, ductile, and malleable. Most of the elements in the periodic table are metals, including familiar ones such as gold, silver, copper, platinum, and iron (see Figure 2.16). (Section 7.6) Metals have a band structure in which the valence electrons are in a partially filled band. In order to conduct electricity, the electrons need to be in empty orbitals; for metals, then, there is almost no energy cost for electrons to jump from the lower, occupied part of

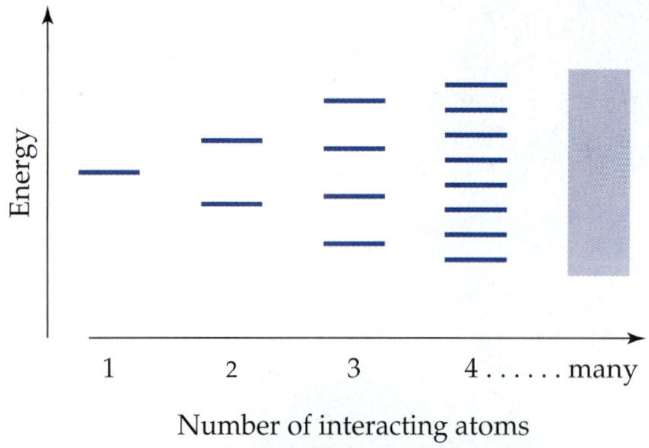

▲ **Figure 12.1 Discrete energy levels in molecules become continuous bands in solids.** Schematic illustration showing how the spacing between energy levels in a molecule decreases as the number of atoms in the molecule increases. In molecules containing very large numbers of atoms, there is essentially no separation between energy levels, and the molecule is characterized as having a continuous band of energy levels. Bands can be either totally filled, empty, or partially filled with electrons, depending on the material. Only one band is shown, but in a material there are other bands that arise from different atomic and molecular orbitals.

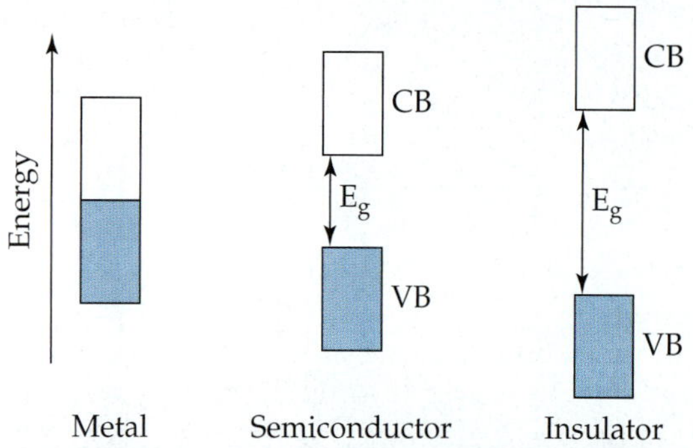

▲ **Figure 12.2 Energy bands in metals, semiconductors, and insulators.** Metals are characterized by the highest-energy electrons occupying a partially filled band. Semiconductors and insulators have an energy gap that separates the totally filled band (shaded in blue) and the empty band (unshaded), known as the band gap and represented by the symbol E_g. The filled band is called the valence band (VB), and the empty band is called the conduction band (CB). Semiconductors have a smaller band gap than insulators.

TABLE 12.1 Electronic Properties of Common Materials[a]

Material	Type	Band Gap Energy, kJ/mol	Band Gap Energy, eV	Conductivity, ohm^{-1} cm^{-1}
C (diamond)	Insulator	~530	~5.5	$<10^{-18}$
C (graphite)	Semiconductor[b]	b	b	10^3—along sheets
Si	Semiconductor	110	1.1	5×10^{-6}
Ge	Semiconductor	65	0.67	0.02
GaP	Semiconductor	220	2.2	5
GaAs	Semiconductor	138	1.43	500
GaP_xAs_{1-x}	Semiconductor (these are a series of solid solutions for x varying between 0 and 1)	Varies between 220 and 138, depending on composition	Varies between 2.2 and 1.4, depending on composition	Varies
CdS	Semiconductor	230	2.4	~0.05
CdSe	Semiconductor	160	1.7	~0.05
CdTe	Semiconductor	140	1.44	~0.001
Al	Metal	—	—	3.8×10^5
Al_2O_3	Insulator	~530	~5.5	$<10^{-14}$
SiO_2	Insulator	>580	>6	$<10^{-18}$
TiO_2	Semiconductor	290	3.0	10^{-12}
Fe	Metal	—	—	1.0×10^5
Cu	Metal	—	—	5.9×10^5
Ag	Metal	—	—	6.3×10^5
Au	Metal	—	—	4.5×10^5

[a]Band gap energies and conductivities are room temperature values. Electron volts (eV) are a commonly used energy unit in the semiconductor industry; 1.602×10^{-19} J = 1 eV.
[b]Graphite does not have a band gap along the sheets, but it does have other characteristics of semiconductors (for example, its conductivity in the bulk has a value like other semiconductors).

the partially filled band to the upper, empty part of the same band. For this reason, metals are very electrically conductive.

Semiconductors, especially silicon, are the heart of the integrated circuits that form the basis for computers and other electronic devices. These materials are characterized by an energy gap, known as the **band gap**, between a filled **valence band** and an empty **conduction band**. The situation is analogous to the energy gap in a molecule between the highest-occupied molecular orbital and the lowest-unoccupied molecular orbital. (Section 9.8 "Organic Dyes") Semiconductors are measurably conductive, but far less so than metals because of the presence of the band gap, which lowers the probability that an electron at any given temperature will have enough energy to jump the gap. The only elements that are semiconductors at room temperature are silicon, germanium, and carbon in its graphite form.*

Inorganic compounds that are semiconductors tend to have chemical formulas that are related to silicon's and germanium's valence electron count of four. For example, gallium, Ga, is in column 3A of the periodic table; arsenic, As, is in column 5A. GaAs, gallium arsenide, has three valence electrons to contribute to the bonding from Ga and five from As, which average out to four, the same number as in silicon or germanium. GaAs is indeed a semiconductor, like silicon and germanium. Similarly, InP is a semiconductor whose valence electron count averages out to four, in which the indium provides three valence electrons and phosphorus contributes five. In cadmium sulfide, CdS, cadmium contributes two valence electrons and sulfur contributes six, which also averages out to four; and therefore CdS is a semiconductor. CdSe and CdTe follow this same trend: They have an average valence electron count of four, and they are semiconductors. The bonding in semiconductors is covalent-network, ionic, or a combination of the two.

* Graphite is unusual. Actually, within the layers graphite is nearly metallic, but it is a semiconductor between the layers. (Section 11.8)

GIVE IT SOME THOUGHT

Do you think GaN is a semiconductor?

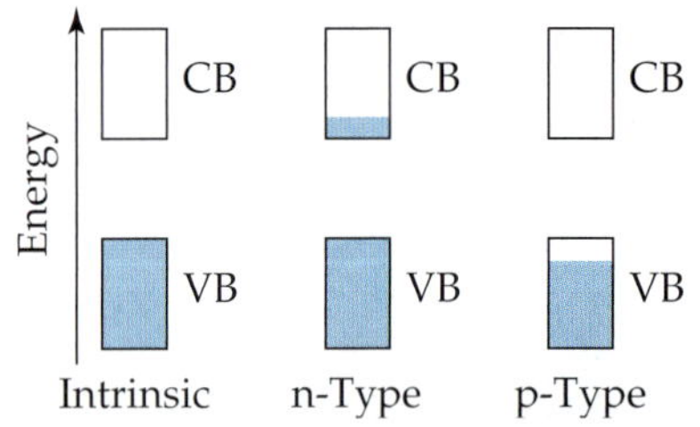

▲ **Figure 12.3 The addition of controlled small amounts of impurities (doping) to a semiconductor changes the electronic makeup of the material.** Left: A pure, intrinsic semiconductor has a filled valence band and an empty conduction band (ideally). Middle: The addition of a dopant atom that has more valence electrons than the host atom adds electrons to the conduction band. An example of this is phosphorus doped into silicon. The resulting material is an *n-type semiconductor*. Right: The addition of a dopant atom that has fewer valence electrons than the host atom leads to fewer electrons in the valence band (or, more holes in the valence band). An example of this is aluminum doped into silicon. The resulting material is a *p-type semiconductor*.

ACTIVITY
Band Structure

The degree to which semiconductors conduct electricity is influenced by the presence of small amounts of impurity atoms—generally at the parts-per-million level. The process of adding controlled amounts of impurity atoms to a material is known as **doping**. Consider what happens when phosphorus replaces a few Si atoms in a silicon crystal. Phosphorus has five valence electrons, and silicon has only four; therefore, the extra electrons that come with the phosphorus atoms are forced to occupy the conduction band, since the valence band is already totally filled (Figure 12.3 ◀, middle). The resulting material is called an *n-type* semiconductor, "n" signifying that the number of negatively charged electrons in the conduction band has increased. These extra electrons can move very easily in the conduction band. Thus, just a few parts per million (ppm) of phosphorus in silicon can increase silicon's intrinsic conductivity by a factor of a million.

It is also possible to dope semiconductors with atoms that have fewer valence electrons than the host material. Consider what happens when aluminum, in group 3A, replaces a few silicon atoms in a silicon crystal. Al has only three valence electrons compared to silicon's four. Thus, there are electron vacancies, known as *holes*, in the valence band when silicon is doped with aluminum. Since the negatively charged electron is not there, the hole can be thought of as having a positive charge. When an adjacent electron jumps into the hole, the electron leaves behind a new hole. Thus, the positive hole moves about in the lattice like a particle itself. This is analogous to watching people changing seats in a classroom; you can watch the people (electrons) move about the seats (atoms), or you can watch the empty seats (holes) "move." Thus, holes can conduct too, and a material that is like this is called a *p-type* semiconductor, "p" signifying that the number of positive holes in the material has increased. Like n-type conductivity, only parts-per-million levels of a p-type dopant can lead to a millionfold increase in the conductivity of the semiconductor—but in this case, the holes in the valence band are doing the conduction (Figure 12.3, right). The junction of an n-type semiconductor with a p-type semiconductor forms the basis for diodes, transistors, and other devices.

SAMPLE EXERCISE 12.1 | Identifying Types of Semiconductors

Which of the following elements, if doped into silicon, would yield an n-type semiconductor? Ga; As; C.

Solution

Analyze: An n-type semiconductor means that the dopant atoms must have more valence electrons than the host material. Silicon is the host material in this case.
Plan: We must look at the periodic table and determine how many valence electrons are in Si, Ga, As, and C by what column they are in. The elements with more valence electrons than silicon are the ones that will produce an n-type material upon doping.
Solve: Si is in column 4A, and so has four valence electrons. Ga is in column 3A, and so has three valence electrons. As is in column 5A, and so has five valence electrons; C is in column 4A, and so has four valence electrons. Therefore, As, if doped into silicon, would yield an n-type semiconductor.

PRACTICE EXERCISE

Suggest an element that could be used to dope silicon to yield a p-type material.
Answer: Because Si is in group 4A, we need to pick an element in group 3A. Boron or aluminum is a good choice—both are in group 3A. In the semiconductor industry boron and aluminum are commonly used dopants for silicon.

Insulators and Ceramics

Semiconductors have band gap energies that range from ~50 kJ/mol to ~300 kJ/mol (Table 12.1). Insulators have a band structure similar to that found in semiconductors, except that in insulators the band gap is larger—more than ~350 kJ/mol (Table 12.1 and Figure 12.2). Because the energy required to promote an electron from the valence band to the conduction band is about the same as the energy required to break chemical bonds in the material, insulators are not electrically conductive. Many ionic solids and complex network solids and most organic compounds are insulators; elements that are solid insulators include carbon in its diamond form and sulfur. (Section 11.8)

Ceramics are inorganic ionic solids that are normally hard and brittle and are stable at high temperatures. They are generally electrical insulators. Ceramic materials include such familiar objects as pottery, china, cement, roof tiles, and spark-plug insulators.

Ceramic materials (Table 12.2 ▼) come in a variety of chemical forms, including *oxides* (oxygen and metals), *carbides* (carbon and metals), *nitrides* (nitrogen and metals), *silicates* (silica, SiO_2, mixed with metal oxides), and *aluminates* (alumina, Al_2O_3, mixed with metal oxides).

Ceramics are highly resistant to heat, corrosion, and wear; do not readily deform under stress; and are less dense than the metals used for high-temperature applications. Some ceramics used in aircraft, missiles, and spacecraft weigh only 40% as much as the metal components they replace. In spite of these many advantages, the use of ceramics as engineering materials has been limited because they are extremely brittle. Whereas a metal component might suffer a dent when struck, a ceramic part typically shatters because the greater extent of ionic bonding in a ceramic prevents the atoms from sliding over one another.

TABLE 12.2 Properties of Some Ceramic and Nonceramic Materials

Material	Melting Point (°C)	Density (g/cm^3)	Hardness (Mohs)[a]	Modulus of Elasticity[b]	Coefficient of Thermal Expansion[c]
		Ceramic materials			
Alumina, Al_2O_3	2050	3.8	9	34	8.1
Silicon carbide, SiC	2800	3.2	9	65	4.3
Zirconia, ZrO_2	2660	5.6	8	24	6.6
Beryllia, BeO	2550	3.0	9	40	10.4
		Nonceramic materials			
Mild steel	1370	7.9	5	17	15
Aluminum	660	2.7	3	7	24

[a] The Mohs scale is a logarithmic scale based on the relative ability of a material to scratch another softer material. Diamond, the hardest material, is assigned a value of 10.
[b] A measure of the stiffness of a material when subjected to a load (MPa $\times 10^4$). The larger the number, the stiffer the material.
[c] In units of ($K^{-1} \times 10^{-6}$). The larger the number, the greater the size change upon heating or cooling.

3-D MODEL Silicon Carbide

Superconductors

Even metals are not infinitely conductive; there is some resistance to electron flow. However, in 1911 the Dutch physicist H. Kamerlingh Onnes discovered that when mercury is cooled below 4.2 K, it loses all resistance to the flow of an electrical current. Since that discovery, scientists have found that many substances exhibit this "frictionless" flow of electrons, known as **superconductivity**. Substances that exhibit superconductivity do so only when cooled below a particular temperature called the **superconducting transition temperature, T_c**.

TABLE 12.3 Superconducting Materials: Dates of Discovery and Transition Temperatures

Substance	Discovery Date	T_c(K)
Hg	1911	4.0
Nb_3Sn	1954	18.0
$SrTiO_3$	1966	0.3
Nb_3Ge	1973	22.3
$BaPb_{1-x}Bi_xO_3$	1975	13.0
$LaBa_2CuO_4$	1986	35.0
$YBa_2Cu_3O_7$	1987	95.0
$BiSrCaCu_2O_x$	1988	100.0
$Tl_2Ba_2Ca_2Cu_3O_{10}$	1988	125.0
$HgBa_2Ca_2Cu_3O_{8+x}$	1993	133.0
$Hg_{0.8}Tl_{0.2}Ba_2Ca_2Cu_3O_{8.33}$	1993	138
Cs_3C_{60}	1995	40
MgB_2	2001	39

The observed values of T_c are generally very low. Table 12.3 ◀ lists numerous superconducting materials, the year of their discovery, and their transition temperature. Some are notable for their relatively high T_c, and others for the mere fact that a material of that composition would superconduct at all. Most scientists looking for superconductivity would examine metals and mixtures of metals first, and not ceramics.

Superconductivity has tremendous economic potential. If electrical power lines or the conductors in a variety of electrical devices were capable of conducting current without resistance, enormous amounts of energy could be saved. The lack of resistance to electron flow could in theory lead to superconductor-based "petaflop" computers capable of performing 10^{15} operations per second. In addition, superconducting materials exhibit a property called the *Meissner effect* (Figure 12.4 ▼), in which they exclude all magnetic fields from their volume. The Meissner effect is being explored for high-speed trains that are magnetically levitated. In November 2003 a magnetically levitated ("maglev") train achieved a top speed of 311 mph in Shanghai, China (Figure 12.5 ▼). Because superconductivity appears in most materials only at very low temperatures, applications of this phenomenon to date have been limited. One important use of superconductors is in the windings of the large magnets that form the magnetic fields needed in magnetic resonance imaging (MRI) instruments, used in medical imaging (Figure 12.6 ▶). The magnet windings, typically formed from Nb_3Sn, must be kept cooled with liquid helium, which boils at about 4 K. The cost of liquid helium is a significant factor in the cost of using an MRI instrument.

Before 1986 the highest value observed for T_c was about 22 K for a niobium-germanium compound (Table 12.3). In 1986, however, J. G. Bednorz and K. A. Müller, working at the IBM research laboratories in Zürich, Switzerland, discovered superconductivity above 30 K in a ceramic oxide containing lanthanum, barium, and copper. This material represents the first **superconducting ceramic**. That discovery, for which Bednorz and Müller received the Nobel Prize in Physics in 1987, set off a flurry of research activity all over the world. Before the end of 1987 scientists had verified the onset of superconductivity at

▲ **Figure 12.4 Magnetic levitation.** A small permanent magnet is levitated by its interaction with a ceramic superconductor that is cooled to liquid nitrogen temperature, 77 K. The magnet floats in space because the superconductor excludes magnetic field lines, a property known as the Meissner effect.

▲ **Figure 12.5 Superconduction in action.** High speed superconductive magnetic levitation (maglev) train. Shown is the ML002 experimental train being tested in Japan.

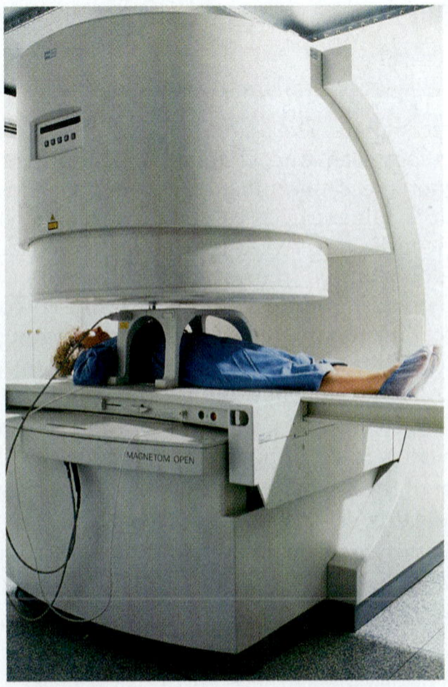

◀ **Figure 12.6 A magnetic resonance imaging (MRI) machine used in medical diagnosis.** The magnetic field needed for the procedure is generated by current flowing in superconducting wires, which must be kept below their superconducting transition temperature, T_c, of 18 K. This low temperature requires liquid helium as a coolant.

95 K in yttrium-barium-copper oxide, $YBa_2Cu_3O_7$. The highest temperature observed to date for onset of zero resistance at 1 atm pressure is 138 K, which was achieved in another complex copper oxide, $Hg_{0.8}Tl_{0.2}Ba_2Ca_2Cu_3O_{8.33}$.

The discovery of so-called **high-temperature** (high-T_c) **superconductivity** is of great significance. Because maintaining extremely low temperatures is costly, many applications of superconductivity will become feasible only with the development of usable high-temperature superconductors. Liquid nitrogen is the favored coolant because of its low cost (it is cheaper than milk), but it can cool objects to only 77 K. The only readily available safe coolant at temperatures below 77 K is liquid helium, which costs as much as fine wine. Thus, practical applications of superconductors will become commercially viable only if T_c is well above 77 K. Alternatively, mechanical cooling devices might be feasible in some applications.

What is the mechanism of superconductivity? The answer to this question is still being vigorously debated. One of the most widely studied ceramic superconductors is $YBa_2Cu_3O_7$, whose structure is shown in Figure 12.7 ▶. Extensive work on modifying this and related copper oxide superconductors by doping other atoms in various atomic positions indicates that the conductivity and superconductivity take place in the copper-oxygen planes. At temperatures above T_c the electrical conductivity parallel to the copper-oxygen planes is 10^4 times greater than in the perpendicular direction. The Cu^{2+} ions have an $[Ar]3d^9$ electron configuration with a single electron in the $3d_{x^2-y^2}$ orbital. (Section 6.6) Although the mechanism of conduction and superconduction is not yet well understood, it is believed to be important that the lobes of the $3d_{x^2-y^2}$ orbital point toward the neighboring O^{2-} ions.

It is still not clear what makes any particular material a superconductor. Superconductivity in metals and metal alloys, such as Nb_3Sn, is reasonably well accounted for by *BCS theory*, named after its inventors John Bardeen, Leon Cooper, and Robert Schrieffer. However, after years of research there is still no satisfactory theory of superconductivity in ceramic materials. Because it seems that superconductivity might arise in many different kinds of materials, much empirical research is devoted to discovering new classes of superconductors. As noted in Table 12.3, it was discovered in 1995 that when C_{60} (see "Closer Look" box on page 474) reacts with an alkali metal, it becomes an electrically conducting material and becomes superconducting at about 40 K. Even more

3-D MODEL
$YBa_2Cu_3O_7$

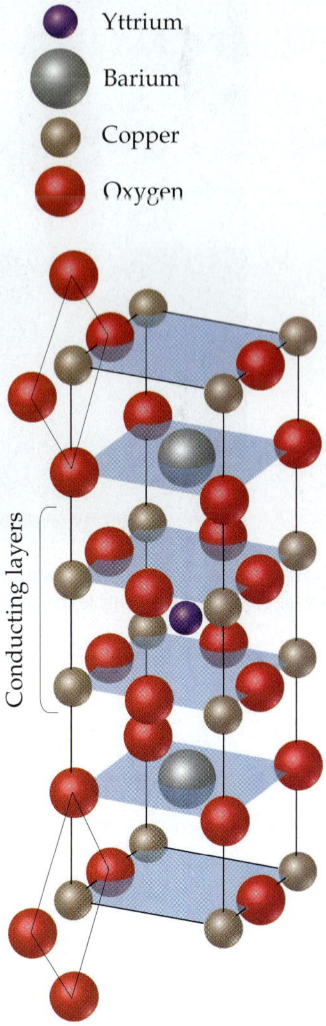

▲ **Figure 12.7 Unit cell of $YBa_2Cu_3O_7$.** A few oxygen atoms that fall outside the unit cell are also shown to illustrate the arrangement of oxygen atoms about each copper atom. The unit cell is defined by the black lines that outline the large rectangular box.

CHEMISTRY AT WORK | Cell Phone Tower Range

Cell phone towers increasingly dot both rural and urban landscapes (Figure 12.8 ▼), yet it can be difficult to maintain contact with a tower in the midst of a telephone conversation. A cell phone communicates with the system by receiving a signal from the tower's transmitter and transmitting signals back to it. Although the tower's transmitter may be quite powerful, the cell phone has very limited power. As the distance from the tower increases or intervening structures interfere, the cell phone signal becomes too weak to be distinguished from general electronic noise.

The amplifiers in the tower's receiver have electronic filters that discriminate between the desired incoming signal and other electronic signals. The sharper the filter, the higher the discrimination between one channel and another, and therefore, the higher the capability of detecting the desired signal with clarity. Filters can be made from a high-temperature superconducting oxide that when cooled below its T_c provides much sharper filtering capabilities than conventional filters. By incorporating such filters in the tower receivers, communications companies can extend the range of the tower by up to a factor of 2, which saves construction costs and improves the reliability of communication.

Superconducting technology is now being deployed in PC-sized boxes located in cell phone base stations (the little building at the foot of the tower). The filters are formed from a ceramic oxide, typically $YBa_2Cu_3O_7$ or $Tl_2Ba_2CaCu_2O_8$. The requisite cooling is provided by a mechanical cooling device, which is a small refrigeration unit capable of cooling the filter below its T_c (Figure 12.9 ▼).

▲ **Figure 12.8 Wireless communications tower.** Towers like this (here, disguised as a tree) enable cell phone communicaton.

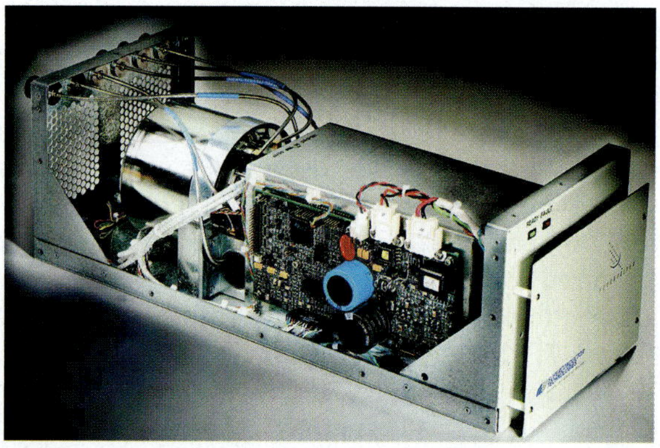

▲ **Figure 12.9 Cryogenic receiver system employing a cryo-cooled superconducting filter and a low-noise amplifier (LNA).** The cylindrical object to the left is the cryogenic refrigerator used to keep the filter and LNA at a temperature below its T_c value.

recently, the simple binary compound magnesium diboride, MgB_2, was found to become superconducting at 39 K. This observation is surprising and potentially quite important, since MgB_2, which is a semiconductor somewhat like graphite, is a relatively cheap material. Other, related compounds might prove to have higher superconducting transition temperatures. The field of superconductivity has a great deal of promise, and the appearance of superconductors in major commercial products is just beginning (see "Chemistry at Work" box).

12.2 | Materials for Structure

Many of the applications of materials, such as plastic bottles, leather dog leashes, concrete roads, and steel girders in buildings, rely on their structural properties. These properties depend a great deal on the fundamental bonding and atomic arrangements of the material at the microscopic level, and different needs will require different materials. Only in rare situations would concrete bottles, steel dog leashes, plastic roads, or leather girders in buildings be desirable.

Soft Materials: Polymers and Plastics

In nature we find many substances of very high molecular mass, running into millions of amu, that make up much of the structure of living organisms and tissues. Some examples include starch and cellulose, which abound in plants, and proteins, which are found in both plants and animals. In 1827 Jons Jakob Berzelius coined the word **polymer** (from the Greek *polys*, "many," and *meros*, "parts") to denote molecular substances of high molecular mass formed by the *polymerization* (joining together) of **monomers**, molecules with low molecular mass.

For a long time humans processed naturally occurring polymers, such as wool, leather, silk, and natural rubber, into usable materials. During the past 70 years or so, chemists have learned to form synthetic polymers by polymerizing monomers through controlled chemical reactions. A great many of these synthetic polymers have a backbone of carbon–carbon bonds because carbon atoms have an exceptional ability to form strong stable bonds with one another.

Plastics are materials that can be formed into various shapes, usually by the application of heat and pressure. **Thermoplastic** materials can be reshaped. For example, plastic milk containers are made from a polymer known as *polyethylene* that has a high molecular mass. These containers can be melted down and the polymer recycled for some other use. In contrast, a **thermosetting plastic** is shaped through irreversible chemical processes and therefore cannot be reshaped readily.

An **elastomer** is a material that exhibits rubbery or elastic behavior. When subjected to stretching or bending, it regains its original shape upon removal of the distorting force, provided that it has not been distorted beyond some elastic limit. Some polymers, such as nylon and polyesters, can also be formed into *fibers* that, like hair, are very long relative to their cross-sectional area and are not elastic. These fibers can be woven into fabrics and cords and fashioned into clothing, tire cord, and other useful objects.

Making Polymers

The simplest example of a polymerization reaction is the formation of polyethylene from ethylene molecules. In this reaction the double bond in each ethylene molecule "opens up," and two of the electrons originally in this bond are used to form new C—C single bonds with two other ethylene molecules:

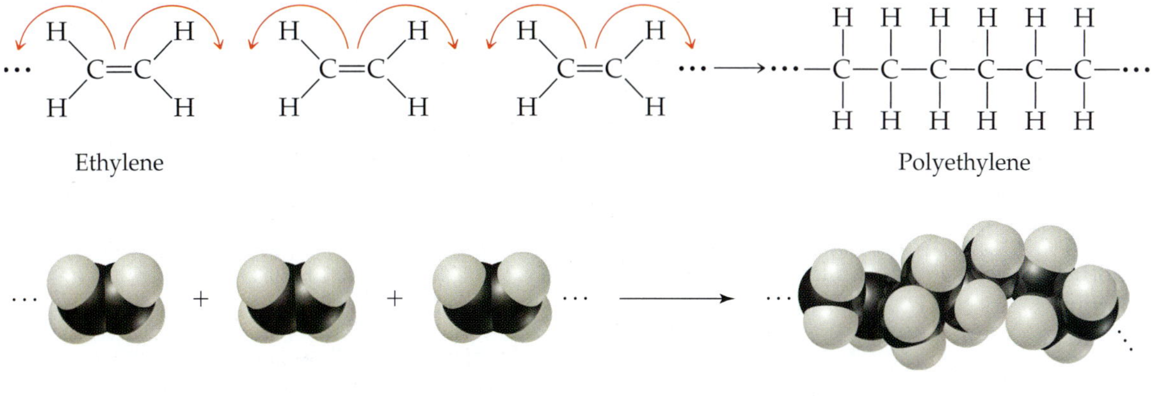

Polymerization that occurs by the coupling of monomers through their multiple bonds is called **addition polymerization**.

3-D MODELS
Polyethylene, Polypropylene, Polystyrene, Polyvinyl Chloride, Nylon 6,6

We can write the equation for the polymerization reaction as follows:

$$n\,\mathrm{CH_2{=}CH_2} \longrightarrow \left[\begin{array}{c} \mathrm{H\quad H} \\ |\quad\ | \\ -\mathrm{C}-\mathrm{C}- \\ |\quad\ | \\ \mathrm{H\quad H} \end{array}\right]_n \qquad [12.1]$$

Here n represents the large number—ranging from hundreds to many thousands—of monomer molecules (ethylene in this case) that react to form one large polymer molecule. Within the polymer a repeat unit (the unit shown in brackets in Equation 12.1) appears along the entire chain. The ends of the chain are capped by carbon–hydrogen bonds or by some other bond, so that the end carbons have four bonds.

Polyethylene is a very important material; more than 20 billion pounds are produced in the United States each year. Although its composition is simple, the polymer is not easy to make. Only after many years of research were the right conditions identified for manufacturing this commercially useful polymer. Today many different forms of polyethylene, varying widely in physical properties, are known. Polymers of other chemical compositions provide still greater variety in physical and chemical properties. Table 12.4 ▼ lists several other common polymers obtained by addition polymerization.

TABLE 12.4 Polymers of Commercial Importance

Polymer	Structure	Uses
Addition Polymers		
Polyethylene	$\mathrm{+\!\!\!-CH_2-CH_2-\!\!\!+}_n$	Films, packaging, bottles
Polypropylene	$\left[-\mathrm{CH_2}-\underset{\mathrm{CH_3}}{\underset{\vert}{\mathrm{CH}}}-\right]_n$	Kitchenware, fibers, appliances
Polystyrene	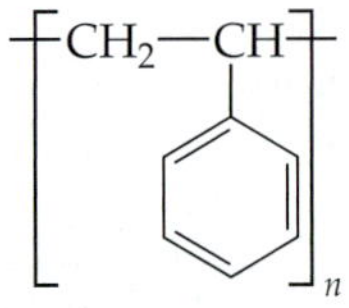	Packaging, disposable food containers, insulation
Polyvinyl chloride (PVC)	$\left[-\mathrm{CH_2}-\underset{\mathrm{Cl}}{\underset{\vert}{\mathrm{CH}}}-\right]_n$	Pipe fittings, clear film for meat packaging
Condensation polymers		
Polyurethane	$\left[-\underset{\mathrm{O}}{\underset{\Vert}{\mathrm{C}}}-\mathrm{NH-R-NH}-\underset{\mathrm{O}}{\underset{\Vert}{\mathrm{C}}}-\mathrm{O-R'-O}-\right]_n$ R, R′ = $\mathrm{-CH_2-CH_2-}$ (for example)	"Foam" furniture stuffing, spray-on insulation, automotive parts, footwear, water-protective coatings
Polyethylene terephthalate (a polyester)	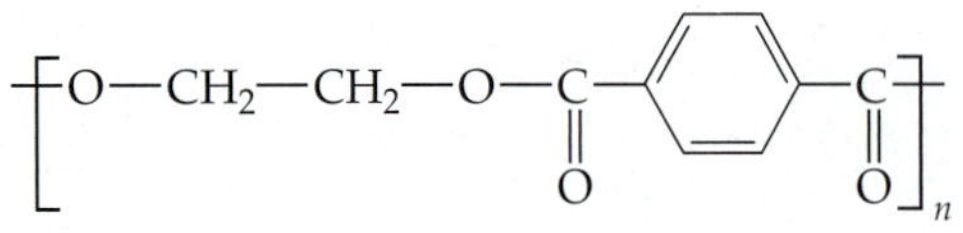	Tire cord, magnetic tape, apparel, soft-drink bottles
Nylon 6,6	$\left[-\mathrm{NH{+\!\!\!-}CH_2{-\!\!\!+}_6NH}-\underset{\mathrm{O}}{\underset{\Vert}{\mathrm{C}}}-\mathrm{(CH_2)_4}-\underset{\mathrm{O}}{\underset{\Vert}{\mathrm{C}}}-\right]_n$	Home furnishings, apparel, carpet, fishing line, toothbrush bristles
Polycarbonate	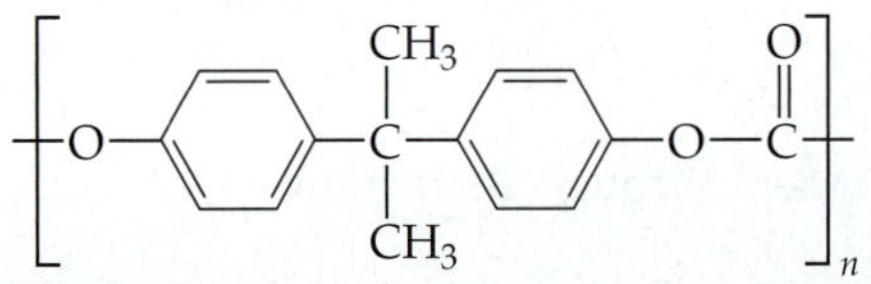	Shatterproof eyeglass lenses, CDs, DVDs, bulletproof windows, greenhouses

CHEMISTRY AT WORK | Recycling Plastics

If you look at the bottom of a plastic container, you are likely to see a recycle symbol containing a number, as seen in Figure 12.10 ▶. The number and the letter abbreviation below it indicate the kind of polymer from which the container is made, as summarized in Table 12.5 ▼. (The chemical structures of these polymers are shown in Table 12.4.) These symbols make it possible to sort containers by composition. In general, the lower the number, the greater the ease with which the material can be recycled.

TABLE 12.5 Categories Used for Recycling Polymeric Materials in the United States

Number	Abbreviation	Polymer
1	PET or PETE	Polyethylene terephthalate
2	HDPE	High-density polyethylene
3	V	Polyvinyl chloride (PVC)
4	LDPE	Low-density polyethylene
5	PP	Polypropylene
6	PS	Polystyrene
7	—	Others

▲ **Figure 12.10 Recycling symbols.** Every plastic container manufactured today has a recycling symbol imprinted on it, indicating the type of polymer used to make the container and the polymer's suitability for recycling.

A second general kind of reaction used to synthesize commercially important polymers is **condensation polymerization**. In a condensation reaction two molecules are joined to form a larger molecule by elimination of a small molecule such as H_2O. For example, an amine (a compound containing the $-NH_2$ group) will react with a carboxylic acid (a compound containing the $-COOH$ group) to form a bond between N and C along with the formation of H_2O.

MOVIE Synthesis of Nylon 6,10

$$-\overset{\displaystyle H}{\underset{}{\overset{|}{N}}}-H + H-O-\overset{\displaystyle O}{\overset{\|}{C}}- \longrightarrow -\overset{\displaystyle H}{\overset{|}{N}}-\overset{\displaystyle O}{\overset{\|}{C}}- + H_2O \qquad [12.2]$$

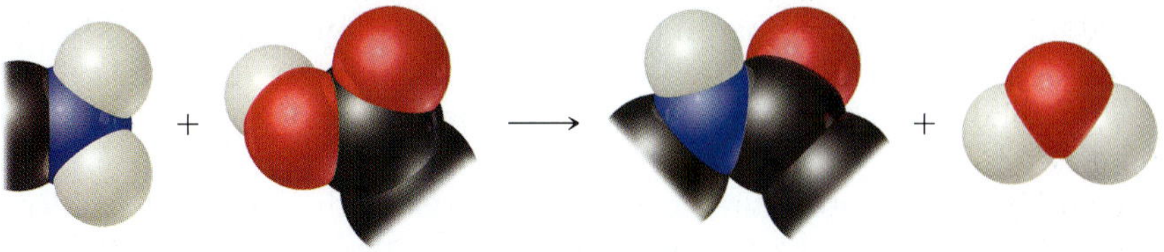

Polymers formed from two different monomers are called **copolymers**. In the formation of many nylons a *diamine*, a compound with an $-NH_2$ group at each end, is reacted with a *diacid*, a compound with a $-COOH$ group at each end. For example, nylon 6,6 is formed when a diamine that has six carbon atoms and an amino group on each end is reacted with adipic acid, which also has six carbon atoms.

$$\underset{\text{Diamine}}{n\, H_2N\!-\!(CH_2)_6\!-\!NH_2} + \underset{\text{Adipic acid}}{n\, HO\overset{\displaystyle O}{\overset{\|}{C}}\!-\!(CH_2)_4\!-\!\overset{\displaystyle O}{\overset{\|}{C}}OH} \longrightarrow \underset{\text{Nylon 6,6}}{\left[NH(CH_2)_6NH-\overset{\displaystyle O}{\overset{\|}{C}}(CH_2)_4\overset{\displaystyle O}{\overset{\|}{C}}\right]_n} + 2n\, H_2O \qquad [12.3]$$

A condensation reaction occurs on each end of the diamine and the acid. The components of H_2O are split out, and N—C bonds are formed between molecules. Table 12.4 lists nylon 6,6 and some other common polymers obtained by condensation polymerization. Notice that these polymers have backbones containing N or O atoms as well as C atoms. In Chapter 25 we will see that proteins, DNA, and RNA are also condensation polymers.

GIVE IT SOME THOUGHT

Would it be possible to make a condensation polymer out of this molecule alone?

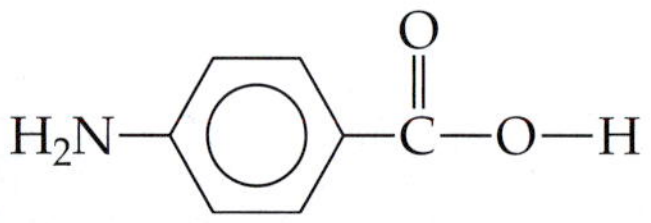

Structure and Physical Properties of Polymers

The simple structural formulas given for polyethylene and other polymers are deceptive. Because each carbon atom in polyethylene is surrounded by four bonds, the atoms are arranged in a tetrahedral fashion, so that the chain is not straight as we have depicted it. Furthermore, the atoms are relatively free to rotate around the C—C single bonds. Rather than being straight and rigid, therefore, the chains are very flexible, folding readily (Figure 12.11 ◀). The flexibility in the molecular chains causes the entire polymer material to be very flexible.

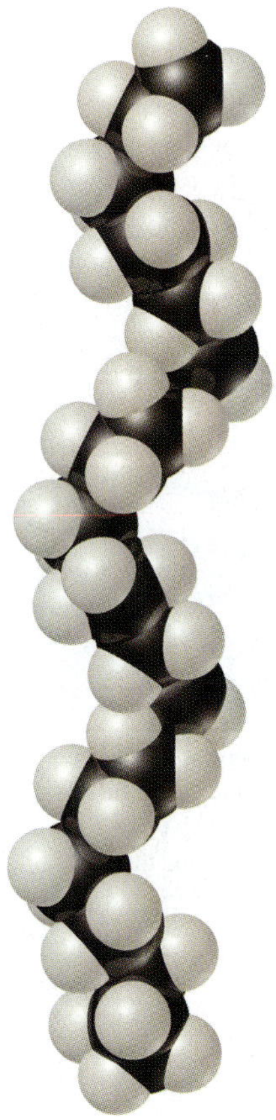

▲ **Figure 12.11 A segment of a polyethylene chain.** The segment shown here consists of 28 carbon atoms. In commercial polyethylenes the chain lengths range from about 10^3 to 10^5 CH_2 units. As this illustration implies, the chains are flexible and can coil and twist in random fashion.

Both synthetic and naturally occurring polymers commonly consist of a collection of *macromolecules* (large molecules) of different molecular weights. Depending on the conditions of formation, the molecular weights may be distributed over a wide range or be closely clustered around an average value. In part because of this distribution in molecular weights, polymers are largely amorphous (noncrystalline) materials. Rather than exhibiting a well-defined crystalline phase with a sharp melting point, they soften over a range of temperatures. They may, however, possess short-range order in some regions of the solid, with chains lined up in regular arrays as shown in Figure 12.12 ◀. The extent of such ordering is indicated by the degree of ***crystallinity*** of the polymer. The crystallinity of a polymer can frequently be enhanced by mechanical stretching or pulling to align the chains as the molten polymer is drawn through small holes. Intermolecular forces between the polymer chains hold the chains together in the ordered crystalline regions, making the polymer denser, harder, less soluble, and more resistant to heat. Table 12.6 ▼ shows how the properties of polyethylene change as the degree of crystallinity increases.

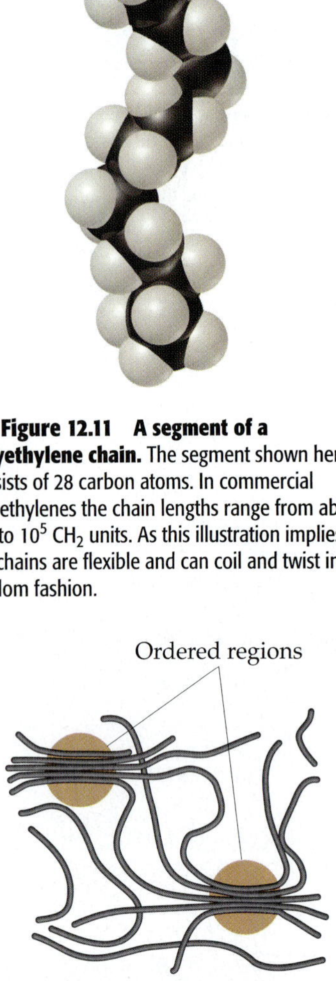

▲ **Figure 12.12 Interactions between polymer chains.** In the circled regions, the forces that operate between adjacent polymer-chain segments lead to ordering analogous to the ordering in crystals, though less regular.

The simple linear structure of polyethylene is conducive to intermolecular interactions that lead to crystallinity. However, the degree of crystallinity

TABLE 12.6 Properties of Polyethylene as a Function of Crystallinity

	Crystallinity				
	55%	62%	70%	77%	85%
Melting point (°C)	109	116	125	130	133
Density (g/cm^3)	0.92	0.93	0.94	0.95	0.96
Stiffness*	25	47	75	120	165
Yield stress*	1700	2500	3300	4200	5100

*These test results show that the mechanical strength of the polymer increases with increased crystallinity. The physical units for the stiffness test are psi × 10^{-3} (psi = pounds per square inch); those for the yield stress test are psi. Discussion of the exact meaning and significance of these tests is beyond the scope of this text.

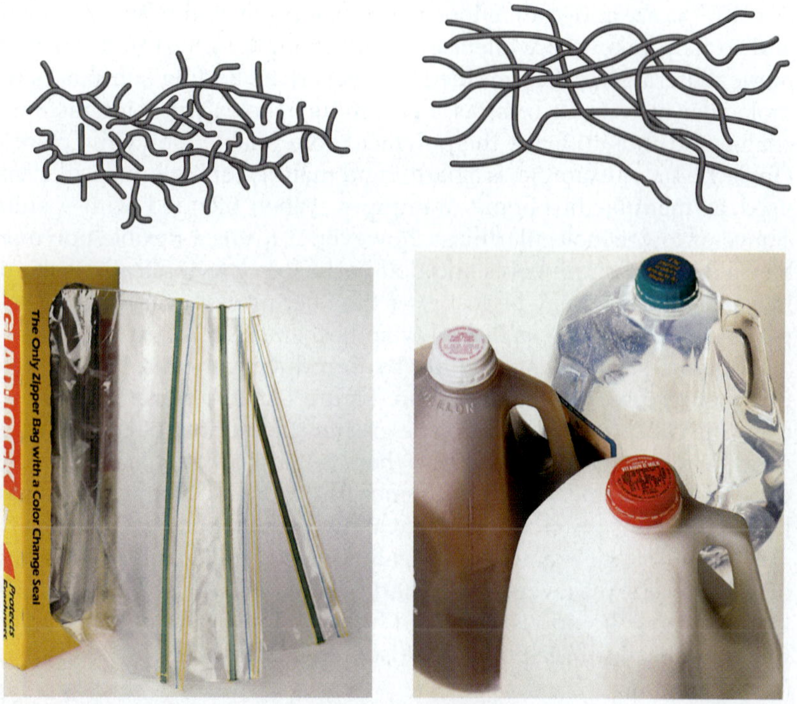

(a) (b)

◀ **Figure 12.13 Two types of polyethylene.** (a) Schematic illustration of the structure of low-density polyethylene (LDPE) and food-storage bags made of LDPE film.
(b) Schematic illustration of the structure of high-density polyethylene (HDPE) and containers formed from HDPE.

in polyethylene strongly depends on the average molecular mass. Polymerization results in a mixture of macromolecules with varying values of n and hence varying molecular masses. So-called low-density polyethylene (LDPE) used in forming films and sheets has an average molecular mass in the range of 10^4 amu and has substantial chain branching. That is, there are side chains off the main chain of the polymer, much like spur lines that branch from a main railway line. These branches inhibit the formation of crystalline regions, reducing the density of the material. High-density polyethylene (HDPE), used to form bottles, drums, and pipes, has an average molecular mass in the range of 10^6 amu. This form has less branching and thus a higher degree of crystallinity. Low-density and high-density polyethylene are illustrated in Figure 12.13 ▲. Thus, the properties of polyethylene can be "tuned" by varying the average length, crystallinity, and branching of the chains, making it a very versatile material.

GIVE IT SOME THOUGHT

Both the melting point and the degree of crystallinity decrease in copolymers made of ethylene and vinyl acetate monomers as the percentage of vinyl acetate increases. Suggest an explanation.

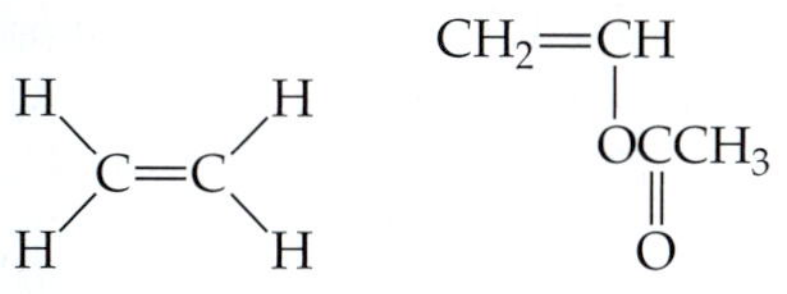

Ethylene Vinyl acetate

Various substances may be added to polymers to provide protection against the effects of sunlight or against degradation by oxidation. For example, manganese(II) salts and copper(I) salts, in concentrations as low as

$5 \times 10^{-4}\%$, are added to nylons to provide protection from light and oxidation and to help maintain whiteness. In addition, the physical properties of polymeric materials can be modified extensively by adding substances with lower molecular mass, called *plasticizers*, to reduce the extent of interactions between chains and thus to make the polymer more pliable. Polyvinyl chloride (PVC) (Table 12.4), for example, is a hard, rigid material of high molecular mass that is used to manufacture home drainpipes. When blended with a suitable substance of lower molecular mass, however, it forms a flexible polymer that can be used to make rain boots and doll parts. In some applications the plasticizer in a plastic object may be lost over time because of evaporation. As this happens, the plastic loses its flexibility and becomes subject to cracking.

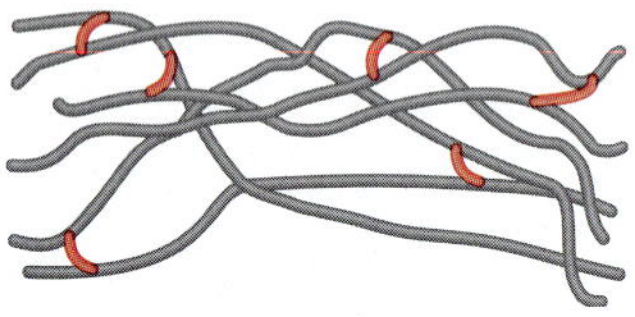

▲ **Figure 12.14 Cross-linking of polymer chains.** The cross-linking groups (red) constrain the relative motions of the polymer chains, making the material harder and less flexible than when the cross-links are not present.

Polymers can be made stiffer by introducing chemical bonds between the polymer chains, as illustrated in Figure 12.14 ◀. Forming bonds between chains is called **cross-linking**. The greater the number of cross-links in a polymer, the more rigid the material. Whereas thermoplastic materials consist of independent polymer chains, thermosetting ones become cross-linked when heated, and it is the cross-links that allow them to hold their shapes.

An important example of cross-linking is the **vulcanization** of natural rubber, a process discovered by Charles Goodyear in 1839. Natural rubber is formed from a liquid resin derived from the inner bark of the *Hevea brasiliensis* tree. Chemically, it is a polymer of isoprene, C_5H_8.

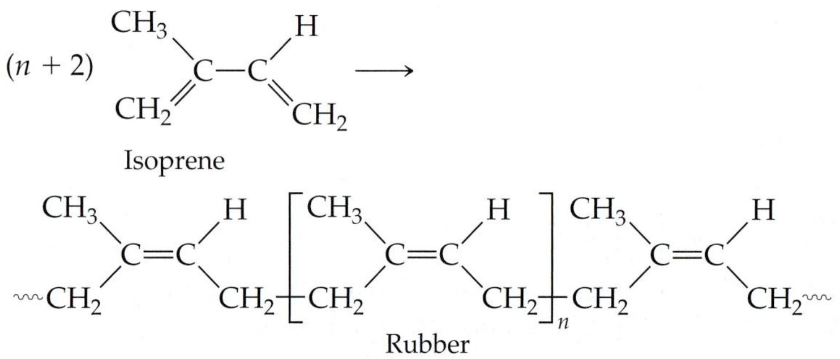

[12.4]

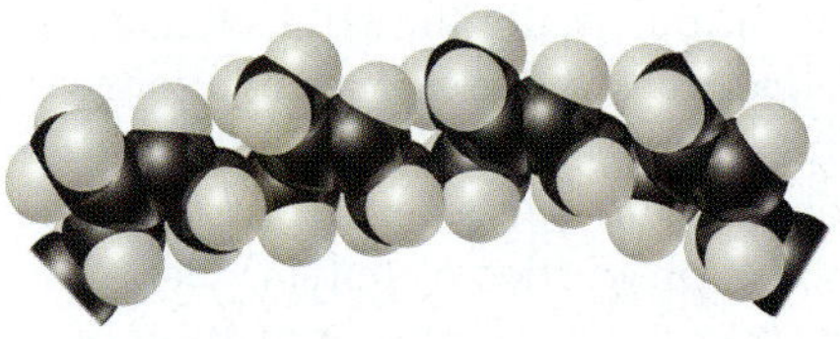

Because rotation about the carbon–carbon double bond does not readily occur, the orientation of the groups bound to the carbons is rigid. In naturally occurring rubber, the chain extensions are on the same side of the double bond, as shown in Equation 12.4. This form is called *cis*-polyisoprene; the prefix cis- is derived from a Latin phrase meaning "on this side."

Natural rubber is not a useful polymer because it is too soft and too chemically reactive. Goodyear accidentally discovered that adding sulfur to rubber and then heating the mixture makes the rubber harder and reduces its susceptibility to oxidation or other chemical attack. The sulfur changes rubber into a thermosetting polymer by cross-linking the polymer chains through reactions at some of the double bonds, as shown schematically in Figure 12.15 ▶. Cross-linking of about 5% of the double bonds creates a flexible, resilient rubber. When the rubber is stretched, the cross-links help prevent the chains from slipping, so that the rubber retains its elasticity.

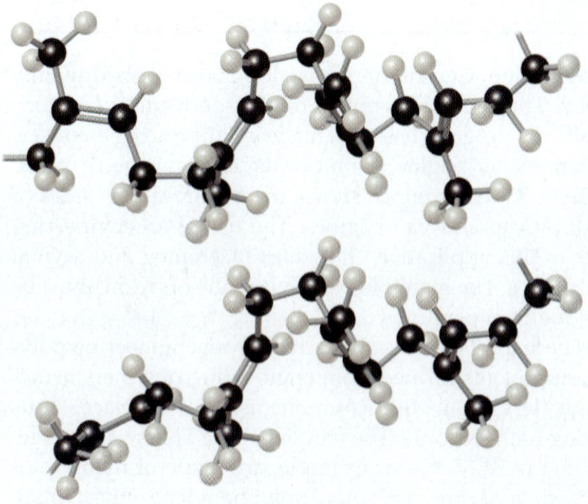
(a)

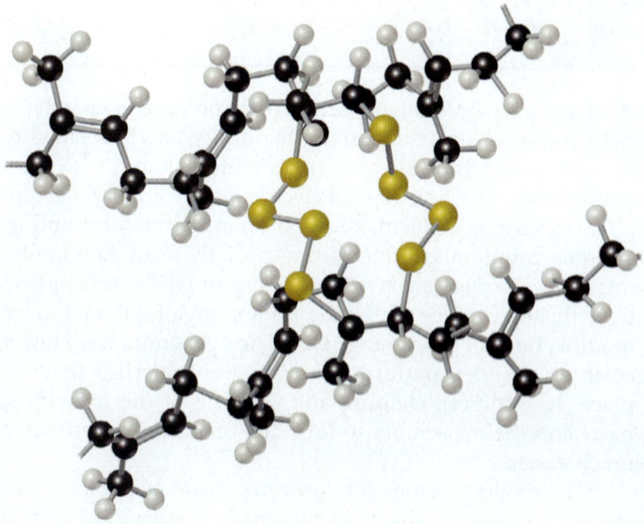
(b)

▲ **Figure 12.15 Vulcanization of rubber.** (a) In polymeric natural rubber, there are carbon–carbon double bonds at regular intervals along the chain, as shown in Equation 12.4. (b) Chains of four sulfur atoms have been added across two polymer chains, by the opening of a carbon–carbon double bond on each chain.

SAMPLE EXERCISE 12.2 | The Stoichiometry of Cross-Linking

If we assume four sulfur atoms per cross-link connection, what mass of sulfur per gram of isoprene, C_5H_8, is required to establish a cross-link as illustrated in Figure 12.15 with 5% of the isoprene units in rubber?

Solution

Analyze: We are asked to calculate the mass of sulfur required per gram of isoprene.
Plan: We need to evaluate the ratio of sulfur atoms to isoprene units based on Figure 12.15, then scale the required mass of sulfur to take account of the 5% cross-linking.
Solve: We can see from the figure that each cross-link involves eight sulfur atoms for every two isoprene units; this means that the ratio of S to C_5H_8 is four. Thus, with 5% (0.05) of the isoprene units cross-linked, we have

$$(1.0 \text{ g } C_5H_8)\left(\frac{1 \text{ mol } C_5H_8}{68.1 \text{ g } C_5H_8}\right)\left(\frac{4 \text{ mol S}}{1 \text{ mol } C_5H_8}\right)\left(\frac{32.1 \text{ g S}}{1 \text{ mol S}}\right)(0.05) = 0.09 \text{ g S}$$

PRACTICE EXERCISE

How would you expect the properties of vulcanized rubber to vary as the percentage of sulfur increases? Explain.
Answer: The rubber would be harder and less flexible as the percentage of sulfur increases, because of an increased degree of cross-linking, which covalently bonds the polymer chains together.

Hard Materials: Metals and Ceramics

Many objects are made of metals: kitchen pots and pans, jewelry, train tracks, electrical wires. Some metals are abundant in nature, especially iron and aluminum. Their atomic structure, in which the metal atoms are held together by a sea of delocalized electrons, allows for metal atoms to slip past one another under force, giving rise to the important properties of malleability and ductility. The properties of metals can be improved by the addition of other substances. Iron, for example, easily rusts. Stainless steel is a mixture of iron with other elements, and it is far more resistant to rust than pure iron.

CHEMISTRY AT WORK | Toward the Plastic Car

Many polymers can be formulated and processed to have sufficient structural strength, rigidity, and heat stability to replace metals, glass, and other materials in a variety of applications. The housings of electric motors and kitchen appliances such as coffeemakers and can openers, for example, are now commonly formed from specially formulated polymers. *Engineering polymers* are tailored to particular applications through choice of polymers, blending of polymers, and modifications of processing steps. They generally have lower costs or superior performance over the materials they replace. In addition, shaping and coloring of the individual parts and their assembly to form the final product are often much easier.

The modern automobile provides many examples of the inroads engineering polymers have made in automobile design and construction. Car interiors have long been formed mainly of plastics. Through development of high-performance materials, significant progress has been made in introducing engineering polymers as engine components and car body parts. Figure 12.16 ▼, for example, shows the manifold in a series of Ford V-8 pickup and van engines. The use of an engineering polymer in this application eliminates machining and several assembly steps. The manifold, which is made of nylon, must be stable at high temperatures.

Car body parts can also be formed from engineering polymers. Components formed from engineering polymers usually weigh less than the components they replace, thus improving fuel economy. The fenders of the Volkswagen New Beetle (Figure 12.17 ▼), for example, are made of nylon reinforced with a second polymer, polyphenylene ether (ppe), which has the following structure:

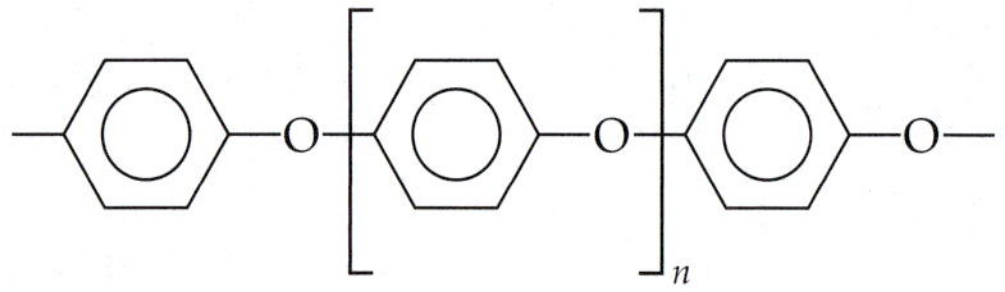

Because the polyphenylene ether polymer is linear and rather rigid, the ppe confers rigidity and shape retention.

A big advantage of most engineering polymers over metals is that they eliminate the need for costly corrosion protection steps in manufacture. In addition, some engineering polymer formulations permit manufacturing with the color built in, thus eliminating painting steps (Figure 12.18 ▼).

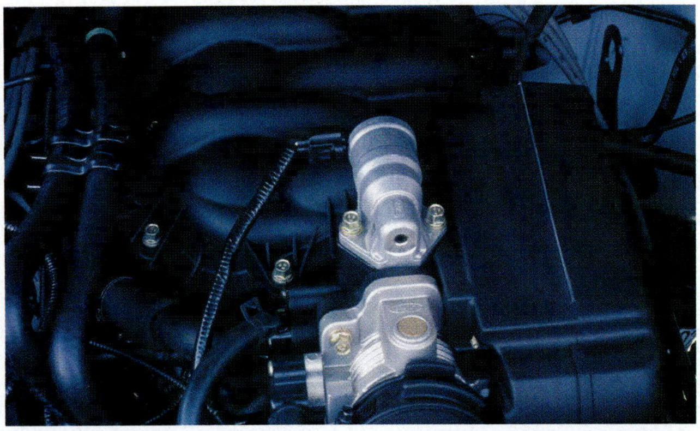

▲ **Figure 12.16 Plastic engines.** The intake manifold of some Ford V-8 engines is formed from nylon.

▲ **Figure 12.17 Plastic fenders.** The fenders of this Volkswagen Beetle are made of General Electric Noryl GTX, a composite of nylon and polyphenylene ether.

▲ **Figure 12.18 A mostly plastic car.** This truck has door handles and bumper supports made of polycarbonate/polybutylene plastic. Coloring pigments are incorporated into the manufacture of the plastic, eliminating the need for painting.

Ceramics are used in structural applications such as bricks and concrete, where the major components are SiO_2, Al_2O_3, Fe_2O_3, and CaO. The bonding in ceramics is either ionic or a mixture of ionic and covalent-network. Therefore, under applied force, the material does not deform easily like a metal and is brittle because planes of atoms of like charge would be required to slip past each other. Ceramic components are also difficult to manufacture free of defects. Ceramic parts often develop random, undetectable microcracks and voids (hollow spaces) during processing. These defects are more susceptible to stress than the rest of the ceramic; thus, they are generally the origin of cracking and fractures. To "toughen" a ceramic—to increase its resistance to fracture—scientists frequently produce very pure uniform particles of the ceramic material that are less than 1 μm (10^{-6} m) in diameter. These are then *sintered* (heated at high temperature under pressure so that the individual particles bond together) to form the desired object.

Ceramic materials are used in the manufacture of ceramic tiles for the surfaces of the Space Shuttle, to protect against overheating on reentry into Earth's atmosphere (Figure 12.19 ▶). The tiles are made of short, high-purity silica fibers reinforced with aluminum borosilicate fibers. The material is formed into blocks, sintered at over 1300°C, and then cut into tiles. The tiles have a density of only 0.2 g/cm^3, yet they are able to keep the shuttle's aluminum skin below 180°C while sustaining a surface temperature of up to 1250°C.

▲ **Figure 12.19 Ceramics in space.** A worker applying thermally insulating ceramic tiles to the body of the Space Shuttle orbiter.

Making Ceramics

Many ceramics are naturally occurring and can be mined from the earth. However, there are inorganic counterparts of the polymerization reactions we discussed earlier that are useful for making ceramic materials with increased purity and controlled particle size. The **sol-gel process** is an important method of forming extremely fine particles of uniform size. A typical sol-gel procedure begins with a *metal alkoxide*. An alkoxide contains organic groups bonded to a metal atom through oxygen atoms. Alkoxides are produced when the metal reacts with an alcohol, which is an organic compound containing an OH group bonded to carbon. (Section 2.9) To illustrate this process, we will use titanium as the metal and ethanol, CH_3CH_2OH, as the alcohol.

$$Ti(s) + 4\,CH_3CH_2OH(l) \longrightarrow Ti(OCH_2CH_3)_4(s) + 2\,H_2(g) \qquad [12.5]$$

The alkoxide product, $Ti(OCH_2CH_3)_4$, is dissolved in an appropriate alcohol solvent. Water is then added, and it reacts with the alkoxide to form Ti—OH groups and to regenerate ethanol.

$$Ti(OCH_2CH_3)_4(soln) + 4\,H_2O(l) \longrightarrow Ti(OH)_4(s) + 4\,CH_3CH_2OH(l) \qquad [12.6]$$

The reaction with ethanol is used, even though the ethanol is simply regenerated, because the direct reaction of Ti(*s*) with $H_2O(l)$ leads to a complex mixture of titanium oxides and titanium hydroxides. The intermediate formation of $Ti(OC_2H_5)_4(s)$ ensures that a uniform suspension of $Ti(OH)_4$ will be formed. The $Ti(OH)_4$ is present at this stage as a *sol*, a suspension of extremely small particles. The acidity or basicity of the sol is adjusted to split out water from between two of the Ti—OH bonds.

$$(HO)_3Ti{-}O{-}H(s) + H{-}O{-}Ti(OH)_3(s) \longrightarrow$$
$$(HO)_3Ti{-}O{-}Ti(OH)_3(s) + H_2O(l) \qquad [12.7]$$

This is another example of a condensation reaction. Condensation also occurs at some of the other OH groups bonded to the central titanium atom, producing a three-dimensional network. The resultant material, called a *gel*, is a

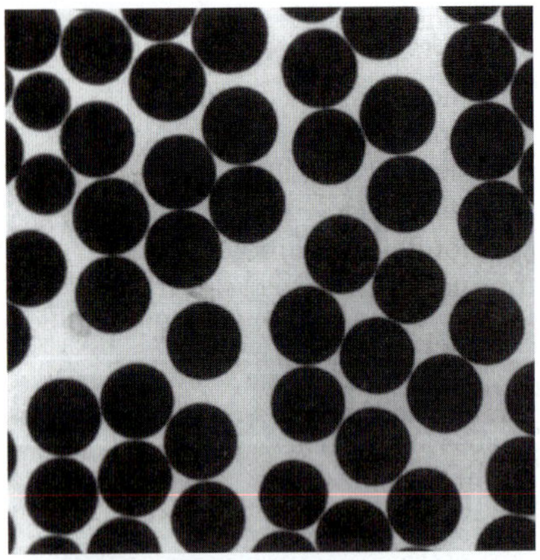

▲ **Figure 12.20 Uniformly sized spheres of amorphous silica, SiO_2.** These are formed by precipitation from a methanol solution of $Si(OCH_3)_4$ upon addition of water and ammonia. The average diameter is 550 nm.

suspension of extremely small particles with the consistency of gelatin. When this material is heated carefully at 200°C to 500°C, all the liquid is removed, and the gel is converted to a finely divided metal oxide powder with particles in the range of 0.003 to 0.1 μm in diameter. Figure 12.20 ◀ shows SiO_2 particles, formed into remarkably uniform spheres by a precipitation process similar to the sol-gel process.

12.3 | Materials for Medicine

Increasingly, modern synthetic materials are being used in medical and biological applications. For our discussion here, a **biomaterial** is any material that has a biomedical application. The material might have a therapeutic use, for example, in the treatment of an injury or a disease. Or it might have a diagnostic use, as part of a system for identifying a disease or for monitoring a quantity such as the glucose level in blood. Whether the use is therapeutic or diagnostic, the biomaterial is in contact with biological fluids, and this material must have properties that meet the demands of that application. For example, a polymer employed to form a disposable contact lens must be soft and have an easily wetted surface, whereas the polymer used to fill a dental cavity must be hard and wear-resistant.

Characteristics of Biomaterials

The most important characteristics that influence the choice of a biomaterial are biocompatibility, physical requirements, and chemical requirements, as illustrated in Figure 12.21 ▼.

Biocompatibility Living systems, particularly the higher animals, have a complex set of protections against invasions of other organisms. Our bodies have an extraordinary ability to determine whether an object is the body's own material or whether it is foreign. Any substance that is foreign to the body has the potential to generate a response from the immune system. Molecular-sized objects are bound by antibodies and rejected, whereas larger objects induce an inflammatory reaction around them. Some materials are more **biocompatible**; that is, they are more readily integrated into the body without inflammatory reactions. The most important determining factors in the biocompatibility of a substance are its chemical nature and the physical texture of its surface.

▼ **Figure 12.21 Schematic illustration of a human-made device implanted in a biological system.** To function successfully, the device must be biocompatible with its surroundings and meet the necessary physical and chemical requirements, some of which are listed for illustrative purposes.

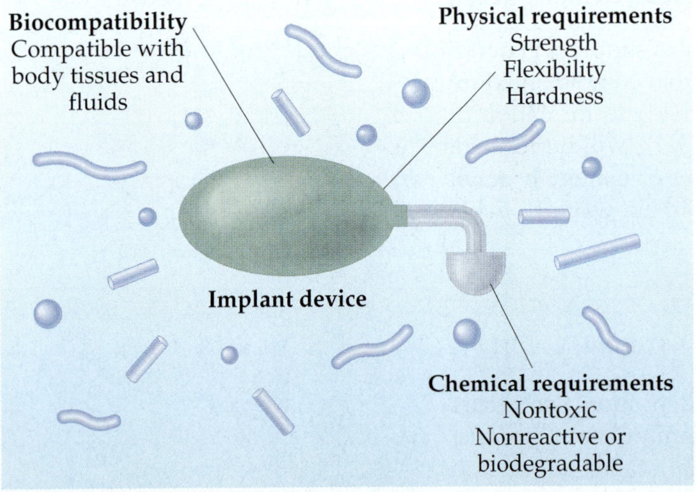

Physical Requirements A biomaterial is often required to meet very exacting demands. Tubing that is to be used to replace a defective blood vessel must be flexible and must not collapse under bending or other distortion. Materials used in joint replacements must be wear-resistant. An artificial heart valve must open and close 70 to 80 times a minute, day after day, for many years. If the valve is to have a life expectancy of twenty years, this means about 750 million cycles of opening and closing! Unlike the failure of the valve in a car engine, failure of the heart valve may have fatal consequences for its owner.

Chemical Requirements Biomaterials must be *medical grade,* which means that they must be approved for use in any particular medical application. Any ingredients present in a medical-grade biomaterial must remain innocuous over the lifetime of the application. Polymers are important biomaterials, but most commercial polymeric materials contain contaminants such as unreacted monomers, fillers or plasticizers, and antioxidants or other stabilizers. The small amounts of foreign materials present in HDPE used in milk jugs (Figure 12.13) pose no threat in that application, but they might if the same plastic material were implanted in the body over a long period.

Polymeric Biomaterials

Specialized polymers have been developed for a variety of biomedical applications. The degree to which the body accepts or rejects the foreign polymer is determined by the nature of the atomic groups along the chain and the possibilities for interactions with the body's own molecules. Our bodies are composed largely of biopolymers such as proteins, polysaccharides (sugars), and polynucleotides (RNA, DNA). We will learn more about these molecules in Chapter 25. For now, we can simply note that the body's own biopolymers have complex structures, with polar groups along the polymer chain. Proteins, for example, are long strings of amino acids (the monomers) that have formed a condensation polymer. The protein chain has the following structure:

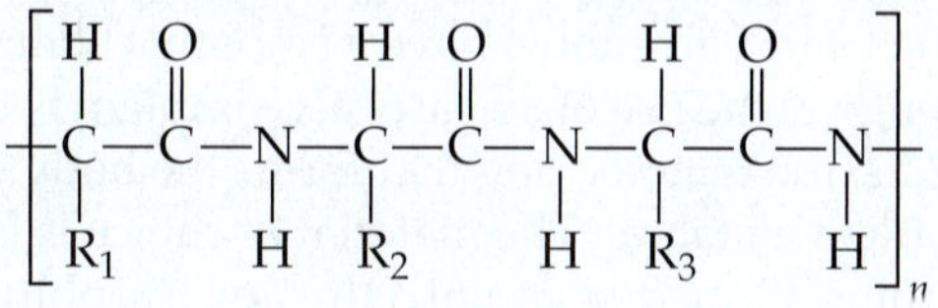

where the R groups vary along the chain [$-CH_3$, $-CH(CH_3)_2$, and so on]. There are 20 different amino acids that are used in different combinations to make up proteins in humans. By contrast, synthetic polymers are simpler, being formed from a single repeating unit or perhaps two different repeating units, as described in Section 12.2. This difference in complexity is one of the reasons that synthetic polymers are identified by the body as foreign objects. Another reason is that there may be few or no polar groups along the chain that can interact with the body's aqueous medium. ∞ (Section 11.2)

We learned in Section 12.2 that polymers can be characterized by their physical properties. Elastomers are used as biomaterials in flexible tubing over leads for implanted heart pacemakers and as catheters (tubes implanted into the body to administer a drug or to drain fluids). Thermoplastics, such as polyethylene or polyesters, are employed as membranes in blood dialysis machines and as replacements for blood vessels. Thermoset plastics find limited, but important uses. Because they are hard, inflexible, and somewhat brittle, they are most often used in dental devices or in orthopedic applications, such as in joint replacements. To fill a cavity, for example, the dentist may pack some material into the cavity, then shine an ultraviolet lamp on it. The light initiates a photochemical reaction that forms a hard, highly cross-linked thermoset polymer.

Examples of Biomaterial Applications

We can best appreciate the kinds of problems encountered in using biomaterials by considering a few specific cases.

Heart Replacement and Repairs The term *cardiovascular* pertains to the heart, blood, and blood vessels. The heart is, of course, an absolutely essential organ. A heart that fails completely may be replaced by a donor organ. About 60,000 people suffer terminal heart failure each year in the United States, yet only about 2500 donor hearts become available for transplant. Many attempts have been made—and continue to be made—to produce an artificial heart that

▲ **Figure 12.22 A disc heart valve.** Known as the St. Jude valve, this device was named for the medical center at which it was developed. The surfaces of the valve are coated with pyrolytic carbon. The valve is secured to the surrounding tissue via a Dacron® sewing ring.

can serve over a long period of time as a replacement for the natural organ. We will not devote attention to these efforts, except to note that recent results are quite promising.

It often happens that only a part of the heart, such as the aortic valve, fails and needs replacement. Repair could be made by using foreign tissue (for example, a pig heart valve) or a mechanical heart valve implant to replace a diseased one. About 250,000 valve replacement procedures are performed annually worldwide. In the United States about 45% of the procedures involve a mechanical valve. The most widely used valve is shown in Figure 12.22 ◄. It has two semicircular discs that move to allow blood to flow in the desired direction as the heart pumps, then they fall back together to form a seal against backflow.

It is vital to minimize fluid disturbance as the blood passes through artificial devices. Surface roughness in a device causes *hemolysis*, the breakdown of red blood cells. Furthermore, surface roughness can serve as a site for invading bacteria to adhere and colonize. Finally, rough surfaces also promote coagulation of the blood, which forms a thrombus, or blood clot. Thus, even though we may have a perfectly fine piece of machinery from a mechanical point of view, the heart valve may not be suitable as a long-term implant. To minimize blood clots, the discs in the heart valve must have a smooth, chemically inert surface.

A second challenge in the use of a heart valve implant is to fix it in place. As shown in Figure 12.22, the retainer ring that forms the body of the valve is covered with a mesh of woven fabric. The material of choice is Dacron®, du Pont's trade name for the fiber formed from polyethylene terephthalate (PET, a polyester; Table 12.4). The mesh acts as a lattice on which the body's tissues can grow, so that the valve becomes incorporated into its surroundings. Tissue grows through the polyester mesh, whereas it does not do so on many other plastics. Apparently the polar, oxygen-containing functional groups along the polyester chain afford attractive interactions to facilitate tissue growth.

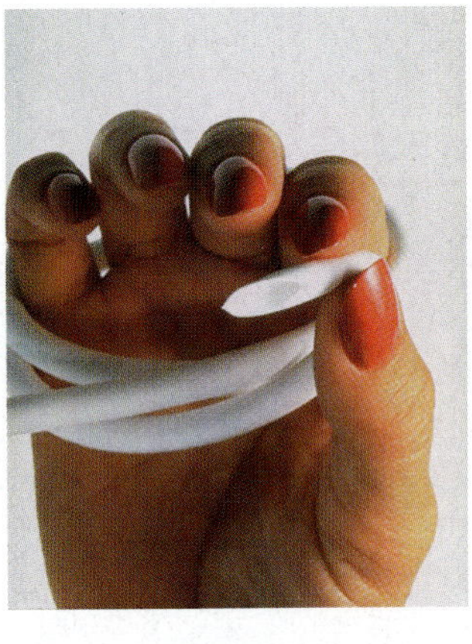

▲ **Figure 12.23 A Dacron® vascular graft.** Grafts like this are used in coronary artery surgery.

Vascular Grafts A vascular graft is a replacement for a segment of diseased arteries. Where possible, diseased blood vessels are replaced with vessels taken from the patient's own body. When this is not feasible, artificial materials must be used. Dacron® is used as replacement for large-diameter arteries around the heart. For this purpose, it is fabricated into a crimped, woven, tubular form, as shown in Figure 12.23 ◄. The tubing is crimped to enable it to bend without serious decrease in its cross-sectional area. The graft must integrate with surrounding tissue after it has been put in place. It must therefore have an open structure, with pores on the order of 10 μm in diameter. During the healing process blood capillaries grow into the graft and new tissues form throughout it. Similarly, polytetrafluoroethylene $[-(CF_2CF_2)_n-]$ is used for the smaller-diameter vascular grafts in the limbs.

Ideally, the inside surface of the graft would become lined with the same sorts of cells that line the native artery, but this process does not occur with materials currently available. Instead, the inside surface of the tubing is recognized as being foreign to the blood. Platelets, which are circulating components of the blood, normally serve the function of sealing up wounds in the blood vessel walls. Unfortunately, they attach to foreign surfaces and cause blood coagulation. The search for a more biocompatible lining for grafts is an area of active research, but at present there is a continuing risk of blood clots. Excess tissue growth at the intersection of the graft with the native artery is also a frequent problem. Because of the possibility of blood clot formation, patients who have received artificial heart valves or vascular grafts are generally required to take anticoagulation drugs on a continuing basis.

Artificial Tissue The treatment of patients who have lost extensive skin tissue—for example, burn patients or those with skin ulcers—is one of the most difficult problems in therapeutic medicine. Today, laboratory-grown skin can be employed to replace grafts in such patients. Ideally, the "artificial" tissue would be grown from cells taken from the patient. When this is not possible, for example, with burn victims, the tissue cells come from another source. When the graft skin is not formed from the patient's own cells, drugs that suppress the body's natural immune defense system must be used or steps must be taken to modify the new cell line to prevent rejection of the tissue.

The challenge in growing artificial tissue is to get the cells to organize themselves as they would in a living system. The first step in accomplishing this objective is to provide a suitable scaffold for the cells to grow on, one that will keep them in contact with each other and allow them to organize. Such a scaffold must be biocompatible; cells must adhere to the scaffold and differentiate (that is, develop into cells of different types) as the culture grows. The scaffolding must also be mechanically strong and biodegradable.

The most successful scaffolds have been lactic acid—glycolic acid copolymers. Formation of the copolymer via a condensation reaction is shown in Equation 12.8.

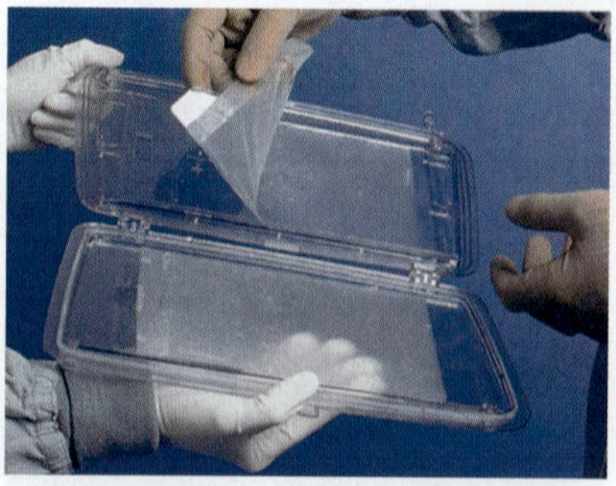

▲ **Figure 12.24 Artificial skin prepared for use as a skin graft.** Burn victims benefit from these biocompatible polymers.

$$\underset{\text{Glycolic acid}}{n\,\mathrm{HOCH_2}\overset{\mathrm{O}}{\overset{\|}{\mathrm{C}}}\mathrm{—OH}} + \underset{\text{Lactic acid}}{n\,\mathrm{HO}\underset{\mathrm{CH_3}}{\underset{|}{\mathrm{CH}}}\overset{\mathrm{O}}{\overset{\|}{\mathrm{C}}}\mathrm{—OH}} \longrightarrow \underset{\text{Copolymer}}{\mathrm{—O—CH_2}\overset{\mathrm{O}}{\overset{\|}{\mathrm{C}}}\mathrm{—O}\!\left[\underset{\mathrm{CH_3}}{\underset{|}{\mathrm{CH}}}\overset{\mathrm{O}}{\overset{\|}{\mathrm{C}}}\mathrm{—O—CH_2}\overset{\mathrm{O}}{\overset{\|}{\mathrm{C}}}\mathrm{—O}\right]_n\!\underset{\mathrm{CH_3}}{\underset{|}{\mathrm{CH}}}\overset{\mathrm{O}}{\overset{\|}{\mathrm{C}}}\mathrm{—}} \qquad [12.8]$$

The copolymer has an abundance of polar carbon–oxygen bonds along the chain, affording many opportunities for hydrogen-bonding interactions. The ester linkages formed in the condensation reaction are susceptible to hydrolysis, which is just the reverse reaction. When the artificial tissue is deployed in the body, the underlying copolymer scaffold slowly hydrolyzes away as the tissue cells continue to develop and merge with adjacent tissue. An example of a graft skin product is shown in Figure 12.24 ▶.

"Smart" Sutures Until the 1970s, if you had to go to the doctor to get stitches, you later had to go back to get the stitches removed. Stitches that are used to hold living tissue together are called *sutures*. Since the 1970s, biodegradable sutures have been available. After the sutures have been applied to the tissue, they slowly dissolve away and do not release any harmful by-products. Today's biodegradable sutures are made out of lactic acid—glycolic acid copolymers, similar to those just described, that slowly hydrolyze over time.

More recently, scientists have developed what are colloquially called "smart" materials for biological applications. In this context, "smart" means a material that will reversibly change its behavior in response to an external stimulus. For example, smart eyeglasses turn dark when you go outside in the sun, then turn light again when you go inside. In the case of smart sutures, scientists have developed a thermoplastic polymer that organizes itself reversibly, via intermolecular forces, as a function of temperature. The material shrinks if it is heated to body temperature or higher. Thus, sutures made from this material can be tied loosely by the surgeon, and then after warming up to body temperature, the sutures tighten up to provide the optimum force (Figure 12.25 ▶). This polymer was first made in 2002, and it is not on the market at the time of this writing. It is expected, however, that it will be very useful to doctors performing surgery in very small spaces (for example, through an endoscope).

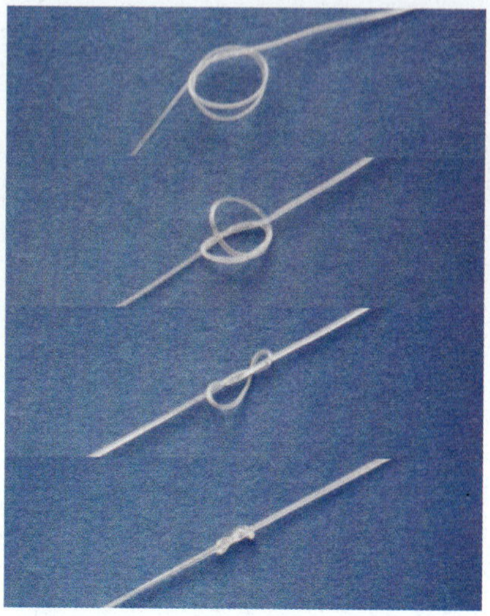

▲ **Figure 12.25 Programmable polymers.** After being stretched about 200% to "program" it, a fiber of thermoplastic polymer was formed into a loose knot, and the ends were fixed (top). The series shows, from top to bottom, how the knot tightens in 20 s when heated from room temperature to 40°C.

12.4 | Materials for Electronics

We can use materials to store information in the form of electrons as "bits," as in the modern integrated circuit. We can also use materials to produce electricity. The latter application will be examined in more detail in Chapter 20, but here we examine the materials that underlie modern electronics.

The Silicon Chip

▲ **Figure 12.26 Photograph of a Pentium 4 processor computer chip.** On this scale, mostly what you can see is the connections between transistors.

Silicon is a semiconductor and one of the most abundant elements on Earth. (Section 1.2) The semiconductor industry that is responsible for the circuitry in computers, cell phones, and a host of other devices relies on silicon wafers called "chips," on which complex patterns of semiconductors, insulators, and metal wires are assembled (Figure 12.26 ◀). Although the details of integrated circuit design are beyond the scope of this book, we can say that the process of making chips relies mostly on chemistry.

Silicon is the semiconductor of choice because it is abundant and cheap in its raw form (obtained from sand), it can be made 99.999999999% pure in highly specialized facilities called "clean rooms" (Figure 12.27 ▼), and it can be grown into enormous crystals that are nearly atomically perfect. (If you recall that only parts-per-million level of impurities can change the conductivity of silicon by a millionfold, it becomes clear why clean rooms are necessary. In addition, the individual units that make up integrated circuits are about 100 nm wide, which makes them about the size of an individual virus particle and quite a bit smaller than ordinary dust particles. Thus, to an integrated circuit, a dust particle looks like a boulder.) Other major advantages of silicon are that it is nontoxic (compared to GaAs, its closest competitor) and its surface can be chemically protected by SiO_2, the natural product of its reaction with air. SiO_2 is an excellent insulator and its atoms are in excellent registry with the underlying silicon substrate, which means it is easy to grow crystalline layers of SiO_2 on top of Si. The basic unit of the integrated circuit, the **transistor**, requires a metal/insulator "gate" between a semiconductor "source" and a semiconductor "drain." Electrons move from the source to the drain when voltage is applied to the gate. That silicon's naturally occurring oxide is a very chemically stable insulator is a huge benefit to silicon technology.

▶ **Figure 12.27 Making silicon chips.** IBM's clean room for the fabrication of 300 mm silicon wafers, in East Fishkill, New York.
Tom Way/Courtesy of International Business Machines Corporation. Unauthorized use not permitted.

A CLOSER LOOK | The Transistor

The transistor is the heart of integrated circuits. By applying a small electric charge, it controls the flow of a substantially larger current. Therefore, the transistor not only controls information flow in the form of electrons, it also amplifies the signal.

One common type of transistor is the MOSFET, which stands for *metal-oxide-semiconductor field-effect transistor.* The design of a MOSFET is shown in Figure 12.28 ▼. A piece of silicon is p-doped to form the substrate, and then patches of n-type doping are created and connected to metal wires. The back of the substrate is also connected to a metal wire. The n-type regions are the *source* and *drain*. They are separated by a p-type *channel*, which is about 60 nm long in a Pentium 4 processor. Deposited on top of the channel is the *gate*, an insulating oxide—usually SiO_2, but it can also be silicon nitride, Si_3N_4, or a mixture of these two. In the Pentium 4 chip, the oxide layer is only about 2 nm, or 20 atoms, thick. A metal contact is made to this oxide gate.

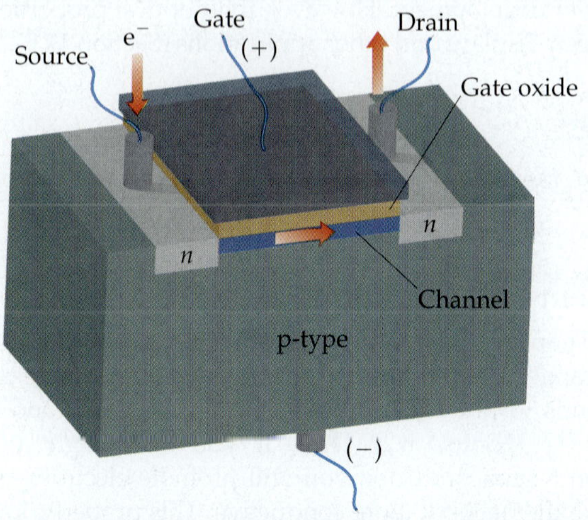

▲ **Figure 12.28 Design of a basic MOSFET.** The substrate is greenish blue, the source and drain are pale gray, the channel is blue, and the gate oxide is pale brown. The blue lines and gray cylinders represent metallic contacts. A small positive voltage applied to the gate creates a positive channel through which electrons (red arrows) from the source are drawn, and go into the drain.

When a tiny positive charge is applied to the gate, electrons from the n-type source are drawn into the channel and flow to the drain. The larger the charge applied, the more the channel is "opened" and the greater the number of electrons that flow.

The reverse situation, in which the source and drain are p-type and the channel is n-type, can also be made. There are many other transistor designs.

At present, the smallest circuit elements in commercial use are 60 nm wide, as in the gate channel length in MOSFETs. This means that a typical integrated circuit board holds in excess of 65 million transistors over an area the size of your fingernail. The patterning of 300-mm commercial silicon wafers, which are about the size of a medium pizza, results from a series of chemical reactions that put down layers of thin films of semiconductor, insulator, and metal in patterns that are defined by a series of masks. Figure 12.29 ▼ shows that the thickness of the layers varies from about 100 nm to 3000 nm. In the research laboratory, film thicknesses have been made as thin as several atoms.

Related Exercises: 12.47, 12.49, 12.81.

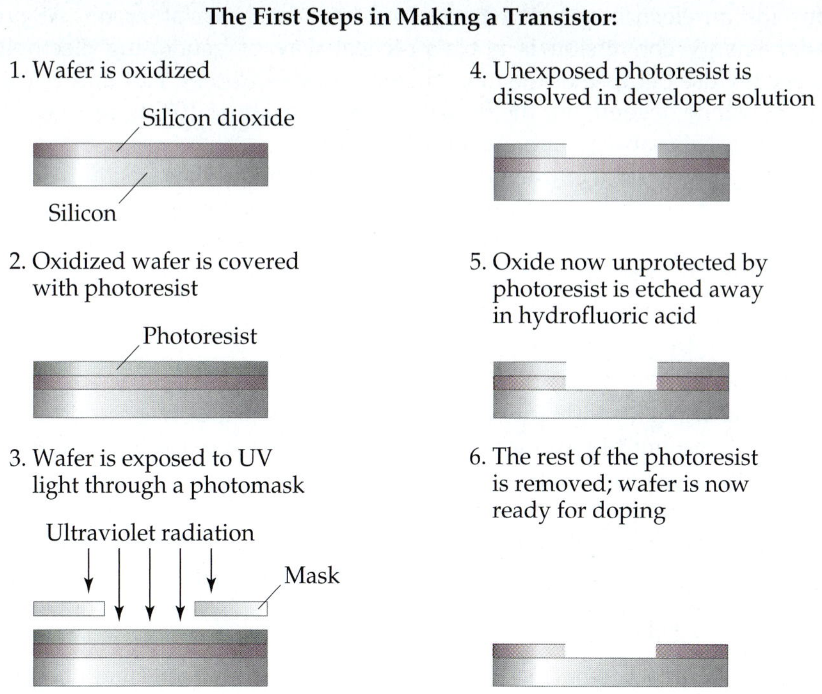

◀ **Figure 12.29 Schematic illustration of the process of photolithography.** Typical film thicknesses are 200–600 μm for the silicon substrate; 80 nm for the silicon dioxide layer; and 0.3–2.5 μm for the photoresist, which is a polymer that cross-links when irradiated with ultraviolet light so it will be chemically resistant to organic solvents in the "developer" solution. The mask protects parts of the photoresist from light and provides patterning to the surface on the ~100-nm scale. Commercial silicon chips can have as many as 30 layers. Other forms of lithography utilize means other than light to pattern the surface (for example, ion-beam lithography and electron-beam lithography use beams of ions or electrons to "write" on the surface).

Plastic Electronics

Silicon's success has not stopped scientists from trying to find other materials to use in electronic devices. Polymers that have structures like those shown in Table 12.4 have somewhat delocalized electrons in their π bonds, and it was found in the 1970s that certain polymers with delocalized electrons are semiconductors. The discoverers of "organic semiconductors"—Alan J. Heeger, Alan G. MacDiarmid, and Hideki Shirakawa—were awarded the Nobel Prize in Chemistry in 2000. Currently, researchers are exploring these and related materials for organic transistors that might be lighter and more physically flexible than silicon-based devices. At present, these polymers are not as robust as silicon, and their use in integrated circuits is still under development. However, their optical properties are very promising for flat-screen displays and other applications (Section 12.5).

GIVE IT SOME THOUGHT

Which do you think is more likely to show semiconducting behavior—a short polymer or a long one?

Solar Energy Conversion

The magnitude of the band gap for semiconductors is ~50 kJ/mol to ~300 kJ/mol (Table 12.1)—just into the range of the bond-dissociation energies for single chemical bonds—and corresponds to photon energies of infrared, visible, and low-energy ultraviolet light. (Section 6.2) Thus, if you shine light of the appropriate wavelength on a semiconductor, you will promote electrons to the conduction band and make the material more conductive. This property, known as *photoconductivity*, is essential for many solar energy conversion applications.

Solar energy conversion became a very active area of research at the time of the energy crisis in the mid-1970s. Silicon in its cheaper noncrystalline form is used to make solar energy conversion panels (Figure 12.30 ◀). Silicon's small band gap is able to capture most photons from the sun, causing electrons to be promoted from the valence band to the conduction band. These electrons then provide an electric current when wired into an electrical circuit. At present, solar energy conversion is not a practical way of generating electricity for everyday use because of the low efficiency of the process. Not only is continuous sunshine needed, but the best efficiencies are about 10% for converting one photon of solar energy into one usable electron.

TiO_2 is another commercially viable solar energy conversion material because it is cheap and abundant. However, its band gap is nearly three times that of silicon (Table 12.1). As a result, it is able to absorb only ultraviolet photons, and its use in solar energy conversion is more limited.

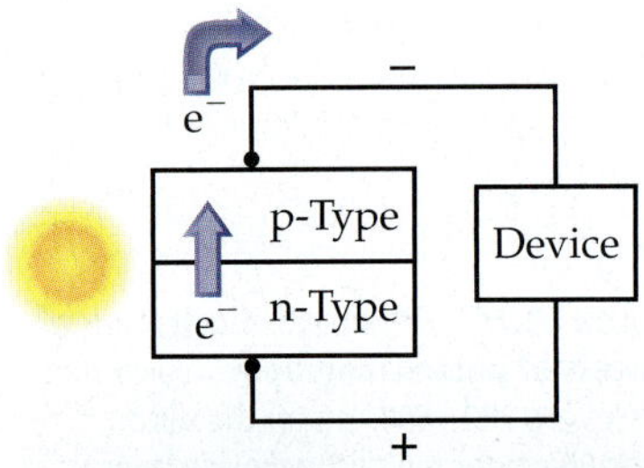

▲ **Figure 12.30 Electricity from sunlight.** Top: A solar cell is formed by the junction of n-type and p-type silicon. Sunlight (which includes photons with a larger energy than silicon's band gap energy) promotes electrons from the n-type side into the p-type side, and the resulting current can be used to power devices. Bottom: Solar panels, made of silicon, are used both as an energy source and as architectural elements in this apartment building in southern California.

12.5 | Materials for Optics

Many materials are valuable for their optical properties; two important examples are light-emitting diodes (LEDs) and liquid crystal displays (LCDs). LEDs are made of semiconductors, and LCDs are made of small organic molecules that arrange themselves in an organized fashion either spontaneously or when a small voltage is applied. In this section we will discuss the molecular principles that give these materials their special optical properties.

Liquid Crystals

When a solid is heated to its melting point, the added thermal energy overcomes the intermolecular attractions that provide molecular order to the solid. (Section 11.1) The liquid that forms is characterized by random molecular orientations and considerable molecular motion, and the molecules are not ordered at all (zero-dimensional order). Some substances, however, exhibit more complex behavior as their solids are heated. Instead of the three-dimensional order that characterizes a crystalline solid, these substances exhibit phases that have one- or two-dimensional order, meaning that the molecules line up so that they point in one direction (one-dimensional order) or are arranged in layers (two-dimensional order).

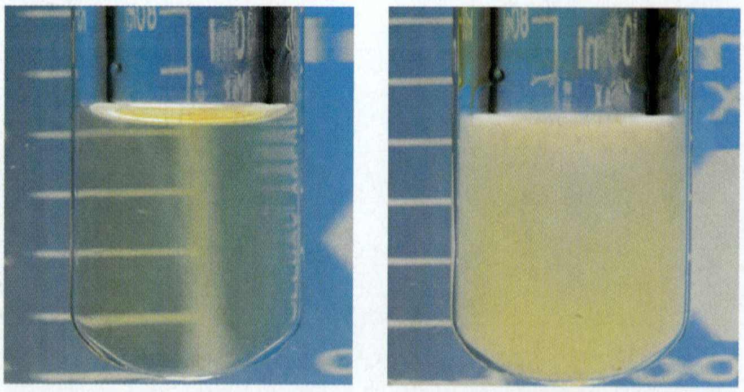

(a) (b)

▲ **Figure 12.31 Liquid and liquid crystal.** (a) Molten cholesteryl benzoate at a temperature above 179°C. In this temperature region the substance is a clear liquid. Note that the printing on the surface of the beaker in back of the sample test tube is readable. (b) Cholesteryl benzoate at a temperature between 179°C and its melting point, 145°C. In this temperature interval cholesteryl benzoate exists as a milky liquid crystalline phase.

In 1888 Frederick Reinitzer, an Austrian botanist, discovered that an organic compound he was studying, cholesteryl benzoate, has interesting and unusual properties. When heated, the substance melts at 145°C to form a viscous milky liquid; at 179°C the milky liquid suddenly becomes clear. When the substance is cooled, the reverse processes occur: The clear liquid turns viscous and milky at 179°C (Figure 12.31 ▶) and the milky liquid solidifies at 145°C. Reinitzer's work represents the first systematic report of what we now call a **liquid crystal**.

Instead of passing directly from the solid to the liquid phase when heated, some substances, such as cholesteryl benzoate, pass through an intermediate liquid crystalline phase that has some of the structure of solids and some of the freedom of motion possessed by liquids. Because of the partial ordering, liquid crystals may be very viscous and possess properties intermediate between those of the solid and liquid phases. The region in which they exhibit these properties is marked by sharp transition temperatures, as in Reinitzer's example.

From the time of their discovery in 1888 until about 30 years ago, liquid crystals were largely a laboratory curiosity. They are now widely used as pressure and temperature sensors and in the displays of electrical devices such as digital watches, calculators, and laptop and handheld computers (Figure 12.32 ▶). Liquid crystals can be used for these applications because the weak intermolecular forces that hold the molecules together in a liquid crystal are easily affected by changes in temperature, pressure, and electric fields.

▲ **Figure 12.32 A handheld wireless device with liquid crystal display.** LCDs can be made very thin and light.

Types of Liquid Crystalline Phases

Substances that form liquid crystals are often composed of long, rodlike molecules. In the normal liquid phase these molecules are oriented randomly [Figure 12.33(a) ▼]. Liquid crystalline phases, by contrast, exhibit some ordering of the molecules. Depending on the nature of the ordering, liquid crystals can be classified as nematic, smectic, or cholesteric.

In the **nematic liquid crystalline phase** the molecules exhibit one-dimensional ordering. The molecules are aligned along their long axes, but there is no ordering with respect to the ends of the molecules [Figure 12.33(b)]. The arrangement of the molecules is like that of a handful of pencils whose

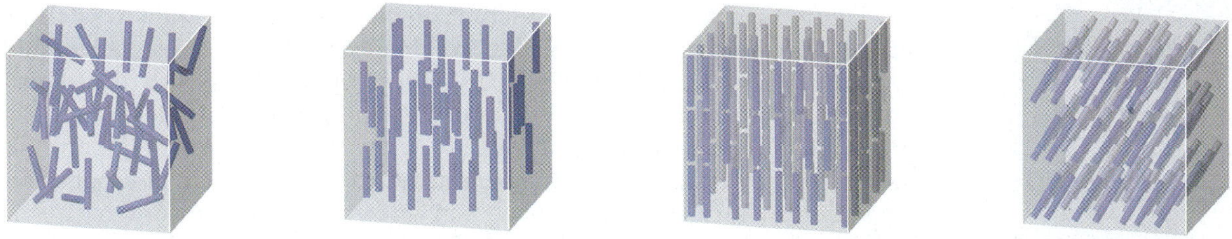

(a) Normal liquid (b) Nematic liquid crystal (c) Smectic A liquid crystal (d) Smectic C liquid crystal

▲ **Figure 12.33 Ordering in liquid crystalline phases and a normal (non–liquid crystalline) liquid.** The ordering arises from intermolecular forces. Smectic A and smectic C phases are both smectic phases; they differ in the angles that the molecules make with respect to the layers.

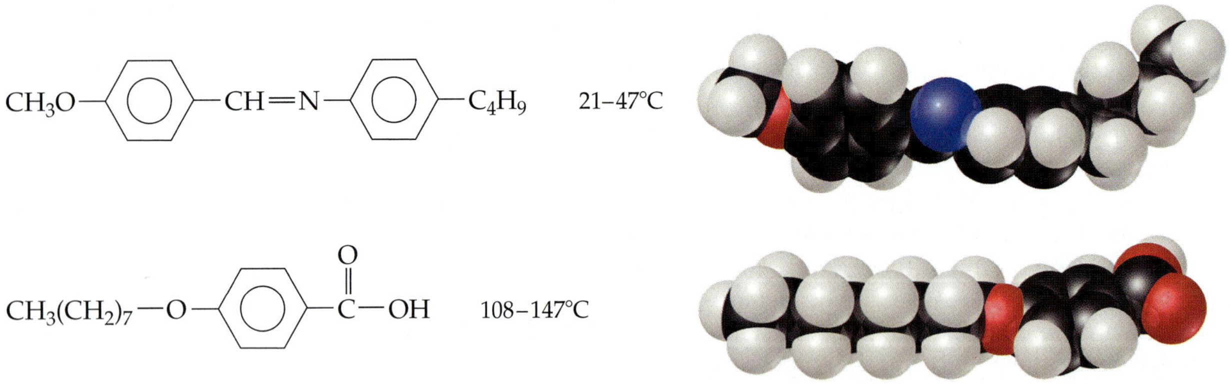

▲ **Figure 12.34 Structures and liquid crystal temperature intervals of two typical liquid crystalline materials.** The temperature interval indicates the temperature range in which the substance exhibits liquid crystalline behavior.

ends are not aligned. In the **smectic liquid crystalline phases** the molecules exhibit two-dimensional ordering; the molecules are aligned along their long axes and in layers [Figure 12.33(c, d)].

Two molecules that exhibit liquid crystalline phases are shown in Figure 12.34 ▲. These molecules are long compared to their thicknesses. The C=N double bond and the benzene rings add rigidity to the chainlike molecules. The flat benzene rings also help the molecules stack with one another. In addition, many liquid crystalline molecules contain polar groups that give rise to dipole-dipole interactions and promote alignment of the molecules. ∞ (Section 11.2) Thus, the molecules order themselves quite naturally along their long axes. They can, however, rotate around their axes and slide parallel to one another. In smectic phases the intermolecular forces between the molecules (such as London dispersion forces, dipole-dipole attractions, and hydrogen bonding) limit the ability of the molecules to slide past one another.

▶ **Figure 12.35 Ordering in a cholesteric liquid crystal.** (a) The molecules in successive layers are oriented at a characteristic angle with respect to those in adjacent layers to avoid repulsive interactions. (b) The result is a screwlike axis.

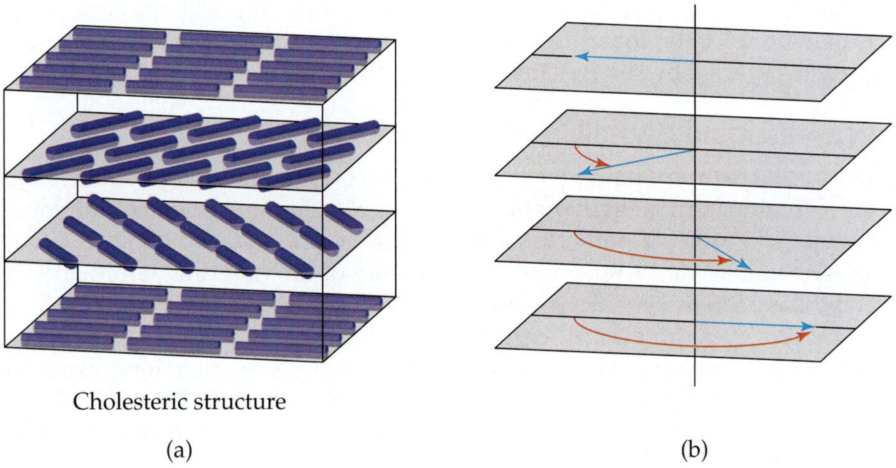

▼ **Figure 12.36 Color change in a cholesteric liquid crystalline material as a function of temperature.**

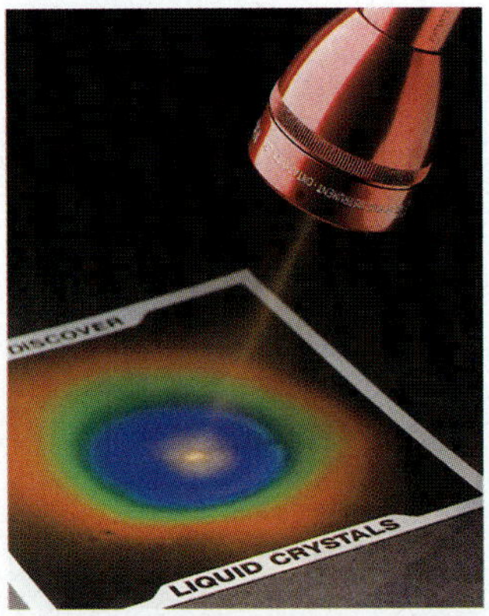

Figure 12.35 ▲ shows the ordering of the **cholesteric liquid crystalline phase**. The molecules are aligned along their long axes as in nematic liquid crystals, but they are arranged in layers with the molecules in each plane twisted slightly in relation to the molecules in the planes above and below. These liquid crystals are so named because many derivatives of cholesterol adopt this structure. The spiral nature of the molecular ordering produces unusual coloring patterns with visible light. Changes in temperature and pressure change the order and hence the color (Figure 12.36 ◀). Cholesteric liquid crystals have been used to monitor temperature changes in situations where conventional methods are not feasible. For example, they can detect hot spots in microelectronic circuits, which may signal the presence of flaws. They can also be fashioned into thermometers for measuring the skin temperature of infants.

SAMPLE EXERCISE 12.3 | Properties of Liquid Crystals

Which of the following substances is most likely to exhibit liquid crystalline behavior?

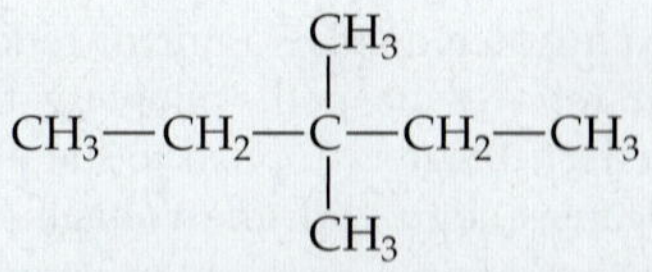

(i)

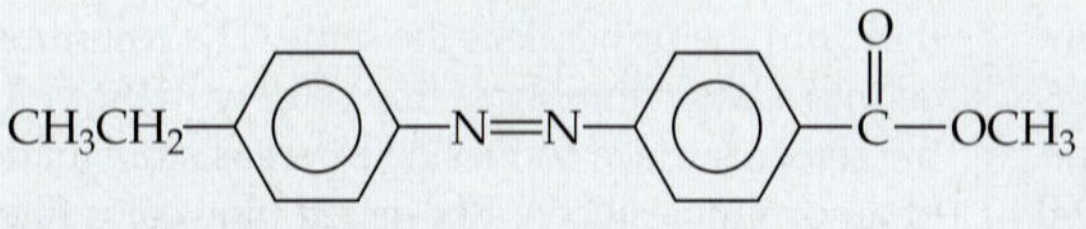

(ii)

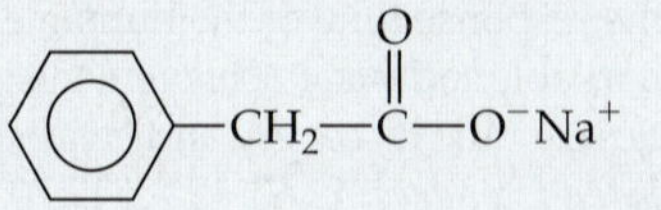

(iii)

Solution

Analyze: We have three molecules of differing molecular structure, and we are asked to determine whether any of them is likely to be a liquid crystalline substance.

Plan: We need to identify the structural features of each case that might induce liquid crystalline behavior.

Solve: Molecule (i) is not likely to be liquid crystalline because it does not have a long axial structure. Molecule (iii) is ionic; the generally high melting points of ionic materials and the absence of a characteristic long axis make it unlikely that this substance will exhibit liquid crystalline behavior. Molecule (ii) possesses the characteristic long axis and the kinds of structural features that are often seen in liquid crystals (Figure 12.34).

PRACTICE EXERCISE

Suggest a reason why the following molecule, decane, does not exhibit liquid crystalline behavior:

$$CH_3CH_2CH_2CH_2CH_2CH_2CH_2CH_2CH_2CH_3$$

Answer: Because rotation can occur about carbon–carbon single bonds, molecules whose backbone consists of C—C single bonds are too flexible; the molecules tend to coil in random ways and thus are not rodlike.

Semiconductor Light-Emitting Diodes

Light-emitting diodes (LEDs) are used as displays in clocks, taillights in some cars, as traffic lights, and in many other places. Most LEDs that are red are made of a mixture of GaAs and GaP. They contain different proportions of P and As to tune the band gap of the material (Table 12.1). Green LEDs are now on the market, and blue ones are beginning to appear in consumer products as well. The mechanism by which LEDs work is the opposite of the mechanism operating in solar energy conversion devices. In an LED, a small voltage is applied to a semiconductor device that has a junction between an n-type semiconductor and a p-type semiconductor. The junction forms an n-p *diode*, in which electrons can flow only one way. When a voltage is applied, electrons in the conduction band from the n-side are forced to combine at the junction with the holes from the p-side to emit light whose photons have energy equal to the band gap.

CHEMISTRY AT WORK | Liquid Crystal Displays

Liquid crystals are widely used in electrically controlled LCD devices in watches, calculators, and computer screens, as illustrated in Figure 12.32. These applications are possible because an applied electrical field changes the orientation of liquid crystal molecules and thus affects the optical properties of the device.

LCDs come in a variety of designs, but the structure illustrated in Figure 12.37 ▼ is typical. A thin layer (5–20 μm) of liquid crystalline material is placed between electrically conducting, transparent glass electrodes. Ordinary light passes through a vertical polarizer that permits light in only the vertical plane to pass. Through a special process the liquid crystal molecules are oriented so that the molecules at the front plate are oriented vertically, those at the bottom plate horizontally. The molecules in between vary in orientation in a regular way, as shown in Figure 12.37. Displays of this kind are called "twisted nematic." The plane of polarization of the light is turned by 90° as it passes through the device, and it is thus in the correct orientation to pass through the horizontal polarizer. In a watch display a mirror reflects the light back, and the light retraces its path, allowing the device to look bright. When a voltage is applied to the plates, the liquid crystalline molecules align with the voltage [Figure 12.37(b)]. The light rays thus are not properly oriented to pass through the horizontal polarizer, and the device appears dark.

Computer screens employ backlighting rather than reflected light, but the principle is the same. The computer screen is divided into a large number of tiny cells, with the voltages at points on the screen surface controlled by transistors made from thin films of amorphous silicon. Red-green-blue color filters are employed to provide full color. The entire display is refreshed at a frequency of about 60 Hz, so the display can change rapidly with respect to the response time of the human eye. Displays of this kind are remarkable technical achievements, based on a combination of basic scientific discovery and creative engineering.

Related Exercises: 12.60, 12.72, 12.76.

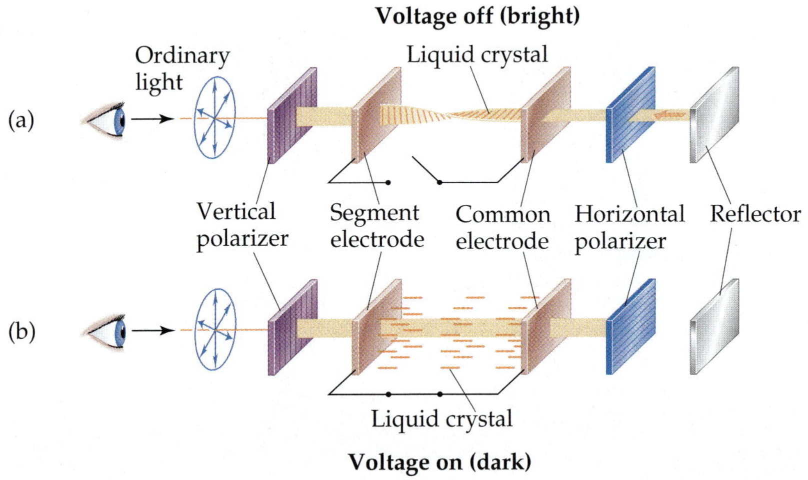

◄ **Figure 12.37 Schematic illustration of the operation of a twisted nematic liquid crystal display (LCD).** (a) Ordinary light, which is unpolarized, passes through the vertical polarizer. The vertically polarized light then passes into the liquid crystalline layer, where the plane of polarization is rotated 90°. It passes through the horizontal polarizer, is reflected, and retraces its path to give a bright display. (b) When a voltage is applied to the segment electrode that covers the small area, the liquid crystal molecules align along the direction of the light path. Thus, the vertically polarized light is not turned 90° and is unable to pass through the horizontal polarizer. The area encompassed by the small transparent segment electrode therefore appears dark. Digital watches typically have such displays.

ANIMATION
Optical Activity

LEDs are manufactured by laying down thin layers of n-type and p-type semiconductors on a transparent substrate (Figure 12.38 ►). The reactants for a GaAs layer, for example, are typically gas-phase $Ga(CH_3)_3$ and AsH_3. In the process known as *chemical vapor deposition* (CVD), $Ga(CH_3)_3$ and AsH_3 in the gas phase are let into a chamber where a hot substrate is located. The gas molecules are deposited on the hot surface (1000°C is typical) and react to form GaAs as a solid thin film on the surface and a gaseous by-product, CH_4.

GIVE IT SOME THOUGHT

Write a balanced chemical equation for the reaction of $Ga(CH_3)_3$ and AsH_3 to make GaAs.

Recently, organic polymers have been used to make an organic version of LEDs, and these are called organic LEDs, or OLEDs. The basic idea is the same as with LEDs: Electrons and holes are forced to combine at a junction region to emit light of an energy that matches the band gap. The types of polymers used

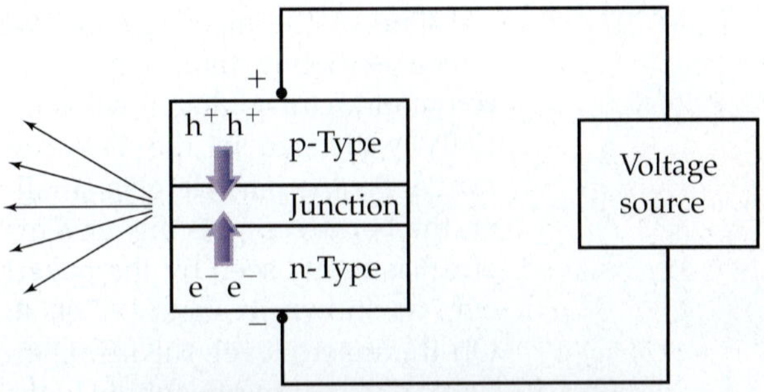

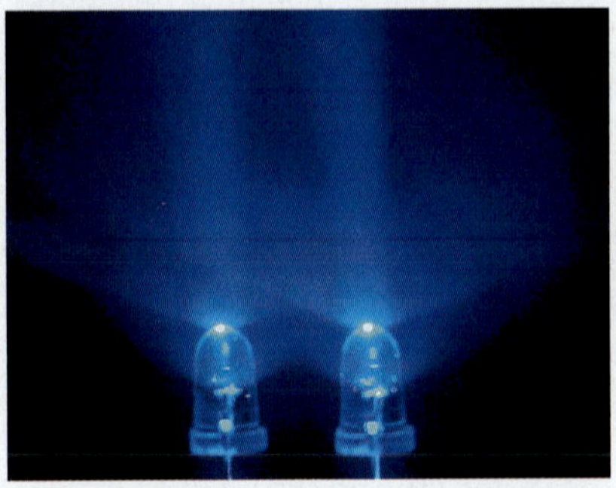

◀ **Figure 12.38 Design of a semiconductor light-emitting diode (LED).** Top: Voltage is applied to a p-n junction. The holes from the p-side are forced to combine with the electrons from the n-side, and light of band gap energy is emitted at the junction. Bottom: Photograph of a blue LED made of indium gallium nitride.

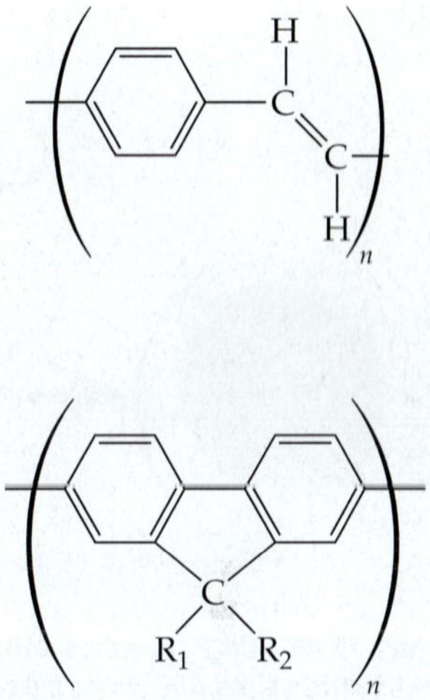

▲ **Figure 12.39 Structures of two polymers commonly used in organic light-emitting diodes (OLEDs), poly (phenylene vinylene) (top) and polyfluorene (bottom).** R_1 and R_2 can be many different organic groups.

in OLEDs are shown in Figure 12.39 ▶. The advantages of OLEDs over traditional LEDs are that OLEDs are lighter and more flexible and they may be brighter and more energy-efficient. Digital cameras that have OLED color displays appeared on the market in 2003.

12.6 | Materials for Nanotechnology

The prefix "nano" means 10^{-9}. ∞ (Section 1.4) When people speak of "nanotechnology," they usually mean making devices that are on the 1–100-nm scale. It turns out that the properties of semiconductors and metals change in this size range. **Nanomaterials**—materials that have dimensions on the 1–100-nm scale—are under intense investigation in the research laboratories of chemists, other scientists, and engineers.

Semiconductors on the Nanoscale

Look back at Figure 12.1. The pictures we have drawn of electron energy levels for molecules and bulk solids are clear: Electrons are in discrete orbitals for individual molecules and in delocalized bands for solids. Let us think about the semiconductor silicon. Atomic silicon has an electron configuration of $1s^22s^22p^63s^23p^2$. ∞ (Section 6.8) We could imagine bonding two silicon atoms together to make an Si_2 molecule. This molecule, then, would have molecular orbitals. ∞ (Section 9.7) We could imagine taking two Si_2 molecules and making an Si_4 molecule, and then making even larger Si_n molecules. At what point does a molecule get so large that it starts behaving as though it has delocalized bands, not localized molecular orbital bonds, for its electrons? For semiconductors, both theory and experiment now tell us that the answer is roughly at 1 to 10 nm (about 10–100 atoms across). The exact number depends on the specific semiconductor material. The equations of quantum mechanics that were used for electrons in atoms can be applied to electrons (and holes) in semiconductors to estimate the size at which unusual effects occur because of the crossover from bonds to bands. Because these effects become important at 1 to 10 nm, semiconductor particles with diameters in this size range are called *quantum dots*.

▲ **Figure 12.40 Cd_3P_2 powders with different particle sizes.** The arrows indicate increasing particle size and a corresponding decrease in the band gap energy, so that the material looks white (smallest nanoparticles), yellow, red, and finally black (the largest nanoparticles).

One of the most spectacular effects is that the band gap of the semiconductor changes substantially with size in the 1–10-nm range. As the particle gets smaller, the band gap gets bigger. This effect is easily seen by the naked eye, as shown in Figure 12.40 ◀. On the macro level, the semiconductor cadmium phosphide looks black because its band gap is small, and it absorbs most wavelengths of visible light. As the particle is made smaller, the material progressively changes color until it looks white! It looks white because now no visible light is absorbed. The band gap is so large that only high-energy ultraviolet light can excite electrons to the conduction band.

Making quantum dots is most easily accomplished using chemical reactions in solution. For example, to make CdS, you can mix $Cd(NO_3)_2$ and Na_2S in water. If you don't do anything else, you will precipitate large crystals of CdS. However, if you add a negatively charged polymer to the water first (such as polyphosphate, $-(OPO_2)_n-$), the Cd^{2+} will associate with the polymer, like tiny "meatballs" in the polymer "spaghetti." When sulfide is added, CdS particles grow, but are kept from forming large crystals by the polymer. A great deal of fine-tuning of reaction conditions is necessary to produce nanocrystals that are of uniform size and shape.

We saw in Section 12.5 that semiconductors can emit light when a voltage is applied. They can also emit light if illuminated by light that has photon energies of the band gap (or higher). An electron from the valence band first absorbs a photon and is promoted to the conduction band. This excited electron can then emit a photon having an energy equal to the band gap energy and fall back down into the hole. In the case of quantum dots, the band gap is tunable with particle size, and thus all the colors of the rainbow can be obtained from just one material (Figure 12.41 ◀). Quantum dots are being explored for applications ranging from electronics to lasers to medical imaging, because they are very bright, very stable, and small enough be taken up by living cells, even after being coated with a biocompatible surface layer.

▲ **Figure 12.41 Semiconductor nanoparticles of CdS of different sizes.** When illuminated with ultraviolet light, these solutions of semiconductor nanoparticles emit light that corresponds to their respective band gap energies. Reprinted with permission from Anal. Chem., October 1, 2002, 74(19), pp. 520A–526A.

Semiconductors do not have to be shrunk down to the nanoscale in all three dimensions to show new properties. They can be laid down in relatively large two-dimensional areas on a substrate, but be only a few nanometers in thickness, or even thinner, to make *quantum wells*. *Quantum wires*, in which the semiconductor wire diameter is only a few nanometers but its length is very long, have also been made by various chemical routes. In both quantum wells and quantum wires, measurements along the nanoscale dimension(s) show quantum behavior, but in the long dimension, the properties seem to be just like those of the bulk material.

Metals on the Nanoscale

Metals also have unusual properties on the 1–100-nm length scale. Fundamentally, this is because the *mean free path* of an electron in a metal at room temperature is typically about 1–100 nm. The mean free path of an electron is the average distance it can move before it bumps into something (another electron, an atom, or the surface of the metal) and "scatters." So, when the particle size of a metal is 100 nm or less, one might expect unusual effects to occur.

In a way, people have known for hundreds of years that metals are different when they are very finely divided. As early as the 1400s, the makers of stained-glass windows knew that gold dispersed in molten glass made the glass a beautiful deep red (Figure 12.42 ▶). Much later, in 1857, Michael Faraday (Chapter 20.5) reported that dispersions of small particles of gold could be made stable and were deeply colored—some of the original colloidal solutions that he made are still in the Royal Institution of Great Britain's Faraday Museum in London (Figure 12.43 ▶). We will examine colloids more closely in Section 13.6. Gold particles less than 20 nm in diameter melt at a far lower temperature than bulk gold, and between 2 and 3 nm gold is no longer a "noble," unreactive metal; in this size range it becomes chemically active.

At nanoscale dimensions, silver has properties analogous to those of gold in its beautiful colors, although it is more reactive than gold. Currently, there is great interest in research laboratories around the world in taking advantage of the unusual optical properties of metal nanoparticles for applications in biomedical imaging and chemical detection.

▲ **Figure 12.42 Stained-glass window from Milan Cathedral in Italy.** The red color is due to gold nanoparticles.
Fototeca della Veneranda Fabbrica del Duomo di Milano.

Carbon Nanotubes

We have already seen that elemental carbon is extraordinarily interesting. In its sp^3-hybridized solid-state form, it is diamond; in its sp^2-hybridized solid-state form, it is graphite; in its molecular form it is C_{60}. (Section 11.8) Soon after the discovery of C_{60}, chemists discovered carbon nanotubes. (Section 1.4) You can think of these as sheets of graphite rolled up and capped at one or both ends by half of a C_{60} molecule (Figure 12.44 ▼). Carbon nanotubes are made in a manner similar to that used to make C_{60}. They can be made in either *multiwall* or *single-walled* forms. Multiwall carbon nanotubes consist of tubes within tubes, nested together; single-walled carbon nanotubes consist of single tubes. Single-walled carbon nanotubes can be 1000 nm long or even longer, but are only about 1 nm in diameter. Depending on how the graphite sheet is rolled up and what its diameter is, carbon nanotubes can behave as either semiconductors or metals.

The fact that carbon nanotubes can be made either semiconducting or metallic without any doping is unique among all solid-state materials, and a great deal of work is ongoing to make carbon-based electronic

▲ **Figure 12.43 The solutions of colloidal gold nanoparticles made by Michael Faraday in the 1850s.** These are on display in the Faraday Museum, London.

3-D MODEL
Fullerene

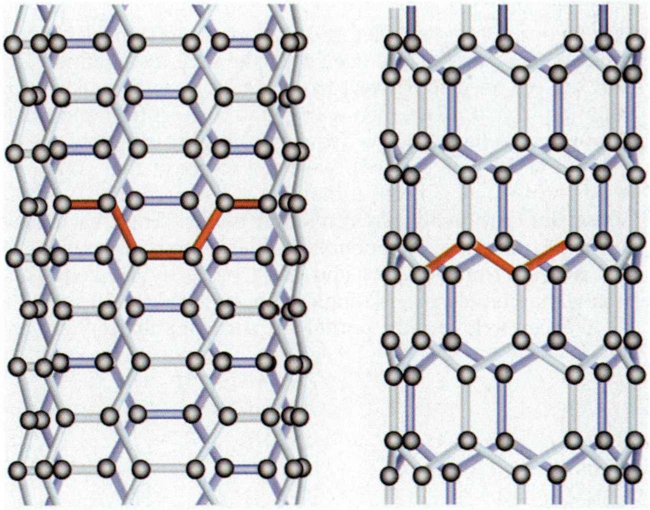

◀ **Figure 12.44 Atomic models of carbon nanotubes.** Left: "Armchair" nanotube, which shows metallic behavior. Right: "Zigzag" nanotube, which can be either semiconducting or metallic, depending on tube diameter.

devices. In 2003 chemists learned how to separate metallic carbon nanotubes from semiconducting ones. Because the basic elements of the transistor are semiconductors and metals, there is great interest in building nanoscale electronic circuits using only carbon nanotubes. The first report of a field effect transistor (FET, related to the MOSFET) using a single carbon nanotube as the channel appeared in 1998.

Carbon nanotubes are also being explored for their mechanical properties. The carbon–carbon bonded framework of the nanotubes means that the imperfections that might appear in a metal nanowire of similar dimensions are nearly absent. Experiments on individual carbon nanotubes suggest that they are stronger than steel if steel were the dimensions of a carbon nanotube. Carbon nanotubes have also been spun into fibers with polymers, adding great strength and toughness to the composite material.

SAMPLE INTEGRATIVE EXERCISE | Putting Concepts Together

A *conducting polymer* is a polymer that can conduct electricity. Some polymers can be made semiconducting; others can be nearly metallic. Polyacetylene is an example of a polymer that is a semiconductor. It can also be doped to increase its conductivity.

Polyacetylene is made from acetylene in a reaction that looks simple but is actually tricky to do:

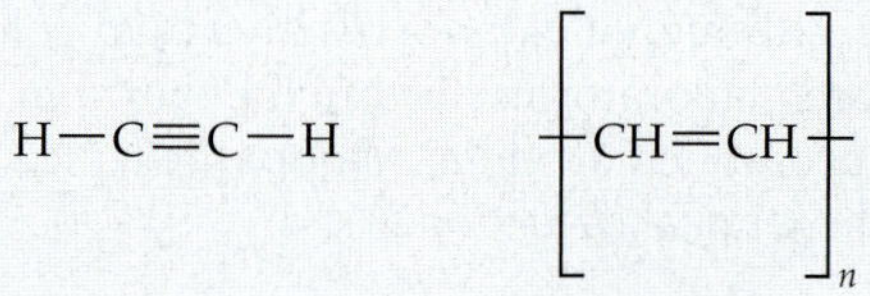

Acetylene Polyacetylene

(a) What is the hybridization of the carbon atoms, and the geometry around those atoms, in acetylene and in polyacetylene?

(b) Write a balanced equation to make polyacetylene from acetylene.

(c) Acetylene is a gas at room temperature and pressure (298 K, 1.00 atm). How many grams of polyacetylene can you make from a 5.00-L vessel of acetylene gas at room temperature and room pressure? Assume acetylene behaves ideally, and that the polymerization reaction occurs with 100% yield.

(d) Using the average bond enthalpies in Table 8.4, predict whether the formation of polyacetylene from acetylene is endothermic or exothermic.

Solution

Analyze: For part (a), we need to recall what we have learned about *sp*, sp^2, and sp^3 hybridization and geometry. (Section 9.5) For part (b), we need to write a balanced equation. For part (c), we need to use the ideal-gas equation. (Section 10.4) For part (d), we need to recall the definitions of endothermic and exothermic, and how bond enthalpies can be used to predict overall reaction enthalpies. (Section 8.8)

Plan: For part (a), we should draw out the chemical structures of the reactant and product. For part (b), we need to make sure the equation is properly balanced. For part (c), we need to convert from liters of gas to moles of gas, using the ideal-gas equation ($PV = nRT$); then we need to convert from moles of acetylene gas to moles of polyacetylene using the answer from part (b); then, we can convert to grams of polyacetylene. For part (d), we need to recall that $\Delta H_{rxn} = \Sigma$(bond enthalpies of bonds broken) $-$ Σ(bond enthalpies of bonds formed).

Solve: **(a)** Carbon always forms four bonds. Thus, each C atom must have a single bond to H and a triple bond to the other C atom in acetylene. As a result, each C atom has two electron domains and must be *sp* hybridized. This *sp* hybridization also means that the H—C—C angles in acetylene are 180°, and the molecule is linear.

We can write out the partial structure of polyacetylene as follows:

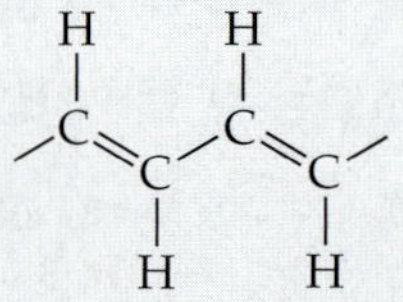

Each carbon is identical but now has three bonding electron domains that surround it. Therefore, the hybridization of each carbon atom is sp^2, and each carbon has local trigonal planar geometry, with 120° angles.

(b) We can write:

$$n\, C_2H_2(g) \longrightarrow \text{-}[CH{=}CH]\text{-}_n$$

Note that all atoms originally present in acetylene end up in the polyacetylene product.

(c) We can use the ideal-gas equation as follows:

$$PV = nRT$$

$$(1.00 \text{ atm})(5.00 \text{ L}) = n(0.08206 \text{ L atm/K mol})(298 \text{ K})$$

$$n = 0.204 \text{ mol}$$

Acetylene has a molar mass of 26.0 g/mol; therefore, the mass of 0.204 mol is

$$(0.204 \text{ mol})(26.0 \text{ g/mol}) = 5.32 \text{ g acetylene}$$

Note that from the answer to part (b), all the atoms in acetylene go into polyacetylene. Due to conservation of mass, then, the mass of polyacetylene produced must also be 5.32 g, if we assume 100% yield.

(d) Let's consider the case for $n = 1$. We note that the reactant side of the equation in part (b) has one C≡C triple bond and two C—H single bonds. The product side of the equation in part (b) has one C=C double bond, one C—C single bond (to link to the adjacent monomer), and two C—H single bonds. Therefore, we are breaking one C≡C triple bond and are forming one C=C double bond and one C—C single bond. Accordingly, the enthalpy change for polyacetylene formation is:

$$\Delta H_{rxn} = (\text{C}{\equiv}\text{C triple bond enthalpy}) - (\text{C}{=}\text{C double bond enthalpy}) - (\text{C}{-}\text{C single bond enthalpy})$$

$$\Delta H_{rxn} = (839 \text{ kJ/mol}) - (614 \text{ kJ/mol}) - (348 \text{ kJ/mol}) = -123 \text{ kJ/mol}$$

Because ΔH is a negative number, the reaction releases heat, and is exothermic.

SUMMARY AND KEY TERMS

Introduction and Section 12.1 In this chapter we consider the classification of materials according to their electronic behavior: **metals**, **semiconductors**, **insulators**, and superconductors. **Ceramics** are usually insulators, but surprisingly can also be superconductors under certain conditions. In extended solids, **bands** rather than bonds are the best way to picture electron energy levels. Semiconductors and insulators have an energy gap known as the **band gap** between the filled **valence band** and empty **conduction band. Doping** semiconductors changes their ability to conduct electricity by orders of magnitude. An n-type semiconductor is one that is doped so that there are excess electrons in the conduction band; a p-type semiconductor is one that is doped so that there are excess **holes** in the valence band.

Superconductivity involves a material that is capable of conducting an electrical current without any apparent resistance when cooled below its **superconducting transition temperature**, T_c. Since discovery of the phenomenon in 1911, the number of known superconducting materials has steadily increased. Until recently, however, all observed T_c values were below about 25 K. An important recent development is the discovery of **high-temperature superconductivity** in certain complex oxides. **Superconducting ceramics** such as $YBa_2Cu_3O_7$ are capable of superconductivity at temperatures higher than that for any nonceramic superconductor. Still other classes of compounds have been shown recently to have relatively high T_c values.

Section 12.2 Materials are examined for their structural properties. Also explored is the synthesis of some materials, both hard and soft. **Polymers** are molecules of high molecular mass formed by joining together large numbers of small molecules, called **monomers**. In an **addition polymerization** reaction, the molecules form new linkages by opening existing bonds. Polyethylene forms, for example, when the carbon–carbon double bonds of ethylene open up. In a **condensation polymerization** reaction, the monomers are joined by splitting out a small molecule from between them. The various kinds of nylon are formed, for example, by removing a water molecule from between an amine and a carboxylic acid. A polymer formed from two different monomers is called a **copolymer**.

Plastics are materials that can be formed into various shapes, usually by the application of heat and pressure. **Thermoplastic** polymers can be reshaped, perhaps through heating, in contrast to **thermosetting plastics**, which are formed into objects through an irreversible chemical process and cannot readily be reshaped. An **elastomer** is a material that exhibits elastic behavior; that is, it returns to its original shape following stretching or bending.

Polymers are largely amorphous, but some materials possess a degree of **crystallinity**. For a given chemical composition, the crystallinity depends on the molecular mass and the degree of branching along the main polymer chain. High-density polyethylene, for example, with little side-chain branching and a high molecular mass, has a higher degree of crystallinity than low-density polyethylene, which has a lower molecular mass and a relatively high degree of branching. Polymer properties are also strongly affected by **cross-linking**, in which short chains of atoms connect the long polymer chains. Rubber is cross-linked by short chains of sulfur atoms in the process called **vulcanization**.

Hard materials such as ceramics are inorganic solids with generally high thermal stability, usually formed through three-dimensional network bonding. The bonding in ceramics may be either covalent or ionic, and ceramics may be crystalline or amorphous. The processing of ceramics generally begins with formation of very small, uniformly sized particles through the **sol-gel process**. The small particles are then compressed and heated at high temperature. They coalesce through a process known as sintering.

Section 12.3 A **biomaterial** is any material that has a biomedical application. Biomaterials are typically in contact with body tissues and fluids. They must be **biocompatible**, which means that they are not toxic, nor do they cause an inflammatory reaction. They must meet physical requirements, such as long-term reliability, strength, and flexibility or hardness, depending on the application. They must also meet chemical requirements of nonreactivity in the biological environment, or of biodegradability. Biomaterials are commonly polymers with special properties matched to the application.

Section 12.4 The reasons for silicon's preeminence in electronics are its abundance, cheapness, and ability to be grown into highly pure crystals, and the ability of its native oxide to grow on it in perfect registry with the atoms underneath. The design of the **transistor**, the heart of integrated circuitry, relies on the junction between n-type and p-type silicon. Semiconductors can absorb sunlight, which includes photon energies larger than its band gap energy, and the resulting electrons promoted to the conduction band can be measured as current. This process is the basis for solar energy conversion devices. Polymers that are semiconducting have been discovered and are being explored as alternatives to silicon for plastic electronics.

Section 12.5 The chemistry behind liquid crystal displays (LCDs) and light-emitting diodes (LEDs) is explored. A **liquid crystal** is a substance that exhibits one or more ordered phases at a temperature above the melting point of the solid. In a **nematic liquid crystalline phase** the molecules are aligned along a common direction, but the ends of the molecules are not lined up. In a **smectic liquid crystalline phase** the ends of the molecules are lined up so that the molecules form sheets. Nematic and smectic liquid crystalline phases are generally composed of molecules with fairly rigid, elongated shapes, with polar groups along the molecules to help retain relative alignment through dipole-dipole interactions. The **cholesteric liquid crystalline phase** is composed of molecules that align as in nematic liquid crystalline phases, but with each molecule twisted with respect to its neighbors, to form a helical structure.

LEDs consist of either inorganic or organic semiconductors that have a junction region between p-type and n-type materials. At the junction, electrons and holes can combine to emit light of band gap energy when a voltage is applied to the device. The process is basically the reverse of that for solar energy conversion.

Section 12.6 **Nanotechnology** is the manipulation of matter, and fabrication of devices, on the 1–100nm scale. The unusual properties that semiconductors, metals, and carbon have on the size scale of 1–100 nm are explored. Quantum dots are semiconductor particles with diameters of 1–10 nm. In this size range the material's band gap energy becomes size-dependent. Metal nanoparticles have different chemical and physical properties in the 1–100-nm size range. Gold, for example, is more reactive and no longer has a golden color. Carbon nanotubes are sheets of graphite rolled up, and they can behave as either semiconductors or metals depending on how the sheet was rolled. Applications of these **nanomaterials** are being developed now for imaging, electronics, and medicine.

VISUALIZING CONCEPTS

12.1 Based on the following band structures, which material is a metal? [Section 12.1]

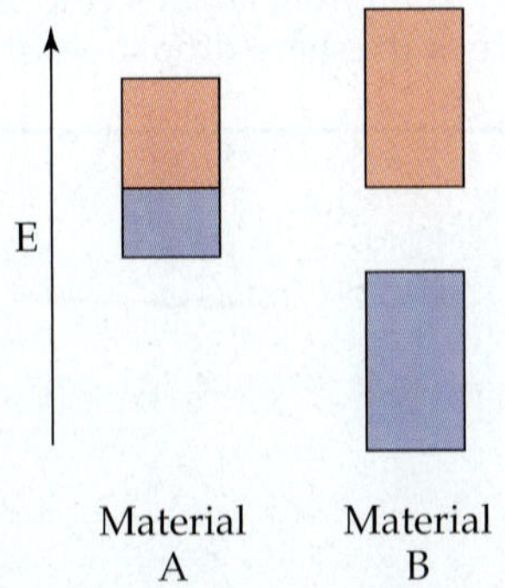

12.2 Both of these materials are semiconductors. Which one is more suitable for solar-energy conversion applications? Why? [Section 12.1 and 12.4]

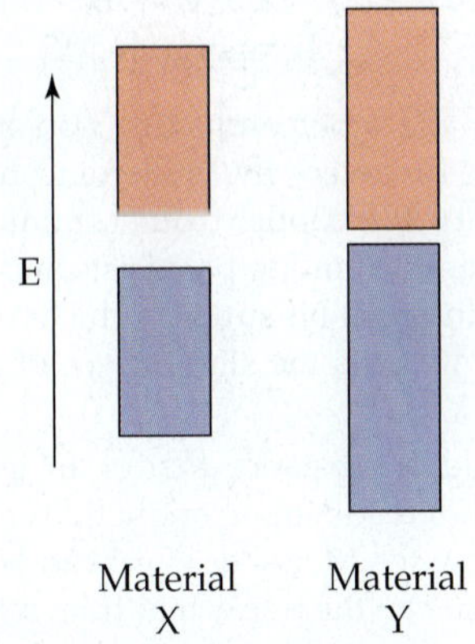

12.3 Shown below are two cartoons of two different polymers. Based on these cartoons, which polymer has stronger intermolecular forces? [Section 12.2]

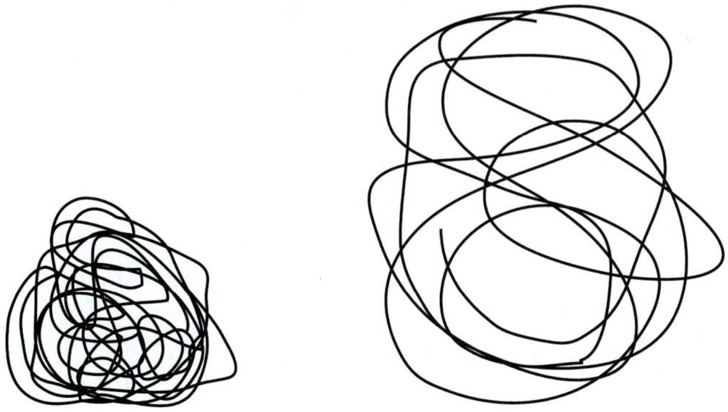

(a) (b)

12.4 Which picture best represents molecules that are in a liquid crystalline phase? [Section 12.5]

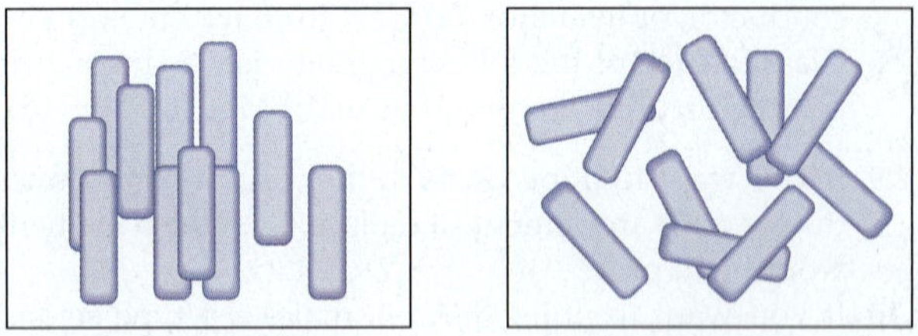

(a) (b)

12.5 Consider the molecules shown below. (a) Which of these would be most likely to form an addition polymer? (b) Which would be most likely to form a condensation polymer? (c) Which would be most likely to form a liquid crystal? [Sections 12.2 and 12.5]

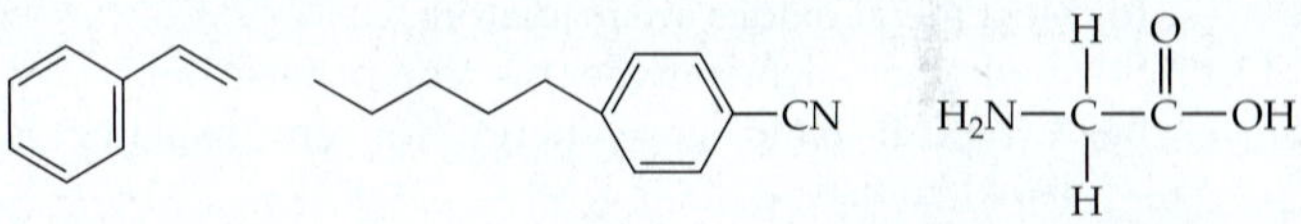

(i) (ii) (iii)

12.6 "Moore's law" is a famous statement made by Gordon Moore, founder of Intel, in 1965. Moore's law says that the complexity of integrated circuits (as judged by the number of components, like transistors, that make up 1 cm^2 of an integrated circuit) doubles every year. Data for Moore's law are shown in the graph below, in which the log of the number of components (per cm^2) is plotted per year:

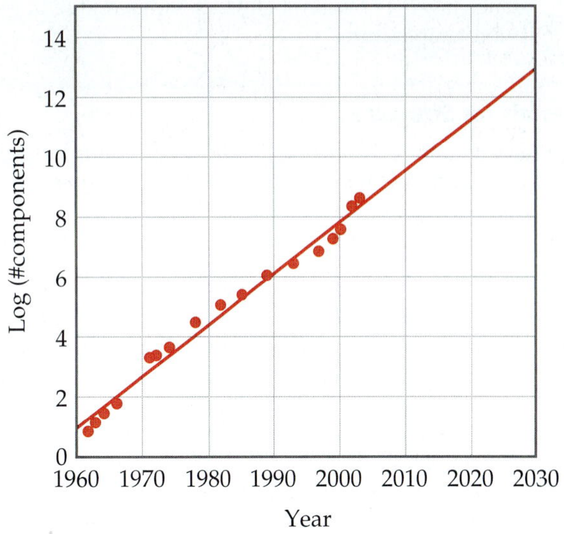

As you can see, Moore had only a few data points to go on in 1965, but his law has held up until now. In 2003, the latest Intel processor was made of 410,000,000 components (and the log of 410,000,000 = 8.6).
It has been estimated that 20 billion components per square centimeter are the most that could be packed into integrated circuits, using current technology. When will this occur if Moore's law holds? [Sections 12.4 and 12.6]

EXERCISES

Classes of Materials

12.7 Classify each of the following materials as metal, semiconductor, or insulator: **(a)** GaN **(b)** B **(c)** ZnO **(d)** Pb

12.8 Classify each of the following materials as metal, semiconductor, or insulator: **(a)** InAs **(b)** MgO **(c)** HgS **(d)** Sn

12.9 If you want to dope GaAs to make an n-type semiconductor with an element to replace Ga, which element(s) would you pick?

12.10 If you want to dope GaAs to make a p-type semiconductor with an element to replace As, which element(s) would you pick?

12.11 State whether each statement is true or false, and why.
(a) Semiconductors have a larger band gap than insulators.
(b) Doping a semiconductor makes it more conductive.
(c) Metals have delocalized electrons.
(d) Most metal oxides are insulators.

12.12 State whether each statement is true or false, and why.
(a) A typical band gap energy for an insulator is 400 kJ/mol.
(b) The conduction band is higher in energy than the valence band.
(c) Electrons can conduct well if they are in a filled valence band.
(d) Holes refer to empty atomic sites in a solid crystal.

12.13 To what does the term *superconductivity* refer? Why might superconductive materials be of value?

12.14 Distinguish between an excellent metallic conductor of electricity (such as silver) and a superconducting substance (such as Nb_3Sn) below its superconducting transition temperature.

12.15 The following graph shows the resistivity (a measure of electrical resistance) of MgB_2 as a function of temperature in the region from about 4 K to 100 K. What is the significance of the sharp drop in resistivity below 40 K?

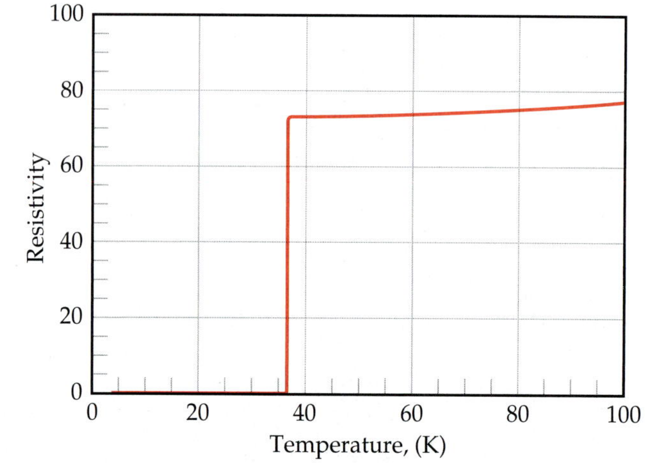

12.16 **(a)** What is the superconducting transition temperature, T_c? **(b)** The discovery by Müller and Bednorz of superconductivity in a copper oxide ceramic at 35 K set off a frantic scramble among physicists and chemists to find materials that exhibit superconductivity at higher temperatures. What is the significance of achieving T_c values above 77 K?

12.17 Why would a substance that is an insulator not be a good candidate for superconductivity?

12.18 Explain how the Meissner effect can be used to levitate trains. What can the tracks and train wheels be made of, and which one would be more likely to be cooled?

Materials for Structure

12.19 The structure of decane is shown in Sample Exercise 12.3. Decane is not considered to be a polymer, whereas polyethylene is. What is the distinction?

12.20 What is a monomer? Give three examples of monomers, taken from the examples given in this chapter.

12.21 An ester is a compound formed by a condensation reaction between a carboxylic acid and an alcohol. Use the index to find the discussion of esters in Chapter 25, and give an example of a reaction forming an ester. How might this kind of reaction be extended to form a polymer (a polyester)?

12.22 Write a chemical equation for formation of a polymer via a condensation reaction from the monomers succinic acid ($HOOCCH_2CH_2COOH$) and ethylenediamine ($H_2NCH_2CH_2NH_2$).

12.23 Draw the structure of the monomer(s) employed to form each of the following polymers shown in Table 12.4: **(a)** polyvinyl chloride, **(b)** nylon 6,6, **(c)** polyethylene terephthalate.

12.24 Write the chemical equation that represents the formation of **(a)** polychloroprene from chloroprene

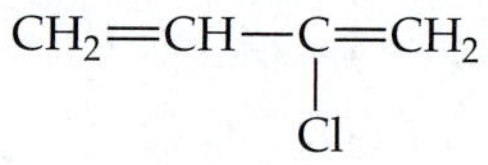

(Polychloroprene is used in highway-pavement seals, expansion joints, conveyor belts, and wire and cable jackets.); **(b)** polyacrylonitrile from acrylonitrile.

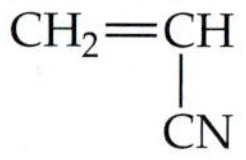

(Polyacrylonitrile is used in home furnishings, craft yarns, clothing, and many other items.)

12.25 The nylon Nomex®, a condensation polymer, has the following structure:

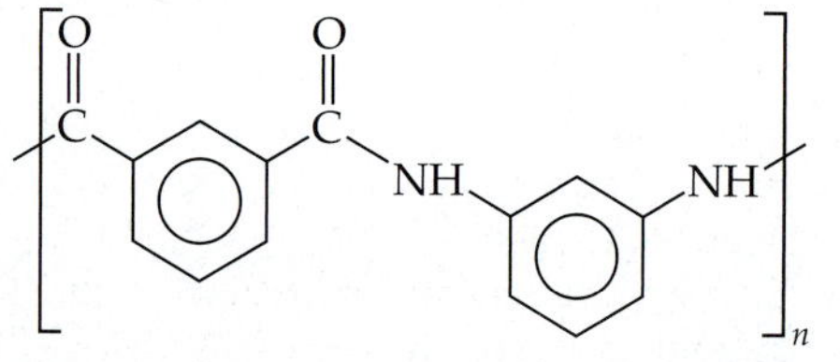

Draw the structures of the two monomers that yield Nomex®.

12.26 Proteins are polymers formed by condensation reactions of amino acids, which have the general structure

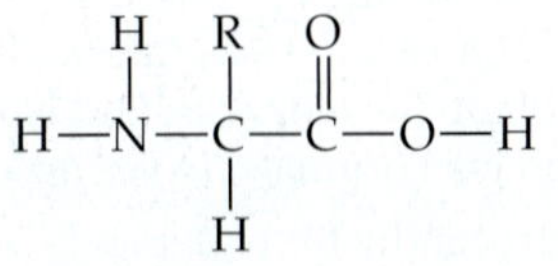

In this structure, R represents —H, —CH_3, or another group of atoms. Draw the general structure for a polyamino acid polymer formed by condensation polymerization of the molecule shown here.

12.27 In addition to the condensation of dicarboxylic acids with diamines, as shown in Equation 12.3, nylons can also be formed by the condensation reactions of aminocarboxylic acids with themselves. Nylon 4, for example, is formed by the polycondensation of 4-aminobutyric acid ($NH_2CH_2CH_2CH_2COOH$). Write a chemical equation to show the formation of nylon 4 from this monomer.

12.28 Kevlar®, a high-performance polymer, has the following structure:

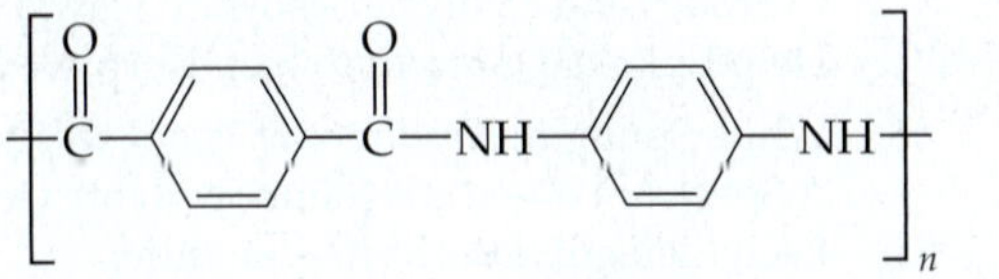

Write the structural formulas for the two substances that are condensed to form this polymer.

12.29 What molecular features make a polymer flexible? Explain how cross-linking affects the chemical and physical properties of the polymer.

12.30 What molecular structural features cause high-density polyethylene to be denser than low-density polyethylene?

12.31 Are high molecular masses and a high degree of crystallinity always desirable properties of a polymer? Explain.

12.32 Briefly describe each of the following: **(a)** elastomer, **(b)** thermoplastic, **(c)** thermosetting plastic, **(d)** plasticizer.

12.33 Metals, such as Al or Fe, and many plastics are recyclable. With the exception of many glasses, such as bottle glass, ceramic materials in general are not recyclable. What characteristics of ceramics make them less readily recyclable?

12.34 You have two solid objects, one formed from a ceramic material and the other from a metal, that look similar to each other. When they are dropped from a height of ten feet onto a concrete surface, one shatters into a variety of pieces, some with sharp edges. The other suffers a big dent and partially splits in two. **(a)** Which one is likely to be the ceramic, and which the metal? **(b)** Why do the two materials differ in their behavior?

12.35 Why is the formation of very small, uniformly sized and shaped particles important for many applications of ceramic materials?

12.36 Describe the general chemical steps in a sol-gel process, beginning with Zr(*s*) and $CH_3CH_2OH(l)$. Indicate whether each step is an oxidation-reduction reaction (refer to Section 4.4), condensation reaction, or other process.

12.37 The hardnesses of several substances according to a scale known as the Knoop value are as follows:

Substance	Knoop Value
Ag	60
$CaCO_3$	135
MgO	370
Soda-lime glass	530
Cr	935
ZrB_2	1550
Al_2O_3	2100
TaC	2000

Which of the materials in this list would you classify as a ceramic? What were your criteria for making this classification? Does classification as a ceramic correlate with Knoop hardness? If you think it does, is hardness alone a sufficient criterion to determine whether a substance is a ceramic? Explain.

12.38 Silicon carbide, SiC, has the three-dimensional structure shown in the figure.

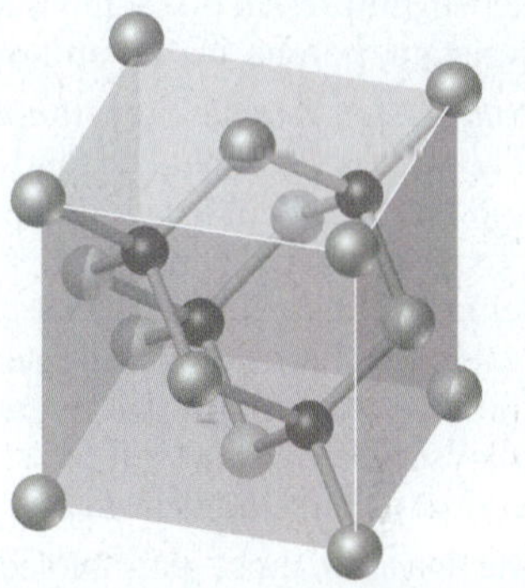

Describe how the bonding and structure of SiC lead to its great thermal stability (to 2700°C) and exceptional hardness.

12.39 Ceramics tend to have high melting points (2000°C or more) and are very hard. Suggest an explanation, based on the structural features of ceramics.

[12.40] The following poem by John Updike is an accurate description of classes of materials. Read the poem, and describe the correlation between what you have learned in this chapter and the verses of the poem.

The Dance of the Solids
by John Updike

All things are Atoms: Earth and Water, Air
And Fire, all, *Democritus* foretold.
Swiss *Paracelsus*, in's alchemic lair,
Saw Sulfur, Salt, and Mercury unfold
Amid Millennial hopes of faking Gold.
Lavoisier dethroned Phlogiston; then
Molecular Analysis made bold
Forays into the gases: Hydrogen
Stood naked in the dazzled sight of Learned Men.

The Solid State, however, kept its grains
Of Microstructure coarsely veiled until
X-ray diffraction pierced the Crystal Planes

That roofed the giddy Dance, the taut Quadrille
Where Silicon and Carbon Atoms will
Link Valencies, four-figured, hand in hand
With common Ions and Rare Earths to fill
The lattices of Matter, Glass or Sand,
With tiny Excitations, quantitatively grand.

The *Metals*, lustrous Monarchs of the Cave,
Are ductile and conductive and opaque
Because each Atom generously gave
Its own Electrons to a mutual Stake,
A Pool that acts as Bond. The Ions take
The stacking shape of Spheres, and slip and flow
When pressed or dented; thusly Metals make
A better Paper Clip than a Window,
Are vulnerable to Shear, and, heated, brightly glow.

Ceramic, muddy Queen of human Arts,
First served as simple Stone. Feldspar supplied
Crude Clay; and Rubies, Porcelain, and Quartz
Came each to light. Aluminum Oxide
Is typical—a Metal is allied
With Oxygen ionically; no free
Electrons form a lubricating tide,
Hence, Empresslike, Ceramics tend to be
Resistant, porous, brittle, and refractory.

Prince Glass, *Ceramic's* son, though crystal-clear,
Is no wise crystalline. The fond Voyeur
And Narcissist alike devoutly peer
Into Disorder, the Disorderer
Being Covalent Bondings that prefer
Prolonged Viscosity and spread loose nets
Photons slip through. The average Polymer
Enjoys a Glassy state, but cools, forgets
To slump, and clouds in closely patterned Minuets.

The *Polymers*, those giant Molecules,
Like Starch and Polyoxymethylene,
Flesh out, as protein serfs and plastic fools,
The Kingdom with Life's Stuff. Our time has seen
The synthesis of Polyisoprene
And many cross-linked Helixes unknown
To *Robert Hooke*; but each primordial Bean
Knew Cellulose by heart: Nature alone
Of Collagen and Apatite compounded Bone.

What happens in these Lattices when Heat
Transports Vibrations through a solid mass?
T = 3Nk is much too neat;
A rigid Crystal's not a fluid Gas.
Debye in 1912 proposed Elas
Tic Waves called phonons which obey *Max Planck's*
Great Quantum Law. Although amorphous Glass,
Umklapp Switchbacks, and Isotopes play pranks
Upon his Formulae, *Debye* deserves warm Thanks.

Electroconductivity depends
On Free Electrons: in Germanium
A touch of Arsenic liberates; in blends
Like Nickel Oxide, Ohms thwart Current. From
Pure Copper threads to wads of Chewing Gum
Resistance varies hugely. Cold and Light
As well as "doping" modify the sum
Of *Fermi* Levels, Ion scatter, site
Proximity, and other factors recondite.

Textbooks and Heaven only are Ideal;
Solidity is an imperfect state.
Within the cracked and dislocated Real
Nonstoichiometric crystals dominate.
Stray Atoms sully and precipitate;
Strange holes, excitons, wander loose; because
Of Dangling Bonds, a chemical Substrate
Corrodes and catalyzes—surface Flaws
Help Epitaxial Growth to fix adsorptive claws.

White Sunlight, *Newton* saw, is not so pure;
A Spectrum bared the Rainbow to his view.
Each Element absorbs its signature:
Go add a negative Electron to
Potassium Chloride; it turns deep blue,
As Chromium incarnadines Sapphire.
Wavelengths, absorbed, are reemitted through
Fluorescence, Phosphorescence, and the higher
Intensities that deadly *Laser Beams* require.

Magnetic Atoms, such as Iron, keep
Unpaired Electrons in their middle shell,
Each one a spinning Magnet that would leap
The *Bloch* Walls whereat antiparallel
Domains converge. Diffuse Material
Becomes Magnetic when another Field
Aligns domains like Seaweed in a swell.
How nicely microscopic forces yield,
In Units growing Visible, the World we wield!

Materials for Medicine

12.41 Neoprene is a polymer of chlorobutadiene.

$$CH_2{=}\underset{\displaystyle H}{\underset{|}{C}}{-}\underset{\displaystyle Cl}{\underset{|}{C}}{=}CH_2$$

The polymer can be used to form flexible tubing that is resistant to chemical attack from a variety of chemical reagents. Suppose it is proposed to use neoprene tubing as a coating for the wires running to the heart from an implanted pacemaker. What questions would you ask to determine whether it might be suitable for such an application?

12.42 On the basis of the structure shown in Table 12.4 for polystyrene and polyurethane, which of these two classes of polymer would you expect to form the most effective interface with biological systems? Explain.

12.43 Patients who receive vascular grafts formed from polymer material such as Dacron® are required to take anticoagulation drugs on a continuing basis to prevent blood clots. Why? What advances in such vascular implants are needed to make this precaution unnecessary?

12.44 Several years ago a biomedical company produced and marketed a new, efficient heart valve implant. It was later withdrawn from the market, however, because patients using it suffered from severe loss of red blood cells. Describe what properties of the valve could have been responsible for this result.

12.45 Skin cells from the body do not differentiate when they are simply placed in a tissue culture medium; that is, they do not organize into the structure of skin, with different layers and different cell types. What is needed to cause such differentiation to occur? Indicate the most important requirements on any material used.

12.46 If you were going to attempt to grow skin cells in a medium that affords an appropriate scaffolding for the cells and you had only two fabrics available, one made from polystyrene and the other from polyethyleneterephthalate (Table 12.4), which would you choose for your experiments? Explain.

Materials for Electronics

12.47 Explain why silicon is the semiconductor of choice for integrated circuits.

12.48 Why is it important for Si crystals to be 99.999999999% pure, as opposed to 99% pure, for silicon chips?

12.49 Look up the diameter of a silicon atom, in Å. The channel length in a Pentium 4 processor chip is 60 nm long; how many silicon atoms does this correspond to?

12.50 Can you think of any applications where you would want electronic circuits to be made of a flexible plastic instead of a relatively hard silicon chip?

12.51 Silicon has a band gap of 1.1 eV at room temperature. What wavelength of light would a photon of this energy correspond to? Draw a vertical line at this wavelength in the figure shown, which shows the light output of the sun as a function of wavelength. Approximately what percent of the solar spectrum can be absorbed by silicon?

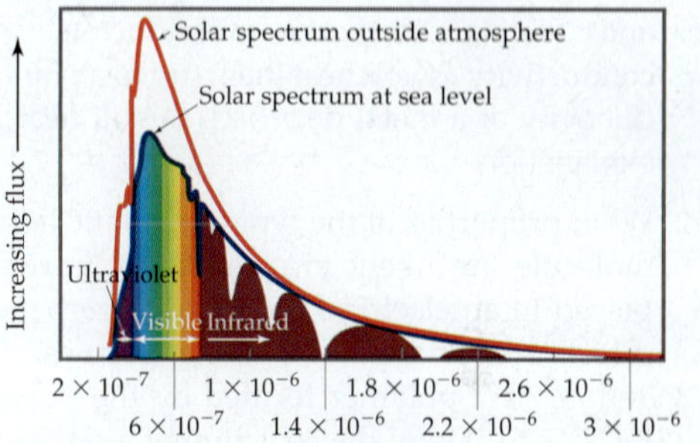

12.52 Repeat Exercise 12.51 for TiO_2 (Table 12.1).

Materials for Optics

12.53 In what ways are a nematic liquid crystalline phase and an ordinary liquid the same, and in what ways do they differ?

12.54 In contrast to ordinary liquids, liquid crystals are said to possess "order." What does this mean?

12.55 Describe what is occurring at the molecular level as a substance passes from the solid to the nematic liquid crystalline to the isotropic (normal) liquid phase upon heating.

12.56 What observations made by Reinitzer on cholesteryl benzoate suggested that this substance possesses a liquid crystalline phase?

12.57 The molecules shown in Figure 12.34 possess polar groups (that is, groupings of atoms that give rise to sizable dipole moments within the molecules). How might the presence of polar groups enhance the tendency toward liquid crystal formation?

12.58 Liquid crystalline phases tend to be more viscous than the isotropic, or normal, liquid phase of the same substance. Why?

12.59 The smectic liquid crystalline phase can be said to be more highly ordered than the nematic. In what sense is this true?

12.60 One of the more effective liquid crystalline substances employed in LCDs is this molecule.

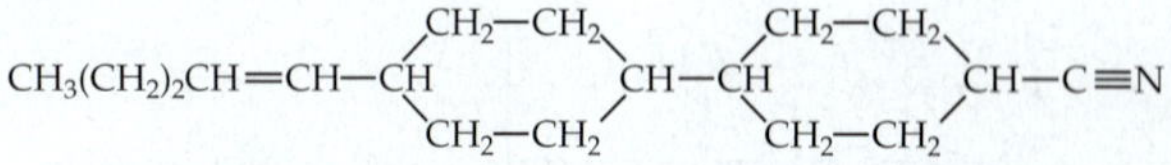

By comparing this structure with the structural formulas and models shown in Figure 12.34, describe the features of the molecule that promote its liquid crystalline behavior.

12.61 Describe how a cholesteric liquid crystal phase differs from a nematic phase.

12.62 It often happens that a substance possessing a smectic liquid crystalline phase just above the melting point passes into a nematic liquid crystalline phase at a higher temperature. Account for this type of behavior in terms of the ideas developed in Chapter 11 relating molecular energies to temperature.

12.63 The semiconductor GaP has a band gap of 2.2 eV. What wavelength of light would be emitted from an LED made from GaP?

12.64 GaAs and GaP make solid solutions that have the same crystal structure as both parent materials, with As and P occupying the same sites. GaP_xAs_{1-x} exists for almost any value of x. If we assume that the band gap varies linearly with composition between $x = 0$ and $x = 1$, estimate the band gap for $GaP_{0.5}As_{0.5}$. What wavelength of light does this correspond to?

Materials for Nanotechnology

12.65 Explain why "bands" may not be the most accurate description of bonding in a solid when the solid has nanoscale dimensions.

12.66 The Nobel Prize–winning physicist Richard Feynman gave a famous speech in 1959, entitled "There's Plenty of Room at the Bottom." Feynman asserted in his speech that if you could write lines as thin as 32 atoms across, then you could write all of the Encyclopedia Britannica on the head of a pin, about $\frac{1}{16}$ of an inch in diameter. Is this true? Look up or estimate the number of pages in the Encyclopedia (over 20 volumes), the area of a page, the average number of words on one page, and the width of the ink lines in the text, and see how close you come to Feynman's calculation.

12.67 True or false:
(a) The band gap of a semiconductor decreases as the particle size decreases, in the 1–10-nm range.
(b) The light that is emitted from a semiconductor, upon external stimulation, is longer and longer in wavelength as the particle size of the semiconductor decreases.

12.68 True or false:
If you want a semiconductor that emits blue light, you could either use a material that has a band gap corresponding to the energy of a blue photon or you could use a material that has a smaller band gap, but make a nanoparticle out of it of the right size.

12.69 Gold is a face-centered cubic structure that has a unit cell edge length of 4.08 Å. There are four gold atoms per unit cell. How many gold atoms are there in a sphere that is 20 nm in diameter? Recall that the volume of a sphere is $\frac{4}{3}\pi r^3$.

12.70 Repeat Exercise 12.69 for a 22-nm sphere.

Additional Exercises

12.71 One major difference in the behavior of semiconductors and metals is that semiconductors increase their conductivity as you heat them (up to a point), but the conductivity of a metal decreases as you heat it. Suggest an explanation.

12.72 What properties of the typical nematic liquid crystalline molecule are likely to cause it to reorient when it is placed in an electrical field that is perpendicular to the direction of orientation of the molecules?

12.73 Teflon® is a polymer formed by the polymerization of $F_2C{=}CF_2$. Draw the structure of a section of this polymer. What type of polymerization reaction is required to form it?

12.74 Classify each of the following as a ceramic, polymer, or liquid crystal.

(a) 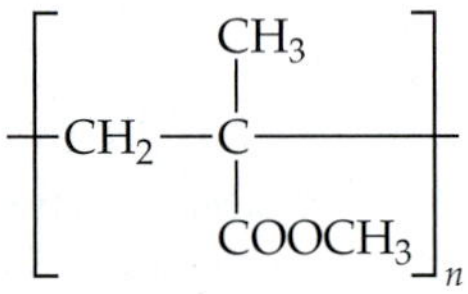

(b) $LiNbO_3$

(c) SiC

(d) 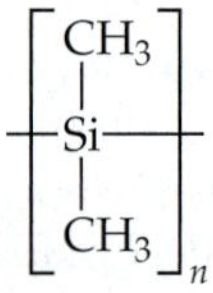

(e) 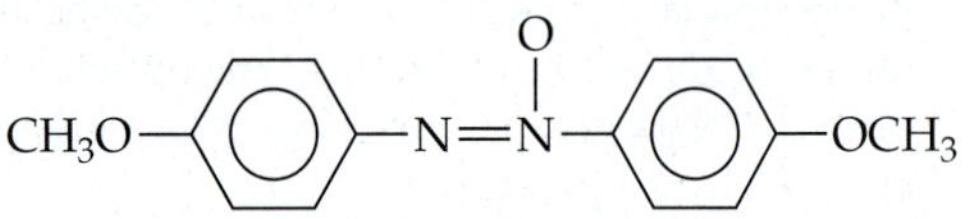

12.75 Ceramics are generally brittle, subject to crack failure, and stable to high temperatures. In contrast, plastics are generally deformable under stress and have limited thermal stability. Discuss these differences in terms of the structures and bonding in the two classes of materials.

12.76 A watch with a liquid crystal display (LCD) does not function properly when it is exposed to low temperatures during a trip to Antarctica. Explain why the LCD might not function well at low temperature.

12.77 The temperature range over which a liquid possesses liquid crystalline behavior is rather narrow (for examples, see Figure 12.34). Why?

12.78 Suppose that a liquid crystalline material such as cholesteryl benzoate is warmed to well above its liquid crystalline range and then cooled. On cooling, the sample unexpectedly remains clear until it reaches a temperature just below the melting point, at which time it solidifies. What explanation can you give for this behavior?

12.79 Hydrogen bonding between polyamide chains plays an important role in determining the properties of a nylon such as nylon 6,6 (Table 12.2). Draw the structural formulas for two adjacent chains of nylon 6,6, and show where hydrogen-bonding interactions could occur between them.

[12.80] A particular liquid crystalline substance has the phase diagram shown in the figure. By analogy with the phase diagram for a non–liquid crystalline substance (Section 11.6), identify the phase present in each area.

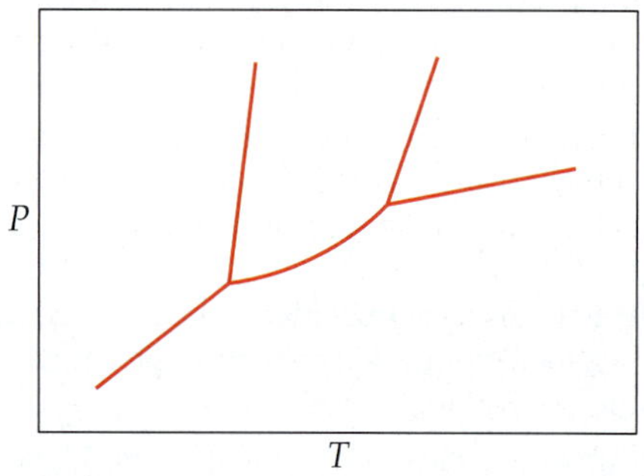

12.81 In fabricating microelectronics circuits, a ceramic conductor such as $TiSi_2$ is employed to connect various regions of a transistor with the outside world, notably aluminum wires. The $TiSi_2$ is deposited as a thin film via chemical vapor deposition, in which $TiCl_4(g)$ and $SiH_4(g)$ are reacted at the Si surface. Write a balanced equation for the reaction, assuming that the other products are H_2 and HCl. Why might $TiSi_2$ behave better as a conducting interconnect on Si than on a metal such as Cu?

12.82 What properties of a material would make it well-suited for solar energy conversion?

Integrative Exercises

[12.83] Employing the bond enthalpy values listed in Table 8.4, estimate the molar enthalpy change occurring upon **(a)** polymerization of ethylene, **(b)** formation of nylon 6,6, **(c)** formation of polyethylene terephthalate (PET).

[12.84] Although polyethylene can twist and turn in random ways, as illustrated in Figure 12.11, the most stable form is a linear one with the carbon backbone oriented as shown in the figure below:

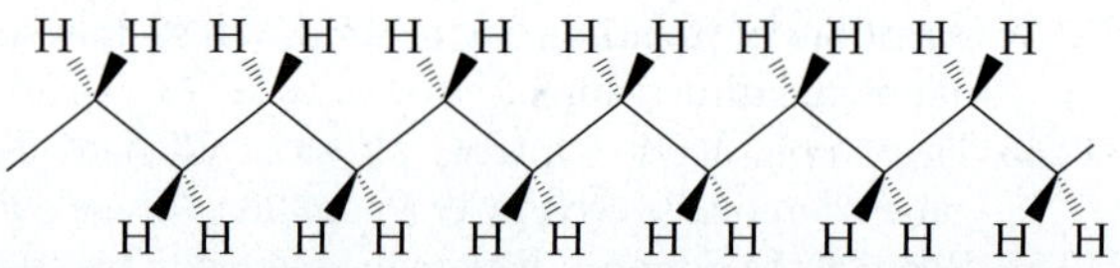

The solid wedges in the figure indicate bonds from carbon that come out of the plane of the page; the dotted wedges indicate bonds that lie behind the plane of the page.

(a) What is the hybridization of orbitals at each carbon atom? What angles do you expect between the bonds?

(b) Now imagine that the polymer is polypropylene rather than polyethylene. Draw structures for polypropylene in which (i) the CH_3 groups all lie on the same side of the plane of the paper (this form is called isotactic polypropylene); (ii) the CH_3 groups lie on alternating sides of the plane (syndiotactic polypropylene); or (iii) the CH_3 groups are randomly distributed on either side (atactic polypropylene). Which of these forms would you expect to have the highest crystallinity and melting point, and which the lowest? Explain in terms of intermolecular interactions and molecular shapes.

(c) Polypropylene fibers have been employed in athletic wear. The product is said to be superior to cotton or polyester clothing in wicking moisture away from the body through the fabric to the outside. Explain the difference between polypropylene and polyester or cotton (which has many —OH groups along the molecular chain), in terms of intermolecular interactions with water.

12.85 In the superconducting ceramic $YBa_2Cu_3O_7$, what is the average oxidation state of copper, assuming that Y and Ba are in their expected oxidation states? Yttrium can be replaced with a rare-earth element such as La, and Ba can be replaced with other similar elements without fundamentally changing the superconducting properties of the material. However, general replacement of copper by any other element leads to a loss of superconductivity. In what respects is the electronic structure of copper different from that of the other two metallic elements in this compound?

[12.86] A sample of the superconducting oxide $HgBa_2Ca_2Cu_3O_{8+x}$ is found to contain 14.99% oxygen by mass. **(a)** Assuming that all other elements are present in the ratios represented by the formula, what is the value of x in the formula? **(b)** Which of the metallic elements in the compound is (are) most likely to have noninteger average charges? Explain your answer. **(c)** Which of the metallic ions in the substance is likely to have the largest ionic radius? Which will have the smallest?

12.87 **(a)** In polyvinyl chloride shown on Table 12.4, which bonds have the lowest average bond enthalpy? **(b)** In the thermal conversion of polyvinyl chloride to diamond, at high pressure, which bonds are most likely to rupture first on heating the material? **(c)** Employing the values of average bond enthalpy in Table 8.4, estimate the overall enthalpy change for converting PVC to diamond.

12.88 Consider *para*-azoxyanisole, which is a nematic liquid crystal in the temperature range of 21°C to 47°C:

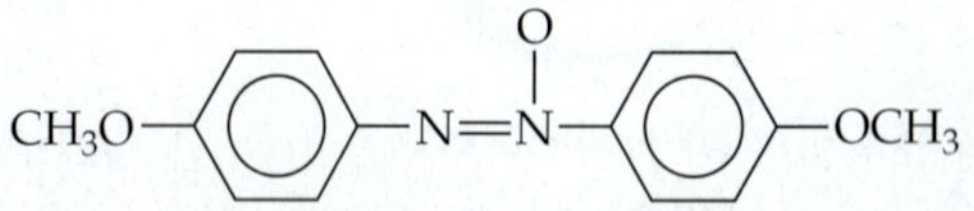

(a) Write out the Lewis structure for this molecule, showing all lone-pair electrons as well as bonds. **(b)** Describe the hybrid orbitals employed by each of the two nitrogens. What bond angles do you anticipate about the nitrogen atom that is bonded to oxygen? **(c)** Replacing one of the $-OCH_3$ groups in *para*-azoxyanisole by a $-CH_2CH_2CH_2CH_3$ group causes the melting point of the substance to drop; the liquid crystal range changes to 19°C to 76°C. Explain why this substitution produces the observed changes in properties. **(d)** How would you expect the density of *para*-azoxyanisole to change upon melting at 117°C? Upon passing from the nematic to the isotropic liquid state at 137°C? Explain.

[12.89] Silicon has a face-centered cubic crystal structure with unit cell edge length of 5.43 Å and four atoms per unit cell. **(a)** How many silicon atoms are there in 1 cm^3 of material? **(b)** Suppose you dope that 1-cm^3 sample of silicon with 1 ppm of phosphorus that will increase the conductivity by a factor of a million. How many milligrams of phosphorus are required?

[12.90] As you go down the periodic table in Group 4A, you see C, Si, Ge, Sn. There are forms of these elements that all have the same structure–the diamond structure. Carbon having the diamond structure is of course diamond; tin can have multiple elemental forms, and the one with the diamond structure is called "gray tin." Si and Ge only have the diamond structure. **(a)** Do you notice a trend in band gap energies as you go down Group 4A for these elements (given the same diamond structure)? Gray tin has a band gap energy of 0.08 eV. **(b)** Does this hold true for other semiconductors, such as the III–V's (such as GaAs) and II–VI's (such as CdS)? Refer to Table 12.1. **(c)** Look up the first ionization energy for each of the Group 4A elements with the diamond structure (use the CRC Handbook of Chemistry and Physics, or another source), and plot them against band gap energy. Suggest an explanation for the trend you observe.

eMEDIA EXERCISES

These exercises make use of the interactive objects available online in OneKey or the Companion Website, and on your Accelerator CD. Access to these resources comes in your MediaPak.

12.91 The **Synthesis of Nylon 6,10** movie (*12.2*) shows the production of a polymer from two organic compounds. The newly formed material is removed from the beaker and wound around a glass rod. If the nylon were not removed, would the same amount be produced? Explain your answer.

12.92 Some of the properties of **Silicon Carbide** (*12.1*) are given in Table 12.2. **(a)** In what ways does silicon carbide resemble diamond? **(b)** In what way is silicon carbide different from most ceramics?

12.93 The superconductor shown by the 3-D model, $\mathbf{YBa_2Cu_3O_7}$ (*12.1*) was one of the first materials shown to exhibit superconductivity at "high" temperature (95 K). What properties of this material prevent its use in many applications where superconductivity would be desirable?

CHAPTER

13

WHAT'S AHEAD

- We begin by considering what happens at a molecular level when a substance dissolves, paying particular attention to the role of *intermolecular forces* in the process. *(Section 13.1)*
- We next examine the changes in *energy* and the changes in how particles are distributed in space that accompany the solution process. *(Section 13.1)*
- We'll see that in *saturated solutions* the dissolved and undissolved solute are in *equilibrium*. *(Section 13.2)*
- The amount of solute in a saturated solution defines its *solubility*, the extent to which a particular solute dissolves in a particular solvent. *(Section 13.2)*
- The nature of the solute and solvent determines the kinds of intermolecular forces between and within them, and thus determines solubility. In addition, pressure affects the solubility of gaseous solutes, and temperature frequently affects solubility in liquid and solid solutions. *(Section 13.3)*
- Because many physical properties of solutions depend on their concentration, we examine several common ways of expressing concentration. *(Section 13.4)*
- Physical properties of solutions that depend only on concentration and not on the identity of the solute are known as *colligative properties*. They include the extent to which the solute lowers the vapor pressure, increases the boiling point, and decreases the freezing point of the solvent. The osmotic pressure of a solution is also a colligative property. *(Section 13.5)*
- We close the chapter by investigating *colloids*, mixtures in which particles larger than molecular size are dispersed in another component. *(Section 13.6)*

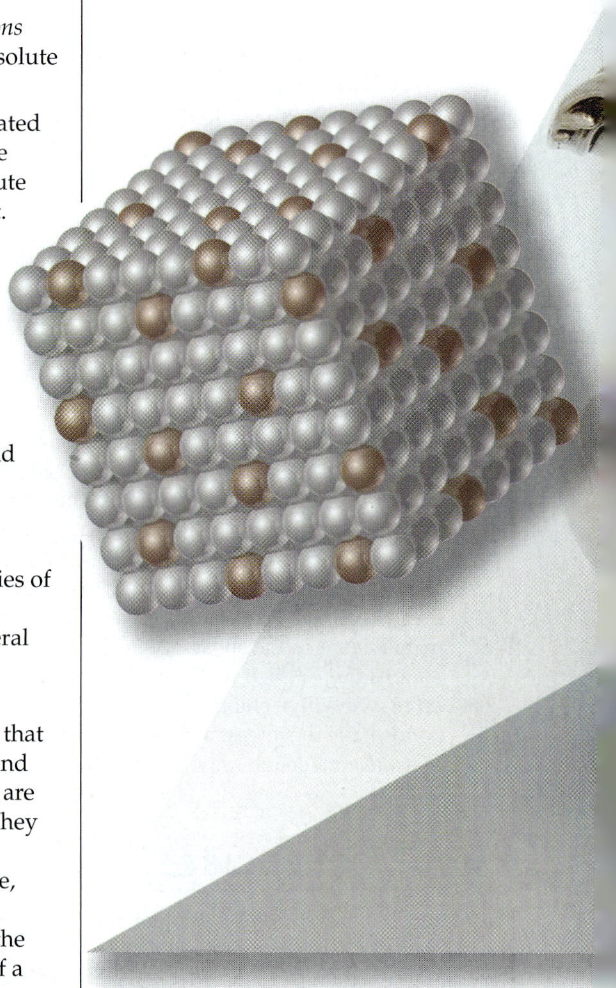

Properties of Solutions

THIS JUG IS MADE OF STERLING SILVER, a solid solution of copper in silver.

WHEN WE THINK OF SOLUTIONS, as described in earlier chapters (Sections 1.2 and 4.1), what comes to mind first are liquids—a solution of salt in water, for example, or a homogeneous mixture of ethanol and water. The beautiful sterling silver art object shown in this chapter's opening photograph is also a solution, however—a homogeneous distribution of about 7% copper in silver. Sterling silver is an example of a solid solution, and other examples of solutions abound in the world around us. To name just three, the air we breathe is a solution of several gases, brass is a solid solution of zinc in copper, and the fluids that run through our bodies are solutions carrying a variety of essential nutrients, salts, and other materials.

Solutions may be gases, liquids, or solids (Table 13.1 ▶). Each of the substances in a solution is called a *component* of the solution. As we saw in Chapter 4, the *solvent* is normally the component present in greatest amount. Other components are called *solutes*. Because liquid solutions are the most common, we will focus our attention on them in this chapter. Our primary goal is to examine the physical properties of solutions, comparing them with the properties of their components. We will be particularly concerned with aqueous solutions of ionic substances because of their central importance in chemistry and in our daily lives.

TABLE 13.1 Examples of Solutions

State of Solution	State of Solvent	State of Solute	Example
Gas	Gas	Gas	Air
Liquid	Liquid	Gas	Oxygen in water
Liquid	Liquid	Liquid	Alcohol in water
Liquid	Liquid	Solid	Salt in water
Solid	Solid	Gas	Hydrogen in palladium
Solid	Solid	Liquid	Mercury in silver
Solid	Solid	Solid	Silver in gold

13.1 | The Solution Process

A solution is formed when one substance disperses uniformly throughout another. With the exception of gas mixtures, all solutions involve substances in a condensed phase. We learned in Chapter 11 that the molecules or ions of substances in the liquid and solid states experience intermolecular attractive forces that hold them together. Intermolecular forces also operate between solute particles and solvent molecules.

ANIMATION
Dissolution of NaCl in Water

Any of the various kinds of intermolecular forces that we discussed in Chapter 11 can operate between solute and solvent particles in a solution. Ion-dipole forces, for example, dominate in solutions of ionic substances in water. Dispersion forces, on the other hand, dominate when a nonpolar substance like C_6H_{14} dissolves in another nonpolar one like CCl_4. Indeed, a major factor determining whether a solution forms is the relative strengths of intermolecular forces between and among the solute and solvent particles.

Solutions form when the attractive forces between solute and solvent particles are comparable in magnitude with those that exist between the solute particles themselves or between the solvent particles themselves. For example, the ionic substance NaCl dissolves readily in water because the attractive interactions between the ions and the polar H_2O molecules overcome the lattice energy of NaCl(*s*). (Section 4.1) Let's examine this solution process more closely, paying attention to these attractive forces.

When NaCl is added to water (Figure 13.1 ▼), the water molecules orient themselves on the surface of the NaCl crystals. The positive end of the water dipole is oriented toward the Cl^- ions, and the negative end of the water dipole is oriented toward the Na^+ ions. The ion-dipole attractions between the ions and water molecules are sufficiently strong to pull the ions from their positions in the crystal.

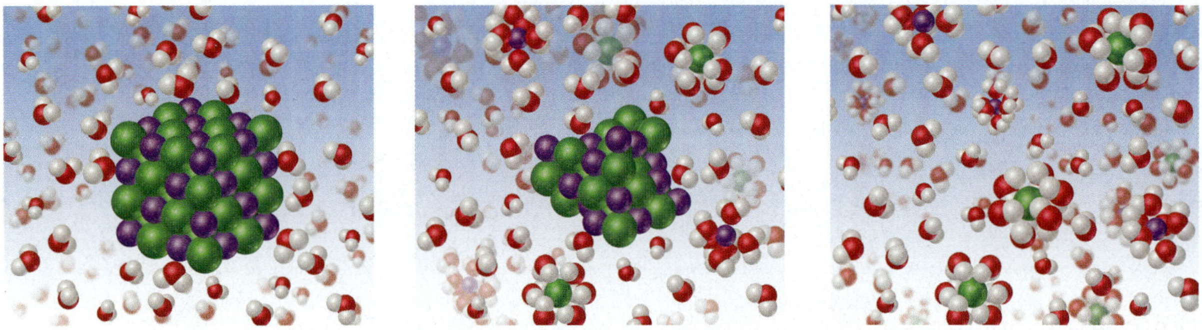

(a) (b) (c)

▲ **Figure 13.1 Dissolution of an ionic solid in water.** (a) A crystal of the ionic solid is hydrated by water molecules, with the oxygen atoms of the water molecules oriented toward the cations (purple) and the hydrogens oriented toward the anions (green). (b, c) As the solid dissolves, the individual ions are removed from the solid surface and become completely separate hydrated species in solution.

Once separated from the crystal, the Na^+ and Cl^- ions are surrounded by water molecules, as shown in Figure 13.1(b and c) and Figure 13.2 ▶. We learned in Section 4.1 that interactions like this between solute and solvent molecules are known as **solvation**. When the solvent is water, the interactions are also referred to as **hydration**.

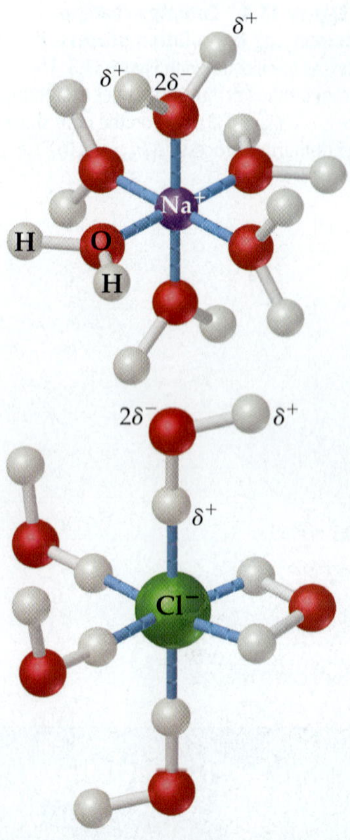

▲ **Figure 13.2 Hydrated Na^+ and Cl^- ions.** The negative ends of the water dipoles point toward the positive ion, and the positive ends point toward the negative ion.

ACTIVITY
Enthalpy of Solution

Energy Changes and Solution Formation

Sodium chloride dissolves in water because the water molecules have a sufficient attraction for the Na^+ and Cl^- ions to overcome the attraction of these two ions for one another in the crystal. To form an aqueous solution of NaCl, water molecules must also separate from one another to form spaces in the solvent that will be occupied by the Na^+ and Cl^- ions. Thus, we can think of the overall energetics of solution formation as having three components, illustrated schematically in Figure 13.3 ▼. The overall enthalpy change in forming a solution, ΔH_{soln}, is the sum of three terms:

$$\Delta H_{soln} = \Delta H_1 + \Delta H_2 + \Delta H_3 \quad [13.1]$$

Figure 13.4 ▶, on the following page, depicts the enthalpy change associated with each of these components. Separation of the solute particles from one another requires an input of energy to overcome their attractive interactions (for example, separating Na^+ and Cl^- ions). The process is therefore endothermic ($\Delta H_1 > 0$). Separation of solvent molecules to accommodate the solute also requires energy ($\Delta H_2 > 0$). The third component arises from the attractive interactions between solute and solvent and is exothermic ($\Delta H_3 < 0$).

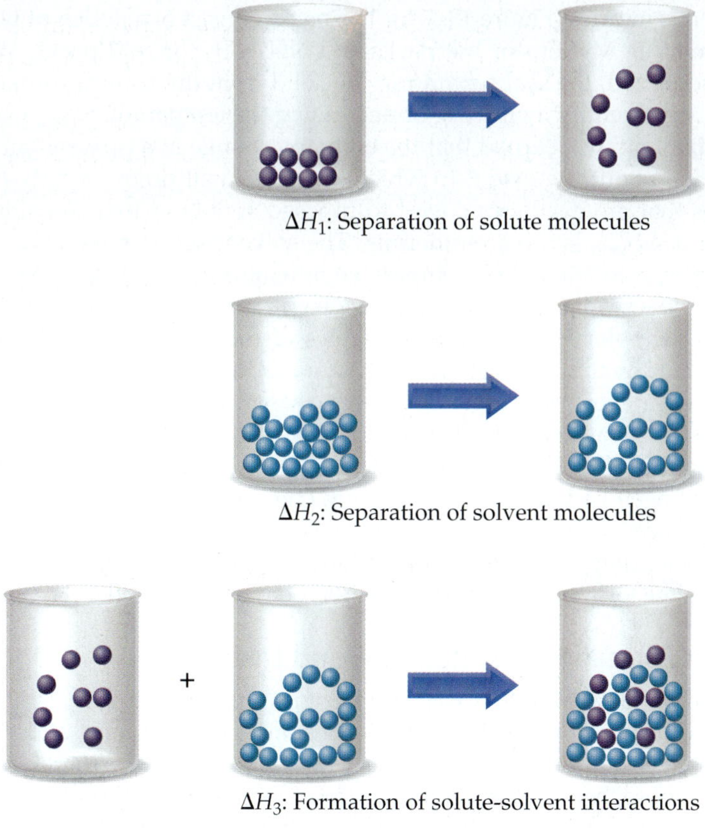

▲ **Figure 13.3 Enthalpy contributions to ΔH_{soln}.** The enthalpy changes ΔH_1 and ΔH_2 represent endothermic processes, requiring an input of energy, whereas ΔH_3 represents an exothermic process.

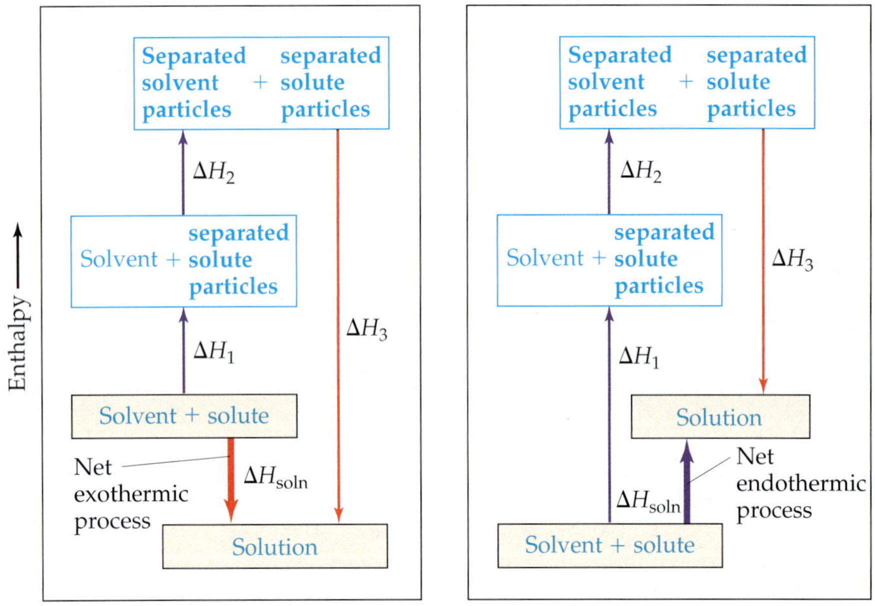

▶ **Figure 13.4 Enthalpy changes accompanying the solution process.** The three processes are illustrated in Figure 13.3. The diagram on the left illustrates a net exothermic process ($\Delta H_{soln} < 0$); that on the right shows a net endothermic process ($\Delta H_{soln} > 0$).

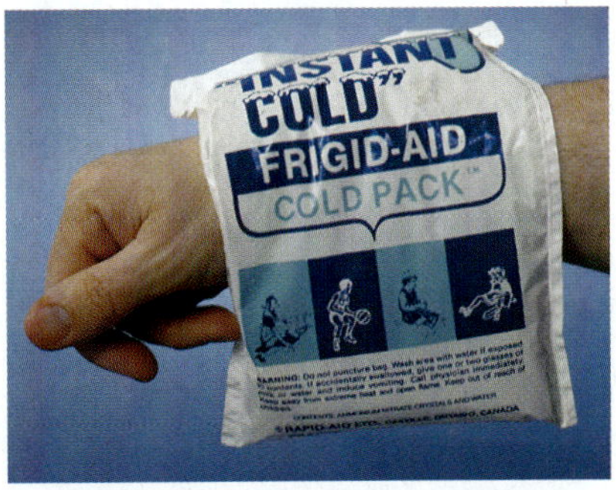

▲ **Figure 13.5 Endothermic dissolution.** Ammonium nitrate instant ice packs are often used to treat athletic injuries. To activate the pack, the container is kneaded, and the kneading breaks an interior seal separating solid NH_4NO_3 from water. The heat of solution of NH_4NO_3 is positive, so the temperature of the solution decreases.

As shown in Figure 13.4, the three enthalpy terms in Equation 13.1 can add together to give either a negative or a positive sum. Thus, the formation of a solution can be either exothermic or endothermic. For example, when magnesium sulfate, $MgSO_4$, is added to water, the resultant solution gets quite warm: $\Delta H_{soln} = -91.2$ kJ/mol. In contrast, the dissolution of ammonium nitrate (NH_4NO_3) is endothermic: $\Delta H_{soln} = 26.4$ kJ/mol. These particular substances have been used to make the instant heat packs and ice packs that are used to treat athletic injuries (Figure 13.5 ◀). The packs consist of a pouch of water and a dry chemical, $MgSO_4$ for hot packs and NH_4NO_3 for cold packs. When the pack is squeezed, the seal separating the solid from the water is broken and a solution forms, either increasing or decreasing the temperature.

In Chapter 5 we learned that the enthalpy change in a process can provide information about the extent to which a process will occur. ∞ (Section 5.4) Processes that are exothermic tend to proceed spontaneously. A solution will not form if ΔH_{soln} is too endothermic. The solvent-solute interaction must be strong enough to make ΔH_3 comparable in magnitude to $\Delta H_1 + \Delta H_2$. This is why ionic solutes like NaCl do not dissolve in nonpolar liquids such as gasoline. The nonpolar hydrocarbon molecules of the gasoline would experience only weak attractive interactions with the ions, and these interactions would not compensate for the energies required to separate the ions from one another.

By similar reasoning, a polar liquid such as water does not form solutions with a nonpolar liquid such as octane (C_8H_{18}). The water molecules experience strong hydrogen-bonding interactions with one another. ∞ (Section 11.2) These attractive forces must be overcome to disperse the water molecules throughout the nonpolar liquid. The energy required to separate the H_2O molecules is not recovered in the form of attractive interactions between H_2O and C_8H_{18} molecules.

Solution Formation, Spontaneity, and Disorder

When carbon tetrachloride (CCl_4) and hexane (C_6H_{14}) are mixed, they dissolve in one another in all proportions. Both substances are nonpolar, and they have similar boiling points (77°C for CCl_4 and 69°C for C_6H_{14}). It is therefore reasonable to suppose that the magnitudes of the attractive forces (London dispersion forces) among molecules in the two substances and in their solution are comparable. When the two are mixed, dissolving occurs spontaneously; that is, it occurs without any extra input of energy from outside the system. Two distinct factors are involved in processes that occur spontaneously. The most obvious is energy; the other is the distribution of each component into a larger volume.

If you let go of a book, it falls to the floor because of gravity. At its initial height, it has a higher potential energy than it has when it is on the floor. Unless it is restrained, the book falls; and as it does, potential energy is converted into kinetic energy. When the book strikes the floor, the kinetic energy is converted largely into heat energy, which is dispersed throughout the surroundings. The book has lost energy to its surroundings in this process. This fact leads us to the first basic principle identifying spontaneous processes and the direction they take: *Processes in which the energy content of the system decreases tend to occur spontaneously.* Spontaneous processes tend to be exothermic. (Section 5.4, "Strategies in Chemistry: Using Enthalpy as a Guide") Change tends to occur in the direction that leads to a lower energy or lower enthalpy for the system.

Some spontaneous processes, however, do not result in lower energy for a system, and even some endothermic processes occur spontaneously. For example, NH_4NO_3 readily dissolves in water, even though the solution process is endothermic. Processes such as this are characterized by a more dispersed state of one or more components, resulting in an overall increase in the randomness of the system. In this example the densely ordered solid, NH_4NO_3, is dispersed throughout the solution as separated NH_4^+ and NO_3^- ions. The mixing of CCl_4 and C_6H_{14} provides another simple example. Suppose that we could suddenly remove a barrier that separates 500 mL of CCl_4 from 500 mL of C_6H_{14}, as in Figure 13.6(a) ▶. Before the barrier is removed, each liquid occupies a volume of 500 mL. All the CCl_4 molecules are in the 500 mL to the left of the barrier and all the C_6H_{14} molecules are in the 500 mL to the right. When equilibrium has been established after the barrier has been removed, the two liquids together occupy a volume of about 1000 mL. Formation of a homogeneous solution has increased the degree of dispersal, or randomness, because the molecules of each substance are now mixed and distributed in a volume twice as large as that which they individually occupied before mixing. The amount of randomness in the system, sometimes referred to as disorder, is given by a thermodynamic quantity called **entropy**. This example illustrates our second basic principle: *Processes occurring at a constant temperature in which the randomness or dispersal in space (entropy) of the system increases tend to occur spontaneously.*

When molecules of different types are brought together, mixing—hence an increased dispersal—occurs spontaneously unless the molecules are restrained by sufficiently strong intermolecular forces or by physical barriers. Thus, gases spontaneously mix and expand unless restrained by their containers; in this case intermolecular forces are too weak to restrain the molecules. However, because strong bonds hold sodium and chloride ions together, sodium chloride does not spontaneously dissolve in gasoline.

We will discuss spontaneous processes again in Chapter 19. At that time we will consider the balance between the tendencies toward lower enthalpy and toward increased entropy in greater detail. For the moment we need to be aware that the solution process involves two factors: a change in enthalpy and a change in entropy. In most cases *formation of solutions is favored by the increase in entropy that accompanies mixing*. Consequently, a solution will form unless solute-solute or solvent-solvent interactions are too strong relative to the solute-solvent interactions.

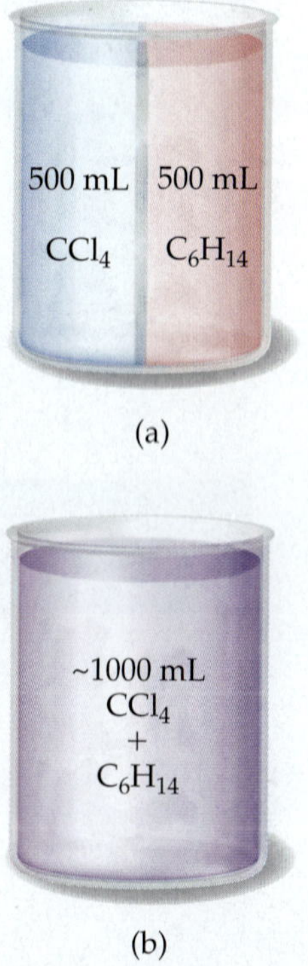

▲ **Figure 13.6 Increasing randomness in a solution process.** A homogeneous solution of CCl_4 and C_6H_{14} forms when a barrier separating the two liquids is removed. Each CCl_4 molecule of the solution in (b) is more dispersed in space than it was in the left compartment in (a), and each C_6H_{14} molecule in (b) is more dispersed than it was in the right compartment in (a).

Solution Formation and Chemical Reactions

In all our discussions of solutions, we must be careful to distinguish the physical process of solution formation from chemical reactions that lead to a solution. For example, nickel metal is dissolved on contact with hydrochloric acid solution because the following chemical reaction occurs:

$$Ni(s) + 2\,HCl(aq) \longrightarrow NiCl_2(aq) + H_2(g) \qquad [13.2]$$

▲ **Figure 13.7 The nickel-acid reaction is *not* a simple dissolution.** (a) Nickel metal and hydrochloric acid. (b) Nickel reacts slowly with hydrochloric acid, forming $NiCl_2(aq)$ and $H_2(g)$. (c) $NiCl_2 \cdot 6\,H_2O$ is obtained when the solution from (b) is evaporated to dryness. Because the residue left after evaporation is chemically different from either reactant, we know what takes place is a chemical reaction rather than merely a solution process.

In this instance the chemical form of the substance being dissolved is changed from Ni to $NiCl_2$. If the solution is evaporated to dryness, $NiCl_2 \cdot 6\,H_2O(s)$, not $Ni(s)$, is recovered (Figure 13.7 ▲). When $NaCl(s)$ is dissolved in water, on the other hand, no chemical reaction occurs. If the solution is evaporated to dryness, NaCl is recovered. Our focus throughout this chapter is on solutions from which the solute can be recovered unchanged from the solution.

GIVE IT SOME THOUGHT

Silver chloride, AgCl, is essentially insoluble in water. Would you expect a significant change in the entropy of the system when 10 g of AgCl is added to 500 mL of water?

A CLOSER LOOK | Hydrates

Frequently, hydrated ions remain in crystalline salts that are obtained by evaporation of water from aqueous solutions. Common examples include $FeCl_3 \cdot 6\,H_2O$ [iron(III) chloride hexahydrate] and $CuSO_4 \cdot 5\,H_2O$ [copper(II) sulfate pentahydrate]. The $FeCl_3 \cdot 6\,H_2O$ consists of $Fe(H_2O)_6{}^{3+}$ and Cl^- ions; the $CuSO_4 \cdot 5\,H_2O$ consists of $Cu(H_2O)_4{}^{2+}$ and $SO_4(H_2O)^{2-}$ ions. Water molecules can also occur in positions in the crystal lattice that are not specifically associated with either a cation or an anion. $BaCl_2 \cdot 2\,H_2O$ (barium chloride dihydrate) is an example. Compounds such as $FeCl_3 \cdot 6\,H_2O$, $CuSO_4 \cdot 5\,H_2O$, and $BaCl_2 \cdot 2\,H_2O$, which contain a salt and water combined in definite proportions, are known as *hydrates*; the water associated with them is called *water of hydration*. Figure 13.8 ▶ shows an example of a hydrate and the corresponding anhydrous (water-free) substance.

Related Exercises: 13.3, 13.102

▲ **Figure 13.8 Hydrates and anhydrous salts.** Hydrated cobalt(II) chloride, $CoCl_2 \cdot 6\,H_2O$ (left), and anhydrous $CoCl_2$ (right).

SAMPLE EXERCISE 13.1 | Assessing Entropy Change

In the process illustrated below, water vapor reacts with excess solid sodium sulfate to form the hydrated form of the salt. The chemical reaction is

$$Na_2SO_4(s) + 10\ H_2O(g) \longrightarrow Na_2SO_4{\cdot}10\ H_2O(s)$$

Essentially all of the water vapor in the closed container is consumed in this reaction. If we consider our system to consist initially of the solid $Na_2SO_4(s)$ and $10\ H_2O(g)$, **(a)** does the system become more or less ordered in this process, and **(b)** does the entropy of the system increase or decrease?

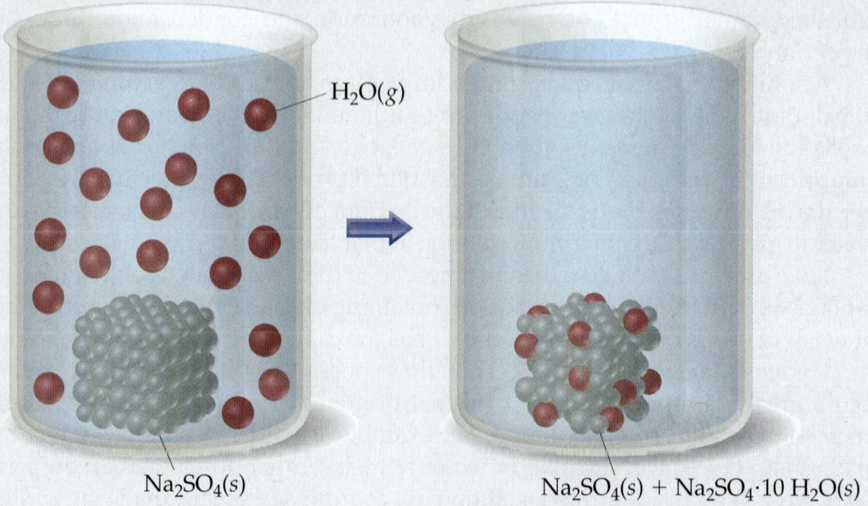

Solution

Analyze: We are asked to determine whether the reaction of water vapor with the solid salt causes the system to become more or less random, or dispersed, and to determine whether the process results in a higher or lower entropy for the system.

Plan: We need to examine the initial and final states and judge whether the process has made the system more or less dispersed. Depending on our answer to that question, we can determine whether the entropy has increased or decreased.

Solve: **(a)** In the course of forming the hydrate of $Na_2SO_4(s)$, the water vapor moves from the vapor state, in which it is dispersed throughout the entire volume of the container, to the solid state, where it is confined to the $Na_2SO_4{\cdot}10\ H_2O(s)$ lattice. This means that the water vapor becomes less dispersed (more ordered). **(b)** When a system becomes more ordered, its entropy is decreased.

PRACTICE EXERCISE

Does the entropy of the system increase or decrease when the stopcock is opened to allow mixing of the two gases in this apparatus?

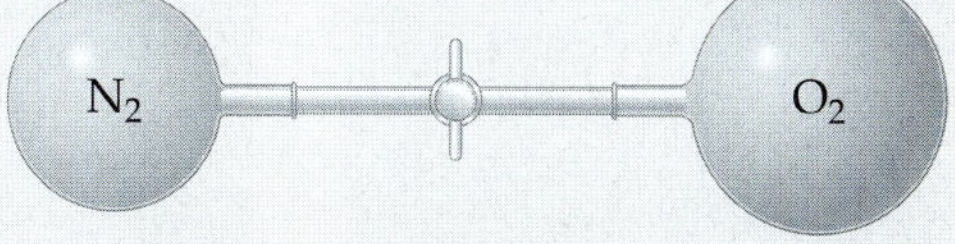

Answer: The entropy increases because each gas eventually becomes dispersed in twice the volume it originally occupied.

13.2 | Saturated Solutions and Solubility

As a solid solute begins to dissolve in a solvent, the concentration of solute particles in solution increases, and so do their chances of colliding with the surface of the solid (Figure 13.9 ▶). Such a collision may result in the solute particle's

▲ **Figure 13.9 Dynamic equilibrium in a saturated solution.** In a solution in which excess ionic solute is present, ions on the surface of the solute are continuously passing into the solution as hydrated species, while hydrated ions from the solution are deposited on the surfaces of the solute. At equilibrium in a saturated solution, the two processes occur at equal rates.

becoming reattached to the solid. This process, which is the opposite of the solution process, is called **crystallization**. Thus, two opposing processes occur in a solution in contact with undissolved solute. This situation is represented in a chemical equation by use of a double arrow:

$$\text{Solute} + \text{solvent} \underset{\text{crystallize}}{\overset{\text{dissolve}}{\rightleftharpoons}} \text{solution} \qquad [13.3]$$

When the rates of these opposing processes become equal, no further net increase in the amount of solute in solution occurs. A dynamic equilibrium is established similar to the one between evaporation and condensation discussed in Section 11.5.

A solution that is in equilibrium with undissolved solute is **saturated**. Additional solute will not dissolve if added to a saturated solution. The amount of solute needed to form a saturated solution in a given quantity of solvent is known as the **solubility** of that solute. For example, the solubility of NaCl in water at 0°C is 35.7 g per 100 mL of water. This is the maximum amount of NaCl that can be dissolved in water to give a stable equilibrium solution at that temperature.

If we dissolve less solute than that needed to form a saturated solution, the solution is **unsaturated**. Thus, a solution containing only 10.0 g of NaCl per 100 mL of water at 0°C is unsaturated because it has the capacity to dissolve more solute.

Under suitable conditions it is sometimes possible to form solutions that contain a greater amount of solute than that needed to form a saturated solution. Such solutions are **supersaturated**. For example, considerably more sodium acetate ($NaC_2H_3O_2$) can dissolve in water at high temperatures than at low temperatures. When a saturated solution of sodium acetate is made at a high temperature and then slowly cooled, all of the solute may remain dissolved even though the solubility decreases as the temperature is reduced. Because the solute in a supersaturated solution is present in a concentration higher than the equilibrium concentration, supersaturated solutions are unstable. Supersaturated solutions result for much the same reason as supercooled liquids (Section 11.4): In order for crystallization to occur, the molecules or ions of solute must arrange themselves properly to form crystals. Addition of a small crystal of the solute (a seed crystal) provides a template for crystallization of the excess solute, leading to a saturated solution in contact with excess solid (Figure 13.10 ▼).

(a) A seed crystal of $NaC_2H_3O_2$ being added to the supersaturated solution.

(b) Excess $NaC_2H_3O_2$ crystallizes from the solution.

(c) The solution arrives at saturation.

▲ **Figure 13.10 Sodium acetate readily forms a supersaturated solution in water.**

GIVE IT SOME THOUGHT

Is a supersaturated solution of sodium acetate a stable equilibrium solution?

13.3 | Factors Affecting Solubility

The extent to which one substance dissolves in another depends on the nature of both the solute and the solvent. It also depends on temperature and, at least for gases, on pressure. Let's consider these factors more closely.

Solute-Solvent Interactions

One factor determining solubility is the natural tendency of substances to mix (the tendency of systems to move toward a more dispersed, or random, state). If this were all that was involved, however, we would expect substances to be completely soluble in one another. This is clearly not the case. So what other factors are involved? As we saw in Section 13.1, the relative forces of attraction among the solute and solvent molecules also play very important roles in the solution process.

Although the tendency toward dispersal and the various interactions among solute and solvent particles are all involved in determining the solubilities, considerable insight can often be gained by focusing on the interaction between the solute and solvent. The data in Table 13.2 ▶ show, for example, that the solubilities of various simple gases in water increase with increasing molecular mass or polarity. The attractive forces between the gas and solvent molecules are mainly of the London dispersion type, which increase with increasing size and mass of the gas molecules. (Section 11.2) Thus, the data indicate that the solubilities of gases in water increase as the attraction between the solute (gas) and solvent (water) increases. In general, when other factors are comparable, *the stronger the attractions between solute and solvent molecules, the greater the solubility.*

TABLE 13.2 Solubilities of Gases in Water at 20°C, with 1 atm Gas Pressure

Gas	Solubility (*M*)
N_2	0.69×10^{-3}
CO	1.04×10^{-3}
O_2	1.38×10^{-3}
Ar	1.50×10^{-3}
Kr	2.79×10^{-3}

As a result of favorable dipole-dipole attractions between solvent molecules and solute molecules, *polar liquids tend to dissolve readily in polar solvents.* Water is not only polar, but also able to form hydrogen bonds. (Section 11.2) Thus, polar molecules, and especially those that can form hydrogen bonds with water molecules, tend to be soluble in water. For example, acetone, a polar molecule whose structural formula is shown in the margin, mixes in all proportions with water. Acetone has a strongly polar C=O bond and pairs of nonbonding electrons on the O atom that can form hydrogen bonds with water.

Acetone

Pairs of liquids such as acetone and water that mix in all proportions are **miscible**, whereas those that do not dissolve in one another are **immiscible**. Gasoline, which is a mixture of hydrocarbons, is immiscible in water. Hydrocarbons are nonpolar substances because of several factors: The C—C bonds are nonpolar, the C—H bonds are nearly nonpolar, and the shapes of the molecules are symmetrical enough to cancel much of the weak C—H bond dipoles. The attraction between the polar water molecules and the nonpolar hydrocarbon molecules is not sufficiently strong to allow the formation of a solution. *Nonpolar liquids tend to be insoluble in polar liquids.* As a result, hexane (C_6H_{14}) does not dissolve in water.

The series of compounds in Table 13.3 ▶ demonstrates that polar liquids tend to dissolve in other polar liquids and nonpolar liquids in nonpolar ones. These organic compounds all contain the OH group attached to a C atom. Organic compounds with this molecular feature are called *alcohols*. The O—H bond is not only polar, but is also able to form hydrogen bonds. For example,

TABLE 13.3 Solubilities of Some Alcohols in Water and in Hexane*

Alcohol	Solubility in H_2O	Solubility in C_6H_{14}
CH_3OH (methanol)	∞	0.12
CH_3CH_2OH (ethanol)	∞	∞
$CH_3CH_2CH_2OH$ (propanol)	∞	∞
$CH_3CH_2CH_2CH_2OH$ (butanol)	0.11	∞
$CH_3CH_2CH_2CH_2CH_2OH$ (pentanol)	0.030	∞
$CH_3CH_2CH_2CH_2CH_2CH_2OH$ (hexanol)	0.0058	∞
$CH_3CH_2CH_2CH_2CH_2CH_2CH_2OH$ (heptanol)	0.0008	∞

*Expressed in mol alcohol/100 g solvent at 20°C. The infinity symbol indicates that the alcohol is completely miscible with the solvent.

CH_3CH_2OH molecules can form hydrogen bonds with water molecules as well as with each other (Figure 13.11 ▼). As a result, the solute-solute, solvent-solvent, and solute-solvent forces are not greatly different within a mixture of CH_3CH_2OH and H_2O. There is no major change in the environments of the molecules as they are mixed. Therefore, the increased dispersal of the two components into a larger combined volume that accompanies mixing plays a significant role in formation of the solution. Ethanol (CH_3CH_2OH), therefore, is completely miscible with water.

The number of carbon atoms in an alcohol affects its solubility in water. As the length of the carbon chain increases, the polar OH group becomes an ever smaller part of the molecule, and the molecule behaves more like a hydrocarbon. The solubility of the alcohol in water decreases correspondingly. On the other hand, the solubility of the alcohol in a nonpolar solvent like hexane (C_6H_{14}) increases as the nonpolar hydrocarbon chain increases in length.

One way to enhance the solubility of a substance in water is to increase the number of polar groups it contains. For example, increasing the number of OH groups along a carbon chain of a solute increases the extent of hydrogen bonding between that solute and water, thereby increasing solubility. Glucose ($C_6H_{12}O_6$) has five OH groups on a six-carbon framework, which makes the molecule very soluble in water (83 g dissolves in 100 mL of water at 17.5°C). The glucose molecule is shown in Figure 13.12 ▶.

Examination of different combinations of solvents and solutes such as those considered in the preceding paragraphs has led to an important generalization: *Substances with similar intermolecular attractive forces tend to be soluble in one another*. This generalization is often simply stated as "*like dissolves like*."

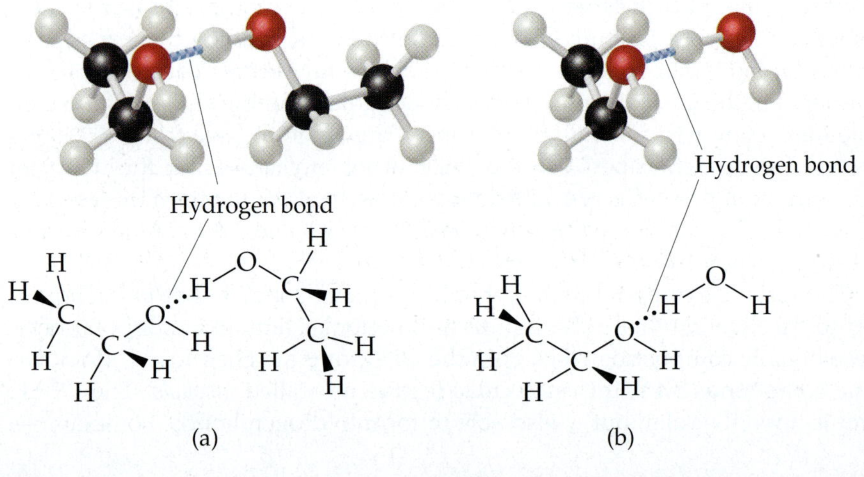

▶ **Figure 13.11 Hydrogen-bonding interactions.** (a) Between two ethanol molecules and (b) between an ethanol molecule and a water molecule.

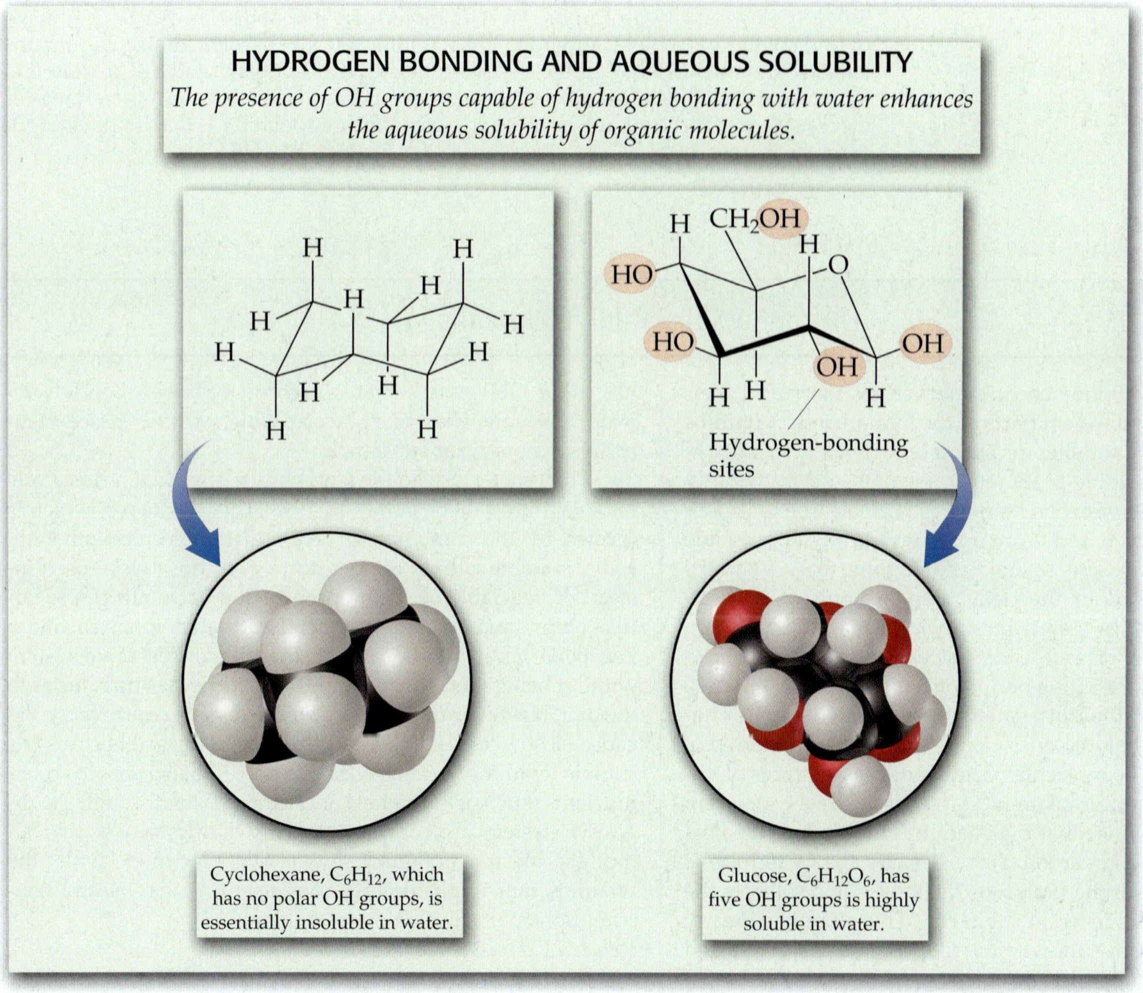

▲ **Figure 13.12 Structure and solubility.**

Nonpolar substances are more likely to be soluble in nonpolar solvents; ionic and polar solutes are more likely to be soluble in polar solvents. Network solids like diamond and quartz are not soluble in either polar or nonpolar solvents because of the strong bonding forces within the solid.

GIVE IT SOME THOUGHT

Suppose the hydrogens on the OH groups in glucose (Figure 13.12) were replaced with methyl groups, CH_3. Would you expect the solubility in water of the resulting molecule to be higher than, lower than, or about the same as the solubility of glucose?

SAMPLE EXERCISE 13.2 | Predicting Solubility Patterns

Predict whether each of the following substances is more likely to dissolve in carbon tetrachloride (CCl_4) or in water: C_7H_{16}, Na_2SO_4, HCl, and I_2.

Solution

Analyze: We are given two solvents, one that is nonpolar (CCl_4) and the other that is polar (H_2O), and asked to determine which will be the best solvent for each solute listed.

Plan: By examining the formulas of the solutes, we can predict whether they are ionic or molecular. For those that are molecular, we can predict whether they are polar or nonpolar. We can then apply the idea that the nonpolar solvent will be best for the nonpolar solutes, whereas the polar solvent will be best for the ionic and polar solutes.

3-D MODELS
Cyclohexane, Glucose

Solve: C_7H_{16} is a hydrocarbon, so it is molecular and nonpolar. Na_2SO_4, a compound containing a metal and nonmetals, is ionic; HCl, a diatomic molecule containing two nonmetals that differ in electronegativity, is polar; and I_2, a diatomic molecule with atoms of equal electronegativity, is nonpolar. We would therefore predict that C_7H_{16} and I_2 would be more soluble in the nonpolar CCl_4 than in polar H_2O, whereas water would be the better solvent for Na_2SO_4 and HCl.

CHEMISTRY AND LIFE | Fat- and Water-Soluble Vitamins

Vitamins have unique chemical structures that affect their solubilities in different parts of the human body. Vitamins B and C are water soluble, for example, whereas vitamins A, D, E, and K are soluble in nonpolar solvents and in the fatty tissue of the body (which is nonpolar). Because of their water solubility, vitamins B and C are not stored to any appreciable extent in the body, and so foods containing these vitamins should be included in the daily diet. In contrast, the fat-soluble vitamins are stored in sufficient quantities to keep vitamin-deficiency diseases from appearing even after a person has subsisted for a long period on a vitamin-deficient diet.

The different solubility patterns of the water-soluble vitamins and the fat-soluble ones can be rationalized in terms of the structures of the molecules. The chemical structures of vitamin A (retinol) and of vitamin C (ascorbic acid) are shown in Figure 13.13 ▼. Notice that the vitamin A molecule is an alcohol with a very long carbon chain. Because the OH group is such a small part of the molecule, the molecule resembles the long-chain alcohols listed in Table 13.3. This vitamin is nearly nonpolar. In contrast, the vitamin C molecule is smaller and has more OH groups that can form hydrogen bonds with water. It is somewhat like glucose, which was discussed earlier. It is a more polar substance.

The Procter & Gamble Company introduced a no-calorie fat substitute called *olestra*® in 1998. This substance, which is formed by the combination of a sugar molecule with fatty acids, is stable at high temperatures and thus can be used instead of vegetable oil in the preparation of potato chips, tortilla chips, and related products. Although it tastes like a vegetable oil, it passes through the human digestive system without being metabolized and thus contributes no calories to the diet. However, its use has generated some controversy. Because olestra consists of large fatlike molecules, it absorbs fat-soluble vitamins (such as A, D, E, and K). It also absorbs other nutrients (such as carotenes) and carries them through the digestive system and out of the body. Critics worry that although olestra-containing foods are fortified with the vitamins that might be lost, long-term, regular olestra consumption could cause dietary problems.

Related Exercises: 13.6, 13.42

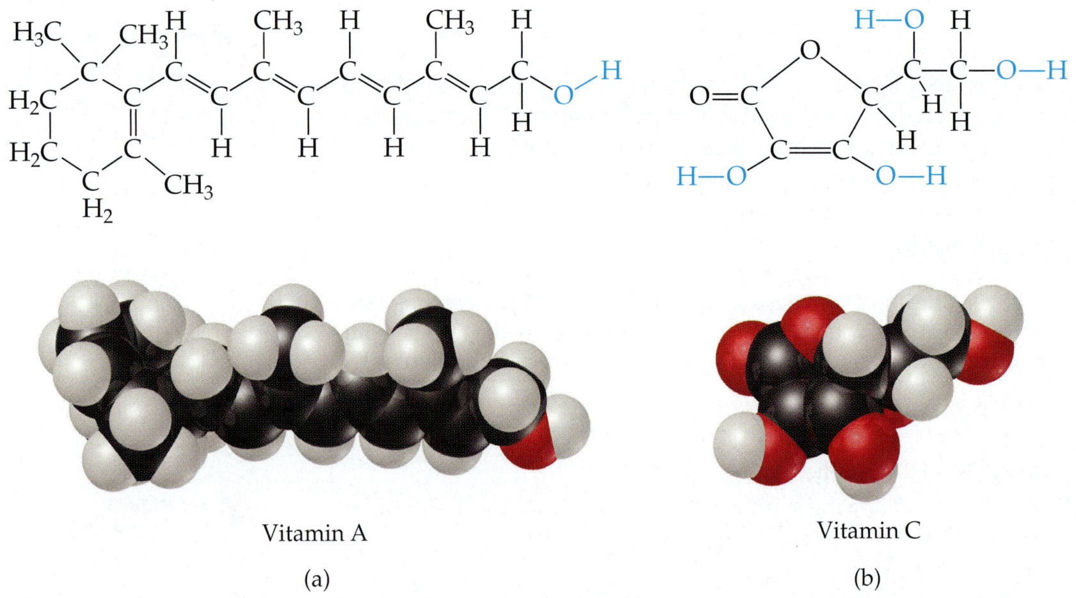

▲ **Figure 13.13 Vitamins A and C.** (a) The molecular structure of vitamin A, a fat-soluble vitamin. The molecule is composed largely of carbon–carbon and carbon–hydrogen bonds, so it is nearly nonpolar. (b) The molecular structure of vitamin C, a water-soluble vitamin. Notice the OH groups and the other oxygen atoms in the molecule, which can interact with water molecules by hydrogen bonding.

PRACTICE EXERCISE

Arrange the following substances in order of increasing solubility in water:

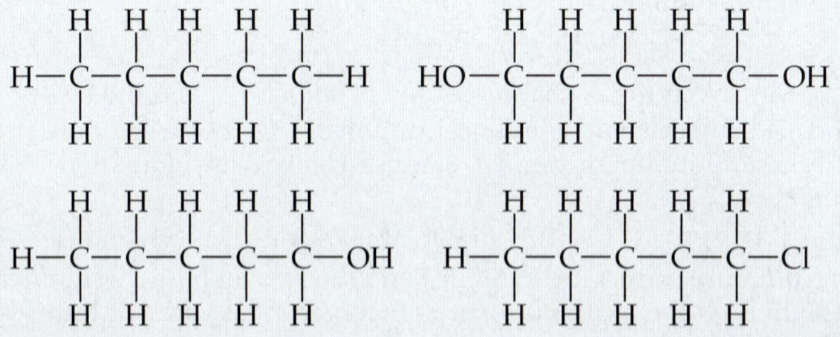

Answer: $C_5H_{12} < C_5H_{11}Cl < C_5H_{11}OH < C_5H_{10}(OH)_2$ (in order of increasing polarity and hydrogen-bonding ability)

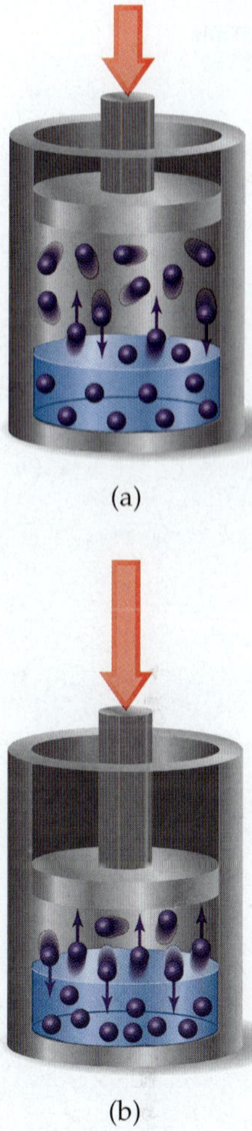

▲ **Figure 13.14 Effect of pressure on gas solubility.** When the pressure is increased, as in (b), the rate at which gas molecules enter the solution increases. The concentration of solute molecules at equilibrium increases in proportion to the pressure.

Pressure Effects

The solubilities of solids and liquids are not appreciably affected by pressure, whereas the solubility of a gas in any solvent is increased as the pressure over the solvent increases. We can understand the effect of pressure on the solubility of a gas by considering the dynamic equilibrium illustrated in Figure 13.14 ▶. Suppose that we have a gaseous substance distributed between the gas and solution phases. When equilibrium is established, the rate at which gas molecules enter the solution equals the rate at which solute molecules escape from the solution to enter the gas phase. The small arrows in Figure 13.14(a) represent the rates of these opposing processes. Now suppose that we exert added pressure on the piston and compress the gas above the solution, as shown in Figure 13.14(b). If we reduced the volume to half its original value, the pressure of the gas would increase to about twice its original value. The rate at which gas molecules strike the surface to enter the solution phase would therefore increase. As a result, the solubility of the gas in the solution would increase until an equilibrium is again established; that is, solubility increases until the rate at which gas molecules enter the solution equals the rate at which solute molecules escape from the solution. Thus, *the solubility of the gas increases in direct proportion to its partial pressure above the solution.*

The relationship between pressure and the solubility of a gas is expressed by a simple equation known as **Henry's law**:

$$S_g = kP_g \qquad [13.4]$$

Here, S_g is the solubility of the gas in the solution phase (usually expressed as molarity), P_g is the partial pressure of the gas over the solution, and k is a proportionality constant known as the *Henry's law constant*. The Henry's law constant is different for each solute-solvent pair. It also varies with temperature. As an example, the solubility of N_2 gas in water at 25°C and 0.78 atm pressure is 5.3×10^{-4} *M*. The Henry's law constant for N_2 in water at 25°C is thus given by $(5.3 \times 10^{-4}\text{ mol/L})/0.78\text{ atm} = 6.8 \times 10^{-4}\text{ mol/L-atm}$. If the partial pressure of N_2 is doubled, Henry's law predicts that the solubility in water at 25°C will also double, to 1.06×10^{-3} *M*.

Bottlers use the effect of pressure on solubility in producing carbonated beverages such as beer and many soft drinks. These are bottled under a carbon dioxide pressure greater than 1 atm. When the bottles are opened to the air, the partial pressure of CO_2 above the solution decreases. Hence, the solubility of CO_2 decreases, and CO_2 bubbles out of the solution (Figure 13.15 ▶).

▲ **Figure 13.15 Solubility decreases as pressure decreases.** CO_2 bubbles out of solution when a carbonated beverage is opened, because the CO_2 partial pressure above the solution is reduced.

ANIMATION
Henry's Law

SAMPLE EXERCISE 13.3 | A Henry's Law Calculation

Calculate the concentration of CO_2 in a soft drink that is bottled with a partial pressure of CO_2 of 4.0 atm over the liquid at 25°C. The Henry's law constant for CO_2 in water at this temperature is 3.1×10^{-2} mol/L-atm.

Solution

Analyze: We are given the partial pressure of CO_2, P_{CO_2}, and the Henry's law constant, k, and asked to calculate the concentration of CO_2 in the solution.

Plan: With the information given, we can use Henry's law, Equation 13.4, to calculate the solubility, S_{CO_2}.

Solve: $S_{CO_2} = kP_{CO_2} = (3.1 \times 10^{-2}\text{ mol/L-atm})(4.0\text{ atm}) = 0.12\text{ mol/L} = 0.12\ M$

Check: The units are correct for solubility, and the answer has two significant figures consistent with both the partial pressure of CO_2 and the value of Henry's constant.

PRACTICE EXERCISE

Calculate the concentration of CO_2 in a soft drink after the bottle is opened and equilibrates at 25°C under a CO_2 partial pressure of 3.0×10^{-4} atm.
Answer: $9.3 \times 10^{-6}\ M$

Temperature Effects

The solubility of most solid solutes in water increases as the temperature of the solution increases. Figure 13.17 ▶ shows this effect for several ionic substances in water. There are exceptions to this rule, however, as seen for $Ce_2(SO_4)_3$, whose solubility curve slopes downward with increasing temperature.

In contrast to solid solutes, *the solubility of gases in water decreases with increasing temperature* (Figure 13.18 ▶). If a glass of cold tap water is warmed, bubbles of air are seen on the inside of the glass. Similarly, carbonated beverages go flat as they are allowed to warm; as the temperature of the solution increases, the solubility of CO_2 decreases and $CO_2(g)$ escapes from the solution. The decreased solubility of O_2 in water as temperature increases is one of the effects of *thermal pollution* of lakes and streams. The effect is particularly

CHEMISTRY AND LIFE | Blood Gases and Deep-Sea Diving

Because the solubility of gases increases with increasing pressure, divers who breathe compressed air (Figure 13.16 ▶) must be concerned about the solubility of gases in their blood. Although the gases are not very soluble at sea level, their solubilities can become appreciable at deep levels where their partial pressures are greater. Thus, deep-sea divers must ascend slowly to prevent dissolved gases from being released rapidly from blood and other fluids in the body. These bubbles affect nerve impulses and give rise to the affliction known as decompression sickness, or "the bends," which is painful and can be fatal. Nitrogen is the main problem because it has the highest partial pressure in air and because it can be removed only through the respiratory system. Oxygen, in contrast, is consumed in metabolism.

Deep-sea divers sometimes substitute helium for nitrogen in the air that they breathe, because helium has a much lower solubility in biological fluids than N_2. For example, divers working at a depth of 100 ft experience a pressure of about 4 atm. At this pressure a mixture of 95% helium and 5% oxygen will give an oxygen partial pressure of about 0.2 atm, which is the partial pressure of oxygen in normal air at 1 atm. If the oxygen partial pressure becomes too great, the urge to breathe is reduced, CO_2 is not removed from the body and CO_2 poisoning occurs. At excessive concentrations in the body, carbon dioxide acts as a neurotoxin, interfering with nerve conduction and transmission.

Related Exercises: 13.53, 13.54, 13.100

▲ **Figure 13.16 Solubility increases as pressure increases.** Divers who use compressed gases must be concerned about the solubility of the gases in their blood.

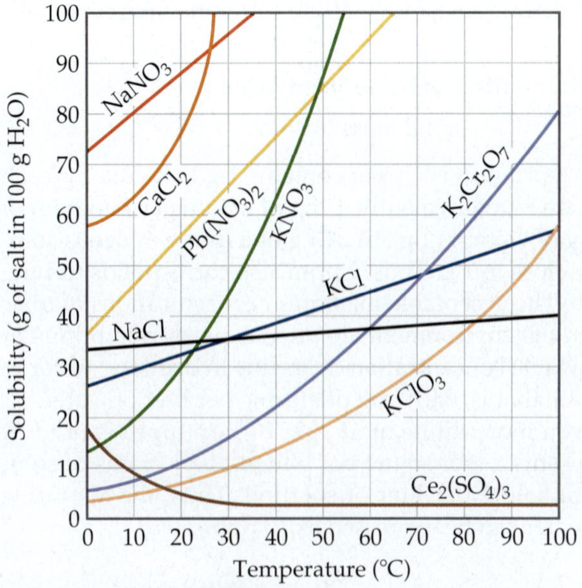

▲ **Figure 13.17 Solubilities of several ionic compounds in water as a function of temperature.**

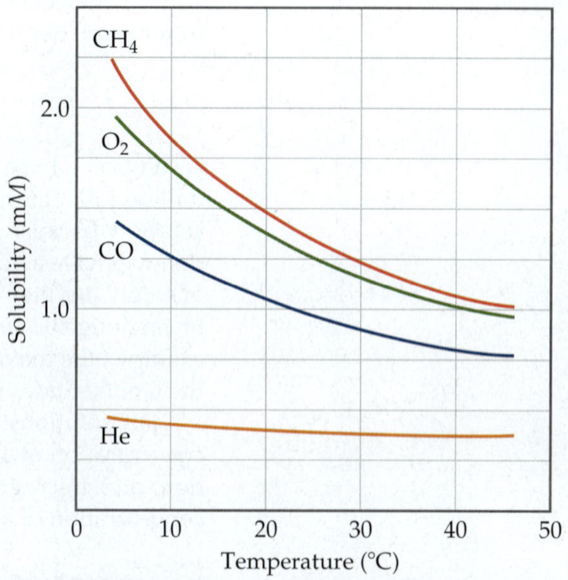

▲ **Figure 13.18 Variation of gas solubility with temperature.** Note that solubilities are in units of millimoles per liter (mmol/L), for a constant total pressure of 1 atm in the gas phase.

serious in deep lakes because warm water is less dense than cold water. It therefore tends to remain on top of cold water, at the surface. This situation impedes the dissolving of oxygen into the deeper layers, thus stifling the respiration of all aquatic life needing oxygen. Fish may suffocate and die under these conditions.

GIVE IT SOME THOUGHT

Why do bubbles form on the inside wall of a cooking pot when water is heated on the stove, even though the temperature is well below the boiling point of water?

13.4 | Ways of Expressing Concentration

The concentration of a solution can be expressed either qualitatively or quantitatively. The terms *dilute* and *concentrated* are used to describe a solution qualitatively. A solution with a relatively small concentration of solute is said to be dilute; one with a large concentration is said to be concentrated. We use several different ways to express concentration in quantitative terms, and we examine four of these in this section: mass percentage, mole fraction, molarity, and molality.

Mass Percentage, ppm, and ppb

One of the simplest quantitative expressions of concentration is the **mass percentage** of a component in a solution, given by

$$\text{Mass \% of component} = \frac{\text{mass of component in soln}}{\text{total mass of soln}} \times 100 \qquad [13.5]$$

where we have abbreviated "solution" as "soln." Thus, a solution of hydrochloric acid that is 36% HCl by mass contains 36 g of HCl for each 100 g of solution.

We often express the concentrations of very dilute solution in **parts per million (ppm)**, defined as

$$\text{ppm of component} = \frac{\text{mass of component in soln}}{\text{total mass of soln}} \times 10^6 \qquad [13.6]$$

A solution whose solute concentration is 1 ppm contains 1 g of solute for each million (10^6) grams of solution or, equivalently, 1 mg of solute per kilogram of solution. Because the density of water is 1 g/mL, 1 kg of a dilute aqueous solution will have a volume very close to 1 L. Thus, 1 ppm also corresponds to 1 mg of solute per liter of solution. The acceptable maximum concentrations of toxic or carcinogenic substances in the environment are often expressed in ppm. For example, the maximum allowable concentration of arsenic in drinking water in the United States is 0.010 ppm; that is, 0.010 mg of arsenic per liter of water.

For solutions that are even more dilute, **parts per billion (ppb)** is used. A concentration of 1 ppb represents 1 g of solute per billion (10^9) grams of solution, or 1 microgram (μg) of solute per liter of solution. Thus, the allowable concentration of arsenic in water can be expressed as 10 ppb.

GIVE IT SOME THOUGHT

A solution of SO_2 in water contains 0.00023 g of SO_2 per liter of solution. What is the concentration of SO_2 in ppm?

SAMPLE EXERCISE 13.4 | Calculation of Mass-Related Concentrations

(a) A solution is made by dissolving 13.5 g of glucose ($C_6H_{12}O_6$) in 0.100 kg of water. What is the mass percentage of solute in this solution? **(b)** A 2.5-g sample of groundwater was found to contain 5.4 μg of Zn^{2+}. What is the concentration of Zn^{2+} in parts per million?

Solution

(a) Analyze: We are given the number of grams of solute (13.5 g) and the number of grams of solvent (0.100 kg = 100 g). From this we must calculate the mass percentage of solute.

Plan: We can calculate the mass percentage by using Equation 13.5. The mass of the solution is the sum of the mass of solute (glucose) and the mass of solvent (water).

Solve:

$$\text{Mass \% of glucose} = \frac{\text{mass glucose}}{\text{mass soln}} \times 100 = \frac{13.5\text{ g}}{13.5\text{ g} + 100\text{ g}} \times 100 = 11.9\%$$

Comment: The mass percentage of water in this solution is (100 − 11.9)% = 88.1%.

(b) Analyze: In this case we are given the number of micrograms of solute. Because 1 μg is 1×10^{-6} g, 5.4 μg = 5.4×10^{-6} g.

Plan: We calculate the parts per million using Equation 13.6.

Solve:

$$\text{ppm} = \frac{\text{mass of solute}}{\text{mass of soln}} \times 10^6 = \frac{5.4 \times 10^{-6}\text{ g}}{2.5\text{ g}} \times 10^6 = 2.2\text{ ppm}$$

PRACTICE EXERCISE

(a) Calculate the mass percentage of NaCl in a solution containing 1.50 g of NaCl in 50.0 g of water. **(b)** A commercial bleaching solution contains 3.62 mass % sodium hypochlorite, NaOCl. What is the mass of NaOCl in a bottle containing 2500 g of bleaching solution?

Answers: **(a)** 2.91%, **(b)** 90.5 g of NaOCl

Mole Fraction, Molarity, and Molality

Concentration expressions are often based on the number of moles of one or more components of the solution. The three most commonly used are mole fraction, molarity, and molality.

Recall from Section 10.6 that the *mole fraction* of a component of a solution is given by

$$\text{Mole fraction of component} = \frac{\text{moles of component}}{\text{total moles of all components}} \qquad [13.7]$$

The symbol X is commonly used for mole fraction, with a subscript to indicate the component of interest. For example, the mole fraction of HCl in a hydrochloric acid solution is represented as X_{HCl}. Thus, a solution containing 1.00 mol of HCl (36.5 g) and 8.00 mol of water (144 g) has a mole fraction of HCl of $X_{HCl} = (1.00 \text{ mol})/(1.00 \text{ mol} + 8.00 \text{ mol}) = 0.111$. Mole fractions have no units because the units in the numerator and the denominator cancel. The sum of the mole fractions of all components of a solution must equal 1. Thus, in the aqueous HCl solution, $X_{H_2O} = 1.000 - 0.111 = 0.889$. Mole fractions are very useful when dealing with gases as we saw in Section 10.6 but have limited use when dealing with liquid solutions.

Recall from Section 4.5 that the *molarity* (*M*) of a solute in a solution is defined as

$$\text{Molarity} = \frac{\text{moles solute}}{\text{liters soln}} \qquad [13.8]$$

For example, if you dissolve 0.500 mol of Na_2CO_3 in enough water to form 0.250 L of solution, then the solution has a concentration of $(0.500 \text{ mol})/(0.250 \text{ L}) = 2.00\ M$ in Na_2CO_3. Molarity is especially useful for relating the volume of a solution to the quantity of solute it contains, as we saw in our discussions of titrations. (Section 4.6)

The **molality** of a solution, denoted *m*, is a unit that we haven't encountered in previous chapters. This concentration unit equals the number of moles of solute per kilogram of solvent:

$$\text{Molality} = \frac{\text{moles of solute}}{\text{kilograms of solvent}} \qquad [13.9]$$

Thus, if you form a solution by mixing 0.200 mol of NaOH (40.0 g) and 0.500 kg of water (500 g), the concentration of the solution is $(0.200 \text{ mol})/(0.500 \text{ kg}) = 0.400\ m$ (that is, 0.400 molal) in NaOH.

The definitions of molarity and molality are similar enough that they can be easily confused. Molarity depends on the *volume* of *solution*, whereas molality depends on the *mass* of *solvent*. When water is the solvent, the molality and molarity of dilute solutions are numerically about the same because 1 kg of solvent is nearly the same as 1 kg of solution, and 1 kg of the solution has a volume of about 1 L.

The molality of a given solution does not vary with temperature because masses do not vary with temperature. Molarity, however, changes with temperature because the expansion or contraction of the solution changes its volume. Thus molality is often the concentration unit of choice when a solution is to be used over a range of temperatures.

GIVE IT SOME THOUGHT

Would it be correct to say that for any aqueous solution under standard conditions of 1 atm and 25°C the molality is always greater than the molarity?

SAMPLE EXERCISE 13.5 | Calculation of Molality

A solution is made by dissolving 4.35 g glucose ($C_6H_{12}O_6$) in 25.0 mL of water at 25°C. Calculate the molality of glucose in the solution.

Solution

Analyze: We are asked to calculate a molality. To do this, we must determine the number of moles of solute (glucose) and the number of kilograms of solvent (water).

Plan: We use the molar mass of $C_6H_{12}O_6$ to convert grams to moles. We use the density of water to convert milliliters to kilograms. The molality equals the number of moles of solute divided by the number of kilograms of solvent (Equation 13.9).

Solve: Use the molar mass of glucose, 180.2 g/mol, to convert grams to moles:

$$\text{Mol } C_6H_{12}O_6 = (4.35 \text{ g } C_6H_{12}O_6)\left(\frac{1 \text{ mol } C_6H_{12}O_6}{180.2 \text{ g } C_6H_{12}O_6}\right) = 0.0241 \text{ mol } C_6H_{12}O_6$$

Because water has a density of 1.00 g/mL, the mass of the solvent is

$$(25.0 \text{ mL})(1.00 \text{ g/mL}) = 25.0 \text{ g} = 0.0250 \text{ kg}$$

Finally, use Equation 13.9 to obtain the molality:

$$\text{Molality of } C_6H_{12}O_6 = \frac{0.0241 \text{ mol } C_6H_{12}O_6}{0.0250 \text{ kg } H_2O} = 0.964\ m$$

PRACTICE EXERCISE

What is the molality of a solution made by dissolving 36.5 g of naphthalene ($C_{10}H_8$) in 425 g of toluene (C_7H_8)?
Answer: 0.670 *m*

Conversion of Concentration Units

Sometimes the concentration of a given solution needs to be known in several different concentration units. It is possible to interconvert concentration units as shown in Sample Exercises 13.6 and 13.7.

SAMPLE EXERCISE 13.6 | Calculation of Mole Fraction and Molality

An aqueous solution of hydrochloric acid contains 36% HCl by mass. **(a)** Calculate the mole fraction of HCl in the solution. **(b)** Calculate the molality of HCl in the solution.

Solution

Analyze: We are asked to calculate the concentration of the solute, HCl, in two related concentration units, given only the percentage by mass of the solute in the solution.

Plan: In converting concentration units based on the mass or moles of solute and solvent (mass percentage, mole fraction, and molality), it is useful to assume a certain total mass of solution. Let's assume that there is exactly 100 g of solution. Because the solution is 36% HCl, it contains 36 g of HCl and (100 − 36) g = 64 g of H_2O. We must convert grams of solute (HCl) to moles in order to calculate either mole fraction or molality. We must convert grams of solvent (H_2O) to moles to calculate mole fractions, and to kilograms to calculate molality.

Solve: (a) To calculate the mole fraction of HCl, we convert the masses of HCl and H_2O to moles and then use Equation 13.7:

$$\text{Moles HCl} = (36 \text{ g HCl})\left(\frac{1 \text{ mol HCl}}{36.5 \text{ g HCl}}\right) = 0.99 \text{ mol HCl}$$

$$\text{Moles } H_2O = (64 \text{ g } H_2O)\left(\frac{1 \text{ mol } H_2O}{18 \text{ g } H_2O}\right) = 3.6 \text{ mol } H_2O$$

$$X_{HCl} = \frac{\text{moles HCl}}{\text{moles } H_2O + \text{moles HCl}} = \frac{0.99}{3.6 + 0.99} = \frac{0.99}{4.6} = 0.22$$

(b) To calculate the molality of HCl in the solution, we use Equation 13.9. We calculated the number of moles of HCl in part (a), and the mass of solvent is 64 g = 0.064 kg:

$$\text{Molality of HCl} = \frac{0.99 \text{ mol HCl}}{0.064 \text{ kg } H_2O} = 15\ m$$

PRACTICE EXERCISE

A commercial bleach solution contains 3.62 mass % NaOCl in water. Calculate **(a)** the molality and **(b)** the mole fraction of NaOCl in the solution.
Answers: **(a)** 0.505 *m*, **(b)** 9.00×10^{-3}

In order to interconvert molality and molarity, we need to know the density of the solution. Figure 13.19 ▶ outlines the calculation of the molarity and molality of a solution from the mass of solute and the mass of solvent. The mass of the solution is the sum of masses of the solvent and solute. The volume of the solution can be calculated from its mass and density.

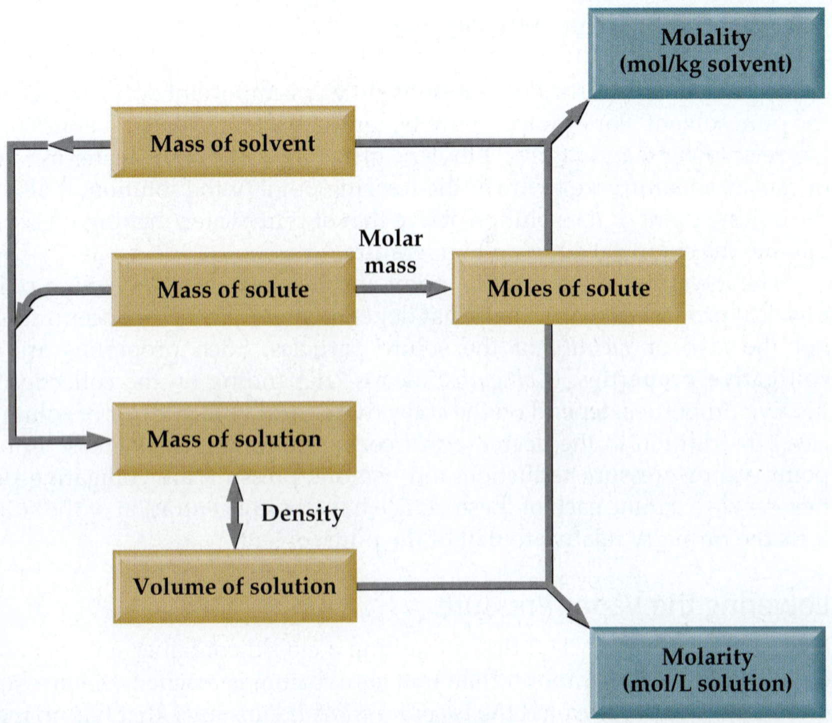

◀ **Figure 13.19 Calculating molality and molarity.** This diagram summarizes the calculation of molality and molarity from the mass of the solute, the mass of the solvent, and the density of the solution.

SAMPLE EXERCISE 13.7 | Calculation of Molarity

A solution contains 5.0 g of toluene (C_7H_8) and 225 g of benzene and has a density of 0.876 g/mL. Calculate the molarity of the solution.

Solution

Analyze: Our goal is to calculate the molarity of a solution, given the masses of solute and solvent and the density of the solution.

Plan: The molarity of a solution is the number of moles of solute divided by the number of liters of solution (Equation 13.8). The number of moles of solute (C_7H_8) is calculated from the number of grams of solute and its molar mass. The volume of the solution is obtained from the mass of the solution (mass of solute + mass of solvent = 5.0 g + 225 g = 230 g) and its density.

Solve: The number of moles of solute is

$$\text{Moles } C_7H_8 = (5.0 \text{ g } C_7H_8)\left(\frac{1 \text{ mol } C_7H_8}{92 \text{ g } C_7H_8}\right) = 0.054 \text{ mol}$$

The density of the solution is used to convert the mass of the solution to its volume:

$$\text{Milliliters soln} = (230 \text{ g})\left(\frac{1 \text{ mL}}{0.876 \text{ g}}\right) = 263 \text{ mL}$$

Molarity is moles of solute per liter of solution:

$$\text{Molarity} = \left(\frac{\text{moles } C_7H_8}{\text{liter soln}}\right) = \left(\frac{0.054 \text{ mol } C_7H_8}{263 \text{ mL soln}}\right)\left(\frac{1000 \text{ mL soln}}{1 \text{ L soln}}\right) = 0.21\,M$$

Check: The magnitude of our answer is reasonable. Rounding moles to 0.05 and liters to 0.25 gives a molarity of

$$(0.05 \text{ mol})/(0.25 \text{ L}) = 0.2\,M$$

The units for our answer (mol/L) are correct, and the answer, 0.21*M*, has two significant figures, corresponding to the number of significant figures in the mass of solute (2).

Comment: Because the mass of the solvent (0.225 kg) and the volume of the solution (0.263 L) are similar in magnitude, the molarity and molality are also similar in magnitude:

$$(0.054 \text{ mol } C_7H_8)/(0.225 \text{ kg solvent}) = 0.24\,m$$

PRACTICE EXERCISE

A solution containing equal masses of glycerol ($C_3H_8O_3$) and water has a density of 1.10 g/mL. Calculate **(a)** the molality of glycerol, **(b)** the mole fraction of glycerol, **(c)** the molarity of glycerol in the solution.

Answers: **(a)** 10.9 *m*, **(b)** $X_{C_3H_8O_3} = 0.163$, **(c)** 5.97 *M*

13.5 | Colligative Properties

Some physical properties of solutions differ in important ways from those of the pure solvent. For example, pure water freezes at 0°C, but aqueous solutions freeze at lower temperatures. Ethylene glycol is added to the water in radiators of cars as an antifreeze to lower the freezing point of the solution. It also raises the boiling point of the solution above that of pure water, making it possible to operate the engine at a higher temperature.

The lowering of the freezing point and the raising of the boiling point are physical properties of solutions that depend on the *quantity* (concentration) but not the *kind* or *identity* of the solute particles. Such properties are called **colligative properties**. (*Colligative* means "depending on the collection"; colligative properties depend on the collective effect of the number of solute particles.) In addition to the decrease in freezing point and the increase in boiling point, vapor-pressure reduction and osmotic pressure are colligative properties. As we examine each of these, notice how the concentration of the solute affects the property relative to that of the pure solvent.

Lowering the Vapor Pressure

ACTIVITY
Raoult's Law

We learned in Section 11.5 that a liquid in a closed container will establish an equilibrium with its vapor. When that equilibrium is reached, the pressure exerted by the vapor is called the *vapor pressure*. A substance that has no measurable vapor pressure is *nonvolatile*, whereas one that exhibits a vapor pressure is *volatile*.

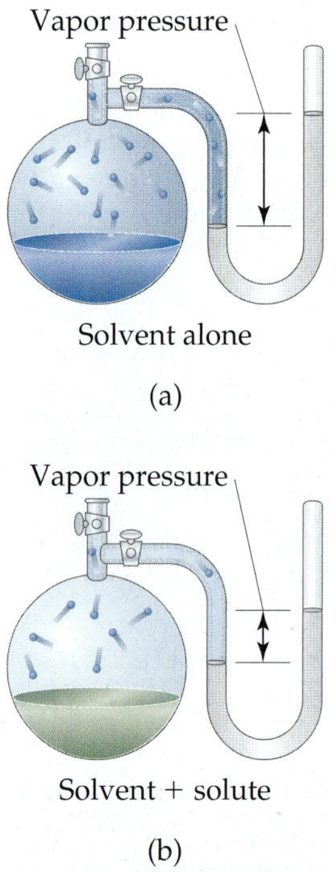

▲ **Figure 13.20 Vapor-pressure lowering.** The vapor pressure over a solution formed by a volatile solvent and a nonvolatile solute (b) is lower than that of the solvent alone (a). The extent of the decrease in the vapor pressure upon addition of the solute depends on the concentration of the solute.

When we compare the vapor pressures of various solvents with those of their solutions, we find that adding a nonvolatile solute to a solvent always lowers the vapor pressure. This effect is illustrated in Figure 13.20 ◀. The extent to which a nonvolatile solute lowers the vapor pressure is proportional to its concentration. This relationship is expressed by **Raoult's law**, which states that the partial pressure exerted by solvent vapor above a solution, P_A, equals the product of the mole fraction of the solvent in the solution, X_A, times the vapor pressure of the pure solvent:

$$P_A = X_A P_A^\circ \qquad [13.10]$$

For example, the vapor pressure of water is 17.5 torr at 20°C. Imagine holding the temperature constant while adding glucose ($C_6H_{12}O_6$) to the water so that the resulting solution has $X_{H_2O} = 0.800$ and $X_{C_6H_{12}O_6} = 0.200$. According to Equation 13.10, the vapor pressure of water over the solution will be 80.0% of that of pure water:

$$P_{H_2O} = (0.800)(17.5 \text{ torr}) = 14.0 \text{ torr}$$

In other words, the presence of the nonvolatile solute lowers the vapor pressure of the volatile solvent by 17.5 torr − 14.0 torr = 3.5 torr.

Raoult's law predicts that when we increase the mole fraction of nonvolatile solute particles in a solution, the vapor pressure over the solution will be reduced. In fact, the reduction in vapor pressure depends on the total concentration of solute particles, regardless of whether they are molecules or ions. Remember that vapor-pressure lowering is a colligative property, so it depends on the concentration of solute particles and not on their kind. The simplest applications of Raoult's law concern solutes that are nonvolatile and nonelectrolytes. We consider the effects of *volatile* substances on vapor pressure in the "Closer Look" box in this section, and we will consider the effects of *electrolytes* in our discussions of freezing points and boiling points.

A CLOSER LOOK | Ideal Solutions with Two or More Volatile Components

Solutions sometimes have two or more volatile components. Gasoline, for example, is a complex solution containing several volatile substances. To gain some understanding of such mixtures, consider an ideal solution containing two components, A and B. The partial pressures of A and B vapors above the solution are given by Raoult's law:

$$P_A = X_A P_A^\circ \quad \text{and} \quad P_B = X_B P_B^\circ$$

The total vapor pressure over the solution is the sum of the partial pressures of each volatile component:

$$P_{total} = P_A + P_B = X_A P_A^\circ + X_B P_B^\circ$$

Consider, for example, a mixture of benzene (C_6H_6) and toluene (C_7H_8) containing 1.0 mol of benzene and 2.0 mol of toluene ($X_{ben} = 0.33$ and $X_{tol} = 0.67$). At 20°C the vapor pressures of the pure substances are

$$\text{Benzene:} \quad P_{ben}^\circ = 75 \text{ torr}$$

$$\text{Toluene:} \quad P_{tol}^\circ = 22 \text{ torr}$$

Thus, the partial pressures of benzene and toluene above the solution are

$$P_{ben} = (0.33)(75 \text{ torr}) = 25 \text{ torr}$$

$$P_{tol} = (0.67)(22 \text{ torr}) = 15 \text{ torr}$$

The total vapor pressure is

$$P_{total} = 25 \text{ torr} + 15 \text{ torr} = 40 \text{ torr}$$

The vapor, therefore, is richer in benzene, the more volatile component. The mole fraction of benzene in the vapor is given by the ratio of its vapor pressure to the total pressure (Equation 10.15):

$$X_{ben} \text{ in vapor} = \frac{P_{ben}}{P_{total}} = \frac{25 \text{ torr}}{40 \text{ torr}} = 0.63$$

Although benzene constitutes only 33% of the molecules in the solution, it makes up 63% of the molecules in the vapor.

When ideal solutions are in equilibrium with their vapor, the more volatile component of the mixture will be relatively richer in the vapor. This fact forms the basis of *distillation*, a technique used to separate (or partially separate) mixtures containing volatile components. Distillation is the procedure by which a moonshiner obtains whiskey using a still and by which petrochemical plants achieve the separation of crude petroleum into gasoline, diesel fuel, lubricating oil, and so forth (Figure 13.21 ▼). It is also used routinely on a small scale in the laboratory. A specially designed *fractional distillation* apparatus can achieve in a single operation a degree of separation that would be equivalent to several successive simple distillations.

Related Exercises: 13.59, 13.61, 13.62

▲ **Figure 13.21 Separating volatile components.** In an industrial distillation tower, such as the ones shown here, the components of a volatile organic mixture are separated according to boiling-point range.

SAMPLE EXERCISE 13.8 | Calculation of Vapor-Pressure Lowering

Glycerin ($C_3H_8O_3$) is a nonvolatile nonelectrolyte with a density of 1.26 g/mL at 25°C. Calculate the vapor pressure at 25°C of a solution made by adding 50.0 mL of glycerin to 500.0 mL of water. The vapor pressure of pure water at 25°C is 23.8 torr (Appendix B).

Solution

Analyze: Our goal is to calculate the vapor pressure of a solution, given the volumes of solute and solvent and the density of the solute.

Plan: We can use Raoult's law (Equation 13.10) to calculate the vapor pressure of a solution. The mole fraction of the solvent in the solution, X_A, is the ratio of the number of moles of solvent (H_2O) to total solution (moles $C_3H_8O_3$ + moles H_2O).

Solve: To calculate the mole fraction of water in the solution, we must determine the number of moles of $C_3H_8O_3$ and H_2O:

$$\text{Moles } C_3H_8O_3 = (50.0 \text{ mL } C_3H_8O_3)\left(\frac{1.26 \text{ g } C_3H_8O_3}{1 \text{ mL } C_3H_8O_3}\right)\left(\frac{1 \text{ mol } C_3H_8O_3}{92.1 \text{ g } C_3H_8O_3}\right) = 0.684 \text{ mol}$$

$$\text{Moles } H_2O = (500.0 \text{ mL } H_2O)\left(\frac{1.00 \text{ g } H_2O}{1 \text{ mL } H_2O}\right)\left(\frac{1 \text{ mol } H_2O}{18.0 \text{ g } H_2O}\right) = 27.8 \text{ mol}$$

$$X_{H_2O} = \frac{\text{mol } H_2O}{\text{mol } H_2O + \text{mol } C_3H_8O_3} = \frac{27.8}{27.8 + 0.684} = 0.976$$

We now use Raoult's law to calculate the vapor pressure of water for the solution:

$$P_{H_2O} = X_{H_2O}P^{\circ}_{H_2O} = (0.976)(23.8 \text{ torr}) = 23.2 \text{ torr}$$

The vapor pressure of the solution has been lowered by 0.6 torr relative to that of pure water.

PRACTICE EXERCISE

The vapor pressure of pure water at 110°C is 1070 torr. A solution of ethylene glycol and water has a vapor pressure of 1.00 atm at 110°C. Assuming that Raoult's law is obeyed, what is the mole fraction of ethylene glycol in the solution?
Answer: 0.290

An ideal gas obeys the ideal-gas equation (Section 10.4), and an **ideal solution** obeys Raoult's law. Real solutions best approximate ideal behavior when the solute concentration is low and when the solute and solvent have similar molecular sizes and similar types of intermolecular attractions.

Many solutions do not obey Raoult's law exactly: They are not ideal solutions. If the intermolecular forces between solvent and solute are weaker than those between solvent and solvent and between solute and solute, then the solvent vapor pressure tends to be greater than predicted by Raoult's law. Conversely, when the interactions between solute and solvent are exceptionally strong, as might be the case when hydrogen bonding exists, the solvent vapor pressure is lower than Raoult's law predicts. Although you should be aware that these departures from ideal solution occur, we will ignore them for the remainder of this chapter.

GIVE IT SOME THOUGHT

In Raoult's law, $P_A = X_A P^{\circ}_A$, what concentration unit is used for X_A?

Boiling-Point Elevation

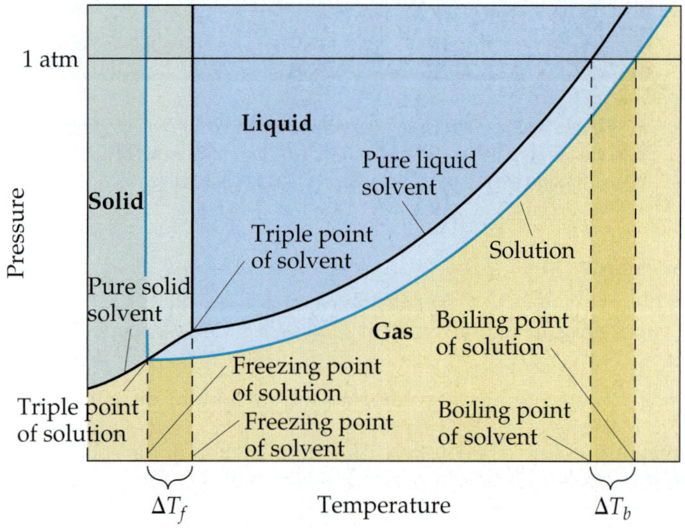

▲ **Figure 13.22 Phase diagrams for a pure solvent and for a solution of a nonvolatile solute.** The vapor pressure of the solid solvent is unaffected by the presence of solute if the solid freezes out without containing a significant concentration of solute, as is usually the case.

In Sections 11.5 and 11.6 we examined the vapor pressures of pure substances and how they can be used to construct phase diagrams. How will the phase diagram of a solution, and hence its boiling and freezing points, differ from those of the pure solvent? The addition of a nonvolatile solute lowers the vapor pressure of the solution. Thus, as shown in Figure 13.22 ◄, the vapor-pressure curve of the solution (blue line) will be shifted downward relative to the vapor-pressure curve of the pure liquid (black line); at any given temperature the vapor pressure of the solution is lower than that of the pure liquid. Recall that the normal boiling point of a liquid is the temperature at which its vapor pressure equals 1 atm. ∞ (Section 11.5) At the normal boiling point of the pure liquid, the vapor pressure of the solution will be less than 1 atm (Figure 13.22). Therefore, a higher temperature is required to attain a vapor pressure of 1 atm. Thus, *the boiling point of the solution is higher than that of the pure liquid.*

The increase in boiling point relative to that of the pure solvent, ΔT_b, is directly proportional to the number of solute particles per mole of solvent molecules. We know that molality expresses the number of moles of solute per 1000 g of solvent, which represents a fixed number of moles of solvent. Thus, ΔT_b is proportional to molality:

$$\Delta T_b = K_b m \qquad [13.11]$$

TABLE 13.4 Molal Boiling-Point-Elevation and Freezing-Point-Depression Constants

Solvent	Normal Boiling Point (°C)	K_b (°C/m)	Normal Freezing Point (°C)	K_f (°C/m)
Water, H_2O	100.0	0.51	0.0	1.86
Benzene, C_6H_6	80.1	2.53	5.5	5.12
Ethanol, C_2H_5OH	78.4	1.22	−114.6	1.99
Carbon tetrachloride, CCl_4	76.8	5.02	−22.3	29.8
Chloroform, $CHCl_3$	61.2	3.63	−63.5	4.68

The magnitude of K_b, which is called the **molal boiling-point-elevation constant**, depends only on the solvent. Some typical values for several common solvents are given in Table 13.4 ▲.

For water, K_b is 0.51°C/m; therefore, a 1 m aqueous solution of sucrose or any other aqueous solution that is 1 m in nonvolatile solute particles will boil 0.51°C higher than pure water. The boiling-point elevation is proportional to the concentration of solute particles, regardless of whether the particles are molecules or ions. When NaCl dissolves in water, 2 mol of solute particles (1 mol of Na^+ and 1 mol of Cl^-) are formed for each mole of NaCl that dissolves. Therefore, a 1 m aqueous solution of NaCl is 1 m in Na^+ and 1 m in Cl^-, making it 2 m in total solute particles. As a result, the boiling-point elevation of a 1 m aqueous solution of NaCl is approximately $(2\ m)(0.51°C/m) = 1°C$, twice as large as a 1 m solution of a nonelectrolyte such as sucrose. Thus, to properly predict the effect of a particular solute on the boiling point (or any other colligative property), it is important to know whether the solute is an electrolyte or a nonelectrolyte. (Sections 4.1 and 4.3)

ACTIVITY Boiling-Point Elevation and Freezing-Point Depression

VCL Boiling-Point Elevation

GIVE IT SOME THOUGHT

An unknown solute dissolved in water causes the boiling point to increase by 0.51°C. Does this mean that the concentration of the solute is 1.0 m?

Freezing-Point Depression

When a solution freezes, crystals of pure solvent usually separate out; the solute molecules are not normally soluble in the solid phase of the solvent. When aqueous solutions are partially frozen, for example, the solid that separates out is almost always pure ice. As a result, the part of the phase diagram in Figure 13.22 that represents the vapor pressure of the solid is the same as that for the pure liquid. The vapor-pressure curves for the liquid and solid phases meet at the triple point. (Section 11.6) In Figure 13.22 we see that the triple point of the solution must be at a lower temperature than that in the pure liquid because the solution has a lower vapor pressure than the pure liquid.

VCL Freezing-Point Depression

The freezing point of a solution is the temperature at which the first crystals of pure solvent begin to form in equilibrium with the solution. Recall from Section 11.6 that the line representing the solid-liquid equilibrium rises nearly vertically from the triple point. Because the triple-point temperature of the solution is lower than that of the pure liquid, *the freezing point of the solution is lower than that of the pure liquid.*

Like the boiling-point elevation, the decrease in freezing point, ΔT_f, is directly proportional to the molality of the solute:

$$\Delta T_f = K_f m \qquad [13.12]$$

The values of K_f, the **molal freezing-point-depression constant**, for several common solvents are given in Table 13.4. For water, K_f is 1.86°C/m; therefore, a 1 m aqueous solution of sucrose or any other aqueous solution that is 1 m in nonvolatile solute particles (such as 0.5 m NaCl) will freeze 1.86°C lower than pure water. The freezing-point lowering caused by solutes explains the use of antifreeze in cars (Sample Exercise 13.8) and the use of calcium chloride ($CaCl_2$) to melt ice on roads during winter.

SAMPLE EXERCISE 13.9 | Calculation of Boiling-Point Elevation and Freezing-Point Lowering

Automotive antifreeze consists of ethylene glycol ($C_2H_6O_2$), a nonvolatile nonelectrolyte. Calculate the boiling point and freezing point of a 25.0 mass % solution of ethylene glycol in water.

Solution

Analyze: We are given that a solution contains 25.0 mass % of a nonvolatile, nonelectrolyte solute and asked to calculate the boiling and freezing points of the solution. To do this, we need to calculate the boiling-point elevation and freezing-point depression.
Plan: In order to calculate the boiling-point elevation and the freezing-point depression using Equations 13.11 and 13.12, we must express the concentration of the solution as molality. Let's assume for convenience that we have 1000 g of solution. Because the solution is 25.0 mass % ethylene glycol, the masses of ethylene glycol and water in the solution are 250 and 750 g, respectively. Using these quantities, we can calculate the molality of the solution, which we use with the molal boiling-point-elevation and freezing-point-depression constants (Table 13.4) to calculate ΔT_b and ΔT_f. We add ΔT_b to the boiling point and subtract ΔT_f from the freezing point of the solvent to obtain the boiling point and freezing point of the solution.
Solve: The molality of the solution is calculated as follows:

$$\text{Molality} = \frac{\text{moles } C_2H_6O_2}{\text{kilograms } H_2O} = \left(\frac{250 \text{ g } C_2H_6O_2}{750 \text{ g } H_2O}\right)\left(\frac{1 \text{ mol } C_2H_6O_2}{62.1 \text{ g } C_2H_6O_2}\right)\left(\frac{1000 \text{ g } H_2O}{1 \text{ kg } H_2O}\right)$$
$$= 5.37\ m$$

We can now use Equations 13.11 and 13.12 to calculate the changes in the boiling and freezing points:

$$\Delta T_b = K_b m = (0.51°\text{C}/m)(5.37\ m) = 2.7°\text{C}$$
$$\Delta T_f = K_f m = (1.86°\text{C}/m)(5.37\ m) = 10.0°\text{C}$$

Hence, the boiling and freezing points of the solution are

$$\text{Boiling point} = (\text{normal bp of solvent}) + \Delta T_b$$
$$= 100.0°\text{C} + 2.7°\text{C} = 102.7°\text{C}$$
$$\text{Freezing point} = (\text{normal fp of solvent}) - \Delta T_f$$
$$= 0.0°\text{C} - 10.0°\text{C} = -10.0°\text{C}$$

Comment: Notice that the solution is a liquid over a larger temperature range than the pure solvent.

PRACTICE EXERCISE

Calculate the freezing point of a solution containing 0.600 kg of $CHCl_3$ and 42.0 g of eucalyptol ($C_{10}H_{18}O$), a fragrant substance found in the leaves of eucalyptus trees. (See Table 13.4.)
Answer: −65.6°C

SAMPLE EXERCISE 13.10 | Freezing-Point Depression in Aqueous Solutions

List the following aqueous solutions in order of their expected freezing point: 0.050 m $CaCl_2$, 0.15 m NaCl, 0.10 m HCl, 0.050 m $HC_2H_3O_2$, 0.10 m $C_{12}H_{22}O_{11}$.

Solution

Analyze: We must order five aqueous solutions according to expected freezing points, based on molalities and the solute formulas.

Plan: The lowest freezing point will correspond to the solution with the greatest concentration of solute particles. To determine the total concentration of solute particles in each case, we must determine whether the substance is a nonelectrolyte or an electrolyte and consider the number of ions formed when it ionizes.
Solve: $CaCl_2$, NaCl, and HCl are strong electrolytes, $HC_2H_3O_2$ is a weak electrolyte, and $C_{12}H_{22}O_{11}$ is a nonelectrolyte. The molality of each solution in total particles is as follows:

$$0.050\ m\ CaCl_2 \Rightarrow 0.050\ m \text{ in } Ca^{2+} \text{ and } 0.10\ m \text{ in } Cl^- \Rightarrow 0.15\ m \text{ in particles}$$

$$0.15\ m\ NaCl \Rightarrow 0.15\ m\ Na^+ \text{ and } 0.15\ m \text{ in } Cl^- \Rightarrow 0.30\ m \text{ in particles}$$

$$0.10\ m\ HCl \Rightarrow 0.10\ m\ H^+ \text{ and } 0.10\ m \text{ in } Cl^- \Rightarrow 0.20\ m \text{ in particles}$$

$$0.050\ m\ HC_2H_3O_2 \Rightarrow \text{weak electrolyte} \Rightarrow \text{between } 0.050\ m \text{ and } 0.10\ m \text{ in particles}$$

$$0.10\ m\ C_{12}H_{22}O_{11} \Rightarrow \text{nonelectrolyte} \Rightarrow 0.10\ m \text{ in particles}$$

Because the freezing points depend on the total molality of particles in solution, the expected ordering is 0.15 *m* NaCl (lowest freezing point), 0.10 *m* HCl, 0.050 *m* $CaCl_2$, 0.10 *m* $C_{12}H_{22}O_{11}$, and 0.050 *m* $HC_2H_3O_2$ (highest freezing point).

PRACTICE EXERCISE

Which of the following solutes will produce the largest increase in boiling point upon addition to 1 kg of water: 1 mol of $Co(NO_3)_2$, 2 mol of KCl, 3 mol of ethylene glycol ($C_2H_6O_2$)?

Answer: 2 mol of KCl because it contains the highest concentration of particles, 2 *m* K^+ and 2 *m* Cl^-, giving 4 *m* in all

Osmosis

Certain materials, including many membranes in biological systems and synthetic substances such as cellophane, are *semipermeable*. When in contact with a solution, they allow some molecules to pass through their network of tiny pores, but not others. Most important, they generally allow small solvent molecules such as water to pass through, but block larger solute molecules or ions. This selectivity gives rise to some interesting and important properties.

Consider a situation in which only solvent molecules are able to pass through a membrane. If such a membrane is placed between two solutions of different concentration, solvent molecules move in both directions through the membrane. The concentration of *solvent* is higher in the solution containing less solute, however, so the rate with which solvent passes from the less concentrated to the more concentrated solution is greater than the rate in the opposite direction. Thus, there is a net movement of solvent molecules from the less concentrated solution into the more concentrated one. In this process, called **osmosis**, *the net movement of solvent is always toward the solution with the higher solute concentration.*

ANIMATION
Osmosis and Osmotic Pressure

Osmosis is illustrated in Figure 13.23 ▶. Let's begin with two solutions of different concentration separated by a semipermeable membrane. Because the solution on the left is more concentrated than the one on the right, there is a net movement of *solvent* through the membrane from right to left, as if the solutions were driven to attain equal concentrations. As a result, the liquid levels in the two arms become unequal. Eventually, the pressure difference resulting from the unequal heights of the liquid in the two arms becomes so large that the net flow of solvent ceases, as shown in the center panel. Alternatively, we may apply pressure to the left arm of the apparatus, as shown in the panel on the right, to halt the net flow of solvent. The pressure required to prevent osmosis by pure solvent is the **osmotic pressure**, π, of the solution. The osmotic pressure obeys a law similar in form to the ideal-gas law, $\pi V = nRT$, where V is the

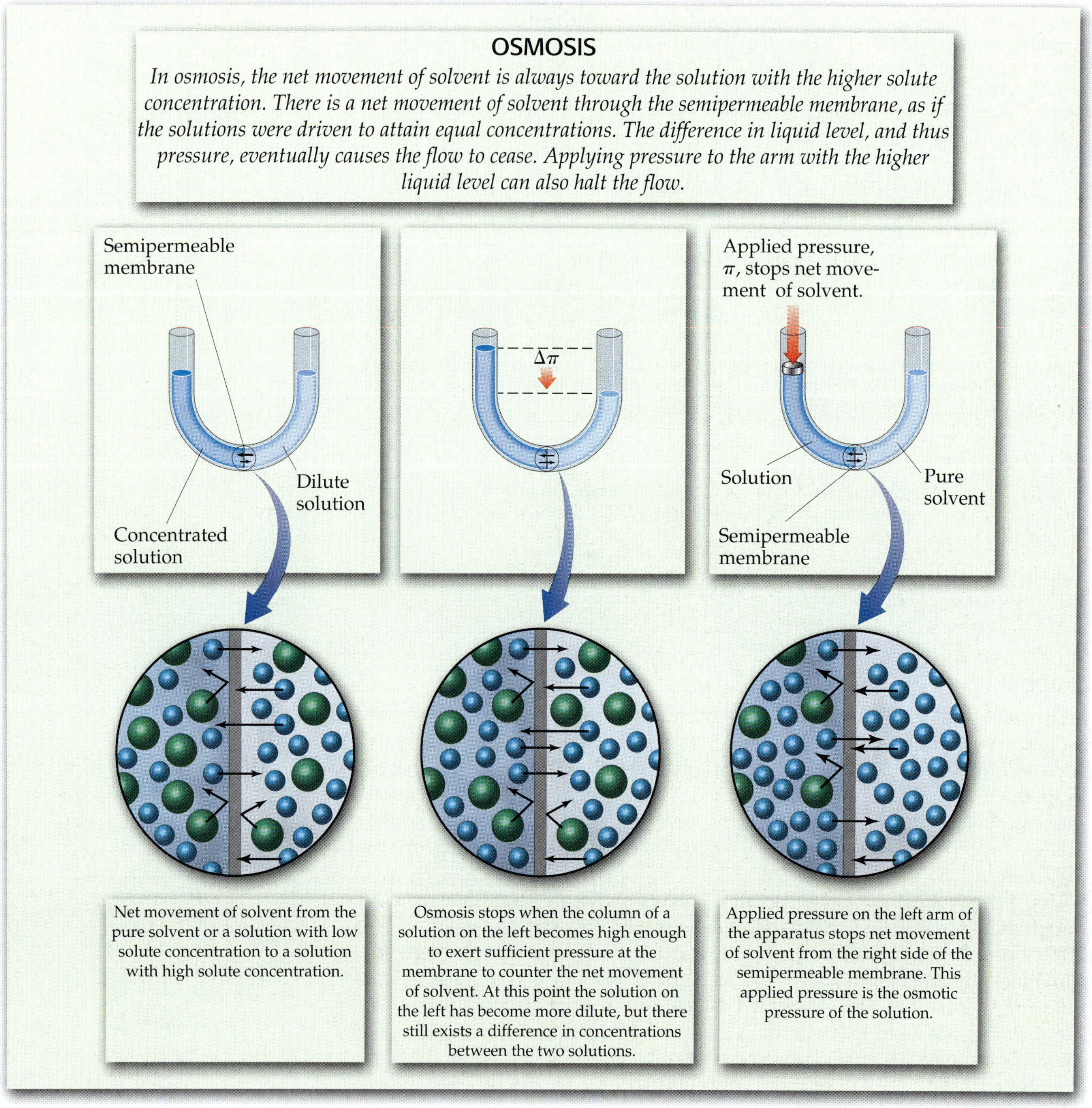

▲ **Figure 13.23 Osmosis.**

volume of the solution, n is the number of moles of solute, R is the ideal-gas constant, and T is the temperature on the Kelvin scale. From this equation, we can write

$$\pi = \left(\frac{n}{V}\right)RT = MRT \qquad [13.13]$$

where M is the molarity of the solution.

If two solutions of identical osmotic pressure are separated by a semipermeable membrane, no osmosis will occur. The two solutions are *isotonic*. If one solution is of lower osmotic pressure, it is *hypotonic* with respect to the more concentrated solution. The more concentrated solution is *hypertonic* with respect to the dilute solution.

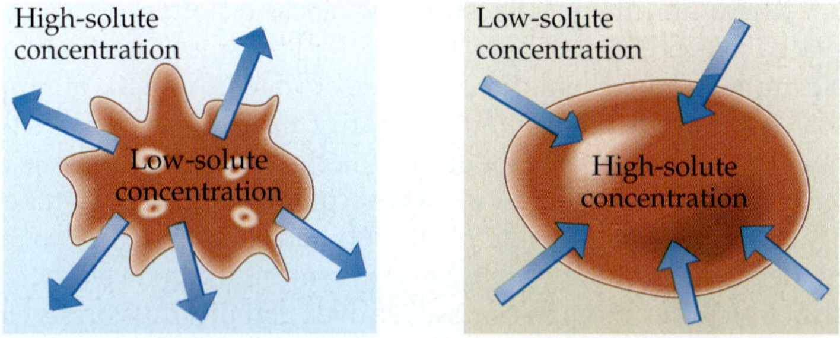

(a) Crenation (b) Hemolysis

◀ **Figure 13.24 Osmosis through red blood cell wall.** The blue arrows represent the net movement of water molecules.

GIVE IT SOME THOUGHT

Of two KBr solutions, one 0.5 *m* and the other 0.20 *m*, which is hypotonic with respect to the other?

Osmosis plays a very important role in living systems. The membranes of red blood cells, for example, are semipermeable. Placing a red blood cell in a solution that is *hyper*tonic relative to the intracellular solution (the solution within the cells) causes water to move out of the cell, as shown in Figure 13.24(a) ▲. This causes the cell to shrivel, a process called *crenation*. Placing the cell in a solution that is *hypo*tonic relative to the intracellular fluid causes water to move into the cell, as illustrated in Figure 13.24(b). This causes the cell to rupture, a process called *hemolysis*. People who need body fluids or nutrients replaced but cannot be fed orally are given solutions by intravenous (IV) infusion, which feeds nutrients directly into the veins. To prevent crenation or hemolysis of red blood cells, the IV solutions must be isotonic with the intracellular fluids of the cells.

SAMPLE EXERCISE 13.11 | Calculations Involving Osmotic Pressure

The average osmotic pressure of blood is 7.7 atm at 25°C. What concentration of glucose ($C_6H_{12}O_6$) will be isotonic with blood?

Solution

Analyze: We are asked to calculate the concentration of glucose in water that would be isotonic with blood, given that the osmotic pressure of blood at 25°C is 7.7 atm.

Plan: Because we are given the osmotic pressure and temperature, we can solve for the concentration, using Equation 13.13.

Solve:

$$\pi = MRT$$

$$M = \frac{\pi}{RT} = \frac{7.7 \text{ atm}}{\left(0.0821 \frac{\text{L-atm}}{\text{mol-K}}\right)(298 \text{ K})} = 0.31 \text{ M}$$

Comment: In clinical situations the concentrations of solutions are generally expressed as mass percentages. The mass percentage of a 0.31 *M* solution of glucose is 5.3%. The concentration of NaCl that is isotonic with blood is 0.16 *M* because NaCl ionizes to form two particles, Na^+ and Cl^- (a 0.155 *M* solution of NaCl is 0.310 *M* in particles). A 0.16 *M* solution of NaCl is 0.9 mass % in NaCl. This kind of solution is known as a physiological saline solution.

PRACTICE EXERCISE

What is the osmotic pressure at 20°C of a 0.0020 *M* sucrose ($C_{12}H_{22}O_{11}$) solution?

Answer: 0.048 atm, or 37 torr

There are many interesting biological examples of osmosis. A cucumber placed in concentrated brine loses water via osmosis and shrivels into a pickle. If a carrot that has become limp because of water loss to the atmosphere is placed in water, the water moves into the carrot through osmosis, making it firm once again. People who eat a lot of salty food retain water in tissue cells and intercellular space because of osmosis. The resultant swelling or puffiness is called *edema*. Water moves from soil into plant roots and subsequently into the upper portions of the plant, at least in part because of osmosis. Bacteria on salted meat or candied fruit lose water through osmosis, shrivel, and die, thus preserving the food.

The movement of a substance from an area where its concentration is high to an area where it is low is spontaneous. Biological cells transport not only water, but also other select materials through their membrane walls. This permits nutrients to enter and waste materials to exit. In some cases substances must be moved across the cell membrane from an area of low concentration to one of high concentration. This movement—called *active transport*—is not spontaneous, so cells must expend energy to do it.

GIVE IT SOME THOUGHT

Which would have the higher osmotic pressure, a 0.10 *M* solution of NaCl or a 0.10 *M* solution of KBr?

ACTIVITY
Determination of Molar Mass

VCL Molar Mass Determination by Freezing-Point Depression

Determination of Molar Mass

The colligative properties of solutions provide a useful means of experimentally determining molar mass. Any of the four colligative properties can be used, as shown in Sample Exercises 13.12 and 13.13.

SAMPLE EXERCISE 13.12 | Molar Mass from Freezing-Point Depression

A solution of an unknown nonvolatile electrolyte was prepared by dissolving 0.250 g of the substance in 40.0 g of CCl_4. The boiling point of the resultant solution was 0.357°C higher than that of the pure solvent. Calculate the molar mass of the solute.

Solution

Analyze: Our goal is to calculate the molar mass of a solute based on knowledge of the boiling-point elevation of its solution in CCl_4, $\Delta T_b = 0.357°C$, and the masses of solute and solvent. Table 13.4 gives K_b for the solvent (CCl_4), $K_b = 5.02°C/m$.

Plan: We can use Equation 13.11, $\Delta T_b = K_b m$, to calculate the molality of the solution. Then we can use molality and the quantity of solvent (40.0 g CCl_4) to calculate the number of moles of solute. Finally, the molar mass of the solute equals the number of grams per mole, so we divide the number of grams of solute (0.250 g) by the number of moles we have just calculated.

Solve: From Equation 13.11 we have

$$\text{Molality} = \frac{\Delta T_b}{K_b} = \frac{0.357°C}{5.02°C/m} = 0.0711\ m$$

Thus, the solution contains 0.0711 mol of solute per kilogram of solvent. The solution was prepared using 40.0 g = 0.0400 kg of solvent (CCl_4). The number of moles of solute in the solution is therefore

$$(0.0400\ \text{kg CCl}_4)\left(0.0711\frac{\text{mol solute}}{\text{kg CCl}_4}\right) = 2.84 \times 10^{-3}\ \text{mol solute}$$

The molar mass of the solute is the number of grams per mole of the substance:

$$\text{Molar mass} = \frac{0.250\ \text{g}}{2.84 \times 10^{-3}\ \text{mol}} = 88.0\ \text{g/mol}$$

PRACTICE EXERCISE

Camphor ($C_{10}H_{16}O$) melts at 179.8°C, and it has a particularly large freezing-point-depression constant, $K_f = 40.0°C/m$. When 0.186 g of an organic substance of unknown molar mass is dissolved in 22.01 g of liquid camphor, the freezing point of the mixture is found to be 176.7°C. What is the molar mass of the solute?

Answer: 110 g/mol

A CLOSER LOOK | Colligative Properties of Electrolyte Solutions

The colligative properties of solutions depend on the total concentration of solute particles, regardless of whether the particles are ions or molecules. Thus, we would expect a 0.100 *m* solution of NaCl to have a freezing-point depression of $(0.200\ m)(1.86°C/m) = 0.372°C$ because it is 0.100 *m* in $Na^+(aq)$ and 0.100 *m* in $Cl^-(aq)$. The measured freezing-point depression is only 0.348°C, however, and the situation is similar for other strong electrolytes. A 0.100 *m* solution of KCl, for example, freezes at −0.344°C.

The difference between the expected and observed colligative properties for strong electrolytes is due to electrostatic attractions between ions. As the ions move about in solution, ions of opposite charge collide and "stick together" for brief moments. While they are together, they behave as a single particle called an *ion pair* (Figure 13.25 ▼). The number of independent particles is thereby reduced, causing a reduction in the freezing-point depression (as well as in the boiling-point elevation, the vapor-pressure reduction, and the osmotic pressure).

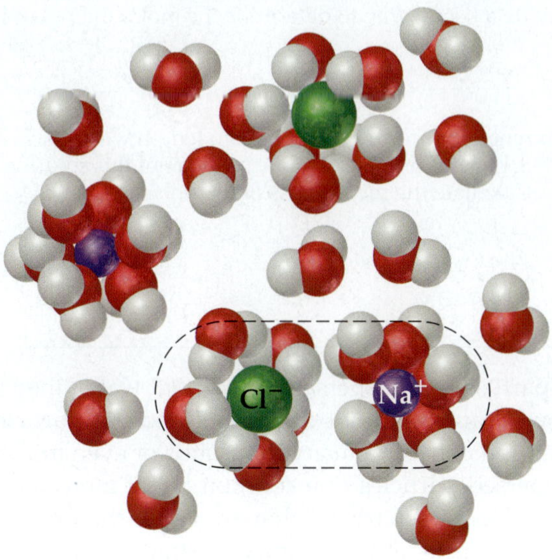

▲ **Figure 13.25 Ion pairing and colligative properties.** A solution of NaCl contains not only separated $Na^+(aq)$ and $Cl^-(aq)$ ions but ion pairs as well. Ion pairing becomes more prevalent as the solution concentration increases and has an effect on all the colligative properties of the solution.

One measure of the extent to which electrolytes dissociate is the *van't Hoff factor, i*. This factor is the ratio of the actual value of a colligative property to the value calculated, assuming the substance to be a nonelectrolyte. Using the freezing-point depression, for example, we have

$$i = \frac{\Delta T_f(\text{measured})}{\Delta T_f(\text{calculated for nonelectrolyte})} \qquad [13.14]$$

The ideal value of *i* can be determined for a salt from the number of ions per formula unit. For NaCl, for example, the ideal van't Hoff factor is 2 because NaCl consists of one Na^+ and one Cl^- per formula unit; for K_2SO_4 it is 3 because K_2SO_4 consists of two K^+ and one SO_4^{2-}. In the absence of any information about the actual value of *i* for a solution, we will use the ideal value in calculations.

Table 13.5 ▼ gives the observed van't Hoff factors for several substances at different dilutions. Two trends are evident in these data. First, dilution affects the value of *i* for electrolytes; the more dilute the solution, the more closely *i* approaches the ideal or limiting value. Thus, the extent of ion pairing in electrolyte solutions decreases upon dilution. Second, the lower the charges on the ions, the less *i* departs from the limiting value because the extent of ion pairing decreases as the ionic charges decrease. Both trends are consistent with simple electrostatics: The force of interaction between charged particles decreases as their separation increases and as their charges decrease.

Related Exercises: 13.75, 13.76, 13.94, 13.97

TABLE 13.5 van't Hoff Factors for Several Substances at 25°C

	Concentration			
Compound	**0.100 *m***	**0.0100 *m***	**0.00100 *m***	**Limiting Value**
Sucrose	1.00	1.00	1.00	1.00
NaCl	1.87	1.94	1.97	2.00
K_2SO_4	2.32	2.70	2.84	3.00
$MgSO_4$	1.21	1.53	1.82	2.00

SAMPLE EXERCISE 13.13 | Molar Mass from Osmotic Pressure

The osmotic pressure of an aqueous solution of a certain protein was measured in order to determine the protein's molar mass. The solution contained 3.50 mg of protein dissolved in sufficient water to form 5.00 mL of solution. The osmotic pressure of the solution at 25°C was found to be 1.54 torr. Calculate the molar mass of the protein.

Solution

Analyze: Our goal is to calculate the molar mass of a high-molecular-mass protein, based on its osmotic pressure and a knowledge of the mass of protein and solution volume.

Plan: The temperature ($T = 25°C$) and osmotic pressure ($\pi = 1.54$ torr) are given, and we know the value of R so we can use Equation 13.13 to calculate the molarity of the solution, M. In doing so, we must convert temperature from °C to K and the osmotic pressure from torr to atm. We then use the molarity and the volume of the solution (5.00 mL) to determine the number of moles of solute. Finally, we obtain the molar mass by dividing the mass of the solute (3.50 mg) by the number of moles of solute.

Solve: Solving Equation 13.13 for molarity gives

$$\text{Molarity} = \frac{\pi}{RT} = \frac{(1.54\ \text{torr})\left(\dfrac{1\ \text{atm}}{760\ \text{torr}}\right)}{\left(0.0821\dfrac{\text{L-atm}}{\text{mol-K}}\right)(298\ \text{K})} = 8.28 \times 10^{-5}\frac{\text{mol}}{\text{L}}$$

Because the volume of the solution is 5.00 ml $= 5.00 \times 10^{-3}$ L, the number of moles of protein must be

$$\text{Moles} = (8.28 \times 10^{-5}\ \text{mol/L})(5.00 \times 10^{-3}\ \text{L}) = 4.14 \times 10^{-7}\ \text{mol}$$

The molar mass is the number of grams per mole of the substance. The sample has a mass of 3.50 mg $= 3.50 \times 10^{-3}$ g. The molar mass is the number of grams divided by the number of moles:

$$\text{Molar mass} = \frac{\text{grams}}{\text{moles}} = \frac{3.50 \times 10^{-3}\ \text{g}}{4.14 \times 10^{-7}\ \text{mol}} = 8.45 \times 10^{3}\ \text{g/mol}$$

Comment: Because small pressures can be measured easily and accurately, osmotic pressure measurements provide a useful way to determine the molar masses of large molecules.

PRACTICE EXERCISE

A sample of 2.05 g of polystyrene of uniform polymer chain length was dissolved in enough toluene to form 0.100 L of solution. The osmotic pressure of this solution was found to be 1.21 kPa at 25°C. Calculate the molar mass of the polystyrene.

Answer: 4.20×10^{4} g/mol

13.6 | Colloids

When finely divided clay particles are dispersed throughout water, they eventually settle out of the water because of gravity. The dispersed clay particles are much larger than molecules and consist of many thousands or even millions of atoms. In contrast, the dispersed particles of a solution are of molecular size. Between these extremes lie dispersed particles that are larger than molecules, but not so large that the components of the mixture separate under the influence of gravity. These intermediate types of dispersions or suspensions are called **colloidal dispersions**, or simply **colloids**. Colloids form the dividing line between solutions and heterogeneous mixtures. Like solutions, colloids can be gases, liquids, or solids. Examples of each are listed in Table 13.6 ▼.

TABLE 13.6 Types of Colloids

Phase of Colloid	Dispersing (solventlike) Substance	Dispersed (solutelike) Substance	Colloid Type	Example
Gas	Gas	Gas	—	None (all are solutions)
Gas	Gas	Liquid	Aerosol	Fog
Gas	Gas	Solid	Aerosol	Smoke
Liquid	Liquid	Gas	Foam	Whipped cream
Liquid	Liquid	Liquid	Emulsion	Milk
Liquid	Liquid	Solid	Sol	Paint
Solid	Solid	Gas	Solid foam	Marshmallow
Solid	Solid	Liquid	Solid emulsion	Butter
Solid	Solid	Solid	Solid sol	Ruby glass

The size of the dispersed particles is used to classify a mixture as a colloid. Colloid particles range in diameter from approximately 5 to 1000 nm. Solute particles are smaller. The colloid particle may consist of many atoms, ions, or molecules, or it may even be a single giant molecule. The hemoglobin molecule, for example, which carries oxygen in blood, has molecular dimensions of 65 Å × 55 Å × 50 Å and a molecular weight of 64,500 amu.

Although colloid particles may be so small that the dispersion appears uniform even under a microscope, they are large enough to scatter light very effectively. Consequently, most colloids appear cloudy or opaque unless they are very dilute. (Homogenized milk is a colloid.) Furthermore, because they scatter light, a light beam can be seen as it passes through a colloidal suspension, as shown in Figure 13.26 ▶. This scattering of light by colloidal particles, known as the **Tyndall effect**, makes it possible to see the light beam of an automobile on a dusty dirt road or the sunlight coming through a forest canopy [Figure 13.27(a) ▼]. Not all wavelengths are scattered to the same extent. As a result, brilliant red sunsets are seen when the sun is near the horizon and the air contains dust, smoke, or other particles of colloidal size [Figure 13.27(b)].

▲ **Figure 13.26 Tyndall effect in the laboratory.** The glass on the left contains a colloidal suspension; that on the right contains a solution. The path of the beam through the colloidal suspension is visible because the light is scattered by the colloidal particles. Light is not scattered by the individual solute molecules in the solution.

◀ **Figure 13.27 Tyndall effect in nature.** (a) Scattering of sunlight by colloidal particles in the misty air of a forest. (b) The scattering of light by smoke or dust particles produces a rich red sunset.

Hydrophilic and Hydrophobic Colloids

The most important colloids are those in which the dispersing medium is water. These colloids may be **hydrophilic** (water loving) or **hydrophobic** (water fearing). Hydrophilic colloids are most like the solutions that we have previously examined. In the human body the extremely large molecules that make up such important substances as enzymes and antibodies are kept in suspension by interaction with surrounding water molecules. The molecules fold in such a way that the hydrophobic groups are away from the water molecules, on the "inside" of the folded molecule, while the hydrophilic, polar groups are on the surface, interacting with the water molecules. These hydrophilic groups generally contain oxygen or nitrogen and often carry a charge. Some examples are shown in Figure 13.28 ▶.

Hydrophobic colloids can be prepared in water only if they are stabilized in some way. Otherwise, their natural lack of affinity for water causes them to separate from the water. Hydrophobic colloids can be stabilized by adsorption of ions on their surface, as shown in Figure 13.29 ▶. (*Adsorption* means to adhere to a surface. It differs from *absorption*, which means to pass into the interior, as when a sponge absorbs water.) These adsorbed ions can interact with water, thereby stabilizing the colloid. At the same time, the mutual repulsion between colloid particles with adsorbed ions of the same charge keeps the particles from colliding and getting larger.

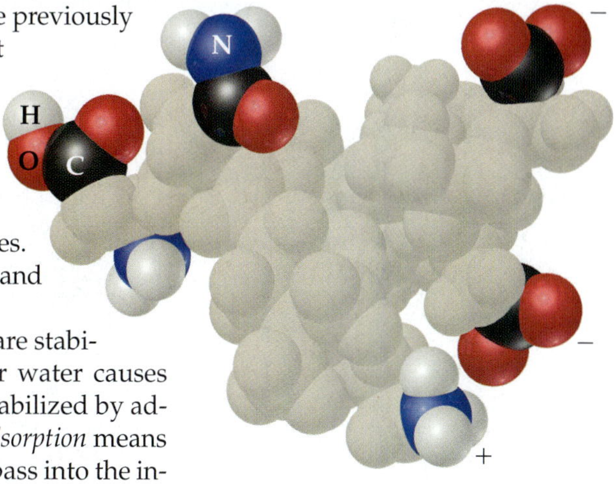

▲ **Figure 13.28 Hydrophilic colloids.** Examples of hydrophilic groups on the surface of a giant molecule (macromolecule) that help to keep the molecule suspended in water.

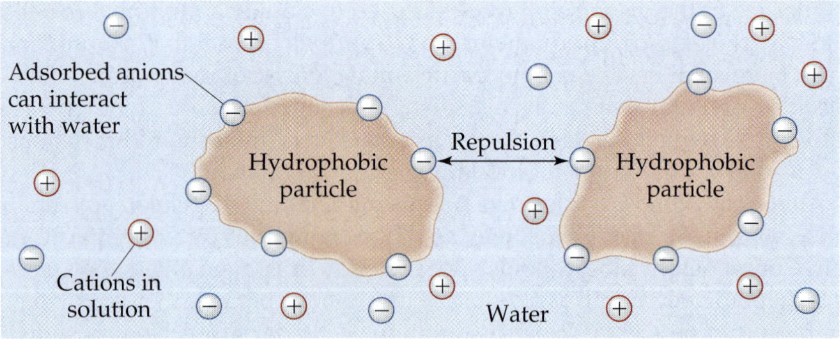

▶ **Figure 13.29 Hydrophobic colloids.** Schematic illustration of how adsorbed ions stabilize a hydrophobic colloid in water.

3-D MODEL Sodium Stearate

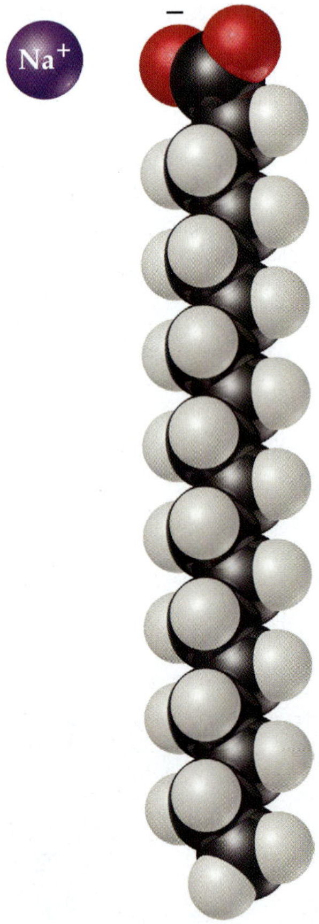

▲ **Figure 13.30 Sodium stearate.**

Hydrophobic colloids can also be stabilized by the presence of hydrophilic groups on their surfaces. Small droplets of oil are hydrophobic, for example, so they do not remain suspended in water. Instead, they aggregate, forming an oil slick on the surface of the water. Sodium stearate (Figure 13.30 ◀), or any similar substance having one end that is hydrophilic (polar, or charged) and one that is hydrophobic (nonpolar), will stabilize a suspension of oil in water. Stabilization results from the interaction of the hydrophobic ends of the stearate ions with the oil droplet and the hydrophilic ends with the water as shown in Figure 13.31 ▼.

GIVE IT SOME THOUGHT

Why don't the oil droplets emulsified by sodium stearate coagulate to form larger oil droplets?

The stabilization of colloids has an interesting application in our own digestive system. When fats in our diet reach the small intestine, a hormone causes the gallbladder to excrete a fluid called bile. Among the components of bile are compounds that have chemical structures similar to sodium stearate; that is, they have a hydrophilic (polar) end and a hydrophobic (nonpolar) end. These compounds emulsify the fats present in the intestine and thus permit digestion and absorption of fat-soluble vitamins through the intestinal wall. The term *emulsify* means "to form an emulsion," a suspension of one liquid in another, as in milk, for example (Table 13.6). A substance that aids in the formation of an emulsion is called an emulsifying agent. If you read the labels on foods and other materials, you will find that a variety of chemicals are used as emulsifying agents. These chemicals typically have a hydrophilic end and a hydrophobic end.

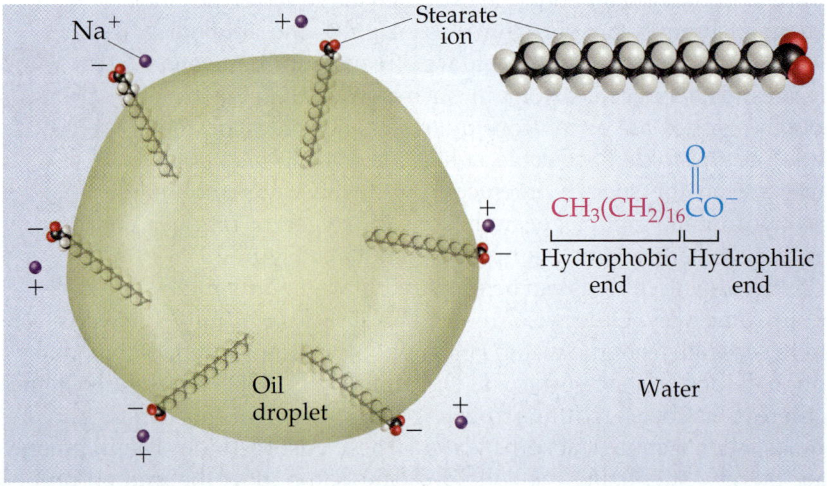

▲ **Figure 13.31 Stabilization of an emulsion of oil in water by stearate ions.**

Removal of Colloidal Particles

Colloidal particles frequently must be removed from a dispersing medium, as in the removal of smoke from stacks or butterfat from milk. Because colloidal particles are so small, they cannot be separated by simple filtration. Instead, the colloidal particles must be enlarged in a process called *coagulation*. The resultant larger particles can then be separated by filtration or merely by allowing them to settle out of the dispersing medium.

CHEMISTRY AND LIFE | Sickle-Cell Anemia

Our blood contains a complex protein called *hemoglobin* that carries oxygen from our lungs to other parts of our body. In the genetic disease known as *sickle-cell anemia*, hemoglobin molecules are abnormal and have a lower solubility, especially in their unoxygenated form. Consequently, as much as 85% of the hemoglobin in red blood cells crystallizes from solution.

The reason for the insolubility of hemoglobin in sickle-cell anemia can be traced to a structural change in one part of an amino acid side chain. Normal hemoglobin molecules have an amino acid in their makeup that has the following side chain protruding from the main body of the molecule:

$$—CH_2—CH_2—\overset{\overset{\large O}{\|}}{C}—OH$$

Normal

Normal

This side chain terminates in a polar group, which contributes to the solubility of the hemoglobin molecule in water. In the hemoglobin molecules of persons suffering from sickle-cell anemia, the side chain is of a different type:

$$\underset{\underset{\large CH_3}{|}}{—CH}—CH_3$$

Abnormal

Abnormal

This abnormal group of atoms is nonpolar (hydrophobic), and its presence leads to the aggregation of this defective form of hemoglobin into particles too large to remain suspended in biological fluids. It also causes the cells to distort into a sickle shape, as shown in Figure 13.32 ▼. The sickled cells tend to clog the capillaries, causing severe pain, physical weakness, and the gradual deterioration of the vital organs. The disease is hereditary, and if both parents carry the defective genes, it is likely that their children will possess only abnormal hemoglobin.

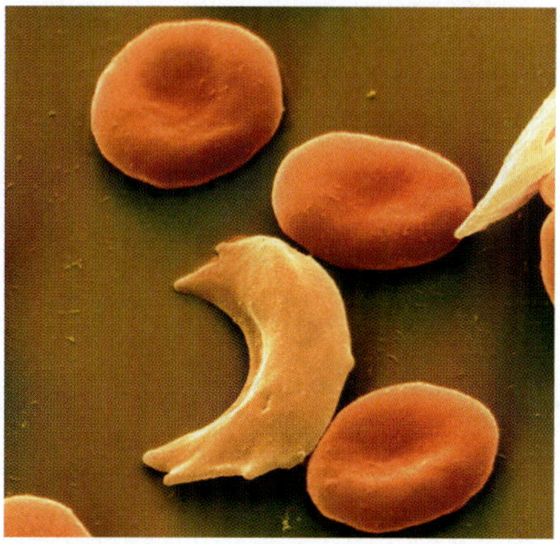

▶ **Figure 13.32 Normal and sickled red blood cells.** Normal red blood cells are about 1×10^{-3}mm in diameter.

Heating the mixture or adding an electrolyte may bring about coagulation. Heating the colloidal dispersion increases the particle motion and so the number of collisions. The particles increase in size as they stick together after colliding. The addition of electrolytes neutralizes the surface charges of the particles, thereby removing the electrostatic repulsions that prevent them from coming together. Wherever rivers empty into oceans or other salty bodies of water, for example, the suspended clay in the river is deposited as a delta when it mixes with the electrolytes in the salt water.

Semipermeable membranes can also be used to separate ions from colloidal particles because the ions can pass through the membrane but the colloidal particles cannot. This type of separation is known as *dialysis* and is used to purify blood in artificial kidney machines. Our kidneys normally remove the waste products of metabolism from blood. In a kidney machine, blood is circulated through a dialyzing tube immersed in a washing solution. The washing solution is isotonic in ions that must be retained by the blood, but is lacking the waste products. Wastes therefore dialyze out of the blood, but the ions do not.

SAMPLE INTEGRATIVE EXERCISE | Putting Concepts Together

A 0.100-L solution is made by dissolving 0.441 g of $CaCl_2(s)$ in water. **(a)** Calculate the osmotic pressure of this solution at 27°C, assuming that it is completely dissociated into its component ions. **(b)** The measured osmotic pressure of this solution is 2.56 atm at 27°C. Explain why it is less than the value calculated in (a), and calculate the van't Hoff factor, *i*, for the solute in this solution. (See the "Closer Look" box in Section 13.5 on page 557.) **(c)** The enthalpy of solution for $CaCl_2$ is $\Delta H = -81.3$ kJ/mol. If the final temperature of the solution was 27.0°C, what was its initial temperature? (Assume that the density of the solution is 1.00 g/mL, that its specific heat is 4.18 J/g-K, and that the solution loses no heat to its surroundings.)

Solution

(a) The osmotic pressure is given by Equation 13.13, $\pi = MRT$. We know the temperature, $T = 27°C = 300$ K and the gas constant, $R = 0.0821$ L-atm/mol-K. We can calculate the molarity of the solution from the mass of $CaCl_2$ and the volume of the solution:

$$\text{Molarity} = \left(\frac{0.441 \text{ g CaCl}_2}{0.100 \text{ L}}\right)\left(\frac{1 \text{ mol CaCl}_2}{111.0 \text{ g CaCl}_2}\right) = 0.0397 \text{ mol CaCl}_2/\text{L}$$

Soluble ionic compounds are strong electrolytes. (Sections 4.1 and 4.3) Thus, $CaCl_2$ consists of metal cations (Ca^{2+}) and nonmetal anions (Cl^-). When completely dissociated, each $CaCl_2$ unit forms three ions (one Ca^{2+} and two Cl^-). Hence the total concentration of ions in the solution is $(3)(0.0397\ M) = 0.119\ M$, and the osmotic pressure is

$$\pi = MRT = (0.119 \text{ mol/L})(0.0821 \text{ L-atm/mol-K})(300 \text{ K}) = 2.93 \text{ atm}$$

(b) The actual values of colligative properties of electrolytes are less than those calculated, because the electrostatic interactions between ions limit their independent movements. In this case the van't Hoff factor, which measures the extent to which electrolytes actually dissociate into ions, is given by

$$i = \frac{\pi(\text{measured})}{\pi(\text{calculated for nonelectrolyte})}$$

$$= \frac{2.56 \text{ atm}}{(0.0397 \text{ mol/L})(0.0821 \text{ L-atm/mol-K})(300 \text{ K})} = 2.62$$

Thus, the solution behaves as if the $CaCl_2$ has dissociated into 2.62 particles instead of the ideal 3.

(c) If the solution is 0.0397 M in $CaCl_2$ and has a total volume of 0.100 L, the number of moles of solute is $(0.100 \text{ L})(0.0397 \text{ mol/L}) = 0.00397$ mol. Hence the quantity of heat generated in forming the solution is $(0.00397 \text{ mol})(-81.3 \text{ kJ/mol}) = -0.323$ kJ. The solution absorbs this heat, causing its temperature to increase. The relationship between temperature change and heat is given by Equation 5.19:

$$q = (\text{specific heat})(\text{grams})(\Delta T)$$

The heat absorbed by the solution is $q = +0.323\ \text{kJ} = 323\ \text{J}$. The mass of the 0.100 L of solution is (100 mL)(1.00 g/mL) = 100 g (to 3 significant figures). Thus the temperature change is

$$\Delta T = \frac{q}{(\text{specific heat of solution})(\text{grams of solution})}$$

$$= \frac{323\ \text{J}}{(4.18\ \text{J/g-K})(100\ \text{g})} = 0.773\ \text{K}$$

A kelvin has the same size as a degree Celsius. (Section 1.4) Because the solution temperature increases by 0.773°C, the initial temperature was 27.0°C − 0.773°C = 26.2°C.

SUMMARY AND KEY TERMS

Section 13.1 Solutions form when one substance disperses uniformly throughout another. The attractive interaction of solvent molecules with solute is called **solvation**. When the solvent is water, the interaction is called **hydration**. The dissolution of ionic substances in water is promoted by hydration of the separated ions by the polar water molecules.

The overall enthalpy change upon solution formation may be either positive or negative. Solution formation is favored both by a negative enthalpy change (exothermic process) and by an increased dispersal in space of the components of the solution, corresponding to a positive **entropy** change.

Section 13.2 The equilibrium between a saturated solution and undissolved solute is dynamic; the process of solution and the reverse process, **crystallization**, occur simultaneously. In a solution in equilibrium with undissolved solute, the two processes occur at equal rates, giving a **saturated** solution. If there is less solute present than is needed to saturate the solution, the solution is **unsaturated**. When solute concentration is greater than the equilibrium concentration value, the solution is **supersaturated**. This is an unstable condition, and separation of some solute from the solution will occur if the process is initiated with a solute seed crystal. The amount of solute needed to form a saturated solution at any particular temperature is the **solubility** of that solute at that temperature.

Section 13.3 The solubility of one substance in another depends on the tendency of systems to become more random, by becoming more dispersed in space, and on the relative intermolecular solute-solute and solvent-solvent energies compared with solute-solvent interactions. Polar and ionic solutes tend to dissolve in polar solvents, and nonpolar solutes tend to dissolve in nonpolar solvents ("like dissolves like"). Liquids that mix in all proportions are **miscible**; those that do not dissolve significantly in one another are **immiscible**. Hydrogen-bonding interactions between solute and solvent often play an important role in determining solubility; for example, ethanol and water, whose molecules form hydrogen bonds with each other, are miscible.

The solubilities of gases in a liquid are generally proportional to the pressure of the gas over the solution, as expressed by **Henry's law**: $S_g = kP_g$. The solubilities of most solid solutes in water increase as the temperature of the solution increases. In contrast, the solubilities of gases in water generally decrease with increasing temperature.

Section 13.4 Concentrations of solutions can be expressed quantitatively by several different measures, including **mass percentage** [(mass solute/mass solution) × 10^2], **parts per million (ppm)** [(mass solute / mass solution) × 10^6], **parts per billion (ppb)** [(mass solute/mass solution) × 10^9], and mole fraction [mass solute/(mol solute + mol solvent)]. Molarity, M, is defined as moles of solute per liter of solution; **molality**, m, is defined as moles of solute per kg of solvent. Molarity can be converted to these other concentration units if the density of the solution is known.

Section 13.5 A physical property of a solution that depends on the concentration of solute particles present, regardless of the nature of the solute, is a **colligative property**. Colligative properties include vapor-pressure lowering, freezing-point lowering, boiling-point elevation, and osmotic pressure. The lowering of vapor pressure is expressed by **Raoult's law**. An **ideal solution** obeys Raoult's law. Differences in solvent-solute as compared with solvent-solvent and solute-solute intermolecular forces cause many solutions to depart from ideal behavior.

A solution containing a nonvolatile solute possesses a higher boiling point than the pure solvent. The **molal boiling-point-elevation constant**, K_b, represents the increase in boiling point for a 1 m solution of solute particles as compared with the pure solvent. Similarly, the **molal freezing-point-depression constant**, K_f, measures the lowering of the freezing point of a solution for a 1 m solution of solute particles. The temperature changes are given by the equations $\Delta T_b = K_b m$ and $\Delta T_f = K_f m$. When NaCl dissolves in water, two moles of solute particles are formed for each mole of dissolved salt. The boiling point or freezing point is thus elevated or depressed, respectively, approximately twice as much as that of a nonelectrolyte solution of the same concentration. Similar considerations apply to other strong electrolytes.

Osmosis is the movement of solvent molecules through a semipermeable membrane from a less concentrated to a more concentrated solution. This net movement of solvent generates an **osmotic pressure**, π, which can be measured in units of gas pressure, such as atm. The osmotic pressure of a solution as compared with pure solvent is proportional to the solution molarity: $\pi = MRT$. Osmosis is a very important process in living systems, in which cell membranes act as semipermeable walls, permitting the passage of water, but restricting the passage of ionic and macromolecular components.

Section 13.6 Particles that are large on the molecular scale but still small enough to remain suspended indefinitely in a solvent system form **colloids**, or **colloidal dispersions**. Colloids, which are intermediate between solutions and heterogeneous mixtures, have many practical applications. One useful physical property of colloids, the scattering of visible light, is referred to as the **Tyndall effect**. Aqueous colloids are classified as **hydrophilic** or **hydrophobic**. Hydrophilic colloids are common in living organisms, in which large molecular aggregates (enzymes, antibodies) remain suspended because they have many polar, or charged, atomic groups on their surfaces that interact with water. Hydrophobic colloids, such as small droplets of oil, may remain in suspension through adsorption of charged particles on their surfaces.

VISUALIZING CONCEPTS

13.1 This figure shows the interaction of a cation with surrounding water molecules.

Would you expect the energy of ion-solvent interaction to be greater for Na^+ or Li^+? Explain. [Section 13.1]

13.2 Referring to the illustration for Exercise 13.1, which ΔH term in Figure 13.4 (either diagram) contains the enthalpy change corresponding to the interaction of the cation with solvent? [Section 13.1]

13.3 A quantity of the pink solid on the left in Figure 13.8 is placed in a warming oven and heated for a time. It slowly turns from pink to the deep blue color of the solid on the right. What has occurred? [Section 13.1]

13.4 Which of the following is the best representation of a saturated solution? Explain your reasoning. [Section 13.2]

(a) (b) (c)

13.5 The solubility of Xe in water at 20°C is approximately 5×10^{-3} *M*. Compare this with the solubilities of Ar and Kr in water, Table 13.2, and explain what properties of the rare gas atoms account for the variation in solubility. [Section 13.3]

13.6 The structures of vitamins E and B_6 are shown below. Predict which is largely water soluble, and which is largely fat soluble. Explain. [Section 13.3]

Vitamin B_6 Vitamin E

13.7 The figure shows two volumetric flasks containing the same solution at two temperatures.

(a) Does the molarity of the solution change with the change in temperature? Explain.

(b) Does the molality of the solution change with the change in temperature? Explain. [Section 13.4]

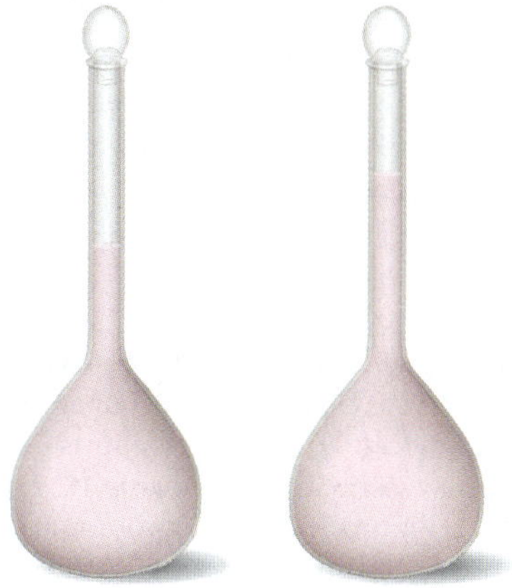

13.8 Suppose you had a balloon made of some highly flexible semipermeable membrane. The balloon is filled completely with a 0.2 M solution of some solute and is submerged in a 0.1 M solution of the same solute:

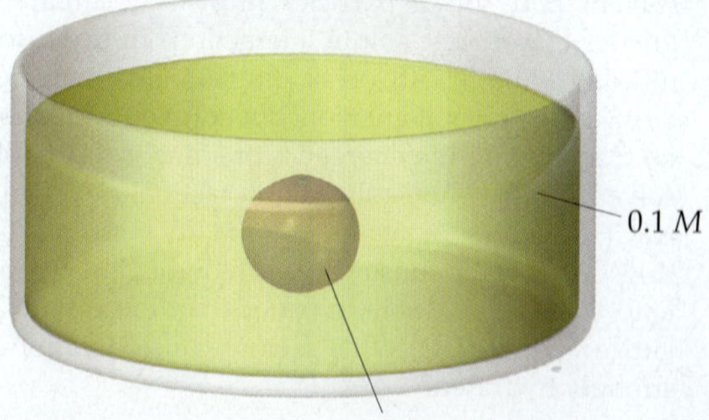

Initially, the volume of solution in the balloon is 0.25 L. Assuming the volume outside the semipermeable membrane is large, as the illustration shows, what would you expect for the solution volume inside the balloon once the system has come to equilibrium through osmosis? [Section 13.5]

13.9 The molecule *n*-octylglucoside, shown here, is widely used in biochemical research as a nonionic detergent for "solubilizing" large hydrophobic protein molecules. What characteristics of this molecule are important for its use in this way? [Section 13.6]

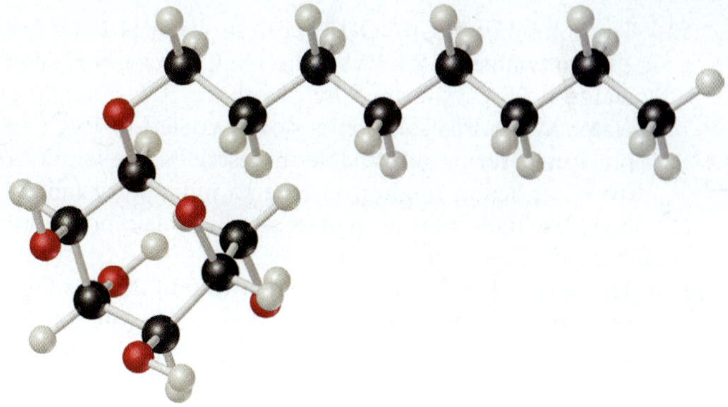

13.10 If you wanted to prepare a solution of CO in water at 25°C in which the CO concentration was 2.5 m*M*, what pressure of CO would you need to use? (See Figure 13.18.) [Section 13.3]

EXERCISES

The Solution Process

13.11 In general, the attractive intermolecular forces between solvent and solute particles must be comparable or greater than solute-solute interactions in order for significant solubility to occur. Explain this statement in terms of the overall energetics of solution formation.

13.12 **(a)** Considering the energetics of solute-solute, solvent-solvent, and solute-solvent interactions, explain why NaCl dissolves in water but not in benzene (C_6H_6). **(b)** Why do ionic substances with higher lattice energies tend to be less soluble in water than those with lower lattice energies? **(c)** What factors cause a cation to be strongly hydrated?

13.13 Indicate the type of solute-solvent interaction (Section 11.2) that should be most important in each of the following solutions: **(a)** CCl_4 in benzene (C_6H_6), **(b)** methanol (CH_3OH) in water, **(c)** KBr in water, **(d)** HCl in acetonitrile (CH_3CN).

13.14 Indicate the principal type of solute-solvent interaction in each of the following solutions, and rank the solutions from weakest to strongest solute-solvent interaction: **(a)** KCl in water, **(b)** CH_2Cl_2 in benzene (C_6H_6), **(c)** methanol (CH_3OH) in water.

13.15 **(a)** In Equation 13.1 which of the energy terms for dissolving an ionic solid would correspond to the lattice energy? **(b)** Which energy terms in this equation are always exothermic?

13.16 The schematic diagram of the solution process as the net sum of three steps in Figure 13.4 does not show the relative magnitudes of the three components because these will vary from case to case. For the dissolution of NH_4NO_3 in water, which of the three enthalpy changes would you expect to be much smaller than the other two? Explain.

13.17 When two nonpolar organic liquids such as hexane (C_6H_{14}) and heptane (C_7H_{16}) are mixed, the enthalpy change that occurs is generally quite small. **(a)** Use the energy diagram in Figure 13.4 to explain why. **(b)** Given that $\Delta H_{soln} \approx 0$, explain why hexane and heptane spontaneously form a solution.

13.18 The enthalpy of solution of KBr in water is about +198 kJ/mol. Nevertheless, the solubility of KBr in water is relatively high. Why does the solution process occur even though it is endothermic?

Saturated Solutions; Factors Affecting Solubility

13.19 The solubility of $Cr(NO_3)_3 \cdot 9\,H_2O$ in water is 208 g per 100 g of water at 15°C. A solution of $Cr(NO_3)_3 \cdot 9\,H_2O$ in water at 35°C is formed by dissolving 324 g in 100 g water. When this solution is slowly cooled to 15°C, no precipitate forms. **(a)** What term describes this solution? **(b)** What action might you take to initiate crystallization? Use molecular-level processes to explain how your suggested procedure works.

13.20 The solubility of $MnSO_4 \cdot H_2O$ in water at 20°C is 70 g per 100 mL of water. **(a)** Is a 1.22 *M* solution of $MnSO_4 \cdot H_2O$ in water at 20°C saturated, supersaturated, or unsaturated? **(b)** Given a solution of $MnSO_4 \cdot H_2O$ of unknown concentration, what experiment could you perform to determine whether the new solution is saturated, supersaturated, or unsaturated?

13.21 By referring to Figure 13.17, determine whether the addition of 40.0 g of each of the following ionic solids to 100 g of water at 40°C will lead to a saturated solution: **(a)** $NaNO_3$, **(b)** KCl, **(c)** $K_2Cr_2O_7$, **(d)** $Pb(NO_3)_2$.

13.22 By referring to Figure 13.17, determine the mass of each of the following salts required to form a saturated solution in 250 g of water at 30°C: **(a)** $KClO_3$, **(b)** $Pb(NO_3)_2$, **(c)** $Ce_2(SO_4)_3$.

13.23 Water and glycerol, $CH_2(OH)CH(OH)CH_2OH$, are miscible in all proportions. What does this mean? How do the OH groups of the alcohol molecule contribute to this miscibility?

13.24 Oil and water are immiscible. What does this mean? Explain in terms of the structural features of their respective molecules and the forces between them.

13.25 **(a)** Would you expect stearic acid, $CH_3(CH_2)_{16}COOH$, to be more soluble in water or in carbon tetrachloride? Explain. **(b)** Which would you expect to be more soluble in water, cyclohexane or dioxane? Explain.

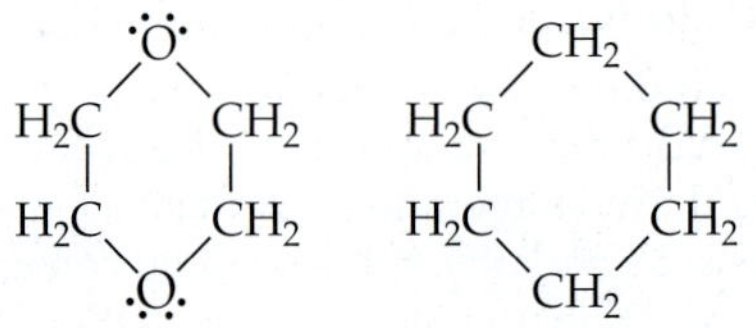

Dioxane Cyclohexane

13.26 Consider a series of carboxylic acids whose general formula is $CH_3(CH_2)_nCOOH$. How would you expect the solubility of these compounds in water and in hexane to change as *n* increases? Explain.

13.27 Which of the following in each pair is likely to be more soluble in hexane, C_6H_{14}: **(a)** CCl_4 or $CaCl_2$; **(b)** benzene (C_6H_6) or glycerol, $CH_2(OH)CH(OH)CH_2OH$; **(c)** octanoic acid, $CH_3CH_2CH_2CH_2CH_2CH_2CH_2COOH$, or acetic acid, CH_3COOH. Explain your answer in each case.

13.28 Which of the following in each pair is likely to be more soluble in water: **(a)** cyclohexane (C_6H_{12}) or glucose ($C_6H_{12}O_6$) (Figure 13.12); **(b)** propionic acid (CH_3CH_2COOH) or sodium propionate (CH_3CH_2COONa); **(c)** HCl or ethyl chloride (CH_3CH_2Cl)? Explain in each case.

13.29 **(a)** Explain why carbonated beverages must be stored in sealed containers. **(b)** Once the beverage has been opened, why does it maintain more carbonation when refrigerated than at room temperature?

13.30 Explain why pressure affects the solubility of O_2 in water, but not the solubility of NaCl in water.

13.31 The Henry's law constant for helium gas in water at 30°C is 3.7×10^{-4} *M*/atm; the constant for N_2 at 30°C is 6.0×10^{-4} *M*/atm. If the two gases are each present at 1.5 atm pressure, calculate the solubility of each gas.

13.32 The partial pressure of O_2 in air at sea level is 0.21 atm. Using the data in Table 13.2, together with Henry's law, calculate the molar concentration of O_2 in the surface water of a mountain lake saturated with air at 20°C and an atmospheric pressure of 665 torr.

Concentrations of Solutions

13.33 **(a)** Calculate the mass percentage of Na_2SO_4 in a solution containing 10.6 g Na_2SO_4 in 483 g water. **(b)** An ore contains 2.86 g of silver per ton of ore. What is the concentration of silver in ppm?

13.34 **(a)** What is the mass percentage of iodine (I_2) in a solution containing 0.045 mol I_2 in 115 g of CCl_4? **(b)** Seawater contains 0.0079 g Sr^{2+} per kilogram of water. What is the concentration of Sr^{2+} measured in ppm?

13.35 A solution is made containing 14.6 g of CH_3OH in 184 g H_2O. Calculate **(a)** the mole fraction of CH_3OH, **(b)** the mass percent of CH_3OH, **(c)** the molality of CH_3OH.

13.36 A solution is made containing 25.5 g phenol (C_6H_5OH) in 495 g ethanol (CH_3CH_2OH). Calculate **(a)** the mole fraction of phenol, **(b)** the mass percent of phenol, **(c)** the molality of phenol.

13.37 Calculate the molarity of the following aqueous solutions: **(a)** 0.540 g $Mg(NO_3)_2$ in 250.0 mL of solution, **(b)** 22.4 g $LiClO_4 \cdot 3\,H_2O$ in 125 mL of solution, **(c)** 25.0 mL of 3.50 M HNO_3 diluted to 0.250 L.

13.38 What is the molarity of each of the following solutions: **(a)** 25.0 g $Al_2(SO_4)_3$ in 0.350 L solution, **(b)** 5.25 g $Mn(NO_3)_2 \cdot 2\,H_2O$ in 175 mL of solution, **(c)** 35.0 mL of 9.00 M H_2SO_4 diluted to 0.500 L?

13.39 Calculate the molality of each of the following solutions: **(a)** 8.66 g benzene (C_6H_6) dissolved in 23.6 g carbon tetrachloride (CCl_4), **(b)** 4.80 g NaCl dissolved in 0.350 L of water.

13.40 **(a)** What is the molality of a solution formed by dissolving 1.50 mol of KCl in 16.0 mol of water? **(b)** How many grams of sulfur (S_8) must be dissolved in 100.0 g naphthalene ($C_{10}H_8$) to make a 0.12 *m* solution?

13.41 A sulfuric acid solution containing 571.6 g of H_2SO_4 per liter of solution has a density of 1.329 g/cm³. Calculate **(a)** the mass percentage, **(b)** the mole fraction, **(c)** the molality, **(d)** the molarity of H_2SO_4 in this solution.

13.42 Ascorbic acid (vitamin C, $C_6H_8O_6$) is a water-soluble vitamin. A solution containing 80.5 g of ascorbic acid dissolved in 210 g of water has a density of 1.22 g/mL at 55°C. Calculate **(a)** the mass percentage, **(b)** the mole fraction, **(c)** the molality, **(d)** the molarity of ascorbic acid in this solution.

13.43 The density of acetonitrile (CH_3CN) is 0.786 g/mL, and the density of methanol (CH_3OH) is 0.791 g/mL. A solution is made by dissolving 22.5 mL CH_3OH in 98.7 mL CH_3CN. **(a)** What is the mole fraction of methanol in the solution? **(b)** What is the molality of the solution? **(c)** Assuming that the volumes are additive, what is the molarity of CH_3OH in the solution?

13.44 The density of toluene (C_7H_8) is 0.867 g/mL, and the density of thiophene (C_4H_4S) is 1.065 g/mL. A solution is made by dissolving 10.0 g of thiophene in 250.0 mL of toluene. **(a)** Calculate the mole fraction of thiophene in the solution. **(b)** Calculate the molality of thiophene in the solution. **(c)** Assuming that the volumes of the solute and solvent are additive, what is the molarity of thiophene in the solution?

13.45 Calculate the number of moles of solute present in each of the following aqueous solutions: **(a)** 600 mL of 0.250 *M* $SrBr_2$, **(b)** 86.4 g of 0.180 *m* KCl, **(c)** 124.0 g of a solution that is 6.45% glucose ($C_6H_{12}O_6$) by mass.

13.46 Calculate the number of moles of solute present in each of the following solutions: **(a)** 245 mL of 1.50 *M* $HNO_3(aq)$, **(b)** 50.0 mg of an aqueous solution that is 1.25 *m* NaCl, **(c)** 75.0 g of an aqueous solution that is 1.50% sucrose ($C_{12}H_{22}O_{11}$) by mass.

13.47 Describe how you would prepare each of the following aqueous solutions, starting with solid KBr: **(a)** 0.75 L of 1.5×10^{-2} *M* KBr, **(b)** 125 g of 0.180 *m* KBr, **(c)** 1.85 L of a solution that is 12.0% KBr by mass (the density of the solution is 1.10 g/mL), **(d)** a 0.150 *M* solution of KBr that contains just enough KBr to precipitate 16.0 g of AgBr from a solution containing 0.480 mol of $AgNO_3$.

13.48 Describe how you would prepare each of the following aqueous solutions: **(a)** 1.50 L of 0.110 *M* $(NH_4)_2SO_4$ solution, starting with solid $(NH_4)_2SO_4$; **(b)** 120 g of a solution that is 0.65 *m* in Na_2CO_3, starting with the solid solute; **(c)** 1.20 L of a solution that is 15.0% of $Pb(NO_3)_2$ by mass (the density of the solution is 1.16 g/mL), starting with solid solute; **(d)** a 0.50 *M* solution of HCl that would just neutralize 5.5 g of $Ba(OH)_2$ starting with 6.0 *M* HCl.

13.49 Commercial aqueous nitric acid has a density of 1.42 g/mL and is 16 *M*. Calculate the percent HNO_3 by mass in the solution.

13.50 Commercial concentrated aqueous ammonia is 28% NH_3 by mass and has a density of 0.90 g/mL. What is the molarity of this solution?

13.51 Brass is a substitutional alloy consisting of a solution of copper and zinc. A particular sample of red brass consisting of 80.0% Cu and 20.0% Zn by mass has a density of 8750 kg/m³. **(a)** What is the molality of Zn in the solid solution? **(b)** What is the molarity of Zn in the solution?

13.52 Caffeine ($C_8H_{10}N_4O_2$) is a stimulant found in coffee and tea. If a solution of caffeine in chloroform ($CHCl_3$) as a solvent has a concentration of 0.0750 *m*, calculate **(a)** the percent caffeine by mass, **(b)** the mole fraction of caffeine.

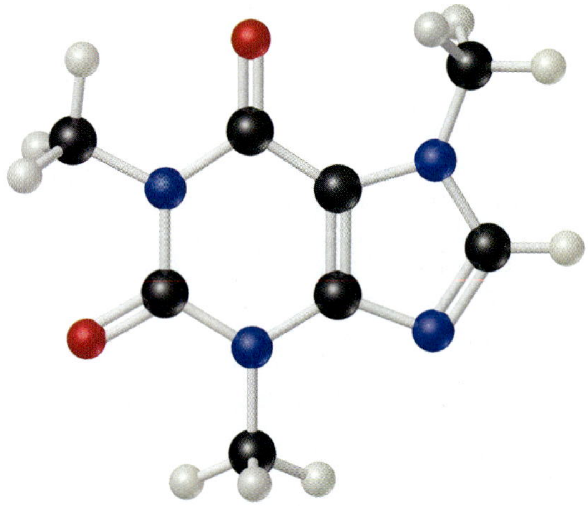

13.53 During a typical breathing cycle the CO_2 concentration in the expired air rises to a peak of 4.6% by volume. Calculate the partial pressure of the CO_2 at this point, assuming 1 atm pressure. What is the molarity of the CO_2 in air at this point, assuming a body temperature of 37°C?

13.54 Breathing air that contains 4.0% by volume CO_2 over a period of time causes rapid breathing, throbbing headache, and nausea, among other symptoms. What is the concentration of CO_2 in such air in terms of **(a)** mol percentage, **(b)** molarity, assuming 1 atm pressure, and a body temperature of 37°C?

Colligative Properties

13.55 List four properties of a solution that depend on the total concentration but not the type of particle or particles present as solute. Write the mathematical expression that describes how each of these properties depends on concentration.

13.56 How does increasing the concentration of a nonvolatile solute in water affect the following properties: **(a)** vapor pressure, **(b)** freezing point, **(c)** boiling point; **(d)** osmotic pressure?

13.57 Consider two solutions, one formed by adding 10 g of glucose ($C_6H_{12}O_6$) to 1 L of water and the other formed by adding 10 g of sucrose ($C_{12}H_{22}O_{11}$) to 1 L of water. Are the vapor pressures over the two solutions the same? Why or why not?

13.58 **(a)** What is an *ideal solution*? **(b)** The vapor pressure of pure water at 60°C is 149 torr. The vapor pressure of water over a solution at 60°C containing equal numbers of moles of water and ethylene glycol (a nonvolatile solute) is 67 torr. Is the solution ideal according to Raoult's law? Explain.

13.59 **(a)** Calculate the vapor pressure of water above a solution prepared by adding 22.5 g of lactose ($C_{12}H_{22}O_{11}$) to 200.0 g of water at 338 K. (Vapor-pressure data for water are given in Appendix B.) **(b)** Calculate the mass of propylene glycol ($C_3H_8O_2$) that must be added to 0.340 kg of water to reduce the vapor pressure by 2.88 torr at 40°C.

[13.60] **(a)** Calculate the vapor pressure of water above a solution prepared by dissolving 35.0 g of glycerin ($C_3H_8O_3$) in 125 g of water at 343 K. (The vapor pressure of water is given in Appendix B.) **(b)** Calculate the mass of ethylene glycol ($C_2H_6O_2$) that must be added to 1.00 kg of ethanol (C_2H_5OH) to reduce its vapor pressure by 10.0 torr at 35°C. The vapor pressure of pure ethanol at 35°C is 1.00×10^2 torr.

[13.61] At 63.5°C the vapor pressure of H_2O is 175 torr, and that of ethanol (C_2H_5OH) is 400 torr. A solution is made by mixing equal masses of H_2O and C_2H_5OH. **(a)** What is the mole fraction of ethanol in the solution? **(b)** Assuming ideal-solution behavior, what is the vapor pressure of the solution at 63.5°C? **(c)** What is the mole fraction of ethanol in the vapor above the solution?

[13.62] At 20°C the vapor pressure of benzene (C_6H_6) is 75 torr, and that of toluene (C_7H_8) is 22 torr. Assume that benzene and toluene form an ideal solution. **(a)** What is the composition in mole fractions of a solution that has a vapor pressure of 35 torr at 20°C? **(b)** What is the mole fraction of benzene in the vapor above the solution described in part (a)?

13.63 **(a)** Why does a 0.10 *m* aqueous solution of NaCl have a higher boiling point than a 0.10 *m* aqueous solution of $C_6H_{12}O_6$? **(b)** Calculate the boiling point of each solution. **(c)** The experimental boiling point of the NaCl solution is lower than that calculated, assuming that NaCl is completely dissociated in solution. Why is this the case?

13.64 Arrange the following aqueous solutions, each 10% by mass in solute, in order of increasing boiling point: glucose ($C_6H_{12}O_6$), sucrose ($C_{12}H_{22}O_{11}$), sodium nitrate ($NaNO_3$).

13.65 List the following aqueous solutions in order of increasing boiling point: 0.120 *M* glucose, 0.050 *M* LiBr, 0.050 *M* $Zn(NO_3)_2$.

13.66 List the following aqueous solutions in order of decreasing freezing point: 0.040 m glycerin ($C_3H_8O_3$), 0.020 m KBr, 0.030 m phenol (C_6H_5OH).

13.67 Using data from Table 13.4, calculate the freezing and boiling points of each of the following solutions: **(a)** 0.22 m glycerol ($C_3H_8O_3$) in ethanol, **(b)** 0.240 mol of naphthalene ($C_{10}H_8$) in 2.45 mol of chloroform, **(c)** 2.04 g KBr and 4.82 g glucose ($C_6H_{12}O_6$) in 188 g of water.

13.68 Using data from Table 13.4, calculate the freezing and boiling points of each of the following solutions: **(a)** 0.40 m glucose in ethanol; **(b)** 20.0 g of decane, $C_{10}H_{22}$, in 45.5 g $CHCl_3$; **(c)** 0.45 mol ethylene glycol and 0.15 mol KBr in 150 g H_2O.

13.69 What is the osmotic pressure of a solution formed by dissolving 44.2 mg of aspirin ($C_9H_8O_4$) in 0.358 L of water at 25°C?

13.70 Seawater contains 3.4 g of salts for every liter of solution. Assuming that the solute consists entirely of NaCl (over 90% is), calculate the osmotic pressure of seawater at 20°C.

13.71 Adrenaline is the hormone that triggers the release of extra glucose molecules in times of stress or emergency. A solution of 0.64 g of adrenaline in 36.0 g of CCl_4 elevates the boiling point by 0.49°C. Is the molar mass of adrenaline calculated from the boiling point elevation in ageement with the following structural formula?

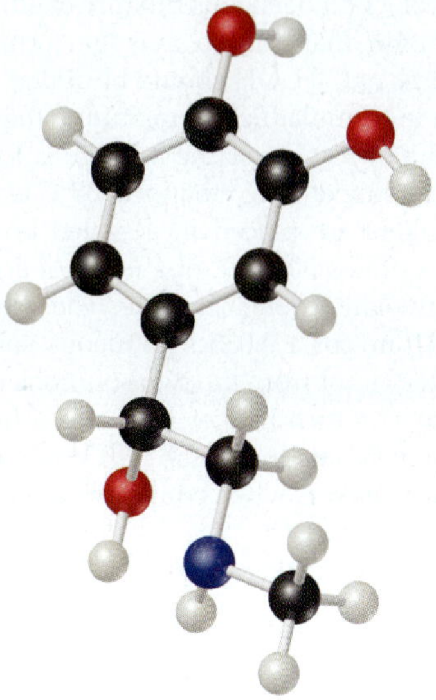

13.72 Lauryl alcohol is obtained from coconut oil and is used to make detergents. A solution of 5.00 g of lauryl alcohol in 0.100 kg of benzene freezes at 4.1°C. What is the approximate molar mass of lauryl alcohol?

13.73 Lysozyme is an enzyme that breaks bacterial cell walls. A solution containing 0.150 g of this enzyme in 210 mL of solution has an osmotic pressure of 0.953 torr at 25°C. What is the molar mass of lysozyme?

13.74 A dilute aqueous solution of an organic compound soluble in water is formed by dissolving 2.35 g of the compound in water to form 0.250 L solution. The resulting solution has an osmotic pressure of 0.605 atm at 25°C. Assuming that the organic compound is a nonelectrolyte, what is its molar mass?

[13.75] The osmotic pressure of 0.010 M aqueous solution of $CaCl_2$ is found to be 0.674 atm at 25°C. **(a)** Calculate the van't Hoff factor, i, for the solution. **(b)** How would you expect the value of i to change as the solution becomes more concentrated? Explain.

[13.76] Based on the data given in Table 13.5, which solution would give the larger freezing-point lowering, a 0.030 m solution of NaCl or a 0.020 m solution of K_2SO_4? How do you explain the departure from ideal behavior and the differences observed between the two salts?

Colloids

13.77 **(a)** Why is there no colloid in which both the dispersed substance and the dispersing substance are gases? **(b)** Michael Faraday first prepared ruby-red colloids of gold particles in water that were stable for indefinite times. (Section 12.6) To the unaided eye these brightly colored colloids are not distinguishable from solutions. How could you determine whether a given colored preparation is a solution or colloid?

13.78 **(a)** Many proteins that remain homogeneously distributed in aqueous medium have molecular masses in the range of 30,000 amu and larger. In what sense is it appropriate to consider such suspensions to be colloids rather than solutions? Explain. **(b)** What general name is given to a colloidal dispersion of one liquid in another? What is an emulsifying agent?

13.79 Indicate whether each of the following is a hydrophilic or a hydrophobic colloid: **(a)** butterfat in homogenized milk, **(b)** hemoglobin in blood, **(c)** vegetable oil in a salad dressing, **(d)** colloidal gold particles in water.

13.80 Explain how each of the following factors helps determine the stability or instability of a colloidal dispersion: **(a)** particulate mass, **(b)** hydrophobic character, **(c)** charges on colloidal particles.

13.81 Colloidal suspensions of proteins, such as a gelatin, can often be caused to separate into two layers by addition of a solution of an electrolyte. Given that protein molecules may carry electrical charges on their outer surface as illustrated in Figure 13.28, what do you believe happens when the electrolyte solution is added?

13.82 Explain how **(a)** a soap such as sodium stearate stabilizes a colloidal dispersion of oil droplets in water; **(b)** milk curdles upon addition of an acid.

Additional Exercises

13.83 Butylated hydroxytoluene (BHT) has the following molecular structure:

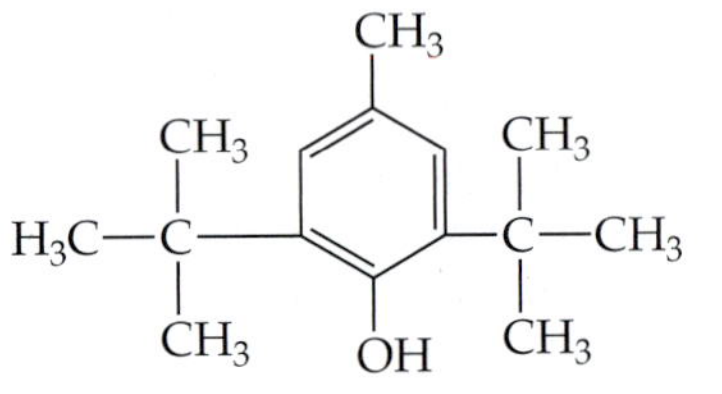

BHT

It is widely used as a preservative in a variety of foods, including dried cereals. Based on its structure, would you expect BHT to be more soluble in water or in hexane (C_6H_{14})? Explain.

13.84 A saturated solution of sucrose ($C_{12}H_{22}O_{11}$) is made by dissolving excess table sugar in a flask of water. There are 50 g of undissolved sucrose crystals at the bottom of the flask, in contact with the saturated solution. The flask is stoppered and set aside. A year later a single large crystal of mass 50 g is at the bottom of the flask. Explain how this experiment provides evidence for a dynamic equilibrium between the saturated solution and the undissolved solute.

13.85 Fish need at least 4 ppm dissolved O_2 for survival. **(a)** What is this concentration in mol/L? **(b)** What partial pressure of O_2 above the water is needed to obtain this concentration at 10°C? (The Henry's law constant for O_2 at this temperature is 1.71×10^{-3} mol/L-atm.)

13.86 The presence of the radioactive gas radon (Rn) in well water obtained from aquifers that lie in rock deposits presents a possible health hazard in parts of the United States. **(a)** Assuming that the solubility of radon in water with 1 atm pressure of the gas over the water at 30°C is 7.27×10^{-3} M, what is the Henry's law constant for radon in water at this temperature? **(b)** A sample consisting of various gases contains 3.5×10^{-6} mole fraction of radon. This gas at a total pressure of 32 atm is shaken with water at 30°C. Calculate the molar concentration of radon in the water.

13.87 Glucose makes up about 0.10% by mass of human blood. Calculate the concentration in **(a)** ppm, **(b)** molality. What further information would you need to determine the molarity of the solution?

13.88 A solution of KCl in water has a concentration of 260 ppm KCl. What is the concentration in ppm of K^+?

13.89 One aqueous solution is 15.0% by weight in propanol ($CH_3CH_2CH_2OH$), and another is 15.0% by weight in ethanol (CH_3CH_2OH). Assuming the two solutions have the same density, which has the higher **(a)** molarity, **(b)** molality, **(c)** mole fraction?

13.90 Acetonitrile (CH_3CN) is a polar organic solvent that dissolves a wide range of solutes, including many salts. The density of a 1.80 M LiBr solution in acetonitrile is 0.826 g/cm^3. Calculate the concentration of the solution in **(a)** molality, **(b)** mole fraction of LiBr, **(c)** mass percentage of CH_3CN.

13.91 Sodium metal dissolves in liquid mercury to form a solution called a sodium amalgam. The densities of Na(s) and Hg(l) are 0.97 g/cm^3 and 13.6 g/cm^3, respectively. A sodium amalgam is made by dissolving 1.0 cm^3 Na(s) in 20.0 cm^3 Hg(l). Assume that the final volume of the solution is 21.0 cm^3. **(a)** Calculate the molality of Na in the solution. **(b)** Calculate the molarity of Na in the solution. **(c)** For dilute aqueous solutions, the molality and molarity are generally nearly equal in value. Is that the case for the sodium amalgam described here? Explain.

13.92 A "canned heat" product used to warm chafing dishes consists of a homogeneous mixture of ethanol (C_2H_5OH) and paraffin that has an average formula of $C_{24}H_{50}$. What mass of C_2H_5OH should be added to 620 kg of the paraffin in formulating the mixture if the vapor pressure of ethanol at 35°C over the mixture is to be 8 torr? The vapor pressure of pure ethanol at 35°C is 100 torr.

[13.93] Two beakers are placed in a sealed box at 25°C. One beaker contains 30.0 mL of a 0.050 M aqueous solution of a nonvolatile nonelectrolyte. The other beaker contains 30.0 mL of a 0.035 M aqueous solution of NaCl. The water vapor from the two solutions reaches equilibrium. **(a)** In which beaker does the solution level rise, and in which one does it fall? **(b)** What are the volumes in the two beakers when equilibrium is attained, assuming ideal behavior?

13.94 Calculate the freezing point of a 0.100 m aqueous solution of K_2SO_4, **(a)** ignoring interionic attractions, and **(b)** taking interionic attractions into consideration by using the van't Hoff factor (Table 13.5).

13.95 When 10.0 g of mercuric nitrate, $Hg(NO_3)_2$, is dissolved in 1.00 kg of water, the freezing point of the solution is −0.162°C. When 10.0 g of mercuric chloride ($HgCl_2$) is dissolved in 1.00 kg of water, the solution freezes at −0.0685°C. Use these data to determine which is the stronger electrolyte, $Hg(NO_3)_2$ or $HgCl_2$.

13.96 Carbon disulfide (CS_2) boils at 46.30°C and has a density of 1.261 g/mL. **(a)** When 0.250 mol of a nondissociating solute is dissolved in 400.0 mL of CS_2, the solution boils at 47.46°C. What is the molal boiling-point-elevation constant for CS_2? **(b)** When 5.39 g of a nondissociating unknown is dissolved in 50.0 mL of CS_2, the solution boils at 47.08°C. What is the molecular weight of the unknown?

[13.97] A 40.0% by weight solution of KSCN in water at 20°C has a density of 1.22 g/mL. **(a)** What is the mole fraction of KSCN in the solution, and what are the molarity and molality? **(b)** Given the calculated mole fraction of salt in the solution, comment on the total number of water molecules available to hydrate each anion and cation. What ion pairing (if any) would you expect to find in the solution? Would you expect the colligative properties of such a solution to be those predicted by the formulas given in this chapter? Explain.

[13.98] A lithium salt used in lubricating grease has the formula $LiC_nH_{2n+1}O_2$. The salt is soluble in water to the extent of 0.036 g per 100 g of water at 25°C. The osmotic pressure of this solution is found to be 57.1 torr. Assuming that molality and molarity in such a dilute solution are the same and that the lithium salt is completely dissociated in the solution, determine an appropriate value of n in the formula for the salt.

Integrative Exercises

13.99 Fluorocarbons (compounds that contain both carbon and fluorine) were, until recently, used as refrigerants. The compounds listed in the following table are all gases at 25°C, and their solubilities in water at 25°C and 1 atm fluorocarbon pressure are given as mass percentages: **(a)** For each fluorocarbon, calculate the molality of a saturated solution. **(b)** Explain why the molarity of each of the solutions should be very close numerically to the molality. **(c)** Based on their molecular structures, account for the differences in solubility of the four fluorocarbons. **(d)** Calculate the Henry's law constant at 25°C for $CHClF_2$, and compare its magnitude to that for N_2 (6.8×10^{-4} mol/L-atm). Can you account for the difference in magnitude?

Fluorocarbon	Solubility (mass %)
CF_4	0.0015
$CClF_3$	0.009
CCl_2F_2	0.028
$CHClF_2$	0.30

[13.100] At ordinary body temperature (37°C) the solubility of N_2 in water in contact with air at ordinary atmospheric pressure (1.0 atm) is 0.015 g/L. Air is approximately 78 mol % N_2. Calculate the number of moles of N_2 dissolved per liter of blood, which is essentially an aqueous solution. At a depth of 100 ft in water, the pressure is 4.0 atm. What is the solubility of N_2 from air in blood at this pressure? If a scuba diver suddenly surfaces from this depth, how many milliliters of N_2 gas, in the form of tiny bubbles, are released into the bloodstream from each liter of blood?

[13.101] Consider the following values for enthalpy of vaporization (kJ/mol) of several organic substances:

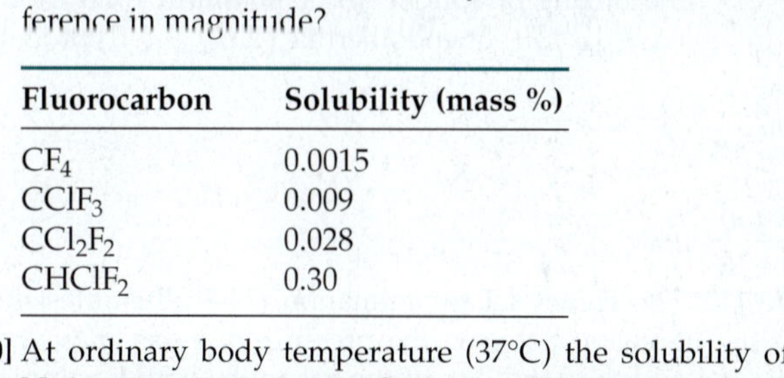

(a) Use variations in the intermolecular forces operating in these organic liquids to account for their variations in heats of vaporization. **(b)** How would you expect the solubilities of these substances to vary in hexane as solvent? In ethanol? Use intermolecular forces, including hydrogen-bonding interactions where applicable, to explain your responses.

[13.102] The enthalpies of solution of hydrated salts are generally more positive than those of anhydrous materials. For example, ΔH of solution for KOH is -57.3 kJ/mol, whereas that for $KOH \cdot H_2O$ is -14.6 kJ/mol. Similarly, ΔH_{soln} for $NaClO_4$ is +13.8 kJ/mol, whereas that for $NaClO_4 \cdot H_2O$ is +22.5 kJ/mol. Use the enthalpy contributions to the solution process depicted in Figure 13.4 to explain this effect.

[13.103] A textbook on chemical thermodynamics states, "The heat of solution represents the difference between the lattice energy of the crystalline solid and the solvation energy of the gaseous ions." **(a)** Draw a simple energy diagram to illustrate this statement. **(b)** A salt such as NaBr is insoluble in most polar nonaqueous solvents such as acetonitrile (CH_3CN) or nitromethane (CH_3NO_2), but salts of large cations, such as tetramethylammonium bromide [$(CH_3)_4NBr$], are generally more soluble. Use the thermochemical cycle you drew in part (a) and the factors that determine the lattice energy (Section 8.2) to explain this fact.

13.104 **(a)** A sample of hydrogen gas is generated in a closed container by reacting 2.050 g of zinc metal with 15.0 mL of 1.00 *M* sulfuric acid. Write the balanced equation for the reaction, and calculate the number of moles of hydrogen formed, assuming that the reaction is complete. **(b)** The volume over the solution is 122 mL. Calculate the partial pressure of the hydrogen gas in this volume at 25°C, ignoring any solubility of the gas in the solution. **(c)** The Henry's law constant for hydrogen in water at 25°C is 7.8×10^{-4} mol/L-atm. Estimate the number of moles of hydrogen gas that remain dissolved in the solution. What fraction of the gas molecules in the system is dissolved in the solution? Was it reasonable to ignore any dissolved hydrogen in part (b)?

[13.105] The following table presents the solubilities of several gases in water at 25°C under a total pressure of gas and water vapor of 1 atm: **(a)** What volume of $CH_4(g)$ under standard conditions of temperature and pressure is contained in 4.0 L of a saturated solution at 25°C? **(b)** Explain the variation in solubility among the hydrocarbons listed (the first three compounds), based on their molecular structures and intermolecular forces. **(c)** Compare the solubilities of O_2, N_2, and NO, and account for the variations based on molecular structures and intermolecular forces. **(d)** Account for the much larger values observed for H_2S and SO_2 as compared with the other gases listed. **(e)** Find several pairs of substances with the same or nearly the same molecular masses (for example, C_2H_4 and N_2), and use intermolecular interactions to explain the differences in their solubilities.

Gas	Solubility (m*M*)
CH_4 (methane)	1.3
C_2H_6 (ethane)	1.8
C_2H_4 (ethylene)	4.7
N_2	0.6
O_2	1.2
NO	1.9
H_2S	99
SO_2	1476

[13.106] Consider the process illustrated in Figure 13.23, the net movement of solvent from the more dilute to the more concentrated solution through the semipermeable membrane. Is there a change in entropy of the system in going from the left to center panels? [*Hint:* Imagine that the more dilute solution on the right hand side of the apparatus is pure solvent.]

[13.107] When 0.55 g of pure benzoic acid ($C_7H_6O_2$) is dissolved in 32.0 g of benzene (C_6H_6), the freezing point of the solution is 0.36°C lower than the freezing point value of 5.5°C for the pure solvent. **(a)** Calculate the molecular weight of benzoic acid in benzene. **(b)** Use the structure of the solute to account for the observed value:

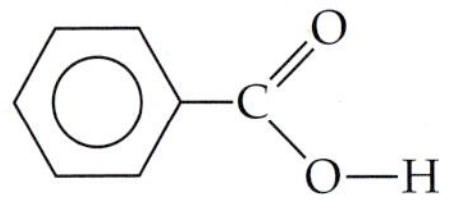

[13.108] At 35°C the vapor pressure of acetone, $(CH_3)_2CO$, is 360 torr, and that of chloroform, $CHCl_3$, is 300 torr. Acetone and chloroform can form weak hydrogen bonds between one another as follows:

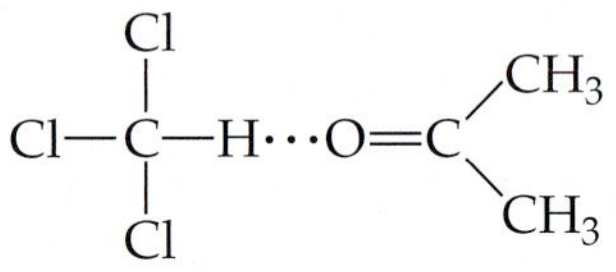

A solution composed of an equal number of moles of acetone and chloroform has a vapor pressure of 250 torr at 35°C. **(a)** What would be the vapor pressure of the solution if it exhibited ideal behavior? **(b)** Use the existence of hydrogen bonds between acetone and chloroform molecules to explain the deviation from ideal behavior. **(c)** Based on the behavior of the solution, predict whether the mixing of acetone and chloroform is an exothermic ($\Delta H_{soln} < 0$) or endothermic ($\Delta H_{soln} > 0$) process.

eMEDIA EXERCISES

These exercises make use of the interactive objects available online in OneKey or the Companion Website, and on your Accelerator CD. Access to these resources comes in your MediaPak.

13.109 **(a)** Of the five solids available in the **Enthalpy of Solution** activity (*13.1*), which would become more soluble and which would become less soluble with increasing temperature? **(b)** Comparing sodium chloride and ammonium nitrate, which would you expect to exhibit the greater change in solubility (in g/L) for a given temperature increase? Explain your reasoning.

13.110 The **Enthalpy of Solution** activity (*13.1*) deals specifically with dissolution of solids. Describe the solution process in terms of three distinct steps. Using the three steps, explain why gases, in contrast to most solids, become *less* soluble as temperature increases.

13.111 The **Enthalpy of Solution** activity (*13.1*) allows you to compare the overall energy changes for the dissolution of various salts. Compare the dissolution of sodium hydroxide with that of sodium chloride. **(a)** How do the energy changes for the overall processes compare? **(b)** Discuss the difference in magnitude of each of the three steps for dissolution of the two compounds.

13.112 The **Henry's Law** animation (*13.3*) illustrates the relationship between the pressure of a gas above a liquid and the solubility of the gas in the liquid. Sample Exercise 13.3 determines the concentration of CO_2 in a carbonated beverage bottled under 4.0 atm of CO_2. **(a)** If the mole fraction of CO_2 in the air is 0.000375, how many times more concentrated is the CO_2 in the unopened bottle than it is in a glass of the beverage that has gone completely flat? **(b)** Use the kinetic-molecular theory of gases to explain why the solubility of a gas is proportional to its pressure over the liquid.

13.113 The **Boiling Point Elevation and Freezing Point Depression** activity (*13.5*) shows how the boiling point and freezing point of water are affected by the addition of various solutes. **(a)** If you were to dissolve 10 g sodium chloride in 500 g water, what would be the boiling point of the solution? **(b)** If you were to heat the solution in part (a), and continue heating it as it boiled, what would happen to the temperature over time? **(c)** In what way is your answer to part (b) contrary to what you learned in Chapter 11 (Section 11.4) regarding constant temperature during phase changes? Explain this apparent contradiction.

eLABORATORY EXERCISES

These exercises allow students to apply the concepts and skills for this chapter in the simulated laboratory of Virtual ChemLab. Worksheets for each of these exercises are found in the Virtual ChemLab Workbook in your MediaPak.

13.114 *(VCL 13-1) Creating a Solution of Known Molality*
In this assignment you will weigh out a sample of solid NH_4Cl and create a solution of known molality.

13.115 *(VCL 13-2) Creating a Solution of Known Molarity*
In this assignment you will weigh out a sample of baking soda ($NaHCO_3$) and create a solution of known molarity.

13.116 *(VCL 13-3) Converting Concentrations to Different Units*
Occasionally, when making solutions in the laboratory, it is convenient to make a solution with a certain concentration unit, such as molarity, and then convert the concentration to a different unit. In this assignment you will make a sodium bicarbonate (baking soda) solution of a certain molarity and then convert that concentration to molality, mass percent, and mole fraction.

13.117 *(VCL 13-4) Boiling-Point Elevation*
If you dissolve a substance such as ordinary table salt (NaCl) in water, the boiling point of the water will increase relative to the boiling point of the pure water. In this assignment you will dissolve a sample of NaCl in water and measure the boiling point elevation for the solution.

13.118 *(VCL 13-5) Freezing-Point Depression*
If you dissolve a substance such as ordinary table salt (NaCl) in water, the freezing point of the water will decrease relative to the freezing point of the pure water. This property is used to melt the snow or ice on roads during the winter or to make homemade ice cream. In this assignment you will dissolve a sample of NaCl in water, add some ice, and then measure the freezing-point depression.

13.119 *(VCL 13-6) Molar Mass Determination by Boiling-Point Elevation*
If you dissolve a substance such as ordinary table salt (NaCl) in water, the boiling point of the water will increase relative to the boiling point of the pure water. You can use this property to calculate the molar mass of an unknown. In this assignment you will dissolve a sample of NaCl in water, measure the boiling-point elevation for the solution, and then calculate the molar mass for NaCl as if it were an unknown.

13.120 *(VCL 13-7) Molar Mass Determination by Freezing-Point Depression*
If you dissolve a substance such as ordinary table salt (NaCl) in water, the boiling point of the water will increase relative to the boiling point of the pure water. You can use this property to calculate the molar mass of an unknown. In this assignment you will dissolve a sample of NaCl in water, measure the boiling point elevation for the solution, and then calculate the molar mass for NaCl as if it were an unknown.

13.121 *(VCL 13-8) Changes in the Boiling Point*
If you dissolve a substance such as ordinary table salt (NaCl) in water, the boiling point of the water will increase relative to the boiling point of the pure water. In this assignment you will dissolve a sample of NaCl in water and observe what happens when you continue to boil the water until the water boils away.

CHAPTER 14

WHAT'S AHEAD

- There are four experimental variables that affect reaction rates: concentration, physical states of reactants, temperature, and catalysts. These factors can be understood in terms of the collisions among reactant molecules that lead to reaction. *(Section 14.1)*
- We consider how to express *reaction rates* and how the rates of disappearance of reactants and the rates of appearance of products are related to the stoichiometry of the reaction. *(Section 14.2)*
- We then examine how the effect of concentration on rate is expressed quantitatively by *rate laws* and how rate laws can be determined experimentally. *(Section 14.3)*
- Rate equations can be written to express how concentrations change with time. We will look at two simple kinds of such rate equations. *(Section 14.4)*
- We next consider the effect of temperature on rate and the fact that reactions require a minimum input of energy called the *activation energy* in order to occur. *(Section 14.5)*
- We then examine the *mechanisms* of reactions, the step-by-step molecular pathways leading from reactants to products. *(Section 14.6)*
- The chapter ends with a discussion of how *catalysts* speed reaction rates, including a discussion of biological catalysts, called *enzymes*. *(Section 14.7)*

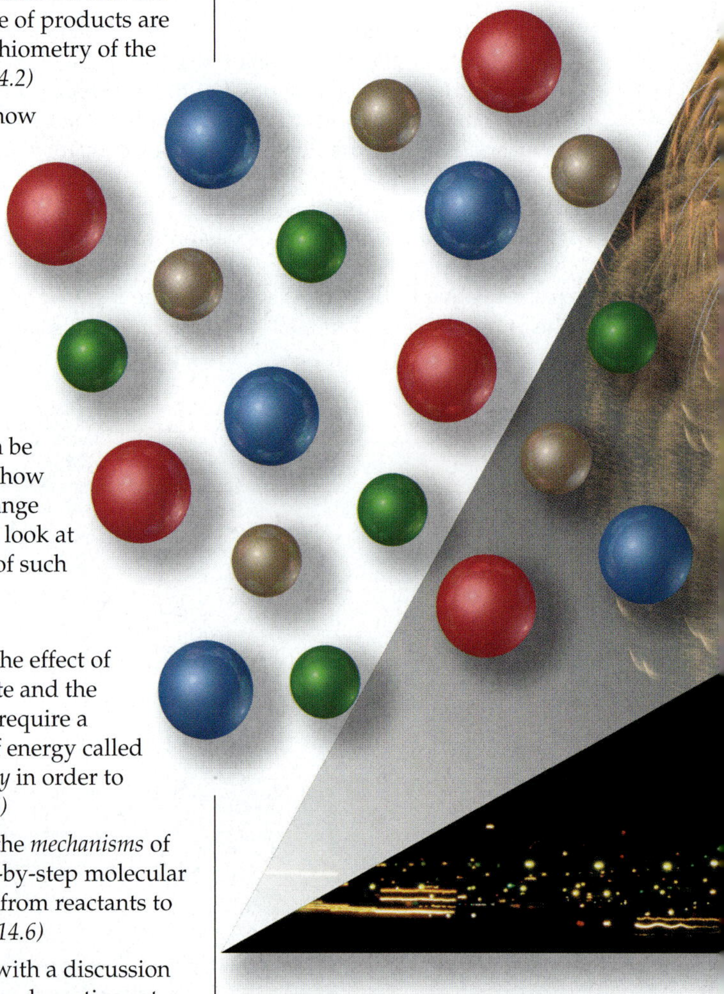

Chemical Kinetics

FIREWORKS RELY ON RAPID CHEMICAL reactions both to propel them skyward and to produce their colorful bursts of light.

CHEMISTRY IS, BY ITS VERY NATURE, CONCERNED WITH CHANGE. Chemical reactions convert substances with well-defined properties into other materials with different properties. Much of our study of chemical reactions is concerned with the formation of new substances from a given set of reactants. However, it is equally important to understand how rapidly chemical reactions occur.

The rates of reactions span an enormous range, from those that are complete within fractions of seconds, such as explosions, to those that take thousands or even millions of years, such as the formation of diamonds or other minerals in Earth's crust (Figure 14.1 ▶). The fireworks shown in this chapter-opening photograph require very rapid reactions, both to propel them skyward and to produce their colorful bursts of light. The chemicals used in the fireworks are chosen both to give the desired colors and to do so very rapidly. The characteristic red, blue, and green colors are produced by salts of strontium, copper and barium, respectively.

The area of chemistry that is concerned with the speeds, or rates, of reactions is called **chemical kinetics**. Chemical kinetics is a subject of broad importance. It relates, for example, to how quickly a medicine is able to work, to whether the formation and depletion of ozone in the upper atmosphere are in balance, and to industrial challenges such as the development of catalysts to synthesize new materials.

Our goal in this chapter is not only to understand how to determine the rates at which reactions occur but also to consider the factors that control these rates.

▲ **Figure 14.1 Reaction rates.** The rates of chemical reactions span a range of time scales. For example, explosions are rapid, occurring in seconds or fractions of seconds; corrosion can take years; and the weathering of rocks takes place over thousands or even millions of years.

For example, what factors determine how rapidly food spoils? How does one design a fast-setting material for dental fillings? What determines the rate at which steel rusts? What controls the rate at which fuel burns in an automobile engine? Although we won't address these specific questions directly, we will see that the rates of all chemical reactions are subject to the same basic principles.

14.1 | Factors that Affect Reaction Rates

Before we examine the quantitative aspects of chemical kinetics, such as how rates are measured, let's examine the key factors that influence the rates of reactions. Because reactions involve the breaking and forming of bonds, their speeds depend on the nature of the reactants themselves. There are, however, four factors that allow us to change the rates at which particular reactions occur:

1. *The physical state of the reactants.* Reactants must come together in order to react. The more readily molecules collide with each other, the more rapidly they react. Most of the reactions we consider are homogeneous, involving either gases or liquid solutions. When reactants are in different phases, such as when one is a gas and another a solid, the reaction is limited to their area of contact. Thus, reactions that involve solids tend to proceed faster if the surface area of the solid is increased. For example, a medicine in the form of a tablet will dissolve in the stomach and enter the bloodstream more slowly than the same medicine in the form of a fine powder.
2. *The concentrations of the reactants.* Most chemical reactions proceed faster if the concentration of one or more of the reactants is increased. For example, steel wool burns with difficulty in air, which contains 20% O_2, but bursts into a brilliant white flame in pure oxygen (Figure 14.2 ◀). As concentration increases, the frequency with which the reactant molecules collide increases, leading to increased rates.
3. *The temperature at which the reaction occurs.* The rates of chemical reactions increase as temperature is increased. It is for this reason that we refrigerate perishable foods such as milk. The bacterial reactions that lead to the spoiling of milk proceed much more rapidly at room temperature than they do at the lower temperature of a refrigerator. Increasing temperature increases the kinetic energies of molecules. ∞ (Section 10.7) As molecules move more rapidly, they collide more frequently and also with higher energy, leading to increased reaction rates.

▼ **Figure 14.2 Effect of concentration on rate.** (a) When heated in air, steel wool glows red-hot but oxidizes slowly. (b) When the red-hot steel wool is placed in an atmosphere of pure oxygen, it burns vigorously, forming Fe_2O_3 at a much faster rate. The difference in behavior is due to the different concentrations of O_2 in the two environments.

(a) (b)

4. *The presence of a catalyst.* Catalysts are agents that increase reaction rates without being used up. They affect the kinds of collisions (the mechanism) that lead to reaction. Catalysts play a crucial role in our lives. The physiology of most living species depends on *enzymes*, protein molecules that act as catalysts, increasing the rates of selected biochemical reactions.

On a molecular level, reaction rates depend on the frequency of collisions between molecules. The greater the frequency of collisions, the greater the rate of reaction. In order for a collision to lead to reaction, however, it must occur with sufficient energy to stretch bonds to a critical length and with suitable orientation for new bonds to form in the proper locations. We will consider these factors as we proceed through this chapter.

GIVE IT SOME THOUGHT

How does increasing the partial pressures of the reactive components of a gaseous mixture affect the rate at which the components react with one another?

14.2 | Reaction Rates

The *speed* of an event is defined as the *change* that occurs in a given interval of *time*: Whenever we talk about speed, we necessarily bring in the notion of time. For example, the speed of a car is expressed as the change in the car's position over a certain period of time. The units of this speed are usually miles per hour (mi/hr)—that is, the quantity that is changing (position, measured in miles) divided by a time interval (hours).

Similarly, the speed of a chemical reaction—its **reaction rate**—is the change in the concentration of reactants or products per unit time. Thus, the units for reaction rate are usually molarity per second (M/s)—that is, the change in concentration (measured in molarity) divided by a time interval (seconds).

ACTIVITY
Progress of Reaction

Let's consider a simple hypothetical reaction, A $\longrightarrow$ B, depicted in Figure 14.3 ▼. Each red sphere represents 0.01 mol of A, and each blue sphere represents 0.01 mol of B. Let's assume that the container has a volume of 1.00 L. At the beginning of the reaction there is 1.00 mol A, so the concentration is 1.00 mol/L = 1.00 M. After 20 s the concentration of A has fallen to 0.54 M, whereas that of B has risen to 0.46 M. The sum of the concentrations is still 1.00 M because 1 mol of B is produced for each mole of A that reacts. After 40 s the concentration of A is 0.30 M and that of B is 0.70 M.

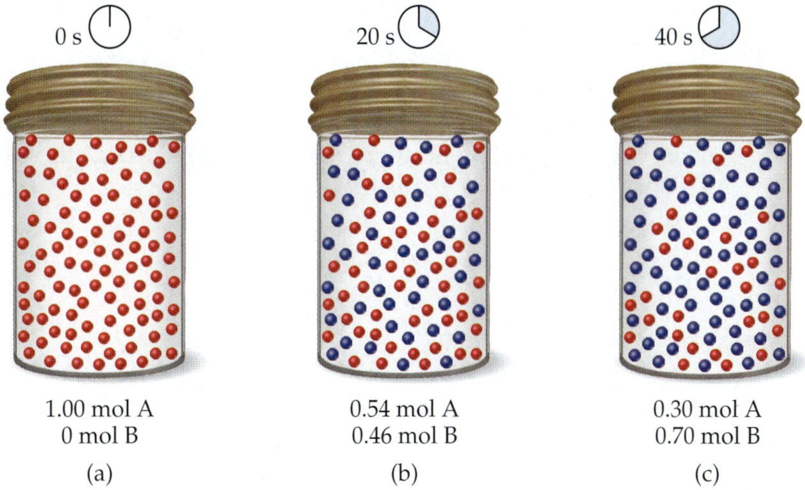

◀ **Figure 14.3 Progress of a hypothetical reaction A → B.** Each red sphere represents 0.01 mol A, each blue sphere represents 0.01 mol B, and the vessel has a volume of 1.00 L. (a) At time zero the vessel contains 1.00 mol A (100 red spheres) and 0 mol B (no blue spheres). (b) After 20 s the vessel contains 0.54 mol A and 0.46 mol B. (c) After 40 s the vessel contains 0.30 mol A and 0.70 mol B.

The rate of this reaction can be expressed either as the rate of disappearance of reactant A or as the rate of appearance of product B. The *average* rate of appearance of B over a particular time interval is given by the change in concentration of B divided by the change in time:

$$\text{Average rate of appearance of B} = \frac{\text{change in concentration of B}}{\text{change in time}}$$

$$= \frac{[\text{B}] \text{ at } t_2 - [\text{B}] \text{ at } t_1}{t_2 - t_1} = \frac{\Delta[\text{B}]}{\Delta t} \qquad [14.1]$$

We use brackets around a chemical formula, as in [B], to indicate the concentration of the substance in molarity. The Greek letter delta, Δ, is read "change in," and it is always equal to the final quantity minus the initial quantity. (Section 5.2) The average rate of appearance of B over the 20-s interval from the beginning of the reaction ($t_1 = 0$ s to $t_2 = 20$ s) is given by

$$\text{Average rate} = \frac{0.46\ M - 0.00\ M}{20\text{ s} - 0\text{ s}} = 2.3 \times 10^{-2}\ M/\text{s}$$

We could equally well express the rate of the reaction with respect to the change of concentration of the reactant, A. In this case we would be describing the rate of disappearance of A, which we express as

$$\text{Average rate of disappearance of A} = -\frac{\Delta[\text{A}]}{\Delta t} \qquad [14.2]$$

Notice the minus sign in this equation. By convention, *rates are always expressed as positive quantities*. Because [A] is decreasing with time, Δ[A] is a negative number. We use the negative sign to convert the negative Δ[A] to a positive rate. Because one molecule of A is consumed for every molecule of B that forms, the average rate of disappearance of A equals the average rate of appearance of B, as the following calculation shows:

$$\text{Average rate} = -\frac{\Delta[\text{A}]}{\Delta t} = -\frac{0.54\ M - 1.00\ M}{20\text{ s} - 0\text{ s}} = 2.3 \times 10^{-2}\ M/\text{s}$$

SAMPLE EXERCISE 14.1 | Calculating an Average Rate of Reaction

From the data given in the caption of Figure 14.3, calculate the average rate at which A disappears over the time interval from 20 s to 40 s.

Solution

Analyze: We are given the concentration of A at 20 s (0.54 *M*) and at 40 s (0.30 *M*) and asked to calculate the average rate of reaction over this time interval.

Plan: The average rate is given by the change in concentration, Δ[A], divided by the corresponding change in time, Δt. Because A is a reactant, a minus sign is used in the calculation to make the rate a positive quantity.

Solve:

$$\text{Average rate} = -\frac{\Delta[\text{A}]}{\Delta t} = -\frac{0.30\ M - 0.54\ M}{40\text{ s} - 20\text{ s}} = 1.2 \times 10^{-2}\ M/\text{s}$$

PRACTICE EXERCISE

For the reaction pictured in Figure 14.3, calculate the average rate of appearance of B over the time interval from 0 to 40 s. (The necessary data are given in the figure caption.)

Answer: $1.8 \times 10^{-2}\ M/\text{s}$

TABLE 14.1 Rate Data for Reaction of C_4H_9Cl with Water

Time, t(s)	$[C_4H_9Cl]$ (M)	Average Rate (M/s)
0.0	0.1000	
		1.9×10^{-4}
50.0	0.0905	
		1.7×10^{-4}
100.0	0.0820	
		1.6×10^{-4}
150.0	0.0741	
		1.4×10^{-4}
200.0	0.0671	
		1.22×10^{-4}
300.0	0.0549	
		1.01×10^{-4}
400.0	0.0448	
		0.80×10^{-4}
500.0	0.0368	
		0.560×10^{-4}
800.0	0.0200	
10,000	0	

Change of Rate with Time

Now, let's consider an actual chemical reaction, one that occurs when butyl chloride (C_4H_9Cl) is placed in water. The products formed are butyl alcohol (C_4H_9OH) and hydrochloric acid:

$$C_4H_9Cl(aq) + H_2O(l) \longrightarrow C_4H_9OH(aq) + HCl(aq) \qquad [14.3]$$

Suppose that we prepare a 0.1000 M aqueous solution of C_4H_9Cl and then measure the concentration of C_4H_9Cl at various times after time zero, collecting the data shown in the first two columns of Table 14.1 ▲. We can use these data to calculate the average rate of disappearance of C_4H_9Cl over the intervals between measurements, and these rates are given in the third column. Notice that the average rate decreases over each 50-s interval for the first several measurements and continues to decrease over even larger intervals through the remaining measurements. It is typical for rates to decrease as a reaction proceeds, because the concentration of reactants decreases. The change in rate as the reaction proceeds is also seen in a graph of the concentration of C_4H_9Cl versus time (Figure 14.4 ▶). Notice how the steepness of the curve decreases with time, indicating a decreasing rate of reaction.

▼ **Figure 14.4 Concentration of butyl chloride (C_4H_9Cl) as a function of time.** The dots represent the experimental data from the first two columns of Table 14.1, and the red curve is drawn to connect the data points smoothly. Lines are drawn that are tangent to the curve at $t = 0$ and $t = 600$ s. The slope of each tangent is defined as the vertical change divided by the horizontal change: $\Delta[C_4H_9Cl]/\Delta t$. The reaction rate at any time is related to the slope of the tangent to the curve at that time. Because C_4H_9Cl is disappearing, the rate is equal to the negative of the slope.

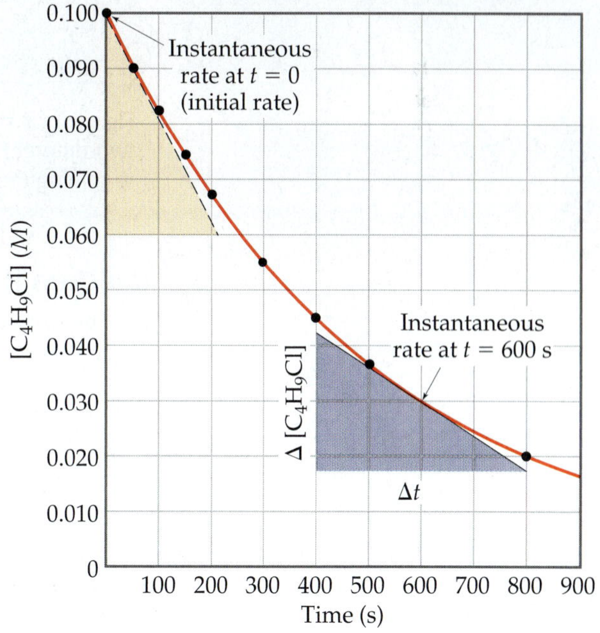

The graph shown in Figure 14.4 is particularly useful because it allows us to evaluate the **instantaneous rate**, the rate at a particular moment in the reaction. The instantaneous rate is determined from the slope (or tangent) of this curve at the point of interest. We have drawn two tangents in Figure 14.4, one at $t = 0$ and the other at $t = 600$ s. The slopes of these tangents give the instantaneous rates at these times.* For example, to determine the instantaneous rate at 600 s, we draw the tangent to the curve at this time, then construct horizontal and vertical lines to form the right triangle shown. The slope is the ratio of the height of the vertical side to the length of the horizontal side:

$$\text{Instantaneous rate} = \frac{\Delta[C_4H_9Cl]}{\Delta t} = -\frac{(0.017 - 0.042)\ M}{(800 - 400)\ \text{s}}$$

$$= 6.2 \times 10^{-5}\ M/\text{s}$$

* You may wish to review briefly the idea of graphical determination of slopes by referring to Appendix A. If you are familiar with calculus, you may recognize that the average rate approaches the instantaneous rate as the time interval approaches zero. This limit, in the notation of calculus, is represented as $-d[C_4H_9Cl]/dt$.

In what follows, the term "rate" means "instantaneous rate," unless indicated otherwise. The instantaneous rate at $t = 0$ is called the *initial rate* of the reaction.

To understand better the difference between average rate and instantaneous rate, imagine that you have just driven 98 mi in 2.0 hr. Your average speed is 49 mi/hr, whereas your instantaneous speed at any moment is the speedometer reading at that time.

SAMPLE EXERCISE 14.2 | Calculating an Instantaneous Rate of Reaction

Using Figure 14.4, calculate the instantaneous rate of disappearance of C_4H_9Cl at $t = 0$ (the initial rate).

Solution

Analyze: We are asked to determine an instantaneous rate from a graph of concentration versus time.

Plan: To obtain the instantaneous rate at $t = 0$, we must determine the slope of the curve at $t = 0$. The tangent is drawn on the graph. The slope of this straight line equals the change in the vertical axis divided by the corresponding change in the horizontal axis (that is, change in molarity over change in time).

Solve: The straight line falls from $[C_4H_9Cl] = 0.100\ M$ to $0.060\ M$ in the time change from 0 s to 200 s, as indicated by the tan triangle shown in Figure 14.4. Thus, the initial rate is

$$\text{Rate} = -\frac{\Delta[C_4H_9Cl]}{\Delta t} = -\frac{(0.060 - 0.100)\ M}{(200 - 0)\ \text{s}} = 2.0 \times 10^{-4}\ M/\text{s}$$

PRACTICE EXERCISE

Using Figure 14.4, determine the instantaneous rate of disappearance of C_4H_9Cl at $t = 300$ s.

Answer: $1.1 \times 10^{-4}\ M/\text{s}$

GIVE IT SOME THOUGHT

Figure 14.4 shows two triangles that are used to determine the slope of the curve at two different times. How do you determine how large a triangle to draw when determining the slope of a curve at a particular point?

Reaction Rates and Stoichiometry

During our earlier discussion of the hypothetical reaction $A \longrightarrow B$, we saw that the stoichiometry requires that the rate of disappearance of A equals the rate of appearance of B. Likewise, the stoichiometry of Equation 14.3 indicates that 1 mol of C_4H_9OH is produced for each mole of C_4H_9Cl consumed. Therefore, the rate of appearance of C_4H_9OH equals the rate of disappearance of C_4H_9Cl:

$$\text{Rate} = -\frac{\Delta[C_4H_9Cl]}{\Delta t} = \frac{\Delta[C_4H_9OH]}{\Delta t}$$

What happens when the stoichiometric relationships are not one to one? For example, consider the following reaction:

$$2\ HI(g) \longrightarrow H_2(g) + I_2(g)$$

We can measure the rate of disappearance of HI or the rate of appearance of either H_2 or I_2. Because 2 mol of HI disappear for each mole of H_2 or I_2 that forms, the rate of disappearance of HI is twice the rate of appearance of either H_2 or I_2. To equate the rates, we must therefore divide the rate of disappearance of HI by 2 (its coefficient in the balanced chemical equation):

$$\text{Rate} = -\frac{1}{2}\frac{\Delta[HI]}{\Delta t} = \frac{\Delta[H_2]}{\Delta t} = \frac{\Delta[I_2]}{\Delta t}$$

A CLOSER LOOK | Using Spectroscopic Methods to Measure Reaction Rates

A variety of techniques can be used to monitor the concentration of a reactant or product during a reaction. Spectroscopic methods, which rely on the ability of substances to absorb (or emit) electromagnetic radiation, are some of the most useful. Spectroscopic kinetic studies are often performed with the reaction mixture in the sample compartment of the spectrometer. The spectrometer is set to measure the light absorbed at a wavelength characteristic of one of the reactants or products. In the decomposition of $HI(g)$ into $H_2(g)$ and $I_2(g)$, for example, both HI and H_2 are colorless, whereas I_2 is violet. During the course of the reaction, the color increases in intensity as I_2 forms. Thus, visible light of appropriate wavelength can be used to monitor the reaction.

Figure 14.5 ▼ shows the basic components of a spectrometer. The spectrometer measures the amount of light absorbed by the sample by comparing the intensity of the light emitted from the light source with the intensity of the light that emerges from the sample. As the concentration of I_2 increases and its color becomes more intense, the amount of light absorbed by the reaction mixture increases, causing less light to reach the detector.

Beer's law relates the amount of light being absorbed to the concentration of the substance absorbing the light:

$$A = abc \qquad [14.5]$$

In this equation, A is the measured absorbance, a is the molar absorptivity constant (a characteristic of the substance being monitored), b is the path length through which the radiation must pass, and c is the molar concentration of the absorbing substance. Thus, the concentration is directly proportional to absorbance.

Related Exercise: 14.88

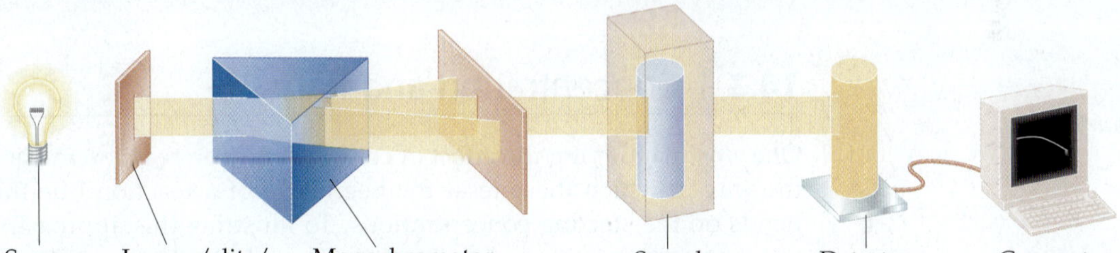

▲ **Figure 14.5 Basic components of a spectrometer.**

In general, for the reaction

$$aA + bB \longrightarrow cC + dD$$

the rate is given by

$$\text{Rate} = -\frac{1}{a}\frac{\Delta[A]}{\Delta t} = -\frac{1}{b}\frac{\Delta[B]}{\Delta t} = \frac{1}{c}\frac{\Delta[C]}{\Delta t} = \frac{1}{d}\frac{\Delta[D]}{\Delta t} \qquad [14.4]$$

When we speak of the rate of a reaction without specifying a particular reactant or product, we will mean it in this sense.*

SAMPLE EXERCISE 14.3 | Relating Rates at Which Products Appear and Reactants Disappear

(a) How is the rate at which ozone disappears related to the rate at which oxygen appears in the reaction $2\,O_3(g) \longrightarrow 3\,O_2(g)$? **(b)** If the rate at which O_2 appears, $\Delta[O_2]/\Delta t$, is $6.0 \times 10^{-5}\ M/s$ at a particular instant, at what rate is O_3 disappearing at this same time, $-\Delta[O_3]/\Delta t$?

Solution

Analyze: We are given a balanced chemical equation and asked to relate the rate of appearance of the product to the rate of disappearance of the reactant.

Plan: We can use the coefficients in the chemical equation as shown in Equation 14.4 to express the relative rates of reactions.

* Equation 14.4 does not hold true if substances other than C and D are formed in significant amounts during the course of the reaction. For example, sometimes intermediate substances build in concentration before forming the final products. In that case the relationship between the rate of disappearance of reactants and the rate of appearance of products will not be given by Equation 14.4. All reactions whose rates we consider in this chapter obey Equation 14.4.

Solve: (a) Using the coefficients in the balanced equation and the relationship given by Equation 14.4, we have:

$$\text{Rate} = -\frac{1}{2}\frac{\Delta[O_3]}{\Delta t} = \frac{1}{3}\frac{\Delta[O_2]}{\Delta t}$$

(b) Solving the equation from part (a) for the rate at which O_3 disappears, $-\Delta[O_3]/\Delta t$, we have:

$$-\frac{\Delta[O_3]}{\Delta t} = \frac{2}{3}\frac{\Delta[O_2]}{\Delta t} = \frac{2}{3}(6.0 \times 10^{-5}\ M/\text{s}) = 4.0 \times 10^{-5}\ M/\text{s}$$

Check: We can directly apply a stoichiometric factor to convert the O_2 formation rate to the rate at which the O_3 disappears:

$$-\frac{\Delta[O_3]}{\Delta t} = \left(6.0 \times 10^{-5}\frac{\text{mol } O_2/\text{L}}{\text{s}}\right)\left(\frac{2\ \text{mol } O_3}{3\ \text{mol } O_2}\right) = 4.0 \times 10^{-5}\frac{\text{mol } O_3/\text{L}}{\text{s}}$$

$$= 4.0 \times 10^{-5}\ M/\text{s}$$

PRACTICE EXERCISE

The decomposition of N_2O_5 proceeds according to the following equation:

$$2\,N_2O_5(g) \longrightarrow 4\,NO_2(g) + O_2(g)$$

If the rate of decomposition of N_2O_5 at a particular instant in a reaction vessel is $4.2 \times 10^{-7}\ M/\text{s}$, what is the rate of appearance of (a) NO_2, (b) O_2?

Answers: **(a)** $8.4 \times 10^{-7}\ M/\text{s}$, **(b)** $2.1 \times 10^{-7}\ M/\text{s}$

ACTIVITY
Decomposition of N_2O_5

14.3 | Concentration and Rate

One way of studying the effect of concentration on reaction rate is to determine the way in which the rate at the beginning of a reaction (the initial rate) depends on the starting concentrations. To illustrate this approach, consider the following reaction:

$$NH_4^+(aq) + NO_2^-(aq) \longrightarrow N_2(g) + 2\,H_2O(l)$$

We might study the rate of this reaction by measuring the concentration of NH_4^+ or NO_2^- as a function of time or by measuring the volume of N_2 collected. Because the stoichiometric coefficients on NH_4^+, NO_2^-, and N_2 are all the same, all of these rates will be equal.

Table 14.2 ▼ shows the initial reaction rate for various starting concentrations of NH_4^+ and NO_2^-. These data indicate that changing either $[NH_4^+]$ or $[NO_2^-]$ changes the reaction rate. Notice that if we double $[NH_4^+]$ while holding $[NO_2^-]$ constant, the rate doubles (compare experiments 1 and 2). If $[NH_4^+]$ is increased by a factor of 4 with $[NO_2^-]$ left unchanged (compare experiments 1 and 3), the rate changes by a factor of 4, and so forth. These results indicate that the rate is proportional to $[NH_4^+]$. When $[NO_2^-]$ is similarly varied while $[NH_4^+]$ is held constant, the rate is affected in the same manner. Thus, the rate is also directly proportional to the concentration of NO_2^-. We can express the

ACTIVITY
Rates of Reaction

TABLE 14.2 Rate Data for the Reaction of Ammonium and Nitrite Ions in Water at 25°C

Experiment Number	Initial NH_4^+ Concentration (M)	Initial NO_2^- Concentration (M)	Observed Initial Rate (M/s)
1	0.0100	0.200	5.4×10^{-7}
2	0.0200	0.200	10.8×10^{-7}
3	0.0400	0.200	21.5×10^{-7}
4	0.0600	0.200	32.3×10^{-7}
5	0.200	0.0202	10.8×10^{-7}
6	0.200	0.0404	21.6×10^{-7}
7	0.200	0.0606	32.4×10^{-7}
8	0.200	0.0808	43.3×10^{-7}

way in which the rate depends on the concentrations of the reactants, NH_4^+ and NO_2^-, in terms of the following equation:

$$\text{Rate} = k[NH_4^+][NO_2^-] \qquad [14.6]$$

An equation such as Equation 14.6, which shows how the rate depends on the concentrations of reactants, is called a **rate law**. For a general reaction,

$$a\,A + b\,B \longrightarrow c\,C + d\,D$$

the rate law generally has the form

$$\text{Rate} = k[A]^m[B]^n \qquad [14.7]$$

The constant k in the rate law is called the **rate constant**. The magnitude of k changes with temperature and therefore determines how temperature affects rate, as we will see in Section 14.5. The exponents m and n are typically small whole numbers (usually 0, 1, or 2). We will consider these exponents more closely very shortly.

If we know the rate law for a reaction and its rate for a set of reactant concentrations, we can calculate the value of the rate constant, k. For example, using the data in Table 14.2 and the results from experiment 1, we can substitute into Equation 14.6

$$5.4 \times 10^{-7}\ M/\text{s} = k(0.0100\ M)(0.200\ M)$$

Solving for k gives

$$k = \frac{5.4 \times 10^{-7}\ M/\text{s}}{(0.0100\ M)(0.200\ M)} = 2.7 \times 10^{-4}\ M^{-1}\,\text{s}^{-1}$$

You may wish to verify that this same value of k is obtained using any of the other experimental results given in Table 14.2.

Once we have both the rate law and the value of the rate constant for a reaction, we can calculate the rate of reaction for any set of concentrations. For example, using Equation 14.6 and $k = 2.7 \times 10^{-4}\ M^{-1}\,\text{s}^{-1}$, we can calculate the rate for $[NH_4^+] = 0.100\ M$ and $[NO_2^-] = 0.100\ M$:

$$\text{Rate} = (2.7 \times 10^{-4}\ M^{-1}\,\text{s}^{-1})(0.100\ M)(0.100\ M) = 2.7 \times 10^{-6}\ M/\text{s}$$

GIVE IT SOME THOUGHT

(a) What is a rate law? (b) What is the name of the quantity k in any rate law?

Exponents in the Rate Law

The rate laws for most reactions have the general form

$$\text{Rate} = k[\text{reactant 1}]^m[\text{reactant 2}]^n \ldots \qquad [14.8]$$

The exponents m and n in a rate law are called **reaction orders**. For example, consider again the rate law for the reaction of NH_4^+ with NO_2^-:

$$\text{Rate} = k[NH_4^+][NO_2^-]$$

Because the exponent of $[NH_4^+]$ is 1, the rate is *first order* in NH_4^+. The rate is also first order in NO_2^-. (The exponent "1" is not shown explicitly in rate laws.) The **overall reaction order** is the sum of the orders with respect to each reactant in the rate law. Thus, the rate law has an overall reaction order of $1 + 1 = 2$, and the reaction is *second order overall*.

The exponents in a rate law indicate how the rate is affected by the concentration of each reactant. Because the rate at which NH_4^+ reacts with NO_2^- depends on $[NH_4^+]$ raised to the first power, the rate doubles when $[NH_4^+]$ doubles, triples when $[NH_4^+]$ triples, and so forth. Doubling or tripling $[NO_2^-]$ likewise doubles or triples the rate. If a rate law is second order with respect to a reactant, $[A]^2$, then doubling the concentration of that substance causes the reaction rate to quadruple ($[2]^2 = 4$), whereas tripling the concentration causes the rate to increase ninefold ($[3]^2 = 9$).

The following are some additional examples of rate laws:

$$2\,N_2O_5(g) \longrightarrow 4\,NO_2(g) + O_2(g) \qquad \text{Rate} = k[N_2O_5] \qquad [14.9]$$

$$CHCl_3(g) + Cl_2(g) \longrightarrow CCl_4(g) + HCl(g) \qquad \text{Rate} = k[CHCl_3][Cl_2]^{1/2} \qquad [14.10]$$

$$H_2(g) + I_2(g) \longrightarrow 2\,HI(g) \qquad \text{Rate} = k[H_2][I_2] \qquad [14.11]$$

Although the exponents in a rate law are sometimes the same as the coefficients in the balanced equation, this is not necessarily the case, as seen in Equations 14.9 and 14.10. *The values of these exponents must be determined experimentally.* In most rate laws, reaction orders are 0, 1, or 2. However, we also occasionally encounter rate laws in which the reaction order is fractional (such as Equation 14.10) or even negative.

GIVE IT SOME THOUGHT

The experimentally determined rate law for the reaction $2\,NO(g) + 2\,H_2(g) \longrightarrow N_2(g) + 2\,H_2O(g)$ is rate $= k[NO]^2[H_2]$. (a) What are the reaction orders in this rate law? (b) Does doubling the concentration of NO have the same effect on rate as doubling the concentration of H_2?

SAMPLE EXERCISE 14.4 | Relating a Rate Law to the Effect of Concentration on Rate

Consider a reaction $A + B \longrightarrow C$ for which rate $= k[A][B]^2$. Each of the following boxes represents a reaction mixture in which A is shown as red spheres and B as purple ones. Rank these mixtures in order of increasing rate of reaction.

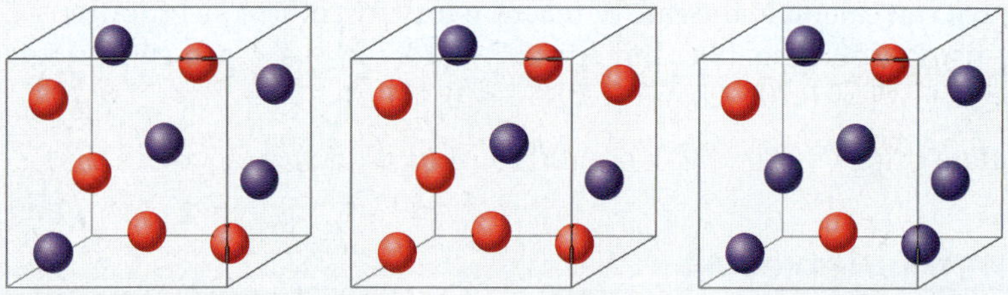

(1) (2) (3)

Solution

Analyze: We are given three boxes containing different numbers of spheres representing mixtures containing different reactant concentrations. We are asked to use the given rate law and the compositions of the boxes to rank the mixtures in order of increasing reaction rates.

Plan: Because all three boxes have the same volume, we can put the number of spheres of each kind into the rate law and calculate the rate for each box.

Solve: Box 1 contains 5 red spheres and 5 purple spheres, giving the following rate:

$$\text{Box 1: Rate} = k(5)(5)^2 = 125k$$

Box 2 contains 7 red spheres and 3 purple spheres:

$$\text{Box 2: Rate} = k(7)(3)^2 = 63k$$

Box 3 contains 3 red spheres and 7 purple spheres:

$$\text{Box 3: Rate} = k(3)(7)^2 = 147k$$

The slowest rate is $63k$ (box 2), and the highest is $147k$ (box 3). Thus, the rates vary in the order $2 < 1 < 3$.

Check: Each box contains 10 spheres. The rate law indicates that in this case [B] has a greater influence on rate than [A] because B has a higher reaction order. Hence, the mixture with the highest concentration of B (most purple spheres) should react fastest. This analysis confirms the order $2 < 1 < 3$.

PRACTICE EXERCISE

Assuming that rate = k[A][B], rank the mixtures represented in this Sample Exercise in order of increasing rate.
Answer: 2 = 3 < 1

Units of Rate Constants

The units of the rate constant depend on the overall reaction order of the rate law. In a reaction that is second order overall, for example, the units of the rate constant must satisfy the equation:

$$\text{Units of rate} = (\text{units of rate constant})(\text{units of concentration})^2$$

Hence, in our usual units of concentration and time

$$\text{Units of rate constant} = \frac{\text{units of rate}}{(\text{units of concentration})^2} = \frac{M/\text{s}}{M^2} = M^{-1}\,\text{s}^{-1}$$

SAMPLE EXERCISE 14.5 | Determining Reaction Orders and Units for Rate Constants

(a) What are the overall reaction orders for the reactions described in Equations 14.9 and 14.10? **(b)** What are the units of the rate constant for the rate law for Equation 14.9?

Solution

Analyze: We are given two rate laws and asked to express **(a)** the overall reaction order for each and **(b)** the units for the rate constant for the first reaction.

Plan: The overall reaction order is the sum of the exponents in the rate law. The units for the rate constant, k, are found by using the normal units for rate (M/s) and concentration (M) in the rate law and applying algebra to solve for k.

Solve: (a) The rate of the reaction in Equation 14.9 is first order in N_2O_5 and first order overall. The reaction in Equation 14.10 is first order in $CHCl_3$ and one-half order in Cl_2. The overall reaction order is three halves.

(b) For the rate law for Equation 14.9, we have

$$\text{Units of rate} = (\text{units of rate constant})(\text{units of concentration})$$

So

$$\text{Units of rate constant} = \frac{\text{units of rate}}{\text{units of concentration}} = \frac{M/\text{s}}{M} = \text{s}^{-1}$$

Notice that the units of the rate constant change as the overall order of the reaction changes.

PRACTICE EXERCISE

(a) What is the reaction order of the reactant H_2 in Equation 14.11? **(b)** What are the units of the rate constant for Equation 14.11?
Answers: **(a)** 1, **(b)** $M^{-1}\,\text{s}^{-1}$

Using Initial Rates to Determine Rate Laws

The rate law for any chemical reaction must be determined experimentally; it cannot be predicted by merely looking at the chemical equation. We often determine the rate law for a reaction by the same method we applied to the data in Table 14.2: We observe the effect of changing the initial concentrations of the reactants on the initial rate of the reaction.

In most reactions the exponents in the rate law are 0, 1, or 2. If a reaction is zero order in a particular reactant, changing its concentration will have no effect on rate (as long as some of the reactant is present) because any concentration raised to the zero power equals 1. On the other hand, we have seen that when a reaction is first order in a reactant, changes in the concentration of that reactant will produce proportional changes in the rate. Thus, doubling the concentration will double the rate, and so forth. Finally, when the rate law is second order in a particular reactant, doubling its concentration increases the rate by a factor of $2^2 = 4$, tripling its concentration causes the rate to increase by a factor of $3^2 = 9$, and so forth.

In working with rate laws, it is important to realize that the *rate* of a reaction depends on concentration, but the *rate constant* does not. As we will see later in this chapter, the rate constant (and hence the reaction rate) is affected by temperature and by the presence of a catalyst.

SAMPLE EXERCISE 14.6 | Determining a Rate Law from Initial Rate Data

The initial rate of a reaction $A + B \longrightarrow C$ was measured for several different starting concentrations of A and B, and the results are as follows:

Experiment Number	[A] (*M*)	[B] (*M*)	Initial Rate (*M*/s)
1	0.100	0.100	4.0×10^{-5}
2	0.100	0.200	4.0×10^{-5}
3	0.200	0.100	16.0×10^{-5}

Using these data, determine **(a)** the rate law for the reaction, **(b)** the magnitude of the rate constant, **(c)** the rate of the reaction when [A] = 0.050 *M* and [B] = 0.100 *M*.

Solution

Analyze: We are given a table of data that relates concentrations of reactants with initial rates of reaction and asked to determine **(a)** the rate law, **(b)** the rate constant, and **(c)** the rate of reaction for a set of concentrations not listed in the table.

Plan: (a) We assume that the rate law has the following form: Rate $= k[A]^m[B]^n$, so we must use the given data to deduce the reaction orders m and n. We do so by determining how changes in the concentration change the rate. **(b)** Once we know m and n, we can use the rate law and one of the sets of data to determine the rate constant k. **(c)** Now that we know both the rate constant and the reaction orders, we can use the rate law with the given concentrations to calculate rate.

Solve: (a) As we move from experiment 1 to experiment 2, [A] is held constant and [B] is doubled. Thus, this pair of experiments shows how [B] affects the rate, allowing us to deduce the order of the rate law with respect to B. Because the rate remains the same when [B] is doubled, the concentration of B has no effect on the reaction rate. The rate law is therefore zero order in B (that is, $n = 0$).

In experiments 1 and 3, [B] is held constant so these data show how [A] affects rate. Holding [B] constant while doubling [A] increases the rate fourfold. This result indicates that rate is proportional to $[A]^2$ (that is, the reaction is second order in A). Hence, the rate law is

$$\text{Rate} = k[A]^2[B]^0 = k[A]^2$$

This rate law could be reached in a more formal way by taking the ratio of the rates from two experiments:

$$\frac{\text{Rate 2}}{\text{Rate 1}} = \frac{4.0 \times 10^{-5}\ M/\text{s}}{4.0 \times 10^{-5}\ M/\text{s}} = 1$$

Using the rate law, we have

$$1 = \frac{\text{rate 2}}{\text{rate 1}} = \frac{k\cancel{[0.100\ M]^m}[0.200\ M]^n}{k\cancel{[0.100\ M]^m}[0.100\ M]^n} = \frac{[0.200]^n}{[0.100]^n} = 2^n$$

2^n equals 1 under only one condition:

$$n = 0$$

We can deduce the value of m in a similar fashion:

$$\frac{\text{Rate 3}}{\text{Rate 1}} = \frac{16.0 \times 10^{-5}\ M/\text{s}}{4.0 \times 10^{-5}\ M/\text{s}} = 4$$

Using the rate law gives

$$4 = \frac{\text{rate 3}}{\text{rate 1}} = \frac{k[0.200\ M]^m\cancel{[0.100\ M]^n}}{k[0.100\ M]^m\cancel{[0.100\ M]^n}} = \frac{[0.200]^m}{[0.100]^m} = 2^m$$

Because $2^m = 4$, we conclude that

$$m = 2$$

(b) Using the rate law and the data from experiment 1, we have

$$k = \frac{\text{rate}}{[\text{A}]^2} = \frac{4.0 \times 10^{-5}\ M/\text{s}}{(0.100\ M)^2} = 4.0 \times 10^{-3}\ M^{-1}\ \text{s}^{-1}$$

(c) Using the rate law from part (a) and the rate constant from part (b), we have

$$\text{Rate} = k[A]^2 = (4.0 \times 10^{-3}\ M^{-1}\ \text{s}^{-1})(0.050\ M)^2 = 1.0 \times 10^{-5}\ M/\text{s}$$

Because [B] is not part of the rate law, it is irrelevant to the rate, provided that there is at least some B present to react with A.
Check: A good way to check our rate law is to use the concentrations in experiment 2 or 3 and see if we can correctly calculate the rate. Using data from experiment 3, we have

$$\text{Rate} = k[\text{A}]^2 = (4.0 \times 10^{-3}\ M^{-1}\ \text{s}^{-1})(0.200\ M)^2 = 1.60 \times 10^{-4}\ M/\text{s}$$

Thus, the rate law correctly reproduces the data, giving both the correct number and the correct units for the rate.

PRACTICE EXERCISE

The following data were measured for the reaction of nitric oxide with hydrogen:

$$2\ \text{NO}(g) + 2\ \text{H}_2(g) \longrightarrow \text{N}_2(g) + 2\ \text{H}_2\text{O}(g)$$

Experiment Number	[NO] (*M*)	[H_2] (*M*)	Initial Rate (*M*/s)
1	0.10	0.10	1.23×10^{-3}
2	0.10	0.20	2.46×10^{-3}
3	0.20	0.10	4.92×10^{-3}

(a) Determine the rate law for this reaction. **(b)** Calculate the rate constant. **(c)** Calculate the rate when [NO] = 0.050 *M* and [H_2] = 0.150 *M*.
Answers: **(a)** rate = $k[\text{NO}]^2[\text{H}_2]$; **(b)** $k = 1.2\ M^{-2}\ \text{s}^{-1}$; **(c)** rate = $4.5 \times 10^{-4}\ M/\text{s}$

14.4 | The Change of Concentration with Time

Rate laws enable us to calculate the rate of a reaction from the rate constant and reactant concentrations. Rate laws can also be converted into equations that tell us what the concentrations of the reactants or products are at any time during the course of a reaction. The mathematics required involves calculus. We don't expect you to be able to perform the calculus operations; however, you should be able to use the resulting equations. We will apply this conversion to two of the simplest rate laws: those that are first order overall and those that are second order overall.

First-Order Reactions

ANIMATION First-Order Process

A **first-order reaction** is one whose rate depends on the concentration of a single reactant raised to the first power. For a reaction of the type A $\longrightarrow$ products, the rate law may be first order:

$$\text{Rate} = -\frac{\Delta[\text{A}]}{\Delta t} = k[\text{A}]$$

This form of a rate law, which expresses how rate depends on concentration, is called the *differential rate law*. Using an operation from calculus called integration, this relationship can be transformed into an equation that relates the concentration of A at the start of the reaction, $[\text{A}]_0$, to its concentration at any other time t, $[\text{A}]_t$:

$$\ln[\text{A}]_t - \ln[\text{A}]_0 = -kt \quad \text{or} \quad \ln\frac{[\text{A}]_t}{[\text{A}]_0} = -kt \qquad [14.12]$$

This form of the rate law is called the *integrated rate law*. The function "ln" in Equation 14.12 is the natural logarithm (Appendix A.2). Equation 14.12 can also be rearranged and written as follows:

$$\ln[A]_t = -kt + \ln[A]_0 \qquad [14.13]$$

Equations 14.12 and 14.13 can be used with any concentration units, as long as the units are the same for both $[A]_t$ and $[A]_0$.

For a first-order reaction, Equation 14.12 or 14.13 can be used in several ways. Given any three of the following quantities, we can solve for the fourth: k, t, $[A]_0$, and $[A]_t$. Thus, these equations can be used, for example, to determine (1) the concentration of a reactant remaining at any time after the reaction has started, (2) the time required for a given fraction of a sample to react, or (3) the time required for a reactant concentration to fall to a certain level.

SAMPLE EXERCISE 14.7 | Using the Integrated First-Order Rate Law

The decomposition of a certain insecticide in water follows first-order kinetics with a rate constant of 1.45 yr^{-1} at 12°C. A quantity of this insecticide is washed into a lake on June 1, leading to a concentration of 5.0×10^{-7} g/cm^3. Assume that the average temperature of the lake is 12°C. **(a)** What is the concentration of the insecticide on June 1 of the following year? **(b)** How long will it take for the concentration of the insecticide to drop to 3.0×10^{-7} g/cm^3?

Solution

Analyze: We are given the rate constant for a reaction that obeys first-order kinetics, as well as information about concentrations and times, and asked to calculate how much reactant (insecticide) remains after one year. We must also determine the time interval needed to reach a particular insecticide concentration. Because the exercise gives time in (a) and asks for time in (b), we know that the integrated rate law, Equation 14.13, is required.

Plan: **(a)** We are given $k = 1.45\ yr^{-1}$, $t = 1.00$ yr, and $[\text{insecticide}]_0 = 5.0 \times 10^{-7} g/cm^3$, and so Equation 14.13 can be solved for $\ln[\text{insecticide}]_t$. **(b)** We have $k = 1.45\ yr^{-1}$, $[\text{insecticide}]_0 = 5.0 \times 10^{-7}\ g/cm^3$, and $[\text{insecticide}]_t = 3.0 \times 10^{-7}\ g/cm^3$, and so we can solve Equation 14.13 for t.

Solve: **(a)** Substituting the known quantities into Equation 14.13, we have

$$\ln[\text{insecticide}]_{t=1\ yr} = -(1.45\ yr^{-1})(1.00\ yr) + \ln(5.0 \times 10^{-7})$$

We use the ln function on a calculator to evaluate the second term on the right, giving

$$\ln[\text{insecticide}]_{t=1\ yr} = -1.45 + (-14.51) = -15.96$$

To obtain $[\text{insecticide}]_{t=1\ yr}$, we use the inverse natural logarithm, or e^x, function on the calculator:

$$[\text{insecticide}]_{t=1\ yr} = e^{-15.96} = 1.2 \times 10^{-7}\ g/cm^3$$

Note that the concentration units for $[A]_t$ and $[A]_0$ must be the same.

(b) Again substituting into Equation 14.13, with $[\text{insecticide}]_t = 3.0 \times 10^{-7}\ g/cm^3$, gives

$$\ln(3.0 \times 10^{-7}) = -(1.45\ yr^{-1})(t) + \ln(5.0 \times 10^{-7})$$

Solving for t gives

$$t = -[\ln(3.0 \times 10^{-7}) - \ln(5.0 \times 10^{-7})]/1.45\ yr^{-1}$$
$$= -(-15.02 + 14.51)/1.45\ yr^{-1} = 0.35\ yr$$

Check: In part (a) the concentration remaining after 1.00 yr (that is, $1.2 \times 10^{-7}\ g/cm^3$) is less than the original concentration ($5.0 \times 10^{-7}\ g/cm^3$), as it should be. In (b) the given concentration ($3.0 \times 10^{-7}\ g/cm^3$) is greater than that remaining after 1.00 yr, indicating that the time must be less than a year. Thus, $t = 0.35$ yr is a reasonable answer.

PRACTICE EXERCISE

The decomposition of dimethyl ether, $(CH_3)_2O$, at 510°C is a first-order process with a rate constant of $6.8 \times 10^{-4}\ s^{-1}$:

$$(CH_3)_2O(g) \longrightarrow CH_4(g) + H_2(g) + CO(g)$$

If the initial pressure of $(CH_3)_2O$ is 135 torr, what is its partial pressure after 1420 s?
Answer: 51 torr

Equation 14.13 can be used to verify whether a reaction is first order and to determine its rate constant. This equation has the form of the general equation for a straight line, $y = mx + b$, in which m is the slope and b is the y-intercept of the line (Appendix A.4):

$$\begin{array}{ccccccc} \ln[\mathrm{A}]_t & = & - & k & \cdot\, t & + & \ln[\mathrm{A}]_0 \\ \updownarrow & & & \updownarrow & \updownarrow & & \updownarrow \\ y & = & & m & \cdot\, x & + & b \end{array}$$

For a first-order reaction, therefore, a graph of $\ln[\mathrm{A}]_t$ versus time gives a straight line with a slope of $-k$ and a y-intercept of $\ln[\mathrm{A}]_0$. A reaction that is not first order will not yield a straight line.

As an example, consider the conversion of methyl isonitrile (CH_3NC) to acetonitrile (CH_3CN) (Figure 14.6 ▶). Because experiments show that the reaction is first order, we can write the rate equation:

$$\ln[\mathrm{CH_3NC}]_t = -kt + \ln[\mathrm{CH_3NC}]_0$$

Figure 14.7(a) ▼ shows how the partial pressure of methyl isonitrile varies with time as it rearranges in the gas phase at 198.9°C. We can use pressure as a unit of concentration for a gas because from the ideal-gas law the pressure is directly proportional to the number of moles per unit volume. Figure 14.7(b) shows a plot of the natural logarithm of the pressure versus time, a plot that yields a straight line. The slope of this line is $-5.1 \times 10^{-5}\ \mathrm{s^{-1}}$. (You should verify this for yourself, remembering that your result may vary slightly from ours because of inaccuracies associated with reading the graph.) Because the slope of the line equals $-k$, the rate constant for this reaction equals $5.1 \times 10^{-5}\ \mathrm{s^{-1}}$.

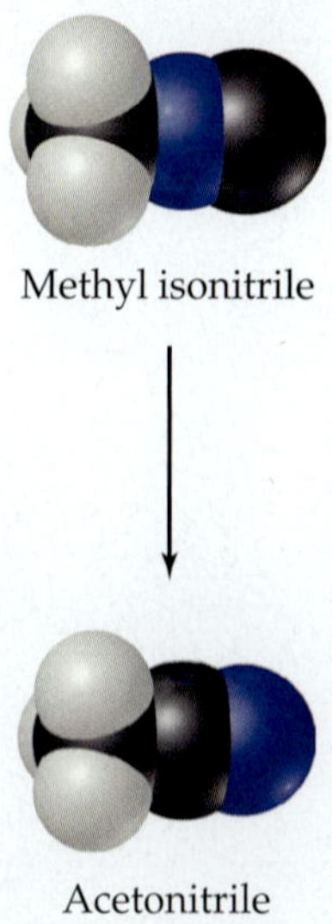

▲ **Figure 14.6 A first-order reaction.** The transformation of methyl isonitrile (CH_3NC) to acetonitrile (CH_3CN) is a first-order process. Methyl isonitrile and acetonitrile are isomers, molecules that have the same atoms arranged differently. This reaction is called an isomerization reaction.

GIVE IT SOME THOUGHT

What do the y-intercepts in Figure 14.7(a) and (b) represent?

Second-Order Reactions

A **second-order reaction** is one whose rate depends on the reactant concentration raised to the second power or on the concentrations of two different reactants, each raised to the first power. For simplicity, let's consider reactions of the

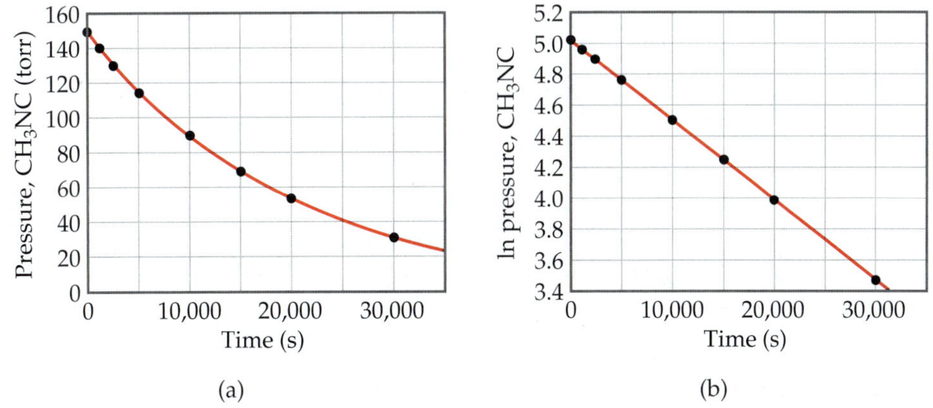

▲ **Figure 14.7 Kinetic data for conversion of methyl isonitrile.** (a) Variation in the partial pressure of methyl isonitrile (CH_3NC) with time during the reaction $CH_3NC \longrightarrow CH_3CN$ at 198.9°C. (b) A plot of the natural logarithm of the CH_3NC pressure as a function of time. The fact that a straight line fits the data confirms that the rate law is first order.

type A $\longrightarrow$ products or A + B $\longrightarrow$ products that are second order in just one reactant, A:

$$\text{Rate} = -\frac{\Delta[\text{A}]}{\Delta t} = k[\text{A}]^2$$

With the use of calculus, this differential rate law can be used to derive the following integrated rate law:

$$\frac{1}{[\text{A}]_t} = kt + \frac{1}{[\text{A}]_0} \qquad [14.14]$$

This equation, like Equation 14.13, has four variables, k, t, $[\text{A}]_0$, and $[\text{A}]_t$, and any one of these can be calculated knowing the other three. Equation 14.14 also has the form of a straight line ($y = mx + b$). If the reaction is second order, a plot of $1/[\text{A}]_t$ versus t will yield a straight line with a slope equal to k and a y-intercept equal to $1/[\text{A}]_0$. One way to distinguish between first- and second-order rate laws is to graph both $\ln[\text{A}]_t$ and $1/[\text{A}]_t$ against t. If the $\ln[\text{A}]_t$ plot is linear, the reaction is first order; if the $1/[\text{A}]_t$ plot is linear, the reaction is second order.

SAMPLE EXERCISE 14.8 | Determining Reaction Order from the Integrated Rate Law

The following data were obtained for the gas-phase decomposition of nitrogen dioxide at 300°C, $NO_2(g) \longrightarrow NO(g) + \frac{1}{2} O_2(g)$:

Time (s)	$[NO_2]$ (M)
0.0	0.01000
50.0	0.00787
100.0	0.00649
200.0	0.00481
300.0	0.00380

Is the reaction first or second order in NO_2?

Solution

Analyze: We are given the concentrations of a reactant at various times during a reaction and asked to determine whether the reaction is first or second order.

Plan: We can plot $\ln[NO_2]$ and $1/[NO_2]$ against time. One or the other will be linear, indicating whether the reaction is first or second order.

Solve: In order to graph $\ln[NO_2]$ and $1/[NO_2]$ against time, we will first prepare the following table from the data given:

Time (s)	$[NO_2]$ (M)	$\ln[NO_2]$	$1/[NO_2]$
0.0	0.01000	−4.610	100
50.0	0.00787	−4.845	127
100.0	0.00649	−5.038	154
200.0	0.00481	−5.337	208
300.0	0.00380	−5.573	263

As Figure 14.8 ▶ shows, only the plot of $1/[NO_2]$ versus time is linear. Thus, the reaction obeys a second-order rate law: Rate = $k[NO_2]^2$. From the slope of this straight-line graph, we determine that $k = 0.543\ M^{-1}\ s^{-1}$ for the disappearance of NO_2.

PRACTICE EXERCISE

Consider again the decomposition of NO_2 discussed in the Sample Exercise. The reaction is second order in NO_2 with $k = 0.543\ M^{-1}\ s^{-1}$. If the initial concentration of NO_2 in a closed vessel is 0.0500 M, what is the remaining concentration after 0.500 h?
Answer: Using Equation 14.14, we find $[NO_2] = 1.00 \times 10^{-3}\ M$

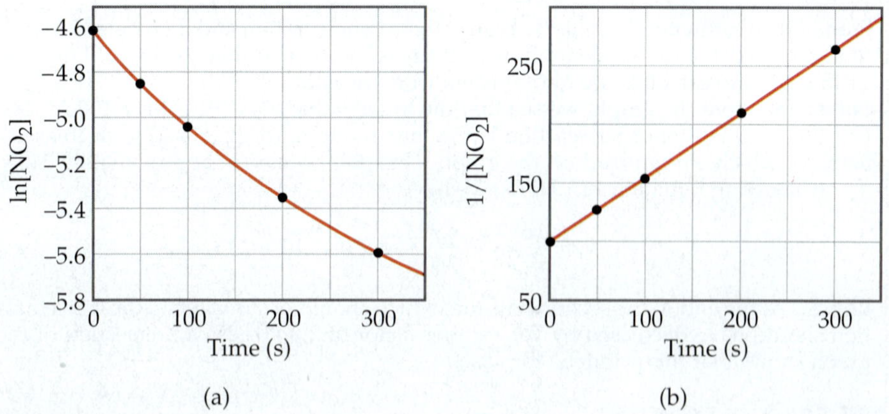

◀ **Figure 14.8 Kinetic data for decomposition of NO_2.** The reaction is $NO_2(g) \longrightarrow NO(g) + \frac{1}{2}O_2(g)$, and the data were collected at 300°C. (a) A plot of $\ln[NO_2]$ versus time is not linear, indicating that the reaction is not first order in NO_2. (b) A plot of $1/[NO_2]$ versus time is linear, indicating that the reaction is second order in NO_2.

ACTIVITY
Rates of Reaction

Half-life

The **half-life** of a reaction, $t_{1/2}$, is the time required for the concentration of a reactant to reach one-half of its initial value, $[A]_{t_{1/2}} = \frac{1}{2}[A]_0$. The half-life is a convenient way to describe how fast a reaction occurs, especially if it is a first-order process. A fast reaction will have a short half-life.

We can determine the half-life of a first-order reaction by substituting $[A]_{t_{1/2}}$ into Equation 14.12:

$$\ln\frac{\frac{1}{2}[A]_0}{[A]_0} = -kt_{1/2}$$

$$\ln\tfrac{1}{2} = -kt_{1/2}$$

$$t_{1/2} = -\frac{\ln\frac{1}{2}}{k} = \frac{0.693}{k} \qquad [14.15]$$

From Equation 14.15, we see that $t_{1/2}$ for a first-order rate law does not depend on the starting concentration. Consequently, the half-life remains constant throughout the reaction. If, for example, the concentration of the reactant is 0.120 *M* at some moment in the reaction, it will be $\frac{1}{2}(0.120\ M) = 0.060\ M$ after one half-life. After one more half-life passes, the concentration will drop to 0.030 *M*, and so on. Equation 14.15 also indicates that we can calculate $t_{1/2}$ for a first-order reaction if k is known, or k if $t_{1/2}$ is known.

The change in concentration over time for the first-order rearrangement of methyl isonitrile at 198.9°C is graphed in Figure 14.9 ▶. The first half-life is shown at 13,600 s (that is, 3.78 h). At a time 13,600 s later, the isonitrile concentration has decreased to one-half of one-half, or one-fourth the original concentration. *In a first-order reaction, the concentration of the reactant decreases by* $\frac{1}{2}$ *in each of a series of regularly spaced time intervals, namely,* $t_{1/2}$. The concept of half-life is widely used in describing radioactive decay, a first-order process that we will discuss in detail in Section 21.4.

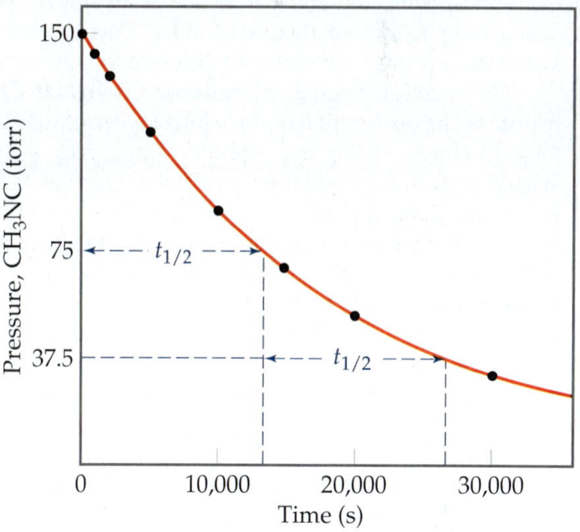

▲ **Figure 14.9 Half-life of a first-order reaction.** Pressure of methyl isonitrile as a function of time showing two successive half-lives of the isomerization reaction depicted in Figure 14.6.

SAMPLE EXERCISE 14.9 | Determining the Half-life of a First-Order Reaction

The reaction of C_4H_9Cl with water is a first-order reaction. Figure 14.4 shows how the concentration of C_4H_9Cl changes with time at a particular temperature. **(a)** From that graph, estimate the half-life for this reaction. **(b)** Use the half-life from (a) to calculate the rate constant.

Solution

Analyze: We are asked to estimate the half-life of a reaction from a graph of concentration versus time and then to use the half-life to calculate the rate constant for the reaction.

Plan: **(a)** To estimate a half-life, we can select a concentration and then determine the time required for the concentration to decrease to half of that value. **(b)** Equation 14.15 is used to calculate the rate constant from the half-life.

Solve: **(a)** From the graph, we see that the initial value of $[C_4H_9Cl]$ is 0.100 *M*. The half-life for this first-order reaction is the time required for $[C_4H_9Cl]$ to decrease to 0.050 *M*, which we can read off the graph. This point occurs at approximately 340 s.

(b) Solving Equation 14.15 for *k*, we have

$$k = \frac{0.693}{t_{1/2}} = \frac{0.693}{340\ \text{s}} = 2.0 \times 10^{-3}\ \text{s}^{-1}$$

Check: At the end of the second half-life, which should occur at 680 s, the concentration should have decreased by yet another factor of 2, to 0.025 *M*. Inspection of the graph shows that this is indeed the case.

PRACTICE EXERCISE

(a) Using Equation 14.15, calculate $t_{1/2}$ for the decomposition of the insecticide described in Sample Exercise 14.7. **(b)** How long does it take for the concentration of the insecticide to reach one-quarter of the initial value?

Answers: **(a)** 0.478 yr $= 1.51 \times 10^7$ s; **(b)** it takes two half-lives, 2(0.478 yr) = 0.956 yr

CHEMISTRY AT WORK | Methyl Bromide in the Atmosphere

Several small molecules containing carbon–chlorine or carbon–bromine bonds, when present in the stratosphere, are capable of reacting with ozone (O_3) and thus contributing to the destruction of Earth's ozone layer. Whether a halogen-containing molecule contributes significantly to destruction of the ozone layer depends in part on the molecule's average lifetime in the atmosphere. It takes quite a long time for molecules formed at Earth's surface to diffuse through the lower atmosphere (called the troposphere) and move into the stratosphere, where the ozone layer is located (Figure 14.10 ▶). Decomposition in the lower atmosphere competes with diffusion into the stratosphere.

The much-discussed chlorofluorocarbons, or CFCs, contribute to the destruction of the ozone layer because they have long lifetimes in the troposphere. Thus, they persist long enough for a substantial fraction of the molecules to find their way to the stratosphere.

Another simple molecule that has the potential to destroy the stratospheric ozone layer is methyl bromide (CH_3Br). This substance has a wide range of uses, including antifungal treatment of plant seeds, and has therefore been produced in large quantities (about 150 million pounds per year). In the stratosphere, the C—Br bond is broken through absorption of short-wavelength radiation. The resultant Br atoms then catalyze decomposition of O_3.

Methyl bromide is removed from the lower atmosphere by a variety of mechanisms, including a slow reaction with ocean water:

$$CH_3Br(g) + H_2O(l) \longrightarrow CH_3OH(aq) + HBr(aq) \quad [14.16]$$

To determine the potential importance of CH_3Br in destruction of the ozone layer, it is important to know how rapidly Equation 14.16 and all other mechanisms together remove CH_3Br from the atmosphere before it can diffuse into the stratosphere.

Scientists have carried out research to estimate the average lifetime of CH_3Br in Earth's atmosphere. Such an estimate is difficult to make. It cannot be done in laboratory-based experiments because the conditions that exist in the atmosphere above the planet are too complex to be simulated in the laboratory. Instead, scientists gathered nearly 4000 samples of the atmosphere during aircraft flights all over the Pacific Ocean and analyzed them for the presence of several trace organic substances, including methyl bromide. From a detailed analysis of the concentrations, it was possible to estimate that the *atmospheric residence time* for CH_3Br is 0.8 ± 0.1 yr.

▲ **Figure 14.10 Distribution and fate of methyl bromide in the atmosphere.** Some CH_3Br is removed from the atmosphere by decomposition, and some diffuses upward into the stratosphere, where it contributes to destruction of the ozone layer. The relative rates of decomposition and diffusion determine how extensively methyl bromide is involved in destruction of the ozone layer.

The atmospheric residence time equals the half-life for CH_3Br in the lower atmosphere, assuming that it decomposes by a first-order process. That is, a collection of CH_3Br molecules present at any given time will, on average, be 50% decomposed after 0.8 years, 75% decomposed after 1.6 years, and so on. A residence time of 0.8 years, while comparatively short, is still sufficiently long so that CH_3Br contributes significantly to the destruction of the ozone layer. In 1997 an international agreement was reached to phase out use of methyl bromide in developed countries by 2005. However, in recent years exemptions for critical agricultural use have been requested and granted.

Related Exercise: 14.105

In contrast to the behavior of first-order reactions, the half-life for second-order and other reactions depends on reactant concentrations and therefore changes as the reaction progresses. Using Equation 14.14, we find that the half-life of a second-order reaction is

$$t_{1/2} = \frac{1}{k[\mathrm{A}]_0} \qquad [14.17]$$

In this case the half-life depends on the initial concentration of reactant—the lower the initial concentration, the greater the half-life.

GIVE IT SOME THOUGHT

How does the half-life of a second-order reaction change as the reaction proceeds?

14.5 | Temperature and Rate

The rates of most chemical reactions increase as the temperature rises. For example, dough rises faster at room temperature than when refrigerated, and plants grow more rapidly in warm weather than in cold. We can literally see the effect of temperature on reaction rate by observing a chemiluminescent reaction (one that produces light). The characteristic glow of fireflies is a familiar example of chemiluminescence. Another is the light produced by Cyalume® light sticks, which contain chemicals that produce chemiluminescence when mixed. As seen in Figure 14.11 ▼, these light sticks produce a brighter light at higher temperature. The amount of light produced is greater because the rate of the reaction is faster at the higher temperature. Although the light stick glows more brightly initially, its luminescence also dies out more rapidly.

How is this experimentally observed temperature effect reflected in the rate expression? The faster rate at higher temperature is due to an increase in the rate constant with increasing temperature. For example, let's reconsider the first-order reaction $CH_3NC \longrightarrow CH_3CN$ (Figure 14.6). Figure 14.12 ▼ shows the rate constant for this reaction as a function of temperature. The rate constant, and hence the rate of the reaction, increases rapidly with temperature, approximately doubling for each 10°C rise.

▲ **Figure 14.11 Temperature affects the rate of the chemiluminescence reaction in Cyalume® light sticks.** The light stick in hot water (left) glows more brightly than the one in cold water (right); at the higher temperature, the reaction is initially faster and produces a brighter light.

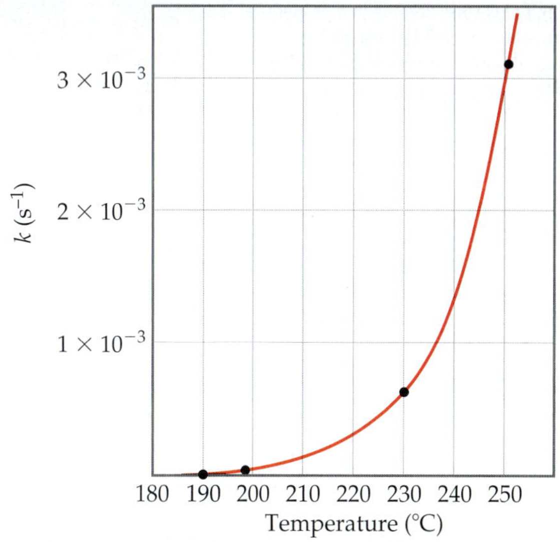

▲ **Figure 14.12 Dependence of rate constant on temperature.** The data show the variation in the first-order rate constant for the rearrangement of methyl isonitrile as a function of temperature. The four points indicated are used in connection with Sample Exercise 14.11.

The Collision Model

ACTIVITY
Arrhenius Model

We have seen that reaction rates are affected both by the concentrations of reactants and by temperature. The **collision model**, which is based on the kinetic-molecular theory (Section 10.7), accounts for both of these effects at the molecular level. The central idea of the collision model is that molecules must collide to react. The greater the number of collisions occurring per second, the greater the reaction rate. As the concentration of reactant molecules increases, therefore, the number of collisions increases, leading to an increase in reaction rate. According to the kinetic-molecular theory of gases, increasing the temperature increases molecular speeds. As molecules move faster, they collide more forcefully (with more energy) and more frequently, increasing reaction rates.

For reaction to occur, though, more is required than simply a collision. For most reactions, only a tiny fraction of the collisions leads to reaction. For example, in a mixture of H_2 and I_2 at ordinary temperatures and pressures, each molecule undergoes about 10^{10} collisions per second. If every collision between H_2 and I_2 resulted in the formation of HI, the reaction would be over in much less than a second. Instead, at room temperature the reaction proceeds very slowly. Only about one in every 10^{13} collisions produces a reaction. What keeps the reaction from occurring more rapidly?

GIVE IT SOME THOUGHT

What is the central idea of the collision model?

The Orientation Factor

In most reactions, molecules must be oriented in a certain way during collisions in order for a reaction to occur. The relative orientations of the molecules during their collisions determine whether the atoms are suitably positioned to form new bonds. For example, consider the reaction of Cl atoms with NOCl:

$$Cl + NOCl \longrightarrow NO + Cl_2$$

The reaction will take place if the collision brings Cl atoms together to form Cl_2, as shown in Figure 14.13(a) ▼. In contrast, the collision shown in Figure 14.13(b) will be ineffective and will not yield products. Indeed, a great many collisions do not lead to reaction, merely because the molecules are not suitably oriented. There is, however, another factor that is usually even more important in determining whether particular collisions result in reaction.

▼ **Figure 14.13 Molecular collisions and chemical reactions.** Two possible ways that Cl atoms and NOCl molecules can collide are shown. (a) If molecules are oriented properly, a sufficiently energetic collision will lead to reaction. (b) If the orientation of the colliding molecules is wrong, no reaction occurs.

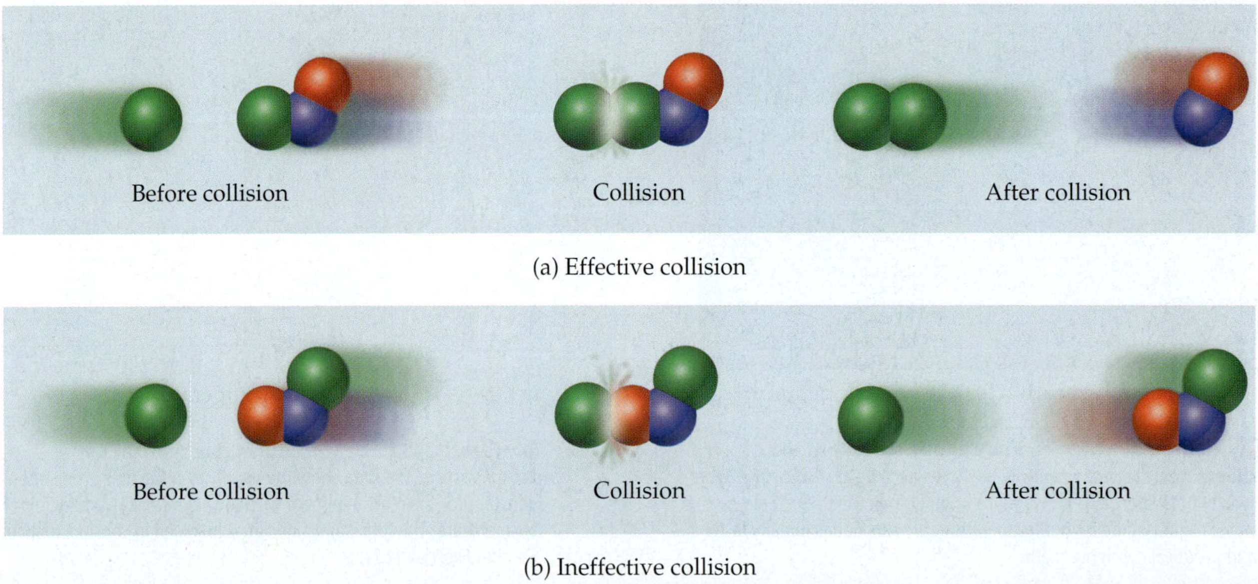

Activation Energy

▲ **Figure 14.14 An energy barrier.** To move the golf ball to the vicinity of the cup, the player must impart enough kinetic energy to the ball to enable it to surmount the barrier represented by the hill. This situation is analogous to a chemical reaction, in which molecules must gain enough energy through collisions to enable them to overcome the barrier to chemical reaction.

In 1888 the Swedish chemist Svante Arrhenius suggested that molecules must possess a certain minimum amount of energy in order to react. According to the collision model, this energy comes from the kinetic energies of the colliding molecules. Upon collision, the kinetic energy of the molecules can be used to stretch, bend, and ultimately break bonds, leading to chemical reactions. That is, the kinetic energy is used to change the potential energy of the molecule. If molecules are moving too slowly, with too little kinetic energy, they merely bounce off one another without changing. In order to react, colliding molecules must have a total kinetic energy equal to or greater than some minimum value. The minimum energy required to initiate a chemical reaction is called the **activation energy**, E_a. The value of E_a varies from reaction to reaction.

The situation during reactions is rather like that shown in Figure 14.14 ▲. The player on the putting green needs to move her ball over the hill to the vicinity of the cup. To do this, she must impart enough kinetic energy with the putter to move the ball to the top of the hill. If she doesn't impart enough energy, the ball will roll partway up the hill and then back down. In the same way, molecules may require a certain minimum energy to break existing bonds during a chemical reaction. In the rearrangement of methyl isonitrile to acetonitrile, for example, we might imagine the reaction passing through an intermediate state in which the N≡C portion of the molecule is sitting sideways:

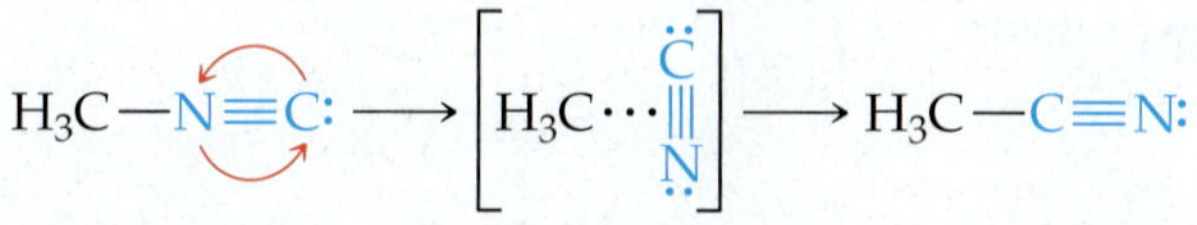

The change in the potential energy of the molecule during the reaction is shown in Figure 14.15 ▼. The diagram shows that energy must be supplied to stretch the bond between the H_3C group and the N≡C group so as to allow

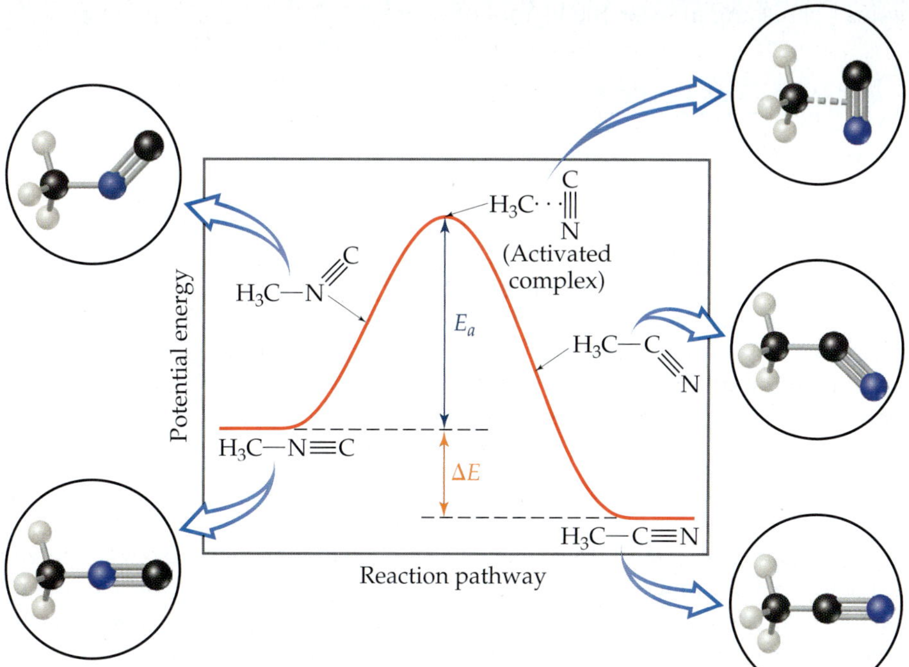

◀ **Figure 14.15 Energy profile for methyl isonitrile isomerization.** The methyl isonitrile molecule must surmount the activation-energy barrier before it can form the product, acetonitrile. The horizontal axis is variously labeled "reaction pathway," as here, or "progress of reaction."

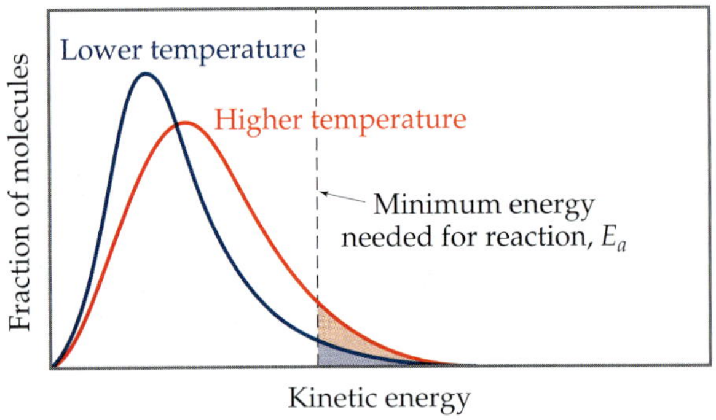

▲ **Figure 14.16 The effect of temperature on the distribution of kinetic energies.** At the higher temperature, a larger number of molecules have higher kinetic energies. Thus, a larger fraction at any one instant will have more than the minimum energy required for reaction.

the N≡C group to rotate. After the N≡C group has twisted sufficiently, the C—C bond begins to form and the energy of the molecule drops. Thus, the barrier represents the energy necessary to force the molecule through the relatively unstable intermediate state to the final product. The energy difference between that of the starting molecule and the highest energy along the reaction pathway is the activation energy, E_a. The particular arrangement of atoms at the top of the barrier is called the **activated complex**, or **transition state**.

The conversion of $H_3C—N≡C$ to $H_3C—C≡N$ is exothermic. Figure 14.15 therefore shows the product as having a lower energy than the reactant. The energy change for the reaction, ΔE, has no effect on the rate of the reaction. The rate depends on the magnitude of E_a; generally, the lower E_a is, the faster the reaction. Notice that the reverse reaction is endothermic. The activation barrier for the reverse reaction is equal to the sum of ΔE and E_a for the forward reaction.

How does any particular methyl isonitrile molecule acquire sufficient energy to overcome the activation barrier? It does so through collisions with other molecules. Recall from the kinetic-molecular theory of gases that, at any given instant, gas molecules are distributed in energy over a wide range. (Section 10.7) Figure 14.16 ▲ shows the distribution of kinetic energies for two different temperatures, comparing them with the minimum energy needed for reaction, E_a. At the higher temperature a much greater fraction of the molecules has kinetic energy greater than E_a, which leads to a much greater rate of reaction.

The fraction of molecules that has an energy equal to or greater than E_a is given by the expression

$$f = e^{-E_a/RT} \qquad [14.18]$$

In this equation R is the gas constant (8.314 J/mol-K) and T is absolute temperature. To get an idea of the magnitude of f, let's suppose that E_a is 100 kJ/mol, a value typical of many reactions, and that T is 300 K, around room temperature. The calculated value of f is 3.8×10^{-18}, an extremely small number! At 310 K the fraction is $f = 1.4 \times 10^{-17}$. Thus, a 10-degree increase in temperature produces a 3.7-fold increase in the fraction of molecules possessing at least 100 kJ/mol of energy.

GIVE IT SOME THOUGHT

Why isn't collision frequency the only factor affecting a reaction rate?

The Arrhenius Equation

Arrhenius noted that for most reactions the increase in rate with increasing temperature is nonlinear, as shown in Figure 14.12. He found that most reaction-rate data obeyed an equation based on three factors: (a) the fraction of molecules possessing an energy of E_a or greater, (b) the number of collisions occurring per second, and (c) the fraction of collisions that have the appropriate orientation. These three factors are incorporated into the **Arrhenius equation**:

$$k = Ae^{-E_a/RT} \qquad [14.19]$$

In this equation k is the rate constant, E_a is the activation energy, R is the gas constant (8.314 J/mol-K), and T is the absolute temperature. The **frequency factor**, A, is constant, or nearly so, as temperature is varied. It is related to the frequency of collisions and the probability that the collisions are favorably oriented for reaction.* As the magnitude of E_a increases, k decreases because the fraction of molecules that possess the required energy is smaller. Thus, *reaction rates decrease as E_a increases.*

SAMPLE EXERCISE 14.10 | Relating Energy Profiles to Activation Energies and Speeds of Reaction

Consider a series of reactions having the following energy profiles:

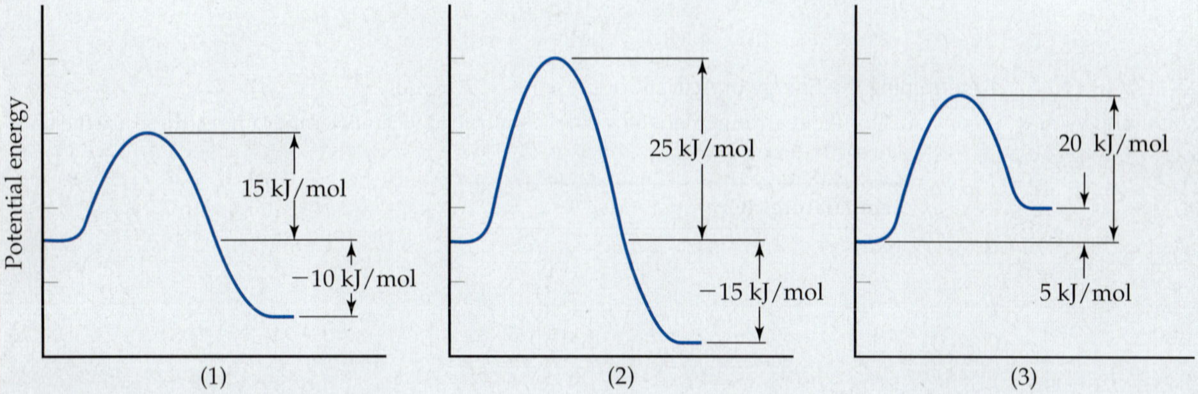

Assuming that all three reactions have nearly the same frequency factors, rank the reactions from slowest to fastest.

Solution The lower the activation energy, the faster the reaction. The value of ΔE does not affect the rate. Hence the order is (2) < (3) < (1).

PRACTICE EXERCISE

Imagine that these reactions are reversed. Rank these reverse reactions from slowest to fastest.

Answer: (2) < (1) < (3) because E_a values are 40, 25, and 15 kJ/mol, respectively

Determining the Activation Energy

Taking the natural log of both sides of Equation 14.19, we have

$$\ln k = -\frac{E_a}{RT} + \ln A \qquad [14.20]$$

Equation 14.20 has the form of a straight line; it predicts that a graph of $\ln k$ versus $1/T$ will be a line with a slope equal to $-E_a/R$ and a y-intercept equal to $\ln A$. Thus, the activation energy can be determined by measuring k at a series of temperatures, graphing $\ln k$ versus $1/T$, and then calculating E_a from the slope of the resultant line.

We can also use Equation 14.20 to evaluate E_a in a nongraphical way if we know the rate constant of a reaction at two or more temperatures. For example, suppose that at two different temperatures, T_1 and T_2, a reaction has rate constants k_1 and k_2. For each condition, we have

ACTIVITY
Rates of Reaction

$$\ln k_1 = -\frac{E_a}{RT_1} + \ln A \quad \text{and} \quad \ln k_2 = -\frac{E_a}{RT_2} + \ln A$$

Subtracting $\ln k_2$ from $\ln k_1$ gives

$$\ln k_1 - \ln k_2 = \left(-\frac{E_a}{RT_1} + \ln A\right) - \left(-\frac{E_a}{RT_2} + \ln A\right)$$

* Because the frequency of collisions increases with temperature, A also has some temperature dependence, but it is small compared to the exponential term. Therefore, A is considered approximately constant.

Simplifying this equation and rearranging it gives

$$\ln\frac{k_1}{k_2} = \frac{E_a}{R}\left(\frac{1}{T_2} - \frac{1}{T_1}\right) \qquad [14.21]$$

Equation 14.21 provides a convenient way to calculate the rate constant, k_1, at some temperature, T_1, when we know the activation energy and the rate constant, k_2, at some other temperature, T_2.

SAMPLE EXERCISE 14.11 | Determining the Energy of Activation

The following table shows the rate constants for the rearrangement of methyl isonitrile at various temperatures (these are the data in Figure 14.12):

Temperature (°C)	$k(s^{-1})$
189.7	2.52×10^{-5}
198.9	5.25×10^{-5}
230.3	6.30×10^{-4}
251.2	3.16×10^{-3}

(a) From these data, calculate the activation energy for the reaction. **(b)** What is the value of the rate constant at 430.0 K?

Solution

Analyze: We are given rate constants, k, measured at several temperatures and asked to determine the activation energy, E_a, and the rate constant, k, at a particular temperature.

Plan: We can obtain E_a from the slope of a graph of $\ln k$ versus $1/T$. Once we know E_a, we can use Equation 14.21 together with the given rate data to calculate the rate constant at 430.0 K.

Solve: (a) We must first convert the temperatures from degrees Celsius to kelvins. We then take the inverse of each temperature, $1/T$, and the natural log of each rate constant, $\ln k$. This gives us the table shown at the right:

T(K)	$1/T\ (K^{-1})$	$\ln k$
462.9	2.160×10^{-3}	−10.589
472.1	2.118×10^{-3}	−9.855
503.5	1.986×10^{-3}	−7.370
524.4	1.907×10^{-3}	−5.757

A graph of $\ln k$ versus $1/T$ results in a straight line, as shown in Figure 14.17 ▶.

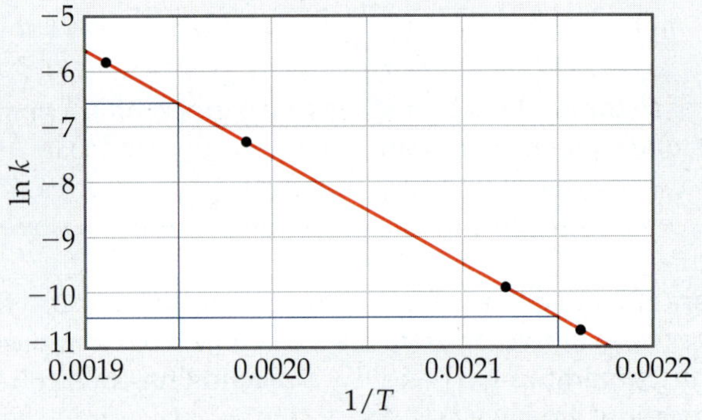

◀ **Figure 14.17 Graphical determination of activation energy.** The natural logarithm of the rate constant for the rearrangement of methyl isonitrile is plotted as a function of $1/T$. The linear relationship is predicted by the Arrhenius equation giving a slope equal to $-E_a/R$.

The slope of the line is obtained by choosing two well-separated points, as shown, and using the coordinates of each:

$$\text{Slope} = \frac{\Delta y}{\Delta x} = \frac{-6.6 - (-10.4)}{0.00195 - 0.00215} = -1.9 \times 10^4$$

Because logarithms have no units, the numerator in this equation is dimensionless. The denominator has the units of $1/T$, namely, K^{-1}. Thus, the overall units for the slope are K. The slope equals $-E_a/R$. We use the value for the molar gas constant R in units of J/mol-K (Table 10.2). We thus obtain

$$\text{Slope} = -\frac{E_a}{R}$$

$$E_a = -(\text{slope})(R) = -(-1.9 \times 10^4\ \text{K})\left(8.314\frac{\text{J}}{\text{mol-K}}\right)\left(\frac{1\ \text{kJ}}{1000\ \text{J}}\right)$$

$$= 1.6 \times 10^2\ \text{kJ/mol} = 160\ \text{kJ/mol}$$

We report the activation energy to only two significant figures because we are limited by the precision with which we can read the graph in Figure 14.17.

(b) To determine the rate constant, k_1, at $T_1 = 430.0$ K, we can use Equation 14.21 with $E_a = 160$ kJ/mol, and one of the rate constants and temperatures from the given data, such as $k_2 = 2.52 \times 10^{-5}\,s^{-1}$ and $T_2 = 462.9$ K:

$$\ln\left(\frac{k_1}{2.52 \times 10^{-5}\,s^{-1}}\right) = \left(\frac{160 \text{ kJ/mol}}{8.314 \text{ J/mol-K}}\right)\left(\frac{1}{462.9 \text{ K}} - \frac{1}{430.0 \text{ K}}\right)\left(\frac{1000 \text{ J}}{1 \text{ kJ}}\right) = -3.18$$

Thus,

$$\frac{k_1}{2.52 \times 10^{-5}\,s^{-1}} = e^{-3.18} = 4.15 \times 10^{-2}$$

$$k_1 = (4.15 \times 10^{-2})(2.52 \times 10^{-5}\,s^{-1}) = 1.0 \times 10^{-6}\,s^{-1}$$

Note that the units of k_1 are the same as those of k_2.

PRACTICE EXERCISE

Using the data in Sample Exercise 14.11, calculate the rate constant for the rearrangement of methyl isonitrile at 280°C.
Answer: $2.2 \times 10^{-2}\,s^{-1}$

14.6 | Reaction Mechanisms

A balanced equation for a chemical reaction indicates the substances present at the start of the reaction and those produced as the reaction proceeds. It provides no information, however, about how the reaction occurs. The process by which a reaction occurs is called the **reaction mechanism.** At the most sophisticated level, a reaction mechanism will describe in great detail the order in which bonds are broken and formed and the changes in relative positions of the atoms in the course of the reaction. We will begin with more rudimentary descriptions of how reactions occur, considering further the nature of the collisions leading to reaction.

Elementary Reactions

We have seen that reactions take place as a result of collisions between reacting molecules. For example, the collisions between molecules of methyl isonitrile (CH_3NC) can provide the energy to allow the CH_3NC to rearrange:

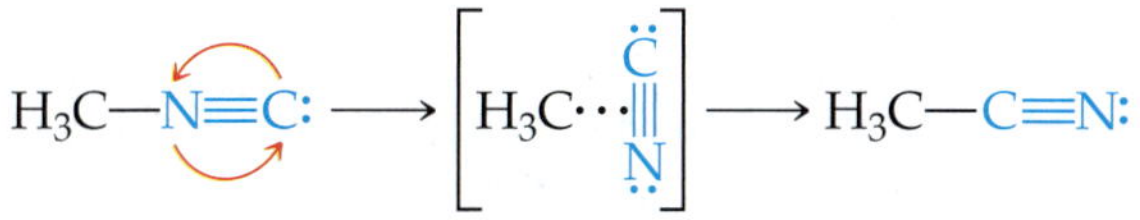

Similarly, the reaction of NO and O_3 to form NO_2 and O_2 appears to occur as a result of a single collision involving suitably oriented and sufficiently energetic NO and O_3 molecules:

$$NO(g) + O_3(g) \longrightarrow NO_2(g) + O_2(g) \qquad [14.22]$$

Both of these processes occur in a single event or step and are called **elementary reactions** (or elementary processes).

The number of molecules that participate as reactants in an elementary reaction defines the **molecularity** of the reaction. If a single molecule is involved, the reaction is **unimolecular**. The rearrangement of methyl isonitrile is a unimolecular process. Elementary reactions involving the collision of two

ANIMATION
Bimolecular Reaction

reactant molecules are **bimolecular**. The reaction between NO and O_3 (Equation 14.22) is bimolecular. Elementary reactions involving the simultaneous collision of three molecules are **termolecular**. Termolecular reactions are far less probable than unimolecular or bimolecular processes and are rarely encountered. The chance that four or more molecules will collide simultaneously with any regularity is even more remote; consequently, such collisions are never proposed as part of a reaction mechanism.

GIVE IT SOME THOUGHT

What is the molecularity of this elementary reaction?

$$NO(g) + Cl_2(g) \longrightarrow NOCl(g) + Cl(g)$$

Multistep Mechanisms

The net change represented by a balanced chemical equation often occurs by a *multistep mechanism*, which consists of a sequence of elementary reactions. For example, consider the reaction of NO_2 and CO:

$$NO_2(g) + CO(g) \longrightarrow NO(g) + CO_2(g) \qquad [14.23]$$

Below 225°C, this reaction appears to proceed in two elementary reactions (or two *elementary steps*), each of which is bimolecular. First, two NO_2 molecules collide, and an oxygen atom is transferred from one to the other. The resultant NO_3 then collides with a CO molecule and transfers an oxygen atom to it:

$$NO_2(g) + NO_2(g) \longrightarrow NO_3(g) + NO(g)$$

$$NO_3(g) + CO(g) \longrightarrow NO_2(g) + CO_2(g)$$

Thus, we say that the reaction occurs by a two-step mechanism.

The chemical equations for the elementary reactions in a multistep mechanism must always add to give the chemical equation of the overall process. In the present example the sum of the two elementary reactions is

$$2\,NO_2(g) + NO_3(g) + CO(g) \longrightarrow NO_2(g) + NO_3(g) + NO(g) + CO_2(g)$$

Simplifying this equation by eliminating substances that appear on both sides of the arrow gives Equation 14.23, the net equation for the process. Because NO_3 is neither a reactant nor a product in the overall reaction—it is formed in one elementary reaction and consumed in the next—it is called an **intermediate**. Multistep mechanisms involve one or more intermediates.

SAMPLE EXERCISE 14.12 | **Determining Molecularity and Identifying Intermediates**

It has been proposed that the conversion of ozone into O_2 proceeds by a two-step mechanism:

$$O_3(g) \longrightarrow O_2(g) + O(g)$$

$$O_3(g) + O(g) \longrightarrow 2\,O_2(g)$$

(a) Describe the molecularity of each elementary reaction in this mechanism. **(b)** Write the equation for the overall reaction. **(c)** Identify the intermediate(s).

Solution

Analyze: We are given a two-step mechanism and asked for **(a)** the molecularities of each of the two elementary reactions, **(b)** the equation for the overall process, and **(c)** the intermediate.

Plan: The molecularity of each elementary reaction depends on the number of reactant molecules in the equation for that reaction. The overall equation is the sum of the equations for the elementary reactions. The intermediate is a substance formed in one step of the mechanism and used in another and therefore not part of the equation for the overall reaction.

Solve: **(a)** The first elementary reaction involves a single reactant and is consequently unimolecular. The second reaction, which involves two reactant molecules, is bimolecular.

(b) Adding the two elementary reactions gives

$$2\,O_3(g) + O(g) \longrightarrow 3\,O_2(g) + O(g)$$

Because $O(g)$ appears in equal amounts on both sides of the equation, it can be eliminated to give the net equation for the chemical process:

$$2\,O_3(g) \longrightarrow 3\,O_2(g)$$

(c) The intermediate is $O(g)$. It is neither an original reactant nor a final product, but is formed in the first step of the mechanism and consumed in the second.

PRACTICE EXERCISE

For the reaction

$$Mo(CO)_6 + P(CH_3)_3 \longrightarrow Mo(CO)_5P(CH_3)_3 + CO$$

the proposed mechanism is

$$Mo(CO)_6 \longrightarrow Mo(CO)_5 + CO$$

$$Mo(CO)_5 + P(CH_3)_3 \longrightarrow Mo(CO)_5P(CH_3)_3$$

(a) Is the proposed mechanism consistent with the equation for the overall reaction? **(b)** What is the molecularity of each step of the mechanism? **(c)** Identify the intermediate(s).

Answers: **(a)** Yes, the two equations add to yield the equation for the reaction. **(b)** The first elementary reaction is unimolecular, and the second one is bimolecular. **(c)** $Mo(CO)_5$

Rate Laws for Elementary Reactions

In Section 14.3 we stressed that rate laws must be determined experimentally; they cannot be predicted from the coefficients of balanced chemical equations. We are now in a position to understand why this is so: Every reaction is made up of a series of one or more elementary steps, and the rate laws and relative speeds of these steps will dictate the overall rate law. Indeed, the rate law for a reaction can be determined from its mechanism, as we will see shortly. Thus, our next challenge in kinetics is to arrive at reaction mechanisms that lead to rate laws that are consistent with those observed experimentally. We will start by examining the rate laws of elementary reactions.

Elementary reactions are significant in a very important way: *If we know that a reaction is an elementary reaction, then we know its rate law.* The rate law of any elementary reaction is based directly on its molecularity. For example, consider the general unimolecular process.

$$A \longrightarrow \text{products}$$

As the number of A molecules increases, the number that decompose in a given interval of time will increase proportionally. Thus, the rate of a unimolecular process will be first order:

$$\text{Rate} = k[A]$$

In the case of bimolecular elementary steps, the rate law is second order, as in the following example:

$$A + B \longrightarrow \text{products} \qquad \text{Rate} = k[A][B]$$

The second-order rate law follows directly from the collision theory. If we double the concentration of A, the number of collisions between molecules of A and B will double; likewise, if we double [B], the number of collisions will double.

TABLE 14.3 Elementary Reactions and Their Rate Laws

Molecularity	Elementary Reaction	Rate Law
*Uni*molecular	$A \longrightarrow$ products	Rate $= k[A]$
*Bi*molecular	$A + A \longrightarrow$ products	Rate $= k[A]^2$
*Bi*molecular	$A + B \longrightarrow$ products	Rate $= k[A][B]$
*Ter*molecular	$A + A + A \longrightarrow$ products	Rate $= k[A]^3$
*Ter*molecular	$A + A + B \longrightarrow$ products	Rate $= k[A]^2[B]$
*Ter*molecular	$A + B + C \longrightarrow$ products	Rate $= k[A][B][C]$

Therefore, the rate law will be first order in both [A] and [B], and second order overall.

The rate laws for all feasible elementary reactions are given in Table 14.3 ▲. Notice how the rate law for each kind of elementary reaction follows directly from the molecularity of that reaction. It is important to remember, however, that we cannot tell by merely looking at a balanced chemical equation whether the reaction involves one or several elementary steps.

SAMPLE EXERCISE 14.13 | Predicting the Rate Law for an Elementary Reaction

If the following reaction occurs in a single elementary reaction, predict the rate law:

$$H_2(g) + Br_2(g) \longrightarrow 2\,HBr(g)$$

Solution

Analyze: We are given the equation and asked for its rate law, assuming that it is an elementary process.

Plan: Because we are assuming that the reaction occurs as a single elementary reaction, we are able to write the rate law using the coefficients for the reactants in the equation as the reaction orders.

Solve: The reaction is bimolecular, involving one molecule of H_2 with one molecule of Br_2. Thus, the rate law is first order in each reactant and second order overall:

$$\text{Rate} = k[H_2][Br_2]$$

Comment: Experimental studies of this reaction show that the reaction actually has a very different rate law:

$$\text{Rate} = k[H_2][Br_2]^{1/2}$$

Because the experimental rate law differs from the one obtained by assuming a single elementary reaction, we can conclude that the mechanism must involve two or more elementary steps.

PRACTICE EXERCISE

Consider the following reaction: $2\,NO(g) + Br_2(g) \longrightarrow 2\,NOBr(g)$. **(a)** Write the rate law for the reaction, assuming it involves a single elementary reaction. **(b)** Is a single-step mechanism likely for this reaction?

Answers: **(a)** Rate $= k[NO]^2[Br_2]$ **(b)** No, because termolecular reactions are very rare

The Rate-Determining Step for a Multistep Mechanism

As with the reaction in Sample Exercise 14.13, most chemical reactions occur by mechanisms that involve two or more elementary reactions. Each of these steps of the mechanism has its own rate constant and activation energy. Often one of the steps is much slower than the others. The overall rate of a reaction cannot exceed the rate of the slowest elementary step of its mechanism. Because the slow step limits the overall reaction rate, it is called the **rate-determining step** (or *rate-limiting step*).

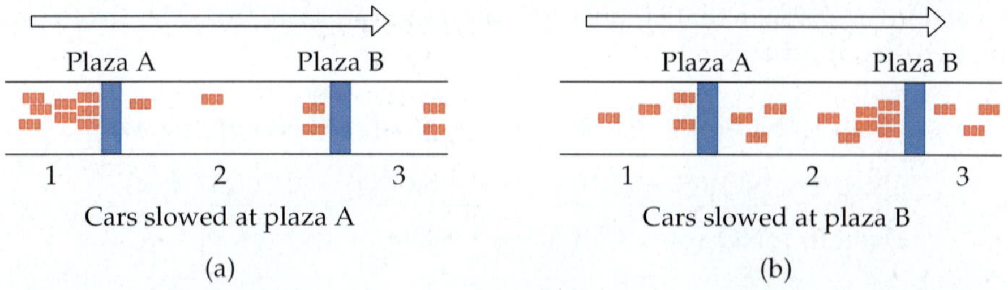

▲ **Figure 14.18 Rate-determining step.** The flow of traffic on a toll road illustrates how a rate-determining step controls reaction rate. The flow is limited by the flow of traffic through the slower toll plaza. In (a) the rate at which cars can reach point 3 is limited by how quickly they can get through plaza A. In this case, getting from point 1 to point 2 is the rate-determining step. In (b), getting from point 2 to point 3 is the rate-determining step.

To understand the concept of a rate-determining step, consider a toll road with two toll plazas (Figure 14.18 ▲). We will measure the rate at which cars exit the toll road. Cars enter the toll road at point 1 and pass through toll plaza A. They then pass an intermediate point 2 before passing through toll plaza B. Upon exiting, they pass point 3. We can therefore envision this trip along the toll road as occurring in two elementary steps:

Step 1: Point 1 $\longrightarrow$ point 2 (through plaza A)

Step 2: Point 2 $\longrightarrow$ point 3 (through plaza B)

Overall: Point 1 $\longrightarrow$ point 3 (through plaza A and B)

Now suppose that several of the gates at toll plaza A are malfunctioning, so that traffic backs up behind it [Figure 14.18(a)]. The rate at which cars can get to point 3 is limited by the rate at which they can get through the traffic jam at plaza A. Thus, step 1 is the rate-determining step of the journey along the toll road. If, however, traffic flows quickly through plaza A, but gets backed up at plaza B [Figure 14.18(b)], there will be a buildup of cars in the intermediate region between the plazas. In this case step 2 is rate determining: The rate at which cars can travel the toll road is limited by the rate at which they can pass through plaza B.

In the same way, *the slowest step in a multistep reaction limits the overall rate.* By analogy to Figure 14.18(a), the rate of a faster step following the rate-determining step does not speed up the overall rate. If the slow step is not the first one, as in Figure 14.18(b), the faster preceding steps produce intermediate products that accumulate before being consumed in the slow step. In either case *the rate-determining step governs the rate law for the overall reaction.*

GIVE IT SOME THOUGHT

Why can't the rate law for a reaction generally be deduced from the balanced equation for the reaction?

Mechanisms with a Slow Initial Step

The relationship between the slow step in a mechanism and the rate law for the overall reaction is most easily seen by considering an example in which the first step in a multistep mechanism is the slow, rate-determining step. As an example, consider the reaction of NO_2 and CO to produce NO and CO_2 (Equation 14.23). Below 225°C, it is found experimentally that the rate law for this reaction is second order in NO_2 and zero order in CO: Rate $= k[NO_2]^2$. Can we propose a reaction mechanism that is consistent with this rate law? Consider the following two-step mechanism:*

$$\text{Step 1:}\quad NO_2(g) + NO_2(g) \xrightarrow{k_1} NO_3(g) + NO(g) \quad \text{(slow)}$$

$$\text{Step 2:}\quad NO_3(g) + CO(g) \xrightarrow{k_2} NO_2(g) + CO_2(g) \quad \text{(fast)}$$

$$\text{Overall:}\quad NO_2(g) + CO(g) \longrightarrow NO(g) + CO_2(g)$$

Step 2 is much faster than step 1; that is, $k_2 \gg k_1$. The intermediate $NO_3(g)$ is slowly produced in step 1 and is immediately consumed in step 2.

Because step 1 is slow and step 2 is fast, step 1 is rate determining. Thus, the rate of the overall reaction equals the rate of step 1, and the rate law of the overall reaction equals the rate law of step 1. Step 1 is a bimolecular process that has the rate law

$$\text{Rate} = k_1[NO_2]^2$$

Thus, the rate law predicted by this mechanism agrees with the one observed experimentally.

Could we propose a one-step mechanism for the preceding reaction? We might suppose that the overall reaction is a single bimolecular elementary process that involves the collision of a molecule of NO_2 with one of CO. However, the rate law predicted by this mechanism would be

$$\text{Rate} = k[NO_2][CO]$$

Because this mechanism predicts a rate law different from that observed experimentally, we can rule it out.

SAMPLE EXERCISE 14.14 | Determining the Rate Law for a Multistep Mechanism

The decomposition of nitrous oxide, N_2O, is believed to occur by a two-step mechanism:

$$N_2O(g) \longrightarrow N_2(g) + O(g) \quad \text{(slow)}$$

$$N_2O(g) + O(g) \longrightarrow N_2(g) + O_2(g) \quad \text{(fast)}$$

(a) Write the equation for the overall reaction. **(b)** Write the rate law for the overall reaction.

Solution

Analyze: Given a multistep mechanism with the relative speeds of the steps, we are asked to write the overall reaction and the rate law for that overall reaction.

Plan: **(a)** The overall reaction is found by adding the elementary steps and eliminating the intermediates. **(b)** The rate law for the overall reaction will be that of the slow, rate-determining step.

Solve: **(a)** Adding the two elementary reactions gives

$$2\,N_2O(g) + O(g) \longrightarrow 2\,N_2(g) + O_2(g) + O(g)$$

Omitting the intermediate, $O(g)$, which occurs on both sides of the equation, gives the overall reaction:

$$2\,N_2O(g) \longrightarrow 2\,N_2(g) + O_2(g)$$

(b) The rate law for the overall reaction is just the rate law for the slow, rate-determining elementary reaction. Because that slow step is a unimolecular elementary reaction, the rate law is first order:

$$\text{Rate} = k[N_2O]$$

* The subscript on the rate constant identifies the elementary step involved. Thus, k_1 is the rate constant for step 1, k_2 is the rate constant for step 2, and so forth. A negative subscript refers to the rate constant for the reverse of an elementary step. For example, k_{-1} is the rate constant for the reverse of the first step.

PRACTICE EXERCISE

Ozone reacts with nitrogen dioxide to produce dinitrogen pentoxide and oxygen:

$$O_3(g) + 2\,NO_2(g) \longrightarrow N_2O_5(g) + O_2(g)$$

The reaction is believed to occur in two steps:

$$O_3(g) + NO_2(g) \longrightarrow NO_3(g) + O_2(g)$$
$$NO_3(g) + NO_2(g) \longrightarrow N_2O_5(g)$$

The experimental rate law is rate $= k[O_3][NO_2]$. What can you say about the relative rates of the two steps of the mechanism?

Answer: Because the rate law conforms to the molecularity of the first step, that must be the rate-determining step. The second step must be much faster than the first one.

Mechanisms with a Fast Initial Step

It is difficult to derive the rate law for a mechanism in which an intermediate is a reactant in the rate-determining step. This situation arises in multistep mechanisms when the first step is *not* rate determining. Let's consider one example: the gas-phase reaction of nitric oxide (NO) with bromine (Br_2).

$$2\,NO(g) + Br_2(g) \longrightarrow 2\,NOBr(g) \qquad [14.24]$$

The experimentally determined rate law for this reaction is second order in NO and first order in Br_2:

$$\text{Rate} = k[NO]^2[Br_2] \qquad [14.25]$$

We seek a reaction mechanism that is consistent with this rate law. One possibility is that the reaction occurs in a single termolecular step:

$$NO(g) + NO(g) + Br_2(g) \longrightarrow 2\,NOBr(g) \qquad \text{Rate} = k[NO]^2[Br_2] \qquad [14.26]$$

As noted in Practice Exercise 14.13, this does not seem likely because termolecular processes are so rare.

Let's consider an alternative mechanism that does not invoke termolecular steps:

$$\text{Step 1:} \quad NO(g) + Br_2(g) \underset{k_{-1}}{\overset{k_1}{\rightleftharpoons}} NOBr_2(g) \quad \text{(fast)}$$

$$\text{Step 2:} \quad NOBr_2(g) + NO(g) \xrightarrow{k_2} 2\,NOBr(g) \quad \text{(slow)}$$

In this mechanism, step 1 actually involves two processes: a forward reaction and its reverse.

Because step 2 is the slow, rate-determining step, the rate of the overall reaction is governed by the rate law for that step:

$$\text{Rate} = k[NOBr_2][NO] \qquad [14.27]$$

However, $NOBr_2$ is an intermediate generated in step 1. Intermediates are usually unstable molecules that have a low, unknown concentration. Thus, our rate law depends on the unknown concentration of an intermediate.

Fortunately, with the aid of some assumptions, we can express the concentration of the intermediate ($NOBr_2$) in terms of the concentrations of the starting reactants (NO and Br_2). We first assume that $NOBr_2$ is intrinsically unstable and that it does not accumulate to a significant extent in the reaction mixture. There are two ways for $NOBr_2$ to be consumed once it is formed: It can either react with NO to form NOBr or fall back apart into NO and Br_2. The first of these possibilities is step 2, a slow process. The second is the reverse of step 1, a unimolecular process:

$$NOBr_2(g) \xrightarrow{k_{-1}} NO(g) + Br_2(g) \qquad [14.28]$$

Because step 2 is slow, we assume that most of the $NOBr_2$ falls apart according to Equation 14.28. Thus, we have both the forward and reverse reactions of step 1 occurring much faster than step 2. Because they occur rapidly with respect to the reaction in step 2, the forward and reverse processes of step 1 establish an equilibrium. We have seen examples of dynamic equilibrium before, in the equilibrium between a liquid and its vapor (Section 11.5) and between a solid solute and its solution. (Section 13.3) As in any dynamic equilibrium, the rates of the forward and reverse reactions are equal. Thus, we can equate the rate expression for the forward reaction in step 1 with the rate expression for the reverse reaction:

$$\underbrace{k_1[NO][Br_2]}_{\text{Rate of forward reaction}} = \underbrace{k_{-1}[NOBr_2]}_{\text{Rate of reverse reaction}}$$

Solving for $[NOBr_2]$, we have

$$[NOBr_2] = \frac{k_1}{k_{-1}}[NO][Br_2]$$

Substituting this relationship into the rate law for the rate-determining step (Equation 14.27), we have

$$\text{Rate} = k_2\frac{k_1}{k_{-1}}[NO][Br_2][NO] = k[NO]^2[Br_2]$$

This is consistent with the experimental rate law (Equation 14.25). The experimental rate constant, k, equals k_2k_1/k_{-1}. This mechanism, which involves only unimolecular and bimolecular processes, is far more probable than the single termolecular step (Equation 14.26).

In general, *whenever a fast step precedes a slow one, we can solve for the concentration of an intermediate by assuming that an equilibrium is established in the fast step.*

SAMPLE EXERCISE 14.15 | Deriving the Rate Law for a Mechanism with a Fast Initial Step

Show that the following mechanism for Equation 14.24 also produces a rate law consistent with the experimentally observed one:

$$\text{Step 1:} \quad NO(g) + NO(g) \underset{k_{-1}}{\overset{k_1}{\rightleftharpoons}} N_2O_2(g) \quad \text{(fast equilibrium)}$$

$$\text{Step 2:} \quad N_2O_2(g) + Br_2(g) \xrightarrow{k_2} 2\,NOBr(g) \quad \text{(slow)}$$

Solution

Analyze: We are given a mechanism with a fast initial step and asked to write the rate law for the overall reaction.

Plan: The rate law of the slow elementary step in a mechanism determines the rate law for the overall reaction. Thus, we first write the rate law based on the molecularity of the slow step. In this case the slow step involves the intermediate N_2O_2 as a reactant. Experimental rate laws, however, do not contain the concentrations of intermediates, but are expressed in terms of the concentrations of starting substances. Thus, we must relate the concentration of N_2O_2 to the concentration of NO by assuming that an equilibrium is established in the first step.

Solve: The second step is rate determining, so the overall rate is

$$\text{Rate} = k_2[N_2O_2][Br_2]$$

We solve for the concentration of the intermediate N_2O_2 by assuming that an equilibrium is established in step 1; thus, the rates of the forward and reverse reactions in step 1 are equal:

$$k_1[\text{NO}]^2 = k_{-1}[\text{N}_2\text{O}_2]$$

$$[\text{N}_2\text{O}_2] = \frac{k_1}{k_{-1}}[\text{NO}]^2$$

Substituting this expression into the rate expression gives

$$\text{Rate} = k_2\frac{k_1}{k_{-1}}[\text{NO}]^2[\text{Br}_2] = k[\text{NO}]^2[\text{Br}_2]$$

Thus, this mechanism also yields a rate law consistent with the experimental one.

PRACTICE EXERCISE

The first step of a mechanism involving the reaction of bromine is

$$\text{Br}_2(g) \underset{k_{-1}}{\overset{k_1}{\rightleftharpoons}} 2\,\text{Br}(g) \quad \text{(fast, equilibrium)}$$

What is the expression relating the concentration of $Br(g)$ to that of $Br_2(g)$?

Answer: $[\text{Br}] = \left(\frac{k_1}{k_{-1}}[\text{Br}_2]\right)^{1/2}$

14.7 | Catalysis

A **catalyst** is a substance that changes the speed of a chemical reaction without undergoing a permanent chemical change itself in the process. Catalysts are very common; most reactions in the body, the atmosphere, and the oceans occur with the help of catalysts. Much industrial chemical research is devoted to the search for new and more effective catalysts for reactions of commercial importance. Extensive research efforts also are devoted to finding means of inhibiting or removing certain catalysts that promote undesirable reactions, such as those that corrode metals, age our bodies, and cause tooth decay.

MOVIE
Catalysis

Homogeneous Catalysis

A catalyst that is present in the same phase as the reacting molecules is called a **homogeneous catalyst**. Examples abound both in solution and in the gas phase. Consider, for example, the decomposition of aqueous hydrogen peroxide, $H_2O_2(aq)$, into water and oxygen:

$$2\,\text{H}_2\text{O}_2(aq) \longrightarrow 2\,\text{H}_2\text{O}(l) + \text{O}_2(g) \qquad [14.29]$$

In the absence of a catalyst, this reaction occurs extremely slowly.

Many different substances are capable of catalyzing the reaction represented by Equation 14.29, including bromide ion, $Br^-(aq)$, as shown in Figure 14.19 ▶. The bromide ion reacts with hydrogen peroxide in acidic solution, forming aqueous bromine and water:

$$2\,\text{Br}^-(aq) + \text{H}_2\text{O}_2(aq) + 2\,\text{H}^+ \longrightarrow \text{Br}_2(aq) + 2\,\text{H}_2\text{O}(l) \qquad [14.30]$$

The brown color observed in the middle photograph of Figure 14.19 indicates the formation of $Br_2(aq)$. If this were the complete reaction, bromide ion would not be a catalyst because it undergoes chemical change during the reaction. However, hydrogen peroxide also reacts with the $Br_2(aq)$ generated in Equation 14.30:

$$\text{Br}_2(aq) + \text{H}_2\text{O}_2(aq) \longrightarrow 2\,\text{Br}^-(aq) + 2\,\text{H}^+(aq) + \text{O}_2(g) \qquad [14.31]$$

The bubbling evident in Figure 14.19(b) is due to the formation of $O_2(g)$.

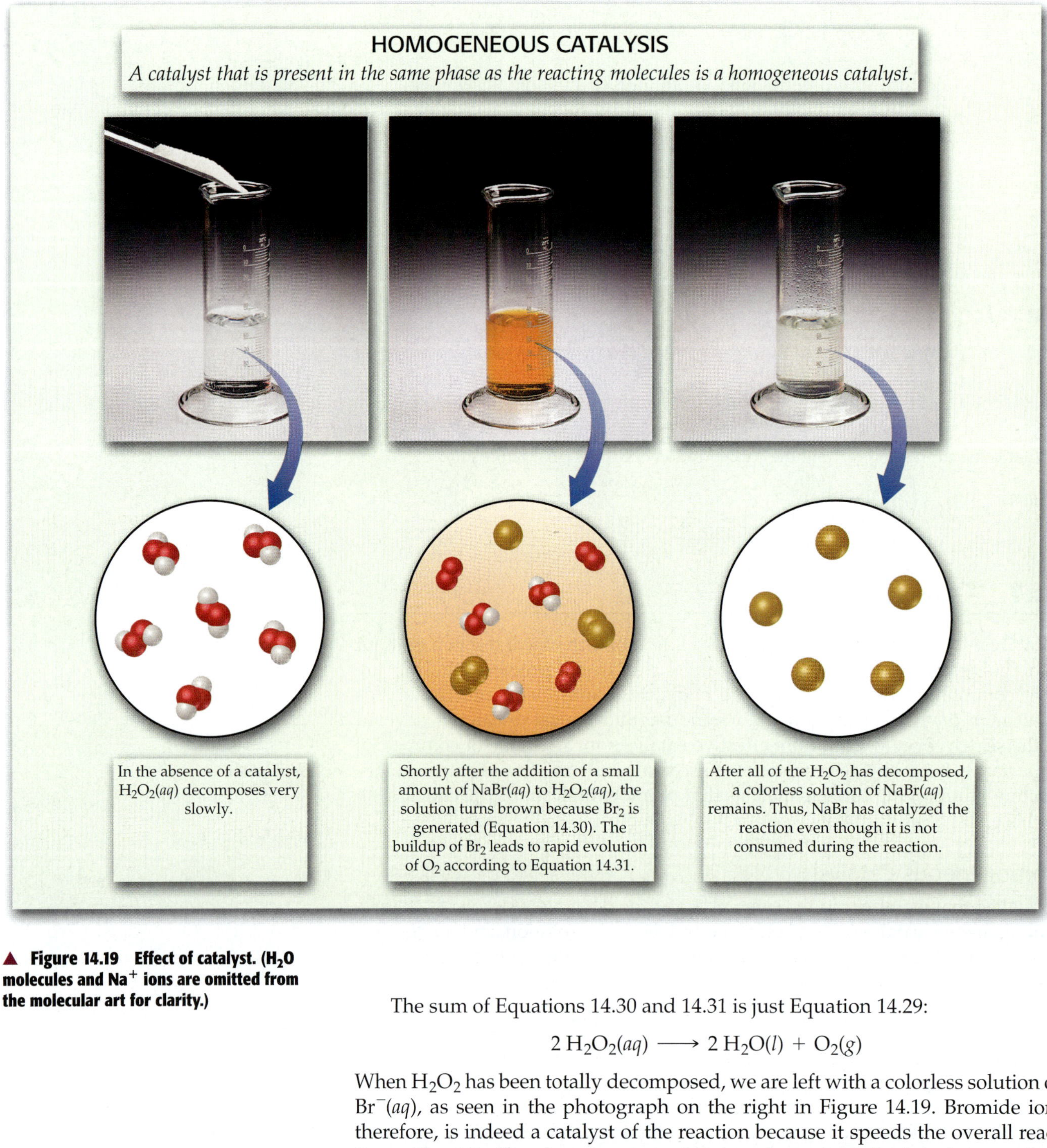

▲ **Figure 14.19 Effect of catalyst. (H_2O molecules and Na^+ ions are omitted from the molecular art for clarity.)**

The sum of Equations 14.30 and 14.31 is just Equation 14.29:

$$2\,H_2O_2(aq) \longrightarrow 2\,H_2O(l) + O_2(g)$$

When H_2O_2 has been totally decomposed, we are left with a colorless solution of $Br^-(aq)$, as seen in the photograph on the right in Figure 14.19. Bromide ion, therefore, is indeed a catalyst of the reaction because it speeds the overall reaction without itself undergoing any net change. In contrast, Br_2 is an intermediate because it is first formed (Equation 14.30) and then consumed (Equation 14.31). Neither the catalyst nor the intermediate appears in the chemical equation for the overall reaction. Notice, however, that the catalyst is there at the start of the reaction, whereas the intermediate is formed during the course of the reaction.

On the basis of the Arrhenius equation (Equation 14.19), the rate constant (k) is determined by the activation energy (E_a) and the frequency factor (A). A catalyst may affect the rate of reaction by altering the value of either E_a or A. The most dramatic catalytic effects come from lowering E_a. As a general rule, *a catalyst lowers the overall activation energy for a chemical reaction.*

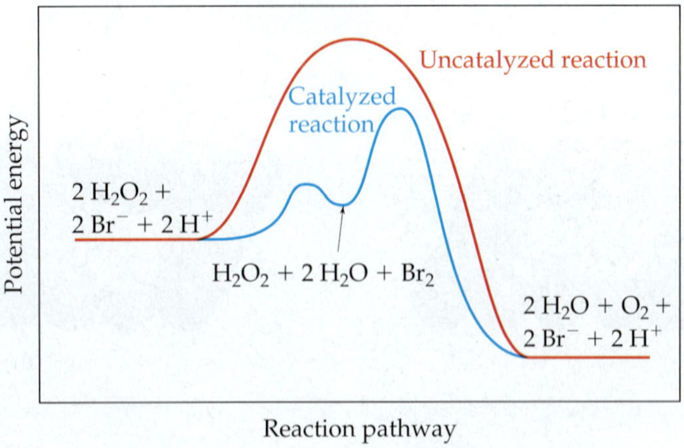

◀ **Figure 14.20 Energy profiles for uncatalyzed and catalyzed reactions.** The energy profiles for the uncatalyzed decomposition of hydrogen peroxide and for the reaction as catalyzed by Br^- are compared. The catalyzed reaction involves two successive steps, each of which has a lower activation energy than the uncatalyzed reaction. Notice that the energies of reactants and products are unchanged by the catalyst.

A catalyst usually lowers the overall activation energy for a reaction by providing a completely different mechanism for the reaction. The examples given previously involve a reversible, cyclic reaction of the catalyst with the reactants. In the decomposition of hydrogen peroxide, for example, two successive reactions of H_2O_2, with bromide and then with bromine, take place. Because these two reactions together serve as a catalytic pathway for hydrogen peroxide decomposition, *both* of them must have significantly lower activation energies than the uncatalyzed decomposition, as shown schematically in Figure 14.20 ▲.

GIVE IT SOME THOUGHT

How does a catalyst increase the rate of a reaction?

Heterogeneous Catalysis

A **heterogeneous catalyst** exists in a different phase from the reactant molecules, usually as a solid in contact with either gaseous reactants or with reactants in a liquid solution. Many industrially important reactions are catalyzed by the surfaces of solids. For example, hydrocarbon molecules are rearranged to form gasoline with the aid of what are called "cracking" catalysts (see the "Chemistry at Work" box in Section 25.3). Heterogeneous catalysts are often composed of metals or metal oxides. Because the catalyzed reaction occurs on the surface, special methods are often used to prepare catalysts so that they have very large surface areas.

The initial step in heterogeneous catalysis is usually **adsorption** of reactants. *Adsorption* refers to the binding of molecules to a surface, whereas *absorption* refers to the uptake of molecules into the interior of another substance. ∞ (Section 13.6) Adsorption occurs because the atoms or ions at the surface of a solid are extremely reactive. Unlike their counterparts in the interior of the substance, surface atoms and ions have unused bonding capacity. This unused bonding capability may be used to bond molecules from the gas or solution phase to the surface of the solid.

The reaction of hydrogen gas with ethylene gas to form ethane gas provides an example of heterogeneous catalysis:

$$\underset{\text{Ethylene}}{C_2H_4(g)} + H_2(g) \longrightarrow \underset{\text{Ethane}}{C_2H_6(g)} \qquad \Delta H^\circ = -137 \text{ kJ/mol} \qquad [14.32]$$

Even though this reaction is exothermic, it occurs very slowly in the absence of a catalyst. In the presence of a finely powdered metal, however, such as nickel, palladium, or platinum, the reaction occurs rather easily at room temperature.

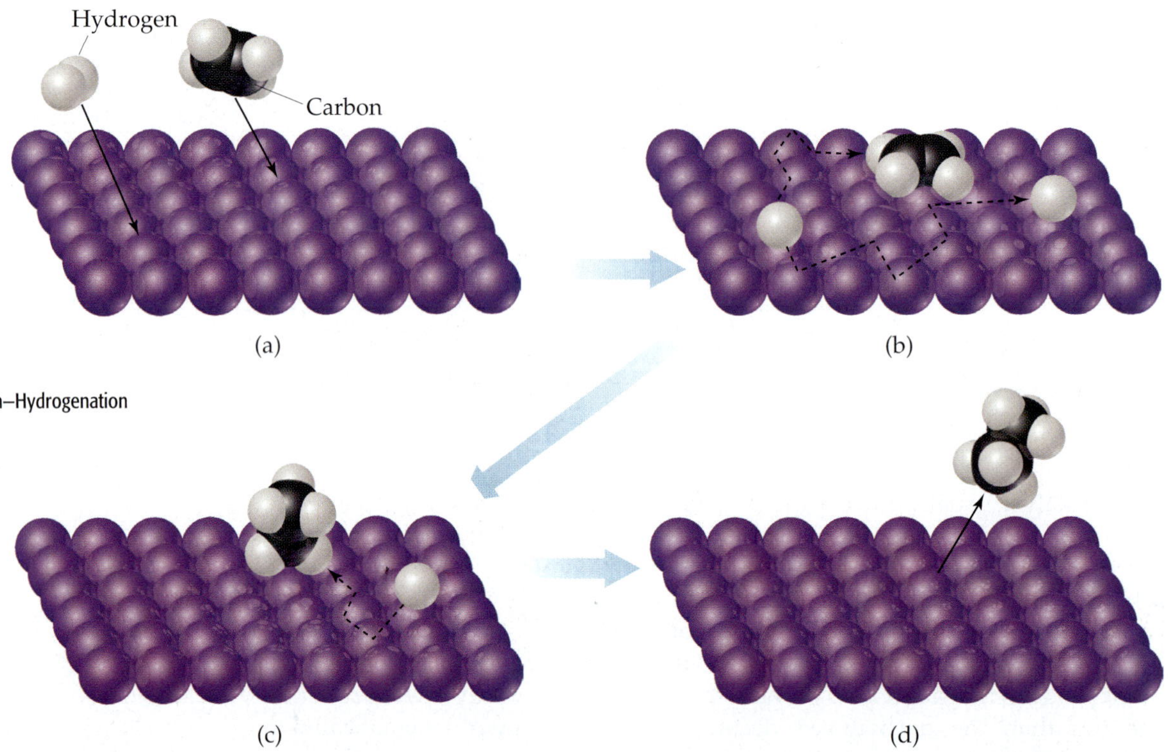

ANIMATION
Surface Reaction–Hydrogenation

▲ **Figure 14.21 Mechanism for reaction of ethylene with hydrogen on a catalytic surface.**
(a) The hydrogen and ethylene are adsorbed at the metal surface. (b) The H—H bond is broken to give adsorbed hydrogen atoms. (c) These migrate to the adsorbed ethylene and bond to the carbon atoms. (d) As C—H bonds are formed, the adsorption of the molecule to the metal surface is decreased and ethane is released.

The mechanism by which the reaction occurs is diagrammed in Figure 14.21 ▲. Both ethylene and hydrogen are adsorbed on the metal surface [Figure 14.21(a)]. Upon adsorption the H—H bond of H_2 breaks, leaving two H atoms that are bonded to the metal surface, as shown in Figure 14.21(b). The hydrogen atoms are relatively free to move about the surface. When a hydrogen encounters an adsorbed ethylene molecule, it can form a σ bond to one of the carbon atoms, effectively destroying the C—C π bond and leaving an *ethyl group* (C_2H_5) bonded to the surface via a metal-to-carbon σ bond [Figure 14.21(c)]. This σ bond is relatively weak, so when the other carbon atom also encounters a hydrogen atom, a sixth C—H σ bond is readily formed and an ethane molecule is released from the metal surface [Figure 14.21(d)]. The site is ready to adsorb another ethylene molecule and thus begin the cycle again.

We can understand the role of the catalyst in this process by considering the bond enthalpies involved. ∞ (Section 8.8) In the course of the reaction, the H—H σ bond and the C—C π bond must be broken, and to do so requires the input of energy, which we can liken to the activation energy of the reaction. The formation of the new C—H σ bonds *releases* an even greater amount of energy, making the reaction exothermic. When H_2 and C_2H_4 are bonded to the surface of the catalyst, less energy is required to break the bonds, lowering the activation energy of the reaction.

Enzymes

Many of the most interesting and important examples of catalysis involve reactions within living systems. The human body is characterized by an extremely complex system of interrelated chemical reactions. All these reactions must

CHEMISTRY AT WORK | Catalytic Converters

Heterogeneous catalysis plays a major role in the fight against urban air pollution. Two components of automobile exhausts that help form photochemical smog are nitrogen oxides and unburned hydrocarbons of various types (Section 18.4). In addition, automobile exhausts may contain considerable quantities of carbon monoxide. Even with the most careful attention to engine design, it is impossible under normal driving conditions to reduce the quantity of these pollutants to an acceptable level in the exhaust gases. It is therefore necessary to remove them from the exhaust before they are vented to the air. This removal is accomplished in the *catalytic converter*.

The catalytic converter, which is part of the exhaust system, must perform two distinct functions: (1) oxidation of CO and unburned hydrocarbons (C_xH_y) to carbon dioxide and water, and (2) reduction of nitrogen oxides to nitrogen gas:

$$CO, C_xH_y \xrightarrow{O_2} CO_2 + H_2O$$

$$NO, NO_2 \longrightarrow N_2$$

These two functions require two distinctly different catalysts, so the development of a successful catalyst system is a difficult challenge. The catalysts must be effective over a wide range of operating temperatures. They must continue to be active in spite of the fact that various components of the exhaust can block the active sites of the catalyst. They must be sufficiently rugged to withstand exhaust gas turbulence and the mechanical shocks of driving under various conditions for thousands of miles.

Catalysts that promote the combustion of CO and hydrocarbons are, in general, the transition-metal oxides and the noble metals, such as platinum. A mixture of two different metal oxides, CuO and Cr_2O_3, might be used, for example. These materials are supported on a structure (Figure 14.22 ▶) that allows the best possible contact between the flowing exhaust gas and the catalyst surface. Either bead or honeycomb structures made from alumina (Al_2O_3) and impregnated with the catalyst may be employed. Such catalysts operate by first adsorbing oxygen gas, also present in the exhaust gas. This adsorption weakens the O—O bond in O_2, so that oxygen atoms are available for reaction with adsorbed CO to form CO_2. Hydrocarbon oxidation probably proceeds somewhat similarly, with the hydrocarbons first being adsorbed followed by rupture of a C—H bond.

The most effective catalysts for reduction of NO to yield N_2 and O_2 are transition-metal oxides and noble metals, the same kinds of materials that catalyze the oxidation of CO and hydrocarbons. The catalysts that are most effective in one reaction, however, are usually much less effective in the other. It is therefore necessary to have two different catalytic components.

Catalytic converters are remarkably efficient heterogeneous catalysts. The automotive exhaust gases are in contact with the catalyst for only 100 to 400 ms. In this very short time, 96% of the hydrocarbons and CO are converted to CO_2 and H_2O, and the emission of nitrogen oxides is reduced by 76%.

There are costs as well as benefits associated with the use of catalytic converters. Some of the metals used in the converters are very expensive. Catalytic converters currently account for about 35% of the platinum, 65% of the palladium, and 95% of the rhodium used annually. All of these metals, which come mainly from Russia and South Africa, are far more expensive than gold.
Related Exercises: 14.54, 14.73, and 14.74

▲ **Figure 14.22 Cross section of a catalytic converter.** Automobiles are equipped with catalytic converters, which are part of their exhaust systems. The exhaust gases contain CO, NO, NO_2, and unburned hydrocarbons which pass over surfaces impregnated with catalysts. The catalysts promote the conversion of the exhaust gases into CO_2, H_2O, and N_2.

occur at carefully controlled rates in order to maintain life. A large number of marvelously efficient biological catalysts known as **enzymes** are necessary for many of these reactions to occur at suitable rates. Most enzymes are large protein molecules with molecular weights ranging from about 10,000 to about 1 million amu. They are very selective in the reactions that they catalyze, and some are absolutely specific, operating for only one substance in only one reaction. The decomposition of hydrogen peroxide, for example, is an important biological process. Because hydrogen peroxide is strongly oxidizing, it can be physiologically harmful. For this reason, the blood and livers of mammals contain an enzyme, *catalase*, which catalyzes the decomposition of hydrogen peroxide into water and oxygen (Equation 14.29). Figure 14.23 ▶ shows the dramatic acceleration of this chemical reaction by the catalase in beef liver.

▶ **Figure 14.23 Effect of an enzyme.** Ground-up beef liver causes hydrogen peroxide to decompose rapidly into water and oxygen. The decomposition is catalyzed by the enzyme *catalase*. Grinding the liver breaks open the cells, so that the reaction takes place more rapidly. The frothing is due to escape of oxygen gas from the reaction mixture.

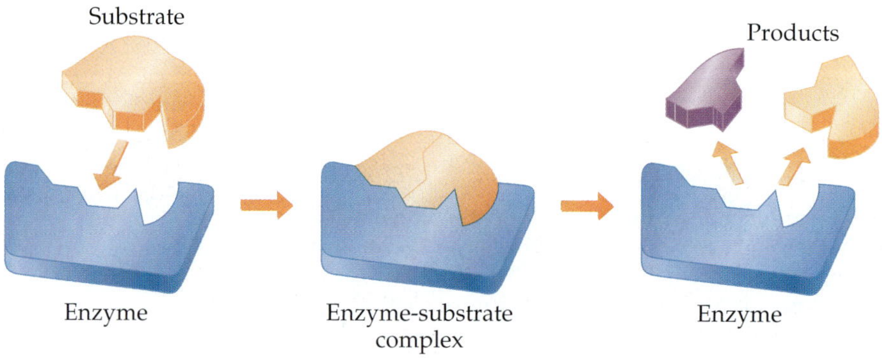

▶ **Figure 14.24 The lock-and-key model for enzyme action.** The correct substrate is recognized by its ability to fit the active site of the enzyme, forming the enzyme-substrate complex. After the reaction of the substrate is complete, the products separate from the enzyme.

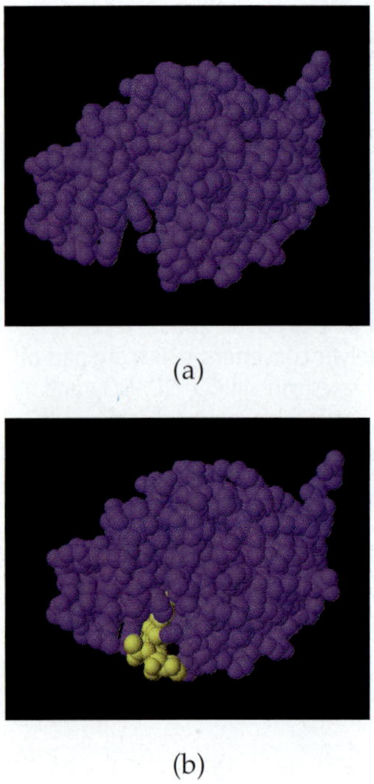

▲ **Figure 14.25 Molecular model of an enzyme.** (a) A molecular model of the enzyme *lysozyme*. Note the characteristic cleft, which is the location of the active site. (b) Lysozyme with a bound substrate molecule.

Although an enzyme is a large molecule, the reaction is catalyzed at a very specific location in the enzyme, called the **active site**. The substances that undergo reaction at this site are called **substrates**. A simple explanation for the specificity of enzymes is provided by the **lock-and-key model**, illustrated in Figure 14.24 ▲. The substrate is pictured as fitting neatly into a special place on the enzyme (the active site), much like a specific key fitting into a lock. The active site is created by coiling and folding of the long protein molecule to form a space, something like a pocket, into which the substrate molecule fits. Figure 14.25 ◀ shows a model of the enzyme *lysozyme* with and without a bound substrate molecule.

The combination of the enzyme and the substrate is called the *enzyme-substrate complex*. Although Figure 14.24 shows both the active site and its complementary substrate as having rigid shapes, there is often a fair amount of flexibility in the active site. Thus, the active site may change shape as it binds the substrate. The binding between the substrate and the active site involves intermolecular forces such as dipole-dipole attractions, hydrogen bonds, and London dispersion forces. ∞ (Section 11.2)

As the substrate molecules enter the active site, they are somehow activated, so that they are capable of extremely rapid reaction. This activation may result from the withdrawal or donation of electron density at a particular bond by the enzyme. In addition, in the process of fitting into the active site, the substrate molecule may be distorted and thus made more reactive. Once the reaction occurs, the products then depart, allowing another substrate molecule to enter.

The activity of an enzyme is destroyed if some molecule in the solution is able to bind strongly to the active site and block the entry of the substrate. Such substances are called *enzyme inhibitors*. Nerve poisons and certain toxic metal ions such as lead and mercury are believed to act in this way to inhibit enzyme activity. Some other poisons act by attaching elsewhere on the enzyme, thereby distorting the active site so that the substrate no longer fits.

Enzymes are enormously more efficient than ordinary nonbiochemical catalysts. The number of individual catalyzed reaction events occurring at a particular active site, called the *turnover number*, is generally in the range of 10^3 to 10^7 per second. Such large turnover numbers correspond to very low activation energies.

GIVE IT SOME THOUGHT

What names are given to the following aspects of enzymes and enzyme catalysis: (a) the place on the enzyme where catalysis occurs, (b) the substances that undergo catalysis?

CHEMISTRY AND LIFE | Nitrogen Fixation and Nitrogenase

Nitrogen is one of the most essential elements in living organisms. It is found in many compounds that are vital to life, including proteins, nucleic acids, vitamins, and hormones. Plants use very simple nitrogen-containing compounds, especially NH_3, NH_4^+, and NO_3^-, as starting materials from which such complex, biologically necessary compounds are formed. Animals are unable to synthesize the complex nitrogen compounds they require from the simple substances used by plants. Instead, they rely on more complicated precursors present in vitamin- and protein-rich foods.

Nitrogen is continually cycling through this biological arena in various forms, as shown in the simplified nitrogen cycle in Figure 14.26 ▼. For example, certain microorganisms convert the nitrogen in animal waste and dead plants and animals into molecular nitrogen, $N_2(g)$, which returns to the atmosphere. In order for the food chain to be sustained, there must be a means of reincorporating this atmospheric N_2 in a form that plants can utilize. The process of converting N_2 into compounds that plants can use is called *nitrogen fixation*. Fixing nitrogen is difficult; N_2 is an exceptionally unreactive molecule, in large part because of its very strong N≡N triple bond. ∞ (Section 8.3) Some fixed nitrogen results from the action of lightning on the atmosphere, and some is produced industrially using a process we will discuss in Chapter 15. About 60% of fixed nitrogen, however, is a consequence of the action of a remarkable and complex enzyme called *nitrogenase*. This enzyme is *not* present in humans or other animals; rather, it is found in bacteria that live in the root nodules of certain plants such as the legumes clover and alfalfa.

Nitrogenase converts N_2 into NH_3, a process that, in the absence of a catalyst, has a very large activation energy. This process is a *reduction* of nitrogen—during the reaction, its oxidation state is reduced from 0 in N_2 to −3 in NH_3. The mechanism by which nitrogenase reduces N_2 is not fully understood. Like many other enzymes, including catalase, the active site of nitrogenase contains transition-metal atoms;

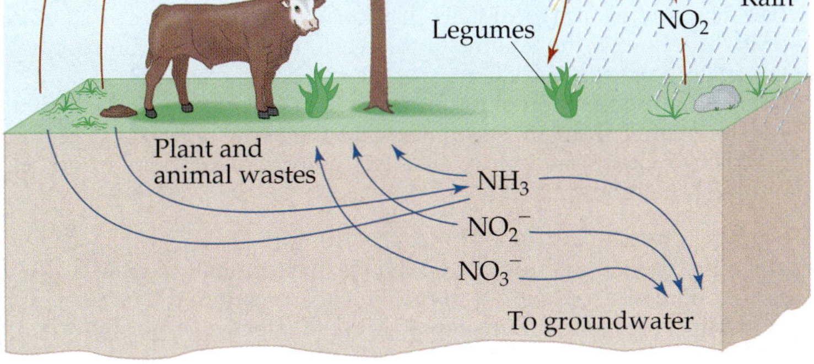

◀ **Figure 14.26 Simplified picture of the nitrogen cycle.** The compounds of nitrogen in the soil are water-soluble species, such as NH_3, NO_2^-, and NO_3^-, which can be washed out of the soil by groundwater. These nitrogen compounds are converted into biomolecules by plants and are incorporated into animals that eat the plants. Animal waste and dead plants and animals are attacked by certain bacteria that release N_2 to the atmosphere. Atmospheric N_2 is fixed in the soil predominantly by the action of certain plants that contain the enzyme nitrogenase, thereby completing the cycle.

continues on next page

continued

such enzymes are called *metalloenzymes*. Because transition metals can readily change oxidation state, metalloenzymes are especially useful for effecting transformations in which substrates are either oxidized or reduced.

It has been known for nearly 20 years that a portion of nitrogenase contains iron and molybdenum atoms. This portion, called the *FeMo-cofactor*, is thought to serve as the active site of the enzyme. The FeMo-cofactor of nitrogenase is a striking cluster of seven Fe atoms and one Mo atom, all linked by sulfur atoms (Figure 14.27 ▼). In 2002, scientists using higher resolution X-ray crystallography discovered a single light atom at the center of the FeMo-cofactor. The specific identity of the atom is not yet known, but it is thought to be a nitrogen atom. The role of this atom in the reduction of N_2 by the cofactor is an area of intense current research.

It is one of the wonders of life that simple bacteria can contain beautifully complex and vitally important enzymes such as nitrogenase. Because of this enzyme, nitrogen is continually cycled between its comparatively inert role in the atmosphere and its critical role in living organisms; without it, life as we know it could not exist on Earth.

Related Exercises: 14.77, 14.103

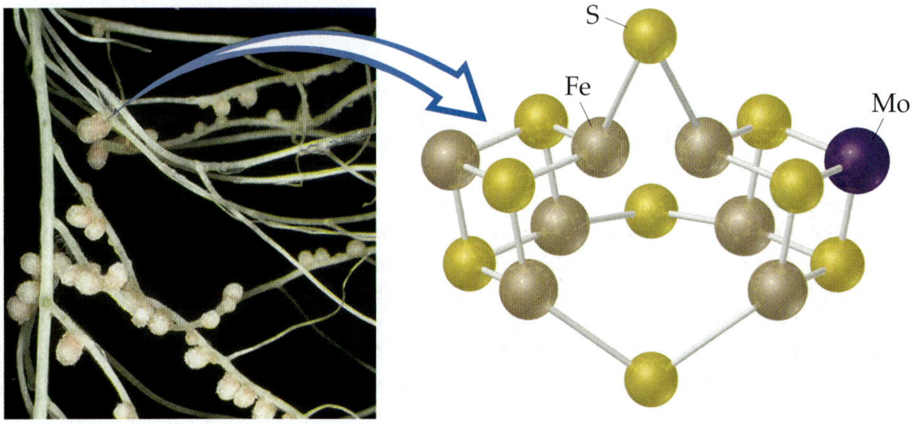

◀ **Figure 14.27 The FeMo-cofactor of nitrogenase.** Nitrogenase is found in nodules in the roots of certain plants, such as the white clover roots shown at the left. The cofactor, which is thought to be the active site of the enzyme, contains seven Fe atoms and one Mo atom, linked by sulfur atoms. The molecules on the outside of the cofactor connect it to the rest of the protein.

SAMPLE INTEGRATIVE EXERCISE | Putting Concepts Together

Formic acid (HCOOH) decomposes in the gas phase at elevated temperatures as follows:

$$HCOOH(g) \longrightarrow CO_2(g) + H_2(g)$$

The decomposition reaction is determined to be first order. A graph of the partial pressure of HCOOH versus time for decomposition at 838 K is shown as the red curve in Figure 14.28 ◀. When a small amount of solid ZnO is added to the reaction chamber, the partial pressure of acid versus time varies as shown by the blue curve in Figure 14.28.

(a) Estimate the half-life and first-order rate constant for formic acid decomposition.

(b) What can you conclude from the effect of added ZnO on the decomposition of formic acid?

(c) The progress of the reaction was followed by measuring the partial pressure of formic acid vapor at selected times. Suppose that, instead, we had plotted the concentration of formic acid in units of mol/L. What effect would this have had on the calculated value of *k*?

(d) The pressure of formic acid vapor at the start of the reaction is 3.00×10^2 torr. Assuming constant temperature and ideal-gas behavior, what is the pressure in the system at the end of the reaction? If the volume of the reaction chamber is 436 cm^3, how many moles of gas occupy the reaction chamber at the end of the reaction?

(e) The standard heat of formation of formic acid vapor is $\Delta H_f^\circ = -378.6$ kJ/mol. Calculate ΔH° for the overall reaction. Assuming that the activation energy (E_a) for the reaction is 184 kJ/mol, sketch an approximate energy profile for the reaction, and label E_a, ΔH°, and the transition state.

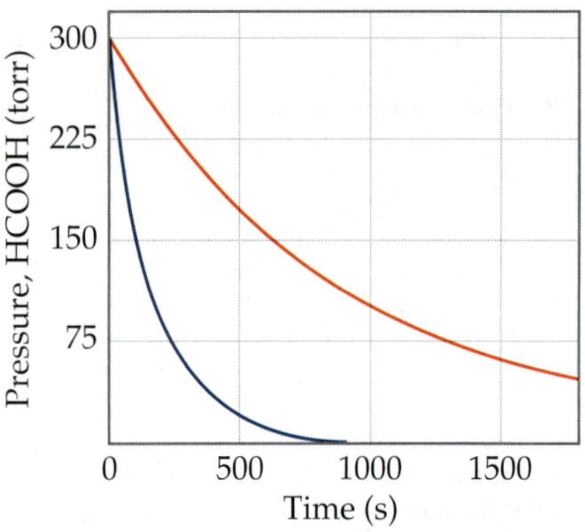

▲ **Figure 14.28 Variation in pressure of HCOOH(*g*) as a function of time at 838 K.** The red line corresponds to decomposition when only gaseous HCOOH is present. The blue line corresponds to decomposition in the presence of added ZnO(*s*).

Solution (a) The initial pressure of HCOOH is 3.00×10^2 torr. On the graph we move to the level at which the partial pressure of HCOOH is 150 torr, half the initial value. This corresponds to a time of about 6.60×10^2 s, which is therefore the half-life. The first-order rate constant is given by Equation 14.15: $k = 0.693/t_{1/2} = 0.693/660 \text{ s} = 1.05 \times 10^{-3} \text{ s}^{-1}$.

(b) The reaction proceeds much more rapidly in the presence of solid ZnO, so the surface of the oxide must be acting as a catalyst for the decomposition of the acid. This is an example of heterogeneous catalysis.

(c) If we had graphed the concentration of formic acid in units of moles per liter, we would still have determined that the half-life for decomposition is 660 seconds, and we would have computed the same value for k. Because the units for k are s^{-1}, the value for k is independent of the units used for concentration.

(d) According to the stoichiometry of the reaction, two moles of product are formed for each mole of reactant. When reaction is completed, therefore, the pressure will be 600 torr, just twice the initial pressure, assuming ideal-gas behavior. (Because we are working at quite high temperature and fairly low gas pressure, assuming ideal-gas behavior is reasonable.) The number of moles of gas present can be calculated using the ideal-gas equation (Section 10.4):

$$n = \frac{PV}{RT} = \frac{(600/760 \text{ atm})(0.436 \text{ L})}{(0.0821 \text{ L-atm/mol-K})(838 \text{ K})} = 5.00 \times 10^{-3} \text{ moles}$$

(e) We first calculate the overall change in energy, ΔH° (Section 5.7 and Appendix C), as in

$$\begin{aligned} \Delta H^\circ &= \Delta H_f^\circ(CO_2(g)) + \Delta H_f^\circ(H_2(g)) - \Delta H_f^\circ(HCOOH(g)) \\ &= -393.5 \text{ kJ/mol} + 0 - (-378.6 \text{ kJ/mol}) \\ &= -14.9 \text{ kJ/mol} \end{aligned}$$

From this and the given value for E_a, we can draw an approximate energy profile for the reaction, in analogy to Figure 14.15.

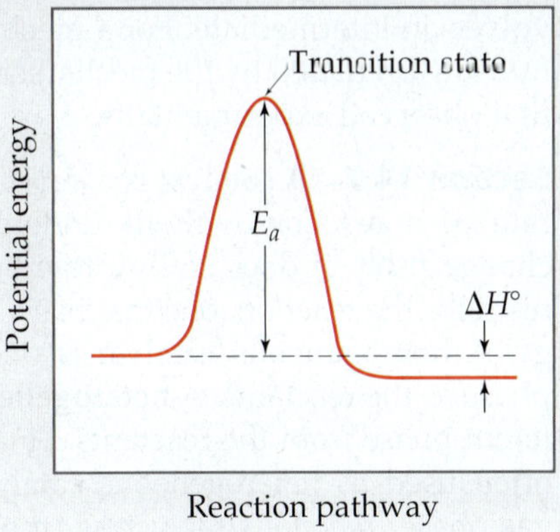

SUMMARY AND KEY TERMS

Introduction and Section 14.1 In this chapter we explored **chemical kinetics**, the area of chemistry that studies the rates of chemical reactions and the factors that affect them, namely, concentration, temperature, and catalysts.

Section 14.2 Reaction rates are usually expressed as changes in concentration per unit time: Typically, for reactions in solution, rates are given in units of molarity per second (M/s). For most reactions, a plot of molarity versus time shows that the rate slows down as the reaction proceeds. The **instantaneous rate** is the slope of a line drawn tangent to the concentration-versus-time curve at a specific time. Rates can be written in terms of the appearance of products or the disappearance of reactants; the stoichiometry of the reaction dictates the relationship between rates of appearance and disappearance. Spectroscopy is one technique that can be used to monitor the course of a reaction. According to **Beer's law**, the absorption of electromagnetic radiation by a substance at a particular wavelength is directly proportional to its concentration.

Section 14.3 The quantitative relationship between rate and concentration is expressed by a **rate law**, which usually has the following form:

$$\text{Rate} = k[\text{reactant 1}]^m[\text{reactant 2}]^n \ldots$$

The constant k in the rate law is called the **rate constant**; the exponents m, n, and so forth are called **reaction orders** for the reactants. The sum of the reaction orders gives the **overall reaction order**. Reaction orders must be determined experimentally. The units of the rate constant depend on the overall reaction order. For a reaction in which the overall reaction order is 1, k has units of s^{-1}; for one in which the overall reaction order is 2, k has units of $M^{-1}\,s^{-1}$.

Section 14.4 Rate laws can be used to determine the concentrations of reactants or products at any time during a reaction. In a **first-order reaction** the rate is proportional to the concentration of a single reactant raised to the first power: Rate $= k[A]$. In such cases the integrated form of the rate law is $\ln[A]_t = -kt + \ln[A]_0$, where $[A]_t$ is the concentration of reactant A at time t, k is the rate constant, and $[A]_0$ is the initial concentration of A. Thus, for a first-order reaction, a graph of $\ln[A]$ versus time yields a straight line of slope $-k$.

A **second-order reaction** is one for which the overall reaction order is 2. If a second-order rate law depends on the concentration of only one reactant, then rate $= k[A]^2$, and the time dependence of [A] is given by the integrated form of the rate law: $1/[A]_t = 1/[A]_0 + kt$. In this case a graph of $1/[A]_t$ versus time yields a straight line.

The **half-life** of a reaction, $t_{1/2}$, is the time required for the concentration of a reactant to drop to one-half of its original value. For a first-order reaction, the half-life depends only on the rate constant and not on the initial concentration: $t_{1/2} = 0.693/k$. The half-life of a second-order reaction depends on both the rate constant and the initial concentration of A: $t_{1/2} = 1/k[A]_0$.

Section 14.5 The **collision model**, which assumes that reactions occur as a result of collisions between molecules, helps explain why the magnitudes of rate constants increase with increasing temperature. The greater the kinetic energy of the colliding molecules, the greater the energy of collision. The minimum energy required for a reaction to occur is called the **activation energy**, E_a. A collision with energy E_a or greater can cause the atoms of the colliding molecules to reach the **activated complex** (or **transition state**), which is the highest energy arrangement in the pathway from reactants to products. Even if a collision is energetic enough, it may not lead to reaction; the reactants must also be correctly oriented relative to one another in order for a collision to be effective.

Because the kinetic energy of molecules depends on temperature, the rate constant of a reaction is very dependent on temperature. The relationship between k and temperature is given by the **Arrhenius equation**: $k = Ae^{-E_a/RT}$. The term A is called the **frequency factor**; it relates to the number of collisions that are favorably oriented for reaction. The Arrhenius equation is often used in logarithmic form: $k = \ln A - E_a/RT$. Thus, a graph of $\ln k$ versus $1/T$ yields a straight line with slope $-E_a/R$.

Section 14.6 A **reaction mechanism** details the individual steps that occur in the course of a reaction. Each of these steps, called **elementary reactions**, has a well-defined rate law that depends on the number of molecules (the **molecularity**) of the step. Elementary reactions are defined as either **unimolecular**, **bimolecular**, or **termolecular**, depending on whether one, two, or three reactant molecules are involved, respectively. Termolecular elementary reactions are very rare. Unimolecular, bimolecular, and termolecular reactions follow rate laws that are first order overall, second order overall, and third order overall, respectively.

Many reactions occur by a multistep mechanism, involving two or more elementary reactions, or steps. An **intermediate** is produced in one elementary step and is consumed in a later elementary step and therefore does not appear in the overall equation for the reaction. When a mechanism has several elementary steps, the overall rate is limited by the slowest elementary step, called the **rate-determining step**. A fast elementary step that follows the rate-determining step will have no effect on the rate law of the reaction. A fast step that precedes the rate-determining step often creates an equilibrium that involves an intermediate. For a mechanism to be valid, the rate law predicted by the mechanism must be the same as that observed experimentally.

Section 14.7 A **catalyst** is a substance that increases the rate of a reaction without undergoing a net chemical change itself. It does so by providing a different mechanism for the reaction, one that has a lower activation energy. A **homogeneous catalyst** is one that is in the same phase as the reactants. A **heterogeneous catalyst** has a different phase from the reactants. Finely divided metals are often used as heterogeneous catalysts for solution- and gas-phase reactions. Reacting molecules can undergo binding, or **adsorption**, at the surface of the catalyst. The adsorption of a reactant at specific sites on the surface makes bond breaking easier, lowering the activation energy. Catalysis in living organisms is achieved by **enzymes**, large protein molecules that usually catalyze a very specific reaction. The specific reactant molecules involved in an enzymatic reaction are called **substrates**. The site of the enzyme where the catalysis occurs is called the **active site**. In the **lock-and-key model** for enzyme catalysis, substrate molecules bind very specifically to the active site of the enzyme, after which they can undergo reaction.

VISUALIZING CONCEPTS

14.1 Consider the following graph of the concentration of a substance over time. **(a)** Is X a reactant or product of the reaction? **(b)** Why is the average rate of the reaction greater between points 1 and 2 than between points 2 and 3? [Section 14.2]

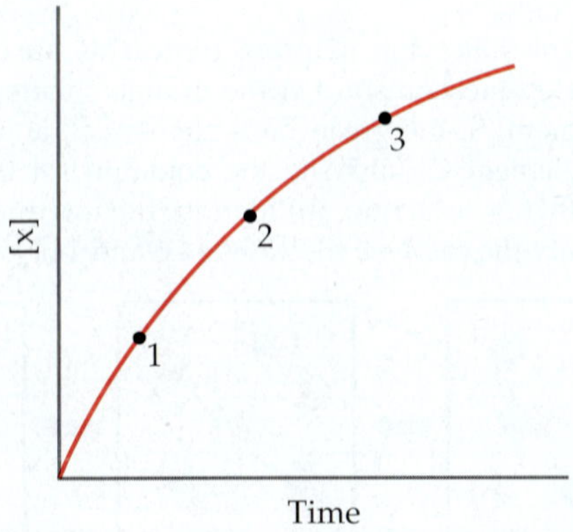

14.2 You study the rate of a reaction, measuring both the concentration of the reactant and the concentration of the product as a function of time, and obtain the following results:

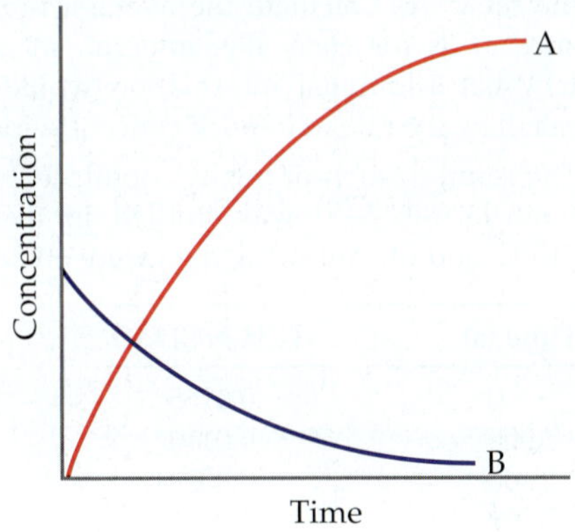

What chemical equation is consistent with this data: **(a)** A $\longrightarrow$ B, **(b)** B $\longrightarrow$ A, **(c)** A $\longrightarrow$ 2 B, **(d)** B $\longrightarrow$ 2 A. Explain your choice. [Section 14.2]

14.3 You perform a series of experiments for the reaction A $\longrightarrow$ B + C and find that the rate law has the form rate $= k[\text{A}]^x$. Determine the value of x in each of the following cases: **(a)** There is no rate change when [A] is tripled. **(b)** The rate increases by a factor of 9 when [A] is tripled. **(c)** When [A] is doubled, the rate increases by a factor of 8. [Section 14.3]

14.4 A friend studies a first-order reaction and obtains the following three graphs for experiments done at two different temperatures. **(a)** Which two lines represent experiments done at the same temperature? What accounts for the difference in these two lines? In what way are they the same? **(b)** Which two lines represent experiments done with the same starting concentration but at different temperatures? Which line probably represents the lower temperature? How do you know? [Section 14.4]

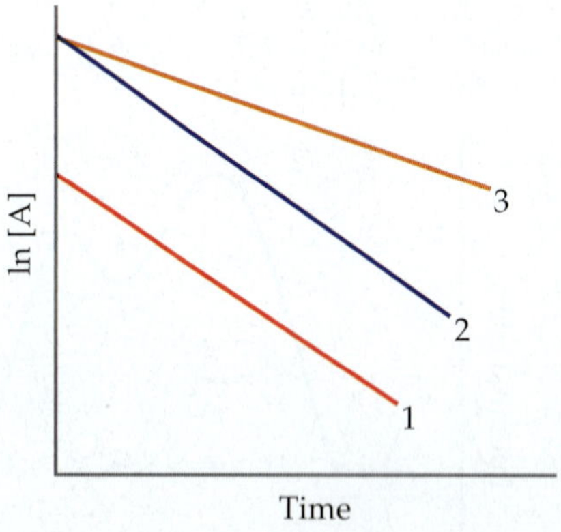

14.5 **(a)** Given the following diagrams at $t = 0$ and $t = 30$ min, what is the half-life of the reaction if it follows first-order kinetics?

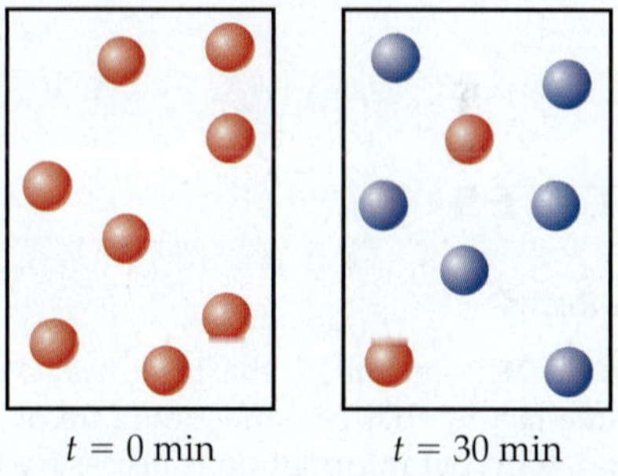

(b) After four half-life periods for a first-order reaction, what fraction of reactant remains? [Section 14.4]

14.6 You study the effect of temperature on the rate of two reactions and graph the natural logarithm of the rate constant for each reaction as a function of $1/T$. How do the two graphs compare **(a)** if the activation energy of the second reaction is higher than the activation energy of the first reaction but the two reactions have the same frequency factor, and **(b)** if the frequency factor of the second reaction is higher than the frequency factor of the first reaction but the two reactions have the same activation energy? [Section 14.5]

14.7 Consider the diagram below, which represents two steps in an overall reaction. The red spheres are oxygen, the blue ones nitrogen, and the green ones fluorine. **(a)** Write the chemical equation for each step in the reaction. **(b)** Write the equation for the overall reaction. **(c)** Identify the intermediate in the mechanism. **(d)** Write the rate law for the overall reaction if the first step is the slow, rate-determining step. [Section 14.6]

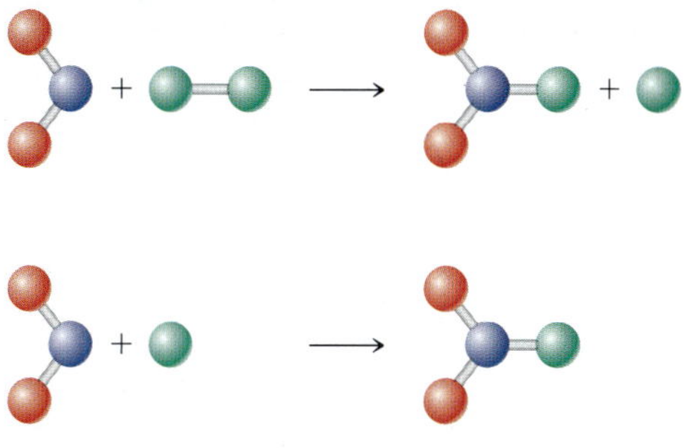

14.8 Based on the following reaction profile, how many intermediates are formed in the reaction A $\longrightarrow$ C? How many transition states are there? Which step is the fastest? Is the reaction A $\longrightarrow$ C exothermic or endothermic? [Section 14.6]

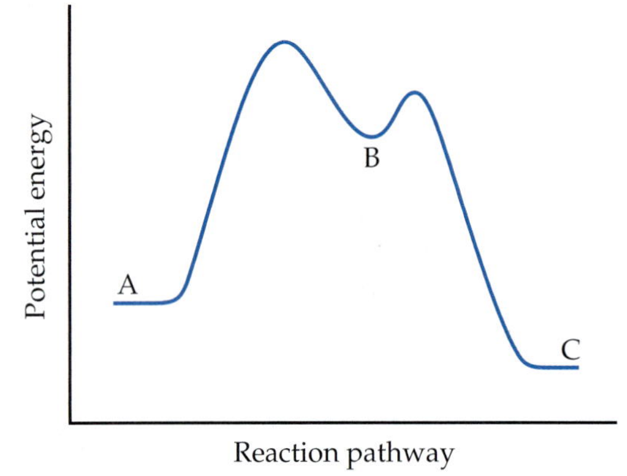

14.9 Draw a possible transition state for the bimolecular reaction depicted below. (The blue spheres are nitrogen atoms, and the red ones are oxygen atoms.) Use dashed lines to represent the bonds that are in the process of being broken or made in the transition state. [Section 14.6]

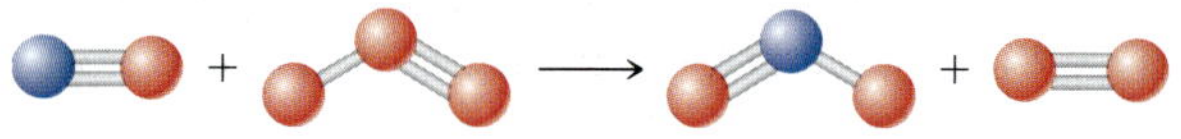

14.10 The following diagram represents an imaginary two-step mechanism. Let the orange spheres represent element A, the green ones element B, and the blue ones element C. **(a)** Write the equation for the net reaction that is occurring. **(b)** Identify the intermediate. **(c)** Identify the catalyst. [Sections 14.6 and 14.7]

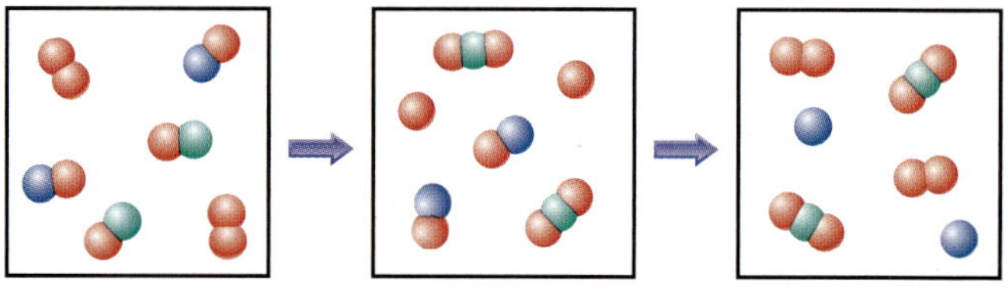

EXERCISES

Reaction Rates

14.11 **(a)** What is meant by the term *reaction rate*? **(b)** Name three factors that can affect the rate of a chemical reaction. **(c)** What information is necessary to relate the rate of disappearance of reactants to the rate of appearance of products?

14.12 **(a)** What are the units usually used to express the rates of reactions occurring in solution? **(b)** From your everyday experience, give two examples of the effects of temperature on the rates of reactions. **(c)** What is the difference between average rate and instantaneous rate?

14.13 Consider the following hypothetical aqueous reaction: $A(aq) \longrightarrow B(aq)$. A flask is charged with 0.065 mol of A in a total volume of 100.0 mL. The following data are collected:

Time (min)	0	10	20	30	40
Moles of A	0.065	0.051	0.042	0.036	0.031

(a) Calculate the number of moles of B at each time in the table, assuming that there are no molecules of B at time zero. **(b)** Calculate the average rate of disappearance of A for each 10-min interval, in units of M/s. **(c)** Between t = 10 min and t = 30 min, what is the average rate of appearance of B in units of M/s? Assume that the volume of the solution is constant.

14.14 A flask is charged with 0.100 mol of A and allowed to react to form B according to the hypothetical gas-phase reaction $A(g) \longrightarrow B(g)$. The following data are collected:

Time (s)	0	40	80	120	160
Moles of A	0.100	0.067	0.045	0.030	0.020

(a) Calculate the number of moles of B at each time in the table. **(b)** Calculate the average rate of disappearance of A for each 40-s interval, in units of mol/s. **(c)** What additional information would be needed to calculate the rate in units of concentration per time?

14.15 The isomerization of methyl isonitrile (CH_3NC) to acetonitrile (CH_3CN) was studied in the gas phase at 215°C, and the following data were obtained:

Time (s)	$[CH_3NC]$ (M)
0	0.0165
2,000	0.0110
5,000	0.00591
8,000	0.00314
12,000	0.00137
15,000	0.00074

(a) Calculate the average rate of reaction, in M/s, for the time interval between each measurement. **(b)** Graph $[CH_3NC]$ versus time, and determine the instantaneous rates in M/s at t = 5000 s and t = 8000 s.

14.16 The rate of disappearance of HCl was measured for the following reaction:

$$CH_3OH(aq) + HCl(aq) \longrightarrow CH_3Cl(aq) + H_2O(l)$$

The following data were collected:

Time (min)	[HCl] (M)
0.0	1.85
54.0	1.58
107.0	1.36
215.0	1.02
430.0	0.580

(a) Calculate the average rate of reaction, in M/s, for the time interval between each measurement.
(b) Graph [HCl] versus time, and determine the instantaneous rates in M/min and M/s at $t = 75.0$ min and $t = 250$ min.

14.17 For each of the following gas-phase reactions, indicate how the rate of disappearance of each reactant is related to the rate of appearance of each product:
(a) $H_2O_2(g) \longrightarrow H_2(g) + O_2(g)$
(b) $2\ N_2O(g) \longrightarrow 2\ N_2(g) + O_2(g)$
(c) $N_2(g) + 3\ H_2(g) \longrightarrow 2\ NH_3(g)$

14.18 For each of the following gas-phase reactions, write the rate expression in terms of the appearance of each product or disappearance of each reactant:
(a) $2\ HBr(g) \longrightarrow H_2(g) + Br_2(g)$
(b) $2\ SO_2(g) + O_2(g) \longrightarrow 2\ SO_3(g)$
(c) $2\ NO(g) + 2\ H_2(g) \longrightarrow N_2(g) + 2\ H_2O(g)$

14.19 (a) Consider the combustion of $H_2(g)$: $2\ H_2(g) + O_2(g) \longrightarrow 2\ H_2O(g)$. If hydrogen is burning at the rate of 0.85 mol/s, what is the rate of consumption of oxygen? What is the rate of formation of water vapor? (b) The reaction $2\ NO(g) + Cl_2(g) \longrightarrow 2\ NOCl(g)$ is carried out in a closed vessel. If the partial pressure of NO is decreasing at the rate of 23 torr/min, what is the rate of change of the total pressure of the vessel?

14.20 (a) Consider the combustion of ethylene, $C_2H_4(g) + 3\ O_2(g) \longrightarrow 2\ CO_2(g) + 2\ H_2O(g)$. If the concentration of C_2H_4 is decreasing at the rate of 0.37 M/s, what are the rates of change in the concentrations of CO_2 and H_2O? (b) The rate of decrease in N_2H_4 partial pressure in a closed reaction vessel from the reaction $N_2H_4(g) + H_2(g) \longrightarrow 2\ NH_3(g)$ is 63 torr/h. What are the rates of change of NH_3 partial pressure and total pressure in the vessel?

Rate Laws

14.21 A reaction $A + B \longrightarrow C$ obeys the following rate law: Rate $= k[B]^2$. (a) If [A] is doubled, how will the rate change? Will the rate constant change? Explain. (b) What are the reaction orders for A and B? What is the overall reaction order? (c) What are the units of the rate constant?

14.22 A reaction $A + B \longrightarrow C$ obeys the following rate law: Rate $= k[A]^2[B]$. (a) If [A] is doubled, how will the rate change? Will the rate constant change? Explain. (b) What are the reaction orders for A and B? What is the overall reaction order? (c) What are the units of the rate constant?

14.23 The decomposition of N_2O_5 in carbon tetrachloride proceeds as follows: $2\ N_2O_5 \longrightarrow 4\ NO_2 + O_2$. The rate law is first order in N_2O_5. At 64°C the rate constant is $4.82 \times 10^{-3}\ s^{-1}$. (a) Write the rate law for the reaction. (b) What is the rate of reaction when $[N_2O_5] = 0.0240\ M$? (c) What happens to the rate when the concentration of N_2O_5 is doubled to 0.0480 M?

14.24 Consider the following reaction:

$$2\ NO(g) + 2\ H_2(g) \longrightarrow N_2(g) + 2\ H_2O(g)$$

(a) The rate law for this reaction is first order in H_2 and second order in NO. Write the rate law. (b) If the rate constant for this reaction at 1000 K is $6.0 \times 10^4\ M^{-2}\ s^{-1}$, what is the reaction rate when $[NO] = 0.025\ M$ and $[H_2] = 0.015\ M$? (c) What is the reaction rate at 1000 K when the concentration of NO is increased to 0.10 M, while the concentration of H_2 is 0.010 M?

14.25 Consider the following reaction:

$$CH_3Br(aq) + OH^-(aq) \longrightarrow CH_3OH(aq) + Br^-(aq)$$

The rate law for this reaction is first order in CH_3Br and first order in OH^-. When $[CH_3Br]$ is $5.0 \times 10^{-3}\ M$ and $[OH^-]$ is 0.050 M, the reaction rate at 298 K is 0.0432 M/s. (a) What is the value of the rate constant? (b) What are the units of the rate constant? (c) What would happen to the rate if the concentration of OH^- were tripled?

14.26 The reaction between ethyl bromide (C_2H_5Br) and hydroxide ion in ethyl alcohol at 330 K, $C_2H_5Br(alc) + OH^-(alc) \longrightarrow C_2H_5OH(l) + Br^-(alc)$, is first order each in ethyl bromide and hydroxide ion. When $[C_2H_5Br]$ is 0.0477 M and $[OH^-]$ is 0.100 M, the rate of disappearance of ethyl bromide is $1.7 \times 10^{-7}\ M/s$. (a) What is the value of the rate constant? (b) What are the units of the rate constant? (c) How would the rate of disappearance of ethyl bromide change if the solution were diluted by adding an equal volume of pure ethyl alcohol to the solution?

14.27 The iodide ion reacts with hypochlorite ion (the active ingredient in chlorine bleaches) in the following way: $OCl^- + I^- \longrightarrow OI^- + Cl^-$. This rapid reaction gives the following rate data:

$[OCl^-]$, M	I^-, M	Rate, M/s
1.5×10^{-3}	1.5×10^{-3}	1.36×10^{-4}
3.0×10^{-3}	1.5×10^{-3}	2.72×10^{-4}
1.5×10^{-3}	3.0×10^{-3}	2.72×10^{-4}

(a) Write the rate law for this reaction. (b) Calculate the rate constant. (c) Calculate the rate when $[OCl^-] = 2.0 \times 10^{-3}\ M$ and $[I^-] = 5.0 \times 10^{-4}\ M$.

14.28 The reaction $2\ ClO_2(aq) + 2\ OH^-(aq) \longrightarrow ClO_3^-(aq) + ClO_2^-(aq) + H_2O(l)$ was studied with the following results:

Experiment	$[ClO_2]$, M	$[OH^-]$, M	Rate, M/s
1	0.060	0.030	0.0248
2	0.020	0.030	0.00276
3	0.020	0.090	0.00828

(a) Determine the rate law for the reaction. (b) Calculate the rate constant. (c) Calculate the rate when $[ClO_2] = 0.010\ M$ and $[OH^-] = 0.025\ M$.

14.29 The following data were measured for the reaction $BF_3(g) + NH_3(g) \longrightarrow F_3BNH_3(g)$:

Experiment	$[BF_3]$ (M)	$[NH_3]$ (M)	Initial Rate (M/s)
1	0.250	0.250	0.2130
2	0.250	0.125	0.1065
3	0.200	0.100	0.0682
4	0.350	0.100	0.1193
5	0.175	0.100	0.0596

(a) What is the rate law for the reaction? **(b)** What is the overall order of the reaction? **(c)** What is the value of the rate constant for the reaction? **(d)** What is the rate when $[BF_3] = 0.100\ M$ and $[NH_3] = 0.500\ M$?

14.30 The following data were collected for the rate of disappearance of NO in the reaction $2\ NO(g) + O_2(g) \longrightarrow 2\ NO_2(g)$:

Experiment	[NO] (*M*)	$[O_2]$ (*M*)	Initial Rate (*M*/s)
1	0.0126	0.0125	1.41×10^{-2}
2	0.0252	0.0250	1.13×10^{-1}
3	0.0252	0.0125	5.64×10^{-2}

(a) What is the rate law for the reaction? **(b)** What are the units of the rate constant? **(c)** What is the average value of the rate constant calculated from the three data sets? **(d)** What is the rate of disappearance of NO when $[NO] = 0.100\ M$ and $[O_2] = 0.0200\ M$? **(e)** What is the rate of disappearance of O_2 at the concentrations given in part (d)?

[14.31] Consider the gas-phase reaction between nitric oxide and bromine at 273°C: $2\ NO(g) + Br_2(g) \longrightarrow 2\ NOBr(g)$. The following data for the initial rate of appearance of NOBr were obtained:

Experiment	[NO] (*M*)	$[Br_2]$ (*M*)	Initial Rate (*M*/s)
1	0.10	0.20	24
2	0.25	0.20	150
3	0.10	0.50	60
4	0.35	0.50	735

(a) Determine the rate law. **(b)** Calculate the average value of the rate constant for the appearance of NOBr from the four data sets. **(c)** How is the rate of appearance of NOBr related to the rate of disappearance of Br_2? **(d)** What is the rate of disappearance of Br_2 when $[NO] = 0.075\ M$ and $[Br_2] = 0.25\ M$?

[14.32] Consider the reaction of peroxydisulfate ion ($S_2O_8^{2-}$) with iodide ion (I^-) in aqueous solution:

$$S_2O_8^{2-}(aq) + 3\ I^-(aq) \longrightarrow 2\ SO_4^{2-}(aq) + I_3^-(aq)$$

At a particular temperature the rate of disappearance of $S_2O_8^{2-}$ varies with reactant concentrations in the following manner:

Experiment	$[S_2O_8^{2-}]$ (*M*)	$[I^-]$ (*M*)	Initial Rate (*M*/s)
1	0.018	0.036	2.6×10^{-6}
2	0.027	0.036	3.9×10^{-6}
3	0.036	0.054	7.8×10^{-6}
4	0.050	0.072	1.4×10^{-5}

(a) Determine the rate law for the reaction. **(b)** What is the average value of the rate constant for the disappearance of $S_2O_8^{2-}$ based on the four sets of data? **(c)** How is the rate of disappearance of $S_2O_8^{2-}$ related to the rate of disappearance of I^-? **(d)** What is the rate of disappearance of I^- when $[S_2O_8^{2-}] = 0.015\ M$ and $[I^-] = 0.040\ M$?

Change of Concentration with Time

14.33 **(a)** Define the following symbols that are encountered in rate equations: $[A]_0$, $t_{1/2}$, $[A]_t$, k. **(b)** What quantity, when graphed versus time, will yield a straight line for a first-order reaction?

14.34 **(a)** For a second-order reaction, what quantity, when graphed versus time, will yield a straight line? **(b)** How do the half-lives of first-order and second-order reactions differ?

14.35 **(a)** The gas-phase decomposition of SO_2Cl_2, $SO_2Cl_2(g) \longrightarrow SO_2(g) + Cl_2(g)$, is first order in SO_2Cl_2. At 600 K the half-life for this process is 2.3×10^5 s. What is the rate constant at this temperature? **(b)** At 320°C the rate constant is $2.2 \times 10^{-5}\ s^{-1}$. What is the half-life at this temperature?

14.36 Americium-241 is used in smoke detectors. It has a rate constant for radioactive decay of $k = 1.6 \times 10^{-3}\ yr^{-1}$. By contrast, iodine-125, which is used to test for thyroid functioning, has a rate constant for radioactive decay of $k = 0.011\ day^{-1}$. **(a)** What are the half-lives of these two isotopes? **(b)** Which one decays at a faster rate? **(c)** How much of a 1.00-mg sample of either isotope remains after three half-lives?

14.37 As described in Exercise 14.35, the decomposition of sulfuryl chloride (SO_2Cl_2) is a first-order process. The rate constant for the decomposition at 660 K is $4.5 \times 10^{-2}\ s^{-1}$. **(a)** If we begin with an initial SO_2Cl_2 pressure of 375 torr, what is the pressure of this substance after 65 s? **(b)** At what time will the pressure of SO_2Cl_2 decline to one-tenth its initial value?

14.38 The first-order rate constant for the decomposition of N_2O_5, $N_2O_5(g) \longrightarrow 2\ NO_2(g) + O_2(g)$, at 70°C is $6.82 \times 10^{-3}\ s^{-1}$. Suppose we start with 0.0250 mol of $N_2O_5(g)$ in a volume of 2.0 L. **(a)** How many moles of N_2O_5 will remain after 2.5 min? **(b)** How many minutes will it take for the quantity of N_2O_5 to drop to 0.010 mol? **(c)** What is the half-life of N_2O_5 at 70°C?

14.39 The reaction

$$SO_2Cl_2(g) \longrightarrow SO_2(g) + Cl_2(g)$$

is first order in SO_2Cl_2. Using the following kinetic data, determine the magnitude of the first-order rate constant:

Time (s)	Pressure SO_2Cl_2 (atm)
0	1.000
2,500	0.947
5,000	0.895
7,500	0.848
10,000	0.803

14.40 From the following data for the first-order gas-phase isomerization of CH_3NC at 215°C, calculate the first-order rate constant and half-life for the reaction:

Time (s)	Pressure CH_3NC (torr)
0	502
2,000	335
5,000	180
8,000	95.5
12,000	41.7
15,000	22.4

14.41 Consider the data presented in Exercise 14.13. **(a)** By using appropriate graphs, determine whether the reaction is first order or second order. **(b)** What is the value of the rate constant for the reaction? **(c)** What is the half-life for the reaction?

14.42 Consider the data presented in Exercise 14.14. **(a)** Determine whether the reaction is first order or second order. **(b)** What is the value of the rate constant? **(c)** What is the half-life?

14.43 The gas-phase decomposition of NO_2, $NO_2(g) \longrightarrow NO(g) + O_2(g)$, is studied at 383°C, giving the following data:

Time (s)	$[NO_2]$ (M)
0.0	0.100
5.0	0.017
10.0	0.0090
15.0	0.0062
20.0	0.0047

(a) Is the reaction first order or second order with respect to the concentration of NO_2? **(b)** What is the value of the rate constant?

14.44 Sucrose ($C_{12}H_{22}O_{11}$), which is commonly known as table sugar, reacts in dilute acid solutions to form two simpler sugars, glucose and fructose, both of which have the formula $C_6H_{12}O_6$: At 23°C and in 0.5 M HCl, the following data were obtained for the disappearance of sucrose:

Time (min)	$[C_{12}H_{22}O_{11}]$ (M)
0	0.316
39	0.274
80	0.238
140	0.190
210	0.146

(a) Is the reaction first order or second order with respect to $[C_{12}H_{22}O_{11}]$? **(b)** What is the value of the rate constant?

Temperature and Rate

14.45 **(a)** What factors determine whether a collision between two molecules will lead to a chemical reaction? **(b)** According to the collision model, why does temperature affect the value of the rate constant?

14.46 **(a)** Explain the rate of a unimolecular (that is, one-molecule) reaction, such as the isomerization of methyl isonitrile (Figure 14.6), in terms of the collision model. **(b)** In a reaction of the form $A(g) + B(g) \longrightarrow$ products, are all collisions of A with B that are sufficiently energetic likely to lead to reaction? Explain. **(c)** How does the kinetic-molecular theory help us understand the temperature dependence of chemical reactions?

14.47 Calculate the fraction of atoms in a sample of argon gas at 400 K that have an energy of 10.0 kJ or greater.

14.48 **(a)** The activation energy for the isomerization of methyl isonitrile (Figure 14.6) is 160 kJ/mol. Calculate the fraction of methyl isonitrile molecules that have an energy of 160.0 kJ or greater at 500 K. **(b)** Calculate this fraction for a temperature of 510 K. What is the ratio of the fraction at 510 K to that at 500 K?

14.49 The gas-phase reaction $Cl(g) + HBr(g) \longrightarrow HCl(g) + Br(g)$ has an overall enthalpy change of −66 kJ. The activation energy for the reaction is 7 kJ. **(a)** Sketch the energy profile for the reaction, and label E_a and ΔE. **(b)** What is the activation energy for the reverse reaction?

14.50 For the elementary process $N_2O_5(g) \longrightarrow NO_2(g) + NO_3(g)$ the activation energy (E_a) and overall ΔE are 154 kJ/mol and 136 kJ/mol, respectively. **(a)** Sketch the energy profile for this reaction, and label E_a and ΔE. **(b)** What is the activation energy for the reverse reaction?

14.51 Based on their activation energies and energy changes and assuming that all collision factors are the same, which of the following reactions would be fastest and which would be slowest? Explain your answer.
(a) $E_a = 45$ kJ/mol; $\Delta E = -25$ kJ/mol;
(b) $E_a = 35$ kJ/mol; $\Delta E = -10$ kJ/mol;
(c) $E_a = 55$ kJ/mol; $\Delta E = 10$ kJ/mol.

14.52 Which of the reactions in Exercise 14.51 will be fastest in the reverse direction? Which will be slowest? Explain.

14.53 A certain first-order reaction has a rate constant of $2.75 \times 10^{-2}\ s^{-1}$ at 20°C. What is the value of k at 60°C if **(a)** $E_a = 75.5$ kJ/mol; **(b)** $E_a = 105$ kJ/mol?

14.54 Understanding the high-temperature behavior of nitrogen oxides is essential for controlling pollution generated in automobile engines. The decomposition of nitric oxide (NO) to N_2 and O_2 is second order with a rate constant of 0.0796 $M^{-1}\ s^{-1}$ at 737°C and 0.0815 $M^{-1}\ s^{-1}$ at 947°C. Calculate the activation energy for the reaction.

14.55 The rate of the reaction

$$CH_3COOC_2H_5(aq) + OH^-(aq) \longrightarrow CH_3COO^-(aq) + C_2H_5OH(aq)$$

was measured at several temperatures, and the following data were collected:

Temperature (°C)	$k(M^{-1}\ s^{-1})$
15	0.0521
25	0.101
35	0.184
45	0.332

Using these data, graph ln k versus $1/T$. Using your graph, determine the value of E_a.

14.56 The temperature dependence of the rate constant for the reaction is tabulated as follows:

Temperature (K)	$k(M^{-1}\,s^{-1})$
600	0.028
650	0.22
700	1.3
750	6.0
800	23

Calculate E_a and A.

[14.57] The activation energy of a certain reaction is 65.7 kJ/mol. How many times faster will the reaction occur at 50°C than at 0°C?

[14.58] The following is a quote from an article in the August 18, 1998, issue of *The New York Times* about the breakdown of cellulose and starch: "A drop of 18 degrees Fahrenheit [from 77°F to 59°F] lowers the reaction rate six times; a 36-degree drop [from 77°F to 41°F] produces a fortyfold decrease in the rate." **(a)** Calculate activation energies for the breakdown process based on the two estimates of the effect of temperature on rate. Are the values consistent? **(b)** Assuming the value of E_a calculated from the 36-degree drop and assuming that the rate of breakdown is first order with a half-life at 25°C of 2.7 years, calculate the half-life for breakdown at a temperature of −15°C.

Reaction Mechanisms

14.59 **(a)** What is meant by the term *elementary reaction?* **(b)** What is the difference between a *unimolecular* and a *bimolecular* elementary reaction? **(c)** What is a *reaction mechanism?*

14.60 **(a)** What is meant by the term *molecularity?* **(b)** Why are termolecular elementary reactions so rare? **(c)** What is an *intermediate* in a mechanism?

14.61 What is the molecularity of each of the following elementary reactions? Write the rate law for each.
(a) $Cl_2(g) \longrightarrow 2\,Cl(g)$
(b) $OCl^-(g) + H_2O(g) \longrightarrow HOCl(g) + OH^-(g)$
(c) $NO(g) + Cl_2(g) \longrightarrow NOCl_2(g)$

14.62 What is the molecularity of each of the following elementary reactions? Write the rate law for each.
(a) $2\,NO(g) \longrightarrow N_2O_2(g)$
(b) $H_2C{-}CH_2(g)$ (ring with CH_2) $\longrightarrow CH_2{=}CH{-}CH_3(g)$
(c) $SO_3(g) \longrightarrow SO_2(g) + O(g)$

14.63 **(a)** Based on the following reaction profile, how many intermediates are formed in the reaction A $\longrightarrow$ D? **(b)** How many transition states are there? **(c)** Which step is the fastest? **(d)** Is the reaction A $\longrightarrow$ D exothermic or endothermic?

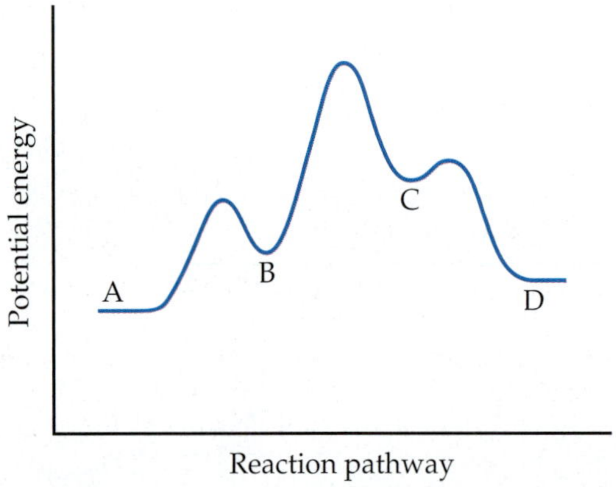

14.64 Consider the following energy profile.

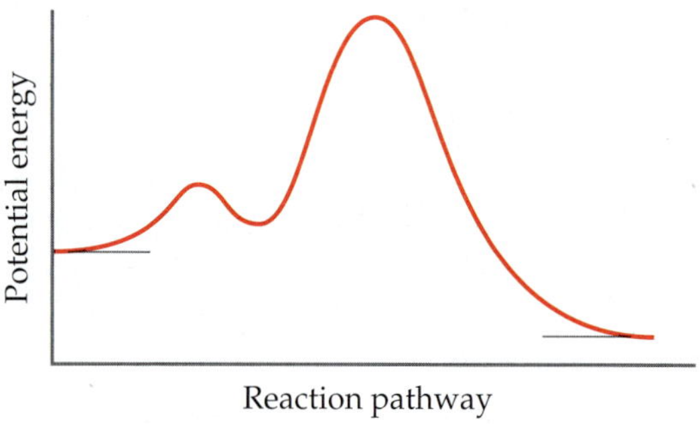

(a) How many elementary reactions are in the reaction mechanism? **(b)** How many intermediates are formed in the reaction? **(c)** Which step is rate limiting? **(d)** Is the overall reaction exothermic or endothermic?

14.65 The following mechanism has been proposed for the gas-phase reaction of H_2 with ICl:

$$H_2(g) + ICl(g) \longrightarrow HI(g) + HCl(g)$$
$$HI(g) + ICl(g) \longrightarrow I_2(g) + HCl(g)$$

(a) Write the balanced equation for the overall reaction. **(b)** Identify any intermediates in the mechanism. **(c)** Write rate laws for each elementary reaction in the mechanism. **(d)** If the first step is slow and the second one is fast, what rate law do you expect to be observed for the overall reaction?

14.66 The decomposition of hydrogen peroxide is catalyzed by iodide ion. The catalyzed reaction is thought to proceed by a two-step mechanism:

$$H_2O_2(aq) + I^-(aq) \longrightarrow H_2O(l) + IO^-(aq) \quad \text{(slow)}$$
$$IO^-(aq) + H_2O_2(aq) \longrightarrow H_2O(l) + O_2(g) + I^-(aq) \quad \text{(fast)}$$

(a) Write the rate law for each of the elementary reactions of the mechanism. **(b)** Write the chemical equation for the overall process. **(c)** Identify the intermediate, if any, in the mechanism. **(d)** Assuming that the first step of the mechanism is rate determining, predict the rate law for the overall process.

14.67 The reaction $2\,NO(g) + Cl_2(g) \longrightarrow 2\,NOCl(g)$ obeys the rate law, rate $= k[NO]^2[Cl_2]$. The following mechanism has been proposed for this reaction:

$$NO(g) + Cl_2(g) \longrightarrow NOCl_2(g)$$

$$NOCl_2(g) + NO(g) \longrightarrow 2\,NOCl(g)$$

(a) What would the rate law be if the first step were rate determining? **(b)** Based on the observed rate law, what can we conclude about the relative rates of the two steps?

14.68 You have studied the gas-phase oxidation of HBr by O_2:

$$4\,HBr(g) + O_2(g) \longrightarrow 2\,H_2O(g) + 2\,Br_2(g)$$

You find the reaction to be first order with respect to HBr and first order with respect to O_2. You propose the following mechanism:

$$HBr(g) + O_2(g) \longrightarrow HOOBr(g)$$

$$HOOBr(g) + HBr(g) \longrightarrow 2\,HOBr(g)$$

$$HOBr(g) + HBr(g) \longrightarrow H_2O(g) + Br_2(g)$$

(a) Indicate how the elementary reactions add to give the overall reaction. [*Hint:* You will need to multiply the coefficients of one of the equations by 2.] **(b)** Based on the rate law, which step is rate determining? **(c)** What are the intermediates in this mechanism? **(d)** If you are unable to detect HOBr or HOOBr among the products, does this disprove your mechanism?

Catalysis

14.69 **(a)** What part of the energy profile of a reaction is affected by a catalyst? **(b)** What is the difference between a homogeneous and a heterogeneous catalyst?

14.70 **(a)** Most heterogeneous catalysts of importance are extremely finely divided solid materials. Why is particle size important? **(b)** What role does adsorption play in the action of a heterogeneous catalyst?

14.71 The oxidation of SO_2 to SO_3 is catalyzed by NO_2 The reaction proceeds as follows:

$$NO_2(g) + SO_2(g) \longrightarrow NO(g) + SO_3(g)$$

$$2\,NO(g) + O_2(g) \longrightarrow 2\,NO_2(g)$$

(a) Show that the two reactions can be summed to give the overall oxidation of SO_2 by O_2 to give SO_3. [*Hint:* The top reaction must be multiplied by a factor so the NO and NO_2 cancel out.] **(b)** Why do we consider NO_2 a catalyst and not an intermediate in this reaction? **(c)** Is this an example of homogeneous catalysis or heterogeneous catalysis?

14.72 NO catalyzes the decomposition of N_2O, possibly by the following mechanism:

$$NO(g) + N_2O(g) \longrightarrow N_2(g) + NO_2(g)$$

$$2\,NO_2(g) \longrightarrow 2\,NO(g) + O_2(g)$$

(a) What is the chemical equation for the overall reaction? Show how the two steps can be added to give the overall equation. **(b)** Why is NO considered a catalyst and not an intermediate? **(c)** If experiments show that during the decomposition of N_2O, NO_2 does not accumulate in measurable quantities, does this rule out the proposed mechanism? If you think not, suggest what might be going on.

14.73 Many metallic catalysts, particularly the precious-metal ones, are often deposited as very thin films on a substance of high surface area per unit mass, such as alumina (Al_2O_3) or silica (SiO_2). Why is this an effective way of utilizing the catalyst material?

14.74 **(a)** If you were going to build a system to check the effectiveness of automobile catalytic converters on cars, what substances would you want to look for in the car exhaust? **(b)** Automobile catalytic converters have to work at high temperatures, as hot exhaust gases stream through them. In what ways could this be an advantage? In what ways a disadvantage? **(c)** Why is the rate of flow of exhaust gases over a catalytic converter important?

14.75 When D_2 reacts with ethylene (C_2H_4) in the presence of a finely divided catalyst, ethane with two deuteriums, $CH_2D{-}CH_2D$, is formed. (Deuterium, D, is an isotope of hydrogen of mass 2.) Very little ethane forms in which two deuteriums are bound to one carbon (for example, $CH_3{-}CHD_2$). Use the sequence of steps involved in the reaction to explain why this is so.

14.76 Heterogeneous catalysts that perform hydrogenation reactions, as illustrated in Figure 14.21, are subject to poisoning, which shuts down their catalytic ability. Compounds of sulfur are often poisons. Suggest a mechanism by which such compounds might act as poisons.

14.77 **(a)** Explain the importance of enzymes in biological systems. **(b)** What chemical transformations are catalyzed (*i*) by the enzyme catalase, (*ii*) by nitrogenase?

14.78 There are literally thousands of enzymes at work in complex living systems such as human beings. What properties of the enzymes give rise to their ability to distinguish one substrate from another?

[14.79] The activation energy of an uncatalyzed reaction is 95 kJ/mol. The addition of a catalyst lowers the activation energy to 55 kJ/mol. Assuming that the collision factor remains the same, by what factor will the catalyst increase the rate of the reaction at **(a)** 25°C, **(b)** 125°C?

[14.80] Suppose that a certain biologically important reaction is quite slow at physiological temperature (37°C) in the absence of a catalyst. Assuming that the collision factor remains the same, by how much must an enzyme lower the activation energy of the reaction in order to achieve a 1×10^5-fold increase in the reaction rate?

Additional Exercises

14.81 Hydrogen sulfide (H_2S) is a common and troublesome pollutant in industrial wastewaters. One way to remove H_2S is to treat the water with chlorine, in which case the following reaction occurs:

$$H_2S(aq) + Cl_2(aq) \longrightarrow S(s) + 2\,H^+(aq) + 2\,Cl^-(aq)$$

The rate of this reaction is first order in each reactant. The rate constant for the disappearance of H_2S at 28°C is $3.5 \times 10^{-2}\ M^{-1}\ s^{-1}$. If at a given time the concentration of H_2S is $2.0 \times 10^{-4}\ M$ and that of Cl_2 is 0.050 M, what is the rate of formation of Cl^-?

14.82 The reaction $2\,NO(g) + O_2(g) \longrightarrow 2\,NO_2(g)$ is second order in NO and first order in O_2. When $[NO] = 0.040\ M$ and $[O_2] = 0.035\ M$, the observed rate of disappearance of NO is $9.3 \times 10^{-5}\ M/s$. **(a)** What is the rate of disappearance of O_2 at this moment? **(b)** What is the value of the rate constant? **(c)** What are the units of the rate constant? **(d)** What would happen to the rate if the concentration of NO were increased by a factor of 1.8?

14.83 For the reaction of iodide ion with hypochorite ion, $I^-(aq) + OCl^-(aq) \longrightarrow OI^-(aq) + Cl^-(aq)$, the reaction is found to be first order each in iodide and hypochlorite ions, and inversely proportional to the concentration of hydroxide ion present in the solution. **(a)** Write the rate law for the reaction. **(b)** By what factor will the rate change if the concentration of iodide ion is tripled? **(c)** By what factor will the rate change if the hydroxide ion concentration is doubled?

14.84 Consider the following reaction between mercury(II) chloride and oxalate ion:

$$2\,HgCl_2(aq) + C_2O_4^{2-}(aq) \longrightarrow 2\,Cl^-(aq) + 2\,CO_2(g) + Hg_2Cl_2(s)$$

The initial rate of this reaction was determined for several concentrations of $HgCl_2$ and $C_2O_4^{2-}$, and the following rate data were obtained for the rate of disappearance of $C_2O_4^{2-}$:

Experiment	$[HgCl_2]$ (*M*)	$[C_2O_4^{2-}]$ (*M*)	Rate (*M*/s)
1	0.164	0.15	3.2×10^{-5}
2	0.164	0.45	2.9×10^{-4}
3	0.082	0.45	1.4×10^{-4}
4	0.246	0.15	4.8×10^{-5}

(a) What is the rate law for this reaction? **(b)** What is the value of the rate constant? **(c)** What is the reaction rate when the concentration of $HgCl_2$ is 0.050 M and that of $C_2O_4^{2-}$ is 0.10 M, if the temperature is the same as that used to obtain the data shown?

14.85 The reaction $2\,NO_2 \longrightarrow 2\,NO + O_2$ has the rate constant $k = 0.63\ M^{-1}\ s^{-1}$. Based on the units for k, is the reaction first or second order in NO_2? If the initial concentration of NO_2 is 0.100 M, how would you determine how long it would take for the concentration to decrease to 0.025 M?

14.86 **(a)** The decomposition of H_2O_2, $H_2O_2(aq) \longrightarrow H_2O(l) + \frac{1}{2}O_2(g)$, is a first-order reaction. At a particular temperature near room temperature, the rate constant equals $7.0 \times 10^{-4}\ s^{-1}$. Calculate the half-life at this temperature. **(b)** At 415°C, $(CH_2)_2O$ decomposes in the gas phase, $(CH_2)_2O(g) \longrightarrow CH_4(g) + CO(g)$. If the reaction is first order with a half-life of 56.3 min at this temperature, calculate the rate constant in s^{-1}.

14.87 Urea (NH_2CONH_2) is the end product in protein metabolism in animals. The decomposition of urea in 0.1 M HCl occurs according to the reaction

$$NH_2CONH_2(aq) + H^+(aq) + 2\,H_2O(l) \longrightarrow 2\,NH_4^+(aq) + HCO_3^-(aq)$$

The reaction is first order in urea and first order overall. When $[NH_2CONH_2] = 0.200\ M$, the rate at 61.05°C is $8.56 \times 10^{-5}\ M/s$. **(a)** What is the value for the rate constant, k? **(b)** What is the concentration of urea in this solution after 4.00×10^3 s if the starting concentration is 0.500 M? **(c)** What is the half-life for this reaction at 61.05°C?

[14.88] The rate of a first-order reaction is followed by spectroscopy, monitoring the absorption of a colored reactant. The reaction occurs in a 1.00-cm sample cell, and the only colored species in the reaction has a molar absorptivity constant of $5.60 \times 10^3\ cm^{-1}\ M^{-1}$. **(a)** Calculate the initial concentration of the colored reactant if the absorbance is 0.605 at the beginning of the reaction. **(b)** The absorbance falls to 0.250 within 30.0 min. Calculate the rate constant in units of s^{-1}. **(c)** Calculate the half-life of the reaction. **(d)** How long does it take for the absorbance to fall to 0.100?

14.89 Cyclopentadiene (C_5H_6) reacts with itself to form dicyclopentadiene ($C_{10}H_{12}$). A 0.0400 M solution of C_5H_6 was monitored as a function of time as the reaction $2\,C_5H_6 \longrightarrow C_{10}H_{12}$ proceeded. The following data were collected:

Time (s)	$[C_5H_6]$ (*M*)
0.0	0.0400
50.0	0.0300
100.0	0.0240
150.0	0.0200
200.0	0.0174

Plot $[C_5H_6]$ versus time, $\ln[C_5H_6]$ versus time, and $1/[C_5H_6]$ versus time. What is the order of the reaction? What is the value of the rate constant?

14.90 **(a)** Two reactions have identical values for E_a. Does this ensure that they will have the same rate constant if run at the same temperature? Explain. **(b)** Two similar reactions have the same rate constant at 25°C, but at 35°C one of the reactions has a higher rate constant than the other. Account for these observations.

14.91 The first-order rate constant for reaction of a particular organic compound with water varies with temperature as follows:

Temperature (K)	Rate Constant (s^{-1})
300	3.2×10^{-11}
320	1.0×10^{-9}
340	3.0×10^{-8}
355	2.4×10^{-7}

From these data, calculate the activation energy in units of kJ/mol.

14.92 The following mechanism has been proposed for the reaction of NO with H_2 to form N_2O and H_2O:

$$NO(g) + NO(g) \longrightarrow N_2O_2(g)$$
$$N_2O_2(g) + H_2(g) \longrightarrow N_2O(g) + H_2O(g)$$

(a) Show that the elementary reactions of the proposed mechanism add to provide a balanced equation for the reaction. **(b)** Write a rate law for each elementary reaction in the mechanism. **(c)** Identify any intermediates in the mechanism. **(d)** The observed rate law is rate = $k[NO]^2[H_2]$. If the proposed mechanism is correct, what can we conclude about the relative speeds of the first and second reactions?

14.93 Using Figure 14.20 as your basis, draw the energy profile for the bromide ion–catalyzed decomposition of hydrogen peroxide. **(a)** Label the curve with the activation energies for reactions [14.30] and [14.31]. **(b)** Notice from Figure 14.19(b) that when $Br^-(aq)$ is added initially, Br_2 accumulates to some extent during the reaction. What does this tell us about the relative rates of reactions [14.30] and [14.31]?

[14.94] The following mechanism has been proposed for the gas-phase reaction of chloroform ($CHCl_3$) and chlorine:

Step 1:	$Cl_2(g) \underset{k_{-1}}{\overset{k_1}{\rightleftharpoons}} 2\,Cl(g)$	(fast)
Step 2:	$Cl(g) + CHCl_3(g) \xrightarrow{k_2} HCl(g) + CCl_3(g)$	(slow)
Step 3:	$Cl(g) + CCl_3(g) \xrightarrow{k_3} CCl_4(g)$	(fast)

(a) What is the overall reaction? **(b)** What are the intermediates in the mechanism? **(c)** What is the molecularity of each of the elementary reactions? **(d)** What is the rate-determining step? **(e)** What is the rate law predicted by this mechanism? [*Hint:* The overall reaction order is not an integer.]

[14.95] In a hydrocarbon solution, the gold compound $(CH_3)_3AuPH_3$ decomposes into ethane (C_2H_6) and a different gold compound, $(CH_3)AuPH_3$. The following mechanism has been proposed for the decomposition of $(CH_3)_3AuPH_3$:

Step 1:	$(CH_3)_3AuPH_3 \underset{k_{-1}}{\overset{k_1}{\rightleftharpoons}} (CH_3)_3Au + PH_3$	(fast)
Step 2:	$(CH_3)_3Au \xrightarrow{k_2} C_2H_6 + (CH_3)Au$	(slow)
Step 3:	$(CH_3)Au + PH_3 \xrightarrow{k_3} (CH_3)AuPH_3$	(fast)

(a) What is the overall reaction? **(b)** What are the intermediates in the mechanism? **(c)** What is the molecularity of each of the elementary steps? **(d)** What is the rate-determining step? **(e)** What is the rate law predicted by this mechanism? **(f)** What would be the effect on the reaction rate of adding PH_3 to the solution of $(CH_3)_3AuPH_3$?

14.96 One of the many remarkable enzymes in the human body is carbonic anhydrase, which catalyzes the interconversion of carbonic acid with carbon dioxide and water. If it were not for this enzyme, the body could not rid itself rapidly enough of the CO_2 accumulated by cell metabolism. The enzyme catalyzes the dehydration (release to air) of up to 10^7 CO_2 molecules per second. Which components of this description correspond to the terms *enzyme*, *substrate*, and *turnover number*?

14.97 The enzyme *invertase* catalyzes the conversion of sucrose, a disaccharide, to invert sugar, a mixture of glucose and fructose. When the concentration of invertase is 4.2×10^{-7} *M* and the concentration of sucrose is 0.0077 *M*, invert sugar is formed at the rate of 1.5×10^{-4} *M*/s. When the sucrose concentration is doubled, the rate of formation of invert sugar is doubled also. **(a)** Assuming that the enzyme-substrate model is operative, is the fraction of enzyme tied up as a complex large or small? Explain. **(b)** Addition of inositol, another sugar, decreases the rate of formation of invert sugar. Suggest a mechanism by which this occurs.

Integrative Exercises

14.98 Dinitrogen pentoxide (N_2O_5) decomposes in chloroform as a solvent to yield NO_2 and O_2. The decomposition is first order with a rate constant at 45°C of $1.0 \times 10^{-5}\ s^{-1}$. Calculate the partial pressure of O_2 produced from 1.00 L of 0.600 *M* N_2O_5 solution at 45°C over a period of 20.0 h if the gas is collected in a 10.0-L container. (Assume that the products do not dissolve in chloroform.)

[14.99] The reaction between ethyl iodide and hydroxide ion in ethanol (C_2H_5OH) solution, $C_2H_5I(alc) + OH^-(alc) \longrightarrow C_2H_5OH(l) + I^-(alc)$, has an activation energy of 86.8 kJ/mol and a frequency factor of $2.10 \times 10^{11}\ M^{-1}\ s^{-1}$. **(a)** Predict the rate constant for the reaction at 35°C. **(b)** A solution of KOH in ethanol is made up by dissolving 0.335 g KOH in ethanol to form 250.0 mL of solution. Similarly, 1.453 g of C_2H_5I is dissolved in ethanol to form 250.0 mL of solution. Equal volumes of the two solutions are mixed. Assuming the reaction is first order in each reactant, what is the initial rate at 35°C? **(c)** Which reagent in the reaction is limiting, assuming the reaction proceeds to completion?

14.100 Zinc metal dissolves in hydrochloric acid according to the reaction

$$Zn(s) + 2\,HCl(aq) \longrightarrow ZnCl_2(aq) + H_2(g)$$

Suppose you are asked to study the kinetics of this reaction by monitoring the rate of production of $H_2(g)$. **(a)** By using a reaction flask, a manometer, and any other common laboratory equipment, design an experimental apparatus that would allow you to monitor the partial pressure of $H_2(g)$ produced as a function of time. **(b)** Explain how you would use the apparatus to determine the rate law of the reaction. **(c)** Explain how you would use the apparatus to determine the reaction order for $[H^+]$ for the reaction. **(d)** How could you use the apparatus to determine the activation energy of the reaction? **(e)** Explain how you would use the apparatus to determine the effects of changing the form of Zn(*s*) from metal strips to granules.

14.101 The gas-phase reaction of NO with F_2 to form NOF and F has an activation energy of $E_a = 6.3$ kJ/mol and a frequency factor of $A = 6.0 \times 10^8\ M^{-1}\,s^{-1}$. The reaction is believed to be bimolecular:

$$NO(g) + F_2(g) \longrightarrow NOF(g) + F(g)$$

(a) Calculate the rate constant at 100°C. **(b)** Draw the Lewis structures for the NO and the NOF molecules, given that the chemical formula for NOF is misleading because the nitrogen atom is actually the central atom in the molecule. **(c)** Predict the structure for the NOF molecule. **(d)** Draw a possible transition state for the formation of NOF, using dashed lines to indicate the weak bonds that are beginning to form. **(e)** Suggest a reason for the low activation energy for the reaction.

14.102 The mechanism for the oxidation of HBr by O_2 to form $2\,H_2O$ and Br_2 is shown in Exercise 14.68. **(a)** Calculate the overall standard enthalpy change for the reaction process. **(b)** HBr does not react with O_2 at a measurable rate at room temperature under ordinary conditions. What can you infer from this about the magnitude of the activation energy for the rate-determining step? **(c)** Draw a plausible Lewis structure for the intermediate HOOBr. To what familiar compound of hydrogen and oxygen does it appear similar?

14.103 Enzymes, the catalysts of biological systems, are high molecular weight protein materials. The active site of the enzyme is formed by the three-dimensional arrangement of the protein in solution. When heated in solution, proteins undergo denaturation, a process in which the three-dimensional structure of the protein unravels or at least partly does so. The accompanying graph shows the variation with temperature of the activity of a typical enzyme. The activity increases with temperature to a point above the usual operating region for the enzyme, then declines rapidly with further temperature increases. What role does denaturation play in determining the shape of this curve? How does your explanation fit in with the lock-and-key model of enzyme action?

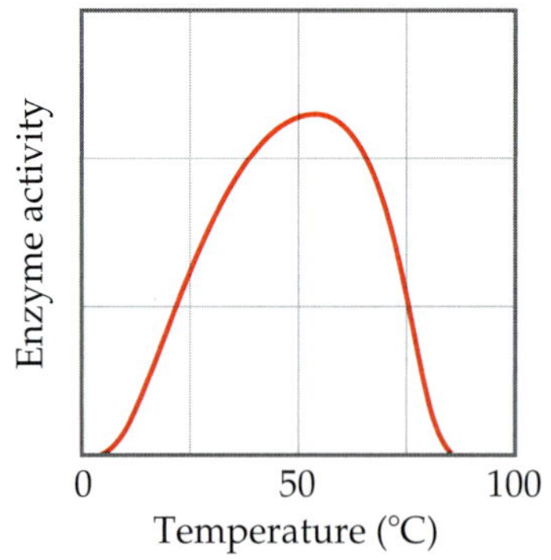

[14.104] Metals often form several cations with different charges. Cerium, for example, forms Ce^{3+} and Ce^{4+} ions, and thallium forms Tl^+ and Tl^{3+} ions. Cerium and thallium ions react as follows:

$$2\,Ce^{4+}(aq) + Tl^{3+}(aq) \longrightarrow 2\,Ce^{3+}(aq) + Tl^{3+}(aq)$$

This reaction is very slow and is thought to occur in a single elementary step. The reaction is catalyzed by the addition of $Mn^{2+}(aq)$, according to the following mechanism:

$$Ce^{4+}(aq) + Mn^{2+}(aq) \longrightarrow Ce^{3+}(aq) + Mn^{3+}(aq)$$
$$Ce^{4+}(aq) + Mn^{3+}(aq) \longrightarrow Ce^{3+}(aq) + Mn^{4+}(aq)$$
$$Mn^{4+}(aq) + Tl^{+}(aq) \longrightarrow Mn^{2+}(aq) + Tl^{3+}(aq)$$

(a) Write the rate law for the uncatalyzed reaction. **(b)** What is unusual about the uncatalyzed reaction? Why might it be a slow reaction? **(c)** The rate for the catalyzed reaction is first order in $[Ce^{4+}]$ and first order in $[Mn^{2+}]$. Based on this rate law, which of the steps in the catalyzed mechanism is rate determining? **(d)** Use the available oxidation states of Mn to comment on its special suitability to catalyze this reaction.

[14.105] The rates of many atmospheric reactions are accelerated by the absorption of light by one of the reactants. For example, consider the reaction between methane and chlorine to produce methyl chloride and hydrogen chloride:

Reaction 1: $CH_4(g) + Cl(g) \longrightarrow CH_3Cl(g) + HCl(g)$

This reaction is very slow in the absence of light. However, $Cl_2(g)$ can absorb light to form Cl atoms:

Reaction 2: $Cl_2(g) + h\nu \longrightarrow 2\,Cl(g)$

Once the Cl atoms are generated, they can catalyze the reaction of CH_4 and Cl_2, according to the following proposed mechanism:

Reaction 3: $CH_4(g) + Cl(g) \longrightarrow CH_3(g) + HCl(g)$

Reaction 4: $CH_3(g) + Cl_2(g) \longrightarrow CH_3Cl(g) + Cl(g)$

The enthalpy changes and activation energies for these two reactions are tabulated as follows:

Reaction	ΔH°_{rxn} (kJ/mol)	E_a (kJ/mol)
3	+4	17
4	−109	4

(a) By using the bond enthalpy for Cl_2 (Table 8.4), determine the longest wavelength of light that is energetic enough to cause reaction 2 to occur. In which portion of the electromagnetic spectrum is this light found? **(b)** By using the data tabulated here, sketch a quantitative energy profile for the catalyzed reaction represented by reactions 3 and 4. **(c)** By using bond enthalpies, estimate where the reactants, $CH_4(g) + Cl_2(g)$, should be placed on your diagram in part (b). Use this result to estimate the value of E_a for the reaction $CH_4(g) + Cl_2(g) \longrightarrow CH_3(g) + HCl(g) + Cl(g)$. **(d)** The species $Cl(g)$ and $CH_3(g)$ in reactions 3 and 4 are radicals, atoms, or molecules with unpaired electrons. Draw a Lewis structure of CH_3, and verify that it is a radical. **(e)** The sequence of reactions 3 and 4 comprise a radical chain mechanism. Why do you think this is called a "chain reaction"? Propose a reaction that will terminate the chain reaction.

eMEDIA EXERCISES

These exercises make use of the interactive objects available online in OneKey or the Companion Website, and on your Accelerator CD. Access to these resources comes in your MediaPak.

14.106 The **Rates of Reaction** activity (*14.3*) allows you to adjust activation energy, overall energy change, temperature, and starting concentration of a reactant to assess the effect of each variable on initial reaction rate. **(a)** If you want to determine the rate law for the reaction using a series of experiments, which variable(s) *must* you vary, which variable(s) should you *not* vary, and for which variable(s) does it not make any difference? **(b)** Determine the rate law for the reaction. What is the overall order of the reaction? **(c)** Determine the value of the rate constant.

14.107 **(a)** Using data from the **Rates of Reaction** activity (*14.3*), select a temperature (T_1) and an activation energy (E_a) and determine the value of the rate constant (k_1). **(b)** Using the (E_a) and (T_1) from part (a), predict the value of the rate constant (k_2) at a higher temperature, $T_2 = 2\,T_1$. **(c)** Again using E_a from part (a) and the temperature from either part (a) or part (b), predict the value of the rate constant (k_3) at a temperature (T_3) that is *lower* than T_1. Use the simulation to check your answers for parts (a), (b), and (c).

14.108 The **First-Order Process** animation (*14.4*) illustrates the change in concentration over time for a process that is first order overall. **(a)** Determine the rate constant for the process shown. **(b)** Assuming that the first-order process is a chemical reaction, what would normally happen to the value of the half-life if the temperature were increased? **(c)** Many enzyme-catalyzed reactions are first order in the concentration of substrate. However, increasing the temperature does not always increase the rate. Why might this be so?

14.109 Using the numerical value of the rate constant you calculated for the reaction in the previous question and using the same "unit" concentration as in the **First-Order Process** animation (*14.4*), generate a curve for a second-order reaction with Equation 14.16. Reproduce the first-order curve, superimpose the second-order curve onto it, and compare them. What can you say about the relative rates of disappearance of the starting material initially and after a significant period of time has passed?

14.110 In Exercise 14.75, you explained why the combination of ethylene and D_2 in the presence of a catalyst predominantly produces $CH_2D{-}CH_2D$. A small amount of product has more than one D bound to a carbon, as in $CH_2D{-}CHD_2$. After watching the **Surface Reaction-Hydrogenation** animation (*14.7*), propose a mechanism by which a product with more than one D per carbon atom could arise.

CHAPTER

15

WHAT'S AHEAD

- We begin by examining the concept of equilibrium. *(Section 15.1)*
- We then define the *equilibrium constant* and learn how to write *equilibrium-constant expressions* for homogeneous reactions. *(Section 15.2)*
- We also learn how to interpret the magnitude of an equilibrium constant and how to determine the way in which its value is affected when the equation is reversed or changed in some other fashion. *(Section 15.3)*
- We then learn how to write equilibrium-constant expressions for heterogeneous reactions. *(Section 15.4)*
- The value of an equilibrium constant can be calculated using equilibrium concentrations of reactants and products. *(Section 15.5)*
- Equilibrium constants can be used to predict the equilibrium concentrations of reactants and products and to determine the direction in which a reaction must proceed in order to achieve equilibrium. *(Section 15.6)*
- The chapter concludes with a discussion of *Le Châtelier's principle*, which predicts how a system at equilibrium responds to changes in concentration, volume, pressure, and temperature. *(Section 15.7)*

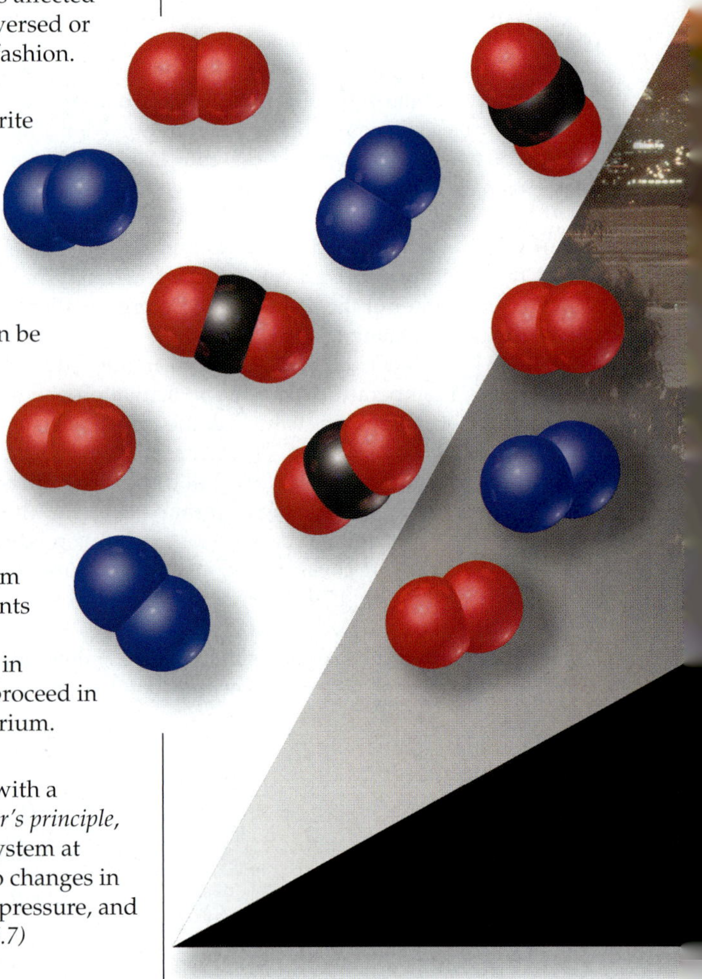

Chemical Equilibrium

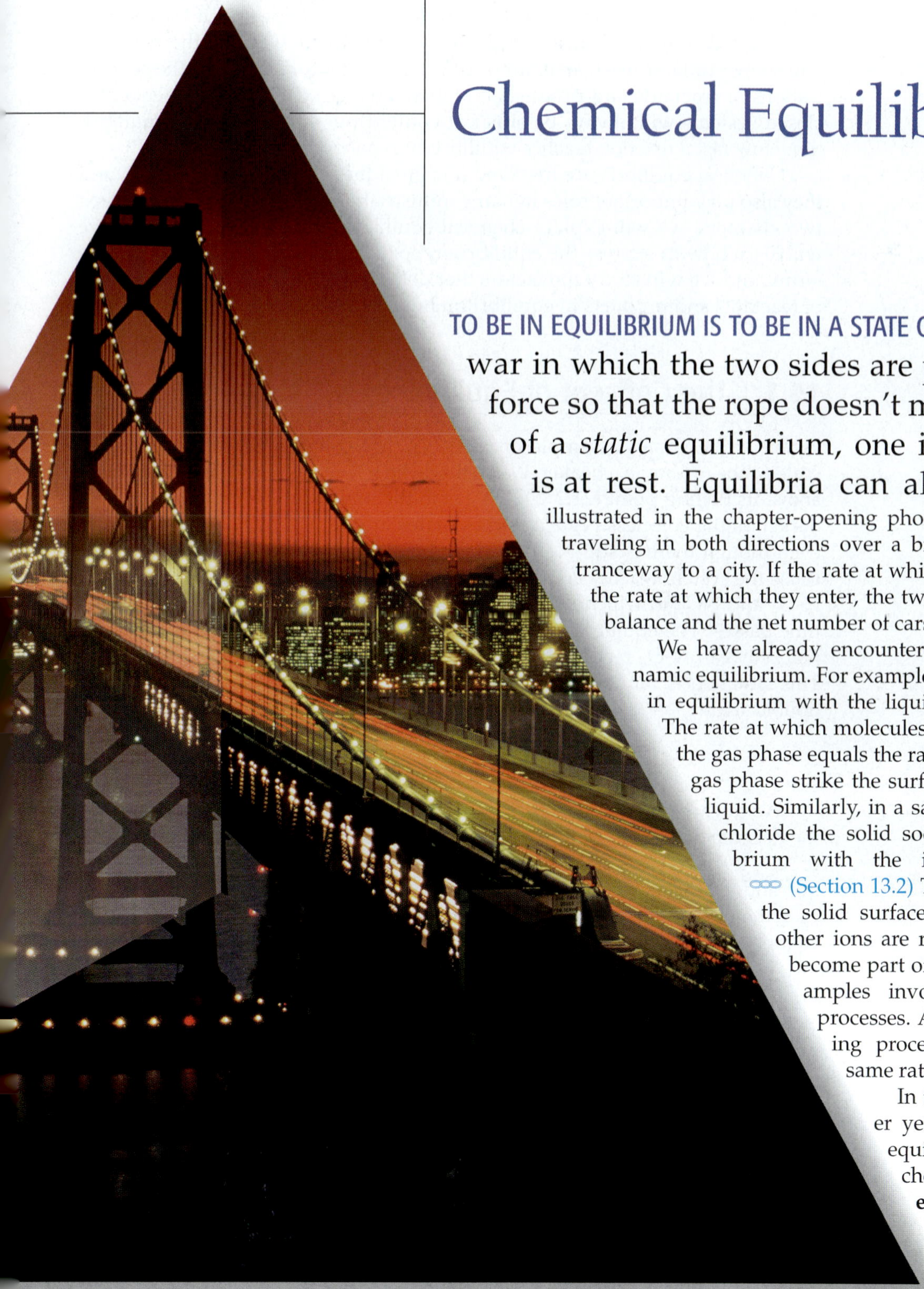

TO BE IN EQUILIBRIUM IS TO BE IN A STATE OF BALANCE. A tug of war in which the two sides are pulling with equal force so that the rope doesn't move is an example of a *static* equilibrium, one in which an object is at rest. Equilibria can also be *dynamic*, as illustrated in the chapter-opening photograph, which shows cars traveling in both directions over a bridge that serves as an entranceway to a city. If the rate at which cars leave the city equals the rate at which they enter, the two opposing processes are in balance and the net number of cars in the city is constant.

We have already encountered several instances of dynamic equilibrium. For example, the vapor above a liquid is in equilibrium with the liquid phase. ∞ (Section 11.5) The rate at which molecules escape from the liquid into the gas phase equals the rate at which molecules in the gas phase strike the surface and become part of the liquid. Similarly, in a saturated solution of sodium chloride the solid sodium chloride is in equilibrium with the ions dispersed in water. ∞ (Section 13.2) The rate at which ions leave the solid surface equals the rate at which other ions are removed from the liquid to become part of the solid. Both of these examples involve a pair of opposing processes. At equilibrium these opposing processes are occurring at the same rate.

In this chapter we will consider yet another type of dynamic equilibrium, one involving chemical reactions. **Chemical equilibrium** *occurs when opposing reactions are proceeding at equal rates*: The rate at which the products are formed from the

TRAFFIC ENTERING AND LEAVING San Francisco over the Bay Bridge at night.

reactants equals the rate at which the reactants are formed from the products. As a result, concentrations cease to change, making it appear as if the reaction has stopped. When chemical reactions occur in closed systems, the reactions will sooner or later reach an equilibrium state—a mixture in which the concentrations of reactants and products no longer change with time. Consequently, closed systems will always be either at equilibrium or approaching equilibrium. How fast a reaction reaches equilibrium is a matter of kinetics.

Chemical equilibria are involved in a great many natural phenomena, and they also play important roles in many industrial processes. In this and the next two chapters, we will explore chemical equilibrium in some detail. Here we will learn how to express the equilibrium position of a reaction in quantitative terms, and we will study the factors that determine the relative concentrations of reactants and products in equilibrium mixtures.

15.1 | The Concept of Equilibrium

Figure 15.1 ▶ shows a sample of solid N_2O_4, a colorless substance, inside a sealed tube resting in a beaker. When this solid is warmed until the substance is above its boiling point (21.2°C), the gas in the sealed tube turns progressively darker as the colorless N_2O_4 gas dissociates into brown NO_2 gas. Eventually, even though there is still N_2O_4 in the tube, the color stops getting darker because the system reaches equilibrium. We are left with an *equilibrium mixture* of N_2O_4 and NO_2 in which the concentrations of the gases no longer change as time passes. The equilibrium mixture results because the reaction is *reversible*. Not only can N_2O_4 react to form NO_2, but NO_2 can also react to form N_2O_4. This situation is represented by writing the equation for the reaction with a double arrow: (Section 4.1)

$$\underset{\text{Colorless}}{N_2O_4(g)} \rightleftharpoons \underset{\text{Brown}}{2\,NO_2(g)} \qquad [15.1]$$

We can analyze this equilibrium using our knowledge of kinetics. Let's call the decomposition of N_2O_4 to form NO_2 the forward reaction and the reaction of NO_2 to re-form N_2O_4 the reverse reaction. In this case both the forward reaction and the reverse reaction are elementary reactions. As we learned in Section 14.6, the rate laws for elementary reactions can be written from their chemical equations:

Forward reaction: $N_2O_4(g) \longrightarrow 2\,NO_2(g) \qquad \text{Rate}_f = k_f[N_2O_4]$ [15.2]

Reverse reaction: $2\,NO_2(g) \longrightarrow N_2O_4(g) \qquad \text{Rate}_r = k_r[NO_2]^2$ [15.3]

where k_f and k_r are the rate constants for the forward and reverse reactions, respectively. At equilibrium the rate at which products are produced from reactants equals the rate at which reactants are produced from products:

$$\underset{\text{Forward reaction}}{k_f[N_2O_4]} = \underset{\text{Reverse reaction}}{k_r[NO_2]^2} \qquad [15.4]$$

Rearranging this equation gives

$$\frac{[NO_2]^2}{[N_2O_4]} = \frac{k_f}{k_r} = \text{a constant} \qquad [15.5]$$

As shown in Equation 15.5, the quotient of two constants, such as k_f and k_r, is itself a constant. Thus, at equilibrium the ratio of the concentration terms involving N_2O_4 and NO_2 equals a constant. (We will consider this constant, called the *equilibrium constant*, in Section 15.2.) It makes no difference whether we start with N_2O_4 or with NO_2, or even with some mixture of the two.

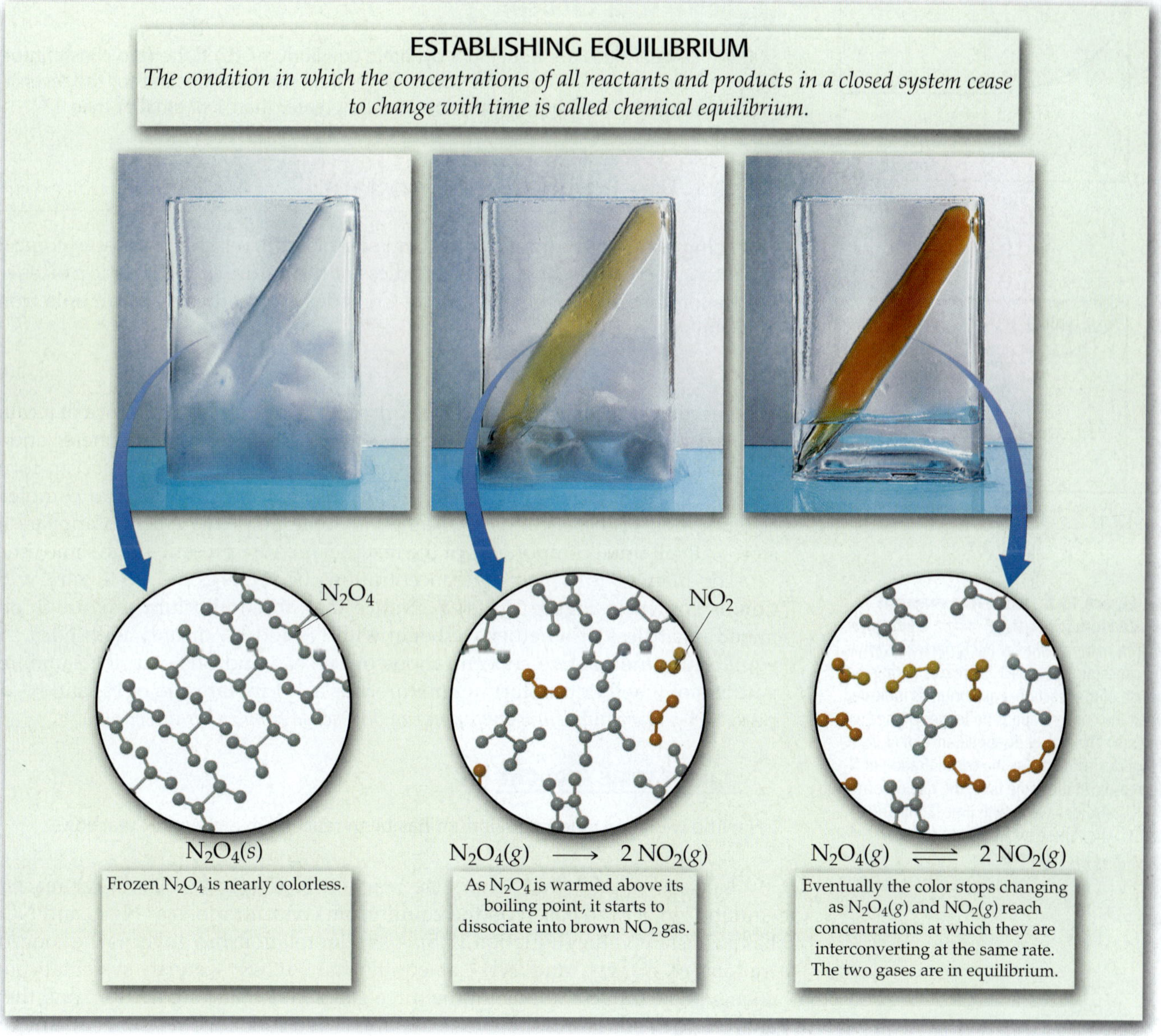

▲ **Figure 15.1 The $N_2O_4(g) \rightleftharpoons 2\ NO_2(g)$ equilibrium.**

At equilibrium the ratio equals a specific value. Thus, there is an important constraint on the proportions of N_2O_4 and NO_2 at equilibrium.

Once equilibrium is established, the concentrations of N_2O_4 and NO_2 no longer change, as shown in Figure 15.2 ▶. Just because the composition of the equilibrium mixture remains constant with time does not mean, however, that N_2O_4 and NO_2 stop reacting. On the contrary, the equilibrium is dynamic—some N_2O_4 is still converting to NO_2, and some NO_2 is still converting to N_2O_4. At equilibrium, however, the two processes occur at the same rate, and so there is no *net* change in their amounts.

ACTIVITY
Chemical Equilibrium

We learn several important lessons about equilibrium from this example. First, the fact that a mixture of reactants and products is formed in which concentrations no longer change with time indicates that the reaction has reached a state of equilibrium. Second, for equilibrium to occur, neither reactants nor products can escape from the system. Third, at equilibrium the particular ratio of concentration terms equals a constant. It is this third fact that we examine in the next section.

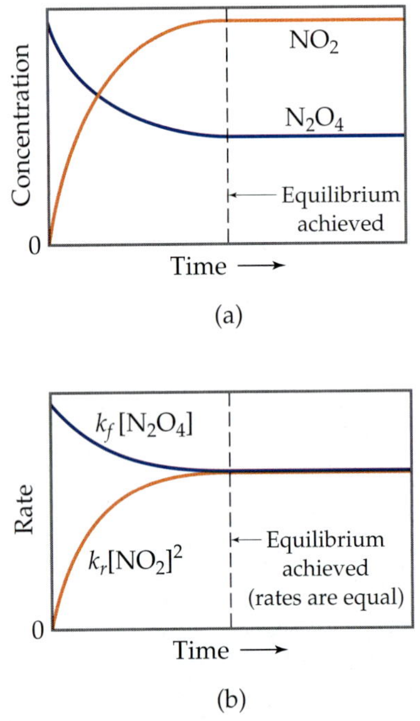

▲ **Figure 15.2 Achieving chemical equilibrium for $N_2O_4(g) \rightleftharpoons 2\,NO_2(g)$.** (a) The concentration of N_2O_4 decreases while the concentration of NO_2 increases during the course of the reaction. Equilibrium is indicated when the concentrations no longer change with time. (b) The rate of disappearance of N_2O_4 decreases with time as the concentration of N_2O_4 decreases. At the same time, the rate of formation of NO_2 also decreases with time. Equilibrium occurs when these two rates are equal.

GIVE IT SOME THOUGHT

(a) Which quantities are equal in a dynamic equilibrium? (b) If the rate constant for the forward reaction in Equation 15.1 is larger than the rate constant for the reverse reaction, will the constant in Equation 15.5 be greater than 1 or smaller than 1?

15.2 | The Equilibrium Constant

Opposing reactions naturally lead to an equilibrium, regardless of how complicated the reaction might be and regardless of the nature of the kinetic processes for the forward and reverse reactions. Consider the synthesis of ammonia from nitrogen and hydrogen:

$$N_2(g) + 3\,H_2(g) \rightleftharpoons 2\,NH_3(g) \qquad [15.6]$$

This reaction is the basis for the **Haber process**, which, in the presence of a catalyst, combines N_2 and H_2 at a pressure of several hundred atmospheres and a temperature of several hundred degrees Celsius. The two gases react to form ammonia under these conditions, but the reaction does not lead to complete consumption of the N_2 and H_2. Rather, at some point the reaction appears to stop, with all three components of the reaction mixture present at the same time.

The manner in which the concentrations of H_2, N_2, and NH_3 vary with time is shown in Figure 15.3(a) ▼. Notice that an equilibrium mixture is obtained regardless of whether we begin with N_2 and H_2 or only with NH_3. At equilibrium the relative concentrations of H_2, N_2, and NH_3 are the same, regardless of whether the starting mixture was a 3:1 molar ratio of H_2 and N_2 or pure NH_3. *The equilibrium condition can be reached from either direction.*

GIVE IT SOME THOUGHT

How do we know when equilibrium has been reached in a chemical reaction?

Earlier, we saw that when the reaction $N_2O_4(g) \rightleftharpoons 2\,NO_2(g)$ reaches equilibrium, a ratio based on the equilibrium concentrations of N_2O_4 and NO_2 has a constant value (Equation 15.5). A similar relationship governs the concentrations of N_2, H_2, and NH_3 at equilibrium. If we were to systematically change the relative amounts of the three gases in the starting mixture and then analyze each equilibrium mixture, we could determine the relationship among the equilibrium concentrations.

Chemists carried out studies of this kind on other chemical systems in the nineteenth century, before Haber's work. In 1864, Cato Maximilian Guldberg (1836–1902) and Peter Waage (1833–1900) postulated their **law of mass action**, which expresses, for any reaction, the relationship between the concentrations of the reactants and products present at equilibrium. Suppose we have the

▶ **Figure 15.3 Concentration changes approaching equilibrium.** (a) Equilibrium for the reaction $N_2 + 3\,H_2 \rightleftharpoons 2\,NH_3$ is approached beginning with H_2 and N_2 present in the ratio 3:1 and no NH_3 present. (b) Equilibrium for the same reaction is approached beginning with only NH_3 in the reaction vessel.

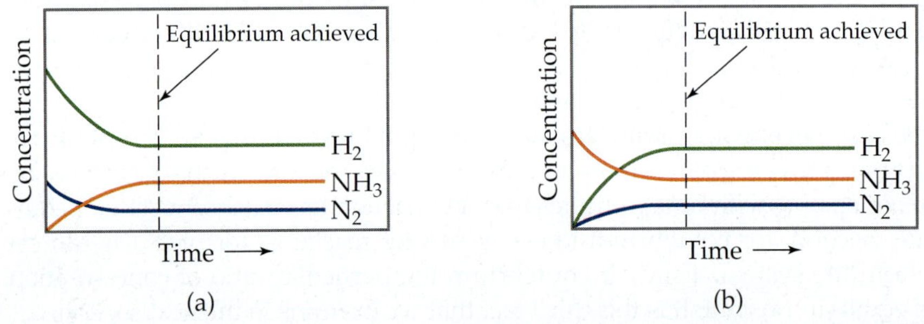

CHEMISTRY AT WORK | The Haber Process

We presented a "Chemistry and Life" box in Section 14.7 that discussed *nitrogen fixation*, the processes that convert N_2 gas into ammonia, which can then be incorporated into living organisms. We learned that the enzyme nitrogenase is responsible for generating most of the fixed nitrogen essential for plant growth. However, the quantity of food required to feed the ever-increasing human population far exceeds that provided by nitrogen-fixing plants, so human agriculture requires substantial amounts of ammonia-based fertilizers that can be applied directly to croplands. Thus, of all the chemical reactions that humans have learned to carry out and control for their own purposes, the synthesis of ammonia from hydrogen and atmospheric nitrogen is one of the most important.

In 1912 the German chemist Fritz Haber (1868–1934) developed a process for synthesizing ammonia directly from nitrogen and hydrogen (Figure 15.4 ▼). The process is sometimes called the *Haber–Bosch process* to also honor Karl Bosch, the engineer who developed the equipment for the industrial production of ammonia. The engineering needed to implement the Haber process requires the use of temperatures and pressures (approximately 500°C and 200 atm) that were difficult to achieve at that time.

The Haber process provides a historically interesting example of the complex impact of chemistry on our lives. At the start of World War I, in 1914, Germany depended on nitrate deposits in Chile for the nitrogen-containing compounds needed to manufacture explosives. During the war the Allied naval blockade of South America cut off this supply. However, by fixing nitrogen from air, Germany was able to continue to produce explosives. Experts have estimated that World War I would have ended before 1918 had it not been for the Haber process.

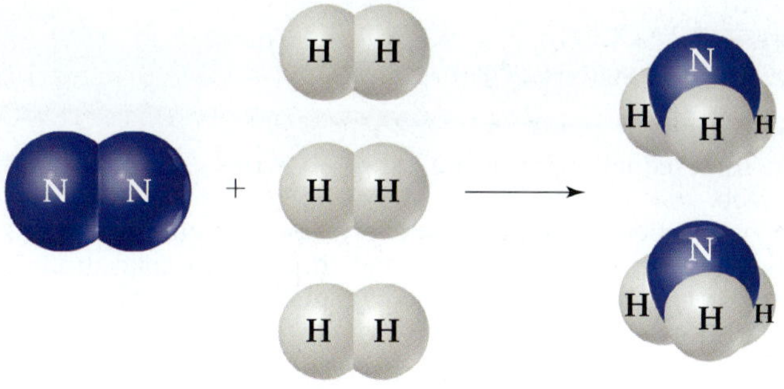

▲ **Figure 15.4 The Haber process.** Used to convert $N_2(g)$ and $H_2(g)$ to $NH_3(g)$, this process, although exothermic, requires breaking the very strong triple bond in N_2.

▲ **Figure 15.5 Liquid ammonia used as fertilizer.** Ammonia, produced by the Haber process, can be added directly to the soil as a fertilizer. Agricultural use is the largest single application of manufactured NH_3.

From these unhappy beginnings as a major factor in international warfare, the Haber process has become the world's principal source of fixed nitrogen. The same process that prolonged World War I has enabled scientists to manufacture fertilizers that have increased crop yields, thereby saving millions of people from starvation. About 40 billion pounds of ammonia are manufactured annually in the United States, mostly by the Haber process. The ammonia can be applied directly to the soil as fertilizer (Figure 15.5 ▲). It can also be converted into ammonium salts—for example, ammonium sulfate, $(NH_4)_2SO_4$, or ammonium hydrogen phosphate, $(NH_4)_2HPO_4$—which, in turn, are used as fertilizers.

Haber was a patriotic German who gave enthusiastic support to his nation's war effort. He served as chief of Germany's Chemical Warfare Service during World War I and developed the use of chlorine as a poison-gas weapon. Consequently, the decision to award him the Nobel Prize for chemistry in 1918 was the subject of considerable controversy and criticism. The ultimate irony, however, came in 1933 when Haber was expelled from Germany because he was Jewish.

Related Exercises: 15.38 and 15.63

following general equilibrium equation:

$$aA + bB \rightleftharpoons dD + eE \qquad [15.7]$$

where A, B, D, and E are the chemical species involved and *a*, *b*, *d*, and *e* are their coefficients in the balanced chemical equation. According to the law of mass action, the equilibrium condition is expressed by the expression

ACTIVITY
Equilibrium Constant

$$K_c = \frac{[D]^d[E]^e}{[A]^a[B]^b} \qquad [15.8]$$

We call this relationship the **equilibrium-constant expression** (or merely the equilibrium expression) for the reaction. The constant K_c, which we call the **equilibrium constant**, is the numerical value obtained when we substitute equilibrium concentrations into the equilibrium-constant expression. The subscript *c* on the *K* indicates that concentrations expressed in molarity are used to evaluate the constant.

In general, the numerator of the equilibrium-constant expression is the product of the concentrations of all substances on the product side of the equilibrium equation, each raised to a power equal to its coefficient in the balanced equation. The denominator is similarly derived from the reactant side of the equilibrium equation. (Remember, the convention is to write the substances on the *product* side in the *numerator* and the substances on the *reactant* side in the *denominator*.) Thus, for the Haber process, $N_2(g) + 3\,H_2(g) \rightleftharpoons 2\,NH_3(g)$, the equilibrium-constant expression is

$$K_c = \frac{[NH_3]^2}{[N_2][H_2]^3} \qquad [15.9]$$

Note that once we know the balanced chemical equation for an equilibrium, we can write the equilibrium-constant expression even if we don't know the reaction mechanism. *The equilibrium-constant expression depends only on the stoichiometry of the reaction, not on its mechanism.*

The value of the equilibrium constant at any given temperature does not depend on the initial amounts of reactants and products. It also does not matter whether other substances are present, as long as they do not react with a reactant or a product. The value of the equilibrium constant depends only on the particular reaction and on the temperature.

SAMPLE EXERCISE 15.1 | Writing Equilibrium-Constant Expressions

Write the equilibrium expression for K_c for the following reactions:

(a) $2\,O_3(g) \rightleftharpoons 3\,O_2(g)$

(b) $2\,NO(g) + Cl_2(g) \rightleftharpoons 2\,NOCl(g)$

(c) $Ag^+(aq) + 2\,NH_3(aq) \rightleftharpoons Ag(NH_3)_2{}^+(aq)$

Solution

Analyze: We are given three equations and are asked to write an equilibrium-constant expression for each.

Plan: Using the law of mass action, we write each expression as a quotient having the product concentration terms in the numerator and the reactant concentration terms in the denominator. Each term is raised to the power of its coefficient in the balanced chemical equation.

Solve: **(a)** $K_c = \dfrac{[O_2]^3}{[O_3]^2}$, **(b)** $K_c = \dfrac{[NOCl]^2}{[NO]^2[Cl_2]}$, **(c)** $K_c = \dfrac{[Ag(NH_3)_2{}^+]}{[Ag^+][NH_3]^2}$

PRACTICE EXERCISE

Write the equilibrium-constant expression, K_c, for **(a)** $H_2(g) + I_2(g) \rightleftharpoons 2\,HI(g)$, **(b)** $Cd^{2+}(aq) + 4\,Br^-(aq) \rightleftharpoons CdBr_4{}^{2-}(aq)$.

Answers: **(a)** $K_c = \dfrac{[HI]^2}{[H_2][I_2]}$; **(b)** $K_c = \dfrac{[CdBr_4{}^{2-}]}{[Cd^{2+}][Br^-]^4}$

We can illustrate how the law of mass action was discovered empirically and demonstrate that the equilibrium constant is independent of starting concentrations by examining some equilibrium concentrations for the gas-phase reaction between dinitrogen tetroxide and nitrogen dioxide:

$$N_2O_4(g) \rightleftharpoons 2\,NO_2(g) \qquad K_c = \frac{[NO_2]^2}{[N_2O_4]} \qquad [15.10]$$

TABLE 15.1 Initial and Equilibrium Concentrations of N_2O_4 and NO_2 in the Gas Phase at 100°C

Experiment	Initial N_2O_4 Concentration (M)	Initial NO_2 Concentration (M)	Equilibrium N_2O_4 Concentration (M)	Equilibrium NO_2 Concentration (M)	K_c
1	0.0	0.0200	0.00140	0.0172	0.211
2	0.0	0.0300	0.00280	0.0243	0.211
3	0.0	0.0400	0.00452	0.0310	0.213
4	0.0200	0.0	0.00452	0.0310	0.213

Figure 15.1 shows the reaction proceeding to equilibrium starting with pure N_2O_4. Because NO_2 is a dark brown gas and N_2O_4 is colorless, the amount of NO_2 in the mixture can be determined by measuring the intensity of the brown color of the gas mixture.

ACTIVITY
NO_2—N_2O_4 Equilibrium

We can determine the numerical value for K_c and verify that it is constant regardless of the starting amounts of NO_2 and N_2O_4 by performing experiments in which we start with several sealed tubes containing different concentrations of NO_2 and N_2O_4, as summarized in Table 15.1 ▲. The tubes are kept at 100°C until no further change in the color of the gas is noted. We then analyze the mixtures and determine the equilibrium concentrations of NO_2 and N_2O_4, as shown in Table 15.1.

To evaluate the equilibrium constant, K_c, we insert the equilibrium concentrations into the equilibrium-constant expression. For example, using the Experiment 1 data, $[NO_2] = 0.0172\ M$ and $[N_2O_4] = 0.00140\ M$, we find

$$K_c = \frac{[NO_2]^2}{[N_2O_4]} = \frac{(0.0172)^2}{0.00140} = 0.211$$

Proceeding in the same way, the values of K_c for the other samples were calculated, as listed in Table 15.1. Note that the value for K_c is constant ($K_c = 0.212$, within the limits of experimental error) even though the initial concentrations vary. Furthermore, the results of Experiment 4 show that equilibrium can be achieved beginning with N_2O_4 rather than with NO_2. That is, equilibrium can be approached from either direction. Figure 15.6 ▶ shows how both Experiments 3 and 4 result in the same equilibrium mixture even though one begins with 0.0400 M NO_2 and the other with 0.0200 M N_2O_4.

Notice that no units are given for the values of K_c either in Table 15.1 or in our calculation using the Experiment 1 data. It is common practice to write equilibrium constants without units for reasons that we address later in this section.

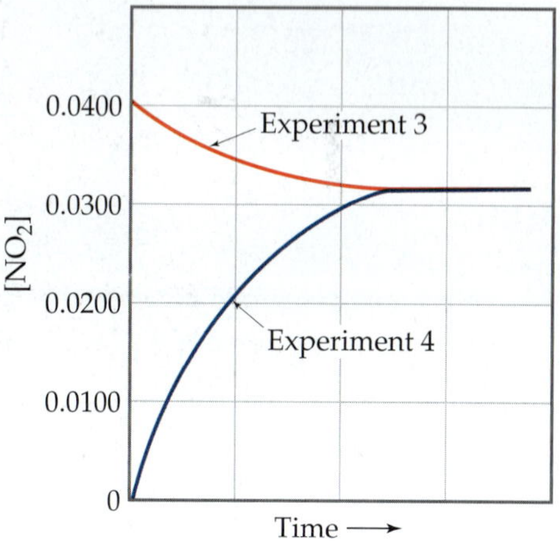

▲ **Figure 15.6 Concentration changes approaching equilibrium.** As seen in Table 15.1, the same equilibrium mixture is produced starting with either 1.22 atm NO_2 (Experiment 3) or 0.612 atm N_2O_4 (Experiment 4).

GIVE IT SOME THOUGHT

How does the value of K_c in Equation 15.10 depend on the starting concentrations of NO_2 and N_2O_4?

Equilibrium Constants in Terms of Pressure, K_p

When the reactants and products in a chemical reaction are gases, we can formulate the equilibrium-constant expression in terms of partial pressures instead of molar concentrations. When partial pressures in atmospheres are used in the equilibrium-constant expression, we can denote the equilibrium constant as K_p (where the subscript p stands for pressure). For the general reaction in Equation 15.7, the expression for K_p is

$$K_p = \frac{(P_D)^d(P_E)^e}{(P_A)^a(P_B)^b} \qquad [15.11]$$

where P_A is the partial pressure of A in atmospheres, and so forth. For example, for $N_2O_4(g) \rightleftharpoons 2\,NO_2(g)$ we have

$$K_p = \frac{(P_{NO_2})^2}{P_{N_2O_4}}$$

GIVE IT SOME THOUGHT

What do the symbols K_c and K_p represent?

For a given reaction, the numerical value of K_c is generally different from the numerical value of K_p. We must therefore take care to indicate, via a subscript c or p, which of these equilibrium constants we are using. It is possible, however, to calculate one from the other using the ideal-gas equation (Section 10.4) to convert between concentration (in molarity, M) and pressure (in atm):

$$PV = nRT, \text{ so } P = \frac{n}{V}RT \qquad [15.12]$$

Using the usual units, n/V has the units of mol/L, and therefore equals molarity, M. For substance A we therefore see that

$$P_A = \frac{n_A}{V}RT = [A]RT \qquad [15.13]$$

When we substitute Equation 15.13 and like expressions for the other gaseous components of the reaction into the expression for K_p (Equation 15.11), we obtain a general expression relating K_p and K_c:

$$K_p = K_c(RT)^{\Delta n} \qquad [15.14]$$

The quantity Δn is the change in the number of moles of gas in the chemical equation for the reaction. It equals the sum of the coefficients of the gaseous products minus the sum of the coefficients of the gaseous reactants:

$$\Delta n = (\text{moles of gaseous product}) - (\text{moles of gaseous reactant}) \qquad [15.15]$$

For example, in the reaction $N_2O_4(g) \rightleftharpoons 2\,NO_2(g)$, there are two moles of the product NO_2 (the coefficient in the balanced equation) and one mole of the reactant N_2O_4. Therefore, $\Delta n = 2 - 1 = 1$, and $K_p = K_c(RT)$ for this reaction. From Equation 15.14, we see that $K_p = K_c$ only when the same number of moles of gas appear on both sides of the balanced chemical equation, which means that $\Delta n = 0$.

SAMPLE EXERCISE 15.2 | Converting Between K_c and K_p

In the synthesis of ammonia from nitrogen and hydrogen,

$$N_2(g) + 3\,H_2(g) \rightleftharpoons 2\,NH_3(g)$$

$K_c = 9.60$ at 300°C. Calculate K_p for this reaction at this temperature.

Solution

Analyze: We are given K_c for a reaction and asked to calculate K_p.

Plan: The relationship between K_c and K_p is given by Equation 15.14. To apply that equation, we must determine Δn by comparing the number of moles of product with the number of moles of reactants (Equation 15.15).

Solve: There are two moles of gaseous products ($2\,NH_3$) and four moles of gaseous reactants ($1\,N_2 + 3\,H_2$). Therefore, $\Delta n = 2 - 4 = -2$. (Remember that Δ functions are always based on products minus reactants.) The temperature, T, is $273 + 300 = 573$ K.

The value for the ideal-gas constant, R, is 0.0821 L-atm/mol-K. Using $K_c = 9.60$, we therefore have

$$K_p = K_c(RT)^{\Delta n} = (9.60)(0.0821 \times 573)^{-2} = \frac{(9.60)}{(0.0821 \times 573)^2} = 4.34 \times 10^{-3}$$

PRACTICE EXERCISE

For the equilibrium $2\,SO_3(g) \rightleftharpoons 2\,SO_2(g) + O_2(g)$, K_c is 4.08×10^{-3} at 1000 K. Calculate the value for K_p.
Answer: 0.335

Equilibrium Constants and Units

The equilibrium constant is related not only to the kinetics of a reaction but also to the thermodynamics of the process. (We will explore this latter connection in Chapter 19.) Equilibrium constants derived from thermodynamic measurements are defined in terms of *activities* rather than concentrations or partial pressures. A full discussion of activities is beyond the scope of this book, but it is useful to explore them very briefly to understand why equilibrium constants are usually reported without units.

The activity of any substance in an ideal mixture is the ratio of the concentration of the substance in molarity to a standard concentration of 1 M or, if the substance is a gas, the ratio of the partial pressure in atmospheres to a standard pressure of 1 atm. For example, if the concentration of a substance in an equilibrium mixture is 0.10 M, its activity is 0.10 M/1 M = 0.10. The units of such ratios always cancel, and consequently, activities have no units. Furthermore, the numerical value of the activity equals the concentration because we have divided by 1. In real systems, activities are not exactly numerically equal to concentrations. In some cases the differences are significant (See the Closer Look box in Section 13.6). However, we will ignore these differences. For pure solids and pure liquids, the situation is even simpler because the activities then merely equal 1 (again with no units). Because activities have no units, the *thermodynamic equilibrium constant* derived from them also has no units. It is therefore common practice to write all types of equilibrium constants without units as well, a practice that we adhere to in this text.

GIVE IT SOME THOUGHT

If the concentration of N_2O_4 in an equilibrium mixture is 0.00140 M, what is its activity?

15.3 | Interpreting and Working with Equilibrium Constants

Before doing calculations with equilibrium constants, it is valuable to understand what the magnitude of an equilibrium constant can tell us about the relative concentrations of reactants and products in an equilibrium mixture. It is also useful to consider how the magnitude of any equilibrium constant depends on how the chemical equation is expressed. We examine these ideas in this section.

The Magnitude of Equilibrium Constants

Equilibrium constants can vary from very large to very small. The magnitude of the constant provides us with important information about the composition of an equilibrium mixture. For example, consider the reaction of carbon monoxide

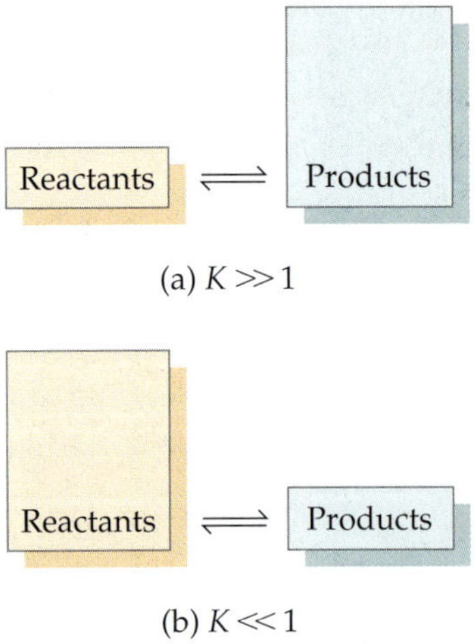

▲ **Figure 15.7** ***K* and the composition of the equilibrium mixture.** The equilibrium expression has products in the numerator and reactants in the denominator. (a) When $K \gg 1$, there are more products than reactants at equilibrium, and the equilibrium is said to lie to the right. (b) When $K \ll 1$, there are more reactants than products at equilibrium, and the equilibrium is said to lie to the left.

gas and chlorine gas at 100°C to form phosgene ($COCl_2$), a toxic gas used in the manufacture of certain polymers and insecticides:

$$CO(g) + Cl_2(g) \rightleftharpoons COCl_2(g) \qquad K_c = \frac{[COCl_2]}{[CO][Cl_2]} = 4.56 \times 10^9$$

In order for the equilibrium constant to be so large, the numerator of the equilibrium-constant expression must be much larger than the denominator. Thus, the equilibrium concentration of $COCl_2$ must be much greater than that of CO or Cl_2, and in fact this is just what we find experimentally. We say that this equilibrium *lies to the right* (that is, toward the product side). Likewise, a very small equilibrium constant indicates that the equilibrium mixture contains mostly reactants. We then say that the equilibrium *lies to the left*. In general,

If $K \gg 1$: Equilibrium lies to the right; products predominate.

If $K \ll 1$: Equilibrium lies to the left; reactants predominate.

These situations are summarized in Figure 15.7 ◀.

SAMPLE EXERCISE 15.3 | Interpreting the Magnitude of an Equilibrium Constant

The reaction of N_2 with O_2 to form NO might be considered a means of "fixing" nitrogen:

$$N_2(g) + O_2(g) \rightleftharpoons 2\,NO(g)$$

The value for the equilibrium constant for this reaction at 25°C is $K_c = 1 \times 10^{-30}$. Describe the feasibility of fixing nitrogen by forming NO at 25°C.

Solution

Analyze: We are asked to comment on the utility of a reaction based on the magnitude of its equilibrium constant.

Plan: We consider the magnitude of the equilibrium constant to determine whether this reaction is feasible for the production of NO.

Solve: Because K_c is so small, very little NO will form at 25°C. The equilibrium lies to the left, favoring the reactants. Consequently, this reaction is an extremely poor choice for nitrogen fixation, at least at 25°C.

PRACTICE EXERCISE

For the reaction $H_2(g) + I_2(g) \rightleftharpoons 2\,HI(g)$, $K_p = 794$ at 298 K and $K_p = 54$ at 700 K. Is the formation of HI favored more at the higher or lower temperature?

Answer: The formation of product, HI, is favored at the lower temperature because K_p is larger at the lower temperature.

The Direction of the Chemical Equation and *K*

Because an equilibrium can be approached from either direction, the direction in which we write the chemical equation for an equilibrium is arbitrary. For example, we have seen that we can represent the $N_2O_4 - NO_2$ equilibrium as

$$N_2O_4(g) \rightleftharpoons 2\,NO_2(g) \quad K_c = \frac{[NO_2]^2}{[N_2O_4]} = 0.212 \quad \text{(at 100°C)} \qquad [15.16]$$

We could equally well consider this same equilibrium in terms of the reverse reaction:

$$2\,NO_2(g) \rightleftharpoons N_2O_4(g)$$

The equilibrium expression is then

$$K_c = \frac{[N_2O_4]}{[NO_2]^2} = \frac{1}{0.212} = 4.72 \quad \text{(at 100°C)} \qquad [15.17]$$

Equation 15.17 is just the reciprocal of the equilibrium-constant expression in Equation 15.16. *The equilibrium-constant expression for a reaction written in one direction is the reciprocal of the one for the reaction written in the reverse direction.* Consequently, the numerical value of the equilibrium constant for the reaction written in one direction is the reciprocal of that for the reverse reaction. Both expressions are equally valid, but it is meaningless to say that the equilibrium constant for the equilibrium between NO_2 and N_2O_4 is 0.212 or 4.72 unless we indicate how the equilibrium reaction is written and also specify the temperature.

SAMPLE EXERCISE 15.4 | Evaluating an Equilibrium Constant When an Equation Is Reversed

(a) Write the equilibrium-constant expression for K_c for the following reaction:

$$2\,NO(g) \rightleftharpoons N_2(g) + O_2(g)$$

(b) Using information in Sample Exercise 15.3, determine the value of this equilibrium constant at 25°C.

Solution

Analyze: We are asked to write the equilibrium-constant expression for a reaction in terms of concentrations and to determine the value of K_c.

Plan: As before, we write the equilibrium-constant expression as a quotient of products over reactants, each raised to a power equal to its coefficient in the balanced equation. We can determine the value of the equilibrium constant by relating the equilibrium-constant expression we write for this equation to the equilibrium-constant expression in Sample Exercise 15.3, where we found $N_2(g) + O_2(g) \rightleftharpoons 2\,NO(g)$, $K_c = 1 \times 10^{-30}$.

Solve: **(a)** Writing products over reactants, we have

$$K_c = \frac{[N_2][O_2]}{[NO]^2}$$

(b) The reaction is just the reverse of the one given in Sample Exercise 15.3. Thus, both the equilibrium-constant expression and the numerical value of the equilibrium constant are the reciprocals of those for the reaction in Sample Exercise 15.3.

$$K_c = \frac{[N_2][O_2]}{[NO]^2} = \frac{1}{1 \times 10^{-30}} = 1 \times 10^{30}$$

Comment: Regardless of the way we express the equilibrium among NO, N_2, and O_2, at 25°C it lies on the side that favors N_2 and O_2.

PRACTICE EXERCISE

For the formation of NH_3 from N_2 and H_2, $N_2(g) + 3\,H_2(g) \rightleftharpoons 2\,NH_3(g)$, $K_p = 4.34 \times 10^{-3}$ at 300°C. What is the value of K_p for the reverse reaction?

Answer: 2.30×10^2

Relating Chemical Equations and Equilibrium Constants

Just as the equilibrium constants of forward and reverse reactions are reciprocals of each other, the equilibrium constants of reactions associated in other ways are also related. For example, if we were to multiply our original $N_2O_4 - NO_2$ equilibrium by 2, we would have

$$2\,N_2O_4(g) \rightleftharpoons 4\,NO_2(g)$$

The equilibrium-constant expression, K_c, for this equation is

$$K_c = \frac{[NO_2]^4}{[N_2O_4]^2}$$

which is simply the square of the equilibrium-constant expression for the original equation, given in Equation 15.10. Because the new equilibrium-constant expression equals the original expression squared, the new equilibrium constant equals the original constant squared: in this case $0.212^2 = 0.0449$ (at 100°C).

GIVE IT SOME THOUGHT

How does the magnitude of the equilibrium constant K_p for the reaction $2\ HI(g) \rightleftharpoons H_2(g) + I_2(g)$ change if the equilibrium is written $6\ HI(g) \rightleftharpoons 3\ H_2(g) + 3\ I_2(g)$?

Sometimes, as in problems that utilize Hess's law (Section 5.6), we must use equations made up of two or more steps in the overall process. We obtain the net equation by adding the individual equations and canceling identical terms. Consider the following two reactions, their equilibrium-constant expressions, and their equilibrium constants at 100°C:

$$2\ NOBr(g) \rightleftharpoons 2\ NO(g) + Br_2(g) \qquad K_c = \frac{[NO]^2[Br_2]}{[NOBr]^2} = 0.014$$

$$Br_2(g) + Cl_2(g) \rightleftharpoons 2\ BrCl(g) \qquad K_c = \frac{[BrCl]^2}{[Br_2][Cl_2]} = 7.2$$

The net sum of these two equations is

$$2\ NOBr(g) + Cl_2(g) \rightleftharpoons 2\ NO(g) + 2\ BrCl(g)$$

and the equilibrium-constant expression for the net equation is the product of the expressions for the individual steps:

$$K_c = \frac{[NO]^2[BrCl]^2}{[NOBr]^2[Cl]_2} = \frac{[NO]^2[Br_2]}{[NOBr]^2} \times \frac{[BrCl]^2}{[Br_2][Cl_2]}$$

Because the equilibrium-constant expression for the net equation is the product of two equilibrium-constant expressions, the equilibrium constant for the net equation is the product of the two individual equilibrium constants: $K_c = 0.014 \times 7.2 = 0.10$.

To summarize:

1. The equilibrium constant of a reaction in the *reverse* direction is the *inverse* of the equilibrium constant of the reaction in the forward direction.
2. The equilibrium constant of a reaction that has been *multiplied* by a number is the equilibrium constant raised to a *power* equal to that number.
3. The equilibrium constant for a net reaction made up of *two or more steps* is the *product* of the equilibrium constants for the individual steps.

SAMPLE EXERCISE 15.5 | Combining Equilibrium Expressions

Given the following information,

$$HF(aq) \rightleftharpoons H^+(aq) + F^-(aq) \qquad K_c = 6.8 \times 10^{-4}$$

$$H_2C_2O_4(aq) \rightleftharpoons 2\ H^+(aq) + C_2O_4^{2-}(aq) \qquad K_c = 3.8 \times 10^{-6}$$

determine the value of K_c for the reaction

$$2\ HF(aq) + C_2O_4^{2-}(aq) \rightleftharpoons 2\ F^-(aq) + H_2C_2O_4(aq)$$

Solution

Analyze: We are given two equilibrium equations and the corresponding equilibrium constants and are asked to determine the equilibrium constant for a third equation, which is related to the first two.

Plan: We cannot simply add the first two equations to get the third. Instead, we need to determine how to manipulate the equations in order to come up with the steps that will add to give us the desired equation.

Solve: If we multiply the first equation by 2 and make the corresponding change to its equilibrium constant (raising to the power 2), we get

$$2\ HF(aq) \rightleftharpoons 2\ H^+(aq) + 2\ F^-(aq) \qquad K_c = (6.8 \times 10^{-4})^2 = 4.6 \times 10^{-7}$$

Reversing the second equation and again making the corresponding change to its equilibrium constant (taking the reciprocal) gives

$$2\,H^+(aq) + C_2O_4^{2-}(aq) \rightleftharpoons H_2C_2O_4(aq) \quad K_c = \frac{1}{3.8 \times 10^{-6}} = 2.6 \times 10^5$$

Now we have two equations that sum to give the net equation, and we can multiply the individual K_c values to get the desired equilibrium constant.

$$2\,HF(aq) \rightleftharpoons 2\,H^+(aq) + 2\,F^-(aq) \quad K_c = 4.6 \times 10^{-7}$$

$$2\,H^+(aq) + C_2O_4^{2-}(aq) \rightleftharpoons H_2C_2O_4(aq) \quad K_c = 2.5 \times 10^5$$

$$2\,HF(aq) + C_2O_4^{2-}(aq) \rightleftharpoons 2\,F^-(aq) + H_2C_2O_4(aq) \quad K_c = (4.6 \times 10^{-7})(2.6 \times 10^5) = 0.12$$

PRACTICE EXERCISE

Given that, at 700 K, $K_p = 54.0$ for the reaction $H_2(g) + I_2(g) \rightleftharpoons 2\,HI(g)$ and $K_p = 1.04 \times 10^{-4}$ for the reaction $N_2(g) + 3\,H_2(g) \rightleftharpoons 2\,NH_3(g)$, determine the value of K_p for the reaction $2\,NH_3(g) + 3\,I_2(g) \rightleftharpoons 6\,HI(g) + N_2(g)$ at 700 K.

Answer: $\dfrac{(54.0)^3}{1.04 \times 10^{-4}} = 1.51 \times 10^9$

15.4 | Heterogeneous Equilibria

Many equilibria, such as the hydrogen-nitrogen-ammonia system, involve substances all in the same phase. Such equilibria are called **homogeneous equilibria**. In other cases the substances in equilibrium are in different phases, giving rise to **heterogeneous equilibria**. As an example, consider the equilibrium that occurs when solid lead(II) chloride ($PbCl_2$) dissolves in water to form a saturated solution:

$$PbCl_2(s) \rightleftharpoons Pb^{2+}(aq) + 2\,Cl^-(aq) \qquad [15.18]$$

This system consists of a solid in equilibrium with two aqueous species. If we write the equilibrium-constant expression for this process, we encounter a problem we have not encountered previously: How do we express the concentration of a solid substance? Although it is possible to express the concentration of a solid in terms of moles per unit volume, it is unnecessary to do so in writing equilibrium-constant expressions. *Whenever a pure solid or a pure liquid is involved in a heterogeneous equilibrium, its concentration is not included in the equilibrium-constant expression for the reaction.* Thus, the equilibrium-constant expression for Equation 15.18 is

$$K_c = [Pb^{2+}][Cl^-]^2 \qquad [15.19]$$

Even though $PbCl_2(s)$ does not appear in the equilibrium-constant expression, it must be present for equilibrium to occur.

The fact that pure solids and pure liquids are excluded from equilibrium-constant expressions can be explained in two ways. First, the concentration of a pure solid or liquid has a constant value. If the mass of a solid is doubled, its volume also doubles. Thus, its concentration, which relates to the ratio of mass to volume, stays the same. Because equilibrium-constant expressions include terms only for reactants and products whose concentrations can change during a chemical reaction, the concentrations of pure solids and pure liquids are omitted.

The omission of pure solids and liquids from equilibrium-constant expressions can also be rationalized in a second way. Recall from the last section that what is substituted into a thermodynamic equilibrium expression is the activity of each substance, which is a ratio of the concentration to a reference value. For a pure substance, the reference value is the concentration of the pure substance itself, so that the activity of any pure solid or liquid is always simply 1.

▶ Figure 15.8 A heterogeneous equilibrium. The equilibrium involving $CaCO_3$, CaO, and CO_2 is a heterogeneous equilibrium. The equilibrium pressure of CO_2 is the same in the two bell jars as long as the two systems are at the same temperature, even though the relative amounts of pure $CaCO_3$ and CaO differ greatly. The equilibrium-constant expression for the reaction is $K_p = P_{CO_2}$.

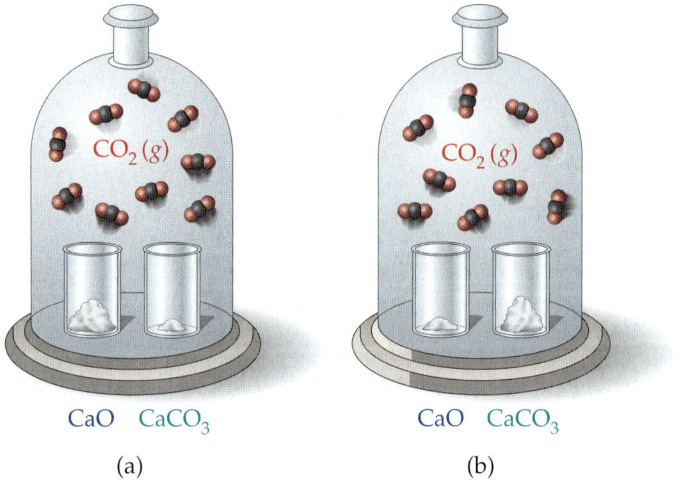

GIVE IT SOME THOUGHT

Write the equilibrium-constant expression for the evaporation of water, $H_2O(l) \rightleftharpoons H_2O(g)$, in terms of partial pressures, K_p.

As a further example of a heterogeneous reaction, consider the decomposition of calcium carbonate:

$$CaCO_3(s) \rightleftharpoons CaO(s) + CO_2(g)$$

Omitting the concentrations of solids from the equilibrium-constant expression gives

$$K_c = [CO_2] \quad \text{and} \quad K_p = P_{CO_2}$$

These equations tell us that at a given temperature, an equilibrium among $CaCO_3$, CaO, and CO_2 will always lead to the same partial pressure of CO_2 as long as all three components are present. As shown in Figure 15.8 ▲, we would have the same pressure of CO_2 regardless of the relative amounts of CaO and $CaCO_3$.

When a solvent is involved as a reactant or product in an equilibrium, its concentration is also excluded from the equilibrium-constant expression, provided the concentrations of reactants and products are low, so that the solvent is essentially a pure substance. Applying this guideline to an equilibrium involving water as a solvent,

$$H_2O(l) + CO_3^{2-}(aq) \rightleftharpoons OH^-(aq) + HCO_3^-(aq) \qquad [15.20]$$

gives an equilibrium-constant expression in which $[H_2O]$ is excluded:

$$K_c = \frac{[OH^-][HCO_3^-]}{[CO_3^{2-}]} \qquad [15.21]$$

SAMPLE EXERCISE 15.6 | Writing Equilibrium-Constant Expressions for Heterogeneous Reactions

Write the equilibrium-constant expression for K_c for each of the following reactions:

(a) $CO_2(g) + H_2(g) \rightleftharpoons CO(g) + H_2O(l)$

(b) $SnO_2(s) + 2\,CO(g) \rightleftharpoons Sn(s) + 2\,CO_2(g)$

Solution

Analyze: We are given two chemical equations, both for heterogeneous equilibria, and asked to write the corresponding equilibrium-constant expressions.

Plan: We use the law of mass action, remembering to omit any pure solids, pure liquids, and solvents from the expressions.

Solve: **(a)** The equilibrium-constant expression is

$$K_c = \frac{[CO]}{[CO_2][H_2]}$$

Because H_2O appears in the reaction as a pure liquid, its concentration does not appear in the equilibrium-constant expression.

(b) The equilibrium-constant expression is

$$K_c = \frac{[CO_2]^2}{[CO]^2}$$

Because SnO_2 and Sn are both pure solids, their concentrations do not appear in the equilibrium-constant expression.

PRACTICE EXERCISE

Write the following equilibrium-constant expressions:

(a) K_c for $Cr(s) + 3\,Ag^+(aq) \rightleftharpoons Cr^{3+}(aq) + 3\,Ag(s)$

(b) K_p for $3\,Fe(s) + 4\,H_2O(g) \rightleftharpoons Fe_3O_4(s) + 4\,H_2(g)$

Answers: **(a)** $K_c = \frac{[Cr^{3+}]}{[Ag^+]^3}$, **(b)** $K_p = \frac{(P_{H_2})^4}{(P_{H_2O})^4}$

SAMPLE EXERCISE 15.7 | Analyzing a Heterogeneous Equilibrium

Each of the following mixtures was placed in a closed container and allowed to stand. Which is capable of attaining the equilibrium $CaCO_3(s) \rightleftharpoons CaO(s) + CO_2(g)$: **(a)** pure $CaCO_3$, **(b)** CaO and a CO_2 pressure greater than the value of K_p, **(c)** some $CaCO_3$ and a CO_2 pressure greater than the value of K_p, **(d)** $CaCO_3$ and CaO?

Solution

Analyze: We are asked which of several combinations of species can establish an equilibrium between calcium carbonate and its decomposition products, calcium oxide and carbon dioxide.

Plan: In order for equilibrium to be achieved, it must be possible for both the forward process and the reverse process to occur. In order for the forward process to occur, there must be some calcium carbonate present. In order for the reverse process to occur, there must be both calcium oxide and carbon dioxide. In both cases, either the necessary compounds may be present initially, or they may be formed by reaction of the other species.

Solve: Equilibrium can be reached in all cases except (c) as long as sufficient quantities of solids are present. **(a)** $CaCO_3$ simply decomposes, forming CaO(*s*) and $CO_2(g)$ until the equilibrium pressure of CO_2 is attained. There must be enough $CaCO_3$, however, to allow the CO_2 pressure to reach equilibrium. **(b)** CO_2 continues to combine with CaO until the partial pressure of the CO_2 decreases to the equilibrium value. **(c)** There is no CaO present, so equilibrium can't be attained because there is no way the CO_2 pressure can decrease to its equilibrium value (which would require some of the CO_2 to react with CaO). **(d)** The situation is essentially the same as in (a): $CaCO_3$ decomposes until equilibrium is attained. The presence of CaO initially makes no difference.

PRACTICE EXERCISE

When added to $Fe_3O_4(s)$ in a closed container, which one of the following substances—$H_2(g)$, $H_2O(g)$, $O_2(g)$—will allow equilibrium to be established in the reaction $3\,Fe(s) + 4\,H_2O(g) \rightleftharpoons Fe_3O_4(s) + 4\,H_2(g)$?

Answer: only $H_2(g)$

15.5 | Calculating Equilibrium Constants

One of the first tasks confronting Haber when he approached the problem of ammonia synthesis was finding the magnitude of the equilibrium constant for the synthesis of NH_3 at various temperatures. If the value of *K* for Equation 15.6 was very small, the amount of NH_3 in an equilibrium mixture would be small relative to the amounts of N_2 and H_2. That is, if the equilibrium lies too far to the left, it would be impossible to develop a satisfactory synthesis process for ammonia.

Haber and his coworkers therefore evaluated the equilibrium constants for this reaction at various temperatures. The method they employed is analogous to that described in constructing Table 15.1: They started with various mixtures of N_2, H_2, and NH_3, allowed the mixtures to achieve equilibrium at a specific temperature, and measured the concentrations of all three gases at equilibrium. Because the equilibrium concentrations of all products and reactants were known, the equilibrium constant could be calculated directly from the equilibrium-constant expression.

ACTIVITY
Using an Equilibrium Table

SAMPLE EXERCISE 15.8 | Calculating *K* When All Equilibrium Concentrations Are Known

A mixture of hydrogen and nitrogen in a reaction vessel is allowed to attain equilibrium at 472°C. The equilibrium mixture of gases was analyzed and found to contain 7.38 atm H_2, 2.46 atm N_2, and 0.166 atm NH_3. From these data, calculate the equilibrium constant K_p for the reaction

$$N_2(g) + 3\,H_2(g) \rightleftharpoons 2\,NH_3(g)$$

Solution

Analyze: We are given a balanced equation and equilibrium partial pressures and are asked to calculate the value of the equilibrium constant.

Plan: Using the balanced equation, we write the equilibrium-constant expression. We then substitute the equilibrium partial pressures into the expression and solve for K_p.

Solve:

$$K_p = \frac{(P_{NH_3})^2}{P_{N_2}(P_{H_2})^3} = \frac{(0.166)^2}{(2.46)(7.38)^3} = 2.79 \times 10^{-5}$$

PRACTICE EXERCISE

An aqueous solution of acetic acid is found to have the following equilibrium concentrations at 25°C: $[HC_2H_3O_2] = 1.65 \times 10^{-2}\,M$; $[H^+] = 5.44 \times 10^{-4}\,M$; and $[C_2H_3O_2^-] = 5.44 \times 10^{-4}\,M$. Calculate the equilibrium constant K_c for the ionization of acetic acid at 25°C. The reaction is

$$HC_2H_3O_2(aq) \rightleftharpoons H^+(aq) + C_2H_3O_2^-(aq)$$

Answer: 1.79×10^{-5}

We often don't know the equilibrium concentrations of all chemical species in an equilibrium mixture. If we know the equilibrium concentration of at least one species, however, we can generally use the stoichiometry of the reaction to deduce the equilibrium concentrations of the others. The following steps outline the procedure we use to do this:

1. Tabulate all the known initial and equilibrium concentrations of the species that appear in the equilibrium-constant expression.
2. For those species for which both the initial and equilibrium concentrations are known, calculate the change in concentration that occurs as the system reaches equilibrium.
3. Use the stoichiometry of the reaction (that is, use the coefficients in the balanced chemical equation) to calculate the changes in concentration for all the other species in the equilibrium.
4. From the initial concentrations and the changes in concentration, calculate the equilibrium concentrations. These are used to evaluate the equilibrium constant.

SAMPLE EXERCISE 15.9 | Calculating *K* from Initial and Equilibrium Concentrations

A closed system initially containing $1.000 \times 10^{-3}\ M\ H_2$ and $2.000 \times 10^{-3}\ M\ I_2$ at 448°C is allowed to reach equilibrium. Analysis of the equilibrium mixture shows that the concentration of HI is $1.87 \times 10^{-3}\ M$. Calculate K_c at 448°C for the reaction taking place, which is

$$H_2(g) + I_2(g) \rightleftharpoons 2\ HI(g)$$

Solution

Analyze: We are given the initial concentrations of H_2 and I_2 and the equilibrium concentration of HI. We are asked to calculate the equilibrium constant K_c for the reaction $H_2(g) + I_2(g) \rightleftharpoons 2\ HI(g)$.

Plan: We construct a table to find equilibrium concentrations of all species and then use the equilibrium concentrations to calculate the equilibrium constant.

Solve: First, we tabulate the initial and equilibrium concentrations of as many species as we can. We also provide space in our table for listing the changes in concentrations. As shown, it is convenient to use the chemical equation as the heading for the table.

	$H_2(g)$ +	$I_2(g)$ $\rightleftharpoons$	$2\ HI(g)$
Initial	$1.000 \times 10^{-3}\ M$	$2.000 \times 10^{-3}\ M$	$0\ M$
Change			
Equilibrium			$1.87 \times 10^{-3}\ M$

Second, we calculate the change in concentration of HI, which is the difference between the equilibrium values and the initial values:

$$\text{Change in [HI]} = 1.87 \times 10^{-3}\ M - 0 = 1.87 \times 10^{-3}\ M$$

Third, we use the coefficients in the balanced equation to relate the change in [HI] to the changes in $[H_2]$ and $[I_2]$:

$$\left(1.87 \times 10^{-3} \frac{\text{mol HI}}{\text{L}}\right)\left(\frac{1\ \text{mol}\ H_2}{2\ \text{mol HI}}\right) = 0.935 \times 10^{-3} \frac{\text{mol}\ H_2}{\text{L}}$$

$$\left(1.87 \times 10^{-3} \frac{\text{mol HI}}{\text{L}}\right)\left(\frac{1\ \text{mol}\ I_2}{2\ \text{mol HI}}\right) = 0.935 \times 10^{-3} \frac{\text{mol}\ I_2}{\text{L}}$$

Fourth, we calculate the equilibrium concentrations of H_2 and I_2, using the initial concentrations and the changes. The equilibrium concentration equals the initial concentration minus that consumed:

$$[H_2] = 1.00 \times 10^{-3}\ M - 0.935 \times 10^{-3}\ M = 0.065 \times 10^{-3}\ M$$

$$[I_2] = 2.000 \times 10^{-3}\ M - 0.935 \times 10^{-3}\ M = 1.065 \times 10^{-3}\ M$$

The completed table now looks like this (with equilibrium concentrations in blue for emphasis):

	$H_2(g)$ +	$I_2(g)$ $\rightleftharpoons$	$2\ HI(g)$
Initial	$1.000 \times 10^{-3}\ M$	$2.000 \times 10^{-3}\ M$	$0\ M$
Change	$-0.935 \times 10^{-3}\ M$	$-0.935 \times 10^{-3}\ M$	$+1.87 \times 10^{-3}\ M$
Equilibrium	$0.065 \times 10^{-3}\ M$	$1.065 \times 10^{-3}\ M$	$1.87 \times 10^{-3}\ M$

Notice that the entries for the changes are negative when a reactant is consumed and positive when a product is formed.

Finally, now that we know the equilibrium concentration of each reactant and product, we can use the equilibrium-constant expression to calculate the equilibrium constant.

$$K_c = \frac{[HI]^2}{[H_2][I_2]} = \frac{(1.87 \times 10^{-3})^2}{(0.065 \times 10^{-3})(1.065 \times 10^{-3})} = 51$$

Comment: The same method can be applied to gaseous equilibrium problems, in which case partial pressures are used as table entries in place of molar concentrations.

PRACTICE EXERCISE

Sulfur trioxide decomposes at high temperature in a sealed container: $2\ SO_3(g) \rightleftharpoons 2\ SO_2(g) + O_2(g)$. Initially, the vessel is charged at 1000 K with $SO_3(g)$ at a partial pressure of 0.500 atm. At equilibrium the SO_3 partial pressure is 0.200 atm. Calculate the value of K_p at 1000 K.

Answer: 0.338

15.6 | Applications of Equilibrium Constants

We have seen that the magnitude of K indicates the extent to which a reaction will proceed. If K is very large, the reaction will tend to proceed far to the right; if K is very small (that is, much less than 1), the equilibrium mixture will contain mainly reactants. The equilibrium constant also allows us to (1) predict the direction in which a reaction mixture will proceed to achieve equilibrium and (2) calculate the concentrations of reactants and products when equilibrium has been reached.

Predicting the Direction of Reaction

For the formation of NH_3 from N_2 and H_2 (Equation 15.6), $K_c = 0.105$ at 472°C. Suppose we place a mixture of 2.00 mol of H_2, 1.00 mol of N_2, and 2.00 mol of NH_3 in a 1.00-L container at 472°C. How will the mixture react to reach equilibrium? Will N_2 and H_2 react to form more NH_3, or will NH_3 decompose to form N_2 and H_2?

To answer this question, we can substitute the starting concentrations of N_2, H_2, and NH_3 into the equilibrium-constant expression and compare its value to the equilibrium constant:

$$\frac{[NH_3]^2}{[N_2][H_2]^3} = \frac{(2.00)^2}{(1.00)(2.00)^3} = 0.500 \quad \text{whereas} \quad K_c = 0.105$$

To reach equilibrium, the quotient $[NH_3]^2/[N_2][H_2]^3$ will need to decrease from the starting value of 0.500 to the equilibrium value of 0.105. This change can happen only if the concentration of NH_3 decreases and the concentrations of N_2 and H_2 increase. Thus, the reaction proceeds toward equilibrium by forming N_2 and H_2 from NH_3; that is, the reaction proceeds from right to left.

The approach we have illustrated can be formalized by defining a quantity called the reaction quotient. The **reaction quotient**, Q, *is a number obtained by substituting starting reactant and product concentrations or partial pressures into an equilibrium-constant expression.* Thus, in our example, when we substituted the starting concentrations into the equilibrium-constant expression, we obtained $Q_c = 0.500$. When we substitute the starting partial pressures into the equilibrium-constant expression, we will label the reaction quotient as Q_p.

To determine the direction in which the reaction will proceed to achieve equilibrium, we compare the values of Q_c and K_c or Q_p and K_p. Three possible situations arise:

- $Q = K$: The reaction quotient will equal the equilibrium constant only if the system is already at equilibrium.
- $Q > K$: The concentration of products is too large and that of reactants too small. Thus, substances on the right side of the chemical equation will react to form substances on the left; the reaction moves from right to left in approaching equilibrium.
- $Q < K$: The concentration of products is too small and that of reactants too large. Thus, the reaction will achieve equilibrium by forming more products; it moves from left to right.

These relationships are summarized in Figure 15.9 ◀.

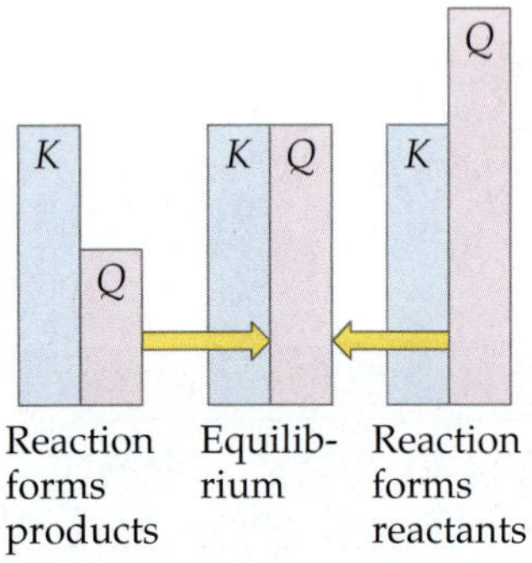

▲ **Figure 15.9 Predicting the direction of a reaction by comparing *Q* and *K*.** The relative magnitudes of the reaction quotient Q and the equilibrium constant K indicate how the reaction mixture changes as it moves toward equilibrium. If Q is smaller than K, the reaction proceeds from left to right until $Q = K$. When $Q = K$, the reaction is at equilibrium and has no tendency to change. If Q is larger than K, the reaction proceeds from right to left until $Q = K$.

SAMPLE EXERCISE 15.10 | Predicting the Direction of Approach to Equilibrium

At 448°C the equilibrium constant K_c for the reaction

$$H_2(g) + I_2(g) \rightleftharpoons 2\,HI(g)$$

is 50.5. Predict in which direction the reaction will proceed to reach equilibrium at 448°C if we start with 2.0×10^{-2} mol of HI, 1.0×10^{-2} mol of H_2, and 3.0×10^{-2} mol of I_2 in a 2.00-L container.

Solution

Analyze: We are given a volume and initial molar amounts of the species in a reaction and asked to determine in which direction the reaction must proceed to achieve equilibrium.

Plan: We can determine the starting concentration of each species in the reaction mixture. We can then substitute the starting concentrations into the equilibrium-constant expression to calculate the reaction quotient, Q_c. Comparing the magnitudes of the equilibrium constant, which is given, and the reaction quotient will tell us in which direction the reaction will proceed.

Solve: The initial concentrations are

$$[HI] = 2.0 \times 10^{-2}\ \text{mol}/2.00\ \text{L} = 1.0 \times 10^{-2}\ M$$

$$[H_2] = 1.0 \times 10^{-2}\ \text{mol}/2.00\ \text{L} = 5.0 \times 10^{-3}\ M$$

$$[I_2] = 3.0 \times 10^{-2}\ \text{mol}/2.00\ \text{L} = 1.5 \times 10^{-2}\ M$$

The reaction quotient is therefore

$$Q_c = \frac{[HI]^2}{[H_2][I_2]} = \frac{(1.0 \times 10^{-2})^2}{(5.0 \times 10^{-3})(1.5 \times 10^{-2})} = 1.3$$

Because $Q_c < K_c$, the concentration of HI must increase and the concentrations of H_2 and I_2 must decrease to reach equilibrium; the reaction will proceed from left to right as it moves toward equilibrium.

PRACTICE EXERCISE

At 1000 K the value of K_p for the reaction $2\ SO_3(g) \rightleftharpoons 2\ SO_2(g) + O_2(g)$ is 0.338. Calculate the value for Q_p, and predict the direction in which the reaction will proceed toward equilibrium if the initial partial pressures are $P_{SO_3} = 0.16$ atm; $P_{SO_2} = 0.41$ atm; $P_{O_2} = 2.5$ atm.

Answer: $Q_p = 16$; $Q_p > K_p$, and so the reaction will proceed from right to left, forming more SO_3.

Calculating Equilibrium Concentrations

Chemists frequently need to calculate the amounts of reactants and products present at equilibrium. Our approach in solving problems of this type is similar to the one we used for evaluating equilibrium constants: We tabulate the initial concentrations or partial pressures, the changes therein, and the final equilibrium concentrations or partial pressures. Usually we end up using the equilibrium-constant expression to derive an equation that must be solved for an unknown quantity, as demonstrated in Sample Exercise 15.11.

SAMPLE EXERCISE 15.11 | Calculating Equilibrium Concentrations

For the Haber process, $N_2(g) + 3\ H_2(g) \rightleftharpoons 2\ NH_3(g)$, $K_p = 1.45 \times 10^{-5}$ at 500°C. In an equilibrium mixture of the three gases at 500°C, the partial pressure of H_2 is 0.928 atm and that of N_2 is 0.432 atm. What is the partial pressure of NH_3 in this equilibrium mixture?

Solution

Analyze: We are given an equilibrium constant, K_p, and the equilibrium partial pressures of two of the three substances in the equation (N_2 and H_2), and we are asked to calculate the equilibrium partial pressure for the third substance (NH_3).

Plan: We can set K_p equal to the equilibrium-constant expression and substitute in the partial pressures that are known. Then we can solve for the only unknown in the equation.

Solve: We tabulate the equilibrium pressures as follows:

	$N_2(g)$	+ $3\ H_2(g)$	$\rightleftharpoons 2\ NH_3(g)$
Equilibrium pressure (atm)	0.432	0.928	x

Because we do not know the equilibrium pressure of NH_3, we represent it with a variable, x. At equilibrium the pressures must satisfy the equilibrium-constant expression:

$$K_p = \frac{(P_{NH_3})^2}{P_{N_2}(P_{H_2})^3} = \frac{x^2}{(0.432)(0.928)^3} = 1.45 \times 10^{-5}$$

We now rearrange the equation to solve for x:

$$x^2 = (1.45 \times 10^{-5})(0.432)(0.928)^3 = 5.01 \times 10^{-6}$$

$$x = \sqrt{5.01 \times 10^{-6}} = 2.24 \times 10^{-3}\ \text{atm} = P_{NH_3}$$

Comment: We can always check our answer by using it to recalculate the value of the equilibrium constant:

$$K_p = \frac{(2.24 \times 10^{-3})^2}{(0.432)(0.928)^3} = 1.45 \times 10^{-5}$$

PRACTICE EXERCISE

At 500 K the reaction $PCl_5(g) \rightleftharpoons PCl_3(g) + Cl_2(g)$ has $K_p = 0.497$. In an equilibrium mixture at 500 K, the partial pressure of PCl_5 is 0.860 atm and that of PCl_3 is 0.350 atm. What is the partial pressure of Cl_2 in the equilibrium mixture?
Answer: 1.22 atm

In many situations we will know the value of the equilibrium constant and the initial amounts of all species. We must then solve for the equilibrium amounts. Solving this type of problem usually entails treating the change in concentration as a variable. The stoichiometry of the reaction gives us the relationship between the changes in the amounts of all the reactants and products, as illustrated in Sample Exercise 15.12.

SAMPLE EXERCISE 15.12 | Calculating Equilibrium Concentrations from Initial Concentrations

A 1.000-L flask is filled with 1.000 mol of H_2 and 2.000 mol of I_2 at 448°C. The value of the equilibrium constant K_c for the reaction

$$H_2(g) + I_2(g) \rightleftharpoons 2\,HI(g)$$

at 448°C is 50.5. What are the equilibrium concentrations of H_2, I_2, and HI in moles per liter?

Solution

Analyze: We are given the volume of a container, an equilibrium constant, and starting amounts of reactants in the container and are asked to calculate the equilibrium concentrations of all species.

Plan: In this case we are not given any of the equilibrium concentrations. We must develop some relationships that relate the initial concentrations to those at equilibrium. The procedure is similar in many regards to that outlined in Sample Exercise 15.9, where we calculated an equilibrium constant using initial concentrations.

Solve: First, we note the initial concentrations of H_2 and I_2 in the 1.000-L flask:

$$[H_2] = 1.000\ M \text{ and } [I_2] = 2.000\ M$$

Second, we construct a table in which we tabulate the initial concentrations:

	$H_2(g)$ +	$I_2(g)$ $\rightleftharpoons$	$2\,HI(g)$
Initial	1.000 *M*	2.000 *M*	0 *M*
Change			
Equilibrium			

Third, we use the stoichiometry of the reaction to determine the changes in concentration that occur as the reaction proceeds to equilibrium. The concentrations of H_2 and I_2 will decrease as equilibrium is established, and that of HI will increase. Let's represent the change in concentration of H_2 by the variable x. The balanced chemical equation tells us the relationship between the changes in the concentrations of the three gases:

For each x mol of H_2 that reacts, x mol of I_2 are consumed and $2x$ mol of HI are produced:

	$H_2(g)$ +	$I_2(g)$ $\rightleftharpoons$	$2\,HI(g)$
Initial	1.000 *M*	2.000 *M*	0 *M*
Change	$-x$	$-x$	$+2x$
Equilibrium			

Fourth, we use the initial concentrations and the changes in concentrations, as dictated by stoichiometry, to express the equilibrium concentrations. With all our entries, our table now looks like this:

	$H_2(g)$ +	$I_2(g)$ $\rightleftharpoons$	$2\,HI(g)$
Initial	1.000 *M*	2.000 *M*	0 *M*
Change	$-x$	$-x$	$+2x$
Equilibrium	$(1.000 - x)\ M$	$(2.000 - x)\ M$	$2x\ M$

Fifth, we substitute the equilibrium concentrations into the equilibrium-constant expression and solve for the single unknown, x:

$$K_c = \frac{[HI]^2}{[H_2][I_2]} = \frac{(2x)^2}{(1.000 - x)(2.000 - x)} = 50.5$$

If you have an equation-solving calculator, you can solve this equation directly for x. If not, expand this expression to obtain a quadratic equation in x:

$$4x^2 = 50.5(x^2 - 3.000x + 2.000)$$
$$46.5x^2 - 151.5x + 101.0 = 0$$

Solving the quadratic equation (Appendix A.3) leads to two solutions for x:

$$x = \frac{-(-151.5) \pm \sqrt{(-151.5)^2 - 4(46.5)(101.0)}}{2(46.5)} = 2.323 \text{ or } 0.935$$

When we substitute $x = 2.323$ into the expressions for the equilibrium concentrations, we find *negative* concentrations of H_2 and I_2. Because a negative concentration is not chemically meaningful, we reject this solution. We then use $x = 0.935$ to find the equilibrium concentrations:

$$[H_2] = 1.000 - x = 0.065\ M$$
$$[I_2] = 2.000 - x = 1.065\ M$$
$$[HI] = 2x = 1.870\ M$$

Check: We can check our solution by putting these numbers into the equilibrium-constant expression:

$$K_c = \frac{[HI]^2}{[H_2][I_2]} = \frac{(1.870)^2}{(0.065)(1.065)} = 51$$

Comment: Whenever you use a quadratic equation to solve an equilibrium problem, one of the solutions will not be chemically meaningful and should be rejected.

PRACTICE EXERCISE

For the equilibrium $PCl_5(g) \rightleftharpoons PCl_3(g) + Cl_2(g)$, the equilibrium constant K_p has the value 0.497 at 500 K. A gas cylinder at 500 K is charged with $PCl_5(g)$ at an initial pressure of 1.66 atm. What are the equilibrium pressures of PCl_5, PCl_3, and Cl_2 at this temperature?

Answer: $P_{PCl_5} = 0.967$ atm; $P_{PCl_3} = P_{Cl_2} = 0.693$ atm

15.7 | Le Châtelier's Principle

In developing his process for making ammonia from N_2 and H_2, Haber sought the factors that might be varied to increase the yield of NH_3. Using the values of the equilibrium constant at various temperatures, he calculated the equilibrium amounts of NH_3 formed under a variety of conditions. Some of his results are shown in Figure 15.10 ▶. Notice that the percent of NH_3 present at equilibrium decreases with increasing temperature and increases with increasing pressure. We can understand these effects in terms of a principle first put forward by Henri-Louis Le Châtelier* (1850–1936), a French industrial chemist. **Le Châtelier's principle** can be stated as follows: *If a system at equilibrium is disturbed by a change in temperature, pressure, or the concentration of one of the components, the system will shift its equilibrium position so as to counteract the effect of the disturbance.*

In this section we will use Le Châtelier's principle to make qualitative predictions about how a system at equilibrium responds to various changes in external conditions. We will consider three

▼ **Figure 15.10 Effect of temperature and pressure on the percentage of NH_3 in an equilibrium mixture of N_2, H_2, and NH_3.** Each mixture was produced by starting with a 3 : 1 molar mixture of H_2 and N_2. The yield of NH_3 is greatest at the lowest temperature and at the highest pressure.

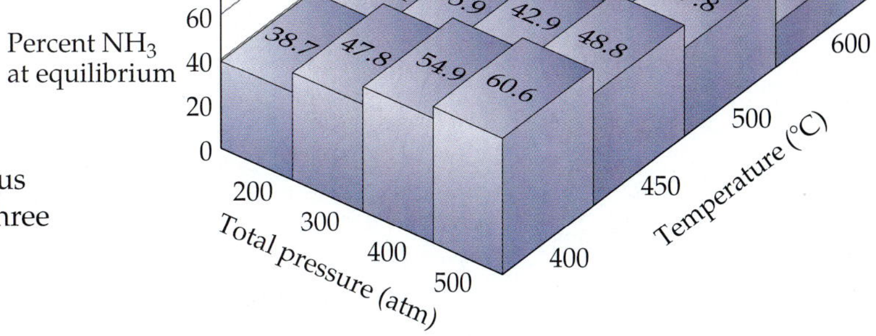

* Pronounced "le-SHOT-lee-ay."

ways that a chemical equilibrium can be disturbed: (1) adding or removing a reactant or product, (2) changing the pressure by changing the volume, and (3) changing the temperature.

Change in Reactant or Product Concentrations

A system at equilibrium is in a dynamic state; the forward and reverse processes are occurring at equal rates, and the system is in a state of balance. Altering the conditions of the system may disturb the state of balance. If this occurs, the equilibrium shifts until a new state of balance is attained. Le Châtelier's principle states that the shift will be in the direction that minimizes or reduces the effect of the change. Therefore, *if a chemical system is at equilibrium and we add a substance (either a reactant or a product), the reaction will shift so as to reestablish equilibrium by consuming part of the added substance. Conversely, removing a substance will cause the reaction to move in the direction that forms more of that substance.*

As an example, consider an equilibrium mixture of N_2, H_2, and NH_3:

$$N_2(g) + 3\,H_2(g) \rightleftharpoons 2\,NH_3(g)$$

Adding H_2 would cause the system to shift so as to reduce the newly increased concentration of H_2. This change can occur only by consuming H_2 and simultaneously consuming N_2 to form more NH_3. This situation is illustrated in Figure 15.11 ▼. Adding more N_2 to the equilibrium mixture would likewise cause the direction of the reaction to shift toward forming more NH_3. Removing NH_3 would also cause a shift toward producing more NH_3, whereas *adding* NH_3 to the system at equilibrium would cause the concentrations to shift in the direction that reduces the newly increased NH_3 concentration; that is, some of the added ammonia would decompose to form N_2 and H_2.

In the Haber reaction, therefore, removing NH_3 from an equilibrium mixture of N_2, H_2, and NH_3 causes the reaction to shift from left to right to form more NH_3. If the NH_3 can be removed continuously, the yield of NH_3 can be increased dramatically. In the industrial production of ammonia, the NH_3 is continuously removed by selectively liquefying it; the boiling point of NH_3 (−33°C) is much higher than that of N_2 (−196°C) and H_2 (−253°C). The liquid NH_3 is removed, and the N_2 and H_2 are recycled to form more NH_3, as diagrammed in Figure 15.12 ▶. By continuously removing the product, the reaction is driven essentially to completion.

▶ **Figure 15.11 Effect of adding H_2 to an equilibrium mixture of N_2, H_2, and NH_3.** When H_2 is added, a portion of the H_2 reacts with N_2 to form NH_3, thereby establishing a new equilibrium position that has the same equilibrium constant. The results shown are in accordance with Le Châtelier's principle.

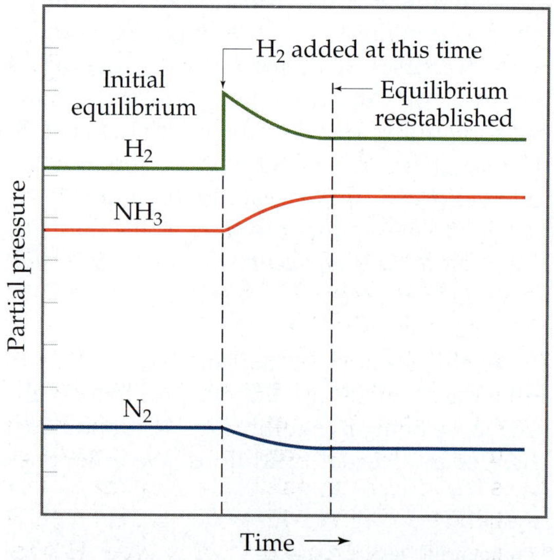

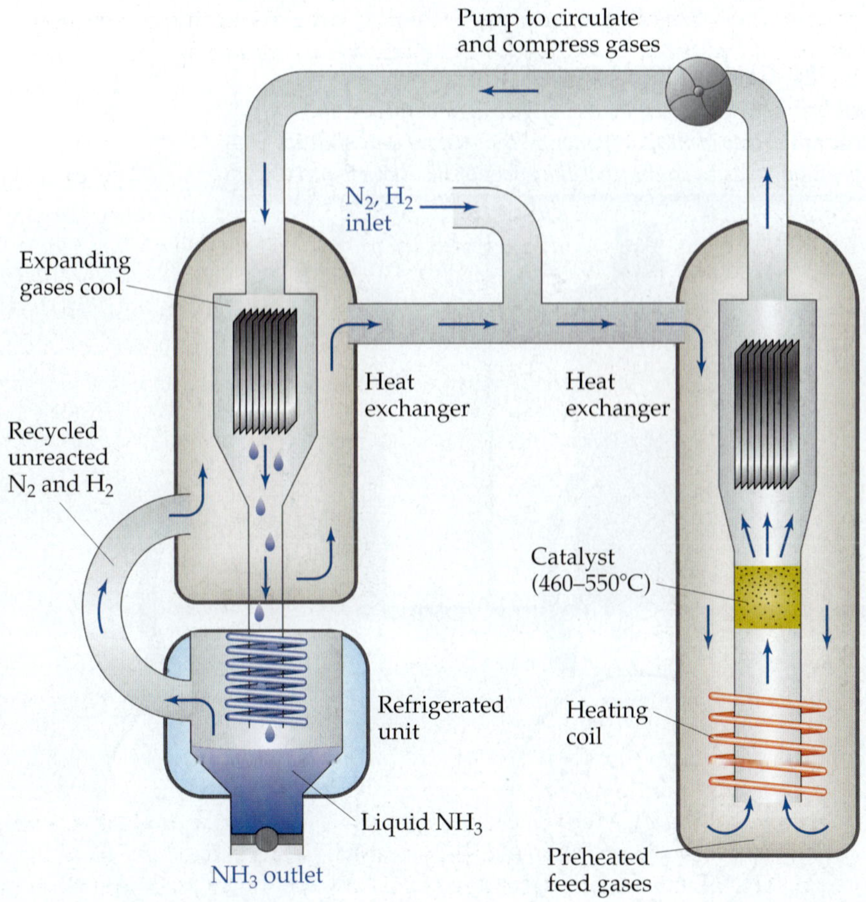

▲ **Figure 15.12 Schematic diagram summarizing the industrial production of ammonia.** Incoming N_2 and H_2 gases are heated to approximately 500°C and passed over a catalyst. The resultant gas mixture is allowed to expand and cool, causing NH_3 to liquefy. Unreacted N_2 and H_2 gases are recycled.

GIVE IT SOME THOUGHT

What happens to the equilibrium $2\ NO(g) + O_2(g) \rightleftharpoons 2\ NO_2(g)$ if (a) O_2 is added to the system, (b) NO is removed?

MOVIE
Nitrogen Dioxide and Dinitrogen Tetroxide

Effects of Volume and Pressure Changes

If a system is at equilibrium and its volume is decreased, thereby increasing its total pressure, Le Châtelier's principle indicates that the system will respond by shifting its equilibrium position to reduce the pressure. A system can reduce its pressure by reducing the total number of gas molecules (fewer molecules of gas exert a lower pressure). Thus, at constant temperature, *reducing the volume of a gaseous equilibrium mixture causes the system to shift in the direction that reduces the number of moles of gas.* Conversely, increasing the volume causes a shift in the direction that produces more gas molecules.

ANIMATIONS
NO_2—N_2O_4 Equilibrium, Temperature Dependence of Equilibrium

For example, let's again consider the equilibrium $N_2O_4(g) \rightleftharpoons 2\ NO_2(g)$. What happens if the total pressure of an equilibrium mixture is increased by decreasing the volume as shown in the sequential photos in Figure 15.13 ▶? According to Le Châtelier's principle, we expect the equilibrium to shift to the side that reduces the total number of moles of gas, which is the reactant side in this case. We therefore expect the equilibrium to shift to the left, so that NO_2 is converted into N_2O_4 as equilibrium is reestablished. In Figure 15.13(a) and Figure 15.13(b), compressing the gas mixture initially causes the color to darken as the concentration of NO_2 increases. The color then fades as equilibrium is reestablished, Figure 15.13(c). The color fades because the pressure increase causes the equilibrium to shift in favor of colorless N_2O_4.

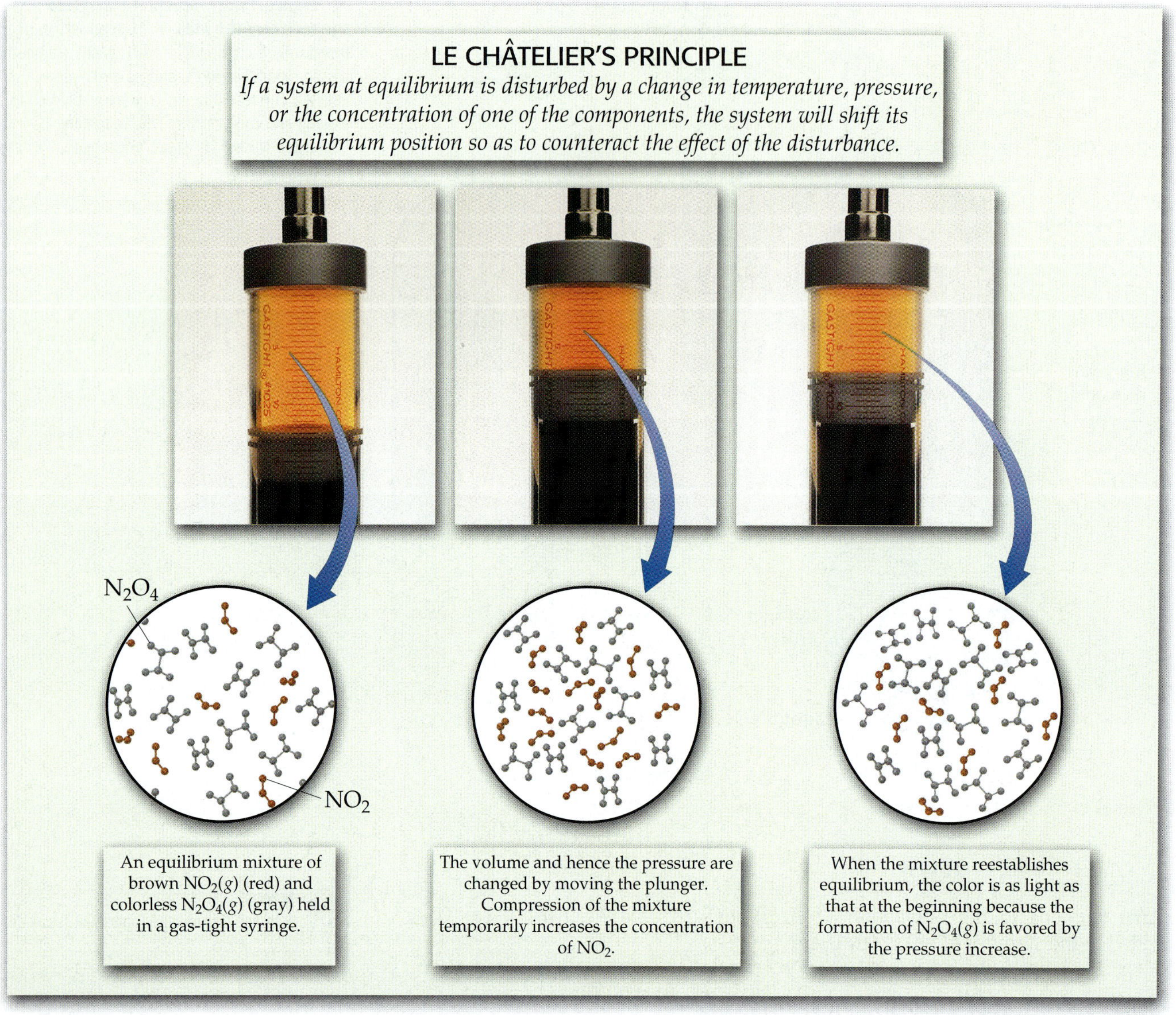

▲ **Figure 15.13 Effect of pressure on an equilibrium.**

GIVE IT SOME THOUGHT

What happens to the equilibrium $2\ SO_2(g) + O_2(g) \rightleftharpoons 2\ SO_3(g)$ if the volume of the system is increased?

For the reaction $N_2(g) + 3\ H_2(g) \rightleftharpoons 2\ NH_3(g)$, there are four molecules of reactant consumed for every two molecules of product produced. Consequently, an increase in pressure (decrease in volume) causes a shift toward the side with fewer gas molecules, which leads to the formation of more NH_3, as indicated in Figure 15.10. In the case of the reaction $H_2(g) + I_2(g) \rightleftharpoons 2\ HI(g)$, the number of molecules of gaseous products (two) equals the number of molecules of gaseous reactants; therefore, changing the pressure will not influence the position of the equilibrium.

Keep in mind that pressure-volume changes do *not* change the value of K as long as the temperature remains constant. Rather, they change the partial pressures of the gaseous substances. In Sample Exercise 15.8 we calculated K_p for an equilibrium mixture at 472°C that contained 7.38 atm H_2, 2.46 atm N_2, and 0.166 atm NH_3. The value of K_p is 2.79×10^{-5}. Consider what happens when we suddenly reduce the volume of the system by one-half. If there were no shift in equilibrium, this volume change would cause the partial pressures of all substances to double, giving $P_{H_2} = 14.76$ atm, $P_{N_2} = 4.92$ atm, and $P_{NH_3} = 0.332$ atm. The reaction quotient would then no longer equal the equilibrium constant.

$$Q_p = \frac{(P_{NH_3})^2}{P_{N_2}(P_{H_2})^3} = \frac{(0.332)^2}{(4.92)(14.76)^3} = 6.97 \times 10^{-6} \neq K_p$$

ACTIVITY
Le Châtelier's Principle

Because $Q_p < K_p$, the system is no longer at equilibrium. Equilibrium will be reestablished by increasing P_{NH_3} and decreasing P_{N_2} and P_{H_2} until $Q_p = K_p = 2.79 \times 10^{-5}$. Therefore, the equilibrium shifts to the right as Le Châtelier's principle predicts.

It is possible to change the total pressure of the system without changing its volume. For example, pressure increases if additional amounts of any of the reacting components are added to the system. We have already seen how to deal with a change in concentration of a reactant or product. The total pressure within the reaction vessel might also be increased by adding a gas that is not involved in the equilibrium. For example, argon might be added to the ammonia equilibrium system. The argon would not alter the partial pressures of any of the reacting components and therefore would not cause a shift in equilibrium.

Effect of Temperature Changes

Changes in concentrations or partial pressures cause shifts in equilibrium without changing the value of the equilibrium constant. In contrast, almost every equilibrium constant changes in value as the temperature changes. For example, consider the equilibrium established when cobalt(II) chloride ($CoCl_2$) is dissolved in hydrochloric acid, HCl(*aq*):

$$\underset{\text{Pale pink}}{Co(H_2O)_6^{2+}(aq)} + 4\,Cl^-(aq) \rightleftharpoons \underset{\text{Deep blue}}{CoCl_4^{2-}(aq)} + 6\,H_2O(l) \qquad \Delta H > 0 \qquad [15.22]$$

The formation of $CoCl_4^{2-}$ from $Co(H_2O)_6^{2+}$ is an endothermic process. We will discuss the significance of this enthalpy change shortly. Because $Co(H_2O)_6^{2+}$ is pink and $CoCl_4^{2-}$ is blue, the position of this equilibrium is readily apparent from the color of the solution. Figure 15.14(a) ▶ shows a room-temperature solution of $CoCl_2$ in HCl(*aq*). Both $Co(H_2O)_6^{2+}$ and $CoCl_4^{2-}$ are present in significant amounts in the solution; the violet color results from the presence of both the pink and blue ions. When the solution is heated [Figure 15.14(b)], it becomes intensely blue in color, indicating that the equilibrium has shifted to form more $CoCl_4^{2-}$. Cooling the solution, as in Figure 15.14(c), leads to a pink solution, indicating that the equilibrium has shifted to produce more $Co(H_2O)_6^{2+}$. How can we explain the dependence of this equilibrium on temperature?

We can deduce the rules for the temperature dependence of the equilibrium constant by applying Le Châtelier's principle. A simple way to do this is to treat heat as if it were a chemical reagent. In an *endothermic* reaction we can consider heat as a *reactant*, whereas in an *exothermic* reaction we can consider heat as a *product*.

Endothermic: Reactants + *heat* $\rightleftharpoons$ products

Exothermic: Reactants $\rightleftharpoons$ products + *heat*

When the temperature of a system at equilibrium is increased, it is as if we have added a reactant to an endothermic reaction or a product to an exothermic reaction. The equilibrium shifts in the direction that consumes the excess reactant (or product), namely heat.

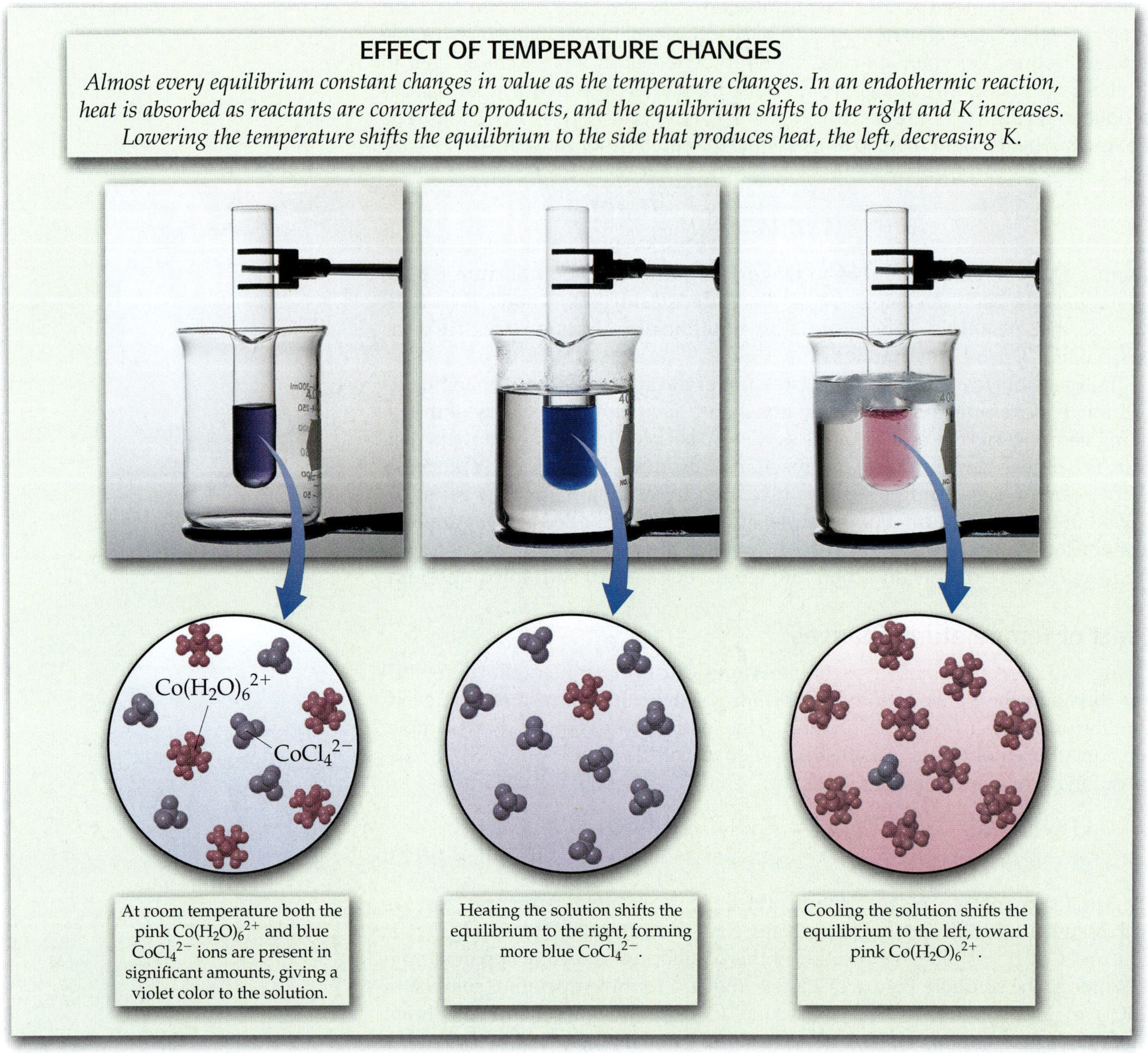

▲ **Figure 15.14 Temperature and equilibrium. (The reaction shown is $Co(H_2O)_6^{2+}(aq) + 4\ Cl^-(aq) \rightleftharpoons$ $4\ Cl^-(aq) \rightleftharpoons CoCl_4^{2-}(aq) + 6\ H_2O(l)$.)**

GIVE IT SOME THOUGHT

Use Le Châtelier's principle to explain why the equilibrium vapor pressure of a liquid increases with increasing temperature.

In an endothermic reaction, such as Equation 15.22, heat is absorbed as reactants are converted to products; thus, increasing the temperature causes the equilibrium to shift to the right, in the direction of products, and K increases. For Equation 15.22, increasing the temperature leads to the formation of more $CoCl_4^{2-}$, as observed in Figure 15.14(b).

In an exothermic reaction the opposite occurs. Heat is absorbed as products are converted to reactants; therefore the equilibrium shifts to the left and K decreases. We can summarize these results as follows:

Endothermic: Increasing T results in an increase in K.

Exothermic: Increasing T results in a decrease in K.

Cooling a reaction has the opposite effect. As we lower the temperature, the equilibrium shifts to the side that produces heat. Thus, cooling an endothermic reaction shifts the equilibrium to the left, decreasing K. We observed this effect in Figure 15.14(c). Cooling an exothermic reaction shifts the equilibrium to the right, increasing K.

SAMPLE EXERCISE 15.13 | Using Le Châtelier's Principle to Predict Shifts in Equilibrium

Consider the equilibrium

$$N_2O_4(g) \rightleftharpoons 2\,NO_2(g) \qquad \Delta H^\circ = 58.0\text{ kJ}$$

In which direction will the equilibrium shift when **(a)** N_2O_4 is added, **(b)** NO_2 is removed, **(c)** the total pressure is increased by addition of $N_2(g)$, **(d)** the volume is increased, **(e)** the temperature is decreased?

Solution

Analyze: We are given a series of changes to be made to a system at equilibrium and are asked to predict what effect each change will have on the position of the equilibrium.

Plan: Le Châtelier's principle can be used to determine the effects of each of these changes.

Solve: **(a)** The system will adjust to decrease the concentration of the added N_2O_4, so the equilibrium shifts to the right, in the direction of products.

(b) The system will adjust to the removal of NO_2 by shifting to the side that produces more NO_2; thus, the equilibrium shifts to the right.

(c) Adding N_2 will increase the total pressure of the system, but N_2 is not involved in the reaction. The partial pressures of NO_2 and N_2O_4 are therefore unchanged, and there is no shift in the position of the equilibrium.

(d) If the volume is increased, the system will shift in the direction that occupies a larger volume (more gas molecules); thus, the equilibrium shifts to the right. (This is the opposite of the effect observed in Figure 15.13, where the volume was decreased.)

(e) The reaction is endothermic, so we can imagine heat as a reagent on the reactant side of the equation. Decreasing the temperature will shift the equilibrium in the direction that produces heat, so the equilibrium shifts to the left, toward the formation of more N_2O_4. Note that only this last change also affects the value of the equilibrium constant, K.

PRACTICE EXERCISE

For the reaction

$$PCl_5(g) \rightleftharpoons PCl_3(g) + Cl_2(g) \qquad \Delta H^\circ = 87.9\text{ kJ}$$

in which direction will the equilibrium shift when **(a)** $Cl_2(g)$ is removed, **(b)** the temperature is decreased, **(c)** the volume of the reaction system is increased, **(d)** $PCl_3(g)$ is added?

Answers: **(a)** right, **(b)** left, **(c)** right, **(d)** left

SAMPLE EXERCISE 15.14 | Predicting the Effect of Temperature on *K*

(a) Using the standard heat of formation data in Appendix C, determine the standard enthalpy change for the reaction

$$N_2(g) + 3\,H_2(g) \rightleftharpoons 2\,NH_3(g)$$

(b) Determine how the equilibrium constant for this reaction should change with temperature.

Solution

Analyze: We are asked to determine the standard enthalpy change of a reaction and how the equilibrium constant for the reaction varies with temperature.

Plan: **(a)** We can use standard enthalpies of formation to calculate $\Delta H°$ for the reaction. **(b)** We can then use Le Châtelier's principle to determine what effect temperature will have on the equilibrium constant.

Solve: **(a)** Recall that the standard enthalpy change for a reaction is given by the sum of the standard molar enthalpies of formation of the products, each multiplied by its coefficient in the balanced chemical equation, less the same quantities for the reactants. At 25°C, $\Delta H_f°$ for $NH_3(g)$ is −46.19 kJ/mol. The $\Delta H_f°$ values for $H_2(g)$ and $N_2(g)$ are zero by definition because the enthalpies of formation of the elements in their normal states at 25°C are defined as zero (Section 5.7). Because 2 mol of NH_3 is formed, the total enthalpy change is

$$(2\ \text{mol})(-46.19\ \text{kJ/mol}) - 0 = -92.38\ \text{kJ}$$

(b) Because the reaction in the forward direction is exothermic, we can consider heat a product of the reaction. An increase in temperature causes the reaction to shift in the direction of less NH_3 and more N_2 and H_2. This effect is seen in the values for K_p presented in Table 15.2 ◀. Notice that K_p changes markedly with changes in temperature and that it is larger at lower temperatures.

Comment: The fact that K_p for the formation of NH_3 from N_2 and H_2 decreases with increasing temperature is a matter of great practical importance. To form NH_3 at a reasonable rate requires higher temperatures. At higher temperatures, however, the equilibrium constant is smaller, and so the percentage conversion to NH_3 is smaller. To compensate for this, higher pressures are needed because high pressure favors NH_3 formation.

TABLE 15.2 Variation in K_p for the Equilibrium $N_2 + 3\ H_2 \rightleftharpoons 2\ NH_3$ as a Function of Temperature

Temperature (°C)	K_p
300	4.34×10^{-3}
400	1.64×10^{-4}
450	4.51×10^{-5}
500	1.45×10^{-5}
550	5.38×10^{-6}
600	2.25×10^{-6}

PRACTICE EXERCISE

Using the thermodynamic data in Appendix C, determine the enthalpy change for the reaction

$$2\ POCl_3(g) \rightleftharpoons 2\ PCl_3(g) + O_2(g)$$

Use this result to determine how the equilibrium constant for the reaction should change with temperature.

Answer: $\Delta H° = 508$ kJ; the equilibrium constant will increase with increasing temperature

The Effect of Catalysts

What happens if we add a catalyst to a chemical system that is at equilibrium? As shown in Figure 15.15 ◀, a catalyst lowers the activation barrier between the reactants and products. The activation energy of the forward reaction is lowered to the same extent as that for the reverse reaction. The catalyst thereby increases the rates of both the forward and reverse reactions. As a result, *a catalyst increases the rate at which equilibrium is achieved, but it does not change the composition of the equilibrium mixture.* The value of the equilibrium constant for a reaction is not affected by the presence of a catalyst.

The rate at which a reaction approaches equilibrium is an important practical consideration. As an example, let's again consider the synthesis of ammonia from N_2 and H_2. In designing a process for ammonia synthesis, Haber had to deal with a rapid decrease in the equilibrium constant with increasing temperature, as shown in Table 15.2. At temperatures sufficiently high to give a satisfactory reaction rate, the amount of ammonia formed was too small. The solution to this dilemma was to develop a catalyst that would produce a reasonably rapid approach to equilibrium at a sufficiently low temperature, so that the equilibrium constant was still reasonably large. The development of a suitable catalyst thus became the focus of Haber's research efforts.

After trying different substances to see which would be most effective, Haber finally settled on iron mixed with metal oxides. Variants of the original catalyst formulations are still used. These catalysts make it possible to obtain a reasonably rapid approach to equilibrium at temperatures around 400°C to 500°C and with gas

▼ **Figure 15.15 Effect of a catalyst on equilibrium.** At equilibrium for the hypothetical reaction A ⇌ B, the forward reaction rate, r_f, equals the reverse reaction rate, r_r. The violet curve represents the path over the transition state in the absence of a catalyst. A catalyst lowers the energy of the transition state, as shown by the green curve. Thus, the activation energy is lowered for both the forward and the reverse reactions. As a result, the rates of forward and reverse reactions in the catalyzed reaction are increased.

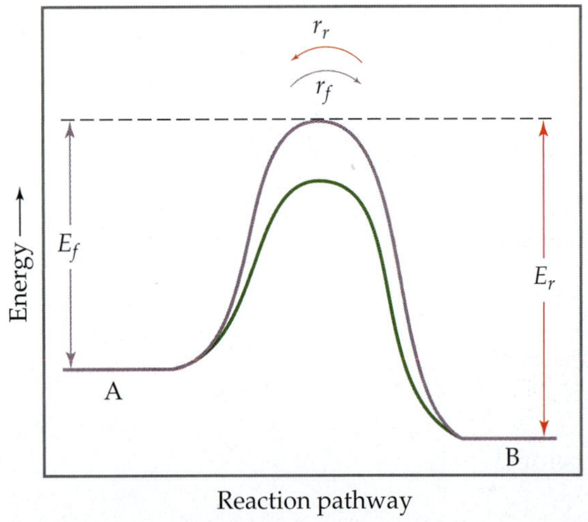

pressures of 200 to 600 atm. The high pressures are needed to obtain a satisfactory degree of conversion at equilibrium. You can see from Figure 15.10 that if an improved catalyst could be found—one that would lead to sufficiently rapid reaction at temperatures lower than 400°C to 500°C—it would be possible to obtain the same degree of equilibrium conversion at much lower pressures. This would result in great savings in the cost of equipment for ammonia synthesis. In view of the growing need for nitrogen as fertilizer, the fixation of nitrogen is a process of ever-increasing importance.

GIVE IT SOME THOUGHT

Does the addition of a catalyst have any effect on the position of an equilibrium?

SAMPLE INTEGRATIVE EXERCISE | Putting Concepts Together

At temperatures near 800°C, steam passed over hot coke (a form of carbon obtained from coal) reacts to form CO and H_2:

$$C(s) + H_2O(g) \rightleftharpoons CO(g) + H_2(g)$$

The mixture of gases that results is an important industrial fuel called *water gas*. **(a)** At 800°C the equilibrium constant for this reaction is $K_p = 14.1$. What are the equilibrium partial pressures of H_2O, CO, and H_2 in the equilibrium mixture at this temperature if we start with solid carbon and 0.100 mol of H_2O in a 1.00-L vessel? **(b)** What is the minimum amount of carbon required to achieve equilibrium under these conditions? **(c)** What is the total pressure in the vessel at equilibrium? **(d)** At 25°C the value of K_p for this reaction is 1.7×10^{-21}. Is the reaction exothermic or endothermic? **(e)** To produce the maximum amount of CO and H_2 at equilibrium, should the pressure of the system be increased or decreased?

Solution (a) To determine the equilibrium partial pressures, we use the ideal gas equation, first determining the starting partial pressure of hydrogen.

$$P_{H_2O} = \frac{n_{H_2O}RT}{V} = \frac{(0.100 \text{ mol})(0.0821 \text{ L-atm/mol-K})(1073 \text{ K})}{1.00 \text{ L}} = 8.81 \text{ atm}$$

We then construct a table of starting partial pressures and their changes as equilibrium is achieved:

	$C(s)$ +	$H_2O(g)$ $\rightleftharpoons$	$CO(g)$ +	$H_2(g)$
Initial		8.81 atm	0 atm	0 atm
Change		$-x$	$+x$	$+x$
Equilibrium		$8.81 - x$ atm	x atm	x atm

There are no entries in the table under C(s) because the reactant, being a solid, does not appear in the equilibrium-constant expression. Substituting the equilibrium partial pressures of the other species into the equilibrium-constant expression for the reaction gives

$$K_p = \frac{P_{CO}P_{H_2}}{P_{H_2O}} = \frac{(x)(x)}{(8.81 - x)} = 14.1$$

Multiplying through by the denominator gives a quadratic equation in x:

$$x^2 = (14.1)(8.81 - x)$$
$$x^2 + 14.1x - 124.22 = 0$$

Solving this equation for x using the quadratic formula yields $x = 6.14$ atm. Hence, the equilibrium partial pressures are $P_{CO} = x = 6.14$ atm, $P_{H_2} = x = 6.14$ atm, and $P_{H_2O} = (8.81 - x) = 2.67$ atm.

(b) Part (a) shows that $x = 6.14$ atm of H_2O must react in order for the system to achieve equilibrium. We can use the ideal-gas equation to convert this partial pressure into a mole amount.

$$n = \frac{PV}{RT} = \frac{(6.14 \text{ atm})(1.00 \text{ L})}{(0.0821 \text{ L-atm/mol-K})(1073 \text{ K})} = 0.0697 \text{ mol}$$

Thus, 0.0697 mol of H_2O and the same amount of C must react to achieve equilibrium. As a result, there must be at least 0.0697 mol of C (0.836 g C) present among the reactants at the start of the reaction.

(c) The total pressure in the vessel at equilibrium is simply the sum of the equilibrium partial pressures:

$$P_{\text{total}} = P_{H_2O} + P_{CO} + P_{H_2} = 2.67 \text{ atm} + 6.14 \text{ atm} + 6.14 \text{ atm} = 14.95 \text{ atm}$$

(d) In discussing Le Châtelier's principle, we saw that endothermic reactions exhibit an increase in K_p with increasing temperature. Because the equilibrium constant for this reaction increases as temperature increases, the reaction must be endothermic. From the enthalpies of formation given in Appendix C, we can verify our prediction by calculating the enthalpy change for the reaction, $\Delta H° = \Delta H_f°(CO) + \Delta H_f°(H_2) - \Delta H°(C) - \Delta H_f°(H_2O) = +131.3$ kJ. The positive sign for $\Delta H°$ indicates that the reaction is endothermic.

(e) According to Le Châtelier's principle, a decrease in the pressure causes a gaseous equilibrium to shift toward the side of the equation with the greater number of moles of gas. In this case there are two moles of gas on the product side and only one on the reactant side. Therefore, the pressure should be reduced to maximize the yield of the CO and H_2.

CHEMISTRY AT WORK | Controlling Nitric Oxide Emissions

The formation of NO from N_2 and O_2 provides another interesting example of the practical importance of changes in the equilibrium constant and reaction rate with temperature. The equilibrium equation and the standard enthalpy change for the reaction are

$$\tfrac{1}{2} N_2(g) + \tfrac{1}{2} O_2(g) \rightleftharpoons NO(g) \qquad \Delta H° = 90.4 \text{ kJ} \qquad [15.23]$$

The reaction is endothermic; that is, heat is absorbed when NO is formed from the elements. By applying Le Châtelier's principle, we deduce that an increase in temperature will shift the equilibrium in the direction of more NO. The equilibrium constant K_p for formation of 1 mol of NO from the elements at 300 K is only about 10^{-15}. In contrast, at a much higher temperature of about 2400 K the equilibrium constant is 10^{13} times as large, about 0.05. The manner in which K_p for Equation 15.23 varies with temperature is shown in Figure 15.16 ▶.

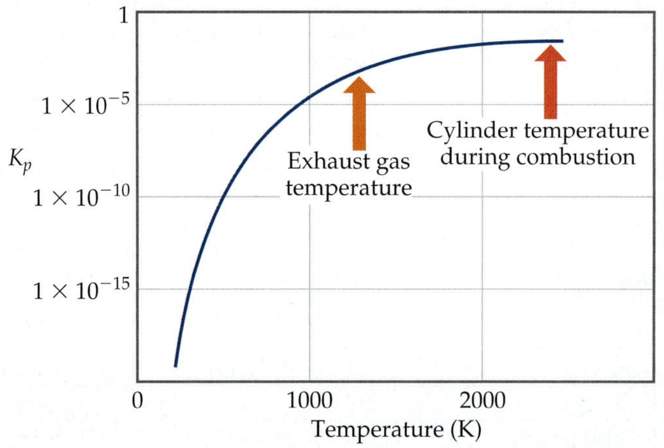

▲ **Figure 15.16 Equilibrium and temperature.** The graph shows how the equilibrium constant for the reaction $\tfrac{1}{2} N_2(g) + \tfrac{1}{2} O_2(g) \rightleftharpoons NO(g)$ varies as a function of temperature. The equilibrium constant increases with increasing temperature because the reaction is endothermic. It is necessary to use a log scale for K_p because the values vary over such a large range.

This graph helps to explain why NO is a pollution problem. In the cylinder of a modern high-compression auto engine, the temperatures during the fuel-burning part of the cycle may be on the order of 2400 K. Also, there is a fairly large excess of air in the cylinder. These conditions favor the formation of some NO. After the combustion, however, the gases are quickly cooled. As the temperature drops, the equilibrium in Equation 15.23 shifts strongly to the left (that is, in the direction of N_2 and O_2). The lower temperatures also mean that the rate of the reaction is decreased, however, so the NO formed at high temperatures is essentially "frozen" in that form as the gas cools.

The gases exhausting from the cylinder are still quite hot, perhaps 1200 K. At this temperature, as shown in Figure 15.16, the equilibrium constant for formation of NO is much smaller. However, the rate of conversion of NO to N_2 and O_2 is too slow to permit much loss of NO before the gases are cooled still further.

As discussed in the "Chemistry at Work" box in Section 14.7, one of the goals of automotive catalytic converters is to achieve the rapid conversion of NO to N_2 and O_2 at the temperature of the exhaust gas. Some catalysts for this reaction have been developed that are reasonably effective under the grueling conditions found in automotive exhaust systems. Nevertheless, scientists and engineers are continually searching for new materials that provide even more effective catalysis of the decomposition of nitrogen oxides.

SUMMARY AND KEY TERMS

Introduction and Section 15.1 A chemical reaction can achieve a state in which the forward and reverse processes are occurring at the same rate. This condition is called **chemical equilibrium**, and it results in the formation of an equilibrium mixture of the reactants and products of the reaction. The composition of an equilibrium mixture does not change with time.

Section 15.2 An equilibrium that is used throughout this chapter is the reaction of $N_2(g)$ with $H_2(g)$ to form $NH_3(g)$: $N_2(g) + 3\,H_2(g) \rightleftharpoons 2\,NH_3(g)$. This reaction is the basis of the **Haber process** for the production of ammonia. The relationship between the concentrations of the reactants and products of a system at equilibrium is given by the **law of mass action**. For a general equilibrium equation of the form $aA + bB \rightleftharpoons dD + eE$, the **equilibrium-constant expression** is written as

$$K_c = \frac{[D]^d[E]^e}{[A]^a[B]^b}$$

The equilibrium-constant expression depends only on the stoichiometry of the reaction. For a system at equilibrium at a given temperature, K_c will be a constant called the **equilibrium constant**. When the equilibrium system of interest consists of gases, it is often convenient to express the concentrations of reactants and products in terms of gas pressures:

$$K_p = \frac{(P_D)^d(P_E)^e}{(P_A)^a(P_B)^b}$$

K_c and K_p are related by the expression $K_p = K_c(RT)^{\Delta n}$.

Section 15.3 The value of the equilibrium constant changes with temperature. A large value of K_c indicates that the equilibrium mixture contains more products than reactants. A small value for the equilibrium constant means that the equilibrium lies toward the reactant side. The equilibrium-constant expression and the equilibrium constant of the reverse of a reaction are the reciprocals of those of the forward reaction. If a reaction is the sum of two or more reactions, its equilibrium constant will be the product of the equilibrium constants for the individual reactions.

Section 15.4 Equilibria for which all substances are in the same phase are called **homogeneous equilibria**; in **heterogeneous equilibria** two or more phases are present. Because the concentrations of pure solids and liquids are constant, these substances are left out of the equilibrium-constant expression for a heterogeneous equilibrium.

Section 15.5 If the concentrations of all species in an equilibrium are known, the equilibrium-constant expression can be used to calculate the value of the equilibrium constant. The changes in the concentrations of reactants and products on the way to achieving equilibrium are governed by the stoichiometry of the reaction.

Section 15.6 The **reaction quotient**, Q, is found by substituting reactant and product partial pressures or concentrations into the equilibrium-constant expression. If the system is at equilibrium, $Q = K$. If $Q \neq K$, however, the system is not at equilibrium. When $Q < K$, the reaction will move toward equilibrium by forming more products (the reaction moves from left to right); when $Q > K$, the reaction will proceed from right to left. Knowing the value of K makes it possible to calculate the equilibrium amounts of reactants and products, often by the solution of an equation in which the unknown is the change in a partial pressure or concentration.

Section 15.7 **Le Châtelier's principle** states that if a system at equilibrium is disturbed, the equilibrium will shift to minimize the disturbing influence. By this principle, if a reactant or product is added to a system at equilibrium, the equilibrium will shift to consume the added substance. The effects of removing reactants or products and of changing the pressure or volume of a reaction can be similarly deduced. For example, if the volume of the system is reduced, the equilibrium will shift in the direction that decreased the number of gas molecules. The enthalpy change for a reaction indicates how an increase in temperature affects the equilibrium: For an endothermic reaction, an increase in temperature shifts the equilibrium to the right; for an exothermic reaction, a temperature increase shifts the equilibrium to the left. Catalysts affect the speed at which equilibrium is reached but do not affect the magnitude of K.

VISUALIZING CONCEPTS

15.1 **(a)** Based on the following energy profile, predict whether $k_f > k_r$ or $k_f < k_r$. **(b)** Using Equation 15.5, predict whether the equilibrium constant for the process is greater than 1 or less than 1. [Section 15.1]

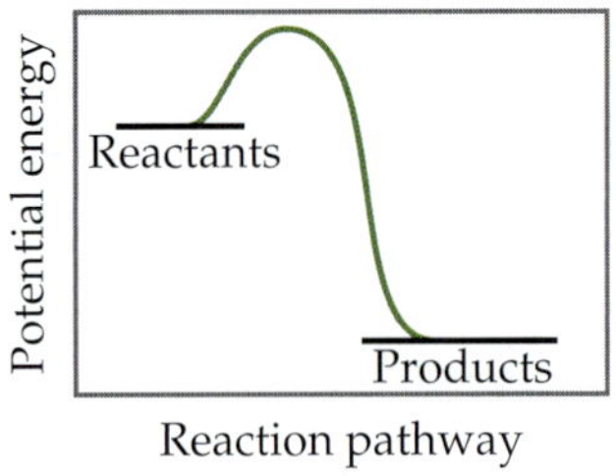

15.2 The following diagrams represent a hypothetical reaction A $\longrightarrow$ B, with A represented by red spheres and B represented by blue spheres. The sequence from left to right represents the system as time passes. Do the diagrams indicate that the system reaches an equilibrium state? Explain. [Sections 15.1 and 15.2]

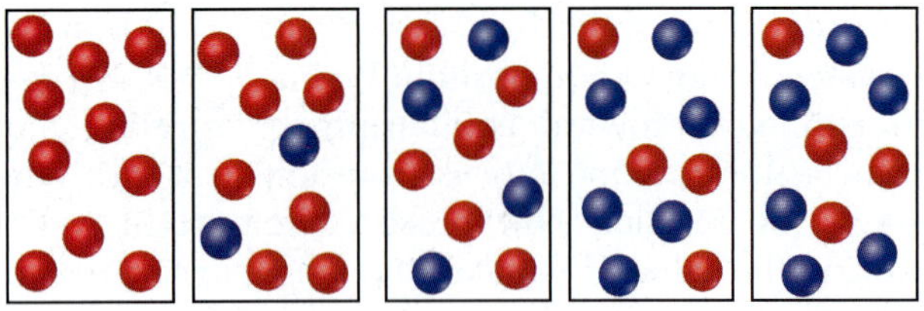

15.3 The following diagram represents an equilibrium mixture produced for a reaction of the type A + X $\rightleftharpoons$ AX. If the volume is 1 L, is K greater or smaller than 1? [Section 15.2]

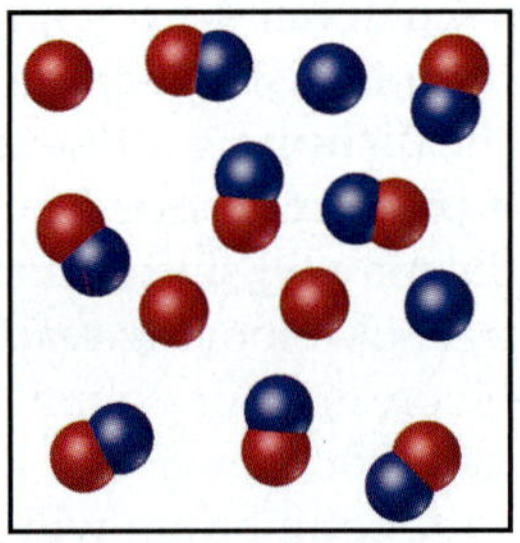

15.4 The following diagram represents a reaction shown going to completion. **(a)** Letting A = red spheres and B = blue spheres, write a balanced equation for the reaction. **(b)** Write the equilibrium-constant expression for the reaction. **(c)** Assuming that all of the molecules are in the gas phase, calculate Δn, the change in the number of gas molecules that accompanies the reaction. **(d)** How can you calculate K_p if you know K_c at a particular temperature? [Section 15.2]

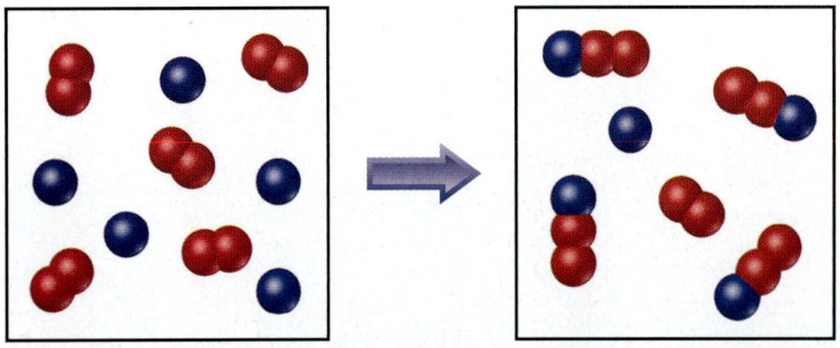

15.5 The reaction $A_2 + B_2 \rightleftharpoons 2\,AB$ has an equilibrium constant $K_c = 1.5$. The following diagrams represent reaction mixtures containing A_2 molecules (red), B_2 molecules (blue) and AB molecules. **(a)** Which reaction mixture is at equilibrium? **(b)** For those mixtures that are not at equilibrium, how will the reaction proceed to reach equilibrium? [Sections 15.5 and 15.6]

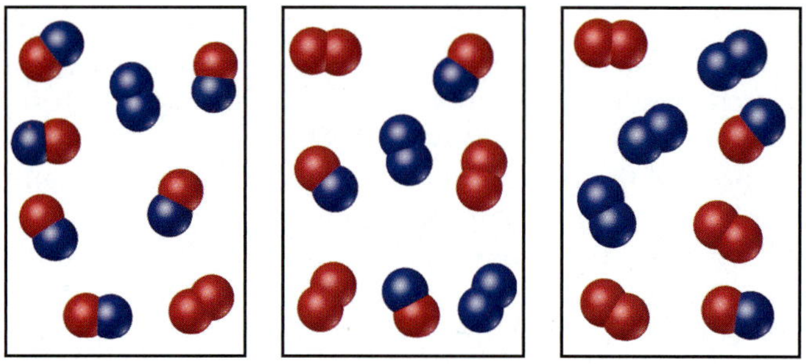

(i) (ii) (iii)

15.6 The reaction $A_2(g) + B(g) \rightleftharpoons A(g) + AB(g)$ has an equilibrium constant of $K_p = 2$. The diagram below shows a mixture containing A atoms (red), A_2 molecules, and AB molecules (red and blue). How many B atoms should be added to the diagram if the system is at equilibrium? [Section 15.6]

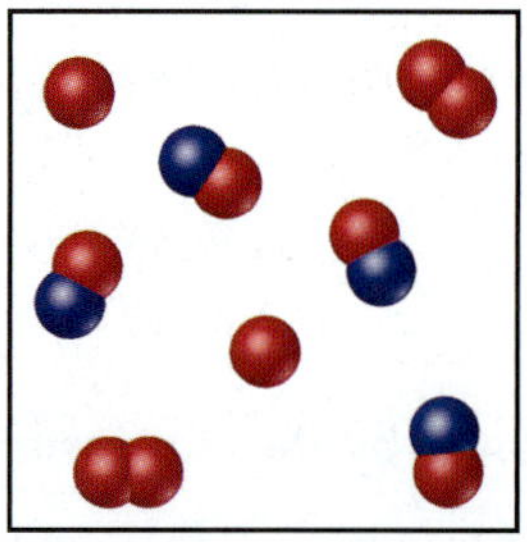

15.7 The following diagram represents the equilibrium state for the reaction $A_2(g) + 2\,B(g) \rightleftharpoons 2\,AB(g)$. **(a)** Assuming the volume is 1 L, calculate the equilibrium constant, K_c, for the reaction. **(b)** If the volume of the equilibrium mixture is decreased, will the number of AB molecules increase or decrease? [Sections 15.5 and 15.7]

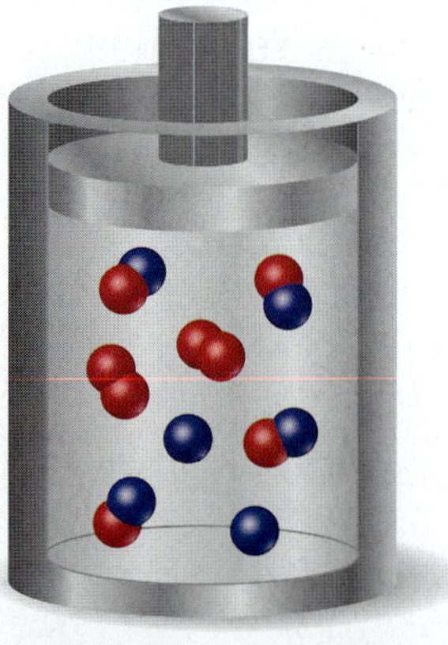

15.8 The following diagrams represent equilibrium mixtures for the reaction $A_2 + B \rightleftharpoons A + AB$ at (1) 300 K and (2) 500 K. The A atoms are red, and the B atoms are blue. Is the reaction exothermic or endothermic? [Section 15.7]

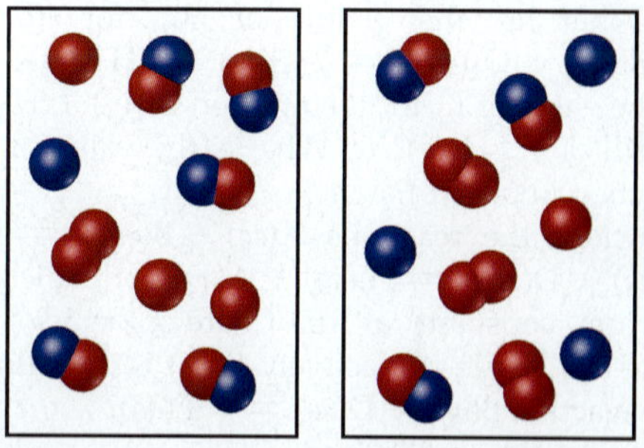

(a) (b)

EXERCISES

Equilibrium; The Equilibrium Constant

15.9 Suppose that the gas-phase reactions $A \longrightarrow B$ and $B \longrightarrow A$ are both elementary processes with rate constants of $3.8 \times 10^{-2}\ s^{-1}$ and $3.1 \times 10^{-1}\ s^{-1}$, respectively. **(a)** What is the value of the equilibrium constant for the equilibrium $A(g) \rightleftharpoons B(g)$? **(b)** Which is greater at equilibrium, the partial pressure of A or the partial pressure of B? Explain.

15.10 Consider the reaction $A + B \rightleftharpoons C + D$. Assume that both the forward reaction and the reverse reaction are elementary processes and that the value of the equilibrium constant is very large. **(a)** Which species predominate at equilibrium, reactants or products? **(b)** Which reaction has the larger rate constant, the forward or the reverse? Explain.

15.11 **(a)** What is the *law of mass action*? Illustrate the law by using the reaction $NO(g) + Br_2(g) \rightleftharpoons NOBr_2(g)$. **(b)** What is the difference between the *equilibrium-constant expression* and the *equilibrium constant* for a given equilibrium? **(c)** Describe an experiment that could be used to determine the value of the equilibrium constant for the reaction in part (a).

15.12 **(a)** The mechanism for a certain reaction $A + B \rightleftharpoons C + D$ is unknown. Is it still possible to apply the law of mass action to the reaction? Explain. **(b)** Write the chemical reaction involved in the Haber process. Why is this reaction important to humanity? **(c)** Write the equilibrium-constant expression for the reaction in part (b).

15.13 Write the expression for K_c for the following reactions. In each case indicate whether the reaction is homogeneous or heterogeneous.

(a) $3\,NO(g) \rightleftharpoons N_2O(g) + NO_2(g)$
(b) $CH_4(g) + 2\,H_2S(g) \rightleftharpoons CS_2(g) + 4\,H_2(g)$
(c) $Ni(CO)_4(g) \rightleftharpoons Ni(s) + 4\,CO(g)$
(d) $HF(aq) \rightleftharpoons H^+(aq) + F^-(aq)$
(e) $2\,Ag(s) + Zn^{2+}(aq) \rightleftharpoons 2\,Ag^+(aq) + Zn(s)$

15.14 Write the expressions for K_c for the following reactions. In each case indicate whether the reaction is homogeneous or heterogeneous.

(a) $N_2(g) + O_2(g) \rightleftharpoons 2\,NO(g)$
(b) $Ti(s) + 2\,Cl_2(g) \rightleftharpoons TiCl_4(l)$
(c) $2\,C_2H_4(g) + 2\,H_2O(g) \rightleftharpoons 2\,C_2H_6(g) + O_2(g)$
(d) $FeO(s) + H_2(g) \rightleftharpoons Fe(s) + H_2O(g)$
(e) $4\,HCl(aq) + O_2(g) \rightleftharpoons 2\,H_2O(l) + 2\,Cl_2(g)$

15.15 When the following reactions come to equilibrium, does the equilibrium mixture contain mostly reactants or mostly products?

(a) $N_2(g) + O_2(g) \rightleftharpoons 2\,NO(g)$; $K_c = 1.5 \times 10^{-10}$
(b) $2\,SO_2(g) + O_2(g) \rightleftharpoons 2\,SO_3(g)$; $K_p = 2.5 \times 10^{9}$

15.16 Which of the following reactions lies to the right, favoring the formation of products, and which lies to the left, favoring formation of reactants?

(a) $2\,NO(g) + O_2(g) \rightleftharpoons 2\,NO_2(g)$; $K_p = 5.0 \times 10^{12}$
(b) $2\,HBr(g) \rightleftharpoons H_2(g) + Br_2(g)$; $K_c = 5.8 \times 10^{-18}$

15.17 If $K_c = 0.042$ for $PCl_3(g) + Cl_2(g) \rightleftharpoons PCl_5(g)$ at 500 K, what is the value of K_p for this reaction at this temperature?

15.18 Calculate K_c at 303 K for $SO_2(g) + Cl_2(g) \rightleftharpoons SO_2Cl_2(g)$ if $K_p = 34.5$ at this temperature.

15.19 The equilibrium constant for the reaction

$$2\,NO(g) + Br_2(g) \rightleftharpoons 2\,NOBr(g)$$

is $K_c = 1.3 \times 10^{-2}$ at 1000 K. **(a)** Calculate K_c for $2\,NOBr(g) \rightleftharpoons 2\,NO(g) + Br_2(g)$. **(b)** At this temperature does the equilibrium favor NO and Br_2, or does it favor NOBr?

15.20 The equilibrium constant for the reaction

$$2\,NO(g) + O_2(g) \rightleftharpoons 2\,NO_2(g)$$

is $K_p = 1.48 \times 10^{4}$ at 184°C. **(a)** Calculate K_p for $2\,NO_2(g) \rightleftharpoons 2\,NO(g) + O_2(g)$. **(b)** Does the equilibrium favor NO and O_2, or does it favor NO_2 at this temperature?

15.21 At 1000 K, $K_p = 1.85$ for the reaction

$$SO_2(g) + \tfrac{1}{2}O_2(g) \rightleftharpoons SO_3(g)$$

(a) What is the value of K_p for the reaction $SO_3(g) \rightleftharpoons SO_2(g) + \frac{1}{2}O_2(g)$? **(b)** What is the value of K_p for the reaction $2\,SO_2(g) + O_2(g) \rightleftharpoons 2\,SO_3(g)$? **(c)** What is the value of K_c for the reaction in part (b)?

15.22 Consider the following equilibrium, for which $K_p = 0.0752$ at 480°C:

$$2\,Cl_2(g) + 2\,H_2O(g) \rightleftharpoons 4\,HCl(g) + O_2(g)$$

(a) What is the value of K_p for the reaction $4\,HCl(g) + O_2(g) \rightleftharpoons 2\,Cl_2(g) + 2\,H_2O(g)$? **(b)** What is the value of K_p for the reaction $Cl_2(g) + H_2O(g) \rightleftharpoons 2\,HCl(g) + \frac{1}{2}O_2(g)$? **(c)** What is the value of K_c for the reaction in part (b)?

15.23 Consider the reactions $A(aq) + B(aq) \rightleftharpoons C(aq)$ and $C(aq) + D(aq) \rightleftharpoons E(aq) + A(aq)$ for which the equilibrium constants at 100°C are $K_c = 1.9 \times 10^{-4}$ and $K_c = 8.5 \times 10^2$, respectively. What is the value of K_c for the reaction $B(aq) + D(aq) \rightleftharpoons E(aq)$?

[15.24] Consider the equilibrium

$$N_2(g) + O_2(g) + Br_2(g) \rightleftharpoons 2\,NOBr(g)$$

Calculate the equilibrium constant K_p for this reaction, given the following information (at 298 K):

$$2\,NO(g) + Br_2(g) \rightleftharpoons 2\,NOBr(g) \qquad K_c = 2.0$$
$$2\,NO(g) \rightleftharpoons N_2(g) + O_2(g) \qquad K_c = 2.1 \times 10^{30}$$

15.25 Mercury(I) oxide decomposes into elemental mercury and elemental oxygen: $2\,Hg_2O(s) \rightleftharpoons 4\,Hg(l) + O_2(g)$. **(a)** Write the equilibrium-constant expression for this reaction in terms of partial pressures. **(b)** Explain why we normally exclude pure solids and liquids from equilibrium-constant expressions.

15.26 Consider the equilibrium $Na_2O(s) + SO_2(g) \rightleftharpoons Na_2SO_3(s)$. **(a)** Write the equilibrium-constant expression for this reaction in terms of partial pressures. **(b)** Why doesn't the concentration of Na_2O appear in the equilibrium-constant expression?

Calculating Equilibrium Constants

15.27 Gaseous hydrogen iodide is placed in a closed container at 425°C, where it partially decomposes to hydrogen and iodine: $2\,HI(g) \rightleftharpoons H_2(g) + I_2(g)$. At equilibrium it is found that $[HI] = 3.53 \times 10^{-3}\,M$, $[H_2] = 4.79 \times 10^{-4}\,M$, and $[I_2] = 4.79 \times 10^{-4}\,M$. What is the value of K_c at this temperature?

15.28 Methanol (CH_3OH) is produced commercially by the catalyzed reaction of carbon monoxide and hydrogen: $CO(g) + 2\,H_2(g) \rightleftharpoons CH_3OH(g)$. An equilibrium mixture in a 2.00-L vessel is found to contain 0.0406 mol CH_3OH, 0.170 mol CO, and 0.302 mol H_2 at 500 K. Calculate K_c at this temperature.

15.29 The equilibrium $2\,NO(g) + Cl_2(g) \rightleftharpoons 2\,NOCl(g)$ is established at 500 K. An equilibrium mixture of the three gases has partial pressures of 0.095 atm, 0.171 atm, and 0.28 atm for NO, Cl_2, and NOCl, respectively. Calculate K_p for this reaction at 500 K.

15.30 Phosphorus trichloride gas and chlorine gas react to form phosphorus pentachloride gas: $PCl_3(g) + Cl_2(g) \rightleftharpoons PCl_5(g)$. A gas vessel is charged with a mixture of $PCl_3(g)$ and $Cl_2(g)$, which is allowed to equilibrate at 450 K. At equilibrium the partial pressures of the three gases are $P_{PCl_3} = 0.124$ atm, $P_{Cl_2} = 0.157$ atm, and $P_{PCl_5} = 1.30$ atm. **(a)** What is the value of K_p at this temperature? **(b)** Does the equilibrium favor reactants or products?

15.31 A mixture of 0.10 mol of NO, 0.050 mol of H_2, and 0.10 mol of H_2O is placed in a 1.0-L vessel at 300 K. The following equilibrium is established:

$$2\,NO(g) + 2\,H_2(g) \rightleftharpoons N_2(g) + 2\,H_2O(g)$$

At equilibrium $[NO] = 0.062\,M$. **(a)** Calculate the equilibrium concentrations of H_2, N_2, and H_2O. **(b)** Calculate K_c.

15.32 A mixture of 1.374 g of H_2 and 70.31 g of Br_2 is heated in a 2.00-L vessel at 700 K. These substances react as follows:

$$H_2(g) + Br_2(g) \rightleftharpoons 2\,HBr(g)$$

At equilibrium the vessel is found to contain 0.566 g of H_2. **(a)** Calculate the equilibrium concentrations of H_2, Br_2, and HBr. **(b)** Calculate K_c.

15.33 A mixture of 0.2000 mol of CO_2, 0.1000 mol of H_2, and 0.1600 mol of H_2O is placed in a 2.000-L vessel. The following equilibrium is established at 500 K:

$$CO_2(g) + H_2(g) \rightleftharpoons CO(g) + H_2O(g)$$

(a) Calculate the initial partial pressures of CO_2, H_2, and H_2O. **(b)** At equilibrium $P_{H_2O} = 3.51$ atm. Calculate the equilibrium partial pressures of CO_2, H_2, and CO. **(c)** Calculate K_p for the reaction.

15.34 A flask is charged with 1.500 atm of $N_2O_4(g)$ and 1.00 atm $NO_2(g)$ at 25°C, and the following equilibrium is achieved:

$$N_2O_4(g) \rightleftharpoons 2\,NO_2(g)$$

After equilibrium is reached, the partial pressure of NO_2 is 0.512 atm. **(a)** What is the equilibrium partial pressure of N_2O_4? **(b)** Calculate the value of K_p for the reaction.

Applications of Equilibrium Constants

15.35 **(a)** How does a reaction quotient differ from an equilibrium constant? **(b)** If $Q_c < K_c$, in which direction will a reaction proceed in order to reach equilibrium? **(c)** What condition must be satisfied so that $Q_c = K_c$?

15.36 **(a)** How is a reaction quotient used to determine whether a system is at equilibrium? **(b)** If $Q_c > K_c$, how must the reaction proceed to reach equilibrium? **(c)** At the start of a certain reaction, only reactants are present; no products have been formed. What is the value of Q_c at this point in the reaction?

15.37 At 100°C the equilibrium constant for the reaction $COCl_2(g) \rightleftharpoons CO(g) + Cl_2(g)$ has the value $K_c = 2.19 \times 10^{-10}$. Are the following mixtures of $COCl_2$, CO, and Cl_2 at 100°C at equilibrium? If not, indicate the direction that the reaction must proceed to achieve equilibrium. **(a)** $[COCl_2] = 2.00 \times 10^{-3}\ M$, $[CO] = 3.3 \times 10^{-6}\ M$, $[Cl_2] = 6.62 \times 10^{-6}\ M$; **(b)** $[COCl_2] = 4.50 \times 10^{-2}\ M$, $[CO] = 1.1 \times 10^{-7}\ M$, $[Cl_2] = 2.25 \times 10^{-6}\ M$; **(c)** $[COCl_2] = 0.0100\ M$, $[CO] = [Cl_2] = 1.48 \times 10^{-6}\ M$.

15.38 As shown in Table 15.2, K_p for the equilibrium

$$N_2(g) + 3\ H_2(g) \rightleftharpoons 2\ NH_3(g)$$

is 4.51×10^{-5} at 450°C. For each of the mixtures listed here, indicate whether the mixture is at equilibrium at 450°C. If it is not at equilibrium, indicate the direction (toward product or toward reactants) in which the mixture must shift to achieve equilibrium. **(a)** 105 atm NH_3, 35 atm N_2, 495 atm H_2; **(b)** 35 atm NH_3, 595 atm H_2, no N_2; **(c)** 26 atm NH_3, 42 atm H_2, 202 atm N_2.

15.39 At 100°C, $K_c = 0.078$ for the reaction

$$SO_2Cl_2(g) \rightleftharpoons SO_2(g) + Cl_2(g)$$

In an equilibrium mixture of the three gases, the concentrations of SO_2Cl_2 and SO_2 are 0.108 M and 0.052 M, respectively. What is the partial pressure of Cl_2 in the equilibrium mixture?

15.40 At 900 K the following reaction has $K_p = 0.345$:

$$2\ SO_2(g) + O_2(g) \rightleftharpoons 2\ SO_3(g)$$

In an equilibrium mixture the partial pressures of SO_2 and O_2 are 0.165 atm and 0.755 atm, respectively. What is the equilibrium partial pressure of SO_3 in the mixture?

15.41 **(a)** At 1285°C the equilibrium constant for the reaction $Br_2(g) \rightleftharpoons 2\ Br(g)$ is $K_c = 1.04 \times 10^{-3}$. A 0.200-L vessel containing an equilibrium mixture of the gases has 0.245 g $Br_2(g)$ in it. What is the mass of $Br(g)$ in the vessel? **(b)** For the reaction $H_2(g) + I_2(g) \rightleftharpoons 2\ HI(g)$, $K_c = 55.3$ at 700 K. In a 2.00-L flask containing an equilibrium mixture of the three gases, there are 0.056 g H_2 and 4.36 g I_2. What is the mass of HI in the flask?

15.42 **(a)** At 800 K the equilibrium constant for $I_2(g) \rightleftharpoons 2\ I(g)$ is $K_c = 3.1 \times 10^{-5}$. If an equilibrium mixture in a 10.0-L vessel contains 2.67×10^{-2} g of $I(g)$, how many grams of I_2 are in the mixture? **(b)** For $2\ SO_2(g) + O_2(g) \rightleftharpoons 2\ SO_3(g)$, $K_p = 3.0 \times 10^4$ at 700 K. In a 2.00-L vessel the equilibrium mixture contains 1.57 g of SO_3 and 0.125 g of O_2. How many grams of SO_2 are in the vessel?

15.43 At 2000°C the equilibrium constant for the reaction

$$2\ NO(g) \rightleftharpoons N_2(g) + O_2(g)$$

is $K_c = 2.4 \times 10^3$. If the initial concentration of NO is 0.200 M, what are the equilibrium concentrations of NO, N_2, and O_2?

15.44 For the equilibrium

$$Br_2(g) + Cl_2(g) \rightleftharpoons 2\ BrCl(g)$$

at 400 K, $K_c = 7.0$. If 0.30 mol of Br_2 and 0.30 mol Cl_2 are introduced into a 1.0-L container at 400 K, what will be the equilibrium concentrations of Br_2, Cl_2, and BrCl?

15.45 At 373 K, $K_p = 0.416$ for the equilibrium

$$2\ NOBr(g) \rightleftharpoons 2\ NO(g) + Br_2(g)$$

If the pressures of $NOBr(g)$ and $NO(g)$ are equal, what is the equilibrium pressure of $Br_2(g)$?

15.46 At 218°C, $K_c = 1.2 \times 10^{-4}$ for the equilibrium

$$NH_4HS(s) \rightleftharpoons NH_3(g) + H_2S(g)$$

Calculate the equilibrium concentrations of NH_3 and H_2S if a sample of solid NH_4HS is placed in a closed vessel and decomposes until equilibrium is reached.

15.47 Consider the reaction

$$CaSO_4(s) \rightleftharpoons Ca^{2+}(aq) + SO_4^{2-}(aq)$$

At 25°C the equilibrium constant is $K_c = 2.4 \times 10^{-5}$ for this reaction. **(a)** If excess $CaSO_4(s)$ is mixed with water at 25°C to produce a saturated solution of $CaSO_4$, what are the equilibrium concentrations of Ca^{2+} and SO_4^{2-}? **(b)** If the resulting solution has a volume of 3.0 L, what is the minimum mass of $CaSO_4(s)$ needed to achieve equilibrium?

15.48 At 80°C, $K_c = 1.87 \times 10^{-3}$ for the reaction

$$PH_3BCl_3(s) \rightleftharpoons PH_3(g) + BCl_3(g)$$

(a) Calculate the equilibrium concentrations of PH_3 and BCl_3 if a solid sample of PH_3BCl_3 is placed in a closed vessel and decomposes until equilibrium is reached. **(b)** If the flask has a volume of 0.500 L, what is the minimum mass of $PH_3BCl_3(s)$ that must be added to the flask to achieve equilibrium?

15.49 For the reaction $I_2(g) + Br_2(g) \rightleftharpoons 2\ IBr(g)$, $K_c = 280$ at 150°C. Suppose that 0.500 mol IBr in a 1.00-L flask is allowed to reach equilibrium at 150°C. What are the equilibrium concentrations of IBr, I_2, and Br_2?

15.50 At 25°C the reaction

$$CaCrO_4(s) \rightleftharpoons Ca^{2+}(aq) + CrO_4^{2-}(aq)$$

has an equilibrium constant $K_c = 7.1 \times 10^{-4}$. What are the equilibrium concentrations of Ca^{2+} and CrO_4^{2-} in a saturated solution of $CaCrO_4$?

Le Châtelier's Principle

15.51 Consider the following equilibrium, for which $\Delta H < 0$:

$$2\ SO_2(g) + O_2(g) \rightleftharpoons 2\ SO_3(g)$$

How will each of the following changes affect an equilibrium mixture of the three gases? **(a)** $O_2(g)$ is added to the system; **(b)** the reaction mixture is heated; **(c)** the volume of the reaction vessel is doubled; **(d)** a catalyst is added to the mixture; **(e)** the total pressure of the system is increased by adding a noble gas; **(f)** $SO_3(g)$ is removed from the system.

15.52 For the following reaction, $\Delta H° = 2816$ kJ:

$$6\ CO_2(g) + 6\ H_2O(l) \rightleftharpoons C_6H_{12}O_6(s) + 6\ O_2(g)$$

How is the equilibrium yield of $C_6H_{12}O_6$ affected by **(a)** increasing P_{CO_2}, **(b)** increasing temperature, **(c)** removing CO_2, **(d)** decreasing the total pressure, **(e)** removing part of the $C_6H_{12}O_6$, **(f)** adding a catalyst?

15.53 How do the following changes affect the value of the equilibrium constant for a gas-phase exothermic reaction: **(a)** removal of a reactant or product, **(b)** decrease in the volume, **(c)** decrease in the temperature, **(d)** addition of a catalyst?

15.54 For a certain gas-phase reaction, the fraction of products in an equilibrium mixture is increased by increasing the temperature and increasing the volume of the reaction vessel. **(a)** What can you conclude about the reaction from the influence of temperature on the equilibrium? **(b)** What can you conclude from the influence of increasing the volume?

15.55 Consider the following equilibrium between oxides of nitrogen

$$3\,NO(g) \rightleftharpoons NO_2(g) + N_2O(g)$$

(a) Use data in Appendix C to calculate $\Delta H°$ for this reaction. **(b)** Will the equilibrium constant for the reaction increase or decrease with increasing temperature? Explain. **(c)** At constant temperature would a change in the volume of the container affect the fraction of products in the equilibrium mixture?

15.56 Methanol (CH_3OH) can be made by the reaction of CO with H_2:

$$CO(g) + 2\,H_2(g) \rightleftharpoons CH_3OH(g)$$

(a) Use thermochemical data in Appendix C to calculate $\Delta H°$ for this reaction. **(b)** In order to maximize the equilibrium yield of methanol, would you use a high or low temperature? **(c)** In order to maximize the equilibrium yield of methanol, would you use a high or low pressure?

Additional Exercises

15.57 Both the forward reaction and the reverse reaction in the following equilibrium are believed to be elementary steps:

$$CO(g) + Cl_2(g) \rightleftharpoons COCl(g) + Cl(g)$$

At 25°C the rate constants for the forward and reverse reactions are $1.4 \times 10^{-28}\ M^{-1}s^{-1}$ and $9.3 \times 10^{10}\ M^{-1}s^{-1}$, respectively. **(a)** What is the value for the equilibrium constant at 25°C? **(b)** Are reactants or products more plentiful at equilibrium?

15.58 A mixture of CH_4 and H_2O is passed over a nickel catalyst at 1000 K. The emerging gas is collected in a 5.00-L flask and is found to contain 8.62 g of CO, 2.60 g of H_2, 43.0 g of CH_4, and 48.4 g of H_2O. Assuming that equilibrium has been reached, calculate K_c and K_p for the reaction.

15.59 When 2.00 mol of SO_2Cl_2 is placed in a 2.00-L flask at 303 K, 56% of the SO_2Cl_2 decomposes to SO_2 and Cl_2:

$$SO_2Cl_2(g) \rightleftharpoons SO_2(g) + Cl_2(g)$$

Calculate K_c for this reaction at this temperature.

15.60 A mixture of H_2, S, and H_2S is held in a 1.0-L vessel at 90°C until the following equilibrium is achieved:

$$H_2(g) + S(s) \rightleftharpoons H_2S(g)$$

At equilibrium the mixture contains 0.46 g of H_2S and 0.40 g H_2. **(a)** Write the equilibrium-constant expression for this reaction. **(b)** What is the value of K_c for the reaction at this temperature? **(c)** Why can we ignore the amount of S when doing the calculation in part (b)?

15.61 A sample of nitrosyl bromide (NOBr) decomposes according to the equation

$$2\,NOBr(g) \rightleftharpoons 2\,NO(g) + Br_2(g)$$

An equilibrium mixture in a 5.00-L vessel at 100°C contains 3.22 g of NOBr, 3.08 g of NO, and 4.19 g of Br_2. **(a)** Calculate K_c. **(b)** What is the total pressure exerted by the mixture of gases?

15.62 Consider the hypothetical reaction $A(g) \rightleftharpoons 2\,B(g)$. A flask is charged with 0.55 atm of pure A, after which it is allowed to reach equilibrium at 0°C. At equilibrium the partial pressure of A is 0.36 atm. **(a)** What is the total pressure in the flask at equilibrium? **(b)** What is the value of K_p?

15.63 As shown in Table 15.2, the equilibrium constant for the reaction $N_2(g) + 3\,H_2(g) \rightleftharpoons 2\,NH_3(g)$ is $K_p = 4.34 \times 10^{-3}$ at 300°C. Pure NH_3 is placed in a 1.00-L flask and allowed to reach equilibrium at this temperature. There are 1.05 g NH_3 in the equilibrium mixture. **(a)** What are the masses of N_2 and H_2 in the equilibrium mixture? **(b)** What was the initial mass of ammonia placed in the vessel? **(c)** What is the total pressure in the vessel?

15.64 For the equilibrium

$$2\,IBr(g) \rightleftharpoons I_2(g) + Br_2(g)$$

$K_p = 8.5 \times 10^{-3}$ at 150°C. If 0.025 atm of IBr is placed in a 2.0-L container, what is the partial pressure of this substance after equilibrium is reached?

15.65 For the equilibrium

$$PH_3BCl_3(s) \rightleftharpoons PH_3(g) + BCl_3(g)$$

$K_p = 0.052$ at 60°C. **(a)** Calculate K_c. **(b)** Some solid PH_3BCl_3 is added to a closed 0.500-L vessel at 60°C; the vessel is then charged with 0.0128 mol of $BCl_3(g)$. What is the equilibrium concentration of PH_3?

[15.66] Solid NH_4HS is introduced into an evacuated flask at 24°C. The following reaction takes place:

$$NH_4HS(s) \rightleftharpoons NH_3(g) + H_2S(g)$$

At equilibrium the total pressure (for NH_3 and H_2S taken together) is 0.614 atm. What is K_p for this equilibrium at 24°C?

[15.67] A 0.831-g sample of SO_3 is placed in a 1.00-L container and heated to 1100 K. The SO_3 decomposes to SO_2 and O_2:

$$2\,SO_3(g) \rightleftharpoons 2\,SO_2(g) + O_2(g)$$

At equilibrium the total pressure in the container is 1.300 atm. Find the values of K_p and K_c for this reaction at 1100 K.

15.68 Nitric oxide (NO) reacts readily with chlorine gas as follows:

$$2\,NO(g) + Cl_2(g) \rightleftharpoons 2\,NOCl(g)$$

At 700 K the equilibrium constant K_p for this reaction is 0.26. Predict the behavior of each of the following mixtures at this temperature: **(a)** $P_{NO} = 0.15$ atm, $P_{Cl_2} = 0.31$ atm, and $P_{NOCl} = 0.11$ atm; **(b)** $P_{NO} = 0.12$ atm, $P_{Cl_2} = 0.10$ atm, and $P_{NOCl} = 0.050$ atm; **(c)** $P_{NO} = 0.15$ atm, $P_{Cl_2} = 0.20$ atm, and $P_{NOCl} = 5.10 \times 10^{-3}$ atm.

15.69 At 900°C, $K_c = 0.0108$ for the reaction

$$CaCO_3(s) \rightleftharpoons CaO(s) + CO_2(g)$$

A mixture of $CaCO_3$, CaO, and CO_2 is placed in a 10.0-L vessel at 900°C. For the following mixtures, will the amount of $CaCO_3$ increase, decrease, or remain the same as the system approaches equilibrium?
(a) 15.0 g $CaCO_3$, 15.0 g CaO, and 4.25 g CO_2
(b) 2.50 g $CaCO_3$, 25.0 g CaO, and 5.66 g CO_2
(c) 30.5 g $CaCO_3$, 25.5 g CaO, and 6.48 g CO_2

15.70 Nickel carbonyl, $Ni(CO)_4$, is an extremely toxic liquid with a low boiling point. Nickel carbonyl results from the reaction of nickel metal with carbon monoxide. For temperatures above the boiling point (42.2°C) of $Ni(CO)_4$, the reaction is

$$Ni(s) + 4\,CO(g) \rightleftharpoons Ni(CO)_4(g)$$

Nickel that is more than 99.9% pure can be produced by the *carbonyl process*: Impure nickel combines with CO at 50°C to produce $Ni(CO)_4(g)$. The $Ni(CO)_4$ is then heated to 200°C, causing it to decompose back into Ni(*s*) and CO(*g*). **(a)** Write the equilibrium-constant expression for the formation of $Ni(CO)_4$. **(b)** Given the temperatures used for the steps in the carbonyl process, do you think the formation reaction is endothermic or exothermic? **(c)** In the early days of automobiles, nickel-plated exhaust pipes were used. Even though the equilibrium constant for the formation of $Ni(CO)_4$ is very small at the temperature of automotive exhaust gases, the exhaust pipes quickly corroded. Explain why this occurred.

15.71 NiO is to be reduced to nickel metal in an industrial process by use of the reaction

$$NiO(s) + CO(g) \rightleftharpoons Ni(s) + CO_2(g)$$

At 1600 K the equilibrium constant for the reaction is $K_p = 6.0 \times 10^2$. If a CO pressure of 150 torr is to be employed in the furnace and total pressure never exceeds 760 torr, will reduction occur?

[15.72] At 700 K the equilibrium constant for the reaction

$$CCl_4(g) \rightleftharpoons C(s) + 2\,Cl_2(g)$$

is $K_p = 0.76$. A flask is charged with 2.00 atm of CCl_4, which then reaches equilibrium at 700 K. **(a)** What fraction of the CCl_4 is converted into C and Cl_2? **(b)** What are the partial pressures of CCl_4 and Cl_2 at equilibrium?

[15.73] The reaction $PCl_3(g) + Cl_2(g) \rightleftharpoons PCl_5(g)$ has $K_p = 0.0870$ at 300°C. A flask is charged with 0.50 atm PCl_3, 0.50 atm Cl_2, and 0.20 atm PCl_5 at this temperature. **(a)** Use the reaction quotient to determine the direction the reaction must proceed in order to reach equilibrium. **(b)** Calculate the equilibrium partial pressures of the gases. **(c)** What effect will increasing the volume of the system have on the mole fraction of Cl_2 in the equilibrium mixture? **(d)** The reaction is exothermic. What effect will increasing the temperature of the system have on the mole fraction of Cl_2 in the equilibrium mixture?

[15.74] An equilibrium mixture of H_2, I_2, and HI at 458°C contains 0.112 mol H_2, 0.112 mol I_2, and 0.775 mol HI in a 5.00-L vessel. What are the equilibrium partial pressures when equilibrium is reestablished following the addition of 0.100 mol of HI?

[15.75] Consider the hypothetical reaction $A(g) + 2\,B(g) \rightleftharpoons 2\,C(g)$, for which $K_c = 0.25$ at some temperature. A 1.00-L reaction vessel is loaded with 1.00 mol of compound C, which is allowed to reach equilibrium. Let the variable x represent the number of mol/L of compound A present at equilibrium. **(a)** In terms of x, what are the equilibrium concentrations of compounds B and C? **(b)** What limits must be placed on the value of x so that all concentrations are positive? **(c)** By putting the equilibrium concentrations (in terms of x) into the equilibrium-constant expression, derive an equation that can be solved for x. **(d)** The equation from part (c) is a cubic equation (one that has the form $ax^3 + bx^2 + cx + d = 0$). In general, cubic equations cannot be solved in closed form. However, you can estimate the solution by plotting the cubic equation in the allowed range of x that you specified in part (b). The point at which the cubic equation crosses the x-axis is the solution. **(e)** From the plot in part (d), estimate the equilibrium concentrations of A, B, and C. [*Hint:* You can check the accuracy of your answer by substituting these concentrations into the equilibrium expression.]

15.76 At 1200 K, the approximate temperature of automobile exhaust gases (Figure 15.16), K_p for the reaction

$$2\,CO_2(g) \rightleftharpoons 2\,CO(g) + O_2(g)$$

is about 1×10^{-13}. Assuming that the exhaust gas (total pressure 1 atm) contains 0.2% CO, 12% CO_2, and 3% O_2 by volume, is the system at equilibrium with respect to the above reaction? Based on your conclusion, would the CO concentration in the exhaust be decreased or increased by a catalyst that speeds up the reaction above?

15.77 Suppose that you worked at the U.S. Patent Office and a patent application came across your desk claiming that a newly developed catalyst was much superior to the Haber catalyst for ammonia synthesis because the catalyst led to much greater equilibrium conversion of N_2 and H_2 into NH_3 than the Haber catalyst under the same conditions. What would be your response?

Integrative Exercises

15.78 Consider the following equilibria in aqueous solution:
(i) $Na(s) + Ag^+(aq) \rightleftharpoons Na^+(aq) + Ag(s)$
(ii) $3\,Hg(l) + 2\,Al^{3+}(aq) \rightleftharpoons 3\,Hg^{2+}(aq) + 2\,Al(s)$
(iii) $Zn(s) + 2\,H^+(aq) \rightleftharpoons Zn^{2+}(aq) + H_2(g)$
(a) For each reaction, write the equilibrium-constant expression for K_c. **(b)** Using information provided in Table 4.5, predict whether K_c is large ($K_c \gg 1$) or small ($K_c \ll 1$). Explain your reasoning. **(c)** At 25°C the reaction $Cd(s) + Fe^{2+}(aq) \rightleftharpoons Cd^{2+}(aq) + Fe(s)$ has $K_c = 6 \times 10^{-2}$. If Cd were added to Table 4.5, would you expect it to be above or below iron? Explain.

15.79 Silver chloride, $AgCl(s)$, is an insoluble strong electrolyte. **(a)** Write the equation for the dissolution of $AgCl(s)$ in $H_2O(l)$. **(b)** Write the expression for K_c for the reaction in part (a). **(c)** Based on the thermochemical data in Appendix C and Le Châtelier's principle, predict whether the solubility of AgCl in H_2O increases or decreases with increasing temperature.

15.80 The hypothetical reaction $A + B \rightleftharpoons C$ occurs in the forward direction in a single step. The energy profile of the reaction is shown in the drawing. **(a)** Is the forward or reverse reaction faster at equilibrium? **(b)** Would you expect the equilibrium to favor reactants or products? **(c)** In general, how would a catalyst affect the energy profile shown? **(d)** How would a catalyst affect the ratio of the rate constants for the forward and reverse reactions? **(e)** How would you expect the equilibrium constant of the reaction to change with increasing temperature?

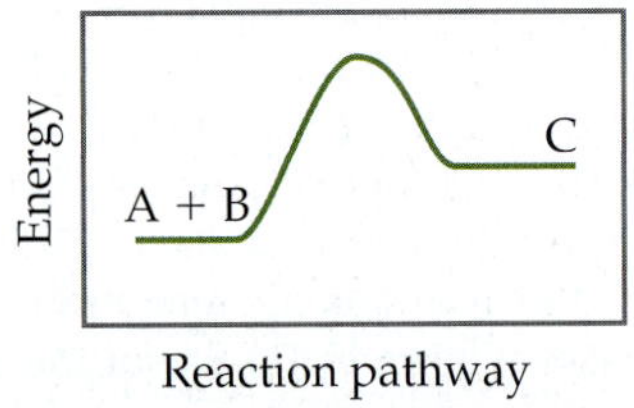

[15.81] Consider the equilibrium $A \rightleftharpoons B$, in which both the forward and reverse reactions are elementary (single-step) reactions. Assume that the only effect of a catalyst on the reaction is to lower the activation energies of the forward and reverse reactions, as shown in Figure 15.15. Using the Arrhenius equation (Section 14.5), prove that the equilibrium constant is the same for the catalyzed reaction as for the uncatalyzed one.

[15.82] At 25°C the reaction

$$NH_4HS(s) \rightleftharpoons NH_3(g) + H_2S(g)$$

has $K_p = 0.120$. A 5.00-L flask is charged with 0.300 g of pure $H_2S(g)$ at 25°C. Solid NH_4HS is then added until there is excess unreacted solid remaining. **(a)** What is the initial pressure of $H_2S(g)$ in the flask? **(b)** Why does no reaction occur until NH_4HS is added? **(c)** What are the partial pressures of NH_3 and H_2S at equilibrium? **(d)** What is the mole fraction of H_2S in the gas mixture at equilibrium? **(e)** What is the minimum mass, in grams, of NH_4HS that must be added to the flask to achieve equilibrium?

[15.83] Write the equilibrium-constant expression for the equilibrium

$$C(s) + CO_2(g) \rightleftharpoons 2\,CO(g)$$

The table included below shows the relative mole percentages of $CO_2(g)$ and $CO(g)$ at a total pressure of 1 atm for several temperatures. Calculate the value of K_p at each temperature. Is the reaction exothermic or endothermic? Explain.

Temperature (°C)	CO_2 (mol %)	CO (mol %)
850	6.23	93.77
950	1.32	98.68
1050	0.37	99.63
1200	0.06	99.94

15.84 In Section 11.5 we defined the vapor pressure of a liquid in terms of an equilibrium. **(a)** Write the equation representing the equilibrium between liquid water and water vapor, and the corresponding expression for K_p. **(b)** By using data in Appendix B, give the value of K_p for this reaction at 30°C. **(c)** What is the value of K_p for any liquid in equilibrium with its vapor at the normal boiling point of the liquid?

[15.85] Polyvinyl chloride (PVC) is one of the most commercially important polymers (Table 12.4). PVC is made by addition polymerization of vinyl chloride (C_2H_3Cl). Vinyl chloride is synthesized from ethylene (C_2H_4) in a two-step process involving the following equilibria:

Equilibrium 1: $C_2H_4(g) + Cl_2(g) \rightleftharpoons C_2H_4Cl_2(g)$

Equilibrium 2: $C_2H_4Cl_2(g) \rightleftharpoons C_2H_3Cl(g) + HCl(g)$

The product of Equilibrium 1 is 1,2-dichloroethane, a compound in which one Cl atom is bonded to each C atom. **(a)** Draw Lewis structures for $C_2H_4Cl_2$ and C_2H_3Cl. What are the C—C bond orders in these two compounds? **(b)** Use average bond enthalpies (Table 8.4) to estimate the enthalpy changes in the two equilibria. **(c)** How would the yield of $C_2H_4Cl_2$ in Equilibrium 1 vary with temperature and volume? **(d)** How would the yield of C_2H_3Cl in Equilibrium 2 vary with temperature and volume? **(e)** Look up the normal boiling points of 1,2-dichloroethane and vinyl chloride in a sourcebook, such as the *CRC Handbook of Chemistry and Physics*. Based on these data, propose a reactor design (analogous to Figure 15.12) that could be used to maximize the amount of C_2H_3Cl produced by using the two equilibria.

eMEDIA EXERCISES

These exercises make use of the interactive objects available online in OneKey or the Companion Website, and on your Accelerator CD. Access to these resources comes in your MediaPak.

15.86 You can choose starting concentrations for reactants in the **Chemical Equilibrium** activity (*15.1*). **(a)** Write the equilibrium-constant expression for the reaction in the simulation. **(b)** Calculate the value of the equilibrium constant. **(c)** Does this reaction lie to the right or to the left?

15.87 Using the equilibrium constant that you calculated in the previous question for the reaction in the **Chemical Equilibrium** activity (*15.1*), predict what the equilibrium concentrations of reactants and products will be if you start with 0.0007 *M* $Fe^{3+}(aq)$ and 0.0004 *M* $SCN^{-}(aq)$. Use the simulation to check your answer.

15.88 The significance of an equilibrium constant's magnitude is illustrated in the **Equilibrium Constant** activity (*15.2*). Select the $A \rightleftharpoons B$ reaction, and carry out several experiments with varying starting concentrations and varying equilibrium constants. Because reactants are favored when K_c is very small, is it possible to enter a very small value for K_c, start with only A (no B), and still have only A at equilibrium? Explain.

15.89 In Exercise 15.34 you calculated K_p for the reaction $N_2O_4(g) \rightleftharpoons 2\,NO_2(g)$. **(a)** Which of the reactions in the **Equilibrium Constant** activity (*15.2*) best represents the equilibrium between $N_2O_4(g)$ and $2\,NO_2(g)$? **(b)** Using the simulation, enter the K_p value from Exercise 15.34 and experiment with various starting concentrations of reactant and products. Can you select starting concentrations of reactants and products such that no change in concentrations is observed? Explain.

15.90 Consider the $A \rightleftharpoons B$ and $A \rightleftharpoons 2\,B$ reactions in the **Equilibrium Constant** activity (*15.2*). Assume that A and B are both gases, that $A \rightleftharpoons B$ is an exothermic reaction, and that $A \rightleftharpoons B$ is an endothermic reaction. For both reactions, state the effect that each of the following changes would have on the value of K_c, the concentration of B at equilibrium, and the rate of the reaction: **(a)** increased temperature, **(b)** increased pressure, and **(c)** addition of a catalyst.

CHAPTER

16

WHAT'S AHEAD

- We start by reviewing the *Arrhenius* definitions of acid and base. *(Section 16.1)*
- We then learn the more general *Brønsted–Lowry* definitions for acid and base. A Brønsted–Lowry acid is a *proton donor* and a Brønsted–Lowry base is a *proton acceptor*. *(Section 16.2)*
- The *conjugate base* of a Brønsted–Lowry acid is what remains after the acid has donated a proton. Similarly, the *conjugate acid* of a Brønsted–Lowry base is the species that results when the base accepts a proton. Two species that differ by the presence or absence of a proton are known as a *conjugate acid-base pair*. *(Section 16.2)*
- *Autoionization* of water produces small concentrations of H_3O^+ and OH^- ions. The *equilibrium constant* for autoionization, K_w, defines the relationship between H_3O^+ and OH^- concentrations in aqueous solutions. *(Section 16.3)*
- The pH scale ($pH = -\log[H^+]$) is used to describe the acidity or basicity of a solution. *(Section 16.4)*
- *Strong* acids and bases are those that ionize or dissociate completely in aqueous solution, whereas *weak* acids and bases ionize only partially. *(Section 16.5)*
- We learn that the ionization of a weak acid in water is an equilibrium process with an equilibrium constant K_a, which can be used to calculate the pH of a weak acid solution. *(Section 16.6)*
- Likewise, ionization of a weak base in water is an equilibrium process with equilibrium constant K_b, which can be used to calculate the pH of a weak base solution. *(Section 16.7)*
- There is a relationship between the K_a and K_b of any conjugate *acid-base* pair: $K_a \times K_b = K_w$. *(Section 16.8)*
- The ions of a salt can serve as Brønsted–Lowry acids or bases. *(Section 16.9)*
- We continue by exploring the relationship between chemical structure and acid-base behavior. *(Section 16.10)*
- Finally, we learn the *Lewis* definitions of acid and base. A Lewis acid is an *electron acceptor*, and a Lewis base is an *electron donor*. *(Section 16.11)*

Acid-Base Equilibria

WHAT'S THE SOUREST FOOD YOU'VE EVER TASTED? Citrus fruit, such as the lemons shown in the chapter-opening photograph? Sour cherries? Rhubarb? The sour taste of foods is due primarily to the presence of acids. Citric acid ($H_3C_6H_5O_7$), malic acid ($H_2C_4H_4O_5$), oxalic acid ($H_2C_2O_4$), and ascorbic acid, also known as vitamin C ($HC_6H_7O_6$), are present in many fruits as well as in certain vegetables such as rhubarb and tomatoes.

Acids and bases are important in numerous chemical processes that occur around us, from industrial processes to biological ones, from reactions in the laboratory to those in our environment. The time required for a metal object immersed in water to corrode, the ability of an aquatic environment to support fish and plant life, the fate of pollutants washed out of the air by rain, and even the rates of reactions that maintain our lives all critically depend upon the acidity or basicity of solutions. Indeed, an enormous amount of chemistry can be understood in terms of acid-base reactions.

We have encountered acids and bases many times in earlier discussions. For example, a portion of Chapter 4 focused on their reactions. But what makes a substance behave as an acid or as a base? In this chapter we reexamine acids and bases, taking a closer look at how they are identified and characterized. In doing so, we will consider their behavior not only in terms of their structure and bonding but also in terms of the chemical equilibria in which they participate.

VARIETIES OF CITRUS FRUITS (oranges, limes, lemons, and tangerines).
FoodPix/Getty Images, Inc. - FoodPix

16.1 | Acids and Bases: A Brief Review

From the earliest days of experimental chemistry, scientists have recognized acids and bases by their characteristic properties. Acids have a sour taste and cause certain dyes to change color (for example, litmus turns red on contact with acids). Indeed, the word *acid* comes from the Latin word *acidus*, meaning sour or tart. Bases, in contrast, have a bitter taste and feel slippery (soap is a good example). The word *base* comes from an old English meaning of the word, which is "to bring low." (We still use the word *debase* in this sense, meaning to lower the value of something.) When bases are added to acids, they lower the amount of acid. Indeed, when acids and bases are mixed in certain proportions, their characteristic properties disappear altogether. ∞ (Section 4.3)

Historically, chemists have sought to relate the properties of acids and bases to their compositions and molecular structures. By 1830 it was evident that all acids contain hydrogen but not all hydrogen-containing substances are acids. In the 1880s the Swedish chemist Svante Arrhenius (1859–1927) linked acid behavior with the presence of H^+ ions, and base behavior with the presence of OH^- ions in aqueous solution. He defined acids as substances that produce H^+ ions in water, and bases as substances that produce OH^- ions in water. Indeed, the properties of aqueous solutions of acids, such as sour taste, are due to $H^+(aq)$, whereas the properties of aqueous solutions of bases are due to $OH^-(aq)$. Over time the Arrhenius concept of acids and bases came to be stated in the following way: *Acids are substances that, when dissolved in water, increase the concentration of H^+ ions*. Likewise, *bases are substances that, when dissolved in water, increase the concentration of OH^- ions*.

Hydrogen chloride is an Arrhenius acid. Hydrogen chloride gas is highly soluble in water because of its chemical reaction with water, which produces hydrated H^+ and Cl^- ions:

$$HCl(g) \xrightarrow{H_2O} H^+(aq) + Cl^-(aq) \quad [16.1]$$

The aqueous solution of HCl is known as hydrochloric acid. Concentrated hydrochloric acid is about 37% HCl by mass and is 12 *M* in HCl. Sodium hydroxide, on the other hand, is an Arrhenius base. Because NaOH is an ionic compound, it dissociates into Na^+ and OH^- ions when it dissolves in water, thereby releasing OH^- ions into the solution.

GIVE IT SOME THOUGHT

What two ions are central to the Arrhenius definitions of acids and bases?

16.2 | Brønsted–Lowry Acids and Bases

The Arrhenius concept of acids and bases, while useful, has limitations. For one thing, it is restricted to aqueous solutions. In 1923 the Danish chemist Johannes Brønsted (1879–1947) and the English chemist Thomas Lowry (1874–1936) independently proposed a more general definition of acids and bases. Their concept is based on the fact that acid-base reactions involve the transfer of H^+ ions from one substance to another.

The H^+ Ion in Water

In Equation 16.1 hydrogen chloride is shown ionizing in water to form $H^+(aq)$. *An H^+ ion is simply a proton with no surrounding valence electron*. This small, positively charged particle interacts strongly with the nonbonding electron pairs of water molecules to form hydrated hydrogen ions. For example, the interaction of a proton with one water molecule forms the **hydronium ion**, $H_3O^+(aq)$:

$$H^+ + \mathrm{:\ddot{O}{-}H} \longrightarrow \left[\mathrm{H{-}\ddot{O}{-}H}\right]^+ \qquad [16.2]$$

(In both structures a second H is bonded below the O.)

The formation of hydronium ions is one of the complex features of the interaction of the H^+ ion with liquid water. In fact, the H_3O^+ ion can form hydrogen bonds to additional H_2O molecules to generate larger clusters of hydrated hydrogen ions, such as $H_5O_2^+$ and $H_9O_4^+$ (Figure 16.1 ▶).

Chemists use $H^+(aq)$ and $H_3O^+(aq)$ interchangeably to represent the same thing—namely the hydrated proton that is responsible for the characteristic properties of aqueous solutions of acids. We often use the $H^+(aq)$ ion for simplicity and convenience, as we did in Equation 16.1. The $H_3O^+(aq)$ ion, however, more closely represents reality.

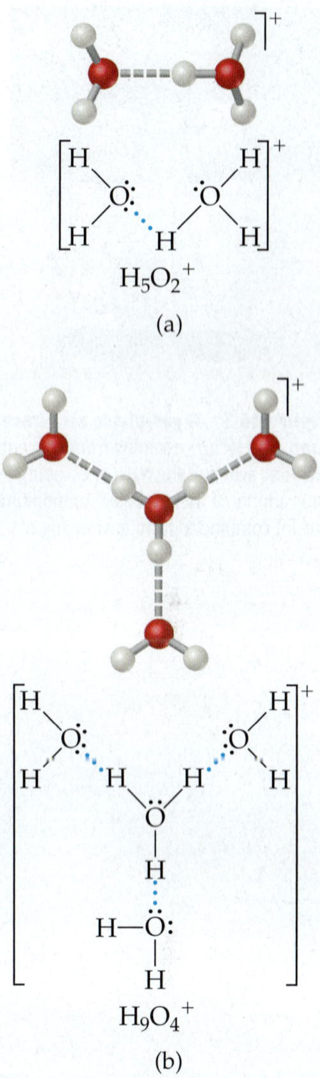

▲ **Figure 16.1 Hydrated hydronium ions.** Lewis structures and molecular models for $H_5O_2^+$ and $H_9O_4^+$. There is good experimental evidence for the existence of both these species.

Proton-Transfer Reactions

When we closely examine the reaction that occurs when HCl dissolves in water, we find that the HCl molecule actually transfers an H^+ ion (a proton) to a water molecule as depicted in Figure 16.2 ▼. Thus, we can represent the reaction as occurring between an HCl molecule and a water molecule to form hydronium and chloride ions:

$$HCl(g) + H_2O(l) \longrightarrow H_3O^+(aq) + Cl^-(aq) \qquad [16.3]$$

The polar H_2O molecule promotes the ionization of acids in water solution by accepting a proton to form H_3O^+.

Brønsted and Lowry proposed defining acids and bases in terms of their ability to transfer protons. According to their definition, *an acid is a substance (molecule or ion) that can donate a proton to another substance.* Likewise, *a base is a substance that can accept a proton.* Thus, when HCl dissolves in water (Equation 16.3), HCl acts as a **Brønsted–Lowry acid** (it donates a proton to H_2O), and H_2O acts as a **Brønsted–Lowry base** (it accepts a proton from HCl).

Because the emphasis in the Brønsted–Lowry concept is on proton transfer, the concept also applies to reactions that do not occur in aqueous solution. In the reaction between HCl and NH_3, for example, a proton is transferred from the acid HCl to the base NH_3:

$$\mathrm{:\ddot{\underset{..}{Cl}}{-}H} + \mathrm{:N{-}H} \longrightarrow \mathrm{:\ddot{\underset{..}{Cl}}:}^- + \left[\mathrm{H{-}N{-}H}\right]^+ \qquad [16.4]$$

(In NH_3 and NH_4^+ the N also carries an H above and an H below.)

▶ **Figure 16.2 A proton-transfer reaction.** When a proton is transferred from HCl to H_2O, HCl acts as the Brønsted–Lowry acid and H_2O acts as the Brønsted–Lowry base.

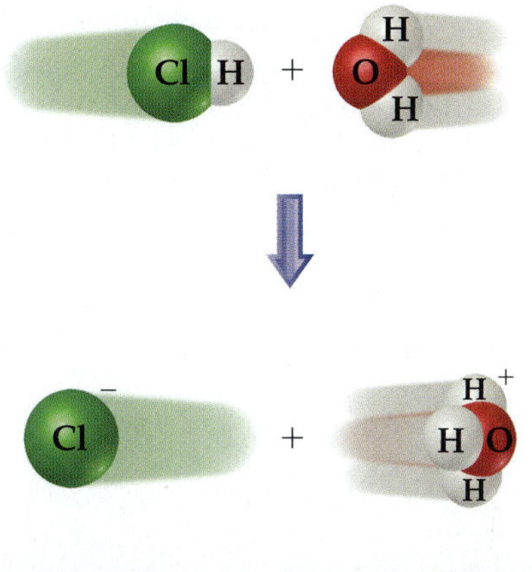

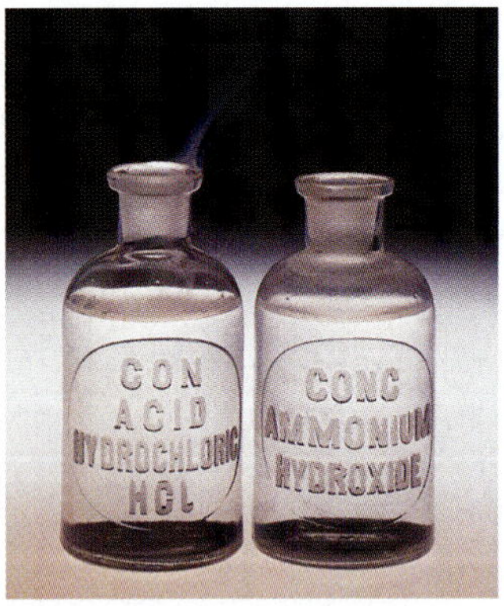

▲ **Figure 16.3 A gas-phase acid-base reaction.** The HCl(*g*) escaping from concentrated hydrochloric acid and the $NH_3(g)$ escaping from aqueous ammonia (here labeled ammonium hydroxide) combine to form a white fog of $NH_4Cl(s)$.

This reaction can occur in the gas phase. The hazy film that forms on the windows of general chemistry laboratories and on glassware in the lab is largely solid NH_4Cl formed by the gas-phase reaction of HCl and NH_3 (Figure 16.3 ◀).

Let's consider another example that compares the relationship between the Arrhenius definitions and the Brønsted–Lowry definitions of acids and bases—an aqueous solution of ammonia, in which the following equilibrium occurs:

$$NH_3(aq) + H_2O(l) \rightleftharpoons NH_4^+(aq) + OH^-(aq) \quad [16.5]$$

Ammonia is an Arrhenius base because adding it to water leads to an increase in the concentration of $OH^-(aq)$. It is a Brønsted–Lowry base because it accepts a proton from H_2O. The H_2O molecule in Equation 16.5 acts as a Brønsted–Lowry acid because it donates a proton to the NH_3 molecule.

An acid and a base always work together to transfer a proton. In other words, a substance can function as an acid only if another substance simultaneously behaves as a base. To be a Brønsted–Lowry acid, a molecule or ion must have a hydrogen atom that it can lose as an H^+ ion. To be a Brønsted–Lowry base, a molecule or ion must have a nonbonding pair of electrons that it can use to bind the H^+ ion.

Some substances can act as an acid in one reaction and as a base in another. For example, H_2O is a Brønsted–Lowry base in its reaction with HCl (Equation 16.3) and a Brønsted–Lowry acid in its reaction with NH_3 (Equation 16.5). A substance that is capable of acting as either an acid or a base is called **amphiprotic**. An amphiprotic substance acts as a base when combined with something more strongly acidic than itself, and as an acid when combined with something more strongly basic than itself.

GIVE IT SOME THOUGHT

In the forward reaction, which substance acts as the Brønsted–Lowry base: $HSO_4^-(aq) + NH_3(aq) \rightleftharpoons SO_4^{2-}(aq) + NH_4^+(aq)$?

Conjugate Acid-Base Pairs

In any acid-base equilibrium both the forward reaction (to the right) and the reverse reaction (to the left) involve proton transfers. For example, consider the reaction of an acid, which we will denote HX, with water.

$$HX(aq) + H_2O(l) \rightleftharpoons X^-(aq) + H_3O^+(aq) \quad [16.6]$$

In the forward reaction HX donates a proton to H_2O. Therefore, HX is the Brønsted–Lowry acid, and H_2O is the Brønsted–Lowry base. In the reverse reaction the H_3O^+ ion donates a proton to the X^- ion, so H_3O^+ is the acid and X^- is the base. When the acid HX donates a proton, it leaves behind a substance, X^-, which can act as a base. Likewise, when H_2O acts as a base, it generates H_3O^+, which can act as an acid.

An acid and a base such as HX and X^- that differ only in the presence or absence of a proton are called a **conjugate acid-base pair**.* Every acid has a **conjugate base**, formed by removing a proton from the acid. For example, OH^- is the conjugate base of H_2O, and X^- is the conjugate base of HX. Similarly, every base has associated with it a **conjugate acid**, formed by adding a proton to the base. Thus, H_3O^+ is the conjugate acid of H_2O, and HX is the conjugate acid of X^-.

* The word *conjugate* means "joined together as a pair."

In any acid-base (proton-transfer) reaction we can identify two sets of conjugate acid-base pairs. For example, consider the reaction between nitrous acid (HNO_2) and water:

remove H^+

$$HNO_2(aq) + H_2O(l) \rightleftharpoons NO_2^-(aq) + H_3O^+(aq) \qquad [16.7]$$

Acid | Base | Conjugate base | Conjugate acid

add H^+

Likewise, for the reaction between NH_3 and H_2O (Equation 16.5), we have

ACTIVITY Conjugate Acids and Bases

add H^+

$$NH_3(aq) + H_2O(l) \rightleftharpoons NH_4^+(aq) + OH^-(aq) \qquad [16.8]$$

Base | Acid | Conjugate acid | Conjugate base

remove H^+

SAMPLE EXERCISE 16.1 | Identifying Conjugate Acids and Bases

(a) What is the conjugate base of each of the following acids: $HClO_4$, H_2S, PH_4^+, HCO_3^-? **(b)** What is the conjugate acid of each of the following bases: CN^-, SO_4^{2-}, H_2O, HCO_3^-?

Solution

Analyze: We are asked to give the conjugate base for each of a series of species and to give the conjugate acid for each of another series of species.

Plan: The conjugate base of a substance is simply the parent substance minus one proton, and the conjugate acid of a substance is the parent substance plus one proton.

Solve: **(a)** $HClO_4$ less one proton (H^+) is ClO_4^-. The other conjugate bases are HS^-, PH_3, and CO_3^{2-}. **(b)** CN^- plus one proton (H^+) is HCN. The other conjugate acids are HSO_4^-, H_3O^+, and H_2CO_3.

Notice that the hydrogen carbonate ion (HCO_3^-) is amphiprotic: It can act as either an acid or a base.

PRACTICE EXERCISE

Write the formula for the conjugate acid of each of the following: HSO_3^-, F^-, PO_4^{3-}, CO.

Answers: H_2SO_3, HF, HPO_4^{2-}, HCO^+

SAMPLE EXERCISE 16.2 | Writing Equations for Proton-Transfer Reactions

The hydrogen sulfite ion (HSO_3^-) is amphiprotic. **(a)** Write an equation for the reaction of HSO_3^- with water, in which the ion acts as an acid. **(b)** Write an equation for the reaction of HSO_3^- with water, in which the ion acts as a base. In both cases identify the conjugate acid-base pairs.

Solution

Analyze and Plan: We are asked to write two equations representing reactions between HSO_3^- and water, one in which HSO_3^- should donate a proton to water, thereby acting as a Brønsted–Lowry acid, and one in which HSO_3^- should accept a proton from water, thereby acting as a base. We are also asked to identify the conjugate pairs in each equation.

Solve: (a)

$$HSO_3^-(aq) + H_2O(l) \rightleftharpoons SO_3^{2-}(aq) + H_3O^+(aq)$$

The conjugate pairs in this equation are HSO_3^- (acid) and SO_3^{2-} (conjugate base); and H_2O (base) and H_3O^+ (conjugate acid).

(b)

$$HSO_3^-(aq) + H_2O(l) \rightleftharpoons H_2SO_3(aq) + OH^-(aq)$$

The conjugate pairs in this equation are H_2O (acid) and OH^- (conjugate base), and HSO_3^- (base) and H_2SO_3 (conjugate acid).

PRACTICE EXERCISE

When lithium oxide (Li_2O) is dissolved in water, the solution turns basic from the reaction of the oxide ion (O^{2-}) with water. Write the reaction that occurs, and identify the conjugate acid-base pairs.

Answer: $O^{2-}(aq) + H_2O(l) \rightleftharpoons OH^-(aq) + OH^-(aq)$. OH^- is the conjugate acid of the base O^{2-}. OH^- is also the conjugate base of the acid H_2O.

Relative Strengths of Acids and Bases

Some acids are better proton donors than others; likewise, some bases are better proton acceptors than others. If we arrange acids in order of their ability to donate a proton, we find that the more easily a substance gives up a proton, the less easily its conjugate base accepts a proton. Similarly, the more easily a base accepts a proton, the less easily its conjugate acid gives up a proton. In other words, *the stronger an acid, the weaker is its conjugate base; the stronger a base, the weaker is its conjugate acid.* Thus, if we know something about the strength of an acid (its ability to donate protons), we also know something about the strength of its conjugate base (its ability to accept protons).

The inverse relationship between the strengths of acids and the strengths of their conjugate bases is illustrated in Figure 16.4 ◀. Here we have grouped acids and bases into three broad categories based on their behavior in water.

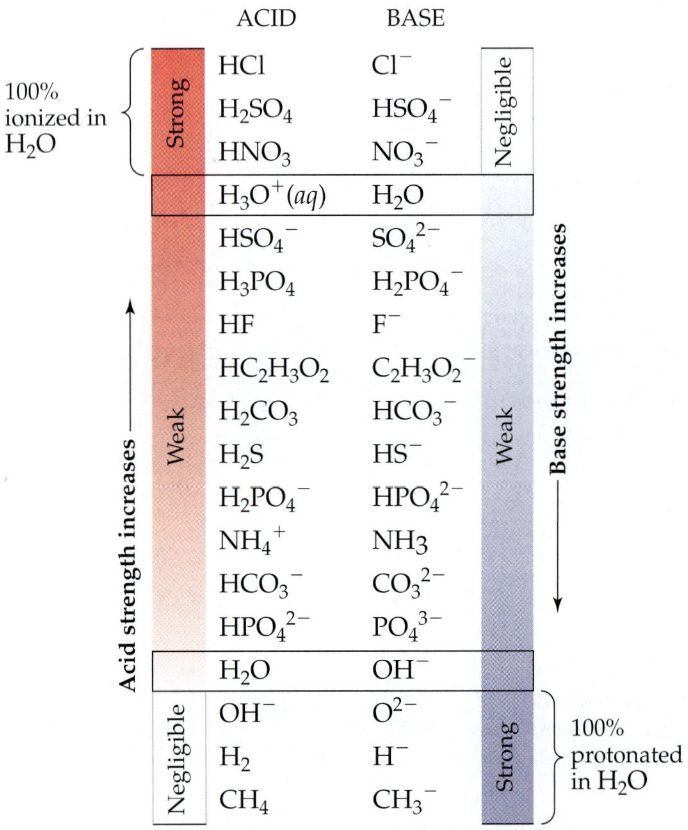

▲ **Figure 16.4 Relative strengths of some conjugate acid-base pairs.** The two members of each pair are listed opposite each other in the two columns. The acids decrease in strength from top to bottom, whereas their conjugate bases increase in strength from top to bottom.

1. The *strong acids* completely transfer their protons to water, leaving no undissociated molecules in solution. ∞ (Section 4.3) Their conjugate bases have a negligible tendency to be protonated (to abstract protons) in aqueous solution.
2. The *weak acids* only partially dissociate in aqueous solution and therefore exist in the solution as a mixture of acid molecules and their constituent ions. The conjugate bases of weak acids show a slight ability to remove protons from water. (*The conjugate bases of weak acids are weak bases.*)
3. The substances with *negligible acidity* are those such as CH_4 that contain hydrogen but do not demonstrate any acidic behavior in water. Their conjugate bases are strong bases, reacting completely with water, abstracting protons to form OH^- ions.

GIVE IT SOME THOUGHT

Using the three categories above, specify the strength of HNO_3 and the strength of its conjugate base, NO_3^-.

We can think of proton-transfer reactions as being governed by the relative abilities of two bases to abstract protons. For example, consider the proton transfer that occurs when an acid HX dissolves in water:

$$HX(aq) + H_2O(l) \rightleftharpoons H_3O^+(aq) + X^-(aq) \qquad [16.9]$$

If H_2O (the base in the forward reaction) is a stronger base than X^- (the conjugate base of HX), then H_2O will abstract the proton from HX to produce H_3O^+ and X^-. As a result, the equilibrium will lie to the right. This describes the behavior of a strong acid in water. For example, when HCl dissolves in water, the solution consists almost entirely of H_3O^+ and Cl^- ions with a negligible concentration of HCl molecules.

$$HCl(g) + H_2O(l) \longrightarrow H_3O^+(aq) + Cl^-(aq) \qquad [16.10]$$

H_2O is a stronger base than Cl^- (Figure 16.4), so H_2O acquires the proton to become the hydronium ion.

When X^- is a stronger base than H_2O, the equilibrium will lie to the left. This situation occurs when HX is a weak acid. For example, an aqueous solution of acetic acid ($HC_2H_3O_2$) consists mainly of $HC_2H_3O_2$ molecules with only a relatively few H_3O^+ and $C_2H_3O_2^-$ ions.

$$HC_2H_3O_2(aq) + H_2O(l) \rightleftharpoons H_3O^+(aq) + C_2H_3O_2^-(aq) \qquad [16.11]$$

$C_2H_3O_2^-$ is a stronger base than H_2O (Figure 16.4) and therefore abstracts the proton from H_3O^+. From these examples, we conclude that *in every acid-base reaction the position of the equilibrium favors transfer of the proton to the stronger base.* That is, the position of the equilibrium favors the reaction of the stronger acid and the stronger base to form the weaker acid and the weaker base. As a result, the equilibrium mixture contains more of the weaker acid and weaker base and less of the stronger acid and stronger base.

SAMPLE EXERCISE 16.3 | Predicting the Position of a Proton-Transfer Equilibrium

For the following proton-transfer reaction, use Figure 16.4 to predict whether the equilibrium lies predominantly to the left (that is, $K_c < 1$) or to the right ($K_c > 1$):

$$HSO_4^-(aq) + CO_3^{2-}(aq) \rightleftharpoons SO_4^{2-}(aq) + HCO_3^-(aq)$$

Solution

Analyze: We are asked to predict whether the equilibrium shown lies to the right, favoring products, or to the left, favoring reactants.

Plan: This is a proton-transfer reaction, and the position of the equilibrium will favor the proton going to the stronger of two bases. The two bases in the equation are CO_3^{2-}, the base in the forward reaction as written, and SO_4^{2-}, the conjugate base of HSO_4^-. We can find the relative positions of these two bases in Figure 16.4 to determine which is the stronger base.

Solve: CO_3^{2-} appears lower in the right-hand column in Figure 16.4 and is therefore a stronger base than SO_4^{2-}. CO_3^{2-}, therefore, will get the proton preferentially to become HCO_3^-, while SO_4^{2-} will remain mostly unprotonated. The resulting equilibrium will lie to the right, favoring products (that is, $K_c > 1$).

$$\underset{\text{Acid}}{HSO_4^-(aq)} + \underset{\text{Base}}{CO_3^{2-}(aq)} \rightleftharpoons \underset{\text{Conjugate base}}{SO_4^{2-}(aq)} + \underset{\text{Conjugate acid}}{HCO_3^-(aq)} \qquad K_c > 1$$

Comment: Of the two acids in the equation, HSO_4^- and HCO_3^-, the stronger one gives up a proton while the weaker one retains its proton. Thus, the equilibrium favors the direction in which the proton moves from the stronger acid and becomes bonded to the stronger base.

PRACTICE EXERCISE

For each of the following reactions, use Figure 16.4 to predict whether the equilibrium lies predominantly to the left or to the right:

(a) $HPO_4^{2-}(aq) + H_2O(l) \rightleftharpoons H_2PO_4^-(aq) + OH^-(aq)$

(b) $NH_4^+(aq) + OH^-(aq) \rightleftharpoons NH_3(aq) + H_2O(l)$

Answers: **(a)** left, **(b)** right

16.3 | The Autoionization of Water

One of the most important chemical properties of water is its ability to act as either a Brønsted acid or a Brønsted base, depending on the circumstances. In the presence of an acid, water acts as a proton acceptor; in the presence of a base, water acts as a proton donor. In fact, one water molecule can donate a proton to another water molecule:

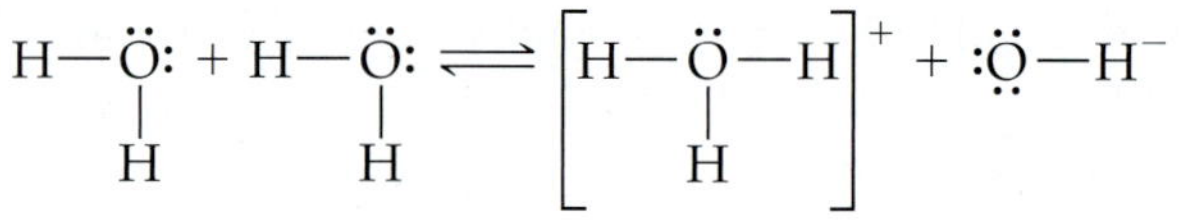
[16.12]

We call this process the **autoionization** of water. No individual molecule remains ionized for long; the reactions are extremely rapid in both directions. At room temperature only about two out of every 10^9 molecules are ionized at any given instant. Thus, pure water consists almost entirely of H_2O molecules and is an extremely poor conductor of electricity. Nevertheless, the autoionization of water is very important, as we will soon see.

ACTIVITY K_w

The Ion Product of Water

Because the autoionization of water (Equation 16.12) is an equilibrium process, we can write the following equilibrium-constant expression for it:

$$K_c = [H_3O^+][OH^-] \qquad [16.13]$$

The term $[H_2O]$ is excluded from the equilibrium-constant expression because we exclude the concentrations of pure solids and liquids. (Section 15.4) Because this equilibrium-constant expression refers specifically to the autoionization of water, we use the symbol K_w to denote the equilibrium constant, which we call the **ion-product constant** for water. At 25°C, K_w equals 1.0×10^{-14}. Thus, we have

$$K_w = [H_3O^+][OH^-] = 1.0 \times 10^{-14} \text{ (at 25°C)} \qquad [16.14]$$

Because we use $H^+(aq)$ and $H_3O(aq)^+$ interchangeably to represent the hydrated proton, the autoionization reaction for water can also be written as

$$H_2O(l) \rightleftharpoons H^+(aq) + OH^-(aq) \qquad [16.15]$$

Likewise, the expression for K_w can be written in terms of either H_3O^+ or H^+, and K_w has the same value in either case:

$$K_w = [H_3O^+][OH^-] = [H^+][OH^-] = 1.0 \times 10^{-14} \text{ (at 25°C)} \qquad [16.16]$$

This equilibrium-constant expression and the value of K_w at 25°C are extremely important, and you should commit them to memory.

What makes Equation 16.16 particularly useful is that it is not only applicable to pure water but also to any aqueous solution. Although the equilibrium between $H^+(aq)$ and $OH^-(aq)$ as well as other ionic equilibria are affected somewhat by the presence of additional ions in solution, it is customary to ignore these ionic effects except in work requiring exceptional accuracy. Thus, Equation 16.16 is taken to be valid for any dilute aqueous solution, and it can be used to calculate either $[H^+]$ (if $[OH^-]$ is known) or $[OH^-]$ (if $[H^+]$ is known).

A solution in which $[H^+] = [OH^-]$ is said to be *neutral*. In most solutions H^+ and OH^- concentrations are not equal. As the concentration of one of these ions increases, the concentration of the other must decrease, so that the product of their concentrations equals 1.0×10^{-14}. In acidic solutions $[H^+]$ exceeds $[OH^-]$. In basic solutions $[OH^-]$ exceeds $[H^+]$.

SAMPLE EXERCISE 16.4 | Calculating $[H^+]$ for Pure Water

Calculate the values of $[H^+]$ and $[OH^-]$ in a neutral solution at 25°C.

Solution

Analyze: We are asked to determine the concentrations of hydronium and hydroxide ions in a neutral solution at 25°C.

Plan: We will use Equation 16.16 and the fact that, by definition, $[H^+] = [OH^-]$ in a neutral solution.

Solve: We will represent the concentration of $[H^+]$ and $[OH^-]$ in neutral solution with x. This gives

$$[H^+][OH^-] = (x)(x) = 1.0 \times 10^{-14}$$

$$x^2 = 1.0 \times 10^{-14}$$

$$x = 1.0 \times 10^{-7}\ M = [H^+] = [OH^-]$$

In an acid solution $[H^+]$ is greater than $1.0 \times 10^{-7}\ M$; in a basic solution $[H^+]$ is less than $1.0 \times 10^{-7}\ M$.

PRACTICE EXERCISE

Indicate whether solutions with each of the following ion concentrations are neutral, acidic, or basic: **(a)** $[H^+] = 4 \times 10^{-9}\ M$; **(b)** $[OH^-] = 1 \times 10^{-7}\ M$; **(c)** $[OH^-] = 7 \times 10^{-13}\ M$.

Answers: **(a)** basic, **(b)** neutral, **(c)** acidic

SAMPLE EXERCISE 16.5 | Calculating $[H^+]$ from $[OH^-]$

Calculate the concentration of $H^+(aq)$ in **(a)** a solution in which $[OH^-]$ is 0.010 M, **(b)** a solution in which $[OH^-]$ is $1.8 \times 10^{-9}\ M$. *Note:* In this problem and all that follow, we assume, unless stated otherwise, that the temperature is 25°C.

Solution

Analyze: We are asked to calculate the hydronium ion concentration in an aqueous solution where the hydroxide concentration is known.

Plan: We can use the equilibrium-constant expression for the autoionization of water and the value of K_w to solve for each unknown concentration.

Solve: **(a)** Using Equation 16.16, we have:

$$[H^+][OH^-] = 1.0 \times 10^{-14}$$

$$[H^+] = \frac{1.0 \times 10^{-14}}{[OH^-]} = \frac{1.0 \times 10^{-14}}{0.010} = 1.0 \times 10^{-12}\ M$$

This solution is basic because

$$[OH^-] > [H^+]$$

(b) In this instance $$[H^+] = \frac{1.0 \times 10^{-14}}{[OH^-]} = \frac{1.0 \times 10^{-14}}{1.8 \times 10^{-9}} = 5.6 \times 10^{-6}\ M$$

This solution is acidic because $[H^+] > [OH^-]$

PRACTICE EXERCISE

Calculate the concentration of $OH^-(aq)$ in a solution in which **(a)** $[H^+] = 2 \times 10^{-6}\ M$; **(b)** $[H^+] = [OH^-]$; **(c)** $[H^+] = 100 \times [OH^-]$.

Answers: **(a)** $5 \times 10^{-9}\ M$, **(b)** $1.0 \times 10^{-7}\ M$, **(c)** $1.0 \times 10^{-8}\ M$

16.4 | The pH Scale

The molar concentration of $H^+(aq)$ in an aqueous solution is usually very small. For convenience, we therefore usually express $[H^+]$ in terms of **pH**, which is the negative logarithm in base 10 of $[H^+]$.*

$$pH = -\log[H^+] \qquad [16.17]$$

If you need to review the use of logs, see Appendix A.

We can use Equation 16.17 to calculate the pH of a neutral solution at 25°C (that is, one in which $[H^+] = 1.0 \times 10^{-7}\ M$):

$$pH = -\log(1.0 \times 10^{-7}) = -(-7.00) = 7.00$$

The pH of a neutral solution is 7.00 at 25°C.

What happens to the pH of a solution as we make the solution acidic? An acidic solution is one in which $[H^+] > 1.0 \times 10^{-7}\ M$. Because of the negative sign in Equation 16.17, *the pH decreases as* $[H^+]$ *increases*. For example, the pH of an acidic solution in which $[H^+] = 1.0 \times 10^{-3}\ M$ is

$$pH = -\log(1.0 \times 10^{-3}) = -(-3.00) = 3.00$$

ACTIVITY
pH Estimation

At 25°C the pH of an acidic solution is less than 7.00.

We can also calculate the pH of a basic solution, one in which $[OH^-] > 1.0 \times 10^{-7}\ M$. Suppose $[OH^-] = 2.0 \times 10^{-3}\ M$. We can use Equation 16.16 to calculate $[H^+]$ for this solution, and Equation 16.17 to calculate the pH:

$$[H^+] = \frac{K_w}{[OH^-]} = \frac{1.0 \times 10^{-14}}{2.0 \times 10^{-3}} = 5.0 \times 10^{-12}\ M$$

$$pH = -\log(5.0 \times 10^{-12}) = 11.30$$

At 25°C the pH of a basic solution is greater than 7.00. The relationships among $[H^+]$, $[OH^-]$, and pH are summarized in Table 16.1 ▼.

TABLE 16.1 Relationships Among $[H^+]$, $[OH^-]$, and pH at 25°C

Solution Type	$[H^+]$ (*M*)	$[OH^-]$ (*M*)	pH Value
Acidic	$>1.0 \times 10^{-7}$	$<1.0 \times 10^{-7}$	<7.00
Neutral	$=1.0 \times 10^{-7}$	$=1.0 \times 10^{-7}$	$=7.00$
Basic	$<1.0 \times 10^{-7}$	$>1.0 \times 10^{-7}$	>7.00

* Because $[H^+]$ and $[H_3O^+]$ are used interchangeably, you might see pH defined as $-\log[H_3O^+]$.

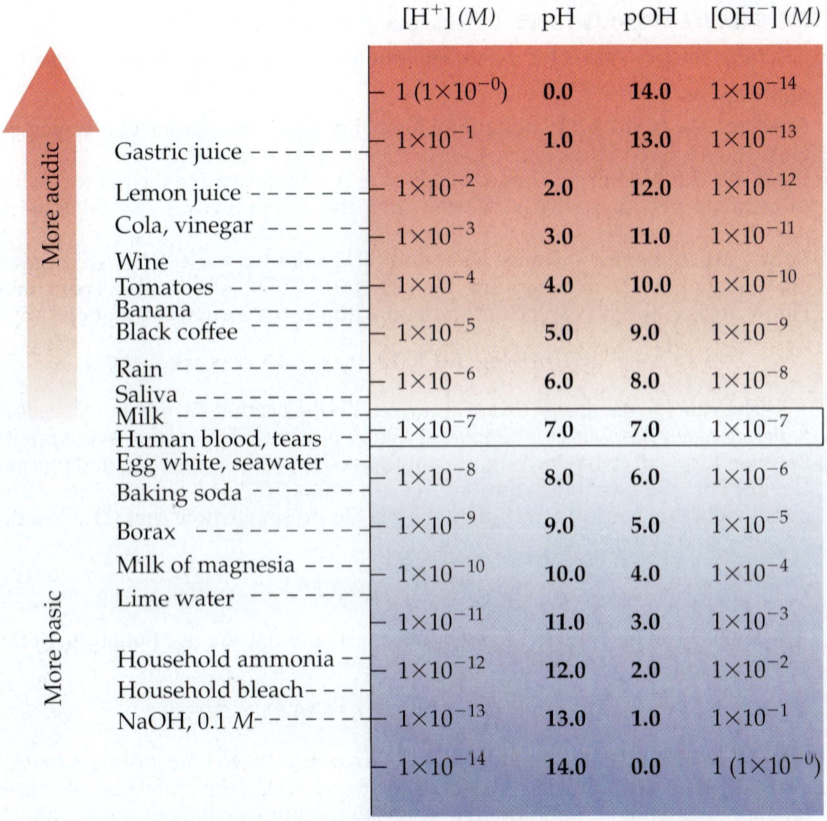

◀ **Figure 16.5 H^+ concentrations and pH values of some common substances at 25°C.** The pH and pOH of a solution can be estimated using the benchmark concentrations of H^+ and OH^- corresponding to whole-number pH values.

ACTIVITY Acids and Bases

The pH values characteristic of several familiar solutions are shown in Figure 16.5 ▲. Notice that a change in $[H^+]$ by a factor of 10 causes the pH to change by 1. Thus, a solution of pH 6 has 10 times the concentration of $H^+(aq)$ as a solution of pH 7.

GIVE IT SOME THOUGHT

(a) What is the significance of pH = 7? **(b)** How does the pH change as OH^- is added to the solution?

You might think that when $[H^+]$ is very small, as it is for some of the examples shown in Figure 16.5, it would be unimportant. Nothing is further from the truth. If $[H^+]$ is part of a kinetic rate law, then changing its concentration will change the rate. (Section 14.3) Thus, if the rate law is first order in $[H^+]$, doubling its concentration will double the rate even if the change is merely from 1×10^{-7} *M* to 2×10^{-7} *M*. In biological systems many reactions involve proton transfers and have rates that depend on $[H^+]$. Because the speeds of these reactions are crucial, the pH of biological fluids must be maintained within narrow limits. For example, human blood has a normal pH range of 7.35 to 7.45. Illness and even death can result if the pH varies much from this narrow range.

A convenient way to estimate pH is to use the "benchmark" H^+ concentrations in Figure 16.5, those for which $[H^+]$ equals 1×10^{-x}, where x is a whole number from 0 to 14. When $[H^+]$ is one of these benchmark concentrations, the pH is simply the corresponding pH value, x. When $[H^+] = 1 \times 10^{-4}$, for example, the pH is simply 4. When $[H^+]$ falls between two benchmark concentrations, the pH will fall between the two corresponding pH values. Consider a solution that is 0.050 *M* in H^+. Because 0.050 (that is, 5.0×10^{-2}) is greater than 1.0×10^{-2} and less than 1.0×10^{-1}, we estimate the pH to be between 2.00 and 1.00. Using Equation 16.17 to calculate the pH gives 1.30.

SAMPLE EXERCISE 16.6 | Calculating pH from $[H^+]$

Calculate the pH values for the two solutions described in Sample Exercise 16.5.

Solution

Analyze: We are asked to determine the pH of aqueous solutions for which we have already calculated $[H^+]$.

Plan: We can use the benchmarks in Figure 16.5 to determine the pH for part (a) and to estimate pH for part (b). We can then use Equation 16.17 to calculate pH for part (b).

Solve: **(a)** In the first instance we found $[H^+]$ to be 1.0×10^{-12} *M*. Although we can use Equation 16.17 to determine the pH, 1.0×10^{-12} is one of the benchmarks in Figure 16.5, so the pH can be determined without any formal calculation.

$$\text{pH} = -\log(1.0 \times 10^{-12}) = -(-12.00) = 12.00$$

The rule for using significant figures with logs is that *the number of decimal places in the log equals the number of significant figures in the original number* (see Appendix A). Because **1.0** $\times 10^{-12}$ has two significant figures, the pH has two decimal places, 12.00.

(b) For the second solution, $[H^+] = 5.6 \times 10^{-6}$ *M*. Before performing the calculation, it is helpful to estimate the pH. To do so, we note that $[H^+]$ lies between 1×10^{-6} and 1×10^{-5}.

$$1 \times 10^{-6} < 5.6 \times 10^{-6} < 1 \times 10^{-5}$$

Thus, we expect the pH to lie between 6.0 and 5.0. We use Equation 16.17 to calculate the pH.

$$\text{pH} = -\log(5.6 \times 10^{-6}) = 5.25$$

Check: After calculating a pH, it is useful to compare it to your prior estimate. In this case the pH, as we predicted, falls between 6 and 5. Had the calculated pH and the estimate not agreed, we should have reconsidered our calculation or estimate or both. Note that although $[H^+]$ lies halfway between the two benchmark concentrations, the calculated pH does not lie halfway between the two corresponding pH values. This is because the pH scale is logarithmic rather than linear.

PRACTICE EXERCISE

(a) In a sample of lemon juice $[H^+]$ is 3.8×10^{-4} *M*. What is the pH? **(b)** A commonly available window-cleaning solution has a $[H^+]$ of 5.3×10^{-9} *M*. What is the pH?

Answers: **(a)** 3.42, **(b)** 8.28

SAMPLE EXERCISE 16.7 | Calculating $[H^+]$ from pH

A sample of freshly pressed apple juice has a pH of 3.76. Calculate $[H^+]$.

Solution

Analyze: We need to calculate $[H^+]$ from pH.

Plan: We will use Equation 16.17, $\text{pH} = -\log[H^+]$, for the calculation.

Solve: From Equation 16.17, we have

$$\text{pH} = -\log[H^+] = 3.76$$

Thus,

$$\log[H^+] = -3.76$$

To find $[H^+]$, we need to determine the *antilog* of −3.76. Scientific calculators have an antilog function (sometimes labeled INV log or 10^x) that allows us to perform the calculation:

$$[H^+] = \text{antilog}(-3.76) = 10^{-3.76} = 1.7 \times 10^{-4}\ M$$

Comment: Consult the user's manual for your calculator to find out how to perform the antilog operation. The number of significant figures in $[H^+]$ is two because the number of decimal places in the pH is two.

Check: Because the pH is between 3.0 and 4.0, we know that $[H^+]$ will be between 1×10^{-3} and 1×10^{-4} *M*. Our calculated $[H^+]$ falls within this estimated range.

PRACTICE EXERCISE

A solution formed by dissolving an antacid tablet has a pH of 9.18. Calculate $[H^+]$.
Answer: $[H^+] = 6.6 \times 10^{-10}$ *M*

Other "p" Scales

The negative log is also a convenient way of expressing the magnitudes of other small quantities. We use the convention that the negative log of a quantity is labeled "p" (quantity). For example, one can express the concentration of OH^- as pOH:

$$\text{pOH} = -\log[\text{OH}^-] \qquad [16.18]$$

By taking the negative log of both sides of Equation 16.16,

$$-\log[\text{H}^+] + (-\log[\text{OH}^-]) = -\log K_w \qquad [16.19]$$

we obtain the following useful expression:

$$\text{pH} + \text{pOH} = 14.00 \text{ (at 25°C)} \qquad [16.20]$$

We will see in Section 16.8 that p scales are also useful when working with equilibrium constants.

GIVE IT SOME THOUGHT

If the pOH for a solution is 3.00, what is the pH of the solution? Is the solution acidic or basic?

Measuring pH

The pH of a solution can be measured quickly and accurately with a *pH meter* (Figure 16.6 ▶). A complete understanding of how this important device works requires a knowledge of electrochemistry, a subject we take up in Chapter 20. In brief, a pH meter consists of a pair of electrodes connected to a meter capable of measuring small voltages, on the order of millivolts. A voltage, which varies with the pH, is generated when the electrodes are placed in a solution. This voltage is read by the meter, which is calibrated to give pH.

The electrodes used with pH meters come in many shapes and sizes, depending on their intended use. Electrodes have even been developed that are so small that they can be inserted into single living cells in order to monitor the pH of the cell medium. Pocket-size pH meters are also available for use in environmental studies, in monitoring industrial effluents, and in agricultural work.

▲ **Figure 16.6 A digital pH meter.** The device is a millivoltmeter, and the electrodes immersed in the solution being tested produce a voltage that depends on the pH of the solution.

Although less precise, acid-base indicators can be used to measure pH. An acid-base indicator is a colored substance that itself can exist in either an acid or a base form. The two forms have different colors. Thus, the indicator turns one color in an acid and another color in a base. If you know the pH at which the indicator turns from one form to the other, you can determine whether a solution has a higher or lower pH than this value. Litmus, for example, changes color in the vicinity of pH 7. The color change, however, is not very sharp. Red litmus indicates a pH of about 5 or lower, and blue litmus indicates a pH of about 8 or higher.

Some of the more common indicators are listed in Figure 16.7 ▶. Methyl orange, for example, changes color over the pH interval from 3.1 to 4.4. Below pH 3.1 it is in the acid form, which is red. In the interval between 3.1 and 4.4, it is gradually converted to its basic form, which has a yellow color. By pH 4.4 the conversion is complete, and the solution is yellow. Paper tape that is

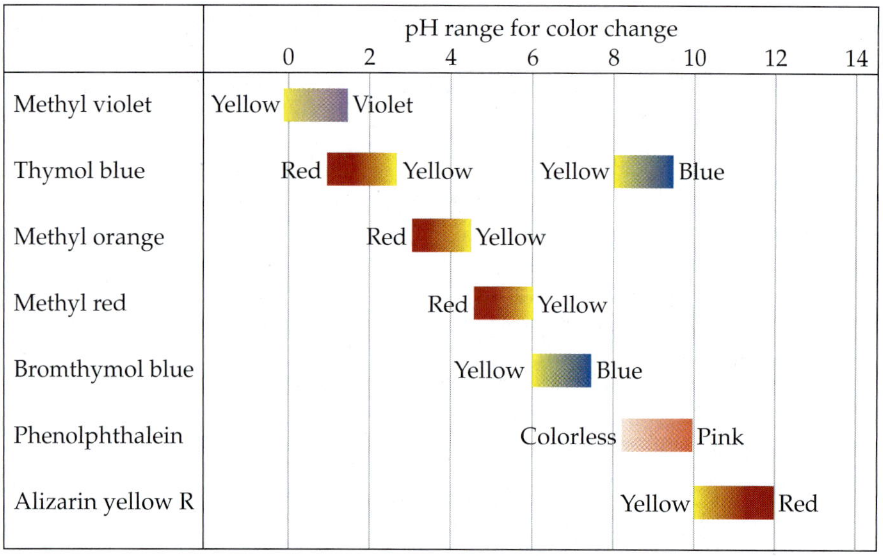

▶ **Figure 16.7 Some common acid-base indicators.** The pH ranges for the color changes of some common acid-base indicators. Most indicators have a useful range of about 2 pH units.

MOVIE
Natural Indicators

impregnated with several indicators and comes complete with a comparator color scale is widely used for approximate determinations of pH.

GIVE IT SOME THOUGHT

If phenolphthalein turns pink when added to a solution, what can we conclude about the pH of the solution?

16.5 | Strong Acids and Bases

The chemistry of an aqueous solution often depends critically on the pH of the solution. It is therefore important to examine how the pH of solutions relate to the concentrations of acids and bases. The simplest cases are those involving strong acids and strong bases. Strong acids and bases are *strong electrolytes*, existing in aqueous solution entirely as ions. There are relatively few common strong acids and bases, and we listed these substances in Table 4.2.

Strong Acids

The seven most common strong acids include six monoprotic acids (HCl, HBr, HI, HNO_3, $HClO_3$, and $HClO_4$), and one diprotic acid (H_2SO_4). Nitric acid (HNO_3) exemplifies the behavior of the monoprotic strong acids. For all practical purposes, an aqueous solution of HNO_3 consists entirely of H_3O^+ and NO_3^- ions.

$$HNO_3(aq) + H_2O(l) \longrightarrow H_3O^+(aq) + NO_3^-(aq) \text{ (complete ionization)} \quad [16.21]$$

We have not used equilibrium arrows for Equation 16.21 because the reaction lies entirely to the right, the side with the ions. (Section 4.1) As noted in Section 16.3, we use $H_3O^+(aq)$ and $H^+(aq)$ interchangeably to represent the hydrated proton in water. Thus, we often simplify the equations for the ionization reactions of acids as follows:

$$HNO_3(aq) \longrightarrow H^+(aq) + NO_3^-(aq)$$

In an aqueous solution of a strong acid, the acid is normally the only significant source of H^+ ions.* As a result, calculating the pH of a solution of a strong

* If the concentration of the acid is 10^{-6} *M* or less, we also need to consider H^+ ions that result from the autoionization of H_2O. Normally, the concentration of H^+ from H_2O is so small that it can be neglected.

monoprotic acid is straightforward because $[H^+]$ equals the original concentration of acid. In a 0.20 M solution of $HNO_3(aq)$, for example, $[H^+] = [NO_3^-] = 0.20\ M$. The situation with the diprotic acid H_2SO_4 is more complex, as we will see in Section 16.6.

SAMPLE EXERCISE 16.8 | Calculating the pH of a Strong Acid

What is the pH of a 0.040 M solution of $HClO_4$?

Solution

Analyze and Plan: We are asked to calculate the pH of a 0.040 M solution of $HClO_4$. Because $HClO_4$ is a strong acid, it is completely ionized, giving $[H^+] = [ClO_4^-] = 0.040\ M$. Because $[H^+]$ lies between benchmarks 1×10^{-2} and 1×10^{-1} in Figure 16.5, we estimate that the pH will be between 2.0 and 1.0.

Solve: The pH of the solution is given by

$$\text{pH} = -\log(0.040) = 1.40.$$

Check: Our calculated pH falls within the estimated range.

PRACTICE EXERCISE

An aqueous solution of HNO_3 has a pH of 2.34. What is the concentration of the acid?
Answer: 0.0046 M

Strong Bases

There are relatively few common strong bases. The most common soluble strong bases are the ionic hydroxides of the alkali metals (group 1A) and the heavier alkaline earth metals (group 2A), such as NaOH, KOH, and $Ca(OH)_2$. These compounds completely dissociate into ions in aqueous solution. Thus, a solution labeled 0.30 M NaOH consists of 0.30 M $Na^+(aq)$ and 0.30 M $OH^-(aq)$; there is essentially no undissociated NaOH.

Because these strong bases dissociate entirely into ions in aqueous solution, calculating the pH of their solutions is also straightforward, as shown in Sample Exercise 16.9.

SAMPLE EXERCISE 16.9 | Calculating the pH of a Strong Base

What is the pH of **(a)** a 0.028 M solution of NaOH, **(b)** a 0.0011 M solution of $Ca(OH)_2$?

Solution

Analyze: We're asked to calculate the pH of two solutions, given the concentration of strong base for each.

Plan: We can calculate each pH by two equivalent methods. First, we could use Equation 16.16 to calculate $[H^+]$ and then use Equation 16.17 to calculate the pH. Alternatively, we could use $[OH^-]$ to calculate pOH and then use Equation 16.20 to calculate the pH.

Solve: **(a)** NaOH dissociates in water to give one OH^- ion per formula unit. Therefore, the OH^- concentration for the solution in (a) equals the stated concentration of NaOH, namely 0.028 M.

Method 1:

$$[H^+] = \frac{1.0 \times 10^{-14}}{0.028} = 3.57 \times 10^{-13}\ M \quad \text{pH} = -\log(3.57 \times 10^{-13}) = 12.45$$

Method 2:

$$\text{pOH} = -\log(0.028) = 1.55 \qquad \text{pH} = 14.00 - \text{pOH} = 12.45$$

(b) $Ca(OH)_2$ is a strong base that dissociates in water to give two OH^- ions per formula unit. Thus, the concentration of $OH^-(aq)$ for the solution in part (b) is $2 \times (0.0011\ M) = 0.0022\ M$.

Method 1:

$$[H^+] = \frac{1.0 \times 10^{-14}}{0.0022} = 4.55 \times 10^{-12}\,M \quad pH = -\log(4.55 \times 10^{-12}) = 11.34$$

Method 2:

$$pOH = -\log(0.0022) = 2.66 \qquad pH = 14.00 - pOH = 11.34$$

PRACTICE EXERCISE

What is the concentration of a solution of **(a)** KOH for which the pH is 11.89; **(b)** $Ca(OH)_2$ for which the pH is 11.68?
Answers: **(a)** $7.8 \times 10^{-3}\,M$, **(b)** $2.4 \times 10^{-3}\,M$

Although all the hydroxides of the alkali metals (group 1A) are strong electrolytes, LiOH, RbOH, and CsOH are not commonly encountered in the laboratory. The hydroxides of the heavier alkaline earth metals, $Ca(OH)_2$, $Sr(OH)_2$, and $Ba(OH)_2$, are also strong electrolytes. They have limited solubilities, however, so they are used only when high solubility is not critical.

Strongly basic solutions are also created by certain substances that react with water to form $OH^-(aq)$. The most common of these contain the oxide ion. Ionic metal oxides, especially Na_2O and CaO, are often used in industry when a strong base is needed. Each mole of O^{2-} reacts with water to form 2 mol of OH^-, leaving virtually no O^{2-} remaining in the solution:

$$O^{2-}(aq) + H_2O(l) \longrightarrow 2\,OH^-(aq) \qquad [16.22]$$

Thus, a solution formed by dissolving 0.010 mol of $Na_2O(s)$ in enough water to form 1.0 L of solution will have $[OH^-] = 0.020\,M$ and a pH of 12.30.

Ionic hydrides and nitrides also react with H_2O to form OH^-:

$$H^-(aq) + H_2O(l) \longrightarrow H_2(g) + OH^-(aq) \qquad [16.23]$$

$$N^{3-}(aq) + 3\,H_2O(l) \longrightarrow NH_3(aq) + 3\,OH^-(aq) \qquad [16.24]$$

Because the anions O^{2-}, H^-, and N^{3-} are stronger bases than OH^- (the conjugate base of H_2O), they are able to remove a proton from H_2O.

GIVE IT SOME THOUGHT

The CH_3^- ion is the conjugate base of CH_4, and CH_4 shows no evidence of being an acid in water. What happens when CH_3^- is added to water?

16.6 | Weak Acids

ANIMATION
Introduction to Aqueous Acids

Most acidic substances are weak acids and are therefore only partially ionized in aqueous solution. We can use the equilibrium constant for the ionization reaction to express the extent to which a weak acid ionizes. If we represent a general weak acid as HA, we can write the equation for its ionization reaction in either of the following ways, depending on whether the hydrated proton is represented as $H_3O^+(aq)$ or $H^+(aq)$:

$$HA(aq) + H_2O(l) \rightleftharpoons H_3O^+(aq) + A^-(aq) \qquad [16.25]$$

or

$$HA(aq) \rightleftharpoons H^+(aq) + A^-(aq) \qquad [16.26]$$

Because $[H_2O]$ is the solvent, it is omitted from the equilibrium-constant expression, which can be written as either

$$K_c = \frac{[H_3O^+][A^-]}{[HA]} \quad \text{or} \quad K_c = \frac{[H^+][A^-]}{[HA]}$$

TABLE 16.2 Some Weak Acids in Water at 25°C*

Acid	Structural Formula	Conjugate Base	Equilibrium Reaction	K_a
Hydrofluoric (HF)	H—F	F^-	$HF(aq) + H_2O(l) \rightleftharpoons H_3O^+(aq) + F^-(aq)$	6.8×10^{-4}
Nitrous (HNO_2)	H—O—N=O	NO_2^-	$HNO_2(aq) + H_2O(l) \rightleftharpoons H_3O^+(aq) + NO_2^-(aq)$	4.5×10^{-4}
Benzoic ($HC_7H_5O_2$)	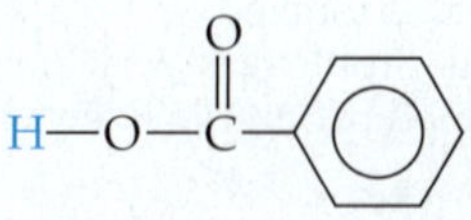	$C_7H_5O_2^-$	$HC_7H_5O_2(aq) + H_2O(l) \rightleftharpoons H_3O^+(aq) + C_7H_5O_2^-(aq)$	6.3×10^{-5}
Acetic ($HC_2H_3O_2$)	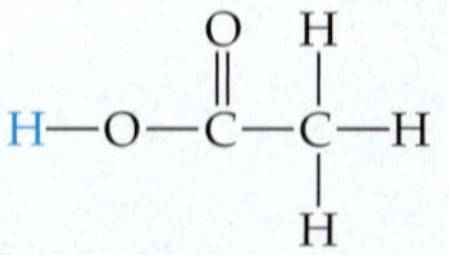	$C_2H_3O_2^-$	$HC_2H_3O_2(aq) + H_2O(l) \rightleftharpoons H_3O^+(aq) + C_2H_3O_2^-(aq)$	1.8×10^{-5}
Hypochlorous (HClO)	H—O—Cl	ClO^-	$HClO(aq) + H_2O(l) \rightleftharpoons H_3O^+(aq) + ClO^-(aq)$	3.0×10^{-8}
Hydrocyanic (HCN)	H—C≡N	CN^-	$HCN(aq) + H_2O(l) \rightleftharpoons H_3O^+(aq) + CN^-(aq)$	4.9×10^{-10}
Phenol (HC_6H_5O)	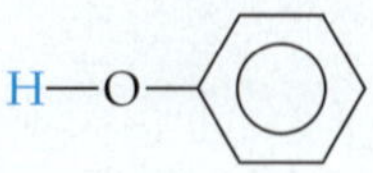	$C_6H_5O^-$	$HC_6H_5O(aq) + H_2O(l) \rightleftharpoons H_3O^+(aq) + C_6H_5O^-(aq)$	1.3×10^{-10}

*The proton that ionizes is shown in blue.

As we did for the ion-product constant for the autoionization of water, we change the subscript on this equilibrium constant to indicate the type of equation to which it corresponds.

$$K_a = \frac{[H_3O^+][A^-]}{[HA]} \quad \text{or} \quad K_a = \frac{[H^+][A^-]}{[HA]} \qquad [16.27]$$

ACTIVITY
Equilibrium Constant

The subscript *a* on K_a denotes that it is an equilibrium constant for the ionization of an acid, so K_a is called the **acid-dissociation constant**.

Table 16.2 ▲ shows the names, structures, and K_a values for several weak acids. A more complete listing is given in Appendix D. Many weak acids are organic compounds composed entirely of carbon, hydrogen, and oxygen. These compounds usually contain some hydrogen atoms bonded to carbon atoms and some bonded to oxygen atoms. In almost all cases the hydrogen atoms bonded to carbon do not ionize in water; instead, the acidic behavior of these compounds is due to the hydrogen atoms attached to oxygen atoms.

The magnitude of K_a indicates the tendency of the acid to ionize in water: *The larger the value of* K_a*, the stronger the acid.* Hydrofluoric acid (HF), for example, is the strongest acid listed in Table 16.2, and phenol (HOC_6H_5) is the weakest. Notice that K_a is typically less than 10^{-3}.

Calculating K_a from pH

In order to calculate either the K_a value for a weak acid or the pH of its solutions, we will use many of the skills for solving equilibrium problems that we developed in Section 15.5. In many cases the small magnitude of K_a allows us to use approximations to simplify the problem. In doing these calculations, it is important to realize that proton-transfer reactions are generally very rapid. As a result, the measured or calculated pH for a weak acid always represents an equilibrium condition.

SAMPLE EXERCISE 16.10 | **Calculating K_a and Percent Ionization from Measured pH**

A student prepared a 0.10 *M* solution of formic acid ($HCHO_2$) and measured its pH using a pH meter of the type illustrated in Figure 16.6. The pH at 25°C was found to be 2.38. **(a)** Calculate K_a for formic acid at this temperature. **(b)** What percentage of the acid is ionized in this 0.10 *M* solution?

Solution

Analyze: We are given the molar concentration of an aqueous solution of weak acid and the pH of the solution at 25°C, and we are asked to determine the value of K_a for the acid and the percentage of the acid that is ionized.

Plan: Although we are dealing specifically with the ionization of a weak acid, this problem is very similar to the equilibrium problems we encountered in Chapter 15. We can solve it using the method first outlined in Sample Exercise 15.8, starting with the chemical reaction and a tabulation of initial and equilibrium concentrations.

Solve: (a) The first step in solving any equilibrium problem is to write the equation for the equilibrium reaction. The ionization equilibrium for formic acid can be written as follows:

$$HCHO_2(aq) \rightleftharpoons H^+(aq) + CHO_2^-(aq)$$

The equilibrium-constant expression is

$$K_a = \frac{[H^+][CHO_2^-]}{[HCHO_2]}$$

From the measured pH, we can calculate $[H^+]$:

$$pH = -\log[H^+] = 2.38$$

$$\log[H^+] = -2.38$$

$$[H^+] = 10^{-2.38} = 4.2 \times 10^{-3}\ M$$

We can do a little accounting to determine the concentrations of the species involved in the equilibrium. We imagine that the solution is initially 0.10 *M* in $HCHO_2$ molecules. We then consider the ionization of the acid into H^+ and CHO_2^-. For each $HCHO_2$ molecule that ionizes, one H^+ ion and one CHO_2^- ion are produced in solution. Because the pH measurement indicates that $[H^+] = 4.2 \times 10^{-3}\ M$ at equilibrium, we can construct the following table:

	$HCHO_2(aq)$ $\rightleftharpoons$	$H^+(aq)$ +	$CHO_2^-(aq)$
Initial	0.10 *M*	0	0
Change	$-4.2 \times 10^{-3}\ M$	$+4.2 \times 10^{-3}\ M$	$+4.2 \times 10^{-3}\ M$
Equilibrium	$(0.10 - 4.2 \times 10^{-3})\ M$	$4.2 \times 10^{-3}\ M$	$4.2 \times 10^{-3}\ M$

Notice that we have neglected the very small concentration of $H^+(aq)$ that is due to the autoionization of H_2O. Notice also that the amount of $HCHO_2$ that ionizes is very small compared with the initial concentration of the acid. To the number of significant figures we are using, the subtraction yields 0.10 *M*:

$$(0.10 - 4.2 \times 10^{-3})\ M \simeq 0.10\ M$$

We can now insert the equilibrium concentrations into the expression for K_a:

$$K_a = \frac{(4.2 \times 10^{-3})(4.2 \times 10^{-3})}{0.10} = 1.8 \times 10^{-4}$$

Check: The magnitude of our answer is reasonable because K_a for a weak acid is usually between 10^{-3} and 10^{-10}.

(b) The percentage of acid that ionizes is given by the concentration of H^+ or CHO_2^- at equilibrium, divided by the initial acid concentration, multiplied by 100%.

$$\text{Percent ionization} = \frac{[H^+]_{\text{equilibrium}}}{[HCHO_2]_{\text{initial}}} \times 100\% = \frac{4.2 \times 10^{-3}}{0.10} \times 100\% = 4.2\%$$

PRACTICE EXERCISE

Niacin, one of the B vitamins, has the following molecular structure:

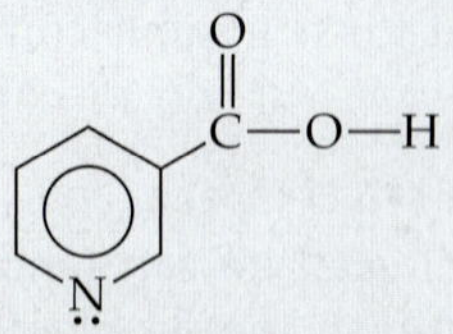

A 0.020 *M* solution of niacin has a pH of 3.26. **(a)** What percentage of the acid is ionized in this solution? **(b)** What is the acid-dissociation constant, K_a, for niacin?
Answers: **(a)** 2.7%, **(b)** 1.5×10^{-5}

Using K_a to Calculate pH

Knowing the value of K_a and the initial concentration of the weak acid, we can calculate the concentration of $H^+(aq)$ in a solution of a weak acid. Let's calculate the pH of a 0.30 *M* solution of acetic acid ($HC_2H_3O_2$), the weak acid responsible for the characteristic odor and acidity of vinegar, at 25°C.

Our *first* step is to write the ionization equilibrium for acetic acid:

$$HC_2H_3O_2(aq) \rightleftharpoons H^+(aq) + C_2H_3O_2^-(aq) \qquad [16.28]$$

VCL Ionization Constants of Weak Acids

According to the structural formula of acetic acid, shown in Table 16.2, the hydrogen that ionizes is the one attached to an oxygen atom. We write this hydrogen separate from the others in the formula to emphasize that only this one hydrogen is readily ionized.

The *second* step is to write the equilibrium-constant expression and the value for the equilibrium constant. From Table 16.2, we have $K_a = 1.8 \times 10^{-5}$. Thus, we can write the following:

$$K_a = \frac{[H^+][C_2H_3O_2^-]}{[HC_2H_3O_2]} = 1.8 \times 10^{-5} \qquad [16.29]$$

As the *third* step, we need to express the concentrations that are involved in the equilibrium reaction. This can be done with a little accounting, as described in Sample Exercise 16.10. Because we want to find the equilibrium value for $[H^+]$, let's call this quantity x. The concentration of acetic acid before any of it ionizes is 0.30 *M*. The chemical equation tells us that for each molecule of $HC_2H_3O_2$ that ionizes, one $H^+(aq)$ and one $C_2H_2O_2^-(aq)$ are formed. Consequently, if x moles per liter of $H^+(aq)$ form at equilibrium, x moles per liter of $C_2H_3O_2^-(aq)$ must also form, and x moles per liter of $HC_2H_3O_2$ must be ionized. This gives rise to the following table with the equilibrium concentrations shown on the last line:

	$HC_2H_3O_2(aq)$ $\rightleftharpoons$	$H^+(aq)$ +	$C_2H_3O_2^-(aq)$
Initial	0.30 *M*	0	0
Change	$-x$ *M*	$+x$ *M*	$+x$ *M*
Equilibrium	$(0.30 - x)$ *M*	x *M*	x *M*

As the *fourth* step of the problem, we need to substitute the equilibrium concentrations into the equilibrium-constant expression. The substitutions give the following equation:

$$K_a = \frac{[H^+][C_2H_3O_2^-]}{[HC_2H_3O_2]} = \frac{(x)(x)}{0.30 - x} = 1.8 \times 10^{-5} \qquad [16.30]$$

This expression leads to a quadratic equation in x, which we can solve by using an equation-solving calculator or by using the quadratic formula. We can also simplify the problem, however, by noting that the value of K_a is quite small. As a result, we anticipate that the equilibrium will lie far to the left and that x will be very small compared to the initial concentration of acetic acid. Thus, we will *assume* that x is negligible compared to 0.30, so that $0.30 - x$ is essentially equal to 0.30.

$$0.30 - x \simeq 0.30$$

As we will see, we can (and should!) check the validity of this assumption when we finish the problem. By using this assumption, Equation 16.30 now becomes

$$K_a = \frac{x^2}{0.30} = 1.8 \times 10^{-5}$$

Solving for x, we have

$$x^2 = (0.30)(1.8 \times 10^{-5}) = 5.4 \times 10^{-6}$$

$$x = \sqrt{5.4 \times 10^{-6}} = 2.3 \times 10^{-3}$$

$$[H^+] = x = 2.3 \times 10^{-3}\ M$$

$$pH = -\log(2.3 \times 10^{-3}) = 2.64$$

We should now go back and check the validity of our simplifying assumption that $0.30 - x \simeq 0.30$. The value of x we determined is so small that, for this number of significant figures, the assumption is entirely valid. We are thus satisfied that the assumption was a reasonable one to make. Because x represents the moles per liter of acetic acid that ionize, we see that, in this particular case, less than 1% of the acetic acid molecules ionize:

$$\text{Percent ionization of } HC_2H_3O_2 = \frac{0.0023\ M}{0.30\ M} \times 100\% = 0.77\%$$

As a general rule, if the quantity x is more than about 5% of the initial value, it is better to use the quadratic formula. You should always check the validity of any simplifying assumptions after you have finished solving a problem.

GIVE IT SOME THOUGHT

Why can we generally assume that the equilibrium concentration of a weak acid equals its initial concentration?

Finally, we can compare the pH value of this weak acid to a solution of a strong acid of the same concentration. The pH of the 0.30 M solution of acetic acid is 2.64. By comparison, the pH of a 0.30 M solution of a strong acid such as HCl is $-\log(0.30) = 0.52$. As expected, the pH of a solution of a weak acid is higher than that of a solution of a strong acid of the same molarity.

SAMPLE EXERCISE 16.11 | Using K_a to Calculate pH

Calculate the pH of a 0.20 M solution of HCN. (Refer to Table 16.2 or Appendix D for the value of K_a.)

Solution

Analyze: We are given the molarity of a weak acid and are asked for the pH. From Table 16.2, K_a for HCN is 4.9×10^{-10}.

Plan: We proceed as in the example just worked in the text, writing the chemical equation and constructing a table of initial and equilibrium concentrations in which the equilibrium concentration of H^+ is our unknown.

Solve: Writing both the chemical equation for the ionization reaction that forms $H^+(aq)$ and the equilibrium-constant (K_a) expression for the reaction:

$$HCN(aq) \rightleftharpoons H^+(aq) + CN^-(aq)$$

$$K_a = \frac{[H^+][CN^-]}{[HCN]} = 4.9 \times 10^{-10}$$

Next, we tabulate the concentration of the species involved in the equilibrium reaction, letting $x = [H^+]$ at equilibrium:

$$HCN(aq) \rightleftharpoons H^+(aq) + CN^-(aq)$$

	$HCN(aq)$	$H^+(aq)$	$CN^-(aq)$
Initial	0.20 *M*	0	0
Change	$-x$ *M*	$+x$ *M*	$+x$ *M*
Equilibrium	$(0.20 - x)$ *M*	x *M*	x *M*

Substituting the equilibrium concentrations from the table into the equilibrium-constant expression yields

$$K_a = \frac{(x)(x)}{0.20 - x} = 4.9 \times 10^{-10}$$

We next make the simplifying approximation that x, the amount of acid that dissociates, is small compared with the initial concentration of acid; that is,

$$0.20 - x \cong 0.20$$

Thus,

$$\frac{x^2}{0.20} = 4.9 \times 10^{-10}$$

Solving for x, we have

$$x^2 = (0.20)(4.9 \times 10^{-10}) = 0.98 \times 10^{-10}$$

$$x = \sqrt{0.98 \times 10^{-10}} = 9.9 \times 10^{-6}\ M = [H^+]$$

A concentration of 9.9×10^{-6} *M* is much smaller than 5% of 0.20, the initial HCN concentration. Our simplifying approximation is therefore appropriate. We now calculate the pH of the solution:

$$pH = -\log[H^+] = -\log(9.9 \times 10^{-6}) = 5.00$$

PRACTICE EXERCISE

The K_a for niacin (Practice Exercise 16.10) is 1.5×10^{-5}. What is the pH of a 0.010 *M* solution of niacin?
Answer: 3.42

The result obtained in Sample Exercise 16.11 is typical of the behavior of weak acids; the concentration of $H^+(aq)$ is only a small fraction of the concentration of the acid in solution. Those properties of the acid solution that relate directly to the concentration of $H^+(aq)$, such as electrical conductivity and rate of reaction with an active metal, are much less evident for a solution of a weak acid than for a solution of a strong acid. Figure 16.8 ▼ presents an experiment that demonstrates the difference in concentration of $H^+(aq)$ in weak and strong acid solutions of the same concentration. The rate of reaction with the metal is much faster for the solution of a strong acid.

(a) (b)

◀ **Figure 16.8 Reaction rates for weak and strong acids.** (a) The flask on the left contains 1 *M* $HC_2H_3O_2$; the one on the right contains 1 *M* HCl. Each balloon contains the same amount of magnesium metal. (b) When the Mg metal is dropped into the acid, H_2 gas is formed. The rate of H_2 formation is higher for the 1 *M* HCl solution on the right as evidenced by more gas in the balloon.

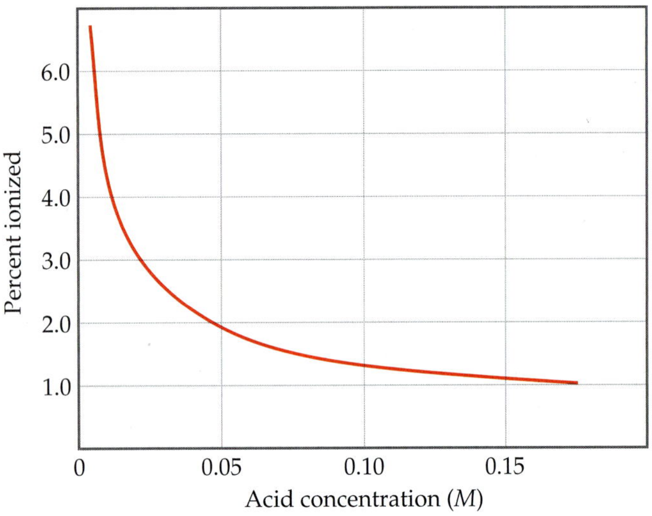

To determine the pH of a solution of a weak acid, you might think that the percent ionization of the acid would be simpler to use than the K_a value. However, the percent ionization at a particular temperature depends not only on the identity of the acid but also on its concentration. As shown in Figure 16.9 ◀, the percent ionization of a weak acid decreases as its concentration increases. This fact is further demonstrated in Sample Exercise 16.12.

◀ **Figure 16.9 The effect of concentration on ionization of a weak acid.** The percent ionization of a weak acid decreases with increasing concentration. The data shown are for acetic acid.

SAMPLE EXERCISE 16.12 | Using K_a to Calculate Percent Ionization

Calculate the percentage of HF molecules ionized in **(a)** a 0.10 *M* HF solution, **(b)** a 0.010 *M* HF solution.

Solution

Analyze: We are asked to calculate the percent ionization of two HF solutions of different concentration.

Plan: We approach this problem as we would previous equilibrium problems. We begin by writing the chemical equation for the equilibrium and tabulating the known and unknown concentrations of all species. We then substitute the equilibrium concentrations into the equilibrium-constant expression and solve for the unknown concentration, that of H^+.

Solve: **(a)** The equilibrium reaction and equilibrium concentrations are as follows:

	$HF(aq)$	$\rightleftharpoons H^+(aq)$	$+ F^-(aq)$
Initial	0.10 *M*	0	0
Change	$-x$ *M*	$+x$ *M*	$+x$ *M*
Equilibrium	$(0.10 - x)$ *M*	x *M*	x *M*

The equilibrium-constant expression is

$$K_a = \frac{[H^+][F^-]}{[HF]} = \frac{(x)(x)}{0.10 - x} = 6.8 \times 10^{-4}$$

When we try solving this equation using the approximation $0.10 - x = 0.10$ (that is, by neglecting the concentration of acid that ionizes in comparison with the initial concentration), we obtain

$$x = 8.2 \times 10^{-3}\ M$$

Because this value is greater than 5% of 0.10 *M*, we should work the problem without the approximation, using an equation-solving calculator or the quadratic formula. Rearranging our equation and writing it in standard quadratic form, we have

$$x^2 = (0.10 - x)(6.8 \times 10^{-4})$$
$$= 6.8 \times 10^{-5} - (6.8 \times 10^{-4})x$$
$$x^2 + (6.8 \times 10^{-4})x - 6.8 \times 10^{-5} = 0$$

This equation can be solved using the standard quadratic formula.

$$x = \frac{-b \pm \sqrt{b^2 - 4ac}}{2a}$$

Substituting the appropriate numbers gives

$$x = \frac{-6.8 \times 10^{-4} \pm \sqrt{(6.8 \times 10^{-4})^2 + 4(6.8 \times 10^{-5})}}{2}$$

$$= \frac{-6.8 \times 10^{-4} \pm 1.6 \times 10^{-2}}{2}$$

Of the two solutions, only the one that gives a positive value for x is chemically reasonable. Thus,

$$x = [H^+] = [F^-] = 7.9 \times 10^{-3}\,M$$

From our result, we can calculate the percent of molecules ionized:

$$\text{Percent ionization of HF} = \frac{\text{concentration ionized}}{\text{original concentration}} \times 100\%$$

$$= \frac{7.9 \times 10^{-3}\,M}{0.10\,M} \times 100\% = 7.9\%$$

(b) Proceeding similarly for the 0.010 M solution, we have

$$\frac{x^2}{0.010 - x} = 6.8 \times 10^{-4}$$

Solving the resultant quadratic expression, we obtain

$$x = [H^+] = [F^-] = 2.3 \times 10^{-3}\,M$$

The percentage of molecules ionized is

$$\frac{0.0023}{0.010} \times 100\% = 23\%$$

Comment: Notice that if we do not use the quadratic formula to solve the problem properly, we calculate 8.2% ionization for (a) and 26% ionization for (b). Notice also that in diluting the solution by a factor of 10, the percentage of molecules ionized increases by a factor of 3. This result is in accord with what we see in Figure 16.9. It is also what we would expect from Le Châtelier's principle. (Section 15.6) There are more "particles" or reaction components on the right side of the equation than on the left. Dilution causes the reaction to shift in the direction of the larger number of particles because this counters the effect of the decreasing concentration of particles.

PRACTICE EXERCISE

In Practice Exercise 16.10, we found that the percent ionization of niacin ($K_a = 1.5 \times 10^{-5}$) in a 0.020 M solution is 2.7%. Calculate the percentage of niacin molecules ionized in a solution that is **(a)** 0.010 M, **(b)** 1.0×10^{-3} M.
Answers: **(a)** 3.8%, **(b)** 12%

Polyprotic Acids

Many acids have more than one ionizable H atom. These acids are known as **polyprotic acids**. For example, each of the H atoms in sulfurous acid (H_2SO_3) can ionize in successive steps:

$$H_2SO_3(aq) \rightleftharpoons H^+(aq) + HSO_3^-(aq) \qquad K_{a1} = 1.7 \times 10^{-2} \qquad [16.31]$$

$$HSO_3^-(aq) \rightleftharpoons H^+(aq) + SO_3^{2-}(aq) \qquad K_{a2} = 6.4 \times 10^{-8} \qquad [16.32]$$

The acid-dissociation constants for these equilibria are labeled K_{a1} and K_{a2}. The numbers on the constants refer to the particular proton of the acid that is ionizing. Thus, K_{a2} always refers to the equilibrium involving removal of the second proton of a polyprotic acid.

In the preceding example K_{a2} is much smaller than K_{a1}. On the basis of electrostatic attractions, we would expect a positively charged proton to be lost more readily from the neutral H_2SO_3 molecule than from the negatively charged HSO_3^- ion. This observation is general: *It is always easier to remove the first proton from a polyprotic acid than to remove the second.* Similarly, for an acid with three ionizable protons, it is easier to remove the second proton than the third. Thus, the K_a values become successively smaller as successive protons are removed.

TABLE 16.3 Acid-Dissociation Constants of Some Common Polyprotic Acids

Name	Formula	K_{a1}	K_{a2}	K_{a3}
Ascorbic	$H_2C_6H_6O_6$	8.0×10^{-5}	1.6×10^{-12}	
Carbonic	H_2CO_3	4.3×10^{-7}	5.6×10^{-11}	
Citric	$H_3C_6H_5O_7$	7.4×10^{-4}	1.7×10^{-5}	4.0×10^{-7}
Oxalic	$H_2C_2O_4$	5.9×10^{-2}	6.4×10^{-5}	
Phosphoric	H_3PO_4	7.5×10^{-3}	6.2×10^{-8}	4.2×10^{-13}
Sulfurous	H_2SO_3	1.7×10^{-2}	6.4×10^{-8}	
Sulfuric	H_2SO_4	Large	1.2×10^{-2}	
Tartaric	$H_2C_4H_4O_6$	1.0×10^{-3}	4.6×10^{-5}	

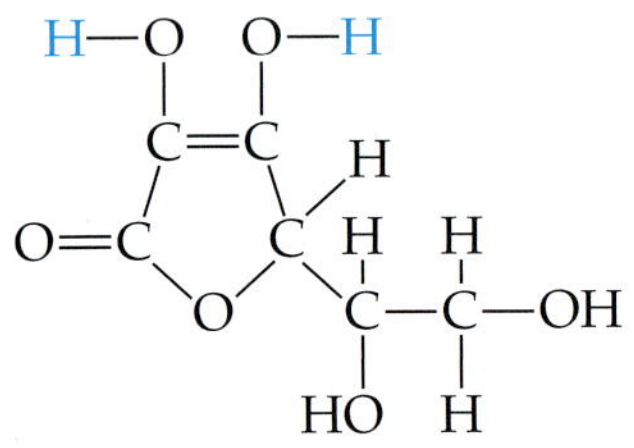

Ascorbic acid (vitamin C)

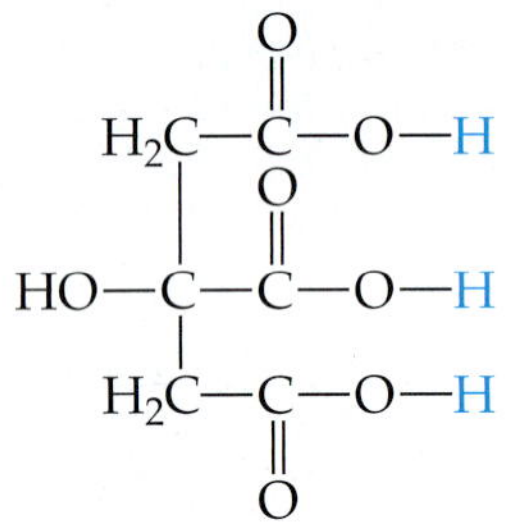

Citric acid

GIVE IT SOME THOUGHT

What is meant by the symbol K_{a3} for H_3PO_4?

The acid-dissociation constants for a few common polyprotic acids are listed in Table 16.3 ▲. A more complete list is given in Appendix D. The structures for ascorbic and citric acids are shown in the margin. Notice that the K_a values for successive losses of protons from these acids usually differ by a factor of at least 10^3. Notice also that the value of K_{a1} for sulfuric acid is listed simply as "large." Sulfuric acid is a strong acid with respect to the removal of the first proton. Thus, the reaction for the first ionization step lies completely to the right:

$$H_2SO_4(aq) \longrightarrow H^+(aq) + HSO_4^-(aq) \quad \text{(complete ionization)}$$

HSO_4^-, on the other hand, is a weak acid for which $K_{a2} = 1.2 \times 10^{-2}$.

Because K_{a1} is so much larger than subsequent dissociation constants for these polyprotic acids, almost all of the $H^+(aq)$ in the solution comes from the first ionization reaction. As long as successive K_a values differ by a factor of 10^3 or more, it is possible to obtain a satisfactory estimate of the pH of polyprotic acid solutions by considering only K_{a1}.

SAMPLE EXERCISE 16.13 | Calculating the pH of a Polyprotic Acid Solution

The solubility of CO_2 in pure water at 25°C and 0.1 atm pressure is 0.0037 *M*. The common practice is to assume that all of the dissolved CO_2 is in the form of carbonic acid (H_2CO_3), which is produced by reaction between the CO_2 and H_2O:

$$CO_2(aq) + H_2O(l) \rightleftharpoons H_2CO_3(aq)$$

What is the pH of a 0.0037 *M* solution of H_2CO_3?

Solution

Analyze: We are asked to determine the pH of a 0.0037 *M* solution of a polyprotic acid.

Plan: H_2CO_3 is a diprotic acid; the two acid-dissociation constants, K_{a1} and K_{a2} (Table 16.3), differ by more than a factor of 10^3. Consequently, the pH can be determined by considering only K_{a1}, thereby treating the acid as if it were a monoprotic acid.

Solve: Proceeding as in Sample Exercises 16.11 and 16.12, we can write the equilibrium reaction and equilibrium concentrations as follows:

	$H_2CO_3(aq)$ ⇌	$H^+(aq)$ +	$HCO_3^-(aq)$
Initial	0.0037 *M*	0	0
Change	$-x$ *M*	$+x$ *M*	$+x$ *M*
Equilibrium	$(0.0037 - x)$ *M*	x *M*	x *M*

The equilibrium-constant expression is as follows:

$$K_{a1} = \frac{[H^+][HCO_3]}{[H_2CO_3]} = \frac{(x)(x)}{0.0037 - x} = 4.3 \times 10^{-7}$$

Solving this equation using an equation-solving calculator, we get

$$x = 4.0 \times 10^{-5}\ M$$

Alternatively, because K_{a1} is small, we can make the simplifying approximation that x is small, so that

$$0.0037 - x \simeq 0.0037$$

Thus,

$$\frac{(x)(x)}{0.0037} = 4.3 \times 10^{-7}$$

Solving for x, we have

$$x^2 = (0.0037)(4.3 \times 10^{-7}) = 1.6 \times 10^{-9}$$

$$x = [H^+] = [HCO_3^-] = \sqrt{1.6 \times 10^{-9}} = 4.0 \times 10^{-5}\ M$$

The small value of x indicates that our simplifying assumption was justified. The pH is therefore

$$pH = -\log[H^+] = -\log(4.0 \times 10^{-5}) = 4.40$$

Comment: If we were asked to solve for $[CO_3^{2-}]$, we would need to use K_{a2}. Let's illustrate that calculation. Using the values of $[HCO_3^-]$ and $[H^+]$ calculated above, and setting $[CO_3^{2-}] = y$, we have the following initial and equilibrium concentration values:

	$HCO_3^-(aq)$ $\rightleftharpoons$	$H^+(aq)$ +	$CO_3^{2-}(aq)$
Initial	$4.0 \times 10^{-5}\ M$	$4.0 \times 10^{-5}\ M$	0
Change	$-y\ M$	$+y\ M$	$+y\ M$
Equilibrium	$(4.0 \times 10^{-5} - y)\ M$	$(4.0 \times 10^{-5} + y)\ M$	$y\ M$

Assuming that y is small compared to 4.0×10^{-5}, we have

$$K_{a2} = \frac{[H^+][CO_3^{2-}]}{[HCO_3^-]} = \frac{(4.0 \times 10^{-5})(y)}{4.0 \times 10^{-5}} = 5.6 \times 10^{-11}$$

$$y = 5.6 \times 10^{-11}\ M = [CO_3^{2-}]$$

The value calculated for y is indeed very small compared to 4.0×10^{-5}, showing that our assumption was justified. It also shows that the ionization of HCO_3^- is negligible compared to that of H_2CO_3, as far as production of H^+ is concerned. However, it is the *only* source of CO_3^{2-}, which has a very low concentration in the solution. Our calculations thus tell us that in a solution of carbon dioxide in water, most of the CO_2 is in the form of CO_2 or H_2CO_3, a small fraction ionizes to form H^+ and HCO_3^-, and an even smaller fraction ionizes to give CO_3^{2-}. Notice also that $[CO_3^{2-}]$ is numerically equal to K_{a2}.

PRACTICE EXERCISE

(a) Calculate the pH of a 0.020 M solution of oxalic acid ($H_2C_2O_4$). (See Table 16.3 for K_{a1} and K_{a2}.) **(b)** Calculate the concentration of oxalate ion, $[C_2O_4^{2-}]$, in this solution.
Answers: **(a)** pH = 1.80, **(b)** $[C_2O_4^{2-}] = 6.4 \times 10^{-5}\ M$

16.7 | Weak Bases

Many substances behave as weak bases in water. Weak bases react with water, abstracting protons from H_2O, thereby forming the conjugate acid of the base and OH^- ions.

ANIMATION
Introduction to Aqueous Bases

$$B(aq) + H_2O \rightleftharpoons HB^+ + OH^-(aq) \quad [16.33]$$

The most commonly encountered weak base is ammonia.

$$NH_3(aq) + H_2O(l) \rightleftharpoons NH_4^+(aq) + OH^-(aq) \quad [16.34]$$

The equilibrium-constant expression for this reaction can be written as

$$K_b = \frac{[NH_4^+][OH^-]}{[NH_3]} \quad [16.35]$$

Water is the solvent, so it is omitted from the equilibrium-constant expression.

As with K_w and K_a, the subscript "b" denotes that this equilibrium constant refers to a particular type of reaction, namely the ionization of a weak base in water. The constant K_b is called the **base-dissociation constant**. *The constant* K_b

TABLE 16.4 Some Weak Bases and Their Aqueous Solution Equilibria

Base	Lewis Structure	Conjugate Acid	Equilibrium Reaction	K_b
Ammonia (NH_3)	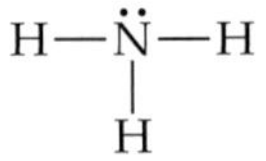	NH_4^+	$NH_3 + H_2O \rightleftharpoons NH_4^+ + OH^-$	1.8×10^{-5}
Pyridine (C_5H_5N)		$C_5H_5NH^+$	$C_5H_5N + H_2O \rightleftharpoons C_5H_5NH^+ + OH^-$	1.7×10^{-9}
Hydroxylamine (H_2NOH)	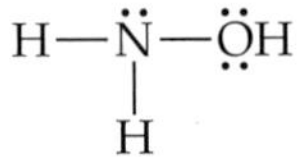	H_3NOH^+	$H_2NOH + H_2O \rightleftharpoons H_3NOH^+ + OH^-$	1.1×10^{-8}
Methylamine (NH_2CH_3)	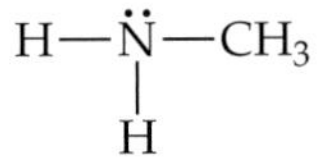	$NH_3CH_3^+$	$NH_2CH_3 + H_2O \rightleftharpoons NH_3CH_3^+ + OH^-$	4.4×10^{-4}
Hydrosulfide ion (HS^-)	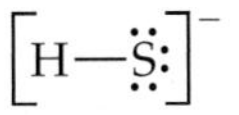	H_2S	$HS^- + H_2O \rightleftharpoons H_2S + OH^-$	1.8×10^{-7}
Carbonate ion (CO_3^{2-})	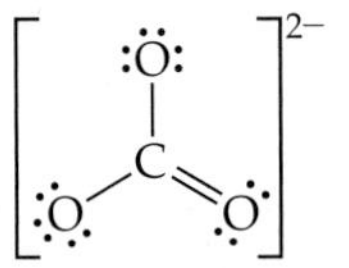	HCO_3^-	$CO_3^{2-} + H_2O \rightleftharpoons HCO_3^- + OH^-$	1.8×10^{-4}
Hypochlorite ion (ClO^-)	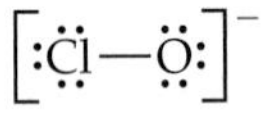	$HClO$	$ClO^- + H_2O \rightleftharpoons HClO + OH^-$	3.3×10^{-7}

always refers to the equilibrium in which a base reacts with H_2O to form the corresponding conjugate acid and OH^-. Table 16.4 ▲ lists the names, formulas, Lewis structures, equilibrium reactions, and values of K_b for several weak bases in water. Appendix D includes a more extensive list. These bases contain one or more lone pairs of electrons because a lone pair is necessary to form the bond with H^+. Notice that in the neutral molecules in Table 16.4, the lone pairs are on nitrogen atoms. The other bases listed are anions derived from weak acids.

SAMPLE EXERCISE 16.14 | Using K_b to Calculate $[OH^-]$

Calculate the concentration of OH^- in a 0.15 *M* solution of NH_3.

Solution

Analyze: We are given the concentration of a weak base and are asked to determine the concentration of OH^-.

Plan: We will use essentially the same procedure here as used in solving problems involving the ionization of weak acids; that is, we write the chemical equation and tabulate initial and equilibrium concentrations.

Solve: We first write the ionization reaction and the corresponding equilibrium-constant (K_b) expression:

$$NH_3(aq) + H_2O(l) \rightleftharpoons NH_4^+(aq) + OH^-(aq)$$

$$K_b = \frac{[NH_4^+][OH^-]}{[NH_3]} = 1.8 \times 10^{-5}$$

We then tabulate the equilibrium concentrations involved in the equilibrium:

	$NH_3(aq)$	+ $H_2O(l)$ ⇌	$NH_4^+(aq)$	+ $OH^-(aq)$
Initial	0.15 *M*	—	0	0
Change	$-x$ *M*	—	$+x$ *M*	$+x$ *M*
Equilibrium	$(0.15 - x)$ *M*	—	x *M*	x *M*

(We ignore the concentration of H_2O because it is not involved in the equilibrium-constant expression.) Inserting these quantities into the equilibrium-constant expression gives the following:

$$K_b = \frac{[NH_4^+][OH^-]}{[NH_3]} = \frac{(x)(x)}{0.15 - x} = 1.8 \times 10^{-5}$$

Because K_b is small, we can neglect the small amount of NH_3 that reacts with water, as compared to the total NH_3 concentration; that is, we can neglect x relative to 0.15 *M*. Then we have

$$\frac{x^2}{0.15} = 1.8 \times 10^{-5}$$

$$x^2 = (0.15)(1.8 \times 10^{-5}) = 2.7 \times 10^{-6}$$

$$x = [NH_4^+] = [OH^-] = \sqrt{2.7 \times 10^{-6}} = 1.6 \times 10^{-3}\ M$$

Check: The value obtained for x is only about 1% of the NH_3 concentration, 0.15 *M*. Therefore, neglecting x relative to 0.15 was justified.

PRACTICE EXERCISE

Which of the following compounds should produce the highest pH as a 0.05 *M* solution: pyridine, methylamine, or nitrous acid?
Answer: methylamine (because it has the largest K_b value)

Types of Weak Bases

How can we recognize from a chemical formula whether a molecule or ion is able to behave as a weak base? Weak bases fall into two general categories. The first category contains neutral substances that have an atom with a nonbonding pair of electrons that can serve as a proton acceptor. Most of these bases, including all of the uncharged bases listed in Table 16.4, contain a nitrogen atom. These substances include ammonia and a related class of compounds called **amines**. In organic amines, one or more of the N—H bonds in NH_3 is replaced with a bond between N and C. Thus, the replacement of one N—H bond in NH_3 with a N—CH_3 bond gives methylamine, NH_2CH_3 (usually written CH_3NH_2). Like NH_3, amines can abstract a proton from a water molecule by forming an additional N—H bond, as shown here for methylamine:

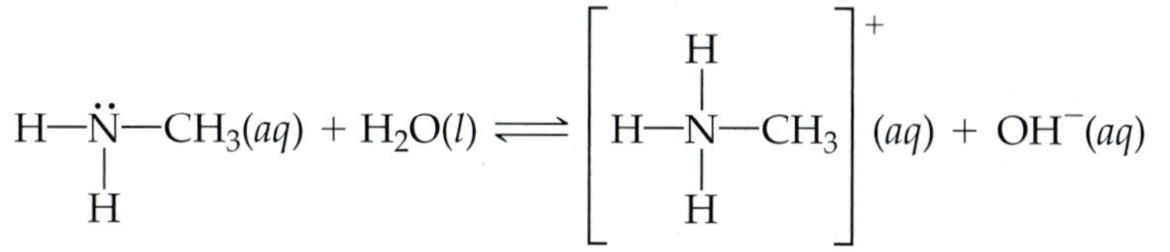
[16.36]

The chemical formula for the conjugate acid of methylamine is usually written $CH_3NH_3^+$.

The second general category of weak bases consists of the anions of weak acids. In an aqueous solution of sodium hypochlorite (NaClO), for example, NaClO dissociates to give Na^+ and ClO^- ions. The Na^+ ion is always a spectator ion in acid-base reactions. (Section 4.3) The ClO^- ion, however, is the conjugate base of a weak acid, hypochlorous acid. Consequently, the ClO^- ion acts as a weak base in water:

$$ClO^-(aq) + H_2O(l) \rightleftharpoons HClO(aq) + OH^-(aq) \qquad K_b = 3.3 \times 10^{-7} \qquad [16.37]$$

SAMPLE EXERCISE 16.15 | Using pH to Determine the Concentration of a Salt

A solution made by adding solid sodium hypochlorite (NaClO) to enough water to make 2.00 L of solution has a pH of 10.50. Using the information in Equation 16.37, calculate the number of moles of NaClO that were added to the water.

Solution

Analyze: We are given the pH of a 2.00-L solution of NaClO and must calculate the number of moles of NaClO needed to raise the pH to 10.50. NaClO is an ionic compound consisting of Na^+ and ClO^- ions. As such, it is a strong electrolyte that completely dissociates in solution into Na^+, which is a spectator ion, and ClO^- ion, which is a weak base with $K_b = 3.33 \times 10^{-7}$ (Equation 16.37).

Plan: From the pH, we can determine the equilibrium concentration of OH^-. We can then construct a table of initial and equilibrium concentrations in which the initial concentration of ClO^- is our unknown. We can calculate $[ClO^-]$ using the equilibrium-constant expression, K_b.

Solve: We can calculate $[OH^-]$ by using either Equation 16.16 or Equation 16.19; we will use the latter method here:

$$pOH = 14.00 - pH = 14.00 - 10.50 = 3.50$$

$$[OH^-] = 10^{-3.50} = 3.16 \times 10^{-4}\ M$$

This concentration is high enough that we can assume that Equation 16.37 is the only source of OH^-; that is, we can neglect any OH^- produced by the autoionization of H_2O. We now assume a value of x for the initial concentration of ClO^- and solve the equilibrium problem in the usual way.

	$ClO^-(aq)$	$+\ H_2O(l) \rightleftharpoons$	$HClO(aq)$	$+\ OH^-(aq)$
Initial	$x\ M$	—	0	0
Change	$-3.16 \times 10^{-4}\ M$	—	$+3.16 \times 10^{-4}\ M$	$+3.16 \times 10^{-4}\ M$
Final	$(x - 3.16 \times 10^{-4})\ M$	—	$3.16 \times 10^{-4}\ M$	$3.16 \times 10^{-4}\ M$

We now use the expression for the base-dissociation constant to solve for x:

$$K_b = \frac{[HClO][OH^-]}{[ClO^-]} = \frac{(3.16 \times 10^{-4})^2}{x - 3.16 \times 10^{-4}} = 3.3 \times 10^{-7}$$

Thus

$$x = \frac{(3.16 \times 10^{-4})^2}{3.3 \times 10^{-7}} + (3.16 \times 10^{-4}) = 0.30\ M$$

We say that the solution is 0.30 *M* in NaClO, even though some of the ClO^- ions have reacted with water. Because the solution is 0.30 *M* in NaClO and the total volume of solution is 2.00 L, 0.60 mol of NaClO is the amount of the salt that was added to the water.

PRACTICE EXERCISE

A solution of NH_3 in water has a pH of 11.17. What is the molarity of the solution?
Answer: 0.12 *M*

16.8 | Relationship Between K_a and K_b

We've seen in a qualitative way that the stronger acids have the weaker conjugate bases. To see if we can find a corresponding *quantitative* relationship, let's consider the NH_4^+ and NH_3 conjugate acid-base pair. Each of these species reacts with water:

$$NH_4^+(aq) \rightleftharpoons NH_3(aq) + H^+(aq) \quad [16.38]$$

$$NH_3(aq) + H_2O(l) \rightleftharpoons NH_4^+(aq) + OH^-(aq) \quad [16.39]$$

Each of these equilibria is expressed by a characteristic dissociation constant:

$$K_a = \frac{[NH_3][H^+]}{[NH_4^+]}$$

$$K_b = \frac{[NH_4^+][OH^-]}{[NH_3]}$$

When Equations 16.38 and 16.39 are added together, the NH_4^+ and NH_3 species cancel and we are left with just the autoionization of water.

$$\begin{array}{rcl} NH_4^+(aq) & \rightleftharpoons & NH_3(aq) + H^+(aq) \\ NH_3(aq) + H_2O(l) & \rightleftharpoons & NH_4^+(aq) + OH^-(aq) \\ \hline H_2O(l) & \rightleftharpoons & H^+(aq) + OH^-(aq) \end{array}$$

Recall that when two equations are added to give a third, the equilibrium constant associated with the third equation equals the product of the equilibrium constants for the two equations added together. (Section 15.2)

Applying this rule to our present example, when we multiply K_a and K_b, we obtain the following:

$$K_a \times K_b = \left(\frac{\cancel{[NH_3]}[H^+]}{\cancel{[NH_4^+]}}\right)\left(\frac{\cancel{[NH_4^+]}[OH^-]}{\cancel{[NH_3]}}\right)$$

$$= [H^+][OH^-] = K_w$$

Thus, the result of multiplying K_a times K_b is just the ion-product constant for water, K_w (Equation 16.16). This is what we would expect, moreover, because adding Equations 16.38 and 16.39 gave us the autoionization equilibrium for water, for which the equilibrium constant is K_w.

This relationship is so important that it should receive special attention: The product of the acid-dissociation constant for an acid and the base-dissociation constant for its conjugate base is the ion-product constant for water.

$$K_a \times K_b = K_w \qquad [16.40]$$

As the strength of an acid increases (larger K_a), the strength of its conjugate base must decrease (smaller K_b) so that the product $K_a \times K_b$ equals 1.0×10^{-14} at 25°C. The K_a and K_b data in Table 16.5 ▼ demonstrate this relationship.

By using Equation 16.40, we can calculate K_b for any weak base if we know K_a for its conjugate acid. Similarly, we can calculate K_a for a weak acid if we know K_b for its conjugate base. As a practical consequence, ionization constants are often listed for only one member of a conjugate acid-base pair. For example, Appendix D does not contain K_b values for the anions of weak acids because these can be readily calculated from the tabulated K_a values for their conjugate acids.

If you look up the values for acid- or base-dissociation constants in a chemistry handbook, you may find them expressed as pK_a or pK_b (that is, as $-\log K_a$ or $-\log K_b$). (Section 16.4) Equation 16.40 can be written in terms of pK_a and pK_b by taking the negative log of both sides:

$$pK_a + pK_b = pK_w = 14.00 \quad \text{at } 25°C \qquad [16.41]$$

TABLE 16.5 Some Conjugate Acid-Base Pairs

Acid	K_a	Base	K_b
HNO_3	(Strong acid)	NO_3^-	(Negligible basicity)
HF	6.8×10^{-4}	F^-	1.5×10^{-11}
$HC_2H_3O_2$	1.8×10^{-5}	$C_2H_3O_2^-$	5.6×10^{-10}
H_2CO_3	4.3×10^{-7}	HCO_3^-	2.3×10^{-8}
NH_4^+	5.6×10^{-10}	NH_3	1.8×10^{-5}
HCO_3^-	5.6×10^{-11}	CO_3^{2-}	1.8×10^{-4}
OH^-	(Negligible acidity)	O^{2-}	(Strong base)

CHEMISTRY AT WORK | Amines and Amine Hydrochlorides

Many amines with low molecular weights have unpleasant "fishy" odors. Amines and NH_3 are produced by the anaerobic (absence of O_2) decomposition of dead animal or plant matter. Two such amines with very disagreeable odors are $H_2N(CH_2)_4NH_2$, known as *putrescine*, and $H_2N(CH_2)_5NH_2$, known as *cadaverine*.

Many drugs, including quinine, codeine, caffeine, and amphetamine (Benzedrine®), are amines. Like other amines, these substances are weak bases; the amine nitrogen is readily protonated upon treatment with an acid. The resulting products are called *acid salts*. If we use A as the abbreviation for an amine, the acid salt formed by reaction with hydrochloric acid can be written as AH^+Cl^-. It is also sometimes written as A·HCl and referred to as a hydrochloride. Amphetamine hydrochloride, for example, is the acid salt formed by treating amphetamine with HCl:

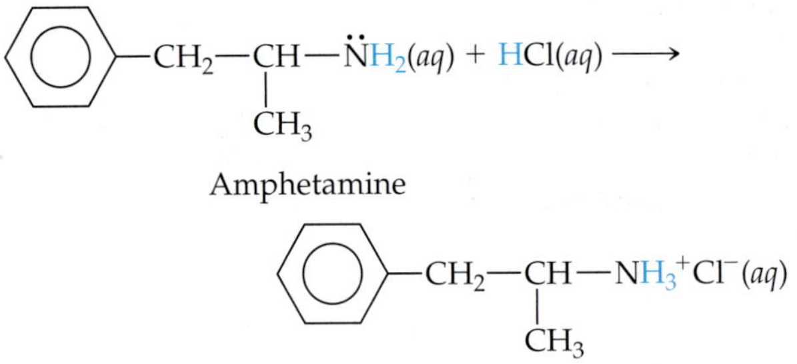

Amphetamine

Amphetamine hydrochloride

Such acid salts are much less volatile, more stable, and generally more water-soluble than the corresponding neutral amines. Many drugs that are amines are sold and administered as acid salts. Some examples of over-the-counter medications that contain amine hydrochlorides as active ingredients are shown in Figure 16.10 ▼.

Related Exercises: 16.77, 16.78, 16.114, and 16.123

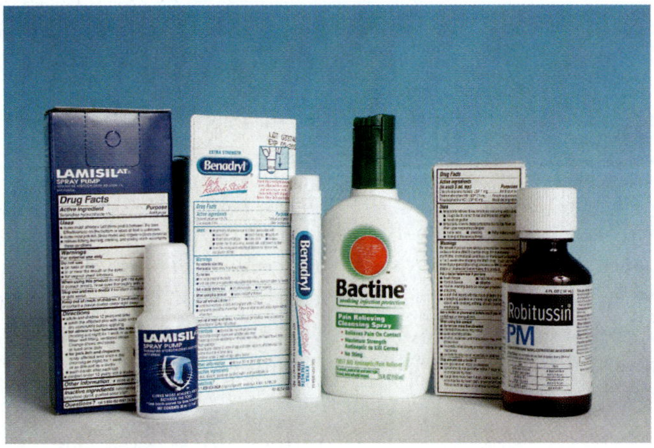

▲ **Figure 16.10** Some over-the-counter medications in which an amine hydrochloride is a major active ingredient.

SAMPLE EXERCISE 16.16 | Calculating K_a or K_b for a Conjugate Acid-Base Pair

Calculate **(a)** the base-dissociation constant, K_b, for the fluoride ion (F^-); **(b)** the acid-dissociation constant, K_a, for the ammonium ion (NH_4^+).

Solution

Analyze: We are asked to determine dissociation constants for F^-, the conjugate base of HF, and NH_4^+, the conjugate acid of NH_3.

Plan: Although neither F^- nor NH_4^+ appears in the tables, we can find the tabulated values for ionization constants for HF and NH_3, and use the relationship between K_a and K_b to calculate the ionization constants for each of the conjugates.

Solve: **(a)** K_a for the weak acid, HF, is given in Table 16.2 and Appendix D as $K_a = 6.8 \times 10^{-4}$. We can use Equation 16.40 to calculate K_b for the conjugate base, F^-:

$$K_b = \frac{K_w}{K_a} = \frac{1.0 \times 10^{-14}}{6.8 \times 10^{-4}} = 1.5 \times 10^{-11}$$

(b) K_b for NH_3 is listed in Table 16.4 and in Appendix D as $K_b = 1.8 \times 10^{-5}$. Using Equation 16.40, we can calculate K_a for the conjugate acid, NH_4^+:

$$K_a = \frac{K_w}{K_b} = \frac{1.0 \times 10^{-14}}{1.8 \times 10^{-5}} = 5.6 \times 10^{-10}$$

PRACTICE EXERCISE

(a) Which of the following anions has the largest base-dissociation constant: NO_2^-, PO_4^{3-}, or N_3^-? **(b)** The base quinoline has the following structure:

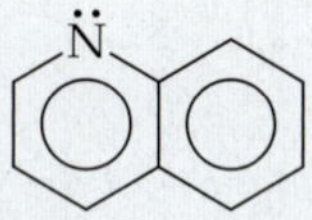

Its conjugate acid is listed in handbooks as having a pK_a of 4.90. What is the base-dissociation constant for quinoline?

Answers: **(a)** PO_4^{3-} ($K_b = 2.4 \times 10^{-2}$), **(b)** 7.9×10^{-10}

16.9 | Acid-Base Properties of Salt Solutions

Even before you began this chapter, you were undoubtedly aware of many substances that are acidic, such as HNO_3, HCl, and H_2SO_4, and others that are basic, such as NaOH and NH_3. However, our recent discussions have indicated that ions can also exhibit acidic or basic properties. For example, we calculated K_a for NH_4^+ and K_b for F^- in Sample Exercise 16.16. Such behavior implies that salt solutions can be acidic or basic. Before proceeding with further discussions of acids and bases, let's examine the way dissolved salts can affect pH.

We can assume that when salts dissolve in water, they are completely dissociated; nearly all salts are strong electrolytes. Consequently, the acid-base properties of salt solutions are due to the behavior of their constituent cations and anions. Many ions are able to react with water to generate $H^+(aq)$ or $OH^-(aq)$. This type of reaction is often called **hydrolysis**. The pH of an aqueous salt solution can be predicted qualitatively by considering the ions of which the salt is composed.

An Anion's Ability to React with Water

In general, an anion, X^-, in solution can be considered the conjugate base of an acid. For example, Cl^- is the conjugate base of HCl and $C_2H_3O_2^-$ is the conjugate base of $HC_2H_3O_2$. Whether an anion reacts with water to produce hydroxide depends upon the strength of the acid to which it is conjugate. To identify the acid and assess its strength, we can simply add a proton to the anion's formula:

$$X^- \text{ plus a proton } (H^+) \text{ gives } HX$$

If the acid determined in this way is one of the strong acids listed at the beginning of Section 16.5, then the anion in question will have a negligible tendency to abstract protons from water (Section 16.2). Consequently, the anion X^- will not affect the pH of the solution. The presence of Cl^- in an aqueous solution, for example, does not result in the production of any OH^- and does not affect the pH. Thus, Cl^- is always a spectator in acid-base chemistry.

Conversely, if HX is *not* one of the seven strong acids, then it is a weak acid. In this case the conjugate base X^- is a weak base. This anion will therefore react to a small extent with water to produce the weak acid and hydroxide ions:

$$X^-(aq) + H_2O(l) \rightleftharpoons HX(aq) + OH^-(aq) \qquad [16.42]$$

The OH^- ion generated in this way increases the pH of the solution, making it basic. Acetate ion ($C_2H_3O_2^-$), for example, being the conjugate base of a weak acid, reacts with water to produce acetic acid and hydroxide ions, thereby increasing the pH of the solution.*

$$C_2H_3O_2^-(aq) + H_2O(l) \rightleftharpoons HC_2H_3O_2(aq) + OH^-(aq) \qquad [16.43]$$

GIVE IT SOME THOUGHT

What effect will each of the following ions have on the pH of a solution: NO_3^- and CO_3^{2-}?

Anions that still have ionizable protons, such as HSO_3^-, are amphiprotic. (Section 16.2) They can act as either acids or bases. Their behavior toward water will be determined by the relative magnitudes of K_a and K_b for the ion, as shown in Sample Exercise 16.17. If $K_a > K_b$, the ion will cause the solution to be acidic. If $K_b > K_a$, the solution will be basic.

A Cation's Ability to React with Water

Polyatomic cations whose formulas contain one or more protons can be considered the conjugate acids of weak bases. NH_4^+, for example, is the conjugate acid of the weak base NH_3. Thus, NH_4^+ is a weak acid and will donate a proton to water, producing hydronium ions and thereby lowering the pH:

$$NH_4^+(aq) + H_2O(l) \rightleftharpoons NH_3(aq) + H_3O^+(aq) \qquad [16.44]$$

Most metal ions can also react with water to decrease the pH of an aqueous solution. The mechanism by which metal ions produce acidic solutions is described in Section 16.11. However, ions of alkali metals and of the heavier alkaline earth metals do not react with water and therefore do not affect pH. Note that these exceptions are the cations found in the strong bases. (Section 16.5)

GIVE IT SOME THOUGHT

Which of the following cations has no effect on the pH of a solution: K^+, Fe^{2+}, or Al^{3+}?

Combined Effect of Cation and Anion in Solution

If an aqueous salt solution contains an anion that does not react with water and a cation that does not react with water, we expect the pH to be neutral. If the solution contains an anion that reacts with water to produce hydroxide and a cation that does not react with water, we expect the pH to be basic. If the solution contains a cation that reacts with water to produce hydronium and an anion that does not react with water, we expect the pH to be acidic. Finally, a solution may contain an anion and a cation *both* capable of reacting with water. In this case both hydroxide and hydronium will be produced. Whether the solution is basic, neutral, or acidic will depend upon the relative abilities of the ions to react with water.

To summarize:

1. An anion that is the conjugate base of a strong acid, for example, Br^-, will not affect the pH of a solution.
2. An anion that is the conjugate base of a weak acid, for example, CN^-, will cause an increase in pH.

* These rules apply to what are called normal salts. These salts are those that contain no ionizable protons on the anion. The pH of an acid salt (such as $NaHCO_3$ or NaH_2PO_4) is affected not only by the hydrolysis of the anion but also by its acid dissociation, as shown in Sample Exercise 16.17.

(a) (b) (c)

◀ **Figure 16.11 Salt solutions can be neutral, acidic, or basic.** These three solutions contain the acid-base indicator bromthymol blue. (a) The NaCl solution is neutral (pH = 7.0); (b) the NH_4Cl solution is acidic (pH = 3.5); (c) the NaClO solution is basic (pH = 9.5).

3. A cation that is the conjugate acid of a weak base, for example, $CH_3NH_3^+$, will cause a decrease in pH.
4. The cations of group 1A and heavier members of group 2A (Ca^{2+}, Sr^{2+}, and Ba^{2+}) will not affect pH. These are the cations of the strong Arrhenius bases.
5. Other metal ions will cause a decrease in pH.
6. When a solution contains both the conjugate base of a weak acid and the conjugate acid of a weak base, the ion with the larger equilibrium constant, K_a or K_b, will have the greater influence on the pH.

Figure 16.11 ▲ demonstrates the influence of several salts on pH.

VCL Acid-Base Classification of Salts

SAMPLE EXERCISE 16.17 | Predicting the Relative Acidity of Salt Solutions

List the following solutions in order of increasing pH: (i) 0.1 *M* $Ba(C_2H_3O_2)_2$, (ii) 0.1 *M* NH_4Cl, (iii) 0.1 *M* NH_3CH_3Br, (iv) 0.1 *M* KNO_3.

Solution

Analyze: We are asked to arrange a series of salt solutions in order of increasing pH (that is, from the most acidic to the most basic).

Plan: We can determine whether the pH of a solution is acidic, basic, or neutral by identifying the ions in solution and by assessing how each ion will affect the pH.

Solve: Solution (i) contains barium ions and acetate ions. Ba^{2+} is an ion of one of the heavy alkaline earth metals and will therefore not affect the pH (summary point 4). The anion, $C_2H_3O_2^-$, is the conjugate base of the weak acid $HC_2H_3O_2$ and will hydrolyze to produce OH^- ions, thereby making the solution basic (summary point 2). Solutions (ii) and (iii) both contain cations that are conjugate acids of weak bases and anions that are conjugate bases of strong acids. Both solutions will therefore be acidic. Solution (i) contains NH_4^+, which is the conjugate acid of NH_3 ($K_b = 1.8 \times 10^{-5}$). Solution (iii) contains $NH_3CH_3^+$, which is the conjugate acid of NH_2CH_3 ($K_b = 4.4 \times 10^{-4}$). Because NH_3 has the smaller K_b and is the weaker of the two bases, NH_4^+ will be the stronger of the two conjugate acids. Solution (ii) will therefore be the more acidic of the two. Solution (iv) contains the K^+ ion, which is the cation of the strong base KOH, and the NO_3^- ion, which is the conjugate base of the strong acid HNO_3. Neither of the ions in solution (iv) will react with water to any appreciable extent, making the solution neutral. Thus, the order of pH is 0.1 *M* NH_4Cl < 0.1 *M* NH_3CH_3Br < 0.1 *M* KNO_3 < 0.1 *M* $Ba(C_2H_3O_2)_2$.

PRACTICE EXERCISE

In each of the following, indicate which salt will form the more acidic (or less basic) 0.010 *M* solution: **(a)** $NaNO_3$, $Fe(NO_3)_3$; **(b)** KBr, KBrO; **(c)** CH_3NH_3Cl, $BaCl_2$, **(d)** NH_4NO_2, NH_4NO_3.

Answers: **(a)** $Fe(NO_3)_3$, **(b)** KBr, **(c)** CH_3NH_3Cl, **(d)** NH_4NO_3

SAMPLE EXERCISE 16.18 | Predicting Whether the Solution of an Amphiprotic Anion is Acidic or Basic

Predict whether the salt Na_2HPO_4 will form an acidic solution or a basic solution on dissolving in water.

Solution

Analyze: We are asked to predict whether a solution of Na_2HPO_4 will be acidic or basic. This substance is an ionic compound composed of Na^+ and HPO_4^{2-} ions.

Plan: We need to evaluate each ion, predicting whether each is acidic or basic. Because Na^+ is the cation of a strong base, NaOH, we know that Na^+ has no influence on pH. It is merely a spectator ion in acid-base chemistry. Thus, our analysis of whether the solution is acidic or basic must focus on the behavior of the HPO_4^{2-} ion. We need to consider the fact that HPO_4^{2-} can act as either an acid or a base.

$$HPO_4^{2-}(aq) \rightleftharpoons H^+(aq) + PO_4^{3-}(aq) \quad [16.45]$$

$$HPO_4^{2-}(aq) + H_2O \rightleftharpoons H_2PO_4^-(aq) + OH^-(aq) \quad [16.46]$$

The reaction with the larger equilibrium constant will determine whether the solution is acidic or basic.

Solve: The value of K_a for Equation 16.45, as shown in Table 16.3, is 4.2×10^{-13}. We must calculate the value of K_b for Equation 16.46 from the value of K_a for its conjugate acid, $H_2PO_4^-$. We make use of the relationship shown in Equation 16.40.

$$K_a \times K_b = K_w$$

We want to know K_b for the base HPO_4^{2-}, knowing the value of K_a for the conjugate acid $H_2PO_4^-$:

$$K_b(HPO_4^{2-}) \times K_a(H_2PO_4^-) = K_w = 1.0 \times 10^{-14}$$

Because K_a for $H_2PO_4^-$ is 6.2×10^{-8} (Table 16.3), we calculate K_b for HPO_4^{2-} to be 1.6×10^{-7}. This is more than 10^5 times larger than K_a for HPO_4^{2-}; thus, the reaction shown in Equation 16.46 predominates over that in Equation 16.45, and the solution will be basic.

PRACTICE EXERCISE

Predict whether the dipotassium salt of citric acid ($K_2HC_6H_5O_7$) will form an acidic or basic solution in water (see Table 16.3 for data).

Answer: acidic

16.10 | Acid-Base Behavior and Chemical Structure

When a substance is dissolved in water, it may behave as an acid, behave as a base, or exhibit no acid-base properties. How does the chemical structure of a substance determine which of these behaviors is exhibited by the substance? For example, why do some substances that contain OH groups behave as bases, releasing OH^- ions into solution, whereas others behave as acids, ionizing to release H^+ ions? Why are some acids stronger than others? In this section we will discuss briefly the effects of chemical structure on acid-base behavior.

Factors that Affect Acid Strength

A molecule containing H will transfer a proton only if the H—X bond is polarized in the following way:

$$\overset{+\!\!\longrightarrow}{\text{H—X}}$$

In ionic hydrides, such as NaH, the reverse is true; the H atom possesses a negative charge and behaves as a proton acceptor (Equation 16.23). Essentially nonpolar H—X bonds, such as the H—C bond in CH_4, produce neither acidic nor basic aqueous solutions.

A second factor that helps determine whether a molecule containing an H—X bond will donate a proton is the strength of the bond. Very strong bonds are less easily dissociated than weaker ones. This factor is important, for example, in the case of the hydrogen halides. The H—F bond is the most polar H—X bond. You therefore might expect that HF would be a very strong acid if the first factor was all that mattered. However, the energy required to dissociate HF into H and F atoms is much higher than it is for the other hydrogen halides, as shown in Table 8.4. As a result, HF is a weak acid, whereas all the other hydrogen halides are strong acids in water.

A third factor that affects the ease with which a hydrogen atom ionizes from HX is the stability of the conjugate base, X^-. In general, the greater the stability of the conjugate base, the stronger is the acid. The strength of an acid is often a combination of all three factors: the polarity of the H—X bond, the strength of the H—X bond, and the stability of the conjugate base, X^-.

Binary Acids

In general, the H—X bond strength is the most important factor determining acid strength among the binary acids (those containing hydrogen and just one other element) in which X is in the same *group* in the periodic table. The strength of an H—X bond tends to decrease as the element X increases in size. As a result, the bond strength decreases and the acidity increases down a group. Thus, HCl is a stronger acid than HF, and H_2S is a stronger acid than H_2O.

Bond strengths change less moving across a row in the periodic table than they do down a group. As a result, bond polarity is the major factor determining acidity for binary acids in the same *row*. Thus, acidity increases as the electronegativity of the element X increases, as it generally does moving from left to right in a row. For example, the acidity of the second-row elements varies in the following order: $CH_4 < NH_3 \ll H_2O < HF$. Because the C—H bond is essentially nonpolar, CH_4 shows no tendency to form H^+ and CH_3^- ions. Although the N—H bond is polar, NH_3 has a nonbonding pair of electrons on the nitrogen atom that dominates its chemistry, so NH_3 acts as a base rather than as an acid. The periodic trends in the acid strengths of binary compounds of hydrogen and the nonmetals of periods 2 and 3 are summarized in Figure 16.12 ▼.

GIVE IT SOME THOUGHT

What is the major factor determining the increase in acidity of binary acids going down a column of the periodic table? What is the major factor going across a period?

	GROUP			
	4A	5A	6A	7A
Period 2	CH_4 No acid or base properties	NH_3 Weak base	H_2O ---	HF Weak acid
Period 3	SiH_4 No acid or base properties	PH_3 Weak base	H_2S Weak acid	HCl Strong acid

Increasing acid strength (down a group; left to right across a period)

Increasing base strength (up a group; right to left across a period)

◀ **Figure 16.12 Trends in acid-base properties of binary hydrides.** The acidity of the binary compounds of hydrogen and nonmetals increases moving left to right across a period and moving top to bottom down a group.

Oxyacids

Many common acids, such as sulfuric acid, contain one or more O—H bonds:

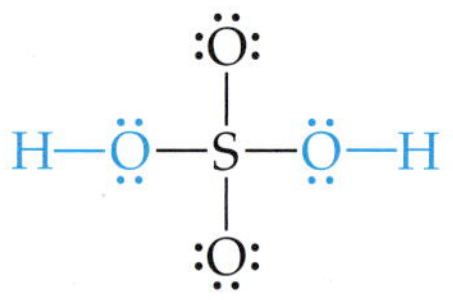

Acids in which OH groups and possibly additional oxygen atoms are bound to a central atom are called **oxyacids**. The OH group is also present in bases. What factors determine whether an OH group will behave as a base or as an acid?

Let's consider an OH group bound to some atom Y, which might in turn have other groups attached to it:

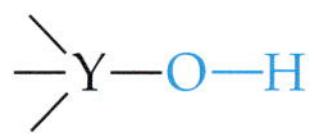

At one extreme, Y might be a metal, such as Na, K, or Mg. Because of their low electronegativities, the pair of electrons shared between Y and O is completely transferred to oxygen, and an ionic compound containing OH^- is formed. Such compounds are therefore sources of OH^- ions and behave as bases.

When Y is a nonmetal, the bond to O is covalent and the substance does not readily lose OH^-. Instead, these compounds are either acidic or neutral. As a general rule, as the electronegativity of Y increases, so will the acidity of the substance. This happens for two reasons: First, as electron density is drawn toward Y, the O—H bond becomes weaker and more polar, thereby favoring loss of H^+ (Figure 16.13 ◀). Second, because the conjugate base is usually an anion, its stability generally increases as the electronegativity of Y increases.

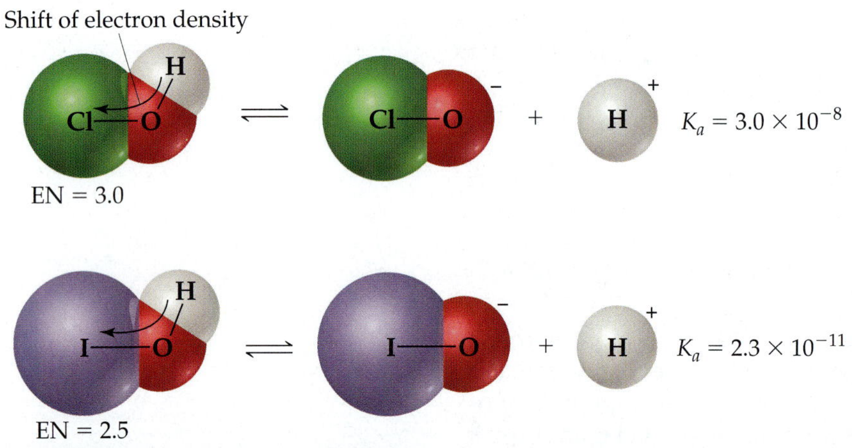

▲ **Figure 16.13 The acidity of oxyacids increases with increasing electronegativity of the central atom.** As the electronegativity of the atom attached to an OH group increases, the ease with which the hydrogen atom is ionized increases. The drift of electron density toward the electronegative atom further polarizes the O—H bond, which favors ionization. In addition, the electronegative atom will help stabilize the conjugate base, which also leads to a stronger acid. Because Cl is more electronegative than I, HClO is a stronger acid than HIO.

Many oxyacids contain additional oxygen atoms bonded to the central atom Y. The additional electronegative oxygen atoms pull electron density from the O—H bond, further increasing its polarity. Increasing the number of oxygen atoms also helps stabilize the conjugate base by increasing its ability to "spread out" its negative charge. Thus, the strength of an acid will increase as additional electronegative atoms bond to the central atom Y.

We can summarize these ideas as two simple rules that relate the acid strength of oxyacids to the electronegativity of Y and to the number of groups attached to Y.

1. For oxyacids that have the same number of OH groups and the same number of O atoms, acid strength increases with increasing electronegativity of the central atom Y. For example, the strength of the hypohalous acids, which have the structure H—O—Y, increases as the electronegativity of Y increases (Table 16.6 ◀).
2. For oxyacids that have the same central atom Y, acid strength increases as the number of oxygen atoms attached to Y increases. For example, the strength of the oxyacids of chlorine steadily increases from hypochlorous acid (HClO) to perchloric acid ($HClO_4$):

TABLE 16.6 Electronegativity Values (EN) of Y and Acid-Dissociation Constants

Acid	EN of Y	K_a
HClO	3.0	3.0×10^{-8}
HBrO	2.8	2.5×10^{-9}
HIO	2.5	2.3×10^{-11}

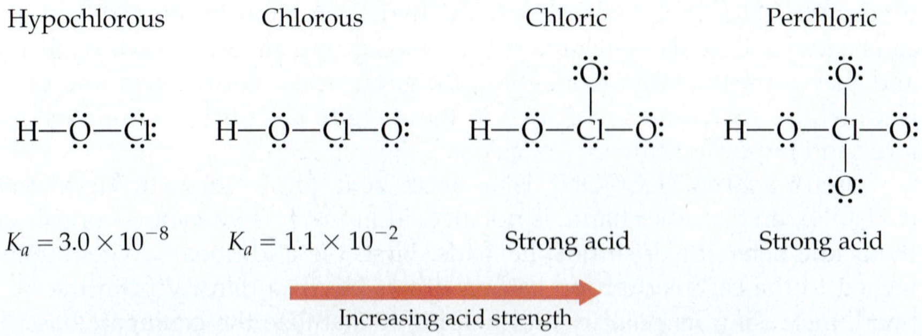

Because the oxidation number of the central atom increases as the number of attached O atoms increases, this correlation can be stated in an equivalent way: In a series of oxyacids, the acidity increases as the oxidation number of the central atom increases.

SAMPLE EXERCISE 16.19 | Predicting Relative Acidities from Composition and Structure

Arrange the compounds in each of the following series in order of increasing acid strength: **(a)** AsH_3, HI, NaH, H_2O; **(b)** H_2SeO_3, H_2SeO_4, H_2O.

Solution

Analyze: We are asked to arrange two sets of compounds in order from weakest acid to strongest acid.

Plan: For the binary acids in part (a), we will consider the electronegativities of As, I, Na, and O, respectively. For the oxyacids in part (b), we will consider the number of oxygen atoms bonded to the central atom and the similarities between the Se-containing compounds and some more familiar acids.

Solve: **(a)** The elements from the left side of the periodic table form the most basic binary hydrogen compounds because the hydrogen in these compounds carries a negative charge. Thus NaH should be the most basic compound on the list. Because arsenic is less electronegative than oxygen, we might expect that AsH_3 would be a weak base toward water. That is also what we would predict by an extension of the trends shown in Figure 16.13. Further, we expect that the binary hydrogen compounds of the halogens, as the most electronegative element in each period, will be acidic relative to water. In fact, HI is one of the strong acids in water. Thus the order of increasing acidity is NaH $<$ AsH_3 $<$ H_2O $<$ HI.

(b) The acidity of oxyacids increases as the number of oxygen atoms bonded to the central atom increases. Thus, H_2SeO_4 will be a stronger acid than H_2SeO_3; in fact, the Se atom in H_2SeO_4 is in its maximum positive oxidation state, and so we expect it to be a comparatively strong acid, much like H_2SO_4. H_2SeO_3 is an oxyacid of a nonmetal that is similar to H_2SO_3. As such, we expect that H_2SeO_3 is able to donate a proton to H_2O, indicating that H_2SeO_3 is a stronger acid than H_2O. Thus, the order of increasing acidity is H_2O $<$ H_2SeO_3 $<$ H_2SeO_4.

PRACTICE EXERCISE

In each of the following pairs choose the compound that leads to the more acidic (or less basic) solution: **(a)** HBr, HF; **(b)** PH_3, H_2S; **(c)** HNO_2, HNO_3; **(d)** H_2SO_3, H_2SeO_3.

Answers: **(a)** HBr, **(b)** H_2S, **(c)** HNO_3, **(d)** H_2SO_3

Carboxylic Acids

Another large group of acids is illustrated by acetic acid ($HC_2H_3O_2$):

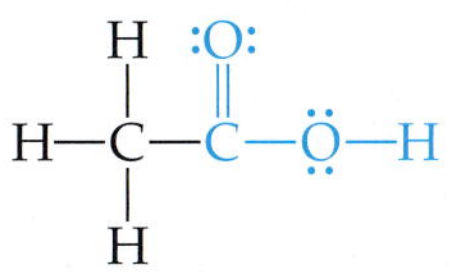

The portion of the structure shown in blue is called the *carboxyl group*, which is often written as COOH. Thus, the chemical formula of acetic acid is often

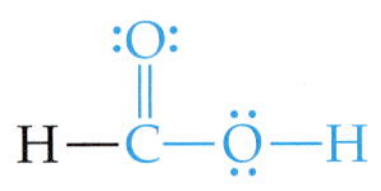

Formic acid

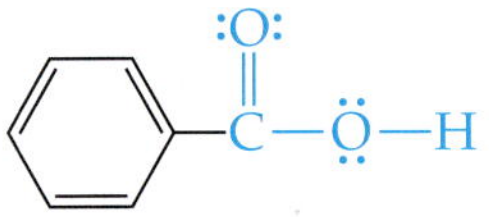

Benzoic acid

written as CH_3COOH, where only the hydrogen atom in the carboxyl group can be ionized. Acids that contain a carboxyl group are called **carboxylic acids**, and they form the largest category of organic acids. Formic acid and benzoic acid, whose structures are drawn in the margin, are further examples of this large and important category of acids.

Acetic acid (CH_3COOH) is a weak acid ($K_a = 1.8 \times 10^{-5}$). Methanol (CH_3OH), on the other hand, is not an acid in water. Two factors contribute to the acidic behavior of carboxylic acids. First, the additional oxygen atom attached to the carboxyl group carbon draws electron density from the O—H bond, increasing its polarity and helping to stabilize the conjugate base. Second, the conjugate base of a carboxylic acid (a *carboxylate anion*) can exhibit resonance (Section 8.6), which contributes further to the stability of the anion by spreading the negative charge over several atoms:

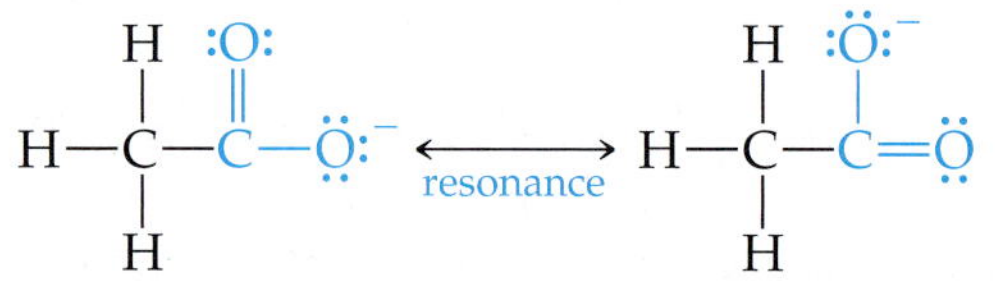

The acid strength of carboxylic acids also increases as the number of electronegative atoms in the acid increases. For example, trifluoroacetic acid (CF_3COOH) has $K_a = 5.0 \times 10^{-1}$; the replacement of three hydrogen atoms of acetic acid with more electronegative fluorine atoms leads to a large increase in acid strength.

GIVE IT SOME THOUGHT

What group of atoms is present in all carboxylic acids?

16.11 | Lewis Acids and Bases

For a substance to be a proton acceptor (a Brønsted–Lowry base), it must have an unshared pair of electrons for binding the proton. NH_3, for example, acts as a proton acceptor. Using Lewis structures, we can write the reaction between H^+ and NH_3 as follows:

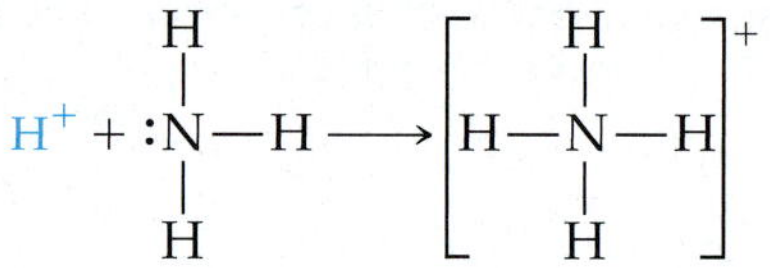

G. N. Lewis was the first to notice this aspect of acid-base reactions. He proposed a definition of acid and base that emphasizes the shared electron pair: A **Lewis acid** is an electron-pair acceptor, and a **Lewis base** is an electron-pair donor.

Every base that we have discussed thus far—whether it be OH^-, H_2O, an amine, or an anion—is an electron-pair donor. Everything that is a base in the Brønsted–Lowry sense (a proton acceptor) is also a base in the Lewis sense (an electron-pair donor). In the Lewis theory, however, a base can donate its electron pair to something other than H^+. The Lewis definition therefore greatly increases the number of species that can be considered acids; H^+ is a Lewis acid, but not the only one. For example, consider the reaction between NH_3 and BF_3. This reaction occurs because BF_3 has a vacant orbital in its valence shell. (Section 8.7) It therefore acts as an electron-pair acceptor (a Lewis acid) toward NH_3, which donates the electron pair. The curved arrow shows the donation of a pair of electrons from N to B to form a covalent bond:

ANIMATION
Lewis Acid-Base Theory

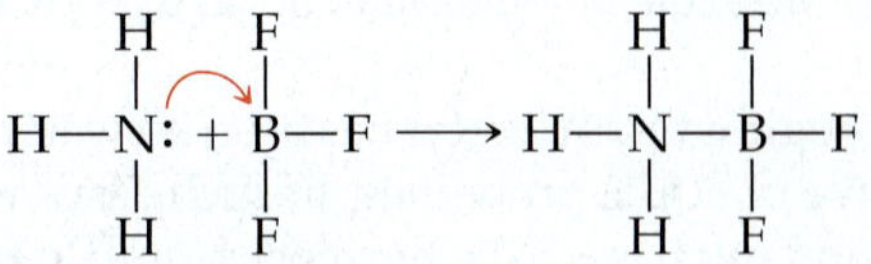

CHEMISTRY AND LIFE | The Amphiprotic Behavior of Amino Acids

Amino acids are the building blocks of proteins. The general structure of amino acids is shown here, where different amino acids have different R groups attached to the central carbon atom:

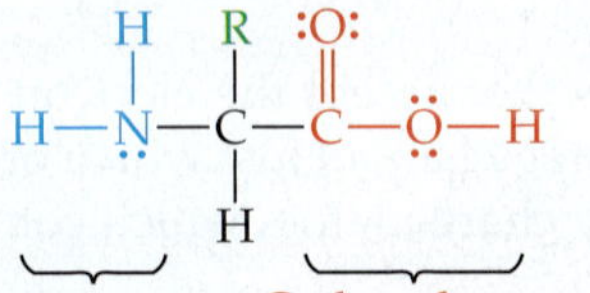

For example, in *glycine*, which is the simplest amino acid, R is a hydrogen atom, whereas in *alanine*, R is a CH_3 group.

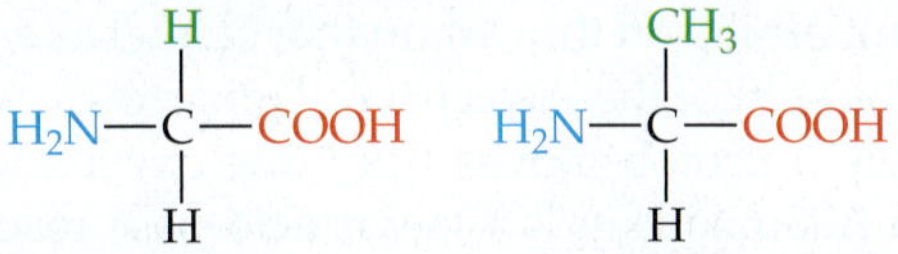

Amino acids contain a carboxyl group and can therefore serve as acids. They also contain an NH_2 group, characteristic of amines (Section 16.7), and thus they can also act as bases. Amino acids, therefore, are amphiprotic. For glycine, we might expect that the acid and the base reactions with water would be as follows:

$$\textit{Acid:}\ H_2N{-}CH_2{-}COOH(aq) + H_2O(l) \rightleftharpoons H_2N{-}CH_2{-}COO^-(aq) + H_3O^+(aq) \quad [16.47]$$

$$\textit{Base:}\ H_2N{-}CH_2{-}COOH(aq) + H_2O(l) \rightleftharpoons {}^+H_3N{-}CH_2{-}COOH(aq) + OH^-(aq) \quad [16.48]$$

The pH of a solution of glycine in water is about 6.0, indicating that it is a slightly stronger acid than a base.

The acid-base chemistry of amino acids is somewhat more complicated than shown in Equations 16.47 and 16.48, however. Because the COOH can act as an acid and the NH_2 group can act as a base, amino acids undergo a "self-contained" Brønsted–Lowry acid-base reaction in which the proton of the carboxyl group is transferred to the basic nitrogen atom:

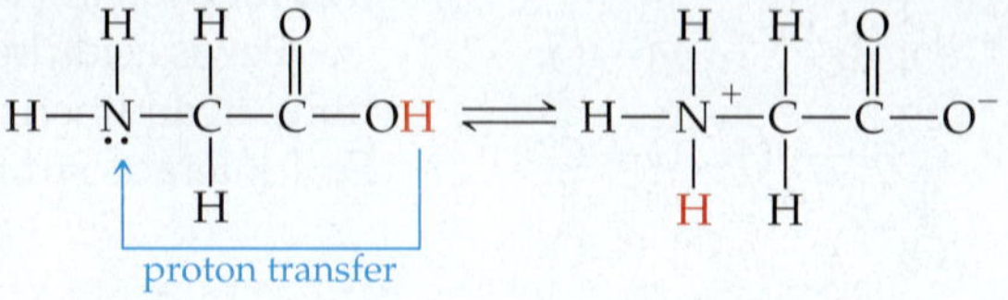

[16.49]

Although the form of the amino acid on the right side of Equation 16.49 is electrically neutral overall, it has a positively charged end and a negatively charged end. A molecule of this type is called a *zwitterion* (German for "hybrid ion").

Do amino acids exhibit any properties indicating that they behave as zwitterions? If so, they should behave similar to ionic substances. (Section 8.2) Crystalline amino acids (Figure 16.14 ▼) have relatively high melting points, usually above 200°C, which is characteristic of ionic solids. Amino acids are far more soluble in water than in nonpolar solvents. In addition, the dipole moments of amino acids are large, consistent with a large separation of charge in the molecule. Thus, the ability of amino acids to act simultaneously as acids and bases has important effects on their properties.

Related Exercise: 16.115

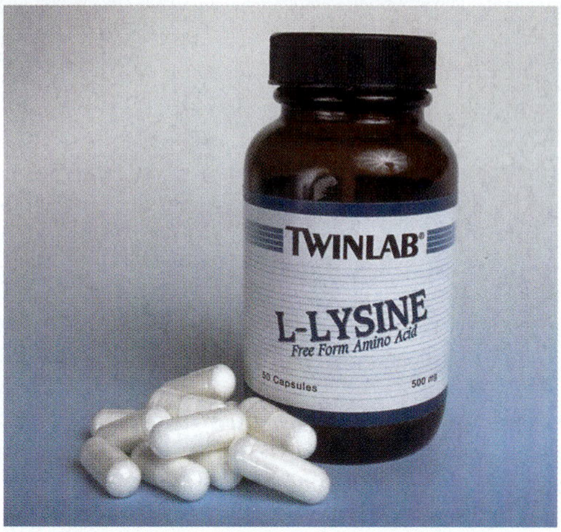

▲ **Figure 16.14 Lysine.** One of the amino acids found in proteins, lysine is available as a dietary supplement. The L on the label refers to a specific arrangement of atoms that is found in naturally occurring amino acids. Molecules with the L arrangement are mirror images of molecules with the D arrangement, much like our left hand is a mirror image of our right hand.

GIVE IT SOME THOUGHT

What feature must any molecule or ion have to act as a Lewis base?

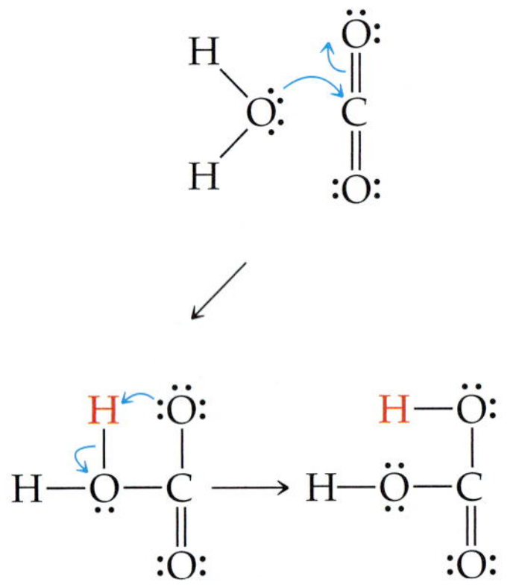

Our emphasis throughout this chapter has been on water as the solvent and on the proton as the source of acidic properties. In such cases we find the Brønsted–Lowry definition of acids and bases to be the most useful. In fact, when we speak of a substance as being acidic or basic, we are usually thinking of aqueous solutions and using these terms in the Arrhenius or Brønsted–Lowry sense. The advantage of the Lewis theory is that it allows us to treat a wider variety of reactions, including those that do not involve proton transfer, as acid-base reactions. To avoid confusion, a substance like BF_3 is rarely called an acid unless it is clear from the context that we are using the term in the sense of the Lewis definition. Instead, substances that function as electron-pair acceptors are referred to explicitly as "Lewis acids."

Lewis acids include molecules that, like BF_3, have an incomplete octet of electrons. In addition, many simple cations can function as Lewis acids. For example, Fe^{3+} interacts strongly with cyanide ions to form the ferricyanide ion, $Fe(CN)_6{}^{3-}$.

$$Fe^{3+} + 6[:C{\equiv}N:]^- \longrightarrow [Fe(C{\equiv}N:)_6]^{3-}$$

The Fe^{3+} ion has vacant orbitals that accept the electron pairs donated by the ions; we will learn more in Chapter 24 about just which orbitals are used by the Fe^{3+} ion. The metal ion is highly charged, too, which contributes to the interaction with CN^- ions.

Some compounds with multiple bonds can behave as Lewis acids. For example, the reaction of carbon dioxide with water to form carbonic acid (H_2CO_3) can be pictured as an attack by a water molecule on CO_2, in which the water acts as an electron-pair donor and the CO_2 as an electron-pair acceptor, as shown in the margin. The electron pair of one of the carbon–oxygen double bonds is moved onto the oxygen, leaving a vacant orbital on the carbon that can act as an electron-pair acceptor. We have shown the shift of these electrons with arrows. After forming the initial acid-base product, a proton moves from one oxygen to another, thereby forming carbonic acid. A similar kind of Lewis acid-base reaction takes place when any oxide of a nonmetal dissolves in water to form an acidic solution.

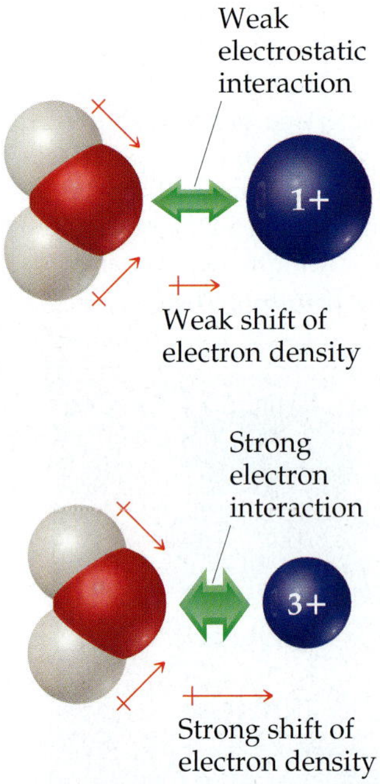

▲ **Figure 16.15 The acidity of a hydrated cation depends on cation charge and size.** The interaction between a water molecule and a cation is much stronger when the cation is a smaller ion of higher charge. The pull of electron density toward the cation weakens the polar O—H bond of the water molecule and allows the transfer of a H^+ ion to a nearby water molecule. As a result, hydrated cations tend to be acidic, with their acidities increasing with increasing charge and decreasing size.

Hydrolysis of Metal Ions

As we have already seen, most metal ions behave as acids in aqueous solution. (Section 16.9) For example, an aqueous solution of $Cr(NO_3)_3$ is quite acidic. An aqueous solution of $ZnCl_2$ is also acidic, though to a lesser extent. The Lewis concept helps explain the interactions between metal ions and water molecules that give rise to this acidic behavior.

Because metal ions are positively charged, they attract the unshared electron pairs of water molecules. It is primarily this interaction, referred to as *hydration*, that causes salts to dissolve in water. (Section 13.1) The process of hydration can be thought of as a Lewis acid-base interaction in which the metal ion acts as a Lewis acid and the water molecules as Lewis bases. When a water molecule interacts with the positively charged metal ion, electron density is drawn from the oxygen, as illustrated in Figure 16.15 ◄. This flow of electron density causes the O—H bond to become more polarized; as a result, water molecules bound to the metal ion are more acidic than those in the bulk solvent.

The hydrated Fe^{3+} ion, $Fe(H_2O)_6{}^{3+}$, which we usually represent simply as $Fe^{3+}(aq)$, acts as a source of protons:

$$Fe(H_2O)_6{}^{3+}(aq) \rightleftharpoons Fe(H_2O)_5(OH)^{2+}(aq) + H^+(aq) \qquad [16.50]$$

The acid-dissociation constant for this hydrolysis reaction has the value $K_a = 2 \times 10^{-3}$, so $Fe^{3+}(aq)$ is a fairly strong acid. Acid-dissociation constants for

▲ **Figure 16.16 Effect of cations on the pH of a solution.** The pH values of 1.0 *M* solutions of a series of nitrate salts are estimated using acid-base indicators.

hydrolysis reactions generally increase with increasing charge and decreasing radius of the ion (Figure 16.15). Thus, the Cu^{2+} ion, which has a smaller charge and a larger radius than Fe^{3+}, forms less acidic solutions than Fe^{3+}: The K_a for $Cu^{2+}(aq)$ is 1×10^{-8}. The acid hydrolysis of a number of salts of metal ions is demonstrated in Figure 16.16 ▲. Note that the Na^+ ion, which is large and has only a 1+ charge (and which we have previously identified as the cation of a strong base), exhibits no acid hydrolysis and yields a neutral solution.

GIVE IT SOME THOUGHT

Which of the following cations will be most acidic and why: Ca^{2+}, Fe^{2+}, Fe^{3+}?

SAMPLE INTEGRATIVE EXERCISE | Putting Concepts Together

Phosphorous acid (H_3PO_3) has the following Lewis structure.

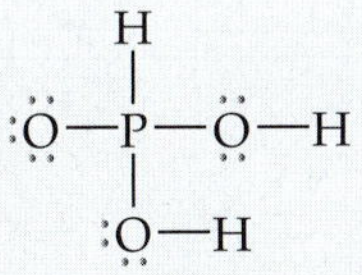

(a) Explain why H_3PO_3 is diprotic and not triprotic. **(b)** A 25.0-mL sample of a solution of H_3PO_3 is titrated with 0.102 *M* NaOH. It requires 23.3 mL of NaOH to neutralize both acidic protons. What is the molarity of the H_3PO_3 solution? **(c)** This solution has a pH of 1.59. Calculate the percent ionization and K_{a1} for H_3PO_3, assuming that $K_{a1} \gg K_{a2}$. **(d)** How does the osmotic pressure of a 0.050 *M* solution of HCl compare with that of a 0.050 *M* solution of H_3PO_3? Explain.

Solution The problem asks us to explain why there are only two ionizable protons in the H_3PO_3 molecule. Further, we are asked to calculate the molarity of a solution of H_3PO_3, given titration-experiment data. We then need to calculate the percent ionization of the H_3PO_3 solution in part (b). Finally, we are asked to compare the osmotic pressure of a 0.050 *M* solution of H_3PO_3 with that of an HCl solution of the same concentration.

We will use what we have learned about molecular structure and its impact on acidic behavior to answer part (a). We will then use stoichiometry and the relationship between pH and $[H^+]$ to answer parts (b) and (c). Finally, we will consider acid strength in order to compare the colligative properties of the two solutions in part (d).

(a) Acids have polar H—X bonds. From Figure 8.6 we see that the electronegativity of H is 2.1 and that of P is also 2.1. Because the two elements have the same electronegativity, the H—P bond is nonpolar. (Section 8.4) Thus, this H cannot be acidic. The other two H atoms, however, are bonded to O, which has an electronegativity of 3.5. The H—O bonds are therefore polar, with H having a partial positive charge. These two H atoms are consequently acidic.

(b) The chemical equation for the neutralization reaction is

$$H_3PO_3(aq) + 2\,NaOH(aq) \longrightarrow Na_2HPO_3(aq) + H_2O(l)$$

From the definition of molarity, $M = \text{mol/L}$, we see that moles $= M \times L$. (Section 4.5) Thus, the number of moles of NaOH added to the solution is $(0.0233\text{ L})(0.102\text{ mol/L}) = 2.377 \times 10^{-3}$ mol NaOH. The balanced equation indicates that 2 mol of NaOH is consumed for each mole of H_3PO_3. Thus, the number of moles of H_3PO_3 in the sample is

$$(2.377 \times 10^{-3}\text{ mol NaOH})\left(\frac{1\text{ mol }H_3PO_3}{2\text{ mol NaOH}}\right) = 1.189 \times 10^{-3}\text{ mol }H_3PO_3$$

The concentration of the H_3PO_3 solution, therefore, equals $(1.189 \times 10^{-3}\text{ mol})/(0.0250\text{ L}) = 0.0475\ M$.

(c) From the pH of the solution, 1.59, we can calculate $[H^+]$ at equilibrium.

$$[H^+] = \text{antilog}(-1.59) = 10^{-1.59} = 0.026\ M \text{ (two significant figures)}$$

Because $K_{a1} \gg K_{a2}$, the vast majority of the ions in solution are from the first ionization step of the acid.

$$H_3PO_3(aq) \rightleftharpoons H^+(aq) + H_2PO_3^-(aq)$$

Because one $H_2PO_3^-$ ion forms for each H^+ ion formed, the equilibrium concentrations of H^+ and $H_2PO_3^-$ are equal: $[H^+] = [H_2PO_3^-] = 0.026\ M$. The equilibrium concentration of H_3PO_3 equals the initial concentration minus the amount that ionizes to form H^+ and $H_2PO_3^-$: $[H_3PO_3] = 0.0475\ M - 0.026\ M = 0.022\ M$ (two significant figures). These results can be tabulated as follows:

	$H_2PO_3^-(aq)$ $\rightleftharpoons$	$H^+(aq)$ +	$H_3PO_3(aq)$
Initial	$0.0475\ M$	0	0
Change	$-0.026\ M$	$+0.026\ M$	$+0.026\ M$
Equilibrium	$0.022\ M$	$0.026\ M$	$0.026\ M$

The percent ionization is

$$\text{Percent ionization} = \frac{[H^+]_{\text{equilibrium}}}{[H_3PO_3]_{\text{initial}}} \times 100\% = \frac{0.026\ M}{0.0475\ M} \times 100\% = 55\%$$

The first acid-dissociation constant is

$$K_{a1} = \frac{[H^+][H_2PO_3^-]}{[H_3PO_3]} = \frac{(0.026)(0.026)}{0.022} = 0.030$$

(d) Osmotic pressure is a colligative property and depends on the total concentration of particles in solution. (Section 13.5) Because HCl is a strong acid, a $0.050\ M$ solution will contain $0.050\ M$ $H^+(aq)$ and $0.050\ M$ $Cl^-(aq)$, or a total of 0.100 mol/L of particles. Because H_3PO_3 is a weak acid, it ionizes to a lesser extent than HCl, and, hence, there are fewer particles in the H_3PO_3 solution. As a result, the H_3PO_3 solution will have the lower osmotic pressure.

SUMMARY AND KEY TERMS

Section 16.1 Acids and bases were first recognized by the properties of their aqueous solutions. For example, acids turn litmus red, whereas bases turn litmus blue. Arrhenius recognized that the properties of acidic solutions are due to $H^+(aq)$ ions and those of basic solutions are due to $OH^-(aq)$ ions.

Section 16.2 The Brønsted–Lowry concept of acids and bases is more general than the Arrhenius concept and emphasizes the transfer of a proton (H^+) from an acid to a base. The H^+ ion, which is merely a proton with no surrounding valence electrons, is strongly bound to water. For this reason, the **hydronium ion**, $H_3O^+(aq)$, is often used to represent the predominant form of H^+ in water instead of the simpler $H^+(aq)$.

A **Brønsted–Lowry acid** is a substance that donates a proton to another substance; a **Brønsted–Lowry base** is a substance that accepts a proton from another substance. Water is an **amphiprotic** substance, one that can function as either a Brønsted–Lowry acid or base, depending on the substance with which it reacts.

The **conjugate base** of a Brønsted–Lowry acid is the species that remains when a proton is removed from the acid. The **conjugate acid** of a Brønsted–Lowry base is the species formed by adding a proton to the base. Together, an acid and its conjugate base (or a base and its conjugate acid) are called a **conjugate acid-base pair**.

The acid-base strengths of conjugate acid-base pairs are related: The stronger an acid, the weaker its conjugate base; the weaker an acid, the stronger its conjugate base. In every acid-base reaction, the position of the equilibrium favors the transfer of the proton from the stronger acid to the stronger base.

Section 16.3 Water ionizes to a slight degree, forming $H^+(aq)$ and $OH^-(aq)$. The extent of this **autoionization** is expressed by the **ion-product constant** for water:

$$K_w = [H^+][OH^-] = 1.0 \times 10^{-14} \ (25°C)$$

This relationship describes both pure water and aqueous solutions. The K_w expression indicates that the product of $[H^+]$ and $[OH^-]$ is a constant. Thus, as $[H^+]$ increases, $[OH^-]$ decreases. Acidic solutions are those that contain more $H^+(aq)$ than $OH^-(aq)$; basic solutions contain more $OH^-(aq)$ than $H^+(aq)$.

Section 16.4 The concentration of $H^+(aq)$ can be expressed in terms of **pH**: $pH = -\log[H^+]$. At 25°C the pH of a neutral solution is 7.00, whereas the pH of an acidic solution is below 7.00, and the pH of a basic solution is above 7.00. The pX notation is also used to represent the negative log of other small quantities, as in pOH and pK_w. The pH of a solution can be measured using a pH meter, or it can be estimated using acid-base indicators.

Section 16.5 Strong acids are strong electrolytes, ionizing completely in aqueous solution. The common strong acids are HCl, HBr, HI, HNO_3, $HClO_3$, $HClO_4$, and H_2SO_4. The conjugate bases of strong acids have negligible basicity.

Common strong bases are the ionic hydroxides of alkali metals and the heavy alkaline earth metals. The cations of strong bases have negligible acidity.

Section 16.6 Weak acids are weak electrolytes; only part of the molecules exist in solution in ionized form. The extent of ionization is expressed by the **acid-dissociation constant**, K_a, which is the equilibrium constant for the reaction $HA(aq) \rightleftharpoons H^+(aq) + A^-(aq)$, which can also be written $HA(aq) + H_2O(l) \rightleftharpoons H_3O^+(aq) + A^-(aq)$. The larger the value of K_a, the stronger the acid. The concentration of a weak acid and its K_a value can be used to calculate the pH of a solution.

Polyprotic acids, such as H_2SO_3, have more than one ionizable proton. These acids have acid-dissociation constants that decrease in magnitude in the order $K_{a1} > K_{a2} > K_{a3}$. Because nearly all the $H^+(aq)$ in a polyprotic acid solution comes from the first dissociation step, the pH can usually be estimated satisfactorily by considering only K_{a1}.

Sections 16.7 and 16.8 Weak bases include NH_3, **amines**, and the anions of weak acids. The extent to which a weak base reacts with water to generate the corresponding conjugate acid and OH^- is measured by the **base-dissociation constant**, K_b. This is the equilibrium constant for the reaction $B(aq) + H_2O(l) \rightleftharpoons HB^+(aq) + OH^-(aq)$, where B is the base.

The relationship between the strength of an acid and the strength of its conjugate base is expressed quantitatively by the equation $K_a \times K_b = K_w$, where K_a and K_b are dissociation constants for conjugate acid-base pairs.

Section 16.9 The acid-base properties of salts can be ascribed to the behavior of their respective cations and anions. The reaction of ions with water, with a resultant change in pH, is called **hydrolysis**. The cations of the alkali metals and the alkaline earth metals and the anions of strong acids do not undergo hydrolysis. They are always spectator ions in acid-base chemistry.

Section 16.10 The tendency of a substance to show acidic or basic characteristics in water can be correlated with its chemical structure. Acid character requires the presence of a highly polar H—X bond. Acidity is also favored when the H—X bond is weak and when the X^- ion is very stable.

For **oxyacids** with the same number of OH groups and the same number of O atoms, acid strength increases with increasing electronegativity of the central atom. For oxyacids with the same central atom, acid strength increases as the number of oxygen atoms attached to the central atom increases. The structures of **carboxylic acids**, which are organic acids containing the COOH group, also helps us to understand their acidity.

Section 16.11 The Lewis concept of acids and bases emphasizes the shared electron pair rather than the proton. A **Lewis acid** is an electron-pair acceptor, and a **Lewis base** is an electron-pair donor. The Lewis concept is more general than the Brønsted–Lowry concept because it can apply to cases in which the acid is some substance other than H^+. The Lewis concept helps to explain why many hydrated metal cations form acidic aqueous solutions. The acidity of these cations generally increases as their charge increases and as the size of the metal ion decreases.

VISUALIZING CONCEPTS

16.1 **(a)** Identify the Brønsted–Lowry acid and the Brønsted–Lowry base in the following reaction:

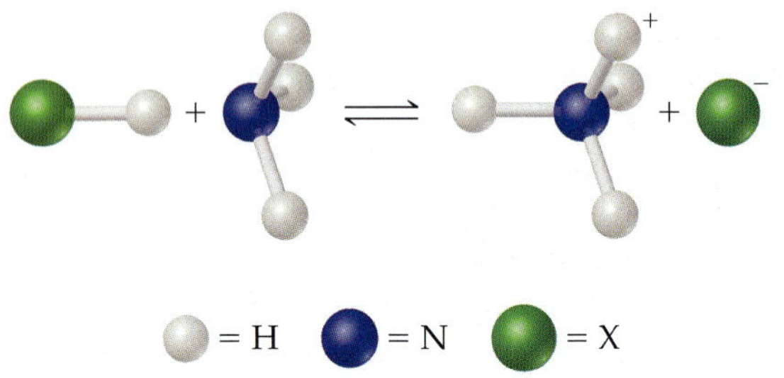

(b) Identify the Lewis acid and the Lewis base in the reaction. [Sections 16.2 and 16.11]

16.2 The following diagrams represent aqueous solutions of two monoprotic acids, HA (A = X or Y). The water molecules have been omitted for clarity. **(a)** Which is the stronger acid, HX or HY? **(b)** Which is the stronger base, X^- or Y^-? **(c)** If you mix equal concentrations of HX and NaY, will the equilibrium

$$HX(aq) + Y^-(aq) \rightleftharpoons HY(aq) + X^-(aq)$$

lie mostly to the right ($K_c > 1$) or to the left ($K_c < 1$)? [Section 16.2]

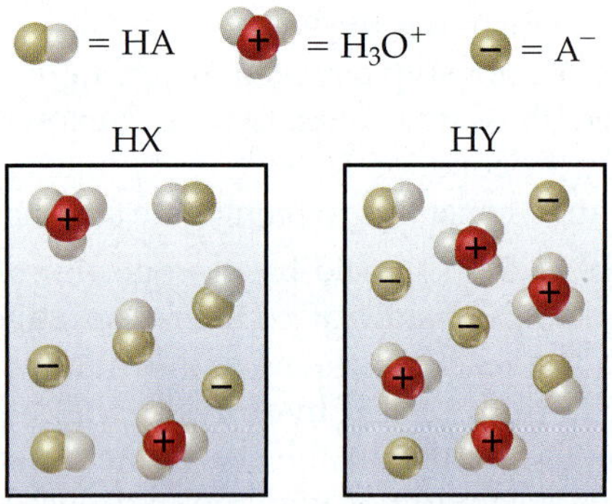

16.3 The following diagrams represent aqueous solutions of three acids, HX, HY, and HZ. The water molecules have been omitted for clarity, and the hydrated proton is represented as a simple sphere rather than as a hydronium ion. **(a)** Which of the acids is a strong acid? Explain. **(b)** Which acid would have the smallest acid-dissociation constant, K_a? **(c)** Which solution would have the highest pH? [Sections 16.5 and 16.6]

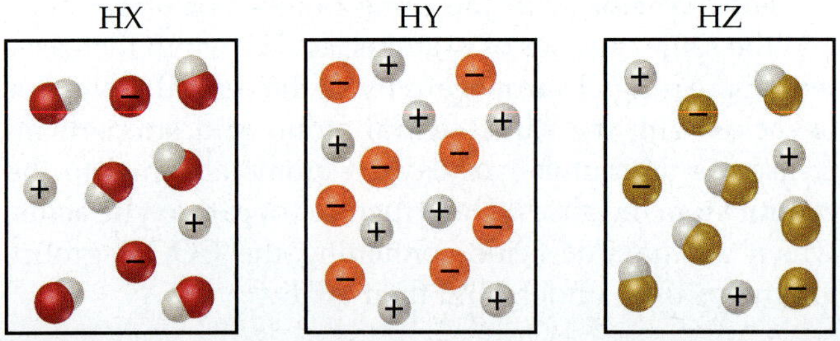

16.4 **(a)** Which of these three lines represents the effect of concentration on the percent ionization of a weak acid? **(b)** Explain in qualitative terms why the curve you choose has the shape it does. [Section 16.6]

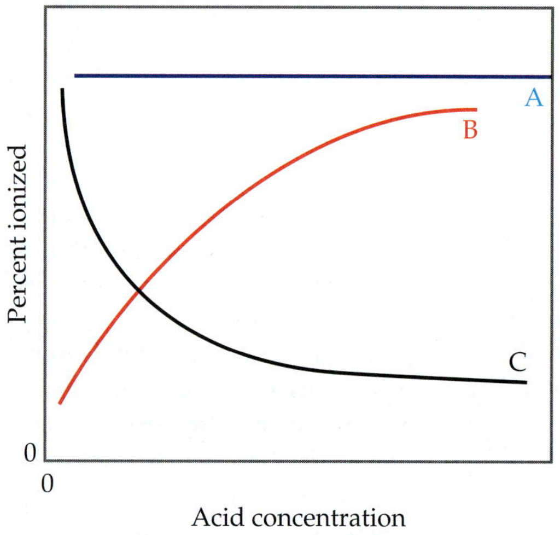

16.5 Refer to the diagrams accompanying Exercise 16.3. **(a)** Rank the anions, X^-, Y^-, and Z^-, in order of increasing basicity. **(b)** Which of the ions would have the largest base-dissociation constant, K_b? [Sections 16.2 and 16.8]

16.6 **(a)** Draw the Lewis structure for the following molecule and explain why it is able to act as a base. **(b)** To what class of organic compounds does this substance belong? [Section 16.7]

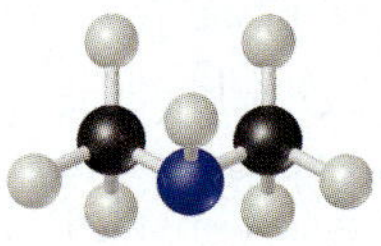

16.7 The following diagram represents an aqueous solution formed by dissolving a sodium salt of a weak acid in water. The diagram shows only the Na^+ ions, the X^- ions, and the HX molecules. What ion is missing from the diagram? If the drawing is completed by drawing all the ions, how many of the missing ions should be shown? [Section 16.9]

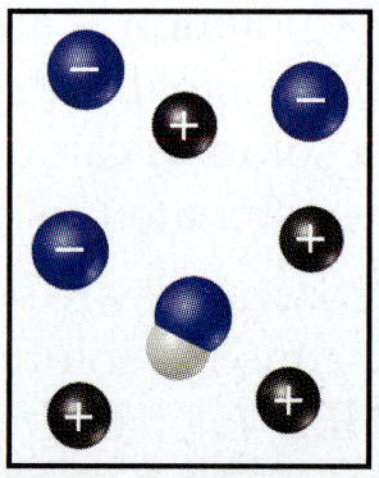

16.8 **(a)** What kinds of acids are represented by the following molecular models? **(b)** Indicate how the acidity of each molecule is affected by increasing the electronegativity of the atom X, and explain the origin of the effect. [Section 16.10]

(a) (b)

16.9 In this model of acetylsalicylic acid (aspirin), identify the carboxyl group in the molecule. [Section 16.10]

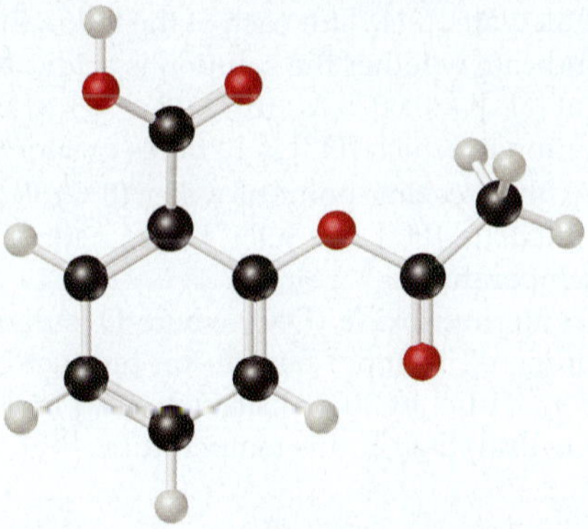

16.10 **(a)** The following diagram represents the reaction of PCl_4^+ with Cl^-. Draw the Lewis structures for the reactants and products, and identify the Lewis acid and the Lewis base in the reaction.

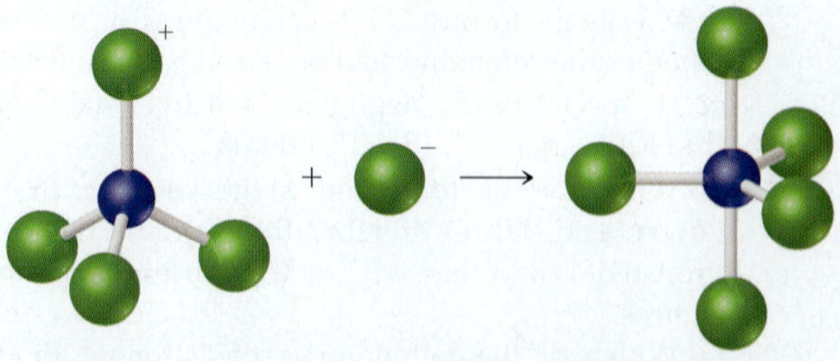

(b) The following reaction represents the acidity of a hydrated cation. How does the equilibrium constant for the reaction change as the charge of the cation increases? [Section 16.11]

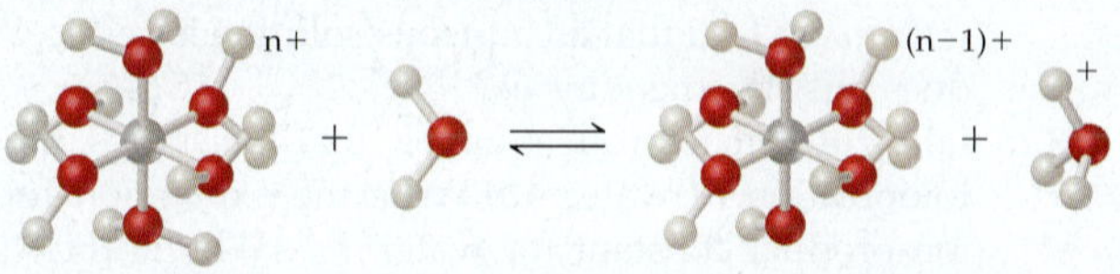

EXERCISES

Arrhenius and Brønsted–Lowry Acids and Bases

16.11 Although HCl and H_2SO_4 have very different properties as pure substances, their aqueous solutions possess many common properties. List some general properties of these solutions, and explain their common behavior in terms of the species present.

16.12 Although pure NaOH and CaO have very different properties, their aqueous solutions possess many common properties. List some general properties of these solutions, and explain their common behavior in terms of the species present.

16.13 **(a)** What is the difference between the Arrhenius and the Brønsted–Lowry definitions of an acid? **(b)** $NH_3(g)$ and $HCl(g)$ react to form the ionic solid $NH_4Cl(s)$ (Figure 16.3). Which substance is the Brønsted–Lowry acid in this reaction? Which is the Brønsted–Lowry base?

16.14 **(a)** What is the difference between the Arrhenius and the Brønsted–Lowry definitions of a base? **(b)** When ammonia is dissolved in water, it behaves both as an Arrhenius base and as a Brønsted–Lowry base. Explain.

16.15 Give the conjugate base of the following Brønsted–Lowry acids: **(a)** HIO_3, **(b)** NH_4^+, **(c)** $H_2PO_4^-$, **(d)** $HC_7H_5O_2$.

16.16 Give the conjugate acid of the following Brønsted–Lowry bases: **(a)** CN^-, **(b)** O^{2-}, **(c)** HPO_4^{2-}, **(d)** $C_2H_5NH_2$.

16.17 Designate the Brønsted–Lowry acid and the Brønsted–Lowry base on the left side of each of the following equations, and also designate the conjugate acid and conjugate base on the right side:

(a) $NH_4^+(aq) + CN^-(aq) \rightleftharpoons HCN(aq) + NH_3(aq)$

(b) $(CH_3)_3N(aq) + H_2O(l) \rightleftharpoons (CH_3)_3NH^+(aq) + OH^-(aq)$

(c) $HCHO_2(aq) + PO_4^{3-}(aq) \rightleftharpoons CHO_2^-(aq) + HPO_4^{2-}(aq)$

16.18 Designate the Brønsted–Lowry acid and the Brønsted–Lowry base on the left side of each equation, and also designate the conjugate acid and conjugate base on the right side.

(a) $HBrO(aq) + H_2O(l) \rightleftharpoons H_3O^+(aq) + BrO^-(aq)$

(b) $HSO_4^-(aq) + HCO_3^-(aq) \rightleftharpoons SO_4^{2-}(aq) + H_2CO_3(aq)$

(c) $HSO_3^-(aq) + H_3O^+(aq) \rightleftharpoons H_2SO_3(aq) + H_2O(l)$

16.19 **(a)** The hydrogen oxalate ion ($HC_2O_4^-$) is amphiprotic. Write a balanced chemical equation showing how it acts as an acid toward water and another equation showing how it acts as a base toward water. **(b)** What is the conjugate acid of $HC_2O_4^-$? What is its conjugate base?

16.20 **(a)** Write an equation for the reaction in which $H_2C_6O_5H_7^-(aq)$ acts as a base in $H_2O(l)$. **(b)** Write an equation for the reaction in which $H_2C_6O_5H_7^-(aq)$ acts as an acid in $H_2O(l)$. **(c)** What is the conjugate acid of $H_2C_6O_5H_7^-$? What is its conjugate base?

16.21 Label each of the following as being a strong acid, a weak acid, or a species with negligible acidity. In each case write the formula of its conjugate base, and indicate whether the conjugate base is a strong base, a weak base, or a species with negligible basicity: **(a)** HNO_2, **(b)** H_2SO_4, **(c)** HPO_4^{2-}, **(d)** CH_4, **(e)** $CH_3NH_3^+$ (an ion related to NH_4^+).

16.22 Label each of the following as being a strong base, a weak base, or a species with negligible basicity. In each case write the formula of its conjugate acid, and indicate whether the conjugate acid is a strong acid, a weak acid, or a species with negligible acidity: **(a)** $C_2H_3O_2^-$, **(b)** HCO_3^-, **(c)** O^{2-}, **(d)** Cl^-, **(e)** NH_3.

16.23 **(a)** Which of the following is the stronger Brønsted–Lowry acid, HBrO or HBr? **(b)** Which is the stronger Brønsted–Lowry base, F^- or Cl^-? Briefly explain your choices.

16.24 **(a)** Which of the following is the stronger Brønsted–Lowry acid, HNO_3 or HNO_2? **(b)** Which is the stronger Brønsted–Lowry base, NH_3 or H_2O? Briefly explain your choices.

16.25 Predict the products of the following acid-base reactions, and also predict whether the equilibrium lies to the left or to the right of the equation:
(a) $O^{2-}(aq) + H_2O(l) \rightleftharpoons$
(b) $HC_2H_3O_2(aq) + HS^-(aq) \rightleftharpoons$
(c) $NO_3^-(aq) + H_2O(l) \rightleftharpoons$

16.26 Predict the products of the following acid-base reactions, and also predict whether the equilibrium lies to the left or to the right of the equation:
(a) $NH_4^+(aq) + OH^-(aq) \rightleftharpoons$
(b) $C_2H_3O_2^-(aq) + H_3O^+(aq) \rightleftharpoons$
(c) $HCO_3^-(aq) + F^-(aq) \rightleftharpoons$

Autoionization of Water

16.27 **(a)** What is meant by the term *autoionization*? **(b)** Explain why pure water is a poor conductor of electricity. **(c)** You are told that an aqueous solution is acidic. What does this statement mean?

16.28 **(a)** Write a chemical equation that illustrates the autoionization of water. **(b)** Write the expression for the ion-product constant for water, K_w. Why is $[H_2O]$ absent from this expression? **(c)** A solution is described as basic. What is meant by this statement?

16.29 Calculate $[H^+]$ for each of the following solutions, and indicate whether the solution is acidic, basic, or neutral: **(a)** $[OH^-] = 0.00045\ M$; **(b)** $[OH^-] = 8.8 \times 10^{-9}\ M$; **(c)** a solution in which $[OH^-]$ is 100 times greater than $[H^+]$.

16.30 Calculate $[OH^-]$ for each of the following solutions, and indicate whether the solution is acidic, basic, or neutral: **(a)** $[H^+] = 0.0075\ M$; **(b)** $[H^+] = 6.5 \times 10^{-10}\ M$; **(c)** a solution in which $[H^+]$ is 10 times greater than $[OH^-]$.

16.31 At the freezing point of water (0°C), $K_w = 1.2 \times 10^{-15}$. Calculate $[H^+]$ and $[OH^-]$ for a neutral solution at this temperature.

16.32 Deuterium oxide (D_2O, where D is deuterium, the hydrogen-2 isotope) has an ion-product constant, K_w, of 8.9×10^{-16} at 20°C. Calculate $[D^+]$ and $[OD^-]$ for pure (neutral) D_2O at this temperature.

The pH Scale

16.33 By what factor does $[H^+]$ change for a pH change of **(a)** 2.00 units, **(b)** 0.50 units?

16.34 Consider two solutions, solution A and solution B. $[H^+]$ in solution A is 500 times greater than that in solution B. What is the difference in the pH values of the two solutions?

16.35 **(a)** If NaOH is added to water, how does $[H^+]$ change? How does pH change? **(b)** Use the pH benchmarks in Figure 16.5 to estimate the pH of a solution with $[H^+] = 0.0006\ M$. Is the solution acidic or basic? **(c)** If the pH of the solution is 5.2, first estimate and then calculate the molar concentrations of $H^+(aq)$ and $OH^-(aq)$ in the solution.

16.36 **(a)** If HNO_3 is added to water, how does $[OH^-]$ change? How does pH change? **(b)** Use the pH benchmarks in Figure 16.5 to estimate the pH of a solution with $[OH^-] = 0.014\ M$. Is the solution acidic or basic? **(c)** If pH = 6.6, first estimate and then calculate the molar concentrations of $H^+(aq)$ and $OH^-(aq)$ in the solution.

16.37 Complete the following table by calculating the missing entries and indicating whether the solution is acidic or basic.

$[H^+]$	$[OH^-]$	pH	pOH	Acidic or basic?
$7.5 \times 10^{-3}\ M$				
	$3.6 \times 10^{-10}\ M$			
		8.25		
			5.70	

16.38 Complete the following table by calculating the missing entries. In each case indicate whether the solution is acidic or basic.

pH	pOH	$[H^+]$	$[OH^-]$	Acidic or basic?
4.75				
	11.89			
		$6.5 \times 10^{-3}\ M$		
			$8.6 \times 10^{-7}\ M$	

16.39 The average pH of normal arterial blood is 7.40. At normal body temperature (37°C), $K_w = 2.4 \times 10^{-14}$. Calculate $[H^+]$, $[OH^-]$, and pOH for blood at this temperature.

16.40 Carbon dioxide in the atmosphere dissolves in raindrops to produce carbonic acid (H_2CO_3), causing the pH of clean, unpolluted rain to range from about 5.2 to 5.6. What are the ranges of $[H^+]$ and $[OH^-]$ in the raindrops?

Strong Acids and Bases

16.41 **(a)** What is a strong acid? **(b)** A solution is labeled 0.500 *M* HCl. What is $[H^+]$ for the solution? **(c)** Which of the following are strong acids: HF, HCl, HBr, HI?

16.42 **(a)** What is a strong base? **(b)** A solution is labeled 0.045 *M* $Sr(OH)_2$. What is $[OH^-]$ for the solution? **(c)** Is the following statement true or false? Because $Mg(OH)_2$ is not very soluble, it cannot be a strong base. Explain.

16.43 Calculate the pH of each of the following strong acid solutions: **(a)** 8.5×10^{-3} *M* HBr, **(b)** 1.52 g of HNO_3 in 575 mL of solution, **(c)** 5.00 mL of 0.250 *M* $HClO_4$ diluted to 50.0 mL, **(d)** a solution formed by mixing 10.0 mL of 0.100 *M* HBr with 20.0 mL of 0.200 *M* HCl.

16.44 Calculate the pH of each of the following strong acid solutions: **(a)** 0.00835 *M* HNO_3, **(b)** 0.525 g of $HClO_4$ in 2.00 L of solution, **(c)** 5.00 mL of 1.00 *M* HCl diluted to 0.500 L, **(d)** a mixture formed by adding 50.0 mL of 0.020 *M* HCl to 150 mL of 0.010 *M* HI.

16.45 Calculate $[OH^-]$ and pH for **(a)** 1.5×10^{-3} *M* $Sr(OH)_2$, **(b)** 2.250 g of LiOH in 250.0 mL of solution, **(c)** 1.00 mL of 0.175 *M* NaOH diluted to 2.00 L, **(d)** a solution formed by adding 5.00 mL of 0.105 *M* KOH to 15.0 mL of 9.5×10^{-2} *M* $Ca(OH)_2$.

16.46 Calculate $[OH^-]$ and pH for each of the following strong base solutions: **(a)** 0.012 *M* KOH, **(b)** 1.565 g of KOH in 500.0 mL of solution, **(c)** 10.0 mL of 0.0105 *M* $Ca(OH)_2$ diluted to 500.0 mL, **(d)** a solution formed by mixing 10.0 mL of 0.015 *M* $Ba(OH)_2$ with 40.0 mL of 7.5×10^{-3} *M* NaOH.

16.47 Calculate the concentration of an aqueous solution of NaOH that has a pH of 11.50.

16.48 Calculate the concentration of an aqueous solution of $Ca(OH)_2$ that has a pH of 12.05.

[16.49] Calculate the pH of a solution made by adding 15.00 g of sodium hydride (NaH) to enough water to make 2.500 L of solution.

[16.50] Calculate the pH of a solution made by adding 2.50 g of lithium oxide (Li_2O) to enough water to make 1.500 L of solution.

Weak Acids

16.51 Write the chemical equation and the K_a expression for the ionization of each of the following acids in aqueous solution. First show the reaction with $H^+(aq)$ as a product and then with the hydronium ion: **(a)** $HBrO_2$, **(b)** $HC_3H_5O_2$.

16.52 Write the chemical equation and the K_a expression for the acid dissociation of each of the following acids in aqueous solution. First show the reaction with $H^+(aq)$ as a product and then with the hydronium ion: **(a)** HC_6H_5O, **(b)** HCO_3^-.

16.53 Lactic acid ($HC_3H_5O_3$) has one acidic hydrogen. A 0.10 *M* solution of lactic acid has a pH of 2.44. Calculate K_a.

16.54 Phenylacetic acid ($HC_8H_7O_2$) is one of the substances that accumulates in the blood of people with phenylketonuria, an inherited disorder that can cause mental retardation or even death. A 0.085 *M* solution of $HC_8H_7O_2$ is found to have a pH of 2.68. Calculate the K_a value for this acid.

16.55 A 0.100 *M* solution of chloroacetic acid ($ClCH_2COOH$) is 11.0% ionized. Using this information, calculate $[ClCH_2COO^-]$, $[H^+]$, $[ClCH_2COOH]$, and K_a for chloroacetic acid.

16.56 A 0.100 *M* solution of bromoacetic acid ($BrCH_2COOH$) is 13.2% ionized. Calculate $[H^+]$, $[BrCH_2COO^-]$, and $[BrCH_2COOH]$.

16.57 A particular sample of vinegar has a pH of 2.90. Assuming that acetic acid is the only acid that vinegar contains ($K_a = 1.8 \times 10^{-5}$), calculate the concentration of acetic acid in the vinegar.

16.58 How many moles of HF ($K_a = 6.8 \times 10^{-4}$) must be present in 0.200 L to form a solution with a pH of 3.25?

16.59 The acid-dissociation constant for benzoic acid ($HC_7H_5O_2$) is 6.3×10^{-5}. Calculate the equilibrium concentrations of H_3O^+, $C_7H_5O_2^-$, and $HC_7H_5O_2$ in the solution if the initial concentration of $HC_7H_5O_2$ is 0.050 *M*.

16.60 The acid-dissociation constant for hypochlorous acid (HClO) is 3.0×10^{-8}. Calculate the concentrations of H_3O^+, ClO^-, and HClO at equilibrium if the initial concentration of HClO is 0.0075 *M*.

16.61 Calculate the pH of each of the following solutions (K_a and K_b values are given in Appendix D): **(a)** 0.095 *M* propionic acid ($HC_3H_5O_2$), **(b)** 0.100 *M* hydrogen chromate ion ($HCrO_4^-$), **(c)** 0.120 *M* pyridine (C_5H_5N).

16.62 Determine the pH of each of the following solutions (K_a and K_b values are given in Appendix D): **(a)** 0.125 *M* hypochlorous acid, **(b)** 0.0085 *M* phenol, **(c)** 0.095 *M* hydroxylamine.

16.63 Saccharin, a sugar substitute, is a weak acid with $pK_a = 2.32$ at 25°C. It ionizes in aqueous solution as follows:

$$HNC_7H_4SO_3(aq) \rightleftharpoons H^+(aq) + NC_7H_4SO_3^-(aq)$$

What is the pH of a 0.10 *M* solution of this substance?

16.64 The active ingredient in aspirin is acetylsalicylic acid ($HC_9H_7O_4$), a monoprotic acid with $K_a = 3.3 \times 10^{-4}$ at 25°C. What is the pH of a solution obtained by dissolving two extra-strength aspirin tablets, containing 500 mg of acetylsalicylic acid each, in 250 mL of water?

16.65 Calculate the percent ionization of hydrazoic acid (HN_3) in solutions of each of the following concentrations (K_a is given in Appendix D): **(a)** 0.400 *M*, **(b)** 0.100 *M*, **(c)** 0.0400 *M*.

16.66 Calculate the percent ionization of propionic acid ($HC_3H_5O_2$) in solutions of each of the following concentrations (K_a is given in Appendix D): **(a)** 0.250 *M*, **(b)** 0.0800 *M*, **(c)** 0.0200 *M*.

[16.67] Show that for a weak acid, the percent ionization should vary as the inverse square root of the acid concentration.

[16.68] For solutions of a weak acid, a graph of pH versus the log of the initial acid concentration should be a straight line. What is the magnitude of the slope of that line?

[16.69] Citric acid, which is present in citrus fruits, is a triprotic acid (Table 16.3). Calculate the pH and the citrate ion ($C_6H_5O_7^{3-}$) concentration for a 0.050 *M* solution of citric acid. Explain any approximations or assumptions that you make in your calculations.

[16.70] Tartaric acid is found in many fruits, including grapes. It is partially responsible for the dry texture of certain wines. Calculate the pH and the tartarate ion ($C_4H_4O_6^{2-}$) concentration for a 0.250 *M* solution of tartaric acid, for which the acid-dissociation constants are listed in Table 16.3. Explain any approximations or assumptions that you make in your calculation.

Weak Bases

16.71 What is the essential structural feature of all Brønsted–Lowry bases?

16.72 What are two kinds of molecules or ions that commonly function as weak bases?

16.73 Write the chemical equation and the K_b expression for the ionization of each of the following bases in aqueous solution: **(a)** dimethylamine, $(CH_3)_2NH$; **(b)** carbonate ion, CO_3^{2-}; **(c)** formate ion, CHO_2^-.

16.74 Write the chemical equation and the K_b expression for the reaction of each of the following bases with water: **(a)** propylamine, $C_3H_7NH_2$; **(b)** monohydrogen phosphate ion, HPO_4^{2-}; **(c)** benzoate ion, $C_6H_5CO_2^-$.

16.75 Calculate the molar concentration of OH^- ions in a 0.075 *M* solution of ethylamine ($C_2H_5NH_2$) ($K_b = 6.4 \times 10^{-4}$). Calculate the pH of this solution.

16.76 Calculate the molar concentration of OH^- ions in a 1.15 *M* solution of hypobromite ion (BrO^-; $K_b = 4.0 \times 10^{-6}$). What is the pH of this solution?

16.77 Ephedrine, a central nervous system stimulant, is used in nasal sprays as a decongestant. This compound is a weak organic base:

$$C_{10}H_{15}ON(aq) + H_2O(l) \rightleftharpoons C_{10}H_{15}ONH^+(aq) + OH^-(aq)$$

A 0.035 *M* solution of ephedrine has a pH of 11.33. **(a)** What are the equilibrium concentrations of $C_{10}H_{15}ON$, $C_{10}H_{15}ONH^+$, and OH^-? **(b)** Calculate K_b for ephedrine.

16.78 Codeine ($C_{18}H_{21}NO_3$) is a weak organic base. A 5.0×10^{-3} *M* solution of codeine has a pH of 9.95. Calculate the value of K_b for this substance. What is the pK_b for this base?

The K_a–K_b Relationship; Acid-Base Properties of Salts

16.79 Although the acid-dissociation constant for phenol (C_6H_5OH) is listed in Appendix D, the base-dissociation constant for the phenolate ion ($C_6H_5O^-$) is not. **(a)** Explain why it is not necessary to list both K_a for phenol and K_b for the phenolate ion. **(b)** Calculate K_b for the phenolate ion. **(c)** Is the phenolate ion a weaker or stronger base than ammonia?

16.80 We can calculate K_b for the carbonate ion if we know the K_a values of carbonic acid (H_2CO_3). **(a)** Is K_{a1} or K_{a2} of carbonic acid used to calculate K_b for the carbonate ion? Explain. **(b)** Calculate K_b for the carbonate ion. **(c)** Is the carbonate ion a weaker or stronger base than ammonia?

16.81 **(a)** Given that K_a for acetic acid is 1.8×10^{-5} and that for hypochlorous acid is 3.0×10^{-8}, which is the stronger acid? **(b)** Which is the stronger base, the acetate ion or the hypochlorite ion? **(c)** Calculate K_b values for $C_2H_3O_2^-$ and ClO^-.

16.82 **(a)** Given that K_b for ammonia is 1.8×10^{-5} and that for hydroxylamine is 1.1×10^{-8}, which is the stronger base? **(b)** Which is the stronger acid, the ammonium ion or the hydroxylammonium ion? **(c)** Calculate K_a values for NH_4^+ and H_3NOH^+.

16.83 Using data from Appendix D, calculate $[OH^-]$ and pH for each of the following solutions: **(a)** 0.10 *M* NaCN, **(b)** 0.080 *M* Na_2CO_3, **(c)** a mixture that is 0.10 *M* in $NaNO_2$ and 0.20 *M* in $Ca(NO_2)_2$.

16.84 Using data from Appendix D, calculate $[OH^-]$ and pH for each of the following solutions: **(a)** 0.085 *M* NaF, **(b)** 0.055 *M* Na_2S, **(c)** a mixture that is 0.045 *M* in $NaC_2H_3O_2$ and 0.055 *M* in $Ba(C_2H_3O_2)_2$.

16.85 Predict whether aqueous solutions of the following compounds are acidic, basic, or neutral: **(a)** NH_4Br, **(b)** $FeCl_3$, **(c)** Na_2CO_3, **(d)** $KClO_4$, **(e)** $NaHC_2O_4$.

16.86 Predict whether aqueous solutions of the following substances are acidic, basic, or neutral: **(a)** $CrBr_3$, **(b)** LiI, **(c)** K_3PO_4, **(d)** $[CH_3NH_3]Cl$, **(e)** $KHSO_4$.

16.87 An unknown salt is either NaF, NaCl, or NaOCl. When 0.050 mol of the salt is dissolved in water to form 0.500 L of solution, the pH of the solution is 8.08. What is the identity of the salt?

16.88 An unknown salt is either KBr, NH_4Cl, KCN, or K_2CO_3. If a 0.100 *M* solution of the salt is neutral, what is the identity of the salt?

16.89 Sorbic acid ($HC_6H_7O_2$) is a weak monoprotic acid with $K_a = 1.7 \times 10^{-5}$. Its salt (potassium sorbate) is added to cheese to inhibit the formation of mold. What is the pH of a solution containing 11.25 g of potassium sorbate in 1.75 L of solution?

16.90 Trisodium phosphate (Na_3PO_4) is available in hardware stores as TSP and is used as a cleaning agent. The label on a box of TSP warns that the substance is very basic (caustic or alkaline). What is the pH of a solution containing 45.0 g of TSP in a liter of solution?

Acid-Base Character and Chemical Structure

16.91 How does the acid strength of an oxyacid depend on **(a)** the electronegativity of the central atom; **(b)** the number of nonprotonated oxygen atoms in the molecule?

16.92 **(a)** How does the strength of an acid vary with the polarity and strength of the H—X bond? **(b)** How does the acidity of the binary acid of an element vary as a function of the electronegativity of the element? How does this relate to the position of the element in the periodic table?

16.93 Explain the following observations: **(a)** HNO_3 is a stronger acid than HNO_2; **(b)** H_2S is a stronger acid than H_2O; **(c)** H_2SO_4 is a stronger acid than HSO_4^-; **(d)** H_2SO_4 is a stronger acid than H_2SeO_4; **(e)** CCl_3COOH is a stronger acid than CH_3COOH.

16.94 Explain the following observations: **(a)** HCl is a stronger acid than H_2S; **(b)** H_3PO_4 is a stronger acid than H_3AsO_4; **(c)** $HBrO_3$ is a stronger acid than $HBrO_2$; **(d)** $H_2C_2O_4$ is a stronger acid than $HC_2O_4^-$; **(e)** benzoic acid (C_6H_5COOH) is a stronger acid than phenol (C_6H_5OH).

16.95 Based on their compositions and structures and on conjugate acid-base relationships, select the stronger base in each of the following pairs: **(a)** BrO^- or ClO^-, **(b)** BrO^- or BrO_2^-, **(c)** HPO_4^{2-} or $H_2PO_4^-$.

16.96 Based on their compositions and structures and on conjugate acid-base relationships, select the stronger base in each of the following pairs: **(a)** NO_3^- or NO_2^-, **(b)** PO_4^{3-} or AsO_4^{3-}, **(c)** HCO_3^- or CO_3^{2-}.

16.97 Indicate whether each of the following statements is true or false. For each statement that is false, correct the statement so that it is true. **(a)** In general, the acidity of binary acids increases from left to right in a given row of the periodic table. **(b)** In a series of acids that have the same central atom, acid strength increases with the number of hydrogen atoms bonded to the central atom. **(c)** Hydrotelluric acid (H_2Te) is a stronger acid than H_2S because Te is more electronegative than S.

16.98 Indicate whether each of the following statements is true or false. For each statement that is false, correct the statement so that it is true. **(a)** Acid strength in a series of H—X molecules increases with increasing size of X. **(b)** For acids of the same general structure but differing electronegativities of the central atoms, acid strength decreases with increasing electronegativity of the central atom. **(c)** The strongest acid known is HF because fluorine is the most electronegative element.

Lewis Acids and Bases

16.99 If a substance is an Arrhenius base, is it necessarily a Brønsted–Lowry base? Is it necessarily a Lewis base? Explain.

16.100 If a substance is a Lewis acid, is it necessarily a Brønsted–Lowry acid? Is it necessarily an Arrhenius acid? Explain.

16.101 Identify the Lewis acid and Lewis base among the reactants in each of the following reactions:

(a) $Fe(ClO_4)_3(s) + 6\,H_2O(l) \rightleftharpoons Fe(H_2O)_6^{3+}(aq) + 3\,ClO_4^-(aq)$

(b) $CN^-(aq) + H_2O(l) \rightleftharpoons HCN(aq) + OH^-(aq)$

(c) $(CH_3)_3N(g) + BF_3(g) \rightleftharpoons (CH_3)_3NBF_3(s)$

(d) $HIO(lq) + NH_2^-(lq) \rightleftharpoons NH_3(lq) + IO^-(lq)$
(*lq* denotes liquid ammonia as solvent)

16.102 Identify the Lewis acid and Lewis base in each of the following reactions:

(a) $HNO_2(aq) + OH^-(aq) \rightleftharpoons NO_2^-(aq) + H_2O(l)$

(b) $FeBr_3(s) + Br^-(aq) \rightleftharpoons FeBr_4^-(aq)$

(c) $Zn^{2+}(aq) + 4\,NH_3(aq) \rightleftharpoons Zn(NH_3)_4^{2+}(aq)$

(d) $SO_2(g) + H_2O(l) \rightleftharpoons H_2SO_3(aq)$

16.103 Predict which member of each pair produces the more acidic aqueous solution: **(a)** K^+ or Cu^{2+}, **(b)** Fe^{2+} or Fe^{3+}, **(c)** Al^{3+} or Ga^{3+}. Explain.

16.104 Which member of each pair produces the more acidic aqueous solution: **(a)** $ZnBr_2$ or $CdCl_2$, **(b)** CuCl or $Cu(NO_3)_2$, **(c)** $Ca(NO_3)_2$ or $NiBr_2$? Explain.

Additional Exercises

16.105 In your own words, define or explain **(a)** K_w, **(b)** K_a, **(c)** pOH.

16.106 Indicate whether each of the following statements is correct or incorrect. For those that are incorrect, explain why they are wrong.

(a) Every Brønsted–Lowry acid is also a Lewis acid.

(b) Every Lewis acid is also a Brønsted–Lowry acid.

(c) Conjugate acids of weak bases produce more acidic solutions than conjugate acids of strong bases.

(d) K^+ ion is acidic in water because it causes hydrating water molecules to become more acidic.

(e) The percent ionization of a weak acid in water increases as the concentration of acid decreases.

16.107 Hemoglobin plays a part in a series of equilibria involving protonation-deprotonation and oxygenation-deoxygenation. The overall reaction is approximately as follows:

$$HbH^+(aq) + O_2(aq) \rightleftharpoons HbO_2(aq) + H^+(aq)$$

where Hb stands for hemoglobin, and HbO_2 for oxyhemoglobin. **(a)** The concentration of O_2 is higher in the lungs and lower in the tissues. What effect does high $[O_2]$ have on the position of this equilibrium? **(b)** The normal pH of blood is 7.4. Is the blood acidic, basic, or neutral? **(c)** If the blood pH is lowered by the presence of large amounts of acidic metabolism products, a condition known as acidosis results. What effect does lowering blood pH have on the ability of hemoglobin to transport O_2?

[16.108] What is the pH of a solution that is 2.5×10^{-9} *M* in NaOH? Does your answer make sense?

16.109 Which of the following solutions has the higher pH? **(a)** a 0.1 *M* solution of a strong acid or a 0.1 *M* solution of a weak acid, **(b)** a 0.1 *M* solution of an acid with $K_a = 2 \times 10^{-3}$ or one with $K_a = 8 \times 10^{-6}$, **(c)** a 0.1 *M* solution of a base with $pK_b = 4.5$ or one with $pK_b = 6.5$.

16.110 Caproic acid ($HC_6H_{11}O_2$) is found in small amounts in coconut and palm oils and is used in making artificial flavors. A saturated solution of the acid contains 11 g/L and has a pH of 2.94. Calculate K_a for the acid.

[16.111] A hypothetical acid H_2X is both a strong acid and a diprotic acid. **(a)** Calculate the pH of a 0.050 *M* solution of H_2X, assuming that only one proton ionizes per acid molecule. **(b)** Calculate the pH of the solution from part (a), now assuming that both protons of each acid molecule completely ionize. **(c)** In an experiment it is observed that the pH of a 0.050 *M* solution of H_2X is 1.27. Comment on the relative acid strengths of H_2X and HX^-. **(d)** Would a solution of the salt NaHX be acidic, basic, or neutral? Explain.

16.112 Arrange the following 0.10 *M* solutions in order of increasing acidity (decreasing pH): (i) NH_4NO_3, (ii) $NaNO_3$, (iii) $NH_4C_2H_3O_2$, (iv) NaF, (v) $NaC_2H_3O_2$.

[16.113] What are the concentrations of H^+, $H_2PO_4^-$, HPO_4^{2-}, and PO_4^{3-} in a 0.0250 *M* solution of H_3PO_4?

[16.114] Many moderately large organic molecules containing basic nitrogen atoms are not very soluble in water as neutral molecules, but they are frequently much more soluble as their acid salts. Assuming that pH in the stomach is 2.5, indicate whether each of the following compounds would be present in the stomach as the neutral base or in the protonated form: nicotine, $K_b = 7 \times 10^{-7}$; caffeine, $K_b = 4 \times 10^{-14}$; strychnine, $K_b = 1 \times 10^{-6}$; quinine, $K_b = 1.1 \times 10^{-6}$.

[16.115] The amino acid glycine (H_2N-CH_2-COOH) can participate in the following equilibria in water:

$$H_2N-CH_2-COOH + H_2O \rightleftharpoons H_2N-CH_2-COO^- + H_3O^+ \quad K_a = 4.3 \times 10^{-3}$$

$$H_2N-CH_2-COOH + H_2O \rightleftharpoons {}^+H_3N-CH_2-COOH + OH^- \quad K_b = 6.0 \times 10^{-5}$$

(a) Use the values of K_a and K_b to estimate the equilibrium constant for the intramolecular proton transfer to form a zwitterion:

$$H_2N-CH_2-COOH \rightleftharpoons {}^+H_3N-CH_2-COO^-$$

What assumptions did you need to make? **(b)** What is the pH of a 0.050 *M* aqueous solution of glycine? **(c)** What would be the predominant form of glycine in a solution with pH 13? With pH 1?

[16.116] The structural formula for acetic acid is shown in Table 16.2. Replacing hydrogen atoms on the carbon with chlorine atoms causes an increase in acidity, as follows:

Acid	Formula	K_a (25°C)
Acetic	CH_3COOH	1.8×10^{-5}
Chloroacetic	$CH_2ClCOOH$	1.4×10^{-3}
Dichloroacetic	$CHCl_2COOH$	3.3×10^{-2}
Trichloroacetic	CCl_3COOH	2×10^{-1}

Using Lewis structures as the basis of your discussion, explain the observed trend in acidities in the series. Calculate the pH of a 0.010 *M* solution of each acid.

Integrative Exercises

16.117 Calculate the number of $H^+(aq)$ ions in 1.0 mL of pure water at 25°C.

16.118 How many milliliters of concentrated hydrochloric acid solution (36.0% HCl by mass, density = 1.18 g/mL) are required to produce 10.0 L of a solution that has a pH of 2.05?

16.119 The volume of an adult's stomach ranges from about 50 mL when empty to 1 L when full. If its volume is 400 mL and its contents have a pH of 2, how many moles of H^+ does it contain? Assuming that all the H^+ comes from HCl, how many grams of sodium hydrogen carbonate will totally neutralize the stomach acid?

16.120 Atmospheric CO_2 levels have risen by nearly 20% over the past 40 years from 315 ppm to 375 ppm. **(a)** Given that the average pH of clean, unpolluted rain today is 5.4, determine the pH of unpolluted rain 40 years ago. Assume that carbonic acid (H_2CO_3) formed by the reaction of CO_2 and water is the only factor influencing pH.

$$CO_2(g) + H_2O(l) \rightleftharpoons H_2CO_3(aq)$$

(b) What volume of CO_2 at 25°C and 1.0 atm is dissolved in a 20.0-L bucket of today's rainwater?

[16.121] In many reactions the addition of $AlCl_3$ produces the same effect as the addition of H^+. **(a)** Draw a Lewis structure for $AlCl_3$ in which no atoms carry formal charges, and determine its structure using the VSEPR method. **(b)** What characteristic is notable about the structure in part (a) that helps us understand the acidic character of $AlCl_3$? **(c)** Predict the result of the reaction between $AlCl_3$ and NH_3 in a solvent that does not participate as a reactant. **(d)** Which acid-base theory is most suitable for discussing the similarities between $AlCl_3$ and H^+?

[16.122] What is the boiling point of a 0.10 *M* solution of $NaHSO_4$ if the solution has a density of 1.002 g/mL?

[16.123] Cocaine is a weak organic base whose molecular formula is $C_{17}H_{21}NO_4$. An aqueous solution of cocaine was found to have a pH of 8.53 and an osmotic pressure of 52.7 torr at 15°C. Calculate K_b for cocaine.

[16.124] The iodate ion is reduced by sulfite according to the following reaction:

$$IO_3^-(aq) + 3\,SO_3^{2-}(aq) \longrightarrow I^-(aq) + 3\,SO_4^{2-}(aq)$$

The rate of this reaction is found to be first order in IO_3^-, first order in SO_3^{2-}, and first order in H^+. **(a)** Write the rate law for the reaction. **(b)** By what factor will the rate of the reaction change if the pH is lowered from 5.00 to 3.50? Does the reaction proceed faster or slower at the lower pH? **(c)** By using the concepts discussed in Section 14.6, explain how the reaction can be pH-dependent even though H^+ does not appear in the overall reaction.

[16.125] **(a)** Using dissociation constants from Appendix D, determine the value for the equilibrium constant for each of the following reactions. (Remember that when reactions are added, the corresponding equilibrium constants are multiplied.)

(i) $HCO_3^-(aq) + OH^-(aq) \rightleftharpoons CO_3^{2-}(aq) + H_2O(l)$

(ii) $NH_4^+(aq) + CO_3^{2-}(aq) \rightleftharpoons NH_3(aq) + HCO_3^-(aq)$

(b) We usually use single arrows for reactions when the forward reaction is appreciable (K much greater than 1) or when products escape from the system, so that equilibrium is never established. If we follow this convention, which of these equilibria might be written with a single arrow?

[16.126] Lactic acid, $CH_3CH(OH)COOH$, received its name because it is present in sour milk as a product of bacterial action. It is also responsible for the soreness in muscles after vigorous exercise. **(a)** The pK_a of lactic acid is 3.85. Compare this with the value for propionic acid (CH_3CH_2COOH, pK_a = 4.89), and explain the difference. **(b)** Calculate the lactate ion concentration in a 0.050 *M* solution of lactic acid. **(c)** When a solution of sodium lactate, $(CH_3CH(OH)COO)Na$, is mixed with an aqueous copper(II) solution, it is possible to obtain a solid salt of copper(II) lactate as a blue-green hydrate, $(CH_3CH(OH)COO)_2\,Cu \cdot xH_2O$. Elemental analysis of the solid tells us that the solid is 22.9% Cu and 26.0% C by mass. What is the value for x in the formula for the hydrate? **(d)** The acid-dissociation constant for the $Cu^{2+}(aq)$ ion is 1.0×10^{-8}. Based on this value and the acid-dissociation constant of lactic acid, predict whether a solution of copper(II) lactate will be acidic, basic, or neutral. Explain your answer.

eMEDIA EXERCISES

These exercises make use of the interactive objects available online in OneKey or the Companion Website, and on your Accelerator CD. Access to these resources comes in your MediaPak.

16.127 You can measure the pH in the **Acids and Bases** activity (*16.4*) for aqueous solutions of 13 different compounds. **(a)** List the compounds available in the simulation, and identify each as a strong acid, a weak acid, a strong base, or a weak base. **(b)** For each compound, measure and record the pH of a 0.05 *M* solution.

16.128 **(a)** Using data from the **Acids and Bases** activity (*16.4*), calculate the K_a of HNO_2. **(b)** Determine the percent ionization of HNO_2 at 2.0 *M*, 0.20 *M*, 0.020 *M*, and 0.0020 *M* concentrations. **(c)** Explain the trend in percent ionization using Le Châtelier's principle.

16.129 The **Introduction to Aqueous Acids** animation (*16.6*) illustrates the ionization in water of two different strong acids and one weak acid. **(a)** Given that all three ionize to produce hydrogen ion in water, what is it about the behavior of the weak acid that makes it different from the strong acids? **(b)** What is the consequence of this difference in terms of pH?

16.130 The **Introduction to Aqueous Bases** animation (*16.7*) illustrates the ionization of a weak base. **(a)** Write the equation that corresponds to the ionization of ammonia in water. **(b)** In the animation the ammonia molecule ionizes, producing aqueous hydroxide. What is it about the behavior of ammonia in water that makes ammonia a *weak* base?

16.131 The K_b of ammonia is 1.8×10^{-5}. **(a)** Calculate the concentration of aqueous ammonia that would have a pH of 8.5. Use the **Acids and Bases** activity (*16.4*) to check your answer. **(b)** What is the percent ionization of ammonia at this concentration?

16.132 Use the **Equilibrium Constant** activity (*16.6*) to experiment with the ionization of a weak acid. Choose the $HA \rightleftharpoons H^+ + A^-$ reaction and enter the K_a of acetic acid (1.8×10^{-5}). **(a)** Enter a starting HA concentration of 0.1 *M* and determine the equilibrium concentration of H^+. **(b)** By what percentage did the concentration of HA change? **(c)** By what percentage did the concentration of H^+ change? **(d)** Calculate pH and pOH of the 0.1 *M* solution of acetic acid.

eLABORATORY EXERCISES

These exercises allow students to apply the concepts and skills for this chapter in the simulated laboratory of Virtual ChemLab. Worksheets for each of these exercises are found in the Virtual ChemLab Workbook in your MediaPak.

16.133 ***(VCL 16-1) Ionization Constants of Weak Acids***
An acid-base indicator is usually a weak acid with a characteristic color in the protonated and deprotonated forms. In this assignment you will monitor the color of an acetic acid solution containing *bromocresol green* as an indicator as the pH is changed, and estimate the ionization constant, K_a, for the indicator.

16.134 ***(VCL 16-2) Acid-Base Classification of Salts***
In this assignment you will be asked to classify aqueous solutions of salts as to whether they are acidic, basic, or neutral. This is most easily done by first identifying how both the cation and anion affect the pH of the solution and then combine the effects. After predicting the acid-base properties of these salts, you will then test your predictions in the laboratory.

16.135 ***(VCL 16-3) Ranking Salt Solutions by pH***
In this assignment you will be asked to rank a set of aqueous solutions of acids, bases, and salts in order of increasing pH of the solutions. After ranking these solutions, you will then test your predictions in the laboratory.

CHAPTER

17

WHAT'S AHEAD

- We begin by considering a specific example of Le Châtelier's principle known as the *common-ion effect*. *(Section 17.1)*
- We then consider the composition of *buffered solutions*, or *buffers*, and learn how they resist pH change upon the addition of small amounts of strong acid or strong base. *(Section 17.2)*
- We continue by examining acid-base titration in detail, and we explore how to determine pH at any point in an acid-base titration. *(Section 17.3)*
- Next, we learn how to use equilibrium constants known as *solubility-product constants* to determine to what extent a sparingly soluble salt will dissolve in water, and we investigate some of the factors that affect solubility. *(Sections 17.4 and 17.5)*
- Continuing the discussion of solubility equilibria, we learn how to precipitate ions selectively. *(Section 17.6)*
- The chapter concludes with an explanation of how the principles of solubility and complexation equilibria can be used to identify ions qualitatively in solution. *(Section 17.7)*

Additional Aspects of Aqueous Equilibria

WATER IS THE MOST COMMON AND MOST IMPORTANT SOLVENT ON EARTH. In a sense, it is the solvent of life. It is difficult to imagine how living matter in all its complexity could exist with any liquid other than water as the solvent. Water occupies its position of importance not only because of its abundance but also because of its exceptional ability to dissolve a wide variety of substances. For example, the chapter-opening photograph shows a hot spring, and this water contains a high concentration of ions (especially Mg^{2+}, Ca^{2+}, Fe^{2+}, CO_3^{2-}, and SO_4^{2-}). The ions are dissolved as the hot water, initially underground, passes through various rocks on its way to the surface and dissolves minerals in the rocks. When the solution reaches the surface and cools, the minerals deposit, making the terracelike formations seen in the photograph.

The various aqueous solutions encountered in nature typically contain many solutes. For example, the aqueous solutions in hot springs and oceans, as well as those in biological fluids, all contain a variety of solutes. Consequently, many equilibria can take place simultaneously in these solutions.

In this chapter we take a step toward understanding such complex solutions by looking first at further applications of acid-base equilibria. The idea is to consider not merely solutions in which there is a single solute but ones containing a mixture of solutes. We then broaden our discussion to include two additional types of aqueous equilibria, those involving slightly soluble salts and those involving the

MAMMOTH HOT SPRING in Yellowstone National Park.

formation of metal complexes in solution. For the most part, the discussions and calculations in this chapter are an extension of those in Chapters 15 and 16.

17.1 | The Common-Ion Effect

In Chapter 16 we examined the equilibrium concentrations of ions in solutions containing a weak acid or a weak base. We now consider solutions that contain not only a weak acid, such as acetic acid ($HC_2H_3O_2$), but also a soluble salt of that acid, such as $NaC_2H_3O_2$. What happens when $NaC_2H_3O_2$ is added to a solution of $HC_2H_3O_2$? Because $C_2H_3O_2^-$ is a weak base, the pH of the solution increases; that is, $[H^+]$ decreases. (Section 16.9) It is instructive, however, to view this effect from the perspective of Le Châtelier's principle. (Section 15.6)

Sodium acetate, $NaC_2H_3O_2$, is a soluble ionic compound and is therefore a strong electrolyte. (Section 4.1) Consequently, it dissociates completely in aqueous solution to form Na^+ and $C_2H_3O_2^-$ ions:

$$NaC_2H_3O_2(aq) \longrightarrow Na^+(aq) + C_2H_3O_2^-(aq)$$

ANIMATION
Common-Ion Effect

In contrast, $HC_2H_3O_2$ is a weak electrolyte that ionizes as follows:

$$HC_2H_3O_2(aq) \rightleftharpoons H^+(aq) + C_2H_3O_2^-(aq) \qquad [17.1]$$

The addition of $C_2H_3O_2^-$, from $NaC_2H_3O_2$, causes this equilibrium to shift to the left, thereby decreasing the equilibrium concentration of $H^+(aq)$.

$$HC_2H_3O_2(aq) \rightleftharpoons H^+(aq) + C_2H_3O_2^-(aq)$$

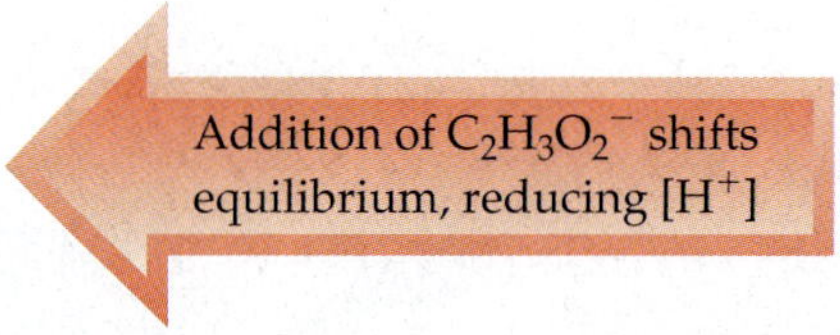

The dissociation of the weak acid $HC_2H_3O_2$ decreases when we add the strong electrolyte $NaC_2H_3O_2$, which has an ion in common with it. We can generalize this observation, which we call the **common-ion effect:** *The extent of ionization of a weak electrolyte is decreased by adding to the solution a strong electrolyte that has an ion in common with the weak electrolyte.* Sample Exercises 17.1 and 17.2 illustrate how equilibrium concentrations may be calculated when a solution contains a mixture of a weak electrolyte and a strong electrolyte that have a common ion. The procedures are similar to those encountered for weak acids and weak bases in Chapter 16.

SAMPLE EXERCISE 17.1 | Calculating the pH When a Common Ion Is Involved

What is the pH of a solution made by adding 0.30 mol of acetic acid ($HC_2H_3O_2$) and 0.30 mol of sodium acetate ($NaC_2H_3O_2$) to enough water to make 1.0 L of solution?

Solution

Analyze: We are asked to determine the pH of a solution of a weak electrolyte ($HC_2H_3O_2$) and a strong electrolyte ($NaC_2H_3O_2$) that share a common ion, $C_2H_3O_2^-$.

Plan: In any problem in which we must determine the pH of a solution containing a mixture of solutes, it is helpful to proceed by a series of logical steps:

1. Identify the major species in solution, and consider their acidity or basicity.
2. Identify the important equilibrium that is the source of H^+ and therefore determines pH.
3. Tabulate the concentrations of ions involved in the equilibrium.
4. Use the equilibrium-constant expression to calculate $[H^+]$ and then pH.

Solve: First, because $HC_2H_3O_2$ is a weak electrolyte and $NaC_2H_3O_2$ is a strong electrolyte, the major species in the solution are $HC_2H_3O_2$ (a weak acid), Na^+ (which is neither acidic nor basic and is therefore a spectator in the acid-base chemistry), and $C_2H_3O_2^-$ (which is the conjugate base of $HC_2H_3O_2$).

Second, $[H^+]$ and, therefore, the pH are controlled by the dissociation equilibrium of $HC_2H_3O_2$:

$$HC_2H_3O_2(aq) \rightleftharpoons H^+(aq) + C_2H_3O_2^-(aq)$$

(We have written the equilibrium using $H^+(aq)$ rather than $H_3O^+(aq)$, but both representations of the hydrated hydrogen ion are equally valid.)

Third, we tabulate the initial and equilibrium concentrations much as we did in solving other equilibrium problems in Chapters 15 and 16:

	$HC_2H_3O_2(aq)$	$\rightleftharpoons$ $H^+(aq)$	+ $C_2H_3O_2^-(aq)$
Initial	$0.30\ M$	0	$0.30\ M$
Change	$-x\ M$	$+x\ M$	$+x\ M$
Equilibrium	$(0.30 - x)\ M$	$x\ M$	$(0.30 + x)\ M$

The equilibrium concentration of $C_2H_3O_2^-$ (the common ion) is the initial concentration that is due to $NaC_2H_3O_2$ (0.30 *M*) plus the change in concentration (x) that is due to the ionization of $HC_2H_3O_2$.

Now we can use the equilibrium-constant expression:

$$K_a = 1.8 \times 10^{-5} = \frac{[H^+][C_2H_3O_2^-]}{[HC_2H_3O_2]}$$

(The dissociation constant for $HC_2H_3O_2$ at 25°C is from Appendix D; addition of $NaC_2H_3O_2$ does not change the value of this constant.)

Substituting the equilibrium-constant concentrations from our table into the equilibrium expression gives

$$K_a = 1.8 \times 10^{-5} = \frac{x(0.30 + x)}{0.30 - x}$$

Because K_a is small, we assume that x is small compared to the original concentrations of $HC_2H_3O_2$ and $C_2H_3O_2^-$ (0.30 *M* each). Thus, we can ignore the very small x relative to 0.30 *M*, giving

$$K_a = 1.8 \times 10^{-5} = \frac{x(0.30)}{0.30}$$

$$x = 1.8 \times 10^{-5}\ M = [H^+]$$

The resulting value of x is indeed small relative to 0.30, justifying the approximation made in simplifying the problem.

Finally, we calculate the pH from the equilibrium concentration of $H^+(aq)$:

$$pH = -\log(1.8 \times 10^{-5}) = 4.74$$

Comment: In Section 16.6 we calculated that a 0.30 *M* solution of $HC_2H_3O_2$ has a pH of 2.64, corresponding to $[H^+] = 2.3 \times 10^{-3}\ M$. Thus, the addition of $NaC_2H_3O_2$ has substantially decreased $[H^+]$, as we would expect from Le Châtelier's principle.

PRACTICE EXERCISE

Calculate the pH of a solution containing 0.085 *M* nitrous acid (HNO_2; $K_a = 4.5 \times 10^{-4}$) and 0.10 *M* potassium nitrite (KNO_2).
Answer: 3.42

SAMPLE EXERCISE 17.2 | Calculating Ion Concentrations When a Common Ion is Involved

Calculate the fluoride ion concentration and pH of a solution that is 0.20 *M* in HF and 0.10 *M* in HCl.

Solution

Analyze: We are asked to determine the concentration of F^- and the pH in a solution containing the weak acid HF and the strong acid HCl. In this case the common ion is H^+.

Plan: We can again use the four steps outlined in Sample Exercise 17.1.

Solve: Because HF is a weak acid and HCl is a strong acid, the major species in solution are HF, H^+, and Cl^-. The Cl^-, which is the conjugate base of a strong acid, is merely a spectator ion in any acid-base chemistry. The problem asks for $[F^-]$, which is formed by ionization of HF. Thus, the important equilibrium is

$$HF(aq) \rightleftharpoons H^+(aq) + F^-(aq)$$

The common ion in this problem is the hydrogen or hydronium ion. Now we can tabulate the initial and equilibrium concentrations of each species involved in this equilibrium:

	$HF(aq)$	$\rightleftharpoons$ $H^+(aq)$	+ $F^-(aq)$
Initial	0.20 M	0.10 M	0
Change	$-x$ M	$+x$ M	$+x$ M
Equilibrium	$(0.20 - x)$ M	$(0.10 + x)$ M	x M

The equilibrium constant for the ionization of HF, from Appendix D, is 6.8×10^{-4}. Substituting the equilibrium-constant concentrations into the equilibrium expression gives

$$K_a = 6.8 \times 10^{-4} = \frac{[H^+][F^-]}{[HF]} = \frac{(0.10 + x)(x)}{0.20 - x}$$

If we assume that x is small relative to 0.10 or 0.20 M, this expression simplifies to

$$\frac{(0.10)(x)}{0.20} = 6.8 \times 10^{-4}$$

$$x = \frac{0.20}{0.10}(6.8 \times 10^{-4}) = 1.4 \times 10^{-3}\ M = [F^-]$$

This F^- concentration is substantially smaller than it would be in a 0.20 M solution of HF with no added HCl. The common ion, H^+, suppresses the ionization of HF. The concentration of $H^+(aq)$ is

$$[H^+] = (0.10 + x)\ M \simeq 0.10\ M$$

Thus,

$$pH = 1.00$$

Comment: Notice that for all practical purposes, $[H^+]$ is due entirely to the HCl; the HF makes a negligible contribution by comparison.

PRACTICE EXERCISE

Calculate the formate ion concentration and pH of a solution that is 0.050 M in formic acid ($HCHO_2$; $K_a = 1.8 \times 10^{-4}$) and 0.10 M in HNO_3.

Answer: $[CHO_2^-] = 9.0 \times 10^{-5}$; pH = 1.00

Sample Exercises 17.1 and 17.2 both involve weak acids. The ionization of a weak base is also decreased by the addition of a common ion. For example, the addition of NH_4^+ (as from the strong electrolyte NH_4Cl) causes the base-dissociation equilibrium of NH_3 to shift to the left, decreasing the equilibrium concentration of OH^- and lowering the pH:

$$NH_3(aq) + H_2O(l) \rightleftharpoons NH_4^+(aq) + OH^-(aq) \quad [17.2]$$

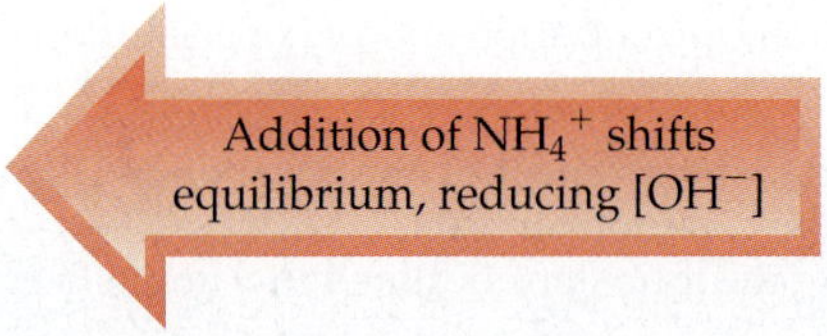

GIVE IT SOME THOUGHT

A mixture of 0.10 mol of NH_4Cl and 0.12 mol of NH_3 is added to enough water to make 1.0 L of solution. (a) What are the initial concentrations of the major species in the solution? (b) Which of the ions in this solution is a spectator ion in any acid-base chemistry occurring in the solution? (c) What equilibrium reaction determines $[OH^-]$ and therefore the pH of the solution?

17.2 | Buffered Solutions

Solutions such as those discussed in Section 17.1, which contain a weak conjugate acid-base pair, can resist drastic changes in pH upon the addition of small amounts of strong acid or strong base. These solutions are called **buffered solutions** (or merely **buffers**). Human blood, for example, is a complex aqueous mixture with a pH buffered at about 7.4 (see the "Chemistry and Life" box near the end of this section). Much of the chemical behavior of seawater is determined by its pH, buffered at about 8.1 to 8.3 near the surface. Buffered solutions find many important applications in the laboratory and in medicine (Figure 17.1 ▶).

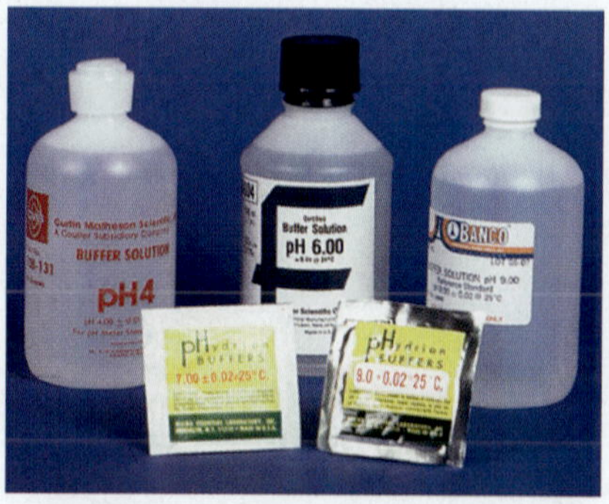

▲ **Figure 17.1 Buffer solutions.** Prepackaged buffer solutions and ingredients for making up buffer solutions of predetermined pH can be purchased.

Composition and Action of Buffered Solutions

A buffer resists changes in pH because it contains both an acidic species to neutralize OH^- ions and a basic species to neutralize H^+ ions. The acidic and basic species that make up the buffer, however, must not consume each other through a neutralization reaction. These requirements are fulfilled by a weak acid-base conjugate pair such as $HC_2H_3O_2$–$C_2H_3O_2^-$ or NH_4^+–NH_3. Thus, buffers are often prepared by mixing a weak acid or a weak base with a salt of that acid or base. The $HC_2H_3O_2$–$C_2H_3O_2^-$ buffer can be prepared, for example, by adding $NaC_2H_3O_2$ to a solution of $HC_2H_3O_2$; the NH_4^+–NH_3 buffer can be prepared by adding NH_4Cl to a solution of NH_3. By choosing appropriate components and adjusting their relative concentrations, we can buffer a solution at virtually any pH.

GIVE IT SOME THOUGHT

Which of the following conjugate acid-base pairs will not function as a buffer: $HCHO_2$ and CHO_2^-; HCO_3^- and CO_3^{2-}; HNO_3 and NO_3^-? Explain.

To understand better how a buffer works, let's consider a buffer composed of a weak acid (HX) and one of its salts (MX, where M^+ could be Na^+, K^+, or another cation). The acid-dissociation equilibrium in this buffered solution involves both the acid and its conjugate base:

$$HX(aq) \rightleftharpoons H^+(aq) + X^-(aq) \qquad [17.3]$$

The corresponding acid-dissociation-constant expression is

$$K_a = \frac{[H^+][X^-]}{[HX]} \qquad [17.4]$$

Solving this expression for $[H^+]$, we have

$$[H^+] = K_a \frac{[HX]}{[X^-]} \qquad [17.5]$$

We see from this expression that $[H^+]$, and thus the pH, is determined by two factors: the value of K_a for the weak-acid component of the buffer and the ratio of the concentrations of the conjugate acid-base pair, $[HX]/[X^-]$.

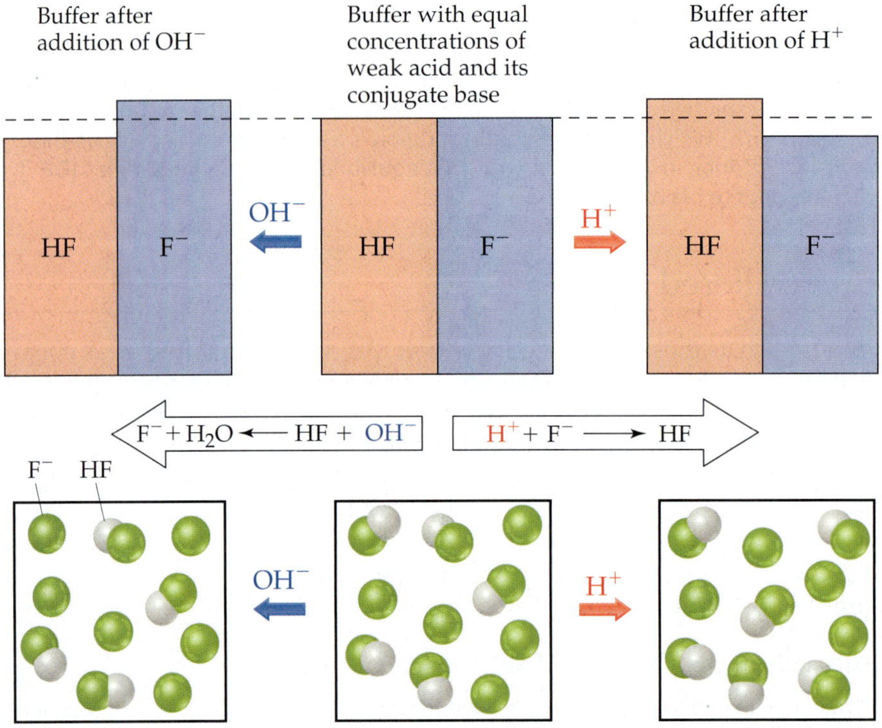

▶ **Figure 17.2 Buffer action.** When a small portion of OH^- is added to a buffer consisting of a mixture of the weak acid HF and its conjugate base (left), the OH^- reacts with the HF, decreasing [HF] and increasing [F^-] in the buffer. Conversely, when a small portion of H^+ is added to the buffer (right), the H^+ reacts with the F^-, decreasing [F^-] and increasing [HF] in the buffer. Because pH depends on the ratio of F^- to HF, the resulting pH change is small.

If OH^- ions are added to the buffered solution, they react with the acid component of the buffer to produce water and the base component (X^-):

$$OH^-(aq) + HX(aq) \longrightarrow H_2O(l) + X^-(aq) \qquad [17.6]$$

This reaction causes [HX] to decrease and [X^-] to increase. As long as the amounts of HX and X^- in the buffer are large compared to the amount of OH^- added, however, the ratio [HX]/[X^-] doesn't change much, and thus the change in pH is small. A specific example of such a buffer, the HF/F^- buffer, is shown in Figure 17.2 ▲.

If H^+ ions are added, they react with the base component of the buffer:

$$H^+(aq) + X^-(aq) \longrightarrow HX(aq) \qquad [17.7]$$

This reaction can also be represented using H_3O^+:

$$H_3O^+(aq) + X^-(aq) \longrightarrow HX(aq) + H_2O(l)$$

Using either equation, the reaction causes [X^-] to decrease and [HX] to increase. As long as the change in the ratio [HX]/[X^-] is small, the change in pH will be small.

Figure 17.2 shows a buffer consisting of equal concentrations of hydrofluoric acid and fluoride ion (center). The addition of OH^- (left) reduces [HF] and increases [F^-]. The addition of H^+ (right) reduces [F^-] and increases [HF].

GIVE IT SOME THOUGHT

(a) What happens when NaOH is added to a buffer composed of $HC_2H_3O_2$ and $C_2H_3O_2^-$? (b) What happens when HCl is added to this buffer?

Calculating the pH of a Buffer

Because conjugate acid-base pairs share a common ion, we can use the same procedures to calculate the pH of a buffer that we used to treat the common-ion effect (see Sample Exercise 17.1). However, an alternate approach is sometimes taken that is based on an equation derived from Equation 17.5. Taking the negative log of both sides of Equation 17.5, we have

$$-\log[H^+] = -\log\left(K_a \frac{[HX]}{[X^-]}\right) = -\log K_a - \log\frac{[HX]}{[X^-]}$$

Because $-\log[H^+] = pH$ and $-\log K_a = pK_a$, we have

$$pH = pK_a - \log\frac{[HX]}{[X^-]} = pK_a + \log\frac{[X^-]}{[HX]} \qquad [17.8]$$

In general,

$$pH = pK_a + \log\frac{[\text{base}]}{[\text{acid}]} \qquad [17.9]$$

where [acid] and [base] refer to the equilibrium concentrations of the conjugate acid-base pair. Note that when [base] = [acid], $pH = pK_a$.

Equation 17.9 is known as the **Henderson–Hasselbalch equation**. Biologists, biochemists, and others who work frequently with buffers often use this equation to calculate the pH of buffers. In doing equilibrium calculations, we have seen that we can normally neglect the amounts of the acid and base of the buffer that ionize. Therefore, we can usually use the starting concentrations of the acid and base components of the buffer directly in Equation 17.9.

ACTIVITIES
Calculating pH Using Henderson–Hasselbalch Equation, Buffer pH

SAMPLE EXERCISE 17.3 | Calculating the pH of a Buffer

What is the pH of a buffer that is 0.12 *M* in lactic acid ($HC_3H_5O_3$) and 0.10 *M* in sodium lactate? For lactic acid, $K_a = 1.4 \times 10^{-4}$.

Solution

Analyze: We are asked to calculate the pH of a buffer containing lactic acid ($HC_3H_5O_3$) and its conjugate base, the lactate ion ($C_3H_5O_3^-$).

Plan: We will first determine the pH using the method described in Section 17.1. The major species in solution are $HC_3H_5O_3$, Na^+, and $C_3H_5O_3^-$. The Na^+ ion is a spectator ion. The $HC_3H_5O_3$–$C_3H_5O_3^-$ conjugate acid-base pair determines $[H^+]$ and thus pH; $[H^+]$ can be determined using the acid-dissociation equilibrium of lactic acid.

Solve: The initial and equilibrium concentrations of the species involved in this equilibrium are

$$HC_3H_5O_3(aq) \rightleftharpoons H^+(aq) + C_3H_5O_3^-(aq)$$

	$HC_3H_5O_3(aq)$	$H^+(aq)$	$C_3H_5O_3^-(aq)$
Initial	0.12 *M*	0	0.10 *M*
Change	$-x$ *M*	$+x$ *M*	$+x$ *M*
Equilibrium	$(0.12 - x)$ *M*	x *M*	$(0.10 + x)$ *M*

The equilibrium concentrations are governed by the equilibrium expression:

$$K_a = 1.4 \times 10^{-4} = \frac{[H^+][C_3H_5O_3^-]}{[HC_3H_5O_3]} = \frac{x(0.10 + x)}{(0.12 - x)}$$

Because K_a is small and a common ion is present, we expect x to be small relative to either 0.12 or 0.10 *M*. Thus, our equation can be simplified to give

$$K_a = 1.4 \times 10^{-4} = \frac{x(0.10)}{0.12}$$

Solving for x gives a value that justifies our approximation:

$$[H^+] = x = \left(\frac{0.12}{0.10}\right)(1.4 \times 10^{-4}) = 1.7 \times 10^{-4}\,M$$

$$pH = -\log(1.7 \times 10^{-4}) = 3.77$$

Alternatively, we could have used the Henderson–Hasselbalch equation to calculate pH directly:

$$pH = pK_a + \log\left(\frac{[\text{base}]}{[\text{acid}]}\right) = 3.85 + \log\left(\frac{0.10}{0.12}\right)$$

$$= 3.85 + (-0.08) = 3.77$$

PRACTICE EXERCISE

Calculate the pH of a buffer composed of 0.12 *M* benzoic acid and 0.20 *M* sodium benzoate. (Refer to Appendix D.)
Answer: 4.42

SAMPLE EXERCISE 17.4 | Preparing a Buffer

How many moles of NH_4Cl must be added to 2.0 L of 0.10 *M* NH_3 to form a buffer whose pH is 9.00? (Assume that the addition of NH_4Cl does not change the volume of the solution.)

Solution

Analyze: Here we are asked to determine the amount of NH_4^+ ion required to prepare a buffer of a specific pH.

Plan: The major species in the solution will be NH_4^+, Cl^-, and NH_3. Of these, the Cl^- ion is a spectator (it is the conjugate base of a strong acid). Thus, the NH_4^+–NH_3 conjugate acid-base pair will determine the pH of the buffer solution. The equilibrium relationship between NH_4^+ and NH_3 is given by the base-dissociation constant for NH_3:

$$NH_3(aq) + H_2O(l) \rightleftharpoons NH_4^+(aq) \qquad K_b = \frac{[NH_4^+][OH^-]}{[NH_3]} = 1.8 \times 10^{-5}$$

The key to this exercise is to use this K_b expression to calculate $[NH_4^+]$.

Solve: We obtain $[OH^-]$ from the given pH:

$$pOH = 14.00 - pH = 14.00 - 9.00 = 5.00$$

and so

$$[OH^-] = 1.0 \times 10^{-5}\ M$$

Because K_b is small and the common ion NH_4^+ is present, the equilibrium concentration of NH_3 will essentially equal its initial concentration:

$$[NH_3] = 0.10\ M$$

We now use the expression for K_b to calculate $[NH_4^+]$:

$$[NH_4^+] = K_b\frac{[NH_3]}{[OH^-]} = (1.8 \times 10^{-5})\frac{(0.10\ M)}{(1.0 \times 10^{-5}\ M)} = 0.18\ M$$

Thus, in order for the solution to have pH = 9.00, $[NH_4^+]$ must equal 0.18 *M*. The number of moles of NH_4Cl needed to produce this concentration is given by the product of the volume of the solution and its molarity:

$$(2.0\ L)(0.18\ mol\ NH_4Cl/L) = 0.36\ mol\ NH_4Cl$$

Comment: Because NH_4^+ and NH_3 are a conjugate acid-base pair, we could use the Henderson–Hasselbalch equation (Equation 17.9) to solve this problem. To do so requires first using Equation 16.41 to calculate pK_a for NH_4^+ from the value of pK_b for NH_3. We suggest you try this approach to convince yourself that you can use the Henderson–Hasselbalch equation for buffers for which you are given K_b for the conjugate base rather than K_a for the conjugate acid.

PRACTICE EXERCISE

Calculate the concentration of sodium benzoate that must be present in a 0.20 *M* solution of benzoic acid ($HC_7H_5O_2$) to produce a pH of 4.00.

Answer: 0.13 *M*

Buffer Capacity and pH Range

Two important characteristics of a buffer are its capacity and its effective pH range. **Buffer capacity** is the amount of acid or base the buffer can neutralize before the pH begins to change to an appreciable degree. The buffer capacity depends on the amount of acid and base from which the buffer is made. The pH of the buffer depends on the K_a for the acid and on the relative concentrations of the acid and base that comprise the buffer. According to Equation 17.5, for example, $[H^+]$ for a 1-L solution that is 1 *M* in $HC_2H_3O_2$ and 1 *M* in $NaC_2H_3O_2$ will be the same as for a 1-L solution that is 0.1 *M* in $HC_2H_3O_2$ and 0.1 *M* in $NaC_2H_3O_2$. The first solution has a greater buffering capacity, however, because it contains more $HC_2H_3O_2$ and $C_2H_3O_2^-$. The greater the amounts of the conjugate acid-base pair, the more resistant the ratio of their concentrations, and hence the pH, is to change.

The **pH range** of any buffer is the pH range over which the buffer acts effectively. Buffers most effectively resist a change in pH in *either* direction when the concentrations of weak acid and conjugate base are about the same. From Equation 17.9 we see that when the concentrations of weak acid and conjugate base are equal, pH = pK_a. This relationship gives the optimal pH of any buffer. Thus, we usually try to select a buffer whose acid form has a pK_a close to the desired pH. In practice, we find that if the concentration of one component of the buffer is more than 10 times the concentration of the other component, the buffering action is poor. Because log 10 = 1, buffers usually have a usable range within ±1 pH unit of pK_a.

GIVE IT SOME THOUGHT

What is the optimal pH buffered by a solution containing $HC_2H_3O_2$ and $NaC_2H_3O_2$? (K_a for $HC_2H_3O_2$ is 1.8×10^{-5}.)

Addition of Strong Acids or Bases to Buffers

Let's now consider in a more quantitative way the response of a buffered solution to the addition of a strong acid or base. In solving these problems, it is important to understand that reactions between strong acids and weak bases proceed essentially to completion, as do those between strong bases and weak acids. Thus, as long as we do not exceed the buffering capacity of the buffer, we can assume that the strong acid or strong base is completely consumed by reaction with the buffer.

Consider a buffer that contains a weak acid HX and its conjugate base X^-. When a strong acid is added to this buffer, the added H^+ is consumed by X^- to produce HX; thus, [HX] increases and $[X^-]$ decreases. When a strong base is added to the buffer, the added OH^- is consumed by HX to produce X^-; in this case [HX] decreases and $[X^-]$ increases. These situations were summarized in Figure 17.2.

To calculate how the pH of the buffer responds to the addition of a strong acid or a strong base, we follow the strategy outlined in Figure 17.3 ▼:

1. Consider the acid-base neutralization reaction, and determine its effect on [HX] and $[X^-]$. This stage of the procedure is a *stoichiometry calculation.*
2. Use K_a and the new concentrations of [HX] and $[X^-]$ from step 1 to calculate $[H^+]$. This second stage of the procedure is a standard *equilibrium calculation* and is most easily done using the Henderson–Hasselbalch equation.

The complete procedure is illustrated in Sample Exercise 17.5.

ACTIVITY
Buffer pH Calculation

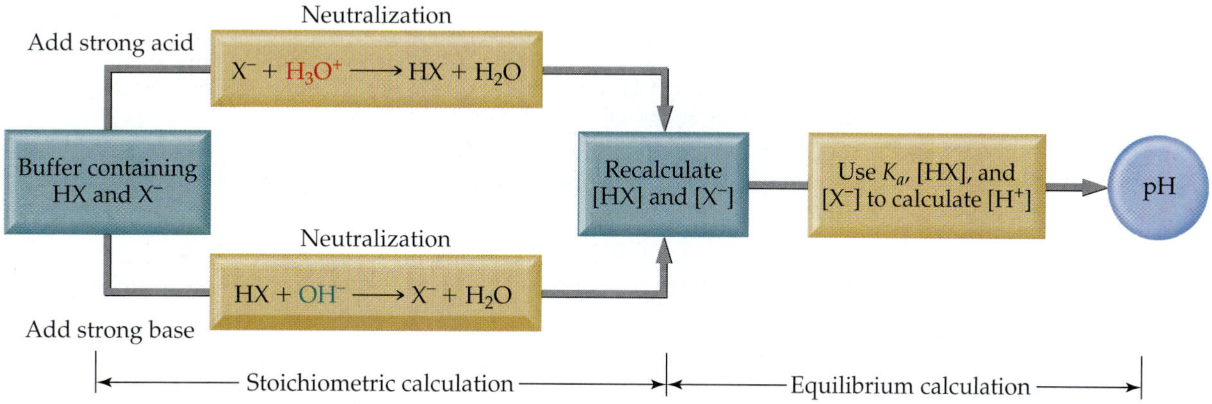

▲ **Figure 17.3 Calculation of the pH of a buffer after the addition of acid or base.** First consider how the neutralization reaction between the added strong acid or strong base and the buffer affects the composition of the buffer (stoichiometry calculation). Then calculate the pH of the remaining buffer (equilibrium calculation). As long as the amount of added acid or base does not exceed the buffer capacity, the Henderson–Hasselbalch equation, Equation 17.9, can be used for the equilibrium calculation.

SAMPLE EXERCISE 17.5 | Calculating pH Changes in Buffers

A buffer is made by adding 0.300 mol $HC_2H_3O_2$ and 0.300 mol $NaC_2H_3O_2$ to enough water to make 1.00 L of solution. The pH of the buffer is 4.74 (Sample Exercise 17.1). **(a)** Calculate the pH of this solution after 0.020 mol of NaOH is added. **(b)** For comparison, calculate the pH that would result if 0.020 mol of NaOH was added to 1.00 L of pure water (neglect any volume changes).

Solution

Analyze: We are asked to determine the pH of a buffer after addition of a small amount of strong base, and to compare the pH change to the pH that would result if we were to add the same amount of strong base to pure water.

Plan: **(a)** Solving this problem involves the two steps outlined in Figure 17.3. Thus, we must first do a stoichiometry calculation to determine how the added OH^- reacts with the buffer and affects its composition. Then we can use the resultant composition of the buffer and either the Henderson–Hasselbalch equation or the equilibrium-constant expression for the buffer to determine the pH.

Solve: *Stoichiometry Calculation:* We assume that the OH^- provided by NaOH is completely consumed by $HC_2H_3O_2$, the weak acid component of the buffer. We can set up a table to see how this reaction changes the composition of the buffer changes. A more compact convention for this type of calculation, however, is to write the number of moles of each species prior to reaction above the equation and the number of moles of each species after reaction below the equation. Prior to the reaction in which the added hydroxide is consumed by acetic acid, there are 0.300 mol each of acetic acid and acetate ion, and 0.020 mol of hydroxide ion.

Before rxn:	0.300 mol	0.020 mol		0.300 mol
	$HC_2H_3O_2(aq)$ +	$OH^-(aq)$ $\longrightarrow$	$H_2O(l)$ +	$C_2H_3O_2^-(aq)$

Because the amount of OH^- added is smaller than the amount of $HC_2H_3O_2$, all the added OH^- will be consumed. An equal amount of $HC_2H_3O_2$ will be consumed, and the same amount of $C_2H_3O_2^-$ will be produced. We write these new, post-reaction amounts below the equation.

Before rxn:	0.300 mol	0.020 mol		0.300 mol
	$HC_2H_3O_2(aq)$ +	$OH^-(aq)$ $\longrightarrow$	$H_2O(l)$ +	$C_2H_3O_2^-(aq)$
After rxn:	0.280 mol	0 mol		0.320 mol

Equilibrium Calculation: We now turn our attention to the equilibrium that will determine the pH of the buffer, namely the ionization of acetic acid.

$$HC_2H_3O_2(aq) \rightleftharpoons H^+(aq) + C_2H_3O_2^-(aq)$$

Using the quantities of $HC_2H_3O_2$ and $C_2H_3O_2^-$ remaining in the buffer, we can determine the pH using the Henderson–Hasselbalch equation.

$$pH = 4.74 + \log\frac{0.320\text{ mol}/1.00\text{ L}}{0.280\text{ mol}/1.00\text{ L}} = 4.80$$

Comment: Notice that we could have used mole amounts in place of concentrations in the Henderson–Hasselbalch equation and gotten the same result. The volumes of the acid and base are equal and cancel.

If 0.020 mol of H^+ was added to the buffer, we would proceed in a similar way to calculate the resulting pH of the buffer. In this case the pH decreases by 0.06 units, giving pH = 4.68, as shown in the figure in the margin.

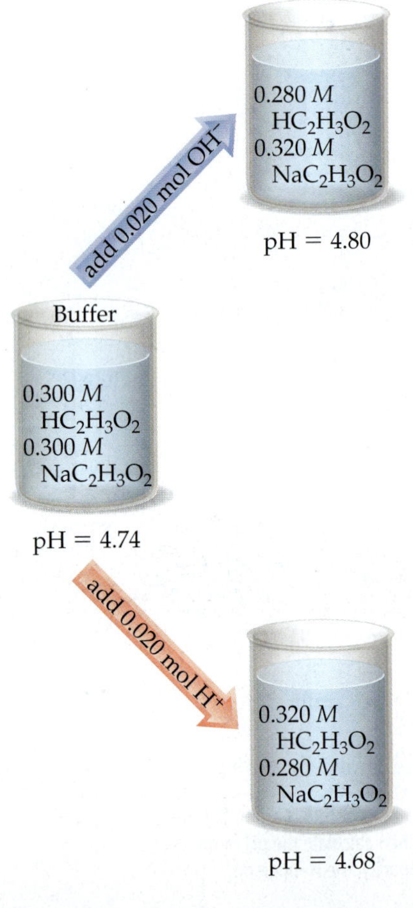

(b) To determine the pH of a solution made by adding 0.020 mol of NaOH to 1.00 L of pure water, we can first determine pOH using Equation 16.18 and subtracting from 14.

$$pH = 14 - (-\log 0.020) = 12.30$$

Note that although the small amount of NaOH is enough to change the pH of water significantly, the pH of the buffer changes very little.

PRACTICE EXERCISE

Determine **(a)** the pH of the original buffer described in Sample Exercise 17.5 after the addition of 0.020 mol HCl, and **(b)** the pH of the solution that would result from the addition of 0.020 mol HCl to 1.00 L of pure water.

Answers: **(a)** 4.68, **(b)** 1.70

CHEMISTRY AND LIFE | Blood as a Buffered Solution

Many of the chemical reactions that occur in living systems are extremely sensitive to pH. Many of the enzymes that catalyze important biochemical reactions, for example, are effective only within a narrow pH range. For this reason the human body maintains a remarkably intricate system of buffers, both within tissue cells and in the fluids that transport cells. Blood, the fluid that transports oxygen to all parts of the body (Figure 17.4 ▼), is one of the most prominent examples of the importance of buffers in living beings.

Human blood is slightly basic with a normal pH of 7.35 to 7.45. Any deviation from this normal pH range can have extremely disruptive effects on the stability of cell membranes, the structures of proteins, and the activities of enzymes. Death may result if the blood pH falls below 6.8 or rises above 7.8. When the pH falls below 7.35, the condition is called *acidosis*; when it rises above 7.45, the condition is called *alkalosis*. Acidosis is the more common tendency because ordinary metabolism generates several acids within the body.

The major buffer system that is used to control the pH of blood is the *carbonic acid–bicarbonate buffer system*. Carbonic acid (H_2CO_3) and bicarbonate ion (HCO_3^-) are a conjugate acid-base pair. In addition, carbonic acid can decompose into carbon dioxide gas and water. The important equilibria in this buffer system are

$$H^+(aq) + HCO_3^-(aq) \rightleftharpoons H_2CO_3(aq) \rightleftharpoons H_2O(l) + CO_2(g) \qquad [17.10]$$

Several aspects of these equilibria are notable. First, although carbonic acid is a diprotic acid, the carbonate ion (CO_3^{2-}) is unimportant in this system. Second, one of the components of this equilibrium, CO_2, is a gas, which provides a mechanism for the body to adjust the equilibria. Removal of CO_2 via exhalation shifts the equilibria to the right, consuming H^+ ions. Third, the buffer system in blood operates at a pH of 7.4, which is fairly far removed from the pK_{a1} value of H_2CO_3 (6.1 at physiological temperatures). In order for the buffer to have a pH of 7.4, the ratio [base]/[acid] must have a value of about 20. In normal blood plasma the concentrations of HCO_3^- and H_2CO_3 are about 0.024 *M* and 0.0012 *M*, respectively. As a consequence, the buffer has a high capacity to neutralize additional acid, but only a low capacity to neutralize additional base.

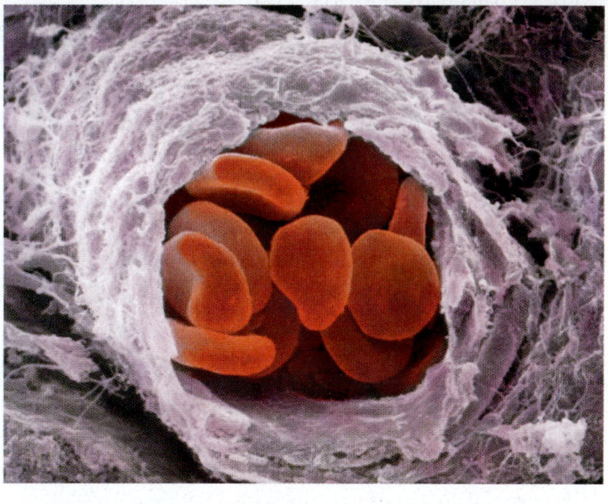

▲ **Figure 17.4 Red blood cells.** A scanning electromicrograph of a group of red blood cells traveling through a small branch of an artery. Blood is a buffered solution whose pH is maintained between 7.35 and 7.45.

The principal organs that regulate the pH of the carbonic acid–bicarbonate buffer system are the lungs and kidneys. Some of the receptors in the brain are sensitive to the concentrations of H^+ and CO_2 in bodily fluids. When the concentration of CO_2 rises, the equilibria in Equation 17.10 shift to the left, which leads to the formation of more H^+. The receptors trigger a reflex to breathe faster and deeper, increasing the rate of elimination of CO_2 from the lungs and shifting the equilibria back to the right. The kidneys absorb or release H^+ and HCO_3^-; much of the excess acid leaves the body in urine, which normally has a pH of 5.0 to 7.0.

The regulation of the pH of blood plasma relates directly to the effective transport of O_2 to bodily tissues. Oxygen is carried by the protein hemoglobin, which is found in red blood cells. Hemoglobin (Hb) reversibly binds both H^+ and O_2. These two substances compete for the Hb, which can be represented approximately by the following equilibrium:

$$HbH^+ + O_2 \rightleftharpoons HbO_2 + H^+ \qquad [17.11]$$

Oxygen enters the blood through the lungs, where it passes into the red blood cells and binds to Hb. When the blood reaches tissue in which the concentration of O_2 is low, the equilibrium in Equation 17.11 shifts to the left and O_2 is released. An increase in H^+ ion concentration (decrease in blood pH) also shifts this equilibrium to the left, as does increasing temperature.

During periods of strenuous exertion, three factors work together to ensure the delivery of O_2 to active tissues: (1) As O_2 is consumed, the equilibrium in Equation 17.11 shifts to the left according to Le Châtelier's principle. (2) Exertion raises the temperature of the body, also shifting the equilibrium to the left. (3) Large amounts of CO_2 are produced by metabolism, which shifts the equilibrium in Equation 17.10 to the left, thus decreasing the pH. Other acids, such as lactic acid, are also produced during strenuous exertion as tissues become starved for oxygen. The decrease in pH shifts the hemoglobin equilibrium to the left, delivering more O_2. In addition, the decrease in pH stimulates an increase in the rate of breathing, which furnishes more O_2 and eliminates CO_2. Without this elaborate arrangement, the O_2 in tissues would be rapidly depleted, making further activity impossible.

Related Exercises: 17.17 and 17.86

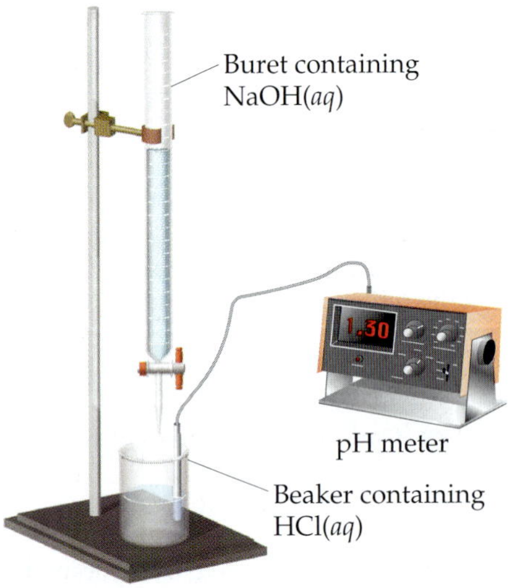

▲ **Figure 17.5 Measuring pH during a titration.** A typical setup for using a pH meter to measure data for a titration curve. In this case a standard solution of NaOH (the titrant) is added by buret to a solution of HCl. The HCl solution is stirred during the titration to ensure uniform composition.

17.3 | Acid-Base Titrations

In Section 4.6 we briefly described *titrations*. In an acid-base titration, a solution containing a known concentration of base is slowly added to an acid (or the acid is added to the base). Acid-base indicators can be used to signal the *equivalence point* of a titration (the point at which stoichiometrically equivalent quantities of acid and base have been brought together). Alternatively, a pH meter can be used to monitor the progress of the reaction producing a **pH titration curve**, a graph of the pH as a function of the volume of the added titrant. The shape of the titration curve makes it possible to determine the equivalence point in the titration. The titration curve can also be used to select suitable indicators and to determine the K_a of the weak acid or the K_b of the weak base being titrated.

A typical apparatus for measuring pH during a titration is illustrated in Figure 17.5 ◀. The titrant is added to the solution from a buret, and the pH is continually monitored using a pH meter. To understand why titration curves have certain characteristic shapes, we will examine the curves for three kinds of titrations: (1) strong acid–strong base, (2) weak acid–strong base, and (3) polyprotic acid–strong base. We will also briefly consider how these curves relate to those involving weak bases.

GIVE IT SOME THOUGHT

For the setup shown in Figure 17.5, will pH increase or decrease as titrant is added?

Strong Acid–Strong Base Titrations

ANIMATION Acid-Base Titration

ACTIVITY Acid-Base Titration

VCL Study of Acid-Base Titrations: Monoprotic Acids

The titration curve produced when a strong base is added to a strong acid has the general shape shown in Figure 17.6 ▶. This curve depicts the pH change that occurs as 0.100 *M* NaOH is added to 50.0 mL of 0.100 *M* HCl. The pH can be calculated at various stages of the titration. To help understand these calculations, we can divide the curve into four regions:

1. *The initial pH:* The pH of the solution before the addition of any base is determined by the initial concentration of the strong acid. For a solution of 0.100 *M* HCl, $[H^+] = 0.100\,M$, and hence $pH = -\log(0.100) = 1.000$. Thus, the initial pH is low.
2. *Between the initial pH and the equivalence point:* As NaOH is added, the pH increases slowly at first and then rapidly in the vicinity of the equivalence point. The pH of the solution before the equivalence point is determined by the concentration of acid that has not yet been neutralized. This calculation is illustrated in Sample Exercise 17.6(a).
3. *The equivalence point:* At the equivalence point an equal number of moles of the NaOH and HCl have reacted, leaving only a solution of their salt, NaCl. The pH of the solution is 7.00 because the cation of a strong base (in this case Na^+) and the anion of a strong acid (in this case Cl^-) do not hydrolyze and therefore have no appreciable effect on pH. (Section 16.9)
4. *After the equivalence point:* The pH of the solution after the equivalence point is determined by the concentration of the excess NaOH in the solution. This calculation is illustrated in Sample Exercise 17.6(b).

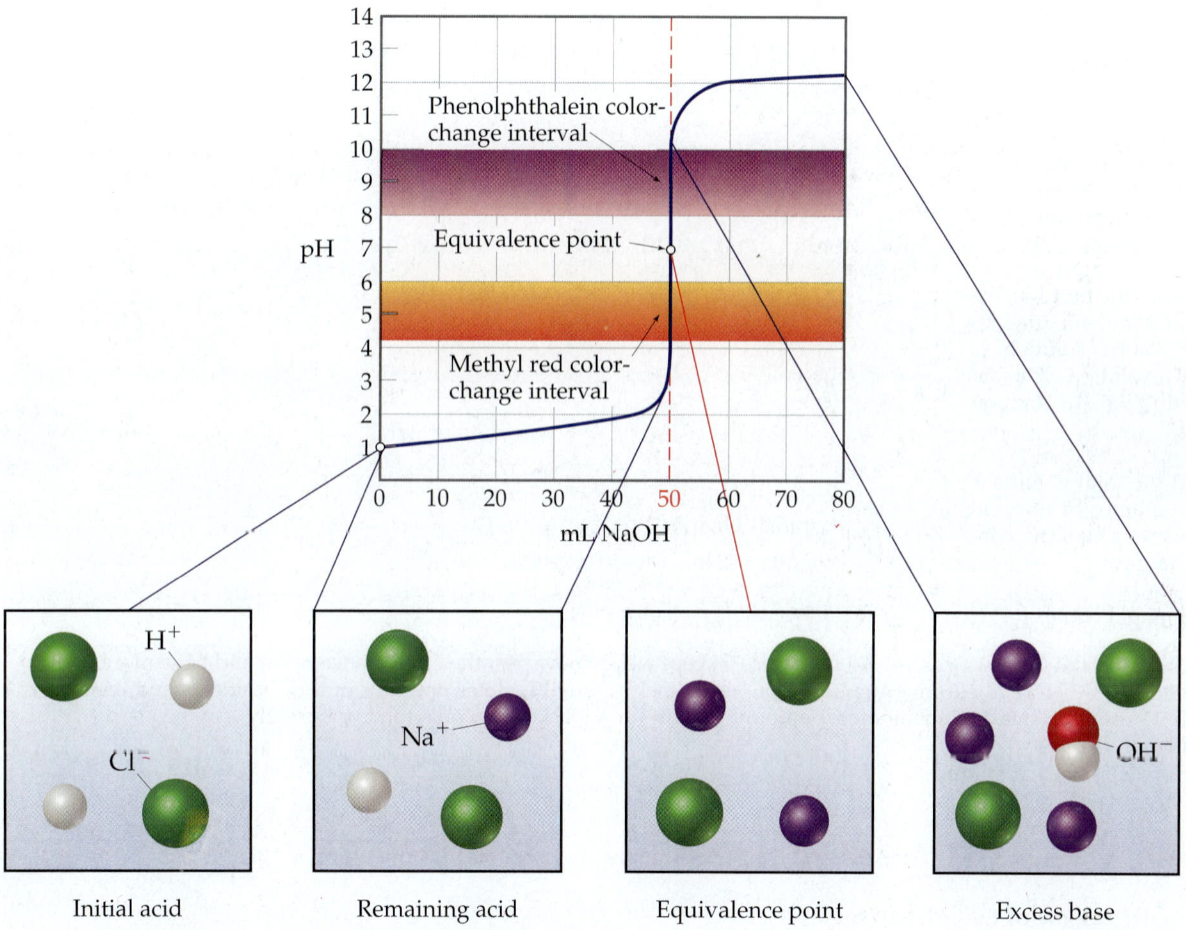

▲ **Figure 17.6 Adding a strong base to a strong acid.** The pH curve for titration of 50.0 mL of a 0.100 *M* solution of a strong acid with a 0.100 *M* solution of a strong base. In this case the acid is HCl and the base is NaOH. The pH starts out at a low value characteristic of the acid and then increases as base is added, rising rapidly at the equivalence point. Both phenolphthalein and methyl red change color at the equivalence point. (For clarity, water molecules have been omitted from the molecular art.)

SAMPLE EXERCISE 17.6 | Calculating pH for a Strong Acid–Strong Base Titration

Calculate the pH when the following quantities of 0.100 *M* NaOH solution have been added to 50.0 mL of 0.100 *M* HCl solution: **(a)** 49.0 mL, **(b)** 51.0 mL.

Solution

Analyze: We are asked to calculate the pH at two points in the titration of a strong acid with a strong base. The first point is just prior to the equivalence point, so we expect the pH to be determined by the small amount of strong acid that has not yet been neutralized. The second point is just after the equivalence point, so we expect this pH to be determined by the small amount of excess strong base.

Plan: **(a)** As the NaOH solution is added to the HCl solution, $H^+(aq)$ reacts with $OH^-(aq)$ to form H_2O. Both Na^+ and Cl^- are spectator ions, having negligible effect on the pH. In order to determine the pH of the solution, we must first determine how many moles of H^+ were originally present and how many moles of OH^- were added. We can then calculate how many moles of each ion remain after the neutralization reaction. In order to calculate $[H^+]$, and hence pH, we must also remember that the volume of the solution increases as we add titrant, thus diluting the concentration of all solutes present.

Solve: The number of moles of H^+ in the original HCl solution is given by the product of the volume of the solution (50.0 mL = 0.0500 L) and its molarity (0.100 *M*):

$$(0.0500\ \text{L soln})\left(\frac{0.100\ \text{mol H}^+}{1\ \text{L soln}}\right) = 5.00 \times 10^{-3}\ \text{mol H}^+$$

Likewise, the number of moles of OH^- in 49.0 mL of 0.100 *M* NaOH is

$$(0.0490 \text{ L soln})\left(\frac{0.100 \text{ mol OH}^-}{1 \text{ L soln}}\right) = 4.90 \times 10^{-3} \text{ mol OH}^-$$

Because we have not yet reached the equivalence point, there are more moles of H^+ present than OH^-. Each mole of OH^- will react with one mole of H^+. Using the convention introduced in Sample Exercise 17.5,

Before rxn:	5.00×10^{-3} mol		4.90×10^{-3} mol		
	$H^+(aq)$	+	$OH^-(aq)$	$\longrightarrow$	$H_2O(l)$
After rxn:	0.10×10^{-3} mol		0.00 mol		

During the course of the titration, the volume of the reaction mixture increases as the NaOH solution is added to the HCl solution. Thus, at this point in the titration, the solution has a volume of 50.0 mL + 49.0 mL = 99.0 mL. (We assume that the total volume is the sum of the volumes of the acid and base solutions.) Thus, the concentration of $H^+(aq)$ is

$$[H^+] = \frac{\text{moles } H^+(aq)}{\text{liters soln}} = \frac{0.10 \times 10^{-3} \text{ mol}}{0.09900 \text{ L}} = 1.0 \times 10^{-3}\ M$$

The corresponding pH equals

$$-\log(1.0 \times 10^{-3}) = 3.00$$

Plan: (b) We proceed in the same way as we did in part (a), except we are now past the equivalence point and have more OH^- in the solution than H^+. As before, the initial number of moles of each reactant is determined from their volumes and concentrations. The reactant present in smaller stoichiometric amount (the limiting reactant) is consumed completely, leaving an excess this time of hydroxide ion.

Solve:

Before rxn:	5.00×10^{-3} mol		5.10×10^{-3} mol		
	$H^+(aq)$	+	$OH^-(aq)$	$\longrightarrow$	$H_2O(l)$
After rxn:	0.0 mol		0.10×10^{-3} mol		

In this case the total volume of the solution is

$$50.0 \text{ mL} + 51.0 \text{ mL} = 101.0 \text{ mL} = 0.1010 \text{ L}$$

Hence, the concentration of $OH^-(aq)$ in the solution is

$$[OH^-] = \frac{\text{moles OH}^-(aq)}{\text{liters soln}} = \frac{0.10 \times 10^{-3} \text{ mol}}{0.1010 \text{ L}} = 1.0 \times 10^{-3}\ M$$

Thus, the pOH of the solution equals

$$\text{pOH} = -\log(1.0 \times 10^{-3}) = 3.00$$

and the pH equals

$$\text{pH} = 14.00 - \text{pOH} = 14.00 - 3.00 = 11.00$$

PRACTICE EXERCISE

Calculate the pH when the following quantities of 0.100 *M* HNO_3 have been added to 25.0 mL of 0.100 *M* KOH solution: **(a)** 24.9 mL, **(b)** 25.1 mL.
Answers: **(a)** 10.30, **(b)** 3.70

Optimally, an indicator would change color at the equivalence point in a titration. In practice, however, that is unnecessary. The pH changes very rapidly near the equivalence point, and in this region merely a drop of titrant can change the pH by several units. Thus, an indicator beginning and ending its color change anywhere on this rapid-rise portion of the titration curve will give a sufficiently accurate measure of the volume of titrant needed to reach the equivalence point. The point in a titration where the indicator changes color is called the *end point* to distinguish it from the actual equivalence point that it closely approximates.

In Figure 17.6 we see that the pH changes very rapidly from about 4 to about 10 near the equivalence point. Consequently, an indicator for this strong

(a) (b)

◀ **Figure 17.7 Methyl red indicator.** Change in appearance of a solution containing methyl red indicator in the pH range 4.2 to 6.3. The characteristic acidic color is shown in (a), and the characteristic basic color in (b).

acid–strong base titration can change color anywhere in this range. Most strong acid–strong base titrations are carried out using phenolphthalein as an indicator (Figure 4.19) because it dramatically changes color in this range. From Figure 16.7, we see that phenolphthalein changes color from pH 8.3 to 10.0. Several other indicators would also be satisfactory, including methyl red, which changes color from pH 4.2 to 6.0 (Figure 17.7 ▲).

Titration of a solution of a strong base with a solution of a strong acid would yield an analogous curve of pH versus added acid. In this case, however, the pH would be high at the outset of the titration and low at its completion, as shown in Figure 17.8 ▶.

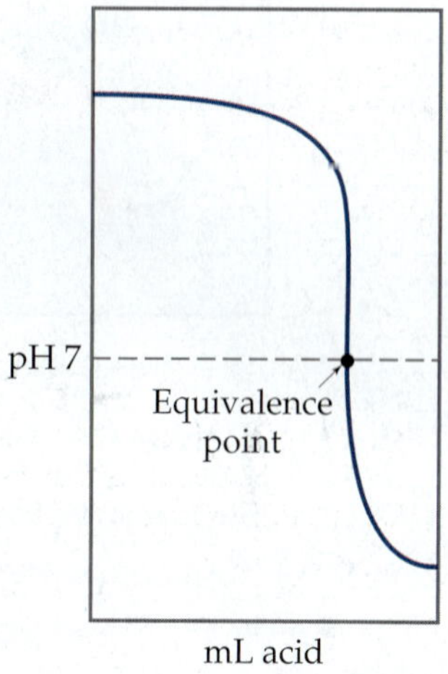

▲ **Figure 17.8 Adding a strong acid to a strong base.** The shape of a pH curve for titration of a strong base with a strong acid. The pH starts out at a high value characteristic of the base and then decreases as acid is added, dropping rapidly at the equivalence point.

VCL Weak Acid-Strong Base Titrations

Weak Acid–Strong Base Titrations

The curve for the titration of a weak acid by a strong base is very similar in shape to that for the titration of a strong acid by a strong base. Consider, for example, the titration curve for the titration of 50.0 mL of 0.100 M acetic acid ($HC_2H_3O_2$) with 0.100 M NaOH shown in Figure 17.9 ▶. We can calculate the pH at points along this curve, using principles we have discussed earlier. As in the case of the titration of a strong acid by a strong base, we can divide the curve into four regions:

1. *The initial pH:* This pH is just the pH of the 0.100 M $HC_2H_3O_2$. We performed calculations of this kind in Section 16.6. The calculated pH of 0.100 M $HC_2H_3O_2$ is 2.89.
2. *Between the initial pH and the equivalence point:* To determine pH in this range, we must consider the neutralization of the acid.

$$HC_2H_3O_2(aq) + OH^-(aq) \longrightarrow C_2H_3O_2^-(aq) + H_2O(l) \qquad [17.12]$$

Prior to reaching the equivalence point, part of the $HC_2H_3O_2$ is neutralized to form $C_2H_3O_2^-$. Thus, the solution contains a mixture of $HC_2H_3O_2$ and $C_2H_3O_2^-$.

The approach we take in calculating the pH in this region of the titration curve involves two main steps. First, we consider the neutralization reaction between $HC_2H_3O_2$ and OH^- to determine the concentrations of $HC_2H_3O_2$ and $C_2H_3O_2^-$ in the solution. Next, we calculate the pH of this buffer pair

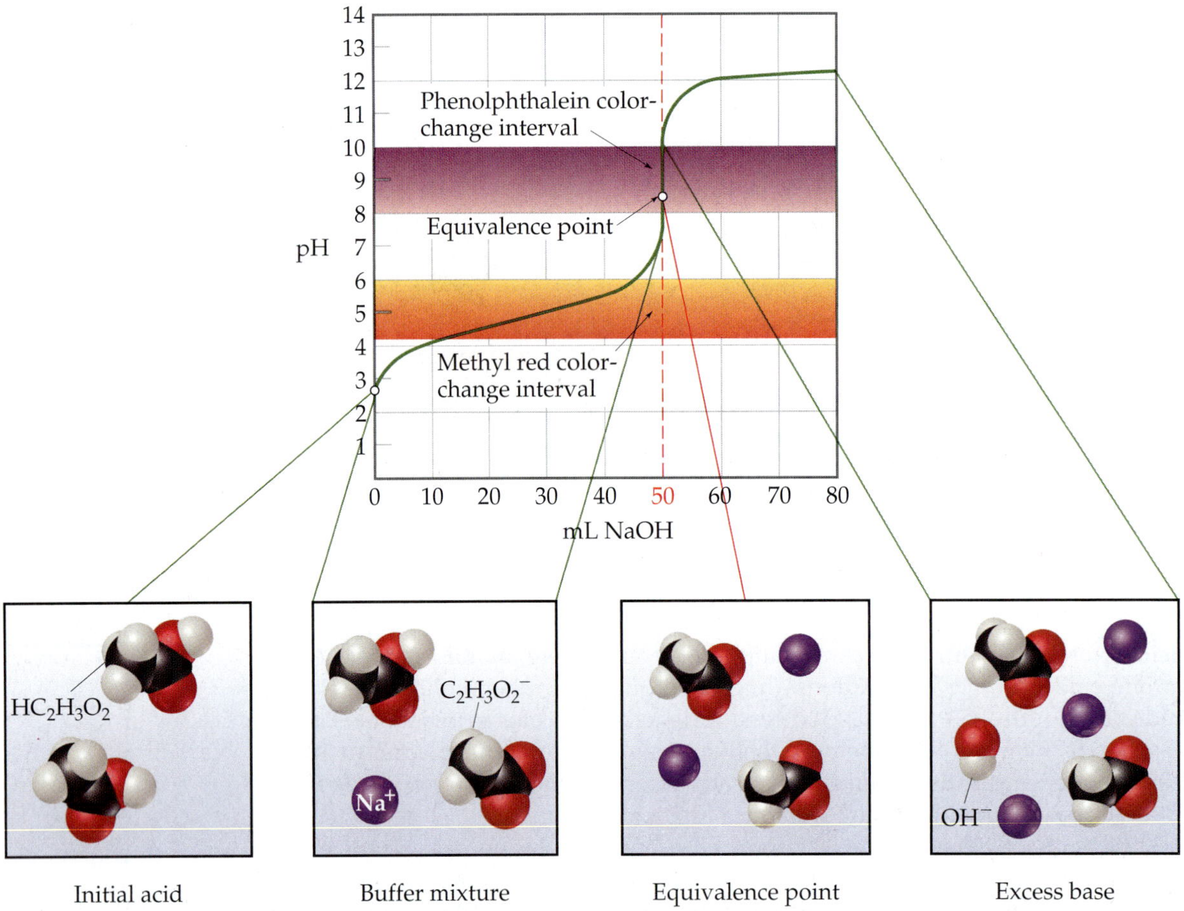

▲ **Figure 17.9 Adding a strong base to a weak acid.** The curve shows the variation in pH as 0.100 *M* NaOH solution is added to 50.0 mL of 0.100 *M* acetic acid solution. Phenolphthalein changes color at the equivalence point, but methyl red does not. (For clarity, water molecules have been omitted from the molecular art.)

using procedures developed in Sections 17.1 and 17.2. The general procedure is diagrammed in Figure 17.10 ▼ and illustrated in Sample Exercise 17.7.

ACTIVITY Titration Calculation

3. *The equivalence point:* The equivalence point is reached after adding 50.0 mL of 0.100 *M* NaOH to the 50.0 mL of 0.100 *M* $HC_2H_3O_2$. At this point the 5.00×10^{-3} mol of NaOH completely reacts with the 5.00×10^{-3} mol of $HC_2H_3O_2$ to form 5.00×10^{-3} mol of their salt, $NaC_2H_3O_2$. The Na^+ ion of this salt has no significant effect on the pH. The $C_2H_3O_2^-$ ion, however, is a weak base, and the pH at the equivalence point is therefore greater than 7. Indeed, the pH at the equivalence point is always above 7 in a weak acid–strong base titration because the anion of the salt formed is a weak base.
4. *After the equivalence point:* In this region of the titration curve, $[OH^-]$ from the reaction of $C_2H_3O_2^-$ with water is negligible compared to $[OH^-]$ from the excess NaOH. Thus, the pH is determined by the concentration of OH^- from the excess NaOH. The method for calculating pH in this region is therefore like that for the strong acid–strong base titration illustrated in

▼ **Figure 17.10 Procedure for calculating the pH when a weak acid is partially neutralized by a strong base.** First consider the effect of the neutralization reaction (stoichiometry calculation). Then determine the pH of the resultant buffer mixture (equilibrium calculation). An analogous procedure can be used for the addition of strong acid to a weak base.

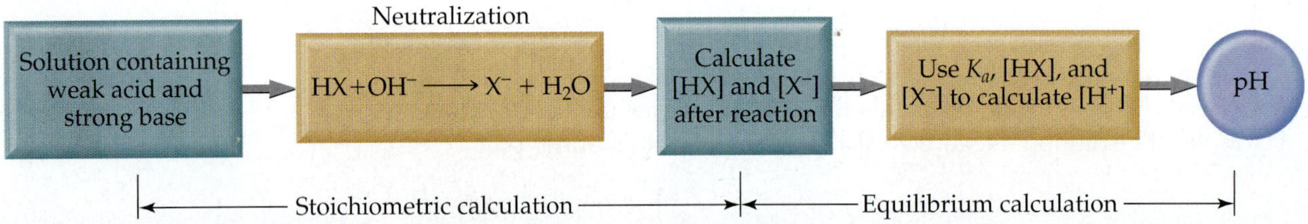

Sample Exercise 17.6(b). Thus, the addition of 51.0 mL of 0.100 *M* NaOH to 50.0 mL of either 0.100 *M* HCl or 0.100 *M* $HC_2H_3O_2$ yields the same pH, 11.00. Notice in Figures 17.6 and 17.9 that the titration curves for the titrations of both the strong acid and the weak acid are the same after the equivalence point.

SAMPLE EXERCISE 17.7 | Calculating pH for a Weak Acid–Strong Base Titration

Calculate the pH of the solution formed when 45.0 mL of 0.100 *M* NaOH is added to 50.0 mL of 0.100 *M* $HC_2H_3O_2$ ($K_a = 1.8 \times 10^{-5}$).

Solution

Analyze: We are asked to calculate the pH prior to the equivalence point of the titration of a weak acid with a strong base.
Plan: We first must determine the number of moles of weak acid and strong base that have been combined. This will tell us how much of the weak acid's conjugate base has been produced, and we can solve for pH using the equilibrium-constant expression.

Solve: *Stoichiometry Calculation:* The product of the volume and concentration of each solution gives the number of moles of each reactant present before the neutralization:

$$(0.0500\text{ L soln})\left(\frac{0.100\text{ mol }HC_2H_3O_2}{1\text{ L soln}}\right) = 5.00 \times 10^{-3}\text{ mol }HC_2H_3O_2$$

$$(0.0450\text{ L soln})\left(\frac{0.100\text{ mol NaOH}}{1\text{ L soln}}\right) = 4.50 \times 10^{-3}\text{ mol NaOH}$$

The 4.50×10^{-3} mol of NaOH consumes 4.50×10^{-3} mol of $HC_2H_3O_2$:

Before rxn:	5.00×10^{-3} mol	4.50×10^{-3} mol	0.0 mol
	$HC_2H_3O_2(aq)$ +	$OH^-(aq) \longrightarrow$	$C_2H_3O_2^-(aq) + H_2O(l)$
After rxn:	0.50×10^{-3} mol	0.0 mol	4.50×10^{-3} mol

The total volume of the solution is

$$45.0\text{ mL} + 50.0\text{ mL} = 95.0\text{ mL} = 0.0950\text{ L}$$

The resulting molarities of $HC_2H_3O_2$ and $C_2H_3O_2^-$ after the reaction are therefore

$$[HC_2H_3O_2] = \frac{0.50 \times 10^{-3}\text{ mol}}{0.0950\text{ L}} = 0.0053\ M$$

$$[C_2H_3O_2^-] = \frac{4.50 \times 10^{-3}\text{ mol}}{0.0950\text{ L}} = 0.0474\ M$$

Equilibrium Calculation: The equilibrium between $HC_2H_3O_2$ and $C_2H_3O_2^-$ must obey the equilibrium-constant expression for $HC_2H_3O_2$:

$$K_a = \frac{[H^+][C_2H_3O_2^-]}{[HC_2H_3O_2]} = 1.8 \times 10^{-5}$$

Solving for $[H^+]$ gives

$$[H^+] = K_a \times \frac{[HC_2H_3O_2]}{[C_2H_3O_2^-]} = (1.8 \times 10^{-5}) \times \left(\frac{0.0053}{0.0474}\right) = 2.0 \times 10^{-6}\ M$$

$$\text{pH} = -\log(2.0 \times 10^{-6}) = 5.70$$

Comment: We could have solved for pH equally well using the Henderson–Hasselbalch equation.

PRACTICE EXERCISE

(a) Calculate the pH in the solution formed by adding 10.0 mL of 0.050 *M* NaOH to 40.0 mL of 0.0250 *M* benzoic acid ($HC_7H_5O_2$, $K_a = 6.3 \times 10^{-5}$). **(b)** Calculate the pH in the solution formed by adding 10.0 mL of 0.100 *M* HCl to 20.0 mL of 0.100 *M* NH_3.
Answers: **(a)** 4.20, **(b)** 9.26

SAMPLE EXERCISE 17.8 | Calculating the pH at the Equivalence Point

Calculate the pH at the equivalence point in the titration of 50.0 mL of 0.100 *M* $HC_2H_3O_2$ with 0.100 *M* NaOH.

Solution

Analyze: We are asked to determine the pH at the equivalence point of the titration of a weak acid with a strong base. Because the neutralization of a weak acid produces its anion, which is a weak base, we expect the pH at the equivalence point to be greater than 7.
Plan: We should first determine how many moles of acetic acid there are initially. This will tell us how many moles of acetate ion there will be at the equivalence point. We then must determine the volume of the solution at the equivalence point and the resultant concentration of acetate ion. Because the acetate ion is a weak base, we can

calculate the pH using K_b and the concentration of acetate as we did for other weak bases in Section 16.7.

Solve: The number of moles of acetic acid in the initial solution is obtained from the volume and molarity of the solution:

$$\text{Moles} = M \times L = (0.100 \text{ mol/L})(0.0500 \text{ L}) = 5.00 \times 10^{-3} \text{ mol } HC_2H_3O_2$$

Hence 5.00×10^{-3} mol of $C_2H_3O_2^-$ is formed. It will take 50.0 mL of NaOH to reach the equivalence point (Figure 17.9). The volume of this salt solution at the equivalence point is the sum of the volumes of the acid and base, 50.0 mL + 50.0 mL = 100.0 mL = 0.1000 L. Thus, the concentration of $C_2H_3O_2^-$ is

$$[C_2H_3O_2^-] = \frac{5.00 \times 10^{-3} \text{ mol}}{0.1000 \text{ L}} = 0.0500\ M$$

The $C_2H_3O_2^-$ ion is a weak base.

$$C_2H_3O_2^-(aq) + H_2O(l) \rightleftharpoons HC_2H_3O_2(aq) + OH^-(aq)$$

The K_b for $C_2H_3O_2^-$ can be calculated from the K_a value of its conjugate acid, $K_b = K_w/K_a = (1.0 \times 10^{-14})/(1.8 \times 10^{-5}) = 5.6 \times 10^{-10}$. Using the K_b expression, we have

$$K_b = \frac{[HC_2H_3O_2][OH^-]}{[C_2H_3O_2^-]} = \frac{(x)(x)}{0.0500 - x} = 5.6 \times 10^{-10}$$

Making the approximation that $0.0500 - x \simeq 0.0500$, and then solving for x, we have $x = [OH^-] = 5.3 \times 10^{-6}\ M$, which gives pOH = 5.28 and pH = 8.72.

Check: The pH is above 7, as expected for the salt of a weak acid and strong base.

PRACTICE EXERCISE

Calculate the pH at the equivalence point when **(a)** 40.0 mL of 0.025 M benzoic acid ($HC_7H_5O_2$, $K_a = 6.3 \times 10^{-5}$) is titrated with 0.050 M NaOH; **(b)** 40.0 mL of 0.100 M NH_3 is titrated with 0.100 M HCl.

Answers: **(a)** 8.21, **(b)** 5.28

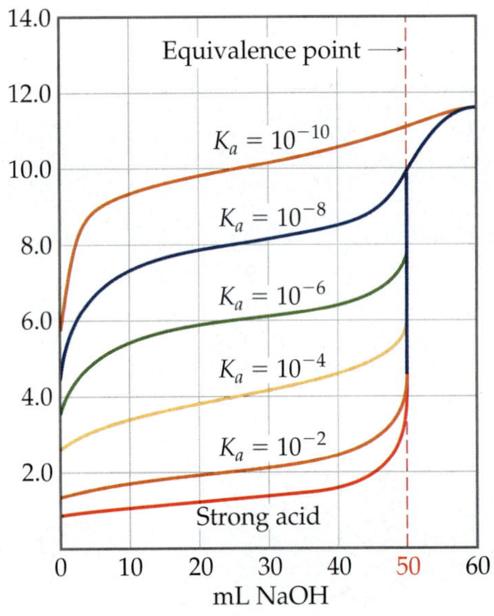

▲ **Figure 17.11 Effect of K_a on titration curves.** This set of curves shows the influence of acid strength (K_a) on the shape of the curve for titration with NaOH. Each curve represents titration of 50.0 mL of 0.10 M acid with 0.10 M NaOH. The weaker the acid, the higher the initial pH and the smaller the pH change at the equivalence point.

The pH titration curves for weak acid–strong base titrations differ from those for strong acid–strong base titrations in three noteworthy ways:

1. The solution of the weak acid has a higher initial pH than a solution of a strong acid of the same concentration.
2. The pH change at the rapid-rise portion of the curve near the equivalence point is smaller for the weak acid than it is for the strong acid.
3. The pH at the equivalence point is above 7.00 for the weak acid–strong base titration.

To illustrate these differences further, consider the family of titration curves shown in Figure 17.11 ◀. As expected, the initial pH of the weak acid solutions is always higher than that of the strong acid solution of the same concentration. Notice also that the pH change near the equivalence point becomes less marked as the acid becomes weaker (that is, as K_a becomes smaller). Finally, the pH at the equivalence point steadily increases as K_a decreases. It is virtually impossible to determine the equivalence point when pK_a is 10 or higher because the pH change is too small and gradual.

Because the pH change near the equivalence point becomes smaller as K_a decreases, the choice of indicator for a weak acid–strong base titration is more critical than it is for a strong acid–strong base titration. When 0.100 M $HC_2H_3O_2$ ($K_a = 1.8 \times 10^{-5}$) is titrated with 0.100 M NaOH, for example, as shown in Figure 17.9, the pH increases rapidly only over the pH range of about 7 to 10. Phenolphthalein is therefore an ideal indicator because it changes color from pH 8.3 to 10.0, close to the pH at the equivalence point. Methyl red is a poor choice, however, because its color change occurs from 4.2 to 6.0, which begins well before the equivalence point is reached.

Titration of a weak base (such as 0.100 *M* NH_3) with a strong acid solution (such as 0.100 *M* HCl) leads to the titration curve shown in Figure 17.12 ▶. In this particular example the equivalence point occurs at pH 5.28. Thus, methyl red would be an ideal indicator, but phenolphthalein would be a poor choice.

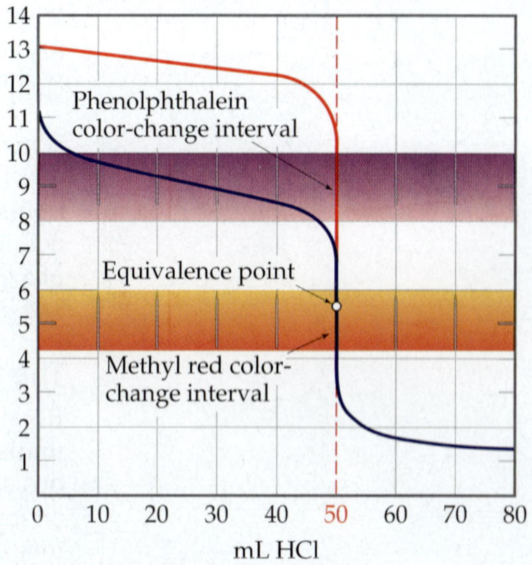

▲ **Figure 17.12 Adding a strong acid to a base.** The blue curve shows pH versus volume of added HCl in the titration of 50.0 mL of 0.10 *M* ammonia (weak base) with 0.10 *M* HCl. The red curve shows pH versus added acid for the titration of 0.10 *M* NaOH (strong base). Both phenolphthalein and methyl red change color at the equivalence point in the titration of the strong base. Phenolphthalein changes color before the equivalence point in the titration of the weak base.

GIVE IT SOME THOUGHT

Why is the choice of indicator more crucial for a weak acid–strong base titration than for a strong acid–strong base titration?

Titrations of Polyprotic Acids

When weak acids contain more than one ionizable H atom, as in phosphorous acid (H_3PO_3), reaction with OH^- occurs in a series of steps. Neutralization of H_3PO_3 proceeds in two stages. (Chapter 16 Sample Integrative Exercise)

$$H_3PO_3(aq) + OH^-(aq) \longrightarrow H_2PO_3^-(aq) + H_2O(l) \quad [17.13]$$

$$H_2PO_3^-(aq) + OH^-(aq) \longrightarrow HPO_3^{2-}(aq) + H_2O(l) \quad [17.14]$$

When the neutralization steps of a polyprotic acid or polybasic base are sufficiently separated, the substance exhibits a titration curve with multiple equivalence points. Figure 17.13 ▼ shows the two distinct equivalence points in the titration curve for the H_3PO_3–$H_2PO_3^-$–HPO_3^{2-} system.

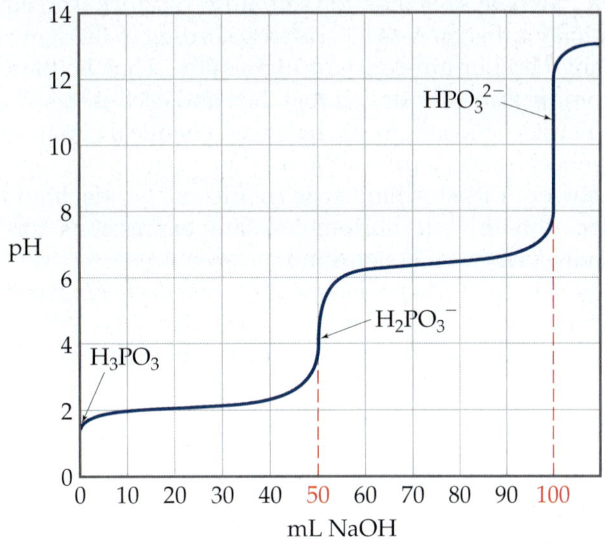

VCL Study of Acid-Base Titrations: Polyprotic Acids

◀ **Figure 17.13 Diprotic acid.** Titration curve for the reaction of 50.0 mL of 0.10 M H_3PO_3 with 0.10 *M* NaOH.

17.4 | Solubility Equilibria

The equilibria that we have considered thus far in this chapter have involved acids and bases. Furthermore, they have been homogeneous; that is, all the species have been in the same phase. Through the rest of this chapter we will consider the equilibria involved in the dissolution or precipitation of ionic compounds. These reactions are heterogeneous.

The dissolving and precipitating of compounds are phenomena that occur both within us and around us. Tooth enamel dissolves in acidic solutions, for example, causing tooth decay. The precipitation of certain salts in our kidneys produces kidney stones. The waters of the earth contain salts dissolved as water passes over and through the ground. Precipitation of $CaCO_3$ from groundwater is responsible for the formation of stalactites and stalagmites within limestone caves (Figure 4.1).

In our earlier discussion of precipitation reactions, we considered some general rules for predicting the solubility of common salts in water. (Section 4.2) These rules give us a qualitative sense of whether a compound will have a low or high solubility in water. By considering solubility equilibria, in contrast, we can make quantitative predictions about the amount of a given compound that will dissolve. We can also use these equilibria to analyze the factors that affect solubility.

The Solubility-Product Constant, K_{sp}

Recall that a *saturated solution* is one in which the solution is in contact with undissolved solute. (Section 13.2) Consider, for example, a saturated aqueous solution of $BaSO_4$ that is in contact with solid $BaSO_4$. Because the solid is an ionic compound, it is a strong electrolyte and yields $Ba^{2+}(aq)$ and $SO_4^{2-}(aq)$ ions upon dissolving. The following equilibrium is readily established between the undissolved solid and hydrated ions in solution:

$$BaSO_4(s) \rightleftharpoons Ba^{2+}(aq) + SO_4^{2-}(aq) \qquad [17.15]$$

ACTIVITY
Solubility Product Constant

As with any other equilibrium, the extent to which this dissolution reaction occurs is expressed by the magnitude of its equilibrium constant. Because this equilibrium equation describes the dissolution of a solid, the equilibrium constant indicates how soluble the solid is in water and is referred to as the **solubility-product constant** (or simply the **solubility product**). It is denoted K_{sp}, where *sp* stands for solubility product. The equilibrium-constant expression for this process is written according to the same rules as those that apply to any equilibrium-constant expression. That is, the concentration terms of the products are multiplied together, and each is raised to the power of its stoichiometric coefficient in the balanced chemical equation. These are divided by the concentration terms of the reactants multiplied together, and each raised to the power of its stoichiometric coefficient. Solids, liquids, and solvents do not appear in the equilibrium-constant expressions for heterogeneous equilibria, however. (Section 15.3)

The solubility product equals the product of the concentration of the ions involved in the equilibrium, each raised to the power of its coefficient in the equilibrium equation.

Thus, the solubility-product expression for the equilibrium expressed in Equation 17.15 is

$$K_{sp} = [Ba^{2+}][SO_4^{2-}] \qquad [17.16]$$

Even though $[BaSO_4]$ is excluded from the equilibrium-constant expression, some undissolved $BaSO_4(s)$ must be present in order for the system to be at equilibrium.

In general, the solubility-product constant (K_{sp}) is the equilibrium constant for the equilibrium that exists between a solid ionic solute and its ions in a saturated aqueous solution. The values of K_{sp} at 25°C for many ionic solids are tabulated in Appendix D. The value of K_{sp} for $BaSO_4$ is 1.1×10^{-10}, a very small number, indicating that only a very small amount of the solid will dissolve in water.

SAMPLE EXERCISE 17.9 | Writing Solubility-Product (K_{sp}) Expressions

Write the expression for the solubility-product constant for CaF_2, and look up the corresponding K_{sp} value in Appendix D.

Solution

Analyze and Plan: We are asked to write an equilibrium-constant expression for the process by which CaF_2 dissolves in water. We apply the same rules for writing any equilibrium-constant expression, making sure to exclude the solid reactant from the expression. We assume that the compound dissociates completely into its component ions.

$$CaF_2(s) \rightleftharpoons Ca^{2+}(aq) + 2\,F^-(aq)$$

Solve: Following the italicized rule stated previously, the expression for K_{sp} is

$$K_{sp} = [Ca^{2+}][F^-]^2$$

In Appendix D we see that this K_{sp} has a value of 3.9×10^{-11}.

PRACTICE EXERCISE

Give the solubility-product-constant expressions and the values of the solubility-product constants (from Appendix D) for the following compounds: **(a)** barium carbonate, **(b)** silver sulfate.

Answers: **(a)** $K_{sp} = [Ba^{2+}][CO_3^{2-}] = 5.0 \times 10^{-9}$; **(b)** $K_{sp} = [Ag^+]^2[SO_4^{2-}] = 1.5 \times 10^{-5}$

Solubility and K_{sp}

It is important to distinguish carefully between solubility and the solubility-product constant. The solubility of a substance is the quantity that dissolves to form a saturated solution. (Section 13.2) Solubility is often expressed as grams of solute per liter of solution (g/L). The molar solubility is the number of moles of the solute that dissolve in forming a liter of saturated solution of the solute (mol/L). The solubility-product constant (K_{sp}) is the equilibrium constant for the equilibrium between an ionic solid and its saturated solution and is a unitless number. Thus, the magnitude of K_{sp} is a measure of how much of the solid dissolves to form a saturated solution.

GIVE IT SOME THOUGHT

Without doing a calculation, predict which of the following compounds will have the greatest molar solubility in water: AgCl ($K_{sp} = 1.8 \times 10^{-10}$), AgBr ($K_{sp} = 5.0 \times 10^{-13}$), or AgI ($K_{sp} = 8.3 \times 10^{-17}$).

The solubility of a substance can change considerably as the concentrations of other solutes change. The solubility of $Mg(OH)_2$, for example, depends highly on pH. The solubility is also affected by the concentrations of other ions in solution, especially Mg^{2+}. In contrast, the solubility-product constant, K_{sp}, has only one value for a given solute at any specific temperature.*

In principle, it is possible to use the K_{sp} value of a salt to calculate solubility under a variety of conditions. In practice, great care must be taken in doing so for the reasons indicated in "A Closer Look" on the limitations of solubility products in Section 17.5. Agreement between measured solubility and that calculated from K_{sp} is usually best for salts whose ions have low charges (1+ and 1−) and do not hydrolyze. Figure 17.14 ▶ summarizes the relationships among various expressions of solubility and K_{sp}.

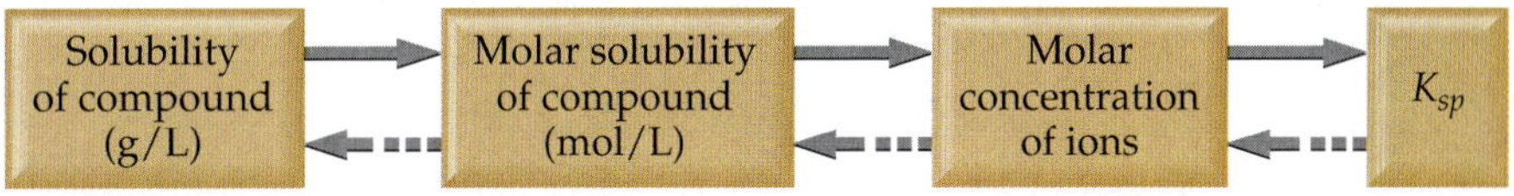

▲ **Figure 17.14 Relationships between solubility and K_{sp}.** The solubility of any compound in grams per liter can be converted to molar solubility. The molar solubility can be used to determine the concentrations of ions in solution. The concentration of ions can be used to calculate K_{sp}. The steps can be reversed, and solubility calculated from K_{sp}.

* This is strictly true only for very dilute solutions. The values of equilibrium constants are somewhat altered when the total concentration of ionic substances in water is increased. However, we will ignore these effects, which are taken into consideration only for work that requires exceptional accuracy.

SAMPLE EXERCISE 17.10 | Calculating K_{sp} from Solubility

Solid silver chromate is added to pure water at 25°C. Some of the solid remains undissolved at the bottom of the flask. The mixture is stirred for several days to ensure that equilibrium is achieved between the undissolved $Ag_2CrO_4(s)$ and the solution. Analysis of the equilibrated solution shows that its silver ion concentration is 1.3×10^{-4} *M*. Assuming that Ag_2CrO_4 dissociates completely in water and that there are no other important equilibria involving the Ag^+ or CrO_4^{2-} ions in the solution, calculate K_{sp} for this compound.

Solution

Analyze: We are given the equilibrium concentration of Ag^+ in a saturated solution of Ag_2CrO_4. From this, we are asked to determine the value of the solubility-product constant for the dissolution of Ag_2CrO_4.

Plan: The equilibrium equation and the expression for K_{sp} are

$$Ag_2CrO_4(s) \rightleftharpoons 2\,Ag^+(aq) + CrO_4^{2-}(aq) \qquad K_{sp} = [Ag^+]^2[CrO_4^{2-}]$$

To calculate K_{sp}, we need the equilibrium concentrations of Ag^+ and CrO_4^{2-}. We know that at equilibrium $[Ag^+] = 1.3 \times 10^{-4}$ *M*. All the Ag^+ and CrO_4^{2-} ions in the solution come from the Ag_2CrO_4 that dissolves. Thus, we can use $[Ag^+]$ to calculate $[CrO_4^{2-}]$.

Solve: From the chemical formula of silver chromate, we know that there must be 2 Ag^+ ions in solution for each CrO_4^{2-} ion in solution. Consequently, the concentration of CrO_4^{2-} is half the concentration of Ag^+.

$$[CrO_4^{2-}] = \left(\frac{1.3 \times 10^{-4}\ \text{mol Ag}^+}{\text{L}}\right)\left(\frac{1\ \text{mol CrO}_4^{2-}}{2\ \text{mol Ag}^+}\right) = 6.5 \times 10^{-5}\ M$$

We can now calculate the value of K_{sp}.

$$K_{sp} = [Ag^+]^2[CrO_4^{2-}] = (1.3 \times 10^{-4})^2(6.5 \times 10^{-5}) = 1.1 \times 10^{-12}$$

Check: We obtain a small value, as expected for a slightly soluble salt. Furthermore, the calculated value agrees well with the one given in Appendix D, 1.2×10^{-12}.

PRACTICE EXERCISE

A saturated solution of $Mg(OH)_2$ in contact with undissolved solid is prepared at 25°C. The pH of the solution is found to be 10.17. Assuming that $Mg(OH)_2$ dissociates completely in water and that there are no other simultaneous equilibria involving the Mg^{2+} or OH^- ions in the solution, calculate K_{sp} for this compound.

Answer: 1.6×10^{-12}

SAMPLE EXERCISE 17.11 | Calculating Solubility from K_{sp}

The K_{sp} for CaF_2 is 3.9×10^{-11} at 25°C. Assuming that CaF_2 dissociates completely upon dissolving and that there are no other important equilibria affecting its solubility, calculate the solubility of CaF_2 in grams per liter.

Solution

Analyze: We are given K_{sp} for CaF_2 and are asked to determine solubility. Recall that the *solubility* of a substance is the quantity that can dissolve in solvent, whereas the *solubility-product constant*, K_{sp}, is an equilibrium constant.

Plan: We can approach this problem by using our standard techniques for solving equilibrium problems. We write the chemical equation for the dissolution process and set up a table of the initial and equilibrium concentrations. We then use the equilibrium-constant expression. In this case we know K_{sp}, and so we solve for the concentrations of the ions in solution.

Solve: Assume initially that none of the salt has dissolved, and then allow *x* moles/liter of CaF_2 to dissociate completely when equilibrium is achieved.

	$CaF_2(s)$	$\rightleftharpoons$ Ca^{2+}	+ $2\,F^-(aq)$
Initial	—	0	0
Change	—	$+x$ *M*	$+2x$ *M*
Equilibrium	—	x *M*	$2x$ *M*

The stoichiometry of the equilibrium dictates that $2x$ moles/liter of F^- are produced for each *x* moles/liter of CaF_2 that dissolve. We now use the expression for K_{sp} and substitute the equilibrium concentrations to solve for the value of *x*:

$$K_{sp} = [Ca^{2+}][F^-]^2 = (x)(2x)^2 = 4x^3 = 3.9 \times 10^{-11}$$

$$x = \sqrt[3]{\frac{3.9 \times 10^{-11}}{4}} = 2.1 \times 10^{-4}\ M$$

(Remember that $\sqrt[3]{y} = y^{1/3}$; to calculate the cube root of a number, you can use the y^x function on your calculator, with $x = \frac{1}{3}$.) Thus, the molar solubility of CaF_2 is 2.1×10^{-4} mol/L. The mass of CaF_2 that dissolves in water to form a liter of solution is

$$\left(\frac{2.1 \times 10^{-4}\ \text{mol CaF}_2}{1\ \text{L soln}}\right)\left(\frac{78.1\ \text{g CaF}_2}{1\ \text{mol CaF}_2}\right) = 1.6 \times 10^{-2}\ \text{g CaF}_2/\text{L soln}$$

Check: We expect a small number for the solubility of a slightly soluble salt. If we reverse the calculation, we should be able to recalculate K_{sp}: $K_{sp} = (2.1 \times 10^{-4})(4.2 \times 10^{-4})^2 = 3.7 \times 10^{-11}$, close to the starting value for K_{sp}, 3.9×10^{-11}.

Comment: Because F^- is the anion of a weak acid, you might expect that the hydrolysis of the ion would affect the solubility of CaF_2. The basicity of F^- is so small ($K_b = 1.5 \times 10^{-11}$), however, that the hydrolysis occurs to only a slight extent and does not significantly influence the solubility. The reported solubility is 0.017 g/L at 25°C, in good agreement with our calculation.

PRACTICE EXERCISE

The K_{sp} for LaF_3 is 2×10^{-19}. What is the solubility of LaF_3 in water in moles per liter?
Answer: 9.28×10^{-6} mol/L

A CLOSER LOOK | Limitations of Solubility Products

The concentrations of ions calculated from K_{sp} sometimes deviate appreciably from those found experimentally. In part, these deviations are due to electrostatic interactions between ions in solution, which can lead to ion pairs. (See "A Closer Look" on colligative properties of electrolyte solutions in Section 13.5.) These interactions increase in magnitude both as the concentrations of the ions increase and as their charges increase. The solubility calculated from K_{sp} tends to be low unless it is corrected to account for these interactions between ions. Chemists have developed procedures for correcting for these "ionic-strength" or "ionic-activity" effects, and these procedures are examined in more advanced chemistry courses. As an example of the effect of these interionic interactions, consider $CaCO_3$ (calcite), whose solubility product, $K_{sp} = 4.5 \times 10^{-9}$, gives a calculated solubility of 6.7×10^{-5} mol/L. Making corrections for the interionic interactions in the solution yields a higher solubility, 7.3×10^{-5} mol/L. The reported solubility, however, is twice as high (1.4×10^{-4} mol/L), so there must be one or more additional factors involved.

Another common source of error in calculating ion concentrations from K_{sp} is ignoring other equilibria that occur simultaneously in the solution. It is also possible, for example, that acid-base equilibria take place simultaneously with solubility equilibria. In particular, both basic anions and cations with high charge-to-size ratios undergo hydrolysis reactions that can measurably increase the solubilities of their salts. For example, $CaCO_3$ contains the basic carbonate ion ($K_b = 1.8 \times 10^{-4}$), which hydrolyzes in water: $CO_3^{2-}(aq) + H_2O(l) \rightleftharpoons HCO_3^-(aq) + OH^-(aq)$. If we consider both the effect of the interionic interactions in the solution and the effect of the simultaneous solubility and hydrolysis equilibria, we calculate a solubility of 1.4×10^{-4} mol/L, in agreement with the measured value.

Finally, we generally assume that ionic compounds dissociate completely into their component ions when they dissolve. This assumption is not always valid. When MgF_2 dissolves, for example, it yields not only Mg^{2+} and F^- ions but also MgF^+ ions in solution. Thus, we see that calculating solubility using K_{sp} can be more complicated than it first appears and it requires considerable knowledge of the equilibria occurring in solution.

17.5 | Factors that Affect Solubility

The solubility of a substance is affected not only by temperature but also by the presence of other solutes. The presence of an acid, for example, can have a major influence on the solubility of a substance. In Section 17.4 we considered the dissolving of ionic compounds in pure water. In this section we examine three factors that affect the solubility of ionic compounds: the presence of common ions, the pH of the solution, and the presence of complexing agents. We will also examine the phenomenon of *amphoterism*, which is related to the effects of both pH and complexing agents.

Common-Ion Effect

The presence of either $Ca^{2+}(aq)$ or $F^-(aq)$ in a solution reduces the solubility of CaF_2, shifting the solubility equilibrium of CaF_2 to the left.

$$CaF_2(s) \rightleftharpoons Ca^{2+}(aq) + 2\,F^-(aq)$$

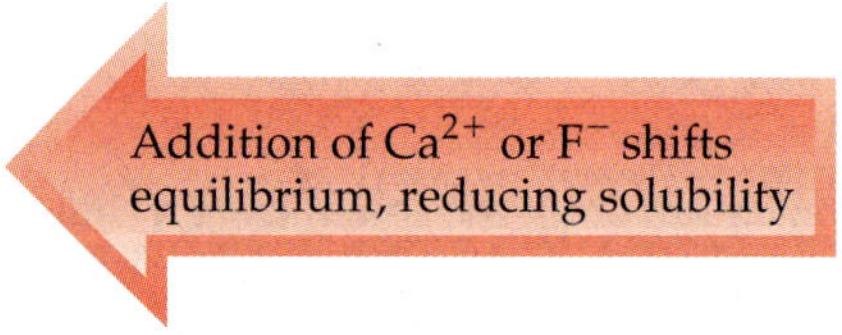

This reduction in solubility is another application of the common-ion effect. (Section 17.1) In general, *the solubility of a slightly soluble salt is decreased by the presence of a second solute that furnishes a common ion.* Figure 17.15 ◀ shows how the solubility of CaF_2 decreases as NaF is added to the solution. Sample Exercise 17.12 shows how the K_{sp} can be used to calculate the solubility of a slightly soluble salt in the presence of a common ion.

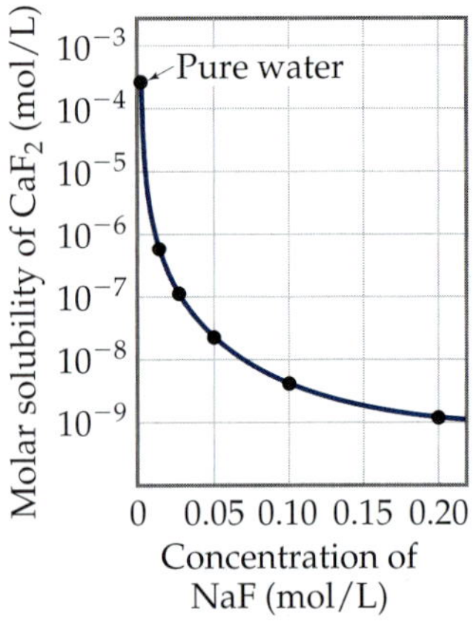

▲ **Figure 17.15 Common-ion effect.** The way in which NaF concentration affects the solubility of CaF_2 demonstrates the common-ion effect. Notice that the CaF_2 solubility is on a logarithmic scale.

SAMPLE EXERCISE 17.12 | Calculating the Effect of a Common Ion on Solubility

Calculate the molar solubility of CaF_2 at 25°C in a solution that is **(a)** 0.010 M in $Ca(NO_3)_2$, **(b)** 0.010 M in NaF.

Solution

Analyze: We are asked to determine the solubility of CaF_2 in the presence of two strong electrolytes, each of which contains an ion common to CaF_2. In (a) the common ion is Ca^{2+} and NO_3^- is a spectator ion. In (b) the common ion is F^- and Na^+ is a spectator ion.

Plan: Because the slightly soluble compound is CaF_2, we need to use the K_{sp} for this compound, which is available in Appendix D:

$$K_{sp} = [Ca^{2+}][F^-]^2 = 3.9 \times 10^{-11}$$

The value of K_{sp} is unchanged by the presence of additional solutes. Because of the common-ion effect, however, the solubility of the salt will decrease in the presence of common ions. We can again use our standard equilibrium techniques of starting with the equation for CaF_2 dissolution, setting up a table of initial and equilibrium concentrations, and using the K_{sp} expression to determine the concentration of the ion that comes only from CaF_2.

Solve: (a) In this instance the initial concentration of Ca^{2+} is 0.010 M because of the dissolved $Ca(NO_3)_2$:

	$CaF_2(s)$ $\rightleftharpoons$	$Ca^{2+}(aq)$ +	$2\,F^-(aq)$
Initial	—	0.010 M	0
Change	$-x$ M	$+x$ M	$+2x$ M
Equilibrium	—	$(0.010 + x)$ M	$2x$ M

Substituting into the solubility-product expression gives

$$K_{sp} = 3.9 \times 10^{-11} = [Ca^{2+}][F^-]^2 = (0.010 + x)(2x)^2$$

This would be a messy problem to solve exactly, but fortunately it is possible to simplify matters greatly. Even without the common-ion effect, the solubility of CaF_2 is very small. Assume that the 0.010 M concentration of Ca^{2+} from $Ca(NO_3)_2$ is very much greater than the small additional concentration resulting from the solubility of CaF_2; that is, x is small compared to 0.010 M, and $0.010 + x \simeq 0.010$. We then have

$$3.9 \times 10^{-11} = (0.010)(2x)^2$$

$$x^2 = \frac{3.9 \times 10^{-11}}{4(0.010)} = 9.8 \times 10^{-10}$$

$$x = \sqrt{9.8 \times 10^{-10}} = 3.1 \times 10^{-5}\ M$$

The very small value for x validates the simplifying assumption we have made. Our calculation indicates that 3.1×10^{-5} mol of solid CaF_2 dissolves per liter of the 0.010 M $Ca(NO_3)_2$ solution.

(b) In this case the common ion is F^-, and at equilibrium we have

$$[Ca^{2+}] = x \quad \text{and} \quad [F^-] = 0.010 + 2x$$

Assuming that $2x$ is small compared to 0.010 M (that is, $0.010 + 2x \simeq 0.010$), we have

$$3.9 \times 10^{-11} = x(0.010)^2$$

$$x = \frac{3.9 \times 10^{-11}}{(0.010)^2} = 3.9 \times 10^{-7}\ M$$

Thus, 3.9×10^{-7} mol of solid CaF_2 should dissolve per liter of 0.010 M NaF solution.

Comment: The molar solubility of CaF_2 in pure water is $2.1 \times 10^{-4}\ M$ (Sample Exercise 17.11). By comparison, our calculations above show that the solubility of CaF_2 in the presence of 0.010 M Ca^{2+} is $3.1 \times 10^{-5}\ M$, and in the presence of 0.010 M F^- ion it is $3.9 \times 10^{-7}\ M$. Thus, the addition of either Ca^{2+} or F^- to a solution of CaF_2 decreases the solubility. However, the effect of F^- on the solubility is more pronounced than that of Ca^{2+} because $[F^-]$ appears to the second power in the K_{sp} expression for CaF_2, whereas Ca^{2+} appears to the first power.

PRACTICE EXERCISE

The value for K_{sp} for manganese(II) hydroxide, $Mn(OH)_2$, is 1.6×10^{-13}. Calculate the molar solubility of $Mn(OH)_2$ in a solution that contains 0.020 M NaOH.

Answer: $4.0 \times 10^{-10}\ M$

Solubility and pH

The solubility of any substance whose anion is basic will be affected to some extent by the pH of the solution. Consider $Mg(OH)_2$, for example, for which the solubility equilibrium is

$$Mg(OH)_2(s) \rightleftharpoons Mg^{2+}(aq) + 2\,OH^-(aq) \quad K_{sp} = 1.8 \times 10^{-11} \quad [17.17]$$

A saturated solution of $Mg(OH)_2$ has a calculated pH of 10.52 and contains $[Mg^{2+}] = 1.7 \times 10^{-4}\ M$. Now suppose that solid $Mg(OH)_2$ is equilibrated with a solution buffered at a more acidic pH of 9.0. The pOH, therefore, is 5.0, so $[OH^-] = 1.0 \times 10^{-5}$. Inserting this value for $[OH^-]$ into the solubility-product expression, we have

$$K_{sp} = [Mg^{2+}][OH^-]^2 = 1.8 \times 10^{-11}$$

$$[Mg^{2+}](1.0 \times 10^{-5})^2 = 1.8 \times 10^{-11}$$

$$[Mg^{2+}] = \frac{1.8 \times 10^{-11}}{(1.0 \times 10^{-5})^2} = 0.18\ M$$

Thus, $Mg(OH)_2$ dissolves in the solution until $[Mg^{2+}] = 0.18\ M$. It is apparent that $Mg(OH)_2$ is quite soluble in this solution. If the concentration of OH^- was reduced even further by making the solution more acidic, the Mg^{2+} concentration would have to increase to maintain the equilibrium condition. Thus, a sample of $Mg(OH)_2$ will dissolve completely if sufficient acid is added (Figure 17.16 ▶).

The solubility of almost any ionic compound is affected if the solution is made sufficiently acidic or basic. The effects are very noticeable, however, only when one or both ions involved are at least moderately acidic or basic. The metal hydroxides, such as $Mg(OH)_2$, are examples of compounds containing a strongly basic ion, the hydroxide ion.

As we have seen, the solubility of $Mg(OH)_2$ greatly increases as the acidity of the solution increases. The solubility of CaF_2 increases as the solution becomes more acidic, too, because the F^- ion is a weak base; it is the conjugate base of the weak acid HF. As a result, the solubility equilibrium of CaF_2 is shifted to the right as the concentration of F^- ions is reduced by protonation to form HF.

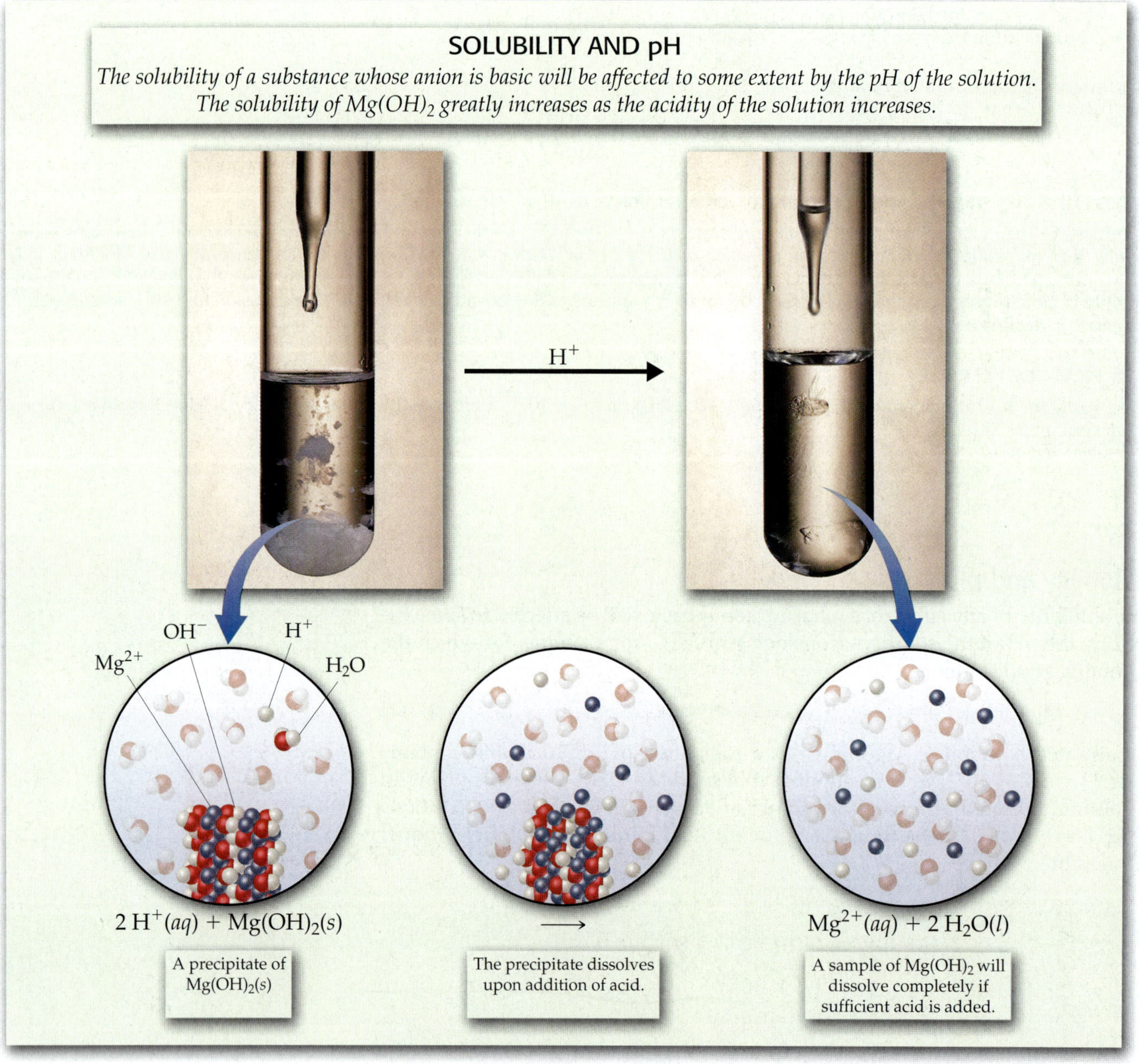

▲ **Figure 17.16 Dissolution of a precipitate in acid. (A white precipitate of $Mg(OH)_2(s)$ in contact with its saturated solution is in the test tube on the left. The dropper poised above the solution surface contains hydrochloric acid. (The anions accompanying the acid have been omitted to simplify the art.))**

ANIMATION
Dissolution of $Mg(OH)_2$ by Acid

Thus, the solution process can be understood in terms of two consecutive reactions.

$$CaF_2(s) \rightleftharpoons Ca^{2+}(aq) + 2\,F^-(aq) \qquad [17.18]$$

$$F^-(aq) + H^+(aq) \rightleftharpoons HF(aq) \qquad [17.19]$$

The equation for the overall process is

$$CaF_2(s) + 2\,H^+(aq) \rightleftharpoons Ca^{2+}(aq) + 2\,HF(aq) \qquad [17.20]$$

CHEMISTRY AND LIFE | Sinkholes

A principal cause of sinkholes is the dissolution of limestone, which is calcium carbonate, by groundwater. Although $CaCO_3$ has a relatively small solubility-product constant, it is quite soluble in the presence of acid.

$$CaCO_3(s) \rightleftharpoons Ca^{2+}(aq) + CO_3^{2-}(aq) \qquad K_{sp} = 4.5 \times 10^{-9}$$

Rainwater is naturally acidic, with a pH range of 5 to 6, and can become more acidic when it comes into contact with decaying plant matter. Because carbonate ion is the conjugate base of the weak acid, hydrogen carbonate ion (HCO_3^-), it readily combines with hydrogen ion.

$$CO_3^{2-}(aq) + H^+(aq) \longrightarrow HCO_3^-(aq)$$

The consumption of carbonate ion shifts the dissolution equilibrium to the right, thus increasing the solubility of $CaCO_3$. This can have profound consequences in areas where the terrain consists of porous calcium carbonate bedrock covered by a relatively thin layer of clay and/or topsoil. As acidic water percolates through and gradually dissolves the limestone, it creates underground voids. A sinkhole results when the overlying ground can no longer be supported by the remaining bedrock and collapses into the underground cavity [Figure 17.17(a) ▼]. Sinkholes are one of a variety of geologic features known as *karst* landforms. Other karst landforms, also caused by the dissolution of bedrock by groundwater, include caves and underground streams. The sudden formation of large sinkholes can pose a serious threat to life and property [Figure 17.17(b)]. The existence of deep sinkholes also increases the risk of contamination of the aquifer.

(a)

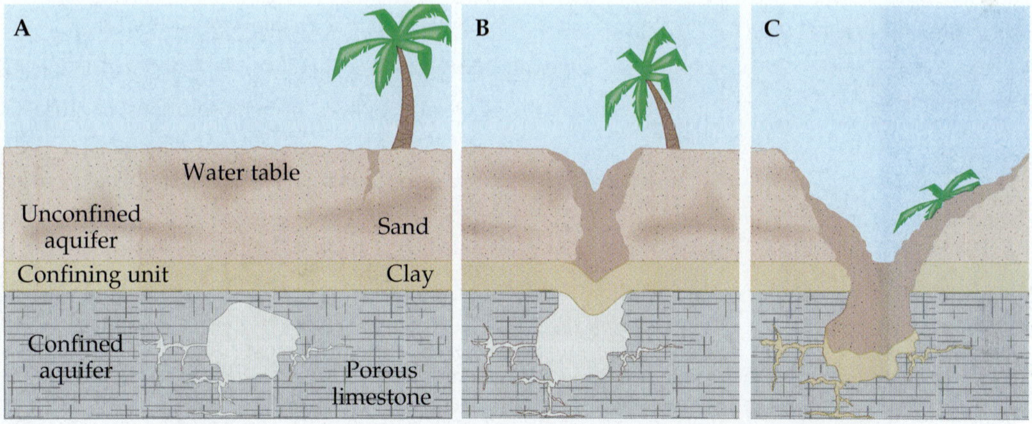

(b)

▲ **Figure 17.17 Sinkhole formation.** (a) The sequence A ⟶ B ⟶ C shows how a sinkhole forms. An underground void develops as limestone, $CaCO_3(s)$, dissolves. Collapse of the overlying ground into an underground cavity causes sinkhole formation. (b) The large sinkhole shown here has destroyed several buildings and part of a highway.

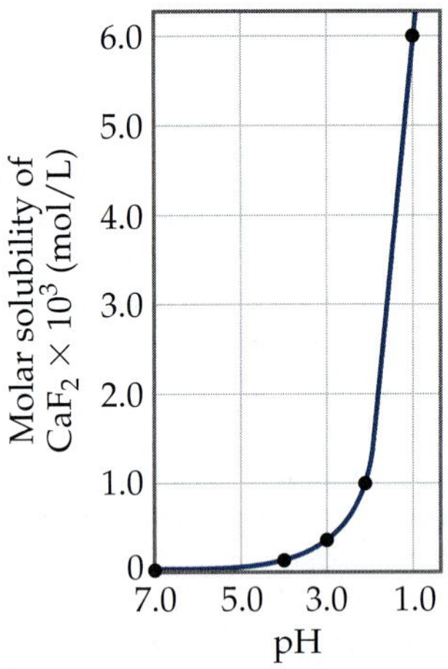

▲ **Figure 17.18 The effect of pH on the solubility of CaF_2.** The solubility increases as the solution becomes more acidic (lower pH). Notice that the vertical scale has been multiplied by 10^3.

Figure 17.18 ◀ shows how the solubility of CaF_2 changes with pH.

Other salts that contain basic anions, such as CO_3^{2-}, PO_4^{3-}, CN^-, or S^{2-}, behave similarly. These examples illustrate a general rule: *The solubility of slightly soluble salts containing basic anions increases as* $[H^+]$ *increases (as pH is lowered).* The more basic the anion, the more the solubility is influenced by pH. Salts with anions of negligible basicity (the anions of strong acids) are unaffected by pH changes.

SAMPLE EXERCISE 17.13 | Predicting the Effect of Acid on Solubility

Which of the following substances will be more soluble in acidic solution than in basic solution: **(a)** $Ni(OH)_2(s)$, **(b)** $CaCO_3(s)$, **(c)** $BaF_2(s)$, **(d)** $AgCl(s)$?

Solution

Analyze: The problem lists four sparingly soluble salts, and we are asked to determine which will be more soluble at low pH than at high pH.

Plan: Ionic compounds that dissociate to produce a basic anion will be more soluble in acid solution.

Solve: (a) $Ni(OH)_2(s)$ will be more soluble in acidic solution because of the basicity of OH^-; the H^+ ion reacts with the OH^- ion, forming water.

$$Ni(OH)_2(s) \rightleftharpoons Ni^{2+}(aq) + 2\,OH^-(aq)$$
$$2\,OH^-(aq) + 2\,H^+(aq) \rightleftharpoons 2\,H_2O(l)$$
$$\text{Overall: } Ni(OH)_2(s) + 2\,H^+(aq) \rightleftharpoons Ni^{2+}(aq) + 2\,H_2O(l)$$

(b) Similarly, $CaCO_3(s)$ dissolves in acid solutions because CO_3^{2-} is a basic anion.

$$CaCO_3(s) \rightleftharpoons Ca^{2+}(aq) + CO_3^{2-}(aq)$$
$$CO_3^{2-}(aq) + 2\,H^+(aq) \rightleftharpoons H_2CO_3(aq)$$
$$H_2CO_3(aq) \longrightarrow CO_2(g) + H_2O(l)$$
$$\text{Overall: } CaCO_3(s) + 2\,H^+(aq) \longrightarrow Ca^{2+}(aq) + CO_2(g) + H_2O(l)$$

The reaction between CO_3^{2-} and H^+ occurs in a stepwise fashion, first forming HCO_3^-. H_2CO_3 forms in appreciable amounts only when the concentration of H^+ is sufficiently high.

(c) The solubility of BaF_2 is also enhanced by lowering the pH, because F^- is a basic anion.

$$BaF_2(s) \rightleftharpoons Ba^{2+}(aq) + 2\,F^-(aq)$$
$$2\,F^-(aq) + 2\,H^+(aq) \rightleftharpoons 2\,HF(aq)$$
$$\text{Overall: } BaF_2(s) + 2\,H^+(aq) \longrightarrow Ba^{2+}(aq) + 2\,HF(aq)$$

(d) The solubility of AgCl is unaffected by changes in pH because Cl^- is the anion of a strong acid and therefore has negligible basicity.

PRACTICE EXERCISE

Write the net ionic equation for the reaction of the following copper(II) compounds with acid: **(a)** CuS, **(b)** $Cu(N_3)_2$.

Answers: **(a)** $CuS(s) + H^+(aq) \rightleftharpoons Cu^{2+}(aq) + HS^-(aq)$
(b) $Cu(N_3)_2(s) + 2\,H^+(aq) \rightleftharpoons Cu^{2+}(aq) + 2\,HN_3(aq)$

Formation of Complex Ions

A characteristic property of metal ions is their ability to act as Lewis acids, or electron-pair acceptors, toward water molecules, which act as Lewis bases, or electron-pair donors. ∞ (Section 16.11) Lewis bases other than water can also interact with metal ions, particularly with transition-metal ions. Such interactions can dramatically affect the solubility of a metal salt. AgCl, for example, which has $K_{sp} = 1.8 \times 10^{-10}$, will dissolve in the presence of aqueous ammonia because Ag^+ interacts with the Lewis base NH_3, as shown in

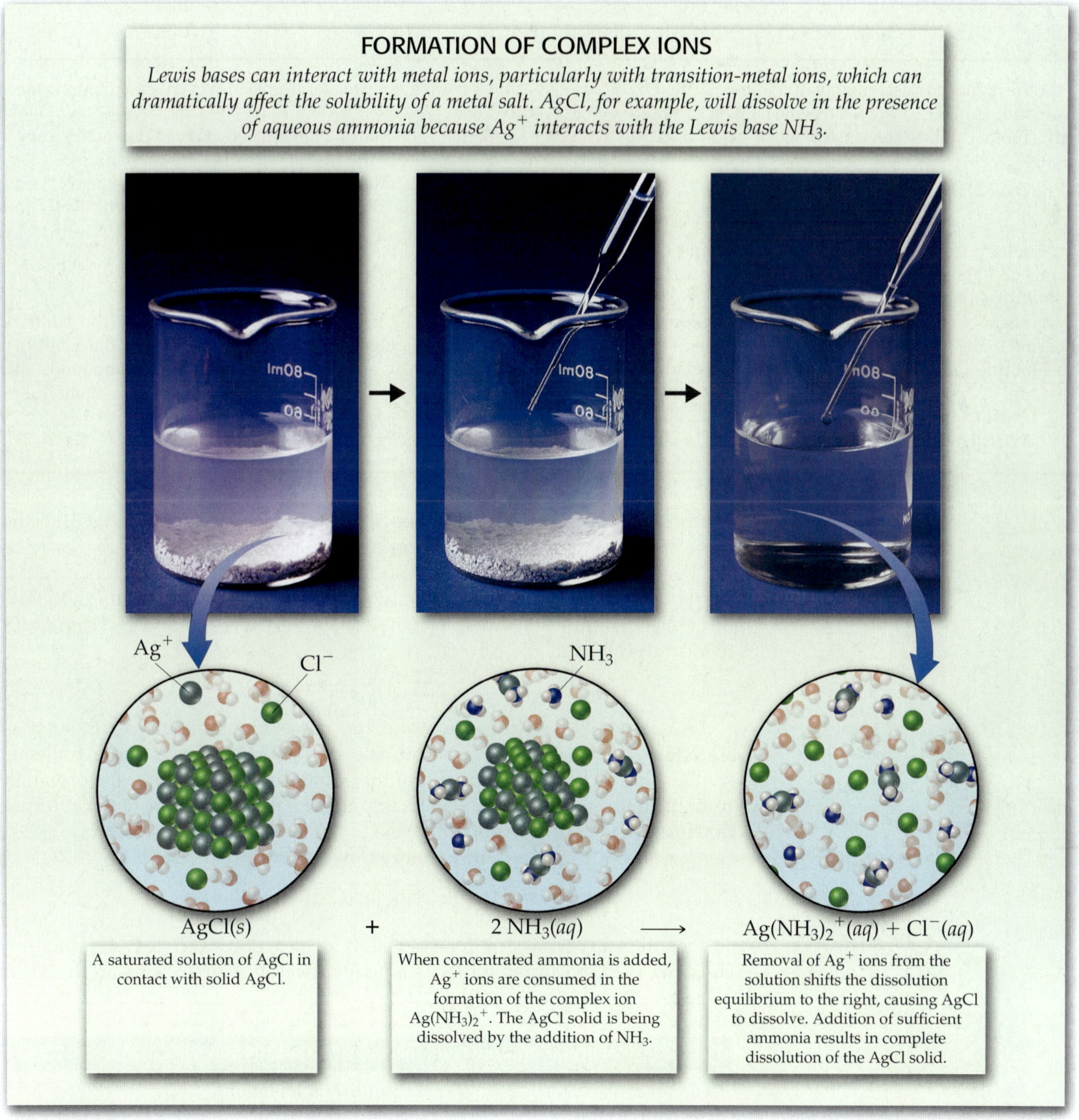

▲ **Figure 17.19 Using $NH_3(aq)$ to dissolve $AgCl(s)$.**

Figure 17.19 ▲. This process can be viewed as the sum of two reactions, the dissolution of AgCl and the Lewis acid-base interaction between Ag^+ and NH_3.

$$AgCl(s) \rightleftharpoons Ag^+(aq) + Cl^-(aq) \qquad [17.21]$$

$$Ag^+(aq) + 2\,NH_3(aq) \rightleftharpoons Ag(NH_3)_2{}^+(aq) \qquad [17.22]$$

$$\text{Overall:}\quad AgCl(s) + 2\,NH_3(aq) \rightleftharpoons Ag(NH_3)_2{}^+(aq) + Cl^-(aq) \qquad [17.23]$$

CHEMISTRY AND LIFE | Tooth Decay and Fluoridation

Tooth enamel consists mainly of a mineral called hydroxyapatite, $Ca_{10}(PO_4)_6(OH)_2$. It is the hardest substance in the body. Tooth cavities are caused when acids dissolve tooth enamel.

$$Ca_{10}(PO_4)_6(OH)_2(s) + 8\,H^+(aq) \longrightarrow 10\,Ca^{2+}(aq) + 6\,HPO_4^{2-}(aq) + 2\,H_2O(l)$$

The resultant Ca^{2+} and HPO_4^{2-} ions diffuse out of the tooth enamel and are washed away by saliva. The acids that attack the hydroxyapatite are formed by the action of specific bacteria on sugars and other carbohydrates present in the plaque adhering to the teeth.

Fluoride ion, present in drinking water, toothpaste, and other sources, can react with hydroxyapatite to form fluoroapatite, $Ca_{10}(PO_4)_6F_2$. This mineral, in which F^- has replaced OH^-, is much more resistant to attack by acids because the fluoride ion is a much weaker Brønsted–Lowry base than the hydroxide ion.

Because the fluoride ion is so effective in preventing cavities, it is added to the public water supply in many places to give a concentration of 1 mg/L (1 ppm). The compound added may be NaF or Na_2SiF_6. Na_2SiF_6 reacts with water to release fluoride ions by the following reaction:

$$SiF_6^{2-}(aq) + 2\,H_2O(l) \longrightarrow 6\,F^-(aq) + 4\,H^+(aq) + SiO_2(s)$$

About 80% of all toothpastes now sold in the United States contain fluoride compounds, usually at the level of 0.1% fluoride by mass. The most common compounds in toothpastes are sodium fluoride (NaF), sodium monofluorophosphate (Na_2PO_3F), and stannous fluoride (SnF_2).

Related Exercise: 17.103

The presence of NH_3 drives the top reaction, the dissolution of AgCl, to the right as $Ag^+(aq)$ is consumed to form $Ag(NH_3)_2^+$.

For a Lewis base such as NH_3 to increase the solubility of a metal salt, it must be able to interact more strongly with the metal ion than water does. The NH_3 must displace solvating H_2O molecules (Sections 13.1 and 16.11) in order to form $Ag(NH_3)_2^+$:

$$Ag^+(aq) + 2\,NH_3(aq) \rightleftharpoons Ag(NH_3)_2^+(aq) \qquad [17.24]$$

An assembly of a metal ion and the Lewis bases bonded to it, such as $Ag(NH_3)_2^+$, is called a **complex ion**. The stability of a complex ion in aqueous solution can be judged by the size of the equilibrium constant for its formation from the hydrated metal ion. For example, the equilibrium constant for formation of $Ag(NH_3)_2^+$ (Equation 17.24) is 1.7×10^7:

$$K_f = \frac{[Ag(NH_3)_2^+]}{[Ag^+][NH_3]^2} = 1.7 \times 10^7 \qquad [17.25]$$

The equilibrium constant for this kind of reaction is called a **formation constant**, K_f. The formation constants for several complex ions are listed in Table 17.1 ▼.

TABLE 17.1 Formation Constants for Some Metal Complex Ions in Water at 25°C

Complex Ion	K_f	Equilibrium Equation
$Ag(NH_3)_2^+$	1.7×10^7	$Ag^+(aq) + 2\,NH_3(aq) \rightleftharpoons Ag(NH_3)_2^+(aq)$
$Ag(CN)_2^-$	1×10^{21}	$Ag^+(aq) + 2\,CN^-(aq) \rightleftharpoons Ag(CN)_2^-(aq)$
$Ag(S_2O_3)_2^{3-}$	2.9×10^{13}	$Ag^+(aq) + 2\,S_2O_3^{2-}(aq) \rightleftharpoons Ag(S_2O_3)_2^{3-}(aq)$
$CdBr_4^{2-}$	5×10^3	$Cd^{2+}(aq) + 4\,Br^-(aq) \rightleftharpoons CdBr_4^{2-}(aq)$
$Cr(OH)_4^-$	8×10^{29}	$Cr^{3+}(aq) + 4\,OH^-(aq) \rightleftharpoons Cr(OH)_4^-(aq)$
$Co(SCN)_4^{2-}$	1×10^3	$Co^{2+}(aq) + 4\,SCN^-(aq) \rightleftharpoons Co(SCN)_4^{2-}(aq)$
$Cu(NH_3)_4^{2+}$	5×10^{12}	$Cu^{2+}(aq) + 4\,NH_3(aq) \rightleftharpoons Cu(NH_3)_4^{2+}(aq)$
$Cu(CN)_4^{2-}$	1×10^{25}	$Cu^{2+}(aq) + 4\,CN^-(aq) \rightleftharpoons Cu(CN)_4^{2-}(aq)$
$Ni(NH_3)_6^{2+}$	1.2×10^9	$Ni^{2+}(aq) + 6\,NH_3(aq) \rightleftharpoons Ni(NH_3)_6^{2+}(aq)$
$Fe(CN)_6^{4-}$	1×10^{35}	$Fe^{2+}(aq) + 6\,CN^-(aq) \rightleftharpoons Fe(CN)_6^{4-}(aq)$
$Fe(CN)_6^{3-}$	1×10^{42}	$Fe^{3+}(aq) + 6\,CN^-(aq) \rightleftharpoons Fe(CN)_6^{3-}(aq)$

SAMPLE EXERCISE 17.14 | Evaluating an Equilibrium Involving a Complex Ion

Calculate the concentration of Ag^+ present in solution at equilibrium when concentrated ammonia is added to a 0.010 M solution of $AgNO_3$ to give an equilibrium concentration of $[NH_3] = 0.20\ M$. Neglect the small volume change that occurs when NH_3 is added.

Solution

Analyze: When $NH_3(aq)$ is added to $Ag^+(aq)$, a reaction occurs forming $Ag(NH_3)_2^+$ as shown in Equation 17.22. We are asked to determine what concentration of $Ag^+(aq)$ will remain uncombined when the NH_3 concentration is brought to 0.20 M in a solution originally 0.010 M in $AgNO_3$.

Plan: We first assume that the $AgNO_3$ is completely dissociated giving 0.10 M Ag^+. Because K_f for the formation of $Ag(NH_3)_2^+$ is quite large, we assume that essentially all the Ag^+ is then converted to $Ag(NH_3)_2^+$ and approach the problem as though we are concerned with the *dissociation* of $Ag(NH_3)_2^+$ rather than its *formation*. In order to facilitate this approach, we will need to reverse the equation to represent the formation of Ag^+ and NH_3 from $Ag(NH_3)_2^+$ and also make the corresponding change to the equilibrium constant.

$$Ag(NH_3)_2^+(aq) \rightleftharpoons Ag^+(aq) + 2\,NH_3(aq)$$

$$\frac{1}{K_f} = \frac{1}{1.7 \times 10^7} = 5.9 \times 10^{-8}$$

Solve: If $[Ag^+]$ is 0.010 M initially, then $[Ag(NH_3)_2^+]$ will be 0.010 M following addition of the NH_3. We now construct a table to solve this equilibrium problem. Note that the NH_3 concentration given in the problem is an *equilibrium* concentration rather than an initial concentration.

	$Ag(NH_3)_2^+(aq)$	$\rightleftharpoons$ $Ag^+(aq)$	+ $2\,NH_3(aq)$
Initial	0.010 M	0 M	
Change	$-x\ M$	$+x\ M$	
Equilibrium	$0.010 - x\ M$	$x\ M$	0.20 M

Because the concentration of Ag^+ is very small, we can ignore x in comparison with 0.010. Thus, $0.010 - x \approx 0.010\ M$. Substituting these values into the equilibrium-constant expression for the dissociation of $Ag(NH_3)_2^+$, we obtain

$$\frac{[Ag^+][NH_3]^2}{[Ag(NH_3)_2^+]} = \frac{(x)(0.20)^2}{0.010} = 5.9 \times 10^{-8}$$

Solving for x, we obtain $x = 1.5 \times 10^{-8}\ M = [Ag^+]$. Thus, formation of the $Ag(NH_3)_2^+$ complex drastically reduces the concentration of free Ag^+ ion in solution.

PRACTICE EXERCISE

Calculate $[Cr^{3+}]$ in equilibrium with $Cr(OH)_4^-$ when 0.010 mol of $Cr(NO_3)_3$ is dissolved in a liter of solution buffered at pH 10.0.

Answer: $1 \times 10^{-16}\ M$

The general rule is that the solubility of metal salts increases in the presence of suitable Lewis bases, such as NH_3, CN^-, or OH^-, if the metal forms a complex with the base. The ability of metal ions to form complexes is an extremely important aspect of their chemistry. In Chapter 24 we will take a much closer look at complex ions. In that chapter and others we will see applications of complex ions to areas such as biochemistry, metallurgy, and photography.

Amphoterism

Some metal oxides and hydroxides that are relatively insoluble in neutral water dissolve in strongly acidic and strongly basic solutions. These substances are soluble in strong acids and bases because they themselves are capable of

MOVIE
Precipitation Reactions

behaving as either an acid or base; they are **amphoteric oxides and bases**. Amphoteric oxides and hydroxides include those of Al^{3+}, Cr^{3+}, Zn^{2+}, and Sn^{2+}. Notice that the term *amphoteric* is applied to the behavior of insoluble oxides and hydroxides that can be made to dissolve in either acidic or basic solutions. The similar term *amphiprotic*, which we encountered in Section 16.2, relates more generally to any molecule or ion that can either gain or lose a proton.

These species dissolve in acidic solutions because they contain basic anions. What makes amphoteric oxides and hydroxides special, though, is that they also dissolve in strongly basic solutions (Figure 17.20 ▼). This behavior

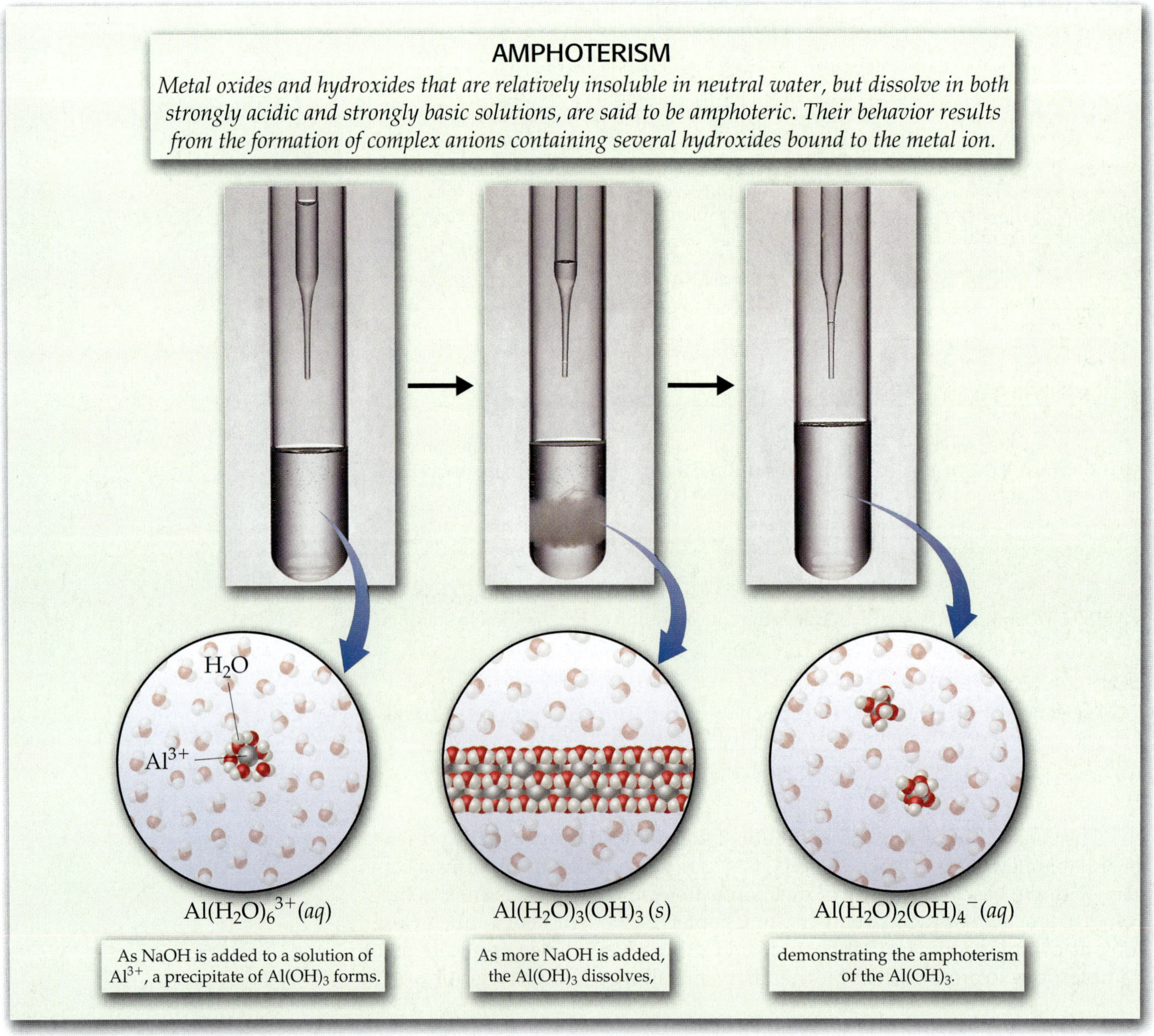

▲ **Figure 17.20 Amphoterism.**

results from the formation of complex anions containing several (typically four) hydroxides bound to the metal ion.

$$Al(OH)_3(s) + OH^-(aq) \rightleftharpoons Al(OH)_4^-(aq) \qquad [17.26]$$

Amphoterism is often explained by the behavior of the water molecules that surround the metal ion and that are bonded to it by Lewis acid-base interactions. (Section. 16.11) For example, $Al^{3+}(aq)$ is more accurately represented as $Al(H_2O)_6^{3+}(aq)$ because six water molecules are bonded to the Al^{3+} in aqueous solution. Recall from Section 16.11 that this hydrated ion is a weak acid. As a strong base is added, $Al(H_2O)_6^{3+}$ loses protons in a stepwise fashion, eventually forming neutral and water-insoluble $Al(H_2O)_3(OH)_3$. This substance then dissolves upon removal of an additional proton to form the anion $Al(H_2O)_2(OH)_4^-$. The reactions that occur are as follows:

$$Al(H_2O)_6^{3+}(aq) + OH^-(aq) \rightleftharpoons Al(H_2O)_5(OH)^{2+}(aq) + H_2O(l)$$

$$Al(H_2O)_5(OH)^{2+}(aq) + OH^-(aq) \rightleftharpoons Al(H_2O)_4(OH)_2^+(aq) + H_2O(l)$$

$$Al(H_2O)_4(OH)_2^+(aq) + OH^-(aq) \rightleftharpoons Al(H_2O)_3(OH)_3(s) + H_2O(l)$$

$$Al(H_2O)_3(OH)_3(s) + OH^-(aq) \rightleftharpoons Al(H_2O)_2(OH)_4^-(aq) + H_2O(l)$$

Removing additional protons is possible, but each successive reaction occurs less readily than the one before. As the charge on the ion becomes more negative, it becomes increasingly difficult to remove a positively charged proton. Addition of an acid reverses these reactions. The proton adds in a stepwise fashion to convert the OH^- groups to H_2O, eventually re-forming $Al(H_2O)_6^{3+}$. The common practice is to simplify the equations for these reactions by excluding the bound H_2O molecules. Thus, we usually write Al^{3+} instead of $Al(H_2O)_6^{3+}$, $Al(OH)_3$ instead of $Al(H_2O)_3(OH)_3$, $Al(OH)_4^-$ instead of $Al(H_2O)_2(OH)_4^-$, and so forth.

The extent to which an insoluble metal hydroxide reacts with either acid or base varies with the particular metal ion involved. Many metal hydroxides—such as $Ca(OH)_2$, $Fe(OH)_2$, and $Fe(OH)_3$—are capable of dissolving in acidic solution but do not react with excess base. These hydroxides are not amphoteric.

The purification of aluminum ore in the manufacture of aluminum metal provides an interesting application of the property of amphoterism. As we have seen, $Al(OH)_3$ is amphoteric, whereas $Fe(OH)_3$ is not. Aluminum occurs in large quantities as the ore *bauxite*, which is essentially Al_2O_3 with additional water molecules. The ore is contaminated with Fe_2O_3 as an impurity. When bauxite is added to a strongly basic solution, the Al_2O_3 dissolves because the aluminum forms complex ions, such as $Al(OH)_4^-$. The Fe_2O_3 impurity, however, is not amphoteric and remains as a solid. The solution is filtered, getting rid of the iron impurity. Aluminum hydroxide is then precipitated by addition of acid. The purified hydroxide receives further treatment and eventually yields aluminum metal. (Section 23.3)

GIVE IT SOME THOUGHT

What kind of behavior characterizes an amphoteric oxide or an amphoteric hydroxide?

17.6 | Precipitation and Separation of Ions

Equilibrium can be achieved starting with the substances on either side of a chemical equation. The equilibrium among $BaSO_4(s)$, $Ba^{2+}(aq)$, and $SO_4^{2-}(aq)$ (Equation 17.15) can be achieved starting with solid $BaSO_4$. It can also be reached starting with solutions of salts containing Ba^{2+} and SO_4^{2-}, say $BaCl_2$ and Na_2SO_4. When these two solutions are mixed, $BaSO_4$ will precipitate if the product of the initial ion concentrations, $Q = [Ba^{2+}][SO_4^{2-}]$, is greater than K_{sp}.

The use of the reaction quotient, Q, to determine the direction in which a reaction must proceed to reach equilibrium was discussed earlier. (Section 15.5) The possible relationships between Q and K_{sp} are summarized as follows:

If $Q > K_{sp}$, precipitation occurs until $Q = K_{sp}$.

If $Q = K_{sp}$, equilibrium exists (saturated solution).

If $Q < K_{sp}$, solid dissolves until $Q = K_{sp}$.

SAMPLE EXERCISE 17.15 | Predicting Whether a Precipitate Will Form

Will a precipitate form when 0.10 L of 8.0×10^{-3} M $Pb(NO_3)_2$ is added to 0.40 L of 5.0×10^{-3} M Na_2SO_4?

Solution

Analyze: The problem asks us to determine whether a precipitate will form when two salt solutions are combined.

Plan: We should determine the concentrations of all ions immediately upon mixing of the solutions and compare the value of the reaction quotient, Q, to the solubility-product constant, K_{sp}, for any potentially insoluble product. The possible metathesis products are $PbSO_4$ and $NaNO_3$. Sodium salts are quite soluble; $PbSO_4$ has a K_{sp} of 6.3×10^{-7} (Appendix D), however, and will precipitate if the Pb^{2+} and SO_4^{2-} ion concentrations are high enough for Q to exceed K_{sp} for the salt.

Solve: When the two solutions are mixed, the total volume becomes 0.10 L + 0.40 L = 0.50 L. The number of moles of Pb^{2+} in 0.10 L of 3.0×10^{-3} M $Pb(NO_3)_2$ is

$$(0.10\ \text{L})\left(8.0 \times 10^{-3}\ \frac{\text{mol}}{\text{L}}\right) = 8.0 \times 10^{-4}\ \text{mol}$$

The concentration of Pb^{2+} in the 0.50-L mixture is therefore

$$[Pb^{2+}] = \frac{8.0 \times 10^{-4}\ \text{mol}}{0.50\ \text{L}} = 1.6 \times 10^{-3}\ M$$

The number of moles of SO_4^{2-} in 0.40 L of 5.0×10^{-3} M Na_2SO_4 is

$$(0.40\ \text{L})\left(5.0 \times 10^{-3}\ \frac{\text{mol}}{\text{L}}\right) = 2.0 \times 10^{-3}\ \text{mol}$$

Therefore, $[SO_4^{2-}]$ in the 0.50-L mixture is

$$[SO_4^{2-}] = \frac{2.0 \times 10^{-3}\ \text{mol}}{0.50\ \text{L}} = 4.0 \times 10^{-3}\ M$$

We then have

$$Q = [Pb^{2+}][SO_4^{2-}] = (1.6 \times 10^{-3})(4.0 \times 10^{-3}) = 6.4 \times 10^{-6}$$

Because $Q > K_{sp}$, $PbSO_4$ will precipitate.

PRACTICE EXERCISE

Will a precipitate form when 0.050 L of 2.0×10^{-2} M NaF is mixed with 0.010 L of 1.0×10^{-2} M $Ca(NO_3)_2$?

Answer: Yes, CaF_2 precipitates because $Q = 4.6 \times 10^{-8}$ is larger than $K_{sp} = 3.9 \times 10^{-11}$

Selective Precipitation of Ions

Ions can be separated from each other based on the solubilities of their salts. Consider a solution containing both Ag^+ and Cu^{2+}. If HCl is added to the solution, AgCl ($K_{sp} = 1.8 \times 10^{-10}$) precipitates, while Cu^{2+} remains in solution because $CuCl_2$ is soluble. Separation of ions in an aqueous solution by using a reagent that forms a precipitate with one or a few of the ions is called *selective precipitation.*

ACTIVITY
Selective Precipitation of Ions

SAMPLE EXERCISE 17.16 | **Calculating Ion Concentrations for Precipitation**

A solution contains 1.0×10^{-2} M Ag^+ and 2.0×10^{-2} M Pb^{2+}. When Cl^- is added to the solution, both AgCl ($K_{sp} = 1.8 \times 10^{-10}$) and $PbCl_2$ ($K_{sp} = 1.7 \times 10^{-5}$) precipitate from the solution. What concentration of Cl^- is necessary to begin the precipitation of each salt? Which salt precipitates first?

Solution

Analyze: We are asked to determine the concentration of Cl^- necessary to begin the precipitation from a solution containing Ag^+ and Pb^{2+} ions, and to predict which metal chloride will begin to precipitate first.

Plan: We are given K_{sp} values for the two possible precipitates. Using these and the metal ion concentrations, we can calculate what concentration of Cl^- ion would be necessary to begin precipitation of each. The salt requiring the lower Cl^- ion concentration will precipitate first.

Solve: For AgCl we have

$$K_{sp} = [Ag^+][Cl^-] = 1.8 \times 10^{-10}$$

Because $[Ag^+] = 1.0 \times 10^{-2}$ M, the greatest concentration of Cl^- that can be present without causing precipitation of AgCl can be calculated from the K_{sp} expression:

$$K_{sp} = [1.0 \times 10^{-2}][Cl^-] = 1.8 \times 10^{-10}$$

$$[Cl^-] = \frac{1.8 \times 10^{-10}}{1.0 \times 10^{-2}} = 1.8 \times 10^{-8}\ M$$

Any Cl^- in excess of this very small concentration will cause AgCl to precipitate from solution. Proceeding similarly for $PbCl_2$, we have

$$K_{sp} = [Pb^{2+}][Cl^-]^2 = 1.7 \times 10^{-5}$$

$$[2.0 \times 10^{-2}][Cl^-]^2 = 1.7 \times 10^{-5}$$

$$[Cl^-]^2 = \frac{1.7 \times 10^{-5}}{2.0 \times 10^{-2}} = 8.5 \times 10^{-4}$$

$$[Cl^-] = \sqrt{8.5 \times 10^{-4}} = 2.9 \times 10^{-2}\ M$$

Thus, a concentration of Cl^- in excess of 2.9×10^{-2} M will cause $PbCl_2$ to precipitate.

Comparing the concentrations of Cl^- required to precipitate each salt, we see that as Cl^- is added to the solution, AgCl will precipitate first because it requires a much smaller concentration of Cl^-. Thus, Ag^+ can be separated from Pb^{2+} by slowly adding Cl^- so $[Cl^-]$ is between 1.8×10^{-8} M and 2.9×10^{-2} M.

PRACTICE EXERCISE

A solution consists of 0.050 M Mg^{2+} and 0.020 M Cu^{2+}. Which ion will precipitate first as OH^- is added to the solution? What concentration of OH^- is necessary to begin the precipitation of each cation? [$K_{sp} = 1.8 \times 10^{-11}$ for $Mg(OH)_2$, and $K_{sp} = 2.2 \times 10^{-20}$ for $Cu(OH)_2$.]

Answer: $Cu(OH)_2$ precipitates first. $Cu(OH)_2$ begins to precipitate when $[OH^-]$ exceeds 1.0×10^{-9} M; $Mg(OH)_2$ begins to precipitate when $[OH^-]$ exceeds 1.9×10^{-5} M.

Sulfide ion is often used to separate metal ions because the solubilities of sulfide salts span a wide range and depend greatly on the pH of the solution. Cu^{2+} and Zn^{2+}, for example, can be separated by bubbling H_2S gas through an acidified solution. Because CuS ($K_{sp} = 6 \times 10^{-37}$) is less soluble than ZnS ($K_{sp} = 2 \times 10^{-25}$), CuS precipitates from an acidified solution (pH = 1) while ZnS does not (Figure 17.21 ▶):

$$Cu^{2+}(aq) + H_2S(aq) \rightleftharpoons CuS(s) + 2\,H^+(aq) \qquad [17.27]$$

The CuS can be separated from the Zn^{2+} solution by filtration. The CuS can then be dissolved by using a high concentration of H^+, shifting the equilibrium shown in Equation 17.27 to the left.

GIVE IT SOME THOUGHT

What experimental conditions will leave the smallest concentration of Cu^{2+} ions in solution according to Equation 17.27?

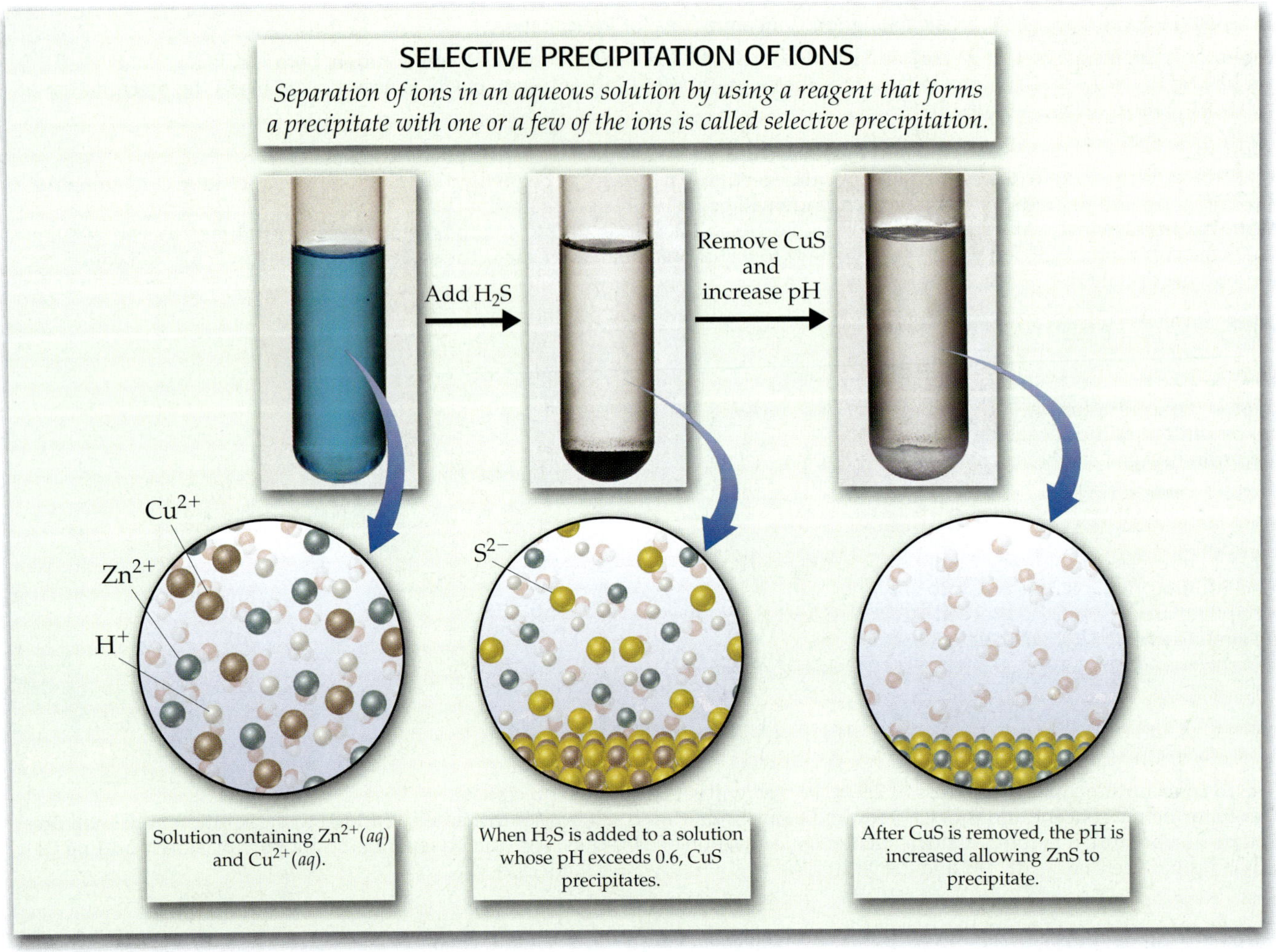

▲ **Figure 17.21 Selective precipitation.**

17.7 | Qualitative Analysis for Metallic Elements

In this chapter we have seen several examples of equilibria involving metal ions in aqueous solution. In this final section we look briefly at how solubility equilibria and complex-ion formation can be used to detect the presence of particular metal ions in solution. Before the development of modern analytical instrumentation, it was necessary to analyze mixtures of metals in a sample by so-called wet chemical methods. For example, a metallic sample that might contain several metallic elements was dissolved in a concentrated acid solution. This solution was then tested in a systematic way for the presence of various metal ions.

Qualitative analysis determines only the presence or absence of a particular metal ion, whereas **quantitative analysis** determines how much of a given substance is present. Wet methods of qualitative analysis have become less important as a means of analysis. They are frequently used in general chemistry laboratory programs, however, to illustrate equilibria, to teach the properties of common metal ions in solution, and to develop laboratory skills. Typically, such analyses proceed in three stages: (1) The ions are separated into broad groups

VCL Identification of Cations in Solution

on the basis of solubility properties. (2) The individual ions within each group are then separated by selectively dissolving members in the group. (3) The ions are then identified by means of specific tests.

A scheme in general use divides the common cations into five groups, as shown in Figure 17.22 ▶. The order of addition of reagents is important. The most selective separations—those that involve the smallest number of ions—are carried out first. The reactions that are used must proceed so far toward completion that any concentration of cations remaining in the solution is too small to interfere with subsequent tests. Let's take a closer look at each of these five groups of cations, briefly examining the logic used in this qualitative analysis scheme.

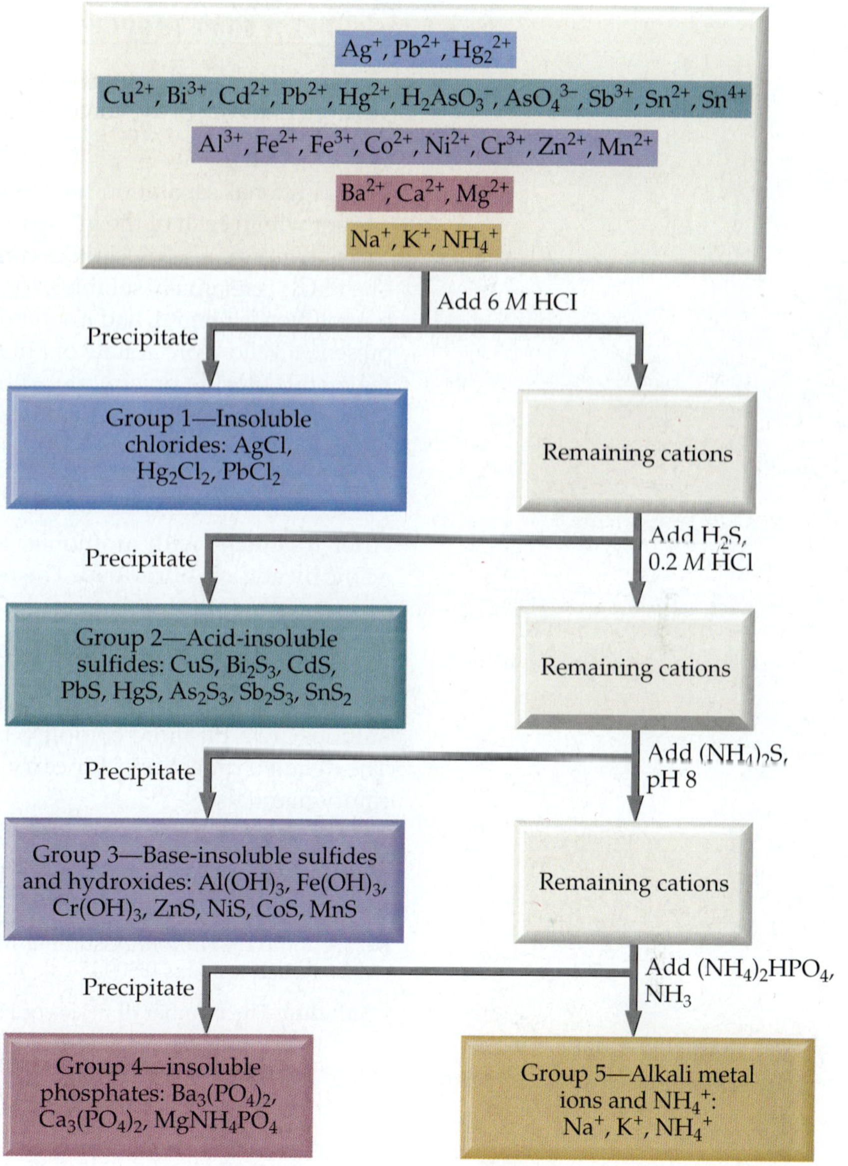

▲ **Figure 17.22 Qualitative analysis.** A flowchart showing the separation of cations into groups as a part of a common scheme for identifying cations.

1. *Insoluble chlorides:* Of the common metal ions, only Ag^+, Hg_2^{2+}, and Pb^{2+} form insoluble chlorides. When dilute HCl is added to a mixture of cations, therefore, only AgCl, Hg_2Cl_2, and $PbCl_2$ will precipitate, leaving the other cations in solution. The absence of a precipitate indicates that the starting solution contains no Ag^+, Hg_2^{2+}, or Pb^{2+}.

2. *Acid-insoluble sulfides:* After any insoluble chlorides have been removed, the remaining solution, now acidic, is treated with H_2S. Only the most insoluble metal sulfides—CuS, Bi_2S_3, CdS, PbS, HgS, As_2S_3, Sb_2S_3, and SnS_2—can precipitate. (Note the very small values of K_{sp} for some of these sulfides in Appendix D.) Those metal ions whose sulfides are somewhat more soluble—for example, ZnS or NiS—remain in solution.

3. *Base-insoluble sulfides and hydroxides:* After the solution is filtered to remove any acid-insoluble sulfides, the remaining solution is made slightly basic, and $(NH_4)_2S$ is added. In basic solutions the concentration of S^{2-} is higher than in acidic solutions. Thus, the ion products for many of the more soluble sulfides are made to exceed their K_{sp} values and precipitation occurs. The metal ions precipitated at this stage are Al^{3+}, Cr^{3+}, Fe^{3+}, Zn^{2+}, Ni^{2+}, Co^{2+}, and Mn^{2+}. (Actually, the Al^{3+}, Fe^{3+}, and Cr^{3+} ions do not form insoluble sulfides; instead they are precipitated as insoluble hydroxides at the same time.)

4. *Insoluble phosphates:* At this point the solution contains only metal ions from periodic table groups 1A and 2A. Adding $(NH_4)_2HPO_4$ to a basic solution precipitates the group 2A elements Mg^{2+}, Ca^{2+}, Sr^{2+}, and Ba^{2+} because these metals form insoluble phosphates.

5. *The alkali metal ions and NH_4^+:* The ions that remain after removing the insoluble phosphates form a small group. We can test for each ion individually. A flame test can be used to determine the presence of K^+, for example, because the flame turns a characteristic violet color if K^+ is present.

ACTIVITY
Qualitative Analysis

GIVE IT SOME THOUGHT

If a precipitate forms upon addition of HCl to an aqueous solution, what conclusions can you draw about the contents of the solution?

Additional separation and testing is necessary to determine which ions are present within each of the groups. Consider, for example, the ions of the insoluble chloride group. The precipitate containing the metal chlorides is boiled in water. The $PbCl_2$ is relatively soluble in hot water, whereas AgCl and Hg_2Cl_2 are not. The hot solution is filtered, and a solution of Na_2CrO_4 is added to the filtrate. If Pb^{2+} is present, a yellow precipitate of $PbCrO_4$ forms. The test for Ag^+ consists of treating the metal chloride precipitate with dilute ammonia. Only Ag^+ forms an ammonia complex. If AgCl is present in the precipitate, it will dissolve in the ammonia solution.

$$AgCl(s) + 2\,NH_3(aq) \rightleftharpoons Ag(NH_3)_2^+(aq) + Cl^-(aq) \qquad [17.28]$$

After treatment with ammonia, the solution is filtered and the filtrate made acidic by adding nitric acid. The nitric acid removes ammonia from solution by forming NH_4^+, thus releasing Ag^+, which re-forms the AgCl precipitate.

$$Ag(NH_3)_2^+(aq) + Cl^-(aq) + 2\,H^+(aq) \rightleftharpoons AgCl(s) + 2\,NH_4^+(aq) \qquad [17.29]$$

The analyses for individual ions in the acid-insoluble and base-insoluble sulfides are a bit more complex, but the same general principles are involved. The detailed procedures for carrying out such analyses are given in many laboratory manuals.

SAMPLE INTEGRATIVE EXERCISE | Putting Concepts Together

A sample of 1.25 L of HCl gas at 21°C and 0.950 atm is bubbled through 0.500 L of 0.150 *M* NH_3 solution. Assuming that all the HCl dissolves and that the volume of the solution remains 0.500 L, calculate the pH of the resulting solution.

Solution The number of moles of HCl gas is calculated from the ideal-gas law.

$$n = \frac{PV}{RT} = \frac{(0.950\ \text{atm})(1.25\ \text{L})}{(0.0821\ \text{L-atm/mol-K})(294\ \text{K})} = 0.0492\ \text{mol HCl}$$

The number of moles of NH_3 in the solution is given by the product of the volume of the solution and its concentration.

$$\text{Moles } NH_3 = (0.500\ \text{L})(0.150\ \text{mol } NH_3/\text{L}) = 0.0750\ \text{mol } NH_3$$

The acid HCl and base NH_3 react, transferring a proton from HCl to NH_3, producing NH_4^+ and Cl^- ions.

$$HCl(g) + NH_3(aq) \longrightarrow NH_4^+(aq) + Cl^-(aq)$$

To determine the pH of the solution, we first calculate the amount of each reactant and each product present at the completion of the reaction.

Before rxn:	0.0492 mol		0.0750 mol		0 mol		0 mol
	$HCl(g)$	+	$NH_3(aq)$	$\longrightarrow$	$NH_4^+(aq)$	+	$Cl^-(aq)$
After rxn:	0 mol		0.0258 mol		0.0492 mol		0.0492 mol

Thus, the reaction produces a solution containing a mixture of NH_3, NH_4^+, and Cl^-. The NH_3 is a weak base ($K_b = 1.8 \times 10^{-5}$), NH_4^+ is its conjugate acid, and Cl^- is neither acidic nor basic. Consequently, the pH depends on $[NH_3]$ and $[NH_4^+]$.

$$[NH_3] = \frac{0.0258\ \text{mol } NH_3}{0.500\ \text{L soln}} = 0.0516\ M$$

$$[NH_4^+] = \frac{0.0492\ \text{mol } NH_4^+}{0.500\ \text{L soln}} = 0.0984\ M$$

We can calculate the pH using either K_b for NH_3 or K_a for NH_4^+. Using the K_b expression, we have

$$NH_3(aq) + H_2O(l) \rightleftharpoons NH_4^+(aq) + OH^-(aq)$$

Initial	0.05156 M	—	0.0984 M	0
Change	$-x$ M	—	$+x$ M	$+x$ M
Equilibrium	$(0.0516 - x)$ M	—	$(0.0984 + x)$ M	x M

$$K_b = \frac{[NH_4^+][OH^-]}{[NH_3]} = \frac{(0.0984 + x)(x)}{(0.0516 - x)} \simeq \frac{(0.0984)x}{0.0516} = 1.8 \times 10^{-5}$$

$$x = [OH^-] = \frac{(0.0516)(1.8 \times 10^{-5})}{0.0984} = 9.4 \times 10^{-6}\ M$$

Hence, $pOH = -\log(9.4 \times 10^{-6}) = 5.03$ and $pH = 14.00 - pOH = 14.00 - 5.03 = 8.97$.

SUMMARY AND KEY TERMS

Section 17.1 In this chapter we've considered several types of important equilibria that occur in aqueous solution. Our primary emphasis has been on acid-base equilibria in solutions containing two or more solutes and on solubility equilibria. The dissociation of a weak acid or weak base is repressed by the presence of a strong electrolyte that provides an ion common to the equilibrium. This phenomenon is called the **common-ion effect.**

Section 17.2 A particularly important type of acid-base mixture is that of a weak conjugate acid-base pair. Such mixtures function as **buffered solutions (buffers)**. Addition of small amounts of a strong acid or a strong base to a buffered solution causes only small changes in pH because the buffer reacts with the added acid or base. (Strong acid–strong base, strong acid–weak base, and weak acid–strong base reactions proceed essentially to completion.) Buffered solutions are usually prepared from a weak acid and a salt of that acid or from a weak base and a salt of that base. Two important characteristics of a buffered solution are its **buffer capacity** and its pH. The pH can be calculated using K_a or K_b. The relationship between pH, pK_a, and the concentrations of an acid and its conjugate base can be expressed by the **Henderson–Hasselbalch equation**: $pH = pK_a + \log\frac{[base]}{[acid]}$.

Section 17.3 The plot of the pH of an acid (or base) as a function of the volume of added base (or acid) is called a **pH titration curve**. Titration curves aid in selecting a proper pH indicator for an acid-base titration. The titration curve of a strong acid–strong base titration exhibits a large change in pH in the immediate vicinity of the equivalence point; at the equivalence point for this titration, $pH = 7$. For strong acid–weak base or weak acid–strong base titrations, the pH change in the vicinity of the equivalence point is not as large. Furthermore, the pH at the equivalence point is not 7 in either of these cases. Rather, it is the pH of the salt solution that results from the neutralization reaction. It is possible to calculate the pH at any point of the titration curve by first considering the stoichiometry of the reaction between the acid and base and then examining equilibria involving remaining solute species.

Section 17.4 The equilibrium between a solid compound and its ions in solution provides an example of heterogeneous equilibrium. The **solubility-product constant** (or simply the **solubility product**), K_{sp}, is an equilibrium constant that expresses quantitatively the extent to which the compound dissolves. The K_{sp} can be used to calculate the solubility of an ionic compound, and the solubility can be used to calculate K_{sp}.

Section 17.5 Several experimental factors, including temperature, affect the solubilities of ionic compounds in water. The solubility of a slightly soluble ionic compound is decreased by the presence of a second solute that furnishes a common ion (the common-ion effect). The solubility of compounds containing basic anions increases as the solution is made more acidic (as pH decreases). Salts with anions of negligible basicity (the anions of strong acids) are unaffected by pH changes.

The solubility of metal salts is also affected by the presence of certain Lewis bases that react with metal ions to form stable **complex ions**. Complex-ion formation in aqueous solution involves the displacement by Lewis bases (such as NH_3 and CN^-) of water molecules attached to the metal ion. The extent to which such complex formation occurs is expressed quantitatively by the **formation constant** for the complex ion. **Amphoteric oxides and hydroxides** are those that are only slightly soluble in water but dissolve on addition of either acid or base. Acid-base reactions involving the

OH^- or H_2O groups bound to the metal ions give rise to the amphoterism.

Section 17.6 Comparison of the ion product, Q, with the value of K_{sp} can be used to judge whether a precipitate will form when solutions are mixed or whether a slightly soluble salt will dissolve under various conditions. Precipitates form when $Q > K_{sp}$. Ions can be separated from each other based on the solubilities of their salts.

Section 17.7 Metallic elements vary a great deal in the solubilities of their salts, in their acid-base behavior, and in their tendencies to form complex ions. These differences can be used to separate and detect the presence of metal ions in mixtures. **Qualitative analysis** determines the presence or absence of species in a sample, whereas **quantitative analysis** determines how much of each species is present. The qualitative analysis of metal ions in solution can be carried out by separating the ions into groups on the basis of precipitation reactions and then analyzing each group for individual metal ions.

VISUALIZING CONCEPTS

17.1 The following boxes represent aqueous solutions containing a weak acid, HX, and its conjugate base, X^-. Water molecules and cations are not shown. Which solution has the highest pH? Explain. [Section 17.1]

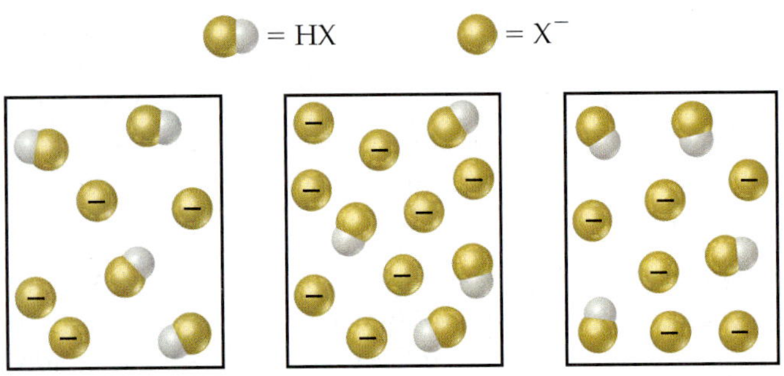

17.2 The beaker on the right contains 0.1 *M* acetic acid solution with methyl orange as an indicator. The beaker on the left contains a mixture of 0.1 *M* acetic acid and 0.1 *M* sodium acetate with methyl orange. **(a)** Using Figure 16.7, estimate the pH of each solution, and explain the difference. **(b)** Which solution is better able to maintain its pH when small amounts of NaOH are added? Explain. [Sections 17.1 and 17.2]

17.3 The drawing on the left represents a buffer composed of equal concentrations of a weak acid, HX, and its conjugate base, X^-. The heights of the columns are proportional to the concentrations of the components of the buffer. **(a)** Which of the three drawings (1), (2), (3) represents the buffer after the addition of a strong acid? **(b)** Which of the three represents the buffer after the addition of a strong base? **(c)** Which of the three represents a situation that cannot arise from the addition of either an acid or a base? [Section 17.2]

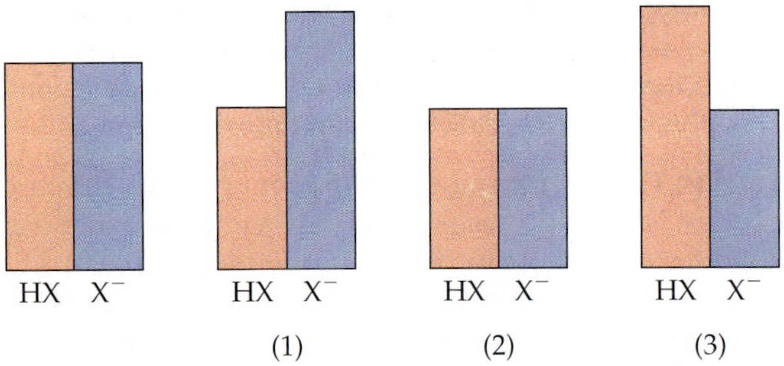

17.4 The following drawings represent solutions at various stages of the titration of a weak acid, HA, with NaOH. (The Na^+ ions and water molecules have been omitted for clarity.) To which of the following regions of the titration curve does each drawing correspond: **(a)** before addition of NaOH, **(b)** after addition of NaOH but before equivalence point, **(c)** at equivalence point, **(d)** after equivalence point? [Section 17.3]

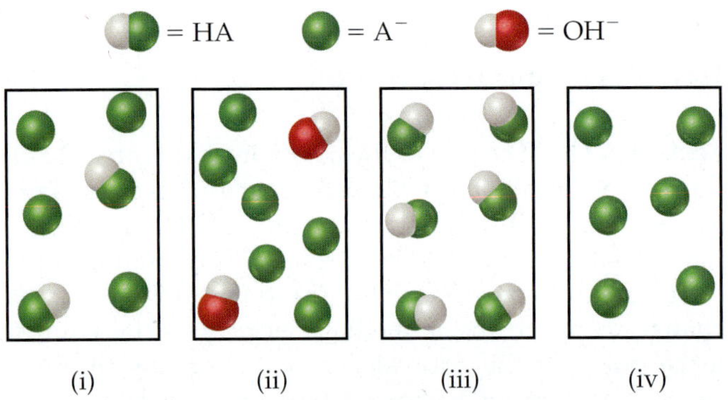

17.5 Match the following descriptions of titration curves with the diagrams: **(a)** strong acid added to strong base,

(b) strong acid added to weak base, **(c)** strong acid added to strong base, **(d)** strong base added to polyprotic acid. [Section 17.3]

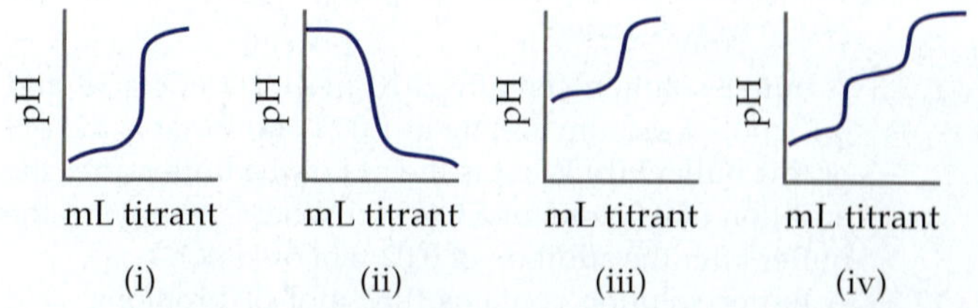

17.6 The following drawings represent saturated solutions of three ionic compounds of silver—AgX, AgY, and AgZ. (Na^+ cations, which might also be present for charge balance, are not shown.) Which compound has the smallest K_{sp}? [Section 17.4]

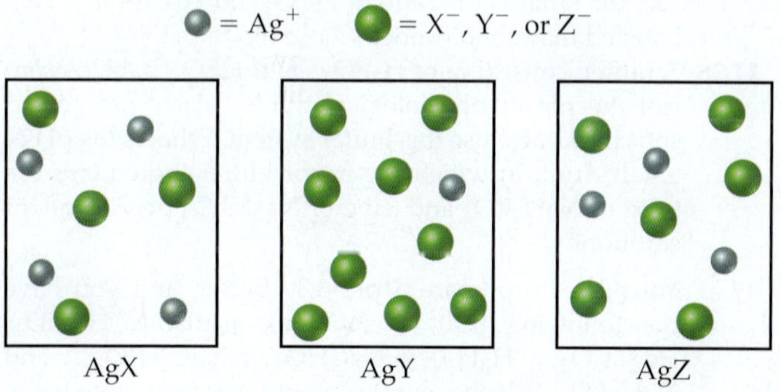

17.7 The following graphs represent the behavior of $BaCO_3$ under different circumstances. In each case the vertical axis indicates the solubility of the $BaCO_3$ and the horizontal axis represents the concentration of some other reagent. **(a)** Which graph represents what happens to the solubility of $BaCO_3$ as HNO_3 is added? **(b)** Which graph represents what happens to the $BaCO_3$ solubility as Na_2CO_3 is added? **(c)** Which represents what happens to the $BaCO_3$ solubility as $NaNO_3$ is added? [Section 17.5]

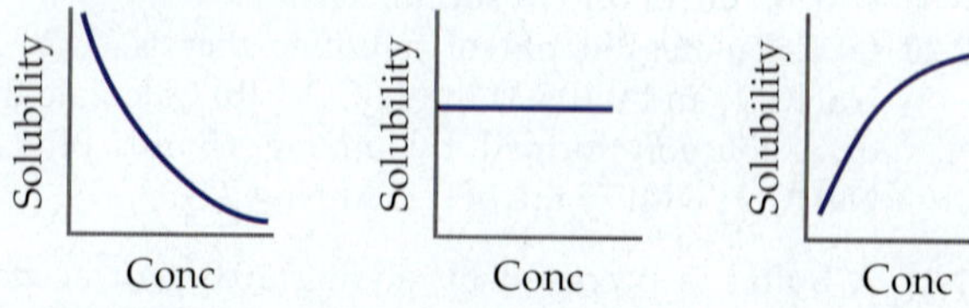

17.8 What is the name given to the kind of behavior demonstrated by a metal hydroxide in this graph? [Section 17.5]

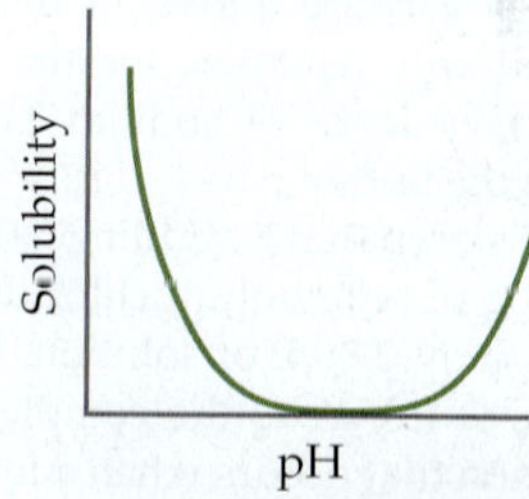

EXERCISES

Common-Ion Effect

17.9 **(a)** What is the common-ion effect? **(b)** Give an example of a salt that can decrease the ionization of HNO_2 in solution.

17.10 **(a)** Consider the equilibrium $B(aq) + H_2O(l) \rightleftharpoons HB^+(aq) + OH^-(aq)$. Using Le Châtelier's principle, explain the effect of the presence of a salt of HB^+ on the ionization of B. **(b)** Give an example of a salt that can decrease the ionization of NH_3 in solution.

17.11 Does the pH increase, decrease, or remain the same when each of the following is added: **(a)** $NaNO_2$ to a solution of HNO_2; **(b)** $(CH_3NH_3)Cl$ to a solution of CH_3NH_2; **(c)** sodium formate to a solution of formic acid; **(d)** potassium bromide to a solution of hydrobromic acid; **(e)** HCl to a solution of $NaC_2H_3O_2$?

17.12 Indicate whether the pH increases, decreases, or remains the same when each of the following is added: **(a)** $Ca(C_7H_5O_2)_2$ to a solution of $HC_7H_5O_2$; **(b)** pyridinium nitrate, $(C_5H_5NH)(NO_3)$, to a solution of pyridine, C_5H_5N; **(c)** ammonia to a solution of hydrochloric acid; **(d)** sodium hydrogen carbonate to a solution of carbonic acid; **(e)** $NaClO_4$ to a solution of NaOH.

17.13 Use information from Appendix D to calculate the pH of **(a)** a solution that is 0.060 *M* in potassium propionate ($KC_3H_5O_2$) and 0.085 *M* in propionic acid ($HC_3H_5O_2$); **(b)** a solution that is 0.075 *M* in trimethylamine, $(CH_3)_3N$, and 0.10 *M* in trimethylammonium chloride, $(CH_3)_3NHCl$; **(c)** a solution that is made by mixing 50.0 mL of 0.15 *M* acetic acid and 50.0 mL of 0.20 *M* sodium acetate.

17.14 Use information from Appendix D to calculate the pH of **(a)** a solution that is 0.160 *M* in sodium formate ($NaCHO_2$) and 0.260 *M* in formic acid ($HCHO_2$); **(b)** a solution that is 0.210 *M* in pyridine (C_5H_5N) and 0.350 *M* in pyridinium chloride (C_5H_5NHCl); **(c)** a solution that is made by combining 125 mL of 0.050 *M* hydrofluoric acid with 50.0 mL of 0.10 *M* sodium fluoride.

17.15 **(a)** Calculate the percent ionization of 0.0075 *M* butanoic acid ($K_a = 1.5 \times 10^{-5}$). **(b)** Calculate the percent ionization of 0.0075 *M* butanoic acid in a solution containing 0.085 *M* sodium butanoate.

17.16 **(a)** Calculate the percent ionization of 0.085 *M* lactic acid ($K_a = 1.4 \times 10^{-4}$). **(b)** Calculate the percent ionization of 0.085 *M* lactic acid in a solution containing 0.050 *M* sodium lactate.

Buffers

17.17 Explain why a mixture of $HC_2H_3O_2$ and $NaC_2H_3O_2$ can act as a buffer while a mixture of HCl and NaCl cannot.

17.18 Explain why a mixture formed by mixing 100 mL of 0.100 *M* $HC_2H_3O_2$ and 50 mL of 0.100 *M* NaOH will act as a buffer.

17.19 **(a)** Calculate the pH of a buffer that is 0.12 *M* in lactic acid and 0.11 *M* in sodium lactate. **(b)** Calculate the pH of a buffer formed by mixing 85 mL of 0.13 *M* lactic acid with 95 mL of 0.15 *M* sodium lactate.

17.20 **(a)** Calculate the pH of a buffer that is 0.120 *M* in $NaHCO_3$ and 0.105 *M* in Na_2CO_3. **(b)** Calculate the pH of a solution formed by mixing 65 mL of 0.20 *M* $NaHCO_3$ with 75 mL of 0.15 M Na_2CO_3.

17.21 A buffer is prepared by adding 20.0 g of acetic acid ($HC_2H_3O_2$) and 20.0 g of sodium acetate ($NaC_2H_3O_2$) to enough water to form 2.00 L of solution. **(a)** Determine the pH of the buffer. **(b)** Write the complete ionic equation for the reaction that occurs when a few drops of hydrochloric acid are added to the buffer. **(c)** Write the complete ionic equation for the reaction that occurs when a few drops of sodium hydroxide solution are added to the buffer.

17.22 A buffer is prepared by adding 5.0 g of ammonia (NH_3) and 20.0 g of ammonium chloride (NH_4Cl) to enough water to form 2.50 L of solution. **(a)** What is the pH of this buffer? **(b)** Write the complete ionic equation for the reaction that occurs when a few drops of nitric acid are added to the buffer. **(c)** Write the complete ionic equation for the reaction that occurs when a few drops of potassium hydroxide solution are added to the buffer.

17.23 How many moles of sodium hypobromite (NaBrO) should be added to 1.00 L of 0.050 *M* hypobromous acid (HBrO) to form a buffer solution of pH 9.15? Assume that no volume change occurs when the NaBrO is added.

17.24 How many grams of sodium lactate ($NaC_3H_5O_3$) should be added to 1.00 L of 0.150 *M* lactic acid ($HC_3H_5O_3$) to form a buffer solution with pH 4.00? Assume that no volume change occurs when the $NaC_3H_5O_3$ is added.

17.25 A buffer solution contains 0.10 mol of acetic acid and 0.13 mol of sodium acetate in 1.00 L. **(a)** What is the pH of this buffer? **(b)** What is the pH of the buffer after the addition of 0.02 mol of KOH? **(c)** What is the pH of the buffer after the addition of 0.02 mol of HNO_3?

17.26 A buffer solution contains 0.12 mol of propionic acid ($HC_3H_5O_2$) and 0.10 mol of sodium propionate ($NaC_3H_5O_2$) in 1.50 L. **(a)** What is the pH of this buffer? **(b)** What is the pH of the buffer after the addition of 0.01 mol of NaOH? **(c)** What is the pH of the buffer after the addition of 0.01 mol of HI?

17.27 **(a)** What is the ratio of HCO_3^- to H_2CO_3 in blood of pH 7.4? **(b)** What is the ratio of HCO_3^- to H_2CO_3 in an exhausted marathon runner whose blood pH is 7.1?

17.28 A buffer, consisting of $H_2PO_4^-$ and HPO_4^{2-}, helps control the pH of physiological fluids. Many carbonated soft drinks also use this buffer system. What is the pH of a soft drink in which the major buffer ingredients are 6.5 g of NaH_2PO_4 and 8.0 g of Na_2HPO_4 per 355 mL of solution?

17.29 You have to prepare a pH 3.50 buffer, and you have the following 0.10 *M* solutions available: $HCHO_2$, $HC_2H_3O_2$, H_3PO_4, $NaCHO_2$, $NaC_2H_3O_2$, and NaH_2PO_4. Which solutions would you use? How many milliliters of each solution would you use to make approximately a liter of the buffer?

17.30 You have to prepare a pH 4.80 buffer, and you have the following 0.10 *M* solutions available: formic acid, sodium formate, propionic acid, sodium propionate, phosphoric acid, and sodium dihydrogen phosphate. Which solutions would you use? How many milliliters of each solution would you use to make approximately a liter of the buffer?

Acid-Base Titrations

17.31 The accompanying graph shows the titration curves for two monoprotic acids. **(a)** Which curve is that of a strong acid? **(b)** What is the approximate pH at the equivalence point of each titration? **(c)** How do the original concentrations of the two acids compare if 40.0 mL of each is titrated to the equivalence point with the same volume of 0.100 *M* base?

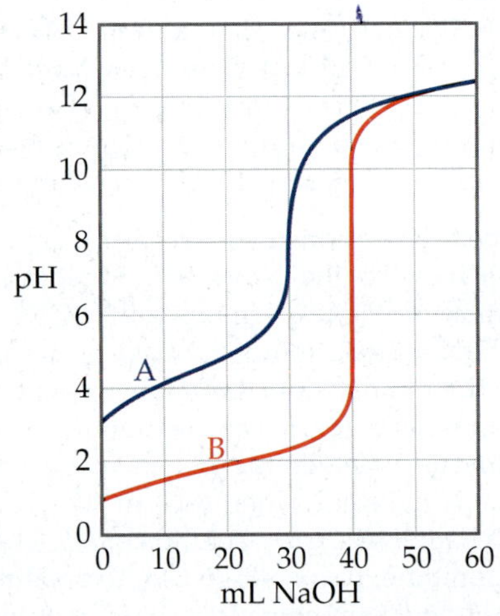

17.32 How does titration of a strong, monoprotic acid with a strong base differ from titration of a weak, monoprotic acid with a strong base with respect to the following: **(a)** quantity of base required to reach the equivalence point, **(b)** pH at the beginning of the titration, **(c)** pH at the equivalence point, **(d)** pH after addition of a slight excess of base, **(e)** choice of indicator for determining the equivalence point?

17.33 Predict whether the equivalence point of each of the following titrations is below, above, or at pH 7: **(a)** $NaHCO_3$ titrated with NaOH, **(b)** NH_3 titrated with HCl, **(c)** KOH titrated with HBr.

17.34 Predict whether the equivalence point of each of the following titrations is below, above, or at pH 7: **(a)** formic acid titrated with NaOH, **(b)** calcium hydroxide titrated with perchloric acid, **(c)** pyridine titrated with nitric acid.

17.35 Two monoprotic acids, both 0.100 *M* in concentration, are titrated with 0.100 *M* NaOH. The pH at the equivalence point for HX is 8.8, and that for HY is 7.9. **(a)** Which is the weaker acid? **(b)** Which indicators in Figure 16.7 could be used to titrate each of these acids?

17.36 Assume that 30.0 mL of a 0.10 *M* solution of a weak base B that accepts one proton is titrated with a 0.10 *M* solution of the monoprotic strong acid HX. **(a)** How many moles of HX have been added at the equivalence point? **(b)** What is the predominant form of B at the equivalence point? **(c)** What factor determines the pH at the equivalence point? **(d)** Which indicator, phenolphthalein or methyl red, is likely to be the better choice for this titration?

17.37 How many milliliters of 0.0850 *M* NaOH are required to titrate each of the following solutions to the equivalence point: **(a)** 40.0 mL of 0.0900 *M* HNO_3, **(b)** 35.0 mL of 0.0850 *M* $HC_2H_3O_2$, **(c)** 50.0 mL of a solution that contains 1.85 g of HCl per liter?

17.38 How many milliliters of 0.105 *M* HCl are needed to titrate each of the following solutions to the equivalence point: **(a)** 55.0 mL of 0.0950 *M* NaOH, **(b)** 22.5 mL of 0.118 *M* NH_3, **(c)** 125.0 mL of a solution that contains 1.35 g of NaOH per liter?

17.39 A 20.0-mL sample of 0.200 *M* HBr solution is titrated with 0.200 *M* NaOH solution. Calculate the pH of the solution after the following volumes of base have been added: **(a)** 15.0 mL, **(b)** 19.9 mL, **(c)** 20.0 mL, **(d)** 20.1 mL, **(e)** 35.0 mL.

17.40 A 30.0-mL sample of 0.150 *M* KOH is titrated with 0.125 *M* $HClO_4$ solution. Calculate the pH after the following volumes of acid have been added: **(a)** 30.0 mL, **(b)** 35.0 mL, **(c)** 36.0 mL, **(d)** 37.0 mL, **(e)** 40.0 mL.

17.41 A 35.0-mL sample of 0.150 *M* acetic acid ($HC_2H_3O_2$) is titrated with 0.150 *M* NaOH solution. Calculate the pH after the following volumes of base have been added: **(a)** 0 mL, **(b)** 17.5 mL, **(c)** 34.5 mL, **(d)** 35.0 mL, **(e)** 35.5 mL, **(f)** 50.0 mL.

17.42 Consider the titration of 30.0 mL of 0.030 *M* NH_3 with 0.025 *M* HCl. Calculate the pH after the following volumes of titrant have been added: **(a)** 0 mL, **(b)** 10.0 mL, **(c)** 20.0 mL, **(d)** 35.0 mL, **(e)** 36.0 mL, **(f)** 37.0 mL.

17.43 Calculate the pH at the equivalence point for titrating 0.200 *M* solutions of each of the following bases with 0.200 *M* HBr: **(a)** sodium hydroxide (NaOH), **(b)** hydroxylamine (NH_2OH), **(c)** aniline ($C_6H_5NH_2$).

17.44 Calculate the pH at the equivalence point in titrating 0.100 *M* solutions of each of the following with 0.080 *M* NaOH: **(a)** hydrobromic acid (HBr), **(b)** lactic acid ($HC_3H_5O_3$), **(c)** sodium hydrogen chromate ($NaHCrO_4$).

Solubility Equilibria and Factors Affecting Solubility

17.45 **(a)** Why is the concentration of undissolved solid not explicitly included in the expression for the solubility-product constant? **(b)** Write the expression for the solubility-product constant for each of the following strong electrolytes: AgI, $SrSO_4$, $Fe(OH)_2$, and Hg_2Br_2.

17.46 **(a)** Explain the difference between solubility and solubility-product constant. **(b)** Write the expression for the solubility-product constant for each of the following ionic compounds: $MnCO_3$, $Hg(OH)_2$, and $Cu_3(PO_4)_2$.

17.47 **(a)** If the molar solubility of CaF_2 at 35°C is 1.24×10^{-3} mol/L, what is K_{sp} at this temperature? **(b)** It is found that 1.1×10^{-2} g of SrF_2 dissolves per 100 mL of aqueous solution at 25°C. Calculate the solubility product for SrF_2. **(c)** The K_{sp} of $Ba(IO_3)_2$ at 25°C is 6.0×10^{-10}. What is the molar solubility of $Ba(IO_3)_2$?

17.48 **(a)** The molar solubility of $PbBr_2$ at 25°C is 1.0×10^{-2} mol/L. Calculate K_{sp}. **(b)** If 0.0490 g of $AgIO_3$ dissolves per liter of solution, calculate the solubility-product constant. **(c)** Using the appropriate K_{sp} value from Appendix D, calculate the solubility of $Cu(OH)_2$ in grams per liter of solution.

17.49 A 1.00-L solution saturated at 25°C with calcium oxalate (CaC_2O_4) contains 0.0061 g of CaC_2O_4. Calculate the solubility-product constant for this salt at 25°C.

17.50 A 1.00-L solution saturated at 25°C with lead(II) iodide contains 0.54 g of PbI_2. Calculate the solubility-product constant for this salt at 25°C.

17.51 Using Appendix D, calculate the molar solubility of AgBr in **(a)** pure water, **(b)** 3.0×10^{-2} *M* $AgNO_3$ solution, **(c)** 0.10 *M* NaBr solution.

17.52 Calculate the solubility of LaF_3 in grams per liter in **(a)** pure water, **(b)** 0.010 *M* KF solution, **(c)** 0.050 *M* $LaCl_3$ solution.

17.53 Calculate the solubility of $Mn(OH)_2$ in grams per liter when buffered at pH **(a)** 7.0, **(b)** 9.5, **(c)** 11.8.

17.54 Calculate the molar solubility of $Fe(OH)_2$ when buffered at pH **(a)** 7.0, **(b)** 10.0, **(c)** 12.0.

17.55 Which of the following salts will be substantially more soluble in acidic solution than in pure water: **(a)** $ZnCO_3$, **(b)** ZnS, **(c)** BiI_3, **(d)** AgCN, **(e)** $Ba_3(PO_4)_2$?

17.56 For each of the following slightly soluble salts, write the net ionic equation, if any, for reaction with acid: **(a)** MnS, **(b)** PbF_2, **(c)** $AuCl_3$, **(d)** $Hg_2C_2O_4$, **(e)** CuBr.

17.57 From the value of K_f listed in Table 17.1, calculate the concentration of Cu^{2+} in 1.0 L of a solution that contains a total of 1×10^{-3} mol of copper(II) ion and that is 0.10 *M* in NH_3.

17.58 To what final concentration of NH_3 must a solution be adjusted to just dissolve 0.020 mol of NiC_2O_4 ($K_{sp} = 4 \times 10^{-10}$) in 1.0 L of solution? [*Hint:* You can neglect the hydrolysis of $C_2O_4^{2-}$ because the solution will be quite basic.]

17.59 By using the values of K_{sp} for AgI and K_f for $Ag(CN)_2^-$, calculate the equilibrium constant for the reaction

$$AgI(s) + 2\,CN^-(aq) \rightleftharpoons Ag(CN)_2^-(aq) + I^-(aq)$$

17.60 Using the value of K_{sp} for Ag_2S, K_{a1} and K_{a2} for H_2S, and $K_f = 1.1 \times 10^5$ for $AgCl_2^-$, calculate the equilibrium constant for the following reaction:

$$Ag_2S(s) + 4\,Cl^-(aq) + 2\,H^+(aq) \rightleftharpoons 2\,AgCl_2^-(aq) + H_2S(aq)$$

Precipitation; Qualitative Analysis

17.61 **(a)** Will $Ca(OH)_2$ precipitate from solution if the pH of a 0.050 *M* solution of $CaCl_2$ is adjusted to 8.0? **(b)** Will Ag_2SO_4 precipitate when 100 mL of 0.050 *M* $AgNO_3$ is mixed with 10 mL of 5.0×10^{-2} *M* Na_2SO_4 solution?

17.62 **(a)** Will $Co(OH)_2$ precipitate from solution if the pH of a 0.020 *M* solution of $Co(NO_3)_2$ is adjusted to 8.5? **(b)** Will $AgIO_3$ precipitate when 100 mL of 0.010 *M* $AgNO_3$ is mixed with 10 mL of 0.015 *M* $NaIO_3$? (K_{sp} of $AgIO_3$ is 3.1×10^{-8}.)

17.63 Calculate the minimum pH needed to precipitate $Mn(OH)_2$ so completely that the concentration of Mn^{2+} is less than 1 μg per liter [1 part per billion (ppb)].

17.64 Suppose that a 10-mL sample of a solution is to be tested for Cl^- ion by addition of 1 drop (0.2 mL) of 0.10 *M* $AgNO_3$. What is the minimum number of grams of Cl^- that must be present in order for AgCl(*s*) to form?

17.65 A solution contains 2.0×10^{-4} *M* Ag^+ and 1.5×10^{-3} *M* Pb^{2+}. If NaI is added, will AgI ($K_{sp} = 8.3 \times 10^{-17}$) or PbI_2 ($K_{sp} = 7.9 \times 10^{-9}$) precipitate first? Specify the concentration of I^- needed to begin precipitation.

17.66 A solution of Na_2SO_4 is added dropwise to a solution that is 0.010 *M* in Ba^{2+} and 0.010 *M* in Sr^{2+}. **(a)** What concentration of SO_4^{2-} is necessary to begin precipitation? (Neglect volume changes. $BaSO_4$: $K_{sp} = 1.1 \times 10^{-10}$; $SrSO_4$: $K_{sp} = 3.2 \times 10^{-7}$.) **(b)** Which cation precipitates first? **(c)** What is the concentration of SO_4^{2-} when the second cation begins to precipitate?

17.67 A solution containing an unknown number of metal ions is treated with dilute HCl; no precipitate forms. The pH is adjusted to about 1, and H_2S is bubbled through. Again, no precipitate forms. The pH of the solution is then adjusted to about 8. Again, H_2S is bubbled through. This time a precipitate forms. The filtrate from this solution is treated with $(NH_4)_2HPO_4$. No precipitate forms. Which metal ions discussed in Section 17.7 are possibly present? Which are definitely absent within the limits of these tests?

17.68 An unknown solid is entirely soluble in water. On addition of dilute HCl, a precipitate forms. After the precipitate is filtered off, the pH is adjusted to about 1 and H_2S is bubbled in; a precipitate again forms. After filtering off this precipitate, the pH is adjusted to 8 and H_2S is again added; no precipitate forms. No precipitate forms upon addition of $(NH_4)_2HPO_4$. The remaining solution shows a yellow color in a flame test. Based on these observations, which of the following compounds might be present, which are definitely present, and which are definitely absent: CdS, $Pb(NO_3)_2$, HgO, $ZnSO_4$, $Cd(NO_3)_2$, and Na_2SO_4?

17.69 In the course of various qualitative analysis procedures, the following mixtures are encountered: **(a)** Zn^{2+} and Cd^{2+}, **(b)** $Cr(OH)_3$ and $Fe(OH)_3$, **(c)** Mg^{2+} and K^+, **(d)** Ag^+ and Mn^{2+}. Suggest how each mixture might be separated.

17.70 Suggest how the cations in each of the following solution mixtures can be separated: **(a)** Na^+ and Cd^{2+}, **(b)** Cu^{2+} and Mg^{2+}, **(c)** Pb^{2+} and Al^{3+}, **(d)** Ag^+ and Hg^{2+}.

17.71 **(a)** Precipitation of the group 4 cations (Figure 17.22) requires a basic medium. Why is this so? **(b)** What is the most significant difference between the sulfides precipitated in group 2 and those precipitated in group 3? **(c)** Suggest a procedure that would serve to redissolve the group 3 cations following their precipitation.

17.72 A student who is in a great hurry to finish his laboratory work decides that his qualitative analysis unknown contains a metal ion from the insoluble phosphate group, group 4 (Figure 17.22). He therefore tests his sample directly with $(NH_4)_2HPO_4$, skipping earlier tests for the metal ions in groups 1, 2, and 3. He observes a precipitate and concludes that a metal ion from group 4 is indeed present. Why is this possibly an erroneous conclusion?

Additional Exercises

17.73 Furoic acid ($HC_5H_3O_3$) has a K_a value of 6.76×10^{-4} at 25°C. Calculate the pH at 25°C of **(a)** a solution formed by adding 35.0 g of furoic acid and 30.0 g of sodium furoate ($NaC_5H_3O_3$) to enough water to form 0.250 L of solution; **(b)** a solution formed by mixing 30.0 mL of 0.250 *M* $HC_5H_3O_3$ and 20.0 mL of 0.22 *M* $NaC_5H_3O_3$ and diluting the total volume to 125 mL; **(c)** a solution prepared by adding 50.0 mL of 1.65 *M* NaOH solution to 0.500 L of 0.0850 *M* $HC_5H_3O_3$.

17.74 The acid-base indicator bromcresol green is a weak acid. The yellow acid and blue base forms of the indicator are present in equal concentrations in a solution when the pH is 4.68. What is the pK_a for bromcresol green?

17.75 Equal quantities of 0.010 M solutions of an acid HA and a base B are mixed. The pH of the resulting solution is 9.2. **(a)** Write the equilibrium equation and equilibrium-constant expression for the reaction between HA and B. **(b)** If K_a for HA is 8.0×10^{-5}, what is the value of the equilibrium constant for the reaction between HA and B? **(c)** What is the value of K_b for B?

17.76 Two buffers are prepared by adding an equal number of moles of formic acid ($HCHO_2$) and sodium formate ($NaCHO_2$) to enough water to make 1.00 L of solution. Buffer A is prepared using 1.00 mol each of formic acid and sodium formate. Buffer B is prepared by using 0.010 mol of each. **(a)** Calculate the pH of each buffer, and explain why they are equal. **(b)** Which buffer will have the greater buffer capacity? Explain. **(c)** Calculate the change in pH for each buffer upon the addition of 1.0 mL of 1.00 M HCl. **(d)** Calculate the change in pH for each buffer upon the addition of 10 mL of 1.00 M HCl. **(e)** Discuss your answers for parts (c) and (d) in light of your response to part (b).

17.77 A biochemist needs 750 mL of an acetic acid–sodium acetate buffer with pH 4.50. Solid sodium acetate ($NaC_2H_3O_2$) and glacial acetic acid ($HC_2H_3O_2$) are available. Glacial acetic acid is 99% $HC_2H_3O_2$ by mass and has a density of 1.05 g/mL. If the buffer is to be 0.20 M in $HC_2H_3O_2$, how many grams of $NaC_2H_3O_2$ and how many milliliters of glacial acetic acid must be used?

17.78 A sample of 0.2140 g of an unknown monoprotic acid was dissolved in 25.0 mL of water and titrated with 0.0950 M NaOH. The acid required 27.4 mL of base to reach the equivalence point. **(a)** What is the molar mass of the acid? **(b)** After 15.0 mL of base had been added in the titration, the pH was found to be 6.50. What is the K_a for the unknown acid?

17.79 Show that the pH at the halfway point of a titration of a weak acid with a strong base (where the volume of added base is half of that needed to reach the equivalence point) is equal to pK_a for the acid.

17.80 Potassium hydrogen phthalate, often abbreviated KHP, can be obtained in high purity and is used to determine the concentrations of solutions of strong bases. Strong bases react with the hydrogen phthalate ion as follows:

$$HP^-(aq) + OH^-(aq) \longrightarrow H_2O(l) + P^{2-}(aq)$$

The molar mass of KHP is 204.2 g/mol and K_a for the HP^- ion is 3.1×10^{-6}. **(a)** If a titration experiment begins with 0.4885 g of KHP and has a final volume of about 100 mL, which indicator from Figure 16.7 would be most appropriate? **(b)** If the titration required 38.55 mL of NaOH solution to reach the end point, what is the concentration of the NaOH solution?

17.81 If 40.00 mL of 0.100 M Na_2CO_3 is titrated with 0.100 M HCl, calculate **(a)** the pH at the start of the titration; **(b)** the volume of HCl required to reach the first equivalence point and the predominant species present at this point; **(c)** the volume of HCl required to reach the second equivalence point and the predominant species present at this point; **(d)** the pH at the second equivalence point.

17.82 A hypothetical weak acid, HA, was combined with NaOH in the following proportions: 0.20 mol of HA, 0.080 mol of NaOH. The mixture was diluted to a total volume of 1.0 L, and the pH measured. **(a)** If pH = 4.80, what is the pK_a of the acid? **(b)** How many additional moles of NaOH should be added to the solution to increase the pH to 5.00?

[17.83] What is the pH of a solution made by mixing 0.30 mol NaOH, 0.25 mol Na_2HPO_4, and 0.20 mol H_3PO_4 with water and diluting to 1.00 L?

[17.84] Suppose you want to do a physiological experiment that calls for a pH 6.5 buffer. You find that the organism with which you are working is not sensitive to the weak acid H_2X ($K_{a1} = 2 \times 10^{-2}$; $K_{a2} = 5.0 \times 10^{-7}$), or its sodium salts. You have available a 1.0 M solution of this acid and a 1.0 M solution of NaOH. How much of the NaOH solution should be added to 1.0 L of the acid to give a buffer at pH 6.50? (Ignore any volume change.)

[17.85] How many microliters of 1.000 M NaOH solution must be added to 25.00 mL of a 0.1000 M solution of lactic acid ($HC_3H_5O_3$) to produce a buffer with pH = 3.75?

17.86 A person suffering from anxiety begins breathing rapidly and as a result suffers alkalosis, an increase in blood pH. **(a)** Using Equation 17.10, explain how rapid breathing can cause the pH of blood to increase. **(b)** One cure for this problem is breathing in a paper bag. Why does this procedure lower blood pH?

17.87 For each pair of compounds, use K_{sp} values to determine which has the greater molar solubility: **(a)** CdS or CuS, **(b)** $PbCO_3$ or $BaCrO_4$, **(c)** $Ni(OH)_2$ or $NiCO_3$, **(d)** AgI or Ag_2SO_4.

17.88 A saturated solution of $Mg(OH)_2$ in water has a pH of 10.38. Estimate the K_{sp} for this compound.

17.89 Excess $Ca(OH)_2$ is shaken with water to produce a saturated solution. The solution is filtered, and a 50.00-mL sample titrated with HCl requires 11.23 mL of 0.0983 M HCl to reach the end point. Calculate K_{sp} for $Ca(OH)_2$. Compare your result with that in Appendix D. Do you think the solution was kept at 25°C?

17.90 The solubility-product constant for barium permanganate, $Ba(MnO_4)_2$, is 2.5×10^{-10}. Suppose that solid $Ba(MnO_4)_2$ is in equilibrium with a solution of $KMnO_4$. What concentration of $KMnO_4$ is required to establish a concentration of 2.0×10^{-8} M for the Ba^{2+} ion in solution?

17.91 Calculate the ratio of $[Ca^{2+}]$ to $[Fe^{2+}]$ in a lake in which the water is in equilibrium with deposits of both $CaCO_3$ and $FeCO_3$, assuming that the water is slightly basic and that the hydrolysis of the carbonate ion can therefore be ignored.

[17.92] The solubility products of $PbSO_4$ and $SrSO_4$ are 6.3×10^{-7} and 3.2×10^{-7}, respectively. What are the values of $[SO_4^{2-}]$, $[Pb^{2+}]$, and $[Sr^{2+}]$ in a solution at equilibrium with both substances?

[17.93] What pH buffer solution is needed to give a Mg^{2+} concentration of 3.0×10^{-2} M in equilibrium with solid magnesium oxalate?

[17.94] The value of K_{sp} for $Mg_3(AsO_4)_2$ is 2.1×10^{-20}. The $AsO_4{}^{3-}$ ion is derived from the weak acid H_3AsO_4 ($pK_{a1} = 2.22$; $pK_{a2} = 6.98$; $pK_{a3} = 11.50$). When asked to calculate the molar solubility of $Mg_3(AsO_4)_2$ in water, a student used the K_{sp} expression and assumed that $[Mg^{2+}] = 1.5[AsO_4{}^{3-}]$. Why was this a mistake?

[17.95] The solubility product for $Zn(OH)_2$ is 3.0×10^{-16}. The formation constant for the hydroxo complex, $Zn(OH)_4{}^{2-}$, is 4.6×10^{17}. What concentration of OH^- is required to dissolve 0.015 mol of $Zn(OH)_2$ in a liter of solution?

Integrative Exercises

17.96 **(a)** Write the net ionic equation for the reaction that occurs when a solution of hydrochloric acid (HCl) is mixed with a solution of sodium formate ($NaCHO_2$). **(b)** Calculate the equilibrium constant for this reaction. **(c)** Calculate the equilibrium concentrations of Na^+, Cl^-, H^+, $CHO_2{}^-$, and $HCHO_2$ when 50.0 mL of 0.15 *M* HCl is mixed with 50.0 mL of 0.15 *M* $NaCHO_2$.

17.97 **(a)** A 0.1044-g sample of an unknown monoprotic acid requires 22.10 mL of 0.0500 *M* NaOH to reach the end point. What is the molecular weight of the unknown? **(b)** As the acid is titrated, the pH of the solution after the addition of 11.05 mL of the base is 4.89. What is the K_a for the acid? **(c)** Using Appendix D, suggest the identity of the acid. Do both the molecular weight and K_a value agree with your choice?

17.98 A sample of 7.5 L of NH_3 gas at 22°C and 735 torr is bubbled into a 0.50-L solution of 0.40 *M* HCl. Assuming that all the NH_3 dissolves and that the volume of the solution remains 0.50 L, calculate the pH of the resulting solution.

17.99 Aspirin has the structural formula

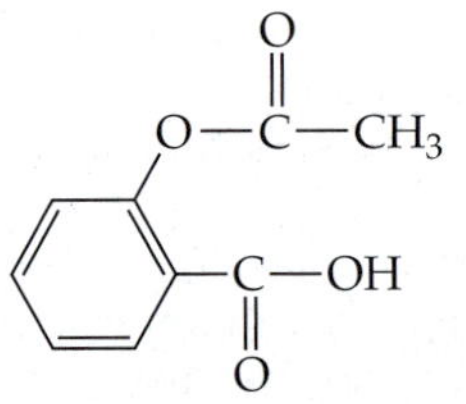

At body temperature (37°C), K_a for aspirin equals 3×10^{-5}. If two aspirin tablets, each having a mass of 325 mg, are dissolved in a full stomach whose volume is 1 L and whose pH is 2, what percent of the aspirin is in the form of neutral molecules?

17.100 What is the pH at 25°C of water saturated with CO_2 at a partial pressure of 1.10 atm? The Henry's law constant for CO_2 at 25°C is 3.1×10^{-2} mol/L-atm. The CO_2 is an acidic oxide, reacting with H_2O to form H_2CO_3.

17.101 The osmotic pressure of a saturated solution of strontium sulfate at 25°C is 21 torr. What is the solubility product of this salt at 25°C?

17.102 A concentration of 10–100 parts per billion (by mass) of Ag^+ is an effective disinfectant in swimming pools. However, if the concentration exceeds this range, the Ag^+ can cause adverse health effects. One way to maintain an appropriate concentration of Ag^+ is to add a slightly soluble salt to the pool. Using K_{sp} values from Appendix D, calculate the equilibrium concentration of Ag^+ in parts per billion that would exist in equilibrium with **(a)** AgCl, **(b)** AgBr, **(c)** AgI.

[17.103] Fluoridation of drinking water is employed in many places to aid in the prevention of dental caries. Typically the F^- ion concentration is adjusted to about 1 ppb. Some water supplies are also "hard"; that is, they contain certain cations such as Ca^{2+} that interfere with the action of soap. Consider a case where the concentration of Ca^{2+} is 8 ppb. Could a precipitate of CaF_2 form under these conditions? (Make any necessary approximations.)

eMEDIA EXERCISES

These exercises make use of the interactive objects available online in OneKey or the Companion Website, and on your Accelerator CD. Access to these resources comes in your MediaPak.

17.104 The **Common-Ion Effect** animation (*17.1*) shows the solubility of an iodide salt being reduced by the addition of sodium iodide. **(a)** How would you expect the solubility of the salt to be affected by the addition of a strong acid? **(b)** How would you expect it to be affected by the addition of a strong base? Explain your reasoning for both answers.

17.105 **(a)** Using data from the **Calculating pH Using Henderson–Hasselbalch Equation** activity (*17.2*), determine the pK_a of benzoic acid ($C_6H_5CO_2H$). **(b)** Using the pK_a that you determined in part (a), calculate the pH of a 0.0015 *M* aqueous solution of benzoic acid. **(c)** Will the pH of the solution in part (b) change if it is diluted with water to twice its original volume? Explain. **(d)** Will the pH of a buffer change when it is diluted with water to twice its original volume? If not, explain why not.

17.106 In Exercise 17.21 you calculated the pH of a buffer prepared by adding 20.0 g each of acetic acid and sodium acetate to enough water to make a 2.00-L solution. **(a)** Use the **Calculating pH Using Henderson–Hasselbalch Equation** activity (*17.2*) to prepare this buffer and verify your answer to Exercise 17.21. **(b)** Describe what is meant by the term *buffer capacity*. **(c)** Of which could you add more to the above buffer without causing a drastic change in pH, strong acid or strong base?

17.107 The **Dissolution of $Mg(OH)_2$ by Acid** animation (*17.5*) shows how the relatively insoluble solid $Mg(OH)_2$ can be made more soluble in water by the addition of acid. **(a)** Write the net ionic equation for the process by which magnesium hydroxide dissolves in neutral water. **(b)** Write the net ionic equation for the combination of two hydronium ions with two hydroxide ions. **(c)** Show that the net ionic equations in parts (a) and (b) add to give the overall net ionic equation shown in the animation. **(d)** Calculate the equilibrium constant for the process represented by the overall net ionic equation (i.e., dissolution of magnesium hydroxide in aqueous acid). **(e)** Determine the solubility of magnesium hydroxide in 0.010 *M* HCl.

eLABORATORY EXERCISES

These exercises allow students to apply the concepts and skills for this chapter in the simulated laboratory of Virtual ChemLab. Worksheets for each of these exercises are found in the Virtual ChemLab Workbook in your MediaPak.

17.108 ***(VCL 17-1) Study of Acid-Base Titrations: Monoprotic Acids***
Titrations provide a method of quantitatively measuring the concentration of an unknown solution. In an acid-base titration this is done by delivering a titrant of known concentration into an analyte of known volume. (The concentration of an unknown titrant can also be determined by titration with an analyte of known concentration and volume.) Titration curves (graphs of volume versus pH) have characteristic shapes. By comparison, the graph can be used to determine the strength or weakness of an acid or base. The equivalence point of the titration, or the point where the analyte has been completely consumed by the titrant, is identified by the point where the pH changes rapidly over a small volume of titrant delivered. In this assignment you will observe this titration curve by titrating the strong acid HCl with the strong base NaOH.

17.109 ***(VCL 17-2) Weak Acid–Strong Base Titrations***
In this assignment you will observe a titration curve associated with a weak acid and strong base by titrating the acid CH_3COOH (acetic acid) with the base NaOH. You will also predict the pH at the equivalence point, validate your prediction experimentally, and calculate the equilibrium constant for the neutralization reaction.

17.110 ***(VCL 17-3) Strong Acid–Weak Base Titrations***
In this assignment you will observe a titration curve associated with a strong acid and weak base by titrating the weak base $NaHCO_3$ (baking soda) with the strong acid HCl. You will also predict the pH at the equivalence point, validate your prediction experimentally, and calculate the equilibrium constant for the neutralization reaction.

17.111 ***(VCL 17-4) Weak Acid–Weak Base Titrations***
In this assignment you will observe a titration curve associated with a weak acid and weak base by titrating the acid CH_3COOH (acetic acid) with the weak base NH_3. You will also predict the pH at the equivalence point, validate your prediction experimentally, and calculate the equilibrium constant for the neutralization reaction.

17.112 ***(VCL 17-5) Study of Acid-Base Titrations: Polyprotic Acids***
Titrations provide a method of quantitatively measuring the concentration of an unknown solution. The equivalence point of the titration, or the point where the analyte has been completely consumed by the titrant, is identified by the point where the pH changes rapidly over a small volume of titrant delivered. For polyprotic acids, there will be multiple equivalence points. In this assignment you will observe a polyprotic acid titration curve by titrating the acid H_2SO_3 with the strong base NaOH.

17.113 ***(VCL 17-6) Acid-Base Standardization***
In this assignment you will determine the molarity of an unknown solution of NaOH by titrating against a primary standard, potassium hydrogen phthalate (KHP).

17.114 ***(VCL 17-7) Analysis of Baking Soda***
In this assignment you will determine the mass percent of an unknown sample of baking soda ($NaHCO_3$) by titrating with an NaOH solution of known concentration.

17.115 ***(VCL 17-8) Identification of Cations in Solution: Flame Tests***
The process of determining the composition of a sample by conducting chemical tests is called qualitative analysis. By using the appropriate tests and applying logic, the identities of the ions present in an unknown solution can be determined. A qualitative analysis scheme is typically made up of a systematic set of chemical reactions where a certain subset of the ions present in the solution are selectively precipitated and removed. The colors of the precipitates and solutions provide the means to identify the ions present. Flame tests are also used to identify certain ions that are difficult to identify chemically. In this assignment you will learn how flame tests are used to identify the cations Na^+ and K^+. As you complete this analysis, remember that careful observation and logical reasoning are the keys to a successful qualitative analysis.

17.116 ***(VCL 17-9) Identification of Cations in Solution: Ag^+, Hg_2^{2+}, Pb^{2+}***
In this assignment you will learn the basics of a qualitative analysis scheme by performing an analysis on a mixture of Ag^+, Hg_2^{2+}, and Pb^{2+}.

17.117 ***(VCL 17-10) Identification of Cations in Solution: Co^{2+}, Cr^{2+}, Cu^{2+}***
In this assignment you will learn the basics of a qualitative analysis scheme by performing an analysis on a mixture of Co^{2+}, Cr^{3+}, and Cu^{2+}.

17.118 ***(VCL 17-11) Identification of Cations in Solution: Ba^{2+}, Sr^{2+}, Ca^{2+}, Mg^{2+}***
In this assignment you will need to develop your own qualitative analysis scheme to separate and identify the group II cations Ba^{2+}, Sr^{2+}, Ca^{2+}, and Mg^{2+}.

17.119 ***(VCL 17-12) Identification of Cations in Solution: Co^{2+}, Cu^{2+}, Ni^{2+}***
In this assignment you will need to develop your own qualitative analysis scheme to separate and identify the transition-metal cations Co^{2+}, Cu^{2+}, and Ni^{2+}.

CHAPTER

18

WHAT'S AHEAD

- This chapter begins with a general look at Earth's atmosphere. We'll study the temperature profile of the atmosphere, its pressure profile, and its chemical composition. *(Section 18.1)*
- The outer regions of the atmosphere, where the pressure is very low, absorb a great deal of high-energy radiation from the Sun through *photoionization* and *photodissociation* reactions. By filtering out high-energy radiation, these processes make it possible for life as we know it to exist on Earth. *(Section 18.2)*
- Ozone in the *stratosphere* acts as a filter of high-energy ultraviolet light. Human activities have contributed to the depletion of the ozone layer by introducing into the stratosphere chemicals that perturb the natural cycle of ozone formation and decomposition. Notable among these are *chlorofluorocarbons*. *(Section 18.3)*
- The innermost region of the atmosphere, the *troposphere*, is the region in which we live. Many minor constituents of the troposphere affect air quality and the acidity of rainwater. The concentrations of many of these minor constituents, including those that cause *acid rain* and *photochemical smog*, have been increased by human activities. *(Section 18.4)*
- Carbon dioxide is an important minor constituent of the atmosphere because it acts as a "greenhouse" gas; that is, it causes a warming of Earth's atmosphere. The combustion of fossil fuels (coal, oil, and natural gas) is projected to cause a doubling of atmospheric CO_2 concentration over the estimated preindustrial level by about 2050. This doubling is expected to cause significant warming of the atmosphere, with attendant climatic changes. *(Section 18.4)*
- Almost all the water on Earth is in the world ocean. Seawater contains many salts that participate in the global cycling of the elements and nutrients. *(Section 18.5)*
- Processing seawater to obtain fresh water is energy-intensive. We count on freshwater sources to supply most of our needs, but these sources often require treatment to render them usable. *(Section 18.6)*
- *Green chemistry* is an international initiative to make all industrial products, processes, and chemical reactions compatible with a sustainable society and environment. We'll examine some reactions and processes in which the goals of green chemistry can be advanced. *(Section 18.7)*

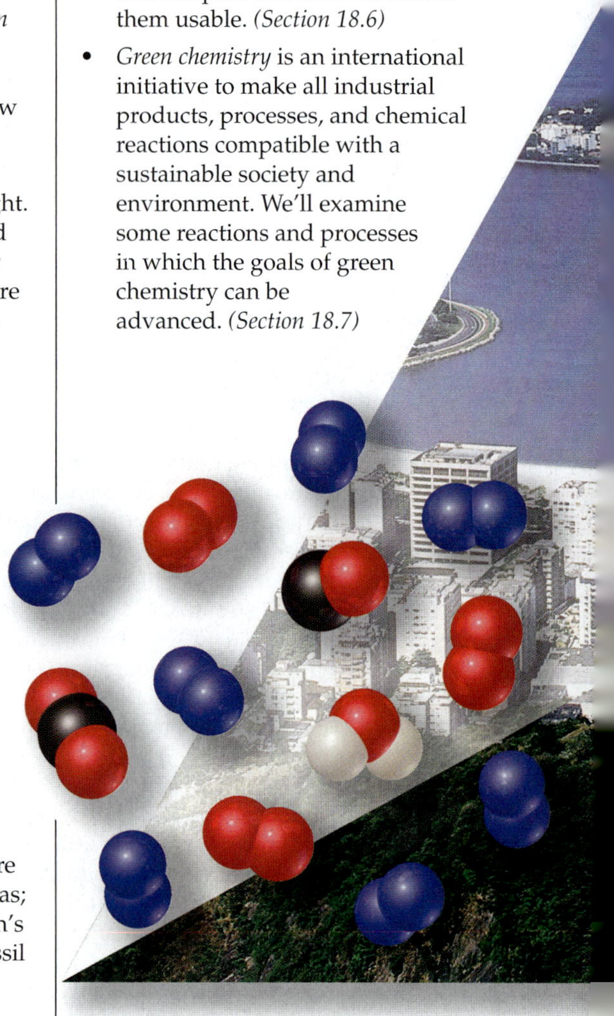

Chemistry of the Environment

RIO DE JANEIRO

IN 1992 REPRESENTATIVES OF 172 COUNTRIES MET IN RIO DE JANEIRO, Brazil, for the United Nations Conference on Environment and Development—a conference that became known as the Earth Summit. Five years later, in December 1997, representatives of 130 nations met in Kyoto, Japan, to discuss the impact of human activities on global warming. Out of that meeting came an initiative to work toward a global treaty that would, among other things, spell out actions to be taken to reduce emissions of gases that cause global warming. In July 2001 in Bonn, Germany, 178 nations signed a treaty based on the so-called Kyoto Protocols.* These efforts to address environmental concerns at the international level indicate that many of the most urgent environmental problems are global in nature.

The economic growth of both developed and underdeveloped nations depends critically on chemical processes. These range from treatment of water supplies to industrial processes, some of which produce products or by-products that are harmful to the environment. We are now in a position to apply the principles we have learned in earlier chapters to an understanding of these processes. In this chapter we consider some aspects of the chemistry of our environment, focusing on Earth's atmosphere and water.

Both the air and water of our planet make life as we know it possible. Water crystallizes to ice high in the atmosphere, and plays an important role in

* The United States, almost alone among nations, refused to sign the treaty.

chemical reactions in the atmosphere. In order to understand and maintain the environment in which we live, we must understand how human-made and natural chemical compounds interact on land, sea, and sky. Our daily decisions as consumers mirror those of the leaders meeting in Bonn and similar international meetings: We must weigh the costs versus the benefits of our actions. Unfortunately, the environmental impacts of our decisions are often very subtle and not immediately evident.

18.1 | Earth's Atmosphere

Because most of us have never been very far from Earth's surface, we tend to take for granted the many ways in which the atmosphere determines the environment in which we live. In this section we will examine some of the important characteristics of our planet's atmosphere.

The temperature of the atmosphere varies in a complex manner as a function of altitude, as shown in Figure 18.1(a) ▼. The atmosphere is divided into four regions based on this temperature profile. Just above the surface, in the **troposphere**, the temperature normally decreases with increasing altitude, reaching a minimum of about 215 K at about 12 km. Nearly all of us live out our entire lives in the troposphere. Howling winds and soft breezes, rain, sunny skies—all that we normally think of as "weather"—occur in this region. Commercial jet aircraft typically fly about 10 km (33,000 ft) above Earth, an altitude that approaches the upper limit of the troposphere, which we call the *tropopause*.

Above the tropopause the temperature increases with altitude, reaching a maximum of about 275 K at about 50 km. The region from 10 km to 50 km is called the **stratosphere**. Beyond the stratosphere are the *mesosphere* and the *thermosphere*. Notice in Figure 18.1 that the temperature extremes that form the boundaries between adjacent regions are denoted by the suffix *-pause*. The boundaries are important because gases mix across them relatively slowly. For example, pollutant gases generated in the troposphere find their way into the stratosphere only very slowly.

Unlike temperature, the pressure of the atmosphere decreases in a regular way with increasing elevation, as shown in Figure 18.1(b). Atmospheric pressure drops off much more rapidly at lower elevations than at higher ones because of the atmosphere's compressibility. Thus, the pressure decreases from an average value of 760 torr (101 kPa) at sea level to 2.3×10^{-3} torr (3.1×10^{-4} kPa) at 100 km, to only 1.0×10^{-6} torr (1.3×10^{-7} kPa) at 200 km. The troposphere and stratosphere together account for 99.9% of the mass of the atmosphere, with 75% of the mass in the troposphere.

▼ **Figure 18.1 Temperature and pressure in the atmosphere.** (a) Temperature variations in the atmosphere at altitudes below 110 km. (b) Variations in atmospheric pressure with altitude. At 80 km the pressure is approximately 0.01 torr.

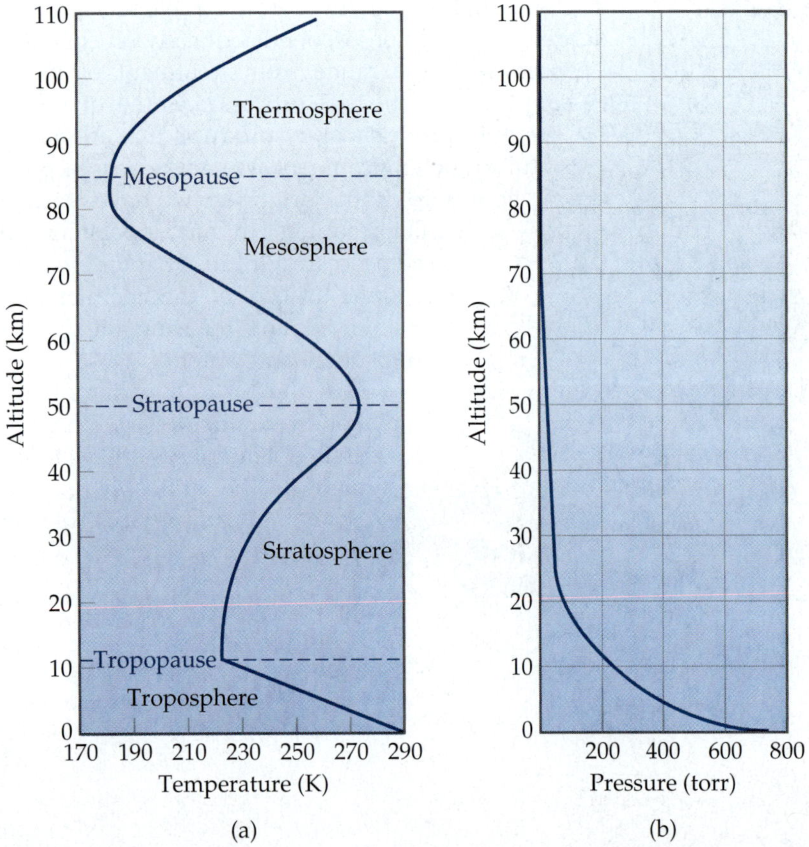

Composition of the Atmosphere

The atmosphere is an extremely complex system. Its temperature and pressure change over a wide range with altitude, as we have just seen. The atmosphere is bombarded by radiation and energetic particles from the Sun. This barrage of energy has profound chemical effects, especially on the outer reaches of the atmosphere (Figure 18.2 ▶). In addition, because of Earth's gravitational field, lighter atoms and molecules tend to rise to the top. This is especially true in the stratosphere and upper layers, which are not as turbulent as the troposphere. As a result of all these factors, the composition of the atmosphere is not uniform.

Table 18.1 ▼ shows the composition by mole fraction of dry air near sea level. Although traces of many substances are present, N_2 and O_2 make up about 99% of the atmosphere. The noble gases and CO_2 make up most of the remainder.

When speaking of trace constituents, we commonly use *parts per million* (ppm) as the unit of concentration. When applied to substances in aqueous solution, parts per million refers to grams of substance per million grams of solution. (Section 13.2) When dealing with gases, however, one part per million refers to one part by *volume* in 1 million volume units of the whole. Because volume (V) is proportional to the number of moles n of gas via the ideal-gas equation ($PV = nRT$), volume fraction and mole fraction are the same. Thus, 1 ppm of a trace constituent of the atmosphere amounts to one mole of that constituent in 1 million moles of total gas; that is, the concentration in ppm is equal to the mole fraction times 10^6. Table 18.1 lists the mole fraction of CO_2 in the atmosphere as 0.000375. Its concentration in ppm is therefore $0.000375 \times 10^6 = 375$ ppm.

Before we consider the chemical processes that occur in the atmosphere, let's review some of the important chemical properties of the two major components, N_2 and O_2. Recall that the N_2 molecule possesses a triple bond between the nitrogen atoms. (Section 8.3) This very strong bond is largely responsible for the very low reactivity of N_2, which undergoes reaction only under extreme conditions. The bond energy in O_2, 495 kJ/mol, is much lower than that in N_2, which is 941 kJ/mol. (Table 8.4) Therefore, O_2 is much more reactive than N_2. For example, oxygen reacts with many substances to form oxides. The oxides of nonmetals, for example SO_2, usually form acidic solutions when dissolved in water. The oxides of active metals, for example CaO, form basic solutions when dissolved in water. (Section 7.6)

▲ **Figure 18.2 The aurora borealis.** This luminous display in the northern sky, also called the northern lights, is produced by collisions of high-speed electrons and protons from the Sun with air molecules. The charged particles are channeled toward the polar regions by Earth's magnetic field.

TABLE 18.1 Composition of Dry Air Near Sea Level

Component*	Content (mole fraction)	Molar Mass
Nitrogen	0.78084	28.013
Oxygen	0.20948	31.998
Argon	0.00934	39.948
Carbon dioxide	0.000375	44.0099
Neon	0.00001818	20.183
Helium	0.00000524	4.003
Methane	0.000002	16.043
Krypton	0.00000114	83.80
Hydrogen	0.0000005	2.0159
Nitrous oxide	0.0000005	44.0128
Xenon	0.000000087	131.30

*Ozone, sulfur dioxide, nitrogen dioxide, ammonia, and carbon monoxide are present as trace gases in variable amounts.

SAMPLE EXERCISE 18.1 | Calculating the Concentration of Water in Air

What is the concentration, in parts per million, of water vapor in a sample of air if the partial pressure of the water is 0.80 torr and the total pressure of the air is 735 torr?

Solution

Analyze: We are given the partial pressure of water vapor and the total pressure of an air sample and asked to determine the water vapor concentration.

Plan: Recall that the partial pressure of a component in a mixture of gases is given by the product of its mole fraction and the total pressure of the mixture (Section 10.6):

$$P_{H_2O} = X_{H_2O}P_t$$

Solve: Solving for the mole fraction of water vapor in the mixture, X_{H_2O}, gives:

$$X_{H_2O} = \frac{P_{H_2O}}{P_t} = \frac{0.80 \text{ torr}}{735 \text{ torr}} = 0.0011$$

The concentration in ppm is the mole fraction times 10^6:

$$0.0011 \times 10^6 = 1100 \text{ ppm}$$

PRACTICE EXERCISE

The concentration of CO in a sample of air is found to be 4.3 ppm. What is the partial pressure of the CO if the total air pressure is 695 torr?

Answer: 3.0×10^{-3} torr

18.2 | Outer Regions of the Atmosphere

Although the outer portion of the atmosphere, beyond the stratosphere, contains only a small fraction of the atmospheric mass, it forms the outer defense against the hail of radiation and high-energy particles that continuously bombard Earth. As the bombarding molecules and atoms pass through the upper atmosphere, they undergo chemical changes.

Photodissociation

The Sun emits radiant energy over a wide range of wavelengths (Figure 18.3 ◀). The shorter-wavelength, higher-energy radiations in the ultraviolet range of the spectrum are sufficiently energetic to cause chemical changes. Recall that electromagnetic radiation can be pictured as a stream of photons. (Section 6.2) The energy of each photon is given by the relationship $E = h\nu$, where h is Planck's constant and ν is the frequency of the radiation. For a chemical change to occur when radiation falls on Earth's atmosphere, two conditions must be met. First, there must be photons with energy sufficient to accomplish whatever chemical process is being considered. Second, molecules must absorb these photons. When these requirements are met, the energy of the photons is converted into some other form of energy within the molecule.

The rupture of a chemical bond resulting from absorption of a photon by a molecule is called **photodissociation**. No ions are formed when the bond between two atoms is cleaved by photodissociation. Instead, half the bonding electrons stay with one of the atoms and half stay with the other atom. The result is two neutral particles.

▼ **Figure 18.3 The solar spectrum.** Shown in this graph is the amount of sunlight (in light energy per area per time, known as flux) at different wavelengths that reaches the top of Earth's atmosphere. For comparison, the corresponding data for the amount of sunlight that reaches sea level is plotted. The atmosphere absorbs much of the ultraviolet and visible light emitted by the Sun.

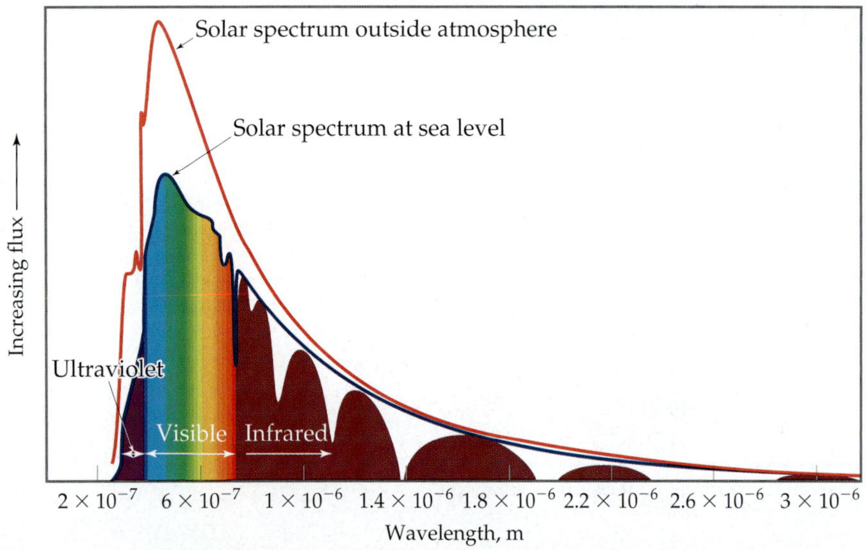

One of the most important processes occurring in the upper atmosphere above about 120-km elevation is the photodissociation of the oxygen molecule:

$$:\ddot{O}=\ddot{O}: + h\nu \longrightarrow :\ddot{O} + \ddot{O}: \qquad [18.1]$$

The minimum energy required to cause this change is determined by the bond energy (or *dissociation energy*) of O_2, 495 kJ/mol. In Sample Exercise 18.2 we calculate the longest wavelength photon having sufficient energy to photodissociate the O_2 molecule.

SAMPLE EXERCISE 18.2 | Calculating the Wavelength Required to Break a Bond

What is the maximum wavelength of light, in nanometers, that has enough energy per photon to dissociate the O_2 molecule?

Solution

Analyze: We are asked to determine the wavelength of a photon that has just sufficient energy to break the O=O double bond in O_2.

Plan: We first need to calculate the energy required to break the O=O double bond in one molecule, then find the wavelength of a photon of this energy.

Solve: The dissociation energy of O_2 is 495 kJ/mol. Using this value and Avogadro's number, we can calculate the amount of energy needed to break the bond in a single O_2 molecule:

$$\left(495 \times 10^3 \frac{\text{J}}{\text{mol}}\right)\left(\frac{1 \text{ mol}}{6.022 \times 10^{23} \text{ molecules}}\right) = 8.22 \times 10^{-19} \frac{\text{J}}{\text{molecule}}$$

We next use the Planck relationship, $E = h\nu$, (Equation 6.2) to calculate the frequency, ν, of a photon that has this amount of energy:

$$\nu = \frac{E}{h} = \frac{8.22 \times 10^{-19} \text{ J}}{6.626 \times 10^{-34} \text{ J-s}} = 1.24 \times 10^{15} \text{ s}^{-1}$$

Finally, we use the relationship between the frequency and wavelength of light (Section 6.1) to calculate the wavelength of the light:

$$\lambda = \frac{c}{\nu} = \left(\frac{3.00 \times 10^8 \text{ m/s}}{1.24 \times 10^{15} \text{ /s}}\right)\left(\frac{10^9 \text{ nm}}{1 \text{ m}}\right) = 242 \text{ nm}$$

Thus, light of wavelength 242 nm, which is in the ultraviolet region of the electromagnetic spectrum, has sufficient energy per photon to photodissociate an O_2 molecule. Because photon energy increases as wavelength *decreases*, any photon of wavelength *shorter* than 242 nm will have sufficient energy to dissociate O_2.

PRACTICE EXERCISE

The bond energy in N_2 is 941 kJ/mol (Table 8.4). What is the longest wavelength a photon can have and still have sufficient energy to dissociate N_2?

Answer: 127 nm

Fortunately for us, O_2 absorbs much of the high-energy, short-wavelength radiation from the solar spectrum before that radiation reaches the lower atmosphere. As it does, atomic oxygen, $:\ddot{O}$, is formed. At higher elevations the dissociation of O_2 is very extensive. At 400 km, for example, only 1% of the oxygen is in the form of O_2; the other 99% is atomic oxygen. At 130 km, O_2 and $:\ddot{O}$ are just about equally abundant. Below 130 km, O_2 is more abundant than atomic oxygen because most of the solar energy has been absorbed in the upper atmosphere.

The dissociation energy of N_2 is very high (Table 8.4). As shown in the Practice Exercise in Sample Exercise 18.2, only photons of very short wavelength possess sufficient energy to dissociate N_2. Furthermore, N_2 does not readily absorb photons, even when they possess sufficient energy. As a result, very little atomic nitrogen is formed in the upper atmosphere by photodissociation of N_2.

TABLE 18.2 Ionization Processes, Ionization Energies, and Maximum Wavelengths Capable of Causing Ionization

Process	Ionization Energy (kJ/mol)	λ_{max}(nm)
$N_2 + h\nu \longrightarrow N_2^+ + e^-$	1495	80.1
$O_2 + h\nu \longrightarrow O_2^+ + e^-$	1205	99.3
$O + h\nu \longrightarrow O^+ + e^-$	1313	91.2
$NO + h\nu \longrightarrow NO^+ + e^-$	890	134.5

Photoionization

In 1901 Guglielmo Marconi received a radio signal in St. John's, Newfoundland, that had been transmitted from Land's End, England, some 2900 km away. Because radio waves were thought to travel in straight lines, it had been assumed that, because of the curvature of Earth's surface, radio communication over large distances on Earth would be impossible. Marconi's successful experiment suggested that Earth's atmosphere in some way substantially affects radio-wave propagation. His discovery led to intensive study of the upper atmosphere. In about 1924 the existence of electrons in the upper atmosphere was established by experimental studies.

For each electron present in the upper atmosphere, there must be a corresponding positively charged particle. The electrons in the upper atmosphere result mainly from the **photoionization** of molecules, caused by solar radiation. Photoionization occurs when a molecule absorbs radiation and the absorbed energy causes an electron to be ejected from the molecule. The molecule then becomes a positively charged ion. For photoionization to occur, therefore, a molecule must absorb a photon and the photon must have enough energy to remove an electron. (Section 7.4)

Some of the more important ionization processes occurring in the atmosphere above about 90 km are shown in Table 18.2 ▲, together with the ionization energies and λ_{max}, the maximum wavelength of a photon capable of causing ionization. Photons with energies sufficient to cause ionization have wavelengths in the high-energy end of the ultraviolet region of the electromagnetic spectrum. These wavelengths are completely filtered out of the radiation reaching Earth, because they are absorbed by the upper atmosphere.

18.3 | Ozone in the Upper Atmosphere

ANIMATION
Stratospheric Ozone

While N_2, O_2, and atomic oxygen absorb photons having wavelengths shorter than 240 nm, ozone, O_3, is the key absorber of photons having wavelengths ranging from 240 to 310 nm. Let's consider how ozone forms in the upper atmosphere and how it absorbs photons.

Below an altitude of 90 km, most of the short-wavelength radiation capable of photoionization has been absorbed. Radiation capable of dissociating the O_2 molecule is sufficiently intense, however, for photodissociation of O_2 (Equation 18.1) to remain important down to an altitude of 30 km. In the region between 30 and 90 km, the concentration of O_2 is much greater than that of atomic oxygen. Therefore, the $:\ddot{O}$ atoms that form in this region undergo frequent collisions with O_2 molecules, resulting in the formation of ozone, O_3:

$$:\ddot{O} + O_2 \longrightarrow O_3^* \qquad [18.2]$$

The asterisk over the O_3 denotes that the ozone molecule contains an excess of energy. The reaction in Equation 18.2 releases 105 kJ/mol. This energy must be transferred away from the O_3^* molecule in a very short time or else the molecule will fly apart again into O_2 and $:\ddot{O}$—a decomposition that is the reverse of the process by which O_3^* is formed.

An energy-rich O_3^* molecule can release its excess energy by colliding with another atom or molecule and transferring some of the excess energy to it. Let's represent the atom or molecule with which O_3^* collides as M. (Usually M is N_2 or O_2 because these are the most abundant molecules in the atmosphere.) The formation of $O_3{}^*$ and the transfer of excess energy to M are summarized by the equations (where O atoms are shown without valence electrons)

$$O(g) + O_2(g) \rightleftharpoons O_3{}^*(g) \quad [18.3]$$

$$O_3{}^*(g) + M(g) \longrightarrow O_3(g) + M^*(g) \quad [18.4]$$

$$O(g) + O_2(g) + M(g) \longrightarrow O_3(g) + M^*(g) \quad [18.5]$$

The rate at which O_3 forms according to Equations 18.3 and 18.4 depends on two factors that vary in opposite directions with increasing altitude. First, the formation of $O_3{}^*$, according to Equation 18.3, depends on the presence of O atoms. At low altitudes most of the radiation energetic enough to dissociate O_2 has been absorbed; thus, the formation of O is favored at higher altitudes. Second, both Equations 18.3 and 18.4 depend on molecular collisions. (Section 14.5) The concentration of molecules is greater at low altitudes, however, and so the frequency of collisions between O and O_2 (Equation 18.3) and between $O_3{}^*$ and M (Equation 18.4) are both greater at lower altitudes. Because these processes vary with altitude in opposite directions, the highest rate of O_3 formation occurs in a band at an altitude of about 50 km, near the stratopause [Figure 18.1(a)]. Overall, roughly 90% of Earth's ozone is found in the stratosphere, between the altitudes of 10 and 50 km.

Once formed, an ozone molecule does not last long. Ozone is capable of absorbing solar radiation, which decomposes it back into O_2 and O. Because only 105 kJ/mol is required for this process, photons of wavelength shorter than 1140 nm are sufficiently energetic to photodissociate O_3. However, most of the output from the Sun is concentrated in the visible and ultraviolet portions of the electromagnetic spectrum (Figure 18.3). Photons with wavelengths shorter than about 300 nm are energetic enough to break many single chemical bonds. If it were not for the layer of ozone in the stratosphere, these high-energy photons would penetrate to Earth's surface. Plant and animal life as we know it could not survive in the presence of this high-energy radiation. The "ozone shield" is therefore essential for our continued well-being. The ozone molecules that form this essential shield against radiation represent only a tiny fraction of the oxygen atoms present in the stratosphere, however, because they are continually destroyed even as they are formed.

GIVE IT SOME THOUGHT

Look at Figure 18.3. Based on the relative areas under the curves, what fraction of the radiation with wavelengths in the ultraviolet portion of the spectrum is absorbed by the upper atmosphere?

The photodecomposition of ozone reverses the reaction that forms it. We thus have a cyclic process of ozone formation and decomposition, summarized as follows:

$$O_2(g) + h\nu \longrightarrow O(g) + O(g)$$

$$O(g) + O_2(g) + M(g) \longrightarrow O_3(g) + M^*(g) \quad \text{(heat released)}$$

$$O_3(g) + h\nu \longrightarrow O_2(g) + O(g)$$

$$O(g) + O(g) + M(g) \longrightarrow O_2(g) + M^*(g) \quad \text{(heat released)}$$

The first and third processes are photochemical; they use a solar photon to initiate a chemical reaction. The second and fourth processes are exothermic chemical reactions. The net result of all four processes is a cycle in which solar radiant energy is converted into thermal energy. The ozone cycle in the stratosphere is responsible for the rise in temperature that reaches its maximum at the stratopause, as illustrated in Figure 18.1(a).

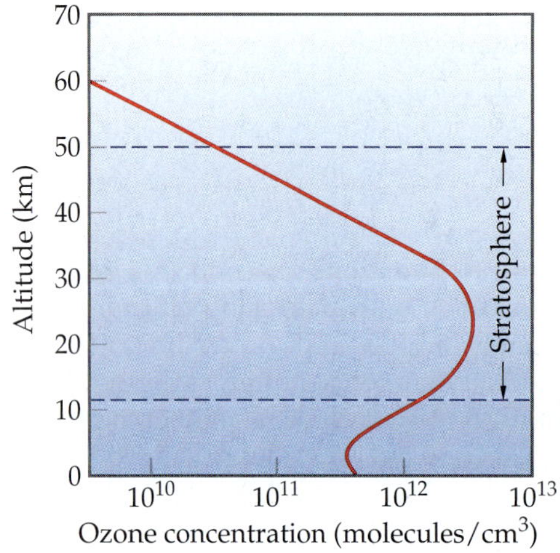

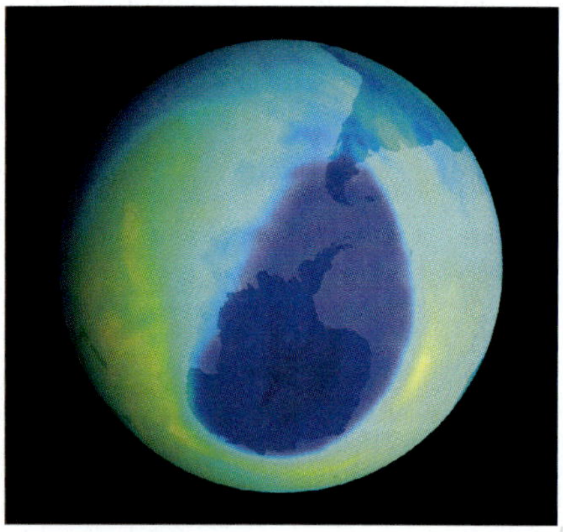

▶ **Figure 18.4 Ozone depletion.**
(Left) Variation in ozone concentration in the atmosphere, as a function of altitude.
(Right) Map of the total ozone present in the Southern Hemisphere, taken in 2004 from an orbiting satellite. The different colors represent different ozone concentrations. The center area, which is over Antarctica, is the area of lowest ozone concentration.

ANIMATIONS
Catalytic Destruction of Stratospheric Ozone, CFCs and Stratospheric Ozone

The scheme described for the formation and decomposition of ozone molecules accounts for some, but not all, of the facts about the ozone layer. Many chemical reactions occur that involve substances other than just oxygen. In addition, the effects of turbulence and winds that mix up the stratosphere must be considered. A very complicated picture results. The overall result of ozone formation and removal reactions, coupled with atmospheric turbulence and other factors, is to produce an ozone profile in the upper atmosphere as shown on the left in Figure 18.4 ▲.

Depletion of the Ozone Layer

In 1995 the Nobel Prize in Chemistry was awarded to F. Sherwood Rowland, Mario Molina, and Paul Crutzen for their studies of ozone depletion in the stratosphere. In 1970 Crutzen showed that naturally occurring nitrogen oxides catalytically destroy ozone. Rowland and Molina recognized in 1974 that chlorine from **chlorofluorocarbons** (CFCs) may deplete the ozone layer that protects Earth's surface from damaging ultraviolet radiation. These substances, principally $CFCl_3$ and CF_2Cl_2, have been widely used as propellants in spray cans, as refrigerant and air-conditioner gases, and as foaming agents for plastics. They are virtually unreactive in the lower atmosphere. Furthermore, they are relatively insoluble in water and are therefore not removed from the atmosphere by rainfall or by dissolution in the oceans. Unfortunately, the lack of reactivity that makes them commercially useful also allows them to survive in the atmosphere and to diffuse eventually into the stratosphere. It is estimated that several million tons of chlorofluorocarbons are now present in the atmosphere.

As CFCs diffuse into the stratosphere, they are exposed to high-energy radiation, which can cause photodissociation. The C—Cl bonds are considerably weaker than the C—F bonds (Table 8.4). As a result, free chlorine atoms are formed readily in the presence of light with wavelengths in the range of 190 to 225 nm, as shown in this equation for CF_2Cl_2:

$$CF_2Cl_2(g) + h\nu \longrightarrow CF_2Cl(g) + Cl(g) \qquad [18.6]$$

Calculations suggest that chlorine atom formation occurs at the greatest rate at an altitude of about 30 km.

Atomic chlorine reacts rapidly with ozone to form chlorine monoxide (ClO) and molecular oxygen (O_2):

$$Cl(g) + O_3(g) \longrightarrow ClO(g) + O_2(g) \qquad [18.7]$$

Equation 18.7 follows a second-order rate law with a very large rate constant:

$$\text{Rate} = k[\text{Cl}][\text{O}_3] \quad k = 7.2 \times 10^9\ M^{-1}\text{s}^{-1} \text{ at } 298\text{ K} \qquad [18.8]$$

Under certain conditions the ClO generated in Equation 18.7 can react to regenerate free Cl atoms. One way that this can happen is by photodissociation of the ClO:

$$\text{ClO}(g) + h\nu \longrightarrow \text{Cl}(g) + \text{O}(g) \qquad [18.9]$$

The Cl atoms generated in Equation 18.9 can react with more O_3, according to Equation 18.7. These two equations form a cycle for the Cl atom–catalyzed decomposition of O_3 to O_2 as we see when we add the equations:

$$2\,\text{Cl}(g) + 2\,\text{O}_3(g) \longrightarrow 2\,\text{ClO}(g) + 2\,\text{O}_2(g)$$

$$2\,\text{ClO}(g) + h\nu \longrightarrow 2\,\text{Cl}(g) + 2\,\text{O}(g)$$

$$\text{O}(g) + \text{O}(g) \longrightarrow \text{O}_2(g)$$

$$2\,\text{Cl}(g) + 2\,\text{O}_3(g) + 2\,\text{ClO}(g) + 2\,\text{O}(g) \longrightarrow 2\,\text{Cl}(g) + 2\,\text{ClO}(g) + 3\,\text{O}_2(g) + 2\,\text{O}(g)$$

The equation can be simplified by eliminating like species from each side to give

$$2\,\text{O}_3(g) \xrightarrow{\text{Cl}} 3\,\text{O}_2(g) \qquad [18.10]$$

Because the rate of Equation 18.7 increases linearly with [Cl], the rate at which ozone is destroyed increases as the quantity of Cl atoms increases. Thus, the greater the amount of CFCs that diffuse into the stratosphere, the faster the destruction of the ozone layer. Rates of diffusion of molecules from the troposphere into the stratosphere are slow. Nevertheless, a thinning of the ozone layer over the South Pole has already been observed, particularly during the months of September and October (Figure 18.4, right).

Because of the environmental problems associated with CFCs, steps have been taken to limit their manufacture and use. A major step was the signing in 1987 of the Montreal Protocol on Substances That Deplete the Ozone Layer, in which participating nations agreed to reduce CFC production. More stringent limits were set in 1992, when representatives of approximately 100 nations agreed to ban the production and use of CFCs by 1996. Nevertheless, because CFCs are unreactive and because they diffuse so slowly into the stratosphere, scientists estimate that ozone depletion will continue for many years to come.

What substances have replaced CFCs? At this time the main alternatives are hydrofluorocarbons, compounds in which C—H bonds replace the C—Cl bonds of CFCs. One such compound in current use is CH_2FCF_3, known as HFC-134a. The U.S. Environmental Protection Agency (EPA) reported that by the year 2000, 45% of existing air-conditioning units in buildings had converted to non-CFC compounds like HFC-134a, with improved energy efficiencies overall.

There are no naturally occurring CFCs, but there are some natural sources that contribute chlorine and bromine to the atmosphere, and, just like halogens from CFC, these naturally occurring Cl and Br atoms can participate in ozone-depleting reactions. The principal sources are methyl bromide and methyl chloride, CH_3Br and CH_3Cl. It is estimated that these molecules contribute less than a third to the total Cl and Br in the atmosphere; the remaining two-thirds is a result of human activities. Volcanoes are a source of HCl, but, generally, the HCl they release reacts with water in the troposphere and does not make it to the upper atmosphere.

TABLE 18.3 Sources and Typical Concentrations of Some Minor Atmospheric Constituents

Minor Constituent	Sources	Typical Concentrations
Carbon dioxide, CO_2	Decomposition of organic matter; release from the oceans; fossil-fuel combustion	375 ppm throughout the troposphere
Carbon monoxide, CO	Decomposition of organic matter; industrial processes; fossil-fuel combustion	0.05 ppm in unpolluted air; 1–50 ppm in urban traffic areas
Methane, CH_4	Decomposition of organic matter; natural-gas seepage	1.77 ppm throughout the troposphere
Nitric oxide, NO	Electrical discharges; internal combustion engines; combustion of organic matter	0.01 ppm in unpolluted air; 0.2 ppm in smog
Ozone, O_3	Electrical discharges; diffusion from the stratosphere; photochemical smog	0 to 0.01 ppm in unpolluted air; 0.5 ppm in photochemical smog
Sulfur dioxide, SO_2	Volcanic gases; forest fires; bacterial action; fossil-fuel combustion; industrial processes	0 to 0.01 ppm in unpolluted air; 0.1–2 ppm in polluted urban environment

18.4 | Chemistry of the Troposphere

The troposphere consists primarily of N_2 and O_2, which together make up 99% of Earth's atmosphere at sea level (Table 18.1). Other gases, although present only at very low concentrations, can have major effects on our environment. Table 18.3 ▲ lists the major sources and typical concentrations of some of the important minor constituents of the troposphere. Many of these substances occur to only a slight extent in the natural environment but exhibit much higher concentrations in certain areas as a result of human activities. In this section we will discuss the most important characteristics of a few of these substances and their chemical roles as air pollutants. As we will see, most form as either a direct or an indirect result of our widespread use of combustion reactions.

TABLE 18.4 Median Concentrations of Atmospheric Pollutants in a Typical Urban Atmosphere

Pollutant	Concentration (ppm)
Carbon monoxide	10
Hydrocarbons	3
Sulfur dioxide	0.08
Nitrogen oxides	0.05
Total oxidants (ozone and others)	0.02

Sulfur Compounds and Acid Rain

Sulfur-containing compounds are present to some extent in the natural, unpolluted atmosphere. They originate in the bacterial decay of organic matter, in volcanic gases, and from other sources listed in Table 18.3. The amount of sulfur-containing compounds released into the atmosphere from natural sources is about 24×10^{12} g per year, which is less than the amount from human activities (about 79×10^{12} g per year). Sulfur compounds, chiefly sulfur dioxide, SO_2, are among the most unpleasant and harmful of the common pollutant gases. Table 18.4 ◀ shows the concentrations of several pollutant gases in a *typical* urban environment (not one that is particularly affected by smog). According to these data, the level of sulfur dioxide is 0.08 ppm or higher about half the time. This concentration is considerably lower than that of other pollutants, notably carbon monoxide. Nevertheless, SO_2 is regarded as the most serious health hazard among the pollutants shown, especially for people with respiratory difficulties.

Combustion of coal and oil accounts for about 80% of the total SO_2 released in the United States. The United States gets 52% of its electricity from coal and another 15% from petroleum and natural gas. The extent to which SO_2 emissions are a problem in the burning of coal and oil depends on the level of their sulfur concentration. Some oil, such as that from the Middle East, is relatively low in sulfur, whereas other oil, such as that from Venezuela, has a higher sulfur content. Because of concern about SO_2 pollution, low-sulfur oil is in greater demand and is thus more expensive.

Coals also vary in their sulfur content. Much of the coal from east of the Mississippi is relatively high in sulfur content, up to 6% by mass. Much of the coal

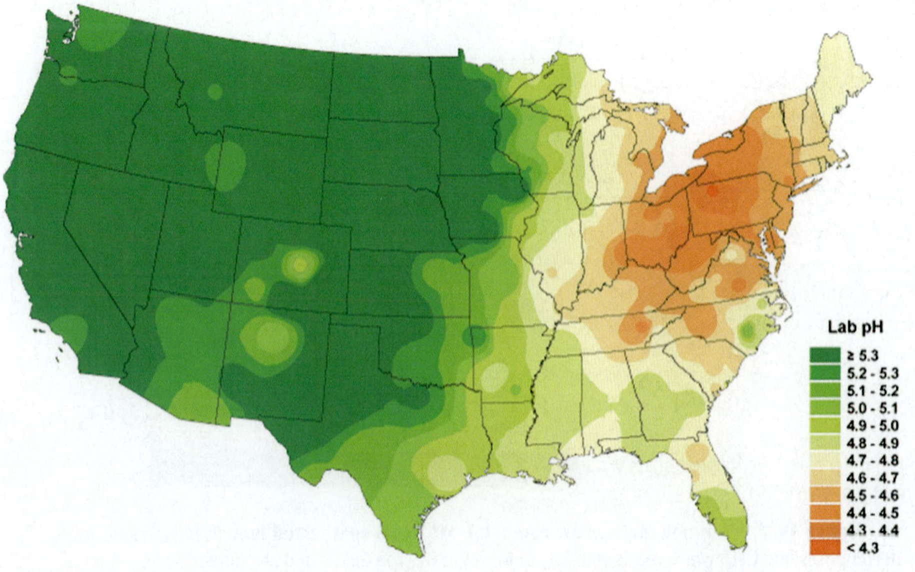

▲ **Figure 18.5 pH values from freshwater sites across the United States, 2001.**

MOVIE
Carbon Dioxide Behaves as an Acid in Water

from the western states has a lower sulfur content. This coal, however, also has a lower heat content per unit mass of coal, so the difference in sulfur content per unit of heat produced is not as large as is often assumed.

Sulfur dioxide is harmful to both human health and property; furthermore, atmospheric SO_2 can be oxidized to SO_3 by several pathways (such as reaction with O_2 or O_3). When SO_3 dissolves in water, it produces sulfuric acid, H_2SO_4:

$$SO_3(g) + H_2O(l) \longrightarrow H_2SO_4(aq)$$

Many of the environmental effects ascribed to SO_2 are actually due to H_2SO_4.

The presence of SO_2 in the atmosphere and the sulfuric acid that it produces result in the phenomenon of **acid rain**. (Nitrogen oxides, which form nitric acid, are also major contributors to acid rain.) Uncontaminated rainwater is naturally acidic and generally has a pH value of about 5.6. The primary source of this natural acidity is CO_2, which reacts with water to form carbonic acid, H_2CO_3. Acid rain typically has a pH value of about 4. This acidity has affected many lakes in the northern Europe, the northern United States, and Canada, reducing fish populations and affecting other parts of the ecological network within the lakes and surrounding forests.

The pH of most natural waters containing living organisms is between 6.5 and 8.5, but as Figure 18.5 ▲ shows, freshwater pH values are far below 6.5 in many parts of the continental United States. At pH levels below 4.0, all vertebrates, most invertebrates, and many microorganisms are destroyed. The lakes that are most susceptible to damage are those with low concentrations of basic ions, such as HCO_3^-, that buffer them against changes in pH. More than 300 lakes in New York State contain no fish, and 140 lakes in Ontario, Canada, are devoid of life. The acid rain that appears to have killed the organisms in these lakes originates hundreds of kilometers upwind in the Ohio Valley and Great Lakes regions. Some of these regions are recovering as sulfur emissions from fossil fuel combustion decrease, in part because of the Clean Air Act of 1990, which required that power plants reduce their sulfur emissions by 80%.

Because acids react with metals and with carbonates, acid rain is corrosive both to metals and to stone building materials. Marble and limestone, for example, whose major constituent is $CaCO_3$, are readily attacked by acid rain (Figure 18.6 ▶). Billions of dollars each year are lost as a result of corrosion that is due to SO_2 pollution.

(a)

(b)

▲ **Figure 18.6 Damage from acid rain.** (a) This statue at the Field Museum in Chicago shows the effects of corrosion from acid rain and atmospheric pollutants. (b) The same statue after restoration.

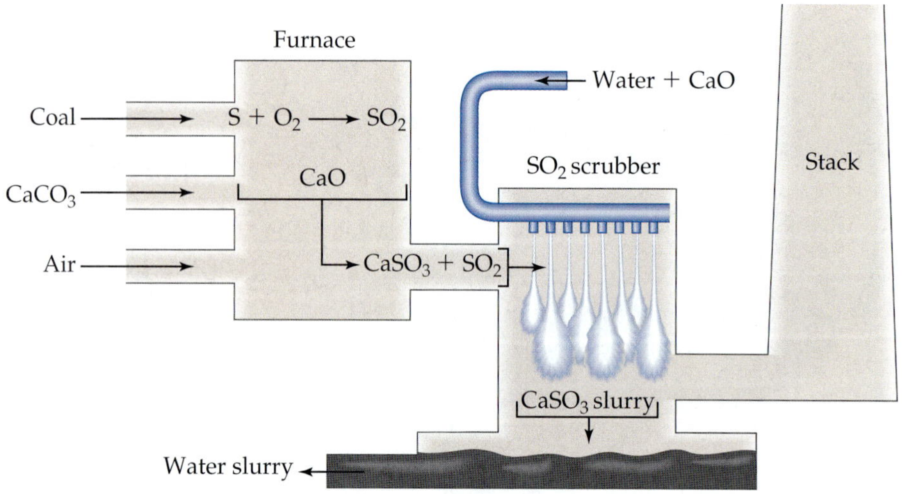

▲ **Figure 18.7 Common method for removing SO_2 from combusted fuel.** Powdered limestone decomposes into CaO, which reacts with SO_2 to form $CaSO_3$. The $CaSO_3$ and any unreacted SO_2 enter a purification chamber called a scrubber, where a shower of CaO and water converts the remaining SO_2 into $CaSO_3$ and precipitates the $CaSO_3$ into a watery residue called a slurry.

One way to reduce the quantity of SO_2 released into the environment is to remove sulfur from coal and oil before it is burned. Although difficult and expensive, several methods have been developed for removing SO_2 from the gases formed when coal and oil are combusted. Powdered limestone ($CaCO_3$), for example, can be injected into the furnace of a power plant, where it decomposes into lime (CaO) and carbon dioxide:

$$CaCO_3(s) \longrightarrow CaO(s) + CO_2(g)$$

The CaO then reacts with SO_2 to form calcium sulfite:

$$CaO(s) + SO_2(g) \longrightarrow CaSO_3(s)$$

The solid particles of $CaSO_3$ as well as much of the unreacted SO_2 can be removed from the furnace gas by passing it through an aqueous suspension of lime (Figure 18.7 ▲). Not all the SO_2 is removed, however, and given the enormous quantities of coal and oil burned worldwide, controllable pollution by SO_2 will probably remain a problem for some time.

Carbon Monoxide

Carbon monoxide is formed by the incomplete combustion of carbon-containing material, such as fossil fuels. In terms of total mass, CO is the most abundant of all the pollutant gases. The level of CO present in unpolluted air is low, probably on the order of 0.05 ppm. The estimated total amount of CO in the atmosphere is about 5.2×10^{14} g. In the United States alone, roughly 1×10^{14} g of CO is produced each year, about two-thirds of which comes from automobiles.

Carbon monoxide is a relatively unreactive molecule and consequently poses no direct threat to vegetation or materials. It does affect humans, however. It has the unusual ability to bind very strongly to **hemoglobin**, the iron-containing protein in red blood cells [Figure 18.8(a) ▶] that transports oxygen in blood. Hemoglobin consists of four protein chains held together by weak intermolecular forces in a cluster [Figure 18.8(b)]. Each protein chain has a heme molecule within its folds. A schematic structure of heme is shown in Figure 18.8(c). Note that iron is situated in the center of a plane of four nitrogen atoms. A hemoglobin molecule in the lungs picks up an O_2 molecule, which reacts with the iron atom to form a species called *oxyhemoglobin*, abbreviated as HbO_2. As the blood circu-

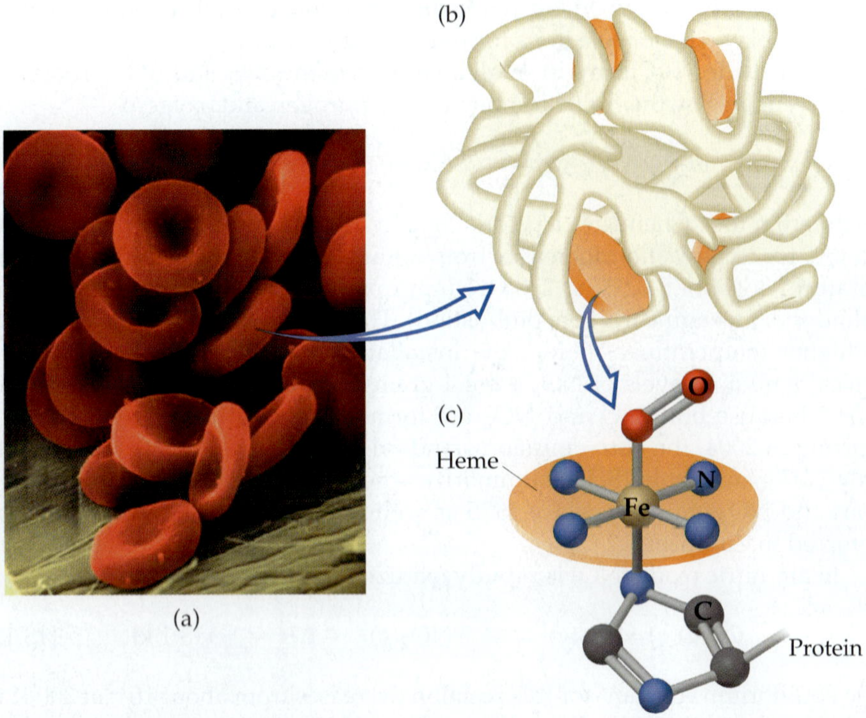

▲ **Figure 18.8 Carbon monoxide harms humans.** Red blood cells (a) contain hemoglobin (b). The hemoglobin contains four heme units, each of which can bind an O_2 molecule (c). When exposed to CO, the heme binds CO in preference to O_2.

lates, the oxygen molecule is released in tissues as needed for cell metabolism, that is, for the chemical processes occurring in the cell. (See the "Chemistry and Life" box on blood as a buffered solution in Section 17.2.)

Like O_2, CO also binds very strongly to the iron in hemoglobin. The complex is called *carboxyhemoglobin* and is represented as COHb. The equilibrium binding constant of human hemoglobin for CO is about 210 times greater than that for O_2. As a result, a relatively small quantity of CO can inactivate a substantial fraction of the hemoglobin in the blood for oxygen transport. For example, a person breathing air that contains only 0.1% of CO takes in enough CO after a few hours of breathing to convert up to 60% of the hemoglobin into COHb, thus reducing the blood's normal oxygen-carrying capacity by 60%.

Under normal conditions a nonsmoker breathing unpolluted air has about 0.3 to 0.5% COHb in the bloodstream. This amount arises mainly from the production of small quantities of CO in the course of normal body chemistry and from the small amount of CO present in clean air. Exposure to higher concentrations of CO causes the COHb level to increase, which in turn leaves fewer Hb sites to which O_2 can bind. If the level of COHb becomes too high, oxygen transport is effectively shut down and death occurs. Because CO is colorless and odorless, CO poisoning occurs with very little warning. Improperly ventilated combustion devices, such as kerosene lanterns and stoves, thus pose a potential health hazard (Figure 18.9 ▶).

▲ **Figure 18.9 Carbon monoxide warnings.** Kerosene lamps and stoves have warning labels concerning use in enclosed spaces, such as an indoor room. Incomplete combustion can produce colorless, odorless carbon monoxide, CO, which is toxic.

Nitrogen Oxides and Photochemical Smog

Nitrogen oxides are primary components of smog, a phenomenon with which city dwellers are all too familiar. The term *smog* refers to a particularly unpleasant condition of pollution in certain urban environments that occurs when weather conditions produce a relatively stagnant air mass. The smog made famous by Los Angeles, but now common in many other urban areas as well, is

▲ **Figure 18.10 Photochemical smog.** Smog is produced largely by the action of sunlight on automobile exhaust gases.

more accurately described as **photochemical smog** because photochemical processes play a major role in its formation (Figure 18.10 ◀).

Nitric oxide, NO, forms in small quantities in the cylinders of internal combustion engines by the direct combination of nitrogen and oxygen:

$$N_2(g) + O_2(g) \rightleftharpoons 2\,NO(g) \qquad \Delta H = 180.8\text{ kJ} \qquad [18.11]$$

As noted in the "Chemistry at Work" box in Section 15.6, the equilibrium constant K for this reaction increases from about 10^{-15} at 300 K (near room temperature) to about 0.05 at 2400 K (approximately the temperature in the cylinder of an engine during combustion). Thus, the reaction is more favorable at higher temperatures. Before the installation of pollution-control devices, typical emission levels of NO_x were 4 grams per mile (g/mi). (The x is either 1 or 2 because both NO and NO_2 are formed, although NO predominates.) Starting in 2004, the auto emission standards for NO_x call for its reduction to only 0.07 g/mi. Table 18.5 ▼ summarizes the federal standards for hydrocarbons and NO_x emissions since 1975 as well as the more restrictive standards enforced in California.

In air nitric oxide (NO) is rapidly oxidized to nitrogen dioxide (NO_2):

$$2\,NO(g) + O_2(g) \rightleftharpoons 2\,NO_2(g) \qquad \Delta H = -113.1\text{ kJ} \qquad [18.12]$$

The equilibrium constant for this reaction decreases from about 10^{12} at 300 K to about 10^{-5} at 2400 K. The photodissociation of NO_2 initiates the reactions associated with photochemical smog. The dissociation of NO_2 into NO and O requires 304 kJ/mol, which corresponds to a photon wavelength of 393 nm. In sunlight, therefore, NO_2 undergoes dissociation to NO and O:

$$NO_2(g) + h\nu \longrightarrow NO(g) + O(g) \qquad [18.13]$$

The atomic oxygen formed undergoes several possible reactions, one of which gives ozone, as described earlier:

$$O(g) + O_2 + M(g) \longrightarrow O_3(g) + M^*(g) \qquad [18.14]$$

Ozone is a key component of photochemical smog. Although it is an essential UV screen in the upper atmosphere, it is an undesirable pollutant in the troposphere. It is extremely reactive and toxic, and breathing air that contains appreciable amounts of ozone can be especially dangerous for asthma sufferers, exercisers, and the elderly. We therefore have two ozone problems: excessive amounts in many urban environments, where it is harmful, and depletion in the stratosphere, where it is vital.

In addition to nitrogen oxides and carbon monoxide, an automobile engine also emits unburned *hydrocarbons* as pollutants. These organic compounds, which are composed entirely of carbon and hydrogen, are the principal components of gasoline (Section 25.1) and are major ingredients of smog. A typical engine without effective emission controls emits about 10 to 15 grams of these compounds per mile. Current standards require that hydrocarbon emissions be less than 0.075 grams per mile. Hydrocarbons are also emitted naturally from living organisms (see "A Closer Look" box later in this section).

Reduction or elimination of smog requires that the essential ingredients for its formation be removed from automobile exhaust. Catalytic converters are designed to drastically reduce the levels of NO_x and hydrocarbons, two of the major ingredients of smog (see the "Chemistry at Work" box in Section 14.6).

TABLE 18.5 National Tailpipe Emission Standards*

Year	Hydrocarbons (g/mi)	Nitrogen Oxides (g/mi)
1975	1.5 (0.9)	3.1 (2.0)
1980	0.41 (0.41)	2.0 (1.0)
1985	0.41 (0.41)	1.0 (0.4)
1990	0.41 (0.41)	1.0 (0.4)
1995	0.25 (0.25)	0.4 (0.4)
2004	0.075 (0.05)	0.07 (0.05)

*California standards in parentheses

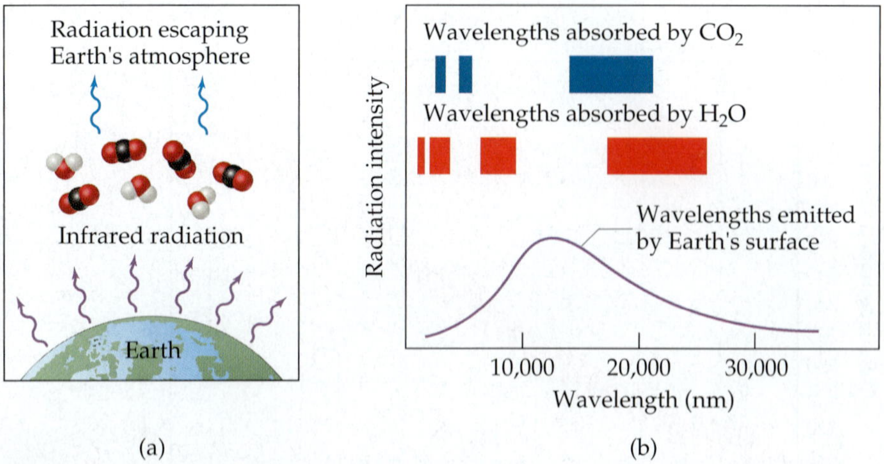

◀ **Figure 18.11 Why Earth appears so cold from outer space.** (a) Carbon dioxide and water absorb certain wavelengths of infrared radiation, which helps keep energy from escaping from Earth's surface. (b) The distribution of the wavelengths absorbed by CO_2 and H_2O compared to the wavelengths emitted by Earth's surface.

Water Vapor, Carbon Dioxide, and Climate

We have seen how the atmosphere makes life as we know it possible on Earth by screening out harmful short-wavelength radiation. In addition, the atmosphere is essential in maintaining a reasonably uniform and moderate temperature on the surface of the planet. The two atmospheric components of greatest importance in maintaining Earth's surface temperature are carbon dioxide and water.

Earth is in overall thermal balance with its surroundings. This means that Earth radiates energy into space at a rate equal to the rate at which it absorbs energy from the Sun. The Sun has a surface temperature of about 6000 K. As seen from outer space, Earth is relatively cold, with a temperature of about 254 K. The distribution of wavelengths in the radiation emitted from an object is determined by its temperature. (Section 6.2) Why does Earth, viewed from outside its atmosphere, appear so much colder than the temperature we usually experience at its surface? The troposphere, transparent to visible light, is not transparent to infrared radiation. Figure 18.11 ▲ shows the distribution of radiation from Earth's surface and the wavelengths absorbed by atmospheric water vapor and carbon dioxide. According to the graph, these atmospheric gases absorb much of the outgoing radiation from Earth's surface. In doing so, they help to maintain a livable uniform temperature at the surface by holding in, as it were, the infrared radiation from the surface, which we feel as heat. The influence of H_2O, CO_2, and certain other atmospheric gases on Earth's temperature is often called the *greenhouse effect* (see the "Chemistry at Work" box in Section 3.7).

The partial pressure of water vapor in the atmosphere varies greatly from place to place and time to time, but it is generally highest near Earth's surface and drops off very sharply with increased elevation. Because water vapor absorbs infrared radiation so strongly, it plays the major role in maintaining the atmospheric temperature at night, when the surface is emitting radiation into space and not receiving energy from the Sun. In very dry desert climates, where the water-vapor concentration is unusually low, it may be extremely hot during the day, but very cold at night. In the absence of an extensive layer of water vapor to absorb and then radiate part of the infrared radiation back to Earth, the surface loses this radiation into space and cools off very rapidly.

Carbon dioxide plays a secondary, but very important, role in maintaining the surface temperature. The worldwide combustion of fossil fuels, principally coal and oil, on a prodigious scale in the modern era has sharply increased the carbon dioxide level of the atmosphere. Measurements carried out over several decades show that the CO_2 concentration in the atmosphere

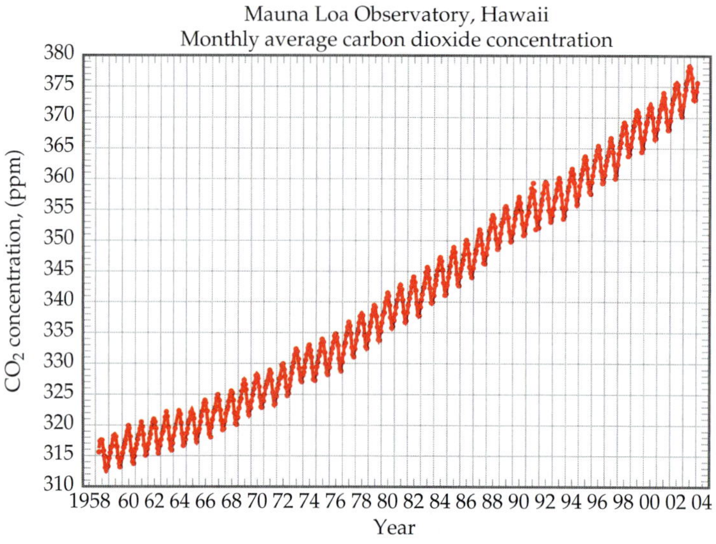

▶ **Figure 18.12 Rising CO_2 levels.** The concentration of atmospheric CO_2 has risen more than 15% since the late 1950s. These data were recorded at the Mauna Loa Observatory in Hawaii by monitoring the absorption of infrared radiation. The sawtooth shape of the graph is due to regular seasonal variations in CO_2 concentration for each year.

is steadily increasing, currently at a high of about 375 ppm (Figure 18.12 ▲). In comparison, it is estimated that CO_2 concentrations in the atmosphere fluctuated between 200 and 300 ppm over the past 150,000 years, based on data from air bubbles trapped in ice cores. A consensus is emerging among scientists that this increase is already perturbing Earth's climate and that it may be responsible for the observed increase in the average global air temperature of 0.3°C to 0.6°C over the past century.

A CLOSER LOOK | Methane as a Greenhouse Gas

Although CO_2 receives most of the attention, other gases in total make an equal contribution to the greenhouse effect. Chief among these is methane, CH_4. Each methane molecule has about 25 times the greenhouse effect of a CO_2 molecule. Studies of atmospheric gas trapped long ago in the Greenland and Antarctic ice sheets show that the concentration of methane in the atmosphere has increased during the industrial age, from preindustrial values in the range of 0.3 to 0.7 ppm to the present value of about 1.8 ppm.

Methane is formed in biological processes that occur in low-oxygen environments. Anaerobic bacteria, which flourish in swamps and landfills, near the roots of rice plants, and in the digestive systems of cows and other ruminant animals, produce methane (Figure 18.13 ▶). It also leaks into the atmosphere during natural-gas extraction and transport (see "Chemistry at Work" box, Section 10.5). It is estimated that about two-thirds of present-day methane emissions, which are increasing by about 1% per year, are related to human activities.

Methane has a half-life in the atmosphere of about 10 years, whereas CO_2 is much longer-lived. This might at first seem a good thing, but there are indirect effects to consider. Some methane is oxidized in the stratosphere, producing water vapor, a powerful greenhouse gas that is otherwise virtually absent from the stratosphere. In the troposphere methane is attacked by reactive species such as OH radicals or nitrogen oxides, eventually producing other greenhouse gases such as O_3. It has been estimated that the climate-changing effects of CH_4 are at least one-third those of CO_2, and perhaps even half as large. Given this large contribution, important reductions of the greenhouse effect could be achieved by reducing methane emissions or capturing the emissions for use as a fuel.

▲ **Figure 18.13 Methane production.** Ruminant animals, such as cows and sheep, produce methane in their digestive systems. In Australia sheep and cattle produce about 14% of the country's total greenhouse emissions.

On the basis of present and expected future rates of fossil-fuel use, the atmospheric CO_2 level is expected to double from its present level sometime between 2050 and 2100. Computer models predict that this increase will result in an average global temperature increase of 1°C to 3°C. Major changes in global climate could result from a temperature change of this magnitude. Because so many factors go into determining climate, we cannot predict with certainty what changes will occur. Clearly, however, humanity has acquired the potential, by changing the concentration of CO_2 and other heat-trapping gases in the atmosphere, to substantially alter the climate of the planet.

18.5 | The World Ocean

Water is the most common liquid on Earth. It covers 72% of Earth's surface and is essential to life. Our bodies are about 65% water by mass. Because of extensive hydrogen bonding, water has unusually high melting and boiling points and a high heat capacity. (Section 11.2) Its highly polar character is responsible for its exceptional ability to dissolve a wide range of ionic and polar-covalent substances. Many reactions occur in water, including reactions in which H_2O itself is a reactant. Recall, for example, that H_2O can participate in acid-base reactions as either a proton donor or a proton acceptor. (Section 16.4) In Chapter 20 we will see that H_2O can also participate in oxidation-reduction reactions as either a donor or an acceptor of electrons. All these properties play a role in our environment.

Seawater

The vast layer of salty water that covers so much of the planet is connected and is generally constant in composition. For this reason, oceanographers speak of a world ocean rather than of the separate oceans we learn about in geography books. The world ocean is huge. Its volume is 1.35×10^9 km^3. Almost all the water on Earth, 97.2%, is in the world ocean. Of the remaining 2.8%, 2.1% is in the form of ice caps and glaciers. All the fresh water—in lakes, rivers, and groundwater—amounts to only 0.6%. Most of the remaining 0.1% is in brackish (salty) water, such as that in the Great Salt Lake in Utah.

Seawater is often referred to as saline water. The **salinity** of seawater is the mass in grams of dry salts present in 1 kg of seawater. In the world ocean the salinity averages about 35. To put it another way, seawater contains about 3.5% dissolved salts by mass. The list of elements present in seawater is very long. Most, however, are present only in very low concentrations. Table 18.6 ▶ lists the 11 ionic species that are most abundant in seawater.

TABLE 18.6 Ionic Constituents of Seawater Present in Concentrations Greater than 0.001 g/kg (1 ppm)

Ionic Constituent	g/kg Seawater	Concentration (*M*)
Chloride, Cl^-	19.35	0.55
Sodium, Na^+	10.76	0.47
Sulfate, SO_4^{2-}	2.71	0.028
Magnesium, Mg^{2+}	1.29	0.054
Calcium, Ca^{2+}	0.412	0.010
Potassium, K^+	0.40	0.010
Carbon dioxide*	0.106	2.3×10^{-3}
Bromide, Br^-	0.067	8.3×10^{-4}
Boric acid, H_3BO_3	0.027	4.3×10^{-4}
Strontium, Sr^{2+}	0.0079	9.1×10^{-5}
Fluoride, F^-	0.0013	7.0×10^{-5}

*CO_2 is present in seawater as HCO_3^- and CO_3^{2-}.

The properties of seawater—its salinity, density, and temperature—vary as a function of depth (Figure 18.14 ▶). Sunlight penetrates well only 200 m into the sea; the region between 200 m and 1000 m deep is the "twilight zone," where visible light is faint. Below 1000 m in depth, the ocean is pitch-black and cold, about 4°C. The transport of heat, salt, and other chemicals throughout the ocean is influenced by these changes in the physical properties of seawater, and in turn the changes in the way heat and substances are transported affects ocean currents and the global climate.

The sea is so vast that if a substance is present in seawater to the extent of only 1 part per billion (ppb, that is, 1×10^{-6} g per kilogram of water), there is still 5×10^9 kg of it in the world ocean. Nevertheless, the ocean

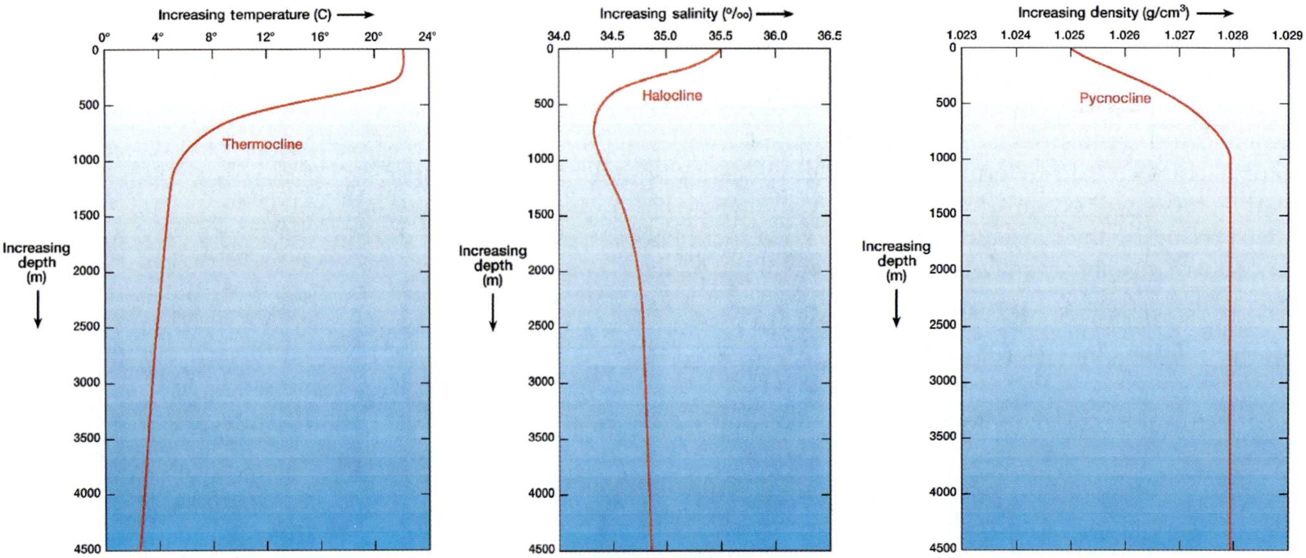

▲ **Figure 18.14 Average temperature, salinity, and density of seawater as a function of depth.**

is rarely used as a source of raw materials because the cost of extracting the desired substances is too high. Only three substances are obtained from seawater in commercially important amounts: sodium chloride, bromine (from bromide salts), and magnesium (from its salts).

Absorption of CO_2 by the ocean plays a large role in global climate. Carbon dioxide reacts with water to form carbonic acid, H_2CO_3 (Section 16.6), and so as CO_2 from the atmosphere is absorbed by the world ocean, the concentration of H_2CO_3 in the ocean increases. Most of the carbon in the ocean, however, is in the form of HCO_3^- and CO_3^{2-} ions. These ions form a buffer system that maintains the ocean's average pH between 8.0 and 8.3. The buffering capacity of the world ocean is predicted to decrease as the concentration of CO_2 in the atmosphere increases, because of the increase in H_2CO_3 concentration. With less buffering capacity, the carbonate ion can precipitate out as $CaCO_3$, one of the main ingredients in seashells. Thus, both acid-base equilibrium reactions and solubility equilibrium reactions (Section 17.4) form a complicated web of interactions that tie the ocean to the atmosphere and to the global climate.

Desalination

Because of its high salt content, seawater is unfit for human consumption and for most of the uses to which we put water. In the United States the salt content of municipal water supplies is restricted by health codes to no more than about 500 ppm. This amount is much lower than the 3.5% dissolved salts present in seawater and the 0.5% or so present in brackish water found underground in some regions. The removal of salts from seawater or brackish water to make the water usable is called **desalination**.

Water can be separated from dissolved salts by *distillation* (described in the "Closer Look" box in Section 13.5) because water is a volatile substance and the salts are nonvolatile. The principle of distillation is simple enough, but carrying out the process on a large scale presents many problems. As water is distilled from a vessel containing seawater, for example, the salts become more and more concentrated and eventually precipitate out.

Seawater can also be desalinated using **reverse osmosis**. Recall that osmosis is the net movement of solvent molecules, but not solute molecules, through a semipermeable membrane. (Section 13.5) In osmosis the solvent passes

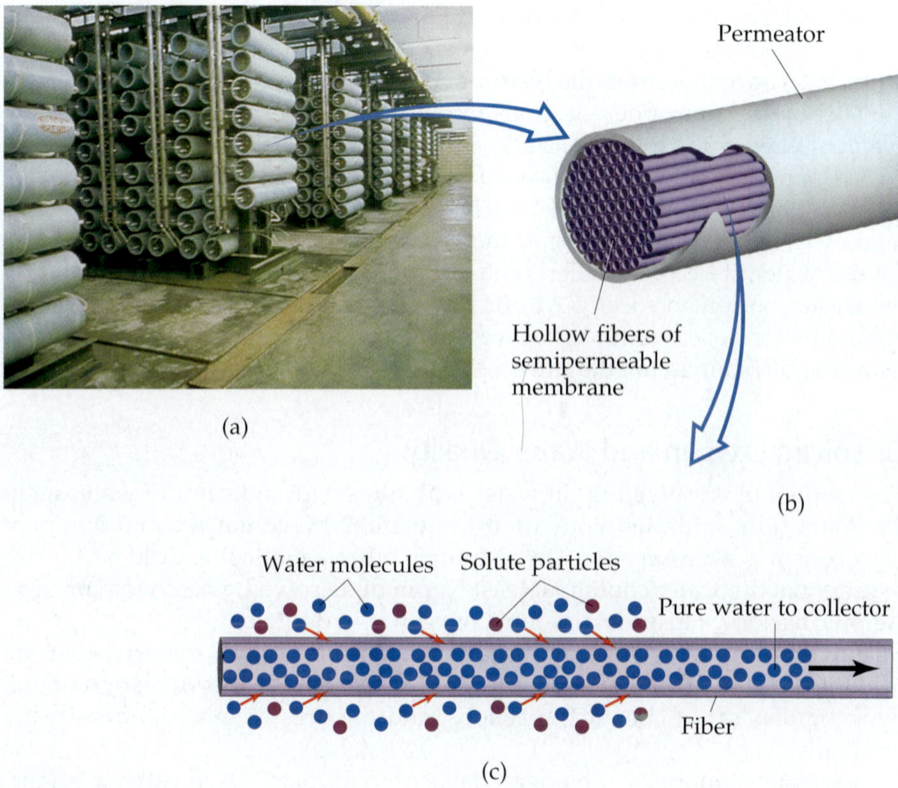

◀ **Figure 18.15 Reverse osmosis.** (a) A room inside a reverse-osmosis desalination plant. (b) Each cylinder shown in (a) is called a permeator and contains several million tiny hollow fibers. (c) When seawater is introduced under pressure into a permeator, water passes through the fiber wall into the fibers and is thereby separated from all the ions initially present in the seawater.

from the more dilute solution into the more concentrated one. However, if sufficient external pressure is applied, osmosis can be stopped and, at still higher pressures, reversed. When this occurs, solvent passes from the more concentrated into the more dilute solution. In a modern reverse-osmosis facility tiny hollow fibers are used as the semipermeable membrane. Water is introduced under pressure into the fibers, and desalinated water is recovered, as illustrated in Figure 18.15 ▲.

The world's largest desalination plant is located in Jubail, Saudi Arabia. This plant provides 50% of that country's drinking water by using reverse osmosis to desalinate seawater from the Persian Gulf. Such plants are becoming increasingly common in the United States. In 1995, for example, the city of Melbourne, Florida, opened a reverse-osmosis plant that can produce 26.5 million gallons of drinking water a day. Small-scale, manually operated reverse-osmosis desalinators are now also available for use in camping and traveling, and at sea (Figure 18.16 ▶).

▲ **Figure 18.16 A portable desalination device.** This hand-operated water desalinator works by reverse osmosis. It can produce 4.5 L (1.2 gal) of pure water from seawater in an hour.

18.6 | Fresh Water

The United States is fortunate in its abundance of fresh water—1.7×10^{15} liters (660 trillion gallons) is the estimated reserve, which is renewed by rainfall. An estimated 9×10^{11} liters of fresh water is used every day in the United States. Most of this is used for agriculture (41%) and hydroelectric power (39%), with small amounts for industry (6%), household needs (6%), and drinking water (1%). An adult needs about 2 liters of water per day for drinking. In the United States our daily use of water per person far exceeds this subsistence level, amounting to an average of about 300 L/day for personal consumption and hygiene. We use about 8 L/person for cooking and drinking, about 120 L/person for cleaning (bathing, laundering, and housecleaning), 80 L/person for flushing toilets, and 80 L/person for watering lawns.

The total amount of fresh water on Earth is not a very large fraction of the total water present. Indeed, fresh water is one of our most precious resources. It forms by evaporation from the oceans and the land. The water vapor that accumulates in the atmosphere is transported by global atmospheric circulation, eventually returning to Earth as rain and snow.

As rain falls and as water runs off the land on its way to the oceans, it dissolves a variety of cations (mainly Na^+, K^+, Mg^{2+}, Ca^{2+}, and Fe^{2+}), anions (mainly Cl^-, SO_4^{2-}, and HCO_3^-), and gases (principally O_2, N_2, and CO_2). As we use water, it becomes laden with additional dissolved material, including the wastes of human society. As our population and output of environmental pollutants increase, we find that we must spend ever-increasing amounts of money and resources to guarantee a supply of fresh water.

Dissolved Oxygen and Water Quality

The amount of dissolved O_2 in water is an important indicator of water quality. Water fully saturated with air at 1 atm and 20°C contains about 9 ppm of O_2. Oxygen is necessary for fish and much other aquatic life. Cold-water fish require that the water contain at least 5 ppm of dissolved oxygen for survival. Aerobic bacteria consume dissolved oxygen in order to oxidize organic materials and so meet their energy requirements. The organic material that the bacteria are able to oxidize is said to be **biodegradable**. This oxidation occurs by a complex set of chemical reactions, and the organic material disappears gradually.

Excessive quantities of biodegradable organic materials in water are detrimental because they deplete the water of the oxygen necessary to sustain normal animal life. Typical sources of these biodegradable materials, which are called *oxygen-demanding wastes*, include sewage, industrial wastes from food-processing plants and paper mills, and effluent (liquid waste) from meatpacking plants.

In the presence of oxygen, the carbon, hydrogen, nitrogen, sulfur, and phosphorus in biodegradable material end up mainly as CO_2, HCO_3^-, H_2O, NO_3^-, SO_4^{2-}, and phosphates. The formation of these oxidation products sometimes reduces the amount of dissolved oxygen to the point where aerobic bacteria can no longer survive. Anaerobic bacteria then take over the decomposition process, forming CH_4, NH_3, H_2S, PH_3, and other products, several of which contribute to the offensive odors of some polluted waters.

▼ **Figure 18.17 Eutrophication.** The growth of algae and duckweed in this pond is due to agricultural wastes. The wastes feed the growth of the algae and weeds, which deplete the oxygen in the water, a process called eutrophication. A eutrophic lake cannot support fish.

Plant nutrients, particularly nitrogen and phosphorus, contribute to water pollution by stimulating excessive growth of aquatic plants. The most visible results of excessive plant growth are floating algae and murky water. More significantly, however, as plant growth becomes excessive, the amount of dead and decaying plant matter increases rapidly, a process called *eutrophication* (Figure 18.17 ◀). The decay of plants consumes O_2 as the plants are biodegraded, leading to the depletion of oxygen in the water. Without sufficient supplies of oxygen, the water, in turn, cannot sustain any form of animal life. The most significant sources of nitrogen and phosphorus compounds in water are domestic

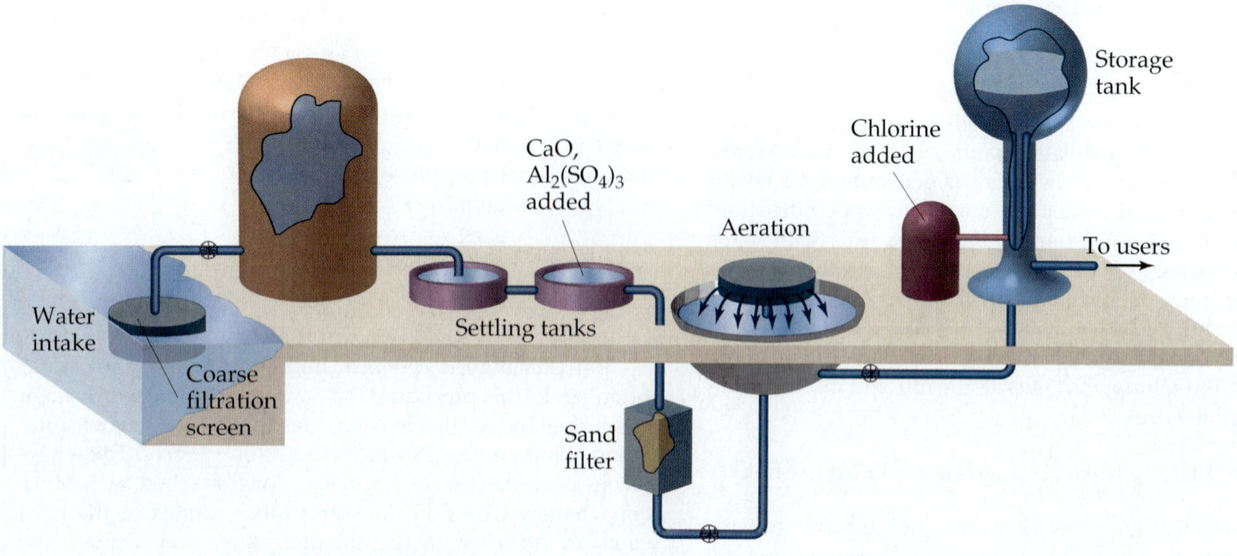

▲ **Figure 18.18 Common steps in treating water for a public water system.**

sewage (phosphate-containing detergents and nitrogen-containing body wastes), runoff from agricultural land (fertilizers containing both nitrogen and phosphorus), and runoff from livestock areas (animal wastes containing nitrogen).

Treatment of Municipal Water Supplies

The water needed for domestic uses, agriculture, and industrial processes is taken either from naturally occurring lakes, rivers, and underground sources or from reservoirs. Much of the water that finds its way into municipal water systems is "used" water: It has already passed through one or more sewage systems or industrial plants. Consequently, this water must be treated before it is distributed to our faucets. Municipal water treatment usually involves five steps: coarse filtration, sedimentation, sand filtration, aeration, and sterilization. Figure 18.18 ▲ shows a typical treatment process.

After coarse filtration through a screen, the water is allowed to stand in large settling tanks in which finely divided sand and other minute particles can settle out. To aid in removing very small particles, the water may first be made slightly basic by adding CaO. Then $Al_2(SO_4)_3$ is added. The aluminum sulfate reacts with OH^- ions to form a spongy, gelatinous precipitate of $Al(OH)_3$ ($K_{sp} = 1.3 \times 10^{-33}$). This precipitate settles slowly, carrying suspended particles down with it, thereby removing nearly all finely divided matter and most bacteria. The water is then filtered through a sand bed. Following filtration, the water may be sprayed into the air to hasten the oxidation of dissolved organic substances.

The final stage of the operation normally involves treating the water with a chemical agent to ensure the destruction of bacteria. Ozone is most effective, but it must be generated at the place where it is used. Chlorine, Cl_2, is therefore more convenient. Chlorine can be shipped in tanks as a liquefied gas and dispensed from the tanks through a metering device directly into the water supply. The amount used depends on the presence of other substances with which the chlorine might react and on the concentrations of bacteria and viruses to be removed. The sterilizing action of chlorine is probably due not to Cl_2 itself, but to hypochlorous acid, which forms when chlorine reacts with water:

$$Cl_2(aq) + H_2O(l) \longrightarrow HClO(aq) + H^+(aq) + Cl^-(aq) \qquad [18.15]$$

A CLOSER LOOK | Water Softening

Water containing a relatively high concentration of Ca^{2+}, Mg^{2+}, and other divalent cations is called **hard water**. Although the presence of these ions is generally not a health threat, they can make water unsuitable for some household and industrial uses. For example, these ions react with soaps to form an insoluble soap scum, the stuff of bathtub rings. In addition, mineral deposits may form when water containing these ions is heated. When water containing calcium ions and bicarbonate ions is heated, some carbon dioxide is driven off. As a result, the solution becomes less acidic and insoluble calcium carbonate forms:

$$Ca^{2+}(aq) + 2\,HCO_3^-(aq) \longrightarrow CaCO_3(s) + CO_2(g) + H_2O(l)$$

The solid $CaCO_3$ coats the surface of hot-water systems and teakettles, thereby reducing heating efficiency. These deposits, called *scale*, can be especially serious in boilers where water is heated under pressure in pipes running through a furnace. Formation of scale reduces the efficiency of heat transfer and reduces the flow of water through pipes (Figure 18.19 ▼).

The removal of the ions that cause hard water is called water *softening*. Not all municipal water supplies require water softening. In those that do, the water is generally taken from underground sources in which it has had considerable contact with limestone, $CaCO_3$, and other minerals containing Ca^{2+}, Mg^{2+}, and Fe^{2+}. The **lime-soda process** is used for large-scale municipal water-softening operations. The water is treated with lime, CaO [or slaked lime, $Ca(OH)_2$], and soda ash, Na_2CO_3. These chemicals precipitate Ca^{2+} as $CaCO_3$ ($K_{sp} = 4.5 \times 10^{-9}$) and Mg^{2+} as $Mg(OH)_2$ ($K_{sp} = 1.6 \times 10^{-12}$):

$$Ca^{2+}(aq) + CO_3^{2-}(aq) \longrightarrow CaCO_3(s)$$

$$Mg^{2+}(aq) + 2\,OH^-(aq) \longrightarrow Mg(OH)_2(s)$$

◀ **Figure 18.19 Scale formation.** A section of water pipe that has been coated on the inside with $CaCO_3$ and other insoluble salts deposited from hard water.

Ion exchange is a typical household method for water softening. In this procedure the hard water is passed through a bed of an ion-exchange resin: plastic beads with covalently bound anion groups such as $-COO^-$ or $-SO_3^-$. These negatively charged groups have Na^+ ions attached to balance their charges. The Ca^{2+} ions and other cations in the hard water are attracted to the anionic groups and displace the lower-charged Na^+ ions into the water. Thus, one type of ion is exchanged for another. To maintain charge balance, 2 Na^+ ions enter the water for each Ca^{2+} removed. If we represent the resin with its anionic site as $R-COO^-$, we can write the equation for the process as follows:

$$2\,Na(R-COO)(s) + Ca^{2+}(aq) \rightleftharpoons Ca(R-COO)_2(s) + 2\,Na^+(aq)$$

Water softened in this way contains an increased concentration of Na^+ ions. Although Na^+ ions do not form precipitates or cause other problems associated with hard-water cations, individuals concerned about their sodium intake, such as those who have high blood pressure (hypertension), should avoid drinking water softened in this way.

When all the available Na^+ ions have been displaced from the ion-exchange resin, the resin is regenerated by flushing it with a concentrated solution of NaCl. Homeowners can do this by charging their units with large amounts of NaCl(*s*), which can be purchased at most grocery stores. The high concentration of Na^+ forces the equilibrium shown in the earlier equation to shift to the left, causing the Na^+ ions to displace the hard-water cations, which are flushed down the drain.

18.7 | Green Chemistry

Since the 1970s, more and more people have realized that the planet on which we live is, to a large extent, a closed system. There is a growing realization that if humankind is to thrive in the future, we must create a *sustainable* society—that is, a society in which all the processes we humans carry out are in balance with Earth's natural processes, a society in which no toxic materials are released to the environment and in which our needs are met with renewable resources. Finally, all of this must be accomplished with the expenditure of the least possible amount of energy.

Although the chemical industry is but a small part of the whole, chemical processes are involved in nearly all aspects of modern life. Chemistry is therefore at the heart of efforts to accomplish these goals. The **green chemistry** initiative promotes the design and application of chemical products and processes that are compatible with human health and that preserve the environment.

Some of the major principles that govern green chemistry are the following:

- It is better to prevent waste than to treat or clean it up after it has been created.
- In synthesizing new substances, the method employed should generate as little waste product as possible. Those substances that are generated should possess little or no toxicity to human health or towards the environment.
- Chemical processes should be designed to be as energy-efficient as possible, avoiding high temperatures and pressures.
- Catalysts that permit the use of common and safe reagents should be employed whenever possible.
- The raw materials for chemical processes should be renewable feedstocks whenever technically and economically feasible.
- Auxiliary substances, such as solvents, should be eliminated or made as innocuous as possible.

Let's consider some of the areas in which green chemistry can operate to improve environmental quality.

Solvents and Reagents

A major area of concern in chemical processes is the use of volatile organic compounds as solvents for reactions. Generally, the solvent in which a reaction is run is not consumed in the reaction, and there are unavoidable releases of solvent into the atmosphere even in the most carefully controlled processes. Further, the solvent may be toxic or may decompose to some extent during the reaction, thus creating waste products. The use of supercritical fluids ("Chemistry at Work" box, Section 11.4) represents a way to replace conventional solvents with other reagents. Recall that a supercritical fluid is an unusual state of matter that has properties of both a gas and a liquid. (Section 11.4) Water and carbon dioxide are the two most popular choices as supercritical fluid solvents. One recently developed industrial process, for example, replaces chlorofluorocarbon solvents with liquid or supercritical CO_2 in the production of polytetrafluoroethylene ($[CF_2CF_2]_n$, sold as Teflon®). The chlorofluorocarbon solvents, aside from their costs, have harmful effects on Earth's ozone layer (Section 18.3). However, CO_2 does participate in global warming processes, as we saw in earlier in this chapter. Therefore, in making choices about "being green," there are trade-offs.

As a further example, *para*-xylene is oxidized to form terephthalic acid, which in turn is used to make polyethylene terephthalate (PET) plastic and polyester fiber (Section 12.2, Table 12.4):

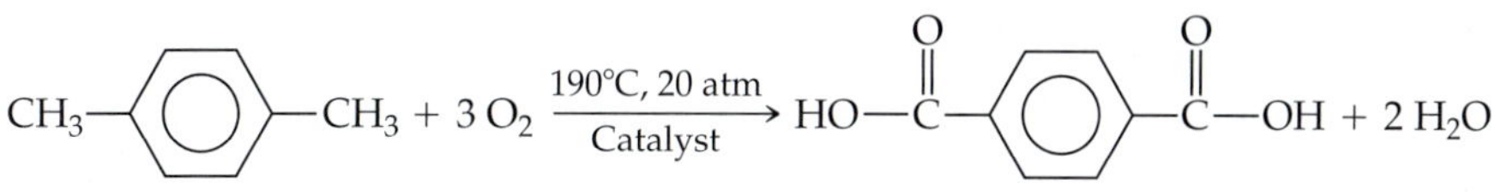

para-Xylene   Terephthalic acid

This commercial process requires pressurization and a relatively high temperature. The catalyst is a manganese/cobalt mixture, oxygen is the oxidizing agent, and the solvent is acetic acid (CH_3COOH). A group at the University of Nottingham in England has developed an alternative route that employs supercritical water as the solvent (Table 11.5) and hydrogen peroxide as the oxidant. This alternative process has several potential advantages, most particularly the elimination of acetic acid as solvent and the use of an innocuous oxidizing agent. Whether it can successfully replace the existing commercial process depends on many factors, which will require further research.

▲ **Figure 18.20 Phosgene use.** These CDs are formed from a polycarbonate plastic manufactured in a process that uses phosgene gas as a starting material.

Another relatively environmentally friendly substance that is a promising candidate as a reagent or solvent is dimethyl carbonate, which has polar character and a relatively low boiling point (90°C). It could replace less environmentally friendly substances such as dimethyl sulfate and methyl halides as a reagent that supplies the methyl (CH_3) group in reactions:

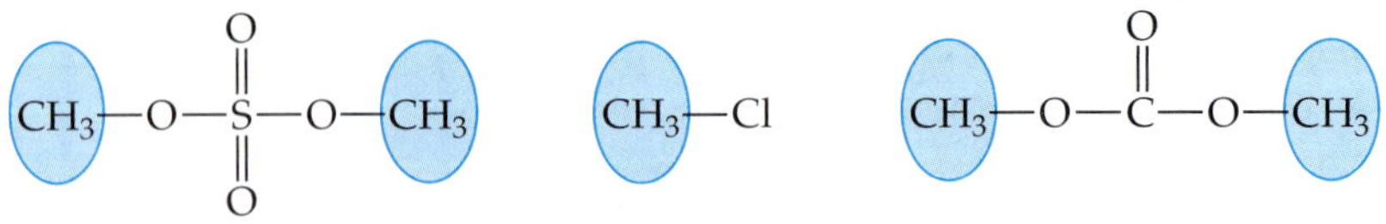

Dimethyl sulfate Methyl chloride Dimethyl carbonate

It might also be used in place of a reagent such as phosgene, Cl—CO—Cl (see the Sample Integrative Exercise in Chapter 8). Not only is phosgene itself toxic, the production of phosgene forms CCl_4 as an undesirable by-product:

$$CO(g) + Cl_2(g) \longrightarrow COCl_2(g) + CCl_4(g) \text{ (by-product)}$$

Phosgene is widely used as a reagent in commercially important reactions, such as formation of polycarbonate plastics (Figure 18.20 ◀) (Section 12.2):

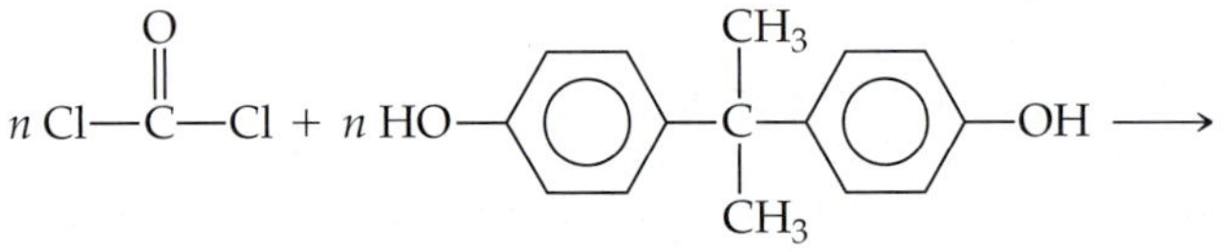

Bisphenol A

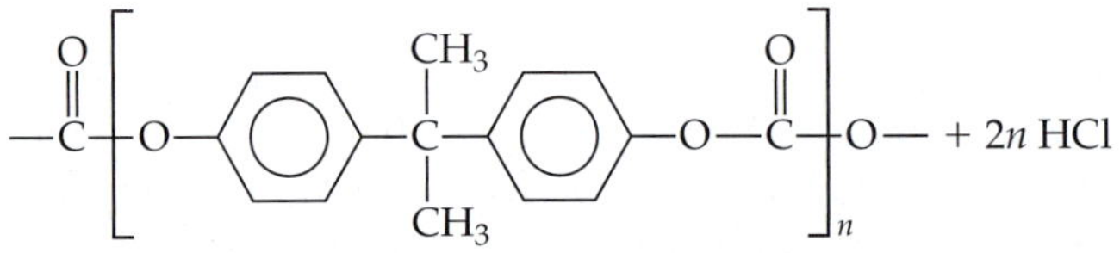

Lexan® polycarbonate

If dimethyl carbonate could be substituted for phosgene in such reactions, the by-product of the reaction would be methanol, CH_3OH, rather than HCl.

Another reaction in which phosgene is used is in the synthesis of urethanes, which can be further reacted to make polyurethanes. The reaction also produces HCl, a strong acid. A "green" synthesis of urethanes uses CO_2 instead of phosgene, and the by-product is water, not HCl:

Traditional synthesis:

1. $RNH_2 + ClC(O)Cl \longrightarrow R{-}N{=}C{=}O + 2\,HCl$
2. $R{-}N{=}C{=}O + R'{-}OH \longrightarrow R{-}NH{-}C({=}O){-}O{-}R'$

Green synthesis:

1. $RNH_2 + CO_2 \longrightarrow R{-}N{=}C{=}O + H_2O$
2. $R{-}N{=}C{=}O + R'{-}OH \longrightarrow R{-}NH{-}C({=}O){-}O{-}R'$

GIVE IT SOME THOUGHT

We noted earlier that increasing carbon dioxide levels contribute to global warming, which seems like a bad thing, but now we are saying that using carbon dioxide in industrial processes is a good thing for the environment. Explain this seeming contradiction.

Other Processes

Many processes that are important in modern society use chemicals not found in nature. Let's briefly examine two of these processes, dry cleaning and the coating of automotive bodies to prevent corrosion, and consider alternatives being developed to reduce harmful environmental impacts.

Dry cleaning of clothing typically uses a chlorinated organic solvent such as tetrachloroethylene ($Cl_2C{=}CCl_2$), which may cause cancer in humans. The widespread use of this and related solvents in dry cleaning, metal cleaning, and other industrial processes has contaminated groundwater in some areas. Alternative dry-cleaning methods that employ liquid CO_2, along with special cleaning agents, are being successfully commercialized (Figure 18.21 ▶).

▲ **Figure 18.21 Green chemistry for your clothes.** This dry cleaning apparatus employs liquid CO_2 as the solvent.

The metal bodies of cars are coated extensively during manufacture to prevent corrosion. One of the key steps is the electrodeposition of a layer of metal ions that creates an interface between the vehicle body and polymeric coatings that serve as the undercoat for painting. In the past, lead was the metal of choice for inclusion in the electrodeposition mixture. Because lead is highly toxic, however, its use in other paints and coatings has been virtually eliminated, and a relatively nontoxic yttrium hydroxide alternative to lead has been developed as an automotive coating (Figure 18.22 ▼). When this coating is subsequently heated, the hydroxide is converted to the oxide, producing an insoluble ceramic-like coating. (Section 12.2)

The Challenges of Water Purification

Access to clean water is essential to the workings of a stable, thriving society. We have seen in the previous section that disinfection of water is an important step in water treatment for human consumption. Water disinfection is one of the greatest public health innovations in human history. It has dramatically decreased the incidences of water-borne bacterial diseases such as cholera and typhus. But this great benefit comes at a price.

◀ **Figure 18.22 Green chemistry for your car.** An automobile body receives a corrosion-protection coating containing yttrium in place of lead.

In 1974 scientists in both Europe and the United States discovered that chlorination of water produces a group of by-products that had previously gone undetected. These by-products are called *trihalomethanes* (THMs) because all have a single carbon atom and three halogen atoms: $CHCl_3$, $CHCl_2Br$, $CHClBr_2$, and $CHBr_3$. These and many other chlorine- and bromine-containing organic substances are produced by the reaction of aqueous chlorine with organic materials present in nearly all natural waters, as well as with substances that are by-products of human activity. Recall that chlorine dissolves in water to form HClO, which is the active oxidizing agent (Section 7.8):

$$Cl_2(g) + H_2O(l) \longrightarrow HClO(aq) + HCl(aq) \qquad [18.16]$$

HClO in turn reacts with organic substances to form the THMs. Bromine enters through the reaction of HClO with dissolved bromide ion:

$$HOCl(aq) + Br^-(aq) \longrightarrow HBrO(aq) + Cl^-(aq) \qquad [18.17]$$

$HBrO(aq)$ halogenates organic substances analogously to $HClO(aq)$.

Some THMs and other halogenated organic substances are suspected carcinogens; others interfere with the body's endocrine system. As a result, the World Health Organization and the EPA have placed concentration limits of 80 μg/L (80 ppb) on the total quantity of such substances in drinking water. The goal is to reduce the levels of THMs and related substances in the drinking water supply while preserving the antibacterial effectiveness of the water treatment. In some cases simply lowering the concentration of chlorine may provide adequate disinfection while reducing the concentrations of THMs formed. Alternative oxidizing agents, such as ozone (O_3) or chlorine dioxide (ClO_2), produce less of the halogenated substances, but they have their own disadvantages. Each is capable of oxidizing aqueous bromide, as shown, for example, for ozone:

$$O_3(aq) + Br^-(aq) + H_2O(l) \longrightarrow HBrO(aq) + O_2(aq) + OH^-(aq) \qquad [18.18]$$

$$HBrO(aq) + 2\,O_3(aq) \longrightarrow BrO_3^-(aq) + 2\,O_2(aq) + H^+(aq) \qquad [18.19]$$

As we have seen, $HBrO(aq)$ is capable of reacting with dissolved organic substances to form halogenated organic compounds. Furthermore, bromate ion has been shown to cause cancer in animal tests.

There seem to be no completely satisfactory alternatives to chlorination at present. The risks of cancer from THMs and related substances in municipal water are very low, however, compared to the risks of cholera, typhus, and gastrointestinal disorders from untreated water. When the water supply is cleaner to begin with, less disinfectant is needed; thus, the danger of contamination through disinfection is reduced. Once the THMs are formed, their concentrations in the water supply can be reduced by aeration because the THMs are more volatile than water. Alternatively, they can be removed by adsorption onto activated charcoal or other adsorbents.

SAMPLE INTEGRATIVE EXERCISE | Putting Concepts Together

(a) Acids from acid rain or other sources are no threat to lakes in areas where the rock is limestone (calcium carbonate), which can neutralize the excess acid. Where the rock is granite, however, no such neutralization occurs. How does the limestone neutralize the acid? **(b)** Acidic water can be treated with basic substances to increase the pH, although such a procedure is usually only a temporary cure. Calculate the minimum mass of lime, CaO, needed to adjust the pH of a small lake ($V = 4 \times 10^9$ L) from 5.0 to 6.5. Why might more lime be needed?

Solution

Analyze: We need to remember what a neutralization reaction is, and calculate the amount of a substance needed to effect a certain change in pH. (Section 4.3)

Plan: For (a), we need to think about how acid can react with calcium carbonate, which evidently does not happen with acid and granite. For (b), we need to think about what reaction would happen with acid and CaO and do stoichiometric calculations. From the proposed change in pH, we can calculate the change in proton concentration needed and then figure out how much CaO it would take to do the reaction.

Solve: **(a)** The carbonate ion, which is the anion of a weak acid, is basic. (Sections 16.2 and 16.7) Thus, the carbonate ion, CO_3^{2-}, reacts with $H^+(aq)$. If the concentration of $H^+(aq)$ is small, the major product is the bicarbonate ion, HCO_3^-. If the concentration of $H^+(aq)$ is higher, however, H_2CO_3 forms and decomposes to CO_2 and H_2O. (Section 4.3)

(b) The initial and final concentrations of $H^+(aq)$ in the lake are obtained from their pH values:

$$[H^+]_{initial} = 10^{-5.0} = 1 \times 10^{-5}\,M \text{ and } [H^+]_{final} = 10^{-6.5} = 3 \times 10^{-7}\,M$$

Using the volume of the lake, we can calculate the number of moles of $H^+(aq)$ at both pH values:

$$(1 \times 10^{-5}\text{ mol/L})(4.0 \times 10^9\text{ L}) = 4 \times 10^4\text{ mol}$$

$$(3 \times 10^{-7}\text{ mol/L})(4.0 \times 10^9\text{ L}) = 1 \times 10^3\text{ mol}$$

Hence, the change in the amount of $H^+(aq)$ is 4×10^4 mol $-$ 1×10^3 mol $\approx$ 4×10^4 mol.

Let's assume that all the acid in the lake is completely ionized, so that only the free $H^+(aq)$ measured by the pH needs to be neutralized. We will need to neutralize at least that much acid, although there may be a great deal more acid in the lake than that.

The oxide ion of CaO is very basic. (Section 16.5) In the neutralization reaction 1 mol of O^{2-} reacts with 2 mol of H^+ to form H_2O. Thus, 4×10^4 mol of H^+ requires the following number of grams of CaO:

$$(4 \times 10^4\text{ mol H}^+)\left(\frac{1\text{ mol CaO}}{2\text{ mol H}^+}\right)\left(\frac{56.1\text{ g CaO}}{1\text{ mol CaO}}\right) = 1 \times 10^6\text{ g CaO}$$

This amounts to slightly more than a ton of CaO. That would not be very costly because CaO is an inexpensive base, selling for less than $100 per ton when purchased in large quantities. The amount of CaO calculated above, however, is the very minimum amount needed because there are likely to be weak acids in the water that must also be neutralized. This liming procedure has been used to adjust the pH of some small lakes to bring their pH into the range necessary for fish to live. The lake in our example would be about a half mile long, a half mile wide, and have an average depth of 20 ft.

SUMMARY AND KEY TERMS

Sections 18.1 and 18.2 In these sections we examined the physical and chemical properties of Earth's atmosphere. The complex temperature variations in the atmosphere give rise to four regions, each with characteristic properties. The lowest of these regions, the **troposphere**, extends from Earth's surface up to an altitude of about 12 km. Above the troposphere, in order of increasing altitude, are the **stratosphere**, mesosphere, and thermosphere. In the upper reaches of the atmosphere, only the simplest chemical species can survive the bombardment of highly energetic particles and radiation from the Sun. The average molecular weight of the atmosphere at high elevations is lower than that at Earth's surface because the lightest atoms and molecules diffuse upward and also because of **photodissociation**, which is the breaking of bonds in molecules because of the absorption of light. Absorption of radiation may also lead to the formation of ions via **photoionization**.

Section 18.3 Ozone is produced in the upper atmosphere from the reaction of atomic oxygen with O_2. Ozone is itself decomposed by absorption of a photon or by reaction with an active species such as NO. **Chlorofluorocarbons** can undergo photodissociation in the stratosphere, introducing atomic chlorine, which is capable of catalytically destroying ozone. A marked reduction in the ozone level in the upper atmosphere would have serious adverse consequences because the ozone layer filters out certain wavelengths of ultraviolet light that are not removed by any other atmospheric component.

Section 18.4 In the troposphere the chemistry of trace atmospheric components is of major importance. Many of these minor components are pollutants. Sulfur dioxide is one of the more noxious and prevalent examples. It is oxidized in air to form sulfur trioxide, which, upon dissolving in water, forms sulfuric acid. The oxides of sulfur are major contributors to **acid rain**. One method of preventing the escape of SO_2 from industrial operations is to react the SO_2 with CaO to form calcium sulfite ($CaSO_3$).

Carbon monoxide (CO) is found in high concentrations in automobile engine exhaust and in cigarette smoke. CO is a health hazard because it can form a strong bond with **hemoglobin** and thus reduce the capacity for blood to transfer oxygen from the lungs.

Photochemical smog is a complex mixture of components in which both nitrogen oxides and ozone play important roles. Smog components are generated mainly in automobile engines, and smog control consists largely of controlling auto emissions.

Carbon dioxide and water vapor are the major components of the atmosphere that strongly absorb infrared radiation. CO_2 and H_2O are therefore critical in maintaining Earth's temperature. The concentrations of CO_2 and other so-called greenhouse gases in the atmosphere are thus important in determining worldwide climate. As a result of the extensive combustion of fossil fuels (coal, oil, and natural gas), the carbon dioxide level of the atmosphere is steadily increasing.

Section 18.5 Seawater contains about 3.5% by mass of dissolved salts and is described as having a **salinity** of 35. Seawater's density and salinity vary with depth. Because most of the world's water is in the oceans, humans may eventually look to the seas for fresh water. **Desalination** is the removal of dissolved salts from seawater or brackish water to make it fit for human consumption. Desalination may be accomplished by distillation or by **reverse osmosis**.

Section 18.6 Fresh water contains many dissolved substances, including dissolved oxygen, which is necessary for fish and other aquatic life. Substances that are decomposed by bacteria are said to be **biodegradable**. Because the oxidation of biodegradable substances by aerobic bacteria consumes dissolved oxygen, these substances are called oxygen-demanding wastes. The presence of an excess amount of oxygen-demanding wastes in water can sufficiently deplete the dissolved oxygen to kill fish and produce offensive odors. Plant nutrients can contribute to the problem by stimulating the growth of plants that become oxygen-demanding wastes when they die.

The water available from freshwater sources may require treatment before it can be used domestically. The several steps generally used in municipal water treatment include coarse filtration, sedimentation, sand filtration, aeration, sterilization, and sometimes water softening. Water softening is required when the water contains ions such as Mg^{2+} and Ca^{2+}, which react with soap to form soap scum. Water containing such ions is called **hard water**. The **lime-soda process**, which involves adding CaO and Na_2CO_3 to hard water, is sometimes used for large-scale municipal water softening. Individual homes usually rely on **ion exchange**, a process by which hard-water ions are exchanged for Na^+ ions.

Section 18.7 The **green chemistry** initiative promotes the design and application of chemical products and processes that are compatible with human health and that preserve the environment. The areas in which the principles of green chemistry can operate to improve environmental quality include choices of solvents and reagents for chemical reactions, development of alternative processes, and improvements in existing systems and practices.

VISUALIZING CONCEPTS

18.1 At room temperature (298 K) and 1 atm pressure (which corresponds to the atmospheric pressure at sea level), one mole of an ideal gas occupies a volume of 22.4 L, according to the ideal-gas equation (Section 10.4). This is approximately the volume of 11 2-liter soda bottles. **(a)** Looking back at Figure 18.1, do you predict that 1 mole of an ideal gas in the middle of the stratosphere would occupy a greater or smaller volume than 22.4 L? **(b)** Looking at Figure 18.1(a), we see that the temperature is lower at 85 km altitude than at 50 km. Does this mean that one mole of an ideal gas would occupy less volume at 85 km than at 50 km? Explain. [Section 18.1]

18.2 Molecules in the upper atmosphere tend to contain double and triple bonds rather than single bonds. Suggest an explanation. [Section 18.2]

18.3 Why does ozone concentration in the atmosphere vary as a function of altitude (see Figure 18.4)? [Section 18.3]

18.4 You are working with an artist who has been commissioned to make a sculpture for a big city in the eastern United States. The artist is wondering what material to make her sculpture out of, since she has heard that acid rain in the eastern U.S. might destroy it over time. You take samples of granite, marble, bronze and other materials, and place them outdoors for a long time in the big city. You periodically examine the appearance of, and measure the mass of, the samples during the course of your study. **(a)** What observations would lead you to conclude that one, or more, of the materials were well-suited for the sculpture? **(b)** In those samples that show erosion through exposure, what chemical process or processes is the most likely source of the erosion? [Section 18.4]

18.5 How does carbon dioxide interact with the world ocean? [Section 18.5]

18.6 The following picture represents an ion-exchange column, in which water containing "hard" ions, like Ca^{2+}, is added to the top of the column, and out the bottom comes "softened" water that has Na^+ instead of Ca^{2+} in it. Explain what is happening in the column. [Section 18.6]

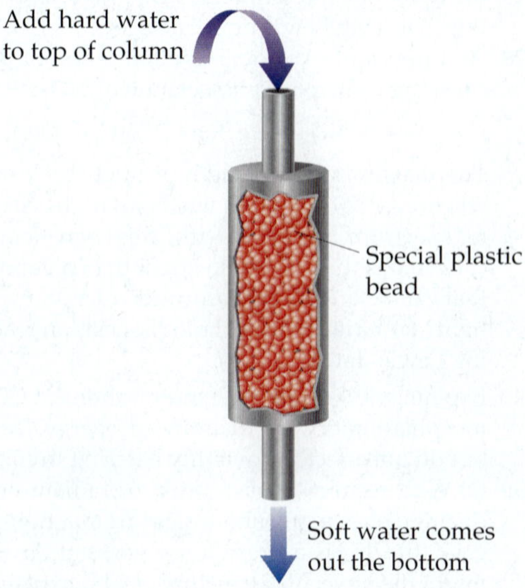

18.7 Describe the basic principles of green chemistry. [Section 18.7]

18.8 One mystery in environmental science is the imbalance in the "carbon dioxide budget." Considering only human activities, scientists have estimated that 1.6 billion metric tons of CO_2 is added to the atmosphere every year because of deforestation (plants use CO_2 to make glucose, and fewer plants will leave more CO_2 in the atmosphere) and another 5.5 billion tons per year is put into the atmosphere as a result of fossil-fuel burning. It is further estimated (again, considering only human activities) that the atmosphere actually has taken up about 3.3 billion tons of CO_2 per year, while the oceans have taken up 2 billion tons per year, leaving about 1.8 billion tons of CO_2 per year unaccounted for. This "missing" CO_2 is assumed to be taken up by the "land." What do you think might be happening?

EXERCISES

Earth's Atmosphere

18.9 **(a)** What is the primary basis for the division of the atmosphere into different regions? **(b)** Name the regions of the atmosphere, indicating the altitude interval for each one.

18.10 **(a)** How are the boundaries between the regions of the atmosphere determined? **(b)** Explain why the stratosphere, which is more than 20 miles thick, has a smaller total mass than the troposphere, which is less than 10 miles thick.

18.11 Mexico has set an air quality standard of 110 ppb (parts per billion) for ozone, but frequently the ozone levels actually measured are much higher. The concentration of ozone in Mexico City has been measured at 441 ppb (0.441 ppm). Mexico City sits at an altitude of 7400 feet, which means its atmospheric pressure is only 0.67 atm. Calculate the partial pressure of ozone at 441 ppb if the atmospheric pressure is 0.67 atm.

18.12 From the data in Table 18.1, calculate the partial pressures of carbon dioxide and argon when the total atmospheric pressure is 98.6 kPa.

18.13 The average concentration of carbon monoxide in air in Dayton, Ohio, was 3.4 ppm in the year 2000. Calculate the number of CO molecules in 1.0 L of this air at a pressure of 755 torr and a temperature of 22°C.

18.14 **(a)** From the data in Table 18.1, what is the concentration of neon in the atmosphere in ppm? **(b)** What is the concentration of neon in the atmosphere in molecules per L, assuming an atmospheric pressure of 743 torr and a temperature of 300 K?

The Upper Atmosphere; Ozone

18.15 The dissociation energy of a carbon–bromine bond is typically about 210 kJ/mol. What is the maximum wavelength of photons that can cause C—Br bond dissociation?

18.16 In CF_3Cl the C—Cl bond-dissociation energy is 339 kJ/mol. In CCl_4 the C—Cl bond-dissociation energy is 293 kJ/mol. What is the range of wavelengths of photons that can cause C—Cl bond rupture in one molecule but not in the other?

18.17 **(a)** Distinguish between *photodissociation* and *photoionization*. **(b)** Use the energy requirements of these two processes to explain why photodissociation of oxygen is more important than photoionization of oxygen at altitudes below about 90 km.

18.18 Why is the photodissociation of N_2 relatively unimportant compared with the photodissociation of O_2?

18.19 What is a hydrofluorocarbon? Why are these compounds potentially less harmful to the ozone layer than CFCs?

18.20 Draw the Lewis structure for the chlorofluorocarbon CFC-11, $CFCl_3$. What chemical characteristics of this substance allow it to effectively deplete stratospheric ozone?

18.21 **(a)** Why is the fluorine present in chlorofluorocarbons not a major contributor to depletion of the ozone layer? **(b)** What are the chemical forms in which chlorine exists in the stratosphere following cleavage of the carbon–chlorine bond?

18.22 Would you expect the substance $CFBr_3$ to be effective in depleting the ozone layer, assuming that it is present in the stratosphere? Explain.

Chemistry of the Troposphere

18.23 For each of the following gases, make a list of known or possible naturally occurring sources: **(a)** CH_4, **(b)** SO_2, **(c)** NO, **(d)** CO.

18.24 Why is rainwater naturally acidic, even in the absence of polluting gases such as SO_2?

18.25 **(a)** Write a chemical equation that describes the attack of acid rain on limestone, $CaCO_3$. **(b)** If a limestone sculpture were treated to form a surface layer of calcium sulfate, would this help to slow down the effects of acid rain? Explain.

18.26 The first stage in corrosion of iron upon exposure to air is oxidation to Fe^{2+}. **(a)** Write a balanced chemical equation to show the reaction of iron with oxygen and protons from acid rain. **(b)** Would you expect the same sort of reaction to occur with a silver surface? Explain.

18.27 Alcohol-based fuels for automobiles lead to the production of formaldehyde (CH_2O) in exhaust gases. Formaldehyde undergoes photodissociation, which contributes to photochemical smog:

$$CH_2O + h\nu \longrightarrow CHO + H$$

The maximum wavelength of light that can cause this reaction is 335 nm. **(a)** In what part of the electromagnetic spectrum is light with this wavelength found? **(b)** What is the maximum strength of a bond, in kJ/mol, that can be broken by absorption of a photon of 335-nm light? **(c)** Compare your answer from part (b) to the appropriate value from Table 8.4. What do you conclude about the C—H bond energy in formaldehyde? **(d)** Write out the formaldehyde photodissociation reaction, showing Lewis-dot structures.

18.28 An important reaction in the formation of photochemical smog is the photodissociation of NO_2:

$$NO_2 + h\nu \longrightarrow NO(g) + O(g)$$

The maximum wavelength of light that can cause this reaction is 420 nm. **(a)** In what part of the electromagnetic spectrum is light with this wavelength found? **(b)** What is the maximum strength of a bond, in kJ/mol, that can be broken by absorption of a photon of 420-nm light? **(c)** Write out the photodissociation reaction showing Lewis-dot structures.

18.29 Explain why increasing concentrations of CO_2 in the atmosphere affect the quantity of energy leaving Earth, but do not affect the quantity entering from the Sun.

18.30 **(a)** With respect to absorption of radiant energy, what distinguishes a greenhouse gas from a non-greenhouse gas? **(b)** CH_4 is a greenhouse gas, but Ar is not. How might the molecular structure of CH_4 explain why it is a greenhouse gas?

The World Ocean

18.31 What is the molarity of Na^+ in a solution of NaCl whose salinity is 5.6 if the solution has a density of 1.03 g/mL?

18.32 Phosphorus is present in seawater to the extent of 0.07 ppm by mass. If the phosphorus is present as phosphate, PO_4^{3-}, calculate the corresponding molar concentration of phosphate in seawater.

18.33 A first-stage recovery of magnesium from seawater is precipitation of $Mg(OH)_2$ with CaO:

$$Mg^{2+}(aq) + CaO(s) + H_2O(l) \longrightarrow Mg(OH)_2(s) + Ca^{2+}(aq)$$

What mass of CaO, in grams, is needed to precipitate 1000 lb of $Mg(OH)_2$?

18.34 Gold is found in seawater at very low levels, about 0.05 ppb by mass. Assuming that gold is worth about $400 per ounce, how many liters of seawater would you have to process to obtain $1,000,000 worth of gold? Assume the density of seawater is 1.03 g/mL and that your gold recovery process is 50% efficient.

18.35 Suppose that one wishes to use reverse osmosis to reduce the salt content of brackish water containing 0.22 *M* total salt concentration to a value of 0.01 *M*, thus rendering it usable for human consumption. What is the minimum pressure that needs to be applied in the permeators (Figure 18.15) to achieve this goal, assuming that the operation occurs at 298 K? [*Hint:* Refer to Section 13.5.]

18.36 Assume that a portable reverse-osmosis apparatus such as that shown in Figure 18.16 operates on seawater, whose concentrations of constituent ions are listed in Table 18.6, and that the desalinated water output has an effective molarity of about 0.02 *M*. What minimum pressure must be applied by hand pumping at 305 K to cause reverse osmosis to occur? [*Hint:* Refer to Section 13.5.]

Fresh Water

18.37 List the common products formed when an organic material containing the elements carbon, hydrogen, oxygen, sulfur, and nitrogen decomposes **(a)** under aerobic conditions, **(b)** under anaerobic conditions.

18.38 **(a)** Explain why the concentration of dissolved oxygen in fresh water is an important indicator of the quality of the water. **(b)** How is the solubility of oxygen in water affected by increasing temperature?

18.39 The organic anion

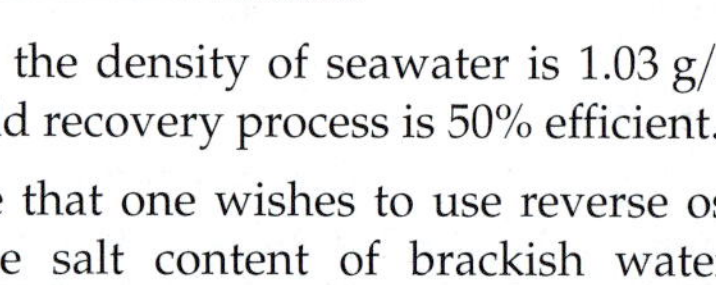

is found in most detergents. Assume that the anion undergoes aerobic decomposition in the following manner:

$$2\,C_{18}H_{29}SO_3^-(aq) + 51\,O_2(aq) \longrightarrow 36\,CO_2(aq) + 28\,H_2O(l) + 2\,H^+(aq) + 2\,SO_4^{2-}(aq)$$

What is the total mass of O_2 required to biodegrade 1.0 g of this substance?

18.40 The average daily mass of O_2 taken up by sewage discharged in the United States is 59 g per person. How many liters of water at 9 ppm O_2 are totally depleted of oxygen in 1 day by a population of 120,000 people?

18.41 Write a balanced chemical equation to describe how magnesium ions are removed in water treatment by the addition of slaked lime, $Ca(OH)_2$.

18.42 **(a)** Which of the following ionic species is, or could be, responsible for hardness in a water supply: Ca^{2+}, K^+, Mg^{2+}, Fe^{2+}, Na^+? **(b)** What properties of an ion determine whether it will contribute to water hardness?

18.43 How many moles of $Ca(OH)_2$ and Na_2CO_3 should be added to soften 1200 L of water in which $[Ca^{2+}] = 5.0 \times 10^{-4}$ *M* and $[HCO_3^-] = 7.0 \times 10^{-4}$ *M*?

18.44 The concentration of Ca^{2+} in a particular water supply is 5.7×10^{-3} *M*. The concentration of bicarbonate ion, HCO_3^-, in the same water is 1.7×10^{-3} *M*. What masses of $Ca(OH)_2$ and Na_2CO_3 must be added to 5.0×10^7 L of this water to reduce the level of Ca^{2+} to 20% of its original level?

18.45 Ferrous sulfate ($FeSO_4$) is often used as a coagulant in water purification. The iron(II) salt is dissolved in the water to be purified, then oxidized to the iron(III) state by dissolved oxygen, at which time gelatinous $Fe(OH)_3$ forms, assuming the pH is above approximately 6. Write balanced chemical equations for the oxidation of Fe^{2+} to Fe^{3+} by dissolved oxygen, and for the formation of $Fe(OH)_3(s)$ by reaction of $Fe^{3+}(aq)$ with $HCO_3^-(aq)$.

18.46 What properties make a substance a good coagulant for water purification?

Green Chemistry

18.47 One of the principles of green chemistry is that it is better to use as few steps as possible in making new chemicals. How, if at all, does this principle relate to energy efficiency?

18.48 Discuss how catalysts can make processes more energy-efficient.

18.49 The Baeyer–Villiger reaction is a classic organic oxidation reaction for converting ketones to lactones, as in this reaction:

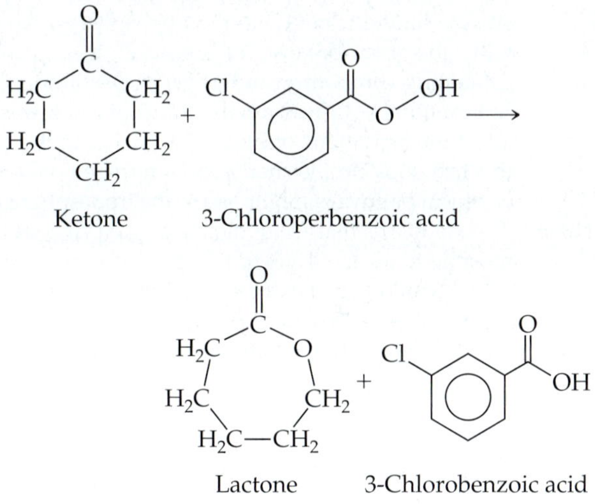

The reaction is used in the manufacture of plastics and pharmaceuticals. The reactant 3-chloroperbenzoic acid is somewhat shock-sensitive, however, and prone to explode. Also, 3-chlorobenzoic acid is a waste product. An alternative process being developed uses hydrogen peroxide and a catalyst consisting of tin deposited within a solid support. The catalyst is readily recovered from the reaction mixture. **(a)** What would you expect to be the other product of oxidation of the ketone to lactone by hydrogen peroxide? **(b)** What principles of green chemistry are addressed by use of the proposed process?

18.50 The reaction shown here was performed with an iridium catalyst, both in supercritical CO_2 ($scCO_2$) and in the chlorinated solvent CH_2Cl_2. The kinetic data for the reaction in both solvents are plotted in the graph. Why is this a good example of a green chemical reaction?

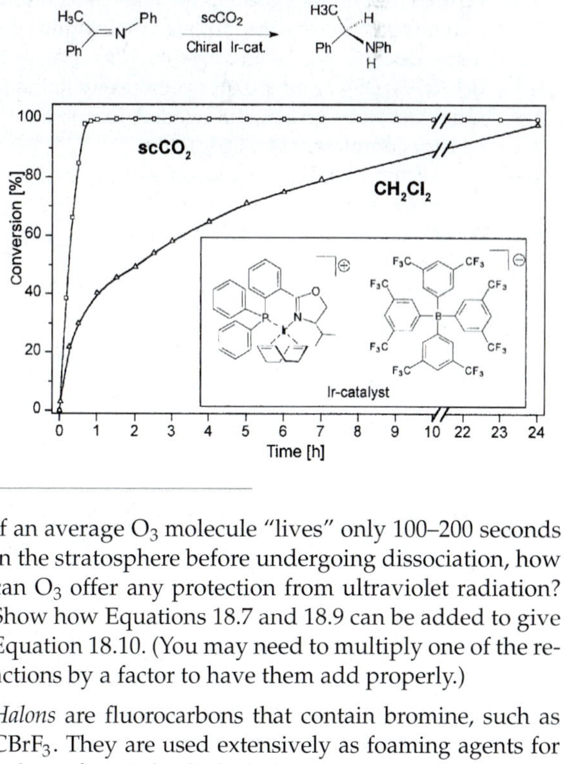

Additional Exercises

18.51 A friend of yours has seen each of the following items in newspaper articles and would like an explanation: **(a)** acid rain, **(b)** greenhouse gas, **(c)** photochemical smog, **(d)** ozone depletion. Give a brief explanation of each term, and identify one or two of the chemicals associated with each.

18.52 Suppose that on another planet the atmosphere consists of 17% Kr, 38% CH_4, and 45% O_2. What is the average molar mass at the surface? What is the average molar mass at an altitude at which all the O_2 is photodissociated?

18.53 If an average O_3 molecule "lives" only 100–200 seconds in the stratosphere before undergoing dissociation, how can O_3 offer any protection from ultraviolet radiation?

18.54 Show how Equations 18.7 and 18.9 can be added to give Equation 18.10. (You may need to multiply one of the reactions by a factor to have them add properly.)

18.55 *Halons* are fluorocarbons that contain bromine, such as $CBrF_3$. They are used extensively as foaming agents for fighting fires. Like CFCs, halons are very unreactive and ultimately can diffuse into the stratosphere. **(a)** Based on the data in Table 8.4, would you expect photodissociation of Br atoms to occur in the stratosphere? **(b)** Propose a mechanism by which the presence of halons in the stratosphere could lead to the depletion of stratospheric ozone.

[18.56] The *hydroxyl radical*, OH, is formed at low altitudes via the reaction of excited oxygen atoms with water:

$$O^*(g) + H_2O(g) \longrightarrow 2\,OH(g)$$

(a) Write the Lewis structure for the hydroxyl radical. [*Hint:* It has one unpaired electron.]
Once produced, the hydroxyl radical is very reactive. Explain why each of the following series of reactions affects the pollution in the troposphere:
(b) $OH + NO_2 \longrightarrow HNO_3$
(c) $OH + CO + O_2 \longrightarrow CO_2 + OOH$
$OOH + NO \longrightarrow OH + NO_2$
(d) $OH + CH_4 \longrightarrow H_2O + CH_3$
$CH_3 + O_2 \longrightarrow OOCH_3$
$OOCH_3 + NO \longrightarrow OCH_3 + NO_2$

18.57 Explain, using Le Châtelier's principle, why the equilibrium constant for the formation of NO from N_2 and O_2 increases with increasing temperature, whereas the equilibrium constant for the formation of NO_2 from NO and O_2 decreases with increasing temperature.

18.58 The affinity of carbon monoxide for hemoglobin is about 210 times that of O_2. Assume a person is inhaling air that contains 125 ppm of CO. If all the hemoglobin leaving the lungs carries either oxygen or CO, calculate the fraction in the form of carboxyhemoglobin.

18.59 Natural gas consists primarily of methane, $CH_4(g)$. **(a)** Write a balanced chemical equation for the complete combustion of methane to produce $CO_2(g)$ as the only carbon-containing product. **(b)** Write a balanced chemical equation for the incomplete combustion of methane to produce $CO(g)$ as the only carbon-containing product. **(c)** At 25°C and 1.0 atm pressure, what is the minimum quantity of dry air needed to combust 1.0 L of $CH_4(g)$ completely to $CO_2(g)$?

18.60 One of the possible consequences of global warming is an increase in the temperature of ocean water. The oceans serve as a "sink" for CO_2 by dissolving large amounts of it. **(a)** How would the solubility of CO_2 in the oceans be affected by an increase in the temperature of the water? **(b)** Discuss the implications of your answer to part (a) for the problem of global warming.

18.61 The solar energy striking Earth averages 169 watts per square meter. The energy radiated from Earth's surface averages 390 watts per square meter. Comparing these numbers, one might expect that the planet would cool quickly, yet it does not. Why not?

18.62 The solar energy striking Earth every day averages 169 watts per square meter, and each day the people of New York City use about 55 watts per square meter. Considering that present technology for solar energy conversion is only about 10% efficient, from how many square meters of land must sunlight be collected in order to power New York City for one day?

18.63 Write balanced chemical equations for each of the following reactions: **(a)** The nitric oxide molecule undergoes photodissociation in the upper atmosphere. **(b)** The nitric oxide molecule undergoes photoionization in the upper atmosphere. **(c)** Nitric oxide undergoes oxidation by ozone in the stratosphere. **(d)** Nitrogen dioxide dissolves in water to form nitric acid and nitric oxide.

18.64 **(a)** Explain why $Mg(OH)_2$ precipitates when CO_3^{2-} ion is added to a solution containing Mg^{2+}. **(b)** Will $Mg(OH)_2$ precipitate when 4.0 g of Na_2CO_3 is added to 1.00 L of a solution containing 125 ppm of Mg^{2+}?

[18.65] It has recently been pointed out that there may be increased amounts of NO in the troposphere as compared with the past because of massive use of nitrogen-containing compounds in fertilizers. Assuming that NO can eventually diffuse into the stratosphere, how might it affect the conditions of life on Earth? Using the index to this text, look up the chemistry of nitrogen oxides. What chemical pathways might NO in the troposphere follow?

[18.66] EPA standards that went into effect in April 2004 limit atmospheric ozone levels to 85 ppb. How many moles of ozone would there be in the air above Los Angeles County (area about 4000 square miles; consider a height of 10 m above the ground) if ozone was at this concentration?

Integrative Exercises

18.67 The estimated average concentration of NO_2 in air in the United States in 2002 was 0.019 ppm. **(a)** Calculate the partial pressure of the NO_2 in a sample of this air when the atmospheric pressure is 755 torr (99.1 kPa). **(b)** How many molecules of NO_2 are present under these conditions at 20°C in a room that measures $15 \times 14 \times 8$ ft?

[18.68] In 1986 an electrical power plant in Taylorsville, Georgia, burned 8,376,726 tons of coal, a national record at that time. **(a)** Assuming that the coal was 83% carbon and 2.5% sulfur and that combustion was complete, calculate the number of tons of carbon dioxide and sulfur dioxide produced by the plant during the year. **(b)** If 55% of the SO_2 could be removed by reaction with powdered CaO to form $CaSO_3$, how many tons of $CaSO_3$ would be produced?

18.69 The water supply for a midwestern city contains the following impurities: coarse sand, finely divided particulates, nitrate ion, trihalomethanes, dissolved phosphorus in the form of phosphates, potentially harmful bacterial strains, dissolved organic substances. Which of the following processes or agents, if any, is effective in removing each of these impurities: coarse sand filtration, activated carbon filtration, aeration, ozonization, precipitation with aluminum hydroxide?

18.70 The concentration of H_2O in the stratosphere is about 5 ppm. It undergoes photodissociation as follows:

$$H_2O(g) \longrightarrow H(g) + OH(g)$$

(a) Write out the Lewis-dot structures for both products and reactant.
(b) Using Table 8.4, calculate the wavelength required to cause this dissociation.
(c) The hydroxyl radicals, OH, can react with ozone, giving the following reactions:

$$OH(g) + O_3(g) \longrightarrow HO_2(g) + O_2(g)$$
$$HO_2(g) + O(g) \longrightarrow OH(g) + O_2(g)$$

What overall reaction results from these two elementary reactions? What is the catalyst in the overall reaction? Explain.

18.71 Bioremediation is the process by which bacteria repair their environment in response, for example, to an oil spill. The efficiency of bacteria for "eating" hydrocarbons depends on the amount of oxygen in the system, pH, temperature, and many other factors. In a certain oil spill, hydrocarbons from the oil disappeared with a first-order rate constant of $2 \times 10^{-6}\ s^{-1}$. How many days did it take for the oil spill to decrease to 10% of its initial value?

18.72 The standard enthalpies of formation of ClO and ClO_2 are 101 and 102 kJ/mol, respectively. Using these data and the thermodynamic data in Appendix C, calculate the overall enthalpy change for each step in the following catalytic cycle:

$$ClO(g) + O_3(g) \longrightarrow ClO_2(g) + O_2(g)$$
$$ClO_2(g) + O(g) \longrightarrow ClO(g) + O_2(g)$$

What is the enthalpy change for the overall reaction that results from these two steps?

18.73 The main reason that distillation is a costly method for purifying water is the high energy required to heat and vaporize water. **(a)** Using the density, specific heat, and heat of vaporization of water from Appendix B, calculate the amount of energy required to vaporize 1.00 gal of water beginning with water at 20°C. **(b)** If the energy is provided by electricity costing \$0.085/kwh, calculate its cost. **(c)** If distilled water sells in the grocery store for \$1.26 per gal, what percentage of the sales price is represented by the cost of the energy?

[18.74] A reaction that contributes to the depletion of ozone in the stratosphere is the direct reaction of oxygen atoms with ozone:

$$O(g) + O_3(g) \longrightarrow 2\,O_2(g)$$

At 298 K the rate constant for this reaction is $4.8 \times 10^5\ M^{-1}\ s^{-1}$. **(a)** Based on the units of the rate constant, write the likely rate law for this reaction. **(b)** Would you expect this reaction to occur via a single elementary process? Explain why or why not. **(c)** From the magnitude of the rate constant, would you expect the activation energy of this reaction to be large or small? Explain. **(d)** Use ΔH_f° values from Appendix C to estimate the enthalpy change for this reaction. Would this reaction raise or lower the temperature of the stratosphere?

18.75 Nitrogen dioxide (NO_2) is the only important gaseous species in the lower atmosphere that absorbs visible light. **(a)** Write the Lewis structure(s) for NO_2. **(b)** How does this structure account for the fact that NO_2 dimerizes to form N_2O_4? Based on what you can find about this dimerization reaction in the text, would you expect to find the NO_2 that forms in an urban environment to be in the form of dimer? Explain. **(c)** What would you expect as products, if any, for the reaction of NO_2 with CO? **(d)** Would you expect NO_2 generated in an urban environment to migrate to the stratosphere? Explain.

[18.76] If the pH of a 1.0-in. rainfall over 1500 mi^2 is 3.5, how many kilograms of H_2SO_4 are present, assuming that it is the only acid contributing to the pH?

[18.77] The Henry's law constant for CO_2 in water at 25°C is $3.1 \times 10^{-2}\ M\ atm^{-1}$. **(a)** What is the solubility of CO_2 in water at this temperature if the solution is in contact with air at normal atmospheric pressure? **(b)** Assume that all of this CO_2 is in the form of H_2CO_3 produced by the reaction between CO_2 and H_2O:

$$CO_2(aq) + H_2O(l) \longrightarrow H_2CO_3(aq)$$

What is the pH of this solution?

[18.78] The precipitation of $Al(OH)_3$ ($K_{sp} = 1.3 \times 10^{-33}$) is sometimes used to purify water. **(a)** Estimate the pH at which precipitation of $Al(OH)_3$ will begin if 5.0 lb of $Al_2(SO_4)_3$ is added to 2000 gal of water. **(b)** Approximately how many pounds of CaO must be added to the water to achieve this pH?

eMEDIA EXERCISES

These exercises make use of the interactive objects available online in OneKey or the Companion Website, and on your Accelerator CD. Access to these resources comes in your MediaPak.

18.79 The formation of ozone molecules in the stratosphere is shown in the **Stratospheric Ozone** animation (*18.3*). **(a)** Describe this process, including equations where appropriate. **(b)** Explain why the process described in part (a) does not occur to the same extent in the troposphere as it does in the stratosphere.

18.80 The **Stratospheric Ozone** animation (*18.3*) shows two ways in which an ozone molecule can dissociate. **(a)** Describe both of these processes. **(b)** Which of these processes protects Earth from harmful radiation?

18.81 The **Catalytic Destruction of Stratospheric Ozone** animation (*18.3*) shows how NO lowers the concentration of ozone by catalyzing its decomposition. Based on the movie, describe how you think a higher NO_2 concentration in the stratosphere would affect ozone concentration. Explain your reasoning, and use equations where necessary.

18.82 The **CFCs and Stratospheric Ozone** animation (*18.3*) portrays chlorofluorocarbons as very stable molecules. **(a)** If CFCs are so stable, why are they considered to be a serious problem in the upper atmosphere? **(b)** If the countries of the world were to stop producing CFCs altogether, their effects in the upper atmosphere would continue for decades. Explain why this is so.

18.83 In Exercise 18.24, you explained why rainwater is acidic even in the absence of pollution, such as SO_2 gas, and the movie **Carbon Dioxide Behaves as an Acid in Water** (*18.4*) shows how CO_2 acidifies water. Compare the typical atmospheric concentrations (Table 18.3) of CO_2 and SO_2, which is oxidized to SO_3 in the troposphere. Which one would you expect to have the greater influence on the pH of rainwater? Explain.

CHAPTER

19

WHAT'S AHEAD

- In this chapter we will see why changes that occur in nature have a directional character: They move *spontaneously* in one direction but not in the reverse direction. *(Section 19.1)*
- The thermodynamic function *entropy* is a state function that is important in determining whether a process is spontaneous. The *second law of thermodynamics* tells us that in any spontaneous process the entropy of the universe (system plus surroundings) increases. *(Section 19.2)*
- On a molecular level we will see that an entropy increase is associated with an increase in the number of accessible *microstates*. Entropy change can also be thought of as a measure of a system's randomness at a stated temperature. *(Section 19.3)*
- The *third law of thermodynamics* states that the entropy of a perfect crystalline solid at 0 K is zero. From this reference point the entropies of pure substances at temperatures above absolute zero can be determined. Using tabulated entropy values, we can find the entropy values of the reactants and products of a reaction and use them to calculate the entropy change of the reaction. *(Sections 19.3 and 19.4)*
- The *free energy* (or *Gibbs free energy*) of a system is a measure of how far removed the system is from equilibrium. It measures the maximum amount of useful work obtainable from a process and yields information on the direction in which a chemical reaction will proceed spontaneously. *(Section 19.5)*
- The relationship among free-energy change, enthalpy change, and entropy change provides important insight into how temperature affects the spontaneity of a process. *(Section 19.6)*
- The standard free-energy change for a chemical reaction can be used to calculate the equilibrium constant for the process. *(Section 19.7)*

Chemical Thermodynamics

THE GAS BURNER SHOWN IN THE CHAPTER-OPENING PHOTOGRAPH uses CH_4 from natural gas and O_2 from air to produce CO_2, H_2O, and heat. The reaction is useful because it occurs rapidly, with virtually complete conversion of the reactants to products, and because it produces energy. Two of the most important questions chemists ask when designing and using chemical reactions are "How fast is the reaction?" and "How far does it proceed?" The first question is addressed by the study of chemical kinetics, which we discussed in Chapter 14. The second question involves the equilibrium constant, which was the focus of Chapter 15.

In Chapter 14 we learned that the rates of chemical reactions are controlled largely by a factor related to energy, namely the activation energy of the reaction. ∞ (Section 14.5) In general, the lower the activation energy, the faster a reaction proceeds. In Chapter 15 we saw that equilibrium depends on the rates of the forward and reverse reactions: Equilibrium is reached when the opposing reactions occur at equal rates. ∞ (Section 15.1) Because reaction rates are closely tied to energy, it is logical that equilibrium also depends in some way on energy.

In this chapter we will explore the connection between energy and the extent of a reaction. Doing so requires us to take a deeper look at *chemical thermodynamics*, the area of chemistry that deals with energy relationships. We first encountered thermodynamics in Chapter 5, where we discussed the nature of energy, the first law of thermodynamics,

THE COMBUSTION OF NATURAL GAS (methane, CH_4) in air, producing carbon dioxide and water, is a spontaneous process. Once lit, natural gas burns spontaneously. The reverse reaction, the combination of CO_2 and H_2O to form CH_4 and O_2, does not occur of its own accord.

and the concept of enthalpy. Recall that the enthalpy change is the heat transferred between the system and its surroundings during a constant-pressure process. (Section 5.3)

Now we will see that reactions involve not only changes in enthalpy but also changes in another important thermodynamic quantity—*entropy*. Our discussion of entropy will lead us to the second law of thermodynamics, which provides insight into why physical and chemical changes tend to favor one direction over another. We drop a brick, for example, and it falls to the ground. We don't expect bricks to spontaneously rise from the ground to our outstretched hand. We light a candle, and it burns down. We don't expect a half-consumed candle to regenerate itself spontaneously, even if we have kept all the gases produced when the candle burned. Thermodynamics helps us understand the significance of this directional character of processes, whether they be exothermic or endothermic.

19.1 | Spontaneous Processes

ANIMATION
Air Bags

The first law of thermodynamics states that *energy is conserved*. (Section 5.2) In other words, energy is neither created nor destroyed in any process, whether it be the falling of a brick, the burning of a candle, or the melting of an ice cube. Energy can be transferred between a system and the surroundings or can be converted from one form to another, but the total energy remains constant. We expressed the first law of thermodynamics mathematically as $\Delta E = q + w$, where ΔE is the change in the internal energy of a system, q is the heat absorbed by the system from the surroundings, and w is the work done on the system by the surroundings.

The first law helps us balance the books, so to speak, on the heat transferred between a system and its surroundings and the work done by a particular process or reaction. But the first law doesn't address another important feature of reactions, the extent to which they occur. As we noted in the introduction, our experience tells us that physical and chemical processes have a directional character. For instance, sodium metal and chlorine gas combine readily to form sodium chloride, which we also know as table salt. We never find table salt decomposing of its own accord to form sodium and chlorine. (Have you ever smelled chlorine gas in the kitchen or seen sodium metal on your table salt?) In both processes—the formation of sodium chloride from sodium and chlorine and the decomposition of sodium chloride into sodium and chlorine—energy is conserved, as it must be according to the first law of thermodynamics. Yet one process occurs, and the other does not. A process that occurs of its own accord without any ongoing outside intervention is said to be spontaneous. A **spontaneous process** is one that proceeds on its own without any outside assistance.

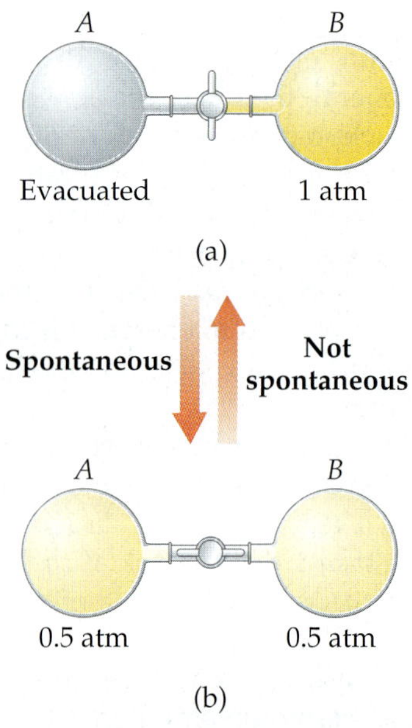

▲ **Figure 19.1 Spontaneous expansion of an ideal gas into an evacuated space.** In (a) flask *B* holds an ideal gas at 1 atm pressure and flask *A* is evacuated. In (b) the stopcock connecting the flasks has been opened. The ideal gas expands to occupy both flasks *A* and *B* at a pressure of 0.5 atm. The reverse process–all the gas molecules moving back into flask *B*–is not spontaneous.

A spontaneous process occurs in a definite direction. For example, as mentioned above, a dropped brick falls downward, toward the ground. Now imagine you were to see a video clip in which a brick rises from the ground, back into your hand. You would conclude that the video is running in reverse—bricks do not magically rise from the ground! A brick falling is a spontaneous process, whereas the reverse process is *nonspontaneous*.

A gas will expand into a vacuum as shown in Figure 19.1 ◀, but the process will never reverse itself. The expansion of the gas is spontaneous. Likewise, a nail left out in the weather will rust (Figure 19.2 ▶). In this process the iron in the nail reacts with oxygen from the air to form an iron oxide. We would never expect the rusty nail to reverse this process and become shiny. The rusting process is spontaneous, whereas the reverse process is nonspontaneous. There are countless other examples we could cite that illustrate the same idea:

Processes that are spontaneous in one direction are nonspontaneous in the opposite direction.

Experimental conditions, such as temperature and pressure, are often important in determining whether a process is spontaneous. Consider, for example, the melting of ice. When the temperature of the surroundings is above 0°C at ordinary atmospheric pressures, ice melts spontaneously and the reverse process—liquid water turning into ice—is not spontaneous. However, when the surroundings are below 0°C, the opposite is true. Liquid water converts into ice spontaneously, and the conversion of ice into water is *not* spontaneous (Figure 19.3 ▼).

What happens at $T = 0°C$, the normal melting point of water, when the flask of Figure 19.3 contains both water and ice? At the normal melting point of a substance, the solid and liquid phases are in equilibrium. ∞ (Section 11.5) At this temperature the two phases are interconverting at the same rate and there is no preferred direction for the process.

It is important to realize that the fact that a process is spontaneous does not necessarily mean that it will occur at an observable rate. A chemical reaction is spontaneous if it occurs on its own accord, regardless of its speed. A spontaneous reaction can be very fast, as in the case of acid-base neutralization, or very slow, as in the rusting of iron. Thermodynamics can tell us the *direction* and *extent* of a reaction but tells us nothing about the *speed* of the reaction.

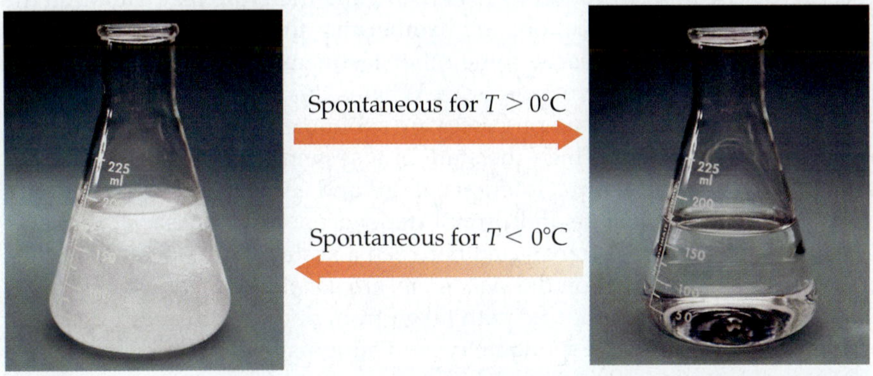

▲ **Figure 19.3 Spontaneity can depend on the temperature.** At $T > 0°C$ ice melts spontaneously to liquid water. At $T < 0°C$ the reverse process, water freezing to ice, is spontaneous. At $T = 0°C$ the two states are in equilibrium and neither conversion occurs spontaneously.

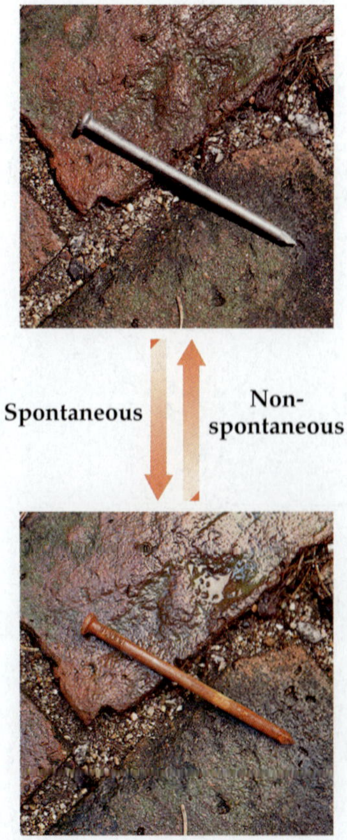

▲ **Figure 19.2 A spontaneous process.** Elemental iron in the shiny nail in the top photograph spontaneously combines with H_2O and O_2 in the surrounding air to form a layer of rust—Fe_2O_3—on the nail surface.

SAMPLE EXERCISE 19.1 | Identifying Spontaneous Processes

Predict whether the following processes are spontaneous as described, spontaneous in the reverse direction, or in equilibrium: **(a)** When a piece of metal heated to 150°C is added to water at 40°C, the water gets hotter. **(b)** Water at room temperature decomposes into $H_2(g)$ and $O_2(g)$. **(c)** Benzene vapor, $C_6H_6(g)$, at a pressure of 1 atm condenses to liquid benzene at the normal boiling point of benzene, 80.1°C.

Solution

Analyze: We are asked to judge whether each process will proceed spontaneously in the direction indicated, in the reverse direction, or in neither direction.

Plan: We need to think about whether each process is consistent with our experience about the natural direction of events or whether it is the reverse process that we expect to occur.

Solve: **(a)** This process is spontaneous. Whenever two objects at different temperatures are brought into contact, heat is transferred from the hotter object to the colder one. In this instance heat is transferred from the hot metal to the cooler water. The final temperature, after the metal and water achieve the same temperature (thermal equilibrium), will be somewhere between the initial temperatures of the metal and the water. **(b)** Experience tells us that this process is not spontaneous—we certainly have never seen hydrogen and oxygen gas bubbling up out of water! Rather, the *reverse* process—the reaction of H_2 and O_2 to form H_2O—is spontaneous once initiated by a spark or flame. **(c)** By definition, the normal boiling point is the temperature at which a vapor at 1 atm is in equilibrium with its liquid. Thus, this is an equilibrium situation. Neither the condensation of benzene vapor nor the reverse process is spontaneous. If the temperature were below 80.1°C, condensation would be spontaneous.

PRACTICE EXERCISE

Under 1 atm pressure $CO_2(s)$ sublimes at −78°C. Is the transformation of $CO_2(s)$ to $CO_2(g)$ a spontaneous process at −100°C and 1 atm pressure?
Answer: No, the reverse process is spontaneous at this temperature.

Seeking a Criterion for Spontaneity

A marble rolling down an incline or a brick falling from your hand loses energy. The loss of energy is a common feature of spontaneous change in mechanical systems. In the 1870s Marcellin Bertholet (1827–1907), a famous chemist of that era, suggested that the direction of spontaneous changes in chemical systems was also determined by the loss of energy. He proposed that all spontaneous chemical and physical changes were exothermic. It takes only a few moments, however, to find exceptions to this generalization. For example, the melting of ice at room temperature is spontaneous even though melting is an endothermic process. Similarly, there are spontaneous solution processes that are endothermic, as we discovered in Section 13.1. We conclude therefore that, although the majority of spontaneous reactions are exothermic, there are spontaneous endothermic ones as well. Clearly, some other factor must be at work in determining the natural direction of processes. What is this factor?

To understand why certain processes are spontaneous, we need to consider more closely the ways in which the state of a system can change. Recall that quantities such as temperature, internal energy, and enthalpy are *state functions*, properties that define a state and do not depend on how we reach that state. ∞ (Section 5.2) The heat transferred between a system and its surroundings, q, and the work done by or on the system, w, are *not* state functions; the values of q and w depend on the specific path taken from one state to another. One of the keys to understanding spontaneity is distinguishing between reversible and irreversible paths between states.

GIVE IT SOME THOUGHT

If a process is nonspontaneous, does that mean that it cannot occur under any circumstances?

Reversible and Irreversible Processes

In 1824 a 28-year-old French engineer by the name of Sadi Carnot (1796–1832) published an analysis of the factors that determine how efficiently a steam engine can convert heat to work. Carnot considered what an ideal engine, one with the highest possible efficiency, would be like. He observed that it is impossible to convert the energy content of a fuel completely to work because a significant amount of heat is always lost to the surroundings. Not only did his analysis give insight into how to build better, more efficient engines, but it also was one of the earliest studies in what has developed into the discipline of thermodynamics.

About forty years later, Rudolph Clausius (1822–1888), a German physicist, extended Carnot's work in an important way. Clausius concluded that a special significance could be ascribed to the ratio of the heat delivered to an ideal engine and the temperature at which it is delivered, q/T. He was so convinced of the importance of this ratio that he gave it a special name, *entropy*. He deliberately selected the name to sound like energy to emphasize his belief that the importance of entropy was comparable to that of energy.

An ideal engine, one with the maximum efficiency, operates under an ideal set of conditions in which all the processes are reversible. In a **reversible process,** a system is changed in such a way that the system and surroundings can be restored to their original state by *exactly* reversing the change. In other words, we can completely restore the system to its original condition with no net change to either the system or its surroundings. An **irreversible process** is one that cannot simply be reversed to restore the system and its surroundings to their original states. What Carnot discovered is that the amount of work we can extract from any spontaneous process depends on the manner in which the process is carried out. *A reversible change produces the maximum amount of work that can be achieved by the system on the surroundings* ($w_{rev} = w_{max}$).

GIVE IT SOME THOUGHT

If you evaporate water and then condense it, have you necessarily performed a reversible process?

Let's examine some examples of reversible and irreversible processes. If two objects are at different temperatures, heat will flow spontaneously from the hotter object to the colder one. Because it is impossible to make heat flow in the opposite direction, the flow of heat is irreversible. Given these facts, can we imagine any conditions under which heat transfer can be made reversible? Consider two objects or a system and its surroundings that are at essentially the same temperature, with just an infinitesimal difference (an extremely small temperature difference, ΔT) to make heat flow in the desired direction (Figure 19.4 ▶). We can then reverse the direction of heat flow by making an infinitesimal change of temperature in the opposite direction. *Reversible processes are those that reverse direction whenever an infinitesimal change is made in some property of the system.*

Now consider another example, the expansion of an ideal gas at constant temperature. A constant-temperature process such as this is said to be **isothermal**. To keep the example simple, consider the gas in the cylinder-and-piston arrangement shown in Figure 19.5 ▼. When the partition is removed, the gas expands spontaneously to fill the evacuated space. Because the gas is expanding into a vacuum with no external pressure, it does no *P-V* work on the surroundings ($w = 0$). (Section 5.3) We can use the piston to compress the gas back to its original state, but doing so requires that the surroundings do work on the system ($w > 0$). That is, reversing the process has produced a change in the surroundings as energy is used to do work on the system. The fact that the system and the surroundings are not both returned to their original conditions indicates that the process is irreversible.

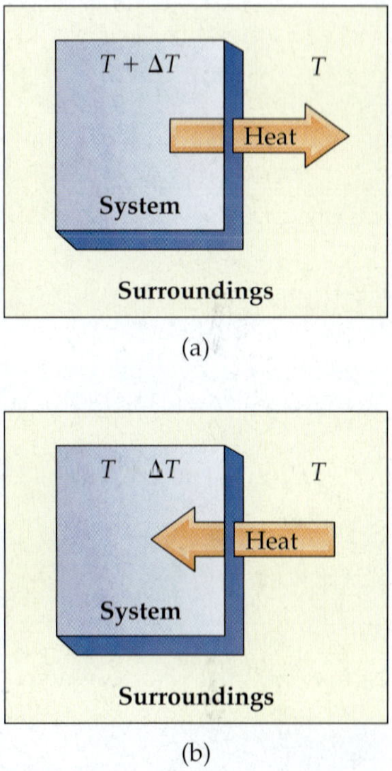

▲ **Figure 19.4 Reversible flow of heat.** Heat can flow reversibly between a system and its surroundings if the two have only an infinitesimally small difference in temperature, ΔT. The direction of heat flow can be changed by increasing or decreasing the temperature of the system by ΔT. (a) Increasing the temperature of the system by ΔT causes heat to flow from the system to the surroundings. (b) Decreasing the temperature of the system by ΔT causes heat to flow from the surroundings into the system.

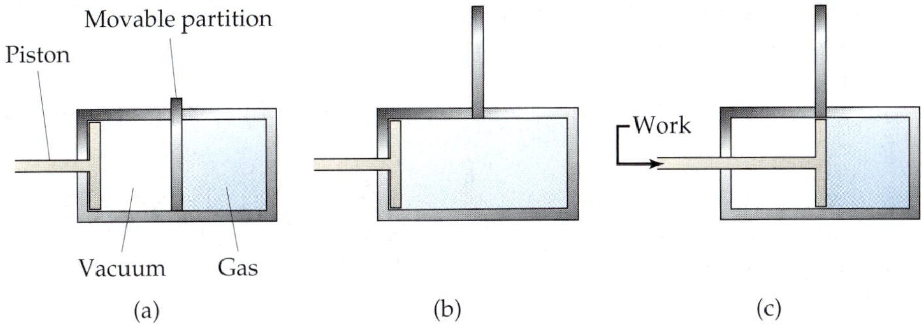

◀ **Figure 19.5 An irreversible process.** Restoring the system to its original state after an irreversible process changes the surroundings. In (a) the gas is confined to the right half of the cylinder by a partition. When the partition is removed (b), the gas spontaneously (irreversibly) expands to fill the whole cylinder. No work is done by the system during this expansion. In (c) we can use the piston to compress the gas back to its original state. Doing so requires that the surroundings do work on the system, which changes the surroundings forever.

What might a reversible, isothermal expansion of an ideal gas be like? It will occur only if the external pressure acting on the piston exactly balances the pressure exerted by the gas. Under these conditions, the piston will not move unless the external pressure is reduced infinitely slowly, allowing the pressure of the confined gas to readjust to maintain a balance in the two pressures. This gradual, infinitely slow process in which the external pressure and internal pressure are always in equilibrium is reversible. If we reverse the process and compress the gas in the same infinitely slow manner, we can return the gas to its original volume. The complete cycle of expansion and compression in this hypothetical process, moreover, is accomplished without any net change to the surroundings.

Because real processes can at best only approximate the slow, ever-in-equilibrium change associated with reversible processes, all real processes are irreversible. Further, the reverse of any spontaneous process is a nonspontaneous process. A nonspontaneous process can occur only if the surroundings do work on the system. Thus, *any spontaneous process is irreversible;* even if we return the system to the original condition, the surroundings will have changed.

19.2 | Entropy and the Second Law of Thermodynamics

We are now closer to understanding spontaneity because we know that any spontaneous process is irreversible. But how can we use this idea to make predictions about the spontaneity of an unfamiliar process? Understanding spontaneity requires us to examine the thermodynamic quantity called **entropy**. Entropy has been variously associated with the extent of *randomness* in a system or with the extent to which energy is distributed among the various motions of the molecules of the system. In fact, entropy is a multifaceted concept whose interpretations are not so quickly summarized by a simple definition. In this section we consider how we can relate entropy changes to heat transfer and temperature. Our analysis will bring us to a profound statement about spontaneity that we call the second law of thermodynamics. In Section 19.3 we examine the molecular significance of entropy.

Entropy Change

The entropy, S, of a system is a state function just like the internal energy, E, and enthalpy, H. As with these other quantities, the value of S is a characteristic of the state of a system. (Section 5.2) Thus, the change in entropy, ΔS, in a system depends only on the initial and final states of the system and not on the path taken from one state to the other:

$$\Delta S = S_{\text{final}} - S_{\text{initial}} \quad [19.1]$$

For the special case of an isothermal process, ΔS is equal to the heat that would be transferred if the process were reversible, q_{rev}, divided by the temperature at which the process occurs:

$$\Delta S = \frac{q_{\text{rev}}}{T} \quad (\text{constant } T) \quad [19.2]$$

Because S is a state function, we can use Equation 19.2 to calculate ΔS for *any* isothermal process, not just those that are reversible. If a change between two states is irreversible, we calculate ΔS by using a reversible path between the states.

GIVE IT SOME THOUGHT

How do we reconcile the fact that S is a state function but that ΔS depends on q, which is not a state function?

ΔS for Phase Changes

The melting of a substance at its melting point and the vaporization of a substance at its boiling point are isothermal processes. Consider the melting of ice. At 1 atm pressure, ice and liquid water are in equilibrium with each other at 0°C. Imagine that we melt one mole of ice at 0°C, 1 atm to form one mole of liquid water at 0°C, 1 atm. We can achieve this change by adding a certain amount of heat to the system from the surroundings: $q = \Delta H_{\text{fusion}}$. Now imagine that we carry out the change by adding the heat infinitely slowly, raising the temperature of the surroundings only infinitesimally above 0°C. When we do it this way, the process is reversible. We can reverse the process simply by infinitely slowly removing the same amount of heat, ΔH_{fusion}, from the system, using immediate surroundings that are infinitesimally below 0°C. Thus, $q_{\text{rev}} = \Delta H_{\text{fusion}}$ and $T = 0°\text{C} = 273\text{ K}$.

The enthalpy of fusion for H_2O is $\Delta H_{\text{fusion}} = 6.01\text{ kJ/mol}$. (The melting is an endothermic process, and so the sign of ΔH is positive.) Thus, we can use Equation 19.2 to calculate ΔS_{fusion} for melting one mole of ice at 273 K:

$$\Delta S_{\text{fusion}} = \frac{q_{\text{rev}}}{T} = \frac{\Delta H_{\text{fusion}}}{T} = \frac{(1\text{ mol})(6.01 \times 10^3\text{ J/mol})}{273\text{ K}} = 22.0\frac{\text{J}}{\text{K}}$$

Notice that the units for ΔS, J/K, are energy divided by absolute temperature, as we expect from Equation 19.2.

SAMPLE EXERCISE 19.2 | Calculating ΔS for a Phase Change

The element mercury, Hg, is a silvery liquid at room temperature. The normal freezing point of mercury is −38.9°C, and its molar enthalpy of fusion is $\Delta H_{\text{fusion}} = 2.29\text{ kJ/mol}$. What is the entropy change of the system when 50.0 g of Hg(l) freezes at the normal freezing point?

Solution

Analyze: We first recognize that freezing is an *exothermic* process; heat is transferred from the system to the surroundings when a liquid freezes ($q < 0$). The enthalpy of fusion is ΔH for the melting process. Because freezing is the reverse of melting, the enthalpy change that accompanies the freezing of 1 mol of Hg is $-\Delta H_{\text{fusion}} = -2.29\text{ kJ/mol}$.

Plan: We can use $-\Delta H_{\text{fusion}}$ and the atomic weight of Hg to calculate q for freezing 50.0 g of Hg:

$$q = (50.0\text{ g Hg})\left(\frac{1\text{ mol Hg}}{200.59\text{ g Hg}}\right)\left(\frac{-2.29\text{ kJ}}{1\text{ mol Hg}}\right)\left(\frac{1000\text{ J}}{1\text{ kJ}}\right) = -571\text{ J}$$

We can use this value of q as q_{rev} in Equation 19.2. We must first, however, convert the temperature to K:

$$-38.9°\text{C} = (-38.9 + 273.15)\text{ K} = 234.3\text{ K}$$

Solve: We can now calculate the value of ΔS_{sys}:

$$\Delta S_{\text{sys}} = \frac{q_{\text{rev}}}{T} = \frac{-571\text{ J}}{234.3\text{ K}} = -2.44\text{ J/K}$$

Check: The entropy change is negative because heat flows from the system making q_{rev} negative.

Comment: The procedure we have used here can be used to calculate ΔS for other isothermal phase changes, such as the vaporization of a liquid at its boiling point.

PRACTICE EXERCISE

The normal boiling point of ethanol, C_2H_5OH, is 78.3°C (Figure 11.24), and its molar enthalpy of vaporization is 38.56 kJ/mol. What is the change in entropy in the system when 68.3 g of $C_2H_5OH(g)$ at 1 atm condenses to liquid at the normal boiling point?

Answer: −163 J/K

A CLOSER LOOK | The Entropy Change when a Gas Expands Isothermally

In general, we will see that if a system becomes more spread out, or more random, the system's entropy increases. Thus, we expect the spontaneous expansion of a gas to result in an increase in entropy. To illustrate how the entropy change associated with an expanding gas can be calculated, consider the expansion of an ideal gas that is initially constrained by a piston, as in Figure 19.5(c). If the gas undergoes a reversible isothermal expansion, the work done on the surroundings by the moving piston can be calculated with the aid of calculus:

$$w_{rev} = -nRT \ln \frac{V_2}{V_1}$$

In this equation, n is the number of moles of gas, R is the gas constant, T is the absolute temperature, V_1 is the initial volume, and V_2 is the final volume. Notice that if $V_2 > V_1$, as it must be in our expansion, then $w_{rev} < 0$, meaning that the expanding gas does work on the surroundings.

One of the characteristics of an ideal gas is that its internal energy depends not on pressure but only on temperature. Thus, when an ideal gas expands at a constant temperature, $\Delta E = 0$. Because $\Delta E = q_{rev} + w_{rev} = 0$, we see that $q_{rev} = -w_{rev} = nRT \ln (V_2/V_1)$. Then, using Equation 19.2, we can calculate the entropy change in the system:

$$\Delta S_{sys} = \frac{q_{rev}}{T} = \frac{nRT \ln \frac{V_2}{V_1}}{T} = nR \ln \frac{V_2}{V_1} \qquad [19.3]$$

For 1.00 L of an ideal gas at 1.00 atm and 0°C, we can calculate the number of moles, $n = 4.46 \times 10^{-2}$ mol. The gas constant, R, can be expressed in units of J/mol-K, 8.314 J/mol-K (Table 10.2). Thus, for the expansion of the gas from 1.00 L to 2.00 L, we have

$$\Delta S_{sys} = (4.46 \times 10^{-2}\ \text{mol})\left(8.314 \frac{\text{J}}{\text{mol-K}}\right)\left(\ln \frac{2.00\ \text{L}}{1.00\ \text{L}}\right)$$
$$= 0.26\ \text{J/K}$$

In Section 19.3 we will see that this increase in entropy is a measure of the increased randomness of the molecules because of the expansion.

Related Exercises: 19.25 and 19.26

The Second Law of Thermodynamics

The key idea of the first law of thermodynamics is that energy is conserved in any process. Thus, the quantity of energy lost by a system equals the quantity gained by its surroundings. (Section 5.1) We will see, however, that entropy is different because it actually increases in any spontaneous process. Thus, the sum of the entropy change of the system and surroundings for any spontaneous process is always greater than zero. Entropy change is like a signpost indicating whether a process is spontaneous. Let's illustrate this generalization by again considering the melting of ice, designating the ice and water as our system.

Let's calculate the entropy change of the system and the entropy change of the surroundings when a mole of ice (a piece roughly the size of an ordinary ice cube) melts in the palm of your hand. The process is not reversible because the system and surroundings are at different temperatures. Nevertheless, because ΔS is a state function, the entropy change of the system is the same regardless of whether the process is reversible or irreversible. We calculated the entropy change of the system just before Sample Exercise 19.2:

$$\Delta S_{system} = \frac{q_{rev}}{T} = \frac{(1\ \text{mol})(6.01 \times 10^3\ \text{J/mol})}{273\ \text{K}} = 22.0 \frac{\text{J}}{\text{K}}$$

The surroundings immediately in contact with the ice are your hand, which we will assume is at body temperature, 37°C = 310 K. The heat lost by your hand is equal in magnitude to the heat gained by the ice but has the opposite sign, -6.01×10^3 J/mol. Hence the entropy change of the surroundings is

$$\Delta S_{surroundings} = \frac{q_{rev}}{T} = \frac{(1\ \text{mol})(-6.01 \times 10^3\ \text{J/mol})}{310\ \text{K}} = -19.4 \frac{\text{J}}{\text{K}}$$

Thus, the total entropy change is positive:

$$\Delta S_{total} = \Delta S_{system} + \Delta S_{surroundings} = \left(22.0\frac{J}{K}\right) + \left(-19.4\frac{J}{K}\right) = 2.6\frac{J}{K}$$

If the temperature of the surroundings were not 310 K but rather some temperature infinitesimally above 273 K, the melting would be reversible instead of irreversible. In that case the entropy change of the surroundings would equal −22.0 J/K and ΔS_{total} would be zero.

In general, any irreversible process results in an overall increase in entropy, whereas a reversible process results in no overall change in entropy. This general statement is known as the **second law of thermodynamics**. The sum of the entropy of a system plus the entropy of the surroundings is everything there is, and so we refer to the total entropy change as the entropy change of the universe, ΔS_{univ}. We can therefore state the second law of thermodynamics in terms of the following equations:

Reversible process: $\Delta S_{univ} = \Delta S_{system} + \Delta S_{surroundings} = 0$

Irreversible process: $\Delta S_{univ} = \Delta S_{system} + \Delta S_{surroundings} > 0$ [19.4]

All real processes that occur of their own accord are irreversible (with reversible processes being a useful idealization). These processes are also spontaneous. Thus, *the total entropy of the universe increases in any spontaneous process.* This profound generalization is yet another way of expressing the second law of thermodynamics.

GIVE IT SOME THOUGHT

The rusting of iron is accompanied by a decrease in the entropy of the system (the iron and oxygen). What can we conclude about the entropy change of the surroundings?

The second law of thermodynamics tells us the essential character of any spontaneous change—it is always accompanied by an overall increase in entropy. We can, in fact, use this criterion to predict whether processes will be spontaneous. Before beginning to use the second law to predict spontaneity, however, we will find it useful to explore further the meaning of entropy from a molecular perspective.

Throughout most of the remainder of this chapter, we will focus mainly on the systems we encounter rather than on their surroundings. To simplify the notation, we will usually refer to the entropy change of the system merely as ΔS rather than explicitly indicating ΔS_{sys}.

19.3 | The Molecular Interpretation of Entropy

As chemists, we are interested in molecules. What does entropy have to do with them and their transformations? What molecular property does entropy reflect? It was Ludwig Boltzmann (1844–1906) who gave conceptual meaning to entropy. To understand Boltzmann's contribution, we need to examine the ways in which molecules can store energy.

Molecular Motions and Energy

When a substance is heated, the motion of its molecules increases. When we studied the kinetic-molecular theory of gases, we found that the average kinetic energy of the molecules of an ideal gas is directly proportional to the absolute temperature of the gas. (Section 10.7) That means the higher the temperature, the faster the molecules are moving and the more kinetic energy they possess. Moreover, hotter systems have a *broader distribution* of molecular speeds, as you can see by referring back to Figure 10.18. The particles of an ideal gas, however, are merely idealized points with no volume and no bonds, points that we visualize as flitting around through space. Real molecules can undergo more complex kinds of motions.

Molecules can undergo three kinds of motion. The entire molecule can move in one direction, as in the motions of the particles of an ideal gas or the motions of larger objects, such as a baseball being thrown around a baseball field. We call such movement **translational motion**. The molecules in a gas have more freedom of translational motion than those in a liquid, which in turn have more translational freedom than the molecules of a solid.

ACTIVITY Molecular Motions

A molecule may also undergo **vibrational motion**, in which the atoms in the molecule move periodically toward and away from one another, much as a tuning fork vibrates about its equilibrium shape. In addition, molecules may possess **rotational motion**, as though they were spinning like tops. Figure 19.6 ▼ shows the vibrational motions and one of the rotational motions possible for the water molecule. These different forms of motion are ways in which a molecule can store energy, and we refer to the various forms collectively as the "motional energy" of the molecule.

GIVE IT SOME THOUGHT

What kinds of motion can a molecule undergo that a single atom cannot?

Boltzmann's Equation and Microstates

The science of thermodynamics developed as a means of describing the properties of matter in our macroscopic world without regard to the microscopic structure of the matter. In fact, thermodynamics was a well-developed field before the modern views of atomic and molecular structure were even known. The thermodynamic properties of water, for example, addressed the behavior of bulk water (or ice or water vapor) as a substance, without considering any specific properties of individual H_2O molecules.

In order to connect the microscopic and macroscopic descriptions of matter, scientists have developed the field of *statistical thermodynamics*, which uses the tools of statistics and probability to provide the link between the microscopic and macroscopic worlds. Here we will show how entropy, which is a property

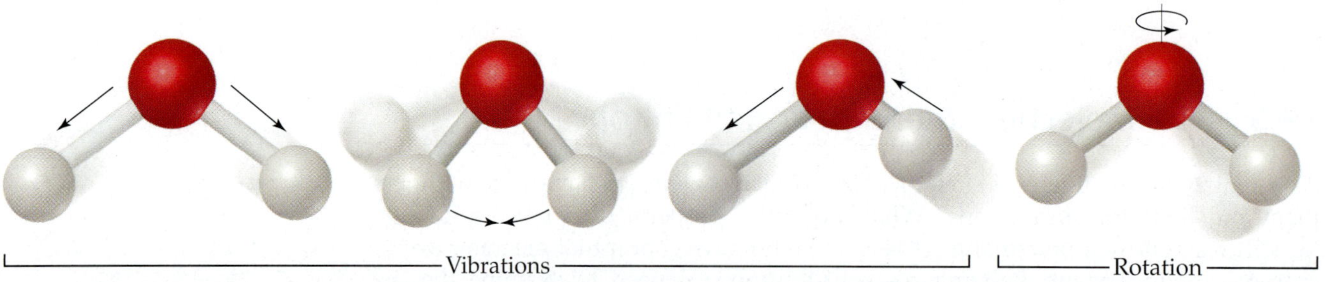

▲ **Figure 19.6 Vibrational and rotational motions in a water molecule.** Vibrational motions in the molecule involve periodic displacements of the atoms with respect to one another. Rotational motions involve the spinning of a molecule about an axis.

of bulk matter, can be connected to the behavior of atoms and molecules. Because the mathematics of statistical thermodynamics is quite complex, our discussion will be largely conceptual.

Let's begin by considering one mole of an ideal gas in a particular thermodynamic state, which we can define by specifying the temperature, T, and volume, V, of the gas. (Remember that the energy, E, of an ideal gas depends only on its temperature and that by fixing the values of n, T, and V, we also fix the value of the pressure, P.) What is happening to our ideal gas sample at the microscopic level, and how does what is going on at the microscopic level relate to the entropy of the sample? To address these questions, we need to consider both the positions of the gas molecules and their individual kinetic energies, which depend on the speeds of the molecules. In our discussion of the kinetic-molecular theory, we considered the gas molecules to be in constant motion within the entire volume of the container. We also saw that the speeds of the gas molecules follow a well-defined distribution at a given temperature, such as that shown in Figure 10.18. ∞ (Section 10.7)

Now, imagine that we could take a "snapshot" of the positions and speeds of all of the molecules at a given instant. That particular set of 6×10^{23} positions and energies of the individual gas molecules is what we call a **microstate** of the thermodynamic system. A microstate is a single possible arrangement of the positions and kinetic energies of the gas molecules when the gas is in a specific thermodynamic state. We could envision continuing to take snapshots of our system to see other possible microstates. In fact, as you no doubt see, there would be such a staggeringly large number of microstates that taking individual snapshots of all of them is not feasible. Because we are examining such a large number of particles, however, we can use the tools of statistics and probability to determine the total number of microstates for the thermodynamic state. (That is where the *statistical* part of statistical thermodynamics comes in.) Each thermodynamic state has a characteristic number of microstates associated with it, and we will use the symbol W for that number.

▲ **Figure 19.7 Ludwig Boltzmann's gravestone.** Boltzmann's gravestone in Vienna is inscribed with his famous relationship between the entropy of a state and the number of available microstates. (In Boltzmann's time, "log" was used to represent the natural logarithm.)

The connection between the number of microstates of a system, W, and its entropy, S, is expressed in a beautifully simple equation developed by Boltzmann:

$$S = k \ln W \qquad [19.5]$$

In this equation, k is Boltzmann's constant, 1.38×10^{-23} J/K. Thus, *entropy is a measure of how many microstates are associated with a particular macroscopic state.* Equation 19.5 appears on Boltzmann's gravestone (Figure 19.7 ▶).

GIVE IT SOME THOUGHT

What is the entropy of a system that has only a single microstate?

The entropy change accompanying any process is

$$\Delta S = k \ln W_{\text{final}} - k \ln W_{\text{initial}} = k \ln \frac{W_{\text{final}}}{W_{\text{initial}}} \qquad [19.6]$$

Thus, any change in the system that leads to an increase in the number of microstates leads to a positive value of ΔS: *Entropy increases with the number of microstates of the system.*

Let's briefly consider two simple changes to our ideal-gas sample and see how the entropy changes in each case. First, suppose we increase the volume of

A CLOSER LOOK | Entropy and Probability

The game of poker is sometimes used as an analogy to explore the idea of the microstates associated with a particular state. There are about 2.6 million different five-card poker hands that can be dealt, and each of these hands can be viewed as a possible "microstate" for the hand dealt to any one player in a game. Table 19.1 ▼ shows two poker hands. The probability that a particular hand will contain five *specific* cards is the same regardless of which five cards are specified. Thus, there is an equal probability of dealing either of the hands shown in Table 19.1. However, the first hand, a royal flush (the ten through ace of a single suit), strikes us as much more highly ordered than the second hand, a "nothing." The reason for this is clear if we compare the number of five-card arrangements that correspond to a royal flush to the number corresponding to a nothing: only 4 hands (microstates) for a royal flush but more than 1.3 million for a nothing hand. The nothing state has a higher probability of being dealt from a shuffled deck than the royal-flush state because there are so many more arrangements of cards that correspond to the nothing state. In other words, the value of W in Boltzmann's equation (Equation 19.5) is much greater for a nothing than for a royal flush.

This example teaches us that there is a connection between probability and entropy. Entropy has a natural tendency to increase because any entropy increase in a system corresponds to the movement of the system from a condition of low probability to a condition of greater probability. Physical and chemical changes are driven by a statistical inevitability formally expressed by the concept of entropy.

We can use similar reasoning to explain the isothermal expansion of a gas, such as that depicted in Figure 19.1. When the stopcock is opened, the gas molecules are less constrained and there are more possible arrangements for them (more microstates) in the larger volume. The various microstates are depicted in a schematic way in Figure 19.8 ▼. In this figure, we make no attempt to describe the motion of the particles, focusing instead only on their locations. The spreading of the molecules over the larger volume represents the more probable state.

When we use the terms randomness and disorder to describe entropy, we have to be careful not to carry an aesthetic sense of what we mean. What we must remember is that the fundamental connection to entropy is not directly with randomness, disorder, or energy dispersal, but with the number of available microstates.

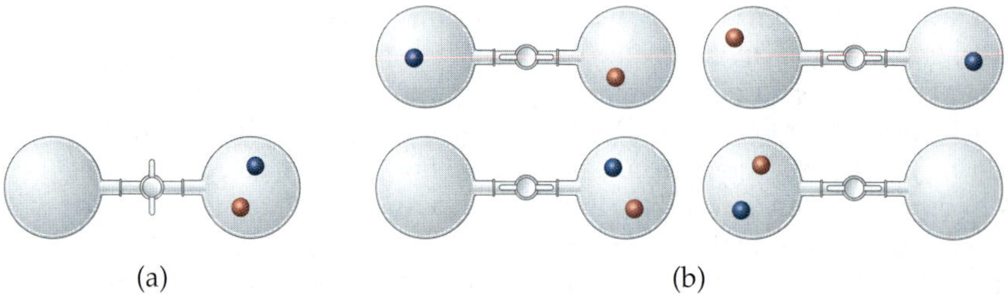

▲ **Figure 19.8 Probability and the locations of gas molecules.** The two molecules are colored red and blue to keep track of them. (a) Before the stopcock is opened, both molecules are in the right-hand flask. (b) After the stopcock is opened, there are four possible arrangements of the two molecules. Only one of the four arrangements corresponds to both molecules' being in the right-hand flask. The greater number of possible arrangements corresponds to greater disorder in the system. In general, the probability that the molecules will stay in the original flask is $\left(\frac{1}{2}\right)^n$, where n is the number of molecules.

TABLE 19.1 A Comparison of the Number of Combinations that Can Lead to a Royal Flush and to a "Nothing" Hand in Poker

Hand	State	Number of Hands that Lead to This State
	Royal flush	4
	"Nothing"	1,302,540

the system, which is analogous to allowing the gas to expand isothermally. A greater volume means that there are a greater number of positions available to the gas atoms. Thus, there will be a greater number of microstates for the system after the increase in volume. The entropy therefore increases as the volume increases, as we saw in the "Closer Look" box in Section 19.2. Second, suppose we keep the volume fixed but increase the temperature. How will this change affect the entropy of the system? Recall the distribution of molecular speeds presented in Figure 10.18. An increase in temperature increases the average (rms) speed of the molecules and broadens the distribution of speeds. Hence, the molecules have a greater number of possible kinetic energies, and the number of microstates will once again increase. The entropy of the system will therefore increase with increasing temperature.

ACTIVITY
Mixing of Gases

If we consider real molecules instead of ideal-gas particles, we must also consider the different amounts of vibrational and rotational energies the molecules have in addition to their kinetic energies. A collection of real molecules therefore has a greater number of microstates available than does the same number of ideal-gas particles. In general, *the number of microstates available to a system increases with an increase in volume, an increase in temperature, or an increase in the number of molecules because any of these changes increases the possible positions and energies of the molecules of the system.*

There are several ways that chemists use to describe an increase in the number of microstates and therefore an increase in the entropy for a system. Each of these ways seeks to capture a sense of the increased freedom of motion that causes molecules to spread out if not restrained by physical barriers or chemical bonds. Some say the increase in entropy represents an increase in the *randomness* or *disorder* of the system as discussed in the preceding "Closer Look" box. Others liken an increase in entropy to an increased *dispersion (spreading out) of energy* because there is an increase in the number of ways the positions and energies of the molecules can be distributed throughout the system. Each of these descriptions (randomness, disorder, and energy dispersal) is conceptually helpful if applied correctly. Indeed, you will find it useful to keep these descriptions in mind as you evaluate entropy changes.

Making Qualitative Predictions About ΔS

It is usually not difficult to construct a mental picture to estimate qualitatively how the entropy of a system changes during a simple process. In most instances, an increase in the number of microstates, and hence an increase in entropy, parallels an increase in

1. temperature
2. volume
3. number of independently moving particles

Thus, we can usually make qualitative predictions about entropy changes by focusing on these factors. For example, when water vaporizes, the molecules spread out into a larger volume. Because they occupy a larger space, there is an increase in their freedom of motion, giving rise to more accessible microstates and hence to an increase in entropy.

In the melting of ice, the rigid structure of the water molecules, shown in Figure 19.9 ▶, restricts motion to only tiny vibrations throughout the crystal. In contrast, the molecules in liquid water are free to move about with

▼ Figure 19.9 Structure of ice. The intermolecular attractions in the three-dimensional lattice restrict the molecules to vibrational motion only.

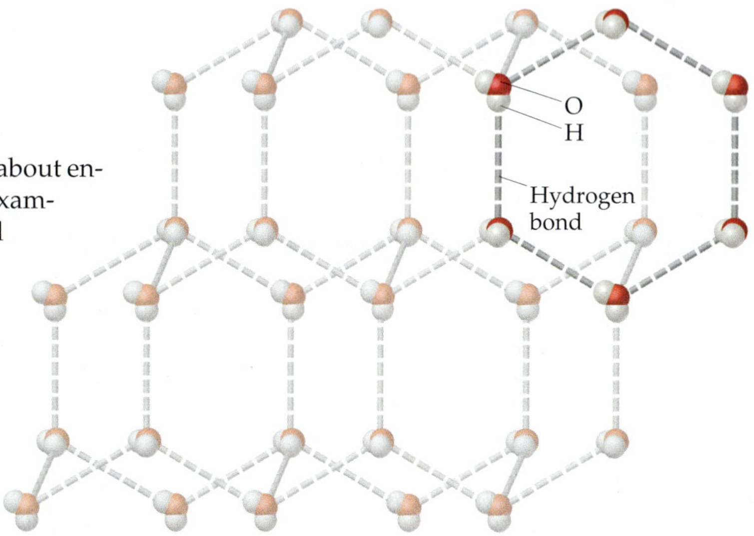

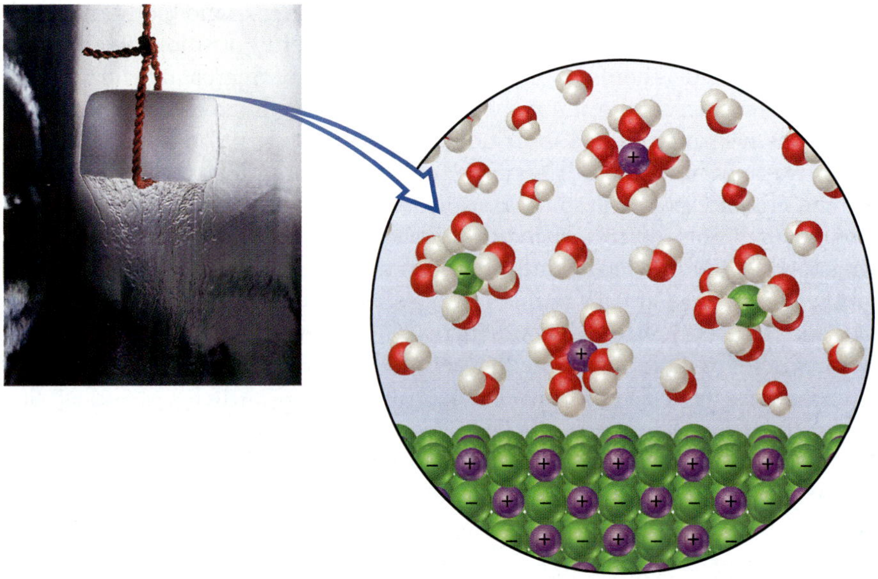

▶ **Figure 19.10 Dissolving an ionic solid in water.** The ions become more spread out and random in their motions, but the water molecules that hydrate the ions become less random.

ACTIVITY
Enthalpy of Solution

Heat of Solution: NH_4NO_3

respect to one another (translation) and to tumble around (rotation) as well as vibrate. During melting, therefore, the number of accessible microstates increases and so does the entropy.

When an ionic solid, such as KCl, dissolves in water, the pure solid and pure water are replaced by a mixture of water and ions (Figure 19.10 ▲). The ions now move in a larger volume and possess more motional energy than in the rigid solid. We have to be careful, however, because water molecules are held around the ions as water of hydration. (Section 13.1) These water molecules have less motional energy than before because they are now confined to the immediate environment of the ions. The greater the charge of an ion, the greater the ion-dipole attractions that hold the ion and the water together, thereby restricting motions. Thus, even though the solution process is normally accompanied by a net increase in entropy, the dissolving of salts with highly charged ions can result in a net *decrease* in entropy.

The same ideas apply to systems involving chemical reactions. Consider the reaction between nitric oxide gas and oxygen gas to form nitrogen dioxide gas:

$$2\,NO(g) + O_2(g) \longrightarrow 2\,NO_2(g) \qquad [19.7]$$

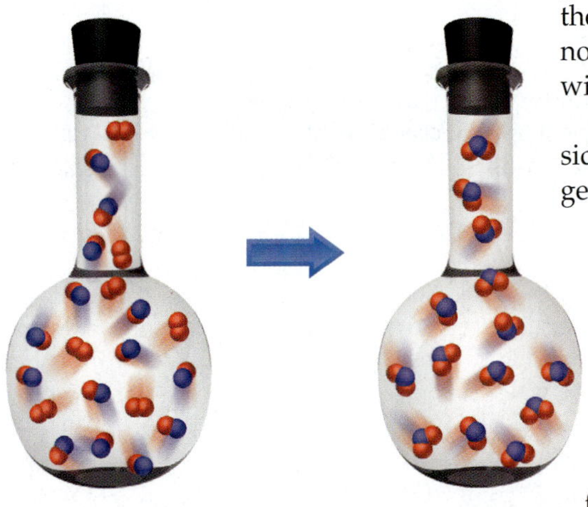

(a) (b)

▲ **Figure 19.11 Entropy change for a reaction.** A decrease in the number of gaseous molecules leads to a decrease in the entropy of the system. When the $NO(g)$ and $O_2(g)$ in (a) react to form the $NO_2(g)$ in (b), the number of gaseous molecules decreases. The atoms have fewer degrees of freedom because new N—O bonds form and the entropy decreases.

In this case the reaction results in a decrease in the number of molecules—three molecules of gaseous reactants form two molecules of gaseous products (Figure 19.11 ◀). The formation of new N—O bonds reduces the motions of the atoms in the system. The formation of new bonds decreases the *number of degrees of freedom*, or forms of motion, available to the atoms. That is, the atoms are less free to move in random fashion because of the formation of new bonds. The decrease in the number of molecules and the resultant decrease in motion result in fewer accessible microstates and therefore a decrease in the entropy of the system.

In summary, we generally expect the entropy of the system to increase for processes in which

1. Gases are formed from either solids or liquids.
2. Liquids or solutions are formed from solids.
3. The number of gas molecules increases during a chemical reaction.

SAMPLE EXERCISE 19.3 | Predicting the Sign of ΔS

Predict whether ΔS is positive or negative for each of the following processes, assuming each occurs at constant temperature:

(a) $H_2O(l) \longrightarrow H_2O(g)$
(b) $Ag^+(aq) + Cl^-(aq) \longrightarrow AgCl(s)$
(c) $4\,Fe(s) + 3\,O_2(g) \longrightarrow 2\,Fe_2O_3(s)$
(d) $N_2(g) + O_2(g) \longrightarrow 2\,NO(g)$

Solution

Analyze: We are given four equations and asked to predict the sign of ΔS for each chemical reaction.

Plan: The sign of ΔS will be positive if there is an increase in temperature, an increase in the volume in which the molecules move, or an increase in the number of gas particles in the reaction. The question states that the temperature is constant. Thus, we need to evaluate each equation with the other two factors in mind.

Solve: **(a)** The evaporation of a liquid is accompanied by a large increase in volume. One mole of water (18 g) occupies about 18 mL as a liquid and 22.4 L as a gas at STP. Because the molecules are distributed throughout a much larger volume in the gaseous state than in the liquid state, an increase in motional freedom accompanies vaporization. Therefore, ΔS is positive.

(b) In this process the ions, which are free to move throughout the volume of the solution, form a solid in which they are confined to a smaller volume and restricted to more highly constrained positions. Thus, ΔS is negative.

(c) The particles of a solid are confined to specific locations and have fewer ways to move (fewer microstates) than do the molecules of a gas. Because O_2 gas is converted into part of the solid product Fe_2O_3, ΔS is negative.

(d) The number of moles of gases is the same on both sides of the equation, and so the entropy change will be small. The sign of ΔS is impossible to predict based on our discussions thus far, but we can predict that ΔS will be close to zero.

PRACTICE EXERCISE

Indicate whether each of the following processes produces an increase or decrease in the entropy of the system:

(a) $CO_2(s) \longrightarrow CO_2(g)$
(b) $CaO(s) + CO_2(g) \longrightarrow CaCO_3(s)$
(c) $HCl(g) + NH_3(g) \longrightarrow NH_4Cl(s)$
(d) $2\,SO_2(g) + O_2(g) \longrightarrow 2\,SO_3(g)$

Answers: **(a)** increase, **(b)** decrease, **(c)** decrease, **(d)** decrease

SAMPLE EXERCISE 19.4 | Predicting Which Sample of Matter Has the Higher Entropy

Choose the sample of matter that has greater entropy in each pair, and explain your choice: **(a)** 1 mol of $NaCl(s)$ or 1 mol of $HCl(g)$ at 25°C, **(b)** 2 mol of $HCl(g)$ or 1 mol of $HCl(g)$ at 25°C, **(c)** 1 mol of $HCl(g)$ or 1 mol of $Ar(g)$ at 298 K.

Solution

Analyze: We need to select the system in each pair that has the greater entropy.

Plan: To do this, we examine the state of the system and the complexity of the molecules it contains.

Solve: **(a)** Gaseous HCl has the higher entropy because gases have more available motions than solids. **(b)** The sample containing 2 mol of HCl has twice the number of molecules as the sample containing 1 mol. Thus, the 2-mol sample has twice the number of microstates and twice the entropy. **(c)** The HCl sample has the higher entropy because the HCl molecule is capable of storing energy in more ways than is Ar. HCl molecules can rotate and vibrate; Ar atoms cannot.

PRACTICE EXERCISE

Choose the substance with the greater entropy in each case: **(a)** 1 mol of $H_2(g)$ at STP or 1 mol of $H_2(g)$ at 100°C and 0.5 atm, **(b)** 1 mol of $H_2O(s)$ at 0°C or 1 mol of $H_2O(l)$ at 25°C, **(c)** 1 mol of $H_2(g)$ at STP or 1 mol of $SO_2(g)$ at STP, **(d)** 1 mol of $N_2O_4(g)$ at STP or 2 mol of $NO_2(g)$ at STP.

Answers: **(a)** 1 mol of $H_2(g)$ at 100°C and 0.5 atm, **(b)** 1 mol of $H_2O(l)$ at 25°C, **(c)** 1 mol of $SO_2(g)$ at STP, **(d)** 2 mol of $NO_2(g)$ at STP

CHEMISTRY AND LIFE | Entropy and Life

The ginkgo leaf shown in Figure 19.12(a) ▶ reveals beautiful patterns of form and color. Both plant systems and animal systems, including ourselves, are incredibly complex structures in which a host of substances come together in organized ways to form cells, tissue, organs, and so on. These various components must operate in synchrony for the organism as a whole to be viable. If even one key system strays far from its optimal state, the organism as a whole may die.

To make a living system from its component molecules—such as a ginkgo leaf from sugar molecules, cellulose molecules, and the other substances present in the leaf—requires a very large reduction in entropy. It would seem, then, that living systems might violate the second law of thermodynamics. They seem spontaneously to become more, not less, organized as they develop. To get the full picture, however, we must take into account the surroundings.

We know that a system can move toward lower entropy if we do work on it. (That is, if we supply energy to the system in a very specific way.) When we do work on a gas, for example, by compressing it isothermally, the entropy of the gas is lowered. The energy for the work done is provided by the surroundings, and in the process the net entropy change in the universe is positive.

The striking thing about living systems is that they are organized to recruit energy from their surroundings spontaneously. Some single-celled organisms, called *autotrophs*, capture energy from sunlight and store it in molecules such as sugars and fats [Figure 19.12(b)]. Others, called *heterotrophs*, absorb food molecules from their surroundings and then break the molecules down to provide needed energy. Whatever their mode of existence, however, living systems gain their order at the expense of the surroundings. Each cell exists at the expense of an increase in the entropy of the universe.

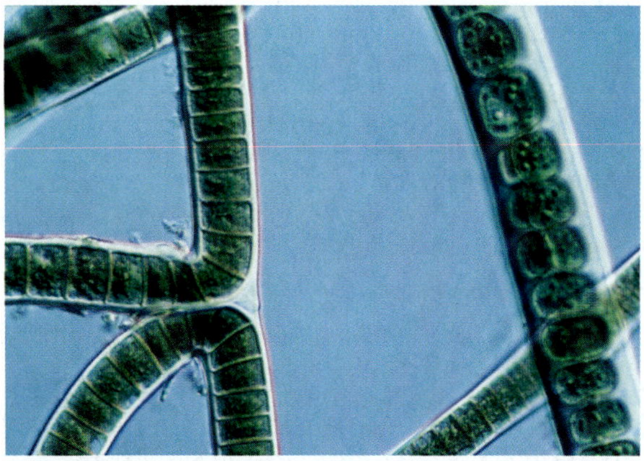

▲ **Figure 19.12 Entropy and life.** (a) This ginkgo leaf represents a highly organized living system. (b) Cyanobacteria absorb light energy and utilize it to synthesize the substances needed for growth.

The Third Law of Thermodynamics

If we decrease the thermal energy of a system by lowering the temperature, the energy stored in translational, vibrational, and rotational forms of motion decreases. As less energy is stored, the entropy of the system decreases. If we keep lowering the temperature, do we reach a state in which these motions are essentially shut down, a point described by a single microstate? This question is addressed by the **third law of thermodynamics**, which states that *the entropy of a pure crystalline substance at absolute zero is zero:* $S(0\text{ K}) = 0$.

Figure 19.13 ▶ shows schematically a pure crystalline solid. At absolute zero all the units of the lattice have no thermal motion. There is, therefore, only one microstate. As a result $S = k \ln W = k \ln 1 = 0$. As the temperature is increased from absolute zero, the atoms or molecules in the crystal gain energy in the form of vibrational motion about their lattice positions. Thus, the degrees of freedom of the crystal increase. The entropy of the lattice therefore increases with temperature because vibrational motion causes the atoms or molecules to have a greater number of accessible microstates.

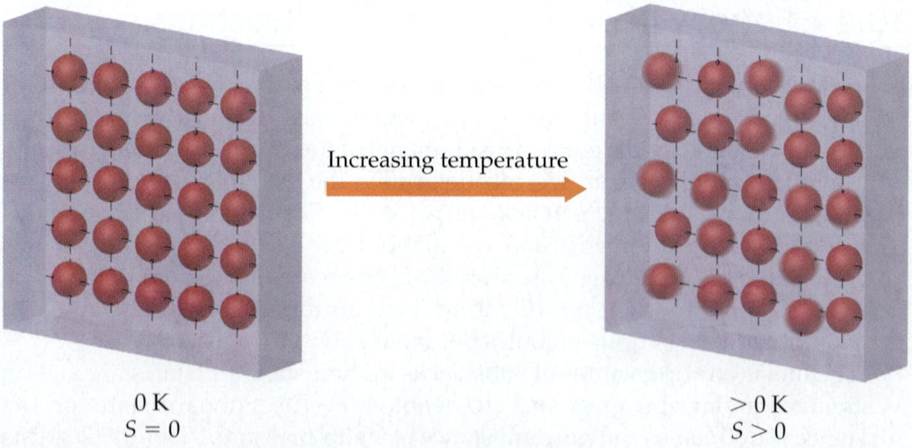

▲ **Figure 19.13 A perfectly ordered crystalline solid at and above 0 K.** At absolute zero (left), all lattice units are in their lattice sites, devoid of thermal motion. As the temperature rises above 0 K (right), the atoms or molecules gain energy and their vibrational motion increases.

ACTIVITY
Temperature Dependence of Entropy

What happens to the entropy of the substance as we continue to heat it? Figure 19.14 ▶ is a plot of how the entropy of a typical substance varies with temperature. We see that the entropy of the solid continues to increase steadily with increasing temperature up to the melting point of the solid. When the solid melts, the bonds holding the atoms or molecules are broken and the particles are free to move about the entire volume of the substance. The added degrees of freedom for the individual molecules allow greater dispersal of the substance's energy thereby increasing its entropy. We therefore see a sharp increase in the entropy at the melting point. After all the solid has melted to liquid, the temperature again increases and with it, the entropy.

At the boiling point of the liquid, another abrupt increase in entropy occurs. We can understand this increase as resulting from the increased volume in which the molecules may be found. When the gas is heated further, the entropy increases steadily as more energy is stored in the translational motion of the gas molecules. At higher temperatures, the distribution of molecular speeds is spread out toward higher values. (Figure 10.18) More of the molecules have speeds that differ greatly from the most probable value. The expansion of the range of speeds of the gas molecules leads to an increased entropy.

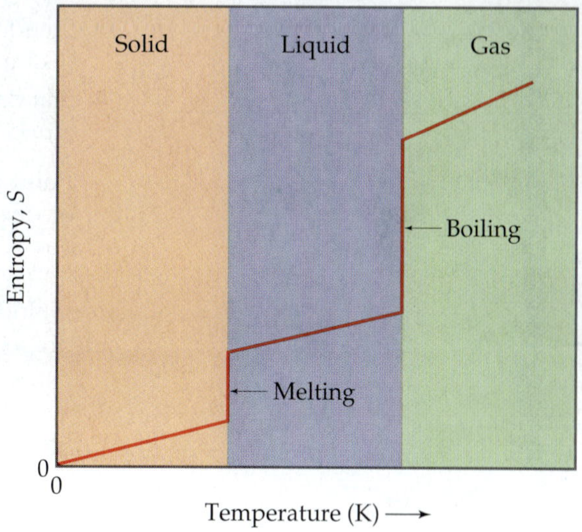

▲ **Figure 19.14 Entropy as a function of temperature.** Entropy increases as the temperature of a crystalline solid is increased from absolute zero. The vertical jumps in entropy correspond to phase changes.

The general conclusions we reach in examining Figure 19.14 are consistent with what we noted earlier: Entropy generally increases with increasing temperature because the increased motional energy can be dispersed in more ways. Further, the entropies of the phases of a given substance follow the order $S_{solid} < S_{liquid} < S_{gas}$. This ordering fits in nicely with our picture of the number of microstates available to solids, liquids, and gases.

GIVE IT SOME THOUGHT

If you are told that the entropy of a certain system is zero, what do you know about the system?

19.4 | Entropy Changes in Chemical Reactions

TABLE 19.2 Standard Molar Entropies of Selected Substances at 298 K

Substance	$S°$, J/mol-K
Gases	
$H_2(g)$	130.6
$N_2(g)$	191.5
$O_2(g)$	205.0
$H_2O(g)$	188.8
$NH_3(g)$	192.5
$CH_3OH(g)$	237.6
$C_6H_6(g)$	269.2
Liquids	
$H_2O(l)$	69.9
$CH_3OH(l)$	126.8
$C_6H_6(l)$	172.8
Solids	
$Li(s)$	29.1
$Na(s)$	51.4
$K(s)$	64.7
$Fe(s)$	27.23
$FeCl_3(s)$	142.3
$NaCl(s)$	72.3

In Section 5.5 we discussed how calorimetry can be used to measure ΔH for chemical reactions. No comparable, easy method exists for measuring ΔS for a reaction. By using experimental measurements of the variation of heat capacity with temperature, however, we can determine the absolute value of the entropy, S, for many substances at any temperature. (The theory and the methods used for these measurements and calculations are beyond the scope of this text.) Absolute entropies are based on the reference point of zero entropy for perfect crystalline solids at 0 K (the third law). Entropies are usually tabulated as molar quantities, in units of joules per mole-kelvin (J/mol-K).

The molar entropy values of substances in their standard states are known as **standard molar entropies** and are denoted $S°$. The standard state for any substance is defined as the pure substance at 1 atm pressure.* Table 19.2 ◀ lists the values of $S°$ for several substances at 298 K; Appendix C gives a more extensive list.

We can make several observations about the $S°$ values in Table 19.2:

1. Unlike enthalpies of formation, the standard molar entropies of elements at the reference temperature of 298 K are *not* zero.
2. The standard molar entropies of gases are greater than those of liquids and solids, consistent with our interpretation of experimental observations, as represented in Figure 19.14.
3. Standard molar entropies generally increase with increasing molar mass. [Compare Li(*s*), Na(*s*), and K(*s*).]
4. Standard molar entropies generally increase with an increasing number of atoms in the formula of a substance.

Point 4 is related to molecular motion (Section 19.3). In general, the number of degrees of freedom for a molecule increases with increasing number of atoms, and thus the number of accessible microstates increases also. Figure 19.15 ▼ compares the standard molar entropies of three hydrocarbons. Notice how the entropy increases as the number of atoms in the molecule increases.

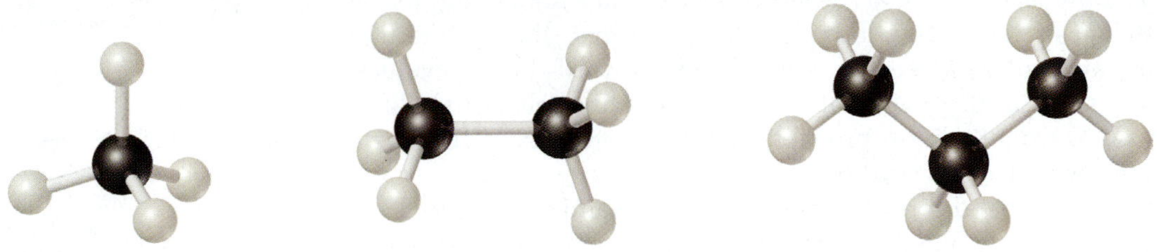

Methane, CH_4 $S° = 186.3\ J\ mol^{-1}\ K^{-1}$ Ethane, C_2H_6 $S° = 229.6\ J\ mol^{-1}\ K^{-1}$ Propane, C_3H_8 $S° = 270.3\ J\ mol^{-1}\ K^{-1}$

▲ **Figure 19.15 Molar entropies.** In general, the more complex a molecule (that is, the greater the number of atoms present), the greater the molar entropy of the substance, as illustrated here by the molar entropies of three simple hydrocarbons.

* The standard pressure used in thermodynamics is no longer 1 atm but is based on the SI unit for pressure, the pascal (Pa). The standard pressure is 10^5 Pa, a quantity known as a *bar*: 1 bar = 10^5 Pa = 0.987 atm. Because 1 bar differs from 1 atm by only 1.3%, we will continue to refer to the standard pressure as 1 atm.

The entropy change in a chemical reaction equals the sum of the entropies of the products less the sum of the entropies of the reactants:

ACTIVITY
Estimation of Entropy Changes

$$\Delta S^\circ = \sum nS^\circ(\text{products}) - \sum mS^\circ(\text{reactants}) \qquad [19.8]$$

As in Equation 5.31, the coefficients n and m are the coefficients in the chemical equation, as illustrated in Sample Exercise 19.5.

SAMPLE EXERCISE 19.5 | Calculating ΔS from Tabulated Entropies

Calculate ΔS° for the synthesis of ammonia from $N_2(g)$ and $H_2(g)$ at 298 K:

$$N_2(g) + 3\,H_2(g) \longrightarrow 2\,NH_3(g)$$

Solution

Analyze: We are asked to calculate the entropy change for the synthesis of $NH_3(g)$ from its constituent elements.

Plan: We can make this calculation using Equation 19.8 and the standard molar entropy values for the reactants and the products that are given in Table 19.2 and in Appendix C.

Solve: Using Equation 19.8, we have

$$\Delta S^\circ = 2S^\circ(NH_3) - [S^\circ(N_2) + 3S^\circ(H_2)]$$

Substituting the appropriate S° values from Table 19.2 yields

$$\Delta S^\circ = (2\text{ mol})(192.5\text{ J/mol-K}) - [(1\text{ mol})(191.5\text{ J/mol-K}) + (3\text{ mol})(130.6\text{ J/mol-K})]$$
$$= -198.3\text{ J/K}$$

Check: The value for ΔS° is negative, in agreement with our qualitative prediction based on the decrease in the number of molecules of gas during the reaction.

PRACTICE EXERCISE

Using the standard entropies in Appendix C, calculate the standard entropy change, ΔS°, for the following reaction at 298 K:

$$Al_2O_3(s) + 3\,H_2(g) \longrightarrow 2\,Al(s) + 3\,H_2O(g)$$

Answer: 180.39 J/K

Entropy Changes in the Surroundings

Tabulated absolute entropy values can be used to calculate the standard entropy change in a system, such as a chemical reaction, as just described, but what about the entropy change in the surroundings? We encountered this situation in Section 19.2, but it is good to revisit it now that we are examining chemical reactions.

We should recognize that the surroundings serve essentially as a large, constant-temperature heat source (or heat sink if the heat flows from the system to the surroundings). The change in entropy of the surroundings will depend on how much heat is absorbed or given off by the system. For an isothermal process, the entropy change of the surroundings is given by

$$\Delta S_{\text{surr}} = \frac{-q_{\text{sys}}}{T} \qquad [19.9]$$

For a reaction occurring at constant pressure, q_{sys} is simply the enthalpy change for the reaction, ΔH. For the reaction in Sample Exercise 19.5, the formation of ammonia from $H_2(g)$ and $N_2(g)$ at 298 K, q_{sys} is the enthalpy change for reaction under standard conditions, ΔH°. (Section 5.7) Using the procedures described in Section 5.7, we have

$$\Delta H^\circ_{\text{rxn}} = 2\Delta H^\circ_f[NH_3(g)] - 3\Delta H^\circ_f[H_2(g)] - \Delta H^\circ_f[N_2(g)]$$
$$= 2(-46.19\text{ kJ}) - 3(0\text{ kJ}) - (0\text{ kJ}) = -92.38\text{ kJ}$$

Thus, at 298 K the formation of ammonia from $H_2(g)$ and $N_2(g)$ is exothermic. Absorption of the heat given off by the system results in an increase in the entropy of the surroundings:

$$\Delta S^\circ_{surr} = \frac{92.38\ \text{kJ}}{298\ \text{K}} = 0.310\ \text{kJ/K} = 310\ \text{J/K}$$

Notice that the magnitude of the entropy gained by the surroundings (310 J/K) is greater than that lost by the system (198.3 J/K, as calculated in Sample Exercise 19.5):

$$\Delta S^\circ_{univ} = \Delta S^\circ_{sys} + \Delta S^\circ_{surr} = -198.3\ \text{J/K} + 310\ \text{J/K} = 112\ \text{J/K}$$

Because ΔS°_{univ} is positive for any spontaneous reaction, this calculation indicates that when $NH_3(g)$, $H_2(g)$, and $N_2(g)$ are together at 298 K in their standard states (each at 1 atm pressure), the reaction system will move spontaneously toward formation of $NH_3(g)$. Keep in mind that while the thermodynamic calculations indicate that formation of ammonia is spontaneous, they do not tell us anything about the rate at which ammonia is formed. Establishing equilibrium in this system within a reasonable period of time requires a catalyst, as discussed in Section 15.6.

GIVE IT SOME THOUGHT

If a process is exothermic, does the entropy of the surroundings (1) always increase, (2) always decrease, or (3) sometimes increase and sometimes decrease, depending on the process?

19.5 | Gibbs Free Energy

We have seen examples of endothermic processes that are spontaneous, such as the dissolution of ammonium nitrate in water. (Section 13.1) We learned in our discussion of the solution process that a spontaneous, endothermic process must be accompanied by an increase in the entropy of the system. However, we have also encountered processes that are spontaneous and yet proceed with a *decrease* in the entropy of the system, such as the highly exothermic formation of sodium chloride from its constituent elements. (Section 8.2) Spontaneous processes that result in a decrease in the system's entropy are always exothermic. Thus, the spontaneity of a reaction seems to involve two thermodynamic concepts, enthalpy and entropy.

There should be a way to use ΔH and ΔS to predict whether a given reaction occurring at constant temperature and pressure will be spontaneous. The means for doing so was first developed by the American mathematician J. Willard Gibbs (1839–1903). Gibbs (Figure 19.16 ◄) proposed a new state function, now called the **Gibbs free energy** (or just **free energy**). The Gibbs free energy, G, of a state is defined as

▲ **Figure 19.16 Josiah Willard Gibbs (1839–1903).** Gibbs was the first person to be awarded a Ph.D. in science from an American university (Yale, 1863). From 1871 until his death, he held the chair of mathematical physics at Yale. He developed much of the theoretical foundation that led to the development of chemical thermodynamics.

$$G = H - TS \qquad [19.10]$$

where T is the absolute temperature. For a process occurring at constant temperature, the change in free energy of the system, ΔG, is given by the expression

$$\Delta G = \Delta H - T\Delta S \qquad [19.11]$$

To see how the state function G relates to reaction spontaneity, recall that for a reaction occurring at constant temperature and pressure

$$\Delta S_{univ} = \Delta S_{sys} + \Delta S_{surr} = \Delta S_{sys} + \left(\frac{-\Delta H_{sys}}{T}\right)$$

Multiplying both sides by $(-T)$ gives us

$$-T\Delta S_{univ} = \Delta H_{sys} - T\Delta S_{sys} \quad [19.12]$$

Comparing Equation 19.12 with Equation 19.11, we see that the free-energy change in a process occurring at constant temperature and pressure, ΔG, is equal to $-T\Delta S_{univ}$. We know that for spontaneous processes, ΔS_{univ} is positive. Thus, the sign of ΔG provides us with extremely valuable information about the spontaneity of processes that occur at constant temperature and pressure. If both T and P are constant, the relationship between the sign of ΔG and the spontaneity of a reaction is as follows:

1. If ΔG is negative, the reaction is spontaneous in the forward direction.
2. If ΔG is zero, the reaction is at equilibrium.
3. If ΔG is positive, the reaction in the forward direction is nonspontaneous; work must be supplied from the surroundings to make it occur. However, the reverse reaction will be spontaneous.

It is more convenient to use ΔG as a criterion for spontaneity than to use ΔS_{univ}, because ΔG relates to the system alone and avoids the complication of having to examine the surroundings.

An analogy is often drawn between the free-energy change during a spontaneous reaction and the potential-energy change when a boulder rolls down a hill. Potential energy in a gravitational field "drives" the boulder until it reaches a state of minimum potential energy in the valley [Figure 19.17(a) ▶]. Similarly, the free energy of a chemical system decreases until it reaches a minimum value [Figure 19.17(b)]. When this minimum is reached, a state of equilibrium exists. *In any spontaneous process at constant temperature and pressure, the free energy always decreases.*

As a specific illustration of these ideas, let's return to the Haber process for the synthesis of ammonia from nitrogen and hydrogen, which we discussed extensively in Chapter 15:

$$N_2(g) + 3\,H_2(g) \rightleftharpoons 2\,NH_3(g)$$

Imagine that we have a reaction vessel that allows us to maintain a constant temperature and pressure and that we have a catalyst that allows the reaction to proceed at a reasonable rate. What will happen if we charge the vessel with a certain number of moles of N_2 and three times that number of moles of H_2? As we saw in Figure 15.3(a), the N_2 and H_2 will react spontaneously to form NH_3 until equilibrium is achieved. Similarly, Figure 15.3(b) demonstrates that if we charge the vessel with pure NH_3, it will decompose spontaneously to form N_2 and H_2 until equilibrium is reached. In each case the free energy of the system is lowered on the way to equilibrium, which represents a minimum in the free energy. We illustrate these cases in Figure 19.18 ▶.

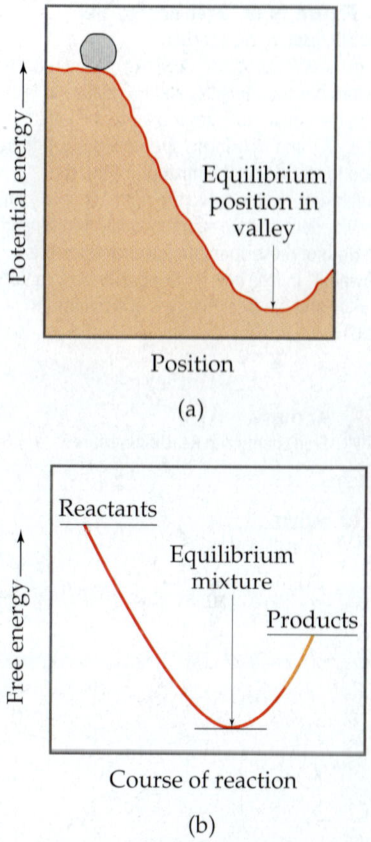

▲ **Figure 19.17 Potential energy and free energy.** An analogy is shown between the gravitational potential-energy change of a boulder rolling down a hill (a) and the free-energy change in a spontaneous reaction (b). The equilibrium position in (a) is given by the minimum gravitational potential energy available to the system. The equilibrium position in (b) is given by the minimum free energy available to the system.

GIVE IT SOME THOUGHT

Give the criterion for spontaneity first in terms of entropy and then in terms of free energy.

VCL The Balance Between Enthalpy and Entropy

This is a good time to remind ourselves of the significance of the reaction quotient, Q, for a system that is not at equilibrium. ∞ (Section 15.6) Recall that when $Q < K$, there is an excess of reactants relative to products. The reaction will proceed spontaneously in the forward direction to reach equilibrium. When $Q > K$, the reaction will proceed spontaneously in the reverse direction. At equilibrium $Q = K$. We have illustrated these points in Figure 19.18. In Section 19.7 we will see how to use the value of Q to calculate the value of ΔG for systems that are not at equilibrium.

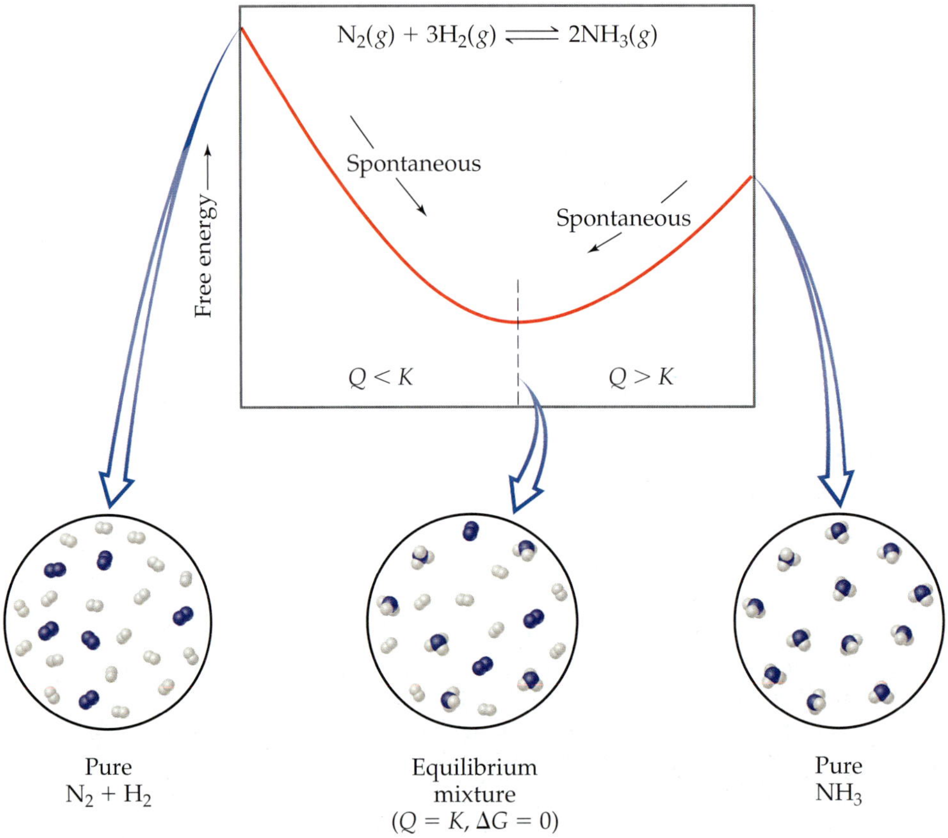

▶ **Figure 19.18 Free energy and equilibrium.** In the reaction $N_2(g) + 3\,H_2(g) \rightleftharpoons 2\,NH_3(g)$, if the reaction mixture has too much N_2 and H_2 relative to NH_3 (left), the equilibrium lies too far to the left $(Q < K)$ and NH_3 forms spontaneously. If there is too much NH_3 in the mixture (right), the equilibrium lies too far to the right $(Q > K)$ and the NH_3 decomposes spontaneously into N_2 and H_2. Both of these spontaneous processes are "downhill" in free energy. At equilibrium (center), $Q = K$ and the free energy is at a minimum $(\Delta G = 0)$.

ACTIVITY
Free Energy and Reaction Mixture

MOVIE
Formation of Water

Standard Free-Energy Changes

Free energy is a state function, like enthalpy. We can tabulate **standard free energies of formation** for substances, just as we can tabulate standard enthalpies of formation. (Section 5.7) It is important to remember that standard values for these functions imply a particular set of conditions, or standard states. The standard state for gaseous substances is 1 atm pressure. For solid substances, the standard state is the pure solid; for liquids, the pure liquid. For substances in solution, the standard state is normally a concentration of 1 *M*. (In very accurate work it may be necessary to make certain corrections, but we need not worry about these.) The temperature usually chosen for purposes of tabulating data is 25°C, but we will calculate $\Delta G°$ at other temperatures as well. Just as for the standard heats of formation, the free energies of elements in their standard states are set to zero. This arbitrary choice of a reference point has no effect on the quantity in which we are really interested, namely, the *difference* in free energy between reactants and products. The rules about standard states are summarized in Table 19.3 ◀. A listing of standard free energies of formation, denoted ΔG_f°, appears in Appendix C.

TABLE 19.3 Conventions Used in Establishing Standard Free Energies

State of Matter	Standard State
Solid	Pure solid
Liquid	Pure liquid
Gas	1 atm pressure
Solution	1 *M* concentration
Elements	Standard free energy of formation of an element in its standard state is defined as zero

GIVE IT SOME THOUGHT

What does the superscript ° indicate when associated with a thermodynamic quantity, as in $\Delta H°$, $\Delta S°$, or $\Delta G°$?

The standard free energies of formation are useful in calculating the *standard free-energy change* for chemical processes. The procedure is analogous to the calculation of ΔH° (Equation 5.31) and ΔS° (Equation 19.8):

$$\Delta G^\circ = \sum n\Delta G_f^\circ(\text{products}) - \sum m\Delta G_f^\circ(\text{reactants}) \qquad [19.13]$$

A CLOSER LOOK | What's "Free" About Free Energy?

The Gibbs free energy is a remarkable thermodynamic quantity. Because so many chemical reactions are carried out under conditions of near-constant pressure and temperature, chemists, biochemists, and engineers use the sign and magnitude of ΔG as exceptionally useful tools in the design and implementation of chemical and biochemical reactions. We will see examples of the usefulness of ΔG throughout the remainder of this chapter and this text.

There are two common questions that often arise when one first learns about the Gibbs free energy: Why does the sign of ΔG tell us about the spontaneity of reactions? What is "free" about free energy? We will address these two questions here by using concepts discussed in Chapter 5 and earlier in this chapter.

In Section 19.2 we saw that the second law of thermodynamics governs the spontaneity of processes. In order to apply the second law (Equation 19.4), however, we must determine ΔS_{univ}, which is often difficult to evaluate. By using the Gibbs free energy under conditions of constant temperature and pressure, we can relate ΔS_{univ} to quantities that depend only on changes in the system, namely ΔH and ΔS (as before, if we do not put a subscript on these quantities, we are referring to the system). Our first step in seeing this relationship is to recall Equation 19.2, which states that at constant temperature ΔS equals the amount of heat that would be transferred to the system in a reversible process, divided by the temperature:

$$\Delta S = q_{rev}/T \qquad \text{(at constant } T\text{)}$$

Remember that because entropy is a state function, this is the entropy change regardless of whether the system changes reversibly or irreversibly. Similarly, if T is constant, the entropy change of the surroundings is given by the heat transferred to the surroundings, q_{surr}, divided by the temperature. Because heat transferred *to* the surroundings must be transferred *from* the system, it follows that $q_{surr} = -q_{sys}$. Combining these ideas allows us to relate ΔS_{surr} to q_{sys}:

$$\Delta S_{surr} = q_{surr}/T = -q_{sys}/T \qquad \text{(constant } T\text{)} \qquad [19.14]$$

If P is also constant, $q_{sys} = q_P = \Delta H$ (Equation 5.10), so

$$\Delta S_{surr} = -q_{sys}/T = -\Delta H/T \qquad \text{(constant } T, P\text{)} \qquad [19.15]$$

We can now use Equation 19.4 to calculate ΔS_{univ} in terms of ΔS and ΔH:

$$\Delta S_{univ} = \Delta S + \Delta S_{surr} = \Delta S - \Delta H/T \qquad \text{(constant } T, P\text{)} \qquad [19.16]$$

Thus, under conditions of constant temperature and pressure, the second law (Equation 19.4) becomes:

Reversible process: $\Delta S - \Delta H/T = 0$

Irreversible process: $\Delta S - \Delta H/T > 0$

$$\text{(constant } T, P\text{)} \qquad [19.17]$$

Now we can see the relationship between ΔG and the second law. If we multiply the preceding equations by $-T$ and rearrange, we reach the following conclusion:

Reversible process: $\Delta G = \Delta H - T\Delta S = 0$

Irreversible process: $\Delta G = \Delta H - T\Delta S < 0$

$$\text{(constant } T, P\text{)} \qquad [19.18]$$

We see that we can use the sign of ΔG to conclude whether a reaction is spontaneous, nonspontaneous, or at equilibrium. The magnitude of ΔG is also significant. A reaction for which ΔG is large and negative, such as the burning of gasoline, is much more capable of doing work on the surroundings than is a reaction for which ΔG is small and negative, such as ice melting at room temperature. In fact, thermodynamics tells us that *the change in free energy for a process, ΔG, equals the maximum useful work that can be done by the system on its surroundings in a spontaneous process occurring at constant temperature and pressure:*

$$\Delta G = -w_{max} \qquad [19.19]$$

This relationship explains why ΔG is called the *free* energy. It is the portion of the energy change of a spontaneous reaction that is free to do useful work. The remainder of the energy enters the environment as heat.

What use can be made of this standard free-energy change for a chemical reaction? The quantity ΔG° tells us whether a mixture of reactants and products, each present under standard conditions, would spontaneously react in the forward direction to produce more products ($\Delta G^\circ < 0$) or in the reverse direction to form more reactants ($\Delta G^\circ > 0$). Because ΔG_f° values are readily available for a large number of substances, the standard free-energy change is easy to calculate for many reactions of interest.

For processes that are not spontaneous ($\Delta G > 0$), the free-energy change is a measure of the *minimum* amount of work that must be done to cause the process to occur. In actual cases we always need to do more than this theoretical minimum amount because of the inefficiencies in the way the changes occur.

SAMPLE EXERCISE 19.6 | Calculating Standard Free-Energy Change from Free Energies of Formation

(a) By using data from Appendix C, calculate the standard free-energy change for the following reaction at 298 K:

$$P_4(g) + 6\,Cl_2(g) \longrightarrow 4\,PCl_3(g)$$

(b) What is ΔG° for the reverse of the above reaction?

Solution

Analyze: We are asked to calculate the free-energy change for the indicated reaction, and then to determine the free-energy change of its reverse.

Plan: To accomplish our task, we look up the free-energy values for the products and reactants and use Equation 19.13: We multiply the molar quantities by the coefficients in the balanced equation, and subtract the total for the reactants from that for the products.

Solve: **(a)** $Cl_2(g)$ is in its standard state, so ΔG_f° is zero for this reactant. $P_4(g)$, however, is not in its standard state, so ΔG_f° is not zero for this reactant. From the balanced equation and using Appendix C, we have:

$$\begin{aligned}\Delta G^\circ_{rxn} &= 4\Delta G_f^\circ[PCl_3(g)] - \Delta G_f^\circ[P_4(g)] - 6\Delta G_f^\circ[Cl_2(g)] \\ &= (4\text{ mol})(-269.6\text{ kJ/mol}) - (1\text{ mol})(24.4\text{ kJ/mol}) - 0 \\ &= -1102.8\text{ kJ}\end{aligned}$$

The fact that ΔG° is negative tells us that a mixture of $P_4(g)$, $Cl_2(g)$, and $PCl_3(g)$ at 25°C, each present at a partial pressure of 1 atm, would react spontaneously in the forward direction to form more PCl_3. Remember, however, that the value of ΔG° tells us nothing about the rate at which the reaction occurs.

(b) Remember that $\Delta G = G\,(\text{products}) - G\,(\text{reactants})$. If we reverse the reaction, we reverse the roles of the reactants and products. Thus, reversing the reaction changes the sign of ΔG, just as reversing the reaction changes the sign of ΔH. (Section 5.4) Hence, using the result from part (a):

$$4\,PCl_3(g) \longrightarrow P_4(g) + 6\,Cl_2(g) \qquad \Delta G^\circ = +1102.8\text{ kJ}$$

PRACTICE EXERCISE

By using data from Appendix C, calculate ΔG° at 298 K for the combustion of methane: $CH_4(g) + 2\,O_2(g) \longrightarrow CO_2(g) + 2\,H_2O(g)$.
Answer: -800.7 kJ

SAMPLE EXERCISE 19.7 | Estimating and Calculating ΔG°

In Section 5.7 we used Hess's law to calculate ΔH° for the combustion of propane gas at 298 K:

$$C_3H_8(g) + 5\,O_2(g) \longrightarrow 3\,CO_2(g) + 4\,H_2O(l) \qquad \Delta H^\circ = -2220\text{ kJ}$$

(a) *Without using data from Appendix C,* predict whether ΔG° for this reaction is more negative or less negative than ΔH°. **(b)** Use data from Appendix C to calculate the standard free-energy change for the reaction at 298 K. Is your prediction from part (a) correct?

Solution

Analyze: In part (a) we must predict the value for ΔG° relative to that for ΔH° on the basis of the balanced equation for the reaction. In part (b) we must calculate the value for ΔG° and compare with our qualitative prediction.

Plan: The free-energy change incorporates both the change in enthalpy and the change in entropy for the reaction (Equation 19.11), so under standard conditions:

$$\Delta G^\circ = \Delta H^\circ - T\Delta S^\circ$$

To determine whether ΔG° is more negative or less negative than ΔH°, we need to determine the sign of the term $T\Delta S^\circ$. T is the absolute temperature, 298 K, so it is a positive number. We can predict the sign of ΔS° by looking at the reaction.

Solve: **(a)** We see that the reactants consist of six molecules of gas, and the products consist of three molecules of gas and four molecules of liquid. Thus, the number of molecules of gas has decreased significantly during the reaction. By using the general rules we discussed in Section 19.3, we would expect a decrease in the number of gas molecules to lead to a decrease in the entropy of the system—the products have fewer accessible microstates than the reactants. We therefore expect $\Delta S°$ and $T\Delta S°$ to be negative numbers. Because we are subtracting $T\Delta S°$, which is a negative number, we would predict that $\Delta G°$ is *less negative* than $\Delta H°$.

(b) Using Equation 19.13 and values from Appendix C, we can calculate the value of $\Delta G°$:

$$\Delta G° = 3\Delta G_f°[CO_2(g)] + 4\Delta G_f°[H_2O(l)] - \Delta G_f°[C_3H_8(g)] - 5\Delta G_f°[O_2(g)]$$

$$= 3\text{ mol}(-394.4\text{ kJ/mol}) + 4\text{ mol}(-237.13\text{ kJ/mol}) -$$

$$1\text{ mol}(-23.47\text{ kJ/mol}) - 5\text{ mol }(0\text{ kJ/mol}) = -2108\text{ kJ}$$

Notice that we have been careful to use the value of $\Delta G_f°$ for $H_2O(l)$; as in the calculation of ΔH values, the phases of the reactants and products are important. As we predicted, $\Delta G°$ is less negative than $\Delta H°$ because of the decrease in entropy during the reaction.

PRACTICE EXERCISE

Consider the combustion of propane to form $CO_2(g)$ and $H_2O(g)$ at 298 K: $C_3H_8(g) + 5\,O_2(g) \longrightarrow 3\,CO_2(g) + 4\,H_2O(g)$. Would you expect $\Delta G°$ to be more negative or less negative than $\Delta H°$?

Answer: more negative

19.6 | Free Energy and Temperature

We have seen that tabulations of $\Delta G_f°$, such as those in Appendix C, make it possible to calculate $\Delta G°$ for reactions at the standard temperature of 25°C. However, we are often interested in examining reactions at other temperatures. How is the change in free energy affected by the change in temperature? Let's look again at Equation 19.11:

$$\Delta G = \Delta H - T\Delta S = \underset{\text{Enthalpy term}}{\Delta H} + \underset{\text{Entropy term}}{(-T\Delta S)}$$

Notice that we have written the expression for ΔG as a sum of two contributions, an enthalpy term, ΔH, and an entropy term, $-T\Delta S$. Because the value of $-T\Delta S$ depends directly on the absolute temperature T, ΔG will vary with temperature. T is a positive number at all temperatures other than absolute zero. We know that the enthalpy term, ΔH, can be positive or negative. The entropy term, $-T\Delta S$, can also be positive or negative. When ΔS is positive, which means that the final state has greater randomness (a greater number of microstates) than the initial state, the term $-T\Delta S$ is negative. When ΔS is negative, the term $-T\Delta S$ is positive.

The sign of ΔG, which tells us whether a process is spontaneous, will depend on the signs and magnitudes of ΔH and $-T\Delta S$. When both ΔH and $-T\Delta S$ are negative, ΔG will always be negative and the process will be spontaneous at all temperatures. Likewise, when both ΔH and $-T\Delta S$ are positive, ΔG will always be positive and the process will be nonspontaneous at all temperatures (the reverse process will be spontaneous at all temperatures). When ΔH and $-T\Delta S$ have opposite signs, however, the sign of ΔG will depend on the magnitudes of these two terms. In these instances temperature is an important consideration. Generally, ΔH and ΔS change very little with temperature. However, the value of T directly affects the magnitude of $-T\Delta S$. As the temperature increases, the magnitude of the term $-T\Delta S$ increases and it will become relatively more important in determining the sign and magnitude of ΔG.

For example, let's consider once more the melting of ice to liquid water at 1 atm pressure:

$$H_2O(s) \longrightarrow H_2O(l) \qquad \Delta H > 0, \Delta S > 0$$

This process is endothermic, which means that ΔH is positive. We also know that the entropy increases during this process, so ΔS is positive and $-T\Delta S$ is negative. At temperatures below 0°C (273K) the magnitude of ΔH is greater than that of $-T\Delta S$. Hence, the positive enthalpy term dominates, leading to a positive value for ΔG. The positive value of ΔG means that the melting of ice is not spontaneous at $T < 0°C$; rather, the reverse process, the freezing of liquid water into ice, is spontaneous at these temperatures.

What happens at temperatures greater than 0°C? As the temperature increases, so does the magnitude of the entropy term $-T\Delta S$. When $T > 0°C$, the magnitude of $-T\Delta S$ is greater than the magnitude of ΔH. At these temperatures the negative entropy term dominates, which leads to a negative value for ΔG. The negative value of ΔG tells us that the melting of ice is spontaneous at $T > 0°C$. At the normal melting point of water, $T = 0°C$, the two phases are in equilibrium. Recall that $\Delta G = 0$ at equilibrium; at $T = 0°C$, ΔH and $-T\Delta S$ are equal in magnitude and opposite in sign, so they cancel one another and give $\Delta G = 0$.

GIVE IT SOME THOUGHT

The normal boiling point of benzene is 80°C. At 100°C and 1 atm, which term is greater for the vaporization of benzene, ΔH or $T\Delta S$?

The possible situations for the relative signs of ΔH and ΔS are given in Table 19.4 ▼, along with examples of each. By applying the concepts we have developed for predicting entropy changes, we often can predict how ΔG will change with temperature.

Our discussion of the temperature dependence of ΔG is also relevant to standard free-energy changes. As we saw in Sample Exercise 19.7, under standard conditions, Equation 19.11 becomes:

$$\Delta G° = \Delta H° - T\Delta S° \qquad [19.20]$$

We can readily calculate the values of $\Delta H°$ and $\Delta S°$ at 298 K from the data tabulated in Appendix C. If we assume that the values of $\Delta H°$ and $\Delta S°$ do not change with temperature, we can use Equation 19.20 to estimate the value of $\Delta G°$ at temperatures other than 298 K.

ACTIVITY
Gibbs Free Energy

TABLE 19.4 Effect of Temperature on the Spontaneity of Reactions

ΔH	ΔS	$-T\Delta S$	$\Delta G = \Delta H - T\Delta S$	Reaction Characteristics	Example
−	+	−	−	Spontaneous at all temperatures	$2\,O_3(g) \longrightarrow 3\,O_2(g)$
+	−	+	+	Nonspontaneous at all temperatures	$3\,O_2(g) \longrightarrow 2\,O_3(g)$
−	−	+	+ or −	Spontaneous at low *T*; nonspontaneous at high *T*	$H_2O(l) \longrightarrow H_2O(s)$
+	+	−	+ or −	Spontaneous at high *T*; nonspontaneous at low T	$H_2O(s) \longrightarrow H_2O(l)$

SAMPLE EXERCISE 19.8 | **Determining the Effect of Temperature on Spontaneity**

The Haber process for the production of ammonia involves the equilibrium

$$N_2(g) + 3\,H_2(g) \rightleftharpoons 2\,NH_3(g)$$

Assume that $\Delta H°$ and $\Delta S°$ for this reaction do not change with temperature. **(a)** Predict the direction in which $\Delta G°$ for this reaction changes with increasing temperature. **(b)** Calculate the values of $\Delta G°$ for the reaction at 25°C and 500°C.

Solution

Analyze: In part (a) we are asked to predict the direction in which $\Delta G°$ for the ammonia synthesis reaction changes as temperature increases. In part (b) we need to determine $\Delta G°$ for the reaction at two different temperatures.

Plan: In part (a) we can make this prediction by determining the sign of ΔS for the reaction and then using that information to analyze Equation 19.20. In part (b) we need to calculate $\Delta H°$ and $\Delta S°$ for the reaction by using the data in Appendix C. We can then use Equation 19.20 to calculate $\Delta G°$.

Solve: **(a)** Equation 19.20 tells us that $\Delta G°$ is the sum of the enthalpy term $\Delta H°$ and the entropy term $-T\Delta S°$. The temperature dependence of $\Delta G°$ comes from the entropy term. We expect $\Delta S°$ for this reaction to be negative because the number of molecules of gas is smaller in the products. Because $\Delta S°$ is negative, the term $-T\Delta S°$ is positive and grows larger with increasing temperature. As a result, $\Delta G°$ becomes less negative (or more positive) with increasing temperature. Thus, the driving force for the production of NH_3 becomes smaller with increasing temperature.

(b) We calculated the value of $\Delta H°$ in Sample Exercise 15.14, and the value of $\Delta S°$ was determined in Sample Exercise 19.5: $\Delta H° = -92.38$ kJ and $\Delta S° = -198.4$ J/K. If we assume that these values don't change with temperature, we can calculate $\Delta G°$ at any temperature by using Equation 19.20. At $T = 298$ K we have:

$$\Delta G° = -92.38 \text{ kJ} - (298 \text{ K})(-198.4 \text{ J/K})\left(\frac{1 \text{ kJ}}{1000 \text{ J}}\right)$$

$$= -92.38 \text{ kJ} + 59.1 \text{ kJ} = -33.3 \text{ kJ}$$

At $T = 500 + 273 = 773$ K we have

$$\Delta G° = -92.38 \text{ kJ} - (773 \text{ K})\left(-198.4\frac{\text{J}}{\text{K}}\right)\left(\frac{1 \text{ kJ}}{1000 \text{ J}}\right)$$

$$= -92.38 \text{ kJ} + 153 \text{ kJ} = 61 \text{ kJ}$$

Notice that we have been careful to convert $-T\Delta S°$ into units of kJ so that it can be added to $\Delta H°$, which has units of kJ.

Comment: Increasing the temperature from 298 K to 773 K changes $\Delta G°$ from -33.3 kJ to $+61$ kJ. Of course, the result at 773 K depends on the assumption that $\Delta H°$ and $\Delta S°$ do not change with temperature. In fact, these values do change slightly with temperature. Nevertheless, the result at 773 K should be a reasonable approximation. The positive increase in $\Delta G°$ with increasing T agrees with our prediction in part (a) of this exercise. Our result indicates that a mixture of $N_2(g)$, $H_2(g)$, and $NH_3(g)$, each present at a partial pressure of 1 atm, will react spontaneously at 298 K to form more $NH_3(g)$. In contrast, at 773 K the positive value of $\Delta G°$ tells us that the reverse reaction is spontaneous. Thus, when the mixture of three gases, each at a partial pressure of 1 atm, is heated to 773 K, some of the $NH_3(g)$ spontaneously decomposes into $N_2(g)$ and $H_2(g)$.

PRACTICE EXERCISE

(a) Using standard enthalpies of formation and standard entropies in Appendix C, calculate $\Delta H°$ and $\Delta S°$ at 298 K for the following reaction: $2\,SO_2(g) + O_2(g) \longrightarrow 2\,SO_3(g)$. **(b)** Using the values obtained in part (a), estimate $\Delta G°$ at 400 K.
Answers: **(a)** $\Delta H° = -196.6$ kJ, $\Delta S° = -189.6$ J/K; **(b)** $\Delta G° = -120.8$ kJ

19.7 | Free Energy and the Equilibrium Constant

In Section 19.5 we saw a special relationship between ΔG and equilibrium: For a system at equilibrium, $\Delta G = 0$. We have also seen how we can use tabulated thermodynamic data, such as those in Appendix C, to calculate values of the standard free-energy change, $\Delta G°$. In this final section of this chapter, we will learn two more ways in which we can use free energy as a powerful tool in our analysis of chemical reactions. First, we will learn how to use the value of $\Delta G°$ to calculate the value of ΔG under *nonstandard* conditions. Second, we will see how we can directly relate the value of $\Delta G°$ for a reaction to the value of the equilibrium constant for the reaction.

The set of standard conditions for which $\Delta G°$ values pertain are given in Table 19.3. Most chemical reactions occur under nonstandard conditions. For any chemical process the general relationship between the standard free-energy change, $\Delta G°$, and the free-energy change under any other conditions, ΔG, is given by the expression

$$\Delta G = \Delta G° + RT \ln Q \qquad [19.21]$$

In this equation R is the ideal-gas constant, 8.314 J/mol-K; T is the absolute temperature; and Q is the reaction quotient that corresponds to the particular reaction mixture of interest. (Section 15.5) Recall that the expression for Q is identical to the equilibrium-constant expression except that the reactants and products need not necessarily be at equilibrium.

Under standard conditions the concentrations of all the reactants and products are equal to 1. Thus, under standard conditions $Q = 1$ and therefore $\ln Q = 0$. We see that Equation 19.21 therefore reduces to $\Delta G = \Delta G°$ under standard conditions, as it should.

SAMPLE EXERCISE 19.9 | Relating ΔG to a Phase Change at Equilibrium

As we saw in Section 11.5, the *normal boiling point* is the temperature at which a pure liquid is in equilibrium with its vapor at a pressure of 1 atm. **(a)** Write the chemical equation that defines the normal boiling point of liquid carbon tetrachloride, $CCl_4(l)$. **(b)** What is the value of $\Delta G°$ for the equilibrium in part (a)? **(c)** Use thermodynamic data in Appendix C and Equation 19.20 to estimate the normal boiling point of CCl_4.

Solution

Analyze: (a) We must write a chemical equation that describes the physical equilibrium between liquid and gaseous CCl_4 at the normal boiling point. **(b)** We must determine the value of $\Delta G°$ for CCl_4 in equilibrium with its vapor at the normal boiling point. **(c)** We must estimate the normal boiling point of CCl_4, based on available thermodynamic data.

Plan: (a) The chemical equation will merely show the change of state of CCl_4 from liquid to solid. **(b)** We need to analyze Equation 19.21 at equilibrium ($\Delta G = 0$). **(c)** We can use Equation 19.20 to calculate T when $\Delta G = 0$.

Solve: (a) The normal boiling point of CCl_4 is the temperature at which pure liquid CCl_4 is in equilibrium with its vapor at a pressure of 1 atm:

$$CCl_4(l) \rightleftharpoons CCl_4(g, 1\text{ atm})$$

(b) At equilibrium $\Delta G = 0$. In any normal boiling-point equilibrium both the liquid and the vapor are in their standard states (Table 19.2). As a consequence, $Q = 1$, $\ln Q = 0$, and $\Delta G = \Delta G°$ for this process. Thus, we conclude that $\Delta G° = 0$ for the equilibrium involved in the normal boiling point of any liquid. We would also find that $\Delta G° = 0$ for the equilibria relevant to normal melting points and normal sublimation points of solids.

(c) Combining Equation 19.20 with the result from part (b), we see that the equality at the normal boiling point, T_b, of $CCl_4(l)$ or any other pure liquid is

$$\Delta G° = \Delta H° - T_b \Delta S° = 0$$

Solving the equation for T_b, we obtain

$$T_b = \Delta H°/\Delta S°$$

Strictly speaking, we would need the values of $\Delta H°$ and $\Delta S°$ for the equilibrium between $CCl_4(l)$ and $CCl_4(g)$ at the normal boiling point to do this calculation. However, we can *estimate* the boiling point by using the values of $\Delta H°$ and $\Delta S°$ for CCl_4 at 298 K, which we can obtain from the data in Appendix C and Equations 5.31 and 19.8:

$$\Delta H° = (1\ \text{mol})(-106.7\ \text{kJ/mol}) - (1\ \text{mol})(-139.3\ \text{kJ/mol}) = +32.6\ \text{kJ}$$
$$\Delta S° = (1\ \text{mol})(309.4\ \text{J/mol-K}) - (1\ \text{mol})(214.4\ \text{J/mol-K}) = +95.0\ \text{J/K}$$

Notice that, as expected, the process is endothermic ($\Delta H > 0$) and produces a gas in which energy can be more spread out ($\Delta S > 0$). We can now use these values to estimate T_b for $CCl_4(l)$:

$$T_b = \frac{\Delta H°}{\Delta S°} = \left(\frac{32.6\ \text{kJ}}{95.0\ \text{J/K}}\right)\left(\frac{1000\ \text{J}}{1\ \text{kJ}}\right) = 343\ \text{K} = 70°\text{C}$$

Note also that we have used the conversion factor between J and kJ to make sure that the units of $\Delta H°$ and $\Delta S°$ match.

Check: The experimental normal boiling point of $CCl_4(l)$ is 76.5°C. The small deviation of our estimate from the experimental value is due to the assumption that $\Delta H°$ and $\Delta S°$ do not change with temperature.

PRACTICE EXERCISE

Use data in Appendix C to estimate the normal boiling point, in K, for elemental bromine, $Br_2(l)$. (The experimental value is given in Table 11.3.)
Answer: 330 K

When the concentrations of reactants and products are nonstandard, we must calculate the value of Q in order to determine the value of ΔG. We illustrate how this is done in Sample Exercise 19.10. At this stage in our discussion, it becomes important to note the units associated with Q in Equation 19.21. The convention used for standard states imposes itself into the way Q is expressed: In Equation 19.21 the concentrations of gases are always expressed by their partial pressures in atmospheres and solutes are expressed by their concentrations in molarity.

SAMPLE EXERCISE 19.10 | Calculating the Free-Energy Change Under Nonstandard Conditions

We will continue to explore the Haber process for the synthesis of ammonia:

$$N_2(g) + 3\ H_2(g) \rightleftharpoons 2\ NH_3(g)$$

Calculate ΔG at 298 K for a reaction mixture that consists of 1.0 atm N_2, 3.0 atm H_2, and 0.50 atm NH_3.

Solution

Analyze: We are asked to calculate ΔG under nonstandard conditions.

Plan: We can use Equation 19.21 to calculate ΔG. Doing so requires that we calculate the value of the reaction quotient Q for the specified partial pressures of the gases and evaluate $\Delta G°$, using a table of standard free energies of formation.

Solve: Solving for the reaction quotient gives:

$$Q = \frac{P_{NH_3}{}^2}{P_{N_2}P_{H_2}{}^3} = \frac{(0.50)^2}{(1.0)(3.0)^3} = 9.3 \times 10^{-3}$$

In Sample Exercise 19.8 we calculated $\Delta G° = -33.3$ kJ for this reaction. We will have to change the units of this quantity in applying Equation 19.21, however. In order for the units in Equation 19.21 to work out, we will use kJ/mol as our units for $\Delta G°$, where "per mole" means "per mole of the reaction as written." Thus, $\Delta G^\circ = -33.3$ kJ/mol implies per 1 mol of N_2, per 3 mol of H_2, and per 2 mol of NH_3.

We can now use Equation 19.21 to calculate ΔG for these nonstandard conditions:

$$\Delta G = \Delta G^\circ + RT \ln Q$$
$$= (-33.3 \text{ kJ/mol}) + (8.314 \text{ J/mol-K})(298 \text{ K})(1 \text{ kJ}/1000 \text{ J}) \ln(9.3 \times 10^{-3})$$
$$= (-33.3 \text{ kJ/mol}) + (-11.6 \text{ kJ/mol}) = -44.9 \text{ kJ/mol}$$

Comment: We see that ΔG becomes more negative, changing from -33.3 kJ/mol to -44.9 kJ/mol, as the pressures of N_2, H_2, and NH_3 are changed from 1.0 atm each (standard conditions, ΔG°) to 1.0 atm, 3.0 atm, and 0.50 atm, respectively. The larger negative value for ΔG indicates a larger "driving force" to produce NH_3.

We would have made the same prediction on the basis of Le Châtelier's principle. (Section 15.6) Relative to standard conditions, we have increased the pressure of a reactant (H_2) and decreased the pressure of the product (NH_3). Le Châtelier's principle predicts that both of these changes should shift the reaction more to the product side, thereby forming more NH_3.

PRACTICE EXERCISE

Calculate ΔG at 298 K for the reaction of nitrogen and hydrogen to form ammonia if the reaction mixture consists of 0.50 atm N_2, 0.75 atm H_2, and 2.0 atm NH_3.
Answer: -26.0 kJ/mol

ACTIVITY
Free Energy and Chemical Equilibrium

We can now use Equation 19.21 to derive the relationship between ΔG° and the equilibrium constant, K. At equilibrium $\Delta G = 0$. Further, recall that the reaction quotient, Q, equals the equilibrium constant, K, when the system is at equilibrium. Thus, at equilibrium, Equation 19.21 transforms as follows:

$$\Delta G = \Delta G^\circ + RT \ln Q$$
$$0 = \Delta G^\circ + RT \ln K$$
$$\Delta G^\circ = -RT \ln K \qquad [19.22]$$

Equation 19.22 also allows us to calculate the value of K if we know the value of ΔG°. If we solve the equation for K, we obtain

$$K = e^{-\Delta G^\circ/RT} \qquad [19.23]$$

As we pointed out in discussing Equation 19.21, some care is necessary in the choice of units. Thus, in Equations 19.22 and 19.23 we again express ΔG° in kJ/mol. For the reactants and products in the equilibrium-constant expression, we use the following conventions: Gas pressures are given in atm; solution concentrations are given in moles per liter (molarity); and solids, liquids, and solvents do not appear in the expression. (Section 15.3) Thus, for gas-phase reactions the equilibrium constant is K_p, whereas for reactions in solution it is K_c. (Section 15.2)

From Equation 19.22 we can see that if ΔG° is negative, then $\ln K$ must be positive. A positive value for $\ln K$ means $K > 1$. Therefore, the more negative ΔG° is, the larger the equilibrium constant, K. Conversely, if ΔG° is positive, then $\ln K$ is negative, which means that $K < 1$. Table 19.5 ◀ summarizes these conclusions by comparing ΔG° and K for both positive and negative values of ΔG°.

TABLE 19.5 Relationship Between ΔG° and K at 298 K

ΔG°, kJ/mol	K
+200	9.1×10^{-36}
+100	3.0×10^{-18}
+50	1.7×10^{-9}
+10	1.8×10^{-2}
+1.0	6.7×10^{-1}
0	1.0
−1.0	1.5
−10	5.6×10^{1}
−50	5.8×10^{8}
−100	3.3×10^{17}
−200	1.1×10^{35}

CHEMISTRY AND LIFE | Driving Nonspontaneous Reactions

Many desirable chemical reactions, including a large number that are central to living systems, are nonspontaneous as written. For example, consider the extraction of copper metal from the mineral *chalcocite*, which contains Cu_2S. The decomposition of Cu_2S to its elements is nonspontaneous:

$$Cu_2S(s) \longrightarrow 2\,Cu(s) + S(s) \qquad \Delta G^\circ = +86.2\ \text{kJ}$$

Because ΔG° is very positive, we cannot obtain Cu(*s*) directly via this reaction. Instead, we must find some way to "do work" on the reaction in order to force it to occur as we wish. We can do this by coupling the reaction to another one so that the overall reaction *is* spontaneous. For example, we can envision the S(*s*) reacting with $O_2(g)$ to form $SO_2(g)$:

$$S(s) + O_2(g) \longrightarrow SO_2(g) \qquad \Delta G^\circ = -300.4\ \text{kJ}$$

By coupling these reactions together, we can extract much of the copper metal via a spontaneous reaction:

$$Cu_2S(s) + O_2(g) \longrightarrow 2\,Cu(s) + SO_2(g)$$
$$\Delta G^\circ = (+86.2\ \text{kJ}) + (-300.4\ \text{kJ}) = -214.2\ \text{kJ}$$

In essence, we have used the spontaneous reaction of S(*s*) with $O_2(g)$ to provide the free energy needed to extract the copper metal from the mineral.

Biological systems employ the same principle of using spontaneous reactions to drive nonspontaneous ones. Many of the biochemical reactions that are essential for the formation and maintenance of highly ordered biological structures are not spontaneous. These necessary reactions are made to occur by coupling them with spontaneous reactions that release energy. The metabolism of food is the usual source of the free energy needed to do the work of maintaining biological systems. For example, complete oxidation of the sugar *glucose*, $C_6H_{12}O_6$, to CO_2 and H_2O yields substantial free energy:

$$C_6H_{12}O_6(s) + 6\,O_2(g) \longrightarrow 6\,CO_2(g) + 6\,H_2O(l)$$
$$\Delta G^\circ = -2880\ \text{kJ}$$

This energy can be used to drive nonspontaneous reactions in the body. However, a means is necessary to transport the energy released by glucose metabolism to the reactions that require energy. One way, shown in Figure 19.19 ▼, involves the interconversion of adenosine triphosphate (ATP) and adenosine diphosphate (ADP), molecules that are related to the building blocks of nucleic acids. The conversion of ATP to ADP releases free energy ($\Delta G^\circ = -30.5$ kJ) that can be used to drive other reactions.

In the human body the metabolism of glucose occurs via a complex series of reactions, most of which release free energy. The free energy released during these steps is used in part to reconvert lower-energy ADP back to higher-energy ATP. Thus, the ATP-ADP interconversions are used to store energy during metabolism and to release it as needed to drive nonspontaneous reactions in the body. If you take a course in biochemistry, you will have the opportunity to learn more about the remarkable sequence of reactions used to transport free energy throughout the human body.

Related Exercises: 19.92 and 19.93

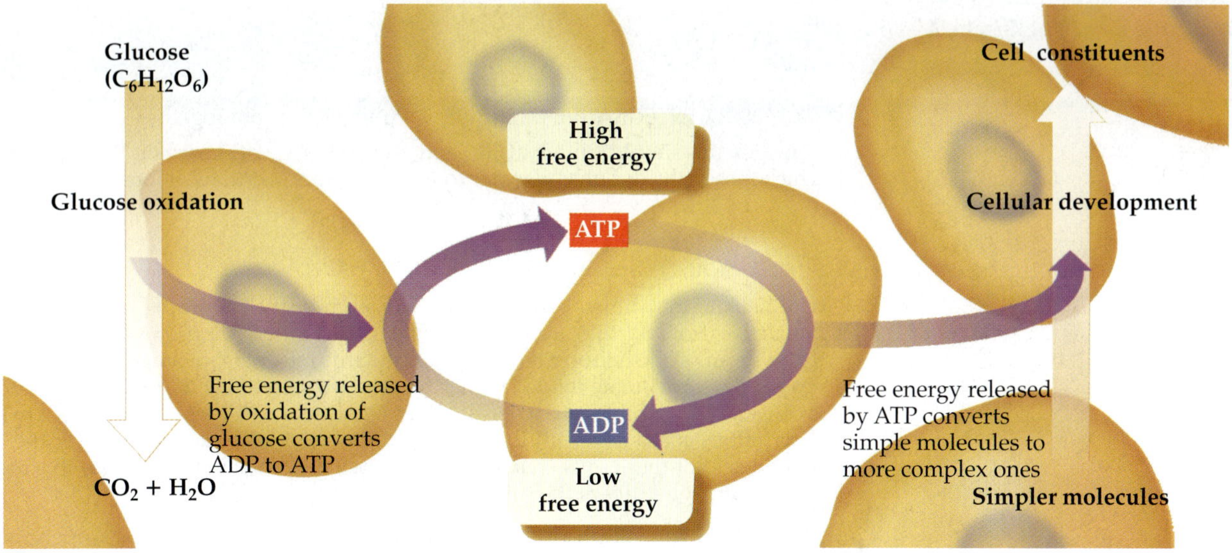

▲ **Figure 19.19 Free energy and cell metabolism.** This schematic representation shows part of the free-energy changes that occur in cell metabolism. The oxidation of glucose to CO_2 and H_2O produces free energy that is then used to convert ADP into the more energetic ATP. The ATP is then used, as needed, as an energy source to convert simple molecules into more complex cell constituents. When it releases its stored free energy, ATP is converted back into ADP.

SAMPLE EXERCISE 19.11 | Calculating an Equilibrium Constant from $\Delta G°$

Use standard free energies of formation to calculate the equilibrium constant, K, at 25°C for the reaction involved in the Haber process:

$$N_2(g) + 3\,H_2(g) \rightleftharpoons 2\,NH_3(g)$$

The standard free-energy change for this reaction was calculated in Sample Exercise 19.8: $\Delta G° = -33.3$ kJ/mol $= -33{,}300$ J/mol.

Solution

Analyze: We are asked to calculate K for a reaction, given $\Delta G°$.

Plan: We can use Equation 19.23 to evaluate the equilibrium constant, which in this case takes the form

$$K = \frac{P_{NH_3}{}^2}{P_{N_2}P_{H_2}{}^3}$$

In this expression the gas pressures are expressed in atmospheres. (Remember that we use kJ/mol as the units of $\Delta G°$ when using Equations 19.21, 19.22, or 19.23).

Solve: Solving Equation 19.23 for the exponent $-\Delta G°/RT$, we have

$$\frac{-\Delta G°}{RT} = \frac{-(-33{,}300\text{ J/mol})}{(8.314\text{ J/mol-K})(298\text{ K})} = 13.4$$

We insert this value into Equation 19.22 to obtain K:

$$K = e^{-\Delta G°/RT} = e^{13.4} = 7 \times 10^5$$

Comment: This is a large equilibrium constant, which indicates that the product, NH_3, is greatly favored in the equilibrium mixture at 25°C. The equilibrium constants for temperatures in the range of 300°C to 600°C, given in Table 15.2, are much smaller than the value at 25°C. Clearly, a low-temperature equilibrium favors the production of ammonia more than a high-temperature one. Nevertheless, the Haber process is carried out at high temperatures because the reaction is extremely slow at room temperature. **Remember:** Thermodynamics can tell us the direction and extent of a reaction, but tells us nothing about the rate at which it will occur. If a catalyst were found that would permit the reaction to proceed at a rapid rate at room temperature, high pressures would not be needed to force the equilibrium toward NH_3.

PRACTICE EXERCISE

Use data from Appendix C to calculate the standard free-energy change, $\Delta G°$, and the equilibrium constant, K, at 298 K for the reaction $H_2(g) + Br_2(l) \rightleftharpoons 2\,HBr(g)$.

Answer: $\Delta G° = -106.4$ kJ/mol, $K = 5 \times 10^{18}$

SAMPLE INTEGRATIVE EXERCISE | Putting Concepts Together

Consider the simple salts NaCl(s) and AgCl(s). We will examine the equilibria in which these salts dissolve in water to form aqueous solutions of ions:

$$NaCl(s) \rightleftharpoons Na^+(aq) + Cl^-(aq)$$

$$AgCl(s) \rightleftharpoons Ag^+(aq) + Cl^-(aq)$$

(a) Calculate the value of $\Delta G°$ at 298 K for each of the preceding reactions. **(b)** The two values from part (a) are very different. Is this difference primarily due to the enthalpy term or the entropy term of the standard free-energy change? **(c)** Use the values of $\Delta G°$ to calculate the K_{sp} values for the two salts at 298 K. **(d)** Sodium chloride is considered a soluble salt, whereas silver chloride is considered insoluble. Are these descriptions consistent with the answers to part (c)? **(e)** How will $\Delta G°$ for the solution process of these salts change with increasing T? What effect should this change have on the solubility of the salts?

Solution (a) We will use Equation 19.13 along with $\Delta G_f°$ values from Appendix C to calculate the $\Delta G°_{soln}$ values for each equilibrium. (As we did in Section 13.1, we use the subscript "soln" to indicate that these are thermodynamic quantities for the formation of a solution.) We find

$$\Delta G°_{soln}(NaCl) = (-261.9\text{ kJ/mol}) + (-131.2\text{ kJ/mol}) - (-384.0\text{ kJ/mol})$$
$$= -9.1\text{ kJ/mol}$$

$$\Delta G°_{soln}(AgCl) = (+77.11\text{ kJ/mol}) + (-131.2\text{ kJ/mol}) - (-109.70\text{ kJ/mol})$$
$$= +55.6\text{ kJ/mol}$$

(b) We can write ΔG°_{soln} as the sum of an enthalpy term, ΔH°_{soln}, and an entropy term, $-T\Delta S^\circ_{soln}$: $\Delta G^\circ_{soln} = \Delta H^\circ_{soln} + (-T\Delta S^\circ_{soln})$. We can calculate the values of ΔH°_{soln} and ΔS°_{soln} by using Equations 5.31 and 19.8. We can then calculate $-T\Delta S^\circ_{soln}$ at $T = 298$ K. All these calculations are now familiar to us. The results are summarized in the following table:

Salt	ΔH°_{soln}	ΔS°_{soln}	$-T\Delta S^\circ_{soln}$
NaCl	+3.6 kJ/mol	+43.2 J/mol-K	−12.9 kJ/mol
AgCl	+65.7 kJ/mol	+34.3 J/mol-K	−10.2 kJ/mol

The entropy terms for the solution of the two salts are very similar. That seems sensible because each solution process should lead to a similar increase in randomness as the salt dissolves, forming hydrated ions. ∞ (Section 13.1) In contrast, we see a very large difference in the enthalpy term for the solution of the two salts. The difference in the values of ΔG°_{soln} is dominated by the difference in the values of ΔH°_{soln}.

(c) The solubility product, K_{sp}, is the equilibrium constant for the solution process. ∞ (Section 17.4) As such, we can relate K_{sp} directly to ΔG°_{soln} by using Equation 19.23:

$$K_{sp} = e^{-\Delta G^\circ_{soln}/RT}$$

We can calculate the K_{sp} values in the same way we applied Equation 19.23 in Sample Exercise 19.13. We use the ΔG°_{soln} values we obtained in part (a), remembering to convert them from to J/mol:

$$\text{NaCl:} \quad K_{sp} = [Na^+(aq)][Cl^-(aq)] = e^{-(-9100)/[(8.314)(298)]} = e^{+3.7} = 40$$

$$\text{AgCl:} \quad K_{sp} = [Ag^+(aq)][Cl^-(aq)] = e^{-(+55,600)/[(8.314)(298)]} = e^{-22.4} = 1.9 \times 10^{-10}$$

The value calculated for the K_{sp} of AgCl is very close to that listed in Appendix D.

(d) A soluble salt is one that dissolves appreciably in water. ∞ (Section 4.2) The K_{sp} value for NaCl is greater than 1, indicating that NaCl dissolves to a great extent. The K_{sp} value for AgCl is very small, indicating that very little dissolves in water. Silver chloride should indeed be considered an insoluble salt.

(e) As we expect, the solution process has a positive value of ΔS for both salts (see the table in part b). As such, the entropy term of the free-energy change, $-T\Delta S^\circ_{soln}$, is negative. If we assume that ΔH°_{soln} and ΔS°_{soln} do not change much with temperature, then an increase in T will serve to make ΔG°_{soln} more negative. Thus, the driving force for dissolution of the salts will increase with increasing T, and we therefore expect the solubility of the salts to increase with increasing T. In Figure 13.17 we see that the solubility of NaCl (and that of nearly any salt) increases with increasing temperature. ∞ (Section 13.3)

SUMMARY AND KEY TERMS

Introduction and Section 19.1 In this chapter we examined some of the aspects of chemical thermodynamics, the area of chemistry that explores energy relationships. Most reactions and chemical processes have an inherent directionality: They are **spontaneous** in one direction and nonspontaneous in the reverse direction. The spontaneity of a process is related to the thermodynamic path the system takes from the initial state to the final state. In a **reversible process**, both the system and its surroundings can be restored to their original state by exactly reversing the change. In an **irreversible process** the system can't return to its original state without there being a change in the surroundings. Any spontaneous process is irreversible. A process that occurs at a constant temperature is said to be **isothermal**.

Section 19.2 The spontaneous nature of processes is related to a thermodynamic state function called **entropy**, denoted S. For a process that occurs at constant temperature, the entropy change of the system is given by the heat absorbed by the system along a reversible path, divided by the temperature: $\Delta S = q_{rev}/T$. The way entropy controls the spontaneity of processes is given by the **second law of thermodynamics**, which governs the change in the entropy of the universe, $\Delta S_{univ} = \Delta S_{sys} + \Delta S_{surr}$. The second law states that in a reversible process $\Delta S_{univ} = 0$; in an irreversible (spontaneous) process $\Delta S_{univ} > 0$. Entropy values are usually expressed in units of joules per kelvin, J/K.

Section 19.3 Molecules can undergo three kinds of motion: In **translational motion** the entire molecule moves in space. Molecules can also undergo **vibrational motion**, in which the atoms of the molecule move toward and away from one another in periodic fashion, and **rotational motion**, in which the entire molecule spins like a top. A particular combination of motions and locations of the atoms and molecules of a system at a particular instant is called a **microstate**. Entropy is a measure of how many microstates are associated with a particular macroscopic state. If the total number of accessible microstates is W, the entropy is given by $S = k \ln W$. The number of available microstates, and therefore the entropy, increases with an increase in volume, temperature, or motion of molecules, because any of these changes increases the possible motions and locations of the molecules. As a result, entropy generally increases when liquids or solutions are formed from solids, gases are formed from either solids or liquids, or the number of molecules of gas increases during a chemical reaction. The **third law of thermodynamics** states that the entropy of a pure crystalline solid at 0 K is zero.

Section 19.4 The third law allows us to assign entropy values for substances at different temperatures. Under standard conditions the entropy of a mole of a substance is called its **standard molar entropy**, denoted $S°$. From tabulated values of $S°$, we can calculate the entropy change for any process under standard conditions. For an isothermal process, the entropy change in the surroundings is equal to $-\Delta H/T$.

Section 19.5 The **Gibbs free energy** (or just **free energy**), G, is a thermodynamic state function that combines the two state functions enthalpy and entropy: $G = H - TS$. For processes that occur at constant temperature, $\Delta G = \Delta H - T\Delta S$. For a process or reaction occurring at constant temperature and pressure, the sign of ΔG relates to the spontaneity of the process. When ΔG is negative, the process is spontaneous. When ΔG is positive, the process is nonspontaneous; the reverse process is spontaneous. At equilibrium the process is reversible and ΔG is zero. The free energy is also a measure of the maximum useful work that can be performed by a system in a spontaneous process. The standard free-energy change, $\Delta G°$, for any process can be calculated from tabulations of **standard free energies of formation**, ΔG_f°, which are defined in a fashion analogous to standard enthalpies of formation, ΔH_f°. The value of ΔG_f° for a pure element in its standard state is defined to be zero.

Sections 19.6 and 19.7 The values of ΔH and ΔS generally do not vary much with temperature. As a consequence, the dependence of ΔG with temperature is governed mainly by the value of T in the expression $\Delta G = \Delta H - T\Delta S$. The entropy term $-T\Delta S$ has the greater effect on the temperature dependence of ΔG and, hence, on the spontaneity of the process. For example, a process for which $\Delta H > 0$ and $\Delta S > 0$, such as the melting of ice, can be nonspontaneous ($\Delta G > 0$) at low temperatures and spontaneous ($\Delta G < 0$) at higher temperatures. Under nonstandard conditions ΔG is related to $\Delta G°$ and the value of the reaction quotient, Q: $\Delta G = \Delta G° + RT \ln Q$. At equilibrium ($\Delta G = 0, Q = K$), $\Delta G° = -RT \ln K$. Thus, the standard free-energy change is directly related to the equilibrium constant for the reaction. This relationship can be used to explain the temperature dependence of equilibrium constants.

VISUALIZING CONCEPTS

19.1 Two different gases occupy two separate bulbs. Consider the process that occurs when the stopcock separating the gases is opened, assuming the gases behave ideally. **(a)** Draw the final (equilibrium) state. **(b)** Predict the signs of ΔH and ΔS for the process. **(c)** Is the process that occurs when the stopcock is opened a reversible one? **(d)** How does the process affect the entropy of the surroundings? [Sections 19.1 and 19.2]

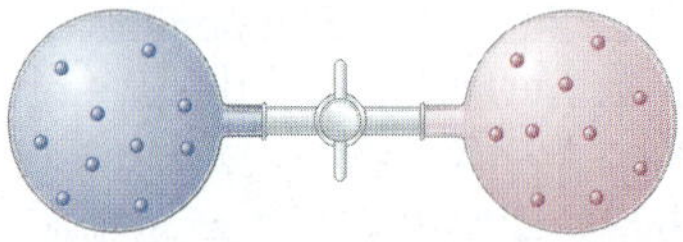

19.2 **(a)** What are the signs of ΔS, ΔH, and ΔG for the process depicted below? **(b)** If energy can flow in and out of the system, what can you say about the entropy change of the surroundings as a result of this process? [Sections 19.2 and 19.5]

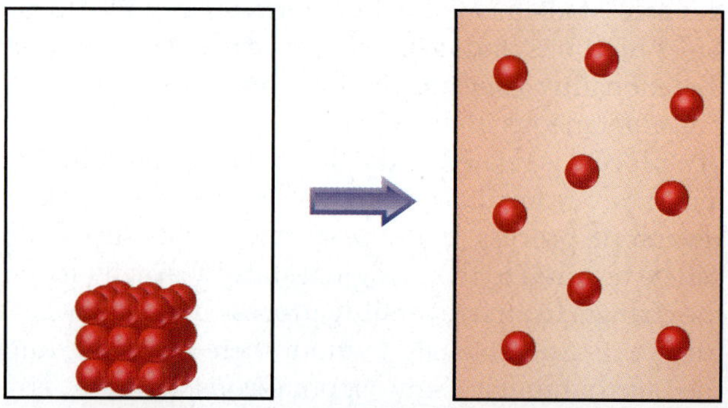

19.3 Predict the sign of ΔS accompanying this reaction. Explain your choice. [Section 19.3]

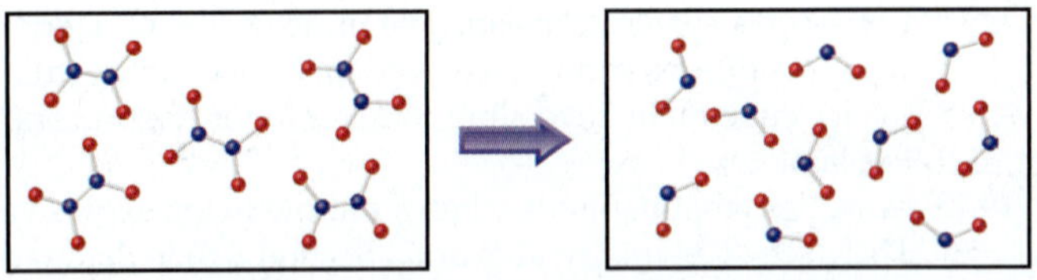

19.4 The diagram below shows how ΔH (red line) and $T\Delta S$ (blue line) change with temperature for a hypothetical reaction. **(a)** What is the significance of the point at 300 K, where ΔH and $T\Delta S$ are equal? **(b)** In what temperature range is this reaction spontaneous? [Section 19.6]

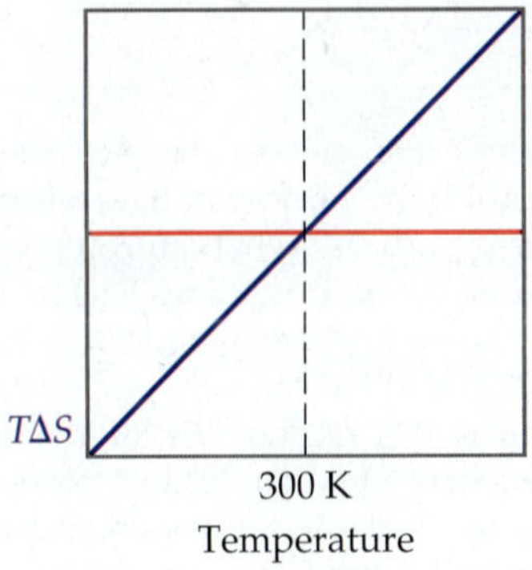

19.5 Consider a reaction $A_2(g) + B_2(g) \rightleftharpoons 2\,AB(g)$, with atoms of A shown in red here and atoms of B shown in blue. **(a)** If box 1 represents an equilibrium mixture, what is the sign of ΔG for the process in which the contents of a reaction vessel change from what is shown in box 2 to what is shown in box 1? **(b)** Rank the boxes in order of increasing magnitude of ΔG for the reaction. [Sections 19.5 and 19.7]

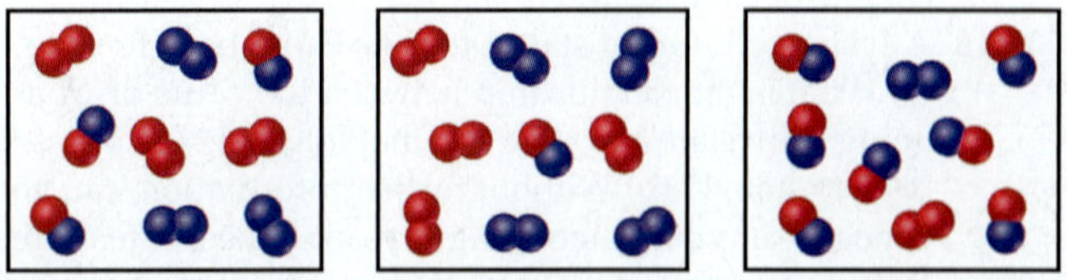

19.6 The diagram below shows how the free energy, G, changes during a hypothetical reaction $A(g) + B(g) \longrightarrow AB(g)$. On the left are pure reactants, each at 1 atm, and on the right is the pure product, also at 1 atm. **(a)** What is the significance of the minimum in the plot? **(b)** What does the quantity x, shown on the right side of the diagram, represent? [Section 19.7]

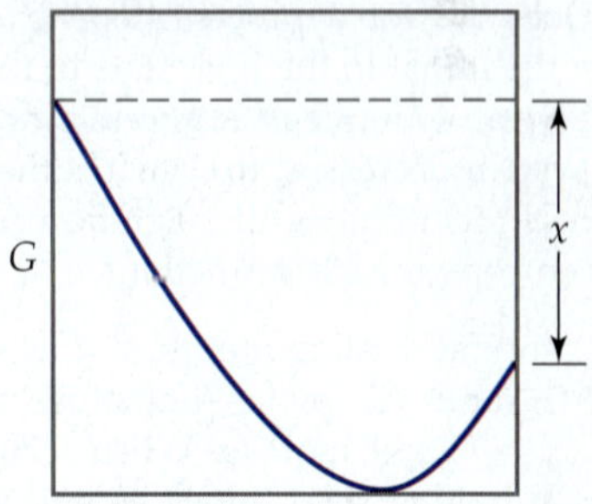

EXERCISES

Spontaneous Processes

19.7 Which of the following processes are spontaneous, and which are nonspontaneous: **(a)** the melting of ice cubes at −5°C and 1 atm pressure, **(b)** dissolution of sugar in a cup of hot coffee, **(c)** the reaction of nitrogen atoms to form N_2 molecules at 25°C and 1 atm, **(d)** alignment of iron filings in a magnetic field, **(e)** formation of CH_4 and O_2 molecules from CO_2 and H_2O at room temperature and 1 atm of pressure?

19.8 Which of the following processes are spontaneous: **(a)** spreading of the fragrance of perfume through a room; **(b)** separating a mixture of N_2 and O_2 into two separate samples, one that is pure N_2 and one that is pure O_2; **(c)** the bursting of an inflated balloon; **(d)** the reaction of sodium metal with chlorine gas to form sodium chloride; **(e)** the dissolution of $HCl(g)$ in water to form concentrated hydrochloric acid?

19.9 **(a)** Give two examples of endothermic processes that are spontaneous. **(b)** Give an example of a process that is spontaneous at one temperature but nonspontaneous at a different temperature.

19.10 A nineteenth-century chemist, Marcellin Berthelot, suggested that all chemical processes that proceed spontaneously are exothermic. Offer some counterexamples that show that Berthelot's idea was too general.

19.11 Consider the vaporization of liquid water to steam at a pressure of 1 atm. **(a)** Is this process endothermic or exothermic? **(b)** In what temperature range is it a spontaneous process? **(c)** In what temperature range is it a nonspontaneous process? **(d)** At what temperature are the two phases in equilibrium?

19.12 The normal freezing point of 1-propanol (C_3H_8O) is −127°C. **(a)** Is the freezing of 1-propanol an endothermic or exothermic process? **(b)** In what temperature range is the freezing of 1-propanol a spontaneous process? **(c)** In what temperature range is it a nonspontaneous process? **(d)** Is there any temperature at which liquid and solid 1-propanol are in equilibrium? Explain.

19.13 **(a)** What is special about a *reversible* process? **(b)** Suppose a reversible process is reversed, restoring the system to its original state. What can be said about the surroundings after the process is reversed? **(c)** Under what circumstances will the vaporization of water to steam be a reversible process?

19.14 **(a)** What is meant by calling a process *irreversible*? **(b)** After an irreversible process the system is restored to its original state. What can be said about the condition of the surroundings after the system is restored to its original state? **(c)** Under what conditions will the condensation of a liquid be an irreversible process?

19.15 Consider a process in which an ideal gas changes from state 1 to state 2 in such a way that its temperature changes from 300 K to 200 K. Does the change in ΔE depend on the particular pathway taken to carry out this change of state? Explain.

19.16 A system goes from state 1 to state 2 and back to state 1. **(a)** What is the relationship between the value of ΔE for going from state 1 to state 2 to that for going from state 2 back to state 1? **(b)** Without further information, can you conclude anything about the amount of heat transferred to the system as it goes from state 1 to state 2 as compared to that upon going from state 2 back to state 1? **(c)** Suppose the changes in state are reversible processes. Can you conclude anything about the work done by the system upon going from state 1 to state 2 as compared to that upon going from state 2 back to state 1?

19.17 Consider a system consisting of an ice cube. **(a)** Under what conditions can the ice cube melt reversibly? **(b)** If the ice cube melts reversibly, is ΔE zero for the process? Explain.

19.18 Consider what happens when a sample of the explosive TNT (see "Chemistry at Work," Section 8.8) is detonated. **(a)** Is the detonation a spontaneous process? **(b)** What is the sign of q for this process? **(c)** Can you determine whether w is positive, negative, or zero for the process? Explain. **(d)** Can you determine the sign of ΔE for the process? Explain.

Entropy and the Second Law of Thermodynamics

19.19 **(a)** How can we calculate ΔS for an isothermal process? **(b)** Does ΔS for a process depend on the path taken from the initial to the final state of the system? Explain.

19.20 **(a)** Give an example of a process in which the entropy of the system decreases. **(b)** What is the sign of ΔS for the process? **(c)** What is the significance of the statement that entropy is a state function?

19.21 The normal boiling point of methanol (CH_3OH) is 64.7°C, and its molar enthalpy of vaporization is $\Delta H_{vap} = 71.8$ kJ/mol. **(a)** When $CH_3OH(l)$ boils at its normal boiling point, does its entropy increase or decrease? **(b)** Calculate the value of ΔS when 1.00 mol of $CH_3OH(l)$ is vaporized at 64.7°C.

19.22 The element cesium (Cs) freezes at 28.4°C, and its molar enthalpy of fusion is $\Delta H_{fus} = 2.09$ kJ/mol **(a)** When molten cesium solidifies to Cs(*s*) at its normal melting point, is ΔS positive or negative? **(b)** Calculate the value of ΔS when 15.0 g of Cs(*l*) solidifies at 28.4°C.

19.23 **(a)** Express the second law of thermodynamics in words. **(b)** If the entropy of the system increases during a reversible process, what can you say about the entropy change of the surroundings? **(c)** In a certain spontaneous process the system undergoes an entropy change, $\Delta S = 42$ J/K. What can you conclude about ΔS_{surr}?

19.24 **(a)** Express the second law of thermodynamics as a mathematical equation. **(b)** In a particular spontaneous process the entropy of the system decreases. What can you conclude about the sign and magnitude of ΔS_{surr}? **(c)** During a certain reversible process, the surroundings undergo an entropy change, $\Delta S_{surr} = -78$ J/K. What is the entropy change of the system for this process?

19.25 The volume of 0.100 mol of helium gas at 27°C is increased isothermally from 2.00 L to 5.00 L. Assuming the gas to be ideal, calculate the entropy change for the process.

19.26 The pressure on 0.500 mol of neon gas is increased from 0.900 atm to 3.00 atm at 100°C. Assuming the gas to be ideal, calculate ΔS for this process.

The Molecular Interpretation of Entropy

19.27 How does each of the following affect the number of microstates available to a system: **(a)** temperature, **(b)** volume, **(c)** state of matter?

19.28 **(a)** Using the heat of vaporization in Appendix B, calculate the entropy change for the vaporization of water at 25°C and at 100°C. **(b)** From your knowledge of microstates and the structure of liquid water, explain the difference in these two values.

19.29 **(a)** What do you expect for the sign of ΔS in a chemical reaction in which two moles of gaseous reactants are converted to three moles of gaseous products? **(b)** For which of the processes in Exercise 19.8 does the entropy of the system increase?

19.30 **(a)** In a chemical reaction two gases combine to form a solid. What do you expect for the sign of ΔS? **(b)** For which of the processes in Exercise 19.7 does the entropy of the system increase?

19.31 How does the entropy of the system change when **(a)** a solid melts, **(b)** a gas liquefies, **(c)** a solid sublimes?

19.32 How does the entropy of the system change when **(a)** the temperature of the system increases, **(b)** the volume of a gas increases, **(c)** a solid dissolves in water?

19.33 **(a)** State the third law of thermodynamics. **(b)** Distinguish between translational motion, vibrational motion, and rotational motion of a molecule. **(c)** Illustrate these three kinds of motion with sketches for the HCl molecule.

19.34 **(a)** The energy of a gas is increased by heating it. Using CO_2 as an example, illustrate the different ways in which additional energy can be distributed among the molecules of the gas. **(b)** You are told that the number of microstates for a system increases. What do you know about the entropy of the system?

19.35 The diagram below shows the variation in entropy with temperature for a substance that is a gas at the highest temperature shown. **(a)** What processes correspond to the entropy increases along the vertical lines labeled 1 and 2 in this diagram? **(b)** Why is the entropy change for 2 larger than that for 1?

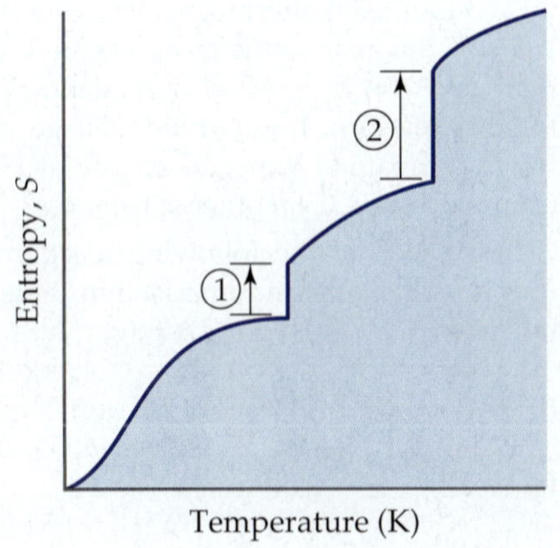

19.36 Propanol (C_3H_7OH) melts at −126.5°C and boils at 97.4°C. Draw a qualitative sketch of how the entropy changes as propanol vapor at 150°C and 1 atm is cooled to solid propanol at −150°C and 1 atm.

19.37 For each of the following pairs, choose the substance with the higher entropy per mole at a given temperature: **(a)** Ar(*l*) or Ar(*g*), **(b)** He(*g*) at 3 atm pressure or He(*g*) at 1.5 atm pressure, **(c)** 1 mol of Ne(*g*) in 15.0 L or 1 mol of Ne(*g*) in 1.50 L, **(d)** $CO_2(g)$ or $CO_2(s)$.

19.38 For each of the following pairs, indicate which substance possesses the larger standard entropy: **(a)** 1 mol of $P_4(g)$ at 300°C, 0.01 atm, or 1 mol of $As_4(g)$ at 300°C, 0.01 atm; **(b)** 1 mol of $H_2O(g)$ at 100°C, 1 atm, or 1 mol of $H_2O(l)$ at 100°C, 1 atm; **(c)** 0.5 mol of $N_2(g)$ at 298 K, 20-L volume, or 0.5 mol $CH_4(g)$ at 298 K, 20-L volume; **(d)** 100 g $Na_2SO_4(s)$ at 30°C or 100 g $Na_2SO_4(aq)$ at 30°C.

19.39 Predict the sign of the entropy change of the system for each of the following reactions:
(a) $2\,SO_2(g) + O_2(g) \longrightarrow 2\,SO_3(g)$
(b) $Ba(OH)_2(s) \longrightarrow BaO(s) + H_2O(g)$
(c) $CO(g) + 2\,H_2(g) \longrightarrow CH_3OH(l)$
(d) $FeCl_2(s) + H_2(g) \longrightarrow Fe(s) + 2\,HCl(g)$

19.40 Predict the sign of ΔS_{sys} for each of the following processes: **(a)** molten Fe solidifies; **(b)** LiCl(*s*) is formed from Li(*s*) and $Cl_2(g)$; **(c)** zinc metal dissolves in hydrochloric acid, forming $ZnCl_2(aq)$ and $H_2(g)$; **(d)** silver bromide precipitates upon mixing $AgNO_3(aq)$ and KBr(*aq*).

Entropy Changes in Chemical Reactions

19.41 In each of the following pairs, which compound would you expect to have the higher standard molar entropy: **(a)** $C_2H_2(g)$ or $C_2H_6(g)$; **(b)** $CO_2(g)$ or CO(*g*)? Explain.

19.42 Cyclopropane and propylene isomers both have the formula C_3H_6. Based on the molecular structures shown, which of these isomers would you expect to have the higher standard molar entropy at 25°C?

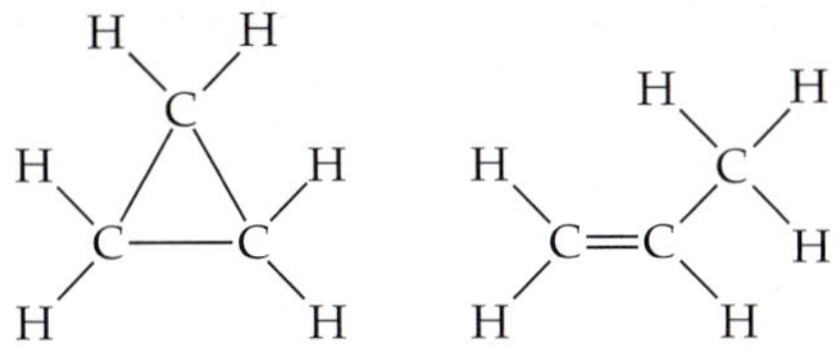

Cyclopropane Propylene

19.43 Use Appendix C to compare the standard entropies at 25°C for the following pairs of substances: **(a)** Sc(*s*) and Sc(*g*), **(b)** $NH_3(g)$ and $NH_3(aq)$, **(c)** 1 mol $P_4(g)$ and 2 mol $P_2(g)$, **(d)** C(graphite) and C(diamond). In each case explain the difference in the entropy values.

19.44 Using Appendix C, compare the standard entropies at 25°C for the following pairs of substances: **(a)** CuO(*s*) and $Cu_2O(s)$; **(b)** 1 mol $N_2O_4(g)$ and 2 mol $NO_2(g)$; **(c)** $CH_3OH(g)$ and $CH_3OH(l)$; **(d)** 1 mol PbO(*s*) plus 1 mol $CO_2(g)$ and 1 mol $PbCO_3(s)$. For each pair, explain the difference in the entropy values.

[19.45] The standard entropies at 298 K for certain of the group 4A elements are as follows: C(*s*, diamond) = 2.43 J/mol-K; Si(*s*) = 18.81 J/mol-K; Ge(*s*) = 31.09 J/mol-K; and Sn(*s*) = 51.18 J/mol-K. All but Sn have the diamond structure. How do you account for the trend in the $S°$ values?

[19.46] Three of the forms of elemental carbon are graphite, diamond, and buckminsterfullerene. The entropies at 298 K for graphite and diamond are listed in Appendix C. **(a)** Account for the difference in the $S°$ values of graphite and diamond in light of their structures (Figure 11.41). **(b)** What would you expect for the $S°$ value of buckminsterfullerene (Figure 11.43) relative to the values for graphite and diamond? Explain.

19.47 Using $S°$ values from Appendix C, calculate $\Delta S°$ values for the following reactions. In each case account for the sign of $\Delta S°$.
(a) $C_2H_4(g) + H_2(g) \longrightarrow C_2H_6(g)$
(b) $N_2O_4(g) \longrightarrow 2\,NO_2(g)$
(c) $Be(OH)_2(s) \longrightarrow BeO(s) + H_2O(g)$
(d) $2\,CH_3OH(g) + 3\,O_2(g) \longrightarrow 2\,CO_2(g) + 4\,H_2O(g)$

19.48 Calculate $\Delta S°$ values for the following reactions by using tabulated $S°$ values from Appendix C. In each case explain the sign of $\Delta S°$.
(a) $N_2H_4(g) + H_2(g) \longrightarrow 2\,NH_3(g)$
(b) $2\,Al(s) + 3\,Cl_2(g) \longrightarrow 2\,AlCl_3(s)$
(c) $Mg(OH)_2(s) + 2\,HCl(g) \longrightarrow MgCl_2(s) + 2\,H_2O(l)$
(d) $2\,CH_4(g) \longrightarrow C_2H_6(g) + H_2(g)$

Gibbs Free Energy

19.49 **(a)** For a process that occurs at constant temperature, express the change in Gibbs free energy in terms of changes in the enthalpy and entropy of the system. **(b)** For a certain process that occurs at constant T and P, the value of ΔG is positive. What can you conclude? **(c)** What is the relationship between ΔG for a process and the rate at which it occurs?

19.50 **(a)** What is the meaning of the standard free-energy change, $\Delta G°$, as compared with ΔG? **(b)** For any process that occurs at constant temperature and pressure, what is the significance of $\Delta G = 0$? **(c)** For a certain process, ΔG is large and negative. Does this mean that the process necessarily occurs rapidly?

19.51 For a certain chemical reaction, $\Delta H° = -35.4$ kJ and $\Delta S° = -85.5$ J/K. **(a)** Is the reaction exothermic or endothermic? **(b)** Does the reaction lead to an increase or decrease in the disorder of the system? **(c)** Calculate $\Delta G°$ for the reaction at 298 K. **(d)** Is the reaction spontaneous at 298 K under standard conditions?

19.52 A certain reaction has $\Delta H° = -19.5$ kJ and $\Delta S° = +42.7$ J/K. **(a)** Is the reaction exothermic or endothermic? **(b)** Does the reaction lead to an increase or decrease in the disorder of the system? **(c)** Calculate $\Delta G°$ for the reaction at 298 K. **(d)** Is the reaction spontaneous at 298 K under standard conditions?

19.53 Using data in Appendix C, calculate $\Delta H°$, $\Delta S°$, and $\Delta G°$ at 298 K for each of the following reactions. In each case show that $\Delta G° = \Delta H° - T\Delta S°$.
(a) $H_2(g) + F_2(g) \longrightarrow 2\,HF(g)$
(b) $C(s, \text{graphite}) + 2\,Cl_2(g) \longrightarrow CCl_4(g)$
(c) $2\,PCl_3(g) + O_2(g) \longrightarrow 2\,POCl_3(g)$
(d) $2\,CH_3OH(g) + H_2(g) \longrightarrow C_2H_6(g) + 2\,H_2O(g)$

19.54 Use data in Appendix C to calculate $\Delta H°$, $\Delta S°$, and $\Delta G°$ at 25°C for each of the following reactions. In each case show that $\Delta G° = \Delta H° - T\Delta S°$.
(a) $Ni(s) + Cl_2(g) \longrightarrow NiCl_2(s)$
(b) $CaCO_3(s, \text{calcite}) \longrightarrow CaO(s) + CO_2(g)$
(c) $P_4O_{10}(s) + 6\,H_2O(l) \longrightarrow 4\,H_3PO_4(aq)$
(d) $2\,CH_3OH(l) + 3\,O_2(g) \longrightarrow 2\,CO_2(g) + 4\,H_2O(l)$

19.55 Using data from Appendix C, calculate $\Delta G°$ for the following reactions. Indicate whether each reaction is spontaneous under standard conditions.
(a) $2\,SO_2(g) + O_2(g) \longrightarrow 2\,SO_3(g)$
(b) $NO_2(g) + N_2O(g) \longrightarrow 3\,NO(g)$
(c) $6\,Cl_2(g) + 2\,Fe_2O_3(s) \longrightarrow 4\,FeCl_3(s) + 3\,O_2(g)$
(d) $SO_2(g) + 2\,H_2(g) \longrightarrow S(s) + 2\,H_2O(g)$

19.56 Using data from Appendix C, calculate the change in Gibbs free energy for each of the following reactions. In each case indicate whether the reaction is spontaneous under standard conditions.
(a) $H_2(g) + Cl_2(g) \longrightarrow 2\,HCl(g)$
(b) $MgCl_2(s) + H_2O(l) \longrightarrow MgO(s) + 2\,HCl(g)$
(c) $2\,NH_3(g) \longrightarrow N_2H_4(g) + H_2(g)$
(d) $2\,NOCl(g) \longrightarrow 2\,NO(g) + Cl_2(g)$

19.57 Cyclohexane (C_6H_{12}) is a liquid hydrocarbon at room temperature. **(a)** Write a balanced equation for the combustion of $C_6H_{12}(l)$ to form $CO_2(g)$ and $H_2O(l)$. **(b)** Without using thermochemical data, predict whether $\Delta G°$ for this reaction is more negative or less negative than $\Delta H°$.

19.58 Sulfur dioxide reacts with strontium oxide as follows:

$$SO_2(g) + SrO(s) \longrightarrow SrSO_3(s)$$

(a) Without using thermochemical data, predict whether $\Delta G°$ for this reaction is more negative or less negative than $\Delta H°$. **(b)** If you had only standard enthalpy data for this reaction, how would you go about making a rough estimate of the value of $\Delta G°$ at 298 K, using data from Appendix C on other substances?

19.59 Classify each of the following reactions as one of the four possible types summarized in Table 19.4:
(a) $N_2(g) + 3\,F_2(g) \longrightarrow 2\,NF_3(g)$
$\Delta H° = -249$ kJ; $\Delta S° = -278$ J/K
(b) $N_2(g) + 3\,Cl_2(g) \longrightarrow 2\,NCl_3(g)$
$\Delta H° = 460$ kJ; $\Delta S° = -275$ J/K
(c) $N_2F_4(g) \longrightarrow 2\,NF_2(g)$
$\Delta H° = 85$ kJ; $\Delta S° = 198$ J/K

19.60 From the values given for $\Delta H°$ and $\Delta S°$, calculate $\Delta G°$ for each of the following reactions at 298 K. If the reaction is not spontaneous under standard conditions at 298 K, at what temperature (if any) would the reaction become spontaneous?
(a) $2\,PbS(s) + 3\,O_2(g) \longrightarrow 2\,PbO(s) + 2\,SO_2(g)$
$\Delta H° = -844$ kJ; $\Delta S° = -165$ J/K
(b) $2\,POCl_3(g) \longrightarrow 2\,PCl_3(g) + O_2(g)$
$\Delta H° = 572$ kJ; $\Delta S° = 179$ J/K

19.61 A particular reaction is spontaneous at 450 K. The enthalpy change for the reaction is +34.5 kJ. What can you conclude about the sign and magnitude of ΔS for the reaction?

19.62 A certain reaction is nonspontaneous at −25°C. The entropy change for the reaction is 95 J/K. What can you conclude about the sign and magnitude of ΔH?

19.63 For a particular reaction, $\Delta H = -32$ kJ and $\Delta S = -98$ J/K. Assume that ΔH and ΔS do not vary with temperature. **(a)** At what temperature will the reaction have $\Delta G = 0$? **(b)** If T is increased from that in part (a), will the reaction be spontaneous or nonspontaneous?

19.64 Reactions in which a substance decomposes by losing CO_2 are called *decarboxylation* reactions. The decarboxylation of acetic acid proceeds as follows:

$$CH_3COOH(l) \longrightarrow CH_4(g) + CO_2(g)$$

By using data from Appendix C, calculate the minimum temperature at which this process will be spontaneous under standard conditions. Assume that $\Delta H°$ and $\Delta S°$ do not vary with temperature.

19.65 Consider the following reaction between oxides of nitrogen:

$$NO_2(g) + N_2O(g) \longrightarrow 3\,NO(g)$$

(a) Use data in Appendix C to predict how $\Delta G°$ for the reaction varies with increasing temperature. **(b)** Calculate $\Delta G°$ at 800 K, assuming that $\Delta H°$ and $\Delta S°$ do not change with temperature. Under standard conditions is the reaction spontaneous at 800 K? **(c)** Calculate ΔG° at 1000 K. Is the reaction spontaneous under standard conditions at this temperature?

19.66 Methanol (CH_3OH) can be made by the controlled oxidation of methane:

$$CH_4(g) + \frac{1}{2}O_2(g) \longrightarrow CH_3OH(g)$$

(a) Use data in Appendix C to calculate ΔH° and ΔS° for this reaction. **(b)** How is ΔG° for the reaction expected to vary with increasing temperature? **(c)** Calculate ΔG° at 298 K. Under standard conditions, is the reaction spontaneous at this temperature? **(d)** Is there a temperature at which the reaction would be at equilibrium under standard conditions and that is low enough so that the compounds involved are likely to be stable?

19.67 **(a)** Use data in Appendix C to estimate the boiling point of benzene, $C_6H_6(l)$. **(b)** Use a reference source, such as the *CRC Handbook of Chemistry and Physics*, to find the experimental boiling point of benzene. How do you explain any deviation between your answer in part (a) and the experimental value?

[19.68] **(a)** Using data in Appendix C, estimate the temperature at which the free-energy change for the transformation from $I_2(s)$ to $I_2(g)$ is zero. What assumptions must you make in arriving at this estimate? **(b)** Use a reference source, such as WebElements (www.webelements.com), to find the experimental melting and boiling points of I_2. **(c)** Which of the values in part (b) is closer to the value you obtained in part (a)? Can you explain why this is so?

19.69 Acetylene gas, $C_2H_2(g)$, is used in welding. **(a)** Write a balanced equation for the combustion of acetylene gas to $CO_2(g)$ and $H_2O(l)$. **(b)** How much heat is produced in burning 1 mol of C_2H_2 under standard conditions if both reactants and products are brought to 298 K? **(c)** What is the maximum amount of useful work that can be accomplished under standard conditions by this reaction?

19.70 **(a)** How much heat is produced in burning 1 mol of ethylene (C_2H_4) under standard conditions if reactants and products are brought to 298 K and $H_2O(l)$ is formed? **(b)** What is the maximum amount of useful work that can be accomplished under standard conditions by this system?

Free Energy and Equilibrium

19.71 Explain qualitatively how ΔG changes for each of the following reactions as the partial pressure of O_2 is increased:
(a) $2\,CO(g) + O_2(g) \longrightarrow 2\,CO_2(g)$
(b) $2\,H_2O_2(l) \longrightarrow 2\,H_2O(l) + O_2(g)$
(c) $2\,KClO_3(s) \longrightarrow 2\,KCl(s) + 3\,O_2(g)$

19.72 Indicate whether ΔG increases, decreases, or does not change when the partial pressure of H_2 is increased in each of the following reactions:
(a) $N_2(g) + 3\,H_2(g) \longrightarrow 2\,NH_3(g)$
(b) $2\,HBr(g) \longrightarrow H_2(g) + Br_2(g)$
(c) $2\,H_2(g) + C_2H_2(g) \longrightarrow C_2H_6(g)$

19.73 Consider the reaction $2\,NO_2(g) \longrightarrow N_2O_4(g)$. **(a)** Using data from Appendix C, calculate ΔG° at 298 K. **(b)** Calculate ΔG at 298 K if the partial pressures of NO_2 and N_2O_4 are 0.40 atm and 1.60 atm, respectively.

19.74 Consider the reaction $H_2(g) + F_2(g) \longrightarrow 2\,HF(g)$. **(a)** Using data from Appendix C, calculate ΔG° at 298 K. **(b)** Calculate ΔG at 298 K if the reaction mixture consists of 8.0 atm of H_2, 4.5 atm of F_2, and 0.36 atm of HF.

19.75 Use data from Appendix C to calculate the equilibrium constant, K, at 298 K for each of the following reactions:
(a) $H_2(g) + I_2(g) \rightleftharpoons 2\,HI(g)$
(b) $C_2H_5OH(g) \rightleftharpoons C_2H_4(g) + H_2O(g)$
(c) $3\,C_2H_2(g) \rightleftharpoons C_6H_6(g)$

19.76 Write the equilibrium-constant expression and calculate the value of the equilibrium constant for each of the following reactions at 298 K, using data from Appendix C:
(a) $NaHCO_3(s) \rightleftharpoons NaOH(s) + CO_2(g)$
(b) $2\,HBr(g) + Cl_2(g) \rightleftharpoons 2\,HCl(g) + Br_2(g)$
(c) $2\,SO_2(g) + O_2(g) \rightleftharpoons 2\,SO_3(g)$

19.77 Consider the decomposition of barium carbonate:

$$BaCO_3(s) \rightleftharpoons BaO(s) + CO_2(g)$$

Using data from Appendix C, calculate the equilibrium pressure of CO_2 at **(a)** 298 K and **(b)** 1100 K.

19.78 Consider the following reaction:

$$PbCO_3(s) \rightleftharpoons PbO(s) + CO_2(g)$$

Using data in Appendix C, calculate the equilibrium pressure of CO_2 in the system at **(a)** 120°C and **(b)** 480°C.

19.79 The value of K_a for nitrous acid (HNO_2) at 25°C is given in Appendix D. **(a)** Write the chemical equation for the equilibrium that corresponds to K_a. **(b)** By using the value of K_a, calculate ΔG° for the dissociation of nitrous acid in aqueous solution. **(c)** What is the value of ΔG at equilibrium? **(d)** What is the value of ΔG when $[H^+] = 5.0 \times 10^{-2}\,M$, $[NO_2^-] = 6.0 \times 10^{-4}\,M$, and $[HNO_2] = 0.20\,M$?

19.80 The K_b for methylamine (CH_3NH_2) at 25°C is given in Appendix D. **(a)** Write the chemical equation for the equilibrium that corresponds to K_b. **(b)** By using the value of K_b, calculate ΔG° for the equilibrium in part (a). **(c)** What is the value of ΔG at equilibrium? **(d)** What is the value of ΔG when $[H^+] = 1.5 \times 10^{-8}\,M$, $[CH_3NH_3^+] = [H^+] = 1.5 \times 10^{-8}\,M$, $[CH_3NH_3^+] = 5.5 \times 10^{-4}\,M$, and $[CH_3NH_2] = 0.120\,M$?

Additional Exercises

19.81 Indicate whether each of the following statements is true or false. If it is false, correct it. **(a)** The feasibility of manufacturing NH_3 from N_2 and H_2 depends entirely on the value of ΔH for the process $N_2(g) + 3\,H_2(g) \longrightarrow 2\,NH_3(g)$. **(b)** The reaction of Na(*s*) with $Cl_2(g)$ to form NaCl(*s*) is a spontaneous process. **(c)** A spontaneous process can in principle be conducted reversibly. **(d)** Spontaneous processes in general require that work be done to force them to proceed. **(e)** Spontaneous processes are those that are exothermic and that lead to a higher degree of order in the system.

19.82 For each of the following processes, indicate whether the signs of ΔS and ΔH are expected to be positive, negative, or about zero. **(a)** A solid sublimes. **(b)** The temperature of a sample of Co(*s*) is lowered from 60°C to 25°C. **(c)** Ethyl alcohol evaporates from a beaker. **(d)** A diatomic molecule dissociates into atoms. **(e)** A piece of charcoal is combusted to form $CO_2(g)$ and $H_2O(g)$.

19.83 The reaction $2\,Mg(s) + O_2(g) \longrightarrow 2\,MgO(s)$ is highly spontaneous and has a negative value for $\Delta S°$. The second law of thermodynamics states that in any spontaneous process there is always an increase in the entropy of the universe. Is there an inconsistency between the above reaction and the second law?

19.84 Ammonium nitrate dissolves spontaneously and endothermally in water at room temperature. What can you deduce about the sign of ΔS for this solution process?

[19.85] *Trouton's rule* states that for many liquids at their normal boiling points, the standard molar entropy of vaporization is about 88 J/mol-K. **(a)** Estimate the normal boiling point of bromine, Br_2, by determining $\Delta H°_{vap}$ for Br_2 using data from Appendix D. Assume that $\Delta H°_{vap}$ remains constant with temperature and that Trouton's rule holds. **(b)** Look up the normal boiling point of Br_2 in a chemistry handbook or at the WebElements Web site (www.webelements.com).

19.86 For the majority of the compounds listed in Appendix C, the value of $\Delta G_f°$ is more positive (or less negative) than the value of $\Delta H_f°$. **(a)** Explain this observation, using $NH_3(g)$, $CCl_4(l)$, and $KNO_3(s)$ as examples. **(b)** An exception to this observation is CO(*g*). Explain the trend in the $\Delta H_f°$ and $\Delta G_f°$ values for this molecule.

19.87 Consider the following three reactions:

(i) $2\,RbCl(s) + 3\,O_2(g) \longrightarrow 2\,RbClO_3(s)$
(ii) $C_2H_2(g) + 4\,Cl_2(g) \longrightarrow 2\,CCl_4(l) + H_2(g)$
(iii) $TiCl_4(l) + 2\,H_2O(l) \longrightarrow TiO_2(s) + 4\,HCl(aq)$

(a) For each of the reactions, use data in Appendix C to calculate $\Delta H°$, $\Delta G°$, and $\Delta S°$ at 25°C. **(b)** Which of these reactions are spontaneous under standard conditions at 25°C? **(c)** For each of the reactions, predict the manner in which the change in free energy varies with an increase in temperature.

19.88 Using the data in Appendix C and given the pressures listed, calculate ΔG for each of the following reactions:

(a) $N_2(g) + 3\,H_2(g) \longrightarrow 2\,NH_3(g)$

$P_{N_2} = 2.6$ atm, $P_{H_2} = 5.9$ atm, $P_{NH_3} = 1.2$ atm

(b) $2\,N_2H_4(g) + 2\,NO_2(g) \longrightarrow 3\,N_2(g) + 4\,H_2O(g)$

$P_{N_2H_4} = P_{NO_2} = 5.0 \times 10^{-2}$ atm,
$P_{N_2} = 0.5$ atm, $P_{H_2O} = 0.3$ atm

(c) $N_2H_4(g) \longrightarrow N_2(g) + 2\,H_2(g)$

$P_{N_2H_4} = 0.5$ atm, $P_{N_2} = 1.5$ atm, $P_{H_2} = 2.5$ atm

19.89 **(a)** For each of the following reactions, predict the sign of $\Delta H°$ and $\Delta S°$ and discuss briefly how these factors determine the magnitude of *K*. **(b)** Based on your general chemical knowledge, predict which of these reactions will have $K > 0$. **(c)** In each case indicate whether *K* should increase or decrease with increasing temperature.

(i) $2\,Mg(s) + O_2(g) \rightleftharpoons 2\,MgO(s)$
(ii) $2\,KI(s) \rightleftharpoons 2\,K(g) + I_2(g)$
(iii) $Na_2(g) \rightleftharpoons 2\,Na(g)$
(iv) $2\,V_2O_5(s) \rightleftharpoons 4\,V(s) + 5\,O_2(g)$

19.90 Acetic acid can be manufactured by combining methanol with carbon monoxide, an example of a *carbonylation* reaction:

$$CH_3OH(l) + CO(g) \longrightarrow CH_3COOH(l)$$

(a) Calculate the equilibrium constant for the reaction at 25°C. **(b)** Industrially, this reaction is run at temperatures above 25°C. Will an increase in temperature produce an increase or decrease in the mole fraction of acetic acid at equilibrium? Why are elevated temperatures used? **(c)** At what temperature will this reaction have an equilibrium constant equal to 1? (You may assume that $\Delta H°$ and $\Delta S°$ are temperature-independent, and you may ignore any phase changes that might occur.)

19.91 The oxidation of glucose ($C_6H_{12}O_6$) in body tissue produces CO_2 and H_2O. In contrast, anaerobic decomposition, which occurs during fermentation, produces ethanol (C_2H_5OH) and CO_2. **(a)** Using data given in Appendix C, compare the equilibrium constants for the following reactions:

$$C_6H_{12}O_6(s) + 6\,O_2(g) \rightleftharpoons 6\,CO_2(g) + 6\,H_2O(l)$$

$$C_6H_{12}O_6(s) \rightleftharpoons 2\,C_2H_5OH(l) + 2\,CO_2(g)$$

(b) Compare the maximum work that can be obtained from these processes under standard conditions.

[19.92] The conversion of natural gas, which is mostly methane, into products that contain two or more carbon atoms, such as ethane (C_2H_6), is a very important industrial chemical process. In principle, methane can be converted into ethane and hydrogen:

$$2\,CH_4(g) \longrightarrow C_2H_6(g) + H_2(g)$$

In practice, this reaction is carried out in the presence of oxygen:

$$2\,CH_4(g) + \tfrac{1}{2}O_2(g) \longrightarrow C_2H_6(g) + H_2O(g)$$

(a) Using the data in Appendix C, calculate *K* for these reactions at 25°C and 500°C. **(b)** Is the difference in $\Delta G°$ for the two reactions due primarily to the enthalpy term (ΔH) or the entropy term ($-T\Delta S$)? **(c)** Explain how the preceding reactions are an example of driving a nonspontaneous reaction, as discussed in the "Chemistry and Life" box in Section 19.7. **(d)** The reaction of CH_4 and O_2 to form C_2H_6 and H_2O must be carried out carefully to avoid a competing reaction. What is the most likely competing reaction?

[19.93] Cells use the hydrolysis of adenosine triphosphate (ATP) as a source of energy (Figure 19.19). The conversion of ATP to ADP has a standard free-energy change of −30.5 kJ/mol. If all the free energy from the metabolism of glucose,

$$C_6H_{12}O_6(s) + 6\,O_2(g) \longrightarrow 6\,CO_2(g) + 6\,H_2O(l)$$

goes into the conversion of ADP to ATP, how many moles of ATP can be produced for each mole of glucose?

[19.94] The potassium-ion concentration in blood plasma is about $5.0 \times 10^{-3}\,M$, whereas the concentration in muscle-cell fluid is much greater (0.15 M). The plasma and intracellular fluid are separated by the cell membrane, which we assume is permeable only to K^+. **(a)** What is ΔG for the transfer of 1 mol of K^+ from blood plasma to the cellular fluid at body temperature (37°C)? **(b)** What is the minimum amount of work that must be used to transfer this K^+?

[19.95] The relationship between the temperature of a reaction, its standard enthalpy change, and the equilibrium constant at that temperature can be expressed as the following linear equation:

$$\ln K = \frac{-\Delta H^\circ}{RT} + \text{constant}$$

(a) Explain how this equation can be used to determine ΔH° experimentally from the equilibrium constants at several different temperatures. **(b)** Derive the preceding equation using relationships given in this chapter. To what is the constant equal?

[19.96] One way to derive Equation19.3 depends on the observation that at constant T the number of ways, W, of arranging m ideal-gas particles in a volume V is proportional to the volume raised to the m power:

$$W \propto V^m$$

Use this relationship and Boltzmann's relationship between entropy and number of arrangements (Equation 19.5) to derive the equation for the entropy change for the isothermal expansion or compression of n moles of an ideal gas.

[19.97] An ad in an auto magazine trumpets a new device that saves on gas. It is called the "Entropy Converter." The ad reads, "Attach the Entropy Converter to your carburetor and see amazing improvement in your gas mileage! Did you know that increases in entropy when the fuel is burned help to drive your car? The Entropy Converter automatically lowers the entropy of the gas molecules as they enter the carburetor. Then, when they are burned in the engine, the entropy increases more than ever, yielding much more energy to drive your car!" Indicate whether you think this device might work. If not, explain why.

Integrative Exercises

19.98 Most liquids follow Trouton's rule, which states that the molar entropy of vaporization lies in the range of 88 ± 5 J/mol-K. The normal boiling points and enthalpies of vaporization of several organic liquids are as follows:

Substance	Normal Boiling Point (°C)	ΔH_{vap} (kJ/mol)
Acetone, $(CH_3)_2CO$	56.1	29.1
Dimethyl ether, $(CH_3)_2O$	−24.8	21.5
Ethanol, C_2H_5OH	78.4	38.6
Octane, C_8H_{18}	125.6	34.4
Pyridine, C_5H_5N	115.3	35.1

(a) Calculate ΔS_{vap} for each of the liquids. Do all of the liquids obey Trouton's rule? **(b)** With reference to intermolecular forces (Section 11.2), can you explain any exceptions to the rule? **(c)** Would you expect water to obey Trouton's rule? By using data in Appendix B, check the accuracy of your conclusion. **(d)** Chlorobenzene (C_6H_5Cl) boils at 131.8°C. Use Trouton's rule to estimate ΔH_{vap} for this substance.

19.99 Consider the polymerization of ethylene to polyethylene. ∞ (Section 12.2) **(a)** What would you predict for the sign of the entropy change during polymerization (ΔS_{poly})? Explain your reasoning. **(b)** The polymerization of ethylene is a spontaneous process at room temperature. What can you conclude about the enthalpy change during polymerization (ΔH_{poly})? **(c)** Use average bond enthalpies (Table 8.4) to estimate the value of ΔH_{poly} per ethylene monomer added. **(d)** Polyethylene is an *addition polymer*. By comparison, Nylon 66 is a *condensation polymer*. How would you expect ΔS_{poly} for a condensation polymer to compare to that for an addition polymer? Explain.

19.100 In chemical kinetics the *entropy of activation* is the entropy change for the process in which the reactants reach the activated complex. The entropy of activation for bimolecular processes is usually negative. Explain this observation with reference to Figure 14.13.

19.101 The following processes were all discussed in Chapter 18, "Chemistry of the Environment." Estimate whether the entropy of the system increases or decreases during each process: **(a)** photodissociation of $O_2(g)$, **(b)** formation of ozone from oxygen molecules and oxygen atoms, **(c)** diffusion of CFCs into the stratosphere, **(d)** desalination of water by reverse osmosis.

19.102 Carbon disulfide (CS_2) is a toxic, highly flammable substance. The following thermodynamic data are available for $CS_2(l)$ and $CS_2(g)$ at 298 K:

	ΔH_f° (kJ/mol)	ΔG_f° (kJ/mol)
$CS_2(l)$	89.7	65.3
$CS_2(g)$	117.4	67.2

(a) Draw the Lewis structure of the molecule. What do you predict for the bond order of the C—S bonds? **(b)** Use the VSEPR method to predict the structure of the CS_2 molecule. **(c)** Liquid CS_2 burns in O_2 with a blue flame, forming $CO_2(g)$ and $SO_2(g)$. Write a balanced equation for this reaction. **(d)** Using the data in the preceding table and in Appendix C, calculate ΔH° and ΔG° for the reaction in part (c). Is the reaction exothermic? Is it spontaneous at 298 K? **(e)** Use the data in the preceding table to calculate ΔS° at 298 K for the vaporization of $CS_2(l)$. Is the sign of ΔS° as you would expect for a vaporization? **(f)** Using data in the preceding table and your answer to part (e), estimate the boiling point of $CS_2(l)$. Do you predict that the substance will be a liquid or a gas at 298 K and 1 atm?

[19.103] The following data compare the standard enthalpies and free energies of formation of some crystalline ionic substances and aqueous solutions of the substances:

Substance	ΔH_f° (kJ/mol)	ΔG_f° (kJ/mol)
$AgNO_3(s)$	−124.4	−33.4
$AgNO_3(aq)$	−101.7	−34.2
$MgSO_4(s)$	−1283.7	−1169.6
$MgSO_4(aq)$	−1374.8	−1198.4

(a) Write the formation reaction for $AgNO_3(s)$. Based on this reaction, do you expect the entropy of the system to increase or decrease upon the formation of $AgNO_3(s)$? **(b)** Use ΔH_f° and ΔG_f° of $AgNO_3(s)$ to determine the entropy change upon formation of the substance. Is your answer consistent with your reasoning in part (a)? **(c)** Is dissolving $AgNO_3$ in water an exothermic or endothermic process? What about dissolving $MgSO_4$ in water? **(d)** For both $AgNO_3$ and $MgSO_4$, use the data to calculate the entropy change when the solid is dissolved in water. **(e)** Discuss the results from part (d) with reference to material presented in this chapter and in the second "Closer Look" box in Section 13.5.

[19.104] Consider the following equilibrium:

$$N_2O_4(g) \rightleftharpoons 2\,NO_2(g)$$

Thermodynamic data on these gases are given in Appendix C. You may assume that ΔH° and ΔS° do not vary with temperature. **(a)** At what temperature will an equilibrium mixture contain equal amounts of the two gases? **(b)** At what temperature will an equilibrium mixture of 1 atm total pressure contain twice as much NO_2 as N_2O_4? **(c)** At what temperature will an equilibrium mixture of 10 atm total pressure contain twice as much NO_2 as N_2O_4? **(d)** Rationalize the results from parts (b) and (c) by using Le Châtelier's principle. (Section 15.6)

[19.105] The reaction

$$SO_2(g) + 2\,H_2S(g) \rightleftharpoons 3\,S(s) + 2\,H_2O(g)$$

is the basis of a suggested method for removal of SO_2 from power-plant stack gases. The standard free energy of each substance is given in Appendix C. **(a)** What is the equilibrium constant for the reaction at 298 K? **(b)** In principle, is this reaction a feasible method of removing SO_2? **(c)** If $P_{SO_2} = P_{H_2S}$ and the vapor pressure of water is 25 torr, calculate the equilibrium SO_2 pressure in the system at 298 K. **(d)** Would you expect the process to be more or less effective at higher temperatures?

19.106 When most elastomeric polymers (e.g., a rubber band) are stretched, the molecules become more ordered, as illustrated here:

Suppose you stretch a rubber band. **(a)** Do you expect the entropy of the system to increase or decrease? **(b)** If the rubber band were stretched isothermally, would heat need to be absorbed or emitted to maintain constant temperature?

eMEDIA EXERCISES

These exercises make use of the interactive objects available online inOneKey or the Companion Website, and on your Accelerator CD. Access to theseresources comes in your MediaPak.

19.107 The **Gibbs Free Energy** activity (*19.6*) allows you to explore the relationship between spontaneity and temperature. **(a)** Of the five reactions available in the simulation, which are spontaneous at all temperatures? **(b)** Which are nonspontaneous at all temperatures? **(c)** Is it possible for a reaction with a positive enthalpy change and a negative entropy change to be spontaneous? Explain.

19.108 Using information from the **Gibbs Free Energy** activity (*19.6*), determine **(a)** the temperature below which the reaction $N_2O_4(g) \longrightarrow 2\,NO_2(g)$ will occur spontaneously; **(b)** the temperature below which the reaction $CaCO_3(s) \longrightarrow CaO(s) + CO_2(g)$ will occur spontaneously; **(c)** the temperature above which the reaction $Si(s) + 2\,Cl_2(g) \longrightarrow SiCl_4(l)$ will occur spontaneously.

19.109 The **Enthalpy of Solution** activity (*19.3*) lets you measure temperature changes for the combination of several different compounds with water. **(a)** Using data from this simulation, determine the signs of $\Delta G°$, $\Delta H°$, and $\Delta S°$ for the dissolution of ammonium nitrate. **(b)** Is the dissolution of ammonium nitrate spontaneous at all temperatures? Explain.

19.110 Watch the **Air Bags** animation again (*19.1*), and answer the following questions. **(a)** What are the signs of $\Delta G°$, $\Delta H°$, and $\Delta S°$ for the decomposition of sodium azide? **(b)** Is there a temperature at which the decomposition of sodium azide is not spontaneous? Explain.

19.111 When a balloon filled with hydrogen is ignited, hydrogen and oxygen react explosively to form water. The **Formation of Water** movie (*19.5*) states, "The equilibrium lies far to the right." **(a)** Explain what this statement means in terms of products and reactants. **(b)** Explain the relationship between the magnitude of an equilibrium constant and the sign of $\Delta G°$.

VCL eLABORATORY EXERCISES

These exercises allow students to apply the concepts and skills for this chapter in the simulated laboratory of Virtual ChemLab. Worksheets for each of these exercises are found in the Virtual ChemLab Workbook in your MediaPak.

19.112 ***(VCL 19-1) The Balance Between Enthalpy and Entropy*** For chemical reactions, we say a reaction proceeds to the right when ΔG is negative and the reaction proceeds to the left when ΔG is positive. At equilibrium, ΔG is zero. The Gibbs–Helmholtz equation specifies that at constant temperature $\Delta G = \Delta H - T\Delta S$ or, in other words, the sign and size of ΔG are governed by the balance between enthalpic (ΔH) and entropic (ΔS) considerations. In this assignment you will dissolve several different salts in water, measure the resulting temperature changes, and then make some deductions about the thermodynamic driving forces behind the dissolving process.

19.113 ***(VCL 19-2) Heat of Solution: NH_4NO_3*** Have you ever used one of those "instant cold packs" that looks like a plastic bag filled with liquid? If you hit the bag and shake it up, it gets extremely cold; but why does it do that? The liquid inside the cold pack is water, and in the water is another plastic bag or tube containing NH_4NO_3. When you hit the cold pack, it breaks the tube so that the water mixes with the NH_4NO_3. The dissolving of a salt, such as NH_4NO_3, in water is called dissolution, and the heat associated with the dissolving process is called the enthalpy of solution. In this problem you will take a sample of NH_4NO_3, dissolve it in water, and from measuring the change in temperature, calculate the enthalpy of solution for NH_4NO_3.

CHAPTER 20

WHAT'S AHEAD

- We briefly review oxidation states and *oxidation-reduction (redox) reactions*. *(Section 20.1)*
- We learn how to balance redox equations. *(Section 20.2)*
- We next consider *voltaic cells*, those that produce electricity from spontaneous redox reactions. In these cells, solid electrodes serve as the surfaces at which oxidation and reduction reactions take place. The electrode where oxidation occurs is the *anode*, and the electrode where reduction occurs is the *cathode*. *(Section 20.3)*
- One of the important characteristics of a voltaic cell is its *emf*, or voltage, which is the difference in the electrical potentials at the two electrodes. Electrode potentials, which are tabulated for reduction reactions under standard conditions (*standard reduction potentials*), are used to calculate cell voltages, to determine the relative strengths of oxidizing agents and reducing agents, and to predict whether or not particular redox reactions are spontaneous. *(Section 20.4)*
- We will relate the Gibbs free energy to the emf of an electrochemical cell. *(Section 20.5)*
- The voltage associated with cells that run under nonstandard conditions can be calculated using standard voltages and the Nernst equation. *(Section 20.6)*
- We describe batteries and fuel cells, which are commercially important energy sources based on electrochemical reactions. *(Section 20.7)*
- Next, we discuss *corrosion*, a spontaneous electrochemical process involving metals. *(Section 20.8)*
- Finally, we focus on nonspontaneous redox reactions, examining *electrolytic cells*, which use electricity to perform chemical reactions. *(Section 20.9)*

Electrochemistry

OXIDATION-REDUCTION (REDOX) REACTIONS ARE AMONG the most common and important chemical reactions. They are involved in a wide variety of important natural processes including the rusting of iron, the browning of foods, and the respiration of animals. As we discussed earlier, *oxidation* refers to the loss of electrons, and *reduction* refers to the gain of electrons. (Section 4.4) Thus, oxidation-reduction reactions occur when electrons are transferred from the atom that is oxidized to the atom that is reduced.

Redox reactions produce new compounds, and can generate heat, just like other kinds of chemical reactions. The transfer of electrons that occurs during oxidation-reduction reactions can also produce energy in the form of electricity. This is a unique feature of spontaneous redox reactions that is used, for example, in the construction of portable energy sources such as batteries and fuel cells. We saw in Chapter 18 that alternative energy sources include solar energy conversion, in which sunlight causes electrons to flow in an external circuit in order to power devices. The Hubble Space Telescope (see photo) uses a combination of solar panels and lithium ion batteries as its power source. In this chapter, we will learn the basic principles behind batteries and fuel cells.

It is also possible to use electrical energy to make certain nonspontaneous chemical processes occur. An everyday example of this is electroplating, in which layers of one metal are coated onto another by the application of a voltage.

THE HUBBLE SPACE TELESCOPE The Hubble Space Telescope orbits the Earth at an altitude of 353 miles, at a speed of 17,500 miles per hour. It is powered by solar energy panels and lithium ion batteries.

Electrochemistry is the study of the relationships between electricity and chemical reactions. We will begin by reviewing how oxidation numbers are used to identify redox reactions.

20.1 | Oxidation States

How do we determine whether a given chemical reaction is an oxidation-reduction reaction? We can do so by keeping track of the oxidation states of all the elements involved in the reaction (Section 4.4). This procedure tells us which elements (if any) are changing oxidation state. For example, when zinc metal is added to a strong acid (Figure 20.1 ▼), electrons are transferred from zinc atoms to hydrogen ions:

$$Zn(s) + 2\,H^+(aq) \longrightarrow Zn^{2+}(aq) + H_2(g) \qquad [20.1]$$

ANIMATIONS
Oxidation-Reduction Reactions Part I
Oxidation-Reduction Reactions Part II

We say that the Zn is oxidized and the H^+ is reduced.

The transfer of electrons that occurs in the reaction in Figure 20.1 produces energy in the form of heat; the reaction is thermodynamically "downhill" and proceeds spontaneously.

The reaction in Equation 20.1 can be written as

$$\underset{0}{Zn(s)} + 2\,\underset{+1}{H^+(aq)} \longrightarrow \underset{+2}{Zn^{2+}(aq)} + \underset{0}{H_2(g)} \qquad [20.2]$$

▲ **Figure 20.1 Redox reaction in acidic solution.** The addition of zinc metal to hydrochloric acid leads to a spontaneous oxidation-reduction reaction: Zinc metal is oxidized to $Zn^{2+}(aq)$, and $H^+(aq)$ is reduced to $H_2(g)$, which produces the vigorous bubbling.

By writing the oxidation state of each element above or below the equation, we can see the oxidation state changes that occur: The oxidation state of Zn changes from 0 to +2, and that of H changes from +1 to 0.

In a reaction such as Equation 20.2, a clear transfer of electrons occurs. Zinc loses electrons as $Zn(s)$ is converted to $Zn^{2+}(aq)$, and hydrogen gains electrons as $H^+(aq)$ is turned into $H_2(g)$. In some reactions, however, the oxidation states change but we can't say that any substance literally gains or loses electrons. For example, consider the combustion of hydrogen gas:

$$2\,H_2(g) + O_2(g) \longrightarrow 2\,H_2O(g) \qquad [20.3]$$

(0) (0) (+1) (−2)

In this reaction, hydrogen is oxidized from the 0 to the +1 oxidation state and oxygen is reduced from the 0 to the −2 oxidation state. Therefore, Equation 20.3 is an oxidation-reduction reaction. Water is not an ionic substance, however, and so there is not a complete transfer of electrons from hydrogen to oxygen as water is formed. Using oxidation states, therefore, is a convenient form of "bookkeeping," but *you should not generally equate the oxidation state of an atom with its actual charge in a chemical compound.* (“A Closer Look” box on oxidation numbers, formal charges, and partial charges in Section 8.5)

GIVE IT SOME THOUGHT

What are the oxidation states of the atoms in cyanide, CN^-?

In any redox reaction, both oxidation and reduction must occur. In other words, if one substance is oxidized, then another must be reduced: The electrons formally have to go from one place to another. The substance that makes it possible for another substance to be oxidized is called either the **oxidizing agent** or the **oxidant**. The oxidizing agent removes electrons from another substance by acquiring them itself; thus, the oxidizing agent is itself reduced. Similarly, a **reducing agent**, or **reductant**, is a substance that gives up electrons, thereby causing another substance to be reduced. The reducing agent is therefore oxidized in the process. In Equation 20.2, $H^+(aq)$ is the oxidizing agent and $Zn(s)$ is the reducing agent.

SAMPLE EXERCISE 20.1 | What Chemical Reactions Occur in a Battery?

The nickel-cadmium (nicad) battery, a rechargeable "dry cell" used in battery-operated devices, uses the following redox reaction to generate electricity:

$$Cd(s) + NiO_2(s) + 2\,H_2O(l) \longrightarrow Cd(OH)_2(s) + Ni(OH)_2(s)$$

Identify the substances that are oxidized and reduced, and indicate which are oxidizing agents and which are reducing agents.

Solution

Analyze: We are given a redox equation and asked to identify the substance oxidized and the substance reduced and to label one as the oxidizing agent and the other as the reducing agent.

Plan: First, we assign oxidation states to all the atoms in the reaction and determine the elements that are changing oxidation state. Second, we apply the definitions of oxidation and reduction.

Solve: $Cd(s) + NiO_2(s) + 2H_2O(l) \longrightarrow Cd(OH)_2(s) + Ni(OH)_2(s)$

(0) (+4)(−2) (+1)(−2) (+2)(−2)(+1) (+2)(−2)(+1)

Cd increases in oxidation state from 0 to +2, and Ni decreases from +4 to +2. Because the Cd atom increases in oxidation state, it is oxidized (loses electrons) and therefore serves as the reducing agent. The Ni atom decreases in oxidation state as NiO_2 is converted into $Ni(OH)_2$. Thus, NiO_2 is reduced (gains electrons) and therefore serves as the oxidizing agent.

Comment: A common mnemonic for remembering oxidation and reduction is "LEO the lion says GER": *l*osing *e*lectrons is *o*xidation; *g*aining *e*lectrons is *r*eduction.

PRACTICE EXERCISE

Identify the oxidizing and reducing agents in the oxidation-reduction reaction

$$2\,H_2O(l) + Al(s) + MnO_4^-(aq) \longrightarrow Al(OH)_4^-(aq) + MnO_2(s)$$

Answer: $Al(s)$ is the reducing agent; $MnO_4^-(aq)$ is the oxidizing agent.

20.2 | Balancing Oxidation-Reduction Equations

Whenever we balance a chemical equation, we must obey the law of conservation of mass: The amount of each element must be the same on both sides of the equation. As we balance oxidation-reduction reactions, there is an additional requirement: The gains and losses of electrons must be balanced. In other words, if a substance loses a certain number of electrons during a reaction, then another substance must gain that same number of electrons. In many simple chemical reactions, such as Equation 20.2, balancing the electrons is handled "automatically"; we can balance the equation without explicitly considering the transfer of electrons. Many redox reactions are more complex than Equation 20.2, however, and cannot be balanced easily without taking into account the number of electrons lost and gained in the course of the reaction. In this section we examine a systematic procedure for balancing redox equations.

Half-Reactions

Although oxidation and reduction must take place simultaneously, it is often convenient to consider them as separate processes. For example, the oxidation of Sn^{2+} by Fe^{3+}:

$$Sn^{2+}(aq) + 2\,Fe^{3+}(aq) \longrightarrow Sn^{4+}(aq) + 2\,Fe^{2+}(aq)$$

can be considered to consist of two processes: (1) the oxidation of Sn^{2+} (Equation 20.4) and (2) the reduction of Fe^{3+} (Equation 20.5):

Oxidation: $$Sn^{2+}(aq) \longrightarrow Sn^{4+}(aq) + 2\,e^- \qquad [20.4]$$

Reduction: $$2\,Fe^{3+}(aq) + 2\,e^- \longrightarrow 2\,Fe^{2+}(aq) \qquad [20.5]$$

Notice that in the oxidation, electrons are shown as products, whereas in the reduction, electrons are shown as reactants.

Equations that show either oxidation or reduction alone, such as Equations 20.4 and 20.5, are called **half-reactions**. In the overall redox reaction, the number of electrons lost in the oxidation half-reaction must equal the number of electrons gained in the reduction half-reaction. When this condition is met and each half-reaction is balanced, the electrons on the two sides cancel when the two half-reactions are added to give the overall balanced oxidation-reduction equation.

Balancing Equations by the Method of Half-Reactions

ACTIVITY
Balancing Redox Equations in Acid

The use of half-reactions provides a general method for balancing oxidation-reduction equations. For balancing a redox reaction that occurs in acidic aqueous solution, the procedure is as follows:

1. Assign oxidation states to see which atoms are gaining electrons and which are losing electrons.
2. Divide the equation into two half-reactions, one for oxidation and the other for reduction.

3. Balance each half-reaction.
   (a) First, balance the elements other than H and O.
   (b) Next, balance the O atoms by adding H_2O as needed.
   (c) Then, balance the H atoms by adding H^+ as needed.
   (d) Finally, balance the charge by adding e^- as needed.
   At this point, you can check whether the number of electrons in each half-reaction corresponds to the changes in oxidation state determined in step 1.
4. Multiply the half-reactions by integers if necessary so that the number of electrons lost in one half-reaction equals the number gained in the other.
5. Add the two half-reactions and wherever possible simplify by canceling species appearing on both sides of the combined equation.
6. Check to make sure that the number of atoms of each kind on the left side of the equation is the same as the number on the right side.
7. Check to make sure that the total charge on the left is the same as the total charge on the right.

As an example, let's consider the reaction between permanganate ion (MnO_4^-) and oxalate ion ($C_2O_4^{2-}$) in acidic aqueous solution. When MnO_4^- is added to an acidified solution of $C_2O_4^{2-}$, the deep purple color of the MnO_4^- ion fades, as illustrated in Figure 20.2 ▼. Bubbles of CO_2 form, and the solution takes on the pale pink color of Mn^{2+}. We can therefore write the unbalanced equation as

$$MnO_4^-(aq) + C_2O_4^{2-}(aq) \longrightarrow Mn^{2+}(aq) + CO_2(aq) \qquad [20.6]$$

Experiments show that H^+ is consumed and H_2O is produced in the reaction. We will see that these facts can be deduced in the course of balancing the equation.

To complete and balance Equation 20.6 by the method of half-reactions, we begin with the unbalanced reaction and assign oxidation states (step 1). Mn g0es from the +7 oxidation state in MnO_4^- to the +2 oxidation state in Mn^{2+}. Therefore, each Mn atom is reduced by five electrons. (In other words, it *gains* five electrons.) Carbon g0es from the +3 oxidation state in $C_2O_4^{2-}$ to the +4 oxidation state in CO_2. Therefore, each C atom is oxidized by one electron. (In other words, it *loses* one electron.)

Next we write the two half-reactions (step 2). The reduction half-reaction must have Mn on both sides of the arrow, and the oxidation half-reaction must have C on both sides of the arrow:

$$MnO_4^-(aq) \longrightarrow Mn^{2+}(aq)$$

$$C_2O_4^{2-}(aq) \longrightarrow CO_2(g)$$

(a) (b) (c)

◀ **Figure 20.2 Titration of an acidic solution of $Na_2C_2O_4$ with $KMnO_4(aq)$.** (a) As it moves from the buret to the reaction flask, the deep purple MnO_4^- is rapidly reduced to extremely pale pink Mn^{2+} by $C_2O_4^{2-}$. (b) Once all the $C_2O_4^{2-}$ in the flask has been consumed, any MnO_4^- added to the flask retains its purple color, and the end point corresponds to the faintest discernible purple color in the solution. (c) Beyond the end point the solution in the flask becomes deep purple because of excess MnO_4^-.

We can now complete and balance the half-reactions. First, let's balance all the atoms except for H and O (step 3a). In the permanganate half-reaction we already have one manganese atom on each side of the equation and so need to do nothing. In the oxalate half-reaction we need to add a coefficient 2 to the right to balance the two carbons on the left:

$$MnO_4^-(aq) \longrightarrow Mn^{2+}(aq)$$

$$C_2O_4^{2-}(aq) \longrightarrow 2\,CO_2(g)$$

The permanganate half-reaction has four oxygens on the left and none on the right; therefore four H_2O molecules are needed as product to balance the four oxygen atoms in MnO_4^- (step 3b):

$$MnO_4^-(aq) \longrightarrow Mn^{2+}(aq) + 4\,H_2O(l)$$

The eight hydrogen atoms now in the products must be balanced by adding $8\,H^+$ to the reactants (step 3c):

$$8\,H^+(aq) + MnO_4^-(aq) \longrightarrow Mn^{2+}(aq) + 4\,H_2O(l)$$

There are now equal numbers of each type of atom on the two sides of the equation, but the charge still needs to be balanced. The total charge of the reactants is $8(1+) + (1-) = 7+$, and that of the products is $(2+) + 4(0) = 2+$. To balance the charge, we must add five electrons to the reactant side (step 3d):

$$5\,e^- + 8\,H^+(aq) + MnO_4^-(aq) \longrightarrow Mn^{2+}(aq) + 4\,H_2O(l)$$

Note that we said earlier that each manganese atom needed to gain five electrons, and now in the balanced half-reaction we can see them.

We balanced the masses in the oxalate half-reaction already in step 3a. We can balance the charge (step 3d) by adding two electrons to the products:

$$C_2O_4^{2-}(aq) \longrightarrow 2\,CO_2(g) + 2\,e^-$$

We noted earlier that each carbon atom needed to be oxidized by one electron. For the two carbon atoms in this chemical equation, there should be two electrons lost, and now in the balanced half-reaction we can see them.

Since we have two balanced half-reactions, we need to add them to obtain an overall balanced reaction. In the final reaction there should not be any free electrons. Therefore, we need to multiply each half-reaction by an appropriate integer so that the number of electrons gained in one half-reaction equals the number of electrons lost in the other (step 4). The MnO_4^- half-reaction must be multiplied by 2, and the $C_2O_4^{2-}$ half-reaction must be multiplied by 5 so that the same number of electrons (10) appears in both equations:

$$10\,e^- + 16\,H^+(aq) + 2\,MnO_4^-(aq) \longrightarrow 2\,Mn^{2+}(aq) + 8\,H_2O(l)$$

$$5\,C_2O_4^{2-}(aq) \longrightarrow 10\,CO_2(g) + 10\,e^-$$

$$16\,H^+(aq) + 2\,MnO_4^-(aq) + 5\,C_2O_4^{2-}(aq) \longrightarrow 2\,Mn^{2+}(aq) + 8\,H_2O(l) + 10\,CO_2(g)$$

The balanced equation is the sum of the balanced half-reactions (step 5). Note that the electrons on the reactant and product sides of the equation cancel each other.

We can check the balanced equation by counting atoms (step 6) and charges (step 7): There are 16 H, 2 Mn, 28 O, 10 C, and a net charge of 4+ on each side of the equation, confirming that the equation is correctly balanced.

SAMPLE EXERCISE 20.2 | Balancing Redox Equations in Acidic Solution

Complete and balance this equation by the method of half-reactions:

$$Cr_2O_7^{2-}(aq) + Cl^-(aq) \longrightarrow Cr^{3+}(aq) + Cl_2(g) \quad \text{(acidic solution)}$$

Solution

Analyze: We are given an incomplete, unbalanced equation for a redox reaction occurring in acidic solution and asked to balance it.

Plan: We use the procedure we just learned.

Solve: Step 1: We check oxidation state changes. Each chromium atom is reduced by three electrons, and each chlorine atom is oxidized by one electron.

Solve: Step 2: We divide the equation into two half-reactions:

$$Cr_2O_7^{2-}(aq) \longrightarrow Cr^{3+}(aq)$$
$$Cl^-(aq) \longrightarrow Cl_2(g)$$

Step 3: We balance each half-reaction. In the first half-reaction the presence of one $Cr_2O_7^{2-}$ among the reactants requires two Cr^{3+} among the products. The seven oxygen atoms in $Cr_2O_7^{2-}$ are balanced by adding seven H_2O to the products. The 14 hydrogen atoms in 7 H_2O are then balanced by adding 14 H^+ to the reactants:

$$14\,H^+(aq) + Cr_2O_7^{2-}(aq) \longrightarrow 2\,Cr^{3+}(aq) + 7\,H_2O(l)$$

Charge is then balanced by adding electrons to the left side of the equation so that the total charge is the same on the two sides:

$$6\,e^- + 14\,H^+(aq) + Cr_2O_7^{2-}(aq) \longrightarrow 2\,Cr^{3+}(aq) + 7\,H_2O(l)$$

Recall that we said earlier that each chromium atom needed to gain three electrons, and now we can see the six electrons needed for two chromium atoms.

In the second half-reaction, two Cl^- are required to balance one Cl_2:

$$2\,Cl^-(aq) \longrightarrow Cl_2(g)$$

We add two electrons to the right side to attain charge balance:

$$2\,Cl^-(aq) \longrightarrow Cl_2(g) + 2\,e^-$$

Recall that we said earlier that each chlorine atom needed to lose one electron, and now we can see the two electrons needed for two chlorine atoms.

Step 4: We equalize the number of electrons transferred in the two half-reactions. To do so, we multiply the Cl half-reaction by 3 so that the number of electrons gained in the Cr half-reaction (six) equals the number lost in the Cl half-reaction, allowing the electrons to cancel when the half-reactions are added:

$$6\,Cl^-(aq) \longrightarrow 3\,Cl_2(g) + 6\,e^-$$

Step 5: The equations are added to give the balanced equation:

$$14\,H^+(aq) + Cr_2O_7^{2-}(aq) + 6\,Cl^-(aq) \longrightarrow 2\,Cr^{3+}(aq) + 7\,H_2O(l) + 3\,Cl_2(g)$$

Steps 6 and 7: There are equal numbers of atoms of each kind on the two sides of the equation (14 H, 2 Cr, 7 O, 6 Cl). In addition, the charge is the same on the two sides (6+). Thus, the equation is balanced.

PRACTICE EXERCISE

Complete and balance the following equations using the method of half-reactions. Both reactions occur in acidic solution.

(a) $Cu(s) + NO_3^-(aq) \longrightarrow Cu^{2+}(aq) + NO_2(g)$

(b) $Mn^{2+}(aq) + NaBiO_3(s) \longrightarrow Bi^{3+}(aq) + MnO_4^-(aq)$

Answers: **(a)** $Cu(s) + 4\,H^+(aq) + 2\,NO_3^-(aq) \longrightarrow Cu^{2+}(aq) + 2\,NO_2(g) + 2\,H_2O(l)$

(b) $2\,Mn^{2+}(aq) + 5\,NaBiO_3(s) + 14\,H^+(aq) \longrightarrow 2\,MnO_4^-(aq) + 5\,Bi^{3+}(aq) + 5\,Na^+(aq) + 7\,H_2O(l)$

GIVE IT SOME THOUGHT

If the equation for a redox reaction is balanced, will free electrons appear in the equation as either reactants or products?

ACTIVITY
Balancing Redox Equations in Base

Balancing Equations for Reactions Occurring in Basic Solution

If a redox reaction occurs in basic solution, the equation must be completed by using OH^- and H_2O rather than H^+ and H_2O. It is difficult to do this using OH^- and H_2O (in parallel with the H^+ and H_2O we used for acidic conditions) because OH^- has both hydrogen and oxygen atoms. Therefore, a good way to balance these reactions is to balance the half-reactions initially as if they occurred in acidic solution. Then, count the H^+ in each half-reaction, and add the same number of OH^- *to each side* of the half-reaction. This way the reaction is still mass-balanced because you are adding the same thing to both sides. In essence, what you are doing is "neutralizing" the protons to form water ($H^+ + OH^- \longrightarrow H_2O$) on the side containing H^+, and the other side ends up with the OH^-. The resulting water molecules can be canceled as needed. This procedure is shown in Sample Exercise 20.3.

SAMPLE EXERCISE 20.3 | Balancing Redox Equations in Basic Solution

Complete and balance this equation for a redox reaction that takes place in basic solution:

$$CN^-(aq) + MnO_4^-(aq) \longrightarrow CNO^-(aq) + MnO_2(s) \text{ (basic solution)}$$

Solution

Analyze: We are given an incomplete equation for a basic redox reaction and asked to balance it.

Plan: We go through the first steps of our procedure as if the reaction were occurring in acidic solution. We then add the appropriate number of OH^- ions to each side of the equation, combining H^+ and OH^- to form H_2O. We complete the process by simplifying the equation.

Solve: Step 1: We assign oxidation states. This is a tricky one!

Mn goes from 7+ to 4+. The sum of the oxidation states of C and N in CN^- must be -1, the overall charge of the ion. In CNO^-, if oxygen has an oxidation state of -2 as usual, the sum of the oxidation states of C and N must be $+1$. So, overall, CN^- is oxidized by two electrons.

Step 2: We write the incomplete, unbalanced half-reactions:

$$CN^-(aq) \longrightarrow CNO^-(aq)$$
$$MnO_4^-(aq) \longrightarrow MnO_2(s)$$

Step 3: We balance each half-reaction as if it took place in acidic solution:

$$CN^-(aq) + H_2O(l) \longrightarrow CNO^-(aq) + 2\,H^+(aq) + 2\,e^-$$
$$3\,e^- + 4\,H^+(aq) + MnO_4^-(aq) \longrightarrow MnO_2(s) + 2\,H_2O(l)$$

Now we need to take into account that the reaction occurs in basic solution, adding OH^- to both sides of both half-reactions to neutralize H^+:

$$CN^-(aq) + H_2O(l) + 2\,OH^-(aq) \longrightarrow CNO^-(aq) + 2\,H^+(aq) + 2\,e^- + 2\,OH^-(aq)$$
$$3\,e^- + 4\,H^+(aq) + MnO_4^-(aq) + 4\,OH^-(aq) \longrightarrow MnO_2(s) + 2\,H_2O(l) + 4\,OH^-(aq)$$

We now "neutralize" H^+ and OH^- by forming H_2O when they are on the same side of either half-reaction:

$$CN^-(aq) + H_2O(l) + 2\,OH^-(aq) \longrightarrow CNO^-(aq) + 2\,H_2O(l) + 2\,e^-$$
$$3\,e^- + 4\,H_2O(l) + MnO_4^-(aq) \longrightarrow MnO_2(s) + 2\,H_2O(l) + 4\,OH^-(aq)$$

Next, we cancel water molecules that appear as both reactants and products:

$$CN^-(aq) + 2\,OH^-(aq) \longrightarrow CNO^-(aq) + H_2O(l) + 2\,e^-$$
$$3\,e^- + 2\,H_2O(l) + MnO_4^-(aq) \longrightarrow MnO_2(s) + 4\,OH^-(aq)$$

Both half-reactions are now balanced—you can check the atoms and the overall charge.

Step 4: Now we multiply the cyanide half-reaction through by 3, which will give six electrons on the product side; and multiply the permanganate half-reaction through by 2, which will give six electrons on the reactant side:

$$3\,CN^-(aq) + 6\,OH^-(aq) \longrightarrow 3\,CNO^-(aq) + 3\,H_2O(l) + 6\,e^-$$
$$6\,e^- + 4\,H_2O(l) + 2\,MnO_4^-(aq) \longrightarrow 2\,MnO_2(s) + 8\,OH^-(aq)$$

Step 5: Now we can add the two half-reactions together and simplify by canceling species that appear as both reactants and products:

$$3\,CN^-(aq) + H_2O(l) + 2\,MnO_4^-(aq) \longrightarrow 3\,CNO^-(aq) + 2\,MnO_2(s) + 2\,OH^-(aq)$$

Steps 6 and 7: Check that the atoms and charges are balanced.

There are 3 C, 3 N, 2 H, 9 O, 2 Mn, and a charge of 5− on both sides of the equation.

PRACTICE EXERCISE

Complete and balance the following equations for oxidation-reduction reactions that occur in basic solution:

(a) $NO_2^-(aq) + Al(s) \longrightarrow NH_3(aq) + Al(OH)_4^-(aq)$

(b) $Cr(OH)_3(s) + ClO^-(aq) \longrightarrow CrO_4^{2-}(aq) + Cl_2(g)$

Answers: **(a)** $NO_2^-(aq) + 2\,Al(s) + 5\,H_2O(l) + OH^-(aq) \longrightarrow NH_3(aq) + 2\,Al(OH)_4^-(aq)$

(b) $2\,Cr(OH)_3(s) + 6\,ClO^-(aq) \longrightarrow 2\,CrO_4^{2-}(aq) + 3\,Cl_2(g) + 2\,OH^-(aq) + 2\,H_2O(l)$

20.3 | Voltaic Cells

The energy released in a spontaneous redox reaction can be used to perform electrical work. This task is accomplished through a **voltaic** (or **galvanic**) **cell**, a device in which the transfer of electrons takes place through an external pathway rather than directly between reactants.

One such spontaneous reaction occurs when a strip of zinc is placed in contact with a solution containing Cu^{2+}. As the reaction proceeds, the blue color of $Cu^{2+}(aq)$ ions fades and copper metal deposits on the zinc. At the same time, the zinc begins to dissolve. These transformations are shown in Figure 20.3 ▶ and are summarized by the equation

ANIMATION
Voltaic Cells I: The Copper-Zinc Cell

$$Zn(s) + Cu^{2+}(aq) \longrightarrow Zn^{2+}(aq) + Cu(s) \qquad [20.7]$$

Figure 20.4 ▶ shows a voltaic cell that uses the redox reaction between Zn and Cu^{2+} given in Equation 20.7. Although the setup shown in Figure 20.4 is more complex than that in Figure 20.3, the reaction is the same in both cases. The significant difference between the two setups is that the Zn metal and $Cu^{2+}(aq)$ are not in direct contact in the voltaic cell. Instead, the Zn metal is placed in contact with $Zn^{2+}(aq)$ in one compartment of the cell, and Cu metal is placed in contact with $Cu^{2+}(aq)$ in another compartment. Consequently, the reduction of the Cu^{2+} can occur only by a flow of electrons through an external circuit, namely, the wire that connects the Zn and Cu strips. In other words, by physically separating the reduction half of a redox reaction from the oxidation half, we create a flow of electrons through an external circuit.

The two solid metals that are connected by the external circuit are called *electrodes*. By definition, the electrode at which oxidation occurs is called the **anode**; the electrode at which reduction occurs is called the **cathode**.* The electrodes can be made of materials that participate in the reaction, as in the present example.

* To help remember these definitions, note that *anode* and *oxidation* both begin with a vowel, and *cathode* and *reduction* both begin with a consonant.

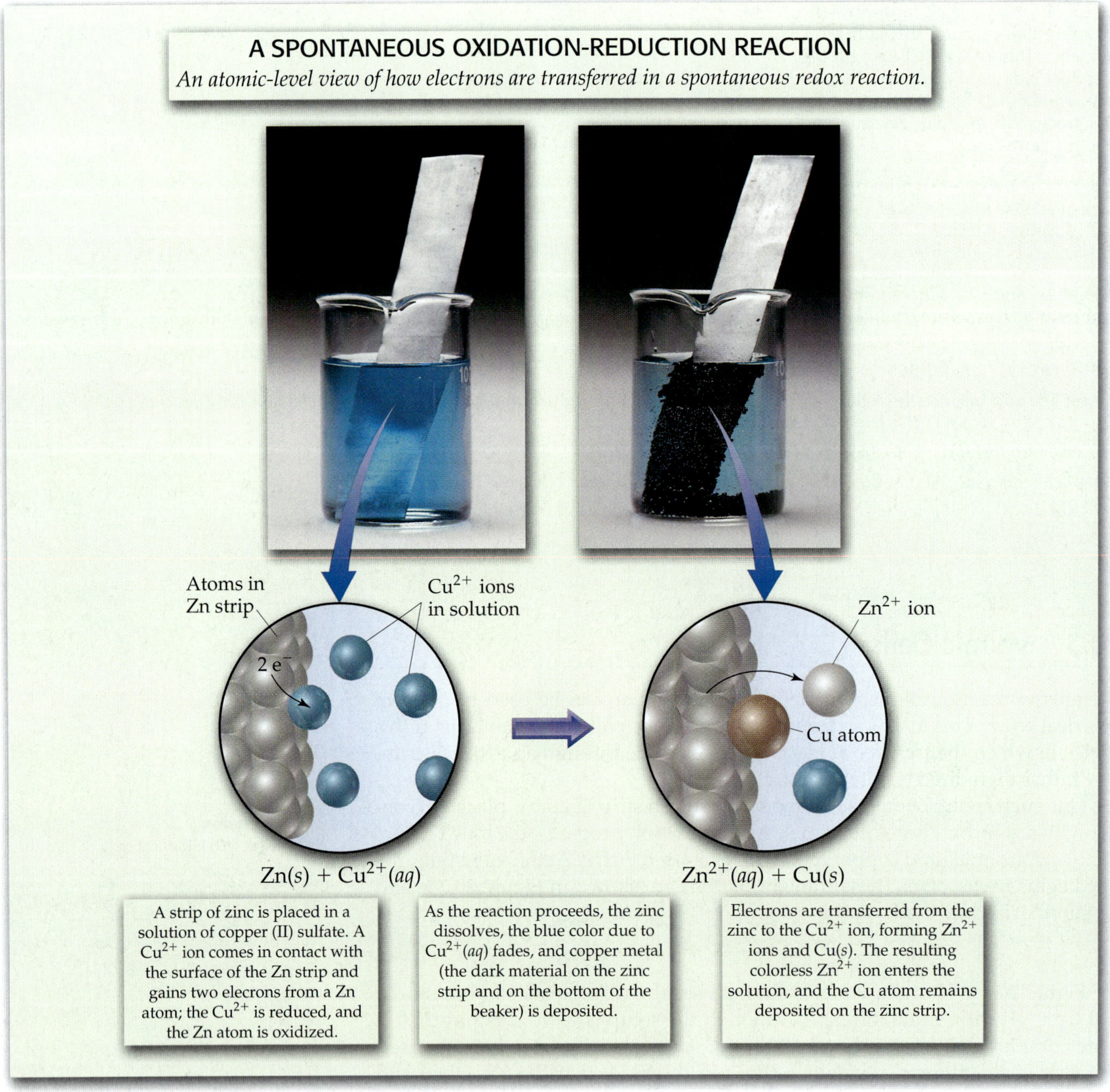

▲ **Figure 20.3 A spontaneous oxidation-reduction reaction.**

MOVIES
Redox Chemistry of Iron and Copper, Formation of Silver Crystals

Over the course of the reaction, the Zn electrode will gradually disappear and the copper electrode will gain mass. More typically, the electrodes are made of a conducting material, such as platinum or graphite, that does not gain or lose mass during the reaction but serves as a surface at which electrons are transferred.

Each of the two compartments of a voltaic cell is called a *half-cell*. One half-cell is the site of the oxidation half-reaction, and the other is the site of the reduction half-reaction. In our present example Zn is oxidized and Cu^{2+} is reduced:

Anode (oxidation half-reaction) $$Zn(s) \longrightarrow Zn^{2+}(aq) + 2\,e^-$$

Cathode (reduction half-reaction) $$Cu^{2+}(aq) + 2\,e^- \longrightarrow Cu(s)$$

Electrons become available as zinc metal is oxidized at the anode. They flow through the external circuit to the cathode, where they are consumed as $Cu^{2+}(aq)$ is reduced. Because $Zn(s)$ is oxidized in the cell, the zinc electrode loses mass, and the concentration of the Zn^{2+} solution increases as the cell operates. Similarly, the Cu electrode gains mass, and the Cu^{2+} solution becomes less concentrated as Cu^{2+} is reduced to $Cu(s)$.

For a voltaic cell to work, the solutions in the two half-cells must remain electrically neutral. As Zn is oxidized in the anode compartment, Zn^{2+} ions enter the solution. Thus, there must be some means for positive ions to migrate out of the anode compartment or for negative ions to migrate in to keep the solution electrically neutral. Similarly, the reduction of Cu^{2+} at the cathode removes positive charge from the solution, leaving an excess of negative charge in that half-cell. Thus, positive ions must migrate into the compartment or negative ions must migrate out. In fact, no measurable electron flow will occur between electrodes unless a means is provided for ions to migrate through the solution from one electrode compartment to the other, thereby completing the circuit.

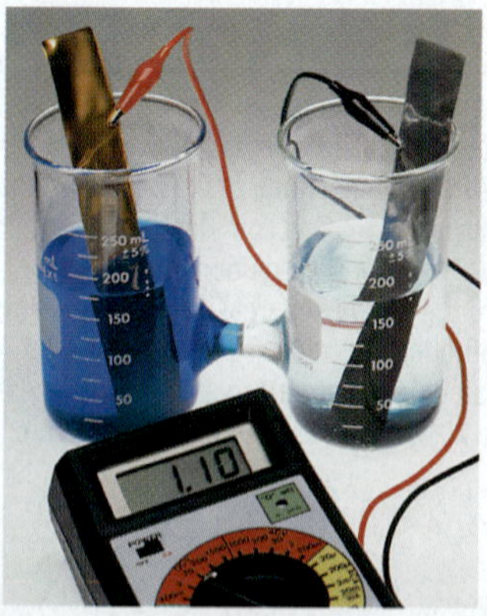

▲ **Figure 20.4 A voltaic cell based on the reaction in Equation 20.7.** The left compartment contains 1 *M* $CuSO_4$ and a copper electrode. The one on the right contains 1 *M* $ZnSO_4$ and a zinc electrode. The solutions are connected by a porous glass disc, which permits contact of the two solutions. The metal electrodes are connected through a voltmeter, which reads the potential of the cell, 1.10 V.

In Figure 20.4, a porous glass disc separating the two compartments allows a migration of ions that maintains the electrical neutrality of the solutions. In Figure 20.5 ▶, a *salt bridge* serves this purpose. A salt bridge consists of a U-shaped tube that contains an electrolyte solution, such as $NaNO_3(aq)$, whose ions will not react with other ions in the cell or with the electrode materials. The electrolyte is often incorporated into a paste or gel so that the electrolyte solution does not pour out when the U-tube is inverted. As oxidation and reduction proceed at the electrodes, ions from the salt bridge migrate to neutralize charge in the cell compartments. Whatever means is used to allow ions to migrate between half-cells, *anions always migrate toward the anode and cations toward the cathode.*

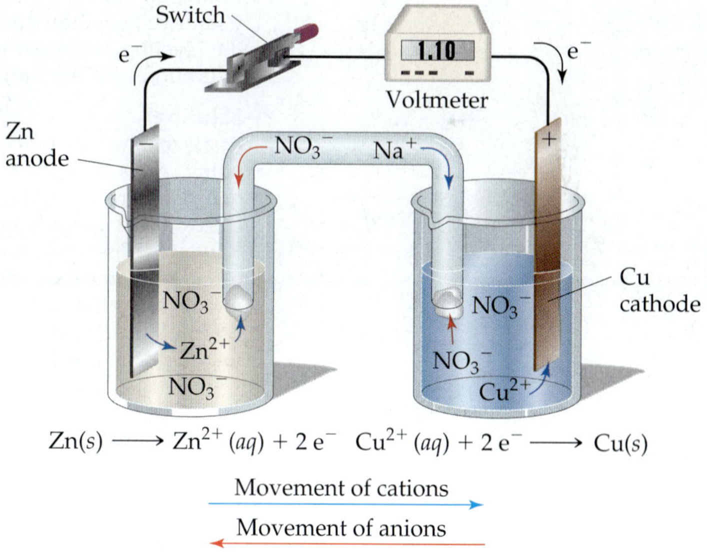

▲ **Figure 20.5 A voltaic cell that uses a salt bridge to complete the electrical circuit.**

GIVE IT SOME THOUGHT

Why do anions in a salt bridge migrate toward the anode?

Figure 20.6 ▶ summarizes the relationships among the anode, the cathode, the chemical process occurring in a voltaic cell, the direction of migration of ions in solution, and the motion of electrons between electrodes in the external circuit. Notice in particular that *in any voltaic cell the electrons flow from the anode through the external circuit to the cathode*. Because the negatively charged electrons flow from the anode to the cathode, the anode in a voltaic cell is labeled with a negative sign and the cathode with a positive sign; we can envision the electrons as being attracted to the positive cathode from the negative anode through the external circuit.*

* Although the anode and cathode are labeled with − and + signs respectively, you should not interpret the labels as charges on the electrodes. The labels simply tell us the electrode at which the electrons are released to the external circuit (the anode) and received from the external circuit (the cathode). The actual charges on the electrodes are essentially zero.

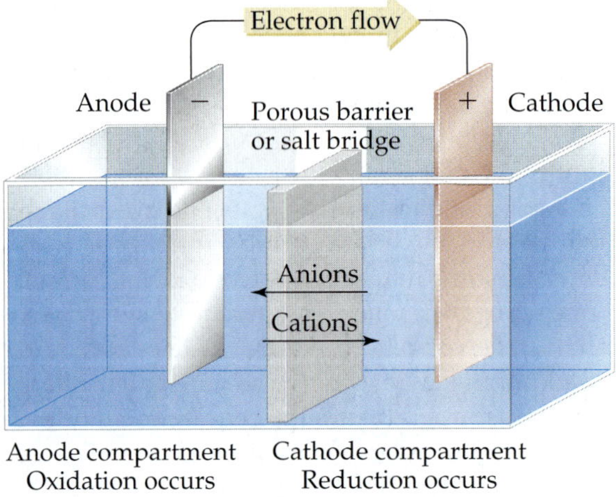

▶ **Figure 20.6 A summary of the terminology used to describe voltaic cells.** Oxidation occurs at the anode; reduction occurs at the cathode. The electrons flow spontaneously from the negative anode to the positive cathode. The electrical circuit is completed by the movement of ions in solution. Anions move toward the anode, whereas cations move toward the cathode. The cell compartments can be separated by either a porous glass barrier (as in Figure 20.4) or by a salt bridge (as in Figure 20.5).

SAMPLE EXERCISE 20.4 | Reactions in a Voltaic Cell

The oxidation-reduction reaction

$$Cr_2O_7^{2-}(aq) + 14\,H^+(aq) + 6\,I^-(aq) \longrightarrow 2\,Cr^{3+}(aq) + 3\,I_2(s) + 7\,H_2O(l)$$

is spontaneous. A solution containing $K_2Cr_2O_7$ and H_2SO_4 is poured into one beaker, and a solution of KI is poured into another. A salt bridge is used to join the beakers. A metallic conductor that will not react with either solution (such as platinum foil) is suspended in each solution, and the two conductors are connected with wires through a voltmeter or some other device to detect an electric current. The resultant voltaic cell generates an electric current. Indicate the reaction occurring at the anode, the reaction at the cathode, the direction of electron migration, the direction of ion migration, and the signs of the electrodes.

Solution

Analyze: We are given the equation for a spontaneous reaction taking place in a voltaic cell and a description of how the cell is constructed. We are asked to write the half-reactions occurring at the anode and at the cathode, as well as the directions of electron and ion movements and the signs assigned to the electrodes.

Plan: Our first step is to divide the chemical equation into half-reactions so that we can identify the oxidation and the reduction processes. We then use the definitions of anode and cathode and the other terminology summarized in Figure 20.6.

Solve: In one half-reaction, $Cr_2O_7^{2-}(aq)$ is converted into $Cr^{3+}(aq)$. Starting with these ions and then completing and balancing the half-reaction, we have

$$Cr_2O_7^{2-}(aq) + 14\,H^+(aq) + 6\,e^- \longrightarrow 2\,Cr^{3+}(aq) + 7\,H_2O(l)$$

In the other half-reaction, $I^-(aq)$ is converted to $I_2(s)$:

$$6\,I^-(aq) \longrightarrow 3\,I_2(s) + 6\,e^-$$

Now we can use the summary in Figure 20.6 to help us describe the voltaic cell. The first half-reaction is the reduction process (electrons shown on the reactant side of the equation), and by definition, this process occurs at the cathode. The second half-reaction is the oxidation (electrons shown on the product side of the equation), which occurs at the anode. The I^- ions are the source of electrons, and the $Cr_2O_7^{2-}$ ions accept the electrons. Hence, the electrons flow through the external circuit from the electrode immersed in the KI solution (the anode) to the electrode immersed in the $K_2Cr_2O_7$–H_2SO_4 solution (the cathode). The electrodes themselves do not react in any way; they merely provide a means of transferring electrons from or to the solutions. The cations move through the solutions toward the cathode, and the anions move toward the anode. The anode (from which the electrons move) is the negative electrode, and the cathode (toward which the electrons move) is the positive electrode.

PRACTICE EXERCISE

The two half-reactions in a voltaic cell are

$$Zn(s) \longrightarrow Zn^{2+}(aq) + 2\,e^-$$

$$ClO_3^-(aq) + 6\,H^+(aq) + 6\,e^- \longrightarrow Cl^-(aq) + 3\,H_2O(l)$$

(a) Indicate which reaction occurs at the anode and which at the cathode. **(b)** Which electrode is consumed in the cell reaction? **(c)** Which electrode is positive?
Answers: **(a)** The first reaction occurs at the anode, the second reaction at the cathode. **(b)** The anode (Zn) is consumed in the cell reaction. **(c)** The cathode is positive.

ANIMATION
Oxidation-Reduction Reactions Part II

A Molecular View of Electrode Processes

To better understand the relationship between voltaic cells and spontaneous redox reactions, let's look at what happens at the atomic or molecular level. The actual processes involved in the transfer of electrons are quite complex; nevertheless, we can learn much by examining these processes in a simplified way.

Let's first consider the spontaneous redox reaction between Zn(*s*) and Cu^{2+}(*aq*), illustrated in Figure 20.3. During the reaction Zn(*s*) is oxidized to Zn^{2+}(*aq*) and Cu^{2+}(*aq*) is reduced to Cu(*s*). Figure 20.7 ▶ shows a schematic diagram of how these processes occur at the atomic level. We can envision a Cu^{2+} ion coming into contact with the strip of Zn metal, as in Figure 20.7(a). Two electrons are transferred directly from a Zn atom to the Cu^{2+} ion, leading to a Zn^{2+} ion and a Cu atom. The Zn^{2+} ion migrates away into the aqueous solution while the Cu atom remains deposited on the metal strip [Figure 20.7(b)]. As the reaction proceeds, we produce more and more Cu(*s*) and deplete the Cu^{2+}(*aq*), as we saw in Figure 20.3(b).

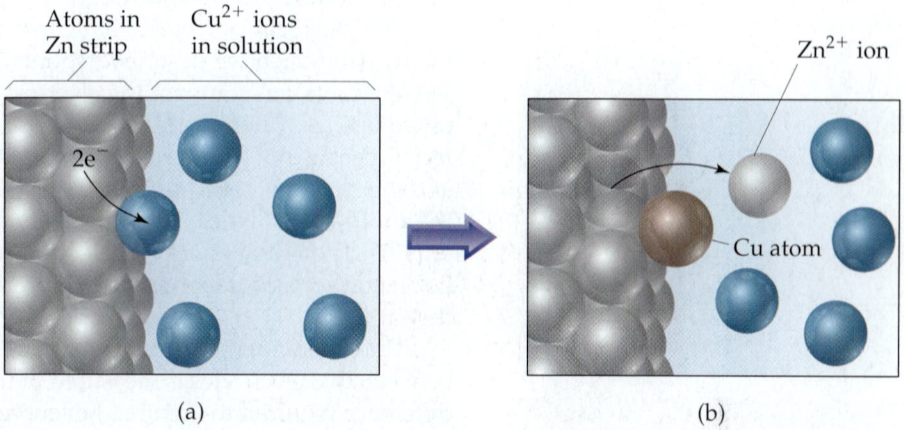

▲ **Figure 20.7 Atomic-level depiction of the Zn(*s*)-Cu^{2+}(*aq*) reaction.** The water molecules and anions in the solution are not shown. (a) A Cu^{2+} ion comes in contact with the surface of the Zn strip and gains two electrons from a Zn atom; the Cu^{2+} ion is reduced, and the Zn atom is oxidized. (b) The resulting Zn^{2+} ion enters the solution, and the Cu atom remains deposited on the strip.

GIVE IT SOME THOUGHT

Which Zn atoms in the strip of Zn metal are primarily involved in the redox reaction with Cu^{2+}(aq), the surface Zn atoms or the interior Zn atoms?

The voltaic cell in Figure 20.5 is also based on the oxidation of Zn(*s*) and the reduction of Cu^{2+}(*aq*). In this case, however, the electrons are not transferred directly from one reacting species to the other. Figure 20.8 ▼ shows qualitatively what happens at each electrode. At the surface of the anode, a Zn atom "loses" two electrons and becomes a Zn^{2+}(*aq*) ion in the anode compartment. We envision the two electrons traveling from the anode through the wire to the cathode. At the surface of the cathode, the two electrons reduce a Cu^{2+} ion to a Cu atom, which is deposited on the cathode. As we noted earlier, the flow of electrons from the anode to the cathode is possible only if ions are transferred through the salt bridge to maintain overall charge balance in each compartment.

The redox reaction between Zn and Cu^{2+} is spontaneous regardless of whether they react directly or in the separate compartments of a voltaic cell. In each case the overall reaction is the same—only the path by which the electrons are transferred from the Zn atom to a Cu^{2+} ion is different. In the next section we will examine *why* this reaction is spontaneous.

▼ **Figure 20.8 Atomic-level depiction of the voltaic cell in Figure 20.5.** At the anode a Zn atom loses two electrons and becomes a Zn^{2+} ion; the Zn atom is oxidized. The electrons travel through the external circuit to the cathode. At the cathode a Cu^{2+} ion gains the two electrons, forming a Cu atom; the Cu^{2+} ion is reduced. Ions migrate through the porous barrier to maintain charge balance between the compartments.

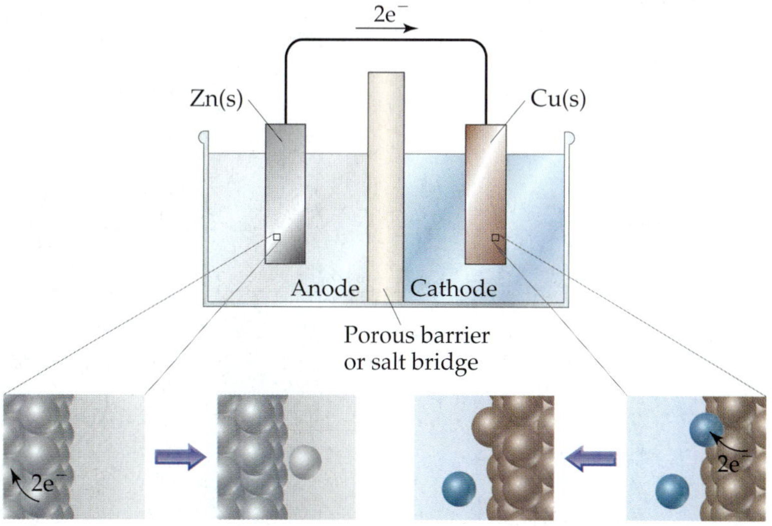

20.4 | Cell EMF Under Standard Conditions

Why do electrons transfer spontaneously from a Zn atom to a Cu^{2+} ion, either directly as in the reaction of Figure 20.3 or through an external circuit as in the voltaic cell of Figure 20.4? In this section we will examine the "driving force" that pushes the electrons through an external circuit in a voltaic cell.

The chemical processes that constitute any voltaic cell are spontaneous in the way in which we described spontaneous processes in Chapter 19. In a simple sense, we can compare the electron flow caused by a voltaic cell to the flow of water in a waterfall (Figure 20.9 ▼). Water flows spontaneously over a waterfall because of a difference in potential energy between the top of the falls and the stream below. (Section 5.1) In a similar fashion, electrons flow from the anode of a voltaic cell to the cathode because of a difference in potential energy. The potential energy of electrons is higher in the anode than in the cathode, and they spontaneously flow through an external circuit from the anode to the cathode.

The difference in potential energy per electrical charge (the *potential difference*) between two electrodes is measured in units of volts. One volt (V) is the potential difference required to impart 1 J of energy to a charge of 1 coulomb (C).

$$1\text{ V} = 1\frac{\text{J}}{\text{C}}$$

Recall that one electron has a charge of 1.60×10^{-19} C. (Section 2.2)

The potential difference between the two electrodes of a voltaic cell provides the driving force that pushes electrons through the external circuit. Therefore, we call this potential difference the **electromotive** ("causing electron motion") **force**, or **emf**. The emf of a cell, denoted E_{cell}, is also called the **cell potential**. Because E_{cell} is measured in volts, we often refer to it as the *cell voltage*. For any cell reaction that proceeds spontaneously, such as that in a voltaic cell, the cell potential will be *positive*.

The emf of a particular voltaic cell depends on the specific reactions that occur at the cathode and anode, the concentrations of reactants and products, and the temperature, which we will assume to be 25°C unless otherwise noted.

▶ **Figure 20.9 Water analogy for electron flow.** The flow of electrons from the anode to the cathode of a voltaic cell can be likened to the flow of water over a waterfall. Water flows over the waterfall because its potential energy is lower at the bottom of the falls than at the top. Likewise, if there is an electrical connection between the anode and cathode of a voltaic cell, electrons flow from the anode to the cathode in order to lower their potential energy.

In this section we will focus on cells that are operated at 25°C under *standard conditions*. Recall from Section 19.5 that standard conditions include 1 *M* concentrations for reactants and products in solution and 1 atm pressure for those that are gases (Table 19.3). Under standard conditions the emf is called the **standard emf**, or the **standard cell potential**, and is denoted E°_{cell}. For the Zn-Cu voltaic cell in Figure 20.5, for example, the standard cell potential at 25°C is +1.10 V.

$$Zn(s) + Cu^{2+}(aq, 1\ M) \longrightarrow Zn^{2+}(aq, 1\ M) + Cu(s) \qquad E^\circ_{cell} = +1.10\ V$$

Recall that the superscript ° indicates standard-state conditions. (Section 5.7)

GIVE IT SOME THOUGHT

A certain voltaic cell has a standard cell potential of +0.85 V at 25°C. Does this mean the redox reaction of the cell is spontaneous?

Standard Reduction (Half-Cell) Potentials

The emf or cell potential of a voltaic cell depends on the particular cathode and anode half-cells involved. We could, in principle, tabulate the standard cell potentials for all possible cathode/anode combinations. However, it is not necessary to undertake this arduous task. Rather, we can assign a standard potential to each individual half-cell and then use these half-cell potentials to determine E°_{cell}.

The cell potential is the difference between two electrode potentials, one associated with the cathode and the other associated with the anode. By convention, the potential associated with each electrode is chosen to be the potential for reduction to occur at that electrode. Thus, standard electrode potentials are tabulated for reduction reactions; they are **standard reduction potentials**, denoted E°_{red}. The cell potential, E°_{cell}, is given by the standard reduction potential of the cathode reaction, E°_{red} (cathode), *minus* the standard reduction potential of the anode reaction, E°_{red} (anode):

$$E^\circ_{cell} = E^\circ_{red}\ (cathode) - E^\circ_{red}\ (anode) \qquad [20.8]$$

We will discuss Equation 20.8 in greater detail shortly.

Because every voltaic cell involves two half-cells, it is not possible to measure the standard reduction potential of a half-reaction directly. If we assign a standard reduction potential to a certain reference half-reaction, however, we can then determine the standard reduction potentials of other half-reactions relative to that reference. The reference half-reaction is the reduction of $H^+(aq)$ to $H_2(g)$ under standard conditions, which is assigned a standard reduction potential of exactly 0 V.

$$2\ H^+(aq, 1\ M) + 2\ e^- \longrightarrow H_2(g, 1\ atm) \qquad E^\circ_{red} = 0\ V \qquad [20.9]$$

An electrode designed to produce this half-reaction is called a **standard hydrogen electrode** (SHE), or the normal hydrogen electrode (NHE). An SHE consists of a platinum wire connected to a piece of platinum foil covered with finely divided platinum that serves as an inert surface for the reaction. The electrode is encased in a glass tube so that hydrogen gas under standard conditions (1 atm) can bubble over the platinum, and the solution contains $H^+(aq)$ under standard (1 *M*) conditions (Figure 20.10 ▶).

Figure 20.11 ▶ shows a voltaic cell using an SHE and a standard Zn^{2+}/Zn electrode. The spontaneous reaction is the one shown in Figure 20.1, namely, the oxidation of Zn and the reduction of H^+.

$$Zn(s) + 2\ H^+(aq) \longrightarrow Zn^{2+}(aq) + H_2(g)$$

Notice that the Zn^{2+}/Zn electrode is the anode and the SHE is the cathode and that the cell voltage is +0.76 V. By using the defined standard reduction

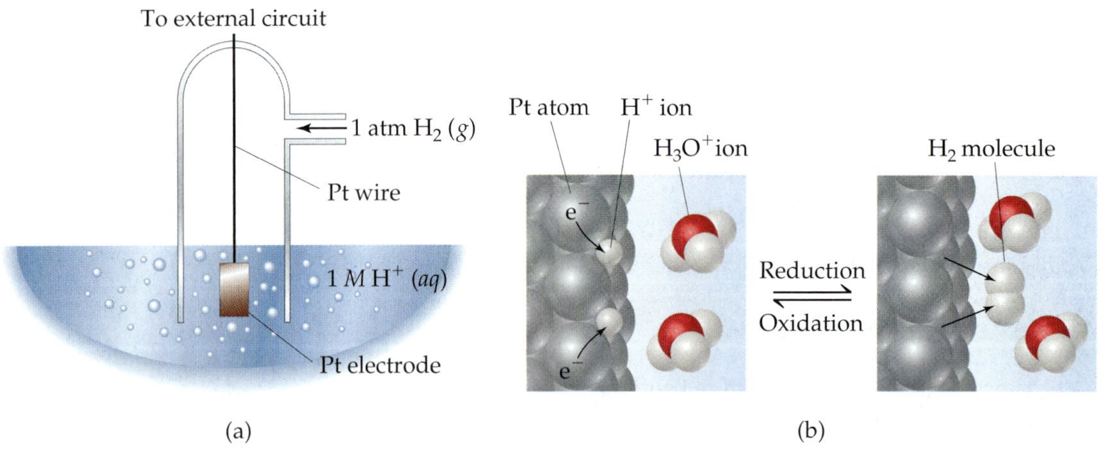

▲ **Figure 20.10 The standard hydrogen electrode (SHE) used as a reference electrode.** (a) An SHE consists of an electrode with finely divided Pt in contact with $H_2(g)$ at 1 atm pressure and an acidic solution with $[H^+] = 1\ M$. (b) Molecular depiction of the processes that occur at the SHE. When the SHE is the cathode of a cell, two H^+ ions each accept an electron from the Pt electrode and are reduced to H atoms. The H atoms bond together to form H_2. When the SHE is the anode of a cell, the reverse process occurs: An H_2 molecule at the electrode surface loses two electrons and is oxidized to H^+. The H^+ ions in solution are hydrated.

potential of $H^+(E^\circ_{red} = 0)$ and Equation 20.8, we can determine the standard reduction potential for the Zn^{2+}/Zn half-reaction:

$$E^\circ_{cell} = E^\circ_{red}\,(\text{cathode}) - E^\circ_{red}\,(\text{anode})$$
$$+0.76\text{ V} = 0\text{ V} - E^\circ_{red}\,(\text{anode})$$
$$E^\circ_{red}\,(\text{anode}) = -0.76\text{ V}$$

Thus, a standard reduction potential of -0.76 V can be assigned to the reduction of Zn^{2+} to Zn.

$$Zn^{2+}(aq, 1\ M) + 2\,e^- \longrightarrow Zn(s) \qquad E^\circ_{red} = -0.76\text{ V}$$

We write the reaction as a reduction even though it is "running in reverse" as an oxidation in the cell in Figure 20.11. *Whenever we assign a potential to a half-reaction, we write the reaction as a reduction.*

The standard reduction potentials for other half-reactions can be established from other cell potentials in a fashion analogous to that used for the Zn^{2+}/Zn half-reaction. Table 20.1 ▶ lists some standard reduction potentials; a more complete list is found in Appendix E. These standard reduction potentials, often called *half-cell potentials*, can be combined to calculate the emfs of a large variety of voltaic cells.

ANIMATION
Voltaic Cells II: The Zinc-Hydrogen Cell

▶ **Figure 20.11 A voltaic cell using a standard hydrogen electrode.**

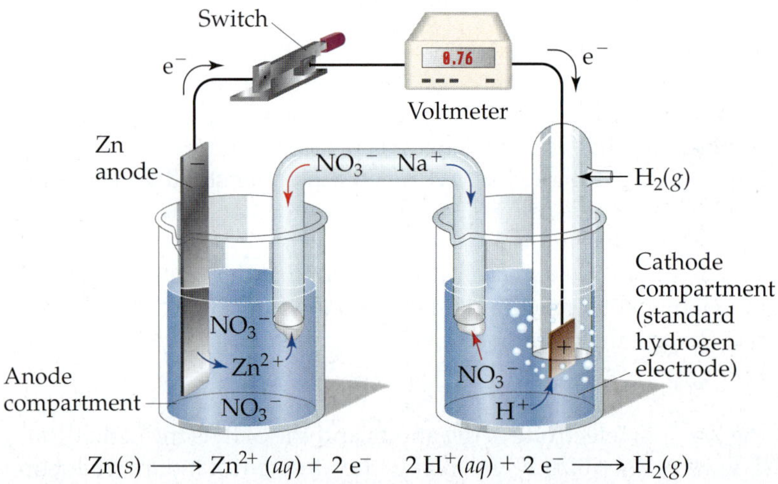

TABLE 20.1 Standard Reduction Potentials in Water at 25°C

Potential (V)	Reduction Half-Reaction
+2.87	$F_2(g) + 2\,e^- \longrightarrow 2\,F^-(aq)$
+1.51	$MnO_4^-(aq) + 8\,H^+(aq) + 5\,e^- \longrightarrow Mn^{2+}(aq) + 4\,H_2O(l)$
+1.36	$Cl_2(g) + 2\,e^- \longrightarrow 2\,Cl^-(aq)$
+1.33	$Cr_2O_7^{2-}(aq) + 14\,H^+(aq) + 6\,e^- \longrightarrow 2\,Cr^{3+}(aq) + 7\,H_2O(l)$
+1.23	$O_2(g) + 4\,H^+(aq) + 4\,e^- \longrightarrow 2\,H_2O(l)$
+1.06	$Br_2(l) + 2\,e^- \longrightarrow 2\,Br^-(aq)$
+0.96	$NO_3^-(aq) + 4\,H^+(aq) + 3\,e^- \longrightarrow NO(g) + 2\,H_2O(l)$
+0.80	$Ag^+(aq) + e^- \longrightarrow Ag(s)$
+0.77	$Fe^{3+}(aq) + e^- \longrightarrow Fe^{2+}(aq)$
+0.68	$O_2(g) + 2\,H^+(aq) + 2\,e^- \longrightarrow H_2O_2(aq)$
+0.59	$MnO_4^-(aq) + 2\,H_2O(l) + 3\,e^- \longrightarrow MnO_2(s) + 4\,OH^-(aq)$
+0.54	$I_2(s) + 2\,e^- \longrightarrow 2\,I^-(aq)$
+0.40	$O_2(g) + 2\,H_2O(l) + 4\,e^- \longrightarrow 4\,OH^-(aq)$
+0.34	$Cu^{2+}(aq) + 2\,e^- \longrightarrow Cu(s)$
0 [defined]	$2\,H^+(aq) + 2\,e^- \longrightarrow H_2(g)$
−0.28	$Ni^{2+}(aq) + 2\,e^- \longrightarrow Ni(s)$
−0.44	$Fe^{2+}(aq) + 2\,e^- \longrightarrow Fe(s)$
−0.76	$Zn^{2+}(aq) + 2\,e^- \longrightarrow Zn(s)$
−0.83	$2\,H_2O(l) + 2\,e^- \longrightarrow H_2(g) + 2\,OH^-(aq)$
−1.66	$Al^{3+}(aq) + 3\,e^- \longrightarrow Al(s)$
−2.71	$Na^+(aq) + e^- \longrightarrow Na(s)$
−3.05	$Li^+(aq) + e^- \longrightarrow Li(s)$

GIVE IT SOME THOUGHT

For the half-reaction $Cl_2(g) + 2\,e^- \longrightarrow 2\,Cl^-(aq)$, what are the standard conditions for the reactant and product?

ANIMATION Standard Reduction Potentials

Because electrical potential measures potential energy per electrical charge, standard reduction potentials are intensive properties. (Section 1.3) In other words, if we increased the amount of substances in a redox reaction, we would increase both the energy and charges involved, but the ratio of energy (joules) to electrical charge (coulombs) would remain constant (V = J/C). Thus, *changing the stoichiometric coefficient in a half-reaction does not affect the value of the standard reduction potential*. For example, E°_{red} for the reduction of 50 Zn^{2+} is the same as that for the reduction of 1 Zn^{2+}:

$$50\,Zn^{2+}(aq, 1\,M) + 100\,e^- \longrightarrow 50\,Zn(s) \qquad E^\circ_{red} = -0.76\text{ V}$$

Earlier, we made an analogy between emf and the flow of water over a waterfall (Figure 20.9). The emf corresponds to the difference in height between the top and bottom of the waterfall. This height remains the same, whether the flow of water is large or small.

SAMPLE EXERCISE 20.5 | Calculating E°_{red} from E°_{cell}

For the Zn-Cu^{2+} voltaic cell shown in Figure 20.5, we have

$$Zn(s) + Cu^{2+}(aq, 1\,M) \longrightarrow Zn^{2+}(aq, 1\,M) + Cu(s) \qquad E^\circ_{cell} = 1.10\text{ V}$$

Given that the standard reduction potential of Zn^{2+} to Zn(*s*) is −0.76 V, calculate the E°_{red} for the reduction of Cu^{2+} to Cu:

$$Cu^{2+}(aq, 1\ M) + 2\ e^- \longrightarrow Cu(s)$$

Solution

Analyze: We are given E°_{cell} and E°_{red} for Zn^{2+} and asked to calculate E°_{red} for Cu^{2+}.

Plan: In the voltaic cell, Zn is oxidized and is therefore the anode. Thus, the given E°_{red} for Zn^{2+} is E°_{red} (anode). Because Cu^{2+} is reduced, it is in the cathode half-cell. Thus, the unknown reduction potential for Cu^{2+} is E°_{red} (cathode). Knowing E°_{cell} and E°_{red} (anode), we can use Equation 20.8 to solve for E°_{red} (cathode).

Solve:

$$E^\circ_{cell} = E^\circ_{red}\,(\text{cathode}) - E^\circ_{red}\,(\text{anode})$$

$$1.10\ V = E^\circ_{red}\,(\text{cathode}) - (-0.76\ V)$$

$$E^\circ_{red}\,(\text{cathode}) = 1.10\ V - 0.76\ V = 0.34\ V$$

Check: This standard reduction potential agrees with the one listed in Table 20.1.

Comment: The standard reduction potential for Cu^{2+} can be represented as $E^\circ_{Cu^{2+}} = 0.34$ V, and that for Zn^{2+} as $E^\circ_{Zn^{2+}} = -0.76$ V. The subscript identifies the ion that is reduced in the reduction half-reaction.

PRACTICE EXERCISE

A voltaic cell is based on the half-reactions

$$In^+(aq) \longrightarrow In^{3+}(aq) + 2\ e^-$$

$$Br_2(l) + 2\ e^- \longrightarrow 2\ Br^-(aq)$$

The standard emf for this cell is 1.46 V. Using the data in Table 20.1, calculate E°_{red} for the reduction of In^{3+} to In^+.

Answer: −0.40 V

SAMPLE EXERCISE 20.6 | Calculating E°_{cell} from E°_{red}

Using the standard reduction potentials listed in Table 20.1, calculate the standard emf for the voltaic cell described in Sample Exercise 20.4, which is based on the reaction

$$Cr_2O_7^{2-}(aq) + 14\ H^+(aq) + 6\ I^-(aq) \longrightarrow 2\ Cr^{3+}(aq) + 3\ I_2(s) + 7\ H_2O(l)$$

Solution

Analyze: We are given the equation for a redox reaction and asked to use data in Table 20.1 to calculate the standard emf (standard potential) for the associated voltaic cell.

Plan: Our first step is to identify the half-reactions that occur at the cathode and the anode, which we did in Sample Exercise 20.4. Then we can use data from Table 20.1 and Equation 20.8 to calculate the standard emf.

Solve: The half-reactions are

Cathode: $Cr_2O_7^{2-}(aq) + 14\ H^+(aq) + 6\ e^- \longrightarrow 2\ Cr^{3+}(aq) + 7\ H_2O(l)$

Anode: $6\ I^-(aq) \longrightarrow 3\ I_2(s) + 6\ e^-$

According to Table 20.1, the standard reduction potential for the reduction of $Cr_2O_7^{2-}$ to Cr^{3+} is +1.33 V, and the standard reduction potential for the reduction of I_2 to I^- (the reverse of the oxidation half-reaction) is +0.54 V. We then use these values in Equation 20.8.

$$E^\circ_{cell} = E^\circ_{red}\,(\text{cathode}) - E^\circ_{red}\,(\text{anode}) = 1.33\ V - 0.54\ V = 0.79\ V$$

Although the iodide half-reaction at the anode must be multiplied by 3 in order to obtain a balanced equation for the reaction, the value of E°_{red} is *not* multiplied by 3. As we have noted, the standard reduction potential is an intensive property, so it is independent of the specific stoichiometric coefficients.

Check: The cell potential, 0.79 V, is a positive number. As noted earlier, a voltaic cell must have a positive emf in order to operate.

PRACTICE EXERCISE

Using data in Table 20.1, calculate the standard emf for a cell that employs the following overall cell reaction:

$$2\,Al(s) + 3\,I_2(s) \longrightarrow 2\,Al^{3+}(aq) + 6\,I^-(aq)$$

Answer: +2.20 V

We are now in a position to discuss Equation 20.8 more fully. First of all, we note that the definition of $E^\circ_{cell} = E^\circ_{red}$ (cathode) $- E^\circ_{red}$ (anode) makes sense from Hess's law. (Section 5.6) Consider again the redox reaction of Figures 20.3 and 20.4:

$$Zn(s) + Cu^{2+}(aq) \longrightarrow Zn^{2+}(aq) + Cu(s)$$

We can easily write this as the sum of the two half-reactions as follows, and add up their respective E°'s:

$$Cu^{2+}(aq) + 2\,e- \longrightarrow Cu(s) \qquad E^\circ = E^\circ_{red} = +0.34\ V$$

$$Zn(s) \longrightarrow Zn^{2+}(aq) + 2\,e^- \qquad E^\circ = -E^\circ_{red} = -(-0.76\ V) = +0.76\ V$$

$$Zn(s) + Cu^{2+}(aq) \longrightarrow Zn^{2+}(aq) + Cu(s) \qquad E^\circ_{cell} = +1.10\ V$$

The copper half-reaction is a reduction, so its reduction potential is the same as that of the standard table. However, the zinc half-reaction is an oxidation, but in the standard table it is written as a reduction. Therefore, when we write the reduction reaction in reverse to make it an oxidation, we have to change the sign of the reduction potential. This is just what we do with ΔH too when we turn a reaction around: We flip the sign of ΔH. Then, finally, when we add the reactions up to get the one we want, we just add up the potentials for the half-reactions that go with the half-reactions as they are written. When you recall that the cathode is where reduction takes place and the anode is where oxidation takes place, you can see that $E^\circ_{cell} = E^\circ_{red}$ (cathode) $+ (-E^\circ_{red}$ (anode)) $= E^\circ_{red}$ (cathode) $- E^\circ_{red}$ (anode).

For each of the half-cells in a voltaic cell, the standard reduction potential provides a measure of the driving force for reduction to occur: *The more positive the value of* E°_{red}*, the greater the driving force for reduction.* In any voltaic cell the reaction at the cathode has a more positive value of E°_{red} than does the reaction at the anode. In essence, the greater driving force of the cathode half-reaction is used to force the anode reaction to occur "in reverse," as an oxidation.

Equation 20.8 tells us that the cell potential, E°_{cell}, is the difference between the standard reduction potential of the cathode half-reaction, E°_{red} (cathode), and that of the anode half-reaction, E°_{red} (anode). We can interpret E°_{cell} as the "net driving force" that pushes the electrons through an external circuit. Equation 20.8 is illustrated graphically in Figure 20.12 ▶, in which the standard reduction potentials are shown on a scale, with more positive E°_{red} values higher on the scale (as in Table 20.1). In any voltaic cell the cathode half-reaction is the one higher on the scale, and the difference between the two standard reduction potentials is the cell potential. Figure 20.13 ▶ shows the specific values of E°_{red} for the Zn-Cu voltaic cell in Figure 20.5.

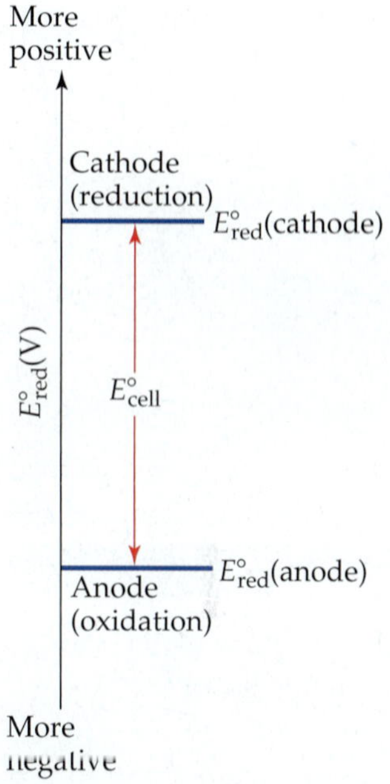

▲ **Figure 20.12 Standard cell potential of a voltaic cell.** The cell potential measures the difference in the standard reduction potentials of the cathode and the anode reactions: $E^\circ_{cell} = E^\circ_{red}$ (cathode) $- E^\circ_{red}$ (anode). In a voltaic cell the cathode reaction is always the one that has the more positive (or less negative) value for E°_{red}.

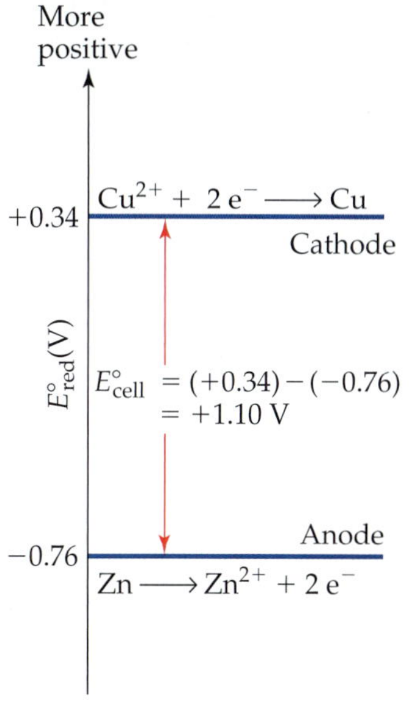

▲ **Figure 20.13 Half-cell potentials.** The half-cell potentials for the voltaic cell in Figure 20.5, diagrammed in the style of Figure 20.12.

GIVE IT SOME THOUGHT

Is the following statement true or false? The smaller the difference between cathode and anode standard reduction potentials, the smaller the driving force for the overall redox reaction.

SAMPLE EXERCISE 20.7 | From Half-Reactions to Cell EMF

A voltaic cell is based on the following two standard half-reactions:

$$Cd^{2+}(aq) + 2\,e^- \longrightarrow Cd(s)$$
$$Sn^{2+}(aq) + 2\,e^- \longrightarrow Sn(s)$$

By using the data in Appendix E, determine **(a)** the half-reactions that occur at the cathode and the anode, and **(b)** the standard cell potential.

Solution

Analyze: We have to look up E°_{red} for two half-reactions and use these values to predict the cathode and anode of the cell and to calculate its standard cell potential, E°_{cell}.

Plan: The cathode will have the reduction with the most positive E°_{red} value. The anode will have the less positive E°_{red}. To write the half-reaction at the anode, we reverse the half-reaction written for the reduction.

Solve: (a) According to Appendix E, $E^\circ_{red}(Cd^{2+}/Cd) = -0.403$ V and $E^\circ_{red}(Sn^{2+}/Sn) = -0.136$ V. The standard reduction potential for Sn^{2+} is more positive (less negative) than that for Cd^{2+}; hence, the reduction of Sn^{2+} is the reaction that occurs at the cathode.

Cathode: $$Sn^{2+}(aq) + 2\,e^- \longrightarrow Sn(s)$$

The anode reaction therefore is the loss of electrons by Cd.

Anode: $$Cd(s) \longrightarrow Cd^{2+}(aq) + 2\,e^-$$

(b) The cell potential is given by Equation 20.8.

$$E^\circ_{cell} = E^\circ_{red}\,(\text{cathode}) - E^\circ_{red}\,(\text{anode}) = (-0.136\text{ V}) - (-0.403\text{ V}) = 0.267\text{ V}$$

Notice that it is unimportant that the E°_{red} values of both half-reactions are negative; the negative values merely indicate how these reductions compare to the reference reaction, the reduction of $H^+(aq)$.

Check: The cell potential is positive, as it must be for a voltaic cell.

PRACTICE EXERCISE

A voltaic cell is based on a Co^{2+}/Co half-cell and an AgCl/Ag half-cell. **(a)** What reaction occurs at the anode? **(b)** What is the standard cell potential?

Answers: **(a)** $Co \longrightarrow Co^{2+} + 2\,e^-$; **(b)** +0.499 V

Strengths of Oxidizing and Reducing Agents

We have thus far used the tabulation of standard reduction potentials to examine voltaic cells. We can also use E°_{red} values to understand aqueous reaction chemistry. Recall, for example, the reaction between Zn(*s*) and $Cu^{2+}(aq)$ shown in Figure 20.3.

$$Zn(s) + Cu^{2+}(aq) \longrightarrow Zn^{2+}(aq) + Cu(s)$$

Zinc metal is oxidized and $Cu^{2+}(aq)$ is reduced in this reaction. These substances are in direct contact, however, so we are not producing usable electrical work; the direct contact essentially "short-circuits" the cell. Nevertheless, the driving force for the reaction is the same as that in a voltaic cell, as in Figure 20.5. Because the E°_{red} value for the reduction of Cu^{2+} (0.34 V) is more positive than the E°_{red} value for the reduction of Zn^{2+}(−0.76 V), the reduction of $Cu^{2+}(aq)$ by Zn(*s*) is a spontaneous process.

We can generalize the relationship between the value of E°_{red} and the spontaneity of redox reactions: *The more positive the* E°_{red} *value for a half-reaction, the greater the tendency for the reactant of the half-reaction to be reduced and, therefore, to oxidize another species.* In Table 20.1, for example, F_2 is the most easily reduced species, so it is the strongest oxidizing agent listed.

$$F_2(g) + 2\,e^- \longrightarrow 2\,F^-(aq) \qquad E^\circ_{red} = 2.87\text{ V}$$

Among the most frequently used oxidizing agents are the halogens, O_2, and oxyanions such as MnO_4^-, $Cr_2O_7^{2-}$, and NO_3^-, whose central atoms have high positive oxidation states. According to Table 20.1, all these species undergo reduction with large positive values of E°_{red}.

Lithium ion (Li^+) is the most difficult species to reduce and is therefore the poorest oxidizing agent:

$$Li^+(aq) + e^- \longrightarrow Li(s) \qquad E^\circ_{red} = -3.05\ V$$

Because Li^+ is so difficult to reduce, the reverse reaction, the oxidation of $Li(s)$ to $Li^+(aq)$, is a highly favorable reaction. *The half-reaction with the smallest reduction potential is most easily reversed as an oxidation.* Thus, lithium metal has a great tendency to transfer electrons to other species. In water, Li is the strongest reducing agent among the substances listed in Table 20.1.

Commonly used reducing agents include H_2 and the active metals such as the alkali metals and the alkaline earth metals. Other metals whose cations have negative E°_{red} values—Zn and Fe, for example—are also used as reducing agents. Solutions of reducing agents are difficult to store for extended periods because of the ubiquitous presence of O_2, a good oxidizing agent. For example, developer solutions used in photography are mild reducing agents; they have only a limited shelf life because they are readily oxidized by O_2 from the air.

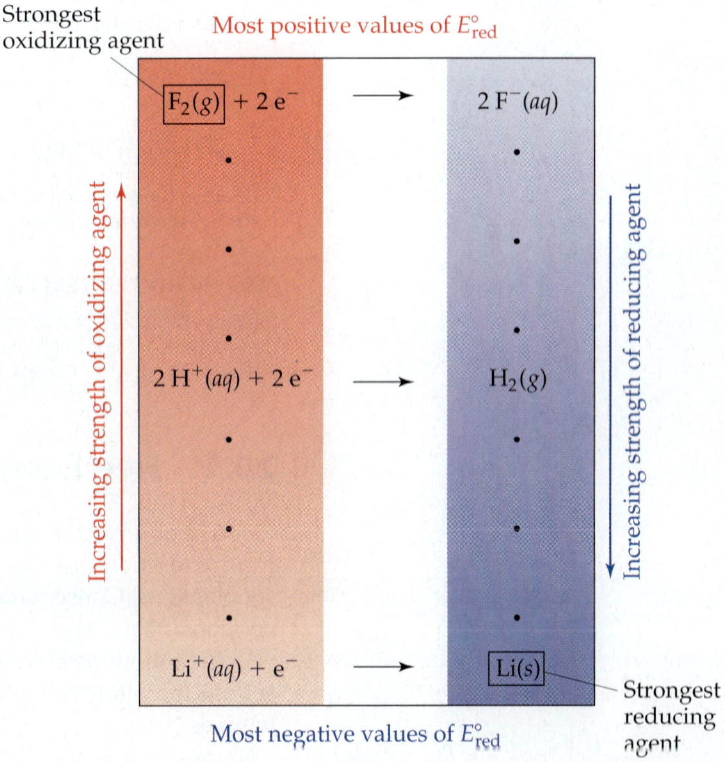

▲ **Figure 20.14 Relative strengths of oxidizing agents.** The standard reduction potentials, E°_{red}, listed in Table 20.1 are related to the ability of substances to serve as oxidizing or reducing agents. Species on the left side of the half-reactions can act as oxidizing agents, and those on the right side can act as reducing agents. As E°_{red} becomes more positive, the species on the left become stronger and stronger oxidizing agents. As E°_{red} becomes more negative, the species on the right become stronger and stronger reducing agents.

The list of E°_{red} values in Table 20.1 orders the ability of substances to act as oxidizing or reducing agents and is summarized in Figure 20.14 ▲. The substances that are most readily reduced (strong oxidizing agents) are the reactants on the top left of the table. Their products, on the top right of the table, are oxidized with difficulty (weak reducing agents). The substances on the bottom left of the table are reduced with difficulty, but their products are readily oxidized. This inverse relationship between oxidizing and reducing strength is similar to the inverse relationship between the strengths of conjugate acids and bases. (Section 16.2 and Figure 16.4)

To help you remember the relationships between the strengths of oxidizing and reducing agents, recall the very exothermic reaction between sodium metal and chlorine gas to form sodium chloride (Figure 8.2). In this reaction $Cl_2(g)$ is reduced (it serves as a strong oxidizing agent) and Na(s) is oxidized (it serves as a strong reducing agent). The products of this reaction—Na^+ and Cl^- ions—are very weak oxidizing and reducing agents, respectively.

SAMPLE EXERCISE 20.8 | Determining the Relative Strengths of Oxidizing Agents

Using Table 20.1, rank the following ions in order of increasing strength as oxidizing agents: $NO_3^-(aq)$, $Ag^+(aq)$, $Cr_2O_7^{2-}(aq)$.

Solution

Analyze: We are given several ions and asked to rank their abilities to act as oxidizing agents.

Plan: The more readily an ion is reduced (the more positive its E°_{red} value), the stronger it is as an oxidizing agent.

Solve: From Table 20.1, we have

$$NO_3^-(aq) + 4\,H^+(aq) + 3\,e^- \longrightarrow NO(g) + 2\,H_2O(l) \qquad E^\circ_{red} = +0.96\text{ V}$$

$$Ag^+(aq) + e^- \longrightarrow Ag(s) \qquad E^\circ_{red} = +0.80\text{ V}$$

$$Cr_2O_7^{2-}(aq) + 14\,H^+(aq) + 6\,e^- \longrightarrow 2\,Cr^{3+}(aq) + 7\,H_2O(l) \qquad E^\circ_{red} = +1.33\text{ V}$$

Because the standard reduction potential of $Cr_2O_7^{2-}$ is the most positive, $Cr_2O_7^{2-}$ is the strongest oxidizing agent of the three. The rank order is $Ag^+ < NO_3^- < Cr_2O_7^{2-}$.

PRACTICE EXERCISE

Using Table 20.1, rank the following species from the strongest to the weakest reducing agent: $I^-(aq)$, Fe(*s*), Al(*s*).
Answer: Al(*s*) > Fe(*s*) > $I^-(aq)$

20.5 | Free Energy and Redox Reactions

We have observed that voltaic cells use redox reactions that proceed spontaneously. Any reaction that can occur in a voltaic cell to produce a positive emf must be spontaneous. Consequently, it is possible to decide whether a redox reaction will be spontaneous by using half-cell potentials to calculate the emf associated with it.

The following discussion will pertain to general redox reactions, not just reactions in voltaic cells. Thus, we will make Equation 20.8 more general by writing it as

$$E^\circ = E^\circ_{red}\text{ (reduction process)} - E^\circ_{red}\text{ (oxidation process)} \qquad [20.10]$$

In modifying Equation 20.8, we have dropped the subscript "cell" to indicate that the calculated emf does not necessarily refer to a voltaic cell. Similarly, we have generalized the standard reduction potentials on the right side of the equation by referring to the reduction and oxidation processes, rather than the cathode and the anode. We can now make a general statement about the spontaneity of a reaction and its associated emf, E: *A positive value of* E *indicates a spontaneous process, and a negative value of* E *indicates a nonspontaneous one.* We will use E to represent the emf under nonstandard conditions, and E° to indicate the standard emf.

SAMPLE EXERCISE 20.9 | Spontaneous or Not?

Using standard reduction potentials (Table 20.1), determine whether the following reactions are spontaneous under standard conditions.

(a) $Cu(s) + 2\,H^+(aq) \longrightarrow Cu^{2+}(aq) + H_2(g)$
(b) $Cl_2(g) + 2\,I^-(aq) \longrightarrow 2\,Cl^-(aq) + I_2(s)$

Solution

Analyze: We are given two equations and must determine whether or not each is spontaneous.

Plan: To determine whether a redox reaction is spontaneous under standard conditions, we first need to write its reduction and oxidation half-reactions. We can then use the standard reduction potentials and Equation 20.10 to calculate the standard emf, E°, for the reaction. If a reaction is spontaneous, its standard emf must be a positive number.

Solve: (a) In this reaction Cu is oxidized to Cu^{2+} and H^+ is reduced to H_2. The corresponding half-reactions and associated standard reduction potentials are

Reduction: $2\,H^+(aq) + 2\,e^- \longrightarrow H_2(g) \qquad E^\circ_{red} = 0\text{ V}$

Oxidation: $Cu(s) \longrightarrow Cu^{2+}(aq) + 2\,e^- \qquad E^\circ_{red} = +0.34\text{ V}$

Notice that for the oxidation, we use the standard reduction potential from Table 20.1 for the reduction of Cu^{2+} to Cu. We now calculate E° by using Equation 20.10:

$$E^\circ = E^\circ_{red}\text{ (reduction process)} - E^\circ_{red}\text{ (oxidation process)}$$
$$= (0\text{ V}) - (0.34\text{ V}) = -0.34\text{ V}$$

Because $E°$ is negative, the reaction is not spontaneous in the direction written. Copper metal does not react with acids in this fashion. The reverse reaction, however, *is* spontaneous and would have an $E°$ of +0.34 V:

$$Cu^{2+}(aq) + H_2(g) \longrightarrow Cu(s) + 2\,H^+(aq) \qquad E° = +0.34\ V$$

Cu^{2+} can be reduced by H_2.

(b) We follow a procedure analogous to that in (a):

Reduction: $Cl_2(g) + 2\,e^- \longrightarrow 2\,Cl^-(aq) \qquad E°_{red} = +1.36\ V$

Oxidation: $2\,I^-(aq) \longrightarrow I_2(s) + 2\,e^- \qquad E°_{red} = +0.54\ V$

In this case

$$E° = (1.36\ V) - (0.54\ V) = +0.82\ V$$

Because the value of $E°$ is positive, this reaction is spontaneous and could be used to build a voltaic cell.

PRACTICE EXERCISE

Using the standard reduction potentials listed in Appendix E, determine which of the following reactions are spontaneous under standard conditions:

(a) $I_2(s) + 5\,Cu^{2+}(aq) + 6\,H_2O(l) \longrightarrow 2\,IO_3^-(aq) + 5\,Cu(s) + 12\,H^+(aq)$
(b) $Hg^{2+}(aq) + 2\,I^-(aq) \longrightarrow Hg(l) + I_2(s)$
(c) $H_2SO_3(aq) + 2\,Mn(s) + 4\,H^+(aq) \longrightarrow S(s) + 2\,Mn^{2+}(aq) + 3\,H_2O(l)$

Answer: Reactions (b) and (c) are spontaneous.

We can use standard reduction potentials to understand the activity series of metals. (Section 4.4) Recall that any metal in the activity series will be oxidized by the ions of any metal below it. We can now recognize the origin of this rule on the basis of standard reduction potentials. The activity series, tabulated in Table 4.5, consists of the oxidation reactions of the metals, ordered from the strongest reducing agent at the top to the weakest reducing agent at the bottom. (Thus, the ordering is "flipped over" relative to that in Table 20.1.) For example, nickel lies above silver in the activity series. We therefore expect nickel to displace silver, according to the net reaction

$$Ni(s) + 2\,Ag^+(aq) \longrightarrow Ni^{2+}(aq) + 2\,Ag(s)$$

In this reaction Ni is oxidized and Ag^+ is reduced. Therefore, using data from Table 20.1, the standard emf for the reaction is

$$E° = E°_{red}(Ag^+/Ag) - E°_{red}(Ni^{2+}/Ni)$$
$$= (+0.80\ V) - (-0.28\ V) = +1.08\ V$$

The positive value of $E°$ indicates that the displacement of silver by nickel is a spontaneous process. Remember that although the silver half-reaction is multiplied by 2, the reduction potential is not.

GIVE IT SOME THOUGHT

Based on Table 4.5, which is the stronger reducing agent, $Hg(l)$ or $Pb(s)$?

EMF and ΔG

The change in Gibbs free energy, ΔG, is a measure of the spontaneity of a process that occurs at constant temperature and pressure. (Section 19.5) Because the emf, E, of a redox reaction indicates whether the reaction is spontaneous, the relationship between emf and the free-energy change is

$$\Delta G = -nFE \qquad [20.11]$$

▲ **Figure 20.15 Michael Faraday.** Faraday (1791–1867) was born in England, a child of a poor blacksmith. At the age of 14 he was apprenticed to a bookbinder who gave Faraday time to read and to attend lectures. In 1812 he became an assistant in Humphry Davy's laboratory at the Royal Institution. He succeeded Davy as the most famous and influential scientist in England, making an amazing number of important discoveries, including his formation of the quantitative relationships between electrical current and the extent of chemical reaction in electrochemical cells.

In this equation n is a positive number without units that represents the number of electrons transferred in the reaction. The constant F is called *Faraday's constant*, named after Michael Faraday (Figure 20.15 ◀). Faraday's constant is the quantity of electrical charge on one mole of electrons. This quantity of charge is called a **faraday** (F).

$$1\ F = 96{,}485\ \text{C/mol} = 96{,}485\ \text{J/V-mol}$$

The units of ΔG calculated by using Equation 20.11 are J/mol; as in Equation 19.21, we use "per mole" to mean "per mole of reaction as written." ∞ (Section 19.7)

Both n and F are positive numbers. Thus, a positive value of E in Equation 20.11 leads to a negative value of ΔG. Remember: *A positive value of* E *and a negative value of* ΔG *both indicate that a reaction is spontaneous.* When the reactants and products are all in their standard states, Equation 20.11 can be modified to relate $\Delta G°$ and $E°$.

$$\Delta G° = -nFE° \qquad [20.12]$$

Equation 20.12 is very important. It relates the standard emf, $E°$, of an electrochemical reaction to the standard free-energy change, $\Delta G°$, for the reaction. Because $\Delta G°$ is related to the equilibrium constant, K, for a reaction by the expression $\Delta G° = -RT \ln K$ ∞ (Section 19.7), we can relate the standard emf to the equilibrium constant for the reaction.

SAMPLE EXERCISE 20.10 | Determining $\Delta G°$ and K

(a) Use the standard reduction potentials listed in Table 20.1 to calculate the standard free-energy change, $\Delta G°$, and the equilibrium constant, K, at room temperature (T = 298 K) for the reaction

$$4\ Ag(s) + O_2(g) + 4\ H^+(aq) \longrightarrow 4\ Ag^+(aq) + 2\ H_2O(l)$$

(b) Suppose the reaction in part (a) was written

$$2\ Ag(s) + \tfrac{1}{2}O_2(g) + 2\ H^+(aq) \longrightarrow 2\ Ag^+(aq) + H_2O(l)$$

What are the values of $E°$, $\Delta G°$, and K when the reaction is written in this way?

Solution

Analyze: We are asked to determine $\Delta G°$ and K for a redox reaction, using standard reduction potentials.

Plan: We use the data in Table 20.1 and Equation 20.10 to determine $E°$ for the reaction and then use $E°$ in Equation 20.12 to calculate $\Delta G°$. We will then use Equation 19.22, $\Delta G° = -RT \ln K$, to calculate K.

Solve: **(a)** We first calculate $E°$ by breaking the equation into two half-reactions, as we did in Sample Exercise 20.9, and then obtain $E°_{red}$ values from Table 20.1 (or Appendix E):

Reduction: $O_2(g) + 4\ H^+(aq) + 4\ e^- \longrightarrow 2\ H_2O(l)$ $\quad E°_{red} = +1.23\ \text{V}$

Oxidation: $4\ Ag(s) \longrightarrow 4\ Ag^+(aq) + 4\ e^-$ $\quad E°_{red} = +0.80\ \text{V}$

Even though the second half-reaction has 4 Ag, we use the $E°_{red}$ value directly from Table 20.1 because emf is an intensive property.

Using Equation 20.10, we have

$$E° = (1.23\ \text{V}) - (0.80\ \text{V}) = 0.43\ \text{V}$$

The half-reactions show the transfer of four electrons. Thus, for this reaction $n = 4$. We now use Equation 20.12 to calculate $\Delta G°$:

$$\begin{aligned}\Delta G° &= -nFE° \\ &= -(4)(96{,}485\ \text{J/V-mol})(+0.43)\ \text{V} \\ &= -1.7 \times 10^5\ \text{J/mol} = -170\ \text{kJ/mol}\end{aligned}$$

The positive value of $E°$ leads to a negative value of $\Delta G°$.

Now we need to calculate the equilibrium constant, K, using $\Delta G^\circ = -RT \ln K$. Because ΔG° is a large negative number, which means the reaction is thermodynamically very favorable, we expect K to be large.

$$\Delta G^\circ = -RT \ln K$$

$$-1.7 \times 10^5 \text{ J/mol} = -(8.314 \text{ J/K mol})(298 \text{ K}) \ln K$$

$$\ln K = \frac{-1.7 \times 10^5 \text{ J/mol}}{-(8.314 \text{ J/K mol})(298 \text{ K})}$$

$$\ln K = 69$$

$$K = 9 \times 10^{29}$$

K is indeed very large! This means that we expect silver metal to oxidize in acidic environments, in air, to Ag^+. Notice that the voltage calculated for the reaction was 0.43 V, which is easy to measure. Directly measuring such a large equilibrium constant by measuring reactant and product concentrations at equilibrium, on the other hand, would be very difficult.

(b) The overall equation is the same as that in part (a), multiplied by $\frac{1}{2}$. The half-reactions are

Reduction: $\frac{1}{2} O_2(g) + 2 H^+(aq) + 2 e^- \longrightarrow H_2O(l)$

Oxidation: $2 Ag(s) \longrightarrow 2 Ag^+(aq) + 2 e^- \quad E^\circ_{red} = +0.80 \text{ V}$

The values of E°_{red} are the same as they were in part (a); they are not changed by multiplying the half-reactions by $\frac{1}{2}$. Thus, E° has the same value as in part (a):

$$E^\circ = +0.43 \text{ V}$$

Notice, though, that the value of n has changed to $n = 2$, which is $\frac{1}{2}$ the value in part (a). Thus, ΔG° is half as large as in part (a).

$$\Delta G^\circ = -(2)(96{,}485 \text{ J/V mol})(+0.43 \text{ V}) = -83 \text{ kJ/mol}$$

Now we can calculate K as before:

$$-8.3 \times 10^4 \text{ J/mol} = -(8.314 \text{ J/K mol})(298 \text{ K}) \ln K$$

$$K = 3.5 \times 10^{14}$$

Comment: E° is an *intensive* quantity, so multiplying a chemical equation by a certain factor will not affect the value of E°. Multiplying an equation will change the value of n, however, and hence the value of ΔG°. The change in free energy, in units of J/mol of reaction as written, is an *extensive* quantity. The equilibrium constant is also an extensive quantity.

PRACTICE EXERCISE

For the reaction

$$3 Ni^{2+}(aq) + 2 Cr(OH)_3(s) + 10 OH^-(aq) \longrightarrow 3 Ni(s) + 2 CrO_4^{2-}(aq) + 8 H_2O(l)$$

(a) What is the value of n? **(b)** Use the data in Appendix E to calculate ΔG°. **(c)** Calculate K at $T = 298$ K.
Answers: **(a)** 6, **(b)** +87 kJ/mol, **(c)** $K = 6 \times 10^{-16}$

20.6 | Cell EMF Under Nonstandard Conditions

We have seen how to calculate the emf of a cell when the reactants and products are under standard conditions. As a voltaic cell is discharged, however, the reactants of the reaction are consumed and the products are generated, so the concentrations of these substances change. The emf progressively drops until $E = 0$, at which point we say the cell is "dead." At that point the concentrations of the reactants and products cease to change; they are at equilibrium. In this section we will examine how the cell emf depends on the concentrations of the reactants and products of the cell reaction. The emf generated under nonstandard conditions can be calculated by using an equation first derived by Walther Nernst (1864–1941), a German chemist who established many of the theoretical foundations of electrochemistry.

The Nernst Equation

The dependence of the cell emf on concentration can be obtained from the dependence of the free-energy change on concentration. ∞ (Section 19.7) Recall that the free-energy change, ΔG, is related to the standard free-energy change, ΔG°:

$$\Delta G = \Delta G^\circ + RT \ln Q \qquad [20.13]$$

The quantity Q is the reaction quotient, which has the form of the equilibrium-constant expression except that the concentrations are those that exist in the reaction mixture at a given moment. ∞ (Section 15.5)

Substituting $\Delta G = -nFE$ (Equation 20.11) into Equation 20.13 gives

$$-nFE = -nFE^\circ + RT \ln Q$$

Solving this equation for E gives the **Nernst equation**:

ACTIVITY
Nernst Equation

$$E = E^\circ - \frac{RT}{nF} \ln Q \qquad [20.14]$$

This equation is customarily expressed in terms of common (base 10) logarithms, which is related to natural logarithms by a factor of 2.303:

$$E = E^\circ - \frac{2.303\ RT}{nF} \log Q \qquad [20.15]$$

At $T = 298$ K the quantity $2.303\ RT/F$ equals 0.0592, with units of volts (V), and so the equation simplifies to

$$E = E^\circ - \frac{0.0592}{n} \log Q \qquad (T = 298\ \text{K}) \qquad [20.16]$$

We can use this equation to find the emf produced by a cell under nonstandard conditions or to determine the concentration of a reactant or product by measuring the emf of the cell.

To show how Equation 20.16 might be used, consider the following reaction, which we discussed earlier:

$$\text{Zn}(s) + \text{Cu}^{2+}(aq) \longrightarrow \text{Zn}^{2+}(aq) + \text{Cu}(s)$$

In this case $n = 2$ (two electrons are transferred from Zn to Cu^{2+}) and the standard emf is +1.10 V. Thus, at 298 K the Nernst equation gives

$$E = 1.10\ \text{V} - \frac{0.0592\ \text{V}}{2} \log \frac{[\text{Zn}^{2+}]}{[\text{Cu}^{2+}]} \qquad [20.17]$$

Recall that pure solids are excluded from the expression for Q. ∞ (Section 15.5) According to Equation 20.17, the emf increases as $[Cu^{2+}]$ increases and as $[Zn^{2+}]$ decreases. For example, when $[Cu^{2+}]$ is 5.0 M and $[Zn^{2+}]$ is 0.050 M, we have

$$E = 1.10\ \text{V} - \frac{0.0592\ \text{V}}{2} \log\left(\frac{0.050}{5.0}\right)$$

$$= 1.10\ \text{V} - \frac{0.0592\ \text{V}}{2}(-2.00) = 1.16\ \text{V}$$

Thus, increasing the concentration of the reactant (Cu^{2+}) and decreasing the concentration of the product (Zn^{2+}) relative to standard conditions increases the emf of the cell ($E = +1.16$ V) relative to standard conditions ($E^\circ = +1.10$ V). We could have anticipated this result by applying Le Châtelier's principle. ∞ (Section 15.6)

In general, if the concentrations of reactants increase relative to those of products, the emf increases. Conversely, if the concentrations of products increase relative to reactants, the emf decreases. As a voltaic cell operates, reactants are converted into products, which increases the value of Q and decreases the emf. The Nernst equation helps us understand why the emf of a voltaic cell drops as the cell discharges: As reactants are converted to products, the value of Q increases, so the value of E decreases, eventually reaching $E = 0$. Because $\Delta G = -nFE$ (Equation 20.11), it follows that $\Delta G = 0$ when $E = 0$. Recall that a system is at equilibrium when $\Delta G = 0$. ∞ (Section 19.7) Thus, when $E = 0$, the cell reaction has reached equilibrium, and no net reaction is occurring.

SAMPLE EXERCISE 20.11 | Voltaic Cell EMF Under Nonstandard Conditions

Calculate the emf at 298 K generated by the cell described in Sample Exercise 20.4 when $[Cr_2O_7^{2-}] = 2.0\ M$, $[H^+] = 1.0\ M$, $[I^-] = 1.0\ M$, and $[Cr^{3+}] = 1.0 \times 10^{-5}\ M$.

$$Cr_2O_7^{2-}(aq) + 14\,H^+(aq) + 6\,I^-(aq) \longrightarrow 2\,Cr^{3+}(aq) + 3\,I_2(s) + 7\,H_2O(l)$$

Solution

Analyze: We are given a chemical equation for a voltaic cell and the concentrations of reactants and products under which it operates. We are asked to calculate the emf of the cell under these nonstandard conditions.

Plan: To calculate the emf of a cell under nonstandard conditions, we use the Nernst equation in the form of Equation 20.16.

Solve: We first calculate $E°$ for the cell from standard reduction potentials (Table 20.1 or Appendix E). The standard emf for this reaction was calculated in Sample Exercise 20.6: $E° = 0.79$ V. As you will see if you refer back to that exercise, the balanced equation shows six electrons transferred from reducing agent to oxidizing agent, so $n = 6$. The reaction quotient, Q, is

$$Q = \frac{[Cr^{3+}]^2}{[Cr_2O_7^{2-}][H^+]^{14}[I^-]^6} = \frac{(1.0 \times 10^{-5})^2}{(2.0)(1.0)^{14}(1.0)^6} = 5.0 \times 10^{-11}$$

Using Equation 20.16, we have

$$E = 0.79\text{ V} - \frac{0.0592\text{ V}}{6}\log(5.0 \times 10^{-11})$$

$$= 0.79\text{ V} - \frac{0.0592\text{ V}}{6}(-10.30)$$

$$= 0.79\text{ V} + 0.10\text{ V} = 0.89\text{ V}$$

Check: This result is qualitatively what we expect: Because the concentration of $Cr_2O_7^{2-}$ (a reactant) is greater than 1 M and the concentration of Cr^{3+} (a product) is less than 1 M, the emf is greater than $E°$. Q is about 10^{-10}, so $\log Q$ is about -10. Thus, the correction to $E°$ is about $0.06 \times (10)/6$, which is 0.1, in agreement with the more detailed calculation.

PRACTICE EXERCISE

Calculate the emf generated by the cell described in the practice exercise accompanying Sample Exercise 20.6 when $[Al^{3+}] = 4.0 \times 10^{-3}\ M$ and $[I^-] = 0.010\ M$.
Answer: $E = +2.36$ V

SAMPLE EXERCISE 20.12 | Concentrations in a Voltaic Cell

If the voltage of a $Zn-H^+$ cell (like that in Figure 20.11) is 0.45 V at 25°C when $[Zn^{2+}] = 1.0\ M$ and $P_{H_2} = 1.0$ atm, what is the concentration of H^+?

Solution

Analyze: We are given a description of a voltaic cell, its emf, and the concentrations of all reactants and products except H^+, which we are asked to calculate.

Plan: First, we write the equation for the cell reaction and use standard reduction potentials from Table 20.1 to calculate $E°$ for the reaction. After determining the value of n from our reaction equation, we solve the Nernst equation for Q. Finally, we use the equation for the cell reaction to write an expression for Q that contains $[H^+]$ to determine $[H^+]$.

Solve: The cell reaction is

$$Zn(s) + 2\,H^+(aq) \longrightarrow Zn^{2+}(aq) + H_2(g)$$

The standard emf is

$$E^\circ = E^\circ_{red}\,(\text{reduction}) - E^\circ_{red}\,(\text{oxidation})$$
$$= 0\text{ V} - (-0.76\text{ V}) = +0.76\text{ V}$$

Because each Zn atom loses two electrons,

$$n = 2$$

Using Equation 20.16, we can solve for Q:

$$0.45\text{ V} = 0.76\text{ V} - \frac{0.0592\text{ V}}{2}\log Q$$

$$\log Q = (0.76\text{ V} - 0.45\text{ V})\left(\frac{2}{0.0592\text{ V}}\right) = 10.47$$

$$Q = 10^{10.47} = 3.0 \times 10^{10}$$

Q has the form of the equilibrium constant for the reaction

$$Q = \frac{[Zn^{2+}]P_{H_2}}{[H^+]^2} = \frac{(1.0)(1.0)}{[H^+]^2} = 3.0 \times 10^{10}$$

Solving for $[H^+]$, we have

$$[H^+]^2 = \frac{1.0}{3.0 \times 10^{10}} = 3.3 \times 10^{-11}$$

$$[H^+] = \sqrt{3.3 \times 10^{-11}} = 5.8 \times 10^{-6}\,M$$

Comment: A voltaic cell whose cell reaction involves H^+ can be used to measure $[H^+]$ or pH. A pH meter is a specially designed voltaic cell with a voltmeter calibrated to read pH directly. (Section 16.4)

PRACTICE EXERCISE

What is the pH of the solution in the cathode compartment of the cell pictured in Figure 20.11 when $P_{H_2} = 1.0$ atm, $[Zn^{2+}]$ in the anode compartment is 0.10 *M*, and cell emf is 0.542 V?
Answer: pH = 4.19

Concentration Cells

In each of the voltaic cells that we have looked at thus far, the reactive species at the anode has been different from the one at the cathode. Cell emf depends on concentration, however, so a voltaic cell can be constructed using the *same* species in both the anode and cathode compartments as long as the concentrations are different. A cell based solely on the emf generated because of a difference in a concentration is called a **concentration cell.**

A diagram of a concentration cell is shown in Figure 20.16(a) ▼. One compartment consists of a strip of nickel metal immersed in a 1.00 *M* solution of $Ni^{2+}(aq)$. The other compartment also has an Ni(*s*) electrode, but it is immersed in a 1.00×10^{-3} *M* solution of $Ni^{2+}(aq)$. The two compartments are connected by

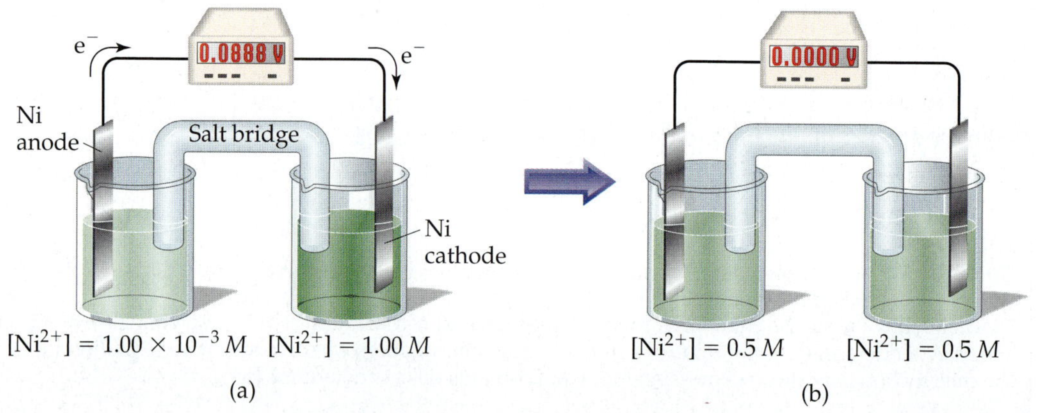

▼ **Figure 20.16 Concentration cell based on the Ni^{2+}-Ni cell reaction.** In (a) the concentrations of $Ni^{2+}(aq)$ in the two compartments are unequal, and the cell generates an electrical current. The cell operates until the concentrations of $Ni^{2+}(aq)$ in the two compartments become equal, (b) at which point the cell has reached equilibrium and is "dead."

a salt bridge and by an external wire running through a voltmeter. The half-cell reactions are the reverse of each other:

Anode: $Ni(s) \longrightarrow Ni^{2+}(aq) + 2\,e^-$ $\quad E^\circ_{red} = -0.28\text{ V}$

Cathode: $Ni^{2+}(aq) + 2\,e^- \longrightarrow Ni(s)$ $\quad E^\circ_{red} = -0.28\text{ V}$

Although the *standard* emf for this cell is zero, $E^\circ_{cell} = E^\circ_{red}$ (cathode) $-$ E°_{red} (anode) $= (-0.28\text{ V}) - (-0.28\text{ V}) = 0\text{ V}$, the cell operates under *nonstandard* conditions because the concentration of $Ni^{2+}(aq)$ is different in the two compartments. In fact, the cell will operate until the concentrations of Ni^{2+} in both compartments are equal. Oxidation of Ni(*s*) occurs in the half-cell containing the more dilute solution, thereby increasing the concentration of $Ni^{2+}(aq)$. It is therefore the anode compartment of the cell. Reduction of $Ni^{2+}(aq)$ occurs in the half-cell containing the more concentrated solution, thereby decreasing the concentration of $Ni^{2+}(aq)$, making it the cathode compartment. The *overall* cell reaction is therefore

Anode: $Ni(s) \longrightarrow Ni^{2+}(aq,\text{ dilute}) + 2\,e^-$

Cathode: $Ni^{2+}(aq,\text{ concentrated}) + 2\,e^- \longrightarrow Ni(s)$

Overall: $Ni^{2+}(aq,\text{ concentrated}) \longrightarrow Ni^{2+}(aq,\text{ dilute})$

We can calculate the emf of a concentration cell by using the Nernst equation. For this particular cell, we see that $n = 2$. The expression for the reaction quotient for the overall reaction is $Q = [Ni^{2+}]_{dilute}/[Ni^{2+}]_{concentrated}$. Thus, the emf at 298 K is

$$E = E^\circ - \frac{0.0592}{n}\log Q$$

$$= 0 - \frac{0.0592}{2}\log\frac{[Ni^{2+}]_{dilute}}{[Ni^{2+}]_{concentrated}} = -\frac{0.0592}{2}\log\frac{1.00 \times 10^{-3}\,M}{1.00\,M}$$

$$= +0.0888\text{ V}$$

This concentration cell generates an emf of nearly 0.09 V even though $E^\circ = 0$. The difference in concentration provides the driving force for the cell. When the concentrations in the two compartments become the same, the value of $Q = 1$ and $E = 0$.

The idea of generating a potential by a difference in concentration is the basis for the operation of pH meters (Figure 16.6). It is also a critical aspect in biology; nerve cells in the brain generate a voltage across the cell membrane by having different concentrations of ions on either side of the membrane. The regulation of the heartbeat in mammals, as discussed in the "Chemistry and Life" box in this section, is another example of the importance of electrochemistry to living organisms.

SAMPLE EXERCISE 20.13 | pH of a Concentration Cell

A voltaic cell is constructed with two hydrogen electrodes. Electrode 1 has $P_{H_2} = 1.00$ atm and an unknown concentration of $H^+(aq)$. Electrode 2 is a standard hydrogen electrode ($[H^+] = 1.00\,M$, $P_{H_2} = 1.00$ atm). At 298 K the measured cell voltage is 0.211 V, and the electrical current is observed to flow from electrode 1 through the external circuit to electrode 2. Calculate $[H^+]$ for the solution at electrode 1. What is its pH?

Solution

Analyze: We are given the voltage of a concentration cell and the direction in which the current flows. We also have the concentrations of all reactants and products except for $[H^+]$ in half-cell 1, which is our unknown.

Plan: We can use the Nernst equation to determine Q and then use Q to calculate the unknown concentration. Because this is a concentration cell, $E^\circ_{cell} = 0\text{ V}$.

Solve: Using the Nernst equation, we have

$$0.211\text{ V} = 0 - \frac{0.0592\text{ V}}{2}\log Q$$

$$\log Q = -(0.211\text{ V})\left(\frac{2}{0.0592\text{ V}}\right) = -7.13$$

$$Q = 10^{-7.13} = 7.4 \times 10^{-8}$$

CHEMISTRY AND LIFE | Heartbeats and Electrocardiography

The human heart is a marvel of efficiency and dependability. In a typical day an adult's heart pumps more than 7000 L of blood through the circulatory system, usually with no maintenance required beyond a sensible diet and lifestyle. We generally think of the heart as a mechanical device, a muscle that circulates blood via regularly spaced muscular contractions. However, more than two centuries ago, two pioneers in electricity, Luigi Galvani (1729–1787) and Alessandro Volta (1745–1827), discovered that the contractions of the heart are controlled by electrical phenomena, as are nerve impulses throughout the body. The pulses of electricity that cause the heart to beat result from a remarkable combination of electrochemistry and the properties of semipermeable membranes. (Section 13.5)

Cell walls are membranes with variable permeability with respect to a number of physiologically important ions (especially Na^+, K^+, and Ca^{2+}). The concentrations of these ions are different for the fluids inside the cells (the *intracellular fluid*, or ICF) and outside the cells (the *extracellular fluid*, or ECF). In cardiac muscle cells, for example, the concentrations of K^+ in the ICF and ECF are typically about 135 millimolar (m*M*) and 4 m*M*, respectively. For Na^+, however, the concentration difference between the ICF and ECF is opposite that for K^+; typically, $[Na^+]_{ICF} = 10\ mM$ and $[Na^+]_{ECF} = 145\ mM$.

The cell membrane is initially permeable to K^+ ions, but much less so to Na^+ and Ca^{2+}. The difference in concentration of K^+ ions between the ICF and ECF generates a concentration cell: Even though the same ions are present on both sides of the membrane, there is a potential difference between the two fluids that we can calculate using the Nernst equation with $E° = 0$. At physiological temperature (37°C) the potential in millivolts for moving K^+ from the ECF to the ICF is

$$E = E° - \frac{2.30\ RT}{nF} \log \frac{[K^+]_{ICF}}{[K^+]_{ECF}}$$

$$= 0 - (61.5\ \text{mV}) \log\left(\frac{135\ mM}{4\ mM}\right) = -94\ \text{mV}$$

In essence, the interior of the cell and the ECF together serve as a voltaic cell. The negative sign for the potential indicates that work is required to move K^+ into the intracellular fluid.

Changes in the relative concentrations of the ions in the ECF and ICF lead to changes in the emf of the voltaic cell. The cells of the heart that govern the rate of heart contraction are called the *pacemaker cells*. The membranes of the cells regulate the concentrations of ions in the ICF, allowing them to change in a systematic way. The concentration changes cause the emf to

---

Because electrons flow from electrode 1 to electrode 2, electrode 1 is the anode of the cell and electrode 2 is the cathode. The electrode reactions are therefore as follows, with the concentration of $H^+(aq)$ in electrode 1 represented with the unknown *x*:

Electrode 1: $H_2(g, 1.00\ \text{atm}) \longrightarrow 2\,H^+(aq, x\ M) + 2\,e^-$ $\quad E°_{red} = 0\ V$

Electrode 2: $2\,H^+(aq; 1.00\ M) + 2\,e^- \longrightarrow H_2(g, 1.00\ \text{atm})$ $\quad E°_{red} = 0\ V$

Thus,

$$Q = \frac{[H^+\ (\text{electrode 1})]^2\, P_{H_2}\ (\text{electrode 2})}{[H^+\ (\text{electrode 2})]^2\, P_{H_2}\ (\text{electrode 1})}$$

$$= \frac{x^2(1.00)}{(1.00)^2(1.00)} = x^2 = 7.4 \times 10^{-8}$$

$$x = \sqrt{7.4 \times 10^{-8}} = 2.7 \times 10^{-4}$$

At electrode 1, therefore,

$$[H^+] = 2.7 \times 10^{-4}\ M$$

and the pH of the solution is

$$pH = -\log[H^+] = -\log(2.7 \times 10^{-4}) = 3.6$$

Comment: The concentration of H^+ at electrode 1 is lower than that in electrode 2, which is why electrode 1 is the anode of the cell: The oxidation of H_2 to $H^+(aq)$ increases $[H^+]$ at electrode 1.

PRACTICE EXERCISE

A concentration cell is constructed with two $Zn(s)$-$Zn^{2+}(aq)$ half-cells. The first half-cell has $[Zn^{2+}] = 1.35\ M$, and the second half-cell has $[Zn^{2+}] = 3.75 \times 10^{-4}\ M$. **(a)** Which half-cell is the anode of the cell? **(b)** What is the emf of the cell?

Answers: **(a)** the second half-cell, **(b)** 0.105 V

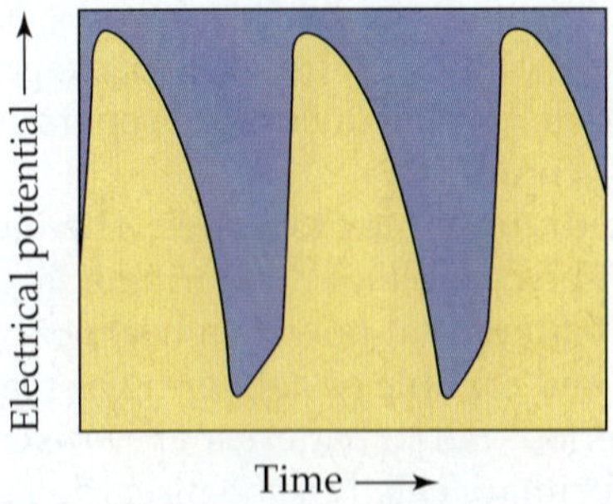

◀ **Figure 20.17 Ion concentration and emf in the human heart.** Variation of the electrical potential caused by changes of ion concentrations in the pacemaker cells of the heart.

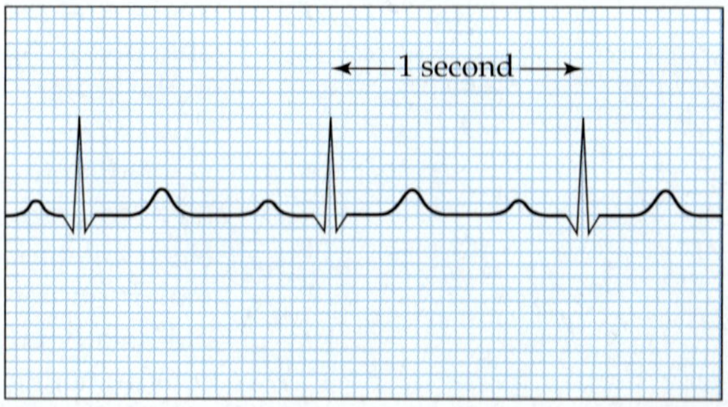

▲ **Figure 20.18 A typical electrocardiogram.** An electrocardiogram (ECG) printout records the electrical events monitored by electrodes attached to the body surface. The horizontal axis is time, and the vertical displacement is the emf.

change in a cyclic fashion, as shown in Figure 20.17 ▲. The emf cycle determines the rate at which the heart beats. If the pacemaker cells malfunction because of disease or injury, an artificial pacemaker can be surgically implanted. The artificial pacemaker is a small battery that generates the electrical pulses needed to trigger the contractions of the heart.

In the late 1800s scientists discovered that the electrical impulses that cause the contraction of the heart muscle are strong enough to be detected at the surface of the body. This observation formed the basis for *electrocardiography*, noninvasive monitoring of the heart by using a complex array of electrodes on the skin to measure voltage changes during heartbeats. A typical electrocardiogram is shown in Figure 20.18 ▲. It is quite striking that, although the heart's major function is the *mechanical* pumping of blood, it is most easily monitored by using the *electrical* impulses generated by tiny voltaic cells.

20.7 | Batteries and Fuel Cells

A **battery** is a portable, self-contained electrochemical power source that consists of one or more voltaic cells (Figure 20.19 ▼). For example, the common 1.5-V batteries used to power flashlights and many consumer electronic devices are single voltaic cells. Greater voltages can be achieved by using multiple voltaic cells in a single battery, as is the case in 12-V automotive batteries. When cells are connected in series (with the cathode of one attached to the anode of another), the battery produces a voltage that is the sum of the emfs of the individual cells. Higher emfs can also be achieved by using multiple batteries in series (Figure 20.20 ▶). The electrodes of batteries are marked following the convention of Figure 20.6—the cathode is labeled with a plus sign and the anode with a minus sign.

◀ **Figure 20.19 A battery of batteries.** Batteries are voltaic cells that serve as portable sources of electricity. Batteries vary markedly in size and in the electrochemical reaction used to generate electricity.

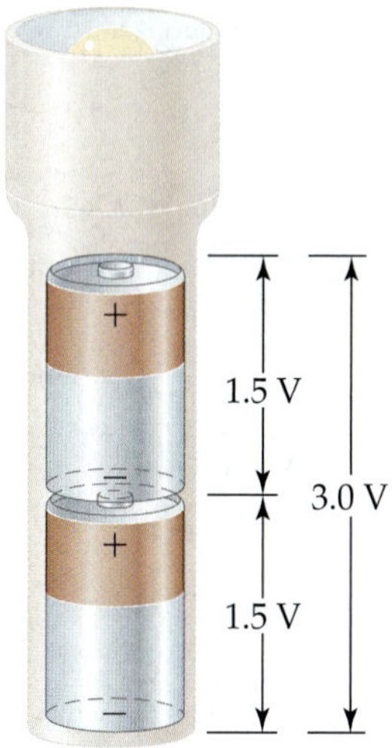

▲ **Figure 20.20 Combining batteries.** When batteries are connected in series, as in most flashlights, the total emf is the sum of the individual emfs.

Although any spontaneous redox reaction can serve as the basis for a voltaic cell, making a commercial battery that has specific performance characteristics can require considerable ingenuity. The substances that are oxidized at the anode and reduced by the cathode determine the emf of a battery, and the usable life of the battery depends on the quantities of these substances packaged in the battery. Usually the anode and the cathode compartments are separated by a barrier analogous to the porous barrier of Figure 20.6.

Different applications require batteries with different properties. The battery required to start a car, for example, must be capable of delivering a large electrical current for a short time period. The battery that powers a heart pacemaker, on the other hand, must be very small and capable of delivering a small but steady current over an extended time period. Some batteries are *primary* cells, meaning that they can't be recharged. A primary cell must be discarded or recycled after its emf drops to zero. A *secondary* cell can be recharged from an external power source after its emf has dropped.

In this section we will briefly discuss some common batteries. As we do so, notice how the principles we have discussed so far in this chapter help us understand these important sources of portable electrical energy.

Lead-Acid Battery

A 12-V lead-acid automotive battery consists of six voltaic cells in series, each producing 2 V. The cathode of each cell consists of lead dioxide (PbO_2) packed on a metal grid. The anode of each cell is composed of lead. Both electrodes are immersed in sulfuric acid. The electrode reactions that occur during discharge are

$$\text{Cathode:} \quad PbO_2(s) + HSO_4^-(aq) + 3\,H^+(aq) + 2\,e^- \longrightarrow PbSO_4(s) + 2\,H_2O(l)$$

$$\text{Anode:} \quad Pb(s) + HSO_4^-(aq) \longrightarrow PbSO_4(s) + H^+(aq) + 2\,e^-$$

$$PbO_2(s) + Pb(s) + 2\,HSO_4^-(aq) + 2\,H^+(aq) \longrightarrow 2\,PbSO_4(s) + 2\,H_2O(l)$$

The standard cell potential can be obtained from the standard reduction potentials in Appendix E:

$$E^\circ_{cell} = E^\circ_{red}\,(\text{cathode}) - E^\circ_{red}\,(\text{anode}) = (+1.685\text{ V}) - (-0.356\text{ V}) = +2.041\text{ V}$$

The reactants Pb and PbO_2 serve as the electrodes. Because the reactants are solids, there is no need to separate the cell into anode and cathode compartments; the Pb and PbO_2 cannot come into direct physical contact unless one electrode plate touches another. To keep the electrodes from touching, wood or glass-fiber spacers are placed between them (Figure 20.21 ◀).

▼ Figure 20.21 A 12-V automotive lead-acid battery. Each anode/cathode pair of electrodes in this schematic cutaway produces a potential of about 2 V. Six pairs of electrodes are connected in series, producing the desired battery voltage.

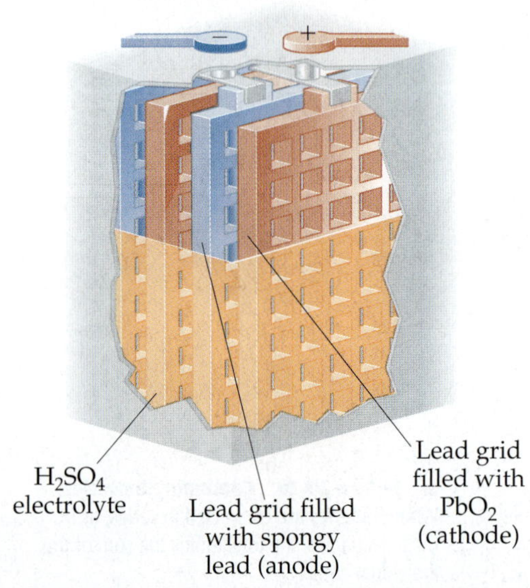

Using a reaction whose reactants and products are solids has another benefit. Because solids are excluded from the reaction quotient Q, the relative amounts of Pb(s), $PbO_2(s)$, and $PbSO_4(s)$ have no effect on the emf of the lead storage battery, helping the battery maintain a relatively constant emf during discharge. The emf does vary somewhat with use because the concentration of H_2SO_4 varies with the extent of cell discharge. As the equation for the overall cell reaction indicates, H_2SO_4 is consumed during the discharge.

One advantage of a lead-acid battery is that it can be recharged. During recharging, an external source of energy is used to reverse the direction of the overall cell reaction, regenerating Pb(s) and $PbO_2(s)$.

$$2\,PbSO_4(s) + 2\,H_2O(l) \longrightarrow PbO_2(s) + Pb(s) + 2\,HSO_4^-(aq) + 2\,H^+(aq)$$

In an automobile the energy necessary for recharging the battery is provided by the alternator, driven by the engine. Recharging is possible because $PbSO_4$ formed during discharge adheres to the electrodes. As the external source forces electrons from one electrode to another, the $PbSO_4$ is converted to Pb at one electrode and to PbO_2 at the other.

Alkaline Battery

The most common primary (nonrechargeable) battery is the alkaline battery. More than 10^{10} alkaline batteries are produced annually. The anode of this battery consists of powdered zinc metal immobilized in a gel in contact with a concentrated solution of KOH (hence the name *alkaline* battery). The cathode is a mixture of $MnO_2(s)$ and graphite, separated from the anode by a porous fabric. The battery is sealed in a steel can to reduce the risk of leakage of the concentrated KOH. A schematic view of an alkaline battery is shown in Figure 20.22 ►. The cell reactions are complex, but can be approximately represented as follows:

Cathode: $2\,MnO_2(s) + 2\,H_2O(l) + 2\,e^- \longrightarrow 2\,MnO(OH)(s) + 2\,OH^-(aq)$

Anode: $Zn(s) + 2\,OH^-(aq) \longrightarrow Zn(OH)_2(s) + 2\,e^-$

The emf of an alkaline battery is 1.55 V at room temperature. The alkaline battery provides far superior performance over the older "dry cells" that were also based on MnO_2 and Zn as the electrochemically active substances.

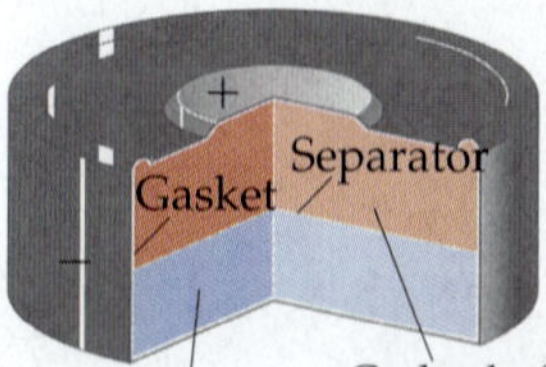

▲ **Figure 20.22 Cutaway view of a miniature alkaline battery.**

ACTIVITY
Batteries

Nickel-Cadmium, Nickel-Metal-Hydride, and Lithium-Ion Batteries

The tremendous growth in high-power-demand portable electronic devices, such as cell phones, notebook computers, and video recorders, has increased the demand for lightweight, readily recharged batteries. One of the most common rechargeable batteries is the nickel-cadmium (nicad) battery. During discharge, cadmium metal is oxidized at the anode of the battery while nickel oxyhydroxide [NiO(OH)(*s*)] is reduced at the cathode.

Cathode: $2\,NiO(OH)(s) + 2\,H_2O(l) + 2\,e^- \longrightarrow 2\,Ni(OH)_2(s) + 2\,OH^-(aq)$

Anode: $Cd(s) + 2\,OH^-(aq) \longrightarrow Cd(OH)_2(s) + 2\,e^-$

As in the lead-acid battery, the solid reaction products adhere to the electrodes, which permits the electrode reactions to be reversed during charging. A single nicad voltaic cell has an emf of 1.30 V. Nicad battery packs typically contain three or more cells in series to produce the higher emfs needed by most electronic devices.

There are drawbacks to nickel-cadmium batteries. Cadmium is a toxic heavy metal. Its use increases the weight of batteries and provides an environmental hazard—roughly 1.5 billion nickel-cadmium batteries are produced annually, and these must eventually be recycled as they lose their ability to be recharged. Some of these problems have been alleviated by the development of nickel-metal-hydride (NiMH) batteries. The cathode reaction of NiMH batteries is the same as that for nickel-cadmium batteries, but the anode reaction is very different. The anode consists of a metal *alloy*, such as $ZrNi_2$, that has the ability to absorb hydrogen atoms (we will discuss alloys in Section 23.6). During the oxidation at the anode, the hydrogen atoms lose electrons, and the resultant H^+ ions react with OH^- ions to form H_2O, a process that is reversed during charging. Hybrid gas-electric automobiles, which are powered by both a gasoline engine and an electric motor, use NiMH batteries to store the electrical power. The batteries are recharged by the electric motor while braking, and as a result, the batteries can last up to 8 years.

The newest rechargeable battery to receive large use in consumer electronic devices is the lithium-ion (Li-ion) battery. This is the battery you will find in cell phones and laptop computers. Because lithium is a very light element, Li-ion batteries achieve a greater *energy density*—the amount of energy stored per unit mass—than nickel-based batteries. The technology of Li-ion batteries is very different from that of the other batteries we have described here, and it is based on the ability of Li^+ ions to be inserted into and removed from certain layered solids. For example, Li^+ ions can be inserted reversibly into layers of graphite (Figure 11.41). In most commercial cells, one electrode is graphite or some other carbon-based material, and the other is usually made of lithium cobalt oxide ($LiCoO_2$). When charged, cobalt ions are oxidized and Li^+ ions migrate into the graphite. During discharge, when the battery is producing electricity for use, the Li^+ ions spontaneously migrate from the graphite anode to the cathode, enabling electrons to flow through the external circuit. An Li-ion battery produces a maximum voltage of 3.7 V, considerably higher than typical 1.5-V alkaline batteries.

Hydrogen Fuel Cells

The thermal energy released by the combustion of fuels can be converted to electrical energy. The heat may convert water to steam, which drives a turbine that in turn drives a generator. Typically, a maximum of only 40% of the energy from combustion is converted to electricity; the remainder is lost as heat. The direct production of electricity from fuels by a voltaic cell could, in principle, yield a higher rate of conversion of the chemical energy of the reaction. Voltaic cells that perform this conversion using conventional fuels, such as H_2 and CH_4, are called **fuel cells**. Strictly speaking, fuel cells are *not* batteries because they are not self-contained systems.

The most promising fuel-cell system involves the reaction of $H_2(g)$ and $O_2(g)$ to form $H_2O(l)$ as the only product. These cells can generate electricity twice as efficiently as the best internal combustion engine. Under acidic conditions, the electrode reactions are

Cathode: $O_2(g) + 4\,H^+ + 4\,e^- \longrightarrow 2\,H_2O(l)$

Anode: $2\,H_2(g) \longrightarrow 4\,H^+ + 4\,e^-$

Overall: $2\,H_2(g) + O_2(g) \longrightarrow 2\,H_2O(l)$

The standard emf of an H_2-O_2 fuel cell is +1.23 V, reflecting the large driving force for the reaction of H_2 and O_2 to form H_2O.

In this fuel cell (known as the PEM fuel cell, for "proton-exchange membrane") the anode and cathode are separated by a thin polymer membrane that is permeable to protons but not electrons. The polymer membrane therefore acts as a salt bridge. The electrodes are typically made from graphite. A PEM cell operates at a temperature of around 80°C. At this low temperature the electrochemical reactions would normally occur very slowly, and so they are catalyzed by a thin layer of platinum on each electrode.

Under basic conditions the electrode reactions in the hydrogen fuel cell are

Cathode: $4\,e^- + O_2(g) + 2\,H_2O(l) \longrightarrow 4\,OH^-(aq)$

Anode: $2\,H_2(g) + 4\,OH^-(aq) \longrightarrow 4\,H_2O(l) + 4\,e^-$

$2\,H_2(g) + O_2(g) \longrightarrow 2\,H_2O(l)$

▼ **Figure 20.23 A low-temperature H_2-O_2 fuel cell.** The porous membrane allows the H^+ ions generated by the oxidation of H_2 at the anode to migrate to the cathode, where H_2O is formed.

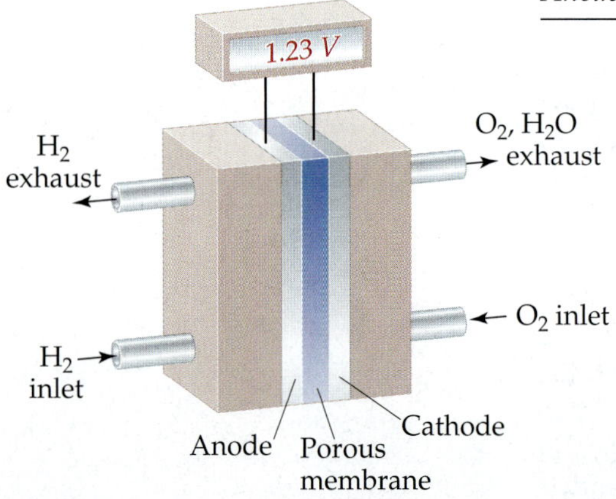

NASA has used this basic hydrogen fuel cell as the energy source for its spacecraft. Liquid hydrogen and oxygen are stored as fuel, and the product of the reaction, water, is drunk by the spacecraft crew.

A schematic drawing of a low-temperature H_2-O_2 fuel cell is shown in Figure 20.23 ◀. This technology is the basis for pollution-free fuel cell–powered vehicles ("Chemistry at Work" box, Section 1.4) that may be possible should the "hydrogen economy" come to pass. Currently a great deal of research is going into improving fuel cells. Much effort is being directed toward developing fuel cells that use conventional fuels such as hydrocarbons and alcohols, which are not as difficult to handle and distribute as hydrogen gas.

Direct Methanol Fuel Cells

The direct methanol fuel cell is similar to the PEM fuel cell, but instead of using hydrogen gas as a reactant , it uses methanol, CH_3OH. The reactions are

Cathode:	$\frac{3}{2}O_2(g) + 6\,H^+ + 6\,e^- \longrightarrow 3\,H_2O(g)$
Anode:	$CH_3OH(l) + H_2O(g) \longrightarrow CO_2(g) + 6\,H^+ + 6\,e^-$
Overall:	$CH_3OH(g) + \frac{3}{2}O_2(g) \longrightarrow CO_2(g) + 2\,H_2O(g)$

These cells operate at around 120°C, which is slightly higher than the temperature at which the standard PEM fuel cell operates. One shortcoming of the methanol cell is that the quantity of platinum catalyst needed is greater than in conventional PEM cells. Also, the carbon dioxide produced in the methanol reaction is not as environmentally friendly as water. However, liquid methanol is a far more attractive fuel to store and transport than hydrogen gas.

20.8 | Corrosion

Batteries are examples of how spontaneous redox reactions can be used productively. In this section we will examine the undesirable redox reactions that lead to the **corrosion** of metals. Corrosion reactions are spontaneous redox reactions in which a metal is attacked by some substance in its environment and converted to an unwanted compound.

For nearly all metals, oxidation is a thermodynamically favorable process in air at room temperature. When the oxidation process is not inhibited in some way, it can be very destructive. Oxidation can also form an insulating protective oxide layer, however, that prevents further reaction of the underlying metal. On the basis of the standard reduction potential for Al^{3+}, for example, we would expect aluminum metal to be very readily oxidized. The many aluminum soft-drink and beer cans that litter the environment are ample evidence, however, that aluminum undergoes only very slow chemical corrosion. The exceptional stability of this active metal in air is due to the formation of a thin protective coat of oxide—a hydrated form of Al_2O_3—on the surface of the metal. The oxide coat is impermeable to O_2 or H_2O and so protects the underlying metal from further corrosion. Magnesium metal is similarly protected. Some metal alloys, such as stainless steel, likewise form protective impervious oxide coats. The semiconductor silicon, as we saw in Chapter 12, also readily forms a protective SiO_2 coating that is important to its use in electronic circuits.

Corrosion of Iron

The rusting of iron (Figure 20.24 ▶) is a familiar corrosion process that carries a significant economic impact. It is estimated that up to 20% of the iron produced annually in the United States is used to replace iron objects that have been discarded because of rust damage.

The rusting of iron requires both oxygen and water. Other factors—such as the pH of the solution, the presence of salts, contact with metals more difficult to oxidize than iron, and stress on the iron—can accelerate rusting.

The corrosion of iron is electrochemical in nature. Not only does the corrosion process involve oxidation and reduction, the metal itself conducts electricity. Thus, electrons can move through the metal from a region where oxidation occurs to another region where reduction occurs, as in voltaic cells.

▲ **Figure 20.24 Corrosion.** The corrosion of iron is an electrochemical process of great economic importance. The annual cost of metallic corrosion in the United States is estimated to be $70 billion.

Because the standard reduction potential for the reduction of $Fe^{2+}(aq)$ is less positive than that for the reduction of O_2, $Fe(s)$ can be oxidized by $O_2(g)$.

Cathode:	$O_2(g) + 4\,H^+(aq) + 4\,e^- \longrightarrow 2\,H_2O(l)$	$E^\circ_{red} = 1.23\ V$
Anode:	$Fe(s) \longrightarrow Fe^{2+}(aq) + 2\,e^-$	$E^\circ_{red} = -0.44V$

A portion of the iron can serve as an anode at which the oxidation of Fe to Fe^{2+} occurs. The electrons produced migrate through the metal to another portion of the surface that serves as the cathode, at which O_2 is reduced. The reduction of O_2 requires H^+, so lowering the concentration of H^+ (increasing the pH) makes the reduction of O_2 less favorable. Iron in contact with a solution whose pH is greater than 9 does not corrode.

The Fe^{2+} formed at the anode is eventually oxidized further to Fe^{3+}, which forms the hydrated iron(III) oxide known as rust*:

ACTIVITY Prevention of Corrosion

$$4\,Fe^{2+}(aq) + O_2(g) + 4\,H_2O(l) + 2\,xH_2O(l) \longrightarrow 2\,Fe_2O_3{\cdot}xH_2O(s) + 8\,H^+(aq)$$

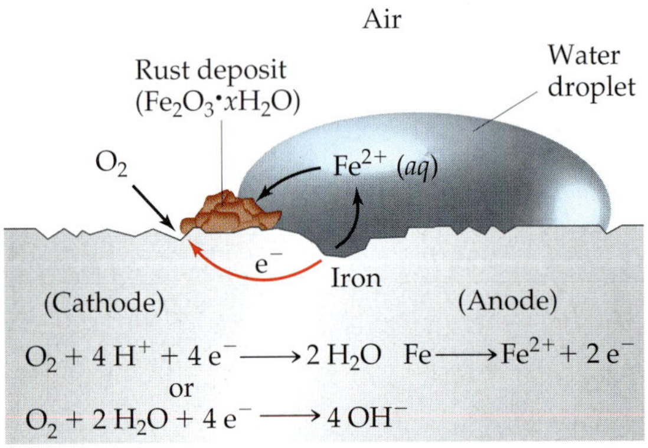

▲ **Figure 20.25 Corrosion of iron in contact with water.**

Because the cathode is generally the area having the largest supply of O_2, rust often deposits there. If you look closely at a shovel after it has stood outside in the moist air with wet dirt adhered to its blade, you may notice that pitting has occurred under the dirt but that rust has formed elsewhere, where O_2 is more readily available. The corrosion process is summarized in Figure 20.25 ◀.

The enhanced corrosion caused by the presence of salts is usually evident on autos in areas where roads are heavily salted during winter. Like a salt bridge in a voltaic cell, the ions of the salt provide the electrolyte necessary to complete the electrical circuit.

Preventing the Corrosion of Iron

Iron is often covered with a coat of paint or another metal such as tin or zinc to protect its surface against corrosion. Covering the surface with paint or tin is simply a means of preventing oxygen and water from reaching the iron surface. If the coating is broken and the iron is exposed to oxygen and water, corrosion will begin.

Galvanized iron, which is iron coated with a thin layer of zinc, uses the principles of electrochemistry to protect the iron from corrosion even after the surface coat is broken. The standard reduction potentials for iron and zinc are

$$Fe^{2+}(aq) + 2\,e^- \longrightarrow Fe(s) \qquad E^\circ_{red} = -0.44\text{ V}$$

$$Zn^{2+}(aq) + 2\,e^- \longrightarrow Zn(s) \qquad E^\circ_{red} = -0.76\text{ V}$$

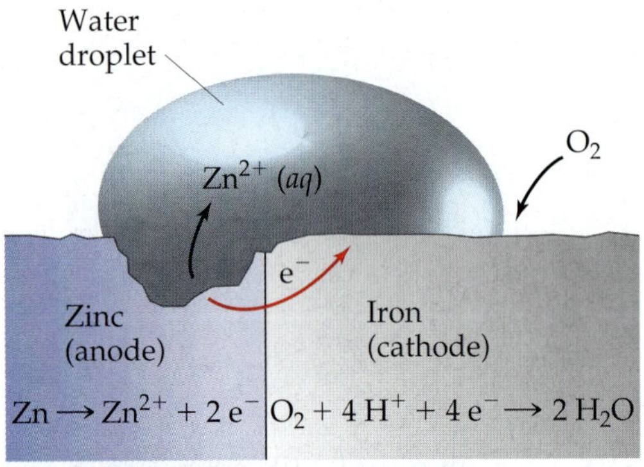

▲ **Figure 20.26 Cathodic protection of iron in contact with zinc.**

Because the E°_{red} value for the reduction of Fe^{2+} is less negative (more positive) than that for the reduction of Zn^{2+}, Fe^{2+} is easier to reduce than Zn^{2+}. Conversely, Zn(*s*) is easier to oxidize than Fe(*s*). Thus, even if the zinc coating is broken and the galvanized iron is exposed to oxygen and water, the zinc, which is most easily oxidized, serves as the anode and is corroded instead of the iron. The iron serves as the cathode at which O_2 is reduced, as shown in Figure 20.26 ◀.

Protecting a metal from corrosion by making it the cathode in an electrochemical cell is known as **cathodic protection**. The metal that is oxidized while protecting the cathode is called the *sacrificial anode*. Underground pipelines are often protected against corrosion by making the pipeline the cathode of a voltaic cell. Pieces of an active metal such as magnesium are buried along the pipeline and connected to it by wire, as shown in Figure 20.27 ▶. In moist soil, where corrosion can occur, the active metal serves as the anode, and the pipe experiences cathodic protection.

* Frequently, metal compounds obtained from aqueous solution have water associated with them. For example, copper(II) sulfate crystallizes from water with five moles of water per mole of $CuSO_4$. We represent this formula as $CuSO_4{\cdot}5\,H_2O$. Such compounds are called hydrates. (Section 13.1) Rust is a hydrate of iron(III) oxide with a variable amount of water of hydration. We represent this variable water content by writing the formula as $Fe_2O_3{\cdot}xH_2O$.

GIVE IT SOME THOUGHT

Based on the standard reduction potentials in Table 20.1, which of the following metals could provide cathodic protection to iron: Al, Cu, Ni, Zn?

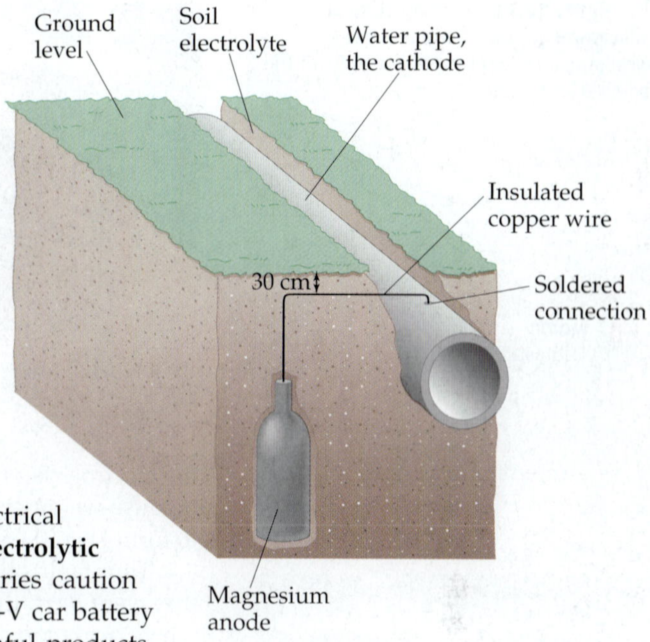

▲ **Figure 20.27 Cathodic protection of an iron water pipe.** The magnesium anode is surrounded by a mixture of gypsum, sodium sulfate, and clay to promote conductivity of ions. The pipe, in effect, is the cathode of a voltaic cell.

20.9 | Electrolysis

Voltaic cells are based on spontaneous oxidation-reduction reactions. Conversely, it is possible to use electrical energy to cause nonspontaneous redox reactions to occur. For example, electricity can be used to decompose molten sodium chloride into its component elements:

$$2\,NaCl(l) \longrightarrow 2\,Na(l) + Cl_2(g)$$

Such processes, which are driven by an outside source of electrical energy, are called **electrolysis reactions** and take place in **electrolytic cells.** This is the reason manufacturers of automotive batteries caution against immersing the battery in salt water—the standard 12-V car battery has more than enough electromotive force to produce harmful products, such as poisonous chlorine gas.

An electrolytic cell consists of two electrodes in a molten salt or a solution. A battery or some other source of direct electrical current acts as an electron pump, pushing electrons into one electrode and pulling them from the other. Just as in voltaic cells, the electrode at which the reduction occurs is called the cathode, and the electrode at which oxidation occurs is called the anode. In the electrolysis of molten NaCl, shown in Figure 20.28 ▼, Na^+ ions pick up electrons and are reduced to Na at the cathode. As the Na^+ ions near the cathode are depleted, additional Na^+ ions migrate in. Similarly, there is net movement of Cl^- ions to the anode, where they are oxidized. The electrode reactions for the electrolysis of molten NaCl are summarized as follows:

Cathode: $2\,Na^+(l) + 2\,e^- \longrightarrow 2\,Na(l)$

Anode: $2\,Cl^-(l) \longrightarrow Cl_2(g) + 2\,e^-$

$$2\,Na^+(l) + 2\,Cl^-(l) \longrightarrow 2\,Na(l) + Cl_2(g)$$

Notice the manner in which the voltage source is connected to the electrodes in Figure 20.28. In a voltaic cell (or any other source of direct current) the electrons move from the negative terminal (Figure 20.6). Thus, the electrode of the electrolytic cell that is connected to the negative terminal of the voltage source is the cathode of the cell; it receives electrons that are used to reduce a substance. The electrons that are removed during the oxidation process at the anode travel to the positive terminal of the voltage source, thus completing the circuit of the cell.

The electrolysis of molten salts is an important industrial process for the production of active metals such as sodium and aluminum. We will have more to say about it in Chapter 23, when we discuss how ores are refined into metals.

Because of the high melting points of ionic substances, the electrolysis of molten salts requires very high temperatures. (Section 11.8) Do we obtain the same products if we electrolyze the aqueous solution of a salt instead of the molten salt? Frequently the answer is no;

▼ **Figure 20.28 Electrolysis of molten sodium chloride.** Cl^- ions are oxidized to $Cl_2(g)$ at the anode, and Na^+ ions are reduced to $Na(l)$ at the cathode.

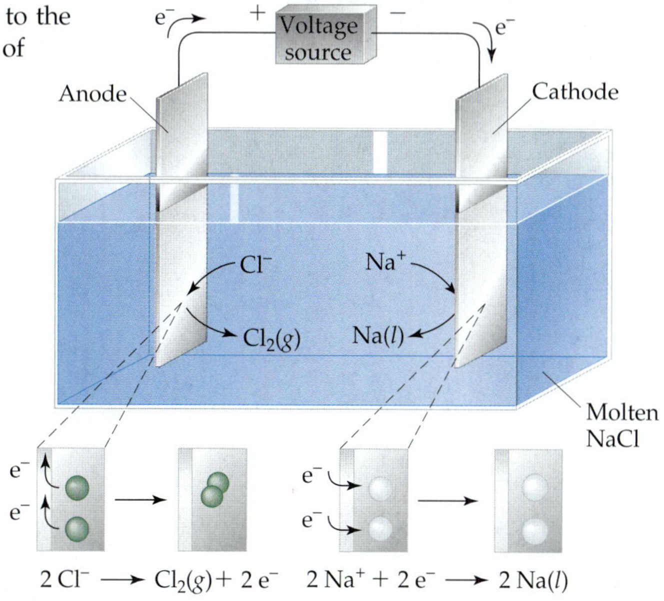

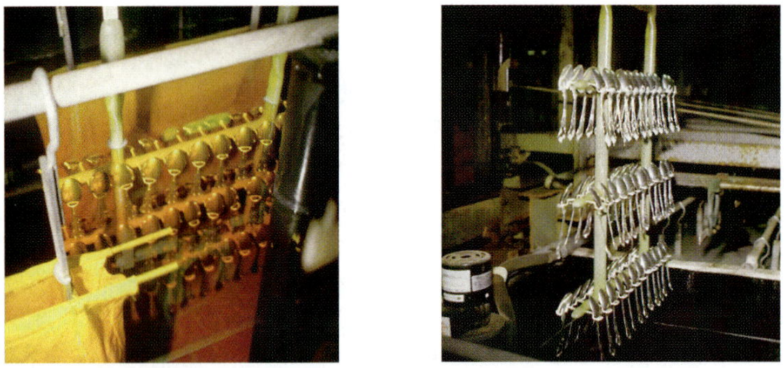

(a) (b)

▶ **Figure 20.29 Electroplating of silverware.** (a) The silverware is being withdrawn from the electroplating bath. (b) The polished final product.

MOVIE
Electroplating

the electrolysis of an aqueous solution is complicated by the presence of water, because we must consider whether the water is oxidized to form O_2 or reduced to form H_2 rather than the ions of the salt. These reactions also depend on pH.

Thus far in our discussion of electrolysis, we have encountered only electrodes that were *inert*; they did not undergo reaction but merely served as the surface where oxidation and reduction occurred. Several practical applications of electrochemistry, however, are based on *active* electrodes—electrodes that participate in the electrolysis process. *Electroplating*, for example, uses electrolysis to deposit a thin layer of one metal on another metal in order to improve beauty or resistance to corrosion (Figure 20.29 ▲). We can illustrate the principles of electrolysis with active electrodes by describing how to electroplate nickel on a piece of steel.

Figure 20.30 ◀ illustrates the electrolytic cell for our electroplating experiment. The anode of the cell is a strip of nickel metal, and the cathode is the piece of steel that will be electroplated. The electrodes are immersed in a solution of $NiSO_4(aq)$. What happens at the electrodes when the external voltage source is turned on? Reduction will occur at the cathode. The standard reduction potential of Ni^{2+} ($E^\circ_{red} = -0.28$ V) is less negative than that of H_2O ($E^\circ_{red} = -0.83$ V), so Ni^{2+} will be preferentially reduced at the cathode.

▲ **Figure 20.30 Electrolytic cell with an active metal electrode.** Nickel dissolves from the anode to form $Ni^{2+}(aq)$. At the cathode $Ni^{2+}(aq)$ is reduced and forms a nickel "plate" on the cathode.

At the anode we need to consider which substances can be oxidized. For the $NiSO_4(aq)$ solution, only the H_2O solvent is readily oxidized because neither Na^+ nor the $SO_4{}^{2-}$ can be oxidized (both already have their elements in their highest possible oxidation state). The Ni atoms in the anode, however, can undergo oxidation. Thus, the two possible oxidation processes are

$$2\,H_2O(l) \longrightarrow O_2(g) + 4\,H^+(aq) + 4\,e^- \qquad E^\circ_{red} = +1.23\text{ V}$$

$$Ni(s) \longrightarrow Ni^{2+}(aq) + 2\,e^- \qquad E^\circ_{red} = -0.28\text{ V}$$

where the potentials are the standard *reduction* potentials of these reactions. Because this is an oxidation, the half-reaction with the more negative value of E°_{red} is favored. We therefore expect $Ni(s)$ to be oxidized at the anode. Another thought process that leads to the same conclusion is to calculate the standard potentials of these two half-reactions as oxidations—which means we just flip the sign of the reduction potentials, as we discussed above. The potentials for the half-reactions as written would be -1.23 V and $+0.28$ V, respectively, and based on $\Delta G = -nFE$ we can see that the nickel oxidation should be spontaneous. We can summarize the electrode reactions as

Cathode (steel strip): $$Ni^{2+}(aq) + 2\,e^- \longrightarrow Ni(s) \qquad E^\circ_{red} = -0.28\text{ V}$$

Anode (nickel strip): $$Ni(s) \longrightarrow Ni^{2+}(aq) + 2\,e^- \qquad E^\circ_{red} = -0.28\text{ V}$$

If we look at the overall reaction, it appears as if nothing has been accomplished. During the electrolysis, however, we are transferring Ni atoms

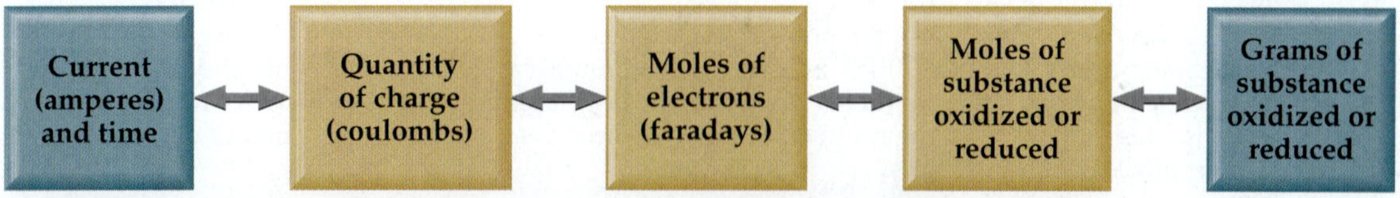

▲ **Figure 20.31 Relationship between charge and amount of reactant and product in electrolysis reactions.** This flowchart shows the steps relating the quantity of electrical charge used in electrolysis to the amounts of substances oxidized or reduced.

from the Ni anode to the steel cathode, plating the steel electrode with a thin layer of nickel atoms. The standard emf for the overall reaction is $E^\circ_{cell} = E^\circ_{red}\text{(cathode)} - E^\circ_{red}\text{(anode)} = 0\text{ V}$. Only a small emf is needed to provide the "push" to transfer the nickel atoms from one electrode to the other. In Chapter 23 we will explore further the utility of electrolysis with active electrodes as a means of purifying crude metals.

Quantitative Aspects of Electrolysis

The stoichiometry of a half-reaction shows how many electrons are needed to achieve an electrolytic process. For example, the reduction of Na^+ to Na is a one-electron process:

ACTIVITIES
Electrolysis, Electrolysis Calculation Example

$$Na^+ + e^- \longrightarrow Na$$

Thus, one mole of electrons will plate out 1 mol of Na metal, two moles of electrons will plate out 2 mol of Na metal, and so forth. Similarly, two moles of electrons are required to produce 1 mol of copper from Cu^{2+}, and three moles of electrons are required to produce 1 mol of aluminum from Al^{3+}.

$$Cu^{2+} + 2\,e^- \longrightarrow Cu$$

$$Al^{3+} + 3\,e^- \longrightarrow Al$$

For any half-reaction, the amount of a substance that is reduced or oxidized in an electrolytic cell is directly proportional to the number of electrons passed into the cell.

The quantity of charge passing through an electrical circuit, such as that in an electrolytic cell, is generally measured in *coulombs*. As noted in Section 20.5, the charge on one mole of electrons is 96,485 C (1 faraday). A coulomb is the quantity of charge passing a point in a circuit in 1 s when the current is 1 ampere (A). Therefore, the number of coulombs passing through a cell can be obtained by multiplying the amperage and the elapsed time in seconds.

$$\text{Coulombs} = \text{amperes} \times \text{seconds} \qquad [20.18]$$

Figure 20.31 ▲ shows how the quantities of substances produced or consumed in electrolysis are related to the quantity of electrical charge that is used. The same relationships can also be applied to voltaic cells. In other words, electrons can be thought of as reagents in electrolysis reactions.

SAMPLE EXERCISE 20.14 | Aluminum Electrolysis

Calculate the number of grams of aluminum produced in 1.00 h by the electrolysis of molten $AlCl_3$ if the electrical current is 10.0 A.

Solution

Analyze: We are told that $AlCl_3$ is electrolyzed to form Al and asked to calculate the number of grams of Al produced in 1.00 h with 10.0 A.

Plan: Figure 20.31 provides a road map of the problem. First, the product of the amperage and the time in seconds gives the number of coulombs of electrical charge being used (Equation 20.18). Second, the coulombs can be converted with the Faraday constant (F = 96,485 C/mole electrons) to tell us the number of moles of electrons being supplied. Third, reduction of 1 mol of Al^{3+} to Al requires three moles of electrons. Hence we can use the number of moles of electrons to calculate the number of moles of Al metal it produces. Finally, we convert moles of Al into grams.

Solve: First, we calculate the coulombs of electrical charge that are passed into the electrolytic cell:

$$\text{Coulombs} = \text{amperes} \times \text{seconds}$$

$$= (10.0\ \text{A})(1.00\ \text{h})\frac{(3600\ \text{s})}{\text{h}} = 3.60 \times 10^4\ \text{C}$$

Second, we calculate the number of moles of electrons that pass into the cell:

$$\text{Moles e}^- = (3.60 \times 10^4\ \text{C})\left(\frac{1\ \text{mol e}^-}{96{,}485\ \text{C}}\right) = 0.373\ \text{mol e}^-$$

Third, we relate the number of moles of electrons to the number of moles of aluminum being formed, using the half-reaction for the reduction of Al^{3+}:

$$Al^{3+} + 3\,e^- \longrightarrow Al$$

Thus, three moles of electrons (3 F of electrical charge) are required to form 1 mol of Al:

$$\text{Moles Al} = (0.373\ \text{mol e}^-)\left(\frac{1\ \text{mol Al}}{3\ \text{mol e}^-}\right) = 0.124\ \text{mol Al}$$

Finally, we convert moles to grams:

$$\text{Grams Al} = (0.124\ \text{mol Al})\left(\frac{27.0\ \text{g Al}}{1\ \text{mol Al}}\right) = 3.36\ \text{g Al}$$

Because each step involves a multiplication by a new factor, the steps can be combined into a single sequence of factors:

$$\text{Grams Al} = (3.60 \times 10^4\ \text{C})\left(\frac{1\ \text{mole e}^-}{96{,}485\ \text{C}}\right)\left(\frac{1\ \text{mol Al}}{3\ \text{mol e}^-}\right)\left(\frac{27.0\ \text{g Al}}{1\ \text{mol Al}}\right) = 3.36\ \text{g Al}$$

PRACTICE EXERCISE

(a) The half-reaction for formation of magnesium metal upon electrolysis of molten $MgCl_2$ is $Mg^{2+} + 2\,e^- \longrightarrow Mg$. Calculate the mass of magnesium formed upon passage of a current of 60.0 A for a period of 4.00×10^3 s. **(b)** How many seconds would be required to produce 50.0 g of Mg from $MgCl_2$ if the current is 100.0 A?
Answers: **(a)** 30.2 g of Mg, **(b)** 3.97×10^3 s

Electrical Work

We have already seen that a positive value of E is associated with a negative value for the free-energy change and, thus, with a spontaneous process. We also know that for any spontaneous process, ΔG is a measure of the maximum useful work, w_{max}, that can be extracted from the process: $\Delta G = w_{max}$. (Section 5.2) Because $\Delta G = -nFE$, the maximum useful electrical work obtainable from a voltaic cell is

$$w_{max} = -nFE \qquad [20.19]$$

The cell emf, E, is a positive number for a voltaic cell, so w_{max} will be a negative number for a voltaic cell. Work done *by* a system *on* its surroundings is indicated by a negative sign for w. (Section 5.2) Thus, the negative value for w_{max} means that a voltaic cell does work on its surroundings.

In an electrolytic cell we use an external source of energy to bring about a nonspontaneous electrochemical process. In this case ΔG is positive and E_{cell} is negative. To force the process to occur, we need to apply an external potential, E_{ext}, which must be larger in magnitude than E_{cell}: $E_{ext} > -E_{cell}$. For example, if a nonspontaneous process has $E_{cell} = -0.9$ V, then the external potential E_{ext} must be greater than 0.9 V in order for the process to occur.

When an external potential E_{ext} is applied to a cell, the surroundings are doing work on the system. The amount of work performed is given by

$$w = nFE_{ext} \qquad [20.20]$$

Unlike Equation 20.19, there is no minus sign in Equation 20.20. The work calculated in Equation 20.20 will be a positive number because the surroundings are doing work on the system. The quantity n in Equation 20.20 is the number of moles of electrons forced into the system by the external potential. The product $n \times F$ is the total electrical charge supplied to the system by the external source of electricity.

Electrical work can be expressed in energy units of watts times time. The **watt** (W) is a unit of electrical power (that is, the rate of energy expenditure).

$$1\ \text{W} = 1\ \text{J/s}$$

Thus, a watt-second is a joule. The unit employed by electric utilities is the kilowatt-hour (kWh), which equals 3.6×10^6 J.

$$1\ \text{kWh} = (1000\ \text{W})(1\ \text{hr})\left(\frac{3600\ \text{s}}{1\ \text{hr}}\right)\left(\frac{1\ \text{J/s}}{1\ \text{W}}\right) = 3.6 \times 10^6\ \text{J} \qquad [20.21]$$

Using these considerations, we can calculate the maximum work obtainable from the voltaic cells and the minimum work required to bring about desired electrolysis reactions.

SAMPLE EXERCISE 20.15 | Calculating Energy in Kilowatt-hours

Calculate the number of kilowatt-hours of electricity required to produce 1.0×10^3 kg of aluminum by electrolysis of Al^{3+} if the applied voltage is 4.50 V.

Solution

Analyze: We are given the mass of Al produced from Al^{3+} and the applied voltage and asked to calculate the energy, in kilowatt-hours, required for the reduction.

Plan: From the mass of Al, we can calculate first the number of moles of Al, then the number of coulombs required to obtain that mass. We can then use Equation 20.20, $w = nFE_{ext}$, where nF is the total charge in coulombs and E_{ext} is the applied potential, 4.50 V.

Solve: First, we need to calculate nF, the number of coulombs required:

$$\text{Coulombs} = (1.00 \times 10^3\ \text{kg Al})\left(\frac{1000\ \text{g Al}}{1\ \text{kg Al}}\right)\left(\frac{1\ \text{mol Al}}{27.0\ \text{g Al}}\right)\left(\frac{3\ \text{mol e}^-}{1\ \text{mol Al}}\right)\left(\frac{96{,}485\ \text{C}}{1\ \text{mole e}^-}\right)$$

$$= 1.07 \times 10^{10}\ \text{C}$$

We can now calculate w. In doing so, we must apply several conversion factors, including Equation 20.21, which gives the conversion between kilowatt-hours and joules:

$$\text{Kilowatt-hours} = (1.07 \times 10^{10}\ \text{C})(4.50\ \text{V})\left(\frac{1\ \text{J}}{1\ \text{C-V}}\right)\left(\frac{1\ \text{kWh}}{3.6 \times 10^6\ \text{J}}\right)$$

$$= 1.34 \times 10^4\ \text{kWh}$$

Comment: This quantity of energy does not include the energy used to mine, transport, and process the aluminum ore, and to keep the electrolysis bath molten during electrolysis. A typical electrolytic cell used to reduce aluminum ore to aluminum metal is only 40% efficient, with 60% of the electrical energy being dissipated as heat. It therefore requires on the order of 33 kWh of electricity to produce 1 kg of aluminum. The aluminum industry consumes about 2% of the electrical energy generated in the United States. Because this is used mainly to reduce aluminum, recycling this metal saves large quantities of energy.

PRACTICE EXERCISE

Calculate the number of kilowatt-hours of electricity required to produce 1.00 kg of Mg from electrolysis of molten $MgCl_2$ if the applied emf is 5.00 V. Assume that the process is 100% efficient.

Answer: 11.0 kWh

SAMPLE INTEGRATIVE EXERCISE | Putting Concepts Together

The K_{sp} at 298 K for iron(II) fluoride is 2.4×10^{-6}. **(a)** Write a half-reaction that gives the likely products of the two-electron reduction of $FeF_2(s)$ in water. **(b)** Use the K_{sp} value and the standard reduction potential of $Fe^{2+}(aq)$ to calculate the standard reduction potential for the half-reaction in part (a). **(c)** Rationalize the difference in the reduction potential for the half-reaction in part (a) with that for $Fe^{2+}(aq)$.

Solution

Analyze:

We are going to have to combine what we know about equilibrium constants and electrochemistry to obtain reduction potentials.

Plan:

For (a) we need to determine which ion, Fe^{2+} or F^-, is more likely to be reduced by 2 electrons and write the overall reaction for $FeF_2 + 2\,e^- \longrightarrow$? For (b) we need to write the K_{sp} reaction and manipulate it to get $E°$ for the reaction in (a). For (c) we need to see what we get for (a) and (b).

Solve: **(a)** Iron(II) fluoride is an ionic substance that consists of Fe^{2+} and F^- ions. We are asked to predict where two electrons could be added to FeF_2. We can't envision adding the electrons to the F^- ions to form F^{2-}, so it seems likely that we could reduce the Fe^{2+} ions to Fe(*s*). We therefore predict the half-reaction

$$FeF_2(s) + 2\,e^- \longrightarrow Fe(s) + 2\,F^-(aq)$$

(b) The K_{sp} value refers to the following equilibrium (Section 17.4):

$$FeF_2(s) \rightleftharpoons Fe^{2+}(aq) + 2\,F^-(aq) \qquad K_{sp} = [Fe^{2+}][F^-]^2 = 2.4 \times 10^{-6}$$

We were also asked to use the standard reduction potential of Fe^{2+}, whose half-reaction and standard voltage are listed in Appendix E:

$$Fe^{2+}(aq) + 2\,e^- \longrightarrow Fe(s) \qquad E° = -0.440\ \text{V}$$

Recall that according to Hess's law, we can add reactions to get the one we want and that we can add thermodynamic quantities like ΔH and ΔG to solve for the enthalpy or free energy of the reaction we want. (Section 5.6) In this case notice that if we add the K_{sp} reaction to the standard reduction half-reaction for Fe^{2+}, we get the half-reaction we want:

$$\begin{array}{lll} & 1. & FeF_2(s) \longrightarrow Fe^{2+}(aq) + 2\,F^-(aq) \\ & 2. & Fe^{2+}(aq) + 2\,e^- \longrightarrow Fe(s) \\ \hline \textit{Overall:} & 3. & FeF_2(s) + 2\,e^- \longrightarrow Fe(s) + 2\,F^-(aq) \end{array}$$

Reaction 3 is still a half-reaction, so we do see the free electrons.

If we knew $\Delta G°$ for reactions 1 and 2, we could add them to get $\Delta G°$ for reaction 3. Recall that we can relate $\Delta G°$ to $E°$ by $\Delta G° = -nFE°$ and to K by $\Delta G° = -RT \ln K$. We know K for reaction 1; it is K_{sp}. We know $E°$ for reaction 2. Therefore we can calculate $\Delta G°$ for reactions 1 and 2:

Reaction 1:

$$\Delta G° = -RT \ln K = -(8.314\ \text{J/K mol})(298\ \text{K}) \ln(2.4 \times 10^{-6}) = 3.2 \times 10^4\ \text{J/mol}$$

Reaction 2:

$$\Delta G° = -nFE° = -(2\ \text{mol})(96{,}485\ \text{C/mol})(-0.440\ \text{J/C}) = 8.49 \times 10^4\ \text{J}$$

(Recall that 1 volt is 1 joule per coulomb.)

Then, $\Delta G°$ for reaction 3, the one we want, is 3.2×10^4 J (for one mole of FeF_2) + $8.49 \times 10^4\ \text{J} = 1.2 \times 10^5$ J. We can convert this to $E°$ easily from the relationship $\Delta G° = -nFE°$:

$$1.2 \times 10^5\ \text{J} = -(2\ \text{mol})(96{,}485\ \text{C/mol})\ E°$$

$$E° = -0.61\ \text{J/C} = -0.61\ \text{V}.$$

(c) The standard reduction potential for FeF_2 (−0.61 V) is more negative than that for Fe^{2+} (−0.440 V), telling us that the reduction of FeF_2 is the less favorable process. When FeF_2 is reduced, we not only reduce the Fe^{2+} ions but also break up the ionic solid. Because this additional energy must be overcome, the reduction of FeF_2 is less favorable than the reduction of Fe^{2+}.

SUMMARY AND KEY TERMS

Introduction and Section 20.1 In this chapter we have focused on **electrochemistry**, the branch of chemistry that relates electricity and chemical reactions. Electrochemistry involves oxidation-reduction reactions, also called redox reactions. These reactions involve a change in the oxidation state of one or more elements. In every oxidation-reduction reaction one substance is oxidized (its oxidation state increases) and one substance is reduced (its oxidation state decreases). The substance that is oxidized is referred to as a **reducing agent**, or **reductant**, because it causes the reduction of some other substance. Similarly, the substance that is reduced is referred to as an **oxidizing agent**, or **oxidant**, because it causes the oxidation of some other substance.

Section 20.2 An oxidization-reduction reaction can be balanced by dividing the reaction into two **half-reactions**, one for oxidation and one for reduction. A half-reaction is a balanced chemical equation that includes electrons. In oxidation half-reactions the electrons are on the product (right) side of the reaction; we can envision that these electrons are transferred from a substance when it is oxidized. In reduction half-reactions the electrons are on the reactant (left) side of the reaction. Each half-reaction is balanced separately, and the two are brought together with proper coefficients to balance the electrons on each side of the equation.

Section 20.3 A **voltaic** (or **galvanic**) **cell** uses a spontaneous oxidation-reduction reaction to generate electricity. In a voltaic cell the oxidation and reduction half-reactions often occur in separate compartments. Each compartment has a solid surface called an electrode, where the half-reaction occurs. The electrode where oxidation occurs is called the **anode**; reduction occurs at the **cathode**. The electrons released at the anode flow through an external circuit (where they do electrical work) to the cathode. Electrical neutrality in the solution is maintained by the migration of ions between the two compartments through a device such as a salt bridge.

Section 20.4 A voltaic cell generates an **electromotive force (emf)** that moves the electrons from the anode to the cathode through the external circuit. The origin of emf is a difference in the electrical potential energy of the two electrodes in the cell. The emf of a cell is called its **cell potential**, E_{cell}, and is measured in volts. The cell potential under standard conditions is called the **standard emf**, or the **standard cell potential**, and is denoted E°_{cell}.

A **standard reduction potential**, E°_{red}, can be assigned for an individual half-reaction. This is achieved by comparing the potential of the half-reaction to that of the **standard hydrogen electrode** (SHE), which is defined to have $E^\circ_{red} = 0$ V and is based on the following half-reaction:

$$2\,H^+(aq, 1\,M) + 2\,e^- \longrightarrow H_2\,(g, 1\text{ atm}) \qquad E^\circ_{red} = 0\text{ V}$$

The standard cell potential of a voltaic cell is the difference between the standard reduction potentials of the half-reactions that occur at the cathode and the anode: $E^\circ_{cell} = E^\circ_{red}\,(\text{cathode}) - E^\circ_{red}\,(\text{anode})$. The value of E°_{cell} is positive for a voltaic cell.

For a reduction half-reaction, E°_{red} is a measure of the tendency of the reduction to occur; the more positive the value for E°_{red}, the greater the tendency of the substance to be reduced. Thus, E°_{red} provides a measure of the oxidizing strength of a substance. Fluorine (F_2) has the most positive value for E°_{red} and is the strongest oxidizing agent. Substances that are strong oxidizing agents produce products that are weak reducing agents, and vice versa.

Section 20.5 The emf, E, is related to the change in Gibbs free energy, ΔG: $\Delta G = -nFE$, where n is the number of electrons transferred during the redox process and F is *Faraday's constant*, defined as the quantity of electrical charge on one mole of electrons. This amount of charge is 1 **faraday** (F): 1 F = 96,485 C/mol. Because E is related to ΔG, the sign of E indicates whether a redox process is spontaneous: $E > 0$ indicates a spontaneous process, and $E < 0$ indicates a nonspontaneous one. Because ΔG is also related to the equilibrium constant for a reaction ($\Delta G = -RT \ln K$), we can relate E to K as well.

Section 20.6 The emf of a redox reaction varies with temperature and with the concentrations of reactants and products. The **Nernst equation** relates the emf under nonstandard conditions to the standard emf and the reaction quotient Q:

$$E = E^\circ - (RT/nF) \ln Q = E^\circ - (0.0592/n) \log Q$$

The factor 0.0592 is valid when $T = 298$ K. A **concentration cell** is a voltaic cell in which the same half-reaction occurs at both the anode and cathode but with different concentrations of reactants in each compartment.

At equilibrium, $Q = K$ and $E = 0$. The standard emf is therefore related to the equilibrium constant.

Section 20.7 A **battery** is a self-contained electrochemical power source that contains one or more voltaic cells. Batteries are based on a variety of different redox reactions. There are several common batteries that were discussed here. The lead-acid battery, the nickel-cadmium battery, the nickel-metal-hydride battery, and the lithium-ion battery are examples of rechargeable batteries. The common alkaline dry cell is not rechargeable. **Fuel cells** are voltaic cells that utilize redox reactions in which reactants such as H_2 have to be continuously supplied to the cell to generate voltage.

Section 20.8 Electrochemical principles help us understand **corrosion**, undesirable redox reactions in which a metal is attacked by some substance in its environment.

The corrosion of iron into rust is caused by the presence of water and oxygen, and it is accelerated by the presence of electrolytes, such as road salt. The protection of a metal by putting it in contact with another metal that more readily undergoes oxidation is called **cathodic protection**. Galvanized iron, for example, is coated with a thin layer of zinc; because zinc is oxidized more readily than iron, the zinc serves as a sacrificial anode in the redox reaction.

Section 20.9 An **electrolysis reaction**, which is carried out in an **electrolytic cell**, employs an external source of electricity to drive a nonspontaneous electrochemical reaction. The negative terminal of the external source is connected to the cathode of the cell, and the positive terminal to the anode. The current-carrying medium within an electrolytic cell may be either a molten salt or an electrolyte solution. The products of electrolysis can generally be predicted by comparing the reduction potentials associated with possible oxidation and reduction processes. The electrodes in an electrolytic cell can be active, meaning that the electrode can be involved in the electrolysis reaction. Active electrodes are important in electroplating and in metallurgical processes.

The quantity of substances formed during electrolysis can be calculated by considering the number of electrons involved in the redox reaction and the amount of electrical charge that passes into the cell. The maximum amount of electrical work produced by a voltaic cell is given by the product of the total charge delivered, nF, and the emf, E: $w_{max} = -nFE$. The work performed in an electrolysis is given by $w = nFE_{ext}$, where E_{ext} is the applied external potential. The **watt** is a unit of power: $1\ W = 1\ J/s$. Electrical work is often measured in kilowatt-hours.

VISUALIZING CONCEPTS

20.1 One way to think about redox reactions is that they are electron-transfer reactions, in the same way that acid-base reactions are proton-transfer reactions. Explain. [Sections 20.1 and 20.2]

20.2 Consider the reaction in Figure 20.3. Describe what would happen if **(a)** the solution contained cadmium(II) sulfate and the metal was zinc, **(b)** the solution contained silver nitrate and the metal was copper. [Section 20.3]

20.3 Redraw Figure 20.5, but use the two following half-reactions to make a voltaic cell:

$$Ni^{2+}(aq) + 2\,e^- \longrightarrow Ni(s)$$

$$Fe^{2+}(aq) + 2\,e^- \longrightarrow Fe(s)$$

You will need to figure out which reaction occurs at the anode, which at the cathode, and what the voltmeter would read under standard conditions. Assume KNO_3 is in the salt bridge. [Sections 20.3 and 20.4]

20.4 Look at Figure 20.14. Indicate where on the diagram you would find **(a)** the chemical species that is the easiest to oxidize, and **(b)** the chemical species that is the easiest to reduce. [Section 20.4]

20.5 For the generic reaction $A(aq) + B(aq) \longrightarrow A^-(aq) + B^+(aq)$ for which $E°$ is a positive number, answer the following questions:
(a) What is being oxidized, and what is being reduced?
(b) If you made a voltaic cell out of this, what half-reaction would be occurring at the cathode, and what half-reaction would be occurring at the anode?
(c) Which half-reaction from (b) is higher in potential energy? [Sections 20.4 and 20.5]

20.6 Draw a generic picture of a fuel cell. What is the main difference between it and a battery, regardless of the redox reactions that occur inside? [Section 20.7]

20.7 How does a zinc coating on iron protect the iron from unwanted oxidation? [Section 20.8]

20.8 You may have heard that "antioxidants" are good for your health. Based on what you have learned in this chapter, what do you deduce an "antioxidant" is? [Sections 20.1 and 20.2]

EXERCISES

Oxidation States

20.9 **(a)** What is meant by the term *oxidation*? **(b)** On which side of an oxidation half-reaction do the electrons appear? **(c)** What is meant by the term *oxidant*? **(d)** What is meant by the term *oxidizing agent*?

20.10 **(a)** What is meant by the term *reduction*? **(b)** On which side of a reduction half-reaction do the electrons appear? **(c)** What is meant by the term *reductant*? **(d)** What is meant by the term *reducing agent*?

20.11 Indicate whether each of the following statements is true or false:
(a) If something is oxidized, it is formally losing electrons.
(b) For the reaction $Fe^{3+}(aq) + Co^{2+}(aq) \longrightarrow Fe^{2+}(aq) + Co^{3+}(aq)$, $Fe^{3+}(aq)$ is the reducing agent and $Co^{2+}(aq)$ is the oxidizing agent.
(c) If there are no changes in the oxidation state of the reactants or products of a particular reaction, that reaction is not a redox reaction.

20.12 Indicate whether each of the following statements is true or false:
(a) If something is reduced, it is formally losing electrons.
(b) A reducing agent gets oxidized as it reacts.
(c) Oxidizing agents can convert CO into CO_2.

20.13 In each of the following balanced oxidation-reduction equations, identify those elements that undergo changes in oxidation number and indicate the magnitude of the change in each case.
(a) $I_2O_5(s) + 5\,CO(g) \longrightarrow I_2(s) + 5\,CO_2(g)$
(b) $2\,Hg^{2+}(aq) + N_2H_4(aq) \longrightarrow 2\,Hg(l) + N_2(g) + 4\,H^+(aq)$
(c) $3\,H_2S(aq) + 2\,H^+(aq) + 2\,NO_3^-(aq) \longrightarrow 3\,S(s) + 2\,NO(g) + 4\,H_2O(l)$
(d) $Ba^{2+}(aq) + 2\,OH^-(aq) + H_2O_2(aq) + 2\,ClO_2(aq) \longrightarrow Ba(ClO_2)_2(s) + 2\,H_2O(l) + O_2(g)$

20.14 Indicate whether the following balanced equations involve oxidation-reduction. If they do, identify the elements that undergo changes in oxidation number.
(a) $PBr_3(l) + 3\,H_2O(l) \longrightarrow H_3PO_3(aq) + 3\,HBr(aq)$
(b) $NaI(aq) + 3\,HOCl(aq) \longrightarrow NaIO_3(aq) + 3\,HCl(aq)$
(c) $3\,SO_2(g) + 2\,HNO_3(aq) + 2\,H_2O(l) \longrightarrow 3\,H_2SO_4(aq) + 2\,NO(g)$
(d) $2\,H_2SO_4(aq) + 2\,NaBr(s) \longrightarrow Br_2(l) + SO_2(g) + Na_2SO_4(aq) + 2\,H_2O(l)$

Balancing Oxidation-Reduction Reactions

20.15 At 900°C titanium tetrachloride vapor reacts with molten magnesium metal to form solid titanium metal and molten magnesium chloride. **(a)** Write a balanced equation for this reaction. **(b)** What is being oxidized, and what is being reduced? **(c)** Which substance is the reductant, and which is the oxidant?

20.16 Hydrazine (N_2H_4) and dinitrogen tetroxide (N_2O_4) form a self-igniting mixture that has been used as a rocket propellant. The reaction products are N_2 and H_2O. **(a)** Write a balanced chemical equation for this reaction. **(b)** What is being oxidized, and what is being reduced? **(c)** Which substance serves as the reducing agent, and which as the oxidizing agent?

20.17 Complete and balance the following half-reactions. In each case indicate whether the half-reaction is an oxidation or a reduction.
(a) $Sn^{2+}(aq) \longrightarrow Sn^{4+}(aq)$ (acidic or basic solution)
(b) $TiO_2(s) \longrightarrow Ti^{2+}(aq)$ (acidic solution)
(c) $ClO_3^-(aq) \longrightarrow Cl^-(aq)$ (acidic solution)
(d) $OH^-(aq) \longrightarrow O_2(g)$ (basic solution)
(e) $SO_3^{2-}(aq) \longrightarrow SO_4^{2-}(aq)$ (basic solution)
(f) $N_2(g) \longrightarrow NH_4^+(aq)$ (acidic solution)
(g) $N_2(g) \longrightarrow NH_3(g)$ (basic solution)

20.18 Complete and balance the following half-reactions. In each case indicate whether the half-reaction is an oxidation or a reduction.
(a) $Mo^{3+}(aq) \longrightarrow Mo(s)$ (acidic or basic solution)
(b) $H_2SO_3(aq) \longrightarrow SO_4^{2-}(aq)$ (acidic solution)
(c) $NO_3^-(aq) \longrightarrow NO(g)$ (acidic solution)
(d) $Mn^{2+}(aq) \longrightarrow MnO_2(s)$ (basic solution)
(e) $Cr(OH)_3(s) \longrightarrow CrO_4^{2-}(aq)$ (basic solution)
(f) $O_2(g) \longrightarrow H_2O(l)$ (acidic solution)
(g) $O_2(g) \longrightarrow H_2O(l)$ (basic solution)

20.19 Complete and balance the following equations, and identify the oxidizing and reducing agents:
(a) $Cr_2O_7^{2-}(aq) + I^-(aq) \longrightarrow Cr^{3+}(aq) + IO_3^-(aq)$ (acidic solution)
(b) $MnO_4^-(aq) + CH_3OH(aq) \longrightarrow Mn^{2+}(aq) + HCO_2H(aq)$ (acidic solution)
(c) $I_2(s) + OCl^-(aq) \longrightarrow IO_3^-(aq) + Cl^-(aq)$ (acidic solution)
(d) $As_2O_3(s) + NO_3^-(aq) \longrightarrow H_3AsO_4(aq) + N_2O_3(aq)$ (acidic solution)
(e) $MnO_4^-(aq) + Br^-(aq) \longrightarrow MnO_2(s) + BrO_3^-(aq)$ (basic solution)
(f) $Pb(OH)_4^{2-}(aq) + ClO^-(aq) \longrightarrow PbO_2(s) + Cl^-(aq)$ (basic solution)

20.20 Complete and balance the following equations, and identify the oxidizing and reducing agents. Recall that the O atoms in hydrogen peroxide, H_2O_2, have an atypical oxidation state. (Table 2.5)
(a) $NO_2^-(aq) + Cr_2O_7^{2-}(aq) \longrightarrow Cr^{3+}(aq) + NO_3^-(aq)$ (acidic solution)
(b) $As(s) + ClO_3^-(aq) \longrightarrow H_3AsO_3(aq) + HClO(aq)$ (acidic solution)
(c) $Cr_2O_7^{2-}(aq) + CH_3OH(aq) \longrightarrow HCO_2H(aq) + Cr^{3+}(aq)$ (acidic solution)
(d) $MnO_4^-(aq) + Cl^-(aq) \longrightarrow Mn^{2+}(aq) + Cl_2(aq)$ (acidic solution)
(e) $H_2O_2(aq) + ClO_2(aq) \longrightarrow ClO_2^-(aq) + O_2(g)$ (basic solution)
(f) $H_2O_2(aq) + Cl_2O_7(aq) \longrightarrow ClO_2^-(aq) + O_2(g)$ (basic solution)

Voltaic Cells

20.21 **(a)** What are the similarities and differences between Figure 20.3 and Figure 20.4? **(b)** Why are Na^+ ions drawn into the cathode compartment as the voltaic cell shown in Figure 20.5 operates?

20.22 **(a)** What is the role of the porous glass disc shown in Figure 20.4? **(b)** Why do NO_3^- ions migrate into the anode compartment as the voltaic cell shown in Figure 20.5 operates?

20.23 A voltaic cell similar to that shown in Figure 20.5 is constructed. One electrode compartment consists of a silver strip placed in a solution of $AgNO_3$, and the other has an iron strip placed in a solution of $FeCl_2$. The overall cell reaction is

$$Fe(s) + 2\,Ag^+(aq) \longrightarrow Fe^{2+}(aq) + 2\,Ag(s)$$

(a) What is being oxidized, and what is being reduced? **(b)** Write the half-reactions that occur in the two electrode compartments. **(c)** Which electrode is the anode, and which is the cathode? **(d)** Indicate the signs of the electrodes. **(e)** Do electrons flow from the silver electrode to the iron electrode, or from the iron to the silver? **(f)** In which directions do the cations and anions migrate through the solution?

20.24 A voltaic cell similar to that shown in Figure 20.5 is constructed. One electrode compartment consists of an aluminum strip placed in a solution of $Al(NO_3)_3$, and the other has a nickel strip placed in a solution of $NiSO_4$. The overall cell reaction is

$$2\,Al(s) + 3\,Ni^{2+}(aq) \longrightarrow 2\,Al^{3+}(aq) + 3\,Ni(s)$$

(a) What is being oxidized, and what is being reduced? **(b)** Write the half-reactions that occur in the two electrode compartments. **(c)** Which electrode is the anode, and which is the cathode? **(d)** Indicate the signs of the electrodes. **(e)** Do electrons flow from the aluminum electrode to the nickel electrode, or from the nickel to the aluminum? **(f)** In which directions do the cations and anions migrate through the solution? Assume the Al is not coated with its oxide.

Cell EMF Under Standard Conditions

20.25 **(a)** What is meant by the term *electromotive force*? **(b)** What is the definition of the *volt*? **(c)** What is meant by the term *cell potential*?

20.26 **(a)** Which electrode of a voltaic cell, the cathode or the anode, corresponds to the higher potential energy for the electrons? **(b)** What are the units for electrical potential? How does this unit relate to energy expressed in joules? **(c)** What is special about a *standard* cell potential?

20.27 **(a)** Write the half-reaction that occurs at a hydrogen electrode in acidic aqueous solution when it serves as the cathode of a voltaic cell. **(b)** What is *standard* about the standard hydrogen electrode? **(c)** What is the role of the platinum foil in a standard hydrogen electrode?

20.28 **(a)** Write the half-reaction that occurs at a hydrogen electrode in acidic aqueous solution when it serves as the anode of a voltaic cell. **(b)** The platinum electrode in a standard hydrogen electrode is specially prepared to have a large surface area. Why is this important? **(c)** Sketch a standard hydrogen electrode.

20.29 **(a)** What is a *standard reduction potential*? **(b)** What is the standard reduction potential of a standard hydrogen electrode? **(c)** Based on the standard reduction potentials listed in Appendix E, which is the more favorable process, the reduction of $Ag^+(aq)$ to $Ag(s)$ or the reduction of $Sn^{2+}(aq)$ to $Sn(s)$?

20.30 **(a)** Why is it impossible to measure the standard reduction potential of a single half-reaction? **(b)** Describe how the standard reduction potential of a half-reaction can be determined. **(c)** By using data in Appendix E, determine which is the more unfavorable reduction, $Cd^{2+}(aq)$ to $Cd(s)$ or $Ca^{2+}(aq)$ to $Ca(s)$.

20.31 A voltaic cell that uses the reaction

$$Tl^{3+}(aq) + 2\,Cr^{2+}(aq) \longrightarrow Tl^{+}(aq) + 2\,Cr^{3+}(aq)$$

has a measured standard cell potential of +1.19 V. **(a)** Write the two half-cell reactions. **(b)** By using data from Appendix E, determine E°_{red} for the reduction of $Tl^{3+}(aq)$ to $Tl^+(aq)$. **(c)** Sketch the voltaic cell, label the anode and cathode, and indicate the direction of electron flow.

20.32 A voltaic cell that uses the reaction

$$PdCl_4^{2-}(aq) + Cd(s) \longrightarrow Pd(s) + 4\,Cl^-(aq) + Cd^{2+}(aq)$$

has a measured standard cell potential of +1.03 V. **(a)** Write the two half-cell reactions. **(b)** By using data from Appendix E, determine E°_{red} for the reaction involving Pd. **(c)** Sketch the voltaic cell, label the anode and cathode, and indicate the direction of electron flow.

20.33 Using standard reduction potentials (Appendix E), calculate the standard emf for each of the following reactions:
(a) $Cl_2(g) + 2\,I^-(aq) \longrightarrow 2\,Cl^-(aq) + I_2(s)$
(b) $Ni(s) + 2\,Ce^{4+}(aq) \longrightarrow Ni^{2+}(aq) + 2\,Ce^{3+}(aq)$
(c) $Fe(s) + 2\,Fe^{3+}(aq) \longrightarrow 3\,Fe^{2+}(aq)$
(d) $2\,Al^{3+}(aq) + 3\,Ca(s) \longrightarrow 2\,Al(s) + 3\,Ca^{2+}(aq)$

20.34 Using data in Appendix E, calculate the standard emf for each of the following reactions:
(a) $H_2(g) + F_2(g) \longrightarrow 2\,H^+(aq) + 2\,F^-(aq)$
(b) $Cu(s) + Ba^{2+}(aq) \longrightarrow Cu^{2+}(aq) + Ba(s)$
(c) $3\,Fe^{2+}(aq) \longrightarrow Fe(s) + 2\,Fe^{3+}(aq)$
(d) $Hg_2^{2+}(aq) + 2\,Cu^+(aq) \longrightarrow 2\,Hg(l) + 2\,Cu^{2+}(aq)$

20.35 The standard reduction potentials of the following half-reactions are given in Appendix E:

$$Ag^+(aq) + e^- \longrightarrow Ag(s)$$
$$Cu^{2+}(aq) + 2\,e^- \longrightarrow Cu(s)$$
$$Ni^{2+}(aq) + 2\,e^- \longrightarrow Ni(s)$$
$$Cr^{3+}(aq) + 3\,e^- \longrightarrow Cr(s)$$

(a) Determine which combination of these half-cell reactions leads to the cell reaction with the largest positive cell emf, and calculate the value. **(b)** Determine which combination of these half-cell reactions leads to the cell reaction with the smallest positive cell emf, and calculate the value.

20.36 Given the following half-reactions and associated standard reduction potentials:

$$AuBr_4^-(aq) + 3\,e^- \longrightarrow Au(s) + 4\,Br^-(aq) \qquad E^\circ_{red} = -0.858\ V$$
$$Eu^{3+}(aq) + e^- \longrightarrow Eu^{2+}(aq) \qquad E^\circ_{red} = -0.43\ V$$
$$IO^-(aq) + H_2O(l) + 2\,e^- \longrightarrow I^-(aq) + 2\,OH^-(aq) \qquad E^\circ_{red} = +0.49\ V$$
$$Sn^{2+}(aq) + 2\,e^- \longrightarrow Sn(s) \qquad E^\circ_{red} = -0.14\ V$$

(a) Write the cell reaction for the combination of these half-cell reactions that leads to the largest positive cell emf, and calculate the value. **(b)** Write the cell reaction for the combination of half-cell reactions that leads to the smallest positive cell emf, and calculate that value.

20.37 A 1 *M* solution of $Cu(NO_3)_2$ is placed in a beaker with a strip of Cu metal. A 1 *M* solution of $SnSO_4$ is placed in a second beaker with a strip of Sn metal. The two beakers are connected by a salt bridge, and the two metal electrodes are linked by wires to a voltmeter. **(a)** Which electrode serves as the anode, and which as the cathode? **(b)** Which electrode gains mass and which loses mass as the cell reaction proceeds? **(c)** Write the equation for the overall cell reaction. **(d)** What is the emf generated by the cell under standard conditions?

20.38 A voltaic cell consists of a strip of lead metal in a solution of $Pb(NO_3)_2$ in one beaker, and in the other beaker a platinum electrode is immersed in a NaCl solution, with Cl_2 gas bubbled around the electrode. The two beakers are connected with a salt bridge. **(a)** Which electrode serves as the anode, and which as the cathode? **(b)** Does the Pb electrode gain or lose mass as the cell reaction proceeds? **(c)** Write the equation for the overall cell reaction. **(d)** What is the emf generated by the cell under standard conditions?

Strengths of Oxidizing and Reducing Agents

20.39 From each of the following pairs of substances, use data in Appendix E to choose the one that is the stronger oxidizing agent:
(a) $Cl_2(g)$ or $Br_2(l)$
(b) $Ni^{2+}(aq)$ or $Cd^{2+}(aq)$
(c) $BrO_3^-(aq)$ or $IO_3^-(aq)$
(d) $H_2O_2(aq)$ or $O_3(g)$

20.40 From each of the following pairs of substances, use data in Appendix E to choose the one that is the stronger reducing agent:
(a) $Fe(s)$ or $Mg(s)$
(b) $Ca(s)$ or $Al(s)$
(c) H_2 (*g*, acidic solution) or $H_2S(g)$
(d) $H_2SO_3(aq)$ or $H_2C_2O_4(aq)$

20.41 By using the data in Appendix E, determine whether each of the following substances is likely to serve as an oxidant or a reductant: **(a)** $Cl_2(g)$, **(b)** MnO_4^- (*aq*, acidic solution), **(c)** $Ba(s)$, **(d)** $Zn(s)$.

20.42 Is each of the following substances likely to serve as an oxidant or a reductant: **(a)** $Na(s)$, **(b)** $O_3(g)$, **(c)** $Ce^{3+}(aq)$, **(d)** $Sn^{2+}(aq)$?

20.43 **(a)** Assuming standard conditions, arrange the following in order of increasing strength as oxidizing agents in acidic solution: $Cr_2O_7^{2-}$, H_2O_2, Cu^{2+}, Cl_2, O_2. **(b)** Arrange the following in order of increasing strength as reducing agents in acidic solution: Zn, I^-, Sn^{2+}, H_2O_2, Al.

20.44 Based on the data in Appendix E, **(a)** which of the following is the strongest oxidizing agent and which is the weakest in acidic solution: Ce^{4+}, Br_2, H_2O_2, Zn? **(b)** Which of the following is the strongest reducing agent, and which is the weakest in acidic solution: F^-, Zn, $N_2H_5^+$, I_2, NO?

20.45 The standard reduction potential for the reduction of $Eu^{3+}(aq)$ to $Eu^{2+}(aq)$ is −0.43 V. Using Appendix E, which of the following substances is capable of reducing $Eu^{3+}(aq)$ to $Eu^{2+}(aq)$ under standard conditions: Al, Co, H_2O_2, $N_2H_5^+$, $H_2C_2O_4$?

20.46 The standard reduction potential for the reduction of $RuO_4^-(aq)$ to $RuO_4^{2-}(aq)$ is +0.59 V. By using Appendix E, which of the following substances can oxidize $RuO_4^{2-}(aq)$ to $RuO_4^-(aq)$ under standard conditions: $Cr_2O_7^{2-}(aq)$, $ClO^-(aq)$, $Pb^{2+}(aq)$, $I_2(s)$, $Ni^{2+}(aq)$?

Free Energy and Redox Reactions

20.47 Given the following reduction half-reactions:

$$Fe^{3+}(aq) + e^- \longrightarrow Fe^{2+}(aq) \qquad E^\circ_{red} = +0.77\text{ V}$$

$$S_2O_6^{2-}(aq) + 4\,H^+(aq) + 2\,e^- \longrightarrow 2\,H_2SO_3(aq) \qquad E^\circ_{red} = +0.60\text{ V}$$

$$N_2O(aq) + 2\,H^+(aq) + 2\,e^- \longrightarrow N_2(g) + H_2O(l) \qquad E^\circ_{red} = -1.77\text{ V}$$

$$VO_2^+(aq) + 2\,H^+(aq) + e^- \longrightarrow VO^{2+}(aq) + H_2O(l) \qquad E^\circ_{red} = +1.00\text{ V}$$

(a) Write balanced chemical equations for the oxidation of $Fe^{2+}(aq)$ by $S_2O_6^{2-}(aq)$, by $N_2O(aq)$, and by $VO_2^+(aq)$. **(b)** Calculate ΔG° for each reaction at 298 K. **(c)** Calculate the equilibrium constant *K* for each reaction at 298 K.

20.48 For each of the following reactions, write a balanced equation, calculate the standard emf, calculate ΔG° at 298 K, and calculate the equilibrium constant *K* at 298 K. **(a)** Aqueous iodide ion is oxidized to $I_2(s)$ by $Hg_2^{2+}(aq)$. **(b)** In acidic solution, copper(I) ion is oxidized to copper(II) ion by nitrate ion. **(c)** In basic solution, $Cr(OH)_3(s)$ is oxidized to $CrO_4^{2-}(aq)$ by $ClO^-(aq)$.

20.49 If the equilibrium constant for a two-electron redox reaction at 298 K is 1.5×10^{-4}, calculate the corresponding ΔG° and E°_{cell} under standard conditions.

20.50 If the equilibrium constant for a one-electron redox reaction at 298 K is 3.7×10^6, calculate the corresponding ΔG° and E°_{cell} under standard conditions.

20.51 Using the standard reduction potentials listed in Appendix E, calculate the equilibrium constant for each of the following reactions at 298 K:
(a) $Fe(s) + Ni^{2+}(aq) \longrightarrow Fe^{2+}(aq) + Ni(s)$
(b) $Co(s) + 2\,H^+(aq) \longrightarrow Co^{2+}(aq) + H_2(g)$
(c) $10\,Br^-(aq) + 2\,MnO_4^-(aq) + 16\,H^+(aq) \longrightarrow 2\,Mn^{2+}(aq) + 8\,H_2O(l) + 5\,Br_2(l)$

20.52 Using the standard reduction potentials listed in Appendix E, calculate the equilibrium constant for each of the following reactions at 298 K:
(a) $2\,VO_2^+(aq) + 4\,H^+(aq) + 2\,Ag(s) \longrightarrow 2\,VO^{2+}(aq) + 2\,H_2O(l) + 2\,Ag^+(aq)$
(b) $3\,Ce^{4+}(aq) + Bi(s) + H_2O(l) \longrightarrow 3\,Ce^{3+}(aq) + BiO^+(aq) + 2\,H^+(aq)$
(c) $N_2H_5^+(aq) + 4\,Fe(CN)_6^{3-}(aq) \longrightarrow N_2(g) + 5\,H^+(aq) + 4\,Fe(CN)_6^{4-}(aq)$

20.53 A cell has a standard emf of +0.177 V at 298 K. What is the value of the equilibrium constant for the cell reaction **(a)** if $n = 1$? **(b)** if $n = 2$? **(c)** if $n = 3$?

20.54 At 298 K a cell reaction has a standard emf of +0.17 V. The equilibrium constant for the cell reaction is 5.5×10^5. What is the value of n for the cell reaction?

Cell EMF Under Nonstandard Conditions

20.55 **(a)** Under what circumstances is the Nernst equation applicable? **(b)** What is the numerical value of the reaction quotient, Q, under standard conditions? **(c)** What happens to the emf of a cell if the concentrations of the reactants are increased?

20.56 **(a)** A voltaic cell is constructed with all reactants and products in their standard states. Will this condition hold as the cell operates? Explain. **(b)** Can the Nernst equation be used at temperatures other than room temperature? Explain. **(c)** What happens to the emf of a cell if the concentrations of the products are increased?

20.57 What is the effect on the emf of the cell shown in Figure 20.11, which has the overall reaction $Zn(s) + 2\,H^+(aq) \longrightarrow Zn^{2+}(aq) + H_2(g)$, for each of the following changes? **(a)** The pressure of the H_2 gas is increased in the cathode compartment. **(b)** Zinc nitrate is added to the anode compartment. **(c)** Sodium hydroxide is added to the cathode compartment, decreasing $[H^+]$. **(d)** The surface area of the anode is doubled.

20.58 A voltaic cell utilizes the following reaction:

$$Al(s) + 3\,Ag^+(aq) \longrightarrow Al^{3+}(aq) + 3\,Ag(s)$$

What is the effect on the cell emf of each of the following changes? **(a)** Water is added to the anode compartment, diluting the solution. **(b)** The size of the aluminum electrode is increased. **(c)** A solution of $AgNO_3$ is added to the cathode compartment, increasing the quantity of Ag^+ but not changing its concentration. **(d)** HCl is added to the $AgNO_3$ solution, precipitating some of the Ag^+ as AgCl.

20.59 A voltaic cell is constructed that uses the following reaction and operates at 298 K:

$$Zn(s) + Ni^{2+}(aq) \longrightarrow Zn^{2+}(aq) + Ni(s)$$

(a) What is the emf of this cell under standard conditions? **(b)** What is the emf of this cell when $[Ni^{2+}] = 3.00\ M$ and $[Zn^{2+}] = 0.100\ M$? **(c)** What is the emf of the cell when $[Ni^{2+}] = 0.200\ M$ and $[Zn^{2+}] = 0.900\ M$?

20.60 A voltaic cell utilizes the following reaction and operates at 298 K:

$$3\,Ce^{4+}(aq) + Cr(s) \longrightarrow 3\,Ce^{3+}(aq) + Cr^{3+}(aq)$$

(a) What is the emf of this cell under standard conditions? **(b)** What is the emf of this cell when $[Ce^{4+}] = 2.0\ M$, $[Ce^{3+}] = 0.010\ M$, and $[Cr^{3+}] = 0.010\ M$? **(c)** What is the emf of the cell when $[Ce^{4+}] = 0.35\ M$, $[Ce^{3+}] = 0.85\ M$, and $[Cr^{3+}] = 1.2\ M$?

20.61 A voltaic cell utilizes the following reaction:

$$4\,Fe^{2+}(aq) + O_2(g) + 4\,H^+(g) \longrightarrow 4\,Fe^{3+}(aq) + 2\,H_2O(l)$$

(a) What is the emf of this cell under standard conditions? **(b)** What is the emf of this cell when $[Fe^{2+}] = 1.3\ M$, $[Fe^{3+}] = 0.010\ M$, $P_{O_2} = 0.50$ atm, and the pH of the solution in the cathode compartment is 3.50?

20.62 A voltaic cell utilizes the following reaction:

$$2\,Fe^{3+}(aq) + H_2(g) \longrightarrow 2\,Fe^{2+}(aq) + 2\,H^+(aq)$$

(a) What is the emf of this cell under standard conditions? **(b)** What is the emf for this cell when $[Fe^{3+}] = 1.50\ M$, $P_{H_2} = 0.50$ atm, $[Fe^{2+}] = 0.0010\ M$, and the pH in both compartments is 5.00?

20.63 A voltaic cell is constructed with two Zn^{2+}-Zn electrodes. The two cell compartments have $[Zn^{2+}] = 1.8\ M$ and $[Zn^{2+}] = 1.00 \times 10^{-2}\ M$, respectively. **(a)** Which electrode is the anode of the cell? **(b)** What is the standard emf of the cell? **(c)** What is the cell emf for the concentrations given? **(d)** For each electrode, predict whether $[Zn^{2+}]$ will increase, decrease, or stay the same as the cell operates.

20.64 A voltaic cell is constructed with two silver–silver chloride electrodes, each of which is based on the following half-reaction:

$$AgCl(s) + e^- \longrightarrow Ag(s) + Cl^-(aq)$$

The two cell compartments have $[Cl^-] = 0.0150\ M$ and $[Cl^-] = 2.55\ M$, respectively. **(a)** Which electrode is the cathode of the cell? **(b)** What is the standard emf of the cell? **(c)** What is the cell emf for the concentrations given? **(d)** For each electrode, predict whether $[Cl^-]$ will increase, decrease, or stay the same as the cell operates.

20.65 The cell in Figure 20.11 could be used to provide a measure of the pH in the cathode compartment. Calculate the pH of the cathode compartment solution if the cell emf at 298 K is measured to be +0.684 V when $[Zn^{2+}] = 0.30\ M$ and $P_{H_2} = 0.90$ atm.

20.66 A voltaic cell is constructed that is based on the following reaction:

$$Sn^{2+}(aq) + Pb(s) \longrightarrow Sn(s) + Pb^{2+}(aq)$$

(a) If the concentration of Sn^{2+} in the cathode compartment is 1.00 M and the cell generates an emf of +0.22 V, what is the concentration of Pb^{2+} in the anode compartment? **(b)** If the anode compartment contains $[SO_4^{2-}] = 1.00\ M$ in equilibrium with $PbSO_4(s)$, what is the K_{sp} of $PbSO_4$?

Batteries and Fuel Cells

20.67 **(a)** What happens to the emf of a battery as it is used? Why does this happen? **(b)** The AA-size and D-size alkaline batteries are both 1.5-V batteries that are based on the same electrode reactions. What is the major difference between the two batteries? What performance feature is most affected by this difference?

20.68 Suggest an explanation for why liquid water is needed in an alkaline battery.

20.69 During a period of discharge of a lead-acid battery, 402 g of Pb from the anode is converted into $PbSO_4(s)$. What mass of $PbO_2(s)$ is reduced at the cathode during this same period?

20.70 During the discharge of an alkaline battery, 12.6 g of Zn are consumed at the anode of the battery. What mass of MnO_2 is reduced at the cathode during this discharge?

20.71 Heart pacemakers are often powered by lithium–silver chromate "button" batteries. The overall cell reaction is:

$$2\,Li(s) + Ag_2CrO_4(s) \longrightarrow Li_2CrO_4(s) + 2\,Ag(s)$$

(a) Lithium metal is the reactant at one of the electrodes of the battery. Is it the anode or the cathode? **(b)** Choose the two half-reactions from Appendix E that *most closely approximate* the reactions that occur in the battery. What standard emf would be generated by a voltaic cell based on these half-reactions? **(c)** The battery generates an emf of +3.5 V. How close is this value to the one calculated in part (b)? **(d)** Calculate the emf that would be generated at body temperature, 37°C. How does this compare to the emf you calculated in part (b)?

20.72 Mercuric oxide dry-cell batteries are often used where a high-energy density is required, such as in watches and cameras. The two half-cell reactions that occur in the battery are

$$HgO(s) + H_2O(l) + 2\,e^- \longrightarrow Hg(l) + 2\,OH^-(aq)$$
$$Zn(s) + 2\,OH^-(aq) \longrightarrow ZnO(s) + H_2O(l) + 2\,e^-$$

(a) Write the overall cell reaction. **(b)** The value of E°_{red} for the cathode reaction is +0.098 V. The overall cell potential is +1.35 V. Assuming that both half-cells operate under standard conditions, what is the standard reduction potential for the anode reaction? **(c)** Why is the potential of the anode reaction different than would be expected if the reaction occurred in an acidic medium?

20.73 **(a)** Suppose that an alkaline battery was manufactured using cadmium metal rather than zinc. What effect would this have on the cell emf? **(b)** What environmental advantage is provided by the use of nickel-metal-hydride batteries over the nickel-cadmium batteries?

20.74 **(a)** The nonrechargeable lithium batteries used for photography use lithium metal as the anode. What advantages might be realized by using lithium rather than zinc, cadmium, lead, or nickel? **(b)** The rechargeable lithium-ion battery does not use lithium metal as an electrode material. Nevertheless, it still has a substantial advantage over nickel-based batteries. Suggest an explanation.

20.75 The hydrogen-oxygen fuel cell has a standard emf of 1.23 V. What advantages and disadvantages are there to using this device as a source of power, compared to a 1.55-V alkaline battery?

20.76 Can the "fuel" of a fuel cell be a solid? Explain.

Corrosion

20.77 **(a)** Write the anode and cathode reactions that cause the corrosion of iron metal to aqueous iron(II). **(b)** Write the balanced half-reactions involved in the air oxidation of $Fe^{2+}(aq)$ to $Fe_2O_3 \cdot 3\,H_2O$.

20.78 **(a)** Based on standard reduction potentials, would you expect copper metal to oxidize under standard conditions in the presence of oxygen and hydrogen ions? **(b)** When the Statue of Liberty was refurbished, Teflon® spacers were placed between the iron skeleton and the copper metal on the surface of the statue. What role do these spacers play?

20.79 **(a)** Magnesium metal is used as a sacrificial anode to protect underground pipes from corrosion. Why is the magnesium referred to as a "sacrificial anode"? **(b)** Looking in Appendix E, suggest what metal the underground pipes could be made from in order for magnesium to be successful as a sacrificial anode.

20.80 An iron object is plated with a coating of cobalt to protect against corrosion. Does the cobalt protect iron by cathodic protection? Explain.

20.81 A plumber's handbook states that you should not connect a brass pipe directly to a galvanized steel pipe because electrochemical reactions between the two metals will cause corrosion. The handbook recommends you use, instead, an insulating fitting to connect them. Brass is a mixture of copper and zinc. What spontaneous redox reaction(s) might cause the corrosion? Justify your answer with standard emf calculations.

20.82 A plumber's handbook states that you should not connect a copper pipe directly to a steel pipe because electrochemical reactions between the two metals will cause corrosion. The handbook recommends you use, instead, an insulating fitting to connect them. What spontaneous redox reaction(s) might cause the corrosion? Justify your answer with standard emf calculations.

Electrolysis; Electrical Work

20.83 **(a)** What is *electrolysis*? **(b)** Are electrolysis reactions thermodynamically spontaneous? Explain. **(c)** What process occurs at the anode in the electrolysis of molten NaCl?

20.84 **(a)** What is an *electrolytic cell*? **(b)** The negative terminal of a voltage source is connected to an electrode of an electrolytic cell. Is the electrode the anode or the cathode of the cell? Explain. **(c)** The electrolysis of water is often done with a small amount of sulfuric acid added to the water. What is the role of the sulfuric acid?

20.85 **(a)** A $Cr^{3+}(aq)$ solution is electrolyzed, using a current of 7.60 A. What mass of $Cr(s)$ is plated out after 2.00 days? **(b)** What amperage is required to plate out 0.250 mol Cr from a Cr^{3+} solution in a period of 8.00 h?

20.86 Metallic magnesium can be made by the electrolysis of molten $MgCl_2$. **(a)** What mass of Mg is formed by passing a current of 5.25 A through molten $MgCl_2$ for 2.50 days? **(b)** How many minutes are needed to plate out 10.00 g Mg from molten $MgCl_2$, using 3.50 A of current?

20.87 A voltaic cell is based on the reaction

$$Sn(s) + I_2(s) \longrightarrow Sn^{2+}(aq) + 2\,I^-(aq)$$

Under standard conditions, what is the maximum electrical work, in joules, that the cell can accomplish if 75.0 g of Sn is consumed?

20.88 Consider the voltaic cell illustrated in Figure 20.5, which is based on the cell reaction

$$Zn(s) + Cu^{2+}(aq) \longrightarrow Zn^{2+}(aq) + Cu(s)$$

Under standard conditions, what is the maximum electrical work, in joules, that the cell can accomplish if 50.0 g of copper is plated out?

20.89 **(a)** Calculate the mass of Li formed by electrolysis of molten LiCl by a current of 7.5×10^4 A flowing for a period of 24 h. Assume the electrolytic cell is 85% efficient. **(b)** What is the energy requirement for this electrolysis per mole of Li formed if the applied emf is +7.5 V?

20.90 Elemental calcium is produced by the electrolysis of molten $CaCl_2$. **(a)** What mass of calcium can be produced by this process if a current of 6.5×10^3 A is applied for 48 h? Assume that the electrolytic cell is 68% efficient. **(b)** What is the total energy requirement for this electrolysis if the applied emf is +5.00 V?

Additional Exercises

20.91 A *disproportionation* reaction is an oxidation-reduction reaction in which the same substance is oxidized and reduced. Complete and balance the following disproportionation reactions:

(a) $Ni^+(aq) \longrightarrow Ni^{2+}(aq) + Ni(s)$ (acidic solution)

(b) $MnO_4^{2-}(aq) \longrightarrow MnO_4^-(aq) + MnO_2(s)$ (acidic solution)

(c) $H_2SO_3(aq) \longrightarrow S(s) + HSO_4^-(aq)$ (acidic solution)

(d) $Cl_2(aq) \longrightarrow Cl^-(aq) + ClO^-(aq)$ (basic solution)

20.92 This oxidation-reduction reaction in acidic solution is spontaneous:

$$5\,Fe^{2+}(aq) + MnO_4^-(aq) + 8\,H^+(aq) \longrightarrow 5\,Fe^{3+}(aq) + Mn^{2+}(aq) + 4\,H_2O(l)$$

A solution containing $KMnO_4$ and H_2SO_4 is poured into one beaker, and a solution of $FeSO_4$ is poured into another. A salt bridge is used to join the beakers. A platinum foil is placed in each solution, and the two solutions are connected by a wire that passes through a voltmeter. **(a)** Sketch the cell, indicating the anode and the cathode, the direction of electron movement through the external circuit, and the direction of ion migrations through the solutions. **(b)** Sketch the process that occurs at the atomic level at the surface of the anode. **(c)** Calculate the emf of the cell under standard conditions. **(d)** Calculate the emf of the cell at 298 K when the concentrations are the following: pH = 0.0, $[Fe^{2+}] = 0.10\,M$, $[MnO_4^-] = 1.50\,M$, $[Fe^{3+}] = 2.5 \times 10^{-4}\,M$, $[Mn^{2+}] = 0.001\,M$.

20.93 A common shorthand way to represent a voltaic cell is to list its components as follows:

anode|anode solution||cathode solution|cathode

A double vertical line represents a salt bridge or a porous barrier. A single vertical line represents a change in phase, such as from solid to solution. **(a)** Write the half-reactions and overall cell reaction represented by $Fe|Fe^{2+}||Ag^+|Ag$; sketch the cell. **(b)** Write the half-reactions and overall cell reaction represented by $Zn|Zn^{2+}||H^+|H_2$; sketch the cell. **(c)** Using the notation just described, represent a cell based on the following reaction:

$$ClO_3^-(aq) + 3\,Cu(s) + 6\,H^+(aq) \longrightarrow Cl^-(aq) + 3\,Cu^{2+}(aq) + 3\,H_2O(l)$$

Pt is used as an inert electrode in contact with the ClO_3^- and Cl^-. Sketch the cell.

20.94 Predict whether the following reactions will be spontaneous in acidic solution under standard conditions: **(a)** oxidation of Sn to Sn^{2+} by I_2 (to form I^-), **(b)** reduction of Ni^{2+} to Ni by I^- (to form I_2), **(c)** reduction of Ce^{4+} to Ce^{3+} by H_2O_2, **(d)** reduction of Cu^{2+} to Cu by Sn^{2+} (to form Sn^{4+}).

[20.95] Gold exists in two common positive oxidation states, +1 and +3. The standard reduction potentials for these oxidation states are

$$Au^+(aq) + e^- \longrightarrow Au(s) \qquad E^\circ_{red} = +1.69\ V$$

$$Au^{3+}(aq) + 3\,e^- \longrightarrow Au(s) \qquad E^\circ_{red} = +1.50\ V$$

(a) Can you use these data to explain why gold does not tarnish in the air? **(b)** Suggest several substances that should be strong enough oxidizing agents to oxidize gold metal. **(c)** Miners obtain gold by soaking gold-containing ores in a aqueous solution of sodium cyanide. A very soluble complex ion of gold forms in the aqueous solution as a result of the redox reaction

$$4\,Au(s) + 8\,NaCN(aq) + 2\,H_2O(l) + O_2(g) \longrightarrow 4\,Na[Au(CN)_2](aq) + 4\,NaOH(aq)$$

What is being oxidized, and what is being reduced, in this reaction? **(d)** Gold miners then react the basic aqueous product solution from part (c) with Zn dust to get gold metal. Write a balanced redox reaction for this process. What is being oxidized, and what is being reduced?

20.96 A voltaic cell is constructed from an $Ni^{2+}(aq)$-Ni(*s*) half-cell and an $Ag^+(aq)$-Ag(*s*) half-cell. The initial concentration of $Ni^{2+}(aq)$ in the Ni^{2+}-Ni half-cell is $[Ni^{2+}] = 0.0100\,M$. The initial cell voltage is +1.12 V. **(a)** By using data in Table 20.1, calculate the standard emf of this voltaic cell. **(b)** Will the concentration of $Ni^{2+}(aq)$ increase or decrease as the cell operates? **(c)** What is the initial concentration of $Ag^+(aq)$ in the Ag^+-Ag half-cell?

[20.97] A voltaic cell is constructed that uses the following half-cell reactions:

$$Cu^{+}(aq) + e^{-} \longrightarrow Cu(s)$$
$$I_2(s) + 2\,e^{-} \longrightarrow 2\,I^{-}(aq)$$

The cell is operated at 298 K with $[Cu^{+}] = 2.5\ M$ and $[I^{-}] = 3.5\ M$. **(a)** Determine E for the cell at these concentrations. **(b)** Which electrode is the anode of the cell? **(c)** Is the answer to part (b) the same as it would be if the cell was operated under standard conditions? **(d)** If $[Cu^{+}]$ was equal to 1.4 M, at what concentration of I^{-} would the cell have zero potential?

20.98 Derive an equation that directly relates the standard emf of a redox reaction to its equilibrium constant.

20.99 Using data from Appendix E, calculate the equilibrium constant for the disproportionation of the copper(I) ion at room temperature: $2\,Cu^{+}(aq) \longrightarrow Cu^{2+}(aq) + Cu(s)$

20.100 Indicate whether each of the following statements is true or false:

(a) A perfectly reasonable number for an aqueous E°_{cell} is 9 V.

(b) If you double the concentrations of reagents in a voltaic cell, you will double the voltage obtainable.

(c) The stronger the oxidizing agent, the larger its reduction potential.

20.101 **(a)** Write the reactions for the discharge and charge of a nickel-cadmium rechargeable battery. **(b)** Given the following reduction potentials, calculate the standard emf of the cell:

$$Cd(OH)_2(s) + 2\,e^{-} \longrightarrow Cd(s) + 2\,OH^{-}(aq)$$
$$E^{\circ}_{red} = -0.76\ V$$
$$NiO(OH)(s) + H_2O(l) + e^{-} \longrightarrow Ni(OH)_2(s) + OH(aq)$$
$$E^{\circ}_{red} = +0.49\ V$$

(c) A typical nicad voltaic cell generates an emf of +1.30 V. Why is there a difference between this value and the one you calculated in part (b)? **(d)** Calculate the equilibrium constant for the overall nicad reaction based on this typical emf value.

20.102 If you were going to apply a small potential to a steel ship resting in the water as a means of inhibiting corrosion, would you apply a negative or a positive charge? Explain.

20.103 The following quotation is taken from an article dealing with corrosion of electronic materials: "Sulfur dioxide, its acidic oxidation products, and moisture are well established as the principal causes of outdoor corrosion of many metals." Using Ni as an example, explain why the factors cited affect the rate of corrosion. Write chemical equations to illustrate your points. [*Note:* NiO(*s*) is soluble in acidic solution.]

20.104 Two wires from a battery are tested with a piece of filter paper moistened with NaCl solution containing phenolphthalein, an acid-base indicator that is colorless in acid and pink in base. When the wires touch the paper about an inch apart, the rightmost wire produces a pink coloration on the filter paper and the leftmost produces none. Which wire is connected to the positive terminal of the battery? Explain.

[20.105] **(a)** How many coulombs are required to plate a layer of chromium metal 0.25 mm thick on an auto bumper with a total area of 0.32 m^2 from a solution containing CrO_4^{2-}? The density of chromium metal is 7.20 g/cm^3. **(b)** What current flow is required for this electroplating if the bumper is to be plated in 10.0 s? **(c)** If the external source has an emf of +6.0 V and the electrolytic cell is 65% efficient, how much electrical power is expended to electroplate the bumper?

20.106 **(a)** What is the maximum amount of work that a 6-V lead-acid battery of a golf cart can accomplish if it is rated at 300 A-hr? **(b)** List some of the reasons why this amount of work is never realized.

[20.107] Some years ago a unique proposal was made to raise the *Titanic*. The plan involved placing pontoons within the ship using a surface-controlled submarine-type vessel. The pontoons would contain cathodes and would be filled with hydrogen gas formed by the electrolysis of water. It has been estimated that it would require about 7×10^8 mol of H_2 to provide the buoyancy to lift the ship [*J. Chem. Educ.*, Vol. *50*, **1973**, 61]. **(a)** How many coulombs of electrical charge would be required? **(b)** What is the minimum voltage required to generate H_2 and O_2 if the pressure on the gases at the depth of the wreckage (2 mi) is 300 atm? **(c)** What is the minimum electrical energy required to raise the *Titanic* by electrolysis? **(d)** What is the minimum cost of the electrical energy required to generate the necessary H_2 if the electricity costs 85 cents per kilowatt-hour to generate at the site?

Integrative Exercises

20.108 Copper dissolves in concentrated nitric acid with the evolution of NO(*g*), which is subsequently oxidized to $NO_2(g)$ in air. In contrast, copper does not dissolve in concentrated hydrochloric acid. Explain these observations by using standard reduction potentials from Table 20.1.

20.109 The Haber process is the principal industrial route for converting nitrogen into ammonia:

$$N_2(g) + 3\,H_2(g) \longrightarrow 2\,NH_3(g)$$

(a) What is being oxidized, and what is being reduced? **(b)** Using the thermodynamic data in Appendix C, calculate the equilibrium constant for the process at room temperature. **(c)** Calculate the standard emf of the Haber process at room temperature.

[20.110] In a galvanic cell the cathode is an Ag^{+} (1.00 M)/Ag(*s*) half-cell. The anode is a standard hydrogen electrode immersed in a buffer solution containing 0.10 M benzoic acid (C_6H_5COOH) and 0.050 M sodium benzoate ($C_6H_5COO^{-}Na^{+}$). The measured cell voltage is 1.030 V. What is the pK_a of benzoic acid?

20.111 Consider the general oxidation of a species A in solution: $A \longrightarrow A^{+} + e^{-}$. The term "oxidation potential" is sometimes used to describe the ease with which species A is oxidized—the easier a species is to oxidize, the greater its oxidation potential. **(a)** What is the relationship between the standard oxidation potential of A and the standard reduction potential of A^+? **(b)** Which of the metals listed in Table 4.5 has the highest standard oxidation potential? Which has the lowest? **(c)** For a series of substances, the trend in oxidation potential is often related to the trend in the first ionization energy. Explain why this relationship makes sense.

[20.112] Gold metal dissolves in aqua regia, a mixture of concentrated hydrochloric acid and concentrated nitric acid. The standard reduction potentials

$$Au^{3+}(aq) + 3\,e^- \longrightarrow Au(s) \qquad E^\circ_{red} = +1.498\text{ V}$$

$$AuCl_4^-(aq) + 3\,e^- \longrightarrow Au(s) + 4\,Cl^-(aq) \qquad E^\circ_{red} = +1.002\text{ V}$$

are important in gold chemistry. **(a)** Use half-reactions to write a balanced equation for the reaction of Au and nitric acid to produce Au^{3+} and $NO(g)$, and calculate the standard emf of this reaction. Is this reaction spontaneous? **(b)** Use half-reactions to write a balanced equation for the reaction of Au and hydrochloric acid to produce $AuCl_4^-(aq)$ and $H_2(g)$, and calculate the standard emf of this reaction. Is this reaction spontaneous? **(c)** Use half-reactions to write a balanced equation for the reaction of Au and aqua regia to produce $AuCl_4^-(aq)$ and $NO(g)$, and calculate the standard emf of this reaction. Is this reaction spontaneous under standard conditions? **(d)** Use the Nernst equation to explain why aqua regia made from *concentrated* hydrochloric and nitric acids is able to dissolve gold.

20.113 A voltaic cell is based on $Ag^+(aq)/Ag(s)$ and $Fe^{3+}(aq)/Fe^{2+}(aq)$ half-cells. **(a)** What is the standard emf of the cell? **(b)** Which reaction occurs at the cathode, and which at the anode of the cell? **(c)** Use S° values in Appendix C and the relationship between cell potential and free-energy change to predict whether the standard cell potential increases or decreases when the temperature is raised above 25°C.

20.114 Hydrogen gas has the potential as a clean fuel in reaction with oxygen. The relevant reaction is

$$2\,H_2(g) + O_2(g) \longrightarrow 2\,H_2O(l)$$

Consider two possible ways of utilizing this reaction as an electrical energy source: (i) Hydrogen and oxygen gases are combusted and used to drive a generator, much as coal is currently used in the electric power industry; (ii) hydrogen and oxygen gases are used to generate electricity directly by using fuel cells that operate at 85°C. **(a)** Use data in Appendix C to calculate ΔH° and ΔS° for the above reaction. We will assume that these values do not change appreciably with temperature. **(b)** Based on the values from part (a), what trend would you expect for the magnitude of ΔG for the above reaction as the temperature increases? **(c)** What is the significance of the change in the magnitude of ΔG with temperature with respect to the utility of hydrogen as a fuel (recall Equation 19.19)? **(d)** Based on the analysis here, would it be more efficient to use the combustion method or the fuel-cell method to generate electrical energy from hydrogen?

20.115 Cytochrome, a complicated molecule that we will represent as $CyFe^{2+}$, reacts with the air we breathe to supply energy required to synthesize adenosine triphosphate (ATP). The body uses ATP as an energy source to drive other reactions. (Section 19.7) At pH 7.0 the following reduction potentials pertain to this oxidation of $CyFe^{2+}$:

$$O_2(g) + 4\,H^+(aq) + 4\,e^- \longrightarrow 2\,H_2O(l) \qquad E^\circ_{red} = +0.82\text{ V}$$

$$CyFe^{3+}(aq) + e^- \longrightarrow CyFe^{2+}(aq) \qquad E^\circ_{red} = +0.22\text{ V}$$

(a) What is ΔG for the oxidation of $CyFe^{2+}$ by air? **(b)** If the synthesis of 1.00 mol of ATP from adenosine diphosphate (ADP) requires a ΔG of 37.7 kJ, how many moles of ATP are synthesized per mole of O_2?

[20.116] The standard potential for the reduction of $AgSCN(s)$ is +0.0895 V.

$$AgSCN(s) + e^- \longrightarrow Ag(s) + SCN^-(aq)$$

Using this value and the electrode potential for $Ag^+(aq)$, calculate the K_{sp} for AgSCN.

[20.117] The K_{sp} value for $PbS(s)$ is 8.0×10^{-28}. By using this value together with an electrode potential from Appendix E, determine the value of the standard reduction potential for the reaction

$$PbS(s) + 2\,e^- \longrightarrow Pb(s) + S^{2-}(aq)$$

[20.118] A pH meter (Figure 16.6) employs a voltaic cell for which the cell potential is very sensitive to pH. A simple (but impractical) pH meter can be constructed by using two hydrogen electrodes: one standard hydrogen electrode (Figure 20.10) and a hydrogen electrode (with 1 atm pressure of H_2 gas) dipped into the solution of unknown pH. The two half-cells are connected by a salt bridge or porous glass disk. **(a)** Sketch the cell described above. **(b)** Write the half-cell reactions for the cell, and calculate the standard emf. **(c)** What is the pH of the solution in the half-cell that has the standard hydrogen electrode? **(d)** What is the cell emf when the pH of the unknown solution is 5.0? **(e)** How precise would a voltmeter have to be in order to detect a change in the pH of 0.01 pH units?

[20.119] A student designs an ammeter (a device that measures electrical current) that is based on the electrolysis of water into hydrogen and oxygen gases. When electrical current of unknown magnitude is run through the device for 2.00 min, 12.3 mL of water-saturated $H_2(g)$ is collected. The temperature of the system is 25.5°C, and the atmospheric pressure is 768 torr. What is the magnitude of the current in amperes?

eMEDIA EXERCISES

These exercises make use of the interactive objects available online in OneKey or the Companion Website, and on your Accelerator CD. Access to these resources comes in your MediaPak.

20.120 The oxidation of zinc by molecular oxygen is shown in the **Oxidation-Reduction Reactions—Part I** animation (*20.1*). **(a)** Write the equation for the reaction between zinc and oxygen. **(b)** Assign oxidation states to each element on each side of the equation. **(c)** Is it possible for oxidation to occur without oxygen? Explain.

20.121 In Exercise 20.35 you chose from a list of half-reactions to construct voltaic cells of specified, relative emf values. The reaction shown in the **Formation of Silver Crystals** movie (*20.3*) can be represented with two of the half-reactions listed in Exercise 20.35. **(a)** Write and balance the equation for the reaction in the movie. **(b)** Which species is the oxidizing agent and which is the reducing agent in this reaction?

20.122 The **Oxidation-Reduction Reactions—Part II** animation (*20.1*) illustrates the reaction that occurs when zinc metal is placed in an aqueous solution containing Cu^{2+} ions. The animation **Voltaic Cells I: The Copper-Zinc Cell** (*20.3*) uses the same reaction for the construction of a voltaic cell. **(a)** Explain why the reaction as depicted in the **Oxidation-Reduction Reactions—Part II** animation cannot be used to generate a voltage. **(b)** Sketch a voltaic cell based on the Zn-Cu reaction. Label and give the purpose of each component of the cell.

20.123 In the **Redox Chemistry of Iron and Copper** movie (*20.3*) an iron nail is placed in an aqueous solution of copper sulfate. **(a)** Write the half-reaction for the formation of the copper coating on the nail. **(b)** Is there another half-reaction occurring in the solution? If so, what is it? **(c)** What is the sign of ΔG for the overall process that occurs?

20.124 Chromium metal can be electroplated onto a copper electrode from an aqueous solution of $H_2Cr_2O_7$, as shown in the **Electroplating** movie (*20.9*). **(a)** Write the half-reaction for the reduction of dichromate ion to chromium metal. **(b)** What current would have to be applied for 15 minutes in order to plate out 0.75 gram of chromium metal from a solution of $H_2Cr_2O_7$?

eLABORATORY EXERCISES

These exercises allow students to apply the concepts and skills for this chapter in the simulated laboratory of Virtual ChemLab. Worksheets for each of these exercises are found in the Virtual ChemLab Workbook in your MediaPak.

20.125 ***(VCL 20-1) Study of Oxidation-Reduction Titrations***
Titrations provide a method of quantitatively measuring the concentration of an unknown solution. This is done by delivering a titrant of known concentration into an analyte of known volume. (The concentration of an unknown titrant can also be determined by titration with an analyte of known concentration and volume.) In oxidation-reduction (redox) titrations, the voltage resulting from the mixture of an oxidant and a reductant can be measured as the titration proceeds. The equivalence point of the titration, or the point where the analyte has been completely consumed by the titrant, is identified as the point where the voltage changes rapidly over a small volume of titrant delivered. In this assignment you will observe this titration curve by titrating ferrous chloride ($FeCl_2$) with potassium permanganate ($KMnO_4$).

20.126 ***(VCL 20-2) Standardization of a Permanganate Solution***
In this assignment you will determine the molarity of an unknown solution of potassium permanganate ($KMnO_4$) by titrating against a primary standard, solid arsenic trioxide (As_2O_3).

20.127 ***(VCL 20-3) Analysis of a Ferrous Chloride Sample***
In this assignment you will determine the mass percent of an unknown sample of ferrous chloride ($FeCl_2$) by titrating with a potassium permanganate ($KMnO_4$) solution of known concentration.

APPENDIX A

Mathematical Operations

A.1 | Exponential Notation

The numbers used in chemistry are often either extremely large or extremely small. Such numbers are conveniently expressed in the form

$$N \times 10^n$$

where N is a number between 1 and 10, and n is the exponent. Some examples of this *exponential notation*, which is also called *scientific notation*, follow.

1,200,000 is 1.2×10^6 (read "one point two times ten to the sixth power")

0.000604 is 6.04×10^{-4} (read "six point zero four times ten to the negative fourth power")

A positive exponent, as in the first example, tells us how many times a number must be multiplied by 10 to give the long form of the number:

$$1.2 \times 10^6 = 1.2 \times 10 \times 10 \times 10 \times 10 \times 10 \times 10 \quad \text{(six tens)}$$
$$= 1{,}200{,}000$$

It is also convenient to think of the *positive exponent* as the number of places the decimal point must be moved to the *left* to obtain a number greater than 1 and less than 10: If we begin with 3450 and move the decimal point three places to the left, we end up with 3.45×10^3.

In a related fashion, a negative exponent tells us how many times we must divide a number by 10 to give the long form of the number.

$$6.04 \times 10^{-4} = \frac{6.04}{10 \times 10 \times 10 \times 10} = 0.000604$$

It is convenient to think of the *negative exponent* as the number of places the decimal point must be moved to the *right* to obtain a number greater than 1 but less than 10: If we begin with 0.0048 and move the decimal point three places to the right, we end up with 4.8×10^{-3}.

In the system of exponential notation, with each shift of the decimal point one place to the right, the exponent *decreases* by 1:

$$4.8 \times 10^{-3} = 48 \times 10^{-4}$$

Similarly, with each shift of the decimal point one place to the left, the exponent *increases* by 1:

$$4.8 \times 10^{-3} = 0.48 \times 10^{-2}$$

Many scientific calculators have a key labeled EXP or EE, which is used to enter numbers in exponential notation. To enter the number 5.8×10^3 on such a calculator, the key sequence is

[5] [·] [8] [EXP] (or [EE]) [3]

On some calculators the display will show 5.8, then a space, followed by 03, the exponent. On other calculators, a small 10 is shown with an exponent 3.

To enter a negative exponent, use the key labeled +/−. For example, to enter the number 8.6×10^{-5}, the key sequence is

[8] [·] [6] [EXP] [+/−] [5]

When entering a number in exponential notation, do not key in the 10 if you use the EXP or EE button.

In working with exponents, it is important to recall that $10^0 = 1$. The following rules are useful for carrying exponents through calculations.

1. **Addition and Subtraction** In order to add or subtract numbers expressed in exponential notation, the powers of 10 must be the same.

$$\begin{aligned}(5.22 \times 10^4) + (3.21 \times 10^2) &= (522 \times 10^2) + (3.21 \times 10^2)\\ &= 525 \times 10^2 \quad \text{(3 significant figures)}\\ &= 5.25 \times 10^4\end{aligned}$$

$$\begin{aligned}(6.25 \times 10^{-2}) - (5.77 \times 10^{-3}) &= (6.25 \times 10^{-2}) - (0.577 \times 10^{-2})\\ &= 5.67 \times 10^{-2} \quad \text{(3 significant figures)}\end{aligned}$$

When you use a calculator to add or subtract, you need not be concerned with having numbers with the same exponents, because the calculator automatically takes care of this matter.

2. **Multiplication and Division** When numbers expressed in exponential notation are multiplied, the exponents are added; when numbers expressed in exponential notation are divided, the exponent of the denominator is subtracted from the exponent of the numerator.

$$\begin{aligned}(5.4 \times 10^2)(2.1 \times 10^3) &= (5.4)(2.1) \times 10^{2+3}\\ &= 11 \times 10^5\\ &= 1.1 \times 10^6\end{aligned}$$

$$(1.2 \times 10^5)(3.22 \times 10^{-3}) = (1.2)(3.22) \times 10^{5-3} = 3.9 \times 10^2$$

$$\frac{3.2 \times 10^5}{6.5 \times 10^2} = \frac{3.2}{6.5} \times 10^{5-2} = 0.49 \times 10^3 = 4.9 \times 10^2$$

$$\frac{5.7 \times 10^7}{8.5 \times 10^{-2}} = \frac{5.7}{8.5} \times 10^{7-(-2)} = 0.67 \times 10^9 = 6.7 \times 10^8$$

3. **Powers and Roots** When numbers expressed in exponential notation are raised to a power, the exponents are multiplied by the power. When the roots of numbers expressed in exponential notation are taken, the exponents are divided by the root.

$$\begin{aligned}(1.2 \times 10^5)^3 &= (1.2)^3 \times 10^{5\times3}\\ &= 1.7 \times 10^{15}\\ \sqrt[3]{2.5 \times 10^6} &= \sqrt[4]{2.5} \times 10^{6/3}\\ &= 1.3 \times 10^2\end{aligned}$$

Scientific calculators usually have keys labeled x^2 and $\sqrt{x}$ for squaring and taking the square root of a number, respectively. To take higher powers or roots, many calculators have y^x and $\sqrt[x]{y}$ (or INV y^x) keys. For example, to perform the operation $\sqrt[3]{7.5 \times 10^{-4}}$ on such a calculator, you would key in 7.5×10^{-4}, press the $\sqrt[x]{y}$ key (or the INV and then the y^x keys), enter the root, 3, and finally press =. The result is 9.1×10^{-2}.

SAMPLE EXERCISE 1 | Using Exponential Notation

Perform each of the following operations, using your calculator where possible:

(a) Write the number 0.0054 in standard exponential notation
(b) $(5.0 \times 10^{-2}) + (4.7 \times 10^{-3})$
(c) $(5.98 \times 10^{12})(2.77 \times 10^{-5})$
(d) $\sqrt[4]{1.75 \times 10^{-12}}$

Solution (a) Because we move the decimal point three places to the right to convert 0.0054 to 5.4, the exponent is −3:

$$5.4 \times 10^{-3}$$

Scientific calculators are generally able to convert numbers to exponential notation using one or two keystrokes. Consult your instruction manual to see how this operation is accomplished on your calculator.

(b) To add these numbers longhand, we must convert them to the same exponent.

$$(5.0 \times 10^{-2}) + (0.47 \times 10^{-2}) = (5.0 + 0.47) \times 10^{-2} = 5.5 \times 10^{-2}$$

(Note that the result has only two significant figures.) To perform this operation on a calculator, we enter the first number, strike the + key, then enter the second number and strike the = key.

(c) Performing this operation longhand, we have

$$(5.98 \times 2.77) \times 10^{12-5} = 16.6 \times 10^{7} = 1.66 \times 10^{8}$$

On a scientific calculator, we enter 5.98×10^{12}, press the × key, enter 2.77×10^{-5}, and press the = key.

(d) To perform this operation on a calculator, we enter the number, press the $\sqrt[x]{y}$ key (or the INV and y^x keys), enter 4, and press the = key. The result is 1.15×10^{-3}.

PRACTICE EXERCISE

Perform the following operations:

(a) Write 67,000 in exponential notation, showing two significant figures;
(b) $(3.378 \times 10^{-3}) - (4.97 \times 10^{-5})$;
(c) $(1.84 \times 10^{15})/(7.45 \times 10^{-2})$;
(d) $(6.67 \times 10^{-8})^3$.

Answers: **(a)** 6.7×10^4, **(b)** 3.328×10^{-3}, **(c)** 2.47×10^{16}, **(d)** 2.97×10^{-22}

A.2 | Logarithms

Common Logarithms

The common, or base-10, logarithm (abbreviated log) of any number is the power to which 10 must be raised to equal the number. For example, the common logarithm of 1000 (written log 1000) is 3 because raising 10 to the third power gives 1000.

$$10^3 = 1000, \text{ therefore, } \log 1000 = 3$$

Further examples are

$$\log 10^5 = 5$$

$$\log 1 = 0 \qquad (\text{Remember that } 10^0 = 1)$$

$$\log 10^{-2} = -2$$

In these examples the common logarithm can be obtained by inspection. However, it is not possible to obtain the logarithm of a number such as 31.25 by inspection. The logarithm of 31.25 is the number x that satisfies the following relationship:

$$10^x = 31.25$$

Most electronic calculators have a key labeled LOG that can be used to obtain logarithms. For example, on many calculators we obtain the value of log 31.25 by entering 31.25 and pressing the LOG key. We obtain the following result:

$$\log 31.25 = 1.4949$$

Notice that 31.25 is greater than 10 (10^1)) and less than 100 (10^2). The value for log 31.25 is accordingly between log 10 and log 100, that is, between 1 and 2.

Significant Figures and Common Logarithms

For the common logarithm of a measured quantity, the number of digits after the decimal point equals the number of significant figures in the original number. For example, if 23.5 is a measured quantity (three significant figures), then log 23.5 = 1.371 (three significant figures after the decimal point).

Antilogarithms

The process of determining the number that corresponds to a certain logarithm is known as obtaining an *antilogarithm*. It is the reverse of taking a logarithm. For example, we saw above that log 23.5 = 1.371. This means that the antilogarithm of 1.371 equals 23.5.

$$\log 23.5 = 1.371$$

$$\text{antilog } 1.371 = 23.5$$

The process of taking the antilog of a number is the same as raising 10 to a power equal to that number.

$$\text{antilog } 1.371 = 10^{1.371} = 23.5$$

Many calculators have a key labeled 10^x that allows you to obtain antilogs directly. On others, it will be necessary to press a key labeled INV (for *inverse*), followed by the LOG key.

Natural Logarithms

Logarithms based on the number *e* are called natural, or base *e*, logarithms (abbreviated ln). The natural log of a number is the power to which *e* (which has the value 2.71828...) must be raised to equal the number. For example, the natural log of 10 equals 2.303.

$$e^{2.303} = 10, \text{ therefore } \ln 10 = 2.303$$

Your calculator probably has a key labeled LN that allows you to obtain natural logarithms. For example, to obtain the natural log of 46.8, you enter 46.8 and press the LN key.

$$\ln 46.8 = 3.846$$

The natural antilog of a number is *e* raised to a power equal to that number. If your calculator can calculate natural logs, it will also be able to calculate natural antilogs. On some calculators there is a key labeled e^x that allows you to calculate natural antilogs directly; on others, it will be necessary to first press the INV key followed by the LN key. For example, the natural antilog of 1.679 is given by

$$\text{Natural antilog } 1.679 = e^{1.679} = 5.36$$

The relation between common and natural logarithms is as follows:

$$\ln a = 2.303 \log a$$

Notice that the factor relating the two, 2.303, is the natural log of 10, which we calculated above.

Mathematical Operations Using Logarithms

Because logarithms are exponents, mathematical operations involving logarithms follow the rules for the use of exponents. For example, the product of z^a and z^b (where z is any number) is given by

$$z^a \cdot z^b = z^{(a+b)}$$

Similarly, the logarithm (either common or natural) of a product equals the *sum* of the logs of the individual numbers.

$$\log ab = \log a + \log b \qquad \ln ab = \ln a + \ln b$$

For the log of a quotient,

$$\log(a/b) = \log a - \log b \qquad \ln(a/b) = \ln a - \ln b$$

Using the properties of exponents, we can also derive the rules for the logarithm of a number raised to a certain power.

$$\log a^n = n \log a \qquad \ln a^n = n \ln a$$

$$\log a^{1/n} = (1/n) \log a \qquad \ln a^{1/n} = (1/n) \ln a$$

pH Problems

One of the most frequent uses for common logarithms in general chemistry is in working pH problems. The pH is defined as $-\log[H^+]$, where $[H^+]$ is the hydrogen ion concentration of a solution (Section 16.4). The following sample exercise illustrates this application.

SAMPLE EXERCISE 2 | Using Logarithms

(a) What is the pH of a solution whose hydrogen ion concentration is 0.015 *M*?
(b) If the pH of a solution is 3.80, what is its hydrogen ion concentration?

Solution **(a)** We are given the value of $[H^+]$. We use the LOG key of our calculator to calculate the value of $\log[H^+]$. The pH is obtained by changing the sign of the value obtained. (Be sure to change the sign *after* taking the logarithm.)

$$[H^+] = 0.015$$

$$\log[H^+] = -1.82 \quad \text{(2 significant figures)}$$

$$pH = -(-1.82) = 1.82$$

(b) To obtain the hydrogen ion concentration when given the pH, we must take the antilog of $-$pH.

$$pH = -\log[H^+] = 3.80$$

$$\log[H^+] = -3.80$$

$$[H^+] = \text{antilog}(-3.80) = 10^{-3.80} = 1.6 \times 10^{-4}\ M$$

PRACTICE EXERCISE

Perform the following operations: **(a)** $\log(2.5 \times 10^{-5})$, **(b)** ln 32.7, **(c)** antilog -3.47, **(d)** $e^{-1.89}$.
Answers: **(a)** -4.60, **(b)** 3.487, **(c)** 3.4×10^{-4}, **(d)** 1.5×10^{-1}

A.3 | Quadratic Equations

An algebraic equation of the form $ax^2 + bx + c = 0$ is called a *quadratic equation*. The two solutions to such an equation are given by the quadratic formula:

$$x = \frac{-b \pm \sqrt{b^2 - 4ac}}{2a}$$

SAMPLE EXERCISE 3 | **Using the Quadratic Equation**

Find the values of x that satisfy the equation $2x^2 + 4x = 1$.

Solution To solve the given equation for x, we must first put it in the form

$$ax^2 + bx + c = 0$$

and then use the quadratic formula. If

$$2x^2 + 4x = 1$$

then

$$2x^2 + 4x - 1 = 0$$

Using the quadratic formula, where $a = 2$, $b = 4$, and $c = -1$, we have

$$x = \frac{-4 \pm \sqrt{(4)(4) - 4(2)(-1)}}{2(2)}$$

$$= \frac{-4 \pm \sqrt{16 + 8}}{4} = \frac{-4 \pm \sqrt{24}}{4} = \frac{-4 \pm 4.899}{4}$$

The two solutions are

$$x = \frac{0.899}{4} = 0.225 \quad \text{and} \quad x = \frac{-8.899}{4} = -2.225$$

Often in chemical problems the negative solution has no physical meaning, and only the positive answer is used.

A.4 | Graphs

Often the clearest way to represent the interrelationship between two variables is to graph them. Usually, the variable that is being experimentally varied, called the *independent variable,* is shown along the horizontal axis (x-axis). The variable that responds to the change in the independent variable, called the *dependent variable,* is then shown along the vertical axis (y-axis). For example, consider an experiment in which we vary the temperature of an enclosed gas and measure its pressure. The independent variable is temperature, and the dependent variable is pressure. The data shown in Table A-1 ▶ can be obtained by means of this experiment. These data are shown graphically in Figure 1 ▶. The relationship between temperature and pressure is linear. The equation for any straight-line graph has the form

$$y = mx + b$$

where m is the slope of the line and b is the intercept with the y-axis. In the case of Figure 1, we could say that the relationship between temperature and pressure takes the form

$$P = mT + b$$

where P is pressure in atm and T is temperature in °C. As shown in Figure 1, the slope is 4.10×10^{-4} atm/°C, and the intercept—the point where the line crosses the y-axis—is 0.112 atm. Therefore, the equation for the line is

$$P = \left(4.10 \times 10^{-4} \frac{\text{atm}}{°\text{C}}\right)T + 0.112 \text{ atm}$$

TABLE A-1 Interrelation Between Pressure and Temperature

Temperature (°C)	Pressure (atm)
20.0	0.120
30.0	0.124
40.0	0.128
50.0	0.132

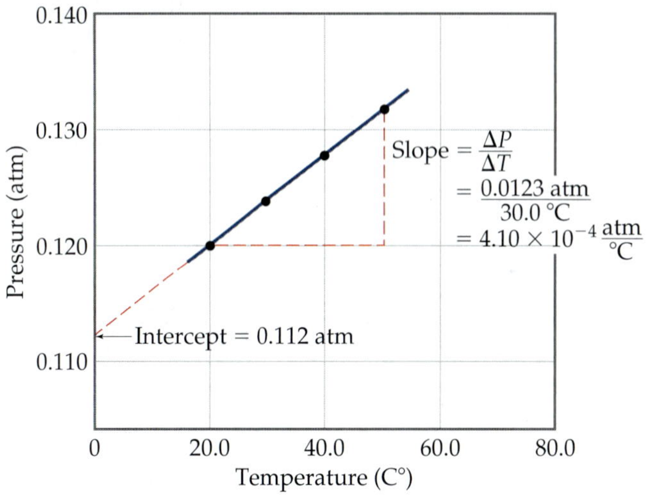

▲ **Figure 1**

APPENDIX B

Properties of Water

Density:	0.99987 g/mL at 0°C 1.00000 g/mL at 4°C 0.99707 g/mol at 25°C 0.95838 g/mol at 100°C
Heat of fusion:	6.008 kJ/mol at 0°C
Heat of vaporization:	44.94 kJ/mol at 0°C 44.02 kJ/mol at 25°C 40.67 kJ/mol at 100°C
Ion-product constant, K_w:	1.14×10^{-15} at 0°C 1.01×10^{-14} at 25°C 5.47×10^{-14} at 50°C
Specific heat:	Ice (at −3°C) 2.092 J/g-K Water (at 14.5°C) 4.184 J/g-K Steam (at 100°C) 1.841 J/g-K

Vapor Pressure (torr)

T(°C)	*P*	*T*(°C)	*P*	*T*(°C)	*P*	*T*(°C)	*P*
0	4.58	21	18.65	35	42.2	92	567.0
5	6.54	22	19.83	40	55.3	94	610.9
10	9.21	23	21.07	45	71.9	96	657.6
12	10.52	24	22.38	50	92.5	98	707.3
14	11.99	25	23.76	55	118.0	100	760.0
16	13.63	26	25.21	60	149.4	102	815.9
17	14.53	27	26.74	65	187.5	104	875.1
18	15.48	28	28.35	70	233.7	106	937.9
19	16.48	29	30.04	80	355.1	108	1004.4
20	17.54	30	31.82	90	525.8	110	1074.6

APPENDIX C

Thermodynamic Quantities for Selected Substances at 298.15 K (25°C)

Substance	ΔH_f° (kJ/mol)	ΔG_f° (kJ/mol)	S° (J/mol-K)
Aluminum			
$Al(s)$	0	0	28.32
$AlCl_3(s)$	−705.6	−630.0	109.3
$Al_2O_3(s)$	−1669.8	−1576.5	51.00
Barium			
$Ba(s)$	0	0	63.2
$BaCO_3(s)$	−1216.3	−1137.6	112.1
$BaO(s)$	−553.5	−525.1	70.42
Beryllium			
$Be(s)$	0	0	9.44
$BeO(s)$	−608.4	−579.1	13.77
$Be(OH)_2(s)$	−905.8	−817.9	50.21
Bromine			
$Br(g)$	111.8	82.38	174.9
$Br^-(aq)$	−120.9	−102.8	80.71
$Br_2(g)$	30.71	3.14	245.3
$Br_2(l)$	0	0	152.3
$HBr(g)$	−36.23	−53.22	198.49
Calcium			
$Ca(g)$	179.3	145.5	154.8
$Ca(s)$	0	0	41.4
$CaCO_3$ (s, calcite)	−1207.1	−1128.76	92.88
$CaCl_2(s)$	−795.8	−748.1	104.6
$CaF_2(s)$	−1219.6	−1167.3	68.87
$CaO(s)$	−635.5	−604.17	39.75
$Ca(OH)_2(s)$	−986.2	−898.5	83.4
$CaSO_4(s)$	−1434.0	−1321.8	106.7
Carbon			
$C(g)$	718.4	672.9	158.0
C(s, diamond)	1.88	2.84	2.43
C(s, graphite)	0	0	5.69
$CCl_4(g)$	−106.7	−64.0	309.4
$CCl_4(l)$	−139.3	−68.6	214.4
$CF_4(g)$	−679.9	−635.1	262.3
$CH_4(g)$	−74.8	−50.8	186.3
$C_2H_2(g)$	226.77	209.2	200.8
$C_2H_4(g)$	52.30	68.11	219.4
$C_2H_6(g)$	−84.68	−32.89	229.5
$C_3H_8(g)$	−103.85	−23.47	269.9
$C_4H_{10}(g)$	−124.73	−15.71	310.0
$C_4H_{10}(l)$	−147.6	−15.0	231.0
$C_6H_6(g)$	82.9	129.7	269.2
$C_6H_6(l)$	49.0	124.5	172.8
$CH_3OH(g)$	−201.2	−161.9	237.6
$CH_3OH(l)$	−238.6	−166.23	126.8
$C_2H_5OH(g)$	−235.1	−168.5	282.7
$C_2H_5OH(l)$	−277.7	−174.76	160.7
$C_6H_{12}O_6(s)$	−1273.02	−910.4	212.1
$CO(g)$	−110.5	−137.2	197.9
$CO_2(g)$	−393.5	−394.4	213.6
$HC_2H_3O_2(l)$	−487.0	−392.4	159.8
Cesium			
$Cs(g)$	76.50	49.53	175.6
$Cs(l)$	2.09	0.03	92.07
$Cs(s)$	0	0	85.15
$CsCl(s)$	−442.8	−414.4	101.2
Chlorine			
$Cl(g)$	121.7	105.7	165.2
$Cl^-(aq)$	−167.2	−131.2	56.5
$Cl_2(g)$	0	0	222.96
$HCl(aq)$	−167.2	−131.2	56.5
$HCl(g)$	−92.30	−95.27	186.69
Chromium			
$Cr(g)$	397.5	352.6	174.2
$Cr(s)$	0	0	23.6
$Cr_2O_3(s)$	−1139.7	−1058.1	81.2
Cobalt			
$Co(g)$	439	393	179
$Co(s)$	0	0	28.4
Copper			
$Cu(g)$	338.4	298.6	166.3
$Cu(s)$	0	0	33.30
$CuCl_2(s)$	−205.9	−161.7	108.1
$CuO(s)$	−156.1	−128.3	42.59
$Cu_2O(s)$	−170.7	−147.9	92.36

Substance	ΔH_f° (kJ/mol)	ΔG_f° (kJ/mol)	S° (J/mol-K)
Fluorine			
$F(g)$	80.0	61.9	158.7
$F^-(aq)$	−332.6	−278.8	−13.8
$F_2(g)$	0	0	202.7
$HF(g)$	−268.61	−270.70	173.51
Hydrogen			
$H(g)$	217.94	203.26	114.60
$H^+(aq)$	0	0	0
$H^+(g)$	1536.2	1517.0	108.9
$H_2(g)$	0	0	130.58
Iodine			
$I(g)$	106.60	70.16	180.66
$I^-(aq)$	−55.19	−51.57	111.3
$I_2(g)$	62.25	19.37	260.57
$I_2(s)$	0	0	116.73
$HI(g)$	25.94	1.30	206.3
Iron			
$Fe(g)$	415.5	369.8	180.5
$Fe(s)$	0	0	27.15
$Fe^{2+}(aq)$	−87.86	−84.93	113.4
$Fe^{3+}(aq)$	−47.69	−10.54	293.3
$FeCl_2(s)$	−341.8	−302.3	117.9
$FeCl_3(s)$	−400	−334	142.3
$FeO(s)$	−271.9	−255.2	60.75
$Fe_2O_3(s)$	−822.16	−740.98	89.96
$Fe_3O_4(s)$	−1117.1	−1014.2	146.4
$FeS_2(s)$	−171.5	−160.1	52.92
Lead			
$Pb(s)$	0	0	68.85
$PbBr_2(s)$	−277.4	−260.7	161
$PbCO_3(s)$	−699.1	−625.5	131.0
$Pb(NO_3)_2(aq)$	−421.3	−246.9	303.3
$Pb(NO_3)_2(s)$	−451.9	—	—
$PbO(s)$	−217.3	−187.9	68.70
Lithium			
$Li(g)$	159.3	126.6	138.8
$Li(s)$	0	0	29.09
$Li^+(aq)$	−278.5	−273.4	12.2
$Li^+(g)$	685.7	648.5	133.0
$LiCl(s)$	−408.3	−384.0	59.30
Magnesium			
$Mg(g)$	147.1	112.5	148.6
$Mg(s)$	0	0	32.51
$MgCl_2(s)$	−641.6	−592.1	89.6
$MgO(s)$	−601.8	−569.6	26.8
$Mg(OH)_2(s)$	−924.7	−833.7	63.24
Manganese			
$Mn(g)$	280.7	238.5	173.6
$Mn(s)$	0	0	32.0
$MnO(s)$	−385.2	−362.9	59.7

Substance	ΔH_f° (kJ/mol)	ΔG_f° (kJ/mol)	S° (J/mol-K)
$MnO_2(s)$	−519.6	−464.8	53.14
$MnO_4^-(aq)$	−541.4	−447.2	191.2
Mercury			
$Hg(g)$	60.83	31.76	174.89
$Hg(l)$	0	0	77.40
$HgCl_2(s)$	−230.1	−184.0	144.5
$Hg_2Cl_2(s)$	−264.9	−210.5	192.5
Nickel			
$Ni(g)$	429.7	384.5	182.1
$Ni(s)$	0	0	29.9
$NiCl_2(s)$	−305.3	−259.0	97.65
$NiO(s)$	−239.7	−211.7	37.99
Nitrogen			
$N(g)$	472.7	455.5	153.3
$N_2(g)$	0	0	191.50
$NH_3(aq)$	−80.29	−26.50	111.3
$NH_3(g)$	−46.19	−16.66	192.5
$NH_4^+(aq)$	−132.5	−79.31	113.4
$N_2H_4(g)$	95.40	159.4	238.5
$NH_4CN(s)$	0.0	—	—
$NH_4Cl(s)$	−314.4	−203.0	94.6
$NH_4NO_3(s)$	−365.6	−184.0	151
$NO(g)$	90.37	86.71	210.62
$NO_2(g)$	33.84	51.84	240.45
$N_2O(g)$	81.6	103.59	220.0
$N_2O_4(g)$	9.66	98.28	304.3
$NOCl(g)$	52.6	66.3	264
$HNO_3(aq)$	−206.6	−110.5	146
$HNO_3(g)$	−134.3	−73.94	266.4
Oxygen			
$O(g)$	247.5	230.1	161.0
$O_2(g)$	0	0	205.0
$O_3(g)$	142.3	163.4	237.6
$OH^-(aq)$	−230.0	−157.3	−10.7
$H_2O(g)$	−241.82	−228.57	188.83
$H_2O(l)$	−285.83	−237.13	69.91
$H_2O_2(g)$	−136.10	−105.48	232.9
$H_2O_2(l)$	−187.8	−120.4	109.6
Phosphorus			
$P(g)$	316.4	280.0	163.2
$P_2(g)$	144.3	103.7	218.1
$P_4(g)$	58.9	24.4	280
$P_4(s, \text{red})$	−17.46	−12.03	22.85
$P_4(s, \text{white})$	0	0	41.08
$PCl_3(g)$	−288.07	−269.6	311.7
$PCl_3(l)$	−319.6	−272.4	217
$PF_5(g)$	−1594.4	−1520.7	300.8
$PH_3(g)$	5.4	13.4	210.2
$P_4O_6(s)$	−1640.1	—	—
$P_4O_{10}(s)$	−2940.1	−2675.2	228.9
$POCl_3(g)$	−542.2	−502.5	325
$POCl_3(l)$	−597.0	−520.9	222
$H_3PO_4(aq)$	−1288.3	−1142.6	158.2

Substance	ΔH_f° (kJ/mol)	ΔG_f° (kJ/mol)	S° (J/mol-K)
Potassium			
$K(g)$	89.99	61.17	160.2
$K(s)$	0	0	64.67
$KCl(s)$	−435.9	−408.3	82.7
$KClO_3(s)$	−391.2	−289.9	143.0
$KClO_3(aq)$	−349.5	−284.9	265.7
$K_2CO_3(s)$	−1150.18	−1064.58	155.44
$KNO_3(s)$	−492.70	−393.13	132.9
$K_2O(s)$	−363.2	−322.1	94.14
$KO_2(s)$	−284.5	−240.6	122.5
$K_2O_2(s)$	−495.8	−429.8	113.0
$KOH(s)$	−424.7	−378.9	78.91
$KOH(aq)$	−482.4	−440.5	91.6
Rubidium			
$Rb(g)$	85.8	55.8	170.0
$Rb(s)$	0	0	76.78
$RbCl(s)$	−430.5	−412.0	92
$RbClO_3(s)$	−392.4	−292.0	152
Scandium			
$Sc(g)$	377.8	336.1	174.7
$Sc(s)$	0	0	34.6
Selenium			
$H_2Se(g)$	29.7	15.9	219.0
Silicon			
$Si(g)$	368.2	323.9	167.8
$Si(s)$	0	0	18.7
$SiC(s)$	−73.22	−70.85	16.61
$SiCl_4(l)$	−640.1	−572.8	239.3
$SiO_2(s, \text{quartz})$	−910.9	−856.5	41.84
Silver			
$Ag(s)$	0	0	42.55
$Ag^+(aq)$	105.90	77.11	73.93
$AgCl(s)$	−127.0	−109.70	96.11
$Ag_2O(s)$	−31.05	−11.20	121.3
$AgNO_3(s)$	−124.4	−33.41	140.9
Sodium			
$Na(g)$	107.7	77.3	153.7
$Na(s)$	0	0	51.45
$Na^+(aq)$	−240.1	−261.9	59.0
$Na^+(g)$	609.3	574.3	148.0
$NaBr(aq)$	−360.6	−364.7	141.00
$NaBr(s)$	−361.4	−349.3	86.82
$Na_2CO_3(s)$	−1130.9	−1047.7	136.0
$NaCl(aq)$	−407.1	−393.0	115.5
$NaCl(g)$	−181.4	−201.3	229.8
$NaCl(s)$	−410.9	−384.0	72.33
$NaHCO_3(s)$	−947.7	−851.8	102.1
$NaNO_3(aq)$	−446.2	−372.4	207
$NaNO_3(s)$	−467.9	−367.0	116.5
$NaOH(aq)$	−469.6	−419.2	49.8
$NaOH(s)$	−425.6	−379.5	64.46
Strontium			
$SrO(s)$	−592.0	−561.9	54.9
$Sr(g)$	164.4	110.0	164.6
Sulfur			
$S(s, \text{rhombic})$	0	0	31.88
$S_8(g)$	102.3	49.7	430.9
$SO_2(g)$	−296.9	−300.4	248.5
$SO_3(g)$	−395.2	−370.4	256.2
$SO_4^{2-}(aq)$	−909.3	−744.5	20.1
$SOCl_2(l)$	−245.6	—	—
$H_2S(g)$	−20.17	−33.01	205.6
$H_2SO_4(aq)$	−909.3	−744.5	20.1
$H_2SO_4(l)$	−814.0	−689.9	156.1
Titanium			
$Ti(g)$	468	422	180.3
$Ti(s)$	0	0	30.76
$TiCl_4(g)$	−763.2	−726.8	354.9
$TiCl_4(l)$	−804.2	−728.1	221.9
$TiO_2(s)$	−944.7	−889.4	50.29
Vanadium			
$V(g)$	514.2	453.1	182.2
$V(s)$	0	0	28.9
Zinc			
$Zn(g)$	130.7	95.2	160.9
$Zn(s)$	0	0	41.63
$ZnCl_2(s)$	−415.1	−369.4	111.5
$ZnO(s)$	−348.0	−318.2	43.9

APPENDIX

D Aqueous Equilibrium Constants

TABLE D-1 Dissociation Constants for Acids at 25°C

Name	Formula	K_{a1}	K_{a2}	K_{a3}
Acetic	$HC_2H_3O_2$	1.8×10^{-5}		
Arsenic	H_3AsO_4	5.6×10^{-3}	1.0×10^{-7}	3.0×10^{-12}
Arsenous	H_3AsO_3	5.1×10^{-10}		
Ascorbic	$HC_6H_7O_6$	8.0×10^{-5}	1.6×10^{-12}	
Benzoic	$HC_7H_5O_2$	6.3×10^{-5}		
Boric	H_3BO_3	5.8×10^{-10}		
Butanoic	$HC_4H_7O_2$	1.5×10^{-5}		
Carbonic	H_2CO_3	4.3×10^{-7}	5.6×10^{-11}	
Chloroacetic	$HC_2H_2O_2Cl$	1.4×10^{-3}		
Chlorous	$HClO_2$	1.1×10^{-2}		
Citric	$H_3C_6H_5O_7$	7.4×10^{-4}	1.7×10^{-5}	4.0×10^{-7}
Cyanic	HCNO	3.5×10^{-4}		
Formic	$HCHO_2$	1.8×10^{-4}		
Hydroazoic	HN_3	1.9×10^{-5}		
Hydrocyanic	HCN	4.9×10^{-10}		
Hydrofluoric	HF	6.8×10^{-4}		
Hydrogen chromate ion	$HCrO_4^-$	3.0×10^{-7}		
Hydrogen peroxide	H_2O_2	2.4×10^{-12}		
Hydrogen selenate ion	$HSeO_4$	2.2×10^{-2}		
Hydrosulfuric acid	H_2S	9.5×10^{-8}	1×10^{-19}	
Hypobromous	HBrO	2.5×10^{-9}		
Hypochlorous	HClO	3.0×10^{-8}		
Hypoiodous	HIO	2.3×10^{-11}		
Iodic	HIO_3	1.7×10^{-1}		
Lactic	$HC_3H_5O_3$	1.4×10^{-4}		
Malonic	$H_2C_3H_2O_4$	1.5×10^{-3}	2.0×10^{-6}	
Nitrous	HNO_2	4.5×10^{-4}		
Oxalic	$H_2C_2O_4$	5.9×10^{-2}	6.4×10^{-5}	
Paraperiodic	H_5IO_6	2.8×10^{-2}	5.3×10^{-9}	
Phenol	HC_6H_5O	1.3×10^{-10}		
Phosphoric	H_3PO_4	7.5×10^{-3}	6.2×10^{-8}	4.2×10^{-13}
Propionic	$HC_3H_5O_2$	1.3×10^{-5}		
Pyrophosphoric	$H_4P_2O_7$	3.0×10^{-2}	4.4×10^{-3}	
Selenous	H_2SeO_3	2.3×10^{-3}	5.3×10^{-9}	
Sulfuric	H_2SO_4	Strong acid	1.2×10^{-2}	
Sulfurous	H_2SO_3	1.7×10^{-2}	6.4×10^{-8}	
Tartaric	$H_2C_4H_4O_6$	1.0×10^{-3}	4.6×10^{-5}	

TABLE D-2 Dissociation Constants for Bases at 25°C

Name	Formula	K_b
Ammonia	NH_3	1.8×10^{-5}
Aniline	$C_6H_5NH_2$	4.3×10^{-10}
Dimethylamine	$(CH_3)_2NH$	5.4×10^{-4}
Ethylamine	$C_2H_5NH_2$	6.4×10^{-4}
Hydrazine	H_2NNH_2	1.3×10^{-6}
Hydroxylamine	$HONH_2$	1.1×10^{-8}
Methylamine	CH_3NH_2	4.4×10^{-4}
Pyridine	C_5H_5N	1.7×10^{-9}
Trimethylamine	$(CH_3)_3N$	6.4×10^{-5}

TABLE D-3 Solubility-Product Constants for Compounds at 25°C

Name	Formula	K_{sp}	Name	Formula	K_{sp}
Barium carbonate	$BaCO_3$	5.0×10^{-9}	Lead(II) fluoride	PbF_2	3.6×10^{-8}
Barium chromate	$BaCrO_4$	2.1×10^{-10}	Lead(II) sulfate	$PbSO_4$	6.3×10^{-7}
Barium fluoride	BaF_2	1.7×10^{-6}	Lead(II) sulfide*	PbS	3×10^{-28}
Barium oxalate	BaC_2O_4	1.6×10^{-6}	Magnesium hydroxide	$Mg(OH)_2$	1.6×10^{-12}
Barium sulfate	$BaSO_4$	1.1×10^{-10}	Magnesium carbonate	$MgCO_3$	3.5×10^{-8}
Cadmium carbonate	$CdCO_3$	1.8×10^{-14}	Magnesium oxalate	MgC_2O_4	8.6×10^{-5}
Cadmium hydroxide	$Cd(OH)_2$	2.5×10^{-14}	Manganese(II) carbonate	$MnCO_3$	5.0×10^{-10}
Cadmium sulfide*	CdS	8×10^{-28}	Manganese(II) hydroxide	$Mn(OH)_2$	1.6×10^{-13}
Calcium carbonate (calcite)	$CaCO_3$	4.5×10^{-9}	Manganese(II) sulfide*	MnS	2×10^{-53}
Calcium chromate	$CaCrO_4$	7.1×10^{-4}	Mercury(I) chloride	Hg_2Cl_2	1.2×10^{-18}
Calcium fluoride	CaF_2	3.9×10^{-11}	Mercury(I) iodide	Hg_2I_2	1.1×10^{-28}
Calcium hydroxide	$Ca(OH)_2$	6.5×10^{-6}	Mercury(II) sulfide*	HgS	2×10^{-53}
Calcium phosphate	$Ca_3(PO_4)_2$	2.0×10^{-29}	Nickel(II) carbonate	$NiCO_3$	1.3×10^{-7}
Calcium sulfate	$CaSO_4$	2.4×10^{-5}	Nickel(II) hydroxide	$Ni(OH)_2$	6.0×10^{-16}
Chromium(III) hydroxide	$Cr(OH)_3$	1.6×10^{-30}	Nickel(II) sulfide*	NiS	3×10^{-20}
Cobalt(II) carbonate	$CoCO_3$	1.0×10^{-10}	Silver bromate	$AgBrO_3$	5.5×10^{-5}
Cobalt(II) hydroxide	$Co(OH)_2$	1.3×10^{-15}	Silver bromide	$AgBr$	5.0×10^{-13}
Cobalt(II) sulfide*	CoS	5×10^{-22}	Silver carbonate	Ag_2CO_3	8.1×10^{-12}
Copper(I) bromide	$CuBr$	5.3×10^{-9}	Silver chloride	$AgCl$	1.8×10^{-10}
Copper(II) carbonate	$CuCO_3$	2.3×10^{-10}	Silver chromate	Ag_2CrO_4	1.2×10^{-12}
Copper(II) hydroxide	$Cu(OH)_2$	4.8×10^{-20}	Silver iodide	AgI	8.3×10^{-17}
Copper(II) sulfide*	CuS	6×10^{-37}	Silver sulfate	Ag_2SO_4	1.5×10^{-5}
Iron(II) carbonate	$FeCO_3$	2.1×10^{-11}	Silver sulfide*	Ag_2S	6×10^{-51}
Iron(II) hydroxide	$Fe(OH)_2$	7.9×10^{-16}	Strontium carbonate	$SrCO_3$	9.3×10^{-10}
Lanthanum fluoride	LaF_3	2×10^{-19}	Tin(II) sulfide*	SnS	1×10^{-26}
Lanthanum iodate	$La(IO_3)_3$	6.1×10^{-12}	Zinc carbonate	$ZnCO_3$	1.0×10^{-10}
Lead(II) carbonate	$PbCO_3$	7.4×10^{-14}	Zinc hydroxide	$Zn(OH)_2$	3.0×10^{-16}
Lead(II) chloride	$PbCl_2$	1.7×10^{-5}	Zinc oxalate	ZnC_2O_4	2.7×10^{-8}
Lead(II) chromate	$PbCrO_4$	2.8×10^{-13}	Zinc sulfide*	ZnS	2×10^{-25}

*For a solubility equilibrium of the type $MS(s) + H_2O(l) \rightleftharpoons M^{2+}(aq) + HS^-(aq) + OH^-(aq)$

APPENDIX E

Standard Reduction Potentials at 25°C

Half-Reaction	$E°$(V)
$Ag^+(aq) + e^- \longrightarrow Ag(s)$	+0.799
$AgBr(s) + e^- \longrightarrow Ag(s) + Br^-(aq)$	+0.095
$AgCl(s) + e^- \longrightarrow Ag(s) + Cl^-(aq)$	+0.222
$Ag(CN)_2^-(aq) + e^- \longrightarrow Ag(s) + 2\,CN^-(aq)$	−0.31
$Ag_2CrO_4(s) + 2\,e^- \longrightarrow 2\,Ag(s) + CrO_4^{2-}(aq)$	+0.446
$AgI(s) + e^- \longrightarrow Ag(s) + I^-(aq)$	−0.151
$Ag(S_2O_3)_2^{3-}(aq) + e^- \longrightarrow Ag(s) + 2\,S_2O_3^{2-}(aq)$	+0.01
$Al^{3+}(aq) + 3\,e^- \longrightarrow Al(s)$	−1.66
$H_3AsO_4(aq) + 2\,H^+(aq) + 2\,e^- \longrightarrow H_3AsO_3(aq) + H_2O(l)$	+0.559
$Ba^{2+}(aq) + 2\,e^- \longrightarrow Ba(s)$	−2.90
$BiO^+(aq) + 2\,H^+(aq) + 3\,e^- \longrightarrow Bi(s) + H_2O(l)$	+0.32
$Br_2(l) + 2\,e^- \longrightarrow 2\,Br^-(aq)$	+1.065
$BrO_3^-(aq) + 6\,H^+(aq) + 5\,e^- \longrightarrow Br_2(l) + 3\,H_2O(l)$	+1.52
$2\,CO_2(g) + 2\,H^+(aq) + 2\,e^- \longrightarrow H_2C_2O_4(aq)$	−0.49
$Ca^{2+}(aq) + 2\,e^- \longrightarrow Ca(s)$	−2.87
$Cd^{2+}(aq) + 2\,e^- \longrightarrow Cd(s)$	−0.403
$Ce^{4+}(aq) + e^- \longrightarrow Ce^{3+}(aq)$	+1.61
$Cl_2(g) + 2\,e^- \longrightarrow 2\,Cl^-(aq)$	+1.359
$HClO(aq) + H^+(aq) + e^- \longrightarrow Cl_2(g) + H_2O(l)$	+1.63
$ClO^-(aq) + H_2O(l) + 2\,e^- \longrightarrow Cl^-(aq) + 2\,OH^-(aq)$	+0.89
$ClO_3^-(aq) + 6\,H^+(aq) + 5\,e^- \longrightarrow Cl_2(g) + 3\,H_2O(l)$	+1.47
$Co^{2+}(aq) + 2\,e^- \longrightarrow Co(s)$	−0.277
$Co^{3+}(aq) + e^- \longrightarrow Co^{2+}(aq)$	+1.842
$Cr^{3+}(aq) + 3\,e^- \longrightarrow Cr(s)$	−0.74
$Cr^{3+}(aq) + e^- \longrightarrow Cr^{2+}(aq)$	−0.41
$Cr_2O_7^{2-}(aq) + 14\,H^+(aq) + 6\,e^- \longrightarrow 2\,Cr^{3+}(aq) + 7\,H_2O(l)$	+1.33
$CrO_4^{2-}(aq) + 4\,H_2O(l) + 3\,e^- \longrightarrow Cr(OH)_3(s) + 5\,OH^-(aq)$	−0.13
$Cu^{2+}(aq) + 2\,e^- \longrightarrow Cu(s)$	+0.337
$Cu^{2+}(aq) + e^- \longrightarrow Cu^+(aq)$	+0.153
$Cu^+(aq) + e^- \longrightarrow Cu(s)$	+0.521
$CuI(s) + e^- \longrightarrow Cu(s) + I^-(aq)$	−0.185
$F_2(g) + 2\,e^- \longrightarrow 2\,F^-(aq)$	+2.87
$Fe^{2+}(aq) + 2\,e^- \longrightarrow Fe(s)$	−0.440
$Fe^{3+}(aq) + e^- \longrightarrow Fe^{2+}(aq)$	+0.771
$Fe(CN)_6^{3-}(aq) + e^- \longrightarrow Fe(CN)_6^{4-}(aq)$	+0.36
$2\,H^+(aq) + 2\,e^- \longrightarrow H_2(g)$	0.000
$2\,H_2O(l) + 2\,e^- \longrightarrow H_2(g) + 2\,OH^-(aq)$	−0.83

Half-Reaction	$E°$(V)
$HO_2^-(aq) + H_2O(l) + 2\,e^- \longrightarrow 3\,OH^-(aq)$	+0.88
$H_2O_2(aq) + 2\,H^+(aq) + 2\,e^- \longrightarrow 2\,H_2O(l)$	+1.776
$Hg_2^{2+}(aq) + 2\,e^- \longrightarrow 2\,Hg(l)$	+0.789
$2\,Hg^{2+}(aq) + 2\,e^- \longrightarrow Hg_2^{2+}(aq)$	+0.920
$Hg^{2+}(aq) + 2\,e^- \longrightarrow Hg(l)$	+0.854
$I_2(s) + 2\,e^- \longrightarrow 2\,I^-(aq)$	+0.536
$IO_3^-(aq) + 6\,H^+(aq) + 5\,e^- \longrightarrow I_2(s) + 3\,H_2O(l)$	+1.195
$K^+(aq) + e^- \longrightarrow K(s)$	−2.925
$Li^+(aq) + e^- \longrightarrow Li(s)$	−3.05
$Mg^{2+}(aq) + 2\,e^- \longrightarrow Mg(s)$	−2.37
$Mn^{2+}(aq) + 2\,e^- \longrightarrow Mn(s)$	−1.18
$MnO_2(s) + 4\,H^+(aq) + 2\,e^- \longrightarrow Mn^{2+}(aq) + 2\,H_2O(l)$	+1.23
$MnO_4^-(aq) + 8\,H^+(aq) + 5\,e^- \longrightarrow Mn^{2+}(aq) + 4\,H_2O(l)$	+1.51
$MnO_4^-(aq) + 2\,H_2O(l) + 3\,e^- \longrightarrow MnO_2(s) + 4\,OH^-(aq)$	+0.59
$HNO_2(aq) + H^+(aq) + e^- \longrightarrow NO(g) + H_2O(l)$	+1.00
$N_2(g) + 4\,H_2O(l) + 4\,e^- \longrightarrow 4\,OH^-(aq) + N_2H_4(aq)$	−1.16
$N_2(g) + 5\,H^+(aq) + 4\,e^- \longrightarrow N_2H_5^+(aq)$	−0.23
$NO_3^-(aq) + 4\,H^+(aq) + 3\,e^- \longrightarrow NO(g) + 2\,H_2O(l)$	+0.96
$Na^+(aq) + e^- \longrightarrow Na(s)$	−2.71
$Ni^{2+}(aq) + 2\,e^- \longrightarrow Ni(s)$	−0.28
$O_2(g) + 4\,H^+(aq) + 4\,e^- \longrightarrow 2\,H_2O(l)$	+1.23
$O_2(g) + 2\,H_2O(l) + 4\,e^- \longrightarrow 4\,OH^-(aq)$	+0.40
$O_2(g) + 2\,H^+(aq) + 2\,e^- \longrightarrow H_2O_2(aq)$	+0.68
$O_3(g) + 2\,H^+(aq) + 2\,e^- \longrightarrow O_2(g) + H_2O(l)$	+2.07
$Pb^{2+}(aq) + 2\,e^- \longrightarrow Pb(s)$	−0.126
$PbO_2(s) + HSO_4^-(aq) + 3\,H^+(aq) + 2\,e^- \longrightarrow PbSO_4(s) + 2\,H_2O(l)$	+1.685
$PbSO_4(s) + H^+(aq) + 2\,e^- \longrightarrow Pb(s) + HSO_4^-(aq)$	−0.356
$PtCl_4^{2-}(aq) + 2\,e^- \longrightarrow Pt(s) + 4\,Cl^-(aq)$	+0.73
$S(s) + 2\,H^+(aq) + 2\,e^- \longrightarrow H_2S(g)$	+0.141
$H_2SO_3(aq) + 4\,H^+(aq) + 4\,e^- \longrightarrow S(s) + 3\,H_2O(l)$	+0.45
$HSO_4^-(aq) + 3\,H^+(aq) + 2\,e^- \longrightarrow H_2SO_3(aq) + H_2O(l)$	+0.17
$Sn^{2+}(aq) + 2\,e^- \longrightarrow Sn(s)$	−0.136
$Sn^{4+}(aq) + 2\,e^- \longrightarrow Sn^{2+}(aq)$	+0.154
$VO_2^+(aq) + 2\,H^+(aq) + e^- \longrightarrow VO^{2+}(aq) + H_2O(l)$	+1.00
$Zn^{2+}(aq) + 2\,e^- \longrightarrow Zn(s)$	−0.763

Answers to Selected Exercises

Chapter 1

1.1 (a) Pure element: i, v (b) mixture of elements: vi (c) pure compound: iv (d) mixture of an element and a compound: ii, iii **1.5** (a) 7.5 cm, 2 significant figures (sig figs) (b) 140°C, 2 sig figs **1.8**

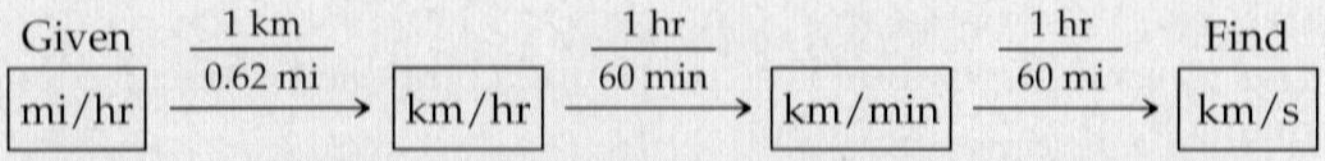

1.9 (a) Heterogeneous mixture (b) homogeneous mixture (heterogeneous if there are undissolved particles) (c) pure substance (d) homogeneous mixture. **1.11** (a) S (b) K (c) Cl (d) Cu (e) Si (f) N (g) Ca (h) He **1.13** (a) Lithium (b) aluminum (c) lead (d) sulfur (e) bromine (f) tin (g) chromium (h) zinc **1.15** C is a compound; it contains carbon and oxygen. A is a compound; it contains at least carbon and oxygen. B is not defined by the data given; it is probably a compound because few elements exist as white solids. **1.17** Physical properties: silvery white; lustrous; melting point = 649°C; boiling point = 1105°C; density at 20°C = 1.738 g/cm^3; pounded into sheets; drawn into wires; good conductor. Chemical properties: burns in air; reacts with Cl_2. **1.19** (a) Chemical (b) physical (c) physical (d) chemical (e) chemical **1.21** (a) Add water to dissolve the sugar; filter this mixture, collecting the sand on filter paper and the sugar water in a flask. Evaporate water from the flask to reproduce solid sugar. (b) Heat the mixture until sulfur melts, then decant the liquid sulfur. **1.23** (a) 1×10^{-1} (b) 1×10^{-2} (c) 1×10^{-15} (d) 1×10^{-6} (e) 1×10^{6} (f) 1×10^{3} (g) 1×10^{-9} (h) 1×10^{-3} (i) 1×10^{-12} **1.25** (a) 2.55×10^{-2} g (b) 0.40 nm (c) 575 μm **1.27** (a) 1.59 g/mL. Carbon tetrachloride, 1.59 g/mL, is more dense than water, 1.00 g/mL; carbon tetrachloride will sink rather than float on water. (b) 1.609 kg (c) 50.35 cm^3 **1.29** (a) Calculated density = 0.86 g/mL. The substance is probably toluene, density = 0.866 g/mL. (b) 40.4 mL ethylene glycol (c) 1.11×10^3 g nickel **1.31** 4.6×10^{-8} m; 46 nm **1.33** (a) 17°C (b) 422.1°F (c) 506 K (d) 108°F (e) 1644 K **1.35** Exact: (c), (d), and (f) **1.37** (a) 3 (b) 2 (c) 5 (d) 3 (e) 5 **1.39** (a) 1.025×10^2 (b) 6.570×10^5 (c) 8.543×10^{-3} (d) 2.579×10^{-4} (e) -3.572×10^{-2} **1.41** (a) 21.11 (b) 237.4 (c) 652 (d) 7.66×10^{-2} **1.43** (a) 2.54 cm = 1 in., 12 in. = 1 ft (b) 1 in.^{3} = $(2.54)^3$ cm^3 **1.45** (a) 76 mL (b) 50 nm (c) 6.88×10^{-4} s (d) 2.3×10^2 g (e) 1.55 g/L (f) 6.151×10^{-3} L/s **1.47** (a) 4.32×10^5 s (b) 88.5 m (c) \$0.499/L (d) 46.6 km/hr (e) 1.420 L/s (f) 707.9 cm^3 **1.49** (a) 1.2×10^2 L (b) 4×10^2 mg (c) 9.64 km/L (d) 26 mL **1.51** 52 kg air **1.53** Use the cm as a unit for comparison (1 in. ≈ 2.5 cm). 57 cm = 57 cm; 14 in. ≈ 35 cm; 1.1 m = 110 cm. The order of length from shortest to longest is 14 in. < 57 cm < 1.1 m. **1.55** (a) \$0.91 (b) Since coins come in integer numbers, 7 coins are required. **1.57** Composition is the contents of a substance; structure is the arrangement of these contents. **1.60** 8.47 g O; the *law of constant composition* **1.63** (a) Volume (b) area (c) volume (d) density (e) time (f) length (g) temperature **1.66** (a) Density of peat = 0.13 g/cm^3, density of topsoil = 2.5 g/cm^3. It is not correct to say that peat is "lighter" than topsoil. Volumes must be specified in order to compare masses. (b) Buy 7 bags of peat (6.9 are needed). **1.69** (a) 4.67 g/cm^3 (b) 2.54 L (c) 1.95×10^6 g. The sphere weighs 1950 kg, or 4300 lb; the thief is unlikely to be able to carry it. **1.73** (a) Average speed = 7.0393 mi/hr (b) pace = 8.5236 min/mi or 511.42 s/mi **1.76** (a) 2.013×10^{-3}/L or 0.2013 ¢/L (b) \$3.397/day **1.79** (a) 61.5% Au by mass (b) 15 carat gold **1.81** The separation is successful if two distinct spots are seen on the paper. To quantify the characteristics of the separation, calculate a reference value for each spot: distance travelled by spot/distance travelled by solvent. If the values for the two spots are fairly different, the separation is successful.

Chapter 2

2.1 (a) The path of the charged particle bends because it is repelled by the negatively charged plate and attracted to the positively charged plate. (b) (−) (c) increase (d) decrease **2.3** The particle is an ion. $^{32}_{16}S^{2-}$ **2.5** Formula: IF_5; name: iodine pentafluoride; the compound is molecular. **2.7** Postulate 4 of the atomic theory states that the relative number and kinds of atoms in a compound are constant, regardless of the source. Therefore, 1.0 g of pure water should always contain the same relative amounts of hydrogen and oxygen, no matter where or how the sample is obtained. **2.9** (a) 0.5711 g O/1 g N; 1.142 g O/1 g N; 2.284 g O/1 g N; 2.855 g O/1 g N (b) The numbers in part (a) obey the *law of multiple proportions*. Multiple proportions arise because atoms are the indivisible entities combining, as stated in Dalton's atomic theory. **2.11** (1) Electric and magnetic fields deflected the rays in the same way they would deflect negatively charged particles. (2) A metal plate exposed to cathode rays acquired a negative charge. **2.13** (a) If the positive plate were lower than the negative plate, the oil drops "coated" with negatively charged electrons would be attracted to the positively charged plate and would descend much more quickly. (b) The more times a measurement is repeated, the better the chance of detecting and compensating for experimental errors. Millikan wanted to demonstrate the validity of his result via its reproducibility. **2.15** (a) 0.19 nm; 1.9×10^2 or 190 pm (b) 2.6×10^6 Kr atoms (c) 2.9×10^{-23} cm^3 **2.17** (a) Proton, neutron, electron (b) proton = +1, neutron = 0, electron = −1 (c) The neutron is most massive, the electron least massive. (The neutron and proton have very similar masses.) **2.19** (a) Atomic number is the number of protons in the nucleus of an atom. Mass number is the total number of nuclear particles, protons plus neutrons, in an atom. (b) mass number **2.21** (a) ^{40}Ar: 18 p, 22 n, 18 e (b) ^{65}Zn: 30 p, 35 n, 30 e (c) ^{70}Ga: 31 p, 39 n, 31 e (d) ^{80}Br: 35 p, 45 n, 35 e (e) ^{184}W: 74 p, 110 n, 74 e (f) ^{243}Am: 95 p, 148 n, 95 e **2.23**

Symbol	^{52}Cr	^{55}Mn	^{112}Cd	^{222}Rn	^{207}Pb
Protons	24	25	48	86	82
Neutrons	28	30	64	136	125
Electrons	24	25	48	86	82
Mass no.	52	55	112	222	207

2.25 (a) $^{196}_{78}$Pt (b) $^{84}_{36}$Kr (c) $^{75}_{33}$As (d) $^{24}_{12}$Mg **2.27** (a) $^{12}_{6}$C (b) Atomic weights are average atomic masses, the sum of the mass of each naturally occurring isotope of an element times its fractional abundance. Each B atom will have the mass of one of the naturally occurring isotopes, while the "atomic

weight" is an average value. **2.29** 63.55 amu **2.31** (a) In Thomson's cathode-ray experiments and in mass spectrometry, a stream of charged particles is passed through the poles of a magnet. The charged particles are deflected by the magnetic field according to their mass and charge. (b) The *x*-axis label is atomic weight, and the *y*-axis label is signal intensity. (c) Uncharged particles are not deflected in a magnetic field. The effect of the magnetic field on charged moving particles is the basis of their separation by mass.
2.33 (a) average atomic mass = 24.31 amu
(b)

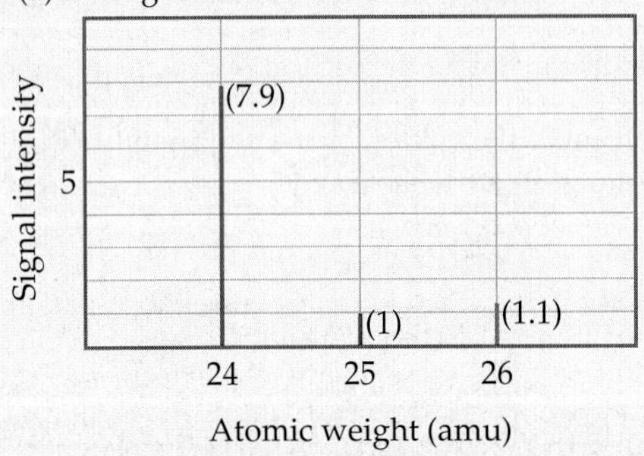

2.35 (a) Cr (metal) (b) He (nonmetal) (c) P (nonmetal) (d) Zn (metal) (e) Mg (metal) (f) Br (nonmetal) (g) As (metalloid) **2.37** (a) K, alkali metals (metal) (b) I, halogens (nonmetal) (c) Mg, alkaline earth metals (metal) (d) Ar, noble gases (nonmetal) (e) S, chalcogens (nonmetal) **2.39** An empirical formula shows the simplest mole ratio of elements in a compound. A molecular formula shows the exact number and kinds of atoms in a molecule. A structural formula shows which atoms are attached to which. **2.41** (a) $AlBr_3$ (b) C_4H_5 (c) C_2H_4O (d) P_2O_5 (e) C_3H_2Cl (f) BNH_2 **2.43** (a) 6 (b) 6 (c) 12

2.45 (a) C_2H_6O,

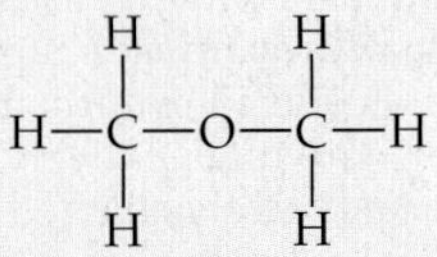

(b) C_2H_6O,

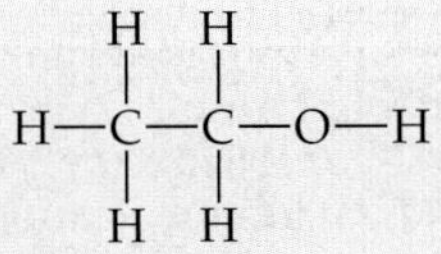

(c) CH_4O,

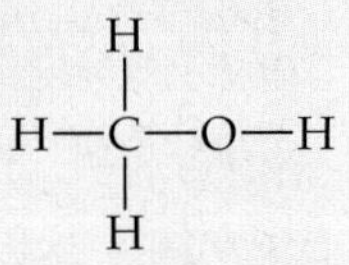

(d) PF_3,

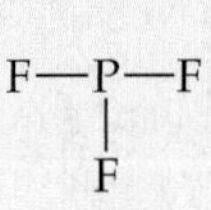

2.47

Symbol	$^{59}Co^{3+}$	$^{70}Se^{2-}$	$^{192}Os^{2+}$	$^{200}Hg^{2+}$
Protons	27	34	76	80
Neutrons	32	46	116	120
Electrons	24	36	74	78
Net Charge	3+	2−	2+	2+

2.49 (a) Mg^{2+} (b) Al^{3+} (c) K^+ (d) S^{2-} (e) F^- **2.51** (a) GaF_3, gallium(III) fluoride (b) LiH, lithium hydride (c) AlI_3, aluminum iodide (d) K_2S, potassium sulfide **2.53** (a) $CaBr_2$ (b) K_2CO_3 (c) $Al(C_2H_3O_2)_3$ (d) $(NH_4)_2SO_4$ (e) $Mg_3(PO_4)_2$ **2.55** Molecular: (a) B_2H_6 (b) CH_3OH (f) NOCl (g) NF_3. Ionic: (c) $LiNO_3$ (d) Sc_2O_3 (e) CsBr (h) Ag_2SO_4 **2.57** (a) ClO_2^- (b) Cl^- (c) ClO_3^- (d) ClO_4^- (e) ClO^- **2.59** (a) Magnesium oxide (b) aluminum chloride (c) lithium phosphate (d) barium perchlorate (e) copper(II) nitrate (cupric nitrate) (f) iron(II) hydroxide (ferrous hydroxide) (g) calcium acetate (h) chromium(III) carbonate (chromic carbonate) (i) potassium chromate (j) ammonium sulfate **2.61** (a) $Al(OH)_3$ (b) K_2SO_4 (c) Cu_2O (d) $Zn(NO_3)_2$ (e) $HgBr_2$ (f) $Fe_2(CO_3)_3$ (g) NaBrO
2.63 (a) Bromic acid (b) hydrobromic acid (c) phosphoric acid (d) HClO (e) HIO_3 (f) H_2SO_3 **2.65** (a) Sulfur hexafluoride (b) iodine pentafluoride (c) xenon trioxide (d) N_2O_4 (e) HCN (f) P_4S_6 **2.67** (a) $ZnCO_3$, ZnO, CO_2 (b) HF, SiO_2, SiF_4, H_2O (c) SO_2, H_2O, H_2SO_3 (d) PH_3 (e) $HClO_4$, Cd, $Cd(ClO_4)_2$ (f) VBr_3
2.69 (a) A hydrocarbon is a compound composed of the elements hydrogen and carbon only.
(b)

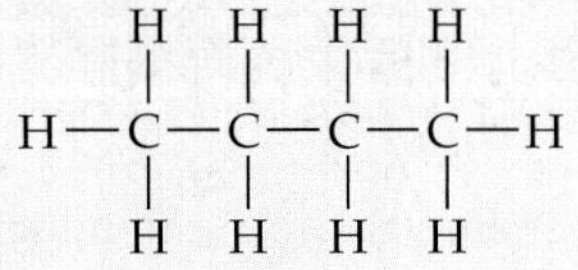

Molecular: C_4H_{10}
Empirical: C_2H_5

2.71 (a) Functional groups are groups of specific atoms that are constant from one molecule to the next. (b) —OH
(c)

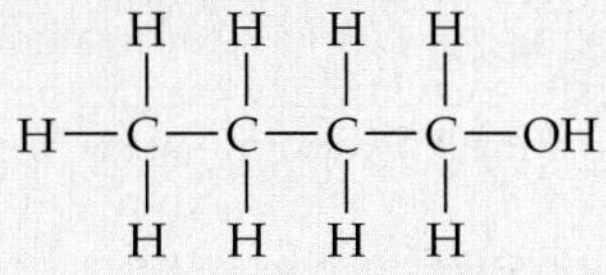

2.74 Radioactivity is the spontaneous emission of radiation from a substance. Becquerel's discovery showed that atoms could decay, or degrade, *implying* that they are not indivisible. However, it wasn't until Rutherford and others characterized the nature of radioactive emissions that the full significance of the discovery was apparent. **2.77** (a) 2 protons, 1 neutron, 2 electrons (b) Tritium, ^3H, is more massive. (c) A precision of 1×10^{-27} g would be required to differentiate between ^3H and ^3He.
2.79 (a) 5.90×10^{22} Au atoms in the cube (b) 2.6 − 6.8 Å (c) Some atomic arrangement must be assumed. The smaller diameter, 2.6 Å, assumes an arrangement with maximum empty space. The larger radius, 6.8 Å, assumes an arrangement with no empty space. **2.82** $^{16}_{8}O$, $^{17}_{8}O$, $^{18}_{8}O$ (b) All isotopes are atoms of the same element, oxygen, with the same atomic number, 8 protons in the nucleus and 8 electrons. We expect their electron arrangements to be the same and their chemical properties to be very similar. Each has a different number of neutrons, a different mass number, and a different atomic mass.
2.86 (a) 157.836 amu, $^{79}Br–^{79}Br$; 159.834 amu, $^{79}Br–^{81}Br$; 161.832 amu, $^{81}Br–^{81}Br$ (b) ^{79}Br, 78.918 amu; ^{81}Br, 80.916 amu (c) average molecular mass of Br_2 = 159.79 amu (d) average atomic mass = 79.895 amu (e) ^{79}Br, 51.10%; ^{81}Br, 48.90%
2.89 (a) An alkali metal: K (b) an alkaline earth metal: Ca (c) a noble gas: Ar (d) a halogen: Br (e) a metalloid: Ge (f) a nonmetal in group 1A: H (g) a metal that forms a 3+ ion: Al (h) a nonmetal that forms a 2− ion: O (i) an element that resembles Al: Ga
2.92 (a) Nickel(II) oxide, 2+ (b) manganese(IV) oxide, 4+ (c) chromium(III) oxide, 3+ (d) molybdenum(VI) oxide, 6+
2.95 (a) Sodium chloride (b) sodium bicarbonate (or sodium hydrogen carbonate) (c) sodium hypochlorite (d) sodium hydroxide (e) ammonium carbonate (f) calcium sulfate **2.97** (a) CaS, $Ca(HS)_2$ (b) HBr, $HBrO_3$ (c) AlN, $Al(NO_2)_3$ (d) FeO, Fe_2O_3 (e) NH_3, NH_4^+ (f) K_2SO_3, $KHSO_3$ (g) Hg_2Cl_2, $HgCl_2$ (h) $HClO_3$, $HClO_4$

Chapter 3

3.1 Equation (a) best fits the diagram. **3.4** The empirical formula of the hydrocarbon is CH_4.
3.7

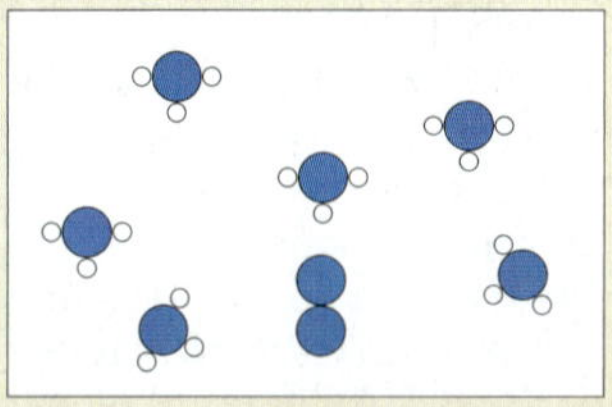

$N_2 + 3\,H_2 \longrightarrow 2\,NH_3$. Eight N atoms (4 N_2 molecules) require 24 H atoms (12 H_2 molecules) for complete reaction. Only 9 H_2 molecules are available, so H_2 is the limiting reactant. Nine H_2 molecules (18 H atoms) determine that 6 NH_3 molecules are produced. One N_2 molecule is in excess.
3.9 (a) Conservation of mass (b) Subscripts in chemical formulas should not be changed when balancing equations, because changing the subscript changes the identity of the compound (*law of constant composition*). (c) (g), (l), (s), (aq)
3.11 (a) $2\,CO(g) + O_2(g) \longrightarrow 2\,CO_2(g)$
(b) $N_2O_5(g) + H_2O(l) \longrightarrow 2\,HNO_3(aq)$
(c) $CH_4(g) + 4\,Cl_2(g) \longrightarrow CCl_4(l) + 4\,HCl(g)$
(d) $Al_4C_3(s) + 12\,H_2O(l) \longrightarrow 4\,Al(OH)_3(s) + 3\,CH_4(g)$
(e) $2\,C_5H_{10}O_2(l) + 13\,O_2(g) \longrightarrow 10\,CO_2(g) + 10\,H_2O(l)$
(f) $2\,Fe(OH)_3(s) + 3\,H_2SO_4(aq) \longrightarrow Fe_2(SO_4)_3(aq) + 6\,H_2O(l)$
(g) $Mg_3N_2(s) + 4\,H_2SO_4(aq) \longrightarrow 3\,MgSO_4(aq) + (NH_4)_2SO_4(aq)$
3.13 (a) $CaC_2(s) + 2\,H_2O(l) \longrightarrow Ca(OH)_2(aq) + C_2H_2(g)$
(b) $2\,KClO_3(s) \xrightarrow{\Delta} 2\,KCl(s) + 3\,O_2(g)$
(c) $Zn(s) + H_2SO_4(aq) \longrightarrow ZnSO_4(aq) + H_2(g)$
(d) $PCl_3(l) + 3\,H_2O(l) \longrightarrow H_3PO_3(aq) + 3\,HCl(aq)$
(e) $3\,H_2S(g) + 2\,Fe(OH)_3(s) \longrightarrow Fe_2S_3(s) + 6\,H_2O(g)$
3.15 (a) Determine the formula by balancing the positive and negative charges in the ionic product. All ionic compounds are solids. $2\,Na(s) + Br_2(l) \longrightarrow 2\,NaBr(s)$ (b) The second reactant is $O_2(g)$. The products are $CO_2(g)$ and $H_2O(l)$.
$2\,C_6H_6(l) + 15\,O_2(g) \longrightarrow 12\,CO_2(g) + 6\,H_2O(l)$
3.17 (a) $Mg(s) + Cl_2(g) \longrightarrow MgCl_2(s)$
(b) $BaCO_3(s) \xrightarrow{\Delta} BaO(s) + CO_2(g)$
(c) $C_8H_8(l) + 10\,O_2(g) \longrightarrow 8\,CO_2(g) + 4\,H_2O(l)$
(d) $C_2H_6O(l) + 3\,O_2(g) \longrightarrow 2\,CO_2(g) + 3\,H_2O(l)$
3.19 (a) $2\,Al(s) + 3\,Cl_2(g) \longrightarrow 2\,AlCl_3(s)$ combination
(b) $C_2H_4(g) + 3\,O_2(g) \longrightarrow 2\,CO_2(g) + 2\,H_2O(l)$ combustion
(c) $6\,Li(s) + N_2(g) \longrightarrow 2\,Li_3N(s)$ combination
(d) $PbCO_3(s) \longrightarrow PbO(s) + CO_2(g)$ decomposition
(e) $C_7H_8O_2(l) + 8\,O_2(g) \longrightarrow 7\,CO_2(g) + 4\,H_2O(l)$ combustion
3.21 (a) 108.0 amu (b) 159.6 amu (c) 149.0 amu (d) 162.1 amu (e) 150.3 amu (f) 399.9 amu (g) 535.6 amu **3.23** (a) 59.9% (b) 43.2% (c) 60.5% (d) 45.0% (e) 60.0% **3.25** (a) 79.2% (b) 63.2% (c) 64.6% **3.27** (a) 6.022×10^{23} (b) The formula weight of a substance in amu has the same numerical value as the molar mass expressed in grams. **3.29** 23 g Na contains 1 mol of atoms; 0.5 mol H_2O contains 1.5 mol atoms; 6.0×10^{23} N_2 molecules contains 2 mol of atoms. **3.31** 4.4×10^{24} kg. One mole of shot-put balls weighs 0.73 times as much as Earth.
3.33 (a) 32.5 g CaH_2 (b) 0.0361 mol $Mg(NO_3)_2$
(c) 1.84×10^{22} CH_3OH molecules (d) 1.41×10^{24} C atoms
3.35 (a) 0.373 g $(NH_4)_3PO_4$ (b) 5.737×10^{-3} mol Cl^- (c) 0.248 g $C_8H_{10}N_4O_2$ (d) 387 g cholesterol/mol **3.37** (a) molar mass = 162.3 g (b) 3.08×10^{-5} mol allicin (c) 1.86×10^{19} allicin molecules (d) 3.71×10^{19} S atoms **3.39** (a) 2.500×10^{21} H atoms (b) 2.083×10^{20} $C_6H_{12}O_6$ molecules (c) 3.460×10^{-4} mol $C_6H_{12}O_6$ (d) 0.06227 g $C_6H_{12}O_6$ **3.41** 3.2×10^{-8} mol C_2H_3Cl/L; 1.9×10^{16} molecules/L **3.43** (a) C_2H_6O (b) Fe_2O_3 (c) CH_2O **3.45** (a) $CSCl_2$ (b) C_3OF_6 (c) Na_3AlF_6
3.47 (a) C_6H_{12} (b) NH_2Cl **3.49** (a) Empirical formula, CH; molecular formula, C_8H_8 (b) empirical formula, $C_4H_5N_2O$; molecular formula, $C_8H_{10}N_4O_2$ (c) empirical formula and molecular formula, $NaC_5H_8O_4N$ **3.51** (a) C_7H_8 (b) The empirical and molecular formulas are $C_{10}H_{20}O$. **3.53** $x = 10$; $Na_2CO_3{\cdot}10\,H_2O$ **3.55** If the equation is not balanced, the mole ratios derived from the coefficients will be incorrect and lead to erroneous calculated amounts of products.
3.57 (a) 2.4 mol HF (b) 5.25 g NaF (c) 0.610 g Na_2SiO_3
3.59 (a) $Al(OH)_3(s) + 3\,HCl(aq) \longrightarrow AlCl_3(aq) + 3\,H_2O(l)$
(b) 0.701 g HCl (c) 0.855 g $AlCl_3$; 0.347 g H_2O (d) Mass of reactants = 0.500 g + 0.701 g = 1.201 g; mass of products = 0.855 g + 0.347 g = 1.202 g. Mass is conserved, within the precision of the data. **3.61** (a) $Al_2S_3(s) + 6\,H_2O(l) \longrightarrow 2\,Al(OH)_3(s) + 3\,H_2S(g)$ (b) 7.01 g $Al(OH)_3$ **3.63** (a) 2.25 mol N_2 (b) 15.5 g NaN_3 (c) 548 g NaN_3 **3.65** (a) 5.50×10^{-3} mol Al (b) 1.47 g $AlBr_3$ **3.67** (a) The *limiting reactant* determines the maximum number of product moles resulting from a chemical reaction; any other reactant is an *excess reactant*. (b) The limiting reactant regulates the amount of products because it is completely used up during the reaction; no more product can be made when one of the reactants is unavailable. **3.69** (a) 2255 bicycles (b) 50 frames left over, 305 wheels left over (c) the handlebars
3.71 NaOH is the limiting reactant; 0.925 mol Na_2CO_3 can be produced; 0.075 mol CO_2 remain. **3.73** (a) $NaHCO_3$ is the limiting reactant. (b) 0.524 g CO_2 (c) 0.238 g citric acid remain
3.75 0.00 g $AgNO_3$ (limiting reactant), 1.94 g Na_2CO_3, 4.06 g Ag_2CO_3, 2.50 g $NaNO_3$ **3.77** (a) The theoretical yield is 60.3 g C_6H_5Br. (b) 94.0% yield **3.79** 7.40 g Li_3N actual yield
3.81 (a) $C_4H_8O_2(l) + 5\,O_2(g) \longrightarrow 4\,CO_2(g) + 4\,H_2O(l)$
(b) $Ni(OH)_2(s) \longrightarrow NiO(s) + H_2O(l)$
(c) $Zn(s) + Cl_2(g) \longrightarrow ZnCl_2(s)$ **3.83** (a) 1.25×10^{22} C atoms (b) 2.77×10^{-3} mol $HC_9H_7O_4$; 1.67×10^{21} $HC_9H_7O_4$ molecules **3.86** $C_{10}H_{12}N_2O$ **3.89** C_6H_5Cl **3.92** (a) In a combustion reaction, all H in the fuel is transformed to H_2O in the products. The reactant with most (mol H/mol fuel) will produce the most H_2O. A 1.5-mol sample of C_3H_8 or $CH_3CH_2COCH_3$ will produce the same amount of H_2O, 1.5 mol C_2H_5OH will produce less H_2O. **3.95** 1.1 kg H_2O
3.98 10.2 g $KClO_3$, 20.0 g $KHCO_3$, 13.8 g K_2CO_3, 56.0 g KCl
3.101 6.46×10^{24} O atoms **3.103** 88 kg CO_2 **3.105** (a) $S(s) + O_2(g) \longrightarrow SO_2(g)$; $SO_2(g) + CaO(s) \longrightarrow CaSO_3(s)$
(b) 1.7×10^5 kg $CaSO_3$/day

Chapter 4

4.1 Diagram (c) represents Li_2SO_4 **4.3** (a) AX is a nonelectrolyte. (b) AY is a weak electrolyte. (c) AZ is a strong electrolyte. **4.5** Solid A is NaOH, solid B is AgBr, and solid C is glucose. **4.7** (b) NO_3^- and (c) NH_4^+ will always be spectator ions. **4.9** Reaction (a) is represented by the diagram.
4.11 No. Electrolyte solutions conduct electricity because the dissolved ions carry charge through the solution from one electrode to the other. **4.13** Although H_2O molecules are electrically neutral, there is an unequal distribution of electrons throughout the molecule. The partially positive ends of H_2O molecules are attracted to the anions in the solid, while the partially negative ends are attracted to the cations. Thus, both cations and anions in an ionic solid are surrounded and separated (dissolved) by H_2O. **4.15** (a) $ZnCl_2(aq) \longrightarrow Zn^{2+}(aq) + 2\,Cl^-(aq)$ (b) $HNO_3(aq) \longrightarrow H^+(aq) + NO_3^-(aq)$
(c) $(NH_4)_2SO_4(aq) \longrightarrow 2\,NH_4^+(aq) + SO_4^{2-}(aq)$
(d) $Ca(OH)_2(aq) \longrightarrow Ca^{2+}(aq) + 2\,OH^-(aq)$ **4.17** $HCHO_2$ molecules, H^+ ions, and CHO_2^- ions; $HCHO_2(aq) \rightleftharpoons H^+(aq)$

$+ CHO_2^-(aq)$ **4.19** (a) Soluble (b) insoluble (c) soluble (d) insoluble (e) insoluble **4.21** (a) $Na_2CO_3(aq) + 2 AgNO_3(aq) \longrightarrow Ag_2CO_3(s) + 2 NaNO_3(aq)$ (b) No precipitate (c) $FeSO_4(aq) + Pb(NO_3)_2(aq) \longrightarrow PbSO_4(s) + Fe(NO_3)_2(aq)$ **4.23** (a) Na^+, SO_4^{2-} (b) Na^+, NO_3^- (c) NH_4^+, Cl^- **4.25** The solution contains Pb^{2+}.
4.27

Compound	$Ba(NO_3)_2$ Result	NaCl Result
$AgNO_3(aq)$	No ppt	AgCl ppt
$CaCl_2(aq)$	No ppt	No ppt
$Al_2(SO_4)_3(aq)$	$BaSO_4$ ppt	No ppt

This sequence of tests would definitely identify the bottle contents. **4.29** LiOH is a strong base, HI is a strong acid, and CH_3OH is a molecular compound and nonelectrolyte. The strong acid HI will have the greatest concentration of solvated protons. **4.31** (a) A monoprotic acid has one ionizable (acidic) H, whereas a diprotic acid has two. (b) A strong acid is completely ionized in aqueous solution, whereas only a fraction of weak acid molecules are ionized. (c) An acid is an H^+ donor, and a base is an H^+ acceptor. **4.33** The strong acid $HClO_4$ is completely ionized; in solution it exists exclusively as $H^+(aq)$ and $ClO_4^-(aq)$. The weak acid $HClO_2$ is only partially ionized; in solution it exists as a mixture of neutral $HClO_2$ molecules, $H^+(aq)$ and $ClO_2^-(aq)$. **4.35** (a) Acid, mixture of ions and molecules (weak electrolyte) (b) none of the above, entirely molecules (nonelectrolyte) (c) salt, entirely ions (strong electrolyte) (d) base, entirely ions (strong electrolyte) **4.37** (a) H_2SO_3, weak electrolyte (b) C_2H_5OH, nonelectrolyte (c) NH_3, weak electrolyte (d) $KClO_3$, strong electrolyte (e) $Cu(NO_3)_2$, strong electrolyte **4.39** (a) $2 HBr(aq) + Ca(OH)_2(aq) \longrightarrow CaBr_2(aq) + 2 H_2O(l)$; $H^+(aq) + OH^-(aq) \longrightarrow H_2O(l)$ (b) $Cu(OH)_2(s) + 2 HClO_4(aq) \longrightarrow Cu(ClO_4)_2(aq) + 2 H_2O(l)$; $Cu(OH)_2(s) + 2 H^+(aq) \longrightarrow 2 H_2O(l) + Cu^{2+}(aq)$ (c) $Al(OH)_3(s) + 3 HNO_3(aq) \longrightarrow Al(NO_3)_3(aq) + 3 H_2O(l)$; $Al(OH)_3(s) + 3 H^+(aq) \longrightarrow 3 H_2O(l) + Al^{3+}(aq)$ **4.41** (a) $CdS(s) + H_2SO_4(aq) \longrightarrow CdSO_4(aq) + H_2S(g)$; $CdS(s) + 2 H^+(aq) \longrightarrow H_2S(g) + Cd^{2+}(aq)$ (b) $MgCO_3(s) + 2 HClO_4(aq) \longrightarrow Mg(ClO_4)_2(aq) + H_2O(l) + CO_2(g)$; $MgCO_3(s) + 2 H^+(aq) \longrightarrow H_2O(l) + CO_2(g) + Mg^{2+}(aq)$ **4.43** (a) $CaCO_3(s) + 2 HNO_3(aq) \longrightarrow Ca(NO_3)_2(aq) + H_2O(l) + CO_2(g)$; $2 H^+(aq) + CaCO_3(s) \longrightarrow Ca^{2+}(aq) + H_2O(l) + CO_2(g)$ (b) $FeS(s) + 2 HBr(aq) \longrightarrow FeBr_2(aq) + H_2S(g)$; $FeS(s) + 2 H^+(aq) \longrightarrow Fe^{2+}(aq) + H_2S(g)$ **4.45** (a) In terms of electron transfer, oxidation is the loss of electrons by a substance and reduction is the gain of electrons (LEO says GER). (b) Relative to oxidation numbers, when a substance is oxidized, its oxidation number increases. When a substance is reduced, its oxidation number decreases. **4.47** Metals in region A are most easily oxidized. Nonmetals in region D are least easily oxidized. **4.49** (a) +4 (b) +4 (c) +7 (d) +1 (e) 0 (f) −1 **4.51** (a) $Ni \longrightarrow Ni^{2+}$, Ni is oxidized; $Cl_2 \longrightarrow 2 Cl^-$, Cl is reduced (b) $Fe^{2+} \longrightarrow Fe$, Fe is reduced; $Al \longrightarrow Al^{3+}$, Al is oxidized (c) $Cl_2 \longrightarrow 2 Cl^-$, Cl is reduced; $2 I^- \longrightarrow I_2$, I is oxidized (d) $S^{2-} \longrightarrow SO_4^{2-}$, S is oxidized; $H_2O_2 \longrightarrow H_2O$; O is reduced **4.53** (a) $Mn(s) + H_2SO_4(aq) \longrightarrow MnSO_4(aq) + H_2(g)$; $Mn(s) + 2 H^+(aq) \longrightarrow Mn^{2+}(aq) + H_2(g)$ (b) $2 Cr(s) + 6 HBr(aq) \longrightarrow 2 CrBr_3(aq) + 3 H_2(g)$; $2 Cr(s) + 6 H^+(aq) \longrightarrow 2 Cr^{3+}(aq) + 3 H_2(g)$ (c) $Sn(s) + 2 HCl(aq) \longrightarrow SnCl_2(aq) + H_2(g)$; $Sn(s) + 2 H^+(aq) \longrightarrow Sn^{2+}(aq) + H_2(g)$ (d) $2 Al(s) + 6 HCHO_2(aq) \longrightarrow 2 Al(CHO_2)_3(aq) + 3 H_2(g)$; $2 Al(s) + 6 HCHO_2(aq) \longrightarrow 2 Al^{3+}(aq) + 6 CHO_2^-(aq) + 3 H_2(g)$ **4.55** (a) $Fe(s) + Cu(NO_3)_2(aq) \longrightarrow Fe(NO_3)_2(aq) + Cu(s)$ (b) NR (c) $Sn(s) + 2 HBr(aq) \longrightarrow SnBr_2(aq) + H_2(g)$ (d) NR (e) $2 Al(s) + 3 CoSO_4(aq) \longrightarrow Al_2(SO_4)_3(aq) + 3 Co(s)$ **4.57** (a) i. $Zn(s) + Cd^{2+}(aq) \longrightarrow Cd(s) + Zn^{2+}(aq)$; ii. $Cd(s) + Ni^{2+}(aq) \longrightarrow Ni(s) + Cd^{2+}(aq)$ (b) Cd is between Zn and Ni on the activity series. (c) Place an iron strip in $CdCl_2(aq)$. If Cd(s) is deposited, Cd is less active than Fe; if there is no reaction, Cd is more active than Fe. Do the same test with Co if Cd is less active than Fe or with Cr if Cd is more active than Fe. **4.59** (a) Intensive; the ratio of amount of solute to total amount of solution is the same, regardless of how much solution is present. (b) The term 0.50 mol HCl defines an amount (~18 g) of the pure substance HCl. The term 0.50 *M* HCl is a ratio; it indicates that there are 0.50 mol of HCl solute in 1.0 liter of solution. **4.61** (a) 0.0863 *M* NH_4Cl (b) 0.0770 mol HNO_3 (c) 83.3 mL of 1.50 *M* KOH **4.63** 16 g $Na^+(aq)$ **4.65** (a) 4.46 g KBr (b) 0.145 *M* $Ca(NO_3)_2$ (c) 20.3 mL of 1.50 *M* Na_3PO_4 **4.67** (a) 0.15 *M* K_2CrO_4 has the highest K^+ concentration. (b) 30.0 mL of 0.15 *M* K_2CrO_4 has more K^+ ions. **4.69** (a) 0.22 *M* Na^+, 0.22 *M* OH^- (b) 0.16 *M* Ca^{2+}, 0.32 *M* Br^- (c) 0.15 *M* CH_3OH (d) 0.080 *M* K^+, 0.080 *M* ClO_3^-, 0.21 *M* Na^+, 0.10 *M* SO_4^{2-} **4.71** (a) 1.69 mL 14.8 *M* NH_3 (b) 0.592 *M* NH_3 **4.73** (a) Add 6.42 g $C_{12}H_{22}O_{11}$ to a 125-mL volumetric flask, dissolve in a small volume of water, and add water to the mark on the neck of the flask. Agitate thoroughly to ensure total mixing. (b) Thoroughly rinse, clean, and fill a 50-mL buret with the 1.50 *M* $C_{12}H_{22}O_{11}$. Dispense 26.7 mL of this solution into a 400-mL volumetric container, add water to the mark, and mix thoroughly. **4.75** 1.398 *M* $HC_2H_3O_2$ **4.77** 0.117 g NaCl **4.79** (a) 38.0 mL of 0.115 *M* $HClO_4$ (b) 769 mL of 0.128 *M* HCl (c) 0.408 *M* $AgNO_3$ (d) 0.275 g KOH **4.81** 27 g $NaHCO_3$ **4.83** 1.22×10^{-2} *M* $Ca(OH)_2$ solution; the solubility of $Ca(OH)_2$ is 0.0904 g in 100 mL solution. **4.85** (a) $NiSO_4(aq) + 2 KOH(aq) \longrightarrow Ni(OH)_2(s) + K_2SO_4(aq)$ (b) $Ni(OH)_2$ (c) KOH is the limiting reactant. (d) 0.927 g $Ni(OH)_2$ (e) 0.0667 *M* $Ni^{2+}(aq)$, 0.0667 *M* $K^+(aq)$, 0.100 *M* $SO_4^{2-}(aq)$ **4.87** 91.40% $Mg(OH)_2$ **4.90** The precipitate is CdS(s). $Na^+(aq)$ and $NO_3^-(aq)$ are spectator ions and remain in solution, along with any excess reactant ions. The net ionic equation is $Cd^{2+}(aq) + S^{2-}(aq) \longrightarrow CdS(s)$. **4.92** (a, b) Expt. 1: NR; Expt. 2: $2 Ag^+(aq) + CrO_4^{2-}(aq) \longrightarrow Ag_2CrO_4(s)$ red precipitate; Expt. 3: NR; Expt. 4: $2 Ag^+(aq) + C_2O_4^{2-}(aq) \longrightarrow Ag_2C_2O_4(s)$ white precipitate; Expt. 5: $Ca^{2+}(aq) + C_2O_4^{2-}(aq) \longrightarrow CaC_2O_4(s)$ white precipitate; Expt. 6: $Ag^+(aq) + Cl^-(aq) \longrightarrow AgCl(s)$ white precipitate. (c) Chromate salts appear to be more soluble than oxalate salts. **4.95** (a) All three reactions are redox reactions. (b) In the first reaction, N is oxidized and O is reduced. In the second reaction, N is oxidized and O is reduced. In the third reaction, N is both oxidized and reduced. **4.98** 1.42 *M* KBr **4.100** (a) 2.2×10^{-9} *M* Na^+ (b) 1.3×10^{12} Na^+ ions/cm^3 **4.103** 0.521 *M* H_2O_2 **4.105** 1.09 g $NaClO_2$ **4.107** 4.793% Ba by mass **4.110** (a) $Na_2SO_4(aq) + Pb(NO_3)_2(s) \longrightarrow PbSO_4(s) + NaNO_3(aq)$ (b) $Pb(NO_3)_2$ is the limiting reactant. (c) 0.200 *M* Na^+, 0.0725 *M* NO_3^-, 0.064 *M* SO_4^{2-} **4.113** 1.766% Cl^- by mass **4.115** 2.8×10^{-5} g Na_3AsO_4 in 1.00 L H_2O

Chapter 5

5.1 As the book falls, potential energy decreases and kinetic energy increases. At the instant before impact, all potential energy has been converted to kinetic energy, so the book's total kinetic energy is 85 J, assuming no transfer of energy as heat. **5.4** (a) No. The distance traveled to the top of a mountain depends on the path taken by the hiker. Distance is a path function, not a state function. (b) Yes. Change in elevation depends only on the location of the base camp and the height of the mountain, not on the path to the top. Change in elevation is a state function, not a path function. **5.6** (a) The temperatures of the system and surroundings will equalize, so the temperature of the hotter system will decrease and the temperature of the colder surroundings will increase. The sign of q is $(-)$; the process is exothermic. (b) If neither volume nor pressure of the system changes, $w = 0$ and $\Delta E = q = \Delta H$. The change in internal energy is equal to the change in enthalpy. **5.9** An object can possess energy by virtue of its motion or position. Kinetic energy depends on the mass of the object and its velocity. Potential energy depends on the position of the object relative to the body with which it interacts. **5.11** (a) 84 J (b) 20 cal (c) As the ball hits the sand, its speed (and hence its kinetic energy) drops to zero. Most of the kinetic energy is transferred to the sand, which deforms when the ball lands. Some energy is released as heat through friction between the ball and the sand. **5.13** 1 Btu = 1054 J **5.15** The energy source of a 100-watt lightbulb is electrical current from household wiring. Energy is radiated in the form of heat and visible light. The energy source for an adult person is food. After a series of complex chemical reactions that release the potential energy stored in chemical bonds, some energy is transferred as electrical impulses that trigger muscle action and become kinetic energy. Some is released as heat. In both cases energy must travel through a network and undergo several changes in form before it is in the correct location and form to be useful. **5.17** (a) The *system* is the well-defined part of the universe whose energy changes are being studied. (b) A *closed system* can exchange heat but not mass with its surroundings. **5.19** (a) Work is a force applied over a distance. (b) The amount of work done is the magnitude of the force times the distance over which it is applied. $w = F \times d$. **5.21** (a) Gravity; work is done because the force of gravity is opposed and the pencil is lifted. (b) Mechanical force; work is done because the force of the coiled spring is opposed as the spring is compressed over a distance. **5.23** (a) In any chemical or physical change, energy can be neither created nor destroyed; energy is conserved. (b) The *internal energy (E)* of a system is the sum of all the kinetic and potential energies of the system components. (c) Internal energy increases when work is done on the system and when heat is transferred to the system. **5.25** (a) $\Delta E = 56$ kJ, endothermic (b) $\Delta E = 0.84$ kJ, endothermic (c) $\Delta E = -71.0$ kJ, exothermic **5.27** (a) Since no work is done by the system in case (2), the gas will absorb most of the energy as heat; the case (2) gas will have the higher temperature. (b) In case (2) $w = 0$ and $q = 100$ J. In case (1) energy will be used to do work on the surroundings $(-w)$, but some will be absorbed as heat $(+q)$. (c) ΔE is greater for case (2) because the entire 100 J increases the internal energy of the system, rather than a part of the energy doing work on the surroundings. **5.29** (a) A *state function* is a property that depends only on the physical state (pressure, temperature, etc.) of the system, not on the route used to get to the current state. (b) Internal energy is a state function; heat is not a state function. (c) Work is not a state function. The amount of work required to move from one state to another depends on the specific series of processes or path used to accomplish the change. **5.31** (a) ΔH is usually easier to measure than ΔE because at constant pressure $\Delta H = q$. The heat flow associated with a process at constant pressure can easily be measured as a change in temperature, while measuring ΔE requires a means to measure both q and w. (b) The process is exothermic. **5.33** (a) $HC_2H_3O_2(l) + 2\,O_2(g) \longrightarrow 2\,H_2O(l) + 2\,CO_2(g)$, $\Delta H = -871.7$ kJ

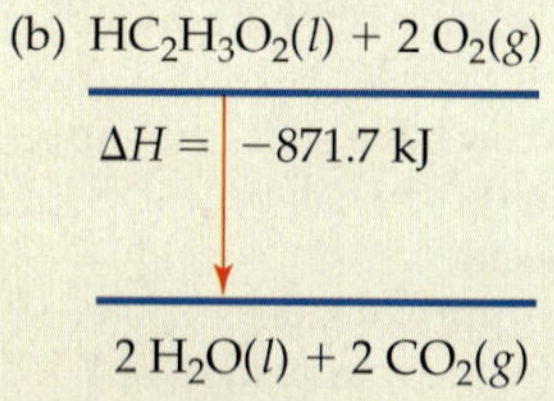

5.35 The reactant, 2 Cl(g), has the higher enthalpy. **5.37** (a) Exothermic (b) −59 kJ heat transferred (c) 6.43 g MgO produced (d) 112 kJ heat absorbed **5.39** (a) −13.1 kJ (b) −1.14 kJ (c) 9.83 J **5.41** At constant pressure, $\Delta E = \Delta H - P\,\Delta V$. The values of either P and ΔV or T and Δn must be known to calculate ΔE from ΔH. **5.43** $\Delta E = -97$ kJ; $\Delta H = -79$ kJ **5.45** (a) $\Delta H = 726.5$ kJ (b) $\Delta H = -1453$ kJ (c) The exothermic forward reaction is more likely to be thermodynamically favored. (d) Vaporization is endothermic. If the product were $H_2O(g)$, the reaction would be more endothermic and would have a smaller negative ΔH. **5.47** (a) J/°C or J/K (b) J/g-°C or J/g-K (c) To calculate heat capacity from specific heat, the mass of the particular piece of copper pipe must be known. **5.49** (a) 4.184 J/g-K (b) 75.40 J/mol-°C (c) 774 J/°C (d) 904 kJ **5.51** 3.00×10^4 J **5.53** $\Delta H = -45.7$ kJ/mol NaOH **5.55** $\Delta E_{rxn} = -25.5$ kJ/g $C_6H_4O_2$ or -2.75×10^3 kJ/mol $C_6H_4O_2$ **5.57** (a) Heat capacity of the complete calorimeter = 14.4 kJ/°C (b) 5.40°C **5.59** Hess's law is a consequence of the fact that enthalpy is a state function. Since ΔH is independent of path, we can describe a process by any series of steps that add up to the overall process. ΔH for the process is the sum of ΔH values for the steps. **5.61** $\Delta H = -1300.0$ kJ **5.63** $\Delta H = -2.49 \times 10^3$ kJ **5.65** (a) *Standard conditions* for enthalpy changes are $P = 1$ atm and some common temperature, usually 298 K. (b) *Enthalpy of formation* is the enthalpy change that occurs when a compound is formed from its component elements. (c) *Standard enthalpy of formation* ΔH_f° is the enthalpy change that accompanies formation of one mole of a substance from elements in their standard states. **5.67** (a) $\frac{1}{2}N_2(g) + \frac{3}{2}H_2(g) \longrightarrow NH_3(g)$, $\Delta H_f^\circ = -46.19$ kJ (b) $S(s) + O_2(g) \longrightarrow SO_2(g)$, $\Delta H_f^\circ = -296.9$ kJ (c) $Rb(s) + \frac{1}{2}Cl_2(g) + \frac{3}{2}O_2(g) \longrightarrow RbClO_3(s)$, $\Delta H_f^\circ = -392.4$ kJ (d) $N_2(g) + 2\,H_2(g) + \frac{3}{2}O_2(g) \longrightarrow NH_4NO_3(s)$, $\Delta H_f^\circ = -365.6$ kJ **5.69** $\Delta H^\circ_{rxn} = -847.6$ kJ **5.71** (a) $\Delta H^\circ_{rxn} = -196.6$ kJ (b) $\Delta H^\circ_{rxn} = 37.1$ kJ (c) $\Delta H^\circ_{rxn} = -433.7$ kJ (d) $\Delta H^\circ_{rxn} = -68.3$ kJ **5.73** $\Delta H_f^\circ = -248$ kJ **5.75** $\Delta H_f^\circ = -924.8$ kJ **5.77** (a) $C_8H_{18}(l) + \frac{25}{2}O_2(g) \longrightarrow 8\,CO_2(g) + 9\,H_2O(g)$, $\Delta H = -5069$ kJ (b) $8\,C(s, g) + 9\,H_2(g) \longrightarrow C_8H_{18}(l)$ (c) $\Delta H_f^\circ = -255$ kJ **5.79** (a) *Fuel value* is the amount of heat produced when 1 g of a substance (fuel) is combusted. (b) 5 g of fat **5.81** 104 or 1×10^2 Cal/serving **5.83** 59.7 Cal

5.85 (a) $\Delta H_{comb} = -1850$ kJ/mol C_3H_4, -1926 kJ/mol C_3H_6, -2044 kJ/mol C_3H_8 (b) $\Delta H_{comb} = -4.616 \times 10^4$ kJ/kg C_3H_4, -4.578×10^4 kJ/kg C_3H_6, -4.635×10^4 kJ/kg C_3H_8 (c) These three substances yield nearly identical quantities of heat per unit mass, but propane is marginally higher than the other two. **5.87** (a) 469.4 m/s (b) 5.124×10^{-21} J (c) 3.086 kJ/mol **5.90** The spontaneous air bag reaction is probably exothermic, with $-\Delta H$ and thus $-q$. When the bag inflates, work is done by the system, so the sign of w is also negative. **5.93** $\Delta H = 38.95$ kJ; $\Delta E = 36.48$ kJ **5.95** 1.8×10^4 or 18,000 bricks. **5.98** 4.90 g CH_4 **5.101** (a) $\Delta H^\circ = -631.1$ kJ (b) 3 mol of acetylene gas has greater enthalpy. (c) Fuel values are 50 kJ/g $C_2H_2(g)$, 42 kJ/g $C_6H_6(l)$. **5.105** If all work is used to increase the man's potential energy, the stair-climbing uses 58 Cal and will not compensate for the extra order of 245 Cal fries. (More than 58 Cal will be required to climb the stairs, because some energy is used to move limbs and some will be lost as heat.) **5.108** (a) 1.479×10^{-18} J/molecule (b) 1×10^{-15} J/photon. The X ray has approximately 1000 times more energy than is produced by the combustion of 1 molecule of $CH_4(g)$. **5.111** (a) $\Delta H°$ for the neutralization of the acids is HNO_3, −55.8 kJ; HCl, −56.1 kJ; NH_4^+, −4.1 kJ. (b) $H^+(aq) + OH^-(aq) \longrightarrow H_2O(l)$ is the net ionic equation for the first two reactions. $NH_4^+(aq) + OH^-(aq) \longrightarrow NH_3(aq) + H_2O(l)$ (c) The $\Delta H°$ values for the first two reactions are nearly identical, −55.9 kJ and −56.2 kJ. Since spectator ions do not change during a reaction and these two reactions have the same net ionic equation, it is not surprising that they have the same $\Delta H°$. (d) Strong acids are more likely than weak acids to donate H^+. The neutralization of the two strong acids is energetically favorable, while the third reaction is barely so. NH_4^+ is likely a weak acid.

Chapter 6

6.2 (a) No. Visible radiation has wavelengths much shorter than 1 cm. (b) Energy and wavelength are inversely proportional. Photons of the longer 1 cm radiation have less energy than visible photons. (c) Radiation with $\lambda = 1$ cm is in the microwave region. The appliance is probably a microwave oven. **6.5** (a) $n = 1, n = 4$ (b) $n = 1, n = 2$ (c) In order of increasing wavelength and decreasing energy: (iii) < (iv) < (ii) < (i) **6.8** (a) In the left-most box, the two electrons cannot have the same spin. (b) Flip one of the arrows in the left-most box, so that one points up and the other down. (c) Group 6A **6.9** (a) Meters (b) 1/second (c) meters/second **6.11** (a) True (b) False. The frequency of radiation decreases as the wavelength increases. (c) False. Ultraviolet light has shorter wavelengths than visible light. (d) False. Electromagnetic radiation and sound waves travel at different speeds. **6.13** Wavelength of X rays < ultraviolet < green light < red light < infrared < radio waves **6.15** (a) 3.14×10^{11} s^{-1} (b) 5.45×10^{-7} m = 545 nm (c) The radiation in (b) is visible, the radiation in (a) is not. (d) 1.50×10^4 m **6.17** 6.88×10^{14} s^{-1}; blue **6.19** (a) Quantization means that energy can only be absorbed or emitted in specific amounts or multiples of these amounts. This minimum amount of energy is equal to a constant times the frequency of the radiation absorbed or emitted; $E = h\nu$. (b) In everyday activities, macroscopic objects such as our bodies gain and lose total amounts of energy much larger than a single quantum, $h\nu$. The gain or loss of the relatively minuscule quantum of energy is unnoticed. **6.21** (a) 4.54×10^{-19} J (b) 4.47×10^{-21} J (c) 6.92×10^{-8} m = 69.2 nm; ultraviolet **6.23** (a) $\lambda = 3.3\ \mu$m, $E = 6.0 \times 10^{-20}$ J; $\lambda = 0.154$ nm, $E = 1.29 \times 10^{-15}$ J (b) The 3.3 μm photon is in the infrared and the 0.154 nm photon is in the X ray region; the X ray photon has the greater energy. **6.25** (a) 6.11×10^{-19} J/photon (b) 368 kJ/mol (c) 1.64×10^{15} photons **6.27** (a) The ~1×10^{-6} m radiation is in the infrared portion of the spectrum. (b) 8.1×10^{16} photons/s **6.29** (a) $E_{min} = 7.22 \times 10^{-19}$ J (b) $\lambda = 275$ nm (c) $E_{120} = 1.66 \times 10^{-18}$ J. The excess energy of the 120 nm photon is converted into the kinetic energy of the emitted electron. $E_k = 9.3 \times 10^{-19}$ J/electron. **6.31** When applied to atoms, the notion of quantized energies means that only certain values of ΔE are allowed. These are represented by the lines in the emission spectra of excited atoms. **6.33** (a) Emitted (b) absorbed (c) emitted **6.35** $E_2 = -5.45 \times 10^{-19}$ J; $E_6 = -0.606 \times 10^{-19}$ J; $\Delta E = 4.84 \times 10^{-19}$ J; $\lambda = 410$ nm; visible, violet. **6.37** (a) Only lines with $n_f = 2$ represent ΔE values and wavelengths that lie in the visible portion of the spectrum. Lines with $n_f = 1$ have shorter wavelengths and lines with $n_f > 2$ have longer wavelengths than visible radiation. (b) $n_i = 3$, $n_f = 2$; $\lambda = 6.56 \times 10^{-7}$ m; this is the red line at 656 nm. $n_i = 4$, $n_f = 2$; $\lambda = 4.86 \times 10^{-7}$ m; this is the blue line at 486 nm. $n_i = 5$, $n_f = 2$; $\lambda = 4.34 \times 10^{-7}$ m; this is the violet line at 434 nm. **6.39** (a) Ultraviolet region (b) $n_i = 6$, $n_f = 1$ **6.41** (a) $\lambda = 5.6 \times 10^{-37}$ m (b) $\lambda = 2.65 \times 10^{-34}$ m (c) $\lambda = 2.3 \times 10^{-13}$ m **6.43** 4.14×10^3 m/s **6.45** (a) $\Delta x \geq 4 \times 10^{-27}$ m (b) $\Delta x \geq 3 \times 10^{-10}$ m **6.47** (a) The uncertainty principle states that there is a limit to how precisely we can simultaneously know the position and momentum (a quantity related to energy) of an electron. The Bohr model states that electrons move about the nucleus in precisely circular orbits of known radius and energy. This violates the uncertainty principle. (b) De Broglie stated that electrons demonstrate the properties of both particles and waves, that each particle has a wave associated with it. A wave function is the mathematical description of the matter wave of an electron. (c) Although we cannot predict the exact location of an electron in an allowed energy state, we can determine the probability of finding an electron at a particular position. This statistical knowledge of electron location is the *probability density* and is a function of Ψ^2, the square of the wave function Ψ. **6.49** (a) $n = 4, l = 3, 2, 1, 0$ (b) $l = 2, m_l = -2, -1, 0, 1, 2$ **6.51** (a) 3*p*: $n = 3, l = 1$ (b) 2*s*: $n = 2, l = 0$ (c) 4*f*: $n = 4, l = 3$ (d) 5*d*: $n = 5, l = 2$ **6.53** (a) impossible, 1*p* (b) possible (c) possible (d) impossible, 2*d* **6.55**

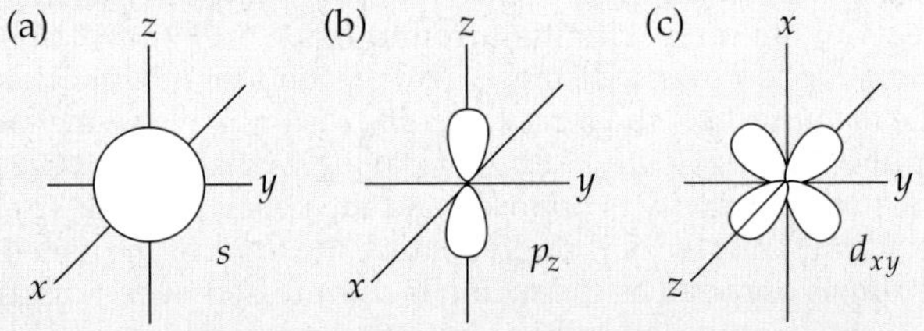

6.57 (a) The hydrogen atom 1*s* and 2*s* orbitals have the same overall spherical shape, but the 2*s* orbital has a larger radial extension and one more node than the 1*s* orbital. (b) A single 2*p* orbital is directional in that its electron density is concentrated along one of the three Cartesian axes of the atom. The $d_{x^2-y^2}$ orbital has electron density along both the *x*- and *y*-axes, while the p_x orbital has density only along the *x*-axis. (c) The average distance of an electron from the nucleus in a 3*s* orbital is greater than for an electron in a 2*s* orbital. (d) 1*s* < 2*p* < 3*d* < 4*f* < 6*s* **6.59** (a) In the hydrogen atom, orbitals with the same principal quantum number, *n*, have the same energy. (b) In a many-electron atom, for a given *n* value, orbital energy increases with increasing *l* value: *s* < *p* < *d* < *f* **6.61** (a) $+\frac{1}{2}, -\frac{1}{2}$ (b) a magnet with a strong inhomogeneous magnetic field (c) they must have different m_s values; the Pauli exclusion principle **6.63** (a) 6 (b) 10 (c) 2 (d) 14 **6.65** (a) Each box represents an orbital. (b) Electron spin is represented by the direction of the half-arrows. (c) No. In Be, there are no electrons in subshells that have degenerate orbitals, so Hund's rule is not used. **6.67** (a) Cs, $[Xe]6s^1$ (b) Ni, $[Ar]4s^23d^8$ (c) Se, $[Ar]4s^23d^{10}4p^4$ (d) Cd, $[Kr]5s^24d^{10}$ (e) Ac, $[Rn]7s^26d^1$ (f) Pb, $[Xe]6s^24f^{14}5d^{10}6p^2$ **6.69** Mt, $[Rn]7s^25f^{14}6d^7$ **6.71** (a) Mg (b) Al (c) Cr (d) Te **6.73** (a) The fifth electron would fill the 2*p* subshell before the 3*s*. (b) Either the core is [He], or the outer electron configuration should be $3s^23p^3$. (c) The 2*p* subshell would fill before the 3*d*. **6.75** (a) $\lambda_A = 3.6 \times 10^{-8}$ m, $\lambda_B = 8.0 \times 10^{-8}$ m

(b) $\nu_A = 8.4 \times 10^{15}\ s^{-1}$, $\nu_B = 3.7 \times 10^{15}\ s^{-1}$ (c) A, ultraviolet; B, ultraviolet **6.77** 66.7 min **6.80** 263 kJ/mol of photons **6.84** (a) The Paschen series lies in the infrared. (b) $n_i = 4$, $\lambda = 1.87 \times 10^{-6}$ m; $n_i = 5$, $\lambda = 1.28 \times 10^{-6}$ m; $n_i = 6$, $\lambda = 1.09 \times 10^{-6}$ m **6.87** $v = 1.02 \times 10^7$ m/s **6.90** (a) l (b) n and l (c) m_s (d) m_l **6.96** If m_s had three allowed values instead of two, each orbital would hold three electrons instead of two. Assuming that there is no change in the n, l, and m_l values, the number of elements in each of the first four rows would be: 1st row, 3 elements; 2nd row, 12 elements; 3rd row, 12 elements; 4th row, 27 elements **6.99** 1.7×10^{28} photons **6.103** (a) Bohr's theory was based on the Rutherford nuclear model of the atom: a dense positive charge at the center and a diffuse negative charge surrounding it. Bohr's theory then specified the nature of the diffuse negative charge. The prevailing theory before the nuclear model was Thomson's plum pudding model: discrete electrons scattered about a diffuse positive charge cloud. Bohr's theory could not have been based on the Thomson model of the atom. (b) De Broglie's hypothesis is that electrons exhibit both particle and wave properties. Thomson's conclusion that electrons have mass is a particle property, while the nature of cathode rays is a wave property. De Broglie's hypothesis actually rationalizes these two seemingly contradictory observations about the properties of electrons.

Chapter 7

7.2 The billiard ball has a definite "hard" boundary, while a quantum mechanical atom has no definitive edge. Billiard balls can be used to model nonbonding interactions, such as Ne atoms colliding, because there is not penetration of electron clouds. They are an inappropriate model for bonding interactions, because bonding electron clouds do penetrate and the bonding atomic radius of an atom is smaller than its nonbonding radius. **7.5** The energy change for the reaction is the ionization energy of A plus the electron affinity of A. **7.7** Mendeleev placed elements with similar chemical and physical properties within a family or column of the table. For undiscovered elements, he left blanks. He predicted properties for the "blanks" based on properties of other elements in the family and on either side. **7.9** (a) *Effective nuclear charge*, Z_{eff}, is a representation of the average electrical field experienced by a single electron. It is the average environment created by the nucleus and the other electrons in the molecule, expressed as a net positive charge at the nucleus. (b) Going from left to right across a period, effective nuclear charge increases. **7.11** (a) $Z = 19$, $S = 18$, $Z_{eff} = Z - S = 1$. (b) The valence electron in K is a 4s electron. Because s electrons have finite probability of being close to the nucleus, shielding of them is never perfect and calculated values of Z_{eff} reflect this. **7.13** The $n = 3$ electrons in Kr experience a greater effective nuclear charge and thus have a greater probability of being closer to the nucleus. **7.15** Atomic radii are determined by distances between atoms in various situations. Bonding radii are calculated from the internuclear separation of two atoms joined by a chemical bond. Nonbonding radii are calculated from the internuclear separation between two gaseous atoms that collide and move apart but do not bond. **7.17** 1.37 Å **7.19** From the sum of the atomic radii, As—I = 2.52 Å. This is very close to the experimental value of 2.55 Å. **7.21** (a) Decrease (b) increase (c) F < S < P < As **7.23** (a) Be < Mg < Ca (b) Br < Ge < Ga (c) Si < Al < Tl **7.25** (a) Electrostatic repulsions are reduced by removing an electron from a neutral atom, effective nuclear charge increases, and the cation is smaller. (b) The additional electrostatic repulsion produced by adding an electron to a neutral atom decreases the effective nuclear charge experienced by the valence electrons, and increases the size of the anion. (c) Going down a column, valence electrons are further from the nucleus, and they experience greater shielding by core electrons. The greater radial extent of the valence electrons outweighs the increase in Z. **7.27** The red sphere is a metal; its size decreases on reaction, characteristic of the change in radius when a metal atom forms a cation. The blue sphere is a nonmetal; its size increases on reaction, characteristic of the change in radius when a nonmetal atom forms an anion. **7.29** (a) An isoelectronic series is a group of atoms or ions that have the same number of electrons. (b) (i) N^{3-}: Ne (ii) Ba^{2+}: Xe (iii) Se^{2-}: Kr (iv) Bi^{3+}: Hg **7.31** (a) Since the number of electrons in an isoelectronic series is the same, repulsion and shielding effects are usually similar for the different particles. As Z increases, the valence electrons are more strongly attracted to the nucleus and the size of the particle decreases. (b) A 2p electron in Na^+ **7.33** (a) $Se < Se^{2-} < Te^{2-}$ (b) $Co^{3+} < Fe^{3+} < Fe^{2+}$ (c) $Ti^{4+} < Sc^{3+} < Ca$ (d) $Be^{2+} < Na^+ < Ne$ **7.35** $B(g) \longrightarrow B^+(g) + e^-$; $B^+(g) \longrightarrow B^{2+}(g) + e^-$; $B^{2+}(g) \longrightarrow B^{3+}(g) + e^-$ **7.37** (a) The electrons in any atom are bound to the atom by the electrostatic attraction to the nucleus. Therefore, energy must be added in order to remove an electron from an atom. The ionization energy, ΔE for this process, is thus positive. (b) F has a greater first ionization energy than O because F has a greater Z_{eff} and the outer electrons in both elements are approximately the same distance from the nucleus. (c) The second ionization energy of an element is greater than the first because more energy is required to overcome the larger Z_{eff} of the 1+ cation than that of the neutral atom. **7.39** (a) The smaller the atom, the larger its first ionization energy. (b) Of the nonradioactive elements, He has the largest, and Cs the smallest first ionization energy. **7.41** (a) Ar (b) Be (c) Co (d) S (e) Te **7.43** (a) Si^{2+}, [Ne]$3s^2$ (b) Bi^{3+}, [Xe]$6s^2 4f^{14} 5d^{10}$ (c) Te^{2-}, [Kr]$5s^2 4d^{10} 5p^6$ or [Xe] (d) V^{3+}, [Ar]$3d^2$ (e) Hg^{2+}, [Xe]$4f^{14} 5d^{10}$ (f) Ni^{2+}, [Ar]$3d^8$ **7.45** (a) Mn^{2+}, [Ar]$3d^5$, 5 unpaired electrons (b) Si^{2-}, [Ne]$3s^2 3p^4$, 2 unpaired electrons **7.47** Ionization energy: $Se(g) \longrightarrow Se^+(g) + e^-$; [Ar]$4s^2 3d^{10} 4p^4 \longrightarrow$ [Ar]$4s^2 3d^{10} 4p^3$; electron affinity: $Se(g) + e^- \longrightarrow Se^-(g)$; [Ar]$4s^2 3d^{10} 4p^4 \longrightarrow$ [Ar]$4s^2 3d^{10} 4p^5$ **7.49** $Li + e^- \longrightarrow Li^-$; [He]$2s^1 \longrightarrow$ [He]$2s^2$; $Be + e^- \longrightarrow Be^-$; [He]$2s^2 \longrightarrow$ [He]$2s^2 2p^1$. Adding an electron to Li completes the 2s subshell. The added electron experiences essentially the same effective nuclear charge as the other valence electron; there is an overall stabilization and ΔE is negative. An extra electron in Be would occupy the higher energy 2p subshell. This electron is shielded from the full nuclear charge by the 2s electrons and does not experience a stabilization in energy; ΔE is positive. **7.51** Ionization energy of F^-: $F^-(g) \longrightarrow F(g) + e^-$; electron affinity of F: $F(g) + e^- \longrightarrow F^-(g)$. The two processes are the reverse of each other. The energies are equal in magnitude but opposite in sign. $I_1(F^-) = -E(F)$. **7.53** The smaller the first ionization energy of an element, the greater the metallic character of that element. **7.55** (a) Li (b) Na (c) Sn (d) Al **7.57** Ionic: MgO, Li_2O, Y_2O_3; molecular: SO_2, P_2O_5, N_2O, XeO_3. Ionic compounds are formed by combining a metal and a nonmetal; molecular compounds are formed by two or more nonmetals. **7.59** (a) An acidic oxide dissolved in water produces an acidic solution; a basic oxide dissolved in water produces a basic solution. (b) Oxides of nonmetals, such as SO_3, are acidic; oxides of metals, such as CaO, are basic. **7.61** (a) Dichlorineseptoxide (b) $2\,Cl_2(g) + 7\,O_2(g) \longrightarrow 2\,Cl_2O_7(l)$ (c) While most nonmetal oxides we have seen, such as CO_2 or SO_2, are gases, a boiling point of 81°C is not totally unexpected for a large molecule like Cl_2O_7. (d) Cl_2O_7 is an acidic oxide, so it will be more reactive to base, OH^-. **7.63** (a) $BaO(s) + H_2O(l) \longrightarrow Ba(OH)_2(aq)$ (b) $FeO(s) + 2\,HClO_4(aq) \longrightarrow Fe(ClO_4)_2(aq) + H_2O(l)$ (c) $SO_3(g) + H_2O(l) \longrightarrow H_2SO_4(aq)$ (d) $CO_2(g) + 2\,NaOH(aq) \longrightarrow Na_2CO_3(aq) + H_2O(l)$ **7.65** (a) Na, $[Ne]3s^1$; Mg, $[Ne]3s^2$ (b) When forming ions, both adopt the stable configuration of Ne; Na loses one electron and Mg two electrons to achieve this configuration. (c) The effective nuclear charge of Mg is greater, so its ionization energy is greater. (d) Mg is less reactive because it has a higher ionization energy. (e) The atomic radius of Mg is smaller because its effective nuclear charge is greater. **7.67** (a) Ca is more reactive because it has a lower ionization energy than Mg. (b) K is more reactive because it has a lower ionization energy than Ca. **7.69** (a) $2\,K(s) + Cl_2(g) \longrightarrow 2\,KCl(s)$ (b) $SrO(s) + H_2O(l) \longrightarrow Sr(OH)_2(aq)$ (c) $4\,Li(s) + O_2(g) \longrightarrow 2\,Li_2O(s)$ (d) $2\,Na(s) + S(l) \longrightarrow Na_2S(s)$ **7.71** H, $1s^1$; Li, $[He]2s^1$; F, $[He]2s^22p^5$. Like Li, H has only one valence electron, and its most common oxidation number is +1. Like F, H needs only one electron to adopt the stable electron configuration of the nearest noble gas; both H and F can exist in the −1 oxidation state. **7.73** (a) F, $[He]2s^22p^5$; Cl, $[Ne]3s^23p^5$ (b) F and Cl are in the same group, and both adopt a 1− ionic charge. (c) The $2p$ valence electrons in F are closer to the nucleus and more tightly held than are the $3p$ electrons of Cl, so the ionization energy of F is greater. (d) The high ionization energy of F coupled with a relatively large exothermic electron affinity makes it more reactive than Cl toward H_2O. (e) While F has approximately the same effective nuclear charge as Cl, its small atomic radius gives rise to large repulsions when an extra electron is added, so the overall electron affinity of F is less exothermic than that of Cl. (f) The $2p$ valence electrons in F are closer to the nucleus so the atomic radius is smaller than that of Cl. **7.75** Under ambient conditions, the group 8A elements are all gases that are extremely unreactive, so the name "inert gases" seemed appropriate. It is inappropriate because both Xe and Kr have been found to react with substances having a strong tendency to remove electrons, such as F_2. **7.77** (a) $2\,O_3(g) \longrightarrow 3\,O_2(g)$ (b) $Xe(g) + F_2(g) \longrightarrow XeF_2(g)$; $Xe(g) + 2\,F_2(g) \longrightarrow XeF_4(s)$; $Xe(g) + 3\,F_2(g) \longrightarrow XeF_6(s)$ (c) $S(s) + H_2(g) \longrightarrow H_2S(g)$ (d) $2\,F_2(g) + 2\,H_2O(l) \longrightarrow 4\,HF(aq) + O_2(g)$ **7.79** Up to $Z = 82$, there are three instances where atomic weights are reversed relative to atomic numbers: Ar and K; Co and Ni; Te and I. In each case the most abundant isotope of the element with the larger atomic number has one more proton, but fewer neutrons than the element with the smaller atomic number. The smaller number of neutrons causes the element with the larger Z to have a smaller than expected atomic weight. **7.81** (a) $4s$ (b) Because s electrons have a greater probability of being close to the nucleus, they experience less shielding than p electrons with the same n-value. $Z_{eff} = Z - S$ and Z is the same for all electrons in As. If S is smaller for $4s$ than $4p$, Z_{eff} will be greater for $4s$ electrons and they will have lower energy. **7.85** Moving across the representative elements, electrons added to ns or np valence orbitals do not effectively screen each other. The increase in Z is not accompanied by a similar increase in S. Z_{eff} increases and atomic size decreases. Moving across the transition elements, electrons are added to $(n-1)d$ orbitals, which do significantly screen the ns valence electrons. The increase in Z is accompanied by a corresponding increase in S. Z_{eff} increases more slowly and atomic size decreases more slowly. **7.88** The completed $4f$ subshell in Hf leads to a much larger change in Z and Z_{eff} going from Zr to Hf than in going from Y to La. This larger increase in Z_{eff} going from Zr to Hf leads to a smaller increase in atomic radius than in going from Y to La. **7.91** I_1 through I_4 represent loss of the $2p$ and $2s$ electrons from the outer shell of the atom. Z is constant while removal of each electron reduces repulsion between the remaining electrons, so Z_{eff} increases and I increases. I_5 and I_6 represent loss of the $1s$ core electrons. These electrons are closer to the nucleus and experience the full nuclear charge, so the values of I_5 and I_6 are significantly greater than $I_1 - I_4$. I_6 is larger than I_5 because all electron-electron repulsion has been eliminated.

7.94

O, $[He]2s^22p^4$ 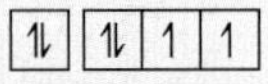

$2s$ $2p$

O^{2-}, $[He]2s^22p^6 = [Ne]$ 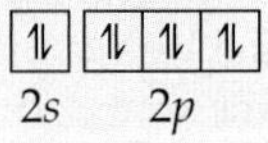

$2s$ $2p$

O^{3-}, $[Ne]3s^1$ The third electron would be added to the $3s$ orbital, which is farther from the nucleus and more strongly shielded by the [Ne] core. The overall attraction of this $3s$ electron for the oxygen nucleus is not large enough for O^{3-} to be a stable particle.

7.96 (a) The group 2B metals have complete $(n-1)d$ subshells. An additional electron would occupy an np subshell and be substantially shielded by both ns and $(n-1)d$ electrons. This is not a lower energy state than the neutral atom and a free electron. (b) Group 1B elements have the generic electron configuration $ns^1(n-1)d^{10}$. An additional electron would complete the ns subshell and experience repulsion with the other ns electron. Going down the group, size of the ns subshell increases and repulsion effects decrease, so effective nuclear charge increases and electron affinities become more negative. **7.99** $O_2 < Br_2 < K < Mg$. O_2 and Br_2 are nonpolar nonmetals. O_2, with the much lower molar mass, should have the lower melting point. K and Mg are metallic solids with higher melting points than the two nonmetals. Since alkaline earth metals are typically harder, more dense, and higher melting than alkali metals, Mg should have the highest melting point of the group. This order of melting points is confirmed by data in Tables 7.4, 7.5, 7.6, and 7.7. **7.103** Chlorine and bromine are much closer in ionization energy and electron affinity to carbon than they are to the metals. Carbon has a much greater tendency than a metal to keep its own electrons and at least some attraction for the electrons of other elements. Carbon is unlikely to form a simple cation, so compounds of carbon and the halogens are molecular, rather than ionic. **7.105** (a) Li, $[He]2s^1$; $Z_{eff} \approx 1+$. (b) $I_1 \approx 5.45 \times 10^{-19}$ J/atom ≈ 328 kJ/mol (c) The estimated value of 328 kJ/mol is less than the Table 7.4 value of 520 kJ/mol. Our estimate for Z_{eff} was a lower limit; the [He] core electrons do not perfectly shield the $2s$ electron from the nuclear charge. (d) Based on the experimental ionization energy, $Z_{eff} = 1.26$. This value is greater than the estimate from part (a), which is consistent with the explanation in part (c). **7.108** (a) Mg_3N_2 (b) $Mg_3N_2(s) + 3\,H_2O(l) \longrightarrow 3\,MgO(s) + 2\,NH_3(g)$; the driving force is the production of $NH_3(g)$ (c) 17% Mg_3N_2 (d) $3\,Mg(s) + 2\,NH_3(g) \longrightarrow Mg_3N_2(s) + 3\,H_2(g)$. NH_3 is the limiting reactant and 0.46 g H_2 are formed. (e) $\Delta H^\circ_{rxn} = -368.70$ kJ

Chapter 8

8.2 (a) No. A_1 and A_2 have like charges and repel each other. (b) A_1Z_1. Lattice energy increases as ionic charge increases and decreases as inter-ionic distance increases. These ions all have the same magnitude of charge; A_1 and Z_1 have the smallest radii and their combination leads to the largest lattice energy. (c) A_2Z_2 has the smallest lattice energy, because it has the largest inter-ionic distance. **8.5** (a) Moving from left to right along the molecule, the first C needs 2 H atoms, the second needs 1, the third needs none and the fourth needs 1. (b) In order of increasing bond length: $3 < 1 < 2$ (c) In order of increasing bond enthalpy: $2 < 1 < 3$ **8.7** (a) Valence electrons are those that take part in chemical bonding. This usually means the electrons beyond the core noble-gas configuration of the atom, although it is sometimes only the outer-shell electrons. (b) A nitrogen atom has 5 valence electrons. (c) The atom (Si) has 4 valence electrons. **8.9** P, $1s^22s^22p^63s^23p^3$. A 3*s* electron is a valence electron; a 2*s* (or 1*s*) electron is a nonvalence electron. The 3*s* valence electron is involved in chemical bonding, while the 2*s* or 1*s* nonvalence electron is not. **8.11** (a) ·Ål· (b) :B̈r: (c) :Är: (d) ·Sr·

8.13 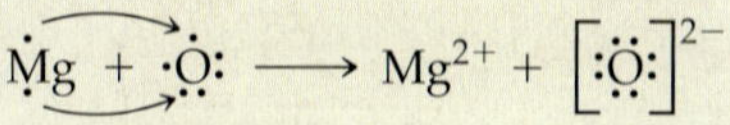

8.15 (a) AlF_3 (b) K_2S (c) Y_2O_3 (d) Mg_3N_2 **8.17** (a) Sr^{2+}, $[Ar]4s^23d^{10}4p^6 = [Kr]$, noble-gas configuration (b) Ti^{2+}, $[Ar]3d^2$ (c) Se^{2-}, $[Ar]4s^23d^{10}4p^6 = [Kr]$, noble-gas configuration (d) Ni^{2+}, $[Ar]3d^8$ (e) Br^-, $[Ar]4s^23d^{10}4p^6 = [Kr]$, noble-gas configuration (f) Mn^{3+}, $[Ar]3d^4$ **8.19** (a) *Lattice energy* is the energy required to totally separate one mole of solid ionic compound into its gaseous ions. (b) The magnitude of the lattice energy depends on the magnitudes of the charges of the two ions, their radii and the arrangement of ions in the lattice. **8.21** KF, 808 kJ/mol; CaO, 3414 kJ/mol; ScN, 7547 kJ/mol. The interionic distances in the three compounds are similar. For compounds with similar ionic separations, the lattice energies should be related as the product of the charges of the ions. The lattice energies above are approximately related as 1 : 4 : 9. Slight variations are due to the small differences in ionic separations. **8.23** Since the ionic charges are the same in the two compounds, the KBr and CsCl separations must be approximately equal. **8.25** The large attractive energy between oppositely charged Ca^{2+} and O^{2-} more than compensates for the energy required to form Ca^{2+} and O^{2-} from the neutral atoms. **8.27** The lattice energy of RbCl(*s*) is +692 kJ/mol. This value is smaller than the lattice energy for NaCl because Rb^+ has a larger ionic radius than Na^+ and therefore cannot approach Cl^- as closely as Na^+ can. **8.29** (a) A *covalent bond* is the bond formed when two atoms share one or more pairs of electrons. (b) Any simple compound whose component atoms are nonmetals, such as H_2, SO_2, and CCl_4, are molecular and have covalent bonds between atoms. (c) Covalent, because it is a gas at room temperature and below.

8.31

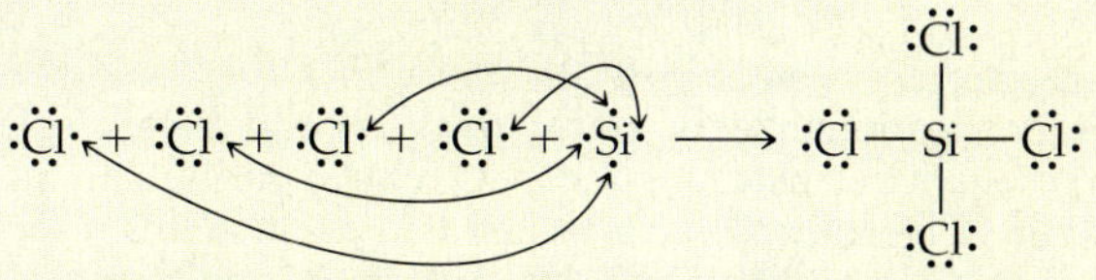

8.33 Ö=Ö (b) A double bond is required because there are not enough electrons to satisfy the octet rule with single bonds and unshared pairs. (c) The greater the number of shared electron pairs between two atoms, the shorter the distance between the atoms. An O=O double bond is shorter than an O—O single bond. **8.35** (a) Electronegativity is the ability of an atom in a molecule to attract electrons to itself. (b) The range of electronegativities on the Pauling scale is 0.7–4.0. (c) Fluorine is the most electronegative element. (d) Cesium is the least electronegative element that is not radioactive. **8.37** (a) Br (b) C (c) P (d) Be **8.39** The bonds in (a), (c), and (d) are polar. The more electronegative element in each polar bond is: (a) F (c) O (d) I **8.41** The calculated charge on H and F is 0.41*e*. **8.43** (a) MnO_2, ionic (b) P_2S_3, covalent (c) CoO, ionic (d) copper(I) sulfide, ionic (e) chlorine trifluoride, covalent (f) vanadium(V) fluoride, ionic

8.45

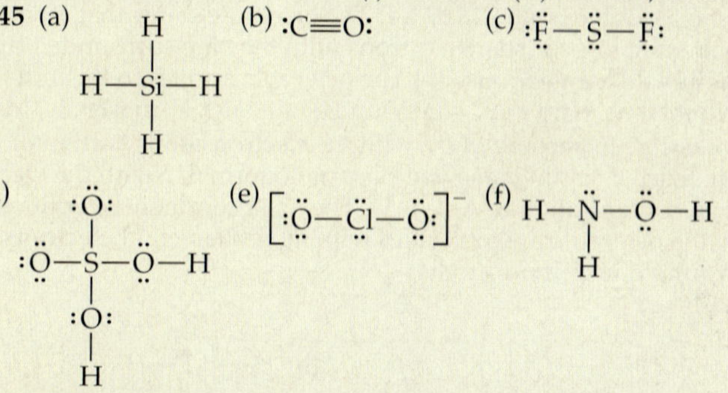

8.47 (a) *Formal charge* is the charge on each atom in a molecule, assuming all atoms have the same electronegativity. (b) Formal charges are not actual charges. They are a bookkeeping system that assumes perfect covalency, one extreme for the possible electron distribution in a molecule. (c) Oxidation numbers are a bookkeeping system that assumes the more electronegative element holds all electrons in a bond. The true electron distribution is some composite of the two extremes. **8.49** Formal charges are shown on the Lewis structures; oxidation numbers are listed below each structure.

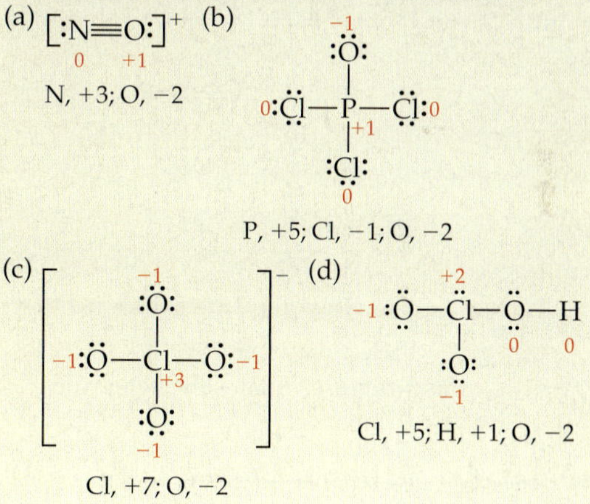

8.51 (a) 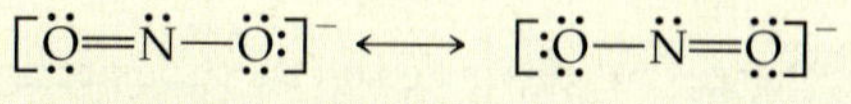

(b) O_3 is isoelectronic with NO_2^-; both have 18 valence electrons. (c) Since each N—O bond has partial double-bond character, the N—O bond length in NO_2^- should be shorter than in species with formal N—O single bonds. **8.53** The more electron pairs shared by two atoms, the shorter the bond. Thus, the C—O bond lengths vary in the order $CO < CO_2 < CO_3^{2-}$. **8.55** (a) Two equally valid Lewis structures can be drawn for benzene.

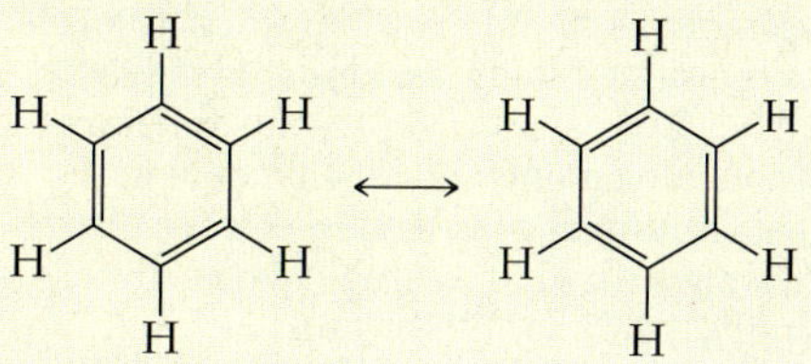

The concept of resonance dictates that the true description of bonding is some hybrid or blend of these two Lewis structures.

The most obvious blend of these two resonance structures is a molecule with six equivalent C—C bonds with equal lengths. (b) In order for the six C—C bonds in benzene to be equivalent, each must have some double-bond character. That is, more than one pair but fewer than two pairs of electrons are involved in each C—C bond. This model predicts a uniform C—C bond length that is shorter than a single bond but longer than a double bond. **8.57** (a) The octet rule states that atoms will gain, lose, or share electrons until they are surrounded by eight valence electrons. (b) The octet rule applies to the individual ions in an ionic compound. For example, in $MgCl_2$, Mg loses 2 e^- to become Mg^{2+} with the electron configuration of Ne. Each Cl atom gains one electron to form Cl^- with the electron configuration of Ar. **8.59** The most common exceptions to the octet rule are molecules with more than eight electrons around one or more atoms.
8.61

(a) 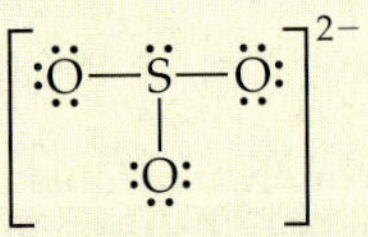

Other resonance structures that violate the octet rule can be drawn, but the single best structure, the one shown here, obeys the octet rule

(b) 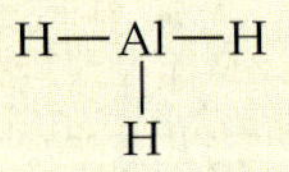

6 electrons around Al

(c) 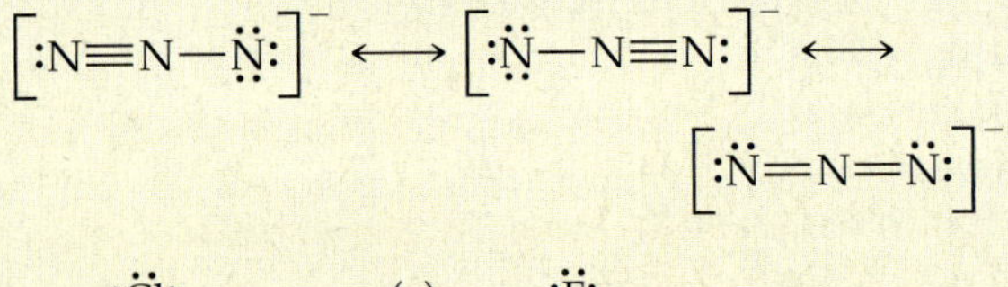

(d) 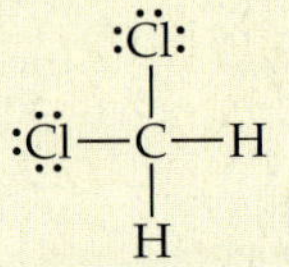

(e) 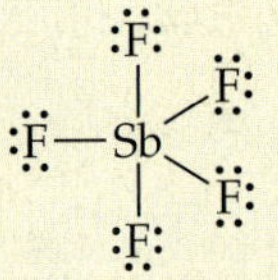

10 electrons around Sb

8.63 (a) 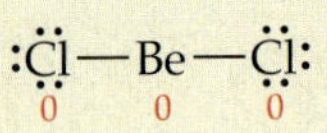 this structure violates the octet rule.

(b) 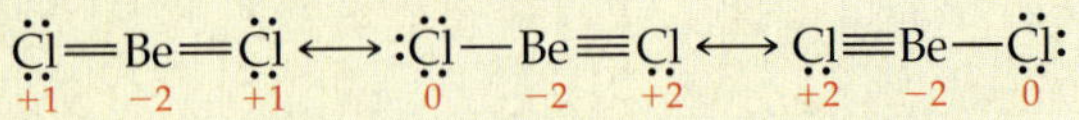

(c) Since formal charges are minimized on the structure that violates the octet rule, this form is probably most important. This is a different conclusion than for molecules that have resonance structures with expanded octets that minimize formal charge.
8.65 (a) $\Delta H = -304$ kJ (b) $\Delta H = -82$ kJ (c) $\Delta H = -467$ kJ
8.67 (a) $\Delta H = -321$ kJ (b) $\Delta H = -103$ kJ (c) $\Delta H = -203$ kJ
8.69 (a) Exothermic (b) The ΔH calculated from bond enthalpies (−97 kJ) is slightly more exothermic (more negative) than that obtained using ΔH_f° values (−92.38 kJ). **8.71** The average Ti—Cl bond enthalpy is 430 kJ/mol. **8.73** (a) Six (nonradioactive) elements. Yes, they are in the same family, assuming H is placed with the alkali metals. The Lewis symbol represents the number of valence electrons of an element, and all elements in the same family have the same number of valence electrons. By definition of a family, all elements with the same Lewis symbol must be in the same family. **8.75** (a) 779 kJ/mol
(b) 627 kJ/mol (c) 2195 kJ/mol **8.78** (a) A polar molecule has a measurable dipole moment while a nonpolar molecule has a zero net dipole moment. (b) Yes. If X and Y have different electronegativities, the electron density around the more electronegative atom will be greater, producing a charge separation or dipole in the molecule. (c) The dipole moment, μ, is the product of the magnitude of the separated charges, Q, and the distance between them, r. $\mu = Qr$
8.81 (a) 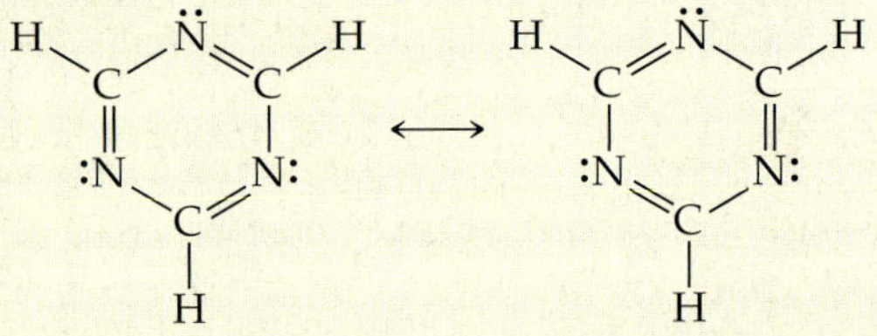

The structures shown here minimize formal charges on the atoms. (b) Carbon–nitrogen bond distances should be in the range 1.40–1.41 Å, intermediate between C—N and C=N bonds. **8.84** (a) +1 (b) −1 (c) +1 (assuming the odd electron is on N) (d) 0 (e) +3 **8.87** ΔH is +42 kJ for the first reaction and −200 kJ for the second. The latter is much more favorable because the formation of 2 mol of O—H bonds is more exothermic than the formation of 1 mol of H—H bonds.
8.90 (a) 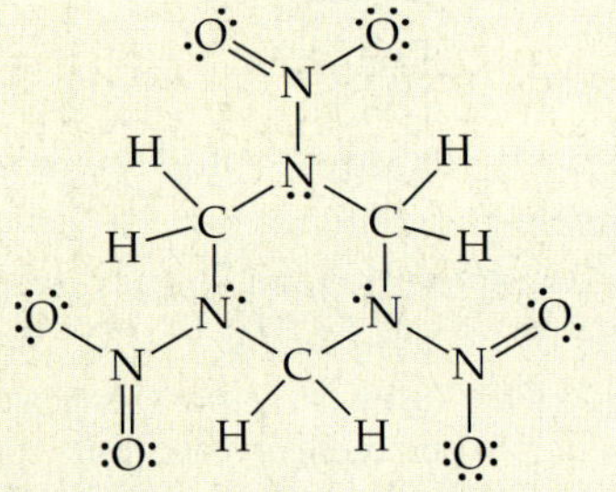

(b) Each terminal O=N—O group has two possible placements for the N=O; this leads to 8 resonance structures with the carbon–nitrogen framework shown in (a). Structures with N=N can be drawn, but these do not minimize formal charge. There are 7 resonance structures with one or more N=N, for a total of 15 resonance structures. (c) $C_3H_6N_6O_6(s) \longrightarrow 3\,CO(g) + 3\,N_2(g) + 3\,H_2O(g)$ (d) According to Table 8.4, N—N single bonds have the smallest bond enthalpy and are weakest.
(e) The enthalpy of decomposition for the resonance structure in (a) is −1668 kJ/mol $C_3H_6N_6O_6$. For the resonance structure with 3 N=N and 6 N—O bonds, $\Delta H = -2121$ kJ/mol. The actual value is somewhere between these two values. The enthalpy change for 5.0 g of RDX is then in the range 38–48 kJ.
8.93 (a) 1.77 Å (b) 1.75–1.76 Å (c) The resonance structures for SO_2 indicate that the sulfur–oxygen bond is intermediate between a double and single bond, so the bond length of 1.43 Å should be significantly shorter than the single bond distance of 1.75 Å.
(d) 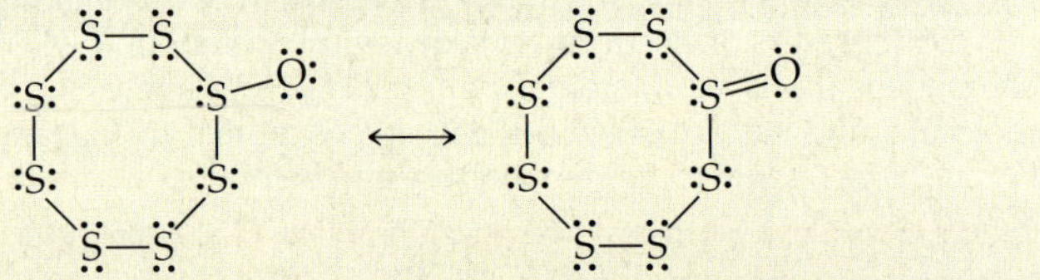

In order to rationalize the sulfur–oxygen bond distance of 1.48 Å, a resonance structure with S=O must contribute to the true structure. The sulfur atom attached to oxygen in this resonance structure has more than an octet of electrons and violates the octet rule. **8.94** (a) Ti^{2+}, [Ar]$3d^2$; Ca, [Ar]$4s^2$. The 2 valence electrons in Ti^{2+} and Ca are in different principal quantum levels and different subshells. (b) In Ca the 4*s* is lower in energy than the 3*d*, while in Ti^{2+} the 3*d* is lower in energy than the 4*s*. (c) There is only one 4*s* orbital, so the 2 valence electrons in Ca are paired; there are 5 degenerate 3*d* orbitals, so the 2 valence electrons in Ti^{2+} are unpaired. **8.97** (a) A, In_2S; B, InS; C, In_2S_3 (b) A, In(I); B, In(II); C, In(III) (c) In(I), [Kr]$5s^24d^{10}$; In(II),

[Kr]$5s^1 4d^{10}$; In(III), [Kr]$4d^{10}$. None of these are noble-gas configurations. (d) In(III) in compound C has the smallest ionic radius. Removing successive electrons, from an atom reduces electron repulsion, increases the effective nuclear charge experienced by the valence electrons and decreases the ionic radius. (e) Lattice energy increases as ionic charge increases and as inter-ionic distance decreases. Only the charge and size of In changes in the three compounds. In(I) in compound A has the lowest charge and largest ionic radius, so compound A has the smallest lattice energy and lowest melting point. Compound C has the largest lattice energy and highest melting point. **8.100** (a) Azide ion is N_3^-. (b) Resonance structures with formal charges are shown.

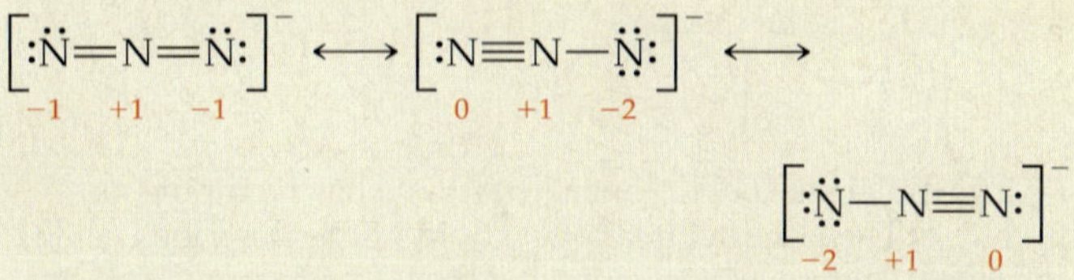

(c) The left structure minimizes formal charges and is probably the main contributor. (d) The N—N distances will be equal, and have the approximate length of a N—N double bond, 1.24 Å. **8.104** (a) $\Delta H = 1551$ kJ (b) $\Delta H = 1394$ kJ (c) $\Delta H = 1353$ kJ

Chapter 9

9.3 (a) Square pyramidal (b) Yes, there is one nonbonding electron domain on A. If there were only the 5 bonding domains, the shape would be trigonal bipyramidal. (c) Yes. If the B atoms are halogens, A is also a halogen. **9.6** 90°: sp^3d or sp^3d^2; 109.5°: sp^3; 120°: sp^2 or sp^3d **9.9** (a) i, Formed by two *s* atomic orbitals; ii, σ-type MO; iii, antibonding MO (b) i, formed by two *p* atomic orbitals overlapping end-to-end; ii, σ-type MO; iii, bonding MO (c) i, formed by two *p* atomic orbitals overlapping side-to-side; ii, π-type MO; iii, antibonding MO **9.11** (a) Yes. The bond angle specifies the shape and the bond length tells the size. (b) Yes. The only possible bond angles in this arrangement are 120° angles. **9.13** (a) An *electron domain* is a region in a molecule where electrons are most likely to be found. (b) Like the balloons in Figure 9.5, each electron domain occupies a finite volume of space, so they also adopt an arrangement where repulsions are minimized. **9.15** (a) Trigonal planar (b) tetrahedral (c) trigonal bipyramidal (d) octahedral **9.17** The electron-domain geometry indicated by VSEPR describes the arrangement of all bonding and nonbonding electron domains. The molecular geometry describes just the atomic positions. In H_2O there are 4 electron domains around oxygen, so the electron-domain geometry is tetrahedral. Because there are 2 bonding and 2 nonbonding domains, the molecular geometry is bent. **9.19** (a) Tetrahedral, tetrahedral (b) trigonal bipyramidal, T-shaped (c) octahedral, square pyramidal **9.21** (a) Tetrahedral, trigonal pyramidal (b) trigonal planar, trigonal planar (c) trigonal bipyramidal, T-shaped (d) tetrahedral, tetrahedral (e) trigonal bipyramidal, linear (f) tetrahedral, bent **9.23** (a) i, trigonal planar; ii, tetrahedral; iii, trigonal bipyramidal (b) i, 0; ii, 1; iii, 2 (c) N and P (d) Cl (or Br or I). This T-shaped molecular geometry arises from a trigonal bipyramidal electron-domain geometry with 2 nonbonding domains. Assuming each F atom has 3 nonbonding domains and forms only single bonds with A, A must have 7 valence electrons and be in or below the third row of the periodic table to produce these electron-domain and molecular geometries. **9.25** (a) 1–109°, 2–109° (b) 3–109°, 4–109° (c) 5–180° (d) 6–120°, 7–109°, 8–109° **9.27** Each species has 4 electron domains, but the number of nonbonding domains decreases from 2 to 0, going from NH_2^- to NH_4^+. Since nonbonding domains exert greater repulsive forces on adjacent domains, the bond angles expand as the number of nonbonding domains decreases. **9.29** (a) Although both ions have 4 bonding electron domains, the 6 total domains around Br require octahedral domain geometry and square-planar molecular geometry, while the 4 total domains about B lead to tetrahedral domain and molecular geometry. (b) CF_4 will have bond angles closest to the value predicted by VSEPR because there are no nonbonding electron domains around C. In SF_4 the single nonbonding domain will enter into more repulsive interactions, "push back" the bonding domains, and lead to bond angles that are nonideal. **9.31** Yes. The dipole moment vector points along the O—S—O angle bisector with the negative end of the dipole pointing away from the S atom. **9.33** (a) In Exercise 9.23, molecules (ii) and (iii) will have nonzero dipole moments. Molecule (i) has no nonbonding electron pairs on A, and the 3 A—F bond dipoles are oriented so that they cancel. Molecules (ii) and (iii) have nonbonding electron pairs on A and their bond dipoles do not cancel. (b) In Exercise 9.24, molecules (i) and (ii) have a zero dipole moment. **9.35** (b) NH_3, (c) SF_4, and (e) CH_3Br are polar. **9.37** The middle isomer has a zero net dipole moment. **9.39** (a) *Orbital overlap* occurs when valence atomic orbitals on two adjacent atoms share the same region of space. (b) In valence-bond theory, orbital overlap allows two bonding electrons to mutually occupy the space between the bonded nuclei. (c) Valence-bond theory is a combination of the atomic orbital concept and the Lewis model of electron-pair bonding. **9.41** (a) H—Mg—H, linear electron domain and molecular geometry (b) The Mg atom has 2 electrons in its 2*s* orbital; these electrons are paired. Without promotion, Mg has no unpaired electrons available to form bonds with H. (c) The linear electron domain geometry in MgH_2 requires *sp* hybridization. (d)

9.43 (a) *sp*, 180° (b) sp^3, 109° (c) sp^2, 120° (d) sp^3d^2, 90° and 180° (e) sp^3d, 90°, 120°, and 180° **9.45** (a) B, [He]$2s^2 2p^1$. One 2*s* electron is promoted to an empty 2*p* orbital. The 2*s* and two 2*p* orbitals that each contain one electron are hybridized to form three equivalent hybrid orbitals in a trigonal planar arrangement. (b) sp^2

(c)

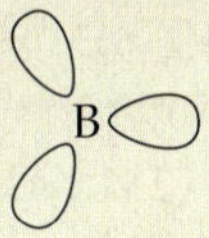

(d) A single 2*p* orbital is unhybridized. It lies perpendicular to the trigonal plane of the sp^2 hybrid orbitals. **9.47** (a) sp^2 (b) sp^3 (c) *sp* (d) sp^3d (e) sp^3d^2

9.49

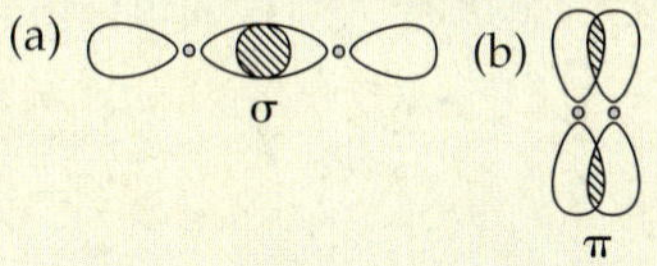

(c) A σ bond is generally stronger than a π bond because there is more extensive orbital overlap.

9.51 (a)

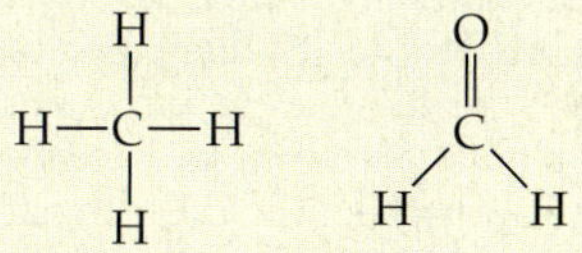

(b) sp^3, sp^2 (c) The C atom in CH_4 is sp^3 hybridized; there are no unhybridized *p* orbitals available for the π overlap required by multiple bonds. In CH_2O the C atom is sp^2 hybridized, with one *p* atomic orbital available to form the π overlap in the C=O double bond. **9.53** (a) 18 valence electrons (b) 16 valence electrons form σ bonds (c) 2 valence electrons form π bonds (d) no valence electrons are nonbonding (e) The left and central C atoms are sp^2 hybridized; the right C atom is sp^3 hybridized.

9.55 (a) ~109° about the leftmost C, sp^3, ~120° about the right-hand C, sp^2 (b) The doubly bonded O can be viewed as sp^2, and the other as sp^3; the nitrogen is sp^3 with approximately 109° bond angles. (c) nine σ bonds, one π bond **9.57** (a) In a localized π bond, the electron density is concentrated between the two atoms forming the bond. In a delocalized π bond, the electron density is spread over all the atoms that contribute p orbitals to the network. (b) The existence of more than one resonance form is a good indication that a molecule will have delocalized π bonding. (c) delocalized **9.59** (a) Both atomic and molecular orbitals have a characteristic energy and shape; each can hold a maximum of two electrons. Atomic orbitals are localized, and their energies are the result of interactions between the subatomic particles in a single atom. Molecular orbitals can be delocalized and their energies are influenced by interactions between electrons on several atoms. (b) There is a net lowering in energy that accompanies bond formation, because the electrons in H_2 are strongly attracted to both H nuclei. (c) 2

9.61 (a)

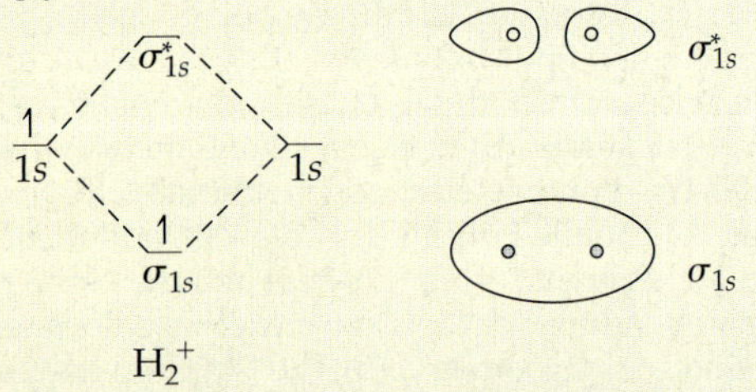

(b) There is one electron in H_2^+. (c) $\sigma_{1s}^{\ 1}$ (d) BO $= \frac{1}{2}$ (e) Fall apart. If the single electron in H_2^+ is excited to the σ^*_{1s} orbital, its energy is higher than the energy of an H 1s atomic orbital and H_2^+ will decompose into a hydrogen atom and a hydrogen ion.

9.63

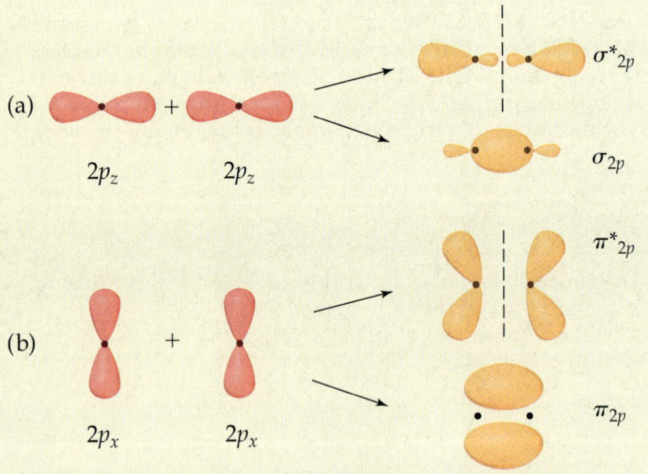

(c) $\sigma_{2p} < \pi_{2p} < \pi^*_{2p} < \sigma^*_{2p}$ **9.65** (a) When comparing the same two bonded atoms, bond order and bond energy are directly related, while bond order and bond length are inversely related. When comparing different bonded nuclei, there are no simple relationships. (b) Be_2 is not expected to exist; it has a bond order of zero and is not energetically favored over isolated Be atoms. Be_2^+ has a bond order of 0.5 and is slightly lower in energy than isolated Be atoms. It will probably exist under special experimental conditions. **9.67** (a, b) Substances with no unpaired electrons are weakly repelled by a magnetic field. This property is called *diamagnetism*. (c) O_2^{2-}, Be_2^{2+} **9.69** (a) B_2^+, $\sigma_{2s}^{\ 2}\sigma^{*\,2}_{2s}\pi_{2p}^{\ 1}$, increase (b) Li_2^+, $\sigma_{1s}^{\ 2}\sigma^{*\,2}_{1s}\sigma_{2s}^{\ 1}$, increase (c) N_2^+, $\sigma_{2s}^{\ 2}\sigma^{*\,2}_{2s}\pi_{2p}^{\ 4}\pi_{2p}^{\ 1}$, increase (d) Ne_2^{2+}, $\sigma_{2s}^{\ 2}\sigma^{*\,2}_{2s}\pi_{2p}^{\ 4}\pi^{*\,4}_{2s}$, decrease **9.71** CN, $\sigma_{2s}^{\ 2}\sigma^{*\,2}_{2s}\pi_{2p}^{\ 4}\sigma_{2p}^{\ 1}$, bond order = 2.5, paramagnetic; CN^+, $\sigma_{2s}^{\ 2}\sigma^{*\,2}_{2s}\pi_{2p}^{\ 4}$, bond order = 2.0, diamagnetic; CN^-, $\sigma_{2s}^{\ 2}\sigma^{*\,2}_{2s}\pi_{2p}^{\ 4}\sigma_{2p}^{\ 2}$, bond order = 3.0, diamagnetic **9.73** (a) 3s, 3p_x, 3p_y, 3p_z (b) π_{3p} (c) 2 (d) If the MO diagram for P_2 is similar to that of N_2, P_2 will have no unpaired electrons and be diamagnetic. **9.76** SiF_4 is tetrahedral, SF_4 is seesaw, XeF_4 is square planar. The shapes are different because the number of nonbonding electron domains is different in each molecule, even though all have four bonding electron domains. Bond angles and thus molecular shape are determined by the total number of electron domains. **9.79** (a) Two sigma, two pi (b) two sigma, two pi (c) three sigma, one pi (d) four sigma, one pi **9.82** The compound on the right has a nonzero dipole moment.

9.85

(a) The molecule is nonplanar. (b) Allene has no dipole moment. (c) The bonding in allene would not be described as delocalized. The π electron clouds of the two adjacent C=C are mutually perpendicular, so there is no overlap and no delocalization of π electrons. **9.91** $\sigma_{2s}^{\ 2}\sigma^{*\,2}_{2s}\pi_{2p}^{\ 4}\sigma_{2p}^{\ 1}\pi^{*\,1}_{2p}$ The bond order of N_2 in the ground state is 3; in the first excited state it has a bond order of 2. Owing to the reduction in bond order, N_2 in the first excited state would be more reactive, have a smaller bond energy and a longer N—N separation. **9.95** (a) :C≡O: (b) The bond order of 3.0 predicted using Figure 9.48 agrees with the triple bond in the Lewis structure. (c) According to Figure 9.48, the highest energy electrons in CO would occupy a π_{2p} molecular orbital. If the highest energy electrons occupy a σ-type orbital, the order of π_{2p} and σ_{2p} orbitals must be reversed. Figure 9.44 shows how the interaction of the 2s orbitals on one atom and the 2p orbitals on the other atom can cause this reversal. (d) The π_{2p} bonding MOs of CO will have a greater contribution from the more electronegative O atom. **9.96** (a) HNO_2 (b) $\ddot{O}=\ddot{N}-\ddot{O}-H$ (c) The geometry around N is trigonal planar. (d) sp^2 hybridization around N (e) three sigma, one pi **9.99** (a) sp^2 hybridization at the two central C atoms (b) Isomerization requires 180° rotation around the C=C double bond. A 90° rotation eliminates all overlap of the p orbitals that form the π bond and it is broken. (c) 4.42×10^{-19} J/molecule (d) λ = 450 nm (e) Yes, 450 nm light is in the visible portion of the spectrum. A cis-trans isomerization in the retinal portion of the large molecule rhodopsin is the first step in a sequence of molecular transformations in the eye that leads to vision. The sequence of events enables the eye to detect visible photons, in other words, to see. **9.101** From bond enthalpies, ΔH = 5364 kJ; according to Hess's law, $\Delta H°$ = 5535 kJ. The difference in the two results, 171 kJ, is due to the resonance stabilization in benzene. The amount of energy actually required to decompose 1 mol of $C_6H_6(g)$ is greater than the sum of the localized bond enthalpies. **9.103** (a) $3d_{z^2}$
(b)

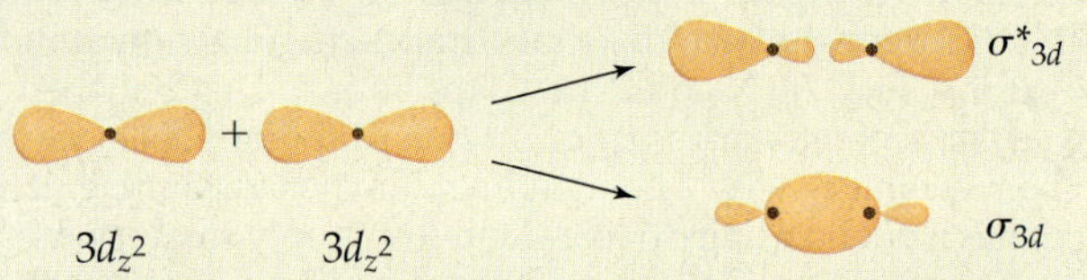

The "donuts" of the $3d_{z^2}$ orbitals have been omitted from the diagram for clarity.
(c) A node is generated in σ^*_{3d} because antibonding MOs are formed when atomic orbital lobes with opposite phases interact. Electron density is excluded from the internuclear region and a node is formed in the MO.

(d)

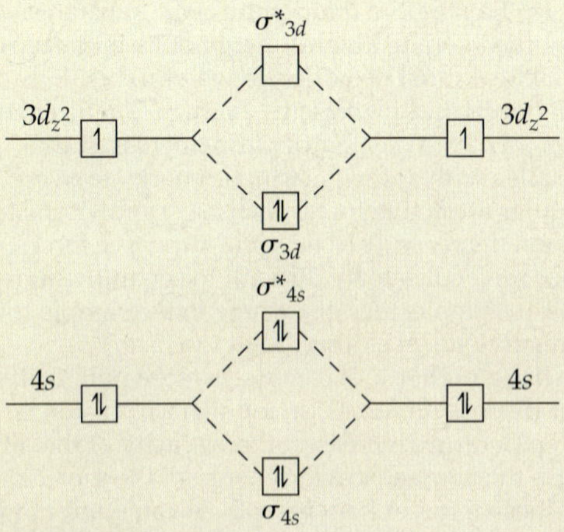

(e) The bond order of Sc_2 is 1.0.

Chapter 10

10.1

(a) $V_2 = \frac{5}{3} V_1$ (b) $V_2 = \frac{1}{2} V_1$

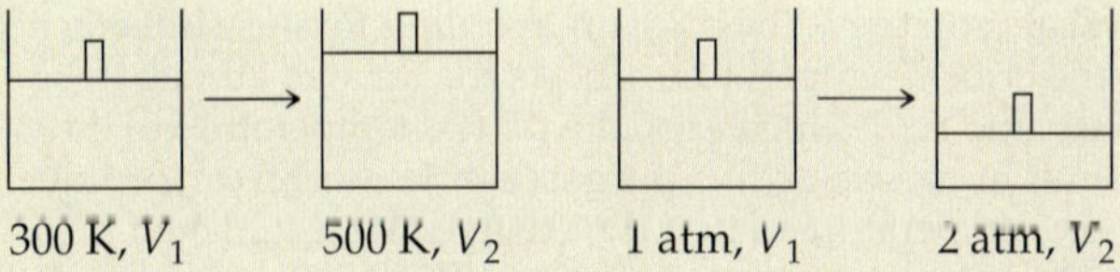

10.4 Over time, the gases will mix perfectly. Each bulb will contain 4 blue and 3 red atoms. The "blue" gas has the greater partial pressure after mixing, because it has the greater number of particles at the same T and V as the "red" gas. **10.7** (a) Curve A is $O_2(g)$ and curve B is He(g). (b) For the same gas at different temperatures, curve A represents the lower temperature and curve B the higher temperature. **10.9** (a) A gas is much less dense than a liquid. (b) A gas is much more compressible than a liquid. (c) All mixtures of gases are homogenous. Similar liquid molecules form homogeneous mixtures, while very dissimilar molecules form heterogeneous mixtures. **10.11** 1.8×10^3 kPa **10.13** (a) 10.3 m (b) 2.0 atm **10.15** (a) The tube can have any cross-sectional area. (b) At equilibrium the force of gravity per unit area acting on the mercury column at the level of the outside mercury is not equal to the force of gravity per unit area acting on the atmosphere. (c) The column of mercury is held up by the pressure of the atmosphere applied to the exterior pool of mercury. **10.17** (a) 0.349 atm (b) 265 mm Hg (c) 3.53×10^4 Pa (d) 0.353 bar **10.19** (a) $P = 773.4$ torr (b) The pressure in Chicago is greater than standard atmospheric pressure, and so it makes sense to classify this weather system as a "high-pressure system." **10.21** (i) 0.30 atm (ii) 1.073 atm (iii) 0.136 atm **10.23** (a) If V decreases by a factor of 4, P increases by a factor of 4. (b) If T decreases by a factor of 2, P decreases by a factor of 2. (c) If n decreases by a factor of 2, P decreases by a factor of 2. **10.25** (a) If equal volumes of gases at the same temperature and pressure contain equal numbers of molecules and molecules react in the ratios of small whole numbers, it follows that the volumes of reacting gases are in the ratios of small whole numbers. (b) Since the two gases are at the same temperature and pressure, the ratio of the numbers of atoms is the same as the ratio of volumes. There are 1.5 times as many Xe atoms as Ne atoms. **10.27** (a) $PV = nRT$; P in atmospheres, V in liters, n in moles, T in kelvins. (b) An ideal gas exhibits pressure, volume, and temperature relationships described by the equation $PV = nRT$. **10.29** Flask A contains the gas with $\mathcal{M} = 30$ g/mol, and flask B contains the gas with $\mathcal{M} = 60$ g/mol,

10.31

P	V	n	T
2.00 atm	1.00 L	0.500 mol	48.7 K
0.300 atm	0.250 L	3.05×10^{-3} mol	27°C
650 torr	11.2 L	0.333 mol	350 K
10.3 atm	585 mL	0.250 mol	295 K

10.33 1.7×10^4 kg H_2 **10.35** 5.72×10^{22} molecules **10.37** (a) 91 atm (b) 2.3×10^2 L **10.39** (a) 29.8 g Cl_2 (b) 9.42 L (c) 501 K (d) 1.90 atm **10.41** (a) $n = 2 \times 10^{-4}$ mol O_2 (b) The roach needs 8×10^{-3}mol O_2 in 48 h, more than 100% of the O_2 in the jar. **10.43** For gas samples at the same conditions, molar mass determines density. Of the three gases listed, (c) Cl_2 has the largest molar mass. **10.45** (c) Because the helium atoms are of lower mass than the average air molecule, the helium gas is less dense than air. The balloon thus weighs less than the air displaced by its volume. **10.47** (a) $d = 1.77$ g/L (b) $\mathcal{M} = 80.1$ g/mol **10.49** $\mathcal{M} = 89.4$ g/mol **10.51** 3.5×10^{-9} g Mg **10.53** 21.4 L CO_2 **10.55** 0.402 g Zn **10.57** (a) When the stopcock is opened, the volume occupied by $N_2(g)$ increases from 2.0 L to 5.0 L. $P_{N_2} = 0.40$ atm (b) When the gases mix, the volume of $O_2(g)$ increases from 3.0 L to 5.0 L. $P_{O_2} = 1.2$ atm (c) $P_t = 1.6$ atm **10.59** (a) $P_{He} = 1.88$ atm, $P_{Ne} = 1.10$ atm, $P_{Ar} = 0.360$ atm (b) $P_t = 3.34$ atm **10.61** $P_{CO_2} = 0.305$ atm, $P_t = 1.232$ atm **10.63** $P_{N_2} = 0.98$ atm, $P_{O_2} = 0.39$ atm, $P_{CO_2} = 0.20$ atm **10.65** 2.5 mole % O_2 **10.67** $P_t = 3.04$ atm **10.69** (a) Increase in temperature at constant volume or decrease in volume or increase in pressure (b) decrease in temperature (c) increase in volume, decrease in pressure (d) increase in temperature **10.71** The fact that gases are readily compressible supports the assumption that most of the volume of a gas sample is empty space. **10.73** (a) Vessel A has more molecules. (b) The density of CO is 1.25 g/L, and the density of SO_2 is 1.33 g/L. Vessel B has more mass. (c) The average kinetic energy of the molecules in vessel B is higher. (d) $u_A/u_B = 1.46$. The molecules in vessel A have the greater rms speed. **10.75** (a) In order of increasing speed and decreasing molar mass: HBr < NF_3 < SO_2 < CO < Ne (b) $u_{NF_3} = 324$ m/s **10.77** The order of increasing rate of effusion is $^2H^{37}Cl < {}^1H^{37}Cl < {}^2H^{35}Cl < {}^1H^{35}Cl$. **10.79** As_4S_6 **10.81** (a) Non-ideal-gas behavior is observed at very high pressures and low temperatures. (b) The real volumes of gas molecules and attractive intermolecular forces between molecules cause gases to behave nonideally. (c) According to the ideal-gas law, the ratio PV/RT should be constant for a given gas sample at all combinations of pressure, volume, and temperature. If this ratio changes with increasing pressure, the gas sample is not behaving ideally. **10.83** Ar ($a = 1.34$, $b = 0.0322$) will behave more like an ideal gas than CO_2 ($a = 3.59$, $b = 0.427$) at high pressures. **10.85** (a) $P = 0.917$ atm (b) $P = 0.896$ atm **10.87** A mercury barometer with water trapped in its tip will not read the correct pressure. Water at the top of the Hg column establishes a vapor pressure that exerts downward pressure in addition to gravity and partially counterbalances the pressure of the atmosphere. **10.90** 742 balloons can be filled completely, with a bit of He left over. **10.93** 5.4×10^3 g O_2 **10.96** (a) 3.78×10^{-3} mol O_2 (b) 0.0345 g C_8H_{18} **10.99** Oxygen is 70.1 mole % of the mixture. **10.102** (a) As a gas is compressed at constant temperature, the number of intermolecular collisions increases. Intermolecular attraction causes some of these collisions to be inelastic, which amplifies the deviation from ideal behavior. (b) As the temperature of a gas is increased at constant volume, a larger fraction of the molecules has sufficient kinetic energy to overcome intermolecular attractions and the effect of intermolecular attraction becomes less significant. **10.105** (a) The molecular formula is C_3H_6. (b) Ar and C_3H_6 are nonpolar, have similar molar masses, and experience London disperson forces. We expect the effect of both volume and intermolecular attractions to be more significant for the structurally more complex C_3H_6. Cyclopropane will deviate more from ideal behavior at the specified conditions. **10.108** $\Delta H = -1.1 \times 10^{14}$ kJ (assuming $H_2O(l)$ is a product) **10.112** (a) The partial pressure of IF_5 is 0.515 atm. (b) The mole fraction of IF_5 is 0.544.

Chapter 11

11.1 The diagram best describes a liquid. The particles are close together, mostly touching, but there is no regular arrangement or order. This rules out a gaseous sample, where the particles are far apart, and a crystalline solid, which has a regular repeating structure in all three directions. **11.4** (a) 385 mm Hg (b) 22°C (c) 47°C **11.7** (a) 3 Nb atoms, 3 O atoms (b) NbO (c) This is primarily an ionic solid, because Nb is a metal and O is a nonmetal. There may be come covalent character to the Nb···O bonds. **11.9** (a) Solid < liquid < gas (b) gas < liquid < solid **11.11** (a) Gases are more compressible than liquids because there is much empty space between gas particles. (b) The solid and liquid forms of a substance are called *condensed phases* because in both states there is very little space between particles; the volume of the substance is condensed. (c) Liquids have greater ability to flow than solids because their average kinetic energy is on the same order of magnitude as the average attractive energy among particles. **11.13** (a) London dispersion forces (b) dipole-dipole and London dispersion forces (c) dipole-dipole forces and in certain cases hydrogen bonding **11.15** (a) Nonpolar covalent molecule; London dispersion forces only (b) polar covalent molecule with O—H bonds; hydrogen bonding, dipole-dipole forces and London dispersion forces (c) polar covalent molecule; dipole-dipole and London dispersion forces (but not hydrogen bonding) **11.17** (a) Polarizability is the ease with which the charge distribution in a molecule can be distorted to produce a transient dipole. (b) Te is most polarizable because its valence electrons are farthest from the nucleus and least tightly held. (c) in order of increasing polarizability: $CH_4 < SiH_4 < SiCl_4 < GeCl_4 < GeBr_4$ (d) The magnitudes of London dispersion forces and thus the boiling points of molecules increase as polarizability increases. The order of increasing boiling points is the order of increasing polarizability given in (c). **11.19** (a) H_2S (b) CO_2 (c) SiH_4 **11.21** Both rodlike butane molecules and spherical 2-methylpropane molecules experience dispersion forces. The larger contact surface between butane molecules facilitates stronger forces and produces a higher boiling point. **11.23** (a) Molecules with N—H, O—H, and F—H bonds form hydrogen bonds with like molecules. (b) CH_3NH_2 and CH_3OH. **11.25** (a) HF has the higher boiling point because hydrogen bonding is stronger than dipole-dipole forces. (b) $CHBr_3$ has the higher boiling point because it has the higher molar mass, indicating greater polarizability and stronger dispersion forces. (c) ICl has the higher boiling point because the molecules have similar molar masses (hence similar dispersion forces), but ICl is polar, giving it dipole-dipole forces that are absent for the nonpolar Br_2. **11.27** (a) Hydrogen bonding occurs in both water and ice. In water, molecules are close together and moving. Hydrogen bonds are constantly broken and new ones formed. In ice, molecules adopt an ordered structure with four hydrogen bonds per molecule. These interactions produce a network structure with more open space between molecules than in the liquid state and a corresponding lower density for the solid than the liquid. (b) Raising the temperature of a substance means increasing the average kinetic energy of its molecules. A large amount of heat must be added to water to break its strong hydrogen bonds and increase its average kinetic energy and temperature. **11.29** (a) As temperature increases, the number of molecules with sufficient kinetic energy to overcome intermolecular attractive forces increases, and viscosity and surface tension decrease. (b) The same attractive forces that cause surface molecules to be difficult to separate (high surface tension) cause molecules elsewhere in the sample to resist movement relative to each other (high viscosity). **11.31** (a) $CHBr_3$ has a higher molar mass, is more polarizable, and has stronger dispersion forces, so the surface tension is greater. (b) As temperature increases, the viscosity of the oil decreases because the average kinetic energy of the molecules increases. (c) Adhesive forces between polar water and nonpolar car wax are weak, so the large surface tension of water draws the liquid into the shape with the smallest surface area, a sphere. **11.33** (a) Freezing, exothermic (b) evaporation, endothermic (c) deposition, exothermic **11.35** Melting does not require separation of molecules, so the energy requirement is smaller than for vaporization, where molecules must be separated. **11.37** 2.2×10^3 g H_2O **11.39** 105 kJ **11.41** (a) The critical pressure is the pressure required to cause liquefaction at the critical temperature. (b) As the force of attraction between molecules increases, the critical temperature of the compound increases. (c) All the gases in Table 11.5 can be liquefied at the temperature of liquid nitrogen, given sufficient pressure. **11.43** (a) No effect (b) no effect (c) Vapor pressure decreases with increasing intermolecular attractive forces because fewer molecules have sufficient kinetic energy to overcome attractive forces and escape to the vapor phase. (d) Vapor pressure increases with increasing temperature because average kinetic energies of molecules increase. **11.45** (a) $CBr_4 < CHBr_3 < CH_2Br_2 < CH_2Cl_2 < CH_3Cl < CH_4$. The trend is dominated by dispersion forces even though four of the molecules are polar. The order of increasing volatility is the order of increasing vapor pressure, decreasing molar mass, and decreasing strength of dispersion forces. (b) Boiling point increases as the strength of intermolecular forces increases; this is the order of decreasing volatility and the reverse of the order in part (a). $CH_4 < CH_3Cl < CH_2Cl_2 < CH_2Br_2 < CHBr_3 < CBr_4$ **11.47** (a) The temperature of the water in the two pans is the same. (b) Vapor pressure does not depend on either volume or surface area of the liquid. At the same temperature, the vapor pressures of water in the two containers are the same. **11.49** (a) Approximately 17°C (b) approximately 510 torr (c) approximately 26°C (d) approximately 105°C **11.51** (a) The critical point is the temperature and pressure beyond which the gas and liquid phases are indistinguishable. (b) The line that separates the gas and liquid phases ends at the critical point because at conditions beyond the critical temperature and pressure, there is no distinction between gas and liquid. In experimental terms a gas cannot be liquefied at temperatures higher than the critical temperature, regardless of pressure. **11.53** (a) $H_2O(g)$ will condense to $H_2O(s)$ at approximately 4 torr; at a higher pressure, perhaps 5 atm or so, $H_2O(s)$ will melt to form $H_2O(l)$. (b) At 100°C and 0.50 atm, water is in the vapor phase. As it cools, water vapor condenses to the liquid at approximately 82°C, the temperature where the vapor pressure of liquid water is 0.50 atm. Further cooling results in freezing at approximately 0°C. The freezing point of water increases with decreasing pressure, so at 0.50 atm the freezing temperature is very slightly above 0°C.

11.55
(a)

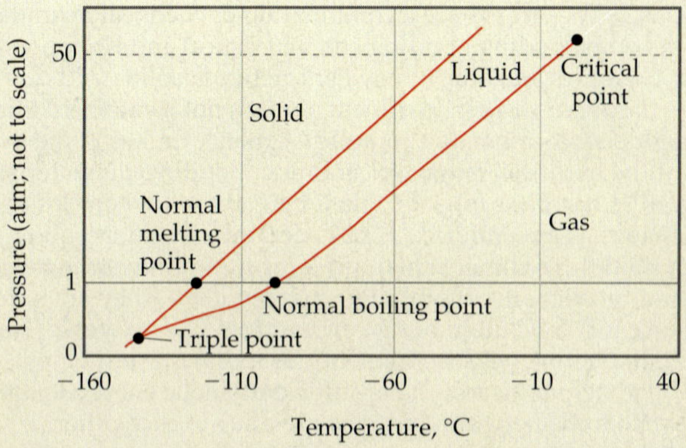

(b) Xe(s) is denser than Xe(l) because the solid-liquid line on the phase diagram is normal. (c) Cooling Xe(g) at 100 torr will cause deposition of the solid because 100 torr is below the pressure of the triple point. **11.57** In a crystalline solid the component particles are arranged in an ordered repeating pattern. In an amorphous solid there is no orderly structure. **11.59** $CaTiO_3$ **11.61** (a) $r = 1.355$ Å (b) density $= 22.67\ g/cm^3$ **11.63** Atomic weight $= 55.8$ g/mol **11.65** (a) 12 (b) 6 (c) 8 **11.67** $a = 6.13$ Å **11.69** (a) The U^{4+} ions in UO_2 are represented by the smaller spheres in Figure 11.42(c). The ratio of large spheres to small spheres matches the O^{2-} to U^{4+} ratio in the chemical formula, so the small ones must represent U^{4+}. (b) density $= 10.97\ g/cm^3$ **11.71** (a) Hydrogen bonding, dipole-dipole forces, London dispersion forces (b) covalent chemical bonds (c) ionic bonds (d) metallic bonds **11.73** In molecular solids relatively weak intermolecular forces bind the molecules in the lattice, so relatively little energy is required to disrupt these forces. In covalent-network solids, covalent bonds join atoms into an extended network. Melting or deforming a covalent-network solid means breaking covalent bonds, which requires a large amount of energy. **11.75** Because of its relatively high melting point and properties as a conducting solution, the solid must be ionic. **11.77** (a) Xe, greater atomic weight, stronger dispersion forces (b) SiO_2, covalent-network lattice versus weak dispersion forces (c) KBr, strong ionic versus weak dispersion forces (d) C_6Cl_6, both are influenced by dispersion forces, C_6Cl_6 has the higher molar mass **11.79** (a) Decrease (b) increase (c) increase (d) increase (e) increase (f) increase (g) increase **11.82** (a) In CH_2Br_2 the dispersion force will be more influential than the dipole-dipole force. Relative to CH_2Cl_2, Br is more polarizable than Cl and the dipole moment of CH_2Br_2 is smaller. (b) Dipole-dipole forces will play a larger role for CH_2F_2 than for CH_2Cl_2 because CH_2F_2 has a larger dipole moment and F is less polarizable than Cl. **11.85** The two O—H groups in ethylene glycol are involved in many hydrogen bonding interactions, leading to its high boiling point and viscosity, relative to pentane, which experiences only dispersion forces. **11.88** (a) Evaporation is an endothermic process. The heat required to vaporize sweat is absorbed from your body, helping to keep it cool. (b) The vacuum pump reduces the pressure of the atmosphere above the water until atmospheric pressure equals the vapor pressure of water and the water boils. Boiling is an endothermic process, and the temperature drops if the system is not able to absorb heat from the surroundings fast enough. As the temperature of the water decreases, the water freezes. **11.92** The large difference in melting points is due to the very different forces imposing atomic order in the solid state. Much more kinetic energy is required to disrupt the delocalized metallic bonding in gold than to overcome the relatively weak London dispersion forces in Xe. **11.95** The most effective diffraction occurs when distances between layers of atoms in the crystal are similar to the wavelength of the light being diffracted. Molybdenum X rays of 0.71 Å are on the same order of magnitude as interlayer distances in crystals and are diffracted. Visible light, 400–700 nm or 4000–7000 Å, is too long to be diffracted effectively. **11.98** 16 Al atoms, 8 Mg atoms, 32 O atoms
11.101

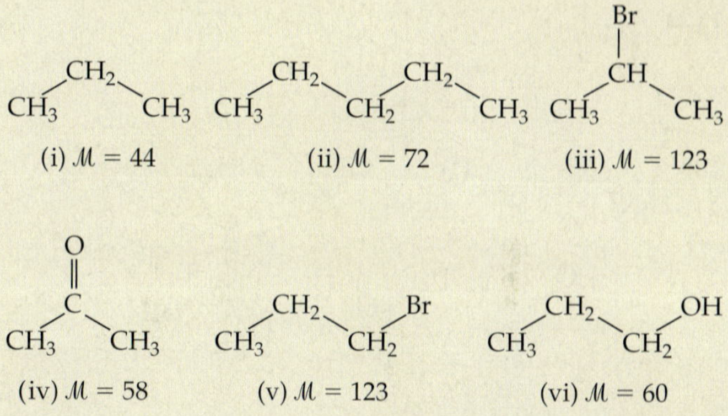

(a) Molar mass: Compounds (i) and (ii) have similar rodlike structures. The longer chain in (ii) leads to greater molar mass, stronger London-dispersion forces, and higher heat of vaporization. (b) Molecular shape: Compounds (iii) and (v) have the same chemical formula and molar mass but different molecular shapes. The more rodlike shape of (v) leads to more contact between molecules, stronger dispersion forces, and higher heat of vaporization. (c) Molecular polarity: Compound (iv) has a smaller molar mass than (ii) but a larger heat of vaporization, which must be due to the presence of dipole-dipole forces. (d) Hydrogen bonding interactions: Molecules (v) and (vi) have similar structures. Even though (v) has larger molar mass and dispersion forces, hydrogen bonding causes (vi) to have the higher heat of vaporization. **11.105** P(benzene vapor) $= 98.6$ torr.

Chapter 12

12.1 The band structure of material A is that of a metal. **12.4** Picture (a) has a one-dimensional order characteristic of a liquid crystalline phase. **12.7** (a) Semiconductor (b) insulator (c) semiconductor (d) metal **12.9** Ge or Si (Ge is closer to Ga in bonding atomic radius.) **12.11** (a) False. Semiconductors have a smaller band gap; they conduct some electricity, while insulators do not. (b) True. Dopants create either more electrons or more "holes," both of which increase the conductivity of the semiconductor. (c) True. Metals conduct electricity because delocalized electrons in the lattice provide a mechanism for charge mobility. (d) True. Metal oxides are ionic substances with essentially localized electrons. This barrier to charge mobility renders them insulators. **12.13** A superconducting material offers no resistance to the flow of electrical current. Superconductive materials could transmit electricity with much greater efficiency than current carriers. **12.15** The sharp drop in resistivity of MgB_2 near 39 K is the superconducting transition temperature, T_c. **12.17** Superconductivity acts on electrons that are conduction electrons. Insulators have no electrons in the conduction band available for conduction. **12.19** n-Decane does not have a sufficiently high chain length or molecular mass to be considered a polymer.

12.21

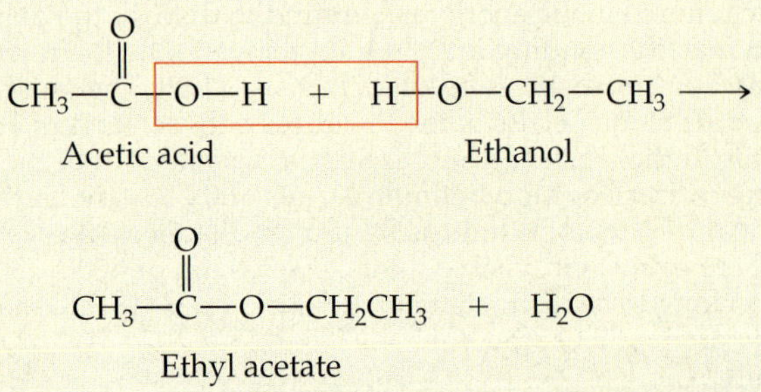

If a dicarboxylic acid and a dialcohol are combined, there is the potential for propagation of the polymer chain at both ends of both monomers.

12.23 (a)

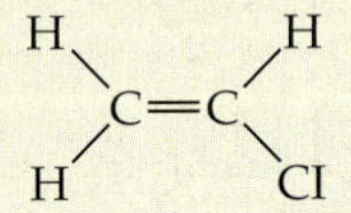

(b)

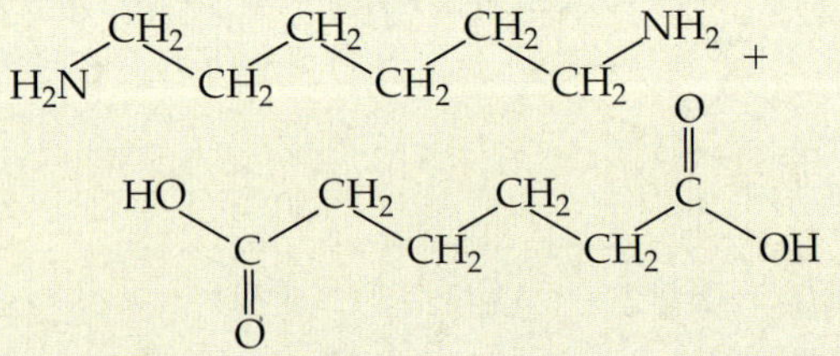

(c)

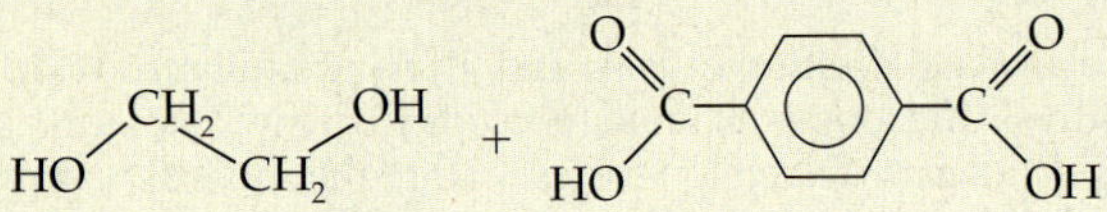

12.25

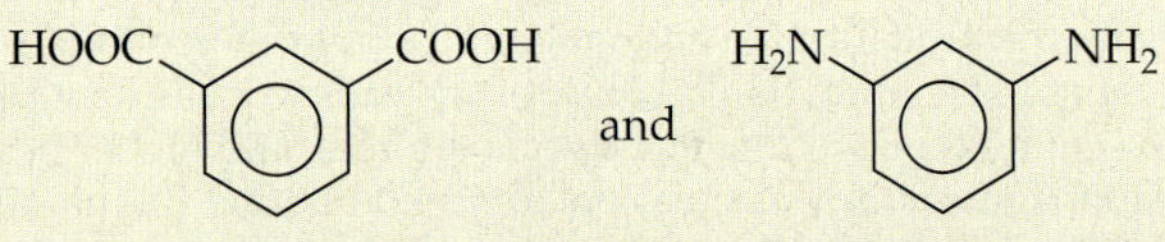

12.27

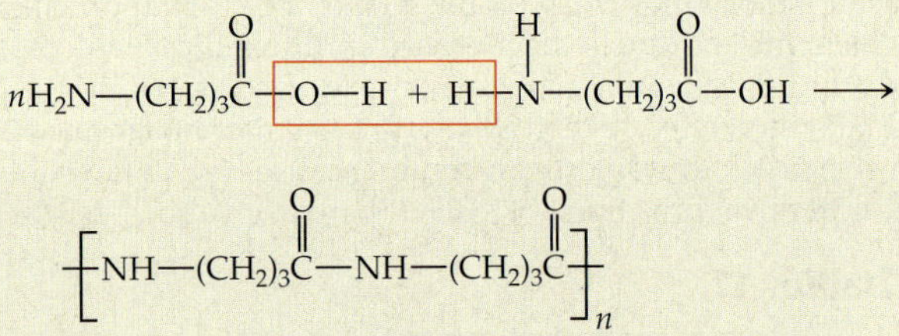

12.29 Flexibility of molecular chains causes flexibility of the bulk polymer. Flexibility is enhanced by molecular features that inhibit order, such as branching, and diminished by features that encourage order, such as cross-linking or delocalized π-electron density. Cross-linking, the formation of chemical bonds between polymer chains, reduces flexibility of the molecular chains, increases the hardness of the material, and decreases the chemical reactivity of the polymer. **12.31** The function of the polymer determines whether high molecular masses and high degree of crystallinity are desirable properties. If the polymer will be used as a flexible wrapping or fiber, rigidity that is due to high molecular mass is an undesirable property. **12.33** Ceramics are not readily recyclable because of their extremely high melting points and rigid ionic or covalent-network structures. **12.35** Very small, uniformly sized and shaped particles are required for the production of a strong ceramic object by sintering. Upon heating to initiate condensation reactions, the more uniform the particle size and the greater the total surface area of the solid, the more chemical bonds are formed and the stronger the ceramic object.

12.37 The ceramics are MgO, soda-lime glass, ZrB_2, Al_2O_3, and TaC. The criteria are a combination of chemical formula (with corresponding bonding characteristics) and Knoop values. Ceramics are ionic or covalent-network solids with fairly large hardness values. Hardness alone is not a sufficient criterion for classification as a ceramic. A metal, Cr, lies in the middle of the hardness range for ceramics. Bonding characteristics as well as hardness must be taken into account when classifying materials as ceramics. **12.39** Ceramics are ionic solids with rigid three-dimensional order resulting from strong electrostatic attractions among fully charged ions. They are hard because it is difficult to budge an ion from this network. They are high melting because tremendous thermal energy must be supplied before the ions have sufficient kinetic energy to break away from the network and move relative to each other.
12.41 Is the neoprene biocompatible? Does it provoke inflammatory reactions? Does neoprene meet the physical requirements of a flexible lead? Will it remain resistant to degradation and maintain elasticity? Can neoprene be prepared in sufficiently pure form so that it can be classified as medical grade?
12.43 Current vascular graft materials cannot be lined with cells similar to those in the native artery. The body detects that the graft is "foreign," and platelets attach to the inside surfaces, causing blood clots. The inside surfaces of future vascular implants need to accommodate a lining of cells that do not attract or attach to platelets. **12.45** In order for skin cells in a culture medium to develop into synthetic skin, a mechanical matrix that holds the cells in contact with one another must be present. The matrix must be strong, biocompatible, and biodegradable. It probably has polar functional groups that form hydrogen bonds with biomolecules in the tissue cells.
12.47 Silicon is the semiconductor of choice because it is abundant, cheap, nontoxic; can be highly purified; and grows nearly perfect enormous crystals. **12.49** The diameter of a Si atom is 2.22 Å; 60 nm $= 2.7 \times 10^2$ Si atoms. **12.51** A 1.1 eV photon corresponds to a wavelength of 1.1×10^{-6} m. According to the figure, Si can absorb approximately 32% of the wavelengths in the solar spectrum and much more than 32% of the flux.
12.53 Both an ordinary liquid and a nematic liquid crystal phase are fluids; they are converted directly to the solid phase upon cooling. The nematic phase is cloudy and more viscous than an ordinary liquid. Upon heating, the nematic phase is converted to an ordinary liquid. **12.55** In the solid state the relative orientation of the molecules is fixed and repeating in all three dimensions. When a substance changes to the nematic liquid crystalline phase, the molecules remain aligned in one dimension; translational motion is allowed, but rotational motion is restricted. Transformation to the isotropic liquid phase destroys the one-dimensional order, resulting in free translational and rotational motion. **12.57** The presence of polar groups or nonbonded electron pairs leads to relatively strong dipole-dipole interactions between molecules. These are a significant part of the orienting forces necessary for liquid crystal formation. **12.59** In the nematic phase there is one-dimensional order, while in a smectic phase there is two-dimensional order. In a smectic phase the long directions of the molecules and the ends of the molecules are aligned. **12.61** A nematic phase is composed of sheets of molecules aligned along their lengths, with no additional order within the sheet or between sheets. A cholesteric phase also contains this kind of sheet, but with some ordering between sheets. **12.63** $\lambda = 560$ nm **12.65** If a solid has nanoscale dimensions of 1–10 nm, there may not be enough atoms contributing atomic orbitals to produce continuous energy bands of molecular orbitals. **12.67** (a) False. As particle size decreases, the band gap increases. (b) False. As particle size

decreases, wavelength decreases. **12.69** 2.47×10^5 Au atoms **12.72** A dipole moment roughly parallel to the long dimension of the molecule would cause the molecules to reorient when an electric field is applied perpendicular to the usual direction of molecular orientation. **12.77** At the temperature where a substance changes from the solid to the liquid crystalline phase, kinetic energy sufficient to overcome most of the long-range order in the solid has been supplied. A relatively small increase in temperature is required to overcome the remaining aligning forces and produce an isotropic liquid.
12.81 $TiCl_4(g) + 2\,SiH_4(g) \longrightarrow TiSi_2(s) + 4\,HCl(g) + 2\,H_2(g)$. As a ceramic, $TiSi_2$ will have a three-dimensional network structure similar to that of Si. At the surface of the thin film, there will be Ti and Si atoms with incomplete valences that will chemically bond with Si atoms on the surface of the substrate. This kind of bonding would not be possible with a Cu thin film. **12.83** (a) $\Delta H = -82$ kJ/mol C_2H_4 (b) $\Delta H = -14$ kJ/mol (of either reactant) (c) $\Delta H = 0$ kJ **12.86** (a) $x = 0.22$ (b) Hg and Cu both have more than one stable oxidation state. If different ions in the solid lattice have different charges, the average charge is a noninteger value. Ca and Ba are stable only in the +2 oxidation state and are unlikely to have noninteger average charge. (c) Ba^{2+} is largest; Cu^{2+} is smallest.

Chapter 13

13.1 The ion-solvent interaction should be greater for Li^+. The smaller ionic radius of Li^+ means that the ion-dipole interactions with polar water molecules are stronger. **13.6** Vitamin B_6 is largely water soluble, owing to its extensive possibilities for hydrogen bonding with water. Vitamin E is largely fat soluble due to the strong dispersion interactions of its long, rod-like carbon chain. **13.9** In *n*-octylglucoside, the eight-carbon *n*-octyl chain has strong dispersion interactions with the hydrophobic protein. The —OH groups on the glycoside ring form strong hydrogen bonds with water. This causes the glycoside to dissolve, dragging the protein along with it. **13.11** If the magnitude of ΔH_3 is small relative to the magnitude of ΔH_1, ΔH_{soln} will be large and endothermic (energetically unfavorable) and not much solute will dissolve. **13.13** (a) Dispersion (b) hydrogen bonding (c) ion-dipole (d) dipole-dipole **13.15** (a) ΔH_1 (b) ΔH_3 **13.17** (a) Since the solute and solvent experience very similar London dispersion forces, the energy required to separate them individually and the energy released when they are mixed are approximately equal. $\Delta H_1 + \Delta H_2 \approx -\Delta H_3$. Thus, ΔH_{soln} is nearly zero. (b) Since no strong intermolecular forces prevent the molecules from mixing, they do so spontaneously because of the increase in randomness. **13.19** (a) Supersaturated (b) Add a seed crystal. A seed crystal provides a nucleus of prealigned molecules, so that ordering of the dissolved particles (crystallization) is more facile.
13.21 (a) Unsaturated (b) saturated (c) saturated (d) unsaturated **13.23** The liquids water and glycerol form homogenous mixtures (solutions) regardless of the relative amounts of the two components. The —OH groups of glycerol facilitate strong hydrogen bonding similar to that in water; like dissolves like.
13.25 (a) Dispersion interactions among nonpolar $CH_3(CH_2)_{16}$— chains dominate the properties of stearic acid, causing it to be more soluble in nonpolar CCl_4. (b) Dioxane can act as a hydrogen bond acceptor, so it will be more soluble than cyclohexane in water. **13.27** (a) CCl_4 is more soluble because dispersion forces among nonpolar CCl_4 molecules are similar to dispersion forces in hexane. (b) C_6H_6 is a nonpolar hydrocarbon and will be more soluble in the similarly nonpolar hexane. (c) The long, rod-like hydrocarbon chain of octanoic acid forms strong dispersion interactions and causes it to be more soluble in hexane.
13.29 (a) A sealed container is required to maintain a partial pressure of $CO_2(g)$ greater than 1 atm above the beverage. (b) Since the solubility of gases increases with decreasing temperature, some $CO_2(g)$ will remain dissolved in the beverage if it is kept cool. **13.31** $S_{He} = 5.6 \times 10^{-4}$ *M*, $S_{N_2} = 9.0 \times 10^{-4}$ *M* **13.33** (a) 2.15% Na_2SO_4 by mass (b) 3.15 ppm Ag **13.35** (a) $X_{CH_3OH} = 0.0427$ (b) 7.35% CH_3OH by mass (c) 2.48 *m* CH_3OH **13.37** (a) 1.46×10^{-2} *M* $Mg(NO_3)_2$ (b) 1.12 *M* $LiClO_4 \cdot 3\,H_2O$ (c) 0.350 *M* HNO_3 **13.39** (a) 4.70 *m* C_6H_6 (b) 0.235 *m* NaCl **13.41** (a) 43.01% H_2SO_4 by mass (b) $X_{H_2SO_4} = 0.122$ (c) 7.69 *m* H_2SO_4 (d) 5.827 *M* H_2SO_4 **13.43** (a) $X_{CH_3OH} = 0.227$ (b) 7.16 *m* CH_3OH (c) 4.58 *M* CH_3OH **13.45** (a) 0.150 mol $SrBr_2$ (b) 1.56×10^{-2} mol KCl (c) 4.44×10^{-2} mol $C_6H_{12}O_6$ **13.47** (a) Weigh out 1.3 g KBr, dissolve in water, dilute with stirring to 0.75 L. (b) Weigh out 2.62 g KBr, dissolve it in 122.38 g H_2O to make exactly 125 g of 0.180 *m* solution. (c) Weigh 244 g KBr and dissolve in enough H_2O to make 1.85 L solution. (d) Weigh 10.1 g KBr, dissolve it in a small amount of water, and dilute to 0.568 L. **13.49** 71% HNO_3 by mass **13.51** (a) 3.82 *m* Zn (b) 26.8 *M* Zn **13.53** 1.8×10^{-3} *M* CO_2 **13.55** Freezing point depression, $\Delta T_f = K_f(m)$; boiling point elevation, $\Delta T_b = K_b(m)$; osmotic pressure, $\pi = MRT$; vapor pressure lowering, $P_A = X_A P_A^{\circ}$ **13.57** (a) Sucrose has a greater molar mass than glucose, so the sucrose solution will contain fewer particles and have a higher vapor pressure. **13.59** (a) $P_{H_2O} = 186.4$ torr (b) 78.9 g $C_3H_8O_2$ **13.61** (a) $X_{Eth} = 0.2812$ (b) $P_{soln} = 238$ torr (c) X_{Eth} in vapor $= 0.472$ **13.63** (a) Because NaCl is a strong electrolyte, one mole of NaCl produces twice as many dissolved particles as one mole of the molecular solute $C_6H_{12}O_6$. Boiling-point elevation is directly related to total moles of dissolved particles, so 0.10 *m* NaCl has the higher boiling point. (b) 0.10 *m* NaCl: $\Delta T_b = 0.101°C$, $T_b = 100.1°C$; 0.10 *m* $C_6H_{12}O_6$: $\Delta T_b = 0.051°C$, $T_b = 100.1°C$ (c) Interactions between ions in solution result in non-ideal behavior. **13.65** 0.050 *M* LiBr < 0.120 *M* glucose < 0.050 *M* $Zn(NO_3)_2$
13.67 (a) $T_f = -115.0°C$, $T_b = 78.7°C$ (b) $T_f = -3.84°C$, $T_b = 64.2°C$ (c) $T_f = -0.604°C$, $T_b = 100.2°C$
13.69 $\pi = 0.0168$ atm $= 12.7$ torr **13.71** Molar mass of adrenaline $= 1.8 \times 10^2$ g **13.73** Molar mass of lysozyme $= 1.39 \times 10^4$ g **13.75** (a) $i = 2.76$ (b) The more concentrated the solution, the greater the ion-pairing and the smaller the measured value of *i*. **13.77** (a) In the gaseous state, particles are far apart and intermolecular attractive forces are small. When two gases combine, all terms in Equation 13.1 are essentially zero and the mixture is always homogeneous. (b) To determine whether Faraday's dispersion is a true solution or a colloid, shine a beam of light on it. If light is scattered, the dispersion is a colloid. **13.79** (a) Hydrophobic (b) hydrophilic (c) hydrophobic (d) hydrophobic (but stabilized by adsorbed charges). **13.81** When electrolytes are added to a suspension of proteins, dissolved ions form ion pairs with the protein surface charges, effectively neutralizing them. The protein's capacity for ion-dipole interactions with water is diminished and the colloid separates into a protein layer and a water layer. **13.83** The periphery of the BHT molecule is mostly hydrocarbon-like groups, such as $—CH_3$. The one —OH group is rather buried inside and probably does little to enhance solubility in water. Thus, BHT is more likely to be soluble in the nonpolar hydrocarbon hexane, C_6H_{14}, than in polar water. **13.85** (a) 1×10^{-4} *M* O_2 (b) $P_{O_2} = 0.07$ atm $= 60$ torr **13.88** 136 ppm K^+ **13.91** (a) 0.16 *m* Na (b) 2.0 *M* Na (c) Clearly, molality and molarity are not the same for this amalgam. Only in the instance that one kg solvent and the mass of one liter of solution are nearly equal do the two concentration units have similar values.

13.95 $Hg(NO_3)_2$, $i = 2.83$, is almost completely dissociated into ions; $HgCl_2$, $i = 1.00$, behaves essentially like a nonelectrolyte. $Hg(NO_3)_2$ is the stronger electrolyte. **13.98** $n \approx 14$. The formula of the salt is $LiC_{14}H_{29}O_2$. **13.99** (a) CF_4, $1.7 \times 10^{-4}\ m$; $CClF_3$, $9 \times 10^{-4}\ m$; CCl_2F_2, $2.3 \times 10^{-2}\ m$; $CHClF_2$, $3.5 \times 10^{-2}\ m$ (b) Molality and molarity are numerically similar when kilograms solvent and liters solution are nearly equal. This is true when solutions are dilute and when the density of the solvent is nearly 1 g/mL, as in this problem. (c) Water is a polar solvent; the solubility of solutes increases as their polarity increases. Nonpolar CF_4 has the lowest solubility and the most polar fluorocarbon, $CHClF_2$, has the greatest solubility in H_2O. (d) The Henry's law constant for $CHClF_2$ is 3.5×10^{-2} mol/L-atm. This value is greater than the Henry's law constant for $N_2(g)$ because $N_2(g)$ is nonpolar and of lower molecular mass than $CHClF_2$.
13.103

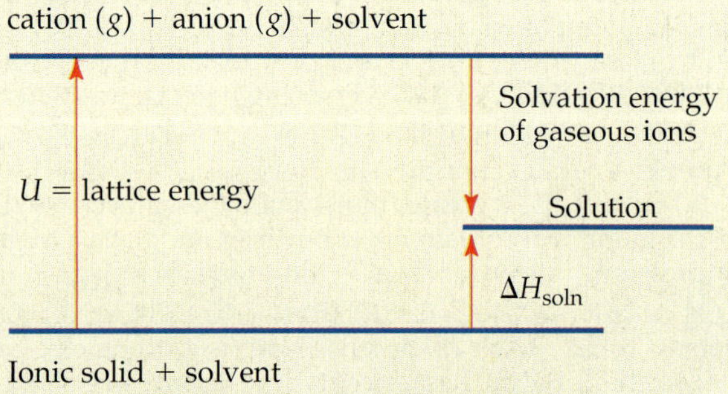

(b) Lattice energy (U) is inversely related to the distance between ions, so salts with large cations like $(CH_3)_4N^+$ have smaller lattice energies than salts with simple cations like Na^+. Also, the —CH_3 groups in the large cation are capable of dispersion interactions with nonpolar groups of the solvent molecules, resulting in a more negative solvation energy of the gaseous ions. Overall, for salts with larger cations, lattice energy is smaller (less positive), the solvation energy of the gaseous ions is more negative, and ΔH_{soln} is less endothermic. These salts are more soluble in polar nonaqueous solvents.
13.106 The process is spontaneous with no significant change in enthalpy, so we suspect that there is an increase in entropy. In general, dilute solutions of the same solute have greater entropy than concentrated ones, because the solute particles are more free to move about the solution. In the limiting case that the more dilute solution is pure solvent, there is a definite increase in entropy as the concentrated solution is diluted. In Figure 13.23, there may be some entropy decrease as the dilute solution loses solvent, but this is more than offset by the entropy increase that accompanies dilution. There is a net increase in entropy.

Chapter 14

14.1 (a) Product (b) The average rate of reaction is greater between points 1 and 2 because they are earlier in the reaction when the concentrations of reactants are greater.
14.3 (a) $x = 0$ (b) $x = 2$ (c) $x = 3$ **14.5** (a) The half-life is 15 min (b) After four half-life periods, 1/16 of the reactant remains. **14.8** There is one intermediate, B, and two transition states, A $\longrightarrow$ B and B $\longrightarrow$ C. The B $\longrightarrow$ C step is faster, and the overall reaction is exothermic. **14.11** (a) *Reaction rate* is the change in the amount of products or reactants in a given amount of time. (b) Rates depend on concentration of reactants, surface area of reactants, temperature, and presence of catalyst. (c) The stoichiometry of the reaction (mole ratios of reactants and products) must be known to relate rate of disappearance of reactants to rate of appearance of products.

14.13

Time (min)	Mol A	(a) Mol B	[A] (mol/L)	Δ[A] (mol/L)	(b) Rate (*M*/s)
0	0.065	0.000	0.65		
10	0.051	0.014	0.51	−0.14	2.3×10^{-4}
20	0.042	0.023	0.42	−0.09	1.5×10^{-4}
30	0.036	0.029	0.36	−0.06	1.0×10^{-4}
40	0.031	0.034	0.31	−0.05	0.8×10^{-4}

(c) $\Delta[B]_{avg}/\Delta t = 1.3 \times 10^{-4}\ M/s$
14.15 (a)

Time (s)	Time Interval (s)	Concentration (*M*)	Δ *M*	Rate (*M*/s)
0		0.0165		
2,000	2,000	0.0110	−0.0055	28×10^{-7}
5,000	3,000	0.00591	−0.0051	17×10^{-7}
8,000	3,000	0.00314	−0.00277	9.23×10^{-7}
12,000	4,000	0.00137	−0.00177	4.43×10^{-7}
15,000	3,000	0.00074	−0.00063	2.1×10^{-7}

(b) From the slopes of the tangents to the graph, the rates are $12 \times 10^{-7}\ M/s$ at 5000 s, $5.8 \times 10^{-7}\ M/s$ at 8000 s.
14.17 (a) $-\Delta[H_2O_2]/\Delta t = \Delta[H_2]/\Delta t = \Delta[O_2]/\Delta t$
(b) $-\frac{1}{2}\Delta[N_2O]/\Delta t = \frac{1}{2}\Delta[N_2]/\Delta t = \Delta[O_2]/\Delta t$
(c) $-\Delta[N_2]/\Delta t = -\frac{1}{3}\Delta[H_2]/\Delta t = \frac{1}{2}\Delta[NH_3]\Delta t$
14.19 (a) $-\Delta[O_2]/\Delta t = 0.43$ mol/s; $\Delta[H_2O]/\Delta t = 0.85$ mol/s (b) P_{total} decreases by 12 torr/min **14.21** (a) If [A] doubles, there is no change in the rate or the rate constant. The overall rate is unchanged because [A] does not appear in the rate law; the rate constant changes only with a change in temperature. (b) The reaction is zero order in A, second order in B, and second order overall. (c) units of $k = M^{-1}s^{-1}$ **14.23** (a) Rate $= k[N_2O_5]$ (b) Rate $= 1.16 \times 10^{-4}\ M/s$ (c) When the concentration of N_2O_5 doubles, the rate of the reaction doubles. **14.25** (a, b) $k = 1.7 \times 10^2\ M^{-1}s^{-1}$ (c) If $[OH^-]$ is tripled, the rate triples. **14.27** (a) Rate $= k[OCl^-][I^-]$ (b) $k = 60\ M^{-1}s^{-1}$ (c) Rate $= 6.0 \times 10^{-5}\ M/s$ **14.29** (a) Rate $= k[BF_3][NH_3]$ (b) The reaction is second order overall. (c) $k_{avg} = 3.41\ M^{-1}s^{-1}$ (d) 0.170 *M/s* **14.31** (a) Rate $= k[NO]^2[Br_2]$ (b) $k_{avg} = 1.2 \times 10^4\ M^{-2}s^{-1}$ (c) $\frac{1}{2}\Delta[NOBr]/\Delta t = -\Delta[Br_2]/\Delta t$ (d) $-\Delta[Br_2]/\Delta t = 8.4\ M/s$ **14.33** (a) $[A]_0$ is the molar concentration of reactant A at time 0. $[A]_t$ is the molar concentration of reactant A at time t. $t_{1/2}$ is the time required to reduce $[A]_0$ by a factor of 2. k is the rate constant for a particular reaction. (b) A graph of ln[A] versus time yields a straight line for a first-order reaction. **14.35** (a) $k = 3.0 \times 10^{-6}\ s^{-1}$ (b) $t_{1/2} = 3.2 \times 10^4$ s **14.37** (a) $P = 20$ torr (b) $t = 51$ s **14.39** Plot ln $P_{SO_2Cl_2}$ versus time, $k = -\text{slope} = 2.19 \times 10^{-5}\ s^{-1}$ **14.41** (a) The plot of 1/[A] versus time is linear, so the reaction is second order in [A]. (b) $k = 0.040\ M^{-1}min^{-1}$ (c) $t_{1/2} = 38$ min **14.43** (a) The plot of $1/[NO_2]$ versus time is linear, so the reaction is second order in NO_2. (b) $k = \text{slope} = 10\ M^{-1}s^{-1}$ **14.45** (a) The energy of the collision and the orientation of the molecules when they collide determine whether a reaction will occur. (b) At a higher temperature, there are more total collisions and each collision is more energetic. **14.47** $f = 4.94 \times 10^{-2}$. At 400 K approximately 1 out of 20 molecules has this kinetic energy.

14.49 (a)

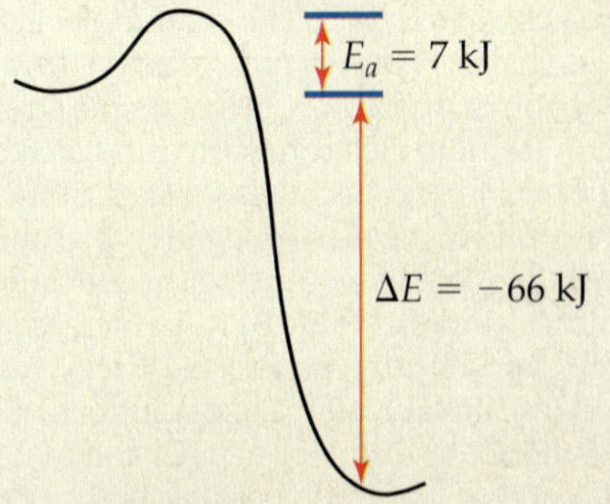

(b) E_a(reverse) = 73 kJ **14.51** Reaction (b) is fastest and reaction (c) is slowest. **14.53** (a) $k = 1.1\ s^{-1}$ (b) $k = 4.9\ s^{-1}$ **14.55** A plot of ln k versus $1/T$ has a slope of -5.71×10^3; $E_a = -R$(slope) = 47.5 kJ/mol. **14.57** The reaction will occur 88 times faster at 50°C, assuming equal initial concentrations. **14.59** (a) An *elementary reaction* is a process that occurs in a single event; the order is given by the coefficients in the balanced equation for the reaction. (b) A *unimolecular* elementary reaction involves only one reactant molecule; a *bimolecular* elementary reaction involves two reactant molecules. (c) A *reaction mechanism* is a series of elementary reactions that describe how an overall reaction occurs and explain the experimentally determined rate law. **14.61** (a) Unimolecular, rate = $k[Cl_2]$ (b) bimolecular, rate = $k[OCl^-][H_2O]$ (c) bimolecular, rate = $k[NO][Cl_2]$ **14.63** (a) Two intermediates, B and C. (b) three transition states (c) C $\longrightarrow$ D is fastest. (d) endothermic **14.65** (a) $H_2(g) + 2\,ICl(g) \longrightarrow I_2(g) + 2\,HCl(g)$ (b) HI is the intermediate. (c) first step: rate = $k_1[H_2][ICl]$, second step: rate = $k_2[HI][ICl]$ (d) If the first step is slow, the observed rate law is rate = $k[H_2][ICl]$. **14.67** (a) Rate = $k[NO][Cl_2]$ (b) The second step must be slow relative to the first step. **14.69** (a) A catalyst increases the rate of reaction by decreasing the activation energy, E_a, or increasing the frequency factor, A. (b) A homogeneous catalyst is in the same phase as the reactants, while a hetereogeneous catalyst is in a different phase. **14.71** (a) Multiply the coefficients in the first reaction by 2 and sum. (b) $NO_2(g)$ is a catalyst because it is consumed and then reproduced in the reaction sequence. (c) This is a homogeneous catalysis. **14.73** Use of chemically stable supports makes it possible to obtain very large surface areas per unit mass of the precious metal catalyst because the metal can be deposited in a very thin, even monomolecular, layer on the surface of the support. **14.75** To put two D atoms on a single carbon, it is necessary that one of the already existing C—H bonds in ethylene be broken while the molecule is adsorbed, so that the H atom moves off as an adsorbed atom and is replaced by a D atom. This requires a larger activation energy than simply adsorbing C_2H_4 and adding one D atom to each carbon. **14.77** (a) Living organisms operate efficiently in a very narrow temperature range; the role of enzymes as homogeneous catalysts that speed up desirable reactions, without heating and undesirable side effects, is crucial for biological systems. (b) catalase: $2\,H_2O_2 \longrightarrow 2\,H_2O + O_2$; nitrogenase: $N_2 \longrightarrow 2\,NH_3$ (nitrogen fixation) **14.79** (a) The catalyzed reaction is approximately 10,000,000 times faster at 25°C. (b) The catalyzed reaction is 180,000 times faster at 125°C.

14.81 $\Delta[Cl^-]/\Delta t = 7.0 \times 10^{-7}\ M/s$

14.84 (a) Rate = $k[HgCl_2][C_2O_4{}^{2-}]^2$ (b) $k = 8.7 \times 10^{-3}\ M^{-2}s^{-1}$ (c) rate = $4.3 \times 10^{-6}\ M/s$ **14.87** (a) $k = 4.28 \times 10^{-4}\ s^{-1}$ (b) [urea] = 0.090 M (c) $t_{1/2} = 1.62 \times 10^3$ s **14.91** A plot of ln k versus $1/T$ is linear with slope = -1.751×10^4. $E_a = -$(slope)$R = 1.5 \times 10^2$ kJ/mol. **14.94** (a) $Cl_2(g) + CHCl_3(g) \longrightarrow HCl(g) + CCl_4(g)$ (b) $Cl(g)$, $CCl_3(g)$ (c) reaction 1, unimolecular; reaction 2, bimolecular; reaction 3, bimolecular (d) Reaction 2, the slow step, is rate determining. (e) Rate = $k[CHCl_3][Cl_2]^{1/2}$ **14.96** *Enzyme*: carbonic anhydrase; *substrate*: carbonic acid (H_2CO_3); *turnover number*: 1×10^7 molecules/s. **14.98** partial pressure of O_2 = 0.402 atm **14.100** (a) Use an open-end manometer, a clock, a ruler, and a constant-temperature bath. Load the flask with $HCl(aq)$, and read the height of the Hg in both arms of the manometer. Quickly add $Zn(s)$ to the flask, and record time = 0 when the $Zn(s)$ contacts the acid. Record the height of the Hg in one arm of the manometer at convenient time intervals such as 5 s. Calculate the pressure of $H_2(g)$ at each time. Since $P = (n/V)RT$, $\Delta P/\Delta t$ at constant temperature is an acceptable measure of reaction rate. (b) Keep the amount of $Zn(s)$ constant, and vary the concentration of $HCl(aq)$ to determine the reaction order for H^+ and Cl^-. Keep the concentration of $HCl(aq)$ constant, and vary the amount of $Zn(s)$ to determine the order for $Zn(s)$. Combine this information to write the rate law. (c) $-\Delta[H^+]/\Delta t = 2\,\Delta[H_2]/\Delta t$; the rate of disappearance of H^+ is twice the rate of appearance of $H_2(g)$. $[H_2] = P/RT$ (d) By changing the temperature of the constant-temperature bath, measure the rate data at several temperatures and calculate the rate constant k at these temperatures. Plot ln k versus $1/T$; the slope of the line is $-E_a/R$. (e) Measure rate data at constant temperature, HCl concentration, and mass of $Zn(s)$, varying only the form of the $Zn(s)$. Compare the rate of reaction for metal strips and granules. **14.103** Changes in temperature change the kinetic energy of the various groups on the enzyme and their tendency to form intermolecular associations or break free from them. At temperatures above the temperature of maximum activity, sufficient kinetic energy has been imparted so that the three-dimensional structure of the enzyme is destroyed. This is the process of denaturation. In the lock-and-key model of enzyme action, the active site is the specific location in the enzyme where the reaction takes place. The precise geometry of the active site both accommodates and activates the substrate. When an enzyme is denatured, its activity is destroyed because the active site has collapsed.

Chapter 15

15.1 $k_f > k_r$ (b) The equilibrium constant is greater than 1. **15.4** (a) $A_2 + B \longrightarrow A_2B$ (b) $K_c = [A_2B]/[A_2][B]$ (c) $\Delta n = -1$ (d) $K_p = K_c(RT)^{\Delta n}$ **15.7** (a) $K_c = 2$ (b) increase **15.9** (a) $K_p = K_c = 2.8 \times 10^{-2}$ (b) Since $k_f < k_r$, in order for the two rates to be equal, [A] must be greater than [B], and the partial pressure of A is greater than the partial pressure of B. **15.11** (a) The *law of mass action* expresses the relationship between the concentrations of reactants and products at equilibrium for any reaction. $K_c = [NOBr_2]/[NO][Br_2]$ (b) The *equilibrium-constant expression* is an algebraic equation where the variables are the equilibrium concentrations of the reactants and products for a specific chemical reaction. The *equilibrium constant* is a number; it is the ratio calculated from the equilibrium expression for a particular chemical reaction. (c) Introduce a known quantity of $NOBr_2(g)$ into a vessel of known volume at constant (known) temperature. After equilibrium has been established, measure the total pressure in the flask. Using an equilibrium table, calculate equilibrium pressures, concentrations, and the value of K_c.

15.13 (a) $K_c = [N_2O][NO_2]/[NO]^3$; homogeneous
(b) $K_c = [CS_2][H_2]^4/[CH_4][H_2S]^2$; homogeneous
(c) $K_c = [CO]^4/[Ni(CO)_4]$; heterogeneous
(d) $K_c = [H^+][F^-]/[HF]$; homogeneous
(e) $K_c = [Ag^+]^2/[Zn^{2+}]$; heterogeneous **15.15** (a) Mostly reactants (b) mostly products **15.17** $K_p = 1.0 \times 10^{-3}$ **15.19** (a) $K_c = 1.3 \times 10^{-3}$ (b) The equilibrium favors NO and Br_2 at this temperature. **15.21** (a) $K_p = 0.541$ (b) $K_p = 3.42$ (c) $K_c = 128$ **15.23** $K_c = 0.16$ **15.25** (a) $K_p = P_{O_2}$ (b) The molar concentration, the ratio of moles of a substance to volume occupied by the substance, is a constant for pure solids and liquids.

15.27 $K_c = 0.0204$ **15.29** $K_p = 51$ **15.31** (a) $[H_2] = 0.012\ M$, $[N_2] = 0.019\ M$, $[H_2O] = 0.138\ M$ (b) $K_c = 653.7 = 7 \times 10^2$ **15.33** (a) $P_{CO_2} = 4.10$ atm, $P_{H_2} = 2.05$ atm, $P_{H_2O} = 3.28$ atm (b) $P_{CO_2} = 3.87$ atm, $P_{H_2} = 1.82$ atm, $P_{H_2O} = 3.51$ atm (c) $K_p = 0.11$ **15.35** (a) A reaction quotient is the result of a general set of concentrations whereas the equilibrium constant requires equilibrium concentrations. (b) to the right (c) The concentrations used to calculate Q must be equilibrium concentrations. **15.37** (a) $Q = 1.1 \times 10^{-8}$, the reaction will proceed to the left. (b) $Q = 5.5 \times 10^{-12}$, the reaction will proceed to the right. (c) $Q = 2.2 \times 10^{-10}$, the mixture is at equilibrium. **15.39** $P_{Cl_2} = 4.96$ atm **15.41** (a) $[Br_2] = 0.00767\ M$, $[Br] = 0.00282\ M$, 0.0451 g Br(*g*) (b) $[H_2] = 0.014\ M$, $[I_2] = 0.00859\ M$, $[HI] = 0.081\ M$, 21 g HI **15.43** $[NO] = 0.002\ M$, $[N_2] = [O_2] = 0.099\ M$ **15.45** The equilibrium pressure of $Br_2(g)$ is 0.416 atm. **15.47** (a) $[Ca^{2+}] = [SO_4^{2-}] = 4.9 \times 10^{-3}\ M$ (b) A bit more than 2.0 g $CaSO_4$ is needed in order to have some undissolved $CaSO_4(s)$ in equilibrium with 3.0 L of saturated solution. **15.49** $[IBr] = 0.447\ M$, $[I_2] = [Br_2] = 0.0267\ M$ **15.51** (a) Shift equilibrium to the right (b) decrease the value of K (c) shift equilibrium to the right (b) decrease the value of K (c) shift equilibrium to the left (d) no effect (e) no effect (f) shift equilibrium to the right **15.53** (a) No effect (b) no effect (c) increase equilibrium constant (d) no effect **15.55** (a) $\Delta H° = -155.7$ kJ (b) The reaction is exothermic, so the equilibrium constant will decrease with increasing temperature. (c) Δn does not equal zero, so a change in volume at constant temperature will affect the fraction of products in the equilibrium mixture. **15.57** (a) $K_c = K_p = 1.5 \times 10^{-39}$ (b) Reactants are much more plentiful than products at equilibrium. **15.59** $K_c = 0.71$ **15.62** (a) $P_t = 0.74$ atm (b) $K_p = 0.40$ **15.65** (a) $K_c = 7.0 \times 10^{-5}$ (b) $[PH_3] = 2.5 \times 10^{-3}\ M$ **15.68** (a) $Q = 1.7; Q > K_p$; the reaction will shift to the left. (b) $Q = 1.7; Q > K_p$; the reaction will shift to the left. (c) $Q = 5.8 \times 10^{-3}; Q < K_p$; the reaction will shift to the right. **15.71** The maximum value of Q is 4.1, which is less than K_p; reduction will occur. **15.74** $P_{H_2} = P_{I_2} = 1.48$ atm; $P_{HI} = 10.22$ atm **15.77** The patent claim is false. A catalyst does not alter the position of equilibrium in a system, only the rate of approach to the equilibrium condition. **15.78** (a) (i) $K_c = [Na^+]/[Ag^+]$ (ii) $K_c = [Hg^{2+}]^3/[Al^{3+}]^2$ (iii) $K_c = [Zn^{2+}]\,P_{H_2}/[H^+]^2$ (b) Using data from Table 4.5: (i) Ag^+ is far below Na, so the reaction will proceed to the right and K_c will be large. (ii) Al^{3+} is above Hg, so the reaction will not proceed to the right and K_c will be small. (iii) H^+ is below Zn, so the reaction will proceed to the right and K_c will be large. (c) $K_c < 1$ for this reaction, so Cd is below Fe. **15.80** (a) At equilibrium, the forward and reverse reactions occur at equal rates. (b) Reactants are favored at equilibrium. (c) A catalyst lowers the activation energy for both the forward and reverse reactions. (d) The ratio of the rate constants remains unchanged. (e) The value of K will increase with increasing temperature. **15.83** At 850°C, $K_p = 14.1$; at 950°C, $K_p = 73.8$; at 1050°C, $K_p = 2.7 \times 10^2$; at 1200°C, $K_p = 1.7 \times 10^3$. Because K increases with increasing temperature, the reaction is endothermic.

Chapter 16

16.1 (a) HX, the H^+ donor, is the Brønsted–Lowry acid. NH_3, the H^+ acceptor, is the Brønsted–Lowry base. (b) HX, the electron pair acceptor, is the Lewis acid. NH_3, the electron pair donor, is the Lewis base. **16.5** (a) The order of base strength is the reverse of the order of acid strength; the diagram of the acid with the most free H^+ has the weakest conjugate base Y^-. The order of increasing basicity is: $Y^- < Z^- < X^-$. (b) The strongest base, X^-, has the largest K_b value. **16.8** (a) Both molecules are oxyacids; the ionizable H atom is attached to O. The right molecule is a carboxylic acid; the ionizable H is part of a carboxyl group. (b) Increasing the electronegativity of X increases the strength of both acids. As X becomes more electronegative and attracts more electron density, the O—H bond becomes weaker, more polar, and more likely to be ionized. An electronegative X group also stabilizes the anionic conjugate base, causing the ionization equilibrium to favor products and the value of K_a to increase. **16.11** Solutions of HCl and H_2SO_4 conduct electricity, taste sour, turn litmus paper red (are acidic), neutralize solutions of bases, and react with active metals to form $H_2(g)$. HCl and H_2SO_4 solutions have these properties in common because both compounds are strong acids. That is, they both dissociate completely in H_2O to form $H^+(aq)$ and an anion. (HSO_4^- is not completely dissociated, but the first dissociation step for H_2SO_4 is complete.) The presence of ions enables the solutions to conduct electricity; the presence of $H^+(aq)$ in excess of $1 \times 10^{-7}\ M$ accounts for all other properties listed. **16.13** (a) The Arrhenius definition of an acid is confined to aqueous solution; the Brønsted–Lowry definition applies to any physical state. (b) HCl is the Brønsted–Lowry acid; NH_3 is the Brønsted–Lowry base. **16.15** (a) IO_3^- (b) NH_3 (c) HPO_4^{2-} (d) $C_7H_5O_2^-$
16.17

Acid	+	Base	⇌	Conjugate Acid	+	Conjugate Base
(a) $NH_4^+(aq)$		$CN^-(aq)$		$HCN(aq)$		$NH_3(aq)$
(b) $H_2O(l)$		$(CH_3)_3N(aq)$		$(CH_3)_3NH^+(aq)$		$OH^-(aq)$
(c) $HCHO_2(aq)$		$PO_4^{3-}(aq)$		$HPO_4^{2-}(aq)$		$CHO_2^-(aq)$

16.19 (a) Acid: $HC_2O_4^-(aq) + H_2O(l) \rightleftharpoons C_2O_4^{2-}(aq) + H_3O^+(aq)$; Base: $HC_2O_4^-(aq) + H_2O(l) \rightleftharpoons H_2C_2O_4(aq) + OH^-(aq)$ (b) $H_2C_2O_4$ is the conjugate acid of $HC_2O_4^-$. $C_2O_4^{2-}$ is the conjugate base of $HC_2O_4^-$. **16.21** (a) HNO_2, weak acid; NO_2^-, weak base; (b) H_2SO_4, strong acid; HSO_4^-, negligible base (c) HPO_4^{2-}, weak acid; PO_4^{3-}, weak base (d) CH_4, negligible acid; CH_3^-, strong base (e) $CH_3NH_3^+$, weak acid; CH_3NH_2, weak base **16.23** (a) HBr. It is one of the seven strong acids. (b) F^-. HCl is a stronger acid than HF, so F^- is the stronger conjugate base. **16.25** (a) $OH^-(aq) + OH^-(aq)$, the equilibrium lies to the right. (b) $H_2S(aq) + C_2H_3O_2^-(aq)$, the equilibrium lies to the right. (c) $HNO_3(aq) + OH^-(aq)$, the equilibrium lies to the left. **16.27** (a) *Autoionization* is the ionization of a neutral molecule into an anion and a cation. The equilibrium expression for the autoionization of water is $H_2O(l) \rightleftharpoons H^+(aq) + OH^-(aq)$. (b) Pure water is a poor conductor of electricity because it contains very few ions. (c) If a solution is acidic, it contains more H^+ than OH^-. **16.29** (a) $[H^+] = 2.2 \times 10^{-11}\ M$, basic (b) $[H^+] = 1.1 \times 10^{-6}\ M$, acidic (c) $[H^+] = 1.0 \times 10^{-8}\ M$, basic **16.31** $[H^+] = [OH^-] = 3.5 \times 10^{-8}\ M$ **16.33** (a) $[H^+]$ changes by a factor of 100. (b) $[H^+]$ changes by a factor of 3.2 **16.35** (a) $[H^+]$ decreases, pH increases (b) The pH is between 3 and 4. By calculation, pH = 3.2; the solution is acidic. (c) pH = 5.2 is between pH 5 and pH 6, closer to pH = 5. A good estimate is $7 \times 10^{-6}\ M\ H^+$ and $3 \times 10^{-9}\ M\ OH^-$. By calculation, $[H^+] = 6 \times 10^{-6}\ M$ and $[OH^-] = 2 \times 10^{-9}\ M$.
16.37

$[H^+]$	$[OH^-]$	pH	pOH	Acidic or Basic
$7.5 \times 10^{-3}\ M$	$1.3 \times 10^{-12}\ M$	2.12	11.88	acidic
$2.8 \times 10^{-5}\ M$	$3.6 \times 10^{-10}\ M$	4.56	9.44	acidic
$5.6 \times 10^{-9}\ M$	$1.8 \times 10^{-6}\ M$	8.25	5.75	basic
$5.0 \times 10^{-9}\ M$	$2.0 \times 10^{-6}\ M$	8.30	5.70	basic

16.39 $[H^+] = 4.0 \times 10^{-8}\ M$, $[OH^-] = 6.0 \times 10^{-7}\ M$, pOH = 6.22 **16.41** (a) A strong acid is completely dissociated into ions in aqueous solution. (b) $[H^+] = 0.500\ M$ (c) HCl, HBr, HI **16.43** (a) $[H^+] = 8.5 \times 10^{-3}\ M$, pH = 2.07 (b) $[H^+] = 0.0419\ M$, pH = 1.377 (c) $[H^+] = 0.0250\ M$, pH = 1.602 (d) $[H^+] = 0.167\ M$, pH = 0.778 **16.45** (a) $[OH^-] = 3.0 \times 10^{-3}\ M$, pH = 11.48 (b) $[OH^-] = 0.3758\ M$, pH = 13.5750 (c) $[OH^-] = 8.75 \times 10^{-5}\ M$, pH = 9.942 (d) $[OH^-] = 0.17\ M$, pH = 13.23 **16.47** $3.2 \times 10^{-3}\ M$ NaOH **16.49** pH = 13.400 **16.51** (a) $HBrO_2(aq) \rightleftharpoons H^+(aq) + BrO_2^-(aq)$, $K_a = [H^+][BrO_2^-]/[HBrO_2]$; $HBrO_2(aq) + H_2O(l) \rightleftharpoons H_3O^+(aq) + BrO_2^-(aq)$, $K_a = [H_3O^+][BrO_2^-]/[HBrO_2]$ (b) $HC_3H_5O_2(aq) \rightleftharpoons H^+(aq) + C_3H_5O_2^-(aq)$, $K_a = [H^+][C_3H_5O_2^-]/[HC_3H_5O_2]$; $HC_3H_5O_2(aq) + H_2O(l) \rightleftharpoons H_3O^+(aq) + C_3H_5O_2^-(aq)$, $K_a = [H_3O^+][C_3H_5O_2^-]/[HC_3H_5O_2]$ **16.53** $K_a = 1.4 \times 10^{-4}$ **16.55** $[H^+] = [ClCH_2COO^-] = 0.0110\ M$, $[ClCH_2COOH] = 0.089\ M$, $K_a = 1.4 \times 10^{-3}$. **16.57** 0.089 *M* $HC_2H_3O_2$ **16.59** $[H^+] = [C_7H_5O_2^-] = 1.8 \times 10^{-3}\ M$, $[HC_7H_5O_2] = 0.048\ M$ **16.61** (a) $[H^+] = 1.1 \times 10^{-3}\ M$, pH = 2.95 (b) $[H^+] = 1.7 \times 10^{-4}\ M$, pH = 3.76 (c) $[OH^-] = 1.4 \times 10^{-5}\ M$, pH = 9.15 **16.63** $[H^+] = 2.0 \times 10^{-2}\ M$, pH = 1.71 **16.65** (a) $[H^+] = 2.8 \times 10^{-3}\ M$, 0.69% ionization (b) $[H^+] = 1.4 \times 10^{-3}\ M$, 1.4% ionization (c) $[H^+] = 8.7 \times 10^{-4}\ M$, 2.2% ionization **16.67** $HX(aq) \rightleftharpoons H^+(aq) + X^-(aq)$; $K_a = [H^+][X^-]/[HX]$. Assume that the percent of acid that ionizes is small. Let $[H^+] = [X^-] = y$. $K_a = y^2/[HX]$; $y = K_a^{1/2}[HX]^{1/2}$. Percent ionization = $y/[HX] \times 100$. Substituting for y, percent ionization = $100\ K_a^{1/2}[HX]^{1/2}/[HX]$ or $100\ K_a^{1/2}/[HX]^{1/2}$. That is, percent ionization varies inversely as the square root of the concentration of HX. **16.69** $[H^+] = 5.72 \times 10^{-3}\ M$, pH = 2.24, $[HC_6H_5O_7^{3-}] = 1.2 \times 10^{-9}\ M$. The approximation that the first ionization is less than 5% of the total acid concentration is not valid; the quadratic equation must be solved. The $[H^+]$ produced from the second and third ionizations is small with respect to that present from the first step; the second and third ionizations can be neglected when calculating the $[H^+]$ and pH. **16.71** All Brønsted–Lowry bases contain at least one unshared (lone) pair of electrons to attract H^+. **16.73** (a) $(CH_3)_2NH(aq) + H_2O(l) \rightleftharpoons (CH_3)_2NH_2^+(aq) + OH^-(aq)$; $K_b = [(CH_3)_2NH_2^+][OH^-]/[(CH_3)_2NH]$ (b) $CO_3^{2-}(aq) + H_2O(l) \rightleftharpoons HCO_3^-(aq) + OH^-(aq)$; $K_b = [HCO_3^-][OH^-]/[CO_3^{2-}]$ (c) $CHO_2^-(aq) + H_2O(l) \rightleftharpoons HCHO_2(aq) + OH^-(aq)$; $K_b = [HCHO_2][OH^-]/[CHO_2^-]$ **16.75** From the quadratic formula, $[OH^-] = 6.6 \times 10^{-3}\ M$, pH = 11.82. **16.77** (a) $[C_{10}H_{15}ON] = 0.033\ M$, $[C_{10}H_{15}ONH^+] = [OH^-] = 2.1 \times 10^{-3}\ M$ (b) $K_b = 1.4 \times 10^{-4}$ **16.79** (a) For a conjugate acid/conjugate base pair such as $C_6H_5OH/C_6H_5O^-$, K_b for the conjugate base can always be calculated from K_a for the conjugate acid, so a separate list of K_b values is not necessary. (b) $K_b = 7.7 \times 10^{-5}$ (c) Phenolate is a stronger base than NH_3. **16.81** (a) Acetic acid is stronger. (b) Hypochlorite ion is the stronger base. (c) K_b for $C_2H_3O_2^- = 5.6 \times 10^{-10}$, K_b for $ClO^- = 3.3 \times 10^{-7}$ **16.83** (a) $[OH^-] = 1.4 \times 10^{-3}\ M$, pH = 11.15 (b) $[OH^-] = 3.8 \times 10^{-3}\ M$, pH = 11.58 (c) $[NO_2^-] = 0.50\ M$, $[OH^-] = 3.3 \times 10^{-6}\ M$, pH = 8.52 **16.85** (a) Acidic (b) acidic (c) basic (d) neutral (e) acidic **16.87** The unknown is NaF. **16.89** $[OH^-] = 5.0 \times 10^{-6}\ M$, pH = 8.70 **16.91** (a) As the electronegativity of the central atom (X) increases, the strength of the oxyacid increases. (b) As the number of nonprotonated oxygen atoms in the molecule increases, the strength of the oxyacid increases. **16.93** (a) HNO_3 is a stronger acid because it has one more nonprotonated oxygen atom and thus a higher oxidation number on N. (b) For binary hydrides, acid strength increases going down a family, so H_2S is a stronger acid than H_2O. (c) H_2SO_4 is a stronger acid because H^+ is much more tightly held by the anion HSO_4^-. (d) For oxyacids, the greater the electronegativity of the central atom, the stronger the acid, so H_2SO_4 is the stronger acid. (e) CCl_3COOH is stronger because the electronegative Cl atoms withdraw electron density from other parts of the molecule, which weakens the O—H bond and stabilizes the anionic conjugate base. Both effects favor increased ionization and acid strength. **16.95** (a) BrO^- (b) BrO^- (c) HPO_4^{2-} **16.97** (a) True (b) False. In a series of acids that have the same central atom, acid strength increases with the number of nonprotonated oxygen atoms bonded to the central atom. (c) False. H_2Te is a stronger acid than H_2S because the H—Te bond is longer, weaker, and more easily ionized than the H—S bond. **16.99** Yes. The Arrhenius definition of a base, an $OH^-(aq)$ donor, is most restrictive; the Brønsted definition, an H^+ acceptor, is more general; and the Lewis definition, an electron-pair donor, is most general. Any substance that fits the narrow Arrhenius definition will fit the broader Brønsted and Lewis definitions. **16.101** (a) Acid, $Fe(ClO_4)_3$ or Fe^{3+}; base, H_2O (b) Acid, H_2O; base, CN^- (c) Acid, BF_3; base, $(CH_3)_3N$ (d) Acid, HIO; base, NH_2^- **16.103** (a) Cu^{2+}, higher cation charge (b) Fe^{3+}, higher cation charge (c) Al^{3+}, smaller cation radius, same charge **16.105** (a) K_w is the equilibrium constant for the reaction of two water molecules to form hydronium ion and hydroxide ion. (b) K_a is the equilibrium constant for the reaction of any acid, neutral or ionic, with water to form hydronium ion and the conjugate base of the acid. (c) pOH is the negative log of hydroxide ion concentration; pOH decreases as hydroxide ion concentration increases. **16.107** (a) A higher O_2 concentration displaces protons from HbH^+, producing a more acidic solution with lower pH in the lungs than in the tissues. (b) basic (c) A high $[H^+]$ shifts the equilibrium toward the proton-bound form HbH^+. This results in a lower concentration of HbO_2 in the blood and impedes the ability of hemoglobin to transport oxygen. **16.110** $K_a = 1.4 \times 10^{-5}$ **16.113** $[H^+] = 0.010\ M$, $[H_2PO_4^-] = 0.010\ M$, $[HPO_4^{2-}] = 6.2 \times 10^{-8}\ M$, $[PO_4^{3-}] = 2.6 \times 10^{-18}\ M$ **16.117** 6.0×10^{13} H^+ ions **16.120** (a) To the precision of the reported data, the pH of rainwater 40 years ago was 5.4, no different from the pH today. With extra significant figures, $[H^+] = 3.63 \times 10^{-6}\ M$, pH = 5.440. (b) A 20.0-L bucket of today's rainwater contains 0.02 L (with extra significant figures, 0.0200 L) of dissolved CO_2. **16.123** [cocaine] = $2.93 \times 10^{-3}\ M$, $[OH^-] = 3.4 \times 10^{-6}\ M$, $K_b = 3.9 \times 10^{-9}$ **16.125** (a) $K(i) = 5.6 \times 10^3$, $K(ii) = 10$ (b) Both (i) and (ii) have $K > 1$, so both could be written with a single arrow.

Chapter 17

17.1 The middle box has the highest pH. For equal amounts of acid HX, the greater the amount of conjugate base X^-, the smaller the amount of H^+ and the higher the pH. **17.3** (a) Drawing 3 (b) Drawing 1 (c) Drawing 2 **17.6** AgY **17.9** (a) The extent of ionization of a weak electrolyte is decreased when a strong electrolyte containing an ion in common with the weak electrolyte is added to it. (b) $NaNO_2$ **17.11** (a) pH increases (b) pH decreases (c) pH increases (d) no change (e) pH decreases **17.13** (a) $[H^+] = 1.8 \times 10^{-5}\ M$, pH = 4.73 (b) $[OH^-] = 4.8 \times 10^{-5}\ M$, pH = 9.68 (c) $[H^+] = 1.4 \times 10^{-5}\ M$, pH = 4.87 **17.15** (a) 4.5% ionization (b) 0.018% ionization **17.17** In a mixture of $HC_2H_3O_2$ and $NaC_2H_3O_2$, $HC_2H_3O_2$ reacts with added base and $C_2H_3O_2^-$

combines with addedacid, leaving $[H^+]$ relatively unchanged. Although HCl and Cl^- are a conjugate acid-base pair, Cl^- has no tendency to combine with added acid to form undissociated HCl. Any added acid simply increases $[H^+]$ in an HCl-NaCl mixture. **17.19** (a) pH = 3.82 (b) pH = 3.96 **17.21** (a) pH = 4.60 (b) $Na^+(aq) + C_2H_3O_2^-(aq) + H^+(aq) + Cl^-(aq) \longrightarrow HC_2H_3O_2(aq) + Na^+(aq) + Cl^-(aq)$ (c) $HC_2H_3O_2(aq) + Na^+(aq) + OH^-(aq) \longrightarrow C_2H_3O_2^-(aq) + H_2O(l) + Na^+(aq)$ **17.23** 0.18 mol NaBrO **17.25** (a) pH = 4.86 (b) pH = 5.0 (c) pH = 4.71 **17.27** (a) $[HCO_3^-]/[H_2CO_3] = 11$ (b) $[HCO_3^-]/[H_2CO_3] = 5.4$ **17.29** 360 mL of 0.10 M $NaCHO_2$, 640 mL of 0.10 M $HCHO_2$ **17.31** (a) Curve B (b) pH at the approximate equivalence point of curve A = 8.0, pH at the approximate equivalence point of curve B = 7.0 (c) For equal volumes of A and B, the concentration of acid B is greater, since it requires a larger volume of base to reach the equivalence point. **17.33** (a) Above pH 7 (b) below pH 7 (c) at pH 7 **17.35** (a) HX is weaker. The higher the pH at the equivalence point, the stronger the conjugate base (X^-) and the weaker the conjugate acid HX. (b) Phenolphthalein, which changes color in the pH 8–10 range, is perfect for HX and probably appropriate for HY. **17.37** (a) 42.4 mL NaOH soln (b) 35.0 mL NaOH soln (c) 29.8 mL NaOH soln **17.39** (a) pH = 1.54 (b) pH = 3.30 (c) pH = 7.00 (d) pH = 10.69 (e) pH = 12.74 **17.41** (a) pH = 2.78 (b) pH = 4.74 (c) pH = 6.58 (d) pH = 8.81 (e) pH = 11.03 (f) pH = 12.42 **17.43** (a) pH = 7.00 (b) $[HONH_3^+] = 0.100\ M$, pH = 3.52 (c) $[C_6H_5NH_3^+] = 0.100\ M$, pH = 2.82 **17.45** (a) The concentration of undissolved solid does not appear in the solubility product expression because it is constant. (b) $K_{sp} = [Ag^+][I^-]$; $K_{sp} = [Sr^{2+}][SO_4^{2-}]$; $K_{sp} = [Fe^{2+}][OH^-]^2$; $K_{sp} = [Hg_2^{2+}][Br^-]^2$; **17.47** (a) $K_{sp} = 7.63 \times 10^{-9}$ (b) $K_{sp} = 2.7 \times 10^{-9}$ (c) 5.3×10^{-4} mol $Ba(IO_3)_2$/L **17.49** $K_{sp} = 2.3 \times 10^{-9}$ **17.51** (a) 7.1×10^{-7} mol AgBr/L (b) 1.7×10^{-11} mol AgBr/L (c) 5.0×10^{-12} mol AgBr/L **17.53** (a) 1.4×10^3 g $Mn(OH)_2$/L (b) 0.014 g/L (c) 3.6×10^{-7} g/L **17.55** More soluble in acid: (a) $ZnCO_3$ (b) ZnS (d) AgCN (e) $Ba_3(PO_4)_2$ **17.57** $[Cu^{2+}] = 2 \times 10^{-12}\ M$ **17.59** $K = K_{sp} \times K_f = 8 \times 10^4$ **17.61** (a) $Q < K_{sp}$; no $Ca(OH)_2$ precipitates (a) $Q < K_{sp}$; no Ag_2SO_4 precipitates **17.63** pH = 11.5 **17.65** AgI will precipitate first, at $[I^-] = 4.2 \times 10^{-13}\ M$. **17.67** The first two experiments eliminate group 1 and 2 ions (Figure 17.22). The absence of insoluble carbonate precipitates in the filtrate from the third experiment rules out group 4 ions. The ions that might be in the sample are those from group 3, Al^{3+}, Fe^{3+}, Zn^{2+}, Cr^{3+}, Ni^{2+}, Co^{2+}, or Mn^{2+}, and from group 5, NH_4^+, Na^+, or K^+. **17.69** (a) Make the solution acidic with 0.5 M HCl; saturate with H_2S. CdS will precipitate; ZnS will not. (b) Add excess base; $Fe(OH)_3(s)$ precipitates, but Cr^{3+} forms the soluble complex $Cr(OH)_4^-$. (c) Add $(NH_4)_2HPO_4$; Mg^{2+} precipitates as $MgNH_4PO_4$; K^+ remains soluble. (d) Add 6 M HCl; precipitate Ag^+ as $AgCl(s)$; Mn^{2+} remains soluble. **17.71** (a) Base is required to increase $[PO_4^{3-}]$ so that the solubility product of the metal phosphates of interest is exceeded and the phosphate salts precipitate. (b) K_{sp} for the cations in group 3 is much larger, and so to exceed K_{sp}, a higher $[S^{2-}]$ is required. (c) They should all redissolve in strongly acidic solution. **17.73** (a) pH = 3.025 (b) pH = 2.938 (c) pH = 12.862 **17.76** (a) pH of buffer A = pH of buffer B = 3.74. For buffers containing the same conjugate acid and base components, the pH is determined by the ratio of concentrations of conjugate acid and base. Buffers A and B have the same ratio of concentrations, so their pH values are equal. (b) Buffer capacity is determined by the absolute amount of buffer components available to absorb strong acid or strong base. Buffer A has the greater capacity because it contains the greater absolute concentrations of $HCHO_2$ and CHO_2^-. (c) Buffer A: pH = 3.74, ΔpH = 0.00; buffer B: pH = 3.66, ΔpH = −0.12 (d) Buffer A: pH = 3.74, ΔpH = 0.00; buffer B: pH = 2.74, ΔpH = −1.00 (e) The results of parts (c) and (d) are quantitative confirmation that buffer A has a significantly greater capacity than buffer B. **17.78** (a) molar mass = 82.2 g/mol (b) $K_a = 3.8 \times 10^{-7}$ **17.81** (a) $[OH^-] = 4.23 \times 10^{-3}\ M$, pH = 11.63 (b) It will require 40.00 mL of 0.100 M HCl to reach the first equivalence point; HCO_3^- is the predominant species at this point. (c) An additional 40.00 mL are required to form H_2CO_3, the predominant species at the second equivalence point.

(d) $[H^+] = 1.2 \times 10^{-4}\ M$, pH = 3.92 **17.84** 1.6 L of 1.0 M NaOH **17.87** (a) CdS (b) $BaCrO_4$ (c) $NiCO_3$ (d) Ag_2SO_4 **17.90** $[KMnO_4] = [MnO_4^-] = 0.11\ M$ **17.93** $[OH^-] = 1.7 \times 10^{-11}\ M$, pH of the buffer = 3.22 **17.96** (a) $H^+(aq) + CHO_2^-(aq) \rightleftharpoons HCHO_2(aq)$ (b) $K = 5.6 \times 10^3$ (c) $[Na^+] = [Cl^-] = 0.075\ M$, $[H^+] = [HCHO_2^-] = 3.7 \times 10^{-3}\ M$, $[HCHO_2] = 0.071\ M$ **17.98** $[NH_4^+] = 0.40\ M$, $[NH_3] = 0.20\ M$, pH = 8.95 **17.101** $[Sr^{2+}] = [SO_4^{2-}] = 5.7 \times 10^{-4}\ M$, $K_{sp} = 3.2 \times 10^{-7}$

Chapter 18

18.1 A greater volume than 22.4 L (b) The gas will occupy more volume at 85 km than at 50 km. **18.3** Ozone concentration varies with altitude because conditions favorable to ozone formation and unfavorable to its decomposition vary with altitude. Above 60 km, there are too few O_2 molecules. Below 30 km, there are too few O atoms. Between 30 km and 60 km, O_3 concentration varies depending on the concentrations of O, O_2 and M*. **18.5** $CO_2(g)$ dissolves in seawater to form $H_2CO_3(aq)$. Carbon is removed from the ocean as $CaCO_3(s)$ in the form of sea shells, coral, and other carbonates. As carbon is removed, more $CO_2(g)$ dissolves to maintain the balance of complex and interacting acid-base and precipitation equilibria. **18.7** The guiding principle of green chemistry is that processes should be designed to minimize or eliminate solvents and waste, generate nontoxic waste, be energy efficient, employ renewable starting materials, and take advantage of catalysts that enable the use of safe and common reagents. **18.9** (a) Its temperature profile (b) troposphere, 0 to 12 km; stratosphere, 12 to 50 km; mesosphere, 50 to 85 km; thermosphere, 85 to 110 km **18.11** The partial pressure of O_3 is 3.0×10^{-7} atm (2.2×10^{-4} torr). **18.13** 8.4×10^{16} CO molecules **18.15** 570 nm **18.17** (a) *Photodissociation* is cleavage of a bond such that two neutral species are produced. *Photoionization* is absorption of a photon with sufficient energy to eject an electron, producing an ion and the ejected electron. (b) Photoionization of O_2 requires 1205 kJ/mol. Photodissociation requires only 495 kJ/mol. At lower elevations, high-energy short-wavelength solar radiation has already been absorbed. Below 90 km, the increased concentration of O_2 and the availability of longer wavelength radiation cause the photodissociation process to dominate. **18.19** A hydrofluorocarbon is a compound that contains hydrogen, fluorine, and carbon; it contains hydrogen in place of chlorine. HFCs are potentially less harmful than CFCs because photodissociation does not produce Cl atoms, which catalyze the destruction of ozone. **18.21** (a) The C—F bond requires more energy for dissociation than the C—Cl bond and is not readily cleaved by the available wavelengths of UV light. (b) Chlorine is present as chlorine atoms and chlorine oxide molecules, Cl and ClO, respectively. **18.23** (a) Methane, CH_4, arises from decomposition of organic matter by certain microorganisms; it also escapes from underground gas deposits. (b) SO_2 is released in volcanic gases and also is produced by bacterial action on decomposing vegetable and animal matter. (c) Nitric oxide, NO, results from oxidation of decomposing organic matter and is formed in lightning flashes. (d) CO is a possible product of some vegetable matter decay. **18.25** (a) $H_2SO_4(aq) + CaCO_3(s) \longrightarrow CaSO_4(s) + H_2O(l) + CO_2(g)$ (b) The $CaSO_4(s)$ would be much less reactive with acidic solution, since it would require a strongly acidic solution to shift the relevant equilibrium to the right. $CaSO_4(s) + 2\,H^+(aq) \rightleftharpoons Ca^{2+}(aq) + 2\,HSO_4^-(aq)$ **18.27** (a) Ultraviolet (b) 357 kJ/mol (c) The average C—H bond energy from Table 8.4 is 413 kJ/mol. The C—H bond energy in CH_2O, 357 kJ/mol, is less than the "average" C—H bond energy. (d)

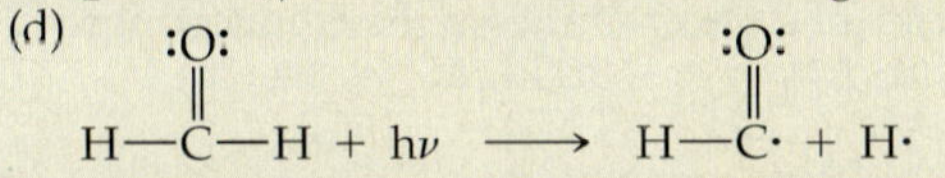

18.29 Incoming and outgoing energies are in different regions of the electromagnetic spectrum. CO_2 is transparent to incoming visible radiation but absorbs outgoing infrared radiation. **18.31** 0.099 M Na^+ **18.33** 4.361×10^5 g CaO **18.35** The minimum pressure required to initiate reverse osmosis is greater than 5.1 atm. **18.37** (a) $CO_2(g)$, HCO_3^-, $H_2O(l)$, SO_4^{2-}, NO_3^-, HPO_4^{2-}, $H_2PO_4^-$ (b) $CH_4(g)$, $H_2S(g)$, $NH_3(g)$, $PH_3(g)$ **18.39** 2.5 g O_2 **18.41** $Mg^{2+}(aq) + Ca(OH)_2(s) \longrightarrow Mg(OH)_2(s) + Ca^{2+}(aq)$ **18.43** 0.42 mol $Ca(OH)_2$, 0.18 mol Na_2CO_3 **18.45** $4\,FeSO_4(aq) + O_2(aq) + 2\,H_2O(l) \longrightarrow 4\,Fe^{3+}(aq) + 4\,OH^- + 4\,SO_4^{2-}(aq)$; $Fe^{3+}(aq) + 3\,HCO_3^-(aq) \longrightarrow Fe(OH)_3(s) + 3\,CO_2(g)$ **18.47** The fewer steps in a process, the less waste is generated. Processes with fewer steps require less energy at the site of the process, and for subsequent cleanup or disposal of waste. **18.49** (a) H_2O (b) It is better to prevent waste than to treat it. Produce as little nontoxic waste as possible. Chemical processes should be efficient. Raw materials should be renewable. **18.51** (a) Acid rain is rain with a larger $[H^+]$ and a lower pH than expected. The additional H^+ is produced by the dissolution of sulfur and nitrogen oxides in rain droplets to form sulfuric and nitric acid. (b) A greenhouse gas such as CO_2 or H_2O absorbs infrared or "heat" radiation emitted from Earth's surface and helps to maintain a relatively constant temperature at the surface. A significant build-up of greenhouse gases in the atmosphere could cause a corresponding increase in the average surface temperature and stimulate global climate change. (c) Photochemical smog is an unpleasant collection of atmospheric pollutants initiated by photochemical dissociation of NO_2 to form NO and O atoms. The major components are $NO(g)$, $NO_2(g)$, $CO(g)$, $O_3(g)$, and unburned hydrocarbons from gasoline. (d) Ozone depletion is the loss of stratospheric ozone due to reactions between O_3 and Cl. Insufficient stratospheric ozone allows damaging ultraviolet radiation to reach our ecosystem. **18.55** (a) Photodissociation of $CBrF_3$ to form Br atoms requires less energy than the production of Cl atoms and should occur readily in the stratosphere. (b) $CBrF_3(g) \xrightarrow{h\nu} CF_3(g) + Br(g)$; $Br(g) + O_3(g) \longrightarrow BrO(g) + O_2(g)$ **18.57** The formation of $NO(g)$ is endothermic, so K increases with increasing temperature. The oxidation of $NO(g)$ to $NO_2(g)$ is exothermic, so the value of K decreases with increasing temperature. **18.59** (a) $CH_4(g) + 2\,O_2(g) \longrightarrow CO_2(g) + 2\,H_2O(g)$ (b) $2\,CH_4(g) + 3\,O_2(g) \longrightarrow 2\,CO(g) + 4\,H_2O(g)$ (c) 9.5 L dry air **18.61** Most of the 390 watts/m^2 radiated from Earth's surface is in the infrared region of the spectrum. Greenhouse gases absorb much of this radiation, which warms the atmosphere close to Earth's surface and makes the planet liveable. **18.63** (a) $NO(g) + h\nu \longrightarrow N(g) + O(g)$ (b) $NO(g) + h\nu \longrightarrow NO^+(g) + e^-$ (c) $NO(g) + O_3(g) \longrightarrow NO_2(g) + O_2(g)$ (d) $3\,NO_2(g) + H_2O(l) \longrightarrow 2\,HNO_3(aq) + NO(g)$
18.65 Because NO has an odd electron, like Cl(g), it could act as a catalyst for decomposition of stratospheric ozone. This would lead to an increase in the amount of damaging ultraviolet radiation that reaches the troposphere. In the troposphere NO is oxidized to NO_2 by O_2. On dissolving in water, NO_2 disproportionates into $NO_3^-(aq)$ and $NO(g)$. Over time the NO in the troposphere will be converted into NO_3^-, which is in turn incorporated into soils. **18.67** (a) $P_{NO_2} = 1.4 \times 10^{-5}$ torr (b) 2×10^{19} NO_2 molecules **18.71** 13.3 days. The value of the rate constant limits the result to 1 significant figure. This implies an uncertainty of ± 1 in the tens place of our answer. Realistically, the remediation could take anywhere from 1 to 30 days. **18.73** (a) The energy required is 9.8×10^3 kJ/gal H_2O. (b) The energy cost is \$0.23/gal H_2O. (c) 18% of the total cost is energy.
18.75 (a) $17e^-$, $8.5e^-$ pairs

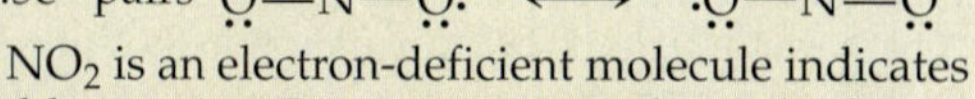

(b) The fact that NO_2 is an electron-deficient molecule indicates that it will be highly reactive. Dimerization results in the formation of a N—N single bond which completes the octet of both N atoms. In a closed system, NO_2 and N_2O_4 exist in equilibrium. In an urban environment, NO_2 is produced from hot automobile combustion. At these temperatures, equilibrium favors the monomer because the reaction is exothermic. (c) NO_2 is an oxidizing agent and CO is a reducing agent, so we expect products to contain N in a more reduced form, NO or N_2, and C in a more oxidized form, CO_2. (d) No. Because NO_2 is a very reactive odd-electron molecule, we expect it to react or photodisssociate before it can migrate to the stratosphere. **18.77** (a) 1.2×10^{-5} M (b) by solving the quadratic formula, $[H^+] = 2.03 \times 10^{-6}$ M, pH = 5.69.

Chapter 19

19.1 (a)

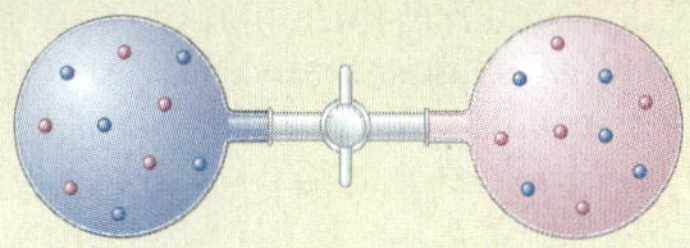

(b) $\Delta H = 0$ for mixing ideal gases. ΔS is positive, because the disorder of the system increases. (c) The process is spontaneous and therefore irreversible. (d) Since $\Delta H = 0$, the process does not affect the entropy of the surroundings. **19.4** (a) At 300 K, $\Delta H = T\Delta S$, $\Delta G = 0$ and the system is at equilibrium. (b) The reaction is spontaneous at temperatures above 300 K.
19.7 Spontaneous: b, c, d; nonspontaneous: a, e
19.9 (a) $NH_4NO_3(s)$ dissolves in water, as in a chemical cold pack. Naphthalene (moth balls) sublimes at room temperature. (b) Melting of a solid is spontaneous above its melting point but nonspontaneous below its melting point. **19.11** (a) Endothermic (b) at or above 100°C (c) below 100°C (d) at 100°C
19.13 (a) For a *reversible* process, the forward and reverse changes occur by the same path. In a reversible process, both the system and the surroundings are restored to their original condition by exactly reversing the change. A reversible change produces the maximum amount of work. (b) There is no net change in the surroundings. (c) The vaporization of water to steam is reversible if it occurs at the boiling temperature of water for a specified external (atmospheric) pressure, and only if the heat is heat needed is

added infinitely slowly. **19.15** No. ΔE is a state function. $\Delta E = q + w$; q and w are not state functions. Their values do depend on path, but their sum, ΔE, does not. **19.17** (a) An ice cube can melt reversibly at the conditions of temperature and pressure where the solid and liquid are in equilibrium. (b) We know that melting is a process that increases the energy of the system even though there is no change in temperature. ΔE is not zero for the process. **19.19** (a) At constant temperature, $\Delta S = q_{rev}/T$, where q_{rev} is the heat that would be transferred if the process were reversible. (b) No. ΔS is a state function, so it is independent of path. **19.21** (a) Entropy increases (b) 213 J/K **19.23** (a) For a spontaneous process, the entropy of the universe increases; for a reversible process, the entropy of the universe does not change. (b) For a reversible process, if the entropy of the system increases, the entropy of the surroundings must decrease by the same amount. (c) For a spontaneous process, the entropy of the universe must increase, so the entropy of the surroundings must decrease by less than 42 J/K. **19.25** $\Delta S = 0.762$ J/K **19.27** (a) The higher the temperature, the more microstates are available to a system. (b) The greater the volume, the more microstates are available to a system. (c) Going from solid to liquid to gas, the number of available microstates increases. **19.29** (a) ΔS is positive. (b) ΔS is positive for Exercise 19.8(a) and (c). **19.31** S increases in (a) and (c); S decreases in (b) **19.33** (a) The entropy of a pure crystalline substance at absolute zero is zero. (b) In translational motion the entire molecule moves in a single direction; in rotational motion the molecule rotates or spins around a fixed axis. In vibrational motion the bonds within a molecule stretch and bend, but the average position of the atoms does not change.

(c) H—Cl $\longrightarrow$ H—Cl

Translational

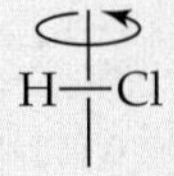

Rotational

$\overleftarrow{\text{H}}$——$\overrightarrow{\text{Cl}}$ $\longleftrightarrow$ $\overrightarrow{\text{H}}$—$\overleftarrow{\text{Cl}}$

Vibrational

19.35 (a) phase changes: 1 = melting (fusion), 2 = vaporization. (b) ΔS for vaporization is always larger than ΔS for fusion because of the large volume increase of the system. **19.37** (a) Ar(g) (b) He(g) at 1.5 atm (c) 1 mol of Ne(g) in 15.0 L (d) $CO_2(g)$ **19.39** (a) $\Delta S < 0$ (b) $\Delta S > 0$ (c) $\Delta S < 0$ (d) $\Delta S > 0$ **19.41** (a) $C_2H_6(g)$ (b) $CO_2(g)$ **19.43** (a) Sc(s), 34.6 J/mol-K; Sc(g), 174.7 J/mol-K. In general, the gas phase of a substance has a larger $S°$ than the solid phase because of the greater volume and motional freedom of the molecules. (b) $NH_3(g)$, 192.5 J/mol-K; $NH_3(aq)$, 111.3 J/mol-K. Molecules in the gas phase have more motional freedom than molecules in solution. (c) 1 mol of $P_4(g)$, 280 J/K; 2 mol of $P_2(g)$, 2(218.1) = 436.2 J/K. More particles have a greater motional energy (more available microstates). (d) C (diamond), 2.43 J/mol-K; C (graphite), 5.69 J/mol-K. The internal entropy in graphite is greater because there is translational freedom among planar sheets of C atoms, while there is very little freedom within the covalent-network diamond lattice. **19.45** For elements with similar structures, the heavier the atoms, the lower the vibrational frequencies at a given temperature. This means that more vibrations can be accessed at a particular temperature, resulting in greater absolute entropy for the heavier elements. **19.47** (a) $\Delta S° = -120.5$ J/K. $\Delta S°$ is negative because there are fewer moles of gas in the products. (b) $\Delta S° = +176.6$ J/K. $\Delta S°$ is positive because there are more moles of gas in the products. (c) $\Delta S° = +152.39$ J/K. $\Delta S°$ is positive because the product contains more total particles and more moles of gas. (d) $\Delta S° = +92.3$ J/K. $\Delta S°$ is positive because there are more moles of gas in the products. **19.49** (a) $\Delta G = \Delta H - T\Delta S$ (b) If ΔG is positive, the process is nonspontaneous, but the reverse process is spontaneous. (c) There is no relationship between ΔG and rate of reaction. **19.51** (a) Exothermic (b) $\Delta S°$ is negative; the reaction leads to a decrease in disorder. (c) $\Delta G° = -9.9$ kJ (d) If all reactants and products are present in their standard states, the reaction is spontaneous in the forward direction at this temperature. **19.53** (a) $\Delta H° = -537.22$ kJ, $\Delta S° = 13.7$ J/K, $\Delta G° = -541.40$ kJ, $\Delta G° = \Delta H° - T\Delta S° = -541.31$ kJ (b) $\Delta H° = -106.7$ kJ, $\Delta S° = -142.2$ J/K, $\Delta G° = -64.0$ kJ, $\Delta G° = \Delta H° - T\Delta S° = -64.3$ kJ (c) $\Delta H° = -508.3$ kJ, $\Delta S° = -178$ J/K, $\Delta G° = -465.8$ kJ, $\Delta G° = \Delta H° - T\Delta S° = -455.1$ kJ. The discrepancy in $\Delta G°$ values is due to experimental uncertainties in the tabulated thermodynamic data. (d) $\Delta H° = -165.9$ kJ, $\Delta S° = 1.4$ J/K, $\Delta G° = -166.2$ kJ, $\Delta G° = \Delta H° - T\Delta S° = -166.3$ kJ **19.55** (a) $\Delta G° = -140.0$ kJ, spontaneous (b) $\Delta G° = +104.70$ kJ, nonspontaneous (c) $\Delta G° = +146$ kJ, nonspontaneous (d) $\Delta G° = -156.7$ kJ, spontaneous **19.57** (a) $C_6H_{12}(l) + 9\,O_2(g) \longrightarrow 6\,CO_2(g) + 6\,H_2O(l)$ (b) Because $\Delta S°$ is negative, $\Delta G°$ is less negative than $\Delta H°$. **19.59** (a) The forward reaction is spontaneous at low temperatures, but becomes nonspontaneous at higher temperatures. (b) The reaction is nonspontaneous in the forward direction at all temperatures. (c) The forward reaction is nonspontaneous at low temperatures, but becomes spontaneous at higher temperatures. **19.61** $\Delta S > +76.7$ J/K **19.63** (a) $T = 330$ K (b) nonspontaneous **19.65** (a) $\Delta H° = 155.7$ kJ, $\Delta S° = 171.4$ J/K. Since $\Delta S°$ is positive, $\Delta G°$ becomes more negative with increasing temperature. (b) $\Delta G° = 19$ kJ. The reaction is not spontaneous under standard conditions at 800 K (c) $\Delta G° = -15.7$ kJ, The reaction is spontaneous under standard conditions at 1000 K. **19.67** (a) $T_b = 79°C$ (b) From the *Handbook of Chemistry and Physics*, 74th Edition, $T_b = 80.1°C$. The values are remarkably close; the small difference is due to deviation from ideal behavior by $C_6H_6(g)$ and experimental uncertainty in the boiling point measurement and the thermodynamic data. **19.69** (a) $C_2H_2(g) + \frac{5}{2}O_2(g) \longrightarrow 2\,CO_2(g) + H_2O(l)$ (b) -1299.5 kJ of heat produced/mol C_2H_2 burned (c) $w_{max} = -1235.1$ kJ/mol C_2H_2 **19.71** (a) ΔG becomes more negative. (b) ΔG becomes more positive. (c) ΔG becomes more positive. **19.73** (a) $\Delta G° = -5.40$ kJ (b) $\Delta G = 0.30$ kJ **19.75** (a) $\Delta G° = -16.77$ kJ, $K = 870$ (b) $\Delta G° = 8.0$ kJ, $K = 0.04$ (c) $\Delta G° = -497.9$ kJ, $K = 2 \times 10^{87}$ **19.77** $\Delta H° = 269.3$ kJ, $\Delta S° = 0.1719$ kJ/K (a) $P_{CO_2} = 6.0 \times 10^{-39}$ atm (b) $P_{CO_2} = 1.6 \times 10^{-4}$ atm **19.79** (a) $HNO_2(aq) \rightleftharpoons H^+(aq) + NO_2^-(aq)$ (b) $\Delta G° = 19.1$ kJ (c) $G = 0$ at equilibrium (d) $\Delta G = -2.72$ kJ **19.82** (a) $\Delta H > 0$, $\Delta S > 0$ (b) $\Delta H < 0$, $\Delta S < 0$ (c) $\Delta H > 0$, $\Delta S > 0$ (d) $\Delta H > 0$, $\Delta S > 0$ (e) $\Delta H < 0$, $\Delta S > 0$ **19.86** (a) For all three compounds listed, there are fewer moles of gaseous products than reactants in the formation reaction, so we expect $\Delta S_f°$ to be negative. If $\Delta G_f° = \Delta H_f° - T\Delta S_f°$ and $\Delta S_f°$ is negative, $-T\Delta S_f°$ is positive and $\Delta G_f°$ is more positive than $\Delta H_f°$. **19.90** (a) $K = 4 \times 10^{15}$ (b) An increase in temperature will decrease the mole fraction of CH_3COOH at equilibrium. Elevated temperatures must be used to increase the speed of the reaction. (c) $K = 1$ at 836 K or 563°C. **19.94** (a) $\Delta G = 8.77$ kJ (b) $w_{min} = 8.77$ kJ. In practice, a larger than minimum amount of work is required. **19.98** (a) Acetone, $\Delta S^\circ_{vap} = 88.4$ J/mol-K; dimethyl ether, $\Delta S^\circ_{vap} = 86.6$ J/mol-K; ethanol, $\Delta S^\circ_{vap} = 110$ J/mol-K; octane, $\Delta S^\circ_{vap} = 86.3$ J/mol-K; pyridine, $\Delta S^\circ_{vap} = 90.4$ J/mol-K. Ethanol does not obey Trouton's rule. (b) Hydrogen bonding (in ethanol and other liquids) leads to more ordering in the liquid state and a greater than usual increase in entropy upon vaporization. Liquids that experience hydrogen bonding are probably exceptions to Trouton's rule. (c) Owing to strong hydrogen bonding interactions, water probably does not obey Trouton's rule. $\Delta S^\circ_{vap} = 109.0$ J/mol-K. (d) ΔH_{vap} for $C_6H_5Cl \approx 36$ kJ/mol **19.101** (a) S increases (b) S decreases (c) S increases (d) S decreases **19.104** (a) For any given total pressure, the condition of equal moles of the two gases can be achieved at some temperature. For individual gas pressures of 1 atm and a total pressure of 2 atm, the mixture is at equilibrium at 328.5 K or 55.5°C. (b) 333.0 K or 60°C (c) 374.2 K or 101.2°C (d) The reaction is endothermic, so an increase in the value of K as calculated in parts (a)–(c) should be accompanied by an increase in T.

Chapter 20

20.1 In a Brønsted-Lowry acid-base reaction, H^+ is transferred from the acid to the base. In a redox reaction, one or more electrons are transferred from the reductant to the oxidant. **20.3** $E^\circ_{cell} = 0.16$ V

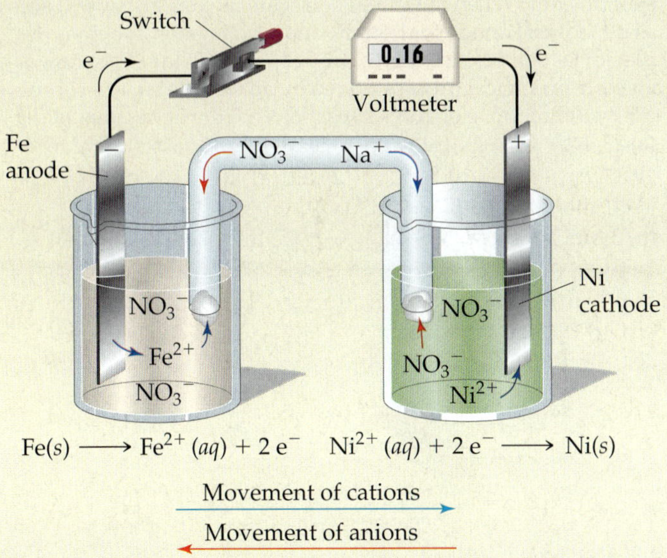

20.5 (a) A is reduced; B is oxidized. (b) Cathode: $A(aq) + 1e^- \longrightarrow A^-(aq)$; anode: $B(aq) \longrightarrow B^+(aq) + 1e^-$ (c) In a cell with a positive E° value, electrons flow spontaneously from the anode to the cathode, so the anode half-reaction, $B(aq) \longrightarrow B^+(aq) + 1e^-$, is at higher potential energy. **20.7** Zinc, with a more negative E°_{red}, is more easily oxidized than iron. If conditions are favorable for oxidation, the zinc coating will be preferentially oxidized, preventing iron from corroding. **20.9** (a) *Oxidation* is the loss of electrons. (b) Electrons appear on the products' side (right side). (c) The *oxidant* is the reactant that is reduced. (d) An *oxidizing agent* is the substance that promotes oxidation; it is the oxidant.

20.11 (a) True (b) false (c) true **20.13** (a) I, +5 to 0; C, +2 to +4 (b) Hg, +2 to 0; N, −2 to 0 (c) N, +5 to +2; S, −2 to 0 (d) Cl, +4 to +3; O, −1 to 0 **20.15** (a) $TiCl_4(g) + 2\,Mg(l) \longrightarrow Ti(s) + 2\,MgCl_2(l)$ (b) $Mg(l)$ is oxidized; $TiCl_4(g)$ is reduced. (c) $Mg(l)$ is the reductant; $TiCl_4(g)$ is the oxidant. **20.17** (a) $Sn^{2+}(aq) \longrightarrow Sn^{4+}(aq) + 2e^-$, oxidation
(b) $TiO_2(s) + 4\,H^+(aq) + 2e^- \longrightarrow Ti^{2+}(aq) + 2\,H_2O(l)$, reduction
(c) $ClO_3^-(aq) + 6\,H^+(aq) + 6e^- \longrightarrow Cl^-(aq) + 3\,H_2O(l)$, reduction
(d) $4\,OH^-(aq) \longrightarrow O_2(g) + 2\,H_2O(l) + 4e^-$, oxidation
(e) $SO_3^{2-}(aq) + 2\,OH^-(aq) \longrightarrow SO_4^{2-}(aq) + H_2O(l) + 2e^-$, oxidation
(f) $N_2(g) + 8\,H^+(aq) + 6e^- \longrightarrow 2\,NH_4^+(aq)$, reduction
(g) $N_2(g) + 6\,H_2O(l) + 2e^- \longrightarrow 2\,NH_3(g) + 6\,OH^-(aq)$, reduction **20.19** (a) $Cr_2O_7^{2-}(aq) + I^-(aq) \longrightarrow 2\,Cr^{3+}(aq) + IO_3^-(aq) + 4\,H_2O(l)$; oxidizing agent, $Cr_2O_7^{2-}$; reducing agent, I^- (b) $4\,MnO_4^-(aq) + 5\,CH_3OH(aq) + 12\,H^+(aq) \longrightarrow 4\,Mn^{2+}(aq) + 5\,HCO_2H(aq) + 11\,H_2O(l)$; oxidizing agent, MnO_4^-; reducing agent, CH_3OH (c) $I_2(s) + 5\,OCl^-(aq) + H_2O(l) \longrightarrow 2\,IO_3^-(aq) + 5\,Cl^-(aq) + 2\,H^+(aq)$; oxidizing agent, OCl^-; reducing agent, I_2 (d) $As_2O_3(s) + 2\,NO_3^-(aq) + 2\,H_2O(l) + 2\,H^+(aq) \longrightarrow 2\,H_3AsO_4(aq) + N_2O_3(aq)$; oxidizing agent, NO_3^-; reducing agent, As_2O_3 (e) $2\,MnO_4^-(aq) + Br^-(aq) + H_2O(l) \longrightarrow 2\,MnO_2(s) + BrO_3^-(aq) + 2\,OH^-(aq)$; oxidizing agent, MnO_4^-; reducing agent, Br^- (f) $Pb(OH)_4^{2-}(aq) + ClO^-(aq) \longrightarrow PbO_2(s) + Cl^-(aq) + 2\,OH^-(aq) + H_2O(l)$; oxidizing agent, ClO^-; reducing agent, $Pb(OH)_4^{2-}$ **20.21** (a) The reaction $Cu^{2+}(aq) + Zn(s) \longrightarrow Cu(s) + Zn^{2+}(aq)$ is occurring in both figures. In Figure 20.3 the reactants are in contact, while in Figure 20.4 the oxidation half-reaction and reduction half-reaction are occurring in separate compartments. In Figure 20.3 the flow of electrons cannot be isolated or utilized; in Figure 20.4 electrical current is isolated and flows through the voltmeter. (b) Na^+ cations are drawn into the cathode compartment to maintain charge balance as Cu^{2+} ions are removed.
20.23 (a) $Fe(s)$ is oxidized, $Ag^+(aq)$ is reduced. (b) $Ag^+(aq) + 1e^- \longrightarrow Ag(s)$; $Fe(s) \longrightarrow Fe^{2+}(aq) + 2e^-$ (c) $Fe(s)$ is the anode, $Ag(s)$ is the cathode. (d) $Fe(s)$ is negative; $Ag(s)$ is positive. (e) Electrons flow from the Fe electrode (−) toward the Ag electrode (+). (f) Cations migrate toward the $Ag(s)$ cathode; anions migrate toward the $Fe(s)$ anode. **20.25** *Electromotive force*, emf, is the potential energy difference between an electron at the anode and an electron at the cathode of a voltaic cell. (b) One *volt* is the potential energy difference required to impart 1 J of energy to a charge of 1 coulomb. (c) *Cell potential*, E_{cell}, is the emf of an electrochemical cell. **20.27** (a) $2\,H^+(aq) + 2e^- \longrightarrow H_2(g)$ (b) A *standard* hydrogen electrode, SHE, has components that are at standard conditions, 1 M $H^+(aq)$ and $H_2(g)$ at 1 atm. (c) The platinum foil in a SHE serves as an inert electron carrier and a solid reaction surface. **20.29** (a) A *standard reduction potential* is the relative potential of a reduction half-reaction measured at standard conditions. (b) $E^\circ_{red} = 0$ (c) The reduction of $Ag^+(aq)$ to $Ag(s)$ is much more energetically favorable. **20.31** (a) $Cr^{2+}(aq) \longrightarrow Cr^{3+}(aq) + e^-$; $Tl^{3+}(aq) + 2e^- \longrightarrow Tl^+(aq)$ (b) $E^\circ_{red} = 0.78$ V

(c)

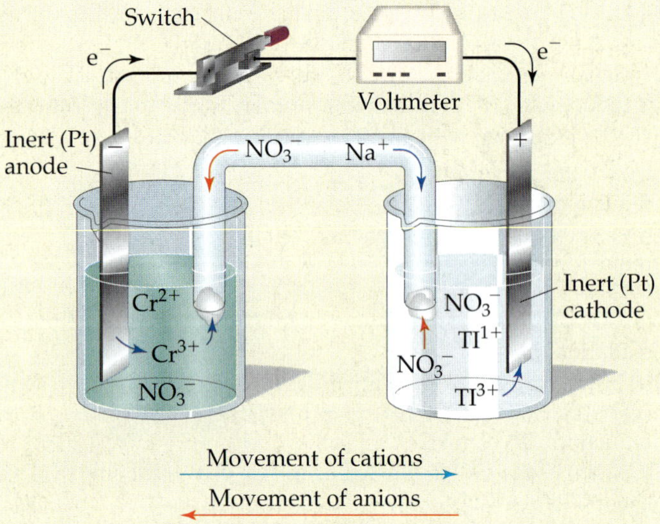

20.33 (a) $E° = 0.823$ V (b) $E° = 1.89$ V (c) $E° = 1.211$ V (d) $E° = 1.21$ V **20.35** (a) $3\,Ag^+(aq) + Cr(s) \longrightarrow 3\,Ag(s) + Cr^{3+}(aq)$, $E° = 1.54$ V (b) Two of the combinations have essentially equal $E°$ values: $2\,Ag^+(aq) + Cu(s) \longrightarrow 2\,Ag(s) + Cu^{2+}(aq)$, $E° = 0.462$ V; $3\,Ni^{2+}(aq) + 2\,Cr(s) \longrightarrow 3\,Ni(s) + 2\,Cr^{3+}(aq)$, $E° = 0.46$ V **20.37** (a) Anode, Sn(*s*); cathode, Cu(*s*). (b) The copper electrode gains mass as Cu is plated out, and the tin electrode loses mass as Sn is oxidized. (c) $Cu^{2+}(aq) + Sn(s) \longrightarrow Cu(s) + Sn^{2+}(aq)$. (d) $E° = 0.473$ V. **20.39** (a) $Cl_2(g)$ (b) $Ni^{2+}(aq)$ (c) $BrO_3^-(aq)$ (d) $O_3(g)$ **20.41** (a) $Cl_2(aq)$, strong oxidant (b) $MnO_4^-(aq)$, acidic, strong oxidant (c) Ba(*s*), strong reductant (d) Zn(*s*), reductant **20.43** (a) $Cu^{2+}(aq) < O_2(g) < Cr_2O_7^{2-}(aq) < Cl_2(g) < H_2O_2(aq)$ (b) $H_2O_2(aq) < I^-(aq) < Sn^{2+}(aq) < Zn(s) < Al(s)$ **20.45** Al and $H_2C_2O_4$ **20.47** (a) $2\,Fe^{2+}(aq) + S_2O_6^{2-}(aq) + 4\,H^+(aq) \longrightarrow 2\,Fe^{3+}(aq) + 2\,H_2SO_3(aq)$; $2\,Fe^{2+}(aq) + N_2O(aq) + 2\,H^+(aq) \longrightarrow 2\,Fe^{3+}(aq) + N_2(g) + H_2O(l)$; $Fe^{2+}(aq) + VO_2^+(aq) + 2\,H^+(aq) \longrightarrow Fe^{3+}(aq) + VO^{2+}(aq) + H_2O(l)$ (b) $E° = -0.17$ V, $\Delta G° = 33$ kJ; $E° = -2.54$ V, $\Delta G° = 4.90 \times 10^2$ kJ; $E° = 0.23$ V, $\Delta G° = -22$ kJ (c) $K = 1.8 \times 10^{-6} = 10^{-6}$; $K = 1.2 \times 10^{-86} = 10^{-86}$; $K = 7.8 \times 10^3 = 8 \times 10^3$ **20.49** $\Delta G° = 21.8$ kJ, $E°_{cell} = -0.113$ V **20.51** (a) $E° = 0.16$ V, $K = 2.54 \times 10^5 = 3 \times 10^5$ (b) $E° = 0.277$ V, $K = 2.3 \times 10^9$ (c) $E° = 0.45$ V, $K = 1.5 \times 10^{75} = 10^{75}$ **20.53** (a) $K = 9.8 \times 10^2$ (b) $K = 9.5 \times 10^5$ (c) $K = 9.3 \times 10^8$ **20.55** (a) The Nernst equation is applicable when the components of an electrochemical cell are at nonstandard conditions. (b) $Q = 1$ (c) Q decreases and E increases **20.57** (a) E decreases (b) E decreases (c) E decreases (d) no effect **20.59** (a) $E° = 0.48$ V (b) $E = 0.53$ V (c) $E = 0.46$ V **20.61** (a) $E° = 0.46$ V (b) $E = 0.37$ V **20.63** (a) The compartment with $[Zn^{2+}] = 1.00 \times 10^{-2}\,M$ is the anode. (b) $E° = 0$ (c) $E = 0.0668$ V (d) In the anode compartment $[Zn^{2+}]$ increases; in the cathode compartment $[Zn^{2+}]$ decreases **20.65** $E° = 0.763$ V, pH = 1.6 **20.67** (a) The emf of a battery decreases as it is used. The concentrations of products increase and the concentrations of reactants decrease, causing Q to increase and E_{cell} to decrease. (b) A D-size battery contains more reactants than a AA, enabling the D to provide power for a longer time. **20.69** 464 g PbO_2 **20.71** (a) The anode (b) $E° = 3.50$ V (c) The emf of the battery, 3.5 V, is exactly the standard cell potential calculated in part (b). (d) At ambient conditions, $E \approx E°$, so $\log Q \approx 1$. Assuming that the value of $E°$ is relatively constant with temperature, the value of the second term in the Nernst equation is approximately zero at 37°C, and $E \approx 3.5$ V. **20.73** (a) The cell emf will have a smaller value. (b) NiMH batteries use an alloy such as $ZrNi_2$ as the anode material. This eliminates the use and disposal problems associated with Cd, a toxic heavy metal. **20.75** The main advantage of a fuel cell is that fuel is continuously supplied, so that it can produce electrical current for a time limited only by the amount of available fuel. For the hydrogen-oxygen fuel cell, this is also a disadvantage because volatile and explosive hydrogen must be acquired and stored. Alkaline batteries are convenient, but they have a short lifetime, and the disposal of their zinc and manganese solids is more problematic than disposal of water produced by the hydrogen-oxygen fuel cell. **20.77** (a) anode: $Fe(s) \longrightarrow Fe^{2+}(aq) + 2e^-$; cathode: $O_2(g) + 4\,H^+(aq) + 4e^- \longrightarrow 2\,H_2O(l)$ (b) $2\,Fe^{2+}(aq) + 3\,H_2O(l) + 3\,H_2O(l) \longrightarrow Fe_2O_3{\cdot}3\,H_2O(s) + 6\,H^+(aq) + 2e^-$; $O_2(g) + 4\,H^+(aq) + 4e^- \longrightarrow 2\,H_2O(l)$ **20.79** (a) Mg is called a "sacrificial anode" because it has a more negative $E°_{red}$ than the pipe metal and is preferentially oxidized when the two are coupled. It is sacrificed to preserve the pipe. (b) $E°_{red}$ for Mg^{2+} is −2.37 V, more negative than most metals present in pipes, including Fe and Zn. **20.81** Under acidic conditions, air (O_2) oxidation of Zn(*s*), 1.99 V, Fe(*s*), 1.67 V, and Cu(*s*), 0.893 V, are all spontaneous. When the three metals are in contact, Zn will act as a sacrificial anode for both Fe and Cu, but after the Zn is depleted, Fe will be oxidized (corroded). **20.83** (a) *Electrolysis* is an electrochemical process driven by an outside energy source. (b) By definition, electrolysis reactions are nonspontaneous. (c) $2\,Cl^-(l) \longrightarrow Cl_2(g) + 2e^-$ **20.85** (a) 236 g Cr(*s*) (b) 2.51 A **20.87** $w_{max} = -8.19 \times 10^4$ J **20.89** (a) 4.0×10^5 g Li (b) 0.24 kWh/mol Li **20.91** (a) $2\,Ni^+(aq) \longrightarrow Ni(s) + Ni^{2+}(aq)$ (b) $3\,MnO_4^{2-}(aq) + 4\,H^+(aq) \longrightarrow 2\,MnO_4^-(aq) + MnO_2(s) + 3\,H_2O(l)$ (c) $3\,H_2SO_3(aq) \longrightarrow S(s) + 2\,HSO_4^-(aq) + 2\,H^+(aq) + H_2O(l)$ (d) $Cl_2(aq) + 2\,OH^-(aq) \longrightarrow Cl^-(aq) + ClO^-(aq) + H_2O(l)$ **20.93** (a) $Fe(s) \longrightarrow Fe^{2+}(aq) + 2e^-$; $2\,Ag^+(aq) + 2e^- \longrightarrow Ag(s)$; $Fe(s) + 2\,Ag^+(aq) \longrightarrow Fe^{2+}(aq) + Ag(s)$

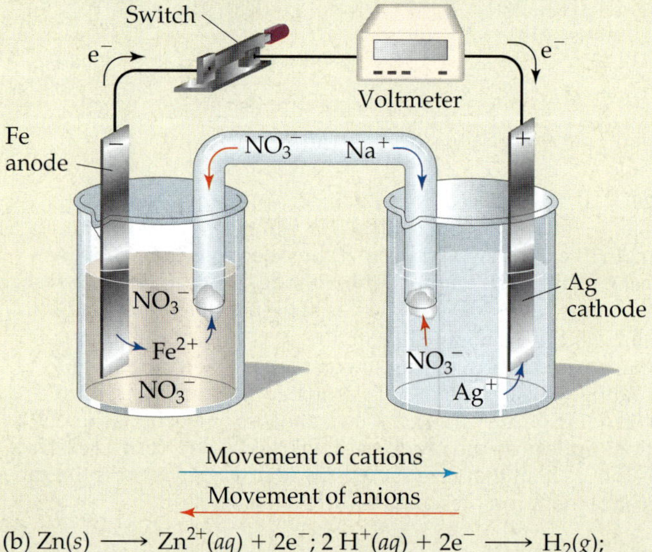

(b) $Zn(s) \longrightarrow Zn^{2+}(aq) + 2e^-$; $2\,H^+(aq) + 2e^- \longrightarrow H_2(g)$; $Zn(s) + 2\,H^+(aq) \longrightarrow Zn^{2+}(aq) + H_2(g)$

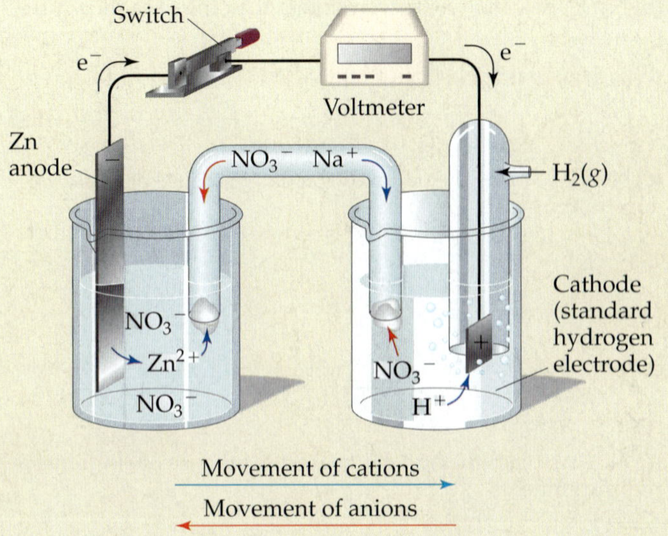

(c) $Cu \,|Cu^{2+}||ClO_3^-, Cl^-|\, Pt$

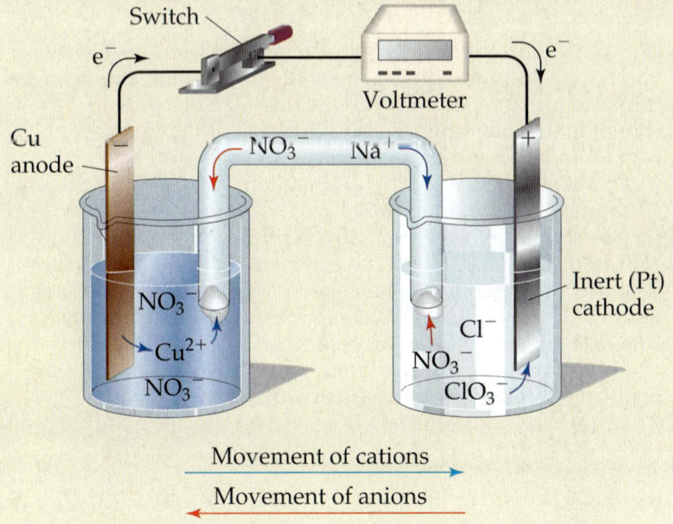

20.95 (a) The reduction potential for $O_2(g)$ in the presence of acid is 1.23 V. $O_2(g)$ cannot oxidize $Au(s)$ to $Au^+(aq)$, even in the presence of acid. (b) Definitely: $Co^{3+}(aq)$, $F_2(g)$, $H_2O_2(aq)$, $O_3(g)$; marginally: $BrO_3^-(aq)$, $Ce^{4+}(aq)$, $HClO(aq)$, $MnO_4^-(aq)$, $PbO_2(s)$ (c) $Au(s)$ is being oxidized; $O_2(g)$ is being reduced. (d) $2\ Na[Au(CN)_2](aq) + Zn(s) \longrightarrow 2\ Au(s) + Zn^{2+}(aq) + 2\ Na^+(aq) + 4\ CN^-(aq)$. $Zn(s)$ is being oxidized; $[Au(CN)_2]^-(aq)$ is being reduced. **20.97** (a) $E = -0.041$ V (b) the inert electrode (c) No, at standard conditions, $Cu(s)$ is the anode. (d) 1.3 M I^- **20.101** (a) In discharge, $Cd(s) + 2\ NiO(OH)(s) + 2\ H_2O(l) \longrightarrow Cd(OH)_2(s) + 2\ Ni(OH)_2(s)$. In charging, the reverse reaction occurs. (b) $E° = 1.25$ V (c) 1.25 V is the standard cell potential, $E°$. The concentrations of reactants and products inside the battery are adjusted so that the cell output is greater than $E°$. (d) $K = 2 \times 10^{42}$ **20.105** (a) 6.4×10^6 C (b) 6.4×10^5 amp (c) 16 kWh **20.109** (a) $H_2(g)$ is being oxidized and $N_2(g)$ is being reduced. (b) $K = 6.9 \times 10^5$ (c) $E° = 0.05755$ V **20.113** (a) $E° = 0.028$ V (b) anode: $Fe^{2+}(aq) \longrightarrow Fe^{3+}(aq) + 1e^-$; cathode: $Ag^+(aq) + e^- \longrightarrow Ag(s)$ (c) $\Delta S° = 148.5$ J. Since $\Delta S°$ is positive, $\Delta G°$ will become more negative and $E°$ will become more positive as temperature is increased. **20.117** $E° = -0.928$ V

Chapter 21

21.1 (a) ^{24}Ne; outside; reduce neutron-to-proton ratio via β-decay (b) ^{32}Cl; outside; increase neutron-to-proton ratio via positron emission or orbital electron capture (c) ^{108}Sn; outside; increase neutron-to-proton ratio via positron emission or orbital electron capture (d) ^{216}Po; outside; nuclei with $Z \geq 84$ usually decay via α-emission. **21.4** (a) 7 min (b) 0.1 min^{-1} (c) 30% (3/10) of the sample remains after 12 min. (d) $^{88}_{41}Nb$
21.7 (a) 25 protons, 30 neutrons (b) 80 protons, 121 neutrons (c) 19 protons, 20 neutrons **21.9** (a) ^1_1p (b) ^0_1e (c) $^0_{-1}\beta$ or $^0_{-1}e$
21.11 (a) $^{214}_{83}Bi \longrightarrow ^{214}_{84}Po + ^0_{-1}e$
(b) $^{195}_{79}Au + ^0_{-1}e$ (orbital electron) $\longrightarrow ^{195}_{78}Pt$
(c) $^{38}_{19}K \longrightarrow ^{38}_{18}Ar + ^0_1e$ (d) $^{242}_{94}Pu \longrightarrow ^{238}_{92}U + ^4_2He$
21.13 (a) $^{211}_{82}Pb \longrightarrow ^{211}_{83}Bi + ^0_{-1}\beta$ (b) $^{50}_{25}Mn \longrightarrow ^{50}_{42}Cr + ^0_1e$
(c) $^{179}_{74}W + ^0_{-1}e \longrightarrow ^{179}_{73}Ta$ (d) $^{230}_{90}Th \longrightarrow ^{226}_{88}Ra + ^4_2He$
21.15 7 alpha emissions, 4 beta emissions **21.17** (a) Positron emission (for low atomic numbers, positron emission is more common than electron capture) (b) beta emission (c) beta emission (d) beta emission **21.19** (a) Stable: $^{39}_{19}K$, 20 neutrons is a magic number (b) stable: $^{209}_{83}Bi$, 126 neutrons is a magic number (c) stable: $^{25}_{12}Mg$; $^{24}_{10}Ne$ has a much higher neutron/proton ratio **21.21** (a) ^4_2He (c) $^{40}_{20}Ca$ (e) $^{126}_{82}Pb$
21.23 Radioactive: $^{14}_8O$—low neutron/proton ratio; $^{115}_{52}Te$—low neutron/proton ratio; $^{208}_{84}Po$—atomic number ≥ 84
21.25 Protons and alpha particles are positively charged and must be moving very fast to overcome electrostatic forces which would repel them from the target nucleus. Neutrons are electrically neutral and not repelled by the nucleus.
21.27 (a) $^{32}_{16}S + ^1_0n \longrightarrow ^1_1p + ^{32}_{15}P$
(b) $^7_4Be + ^0_{-1}e$ (orbital electron) $\longrightarrow ^7_3Li$
(c) $^{187}_{75}Re \longrightarrow ^{187}_{76}Os + ^0_{-1}e$ (d) $^{98}_{42}Mo + ^2_1H \longrightarrow ^1_0n + ^{99}_{43}Tc$
(e) $^{235}_{92}U + ^1_0n \longrightarrow ^{135}_{54}Xe + ^{99}_{38}Sr + 2\ ^1_0n$
21.29 (a) $^{238}_{92}U + ^1_0n \longrightarrow ^{239}_{92}U + ^0_0\gamma$
(b) $^{14}_7N + ^1_1H \longrightarrow ^{11}_6C + ^4_2He$ (c) $^{18}_8O + ^1_0n \longrightarrow ^{19}_9F + ^0_{-1}e$
21.31 The energy changes involved in chemical reactions are much too small to allow us to alter nuclear properties via chemical processes. The nuclei that are formed in a nuclear reaction will continue to be radioactive regardless of any chemical changes we bring to bear. **21.33** 24.0 mg tritium remain after 12.3 yr, 3.0 mg after 49.2 yr
21.35 $k = 0.0249\ d^{-1}$, $t = 53.9$ d **21.37** (a) 4.1×10^{-12} g ^{226}Ra decays in 1.0 min, 1.1×10^{10} alpha particles emitted (b) 4.9 mCi **21.39** $k = 1.21 \times 10^{-4}\ yr^{-1}$, $t = 2.20 \times 10^3$ yr
21.41 $k = 1.5 \times 10^{-10}\ yr^{-1}$; the original rock contained 66.2 mg ^{238}U and is 1.8×10^9 yr old. **21.43** $\Delta m = 4.378 \times 10^{-9}$ g
21.45 $\Delta m = 0.200287$ amu, $\Delta E = 2.98912 \times 10^{-11}$ J/^{23}Na nucleus required, 1.80009×10^{13} J/mol ^{23}Na
21.47 (a) mass defect = 0.098940 amu,
binding energy/nucleon = 1.2305×10^{-12} J
(b) mass defect = 0.340423 amu,
binding energy/nucleon = 1.37312×10^{-12} J
(c) mass defect = 1.234249 amu,
binding energy/nucleon = 1.344536×10^{-12} J
21.49 (a) 1.71×10^5 kg/d (b) 2.1×10^8 g ^{235}U **21.51** (a) ^{59}Co; it has the largest binding energy per nucleon, and binding energy gives rise to mass defect. **21.53** The ^{59}Fe is incorporated into the diet component and fed to the rabbits. Blood samples are removed from the animals, the red blood cells separated, and the radioactivity of the sample measured. If the iron in the dietary compounds has been incorporated into blood hemoglobin,

the blood cell sample should show beta emission. Samples can be taken at various times to determine the rate of iron uptake, rate of loss of iron from the blood, and so forth. **21.55** (a) Control rods control neutron flux so that there are enough neutrons to sustain the chain reaction but not so many that the core overheats. (b) A moderator slows neutrons so that they are more easily captured by fissioning nuclei. **21.57** (a) 4 $^{1}_{0}n$ (b) $^{94}_{36}Kr$ **21.59** The extremely high temperature is required to overcome electrostatic charge repulsions between the nuclei so that they can come together to react. **21.61** ·OH is a free radical; it can react with almost any particle (atom, molecule, ion) to acquire an electron and become OH^-. This often starts a disruptive chain of reactions, each producing a different free radical. Hydroxide ion, OH^-, on the other hand, will be readily neutralized in the buffered cell environment. The acid-base reactions of OH^- are usually much less disruptive to the organism than the chain of redox reactions initiated by ·OH radical.
21.63 (a) 3.2×10^8 dis/s, 3.2×10^8 Bq (b) 1.5×10^2 mrad, 1.5×10^{-3} Gy (c) 1.5×10^3 mrem, 1.5×10^{-2} Sv **21.65** $^{210}_{82}Pb$
21.67 The most massive radionuclides will have the highest neutron/proton ratios, so they will decay by a process that lowers this ratio, beta emission. The least massive nuclides will decay by a process that increases the neutron/proton ratio, positron emission or orbital electron capture.
21.70

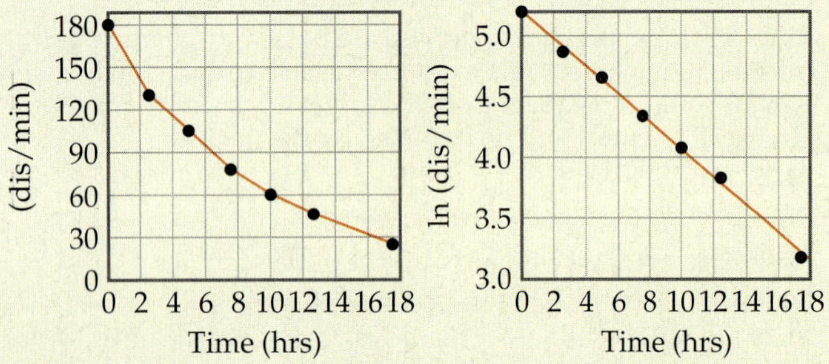

From the curve, $t_{1/2} \approx 6.0$ hr From the slope of the line, $k = 0.11$, $t_{1/2} \approx 6.3$ hr **21.71** 3.7×10^4 disintegration/s; 4.8×10^{13} ^{90}Sr nuclei or 7.2×10^{-9} g ^{90}Sr **21.74** If there is no depletion of iodide due to plant uptake, the calculated activity after 30 days is 13.9 counts/min (to the appropriate numbers of figures, 1×10^1 counts/min). The observed activity, 13.5 counts/min, is slightly lower than the calculated value. The uncertainty in the calculated value is larger than the difference in the two values, so we conclude that little, if any, iodide is absorbed by the plant. **21.76** ^7Be, 8.612×10^{-13} J/nucleon; ^9Be, 1.035×10^{-12} J/nucleon; ^{10}Be: 1.042×10^{-12} J/nucleon. The binding energies/nucleon for ^9Be and ^{10}Be are very similar; that for ^{10}Be is slightly higher. **21.81** 1.3×10^4 kg $C_3H_8(g)$ **21.84** $\lambda = 2.426 \times 10^{-12}$ m. These are short -wavelength gamma ray photons.

Chapter 22

22.2 (a) Acid-base (Brønsted) (b) From left to right in the reaction, the charges are: 0, 0, 1+, 1−
(c) $NH_3(aq) + H_2O(l) \rightleftharpoons NH_4^+(aq) + OH^-(aq)$
22.5 (a) N_2O_5. Resonance structures other than the ones shown are possible. Those with double bonds to the central oxygen do not minimize formal charge and make a smaller contribution to the net bonding model.

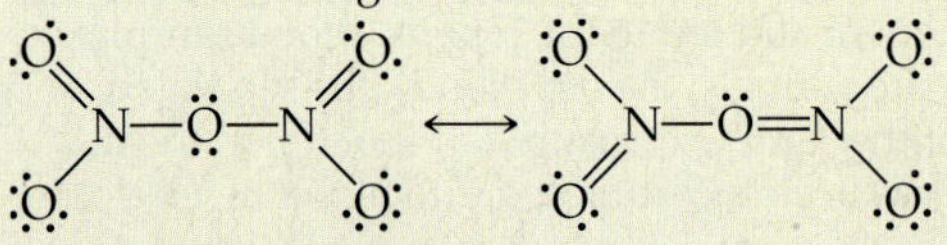

(b) N_2O_4. Resonance structures other than the ones shown are possible.

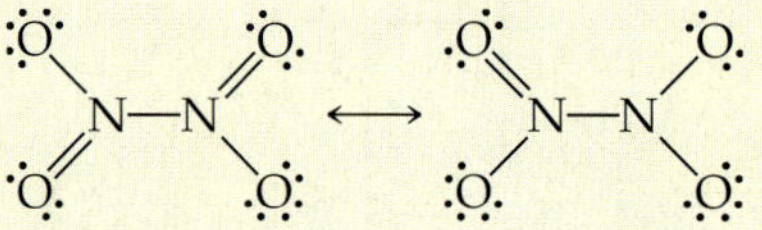

(c) NO_2. We place the odd electron on N because of electronegativity arguments.

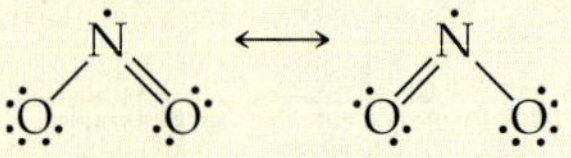

(d) N_2O_3.

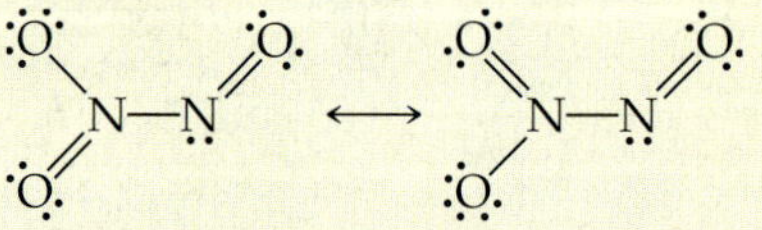

(e) NO. We place the odd electron on N because of electronegativity arguments.

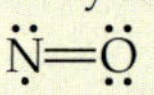

(f) N_2O. The right-most structure does not minimize formal charge and makes a smaller contribution to the net bonding model.

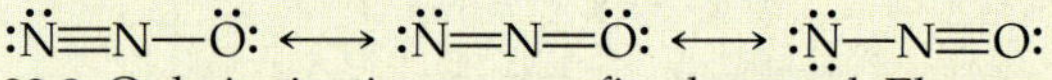

22.8 Only ionization energy fits the trend. Electronegativity varies smoothly, and there is no value for Ar. Both atomic radius and melting point vary in the direction opposite that shown in the diagram. X–X single bond enthalpies show no consistent trend and there should be no value for Ar.
22.11 Metals: (b) Sr, (c) Mn, (e) Rh; nonmetals: (a) P, (d) Se, (f) Kr; metalloids: none. **22.13** (a) O (b) Br (c) Ba (d) O (e) Co **22.15** (a) N is too small a central atom to fit five fluorine atoms, and it does not have available *d* orbitals, which can help accommodate more than eight electrons. (b) Si does not readily form π bonds, which are necessary to satisfy the octet rule for both atoms in the molecule. (c) As has a lower electronegativity than N; that is, it more readily gives up electrons to an acceptor and is more easily oxidized.
22.17 (a) $Mg_3N_2(s) + 6\,H_2O(l) \longrightarrow 2\,NH_3(aq) + 3\,Mg(OH)_2(s)$
(b) $2\,C_3H_7OH(l) + 9\,O_2(g) \longrightarrow 6\,CO_2(g) + 8\,H_2O(l)$
(c) $MnO_2(s) + C(s) \xrightarrow{\Delta} CO(g) + MnO(s)$ or
$MnO_2(s) + 2\,C(s) \xrightarrow{\Delta} 2\,CO(g) + Mn(s)$ or
$MnO_2(s) + C(s) \xrightarrow{\Delta} CO_2(g) + Mn(s)$
(d) $AlP(s) + 3\,H_2O(l) \longrightarrow PH_3(g) + Al(OH)_3(s)$
(e) $Na_2S(s) + 2\,HCl(aq) \longrightarrow H_2S(g) + 2\,NaCl(aq)$
22.19 (a) ^1_1H, protium; ^2_1H, deuterium; ^3_1H, tritium (b) in order of decreasing natural abundance: protium > deuterium > tritium **22.21** Like other elements in group 1A, hydrogen has only one valence electron and its most common oxidation number is +1.
22.23 (a) $Mg(s) + 2\,H^+(aq) \longrightarrow Mg^{2+}(aq) + H_2(g)$
(b) $C(s) + H_2O(g) \xrightarrow{1000°C} CO(g) + H_2(g)$
(c) $CH_4(g) + H_2O(g) \xrightarrow{1100°C} CO(g) + 3\,H_2(g)$
22.25 (a) $NaH(s) + H_2O(l) \longrightarrow NaOH(aq) + H_2(g)$
(b) $Fe(s) + H_2SO_4(aq) \longrightarrow Fe^{2+}(aq) + H_2(g) + SO_4^{2-}(aq)$
(c) $H_2(g) + Br_2(g) \longrightarrow 2\,HBr(g)$
(d) $2\,Na(l) + H_2(g) \longrightarrow 2\,NaH(s)$
(e) $PbO(s) + H_2(g) \xrightarrow{\Delta} Pb(s) + H_2O(g)$
22.27 (a) Ionic (b) molecular (c) metallic **22.29** Xenon has a lower ionization energy than argon; because the valence electrons are not as strongly attracted to the nucleus, they are more readily promoted to a state in which the atom can form bonds

with fluorine. Also, Xe is larger and can more easily accommodate an expanded octet of electrons. **22.31** (a) ClO_3^-, +5 (b) HI, −1 (c) ICl_3; I, +3, Cl, −1 (d) NaOCl, +1 (e) $HClO_4$, +7 (f) XeF_4; Xe, +4, F, −1 **22.33** (a) iron(III) chlorate (b) chlorous acid (c) xenon hexafluoride (d) bromine pentafluoride (e) xenon oxide tetrafluoride (f) iodic acid **22.35** (a) Van der Waals intermolecular attractive forces increase with increasing number of electrons in the atoms. (b) F_2 reacts with water. $F_2(g) + H_2O(l) \longrightarrow 2\,HF(g) + O_2(g)$. That is, fluorine is too strong an oxidizing agent to exist in water. (c) HF has extensive hydrogen bonding. (d) Oxidizing power is related to electronegativity. Electronegativity and oxidizing power decrease in the order given. **22.37** Vehicle fuels produce energy via combustion reactions. The combustion of hydrogen is very exothermic and its only product, H_2O, is a nonpollutant. **22.39** Perchlorate anion, ClO_4^-, must be extremely unreactive in aqueous solutions and aerobic environments. Although chlorine is in a very high oxidation state, it is not readily reduced, because the ion has a stable, symmetric structure that protects it against reactions.

22.41 (a) As an oxidizing agent in steelmaking; to bleach pulp and paper; in oxyacetylene torches; in medicine to assist in breathing (b) synthesis of pharmaceuticals, lubricants, and other organic compounds where C=C bonds are cleaved; in water treatment.

22.43 (a) $2\,HgO(s) \xrightarrow{\Delta} 2\,Hg(l) + O_2(g)$

(b) $2\,Cu(NO_3)_2(s) \xrightarrow{\Delta} 2\,CuO(s) + 4\,NO_2(g) + O_2(g)$

(c) $PbS(s) + 4\,O_3(g) \longrightarrow PbSO_4(s) + 4\,O_2(g)$

(d) $2\,ZnS(s) + 3\,O_2(g) \longrightarrow 2\,ZnO(s) + 2\,SO_2(g)$

(e) $2\,K_2O_2(s) + 2\,CO_2(g) \longrightarrow 2\,K_2CO_3(s) + O_2(g)$

22.45 (a) acidic (b) acidic (c) basic (d) amphoteric **22.47** (a) H_2SeO_3, +4 (b) $KHSO_3$, +4 (c) H_2Te, −2 (d) CS_2, −2 (e) $CaSO_4$, +6

22.49 (a) $2\,Fe^{3+}(aq) + H_2S(aq) \longrightarrow 2\,Fe^{2+}(aq) + S(s) + 2\,H^+(aq)$

(b) $Br_2(l) + H_2S(aq) \longrightarrow 2\,Br^-(aq) + S(s) + 2\,H^+(aq)$

(c) $2\,MnO_4^-(aq) + 6\,H^+(aq) + 5\,H_2S(aq) \longrightarrow 2\,Mn^{2+}(aq) + 5\,S(s) + 8\,H_2O(l)$

(d) $2\,NO_3^-(aq) + H_2S(aq) + 2\,H^+(aq) \longrightarrow 2\,NO_2(aq) + S(s) + 2\,H_2O(l)$

22.51 (a)

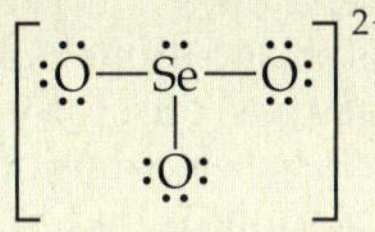

Trigonal pyramidal

(b)

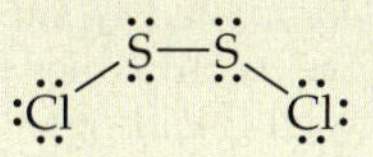

Bent (free rotation around S–S bond)

(c)

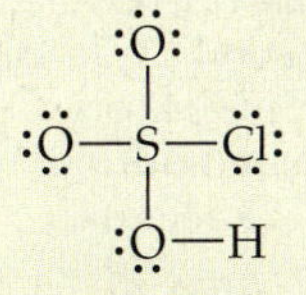

Tetrahedral

22.53 (a) $SO_2(s) + H_2O(l) \rightleftharpoons H_2SO_3(aq) \rightleftharpoons H^+(aq) + HSO_3^-(aq)$

(b) $ZnS(s) + 2\,HCl(aq) \longrightarrow ZnCl_2(aq) + H_2S(g)$

(c) $8\,SO_3^{2-}(aq) + S_8(s) \longrightarrow 8\,S_2O_3^{2-}(aq)$

(d) $SO_3(aq) + H_2SO_4(l) \longrightarrow H_2S_2O_7(l)$

22.55 (a) $NaNO_2$, +3 (b) NH_3, −3 (c) N_2O, +1 (d) NaCN, −3 (e) HNO_3, +5 (f) NO_2, +4

22.57 (a) :Ö=N̈—Ö—H ⟷ :Ö—N̈=Ö—H

The molecule is bent around the central oxygen and nitrogen atoms; the four atoms need not be coplanar. The right-most form does not minimize formal charges and is less important in the actual bonding model.

(b)

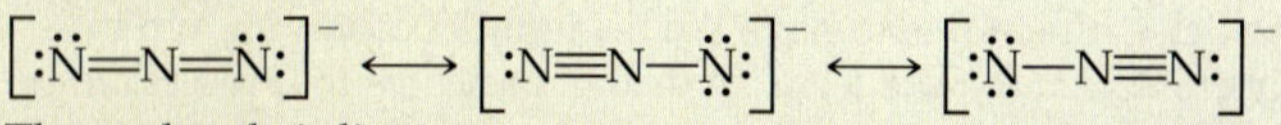

The molecule is linear.

(c)

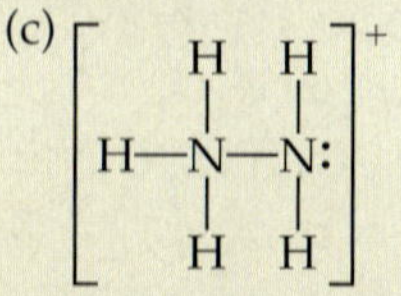

The geometry is tetrahedral around the left nitrogen, trigonal pyramidal around the right.

(d)

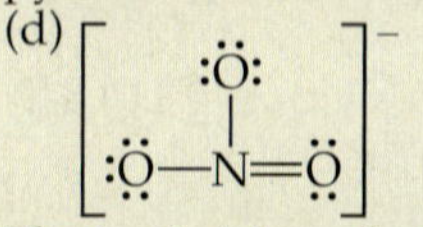

The ion is trigonal planar; it has three equivalent resonance forms.

22.59 (a) $Mg_3N_2(s) + 6\,H_2O(l) \longrightarrow 3\,Mg(OH)_2(s) + 2\,NH_3(aq)$

(b) $2\,NO(g) + O_2(g) \longrightarrow 2\,NO_2(g)$

(c) $N_2O_5(g) + H_2O(l) \longrightarrow 2\,H^+(aq) + 2\,NO_3^-(aq)$

(d) $NH_3(aq) + H^+(aq) \longrightarrow NH_4^+(aq)$

(e) $N_2H_4(l) + O_2(g) \longrightarrow N_2(g) + 2\,H_2O(g)$

22.61 (a) $HNO_2(aq) + H_2O(l) \longrightarrow NO_3^-(aq) + 3\,H^+(aq) + 2e^-,\ E^\circ_{red} = 0.96\ V$

(b) $N_2(g) + H_2O(l) \longrightarrow N_2O(aq) + 2\,H^+(aq) + 2e^-,\ E^\circ_{red} = 1.77\ V$

22.63 (a) H_3PO_3, +3 (b) $H_4P_2O_7$, +5 (c) $SbCl_3$, +3 (d) Mg_3As_2, +5 (e) P_2O_5, +5 **22.65** (a) Phosphorus is a larger atom than nitrogen, and P has energetically available 3*d* orbitals, which participate in the bonding, but nitrogen does not. (b) Only one of the three hydrogens in H_3PO_2 is bonded to oxygen. The other two are bonded directly to phosphorus and are not easily ionized. (c) PH_3 is a weaker base than H_2O so any attempt to add H^+ to PH_3 in the presence of H_2O causes protonation of H_2O. (d) Because of the severely strained bond angles in P_4 molecules, white phosphorus is highly reactive.

22.67 (a) $2\,Ca_3PO_4(s) + 6\,SiO_2(s) + 10\,C(s) \longrightarrow P_4(g) + 6\,CaSiO_3(l) + 10\,CO(g)$

(b) $PBr_3(l) + 3\,H_2O(l) \longrightarrow H_3PO_3(aq) + 3\,HBr(aq)$

(c) $4\,PBr_3(g) + 6\,H_2(g) \longrightarrow P_4(g) + 12\,HBr(g)$

22.69 (a) HCN (b) $Ni(CO)_4$ (c) $Ba(HCO_3)_2$ (d) CaC_2

22.71 (a) $ZnCO_3(s) \xrightarrow{\Delta} ZnO(s) + CO_2(g)$

(b) $BaC_2(s) + 2\,H_2O(l) \longrightarrow Ba^{2+}(aq) + 2\,OH^-(aq) + C_2H_2(g)$

(c) $2\,C_2H_2(g) + 5\,O_2(g) \longrightarrow 4\,CO_2(g) + 2\,H_2O(g)$

(d) $CS_2(g) + 3\,O_2(g) \longrightarrow CO_2(g) + 2\,SO_2(g)$

(e) $Ca(CN)_2(s) + 2\,HBr(aq) \longrightarrow CaBr_2(aq) + 2\,HCN(aq)$

22.73 (a)

$2\,CH_4(g) + 2\,NH_3(g) + 3\,O_2(g) \xrightarrow[cat]{800°C} 2\,HCN(g) + 6\,H_2O(g)$

(b) $NaHCO_3(s) + H^+(aq) \longrightarrow CO_2(g) + H_2O(l) + Na^+(aq)$

(c) $2\,BaCO_3(s) + O_2(g) + 2\,SO_2(g) \longrightarrow 2\,BaSO_4(s) + 2\,CO_2(g)$

22.75 (a) H_3BO_3, +3 (b) $SiBr_4$, +4 (c) $PbCl_2$, +2 (d) $Na_2B_4O_7 \cdot 10\,H_2O$, +3 (e) B_2O_3, +3 **22.77** (a) Lead (b) carbon, silicon, and germanium (c) silicon **22.79** (a) Tetrahedral (b) Metasilicic acid will probably adopt the single-strand silicate chain structure shown in Figure 22.48(a). The Si to O ratio is correct and there are two terminal O atoms per Si that can accomodate the two H atoms associated with each Si atom of the acid. **22.81** (a) Diborane has bridging H atoms linking the two B atoms. The structure of ethane has the C

atoms bound directly, with no bridging atoms. (b) B_2H_6 is an electron-deficient molecule. The 6 valence electron pairs are all involved in B—H sigma bonding, so the only way to satisfy the octet rule at B is to have the bridging H atoms shown in Figure 22.50. (c) The term 'hydridic' indicates that the H atoms in B_2H_6 have more than the usual amount of electron density for a covalently bound H atom. **22.84** (a) 88.2 g H_2 (b) 980 L **22.87** SiH_4, CO, Mg **22.89** (a) SO_3 (b) Cl_2O_5 (c) N_2O_3 (d) CO_2 (e) P_2O_5 **22.95** $GeO_2(s) + C(s) \xrightarrow{\Delta} Ge(l) + CO_2(g)$
$Ge(l) + 2\,Cl_2(g) \longrightarrow GeCl_4(l)$
$GeCl_4(l) + 2\,H_2O(l) \longrightarrow GeO_2(s) + 4\,HCl(g)$
$GeO_2(s) + 2\,H_2(g) \longrightarrow Ge(s) + 2\,H_2O(l)$
22.97 (a) $N_2H_4(aq) + 2\,H_2O_2(aq) \longrightarrow N_2(g) + 4\,H_2O(l)$
(b)
$SO_2(g) + 2\,OH^-(aq) + 2\,H_2O_2(aq) \longrightarrow SO_4^{2-}(aq) + 2\,H_2O(l)$
(c) $NO_2^-(aq) + H_2O_2(aq) \longrightarrow NO_3^-(aq) + H_2O(l)$
(d) $H_2S(g) + H_2O_2(aq) \longrightarrow S(s) + 2\,H_2O(l)$
(e)
$2\,Fe^{2+}(aq) + H_2O_2(aq) + 2\,H^+(aq) \longrightarrow 2\,Fe^{3+}(aq) + 2\,H_2O(l)$
22.99 The average Xe–F bond enthalpy in XeF_2 and XeF_4 is 134 kJ; in XeF_6 it is 130 kJ. These values are remarkably constant in the series. **22.101** [HClO] = 0.036 *M* **22.104** (a) $SO_2(g) + 2\,H_2S(aq) \longrightarrow 3\,S(s) + 2\,H_2O(l)$ or $8\,SO_2(g) + 16\,H_2S(aq) \longrightarrow 3\,S_8(s) + 16\,H_2O(l)$ (b) 2.0×10^3 mol = 5.01×10^4 L $H_2S(g)$ (c) 9.5×10^4 g S = 210 lb S per ton of coal combusted. **22.106** $D(H—O) = 463$ kJ, $D(H—S) = 367$ kJ, $D(H—Se) = 316$ kJ, $D(H—Te) = 266$ kJ. The H—X bond energy decreases steadily in the series, probably due to the increasing size of the orbital from X with which the hydrogen 1*s* orbital must overlap. **22.109** 12 tons N_2O_4

Chapter 23

23.1 (c) Al_2O_3 and Fe_2O_3. The compounds have the same stoichiometry and both metals are in high oxidation states, so they could have been formed in the same environment. Both compounds are inert solids that would survive environmental degradation. **23.5** The most common solid-state structure for metals is one of the close-packed forms because they result in maximum metal-metal contacts, orbital interactions, and delocalization in metallic bonding. **23.9** Iron: hematite, Fe_2O_3; magnetite, Fe_3O_4. Aluminum: bauxite, $Al_2O_3{\cdot}xH_2O$. In ores, iron is present as the 3+ ion or as both the 2+ and 3+ ions as in magnetite. Aluminum is always present in the +3 oxidation state. **23.11** An ore consists of a little bit of the stuff we want (chalcopyrite, $CuFeS_2$) and lots of other junk (gangue).
23.13 (a) $Cr_2O_3(s) + 6\,Na(l) \longrightarrow 2\,Cr(s) + 3\,Na_2O(s)$
(b) $PbCO_3(s) \xrightarrow{\Delta} PbO(s) + CO_2(g)$
(c) $2\,CdS(s) + 3\,O_2(g) \xrightarrow{\Delta} 2\,CdO(s) + 2\,SO_2(g)$
(d) $ZnO(s) + CO(g) \xrightarrow{\Delta} Zn(l) + CO_2(g)$
23.15 (a) $SO_3(g)$ (b) $CO(g)$ provides a reducing environment for the transformation of Pb^{2+} to Pb. (c) $PbSO_4(s) \longrightarrow PbO(s) + SO_3(g)$; $PbO(s) + CO(g) \longrightarrow Pb(s) + CO_2(g)$
23.17 The major reducing agent is CO, formed by partial oxidation of the coke (C) with which the furnace is charged. $Fe_2O_3(s) + 3\,CO(g) \longrightarrow 2\,Fe(l) + 3\,CO_2(g)$; $Fe_3O_4(s) + 4\,CO(g) \longrightarrow 3\,Fe(l) + CO_2(g)$ **23.19** (a) Air serves mainly to oxidize coke to CO; this exothermic reaction also provides heat for the furnace: $2\,C(s) + O_2(g) \longrightarrow 2\,CO(g)$, $\Delta H = -221$ kJ. (b) Limestone, $CaCO_3$, is the source of basic oxide for slag formation: $CaCO_3(s) \xrightarrow{\Delta} CaO(s) + CO_2(g)$; $CaO(l) + SiO_2(l) \longrightarrow CaSiO_3(l)$. (c) Coke is the fuel for the blast furnace and the source of CO, the major reducing agent in the furnace. $2\,C(s) + O_2(g) \longrightarrow 2\,CO(g)$; $4\,CO(g) + Fe_3O_4(s) \longrightarrow 4\,CO_2(g) + 3\,Fe(l)$ (d) Water acts as a source of hydrogen and as a means of controlling temperature. $C(s) + H_2O(g) \longrightarrow CO(g) + H_2(g)$, $\Delta H = +131$ kJ **23.21** To purify crude cobalt electrochemically, use an electrolysis cell in which the crude metal is the anode, a thin sheet of pure cobalt is the cathode, and the electrolyte is an aqueous solution of a soluble cobalt salt such as $CoSO_4{\cdot}7\,H_2O$. Reduction of water doesn't occur, because of kinetic effects. Anode reaction: $Co(s) \longrightarrow Co^{2+}(aq) + 2e^-$; cathode reaction: $Co^{2+}(aq) + 2e^- \longrightarrow Co(s)$. **23.23** Sodium is metallic; each atom is bonded to many others. When the metal lattice is distorted, many bonds remain intact. In NaCl the ionic forces are strong, and the ions are arranged in very regular arrays. The ionic forces tend to be broken along certain cleavage planes in the solid, and the substance does not tolerate much distortion before cleaving. **23.25** In the electron-sea model, valence electrons move about the metallic lattice, while metal atoms remain more or less fixed in position. Under the influence of an applied potential, the electrons are free to move throughout the structure, giving rise to thermal and electrical conductivity. **23.27** The variation in densities indicates shorter metal-metal bond distances moving from left to right in the series. It seems that the extent of metal-metal bonding increases in the series, which is consistent with greater occupancy of the bonding band as the number of valence electrons increases up to 6. **23.29** (a) K (b) Zn. Ductility would be lower as the number of bonding electrons per atom increases, producing a stiffer lattice, less susceptible to distortion. **23.31** White tin is more metallic in character; it has a higher conductivity and a longer Sn–Sn distance (3.02 Å) than gray tin (2.81 Å). **23.33** An *alloy* contains atoms of more than one element and has the properties of a metal. In a solution alloy the components are randomly dispersed. In a heterogeneous alloy the components are not evenly dispersed and can be distinguished at a macroscopic level. In an intermetallic compound the components have interacted to form a compound substance, as in Cu_3As. **23.35** Isolated atoms: (b) and (f); bulk metal: (a), (c), (d), and (e). Although it seems that atomic radius is a property of isolated atoms, it can only be measured on a bulk sample. **23.37** Hf has a completed 4*f* subshell, while Zr does not. The build-up in Z that accompanies the filling of the 4*f* orbitals, along with additional shielding of valence electrons, offsets the typical effect of the larger *n* value for the valence electrons of Hf. Thus, the atomic radius of Hf is about the same as that of Zr, the element above it in Group 4B. **23.39** (a) ScF_3 (b) CoF_3 (c) ZnF_2 (d) MoF_6 **23.41** Chromium, $[Ar]4s^13d^5$, has six valence electrons, some or all of which can be involved in bonding, leading to multiple stable oxidation states. Al, $[Ne]3s^23p^1$, has only three valence electrons, which are all lost or shared during bonding, producing the +3 state exclusively. **23.43** (a) Cr^{3+}, $[Ar]3d^3$ (b) Au^{3+}, $[Xe]4f^{14}5d^8$ (c) Ru^{2+}, $[Kr]4d^6$ (d) Cu^+, $[Ar]3d^{10}$ (e) Mn^{4+}, $[Ar]3d^3$ (f) Ir^+, $[Xe]4f^{14}5d^8$ **23.45** Ti^{2+}
23.47 Fe^{2+} is a reducing agent that is readily oxidized to Fe^{3+} in the presence of O_2 from air.
23.49 (a) $Fe(s) + 2\,HCl(aq) \longrightarrow FeCl_2(aq) + H_2(g)$
(b) $Fe(s) + 4\,HNO_3(g) \longrightarrow Fe(NO_3)_3(aq) + NO(g) + 2\,H_2O(l)$
23.51 The unpaired electrons in a paramagnetic material cause it to be weakly attracted into a magnetic field. A diamagnetic material, where all electrons are paired, is very weakly repelled by a magnetic field. **23.53** $PbS(s) + O_2(g) \longrightarrow Pb(l) + SO_2(g)$. $SO_2(g)$ is a product of roasting sulfide ores. In an oxygen-rich environment, $SO_2(g)$ is oxidized to $SO_3(g)$, which dissolves in

$H_2O(l)$ to form sulfuric acid, $H_2SO_4(aq)$. A sulfuric acid plant near a roasting plant would provide a means for disposing of hazardous $SO_2(g)$ that would also generate a profit. **23.55** $CO(g)$: $Pb(s)$; $H_2(g)$: $Fe(s)$; $Zn(s)$: $Au(s)$ **23.58** Because selenium and tellurium are both nonmetals, we expect them to be difficult to oxidize. Both Se and Te are likely to accumulate as the free elements in the so-called anode slime, along with noble metals that are not oxidized. **23.61** (a) The low boiling point indicates a covalent molecule. Nonbonded electron pairs on a metal are unlikely, so VSEPR predicts a tetrahedral structure. (b) The +8 oxidation state required for a tetraoxide is not common for Fe. Also, in a reversal of trend, the electronegativity difference between Fe and O is greater than that between Os and O, so the iron oxides are more likely to be ionic. **23.64** $E°$ will become more negative as the stability (K_f value) of the complex increases.
23.67 (a) $Mn(s) + 2\,HNO_3(aq) \longrightarrow Mn(NO_3)_2(aq) + H_2(g)$
(b) $Mn(NO_3)_2(s) \xrightarrow{\Delta} MnO_2(s) + 2\,NO_2(g)$
(c) $MnO_2(s) \xrightarrow{\Delta} Mn_3O_4(s) + O_2(g)$
(d) $2\,MnCl_2(s) + 9\,F_2(g) \longrightarrow 2\,MnF_3(s) + 4\,ClF_3(g)$
23.70 (a) Insulator (b) semiconductor (c) metallic conductor (d) metallic conductor (e) insulator (f) metallic conductor **23.71** (a) 6.7×10^8 g SO_2 (b) 2.3 mol Cu/mol Fe (c) CuO and Fe_2O_3 (d) $Cu_2S(s) + 2\,O_2(g) \longrightarrow 2\,CuO(s) + SO_2(g)$; $4\,FeS(s) + 7\,O_2(g) \longrightarrow 2\,Fe_2O_3(s) + 4\,SO_2(g)$ **23.73** 7.3×10^2 g $Ni^{2+}(aq)$ **23.76** $K = 1.6 \times 10^6$ **23.78** (a) The standard reduction potential for $H_2O(l)$ is much greater than that of $Mg^{2+}(aq)$ (−0.83 V vs. −2.37 V). In aqueous solution $H_2O(l)$ would be preferentially reduced and no $Mg(s)$ would be obtained. (b) 1.0×10^3 kg Mg

Chapter 24

24.1 (a)

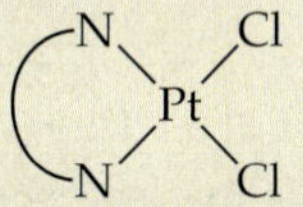

(b) coordination number = 4, coordination geometry = square planar (c) oxidation number = +2 **24.4** Structures (3) and (4) are identical to (1); (2) and (5) are geometric isomers of (1). **24.6** The yellow-orange solution absorbs blue-violet; the blue-green solution absorbs orange-red. **24.9** (a) A *metal complex* consists of a central metal ion bonded to a number of surrounding molecules or ions. The number of bonds formed by the central metal ion is the *coordination number*. The surrounding molecules or ions are the *ligands*. (b) Metal ions, by virtue of their positive charge and empty *d*, *s*, or *p* orbitals, act as electron-pair acceptors, or Lewis acids. Ligands, which have at least one unshared electron pair, act as electron-pair donors, or Lewis acids. **24.11** (a) +2 (b) 6 (c) 2 mol AgBr(*s*) will precipitate per mole of complex. **24.13** (a) Coordination number = 4, oxidation number = +2 (b) 5, +4 (c) 6, +3 (d) 5, +2 (e) 6, +3 (f) 4, +2 **24.15** (a) 4 Cl^- (b) 4 Cl^-, 1 O^{2-} (c) 4 N, 2 Cl^- (d) 5 C (e) 6 O (f) 4 N **24.17** (a) A monodentate ligand binds to a metal via one atom, a bidentate ligand through two atoms. (b) Three bidentate ligands fill the coordination sphere of a six-coordinate complex. (c) A tridentate ligand has at least three atoms with unshared electron pairs in the correct orientation to simultaneously bind one or more metal ions. **24.19** (a) *Ortho*-phenanthroline, *o*-phen, is bidentate (b) oxalate, $C_2O_4^{2-}$, is bidentate (c) ethylenediaminetetraacetate, EDTA, is pentadentate (d) ethylenediamine, en, is bidentate **24.21** (a) The term *chelate effect* refers to the special stability associated with formation of a metal complex containing a polydentate (chelate) ligand relative to a complex containing only monodentate ligands. (b) The increase in entropy, $+\Delta S$, associated with the substitution of a chelating ligand for two or more monodentate ligands generally gives rise to the *chelate effect*. Chemical reactions with $+\Delta S$ tend to be spontaneous, have negative ΔG and large positive values of K. (c) Polydentate ligands are used as *sequestering agents* to bind metal ions and prevent them from undergoing unwanted chemical reactions without removing them from solution.
24.23 (a) $[Cr(NH_3)_6](NO_3)_3$ (b) $[Co(NH_3)_4CO_3]_2SO_4$ (c) $[Pt(en)_2Cl_2]Br_2$ (d) $K[V(H_2O)_2Br_4]$ (e) $[Zn(en)_2][HgI_4]$
24.25 (a) tetraaminedichlororhodium(III) chloride (b) potassium hexachlorotitanate(IV) (c) tetrachlorooxomolybdenum(VI) (d) tetraaqua(oxalato)platinum(IV) bromide
24.27 (a)

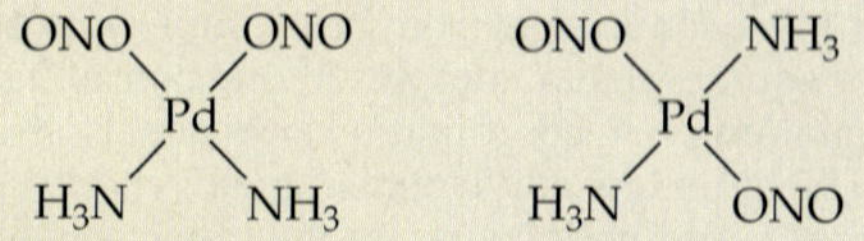

cis trans

(b) $[Pd(NH_3)_2(ONO)_2]$, $[Pd(NH_3)_2(NO_2)_2]$
(c)

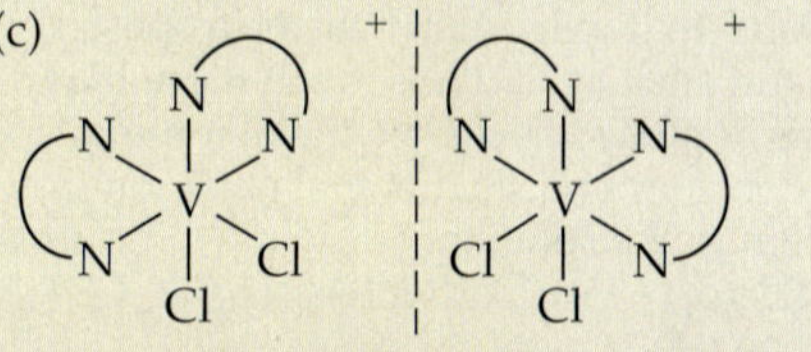

(d) $[Co(NH_3)_4Br_2]Cl$, $[Co(NH_3)_4BrCl]Br$ **24.29** Yes. No structural or stereoisomers are possible for a tetrahedral complex of the form MA_2B_2. The complex must be square planar with cis and trans geometric isomers. **24.31** In the trans isomer the Cl—Co—NH_3 bond angle is 180°; in the cis isomer this bond angle is 90°. The cis isomer is chiral. **24.33** (a) One isomer (b) trans and cis isomers with 180° and 90° Cl—Ir—Cl angles, respectively (c) trans and cis isomers with 180° and 90° Cl—Fe—Cl angles, respectively. The cis isomer is optically active. **24.35** (a) Visible light has wavelengths between 400 and 700 nm. (b) *Complementary* colors are opposite each other on an artist's color wheel. (c) A colored metal complex absorbs visible light of its complementary color. (d) 196 kJ/mol
24.37 Most of the attraction between a metal ion and a ligand is electrostatic. Whether the interaction is ion-ion or ion-dipole, the ligand is strongly attracted to the metal center and can be modeled as a point negative charge.
24.39 (a)

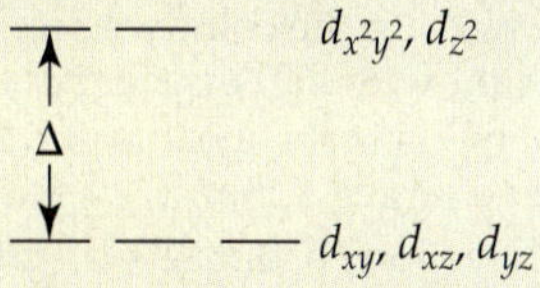

(b) The magnitude of Δ and the energy of the *d-d* transition for a d^1 complex are equal. (c) Δ = 203 kJ/mol **24.41** A yellow color is due to absorption of light around 400 to 430 nm, a blue color to absorption near 620 nm. The shorter wavelength corresponds to a higher-energy electron transition and larger Δ value. Cyanide is a stronger-field ligand, and its complexes are expected to have larger Δ values than aqua complexes.
24.43 (a) Ti^{3+}, d^1 (b) Co^{3+}, d^6 (c) Ru^{3+}, d^5 (d) Mo^{5+}, d^1 (e) Re^{3+}, d^4 **24.45** (a) Mn, $[Ar]4s^23d^5$; Mn^{2+}, $[Ar]3d^5$; 1 unpaired electron (b) Ru, $[Kr]5s^14d^7$; Ru^{2+}, $[Kr]4d^6$; 0 unpaired electrons (c) Rh, $[Kr]5s^14d^8$; Rh^{2+}, $[Kr]4d^7$; 1 unpaired electron **24.47** All complexes in this exercise are six-coordinate octahedral.

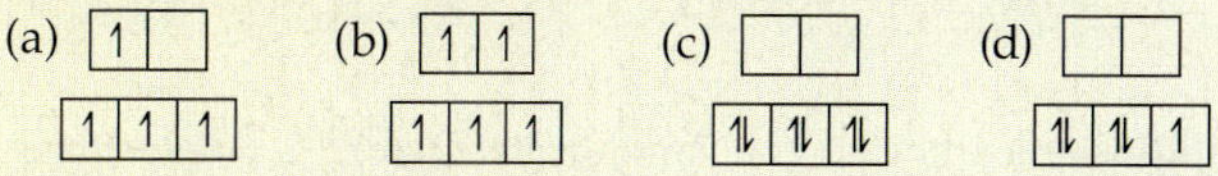

d^4, high spin d^5, high spin d^6, low spin d^5, low spin

(e) (f) 24.49

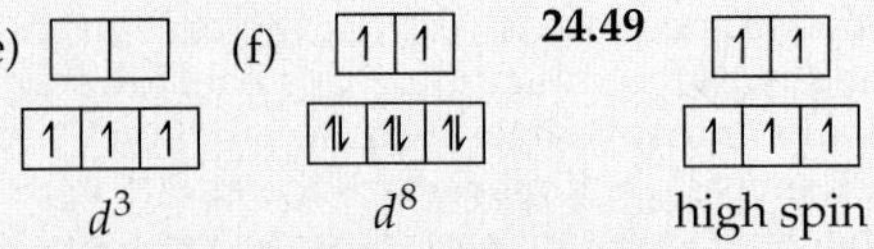

24.51 $[Pt(NH_3)_6]Cl_4$; $[Pt(NH_3)_4Cl_2]Cl_2$; $[Pt(NH_3)_3Cl_3]Cl$; $[Pt(NH_3)_2Cl_4]$; $K[Pt(NH_3)Cl_5]$ **24.54** (a) [24.53(a)] *cis*-tetraamminediaquacobalt(II) nitrate; [24.53(b)] sodium aquapentachlororuthenate(III); [24.53(c)] ammonium *trans*-diaquabisoxalatocobaltate(III); [24.53(d)] *cis*-dichlorobisethylenediamineruthenium(II) (b) Only the complex in 24.53(d) is optically active. The mirror images of (a)–(c) can be superimposed on the original structure. The chelating ligands in (d) prevent its enantiomers from being superimposable. **24.56** (a) In a square-planar complex, if one pair of ligands is trans, the remaining two coordination sites are also trans to each other. The bidentate ethylenediamine ligand is too short to occupy trans coordination sites, so the trans isomer of $[Pt(en)Cl_2]$ is unknown. (b) The minimum steric requirement for a trans-bidentate ligand is a medium-length chain between the two coordinating atoms that will occupy the trans positions. A polydentate ligand such as EDTA is much more likely to occupy trans positions because it locks the metal ion in place with multiple coordination sites and shields the metal ion from competing ligands present in the solution.
24.59 (a) $AgCl(s) + 2\,NH_3(aq) \longrightarrow [Ag(NH_3)_2]^+(aq) + Cl^-(aq)$
(b) $[Cr(en)_2Cl_2]Cl(aq) + 2\,H_2O(l) \longrightarrow [Cr(en)_2(H_2O)_2]^{3+}(aq) + 3\,Cl^-(aq)$;
$3\,Ag^+(aq) + 3\,Cl^-(aq) \longrightarrow 3\,AgCl(s)$ (c)
$Zn(OH_3)_2(aq) + 2\,NaOH(aq) \longrightarrow Zn(OH)_2(s) + 2\,NaNO_3(aq)$;
$Zn(OH)_2(s) + 2\,NaOH(aq) \longrightarrow [Zn(OH)_4]^{2-}(aq) + 2\,Na^+(aq)$
(d) $Co^{2+}(aq) + 4\,Cl^-(aq) \longrightarrow [CoCl_4]^{2-}(aq)$
24.62 (a)

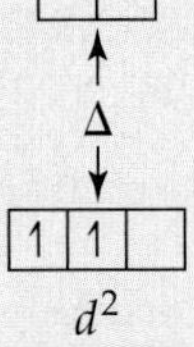

(b) Visible light with $\lambda = hc/\Delta$ is absorbed by the complex, promoting one of the *d* electrons into a higher energy *d* orbital. The remaining wavelengths are reflected or transmitted; the combination of these wavelengths is the color we see. (c) $[V(H_2O)_6]^{3+}$ will absorb light with higher energy because it has a larger Δ than $[VF_6]^{3-}$. H_2O is in the middle of the spectrochemical series and causes a larger Δ than F^-, a weak-field ligand. **24.64** $[Co(NH_3)_6]^{3+}$, yellow; $[Co(H_2O)_6]^{2+}$, pink; $[CoCl_4]^{2-}$, blue **24.66** (a) $[FeF_6]^{4-}$. F^- is a weak-field ligand that imposes a smaller Δ and longer λ for the complex ion. (b) $[V(H_2O)_6]^{2+}$. V^{2+} has a lower charge, so the interaction with the ligand will produce a weaker field, a smaller Δ, and a longer absorbed wavelength. (c) $[CoCl_4]^{2-}$. Cl^- is a weak-field ligand that imposes a smaller Δ and a longer λ for the complex ion.

24.69 (a)

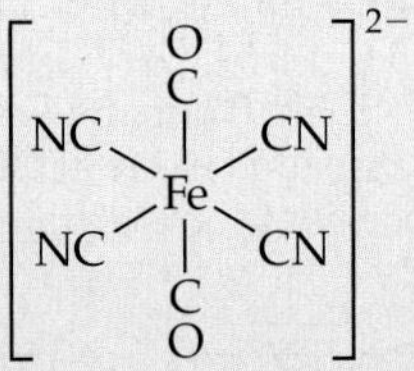

(b) sodium dicarbonyltetracyanoferrate(II) (c) +2, 6*d* electrons (d) We expect the complex to be low spin. Cyanide (and carbonyl) are high on the spectrochemical series, which means the complex will have a large Δ splitting, characteristic of low-spin complexes. **24.73** In carbonic anhydrase, Zn^{2+} is the Lewis acid, withdrawing electron density from H_2O, the Lewis base. The O—H bond is polarized and H becomes more ionizable, more acidic than the bulk solvent. **24.75** $K_4[Mn(ox)_2Br_2]$
24.78 47.3 mg Mg^{2+}/L, 53.4 mg Ca^{2+}/L
24.81 $\Delta E = 3.02 \times 10^{-19}$ J/photon, $\lambda = 657$ nm. The complex will absorb in the visible around 660 nm and appear blue-green.

Chapter 25

25.2 Molecule (c) **25.5** (a) Molecule (i), disaccharide (b) molecule (iii), amino acid (c) molecule (ii), organic base (d) molecule (iv), alcohol **25.7** The carbon atoms with two and three hydrogens both have tetrahedral electron-domain geometry, 109° bond angles, and sp^3 hybridization. The carbon atom with one hydrogen has trigonal-planar electron-domain geometry, 120° bond angles, and sp^2 hybridization. **25.9** Carbon, hydrogen, oxygen, nitrogen, sulfur, phosphorus, fluorine, and chlorine. Oxygen, nitrogen, fluorine, and chlorine are more electronegative than carbon; sulfur has the same electronegativity as carbon.
25.11 (a) A straight-chain hydrocarbon has all carbon atoms connected in a continuous chain. A branched-chain hydrocarbon has a branch; at least one carbon atom is bound to three or more carbon atoms. (b) An alkane is a complete molecule composed of carbon and hydrogen in which all bonds are σ bonds. An alkyl group is a substituent formed by removing a hydrogen atom from an alkane. (c) Alkanes are said to be saturated because they cannot undergo addition reactions, such as those characteristic of carbon-carbon double bonds. **25.13** (a) C_5H_{12} (b) C_5H_{10} (c) C_5H_{10} (d) C_5H_8; saturated: (a), (b); unsaturated: (c), (d) **25.15** C_nH_{2n-2}
25.17

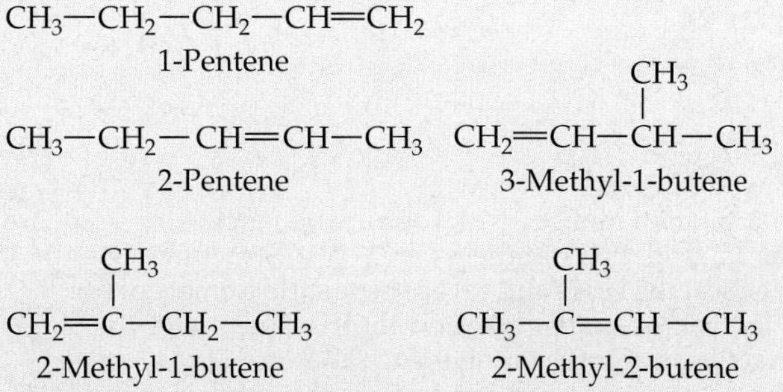

25.19 (a) 109° (b) 120° (c) 180° **25.21** (a) 2-methylhexane (b) 4-ethyl-2,4-dimethyldecane

(c) 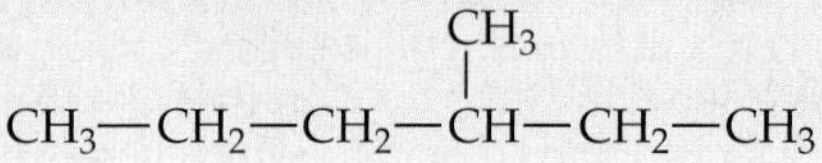

(d) 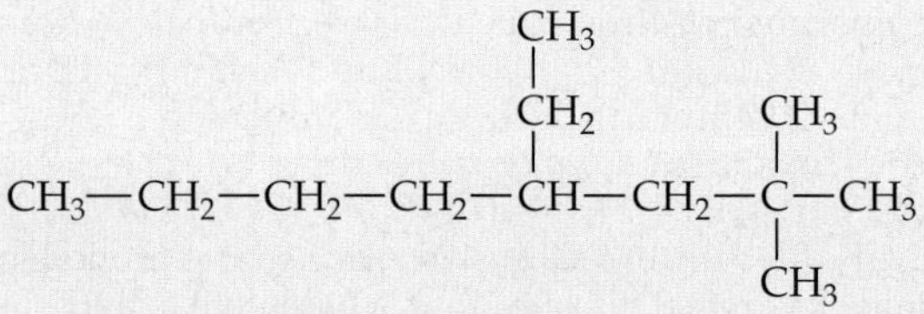

(e) 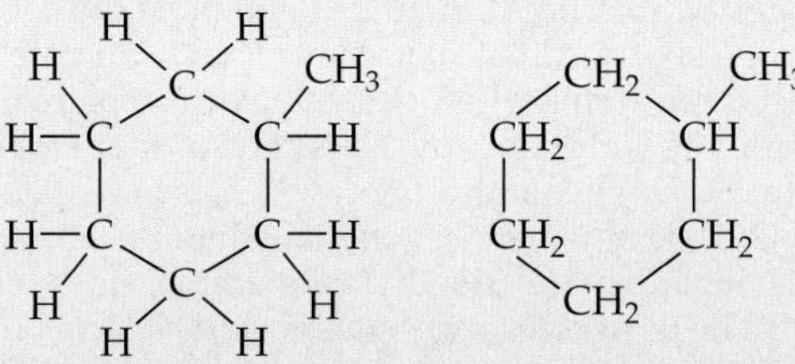

25.23 (a) 2,3-dimethylheptane (b) *cis*-6-methyl-3-octene (c) *para*-dibromobenzene (d) 4,4-dimethyl-1-hexyne

(e) methylcyclobutane **25.25** Geometric isomerism in alkenes is the result of restricted rotation about the double bond. In alkanes bonding sites are interchangeable by free rotation about the C—C single bonds. In alkynes there is only one additional bonding site on a triply bound carbon, so no isomerism results. **25.27** (a) No

(b)

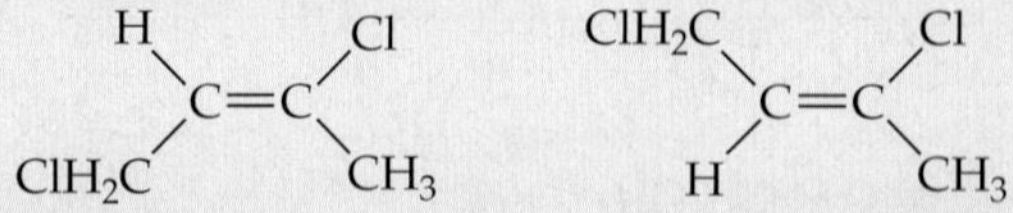

(c) no (d) no **25.29** 65 **25.31** (a) An addition reaction is the addition of some reagent to the two atoms that form a multiple bond. In a substitution reaction one atom or group of atoms replaces another atom. Alkenes typically undergo addition, while aromatic hydrocarbons usually undergo substitution.

(b)

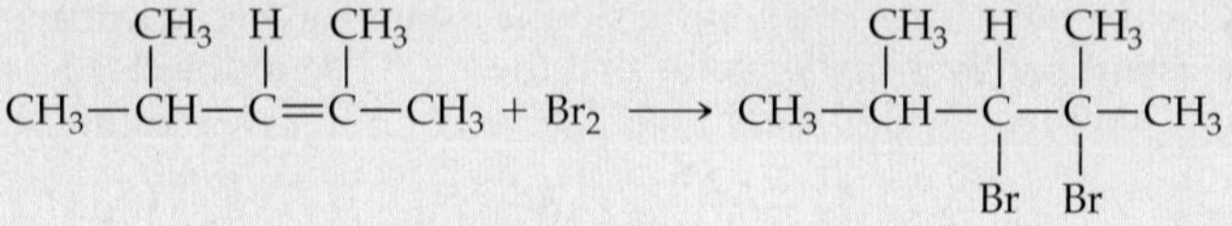

(c)

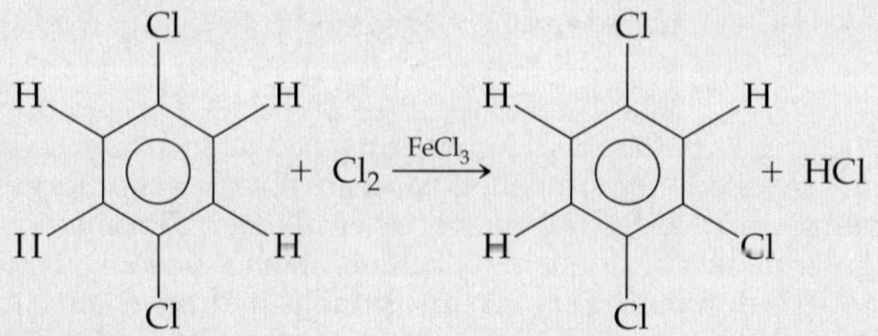

25.33 (a) The 60° C—C—C angles in the cyclopropane ring cause strain that provides a driving force for reactions that result in ring opening. There is no comparable strain in the five- or six-membered rings.

(b) $C_2H_4(g) + HBr(g) \longrightarrow CH_3CH_2Br(l)$;

$C_6H_6(l) + CH_3CH_2Br(l) \xrightarrow{AlCl_3} C_6H_5CH_2CH_3(l) + HBr(g)$

25.35 The intermediate is a carbocation. One of the C atoms formerly involved in the π bond is now bound to a second H atom from the hydrogen halide. The other C atom formerly involved in the π bond carries a full positive charge and forms only three sigma bonds, two to adjacent C atoms and one to H. **25.37** ΔH_{comb}/mol CH_2 for cyclopropane = 696.3 kJ, for cyclopentane = 663.4 kJ. $\Delta H_{comb}/CH_2$ group for cyclopropane is greater because C_3H_6 contains a strained ring. When combustion occurs, the strain is relieved and the stored energy is released. **25.39** (a) Ketone (b) carboxylic acid (c) alcohol (d) ester (e) amide (f) amine **25.41** (a) Propionaldehyde (or propanal):

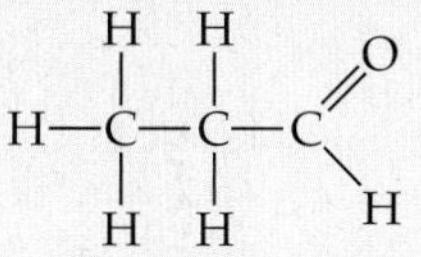

(b) ethylmethyl ether:

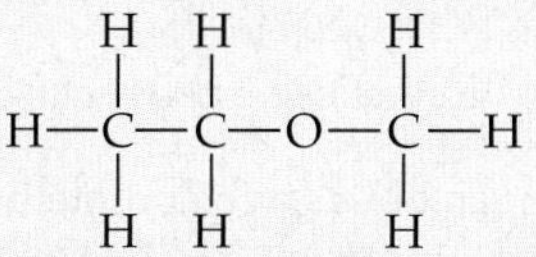

25.43 (a) Methanoic acid (b) butanoic acid (c) 3-methylpentanoic acid

25.45 (a)

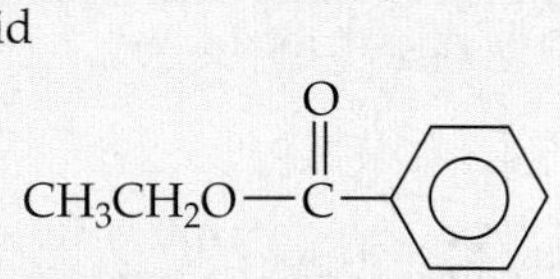

Ethylbenzoate

(b)

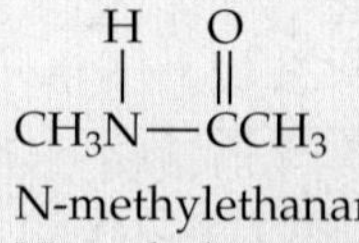

N-methylethanamide or N-methylacetamide

(c)

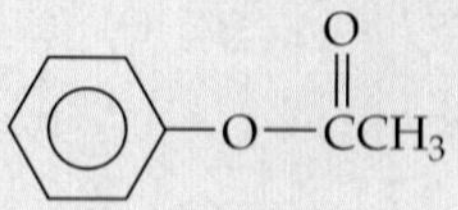

Phenylacetate

25.47

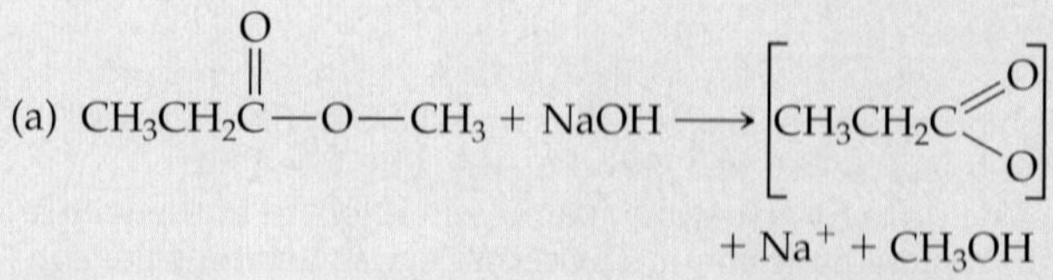

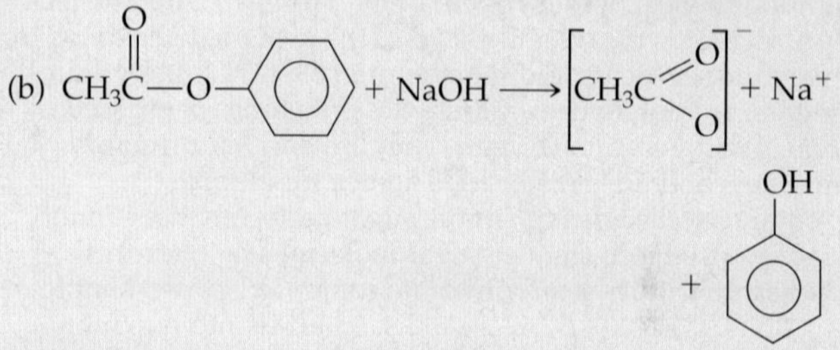

25.49 The presence of both —OH and —C=O groups in pure acetic acid leads us to conclude that it will be a strongly hydrogen-bonded substance. That the melting and boiling points of pure acetic acid are both higher than those of water, a substance we know to be strongly hydrogen-bonded, supports this conclusion.

25.51

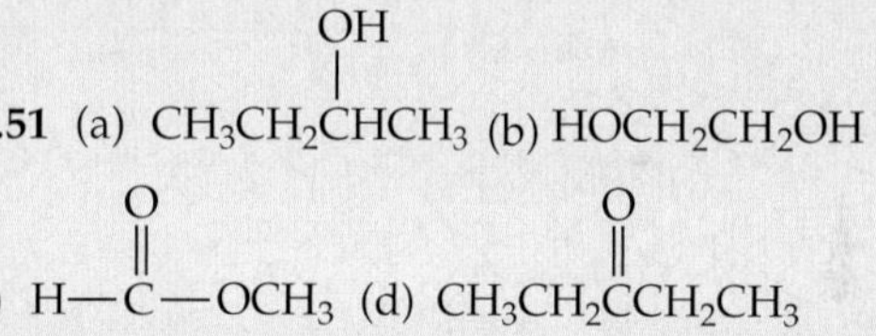

(e) $CH_3CH_2OCH_2CH_3$

25.53

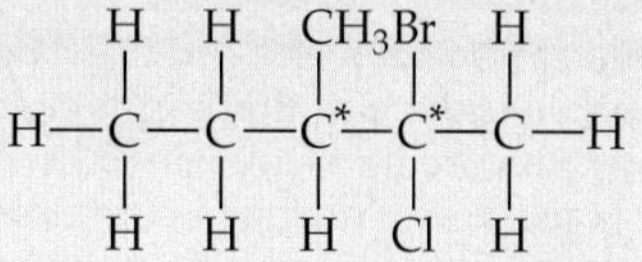

*chiral C atoms

25.55 (a) An α-amino acid contains an NH_2 group attached to the carbon adjacent to the carboxcylic acid function. (b) In protein formation, amino acids undergo a condensation reaction between the amino group of one molecule and the carboxylic acid group of another to form the amide linkage.

25.57

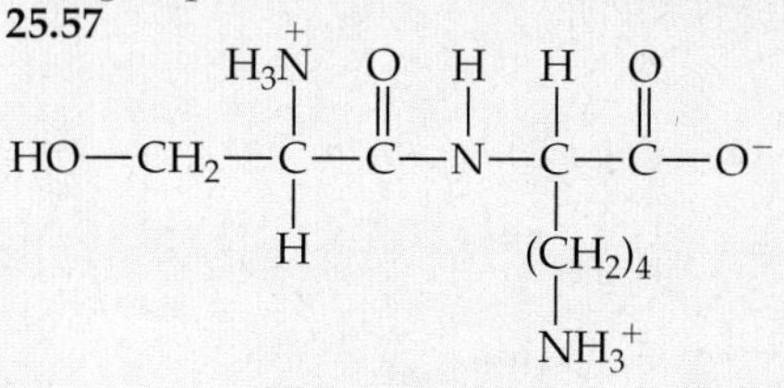

Seryllysine

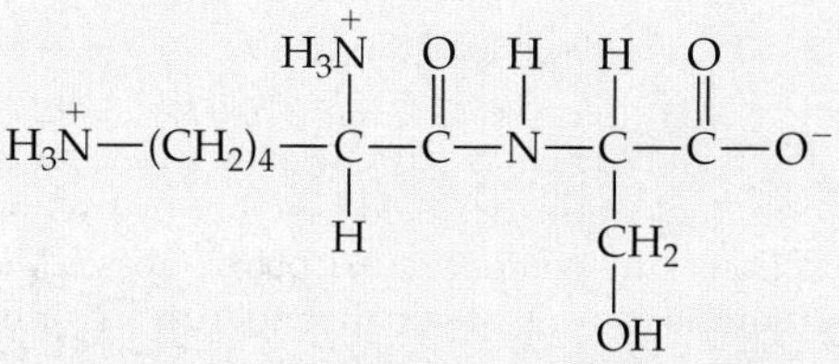

Lysylserine

25.59 (a)

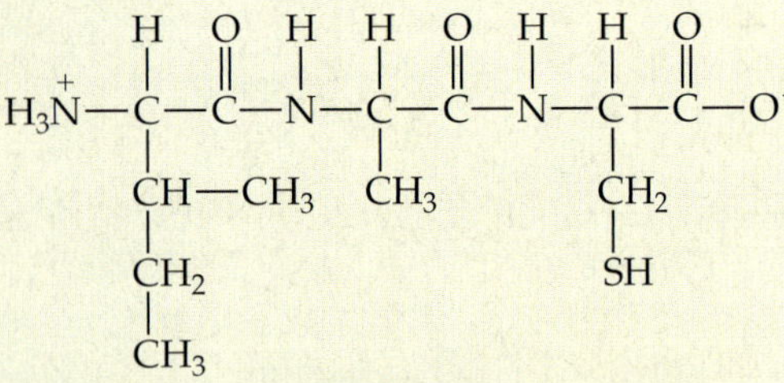

Ile-Ala-Cys

(b) Eight: Ser-Ser-Ser; Ser-Ser-Phe; Ser-Phe-Ser; Phe-Ser-Ser; Ser-Phe-Phe; Phe-Ser-Phe; Phe-Phe-Ser; Phe-Phe-Phe **25.61** The primary structure of a protein refers to the sequence of amino acids in the chain. The secondary structure is the configuration (helical, folded, open) of the protein chain. The tertiary structure is the overall shape of the protein determined by the way the segments fold together **25.63** (a) Carbohydrates, or sugars, are polyhydroxyaldehydes or ketones composed of carbon, hydrogen, and oxygen. They are derived primarily from plants and are a major food source for animals.
(b) A monosaccharide is a simple sugar molecule that cannot be decomposed into smaller sugar molecules by hydrolysis.
(c) A disaccharide is a carbohydrate composed of two simple sugar units. Hydrolysis breaks a disaccharide into two monosaccharides. **25.65** (a) In the linear form of galactose, the aldehydic carbon is C1. Carbon atoms 2, 3, 4, and 5 are chiral because they each carry four different groups. (b) Both the β form (shown here) and the α form (OH on carbon 1 on same side of ring as OH on carbon 2) are possible.

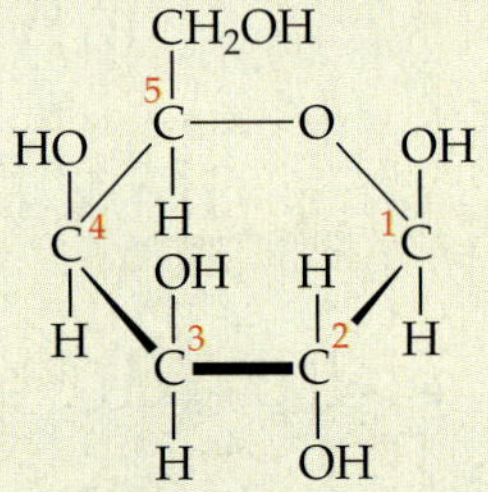

Galactose

25.67 The empirical formula of glycogen is $C_6H_{10}O_5$. The six-membered ring form of glucose is the unit that forms the basis of glycogen. The monomeric glucose units are joined by α linkages. **25.69** A nucleotide consists of a nitrogen-containing aromatic compound, a sugar in the furanose (five-membered) ring form, and a phosphoric acid group. The structure of deoxycytidine monophosphate is

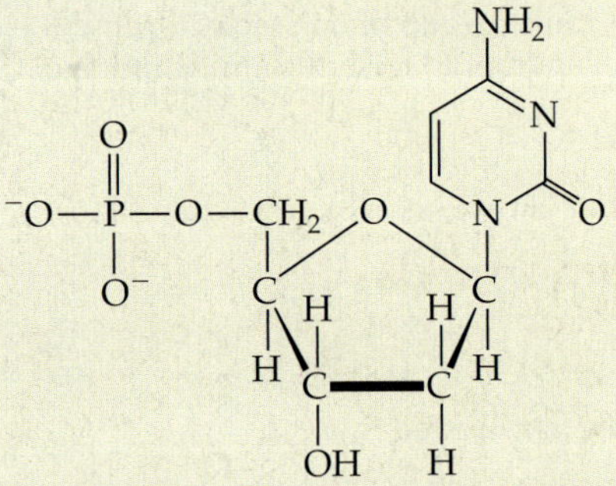

25.71 $C_4H_7O_3CH_2OH + HPO_4^{2-} \longrightarrow C_4H_7O_3CH_2{-}O{-}PO_3^{2-} + H_2O$

25.73 In the helical structure for DNA, the strands of the polynucleotides are held together by hydrogen-bonding interactions between particular pairs of bases. Adenine and thymine are one base pair, and guanine and cytosine are another. For each adenine in one strand, there is a cytosine in the other, so that total adenine equals total thymine, and total guanine equals total cytosine.

25.75

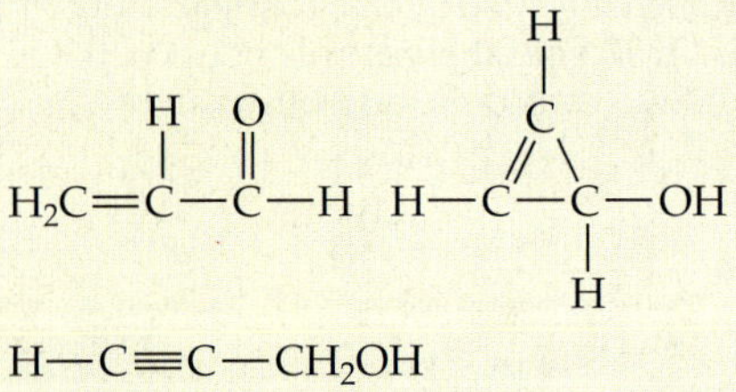

25.78

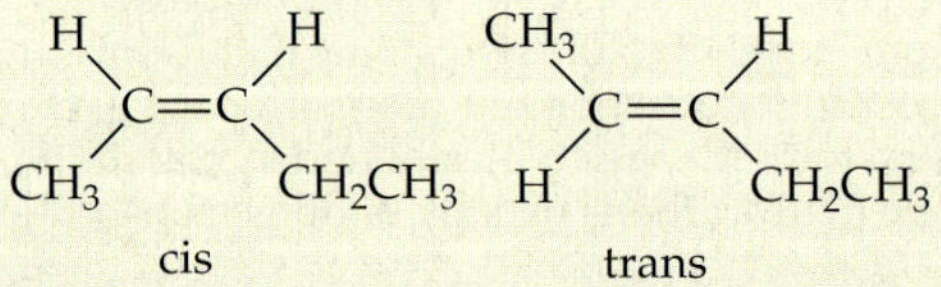

Cyclopentene does not show cis-trans isomerism because the existence of the ring demands that the C—C bonds be cis to one another.

25.81

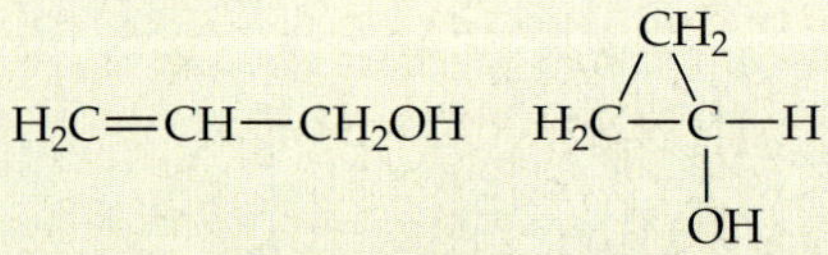

(Structures with the —OH group attached to an alkene carbon atom are called "vinyl alcohols" and are not the major form at equilibrium.) **25.84** (a) Alkene, ether, alkene (b) aromatic ring, carboxylic acid, ester (c) ketone, alkene, alcohol
25.86 The difference between an alcoholic hydrogen and a carboxylic acid hydrogen is two-fold. In a carboxylic acid the electronegative carbonyl oxygen withdraws electron density from the O—H bond, rendering the bond more polar and the H more ionizable. Also, the conjugate base of a carboxylic acid is stabilized by resonance, which encourages ionization of the carboxylic acid. In an alcohol no electronegative atoms are bound to the carbon that holds the —OH group, there is no resonance stabilization of the conjugate base, and H is tightly bound to O. **25.89** (a) None (b) The carbon bearing the secondary —OH has four different groups attached, and is thus chiral. (c) The carbon bearing the $—NH_2$ group and the carbon bearing the CH_3 group are both chiral. **25.91** Glu-Cys-Gly is the only possible order. Glutamic acid has two carboxyl groups that can form a peptide bond with cysteine, so there are two possible structures for glutathione.

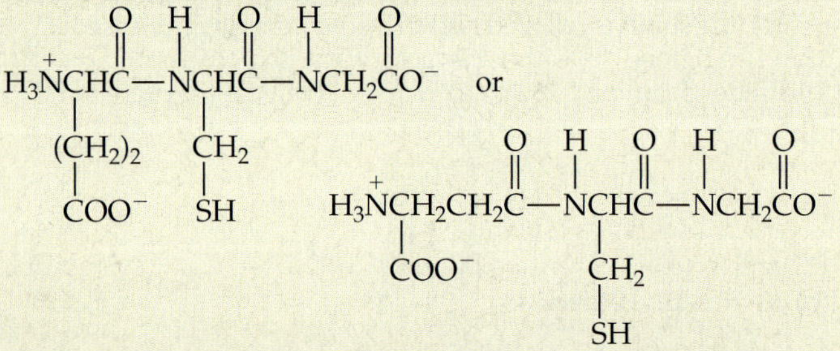

25.93 Glucose is an aldohexose, and fructose is a ketohexose.
25.95 Ethanol contains —O—H bonds, which form strong intermolecular hydrogen bonds, while dimethyl ether experiences only weak dipole-dipole and dispersion forces. Polar CH_2F_2 experiences dipole-dipole and dispersion forces, while nonpolar CF_4 experiences only dispersion forces.

25.97

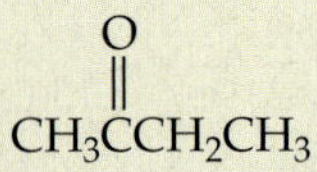

25.99 $\Delta G° = 13$ kJ **25.103** $[AMPOH^-]/[AMPO^{2-}] = 0.65$

Answers to "Give It Some Thought" Questions

Chapter 1

page 3 (a) about 100 elements, (b) atoms and molecules.

page 6 oxygen, O

page 8 The water molecule contains the atoms of two different elements.

page 11 (a) is a chemical change because a new substance is being formed. (b) is a physical change because the water merely changes its physical state and not its composition.

page 15 1 pg, which equals 10^{-12} g.

page 21 (b) is inexact because it is a measured quantity. Both (a) and (c) are exact; (a) involves counting, and (c) is a defined value.

page 26 Whenever possible, we must avoid using a conversion factor that has fewer significant figures than the data whose units are being converted. It is best to use at least one more significant figure in the conversion factor than in the data, which is what was done in Sample Exercise 1.9.

Chapter 2

page 39 (a) the law of multiple proportions. (b) The second compound must contain two oxygen atoms for each carbon atom (that is, twice as many carbon atoms as the first compound.)

page 43 (top) Most α particles pass through the foil without being deflected because most of the volume of the atoms that comprise the foil is empty space.

page 43 (bottom) (a) The atom has 15 electrons because atoms have equal numbers of electrons and protons. (b) The protons reside in the nucleus of the atom.

page 47 Any single atom of chromium must be one of the isotopes of that element. The isotope mentioned has a mass of 52.94 amu and is probably ^{53}Cr. The atomic weight differs from the mass of any particular atom because it is the average atomic mass of the naturally occurring isotopes of the element.

page 51 (a) Cl. (b) third period and group 7A. (c) 17. (d) nonmetal.

page 55 (a) C_2H_6. (b) CH_3. (c) Probably the ball-and-stick model because the angles between the sticks indicate the angles between the atoms.

page 58 We write the empirical formulas for ionic compounds. Thus, the formula is CaO.

page 61 The transition metals can form more than one type of cation, and the charges of these ions are therefore indicated explicitly with Roman numerals: Chromium(II) ion is Cr^{2+}. Calcium, on the other hand, always forms the Ca^{2+} ion, so there is no need to distinguish it from other calcium ions with different charges.

page 63 (top) An *-ide* ending usually means a monatomic anion, although there are some anions with two atoms that are also named this way. An *-ate* ending indicates an oxyanion. The most common oxyanions have the *-ate* ending. An *-ite* ending also indicates an oxyanion, but one having less O than the anion whose name ends in *-ate*.

page 63 (bottom) BO_3^{3-} and SiO_4^{4-}. The borate has three O atoms, like the other oxyanions of the second period in Figure 2.27, and its charge is 3−, following the trend of increasing negative charge as you move to the left in the period. The silicate has four O atoms, as do the other oxyanions in the third period in Figure 2.27, and its charge is 4−, also following the trend of increasing charge moving to the left.

Chapter 3

page 80 Each $Mg(OH)_2$ has 1 Mg, 2 O, and 2 H; thus, 3 $Mg(OH)_2$ represents 3 Mg, 6 O, and 6 H.

page 85 The product is an ionic compound involving Na^+ and S^{2-}, and its chemical formula is therefore Na_2S.

page 100 There are experimental uncertainties in the measurements.

page 101 3.14 mol because 2 mol $H_2 \simeq$ 1 mol O_2.

Chapter 4

page 123 (a) $K^+(aq)$ and $CN^-(aq)$, (b) $Na^+(aq)$ and $ClO_4^-(aq)$.

page 125 $MgBr_2$, because it leads to ions in solution.

page 132 HBr.

page 137 $SO_2(g)$.

page 139 (a) Ne, (b) 0.

page 144 $Ni^{2+}(aq)$

page 150 Molarity is reduced to 0.25 *M*

page 152 12.50 mL

Chapter 5

page 169 (a) kinetic energy, (b) potential energy, (c) heat, (d) work.

page 176 The balance (current state) doesn't depend on the ways the money may have been transferred into the account or on the particular expenditures made in withdrawing money from the account. It depends only on the net total of all the transactions.

page 178 It provides us with a state function that allows us to focus on heat flow, which is easier to measure than the work that accompanies a process.

page 180 The coefficients indicate the numbers of moles of reactants and products that give rise to the stated enthalpy change.

page 185 (a) The energy lost by a system is gained by its surroundings. (b) $q_{system} = -q_{surroundings}$

page 189 (a) the sign of ΔH changes, (b) the magnitude of ΔH doubles.

page 192 $2\,C(s) + H_2(g) \longrightarrow C_2H_2(g) \qquad \Delta H_f^\circ = 226.7$ kJ

page 197 fats

Chapter 6

page 218 Visible light is one form of electromagnetic radiation; all visible light is electromagnetic radiation, but not all electromagnetic radiation is visible.

page 220 Both visible light and X rays are forms of electromagnetic radiation. They therefore both travel at the speed of light, *c*. The differing ability to penetrate skin is due to the different energies of visible light and X rays, which we will discuss in the next section.

page 222 As temperature increases, the average energy of the emitted radiation increases. Blue-white light is at the short end of the visible spectrum (at about 400 nm) whereas red light is closer to the other end of the visible spectrum (about 700 nm). Thus, the blue-white light has a higher frequency, is more energetic, and is consistent with higher temperatures.

page 223 Electrons would be ejected, and the ejected electrons would have greater kinetic energy than they did when ejected by yellow light.

page 226 According to the third postulate, photons of only certain allowed frequencies can be absorbed or emitted as the electron changes energy state. The lines in the spectrum correspond to the allowed frequencies.

page 227 Absorb, because it is moving from a lower-energy state ($n = 3$) to a higher-energy state ($n = 7$).

page 229 Yes, all moving objects produce matter waves, but the wavelengths associated with macroscopic objects, such as the baseball, are too small to allow for any way of observing them.

page 231 The small size and mass of a subatomic particle. The term $h/4\pi$ in the uncertainty principle is a very small number that becomes important only when considering extremely small objects, such as electrons.

page 232 (top) Yes, there is a difference. The first statement says that the electron's position is known exactly, which violates the uncertainty principle. The second statement says that there is a high probability of where the electron is, but there is still uncertainty in its position.

page 232 (bottom) Bohr proposed that the electron in the hydrogen atom moves in a well-defined circular orbit about the nucleus, which violates the uncertainty principle. An orbital is a wave function that gives the probability of finding the electron at any point in space, in accord with the uncertainty principle.

page 233 The energy of an orbital is proportional to $-1/n^2$. The difference between $-1/(2)^2$ and $-1/(1)^2$ is much greater than the difference between $-1/(3)^2$ and $-1/(2)^2$.

page 237 The radial probability function for the 4*s* orbital will have four maxima and three nodes.

page 238 The probability of finding the electron when it is in this *p* orbital is greater in the interior of the lobe than on the edges.

page 239 No. We know that the 4*s* orbital is higher in energy than the 3*s* orbital. Likewise, we know that the 3*d* orbitals are higher in energy than the 3*s* orbital. But, without more information, we do not know whether the 4*s* orbital is higher or lower in energy than the 3*d* orbitals.

page 245 The 6*s* orbital, which starts to hold electrons at element 55, Cs.

page 249 We can't conclude anything! Each of the three elements has a different valence electron configuration for its *nd* and $(n + 1)s$ subshells: For Ni, $3d^8 4s^2$; for Pd, $4d^{10}$; and for Pt, $5d^9 6s^1$.

Chapter 7

page 263 The atomic number of an element depends on the number of protons in the nucleus, whereas the atomic weight depends (mainly) on the number of protons *and* the number of neutrons in the nucleus.

page 265 The 2*p* electron of an Ne atom has a greater Z_{eff}. The atomic number of Na is greater than that of Ne, but an electron in the 3*s* orbital of Na is farther from the nucleus and therefore more screened than is an electron in the 2*p* orbital of Ne.

page 267 No. Atomic weight is determined almost entirely by the composition of the nucleus. Atomic radius is determined by the distribution of the electrons.

page 271 The photoelectric effect, first explained by Einstein, involves the ejection of electrons using the energy of light. If the energy of a photon, $h\nu$, is greater than the ionization energy of an atom, then the light can, in principle, be used to ionize the atom.

page 272 I_2 for a carbon atom. In each process an electron is being removed from an atom or ion with five electrons, either $B(g)$ or $C^+(g)$. The higher nuclear charge of C makes I_2 for carbon greater than I_1 for boron.

page 274 The same: $[Ar]3d^3$

page 276 The first ionization energy of $Cl^-(g)$ is the energy needed to remove an electron from Cl^-, forming $Cl(g) + e^-$. That is the reverse process of Equation 7.6, so the first ionization energy of $Cl^-(g)$ is +349 kJ/mol.

page 278 In general, increasing ionization energy correlates with decreasing metallic character.

page 280 Molecular; P

page 283 Cs has the lowest ionization energy among the alkali metals.

page 286 Gastric fluids in the stomach are very acidic (see the "Chemistry at Work" box on antacids in Section 4.3). Metal carbonates are soluble in acidic solution, where they react with the acid to form $CO_2(g)$ and soluble salts, as in Equations 4.19 and 4.20.

page 288 We can extrapolate the data in the table to make intelligent guesses for these numbers. Notice that the atomic radii increase by 0.15 and 0.19 Å, respectively, from Cl to Br and from Br to I. We might therefore expect an increase of 0.15–0.20 Å from I to At, leading to an estimate of roughly 1.5 Å for the atomic radius of At. Similarly, we might expect I_1 for At to be about 900 kJ/mol.

Chapter 8

page 302 Cl has seven valence electrons. The first and second Lewis symbols are both correct—they both show seven valence electrons, and it doesn't matter which of the four sides has the single electron. The third symbol shows only five electrons and is incorrect.

page 304 Magnesium fluoride is an ionic substance, MgF_2, which consists of Mg^{2+} and F^- ions (Section 2.7). Each magnesium atom loses two electrons, and each fluorine atom gains one electron. Thus, we can say that each Mg atom transfers one electron to each of two F atoms.

page 309 Rhodium, Rh.

page 310 Attractive forces are between each electron and either nucleus. Repulsive forces are those between the two nuclei and those between any two electrons. Because He_2 doesn't exist, it is likely that the repulsions are greater than the attractions.

page 312 CO_2 has C—O double bonds. Because the C—O bond in carbon monoxide is shorter, it is likely to be a triple bond.

page 313 Electron affinity measures the energy released when an isolated atom gains an electron to form a 1− ion. The electronegativity measures the ability of the atom to hold on to its own electrons and attract electrons from other atoms in compounds.

page 314 Polar covalent. Based on the examples of F_2, HF, and LiF, the difference in electronegativity is great enough to introduce some polarity to the bond, but not great enough to cause a complete electron transfer from one atom to the other.

page 315 IF. Because the difference in electronegativity between I and F is greater than that between Cl and F, the magnitude of Q should be greater for IF. In addition, because I has a larger atomic radius than Cl, the bond length in IF is longer than that in ClF. Thus, both Q and r are larger for IF, and therefore $\mu = Qr$ will be larger for IF.

page 316 Red indicates very high electron density on one end of the molecule, which means a large separation of positive and negative charges. In HBr and HI, the electronegativity differences are too small to lead to large charge separations in the molecules.

page 317 MoO_3 is expected to have the higher melting point. It is named using the conventions for ionic compounds, whereas OsO_4 is named using the conventions for molecular compounds, and ionic compounds generally have much higher melting points than molecular compounds. Experimental data support this conclusion: MoO_3 melts at 795°C, and OsO_4 melts at 40°C.

page 321 There is probably a better choice of Lewis structure than the one chosen. Because the formal charges must add up to 0 and the formal charge on the F atom is +1, there must be an atom that has a formal charge of −1. Because F is the most electronegative element, we don't expect it to carry a positive formal charge.

page 323 Yes. There are two resonance structures for ozone that each contribute equally to the overall description of the molecule. Each O—O bond is therefore an average of a single bond and a double bond, which is a "one-and-a-half" bond.

page 324 As "one-and-a-third" bonds. There are three resonance structures, and each of the three N—O bonds is single in two of those structures and double in the third. Each bond in the actual ion is an average of these: $(1 + 1 + 2)/3 = 1^1/_3$.

page 325 No, it will not have multiple resonance structures. We can't "move" the double bonds, as we did in benzene, because the positions of the hydrogen atoms dictate specific positions for the double bonds. We can't write any other reasonable Lewis structures for the molecule.

page 329 The atomization of ethane produces $2\,C(g) + 6\,H(g)$. In this process, six C—H bonds and one C—C bond are broken. We can use $6D(C—H)$ to estimate the amount of enthalpy needed to break the six C—H bonds. The difference between that number and the enthalpy of atomization is an estimate of the bond enthalpy of the C—C bond, $D(C—C)$.

Chapter 9

page 348 (top) Octahedral. Removing two atoms that are opposite each other leads to a square-planar geometry.

page 348 (bottom) The molecule does not follow the octet rule because it has ten electrons around the central A atom. There are four electron domains around A: two single bonds, one double bond, and one nonbonding pair.

page 353 Yes. Based on one resonance structure, we might expect the electron domain that is due to the double bond to "push" the domains that are due to the single bonds, leading to angles slightly different from 120°. However, we must remember that there are two other equivalent resonance structures—each of the three O atoms has a double bond to N in one of the three resonance structures. (Section 8.6) Because of resonance, all three O atoms are equivalent, and they will experience the same amount of repulsion, which leads to bond angles equal to 120°.

page 358 No. The C—O and C—S bond dipoles exactly oppose each other, like in CO_2, but because O and S have different electronegativities, the magnitudes of the bond dipoles will be different. As a consequence, the bond dipoles will not cancel each other and the OCS molecule has a nonzero dipole moment.

page 361 Yes. In both F_2 and Cl_2 the bond is formed by the overlap of singly occupied p orbitals, as in Figure 9.14(c). In F_2, however, the overlapping orbitals are $2p$ orbitals, whereas in Cl_2 they are $3p$ orbitals. Because $3p$ orbitals extend farther in space, the optimum overlap in Cl_2 is achieved at a longer distance than in F_2. Thus, the bond length in Cl_2 is longer than that in F_2.

page 363 The three $2p$ orbitals are equivalent to one another; they differ only in their orientation. Thus, the two Be—F bonds would be equivalent to each other. Because p orbitals are perpendicular to one another, we would expect an F—Be—F bond angle of 90°. Experimentally, the bond angle is 180°.

page 364 The unhybridized p orbital is oriented perpendicular to the plane defined by the three sp^2 hybrids, with one lobe on each side of the plane.

page 368 The molecule should not be linear. Because there are three electron domains around each N atom, we expect sp^2 hybridization and H—N—N angles of approximately 120°. The molecule is expected to be planar; the unhybridized $2p$ orbitals on the N atoms can form a π bond only if all four atoms lie in the same plane. You might notice that there are two ways in which the H atoms can be arranged: They can be both on the same side of the N═N bond or on opposite sides of the N═N bond.

page 375 (top) The molecule would fall apart. With one electron in the bonding MO and one in the antibonding MO, there is no net stabilization of the electrons relative to two separate H atoms.

page 375 (bottom) Zero, because there is one electron in the bonding MO and one electron in the antibonding MO.

page 377 Yes. In Be_2^+ there would be two electrons in the σ_{2s} MO but only one electron in the σ_{2s}^* MO, and so the ion is predicted to have a bond order of $\frac{1}{2}$. It should (and does) exist.

page 382 No. If the σ_{2p} MO were lower in energy than the π_{2p} MOs, we would expect the σ_{2p} MO to hold two electrons and the π_{2p} MOs to hold one electron each, with the same spin. The molecule would therefore be paramagnetic.

Chapter 10

page 401 The major reason is the relatively large distance between molecules.

page 402 The height of the column decreases because atmospheric pressure decreases with increasing altitude.

page 406 As the pressure increases, the volume decreases. Doubling the pressure causes the volume to decrease to half its original value.

page 407 The volume decreases, but it doesn't decrease to half because the volume is proportional to the temperature on the Kelvin scale but not on the Celsius scale.

page 409 Because 22.41 L is the volume of one mole of the gas at STP, it contains Avogadro's number of molecules, 6.022×10^{23}.

page 414 Because water has a lower molar mass (18.0 g/mol) than N_2 (28.0 g/mol), the water vapor is less dense.

page 417 According to Dalton's law of partial pressures, the pressure that is due to N_2 (its partial pressure) doesn't change. However, the total pressure that is due to the partial pressures of both N_2 and O_2 increases.

page 422 The average kinetic energies depend only on temperature and not on the identity of the gas. Thus, the trend in average kinetic energies is $HCl\ (298\ K) = H_2\ (298\ K) < O_2\ (350\ K)$.

page 427 (a) Mean free path decreases because the molecules are crowded closer together. (b) There is no effect. Although the molecules are moving faster at the higher temperature, they are not crowded any closer together.

page 428 (b) Gases deviate from ideality most at low temperatures and high pressures. Thus, the helium gas would deviate most from ideal behavior at 100 K (the lowest temperature listed) and 5 atm (the highest pressure listed).

page 429 The fact that real gases deviate from ideal behavior can be attributed to the finite sizes of the molecules and to the attractions that exist between molecules.

Chapter 11

page 445 (a) In a gas the energy of attraction is less than the average kinetic energy. (b) In a solid, the energy of attraction is greater than the average kinetic energy.

page 446 $Ca(NO_3)_2$ in water. CH_3OH is a molecular substance and a nonelectrolyte. When it dissolves in water, no ions are present. $Ca(NO_3)_2$ is an ionic compound and a strong electrolyte. When it dissolves in water, the Ca^{2+} ions and the NO_3^- ions interact with the polar water molecules by ion-dipole attractive forces.

page 447 The magnitude of the dipole-dipole force depends on both the magnitude of the dipoles and the distance between the dipoles. We cannot judge the distance from Table 11.2, but we see that acetonitrile has the largest dipole moment and the highest boiling point, suggesting that the dipole-dipole attractions are greatest for that substance.

page 448 (a) The polarizability increases in order of increasing size and molecular weight: $CH_4 < CCl_4 < CI_4$. (b) The strength of the dispersion forces follows the same order: $CH_4 < CCl_4 < CI_4$.

page 452 For nearly all substances, the solid phase is more dense than the liquid phase. For water, however, the solid is less dense than the liquid.

page 454 (top) (a) Both viscosity and surface tension decrease with increasing temperature because of the increased molecular motion. (b) Both properties increase as the strength of intermolecular forces increases.

page 454 (bottom) Cohesive. The properties reflect the attraction of the molecules of a substance for one another.

page 456 Melting (or fusion), endothermic.

page 461 Both compounds are nonpolar. Consequently, only dispersion forces exist between the molecules. (a) Because dispersion forces are stronger for the larger, heavier CBr_4 than for CCl_4, CBr_4 has a lower vapor pressure than CCl_4. (b) The substance with the lower vapor pressure is less volatile. Thus, CBr_4 is less volatile than CCl_4.

page 463 The line slopes to the right with increasing pressure because the solid is usually more dense than the liquid.

page 466 Crystalline solids melt at a specific temperature, whereas amorphous ones tend to melt over a temperature range.

page 467 As summarized in Table 11.6 and shown in Figure 11.34, one-eighth of an atom is at each of the 8 corners, and a whole atom is at the center. Thus, the total number of atoms is $8\left(\frac{1}{8}\right) + 1 = 2$.

page 470 The higher the coordination number of the particles in a crystal, the greater the packing efficiency.

page 472 Molecular solids are composed of molecules or nonmetal atoms. Because Co is a metal and K_2O is an ionic substance, they do not form molecular solids. C_6H_6, however, is a molecular substance and forms a molecular solid.

Chapter 12

page 490 Yes. Its elements are from columns 3A and 5A.

page 498 Yes, the molecule has both the $-NH_2$ and $-COOH$ groups in it, which can react as in nylon to make a polymer.

page 499 Vinyl acetate interferes with intermolecular interactions between adjacent ethylene chains, and so reduces crystallinity and melting point.

page 510 Long. The longer the polymer, the more atoms that can participate in delocalized π bonding.

page 514 $Ga(CH_3)_3 + AsH_3 \longrightarrow GaAs + 3\,CH_4$

Chapter 13

page 534 No, because the AgCl is not dispersed throughout the liquid phase.

page 537 No. The concentration of sodium acetate is higher than the stable equilibrium value, so some of the dissolved solute comes out of solution when a seed crystal initiates the process.

page 539 Considerably lower because there would no longer be hydrogen bonding with water.

page 543 Dissolved gases are less soluble as temperature increases.

page 544 230 ppm

page 545 Yes. Molarity is the number of moles of solute per liter of solution. In a liter of solution, however, there will generally be less than 1 kg of water because the solute takes up some of the volume. Of course, for dilute solutions the two concentration units are approximately the same numerically.

page 550 Mole fraction

page 551 Not necessarily; if the solute dissociates into particles, it could have a lower molality and still cause an increase of 0.51°C. The total molality of all the particles in the solution is 1 *m*.

page 555 The 0.20 *m* solution is hypotonic with respect to the 0.5 *m* solution.

page 556 They would have the same osmotic pressure because they would have the same total concentration of dissolved particles.

page 560 The smaller droplets carry negative charges because of the embedded stearate ions and thus repel one another.

Chapter 14

page 577 Increasing the partial pressure of a gas increases the number of molecules in a given volume and therefore increases the concentration of the gas. Rates normally increase with increasing reactant concentrations.

page 580 The size of the triangle is mostly a matter of convenience. Drawing a larger triangle will give larger values of both $\Delta[C_4H_9Cl]$ and Δt, but their ratio, $\Delta[C_4H_9Cl]/\Delta t$, remains constant. Thus, the calculated slope is independent of the size of the triangle.

page 583 (a) The rate law for any chemical reaction is the equation that relates the concentrations of reactants to the rate of the reaction. The general form for a rate law is given in Equation 14.7. (b) The quantity k in any rate law is the rate constant.

page 584 (a) The rate law is second order in NO, first order in H_2, and third order overall. (b) No. Doubling [NO] will cause the rate to increase fourfold, whereas doubling $[H_2]$ will merely double the rate.

page 589 In (a) the intercept is the initial partial pressure of the CH_3NC, 150 torr. In (b) it is the natural logarithm of this pressure, $\ln(150) = 5.01$.

page 593 As the reaction proceeds, the concentration of the reactant decreases. Thus, according to Equation 14.16, the half-life increases during the course of the reaction.

page 594 Molecules must collide in order to react.

page 596 The molecules must not merely collide in order to react, but they must collide in the proper orientation and with an energy greater than the activation energy for the reaction.

page 600 Because the elementary reaction involves two molecules, it is bimolecular.

page 603 The rate law depends not on the overall reaction, but on the slowest step in the mechanism.

page 609 Generally, a catalyst lowers the activation energy by providing a different, lower-energy pathway (different mechanism) for a reaction.

page 613 (a) active site, (b) substrate.

Chapter 15

page 632 (top) (a) The rates of opposing reactions are equal. (b) If the forward step is faster than the reverse step, $k_f > k_r$ and the constant, which equals k_f/k_r, will be greater than 1.

page 632 (bottom) The fact that concentrations no longer change with time indicates that equilibrium has been reached.

page 635 It is independent of the starting concentrations of reactants and products.

page 636 They represent equilibrium constants. K_c is obtained when equilibrium concentrations expressed in molarity are substituted into the equilibrium-constant expression. K_p is obtained when equilibrium partial pressures expressed in atmospheres are substituted into the expression.

page 637 $0.00140\ M/1\ M = 0.00140$ (no units)

page 640 Because the coefficients in the equation have been multiplied by 3, the exponents in K_p are also multiplied by 3, so that the magnitude of the equilibrium constant will be $(K_p)^3$.

page 642 Because the pure H_2O liquid is omitted from the equilibrium-constant expression, $K_p = P_{H_2O}$. Thus, at any particular temperature, the equilibrium vapor pressure is a constant.

page 651 (a) The equilibrium shifts to the right, using up some of the added O_2 and forming NO_2. (b) The equilibrium shifts to the left, forming more NO to replace some of the NO that was removed.

page 652 Increasing the volume decreases the pressure, and so equilibrium shifts in the direction that causes the pressure to increase. Because there are three gas molecules on the left in the equation and only two on the right, the equilibrium shifts to the left, increasing the number of gas molecules by converting SO_3 into SO_2 and O_2.

page 654 Evaporation is an endothermic process. Increasing the temperature of an endothermic process shifts the equilibrium to the right, forming more product. Because the product of the evaporation is the vapor, the vapor pressure increases.

page 657 No; catalysts have no effect on the position of an equilibrium, although they do affect how quickly equilibrium is reached.

Chapter 16

page 670 The H^+ ion for acids and the OH^- ion for bases.

page 672 Because it adds H^+ as it moves from the left side of the equation to the right side, NH_3 is the base.

page 675 HNO_3 is a strong acid, which means that NO_3^- has negligible basicity.

page 679 (a) The solution is neutral, $[H^+] = [OH^-]$. (b) The pH increases as $[OH^-]$ increases.

page 681 pH = 14.00 − 3.00 = 11.00. The solution is basic because pH > 7.

page 682 From Figure 16.7, we see that the pH must be above about 8, meaning that the solution is basic.

page 684 Because it is the conjugate base of a substance that has negligible acidity, CH_3^- must be a strong base. Bases stronger than OH^- abstract H^+ from water molecules: $CH_3^- + H_2O \longrightarrow CH_4 + OH^-$.

page 688 Because weak acids typically undergo very little ionization, often less than 1%.

page 692 This is the acid-dissociation constant for the loss of the third and final proton from H_3PO_4: $HPO_4^{2-}(aq) \rightleftharpoons H^+(aq) + PO_4^{3-}(aq)$.

page 700 (top) Because the NO_3^- ion is the conjugate base of a strong acid, it will have no effect on pH. (NO_3^- has negligible basicity.) Because the CO_3^{2-} ion is the conjugate base of a weak acid, it will have an effect on pH, increasing the pH.

page 700 (bottom) K^+, an alkali metal cation, does not affect pH.

page 703 The increasing acidity down a column is due mainly to the decrease in H—X bond strength. The trend across a period is due mainly to the increasing electronegativity.

page 706 The carboxyl group, —COOH.

page 708 It must have an unshared pair of electrons.

page 709 Fe^{3+}, because it has the highest charge.

Chapter 17

page 725 (top) (a) $[NH_4^+] = 0.10\ M$; $[Cl^-] = 0.10\ M$; $[NH_3] = 0.12\ M$, (b) Cl^-, (c) Equation 17.2.

page 725 (bottom) HNO_3 and NO_3^-. This is a strong acid and its conjugate base. Buffers are composed of weak acids and their conjugate bases. The NO_3^- ion is merely a spectator in any acid-base chemistry and is therefore ineffective in helping control the pH of a solution.

page 726 (a) The NaOH (a strong base) reacts with the acid member of the buffer ($HC_2H_3O_2$), abstracting a proton. Thus, [$HC_2H_3O_2$] decreases and [$C_2H_3O_2^-$] increases. (b) The HCl (a strong acid) reacts with the base member of the buffer ($C_2H_3O_2^-$). Thus, [$C_2H_3O_2^-$] decreases and [$HC_2H_3O_2$] increases.

page 729 The solution will most effectively resist a change in either direction when its pH = pK_a. Thus, the optimal pH is pH = pK_a = $-\log(1.8 \times 10^{-5})$ = 4.74, and the pH range of the buffer is about 4.7 ± 1.

page 732 Increase because the added base makes the solution less acidic (more basic).

page 739 The nearly vertical part of the titration curve at the equivalence point is smaller for the weak acid–strong base titration, and fewer indicators undergo their color change within this narrow range.

page 741 Because all three compounds produce the same number of ions, their relative solubilities correspond directly to their K_{sp} values, with the compound with the largest K_{sp} value being most soluble, AgCl.

page 753 They are insoluble in water but dissolve in the presence of either an acid or a base.

page 755 A high concentration of H_2S and a low concentration of H^+ (that is, high pH) will shift the equilibrium to the right, reducing [Cu^{2+}].

page 758 The solution must contain one or more of the cations in group 1 of the qualitative analysis scheme shown in Figure 17.22: Ag^+, Pb^{2+}, or Hg_2^{2+}.

Chapter 18

page 775 It looks like the area of the upper curve to the left of the visible portion is about twice as large as the area of the lower curve. The upper curve corresponds to radiation at the "top" of the atmosphere, and the bottom curve corresponds to radiation at sea level; therefore, we estimate about half of the ultraviolet light that arrives at Earth from the Sun gets absorbed by the upper atmosphere and does not make it to the ground.

page 793 As with so many other things, there are trade-offs in selecting processes for green chemistry. Yes, increasing CO_2 is expected to lead to global warming; however, the main source of CO_2 is from combustion of fossil fuels. If we can reduce the amount of CO_2 entering the atmosphere from that source, perhaps we can afford to use supercritical or liquid CO_2 in industrial processes, which are safer for workers than other chemicals and form less harmful by-products. Supercritical (or just liquid) water would be a better choice environmentally if the industrial process you want to change still works with water as a solvent.

Chapter 19

page 806 No, nonspontaneous processes can occur so long as they receive some continuous outside assistance. Examples of nonspontaneous processes with which we may be familiar include the building of a brick wall and the electrolysis of water to form hydrogen gas and oxygen gas.

page 807 No. Just because the system is restored to its original condition doesn't mean that the surroundings have likewise been restored to their original condition.

page 809 ΔS depends not merely on q but on q_{rev}. Although there are many possible paths that could take a system from its initial to final state, there is always only one reversible isothermal path between two states. Thus, ΔS has only one particular value regardless of the path taken between states.

page 811 Because rusting is a spontaneous process, ΔS_{univ} must be positive. Therefore the entropy of the surroundings must increase, and that increase must be larger than the entropy decrease of the system.

page 812 A molecule can vibrate (atoms moving relative to one another) and rotate (tumble), whereas a single atom cannot undergo these motions.

page 813 $S = 0$, based on Equation 19.5 and the fact that $\ln 1 = 0$.

page 819 It must be a perfect crystal at 0 K (third law of thermodynamics), which means it has only a single accessible microstate.

page 822 Always increase (positive ΔS).

page 823 In any spontaneous process the entropy of the universe increases. In any spontaneous process operating at constant temperature, the free energy of the system decreases.

page 824 It indicates that the process to which the thermodynamic quantity refers has taken place under standard conditions, as summarized in Table 19.3.

page 828 Above the boiling point, vaporization is spontaneous, so that $\Delta G > 0$. Because $\Delta H - T\Delta S > 0$, $\Delta H > T\Delta S$.

Chapter 20

page 849 This is tricky. If we think of C as +4 (as it would be in CO_2), then N has to be −5. However, if we think of C as +2 (as it would be in CO), then N would have to be −3. The best thing to do is to say that the *sum* of the oxidation states of the carbon and nitrogen in cyanide is −1, and leave the individual atoms unassigned.

page 854 No. Electrons should appear in the half-reactions but cancel if the half-reactions are adjusted properly. So, when you are balancing a redox equation and find yourself with e^- on either side of the reaction arrow when you are done, you know you should go back and check your work.

page 857 The anode is where oxidation takes place. Because electrons are being removed from the anode, negatively charged anions must migrate to the anode to maintain charge balance.

page 859 The surface atoms. They are the ones available to make contact with the copper ions.

page 861 Yes.

page 863 1 atm pressure of $Cl_2(g)$ and 1 M solution of $Cl^-(aq)$.

page 865 True, since $E^\circ_{cell} = E^\circ_{red}$ (cathode) $- E^\circ_{red}$ (anode), and smaller E°_{cell} means a smaller net driving force for the overall redox reaction.

page 869 Pb(s) is stronger because it lies above Hg(l) in the activity series, which means that it more readily loses electrons.

page 883 Al, Zn

Chapter 21

page 902 The mass number decreases by 4.

page 905 Only the neutron, as it is the only neutral particle listed.

page 906 From Figure 21.2, we see that the center of the belt of stability for a nucleus containing 60 protons lies at about 85 neutrons.

page 908 24. Having 30 neutrons would take it outside the belt of stability; 25 would be an odd number, which is generally less stable than an even number.

page 910 The reactors produce lots of neutrons as by-products, and neutrons are the most generally useful particles to effect nuclear reactions.

page 913 (a) Yes; doubling the mass would double the amount of radioactivity of the sample. (b) No; changing the mass would not change the half-life.

page 916 No. Alpha particles are more readily absorbed by matter than beta or gamma rays. Geiger counters must be calibrated for the radiation they are being used to detect.

page 920 No. Stable nuclei having mass numbers around 100 are the most stable nuclei. They could not form a still more stable nucleus with an accompanying release of energy.

Chapter 22

page 939 No. phosphorus, being a third-row element, does not have a capacity for forming strong π bonds. Instead, it exists as a solid in which the phosphorus atoms are singly bonded to one another.

page 945 No. As shown in Figure 22.7, the free energy of formation of $H_2Se(g)$ is positive, indicating that the equilibrium constant for the reaction will be small. (Section 19.7)

page 952 Yes; the bromine is in the +5 oxidation state and readily reducible to a lower oxidation state such as 0 or −1.

page 955 HIO_3. Note that the iodine remains in the +5 oxidation state.

page 961 No; the oxidation number of sulfur is the same in both reactants and products.

page 967 The nitrogen atom in nitric acid has an oxidation number of +5 and that in nitrous acid has an oxidation number of +3.

page 975 $MgCO_3(s)$. The bicarbonate ion of washing soda dissociates a proton to form CO_3^{2-} ion.

Chapter 23

page 996 Not CaO, because Ca is too active a metal. Ag_2O should reduce readily with CO because Ag is not an active metal, and Ag_2O is therefore not so stable.

page 1000 (top) Gold is the reducing agent, oxidized from 0 to +1 oxidation state. Dissolved oxygen is the oxidizing agent, reduced from 0 to −2 oxidation state.

page 1000 (bottom) Al_2O_3

page 1001 Na^+ and Cl^-

page 1003 Cu^{2+} and H^+ as cations, and SO_4^{2-} as the anion.

page 1006 Pd

page 1008 Interstitial, because B is so much smaller than Pd.

page 1011 From Figure 23.18, we see that in the group 8 metals, the elements of the second series have higher bonding atomic radii than those of the first series. So, Pd, in the second series, is larger than Ni, which is in the first series.

page 1014 Yes. From Section 6.9 we would expect Cr^{3+} to have an outer $4s^2 3d^1$ or $3d^3$ configuration. In either case there is at least one unpaired electron present.

Chapter 24

page 1028 $[Fe(H_2O)_6]^{3+}(aq) + SCN^-(aq) \rightarrow [Fe(H_2O)_5NCS]^{2+}(aq) + H_2O(l)$

page 1030 (a) tetrahedral and square planar; (b) octahedral

page 1033 Each NH_3 has one donor atom. Consequently, the CO_3^{2-} ion must have two donor atoms to give the cobalt atom a coordination number of 6. Thus, CO_3^{2-} is acting as a bidentate ligand.

page 1042 The two compounds have the same composition and bonds, but they are optical isomers of each other (nonsuperimposable mirror images). The *d* isomer rotates plane-polarized light to the right (dextrorotatory), whereas the *l* isomer rotates plane-polarized light to the left (levorotatory).

page 1045 (a) $[Ar]4s^2 3d^7$, three unpaired electrons; (b) $[Ar]3d^6$, four unpaired electrons.

page 1047 The d_{z^2} and $d_{x^2-y^2}$ orbitals.

page 1048 The Ti(IV) ion has no *d* electrons, so there can be no *d*-*d* transitions, which are the ones usually responsible for the color of transition-metal compounds.

Chapter 25

page 1065 C=N, because it is a polar double bond.

page 1068 Two C—H bonds and two C—C bonds.

page 1071 C_3H_7

page 1075 Only one.

page 1081

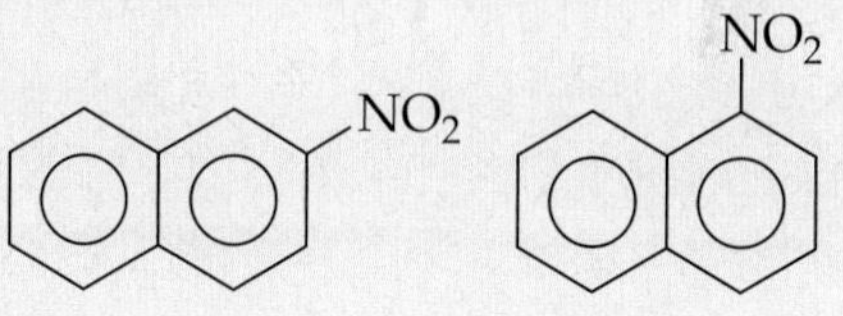

page 1086

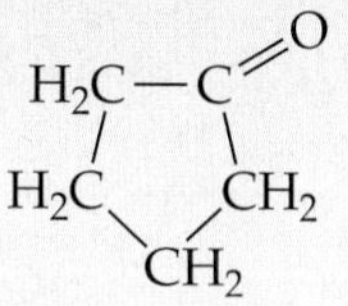

page 1093 All four groups must be different from one another.

page 1098 The hydrogen bond between an N—H and a carbonyl (C=O) group on the adjacent turn of the helix.

page 1103 The α form of the C—O—C linkage. Glycogen serves as a source of energy in the body, which means that the body's enzymes must be able to hydrolyze it to sugars. The enzymes work only on polysaccharides having the α linkage.

Glossary

absorption spectrum The amount of light absorbed by a sample as a function of wavelength. (Section 24.5)

accuracy A measure of how closely individual measurements agree with the correct value. (Section 1.5)

acid A substance that is able to donate a H^+ ion (a proton) and hence increases the concentration of $H^+(aq)$ when it dissolves in water. (Section 4.3)

acid-dissociation constant (K_a) An equilibrium constant that expresses the extent to which an acid transfers a proton to solvent water. (Section 16.6)

acidic anhydride (acidic oxide) An oxide that forms an acid when added to water; soluble nonmetal oxides are acidic anhydrides. (Section 22.5)

acidic oxide (acidic anhydride) An oxide that either reacts with a base to form a salt or with water to form an acid. (Section 22.5)

acid rain Rainwater that has become excessively acidic because of absorption of pollutant oxides, notably SO_3, produced by human activities. (Section 18.4)

actinide element Element in which the 5*f* orbitals are only partially occupied. (Section 6.8)

activated complex (transition state) The particular arrangement of atoms found at the top of the potential-energy barrier as a reaction proceeds from reactants to products. (Section 14.5)

activation energy (E_a) The minimum energy needed for reaction; the height of the energy barrier to formation of products. (Section 14.5)

active site Specific site on a heterogeneous catalyst or an enzyme where catalysis occurs. (Section 14.7)

activity The decay rate of a radioactive material, generally expressed as the number of disintegrations per unit time. (Section 21.4)

activity series A list of metals in order of decreasing ease of oxidation. (Section 4.4)

addition polymerization Polymerization that occurs through coupling of monomers with one another, with no other products formed in the reaction. (Section 12.2)

addition reaction A reaction in which a reagent adds to the two carbon atoms of a carbon–carbon multiple bond. (Section 25.4)

adsorption The binding of molecules to a surface. (Section 14.7)

alcohol An organic compound obtained by substituting a hydroxyl group ($-OH$) for a hydrogen on a hydrocarbon. (Sections 2.9 and 25.5)

aldehyde An organic compound that contains a carbonyl group to which at least one hydrogen atom is attached. (Section 25.6)

alkali metals Members of group 1A in the periodic table. (Section 7.7)

alkaline earth metals Members of group 2A in the periodic table. (Section 7.7)

alkanes Compounds of carbon and hydrogen containing only carbon–carbon single bonds. (Sections 2.9 and 25.2)

alkenes Hydrocarbons containing one or more carbon–carbon double bonds. (Section 25.2)

alkyl group A group that is formed by removing a hydrogen atom from an alkane. (Section 25.3)

alkynes Hydrocarbons containing one or more carbon–carbon triple bonds. (Section 25.2)

alloy A substance that has the characteristic properties of a metal and contains more than one element. Often there is one principal metallic component, with other elements present in smaller amounts. Alloys may be homogeneous or heterogeneous in nature. (Section 23.6)

alpha (α) helix A protein structure in which the protein is coiled in the form of a helix, with hydrogen bonds between $C{=}O$ and $N{-}H$ groups on adjacent turns. (Section 25.9)

alpha particles Particles that are identical to helium-4 nuclei, consisting of two protons and two neutrons, symbol ${}^4_2He$ or ${}^4_2\alpha$. (Section 21.1)

amide An organic compound that has an NR_2 group attached to a carbonyl. (Section 25.6)

amine A compound that has the general formula R_3N, where R may be H or a hydrocarbon group. (Section 16.7)

amino acid A carboxylic acid that contains an amino ($-NH_2$) group attached to the carbon atom adjacent to the carboxylic acid ($-COOH$) functional group. (Section 25.9)

amorphous solid A solid whose molecular arrangement lacks a regular, long-range pattern. (Section 11.7)

amphiprotic Refers to the capacity of a substance to act as either an acid or a base. (Section 16.2)

amphoteric Capable of behaving as either an acid or a base. (Sections 16.2 and 17.5)

angstrom A common non-SI unit of length, denoted Å, that is used to measure atomic dimensions: $1\ \text{Å} = 10^{-10}$ m. (Section 2.3)

anion A negatively charged ion. (Section 2.7)

anode An electrode at which oxidation occurs. (Section 20.3)

antibonding molecular orbital A molecular orbital in which electron density is concentrated outside the region between the two nuclei of bonded atoms. Such orbitals, designated as σ^* or π^*, are less stable (of higher energy) than bonding molecular orbitals. (Section 9.7)

aqueous solution A solution in which water is the solvent. (Chapter 4: Introduction)

aromatic hydrocarbons Hydrocarbon compounds that contain a planar, cyclic arrangement of carbon atoms linked by both σ and delocalized π bonds. (Section 25.2)

Arrhenius equation An equation that relates the rate constant for a reaction to the frequency factor, *A*, the activation energy, E_a, and the temperature, T: $k = Ae^{-E_a/RT}$. In its logarithmic form it is written $\ln k = -E_a/RT + \ln A$. (Section 14.5)

atmosphere (atm) A unit of pressure equal to 760 torr; 1 atm = 101.325 kPa. (Section 10.2)

atom The smallest representative particle of an element. (Sections 1.1 and 2.1)

atomic mass unit (amu) A unit based on the value of exactly 12 amu for the mass of the isotope of carbon that has six protons and six neutrons in the nucleus. (Sections 2.3 and 3.3)

atomic number The number of protons in the nucleus of an atom of an element. (Section 2.3)

atomic radius An estimate of the size of an atom. See **bonding atomic radius**. (Section 7.3)

atomic weight The average mass of the atoms of an element in atomic mass units (amu); it is numerically equal to the mass in grams of one mole of the element. (Section 2.4)

autoionization The process whereby water spontaneously forms low concentrations of $H^+(aq)$ and $OH^-(aq)$ ions by proton transfer from one water molecule to another. (Section 16.3)

Avogadro's hypothesis A statement that equal volumes of gases at the same temperature and pressure contain equal numbers of molecules. (Section 10.3)

Avogadro's law A statement that the volume of a gas maintained at constant temperature and pressure is directly proportional to the number of moles of the gas. (Section 10.3)

Avogadro's number The number of ${}^{12}C$ atoms in exactly 12 g of ${}^{12}C$; it equals 6.022×10^{23}. (Section 3.4)

band An array of closely spaced molecular orbitals occupying a discrete range of energy. (Section 12.1)

band gap The energy gap between an occupied valence band and a vacant band called the conduction band. (Section 12.1)

bar A unit of pressure equal to 10^5 Pa. (Section 10.2)

base A substance that is an H^+ acceptor; a base produces an excess of $OH^-(aq)$ ions when it dissolves in water. (Section 4.3)

base-dissociation constant (K_b) An equilibrium constant that expresses the extent to which a base reacts with solvent water, accepting a proton and forming $OH^-(aq)$. (Section 16.7)

basic anhydride (basic oxide) An oxide that forms a base when added to water; soluble metal oxides are basic anhydrides. (Section 22.5)

basic oxide (basic anhydride) An oxide that either reacts with water to form a base or reacts with an acid to form a salt and water. (Section 22.5)

battery A self-contained electrochemical power source that contains one or more voltaic cells. (Section 20.7)

Bayer process A hydrometallurgical procedure for purifying bauxite in the recovery of aluminum from bauxite-containing ores. (Section 23.3)

becquerel The SI unit of radioactivity. It corresponds to one nuclear disintegration per second. (Section 21.4)

Beer's law The light absorbed by a substance (*A*) equals the product of its molar absorptivity constant (*a*), the path length through which the light passes (*b*), and the molar concentration of the substance (*c*): $A = abc$. (Section 14.2)

beta particles Energetic electrons emitted from the nucleus, symbol $^{0}_{-1}e$. (Section 21.1)

bidentate ligand A ligand in which two linked coordinating atoms are bound to a metal. (Section 24.2)

bimolecular reaction An elementary reaction that involves two molecules. (Section 14.6)

biochemistry The study of the chemistry of living systems. (Chapter 25: Introduction)

biocompatible Any substance or material that can be compatibly placed within living systems. (Section 12.3)

biodegradable Organic material that bacteria are able to oxidize. (Section 18.6)

biomaterial Any material that has a biomedical application. (Section 12.3)

biopolymer A polymeric molecule of high molecular weight found in living systems. The three major classes of biopolymers are proteins, carbohydrates, and nucleic acids. (Section 25.8)

body-centered cubic cell A cubic unit cell in which the lattice points occur at the corners and at the center. (Section 11.7)

bomb calorimeter A device for measuring the heat evolved in the combustion of a substance under constant-volume conditions. (Section 5.5)

bond angles The angles made by the lines joining the nuclei of the atoms in a molecule. (Section 9.1)

bond dipole The dipole moment that is due to unequal electron sharing between two atoms in a covalent bond. (Section 9.3)

bond enthalpy The enthalpy change, ΔH, required to break a particular bond when the substance is in the gas phase. (Section 8.8)

bonding atomic radius The radius of an atom as defined by the distances separating it from other atoms to which it is chemically bonded. (Section 7.3)

bonding molecular orbital A molecular orbital in which the electron density is concentrated in the internuclear region. The energy of a bonding molecular orbital is lower than the energy of the separate atomic orbitals from which it forms. (Section 9.7)

bonding pair In a Lewis structure a pair of electrons that is shared by two atoms. (Section 9.2)

bond length The distance between the centers of two bonded atoms. (Section 8.8)

bond order The number of bonding electron pairs shared between two atoms, less the number of antibonding electron pairs: bond order = (number of bonding electrons − number of antibonding electrons)/2. (Section 9.7)

bond polarity A measure of the degree to which the electrons are shared unequally between two atoms in a chemical bond. (Section 8.4)

boranes Covalent hydrides of boron. (Section 22.11)

Born–Haber cycle A thermodynamic cycle based on Hess's law that relates the lattice energy of an ionic substance to its enthalpy of formation and to other measurable quantities. (Section 8.2)

Boyle's law A law stating that at constant temperature, the product of the volume and pressure of a given amount of gas is a constant. (Section 10.3)

Brønsted–Lowry acid A substance (molecule or ion) that acts as a proton donor. (Section 16.2)

Brønsted–Lowry base A substance (molecule or ion) that acts as a proton acceptor. (Section 16.2)

buffer capacity The amount of acid or base a buffer can neutralize before the pH begins to change appreciably. (Section 17.2)

buffered solution (buffer) A solution that undergoes a limited change in pH upon addition of a small amount of acid or base. (Section 17.2)

calcination The heating of an ore to bring about its decomposition and the elimination of a volatile product. For example, a carbonate ore might be calcined to drive off CO_2. (Section 23.2)

calorie A unit of energy, it is the amount of energy needed to raise the temperature of 1 g of water by 1°C from 14.5°C to 15.5°C. A related unit is the joule: 1 cal = 4.184 J. (Section 5.1)

calorimeter An apparatus that measures the evolution of heat. (Section 5.5)

calorimetry The experimental measurement of heat produced in chemical and physical processes. (Section 5.5)

capillary action The process by which a liquid rises in a tube because of a combination of adhesion to the walls of the tube and cohesion between liquid particles. (Section 11.3)

carbide A binary compound of carbon with a metal or metalloid. (Section 22.9)

carbohydrates A class of substances formed from polyhydroxy aldehydes or ketones. (Section 25.10)

carbon black A microcrystalline form of carbon. (Section 22.9)

carbonyl group The C=O double bond, a characteristic feature of several organic functional groups, such as ketones and aldehydes. (Section 25.6)

carboxylic acid A compound that contains the —COOH functional group. (Sections 16.10 and 25.6)

catalyst A substance that changes the speed of a chemical reaction without itself undergoing a permanent chemical change in the process. (Section 14.7)

cathode An electrode at which reduction occurs. (Section 20.3)

cathode rays Streams of electrons that are produced when a high voltage is applied to electrodes in an evacuated tube. (Section 2.2)

cathodic protection A means of protecting a metal against corrosion by making it the cathode in a voltaic cell. This can be achieved by attaching a more easily oxidized metal, which serves as an anode, to the metal to be protected. (Section 20.8)

cation A positively charged ion. (Section 2.7)

cell potential A measure of the driving force, or "electrical pressure," for an electrochemical reaction; it is measured in volts: 1 V = 1 J/C. Also called electromotive force. (Section 20.4)

cellulose A polysaccharide of glucose; it is the major structural element in plant matter. (Section 25.10)

Celsius scale A temperature scale on which water freezes at 0° and boils at 100° at sea level. (Section 1.4)

ceramic A solid inorganic material, either crystalline (oxides, carbides, silicates) or amorphous (glasses). Most ceramics melt at high temperatures. (Section 12.1)

chain reaction A series of reactions in which one reaction initiates the next. (Section 21.7)

changes of state Transformations of matter from one state to a different one, for example, from a gas to a liquid. (Section 1.3)

charcoal A form of carbon produced when wood is heated strongly in a deficiency of air. (Section 22.9)

Charles's law A law stating that at constant pressure, the volume of a given quantity of gas is proportional to absolute temperature. (Section 10.3)

chelate effect The generally larger formation constants for polydentate ligands as compared with the corresponding *monodentate* ligands. (Section 24.2)

chelating agent A polydentate ligand that is capable of occupying two or more sites in the coordination sphere. (Section 24.2)

chemical bond A strong attractive force that exists between atoms in a molecule. (Section 8.1)

chemical changes Processes in which one or more substances are converted into other substances; also called **chemical reactions**. (Section 1.3)

chemical equation A representation of a chemical reaction using the chemical formulas of the reactants and products; a balanced chemical equation contains equal numbers of atoms of each element on both sides of the equation. (Section 3.1)

chemical equilibrium A state of dynamic balance in which the rate of formation of the products of a reaction from the reactants equals the rate of formation of the reactants from the products; at equilibrium the concentrations of the reactants and products remain constant. (Section 4.1; Chapter 15: Introduction)

chemical formula A notation that uses chemical symbols with numerical subscripts to convey the relative proportions of atoms of the different elements in a substance. (Section 2.6)

chemical kinetics The area of chemistry concerned with the speeds, or rates, at which chemical reactions occur. (Chapter 14: Introduction)

chemical nomenclature The rules used in naming substances. (Section 2.8)

chemical properties Properties that describe a substance's composition and its reactivity; how the substance reacts or changes into other substances. (Section 1.3)

chemical reactions Processes in which one or more substances are converted into other substances; also called **chemical changes**. (Section 1.3)

chemistry The scientific discipline that treats the composition, properties, and transformations of matter. (Chapter 1: Introduction)

chiral A term describing a molecule or an ion that cannot be superimposed on its mirror image. (Sections 24.4 and 25.7)

chlorofluorocarbons Compounds composed entirely of chlorine, fluorine, and carbon. (Section 18.3)

chlorophyll A plant pigment that plays a major role in conversion of solar energy to chemical energy in photosynthesis. (Section 24.2)

cholesteric liquid crystalline phase A liquid crystal formed from flat, disc-shaped molecules that align through a stacking of the molecular discs. (Section 12.5)

coal A naturally occurring solid containing hydrocarbons of high molecular weight, as well as compounds containing sulfur, oxygen, and nitrogen. (Section 5.8)

coke An impure form of carbon, formed when coal is heated strongly in the absence of air. (Section 22.9)

colligative properties Those properties of a solvent (vapor-pressure lowering, freezing-point lowering, boiling-point elevation, osmotic pressure) that depend on the total concentration of solute particles present. (Section 13.5)

collision model A model of reaction rates based on the idea that molecules must collide to react; it explains the factors influencing reaction rates in terms of the frequency of collisions, the number of collisions with energies exceeding the activation energy, and the probability that the collisions occur with suitable orientations. (Section 14.5)

colloids (colloidal dispersions) Mixtures containing particles larger than normal solutes but small enough to remain suspended in the dispersing medium. (Section 13.6)

combination reaction A chemical reaction in which two or more substances combine to form a single product. (Section 3.2)

combustion reaction A chemical reaction that proceeds with evolution of heat and usually also a flame; most combustion involves reaction with oxygen, as in the burning of a match. (Section 3.2)

common-ion effect A shift of an equilibrium induced by an ion common to the equilibrium. For example, added Na_2SO_4 decreases the solubility of the slightly soluble salt $BaSO_4$, or added $NaC_2H_3O_2$ decreases the percent ionization of $HC_2H_3O_2$. (Section 17.1)

complementary colors Colors that, when mixed in proper proportions, appear white or colorless. (Section 24.5)

complete ionic equation A chemical equation in which dissolved strong electrolytes (such as dissolved ionic compounds) are written as separate ions. (Section 4.2)

complex ion (complex) An assembly of a metal ion and the Lewis bases (ligands) bonded to it. (Sections 17.5 and 24.1)

compound A substance composed of two or more elements united chemically in definite proportions. (Section 1.2)

concentration The quantity of solute present in a given quantity of solvent or solution. (Section 4.5)

concentration cell A voltaic cell containing the same electrolyte and the same electrode materials in both the anode and cathode compartments. The emf of the cell is derived from a difference in the concentrations of the same electrolyte solutions in the compartments. (Section 20.6)

condensation polymerization Polymerization in which molecules are joined together through condensation reactions. (Section 12.2)

condensation reaction A chemical reaction in which a small molecule (such as a molecule of water) is split out from between two reacting molecules. (Section 22.8)

conduction band A band of molecular orbitals lying higher in energy than the occupied valence band, and distinctly separated from it. (Section 12.1)

conjugate acid A substance formed by addition of a proton to a Brønsted–Lowry base. (Section 16.2)

conjugate acid-base pair An acid and a base, such as H_2O and OH^-, that differ only in the presence or absence of a proton. (Section 16.2)

conjugate base A substance formed by the loss of a proton from a Brønsted–Lowry acid. (Section 16.2)

continuous spectrum A spectrum that contains radiation distributed over all wavelengths. (Section 6.3)

conversion factor A ratio relating the same quantity in two systems of units that is used to convert the units of measurement. (Section 1.6)

coordination compound or complex A compound containing a metal ion bonded to a group of surrounding molecules or ions that act as ligands. (Section 24.1)

coordination number The number of adjacent atoms to which an atom is directly bonded. In a complex the coordination number of the metal ion is the number of donor atoms to which it is bonded. (Sections 11.7 and 24.1)

coordination sphere The metal ion and its surrounding ligands. (Section 24.1)

coordination-sphere isomers Structural isomers of coordination compounds in which the ligands within the coordination sphere differ. (Section 24.4)

copolymer A complex polymer resulting from the polymerization of two or more chemically different monomers. (Section 12.2)

core electrons The electrons that are not in the outermost shell of an atom. (Section 6.8)

corrosion The process by which a metal is oxidized by substances in its environment. (Section 20.8)

covalent bond A bond formed between two or more atoms by a sharing of electrons. (Section 8.1)

covalent-network solids Solids in which the units that make up the three-dimensional network are joined by covalent bonds. (Section 11.8)

critical mass The amount of fissionable material necessary to maintain a chain reaction. (Section 21.7)

critical pressure The pressure at which a gas at its critical temperature is converted to a liquid state. (Section 11.4)

critical temperature The highest temperature at which it is possible to convert the gaseous form of a substance to a liquid. The critical temperature increases with an increase in the magnitude of intermolecular forces. (Section 11.4)

cross-linking The formation of bonds between polymer chains. (Section 12.2)

crystal-field theory A theory that accounts for the colors and the magnetic and other properties of transition-metal complexes in terms of the splitting of the energies of metal ion *d* orbitals by the electrostatic interaction with the ligands. (Section 24.6)

crystal lattice An imaginary network of points on which the repeating unit of the structure of a solid (the contents of the unit cell) may be imagined to be laid down so that the structure of the crystal is obtained. Each point represents an identical environment in the crystal. (Section 11.7)

crystalline solid (crystal) A solid whose internal arrangement of atoms, molecules, or ions shows a regular repetition in any direction through the solid. (Section 11.7)

crystallinity A measure of the extent of crystalline character (order) in a polymer. (Section 12.2)

crystallization The process in which a dissolved solute comes out of solution and forms a crystalline solid. (Section 13.2)

cubic close packing A close-packing arrangement in which the atoms of the third layer of a solid are not directly over those in the first layer. (Section 11.7)

curie A measure of radioactivity: 1 curie = 3.7×10^{10} nuclear disintegrations per second. (Section 21.4)

cycloalkanes Saturated hydrocarbons of general formula C_nH_{2n} in which the carbon atoms form a closed ring. (Section 25.3)

Dalton's law of partial pressures A law stating that the total pressure of a mixture of gases is the sum of the pressures that each gas would exert if it were present alone. (Section 10.6)

***d-d* transition** The transition of an electron from a lower-energy *d* orbital to a higher-energy *d* orbital. (Section 24.6)

decomposition reaction A chemical reaction in which a single compound reacts to give two or more products. (Section 3.2)

degenerate A situation in which two or more orbitals have the same energy. (Section 6.7)

delocalized electrons Electrons that are spread over a number of atoms in a molecule rather than localized between a pair of atoms. (Section 9.6)

density The ratio of an object's mass to its volume. (Section 1.4)

deoxyribonucleic acid (DNA) A polynucleotide in which the sugar component is deoxyribose. (Section 25.11)

desalination The removal of salts from seawater, brine, or brackish water to make it fit for human consumption. (Section 18.5)

deuterium The isotope of hydrogen whose nucleus contains a proton and a neutron: ${}^{2}_{1}H$. (Section 22.2)

dextrorotatory, or merely dextro or *d*. A term used to label a chiral molecule that rotates the plane of polarization of plane-polarized light to the right (clockwise). (Section 24.4)

diamagnetism A type of magnetism that causes a substance with no unpaired electrons to be weakly repelled from a magnetic field. (Section 9.8)

diatomic molecule A molecule composed of only two atoms. (Section 2.6)

diffusion The spreading of one substance through a space occupied by one or more other substances. (Section 10.8)

dilution The process of preparing a less concentrated solution from a more concentrated one by adding solvent. (Section 4.5)

dimensional analysis A method of problem solving in which units are carried through all calculations. Dimensional analysis ensures that the final answer of a calculation has the desired units. (Section 1.6)

dipole A molecule with one end having a slight negative charge and the other end having a slight positive charge; a polar molecule. (Section 8.4)

dipole-dipole force The force that exists because of the interactions of dipoles on polar molecules in close contact. (Section 11.2)

dipole moment A measure of the separation and magnitude of the positive and negative charges in polar molecules. (Section 8.4)

displacement reaction A reaction in which an element reacts with a compound, displacing an element from it. (Section 4.4)

disproportionation A reaction in which a species undergoes simultaneous oxidation and reduction [as in $N_2O_3(g) \longrightarrow NO(g) + NO_2(g)$.] (Section 22.5)

donor atom The atom of a ligand that bonds to the metal. (Section 24.1)

doping Incorporation of a hetero atom into a solid to change its electrical properties. For example, incorporation of P into Si. (Section 12.1)

double bond A covalent bond involving two electron pairs. (Section 8.3)

double helix The structure for DNA that involves the winding of two DNA polynucleotide chains together in a helical arrangement. The two strands of the double helix are complementary in that the organic bases on the two strands are paired for optimal hydrogen bond interaction. (Section 25.11)

Downs cell A cell used to obtain sodium metal by electrolysis of molten NaCl. (Section 23.4)

dynamic equilibrium A state of balance in which opposing processes occur at the same rate. (Section 11.5)

effective nuclear charge The net positive charge experienced by an electron in a many-electron atom; this charge is not the full nuclear charge because there is some shielding of the nucleus by the other electrons in the atom. (Section 7.2)

effusion The escape of a gas through an orifice or hole. (Section 10.8)

elastomer A material that can undergo a substantial change in shape via stretching, bending, or compression and return to its original shape upon release of the distorting force. (Section 12.2)

electrochemistry The branch of chemistry that deals with the relationships between electricity and chemical reactions. (Chapter 20: Introduction)

electrolysis reaction A reaction in which a nonspontaneous redox reaction is brought about by the passage of current under a sufficient external electrical potential. The devices in which electrolysis reactions occur are called electrolytic cells. (Section 20.9)

electrolyte A solute that produces ions in solution; an electrolytic solution conducts an electric current. (Section 4.1)

electrolytic cell A device in which a nonspontaneous oxidation-reduction reaction is caused to occur by passage of current under a sufficient external electrical potential. (Section 20.9)

electromagnetic radiation (radiant energy) A form of energy that has wave characteristics and that propagates through a vacuum at the characteristic speed of 3.00×10^8 m/s. (Section 6.1)

electrometallurgy The use of electrolysis to reduce or refine metals. (Section 23.4)

electromotive force (emf) A measure of the driving force, or *electrical pressure*, for the completion of an electrochemical reaction. Electromotive force is measured in volts: $1\ V = 1\ J/C$. Also called the cell potential. (Section 20.4)

electron A negatively charged subatomic particle found outside the atomic nucleus; it is a part of all atoms. An electron has a mass 1/1836 times that of a proton. (Section 2.3)

electron affinity The energy change that occurs when an electron is added to a gaseous atom or ion. (Section 7.5)

electron capture A mode of radioactive decay in which an inner-shell orbital electron is captured by the nucleus. (Section 21.1)

electron configuration A particular arrangement of electrons in the orbitals of an atom. (Section 6.8)

electron density The probability of finding an electron at any particular point in an atom; this probability is equal to ψ^2, the square of the wave function. (Section 6.5)

electron domain In the VSEPR model, regions about a central atom in which electrons are concentrated. (Section 9.2)

electron-domain geometry The three-dimensional arrangement of the electron domains around an atom according to the VSEPR model. (Section 9.2)

electronegativity A measure of the ability of an atom that is bonded to another atom to attract electrons to itself. (Section 8.4)

electronic charge The negative charge carried by an electron; it has a magnitude of 1.602×10^{-19} C. (Section 2.3)

electronic structure The arrangement of electrons of an atom or molecule. (Chapter 6: Introduction)

electron-sea model A model for the behavior of electrons in metals. (Section 23.5)

electron shell A collection of orbitals that have the same value of n. For example, the orbitals with $n = 3$ (the 3s, 3p, and 3d orbitals) comprise the third shell. (Section 6.5)

electron spin A property of the electron that makes it behave as though it were a tiny magnet. The electron behaves as if it were spinning on its axis; electron spin is quantized. (Section 6.7)

element A substance that cannot be separated into simpler substances by chemical means. (Sections 1.1 and 1.2)

elementary reaction A process in a chemical reaction that occurs in a single event or step. An overall chemical reaction consists of one or more elementary reactions or steps. (Section 14.6)

empirical formula (simplest formula) A chemical formula that shows the kinds of atoms and their relative numbers in a substance in the smallest possible whole-number ratios. (Section 2.6)

enantiomers Two mirror-image molecules of a chiral substance. The enantiomers are nonsuperimposable. (Section 24.4)

endothermic process A process in which a system absorbs heat from its surroundings. (Section 5.2)

energy The capacity to do work or to transfer heat. (Section 5.1)

energy-level diagram A diagram that shows the energies of molecular orbitals relative to the atomic orbitals from which they are derived. Also called a **molecular-orbital diagram**. (Section 9.7)

enthalpy A quantity defined by the relationship $H = E + PV$; the enthalpy change, ΔH, for a reaction that occurs at constant pressure is the heat evolved or absorbed in the reaction: $\Delta H = q_p$. (Section 5.3)

enthalpy (heat) of formation The enthalpy change that accompanies the formation of a substance from the most stable forms of its component elements. (Section 5.7)

enthalpy of reaction The enthalpy change associated with a chemical reaction. (Section 5.4)

entropy A thermodynamic function associated with the number of different equivalent energy states or spatial arrangements in which a system may be found. It is a thermodynamic state function, which means that once we specify the conditions for a system—that is, the temperature, pressure, and so on—the entropy is defined. (Sections 13.1 and 19.2)

enzyme A protein molecule that acts to catalyze specific biochemical reactions. (Section 14.7)

equilibrium constant The numerical value of the equilibrium-constant expression for a system at equilibrium. The equilibrium constant is most usually denoted by K_p for gas-phase systems or K_c for solution-phase systems. (Section 15.2)

equilibrium-constant expression The expression that describes the relationship among the concentrations (or partial pressures) of the substances present in a system at equilibrium. The numerator is obtained by multiplying the concentrations of the substances on the product side of the equation, each raised to a power equal to its coefficient in the chemical equation. The denominator similarly contains the concentrations of the substances on the reactant side of the equation. (Section 15.2)

equivalence point The point in a titration at which the added solute reacts completely with the solute present in the solution. (Section 4.6)

ester An organic compound that has an OR group attached to a carbonyl; it is the product of a reaction between a carboxylic acid and an alcohol. (Section 25.6)

ether A compound in which two hydrocarbon groups are bonded to one oxygen. (Section 25.5)

exchange (metathesis) reaction A reaction between compounds that when written as a molecular equation appears to involve the exchange of ions between the two reactants. (Section 4.2)

excited state A higher energy state than the ground state. (Section 6.3)

exothermic process A process in which a system releases heat to its surroundings. (Section 5.2)

extensive property A property that depends on the amount of material considered; for example, mass or volume. (Section 1.3)

face-centered cubic cell A cubic unit cell that has lattice points at each corner as well as at the center of each face. (Section 11.7)

faraday A unit of charge that equals the total charge of one mole of electrons: $1\ F = 96{,}500$ C. (Section 20.5)

***f*-block metals** Lanthanide and actinide elements in which the $4f$ or $5f$ orbitals are partially occupied. (Section 6.9)

ferromagnetism The ability of some substances to become permanently magnetized through alignment of unpaired electron spins. (Section 23.7)

first law of thermodynamics A statement of our experience that energy is conserved in any process. We can express the law in many ways. One of the more useful expressions is that the change in internal energy, ΔE, of a system in any process is equal to the heat, q, added to the system, plus the work, w, done on the system by its surroundings: $\Delta E = q + w$. (Section 5.2)

first-order reaction A reaction in which the reaction rate is proportional to the concentration of a single reactant, raised to the first power. (Section 14.4)

fission The splitting of a large nucleus into two smaller ones. (Section 21.6)

force A push or a pull. (Section 5.1)

formal charge The number of valence electrons in an isolated atom minus the number of electrons assigned to the atom in the Lewis structure. (Section 8.5)

formation constant For a metal ion complex, the equilibrium constant for formation of the complex from the metal ion and base species present in solution. It is a measure of the tendency of the complex to form. (Section 17.5)

formula weight The mass of the collection of atoms represented by a chemical formula. For example, the formula weight of NO_2 (46.0 amu) is the sum of the masses of one nitrogen atom and two oxygen atoms. (Section 3.3)

fossil fuels Coal, oil, and natural gas, which are presently our major sources of energy. (Section 5.8)

free energy (Gibbs free energy, *G*) A thermodynamic state function that gives a criterion for spontaneous change in terms of enthalpy and entropy: $G = H - TS$. (Section 19.5)

free radical A substance with one or more unpaired electrons. (Section 21.9)

frequency The number of times per second that one complete wavelength passes a given point. (Section 6.1)

frequency factor (*A*) A term in the Arrhenius equation that is related to the frequency of collision and the probability that the collisions are favorably oriented for reaction. (Section 14.5)

fuel cell A voltaic cell that utilizes the oxidation of a conventional fuel, such as H_2 or CH_4, in the cell reaction. (Section 20.7)

fuel value The energy released when 1 g of a substance is combusted. (Section 5.8)

functional group An atom or group of atoms that imparts characteristic chemical properties to an organic compound. (Section 25.1)

fusion The joining of two light nuclei to form a more massive one. (Section 21.6)

galvanic cell See **voltaic cell**. (Section 20.3)

gamma radiation Energetic electromagnetic radiation emanating from the nucleus of a radioactive atom. (Section 21.1)

gas Matter that has no fixed volume or shape; it conforms to the volume and shape of its container. (Section 1.2)

gas constant (*R*) The constant of proportionality in the ideal-gas equation. (Section 10.4)

Geiger counter A device that can detect and measure radioactivity. (Section 21.5)

geometric isomers Compounds with the same type and number of atoms and the same chemical bonds but different spatial arrangements of these atoms and bonds. (Sections 24.4 and 25.4)

Gibbs free energy A thermodynamic state function that combines enthalpy and entropy, in the form $G = H - TS$. For a change occurring at constant temperature and pressure, the change in free energy is $\Delta G = \Delta H - T\Delta S$. (Section 19.5)

glass An amorphous solid formed by fusion of SiO_2 CaO, and Na_2O. Other oxides may also be used to form glasses with differing characteristics. (Section 22.10)

glucose A polyhydroxy aldehyde whose formula is $CH_2OH(CHOH)_4CHO$; it is the most important of the monosaccharides. (Section 25.10)

glycogen The general name given to a group of polysaccharides of glucose that are synthesized in mammals and used to store energy from carbohydrates. (Section 25.10)

Graham's law A law stating that the rate of effusion of a gas is inversely proportional to the square root of its molecular weight. (Section 10.8)

gray (Gy) The SI unit for radiation dose corresponding to the absorption of 1 J of energy per kilogram of tissue; 1 Gy = 100 rads. (Section 21.9)

green chemistry Chemistry that promotes the design and application of chemical products and processes that are compatible with human health and that preserve the environment. (Section 18.7)

ground state The lowest-energy, or most stable, state. (Section 6.3)

group Elements that are in the same column of the periodic table; elements within the same group or family exhibit similarities in their chemical behavior. (Section 2.5)

Haber process The catalyst system and conditions of temperature and pressure developed by Fritz Haber and coworkers for the formation of NH_3 from H_2 and N_2 (Section 15.1)

half-life The time required for the concentration of a reactant substance to decrease to half its initial value; the time required for half of a sample of a particular radioisotope to decay. (Sections 14.4 and 21.4)

half-reaction An equation for either an oxidation or a reduction that explicitly shows the electrons involved, for example, $Zn^{2+}(aq) + 2e^- \longrightarrow Zn(s)$. (Section 20.2)

Hall process A process used to obtain aluminum by electrolysis of Al_2O_3 dissolved in molten cryolite, Na_3AlF_6. (Section 23.4)

halogens Members of group 7A in the periodic table. (Section 7.8)

hard water Water that contains appreciable concentrations of Ca^{2+} and Mg^{2+}; these ions react with soaps to form an insoluble material. (Section 18.6)

heat The flow of energy from a body at higher temperature to one at lower temperature when they are placed in thermal contact. (Section 5.1)

heat capacity The quantity of heat required to raise the temperature of a sample of matter by 1°C (or 1 K). (Section 5.5)

heat of fusion The enthalpy change, ΔH, for melting a solid. (Section 11.4)

heat of sublimation The enthalpy change, ΔH, for vaporization of a solid. (Section 11.4)

heat of vaporization The enthalpy change, ΔH, for vaporization of a liquid. (Section 11.4)

hemoglobin An iron-containing protein responsible for oxygen transport in the blood. (Section 18.4)

Henderson–Hasselbalch equation The relationship among the pH, pK_a, and the concentrations of acid and conjugate base in an aqueous solution: $pH = pK_a + \log\frac{[base]}{[acid]}$. (Section 17.2)

Henry's law A law stating that the concentration of a gas in a solution, C_g, is proportional to the pressure of gas over the solution: $C_g = kP_g$. (Section 13.3)

Hess's law The heat evolved in a given process can be expressed as the sum of the heats of several processes that, when added, yield the process of interest. (Section 5.6)

heterogeneous alloy An alloy in which the components are not distributed uniformly; instead, two or more distinct phases with characteristic compositions are present. (Section 23.6)

heterogeneous catalyst A catalyst that is in a different phase from that of the reactant substances. (Section 14.7)

heterogeneous equilibrium The equilibrium established between substances in two or more different phases, for example, between a gas and a solid or between a solid and a liquid. (Section 15.4)

hexagonal close packing A close-packing arrangement in which the atoms of the third layer of a solid lie directly over those in the first layer. (Section 11.7)

high-spin complex A complex whose electrons populate the *d* orbitals to give the maximum number of unpaired electrons. (Section 24.6)

high-temperature superconductivity The "frictionless" flow of electrical current (superconductivity) at temperatures above 30 K in certain complex metal oxides. (Section 12.1)

homogeneous catalyst A catalyst that is in the same phase as the reactant substances. (Section 14.7)

homogeneous equilibrium The equilibrium established between reactant and product substances that are all in the same phase. (Section 15.3)

Hund's rule A rule stating that electrons occupy degenerate orbitals in such a way as to maximize the number of electrons with the same spin. In other words, each orbital has one electron placed in it before pairing of electrons in orbitals occurs. Note that this rule applies only to orbitals that are degenerate, which means that they have the same energy. (Section 6.8)

hybridization The mixing of different types of atomic orbitals to produce a set of equivalent hybrid orbitals. (Section 9.5)

hybrid orbital An orbital that results from the mixing of different kinds of atomic orbitals on the same atom. For example, an sp^3 hybrid results from the mixing, or hybridizing, of one *s* orbital and three *p* orbitals. (Section 9.5)

hydration Solvation when the solvent is water. (Section 13.1)

hydride ion An ion formed by the addition of an electron to a hydrogen atom: H^-. (Section 7.7)

hydrocarbons Compounds composed of only carbon and hydrogen. (Section 2.9)

hydrogen bonding Bonding that results from intermolecular attractions between molecules containing hydrogen bonded to an electronegative element. The most important examples involve OH, NH, and HF. (Section 11.2)

hydrolysis A reaction with water. When a cation or anion reacts with water, it changes the pH. (Section 16.9)

hydrometallurgy Aqueous chemical processes for recovery of a metal from an ore. (Section 23.3)

hydronium ion (H_3O^+) The predominant form of the proton in aqueous solution. (Section 16.2)

hydrophilic Water attracting. The term is often used to describe a type of colloid. (Section 13.6)

hydrophobic Water repelling. The term is often used to describe a type of colloid. (Section 13.6)

hypothesis A tentative explanation of a series of observations or of a natural law. (Section 1.3)

ideal gas A hypothetical gas whose pressure, volume, and temperature behavior is completely described by the ideal-gas equation. (Section 10.4)

ideal-gas equation An equation of state for gases that embodies Boyle's law, Charles's law, and Avogadro's hypothesis in the form $PV = nRT$. (Section 10.4)

ideal solution A solution that obeys Raoult's law. (Section 13.5)

immiscible liquids Liquids that do not dissolve in one another to a significant extent. (Section 13.3)

indicator A substance added to a solution to indicate by a color change the point at which the added solute has reacted with all the solute present in solution. (Section 4.6)

instantaneous rate The reaction rate at a particular time as opposed to the average rate over an interval of time. (Section 14.2)

insulator A solid with extremely low electrical conductivity. (Section 12.1)

intensive property A property that is independent of the amount of material considered, for example, density. (Section 1.3)

interhalogens Compounds formed between two different halogen elements. Examples include IBr and BrF_3. (Section 22.4)

intermediate A substance formed in one elementary step of a multistep mechanism and consumed in another; it is neither a reactant nor an ultimate product of the overall reaction. (Section 14.6)

intermetallic compound A homogeneous alloy with definite properties and composition. Intermetallic compounds are stoichiometric compounds, but their compositions are not readily explained in terms of ordinary chemical bonding theory. (Section 23.6)

intermolecular forces The short-range attractive forces operating between the particles that make up the units of a liquid or solid substance. These same forces also cause gases to liquefy or solidify at low temperatures and high pressures. (Chapter 11: Introduction)

internal energy The total energy possessed by a system. When a system undergoes a change, the change in internal energy, ΔE, is defined as the heat, q, added to the system, plus the work, w, done on the system by its surroundings: $\Delta E = q + w$. (Section 5.2)

ion Electrically charged atom or group of atoms (polyatomic ion); ions can be positively or negatively charged, depending on whether electrons are lost (positive) or gained (negative) by the atoms. (Section 2.7)

ion-dipole force The force that exists between an ion and a neutral polar molecule that possesses a permanent dipole moment. (Section 11.2)

ion exchange A process in which ions in solution are exchanged for other ions held on the surface of an ion-exchange resin; the exchange of a hard-water cation such as Ca^{2+} for a soft-water cation such as Na^+ is used to soften water. (Section 18.6)

ionic bond A bond between oppositely charged ions. The ions are formed from atoms by transfer of one or more electrons. (Section 8.1)

ionic compound A compound composed of cations and anions. (Section 2.7)

ionic hydrides Compounds formed when hydrogen reacts with alkali metals and also the heavier alkaline earths (Ca, Sr, and Ba); these compounds contain the hydride ion, H^-. (Section 22.2)

ionic solids Solids that are composed of ions. (Section 11.8)

ionization energy The energy required to remove an electron from a gaseous atom when the atom is in its ground state. (Section 7.4)

ionizing radiation Radiation that has sufficient energy to remove an electron from a molecule, thereby ionizing it. (Section 21.9)

ion-product constant For water, K_w is the product of the aquated hydrogen ion and hydroxide ion concentrations: $[H^+][OH^-] = K_w = 1.0 \times 10^{-14}$ at 25°C. (Section 16.3)

irreversible process A process that is not reversible; and as a result, some of its potential for accomplishing work is dissipated as heat. Any spontaneous process is irreversible in practice. (Section 19.1)

isoelectronic series A series of atoms, ions, or molecules having the same number of electrons. (Section 7.3)

isomers Compounds whose molecules have the same overall composition but different structures. (Section 24.4)

isothermal process One that occurs at constant temperature. (Section 19.1)

isotopes Atoms of the same element containing different numbers of neutrons and therefore having different masses. (Section 2.3)

joule (J) The SI unit of energy, 1 kg-m^2/s^2. A related unit is the calorie: 4.184 J = cal. (Section 5.1)

Kelvin scale The absolute temperature scale; the SI unit for temperature is the kelvin. Zero on the Kelvin scale corresponds to −273.15°C; therefore, K = °C + 273.15. (Section 1.4)

ketone A compound in which the carbonyl group occurs at the interior of a carbon chain and is therefore flanked by carbon atoms. (Section 25.6)

kinetic energy The energy that an object possesses by virtue of its motion. (Section 5.1)

kinetic-molecular theory A set of assumptions about the nature of gases. These assumptions, when translated into mathematical form, yield the ideal-gas equation. (Section 10.7)

lanthanide contraction The gradual decrease in atomic and ionic radii with increasing atomic number among the lanthanide elements, atomic numbers 57 through 70. The decrease arises because of a gradual increase in effective nuclear charge through the lanthanide series. (Section 23.7)

lanthanide (rare earth) element Element in which the 4*f* subshell is only partially occupied. (Sections 6.8 and 6.9)

lattice energy The energy required to separate completely the ions in an ionic solid. (Section 8.2)

law of conservation of mass A scientific law stating that the total mass of the products of a chemical reaction is the same as the total mass of the reactants, so that mass remains constant during the reaction. (Section 3.1)

law of constant composition A law that states that the elemental composition of a pure compound is always the same, regardless of its source; also called the **law of definite proportions**. (Section 1.2)

law of definite proportions A law that states that the elemental composition of a pure substance is always the same, regardless of its source; also called the **law of constant composition**. (Section 1.2)

law of mass action The rules by which the equilibrium constant is expressed in terms of the concentrations of reactants and products, in accordance with the balanced chemical equation for the reaction. (Section 15.2)

leaching The selective dissolution of a desired mineral by passing an aqueous reagent solution through an ore. (Section 23.3)

Le Châtelier's principle A principle stating that when we disturb a system at chemical equilibrium, the relative concentrations of reactants and products shift so as to partially undo the effects of the disturbance. (Section 15.7)

levorotatory, or merely levo or *l*. A term used to label a chiral molecule that rotates the plane of polarization of plane-polarized light to the left (counterclockwise). (Section 24.4)

Lewis acid An electron-pair acceptor. (Section 16.11)

Lewis base An electron-pair donor. (Section 16.11)

Lewis structure A representation of covalent bonding in a molecule that is drawn using Lewis symbols. Shared electron pairs are shown as lines, and unshared electron pairs are shown as pairs of dots. Only the valence-shell electrons are shown. (Section 8.3)

Lewis symbol (electron-dot symbol) The chemical symbol for an element, with a dot for each valence electron. (Section 8.1)

ligand An ion or molecule that coordinates to a metal atom or to a metal ion to form a complex. (Section 24.1)

lime-soda process A method used in large-scale water treatment to reduce water hardness by removing Mg^{2+} and Ca^{2+}. The substances added to the water are lime, CaO [or slaked lime, $Ca(OH)_2$], and soda ash, Na_2CO_3, in amounts determined by the concentrations of the undesired ions. (Section 18.6)

limiting reactant (limiting reagent) The reactant present in the smallest stoichiometric quantity in a mixture of reactants; the amount of product that can form is limited by the complete consumption of the limiting reactant. (Section 3.7)

line spectrum A spectrum that contains radiation at only certain specific wavelengths. (Section 6.3)

linkage isomers Structural isomers of coordination compounds in which a ligand differs in its mode of attachment to a metal ion. (Section 24.4)

liquid Matter that has a distinct volume but no specific shape. (Section 1.2)

liquid crystal A substance that exhibits one or more partially ordered liquid phases above the melting point of the solid form. By contrast, in nonliquid crystalline substances the liquid phase that forms upon melting is completely unordered. (Section 12.5)

lithosphere That portion of our environment consisting of the solid Earth. (Section 23.1)

lock-and-key model A model of enzyme action in which the substrate molecule is pictured as fitting rather specifically into the active site on the enzyme. It is assumed that in being bound to the active site, the substrate is somehow activated for reaction. (Section 14.7)

London dispersion forces Intermolecular forces resulting from attractions between induced dipoles. (Section 11.2)

low-spin complex A metal complex in which the electrons are paired in lower-energy orbitals. (Section 24.6)

magic numbers Numbers of protons and neutrons that result in very stable nuclei. (Section 21.2)

main-group elements Elements in the *s* and *p* blocks of the periodic table. (Section 6.9)

mass A measure of the amount of material in an object. It measures the resistance of an object to being moved. In SI units, mass is measured in kilograms. (Section 1.4)

mass defect The difference between the mass of a nucleus and the total masses of the individual nucleons that it contains. (Section 21.6)

mass number The sum of the number of protons and neutrons in the nucleus of a particular atom. (Section 2.3)

mass percentage The number of grams of solute in each 100 g of solution. (Section 13.4)

mass spectrometer An instrument used to measure the precise masses and relative amounts of atomic and molecular ions. (Section 2.4)

matter Anything that occupies space and has mass; the physical material of the universe. (Section 1.1)

matter waves The term used to describe the wave characteristics of a particle. (Section 6.4)

mean free path The average distance traveled by a gas molecule between collisions. (Section 10.8)

metal complex (complex ion or complex) An assembly of a metal ion and the Lewis bases bonded to it. (Section 24.1)

metallic bond Bonding, usually in solid metals, in which the bonding electrons are relatively free to move throughout the three-dimensional structure. (Section 8.1)

metallic character The extent to which an element exhibits the physical and chemical properties characteristic of metals, for example, luster, malleability, ductility, and good thermal and electrical conductivity. (Section 7.6)

metallic elements (metals) Elements that are usually solids at room temperature, exhibit high electrical and heat conductivity, and appear lustrous. Most of the elements in the periodic table are metals. (Section 2.5)

metallic hydrides Compounds formed when hydrogen reacts with transition metals; these compounds contain the hydride ion, H^-. (Section 22.2)

metallic solids Solids that are composed of metal atoms. (Section 11.8)

metalloids Elements that lie along the diagonal line separating the metals from the nonmetals in the periodic table; the properties of metalloids are intermediate between those of metals and nonmetals. (Section 2.5)

metallurgy The science of extracting metals from their natural sources by a combination of chemical and physical processes. It is also concerned with the properties and structures of metals and alloys. (Section 23.1)

metathesis (exchange) reaction A reaction in which two substances react through an exchange of their component ions: $AX + BY \longrightarrow AY + BX$. Precipitation and acid-base neutralization reactions are examples of metathesis reactions. (Section 4.2)

metric system A system of measurement used in science and in most countries. The meter and the gram are examples of metric units. (Section 1.4)

microstate The state of a system at a particular instant; one of many possible states of the system. (Section 19.3)

mineral A solid, inorganic substance occurring in nature, such as calcium carbonate, which occurs as calcite. (Section 23.1)

miscible Liquids that mix in all proportions. (Section 13.3)

mixture A combination of two or more substances in which each substance retains its own chemical identity. (Section 1.2)

molal boiling-point-elevation constant (K_b) A constant characteristic of a particular solvent that gives the increase in boiling point as a function of solution molality: $\Delta T_b = K_b m$. (Section 13.5)

molal freezing-point-depression constant (K_f) A constant characteristic of a particular solvent that gives the decrease in freezing point as a function of solution molality: $\Delta T_f = K_f m$. (Section 13.5)

molality The concentration of a solution expressed as moles of solute per kilogram of solvent; abbreviated m. (Section 13.4)

molar heat capacity The heat required to raise the temperature of one mole of a substance by 1°C. (Section 5.5)

molarity The concentration of a solution expressed as moles of solute per liter of solution; abbreviated M. (Section 4.5)

molar mass The mass of one mole of a substance in grams; it is numerically equal to the formula weight in atomic mass units. (Section 3.4)

mole A collection of Avogadro's number (6.022×10^{23}) of objects; for example, a mole of H_2O is 6.022×10^{23} H_2O molecules. (Section 3.4)

molecular compound A compound that consists of molecules. (Section 2.6)

molecular equation A chemical equation in which the formula for each substance is written without regard for whether it is an electrolyte or a nonelectrolyte. (Section 4.2)

molecular formula A chemical formula that indicates the actual number of atoms of each element in one molecule of a substance. (Section 2.6)

molecular geometry The arrangement in space of the atoms of a molecule. (Section 9.2)

molecular hydrides Compounds formed when hydrogen reacts with nonmetals and metalloids. (Section 22.2)

molecularity The number of molecules that participate as reactants in an elementary reaction. (Section 14.6)

molecular orbital (MO) An allowed state for an electron in a molecule. According to molecular-orbital theory, a molecular orbital is entirely analogous to an atomic orbital, which is an allowed state for an electron in an atom. A molecular orbital may be classified as σ or π, depending on the disposition of electron density with respect to the internuclear axis. (Section 9.7)

molecular-orbital diagram A diagram that shows the energies of molecular orbitals relative to the atomic orbitals from which they are derived; also called an **energy-level diagram**. (Section 9.7)

molecular-orbital theory A theory that accounts for the allowed states for electrons in molecules. (Section 9.7)

molecular solids Solids that are composed of molecules. (Section 11.8)

molecular weight The mass of the collection of atoms represented by the chemical formula for a molecule. (Section 3.3)

molecule A chemical combination of two or more atoms. (Sections 1.1 and 2.6)

mole fraction The ratio of the number of moles of one component of a mixture to the total moles of all components; abbreviated X, with a subscript to identify the component. (Section 10.6)

momentum The product of the mass, m, and velocity, v, of a particle. (Section 6.4)

monodentate ligand A ligand that binds to the metal ion via a single donor atom. It occupies one position in the coordination sphere. (Section 24.2)

monomers Molecules with low molecular weights, which can be joined together (polymerized) to form a polymer. (Section 12.2)

monosaccharide A simple sugar, most commonly containing six carbon atoms. The joining together of monosaccharide units by condensation reactions results in formation of polysaccharides. (Section 25.10)

multiple bonding Bonding involving two or more electron pairs. (Section 8.3)

nanomaterial A material whose useful characteristics are the result of features in the range from 1 to 100 nm. (Section 12.6)

nanotechnology Technology that relies on the properties of matter at the nanoscale, that is, in the range from 1 to 100 nm. (Section 12.6)

natural gas A naturally occurring mixture of gaseous hydrocarbon compounds composed of hydrogen and carbon. (Section 5.8)

nematic liquid crystalline phase A liquid crystal in which the molecules are aligned in the same general direction, along their long axes, but in which the ends of the molecules are not aligned. (Section 12.5)

Nernst equation An equation that relates the cell emf, E, to the standard emf, $E°$, and the reaction quotient, $Q: E = E° - (RT/nF)\ln Q$. (Section 20.6)

net ionic equation A chemical equation for a solution reaction in which soluble strong electrolytes are written as ions and spectator ions are omitted. (Section 4.2)

neutralization reaction A reaction in which an acid and a base react in stoichiometrically equivalent amounts; the neutralization reaction between an acid and a metal hydroxide produces water and a salt. (Section 4.3)

neutron An electrically neutral particle found in the nucleus of an atom; it has approximately the same mass as a proton. (Section 2.3)

noble gases Members of group 8A in the periodic table. (Section 7.8)

node A locus of points in an atom at which the electron density is zero. For example, the node in a 2*s* orbital is a spherical surface. (Section 6.6)

nonbonding pair In a Lewis structure a pair of electrons assigned completely to one atom; also called a lone pair. (Section 9.2)

nonelectrolyte A substance that does not ionize in water and consequently gives a nonconducting solution. (Section 4.1)

nonionizing radiation Radiation that does not have sufficient energy to remove an electron from a molecule. (Section 21.9)

nonmetallic elements (nonmetals) Elements in the upper right corner of the periodic table; nonmetals differ from metals in their physical and chemical properties. (Section 2.5)

nonpolar covalent bond A covalent bond in which the electrons are shared equally. (Section 8.4)

normal boiling point The boiling point at 1 atm pressure. (Section 11.5)

normal melting point The melting point at 1 atm pressure. (Section 11.6)

nuclear binding energy The energy required to decompose an atomic nucleus into its component protons and neutrons. (Section 21.6)

nuclear disintegration series A series of nuclear reactions that begins with an unstable nucleus and terminates with a stable one; also called a **radioactive series.** (Section 21.2)

nuclear transmutation A conversion of one kind of nucleus to another. (Section 21.3)

nucleic acids Polymers of high molecular weight that carry genetic information and control protein synthesis. (Section 25.11)

nucleon A particle found in the nucleus of an atom. (Section 21.1)

nucleotide Compounds formed from a molecule of phosphoric acid, a sugar molecule, and an organic nitrogen base. Nucleotides form linear polymers called DNA and RNA, which are involved in protein synthesis and cell reproduction. (Section 25.11)

nucleus The very small, very dense, positively charged portion of an atom; it is composed of protons and neutrons. (Section 2.2)

nuclide A nucleus of a specific isotope of an element. (Section 2.3)

octet rule A rule stating that bonded atoms tend to possess or share a total of eight valence-shell electrons. (Section 8.1)

optical isomers Stereoisomers in which the two forms of the compound are nonsuperimposable mirror images. (Section 24.4)

optically active Possessing the ability to rotate the plane of polarized light. (Section 24.4)

orbital An allowed energy state of an electron in the quantum mechanical model of the atom; the term orbital is also used to describe the spatial distribution of the electron. An orbital is defined by the values of three quantum numbers: n, l, and m_l. (Section 6.5)

ore A source of a desired element or mineral, usually accompanied by large quantities of other materials such as sand and clay. (Section 23.1)

organic chemistry The study of carbon-containing compounds, typically containing carbon–carbon bonds. (Section 2.9; Chapter 25: Introduction)

osmosis The net movement of solvent through a semipermeable membrane toward the solution with greater solute concentration. (Section 13.5)

osmotic pressure The pressure that must be applied to a solution to stop osmosis from pure solvent into the solution. (Section 13.5)

Ostwald process An industrial process used to make nitric acid from ammonia. The NH_3 is catalytically oxidized by O_2 to form NO; NO in air is oxidized to NO_2; HNO_3 is formed in a disproportionation reaction when NO_2 dissolves in water. (Section 22.7)

overall reaction order The sum of the reaction orders of all the reactants appearing in the rate expression when the rate can be expressed as $rate = k[A]^a[B]^b \ldots$. (Section 14.3)

overlap The extent to which atomic orbitals on different atoms share the same region of space. When the overlap between two orbitals is large, a strong bond may be formed. (Section 9.4)

oxidation A process in which a substance loses one or more electrons. (Section 4.4)

oxidation number (oxidation state) A positive or negative whole number assigned to an element in a molecule or ion on the basis of a set of formal rules; to some degree it reflects the positive or negative character of that atom. (Section 4.4)

oxidation-reduction (redox) reaction A chemical reaction in which the oxidation states of certain atoms change. (Chapter 20: Introduction)

oxidizing agent, or oxidant The substance that is reduced and thereby causes the oxidation of some other substance in an oxidation-reduction reaction. (Section 20.1)

oxyacid A compound in which one or more OH groups, and possibly additional oxygen atoms, are bonded to a central atom. (Section 16.10)

oxyanion A polyatomic ion that contains one or more oxygen atoms. (Section 2.8)

ozone The name given to O_3, an allotrope of oxygen. (Section 7.8)

paramagnetism A property that a substance possesses if it contains one or more unpaired electrons. A paramagnetic substance is drawn into a magnetic field. (Section 9.8)

partial pressure The pressure exerted by a particular gas in a mixture. (Section 10.6)

particle accelerator A device that uses strong magnetic and electrostatic fields to accelerate charged particles. (Section 21.3)

parts per billion (ppb) The concentration of a solution in grams of solute per 10^9 (billion) grams of solution; equals micrograms of solute per liter of solution for aqueous solutions. (Section 13.4)

parts per million (ppm) The concentration of a solution in grams of solute per 10^6 (million) grams of solution; equals milligrams of solute per liter of solution for aqueous solutions. (Section 13.4)

pascal (Pa) The SI unit of pressure: $1\ \text{Pa} = \text{N/m}^2$. (Section 10.2)

Pauli exclusion principle A rule stating that no two electrons in an atom may have the same four quantum numbers (n, l, m_l, and m_s). As a reflection of this principle, there can be no more than two electrons in any one atomic orbital. (Section 6.7)

peptide bond A bond formed between two amino acids. (Section 25.9)

percent yield The ratio of the actual (experimental) yield of a product to its theoretical (calculated) yield, multiplied by 100. (Section 3.7)

period The row of elements that lie in a horizontal row in the periodic table. (Section 2.5)

periodic table The arrangement of elements in order of increasing atomic number, with elements having similar properties placed in vertical columns. (Section 2.5)

petroleum A naturally occurring combustible liquid composed of hundreds of hydrocarbons and other organic compounds. (Section 5.8)

pH The negative log in base 10 of the aquated hydrogen ion concentration: $pH = -\log[H^+]$. (Section 16.4)

pH titration curve See **titration curve.** (Section 17.3)

phase change The conversion of a substance from one state of matter to another. The phase changes we consider are melting and freezing (solid ↔ liquid), sublimation and deposition (solid ↔ gas), and vaporization and condensation (liquid ↔ gas). (Section 11.4)

phase diagram A graphic representation of the equilibria among the solid, liquid, and gaseous phases of a substance as a function of temperature and pressure. (Section 11.6)

photochemical smog A complex mixture of undesirable substances produced by the action of sunlight on an urban atmosphere polluted with automobile emissions. The major starting ingredients are nitrogen oxides and organic substances, notably olefins and aldehydes. (Section 18.4)

photodissociation The breaking of a molecule into two or more neutral fragments as a result of absorption of light. (Section 18.2)

photoelectric effect The emission of electrons from a metal surface induced by light. (Section 6.2)

photoionization The removal of an electron from an atom or molecule by absorption of light. (Section 18.2)

photon The smallest increment (a quantum) of radiant energy; a photon of light with frequency v has an energy equal to hv. (Section 6.2)

photosynthesis The process that occurs in plant leaves by which light energy is used to convert carbon dioxide and water to carbohydrates and oxygen. (Section 24.2)

physical changes Changes (such as a phase change) that occur with no change in chemical composition. (Section 1.3)

physical properties Properties that can be measured without changing the composition of a substance, for example, color and freezing point. (Section 1.3)

pi (π) bond A covalent bond in which electron density is concentrated above and below the line joining the bonded atoms. (Section 9.6)

pi (π) molecular orbital A molecular orbital that concentrates the electron density on opposite sides of a line that passes through the nuclei. (Section 9.8)

Planck's constant (h) The constant that relates the energy and frequency of a photon, $E = hv$. Its value is 6.626×10^{-34} J-s. (Section 6.2)

plastic A material that can be formed into particular shapes by application of heat and pressure. (Section 12.2)

polar covalent bond A covalent bond in which the electrons are not shared equally. (Section 8.4)

polarizability The ease with which the electron cloud of an atom or a molecule is distorted by an outside influence, thereby inducing a dipole moment. (Section 11.2)

polar molecule A molecule that possesses a nonzero dipole moment. (Section 8.4)

polyatomic ion An electrically charged group of two or more atoms. (Section 2.7)

polydentate ligand A ligand in which two or more donor atoms can coordinate to the same metal ion. (Section 24.2)

polymer A large molecule of high molecular mass, formed by the joining together, or polymerization, of a large number of molecules of low molecular mass. The individual molecules forming the polymer are called monomers. (Section 12.2)

polypeptide A polymer of amino acids that has a molecular weight of less than 10,000. (Section 25.9)

polyprotic acid A substance capable of dissociating more than one proton in water; H_2SO_4 is an example. (Section 16.6)

polysaccharide A substance made up of several monosaccharide units joined together. (Section 25.10)

porphyrin A complex derived from the porphine molecule. (Section 24.2)

positron A particle with the same mass as an electron but with a positive charge, symbol ${}^{0}_{1}e$. (Section 21.1)

potential energy The energy that an object possesses as a result of its composition or its position with respect to another object. (Section 5.1)

precipitate An insoluble substance that forms in, and separates from, a solution. (Section 4.2)

precipitation reaction A reaction that occurs between substances in solution in which one of the products is insoluble. (Section 4.2)

precision The closeness of agreement among several measurements of the same quantity; the reproducibility of a measurement. (Section 1.5)

pressure A measure of the force exerted on a unit area. In chemistry, pressure is often expressed in units of atmospheres (atm) or torr: 760 torr = 1 atm; in SI units pressure is expressed in pascals (Pa). (Section 10.2)

pressure-volume (PV) work Work performed by expansion of a gas against a resisting pressure. (Section 5.3)

primary structure The sequence of amino acids along a protein chain. (Section 25.9)

primitive cubic cell A cubic unit cell in which the lattice points are at the corners only. (Section 11.7)

probability density (ψ^2) A value that represents the probability that an electron will be found at a given point in space. (Section 6.5)

product A substance produced in a chemical reaction; it appears to the right of the arrow in a chemical equation. (Section 3.1)

property A characteristic that gives a sample of matter its unique identity. (Section 1.1)

protein A biopolymer formed from amino acids. (Section 25.9)

protium The most common isotope of hydrogen. (Section 22.2)

proton A positively charged subatomic particle found in the nucleus of an atom. (Section 2.3)

pure substance Matter that has a fixed composition and distinct properties. (Section 1.2)

pyrometallurgy A process in which heat converts a mineral in an ore from one chemical form to another and eventually to the free metal. (Section 23.2)

qualitative analysis The determination of the presence or absence of a particular substance in a mixture. (Section 17.7)

quantitative analysis The determination of the amount of a given substance that is present in a sample. (Section 17.7)

quantum The smallest increment of radiant energy that may be absorbed or emitted; the magnitude of radiant energy is $h\nu$. (Section 6.2)

racemic mixture A mixture of equal amounts of the dextrorotatory and levorotatory forms of a chiral molecule. A racemic mixture will not rotate polarized light. (Section 24.4)

rad A measure of the energy absorbed from radiation by tissue or other biological material; 1 rad = transfer of 1×10^{-2} J of energy per kilogram of material. (Section 21.9)

radial probability function The probability that the electron will be found at a certain distance from the nucleus. (Section 6.6)

radioactive Possessing radioactivity, the spontaneous disintegration of an unstable atomic nucleus with accompanying emission of radiation. (Section 2.2; Chapter 21: Introduction)

radioactive series A series of nuclear reactions that begins with an unstable nucleus and terminates with a stable one. Also called **nuclear disintegration series.** (Section 21.2)

radioisotope An isotope that is radioactive; that is, it is undergoing nuclear changes with emission of radiation. (Section 21.1)

radionuclide A radioactive nuclide. (Section 21.1)

radiotracer A radioisotope that can be used to trace the path of an element in a chemical system. (Section 21.5)

Raoult's law A law stating that the partial pressure of a solvent over a solution, P_A, is given by the vapor pressure of the pure solvent, P_A°, times the mole fraction of a solvent in the solution, $X_A : P_A = X_A P_A^\circ$. (Section 13.5)

rare earth element See **lanthanide element.** (Sections 6.8 and 6.9)

rate constant A constant of proportionality between the reaction rate and the concentrations of reactants that appear in the rate law. (Section 14.3)

rate-determining step The slowest elementary step in a reaction mechanism. (Section 14.6)

rate law An equation that relates the reaction rate to the concentrations of reactants (and sometimes of products also). (Section 14.3)

reactant A starting substance in a chemical reaction; it appears to the left of the arrow in a chemical equation. (Section 3.1)

reaction mechanism A detailed picture, or model, of how the reaction occurs; that is, the order in which bonds are broken and formed and the changes in relative positions of the atoms as the reaction proceeds. (Section 14.6)

reaction order The power to which the concentration of a reactant is raised in a rate law. (Section 14.3)

reaction quotient (Q) The value that is obtained when concentrations of reactants and products are inserted into the equilibrium expression. If the concentrations are equilibrium concentrations, $Q = K$; otherwise, $Q \neq K$. (Section 15.6)

reaction rate The decrease in concentration of a reactant or the increase in concentration of a product with time. (Section 14.2)

redox (oxidation-reduction) reaction A reaction in which certain atoms undergo changes in oxidation states. The substance increasing in oxidation state is oxidized; the substance decreasing in oxidation state is reduced. (Chapter 20: Introduction)

reducing agent, or reductant The substance that is oxidized and thereby causes the reduction of some other substance in an oxidation-reduction reaction. (Section 20.1)

reduction A process in which a substance gains one or more electrons. (Section 4.4)

refining The process of converting an impure form of a metal into a more usable substance of well-defined composition. For example, crude pig iron from the blast furnace is refined in a converter to produce steels of desired compositions. (Section 23.2)

rem A measure of the biological damage caused by radiation; rems = rads × RBE. (Section 21.9)

renewable energy Energy such as solar energy, wind energy, and hydroelectric energy derived from essentially inexhaustible sources. (Section 5.8)

representative (main-group) element An element from within the *s* and *p* blocks of the periodic table (Figure 6.29). (Section 6.9)

resonance structures (resonance forms) Individual Lewis structures in cases where two or more Lewis structures are equally good descriptions of a single molecule. The resonance structures in such an instance are "averaged" to give a more accurate description of the real molecule. (Section 8.6)

reverse osmosis The process by which water molecules move under high pressure through a semipermeable membrane from the more concentrated to the less concentrated solution. (Section 18.5)

reversible process A process that can go back and forth between states along exactly the same path; a system at equilibrium is reversible if equilibrium can be shifted by an infinitesimal modification of a variable such as temperature. (Section 19.1)

ribonucleic acid (RNA) A polynucleotide in which ribose is the sugar component. (Section 25.11)

roasting Thermal treatment of an ore to bring about chemical reactions involving the furnace atmosphere. For example, a sulfide ore might be roasted in air to form a metal oxide and SO_2. (Section 23.2)

root-mean-square (rms) speed (μ) The square root of the average of the squared speeds of the gas molecules in a gas sample. (Section 10.7)

rotational motion Movement of a molecule as though it is spinning like a top. (Section 19.3)

salinity A measure of the salt content of seawater, brine, or brackish water. It is equal to the mass in grams of dissolved salts present in 1 kg of seawater. (Section 18.5)

salt An ionic compound formed by replacing one or more H^+ of an acid by other cations. (Section 4.3)

saponification Hydrolysis of an ester in the presence of a base. (Section 25.6)

saturated solution A solution in which undissolved solute and dissolved solute are in equilibrium. (Section 13.2)

scientific law A concise verbal statement or a mathematical equation that summarizes a wide range of observations and experiences. (Section 1.3)

scientific method The general process of advancing scientific knowledge by making experimental observations and by formulating hypotheses, theories, and laws. (Section 1.3)

scintillation counter An instrument that is used to detect and measure radiation by the fluorescence it produces in a fluorescing medium. (Section 21.5)

secondary structure The manner in which a protein is coiled or stretched. (Section 25.9)

second law of thermodynamics A statement of our experience that there is a direction to the way events occur in nature. When a process occurs spontaneously in one direction, it is nonspontaneous in the reverse direction. It is possible to state the second law in many different forms, but they all relate back to the same idea about spontaneity. One of the most common statements found in chemical contexts is that in any spontaneous process the entropy of the universe increases. (Section 19.2)

second-order reaction A reaction in which the overall reaction order (the sum of the concentration-term exponents) in the rate law is 2. (Section 14.4)

semiconductor A solid with limited electrical conductivity. (Section 12.1)

sigma (σ) bond A covalent bond in which electron density is concentrated along the internuclear axis. (Section 9.6)

sigma (σ) molecular orbital A molecular orbital that centers the electron density about an imaginary line passing through two nuclei. (Section 9.7)

significant figures The digits that indicate the precision with which a measurement is made; all digits of a measured quantity are significant, including the last digit, which is uncertain. (Section 1.5)

silicates Compounds containing silicon and oxygen, structurally based on SiO_4 tetrahedra. (Section 22.10)

single bond A covalent bond involving one electron pair. (Section 8.3)

SI units The preferred metric units for use in science. (Section 1.4)

slag A mixture of molten silicate minerals. Slags may be acidic or basic, according to the acidity or basicity of the oxide added to silica. (Section 23.2)

smectic liquid crystalline phase A liquid crystal in which the molecules are aligned along their long axes and arranged in sheets, with the ends of the molecules aligned. There are several different kinds of smectic phases. (Section 12.5)

smelting A melting process in which the materials formed in the course of the chemical reactions that occur separate into two or more layers. For example, the layers might be slag and molten metal. (Section 23.2)

sol-gel process A process in which extremely small particles (0.003 to 0.1 μm in diameter) of uniform size are produced in a series of chemical steps, followed by controlled heating. (Section 12.2)

solid Matter that has both a definite shape and a definite volume. (Section 1.2)

solubility The amount of a substance that dissolves in a given quantity of solvent at a given temperature to form a saturated solution. (Sections 4.2 and 13.2)

solubility-product constant (solubility product) (K_{sp}) An equilibrium constant related to the equilibrium between a solid salt and its ions in solution. It provides a quantitative measure of the solubility of a slightly soluble salt. (Section 17.4)

solute A substance dissolved in a solvent to form a solution; it is normally the component of a solution present in the smaller amount. (Section 4.1)

solution A mixture of substances that has a uniform composition; a homogeneous mixture. (Section 1.2)

solution alloy A homogeneous alloy, with the components distributed uniformly throughout. (Section 23.6)

solvation The clustering of solvent molecules around a solute particle. (Section 13.1)

solvent The dissolving medium of a solution; it is normally the component of a solution present in the greater amount. (Section 4.1)

specific heat The heat capacity of 1 g of a substance; the heat required to raise the temperature of 1 g of a substance by 1°C. (Section 5.5)

spectator ions Ions that go through a reaction unchanged and that appear on both sides of the complete ionic equation. (Section 4.2)

spectrochemical series A list of ligands arranged in order of their abilities to split the *d*-orbital energies (using the terminology of the crystal-field model). (Section 24.6)

spectrum The distribution among various wavelengths of the radiant energy emitted or absorbed by an object. (Section 6.3)

spin magnetic quantum number (m_s) A quantum number associated with the electron spin; it may have values of $+\frac{1}{2}$ or $-\frac{1}{2}$. (Section 6.7)

spin-pairing energy The energy required to pair an electron with another electron occupying an orbital. (Section 24.6)

spontaneous process A process that is capable of proceeding in a given direction, as written or described, without needing to be driven by an outside source of energy. A process may be spontaneous even though it is very slow. (Section 19.1)

standard atmospheric pressure Defined as 760 torr or, in SI units, 101.325 kPa. (Section 10.2)

standard electrode potential See **standard reduction potential**. (Section 20.4)

standard emf, also called the standard cell potential ($E°$) The emf of a cell when all reagents are at standard conditions. (Section 20.4)

standard enthalpy change ($\Delta H°$) The change in enthalpy in a process when all reactants and products are in their stable forms at 1 atm pressure and a specified temperature, commonly 25°C. (Section 5.7)

standard enthalpy of formation ($\Delta H_f°$) The change in enthalpy that accompanies the formation of one mole of a substance from its elements, with all substances in their standard states. (Section 5.7)

standard free energy of formation ($\Delta G_f°$) The change in free energy associated with the formation of a substance from its elements under standard conditions. (Section 19.5)

standard hydrogen electrode An electrode based on the half-reaction $2 H^+(1 M) + 2 e^- \longrightarrow H_2(1 \text{ atm})$. The standard electrode potential of the standard hydrogen electrode is defined as 0 V. (Section 20.4)

standard molar entropy ($S°$) The entropy value for a mole of a substance in its standard state. (Section 19.4)

standard reduction potential ($E°_{red}$) The potential of a reduction half-reaction under standard conditions, measured relative to the standard hydrogen electrode. A standard reduction potential is also called a **standard electrode potential**. (Section 20.4)

standard solution A solution of known concentration. (Section 4.6)

standard temperature and pressure (STP) Defined as 0°C and 1 atm pressure; frequently used as reference conditions for a gas. (Section 10.4)

starch The general name given to a group of polysaccharides that acts as energy-storage substances in plants. (Section 25.10)

state function A property of a system that is determined by the state or condition of the system and not by how it got to that state; its value is fixed when temperature, pressure, composition, and physical form are specified; *P*, *V*, *T*, *E*, and *H* are state functions. (Section 5.2)

states of matter The three forms that matter can assume: solid, liquid, and gas. (Section 1.2)

stereoisomers Compounds possessing the same formula and bonding arrangement but differing in the spatial arrangements of the atoms. (Section 24.4)

stoichiometry The relationships among the quantities of reactants and products involved in chemical reactions. (Chapter 3: Introduction)

stratosphere The region of the atmosphere directly above the troposphere. (Section 18.1)

strong acid An acid that ionizes completely in water. (Section 4.3)

strong base A base that ionizes completely in water. (Section 4.3)

strong electrolyte A substance (strong acids, strong bases, and most salts) that is completely ionized in solution. (Section 4.1)

structural formula A formula that shows not only the number and kinds of atoms in the molecule but also the arrangement (connections) of the atoms. (Section 2.6)

structural isomers Compounds possessing the same formula but differing in the bonding arrangements of the atoms. (Sections 24.4 and 25.3)

subatomic particles Particles such as protons, neutrons, and electrons that are smaller than an atom. (Section 2.2)

subshell One or more orbitals with the same set of quantum numbers *n* and *l*. For example, we speak of the 2*p* subshell ($n = 2, l = 1$), which is composed of three orbitals ($2p_x$, $2p_y$, and $2p_z$). (Section 6.5)

substitution reactions Reactions in which one atom (or group of atoms) replaces another atom (or group) within a molecule; substitution reactions are typical for alkanes and aromatic hydrocarbons. (Section 25.4)

substrate A substance that undergoes a reaction at the active site in an enzyme. (Section 14.7)

superconducting ceramic A complex metal oxide that undergoes a transition to a superconducting state at a low temperature. (Section 12.1)

superconducting transition temperature (T_c) The temperature below which a substance exhibits superconductivity. (Section 12.1)

superconductivity The "frictionless" flow of electrons that occurs when a substance loses all resistance to the flow of electrical current. (Section 12.1)

supercritical mass An amount of fissionable material larger than the critical mass. (Section 21.7)

supersaturated solutions Solutions containing more solute than an equivalent saturated solution. (Section 13.2)

surface tension The intermolecular, cohesive attraction that causes a liquid to minimize its surface area. (Section 11.3)

surroundings In thermodynamics, everything that lies outside the system that we study. (Section 5.1)

system In thermodynamics, the portion of the universe that we single out for study. We must be careful to state exactly what the system contains and what transfers of energy it may have with its surroundings. (Section 5.1)

termolecular reaction An elementary reaction that involves three molecules. Termolecular reactions are rare. (Section 14.6)

tertiary structure The overall shape of a large protein, specifically, the manner in which sections of the protein fold back upon themselves or intertwine. (Section 25.9)

theoretical yield The quantity of product that is calculated to form when all of the limiting reagent reacts. (Section 3.7)

theory A tested model or explanation that satisfactorily accounts for a certain set of phenomena. (Section 1.3)

thermochemistry The relationship between chemical reactions and energy changes. (Chapter 5: Introduction)

thermodynamics The study of energy and its transformation. (Chapter 5: Introduction)

thermonuclear reaction Another name for fusion reactions; reactions in which two light nuclei are joined to form a more massive one. (Section 21.8)

thermoplastic A polymeric material that can be readily reshaped by application of heat and pressure. (Section 12.2)

thermosetting plastic A plastic that, once formed in a particular mold, is not readily reshaped by application of heat and pressure. (Section 12.2)

third law of thermodynamics A law stating that the entropy of a pure, crystalline solid at absolute zero temperature is zero: $S(0\text{ K}) = 0$. (Section 19.3)

titration The process of reacting a solution of unknown concentration with one of known concentration (a standard solution). (Section 4.6)

titration curve A graph of pH as a function of added titrant. (Section 17.3)

torr A unit of pressure (1 torr = 1 mm Hg). (Section 10.2)

transistor An electrical device that forms the heart of an integrated circuit. (Section 12.4)

transition elements (transition metals) Elements in which the *d* orbitals are partially occupied. (Section 6.8)

transition state (activated complex) The particular arrangement of reactant and product molecules at the point of maximum energy in the rate-determining step of a reaction. (Section 14.5)

translational motion Movement in which an entire molecule moves in a definite direction. (Section 19.3)

transuranium elements Elements that follow uranium in the periodic table. (Section 21.3)

triple bond A covalent bond involving three electron pairs. (Section 8.3)

triple point The temperature at which solid, liquid, and gas phases coexist in equilibrium. (Section 11.6)

tritium The isotope of hydrogen whose nucleus contains a proton and two neutrons. (Section 22.2)

troposphere The region of Earth's atmosphere extending from the surface to about 12 km altitude. (Section 18.1)

Tyndall effect The scattering of a beam of visible light by the particles in a colloidal dispersion. (Section 13.6)

uncertainty principle A principle stating there is an inherent uncertainty in the precision with which we can simultaneously specify the position and momentum of a particle. This uncertainty is significant only for extremely small particles, such as electrons. (Section 6.4)

unimolecular reaction An elementary reaction that involves a single molecule. (Section 14.6)

unit cell The smallest portion of a crystal that reproduces the structure of the entire crystal when repeated in different directions in space. It is the repeating unit or building block of the crystal lattice. (Section 11.7)

unsaturated solutions Solutions containing less solute than a saturated solution. (Section 13.2)

valence band A band of closely spaced molecular orbitals that is essentially fully occupied by electrons. (Section 12.1)

valence-bond theory A model of chemical bonding in which an electron-pair bond is formed between two atoms by the overlap of orbitals on the two atoms. (Section 9.4)

valence electrons The outermost electrons of an atom; those that occupy orbitals not occupied in the nearest noble-gas element of lower atomic number. The valence electrons are the ones the atom uses in bonding. (Section 6.8)

valence orbitals Orbitals that contain the outer-shell electrons of an atom. (Chapter 7: Introduction)

valence-shell electron-pair repulsion (VSEPR) model A model that accounts for the geometric arrangements of shared and unshared electron pairs around a central atom in terms of the repulsions between electron pairs. (Section 9.2)

van der Waals equation An equation of state for nonideal gases that is based on adding corrections to the ideal-gas equation. The correction terms account for intermolecular forces of attraction and for the volumes occupied by the gas molecules themselves. (Section 10.9)

vapor Gaseous state of any substance that normally exists as a liquid or solid. (Section 10.1)

vapor pressure The pressure exerted by a vapor in equilibrium with its liquid or solid phase. (Section 11.5)

vibrational motion Movement of the atoms within a molecule in which they move periodically toward and away from one another. (Section 19.3)

viscosity A measure of the resistance of fluids to flow. (Section 11.3)

volatile Tending to evaporate readily. (Section 11.5)

voltaic (galvanic) cell A device in which a spontaneous oxidation-reduction reaction occurs with the passage of electrons through an external circuit. (Section 20.3)

vulcanization The process of cross-linking polymer chains in rubber. (Section 12.2)

watt A unit of power; 1 W = 1 J/s. (Section 20.9)

wave function A mathematical description of an allowed energy state (an orbital) for an electron in the quantum mechanical model of the atom; it is usually symbolized by the Greek letter ψ. (Section 6.5)

wavelength The distance between identical points on successive waves. (Section 6.1)

weak acid An acid that only partly ionizes in water. (Section 4.3)

weak base A base that only partly ionizes in water. (Section 4.3)

weak electrolyte A substance that only partly ionizes in solution. (Section 4.1)

work The movement of an object against some force. (Section 5.1)

Photo/Art Credits

FM.1 Dr. Theodore L. Brown **FM.2** H. Eugene LeMay, Jr. **FM.3** Bruce E. Bursten

Chapter 1: CO01 NASA, H. Ford (JHU), G. Illingworth (USCS/LO), M. Clampin (STScI), G. Hartig (STScI), the ACS Science Team and ESA **1.2a** Andrew Syred/SPL/Photo Researchers, Inc. **1.2b** Francis G. Mayer/Corbis/Bettmann **1.2c** G. Murti/Photo Researchers, Inc. **1.3** Richard Megna/Fundamental Photographs, NYC **1.4** Dale Wilson/Green Stock/Corbis/Bettmann **1.7** Charles D. Winters/Photo Researchers, Inc. **1.8a** M. Angelo/Corbis/Bettmann **1.8b** Richard Megna/Fundamental Photographs, NYC **1.11a–c** Donald Clegg and Roxy Wilson/Pearson Education/PH College **1.12a–b** Donald Clegg and Roxy Wilson/Pearson Education/PH College **1.14a–c** Richard Megna/Fundamental Photographs, NYC **1.16** © Rachael Epstein/PhotoEdit Inc **1.17** Australian Postal Service/National Standards Commission **1.21** Copyright 2004 General Motors Corp. Used with permission, GM Media Archive. **1.25.1UN** ©2006 by Sidney Harris

Chapter 2: CO02 Courtesy of IBM Archives. Unauthorized use not permitted. **2.1** Corbis/Bettmann **2.2** IBMRL/Visuals Unlimited **2.3b–c** Richard Megna/Fundamental Photographs, NYC **2.6** Radium Institute, Courtesy AIP Emilio Segre Visual Archives **2.7** G.R. "Dick" Roberts Photo Library/The Natural Sciences Image Library (NSIL) **2.12** Stephen Frisch/Stock Boston **2.17** Richard Megna/Fundamental Photographs, NYC **2.18** Ernest Orlando Lawrence Berkeley National Laboratory, Courtesy AIP Emilio Segre Visual Archives **2.23c** Andrew Syred/Science Photo Library/Photo Researchers, Inc. **2.25** Richard Megna/Fundamental Photographs, NYC

Chapter 3: CO03 Richard Megna/Fundamental Photographs, NYC **3.1** Jean-Loup Charmet/Science Photo Library/Photo Researchers, Inc. **3.3** Charles D. Winters/Photo Researchers, Inc. **3.3.1UN** Dave Carpenter **3.5a–c** Richard Megna/Fundamental Photographs, NYC **3.6** Donald Johnston/Getty Images Inc. - Stone Allstock **3.7** Richard Megna/Fundamental Photographs, NYC **3.9** Richard Megna/Fundamental Photographs, NYC **3E.65** Carey B. Van Loon/Carey B. Van Loon **3E.73** Paul Silverman/Fundamental Photographs, NYC

Chapter 4: CO04 Everton, Macduff/Getty Images Inc. - Image Bank **4.1** Richard Megna/Fundamental Photographs, NYC **4.2a–c** Richard Megna/Fundamental Photographs, NYC **4.4.1–3** Richard Megna/Fundamental Photographs, NYC **4.5** Robert Mathena/Fundamental Photographs, NYC **4.7** Richard Megna/Fundamental Photographs, NYC **4.8a–c** Richard Megna/Fundamental Photographs, NYC **4.9** Richard Megna/Fundamental Photographs, NYC **4.10** Robert Mathena/Fundamental Photographs, NYC **4.11** Jorg Heimann/Bilderberg/Peter Arnold, Inc. **4.12** Richard Megna/Fundamental Photographs, NYC **4.13** Richard Megna/Fundamental Photographs, NYC **4.14a–c** Peticolas/Megna/Fundamental Photographs, NYC **4.15** Erich Lessing/Art Resource, N.Y. **4.16a–d** Donald Clegg and Roxy Wilson/Pearson Education/PH College **4.17** Al Bello/Getty Images, Inc - Liaison **4.18a–c** Richard Megna/Fundamental Photographs, NYC **4.20a–c** Richard Megna/Fundamental Photographs, NYC **4.21.6UN** Richard Megna/Fundamental Photographs, NYC

Chapter 5: CO05 H Richard Johnston/Getty Images Inc. - Stone Allstock **5.1a** Amoz Eckerson/Visuals Unlimited **5.1b** Tom Pantages/Tom Pantages **5.8a–b** Richard Megna/Fundamental Photographs, NYC **5.14a–b** Donald Clegg and Roxy Wilson/Pearson Education/PH College **5.15** UPI/Corbis/Bettmann **5.20** Gerard Vandystadt/Photo Researchers, Inc. **5.23** Robb Gregg/PhotoEdit **5.25** AP Wide World Photos

Chapter 6: CO06 Richard Cummins/The Viesti Collection, Inc. **6.1** Pal Hermansen/Getty Images Inc. - Stone Allstock **6.5** AGE/Peter Arnold, Inc. **6.6a** Laura Martin/Visuals Unlimited **6.6b** PhotoDisc/Getty Images, Inc. - PhotoDisc **6.8** Photograph by Paul Ehrenfest, courtesy AIP Emilio Segre Visual Archives, Ehrenfest Collection **6.9** Digital Vision Ltd. **6.11a–b** Tom Pantages/Tom Pantages **6.14** National Institute for Biological Standards and Control (U.K.)/Science Photo Library/Photo Researchers, Inc. **6.15** AIP Emilio Segre Visual Archives **6.28** Alfred Pasieka/Science Photo Library/Photo Researchers, Inc. **6.30.2UN** E.R. Degginger/Color-Pic, Inc. **6.30.3UN** Getty Images - Photodisc **6.30.4UN** E.R. Degginger/Color-Pic, Inc.

Chapter 7: CO07 Claude Monet (1840–1926) "Rue Montorgueil in Paris, Festival of 30 June 1878" 1878. Herve Lewandowski/Reunion des Musees Nationaux/Art Resource, NY. **7.1** Richard Megna/Fundamental Photographs, NYC **7.9.1** 1ca2: A.E. Eriksson, T.A. Jones, A. Liljas: Refined structure of human anhydrase II at 2.0 Angstrom resolution. Proteins 4 pp. 274 (1988). PDB: H.M. Berman, J. Westbrook, Z. Feng, G. Gilliland, T.N. Bhat, H. Weissig, I.N. Shindyalov, P.E. Bourne: The Protein Data Bank. Nucleic Acids Research, 28 pp. 235–242 (2000). Protein Data Bank (PDB) ID 1ca2 from *www.pdb.org*. **7.14** © Judith Miller/Dorling Kindersley/Freeman's **7.16a–b** Richard Megna/Fundamental Photographs, NYC **7.17** Ed Degginger/ Color-Pic, Inc. **7.18a–b** Richard Megna/Fundamental Photographs, NYC **7.19** Richard Megna/Fundamental Photographs, NYC **7.20** Richard Megna/Fundamental Photographs, NYC **7.21** Jeff Daly/Visuals Unlimited **7.22a–c** Richard Megna/Fundamental Photographs, NYC **7.23a–c** H. Eugene LeMay, Jr./H. Eugene LeMay, Jr. **7.24** Phil Degginger/Color-Pic, Inc. **7.25** Paparazzi Photography Studio - Dr. Pepper/Seven Up and Jeffrey L. Rodengen, Write Stuff Syndicte, Inc. **7.26** Tom Pantages **7.27** RNHRD NHS Trust/Getty Images Inc. - Stone Allstock **7.28** JPL/Space Science Institute/NASA Headquarters **7.29** Ed Degginger/Color-Pic, Inc. **7.31** Richard Megna/Fundamental Photographs, NYC **7.32** Ed Degginger/Color-Pic, Inc. **7.32.2UN** Phil Degginger/ Color-Pic, Inc.

Chapter 8: CO08 Richard Megna/Fundamental Photographs, NYC **8.1a–c** Richard Megna/Fundamental Photographs, NYC **8.2a–c** Donald Clegg and Roxy Wilson/Pearson Education/PH College **8.13a** Tom Pantages **8.15** Corbis/Bettmann

Chapter 9: CO09 Inga Spence/Visuals Unlimited **9.5a–c** Kristen Brochmann/Fundamental Photographs, NYC **9.31** Bill Longcore/Photo Researchers, Inc. **9.47** Richard Megna/Fundamental Photographs, NYC **9.49a–b** Phil Degginger/Color-Pic, Inc.

Chapter 10: CO10 Steve Bloom/Taxi/Getty Images **10.4** Yoav Levy/Phototake NYC **10.5** Roland Seitre/Peter Arnold, Inc. **10.8a–b** Richard Megna/Fundamental Photographs, NYC **10.14** Color-Pic, Inc. **10.15** Lowell Georgia/Corbis/Bettmann **10.21a–b** Richard Megna/Fundamental Photographs, NYC

Chapter 11: CO11 Chase Swift/Corbis/Bettmann **11.9** Richard Megna/Fundamental Photographs, NYC **11.10.1UN** Calvin and Hobbes © Watterson. Dist. by Universal Press Syndicate. Reprinted with permission. All rights reserved. **11.10c** Astrid and Hanns-Frieder Michler/Peter Arnold, Inc. **11.11** Richard Megna/Fundamental Photographs, NYC **11.13** Kristen Brochmann/Fundamental Photographs, NYC **11.14** Hermann Eisenbeiss/Photo Researchers, Inc. **11.16** Richard Megna/Fundamental Photographs, NYC **11.29a** Dan McCoy/Rainbow **11.29b** Herve Berthoule/Jacana Scientific Control/Photo Researchers, Inc. **11.29c** Michael Dalton/Fundamental Photographs, NYC **11.39** Science Source/Photo Researchers, Inc. **11.43** Phil Degginger/Merck/Color-Pic, Inc. **11.44** Robert L. Whetten, University of California at Los Angeles **11.45.8** Tony Mendoza/Stock Boston

Chapter 12: CO12 Corbis Royalty Free **12.4** David Parker/IMI/University of Birmingham High TC Consortium/Science Photo Library/Photo Researchers, Inc. **12.5** National Railway Japan/Phototake NYC **12.6** Mauro Fermariello/Science Photo Library/Photo Researchers, Inc. **12.8** Darren McCollester/Getty Images, Inc - Liaison **12.9** © Copyright and courtesy Superconductor Technologies Inc. **12.10** Richard Megna/Fundamental Photographs, NYC **12.13a–b** Richard Megna/Fundamental Photographs, NYC **12.16** Tom Pantages **12.17** Leonard Lessin/Peter Arnold, Inc. **12.18** Copyright 2004 General Motors Corp. Used with permission of GM Media Archives. **12.19** James L. Amos/Peter Arnold, Inc. **12.20** Professor Charles Zukoski, Department of Chemical Engineering, University of Illinois/Urbana-Champaign, Illinois **12.22** SJM is a registered trademark of St. Jude Medical, Inc. Copyright St. Jude Medical, Inc. 2002. This image is provided courtesy of St. Jude Medical, Inc. All rights reserved. **12.23** Southern Illinois University/Photo Researchers, Inc. **12.24** Advanced Tissue Sciences, Inc. **12.25** Reproduced with permission. From Fig. 5, Science 2002, 296, 1673–1676 "Biodegradable, Elastic Shape-memory Polymers for Potential Biomedical Applications" by A. Lendlein and R. Langer. Copyright American Association for the Advancement of Science. **12.26** Intel Corporation Pressroom Photo Archives **12.27** Tom Way/ Courtesy of International Business Machines Corporation. Unauthorized use not permitted. **12.30** Marvin Rand/PUGH + SCARPA Architecture **12.31a–b** Richard Megna/Fundamental Photographs, NYC **12.32** Siemens AG **12.36** Richard Megna/Fundamental Photographs, NYC **12.38** Dr. Alois Krost **12.40** Prof. Dr. Horst Weller from H. Weller, Angew.Chem. Int. Ed. Engl. 1993, 32, 41–53, Fig. 1. **12.41** Reprinted with permission from Anal. Chem., October 1, 2002, 74(19), pp. 520A–526A. ©Copyright 2002 American Chemical Society. **12.42** Fototeca della Veneranda Fabbrica del Duomo di Milano **12.43** Reproduced by courtesy of The Royal Instiution of Great Britain. **12.44** Erik T. Thostenson, "Carbon Nanotube-Reinforced Composites: Processing, Characterization and Modeling" Ph.D. Dissertation, University of Delaware, 2004. **12.44.9UN** Courtesy Earth Observatory/NASA and Nicholas M. Short, Sr.

Chapter 13: CO13 © Judith Miller/Dorling Kindersley/John Bull Silver **13.5** Tom Bochsler/Pearson Education/PH College **13.7a–c** Richard Megna/Fundamental Photographs, NYC **13.8** Ed Degginger/Color-Pic, Inc. **13.10a–c** Richard Megna/

Fundamental Photographs, NYC **13.15** Charles D. Winters/Photo Researchers, Inc. **13.16** Joan Richardson/Visuals Unlimited **13.21** Grant Heilman/Grant Heilman Photography, Inc. **13.26** Leonard Lessin/Peter Arnold, Inc. **13.27a** E.R. Degginger/Color-Pic, Inc. **13.27b** Gene Rhoden/Visuals Unlimited **13.32** Oliver Meckes & Nicole Ottawa/Photo Researchers, Inc.

Chapter 14: CO14 Mark Green/Getty Images Inc. - Stone Allstock **14.1a** Michael S. Yamashita/Corbis/Bettmann **14.1b** S.C. Fried/Photo Researchers, Inc. **14.1c** David N. Davis/Photo Researchers, Inc. **14.2a** Michael Dalton/Fundamental Photographs, NYC **14.2b** Richard Megna/Fundamental Photographs, NYC **14.11** Richard Megna/Fundamental Photographs, NYC **14.19a–c** Richard Megna/Fundamental Photographs, NYC **14.22** Delphi Energy & Chassis, Troy, Michigan **14.23** Richard Megna/Fundamental Photographs, NYC **14.27a** Science Photo Library/Photo Researchers, Inc.

Chapter 15: CO15 Roger Ressmeyer/Corbis/Bettmann **15.1a–c** Richard Megna/Fundamental Photographs, NYC **15.5** Ed Degginger/Color Pic, Inc. Courtesy Farmland Industries, Inc. **15.13a–c** Richard Megna/Fundamental Photographs, NYC **15.14a–c** Richard Megna/Fundamental Photographs, NYC

Chapter 16: CO16 FoodPix/Getty Images, Inc. - FoodPix **16.3** Richard Megna/Fundamental Photographs, NYC **16.6** Yoav Levy/Phototake NYC **16.8a–b** Donald Clegg and Roxy Wilson/Pearson Education/PH College **16.10** Michael Dalton/Fundamental Photographs, NYC **16.11a–c** Richard Megna/Fundamental Photographs, NYC **16.14** Frank LaBua/Pearson Education/PH College **16.16** Tom Pantages

Chapter 17: CO17 Larry Ulrich/DRK Photo **17.1** Donald Clegg & Roxy Wilson/Pearson Education/PH College **17.4** Profs. P. Motta and S. Correr/Science Photo Library/Photo Researchers, Inc. **17.7a–b** Richard Megna/Fundamental Photographs, NYC **17.16a–b** Richard Megna/Fundamental Photographs, NYC **17.17b** Gerry Davis/Phototake NYC **17.19a–c** Richard Megna/Fundamental Photographs, NYC **17.20a–c** Richard Megna/Fundamental Photographs, NYC **17.21a–c** Richard Megna/Fundamental Photographs, NYC **17.22.2** Richard Megna/Fundamental Photographs, NYC

Chapter 18: CO18 Corbis Digital Stock **18.2** Pekka Parviainen/Science Photo Library/Photo Researchers, Inc. **18.3** Courtesy Earth Observatory/NASA and Nicholas M. Short, Sr. **18.4** NASA Headquarters **18.5** National Atmospheric Deposition Program (NRSP-3) 2004. NADP Program Office, Illinois State Water Survey. **18.6a–b** Don and Pat Valenti **18.8a** Dennis Kunkel/Phototake NYC **18.9** Corbis Royalty Free **18.10** Ulf E. Wallin/Ulf Wallin Photography **18.12** C.D. Keeling and T.P Whorf, Carbon Dioxide Research Group, Scripps Institution of Oceanography, University of California, La Jolla, CA 92093-0444. **18.13** Australia Picture Library/Corbis/Bettmann **18.14L-R** Windows to the Universe, of the University Corporation for Atmospheric Research. Copyright ©2004 University Corporation for Atmospheric Research. All rights reserved. **18.15a** DuPont **18.16** © Katadyn North America. Used by permission. **18.17** Robert T. Zappalorti/Photo Researchers, Inc. **18.19** Sheila Terry/Science Photo Library/Photo Researchers, Inc. **18.20** Courtesy of GE Plastics **18.21** Kim Fennema/Visuals Unlimited **18.22** PPG Industries Inc. **18.22.3UN** Reprinted with permission from Walter Leitner, "Supercritical Carbon Dioxide as a Green Reaction Medium for Catalysis", Acc. Chem Res., 35 (9), 746–756, 2002. Copyright 2002 American Chemical Society.

Chapter 19: CO19 PNNL/RDF/Visuals Unlimited **19.2a–b** Richard Megna/Fundamental Photographs, NYC **19.3.1–2** Michael Dalton/Fundamental Photographs, NYC **19.7** Austrian Central Library for Physics, Vienna, Austria **19.10.1** Richard Megna/Fundamental Photographs, NYC **19.12a** Klaus Pavsan/ Peter Arnold, Inc. **19.12b** Biophoto Associates/Photo Researchers, Inc. **19.16** Library of Congress

Chapter 20: CO20 NASA Headquarters **20.1a–b** Richard Megna/Fundamental Photographs, NYC **20.2a–c** Richard Megna/Fundamental Photographs, NYC **20.3a–b** Richard Megna/Fundamental Photographs, NYC **20.4** Richard Megna/Fundamental Photographs, NYC **20.9a** Jeff Gnass/Corbis/Stock Market **20.15** The Burndy Library, Dibner Institute for the History of Science and Technology, Cambridge, Massachusetts **20.19** Kaj R. Svenson/Science Photo Library/Photo Researchers, Inc. **20.24** Erich Schrempp/Photo Researchers, Inc. **20.29a–b** Reed Barton/Tom Pantages

Chapter 21: CO21 Martin Dohrn/Science Photo Library/Photo Researchers, Inc. **21.1** CNRI/Photo Researchers, Inc. **21.6** Fermi National Accelerator Laboratory/Science Photo Library/Photo Researchers, Inc. **21.8** James King-Holmes/SPL/Photo Researchers, Inc. **21.9** Yoav Levy/Phototake NYC **21.11** Kevin Schafer/Peter Arnold, Inc. **21.12a** Hank Morgan/Rainbow **21.12b** Wellcome Dept. of Cognitive Neurology/Science Photo Library/Photo Researchers, Inc. **21.18** Gary Sheahan "Birth of the Atomic Age" Chicago (Illinois); 1957. Chicago Historical Society, ICHi-33305. **21.20b** Ed Degginger/Color-Pic, Inc. **21.24** Earl Roberge/Photo Researchers, Inc. **21.25** Richard Megna/Fundamental Photographs, NYC

Chapter 22: CO22 Philip Wegener-Kantor/Index Stock Imagery, Inc. **22.5** Russell Curtis/Photo Researchers, Inc. **22.6a–b** Richard Megna/Fundamental Photographs, NYC **22.9** Donald Clegg and Roxy Wilson/Pearson Education/PH College **22.11** Paul Silverman/Fundamental Photographs, NYC **22.12a–b** Richard Megna/Fundamental Photographs, NYC **22.13** Robert and Beth Plowes Photography **22.14** NASA/Johnson Space Center **22.16** Joseph Priestley (1733–1804): colored English engraving, 19th Century. The Granger Collection, New York. **22.17** John Hill/Getty Images Inc. - Image Bank **22.19a–c** Richard Megna/Fundamental Photographs, NYC **22.20** Courtesy of DuPont Nomex® **22.22a** Jeffrey A. Scovil/Jeffrey A. Scovil **22.23** Lawrence Migdale/Science Source/Photo Researchers, Inc. **22.25** Dan McCoy/Rainbow **22.26a–c** Kristen Brochmann/Fundamental Photographs, NYC **22.31a–c** Donald Clegg and Roxy Wilson/Pearson Education/PH College **22.33** Kristen Brochmann/Fundamental Photographs, NYC **22.33.2UN** Michael Dalton/Fundamental Photographs, NYC **22.35** Richard Megna/Fundamental Photographs, NYC **22.40** General Electric Corporate Research & Development Center **22.41** AP Wide World Photos **22.42** E.R. Degginger/Color-Pic, Inc. **22.43** Chad Ehlers/Getty Images Inc. - Stone Allstock **22.45** Photo Courtesy of Texas Instruments Incorporated **22.49** National Institute for Occupational Safety & Health

Chapter 23: CO23 Frank Micelotta/Getty Images, Inc - Liaison **23.1b** Brownie Harris/Corbis/Stock Market **23.2a–b** Karl Hartmann/Traudel Sachs/Phototake NYC **23.2c** Jeffrey A. Scovil/Jeffrey A. Scovil **23.3** The Cleveland-Cliffs Iron Company **23.5** Robin Smith/Getty Images Inc. - Stone Allstock **23.10** Oberlin College Archives, Oberlin, Ohio **23.12** Richard Megna/Fundamental Photographs, NYC **23.14a** Peter Christopher/Masterfile Corporation **23.14b** IT Stock International/Index Stock Imagery, Inc. **23.15** Magneti Ljubljana, d.d. **23.19** Donald Clegg and Roxy Wilson/Pearson Education/PH College **23.22** Michael Dalton/Fundamental Photographs, NYC **23.23** Donald Clegg and Roxy Wilson/Pearson Education/PH College **23.24** Richard Megna/Fundamental Photographs, NYC **23.25** Donald Clegg and Roxy Wilson/Pearson Education/PH College **23.26** Paul Silverman/Fundamental Photographs, NYC **23.27** Donald Clegg and Roxy Wilson/Pearson Education/PH College

Chapter 24: CO24 Robert Garvey/Corbis/Bettmann **24.2a–b** Richard Megna/Fundamental Photographs, NYC **24.7** Gary C. Will/Visuals Unlimited **24.14** Kim Heacox Photography/DRK Photo **24.18L–R** Richard Megna/Fundamental Photographs, NYC **24.22** Richard Megna/Fundamental Photographs, NYC **24.26** Richard Megna/Fundamental Photographs, NYC **24.36a–c** Tom Pantages **24.37.8c** Carey B. Van Loon/Carey B. Van Loon

Chapter 25: CO25 Jack Anderson/PictureArts/Food Pix/PictureArts Corporation **25.8** Ed Degginger/Color-Pic, Inc. **25.9** Wes Thompson/Corbis/Stock Market **25.15** Richard Megna/Fundamental Photographs, NYC **25.16a** Richard Megna/Fundamental Photographs, NYC **25.16b** Robert and Beth Plowes Photography **25.18** Richard Megna/Fundamental Photographs, NYC **25.19** Culver Pictures, Inc. **25.25** Richard Megna/Fundamental Photographs, NYC **25.27** Oxford Molecular Biophysics Laboratory/Science Photo Library/Photo Researchers, Inc.

Index

COMMON IONS

Positive Ions (Cations)

1+

Ammonium (NH_4^+)
Cesium (Cs^+)
Copper(I) or cuprous (Cu^+)
Hydrogen (H^+)
Lithium (Li^+)
Potassium (K^+)
Silver (Ag^+)
Sodium (Na^+)

2+

Barium (Ba^{2+})
Cadmium (Cd^{2+})
Calcium (Ca^{2+})
Chromium(II) or chromous (Cr^{2+})
Cobalt(II) or cobaltous (Co^{2+})
Copper(II) or cupric (Cu^{2+})
Iron(II) or ferrous (Fe^{2+})
Lead(II) or plumbous (Pb^{2+})
Magnesium (Mg^{2+})
Manganese(II) or manganous (Mn^{2+})
Mercury(I) or mercurous (Hg_2^{2+})
Mercury(II) or mercuric (Hg^{2+})
Strontium (Sr^{2+})
Nickel(II) (Ni^{2+})
Tin(II) or stannous (Sn^{2+})
Zinc (Zn^{2+})

3+

Aluminum (Al^{3+})
Chromium(III) or chromic (Cr^{3+})
Iron(III) or ferric (Fe^{3+})

Negative Ions (Anions)

1−

Acetate ($C_2H_3O_2^-$)
Bromide (Br^-)
Chlorate (ClO_3^-)
Chloride (Cl^-)
Cyanide (CN^-)
Dihydrogen phosphate ($H_2PO_4^-$)
Fluoride (F^-)
Hydride (H^-)
Hydrogen carbonate or bicarbonate (HCO_3^-)
Hydrogen sulfite or bisulfite (HSO_3^-)
Hydroxide (OH^-)
Iodide (I^-)
Nitrate (NO_3^-)
Nitrite (NO_2^-)
Perchlorate (ClO_4^-)
Permanganate (MnO_4^-)
Thiocyanate (SCN^-)

2−

Carbonate (CO_3^{2-})
Chromate (CrO_4^{2-})
Dichromate ($Cr_2O_7^{2-}$)
Hydrogen phosphate (HPO_4^{2-})
Oxide (O^{2-})
Peroxide (O_2^{2-})
Sulfate (SO_4^{2-})
Sulfide (S^{2-})
Sulfite (SO_3^{2-})

3−

Arsenate (AsO_4^{3-})
Phosphate (PO_4^{3-})

FUNDAMENTAL CONSTANTS*

Atomic mass unit	1 amu	$= 1.66053873 \times 10^{-24}$ g
	1 g	$= 6.02214199 \times 10^{23}$ amu
Avogadro's number	N	$= 6.02214199 \times 10^{23}$/mol
Boltzmann's constant	k	$= 1.3806503 \times 10^{-23}$ J/K
Electron charge	e	$= 1.602176462 \times 10^{-19}$ C
Faraday's constant	F	$= 9.64853415 \times 10^{4}$ C/mol
Gas constant	R	= 0.082058205 L-atm/mol-K
Mass of electron	m_e	$= 5.485799 \times 10^{-4}$ amu
		$= 9.10938188 \times 10^{-28}$ g
Mass of neutron	m_n	= 1.0086649 amu
		$= 1.67492716 \times 10^{-24}$ g
Mass of proton	m_p	= 1.0072765 amu
		$= 1.67262158 \times 10^{-24}$ g
Pi	π	= 3.1415927
Planck's constant	h	$= 6.62606876 \times 10^{-34}$ J-s
Speed of light	c	$= 2.99792458 \times 10^{8}$ m/s

*Fundamental constants are listed at the National Institute of Standards and Technology website: http://physics.nist.gov/PhysRefData/contents.html